Handbook of the
Birds of Europe
the Middle East and
North Africa
The Birds of the
Western Palearctic

Volume VIII

# Handbook of the
# Birds of Europe
# the Middle East and
# North Africa

## The Birds of the Western Palearctic

### Volume VIII · Crows to Finches

†Stanley Cramp
C M Perrins   *Senior Editor*
Duncan J Brooks   *Executive Editor*

| | | |
|---|---|---|
| Euan Dunn | Robert Gillmor | Joan Hall-Craggs |
| Brian Hillcoat | P A D Hollom | E M Nicholson |
| C S Roselaar | W T C Seale | P J Sellar |
| K E L Simmons | D W Snow | Dorothy Vincent |
| K H Voous | D I M Wallace | M G Wilson |

OXFORD   NEW YORK
OXFORD UNIVERSITY PRESS · 1994

DEDICATED TO THE MEMORY OF

## H F WITHERBY

(1873–1943)

EDITOR OF *THE HANDBOOK OF BRITISH BIRDS*
(1938–41)

*Oxford University Press, Walton Street, Oxford OX2 6DP*
*Oxford New York Toronto*
*Delhi Bombay Calcutta Madras Karachi*
*Kuala Lumpur Singapore Hong Kong Tokyo*
*Nairobi Dar es Salaam Cape Town*
*Melbourne Auckland Madrid*
*and associated companies in*
*Berlin Ibadan*

*Oxford is a trade mark of Oxford University Press*

*Published in the United States by*
*Oxford University Press Inc., New York*

*A catalogue record for this book is available from the British Library*

*Library of Congress Cataloging in Publication Data*
*Data available*

*ISBN 0–19–854679–3*

*Typeset by Latimer Trend Ltd., Plymouth*
*Printed in Hong Kong*

# CONTENTS

INTRODUCTION      *Page*   1

ACKNOWLEDGEMENTS      2

**PASSERIFORMES** (*continued*)
CORVIDAE crows and allies      5

*Garrulus glandarius* **Jay**      7
*Perisoreus infaustus* **Siberian Jay**      31
*Cyanopica cyanus* **Azure-winged Magpie**      42
*Pica pica* **Magpie**      54
*Nucifraga caryocatactes* **Nutcracker**      76
*Pyrrhocorax graculus* **Alpine Chough**      95
*Pyrrhocorax pyrrhocorax* **Chough**      105
*Corvus monedula* **Jackdaw**      120
*Corvus dauuricus* **Daurian Jackdaw**      140
*Corvus splendens* **House Crow**      143
*Corvus frugilegus* **Rook**      151
*Corvus corone* **Carrion Crow**      172
*Corvus albus* **Pied Crow**      195
*Corvus ruficollis* **Brown-necked Raven**      197
*Corvus corax* **Raven**      206
*Corvus rhipidurus* **Fan-tailed Raven**      223

STURNIDAE starlings      228

*Onychognathus tristramii* **Tristram's Grackle**      229
*Sturnus sturninus* **Daurian Starling**      234
*Sturnus sinensis* **Grey-backed Starling**      237
*Sturnus vulgaris* **Starling**      238
*Sturnus unicolor* **Spotless Starling**      260
*Sturnus roseus* **Rose-coloured Starling**      269
*Acridotheres tristis* **Common Myna**      280

PASSERIDAE sparrows, rock sparrows, snow finches      288

*Passer domesticus* **House Sparrow**      289
*Passer hispaniolensis* **Spanish Sparrow**      308
*Passer moabiticus* **Dead Sea Sparrow**      321
*Passer iagoensis* **Iago Sparrow**      327
*Passer simplex* **Desert Sparrow**      331
*Passer montanus* **Tree Sparrow**      336
*Passer luteus* **Sudan Golden Sparrow**      351
*Carpospiza brachydactyla* **Pale Rock Sparrow**      357

Contents

*Petronia xanthocollis* **Yellow-throated Sparrow**    365

*Petronia petronia* **Rock Sparrow**    371

*Montifringilla nivalis* **Snow Finch**    386

PLOCEIDAE weavers and allies    400

*Ploceus cucullatus* **Village Weaver**    401

*Ploceus manyar* **Streaked Weaver**    401

*Quelea quelea* **Red-billed Quelea**    409

ESTRILDIDAE waxbills, grassfinches, mannikins    409

*Lagonosticta senegala* **Red-billed Firefinch**    411

*Uraeginthus bengalus* **Red-cheeked Cordon-bleu**    420

*Estrilda astrild* **Common Waxbill**    420

*Amandava amandava* **Red Avadavat**    427

*Euodice malabarica* **Indian Silverbill**    437

*Euodice cantans* **African Silverbill**    437

VIREONIDAE vireos    439

*Vireo flavifrons* **Yellow-throated Vireo**    440

*Vireo philadelphicus* **Philadelphia Vireo**    442

*Vireo olivaceus* **Red-eyed Vireo**    444

FRINGILLIDAE finches    447

FRINGILLINAE chaffinches    447

*Fringilla coelebs* **Chaffinch**    448

*Fringilla teydea* **Blue Chaffinch**    474

*Fringilla montifringilla* **Brambling**    479

CARDUELINAE typical finches    497

*Serinus pusillus* **Red-fronted Serin**    499

*Serinus serinus* **Serin**    508

*Serinus syriacus* **Syrian Serin**    521

*Serinus canaria* **Canary**    528

*Serinus citrinella* **Citril Finch**    536

*Carduelis chloris* **Greenfinch**    548

*Carduelis carduelis* **Goldfinch**    568

*Carduelis spinus* **Siskin**    587

*Carduelis pinus* **Pine Siskin**    604

*Carduelis cannabina* **Linnet**    604

*Carduelis flavirostris* **Twite**    625

*Carduelis flammea* **Redpoll**    639

*Carduelis hornemanni* **Arctic Redpoll**    661

*Loxia leucoptera* **Two-barred Crossbill**    672

*Loxia curvirostra* **Crossbill**    686

*Loxia scotica* **Scottish Crossbill** 707

*Loxia pytyopsittacus* **Parrot Crossbill** 717

*Rhodopechys sanguinea* **Crimson-winged Finch** 729

*Rhodospiza obsoleta* **Desert Finch** 739

*Bucanetes mongolicus* **Mongolian Trumpeter Finch** 749

*Bucanetes githagineus* **Trumpeter Finch** 754

*Carpodacus erythrinus* **Scarlet Rosefinch** 764

*Carpodacus synoicus* **Sinai Rosefinch** 783

*Carpodacus roseus* **Pallas's Rosefinch** 789

*Carpodacus rubicilla* **Great Rosefinch** 792

*Pinicola enucleator* **Pine Grosbeak** 802

*Uragus sibiricus* **Long-tailed Rosefinch** 814

*Pyrrhula pyrrhula* **Bullfinch** 815

*Eophona migratoria* **Yellow-billed Grosbeak** 832

*Eophona personata* **Japanese Grosbeak** 832

*Coccothraustes coccothraustes* **Hawfinch** 832

*Hesperiphona vespertina* **Evening Grosbeak** 847

REFERENCES 851

CORRECTIONS 895

INDEXES   Scientific names 896

English names 897

Noms français 897

Deutsche Namen 898

# INTRODUCTION

History has repeated itself! As readers already know, when the editors completed the work originally planned for Volume VI, there was so much material that it had to be split into two volumes, VI and VII. Exactly the same thing has happened with what was to have been the final volume; when the material was put together, we estimated that there were almost 1500 pages (not including the plates). Apart from the difficulties for the reader in handling a volume of this size, it would have been physically impossible to bind it with normal binding machinery. So, the volume originally planned as Volume VII now appears as Volumes VIII and IX.

The editors apologize for the resulting inconvenience to readers. But, at one level, apologies are perhaps unnecessary. The volumes have become so large only because there is so much material to present. The long interval since the start of this immense project has been accompanied by an explosion of ornithological literature. It was inevitable that much more would be known about the species treated at the end of the work than was known about those covered at the beginning, two decades ago. In addition, the passerines have some of the most complex (and best known) displays and by far the most elaborate songs; there are more than a thousand sonagrams in the last two volumes.

The need to divide the material into two volumes has caused a further problem. Ideally, we would have liked to produce two volumes of equal size. Unfortunately, the Fringillinae fall in the middle of the species to be covered and take up some 400 pages. Putting them into Volume IX would have created a similar imbalance to the one we now have. Splitting the group or changing the order of the families seemed even more undesirable than the solution that we have adopted.

Scientific nomenclature in this volume deviates from the *List of Recent Holarctic Bird Species* (Voous 1977) in using the names *Cyanopica cyanus* (Azure-winged Magpie) rather than *C. cyana*, and *Carpospiza brachydactyla* (Pale Rock Sparrow) rather than *Petronia brachydactyla*.

For a full introduction, detailing the scope of the work as a whole and of the individual sections (including glossaries of terminology), see Volume V.

The editors with special responsibility for the various sections of the species accounts in this volume are as follows:

*Field Characters*  D I M Wallace
*Habitat*  E M Nicholson
*Distribution* and *Population*  Dr D W Snow
*Movements*  D F Vincent
*Food*  B Hillcoat
*Social Pattern and Behaviour*  Dr E K Dunn and M G Wilson
*Voice*  Dr E K Dunn, M G Wilson, J Hall-Craggs, W T C Seale, and P J Sellar
*Breeding*  B Hillcoat
*Plumages*, *Bare Parts*, *Moults*, *Measurements*, *Weights*, *Structure*, and *Geographical Variation*  Drs C S Roselaar
*Family Treatments*  Dr K E L Simmons and Drs C S Roselaar

D J Brooks is responsible for the editing of the entire volume.

The paintings are the work of Norman Arlott, Trevor Boyer, Hilary Burn, Dr P J K Burton, Ian Lewington, Darren Rees, Chris Rose, and C E Talbot-Kelly; their initials appear at the end of the caption for each plate. Euan Dunn, Robert Gillmor, Julian Hough, David Nurney, and David Quinn have prepared the line drawings for the Social Pattern and Behaviour sections. Robert Gillmor is the editor with general responsibility for artwork. R J Connor and A C Parker have given most generously of their time and expertise in providing photographs of eggs in the collection of the Natural History Museum at Tring where they were joined in this work by M P Walters of the Museum staff.

For purely practical reasons it has not always been possible to incorporate in our treatment the many recent changes in political boundaries and names of states both within and outside the west Palearctic. In general, the European part of the former USSR has been treated as an entity for distributional purposes, as in previous volumes. Other data from the former USSR have been linked to the various new republics individually wherever possible, but there are many instances where this could not be done conveniently and in such cases the term 'USSR' has been used for simplicity. The names Czechoslovakia and Yugoslavia have been retained throughout, and it has sometimes been necessary to refer to West and East Germany.

# ACKNOWLEDGEMENTS

In the preparation of this volume the continued substantial financial support of the Delegates of the Oxford University Press has been fundamental. Their generous help and understanding, together with the patience of other benefactors, particularly the Royal Society for the Protection of Birds, the Pilgrim Trust, and the World Wide Fund for Nature, who advanced loans long ago, are gratefully acknowledged.

Time has been generously given by many ornithologists and others throughout the world. For facilitating their labours, the Editorial Board are grateful to the Institute of Taxonomic Zoology (University of Amsterdam), and to the Edward Grey Institute of Field Ornithology (University of Oxford) where Dr L Birch of the Alexander Library has provided invaluable back-up over many years.

At the University of Cambridge the Directors of the Sub-Department of Animal Behaviour—Prof. P P G Bateson and his successor Dr E B Keverne—and Prof. G Horn, Head of the Department of Zoology, have afforded long-term and generous support through providing a working environment and all of the apparatus needed for preparation of the sonagrams and analysis of these and many other recordings. J Hall-Craggs, who has enjoyed free access to these facilities throughout the life of the project, is joined in this work by W T C Seale who has been solely responsible for voice illustration of twenty species. Dr P K McGregor and L M McGregor of the Behaviour and Ecology Research Group of the Department of Zoology, University of Nottingham, were responsible for the preparation of sonagrams for many species, and Prof. P J Slater for two. P J Sellar continued to give his time and expertise to amassing, indexing, and re-recording the quantities of material required for each volume, along with the use of his studio and much of his own equipment. R Ranft, Curator of the British Library of Wildlife Sounds, has been always ready and willing to give help and to seek out the best examples of items needed. Gratitude and keen appreciation are due to the members of the Oxford University Press who patiently guided the many sonagrams through the processing that transforms them into intelligible figures. The Voice text for many species has benefited from comments by L Svensson.

Where recordings used for sonagrams are available as published gramophone records or cassettes, references are given in the captions as follows:

Ferdinand, L (1991) *Bird voices in the North Atlantic*. Tórshavn.

Mild, K (1987) *Soviet bird songs*. Stockholm.

Mild, K (1990) *Bird songs of Israel and the Middle East*. Stockholm.

Roché, J-C (1964) *Guide sonore des oiseaux d'Europe* 1; (1966) 2; (1970) 3. Institut Echo, Aubenas-les-Alpes, France.

Roché, J-C (1968) *A sound guide to the birds of north-west Africa*. Institut Echo, Aubenas-les-Alpes, France.

Roché, J-C (1990) *All the bird songs of Europe*. Sittelle, La Mure.

Sveriges Radio (1972-80) *A field guide to the bird songs of Britain and Europe* by S Palmér and J Boswall 1-15 (discs).

Swedish Radio Company (1981) *A field guide to the bird songs of Britain and Europe* by S Palmér and J Boswall 1-16 (cassettes).

For recordings which have not been published commercially, assistance in contacting the original recordists may often be obtained from the Curator, British Library of Wildlife Sounds, National Sound Archive, 29 Exhibition Road, London SW7 2AS.

A vital part was played by the correspondents who provided much of the basic data on status, distribution, and populations for species occurring in the following regions:

ALBANIA  Dr E Nowak
ALGERIA  E D H Johnson
AUSTRIA  Dr H Schifter
AZORES  Dr G Le Grand
BRITAIN  R Spencer
BULGARIA  T Michev
CAPE VERDE ISLANDS  C J Hazevoet
CHAD  C Erard
CYPRUS  P R Flint, P F Stewart
CZECHOSLOVAKIA  Dr K Hudec
DENMARK  U Gjøl Sørensen
EGYPT  P L Meininger, W C Mullié, S M Goodman
FAEROES  Dr D Bloch, S Sørensen
FINLAND  Dr O Hildén
FRANCE  R Cruon
GERMANY  A Hill, S Schnabel
GREECE  G I Handrinos
HUNGARY  G Magyar
ICELAND  Dr Æ Petersen
IRAQ  H Y Siman
IRELAND  C D Hutchinson
ISRAEL  H Shirihai
ITALY  P Brichetti, B Massa
JORDAN  I Andrews, P A D Hollom, D I M Wallace
KUWAIT  Prof. C W T Pilcher
LEBANON  Lt-Col A M Macfarlane
LIBYA  G Bundy
MADEIRA  P A Zino, G Maul
MALI  Dr J M Thiollay, B Lamarche

MALTA  J Sultana, C Gauci
MAURITANIA  R A Williams, J Trotignon, B Lamarche
MOROCCO  Dr M Thévenot, J D R Vernon
NETHERLANDS  Drs C S Roselaar
NIGER  Dr J M Thiollay
NORWAY  V Ree
POLAND  Dr A Dyrcz, Dr L Tomiałojć
PORTUGAL  R Rufino, G A Vowles
RUMANIA  P Weber
SAUDI ARABIA  M C Jennings, *Atlas of the Breeding Birds of Arabia* project
SPAIN  A Noval
SWEDEN  L Risberg
SWITZERLAND  R Winkler
SYRIA  Lt-Col A M Macfarlane
TURKEY  R P Martins
USSR  Prof. L S Stepanyan, Prof. A F Kovshar', V Konstantinov
YUGOSLAVIA  V F Vasić

We also wish to thank all those who made available photographs, sketches, and published material on which the drawings illustrating the Social Pattern and Behaviour section were based; their names are given at the end of the relevant accounts.

F E Warr carried out much of the essential basic literature research, and L Cruickshank gave valuable assistance with typing. For help with translations we are grateful to Dr P Ahlberg, J E Arévalo, B Arroyo, M Cellier, Dr S Haywood, N Hillcoat-Kayser, Dr V Karpov, J King, T Köhler, M Kohlhaas, Pak Fook Chinese Restaurant (Oxford), E R Potapov, L Rode, P Stephenson, Dr E Syroechkovski, Dr T Székely, J C Yoo, L Zadorina, and M Zernicka-Goetz.

Finally, we are greatly indebted to the following, who assisted in many ways too diverse to specify in detail though credits are given in the text where appropriate:

G H Acklam, Dr V C Ambedkar, I J Andrews, Dr J S Ash, G Åström, P Baldwin, L Batt, W R R de Batz, M Beaman, Dr G S Bel'skaya, Dr C W Benkman, P Bennett, Prof. H-H Bergmann, H Biebach, H van den Bijtel, J Bjørn Andersen, B Bjørnsen, W J Bock, J H R Boswall, J Bowler, P Brichetti, B de Bruin, C de Bruyn, Dr I D Bullock, Dr I Byrkjedal, M Cellier, A E Chapman, K Colcomb-Heiliger, Dr N J Collar, P R Colston, P J Conder, Dr C de la Cruz Solis, A Dawson, Dr R W R J Dekker, H Delin, P von Dom, P J Dubois, Dr A Dyrcz, E B Ebels, S Eck, Dr C C Eley, Dr S Elliott, J Elmberg, Dr R van den Elzen, R E Emmett, P Enggist-Düblin, Dr A Evans, J Evans, Dr P G H Evans, Dr C J Feare, G D Field, Prof. V E Flint, Free University of Amsterdam, R J Fuller, D Goodwin, D Gosney, H Göttgens, A Grabher Meyer, M Grahn, G Le Grand, A Gretton, G I Handrinos, P S Hansen, Dr D G C Harper, S Harrap, Dr I R Hartley, C J Hazevoet, M Herremans, I and M Hills, Dr O Hogstad, Dr G Högstedt, D J Holman, P R Holmes, Dr D T Holyoak, P D Housley, Prof. V D Ilyichev, M P S Irwin, Dr V V Ivanitski, Dr B Ivanov, J Jackson, H Jännes, D C Jardine, Dr P J Jones, J Jukema, E I Khlebosolov, the late B King, J King, J L Kitwood, Dr A G Knox, Dr A J Knystautas, J Langer, P A Lassey, P G Lee, Dr G Le Grand, L R Lewis, C M Liebregts-Hooker, M Limbert, F Lindgren, M L Long, F de Lope Rebollo, Dr V M Loskot, F Lovaty, S C Madge, Dr W Mann, Dr A Martín, R P Martins, J A McGeoch, D McGinn, Dr G Mauersberger, C J Mead, B S Meadows, Dr G F Mees, Prof. H Mendelssohn, T Michev, Dr P Mierauskas, Dr A Mikkonen, K Mild, Dr A P Møller, N C Moore, M-Y Morel, A Motis, National University of Singapore, E Nemeth, Dr I A Neufeldt, Dr I Newton, Dr B Nicolai, Dr E Nieboer, Dr I C T Nisbet, the late M E W North, T B Oatley, E Olafsson, Dr G Olioso, U Olsson, Dr S L C O'Malley, Prof. D F Owen, J Palfery, Dr E N Panov, Dr I J Patterson, J Paul, Dr S J Peris, Dr Æ Petersen, G Pétursson, F Pieters, R Pinxten, E R Potapov, Prof. R L Potapov, Dr G R Potts, T Prins, L Profirov, J A Ramos, R Reijnders, A Renard, Dr G Rheinwald, S Rick, Dr D Robel, T J Roberts, V A D Sales, Dr D A Scott, R Scudamore, Y M Shchadilov, B C Sheldon, M Shepherd, H Shirihai, R D Smith, Dr T Stawarczyk, Prof. L S Stepanyan, Prof. B Stephan, J D Summers-Smith, L Svensson, Dr T Székely, P Tatner, C Thomas, J Tigner, L M Tuck, M Ullman, the late Prof. B N Veprintsev, B D Waite, Dr K Walasz, L K Wang, J M Warnes, Dr A Watson, F E Warr, R B Warren, Dr J Wattel, T C White, D Yakutiel, Dr Y Yom-Tov, Dr V A Zubakin, R L Zusi, E Zwart.

## CITATION

The editors recommend that for references to this volume in scientific publications the following citation should be used: Cramp, S and Perrins, C M (eds) (1994) *The Birds of the Western Palearctic* Vol. VIII.

# Order PASSERIFORMES (continued)

# Family CORVIDAE crows and allies

Medium to very large oscine passerines (suborder Passeres), including some of most adaptable and successful birds of entire order. Raven *Corvus corax* is largest of all passerines. Corvids found in many habitats, from forest, woodland, and steppe to tundra and desert. Many wholly or partially arboreal but some terrestrial. Most have generalized diet, taking wide variety of animal and vegetable food, often by scavenging, in trees and/or on ground. Some also specialize to greater or lesser extent in eating large seeds (e.g. those of oaks *Quercus*) and storing them for later use; concealment of surplus food, however, is characteristic of family as a whole. Of almost worldwide distribution but absent from Arctic, Antarctic, and most oceanic islands. Mainly sedentary or partially migratory; some irruptive when main food supply fails. 115 species (Roselaar 1991; see also Goodwin 1986 for monograph); 16 in west Palearctic, 14 breeding (including introduced House Crow *C. splendens*).

Corvids form well-defined, monophyletic group (here given its traditional family status) comprising jays, magpies, nutcrackers, choughs, etc., as well as typical crows. 24 genera, including (e.g.) (1) *Garrulus* (typical jays), 3 species—Eurasia (one confined to Ryukyu Islands south of Japan); (2) *Perisoreus* (grey jays), 3 species—Eurasia and North America: (3) *Urocissa* (blue-magpies, etc.), 5 species—Asia; (4) *Cissa* (green magpies, etc.), 5 species—Asia; (5) *Cyanopica* (Azure-winged Magpie *C. cyanus*), monotypic—discontinuously south-west Europe and eastern Asia; (6) *Dendrocitta* (tree-pies), 6 species—Asia to Indonesia; (7) *Pica* (typical magpies), 2 species—Eurasia and North America; (8) *Podoces* (ground-jays), 4 species—Asia from Iran to China; (9) *Nucifraga* (nutcrackers), 2 species—Eurasia and North America; (10) *Pyrrhocorax* (choughs), 2 species—Eurasia; (11) *Corvus* (typical crows), 42 species—nearly worldwide, including the only corvids (*sensu stricto*) occurring in Australia, but absent from South America and New Zealand (where, however, Rook *C. frugilegus* is introduced). See also below for 6 of the 7 genera of New World jays.

Sexes generally of similar size. Bill of most species stout, strong, and fairly long with slight hook, but (e.g.) rather longer and pointed in *Nucifraga* and long and decurved in *Podoces* and *Pyrrhocorax*. Used to hammer open hard food, crack seeds, probe, flick aside debris and earth, etc. (see Goodwin 1986 for these and other behavioural characters); also for prying into holes (etc.) using open-bill method (see Sturnidae). Nostrils usually rounded; open but closely covered by dense bristles. Rictal bristles also present. Special sub-lingual pouch found in some species;

used to transport food (e.g. for storage or feeding young). Wing shape variable but often broad; 10 primaries, p10 somewhat reduced (length 35–65% of longest), relatively longer in juveniles than in adults, relatively shortest in adult *Pica*, relatively longest in juvenile *Corvus corax*. Flight typically strong and straight with rather deliberate wing-beats but more laboured-looking in shorter-winged, longer-tailed species (e.g. Magpie *P. pica*); some species (e.g. Carrion Crow *C. corone*) perform aerial displays, with deep slow wing-beats, and others (most notably *C. corax*) glide, soar, tumble, and turn aerial somersaults. Tail variable, short to long, often graduated, central feathers much elongated in some species; 12 feathers. Leg and toes sturdy; leg quite long in most species. Tarsus usually strongly scutellate, booted at rear. Gait often a hop or bounding gallop (so-called 'polka step' in which legs leave ground one after the other), but most species also walk and run, some habitually. Foot employed to hold down food items by 'clamping', using one or both feet (see, e.g., Timaliidae in Volume VII); in *Corvus* at times also used to carry objects, though bill more usually employed for this task, including dropping of shellfish (etc.) from height on to hard surface in order to break them.

Shell-dropping is typical example of often remarked 'intelligence' (especially in opportunistic feeders of genus *Corvus*), leading many systematists to consider them the 'highest' of all birds and place them last (or first) in sequence. This ability extends to other areas of foraging behaviour, including hiding and recovering of stored food, and indicates exceptionally good powers of memory and learning. Linked (e.g.) with skill and dexterity in use of bill and foot, curious and cautious nature, and (in larger species at least) prolonged period of dependence and semi-dependence when young (see below) and relatively long life. Manifest, too, by tendency of some corvids to use objects as tools (e.g. Boswall 1985*b*, Rolando and Zunino 1992) and to indulge in 'play' behaviour, including (in *Corvus*) aerobatics, dropping and catching twigs (etc.), and, perhaps, hanging upside-down from perch.

As in majority of passerines, head-scratching typically by indirect method though some individuals of genus *Corvus* reported doing so directly on occasion (see Goodwin 1986), sometimes in flight (e.g. Simmons 1974), and *Pyrrhocorax* will occasionally scratch bill directly (Holyoak 1972*b*). Most species drink by dipping bill in water then raising head, but some suction probably also involved (Goodwin 1965*a*); Nutcracker *N. caryocatactes*, however, seems habitually to drink by continuous suction ('pumping'), e.g. in manner of pigeons (Columbidae) (see

Hollyer 1970). Usually bathe in typical 'stand-in' manner, but, if water too deep, will do so by 'standing out' (see Simmons 1985); same movements used in snow by individuals of some species, e.g. *P. pica* (Coombs 1978) and *N. caryocatactes* (Pfeifer 1956). Sunning frequently observed, especially in moulting birds (Goodwin 1986); of usual 2 main passerine types, but lateral posture more prevalent, full spreadeagle being comparatively rare (see also Simmons 1986a). Anting reported from many species (see also Goodwin 1953b; Simmons 1957a, 1961b, 1966); probably occurs throughout family but apparently not yet reliably recorded from such well-known birds as *C. corax* and Jackdaw *C. monedula*. Both direct active method (ant-application) and indirect passive method (ant-exposure) used by some species (e.g. *C. corone*), often at same time, but others (e.g. *P. pica*) appear to apply ants with bill only, some collecting several ants in ball or wad (see Sturnidae). Jay *Garrulus glandarius* and a few extralimital corvids ant-expose only while stirring up ants with forward-posturing wings and/or (in case of Red-billed Blue Magpie *Urocissa erythrorhyncha*) fanned tail; these species never actually apply ants to plumage though make token movements of so doing. Anting with substitutes, including smoke and flames, also recorded (mainly from captive birds). Dusting not recorded, even in desert species.

Voice unspecialized; calls mostly simple, loud, and harsh. Typical advertising song of most other passerine groups lacking, but quiet song (really subsong), usually uttered by lone birds, occurs in all or most species; mimicry (by both sexes) common. Oten gregarious at times, mainly while feeding and roosting, but most species solitary and territorial when breeding. *C. frugilegus*, *C. monedula*, and a few other species nest colonially, however, associating in flocks throughout year, and *C. cyana* and many American jays (in genera *Aphelocoma*, *Cyanocorax*, *Cissilopha*, *Psilorhinus*, *Calocitta*, and *Gymnorhinus*) often nest communally, i.e. with helpers (usually earlier offspring of dominant pair) assisting in rearing of young (see also Brown 1987). Pair-bond strong; usually sustained all year and probably often life-long. Like most passerines, some corvids maintain clear 'individual distance' and avoid physical contact with others of same species (even mate); some, however, associate at times in close bodily contact with mate and other members of same social group, especially when loafing or roosting, often allopreening. Nest an open cup or rough domed structure; built by both sexes. Incubation usually by ♀ only (fed by ♂); young fed by both

sexes. Family bonds strong; post-fledging care, and association between adults and young and between siblings, prolonged in some species, especially larger *Corvus* (up to 6 months in *C. corax*).

Plumage colour variable but often wholly black, black and white, black and grey, or brown; some magpies and many jays, however, are highly colourful, often with contrasting blue, purple, or green patterns on head, throat, breast, wing, or tail. Little or no seasonal variation. Some species crested. Sexes usually closely similar. Juvenile plumage like that of adult in most species but usually duller, looser, and fluffier. Nestlings of most species naked or with scanty down. Adults have single, complete, post-breeding moult annually. Progress sometimes rapid (e.g. in *N. caryocatactes*), sometimes slow (e.g. in *C. corax*). Moult of wing feathers usually started after fledging of young but sometimes during nestling period or, in *C. corax*, sometimes even during incubation. Post-juvenile moult incomplete: body starts within 1–2 months after fledging; wing-feathers, and usually tail, retained until 1 year old.

Relationship of corvids to other passerine groups much debated but in most classifications thought to be closest to bowerbirds (Ptilonorhynchidae), birds-of-paradise (Paradisaeidae), and some other Australasian groups such as bell-magpies and allies (Cracticidae) and magpie-larks (Grallinidae); other families suggested include orioles (Oriolidae), drongos (Dicruridae), starlings (Sturnidae), tits (Paridae), and true shrikes (Laniidae), especially latter from egg-white protein data (see summary by Sibley 1970). Cuckoo-shrikes (Campephagidae) and wood-swallows (Artamidae) not previously considered at all close, but DNA data (Sibley and Ahlquist 1990) now indicate that they, together with birds-of-paradise, bell-magpies and allies, and orioles, are closest relatives—close enough, in fact, for all to be placed in same subfamily within greatly enlarged and diverse 'Corvidae' which also includes drongos, whistlers (Pachycephalidae), ioras (Aegithinidae), and bush-shrikes (Malaconotinae), together with certain other indigenous birds (some crow-like, others not) of Australasia. True shrikes, vireos (Vireonidae), leafbirds (Irenidae), and an even larger and diverse collection of Australasian groups (but not bowerbirds) are placed in same superfamily as corvids (etc.). Ancestors of endemic Australasian assemblage thought to have given rise to corvids elsewhere, present-day Australian *Corvus* being derived from much more recent invasion of that derived stock from north.

# *Garrulus glandarius* **Jay**

Du. Gaai    Fr. Geai des chênes    Ge. Eichelhäher
Ru. Сойка    Sp. Arrendajo    Sw. Nötskrika

*Corvus glandarius* Linnaeus, 1758

Polytypic. GLANDARIUS GROUP. Nominate *glandarius* (Linnaeus, 1758), Fenno-Scandia and European USSR east to Pechora basin, Kirov, Kazan', and Ulyanovsk, south to Ukraine, Rumania, north-west Bulgaria, northern Yugoslavia (except Dalmatian coast), Alps, and France (except Bretagne) to Pyrénées, grading into *brandtii* in Pechora basin and from Kirov, Kazan', and Ulyanovsk east to western slopes of central and southern Urals (intermediates have been named *severtzowi* Bogdanov, 1871); *rufitergum* Hartert, 1903, England, Wales, and (showing some influence of nominate *glandarius*) Scotland ('*caledonicus*' Hazelwood and Gorton, 1953) and Bretagne ('*armoricanus*' Lebeurier and Rapine 1939); *hibernicus* Witherby and Hartert, 1911, Ireland; *lusitanicus* Voous, 1953, northern Portugal and Salamanca (Spain), grading into nominate *glandarius* in Galicia and western Pyrénées; *fasciatus* A E Brehm, 1857, southern and eastern Spain, grading into *lusitanicus* in southern Portugal and probably in central Spain, and probably into nominate *glandarius* in north-east Spain; *corsicanus* Laubmann, 1912, Corsica; *ichnusae* Kleinschmidt, 1903, Sardinia; *albipectus* Kleinschmidt, 1920, mainland Italy, Dalmatian coast of Yugoslavia, Albania, and Ionian islands (western Greece), grading into nominate *glandarius* in northern Italy, in mountains of western Yugoslavia ('*yugoslavicus*' Voous, 1953), and probably in eastern Albania; *jordansi* Keve, 1966, Sicily; *graecus* Kleiner, 1939, southern Yugoslavia south from Kosovo and southern Serbia, southern Bulgaria south from Rila and Rodopi mountains, and mainland Greece south to Pelopónnisos; *cretorum* Meinertzhagen, 1920, Crete; *ferdinandi* Keve, 1943, south-east Bulgaria and Istranca mountains in northern European Turkey. BRANDTII GROUP. *G. g. brandtii* Eversmann, 1842, from northern Urals and eastern slopes of central and southern Urals east through Siberia to Lake Baykal area, south to Altai and Sayan mountains; *bambergi* Lönnberg, 1909, Mongolia and Transbaykalia east to Sakhalin, southern Kuril islands, Hokkaido (Japan), and Korea, grading into *sinensis* of *bispecularis* group in Hopeh (China) ('*pekingensis*' Reichenow, 1905) (extralimital); *kansuensis* Stresemann, 1928, eastern Tien Shan and Kansu, China (extralimital). ATRICAPILLUS GROUP. *G. g. krynicki* Kaleniczenko, 1839, Kuban' area, Caucasus, Transcaucasia (except Lenkoran' area), and northern Asia Minor west to about Samsun; *iphigenia* Sushkin and Ptushenko, 1914, Crimea; *anatoliae* Seebohm, 1883, Istanbul area (European Turkey), western Asia Minor (in north, east to at least Kastamonu), Lesbos, Khios, Rhodes, and perhaps Kos (eastern Aegean Sea, Greece), southern Asia Minor east to Van Gölü, south to Latakia (north-west Syria), northern Iraq, and Zagros mountains (south-west Iran); *samios* Kleiner, 1939, Samos and Ikaria (eastern Aegean Sea); *atricapillus* Geoffroy St Hilaire, 1832, Lebanon, southern Syria, Israel, and western Jordan; *cervicalis* Bonaparte, 1853, Tunisia and north-east Algeria (west to Algiers and Médéa, south to Batna area); *whitakeri* Hartert, 1903, Tanger area and Er Rif (northern Morocco) and Tlemcen area (north-west Algeria). MINOR GROUP. *G. g. minor* Verreaux, 1857, Atlas Saharien (Algeria) from Djelfa area west to Haut Atlas of Morocco, grading into *whitakeri* in Moyen Atlas ('*theresae*' Meinertzhagen, 1939); *glaszneri* Madarász, 1902, Cyprus; *hyrcanus* Blanford, 1873, Elburz mountains and southern shore of Caspian Sea (northern Iran), grading into *krynicki* in Lenkoran' area of south-east Transcaucasia ('*caspius*' Seebohm, 1883). Extralimital: *japonicus* group with 4–6 races Japan and surrounding islands (except Hokkaido), *bispecularis* group with 7–11 races Himalayas, China (except southern Yunnan, Kansu, Hopeh, and Manchuria), and Taiwan, south to western Burma and northern Vietnam, *leucotis* group with 1 race in eastern Burma, southern Yunnan, and Thailand to central Vietnam.

**Field characters.** 34–35 cm; wing-span 52–58 cm. Somewhat larger than Jackdaw *Corvus monedula*, with deeper body and broader wings. Rather small corvid, most colourful of family in west Palearctic, with short bill, domed head, broad wings, chesty body, and rather long tail. West and central European races pink- to grey-brown with black 'moustache', black, white-barred, and blue-splashed wings, bold white rump, and black tail; other races differ especially in head pattern and overall colour. Flight jerky and weak-looking. Screeching call distinctive. Sexes similar; little seasonal variation. Juvenile separable at close range. 23 races in west Palearctic, falling into 4 groups; only north-west European races described here (for others, see Geographical Variation).

ADULT. Complete moult after breeding (mostly June–October). (1) European and north Mediterranean races, ranging east to Crete (*glandarius* group). Commonly show vinous-grey head and body and streaked crowns. (1a) Scandinavian, north and central European, and southwest Russian race, nominate *glandarius*. Forehead and crown dull to vinous-white, streaked black; crown feathers lanceolate and often raised into short crest. Face vinous-buff, warmer-toned than crown; encloses whitish patch around staring bluish-white eye and is punctuated by broad black moustache which extends back to behind eye. Moustache and black-horn bill contrast with white throat, often puffed-out. Side of neck, back, breast, and flank warm vinous-pink, with obvious grey-brown wash on back and scapulars making these areas darker than underbody. Wings have smaller coverts concolorous with scapulars but other feathers are at least partly black, boldly patterned with white on bases of at least 5 outermost secondaries, narrow but sheeny blue bars on outer greater coverts, primary coverts, bastard wing, and (visible only at close range) whitish outer webs of primaries. Tail appears wholly black in most views, contrasting strongly with obvious white rump and upper tail-coverts and white to vinous-white on under tail-coverts and centre of underbody. With wear, body plumage becomes greyer. Legs pale brown-flesh. (1b) Irish race, *hibernicus*. Distinctly more reddish-brown, particularly below, than nominate *glandarius*. (1c) South Scottish, Welsh, and English race, *rufitergum*. Intermediate between nominate *glandarius* and *hibernicus*; usually warmer pink and less grey above than nominate *glandarius*. (2) North Russian and Siberian races (*brandtii* group). Mostly show chestnut head and underbody, contrasting with rather clean grey back; crown streaked, as in *glandarius* group, but lore and streak behind eye also black. (3) Mediterranean, Crimean, Turkish, and Tunisian races (*atricapillus* group). Mostly show monotoned back, hindneck, and underbody and wholly black crown. (4) Moroccan, Algerian, and Cyprus races (*minor* group). Small in size. Mostly show dark, 2-toned body, dark vinous-brown side of head, and dark crown. JUVENILE. Characteristically fluffy underparts. *Glandarius* and *minor* groups resemble adult but streaks on forehead and crown smaller and ground-colour of head and body distinctly redder inviting confusion with *brandtii* group. Other groups closely resemble adult (see Geographical Variation).

Unmistakable when seen well but subject to confusion with Nutcracker *Nucifraga caryocatactes* when only a dark silhouette flitting through trees and even Hoopoe *Upupa epops* when briefly glimpsed in sun and shadow. General character rather graceless, not improved by harsh voice. Flight lacks freedom and relative grace of crows *Corvus*, having characteristic action (again partly exhibited by *N. caryocatactes* and *U. epops*) of rather quick, somewhat mechanical, uneven, half-depressed beats and full flaps of much extended and spread, rounded wings. Over open country, flight laborious, with hints of stall followed by exaggerated bursts of wing-beats, occasional floating and terminal dipping; makes flight at first level and flapping, then distinctively rather undulating. On eruption, flight still appears unskilled but develops into more marked and longer undulations, with bursts of wing-beats more even and wings more closed. Within trees, flight appears more accomplished, with more even wing-beats. Gait rather heavy but confident, with muscular hop and bounce along branches and ground; accompanied by frequent twitching and jerking of tail either up and down or side to side; also flicks out wings, exposing blue 'shoulder' more fully. Usually seen singly in open country, but often in pairs or small noisy groups in woods.

Usual call a loud, harsh, unattractive screech which carries far, even through dense woods. Voice also includes weak imperfect crooning gurgle (approaching song) and some mimicry.

**Habitat.** Breeds across wooded middle and lower middle latitudes of west Palearctic to July isotherm of 14°C, mainly in continental temperate and Mediterranean climates, but marginally in oceanic, boreal, and wooded steppe zones. Predominantly lowland, but in Switzerland (infrequently) to 1200 m and in Carpathians to treeline at 1600 m. Strongly arboreal, and at home in fairly dense cover of trees, scrub, and woody undergrowth, especially in woodlands of oak *Quercus*, beech *Fagus*, and hornbeam *Carpinus*, but also inhabits other broad-leaved and, in parts of range, coniferous forests. In some regions has spread into smaller outlying woodlands, spinneys, and copses, and even to urban and rural parkland, and to large gardens, where sometimes uses walls and stone ledges. Tends to avoid situations not affording instant retreat to cover, especially when attracted to feeding on ground. In contrast to many woodland birds is not generally attracted to edges, glades, or clearings (Yapp 1962), although an exception appears to occur in Swiss coniferous forest (Glutz von Blotzheim 1962). Habit of burying food, especially acorns, indicates particularly strong and ancient association with oaks, and probable role in enabling them to spread uphill and over open country (Goodwin 1951). After breeding season, tends to occur also in orchards, vineyards, and gardens, and in groups of trees scattered in steppes and semi-deserts, as well as around human settlements and along farm hedgerows. *G. g. brandtii* in Russia also resorts to wooded steppes, mainly when not breeding (Goodwin 1986). Highly sensitive to unfavourable and favourable environmental factors, including human persecution or toleration, and will adapt quickly to changes in these. Although normally reluctant to break cover will on occasion make long flights well above lower airspace.

**Distribution.** No evidence of major changes in recent years, except for range expansion in Scotland, Russia, and Israel, and possible contraction in Algeria. Following decrease in human persecution, has spread into suburban and urban habitats in Britain (Sharrock 1976), Poland (Tomiałojć 1990), Russia (Idzelis 1986), and probably elsewhere.

BRITAIN. Considerable expansion in Scotland, especially since early 1970s, associated with colonization of maturing conifer plantations (Thom 1986). IRELAND. Expansion throughout 20th century (Hutchinson 1989). RUSSIA. Caucasus race, *krynicki*, has spread north in recent decades with increase and maturing of shelter belts (Taranenko 1979). ISRAEL. Until 1930s confined to areas with natural woodland; expansion from 1950s, following development of settlements, agriculture, and afforestation (Shi-

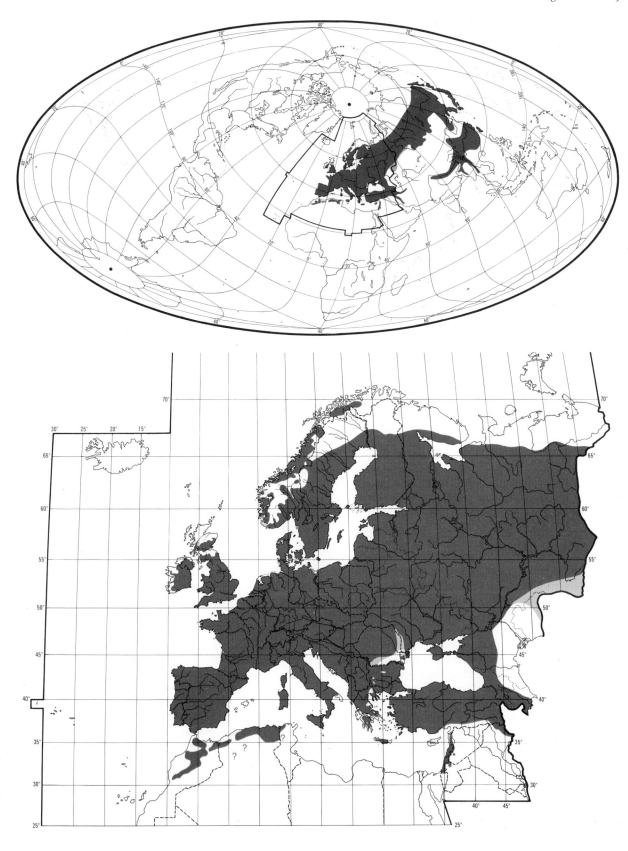

rihai in press). ALGERIA. Not recorded in recent years from Atlas Saharien (EDHJ).

**Population.** Few changes reported.

BRITAIN, IRELAND. About 100 000 pairs (Sharrock 1976). Considerable increase in Scotland in recent decades (Thom 1986). Increased in Ireland throughout 20th century (Hutchinson 1989). FRANCE. 100 000 to 1 million pairs; has probably increased (Yeatman 1976). Numbers in winter fluctuate according to irruptions of eastern and northern birds; last big invasion 1977 (RC). BELGIUM. Population probably stable in recent decades; estimated 22 000 pairs (Devillers *et al.* 1988). NETHERLANDS. Estimated 30 000-60 000 pairs (SOVON 1987). GERMANY. 835 000 pairs (G Rheinwald). East Germany: 100 000 ± 45 000 pairs (Nicolai 1993). SWEDEN. Estimated 450 000 pairs (Lundberg *et al.* 1980). FINLAND. May have increased in north in last few decades, otherwise stable; estimated 100 000-200 000 pairs (Koskimies 1989). ISRAEL. A few tens of thousands of pairs (Shirihai in press).

Survival. Britain: mortality in 1st calendar year 40%, in 2nd year 55%, and in 3rd–5th years 41% (Holyoak 1971). Europe: mortality in 1st year of life 60·7% (Busse 1969). Oldest ringed bird 17 years 11 months (Rydzewski 1978).

**Movements.** Sedentary in west and south of range, eruptive migrant in east and north. For reviews, see John and Roskell (1985) for Britain, Berndt and Dancker (1960), and especially Zink (1981).

British race *rufitergum* mostly sedentary; 98% of 276 recoveries 1972-81 were within 50 km of ringing site (Mead and Hudson 1984); for exceptional westward dispersal during 1983 irruption of nominate *glandarius*, see below. In Ireland, *hibernicus* also fairly sedentary, with evidence of westward dispersal in winter, some birds probably staying to breed (Hutchinson 1989).

Eruptive (diurnal) migration of north and central European race, nominate *glandarius*, probably chiefly due to failure of acorn *Quercus* crop; notable years have included 1955, 1977, and 1983. Populations involved and extent of movement vary greatly. Heading usually ranges between west and SSW, with strong westerly component for eastern (Russian) birds; movement in reverse of normal direction not uncommon, mostly influenced by leading-line effect of coastline or by weather. Short-distance movements in search of alternative food source may be mistaken for eruptions. Most migrants are juveniles, and considerable proportion returns to area of origin in spring.

Autumn movement mid- or late September to early or mid-November, spring movement (in smaller numbers than autumn) March–June, thus continuing markedly late, when breeding season well under way (for discussion of cause and effect, see Gatter 1977). Birds are reluctant to cross sea; thus, passage migrants rare on Helgoland (western Germany) and Ottenby (Öland island, Sweden); no

records of sea crossings from Norway or Finland, and small number of Swedish records are mostly to Danish islands; observers in southern Scandinavia report unsuccessful attempts to leave land, or departure after several false attempts. However, British records and observations, especially in southern and eastern coastal areas, show that in some years continental birds reach Britain. Longest distance recorded *c.* 3000 km: bird ringed Hessen (western Germany), winter, recovered on Ural river in April of following year. (Berndt and Dancker 1960; Paevski 1971; Zink 1981; Andrén 1985; John and Roskell 1985; Goodwin 1986; Klafs and Stübs 1987.)

In Scandinavia, seasonal movements occur each year, notably from north of distribution of oak *Quercus*, with eruptions every few years. At Falsterbo (south-west Sweden), 8 large-scale movements since early 1940s, all coinciding with poor acorn crop; for 1955, see below; major 1983 irruption (see below) reached northern Skåne in southern Sweden, but was scarcely perceptible as far south as Falsterbo, perhaps due to abundance of beechmast *Fagus* crop further north. In 1977, movement began mid-September, peaking end September in central Sweden and early October in southern Sweden, Denmark, and Norway (few migrants in Finland); many birds remained in breeding areas, however, and most migrants ringed were juveniles; *c.* 16 000 passed through Falsterbo, of which 82% on 3 and 12 October, and 3 were recovered in eastern Denmark. In Norway, strong reluctance to cross open sea caused birds to follow coast northward after reaching extreme south. (Gyllin 1965*b*; Svensson 1973; Roos 1978; Løfaldli 1983; Andrén 1985.) 1977 movement conspicuous also in Belgium and Switzerland and on southern Baltic coast; in Schwäbische Alb (southern Germany), *c.* 6000 migrants recorded (*Br. Birds* 1978, 71, 257; Gatter *et al.* 1979; Zink 1981).

In Schwäbische Alb, 1954-73, irruptions occurred every 4-5 years. In 1972, first record 25 August, regular from 1 September, peaking late September and continuing to end October or early November; movement was in several waves, with headings from north-west to south, but simultaneously south or south-west and north or north-west, apparently reflecting differing provenance; passage at 06.00-18.00 hrs, peaking 08.00-10.00 hrs; numbers remained high throughout winter, and following heavy snowfall westward movement noted 28 January to 27 February 1973; return movement in smaller numbers recorded 18 April to 5 June, chiefly 23 April to 12 May, peaking at 05.30-07.30 hrs; heading north to north-east 23 April to 1 May, east to ESE 2-12 May. (Gatter 1974.) Conspicuous autumn movement 1972 noted also at Warnemünde (Baltic coast, Germany): on 1 October, 8230 birds flying between south and WSW in 4 hrs (Müller 1977). In Switzerland, 1959-82, passage of varying intensity recorded in 14 years, of which 1959, 1966, 1972, and 1977 were invasion years; at Col de Bretolet in west, peak passage 28 September to 7 October; return movement chiefly in springs following irruptions, sometimes from March, usually from mid-April, continuing to beginning of June (Winkler 1984). In France, few records of distant movements of local breeders: 8 recoveries at 30–90 km, 2 at slightly more than 100 km. In Rhône-Alpes in south-east, invasions occurred 1947, 1959, 1964, and 1972; 15 recoveries involve birds ringed in Switzerland, chiefly on passage. (Lebreton 1977; Zink 1981.)

At Pape (Latvia), records suggest tendency to move every year, but in greatly fluctuating numbers; in exceptional 1981 passage, recorded from 12 September with rapid increase to 25 September, and sharp decline thereafter to late October; in all, probably 13 000–16 000 birds involved (Rute 1984). In St Petersburg region (north-west Russia), short-distance autumn movements reported almost every year, and larger-scale movements in some years; on southern shore of Gulf of Finland, passage both east and west observed; in 1975, near Pskov-Chudskoe reservoir, more than 34 000 reported; in 1961, large numbers seen travelling east and south-east at Lake Il'men', Lake Chudskoe, and in Lithuania (Mal'chevski and Pukinski 1983). Recovery of passage birds ringed on southern Baltic coast (northern Poland and Kaliningrad region) exemplifies varying distance of movement; most remain north of *c*. 49°N; 2 recovered as far as west coast of France and north-east coast of Italy, but not in winter following ringing, and movement in intervening period not known; in 1964, heading WSW and south-west, most moved only to north-east Germany (Mecklenburg and Brandenburg), with 1 record in Bayern (southern Germany); in 1961, migrating south-west, they reached upper Elbe, with 1 further afield on lower Rhine; in 1965, movement was yet further, mostly WSW to north-west Germany and Belgium, with a few birds reaching Denmark, north-east France, and central and southern Germany. Recoveries in direction of origin are chiefly north-east, extending north to Gulf of Finland, east to upper Volga. Bird ringed Kaliningrad, September, recovered south-east in Ukraine the following September, had apparently taken different autumn routes in successive years. (Paevski 1971; Zink 1981.)

In Kaliningrad region, 1959–84, intensive migration (involving 60% of all birds caught) in 1961, 1964, 1966, and 1981; no migration observed in 1963, 1967, 1971, 1976, 1978, and 1982; return movement always noticed in springs following autumn movement, and numbers in spring relative to autumn are higher than in other 'invasion' species (Shapoval 1989).

Major 1955 irruption in north-central Europe apparently had no distant origin, but involved birds from Belgium east to western Poland, and southern Sweden south to Erzgebirge (*c*. 51°30′N, eastern Germany); heading ranged between north-west and south-west, and invasion reached southern England and north-east France in north, southern Germany and Switzerland in south; occurred in several waves, beginning progressively later further north

and east, and averaging 60–80 km per day; passage through Low Countries reported from 10 September, strong by late September; 2 Belgian-ringed nestlings were retrapped 47 and 171 km WSW of ringing site, indicating involvement of local birds; in north-central Germany, thousands reported at Braunschweig in late September, *c*. 50 000 at Magdeburg on 28 September. In Sweden, little evidence of eruption north of *c*. 59°N, but movement south-west reported at Falsterbo from 15 September, with 5 waves involving 10 340 between 23 September and 31 October; probably, 1st wave reached England, 2nd and 3rd waves northern Germany and Low Countries, 4th and 5th only eastern Denmark and southern Sweden. Recovery 2 June 1956, Netherlands, of bird ringed 12 October 1955, Dungeness (Kent, south-east England), provided sole ringing evidence for sea-crossing to Britain. (Berndt and Dancker 1960; Keve 1985.)

In 1983, largest recorded influx into Britain, involving thousands of birds, began 24 September, chiefly October, continuing to 9 November. Especially noticeable in East Anglia (where birds seen flying in over sea) and southern counties, but increased numbers reported as far north as Perthshire (central Scotland). Heading predominantly westward, peaking Kent in early October (1500 on 2 October), Cornwall in south-west mid-October (6000 on 17 October). Acorn crop failed in both Britain and continental Europe, and location and timing of movements suggest that many British birds participated, presumably in search of food. Of 39 recoveries October 1983 to March 1984, 7 (of which at least 5 were British birds) had moved more than 50 km (up to 285 km). Several reports of unusually high numbers in spring 1984, some presumably relating to return passage. Elsewhere in Europe, 1983 invasion reported from Switzerland and Poland; in southern Sweden, high numbers included *c*. 10 000 migrating west to north-west at Vänersborgsviken late September to early October; no sizeable movement noted in France or Belgium. (John and Roskell 1985.)

Notable instance of short-term reversed migration in western Germany, late September 1977: passage eastwards, involving a region nearly 1000 km across, from Schwäbische Alb north to Schleswig-Holstein, entailed movement from low to high pressure area, and also coincided with displacement of warm sea air ending cold weather in reversed migration area; observations in Schwäbische Alb showed that direction was between SSW and WNW 21–23 September, and between south-east and north-east 25–29 September; thereafter heading between south and west predominated (Gatter *et al.* 1979).

North Asian race *brandtii* partially resident, but sometimes undertakes movements of considerable extent. In Kazakhstan, breeds only in north-east, but migrants occur in north and west, e.g. in Orenburg region (Korelov *et al.* 1974). In north-east Altai, August to beginning of September, numbers increase considerably with arrival of southward-dispersing birds (Ravkin 1973). At north-west of Teletskoe lake (north-east Altai), 1968–78, heavy west-ward passage occurred from August to 2nd half of September, more prolonged and intensive in warm, dry autumns, continuing from 07.00 or 07.30 hrs to 19.00 or 20.00 hrs, and peaking at 11.00–12.00 hrs and 16.00–17.00 hrs; groups of 5–6(–20) birds moved through at height of 30–40 m; in September 1977, perhaps up to 150 000 birds involved (Sobanski 1979). In Ussuriland, regular passage reported September, and May to early June (Panov 1973a); in extreme south, at Sudzukhe reserve, large-scale movement observed in 1944 (when acorns scarce); from 2nd third of August until early September flocks flew along coast northeastwards, then changed heading to south-west, continuing to end of September; from early October, random dispersal widespread (Dementiev and Gladkov 1954).

Some of Caucasian race *krynicki* descend from mountains, though others are sedentary birds; autumn passage through Klukhorskiy pass recorded, and numbers in lowlands increase in winter (Dementiev and Gladkov 1954; Polivanov and Polivanova 1986). Italian and Yugoslavian race *albipectus* disperses north-east to Hungary (reported east to Budapest) in severe winters; single records 1939–40 and 1959–60, and 25 birds 4 October to 20 March 1962–3 (Keve 1985). Other southern races sedentary, with some local dispersal; on Cyprus, 3 records away from breeding range may involve vagrants crossing from Turkey (Flint and Stewart 1992).                          DFV

**Food.** Invertebrates, especially beetles (Coleoptera) and Lepidoptera larvae, fruits, and seeds, especially acorns *Quercus*; small vertebrates occasionally taken, most often in winter or when feeding young; also carrion and domestic scraps (Keve 1985; Goodwin 1986). During breeding season most food collected from leaves of trees, mainly caterpillars (Lepidoptera) in oak, but otherwise forages principally on ground except when collecting acorns in autumn for storing (Kadochnikov and Eygelis 1954; Owen 1956). On ground, digs in leaf litter and soil surface with swinging movement of bill; otherwise inserts and opens bill ('prying') in holes, cracks, rolled leaves (where it finds Tortricidae caterpillars), etc., pulling or levering away loose bark on trunk and lower branches (Kadochnikov and Eygelis 1954; Goodwin 1986). Feeds readily on cereals, both during sowing and at harvest time (Fuye 1911; Keve and Sterbetz 1968; Bossema 1979), and may hover for 1–2 s to pull off seed-heads, returning to perch for consumption (Sandberg and Wallengaard 1987; Dahlén 1988). Will pursue insects in the air, especially swarming flies (Diptera), 'in manner of shrike *Lanius*' (Stahlbaum 1967), and will catch cicadas *Cicada orni* in flight, though more usually pounces on them from perch on tree or ground (Patterson *et al.* 1991). Pulls open wasps' nests to get at larvae inside (Béthune 1986); when eating adult wasp, first crushes it in bill then bites repeatedly at tip of abdomen, though sting not removed; pulls large

insects apart while holding them under both feet, more rarely under one (Goodwin 1952, 1986). Pulls unripe acorns from twig with cup and stalk; loosens ripe ones from cup and swallows them whole if to be cached; prefers largest acorns which can still be swallowed for caching, larger ones carried in bill or eaten immediately (Bossema 1979). When eating acorn grips it in both feet then bites and levers (rarely hammers) until shell is pierced, rotating acorn continuously; pieces either torn off or rasped by lower mandible, holding head upside-down (Goodwin 1986). Recorded in Slovakia and Rumania feeding on beech *Fagus* nuts and apparently ignoring acorns; seen eating resinous conifer bark (Keve and Sterbetz 1968). In Israel, robs Syrian Woodpecker *Dendrocopos syriacus* of pecan *Carya* nuts after it has opened them as unable to do so alone (Mann and Hochberg 1982). Bird killed newly-fledged House Sparrow *Passer domesticus* by blows to head while holding it under foot; head and intestines discarded (Heuer 1986). Recorded killing adult *P. domesticus*, returning at same time next day to take another (Guex 1986). In 40-year study of 5280 fledged broods in a forest songbird population in eastern Germany, said to take up to 85% of newly-fledged young, though density of *G. glandarius* here artificially high due to supply of grain for gamebirds (Henze 1979). On several occasions seen to rush at newly-fledged tits (Paridae) with beating wings, trying to knock them to ground (C M Perrins). In northern Rumania took 16% of eggs and 5·5% of nestlings from 91 nests of Song Thrush *Turdus philomelos*, and 7·9% and 2·8% from 77 nests of Blackbird *T. merula*; most important predator of *T. philomelos* in this study (Korodi Gál 1968). Takes eggs and young from nest-boxes of Pied Flycatcher *Ficedula hypoleuca* in Wales (Playford 1985). See Kolbe (1982) for review of effect on songbird populations, and Vercautern (1984) for study in Belgium. Captive birds ignored small passerines that had got into their aviaries, attacking them only when they had well-grown nestlings to feed; wild *G. glandarius* mobbed by passerines will immediately enter nearest tree or bush to search for nests (Goodwin 1986). Once seen to eat conspecific bird *c.* 16 days old (Arff 1962). Recorded taking bats (Chiroptera) from tree-holes (Henze 1975). Bird seen to stall in flight above river, drop to surface, and take live fish in bill (Sharrock 1963); one bird returned over several days to low branch over river where it leaned over to snap up smooth newts *Triturus vulgaris* from water (Renglin 1975). Often at all kinds of carrion and food scraps (Vásárhelyi 1967; Görner 1971, which see for feeding technique at dead mouse *Apodemus*; Polivanov and Polivanova 1986); very partial to peanuts in gardens, hanging upside-down from mesh bags or manipulating feeders to get at them (Salfeld 1963; Goodwin 1986; Hughes 1986). For further details of diet and comparison of feeding ecology with Siberian Jay *Perisoreus infaustus* in southern Lake Baykal area (eastern Russia) see Durnev *et al.* (1991).

During September–October in central and western

Europe intense collection of acorns takes place, which are then cached, generally in ground. Nuts of beech and hazel *Corylus* stored to much lesser extent; see Turček and Kelso (1968) for list of seeds and nuts stored. Birds arrive at oak stands from all directions and carry acorns away to own home-range, often covering considerable distances. At one wood in eastern Germany, over 1000 birds per hr arrived, though around 250 individuals involved, taking *c.* 5 min to fill crop and *c.* 15 min to cache acorns, flying average 4 km to store; estimated 3000 kg of acorns removed by all birds in 20 days (Wadewitz 1976). In one year in south-east England, caching began in 1st week of September, increasing to peak in mid-October when *c.* 40 birds taking part at one site, probably whole local population as no birds observed in usual locations elsewhere; last seen transporting acorns in mid-November when apparently none left on trees. Activity occupied *c.* 10 hr per day at peak intensity, when average time to collect, carry, and store acorn load was *c.* 10 min, with average 1·2 km to cache; each bird stored *c.* 5000 acorns during autumn. (Chettleburgh 1952.) In another German study, each bird stored *c.* 4500, carrying them up to 4 km and travelling up to 175 km per day (Schuster 1950). In Poland, 6 km recorded between source and store (Turček and Kelso 1968). Acorns often carried to other stands of trees for burial; in Sweden (where birds cache grain before acorns ripe), taken 2–3 km to conifer forest (Swanberg 1969), and in Italy carried from oak scrub to pine wood (Patterson *et al.* 1991). Up to 9 acorns can be transported in gullet (90 large pine seeds, 15 beech nuts, or 10 hazel nuts), though 1–3 more usual with generally 1 in bill; heavier loads are usually taken to more distant caches (Schuster 1950; Chettleburgh 1952; Turček 1961; Bossema 1979). Often approaches storage site moving from tree to tree, probably to reduce detection (Schuster 1950; Chettleburgh 1952). Usual sites are under leaf litter, moss, in roots, etc., and natural holes are preferred, hardly any birds in English study digging own holes; caches sometimes in crevices in tree bark (Chettleburgh 1952; Bossema 1979; Goodwin 1986). Usually 1, sometimes 2 or more, acorns per store; pushed into soil at *c.* 45° and hammered a few times if still visible, hole then filled in with sideways movements of bill, and covered with leaves, stick, small stone, etc. (though covering not noted by Chettleburgh 1952); more rarely, closed bill pushed into ground, opened, and food item dropped. Caches are generally in open areas within woodland (clearings, slopes, edges, etc.), a tendency which helps oak to spread since many acorns are not retrieved (Bossema 1979, which see for other evidence of symbiotic relationship between oak and *G. glandarius*). In other places, caches recorded close to trees and bushes (Chettleburgh 1952). Visual memory, and cues from near and distant landmarks employed in finding stores (Bossema and Pot 1974; Bossema 1979). Most acorns eaten in winter and spring come from caches; bird generally goes directly to site and digs up acorn, even in 40 cm of snow, when digs at an angle with side-to-side bill movement, giving up after *c.* 1 min if unsuccessful; may re-bury at another site, especially if conspecifics around (Chettleburgh 1952; Swanberg 1969; Bossema 1979). According to Swanberg (1969) finds store in snow 'almost every time'; Bossema (1979, which see for retrieval experiments using captive birds) found success rate under snow in the wild of 16%, though actual rate 'probably much higher'. Will also cache peanuts and food scraps, scraps perhaps more often in trees where they will keep longer (Salfeld 1969; Goodwin 1986).

Diet in west Palearctic includes the following. Invertebrates: damsel flies and dragonflies (Odonata: Aeshnidae, Gomphidae, Libellulidae), grasshoppers, etc., (Orthoptera: Gryllidae, Tettigoniidae, Gryllotalpidae, Acrididae, Tetrigidae), earwigs (Dermaptera: Forficulidae), bugs (Hemiptera: Pentatomidae, Reduviidae, Cicadidae), adult and larval Lepidoptera (Lycaenidae, Pieridae, Hepialidae, Tortricidae, Tineidae, Notodontidae, Arctiidae, Noctuidae, Lymantriidae, Lasiocampidae, Sphingidae, Geometridae, Papilionidae), adult and larval flies (Diptera: Tipulidae, Bibionidae, Stratiomyidae, Tabanidae, Empididae, Asilidae, Syrphidae, Otitidae, Oestridae, Tachinidae, Calliphoridae, Muscidae), Hymenoptera (Siricidae, Tenthredinidae, Diprionidae, Cynipidae, Ichneumonidae, ants Formicidae, wasps Vespidae, bees Apoidea), beetles (Coleoptera: Cicindelidae, Carabidae, Dytiscidae, Histeridae, Silphidae, Staphylinidae, Lucanidae, Geotrupidae, Scarabaeidae, Elateridae, Cantharidae, Meloidae, Cerambycidae, Chrysomelidae, Curculionidae), spiders (Araneae: Theridiidae, Araneidae, Tetragnathidae, Lycosidae, Pisauridae, Clubionidae, Thomisidae, Salticidae), harvestmen (Opiliones: Phalangiidae), woodlice (Isopoda: Asellota, Oniscidae, Porcellionidae, Armadillidiidae), millipedes (Diplopoda: Julidae), centipedes (Chilopoda: Lithobiidae), earthworms (Lumbricidae), snails (Gastropoda: Arionidae, Limacidae, Helicidae). Vertebrates: bat (Chiroptera), hazel dormouse *Muscardinus avellanarius*, mouse *Mus*, yellow-necked mouse *Apodemus flavicollis*, vole *Arvicola*; fish (Pisces); frogs *Rana temporaria*, *Hyla arborea*, smooth newt *Triturus vulgaris*; lizards *Lacerta agilis*, *L. vivipara*, slow worm *Anguis fragilis*, grass snake *Natrix*, adder *Vipera berus*; adults, nestlings, and eggs of many birds, great majority songbirds but including eggs of Sparrowhawk *Accipiter nisus*, Phasianidae, and Columbidae, and nestlings of Phasianidae, probable Turtle Dove *Streptopelia turtur*, and Red-backed Shrike *Lanius collurio*. Plant material: seeds and fruits of pine *Pinus*, spruce *Picea*, yew *Taxus*, juniper *Juniperus*, oak *Quercus*, beech *Fagus*, hornbeam *Carpinus*, maple and sycamore *Acer*, birch *Betula*, whitebeam and rowan *Sorbus*, chestnut *Castanea*, horse chestnut *Aesculus*, hazel *Corylus*, walnut *Juglans*, pecan *Carya*, lime *Tilia*, willow *Salix*, elder *Sambucus*, olive *Olea*, strawberry tree *Arbutus*, spindle *Euonymus*, dogwood *Cornus*, holly *Ilex*, ivy *Hedera*, mistletoe *Viscum*,

*Loranthus*, hawthorn *Crataegus*, mulberry *Morus*, privet *Ligustrum*, buckthorn *Rhamnus*, sea-buckthorn *Hippophae*, grape *Vitis*, currant and gooseberry *Ribes*, rose *Rosa*, raspberry and bramble *Rubus*, cherry *Prunus*, pear *Pyrus*, apple *Malus*, strawberry *Fragaria*, mespil *Amelanchier*, quince *Cydonia*, *Cotoneaster*, crowberry *Empetrum*, various *Vaccinium* species, barberry *Berberis*, Oregon grape *Mahonia*, locust *Gleditsia*, *Robinia*, *Sophora*, jujube *Zizyphus*, oleaster *Elaeagnus*, *Catalpa*, honeysuckle *Lonicera*, spurge laurel *Daphne*, *Clematis*, *Viburnum*, bittersweet and potato *Solanum*, pea *Pisum*, broad bean *Vicia*, *Asparagus*, maize *Zea*, wheat *Triticum*, barley *Hordeum*, rye *Secale*, fungus *Boletus*. (Jäckel 1891; Newstead 1908; Baer 1910; Rey 1910; Fuye 1911; Csiki 1913; Madon 1928a; Collinge 1930; Schuster 1930; Ssokolow 1932; Campbell 1936; Jenner 1947; Wortelaers 1950; Kovačević and Danon 1952, 1959; Kadochnikov and Eygelis 1954; Gebhardt 1955; Owen 1956; Szemere 1957; Neufeldt 1961; Turček 1961; Arff 1962; Bossema 1967, 1979; Holyoak 1968; Keve and Sterbetz 1968; Korodi Gál 1968, 1972; Trommer 1971; Henze 1975, 1979; Renglin 1975; Vieweg 1981; Kolbe 1982; Mann and Hochberg 1982; Vercauteren 1984; John and Roskell 1985; Guex 1986; Polivanov and Polivanova 1986; Guitián 1987, 1989; Snow and Snow 1988; E K Dunn.)

More plant material eaten in winter, particularly cached acorns, though where these absent will store beech nuts or conifer seeds (Fuye 1911; Keve and Sterbetz 1968; Bossema 1979). In winter, Czechoslovakia, plant material 93·6% by weight of diet, $n = 43$ stomachs (Hell and Soviš 1958); in Hungary, present in 85% of 655 stomachs November–February, 27% of 769 March–October (Keve and Sterbetz 1968, which see for comprehensive comparison of all food studies); over the year in England, 28·5% by volume, sample size unknown (Collinge 1930). In warmer habitats seems to take less plant material (Kovačević and Danon 1952; Keve 1985). For diet of adult, see Tables A–B. In study of 301 stomachs in Netherlands by Bossema (1979), few invertebrates taken in winter but increased from February and present in 100% of stomachs in May, remaining at *c.* 85% until November then declining sharply; invertebrates taken according to abundance, especially caterpillars, great majority oak-dwelling species. During December–March, only invertebrates found in samples were earwigs, beetles, and spiders; in April, beetles 82% by number and caterpillars 18% ($n = 22$ invertebrates), increasing to 61% (including a few adult Lepidoptera) in June ($n = 216$); Hymenoptera appeared in diet in May with 16% by number ($n = 784$) rising to maximum of 68% in August ($n = 25$), while Diptera made up 46% of invertebrates in July ($n = 45$). Proportion of stomachs containing seeds and fruits (excluding cereals and acorns) rose from 0 in April to 50% in November, though proportion by dry weight was highest in June at 35%, only 1–7% in all other months; succulent fruits comprised most plant material of this sort in summer and

Table A   Food of adult Jay *Garrulus glandarius* from stomach analysis; figures are percentages, animal and plant material treated separately.

| | Hungary | | St Petersburg region (Russia), Aug | Netherlands, all year |
|---|---|---|---|---|
| | Mar–Oct | Nov–Feb | | |
| *Animal material (% by no.)* | | | | |
| Vertebrates | 3·7 | 1·7 | 0·8 | 1·6 |
| (mammals) | (3·7) | (1·7) | (0·8) | (0·7) |
| (birds) | (0) | (0) | (0) | (0·6) |
| Odonata | 0·8 | 0·9 | — | 0 |
| Orthoptera | 5·7 | 17·9 | 2·3 | 0 |
| (Gryllidae) | (3·5) | (17·0) | — | (0) |
| Dermaptera | 1·8 | 21·2 | — | 8·0 |
| Hemiptera | 7·8 | 1·7 | fragments | 0 |
| (Pentatomidae) | (5·3) | (1·7) | — | (0) |
| Lepidoptera | 1·3 | — | 6·3 | 44·8 |
| Diptera | 0 | — | — | 7·1 |
| Hymenoptera | 8·9 | 12·7 | 28·1 | 18·5 |
| (Vespidae) | (5·9) | (7·6) | (3·9) | — |
| (Formicidae) | (2·9) | (5·1) | (19·5) | — |
| Coleoptera | 69·0 | 40·7 | 47·7 | 12·3 |
| (Scarabaeidae) | (21·6) | (4·2) | (2·3) | — |
| (Carabidae) | (21·7) | (6·8) | (25·0) | — |
| (Curculionidae) | (16·2) | (20·3) | (13·3) | — |
| Spiders | 0·8 | — | 3·9 | 7·2 |
| Total no. of items | 869 | 118 | 128 | 1419 |
| *Plant material* | *(% of total occurrences)* | | *(% by number)* | *(% by dry weight)* |
| Acorns | 36·7 | 31·1 | 0 | 71·2 |
| Cereals | 12·1 | 31·8 | 2·5 | 22·2 |
| Fruits and seeds | 8·2 | 3·7 | 97·5 | 5·0 |
| Scraps | 43·1 | 30·6 | 0 | — |
| Other | 0 | 2·8 | 0 | 1·6 |
| Sample | 207 occurrences | 559 occurrences | 1348 items | 390 stomachs |
| Source | Keve and Sterbetz (1968) | | Ssokolow (1932) | Bossema (1979) |

autumn, beech nuts in winter. Consumption of cereal grain highest July–August, when 90% by weight of plant material in 80% of stomachs. Acorns present in stomachs all year: in 14–100% of stomachs in different months, making up 9–98% by weight of plant material; from November to March 60–96% by weight, all from caches. Vertebrate prey absent July–November, mostly small mammals December–May, and mainly nestlings and eggs May–July. Of 74 stomachs from all of England (only 14 for period July–December), 95% from September–February contained acorns, declining quickly to 0 in July–August; beetles present all year, in 14–100% of stomachs in different months; grain in 6–75%, absent only September–October (Holyoak 1968). For data from northern France (75 stomachs), see Fuye (1911). Average daily consumption in the wild *c.* 400–500 *Picea* seeds; 1 captive

Table B  Food of adult Jay *Garrulus glandarius* from stomach analysis. Figures are percentages by volume, animal and plant material treated together.

| | England, all year | SE England May–Nov |
|---|---|---|
| Vertebrates | 21·0 | 7·2 |
| (mammals) | (9·5) | (0) |
| (birds and eggs) | (6·5) | (7·2) |
| Lepidoptera (almost all larvae) | — | 14·4 |
| Hymenoptera (Cynipidae galls) | — | 4·7 |
| Other insects (including spiders) | 38·5 | 1·6 |
| Snails | 3·5 | 0 |
| Earthworms | 8·5 | 0 |
| Acorns | — | 50·3 |
| Cereals and peas | 4·5 | 10·2 |
| Other fruits and seeds | 22·0 | 11·6 |
| Other | 2·0 | 0 |
| Sample | unknown | 23 stomachs, unknown no. of items |
| Source | Collinge (1930) | Campbell (1936) |

Table C  Food of nestling Jay *Garrulus glandarius* from collar-samples. Figures are percentages by number.

| | S England | N Rumania | SW USSR | SW USSR |
|---|---|---|---|---|
| Vertebrates | 0·08 | 0·2 | <1·0 | 5·1 |
| Odonata | 0·1 | 0·02 | 0·2 | 5·1 |
| Orthoptera | 0 | 0·1 | 1·5 | 0 |
| Dermaptera | 0·1 | 0 | 0 | 0 |
| Hemiptera | 0·4 | 0 | 0·2 | 0 |
| Lepidoptera | 86·6 | 60·0 | 70·4 | 68·8 |
| | (0·1 adults) | (0·1 adults) | (0·4 adults) | |
| (Tortricidae) | (57·0) | — | (7·5) | (0) |
| (Geometridae) | (20·0) | (41·6) | (5·5) | (24·2) |
| (Noctuidae) | (8·6) | (3·1) | (2·3) | (44·6) |
| (Notodontidae) | (0·4) | (11·3) | — | (0) |
| (Lasiocampidae) | (0·04) | (1·5) | (35·7) | (0) |
| (Lymantriidae) | (0) | (0·6) | (10·6) | (0) |
| Diptera | 1·9 | 1·3 | 1·5 | 1·9 |
| (Empididae) | (1·0) | — | — | (0) |
| (Tipulidae) | — | — | — | (1·9) |
| Hymenoptera | 0·6 | 0·5 | 0·9 | 0 |
| (Formicidae) | (0) | (0·4) | (0·3) | (0) |
| Coleoptera | 6·0 | 8·5 | 2·7 | 15·3 |
| (Scarabaeidae) | (2·0) | (1·3) | (0·8) | (5·1) |
| (Carabidae) | (0·4) | (1·4) | (0·9) | (1·3) |
| (Tenebrionidae) | (0) | (2·5) | — | (0) |
| (Elateridae) | (0) | (1·5) | (0·1) | (2·5) |
| Spiders | 2·7 | 27·2 | 20·7 | 3·8 |
| Snails | 1·6 | 0·8 | 1·5 | 0 |
| Total no. of items | 2566 | 11 016 | 913 | 157 |
| Source | Owen (1956) | Korodi Gál (1972) | Kadochnikov and Eygelis (1954) | Polivanov and Polivanova (1986) |

bird ate 17 acorns per day with total weight of 35 g, 24% of body weight (Turček 1961).

Diet of nestlings principally caterpillars (Table C), but if these become unavailable chicks of other birds are sometimes brought as food (Goodwin 1951). In Netherlands, at 8–16 days, 95·5% of diet by volume invertebrates, 4% plant material, and 0·5% vertebrates (eggs), $n = 17$ collar-samples; at 16–24 days, respectively 74%, 6%, and 20% (nestlings), $n = 31$; in May–June diet very like adults', but no Hymenoptera, though later in season fledglings consumed proportionately more acorns and nestlings than adults, most acorns in June probably coming from pulled-up seedlings. Of caterpillars fed to nestlings, 23% (by number) *Erannis leucophaearia*, 13·6% *E. aurentiaria*, 20·5% *Orthosia stabilis*, and c. 13% *Tortrix viridana*, all oak-dwelling species. (Bossema 1967, 1979.) In southern England, caterpillars from oak were 83% of diet and commonest beetle was cockchafer *Melolontha melolontha*, an important element since large size means equivalent of 10 caterpillars (wing-cases, legs, and head removed before feeding); only 0·6% of 2583 items were fragments of acorn which was only plant material in diet (Owen 1956). In northern Rumania, average consumption at 1–5 days was 7 g per day (17% of body weight, 71% of which used for growth; $n = 76$ birds); at 21–22 days, 77 g per day (60% and 3·0%; $n = 52$); overall average consumption 43·4 g per day (41% and 25·1%; $n = 748$); in this study plant material (acorns) accounted for 0·5% by number of 11 499 food items (Korodi Gál 1972). In Voronezh region of south European Russia, plant material was 0·1% of 914 items; small pieces of vertebrate meat fed without bones (Kadochnikov and Eygelis 1954). In this study total average intake over 16 days was 320 g; maximum amount brought per visit by adult was 14 g and maximum fed to one nestling 4·2 g, $n = 790$ collar-samples (Eygelis 1965, which see for average daily consumption during this period). Adult bird fills crop with prey items before returning to nest to feed young; feeding is by regurgitation, bill of adult being inserted deep into nestling's throat (Owen 1956; Goodwin 1986).                    BH

**Social pattern and behaviour.** Extensive studies in southern England by Goodwin (1949, 1951, 1952*a*, 1956), mostly on captive tame birds, but much also applicable to the wild. For reviews see Coombs (1978), Keve (1985), and, especially, Goodwin (1986).

1. Breeding pairs dispersed at low density on home-ranges throughout the year; in contrast to (e.g.) Magpie *Pica pica* and Carrion Crow *Corvus corone*, non-breeders do not typically flock (Bossema 1979; Keve 1985; Goodwin 1986; Grahn 1990), although immatures tend to do so in Israel, June–September, e.g. 25 seen together (Shirihai in press). As in Nutcracker *Nucifraga caryocatactes*, winter territoriality seems to be weak or non-existent, but in spring residents become territorial and no longer tolerate 1-year-olds which had coexisted on their home-ranges outside breeding season (these perhaps not offspring, which typically leave parental territory in September: Bossema 1979); 1-year-olds respond by dispersing to seek breeding vacancies

elsewhere, becoming 'floaters' if they fail to settle. Average size of home-range outside breeding season 12·7 ha ($n = 24$). (Grahn 1990.) In Toscana (northern Italy) dispersion outside breeding season, and perhaps even social system, clearly different from northern Europe: radio-tracking indicated that individuals (status unknown) shifted home-range c. 1–2 km from summer (pine *Pinus* woods) to autumn (scrub and olive *Olea*), so year-round area used is likely to be much larger than seasonal home-ranges found by tracking 2 individuals (varied from exceptionally large 359 ha in July to 29·6 ha in October). Overall densities extremely high in Toscana: mean 484 birds per km² (327–676) in July, 584 (544–612) in late October and early November. (Patterson *et al.* 1991.) Throughout much of range, dispersion outside breeding season complicated by lengthy flights to collect acorns for hoarding (see Food), e.g. birds in conifer woods seen flying up to 18 km to collect acorns (Wadewitz 1976). In occasional years, failure of acorn crop results in mass evacuation of breeding areas in autumn (see Movements) and sometimes large eruptive flocks. During irruptions, birds can pass in steady stream of small groups, e.g. in Plymouth area (south-west England), October, c. 1000 passed through in c. 75 min, in parties of 3–10(–600) (John and Roskell 1985); for similar passage in northern Germany, see Sanden (1956). Movements rarely associated with other Corvidae, but flocks in British irruptions have exceptionally included *N. caryocatactes* (Mead 1983). For flocking of *G. glandarius* in spring, see Flock Behaviour (below). BONDS. Mating system monogamous, pair-bond lifelong (M Grahn). In Wytham (Oxford, England), probable case of bigamy (1 ♂, 2 ♀♀) (M Cellier), and further study needed in light of exceptionally close nesting recorded in Buckinghamshire (see Breeding Dispersion, below). Both members of pair build nest. Incubation by ♀ only; brooding by ♀, though occasional records (e.g. Tutt 1952) of ♂ brooding for short periods. Both sexes feed young, at first mostly with food brought by ♂. Latter stages of post-fledging care, at least in captivity, mainly by ♂, and frequent occurrence in the wild of family parties containing only 1 parent perhaps suggests ♂'s parental role typically lasts longer than ♀'s. (Goodwin 1951, 1986.) Young largely independent at 7–8 weeks (i.e. 4–5 weeks after leaving nest) though may remain in parental territory for 1 month further (Bossema 1979). Age of first breeding typically 2 years; competition for space excludes all but a few 1-year-olds from breeding (M Grahn). In Wytham, minority of 1-year-old ♀♀ breed, but no 1-year-old ♂♂ (Cellier 1992). BREEDING DISPERSION. Typically solitary · and territorial, although evidence from one study (Buckinghamshire, England) of clumped nesting, with active nests as little as 35 m apart (R J Fuller). In Wytham, assuming territories do not overlap, mean maximum size of breeding territory 6·7 ha (Cellier 1992), in southern Sweden 14 ha ($n = 24$), not significantly larger than in winter; although territory size often changed markedly from year to year (according to density of breeders) residents highly site-faithful and none known to change location (core) of home-range more than a few hundred metres in consecutive years (Grahn 1990). Territory serves for courtship, breeding, and feeding, but birds may also leave territory to join Spring Gatherings (see below) and to collect food (although all food-caching by breeders is inside territory) (Bossema 1979; Goodwin 1951, 1986; M Grahn). Resident pairs build new nest each year, and survival of old nests from year to year (Eygelis 1970) means caution needed to establish true density of pairs. Kolbe (1982) gave comprehensive review of density based on 71 central European studies where study area at least 9 ha (smaller areas bias results): average density 9·4 pairs per km² (0·4–27·7), rather little variation with habitat. Other densities as follows: in Britain, 4·2 pairs per km² on Suffolk farmland (Benson and Williamson 1972), c. 15

pairs per km² in mixed deciduous woodland (M Cellier). In Åland (Finland) and central Sweden, not more than 3 pairs per km² overall (Palmgren 1930; Ulfstrand and Hogstedt 1976). In Voronezh (south European Russia), 8–12 pairs per km², occupied nests 200–250 m apart (Eygelis 1970). In Bulgaria, maximum 50 pairs per km² in stands dominated by lime *Tilia*, c. 10–20 in other woodland (Simeonov 1975; Simeonov and Petrov 1977; Petrov 1988). In Vosges du Nord (France), 2 pairs per km² in Scots pine *P. sylvestris*, 4 in beech *Fagus sylvatica* (Muller 1987); in Bourgogne, 0·5–8 pairs per km² in shelterwood, 8–11 in coppice-with-standards (Ferry and Frochot 1970). In Provence (France), 6 pairs per km², in Corsica 10 pairs per km² (Blondel 1979). In mixed woodland in Toscana, 5 territories per km² (Lambertini 1981). In north-west Spain, 3 pairs per km² (Carrascal 1987). In Morocco, 6 pairs per km² in semi-arid *Tetraclinis* scrub, slightly fewer in maquis and woodland (Thévenot 1982). ROOSTING. More information needed. Typically in territory but no details of sites used. During breeding season ♀ sometimes roosts on nest at night before laying 1st egg. Captive young recently out of nest often roosted in contact with one another, this ceasing when they became fully independent. (Goodwin 1956, 1986.) In Toscana in July, birds left pinewood feeding grounds to roost in scrub on limestone ridge, with pronounced roosting flights from just after sunset to 90 min after sunset; morning exodus apparently from sunrise to c. 90 min after (Patterson *et al.* 1991). Temporary communal woodland roost (no details of spacing between individuals) in mid-October in Midlothian (Scotland) held c. 320 birds, though dwindled rapidly—empty 3 days after discovery (Young 1984; see also Flock Behaviour, below). For details of anting behaviour, see Goodwin (1951, 1952a).

2. Timid and wary of man where persecuted but readily becomes tame or indifferent in towns where not molested; less suspicious than *Corvus* of strange or new objects and thus easily caught in traps. For behaviour of wild and hand-reared birds when handled, see Goodwin (1956), also 8 in Voice. Parties crossing open ground always go one by one at intervals (Goodwin 1951). On point of fleeing, sleeks most plumage but raises crest and belly feathers. Flight-intention movements consist of flicking tail and (at high intensity) closed wings above line of back, especially vigorous during confrontations (see Coombs 1978 for drawing): Tail-flicking seems to consist of quick downward flick, then upward movement in which feathers partly spread then closed again. If perched while mobbing predator, swings head and body quickly downwards and sideways with strong bending of legs and upward swing of tail; then jerks body upright again with downward flick of tail (Fig A: Goodwin 1956). On first catching sight of intimidating bird, no uncommonly mimics its call, e.g. Grey Heron *Ardea cinerea*, Goshawk

A

*Accipiter gentilis* ('kikikiki. . .'), Tawny Owl *Strix aluco* (hooting or 'ke-wick' call), *C. corone* (Goodwin 1951, 1956, 1986; Paquet 1979). Of 7 *G. glandarius*, 2 gave 'ke-wick' call before venturing closer to stuffed *S. aluco* (Löhrl 1980). For interpretation of such mimicry, see Goodwin (1956), Crombrugghe (1980), and Löhrl (1980). For mimicking calls of various predatory birds, or alarm-calls of non-predatory birds, when mobbing humans at nest, see Parental Anti-predator Strategies (below) and Voice. On spotting flying Sparrowhawk *Accipiter nisus*, Kestrel *Falco tinnunculus*, or Hobby *F. subbuteo*, tame *G. glandarius* typically froze in crouched posture, watched raptor intently, and gave Hawk-alarm call (Goodwin 1951: see 7d in Voice). Response to hawk-alarm calls of other passerines varies from freezing to diving into cover or giving Screech-calls (7a in Voice). If attacked by *Accipiter*, dashes from branch to branch within tree or bush, ducking, dodging, and sometimes giving low-intensity Grating-call (7c in Voice); fleeing thus similar to 'play' behaviour, seen mostly in young birds. However, perched *A. nisus* may be violently and noisily mobbed (see 7a, c in Voice), and even struck, sometimes collectively by (e.g.) 4-5 birds. (Goodwin 1949, 1951, which see for details of mobbing and other alarm-responses to various other birds and mammals; see also Parental Anti-predator Strategies, below.) FLOCK BEHAVIOUR. Small parties never compact, and even pair-members in flight usually separated by several metres (Goodwin 1986). Some irruptive flocks (including one of 500-600) in Britain, however, quite compact and members seem to call continuously (no details); at Porthgwarra (Cornwall), where *c.* 200 present, flocks flew up, climbing for several hundred metres as if heading out to sea, but always returned to land, often 'whiffling' down like Curlews *Numenius arquata* (John and Roskell 1985). In exceptional pre-roost assembly in October, Scotland (see Roosting), small groups of up to 10 or more flew in to join flock in field until 320 assembled; these very active and vocal, eventually all flying off simultaneously to roost-site, in contrast with morning exodus which was protracted dispersal of single birds (Young 1984). Communal display known as Spring Gathering first described by Geyr von Schweppenburg (1939) and subsequently by Goodwin (1951, 1986), Charvoz (1953), and Kipp (1978, which see for comparison with gatherings in other Corvidae). Following description based mostly on Goodwin (1951, 1986). Spring Gatherings (hereafter 'Gatherings') last up to 20-30 min and occur at no set time or place, though mostly on fine mornings from late February until late April or early May (Kipp 1978), sometimes even mid-May to early June if one or more nests in an area have recently failed. 3-20 or more birds may participate; some smaller Gatherings seem to consist of 1 ♀ attended by 2-3 ♂♂, but larger ones probably include more than 1 ♀ (Charvoz 1953; Kipp 1978). Although there are frequent short lulls in activity and although some birds are apparently passive spectators, main impression of Gathering is of intense excitement and noise; participants, often including several apparent pairs, may be spread over several trees and continually change position over wide area, one bird taking flight, the rest following. Display of perched birds is accompanied by great diversity of vocalizations (see Goodwin 1951, Charvoz 1953, Kipp 1978, and Voice) including mimicry, but most notably Flight-appeal call and Kraah-call (see 2d and 4 in Voice). Kraah-calls given mainly by ♀♀, perched in Advertising-posture (see Fig D, and Antagonistic Behaviour, below). Gathering apparently starts with single bird giving Kraah-calls (Hudson 1915; Goodwin 1986), inducing others (apparently ♂♂) to pursue giving Flight-appeal calls; followers display in leader's vicinity wherever it alights, including sometimes mutual Advertising-display between apparent ♂ and ♀. Brief fights not uncommon in southern England, although none seen in German

study by Kipp (1978). Gatherings hard to interpret but thought mainly to bring together unpaired birds (though not necessarily a prerequisite of pairing: Kipp 1978); thus when birds in aviary detected nearby Gathering in the wild, only unpaired captives tried to escape, ♂♂ with Flight-appeal call, ♀♀ with Kraah-call; moreover, Gatherings can be stimulated by killing a member of an established pair (Raspail 1901). Gatherings might also mediate readjustment of territorial ownership or boundaries (Goodwin 1986: see *P. pica* for similar interpretation of Ceremonial Gatherings in thta species). SONG-DISPLAY. Song (see 1a in Voice) heard mostly from unpaired 1st-year birds (sometimes older), usually when alone and at ease, e.g. sitting quietly or foraging casually (Goodwin 1951, 1952a, 1986). In captivity, ♀♀ also sing, though less persistently than ♂♂ (Goodwin 1952a). In the wild, singing heard from late January (Desfayes 1951), more usually late March and April when apparently directed by ♂♂ at unpaired ♀♀ (Lebeurier and Rapine 1939: see Heterosexual Behaviour, below, and 1b in Voice). ANTAGONISTIC BEHAVIOUR. (1) General. Intraspecific aggression apparently not marked in the wild, but varies seasonally. Dominance hierarchy evident whenever 2 or more birds kept in same aviary. Usually ♂♂ dominate ♀♀, and adults dominate juveniles. (Goodwin 1951.) Pair dominant in their territory although conspecific intruders not always expelled (see introduction to part 1, above). In early part of breeding season, intolerance increases, especially in nest-area. (Bossema 1979.) Intrusions by breeders in neighbouring territories are reduced at this time, and resistance to non-breeders even greater (Grahn 1990, M Grahn). Later, in post-fledging period, breeding adults gradually again become more tolerant (Bossema 1979). (2) Threat and fighting. Following account based on Goodwin (1949, 1951, 1952a, 1956, 1986). Display varies considerably with degree of excitement, individual involved, and its sex. Dominant bird typically supplants subordinate (e.g. at food source) as follows: hops up to subordinate, stretches head slightly towards it, and Bill-snaps, usually inducing immediate retreat. With intransigence, however, threat may develop into attack, or threatening bird may somewhat retract its head and partially erect its plumage (low-intensity Defensive-threat display). In Defensive-threat, bird faces rival, ruffles body plumage (including conspicuously its crest), slightly spreads and droops wings, and may slightly spread tail (Fig B); at highest

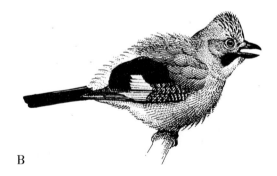

B

intensity, spreads wings widely. Defensive-threat used by either sex mainly in defence of nest or young and may precede attack accompanied by Grating-call, Screech-call, or Panting-call (see 7b in Voice). Advertising-display similar to Defensive-threat but differs in that some parts of plumage (rump, belly, flanks) are ruffled and others sleeked, and performer (especially ♂) usually orientates itself side-on to object of threat (hence 'Lateral display' in Goodwin 1952a); wings are drooped somewhat to

C

accentuate bold patterning. In display of ♂ (Fig C), higher crest indicates greater attack-flee conflict: performer reaches forward (Fig C, right, with crest lowered) and gives series ('display phrase') of Hissing-or mimicked calls (see 5 in Voice). If on ground, may totter a few steps, often tilting itself towards threatened bird. Advertising-display of ♀ similar but less exaggerated: mantle somewhat ruffled, crest seldom so, and, as performer reaches forward and calls (usually Castanet-call: see 6a in Voice), she moves from upright posture (Fig D, left) to head-down (Fig D, right). For use of Advertising-display in sexual

D

context, see Heterosexual Behaviour (below). In boundary disputes between ♂♂, and other cases of 'thwarted aggression' (including between mates), display by either sex accompanied variously by bill-wiping, pecking, tugging, or tearing at ground, leaves, twigs, etc. Territorial disputes rarely fierce: in Swedish study, rivals seen fighting over vacant territory at start of breeding season, rolling on ground 'like cats' (M Grahn). For appeasement between mates, see Heterosexual Behaviour (below). For mobbing of other birds, see above and also Parental Anti-predator Strategies (below). Subordinate in disputes with *C. corone* and *P. pica* (Bossema 1979). HETEROSEXUAL BEHAVIOUR. (1) General. Except where stated, following from Goodwin (1949, 1951, 1952a, 1956, 1986). (2) Pair-bonding behaviour. No precise information on when pairing takes place or sequence of events in birds pairing anew, but see Flock Behaviour for likely pairing display during Spring Gatherings. In Voronezh, flock-display (presumably Spring Gatherings) began in January, but pair-display not seen until beginning of March (Eygelis 1970). Display between mates or prospective mates well studied, especially in captive birds. Apart from frequent pursuit-flights, the following more ritualized displays occur. Particularly in late winter and early spring, ♀ commonly greets mate with Advertising-display (as described above). ♂ also displays thus to mate,

especially in presence of potential ♂-rivals. In heightened sexual mood, singing ♂ (see 1b in Voice) hops around ♀; such display, seen in both captive and wild birds, may precede Courtship-feeding and/or copulation (see below). Whether ♂ or ♀ is performer, Advertising-display often occurs during or immediately after hostile encounters with rivals when it may serve to reinforce pair-bond. ♀ also responds to ♂'s song with Chirruping-calls (3 in Voice) and other sounds, and often appears to greet him with Chin-up display (Fig E): makes typically excited jerking

E

movements of body and tail and tends to lift head and jerk it slightly from side to side, thus presenting her white throat (white feathers sleeked in contrast to ruffled black malar stripes). Chirruping-calls, accompanied by slight jerking of body and tail, often used by either mate to appease other; pairs also frequently hold duets of Chirruping-calls. In captivity, throughout the year, ♂♂ will also signal appeasement to keeper (and in one case to dominant mate) with Submissive-display (see subsection 4, below); not known if this display used in context of appeasing dominant in the wild. (3) Courtship-feeding. Mainly from Goodwin (1951, 1956, 1986). From before nest-building through to brooding, ♂ feeds ♀ (in captivity, reverse also occasionally occurs). ♂ usually proffers ♀ mandibulated food which he holds in his mouth or gullet, tilting his head sideways to let ♀ remove it with her bill at right angles to his. Most often, food transfer is simple exchange accompanied by Food-offering call (see 2b in Voice) from both birds; no other special display or preliminaries, but feathers of rump and malar stripes usually somewhat erect. Either sex may initiate, but mostly ♂. When ♂ takes initiative, it is sometimes culmination of hopping approach through branches to ♀ in which he performs Advertising-display and sings (see subsection 2, above). Before nest-building, ♂ may display thus but withhold food or relinquish only some of it at a time; this seen in both captive and wild birds. Before nest-building, ♀ sometimes begs with low-intensity Submissive-display (for full display see subsection 4, below) in which wings only slightly extended and not necessarily quivered. Especially when ovulating, laying, or incubating, ♀ also often solicits and/or receives food with juvenile-type flapping or fluttering of wings and Food-begging-call (see 2c in Voice). Courtship-feeding seen in the wild from February onwards. ♀ almost entirely dependent on ♂ for food during incubation and 1st week of nestling care (Bossema 1979). In the wild, ♂ fed ♀ on nest about every 2 hrs, although ♀ also occasionally left nest to feed herself or sometimes to be fed by ♂. (4) Mating and Mate-guarding. ♀ solicits with Submissive-display which varies in intensity. In full display (Fig F), adopts rather horizontal posture, slightly ruffles rump and raises tail, and jerks and quivers them violently; wings spread, slightly arched, and quivered. At lower intensities (e.g. Courtship-feeding), wings less extended, and quivered less or not at all, and quivering then restricted to tail (see Goodwin 1956 for drawings of various intensities). Copulation or attempted

F

copulation may follow courtship sequence in which ♂ makes hopping, singing approach in Advertising-display, and may then feed ♀ before mounting. Sequence in one captive pair as follows (Goodwin 1949): ♀ in mild Advertising-display was watched by nearby ♂; ♀ suddenly switched to intense Submissive-display with body tilted forward and wings outstretched so as to present them somewhat frontally; ♂ then flew onto her back and copulated with fluttering wings. ♀ maintained wings-spread posture throughout. In Swedish study, intensity of Mate-guarding, as measured by average distance between pair-members, increased markedly between pre-breeding period (January to late April) and breeding season (late April to July); egg-laying preceded by 3–9 days of intense guarding (Fröding 1987; M Grahn). (5) Nest-site selection. ♂ tends to take initiative but information scanty and mostly from captive birds. Before building, ♂ often enters suitable cover and gives Nest-call (Goodwin 1956: see 2e in Voice); according to Goodwin (1986) either sex may call thus on finding suitable site. ♂ in the wild seen Nest-calling from entrance to tree-hole (where pair subsequently nested), looking towards his mate and slightly fluttering his wings; ♀ sometimes attracted by this invitation, sometimes not (Goodwin 1956). (6) Behaviour at nest. ♂ approaching nest to feed ♀ does so with great caution. ♀ leaves nest every *c.* 3 hrs for periods of *c.* 5–15 min, rarely longer. Near end of incubation stint, ♀ commonly fidgets and probes nest-lining. (Goodwin 1956, 1986.) Goodwin (1951) suggested that some sudden losses of eggs and newly hatched young were attributable to predation by conspecifics, but subsequently (Goodwin 1956) considered that, at least in the case of eggs, owners of nest more likely responsible. RELATIONS WITHIN FAMILY GROUP. Based mainly on Goodwin (1951, 1956, 1986) for captive rearing of nestlings. After hatching, ♀ (both wild and captive) eats eggshells and cleans plumage of young if soiled. Both sexes feed young, at first mostly with food brought by ♂. ♂ regurgitates food, typically giving some to ♀ (1st week of chick-rearing) and both feed brood together. Rarely ♂ refuses to share food with mate and feeds young himself. ♂ or ♀ induces young to gape with Food-offering call and inserts bill deeply into nestling's throat before regurgitating. Both sexes perform nest-sanitation; at least after 1st week, ♂ typically discards faecal sacs away from nest but ♀ evidently eats those she collects, at least during brooding period. In the wild, probing in and pulling at nest-lining follows apparently unsuccessful attempts to catch and eat ectoparasites. For behavioural development of young, see Goodwin (1951). Leave nest at 21–23 days, initially capable only of rudimentary flight, but well fledged after *c.* 1 week out of nest and gradually start to follow parents (Bossema 1979). According to Goodwin (1951), young appear to start following when parents start neglecting them. At this stage young still fully dependent, but largely self-feeding *c.* 4–5 weeks after leaving nest (Bossema 1979). For details of transition to inde-

pendence, see Goodwin (1951). In 2 captive pairs, ♀ stopped feeding young *c.* 1 week before ♂, and this difference perhaps common, even in the wild. ANTI-PREDATOR RESPONSES OF YOUNG. After 9–12 days, young disturbed at nest cower and gape with crests raised. From day 9, though usually later, express extreme alarm by calling noisily (see Voice) which immediately summons parents. Young will jump out of nest as early as 17 days (i.e. 4–6 days early) if badly frightened. (Goodwin 1951, 1986.) Juveniles defend themselves with hissing sounds and Bill-snapping (Bergmann and Helb 1982: see Voice). PARENTAL ANTI-PREDATOR STRATEGIES. (1) Passive measures. ♀ disturbed on nest crouches low, with plumage sleeked, and bill slightly open. Usually flees to cover, silently or with screeching if (e.g.) man continues approach (but see below for rarer hostility). Some sit tight, not flushing until nest-tree climbed, but exceptionally flush at *c.* 50 m. Sudden jarring of nest causes ♀ to flush instantly and fly some distance before alighting. (Goodwin 1951, 1956, 1986.) Occasionally, bird thus disturbed abandons nest completely, exceptionally even when caring for young (Schuster 1921–3). (2) Active measures: against birds. For mimicry of calls of species being mobbed, see introduction to part 2 (above). Parents with young in or out of nest will fly some distance to launch vigorous flying attacks on *Strix aluco* (which is mobbed with distinctive screech, see 7a in Voice, as well as with mimicry), *Accipiter nisus*, and any corvid (notably *C. corone* and *C. monedula*). Neighbouring *G. glandarius* may join in mobbing attacks, which are accompanied by Grating-calls and sometimes Panting-calls. (Goodwin 1951, 1986.) In experiment in which dead *P. pica* placed above nest containing well-grown young, arriving parent dived silently through branches and struck *P. pica* with bill and claws, then flew up and repeated attack 4 times (Tutt 1952). (3) Active measures: against man. When brooding, response varies from passive flushing (see above) to more demonstrative, especially with well-grown young. Parents often scold from cover with Screech-calls and mimicry of various predatory birds (see Voice). Not infrequently venture near intruder and more rarely (notably if young screech in alarm) mob him with violent screeching and usually alarm-calls of other species (including passerines); also break off twigs and fling them down. (Goodwin 1951, 1956.) In Ireland, mobbing birds similarly recorded dropping twigs on heads of intruders and using mimicry at stage when offspring not long out of nest; these considered typical responses to threat by man (Humphreys 1928). In mobbing by wild birds, physical contact rare and slight; commoner in captivity as nest is approached but ceases when keeper right beside nest (Goodwin 1951, 1956). (4) Active measures: against other animals. Squirrel *Sciurus* which came near nest was driven off by both parents (Hulten 1967). 2 reports of weasel or stoat *Mustela* seizing and killing incubating bird (Goodwin 1956) but no information on any counteractive measures.

(Figs by D Quinn: A from drawing in Coombs 1978; B–D from drawings in Goodwin 1956; E from drawing in Goodwin 1952*a*; F from drawing in Goodwin 1951.)                    EKD

**Voice.** Extensive repertoire determined by Goodwin (1949)—largely from tame captive birds, also by cross-checking with wild birds (Goodwin 1949). Call 7a is best-known call. Innate repertoire supplemented by highly developed and widely used capacity of both sexes for accurate mimicry of various sounds, e.g. in Germany, wild bird (probably ♂) mimicked Buzzard *Buteo buteo*, Kestrel *Falco tinnunculus*, Grey Partridge *Perdix perdix*, Crane *Grus grus*, Magpie *Pica pica*, Carrion Crow *Corvus corone*,

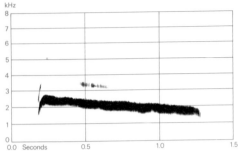

I  J-C Roché  Morocco  May 1983

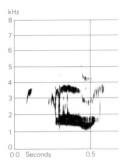

II  K Turner  France  April 1988

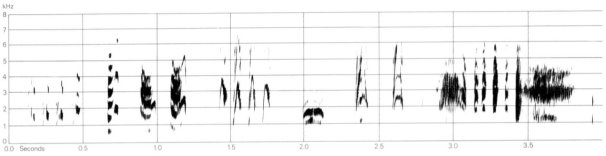

III  J-C Roché  Morocco  May 1966

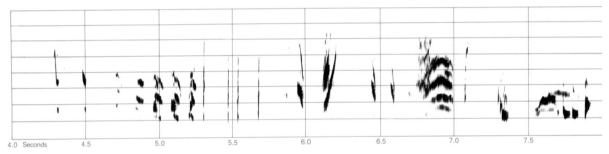

III  *cont.*

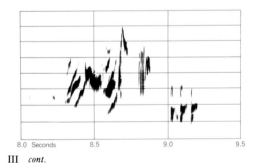

III  *cont.*

and Starling *Sturnus vulgaris* (Schuster 1921–3). Captive birds imitate sounds such as motorbike horn, human voice, whistled songs (Creutz 1970), barking dog, and (probably) lawnmower (Goodwin 1956). Fig I depicts mimicry of mewing call of *B. buteo* (one of the most commonly mimicked species) and sound in Fig II is probably imitated component of Whistling-song of Golden Oriole *Oriolus oriolus* (W T C Seale). Frequently mimics calls (commonly interspersed with low-pitched conversational variants of call 2) when foraging or resting, also when excited (e.g. during Spring Gathering) or anxious, e.g. when mobbing human at nest, typically gives alarm-calls of other (predatory and non-predatory) birds; also often mimics the bird of prey confronted, e.g. gives hoot or 'ke-wick' of Tawny Owl *Strix aluco* when, or just before, mobbing it (Goodwin 1951, 1956, which see for discussion). For further mimicry, see below. Many innate calls husky, noisy, and hard to render. Unless otherwise attributed, following descriptions and names based largely on Goodwin (1949, 1951, 1952a, 1986). For additional sonagrams, see Bergmann and Helb (1982). Geographical variation little studied, though apparently not marked in call 7a in the few races studied. Adults and young commonly intersperse calls with Bill-snapping in aggressive context (Goodwin 1949, 1951; Bergmann and Helb 1982).

CALLS OF ADULTS. (1) Song. (1a) Warbling-song. Medley

of mimicked calls and phrases (e.g. song of Bullfinch *Pyrrhula pyrrhula*) and quiet variants of call 2, given mainly by young unpaired birds and more often by ♂♂. One captive bird sang compilation of human whistles, also beautiful, subdued, liquid composition of calls copied from song of various passerines. In Switzerland, late January, song of one bird somewhat recalled Blackcap *Sylvia atricapilla*, even more so subsong of Mistle Thrush *Turdus viscivorus* though more varied (Desfayes 1951). For another description, including suggestion that song has ventriloquial quality, see Bergmann and Helb (1982). Song depicted in Fig III contains rich variety of fairly subdued sounds, including chirruping, chattering, clicking, knocking (and other percussive sounds), chuckling, buzzes, and wheezes (W T C Seale). (1b) Courtship-song of ♂. Low-pitched disconnected song of squeaking, liquid bubbling and metallic chinking sounds (including mimicry) given by ♂ in heightened sexual display toward ♀; courtship-feeding and/or copulation may follow. Recording of Courtship-song described as typical (D Goodwin) includes clicking sounds (probably vocal in original: D Goodwin) remarkable for patterned delivery: e.g. Fig IV (double speed to clarify pattern) shows 3 identical click motifs, each consisting of 7 clicks and 1 non-click unit; Fig V (normal speed) shows single motif (from *c.* 0.45 to 1.5 s on sonagram), identical to those in Fig IV, preceded by subdued chuckle (descending glissando) (W T C Seale). (2) Appeal-call. Difficult to render but basically 'aaa' or 'aar'; 'oor' or 'choork' for quieter variants which have muted 'kissing' quality. Appeal-calls are used for expressing almost any need; numerous variants (according to need) as follows. (2a) Mewing-call. Querulous mewing, often given when hungry (see Calls of Young) and also as contact-call.

(2b) Food-offering call. Husky, low-pitched, throaty sound used when offering food to mate or (less huskily and more tenderly) to young. Also heard when pair pass objects to each other at nest-site, and sometimes apparently to conciliate mate. In recording at nest containing eggs, calls apparently of this type rendered a croaking 'rrar rrhar' (Fig VI). Wheezy, wet, rasping sounds during nest-building are perhaps related, consisting of train of pulses: Fig VII shows 2 bouts, in each of which pulse rate progressively slows

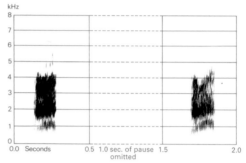

VI   P A D Hollom   England   May 1987

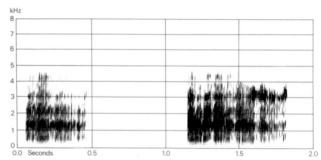

VII   P A D Hollom   England   April 1979

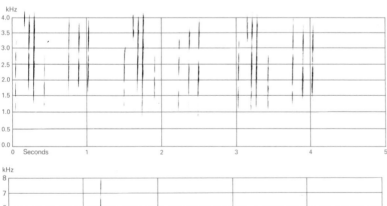

IV   P A D Hollom   England   March 1979

V   P A D Hollom   England   March 1979

(e.g. in 2nd bout from *c.* 150 per s to *c.* 110 per s) (W T C Seale). (2c) Food-begging call ('Hunger-appeal'). Urgent 'aa aaa', very like juvenile food-call (see below) given to solicit food, notably by incubating ♀ to mate. (2d) Flight-appeal call. Rather high pitched, usually not loud, with pathetic yearning tone, heard at Spring Gatherings and during pursuit-flights, also when seeking hiding place for food or when foraging for offspring. (2e) Nest-call. Similar to 2d but quieter, shorter, and lower pitched, and used by both sexes on finding (selecting) nest-site, probably to summon mate. (3) Chirruping-calls. Series of highly variable but generally quiet stammering sounds, usually interspersed with, apparently, quiet variants of call 2 (especially 2b). Chirruping-calls seem to be for intimate appeasement, and typically given by subordinate pair-member, although pairs also hold chirruping duets. Chirruping-calls also heard when apparently perplexed, or relieved from tension or fear. (4) Kraah-call. Loud, resonant, usually disyllabic 'kraa-aah' or 'kraah' as loud as, but less harsh than, intense variants of call 7a, heard mainly during Spring Gatherings, perhaps only from ♀♀ (for details of associated display, see Social Pattern and Behaviour). Fig VIII shows typical call (D Goodwin), roughly 'deeJARRR', a harsh rasping 'JARRR' with more tonal ascending introduction (W T C Seale). Numerous variants, including commonly somewhat nasal 'k' yankh' or 'k' ya-arnkh'; Fig IX depicts variant (D Goodwin) with rather strangled but penetrating bleating sound (W T C Seale). (5) Hissing-call of ♂. Peculiar hissing, or hissing-crashing sound, suggestive of load of gravel being tipped onto road, given repeatedly (to form 'display-phrases') by ♂♂ performing Advertising-display. ♂ usually has 2 or more such phrases which may consist partly (sometimes entirely) of mimicked sounds, although hissing seems not to be mimicry. Typical call (D Goodwin) depicted in Fig X a crescendo, with diadic start to noisy wavering hiss (W T C Seale). (6) Clicking-calls. (6a) Castanet-call. Repeated, individually variable, guttural clicking or grating sounds given by ♀ in display-phrases during her Advertising-display, sometimes combined with mimicry. In example (not very typical: D Goodwin) shown in Fig XI, the 2nd unit, and unit between the 2 short series of clicks, are low-pitched creaks ('frrarr'); 1st unit in Fig XI is brief quiet 'pu' (W T C Seale). Recurrent pattern of clicks comparable to those in Fig X of *P. pica* (J Hall-Craggs, W T C Seale). (6b) Clicking sounds given by ♂ in display differ from 6a in being always lighter and more suggestive of scissors snipping. Calls in Fig XII resemble sound made by garden shears (P J Sellar) or 'schp' by flicking tongue off roof of mouth (W T C Seale). (7) Alarm- and Attack-calls. (7a) Screech-call. Strident harsh rasping screech, given singly or more often twice in quick succession, then pause before 2 more; 'kshehr' (Bruun *et al.* 1986). Best-known call, typically given in alarm, most intensely when mobbing predator (e.g. nagging and drawn-out, as in Fig XIII, when mobbing *S. aluco*), but highly variable and also heard in less intense moods of annoyance or threat; Fig XIV shows variant typical of (e.g.) bird startled by man (D Goodwin). See also call 4. For comparison of *corsicanus* and *cervicalis*, see Chappuis (1976). (7b) Panting-call ('Anger-appeal'). Short, husky, rapidly repeated panting sound,

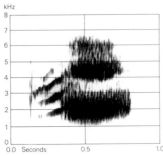

VIII   K Turner   France   April 1988

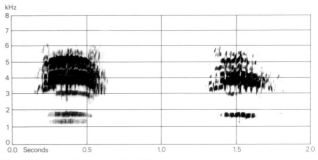

IX   K Turner   France   April 1988

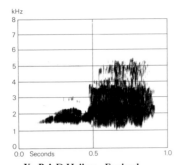

X   P A D Hollom   England April 1991

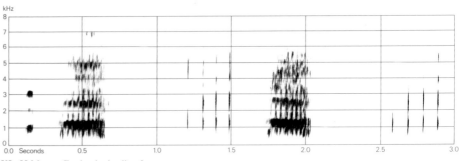

XI   H Myers   England   April 1981

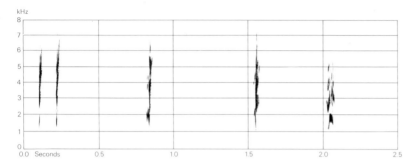

XII  K Turner  France  April 1988

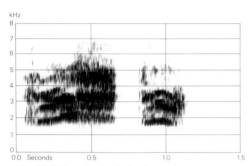

XIII  P Riddett  England  April 1990

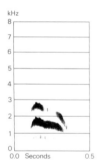

XV  W T C Seale  England  June 1991

XIV  J-C Roché  France  May 1983

chippering much like young *P. pica* but much quieter. Nestlings often receive food with faint squeaking gabble (apparently calling while swallowing) and chirrup contentedly after being fed. Begin to utter querulous food-call, like adult call 2b, shortly before leaving nest, this becoming louder after fledging. When hungry, young give frantic mewing scream, like intense call 2a. If highly alarmed, deliver continuous sound intermediate between adult screech (call 7a) and sound of kettle boiling over. (Goodwin 1951, 1986.)

EKD

suggestive of call 2, often given by tame birds launching flying-attacks in response to alarm-calls of handled young (see below). Also once heard from breeding pair chasing Sparrowhawk *Accipiter nisus*. (7c) Grating-call. Short, hard, grating rattle, rather like *T. viscivorus* (Goodwin 1956); 'krr-kr-kr' (Bergmann and Helb 1982). Often given when making darting attack at corvid or other predator, or when turning in defence against mobbing by other birds. Grades into calls 7a–b. Less intense form heard when 'playing with' or threatening conspecifics. (7d) Hawk-alarm call. Very quiet, low-pitched, melancholy monosyllable with ventriloquial quality, given when crouched at sight of flying bird of prey, e.g. *A. nisus*. (7e) Subdued muffled 'meeoo' (Fig XV) once heard at end of series of call 7a when mobbing bird of prey (W T C Seale); different from call 7d (D Goodwin) but not known if mimicry. (8) Distress-call. Loud drawn-out scream, sounding like call 7a but much longer and more continuous, from trapped or handled bird (Goodwin 1956).

CALLS OF YOUNG. Food-call of small nestlings a very faint

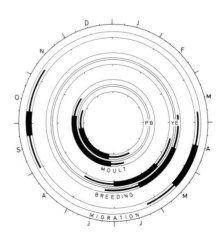

**Breeding.** SEASON. Britain: eggs laid mid-April to early June; average date of 1st egg in Wales (below 160 m altitude) 3 May ($n=9$), southern England 7 May ($n=168$), northern England 14 May ($n=78$) (Holyoak 1967b). Finland: in central latitudes eggs laid end of April to early June, in south to end of May (Haartman 1969). Germany: see diagram; in Rheinland, egg-laying from early April; end of breeding season 1st half of July (Mildenberger 1984). Switzerland: eggs laid mid-April to end of May; varies 2–3 weeks depending on foliage cover (Glutz von Blotzheim 1962). Czechoslovakia: c. 55% of 85 clutches laid in April (Hudec 1983). Rumania: eggs laid mid-April to early May (Korodi Gál 1972). Turkey and North Africa: laying starts early April (Goodwin 1986). For detailed review, see Keve (1985). SITE. In fork or on branch of tree or bush, often thorny, usually close to or against trunk in middle of lower crown, or high in crown of young tree or of conifer; generally supported by 1–2 thick branches or many twigs; frequently in creeper (e.g. ivy *Hedera*, honeysuckle *Lonicera*); perhaps increasingly on buildings, and rarely in tree-hole, large nest-box, or rock-face crevice (Goodwin 1953a, 1956; Tutt 1953; Piechocki 1956; Glutz von Blotzheim 1962; Hulten 1967; Peter 1968; Wittenberg 1970; Kulczycki 1973; Mildenberger 1984). Of 165 nests in western Germany, 47% in conifers, 30% in broad-leaved trees, 11% in creepers, 8% in thorn bushes, and 4% on buildings (Mildenberger 1984, which see for height distribution of 159 nests). Of 43 nests in Poland, 5% 1–2 m above ground, 21% 2–3 m, 21% 3–4 m, 12% 4–5 m, 23% 5–8 m, and 19% over 8 m, average height 5·3 m; of 41 nests, 17% in pine *Pinus*, 15% in oak *Quercus*, 10% in larch *Larix*, and remainder in 12 other tree species (Kulczycki 1973, which see for discussion of nest-sites in Palearctic). Average height of 41 nests in Rumania 4·1 m (1·2–8·5); 30% of nests in pear *Pyrus*, remainder in 8 other trees and bushes; 50% in branches, 30% in main fork of trunk (Korodi Gál 1972). In Luxembourg, 46% of 24 nests directly against trunk; average height of 22 nests 4·7 m (2–11) (Hulten 1967); 98% of 111 nests in Finland less than 8 m above ground and most against trunk of spruce *Picea* (Haartman 1969). Apparently important for nest to be concealed from above by foliage, though some easily visible (Goodwin 1956). Nest recorded inside roof of shed 2·5 m above human activity (Vogt 1974). Nest: rough foundation of twigs c. 0·3–1·5 cm in diameter, with inside layer of soft, thinner twigs, roots, stalks, etc., lined with rootlets, bast, grass, moss, leaves, hair, and rarely feathers; well-supported nests in ivy, etc., may have very little foundation; average outer diameter of 31 nests, Poland, 23·5 cm (16·0–33·5), inner diameter 12·6 cm (10·5–15·7), overall height 15·8 cm (8·5–26·0), depth of cup 6·5 cm (5·0–9·5) (Goodwin 1956; Kulczycki 1973, which see for discussion of materials used in Palearctic and Japan, where e.g. earth used in building). Average weight of 41 nests in Rumania 153 g (Korodi Gál 1972, which see for dimensions, etc.). For description of nests in Luxembourg, see Hulten (1967). Building: by both sexes, each placing own material in nest; lining per-

haps mostly by ♂; thicker twigs bitten off trees and shrubs, lining gathered from ground; takes (5–)6–9 days, mostly in early morning and afternoon (Goodwin 1951; Haartman 1969; Korodi Gál 1972). EGGS. See Plate 57. Sub-elliptical, sometimes short sub-elliptical or oval, smooth and slightly glossy or matt; pale green to blue-green, olive, or olive-buff, occasionally sand-coloured, finely speckled all over (sometimes very densely) with olive, light brown, or greyish-green; sometimes black hair-streaks at broad end where speckling may also be concentrated (Harrison 1975; Makatsch 1976; Goodwin 1986). Nominate *glandarius*: $31·3 \times 23·0$ mm (27·5–36·0 × 20·6–25·9), $n=781$; calculated weight 8·5 g. *G. g. rufitergum*: $31·5 \times 22·9$ mm (28·2–34·3 × 21·1–24·6), $n=123$; calculated weight 8·6 g. *G. g. hibernicus*: $31·7 \times 23·5$ mm (29·6–34·4 × 22·5–23·8), $n=19$; calculated weight 9·0 g. *G. g. fasciatus*: $33·1 \times 23·6$ mm (32·4–34·0 × 23·5–23·8), $n=4$; calculated weight 9·5 g. *G. g. atricapillus*: $32·8 \times 23·3$ mm (30·0–35·0 × 21·9–23·9), $n=30$; calculated weight 9·0 g. *G. g. brandtii*: $30·4 \times 22·9$ mm (27·5–32·7 × 21·5–23·6), $n=44$; calculated weight 8·2 g. *G. g. cervicalis*: $31·3 \times 22·8$ (29·6–33·6 × 22·0–24·4), $n=52$; calculated weight 8·4 g. (Schönwetter 1984.) Clutch: 5–7 (3–10). In Britain, average size on south coast of England 4·3 ($n=52$), in northern England 4·7 ($n=38$), in southern England 4·8 ($n=68$), and in Wales 4·9 ($n=10$). Of 121 British clutches: 3 eggs, 9%; 4, 26%; 5, 50%; 6, 16%. (Holyoak 1967b.) For Ireland, see Humphreys (1928). In Finland, average 6·3, $n=35$ (Haartman 1969). In western Germany, of 90 clutches: 4 eggs, 7%; 5, 27%; 6, 57%; 7, 10%; average 5·7 (Mildenberger 1984). In Czechoslovakia, of 59 clutches: 4 eggs, 12%; 5, 19%; 6, 44%; 7, 21%; 8, 4%; average 6·9 (Hudec 1983). In Rumania, average 5·7, $n=19$ (Korodi Gál 1972). Eggs laid daily; in early morning, more rarely in evening (Holyoak 1967b; Korodi Gál 1972; Goodwin 1986). One brood. In England, eggs laid after mid-June are replacement clutches (Goodwin 1986); record in Germany of replacement clutch laid 1 week after loss of newly-hatched young, and of 2nd replacement (Schuster 1921–3; Mildenberger 1984). INCUBATION. Takes 16–17(–19) days; by ♀ only; starts properly with 2nd–4th egg, although eggs covered for some time from outset (Korodi Gál 1972; Goodwin 1986). In Rumania, 16 days at 92% of 41 nests, 17 days at 8%; eggs hatch over 1–2 days depending on clutch size (Stein 1929; Goodwin 1951; Korodi Gál 1972). Incubating ♀ leaves nest for 5–15 min every 3 hrs (Goodwin 1956). YOUNG. Fed and cared for by both parents; ♂ brings food to ♀ and also feeds young (Goodwin 1986). Brooded for c. 14–15 days (Eygelis 1970), by ♀, but both birds once seen on nest with young (Tutt 1952). FLEDGING TO MATURITY. Fledging period 21–22 (19–23) days (Hulten 1967; Haartman 1969; Korodi Gál 1972; Goodwin 1986); young cease to be fed by parents at 6–8 weeks (Goodwin 1986). First breeding generally at 2 years old (Niethammer 1937), though some at 1 year (D Goodwin; M Grahn);

see also Dementiev and Gladkov (1954) and Berndt and Dancker (1960). BREEDING SUCCESS. In Britain, of 571 eggs, 60% hatched, highest success rates in clutches of 4 (64%, $n=31$) and 5 (65%, $n=60$), lowest in those of 3 (33%, $n=11$); hatching success if 1st egg laid in April was $c.$ 45% including total failures, $c.$ 80% excluding ($n=20$); for late April and early May, $c.$ 62% and 84% ($n=27$); for mid-May, $c.$ 56% and 90% ($n=30$); for late May and June, $c.$ 84% and 86% ($n=16$); this perhaps reflects lower predation rate because of increasing foliage cover; most nestling deaths occurred in middle third of nestling period (Holyoak 1967b). In south European USSR, of 181 eggs, 67% hatched and 38% produced fledged young; most nests destroyed by man, other predators were corvids, Sparrowhawk *Accipiter nisus*, and Buzzard *Buteo buteo* (Eygelis 1970). In north European USSR, of 96 eggs, 64% hatched and 41% produced fledged young; 1 clutch of 7 ejected from nest by parent (Mal'chevski 1959). In Rumania, of 41 clutches, 54% destroyed (36% of losses due to squirrel *Sciurus vulgaris*, 32% human interference, 14% dormouse *Eliomys quercinus*, 9% Magpie *Pica pica*, and 9% conspecifics); of 109 eggs in remaining 19 clutches, 94% hatched, resulting in 5·4 nestlings per successful nest, or 2·5 per nest overall (Korodi Gál 1972). Average brood size in Czechoslovakia 4·75 ($n=24$), 69% of average clutch size (Hudec 1983). Very readily deserts eggs when disturbed (Schuster 1923; Holyoak 1967b), and possibly eats own eggs or young, probably when already dead, since 40-80% of nests examined in southern England had lost eggs or nestlings and not all losses explicable by predation; in Britain most losses due to *P. pica*, Carrion Crow *Corvus corone*, squirrel *S. carolinensis*, and human interference (Goodwin 1956, 1986). For predation by other Corvidae in Belgium, see Vercauteren (1984). See also Keve (1985).      BH

**Plumages.** (nominate *glandarius*). ADULT. Nasal bristles, forehead, forecrown, and stripe above eye white or isabelline-white, often slightly tinged brown on tips of nasal bristles and above lore, merging into greyish-vinous-brown of hindcrown and hindneck; on hindcrown, each feather minutely barred vinous and light grey at tip, uniform rufous-brown subterminally; forehead and forecrown with bold black spots and streaks 2-3(-4) mm wide, borders of black sometimes tinged pink-brown. Lore, ear-coverts, and lower side of neck vinous-cinnamon, sometimes mottled dark grey or isabelline on lore, often tinged vinous-grey on rear of ear-coverts and side of neck; patch round eye mottled isabelline-white and dull black. Broad velvet-black malar stripe from base of lower mandible over lower cheek. Mantle, scapulars, and back medium grey, often distinctly vinous on upper mantle and pink-buff on lower scapulars, more vinous-pink on back, and sometimes slightly vinous on remainder also. Rump and upper tail-coverts white. Chin and upper throat dirty white, sometimes tinged isabelline, forming sharply demarcated patch; shafts often purer white, some grey of feather-bases showing through. Vent and under tail-coverts white; remainder of underparts saturated vinous-pink, darkest across lower throat (where sometimes blotched dull black), tinged grey on chest, breast, and upper flank, sometimes variegated isabelline-white on mid-belly.

Distal half of tail black, more sooty-grey on outermost feathers and on inner borders of others; basal half of tail grey, incompletely barred grey and black towards middle portion of feathers; extent of grey barring variable, sometimes extending well beyond tail-coverts; grey sometimes tinged blue. Primaries sooty-grey or dull black, outer web of (p1-)p2-p9 fringed silvery-greyish-white (sometimes incomplete on p1-p3), outer fringes of p1—p2(-p3) sometimes tinged cerulean-blue and with more or less complete black bars or notches, p10 entirely deep black. Secondaries (s1-s6) and outer tertials (s7-s8) deep velvety-black, outer web of s1-s5(-s6) with white or bluish-white middle portion (most extensive on s1) and with base barred blue and black (blue of bases sometimes largely hidden beneath greater coverts, especially on outer secondaries); inner tertials (s9-s10) rufous-chestnut, s9 with broad black tip, base of s8 often also with (hidden) chestnut. Inner greater upper wing-coverts and inner webs of outer greater coverts and greater upper primary coverts black; outer webs of outer 5 greater coverts and of primary coverts, as well as virtually whole bastard wing contrastingly barred blue and black (each blue bar paler cerulean basally, deeper azure distally). Median upper wing-coverts chestnut- or cinnamon-rufous, tips of outer with fine light grey bars, bases dull brown-grey; lesser upper wing-coverts vinous brown-grey. Under wing-coverts and axillaries vinous-pink, longer coverts light grey. *In worn plumage* (from June onwards), feather-tips of mantle, scapulars, and underparts wear to dirty white, upperparts appearing blotched isabelline and vinous-grey, underparts dirty isabelline-grey, some vinous-pink remaining on side of breast and flank only. Throughout October–May, individual variation in Norwegian and Swedish birds (described above) is limited, all with distinctly grey lower mantle and scapulars and dark greyish-vinous chest, breast, and flank; only occasionally are both upperparts and underparts tinged vinous-brown, less grey (see Voous 1953); throughout central Europe south to Alps, northern Yugoslavia, and western Rumania populations are similar, but further west in westernmost Germany, Netherlands, and north-east France lower mantle and scapulars are less uniform, grey paler and more restricted, more extensively tinged vinous-pink or vinous-drab, chest, breast, and upper flank less grey, underparts slightly paler and more uniform vinous-pink, sometimes with more pink-white or isabelline showing on mid-belly. Sexes similar (but see Structure). NESTLING. Naked. For development, see Heinroth and Heinroth (1924-6) and Keve (1969). JUVENILE. Forehead, nasal bristles, lore, and region round eye isabelline-white to pale buff, grading into rufous-cinnamon of crown, hindneck, ear-coverts, and side of neck; forehead speckled dull black, lore and region round eye speckled dark grey, crown marked with narrow rather ill-defined dull black spots or streaks; hindcrown sometimes slightly tinged vinous-grey. Black malar stripe as in adult. Mantle, scapulars, back, and lesser and median upper wing-coverts rufous-brown, drab-brown, or brown-grey, often with grey of feather-bases visible, brighter rufous on tips of longer coverts, outer scapulars, and back feathers; rump and upper tail-coverts white, often slightly pink-buff. Chin, upper throat, vent, and under tail-coverts white or cream, remainder of underparts buff-brown or pale rufous-brown (brightest on chest and flank) with much pale grey of feather-bases visible and often with extensive isabelline-white wash on mid-belly; feathering of entire underparts markedly short and loose. Tertials as adult, but chestnut of innermost often less bright and not sharply demarcated from black of tip. Flight-feathers, primary coverts, and tail like adult, but see 1st adult (below). Thus, easily distinguished from adult by largely rufous head and body, reduced white on face, reduced and ill-defined black streaks on cap, absence of fine vinous and grey bars on hindcrown, and fluffy underparts; however, adult *brandtii* are rather similar to this, differing mainly in sharp black streaks on cap, more uniform grey mantle and scapulars, and saturated greyish pink-brown underparts. FIRST ADULT. Like adult, but juvenile flight-feathers, greater upper primary coverts, tail, and variable number of tertials, greater upper secondary coverts, and feathers of bastard wing retained. Retained juvenile feathers sometimes indistinguishable from adult feathers, and not all birds can be aged reliably; even between brood siblings, great variation occurs in juvenile character of retained feathers. Birds are 1st adult if: (1) a difference in colour, pattern, shape, or degree of abrasion occurs between (new) inner and (retained juvenile) outer greater coverts, bastard wing, or (rarely) tail-feathers (in adult and some 1st adult, all feathers similar); (2) tail-feathers narrow (t5 20–25 mm wide at *c.* 40 mm from tip: Svensson 1992), tips rounded or slightly pointed and sometimes washed grey (in adult and some 1st adults, 25–30 mm wide, tip truncate, uniform black or with faint grey edge); (3) any irregularity in barring of greater primary coverts (and, if retained, greater secondary coverts or feathers of bastard wing) appears at about same distance from tip of each feather, as feathers in juvenile grow all at same time, unlike adult; see Svensson (1992) for figure (in adult and some 1st adults, pattern very regular on all blue-and-black feathers or, if irregular, irregularities fall at different positions on feathers); (4) outer web of longest feather of bastard wing shows 10–11 black bars (10–14 in adult), that of longest greater primary covert 7–8 bars (7–10 in adult); distance between distal 5 bars 12–16·5 mm on longest feathers of bastard wing (11–14 in adult), 19–22 mm on longest primary covert (16–19 in adult) (Mayaud 1948; see also Svensson 1992). In summer, shortly before moult, generally much more heavily worn than adult, especially tail and outer primaries (Bährmann 1958).

**Bare parts.** *Glandarius* group. ADULT. Iris pearl-grey, blue-white, blue-grey, or light sky-blue, usually slightly mixed with reddish and with narrow brownish, pale red-brown, purplish-red, or brown-pink outer ring, averaging wider in ♀; some ♀♀ have iris all-brown or brown-pink, some ♂♂ have outer ring very narrow; iris often suffused red when bird in stress. Bill black with slight blue or brown tinge. Mouth greyish-flesh, tinged flesh-grey on tongue and palate; inside of mandibles grey or (in older birds) black. Leg pink-flesh with brown or grey tinge, foot flesh-grey to brownish-grey. NESTLING. At hatching, bare skin pink-yellow or yellow-flesh, bill and leg flesh-pink or pale mauve-grey, mouth bright pink-red, gape-flanges and cutting edges of bill pinkish-white. In 1st week bright olive-yellow or bronze-green tinge develops on upperparts (depending on diet), and culmen, bill-tip, and scutes of tarsus and toes darken to grey. At fledging, iris uniform clear pale blue or greyish-blue, without brown outer ring; rarely, uniform brown; bill dark grey or slate-grey with blackish tip, mauve tinge at base, flesh-pink cutting edges, and narrow flesh-coloured flanges; mouth bright red to pale mauve-pink, sometimes tinged yellow, sometimes largely dull flesh-yellow; inside of mandibles pink-white; leg and foot as adult, but slightly tinged yellow. JUVENILE, FIRST ADULT. Iris uniform bluish-white to grey-blue; rarely, pink, brown-grey, or hazel; brown ring (see Adult) gradually develops; Bill plumbeous-grey with black tip; inside of mandibles grey with large white tip; leg and foot greyish-flesh. Adult colours obtained during post-juvenile moult. (Heinroth and Heinroth 1924–6; Witherby *et al.* 1938; Keve 1969; Goodwin 1986; Cellier 1992; RMNH, ZMA.)

In other groups, bare parts mainly as in *glandarius* group, but iris of *atricapillus* dark purplish-grey with dark inner ring, in

*bispecularis* group grey-yellow, pink-grey, brown-pink, greyish, dark bluish-brown, dark brown, or blackish (depending on race), in *leucotis* group dark brown (Goodwin 1986, which see for other details).

**Moults.** ADULT POST-BREEDING. Complete; primaries descendent (exceptionally, moult from a focus on p3: Bährmann 1971*b*). In Britain, *rufitergum* starts with p1 late May to mid-July, completed with re-growth of p9–p10 after *c.* 92 days mid-August to mid-October (Holyoak 1974*b*; Seel 1976; Ginn and Melville 1983). On average, both ♂ and ♀ start (with p1) on 9 July; tail starts on average *c.* 22 July, sequence centrifugal, completed after *c.* 66 days; tertials start *c.* 2 August, completed after *c.* 53 days; secondaries *c.* 4 August, completed after *c.* 74 days, on average *c.* 1 week after completion of primaries. Body moult starts with mantle from *c.* 4 July (thus slightly before p1), scapulars, breast, and belly from *c.* 13 July, completed at about same time as secondaries but slight moult continues during winter; head moulted last, with mean start on 7 August and completion after *c.* 65 days. Wing-coverts start 2nd half of July, completed by late September. (Seel 1976, which see for further details.) Tertials moult inwards and outwards from s8–s9 (Mayaud 1948). Small samples from Netherlands, Sweden, Germany, Yugoslavia, Greece, and Turkey point to start of primary moult between early June and mid-July, with duration and sequence apparently as in Britain; some from Sardinia and Algeria start mid-May (Stresemann 1920; RMNH, ZFMK, ZMA). In western France, moult early July to late September (Mayaud 1948); in eastern Germany, starts early June (non-breeders) or 2nd half of June (successful breeders), completed after 11–12 weeks early September to early October (Bährmann 1958). On Cyprus, singles in moult on 7 and 11 September, in north-west Turkey singles in August and on 15 September (Kleiner 1939*a*). In Algeria, *minor* still in moult on 17 October (Ticehurst and Whistler 1938). In Mongolia, 2 *bambergi* had just started moult on 11 June (Piechocki and Bolod 1972); moult completed in 2 Manchurian birds from early September (Piechocki 1959). ADULT PRE-BREEDING. No complete body moult; often slight moult of body feathers October–March, but this probably a continuation of post-breeding or a replacement of feathers after accidental loss. POST-JUVENILE. Partial. Starts shortly after fledging, at age of 7–9 weeks, sometimes before flight- or tail-feathers full-grown (Stresemann 1920; Dementiev and Gladkov 1954; Bährmann 1958). Earliest birds start from early June, but a few (especially in Mediterranean basin) from late May, some not until early August. Moult slow, completed after *c.* 3 months, mid-August to October (RMNH, ZFMK, ZMA). In western France, moult from 2nd half of July to late September or October (Mayaud 1948); in eastern Germany, starts July, completed after 9–10 weeks (Bährmann 1958); in Algeria, some *cervicalis* still in moult on 12 October (Ticehurst and Whistler 1938). Moult includes head, body, and lesser and median upper wing-coverts. Depending on locality, sometimes a number of greater upper wing-coverts or feathers of bastard wing replaced; in Sweden, no coverts or bastard wing replaced, but in Britain up to *c.* 6 greater coverts new; in Switzerland, 50% of 50 birds replaced 1 or more greater coverts and feathers of bastard wing (6% replaced all) (Svensson 1992); in Netherlands, inner greater coverts and shorter 2 feathers of bastard wing often new, but outer feathers rarely (RMNH, ZMA); in western France, moult includes all greater coverts and 2 shorter feathers of bastard wing (Mayaud 1948). In eastern Germany, usually includes all greater coverts also (Bährmann 1958). No 1st pre-breeding, but, as in adult, often a few feathers replaced November–March, probably as continuation of post-juvenile; in North Africa and Middle East, in particular, some birds then replace outer greater coverts, some tertials, 1–4 pairs of central tail-feathers, or (rarely) some inner secondaries or primaries (ZFMK, ZMA).

**Measurements.** ADULT, FIRST ADULT. Nominate *glandarius*. Scandinavia and Germany south to northern Switzerland, Austria, and Hungary, all year, ages combined; skins (RMNH, ZFMK, ZMA). Bill (S) to skull, bill (N) to distal corner of nostril; exposed culmen on average *c.* 3·8 less than bill (S).

| | ♂ | ♀ |
|---|---|---|
| WING | 186·5 (4·65; 34) 178–196 | 181·0 (3·98; 21) 175–188 |
| TAIL | 151·8 (5·03; 24) 143–160 | 146·5 (4·53; 17) 140–155 |
| BILL (S) | 33·4 (1·36; 33) 31·0–35·9 | 32·3 (1·73; 21) 29·5–34·7 |
| BILL (N) | 20·8 (0·86; 33) 19·3–22·7 | 19·7 (0·90; 21) 18·4–21·2 |
| TARSUS | 43·4 (1·17; 26) 41·5–45·2 | 42·8 (1·32; 17) 40·6–44·8 |

Sex differences significant, except for tarsus.

Netherlands, Belgium, German Rheinland, and north-east France, all year; skins (RMNH, ZFMK, ZMA). Juvenile wing and tail include 1st adult.

| | ♂ | ♀ |
|---|---|---|
| WING AD | 183·8 (4·14; 23) 176–192 | 178·9 (3·27; 19) 175–185 |
| JUV | 181·2 (4·83; 40) 172–194 | 177·2 (3·75; 29) 171–186 |
| TAIL AD | 145·6 (4·09; 23) 139–153 | 141·8 (3·54; 18) 137–148 |
| JUV | 143·1 (5·90; 41) 133–152 | 140·0 (4·65; 26) 132–147 |
| BILL (S) | 33·5 (1·33; 62) 31·2–36·2 | 32·9 (1·32; 44) 30·6–36·5 |
| BILL (N) | 20·6 (0·90; 58) 18·9–22·0 | 20·0 (0·97; 38) 18·5–22·5 |
| TARSUS | 43·3 (1·23; 47) 40·8–45·2 | 42·4 (1·26; 46) 40·1–45·0 |

Sex differences significant, except for juvenile tail.

Wing, ages combined; skins. (1) Finland, Poland, and European USSR (Kleiner 1939*a*). (2) Finland, (3) eastern Poland, (4) Norway, (5) Sweden (Voous 1953). (6) East Germany (Eck 1984). (7) West Germany (Voous 1953). (8) Hungary (Kleiner 1939*a*). (9) Netherlands, breeding, May–August (RMNH, ZMA).

| | ♂ | ♀ |
|---|---|---|
| (1) | 183·3 (4·82; 11) 175–193 | 178·6 (4·98; 12) 170–186 |
| (2) | 184·1 (5·10; 7) 179–191 | 183·0 ( — ; 3) 180–187 |
| (3) | 185·7 (4·47; 15) 178–194 | 184·8 (2·98; 8) 180–191 |
| (4) | 186·8 (4·24; 13) 180–198 | 181·7 (2·31; 9) 177–185 |
| (5) | 186·3 (4·68; 58) 176–198 | 181·9 (3·59; 41) 173–189 |
| (6) | 186·8 (4·52; 153) 176–197 | 181·7 (4·50; 139) 172–192 |
| (7) | 185·5 (5·82; 11) 175–198 | 179·2 (2·64; 5) 175–183 |
| (8) | 180·7 (4·97; 107) 168–194 | 177·4 (4·08; 61) 162–185 |
| (9) | 181·3 (3·32; 21) 175–185 | 177·6 (3·71; 15) 171–184 |

For other measurements or other localities, see Kleiner (1939*a*, *b*), Voous (1944, 1953), Hell and Soviš (1959), Bährmann (1963, 1976), Ekelöf and Kuschert (1979), and Eck (1984). In all races, bill length somewhat variable throughout year, longest in autumn (Keve 1969).

*G. g. rufitergum*. England, all year, skins (RMNH, ZFMK, ZMA).

| | ♂ | ♀ |
|---|---|---|
| WING | 182·2 (3·58; 24) 176–190 | 177·5 (2·86; 18) 173–183 |
| TAIL | 145·2 (4·18; 14) 137–152 | 141·3 (3·74; 17) 135–149 |
| BILL (S) | 33·8 (1·42; 14) 31·7–35·6 | 33·5 (1·15; 17) 31·2–35·2 |
| BILL (N) | 20·1 (0·62; 14) 19·1–21·3 | 19·9 (1·02; 17) 18·4–21·3 |
| TARSUS | 44·0 (0·94; 15) 42·0–45·5 | 42·9 (0·83; 17) 42·1–44·7 |

Sex differences significant, except for bill. Finistere (France), wing: ♂ 176·9 (2·87; 15) 172–180, ♀ 172·5 (4·17; 15) 165–178 (Lebeurier and Rapine 1939). Live birds, Oxfordshire (England): wing, ♂ 188·9 (3·6; 25), ♀ 178·0 (3·1; 30) (Cellier 1992). Perthshire, Scotland: wing 184–201 (*n* = 14) (Hazelwood and Gorton 1953).

*G. g. hibernicus*. Ireland, wing: ♂ 183·0 (4·15; 17) 176–190, ♀ 179·5 (4·09; 8) 174–188 (Voous 1953); 182 (27) 175–190 (Vaurie 1954*b*).

*G. g. lusitanicus*. Salamanca, Spain, all year; skins (RMNH, ZFMK, ZMA).

| | ♂ | ♀ |
|---|---|---|
| WING | 183·0 (4·62; 11) 175–189 | 174·3 (3·58; 12) 168–178 |
| TAIL | 147·2 (6·71; 7) 139–155 | 138·8 (3·46; 8) 133–144 |
| BILL (S) | 33·5 (1·12; 11) 31·8–35·4 | 32·2 (1·35; 12) 30·2–34·1 |
| BILL (N) | 20·0 (0·58; 11) 19·1–20·9 | 19·0 (0·61; 12) 18·2–19·9 |
| TARSUS | 42·6 (0·48; 7) 41·8–43·0 | 42·0 (0·67; 8) 41·2–43·0 |

Sex differences significant, except for tarsus.

*G. g. fasciatus*. Southern Portugal and southern Spain, all year; skins (RMNH, ZFMK, ZMA).

| | ♂ | ♀ |
|---|---|---|
| WING | 184·8 (4·17; 8) 177–191 | 179·2 (6·54; 8) 170–189 |
| BILL (S) | 34·2 (1·19; 8) 32·2–35·8 | 33·1 (1·76; 7) 30·7–35·2 |
| BILL (N) | 20·8 (0·69; 8) 19·8–21·7 | 20·1 (0·89; 7) 19·0–21·2 |

Sex differences not significant.

*G. g. corsicanus*. Wing, Corsica, skins: ♂ 179·2 (6·71; 8) 170–189, ♀ 176·8 (2·58; 5) 174–181 (Kleiner 1939*a*); ♂ 182–191 (*n* = 4), ♀♀ 178, 179 (Laubmann 1912*b*).

*G. g. ichnusae*. Sardinia, all year; skins (RMNH, ZMA).

| | ♂ | ♀ |
|---|---|---|
| WING | 178·6 (3·06; 15) 175–184 | 173·0 (3·96; 24) 166–181 |
| TAIL | 147·4 (4·21; 15) 140–153 | 142·5 (3·96; 24) 136–152 |
| BILL (S) | 33·9 (1·23; 14) 32·5–36·2 | 33·1 (1·22; 24) 31·2–35·2 |
| BILL (N) | 20·5 (0·71; 14) 19·5–21·4 | 20·1 (0·90; 22) 19·0–22·2 |
| TARSUS | 42·9 (1·41; 7) 41·4–45·0 | 41·8 (1·03; 15) 40·5–43·5 |

Sex differences significant for wing and tail. Wing, skins: ♂ 174·6 (5·23; 17) 165–187, ♀ 171·9 (4·54; 12) 163–179 (Keve 1966*b*); in 18 birds, ♂ 166–173, ♀ 167–171 (Bezzel 1957); 177·1 (5·16; 12) 168–182 (once 188) (Vaurie 1954*b*).

*G. g. albipectus*. Mainland Italy and western Yugoslavia, all year; skins (RMNH, ZFMK, ZMA).

| | ♂ | ♀ |
|---|---|---|
| WING | 184·6 (5·69; 14) 178–194 | 179·9 (6·30; 14) 172–191 |
| TAIL | 150·0 (6·52; 10) 141–160 | 145·1 (7·00; 12) 135–157 |
| BILL (S) | 34·5 (0·91; 9) 33·2–35·8 | 34·1 (1·71; 12) 31·7–36·8 |
| BILL (N) | 21·7 (1·01; 9) 20·2–23·2 | 20·7 (1·10; 12) 19·3–22·5 |
| TARSUS | 44·0 (1·50; 10) 42·2–46·2 | 43·3 (1·04; 12) 41·5–45·0 |

Sex differences significant for wing and bill (N).

*G. g. jordansi*. Sicily, wing; skins: ♂ 179·4 (7·83; 11) 162–194, ♀ 173·8 (6·44; 9) 168–188 (Keve 1966*b*).

*G. g. graecus*. Wing, skins: southern Yugoslavia, ♂ 182·7 (4·55; 13) 177–192, ♀ 177·2 (4·76; 12) 169–186 (Stresemann 1920); mainland Greece and Peloponnísos, 179·9 (5·11; 11) 174–188 (Kleiner 1939*b*; Niethammer 1943; Makatsch 1950; RMNH).

*G. g. cretorum*. Crete, wing; skins: ♂♂ 170, 171, unsexed 169 (Kleiner 1939*a*; Niethammer 1942; ZMA).

*G. g. ferdinandi*. South-east Bulgaria, wing; skins: ♂ 179·7 (3·09; 7) 174–183, ♀ 174·5 (2) 169–180 (Keve-Kleiner 1943).

*Brandti* group. Wing; skins. *G. g. brandti*, Ural mountains to Lake Baykal (USSR): (1) (Stresemann 1928*b*; RMNH, ZFMK, ZMA), (2) (Kleiner 1939*a*). *G. g. kansuensis*, Sinkiang (K'u-erh-le) and northern Kansu (China): (3) (Streseman 1928*b*; Kleiner 1939*a*; RMNH, ZMA). *G. g. bambergi*, northern Mongolia and Transbaykalia (USSR) to Hokkaido (Japan): (4) (Piechocki and Bolod 1972; RMNH, ZFMK), (5) (Kleiner 1939*a*).

| | ♂ | ♀ |
|---|---|---|
| (1) | 174·2 (2·48; 6) 170–177 | 171·0 (2·41; 11) 169–176 |
| (2) | 174·8 (4·63; 25) 164–181 | 168·9 (4·10; 15) 162–175 |
| (3) | 181·1 (5·24; 5) 175–188 | 177·6 (3·09; 8) 173–183 |
| (4) | 180·8 (4·34; 7) 176–187 | 182·5 ( — ; 3) 180–185 |
| (5) | 174·7 (4·78; 16) 169–189 | 175·9 (5·65; 11) 167–185 |

Other measurements. Sexes combined: (A) *brandti*. (B) *kansuensis* and *bambergi* (RMNH, ZFMK, ZMA).

| | (A) | (B) |
|---|---|---|
| TAIL | 141·3 ( — ; 3) 136–147 | 152·2 (2·30; 12) 150–157 |
| BILL (S) | 31·0 (1·28; 10) 29·1–32·9 | 33·1 (1·22; 15) 30·9–35·6 |
| BILL (N) | 19·0 (0·71; 10) 18·1–20·1 | 20·8 (0·68; 15) 19·7–22·1 |
| TARSUS | 40·8 ( — ; 3) 39·7–40·1 | 41·8 (0·83; 10) 40·4–43·7 |

*G. g. krynicki*. Caucasus and north-east Turkey, all year; skins (BMNH, RMNH, ZFMK, ZMA).

| | ♂ | ♀ |
|---|---|---|
| WING | 189·0 (3·15; 13) 185–194 | 189·9 (4·25; 10) 183–197 |

| | | |
|---|---|---|
| TAIL | 151·8 (5·13; 13) 146–162 | 151·5 (4·87; 10) 145–159 |
| BILL (S) | 34·7 (1·59; 13) 33·3–37·3 | 34·0 (2·09; 10) 30·5–36·5 |
| BILL (N) | 21·6 (0·90; 13) 20·1–22·8 | 21·0 (0·61; 10) 19·8–21·8 |
| TARSUS | 45·5 (1·14; 13) 44·0–47·5 | 44·5 (1·99; 10) 42·0–47·4 |

Sex differences not significant. Wing, skins: Caucasus area, ♂ 190·9 (5·44; 32) 179–200, ♀ 187·0 (7·07; 16) 174–204, unsexed 187·1 (5·81; 21) 178–200; northern Asia Minor, ♂ 187·3 (7·97; 13) 180–192 (once 210), ♀ 183·8 (5·39; 13) 176–195 (Keve 1973).

*G. g. iphigenia.* Crimea, skins: wing, ♂ 188·5 (5·30; 13) 175–196, ♀ 180·8 (4·02; 5) 176–189 (Keve 1973).

*G. g. samios.* Samos and Ikaria (Aegean, Greece), sexes combined: wing 183·6 (2·61; 5) 179–185 (Kleiner 1939b; Keve 1973).

*G. g. anatoliae.* Western and southern Asia Minor and north-west Syria, all year; skins (BMNH, RMNH, ZFMK).

| | | |
|---|---|---|
| WING | ♂ 189·2 (5·98; 16) 177–200 | ♀ 182·4 (3·98; 7) 178–190 |
| TAIL | 151·0 (6·17; 7) 145–157 | 146·3 (6·22; 7) 138–154 |
| BILL (S) | 34·8 (1·90; 16) 31·4–37·5 | 36·0 (1·86; 7) 33·0–37·9 |
| BILL (N) | 21·6 (1·05; 16) 19·5–23·0 | 21·4 (0·45; 7) 20·8–21·9 |

Sex differences significant for wing. Wing, sexes combined: Bosporus area 183·5 (4·20; 11) 176–190; Lesbos and Khios 178·8 (5·65; 13) 168–185; Rodhos 185, 189; Izmir 177·6 (2·70; 5) 173–180; Cilician Taurus 183·9 (5·64; 9) 176–193 (Kleiner 1939a, b; Keve 1973); Zagros mountains (Iran) 184·6 (6·46; 9) 173–192 (Paludan 1938; Keve 1973).

Zagros mountains: wing 178·0 (3·46; 5) 173–181, bill (S) 33·4 (0·81; 5) 32·0–34·2 (BMNH).

*G. g. atricapilla.* Lebanon, Syria, Israel, and Jordan, all year; skins (BMNH, RMNH, ZFMK, ZMA).

| | | |
|---|---|---|
| WING | ♂ 187·3 (5·23; 15) 176–192 | ♀ 177·9 (4·45; 8) 173–187 |
| TAIL | 149·9 (4·31; 10) 143–156 | 141·6 (4·68; 8) 136–149 |
| BILL (S) | 35·0 (1·48; 9) 32·3–36·7 | 34·3 (1·75; 8) 31·3–36·2 |
| BILL (N) | 21·7 (0·73; 9) 20·5–22·5 | 21·4 (1·58; 8) 19·2–23·9 |
| TARSUS | 44·3 (2·36; 8) 41·5–47·5 | 43·3 (1·27; 7) 41·0–45·0 |

Sex differences significant for wing and tail. Levant, wing, sexes combined: 178·0 (5·26; 12) 165–186 (Keve 1973).

*G. g. cervicalis.* Northern Algeria (east from Blida and Médéa) and northern Tunisia, all year; skins (RMNH, ZFMK, ZMA).

| | | |
|---|---|---|
| WING | ♂ 183·0 (5·63; 16) 174–190 | ♀ 178·2 (3·27; 11) 174–182 |
| TAIL | 157·2 (5·38; 16) 149–167 | 153·3 (2·62; 10) 150–158 |
| BILL (S) | 33·7 (1·68; 16) 31·8–36·0 | 32·5 (1·38; 11) 30·5–34·6 |
| BILL (N) | 20·7 (1·02; 15) 19·4–22·4 | 20·3 (1·14; 11) 18·6–21·7 |
| TARSUS | 43·2 (1·39; 16) 40·8–45·4 | 42·9 (0·92; 11) 40·8–44·0 |

Sex differences significant for wing. North-east Algeria and Tunisia, wing, sexes combined: 178·4 (5·22; 16) 166–185 (Kleiner 1939a).

*G. g. whitakeri.* Wing, sexes combined: Tanger area and Rif (Morocco), 182·6 (4·64; 9) 175–190; Tlemcen (north-west Algeria), 173·9 (2·27; 7) 170–177 (Vaurie 1954b).

*G. g. minor.* Wing of (1) Atlas Saharien (Algeria), (2) Haut Atlas, (3) Moyen Atlas (Morocco); other measurements Atlas Saharien only, all year; skins (BMNH, RMNH, ZFMK, ZMA).

| | | |
|---|---|---|
| WING (1) | ♂ 163·3 (3·28; 6) 158–167 | ♀ 160·6 (2·54; 9) 157–165 |
| (2) | 164·7 (1·63; 6) 163–167 | 158·6 (5·04; 5) 153–165 |
| (3) | 170·2 (4·44; 8) 163–175 | 165·3 (1·53; 3) 164–167 |
| TAIL | 137·5 (3·56; 6) 132–142 | 135·6 (3·50; 9) 132–142 |
| BILL (S) | 29·5 (1·27; 5) 27·3–30·5 | 28·9 (1·10; 8) 28·0–30·5 |
| BILL (N) | 18·5 (0·61; 4) 17·6–18·9 | 17·5 (0·50; 5) 16·8–18·3 |
| TARSUS | 40·3 (0·42; 5) 39·8–40·9 | 39·2 (0·47; 7) 38·5–39·7 |

Sex differences significant for wing (2), bill (N), and tarsus. See also Ticehurst and Whistler (1938), Kleiner (1939a), Meinertzhagen (1939, 1940), and Vaurie (1954b).

*G. g. glaszneri.* Cyprus, all year; skins (BMNH, RMNH, ZFMK, ZMA).

| | | |
|---|---|---|
| WING | ♂ 171·7 (3·95; 13) 165–176 | ♀ 168·4 (3·94; 10) 159–173 |

| | | |
|---|---|---|
| TAIL | 148·8 (4·76; 12) 140–151 | 143·1 (5·08; 9) 132–151 |
| BILL (S) | 31·6 (0·98; 12) 30·4–33·9 | 32·0 (1·01; 9) 30·3–33·4 |
| BILL (N) | 18·6 (0·86; 12) 17·4–20·6 | 18·8 (0·56; 9) 17·9–19·7 |
| TARSUS | 40·4 (1·47; 12) 38·2–42·3 | 40·2 (0·90; 9) 39·2–41·6 |

Sex differences significant for tail. Wing, skins: ♂ 169·6 (3·90; 16) 161–178, ♀ 163·9 (4·05; 14) 154–168 (Kleiner 1939a).

*G. g. hyrcanus.* Northern Iran, wing, skins: ♂ 171·4 (3·52; 10) 168–176, ♀ 166·0 (1·55; 6) 164–168 (Stresemann 1928a; Paludan 1940; Schüz 1959; ZMA). Lenkoran area (south-east Transcaucasia, USSR), wing; skins, sexes combined: 177·3 (8·76; 18) 162–195 (Kleiner 1939a; Keve 1973; ZMA).

**Weights.** Nominate *glandarius*. (1) Norway, September–April (Haftorn 1971). (2) Sweden, October–April (RMNH, ZMA). (3) East Germany, September–June (Eck 1984, which see for monthly data). (4) Netherlands, all year (RMNH, ZMA). (5) North Balkan region (Kleiner 1939a). (6) European USSR (Dementiev and Gladkov 1954).

| | | |
|---|---|---|
| (1) | ♂ 167·1 ( — ; 18) 150–180 | ♀ 156·3 ( — ; 18) 138–180 |
| (2) | 171·9 (6·23; 7) 162–182 | 171·8 (8·98; 5) 164–185 |
| (3) | 171·6 (8·75; 107) 153–197 | 163·5 (11·1; 101) 129–188 |
| (4) | 167·4 (12·1; 20) 152–192 | 161·4 (16·4; 14) 135–205 |
| (5) | 164·0 ( — ; 61) 145–190 | 158·7 ( — ; 46) 142–183 |
| (6) | 158·8 ( — ; 11) 145–197 | 165·3 ( — ; 3) 149–168 |

Norway, all year: unsexed, 161·1 (24) 141–177 (Haftorn 1971). West Germany, autumn migrants, during invasion: 166 (16·2; 22) 125–200 (Ekelöf and Kuschert 1979). French Alps: 172·4 (18·39; 10) 145–198 (Keve 1969). European Turkey, May: ♂ 178 (Rokitansky and Schifter 1971).

*G. g. rufitergum.* Britain: average, whole year, ♂ 168·5 (232), ♀ 158·5 (198); average of ♂ reaching peak in October, 180 (22), low in July, 162 (35); average of series of ♀♀ reaches peaks in May and October–December, 163–166, more or less constant low of 154–158 during rest of year (Seel 1976, which see for details). Oxfordshire (England), March–November: ♂ 171·8 (7·0; 25), ♀ 163·6 (8·6; 29) (Cellier 1992). Finistere (France), October–June: average, ♂ 173·5 (12), ♀ 168·7 (18) (Lebeurier and Rapine 1939).

*G. g. ichnusae.* Sardinia, October–February: ♂ 151·7 (3) 147–158 (Demartis 1987).

*G. g. albipectus.* Western Yugoslavia, September–November: ♂♂ 170, 172, ♀ 164·3 (3) 163–167 (ZMA).

*G. g. graecus.* Crete, November: ♂ 130 (ZMA).

*G. g. brandti* and *bambergi.* USSR, Mongolia, and Manchuria (China), mainly June–September: ♂ 136·6 (11·48; 9) 120–152, ♀ 137·6 (11·86; 6) 115–149 (Piechocki 1959; Nechaev 1969; Piechocki and Bolod 1972; Korelov *et al.* 1974). In USSR, 2 birds 162, 163, but down to 123 if food scarce (Dementiev and Gladkov 1954).

*G. g. hyrcanus.* Northern Iran, July: juvenile ♀, 134·2 (Paludan 1940).

*G. g. krynicki.* North-east Turkey, October–November: ♂♂ 193, 204 (Kumerloeve 1967a).

*G. g. anatoliae.* Western and southern Asia Minor, July: adult, ♂ 180, ♀ 174, 178; juvenile, ♂ 157, ♀ 154·7 (3) 147–165 (Rokitansky and Schifter 1971; ZFMK). Zagros mountains (south-west Iran), May: ♂ 166·7 (3) 158–178, ♀ 163·5 (Paludan 1938).

*G. g. atricapillus.* Israel: 160–191 (Paz 1987).

*G. g. cervicalis.* Northern Algeria, October: ♂ 183, ♀ 168 (3) 164–171 (ZFMK).

**Structure.** Wing short and broad; tip rounded. 10 primaries: p6 usually longest and p5 up to 4 shorter, but occasionally p5 longest and p6 0–2 shorter; p7 2–7 shorter than longest, p8 9–17, p4 2–7; in *glandarius*, *brandtii*, and *atricapillus* groups, p9 36 (20) 30–42 shorter, p10 78 (20) 72–84, p1 31 (20) 26–40, but in *minor* group p9 32 (10) 27–37 shorter, p10 70 (10) 62–78, p1 26 (10) 22–33; in all groups, p10 29.4 (30) 25–36 longer than longest upper primary covert. Outer web of p3—p8 slightly emarginated (most distinct on p4–p6), inner web of (p4–)p6–p9 with notch near base. Tertials short; outermost secondary (s1) 1 longer to 10 shorter than p1 (perhaps dependent on age, distance larger in adults). Tail long (relatively longest in races *minor*, *glaszneri*, *brandtii*, and *cervicalis*), tip slightly rounded; 12 feathers, t2 longest, t6 5–10 shorter, t1 often slightly shorter; in adult and some 1st adults, feathers markedly broad towards truncate tip, in many other 1st adults narrower and more even in width, tip rounded. Bill strong, about half of head length, deep and broad at base, laterally compressed at tip; tip of upper mandible often with distinct hook, but this sometimes absent (shape of bill subject to individual variation and abrasion); bill depth at middle of nostril on average 11·0 in *minor* and *japonicus* group, 11·5 in *brandtii* group, 12·2 in *glandarius* group, 13·0 in *atricapillus* group; in *glandarius* group, depth at middle of nostril in ♂ 12·4 (103) 11·5–13·3 (12·1 or less in 30 birds), in ♀ 11·9 (76) 11·0–13·0 (12·2 or more in 20 birds) (ZMA). Nostril rounded, covered by bristle-like feathers projecting forward from base of upper mandible; *c.* 6 stiff bristles projecting over gape at side at upper mandible. Plumage of head and body soft and full, crown feathers slightly elongated, especially in *atricapillus* group. Tarsus and toes rather long and strong, distinctly scutellate; middle toe with claw 32·7 (20) 30–36 in *glandarius* and *atricapillus* group, 30·5 (8) 28–33 in *minor* group, 29·0 (5) 28–31 in *brandtii* group; outer toe with claw *c.* 77% of middle toe with claw, inner *c.* 70%, hind *c.* 81%.

**Geographical variation.** Marked and complex. 30–60 races have been recognized, combined for convenience into 6–8 subspecies-groups; for grouping arrangements, see Stresemann (1940), Vaurie (1954*b*, 1959), and Goodwin (1986). Recognition of races here follows work by A Keve (also known as Keve-Kleiner or Kleiner): Kleiner (1938, 1939*b–e*), Keve-Kleiner (1943), Keve (1958*b*, 1966*a*, *b*, 1967*a*, *b*, 1969, 1973), and Keve and Dončev (1967), with adaptations after examination of specimens in (especially) BMNH and ZFMK, and with somewhat different interpretation of subspecies-groups. 6 groups recognized. *Bispecularis* group and *leucotis* group show largely black greater coverts and tertials, blue-and-black barred secondaries (except tip), primary coverts, and bastard wing, and restricted white along outer webs of primaries—*leucotis* group (1 race only, south-east Asia) with black cap, *bispecularis* group (7–11 races, Himalaya to Taiwan) with head almost uniform pink or rufous (similar in colour to body), except for black malar patch. In remaining 4 groups, basal ♂ of secondaries white, greater coverts barred blue and black, and outer webs of primaries with much white: *japonicus* group (4–6 races, Japan, except Hokkaido but including surrounding islands of Sado, Tsushima, and Yaku-jima) has cap white with contrasting black streaks, lore and region round eye black, joining black malar patch, body saturated vinous-brown with contrasting white chin and greyish breast; *brandtii* group has head mainly rufous, like body, thus as in *bispecularis* group, but forehead and crown narrowly streaked black; *atricapillus* group has contrasting black cap of rather elongated feathers, size large; *minor* group has cap broadly streaked black, showing fine whitish and vinous streaks only, dark cap extending not as far to rear as in *atricapillus* group, forehead largely black, side of head largely vinous, body colour saturated vinous, size small; *glandarius* group has white or pale buff forehead and forecrown with rather narrow streaks (intermediate between those of *brandtii* and *minor* groups), pale colour of face gradually merging into vinous of body. Previous authors have considered each member of *minor* group to be either aberrant and without close relationships (with each other or with other groups), or to be hybrids between members of black-headed *atricapillus* group and streaky-headed *glandarius* group. However, they have much in common (e.g. saturated colour, and smaller size than any other Palearctic group except *japonicus*) and each differs greatly from nearest race of other groups, pointing to long isolation; intergradation with other groups, if any, appears to be secondary in character (with narrow highly variable hybrid zone); *minor* group, though rather near *glandarius* group, appears to be a relict group of an old *G. glandarius* stock, fragmented in range by intrusion from more recent *glandarius* group from north and by *atricapillus* group from Levant. (C S Roselaar.) For geographical range of individual races in the following account, see heading for this species.

Within *glandarius* group, variation largely clinal and most races poorly defined. In general, colour gradually more pink (less grey) from central Europe westwards, leading through vinous-pink *rufitergum* to saturated rufous-pink *hibernicus*; also, slightly more rufous on head and neck toward east. Towards south, either paler, light vinous-pink above and whitish below (*lusitanicus*, *jordansi*, *ferdinandi*), or darker, either deeper vinous-brown (*corsicanus*) or greyer all over (*fasciatus*, *ichnusae*, *cretorum*); within these paler and darker southern races, differences often slight, and recognition of intermediate races between nominate *glandarius* and any of these paler or darker races is doubtful—e.g. *graecus* (intermediate between nominate *glandarius* and *cretorum*) and *albipectus* (between nominate *glandarius* and *jordansi*)—but these intermediates maintained here, as differences just visible in series of skins. *G. g. rufitergum* from England and Wales is more saturated vinous-pink on upperparts, chest, breast, and flank than nominate *glandarius*, hardly grey, ear-coverts deeper rufous and with vinous tinge at rear, fine grey and vinous bars on nape often more pronounced (blue-grey bars frequently invading black of crown-streaks); birds from Scotland sometimes separated as '*caledonicus*', larger and more vinous than *rufitergum*, barring of cap and saturated vinous-pink of upperparts, chest, breast, and flank as in *rufitergum*, but grey of body slightly more pronounced and size large, tending slightly towards Scandinavian nominate *glandarius*, but nearer *rufitergum*. Irish race *hibernicus* more saturated vinous-brown on body, vinous tinge extending on to forecrown; black streaks of crown often with extensive fine blue-grey bars; lore and ear-coverts vinous-rufous; forehead buffish, sometimes heavily spotted black; buff-white chin contrasting more with vinous-brown belly; mid-belly tinged cream-buff, less whitish. Birds of Bretagne (sometimes separated as '*armoricanus*') rather saturated vinous-brown on upperparts, chest, and flank, intermediate between *hibernicus* and *rufitergum*, but breast and belly rather pale; here included in *rufitergum*. Birds from Netherlands, western Germany, and north-east France intermediate between *rufitergum* and nominate *glandarius*, but mainly nearer latter (see, e.g. Voous 1944, 1953); birds from Belgium, central and southern France, and western Switzerland (north of Alps) intermediate to varying extent between those of Britain, Italy, and central Europe but included here in nominate *glandarius* (Kleiner 1938; Voous 1945, 1953; Keve 1958*b*). In Finland, eastern Poland, eastern Rumania, and European USSR, populations occur which closely resemble nominate *glandarius*, but which on average show more buffish-isabelline ground-colour of forehead and forecrown, deeper vinous-rufous nape, a distinct rufous band across hindneck, more extensive rufous ear-coverts, and slightly browner-grey mantle and chest; these sometimes separated as 'septentrionalis' Brehm, 1831, but differences slight and not all birds separable (Kleiner 1939b; Voous 1953; Keve 1966a; Keve and Dončev 1967); slightly more rufous-brown tone perhaps point to introgression of brandtii characters, but size large (as nominate glandarius), not as small as brandtii (C S Roselaar). True intermediates between nominate glandarius and brandtii (sometimes separated as 'severtzowi') occur in relatively narrow zone through east European USSR (Johansen 1944; Dementiev and Gladkov 1954; Keve 1966a), where individuals highly variable. In south, lusitanicus from Salamanca (Spain) has ground-colour of cap whiter than nominate glandarius, extending further to rear; black crown-streaks highly variable in width, grey of mantle and scapulars lighter and less saturated, tinged light vinous-pink, chin and upper throat whiter; chest, side of breast, and flank paler vinous (hardly grey), lower throat in some birds (as in other races) sometimes with dark grey dots or suffusion; breast and belly distinctly paler, vinous-pink with white wash. Crown of jordansi from Sicily pale as lusitanicus, remainder of upperparts almost equally pale, but slightly greyer, less diluted vinous-pink; underparts pale as lusitanicus, but chest whiter, rather like belly, gradually merging into white of throat, and flank slightly greyer vinous. G. a. albipectus of mainland Italy and coastal Yugoslavia north to southern Switzerland (Kleiner 1939e) and south to Ionian islands (Kleiner 1939b; Keve and Dončev 1967) rather variable: some birds as jordansi, others with more vinous-grey upperparts and more saturated vinous chest and flank, nearer nominate glandarius of (e.g.) Netherlands, but belly usually still whitish vinous-pink; see also Kleiner (1939b) and Keve (1966b). G. g. ferdinandi from south-east Bulgaria and neighbouring northern European Turkey near lusitanicus, less grey than neighbouring graecus; more vinous-pink on upperparts than lusitanicus, less pallid vinous-grey (Keve-Kleiner 1943; Keve 1958b; Keve and Dončev 1967). Of dark greyish races of glandarius group, fasciatus of eastern and southern Spain has upperparts extensively grey, similar to nominate glandarius from Sweden; underparts distinctly greyer, however, especially on chest and flank, breast rather dark vinous; black crown-streaks on average wider than in nominate glandarius. Grey upperparts of ichnusae from Sardinia slightly purer and more extensive than in both fasciatus and nominate glandarius, underparts as grey as in fasciatus or grey slightly more extensive; black streaks on crown slightly narrower than in fasciatus. G. g. corsicanus from Corsica dark, but mainly vinous, not grey as in ichnusae (Keve 1958b); rather similar to nominate glandarius from western Europe, but body darker and colours more saturated. In cretorum from Crete, extent of grey on upperparts as in Scandinavian nominate glandarius or fasciatus, but grey slightly more pure, less tinged vinous or drab, contrasting more with vinous of hindneck; underparts extensively tinged grey, often even more so than in ichnusae; graecus from mainland Greece, south-east Yugoslavia (Dimovski and Matvejev 1955; Keve 1958b; Keve and Dončev 1967; but see Stresemann 1920) and central and southern Bulgaria (Niethammer 1943, 1950) is similar to cretorum, but grey on underparts less extensive and belly whiter, contrasting rather with vinous-grey breast and flank; also rather like Scandinavian nominate glandarius, but mantle, scapulars, breast, and flank slightly purer grey and belly whiter.

In brandtii group, brandtii from Ural mountains east to Lake Baykal area characterized by bright rufous-cinnamon head and neck (chin buffish and malar patch black, as in other races); forehead, nasal bristles, lore, and area round eye more buffish-cinnamon, heavily spotted black (spots sometimes almost coalescing); crown rather narrowly streaked black, streaks c. 2–3 mm wide on forecrown but reduced to c. 1 mm wide towards rear.

Upper mantle buff-brown, gradually merging into pure medium grey of lower mantle, scapulars, and back. Underparts buff-brown, tinged rufous on lower throat, slightly vinous on side of breast and flank, slightly grey on chest, pale salmon on mid-belly; tail-coverts, wing, and tail as in other races; smaller than nominate glandarius. G. g. kansuensis from Chinese eastern Tien Shan and northern Kansu like brandtii, but larger and on average many fewer black spots on forehead, lore and round eye; bambergi from Mongolia and Transbaykalia eastward intermediate in size, slightly paler on head and body, intermediate in amount of black on forehead, lore, and round eye, but much local variation in body colour and in amount of black. See also Lönnberg (1909), Stresemann (1928b), Stegmann (1931a), Stresemann et al. (1937), Voous (1945), Dementiev and Gladkov (1954), and Vaurie (1954b) for differing opinions about characters of races of brandtii group and their distribution.

In atricapillus group, krynicki from Caucasus area south to Iranian Azarbaijan and west to north-east Turkey has nasal bristles and forehead white, spotted black to variable extent (rarely, all black); cap contrastingly uniform black, feathers at rear distinctly elongated, tips sometimes faintly barred grey. Side of head from lore to ear-coverts vinous-pink, white of feather-bases showing through on shorter feathers; hindneck and side of neck vinous; mantle and scapulars dark vinous-grey; chest, breast, and flank dark greyish-vinous, chest in particular often distinctly grey, lower throat frequently blotched dark grey; general tone of body nearest ichnusae, but chest grey; also, more streaked vinous and white behind eye. Slightly paler populations occur in westernmost Caucasus area (Keve 1973). In anatoliae from western and southern Asia Minor and south-west Iran, forehead and ear-coverts whiter than krynicki, ear-coverts slightly streaked vinous only; body distinctly less saturated vinous and grey, belly more vinous-pink; forehead and ear-coverts less uniform white than atricapillus, upperparts more greyish-vinous, less vinous-pink, and remainder of underparts less white. Opinions about distribution of races in Asia Minor differ between authors: according to Keve (1973), krynicki breeds west to Ulu Dag; throughout northern Asia Minor, colour gradually darker towards east (Kleiner 1939b; Kumerloeve 1961), but only in north-east are birds as dark as those of central Caucasus; birds from southern Taurus attributed mainly to atricapillus or intermediate (Vaurie 1959; Kumerloeve 1961); populations of western Turkey (including Istanbul) considered to be atricapillus (Watson in Keve 1973). According to large series examined (BMNH, ZFMK), birds from Ulu Dag, Bolu, and Kastamonu (north-west Asia Minor) are similar to anatoliae of western Asia Minor and Taurus mountains, while those of south-west Iran (sometimes separated as 'susianae' Keve, 1973) are also similar or very slightly paler below (Paludan 1938; C S Roselaar); all these slightly darker and greyer still than atricapillus. Slight colour variations occur between several populations of western Turkey and neighbouring islands of Greece, but these insufficient to split off additional races from anatoliae, except for samios of Samos and Ikaria (as dark as krynicki, but mantle and scapulars rufous dark grey, hindneck violet-rusty, and underparts dark, rather rufous: Kleiner 1939c; Keve 1973). Included here in anatoliae is 'hansguentheri' Keve, 1967, restricted to small area on both sides of Bosporus, which is pale, near anatoliae, but shows rather variable extent of black on crown, latter being slightly streaked white in some birds, apparently due to introgression of characters from glandarius group, and mantle either greyish (near graecus) or more reddish-vinous (near ferdinandi) (Keve 1967a); 'zervasi' Keve, 1939 (Lesbos), 'chiou' Keve, 1939 (Khios), and birds from Izmir (RMNH, ZFMK) paler than typical anatoliae from Taurus mountains, and 'rhodius' Salvadori and Festa, 1913 (Rhodes and perhaps Kos) paler still, almost as pale as *atricapillus*, but greyer. *G. g. atricapillus* from Levant north to Lebanon and southern Syria has broad white forehead; lore, area round eye, and ear-coverts white, apart from occasional slight vinous-pink suffusion; hindneck, mantle, scapulars, back, and flank pale vinous-pink (darkest on hindneck), hardly grey, chest pale vinous with slight grey tinge, hardly contrasting with white throat (but occasionally mottled grey on lower throat), breast and belly cream-pink to whitish. *G. g. iphigenia* from Crimea close to *anatoliae*, distinctly paler than *krynicki*; forehead white with black spots; black of cap mixed with vinous at feather-bases, finely barred grey on longer feathers, less uniform than in *anatoliae*, and often appearing streaked; upperparts grey with slight cinnamon cast, underparts pale vinous-pink with whitish mid-belly; close to *anatoliae* (especially to some '*hansguentheri*'), but separated in view of isolated breeding range. Nasal bristles and forehead of *cervicalis* from northern Tunisia and north-east Algeria white with blackish tips, forehead sometimes largely black; cap black, feather-sides of hindcrown partly tinged chestnut (contrast between hindneck and black cap not as marked as in *atricapillus* and *anatoliae*); broad chestnut band across hindneck down sides of neck, often joining black malar patch at rear, contrasting strongly with white side of head; mantle and scapulars grey, tinged vinous on upper mantle, outer scapulars and towards back; chest, side of breast, and flank light vinous-grey, gradually merging into extensively whitish belly; lower throat often with dark grey dots; a distinct race in fresh plumage, but somewhat resembles *atricapillus* when worn, as chestnut half-collar then paler, more vinous, indistinctly extending down to side of neck, grey of upperparts lighter, tinged vinous-pink, and underparts paler, greyish-white with vinous-pink flanks. *G. g. whitakeri* of northern Morocco and extreme north-west Algeria has traces of white and vinous streaks in black of cap; half-collar more vinous-brown, less chestnut; rear of ear-coverts tinged vinous; upperparts light grey, without vinous tinge; underparts as *cervicalis* or slightly darker grey; depth and extent of grey on body rather like *fasciatus*. Birds from extreme north-west Algeria on average smaller and greyer than those of northern Morocco; those from Tanger paler than those from Er Rif (Kleiner 1939*b*; Vaurie 1954*b*).

In *minor* group, *minor* from Atlas Saharien in Algeria has forehead and crown black, faintly streaked white on forehead and narrowly streaked vinous-grey on crown; black of cap not extending as far to rear as *cervicalis*; lore and area round eye buff-white with some grey mottling; ear-coverts, side of neck, and hindneck deep vinous-brown, slightly tinged grey on hindneck; mantle, scapulars, and back dark grey; underparts saturated vinous-grey, slightly purer vinous towards lower throat and lower flank, white restricted to vent and under tail-coverts; thus, birds from Djelfa area in eastern Atlas Saharien quite different

from nearby *cervicalis* from Batna area, underparts and side of head almost similar to those of Scandinavian nominate *glandarius*, but cap much blacker than nominate *glandarius*, mantle and scapulars darker and greyer, and size much smaller; if worn, cap of *minor* sometimes more streaky, mantle and scapulars brownish-grey, underparts dirty buff-grey with vinous-pink cast. Birds from Haut Atlas (Morocco), sometimes separated as '*oenops*' Whitaker, 1897, but inseparable from typical *minor* from Djelfa area (BMNH). Birds of Moyen Atlas, sometimes separated as '*theresae*' Meinertzhagen, 1939, larger than *minor*, intermediate in size and colour between *minor* and *whitakeri*, width of black crown-streaks variable (Meinertzhagen 1939, 1940; Vaurie 1954*b*). Upperparts of *glaszneri* from Cyprus similar to *minor*, but black streaks on forehead and crown slightly narrower, feather-sides more extensively rufous-brown, hardly white and grey; outer scapulars and rump more strongly vinous, mantle and scapulars less uniform dark grey than *minor*; side of head deep vinous-brown; lower throat to lower belly deep vinous, less grey than *minor*. In *hyrcanus*, black streaks of crown narrower still (though on average wider than in nominate *glandarius*), but forehead sometimes almost black; black of crown sometimes finely streaked grey; ground-colour of cap buff-white (forehead) or vinous-pink (crown), mantle and scapulars saturated vinous-brown, hardly tinged grey; underparts saturated deep vinous, like *glaszneri* or even deeper and less grey. In Lenkoran' area (south-east Azerbaydzhan, USSR), highly variable hybrids between *hyrcanus* and *krynicki* occur (sometimes separated as '*caspius*'): either near *hyrcanus* in colour but having largely black forehead and crown with limited vinous-brown streaking, or near *krynicki* but having deeper vinous-brown side of head, hindneck, and underparts (Keve 1973; ZMA).

Juveniles of *atricapillus* and *brandtii* groups closely resemble respective adults, differing mainly in looser and shorter plumage of head and underparts; upperparts often tinged rufous, some juveniles of *atricapillus* group showing rufous half-collar, and those of *atricapillus* in particular then resembling *cervicalis*. See also Keve (1973). Juveniles of *glandarius* and *minor* groups have rufous head and neck with rather narrowly streaked cap, resembling adults of *brandtii* group; depth of grey or rufous-brown of mantle and scapulars and depth and extent of greyish-buff to rufous on chest and flank highly variable, individually rather than geographically; only a slight tendency of juveniles of rufous races to be more rufous than juvenile nominate *glandarius* and of dark grey races to be dark (see, e.g. Keve 1967*b*), with many birds indistinguishable.

For extralimital races, see Kleiner (1939*b*) and Vaurie (1954*b*, 1959). For evolutionary history of distribution, see Voous (1953) and Keve (1969). Nearest relative is Lanceolated Jay *G. lanceolatus* of Himalayas (Horváth 1976).                    CSR

---

## *Perisoreus infaustus*  Siberian Jay

PLATES 2 and 5 (flight)
[between pages 136 and 137]

Du. Taigagaai      FR. Mésangeai imitateur      GE. Unglückshäher
RU. Кукша      SP. Arrendajo funesto      Sw. Lavskrika

*Corvus infaustus* Linnaeus, 1758. Synonym: *Cractes infaustus*.

Polytypic. Nominate *infaustus* (Linnaeus, 1758), Fenno-Scandia and Kola peninsula (Russia), grading into *ostjakorum* in north European Russia and into *ruthenus* in Kareliya; *ostjakorum* Suschkin and Stegmann, 1929, Russia north of *c.* 64°N from northern Urals east to Olenek basin; *ruthenus* Buturlin, 1916, Russia from Lake Onega, St Petersburg area, and Moscow east to Barnaul, Tomsk, and Vakh basin, grading into *ostjakorum* in north. Extralimital: 3–13 races from Siberia south to Mongolia and Manchuria.

**Field characters.** 30–31 cm; wing-span 40–46 cm. Noticeably smaller, slighter, and proportionately longer-tailed than Jay *Garrulus glandarius*, with much shorter, more pointed bill. Smallest and most delicate of west Palearctic Corvidae; rather drab brown-grey, with rufous-chestnut near wing-bend and on under wing-coverts, rump, and sides of tail setting bird 'on fire' in flight. Sexes similar; no seasonal variation. Juvenile separable at close range. 3 races in west Palearctic; only Lapland race, nominate *infaustus*, described here; other races differ in plumage tone and size of wing-patch (see Geographical Variation).

Adult. Complete moult, May–July. Crown, nape, and upper face sooty-brown, with dense, pale buff bristles over base of bill clearly visible. Lower hindneck, back, and underparts dull brown, more or less suffused grey, becoming rufous on belly, rear flank, and under tail-coverts. Wings sooty-brown, on which fox-red outer greater coverts, greater primary coverts, and bases to inner primaries and outer secondaries form obvious bright warm patch on upperwing when folded or short bar in flight. Underwing shows even larger and much paler chestnut panel in similar position to that on upperwing, rufous axillaries and coverts contrasting with buff-grey bases and sooty rim to flight-feathers. Rump, upper and under tail-coverts, and tail chestnut-brown to fox-red, but bright uniform colour of uppertail interrupted by lead-grey central pair of feathers; grey tips to other tail-feathers rarely show in the field. Bill and legs black. Juvenile. Similar to adult but ground-colours darker above and paler and duller below, and rufous areas less bright.

Unmistakable. Chief problem in identification is marked shyness during breeding season in distinct contrast to often tame behaviour in winter or time of food shortage. Flight differs from *G. glandarius*: quite accomplished, but deeply undulating due to bird closing wings between burst of quick, easy, loose wing-beats. Gait a brisk, light hop, varied by sidling and jumping. Able to cling upside down like tit *Parus*. Small groups often move in follow-my-leader line.

Voice includes: somewhat harsh, melancholy 'chair' or 'tchair' recalling angry tit or *G. glandarius*; brisk, cheerful 'kook kook'; disyllabic 'whisk-ee' or 'kij kij'; mew 'geeah', suggesting buzzard *Buteo*.

**Habitat.** Breeds across higher latitudes of west Palearctic, and is predominantly at all seasons a bird of coniferous forest, mainly of Norway spruce *Picea abies*, and Scots pine *Pinus sylvestris* but also of larch *Larix decidua* and downy birch *Betula pubescens*. Favours dense stands of forest rather than open growth lowlands, but in south of range extralimitally ascends to 2200 m in Altai (Dementiev

and Gladkov 1954). Occupies natural habitat which has suffered only locally from modification by man, but readily attaches itself to human travellers and their living quarters. Quite at home on ground; flies only in lowest airspace and usually for short distances. Contrasts with Jay *Garrulus glandarius* in lack of fear of man, but this has little effect on choice of habitat since normal range is largely uninhabited by people.

**Distribution.** Marked range contraction in Fenno-Scandia in recent decades.

Sweden. Probably bred as far south as 60°15′N in mid-19th century; has since withdrawn from southern parts of range (LR). Finland. Striking contraction of breeding range in recent decades, probably due to modern forestry practice and spread of cultivation and human habitation; scattered populations in south in danger of extinction (Koskimies 1989; Kemppainen and Kemppainen 1991; OH). Latvia. Only one definite record in 20th century; recorded more often in past (Vīksne 1983).

Accidental. Poland, Czechoslovakia.

**Population.** Has decreased in Fenno-Scandia in recent decades.

Sweden. Rough estimate of 150 000 pairs (Ulfstrand and Högstedt 1976); decreasing (LR). Finland. Striking decrease (OH); perhaps 50 000–100 000 pairs (Koskimies 1989).

Survival. Oldest ringed bird 11 years 5 months (Staav 1983).

**Movements.** Resident, with all-year territory, but winter movements reported from outside Europe for northern population in eastern Siberia.

No movements observed in northern Scandinavia or on Kola peninsula (north-west Russia), with territory occupancy by same bird for up to 12 years in Scandinavia (Blomgren 1964) and 7 years in Russia (Kokhanov 1982). For other evidence of marked site-fidelity, see Social Pattern and Behaviour.

Rare but regular vagrant outside breeding range, mainly in autumn and winter. Occasional observations from southern and central Scandinavia in autumn (Wiman 1943; Moberg 1949; Ekström 1952; Wahlmino and Petersson 1956; Nilsson 1957). Rare migrant in Moscow region with only 2 observations (October) in 60 years (Ptushenko and Inozemtsev 1968); in St Petersburg region, occasionally recorded in autumn and winter in central and western areas where none breed (Mal'chevski and Pukinski 1983); no records south of breeding range reported in 30 years for Volga-Kama region (Popov 1978); first observation in Ukraine in January 1977 (*Br. Birds* 1988, 81, 338).

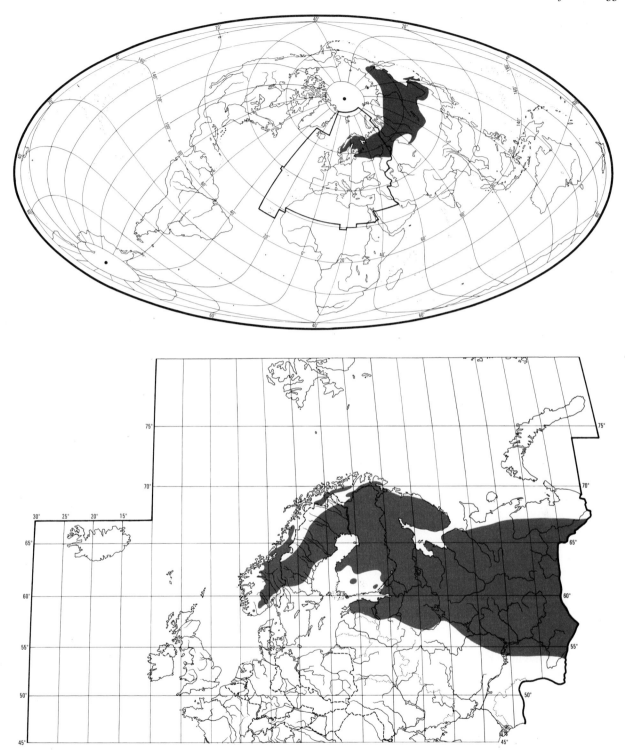

In northern Siberia, birds reported to move south in winter (Rogacheva 1988). Influx of birds to Yakutsk region reported January-February, and interpreted as migrating birds coming from north (Larionov 1959). JE

**Food.** Omnivorous all year. Captures and kills small mammals and small passerines up to size of tit *Parus*; also plunders nests for eggs and nestlings. Feeds from all kinds of carrion. Takes variety of arthropods. Plant material forms substantial portion of diet, especially various berries occurring in coniferous forest. Andreev (1982a) reported preference for animal food. Remarkably tame and regularly searches out humans in winter (Blomgren 1964), taking discarded food. One report of stealing fish left to dry outdoors (Krechmar 1966). Attracted by noise and comes to camp fires where food offered is readily taken. Can easily become hand-tame if regularly offered food. Foraging bird carefully examines conifer branches and trunks, moving up and down trunk, fluttering from branch to branch and clinging to bark; examines crevices under bark, prises off bark, and searches exposed area. Feeds on ground especially during breeding and summer. Attacks voles *Clethrionomys* in spring and one report of unsuccessful attack on chipmunk *Tamias sibiricus* by a pair. (Andreev 1982a.) In northern Sweden, 2 birds jointly attacked small rodent, pecking at it alternately until dead, and 2 others pursued Willow Tit *Parus montanus* carrying food, forcing it onto snow and killing it. Flying insects sometimes caught by sallying; bird reported standing on ground leaping 50 cm into the air to catch flying insect. Larger beetles (Coleoptera), caterpillars (Lepidoptera), lizards (Lacertidae), and nestlings normally carried away and repeatedly beaten against branch until dead. Smaller prey sometimes swallowed whole and larger prey torn to pieces before being eaten. Eggs eaten when small enough to be swallowed whole; larger eggs reported to be ignored. (Blomgren 1964.) Berries swallowed whole or crushed in bill. In spring, searches out first berries to be exposed by snow melting on southern slopes (J Ekman). Conifer seeds never seen taken from cones, but collected on snow (Blomgren 1964). Fur of small rodents and skin of berries regurgitated in pellets (J Ekman).

Marked tendency to store food. Food-storing so intensive that it has been suggested that *P. infaustus* overwinters successfully largely because of stores, and uses them to prepare for breeding and rearing young (Andreev 1982a). Storing occurs in spring (Pravosudov 1984) as well as in autumn and winter (Blomgren 1971; Andreev 1982a; Semenov-Tyan-Shanski and Gilyazov 1991). Food stored in trees: in Murmansk region (north-west Russia), mainly in conifers (pine *Pinus sylvestris*, spruce *Picea obovata*, *P. fennica*: Pravosudov 1984); in Scandinavia, also reported in birch *Betula* (Haftorn 1952); east of Urals, in larch *Larix*, less commonly in Korean willow *Chosenia*, and still less in willow *Salix* (Krechmar *et al.* 1978; Andreev 1982a). Typical hoarding sites are bark crevices, in lichens *Usnea*, or among needles (Blomgren 1964; Andreev 1982a; Pravosudov 1984; Semenov-Tyan-Shanski and Gilyazov 1991). Food carefully concealed and rendered almost invisible. Small pieces of bark and lichens used to cover hoarded items (Haftorn 1952). In autumn, food carried in sublingual pouch and regurgitated at hoarding site (Blomgren 1964). Sublingual salivary gland enlarged to produce saliva to form food balls and to make them stick to hiding places (Andreev 1982a; Pravosudov 1984). Captive birds soon started to hide pieces of food in aviary after enveloping them in saliva (Andreev 1982a). In spring, food carried in bill and not treated with saliva (Pravosudov 1984). Hoarding behaviour highly stereotyped: prey first killed, then taken to hoarding site at mean distance of 10 m (3–25) from where prey found; hoarder looks about quickly, flies to suitable site, pushes item into store with bill, then flies off; storing usually followed by bill-wiping. Stores dispersed over whole winter territory and possibly recovered by memory. (Pravosudov 1984.) Hoarding sites used only once (Blomgren 1964; Pravosudov 1984). From Murmansk, Pravosudov (1984) reported only 1 item per site, but multiple items in one site observed in Sweden; 1–3 bilberries *Vaccinium* per site (usually *Usnea* or *Alectoris* lichens). Blomgren (1964) reported concentration of hoarding in single trees, these containing large numbers of (mainly) bilberries. Re-hoarding occurs (Haftorn 1952). Berries dominate among hoarded food items in winter, but wasps (*Vespidae*) occur (Blomgren 1964). No data on hoarding frequencies in autumn and winter. Storing studied during breeding season (21 May to 2 June) in Murmansk (Pravosudov 1984): hoarding frequency between 4 items per hr (breeding pair with nestlings) and 12 items per hr (non-breeding birds); 80–200 items stored per day.

Forages on ground and in trees. In breeding season, often on ground (Vorobiev 1963; Reymers 1966); 55% of foraging time on ground in spring (Pravosudov 1984). All parts of tree used (Pravosudov 1984); insects taken under pieces of loose bark (Vorobiev 1963). Feeds lower in trees in autumn and winter, in crowns in early spring (Reymers 1966). Feeding rates: in spring, 5–20 s between prey items for breeding pair (1 pair), and 30–60 s between prey items for non-breeding birds (1 solitary bird and 1 non-breeding pair). Breeding bird calculated to capture 1300 items per day. (Pravosudov 1984.)

Diet includes the following. Vertebrates: carrion including deer (probably reindeer *Rangifer*), moose *Alces*, grouse (Tetraonidae). Live prey: voles *Clethrionomys*, shrews *Sorex*, small passerines, lizard *Lacerta vivipara*, and fish fry (Salmonidae). Invertebrates: damsel flies and dragonflies (Odonata), stoneflies (Plecoptera), grasshoppers (Orthoptera: Acrididae), bugs (Hemiptera: Pentatomidae, Cicadidae), adult lacewings (Neuroptera), adult and larval Lepidoptera (Tortricidae), adult flies (Diptera: Tipulidae), adult Hymenoptera (Siricidae, Ichneumonoidea, ants Formicidae, wasps Vespidae), adult beetles (Coleoptera: Carabidae, Geotrupidae, Buprestidae, Elateridae, Tenebrionidae, Cerambycidae, Chrysomelidae, Curculionidae, Scolytidae), spiders (Araneae), centipedes (Chilopoda), earthworms (Lumbricidae), slugs (Gastropoda). Plant material includes mainly seeds and berries (much less often, needles, bark, and shoots) of spruce *Picea*, juniper *Juniperus*, birch *Betula*, apple *Malus*, rowan *Sorbus*, cloudberry *Rubus*, bird cherry *Prunus*, red currant *Ribes*, bearberry *Arctostaphylos*, bilberry, etc.

*Vaccinium*, crowberry *Empetrum*, violet *Viola*, cow-wheat *Melampyrum*, may-lily *Maianthemum*, sedge *Carex*, grass *Melica*, and various fungi (Portenko 1939; Novikov 1952, 1956; Vorobiev 1954, 1963; Neufeldt 1961; Blomgren 1964; Reymers 1966; Krechmar *et al.* 1978; Zonov 1978; Andreev 1982*a*; Germogenov 1982; Kokhanov 1982; Pravosudov 1984; Durnev *et al.* 1991).

Adult diet largely insects in spring and summer; plant material makes up large portion in winter. Limited information from west Palearctic, and account includes extralimital data from east of Urals. Of 18 stomachs collected May–September in Karel'skaya ASSR (north-west Russia), all contained animal food, chiefly beetles (88% of stomachs), also Hymenoptera (60·5%; *Sirex gigas* in 22%), lizards (*Lacerta vivipara*; 17·5%), and vole (*Clethrionomys*; 5·5%); plant material in 55%; large caterpillars of *Celerio galii* (11% of stomachs) suggest feeding in clear-felled areas (Neufeldt 1961). In direct observations and 3 stomachs from Kola peninsula (north-west Russia), May–October, waste near human settlement comprised 48% of 82 items, carrion 6%, small vertebrates 17%, arthropods 8%, berries and seeds 15%, fungi 5%, and lichens 1% (Semenov-Tyan-Shanski and Gilyazov 1991). 3 stomachs collected in Anadyr' region (north-east Russia) on 11 May (1) and 15 May (2) contained 2 small mammals, 2 beetles, 1 spider, 4 insect larvae, *c.* 100 seeds, *c.* 25 berries, shoots, and bark fragments (Portenko 1939). Of 10 stomachs collected in Irkutsk (southern Lake Baykal), spring and summer, all contained animal food (spiders, bugs, beetles, Lepidoptera larvae, Hymenoptera, ants, flies) and half contained conifer seeds (Durnev *et al.* 1991). In stomachs (sample size not known) from Lena valley (eastern Russia) in summer, insects in 95·5%, notably Curculionidae (in 20·4%), Cerambycidae (11·3%), larval Lepidoptera (9·8%), and Acrididae (8·3%) (Germogenov 1982). In direct observations and 3 stomachs collected on Kola peninsula, November–April, waste near human settlement comprised 54% of 56 items, carrion (deer, moose, grouse) 30%, small vertebrates 2%, arthropods 7%, berries and seeds 5%, and lichens 2% (Semenov-Tyan-Shanski and Gilyazov 1991). In Omolon valley (north-east Russia), winter, of 30 stomachs, rodents in 57%, beetles in 20%, Symphyta (Hymenoptera) 17%, fish 10%, larval Trichoptera 7%, bilberries 43%, *Rosa* berries 37%, fungi 20% (Andreev 1982*a*). In Irkutsk in autumn and winter, both plant and animal food found in all of 20 stomachs; rodents (Muridae) in 45% of stomachs, spiders 15%, grasshoppers 30%, bugs 35%, beetles 55%, Lepidoptera larvae 35%, Hymenoptera 25%; plant material chiefly may-lily berries (65% of stomachs), *Pinus sibirica* seeds (20%), and *Vaccinium* berries (55%) (Durnev *et al.* 1991). Another 22 stomachs collected in winter in Irkutsk area contained by volume/frequency of occurrence 8/15% small mammals, 14/50% beetles, 9/23% grasshoppers (Acrididae), 21/23% insect larvae, 8/18% cereal grain, 15/50% rowan berries, 5/20% bird cherry, and 20/80% probable

may-lily berries (Zonov 1978). Based on stomachs (sample size and time of year unknown) from Kola peninsula, Novikov (1956) reported small rodents in 41% of stomachs, spiders in 43%, also shrews, Vespidae larvae and pupae, and various seeds and berries, etc. (spruce, birch, rowan, cow-wheat, crowberry, various *Vaccinium*, *Rubus*, fungi).

Energy consumption calculated from evening stomach contents and photoperiod mid-winter, Omolon valley: maximum dry weight of stomach contents of birds shot on roost or shortly before roosting 23·8–29·2 g; estimated consumption rate 0·56 g dry matter per hr (0·46–0·66); daily food requirement (3·5 hrs daytime activity) 12·7–18·1 g dry matter. Captive bird consumed 12·6–13·5 g dry matter (meat and berries) per day. (Andreev 1982*a*, *b*.) ♀ brooding chicks receives 12–14 g food per day (Kokhanov 1982).

Nestling diet poorly known. Fed with half-digested and regurgitated food. Animal food 64·8% by weight in 19 collar-samples from Murmansk (chicks 6–20 days old); spiders 47·8% by weight and absent in only 4 samples out of 19; mean number of spiders per portion 9 (maximum 28); other animal food 17% by weight (adult and larval insects, voles); plant material 4·4% by weight, remainder unidentified (Kokhanov 1982). Nestling diet studied by collar-samples and collection of pellets and food remains by nest in Lena valley: of 639 items, spiders 58·5% by number, 89·1% by frequency of occurrence (chiefly Lycosidae and Gnaphosidae: 18/20·4%), insects 37·8% (Lepidoptera 15·6%, pupal Hymenoptera 12·1%, beetles 5·6%), earthworms 12·5% by frequency of occurrence, 1·3% by weight, Myriapoda 1·6/0·1%, rodents 20·3/0·5%, and *Vaccinium* berries 9·4/1·7% (Germogenov 1982). One pair fed 2 nestlings with portions of average 2·14 g (1·12–2·94, *n* = 12); brood of 3 (10–17 days old) received 28–29 portions per day (20 g per chick per day) (Kokhanov 1982). JE

**Social pattern and behaviour.** Little information, owing to combination of secretive behaviour, sparse populations, and snowbound terrain. 2 major studies on social behaviour in northern Sweden (Blomgren 1964, 1971; Lindgren 1975), and one in Murmansk region (north-west Russia) (Kokhanov 1982).

1. Adults in resident territorial flocks all year. Flock size at Arvidsjaur (northern Sweden) 2·8 ± SD0·2 birds (*n* = 15) (J Ekman) but mean 4–5 some years, probably reflecting reproductive output (F Lindgren). On Kola peninsula (north-west Russia), usually in pairs but occasionally 3–6 birds in winter (Semenov-Tyan-Shanski and Gilyazov 1991). Adults accompanied in winter flocks by retained offspring and juvenile immigrants (Blomgren 1964; Kokhanov 1982; J Ekman). In 10 winter flocks at Arvidsjaur, relatedness measured through ringing of nestlings and DNA fingerprinting (J Ekman); adults accompanied by 1 retained offspring (3 flocks), 1 immigrant juvenile (3 flocks), 1 retained offspring plus 1 immigrant juvenile (3 flocks), and 2 retained siblings (1 flock). Rarely whole flock shifts to new (adjacent) territory (Blomgren 1964; Lindgren 1975). Most information on winter densities extralimital, but in Arvidsjaur 5·6 birds per km² (J Ekman). In north-east Altai, 0·3 birds per

km² in taiga, 3 birds per km² in sparse forest (Ravkin 1973); 0·6 birds per km² in foothill taiga of northern Urals (Bobretsov and Neufeld 1986); 12–16 birds per km² in spruce *Picea* forest in central Siberia (Rogacheva 1988); 12·4 birds per km² in foothill forests and 7·4 birds per km² in montane taiga in southern Urals (Podol'ski and Sadykov 1983). Winter territory large: 2 territories at Boden (northern Sweden) 1 km² and 1·5 km² (Blomgren 1964); at Arvidsjaur, Lindgren (1975) gave 0·5–1 km² (sample size not known). BONDS. Mating system monogamous. Close association between mates, but attempted copulation by extra-pair ♂ with sitting ♀ observed (Lindgren 1975). Pair-bond probably life-long. No divorces observed (Blomgren 1964, 1971; Lindgren 1975), but for re-pairing after apparent loss of mate, see Breeding Dispersion (below). Both sexes participate in nest-building (Blomgren 1964; Lindgren 1975). ♂ lays foundation and ♀ completes (Kokhanov 1982). Only ♀ incubates and broods nestlings (Blomgren 1964; Lindgren 1975). Young fed by both sexes, for first few days only by ♂ but ♀ gradually contributes more (Blomgren 1964; Kokhanov 1982). Young feed independently after c. 1 month, but some observed to beg throughout 1st year (Blomgren 1964). Retained offspring (ringed) observed being fed by parent after begging at 13 months (J Ekman). Common to find extra birds associating with breeding pair (Lindgren 1975); at 30% of 23 nests at Arvidsjaur (J Ekman). Role of extra birds unclear, but joint incubation found in 1 out of 91 nests (Lindgren 1975) and observations suggesting more than 2 birds feeding young occur (Carlson 1946). Unrelated ♂♂ chased by paired ♂ (J Ekman). In Arvidsjaur, 1st-year birds in possession of territory breed (J Ekman). In Murmansk, Kokhanov (1982) reported first breeding at 3 years (2 cases) after 1st year in parental territory and another 2 years as territory-owner. BREEDING DISPERSION. Territorial and solitary or in small groups. Shift from established territory rare and typically involves movement to adjacent vacant territory (Blomgren 1964; F Lindgren). Territory all-purpose. Territory size difficult to estimate, but less during breeding than at other times; nests placed within restricted part of territory (Blomgren 1964; Lindgren 1975). Minimum 10 m reported between nest-sites of same pair in consecutive years (Lindgren 1975). Mean size of breeding territory 52 ha (45–57) for 5 pairs in Murmansk (Kokhanov 1982). Neighbouring nests 500–1000 m apart in Arvidsjaur (J Ekman). Trespassing uncommon but occurs (Lindgren 1975). Site-fidelity marked. In Sweden, 1 ♀ held same territory for 12 years, and several pairs held same territory for 2 consecutive years (Blomgren 1964). 13 cases where adults have stayed in same territory for 2–12 years; among these, 8 cases including both ♂♂ and ♀♀ where birds have re-paired within territory after apparent loss of mate (Lindgren 1975). Another 14 birds re-observed 1 year after ringing at Arvidsjaur all remained in territory where they had been ringed (J Ekman). 3 birds ringed on Kola peninsula all retrapped within same territory 13 months later (Semenov-Tyan-Shanski and Gilyazov 1991). On Kola peninsula, territory occupancy by same bird for mean of 4 (2–7) years; sample size unknown (Kokhanov 1982). Little information on breeding densities: 1–19 birds per km² in taiga of western and central Siberia (Ravkin 1984). ROOSTING. Observations from Omolon valley (north-east Russia) in mid-winter where birds confined to roost for all but 3·5 hrs (active foraging period) of day (Andreev 1982a): same roost in tree used over many days (accumulation of faeces below tree). Bird found in roost at 15.15 hrs on 27 December (−47°C) was pressed close to trunk of larch *Larix* 3·5 m above ground, with maximum ruffling of feathers, head under wing. Roost-site protected from above by roof of mass of twigs capped with snow. Captive birds chose similar site in winter when frosty: in 2 out of 5 cases, roost used by

pair-members sitting on separate branches 2–4 m above ground.

2. Communication mainly vocal, sometimes associated with displays. Frequently approaches humans, e.g. at camp fires, and readily becomes hand-tame if regularly offered food. Otherwise secretive and hard to see. Flocks jointly mob predators, e.g. owls (Strigidae), other Corvidae, squirrels (Sciuridae). (Blomgren 1964, J Ekman.) FLOCK BEHAVIOUR. Movement of flock-members between trees is silent and unsynchronized. Unrelated flock-members are met with aggression (especially at food), including supplanting and chasing accompanied by calling. Dominance hierarchy exists whereby adult resident ♂♂ dominate all immigrants of both sexes, while adult resident ♀♀ dominate some but not all immigrants. (J Ekman.) SONG-DISPLAY. Singer has no special song-perch, but may sing from anywhere within territory, primarily at dawn and hardly ever in afternoon. Sings in all kinds of weather, including in snowfall at −8°C. Singing bout usually not more than a couple of seconds, but has been heard to sing up to 15 min with only short intervals. Earliest song in northern Sweden 22 February, then continues with decreasing intensity until ♀ lays in mid-April, but also heard at low intensity throughout summer. (Blomgren 1964.) ANTAGONISTIC BEHAVIOUR. (1) General. Little aggression. Trespassing neighbours tolerated except at feeding sites and around nest. Retained offspring also tolerated, including at feeding sites (J Ekman). (2) Threat and fighting. Territorial advertisement and threat signalled by postures and vocalization. Threat-posture comprises low body and spread hanging wings, accompanied by sharp Viv-call (see 2b in Voice), also often by Bill-snapping. Chasing of individuals accompanied by Viv-call. Appeasement-display signalled by lowering body and shivering and hanging wings, in association with Hiss-call (see 2g in Voice). Appeasement-display resembles that of right-hand bird in Fig A (see Heterosexual Behaviour, below,

A

for details); used by (e.g.) young birds begging from territorial birds during dispersal phase. (J Ekman; see also Bonds, above). HETEROSEXUAL BEHAVIOUR. (1) General. Life-long pair-bonds make pair-formation difficult to observe; can take place at any time of year. Lost partners replaced within 1 year (Blomgren 1964; Lindgren 1975), and one ♀ losing her mate in early September replaced him within 3 weeks with immigrant ♂ (J Ekman). (2) Pair-bonding behaviour. Blomgren (1964) and Lindgren (1975) described 2 main displays. ♂ may 'dance' around ♀ by swaying sideways like compass needle with stiff body and slightly spread wings and tail, simultaneously jumping to and fro sideways and twittering. Fig A shows ♂ (right) displaying thus to ♀; arrows indicate distinct sideways and bobbing movements (head up with tail down and vice versa) of otherwise stiffly lowered body, with legs as pivot; movements can be combined into see-saw motion. ♀ passive during dance or she may retreat slowly and intermittently up tree with small jumps. ♂ follows ♀ up tree (Fig B), dancing with her on branches, before finally breaking

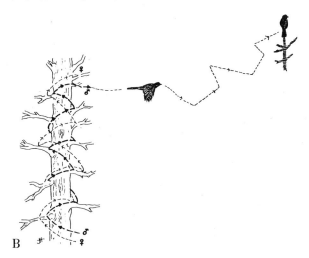

B

off and making zigzag flight to nearby tree-top. (F Lindgren.) This display occurs throughout year. In another display called 'spiral-dance' by Lindgren (1975), ♂ dives from tree-tops (con-ifers) with wings folded, then spreads wings close to ground to turn into branches and fly up trunk in tight spiral (Fig C) with

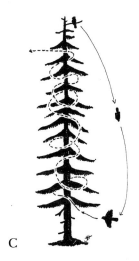

C

vigorous wing-beats. Blomgren (1964) reported this display from spruces only, while according to Lindgren (1975) it occurs in both spruce and pine *Pinus*. (3) Courtship-feeding. ♂ feeds ♀ during nest-building, incubation, and brooding. ♀ begs by gap-ing and ♂ then regurgitates, but ♀ can also feed herself off nest (Blomgren 1964; J Ekman). ♀ perhaps fed only once a day by ♂ (Blomgren 1964). (4) Mating. Takes place in trees, without being preceded by any apparent display. Sometimes preceded by ♂ feeding ♀, but bills can also meet without transfer of food. (Lindgren 1975.) (5) Nest-site selection. Chosen by 'test-brooding' possible sites with raised tail; both sexes participate, but ♂ seems to initiate. Nest-building often discontinued at early

stage, and birds move on to test another site. Even complete nests dismantled and moved (Lindgren 1975). ♀ performs about two-thirds of building, ♂ one-third, according to Blomgren (1964). (6) Behaviour at nest. ♀ alone incubates. Both sexes secretive around nest. ♂ spends much time silent in tree-top look-out near nest and makes silent approach to feed mate, his only vocalization a low contact-call close to nest (see 3 in Voice). RELATIONS WITHIN FAMILY GROUP. Only ♀ broods; up to *c.* 9-10 days, thereafter only in bad weather. ♂ bringing food for nestlings transfers it to ♀ during first few days after hatching but then switches to feeding young directly. Parents feed nestlings by thrusting bill into throat and regurgitating; sticky threads of half-digested food visible when parent withdraws bill. Faecal sacs swallowed by parents. Penetration of nest floor with bill by ♀ suggested to be for sanitation. Eyes fully open at *c.* 12 days. Nestlings gape and hiss on arrival of parents, and flutter wings when fed. Adult stimulates nestlings to beg with Koi-call (see 4 in Voice); no similar response to other calls. (Blomgren 1964.) Nestlings fledge at *c.* 20-23 days (Blomgren 1964; Kokhanov 1982), and spend following week in nearby trees (J Ekman). See Bonds (above) for subsequent dependence. ANTI-PREDATOR RESPONSES OF YOUNG. Little information. Call when handled (J Ekman: see Voice). PARENTAL ANTI-PREDATOR STRATEGIES. (1) Passive measures. ♀ a tight sitter. 9 out of 13 ♀♀ tolerated being lifted partly from nest without leaving (Kokhanov 1982). ♀ crouched when Raven *Corvus corax* heard *c.* 500 m away (J Ekman). (2) Active measures: general. Flutters back and forth, swoops down over intruder and calls. (3) Active measures: against birds. Pair attacked and drove off Jay *Garrulus glandarius*. Confronted stuffed Pygmy Owl *Glaucidium passerinum* with threat-postures, while only excited calls were directed at stuffed Ural Owl *Strix uralensis* (Blomgren 1964). (4) Active measures: against man and other animals. Circles and calls excitedly, some-times chasing squirrels *Sciurus vulgaris*. Swoops down over domestic cats (Blomgren 1964; J Ekman). When human intruders at nest, especially with newly fledged young, performs dive-attacks and calls (Blomgren 1964).

(Figs by J Hough, based on drawings by F Lindgren.)   JE

**Voice.** Not highly vocal, but calls heard all year. Rep-ertoire large and not well studied, so function of calls poorly known. Many calls flexible in pitch and speed of delivery, making exact description difficult and giving rise to confusion and multiplicity of renderings. Bill-snapping a short harsh 'ptt'. Considerable mimetic ability. For addi-tional sonagrams, see Bergmann and Helb (1982).

CALLS OF ADULTS. (1) Song. Gentle flow of subdued phrases composed of low twittering and chattering mixed with melodious whistling and mewing (Blomgren 1965; Lindgren 1975). Contains mimicry of (e.g.) Redpoll *Acan-this flammea*, Willow Tit *Parus montanus*, Bullfinch *Pyrrhula pyrrhula*, and thrushes *Turdus*. Song low and audible only at close range; at *c.* 20 m according to Lindgren (1975). (2) Excitement-calls. (2a) Crow-call. Harsh and usually loud 'eee', 'eeer', or 'kreee', given either singly or several in rapid sequence. Resembles Screech-call of Jay *Garrulus glandarius* but softer (Blomgren 1964). Signals excitement, fear, and anxiety, but when soft also used as greeting between pair at nest (Blomgren 1964). Also observed in connection with food-begging by ♀ (Kok-hanov 1982). Variety of transcriptions: sequence in Fig I

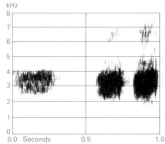

I   S Palmér/Sveriges Radio (1972-80)   Sweden   March 1961

similar to that in Bergmann and Helb (1982) rendered 'chräh chrä chrät'. (2b) Viv-call. Soft, and usually repeated many times in sequence. Used in different contexts. Signals alarm or strong excitement, as when humans approach nest. Also used in aggressive interactions with (especially) intruders. Delivered at different speeds, with higher speed indicating greater excitement. Can be given alone or in combination with other calls. Rendered as 'fu-fee', 'feel-kleeu', 'ful-ful', 'flyu-flyu-flyu' (Kokhanov 1982). (2c) Hui-call. Low, plaintive. Given rarely and usually in connection with Viv-call. Rendered 'tyu' by Kokhanov (1982). (2d) Twee-call. Low, and usually (though rarely) given in connection with 2b and 2c, thus 'twee-hui-viv-viv-viv'. Same sequence rendered 'fee-tyu-flyu-flyu-flyu' by Kokhanov (1982). This combination signals anxiety and is heard from ♂ approaching nest (Kokhanov 1982; J Ekman). (2e) Woodpecker-call. A 'ke' or higher-pitched 'bi', also rendered 'kjik', 'jak', etc. (Bergmann and Helb 1982). Given several times in sequence (at least 3 times but often more), and with varying loudness; Fig II shows rapid series of 5 calls. Signals

anxiety and often used while inspecting novel situations like new feeding sites, but also heard in encounters with conspecific intruders. (2f) Hawk-call. Harsh 'ka' ('kä': Bergmann and Helb 1982) resembling Goshawk *Accipiter gentilis*, or softer more goose-like 'ga'. Loud with several calls (usually 3-4) in sequence; Fig III shows evenly-shaped series of calls. Often given during displays directed towards intruders, and then normally in connection with 2b and 2e. Kokhanov (1982) described this call by ♀ threatening strange conspecific as a guttural not very loud 'kokur-kokur-kokur'. Rendering by Kokhanov (1982) of quiet extended 'o-o-o-ee-ee-ee' given by ♀ on nest hearing call of strange conspecific may also refer to this call. (2g) Hiss-call. A 'dzhee-ee' or 'vzheet-vzheet' (Fig IV) with trembling timbre and varying strength, given either singly or several times in sequence (Kokhanov 1982). Has several functions. Once heard from pair-members very agitated after loss of adult ♂ offspring to predator (J Ekman). Also used by ♀♀ when begging food from ♂♂. Appeasement-call signalling submission along with Appeasement-display, often used by dispersing juveniles when trespassing on territories of established birds (J Ekman). (2h) Balloon-call. Hoarse 'ooork' resembling sound produced of wet fingers drawn over balloon. Rare. Loud, delivered singly or in short sequence (typically repeated 2-4 times). Signals submission and noted only from dispersing juveniles trying to appease established resident birds. (3) Mew-call. Plaintive and high-pitched 'pieeh' (Fig V) of varying strength, mimicking Buzzard *Buteo buteo*. Frequently heard in all seasons. Can be used as contact-call and, when soft, in 'conversation' within pair, but also as warning against both avian and mammalian predators (Kokhanov 1982; Blomgren 1964). Multiplicity of transcriptions, e.g.

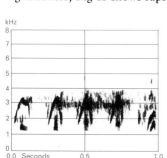

II   J-C Roché   Finland   June 1968

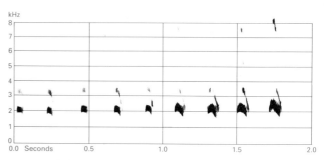

III   S Palmér/Sveriges Radio (1972-80)   Sweden   March 1961

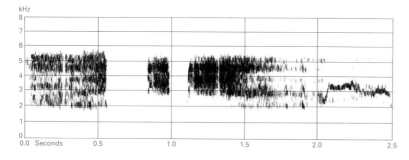

IV   J-C Roché   Finland   June 1968

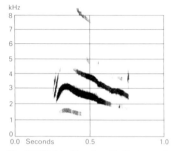

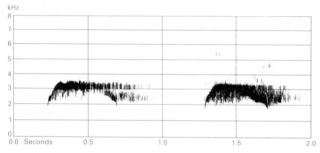

V  G Thielcke/Sveriges Radio (1972-80)  Sweden  July 1964

'kee-ee-ee' (Kokhanov 1982), 'hijäh' (Bergmann and Helb 1982). (4) Koi-call. Normally delivered singly. Can be delivered loudly as advertisement-call. Also often quietly in conversation within pair, especially at nest (Blomgren 1965). (5) Kee-call. Loud and resounding, often given several (2-4) times in sequence. Advertisement-call, sometimes described as 'whis-kee' call, but typical form is clear 'kee-kee-kee' (Kokhanov 1982). (6) Begging-calls of breeding ♀. Variety of calls, including weak whistling 'ufee-ee-t' resembling Spotted Crake *Porzana porzana*; weak sibilant 'chzhee-ee-kee' from ♀ being fed by ♂ (Kokhanov 1982: Fig VI). (7) Courtship-call of ♂. Low guttural 'krootoo-krootoo-krootoo' (Kokhanov 1982).

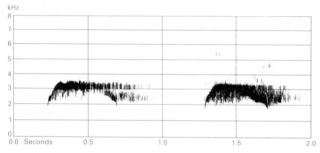

VI  S Palmér/Sveriges Radio (1972-80)  Sweden  March 1961

CALLS OF YOUNG. Nestlings quiet, calling only when being fed or handled. On arrival of parents at nest, small (unfeathered) nestlings give faint cheepings and squawks. From about day 5, start to mix faint hiss-sounds (call 2g) with cheepings (Blomgren 1982). Hiss later becomes more prominent, and feathered nestlings beg with weak 'zhee-ee-ee' (Kokhanov 1982). Same call given in strong trembling voice signals fear (e.g. when handled).    JE

**Breeding.** SEASON. Regularly lays while terrain still snow-bound, normally starts in first half of April. Sweden: median dates of 1st egg for Boden area 13 (3-22) April (Blomgren 1964), for Arvidsjaur area 10 April (31 March to 14 April); clutches started as late as 9 May are presumed replacements (Lindgren 1975). North-west Russia: in Murmansk area, mean date of 1st egg 12 (8-24) April for 14 nests (Kokhanov 1982). SITE. Almost always in conifer,

close to trunk. 82% of nests in pine *Pinus* and 18% in spruce *Picea* in Arvidsjaur (*n* = 58, Lindgren 1975); of 53 nests, 85% in spruce, 15% in pine (Blomgren 1964). Of 36 nests, 50% in spruce, 50% in pine (Kokhanov 1982). Mean height of nests above ground: 6 m (2-13, *n* = 58) (Lindgren 1975); 7 m (4-10·5) in pine and 4·1 m (2·3-7) in spruce (Blomgren 1964); 11 m (3·5-18) in pine and 6 m (1·7-15) in spruce (Kokhanov 1982). Only 2 out of 58 nests oriented to north at Arvidsjaur (Lindgren 1975). Nest: Rather loosely constructed platform of twigs supporting deep and well-insulated cup built mainly of lichens *Alectoris*. Platform built of dead twigs (maximum length 45 cm, maximum diameter 5 mm: Blomgren 1964) mainly from spruce but also pine, broken directly from trees. Twigs interwoven with strips of birch *Betula* bark and bound with spider cocoons. Cup lined mainly with feathers and reindeer *Rangifer* hair. Dimensions of 25 nests in Murmansk (Kokhanov 1982): outer diameter 204·4 mm (140-290), inner diameter 96·4 mm (84-112), overall height 150·3 mm (112-220), depth of cup 70·5 (59-89). Building: by both sexes. Mean time for construction 16 days (11-26) for 12 nests in Boden (Blomgren 1964), 14-24 days (sample size unknown) in Murmansk (Kokhanov 1982). Most material brought from within 25-200 m of nest, twigs sometimes from up to 300 m. Feathers cached for use as lining. (Blomgren 1964.) EGGS. See Plate 57. Sub-elliptical, smooth and glossy. Very pale bluish-green, blue, or bluish-grey; spotted and finely blotched with olive-brown, grey, and lilac-grey, usually more heavily mottled at broad end (Harrison 1975). Nominate *infaustus*: 30·4 × 21·9 mm (27·0-36·0 × 19·5-26·0), *n* = 331; calculated weight 7·5 g. *P. i. ruthenus*: 28·7 × 21·0 mm (27·5-29·4 × 20·8-21·2), *n* = 4; calculated weight 6·5 g. (Schönwetter 1984.) Clutch: 3-4 (2-5). Clutch of 6 reported for nest with 2 incubating ♀♀ (Lindgren 1975). Mean clutch in Sweden 3·7 eggs (3-4, *n* = 63) (Rosenius 1929); 3·9 (3-4, *n* = 23) for Boden (Blomgren 1964), 3·8 (2-5, *n* = 19) for Arvidsjaur (Lindgren 1975; J Ekman). Combined material for Sweden and Finland, 4·2 (4-5, *n* = 13) (Makatsch 1971); Russia: 3·8 (maximum 4, *n* = 17) in Murmansk (Kokhanov 1982); in central Siberia, clutch size normally 3 and rarely 4 (Reymers 1966; sample size unknown). INCUBATION. In Sweden, 19·5 days (*c.* 19 days 1 hr to 19 days 21 hrs, data for 8 years but sample size unknown) (Blomgren 1964). In Russia, 19 days, sometimes 20 (5 of 25 nests: Kokhanov 1982). Laying interval 24 hrs. Incubation by ♀ alone; can start after any egg. Hatching over 2-3 days (rarely 1-4: Kokhanov 1982). YOUNG. Fed and cared for by both parents. Brooded by ♀ alone up to *c.* 10 days, thereafter only in bad weather (Blomgren 1964; Kokhanov 1982). FLEDGING TO MATURITY. Fledging period *c.* 3 weeks. In Sweden, 21-23 days (maximum 24; sample size unknown) (Blomgren 1964). In Russia, 21 days (20-23, *n* = 14) (Kokhanov 1982). See Social Pattern and Behaviour for age at independence and age of first breeding. BREEDING SUCCESS. Nest failure due to predation sometimes high. 80% of 52 nests plundered at Arvidsjaur (Lindgren 1975), but large local variation. In Boden, only 1 out of 20 nests failed entirely (Blomgren 1964). In Murmansk, 35% of 17 nests predated (Kokhanov 1982). Of 73 eggs in Boden, 19% did not hatch (Blomgren 1964), and in Murmansk 7·7% (sample size unknown) (Kokhanov 1982). In Murmansk, of 37 chicks hatched, 59% survived to fledge; Carrion Crow *Corvus corone* took 19% of nestlings lost and 22% died in nest; overall 34% of eggs laid produced fledged young (Kokhanov 1982). On Kola peninsula (north-west Russia), mean 2·8 young fledged per brood ($n = 52$) (Semenov-Tyan-Shanski and Gilyazov 1991). JE

**Plumages.** (nominate *infaustus*). ADULT. Dense tuft of nasal feathers buff with dark brown tips. Cap formed by forehead, crown, hindneck, and side of head down to lore, just below eye, and ear-coverts dark fuscous-brown; feather-bases on forehead sometimes slightly paler grey-brown, ear-coverts slightly grey-brown or olive, lore sometimes (depending on light) sooty-brown; cap not sharply demarcated from remainder of body. Mantle, scapulars, and back buff-brown, feather-bases and (sometimes) feather-tips medium grey, giving greyish tone; often an almost uniform grey feather band across upper mantle and down side of neck; buff-brown often more cinnamon towards lower scapulars. Rump cinnamon-brown, merging into bright cinnamon-rufous of upper tail-coverts. Patch at base of lower mandible buff-white, remainder of lower cheek medium grey with slightly browner feather-tips and paler shafts; upper chin buff-white, feathers with black hair-like tips, remainder of chin and throat light grey. Chest and side of breast light grey with slight buff tinge, grading to pale rufous-cinnamon on flank and belly and this in turn to purer cinnamon on vent and under tail-coverts; some feathers of breast often with light grey tips, giving grey tinge to pale cinnamon-buff breast. Central pair of tail-feathers medium grey with dark horn shafts, other tail-feathers bright cinnamon-rufous above, paler rufous-cinnamon below; outer web of t6 and distal end of outer webs of (t2-)t4-t5 medium grey or brown-grey. Flight-feathers, tertials, inner greater upper wing-coverts, and bastard wing medium grey, inner web of secondaries and inner primaries sooty-black, of outer primaries sooty-grey, fringe of outer web of flight-feathers tinged olive or brown; bases of outer 2-3 secondaries and of inner and middle primaries bright cinnamon-rufous, extending 1-2 cm beyond greater coverts on upperwing, appearing as bright rufous patch, showing as much larger pink-cinnamon or salmon patch on underwing. Outer 4-5 greater upper wing-coverts and all greater upper primary coverts brighter cinnamon-rufous, distal part of inner web contrastingly black. Lesser and median upper wing-coverts grey-brown or buff-brown. Under wing-coverts and axillaries rufous-cinnamon, more rufous than belly and flank, but less deep and duller rufous than rufous on upper wing and upperside of tail. *In worn plumage* (early summer), cap browner, less dark fuscous; body greyer, cinnamon not extending as far up towards upper belly or breast; rufous of wing and tail less bright; grey tail-tips partly worn off. Sexes similar, but see Measurements. NESTLING. Naked (Blomgren 1971), or naked, except for some sparse down on upperparts (Harrison 1975). JUVENILE. Rather like adult, differing mainly in much shorter and looser body plumage, especially on underparts, with much grey of feather-bases showing through. Cap as adult, but buffier on forehead and lower cheek, less sharply defined on hindneck; upperparts (including lesser

and median upper wing-coverts) dark and duller, tinged olive-brown or grey-brown, less buff, rufous-cinnamon of lower rump and upper tail-coverts more contrasting; entire underparts greyish-buff, more grey on flank, and with paler shaft-streaks on chest and breast; chin and throat more buff (less grey) than adult, flank, belly, and vent mainly buff (less rufous-cinnamon), under tail-coverts less bright rufous. T1, outer web of t6, flight-feathers, bastard wing, and tertials often slightly tinged brown or olive, less pure grey than in adult, grey borders of distal outer webs of t2-t5 tinged olive, contrasting less with rufous remainder of feather; black of inner web of greater upper wing-coverts and greater upper primary coverts tinged olive-brown, rufous of outer webs often less bright and deep. FIRST ADULT. Like adult, but juvenile flight-feathers, tail, tertials, greater upper wing-coverts, greater upper primary coverts, and bastard wing retained. On average, grey and black parts of these feathers slightly more olive or brown than in adult, rufous less bright; tips of primary coverts and of feathers of bastard wing and tail often more pointed than in adult, less broadly rounded, tips of these feathers relatively more worn than those of adult at same time of year. However, some birds probably inseparable from adult once last juvenile feathers on head and neck replaced. See also Svensson (1992) and Structure.

**Bare parts.** ADULT, FIRST ADULT. Iris dark brown to black-brown. Bill, leg, and foot black. (Hartert 1903-10; Haftorn 1971.) NESTLING, JUVENILE. No information.

**Moults.** ADULT POST-BREEDING. Complete; primaries descendent. In Sweden, late May to early August (Svensson 1992); primary score *c.* 10 on 11 June (RMNH), *c.* 30 in June (*Br. Birds* 1971, **64**, plate 8); in Siberia, late May to late August (Johansen 1944; Goodwin 1986); in European Russia from about mid-June (Dementiev and Gladkov 1954). Moult finished in birds examined from late August and September (RMNH, ZFMK). Tail moult centrifugal; starts with shedding of t1 at about same time as p4, followed by body and secondaries from about shedding of p5 (RMNH, ZFMK). POST-JUVENILE. Partial; involves head, body, and lesser and median upper wing-coverts; in 20 examined, no tertials, tertial coverts, greater upper wing-coverts, bastard wing, or tail (RMNH, ZFMK, ZMA). In Sweden, July-September (Svensson 1992); in birds examined, Sweden, Finland, and Russia, fully juvenile birds encountered from early June to mid-July, moulting birds early July to early September, birds with moult completed from mid-August onwards (RMNH, ZFMK, ZMA).

**Measurements.** ADULT, FIRST ADULT. Nominate *infaustus*. Fenno-Scandia, all year; skins (RMNH, ZFMK, ZMA). Bill (S) to skull, bill (N) to distal corner of nostril; exposed culmen on average 4·5 shorter than bill (S).

| | ♂ | ♀ |
|---|---|---|
| WING | 144·3 (3·29; 18) 139-152 | 140·8 (2·58; 20) 135-145 |
| TAIL | 134·3 (3·96; 18) 126-143 | 131·5 (4·25; 20) 124-139 |
| BILL (S) | 28·0 (0·92; 15) 26·7-29·6 | 27·0 (1·05; 18) 25·7-28·9 |
| BILL (N) | 16·4 (0·75; 14) 15·0-17·4 | 15·6 (0·71; 18) 14·5-17·0 |
| TARSUS | 38·2 (0·85; 17) 36·5-39·3 | 37·1 (0·99; 20) 35·8-39·0 |

Sex differences significant. Wing and tail of adult on average slightly longer than retained juvenile wing and tail of 1st adult, thus, from sample above: wing, adult ♂ 145·3 (9) 142-152, 1st adult ♂ 144·8 (6) 141-148; tail, adult ♂ 136·4 (9) 134-143, 1st adult ♂ 134·5 (6) 132-138.

Wing: Sweden, ♂ 140-147(-153) (22), ♀ 134-145 (25) (Svensson 1992); Norway, ♂ 137-150, ♀ 137-145 (Haftorn 1971); European Russia, ♂ 138·8 (48) 130·5-146, ♀ 133·5 (46) 128-142 (Dementiev and Gladkov 1954).

Range of wing, tail, bill (N), and tarsus of (1) nominate *infaustus*, (2) *ostjakorum* (including *monjerensis*), (3) *jakutensis* (including *bungei* and Anadyr' birds), (4) *ruthenus*, (5) *sibericus* (including *rogosowi*), (6) *tkachenkoi* (including *varnak*), (7) *opicus*, (8) *caudatus*, (9) *maritimus* (including *sakhalinensis*) (Suschkin and Stegmann 1929; see Geographical Variation for distribution of these races).

|     | WING    | TAIL    | BILL (N) | TARSUS | n  |
|-----|---------|---------|----------|--------|----|
| (1) | 132-141 | 128-134 | 14-16·2  | 35-38  | 20 |
| (2) | 135-144 | 132-141 | 15-17    | 35-39  | 15 |
| (3) | 139-152 | 131-150 | 15-18    | 35-38  | 21 |
| (4) | 136-146 | 133-143 | 16-18    | 36-39  | 52 |
| (5) | 138-148 | 132-141 | 15-18    | 35-39  | 59 |
| (6) | 139-151 | 133-147 | 15-18·2  | 35-40  | 43 |
| (7) | 135-145 | 134-142 | 15-17    | 36-39  | 32 |
| (8) | 138-141 | 141-142 | 15-17    | 35-36  | 3  |
| (9) | 137-152 | 135-146 | 15-17·5  | 36-39  | 17 |

**Weights.** ADULT, FIRST ADULT. Nominate *infaustus*. Norway, November–February: ♂ 88-101 (7), ♀ 81-93 (4) (Haftorn 1971). European Russia: ♂ 87·6 (22) 81-97, ♀ 81·8 (22) 73-89 (Dementiev and Gladkov 1954).

Races combined. Siberia: ♂ 83·4 (7·01; 8) 72-94, ♀♀ 73, 83 (Dementiev and Gladkov 1954). For energy requirement in relation to weight and temperature, see Andreev (1982*b*).

**Structure.** Wing short, broad; tip rounded. 10 primaries: p5-p6 longest, either one 0-1 shorter than other; p7 1-7 shorter than longest, p8 5-13, p9 25-33 (adult, $n = 10$) or 20-25 (juvenile, $n = 7$), p4 2-5, p3 7-12, p2 14-22, p1 19-34; p10 60·5 (10) 54-68 (adult) or 55·0 (7) 48-59 (juvenile) shorter than longest, 26·2 (10) 18-31 (adult) or 33·0 (7) 30-37 (juvenile) longer than longest upper primary covert; in larger samples probably more overlap between ages. Outer web of (p3-)p4-p8 emarginated, inner web of p5-p8 (p4-p10) with rather indistinct notch. Tertials short. Tail long, tip rounded; 12 feathers, t6 28·5 (10) 23-34 (adult) or 24·3 (7) 20-32 (juvenile) shorter than t1; feathers of adult broad, bluntly rounded at tip, those of juvenile slightly narrower and more pointed. Bill short; broad but not very deep at base, gradually tapering towards tip; tip of upper mandible with slight hook, middle part of culmen slightly concave and flattened, tip not laterally compressed. Basal half of upper mandible covered by dense tuft of bristle-like feathers at each side, concealing oval nostril. About 5 stiff bristles above gape, projecting forwards; many fine hairs on chin, lower cheek, and throat. Feathers of body long, soft, and dense. Tarsus and toes rather short and slender; middle toe with claw 26·8 (6) 25·5-28; outer toe with claw *c.* 71% of middle with claw, inner *c.* 72%, hind *c.* 81%.

**Geographical variation.** Marked, but clinal throughout. Involves relative amount and depth of grey and rufous of body plumage, colour of cap, size and intensity of rufous of wing-patch, amount of grey on tips of tail-feathers, and size (expressed in wing and tail length). Early survey of variation by Suschkin and Stegmann (1929), who recognized 14 races: *infaustus* (Scandinavia to Kola peninsula), *ruthenus* (Mezen river, Kareliya, St Petersburg, and Moscow east through central and southern Urals, and south-west Siberia east to Tomsk, Barnaul, and Biysk); *ostjakorum* (northern Ural mountains and north-west Siberia from basin of lower Ob' to basin of Taz river), *monjerensis* (lower Yenisey to Olenek basin, north of Nizhnyaya Tunguska

and Vilyuy rivers), *rogosowi* (middle Yenisey valley and central Siberia from Nizhnyaya Tunguska south to Angara basin, Krasnoyarsk, and northern Sayan mountains), *opicus* (Abakan basin, and Altai and Sayan mountains), *caudatus* (north-central Mongolia and Russia south of Lake Baykal), *sibericus* (Transbaykalia and neighbouring north-east Mongolia north to basins of upper Lena and Vilyuy rivers), *bungei* (northern fringe of the species distribution between lower Lena and lower Kolyma rivers), *tkachenkoi* (Lena valley from Zhigansk to Yakutsk and basin of Aldan river), *varnak* (Stanovoy mountains south to north-west Manchuria, basins of Zeya and middle Amur), *yakutensis* (north-east Siberia, except extreme north and east, in valleys of Yana, Indigirka, and Kolyma rivers south to Verkhoyansk mountains), *maritimus* (south-east Siberia from Uda river and Shantar islands to Sikhote Alin mountains), and *sakhalinensis* (Sakhalin). 2 further races subsequently described: *sokolnikowi* Démentieff, 1935 (Chukotsk and Koryak mountains through Anadyr' basin to northern shores of Sea of Okhotsk) and *manteufeli* Stachanow, 1928 (basin of upper Severnaya Dvina in north-west European Russia). Clinal character of these races clearly summarized by Johansen (1944), who recognized 4 main lines in variation, as follows. (1) Northern races, showing relatively pale colour, small rufous primary-patch, and distinct grey patches on tail-tip; within these races, size gradually greater, cap blacker, and body greyer towards east, from mainly buff nominate *infaustus* through intermediate *ostjakorum*, *monjerensis*, and *bungei* to pale grey *jakutensis*. (2) Races of mid-latitudes, characterized by relatively dark colour, larger and deeper rufous primary-patch, and less developed grey on tail-tip; within group, again smaller and more rufous in west and gradually larger, greyer, and with darker cap in east, from rufous-buff *ruthenus* through intermediate *rogosowi* and *sibericus* to pure medium grey *tkachenkoi*. (3) In southern races, body colour highly saturated, primaries extensively rufous, and tail hardly grey on tip; body of western race *opicus* bright rufous, of *caudatus* more brown-grey, of *varnak* almost uniform dark grey. (4) Towards Pacific, colours more rufous again than in neighbouring grey extremes of other groups; northern *sokolnikowi* light grey with slight ochre tinge, mid-latitude *sakhalinensis* buff-grey, southern *maritimus* buff (like nominate *infaustus*, but larger). This opinion not quite in agreement with Suschkin and Stegmann (1929), as (e.g.) *jakutensis* considered by them to have much rufous on primaries instead of little. Later authors variously combined races. Thus, Dementiev and Gladkov (1954) largely ignored north-south variation and combined all European birds in nominate *infaustus*, all Ural and west Siberian populations in *rogosowi*, south-central Siberian birds in *sibericus*, and north-central and all east Siberian birds in *jakutensis* (apart from *opicus* of Altai-Sayan, *maritimus*, and *sakhalinensis*), thus combining very dark and very pale birds and those with much rufous in wing with those with little in a single race; Stepanyan (1978, 1990) recognized nominate *infaustus* in west and *maritimus* (including *sakhalinensis*) in east, but combined all birds from south European Russia and south-west Siberia east to Sayan mountains into *opicus*, those of north European Russia and north-west and north-central Siberia into *rogosowi*; and all remaining populations of northern and eastern Siberia in *sibericus*. Vaurie (1959) was more cautious and largely followed Suschkin and Stegmann (1929), but included *monjerensis*, *bungei*, and *sokolnikowi* in *jakutensis*, and *caudatus* and *tkachenkoi* in *sibericus*; ranges of races considered in part slightly different from those of Suschkin and Stegmann (1929), with *opicus* supposed to extend west to Tomsk, and *ruthenus* west to southern Fenno-Scandia. Clearly, boundaries impossible to define in a species with such strongly clinal variation. Following (limited) number of specimens examined, separation favoured of:

(1) in north, nominate *infaustus* (including *manteufeli*; buffish body, little rufous on wing), *ostjakorum* (including *monjerensis*; light grey body, hardly rufous in wing), and *jakutensis* (including *bungei* and *sokolnikowi*; light grey body, much rufous in wing); (2) in mid-latitudes, *ruthenus* (rufous-buff body, little rufous in wing), *sibericus* (including *rogosowi*; buffish-grey body, much rufous in wing), and *tkachenkoi* (including *varnak*; pure grey body, intermediate rufous in wing); (3) in south, *opicus* (rufous body, intermediate rufous in wing), *caudatus* (dusky grey-brown body, little rufous in wing), and *maritimus* (including *sakhalinensis*; buffish body, intermediate amount of deep rufous on wing). In Europe, no marked difference between birds of north and south Fenno-Scandia (contra Vaurie 1959), but birds from north-east Finland and north-west European Russia tend to be paler and greyer due to influence of *ostjakorum*.

Forms superspecies with Szechwan Grey Jay *P. internigrans* of western China and Gray Jay *P. canadensis* of North America (Mayr and Short 1970; Eck 1984).                                CSR

---

## *Cyanopica cyanus*  Azure-winged Magpie

PLATES 2 and 5 (flight)
[between pages 136 and 137]

DU. Blauwe Ekster      FR. Pie bleue      GE. Blauelster
RU. Голубая сорока     SP. Rabilargo      SW. Blåskata

*Corvus Cyanus* Pallas, 1776

Polytypic. *C. c. cooki* Bonaparte, 1850, Spain and Portugal. Extralimital: nominate *cyanus* (Pallas, 1776), Irkut valley and southern Transbaykalia east to Nerchinsk (Russia) and neighbouring northern Mongolia from Uliastay region and Selenga basin east to Hentiy, grading into *pallescens* between Sretensk and Dzhalinda (Shilka and Amur valleys) and probably in Arghun valley; *pallescens* Stegmann, 1931, middle Amur valley down to about Amgun' mouth, south to Ussuri valley and Sikhote-Alin' mountains; *stegmanni* Meise, 1932, Manchuria (China) from Greater Khingan range to Liaoning, grading into *swinhoei* in Hopeh; *swinhoei* Hartert, 1903, China from Shantung south to Fukien, and through Yangtze valley west to Szechwan; *japonica* Parrot, 1905, Honshu and Kyushu (Japan); 1–5 further races in Korea, Shansi, Shensi, and Kansu (China).

**Field characters.** 34–35 cm (of which nearly half tail); wing-span 38–40 cm. About 75% size of Magpie *Pica pica*. Small, elegant corvid with rather attenuated form ending in long tail shaped like *P. pica* but looking narrower in flight. Body dove-brown, with black cap, white throat, and mainly azure-blue wings and tail. Flight recalls *P. pica*; other behaviour more like Jay *Garrulus glandarius*. Sexes similar; no seasonal variation. Juvenile separable at close range.

ADULT. Complete moult in autumn. Crown, nape, and side of head to well below eye velvet-black, forming long cap unique in west Palearctic Corvidae. Back and rump rather pale grey-brown, similar in tone to Collared Dove *Streptopelia decaocto*; upper tail-coverts grey-brown. Wings rich azure-blue; blackish inner webs of tertials, greater coverts, and primaries can show in flight, black on inner part of wing also at rest. In fresh plumage, pale whitish outer webs of flight-feathers form panel on edge of closed wing and may even show as soft lines on spread wing in flight. Underwing pale dove-brown, with dark outer rim formed by dusky tips to flight-feathers. Tail markedly graduated; azure-blue above, dusky below. Underbody off-white, washed vinous-grey to ashy-brown on flank and under tail-coverts. Bill and legs black. JUVENILE. As adult but duller, with buff bar across upper mantle, indistinct whitish bar across median coverts, and whitish fringes to tail-feathers.

Unmistakable in good view but may suggest *P. pica* if seen only as silhouette, and even Great Spotted Cuckoo *Clamator glandarius* in brief glimpse. Flight recalls *P. pica* but looser, with far less erratic and staccato bursts of wing-beats; progress thus more sustained, even dashing at times. Gait a light, bouncy hop; also clambers and jumps. Flicks wings and tail in excitement. Secretive rather than shy in breeding season, keeping within cover and rarely crossing wide open spaces; bold and confident at other times, roving widely in noisy groups.

Main call a distinctive husky rising whistle 'zhreee'.

**Habitat.** Breeds in west Palearctic only in warm lower middle latitudes in Iberia, locally in mountain gorges at *c.* 700 m but also at sea-level on sand-dunes overgrown with planted stone pine *Pinus pinea* and introduced eucalyptus. In Coto Doñana (Spain) such modern woodlands are fully colonized, to virtual exclusion of the locally abundant Magpie *Pica pica*, which nests there chiefly in low bushes outside woods. Arboreal, but forages freely on ground under canopy (Mountfort 1958). In contrast to far separated Asian population, shows no marked preference for broad-leaved trees, or for river banks and islands (Dementiev and Gladkov 1954). Also inhabits open cultivated or grass country with groups of trees, scrub, or hedgerows, and orchards or groves of olive *Olea* and cork oak *Quercus suber*. Sometimes comes to scavenge round houses and refuse tips (Goodwin 1986). See also Food.

**Distribution.** No firm evidence of recent changes in distribution.

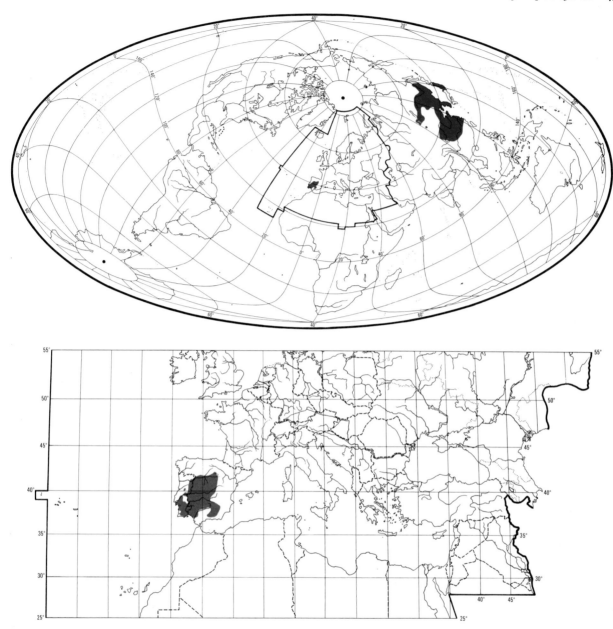

FRANCE. Alleged breeding attempt near Collioure (Pyrénées Orientales) in spring 1956 (Ferrer 1987) not sufficiently well documented to be acceptable (RC). SPAIN. Occurrence in north-east Spain occasional, but no evidence of breeding (Ferrer 1987, which see for map of records in north-east). Disperses in autumn into areas adjacent to breeding range (AN).

**Population.** No information.

**Movements.** Chiefly sedentary; after breeding, moves around in flocks in group-territory; evidence of more movement in east Palearctic.

In Spain, present in same localities all year (Sacarrão 1972). Occasional records of presumed vagrants (mostly in spring) in north-east Spain and Pyrénées-Orientales (southern France) (Ferrer 1987, which see for map). At Polientes (42°48′N 3°56′W, northern Spain), 26 November 1967, report of at least 3 birds: not known whether vagrants or members of undiscovered local breeding colony (Morales 1969).

Resident in China, with some local movements; varying reports over the years from Peking area indicate erratic movements, and presence in non-breeding areas in winter (Hemmingsen and Guildal 1968; Schauensee 1984). In South Korea, apparently occurs chiefly in winter (Austin

1948). Study in Nagano region (central Japan) showed that area of movement of fledged young was limited to diurnal activity range of flock they belonged to; birds ringed as nestlings were recovered after 0·5-20·5 months, and no correlation between dispersal distance and time elapsed (Hosono 1969). At Hikage village (Nagano), resident throughout year, even in winters with 1-1·5 m snow December–April (Hosono 1983). Some short-distance movements reported—one of 3 breeding groups studied in Nagano wintered at foot of mountains c. 2 km from study area (Komeda et al. 1987).

In Baykal region (south-east Russia), present at breeding grounds in Ulan-Ude all year; in south-east Transbaykalia, recorded in non-breeding areas in some winters (Prokofiev 1962; Belik 1981); nearby in extreme northwest Heilungkiang (northern China), where stegmanni resident, nominate cyanus reported only in winter (Schauensee 1984). Further east, in lower Amur region (Russia), winter records (e.g. at Lake Orel') indicate that proportion of population regularly moves north of breeding range (i.e. north of Komsomol'sk) (Babenko 1984b). In Ussuriland (eastern Russia), many birds apparently nomadic in winter, preferring wooded valleys, but occurring in open spaces, and on coasts and islands near Vladivostok; in some breeding areas, rarely recorded December–January, with main return late March to April (Panov 1973a; Nechaev 1974). DFV

**Food.** Invertebrates, especially beetles (Coleoptera), seeds, fruits, and more rarely small vertebrates; also carrion, scraps, and refuse. Forages generally in flocks, very often on ground; scavenges readily around houses, particularly in east Palearctic (Zieger 1967; Goodwin 1975). Feeding habits in public area of zoo in Peking (northern China) described as like Collared Dove Streptopelia decaocto, feeding from animals' troughs, etc. (Fiebig 1983), and in parks in Ulan Bator (Mongolia) forages in litter bins (Mauersberger et al. 1982). In Peking, flock moved around while feeding in follow-the-leader fashion; arrived at sites in trees and bushes but quickly dropped to ground where foraged under leaves, between roots, and in soil, and searched behind bark of dead branches; also on buildings, investigating window-ledges, tops of walls, etc. (J Palfery). In Portugal, some members of flock examined tree roots and undergrowth, while others in tree-tops investigated conifer cones for insect larvae (Mathews 1864). Searches for larvae in ploughed fields (Coverley 1933), and said to damage maize Zea crops in Portugal by eating grain on cobs (Tait 1924). For other agricultural produce eaten, see below and Santos Júnior (1968). Flocks in Peking systematically consumed unripe grapes Vitis vinifera and apples Malus (Fiebig 1983). In Ussuriland (eastern Russia) in winter, feeds by unfrozen streams, taking caddis fly (Trichoptera) larvae and hibernating frogs Rana from under stones (Panov 1973a), and often caught in hunters' traps while taking fish bait (Vorobiev

1954). Recorded feeding at dead donkey Equus asinus (Santos Júnior 1968). May hang upside-down from twig to reach fruit (Nechaev 1974). When feeding, large objects held under foot and pulled apart with bill (Goodwin 1975), and may be taken up into tree for consumption (Restall 1975a). Hairy caterpillars (Lymantriidae) dealt with in same manner as by Magpie Pica pica: squeezed in bill then beaten on substrate several times before being swallowed (Litun 1986). Captive bird tried to attack small birds which seemed injured or helpless (Harrison 1976), though in eastern Russia songbirds nest successfully close to C. cyanus colonies (Prokofiev 1962). Stores food in caches, presumably for later consumption; in Spain, stores acorns Quercus, olives Olea, and pine seeds Pinus in ground; sites often robbed by conspecifics (Turček and Kelso 1968; Kübel and Ullrich 1975). Flock of c. 50 in Peking apparently buried items of plant material in bank of loose soil (J Palfery). Often hides food in captivity, digging holes then covering item with leaf, often retrieving and then hiding it again (Porter 1941; Goodwin 1975; Restall 1975a). Of c. 550 items from 98 stomachs collected in pine forest, southern Spain, 59% by number (46% by weight) animal material, 41% (51%) plant material (remainder grit); 53% of animal material taken in or on soil or in leaf litter, 27% on vegetation, 14% in air and other places, 3% by water, and 1% in water; 56% of plant material from understorey, 28% from other vegetation, and 16% from cultivated areas (Cónsul and Alvarez 1978). No significant difference between diet of ♂ (n = 45) and ♀ (n = 26) in southern Spain (Alvarez and Aguilera 1988).

Diet in west Palearctic includes the following. Invertebrates: mayflies (Ephemeroptera: Ephemeridae), dragonflies (Odonata), grasshoppers, etc. (Orthoptera: Gryllotalpidae, Acrididae), earwigs (Dermaptera: Forficulidae), mantises (Dictyoptera: Mantidae), bugs (Hemiptera: Pentatomidae, Cicadidae), ant-lions (Neuroptera: Myrmeleontidae), adult and larval Lepidoptera (Noctuidae), larval caddis flies (Trichoptera: Hydroptilidae), adult and larval flies (Diptera: Culicidae, Muscidae), Hymenoptera (Ichneumonidae, Cynipidae, Proctotrupoidea, ants Formicidae, wasps Vespidae), adult and larval beetles (Coleoptera: Carabidae, Staphylinidae, Scarabaeidae, Tenebrionidae, Coccinellidae, Chrysomelidae, Cerambycidae, Curculionidae), adult and eggs of spiders (Araneae: Araneidae), mites (Acari: Ixodidae), millipedes (Diplopoda: Julidae), centipedes (Chilopoda: Scolopendridae), snails (Gastropoda: Helicidae, Planorbidae), leeches (Hirudinea). Plant material: fruits and seeds of pine Pinus pinea, juniper Juniperus, oaks Quercus pyrenaica, Q. robur, Q. ilex, elm Ulmus, cherry, plum, peach, and almond Prunus, hawthorn Crataegus, mistletoe Viscum, myrtle Myrtus, mulberry Morus, fig Ficus, grape Vitis, tomato Solanum, orange Citrus, pomegranate Punica, olive Olea, mastic (lentisk) Pistacia lentiscus, strawberry Fragaria, asparagus Asparagus officinalis, daphne Daphne gnidium, maize Zea, fan-palm Chamaerops (Mathews 1864;

Tait 1924; Gil 1927; Turček 1961; Valverde 1967; Santos Júnior 1968; Goodwin 1975; Cónsul and Alvarez 1978; Alvarez and Aguilera 1988.)

In south-west Spain, of 330 invertebrates (excluding eggs) from 98 stomachs, 32·7% by number (49·5% by weight) beetles (including 17% Curculionidae, 1% Coccinellidae, 1% larvae), 21·8% (6·5%) Hymenoptera (including 15% ants and 6% wasps), 13·9% (14·5%) millipedes, 9·7% (3·3%) Diptera (including 3% Muscidae, 2% larvae), 6·1% (5·6%) Hemiptera (including 2% Pentatomidae), 3·6% (4·4%) Lepidoptera (including 3% larvae), 3·3% (6·7%) Orthoptera (including 2% Acrididae), 3% (3%) dragonflies, 1·8% (0·8%) snails, 1·5% (0·7%) leeches; of total animal material, 24·6% by number (0·5% by weight) arthropod eggs. Of *c.* 300 fragments of plant material, 33·7% (9·8%) fruits of mastic (lentisk) tree, 27% (22·8%) pine seeds, 16·9% (5·9%) fruits of *Daphne gnidium*, 12·7% (21·7%) grapes, 2·8% (1·6%) mulberries, 1·7% (3·0%) olives, 1·5% (0·3%) myrtle berries, 1·5% (0·2%) asparagus berries; also 12·5% by weight figs. 96% of stomachs contained animal material (including beetles in 79%, millipedes in 40%, Hymenoptera in 38%, and Diptera in 16%), and 97% contained plant material (including olives in 82%, grapes in 41%, asparagus in 31%, and pine seeds in 31%). (Cónsul and Alvarez 1978.) Contents of 10 stomachs from Valladolid, northern Spain, as follows: 4 from January–February contained many millipedes, 6 beetles, 6 insect larvae, 2 Ichneumonidae, 1 Acrididae, and 12 juniper berries; 2 from June contained many spiders and eggs, 2 beetles *Melolontha* (Scarabaeidae), and 1 Cicadidae; 2 from September contained many beetles, 1 Acrididae, 6 grapes, and 1 pine seed; 2 from November contained 9 beetles, 3 wasps, 2 Acrididae, 1 Pentatomidae, and many grapes (Valverde 1967). See also Alvarez and Aguilera (1988). In South Korea, of 146 items from 15 stomachs, 82% (by number) plant material (36% of which berries of *Smilax*, 34% acorns, and 28% fruits of rose *Rosa*), and 18% animal material (85% of which beetles, 8% ants, and 4% frog). In Japan, items from 408 stomachs comprised 76% (by number) plant material and 24% animal material, which included crustaceans, fish, amphibians, birds, and mammals. (Won 1961.) For eastern Russia, see Vorobiev (1954), Prokofiev (1962), and Nechaev (1974). 60 stomachs from Portugal contained only Colorado beetles *Leptinotarsa decemlineata* (Chrysomelidae) (Restall 1975a). In spring, southern Spain, 70% by weight of diet animal material, in summer 60%, in autumn 47%, and in winter 49%; average weight of food item 1·1 g (0·1–3·1 g) (Cónsul and Alvarez 1978). In Ussuriland, diet over year changes as follows: in spring, insects, spiders, and fruit; summer, insects and spiders; autumn, fruits and berries; winter, plant material, invertebrates from leaf litter, bark, galls, and from water, also carrion, and bait (Nechaev 1974).

Nestling diet studied only in east Palearctic. Of 423 items in 115 collar-samples, Ussuriland, June–July, 48% by number spiders, harvestmen (Opiliones), and their eggs, 29% Lepidoptera larvae and pupae (including Noctuidae, Lasiocampidae, Sphingidae, and Geometridae), 7% adult and larval beetles (including Carabidae, Elateridae, Cerambycidae, Scarabaeidae, and Silphidae), 6% Orthoptera (including Tettigoniidae and Acrididae), 3% snails, 2% Hemiptera, 2% Diptera, 1% frog *Rana*; in first few days fed soft insect bodies and at 10–14 days whole and larger insects, e.g. Orthoptera (Nechaev 1974). In Chita region of Transbaykalia (eastern Russia), 50 collar-samples included 'pond snails' and frog tadpoles (Ogorodnikova and Mironova 1989). On Honshu (Japan), of 105 arthropods in 59 collar-samples, 30% by number Diptera (14% adults, 11% larvae, including 24% Syrphidae), 23% Lepidoptera (13% adults, 6% larvae, 4% pupae, including 13% Tortricidae, 4% Noctuidae), 14% beetles (7% adults, 7% larvae, including 7% Carabidae, 4% Scarabaeidae, 2% Elateridae), 9% adult Orthoptera (Gryllotalpidae), 7% woodlice (Isopoda: Armadillidiidae), 6% Hemiptera, 6% spiders, 2% earwigs, 1% stonefly larvae (Plecoptera), 1% cockroach larvae (Dictyoptera: Blattidae); of total sample of 141 items, 75% by number arthropods, 9% nestling birds, 8% other animal material, and 9% mulberries; average consumption per nestling in 1 hr was 0·2 g on day 3, 1·3 g on day 11, falling to 0·8 g on day 17 (Hosono 1966b). In South Korea, of *c.* 310 items from 180 collar-samples, 61% adult insects, 12% tree frog *Hyla*, 10% spiders, 7% insect pupae, 6% insect larvae, 1% plant material (Won *et al.* 1966). Nestlings fed by regurgitation (Mountfort 1958).                     BH

**Social pattern and behaviour.** Extensive study of *japonica* in Nagano region (central Japan) by Hosono (see below for references), also by Komeda *et al.* (1987) has revealed fine details of strongly developed flock structure—including occurrence of helpers at nest. Much less information for west Palearctic (*cooki*) but, since dispersion and behaviour apparently show strong similarities throughout range, following account draws heavily on *japonica*, also on *pallescens*.

1. Highly gregarious throughout the year. Dispersed in stable flocks, each of which, after breeding season, comprises family parties originating from single breeding colony (see Breeding Dispersion, below). Each flock defends extensive territory throughout year against other flocks, although outside breeding season different flocks, at least in Japan, sometimes roost communally (Komeda *et al.* 1987; Hosono 1989; see Roosting, below). Flocks thus typically comprise adults (breeders and non-breeders) and young, with family groupings evident within post-breeding flocks (no free mixing of young); some failed breeders joined families (Hosono 1967b, 1989). However, flock composition perhaps to some extent dependent on sedentariness (or otherwise) of population. Thus in Ussuriland (eastern Russia) where *pallescens* partially dispersive, juvenile flocks recorded, also in Baykal area (Russia); in Ussuriland, young leaving nest initially form discrete flocks of 15–20 birds around colony before dispersing more widely; in September–October, flocks (presumably now all ages) sometimes up to 35–40 or more (Nechaev 1974); in Baykal area, juvenile flock of at least 30–40 recorded in August (Bogorodski 1981). In Iberia, outside breeding season, generally found in small flocks comprising several family groups

(Goodwin 1975, 1986; Neves 1984), although much larger assemblages reported in earlier years, e.g. in 1935–40 some flocks of 200–300, always with 40–50 Spotless Starlings *Sturnus unicolor* and often 3–4 Hoopoes *Upupa epops* (Neves 1984); earliest report of such a large flock (*c.* 200) in Portugal by Mathews (1865). In southern Ussuriland, April–May, flock size 12–20 (Panov 1973*a*); up to 80–100 (presumably outside breeding season) (Dementiev and Gladkov 1954). In Japan, average flock 22·9 (9–45, *n* = 45) and average territory 0·22 km² (0·11–0·42) (Hosono 1967*c*). For additional information on flock sizes outside breeding season, see Roosting (below). As breeding season approaches, pairs evident within flocks (Panov 1973*a*; Nechaev 1974) and, once breeding under way, non-breeders continue to flock in neighbourhood of colony (Nechaev 1974). BONDS. Mating system essentially monogamous but pair may receive help from other flock members. No information on duration of pair-bond or mate-fidelity between years. In breeding pair, only ♀ incubates and broods; ♂ (mate) feeds ♀ on nest and both sexes feed young (Goodwin 1986; Komeda *et al.* 1987). Nest-help recorded both in Iberia (Araujo 1975) and in Japan (see below for details), but more information needed for west Palearctic. In study in Avila (central Spain), 1 adult simultaneously fed 2 young in 2 nests, and also defended a 3rd nest with young; also, 4 or more birds participated in feeding young at 1 nest (Araujo 1975). Only detailed study of nest-help is in Japan as follows. On average, from flock of *c.* 20 birds, 4–5 pairs bred each year. No active breeders helped at other nests, and amount of help they received was generally small, both in terms of number of helpers and frequency of their visits (although 2 helpers visited 3 nests as often as did parents). Of 14 nests, 6 had 1–2 helpers, 1 had no helpers, and incidence of help (if any) at remaining 7 not known; 1 helper visited 2 nests, 2 visited 4. Helpers participated in carrying nest-material and building, feeding the incubating or brooding ♀, feeding nestlings and fledglings, and removal of faecal sacs; did not participate in egg-laying, incubation, or brooding. Helpers fell into 2 categories: (a) frequent (confident) visitors to nest, contributing significantly to feeding ♀ and young; (b) infrequent (subordinate and nervous) visitors which seemed not to contribute food; one helper filled 1st role at one nest and 2nd role at another. Age of first breeding 1 year. That 3 (out of 6 marked) 1-year-olds attempted to breed and 1 adult acted as a helper, suggests that factors other than age affect individual status. (Komeda *et al.* 1987.) Little known about status, kin relationships, etc., of helpers, but in study by Hosono (1983) one frequent helper at nest was ringed 1-year-old from same colony. For other behaviour of helpers at nest see part 2 (below). Hosono (1971) recorded young being fed at 27 days and more after leaving nest. BREEDING DISPERSION. Colonies formed by members of winter flock, with extensive group-territories which overlap little (at least in Nagano), creating discontinuous distribution of colonies, e.g. in Sevilla (south-west Spain) 2 colonies (6 nests, 13 nests) *c.* 1 km apart (Pacheco *et al.* 1977). Differences exist in flock size and therefore in size of colonies (see Bonds, above, for proportion of flock that breeds). Colonies of 3–16 pairs recorded, occasionally larger (Redondo *et al.* 1989). Dispersion within colony loose, with each nest usually in separate tree or bush, often 40–100 m apart (Goodwin 1986). In Tiétar valley (central Spain), more than 1 nest per tree, mean distance between nests 48·7 ± 4·4 m (*n* = 107) (Muñoz-Pulido *et al.* 1990). In study in southern Spain, 4 colonies (2 perhaps sub-colonies) contained 32 pairs (3, 5, 8, 16) in 8·5 km², average 3·8 pairs per km² (Redondo *et al.* 1989). In Nagano, 3 groups in 6·25 km² (each territory *c.* 1 km²), mean distance between nests within colony 85·3 ± SD89·4 m (15–400, *n* = 33 distances: Komeda *et al.* 1987). However at another

Nagano colony, nests only 10–20 m apart (Hosono 1989). In Peking (China), breeds in avenues of poplar *Populus*, maximum 10 pairs per colony (Fiebig 1983). In study at Ulan-Ude (Baykal area), 1 flock recorded nesting annually over 6 years in deciduous wood of *c.* 1 ha; 2–7 nests each year, 5–15 m apart (once, 2 nests 1·5 m apart in same tree) (Prokofiev 1962), but, in Ussuriland study, colony sites tend to change from year to year; usually 3–10 pairs per colony, nests 5–80 m apart but occasionally solitary (solitary nests tend to be predated) (Nechaev 1974). No records of nests being re-used in subsequent years in Spain, and exceptional in Russia (see Breeding). After nest-failure, however, will desert and build new nest (Prokofiev 1962); in Mongolia, 2 cases of rebuilding 30 m and 120 m away after nest-failure (Mauersberger *et al.* 1982). Rather strict confinement of ♀ to nest area during egg-laying and incubation reduces her home-range relative to ♂'s: in 1 pair, ♀'s home-range during egg-laying *c.* 0·2 ha, ♂'s during incubation *c.* 3·5 ha, home-range of ♂ and ♀ while feeding young *c.* 8·5 ha (rough calculation from Hosono 1971). ROOSTING. Mainly communal throughout year, especially in winter (when different local flocks may join). Main study by Hosono (1967*a*, *c*, 1973, 1989) as follows. Favoured sites were apple *Malus* orchards and garden trees in summer, chiefly evergreens (2–20 m high) in winter; evergreens similarly sought in west Palearctic, e.g. eucalyptus in northern Portugal (Santos Júnior 1968). Site usually in dense grove but sometimes solitary tree near house. Sites outside breeding season are often traditional. In study territory, roost at maximum size (*c.* 50 birds) December–January, declining to *c.* 10–20 in April, less than 10 in June–September. Similar seasonal pattern observed in Peking (J Palfery). Gregariousness emphasized by some communal roosting even in breeding season, not just of non-breeders but also including some pairs. From start of breeding season, typically pair, then ♂ (once ♀ is incubating), and latterly family, roost inside flock territory. From autumn to spring, flock (composed of family groups, etc.) roosts communally, usually inside flock territory but sometimes outside; varying stability of roosting flocks results in variations in winter-spring dispersion of roosts (see Hosono 1967*c*, 1973 for details). In Peking in winter, roosts communally with Magpies *Pica pica* near buildings (Fiebig 1983); for details of winter roosting there, including assembly times, see Flock Behaviour (below). For seasonal variation in roosting times in Nagano, see Hosono (1967*c*).

2. Generally wary and shy, though in places where habituated to man may be quite approachable (Jonsson 1982; J Palfery). Thus, in Ulan-Ude where colony near research station, not shy, and approachable to within a few metres when feeding on rubbish tips (Prokofiev 1962). For responses to predators, including communal mobbing, see Antagonistic Behaviour and Parental Anti-predator Strategies (below). FLOCK BEHAVIOUR. In keeping with flock stability throughout year, cohesion maintained at all times for feeding, roosting, bathing (water-bathing and sun-bathing) and colony defence. Flock on the move commonly adopts single file, likened to babblers *Turdoides* (Swinhoe 1861; Ticehurst and Whistler 1933), with members communicating constantly and noisily with Contact-calls (see 2 in Voice). In Nagano, flock recorded moving 3·5–5·0 km per day (equivalent to 10–17 ha) outside breeding season (Hosono 1967*b*, which see for detailed flock movements and seasonal variation in activity). Pre-roosting sequence in Nagano as follows: after final bout of feeding, flock gathers first in tree then flies to roost-site, and enters, individuals selecting sleeping perches before settling for night (Hosono 1973, which see for influence of light intensity, distance from roost, weather, and food intake, on roosting behaviour). In Peking, December–January, pre-roost assembly occurred *c.* 30 min before sunset, parties of up to 20 birds

arriving excitedly in tree-tops until whole flock (40–50) assembled; whole flock then flew into evergreen roost-site around sunset (*c.* 1 hr before complete darkness); once in roost, flock members continued moving from branch to branch and from tree to tree for *c.* 30 min before settling; some birds seemed to be displaced from perch-sites but with little hostility. Flock appeared to seek denser foliage but no sign of huddling; once, before nightfall, flock disturbed at roost moved *en masse* to (presumably) another site. (J Palfery.) In study at eucalyptus roost in Baca d'Alva (northern Portugal) at end of December, birds sought roost directly in small parties rather than making mass arrival from pre-roost assembly: 67 arrived over 5 min (*c.* 17.00 hrs) in 6 flocks of 3–30; flock left very early the following morning and did not return to roost until evening; up to 35 seen together drinking and bathing (Santos Júnior 1968). In Peking, flock of *c.* 30 came to drink in group of 7–8 at small patch of open water (J Palfery); also in Peking, family party seen sunbathing on sand with wings outspread (Fiebig 1983) and 2 captive young sunbathed thus in Portugal (Tait 1924). SONG-DISPLAY. Thought to occur (see 1 in Voice), but no information on use. ANTAGONISTIC BEHAVIOUR. (1) General. Markedly tolerant of fellow flock members but hostile to alien flocks and to predators. Even in breeding season, all members of flock tolerate each other, with non-breeders usually allowed close access to nest (see Bonds, above, for helpers) except around time of building and laying; also, one incubating pair seen chasing another pair which intended building close to them (Hosono 1971). Breeders sometimes hostile to helpers; at 2 nests, helpers deferred to ♂ breeder over courtship-feeding, avoiding visiting when ♂ was feeding mate; if visits accidentally coincided, ♂ temporarily expelled helper. Occasionally, bird visited without offering any help and, in one such case, incubating ♀ chased 'helper' for *c.* 10 m and physical contact occurred. (Komeda *et al.* 1987.) After fledging, no antagonism between families of same flock, but chases between birds from different flocks seen along common boundary between their territories (Hosono 1989). Outside breeding season, response to 13 species of predator included alarm-calls (see 4 in Voice), attack, mobbing, chasing, and fleeing (Hosono 1975, which see for variation in response according to predator). (2) Threat and fighting. The only good information comes from captive birds. When hostile to keeper, bird performed Forward-threat display (Fig A): leaned towards keeper, slightly lifted and fluttered fol-

A

ded wings (wing nearer object of threat raised less than other), tail partly spread on both sides (*P. pica* spreads tail only on side facing danger); threat-display accompanied by distinctive whistling call (Goodwin 1975, 1986: see 4a in Voice), although others perhaps more typical in the wild, according to context. In obviously same (or similar) display described by Harrison (1983), bird crouched and made jabbing pecks with head lowered, plumage ruffled, wings drooped and partly spread, and tail spread. HETEROSEXUAL BEHAVIOUR. (1) General. More information needed on behaviour in the wild, and studies on

captive birds contribute significantly to following account. (2) Pair-bonding behaviour. Few details, but includes chasing, Courtship-feeding (see 3, below) and probably Allopreening. From shortly before start of breeding season, flocks include pairs (Panov 1973*a*; Nechaev 1974; Hosono 1989), but not known if these are newly formed or involve renewal of former bonds. In Ussuriland, main indication of courtship was ♂ chasing ♀ in low flight, pair twisting through trees before ♂ landed, calling loudly (Nechaev 1974). Allopreening probably occurs, since, when keeper extended his finger towards captive bird, latter solicited Allopreening by adopting stiff upright posture, or tilting and lowering head towards finger, body plumage sleeked but head and neck feathers ruffled (Goodwin 1975, 1986; Restall 1975*a*; Harrison 1983). In captive pair, although no Allopreening witnessed, ♂ sometimes erected head feathers thus (effecting appearance of black busby) when approaching or standing near mate (Goodwin 1986; see also below). (3) Courtship-feeding. In Ussuriland, begins early May when pairs conspicuous in pre-breeding flocks; ♀, with loud Begging-calls (see 3a in Voice) approaches ♂ to be fed (Panov 1973*a*). ♂ continues feeding mate after laying, at rate of 1–2 visits per hr, so that ♀ seldom leaves nest for first 10 days of incubation (Nechaev 1974). Typically, on ♂'s approach, ♀ leaves nest and flies to nearby tree where she crouches, wing-shivering and calling, with gaping bill into which ♂ regurgitates (Nechaev 1974). When fed on nest, ♀ wing-shivers and upstretches neck (Araujo 1975). In captivity, ♀ attracted ♂ with similar display (Fig B): stood upright with

B

bill raised, wings slightly opened and sometimes shivered; gave contact-call. Head-ruffling (inviting Allopreening) by ♂ sometimes associated with Courtship-feeding, although often not. Just after Courtship-feeding, ♂ seen hopping around ♀ with head slightly lowered, tail tilted somewhat towards her, wings partly open and fluttering; display apparently related to Forward-threat (see above). Exceptionally, ♀ also fed ♂. (Goodwin 1975, 1986.) (4) Mating. Pair may evidently copulate repeatedly. In Spain, one pair copulated 3 times in 5 min, firstly in alder *Alnus*, then 10 m up in pine *Pinus*; each mating lasted only a few seconds; in one observation, ♀ perched on dead branch, and leaned forward while ♂ fought to keep balance on her back; ♀ then raised her tail to right, ♂ bringing his down and somewhat to left; mating was accompanied by calls (Araujo 1975). No further information on soliciting-displays from the wild, but captive bird, in state of high excitement with keeper, slightly drooped and spread wings, half-spread tail, arched neck, and, as calling (see 5a in Voice) accelerated, lowered head and bill (pointing down) with pronounced shudder; similar calls heard during mock-copulation (♂-role) when bird perched on hand (Harrison 1983). (5) Behaviour at nest. Nest-site selection and nest-building take *c.* 14 days (Hosono 1971), but no information on role of sexes in site-selection. Both sexes bring nest-material, but only ♀ builds; nest-building pairs remain in flock, spending only some of day building (Hosono 1971). ♀ leaves nest quite often when clutch incomplete, covering eggs with wool or moss on departure. In addition to receiving food from ♂ (see above) incubating ♀ also forages for herself near nest while ♂ guards nearby. (Nechaev 1974.) ♂ also recorded apparently removing ♀'s faeces, ♀ indicating readiness to defecate and attracting ♂ by calling, raising her tail, and waving it persistently from side to side; during 7 hrs observation at one nest, 4 cases of Courtship-feeding and 3 of faeces-removal (Araujo 1975). However, faeces-removal and associated display never seen in study by C de la Cruz, and need confirmation. RELATIONS WITHIN FAMILY GROUP. Eyes of young open (not fully) by day 6. For first few days, young beg by stretching up, calling little; later in nestling period, however, beg loudly. At 12 days, start hopping on to nest-rim, and fledge at 15–16 days. (Nechaev 1974; see also Breeding.) Apparently ♀ broods young alone, and only contributes significantly to feeding once this role subsides. ♂ thus main provider, both for ♀ and nestlings. For first 2–3 days after hatching, ♂ feeds young via ♀, thereafter directly. ♂ will feed (always by regurgitation) 3–4(–5) young at each visit. Most nest-sanitation also by ♂ who will even push ♀ aside when she is trying to collect faecal sacs. (Nechaev 1974.) For role of nest-helpers, see Bonds (above). On fledging, young (unable to fly) keep hidden a few metres apart in bushes, and together with other fledglings in colony make loud commotion when parents arrive with food (Prokofiev 1962). ANTIPREDATOR RESPONSES OF YOUNG. Will fledge 1–2 days prematurely if disturbed (Nechaev 1974). PARENTAL ANTIPREDATOR STRATEGIES. (1) Passive measures. For covering of eggs with nest-material, see above. Sitting ♀ very quiet and wary, slipping off nest unobtrusively when disturbed early in incubation (Lilford 1866; Panov 1973a; Nechaev 1974). From day 6–7, however, sits tightly, allows approach of man to *c.* 0·5–1 m, and even then flies only a few metres away (Vorobiev 1954; Nechaev 1974), sometimes returning to incubate when intruder still present (Panov 1973a). (2) Active measures: general. Most nest defence by ♀ (Nechaev 1974), becoming bolder as incubation progresses (Panov 1973a). In Spain, as elsewhere, colony defence communal, neighbours joining in fierce defence of any endangered nest (Grande 1986; see also below). (3) Active measures: against birds. In Ussuriland, if crows (and notably *P. pica*) encroach on nest, all birds, including non-breeders (and, later in season, failed breeders) will combine to chase them away; several records of chasing Jungle Crow *Corvus macrorhynchos*, *P. pica*, and Jay *Garrulus glandarius* (Nechaev 1974). (4) Active measures: against man. If man touches nest or handles eggs, ♀ will attack, pecking at head and hands. Some incubating and brooding ♀♀ particularly aggressive, attacking anybody and anything; her alarm-calls summon mate who (like neighbours) tends to stand off, however, not joining attack (Nechaev 1974). In various studies, attack included the following: flying at intruder, pecking head and even drawing blood (Santos Júnior 1968) or veering off at last moment, also Bill-snapping (Panov 1973a) and wing-cuffing (Prokofiev 1962); in aviary, fast flying attacks from behind culminated in pecking crown of head and tweaking ears (Porter 1941); also, in Japan, recorded making dive-attacks and throwing twigs down on boys at nest (Kiuchi 1988). (5) Active measures: against other animals. Recorded attacking weasel *Mustela sibirica*, snake (Nechaev 1974), and cat (Hosono 1989).

(Figs by D Quinn from drawings in Goodwin 1986.)    EKD

**Voice.** In keeping with highly gregarious lifestyle, repertoire complex, though not fully known. Calls often intergrade and may also be combined in rapid succession. Homologous calls readily identifiable between races but niceties of geographical variation not known, and apparent phonetic differences in literature hard to evaluate. Unless otherwise specified, descriptions refer to *cooki* of Iberia, compiled partly from captive birds (Goodwin 1975, 1986; Harrison 1983); for other sonagrams of (Spanish) *cooki* see Bergmann and Helb (1982). For additional calls of extralimital nominate *cyanus*, see Mauersberger *et al.* (1982); for *interposita*, Hemmingsen and Guildal (1968).

CALLS OF ADULTS. (1) Song. Probably a quiet chattering but not well known (Bergmann and Helb 1982). Soft high-pitched chattering from ♂ displaying to ♀ (Goodwin 1975, 1986) perhaps song. (2) Contact-call. Constantly heard from flocks on the move. All descriptions, irrespective of race, mention ascending pitch. In *cooki*, hoarse 'zhree(eh)' (Mountfort 1958; Harrison 1983); drawn-out, rather husky, sibilant 'tschrreeeeeh' given by ♀ to attract ♂ (Goodwin 1975, 1986); screaming 'kschrrrie' (Jonsson 1982). Fig I shows a 'schrie' with short rapid initial ascent

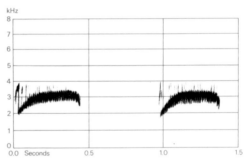

I   E D H Johnson   Spain   June 1959

in pitch and even shorter descent at end (J Hall-Craggs); for sonagrams of same recording, but different phonetic rendering, see Bergmann and Helb (1982). Similar in nominate *cyanus* (Mauersberger 1980), and in *swinhoei* harsh ascending high-pitched 'jwee' or 'zwee' (J Palfery). (3) Tonal calls. (3a) Begging-call. Given at least by ♀, perhaps also ♂. High-pitched call given by ♀ inviting ♂ to courtship-feed, recalling food-call of young (see below) (Araujo 1975), and described as soft, slightly plaintive,

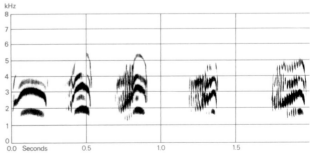

II   J-C Roché   Spain   April 1964

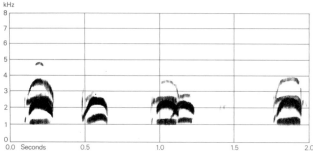

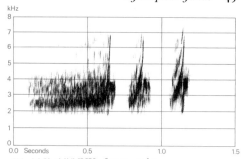

III  J-C Roché  Spain  April 1964

IV  M Yoshii/NHK  Japan  1967

piping, ascending 'quee' or 'quee-u' (Harrison 1983). Recording includes calls apparently of this kind, though context unknown; Fig II shows 10 calls starting tonal ('quee') increasing overall in pitch to 10th ('quii') and also increasing in frequency modulation (effectively rolled 'r' sound) to 6th before reverting gradually to tonal (J Hall-Craggs). In *japonica*, 'qui' or 'shui' (Hosono 1966a). (3b) Recording from which Fig II taken also includes other tonal calls (similar to 1st call in Fig II), roughly 'kwel' or 'kwil' (Fig III, which shows each to be a diad); 3rd unit in Fig III represents overlapping calls of 2 birds (J Hall-Craggs). Function of call not known. (4) Threat-, excitement-, and alarm-calls. (4a) Soft, sweet, high-pitched 'wee-we-we-wee-u', with 'u' sharply ascending in pitch, given in apparent self-assertion or threat to keeper (Goodwin 1975, 1986). High-pitched urgent squeaky 'quer' in response to stuffed Eagle Owl *Bubo bubo* (Mountfort 1958) is perhaps the same. (4b) Rapid series of short metallic ascending 'kwink' calls from captive birds confronted with object arousing fear or curiosity (Goodwin 1975, 1986). Renderings from elsewhere indicate same or similar call throughout geographic range, typically preceded by call 4d (or variant): in *japonica* a sharp 'wit wit wit' (Jahn 1942); Fig IV shows 'krarrah-wit wit wit' (J Hall-Craggs); in *swinhoei* 'wit wit twit-twit' (Swinhoe 1861); in nominate *cyanus*, wader-like 'tu-ittitt' (Mauersberger *et al.* 1982). (4c) Series of 'grid' sounds when agitated (Bergmann and Helb 1982). Low grating 'zhre', shorter than call 2, in response to stuffed *B. bubo* (Mountfort 1958) is perhaps also this call. Although described as high-pitched, the following sequence in response to hawk *Accipiter* and various other birds may be intense version of same: loud

repeated 'kree-kree-kree-kree', recalling mixture of shrieking alarm of Blackbird *Turdus merula* and 'laughing' of Green Woodpecker *Picus viridis* (Harrison 1983). (4d) Strong 'krrrah' in anger, sometimes approaching rattling churr (perhaps call 4f), including when strange cat nearby (Harrison 1983). Also rendered a drawn-out, even-pitched 'grääe' (Bergmann and Helb 1982). Call which typically precedes series of call 4b is evidently this call: loud harsh 'krarraah' (Goodwin 1975, 1986: Fig IV, 1st unit), screeching 'rääit' (Jahn 1942). Also, at low intensity, frequent subdued low-pitched 'kaah' and 'kah' as response to soaring raptor (Harrison 1983). (4e) Rattle-calls. So prominent and variable in recordings of wild birds as to suggest these calls individually recognizable and important in pair (or flock) communication; perhaps serve for long-distance contact since apparently not described from captive birds. Rattles in recording of 2 sorts: (i) slow rattles (*c.* 40 units per s), quite variable in timing, timbre, and amplitude, and typically with tonal coda; Fig V shows double rattle with weak tonal coda; Fig VI shows 4th of quadruple rattle, with clear (6-tone) coda; (ii) fast rattles (*c.* 60 units per s), sound like winding clock and with few tonal codas (e.g. Fig VII depicts complete double rattle, 2nd of which shows crescendo almost to end). (J Hall-Craggs.) Although buzzing churring (rattle) in *cooki* said to resemble Mistle Thrush *Turdus viscivorus* (Mountfort 1958), structure differs significantly from that species (compare Figs V–VII of *C. cyanus* with Figs IV–V of *T. viscivorus*, Volume V, p. 1020). Nominate *cyanus* and *swinhoei* similar to *cooki* in overall pattern of rattle-plus-coda, though following descriptions indicate differences; in nominate *cyanus* 't-k-k-k-k-k-k-k-k-dschrii' like

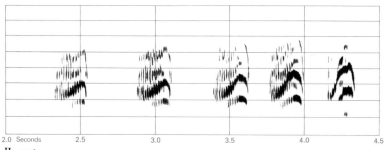

II  *cont.*

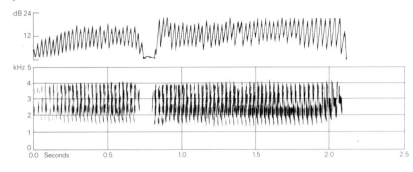

V   J-C Roché   Spain   April 1964

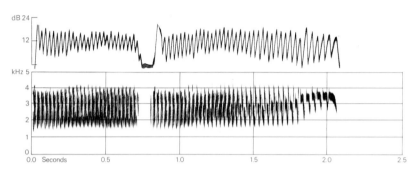

VI   J-C Roché   Spain   April 1964

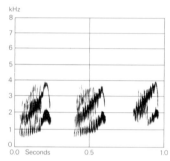

VIII   E R Parrinder/Sveriges Radio (1972–80)
Spain  May–June 1952

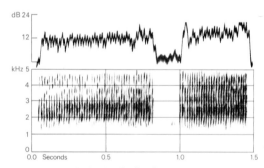

VII   J-C Roché   Spain   April 1964

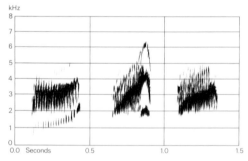

IX   J Burton and D J Tombs/BBC   Spain   May 1982

loud alarm-clock being wound up, ending in quiet whistling screech (Mauersberger 1980). In *swinhoei* very distinctive, slow, somewhat creaky rattle, often ending in rather high-pitched 'keeow' like Jackdaw *Corvus monedula* (J Palfery). Quiet 'ker' given by parents nearing nest, apparently to stimulate young to beg (Mountfort 1958), is perhaps short variant. (4f) A 'woid' reminiscent of Whitethroat *Sylvia communis* but louder (Mauersberger 1980). In Portugal, Mathews (1885) also reported alarm-calls much like loud *S. communis*. (5) Other calls. (5a) Long shivering trill given by bird apparently inviting copulation (Harrison 1983). (5b) Rather sharp chirruping from captive ♂ making nest-building movements (Goodwin 1975, 1986).

CALLS OF YOUNG. Main food-call in nest a high-pitched 'wee' or 'quee' (Harrison 1983: Fig VIII) resembling adult Begging-call (compare with call 3a and Fig II); see also sonagram in Redondo and Arias de Reyna (1988). In fledglings a more rattling 'krrrre' or 'krree' (J Hall-Craggs: Fig IX).                    EKD

**Breeding.** SEASON. Central Spain: at 300 m altitude, eggs laid early April to late May, peak in one year 15–25 April, *n* = 45 clutches (Muñoz-Pulido *et al.* 1990); at 1250 m, eggs laid early June to late July, peak mid-June (Araujo 1975); laying started earlier in nests in evergreen oaks *Quercus ilex* and *Q. suber* (average date 22 April, *n* = 20) than in deciduous oak and ash *Fraxinus* in same area (29 April, *n* = 19) (Alonso *et al.* 1991). Southern Spain: eggs laid end of April to mid-May (Alvarez 1975; Pacheco *et* *al.* 1977). Eastern Portugal: eggs laid late April to early May (Santos Júnior 1968). Record in Iberia of eggs in March (Goodwin 1986). Ussuriland (eastern Russia): eggs laid first half of May to first half of June, peak in second half of May (Vorobiev 1954; Panov 1973*a*; Nechaev 1974). For Japan, see Hosono (1966*a*, 1989). SITE. Generally in fork of branch towards edge of tree crown, top often covered by leaves of supporting branch (Alvarez 1975). Preferred position is in mid-crown vertically but as far from trunk as possible without being visible in outer foliage; apparently best position for avoiding predators (Redondo *et al.* 1989, which see for comparison of 6 nest-site studies; Alonso *et al.* 1991). 2 nests may be in same tree (Makatsch 1976), though only 1 nest per tree found in study of 163 nests in central Spain (Muñoz-Pulido *et al.* 1990). Average diameter of supporting branch *c.* 5 cm (Redondo *et al.* 1989; Alonso *et al.* 1991). In one area of central Spain, preferred nest-tree species (relative to abundance) was ash (33% of 163 nests), though most nests in *Q. ilex*; average height above ground 5·7 m, *n* = 107 (Alonso *et al.* 1991). Also in central Spain, 46% of nests in pine *Pinus pinaster*, 10% in chestnut *Castanea*, 8% in walnut *Juglans*, and 8% in *Robinia* (*n* = 61); average height 6·6 m (2·5–18), *n* = 55, and (exceptionally) 49% of nests in this study against trunk (Araujo 1975). All of 32 nests in southern spain in *Q. ilex*, at average height 4·4 m (2·2–6·3) and 2·1 m from trunk; often near water for mud collection (Redondo *et al.* 1989), and also commonly in stone pine *Pinus pinea* (Alvarez 1975). Nests recorded in Portugal in *Q. ilex*, *Q. suber*, ash, olive *Olea*, orange *Citrus aurantium*, almond *Prunus triloba*, mulberry *Morus*, and *Eucalyptus* (Santos Júnior 1968; Sacarrão and Soares 1976). In Ussuriland, among dense twigs in low trees, especially fruit trees and thorn bushes, more rarely in rock cavities and hollow trees, and record of nest on ground; sometimes under roof of mass of vegetation, sticks, etc., left by floodwater in dense, low bushes less than 1 m above ground (Dementiev and Gladkov 1954; Vorobiev 1954; Nechaev 1974). For Japan, see Hosono (1966*a*, 1971). Nest: rough, loose foundation of twigs of pine, *Eucalyptus*, rockrose *Halimium*, gorse *Ulex*, etc., inside which layer of compacted pellets of earth, mud, or dung in form of bowl 2–5 cm thick, in which is embedded rough lower part of lining (roots, pine needles, small twigs) covered by rootlets, plant down, fibres, moss, lichen, animal hair, sheep's wool, feathers, etc., whole lining *c.* 1 cm thick (Santos Júnior 1968; Alvarez 1975; Araujo 1975; Makatsch 1976; Sacarrão and Soares 1976; Pacheco *et al.* 1977). Fresh and aromatic stalks and flowers may be incorporated (Kulczycki 1973; Neves 1984). Paper and rags found in nests in Russia and Mongolia (Mauersberger *et al.* 1982; Ogorodnikova and Mironova 1989). For Ussuriland see Vorobiev (1954), Panov (1973*a*), and Nechaev (1974). 19 nests in southern Spain had average outer diameter 15·4 cm (12·2–18·3), inner diameter 10·5 cm (8·9–13·0), overall height 11·6 cm (8·2–13·5), and depth of cup

7·8 cm (6·1–9·1) (Pacheco *et al.* 1977). 10 nests from central Spain had average weight 221 g, 22% of which mud (Araujo 1975). For nests in Japan see Hosono (1971, 1983). Building: takes 10–18 days (Hosono 1971; Nechaev 1974; Araujo 1975); by both sexes, apparently mainly by ♀, and ♂ helps to bring material (Nechaev 1974; Komeda *et al.* 1987); captive ♂ tore up stems and roots of grass, then took damp sand in same billful and tried to stick it down on nest, whole action like Swallow *Hirundo rustica* (Goodwin 1986). In Spain new nest built each year (Alvarez 1975; Araujo 1975), but in Russia old nests may be refurbished or old lining re-used in new nests (Prokofiev 1962; Nechaev 1974). Other birds (helpers) may bring material and build (see Social Pattern and Behaviour). EGGS. See Plate 57. Sub-elliptical (short sub-elliptical to oval), smooth and faintly silky; from pale cream or olive-cream, rarely with bluish tinge, to pale yellowish-brown or brown-olive, those of nominate *cyanus* sometimes greenish, with sparse small brown or olive-brown spots and greyish blotches, often concentrated at broad end, sometimes in circle (Harrison 1975; Makatsch 1976; Goodwin 1986). *C. c. cooki*: 26·7 × 20·0 mm (24·0–30·2 × 18·0–22·0), *n* = 451; calculated weight 5·5 g (Schönwetter 1984). Nominate *cyanus*: 26·7 × 19·5 mm (25·6–28·7 × 18·5–20·1), *n* = 18; calculated weight 5·25 g (Makatsch 1976). See Alvarez (1975) for 80 eggs from southern Spain, and Araujo (1975) for 94 from central Spain. For Ussuriland, see Panov (1973*a*). Clutch. 5–7 (4–9). Of 182 clutches, western Spain: 4 eggs, 2·2%; 5, 6·6%; 6, 35·7%; 7, 41·2%; 8, 13·2%; 9, 1·1%; average size of 226 clutches over 3 years 6·58 ± SD0·89, with decline in size of *c.* 1·4 over season (Cruz *et al.* 1990). Of 81 clutches, central Spain: 4 eggs, 7·4%; 5, 12·3%; 6, 35·8%; 7, 39·5%; 8, 4·9%; average size 6·2 ± SE0·11 (Muñoz-Pulido *et al.* 1990). Of 21 clutches, Japan: 5 eggs, 14%; 6, 38%; 7, 24%; 8, 24%; average 6·6; decrease from average of 7·0 in April, *n* = 11, to 5·3 in June, *n* = 3 (Hosono 1971). Average of 7 clutches in Ussuriland 8·4 (Nechaev 1974). Eggs laid daily in early morning (Vorobiev 1954; Prokofiev 1962; Hosono 1971; Muñoz-Pulido *et al.* 1990), though possibility of last 2 or 3 laid on alternate days (Nechaev 1974). In study of 136 nests in central Spain, no replacement or 2nd broods found (Muñoz-Pulido *et al.* 1990), but in Ussuriland replacement clutches of 4–5 eggs laid *c.* 2 weeks after loss of 1st clutch (Nechaev 1974), and observation in Mongolia of replacement nest built *c.* 1 week after loss of clutch (Mauersberger *et al.* 1982). Captive Chinese birds were double-brooded: 2nd nest built when young *c.* 3 weeks old, 1st brood still being fed during incubation of 2nd clutch (Porter 1941). INCUBATION. 15–16 (14–17) days; by ♀ only (Hosono 1971; Pacheco *et al.* 1977; Muñoz-Pulido *et al.* 1990). Starts after laying of 3rd egg (Porter 1941; Hosono 1971), 4th egg (Nechaev 1974), or penultimate egg (Panov 1973*a*). During 3·25 hrs of observation after laying of 3rd egg, average stint on eggs of one ♀ 7·7 min, average break 23 min, and at 3 nests incubation of completed clutches occupied 66–95% of ♀'s time (Hosono 1971). YOUNG. Fed and cared for by both parents, though fed mostly by ♂ via ♀ for first few days after hatching (Porter 1941; Nechaev 1974; Ogorodnikova and Mironova 1989). See Hosono (1983, 1989) and Komeda *et al.* (1987) for helpers at nest, and Araujo (1975) for bird feeding young at 2 nests; see also Social Pattern and Behaviour. For brooding, feeding, and development of young see Hosono (1966a, 1971), Santos Júnior (1968), Alvarez (1975), and Araujo (1975). FLEDGING TO MATURITY. Fledging period in central Spain 14–16 days, *n* = 7 nests (Muñoz-Pulido *et al.* 1990). In Japan, 1962–5, 16·2 days, *n* = 19; 1966–9, 17·9 days, *n* = 21; 19 days at one nest (Hosono 1983). First breeding at 1 year (Komeda *et al.* 1987). Young may be fed by parents for 27 days or more after fledging (Hosono 1971). BREEDING SUCCESS. In central Spain, of 136 nests, 31% abandoned before laying, and clutches lost in 9·6% before completion, so full clutches laid in 81 nests; of 60 for which outcome known, 46% failed completely and 54% fledged young (of which 59% fledged all, 41% only some); average number of fledged young per nest 2·7 (*n* = 60), average per successful nest 5·1 (*n* = 32). Number of young per successful nest decreased over season probably because of predation since food supply increases; although average success low, successful nests show high productivity, again suggesting predation pressure. (Muñoz-Pulido *et al.* 1990.) Using same data, Alonso *et al.* (1991) found nests in most favoured part of crown (see Site, above) were most successful, probably because of reduced predation by nocturnal mammals; potential predators are dormouse *Eliomys*, genet *Genetta*, snake *Malpolon*, lizard *Lacerta*, and Magpie *Pica pica*; nests may also be parasitized by Cuckoo *Cuculus canorus* and Great Spotted Cuckoo *Clamator glandarius*. In western Spain, of eggs in 214 nests, 57% hatched and 40·3% produced fledged young, giving 2·65 young per nest or 4·04 per successful nest (*n* = 125); 42·5% of clutches failed totally while all eggs produced fledged young in 13·6%; breeding success highest in clutches of 8 (47·3%). Of 666 unsuccessful eggs, 62% predated, 19% infertile, 12% deserted, 3% disappeared, and 4% lost to other causes. (Cruz *et al.* 1990, which see for more details.) In one season of adverse weather, southern Spain, 1 nest out of 32 produced fledged young, 20 were deserted before or during incubation, and 11 found to be empty; many losses due to parents leaving nest to find food because insect supply low (Redondo *et al.* 1989). In another study in southern Spain, 43% of 77 eggs produced fledged young, 2·5 per nest (*n* = 13); 23% of clutches abandoned (Pacheco *et al.* 1977). For Japan, see Hosono (1971, 1983).　　BH

**Plumages.** (*C. c. cooki*). ADULT. Nasal bristles, forehead, upper cheek, and ear-coverts to side of neck black, rather velvety on forehead, slightly tinged purplish from eye backwards; crown to hindneck black with strong purple gloss. Mantle, scapulars, back, and rump greyish-drab with slight vinous tinge (feather-tips more drab, vinous-pink bases showing through), often more distinctly vinous-pink on upper mantle at border with black cap. Upper tail-coverts pale drab-grey with blue tinge. Chin, lower cheek, and throat white, slightly tinged pale vinous-pink on rear of cheek and on lower throat when plumage fresh. Side of breast and flank vinous-drab, like mantle and scapulars but slightly paler and less saturated, remainder of underparts vinous-grey, often tinged drab on upper chest and sometimes almost pink-white on mid-belly and vent. Tail uniform azure- or caerulean-blue; shafts light horn. Outer web of p9–p10 and inner web of all flight-feathers dull black; outer web of secondaries and p1–p8 azure-blue, brightest on primaries; terminal $\frac{2}{3}$ (p8) to $\frac{1}{3}$ (p3) of outer webs of primaries contrastingly white, and sometimes a white fringe along terminal outer web of p2. Tertials, bastard wing, and all upper wing-coverts azure-blue, lesser coverts tinged drab-grey. Axillaries and under wing-coverts dark grey with vinous tinge, longest coverts tipped white. *In worn plumage*, about May–July, cap duller black, gloss more bluish; vinous-pink collar at rear of cap more pronounced; mantle and scapulars either darker drab-brown or (if heavily worn) less uniform deep vinous-drab, feather-tips worn to pale drab-grey; underparts paler, more white of feather-bases showing through (occasionally, breast and belly entirely dirty vinous-white); white of throat contrasts less with remainder of underparts; azure-blue of flight-feathers and tail often more glistening, but tips of tail-feathers often partly freckled dull dark grey. Sexes similar; but see Measurements. NESTLING. Naked; feather-pins appear on day 7 (Witherby 1928; Porter 1941). JUVENILE. Like adult, but cap dull black, not glossy, sometimes slightly tinged grey; tips of feathers of cap with dark brown-grey fringe. Bar across upper mantle dirty buff-white; remainder of mantle, scapulars, back, and rump dull drab-brown (without delicate vinous tinge of adult), feathers with greyish-drab base and faint brown-grey fringe along tip. Underparts entirely dirty white, feathers of side of breast, chest, and flank with broad buff-brown tips (sometimes slightly vinous). Tail quite different from adult: colour as adult, but tips of feathers rounded (truncate in adult), bordered by white or isabelline fringe (often faint or absent on t1 and t6); t1 much shorter than t2. Flight-feathers as adult, but tertials tinged drab or brown; greater upper primary coverts as adult, but tinged drab-grey on tip and inner web and often with faint trace of dull grey and white fringe along tip (most pronounced on innermost); lesser and median upper wing-coverts dull buffish-drab with contrasting dirty white tips (tending to form wing-bar *c.* 2 mm wide on median coverts), greater coverts drab-grey with bluish outer web and narrow off-white tip. FIRST ADULT. Like adult, but retaining juvenile flight-feathers (frequently except outermost primaries and innermost secondaries), usually (t2–)t3–t6 and greater upper primary coverts, and rarely a few tertials, outer greater coverts, or feathers of bastard wing. If outer primaries, inner secondaries, and/or part of upper primary coverts new and remainder of flight-feathers and coverts old, ageing generally easy due to contrast in colour and abrasion between these feather generations. If all flight-feathers juvenile, ageing on wing not always easy, except in a few birds which retain some feathers of bastard wing, tertials, or greater coverts (these distinctly browner and more worn on tip than neighbouring fresh feathers); if all tertials and greater coverts new and primary coverts all new or all old, the old feathers are often (not always) browner and more worn than neighbouring new ones. However, in birds with all flight-feathers juvenile, ageing generally possible by tail: t1(–t2) relatively new, 1st adult, outer feathers older, juvenile; tips of juvenile outer feathers more heavily abraded and less truncate than adult at same time of year, distinctly fringed white or with traces of white; 1st adult t1(–t2) new, truncate, contrasting in shape, colour, and degree of abrasion with juvenile outer feathers. Rarely, independent of age, terminal 2–6 mm of t1 white (5% of 60 adults and 1st adults examined). *In worn plumage*, May–July, more strongly abraded than adult at same time of year, especially flight-feathers and tail; adult t1 often also sometimes heavily abraded, but adult t2–t6 still with fairly smooth truncate tip (in 1st adult, all feathers heavily abraded; if retained, juvenile outer feathers have pointed abraded tips, often with traces of white fringes).

**Bare parts.** ADULT, FIRST ADULT. Iris dark brown to brownish-black. Bill, leg, and foot black. NESTLING. Bare skin bright pink, turning to dark leaden-grey from day 1. Bill greyish-flesh, cutting edges and narrow gape-flanges pink-white. Mouth reddish-pink to bright crimson, pinkish at gape; no spots on tongue. (Witherby 1928; Porter 1941; Mountfort and Ferguson-Lees 1961; Goodwin 1986; ZMA.) JUVENILE. No information.

**Moults.** ADULT POST-BREEDING. Complete; primaries descendent. In *cooki* from Badajoz (south-west Spain), moult starts with shedding of p1 between late May and late June, mainly from mid-June; in 5 captive birds, primary moult completed *c.* 118 days after loss of p1 with shedding of p10; p10 full-grown again after *c.* 32 days, giving primary moult duration of *c.* 150 days; in the wild, duration of primary moult *c.* 141 days, mid-September to early November (average *c.* 10 October); some evidence that 1-year-olds start earlier and complete later than older birds, but samples small and ageing impossible when moult advanced (Cruz Solis 1989). Secondary moult starts with loss of s1 from *c.* 20 July (primary moult score *c.* 20), completed with regrowth of s6 on *c.* 25 October, 0–25 days after completion of primaries (score 50); tail starts with t1 on *c.* 10 July, completed with t6 on *c.* 10 October at about same time as completion of primaries; tertial moult mid-July to end of September, moult of bastard wing mainly August–September, body July to October or November. (Cruz Solis 1989; Cruz *et al.* 1991a; see both these sources for further detail). In extralimital *pallescens*, Russia, ♀ starts before ♂; single ♀ in primary and body moult on 29 June, single ♂ just started 4 July (Dementiev and Gladkov 1954). POST-JUVENILE. Partial. In birds examined (*n*=17), involved head, body, lesser and median upper wing-coverts, greater upper wing-coverts and bastard wing (rarely, 1–2 feathers in each retained), tertials (rarely innermost or outermost excluded), occasionally 1–2 innermost secondaries or 1–2 outermost primaries, and always t1 (occasionally, t2 also, rarely t3). In larger sample from Badajoz, moult starts with upper wing-coverts from 12 July onward, at age of 2–3 months, followed by tertials (on average, starts 28 July, sequence s8–s9–s7, lasting from average age of *c.* 90 days to *c.* 175 days, independent of hatching date); bastard wing moulted last; in captive birds, body starts with upperparts from late July, completed with chin late November, duration *c.* 122 days; in 2nd half of September all birds show moult of tertials and tail-feathers, and greater upper wing-coverts new in many. Of 132 birds, 50% replaced t1 only, 34% all tail, remainder t1–t2 and sometimes a few other feathers; following moult of tertials (s7–s9), 69% replaced innermost secondary (s6), 33% s5 also, 11% s4, 2% s3; 13% replaced some outer primaries, mainly p9 and/or p10, in a few birds (3%) 5–8 outer feathers; 17% replaced all greater upper primary coverts, others scattered feathers or none. (Cruz Solis 1989; Cruz Solis *et al.* 1992.) Moult never complete (Cruz *et al.* 1991a, contra Goodwin 1975, 1976), bird retaining at least juvenile inner primaries and outer secondaries. Pronounced sexual difference in retained number of juvenile flight-feathers, tail-feathers, and primary coverts: moult of ♂ more extensive than in ♀ (Cruz Solis *et al.* 1991b, which see for details).

**Measurements.** *C. c. cooki*. Spain and Portugal, October–June; skins (RMNH, ZFMK, ZMA). Juvenile wing includes retained juvenile wing of 1st adult; 1st adult tail is 1st adult with at least t1 new. Bill (S) to skull, bill (N) to distal corner of nostril; exposed culmen on average *c.* 3·5 less than bill (S).

| | | | |
|---|---|---|---|
| WING AD | ♂ | 136·7 (3·27; 21) 131–145 | ♀ 130·1 (3·31; 15) 126–136 |
| JUV | | 134·2 (3·96; 8) 130–142 | 129·3 (2·87; 9) 126–134 |
| TAIL AD | | 187·1 (5·11 21) 178–197 | 173·4 (8·19; 12) 162–185 |
| 1ST AD | | 188·6 (5·06; 6) 183–195 | 171·2 (8·92; 10) 157–183 |
| BILL (S) | | 27·5 (0·99; 14) 25·7–28·6 | 26·7 (1·62; 9) 24·5–28·7 |
| BILL (N) | | 16·3 (0·59; 14) 15·2–17·1 | 15·5 (0·91; 8) 14·2–16·9 |
| TARSUS | | 35·7 (0·94; 15) 34·3–37·4 | 34·8 (0·75; 9) 33·8–35·8 |

Sex differences significant, except for bill (S). Tail below 181 in only 1 of 27 ♂♂, over 180 in only 4 of 22 ♀♀. North-east Portugal: wing, ♂ 132·9 (2·87; 14) 126–137, ♀ 129·8 (3·62; 8) 122–133; tail, ♂ 191·6 (6·90; 14) 178–200, ♀ 183·6 (14·29; 8) 152–200 (Santos Júnior 1968, which see for other measurements). Spain: wing, ♂ 134·5 (8·73; 27) (fresh) or 135·3 (4·75; 25) (skins), ♀ 127·2 (6·19; 18) (fresh) or 131·5 (4·02; 24) (skins); tail, ♂ 179·4 (26·62; 26) (fresh) or 187·9 (24·89; 24) (skins), ♀ 172·6 (17·48; 16) (fresh) or 173·6 (16·40; 23) (skins) (Alvarez and Aguilera 1988, which see for sexual differences in other measurements). See also Cruz Solis (1989) for Badajoz (south-west Spain).

For various measurements throughout Iberia, see Sacarrão (1968). In general, birds from northern Iberia slightly larger than those from south, as follows. Wing: Salamanca, Caceres, and Madrid, ♂ 136·5 (3·55; 19) 130–143, ♀ 130·9 (3·07; 15) 126–136; southern Portugal (Setubal, Algarve) and southern Spain (Andalucia), ♂ 133·4 (2·46; 7) 130–137, ♀ 127·2 (1·89; 5) 126–131 (RMNH, ZFMK, ZMA).

Extralimital races. Nominate *cyanus*: wing 127–151, tail 202–255 (Stegmann 1931a, b); wing, ♂ 141·9 (7) 136–147·5, ♀ 134·7 (4) 125–137·5 (Dementiev and Gladkov 1954). *C. c. pallescens*: wing 132–151, tail 193–254 (Stegmann 1931a, b; Meise 1934a); wing, ♂ 141·1 (15) 133–148, ♀ 137·7 (15) 134–148 (Dementiev and Gladkov 1954). *C. c. stegmanni*: wing 131–154 (69), tail 191–246 (Meise 1934a; Piechocki 1959). *C. c. interposita*, Ching-Ling mountains (Shensi, China): wing 132–152 (51), tail 215–235 (6) (Hartert 1921–2; Hartert and Steinbacher 1932–8; Meise 1934a). *C. c. swinhoei*: wing 131–142(–145) (47), tail 200–222 (4) (Meise 1934a). *C. c. kansuensis*, Kansu (China): wing, ♂ 139–145 (5), ♀ 136–137 (3); tail, ♂ 221·8 (7·08, 5) 213–226, ♀ 210·7 (3) 201–225 (Stresemann *et al.* 1937). *C. c. japonica*: wing 125–140, tail 190–224 (*n*=13 or more) (Meise 1934a); wing 135 (10) 133–139, tail 203 (10) 190–220 (Vaurie 1959). *C. c. pallescens, stegmanni,* and *swinhoei*, sexes combined: wing 138·7 (13) 130–149, tail 213·7 (10) 204–219, bill (S) 30·2 (9) 29·0–31·7 (RMNH, ZFMK, ZMA).

**Weights.** *C. c. cooki*. North-east Portugal, May: ♂ 73·0 (2·74; 5) 70–75, ♀ 69·0 (3·37; 4) 65–73 (Santos Júnior 1968). Spain: ♂ 78·1 (6·08; 29), ♀ 71·1 (6·92; 16) (Alvarez and Aguilera 1988). Caceres (north-west Spain), November–December: ♂ 75·7 (2·39; 5) 72–78, ♀ 72·8 (5·67; 5) 65–79. Southern Spain, May: ♀ 78. (ZFMK.) Badajoz (south-west Spain): average of juvenile 68·1 (93), 1st adult ♂ 73·4 (30), 1st adult ♀ 71·7 (37), adult ♂ 73·8 (24), adult ♀ 70·7 (24) (Cruz Solis 1989). Nestlings reach average of *c.* 75 on day 17 (Araujo 1975, which see for growth curves).

Extralimital races. Nominate *cyanus*: ♀♀ 62, 82 (Dementiev and Gladkov 1954). *C. c. stegmanni*, July–August: adult ♂ 84, juveniles 79 (6) 74–83 (Piechocki 1959).

**Structure.** Wing short, broad at base, tip rounded. 10 primaries: p6 longest, p7 0–3 shorter, p8 4–11, p9 20–29, p5 0–2, p4 3–5, p3 9–13, p2 15–20, p1 17–24; p10 somewhat reduced, 48–60 shorter than p6, 23–29 longer than longest upper primary covert. Outer web of (p4–)p5–p8 emarginated, inner of (p5–)p6–p9 with notch. Tertials short. Tail long, strongly graduated; 12 feathers, t6 94·9 (20) 85–103 shorter than t1 in adult and 1st adult; in juvenile, t1 reduced, much shorter than t2 (Stegmann 1931a; RMNH). Bill, leg, and foot as in Magpie *Pica pica*, but finer and more slender; however, culmen less decurved than in *P. pica*, bill tip more elongated (especially in eastern races) and more sharply pointed. Depth of bill at base 9·2 (8·5–10) in ♂, 8·5 (8–9·5) in ♀ (Santos Júnior 1968) or average 10·2–10·5 in ♂, 9·4–10·0 in ♀ (Alvarez and Aguilera 1988). Middle toe with claw 25·0 (10) 24–26; outer and inner toe with claw both *c.* 72% of middle, hind *c.* 75%.

**Geographical variation.** Rather slight in size and general body colour, mainly clinal in eastern races, but western isolate *cooki* distinct, lacking white tip of t1 usually present in east. Colour variation obscured by influence of bleaching and wear on plumage and by use of old specimens which are much liable to fading (Stegmann 1931b, Fiebig 1983). Stegmann (1931a, b), Meise (1934a), and Vaurie (1959) followed here, though these authors perhaps recognize too many races. Variation within Iberia slight; birds from northern and inland localities (Valladolid, Salamanca, Avila, Madrid, Caceres, Extremadura) on average slightly larger and greyer than those from coastal and southern localities (smal-lest and brownest in Algarve, Sevilla, Jaén, Malaga, and Almeria), but overlap in size too great to warrant recognition of more than one race and difference in colour very slight in fresh plumage (Vaurie 1954b, 1955; Sacarrão 1968; C S Roselaar). East Asian populations differ from those of Spain and Portugal in broad white tip on t1 in adult and 1st adult (16–30 mm long, rarely absent; in *cooki*, rarely up to 5 mm of tip of t1 white), in relatively longer tail and in generally greyer body colour. As in Iberia, paler birds in north and inland, darker birds in south and near coasts; nominate *cyanus* from Transbaykalia and neighbouring Mongolia grey with slight vinous cast on upperparts, white with slight grey tinge on underparts; *pallescens* from far-eastern Russia paler still, but upperparts slightly more olivaceous; birds from Kansu (*kansuensis*), Manchuria (*stegmanni*), and Korea (*koreensis*) slightly darker and browner than *pallescens* on upperparts and greyer on underparts, but doubtfully separable from each other, from *pallescens*, or from *interposita* of Ching-Ling mountains in Shensi, which is slightly darker and browner still; *swinhoei* from Peking south to Fukien and west to Szechwan dark, saturated vinous-drab above, pale drab-grey below; *japonica* from Japan dark also, but greyer than *swinhoei*, rather like nominate *cyanus* on upperparts but colours more saturated. In both colour and size, *japonica* intermediate between remaining Asiatic races and *cooki*.

N.B. Name *C. cyanus* conforms with International Code of Zoological Nomenclature and is thus correct (i.e. not *C. cyana*, contra Voous 1977).                                                    CSR

## *Pica pica* Magpie

PLATES 3 and 5 (flight)
[between pages 136 and 137]

Du. Ekster     Fr. Pie bavarde     Ge. Elster
Ru. Сорока     Sp. Urraca     Sw. Skata     N. Am. Black-billed Magpie

*Corvus Pica* Linnaeus, 1758

Polytypic. Nominate *pica* (Linnaeus, 1758), southern Norway and Sweden (south of a line from Trondheim to Gävle) south through western and central Europe to southern France, mainland Italy, Sicily, Greece, Asia Minor, Cyprus, and Transcaucasia, east to eastern Poland, Soviet and Rumanian Carpathians, and Bulgaria; grades into *fennorum* in Väster Norrland and northern Gävleborg (Sweden), probably in Nord Trøndelag (Norway), and in Lithuania, Belorussiya, and western Ukraine; into *bactriana* in central Ukraine, Moldavia, eastern Rumania, Crimea, Araks valley (Transcaucasia–Iran border), and south-east Turkey (Euphrates to Van area); into *melanotos* in Pyrénées and Roussillon (France); *fennorum* Lönnberg, 1927, northern Norway, northern Sweden (north from Jämtland and Västerbotten), Finland, and from Latvia and Estonia east to Kostroma, Severnaya Dvina basin, and upper Pechora (Russia), grading into *bactriana* in regions of Kursk, Tambov, Moskow, and Gor'kiy, and in upper Vyatka and Kama basins; *bactriana* Bonaparte, 1850, European Russia west to Perm', Kazan', Tambov, Khar'kov, and Dnepropetrovsk, south to Sea of Azov and northern slopes of greater Caucasus; Siberia east to upper Lena basin and Lake Baykal, south through mountains of central Asia and north-west Mongolia to Himalayas, Pakistan, Iran (except north-west), and Iraq, grading into *anderssoni* in Zeya basin (probably) and into *leucoptera* in northern Transbaykalia and in Irkutsk region; *melanotos* A E Brehm, 1857, Iberia; *mauritanica* Malherbe, 1845, north-west Africa. Extralimital: *asirensis* Bates, 1936, south-west Arabia; *bottanensis* Delessert, 1840, eastern Himalayas to east and north-east Tibet; *leucoptera* Gould, 1862, southern Transbaykalia, north-east Mongolia, and north-west Manchuria in Argun basin; *anderssoni* Lönnberg, 1923, northern China from Kansu, Shensi, Shansi, and Hopeh to Manchuria (except north-west), Korea, and Russia from lower Amur valley to Sikhote-Alin'; *sericea* Gould, 1845, eastern and southern China and neighbouring parts of south-east Asia, Taiwan, and southern Japan; *camtschatica* Stejneger, 1884, northern shore of Sea of Okhotsk, Kamchatka, and Anadyrland; *hudsonia* (Sabine, 1823), North America.

**Field characters.** 44–46 cm (of which over 50% is tail in adult); wing-span 52–60 cm. Bill, head, and body size close to Jackdaw *Corvus monedula* but wings rather short, appearing fan-like, and tail very long and graduated. Medium-sized and markedly attenuated crow, with rather flat crown, apparently deep chest and short body, rather long legs, and long graduated tail. Boldly pied: black, with white belly and large white wing-panels. Flight laborious. Diagnostic chattering call. Sexes similar; no seasonal variation. Juvenile separable. 5 races in west Palearctic, of which north-west African most easily distinguished from rest.

ADULT. Moults: May–October (complete). (1) Races of Europe (except Iberia) and Turkey, nominate *pica* and *fennorum*. Plumage basically black, with striking white outer scapulars, breast, flank, and central belly, and variable amount of white on rump and basal inner webs of primaries (visible on upper- and underwing but usually only in flight). White rump most obvious in northernmost populations in which birds are also somewhat larger. Black plumage iridescent, colours clearly visible at close range when sunlit: blue-purple on most of head and body, faint green on crown and inner scapulars, brilliant bronze-green and terminal bands of red-purple and blue- and green-purple on tail, and blue-green on wings. Bill and legs black. (2) Iberian race, *melanotos*. Somewhat smaller than central European birds, with black rump and sometimes a small bare bluish patch behind eye. (3) North-west African race, *mauritanica*. 20% shorter-winged than northernmost birds, with black rump and (invariably) obvious bare bright blue patch behind eye. Iridescence essentially bronze-toned, especially on tail. JUVENILE. All races. Resembles adult but duller, with sooty and dirty white tones producing less sharply pied pattern; iridescence confined to wings and tail and less brilliant. Tail often shorter.

Unmistakable. Flight lacks grace and freedom of black crows; over long distance, action alternates slightly 'desperate' bursts of rapid wing-beats with stalling glides, bird appearing to drag long straight tail. In spite of rather slow, hesitant, laborious progress, often ascends to well above surrounding trees. Over short distance and in and around cover, action more confident, dashing and sweeping with tail often spread; able to follow prey through thick foliage. Gait confident, with high-stepping walk and characteristic brisk, sidling hops and jumps, usually with tail held up; movements far less accomplished among branches, made seemingly awkward by trailing tail. Raises and fans tail when excited. Often visible as sentinel on tree or bush, sitting upright with tail drooped.

Voice quite varied, with babble (approaching song), at least 2 relatively soft or liquid notes, and common, harsh, staccato chatter, 'chatchatchatchatchack' or 'chacker chacker chacker chacker' and delivered with increasingly peevish tone when bird disturbed.

**Habitat.** Breeds in west Palearctic from upper to lower middle continental and oceanic latitudes, from boreal taiga through temperate to Mediterranean, steppe, and semi-desert zones, avoiding both densely forested and treeless regions, precipitous rocky terrain, and extensive wetlands. Occurs up to considerable altitudes, especially towards south of range, where ecological conditions are suitable, but is predominantly a lowland bird of open or lightly wooded country offering good opportunities for foraging on ground and for nesting, roosting, and taking cover in trees or shrubs. Inhabits woods of many different types, both broad-leaved and coniferous, wherever glades, clearings, or more open stands occur, and especially near margins of natural or cultivated grasslands and croplands. Prefers to perch below tallest parts of tree canopy. Intensive cultivation with few trees, tall bushes, or hedgerows is unattractive, as are treeless natural grasslands and wetlands, although exceptionally isolated buildings or bushy thickets may be accepted in place of more suitable cover. Requirement for immediate access to cover may be relaxed or even waived in situations where there is no threat to security. Spread into built-up areas and even major cities is increasingly apparent as result of overall population increase and as need for wariness diminishes; here will sometimes perch freely on artefacts including tall buildings. Adaptability and opportunism together with effects of human intervention in spreading suitable mixed habitats have evidently contributed to historical advance of this species, in common with other Corvidae. Earlier habitat occupancy is perhaps indicated by that typical in North America where *hudsonia* occupies mountainous country, favouring canyons and streamsides where tall thickets and scattered trees provide cover, nest-sites, and foraging grounds. Its range coincides almost exactly with climatic region defined as Cold Type Steppe Dry Climate, in which low mean January temperatures appear to play critical role in distribution. Association with herds of large herbivores, and nesting close to eagles (Accipitridae) and other large predators, are persisting features in North America, which may well have formerly been significant also for Palearctic population. Differs from most Corvidae in not normally flying in any but lowest airspace, but occasionally up to *c.* 70 m on breeding grounds and on longer journeys even higher.

**Distribution.** Few changes in recent decades, except for colonization of suburban and urban habitats (Britain, Belgium, Netherlands, Finland, Poland, Russia, and probably elsewhere), and reoccupation of areas of former occurrence following decrease in human persecution, and some spread in extreme north.

BRITAIN. Much of southern and eastern Scotland reoccupied since 1938, with decrease in persecution and more recently colonization of upland forest plantations (Thom 1986). FRANCE. Recent spread in south-east. Records east of river Var in Alpes-Maritimes no longer uncommon; pair bred north of Menton, near Italian border, in 1987. (RC.)

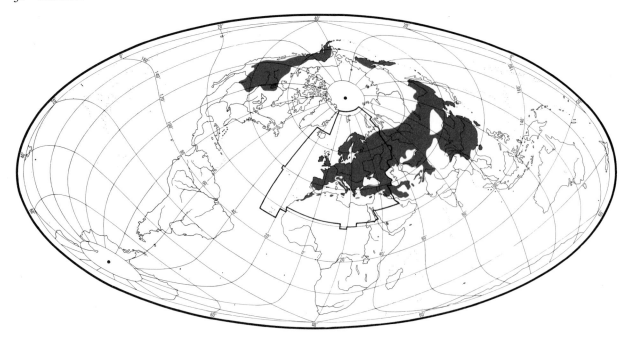

CZECHOSLOVAKIA. Has spread to higher elevations in last 20 years (KH). RUSSIA. Recent expansion east and to lesser extent north in Kola peninsula (Mikhaylov and Fil'chagov 1984). LEBANON. Several reported in 1967–8, one in 1986; apparently accidental (Vere Benson 1970; Khairallah 1986). TUNISIA. Status uncertain; apparently rare resident breeder, but no recent records (Thomsen and Jacobsen 1979).

**Population.** Has increased in some parts of Europe, following decrease in human persecution.

BRITAIN. Probably more than 250 000 pairs; marked increase in 1940s, and more recently with spread into suburban and urban habitats (Sharrock 1976). FRANCE. 100 000 to 1 million pairs (Yeatman 1976). BELGIUM. Increased during World War II due to prohibition of hunting, and more recently through spread into towns; estimated 19 000 pairs (Devillers *et al.* 1988). NETHERLANDS. Estimated 60 000–120 000 pairs (SOVON 1987). GERMANY. 500 000 pairs (G Rheinwald). East Germany: 70 000 ± 30 000 pairs (Nicolai 1993). SWEDEN. Estimated 350 000 pairs (Ulfstrand and Högstedt 1976). FINLAND. Has probably increased, especially in north, due to disruption of forests; perhaps 100 000–200 000 pairs (Koskimies 1989). POLAND. Has increased, probably due to cessation of human persecution (Tomiałojć 1990). HUNGARY. Estimated *c.* 100 000 pairs (Kalotás 1986a).

Survival. Britain and Finland: mortality in 1st calendar year 46%, in 2nd year 58%, in 3rd–5th years 55% (Holyoak 1971). Urban population, Britain: mortality in 1st year after leaving nest 44%, in successive years 30%, 24%, 32%, 46%, and 86% (Tatner 1986). Finland: annual mortality based on all recoveries 61 ± SE4·3%, probably too high; for breeding birds 47 ± SE7·9% probably a good estimate (Haukioja 1969). Europe: mortality in 1st year of life 69·0% (Busse 1969). Oldest ringed bird 15 years 1 month (British Trust for Ornithology).

**Movements.** Sedentary, with limited dispersal, chiefly in north of range. Most ringing recoveries over 30–40 km involve birds from northern Europe, and show no preferred direction. Reluctant to cross sea.

In Britain, adults markedly sedentary, and dispersal distances of 1st-year birds are usually small; of nestlings and juveniles ringed up to 1965, only 2 recoveries over 32 km (one moved 86 km), and only *c.* 20% moved more than 8 km (Holyoak 1971). In Sheffield area (northern England), young birds dispersed chiefly September–October after gaining independence, and most bred close to where they spent first winter; median distance between natal site and first breeding site 425 m (0–798 m), or *c.* 2 territories away (*n* = 48 birds); ♀♀ tended to move slightly further than ♂♂, but difference not significant (Birkhead 1991).

Mostly sedentary also in Scandinavia. In Norway, only 7 of 189 recoveries were more than 40 km (up to 110 km) (Haftorn 1971). In Sweden, dispersive movements are mostly within 50 km radius; of 74 recoveries, 1972–83, 58 at less than 10 km, 14 at 10–32 km, 1 at 57 km, 1 at 65 km; of 210 recoveries of ringed nestlings, 7 over 50 km, of which 3 over 100 km (Zink 1981; *Rep. Swedish Bird-ringing*). At Falsterbo (south-west tip of Sweden), however, incipient migratory movement observed in some autumns, coinciding with end of spells of fine weather, and sometimes associated with movements of Jay *Garrulus glandarius* and Nutcracker *Nucifraga caryocatactes*: e.g. in

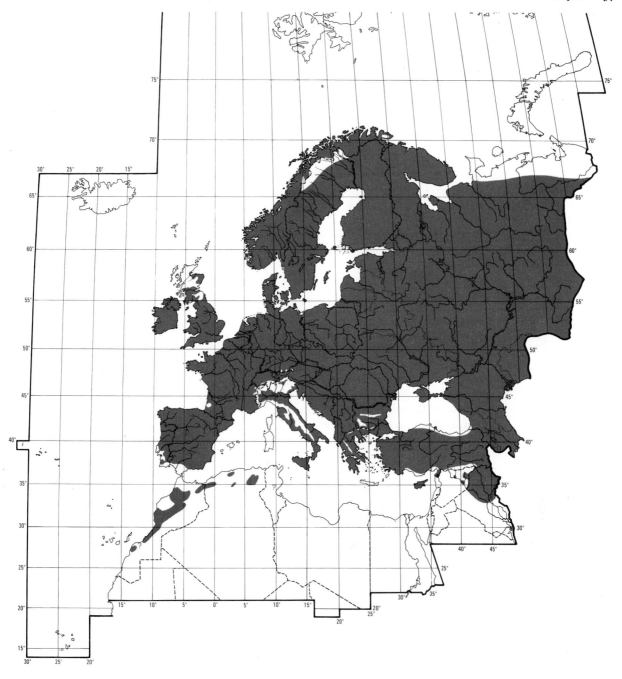

autumn 1945, numbers far exceeded local population, and birds were seen approaching from east, flying low; groups took off from coast many times to head for Denmark 24 km west, and rose to high altitude, but always returned, often 'in full panic'; no successful departures observed. Also in September–October 1943, numbers increased to 225, and southward passage was observed over coastal meadows further north. (Rudebeck 1950, which see for discussion; Ulfstrand 1959.) Similarly, flocks assemble at Skagen (northern tip of Denmark) in spring (usually up to 80, once 150 birds), and often make several attempts to cross sea, occasionally succeeding; bird recovered 70 km north-east in Sweden, another 200 km NNE in Norway; most birds turn back, with reverse movements of 18–210 km recorded. Several recoveries over 100 km in Finland, including bird 240 km north-east of natal site, and another on south coast 450 km south-west of ringing site. (Zink 1981.) Of 20 recoveries in Poland, only 1 over 35 km (80 km north-east) (Busse 1963). In Latvia, 18 of 20 nestlings recovered up to 10 km from ringing site, others at 12 and

55 km (Baltvilks 1970). Very few birds move further than 30 km in central Europe; exceptional recovery of bird ringed as nestling in Doubs (eastern France) found 330 km SSW in Rhône valley $1\frac{1}{2}$ years later (Zink 1981). Swiss birds occasionally move up to 50 km; sporadically reported at Alpine passes, and in winter sometimes recorded above breeding range at 2000–2200 m (Glutz von Blotzheim 1962; Winkler 1984).

Sea-crossing reported from Cape Andreas at north-east tip of Cyprus, where numbers fluctuate markedly in spring; in April 1973, groups seen trying to head for Turkey; attempts always abandoned at slightest freshening of wind, but 3 apparently successful exits were observed in light, variable wind; birds climbed to c. 900 m and moved steadily NNW (Stagg 1974).

In USSR, chiefly sedentary, but some birds move south from north of range in colder years, and in some areas altitudinal movements occur; birds tend to move towards inhabited areas in winter, especially in Siberia (Dementiev and Gladkov 1954; Korelov *et al.* 1974).

In study in Kyushu (Japan), 80% of birds ringed as nestlings dispersed within 1 km (median 585 m) of natal site during first winter, then moved to villages to establish territories in spring; nests were built (usually not until following spring) at 450 m (median) from natal site (Takeishi 1985).

In North America, regularly makes local movements (chiefly in September), and in some years flocks reported east and south of breeding range; autumn movements from interior to coast recorded in Alaska and northern British Columbia; in mountainous areas (e.g. Montana), birds often move upward above breeding range in autumn. Bird ringed Wyoming, May, recovered c. 350 km south in Colorado, January. (Bent 1946.)                    DFV

**Food.** Invertebrates, especially beetles (Coleoptera), fruits, and seeds; occasionally small vertebrates, and all kinds of carrion, refuse, and domestic scraps. Very opportunistic feeder, diet varying considerably according to habitat and local food sources: broadly, consumption of invertebrates highest in spring and summer, vertebrates and plant material in autumn and winter (Sterbetz 1964; Bigot 1966; Bährmann 1968a; Tatner 1983; Fasola *et al.* 1986); if insects available all year then can comprise very high proportion of total diet (Deckert 1980). Feeds mostly on ground, sometimes in trees and shrubs; on agricultural land in central England, winter (243 observations), grassland significantly preferred, particularly pasture, while arable fields with stubble were frequented according to abundance, and bare soil avoided (Waite 1984a, which see for foraging comparisons with other Corvidae); in study in southern England (866 observations), rough grassland beside hedgerows was favoured foraging site, stubble fields avoided more as autumn progressed (Holyoak 1974b). Near Paris, northern France, in mixed farming landscape, greatly preferred roads, paths, and their grassy

verges, followed by meadows and lawns, and arable fields visited least; preferences changed over year depending on crop height; of 1184 foraging observations, 53% in vegetation lower than 10 cm, 20% 10–20 cm, 10% 20–40 cm, and 13% no vegetation (Balança 1984a, which see for many details of foraging differences between adults and juveniles). In general, permanent pasture highly favoured, especially with high water-table; long grass and bare soil avoided except when following plough, etc. (Birkhead 1991). For foraging sites in Sweden, and comparison with other Corvidae, see Högstedt (1980a) and Loman (1980a); see Deckert (1980) for eastern Germany, Denneman (1981) for Netherlands, and Møller (1982a, 1983a) for Denmark. Apparently some may be forced towards houses and cover by competition from Carrion Crow *Corvus corone* (Møller 1983a; Eden 1987b). Common frequenter of farmyards; in suburban and urban areas feeds in parks, on lawns, flower-beds, bird-tables, and refuse tips (Deckert 1980; Birkhead 1991). On grass, walks, stops, looks around, and when prey spotted walks quickly or hops towards it to seize it in bill; jumps up to glean prey from tall plants or snatch flying insect; leaf-litter thrown aside, or scraped with foot, in search of beetles, spiders (Araneae), etc., and dung, stones, or earth overturned; pries in soil (bill inserted then opened) for invertebrates and plucks berries from shrubs and trees, though most caterpillars probably taken from ground; frequents edges of shallow water and often searches roofs; sometimes dives into long grass from post or telephone wire to seize large invertebrates or small vertebrates, but more rarely flies low over meadows searching for prey or hovers briefly; sometimes pursues flying insects (Holyoak 1974b; Deckert 1980). Able to take material from water surface in feet while flying (Glutz von Blotzheim 1962). In winter, 56% of 282 feeding actions were surface picking, 12% dung turning, 11% turning of clods or stones, 9% surface probing, 9% pouncing, 3% deep probing and digging (Waite 1984a); in Swedish study on grass heath and rough grassland, 81% of 1316 actions were picking from ground and vegetation (Högstedt 1980a). In southern England, January–March, c. 50% of 247 daylight observations were of foraging birds, with constant decrease though year to c. 30% (n = 265) by mid-November to December (Holyoak 1974b). In Denmark, fed for c. 10 hrs per day, June–August, and c. 3–4 hrs December–March; peak of 60% of daylight time spent feeding January and August, minimum of c. 35% in March, n = 600 hrs of observation (Møller 1983a). Probably considerable individual differences in feeding, and food sources once discovered will be thoroughly exploited; at 5 nests fairly close to each other, northern France, 3 times more caterpillars were brought to one than to any other, at another 4 times more earthworms (Lumbricidae), while at a third 84% of items brought were of 1 species of beetle; some birds may specialize in vertebrates (Balança 1984b). In one study in Germany, 65% of plant material eaten were barberries *Berberis*, although plant was scarce in vicinity, and 52% of animal prey were earwigs (Forficulidae) (Grimm 1989). Some birds steal conspecifics' caches and hardly store own food (Birkhead 1991). Apparent specialization in robbing caches of Great Grey Shrike *Lanius excubitor* recorded, small groups systematically searching likely sites (Kübel and Ullrich 1975; Olsson 1985). May kleptoparasitize other birds, sometimes in pairs, attacking raptors from behind and forcing them to drop prey (Goodwin 1986); one bird observed flying onto back of Kestrel *Falco tinnunculus* on ground and pecking at head until prey released (Sage 1957). Recorded opening milk bottles and egg-cartons on doorsteps (Åbro 1964; Sharrock 1976), and often eats animal droppings (Summers-Smith 1983; Wassmann 1990). Walnut *Juglans* held under foot and hammered with bill (Creutz 1953); seen to take whole pears *Pyrus* from tree and fly off with them (Birkhead 1991), and dug *c.* 10 cm into soil to get at potatoes (Buzzard 1989). Hairy caterpillars rubbed on ground before eating (Litun 1986); recorded smashing snails (Gastropoda) on hard substrate in manner of Song Thrush *Turdus philomelos* (Deckert 1980); turning over water-lilies (Nymphaceae) systematically to get at snails and eggs underneath (Thesing 1987), and digging with feet in sand to retrieve mussel *Mytilus*, perhaps previously buried, which was then held under feet and prised open with bill (Young 1990). Many observations of killing adult birds of other species in unusual circumstances such as cold weather or when victim weakened; sometimes hunts in small groups. However, can catch healthy birds in flight, on ground, or on nest (Codd 1947; Nein 1982; Williams 1989); 4–5 watched same hedge every day to capture House Sparrows *Passer domesticus*, which were seized in feet, carried to ground where killed and eaten, intestines first, or taken to perch (Schnell 1950); another *P. domesticus* captured after lengthy pursuit-flight (Thomas 1982); healthy adult Swift *Apus apus* beaten to ground and killed just after leaving nest (Pulman 1978). One bird returned over 2 days to Sand Martin *Riparia riparia* colony to dig out nestlings with bill, carrying them away to nest (Brennecke 1953); at reedbed roost of Tree Sparrows *P. montanus*, northern Spain, 6–7 birds at a time (up to 60 involved each evening out of *c.* 250 present during sparrows' arrival) plunged into reeds to seize many sparrows which were transported to rocks or trees to be eaten (Rolfe 1965). See also Liederkerke (1979), Deckert (1980), and Bjordal (1983*b*). Nestlings killed by crushing in bill or beating head against hard surface, larger birds dealt with also by crushing or by heavy blows of bill while held under one or both feet; seems unskilled at times when tackling adult birds, often striking at breast and body rather than head, so that many intended victims escape, e.g. Starling *Sturnus vulgaris* attacked for *c.* 4 min (Hume 1980), and *P. domesticus* surviving attempt by 10 birds to kill it (Bub 1953). See also Schölzel (1981) and Schroeter (1982*a*). Heavy prey probably always transported in bill

(Schroeter 1982*b*). No evidence of any long-term effect on songbird populations in England, including in urban areas (Wilkinson 1988; Gooch *et al.* 1991), and in Berlin and Osnabrück (Germany) no discernible decline in their numbers due to *P. pica* (Witt 1989; Kooiker 1991). However, in 5-year survey in suburban area of Braunschweig (northern Germany), took up to 100% of eggs and young of some species in 10-ha area; total of 64% of open nesters and 1·3% of hole nesters taken, *n* = 209 nests (Wiehe 1990). In rural area of Belgium, accounted for only *c.* 6% of predation by Corvidae on songbirds though comprising 50% of corvid population (Vercauteren 1984); see Birkhead (1991) for review of European studies. Captured rodent held under foot and killed with blows of bill (Birkhead 1991); live snakes carried in bill to perch where killed by blows to head (Carr 1969; Källander and Sylvén 1976); snatched young trout *Trutta* from pond and threw them on grass, eating only intestines (Kumerloeve 1967); bat (Chiroptera) caught in flight and eaten (Alexander 1952). Abundance of vertebrate carrion on motorways apparently leads to locally increased population density (Kummer 1967). Takes ticks (Ixodidae), etc., from backs of sheep (Balança 1984*b*), but can be serious nuisance to large mammals, especially stock, feeding at existing wounds (probably initially on larvae) and often enlarging them by eating flesh and skin causing severe injury or death; see Espmark (1972) for such activity on reindeer *Rangifer*, and Birkhead (1991) for many examples from North America, and also predation on poultry and gamebirds. Birds recorded eating putty from windows; up to *c.* 350 g from one window (Simpson 1970*a*).

Stores food like other Corvidae, but caches seem to be short-term, retrieved within 1–2 weeks at most and majority of stored items perishable (Clarkson *et al.* 1986; Birkhead 1991). In experiment in south-west Spain, of 148 pieces of bread 0·5 cm square, all were eaten on the spot and none carried away, but of 41 pieces 2 cm square, 8% eaten and 92% carried away; of 17 pieces carried away and where outcome definitely known, 15 stored and 2 eaten; pieces 1 cm square were equally likely to be eaten or carried away (Henty 1975). Of 3184 caches in central England, only 1 not in ground; bill and buccal pouch filled with food (up to 120 grains of wheat recorded), hole made in ground with bill then food ejected into it, usually without withdrawing bill, then hole covered with leaf, stone, etc.; caches of established breeding birds averaged 8·3 m apart and were at median distance 28 m from source of food, those of non-breeders 18·5 m and 65 m; overall average distance between caches *c.* 13 m. Food stored all year except July, with strong peak September–December; 44% of 305 items were domestic scraps, 16% dog faeces and kennel waste, 13% grain, 11% acorns *Quercus*, 10% berries, 3% carrion, and 1% pears. (Birkhead 1991.) See Clarkson (1984) and Clarkson *et al.* (1986) for details. In eastern Germany, most caches are by stones, posts, trees, under roof slates, in loose bark, etc.; birds partially eaten and rest cached, some small mammals cached whole; retrieval of rabbit carrion observed after 10 days; caches also made in snow, where head can be pushed in about 10 cm, and others recovered from under 20 cm of snow (Conrad 1979; Deckert 1980). Caching in snow 15 cm deep seen in England (Summers-Smith 1984). In built-up areas stores food in places such as roofs, probably collecting it in early-morning visits to gardens, etc. hence able to maintain regular supply in urban surroundings (Birkhead 1991). Often seen to go straight to caches without hesitation; probably uses visual memory to find location, possibly pin-pointing item by smell (Hayman 1958; Birkhead 1991). Contrary to popular belief, wild birds never seen to hoard anything inedible (Goodwin 1986; Birkhead 1991).

Diet in west Palearctic includes the following. Vertebrates (many as carrion): trout *Trutta*, carp *Cyprinus*, *Carassius*, bream *Abramis*, toad *Bombina*, frog *Rana ridibunda*, newt *Triturus*, lizard *Lacerta agilis*, snakes *Natrix natrix*, *Vipera berus*, hedgehog *Erinaceus europaeus*, mole *Talpa europaea*, shrews *Sorex araneus*, *Crocidura leucodon*, *C. russula*, bats (Chiroptera), suslik *Spermophilus citellus*, voles *Microtus arvalis*, *M. agrestis*, *Clethrionomys glareolus*, rat *Rattus*, mice *Apodemus sylvaticus*, *A. agrarius*, *Micromys minutus*, *Mus musculus*, hare *Lepus capensis*, rabbit *Oryctolagus cuniculus*, eggs, young, and adults of many bird species, particularly sparrows (Passeridae), finches (Fringillidae), thrushes (Turdidae), etc., but also including adult Water Rail *Rallus aquaticus*, Swift *Apus apus*, and Feral Pigeon *Columba livia*. Invertebrates: bristle-tails (Thysanura: Lepismatidae), springtails (Collembola: Entomobryoidea), damsel flies and dragonflies (Odonata: Agriidae, Coenagriidae, Aeshnidae, Libellulidae), grasshoppers, etc. (Orthoptera: Gryllidae, Tettigoniidae, Gryllotalpidae, Acrididae), earwigs (Dermaptera: Labiduridae, Forficulidae), cockroaches (Dictyoptera: Blattidae), bugs (Hemiptera: Scutelleridae, Pentatomidae, Reduviidae, Stenocephalidae, Nabiidae, Miridae, Gerridae, Naucoridae, Corixidae, Cercopidae, Cicadidae, Coccoidea), lacewings, etc. (Neuroptera: ant-lions Myrmeleontidae, Chrysopidae), adult and (mostly) larval Lepidoptera (Nymphalidae, Lycaenidae, Pieridae, Cossidae, Zygaenidae, Tortricidae, Psychidae, Pyralidae, Gelechiidae, Plutellidae, Tineidae, Incurvarioidea, Arctiidae, Noctuidae, Lymantriidae, Lasiocampidae, Sphingidae, Thyatiridae, Geometridae), adult and larval caddis flies (Trichoptera: Phryganeidae, Sericostomatidae), adult and larval flies (Diptera: Tipulidae, Chironomidae, Bibionidae, Tabanidae, Therevidae, Empididae, Asilidae, Syrphidae, Tachinidae, Calliphoridae, Muscidae), Hymenoptera (mostly ants Formicidae, also Pamphilidae, Ichneumonoidea, Scoliidae, Chrysididae, wasps Vespidae, bees Apoidea), adult and larval beetles (Coleoptera: Cicindelidae, Carabidae, Dytiscidae, Hydrophilidae, Histeridae, Silphidae, Staphylinidae, Lucanidae, Geotrupidae, Scarabaeidae, Byrrhidae, Buprestidae, Elateridae, Cantharidae, Dermestidae, Meloidae, Cler-

idae, Anthicidae, Tenebrionidae, Nitidulidae, Coccinellidae, Cerambycidae, Chrysomelidae, Bruchidae, Curculionidae), spiders (Araneae: Theridiidae, Araneidae, Tetragnathidae, Lycosidae, Pisauridae, Eusparassidae, Agelenidae, Thomisidae, Salticidae), harvestmen (Opiliones: Phalangiidae), ticks (Acari: Ixodidae), woodlice (Isopoda: Porcellionidae, Armadillidiidae), shrimps (Amphipoda: Gammaridae), millipedes (Diplopoda: Julidae), centipedes (Chilopoda: Lithobiidae), earthworms (Lumbricidae), snails (Gastropoda: Chondrinidae, Clausiliidae, Valloniidae, Helicidae, Enidae, Bulimidae, Physidae, Dreissenidae, Planorbidae), bivalves (Bivalvia: Mytilidae). Plants: fruits and seeds of pine *Pinus*, spruce *Picea*, juniper *Juniperus*, maple *Acer*, oak *Quercus*, beech *Fagus*, birch *Betula*, chestnut *Castanea*, walnut *Juglans*, rose *Rosa*, various *Rubus* species, rowan and whitebeam *Sorbus*, apple *Malus*, pear *Pyrus*, various *Prunus* species, *Cotoneaster*, mespil *Amelanchier*, *Robinia*, pea tree *Caragana*, holly *Ilex*, spindle *Euonymus*, dogwood *Cornus*, elder

Table A   Food of adult Magpie *Pica pica* from analysis of faeces (Tatner 1983) and stomach contents (other studies); figures are percentages by number, animal and plant material treated separately.

| | NW England, urban, breeding season | SE France, all year | NW Italy, Mar–Jun | SW USSR, Apr–Jul |
|---|---|---|---|---|
| *Animal material* | | | | |
| Vertebrates | 1·3 | 0·5 | 13 | 1·4 |
| Orthoptera | 0 | 5·7 | 4 | 21·6 |
| Hemiptera | 2·4 | 2·6 | 9 | 2·6 |
| Lepidoptera | 6·2 | 6·4 | 2 | 1·0 |
| Diptera | 18·0 | 2·3 | 0 | 5·1 |
| Hymenoptera | 23·5 | 6·9 | 4 | 2·6 |
| Coleoptera | 41·5 | 49·3 | 63 | 64·6 |
| (Curculionidae) | (7·4) | (—) | (13) | (7·7) |
| (Carabidae) | (1·9) | (—) | (22) | (1·0) |
| (Scarabaeidae) | (—) | (—) | (7) | (44·3) |
| (Elateridae) | (3·7) | (—) | (6) | (1·5) |
| Woodlice | 0·6 | 5·7 | 0 | 0 |
| Spiders, harvestmen | 5·0 | 18·3 | 4 | 0·5 |
| Earthworms | 0 | 0 | 0 | 0 |
| Snails | 0·6 | 0·6 | 0 | 1·1 |
| Total no. of items | 162 | 612 | 54 | 277 |
| *Plant material* | | | | |
| Fruits | 0 | 5·2 | 40 | 34·1 |
| Herb seeds | 0 | 0 | 0 | 44·6 |
| Cereal grain | 100 | 94·8 | 60 | 21·3 |
| Total no. of items | 9 | 58 | 5 | 920 |
| Source | Tatner (1983) | Bigot (1966) | Fasola *et al.* (1986) | Eygelis (1964) |

*Sambucus*, mulberry *Morus*, hawthorn *Crataegus*, buckthorn *Rhamnus*, alder buckthorn *Frangula*, sea buckthorn *Hippophae*, privet *Ligustrum*, *Vaccinium*, crowberry *Empetrum*, strawberry tree *Arbutus*, *Ricinus*, *Viburnum*, currants, etc. *Ribes*, barberry *Berberis*, mistletoe *Viscum*, Virginia creeper *Parthenocissus*, grape *Vitis*, olive *Olea*, fig *Ficus*, amaranth *Amaranthus*, knotgrass, etc. *Polygonum*, love-in-a-mist *Nigella*, hemlock *Conium*, peas *Pisum*, beans *Phaseolus*, vetch *Vicia*, medick *Medicago*, plantain *Plantago*, mustard *Sinapis*, pumpkin *Cucurbita*, potato and nightshade *Solanum*, sunflower *Helianthus*, bur-reed *Sparganium*, grasses (Gramineae), sedges (Cyperaceae), oats *Avena*, barley *Hordeum*, wheat *Triticum*, rye *Secale*, millet *Panicum*, maize *Zea*, rice *Oryza*, fungus *Mycena*, *Boleta*. (Csiki 1919; Madon 1928; Schuster 1930; Codd 1947; Alexander 1952; Owen 1956; Altner 1957; Turček 1948, 1961; Eygelis 1964; Sterbetz 1964; Bigot 1966; Bohac 1967; Bährmann 1968a; Sudhaus 1969a; Källander and Sylvén 1976; Pulman 1978; Deckert 1980; Högstedt 1980a; Hume 1980; Denneman 1981; Schroeter 1982a; Spaans *et al.* 1982; Tatner 1983; Balança 1984b; Fraticelli 1984; Jacob 1984; Kiss and Rékási 1986; Krištín 1988; Snow and Snow 1988; Grimm 1989; Wiehe 1990; Young 1990.)

For food of adult see Table A. Adult's diet in spring very like that of nestlings (see below), though probably with fewer Diptera and caterpillars, but in late summer and autumn takes increasing numbers of Orthoptera as well as cereal grains in stubble fields; frequents especially maize fields before and after harvest, taking grain from cobs and ground into November, when there is intense competition among Corvidae for this food; much less often on stubble fields as winter progresses (Holyoak 1974b; Deckert 1980; Balança 1984b). Diet composition in general determined by seasonal availability of invertebrates and plants (Bigot 1966). In Bulgaria, no significant change in proportion of plant material in diet in spring, summer, and autumn, with average of *c.* 33% by number ($n = 773$ items), but proportion in winter was 68% ($n = 108$) (Bährmann 1968a). In Czechoslovakia, winter, 25 stomachs contained 76% by weight plant material (71% cereals) and only 8% insects (Hell and Soviš 1958); 250 pellets were 88% by

Table B  Food of nestling Magpie *Pica pica* from stomach contents and faeces (Tatner 1983) and collar-samples (other studies); figures in italics are percentages by weight, others are percentages by number; animal and plant material taken together.

| | NW England (urban) | | S England | N France | Netherlands | S Sweden | E Czecho-slovakia | SW USSR |
|---|---|---|---|---|---|---|---|---|
| Vertebrates | 1·3 | *62·3* | 0 | 1·2 | 16·5 | 0·9 | *19·7* | 1·8 | 2·3 |
| Orthoptera | 0 | *0* | 0 | 0·2 | | 0 | *0* | 0·2 | 6·3 |
| Hemiptera | 0·1 | *<0·1* | 0·3 | 0·4 | | 0 | 0·2 | 0·4 | 4·6 |
| Lepidoptera | 16·0 | *3·1* | 47·9 | 1·2 | | 11·6 | 7·8 | 17·6 | 30·3 |
| (Noctuidae) | (—) | (—) | (13) | (—) | | (—) | (*6·4*) | (9·8)ʹ | (5·5) |
| (Geometridae) | (—) | (—) | (6) | (—) | | (—) | (*0·1*) | (3·7) | (3·6) |
| (Tortricidae) | (—) | (—) | (24) | (—) | 4·3 | (—) | (—) | (—) | (1·2) |
| Diptera | 12·3 | *10·9* | 19·0 | 4·4 | | 4·0 | 2·1 | 30·7 | 5·7 |
| (Syrphidae) | (—) | (—) | (<1) | (—) | | (0) | 0 | (11·3) | (—) |
| (Bibionidae) | (—) | (—) | (—) | (—) | | (0·2) | 0 | (9·9) | (—) |
| (Tipulidae) | (5·4) | (*10·4*) | (3) | (—) | | (3·8) | (*0·1*) | (—) | (—) |
| Hymenoptera | 5·8 | *1·0* | 0·7 | 0·9 | | 0·2 | 1·7 | 0·6 | 5·9 |
| Coleoptera | 54·4 | *7·8* | 21·0 | 62·4 | 25·9 | 43·0 | 23·7 | 39·8 | 30·4 |
| (Curculionidae) | (37·3) | (*1·1*) | (—) | (1·1) | (—) | (4·4) | (*14·9*) | (15·6) | (2·6) |
| (Carabidae) | (4·9) | (*4·0*) | (10) | (1·9) | (—) | (27·7) | (2·3) | (10·2) | (7·4) |
| (Scarabaeidae) | (—) | (—) | (1) | (50·2) | (—) | (3·3) | (*0·1*) | (4·9) | (9·6) |
| (Elateridae) | (8·7) | (*1·2*) | (5) | (6·7) | (—) | (0·1) | (*5·1*) | (3·0) | (5·1) |
| Woodlice | 0·1 | *<0·1* | 1·7 | 1·6 | <1 | 1·6 | 0 | 0·04 | 0 |
| Centipedes, millipedes | 0·4 | *0·2* | 1·4 | 3·0 | 1·6 | 0·7 | *0·1* | 0 | 0·2 |
| Spiders, harvestmen | 7·4 | *0·6* | 3·4 | 3·5 | *1·0* | 1·6 | 1·2 | 1·1 | 6·0 |
| Earthworms | 0·1 | *4·6* | 0 | 4·9 | 8·7 | 11·5 | *24·9* | 2·5 | 0·4 |
| Snails | 1·1 | *0·1* | 0·7 | 0·2 | 0·6 | 0·1 | 0 | 1·1 | 1·0 |
| Plants | — | *9·2* | 0 | 6·0 | 41·5 (inc scraps and refuse) | 20·6 (calculated separately) | *19·3* | 4·2 | 4·4 |
| Other | 0 | *0* | 3·9 | 10·1 | 0 | 24·8 | *0* | 0 | 2·5 |
| Total no. of items | 4092 | | 357 | 571 | | 811 | | 2561 | 3656 |
| Source | Tatner (1983) | | Owen (1956) | Balança (1984b) | | Spaans *et al.* (1982) | Högstedt (1980a) | Krištín (1988) | Eygelis (1964) |

weight plant and 5% invertebrate material (Turček 1948). For 127 stomachs taken throughout year, northern Rumania, see Kiss and Rékási (1986). Diet of urban birds in north-west England (by volume, from 35 stomachs) was 73% plant material in February-March, 8% invertebrates, and 2% vertebrates, remainder unidentified; particular increase in this study in number of leatherjackets (Tipulidae larvae) consumed from March onwards, rising to 70% of invertebrates by weight (from faecal material), though 80% of total diet was vertebrate material; other percentages (of total) by weight included 2·6% beetles, 0·2% spiders and harvestmen, and 0·1% ants (compare percentages by number in Table A). Plant material in urban winter diet mainly grain and roots, and low animal component probably indicates food shortage at this time of year so birds dependent on scraps and refuse. (Tatner 1983.) For diet by weight in Netherlands see Denneman (1981); for diet by weight and volume in Czechoslovakia see Turček (1948) and Bohac (1967). For comparison with other Corvidae in Britain, see Holyoak (1968). In Camargue (south-east France), 85% of 612 prey items were 0·6-1·5 cm long, 3% 0·4-0·5 cm, 5% 1·6-2·5 cm, and 7% 2·6-7·0 cm (Bigot 1966).

In eastern Germany, February, took 3-6 invertebrates per min on wet meadow, in spring 7-9 per min in winter cereal (Deckert 1980). In central England, winter, 70 birds had average energy intake on grassland of from 8·06 kJ per hr (from medium- sized invertebrates above soil surface) to 0·03 kJ per hr (from small ones beneath surface) (Waite 1984a). In Swedish study, wild bird during breeding season, but before hatching of young, estimated to require c. 630 kJ per day, of which c. 500 kJ expended in normal activity (Högstedt 1981).

For food of nestlings see Table B. In north-west England, most plant material in diet of urban nestlings (moss, grass, roots, etc.) is apparently picked up incidentally by adults when foraging for invertebrates. Up to c. 9 days old, smaller, softer items like spiders, flies, and caterpillars brought, then later on worms, leatherjackets, and beetles; invertebrates probably collected according to grassland abundance, and vertebrate material likely to be mostly carrion. Since 1 vole is equivalent to 54 leatherjackets or 3500 Curculionidae, this suggests live vertebrates taken very rarely; diet in this urban area not dissimilar to rural studies. (Tatner 1983, which see for discussion of methods and difficulties involved in estimating composition of diet, and merits of analysing by weight, volume, or number of items.) In nests near houses, over 40% of diet can be composed of refuse and scraps (Balança 1984b). In south European USSR, spiders and caterpillars also brought in first few days; proportion of beetles peaks around days 5-15, though some fed from day 1; some grasshoppers (Acrididae) brought from day 1 and every day for 3 weeks; no snails before day 7 (Eygelis 1964). In northern France, 3% of 571 items less than 5 mm long, 62% 5-10 mm, 20% 10-15 mm, 11% 15-20 mm,

and 4% greater than 20 mm (Balança 1984b). In southern Sweden, median size of prey 7·0 mm and median weight 5·2 mg, with 89% of items less than 20 mg dry weight (Högstedt 1980a). Daily consumption in Czechoslovakia at 10 days old was 16 g, 21% of body weight, (n=223 collar-samples from 19 nests (Krištín 1988); in USSR, average over first 20 days was 27·5 g per day, n=641 collar-samples (Eygelis 1964).                    BH

Social pattern and behaviour. British, Irish, and continental European populations well known; main studies in Netherlands by Baeyens (1979, 1981a, 1981c), in Sheffield area (northern England) by Birkhead (1979, 1982, 1989, 1991), Birkhead and Clarkson (1985), Birkhead et al. (1986), and Eden (1987a, 1987b, 1989), and in southern England by Vines (1981). For other reviews, see Goodwin (1952b, 1986) and Bährmann (1968a). Asian and Arabian races poorly known, but for useful study of bactriana in northern Kazakhstan, see Smetana (1980). North American hudsonia well known, mainly through studies by Buitron (1983a, b, 1988).

1. Can be either solitary or gregarious throughout year. Outside breeding season, breeding birds may remain as pairs in their territories or abandon them (Baeyens 1981a, c; Birkhead et al. 1986), depending on quality of territory. Site-fidelity of residents is high: some birds in northern England remained on same territory for 4 years, duration of their breeding life. Birds which abandon territory at end of breeding season live either alone or as pairs and often spend time as part of flock with immature non-breeding birds. Non-breeders can make up over half total population in some areas, depending on population density: 40% in northern England, 20% in Netherlands, 60% in southern Sweden (Birkhead et al. 1986; Baeyens 1979, 1981a, c; G Högstedt). In Denmark, young non-breeders formed only 5% of population (A P Møller). Non-breeding flocks very loose and unstructured: members have similar and overlapping home-ranges (c. 18 ha per bird) which also overlap breeding territories of adults (Birkhead et al. 1986). However, flock-members will coalesce into groups of 20-30(-50) where food locally abundant (Birkhead 1991); for other aggregations, see Roosting and Flock Behaviour (below). Age structure of non-breeding flocks in northern England (n = 227 birds): 81% 1st-year, 17% 2nd-year, 2% older (ex-breeders) which had been ousted from territories. Sex-ratio in northern England slightly biased towards ♂♂: of 107 birds sexed from non-breeding flock, 58% ♂. (T R Birkhead.) Non-breeding flocks can form from nucleus of young birds from one or several nearby territories (Eden 1989), possibly because food abundant there. BONDS. Mating system essentially monogamous although ♂♂ commonly promiscuous. In northern England, 2 cases of simultaneous bigamy out of 127 pairs over 8 years. In each case only one nest built and only one ♀ laid. (Birkhead et al. 1986.) Pairs can remain intact for several years, but better data needed. In Netherlands study, divorce common (c. 30% of pairs changed partners between years), but sample sizes small. Typically, birds divorced their mates in order to pair with a bird in better quality territory whose mate had disappeared: in the 11 recorded cases a single pair-member thus switched to neighbouring territory, deserting its previous mate and territory; ♂ initiated divorce in 7 cases, ♀ in 3, and in 1 other case a non-breeding ♀ evicted territorial ♀ and paired up with ♂ (Baeyens 1981c). Both sexes build nest but only ♀ incubates and broods (Birkhead 1991). Neither sex able to rear young alone after losing mate in incubation or chick-rearing (Dunn and Hannon 1989 for hudsonia). Both sexes feed young in nest and

for *c.* 6 weeks after fledging, young becoming independent *c.* 8 weeks after fledging (i.e. at *c.* 80 days old) (Husby 1986; Birkhead 1991). Age of first breeding 1–2 years but some regional variation, e.g. 1–2 in northern England (Birkhead 1991), mostly 2 or older in southern Sweden (G Högstedt). BREEDING DISPERSION. Solitary and strongly territorial. In central and northern Europe defends all-purpose territory. In southern Europe (e.g. Spain) territorial system during breeding season may differ (see below), but needs to be investigated. Territory during breeding season averages 5 ha in northern England (Birkhead *et al.* 1986), 5·8 ha in Haren (Netherlands) (Baeyens 1981*a*), 5–7 ha in southern England (Vines 1981). In 2 studies, foraging (unusually) extended outside territorial boundaries; territories averaged 3·25 ha in Zossen (Germany) (Deckert 1968*a*) and 7·5 ha in Fochteloo (Netherlands); in Fochteloo, extended foraging area created average home-range per pair of *c.* 15 ha (Baeyens 1981*a*). After hatching, parents rarely feed more than 75 m from nest (Högstedt 1980*a*). Young birds can establish territories at any time of year, but 1st-year birds not until September at earliest. Young birds obtain territories in several ways. They may replace dead bird: if breeding ♂ dies, ♀ unable to defend territory alone, and territory then taken over by new pair; if breeding ♀ dies, ♂ remains on territory and re-pairs, either with ♀ from non-breeding flock or with established breeder from another territory (Baeyens 1981*c*; T R Birkhead). Some territories acquired by non-breeders through Ceremonial Gatherings (see Flock Behaviour, below), others by 'squeezing in' whereby non-breeders establish small territory late in breeding season when territorial aggression low, and gradually expand boundaries, until territory of suitable size for breeding. Squeezing in always occurs too late for birds to breed in that season. In northern England study, estimated that ⅓ of territories established by mate-replacement, ⅓ by squeezing in, and ⅓ through Ceremonial Gatherings. (Birkhead and Clarkson 1985.) In most of northern and central Europe, nests are spaced evenly in habitat, but tend to be more aggregated in southern Europe. Note that basing density on nests exaggerates true breeding density due to presence of old nests. Maximum density recorded in northern England, 32 pairs per km², with average *c.* 100 m between nests (Birkhead *et al.* 1986; T R Birkhead). In Netherlands, 7·3 pairs per km², southern England 3·6 pairs per km², southern Sweden 4·0–7·1 pairs per km² (Baeyens 1981*a*; Vines 1981; G Högstedt). In Switzerland, 0·2–10 pairs per km² in fields and meadows, neighbouring groups of up to 12 nests per 10 ha in riverine woods and wooded fens (Schifferli *et al.* 1982). In Germany: 0·18 nests per km² in northern Bayern, up to 4·5 in villages (Dittrich1981); in Heckengäu (Baden-Württemberg), up to 3·5 pairs per km² (Kroymann and Girod 1980); in Tierpark Berlin, density increased from 1·2 pairs per km² in 1955 to 14 in 1969 (Deckert 1980, which see for other densities). For Ukraine, see Vakarenko and Mikhalevich (1986). Increase in density in urban areas since *c.* 1940, e.g. in Sheffield from 1·3 pairs per km² in 1946 to 8·1 pairs per km² in in 1986 (Clarkson and Birkhead 1987). In Manchester (northern England), 1978, 6·9 pairs per km² (Tatner 1982*c*); in Dublin, 1980–3, 16·6 pairs per km², almost 4 times average density recorded on British farmland and twice that in British woodland (Kavanagh 1987*a*). See also Berlin data (above). In high-density area, northern England, no unoccupied space between territories; in adjacent low-density high-altitude area, territories separated by unoccupied land with no suitable nesting trees (Birkhead *et al.* 1986; Birkhead 1991). In Camargue (southern France) and southern Spain, nests tend to be loosely aggregated (Arias de Reyna *et al.* 1984; Birkhead 1991), perhaps associated with shortage of suitable breeding areas or spatial dispersion of food. In North Africa, nests patchily distributed;

within patches, nests evenly spaced, separated by average 264 m (D Gosney), contra 'colonial' distribution stated by Etchécopar and Hüe (1967). No information on territory size or utilization for southern Europe or North Africa. *P. p. hudsonia* typically shows aggregated (but not colonial) nest dispersion. For high degree of site-fidelity, see introduction to part 1 (above). In Sheffield area, median 25 m moved between breeding attempts in successive years; pairs may build new nest or re-use old one and, if 1st clutch lost, almost always build replacement nest, always within territory. (Birkhead 1991.) ROOSTING. Nocturnal. Breeding birds roost in territory in trees, or in nest (♀) if eggs or young present. Outside breeding season, may continue to roost in territory with partner, or may join communal roost of up to 200 birds. In southern Sweden, pairs sometimes roost in nest in winter (A P Møller). Communal roosting usually only with conspecifics, but for case of 30 *P. pica* with 150 Jackdaws *Corvus monedula*, see Diesing (1984). Communal roosts often over water in willows *Salix*, in reedbeds, or in thorny scrub, e.g. hawthorn *Crataegus*. In Denmark, 75% of 55 roosts in deciduous trees; 47% had water beneath trees; most roost-trees 2·5–9 m high (Møller 1985, which see for parts of tree utilized, and variation with weather). In Sheffield area, throughout year, non-breeders roosted in groups of 10–20 within their normal home-range, although in winter communal roost containing up to 150 birds formed *c.* 1 km from study area (Birkhead 1991). In Sweden, study birds flew maximum 4 km to reach communal roost (Gyllin and Källander 1977). In Denmark, average 1·9 km between adjacent communal roosts, which occur mainly November–March (Møller 1985). For roosting behaviour, see Flock Behaviour (below).

2. Furtive and wary wherever in conflict with man, more confiding elsewhere, e.g. Darwin (1872) contrasted wariness in England (where persecuted) with tameness in Norway (where tolerated). See Díaz and Asensio (1991) for vigilance behaviour (e.g. scanning) in relation to group size and distance to protective cover, and see Dhindsa and Boag (1989*b*) for flushing distance of *hudsonia* in relation to age (etc.). For reaction to conspecific Distress-calls (11 in Voice), see Brémond (1962). For mobbing, see Parental Anti-predator Strategies (below). Puzzling response is 'mobbing' of dead conspecific or even of 1–2 isolated tail-feathers; function of these 'funerals' remains obscure (e.g. Verbeek 1972, Birkhead 1991). FLOCK BEHAVIOUR. Within non-breeding flock, dominance hierarchy exists, ♂♂ generally dominant over ♀♀ (Birkhead and Clarkson 1989; see also Komers 1989): high-ranking birds initiate Ceremonial Gatherings by flying into centre of established territories to provoke territorial response from owners. Noise and activity attracts all nearby breeders and non-breeders as spectators. Average number of participants in 225 such gatherings in northern England was 9 (3–24), average duration 10 min (0·5–70). In 95% of 225 gatherings initiators were evicted. In 3 cases initiators fought with and ousted territory owners, taking over territory. In 7 instances birds returned to and initiated gatherings at same location on several consecutive days, eventually establishing small territory there. (Birkhead and Clarkson 1985.) For details of aggression in Ceremonial Gatherings, see Baeyens (1979); see also Antagonistic Behaviour (below). Forms pre-roost gatherings, in northern England from 2–3 hrs before sunset, birds usually arriving in small parties to congregate quietly in fields, or in bushes and trees a few hundred metres from roost; can number 100–200 birds and in early literature were confused with (much smaller) Ceremonial Gatherings (T R Birkhead). Birds move to roost at dusk, and sometimes much aggressive display (chasing) and circling flights prior to roosting. Departure in morning usually before sunrise (Møller 1985). SONG-DISPLAY. 3 types of song

distinguished (see 1 in Voice), but mainly Babble-song which is given by ♂ (less often by ♀) in company of mate (e.g. at prospective nest-site: see Heterosexual Behaviour, below), perhaps to stimulate her in days prior to laying, but no detailed study (Birkhead 1991). Baeyens (1979) noted Babble-song from ♂ in association with the following displays and contexts, suggesting wider role of self-advertisement: fighting, Tree-top Sitting (see Antagonistic Behaviour, below), Circling, and Chase-hopping (see Heterosexual Behaviour, below); pairs also Babble-sing together, notably during Allopreening (see Heterosexual Behaviour, below), and also heard from soliciting ♂♂ in Ceremonial Gatherings. In northern Kazakhstan (*bactriana*), main song-period (presumably Babble-song) 1 March–10 April, ceasing with start of nest-building although most ♂♂ continue singing on feeding grounds; at this time, also groups of up to 40 unpaired ♂♂ gathered in late morning and sang for up to 2½ hrs; no song heard during incubation or chick-rearing but some (probably from juveniles: see Voice) mid-June to early July after fledging, again after break-up of family parties in late July, and also regularly heard from two-thirds of pairs near old nests in autumn (up to early or mid-September) (Smetana 1980). Bähr-mann (1968a) also reported Babble-song from pairs visiting old nests in late winter to early spring. ANTAGONISTIC BEHAVIOUR. (1) General. Assertive and confident, often appearing aggressive. Breeding birds aggressive mainly in territorial defence, but also in defence of mate. In central and northern Europe, territories defended most vigorously just before and during egg-laying (T R Birkhead). For rest of year, if birds remain on territory, defence is weak. (2) Threat and fighting. Territory defended mainly by Tree-top Sitting, equivalent to song in other passerines: one or both members of pair sit in uppermost branches of tall tree in territory (Birkhead 1991). Tree-top Sitting occurs throughout the year but most frequently during breeding season. ♂♂ Tree-top Sit more frequently than ♀♀, and perform alone more often (46% of 54 cases) than ♀♀ (9%). (Baeyens 1981a.) Tree-top Sitting occurs spontaneously, but also in response to territorial intrusions, when it often precedes and follows chases of intruder (Birkhead 1991). Both sexes chase intruders on territory, but ♂♂ more likely to initiate chase and more aggressive than ♀♀ towards intruders. When territorial pair faces another pair, ♂ usually threatens ♂, and ♀ threatens ♀. Fight occurs if intruder on territory does not leave immediately on being threatened or chased. Baeyens (1981a) recorded fights only between ♂♂, but sometimes all members of 2 pairs will fight (T R Birkhead), ♂ with ♂, ♀ with ♀. Fighting consists of kicking while jumping up, grappling with feet interlocked, and thrusting and delivering heavy blows with bill. Usually one bird stands over and pecks at loser who often shrieks (9 in Voice). On one occasion defeated bird found to have suffered broken primaries and unable to fly (P Tatner); once, defeated bird was killed and eaten (T R Birkhead). Sex-specific response to intruder matches threat posed by sex of intruder: ♂ intruders pose greatest threat to ♂ territory-owner since intruder may copulate with his mate. ♀ intruder may pair with territorial ♂ and oust original ♀. (Baeyens 1981a.) Response of territorial birds to caged decoy depends on decoy's sex. If ♀ decoy used and only ♂ resident present, his response is usually sexual—attempts to court and mount. If joined by his partner, his response changes to aggressive, as does his partner's. If decoy is ♂, territorial ♀♀ rarely court, and both partners attack. (Birkhead 1979; Baeyens 1981a.) Territorial neighbours also perform parallel walks along boundaries. Usually performed by 2 pairs together, occasionally 3 where 3 boundaries abut (can look like type of communal display if territory boundaries not known). Birds adopt upright posture, with white feathers ruffled and tail 60° above horizontal. Aggressive displays

(including Tree-top Sitting) also include the following: Wing-spreading—primaries spread while remaining in contact with body, and may be held thus motionless for several seconds; Wing-flickering—primaries of outspread wings are moved quickly in and out, producing flickering effect of black and white markings; Wing-quivering—primaries of horizontally or vertically outspread wings vibrated vigorously; Wing-flirting—quick in-out flick of wings, horizontally (Fig A) or vertically (Fig B); Tail-flirting—tail jerked vigorously up (Fig C), sometimes

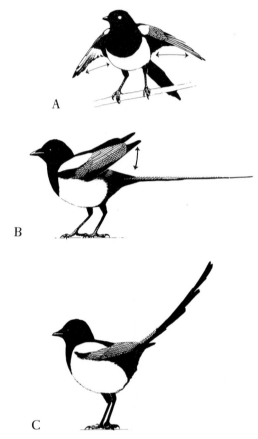

combined with rapid down-up bow of head; Tail-quivering—tail lifted 30–40° above horizontal and tail-feathers vibrated; Tail-spreading—feathers slowly spread and folded again (sometimes on one side only: Goodwin 1952b); for full descriptions of all these, and associated vocalizations, see Baeyens (1979, 1981a) who emphasised gradation between displays and described variable sequences and combinations in which they occur, according to context; often difficult to interpret since same displays also used in heterosexual encounters. Aggression between pair-members occurs if ♀ involved in extra-pair copulation. On one occasion, ♀ was mounted by ♂ from adjacent territory, and her partner chased off intruder and then chased ♀ back to nest with much calling (Birkhead 1991). Rarely aggressive towards birds other than conspecifics in territory, except potential predators. In northern and central Europe, most interspecific aggression occurs with Carrion Crow *Corvus corone* (e.g. Baeyens 1981a, b, c, Vines 1981, Eden 1987a, b); *C. corone* dominant and has been seen to catch 1st-year *P. pica* in flight and kill it (T R Birkhead). For predation by *C. corone*, see Breeding. For experimental demonstration of competition for food with Jackdaw *C. monedula* during breeding season, see Högstedt (1980*b*). In non-breeding flocks, aggression usually takes form of supplanting: one bird approaches another, who simply moves away. After being attacked or when approaching a dominant, either sex may perform Begging-display (see also Heterosexual Behaviour, below) in which bird crouches, Wing-quivers, and usually gives Begging-call (Baeyens 1979: see 5 in Voice). If birds equally matched in size, may perform Chin-up display (head jerked up and held thus for 5–10 s: Fig D, left) or more extreme Stretch-display (Fig D, right). Sometimes, nictitating membrane drawn

D

rapidly and frequently over eye in these encounters ('Blinking': Baeyens 1979); membrane is white with orange spot and serves as signal, but precise function unknown: also used in heterosexual context (Birkhead 1991). HETEROSEXUAL BEHAVIOUR. (1) Pair-bonding behaviour. Subtle, with few specific displays (see Antagonistic Behaviour, above), but not well known. Pair-formation of non-breeders probably occurs in non-breeding flock. Studies of mate choice in captive *hudsonia* indicated that ♀♀ prefer older ♂♂ (Komers and Dhindsa 1989). Meetings between pair-members (or prospective pair-members) are accompanied variously by Wing-flirting, Tail-flirting, Begging by ♀ (see subsection 2, below), and Circling by ♂ (see subsection 3, below). Circling occurs usually on ground but also in trees and bushes as ♀ hops away, climbing ever higher, with ♂ (calling or Babble-singing) spiralling in pursuit; if ♀ flies or hops away to other trees, Circling becomes Chase-hopping, i.e. alternate bouts of hopping and flying, then landing close to other bird which immediately moves away, and so on. (Baeyens 1979.) Paired birds spend more time together than other birds and often cooperate to evict intruders from territory (Baeyens 1981*a*, *b*, *c*). After evicting intruders, pair often sit in body contact and Allopreen (mostly neck, nape, and breast), with ♀ preening ♂ more than vice versa. Allopreening also occurs apparently spontaneously, but less often. Mutual Billing (nibbling at bill) also occurs, usually after Begging-display of ♀. Bond appears most strong during time when ♀ fertile and ♂ guards her by constant following (Birkhead 1991; see also subsection 3, below). (2) Courtship-feeding. In few days before start of egg-laying, ♀ (Fig E, left) begs from ♂: crouches in front of ♂, lifts (often high over back) or flaps wings as if 'loose at the shoulder' (Goodwin 1952*b*, 1986), and gives Begging-call. Initially ♂ usually ignores

♀ and only begins feeding her once egg-laying started. This aspect not studied in detail in nominate *pica* which does not start begging until just after 1st egg laid. Probably differs from that in *hudsonia* (Buitron 1988): 1 ♀ (out of 36) started begging 7 days before laying 1st egg, but most (11) on day before 1st egg; 8 not until day 1st egg laid. ♂ first fed ♀ 2 days before 1st egg (3 of 27 ♂♂) but most ♂♂ (12 of 27) did not begin until day 1st egg laid. ♀♀ begged most frequently on day before laying 1st egg and next 2 days, begging 40–45% of observation time. Some ♂♂ responded to ♀ begging with courtship-display other than feeding. Once incubation starts (usually on day 3rd egg laid), ♂ routinely feeds ♀ on nest. Courtship-feeding starts too late to contribute much to egg-formation; rather, seems to be associated with ♀ being fed on nest during incubation. ♀ performs all incubation (and brooding), and, in North America at least, ♂ provides almost all ♀'s food. (3) Mating. Main associated display by ♂ is Circling and Tilting, often accompanied by Wing-quivering and Babble-singing. ♂ approaches ♀, who is often Begging, and starts to walk round her in tight circles (Fig F, ♂

F

on left), *c*. 0·2–0·5 m away; Circling ♂ adopts upright posture, white plumage ruffled to give almost spherical appearance, and tilts tail inwards towards ♀. As ♂'s excitement increases he flaps or quivers wings (drooped or stiffly open) and Babble-sings, eventually placing one foot on ♀'s back and then mounting (Clegg 1962; Baeyens 1979; Goodwin 1986; Birkhead 1991). Copulation usually on ground; once seen on wall. Copulation infrequent in all studies; in northern England, estimated to occur only 3 times per clutch (Birkhead *et al.* 1987). Lasts up to 5 s; single cloacal contact. Of 9 copulations, all occurred between 4 days before start of laying and day 2nd egg laid (T R Birkhead). Attempted extra-pair copulation, usually between breeders in neighbouring territories, more frequent than copulation between pair-members; ♂♂ sneak into adjacent territory when ♀ neighbour fertile (precise timing of this period not known, but ends on day before ♀ lays penultimate egg; ♀♀ have sperm storage tubules in standard location, utero-vaginal junction of oviduct); sneaking ♂ uses vegetation, walls, etc., to approach ♀, then when close performs abbreviated courtship and attempts to mount. Usually intercepted by ♀'s mate. Once, extra-pair copulation attempted when resident ♂ momentarily asleep. (Birkhead 1979, 1991.) ♂ guards mate by close following from 4 days before 1st egg until 3rd egg laid. ♂ rarely initiates movements away from ♀ during this time; ♀ makes most moves and ♂ follows. For first 3

E

eggs, ♂ sits outside nest while ♀ lays (early morning), and then continues guarding; remains close enough to ♀ to intercept intruding ♂♂. (Birkhead 1979, 1982.) (4) Nest-site selection and behaviour at nest. Not clear which sex chooses nest-site: may be mutual decision. ♂ may attract mate to possible nest or nest-site by Hover-flying: slow, rhythmic, undulating flight to and fro in front of her (Baeyens 1979)—though this not seen by T R Birkhead and may not be ritualized display. Long-standing pairs may start building in early January in northern England. First sign of building is placement of twigs or mud in fork of tree. (Birkhead 1991.) Pair displays (Fig G) at potential nest-site by

G

Wing-spreading, Wing-flirting, and quiet complex vocalizations (including 1, 2a, and 7 in Voice); such display may indicate decision about nest location (Baeyens 1979; Birkhead 1991). Both sexes build (see Breeding); completed 10 days before start of laying (T R Birkhead). Between completion and laying, ♀ feeds intensively (Birkhead 1979). RELATIONS WITHIN FAMILY GROUP. Only ♀ broods. Studied in detail in *hudsonia* (Buitron 1988). Brooding by ♀ decreases from about 90% at hatching to 10% at day 16. ♀ will brood large chicks during heavy rain or in cold weather. Both sexes feed young (by regurgitation), ♂ doing more, partly because in early stages ♀ is brooding, but even does more after chicks 15 days old; this pattern continues after fledging. (Buitron 1988.) Food carried in gular pouch. Faecal sacs not removed by parents: probably eaten when chicks young; later on chicks defecate on or over rim of nest. Eyes open at 7–8 days. (Birkhead 1991.) Flying ability poor at fledging, but young have full-sized legs and feet and grip strongly; spend most of their time in branches of nest-tree. Post-fledging care lasts 6 weeks (Husby 1986), longer than most European Corvidae (Goodwin 1986). During this time parents continue to feed chicks, but to decreasing extent. In Netherlands study, fledged young stayed at least 8 weeks in parental territory, gradually straying more widely (Baeyens 1981a). As young become independent 2 or more family groups coalesce, forming basis for non-breeding flock (Birkhead 1991). ANTI-PREDATOR RESPONSES OF YOUNG. Chicks younger than 13 days respond to human at nest by food-begging. Older chicks show fear response, crouching in nest and grasping lining with feet. Chicks older than 18 days actively try to climb out of nest, moving towards thin branches away from predator, but returning to nest once danger passed (Redondo and Carranza 1989). From *c.* 20 days, disturbed chicks very likely to explode from nest with Chatter-calls similar

to adult (see 10 in Voice). Distress-calls (see 11 in Voice) given when handled from 17 days (Redondo 1991.) PARENTAL ANTI-PREDATOR STRATEGIES. (1) Passive measures. ♀ leaves nest at slightest disturbance during egg-laying, before start of incubation. Incubating ♀♀ often remain on nest as observer climbs tree, particularly at hatching time. (Birkhead 1991.) (2) Active measures. Once chicks hatched and also after fledging, both parents mob predators, calling loudly (see 10 in Voice); also swoop at them, approaching closely. In *hudsonia*, ♂♂ approached potential predator (cat) much more closely than did ♀♀ (Komers and Boag 1988). Response to humans at nest similar, but often hammer at or tear off twigs; studied in detail in *hudsonia* (Buitron 1983b) and in Spain (Redondo and Carranza 1989). Parental nest defence increases from hatching, markedly so once chicks reach 18 days (Redondo and Carranza 1989). Regularly mobs domestic cat, stoat *Mustela*, fox *Vulpes*, owls (Strigiformes) and raptors. Mobbing calls often attract other *P. pica* (up to 10 seen together) and entire group call and swoop at predator; sometimes make physical contact with victim, pulling fur or feathers (Buitron 1983b; Birkhead 1991).

(Figs by D Quinn, first published in Birkhead 1991.)

TRB, EKD

**Voice.** Complex repertoire with several calls intergrading, and some divisions thus rather arbitrary. Following scheme based on study by Enggist-Düblin (1988) which, in turn, drew especially on Goodwin (1952b) and Baeyens (1979). For review of possible differences between races, see Goodwin (1986); recordings indicate Arabian *asirensis* very different from nominate *pica* (T R Birkhead). Captive birds often imitate human voice and other sounds but although some vocalizations in the wild suggest mimicry (see 1, below), no unequivocal evidence of it (Goodwin 1952b, 1986). For additional sonagrams, see Bergmann and Helb (1982), Enggist-Düblin (1988), and Redondo (1991). Bill-snapping probably occurs (Bergmann and Helb 1982).

CALLS OF ADULTS. (1) Song. 3 types distinguished by Enggist-Düblin (1988) all based on amalgamation of calls, relatively quiet, and resembling subsong. Also, very different from these, a Whisper-song (see 1d, below), not previously described. (1a) Babble-song of ♂. Series of soft warbling sounds interspersed with higher-pitched units (Birkhead 1991), usually not audible to observer more than 75 m away. Apart from warbling sounds, song incorporates calls 2–5 (below) and sometimes apparent mimicry, e.g. of Starling *Sturnus vulgaris* and Song Thrush *Turdus philomelos*; structure of song and inclusion of calls show marked individual variation (Baeyens 1979). Fig I shows song well endowed with call-type units, Fig II a harsher example. Apart from juveniles (see below), Babble-song heard only from unpaired adults (P Enggist-Düblin), mainly during early part of breeding season, in variety of antagonistic and heterosexual contexts (see Social Pattern and Behaviour) suggesting general purpose of self-advertisement. (1b) Soft song. Much more 'uniform' and lower pitched than basic Babble-song, heard from either sex during intimate pair-contact and thus ascribed courtship function, e.g. before and after bouts of Allopreening (Baeyens 1979); song given by ♂ Circling ♀ just before mounting (T R Birkhead)

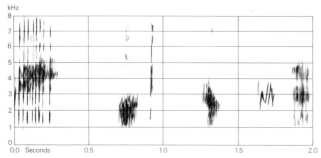

I   P A D Hollom   England   August 1984

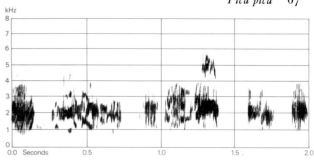

II   S Palmér/Sveriges Radio (1972–80)   Sweden   March 1958

is possibly this type. (1c) Enggist-Düblin (1988) distinguished 'Rhythmic' song of ♂. No detailed description
but repetition of units imparts more structure than in
song-types 1a–b (P Enggist-Düblin); see sonagram in
Enggist-Düblin (1988). Given only by adult ♂ shortly
before and during breeding season, and thus probably a
courtship-song (Enggist-Düblin 1988; Birkhead 1991).
(1d) Whisper-song. Very quiet series of beautiful, musical,
mainly descending 'prrrrrrrr' sounds (units 2–4 in Fig
III); final unit (coda) in Fig III a very faint ascent of 4
sub-units, balancing the 1st descending 'purr' unit of
song; 1st unit in Fig III is call 2 (J Hall-Craggs). This
song heard during nest-building. (2) Tchurch-call. (2a)
Peaceable variant. Soft gentle 'tchurch' frequently heard
during nest-site selection (Birkhead 1991). 1st call in Fig
III (from recording during nest-building) is probably of
this type (D Goodwin). (2b) Aggressive variant. Throaty,
explosive, almost snorting 'tchurch' (Goodwin 1952b),
given almost continuously when attacking, together with

Wing- and Tail-flirting; also heard from ♂ when 'driving'
(Circling and Chase-hopping) prospective mates (Baeyens
1979). (3) Yelping-call. Monosyllabic sound, much like
call 2b but carrying less far and slightly higher pitched;
remarkably like small dog barking. Apparently associated
with self-advertisement, e.g. during Tree-top Sitting.
(Baeyens 1979.) (4) Shrill-call of ♀. The only call exclusive
to ♀. Short, soft (inaudible beyond 25 m), very high-
pitched 'trirr' (Baeyens 1979) or 'irrr-irrr' (Steinfatt 1943),
given by ♀ in response to mate returning with or without
food (Baeyens 1979; Birkhead 1991). Rarely, accompanies
Begging-display to appease attack (Baeyens 1979). (5)
Begging-call. Very variable, but basically an eager-
sounding disyllabic 'cheeuch', or 'cheeuch-uch', based on
juvenile food-call (see below) (Goodwin 1952b, 1986).
Variability evident in the following sonagrams: Fig IV
shows sequence sounding like 'zeer zut-uk', Fig V a
sharper 'keeack-kak' (J Hall-Craggs: compare 'ki-jak' of
Bergmann and Helb 1982, and 'pee-ak' of Holyoak 1967a);

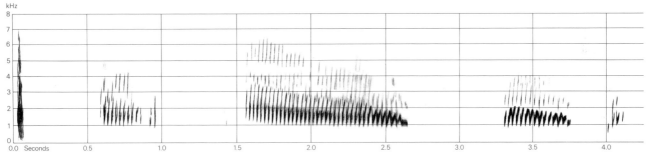

III   R Boughton   England   March 1988

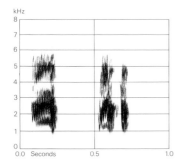

IV   P J Sellar   England   August 1982

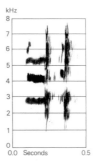

V   J Hall-Craggs   Channel Islands   April 1977

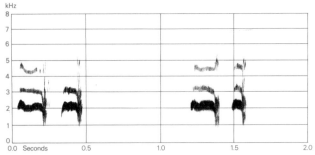

VI  P J Sellar  England  March 1982

♀ soliciting food from ♂ gives disyllabic variant with distinctive pleading squealing quality (Fig VI) which, as excitement increases, is reduced to repetition of only 2nd syllable at rate of up to 4–5 per s (J Hall-Craggs). Also commonly heard is shorter, lower-pitched, more-or-less monosyllabic variant (Goodwin 1952b, 1986). Apart from food-begging by ♀, Begging-calls are used as greeting between mates (Goodwin 1952b, 1986), during allopreening (Holyoak 1974), and by either sex appeasing dominant conspecific (Baeyens 1979). (6) Purr-call. Low guttural or crooning sound (Goodwin 1986), also heard as rather soft cough (Birkhead 1991) from food-bearing parents returning to nest to stimulate young to beg (Goodwin 1986). Different from call 12f (below) (D Goodwin). (7) Nest-call. Prolonged, loud, hoarse, throaty (but not screeching) call, often given for c. 30 s or more without pause (Goodwin 1952b). Also described as loud nasal buzzing, similar to sustained call 2a (Baeyens 1979). Given by either sex when pair are choosing nest-site or about to start building, and serves to summon mate to site; at nest, when pair may call thus together, is usually preceded and followed by bouts of call 10 (Goodwin 1952b; Baeyens 1979). (8) Protest-call. Quiet drawn-out 'tsrae(e)' with pathetic-sounding diminuendo, given presumably in appeasement, by bird under threat or attack by conspecific; also given by captive birds when generally thwarted (Goodwin 1952b, 1986; Baeyens 1979). Develops from food-call of older nestlings (Enggist-Düblin 1988: see below). Under greater threat, gives way to call 9. (9) Shriek-call. Sustained, harsh, relatively high-pitched, far-carrying 'cheerk' when attacked by conspecific (Baeyens 1979; Enggist-Düblin 1988), representing more extreme response than calls 2b and 8. (10) Chatter-call.

Best-known call, a harsh, rattling, far-carrying 'cha-cha-cha-cha', often preceded by 2 longer sounds (Goodwin 1952b); in Fig VII, 16-unit rattle preceded by 4 'zuh' or 'zerh' sounds (J Hall-Craggs). Also rendered 'skrak-ak-ak-ak-ak-ak' (Baeyens 1979), 'tsche-tsche-tsche' or 'gek-gek-gek', sometimes developing, when run together, into single hard rattling 'trre' (Bergmann and Helb 1982); first (5-unit) rattle in Fig VIII is relatively fast (15 units

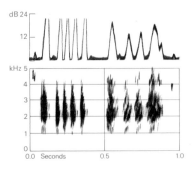

VIII  V C Lewis
England  June 1978

per s compared with c. 10–11 in Fig VII: J Hall-Craggs); amplitude trace shows crescendo in 2nd rattle. For other renderings see Goodwin (1986). Chatter signals alarm, annoyance, and attack-flee conflict (Goodwin 1952b), e.g. when nest threatened or when mobbing predators. Playback attracted P. pica to perch together and chatter at loudspeaker (Baeyens 1979). At lower intensity, commonly a harsh 'shrak-ak', often repeated at brief intervals but never in continuous chatter (Goodwin 1952b); single 'kak' (Fig IX) like Jackdaw Corvus monedula is probably of this

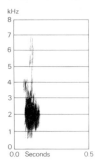

IX  V C Lewis  England  May 1978

low-intensity type (D Goodwin). (11) Distress-call. Very loud, hoarse, sustained screech, given when handled, or seized by conspecific; heard more often from hand-raised

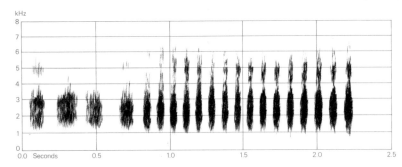

VII  P J Sellar  England  August 1982

than wild birds (Goodwin 1952*b*; Baeyens 1979). (12) Other calls. Calls 12c-f (below) have not been previously reported; based on recordings analysed by J Hall-Craggs. (12a) Various clicking sounds, e.g. 'k' tk', 'kittik', or short explosive 'tchuk' given by tame captive ♀♀ addressed by keeper to which they show affection, accompanied by Wing- and Tail-flirting and down-up head bow; function not clear but also heard at Ceremonial Gatherings in the wild (Goodwin 1952*b*, 1986). (12b) Various low-pitched sounds with submissive tone (perhaps low-intensity variants of call 5) given by captive ♀ in same context as 12a; similar sounds exchanged in the wild between mates, especially at or near nest (Goodwin 1952*b*, 1986). (12c) In recording of small bickering group, some chatters and short rattles (call 10) interspersed with extremely short (10-20 ms) tonal 'pip' whistles; Fig VIII shows 2 pips, one before (consisting of 2 parts) and one after 2 rattles (amplitude trace facilitates location of pips). (12d) Fig X

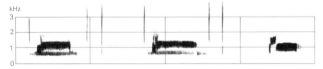

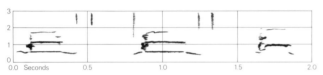

X V C Lewis England May 1978

shows 3 notes resembling lazy crooning sounds of domestic fowl foraging at ease, interspersed with clicks similar (also in respect of recurrent pattern) to those of Jay *Garrulus glandarius* (see Fig XI of that species) (J Hall-Craggs);

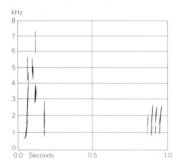

XI V C Lewis England May 1978

context not known. (12e) Fig XI shows liquid 'tlip' followed by brief purr (apparently of Whisper-song type), heard when adults together at nest.

CALLS OF YOUNG. Repertoire and its development described by Enggist-Düblin (1988), Redondo *et al.* (1988), Redondo and Exposito (1990), and Redondo (1991). Of 15 different adult calls found by Enggist-Düblin (1988), 9 develop from weeks 2-7, rest after 8-9

months; unique to nestlings is the first food-call (see below), other nestling calls being forerunners of adult calls (Enggist-Düblin 1988), although quiet whistle-call (serving as a comfort-call) described by Redondo (1991) was not heard after 17 days and may also have no adult counterpart. First food-call is quiet cheeping ('begging trill': Redondo 1991) which, after 2 weeks, gives way to louder, more sustained call, which, in turn, develops into call 8 of adult. At fledging, young also beg with call very like adult call 5, and start giving adult calls 2, 10, and 11. (Enggist-Düblin 1988; Redondo 1991.) Babble-song heard from *c.* 6-20 weeks (P Enggist-Düblin) until autumn (Goodwin 1986), then tends to disappear, reappearing in spring in normal adult context (Baeyens 1979).　　　EKD

Breeding. SEASON. Northern England: median date for laying of first egg over 8 years, 12-24 April (*n* = 427); laying delayed average 3 days per 100 m increase in altitude; nest-building may occur on mild days in late December though eggs not laid until late March (Birkhead 1991); urban birds started laying *c.* 5 days earlier (10 April, *n* = 35) than rural (*n* = 50), probably due to 'heat-island' effect and abundance of food in form of scraps (Eden 1985). For Manchester (north-west England), see Tatner (1982*a*); for Anglesey (north-west Wales), see Seel (1983); for north-east Scotland, see Love and Summers (1973); for major study of variation throughout Britain and Ireland, see Holyoak (1967*c*). Netherlands: mean date of 1st egg over 6 years, 15-20 April, *n* = 178 (Walters 1988); see also Baeyens (1981*b*). Throughout Europe, peak laying period mid- to late April: in northern Finland, earliest laying *c.* 20 April (Haartman 1969); in Coto Doñana, south-west Spain, peak laying mid-April (Alvarez and Arias de Reyna 1974). For Denmark, see Henrikson (1989); for France, see Labitte (1953) and Balanç a (1984*c*); for south-west Germany, see Hund and Prinzinger (1981); for south-west Spain, see Arias de Reyna *et al.* (1984). SITE. Usually in tree, often thorny, but, where unavailable, in low bush or even on ground; on treeless island off Wales, all nests in or under bramble *Rubus* bushes less than 1 m above ground; in built-up areas, sometimes on man-made structures like electricity pylons or even inside large shed or factory (Bährmann 1968*a*; Birkhead 1991). In tree, nest placed in top of crown, fork of trunk, near trunk, or in outer twigs; supporting branches up to 4 cm thick may be built into nest, and weight often borne by live lateral twigs; in tall trees can be highly visible (Bährmann 1968*a*; Kulczycki 1973; Jerzak 1988). In Manchester, preferred species (allowing for abundance of each) included holly *Ilex*, Wheatley elm *Ulmus wheatleyi*, Lombardy poplar *Populus nigra* 'Italica', black poplar *P. nigra*, and hawthorn *Crataegus*; species avoided included birch *Betula*, oak *Quercus*, and sycamore *Acer pseudoplatanus*; no correlation between choice and early leaf cover, though canopies denser in preferred trees; average nest height 13·8 m (Tatner 1982*b*). Of 148 nests in Poland, 20% in various *Prunus* species, 15% in willow *Salix*, and 12% in poplar; only 5% in conifers (Kulczycki 1973); in another Polish study, of 657 nests, only 7% in pine *Pinus* (Jerzak 1988), and conifers avoided in south-west Germany, where favoured old, untended pear *Pyrus* trees (Prinzinger and Hund 1981). However, 89% of 398 nests in Finland in pine or spruce *Picea* (Haartman 1969). Height of nest above ground depends on tree species (Prinzinger and Hund 1981; Birkhead 1991), but higher near houses: average height in villages in Poland 12·3 m ($n=232$), in countryside 4·1 m ($n=354$) (Jerzak 1988); on Lolland (Denmark), 5·7 m near houses ($n=125$), 2·7 m ($n=175$) elsewhere (Hansen 1950); in south-west Germany, nests placed in very tops of trees in urban areas, and as height increases position of nest moves towards trunk (Prinzinger andHund 1981). For discussion of nest height in Canada, see Dhindsa *et al.* (1989). For north-east Germany, see Plath (1988a); for Denmark, see also Møller (1978c) and Fjeldså (1981); for Italy, see Prigioni *et al.* (1985); for south-west Spain, see Alvarez and Arias de Reyna (1974) and Arias de Reyna *et al.* (1984). Records of nests on old quarry ledges in England (Fairhurst 1970), on ground in sand-dunes and on suburban lawn surrounded by trees (Felton 1969b; Bangjord 1986), in low bushes just above water surface in Polish reedbeds (Jedraszko-Dabrowska and Szepietowska 1987), and increasingly on telephone poles, even near suitable trees, in Dublin (Ireland) (Kavanagh 1987). Nest: loose, bulky outer layer of twigs and sticks usually extended to form roof (thus globular), often of thorny twigs, those with roof sometimes having 2 entrances; inside this, main part of nest is mud (occasionally dung) bowl incorporating twigs, roots, etc., this layer often forming distinct rim on top of nest; bound into this bowl is layer of more twigs, grass, etc., above which is lining of grass, hair, feathers, leaves, and other soft material; birds near human settlements often ignore man-made materials (Kulczycki 1973; Alvarez and Arias de Reyna 1974; Birkhead 1991). 45 nests in Poland had average outer diameter 24·1 cm (18·5-35·0), inner diameter 17·3 cm (15·0-21·5), overall height 23·5 cm (18·0-29·0), and depth of cup 12·2 cm (9·0-16·0) (Kulczycki 1973, which see for many details). In northern Norway, in absence of trees, nests recorded built of wire lined with pieces of fishing net, lichen, and grass, some 1 m in diameter (Nagy 1943); increasing tendency for nests in urban areas in Poland to be unroofed and to contain pieces of metal, some consisting largely of wire (Jerzak and Kavanagh 1991). Nests in rock holes or on cliff ledges may only be of twigs or roots and lack roof (Holyoak 1967c). In northern England, *c.* 25% of both urban ($n=35$) and rural nests ($n=50$) had no roof (Eden 1985), while 35% of 26 nests built by 1st-years were open compared to 10% of 137 built by older birds (Birkhead 1991). All 247 nests in Coto Doñana study were roofed (Alvarez and Arias de Reyna 1974, which see for dimensions and materials of 31 nests). Study in Netherlands found that some nests may

remain unroofed because of continual harassment during building by Carrion Crow *Corvus corone* (Baeyens 1981b). For USA, see Erpino (1968b). Building: by both sexes; in northern England, takes 1-8 weeks depending on length of territory occupation; ♂♂ collected more material than ♀♀, who built more; 24% of 460 nests were re-used from previous years, 37% of those above 200 m altitude and 19% of those below, probably reflecting availability of nest-material (Birkhead 1991); similarly, 36% of 94 pairs in urban Manchester re-used old nests (Tatner 1982a). In Coto Doñana, said to take average 8·6 days (7-12) (Alvarez and Arias de Reyna 1974). Outer layer plus roof built first, then mud bowl, followed by lining (Erpino 1968b; Alvarez and Arias de Reyna 1974); in study in South Dakota (USA), great individual variation in roles of ♂ and ♀, but in general ♂ worked mostly on base and roof, ♀ on lining (Buitron 1988, which see for many details of building). Material gathered close to nest, from ground or pulled from bushes and trees or old nests, though mud or thorny twigs for roof may be collected some distance away (Bährmann 1968a; Birkhead 1991). New nest sometimes built very close to a previous year's, even on top, in other cases hundreds of metres away (Oeser 1975; Schroeter 1982c; Birkhead 1989). In general, each pair builds only 1 nest per breeding attempt (Birkhead 1991), though records of ♂ building complete nest while ♀ incubated in neighbouring tree (Bährmann 1968a). EGGS. See Plate 57. Sub-elliptical, sometimes short or long sub-elliptical or even long oval, smooth and glossy; very variable in colour, also within clutch; pale or greenish-blue, or light to dark olive-brown, heavily speckled or blotched olive-brown and grey, only sometimes concentrated at broad end (Harrison 1975; Makatsch 1976). Nominate *pica*: 33·8 × 23·9 mm (27·7-40·1 × 20·7-28·0), $n=723$; calculated weight 9·9 g. *P. p. fennorum*: 33·5 × 24·6 mm (31·0-39·0 × 22·0-27·0), $n=100$; calculated weight 10·5 g. *P. p. melanotos*: 33·8 × 23·9 mm (29·8-38·5 × 23·0-26·0), $n=60$; calculated weight 10·0 g. *P. p. mauretanica*: 33·2 × 23·5 mm (30·0-38·1 × 22·0-25·2), $n=84$; calculated weight 9·4 g. *P. p. bactriana*: 35·5 × 24·4 (29·5-43·1 × 22·0-27·1), $n=238$; calculated weight 10·6 g. (Schönwetter 1984.) For 38 clutches from southern England, see Connor (1965); for 144 eggs, with weight 8·2 g (6·8-10·2), from south-west Spain, see Alvarez and Arias de Reyna (1974); for 125 eggs from Poland, see Keller (1979). Clutch: 5-7 (3-10); in northern England over 8 years, annual average size of 1st clutches varied from 5·5 ($n=18$) to 6·3 ($n=81$); no significant difference between urban and rural nests or between nests at different altitudes (Eden 1985; Birkhead 1991). Of 267 1st clutches from Anglesey: 2 eggs, 1%; 3, 2%; 4, 7%; 5, 21%; 6, 46%; 7, 19%; 8, 2%; average 5·8; decrease of *c.* 1 egg between late March and early May (Seel 1983). For variation throughout Britain and Ireland, see Holyoak (1967c); for north-east Scotland, see Love and Summers (1973). Of 1st clutches from 2 areas in Netherlands: 3 eggs, 1%; 4, 5%; 5, 22%; 6, 43%; 7, 27%; 8,

3%; averages 5·7 (*n* = 107) and 6·2 (*n* = 122) (Walters 1988). Little variation in clutch size throughout western Europe; range in studies above was from 5·4 (*n* = 132) in southern England (Holyoak 1967c) to 6·7 (*n* = 45) in south-west Germany (Hund and Prinzinger 1981); average in Finland 6·4 (*n* = 127) (Haartman 1969) and in Coto Doñana 6·1 (*n* = 108) (Alvarez and Arias de Reyna 1974). For USA, see Erpino (1968b) and Buitron (1988). Replacement clutch laid *c.* 2 weeks after loss of 1st clutch (Makatsch 1976; Tatner 1982a), new nest being built much more quickly than original; in northern England, 7% of 155 replacements laid in original nest (Birkhead 1991); in urban Manchester, 21% of 29 used original nest; 27% of 22 pairs which lost nestlings also relaid (Tatner 1982a). Up to 3 replacement clutches recorded in northern England; average size of 1st replacements by 14 pairs was 6·3 (1st clutches, 6·1), but 2nd replacements significantly smaller at 5·4 (Birkhead 1991). Eggs usually laid daily, starting *c.* 10 days after nest completed, though gaps between eggs of up to 3 days not unusual, especially early in laying. One brood. (Holyoak 1967c; Birkhead 1991; T R Birkhead.) INCUBATION. 21–22 days; 23·7 days (*n* = 52) between laying of 1st egg and hatching of 1st nestling; incubation proper starts with 2nd or 3rd egg, or 2–3 days after laying of 1st (Tatner 1982a). According to Holyoak (1967c), can begin after laying of 1st egg, or last, or any in between, and according to Alvarez and Arias de Reyna (1974) usually starts with last or penultimate egg. By ♀ only. In South Dakota study, ♀ spent 45% of observation time on nest 2 days after laying 1st egg, rising to 95% when clutch complete; hatching asynchronous, over 2–3 days in clutch of 6 or more (Buitron 1988). YOUNG. Fed and cared for by both parents, probably fed mostly by ♂ (Birkhead 1991). Brooded only by ♀ (Buitron 1988, which see for details). FLEDGING TO MATURITY. Fledging period variously estimated at 24–30 days; in South Dakota, 27·2 days (23–32, *n* = 20) (Buitron 1988). See also Bährmann (1968a), Balança (1984c), and Birkhead (1989). Young independent at 70–80 days old (Buitron 1988; Birkhead 1991). Age of first breeding 1–2 years (see Social Pattern and Behaviour). BREEDING SUCCESS. In Netherlands, of 73 pairs, 79% hatched eggs and 55% produced fledged young; 1·4 fledged young per pair, 2·6 per successful pair; main cause of failure was interference by *C. corone*, which stole nest-material, usurped nests, and took eggs and young; roofed nests near houses were most successful (1·8 fledged young per pair, *n* = 46), unroofed ones furthest away fledged no young (*n* = 8) (Baeyens 1981b). Also in Netherlands, 33% of 1124 eggs produced fledged young (Walters 1988). For breeding success and *C. corone* interference in England, Denmark, and France, see Vines (1981), Møller (1982), and Balança (1984c). In northern England, 46% of 430 eggs hatched and 36% produced 14-day-old nestlings; 83% of nestlings survived to 14 days in urban areas (*n* = 80), only 74% in rural (*n* = 116), because of increased predation by *C. corone*, squirrel *Sci-*

*urus carolinensis*, and human interference, as well as more adverse climate; however, no significant difference in overall success, at 2·0 nestlings per pair (*n* = 35) in urban population, 1·7 (*n* = 50) in rural (Eden 1985). In rural part of same area, 44% of 409 pairs succeeded in 1st breeding attempt, 31% abandoned breeding, and 25% laid replacement clutch, 13% successfully, to give 3·2 14-day-old young per successful pair (*n* = 233), or 1·8 per pair overall; no 2nd replacements successful; of 236 failures, 88% occurred at egg stage (desertion, predation, egg disappearance, human interference), and 12% at nestling stage (mostly predation and starvation); most predation by *C. corone* (Birkhead 1991, which see for further data). In Manchester, 2·8 fledged young per successful pair, *n* = 62 (Tatner 1982a). On Anglesey, 57% of 463 nests produced 24-day-old young; of 156 failures where stage definitely known, 41% at egg stage, 28% at nestling stage, and in 31% no eggs laid; predation, weather, and human disturbance main causes of failure (Seel 1983). For north-east Scotland, see Love and Summers (1973). In south-west Spain, *c.* 50–70% of eggs failed to produce fledged young; of 43 nests, 81% of those below 0·8 m predated, 26% of those above; main predators are rodents *Eliomys*, *Glis*, *Mus*, *Apodemus*, *Rattus*, lizards *Lacerta*, and birds (Imperial Eagle *Aquila heliaca*, Buzzard *Buteo buteo*, Red Kite *Milvus milvus*, and especially Black Kite *M. migrans*); *c.* 20% of broods in one study parasitized by Great Spotted Cuckoo *Clamator glandarius* (Alvarez and Arias de Reyna 1974; Arias de Reyna *et al.* 1984). For USA, see Buitron (1988). See Högstedt (1981) and Dhindsa and Boag (1989a) for experiments in improving breeding success by supplementary feeding. For lifetime reproductive success, see Birkhead and Goodburn (1989).        BH

**Plumages.** (nominate *pica*). ADULT. Upperparts and side of head from nasal bristles to back black, slightly glossed metallic green on crown and mantle, faintly green elsewhere. Outer and longer scapulars white, forming bold and contrasting patch above wing. Rump variable: of 20 birds from southern Norway and Sweden, 65% had rump marked by band of light grey *c.* 2 cm wide, 20% had light grey mixed with white, 10% had rump grey or dark grey, and 5% grey mixed with some black (RMNH, ZFMK, ZMA); in eastern Germany, whitish in 9% of 195 birds, light grey and white in 32%, light grey in 45%, dark grey in 14% (Bährmann 1968a); in Netherlands, whitish in 10% of 42 birds, mixed light grey and white in 21%, light grey in 48%, dark grey in 21% (RMNH, ZMA). Upper tail-coverts black. Chin to chest black, slightly glossed purple or green (depending on light); shafts of feathers of chin and throat rather stiff, forming glossy black spikes, contrasting with remainder of feathers in some lights. Belly and flank white, sharply demarcated from black of chest; some restricted dark grey on upper flank hidden beneath wing; lower mid-belly, vent, thigh, and under tail-coverts black. Tail black, central pair of feathers (t1) and outer webs of others strongly glossed metallic green over much of length, but 2–3 cm of tip less strongly glossed greenish-blue, this bordered subterminally by ill-defined band 1–2 cm wide with strong purple or violet-blue lustre, grading into bronze-green at border of metallic green; inner webs of t2–t6 dull black or black with slight purple or green gloss. Outer webs of primaries black, slightly glossed blue-green, especially on basal half; inner webs contrastingly white, but extreme base and $c.$ 1 cm of tip black (on average 12·7 mm on p9, 13·1 on p7–p8, 8·8 on p1), tips sometimes slightly glossed green; black of primary-tips extends into black border ending with sharp tip on about middle of inner web of p6–p9, less far on p4–p5; black on tip of inner web of p10 of ♂ 5·9 (35) 4–9 mm long (exceptionally, 0·5–11), in ♀ 8·0 (30) 5–10 (exceptionally 4–19) (sample from southern Sweden, Germany, Netherlands, and Balkan region: RMNH, ZMA). Secondaries and tertials deep black, outer webs strongly glossed metallic blue or purple-blue, basal half of outer web of secondaries with metallic green stripe in middle; inner webs of tertials less strongly glossed greenish-blue, of secondaries faintly green, mainly on tips; in $c.$ 30% of birds, inner web of outermost secondary (s1) has irregular white subterminal spot, exceptionally also s2. Greater upper wing-coverts, greater upper primary coverts, and bastard wing black with blue-green gloss, inner greater upper wing-coverts with much purple-blue on base. Lesser and median upper wing-coverts black with slight green gloss (strongest on longer lesser and on median coverts). Under wing-coverts and axillaries black. *In fresh plumage*, green of tail lighter, tinged brass colour, tertials slightly tinged green; *in worn plumage*, gloss on black of body slightly more purplish, of tail more uniform dull metallic green. NESTLING. Naked. For development, see Brown (1924), Heinroth and Heinroth (1924–6), Steinfatt (1943), and Bährmann (1968*a*). JUVENILE. Like adult, but black of head down to back and chest as well as that of vent and tail-coverts dull, sooty, and unglossed; bare patch below and behind eye and bare stripe along side of chin; chin and throat without glossy black shafts; some white of feather-bases frequently visible on throat and sometimes on chest. White of scapulars and rump partly suffused pale buff or isabelline, some feather-tips tinged grey, scapular patch less large and less uniform than adult, rump patch narrower and more often appearing (buffish-)grey. White of underparts slightly isabelline and with grey of feather-bases showing through, less pure white and less sharply demarcated from black than in adult; feathers of breast, upper flank, and lower belly sometimes tipped brown-grey. Tail as adult, but feathers narrower, $c.$ 2 cm rather than 2·5–3·5 cm; feather-tips more rounded, distinctly less square; tail not full-grown until 4–8 weeks after fledging. Flight-feathers as adult, but gloss slightly less intense and primaries on average more extensively black on tips, especially outermost: minimum extent of black of p9 31·5 (69) 24–40 mm (exceptionally 16–47), on p10 21·2 (69) 18–25 (exceptionally 10–31), on p1 12·6 (18) 8–16 (–24); in adult, 12·7 (70) 5–16(–21) on p9, 6·9 (65) 4–10 on p10 (exceptionally 0·5–19), 9·1 (10) 8–11 on p1 (RMNH, ZFMK, ZMA; for p7–p8 in eastern Germany, see Bährmann 1958); also, border between black tip and white centre less sharp, often faintly grizzled grey or tinted brown, and p10 tapers more gradually to rounded tip, tip less narrow and inner web less strongly emarginated (see, e.g., Parrot 1907, Bährmann 1958, Erpino 1968*a*, Svensson 1992); outer secondary (s1) has white spot in 82% of 172 birds, s2 in 14% (Bährmann 1968*a*). Upper wing-coverts less glossy than adult, lesser and median coverts mainly dull black, greater and primary coverts as well as bastard wing less intensely glossy, tips metallic green rather than strongly greenish-blue. FIRST ADULT. Like adult, but juvenile tail (sometimes except central pair), flight-feathers, bastard wing, some or all tertials, and sometimes a variable number of greater upper wing-coverts retained; these average less glossy and more worn than those of adult at same time of year, sometimes contrasting in amount of gloss and in shape with neighbouring new feathers. Usually easy to distinguish from adult by greater amount of black

on outer primaries (see above), shape of p10, and shape and relatively greater abrasion of tail-tips.

**Bare parts.** ADULT, FIRST ADULT. Iris dark brown. Bill, leg, and foot black. Nictitating membrane white with elliptical bright orange patch. Bare patch behind eye in *mauritanica* and some *melanotos* bright cobalt-blue (Meinertzhagen 1940; P S Hansen), exceptionally in central Europe also (Eck 1984). Inside of bill blackish-grey, mouth flesh-red. Exceptionally, skin of body (including eye-patch), bill, or soles of feet yellow (Kleiner 1939*a*; Eck 1984). NESTLING. At hatching, bare skin, including leg, foot, and bill, pink-flesh or yellowish; gape-flanges pink-white, mouth dark red. On day 6, bare skin orange-yellow; on day 9–11, black feather-pins appear, bare skin pink-yellow, bill, leg, and foot purple-flesh, iris dark blue-grey. On day 17 (when feather-pins open), bill dusky blue-grey, darkest on culmen; base of bill, cutting edges, and gape-flanges pink-white; leg and foot greyish-black with flesh-yellow grooves between scutes. JUVENILE. At fledging, iris pale grey or greyish-blue, bill greyish-black, cutting edges and flanges pale yellow or whitish; inside of bill pale red to cream-white, darkening to blackish-grey when bill full-grown, mouth deep pink, purple-pink, or red. Bare skin behind eye bluish-slate, inconspicuous; sometimes (especially in Balkan area) yellow. Leg and foot greyish-black. (Heinroth and Heinroth 1924–6; Kleiner 1939*a*; Bährmann 1958; Goodwin 1986; RMNH, ZMA.)

**Moults.** ADULT POST-BREEDING. Complete; primaries descendent. In Britain, starts with p1 mid-May to mid-July (average $c.$ 21 June), completed with regrowth of p10 after $c.$ 105 days (Holyoak 1971; Ginn and Melville 1983) or 131 days (Seel 1976), mid-August to mid-October, rarely November (average $c.$ 1 October); in 1-year-olds, starts late April to mid-June (on average, early May), completed after $c.$ 115 days between early August and late September (on average, late August) (Holyoak 1974*a*; Ginn and Melville 1983). In each age-class, ♂ starts before ♀: on average, adult ♂ 20 June, ♀ 26 June, 1st adult ♂ 11 June, ♀ 22 June. Overall, average start of secondary moult 12 August (secondary replacement takes 113 days), of tertial moult 14 August (duration 69 days), of tail moult 17 July (duration 88 days); tail moult centrifugal, each feather shed 6–11 days after previous one, duration of growth of each feather 63 (t1) to 43 (t5–t6) days (Seel 1976, which see for many other details). In adult, secondaries start at primary moult score $c.$ 10–22; in 1st adult, from score $c.$ 15–33 (Holyoak 1974). In eastern Germany, adults start with p1 from 2nd half of June, 1st adults from early June (Bährmann 1958). In small samples of birds from Europe (Sweden, Netherlands, Spain), start early May (1st adult) to early July (adult), completion late August to early October; single bird from Algeria in last stage of moult 30 October (RMNH, ZMA). In France, starts July, sometimes late June, completed mid-September to October (Mayaud 1933*b*). In *bactriana* (Russia), starts mid-June to early July, in full moult July–August, completed late September (Dementiev and Gladkov 1954). Birds in small Afghanistan sample, 11 August to 8 October, all in moult (Paludan 1959). No pre-breeding moult, but scattered feathers of body sometimes replaced January–April, perhaps after accidental loss (ZMA). POST-JUVENILE. Partial: head, body, lesser and median upper wing-coverts, inner or (usually) all greater upper wing-coverts, occasionally (especially in Spain and north-west Africa, sometimes in Netherlands and France) some tertials or 1(–2) pairs of central tail-feathers (Mayaud 1939; RMNH, ZMA). Starts 4–6 weeks after fledging, often before central tail-feathers full-grown (Bährmann 1958; RMNH, ZMA). Start of moult varies with fledging date: fully juvenile birds encountered April (north-west Africa) to late August (Sweden, Netherlands, Karel'skaya ASSR, Greece, Algeria), birds in full body moult July to mid-October, birds with moult completed from late August onwards (RMNH, ZMA). In Britain, moult June-September (Ginn and Melville 1983), in eastern Germany (July-)August-September (Bährmann 1958), in USSR late July and August (Dementiev and Gladkov 1954), in France July-October (Mayaud 1933*b*).

**Measurements.** ADULT, FIRST ADULT. Nominate *pica*. Central Norway and Sweden (Uppsala and southern Gävleborg to Trondheim); skins (BMNH, RMNH, ZFMK, ZMA). Bill (S) to skull, bill (N) to distal corner of nostril; exposed culmen on average 5·9 less than bill (S).

| | | |
|---|---|---|
| WING AD | ♂ 204·6 ( 4·38; 12) 195–213 | ♀ 192·4 ( 7·71; 8) 183–203 |
| TAIL AD | 260·3 (10·48; 12) 240–277 | 244·9 (10·83; 8) 233–259 |
| BILL (S) | 41·1 ( 1·32; 13) 39·0–43·2 | 39·5 ( 2·65; 13) 36·9–44·4 |
| BILL (N) | 26·2 ( 1·29; 13) 24·4–28·0 | 24·4 ( 1·88; 13) 22·0–27·5 |
| TARSUS | 52·4 ( 1·78; 13) 49·4–53·8 | 49·8 ( 2·68; 13) 46·0–53·4 |

Sex differences significant, except bill (S).

Netherlands; skins (RMNH, ZMA). Juvenile wing and tail include retained juvenile wing and tail of 1st adult.

| | | |
|---|---|---|
| WING AD | ♂ 197·3 (3·82; 32) 190–206 | ♀ 186·1 ( 4·73; 28) 177–195 |
| JUV | 192·8 (5·07; 32) 187–206 | 181·6 ( 4·69; 33) 174–191 |
| TAIL AD | 249·1 (8·32; 31) 234–266 | 231·0 (10·22; 24) 211–246 |
| JUV | 228·2 (8·35; 22) 214–243 | 210·8 ( 7·67; 29) 194–224 |
| BILL (S) | 40·0 (1·46; 28) 38·1–42·4 | 37·5 ( 1·78; 41) 35·2–41·1 |
| BILL (N) | 25·3 (1·06; 28) 23·8–27·0 | 24·0 ( 1·09; 28) 22·2–26·0 |
| TARSUS | 51·4 (1·95; 35) 48·4–54·2 | 48·3 ( 2·04; 30) 45·3–51·5 |

Sex differences significant.

Wing and tail, adults only (except as stated); skins. (1) Sweden, south of *c.* 62°N in west and *c.* 61°N in east (Svensson 1992). (2) Blekinge (southern Sweden) (Kelm and Eck 1985). Germany: (3) Fehmarn (north-east), (4) Lausitz (east), (5) Leipzig and Halle (centre), (6) Emsland (north-west), (7) Baden-Württemberg (south-west) (Eck 1984; Kelm and Eck 1985; Eck and Piechocki 1988; see also Bährmann 1968*a*, 1976). (8) Austria (Kleiner 1939*a*). (9) Ireland (Kavanagh 1988). (10) Rheinland-Pfalz (westernmost Germany) and northern Switzerland (Kleiner 1939*a*; ZFMK). (11) France (Bacmeister and Kleinschmidt 1920; Mayaud 1933*b*; Kleiner 1939*a*; Eck 1984; RMNH, ZMA). (12) Italy, (13) Sicily, adult and unaged birds combined (Kleiner 1939*a*; Eck 1984; Londei and Gnisci 1988; RMNH, ZMA). (14) Carpathian basin (Hungary and Rumania) (Kleiner 1939*a*). (15) Southern Yugoslavia (Stresemann 1920). (16) Western and northern Turkey, (17) south-central and south-east Turkey, adult and unaged birds combined (Kleiner 1939*a*; Jordans and Steinbacher 1948; Kumerloeve 1961, 1967*a*, 1969*a*; Rokitansky and Schifter 1971; ZFMK). (18) Cyprus (Kleiner 1939*a*).

| | | |
|---|---|---|
| WING (1) | ♂ 197·6 ( — ; 43) 186–209 | ♀ 190·5 ( — ; 33) 182–202 |
| (2) | 199·5 ( 5·4 ; 78) 185–215 | 188·2 ( 4·9 ; 38) 177–196 |
| (3) | 195·4 ( 4·5 ; 33) 190–207 | 185·3 ( 4·3 ; 80) 176–194 |
| (4) | 194·6 ( 4·31; 100) 186–204 | 184·1 ( 4·75; 83) 175–198 |
| (5) | 195·7 ( 4·6 ; 74) 185–206 | 184·0 ( 4·8 ; 56) 173–198 |
| (6) | 198·0 ( 4·6 ; 53) 189–207 | 186·8 ( 4·1 ; 75) 173–196 |
| (7) | 194·5 ( 4·1 ; 37) 186–201 | 184·7 ( 3·6 ; 61) 175–194 |
| (8) | 192·8 ( 5·53; 6) 185–201 | — ( — ; — ) — |
| (9) | 192·5 ( 4·81; 110) 179–206 | 182·0 ( 4·39; 122) 172–196 |
| (10) | 191·5 ( 3·63; 10) 186–198 | 184·1 ( 7·08; 7) 172–192 |
| (11) | 195·1 ( 3·94; 14) 189–202 | 184·7 ( 4·60; 23) 179–195 |
| (12) | 194·4 ( 4·41; 13) 185–200 | 184·4 ( 7·25; 7) 176–193 |
| (13) | 194·9 ( 3·13; 7) 191–200 | 182·7 ( 6·12; 6) 176–190 |
| (14) | 194·2 ( 4·37; 43) 185–205 | 181·6 ( 3·49; 25) 175–187 |
| (15) | 195·0 ( 5·23; 4) 190–200 | 185·8 ( 3·86; 4) 182–191 |
| (16) | 199·1 ( 3·95; 11) 189–204 | 191·4 ( 3·97; 8) 185–197 |
| (17) | 201·5 ( 9·30; 5) 185–208 | 196·2 ( 4·02; 7) 190–199 |
| (18) | 197·6 ( 4·44; 8) 191–203 | 179·0 ( — ; 1) — |
| TAIL (1) | — ( — ; 43) 236–269 | — ( — ; 33) 217–268 |
| (2) | 252·4 (12·5 ; 70) 219–278 | 233·0 (10·1 ; 38) 209–252 |
| (3) | 249·0 ( 7·0 ; 31) 233–260 | 232·1 ( 8·8 ; 64) 214–252 |
| (4) | 247·7 ( 9·72; 99) 228–269 | 230·7 (10·1 ; 81) 206–255 |
| (5) | 251·0 (12·8 ; 68) 227–285 | 231·5 (10·5 ; 50) 213–254 |
| (6) | 252·0 ( 9·7 ; 47) 232–280 | 233·5 ( 8·7 ; 69) 211–250 |
| (7) | 245·0 ( 7·1 ; 30) 235–258 | 228·0 ( 8·0 ; 43) 213–247 |
| (8) | 247·2 (14·18; 6) 220–260 | — ( — ; — ) — |
| (9) | 243·9 (12·43; 100) 204–270 | 231·6 ( 8·96; 111) 209–254 |
| (10) | 246·6 (16·67; 10) 207–268 | 236·6 ( 7·50; 6) 227–245 |
| (11) | 248·1 (11·87; 16) 212–265 | 234·8 (11·48; 18) 216–265 |
| (12) | 250·1 (10·54; 9) 238–264 | 231·3 (11·79; 7) 220–248 |
| (13) | 244·2 (15·24; 4) 225–260 | 226·2 ( 8·23; 5) 217–235 |
| (14) | 248·9 (13·91; 43) 213–275 | 225·3 (18·89; 23) 170–255 |
| (15) | 251·0 ( 7·44; 4) 246–262 | 234·0 (11·50; 3) 225–247 |
| (18) | 267·5 (11·65; 8) 250–285 | 228·0 ( — ; 1) — |

Tail data of Kleiner (1939*a*) 10 mm too long due to different method of measurement; corrected in table above (see also Eck 1984). For wing and tail of juveniles and other measurements of samples above, see Stresemann (1920), Kleiner (1939*a*), Bährmann (1968*a*), Eck (1984), Kelm and Eck (1985), Eck and Piechocki (1988), and Svensson (1992). For Czechoslovakia, see Hell and Soviš (1959). Adult ♂, Transcaucasia: wing 193 (14) 182–213, tail 254 (10) 237–288 (Dementiev and Ptushenko 1939). For sexing with discriminant function in Ireland, see Kavanagh (1988).

*P. p. fennorum*. Finland and Karel'skaya ASSR (north-west Russia), all year; skins (Dunajewski 1938; Kleiner 1939*a*; RMNH, ZFMK, ZMA).

| | | |
|---|---|---|
| WING AD | ♂ 207·0 ( 5·54; 13) 200–219 | ♀ 196·0 ( 4·55; 4) 190–200 |
| JUV | 205·7 ( 1·53; 3) 204–207 | 197·0 ( 9·39; 11) 185–211 |
| TAIL AD | 275·6 (22·90; 7) 225–290 | 256·5 (19·64; 4) 230–276 |
| JUV | 255·7 (12·10; 3) 242–265 | 237·5 (12·85; 11) 220–267 |

Sex differences significant for adult wing.

West European USSR: wing, ♂ 203·6 (70) 191–215, ♀ 192·9 (63) 182–210 (Dementiev and Gladkov 1954). Wing: adult 204 (10) 190–221 (Vaurie 1959). Sweden, north of *c.* 62°N in west and *c.* 64°N in east, adult: wing, ♂ 206·5 (11) 202–211, ♀ 194·1 (11) 189–200; tail, ♂ 245–277 (11), ♀ 239–259 (11) (Svensson 1992; L Svensson).

*P. p. melanotos*. Spain and Portugal, all year; ages combined (mainly adult), skins (BMNH, ZFMK, ZFMK, ZMA).

| | | |
|---|---|---|
| WING | ♂ 192·7 ( 4·90; 12) 187–204 | ♀ 181·1 ( 6·04; 7) 172–190 |
| TAIL | 244·5 (15·43; 12) 223–265 | 229·4 (18·14; 8) 200–258 |
| BILL (S) | 39·2 ( 1·83; 11) 37·2–42·3 | 36·5 ( 1·42; 8) 34·9–38·6 |
| BILL (N) | 24·7 ( 0·94; 11) 23·2–26·4 | 23·6 ( 1·02; 8) 23·2–25·0 |
| TARSUS | 51·2 ( 1·80; 11) 48·3–54·2 | 49·5 ( 2·27; 7) 46·5–53·3 |

Sex differences significant for wing and bill.

Salamanca (west-central Spain), adult: wing, ♂ 190·6 (4·6; 40) 181–197, ♀ 181·7 (4·3; 17) 172–189; tail ♂ 245·1 (12·7; 36) 228–266 (282 in one bird), ♀ 231·8 (8·3; 17) 215–243 (Kelm and Eck 1985, which see for other measurements and juveniles).

*P. p. mauritanica*. North-west Africa, all year; ages combined (mainly adult), skins (BMNH, RMNH, ZFMK, ZMA).

| | | |
|---|---|---|
| WING | ♂ 169·5 ( 2·29; 15) 165–172 | ♀ 157·2 ( 3·56; 16) 152–165 |
| TAIL | 268·4 (13·21; 14) 247–298 | 235·1 (14·51; 16) 210–270 |
| BILL (S) | 40·6 ( 1·58; 12) 37·9–42·6 | 37·8 ( 0·71; 12) 36·3–38·6 |
| BILL (N) | 26·0 ( 1·03; 12) 24·4–27·2 | 24·3 ( 0·65; 12) 23·5–26·0 |
| TARSUS | 51·5 ( 1·91; 12) 49·2–53·9 | 47·4 ( 1·74; 12) 44·6–50·2 |

Sex differences significant.

Wing, adult: 163 (20) 152–172 (Vaurie 1959). Morocco and

Algeria, $n=11$: wing 155–172, culmen 34·5–42 (Meinertzhagen 1940).

*P. p. bactriana*. Tomsk, Tien Shan, Tarbagatay, Altai, and western Himalayas; skins (Kleiner 1939a; RMNH, ZFMK, ZMA).

| | | | | | | |
|---|---|---|---|---|---|---|
| WING AD | ♂ | 219·9 ( 5·32; 9) | 204–220 | ♀ | 204·5 ( 8·17; 6) | 195–216 |
| JUV | | 211·7 ( 6·86; 6) | 200–220 | | 205·5 ( 3·39; 6) | 203–212 |
| TAIL AD | | 287·6 (11·88; 9) | 266–302 | | 262·7 (19·39; 6) | 226–279 |
| JUV | | 254·0 (17·68; 4) | 237–275 | | 248·3 (10·89; 5) | 238–262 |
| BILL (S) | | 40·8 ( 2·35; 9) | 37·5–43·6 | | 39·1 ( 2·69; 9) | 36·3–42·8 |
| BILL (N) | | 26·3 ( 1·02; 9) | 24·7–27·6 | | 24·5 ( 1·56; 9) | 23·4–27·2 |
| TARSUS | | 50·0 ( 2·17; 10) | 47·2–53·3 | | 48·3 ( 1·97; 9) | 46·3–51·5 |

Sex differences significant for adult wing, adult tail, and bill (N).

Iran (in part intermediate between nominate *pica* and *bactriana*): wing, ♂ 199·3 (3) 193–204, ♀ 185·0 (7·83; 4) 175–193, sex unknown 199·0 (5·89; 4) 193–205 (Stresemann 1928b; Schüz 1959; Diesselhorst 1962). Turkmeniya, sexes combined, adult: wing 197·1 (3·83; 6) 192–202, tail 258·4 (14·26; 6) 234–267 (Kleiner 1939a; ZMA). Afghanistan, ages combined: wing, ♂ 212·6 (5·74; 10) 201–221, ♀ 199·6 (6·59; 8) 188–209; tail, ♂ 268·8 (18·41; 10) 230–300, ♀ 246·0 (21·05; 7) 224–285 (Paludan 1959). Wing, east European Russia to Altai, ♂ 204·2 (44) 190–218, ♀ 200·3 (37) 183–211 (Dementiev and Gladkov 1954). Kazakhstan, wing, ♂ 199–218 (8), ♀ 188–207 (13); tail, ♂ 222–309 (8), ♀ 242–291 (13) (Korelov *et al.* 1974). Voronezh area, adult wing ♂ 207·0 (11) (Kelm and Eck 1984). Northern slopes of Caucasus, adult ♂, wing 195–208, tail 264–304; eastern Ukraine and regions round Kursk and Rostov-on-Don, adult ♂, wing 204 (199–207), tail 237–292 (Dementiev and Ptushenko 1939). Western Himalayas, wing: ♂ 217·6 (9) 211–225, ♀ 208·2 (10) 200–225 (Vaurie 1972).

*P. p. leucoptera*. Southern end of Lake Baykal (USSR), sexes combined, 1st adult; wing 211·5 (6·62; 10) 203–221, tail 271·8 (13·67; 10) 253–295 (Kleiner 1939a; ZMA). North-east Mongolia, adult: wing, ♂ 220·0 (7·31; 5) 212–230, ♀ 212·6 (5·40; 8) 203–220 (Piechocki and Bolod 1972; Piechocki *et al.* 1982).

*P. p. bottanensis*. Tibet: wing, ♂ 256·1 (22) 242–270, ♀ 246·5 (14) 240–256 (Vaurie 1972); tail, sexes combined, 257·2 (11·5; 10) 242–275 (Vaurie 1955).

**Weights.** ADULT, FIRST ADULT. Nominate *pica*. In Britain, peaks December–January (♂ average 247, $n=9$; ♀ 206·5, $n=26$), decreasing to low in March (♂ 224, $n=9$; ♀ 193, $n=16$), followed by peak again during laying, April–May, especially in ♀ (♂ 236, $n=64$; ♀ 220, $n=141$); generally low June–August (♂ 230, $n=40$; ♀ 196, $n=52$), slightly heavier September–October (♂ 237, $n=18$; ♀ 203, $n=37$) (Seel 1976). In Sweden, monthly average fluctuated between 233 (April) and 250 (December) in ♂, 180 (May) and 222 (December) in ♀; ♀ also rather heavy in April (average 220) (Zedlitz 1926). In eastern Germany, peak April–May, ♂ 223·9 (2·46; 29), ♀ 199·5 (3·08; 27); low when feeding young, June–July, increasing during moult to average of 228 ($n=12$) in ♂, 191·1 ($n=19$) in ♀, low again December, increasing January–March (Bährmann 1968a, 1972, which see for other details).

Ages combined; data from whole year combined or time of year unknown (except as stated): (1) Britain (Seel 1976); (2) Ireland, February–March (Kavanagh 1988); (3) Norway, August–March (Haftorn 1971); (4) southern Sweden (Svensson 1992); (5) USSR (Dementiev and Gladkov 1954); (6) south-east Germany, adult only (Bährmann 1972); central Germany, (7) adult, (8) juvenile and 1st adult (Eck and Piechocki 1988); (9) Netherlands (ZMA, RMNH); (10) north-east France, November–February (Bacmeister and Kleinschmidt 1920); (11) Hun-gary and Rumania (Kleiner 1939a); (12) Turkey, May–October (Kumerloeve 1967a, 1969a; Rokitansky and Schifter 1971; ZFMK).

| | | | | |
|---|---|---|---|---|
| (1) | ♂ 234·6 ( — ; 148) | — | ♀ 210·4 ( — ; 271) | — |
| (2) | 239·1 (18·8; 137) | 171–288 | 201·3 (15·8; 177) | 160–244 |
| (3) | 236 ( — ; 9) | 185–290 | 210 ( — ; 4) | 190–253 |
| (4) | — ( — ; 85) | 200–272 | — ( — ; 87) | 171–240 |
| (5) | 229·4 ( — ; 5) | 203–261 | 168·5 ( — ; 2) | 165–172 |
| (6) | 221·2 (13·6; 143) | 185–247 | 191·1 (16·0; 136) | 161–240 |
| (7) | 221·6 (14·8; 73) | 192–260 | 185·4 (15·6; 56) | 142–212 |
| (8) | 201·2 (31·8; 37) | 127–250 | 188·7 (13·4; 41) | 150–223 |
| (9) | 237·4 (18·4; 16) | 210–272 | 197·8 ( 9·7; 11) | 182–214 |
| (10) | 241·5 (22·4; 6) | 220–275 | 224·0 ( — ; 3) | 190–268 |
| (11) | 215·2 ( — ; 56) | 171–252 | 182·3 ( — ; 39) | 133–215 |
| (12) | 228·3 (27·7; 7) | 183–260 | 206·5 ( 9·3; 4) | 199–220 |

Eastern Czechoslovakia, winter: 193·7 (21·2; 35) 155–245 (Hell and Sovíš 1959).

*P. p. fennorum*. Northern Sweden, ages combined: ♂ 203–275 ($n=37$), ♀ 184–248 ($n=36$) (Svensson 1992). USSR: ♂ 253·7 (8) 192–300, ♀ 218·7 (7) 200–234 (Dementiev and Gladkov 1954).

*P. p. bactriana*. Northern Iran, February–April: ♂♂ 230, 240; ♀ 188·3 (3) 175–198 (Schüz 1959). Afghanistan: May–August, ♂♂ 220, 220, ♀♀ 182, 216; September–October, ♂ 237·8 (17·1; 6) 212–264, ♀ 202·6 (4·2; 5) 199–209 (Paludan 1959). Kazakhstan: ♂ 255·1 (19) 195–314, ♀ 214·1 (11) 196–224 (Korelov *et al.* 1974).

*P. p. leucoptera*. Mongolia, May–August: ♂ 242·7 (17·6; 6) 214–268, ♀ 229·0 (19·3; 15) 204–250 (Piechocki and Bolod 1972; Piechocki *et al.* 1982).

*P. p. mauritanica*. Algeria, November: ♀ 180 (ZFMK).

**Structure.** Wing rather short, broad at base, tip rounded; 10 primaries: p6 longest, p7 0–3 shorter, p8 4–13, p9 26–45, p5 1–4, p4 5–13, p3 14–25, p2 21–37, p1 26–43; p10 reduced, 16·3 (15) 10–21 longer than longest upper primary covert in adult, 21·2 (10) 11–28 in juvenile, 82–99 shorter than p6 in adult, 74–79 in juvenile (all races: RMNH, ZMA). In France, length of p10 46–58 in 25 adults, 53–66 in 16 juveniles (Mayaud 1933b). Outer web of (p3-)p4–p8 emarginated, inner web of (p5-)p6–p9 with notch; in adult, inner web of p10 strongly emarginated, tip narrow; in juvenile, p10 tapers more gradually to tip. Tertials short, longest (s7) slightly shorter than innermost secondary (s6). Tail very long, tip graduated; 12 feathers, t6 119 (106–137) shorter than t1 in 15 adult ♂♂, 109 (97–116) in 15 adult ♀♀, 98 (86–106) in 15 1st adult ♂♂ (with retained juvenile tail), 91 (80–106) in 15 1st adult ♀♀. Bill strong, about equal to head length; deep and wide at base; distal half of culmen strongly arched towards bluntly pointed bill-tip, cutting edges slightly decurved, gonys slightly curved upwards. Bill depth at nostril: nominate *pica*, *fennorum*, and *bactriana*, ♂ 13·8 (30) 13·0–14·5, ♀ 12·5 (35) 11·5–13·4; in *melanotos* and *mauritanica*, ♂ 12·5 (10) 11·9–13·2, ♀ 11·8 (11) 11·3–12·3 (see also Kelm and Eck 1986; Eck and Piechocki 1988). Nostril rounded, covered by long and stiff bunch of bristle-like feathers projecting from lore. Some short bristles at side of gape; many hair-like feather-tips on chin and throat. Tarsus and toes rather short, but strong. Middle toe with claw 35·7 (30) 33–40; outer and inner toe with claw both *c.* 72% of middle with claw, hind *c.* 80%.

**Geographical variation.** Marked, but mainly clinal. Involves size (as expressed in wing, bill, or tarsus length), relative tail length, colour of gloss of wing and tail, size of black tips on white primaries, and amount of white, grey, or black on rump;

also, a bare spot behind eye in some populations. On average, juveniles have shorter wing, relatively shorter tail, somewhat less intense gloss on wing and tail, and more black on primaries, and thus only adults should be used for racial identification. In general, birds darkest (more black on primary tips, rump black) in southern races (*mauritanica*, *asirensis*, *bottanensis*), which are more or less isolated from others and which have relatively much longer tail (*mauritanica*) or much shorter tail (*asirensis*, *bottanensis*) than all other races; palest birds are in northern Asia. A cline runs from north-east to south-west in Europe: size gradually decreasing to south-west, amount of black on primaries increasing slightly, and rump gradually changing from pure white through variable mixture of grey, white, and black to pure black. Another cline runs from northern and central Europe east to central Asia: size gradually increasing eastwards, amount of white on wing and rump increasing, gloss on secondaries changing from blue to green, and on tail from green to brass-yellow (Johansen 1944). In eastern Asia, isolated *camtschatica* from Kamchatka to Anadyrland has secondaries greenest and primaries whitest of all races, but birds smaller than those of interior northern Asia. Further south-east, *sericea* of eastern China, Taiwan, Japan, and south-east Asia much smaller and darker than large and pale birds of northern Asia, but connected with them by series of intermediate populations (here separated as *anderssoni*) in northern China (from Kansu through Hopeh to Manchuria), Korea, Ussuriland, and Amurland. Typical *sericea* closely similar to nominate *pica*, but t1 sometimes more bluish, secondaries darker green, and (especially) tail relatively shorter, 115% of wing length (130% in nominate *pica*: Vaurie 1954, 1959). In north-west North America, *hudsonia* also close to nominate *pica*, but gloss of upperwing and tertials dark bluish-green, less deep blue on secondaries and tertials, gloss on crown slightly more bronze-green.

Due to clinal character of variation in Eurasia, boundaries between races hard to define. In nominate *pica* from southern Sweden, rump mainly light grey, but white or whitish in 4 of 22 adult and 1st adult birds and dark grey in 1 (RMNH, ZFMK, ZMA), rarely blackish (Hartert and Steinbacher 1932–8); in Netherlands, white or whitish in 5% of 90 adult and 1st adult birds, dark grey to blackish in 26%; in mainland Italy, Balkans, and Greece, white or whitish in 6 of 20 birds, dark grey in 2 (RMNH, ZMA); in eastern Germany, 16·2% of 277 birds white, 1·4% black (Kelm and Eck 1985; see also Plumages); in southern Yugoslavia, mainly light grey, sometimes white, exceptionally black (Stresemann 1920); on Sicily, whitish, grey, or black (Eck 1984; RMNH, ZMA). In north-east France and locally in Belgium, rump relatively often blackish (e.g. dark grey or blackish in 5 of 15 birds: Bacmeister and Kleinschmidt 1920), and birds from north-east France therefore sometimes separated as *galliae* Kleinschmidt, 1917, but most birds inseparable from those of southern Sweden in both colour and size, and rump lighter elsewhere in France, becoming darker again only in Roussillon (foot of eastern Pyrénées), there grading into *melanotos* (Mayaud

1933*b*). In Iberian *melanotos*, rump black, occasionally with traces of grey or some (mainly) hidden white on feather-bases; amount of black on primary tips about similar to nominate *pica* (on average, 12·2 mm on p1, 8·0 on adult p10, 21·3 on juvenile p10: RMNH, ZFMK, ZMA); from south of Madrid, some birds have small bare bluish spot behind eye, and this spot present in many or all birds from Extremadura southwards (P S Hansen); such a spot occurs exceptionally in central Europe also (Bährmann 1958; Eck 1984). Much larger cobalt-blue bare spot is present behind eye in all birds of *mauritanica* from north-west Africa; rump in this race always black, wing markedly short but tail long; wing-tips with much black, 22·6 (15) 16–27 mm on p1, 23·0 (6) 15–29 on adult p10, 26·8 (9) 16–32 on juvenile p10; with wear, gloss on tail changes rather rapidly from dull metallic green through purplish-green to dull bronzy-purple. *P. p. fennorum* from northern Norway, northern Sweden, and Finland east to north-west European Russia is similar to nominate *pica*, but larger, and rump either fully white or (sometimes) light grey mixed white. In adults of *bactriana* and *leucoptera*, rump always white and primaries extensively white, white on tip of p3–p4(–p5) visible in closed wing; in juvenile, white usually not visible, but black still distinctly more restricted than that on juvenile wing of otherwise rather similar *fennorum*; in adult *bactriana*, minimum length of black on tip of p1 3·9 (11) 1–6 mm, on p9 3·2 (10) 1–6, on p10 0·7 (8) 0–2; in juvenile, 9·2 (10) 7–12 mm on p1, 23·1 (9) 11–31 on p9, 12·9 (10) 6–16 on p10. Tips and inner webs of tertials metallic green (blue in *fennorum* and nominate *pica*), secondaries green except for narrow bluish outer fringe (not largely deep blue), remainder of upper wing-coverts green (less blue-green or green-blue); tail lighter metallic green or brass-green than in nominate *pica*. Within *bactriana*, birds from plains of Transcaspia east to Lake Zaysan and south to Iraq and Pakistan on average slightly smaller and darker, those of Siberia from Urals east to Lena basin as well as mountains from west-central Asia slightly larger and paler (Johansen 1944), but difference too slight and overlap too large to warrant separation of latter birds as *hemileucoptera* Stegmann, 1928. Boundaries of *fennorum* and nominate *pica* with *bactriana* hard to define; limits given on p. 54 are compilation of data from Stegmann (1928), Dementiev and Ptushenko (1939), and Dementiev and Gladkov (1954) in USSR, and of Stresemann (1928*b*), Kleiner (1939*a*), Vaurie (1954*b*, 1955, 1959), Schüz (1959), Kumerloeve (1961, 1967*a*, 1969*b*), Nicht (1961), and specimens examined (ZFMK) in Turkey and Iran. Size of *leucoptera* from southern Transbaykalia and north-east Mongolia is larger still than *bactriana*, primaries whiter, and tail more yellowish-brass-green. Very large and extensively dull black *bottanensis* of eastern Himalayas and China north to Tsinghai may be separate species (Stegmann 1928; Vaurie 1972), but intermediates with *anderssoni* occur in north-east of range (Stresemann *et al.* 1937) and characters in general similar to those of *P. pica* (Eck 1984).

Forms superspecies with Yellow-billed Magpie *P. nuttalli* of western USA. CSR

# *Nucifraga caryocatactes* Nutcracker

PLATES 4 and 5 (flight)
[between pages 136 and 137]

Du. Notenkraker     Fr. Cassenoix moucheté     Ge. Tannenhäher
Ru. Кедровка     Sp. Cascanueces     Sw. Nötkråka

*Corvus Caryocatactes* Linnaeus, 1758

Polytypic. CARYOCATACTES GROUP. Nominate *caryocatactes* (Linnaeus, 1758), south-east, central, and northern Europe, east to upper Pechora, Vychegda, and Kama basins (east European Russia), where it grades into *macrorhynchos*; *macrorhynchos* C L Brehm, 1823, from western slopes of Urals east to Anadyrland, Kamchatka, Sea of Okhotsk, Ussuriland, and (perhaps) northern Korea, south to Tarbagatay, Altai, northern Mongolia, and Manchuria. Extralimital: *rothschildi* Hartert, 1903, Tien Shan from Talasskiy Alatau in west to Dzhungarskiy Alatau in north-east; *japonicus* Hartert, 1897, Japan and central and southern Kuril islands. MULTIPUNCTATA GROUP. 1 race in north-west Himalayas (extralimital). HEMISPILA GROUP. 3–4 races from western Himalayas (south of *multipunctata*) east to northern Yunnan and north to Shensi and Hopeh (China), also Taiwan (extralimital).

**Field characters.** 22–33 cm; wing-span 52–58 cm. Close in size to Jay *Garrulus glandarius*, with structure differing most in 10% longer, more pointed bill and 10–15% shorter tail, but with rather similar broad rounded wings. Rather small, long-billed, compact, and short-tailed corvid, with (uniquely in Corvidae) pale spotted face and body and bold white vent and tail-rim all obvious against otherwise dark chocolate-brown plumage. Flight and behaviour recall *G. glandarius* but less secretive. Sexes similar; some seasonal variation. Juvenile and 1st-winter separable. 2 races in west Palearctic not distinguishable in the field.

ADULT. Moults: May–August (complete). Nasal bristles (forming pale patch over base of upper mandible) and lore whitish, tending to continue as pale line over eye and contrasting with forepart of dark chocolate-brown cap. Face, throat, back (to edge of rump), and underbody (to vent) mid-chocolate brown, each feather with pear-shaped dull white spot (enlarging with increasing feather size towards rear of back and underbody). Rump and upper tail-coverts dark chocolate-brown, virtually unspotted but glossed green near tail. Wings brown-black, not visibly spotted except on tips of lesser and median coverts but glossed bluish-green overall. Underwing dusky, with rows of white spots. Tail brown-black with broadening white tips from centre to outer feathers (those showing as full terminal band below, in manner of Collared Dove *Streptopelia decaocto*) and bluish-green gloss on upper surface. Vent and under tail-coverts clean white, contrasting strongly with black base to tail and spotted brown underbody. With wear, ground-colours become paler and white spots are lost from most wing-coverts. Bill noticeably long and pointed, black; legs black. JUVENILE. Resembles adult but duller and paler in ground-colour, with pale streaks on crown, less extensive and less defined spotting on body, but with noticeable white tips on greater coverts. FIRST-WINTER. Closely resembles adult but retains most juvenile wing-feathers, especially white-tipped greater coverts.

Unmistakable, with white–black–white bands under tail conspicuous from below or behind even at distances where pale spotting on body becomes invisible. Sharp-billed,

short-tailed silhouette quickly apparent as distinct from that of short-billed, rather long-tailed *G. glandarius*. Flight recalls *G. glandarius* but rather more steady, with flapping, not so jerky wing-beats used during both level and undulating progress. Gait strong but rather heavy, essentially a hop varied by sidling jumps and occasional bounces. Shy when breeding but behaviour overt and confident in winter; sedentary birds often leave cover to hammer at fallen cones or hack out seeds from others still attached to outer branches, and frequently perch on topmost sprays; migrants most regularly seen on ground, pecking at fallen fruit or probing for invertebrates, and may shun woods altogether. Sociable, even gregarious during eruptions during which birds often become tame.

Voice varied, with babble (approaching song) not unlike Starling *Sturnus vulgaris*. Most distinctive calls are far-carrying, rather high-pitched, rasping 'kraak' or 'kreak' (less strident than screech of *G. glandarius*), loud disyllabic 'kerrr kerrr' (from pairs in spring), and rattling churr in alarm recalling Nightjar *Caprimulgus europaeus* and Mistle Thrush *Turdus viscivorus*.

**Habitat.** Breeds in boreal or montane upper and middle latitudes of west Palearctic in cool continental forest lowlands and on mountains up to treeline, wherever essential requirements of coniferous forest and food are fulfilled. At home on ground as well as in trees. Within west Palearctic, resorts mainly to stands of Norway spruce *Picea abies* in northern taiga zone or preferably Arolla pine *Pinus cembra*, but sometimes larch *Larix* or silver fir *Abies alba*. Where *P. cembra* absent, alternative reliance on storage for winter of hazel nuts *Corylus avellana* renders access to these of vital importance, even if this involves repeated flights of several km; in Sweden at least, this factor is apparently indispensable (Swanberg 1951*b*). Unrecovered seeds often grow and thus extend habitat of (e.g.) *P. cembra* (Holtmeier 1966; Turček and Kelso 1968). Seeds must be buried in ground, or under lichen or moss on a rock, and be safely recoverable later despite deep winter snow cover; this involves further habitat demands, including memorising

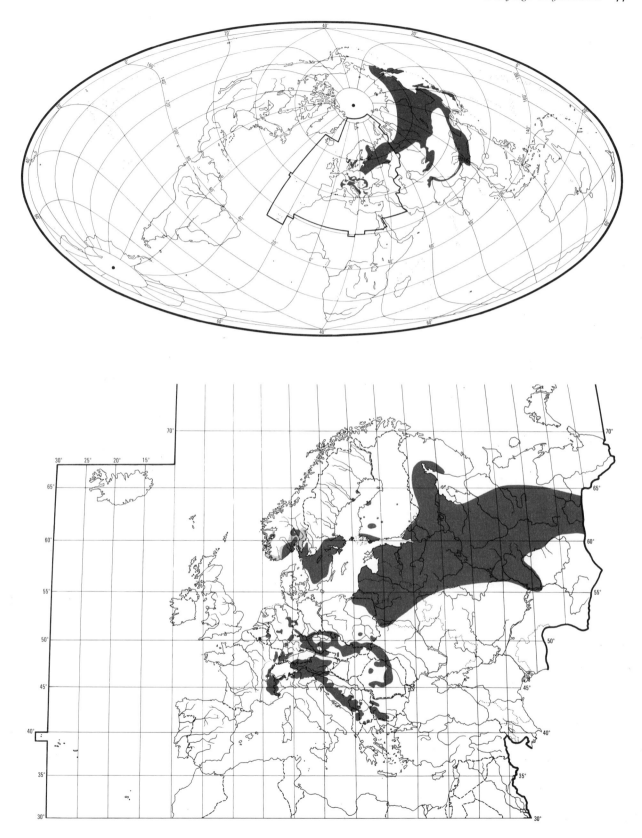

and locating hiding places and ensuring that they are accessible to winter range, even though supplies must be brought to them from a distance. This habit demands sometimes long flights, often at some height, over exposed terrain, contrasting with concealment during breeding season in dense forest. Liking for conspicuous tree-top perches also contrasts with frequently secretive habits, like Jay *Garrulus glandarius*. Successful in limited adaptation to resources of differing habitats in various parts of range, but special needs constrain it to a lifestyle which is vulnerable to shortages of particular foods, in contrast to more robust and flexible pattern of omnivorous relatives. Apart from enforced long-distance displacements of Siberian birds when conifer crops fail, short-distance movements occur elsewhere, involving in Switzerland dispersal not only downwards to lowlands but sometimes upwards above treeline (Lüps *et al.* 1978). In such a situation will fly across a sizeable town (E M Nicholson).

**Distribution.** Recent range expansions in Fenno-Scandia and Belgium.

BELGIUM. Small population established in Ardennes 1968, following major invasion from Siberia; breeding suspected in 1969, proved in 1975 (Devillers *et al.* 1988). At least 60 pairs 1989-91 (Clesse *et al.* 1991). DENMARK. Occasional breeder after invasion years, both nominate *caryocatactes* and *macrorhynchos* (TB). NORWAY. Recently spread to west, where now breeding in several localities (VR). SWEDEN. Recent discovery of small populations north of 60°15′N (limit of hazel *Corylus*, with which species is closely associated), which seem to depend on plantations of Arolla pine *Pinus cembra*; appeared after influx of *macrorhynchos*. In 1980s, *c.* 30 birds at Umeå (*c.* 64°N), *c.* 10 at Skellefteå (*c.* 65°N), and fewer at 2 other places. (LR.) FINLAND. Scattered records of breeding outside main range in south-west refer to *macrorhynchos*, which has established small isolated breeding populations and is increasing (OH). CZECHOSLOVAKIA. Spread to lower elevations noted after 1960 (KH). HUNGARY. No proof of breeding, despite summer records in suitable habitat in north (G Magyar).

Accidental. Turkey. Also occurs irregularly in many other countries as occasional irruptive migrant (see Movements).

**Population.** Some local increases after expansion to new areas (see Distribution); no major changes reported.

FRANCE. 100-1000 pairs (Yeatman 1976). BELGIUM. Estimated 60 pairs; population apparently stable (Clesse *et al.* 1991). GERMANY. Bayern: 5000-15 000 pairs (Bezzel *et al.* 1980). East Germany: 800 ± 200 pairs (Nicolai 1993). SWEDEN. Estimated 3000 pairs (Ulfstrand and Högstedt 1976). FINLAND. Nominate *caryocatactes* estimated *c.* 1000 pairs, perhaps slowly increasing; *macrorhynchos* perhaps 50 pairs (Koskimies 1989).

Survival. Oldest ringed bird 8 years (Rydzewski 1978).

**Movements.** Western race nominate *caryocatactes* and most Asian races chiefly resident, Siberian race *macrorhynchos* eruptive migrant. For general review, see Zink (1981); for review of 1968 eruption, see Kirsch (1992).

Nominate *caryocatactes* resident and dispersive, with weak invasions occasionally reported, e.g. records on Polish coast in extremely cold winters of 1929 and 1940 were presumably from Russian breeding area. Southern Swedish birds sometimes reach (chiefly eastern) Denmark, e.g. 1949 and 1961, but rarely go beyond; at Falsterbo (extreme south of Sweden), unprecedented total of 1530 birds recorded end of August to beginning of October 1975, peaking 5-17 September. (Andersen-Harild *et al.* 1966; Roos 1975; Zink 1981.) More extensive movement in 1977 apparently originated in Komi ASSR and Arkhangel'sk region (Russia), heading between WSW and south-east (chiefly south-west) and spreading through much of European USSR, south to *c.* 48°N in southern Ukraine; main departure late August, with most records September-October (Litun and Plesski 1983); not known to what extent this movement reached other parts of eastern and northern Europe, as invasion of *macrorhynchos* coincided. Considerable movement of Russian nominate *caryocatactes* as well as *macrorhynchos* in 1885 (Formosof 1933; Dementiev and Gladkov 1954). In central European Alpine area, adult nominate *caryocatactes* highly sedentary and movements concern mostly 1st-years; at Col de Bretolet in western Switzerland, passage north-east and south-west equally common in July, but by mid-September entirely to south-west, probably reflecting change to true (short-distance) migration; recoveries (up to 372 km) are limited to region of Alps and Jura (Mattes and Jenni 1984). A few birds with characteristics of nominate *caryocatactes* occur during invasions of Siberian race *macrorhynchos*; these may come from border area of the 2 races, or be caught up on the way (Zink 1981). Only 2 British records, 1860 and 1900 (Dymond *et al.* 1989).

Eruptive (diurnal) migration of Siberian race *macrorhynchos* associated with failure of pine seed crop, especially Siberian stone pine *Pinus sibirica* (see Food), e.g. 1885 irruption probably due to poor crop throughout western Siberia in 1884 and 1885; early start to irruption (notably in 1968), before full dependence on *P. sibirica* begins, thought to result from lack of seeds to cache in post-breeding season. Migration exceptional in that little or no return movement occurs in following spring, most birds failing to survive, though a few breed in wintering areas in subsequent year(s); some return towards area of origin a few weeks after emigrating, however. Eruptions reach Britain, southern France, and northern Italy in exceptional years. Coastlines act as leading lines, and expanses of water as barriers. (Formosof 1933; Kumari 1972; Zink 1981.) Study of museum specimens shows that, unlike in most other eruptive species, irruptions of *macrorhynchos* sometimes consist entirely of birds over 1 year old (1864, 1911, 1968); *c.* 50% of adult ♀♀ collected in major 1968 irruption had never laid eggs, suggesting that at least this irruption took place in a non-breeding year, perhaps following year of poor harvest; in 1885, 1913, and 1954 irruptions, only 1st-years appeared in western Europe; in other irruptions, birds of all ages occurred (Van der Plas and Wattel 1986; C S Roselaar).

20th century irruptions occurred 1911, 1933, 1954, and 1968; to lesser extent in 1900, 1907, 1913, 1917, 1947, 1971, 1977, and 1985 (Zink 1981; *Br. Birds* 1986, **79**, 291); for dates of earlier irruptions, see Formosof (1933). Most widespread and large-scale irruption was 1968. Conspicuous movement reported in area of origin, Siberia: birds flew south along shore of Lake Baykal in continuous stream throughout most of daylight hours, 28 July to 8 August (few after mid-August), stopping only during heavy rain; those taken were extremely fat; flocks also recorded in Mongolia at end of August. In Kazakhstan, passage involved only northern areas. (Piechocki 1971*b*; Korelov *et al.* 1974.) Movement west was via north-east Europe, conspicuous also in north-central and north-west Europe; a few reached southern France and northern Italy, with 1 record each from Portugal and North Africa. Began earlier than other invasions, with records late June as far west as Netherlands, and peak in August, declining dramatically throughout northern Europe from end of October; many birds seen were in poor condition, apparently due to inability to sustain themselves on available food sources. In St Petersburg region (north-west Russia), passage reported from end of July, chiefly in August, with individuals still encountered up to mid-November after snowfall. Thousands present throughout much of Baltic states in August, decreasing in September, with only small numbers later. Widespread and numerous also in Poland. In Finland, recorded end of July to early September; numbers in different localities varied markedly, with most at focal points of leading lines, notably at Vaasa on west coast, where *c.* 3000 reported on 15 August. In Sweden, *c.* 17 000 observed; earliest 22–23 June in Lapland in north, and 50 birds 8 July at Blekinge in south-east; numbers decreased sharply from mid-September, though occasionally reported to May 1969. Recorded in nearly all provinces of Norway. Peak movement 1st week of August in Netherlands (*c.* 6000 records in all), 2nd half of August in Belgium (*c.* 800). In Britain, *c.* 315 immigrants (far outnumbering total in all other years), arrivals coinciding with periods of high pressure over wide area of northern Europe (including Britain). 2 influxes: *c.* 27 birds 6–17 August, restricted to south-east (Norfolk, Suffolk, Kent); main influx from 21 August to mid-September in east (1 on Shetland), spreading west and north to reach Scilly Isles, southern Wales, and Yorkshire (northern England), but not Ireland (where not yet recorded: Dymond *et al.* 1989); small additional arrival 4–9 October in east; minimum total 138 in September, 67 in October, and 17 in December, with 22 reported in 1969 at various dates until early autumn. Entry into France via north-east, thereafter

fanning out with no clear direction; majority of records were in northern half, with highest numbers in September, and isolated reports continuing through winter. In Germany, number of records decreased sharply southward: in Mecklenburg region in north-east, *c.* 10 800, with spectacular passage on coast in early August; in eastern Germany, *c.* 2500 in Elbe-Saale and Brandenburg regions, and *c.* 2000 in Sachsen and Thüringen, mostly in Erzgebirge which acted as barrier; apparently only very weak south of Danube. Similarly, recorded only in small numbers in Austria and Czechoslovakia, with none in Balkans, though some reported from north-west Black Sea area. A few reached northern Switzerland, but arrivals here and in adjoining areas of France masked by local nominate *caryocatactes*. (Hildén 1969*b*; Erard 1970; Hollyer 1970, 1971; Piechocki 1971*b*; Stravinski and Shchepski 1972; Eriksson and Hansson 1973; Zink 1981; Mal'chevski and Pukinski 1983.)

Birds ringed Kaliningrad coast (western Russia) 1968 recovered same autumn between north-west and south-west in Denmark (2), Belgium (1), north-west France (1), and Czechoslovakia (1), and birds ringed Low Countries recovered same autumn in northern France, with one SSW in Massif Central (southern France). Many birds reversed migration within same autumn: northward or eastward heading observed not infrequently in Fenno-Scandia and Baltic states (in part due to leading-line effect of coasts), and predominated in Finland after 20 August. Recoveries (mostly September–October) of 13 birds ringed on Åland islands (south-west Finland), 12 August to 7 September, and of bird ringed 19 September on south coast of Finland, were all between east and SSE of ringing sites, some showing rapid movement; 3 were 2200–3300 km east, thus in area of origin. Also bird ringed Kaliningrad coast 1 August was 2500 km east 2 weeks later. No evidence of return movement in following spring. (Hildén 1969*b*; Zink 1981.)

Invasions of 1911, 1933, and 1954 were later in autumn than 1968, and also relatively stronger inland in central Europe (Zink 1981). 1954 irruption probably originated from Urals and adjacent areas; marked movement there from mid-August, numbers declining in September, with few in October. Remains of bird found in pellet of Eagle Owl *Bubo bubo* in northern Ustyurt, *c.* 44°N in western Kazakhstan, indicating that some moved further south than main direction. In central European Russia, dispersal was over wide area and large numbers not observed; regular in small numbers September–October in Moscow region, and locally in higher numbers in Belorussiya (latest 13 November); reported south to Kiev 50°30′N; none as far north as Karel'skaya ASSR *c.* 61°N. Many thousands (including some nominate *caryocatactes*) reported in Baltic states, especially in coastal regions, with general direction south-west, mostly from mid-August. Reached Czechoslovakia 10 September. (Kumari 1960.) In Germany, *c.* 4000 birds recorded; arrival apparently in 2 waves, 1st via north-east coast mid-September, 2nd slightly further south, heading west over Harz mountains *c.* 22 September and reaching southern Germany; peaked 1-12 October in north, 22-28 October in south; some reached Netherlands, central France, and northern Italy (Huckriede 1969; Zink 1981). Many wintered in central Europe, and slight return movement towards north-east reported midwinter, with records until end of May (Kumari 1960). 1985 invasion only reached northern Europe, and reports involve some nominate *caryocatactes*. In Finland, several thousand recorded, but movement much weaker than 1954 or 1968. In southern and central Sweden, 121 ringed mid-September to mid-October, including some nominate *caryocatactes* but mostly *macrorhynchos*; 5 recoveries of *macrorhynchos* in same autumn showed movement in various directions: east to Finland (2), NNE (1) and WNW (1) within Sweden, and SSW (1) to Denmark. 1500 recorded 9 September to 7 October in Denmark, with peak 252 on 14 September. More than 150 reported from Netherlands, chiefly in 2nd half of October, with isolated sightings continuing to June 1986. 3 records in Britain, 2 in October, 1 from 2 November to 7 December. (*Br. Birds* 1986, **79**, 291; Hildén and Nikander 1986; Rogers *et al.* 1986*b*; Berg 1988; *Rep. Swedish Bird-ringing 1989.*)

Occasionally, birds remain to breed in invasion areas; thus, small colonies established in eastern Belgium following 1968 invasion, and Finland and Sweden following 1977 invasion; breeding also reported from Denmark, Poland, Germany, Netherlands, and France (Zink 1981; Ylimaunu *et al.* 1986; Devillers *et al.* 1988; Crocq 1990).

Himalayan races resident, with some altitudinal movements (Ali and Ripley 1972; Inskipp and Inskipp 1985). Tien Shan race *rothschildi* makes only local movements, according to varying time of seed ripening in different parts of mountains (Kumari 1960; Korelov *et al.* 1974). Japanese race *japonicus*, breeding in high Japanese Alps, usually descends to coniferous forest in lower mountains and foothills, rarely straying to lowlands and plains, exceptionally to coast; a number recorded October 1988 to June 1989 in central and western Hokkaido (Brazil 1991). DFV

**Food.** On breeding grounds, restricted to seeds, nuts, invertebrates, and sometimes small vertebrates, but during irruptions birds virtually omnivorous; winter diet of sedentary birds almost wholly seeds and nuts from supplies stored in autumn. European nominate *caryocatactes* highly dependent on seeds of Arolla pine *Pinus cembra* and nuts of hazel *Corylus avellana*, while Siberian race *macrorhynchos* specializes on seeds of Siberian stone pine *P. sibirica*, other far-eastern pines, and spruce *Picea* (Grote 1940, 1947; Mezhennyi 1964; Goodwin 1986; Crocq 1990; Lanner and Nikkanen 1990). Relationship with *P. cembra* and *P. sibirica* so strong that many authors consider it an example of symbiosis or mutualism, since these seeds are heavy and unwinged, depending on agency other than

wind for dispersal (Formosof 1933; Holtmeier 1966; Crocq 1990; Lanner and Nikkanen 1990). For study of symbiosis in Japan, see Saito (1983). In central Sweden, birds breeding in spruce flew 6 km to feed on hazel (Swanberg 1956*a*); in eastern Siberia up to 10 km travelled in winter to feed on *Pinus pumila* seeds (Andreev 1982*a*). During irruption in Finland, birds of race *macrorhynchos* found planted stands *c.* 6 ha in extent of *P. cembra* and *P. sibirica* and remained there to breed, keeping population of trees viable by seed-caching (Lanner and Nikkanen 1990). Small *macrorhynchos* breeding population in spruce plantations, eastern Belgium, probably feeds mainly on hazel nuts (Devillers *et al.* 1988). Shows remarkable tameness during westward eruptions, often feeding on kitchen scraps, compost heaps, dung, carrion, agricultural crops, and at bird-tables and picnic places, approaching humans close enough to be hand-fed and even entering houses; often close to agricultural or forestry workers to take disturbed insects (Creutz and Flössner 1958; Tricot 1968; Huckriede 1969; Boecker 1970; Hollyer 1970; Piechocki 1971*a*; Mauersberger 1973). Nominate *caryocatactes* can be equally tame within breeding range in times of winter food shortage, or near tourist resorts (Glutz von Blotzheim 1962; Pannach 1983). *P. sibirica* seeds have high energy content (*c.* 28 kJ per g) compared to other conifers (*P. sylvestris c.* 20 kJ per g, larch *Larix c.* 11 kJ per g; hazel nut has *c.* 29 kJ per g), they are much heavier (100 seeds weigh *c.* 30 g; 100 wingless seeds of *P. sylvestris*, *Larix*, *Picea*, and fir *Abies* all weigh less than 1 g), and dry weight comprises 60-75% fat; hence birds probably unable to sustain minimum energy requirement if *P. sibirica* seeds unavailable, and westward invasions probably due to poor crop of these seeds in Siberia (Formosof 1933; Turček 1961; Crocq 1990). Although acorns *Quercus* are eaten, taken much less frequently than hazel and often rejected when offered (Kumari 1960; Plucinski 1970). *P. cembra* seeds may be taken before cones ripen, birds often tackling them while still on tree, but after late July cones are brought to anvil (thick branch, stump, rock, etc.) where held in crevice or fork, usually under foot for additional grip; unripe seeds extracted by cone being broken up by sharp blows with point of bill and soft parts of cone may also be eaten; such seeds not cached, as husks remain fixed to cone. At low altitude in Alps of south-east France, cones begin to ripen from early August and seeds can be extracted for consumption and caching (average 40-60 seeds per cone); mature cones may have seeds extracted *in situ* but this unusual; at anvil, strikes off scales of cone using lower mandible with vigorous blows of slightly open bill, pulling head right up before striking, usually starting at base of cone. Around 12 blows sufficient to expose 15-20 seeds which are eaten or diverted into special sublingual pouch for transportation to cache; then turned and fresh area attacked, whole sequence of behaviour apparently innate. Late in season easily extracts seeds with point of bill from old cones on ground using prying action (opening bill after insertion); if bill resin-covered, will wipe it on moss, etc.; cracks large pine seeds on hard ridge inside base of lower mandible, and unusual forked tongue may be for separating seeds from cone; husk, if removed, often swallowed as grit. (Boie 1866; Glutz von Blotzheim 1956; Mezhennyi 1964; Jung 1968; Löhrl 1970; Piechocki 1971*a*; Crocq 1990, which see for detailed study.) Cones may be picked up and smashed on anvil to release seeds (Knystautas 1987). Hazel nuts similarly tackled, held under foot and hammered between toes; shell opened by 4–5 blows taking 20–30 s, sometimes 2 holes made; contents eaten piecemeal through hole, 2–8 min per nut; one bird had apparent difficulty swallowing pieces of kernel *c.* 6 mm in size; another bird having difficulty opening hazel nut, swallowed and regurgitated it 3 times, keeping it in crop each time for *c.* 1 min before (unsuccessfully) hammering it; hazel nuts can be cracked in bill but this unusual. (Creutz and Flössner 1958; Huckriede 1969; Uloth 1977.) Whole hazel twigs may be pulled off, held under foot, and nuts hammered out of cups (Glutz von Blotzheim 1962; Piechocki 1971*a*); see also Knolle (1990, 1991) for detailed description of treatment of hazel nuts and walnuts *Juglans*. Cereal grains, berries, and fruits may also be taken to anvil for consumption; fruits with large stones generally taken for kernel rather than flesh (Boecker 1970; Hollyer 1971). May jump up from ground to seize hazel nuts on lowest branches (Wernli 1970), or pluck them while hovering (Bannerman 1953*a*). Recorded carrying hazel nut in foot from tree to ground (Meiklejohn and Meiklejohn 1938). See Löhrl (1970) for suggestion that thinner bill of *macrorhynchos* adapted for taking seeds of *P. sibirica* while thicker bill of nominate *caryocatactes* more suitable for hazel nuts. Often recorded at nests of bees (Apoidea) and wasps (Vespidae), taking especially larvae, both on breeding grounds and during invasions; observations of birds working at nests in ground for up to 5·5 hrs, making holes 25 cm deep, throwing soil around vigorously while digging; 8 birds recorded together at a nest; often seen catching emerging bees and wasps, which are rubbed on branch, etc. to remove sting before being swallowed; hornets *Vespa crabro* held under foot and squeezed in bill for up to 15 min; nests of ants (Formicidae) similarly plundered (Sylvester 1968; Glause 1969; Röthing 1969; Piechocki 1971*a*). Searches for insects, etc., behind bark, in crevices in timber, under stones, etc., often hacking at wood like woodpecker (Picinae); recorded turning over fist-sized stones to get at invertebrates; will carry large pieces of bark to perch and break them open; insects, including hairy caterpillars, may be held under foot and killed by blows of bill, then extremities removed; snails with shells beaten on substrate before being eaten; bird recorded hanging upside-down from branch to reach caterpillars (Boecker 1970; Hollyer 1970; Piechocki 1971*a*; Tietze 1971; Rammner 1977). Many reports of hunting technique like that of shrike *Lanius*, using exposed perch and either taking flying insects in sallying flights or dropping on ground insects, returning to same perch; success rate seemingly very high (Pfeifer 1954; Latzel 1968; Hollyer 1970; Crocq 1990). Siberian birds during irruptions recorded following plough for invertebrates (Heidemann and Schüz 1936; Boecker 1970). Both breeding and invasion birds not uncommonly hunt in or over open fields and meadows, sometimes hovering while searching for prey, particularly grasshoppers (Orthoptera) (Heidemann and Schüz 1936; Grote 1940; Jung 1966; Latzel 1968; Litun and Plesski 1983; Crocq 1990). Reported knocking flying or flushed insects to ground with sideways swipe of bill (Bettmann 1969), and also flying up from ground to capture flying insects (Latzel 1968). Captive bird held rodent by nape, shook it violently, and crushed it 2–3 times along spine with bill until dead then opened body by prying action, discarding skin and tail; this bird refused to eat small birds or eggs (Rammner 1977). Another captive bird totally consumed dead sparrow *Passer*, apart from a few flight-feathers, in *c.* 15 min then immediately after killed and ate live sparrow in *c.* 30 min (Boie 1866). Observation in England of House Sparrow *P. domesticus* decapitated and skull split 'like nut'; another bird came 3 times to remove dead sparrows from thatched roof; rodent caught and taken to tree where wedged in fork and hammered until dead, then flesh pulled off in strips (Hollyer 1970). Will take passerines from mist-nets (Piechocki 1971*b*). Recorded removing stored prey of Pygmy Owl *Glaucidium passerinum* on several occasions near owl's nest (Thönen 1965). Regularly takes bait such as fish heads from hunters' traps in Siberia and fish remains on ice of Caspian Sea (Grote 1940, 1947). Frequent reports of drinking water or eating snow (Kollibay 1913; Glutz von Blotzheim 1962; Boecker 1970; Rammner 1977). For many details of foraging behaviour during invasion in northern Germany, see Kirsch (1992).

On breeding grounds in winter, completely dependent on supply of cached seeds and nuts. These transported in sublingual pouch, which expands to maximum size of *c.* 15 cm$^3$ in autumn when storing activity at peak (Jung 1968; Turček and Kelso 1968). Caches are generally on ground in soft soil, under leaf litter, moss, etc., or on trunks of fallen trees or moss-covered rocks; always covered after deposition with leaf, stick, stone, etc. Often placed near object like tree trunk, stone, seedling tree, or tuft of grass, generally in more open areas in clearings, at forest edge, or just above treeline, and often under shelter of branches where snow cover thinnest in winter; rarely above ground, though 7 m up in tree recorded as well as on roof. (Dementiev and Gladkov 1954; Creutz and Flössner 1958; Mezhennyi 1964; Jung 1966; Blana 1970; Löhrl 1970; Crocq 1990, which see for major study of caching; Knolle 1991.) In Siberia, starts to take *P. sibirica* seeds for storing as soon as cones begin to open in August; whole areas of forest suddenly stripped of cones in about a week; up to 167 seeds (218 of *P. pumila*) recorded in pouch; to ensure sufficient supply for September–May, including feeding young and allowing for losses to rodents, etc., each bird probably caches *c.* 100 000 seeds (*c.* 27 000 required by 1 adult); see also below for daily food intake; 40 000–250 000 have been found stored per ha (Jung 1966; Turček and Kelso 1968; Crocq 1990). *P. sibirica* seeds carried up to 16 km to cache (Mezhennyi 1964), and *P. cembra* in Switzerland up to 12 km (Sutter and Amann 1953). Seeds often taken to temporary store where 40–60 cached then distributed among final stores which contain 2–50; average 11·7 in 39 caches in Siberia (Mezhennyi 1964; Jung 1966; Turček and Kelso 1968). Caches in Siberian forest made in straight line at intervals of 25–100 cm, 0·5–4 cm deep (Mezhennyi 1964), those by Siberian birds in Germany storing hazel nuts 15–20 cm apart in 8–10 cm of snow (Wernli 1970). In southern French Alps, birds made up to 17 caches in 3·5 min from 1 pouch-load, dropping 2–12 seeds into each hole (Crocq 1990). Study of captive bird showed it probably carried seeds from only 1 cone at a time, caching them in several stores; bill inserted in soil then opened, allowing some seeds to fall, accompanied by regurgitation actions, and hole then covered; apparently unable to empty full pouch in one action; unripe cones sometimes buried since seed easier to crack open later (Löhrl 1970). Probable that each bird has, and knows, its own stores (Andreev 1982a; Crocq 1990), but see Goodwin (1986) for reference to possible communal stores. Siberian birds in Germany observed making caches in autumn which they used in winter (Boecker 1970). See Turček and Kelso (1968) for estimates from literature of proportion of seed crop taken and stored, number of unretrieved seeds which sprout, and many other aspects of food transportation and storage. In one study, 65% of birds in 125 observations went straight to exact spot where food cached, and 78% of caches found eventually; other studies gave retrieval rate of 82–90%; some retrievals made after 17 months, and caches often found under snow well over 1 m deep (Swanberg 1951b, 1981; Crocq 1990, which see for major discussion). Crocq (1990) suggested that, after finding approximate area of cache by visual orientation among trees and other objects at cache site, birds may be able to pinpoint cache by taste as they dig through snow with open bill. In winter, north-east Siberia, one bird found 8 stores in 15 min (Andreev 1982a). See Balda (1980) and Vorobiev and Kaganova (1980) for retrieval experiments with captive birds. Storing of hazel nuts begins in August, peaking mid-September; record of bird with 26 nuts in pouch but 10–16 more usual, and can carry up to 6 in bill; will fly up to 12 km from trees to store (Swanberg 1951b; Sutter and Amann 1953; Turček and Kelso 1968; Wernli 1970). Generally *c.* 3–18 nuts per store; one Siberian bird in Germany had 62 in 1 store; nuts stored in shell, at depth of bill in soil or pushed under moss (etc.) like pine seeds (Swanberg 1951b; Creutz and Flössner 1958; Piechocki 1971a). In Sweden, food collected either inside or outside nesting territory, but always stored within it (Swanberg 1956a). See also Crocq (1990) for discussion of territory and stores. Rarely, will cache mammals and birds, or parts of them (Kollibay 1913; Piechocki 1971a).

Diet in west Palearctic includes the following. Vertebrates: mole *Talpa europaea*, shrew *Sorex araneus*, squirrel *Sciurus vulgaris*, voles *Microtus agrestis*, *M. brandtii*, harvest mouse *Micromys minutus*, mouse *Mus musculus*, rat *Rattus norvegicus*, frog *Rana*, toad *Bufo*, slow-worm *Anguis fragilis*, lizard *Lacerta vivipara*, passerines up to size of thrush *Turdus*. Invertebrates: damsel flies and dragonflies (Odonata); grasshoppers, etc. (Orthoptera: Gryllidae, Tettigoniidae, Gryllotalpidae, Acrididae), earwigs (Dermaptera: Forficulidae), cockroaches (Dictyoptera: Blattidae), bugs (Hemiptera: Psyllidae), adult and larval Lepidoptera (Pieridae, Noctuidae, Geometridae), adult and larval flies (Diptera: Tipulidae, Calliphoridae, Muscidae), adult and larval Hymenoptera (Tenthredinidae, ants Formicidae, wasps Vespidae, bees Apoidea), adult and larval beetles (Coleoptera: Carabidae, Silphidae, Staphylinidae, Geotrupidae, Scarabaeidae, Byrrhidae, Elateridae, Cantharidae, Tenebrionidae, Coccinellidae, Cerambycidae, Chrysomelidae, Curculionidae, Scolytidae), spiders (Araneae: Araneidae, Linyphiidae, Lycosidae), woodlice (Isopoda), earthworms (Lumbricidae), slugs and snails (Gastropoda: Limacidae, Arionidae, Helicidae). Plant material includes exotic and agricultural species eaten during invasions: seeds, nuts, fruit, and occasionally buds of *Pinus cembra*, *P. sibirica*, *P. peuce*, *P. mugo*, *P. strobus*, *P. nigra*, *P. sylvestris*, *Picea abies*, *P. pungens*, *P. orientalis*, *P. engelmannii*, *Abies alba*, *A. nordmanniana*, juniper *Juniperus*, yew *Taxus*, hemlock *Tsuga*, larch *Larix*, Lawson cypress *Chamaecyparis*, oak *Quercus*, beech *Fagus*, hazel *Corylus*, hornbeam *Carpinus*, sweet chestnut *Castanea*, horse chestnut *Aesculus*, lime *Tilia*, sycamore *Acer*, birch *Betula*, alder *Alnus*, ash *Fraxinus*, box *Buxus*, holly *Ilex*, walnut *Juglans*, elder *Sambucus*, honeysuckle *Lonicera*, rowan *Sorbus*, apple *Malus*, pear *Pyrus*, blackthorn, cherry, plum, and apricot *Prunus*, hawthorn *Crataegus*, *Cotoneaster*, raspberry, bramble, and cloudberry *Rubus*, rose *Rosa*, strawberry *Fragaria*, spindle *Euonymus*, mulberry *Morus*, buckthorn *Rhamnus*, seabuckthorn *Hippophae*, alder buckthorn *Frangula alnus*, privet *Ligustrum*, heather *Calluna*, various species of *Vaccinium*, crowberry *Empetrum*, snowberry *Symphoricarpos*, mistletoe *Viscum*, *Laburnum*, currant *Ribes*, guelder rose *Viburnum*, grape *Vitis*, nightshade and potato *Solanum*, hemp *Cannabis*, sunflower *Helianthus*, lupin *Lupinus*, beans *Phaseolus*, peas *Pisum*, mallow *Malva*, melon *Cucumis*, turnip *Brassica*, *Asparagus*, maize *Zea*, rye *Secale*, wheat *Triticum*, various fungi. (Csiki 1913; Madon 1928b; Schuster 1930; Grote 1947; Pfeifer 1954; Taapken *et al.* 1955a; Creutz and Flössner 1958; Kumari 1960; Turček 1961; Tricot 1968; Boecker 1970; Erard 1970; Hollyer 1970, 1971; Piechocki 1971a, b; Tietze 1971; Uloth 1977; Litun and Plesski 1983; Goodwin 1986; Lanner and Nikkanen 1990.)

Table A Food of adult Nutcracker *Nucifraga caryocatactes macrorhynchos* during irruption in Germany, autumn and winter 1968–9. Figures are percentages by number of feeding observations; animal and plant material (seeds, fruit) treated separately.

| | E Germany | W Germany (Nordrhein–Westfalen) |
|---|---|---|
| Vertebrates | 11·1 | 6·7 |
| Orthoptera | 5·1 | 2·7 |
| Forficulidae | 2·2 | 1·0 |
| Lepidoptera | 5·9 | 1·8 |
| Hymenoptera | 27·5 | 25·0 |
| Coleoptera | 22·6 | 12·9 |
| Other insects | 18·3 | 21·4 |
| Snails | 2·9 | 4·0 |
| Earthworms | 2·2 | 11·2 |
| Total no. of observations | 137 | 224 |
| *Corylus* | 47·5 | 30·1 |
| *Prunus* | 13·4 | 11·3 |
| *Sambucus* | 6·4 | 6·0 |
| *Malus* | 4·4 | 4·5 |
| *Vaccinium* | 3·9 | 1·5 |
| *Pyrus* | 3·6 | 1·9 |
| *Fagus* | 3·3 | 1·1 |
| *Quercus* | 3·0 | 2·6 |
| *Rubus* | 3·0 | 4·1 |
| *Sorbus* | 2·6 | 3·0 |
| *Hippophae* | 2·6 | — |
| Conifers | 2·3 | 6·0 |
| *Rosa* | — | 1·1 |
| Agricultural crops | — | 3·8 |
| Total no. of observations | 387 | 266 |
| Source | Piechocki (1971*a*, *b*) | Boecker (1970) |

See Table A for feeding observations during irruption in Germany. Of 742 animal prey items in 54 stomachs of *macrorhynchos* during irruption in eastern Germany, August–September, 52·7% by number adult and larval Hymenoptera (including 30% wasps, 13% ants), 34·1% adult and larval beetles (including 13% Carabidae, 10% Scarabaeidae), 6·3% Lepidoptera (including 6% Noctuidae caterpillars), 2·7% earwigs, 2·2% spiders, 1% Diptera, and 1% voles *Microtus*; birds also took many seeds from berries (note absence of hazel nuts, acorns, and beechmast in this study); almost all prey from soil, ground layer, herb layer, and decomposing wood; Siberian birds erupting westwards have little chance of survival if restricted to such a diet. (Tietze 1971.) Of 342 items of animal prey in 59 stomachs of nominate *caryocatactes* collected throughout year in Hungary, 67·5% by number beetles (45% Curculionidae, 13% Scarabaeidae), 18·7% Orthoptera (16% Gryllidae), 9·6% Hymenoptera (6% ants, 4% wasps), 2·6% larval and pupal Lepidoptera, 0·6% Hemiptera (Pentatomidae), 0·6% earwigs, and 0·3% vole; 17% of stomachs contained seeds and nuts only, and 20% contained animal material only; proportion of plant material rose quickly after August, as did number of Orthoptera and Carabidae taken September–October (Csiki 1913, which also see for 40 autumn stomachs of *macrorhynchos* in Hungary). Of 37 stomachs from 2 separate studies of *macrorhynchos* during irruptions in west European USSR, 43% contained Scarabaeidae (1 stomach with 50 *Aphodius*), 27% Geotrupidae, 30% Curculionidae, 19% Carabidae, 3% Silphidae; 27% contained wasps, 8% bees; 16% contained Lepidoptera, 14% Acrididae, 11% Diptera, 3% earwigs, 3% Odonata, 19% spiders, 3% snails, 3% birds, 3% frogs, 49% seeds of *Rosa*, 38% hazel nuts, 32% rowan berries, 11% acorns, 5% *Rubus*, 5% oats, and 3% *Vaccinium*; plant material in one study 58% by volume, in other 48% (Kumari 1960; Litun and Plesski 1983). See Reymers (1954) and Mezhennyi (1964) for extralimital Siberia, and Andreev (1982*a*) for 49 stomachs from north-east Siberia, mid-winter, all of which contained *Pinus pumila* seeds and 10% contained voles. In France, of 53 irrupted birds, 75% contained earwigs, 34% beetles (especially Geotrupidae), 17% Hymenoptera, 55% contained hazel nuts, and 45% berries (Erard 1970); see also Madon (1928*b*) and Tricot (1968). For Germany, see Rinne and Bauch (1970) and Kirsch (1992); for Britain, see Hollyer (1970, 1971). In eastern Siberia often feeds on spruce: 87% by volume of items in 23 stomachs were seeds of *Picea schrenkiana* (Grote 1940). In southern French Alps, dietary variation has 3 phases over year: in winter hardly any animal prey taken, relies on stored seeds; with start of breeding season towards late February increase in amount of animal material, which is also fed to nestlings, though proportion depends on weather, extent of stores, etc., with peak of *c*. 80% of diet during June–July; after young fledged in late summer, decline in animal material as storing begins, falling to zero around November (Crocq 1990, which see for change in foraging habitats over year). In Siberia, June–July, larval and adult insects account for over 50% of diet (Mezhennyi 1964).

In French Alps average individual consumption 113 *Pinus cembra* seeds per day in winter (78–140, $n = 6$ estimates) weighing *c*. 22 g or 13% of body weight (Crocq 1990). Captive bird in Siberia ate 170–240 *P. pumila* seeds per day, equivalent to 70–90 *P. sibirica* seeds (Mezhennyi 1964). These figures very similar to those of Andreev (1982*a*) for mid-winter diet in north-east Siberia from 5 birds in wild; energy expenditure in these very low temperatures much less than theoretically expected in seed-eating bird of this size, so needs relatively little food (Andreev 1977).

Little information on nestling diet; collar-sampling made difficult by parents' habit of inserting bill deep into nestling's throat during feeding by regurgitation to avoid sublingual pouch. 31 samples taken from throats of young (some material probably swallowed) in 8 nests over year in southern French Alps gave *c.* 70 *P. cembra* seeds, *c.* 12 spiders, 8 beetles (mainly Carabidae and including 1 larva), 1 Diptera, 1 earthworm, remains of vole *Microtus agrestis* and squirrel *Sciurus vulgaris*; pine seeds brought with husks removed in all months April–June, in first week broken up but later fed whole; animal food probably indispensable in nestling diet, though proportion varies with weather, success of autumn caching, etc. Each young at 10–20 days receives *c.* 60 *P. cembra* seeds daily (*c.* 12 g) (Crocq 1990, which see for development and feeding of young). In one study in Siberia, contents of nestling stomachs 90–100% by number arthropods; at 7–10 days, less than half of seeds in diet from new growth, remainder from caches (Mezhennyi 1964). According to Reymers (1959), however, diet in Siberia 100% *P. sibirica* seeds. In Sweden, nestlings and fledglings almost wholly dependent on hazel nuts (Swanberg 1981). One nestling fallen from nest in Switzerland had crop full of *P. cembra* seeds (Kunz-Plüss 1969).                           BH

**Social pattern and behaviour.** Early breeding in difficult, often snow-bound terrain and generally secretive habits mean many aspects little known. Most comprehensive study and review of literature by Crocq (1990) for nominate *caryocatactes*; account includes data (some extralimital) for *macrorhynchos*.

1. Flocks occur during autumn seed-gathering period and for migration, but basically solitary or in pairs and sedentary on territory throughout year; even birds dispersing in winter often in pairs. Territory (see Breeding Dispersion, below) contains food stores (see Food) vital for winter survival and important also for rearing of young. (Steinfatt 1944; Swanberg 1956a, 1981; Rudat and Rudat 1978; Crocq 1990.) Adults normally faithful to mate and territory throughout year and (probably) life (see Bonds and Breeding Dispersion, below). After attaining independence around early July, young have *c.* 6–8 weeks in which to establish territory (food-storing and wintering site). Birds ringed after seeds stored also site-faithful, moving up to *c.* 2 km from regular roost in winter. (Mattes and Jenni 1984.) In Omolon valley (66°30′N, north-east Russia), some birds overwinter successfully (probably adults which have already done so) in valley territories with dense shrubs and trees; juveniles, and adults whose food stores exhausted, stay initially near mountains, but sooner or later disperse, timing of dispersal and proportions in each group evidently determined by seed crop (Andreev 1982a). Juveniles may spend winter up to early spring on edge of adults' territory and (in years when seeds plentiful) tend to settle eventually where they have laid up stores (Crocq 1990). From July or August and through autumn, adults and (increasingly) juveniles spend most of day collecting nuts or seeds for storing. Collecting sites generally outside territory for hazel *Corylus* nuts (Swanberg 1951b, 1956a; Lefranc and Pfeffer 1975); in Alps (seeds of pine *Pinus cembra* main food), some overlap between collecting site and territory (Crocq 1990). Birds may commute several km (maximum 15 km, with 700 m altitudinal difference, in Switzerland) between roost or larders and collecting sites (Steinfatt 1944; Swanberg 1951b, 1956a; Sutter and Amann 1953; Lefranc and Pfeffer 1975; Andreev 1982a; Mattes and Jenni 1984). Family parties (sometimes several united together) and flocks of varying size (mainly up to *c.* 20 in Europe, up to several hundred recorded in USSR) occur in various parts of range from June or July and through autumn, i.e. during seed-gathering and -storing period; mostly solitary or in pairs again with approach of winter;

flocks of juveniles (e.g. *c.* 10–20, especially after good breeding season and seed-crop), perhaps of non-breeders, also reported (Grote 1947; Sutter and Amann 1953; Dementiev and Gladkov 1954; Glutz von Blotzheim 1962; Mezhennyi 1964; Lefranc and Pfeffer 1975; Rudat 1984; Crocq 1990). During periodic irruptions (principally of Siberian *macrorhynchos*: see Movements), typically loose-knit flocks tend to be largest (up to 300) at start of invasion, subsequently breaking up (Černý 1946; Taapken *et al.* 1955b, 1957; Kumari 1960, 1972; Jung 1966; Erard 1970; Piechocki 1971a). Then generally solitary or in small groups: e.g. in northern Germany, 6–10, but birds several tens of metres apart; some 'pairs' (or adults with young) (Boecker 1970). Of 4500 passing in 160 min, east coast of Sweden, early August, most in compact flocks of 100 or more (Eriksson and Hansson 1973). On migration elsewhere (e.g. Col de Bretolet, Switzerland), usually solitary or in twos, though groups of 3–25 recorded (Mattes and Jenni 1984); compact flock of 121 in October (Blasius 1886). Occasionally associates temporarily (including for migration) with Jay *Garrulus glandarius*, e.g. during invasions of *macrorhynchos* which may also then mingle with nominate *caryocatactes* (Heidemann and Schüz 1936b; Creutz and Flössner 1958; Berndt and Moeller 1960; Boecker 1970; Piechocki 1971a); see also Antagonistic Behaviour (below). BONDS. Mating system monogamous and, though difficult to determine, pair-bond apparently long-term, perhaps life-long (Steinfatt 1944; Swanberg 1951b, 1956a). For ♂♂ re-pairing on territory, and record of ♀ effecting divorce by fighting to regain her mate, see Breeding Dispersion (below). Study of dominance hierarchy at feeding station in Austria showed that ♂ of pair dominated ♀ (A Grabher Meyer). Unlike in other Corvidae, incubation is by both sexes (Swanberg 1956b; Reymers 1959; Crocq 1990); average in one study 32% by ♂ (A Grabher Meyer). Both sexes also brood and feed young, ♂ brooding more than incubating, and also feeding young more during first 7 days than later (Steinfatt 1944; Reymers 1959; A Grabher Meyer). Period of dependence conspicuously long: e.g. young recorded being fed at 105–114 days, then dispersing during food-storing period (Swanberg 1956a); estimates in other studies 60–90 (Simon *et al.* 1983) and 105–119 days (Rudat 1984), with juveniles from late broods recorded still begging close to onset of winter (Crocq 1990). On Öland island (Sweden), birds perhaps paired up at 1 year but did not establish territories or attempt to breed owing to lack of suitable habitat. Probable age of first breeding 2 years or older; no further information on 1-year-olds nor on proportion of non-breeders in population. (Swanberg 1956a; Rudat 1984.) BREEDING DISPERSION. Solitary and territorial. In *Pinus cembra* forest, French Alps, mostly 300–350 m (80–410) between neighbouring nests, or 250–300 m in another plot of *P. cembra*, larch *Larix*, and spruce *Picea* (Crocq 1990); often barely 50 m apart in optimal habitat, Austria (A Grabher Meyer); in Thüringen (eastern Germany), 2 territories *c.* 500 m apart (Uloth 1977). See also Mezhennyi (1964) for USSR. Austrian study showed birds basically resident within home-range of *c.* 5–10 ha (up to 50% overlap with neighbours), and much smaller territory (radius of *c.* 25–40 m from nest-tree) effective only for breeding season and defended against all intruders, though immediate neighbours are often only threatened and not so consistently evicted as strangers which are always vigorously attacked and usually chased beyond territory limits (A Grabher Meyer). Reports of up to 4 different birds within less than 50 m of occupied nest (Lefranc and Pfeffer 1975) may thus refer to owners and neighbours. Further suggestions that only small nest-area territory is defended in Schönbeck (1956) and Crocq (1974); *G. glandarius* also attacked only if ventures near nest, and not in all parts of territory (Swanberg 1956a). In French

Alps, territories generally smaller when more clustered in favoured habitat on slope (Crocq 1990). In Sweden, average size of home-range in spruce forest 13·2 ha (11·2-14·9, *n* = 4). Defended (to limited extent) by both sexes, perhaps mainly when pairing up and establishing new home-range. Proclaimed for greater part of year and thus presumably well known to neighbours which respect ownership with or without encounters. Trespassing not unusual, but intruders not attacked when resting in another's home-range. Tolerated also during 'friendly visits' made, especially in early morning before or during nest-building, for Ceremonial Gatherings (see Flock Behaviour, below). Trespass thus more for social contact and represents no threat to food-stores whose location known anyway only to owners. When nuts placed experimentally in home-range, intruders are attacked but not driven off. (Swanberg 1951*b*, 1956*a*; Mattes 1978; Rudat 1984; Crocq 1990.) See Antagonistic Behaviour (below). After fledging, boundaries less strict and (e.g.) flocks of juveniles may move through home-ranges where adults still feeding late broods, and in Thüringen all members of 2 families moved freely over their adjacent home-ranges (Uloth 1977; Mattes 1978; Crocq 1990). Home-range serves most conspicuously for food-storing, but area suitable for breeding more important. Birds sing, copulate, and nest in home-range, but considerable amount of food gathered outside it: nuts and seeds collected in neutral feeding area (Swanberg 1956*a*, which see for advantages of food-stores in home-range). In Rominter Heide (Poland/Russia), birds forage within *c*. 50-80 m of nest in breeding season (Steinfatt 1944); in Vosges (France), within *c*. 1 km (Lefranc and Pfeffer 1975), and larders also sometimes up to *c*. 1 km from nest (Rudat and Rudat 1978). Sole report of distance moved following nest-failure refers to Harz mountains (Germany) where replacement nest *c*. 150 m from 1st (Kleinschmidt 1909-11). Home-range occupied throughout year and for many years in succession: Swedish pair held same home-range for at least 14 years, and no attempts by other birds to usurp even part of it; ♂ of same pair site-faithful (with new mate) also in following year (Swanberg 1951*b*, 1956*a*); see also Mezhennyi (1964), Crocq (1974), Mattes (1978), and Rudat (1984). 3 cases of ♂♂ retaining territory after losing mate, then re-pairing (Swanberg 1956*a*). In Oberengadin (Switzerland), 6 birds site-faithful for up to 9 years, and no case of adult changing home-range; in one case, ♀ absent from feeding place and home-range during winter, but returned and fought successfully to win back home-range and mate (Mattes 1978; Mattes and Jenni 1984); see also Antagonistic Behaviour (below). In Austria, if 2 or more successive failures of *P. cembra* cone crop (stored food insufficient to raise young), birds sometimes quit home-range, moving up to 120 km; all home-ranges re-occupied in following summer if food-supply adequate (A Grabher Meyer). Unpaired birds (e.g. ♀ before she replaced widowed ♂'s mate in following year) apparently non-territorial (Swanberg 1956*a*). Density difficult to assess owing to difficult terrain (snow, etc.), unobtrusive habits when breeding, difficulty in locating nests, and inadequate knowledge of territorial calls; also, birds travel well beyond limits of home-range in search of seeds; best times for census are before incubation and after fledging (Glutz von Blotzheim 1962; Rudat 1984). Highest density in Switzerland and France in mixed *P. cembra*-*Larix* forests of middle and upper coniferous forest belt, high in central *P. cembra* forest, low in peripheral spruce forest: 2-10 pairs per km² in French Alps, 15-23 pairs per km² in Oberengadin (Glutz von Blotzheim 1962; Mattes and Jenni 1984; Crocq 1990); in Vosges (France), 4 pairs recorded along 800 m and 4 along 1 km (Lefranc and Pfeffer 1975). Values from Germany include 0·20-0·25 pairs per km² in Harz (Haensel 1970), and 1 pair per km² in Thüringen (Rudat and Rudat 1971, 1978).

In Bulgaria, density ranges from 3 pairs per km² in predominantly Scots pine *Pinus sylvestris* forest to 74 pairs per km² in predominantly Himalayan pine *P. excelsa* (Simeonov 1975). On Åland islands (Finland), 1 pair per km² in spruce woods, less in mixed conifers (Palmgren 1930); on Sottunga (Åland islands) food-stores of 40 pairs widely dispersed over *c*. 360 ha, so density presumably 11 pairs per km²; strong evidence that in areas good for breeding but with limited hazel, number of home-ranges restricted by nut supply (Swanberg 1956*a*). For densities (birds per km²) in extralimital Russia, see (e.g.) Reymers (1966), Ravkin (1984), Vartapetov (1984), and Rogacheva (1988). ROOSTING. Nocturnal and apparently mostly solitary; group of 12 non-breeding 1-year-olds, Öland (Sweden), perhaps roosted communally (Swanberg 1956*a*). In Omolon valley, April, pairs recorded roosting close together, but on separate branches of same tree (Andreev 1982*a*); ♂ not in immediate vicinity when ♀ roosting on nest at night (Steinfatt 1944). In eastern Germany, when immigrants stayed long in one area, regularly roosted in wood after feeding in town (Creutz and Flössner 1958). Site chosen typically in dense conifer or other tree (e.g. Creutz and Flössner 1958, Hollyer 1970); in Anadyr' (eastern Russia), early winter, used nest of Magpie *Pica pica* (Portenko 1939). Of 19 birds in Omolon valley, midwinter, 10 roosted in larch, 6 in crown of willow *Salix*, 1 in dense matted witch's broom of willow *Chosenia*, 2 in accumulated jetsam under river bank, same site being used for many days; site typically protected by branches and snow roof and bird pressed close to trunk (Andreev 1982*a*); presumably adopts posture with ruffled plumage and bent tarsi as described by Blana (1970) and Portenko (1939) for daytime rest periods of 10-15 min (usually around midday when gathering seeds: Jung 1966) and nocturnal roosting. In Omolon valley, 2½-3 hrs per day for feeding in winter; come to roost with stomach and gullet full of pine seeds (some also in bill); before roosting and in morning, rest on tree-tops and call (total 25-30 min); roosting proper 18½-20½ hrs; 10-15 min taken up by flights to feeding grounds (Andreev 1982*a*). In Scandinavia, autumn, birds collecting hazel nuts generally fly to food-source before sunrise (Swanberg 1951*b*).

2. Expresses excitement by rapidly flicking tail open and shut, flashing white feathers, and creating curious effect if bird hidden amongst dark foliage; often combined with Wing-flicking, and may also give Rasp-call (see 2a in Voice) with head stretched forward (Naumann 1905; Stübs 1958; Wernli 1970; Ali and Ripley 1972). Throughout range, typically quiet and secretive when breeding (e.g. Donner 1908, Grote 1947, Mezhennyi 1964); nest-building birds fairly confiding (Selous 1907), but may fly about excitedly, call and Bill-snap (Kleinschmidt 1909-11: see Voice). During invasions, Siberian *macrorhynchos* remarkably tame (e.g. eating out of hand), more so than nominate *caryocatactes* and *G. glandarius* (Heidemann and Schüz 1936*b*; Taapken *et al.* 1955*b*). In Germany, 72% of records of *macrorhynchos* involved birds allowing approach to 5·5 m or less, 36% to 2·5 m or less (Boecker 1970); in Switzerland, more cautious and alert only when joined by nominate *caryocatactes* which never allowed approach closer than 8-10 m (Wernli 1970). In European USSR, invading nominate *caryocatactes* could be approached within *c*. 3 m in forest, 7-8 m in the open (Litun and Plesski 1983); in Poland, birds scavenging at tourist facilities in mountains tolerated man within 2-3 m (Pannach 1983). Astonishingly active and noisy when gathering seeds (Jung 1966); birds disturbed while feeding will fly off, calling loudly (Pfeifer 1954; Holtmeier 1966). *N. c. macrorhynchos* may show little or no reaction to cats and dogs (Blana 1970; Boecker 1970), but reported sometimes to give loud and long series of Rasp-calls and to attack these, also crows *Corvus*, Buzzard *Buteo buteo*,

Kestrel *Falco tinnunculus*, Pygmy Owl *Glaucidium passerinum* (but fled screeching when attacked by this species whose larder it was raiding), and Little Owl *Athene noctua*; also chased Blackbird *Turdus merula*, and Great Spotted Woodpecker *Dendrocopos major* from its anvil (Taapken *et al.* 1955b; Creutz and Flössner 1958; Thönen 1965; Sadlik and Haferland 1981); for further details, including attack on man, see Piechocki (1971a). Bird flying at *c.* 100–300 m with seeds will dive down into forest if threatened (Mezhennyi 1964). FLOCK BEHAVIOUR. Most conspicuous feature is Ceremonial Gatherings which perhaps facilitate meetings of unpaired birds: take place immediately before and during nest-building and also in late summer. Up to 10–12 birds assemble in a home-range, owners participating or watching and showing no aggression (though one record of fight). Few details of displays, but in glades and on frozen, snow-covered river in USSR in 2nd half of April, 5–12 birds seen performing bowing movements apparently as form of courtship (Mezhennyi 1964); see also subsection 2 in Heterosexual Behaviour (below). Wide variety of calls (see 7 in Voice) accompanies display. (Swanberg 1951b, 1956a; Mezhennyi 1964; Kipp 1978.) In northern Baykal region (eastern Russia), flocks often fly high and perform aerial evolutions ('play') before diving down with closed wings; also approach roost at some height and make rapid angled descent, with rushing noise from wings (Stegmann 1936). SONG-DISPLAY. No loud territorial song. Function probably fulfilled by harsh and far-carrying Rasp-call, given from tree-top, especially at dawn and dusk in calm conditions, with head raised, bill wide open, tail and wings trembling (Swanberg 1951b; Géroudet 1961; Ptushenko and Inozemtsev 1968). ♂ sings (see 1 in Voice) usually from low, concealed perch; bill pointed up, with head and neck extended and sometimes moved side to side or up and down; rear body lowered while legs bent and wings may be drooped; throat feathers ruffled (Naumann 1905; Swanberg 1951b; Creutz and Flössner 1958; Stübs 1958; Rammner 1977). In Swedish study, sang during few weeks before and during building; apparently functions as courtship-song, perhaps helping to synchronize ♂-♀ sexual cycles (Swanberg 1951b). Captive *macrorhynchos* gave rudimentary song during January–February, with peak (especially when sunny) March–April, waning May and ceasing (after moult) June (Rammner 1977). Similar song-period indicated by Creutz and Flössner (1958) and Berndt and Moeller (1960). ANTAGONISTIC BEHAVIOUR. (1) General. Conflicts rare, especially between established neighbours (Crocq 1990). (2) Territorial advertisement, threat, and fighting. Rasp-call given apparently to advertise home-range, birds starting to call simultaneously in morning, with peak just before nest-building; most calling along boundary facing favourite places of other conspecific birds; also call when other *N. caryocatactes* passing over (during period when newcomers seeking to establish home-range), and chases occur at this time (Swanberg 1956a; Crocq 1990). According to Crocq (1990), Rasp-call used more for intrusions by man or another animal than for conspecific birds. Playback of Rasp-call elicits immediate approach, silent or accompanied by same call (Lefranc and Pfeffer 1975). In apparent territory-marking display, sometimes triggered by rival on boundary, ♂ ruffles breast and belly feathers (bird becoming pear-shaped), droops wings, and hides bill in belly feathers; moves in circle or wide loop across snow, thus forming territorial markers (A Grabher Meyer). In rare fight, ♀ regained home-range and mate usurped by 2nd ♀ during her absence; 2 ♀♀ on ground *c.* 12 min, grappling and pecking (Mattes 1978; Crocq 1990). When nuts placed experimentally in home-range, intruders are attacked (usually silently), but owners make no attempt to evict them (Swanberg 1956a). In Omolon valley, winter, bird uncovering food-store chased away another which

approached (Andreev 1982a). Study at feeding station in Austria revealed existence of dominance hierarchy: lowest-ranking bird still managed to feed successfully by never moving far from food and by distracting others with alarm-calls so as to reach food unhindered; high-ranking individuals attacked especially often by dominant birds would then chase lower-ranking birds and Nuthatches *Sitta europaea* until these dropped food (A Grabher Meyer). Apparently dominant over *G. glandarius* (e.g. Boecker 1970), which is attacked near nest, and generally ignored outside breeding season (Swanberg 1956a). Seen chasing Siberian Jay *Perisoreus infaustus* in eastern Russia (Reymers 1954). 2–3 birds recorded persistently harrying squirrel *Sciurus* with pine cone, not quite pressing home attack (Grote 1947). HETEROSEXUAL BEHAVIOUR. (1) General. Difficult to observe pair-formation in view of apparent life-long pair-bond and long-term retention of home-range (Swanberg 1956a). In French Alps, birds more often together in home-range from February, giving quiet contact-calls; rarely apart during nest-building (Crocq 1990). In Baykal area, pair-formation said by Reymers (1954) to take place in March, and most birds in pairs by mid-month. (2) Pair-bonding behaviour. Pair seen in chases (e.g. late April and early June in Germany), pursuer (not known which sex) copying all manoeuvres (fast descents, sharp twists and turns) of first, with noise created by wings (Creutz and Flössner 1958; Crocq 1990). Pair-bond said by A Grabher Meyer to be strengthened by foraging and collecting nuts rather than by ritualized allopreening; also one watching while other hides food or forages, visual and acoustic recognition of mate even at some distance being important for contact. Other brief descriptions of displays (probably some overlap) as follows. At Lake Baykal, mid- to late March, apparently displaying birds hopped in straight line or spiral on flat, open, snow-covered surface, sometimes around small sapling; also perched on tall dead stumps, pecking and tossing away wood chips (Reymers 1959). In German study, gave great variety of different sounds (perhaps song), 'shrugged shoulders', raised wings slightly, and ruffled crop feathers (Creutz and Flössner 1958). Pattering on branch and mutual bowing noted between paired birds in April by Jung (1966). 'Tree-display' lasting *c.* 3 min and with all neighbours present and following events, calling loudly, mentioned (without further details) by A Grabher Meyer (see also subsection 4, below). Captive ♂, feathers on head and neck sleeked, those on underparts ruffled, tail fanned and brushing ground, seen circling apparently indifferent ♀ while calling; another captive ♂ faced ♀ with body feathers ruffled, head raised and tail horizontal (Noll 1956; Crocq 1990). (3) Courtship-feeding. Unlike in other Corvidae, ♂ shares incubation, allowing ♀ to feed herself (Swanberg 1956b; Reymers 1959; Crocq 1990); for claims or suggestions (presumably exceptional cases) that ♂ nevertheless feeds ♀, see Bartels and Bartels (1929), Mezhennyi (1964), Jung (1966), and Crocq (1990). (4) Mating. Regularly takes place in tree following display; ♀ immediately afterwards flies into nearby tree and calls (probably Rasp-call), both birds then preening vigorously; copulation rarely follows display on ground (A Grabher Meyer). Hoarse croaking crescendo (probably Rasp-call) given by ♂ apparently attempting to copulate with inanimate object (Rammner 1977). (5) Nest-site selection. Not clear how site selected. Nest built by both sexes (Bartels and Bartels 1929; Reymers 1959); lined almost exclusively by ♀ in one study (A Grabher Meyer). Most material collected near nest, sometimes up to *c.* 200 m away; when bad weather interrupted building of one nest, new one built directly below it (Crocq 1990). (6) Behaviour at nest. ♀ (presumably laying) gave harsh call, not allowing ♂ near (Reymers 1959). Both sexes incubate. ♂ may guard nest while ♀, usually very low in nest (Fig A), incubating (Mezhennyi 1964;

A

Jung 1966). Inconspicuous arrival and departure typical (e.g. Matz 1967). ♀ leaves for long periods to feed (she alone knows location of her own caches: see Food), ♂ then taking over; nest-relief ceremony may involve bill-movements by sitting bird (Swanberg 1956*b*: see Fig B), or change-over very rapid (espe-

B

cially in dull conditions) and without ceremony, though incoming bird will give quiet calls (Steinfatt 1944; Reymers 1959; Crocq 1990). RELATIONS WITHIN FAMILY GROUP. Young brooded (by both sexes) up to *c*. 14–15 days, not normally thereafter except in bad weather; night-time brooding longer than 14 days according to Steinfatt (1944); left uncovered for short spells from 7–14 days during which thermoregulation develops; by end of this period also generally more active, but no longer gape when nest-tree shaken, etc. (A Grabher Meyer; also Bartels and Bartels 1929, Reymers 1959, Rudat and Rudat 1971, Crocq 1990). Eyes of chicks fully open at 12 days (Reymers 1959, which see for physical and behavioural development; see also Crocq 1990). Young fed at change-over. Stretch up neck and give quiet food-calls. Adult may also encourage gaping by giving Purring-call (see 6 in Voice). Feeding quite a lengthy process, parent making shivering-swallowing movements, transferring seeds separately from sublingual pouch to bill-tip and thrusting bill well down into chick's throat (Fig C), ever deeper as young grow—this to

C

avoid food going into chick's sublingual pouch. Nestling faeces and other detritus swallowed or carried away by adults. (Bartels and Bartels 1929; Steinfatt 1944; Reymers 1959; Matz 1967; Crocq 1974, 1990.) In eastern Germany, pair usually fed young together, though one apparently dominant (Rudat and Rudat 1971); in another study, pair maintained contact with soft croaking calls, but never seen at nest together (Matz 1967). Young fledge at *c*. 23–28 days (Steinfatt 1944; Reymers 1959; Crocq 1974; Rudat 1984). According to Korodi Gál (1969), spend 20 days in nest, then 5–6 in tree nearby. Austrian study similarly found young to move onto nest-rim at 21 days, parents then attempting to lure them from nest from 25 days, successfully at 28 days; not able to fly well at first and spend several days in nearby trees, gradually moving away (A Grabher Meyer), though some reports of family remaining in general vicinity of nest for up to 3–4 weeks (Rudat and Rudat 1978; Rudat 1984). Wing-flap and give increasingly loud food-calls in a crouched posture when begging, e.g. on ground as parent uncovers cached seeds (Schüz 1932; Pfeifer 1954; Rudat and Rudat 1978; Crocq 1990). Adult recorded carrying chick *c*. 10 days old in feet in flight; reason not known (Beaud 1991). ANTI-PREDATOR RESPONSES OF YOUNG. After begging from man at *c*. 10–11 days, crouched (as also on hearing adult's alarm-call: see 2 in Voice), then half-gaped apparently in threat at *c*. 15 days. Give distress-call when handled; in 3 out of 11 broods ringed, other brood-members joined in when one gave this call. (Steinfatt 1944; Mezhennyi 1964; Rudat 1976; Crocq 1990.) Brood vacated nest at *c*. 18–20 days when tree shaken by observer checking nest (Matz 1967). Ruffled plumage of fledgling perched on bare branch breaks up outline and may function as camouflage-posture (Portenko 1937). PARENTAL ANTI-PREDATOR STRATEGIES. (1) Passive measures. A tight sitter, e.g. leaving only when observer had climbed up level with nest in neighbouring tree, or even tolerating hand on nest (Selous 1907; Grote 1947; Reymers 1959). ♂ on tree-top look-out will drop down and disappear in forest if disturbed (Mezhennyi 1964). (2) Active measures: general. Especially when nestlings threatened, will flutter back and forth near nest, breaking off twigs and calling excitedly from tree-tops (Kleinschmidt 1909–11). (3) Active measures: against birds. Alarm-calls given persistently for perched Sparrowhawk *Accipiter nisus* which was then dive-attacked until it flew off (Rudat and Rudat 1971); dive-attacks, followed by aerial tussle, also reported for *G. passerinum* (Seifert and Schönfuss 1959). In eastern Russia, pair chased White-tailed Eagle *Haliaeetus albicilla*, giving alarm-calls (Mezhennyi 1964). (4) Active measures: against man (no information for other animals). Bird disturbed from nest returned in low flight then swayed in curious fashion while looking at observer (Bartels and Bartels 1929); perhaps form of distraction-lure display. Normally gives alarm-calls when man near nest and until he moves away, or birds call from greater distance, returning stealthily to check (Rudat and Rudat 1971). One bird approached to within *c*. 5 m of man handling chick and gave distress-call (presumably call 2a or 2b, or similar: see Voice), then stayed *c*. 20 m away until young calmed down (Steinfatt 1944).

(Figs by D Quinn: A–B from photographs in Swanberg 1956*b*; C from photograph in Swanberg 1952.) MGW

**Voice.** Generally silent during winter, but vocal with start of breeding (using varied repertoire of calls, especially in courtship), then silent again after laying; noisy in summer when adults with fledged young in territory and (stimulated by newcomers in area) at start of food-storing (Kleinschmidt 1909–11; Bartels and Bartels 1929; Taapken *et al.* 1955*b*; Swanberg 1956*a*; Mezhennyi 1964; Lefranc and Pfeffer 1975; Rudat 1984; Crocq 1990). Repertoire apparently large, but many calls not well known (especially function) and confusion arises through multiplicity of renderings, though these reflect genuine variation in best-known call 2a which serves (rather than song) to advertise territory (Swanberg 1951*b*). Voice of *japonicus* does not differ from European birds (Jahn 1942); for *macella*, see call 2a. Bill-snapping a short, sharp 'ptt' (Rammner 1977), softer than in owl (Strigiformes) (Creutz and Flössner 1958). Will also hammer with bill against tree as displacement activity in antagonistic context (Crocq 1990). Considerable mimetic ability: for bird, animal, and mechanical sounds mimicked by captive individual, see Rammner (1977) and below. For additional sonagrams, see Bergmann and Helb (1982) and Crocq (1990). For musical notation and comparison with other Corvidae, see Stadler (1927).

CALLS OF ADULTS. (1) Song. Long, varied phrase of subdued, at times ventriloquial, harsh chattering, warbling, or babbling, with sweeter (even melodious) more tonal piping, whistling, whining, mewing, or bleating, interspersed with gurgling and clicking sounds. Not unlike song of Magpie *Pica pica* or Jay *Garrulus glandarius*. (Mojsisovics 1886; Simon 1921; Beaufort 1947; Swanberg 1951*b*; Rammner 1977; Crocq 1990.) May begin with quiet guttural 'kroi kroi' (Creutz and Flössner 1958), and then include gentle cheeping, descending bleating tremolo, twittering, or chattering like Starling *Sturnus vulgaris* (rising and falling motifs, many guttural or squeaking but few fluted notes), ending in loud 'öck' (up to 11 units) or loud squeaking 'iäääh-iäääh-iäääh'; several units or motifs also function as separate calls (Rammner 1977, which see for renderings of other parts of song). Main sounds in song described by Stübs (1958) were probably calls 2a, 3, and 5. Song contains mimicry of various small passerines (Crocq 1990), e.g. Swallow *Hirundo rustica* and Turdidae (Mojsisovics 1886; Berndt and Moeller 1960). Delivery rapid and irregular (Stadler 1927), but with definite temporal pattern according to Swanberg (1951*b*). Song apparently given by ♂ alone and has no territorial significance; especially associated with high-intensity courtship in spring (Rammner 1977; Bergmann and Helb 1982), though not confined to breeding season, and perhaps also expresses well-being (Donner 1908). (2) Excitement-calls. (2a) Rasp-call. Harsh and (usually) loud rasping sound with characteristic dry buzzing timbre (Stadler 1927; Bruun *et al.* 1986). Given singly or in series, usually 2–4 successive units (often in pairs), though length of individual units and series varies, as does speed of delivery; long, faster series of short units signal increased excitement or alarm (Naumann 1905; Rammner 1977; Crocq 1990); also serves as advertising-call (Crocq 1990), and allegedly given in contact (Stadler 1927; Stübs 1958). Variety of transcriptions referring to both nominate *caryocatactes* and *macrorhynchos* (according to Wernli 1970, *macrorhynchos* slightly finer, less hard): 'krärr', 'krrre', or 'kkrä' (Taapken *et al.* 1955*b*; Creutz and Flössner 1958; Bergmann and Helb 1982); 'karr' or 'kerr' (Stadler 1927); 'kräk-kräk', 'rät-rät', or 'käärrk käärrk' (Heinroth and Heinroth 1924–6; Stadler 1927; Bartels and Bartels 1929; Stübs 1958). Also 'kraak', 'krark', 'kra', or longer 'kraaaa', like screech of *G. glandarius*, but less incisive and strident (Beaufort 1947; Hollyer 1970; Rammner 1977; Goodwin 1986); according to Stadler (1927) and Rammner (1977), squawk or screech like *G. glandarius* is rare variant, serving as alarm-call; 'chät' (probably fright-call) mentioned by Bartels and Bartels (1929) is presumably the same. Renderings of Rasp-call with other vowel sounds include 'kroi' (Pfeifer 1953, 1954) and 'krrü' resembling Carrion Crow *Corvus corone*, but higher pitched (Taapken *et al.* 1955*b*); see also Stadler (1927). In recording (Fig I), quite long 'krraar' followed by shorter 'krrar', each unit of pair rising in pitch, as do uniformly shorter, higher-pitched 'krrair' units (3 followed by pair) in Fig II. Relatively long rasps ('krrair krrair') in Fig III rise in pitch at start and fall slightly at end, recalling in timbre both *G. glandarius* and rasping screech of Black-headed Gull *Larus ridibundus*; also suggest (reinforced by arrangement in pairs) Craking-call of Corncrake *Crex crex* (which albeit is more clicking or pulsed in timbre). Regional variation suggested by recording of *macella* (Fig IV) in which 'krrarr' rasps are more pulsating or vibrant due to relatively coarse amplitude modulation. Paired units in Fig V comprise short rising 'kra', 'kre', or 'ka' and long, markedly descending, thus disyllabic

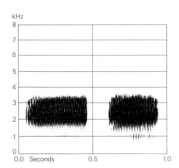

I  Roché (1970)  France  May 1963

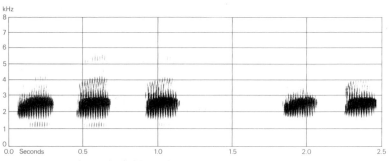

II  J-C Roché  Switzerland  May 1967

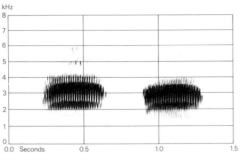

III  J-C Roché  France  May 1983

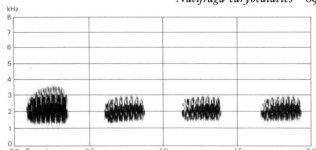

IV  J G Corbett  Nepal  May 1984

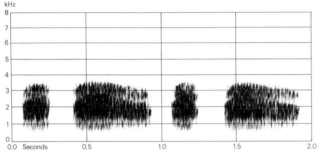

V  S Palmér/Sveriges Radio (1972–80)  Sweden  April 1948

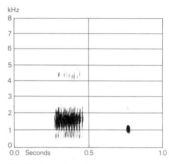

VI  F Jüssi  Estonia  May 1981

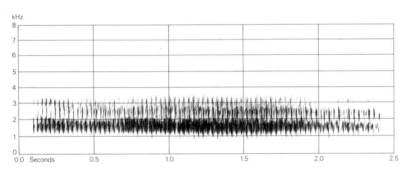

VII  S Palmér/Sveriges Radio (1972–80)
Sweden  April 1948

'krraaoo', 'krEhe', or 'kArre'. In Fig VI, rather short, subdued rasp 'kar' followed by highly tonal, staccato 'oo' (of unknown significance). (W T C Seale, M G Wilson.) Rasp-call sometimes given in long bouts and probably audible up to *c*. 1 km away (Bergmann and Helb 1982). (2b) Churr-call. Drawn-out, harsh, but not especially loud, low-pitched, apparently ventriloquial rattle, variously rendered 'kräärr' (Kinzelbach 1962; Bergmann and Helb 1982), 'rärrkrärrh' (Steinfatt 1944), 'krooooooorrrrr' (Wernli 1970), 'rrrrrrr' (Crocq 1990), 'rurrr', or 'büirrrr' (Stübs 1958). Strikingly similar to churring Advertising-call of Nightjar *Caprimulgus europaeus*; given in bursts of 2–8 s separated by pauses of 6–8 s; apparently serves as alarm-call and used (less commonly than 2a) throughout year (Brandt 1962; Uloth 1977; Palmér and Boswall 1981). Churr in recording (Fig VII) indeed very like *C. europaeus*, with similar pitch, though less wooden and vibrant; pitch undulates slightly and final descent (as reported by Brandt 1962) rather subtle (W T C Seale).

Churr-call derived from call 2a (e.g. Ali and Ripley 1972, Crocq 1990); perhaps homologous with rattling calls of other Corvidae (Goodwin 1986). See also Stadler (1927) and Taapken *et al.* (1955*b*) for this or perhaps related calls. (3) Series of 'jäk', 'jak', 'jaaa', or 'gja' sounds resembling Jackdaw *C. monedula*, though quieter (Stadler 1927; Creutz and Flössner 1958; Kipp 1978; Bergmann and Helb 1982); also rendered 'tjakke-tjakke-tjakketjak' (Taapken *et al.* 1955*b*), while 'karrrjakökjakökjakökja' noted in Germany, February (Brandt 1962), was probably this call combined with 2a. Given during Ceremonial Gatherings (Kipp 1978), and sometimes in song (see above); no other information on function. (4) Nasal 'wäk' (Bergmann and Helb 1982, which see for sonagram); 'oued' with chattering quality (Crocq 1990), and quiet 'gäd' from bird preening in sun (Creutz and Flössner 1958) are perhaps the same or related. (5) Mewing-call. Quiet, drawn-out, plaintive, delicate, slightly fluted, descending tonal 'piou', 'dijüu', 'siep', glissando 'iiea', etc.,

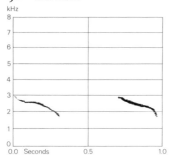

VIII   S Palmér/Sveriges Radio (1972–80) Sweden April 1948

reminiscent of mewing of Buzzard *Buteo buteo*, though also described as (nasal) whistling or whining. In recording (Fig VIII), highly tonal, plaintive, irregularly descending portamento mew with relatively abrupt offset, suggesting 'peeoop' (W T C Seale). Mew-call one of most frequently uttered contact-calls, given throughout year, especially during autumn food-storing, including between adults and fledglings. (Stadler 1927; Portenko 1939; Taapken *et al.* 1955*b*; Creutz and Flössner 1958; Stübs 1958; Mezhennyi 1964; Rammner 1977; Kipp 1978; Bergmann and Helb 1982; Crocq 1990.) Sometimes given in fast series, and rendered 'wiä', 'büeb', or 'miauw' (Taapken *et al.* 1955*b*; Creutz and Flössner 1958); 'A-in' and mewing 'mi-in' given (with Rasp-call) in response to playback of Rasp-call (Lefranc and Pfeffer 1975). For further renderings of calls probably in this category, see Boecker (1970), Wernli (1970), and Crocq (1990). (6) Purring-call. Quiet, voiced purring, sometimes with brief trilling onset, given to encourage young to gape, and in contact with fledglings (Rudat and Rudat 1971; also Reymers 1959). Other calls perhaps belonging here are quiet, gentle 'krrüw' given by ♂ apparently calling ♀ off nest (Bartels and Bartels 1929) and 'grrr-grrr' used by ♂ in courtship (Rammner 1977). Renderings suggest relationship with calls 2a–b. (7) Great variety of calls (changes in pitch and emphasis creating even greater variety) noted during Ceremonial Gatherings in March, also in autumn and winter and from fledglings (though less obvious then). Includes many variants of Rasp-call, wooden cracking noise, clicking (see 1 and 8a), calls 3 and 5, hissing like *G. glandarius*, squeaking 'äääujk' and discordant 'tljau', also various undescribed begging-calls. (Kipp 1978.) (8) Other

calls. (8a) Subdued guttural 'k' sounds illustrated by Bergmann and Helb (1982). A 'gggw-gggw' noted from *macrorhynchos* in Netherlands (Taapken *et al.* 1955*b*) is perhaps the same, or related to call 2a. (8b) A 'jäb-jäb-jäb' given by captive *macrorhynchos* apparently as greeting (Rammner 1977); perhaps related to calls 3–4. (8c) An 'ari' resembling *G. glandarius* (Stadler 1927). (8d) Nasal 'ââ' (Stadler 1927). (8e) Quiet sounds resembling Partridge *Perdix perdix* (Heinroth and Heinroth 1924–6).

CALLS OF YOUNG. Nestlings generally quiet during first 2 weeks, giving only faint cheeping when being fed (Crocq 1990). Most descriptions refer to fledglings (or at least well-grown young): soft, short plaintive note of varying pitch (Beaufort 1947); curious cheeping-whinnying (Bartels and Bartels 1929) or (in *hemispila*) nasal bleating (Ali and Ripley 1972). Recording of *macella* fledglings (Fig IX) similarly contains querulous bleating or urgent wailing squawks, with (before last 2 units) series of short non-discrete bleats like wavering cackle (W T C Seale, M G Wilson). Further descriptions: gentle 'krüik' or 'kroik' (Bartels and Bartels 1929); chattering resembling *P. pica* or Fieldfare *Turdus pilaris* (Steinfatt 1944; Pfeifer 1953); loud harsh Rasp-calls (like adult call 2a) given especially from *c.* 7 weeks (Schüz 1932; Rudat 1984), also presumably similar loud 'rärrk rärrk' when handled (Steinfatt 1944; for sonagrams, see Rudat 1976). In Austrian study, quiet calls said by A Grabher Meyer to be gradually replaced by 'kind of subsong' from *c.* 14 days; captive bird first sang (including quiet 'rärrk') at *c.* 7–8 weeks (Steinfatt 1944).                                                    MGW

**Breeding.** SEASON. Early, so in many areas throughout range whole breeding cycle (nest-building to fledging) can be in cold conditions with deep snow on ground. Sweden: egg-laying peaks last third of March to first third of April (Swanberg 1956*b*). Finland: laying peaks early to mid-April (Haartman 1969). Germany: in Bayern, eggs recorded first week of March, and observations of young still in nest at beginning of July (Wüst 1986); for Thüringen, see Matz (1967) and Rudat and Rudat (1971). In Switzerland, nest-building usually from end of February, but can be delayed 3–4 weeks by weather; newly-fledged young recorded in 1st half of August (Bartels 1931; Glutz von Blotzheim 1962); 2 young seen on 23 January in

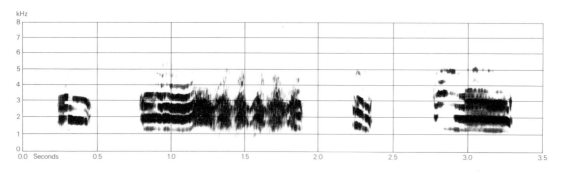

Switzerland would have hatched in last week of December (Saunier 1971). South-east France: during 1967–88, earliest laying 3rd week of March, latest 3rd week of May, peak in 1st week of April; laying is late when snow persists or preceding year's Arolla pine *Pinus cembra* seed crop poor (Crocq 1990). Balkans: laying peaks end of March to beginning of April (Makatsch 1976). Siberia (eastern Russia): 2 nests started at beginning of April (Reymers 1959). SITE. Almost always in conifer, generally against trunk (Rudat 1984; Goodwin 1986), though records from eastern Russia of nests in broad-leaved trees, between pine roots, in rock crevice, and in abandoned building; nests often poorly attached (Kulczycki 1973). Of 44 nests, southern French Alps, 68% in *P. cembra*, 18% in *P. uncinata*, 9% in larch *Larix*, and 5% in spruce *Picea*; Arolla pine apparently preferred because of configuration of branches; 72% of nests against trunk on lateral branches, 17% in cluster of branches near top of tree, and 11% in lateral branches away from trunk, with some in tangle of twigs between 2 trees (Crocq 1990, which see for comparison with other studies). Of 19 nests (7 from Poland, 12 from literature) of nominate *caryocatactes*, 8 in spruce, 8 in fir *Abies*, and 3 in pine *P. sylvestris*; of 10 nests of *macrorhynchos* from literature, 3 in pine *P. sibirica*, 3 in fir, 2 in larch, 1 in spruce, and 1 in alder *Alnus* (Kulczycki 1973). Average height above ground in south-east France 6·7 m, average height of nest-tree 9·6 m ($n = 37$); average height in Poland 5·5 m ($n = 7$), with 6 nests 3–7 m; no significant difference in nest height between nominate *caryocatactes* and *macrorhynchos* (Kulczycki 1973; Crocq 1990). Can be up to 18 m above ground (Makatsch 1976). Nests probably orientated to be sheltered from wind (Crocq 1990). Nest: compact, well-made structure containing 3–5 distinct layers providing good insulation; outer foundation of twigs generally of conifer (often green), though sometimes of other species such as birch *Betula*, beech *Fagus*, or bramble *Rubus fruticosus*, with roots, grass, and lichen, inside which layer of compressed beard-lichen *c*. 3 cm thick, then layer of fragments of decayed wood and moss *c*. 8 cm thick, followed by some earth mixed with various fibres, etc., and lastly lining of grass, rootlets, lichen, hair, etc., rarely feathers (Reiser 1921; Bartels and Bartels 1929; Matz 1967; Haartman 1969; Kulczycki 1973; Goodwin 1986). According to Crocq (1990) positions of compacted lichen and earth layers are reverse of above, but earth layer often lacking depending on local conditions and earth can be brought incidentally on roots, etc. In Poland, 7 nests had average outer diameter of 26 cm (22–38), inner diameter 14 cm (13–16), overall height 15 cm (12–16), and depth of cup 8 cm (7–10) (Kulczycki 1973, which see for comparison of various studies). Weights of 8 nests from French Alps *c*. 400–700 g (Crocq 1990, which see for dimensions of 6 nests and weights of separate layers). For description of nests of recently established *macrorhynchos* population in eastern Belgium, see Simon *et al.* (1975, 1983). Building:

by both sexes, lining mostly by ♀; takes 5–12 days (Haartman 1969; Crocq 1990; A Grabher Meyer). Most material gathered close to nest, though some from up to *c*. 200 m away; dead twigs broken off trees. If snow fills nest before incubation starts new nest may be built nearby; record of 2nd nest 50 cm below 1st. (Reymers 1959; Crocq 1990.) EGGS. See Plate 57. Sub-elliptical, sometimes short sub-elliptical, smooth and matt-glossy, shape variable even within clutch; ground-colour ranges from whitish to deep green but generally pale or greenish-blue, fine olive-brown and grey spots and speckles can be almost invisible or dense, sometimes concentrated at broad end, rarely in ring (Harrison 1975; Makatsch 1976). Nominate *caryocatactes*: $33·7 \times 24·9$ mm ($29·6–38·0 \times 21·5–26·0$), $n = 595$; calculated weight 10·7 g. *N. c. macrorhynchos*: $33·0 \times 25·1$ mm ($32·0–34·9 \times 25·0–25·2$), $n = 3$; calculated weight 10·7 g. (Schönwetter 1984.) Clutch: 3–4 (2–5); in southern Sweden, of 12 clutches following poor or average hazel crop, all were of 3; of 57 clutches following good crop, 14% of 3 and 86% of 4–5; artificial provisioning with food raised average clutch size. Of 123 clutches, 4·1% of 5 eggs, and of 144 none was of 2 eggs; average of 33 clutches was 3·5. Upper limit of clutch perhaps determined by ability to keep all eggs warm in very low ambient temperature; in one case, 55°C difference between them. (Swanberg 1981.) Of 26 clutches in southern French Alps: 2 eggs, 4%; 3, 42%; 4, 50%; 5, 4%; average 3·53 (Crocq 1990). Of 42 clutches in Balkans: 2 eggs, 5%; 3, 52%; 4, 43%; average 3·38; replacement clutch laid 14 days after loss of 1st clutch (Makatsch 1976). Eggs laid immediately on completion of nest (Crocq 1990). In 94% of 86 observations of egg-laying in Sweden, interval between eggs was 1 day, in 6% 2 days; 1st egg laid in early morning then subsequently *c*. 1 hr later each day, $n = 61$ eggs (Swanberg 1956b). No replacement clutches found in 22-year French study (Crocq 1990). One brood (Makatsch 1976). INCUBATION. In Sweden, 18·0 days (17·7–18·5) for incubation of last egg in 17 clutches (13 of 4 eggs, 4 of 3) (Swanberg 1956b). In French Alps, 17·5–18·5 days; by both sexes (♂ has brood-patch as large as ♀); sits from 1st egg but incubation proper only starts with last egg; hatching synchronous. (Bartels and Bartels 1929; Swanberg 1956b; Crocq 1990.) In Sweden, only ♀ incubated at night; relieved by ♂ around sunrise for 20–80 min, then they alternated through day until ♀ remained on eggs from several hours before sunset. ♀ sat for 66% of time at 2 nests, ♂ for 34%, $n = 30$ hrs of daytime observation; often incubates successfully with nest covered by snow around sitting bird. (Swanberg 1956b; which see for more details.) In Austrian study, ♂ incubated for *c*. 32% of time and sat longer when ambient temperature higher; maximum stint of ♂ 3 hrs, of ♀ 6·5 hrs; probably necessary that ♂ incubates since ♀ can then exploit her own food caches (A Grabher Meyer). In Siberia, ♀'s share rose from 6·5 hrs per day on day 2 to 11 hrs on day 16, ♂'s share fell from 9·5 hrs to 5 hrs per day (Reymers 1959). YOUNG. Fed and cared for by both parents; fed by regurgitation, parent inserting bill deep into throat of young to avoid sublingual pouch. Brooded by both parents continuously until about day 7 at least, then irregularly until about day 15; ♂ takes greater share of brooding than of incubation. (Steinfatt 1944; Reymers 1959; Rudat and Rudat 1971; Crocq 1990; A Grabher Meyer.) FLEDGING TO MATURITY. Fledging period 24–25 days (23–28) (Reymers 1959; Rudat 1984; Crocq 1990; A Grabher Meyer). Young can still be fed by parents at 15 weeks; probably breeds first at 2 years or older (Rudat 1984). BREEDING SUCCESS. In southern French Alps, of 97 eggs in 28 clutches, 62% hatched and 44% produced fledged young, giving 1·53 fledged young per pair. Of 38 nests, 14 failed completely, 8 before hatching, 6 afterwards; main predators squirrels *Sciurus*, martens *Martes*, stoat *Mustela erminea*, other Corvidae, owls, especially Eagle Owl *Bubo bubo* and Tengmalm's Owl *Aegolius funereus*, and diurnal raptors, including Golden Eagle *Aquila chrysaetos*, Goshawk *Accipiter gentilis*, and Sparrowhawk *Accipiter nisus*; newly-fledged young also taken by fox *Vulpes*; high losses of eggs and nestlings caused by adverse weather (Crocq 1990). Nestling mortality in Sweden high if insufficient hazel nuts cached (Swanberg 1981). BH

**Plumages.** (nominate *caryocatactes*). ADULT. In fresh plumage (August–September), ground-colour of almost entire head and body deep umber-brown, darkest (near dark fuscous-brown) on cap, upper tail-coverts, side of head, throat, and thigh, paler (near cinnamon-brown) on lower scapulars, back, and rump; under tail-coverts contrastingly white; all umber-brown feathers, except for those of cap, hindneck, rump, and upper tail-coverts, with contrastingly white triangular round-ended drop of varying width on tip, each drop surrounded by narrow fuscous or sooty margin. Short bristles projecting over nostrils closely streaked dull black and white (shorter ones) or almost fully white (longer ones at side of bill). Dark fuscous-brown feathers of cap and hindneck have slightly darker fringes, which may show faint purple-violet gloss in some lights. Drops on mantle and shorter scapulars *c*. 5–7 mm long and *c*. 3–5 mm wide, those on longer scapulars slightly narrower, those on back much smaller, often a narrow speck only. Tips of longer upper tail-coverts black with slight oil-green or blue-green gloss. Lore fully white or cream-white; remainder of side of head and neck streaked with rather equal amounts of brown and white; white often confluent, forming streaky supercilium (broken above eye) and small crescent below eye. Drops on chin and throat small, narrow, and elongate; upper chin largely white, centre of throat mainly brown with limited number of small white specks; drops gradually larger towards side of head and towards lower throat. Spots on chest, side of breast, and upper belly large, *c*. 8–10 mm long, 5–8 mm broad, white predominant over brown, often confluent into long and broad white stripes; spots gradually smaller on flank, lower belly, and vent, where brown predominant over white. Thigh unmarked, except for row of white triangles at rear and some traces of white feather-fringes. Tail mainly black, both webs of central pair (t1) and outer webs of t2–t5 rather faintly glossed oil-green to green-blue; t1 with narrow square-cut white tip *c*. 3–7 mm wide; white tips gradually longer towards outer feathers, largest on t5–t6, with maximum extent of white on middle of inner web of t6 17–24(–27) mm (on average, 21·3, n=15)

(RMNH, ZMA), 22·7 (10) 14–29 mm (Voous 1947), or (measured along shaft) 12–23 mm (Svensson 1992) or 17·4 (2·87; 37) 11–24 mm (Erard 1970). Flight-feathers, tertials, greater upper primary and secondary coverts, and bastard wing black, glossed green-blue or blue with slight purple tinge, black of wing contrasting with brown of body; feathers occasionally entirely unmarked on tip but usually a tiny white speck or short fringe on tip of secondaries, on p3–p4 (p1–p5), on outer primary coverts, or on shortest feather of bastard wing, or a small triangular spot or a trace of a fringe on outer greater coverts; occasionally, tips of secondaries and of p3–p4 show a longer white crescent of up to 2 cm long and 1–3 mm wide; more rarely, tips of secondaries, p1–p6, outer greater coverts, and all primary coverts and feathers of bastard wing with white triangular spot of 2–3 mm wide and 2–5 mm long. Inner border of p4–p5 with large white blob near base, well-visible from below; p3 and p6 sometimes with traces of white at base; blob occasionally present on p4 only or entirely absent. Shorter and inner lesser upper wing-coverts dark umber-brown, longer and outer fuscous-brown to black, in part with oil-green gloss, all with elongate white drop; median coverts black with green-blue gloss, outer with white drops or triangular wedges on tips. Shorter primary coverts mainly white, showing as white patch below bastard wing. Under wing-coverts and axillaries sooty grey with short and broad white blobs on tips, latter showing as rows of parallel white bars on underwing. Sexes similar. *In worn plumage* (March–April), umber-brown of head and body slightly paler, more russet-brown, especially on mantle, scapulars, and belly; narrow dusky margin along tip of each white drop worn off; white drops on body partly worn off, drops slightly smaller, less confluent; white tips of t1, flight-feathers, and coverts (if any) worn off; flight-feathers slightly duller and browner, especially tips of primaries, but bases and coverts still distinctly blacker than in 1st adult. NESTLING. Naked during 1st week of life (Bartels and Bartels 1929); cap, stripe on mid-back, and upper wing covered with grey-white down (Kleinschmidt 1909–11). JUVENILE. Ground-colour of head, body, and lesser and median upper wing-coverts paler than adult, dark brown (when fresh) to dull milk-chocolate brown or brown-grey (when worn); number of marks as in adult, but each mark a narrow streak, widening slightly on feather-tip, rather than a broader triangular drop, less sharply-defined than in adult, head and body appearing narrowly streaked and with brown more predominant, pale marks on side of head mere shaft-streaks only; cap sometimes with faint spots or shaft-streaks, scapulars sometimes partly without spots. Chin, mid-belly, and vent dull grey-brown with large but ill-defined off-white feather-tips. Tail as adult, but feathers on average narrower, tips more rounded, less truncate; black tinged brown, hardly glossy; white on tips of tail-feathers less square-cut, forming shallow notch at shaft on t1 (in adult, a more sharply-defined tip of even width); maximum extent of white on inner web of t6 25·4 (11) 18–28(–32) mm. Flight-feathers, tertials, bastard wing, and greater upper primary and secondary coverts sooty-black, inner webs tinged grey when fresh (distinctly duller and less glossy than in adult), black-brown if worn (less contrasting in colour with body than adult, especially tertials); inner primaries usually with at least trace of narrow white fringe along tip (but often less extensive and clear-cut white than in some adults), p10 pointed and usually with small white spot on tip (in adult, broadly rounded, uniform black), white blobs on bases of inner webs of p4–p5 sometimes absent; coverts and feathers of bastard wing with white or pale buff fringe along tip, pale fringe along greater coverts often forming wing-bar (in adult, usually without white marks, especially inner greater coverts and bastard wing of adult virtually always uniform black on tip; in those adults which show white spots on tips of coverts and bastard wing, these narrow, triangular, not broad and shallow, not forming bar over greater coverts). For pattern of coverts, see also Jenni (1983). FIRST ADULT. Like adult, but juvenile tail, flight-feathers, tertials, all greater upper primary and secondary coverts, and variable number of median upper wing-coverts retained; for characters, see Juvenile, above. In autumn, juvenile tertials, greater upper wing-coverts, bastard wing, and (if any) median coverts distinctly browner and less glossy than in adult, and virtually always more extensively tipped white; in winter and spring, white of tips sometimes largely worn off, but traces usually remain, and tail, tertials, and coverts much browner than in adult at same time of year, tips distinctly frayed. Birds with more limited white on tips of coverts and bastard wing than usual sometimes hard to distinguish from those adults which show more white on tips than is usual for average adult birds; check for presence of white fringe on tip of innermost greater coverts and median coverts and for presence of white spot on tip of p10 (these absent in adult) and tail- and tertial-shape (narrower and with more rounded and more frayed tip in 1st adult). Width of t1 (at 20 mm from tip) 20-27 mm (in adult, 25-28 mm) (Svensson 1992).

**Bare parts.** ADULT, FIRST ADULT. Iris brown. Bill black or plumbeous-black, base slightly more plumbeous-blue. Mouth dark grey to slate-black. Leg and foot black or plumbeous-black, scutes black, soles yellow-horn to greyish-black. (ZMA.) NESTLING. Shortly after hatching, skin flesh-pink; bill flesh with grey tinge, cutting edges and narrow gape-flanges white; mouth dark flesh-red, tinged pink towards flanges, more purplish inwards. Leg and foot dull flesh, scutes gradually darkening to grey. At 2 weeks, bill grey with flesh tinge, cutting edges white or pale yellow; leg and foot grey, fissures between scutes, rear of tarsus, and soles flesh. JUVENILE. Iris grey. Bill bluish-grey, traces of gape-flanges and narrow cutting edges pink-white; mouth red. Leg and foot grey with flesh tinge or horn-brown (Niethammer 1937; Steinfatt 1944; Goodwin 1986; Sorbi *et al.* 1990).

**Moults.** ADULT POST-BREEDING. Complete; primaries descendent. In Pechora basin (north European Russia), moult mid-June to about early or mid-August, duration 40-45 days; in Kirgiziya, starts 2nd half of May, almost completed late June; in Russian Altai, mid-June to late August; elsewhere in Russia, moult between 2nd half of May and 1st days of August; flight-feathers and tail start 20-25 days before body, ♂ slightly before ♀ (Dementiev and Gladkov 1954). In western Siberia, moult from June (Johansen 1944); in Russian Altai, from mid-June, occasionally late May (Grote 1947); in Afghanistan, sometimes from early May (Paludan 1959); in Nepal, in full moult mid-May (Diesselhorst 1962). In birds in moult examined ($n = 13$), mainly from central Europe and Bulgaria, moult starts with p1 between late April and early June, completed with regrowth of p9-p10 mid-July to mid-August; mainly May to July; during large invasion of adult *macrorhynchos* in Netherlands, early August 1968, all birds had primary moult complete, but some feathering of head, neck, or tail occasionally still growing (RMNH, ZFMK, ZMA). One-year-olds apparently moult *c.* 1 month earlier than adults (Kleinschmidt 1909-11). Non-breeding captive birds may start late March (Rammner 1977, which see for sequence of body moult) and complete by early June (ZMA). POST-JUVENILE. Partial: head, body, lesser and many or all median upper wing-coverts, exceptionally innermost greater coverts or central tail-feathers (RMNH, ZMA). In 17 birds from Alps, 4 retained all median coverts, but most birds had 1-6 median coverts new, one had all new; 2 birds had 4-6 innermost greater coverts new

(Jenni 1983). Starts at age of *c.* 5.5-6.5 weeks; in north-east Poland, late May to mid-July; in full moult at age of 2 months, completed early July (Heinroth and Heinroth 1931; Steinfatt 1944). In Latvia, one had moult almost completed late June (Kleinschmidt 1909-11). In USSR, moult early July to 1st half of August (Dementiev and Gladkov 1954); in Amur area, bird half-way through moult 19 July (Stegmann 1931*a*); in Russian Altai, moult sometimes completed by late June (Grote 1947). In Afghanistan, moult May and early June (Paludan 1959); in Nepal, early June to mid-July (Diesselhorst 1962).

**Measurements.** ADULT, FIRST ADULT. Nominate *caryocatactes*. Wing, bill, and bill depth of (1) Scandinavia and northern Poland east to western European Russia; (2) southern Poland, south-east Germany, Czechoslovakia, and northern Carpathian mountains; (3) Alps of southern Germany, Switzerland, Austria, and northern Italy, (4) Bulgaria; other measurements combined; skins (RMNH, SMTD, ZFMK, ZMA). Juvenile wing and tail refer to those of 1st adult. Bill (S) to skull, bill (N) to distal corner of nostril; exposed culmen on average *c.* 8.9 less than bill (S). Depth is bill depth at middle of nostril; at forehead, depth on average 1.41 greater ($n = 26$), at gonys bulge, depth on average 1.25 less ($n = 37$).

| | ♂ | ♀ |
|---|---|---|
| WING JUV (1) | 184.2 (2.64; 6) 181-189 | 179.0 ( — ; 2) 176-182 |
| JUV (2) | 185.8 (4.36; 6) 182-191 | 183.0 ( — ; 2) 182-184 |
| AD (3) | 193.2 (4.00; 12) 189-198 | 185.0 (3.95; 6) 178-193 |
| JUV (3) | 185.5 (3.27; 16) 180-192 | 180.8 (3.01; 10) 177-185 |
| AD (4) | 191.7 (2.38; 6) 188-194 | 186.1 (5.98; 4) 180-192 |
| TAIL AD | 124.6 (2.33; 14) 121-126 | 120.0 (4.10; 7) 115-125 |
| JUV | 120.3 (3.05; 19) 116-126 | 117.9 (4.54; 9) 113-125 |
| BILL (S) (1) | 49.8 (1.93; 6) 46.3-51.8 | 48.4 ( — ; 2) 46.7-50.1 |
| (2) | 53.0 (1.55; 6) 50.6-54.8 | 48.8 ( — ; 2) 48.2-49.5 |
| (3) | 52.1 (1.88; 6) 49.1-54.7 | 48.1 (1.80; 16) 45.5-51.0 |
| (4) | 49.8 (2.97; 6) 46.3-52.2 | 46.6 (2.19; 4) 45.1-48.4 |
| BILL (N) (1) | 36.3 (0.64; 6) 35.4-37.2 | 34.4 ( — ; 2) 33.3-35.5 |
| (2) | 39.1 (1.90; 6) 36.0-40.9 | 35.8 ( — ; 2) 35.0-36.7 |
| (3) | 38.7 (1.64; 28) 35.5-41.0 | 36.1 (2.14; 16) 33.5-39.5 |
| (4) | 37.8 (2.07; 6) 35.3-39.8 | 34.0 (1.52; 4) 32.8-35.2 |
| DEPTH (1) | 15.8 (0.63; 6) 15.4-17.0 | 15.3 ( — ; 2) 15.1-15.5 |
| (2) | 15.6 (0.34; 6) 15.0-16.0 | 15.0 ( — ; 2) 14.2-15.7 |
| (3) | 15.1 (0.56; 26) 14.3-16.4 | 14.6 (0.67; 16) 13.8-16.0 |
| (4) | 14.9 (0.37; 6) 14.5-15.5 | 14.2 (0.60; 4) 13.8-14.8 |
| TARSUS | 42.1 (1.28; 26) 40.0-44.4 | 41.4 (0.83; 12) 40.2-42.4 |

Sex differences significant for wing (3), adult tail, bill (3), and depth.

USSR: wing, ♂ 182.0 (25) 176-195, ♀ 179.6 (14) 174-189; bill to nostril, ♂ 38.6 (26) 34-42, ♀ 37.0 (16) 33.0-39.0 (Dementiev and Gladkov 1954). Alps: wing 185.4 (5.51; 25) (Kleinschmidt 1909-11). Northern Italy: wing, ♂ 189.3 (4.59; 6) 183-195, ♀ 186.2 (3.97; 6) 180-190; bill to nostril, ♂ 39.2 (3.37; 6) 35-40, ♀ 38.3 (1.97; 6) 36-41 (Londei and Gnissci 198). Width of bill at base of lower mandible: Sweden and Baltic 12.6 (0.52; 26) 12.0-13.5, Switzerland 12.0 (0.59; 24) 11.0-13.0 (Kleinschmidt 1909-11).

*N. c. macrorhynchos.* Netherlands: wing, bill, and bill depth of birds in (1) 1968-9 invasion (August-March), (2) other invasions 1850-1979 (late September to December); other measurements combined; skins (RMNH, ZMA).

| | ♂ | ♀ |
|---|---|---|
| WING AD (1) | 187.6 (3.15; 28) 183-190 | 182.3 (3.35; 25) 177-189 |
| AD (2) | 185.1 (4.10; 19) 180-194 | 178.5 (3.27; 6) 175-182 |
| JUV (2) | 182.4 (4.12; 18) 175-191 | 179.2 (3.44; 17) 171-184 |
| TAIL AD | 123.9 (4.03; 24) 118-130 | 121.5 (4.09; 23) 114-128 |
| JUV | 118.2 (2.87; 8) 114-122 | 116.8 (5.38; 8) 110-123 |
| BILL (S) (1) | 55.8 (2.03; 25) 52.3-59.9 | 53.6 (3.30; 25) 48.4-59.0 |
| (2) | 53.6 (1.95; 42) 50.5-56.6 | 50.4 (1.70; 21) 47.3-52.6 |

| | ♂ | ♀ |
|---|---|---|
| BILL (N) (1) | 42·5 (1·94; 27) 40·3-45·6 | 41·1 (2·83; 25) 37·0-44·3 |
| (2) | 40·5 (2·15; 38) 37·0-44·5 | 37·6 (1·47; 22) 35·5-39·8 |
| DEPTH (1) | 14·3 (0·52; 28) 13·5-15·2 | 14·0 (0·52; 25) 13·3-14·9 |
| (2) | 14·1 (0·60; 37) 13·0-15·3 | 13·7 (0·54; 22) 12·9-14·5 |
| TARSUS | 41·3 (1·21; 24) 39·0-43·1 | 40·6 (0·89; 8) 38·9-42·1 |

Sex differences significant, except juvenile tail, bill (N) (1), and tarsus.

During 1968-9 invasion in Netherlands, bill gradually shorter in autumn probably due to change in diet causing greater wear; average of bill to nostril: August, ♂ 43·3 (11), ♀ 41·7 (13); September, ♂ 42·2 (9), ♀ 41·2 (7); October-December, ♂ 40·8 (5), ♀ 37·9 (4). Wing and tail length increase with age of bird: in captives of 1968 invasion (which involved disproportionately large number of 1-year-olds with 1st complete post-breeding moult just finished after failed breeding season), wing of ♂ 194 (3) 193-195 1-7 years later, ♀♀ 187, 190; tail of ♂ 128·5 (3) 123-133, ♀♀ 122, 128 (RMNH, ZMA; C S Roselaar).

Birds from 1968 invasion (virtually all adult): England, wing 182·3 (6·41; 6) 173-190 (Hollyer 1970); Belgium, wing 173-193 (40), bill (N) 37·3-45·0 (45) (Herroelen 1974); France, bill (N) 40·2 (2·16; 38) 36·5-44 (Erard 1970); Helgoland (Germany), wing 183·0 (5·72; 19) 175-191, bill (N) 43·0 (2·49; 19) 39-47·5 (Vauk 1970); eastern Germany, wing ♂ 188·0 (50) 178-197, ♀ 182·0 (20) 174-189 (Piechocki 1971a); Poland, wing 181·4 (4·85; 50) 169-191, exposed culmen 46·3 (3·85; 41) 38-53 (Stravinski and Shchepski 1972). Finland, 1977 invasion: wing 180·5 (4·36; 38) 172-189, exposed culmen 45·6 (2·10; 38) 40·5-51·0 (Lehikoinen 1979, which see for influence of age). Siberia: wing, ♂ 181·0 (139) 170-191, ♀ 177·3 (113) 161-190; bill (N), ♂ 44·7 (47) 39-49, ♀ 40·9 (20) 38-45 (Dementiev and Gladkov 1954). Width of bill at base of lower mandible 10·2 (0·51; 24) 9·5-11 (Kleinschmidt 1909-11).

*N. c. rothschildi.* Tien Shan (central Asia): wing, ♂ 202 (10) 194-212, ♀ 196 (10) 185-206 (Dementiev and Gladkov 1954); sexes combined (mainly ♀), wing 198·4 (3·23; 9) 193-202, tail 135·6 (5·82; 8) 128-145, bill (S) 49·6 (3·24; 9) 44·9-54·3, bill (N) 35·6 (2·52; 9) 32·0-39·2, bill depth at nostril 14·8 (0·41; 9) 14·3-15·6, tarsus 42·5 (0·95; 9) 40·8-44·1 (RMNH, ZMA).

*N. c. multipunctata.* Nuristan (Afghanistan): wing, ♂ 203·7 (4·32; 6) 200-212, ♀ 200·3 (4·46; 6) 195-206 (Paludan 1959).

**Weights.** ADULT, FIRST ADULT. Nominate *caryocatactes*. Sweden (n = 13), average ♂ 198, ♀ 188; Norway, July, ♂ 125; August-January, ♂ 160-190 (3); September-November, ♀ 140-195 (4) (Haftorn 1971). Netherlands, September: ♀ 153 (ZMA). Southern Germany, October: ♀ 155 (SMTD). Austria, August: ♂ 191 (ZFMK). Bulgaria, July: ♀ 172 (Niethammer 1950; ZFMK). North-east Poland, October: 165 (ZFMK). USSR: ♂ 190, ♀♀ 200 (n = 3) (Dementiev and Gladkov 1954).

*N. c. macrorhynchos.* Netherlands, 1968 invasion: (1) dead or dying, mainly August-September; (2) apparently healthy, mainly September-December (RMNH, ZMA). (3) Nordrhein-Westfalen (Germany), August-September 1968 (Boecker 1970). Eastern Germany, 1968 invasion: (4) found exhausted, (5) shot (Piechocki 1971a). (6) USSR (Dementiev and Gladkov 1954). (7) Russian Altai, March-September (n = 21) (Grote 1947).

| | ♂ | | ♀ | |
|---|---|---|---|---|
| (1) | 133 (18·3; 18) | 100-157 | 123 (12·8; 18) | 101-150 |
| (2) | 175 (11·1; 7) | 164-191 | 154 (14·7; 4) | 140-176 |
| (3) | 161 ( — ; 9) | 117-183 | 150 ( — ; 15) | 110-178 |
| (4) | 128 ( — ; 5) | 125-134 | 125 ( — ; 5) | 110-130 |
| (5) | 166 ( — ; 43) | 137-197 | 156 ( — ; 24) | 134-182 |
| (6) | 176 ( — ; 40) | 153-190 | 169 ( — ; 28) | 124-184 |
| (7) | 166 ( — ; —) | 148-190 | 159 ( — ; —) | 143-176 |

Baykal area (Russia), early August: 170 (23) 154-190 (Pie-chocki 1971a). 1954-55 invasion: Sachsen (Germany), ♂♂ 119, 147, ♀ 172 (3) 145-196 (Creutz and Flössner 1958). 1968 invasion: north-west France, 120 (1) (Hollyer 1970); Belgium, 113-182 (24), September 159·5 (9·33; 4) 149-170 (Geuens 1968; Van Winkel 1968); Helgoland (Germany), August and early September, 141·4 (17·1; 19) 105-165 (Vauk 1970); Finland, average 148 (Lehikoinen 1979); Mongolia, August, ♂ 175 (Piechocki and Bolod 1972). See also Stravinski and Shchepski (1972) and Busche (1970). 1977 invasion: Finland, October, 163·0 (12·7; 38) 145-192 (Lehikoinen 1979).

*N. c. rothschildii.* Kirgiziya: ♂♂ 152, 175, ♀♀ 159, 166 (Yanushevich *et al.* 1960).

*N. c. hemispila.* South-east Afghanistan, May-June: ♂ 170·0 (5·48; 6) 164-177, ♀ 160·7 (7·23; 6) 153-173 (Paludan 1959).

**Structure.** Wing rather short, broad at base, tip rounded. 10 primaries: p6 longest, p7 0-3 shorter (rarely, p7 slightly longer than p6), p8 3-11, p9 23-33, p5 1-5, p4 12-24, p3 27-37, p2 35-46, p1 38-52; tip of p9 between tips of p3 and p4, sometimes equal to p3 or p4 or slightly shorter than p3. P10 somewhat reduced, 67-80 (adult) or 53-74 (juvenile) shorter than p6, 30·6 (20) 22-40 (adult) or 35·2 (10) 30-40 (juvenile) longer than longest upper primary covert (RMNH, ZMA). P9 of *macrorhynchos* relatively longer than in nominate *caryocatactes* (Kleinschmidt 1909-11). Outer web of p5-p8 clearly emarginated, inner web of p6-p9 with slight notch. Tip of longest tertial reaches to tip of p1. Tail rather short, tip slightly rounded or almost square; 12 feathers, t6 2-9 shorter than t1. Bill about equal to head-length (nominate *caryocatactes*) or slightly longer (*macrorhynchos*); rather broad and deep at base, gradually tapering to tip in *macrorhynchos*, with marked bulge at gonys in nominate *caryocatactes*, virtually straight (nominate *caryocatactes*) or slightly decurved (*macrorhynchos*), somewhat compressed laterally in middle and tip in *macrorhynchos*. Nostril rather small, rounded, covered by short tuft of bristle-like feathers. Large pouch, opening below tongue (Johansen 1944). Plumage soft and full. Leg and foot rather short and strong. Middle toe with claw 32·7 (15) 30-35; outer toe with claw *c.* 76% of middle with claw, inner *c.* 80%, hind *c.* 89%.

**Geographical variation.** Rather slight in size, more marked in depth and width of bill, in size and extent of white spots, and in amount of white on tail. 3 main groups recognized. (1) *Caryocatactes* group, with 4 races, range extending from Europe to Kamchatka and Japan, south to Balkans and Tien Shan; shows profuse white spotting on body and broad white tips to tail-feathers. (2) *Multipunctata* group, with one race only, in Afghanistan, northern Pakistan, and north-west India east through Kashmir to Lahul; spots on head and body expanded to large and extensive white streaks, almost coalescing; outer tail-feathers extensively white (but central pair blacker than in nominate *caryocatactes* group); bill slender. (3) *Hemispila* group, with 3-4 races, occurring through Himalayas (west to Kashmir and Murree, south of *multipunctata*) east to Shensi and Hopeh (eastern China) and Taiwan; white on tail as in *multipunctata* group, but bill heavy at base, and white spots on head and body markedly reduced in size and extent, mainly restricted to small spots on mantle, side of head and neck, and chest. *Multipunctata* and *hemispila* groups sometimes considered separate species, as colour and proportions differ greatly (e.g. Kleinschmidt 1909-11, Goodwin 1986); breeding ranges generally separated by Pir Panjal range but may overlap in Simla area (Meinertzhagen 1927); abrupt replacement of one form by other without intermediates points to competitive exclusion (Mayr 1963), and this is usually a reason to consider such forms full species but, as some hybrids are reported, traditional view of single species (Vaurie 1954*b*) followed here. For races within *hemispila* group, see Hartert (1903–10), Biswas (1950), and Vaurie (1954*b*, 1959).

Within *caryocatactes* group, races differ mainly in bill shape; also much individual variation in depth of ground-colour, but this mainly related to age of birds and influence of bleaching and abrasion; birds from southern Balkans and Tien Shan often said to be darker than others, but freshly-moulted birds from elsewhere equally dark, all rapidly fading in course of year (Stegmann 1931*a*; Piechocki 1971*a*; C S Roselaar). *N. c. rothschildi* from Tien Shan averages larger than other races of *caryocatactes* group (see Measurements); bill short, heavy at base. Nominate *caryocatactes* from central and northern Europe east to about upper Pechora, Vychegda, and Kama basins has bill short and heavy (see Measurements) and amount of white on tail-tip rather restricted (see Plumages); *macrorhynchos* from east European Russia eastward through Siberia has longer and more slender bill and more white in tail: length of white measured in middle of outer web of t6 27·0 (76) 22–35 mm in adult (rarely 20–37), 29·2 (35) 25–33(–38) in juvenile (RMNH, ZMA), 27·4 (47) 22·5–35 (Voous 1947), or (measured along shaft) 24·2 (2·63; 38) 19–29 (Erard 1970), 24·4 (38) (Lehikoinen 1979), 24·4 (27) 19–29·5 (Piechocki 1971*a*), 18–32 (Herroelen 1974), 19–32 (Svensson 1992), or 23·4 (36) 19–28·5 (Vauk 1970). For summary of differences, see Table A. Within nominate *caryocatactes*, sometimes several races described: (1) typical nominate *caryocatactes* from Scandinavia and north-east Poland eastwards to central European Russia, (2) *relicta* Reichenow, 1889, from Germany and southern Poland south to Alps and north and central Yugoslavia, and (3) *wolfi* Von Jordans, 1940, in southern Yugoslavia, Bulgaria, and Greece; white spots on upperparts of latter 2 said to be smaller than in birds from further north, and bill of *wolfi* said to be narrower at base and more laterally compressed at tip (Matvejev 1948; Matvejev and Vasić 1973; Przygodda 1969). When series of skins compared, no constant difference in spot size discernible; bill of birds from Alps south to Greece indeed more slender at base than those from Scandinavia, but those from Erzgebirge to northern Carpathians intermediate, and overlap between all populations is large (see also Measurements). Also, some local variation in bill depth and length within *macrorhynchos*, e.g. bill long and slender in north-east European Russia and western Siberia, short but slender in Yakutia and Anad-

Table A   Characters of races of Nutcracker *Nucifraga caryocatactes*, according to data from Kleinschmidt (1909–11), Niethammer (1937), Voous (1947), Matvejev (1948), Przygodda (1969), Erard (1970), Vauk (1970), Haftorn (1971), Piechocki (1971*a*), Herroelen (1974), Svensson (1992), and C S Roselaar.

| | Nominate *caryocatactes* | *N. c. macrorhynchos* |
|---|---|---|
| Bill to skull | 45·1–54·8 | 47·3–59·9 |
| Exposed culmen | 39–49(–52) | 41–54 |
| Bill to nostril | 32·5–43 | 35·5–46 |
| Depth of bill (mid-nostril) | 13·8–17·1 | 12·9–15·2 |
| Depth of bill (gonys bulge) | 13·0–16·3 | 11·5–14·0 |
| Ratio of depth (gonys bulge) to exposed culmen | 0·307–0·373 | 0·232–0·307 |
| Ratio of bill (to skull) to depth (nostril) | 2·98–3·60 | 3·44–4·32 |
| Bill width at base of lower mandible | 11–15 | 9–13 |
| White on t6 (at shaft) | 11–24 | 19–32 |
| White on t6 (on middle of inner web) | 14–29(–32) | (20–)22–38 |

yrland, while some thick-billed birds occasionally occur within range of *macrorhynchos* (e.g. in north-east Europe, in Turukhansk area, and locally in eastern Siberia); boundary between *macrorhynchos* and nominate *caryocatactes* not sharp (Grote 1947). *N. c. japonicus* from Japan rather a poor race; bill and amount of white in tail intermediate between *macrorhynchos* and nominate *caryocatactes*, but nearest to latter; spots on underparts on average larger than both these races. Birds from Kamchatka and northern Kuril Islands ('*kamchatkensis*' Barrett-Hamilton, 1898) intermediate between *japonicus* and *macrorhynchos*, but closer to *macrorhynchos* (Vaurie 1954*b*, 1955, 1959). *N. c. macrorhynchos* frequently involved in large invasions to west, south, and east when food suitable for storing is scarce (Piechocki 1971*b*; Kumari 1972); some breed in northern and central Europe after such invasions, sometimes maintaining themselves for considerable number of years (e.g. in Finland, Sweden, Germany, and Netherlands). For possible evolutionary history of races, see Simroth (1908) and Kleinschmidt (1909–11).          CSR

---

## *Pyrrhocorax graculus*  Alpine Chough

PLATES 6 and 9 (flight)<br>[between pages 136 and 137]

Du. Alpenkauw          Fr. Chocard des Alpes          Ge. Alpendohle
Ru. Альпийская галка          Sp. Chova piquigualda          Sw. Alpkaja

*Corvus Graculus* Linnaeus, 1766

Polytypic. Nominate *graculus* (Linnaeus, 1766), southern Europe from central Spain, Pyrénées, Alps, Corsica, and Yugoslavia to Greece; perhaps this race Morocco, Caucasus, and northern Iran; *digitatus* Hemprich and Ehrenberg, 1833, southern Turkey and Levant to Zagros mountains in south-west Iran. Extralimital: *forsythi* Stolickza, 1874, central Asia from Afghanistan through Himalayas and Tien Shan to Sayan mountains and central China.

**Field characters.** 38 cm (tail 12–14 cm); wing-span 75–85 cm. Close in size to Chough *P. pyrrhocorax* but with 40% shorter bill (decurved only towards tip), less broad and 'fingered' wings and 15–20% longer tail; 15% larger

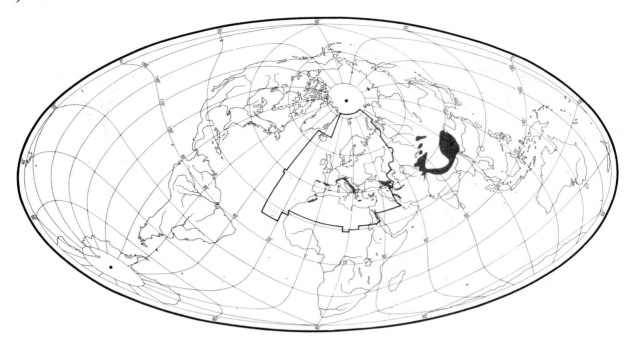

and less compact than Jackdaw *Corvus monedula*. Medium-sized, rather small-headed, graceful crow. Black, with relatively small, short, decurved yellow bill and red legs. Sexes similar; no seasonal variation. Juvenile separable. 2 races in west Palearctic, differing only in size.

ADULT. Moults: June–October (complete). Entire plumage black, glossed with purple and bottle-green but less brilliantly than in *P. pyrrhocorax* so that bird looks drabber. On underwing, flight-feathers appear greyer than coverts. Bill yellow, tinged orange to green at base. Legs usually vermilion. JUVENILE. Brownish-black, with little if any gloss. Bill pale horn at base, dusky or black at tip. Legs dull black with hint of red at joints.

Unless at close range, or calling (see below), difficult to distinguish from *P. pyrrhocorax*, though lacks full aerial grace and quite broad build of that species. Shortness of yellow bill diagnostic, but note that juvenile *P. pyrrhocorax* has initially short, orange bill. Separation from *C. monedula* in flight assisted by 2-toned (not uniform) underwing. Flight swift and skilful, with loose, deep wing-beats, folded wing attitudes, and fanning of full tail allowing high manoeuvrability at cliffs and easy sailing in upcurrents; even so, action slightly stilted compared to *P. pyrrhocorax* and landing often a rather untidy collapse. Flight silhouette differs from *P. pyrrhocorax* due to rather narrower wings with only 4 separated primaries (5–6 in *P. pyrrhocorax*) and longer tail. Gait combines shuffling walk and restricted hop, usually accomplished with tail trailing downwards; flicks wings and bows. Often tame, scavenging near man. Sociable.

Voice includes: frequent, rather thin but musical 'chirrish' or 'keerreesh'; clear, piercing, whining 'tsi-eh' and combining into rather metallic whistling chorus when uttered by several birds; rolled 'krrrree'; in threat, rather short and more punctuated 'tchiupp' recalling *P. pyrrhocorax*.

**Habitat.** Strictly montane, breeding in middle latitudes of west Palearctic in generally colder climates and at greater altitudes than any other bird species, in Switzerland up to 3000 m and only exceptionally below 1500 m, and in Morocco up to 3900 m. While demanding inaccessible nest-sites in steep rock-faces, often in caverns or fissures, is enabled by mastery of air to range over wide variety of foraging habitats, from snowline to treeline or lower, tending however, to avoid snow cover and to favour alpine meadows, newly mown grasslands, and boulder slopes. Especially in winter, favours immediate neighbourhood of huts, hotels, settlements, ski lifts, and other tourist facilities, showing little aversion to human presence in small or large numbers. Only exceptionally, however, descends to cultivated areas below treeline, though has bred in a railway yard in Switzerland as low as 563 m.

Winter movements often involve regular daily foraging excursions to favourable sources of food, often provided by human activities; return to higher ground is usual not long after midday (Glutz von Blotzheim 1962; Lüps *et al.* 1978). In Himalayas, recorded following mountaineers for scraps at 8100 m (Goodwin 1986). In European USSR, occurs during breeding season at 2000–3600 m, descending in cold season to alpine meadows not below 1800 m; extralimitally, ranges somewhat higher and lower, wintering locally close to human habitations around rubbish heaps (Dementiev and Gladkov 1954). In Kashmir in sum-

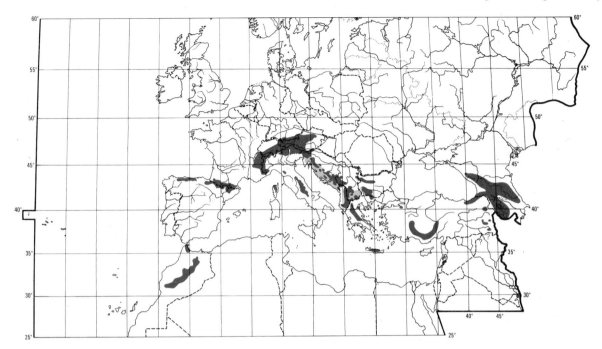

mer lives mainly at *c.* 3500–5000 m, feeding on ground often in company with Chough *P. pyrrhocorax*, but takes much of food on the wing in flocks, airspace forming an important part of regular habitat; trees, bushes, wetlands, and crops normally avoided (Bates and Lowther 1952), though sometimes feeds while perching in trees, shrubs, or bushes. Flies back to roost in mountains when using lower level feeding grounds by day. (Goodwin 1986.)

**Distribution.** No recent changes recorded.

POLAND. Former breeder in Tatra mountains (1850); since recorded only as accidental visitor (Tomiałojć 1990). SYRIA. Breeds Mt Hermon (Shirihai in press).

Accidental. Czechoslovakia, Poland, Hungary, Cyprus.

**Population.** No changes recorded.

FRANCE. 1000–10 000 pairs; some recent increase in Haute-Savoie (Yeatman 1976). GERMANY. Estimated 6000 pairs (G Rheinwald).

Survival. Oldest ringed bird 11 years (Rydzewski 1978).

**Movements.** Mainly sedentary, except for altitudinal movements. Few ringing data.

Alpine populations (nominate *graculus*) often make diurnal altitudinal movements of up to several km between feeding and roosting sites (see Social Pattern and Behaviour) and may make overall movements to lower elevations in winter (e.g. Glutz von Blotzheim 1962, Schifferli *et al.* 1982, Büchel 1983). In Switzerland, daily movements may exceed 20 km in length and cover 1600 m altitude. More birds than formerly may remain high in Alps in winter due

to development of skiing above 3000 m and consequent artificial food sources. (Schifferli *et al.* 1982.) No information on altitudinal movements of other populations of nominate *graculus*. Occasionally wanders outside normal range in winter (e.g. in France to Hérault and Var: R Cruon), but records far outside range (e.g. Belgium: Smets and Draulans 1982; near Leipzig: Jost 1951) may be escaped captive birds; an old record from Helgoland (Vauk 1972) is of uncertain significance. The few ringing recoveries of alpine birds mainly show only local movements (less than 50 km), but there are a few more distant recoveries at 85 km (Glutz von Blotzheim 1962), 53 km (2) and 155 km; most if not all movements appear to be of young birds (Büchel 1983).

*P. g. digitatus* (Lebanon, parts of Middle East, Himalayas to China) also mainly sedentary but with marked altitudinal movements in higher montane regions. Thus, seasonal altitudinal movement sometimes well marked in Himalayas, with records from lowest elevations in winter (Ali and Ripley 1972; Schauensee 1984; Inskipp and Inskipp 1985). Occurrences in Israel regular only on flanks of Mount Hermon, restricted almost entirely to winter (December–April); 2 records in January elsewhere (Shirihai in press). DTH

**Food.** From spring to autumn favours insects, particularly grasshoppers (Orthoptera), Tipulidae larvae (Diptera), and beetles (Coleoptera); in autumn and winter, berries; diet often includes refuse or scraps, particularly in winter. Feeds in pairs or flocks on open ground, often frequenting favoured sites (Rothschild 1957; Tohmé and Neuschwander 1978; Büchel 1983). In summer, usually above treeline on short grass, rocks, scree, and cliffs; searches in crevices, under stones, among vegetation, in dung, and on buildings and walls; often feeds along receding snowline in early summer and occasionally picks items from snow surface; avoids long grass but visits recently mown alpine meadows, especially damper areas; sometimes visits arable fields after sowing or harvest (Strahm 1960; Glutz von Blotzheim 1962; Holyoak 1972b; García Dory 1983; Goodwin 1986; Raboud 1988). In winter, scavenges around ski resorts (etc.), sometimes taking food from hand or even catching thrown food in flight; otherwise may descend 1600 m each day to valley bottoms to feed in towns, around buildings, in orchards, or from fruiting bushes (Murr 1957; Rothschild 1957; Strahm 1958; Lovari 1978; Mattes and Bürkli 1979; Büchel 1983). Feeds by picking items from surface of ground with action similar to Starling *Sturnus vulgaris* (Glutz von Blotzheim 1962). Takes advantage of temporary abundances of prey, e.g. wind-blown insects on glacier (Epprecht 1965; Holyoak 1972b). Overturns stones with bill, or sometimes tears up vegetation, to reveal prey (Strahm 1958; Glutz von Blotzheim 1962). Searches in crevices in cliffs and walls and will dig less well-covered grain from freshly sown field, though tends not to probe or pry with bill like Chough *P. pyrrhocorax*, and in Picos de Europa (northern Spain) tends only to feed on dung previously broken open by *P. pyrrhocorax* (Strahm 1958; Abdusalyamov 1973; García Dory 1983; Sitasuwan and Thaler 1984). Recorded levering old putty from windows in search of invertebrates and sometimes eats fresh putty (Glutz von Blotzheim 1962; Harrison 1970). In Pamirs, June, flocks recorded catching insects in flight (Ivanov 1969). Flies slowly against wind close to rock faces or recently mown hay meadows, dipping down in flight to take individual prey items such as grasshoppers (García Dory 1983; Dendaletche and Saint-Lebe 1988). Kills small vertebrates and recorded following hunting stoat *Mustela erminea* (Comte 1926; Lane 1957; Goodwin 1986). Takes unguarded birds' eggs and nestlings; eggs up to size of Ptarmigan *Lagopus mutus* (Lane 1957; Chappatte 1980). When feeding on berries, perches on or clings to shrubs and bushes, maintaining balance on thin branches with wing movements; flocks may strip bushes completely (Korelov *et al.* 1974; García Dory 1983; Goodwin 1986). Prey usually swallowed whole without manipulation, though holds large prey under one foot, possibly both; foot held well forward of body, tarsus on ground (Sitasuwan and Thaler 1984; Goodwin 1986). Recorded drinking after swallowing sticky food (Holyoak 1972b).

Frequently caches prey, especially in winter. In Switzerland, recorded retrieving and re-caching a segment of orange, with peel, from one deep rock crevice to another 10 m away; in both, orange concealed with 8–10 stones placed on top (Fitzpatrick 1978). At Chateaux d'Oex (Switzerland), one bird used lichen to conceal bread cached in crevice in wooden fence (Strahm 1960). In cap-

tivity, food concealed carefully with rapid to and fro movements of bill; if crevice in vertical wall, food covered with dirt and spiders' webs removed from wall with bill-scissoring movement; on ground, food hidden under stones up to 5 cm in diameter (Goodwin 1986).

Diet in west Palearctic includes the following. Invertebrates: dragonflies (Odonata), grasshoppers, etc. (Orthoptera: Gryllotalpidae, Gryllidae, Acrididae), bugs (Hemiptera: Reduviidae), larval Lepidoptera, flies (Diptera: Tipulidae), Hymenoptera (ants Formicidae), beetles (Coleoptera: Carabidae, Scarabaeidae, Elateridae, Curculionidae), spiders (Araneae), small molluscs (Mollusca), earthworms (Lumbricidae). Also small amphibians, reptiles, birds eggs and nestlings, small rodents, and carrion. Plant material mainly fruits, seeds, and shoots of pine *Pinus*, juniper *Juniperus*, rowan *Sorbus*, pear *Pyrus*, cherry and blackthorn *Prunus*, rose *Rosa*, holly *Ilex*, ivy *Hedera*, olive *Olea*, sea buckthorn *Hippophae*, barberry *Berberis*, hackberry *Celtis australis*, mulberry *Morus*, grape *Vitis*, orange *Citrus*, dogwood *Cornus*, virginia creeper *Parthenocissus*, *Vaccinium*, hemp *Cannabis*, and grasses (Gramineae) including grain. Also range of human-provided foods such as bread, cheese, cooked fruits, sultanas, and potato. (Reiser 1926; Whistler and Harrison 1930; Lane 1957; Turček 1961; Glutz von Blotzheim 1962; Baumgart 1967; Warncke 1968; Chappatte 1980; Dejonghe 1984; Goodwin 1986; Dendaletche and Saint-Lebe 1988; Raboud 1988.)

In Abruzzo (central Italy), late autumn and winter diet mainly plant material, especially *Juniperus* berries and Rosaceae fruits (Lovari 1978). In western Bulgaria, pellets (date unknown) contained remains of fruits of rose, juniper, and dogwood (Baumgart 1967). In winter, Switzerland, one pellet collected from roost at 1700 m contained *c.* 20 beetles, mainly Curculionidae (Warncke 1968). In October, in Pyrénées (southern France), gut of one bird contained one Reduviidae and 40 Acrididae (mainly *Stenobothrus* and *Omocestus*) (Whistler and Harrison 1930). In winter in central Caucasus, stomachs crammed with *Hippophae* berries (Boehme 1958). In winter, Wengen (Switzerland), crop of one bird contained bread, apples, mashed potato, and 20 *Clausilia* snails (Rothschild 1957). In winter in Switzerland, stomachs contained *Sorbus* berries, hemp seed, and young pine *Pinus* shoots; summer stomachs contained insects and small snails (Fatio and Studer 1889). In Picos de Europa, insects taken May–November, with beetles and grasshoppers important from late autumn to early winter; between December and April diet consists mainly of vegetable material (García Dory 1983). For diet outside west Palearctic, see Ludlow (1928), Pek and Fedyanina (1961), Kovshar' (1966), Salikhbaev and Bogdanov (1967), Abdusalyamov (1973), and Korelov *et al.* (1974).

Diet of young little known, though old nest in Alps contained old snail-shells and remains of beetles, grasshoppers, and other insects (Comte 1926). In western

Pyrénées, regurgitations of nestlings being ringed were all Orthoptera, and these also abundant in faeces (Dendaletche 1988). WGM

**Social pattern and behaviour.** Not well known; the only intensive and long-term study of marked birds is by Büchel (1983) in Switzerland. Data on non-breeding birds in Alps are given by Rothschild (1955, 1957, 1960), Strahm (1958, 1959, 1962), Voisin (1963), and Holyoak (1972b); observations at nests by Schifferli and Lang (1946), Ferguson-Lees (1958), Voisin (1968), Warncke (1968), and Büchel (1974). Other studies by Lovari (1978) (Italy), García Dory (1983) (northern Spain), Dendaletche (1988, 1991), and Dendaletche and Saint-Lebe (1991) (French Pyrénées). For general review, see Goodwin (1986).
  1. Highly gregarious throughout year, commonly in small or large flocks, but sometimes in pairs or family parties. In Swiss Alps, often in flocks of up to 300, but over 1000 at times (e.g. Glutz von Blotzheim 1962). Voisin (1965) and Delestrade (1991) recorded winter flocks of 920 and over 1000 at Monthey (Switzerland); flocks there were much smaller May-September (Delestrade 1991). Lovari (1978) found less seasonal variation in flock size in Italian Apennines, where feeding flocks smallest in spring. Pair often appears to be basic unit within flocks (e.g. Bonham 1970a, Holyoak 1972b), but many flocking birds unpaired (Büchel 1983). Highly mobile, often undertaking daily movements over distances of several km. When travelling to or from winter feeding areas, usually does so in flocks which may split up into smaller groups at or near destination. Flocks often associate with Chough *P. pyrrhocorax* (e.g. Lovari 1978, García Dory 1983, Dendaletche 1991). Rothschild (1957) observed that at a Swiss village *P. graculus* and Carrion Crow *Corvus corone* both appeared to avoid areas used by the other species, *P. graculus* keeping to open areas and *C. corone* to vicinity of trees; in summer, when *P. graculus* no longer visited the place, *C. corone* foraged everywhere. BONDS. Essentially monogamous. Many birds in flocks evidently paired throughout year (Büchel 1983). One ♂ feeding incubating ♀ was reported to be accompanied by 3rd bird (Voisin 1968: see below), and fledged young sometimes fed by birds other than parents (Rothschild 1960), suggesting occasional occurrence of helpers in breeding season. Dendaletche and Saint-Lebe (1991) recorded one instance of bigamy: 2 ♀♀ laid total of 8 eggs in one nest. Pair-bond of breeding adults of long duration; 44 marked pairs studied over 10 years by Büchel (1983) were recorded staying together for 1-8 years (average 2·5). Nest-building apparently by ♀, using material brought by ♂. Incubation by ♀, fed at nest by ♂ (see Heterosexual Behaviour, below). ♀ broods small young and feeds them on food brought by ♂; larger young fed by both sexes. Post-breeding dependence of juveniles prolonged. Age at first breeding probably 2 years or more but not reliably known. High proportion of birds are non-breeders (Büchel 1983): of 300 birds in whole of local Swiss population, no more than 13-17% were paired and performing Courtship-feeding, and not all of these necessarily bred, implying non-breeders made up at least 83-87%. In Grisons (Switzerland), Raboud (1988) estimated 40% of non-breeders in population of *c.* 140; near Chamonix (France), feeding birds comprised *c.* 77% non-breeders (Delestrade 1989). Hybridization in the wild with *P. pyrrhocorax* recorded several times; such hybrids may occur when *P. pyrrhocorax* has no conspecific mate available and may be facilitated by dominance of *P. pyrrhocorax* over *P. graculus* (Mattes and Bürkli 1979). BREEDING DISPERSION. Poorly understood. Pairs often solitary but also in colonies of up to 20 or more (Glutz von Blotzheim 1962; Sch-

ifferli 1982). Reports of colonial nesting much repeated in literature (e.g. Etchécopar and Hüe 1967, Hüe and Etchécopar 1970, Flint *et al.* 1984), but few are well documented: Baumgart (1967) recorded colonies in deep shafts in ground in Bulgaria, 2 of which had 15-20 and 8-10 breeding pairs; Dendaletche (1991) recorded 3 pairs in cave in Pyrénées. Schifferli *et al.* (1982) recorded 20-25 pairs along 7·5 km of crags in Switzerland, with closest nests 70 m apart; Raboud (1988) found distances between nests varied from 600 m to 4·8 km in Grisons. Territory may be defended, since Büchel (1983) found that intruding conspecifics were attacked within *c.* 50 m of nest of solitary pair. Breeding densities not well documented, but generally appear to be low in west Palearctic: 26 breeding pairs in 16 km² (1·6 pairs per km²) in Grisons (Raboud 1988). Other estimates of total density in summer doubtless include non-breeders: Lovari (1978) found 250 ± 10 birds in *c.* 400 km² of Apennines; in France, 285 ± SD114 birds per 20 km² (*n*=6) in Corsica, 220 ± 93 (*n*=18) near Vercours, and 120 ± 112 (*n*=72) near Chamonix (Delestrade 1991). Some nest-sites are re-used regularly, otherwise no information on site-fidelity from year to year (Büchel 1983). ROOSTING. Roosts in caves, large rock crevices, or holes in crags, usually communally, at least outside breeding season (in Apennines, such communal sites also used for breeding: Lovari 1978). Sometimes nests in chalets or other buildings in Alps, roosting at same sites during breeding season. Flies back up into mountains to roost even when feeding by day at much lower elevations (Rothschild 1955; Büchel 1983; Goodwin 1986). However, winter roosts are sometimes at lower elevations than usual breeding sites. In Switzerland, communal roost of 25-50 birds was occupied in July as well as winter (Rothschild 1955). In Bulgaria, up to 500 recorded at communal roost in July (Baumgart 1967); in Pyrénées, *c.* 100 at roost (Dendaletche 1991). Sometimes shares communal roosts with *P. pyrrhocorax* (Mayaud 1933b; Lovari 1978; Dendaletche 1991).
  2. Often quite tame where fed by tourists in mountain regions such as Swiss and Austrian Alps, approaching to 2 m or less for food (e.g. Bonham 1970a, Büchel 1983); also tolerates human disturbance at some nest-sites in buildings. Nonetheless, although habitually seen taking food from humans, Delestrade (1989) found that in French Alps they prefer to feed on food scraps after picnickers leave. In Himalayas also tame, entering villages and encampments for food, but not known to roost or nest in buildings there, unlike otherwise shyer *P. pyrrhocorax* (Ali and Ripley 1972). Frequently vocal at all times of year, giving Advertising-calls (see 2 in Voice) most commonly. This call often but not always accompanied by wing-flirting movements very similar to those of *P. pyrrhocorax* (see Fig A in that species) and apparently given in similar contexts. Flocking birds give 3 types of call as response to disturbance or threat (Büchel 1983): Take-off call (see 3 in Voice), given in many contexts; Ground-predator alarm-call (4a in Voice) reported as response to dog, cat, and (once each) marten *Martes foina*, Eagle Owl *Bubo bubo*, and Raven *C. corax*; Flying-predator alarm-call (4b in Voice), in response to mobbing *C. corone*, *C. corax* (see also Holyoak 1972b), paper kite shaped like raptor, and helicopter (Büchel 1983). High-flying birds attacked by raptor were seen to plunge down at great speed (Strahm 1958). Flocks also recorded mobbing Golden Eagle *Aquila chrysaetos* in flight (Nigg 1974; Gubler 1978). In Elburz (northern Iran), several birds dive-attacked human observers climbing ridge and one seen to mob bear *Ursus arctos* (Norton 1958). Lane (1957) described pair closely following hunting stoat *Mustela erminea* without mobbing it, perhaps in hope of catching mice disturbed. FLOCK BEHAVIOUR. Flocks are highly vocal and habitually indulge in spectacular aerobatics, perhaps even more commonly than *P.*

*pyrrhocorax* and Jackdaw *Corvus monedula*. Habitually glides and soars, flocks sometimes rising to great heights on strong updraughts, then often diving in apparent play or as prelude to landing (e.g. Rothschild 1955). Members of flock descending to seek food will often dive steeply, checking and then dashing about just above ground before landing (Rothschild 1955; Goodwin 1986). Members of feeding flocks sometimes rather loosely dispersed on grassland or rocky slopes, at other times clustered, e.g. at artificial food sources. In flocks, individual-distance of rather less than one body length normally maintained (e.g. Büchel 1983). Detailed study of aggressive interactions within groups taking artificial food established existence of well-marked feeding hierarchy. Paired ♂♂ in highest positions, ♀♀ lowest. Unlike ♀♀, unpaired ♂♂ could improve rank by pairing. Observations following experimental dyeing of bill and feet to simulate immature coloration indicated this might influence rank. One high-ranking ♂ adopted new call (Dominance-call: see 6a in Voice) which he used regularly thereafter, and gradually other high-ranking ♂♂ adopted same call; 2 ♂♂ used it only as long as they were paired and thus high-ranked, but not thereafter; call thus apparently used to manifest an already established ranking system rather than to improve position of individual birds (Büchel 1983). Clearly subordinate to *P. pyrrhocorax* in mixed-species flocks (Mattes and Bürkli 1979). In Grisons, Raboud (1988) found that non-breeders did not flock with breeding birds feeding on pasture in breeding season, but from time of fledging through July and August non-breeders did form flocks with breeding birds and their fledged juveniles; such flocks largest around midday; in September, flocks small again, as in breeding season. Swiss birds studied by Rothschild (1955) left communal roost (see above) from *c.* 1 hr after dawn, flying rapidly to feeding areas in pairs, small parties, or flocks. For call given when flocks approach feeding grounds, see 2b in Voice. Time at which birds feeding at lower levels return to roosting area varies from midday until shortly before dusk. Early return apparently encouraged by fine weather, visibility of roosting place from feeding grounds, and probably by possibility of obtaining food near roost. Birds often leave feeding areas in large flock which later subdivides; for call given when leaving feeding grounds for roost, see 3 in Voice. (Rothschild 1955; Goodwin 1986.) Büchel (1983) also found very rapid movements from roost-site to feeding areas: 8·3 km travelled in 2·5 min of extremely fast downward-gliding flight. In Italian Apennines, Lovari (1978) found movement from communal roost to feeding grounds 30–45 min after sunrise; some dispersal of flocks occurred during day, then flocks formed again before return to roost. At mixed roost, left later in morning than *P. pyrrhocorax* and returned earlier in evening. Unlike *P. pyrrhocorax*, returned to roost in large flock. On clear days, pre-roosting birds perched in sun and called almost continually, calling reaching peak immediately after entering roost crevice. (Lovari 1978.) See Dendaletche and Saint-Lebe (1988) for additional data on timing of entering and leaving roosting site. Birds in flocks and pairs give flight-intention movement consisting of quick upward flick of wing-tips in which they are lifted about 1 cm and slightly opened (apparently identical to flight-intention movement of *P. pyrrhocorax*: see Fig B in that species, p. 112) (Holyoak 1972b). SONG-DISPLAY. No ritualized display described, but see 1 in Voice for so-called song. ANTAGONISTIC BEHAVIOUR. Aggression over scraps provided by man reported in French Alps (Holyoak 1972b; Delestrade 1991), although considered by Goodwin (1986) to be rather unaggressive with little or no quarrelling over food, even in winter at artificial food supplies. Much aggression reported in Italian Apennines, with birds often supplanting each other in feeding flocks (successfully in 90% of 153 observations), although overt threats less common

(9·8%) and fighting seen only once (at nest-site: see below) (Lovari 1978). Study by Büchel (1983) revealed frequent aggressive interactions when artificial food provided, serving to maintain hierarchy (see above). Threat-postures used were Bill-thrusting threat (19·1% of 1412 aggressive acts: aggressor thrusts bill at another to displace it from food), Wing-spreading threat (16·2%: aggressor stands upright with head raised and lifts closed- or partly-opened wings high, Fig A; tail held low and may be more or less widely fanned, Fig B, which shows intense display), Head-up threat (14·8%: aggressor stands upright, often with bill pointing upwards, Fig C), Forward-threat (5·3%: aggressor crouches low with body near-horizontal, plumage more or less ruffled, bill pointing forward, wings drooped, and tail more or less spread, Fig D; used less often

than Head-up threat and signals determination); other elements of threat behaviour were Aggressive-approach (14·0%), Bill-snapping (5·9%), stepping on opponent, e.g. on tail-tip (5·5%), Threat-call (see 2e in Voice: 4·3%), Dominance-call (see above and 6a in Voice: 4·2%), and Threat-chatter (see 2f in Voice: 3·1%); actual fighting infrequent, usually involving Tail-pecking (2·4%); at higher intensity, birds fought breast-to-breast, on ground (2·6%) or in air (1·3%); aerial pursuit 1·1%. Response to threat was usually retreat (68·2% of 403 defensive acts); less often, threatened bird adopted Crouch-low posture with head lowered (29·3%) or turned head away (2·5%). Bill-thrusting, Head-up, Forward-threat, Aggressive-approach, retreat of threatened bird, and fighting at food also described by Holyoak (1972b), who noted that captives directed Forward-threat and Head-up at both conspecifics and other birds, often in contexts where food not in dispute. Fighting for 3 min recorded at nest-site in May (Lovari 1978). HETEROSEXUAL BEHAVIOUR. (1) Pair-bonding behaviour. Apparent sexual displays recorded only in groups of birds. These Group-displays, described by Voisin (1966) and investigated in detail by Büchel (1983), probably have function in pair-formation. Usually involve 2–8 birds (2–22, average 5·4, n = 106) mostly unpaired birds of high or moderately high rank (71·7%, n = 205); paired birds relatively scarce (7·2%), young birds absent. Group-display may occur in all months of year, and in various places, e.g. on road, snow-patch, or when perched on fence. Within group, birds perch closer together than normal, individuals almost touching; one bird gives Group-display call (see 2g in Voice) in Group-display posture (Fig E: head raised, bill pointing up at steep angle, wings slightly

E

lowered and slightly spread, plumage rather ruffled, especially on throat, and tail lowered and often fanned). This provokes one or several others to use same call and posture, so that all or part of group may display noisily at same time (Fig F, showing 8 birds from group of 13). Function of Group-display needs confirmation; copulation not seen to follow it and precise mechanism of mate-selection (if that is function) remains unclear. Behaviour used in maintenance of pair-bond as follows (also Courtship-feeding: see subsection 2, below). Pair-call (see 2c in Voice) used to maintain contact and often characteristic of individuals; Greeting-call (see 2d in Voice) also used only by paired birds. Characteristic wing-flicking movements in flight, and flights close together, are also restricted to paired birds. Close proximity when perched is typical of pairs. (Büchel 1983.) Allopreening (head, neck, and mantle) occurs regularly with paired birds (Holyoak 1972b). At least in captive birds, Allopreening may not always be confined to members of pair (Goodwin 1986). Warncke (1968) reported what may have been courtship-display at communal roost, where ♂♂ made nest-shaping movements with pieces of stem while uttering unceasing

calls. Other than this, courtship-display not reported from lone pairs and perhaps restricted to Group-display (Büchel 1983). (2) Courtship-feeding. Occurs all year, most frequently in breeding season: ♂ gives Feeding-call (see 5 in Voice) and regurgitates food from throat; ♀ approaches and begs by fluttering partly opened wings and raising open bill, like begging juvenile. Courtship-feeding restricted to paired ♂♂; high-ranking ♂♂ feed mates more often than lower-ranking ♂♂. (Büchel 1983.) For ♂ feeding ♀ on nest, see subsection 4 (below). (3) Mating. Copulation not described adequately; Büchel's (1983) study and earlier reports led to belief that mating occurs in flight. 8 observations (all in late March or April) supported idea of copulation occurring while pair flying in very close proximity, with body-to-body contact for c. 30 s, before resumption of normal 'close-flight' with wing-flicking; Courtship-feeding sometimes followed. However, possibility that mating occurs at nest is not discounted, and Dendaletche and Saint-Lebe (1988) saw copulation on ground; breeding in captivity (e.g. Sitasuwan and Thaler 1984) and hybridization with ♂ *P. pyrrhocorax* (which mates on ground) in the wild and captivity indicate that aerial copulation (if it occurs) is not indispensable. (4) Nest-site selection and behaviour at nest. Warncke (1968) observed apparent nest-site selection, in which ♂ flew to future nest-site, calling loudly as he landed; ♀ repeatedly flew past him without landing but site was later used. In captivity, ♂ brings nest-material and ♀ builds (Sitasuwan and Thaler 1984). Incubation by both sexes claimed in old literature (Sharpe and Dresser 1871–81) but this probably exceptional or wrong; normally by ♀ only (Voisin 1968; Blaser 1970; Büchel 1974, 1983; Sitasuwan and Thaler 1984), fed at nest by ♂ on regurgitated food (Büchel 1974, 1983), who may come in company with other birds on feeding visits (Ferguson-Lees 1958). At nest watched by Voisin (1968), ♂ usually accompanied by 3rd bird, which did not feed ♀ and only once went into building where nest was. RELATIONS WITHIN FAMILY GROUP. Eggshells are carried out of nesting cave (García Dory 1983). At one nest, young brooded by ♀ for c. 80% of time during first 3 days (Voisin 1968). ♀ feeds small young on food regurgitated by ♂; larger young fed by both sexes (Voisin 1968; Büchel 1983; Sitasuwan and Thaler 1984). Droppings of nestlings removed by ♀ (Voisin 1968). For development of young, see Voisin (1968), Büchel (1974), and (for *P. graculus* × *P. pyrrhocorax* hybrid in captivity) Thaler (1977). For fledging behaviour, see Dendaletche and Saint-Lebe (1988). Young may leave nest but remain in nest-cavity for c. 10 days before flying (Glutz von Blotzheim 1962; Sitasuwan and Thaler 1984). Fledged young fed for some time by parents, from which they beg with calls (see Voice) and movements of partly opened wings. Juveniles receive food well into August in Switzerland (Glutz von Blotzheim 1962; Goodwin 1986); sometimes fed by birds other than parents (Rothschild 1960). ANTI-PREDATOR RESPONSES OF YOUNG. García Dory (1983) noted that recently fledged young entered nest-cave immediately they detected danger. No other details. PARENTAL ANTI-PREDATOR STRATEGIES. Apparently undescribed.

F

(Figs by D Nurney: A–C from drawings in Sitasuwan and Thaler 1984; D from drawings in Coombs 1978 and Sitasuwan and Thaler 1984; E–F from photographs in Büchel 1983.)                                        DTH

**Voice.** Conspicuously vocal throughout year. Most information on nominate *graculus*; the few data and recordings of *digitatus* suggest general similarity. Studied (with sonagrams) in Switzerland by Büchel (1983) and in captivity by Sitasuwan and Thaler (1985), latter including comparisons with Chough *P. pyrrhocorax* and hybrids. Both studies show that calls which are only slightly different may be used in widely different contexts and that calls of conspecifics are often copied (sometimes for deliberate deception: Sitasuwan and Thaler 1985). Thus, function of many calls and significance of their variants remain poorly understood, so following account provisional and attempts only to distinguish main types. For further sonagrams, see Bergmann and Helb (1982). Stadler (1927) described calls using musical notation. High-pitched calls of *P. graculus*, and to lesser extent those of *P. pyrrhocorax*, may be adaptations to montane habitats (Lovari 1978). Bill-snapping used in threat (Büchel 1983, which see for sonagram): single hard snap, short clacking series, or occasionally a longer 'chattering' burst.

CALLS OF ADULTS. (1) Song. Succession of warbling, squeaky, chittering, and churring sounds reported from birds apparently at ease and in flocks or pairs feeding together (Holyoak 1972*b*). 'Subsong' of Sitasuwan and Thaler (1985, which see for sonagrams) evidently similar. Not known which sex makes these sounds. Apparently similar to 'song' in *P. pyrrhocorax* and likewise of obscure function. (2) Advertising-calls. Much the commonest call is high-pitched, penetrating 'chree' or 'tree', given singly or repeated, both with and without Wing-flirting movements (see Social Pattern and Behaviour); musical, whistling 'trree' or 'sree', rather high-pitched and suggestive of football whistle (Goodwin 1986). Heard from flocks in flight or about to take off (and by bird apparently looking for its mate or young: Goodwin 1986); apparently serves as advertising- and contact-call, besides having variants with other functions (see below). Call suggested by Goodwin (1986) to permit instant individual recognition of bird's own mate or young. Varies considerably, with extremes sounding like high, squeaky 'kee', 'squee', or 'skweea' (the last very like squeaky version of call 2 of *P. pyrrhocorax*), usually given without wing movements, apparently as contact-call. The more aggressive the bird, the nearer the call is to rippling 'chree', and such calls always accompanied by Wing-flirting (Holyoak 1972*b*). Several variants of this call or apparently similar calls with other functions have been recognized (Rothschild 1955, 1957; Lovari 1978; Büchel 1983; Sitasuwan and Thaler 1985); most distinctive are as follows. (2a) Contact-call. A 'cher', produced most frequently when in flocks (Lovari 1978); churring 'rrrr' or 'rreee' (P J Sellar: units 1 and 3

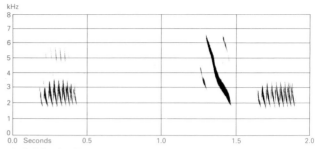

I  J-C Roché  France  July 1987

in Fig I). Also described as sharp 'dschirr' (heard as 'prri' from distance) and not uncommonly combined (as in Fig I) in loose series with alarm-whistle (see call 4b) (Bergmann and Helb 1982). Flight-call of Sitasuwan and Thaler (1985) apparently the same. (2b) Excitement-call. A 'cheew' or 'cheew-cheew' when mobbing predator, when pair in sight of nest-hole, when flock about to land, or when birds take off due to disturbance (Lovari 1978). See Sitasuwan and Thaler (1985) for sonagrams. May be same as 'arrival cry' described by Rothschild (1955, 1957) as series of shrill, double, evenly pitched sounds, given by birds arriving at feeding grounds. (2c) Pair-call. A 'wre', 'wrii', 'wro', or 'wrü', often individually distinctive (Büchel 1983). Fig II shows calls apparently of this type.

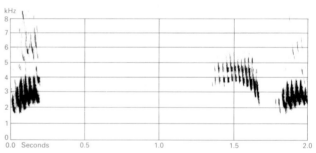

II  A Ausobsky/Sveriges Radio (1972–80)  Austria  June 1964

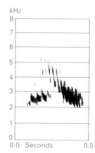

III  J-C Roché  France  June 1962

(2d) Greeting-call of paired birds. A 'wria-wree', 'wria-wrii', or 'wrüi-wree' (Büchel 1983). Call rendered 'prrrEEEaa' (Fig III) is perhaps 1st part of this sequence; see Bergmann and Helb (1982) for similar 'psria'. (2e) Threat-call. A 'diüpp' or 'drupp' (Büchel 1983). 'Aggressionslaut' of Sitasuwan and Thaler (1985) is apparently the same. (2f) Threat-chatter. A 'wrrrl' or 'wirr-wierl-wiirl' (Büchel 1983). (2g) Group-display call. Repeated 'ziupp' by birds giving Group-display (Büchel 1983). Called 'Balzgesang' by Sitasuwan and Thaler (1985). (3) Take-off call. Series of short sounds, 'ke-ke-ke-ka-ka' (Büchel 1983). Alarm-calls of Sitasuwan and Thaler (1985) apparently similar. May also be same as 'departure cries' described by Rothschild (1955) as somewhat like hurried chattering of Jackdaws *C. monedula* in flight but louder and more sustained, given by birds returning to roost or preparing to do so. (4) Alarm-calls. (4a) Ground-predator alarm-call. Harsh, churring 'chrrr' in response to predator or other threat on ground (Büchel 1983: see Social Pattern and Behaviour). In Fig IV, harsh

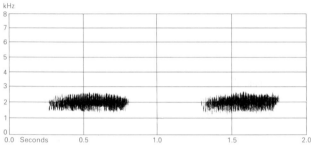

IV A Ausobsky/Sveriges Radio (1972–80) Austria June 1964

'karrr' like crow *Corvus* (P J Sellar); in recording by J-C Roché, similar calls are a more buzzy 'shri' or 'zhree'. This appears to be the unspecific alarm-call of Sitasuwan and Thaler (1985), who gave additional sonagrams. Reported by Holyoak (1972*b*) when flocks threatened by Raven *C. corax*, suddenly disturbed by man, or when large falling rock landed nearby. Lovari (1978) rendered this call 'krrrr' and noted that it was never heard from young less than 6 months old. Goodwin (1986) described what appears to be same call, given in apparent alarm, as harder-sounding, less musical version of call 2, rendered 'chrrurr'. (4b) Flying-predator alarm-calls. High-pitched whistling 'wiii' (Büchel 1983), 'pija', 'psia', or short 'dju' (Bergmann and Helb 1982), combined with low-pitched rattle (Büchel 1983), when flying predator or other aerial threat noticed (see Social Pattern and Behaviour). 1st call in this combination, a liquid descending whistle, 'sweea' or 'pseee' (P J Sellar), is shown in Fig I (middle unit) in series with call 2a (see Bergmann and Helb 1982 for same combination). (5) Feeding-call of ♂. Series of sounds rendered 'wrääg-wrr-wierl-wägagaga' given prior to Courtship-feeding by paired ♂ as invitation to ♀ (Büchel 1983). 'Fütterlaut' of Sitasuwan and Thaler (1985) is probably the same. (6) Other calls. (6a) Dominance-call. Rendered 'dipperiudidipp' (Büchel 1983); see Social Pattern and Behaviour. (6b) Fear-call listed by Sitasuwan and Thaler (1985) without description.

CALLS OF YOUNG. Nestlings give thin high-pitched 'zi' or 'zi-zi-' etc. Fledged young waiting near parents gave rather squeaking, nagging 'pee(a) pee(a)', suggestive both of squeaking of young pigeon *Columba* and of mewing call of Black-headed Gull *Larus ridibundus* (Goodwin 1986); calls repeated more quickly and with increased emphasis when directly soliciting food from parents. See Sitasuwan and Thaler (1985) for sonagrams. Apparently the same call (from *digitatus*) described as curious persistent mewing like 'cheep' of half-grown chicken (Ali and Ripley 1972). Fledged young give squeakier versions of adult Contact-call (2a) and Excitement-call (2b) (Lovari 1978).    DTH

**Breeding.** SEASON. Alps: eggs laid early May to mid-June, rarely April (Glutz von Blotzheim 1962; Codourey 1966; Beaud and Manuel 1983). Morocco: nests mid-May to July; young recorded early June at 2000 m and early July at 2550 m (Roux 1990). Lebanon: probably June–July; fledged young mid-July (Tohmé and Neuschwander 1978). SITE. Ledge or crevice in cave, cliff, tunnel, shaft, or building (Glutz von Blotzheim 1962; Géroudet 1964; Büchel 1983). Access sometimes through small (e.g. 50 cm) entrance hole; site sometimes more than 10 m from entrance and often in darkness (Glutz von Blotzheim 1962; Goodwin 1986); in study in western Pyrénées, 90% of entrances sheltered by whitebeam *Sorbus* (Dendaletche and Saint-Lebe 1991). In building, often on wide beam as little as 6 cm from roof; tolerates human disturbance in factories, etc. (Tintori 1964; Codourey 1966; Voisin 1968; Blaser 1970; Beaud and Manuel 1983). 20–30 nests in Bulgaria 15–18 m down rock shafts (Baumgart 1967); in Picos de Europa (northern Spain), 2 nests 10 and 12·5 m from cave entrances, on ledges 2·5 and 3·5 m from cave floor (García Dory 1983); nests in buildings in Alps 2–9·5 m from ground (average 4·4 m, *n*=6) (Tintori 1964; Codourey 1966, 1968; Voisin 1968; Blaser 1970; Beaud and Manuel 1983). Same site may be used in successive years (Voisin 1968). Nest: bulky mass of sticks (up to 50 cm long), roots, twigs, moss, and plant stems; lined with neat, compact cup of grass, fine twigs, rootlets, hair, and some feathers (Schifferli and Lang 1946; Tintori 1964; Codourey 1966; Voisin 1968); 7 nests in Alps had average outer diameter 34·6 × 42·8 cm (28–50 × 35–60), inner diameter 13·5 cm (12–15), overall height 18 cm (12–30), depth of cup 6·25 cm (5–7) (Schifferli and Lang 1946; Tintori 1964; Codourey 1966; Voisin 1968; Beaud and Manuel 1983); 2 nests weighed 590 g and 600 g, lining 90 g and 110 g (Voisin 1968). Building: in captivity, by both sexes, ♂ mainly bringing material and ♀ mainly building (Sitasuwan and Thaler 1984). EGGS. See Plate 57. Sub-elliptical, smooth and glossy; whitish, tinged creamy or faintly buff, rarely greenish, profusely marked overall with small blotches, spots, and specks of dark brown, reddish-brown or olive-brown, with grey or lilac underlying markings; very like those of *P. pyrrhocorax* (Harrison 1975; Makatsch 1976; Goodwin 1986). Nominate *graculus*: 38·2 × 26·8 mm (34·2–42·3 × 23–28·5), *n*=111; calculated weight 14·1 g (Schönwetter 1984). Clutch: 3–5(–6) (Reiser

1926; Schifferli and Lang 1946; Glutz von Blotzheim 1962). Of 15 clutches in Alps: 3 eggs, 13·3%; 4, 53·3%; 5, 33·3%; average 4·2 (Glutz von Blotzheim 1962; Codourey 1968; Blaser 1970; Büchel 1983). Eggs laid at intervals of 1–2 days in captivity (Sitasuwan and Thaler 1984); 2-day intervals at one nest in Switzerland (Büchel 1983). One brood, though replacement clutches possible (Glutz von Blotzheim 1962; Wüst 1986). INCUBATION. 18–21 days (Schifferli and Lang 1946; Blaser 1970; Büchel 1983). By ♀ only; at one nest in Switzerland ♀ left nest for only 3–6 min at a time (Blaser 1970); at another, sat for 92–95% of time during last 3 days of incubation, sitting for average periods of 40–56 min, off nest for periods of 4–6 min (Voisin 1968). Begins before clutch complete, hatching asynchronous (Glutz von Blotzheim 1962; Voisin 1968; Büchel 1983); in captivity, began with last egg, young hatching over 1–2 days (Sitasuwan and Thaler 1984). YOUNG. Fed and cared for by both parents; brooded by ♀ 80% of time during first 3 days; fed regurgitated food directly by both parents or indirectly via ♀ (Voisin 1968, which see for development of young; Büchel 1983; Sitasuwan and Thaler 1984). FLEDGING TO MATURITY. Fledging period 29–31 days (Glutz von Blotzheim 1962; Voisin 1968; Büchel 1983); 33–38 days in captivity; may leave nest but remain in nest cavity for c. 10 days before flying (Glutz von Blotzheim 1962; Sitasuwan and Thaler 1984). Age of first breeding probably 2–3 years (see Social Pattern and Behaviour). Young fed by parents, and perhaps sometimes by other conspecifics well into August in Switzerland (Glutz von Blotzheim 1962; Goodwin 1986). BREEDING SUCCESS. In captivity, higher than P. pyrrhocorax (Sitasuwan and Thaler 1984). Of 17 eggs laid in 4 nests in France, Spain, and Switzerland, 14 hatched; of 17 nestlings in 5 nests, 9 fledged; overall egg to nestling survival 43% (Tintori 1964; Codourey 1968; Voisin 1968; Beaud and Manuel 1983; García Dory 1983). Of 180 eggs in 36 nests over 3 years, western Pyrénées, 39% produced 3-week-old young, all of which fledged, average 1·94 per pair (Dendaletche and Saint-Lebe 1991).                    WGM

**Plumages.** (nominate *graculus*). ADULT. Entirely black, slightly glossy on head, body, and lesser wing-coverts (gloss not as strongly purplish or greenish as in Chough P. pyrrhocorax, faintly metallic blue-black at most), less glossy and more sooty-black on tail-coverts, tail, flight-feathers, tertials, and greater wing-coverts; hidden feather-bases of head and body with restricted amount of dark grey. *In worn plumage* (late spring and summer), entire plumage still duller black; some dark grey of feather-bases shows through on upper mantle, breast, or belly. Sexes similar, but (when age properly determined) many separable by length of wing or tail (see Measurements) or depth of bill (see Structure). NESTLING. Down short, sparse, on upper-parts and upperwing only (Voisin 1968); grey (Schifferli and Lang 1946). JUVENILE. As adult, but head, body, and lesser and median wing-coverts dull sooty-black, without gloss, feathers much shorter and looser in structure than in adult, appearing rather woolly on rump and underparts. Tail and remainder of wing as adult, but fresh and faintly glossy when flight-feathers

and tail of adult usually worn or in moult, somewhat browner than freshly moulted adult. As in *P. pyrrhocorax* (see p. 117), many (not all) birds have (p9-)p10 and tail-feathers rather pointed at tip, not as rounded as in adult. FIRST ADULT. As adult, but juvenile tail, flight-feathers, tertials, greater upper primary and secondary coverts, and bastard wing retained. Tips of (p9-)p10 and tail-feathers often more pointed than in adult (see above and Svensson 1992); tail-feathers, primaries, and bastard wing often browner than adult, contrasting with black of remainder of plumage, tips more worn than in adult at same time of year. *In worn plumage*, body has more grey of feather-bases showing than in adult, body often appearing tinged grey-brown, less uniform black, especially underparts. Less often reliably aged than *P. pyrrhocorax*; in particular, 1st adult ♂ sometimes hard to distinguish from adult ♀.

**Bare parts.** ADULT, FIRST ADULT. Iris dark brown or dark greyish-brown. Bill yellow, often tinged green (especially at base), golden, or slightly orange; in captivity (and perhaps in the wild), sometimes almost white. Mouth black. Leg and foot orange-red, light orange, salmon-orange, yellow-orange, or coral-red, in captivity sometimes pale yellow; occasionally deep red. (Goodwin 1986.) NESTLING. Mouth flesh-pink (Goodwin 1986). JUVENILE. Iris brown. At fledging, bill dusky horn or pale reddish-white with dark horn tip and black culmen ridge; gape-flanges white; mouth pink-red or reddish-pink; leg and foot olive-brown, grey, or dirty yellow, scutes partly or entirely dark brown or blackish; soles dirty yellow, brown, or grey. Bill bleaches to yellow from early July onwards. In 1st winter and early spring, leg more or less blackish with some orange or red marks, mouth yellow; later in spring, leg red or orange, but mouth still yellow. (Lang 1946; Schifferli and Lang 1946; Voisin 1968; Goodwin 1986; ZFMK.)

**Moults.** ADULT POST-BREEDING. Complete; primaries descendent. Starts with shedding of innermost primary (p1) mid-June to mid-July, completed with regrowth of innermost secondary (s6) mid-September to late October, both up to c. 1 month earlier in 1-year-olds. In captive Austrian bird, p10 shed c. 80 days after p1, duration of growth of each primary 29–40 days, depending on length; tail moult centrifugal, but t2–t4 shed almost simultaneously; t1 shed at about same time as p3, t6 with p7–p8, each feather full-grown after c. 32 days; tertials moult in sequence s8–s9–s7, s8 shed with p6, s7 with p7–p8; secondaries start with s1 at about same time as shedding of p6; s4 completed at about same time as last primary (p10), but s5–s6 not until 2–3 weeks later (Winkler *et al.* 1988). Birds in moult examined (Alps, Crete, Turkey, Iran, Tibet: RMNH, ZMA, ZMB) as well as data from literature (e.g. Nepal, Diesselhorst 1968; USSR, Dementiev and Gladkov 1954; China, Stresemann *et al.* 1937) fit into above scheme, but 2 birds from Tien Shan in heavily worn plumage in early September and early October (Keve and Rokitansky 1966) were perhaps aberrant. POST-JUVENILE. Partial: head, body, and lesser upper wing-coverts plus variable number of median coverts. Fledged birds in fully juvenile plumage appear from late July (Winkler *et al.* 1988); fully juvenile birds examined from 6 July (Crete), late July (Croatia, Nepal), and mid-August (Tien Shan, Russia); birds in moult 28 July (USSR: Dementiev and Gladkov 1954), early August (Nepal: Diesselhorst 1968), and up to mid-October (Alps: SMTD, ZMB); in 1st adult plumage from late September or early October (China and Alps: RMNH, ZMB).

**Measurements.** Nominate *graculus*. Alps, all year; skins

(RMNH, SMTD, ZFMK, ZMA, ZMB). Bill (S) to skull, bill (N) to distal corner of nostril; exposed culmen on average 4·8 less than bill (S). Juvenile wing and tail include retained juvenile wing and tail of 1st adult.

| | | | | |
|---|---|---|---|---|
| WING AD | ♂ | 277·4 (5·18; 10) 270–284 | ♀ | 258·6 (6·30; 7) 251–267 |
| JUV | | 264·2 (6·32; 9) 255–274 | | 253·2 (3·31; 9) 248–259 |
| TAIL AD | | 168·8 (5·78; 10) 161–179 | | 161·8 (3·49; 6) 157–165 |
| JUV | | 161·4 (6·73; 9) 152–168 | | 156·5 (5·60; 6) 147–162 |
| BILL (S) | | 34·9 (1·28; 12) 33·2–37·2 | | 34·0 (1·48; 11) 32·2–36·4 |
| BILL (N) | | 22·4 (1·26; 12) 20·7–24·0 | | 21·4 (0·30; 11) 19·7–23·3 |
| TARSUS | | 46·9 (1·17; 13) 45·3–48·5 | | 45·1 (1·20; 11) 43·4–46·8 |

**Weights.** Nominate *graculus*. Switzerland: December–April, adult 210 (4) 188–240, juvenile 195 (11) 168–246; March, ♂ 251·5 (Lang 1946); January, ♀ in poor condition 168 (ZMA).
  *P. g. digitatus*. Afghanistan, June: ♂ 212·8 (16·46; 4) 191–231 (Paludan 1959). Kazakhstan: ♂ 248·2 (5) 202–280, ♀♀ 210, 215 (Korelov *et al.* 1974). Kirgiziya: 228–250 (3) (Yanushevich *et al.* 1960). Nepal: July, ♂ 228·3 (3) 223–239, ♀ 203; September, 206, 213 (Diesselhorst 1968).

**Structure.** Wing rather long, fairly broad at base, rounded at tip. 10 primaries: p7 longest, p8 (1–)3–12 shorter, p9 30–42, p6 (1–)4–12 (adult) or 0–6 (juvenile), p5 26–38 (adult) or 20–32 (juvenile), p4 44–55 (adult) or 42–52 (juvenile), p1 78–93 (adult) or 79–105 (juvenile). P10 somewhat reduced, 92–120 shorter than p7, 46–57 longer than longest upper primary covert. Outer web of p6–p8 emarginated, inner web of p7–p10 with fairly distinct notch. Tertials short, longest almost equal in length to inner secondaries. Tail rather long, tip rounded; 12 feathers, t6 15–23 (adult) or 10–16 (juvenile) shorter than t1; at rest, tip extends distinctly beyond wing-tip. Bill short, just over half head length; straight, but tip of culmen distinctly decurved (not as gradual and gently as *P. pyrrhocorax*); tip of upper mandible with slight hook; bill markedly small and slender compared with body size, scarcely larger than that of Blackbird *Turdus merula*. Depth of bill at middle of nostril 9·7 (10) 9·1–10·6 in ♂, 9·2 (10) 8·5–9·7 in ♀. Base of upper mandible with dense and rather long tuft of bristle-like feathers, projecting from side forward, covering small oval nostril. Tarsus and toes rather strong; tarsus booted (traces of scutes in juvenile); middle toe with claw 35·0 (5) 32–40; outer toe with claw *c.* 80% of middle with claw, inner *c.* 76%, hind *c.* 85%.

**Geographical variation.** Slight, involving size (length of wing, tail, and bill) and relative length of tarsus and toes. Usually 2 races recognized, smaller nominate *graculus* in west, larger *digitatus* in east, with wing length of adult ♂ nominate *graculus* assumed to be below 280, *digitatus* over 280 (Hartert and Steinbacher 1932–8; Vaurie 1954*a*). However, wing in 33% of sample of 15 adult ♂♂ from Alps over 280, and overlap in size with eastern birds apparently large (see Measurements), though range in size of eastern birds artificially large as most authors have combined age-groups. Boundary between races as defined on p. 95 follows Vaurie (1954*a*), typical *digitatus* described from Lebanon, where now apparently extinct; very small samples from here and from south-west and north-east Iran assumed to be similar in size to eastern birds by Vaurie (1954*a*), but, now size variation of western birds better known, Lebanon and Iran birds appear to be close to nominate *graculus* or intermediate between western and eastern birds. If birds from central Asia are consistently large, and *digitatus* is included in nominate *graculus*, name *forsythi* Stolickza, 1874, is available for central Asian birds. Wing and tail of central Asian populations probably averages *c.* 13 longer than in Alps, bill *c.* 1·5 longer. Data on tarsus conflicting: in Tien Shan, tarsus averages 1·5 shorter (see Measurements), and leg and foot apparently slender also in Szechwan (China) (Mayaud 1933*b*), but tarsus longer in Zagros (Iran) and Himalayas (Vaurie 1954*a*). Thus, as in *P. pyrrhocorax*, birds in southern part of Asian range have tarsus of normal proportion compared with body size, those in northern part of Asian range have leg and foot relatively short and slender. Further study needed.

**Recognition.** For difference from *P. pyrrhocorax*, see that species.                    CSR

---

## *Pyrrhocorax pyrrhocorax*   Chough

PLATES 6 and 9 (flight)
[between pages 136 and 137]

| | | |
|---|---|---|
| DU. Alpenkraai | FR. Crave à bec rouge | GE. Alpenkrähe |
| RU. Клушица | SP. Chova piquirroja | SW. Alpkråka |

*Upupa Pyrrhocorax* Linnaeus, 1758

Polytypic. Nominate *pyrrhocorax* (Linnaeus, 1758), Britain and Ireland; *erythrorhamphus* (Vieillot, 1817), Iberia and southern France through Alps to Austria and perhaps northernmost Yugoslavia, south through Italy to Sicily and Sardinia; birds from Bretagne (western France) intermediate between *erythrorhamphus* and nominate *pyrrhocorax*, but nearer *erythrorhamphus*; *barbarus* Vaurie, 1954, La Palma (Canary Islands) and north-west Africa; *docilis* (S G Gmelin, 1774), southern Yugoslavia and Greece to Crete, east through Turkey, Levant, Caucasus, Transcaucasia, and Iran to southern Turkmeniya and Afghanistan. Extralimital: *centralis* Stresemann, 1928, mountains of Soviet Central Asia east through Altai to Mongolia and Transbaykalia, south to Kashmir and Ladakh; *himalayanus* (Gould, 1862), Himalayas from Nepal east to northern Yunnan, north through Szechwan to central Kansu; *brachypus* (Swinhoe, 1871), Ningsia and Shensi to Hopeh (northern China); *baileyi* Rand and Vaurie, 1955, Simien and Bale mountains of Ethiopia.

**Field characters.** 39–40 cm; wing-span 73–90 cm. Smaller-headed but broader-winged than Jackdaw *Corvus* *monedula*; close in size to Alpine Chough *P. graculus* but with 70% longer, more distinctly and evenly decurved bill, more oval head, somewhat longer, broader, and more 'fingered' wings, shorter tail, and slightly longer legs. Medium-sized, dashing, and graceful crow, with long, thin, decurved red bill (duller in juvenile). Plumage brilliantly glossy black. Flight most aerobatic of Corvidae. Sexes similar; no seasonal variation. Juvenile separable. 4 races in west Palearctic, differing mainly in size.

ADULT. Moults: June–September (complete). Entire body plumage black, glossed with brilliant blue and purple; wings and tail black, glossed with green above and silver-grey on underside of flight-feathers (thus contrasting with coverts). Sheen noticeable at quite long range, creating strangely vivid appearance. Bill and legs bright red, legs tending to vermilion. JUVENILE. Body dull brownish-black, with purplish gloss only on feather-tips. Wings and tail as adult. Bill shorter and less decurved than in adult, pinkish-yellow through orange to pale red. Legs orange to red.

Not confined to rocky mountains and thus more widespread than *P. graculus*, and needing to be distinguished also from *C. monedula* which frequently shares coastal habitats. Distinction from *C. monedula* not difficult except at long range: *P. pyrrhocorax* unmarked except for colourful iridescence, red bare parts, and silvery undersurface to flight-feathers, and is an altogether more aristocratic and graceful bird (see below). Even dullest (British and Irish) *C. monedula* shows distinctly grey rear of head, only silvery sheen to black plumage, stubby black bill, and black legs; altogether more plebeian in general character. Distinction (especially of short-billed juvenile) from *P. graculus* much more difficult (see p. 96). In flight, shows quite broad and well-fingered wings and rather square tail; action buoyant and marvellously accomplished, including gliding and soaring, easy, almost leisurely acceleration into fast direct progress, sweeping dives, and even hurtling rolls and tumbling manoeuvres. No other crow, not even *P. graculus*, gives such impression of flight mastery and enjoyment, with flocks as well as individuals and pairs indulging in aerobatics up and down cliffs. Primary tips turn up on downstrokes and in soaring, suggesting small raptor (as do wing and tail attitudes), but this can also be shown by *P. graculus*. Gait confident: walks and runs, in rather balanced, horizontal posture, or hops with body more upright; also has sideways jump, and will clamber in tight spaces. Generally alert, even excitable; flicks wing-tips and flirts tail, particularly when calling from perch or ground. Sociable. Less often over permanent snow than *P. graculus*.

Voice generally distinctive. Commonest call has yelping, almost merry quality: clear, rather musical and high-pitched, drawn-out 'kjaa' or 'kyeow'. Also a deeper, rather gull-like 'kaah' and 'kwuk-uk-uk'; raptor-like 'kree-aw' or loud harsh 'kwarr' in alarm. Voice less gruff than *C. monedula*, lacking cackling quality, but beware confusion with 'kyow' of juvenile *C. monedula*.

**Habitat.** In west Palearctic, breeds in temperate middle latitudes, either on coastal cliffs or inland crags, and locally on buildings or ruins, especially of stone. Elsewhere in continental mid-latitudes mainly in montane regions. In Switzerland, nests at 1200–1500 m; in Haut Atlas (Morocco) mainly 2000–2500 m, and in southern USSR inhabits mountains at 1200–3600 m, but sometimes occurs on crags at lower levels, nesting in crevices and caves usually near water. Normally breeds at lower altitudes than Alpine Chough *P. graculus*. (Voous 1960; Dementiev and Gladkov 1954). Requirement for such localized nest-sites, which are often tenanted for many years, restricts foraging to normally poor soils and areas marginal for cultivation, excluding fertile intensively farmed lowlands and also woods, groves, bushy or tree-grown areas, and extensive wetlands. Differs from congeners in rarely perching on trees, posts, or bushes. Feeds almost entirely on ground, in Britain and Ireland mainly along coasts where low-intensity farming combines patchy cereal cultivation with plenty of short grass grazed by sheep and cattle or wild rabbits, or kept down by high winds and salt spray, thus permitting essential access to invertebrate prey in soil. Will use small turf patches among gorse *Ulex*, bracken *Pteridium*, or heather *Calluna*; is displaced when such vegetation grows too tall and forms continuous cover over accessible foraging sites, as also when intensive cultivation intrudes on them. Will, however, sometimes forage on stubbles, fallows, and ploughland, and in southern Europe may abound on plateaux with extensive cornfields, as on the Hoya de Guadix (Andalucia, Spain) at 900–1200 m. Sand-dunes, machairs, fields rich in dung deposits from livestock, and even beaches, are also used for foraging. Highly aerial, flying up to great heights, although rarely for great linear distances.

Restricted and fragmented distribution suggests strictly specialized needs, as does widespread susceptibility to declines in numbers and range and even frequent local extinctions. Various speculative explanations for these have not withstood recent critical investigations, which tend to eliminate most alternatives to adverse changes in carrying capacity of much local habitat (Bignal and Curtis 1989).

**Distribution.** Range has contracted markedly in north-west and in Alps; now apparently stabilized at least in Britain and Ireland.

BRITAIN, IRELAND. Range formerly much wider. In Scotland, formerly occurred both inland and on east and west coasts; by early 19th century had vanished from inland areas and was declining on east coast (Thom 1986). In England, bred Devon until 1910, and Cornwall until 1952. In Ireland, formerly bred on east coast. (Sharrock 1976.) FRANCE. Long-term contraction of breeding range. Extinct Normandie from 1910. (Yeatman 1976.) PORTUGAL. Has probably disappeared from some areas where formerly occurred, but evidence not conclusive (Rufino

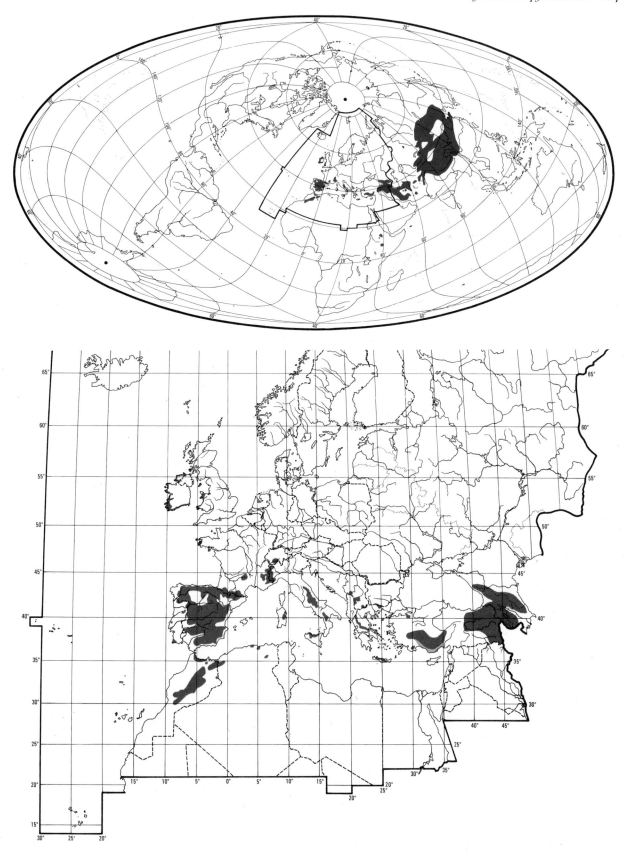

1989). AUSTRIA. Probably bred southern Carinthia in 19th century; now accidental visitor (H-MB). SWITZERLAND. Disappeared as breeding bird in Grisons in 1967 (Winkler 1984); only breeding area now Valais (RW). ITALY. Has become extinct in central and eastern Alps; now only in western Alps (Mingozzi 1982). ALBANIA. Probably breeds in south, but no proof (EN). LEBANON. Status uncertain; flock near Faraya, 1969 (Vere Benson 1970; AMM). IRAQ. Status uncertain; may breed (HYS). TUNISIA. Formerly bred; no records in 20th century (Thomsen and Jacobsen 1979).

Accidental. Austria, Hungary, Israel.

**Population.** Long-term decrease in many parts of range, perhaps now halted in some north-west coastal areas.

BRITAIN, IRELAND. 700–800 pairs in 1963, with perhaps 400 non-breeders (Sharrock 1976); c. 1000 pairs maximum 1982, with c. 860 maximum non-breeders; probably no real increase since 1963 (Bullock et al. 1983a). FRANCE. 100–1000 pairs; long-term decline (Yeatman 1976). Bretagne: estimated 28–37 pairs in 1988 (Thomas 1989); declined at least until 1982, except in Ouessant. Alps: more than 30 pairs in Verdon canyon and mountains north of it, Haute-Provence (Gallardo 1986); c. 50 km further east, c. 20 pairs on Cheiron mountain, north of Grasse (Alpes-Maritimes) (D Siméon). (RC.) SPAIN. Censuses in limited area of central Spain show increase in 1975–90; 324 pairs in 1990 (Blanco et al. 1991). PORTUGAL. Decreasing, at least locally (RR). SWITZERLAND. About 40 pairs; no marked recent change (RW).

Survival. Scotland: mortality in 1st year after fledging 29%, in 2nd year 26%. Earlier estimates of mortality almost certainly too high. (Bignal et al. 1987.)

**Movements.** Mainly sedentary. Recorded far from breeding areas only exceptionally. Ringing data scanty.

Nominate pyrrhocorax in Britain and Ireland essentially sedentary but often makes feeding movements of several km. Recoveries of ringed birds, mainly in winter, show dispersal mainly of less than 10 km but not infrequently up to c. 45 km from natal areas in Wales, Isle of Man, and Ireland (Holyoak 1971; Roberts 1985). Most immatures disperse from Bardsey island (Wales) in 1st autumn and mainland birds disperse to Bardsey at this time; adults markedly sedentary (Roberts 1985, 1989). On Islay (Scotland), Bignal et al. (1989) found local movements of marked birds tended to be longer for ♀♀ (median 9·9 km, n = 21) than ♂♂ (median 3·4 km, n = 17). Occasional longer movements of ringed birds recorded from Bardsey to Liverpool (142 km), of 110 km and 150 km within western Wales, and from Isle of Man to Down (Northern Ireland) (Roberts 1985, 1989; Sapsford 1991). Sight records outside normal range in Britain imply that yet longer movements occur rarely: Orkney (at least 360 km from nearest breeding area), eastern England (210 km), Devon (100 km); several such vagrant records involve lengthy sea crossings, e.g. to Barra (Outer Hebrides) and perhaps across Irish Sea to Dublin (Roberts 1985).

Alpine and Iberian race, erythrorhamphus, also appears to be essentially sedentary (e.g. Glutz von Blotzheim 1962) although commonly making feeding movements of several km. Many but not all alpine birds probably leave high elevations in severe winter weather, mostly moving less than 10 km (Praz and Oggier 1976), but a few wander outside breeding range; ringing recoveries of longer movements are 25, 40, 50, and 60 km, with one at 130 km (Busse 1969; Praz 1971; Praz and Oggier 1976). There is evidence that movements to areas where it does not breed occur mainly in autumn (Praz and Oggier 1976). In northern Spain, also leaves high mountains in hard winter weather, even occurring on coast when ground snow-covered (García Dory 1983).

Race in north-west Africa and Canary Islands, barbarus, mainly sedentary. In some regions of Morocco (including central Haut Atlas), flocks reported to descend in winter from high mountains to lower slopes and nearby plains (Heim de Balsac and Mayaud 1962; Destre 1984).

Race in south-east Europe and Middle East, docilis, mainly sedentary, but in Yugoslavia descends from high mountains in severe winter weather (Matvejev 1955), and altitudinal movements also occur in Turkey (Beaman 1975). In Afghanistan, flocks move down into plains of south in winter (S C Madge). Extralimital races all appear to be essentially sedentary, but Himalayan and Chinese forms show appreciable winter-summer altitudinal movements (Ali and Ripley 1972; Schauensee 1984). Thus in Nepal mainly occurs above 2440 m but recorded down to 1450 m February-March and as high as 7950 m in May (Inskipp and Inskipp 1985).　　　　　DTH

**Food.** Soil-living insects and other invertebrates, with grain and berries taken especially in winter or by upland populations. Forages generally in pairs or flocks on open ground with short vegetation (2–4 cm) or on bare, burnt, rocky, or disturbed ground, often frequenting certain favoured areas (Cowdy 1962; Haycock and Bullock 1982; Bullock and del-Nevo 1983; Roberts 1983; Sharrock 1984). Avoids taller vegetation and scrub; in alpine meadows in Picos de Europa (northern Spain), feeds in recently mown grass; does not do so on Ramsey (Wales) (Cowdy 1973; García Dory 1983; Meyer 1990). Throughout year in Britain, favours grazed or ungrazed rough grassland over limestone, sand, or clifftops; dung important in grazed areas, e.g. in Cornwall (south-west England) dung used for 70% of foraging time in 3 hrs of observation. Cliffs and beaches sometimes important in winter when other ground frozen, and at this time of year sometimes exploits improved grassland, stock-feeding areas, manure heaps, rubbish tips, animal carcasses, and even streets and town gardens. (Rolfe 1966; Bonham 1970a; Bullock et al. 1983a, b; Gamble and Haycock 1989; McCracken 1989; Meyer 1990.) In Mongolia, feeds on balconies and window-sills of buildings (Mey 1988). Of 103 feeding observations on Islay (western Scotland), throughout year, 64% on pasture, 21% on moorland, 14% on arable, and 1% on beach (Warnes and Stroud 1988); of over 8000 bird-minutes of observation in Cornwall in winter, 51% on cliffs and 38% on disturbed ground with much cow-dung, even though these together made up only 12% of available habitat; improved pasture and scrub virtually ignored (Meyer 1990). Feeding efforts often concentrated on inter-faces, e.g. between vegetation and rock outcrops, between stones and earth, at the bases of shrubs and grass tussocks, and along edges of snow patches (Cowdy 1962; Praz and Oggier 1976; Haycock and Bullock 1982; Roberts 1983; Neufeldt 1986; Meyer 1990). In Pamirs, sometimes takes contents of passerine nests; will pick parasites from backs of wild and domestic animals; also takes prey from mud at water's edge (Potapov 1966). Rarely perches on veget-ation, though in south-west Portugal 20% of unknown number of feeding observations were on flower-heads of artichoke *Cynara cardunculus* (Farinha 1989), and else-where noted eating fruit, berries, and, very rarely, insects from trees (Cullen *et al.* 1952; García Dory 1983; Smiddy 1986; Piersma and Bloksma 1987). Food sought mainly by deep-probing in soil with bill; probes substrate by rapid succession of movements at all angles, side of head some-times parallel with and almost lying on ground (Cowdy 1973). Subsequent behaviour seems to depend on prey choice: birds feeding on subterranean ants (Formicidae) and their larvae make up to 8 quick jabs with bill slightly open, moving head from side to side (Cowdy 1973); birds feeding on Tipulidae larvae (Diptera) suddenly probe, jump half-round, probe more deeply, and, with some lever-ing, withdraw bill (Goodwin 1986). Vegetation sometimes pushed or pulled back to facilitate probing, and bill may be inserted to base (Cowdy 1962; Holyoak 1967b; Good-win 1986; Piersma and Bloksma 1987; Meyer 1990). Holes made quickly with half-open bill which is alternately moved backwards and forwards and from side to side in ground (Mey 1988). Depth of probing, measured from holes in turf, up to 7 cm, though mostly less: 5 cm on Ramsey (Cowdy 1973); 60% 2–3 cm on Bardsey (Wales) (Roberts 1983). Prey on surface of substrate taken with picking action: slow-picking, possibly for spiders (Araneae), beetles (Coleoptera), etc., or rapid 'sewing-machine' picking with bill held vertically, for ants (Holyoak 1967b; Cowdy 1973; Haycock and Bullock 1982; Bullock *et al.* 1983b). Overturns stones to uncover prey; reaches over larger stones and pulls them towards body or pushes them to one side with partly open, down-pointed bill; flicks small stones aside (Holyoak 1972b; Goodwin 1986; Neufeldt 1986). Digs in sandy beaches for sand-hoppers (Talitridae) by throwing sand over the head, or to one side, pausing every few strokes to call and look around; digging birds may disappear into holes (10–20 cm deep); may leave hole and return to it later, though all digging possibly ceases if tidewrack seaweed removed by

gales; birds also chip away base of soft sand and clay cliffs to expose sandhoppers (Morgan 1971; Roberts 1982, 1983). Searches dried dung of cow, sheep, or horse for invertebrates and grain fragments; breaks it open and turns it over, searching damp underside and ground beneath for prey; probes fresher dung (Whistler 1941; Bonham 1970a; Gatehouse and Morgan 1973; Dawson 1975; Bullock *et al.* 1983b; Meyer 1990). Takes berries and Lepidoptera larvae from trees and bushes with much fluttering and wing-flapping for balance (García Dory 1983; Smiddy 1986; Piersma and Bloksma 1987); in spring on La Palma (Canary Islands), 350–400 recorded perch-ing, hanging, and hovering in pines *Pinus canariensis*, exploiting infestation of Lepidoptera larvae and pupae; some birds clung to rough pine trunks and large side branches like woodpeckers (Picinae), pushing and twisting off bark with bills to uncover prey (Piersma and Bloksma 1987). In India, recorded taking heads of barley *Hordeum*, thrashing grains out by repeated blows (Ali and Ripley 1972). One old report (Ireland, 1924) of birds catching flying ants in flight (Rolfe 1966). Food items usually swal-lowed whole, rapidly, without manipulation (García Dory 1983; Mey 1988), though larger prey sometimes held under one or both feet (Holyoak 1972b). On La Palma, cat-erpillars or pupae caught in trees were always brought to ground, manipulated in bill and then beaten against stone or piece of wood before ingestion, process taking 10–40 s (Piersma and Bloksma 1987). Sometimes drinks imme-diately after swallowing sticky food (Holyoak 1972b). Bill-wiping (against wood, stone, or vegetation) commonly follows feeding, particularly when probing (e.g. Piersma and Bloksma 1987, Holyoak 1967b). Food collected and brought to young in sublingual pouch and nestlings fed by regurgitation (Matvejev 1955; Cowdy 1962). In captivity, excess food sometimes regurgitated for caching (Turner 1959a, b). Insect exoskeletons, mammal fur and bones, and some plant material are formed into pellets (e.g. Cowdy 1962, Jong and Schilthuizen 1987, Warnes and Stroud 1988, Meyer 1990). Little information on prey capture rates, but in Cornwall in 3 hrs of observation, 60% of time was spent actively foraging, with head down; in December, 346 items taken in 2 hrs by one individual; foraging success significantly higher on cliffs than agri-cultural land, though prey items smaller (Meyer 1990).

Diet in west Palearctic includes the following. Inver-tebrates: grasshoppers, etc. (Orthoptera: Gryllidae, Acrid-idae), earwigs (Dermaptera: Forficulidae), bugs (Hemiptera), adult and larval Lepidoptera (Hepialidae, Pyralidae, Noctuidae, Lymantriidae), adult and larval Diptera (Tipulidae, Bibionidae, Tabanidae, Coelopidae, Calliphoridae), adult and larval Hymenoptera (ants Form-icidae, bees Apoidea), adult and larval beetles (Coleoptera: Carabidae, Staphylinidae, Geotrupidae, Scarabaeidae, Elateridae, Tenebrionidae, Cerambycidae, Chrysomel-idae, Curculionidae), spiders (Araneae), harvestmen (Opiliones), scorpions (Scorpionida), woodlice (Isopoda), sandhoppers (Amphipoda: Talitridae), millipedes (Diplopoda), centipedes (Chilopoda), small molluscs (Gastropoda), earthworms (Lumbricidae). Also lizards *Lacerta*, skink *Chalcides*, shrew (Soricidae), mouse *Mus*, and, as carrion, chamois *Capra*, rabbit *Oryctolagus* intestine, sheep, and wool. Plant material mainly fruits and seeds, of juniper *Juniperus*, rowan, etc. *Sorbus*, pear *Pyrus*, various *Prunus* species, hawthorn *Crataegus*, rose *Rosa*, holly *Ilex*, fig *Ficus*, olive *Olea*, sea buckthorn *Hippophae*, grape *Vitis*, orange *Citrus*, *Vaccinium*, and grasses (Gramineae) including grain. (Schifferli and Lang 1940a; Cullen *et al.* 1952; Matvejev 1955; Turček 1961; Rolfe 1966; Sorci *et al.* 1971; Dawson 1975; Praz and Oggier 1976; Roberts 1982; Bullock and del-Nevo 1983; Bullock *et al.* 1983a; García Dory 1983; Warnes and Stroud 1988.)

Coleoptera and Diptera important all year, spiders and earwigs less so. Tipulidae and Lepidoptera larvae important in breeding season; ants sometimes taken extensively in midsummer, perhaps if larger prey become scarce in dry soil; in winter, invertebrate diet supplemented by berries, grain, and carrion, though earthworms and sandhoppers still important locally (Roberts 1982; García Dory 1983; Warnes and Stroud 1988; Meyer 1990). In Cornwall, of 103 faeces in winter, 75% contained earthworms, 54% adult beetles, 27% Diptera, 7% seeds, 6% earwigs, and less than 2% each of ants, spiders, beetle larvae, and millipedes; 58% of 13 pellets contained adult beetles, 39% earthworms, 23% Lepidoptera larvae, 19% beetle larvae, 8% ants, 7% earwigs, and 7% seeds; Scarabaeidae, Geotrupidae, and Carabidae most frequent beetle groups (Meyer 1990). On Islay, of 68 faeces in winter, *c.* 85% contained cereal grains, *c.* 40% *Aphodius* (Scarabaeidae) larvae, and *c.* 35% *Aphodius* adults, Diptera larvae, and Carabidae larvae; of unknown number of pellets in winter, *c.* 100% contained cereal grains, *c.* 100% beetles, *c.* 60% *Aphodius* adults, *c.* 65% earwigs, and *c.* 10% Carabidae larvae; of 12 faeces in spring, *c.* 65% contained *Aphodius* adults, *c.* 65% *Aphodius* larvae, *c.* 30% Diptera larvae, *c.* 30% millipedes, and *c.* 30% Carabidae larvae; of 13 faeces in summer, *c.* 70% contained *Aphodius* adults, *c.* 70% earwigs, *c.* 40% cereal grains, and *c.* 30% *Aphodius* larvae; of 38 faeces in autumn, *c.* 100% contained cereal grains, *c.* 80% *Aphodius* adults, *c.* 40% earwigs, and *c.* 25% each of Carabidae adults and *Aphodius* larvae, Diptera larvae, millipedes, and Staphylinidae larvae (Warnes and Stroud 1988). On Bardsey, of 328 faeces, late April to late November, beetles in 75-100% of samples late April to early June and late September to October; ants, Tipulidae larvae, Lepidoptera larvae, spiders, and harvestmen all peak at 75-100% late June to late July; sandhoppers and kelp fly *Coelopa* larvae more frequent early November (Roberts 1982). Of 3 stomachs in Makedonija (Yugoslavia) in June, one contained 90% Tipulidae larvae by weight, one 83% Scarabaeidae, and the other 70% Carabidae and 16% Tipulidae larvae (Matvejev 1955). All items in 3 pellets from Spanish Pyrénées, July,

were beetles (mostly Carabidae) (Jong and Schilthuizen 1987). Of 25 gizzards in Hoya de Guadix (southern Spain), July and September, grain and wild seed 69% by volume (Soler 1989c).

Diet of young less well known. On Islay, of 22 faecal samples from young at least 3 weeks old in 6 nests, May-June, 90% contained remains of *Aphodius* adults, 36% Carabidae larvae, 32% Diptera larvae, 27% Carabidae adults, 27% Elateridae adults, 22% earwigs, 18% Elateridae larvae, 13% Lepidoptera larvae, and less than 10% each of cereal grain, Diptera pupae, spiders, and bees (Warnes and Stroud 1988). In Yugoslavia, Noctuidae caterpillars formed 85-90% (probably of weight) of food in sublingual pouch of 2 adults (Matvejev 1955), and similar proportions of prey items recorded in Tarasp (Switzerland) (Schifferli and Lang 1940a).                              WGM

**Social pattern and behaviour.** General aspects fairly well known, but more work with marked birds needed. Main studies by Schifferli and Lang (1940a, b) in western Alps and Cowdy (1962), Holyoak (1972b), Owen (1985), and Roberts (1985) in Wales and Isle of Man. For review, see Goodwin (1986).

1. Throughout year, typically in small or (often temporary) large flocks, up to several hundred birds (Etchécopar and Hüe 1967; Bonham 1970a; Soler 1989c); often also in pairs, sometimes singly. Many flocking birds appear to be paired at all times of year and pairs commonly join and leave flocks. Many years of observation on Bardsey (North Wales) showed maximum annual flock size may occur in any month, but with flocking usually pronounced September-October (Roberts 1985). In Italy, Lovari (1976b, 1978) found largest flocks in summer; feeding flocks tended to be larger than roosting and other flocks at all seasons. For seasonal variation in flock size in southern Spain, see Soler (1989c). Flocks commonly associate with flocks of Alpine Chough *P. graculus* (see Lovari 1978 and Dendaletche and Saint-Lebe 1991); less often and apparently only casually with Jackdaw *Corvus monedula* (e.g. Praz and Oggier 1976, Owen 1985), Rook *C. frugilegus*, or Carrion Crow *C. corone* (e.g. Cullen 1989, Moore 1991). Feeding birds dominant over *C. monedula* (Monaghan 1989) and *P. graculus* (Mattes and Bürkli 1979). No evidence of territorial defence outside breeding season, but paired birds frequently occur in nesting areas. BONDS. Essentially monogamous. Pair-bond in adults almost certainly of long duration; pair-members remain together all year. Several examples of 1- and 2-year-old non-breeders that appeared to be paired (judged from Allopreening and Courtship-feeding) but which changed 'mates' in later weeks, sometimes more than once. (Roberts 1985.) Both sexes involved in nest-building (see below). Incubation by ♀ alone (fed regularly at nest by ♂); young brooded at first by ♀ (fed by ♂), later fed by both parents. Fledged young self-feeding by *c.* 3 weeks after fledging, but remain with parents for further 1-2 weeks. (Holyoak 1972b.) Age at first breeding 2-4 years or later (Roberts 1985; J M Warnes); many may not breed until 4 years or older on Bardsey (Roberts 1985), but this possibly varies between populations depending on density, etc. (Bignal and Curtis 1989). For hybridization with *P. graculus*, see that species (p. 99). Considerable proportion of non-breeders noted in late spring and early summer; estimates for British and Irish populations *c.* 30% (Rolfe 1966; Bullock *et al.* 1983a); in smaller populations, estimated 20% in Italy (Lovari 1976a), 34-43% in southern Isle of Man, *c.* 40% in North Wales (Holyoak 1972b), and varying annually from 20 to 60% on Bardsey (Roberts 1985,

which see for discussion). Most non-breeding birds are paired and (at least) some hold territories (e.g. in 1966 study on Isle of Man, 4 territories held by breeding pairs, 5 by non-breeding pairs: Holyoak 1972*b*). Although not apparently usual (Monaghan 1989; D T Holyoak) there are several records of helpers at nests, which may feed young; in one case, helper was 2-year-old offspring of breeding pair (Bullock *et al.* 1983*a*; Warnes 1983; Jennings 1984). Owen (1985) recorded immature nest-helper that was almost certainly unrelated to pair it assisted. On Bardsey, some adult pairs do not breed every year (Roberts 1985). BREEDING DISPERSION. Solitary, occasionally in small loose colonies. In Britain, nests usually several hundred metres or more apart where densities relatively high, as in Isle of Man; average 1·4 km (*n* = 32) between nests in best areas in Wales, 2·6 km (*n* = 27) in poorer areas (Bullock *et al.* 1983*a*); only rarely close together (2 nests within 180 m on cliff and 2 on same ruined castle: Bullock *et al.* 1983*a*). Restricted availability of sites and low overall population density may contribute to spacing, but some evidence for active separation (Holyoak 1972*b*; Bignal and Curtis 1989). Where birds are numerous or suitable nest-sites are few and concentrated (e.g. parts of Spain, Tibet) several pairs may breed close together (Goodwin 1986; Bignal and Curtis 1989). Territory around nest apparently defended, as one or both of pair frequently seen to drive conspecifics away when they intruded within several hundred metres of nest-site, though wandering birds seem to show little fear of passing close to occupied nests (Holyoak 1972*b*). Much but by no means all feeding carried out within home-range away from nesting territory. Inland in North Wales, several pairs regularly seen flying from nests to collect food 2–3 km away (Holyoak 1972*b*), although in Welsh coastal areas distances between nests and observed feeding places averaged 0·7 km (*n* = 58) (Bullock *et al.* 1983*a*). In Valais (Switzerland), breeding birds fed up to 2 km from nests (Praz and Oggier 1976). For review of breeding densities in Britain, see Bullock *et al.* (1983*a*). Highest British density probably on Islay (Scotland) with up to 0·33 nesting pairs per km² (Monaghan *et al.* 1989), much lower in mountain area of North Wales and in French Alps (Holyoak 1972*b*). In south-east Madrid province (central Spain), average 3·9 pairs per km of cliff (*n* = 324 pairs), maximum locally 12·3 pairs per km (Blanco *et al.* 1991). Nest-sites are occupied by same pairs year after year (Roberts 1985). Some sites are occupied regularly beyond possible lifetime of individuals, others only used occasionally or irregularly. Lack of close fidelity to natal site suggested from Bardsey where none of breeding population likely to have been reared on the island; birds reared there are known to have bred on mainland (Owen 1985; Roberts 1985, 1989). Sometimes breeds among *C. monedula* (e.g. Bullock 1985, Soler 1989*c*) or *P. graculus* (Goodwin 1986; Bignal and Curtis 1989; Dendaletche and Saint-Lebe 1991). Other nesting associates locally include Rock Dove *Columba livia* and Bald Ibis *Geronticus eremita* (Etchécopar and Hüe 1967). ROOSTING. Sites are typically rock holes and sheltered ledges or equivalent sites on buildings. Open cliffs are used locally (Roberts 1985; Still *et al.* 1987). Breeding ♀ roosts on nest, ♂ nearby (usually on same ledge or another within a few metres of nest) until young finally leave nest area. In study of family party in North Wales, brood continued to return to nest-area for *c.* 2 months after fledging, but parents roosted separately from them (in cave *c.* 100 m away) from *c.* 10 days after brood fledged (Cowdy 1962). On Islay, pair and fledged young initially roosted at nest-site; later used various sites over their 'extended home range'; pair then took offspring to main communal roost (see below) where family roosted together, but eventually pair returned to nest-site or general vicinity to roost; thereafter some 1st-winter birds roosted with parents but most used communal

roosts (see below) (Bignal and Curtis 1989). In spring and summer, some paired non-breeding birds apparently roost on territories (D T Holyoak). On Bardsey, non-breeders used 'variety of loose summer roosts' (Roberts 1985). Roosts communally outside (sometimes also during) breeding season. In south-east Madrid province, large numbers of non-breeders roost communally throughout the year (Blanco *et al.* 1991). In Alquife (southern Spain), regular winter roost of 450 birds (Soler 1989). On Bardsey, communal roosts of up to 50 birds used regularly October–March, sometimes as early as August (Roberts 1985). Still *et al.* (1987) described communal roost (cliff) on Islay used regularly by 50–130 birds throughout year, as follows. During July–September (period of study), different age-classes were spatially segregated: 3rd-year birds comprised densest part of roost and were significantly more likely to attack conspecifics than were 1st-years, which roosted significantly lower down and on periphery of flock. At this roost individuals tended to use particular site repeatedly. (Monaghan 1989.) Individual pairs may roost by themselves even in winter (Goodwin 1986; Monaghan 1989). Some pairs which breed in areas with apparently limited suitability for feeding outside breeding season are known to use communal roosts away from their nesting areas, but may continue to maintain daytime contact with nest-site (Bignal and Curtis 1989). In severe winter weather some adults that are usually resident in their nesting area have been observed to move away and join feeding flocks (Bignal and Curtis 1989). Regularly shares roost-sites with *P. graculus* in some regions (Mayaud 1933*b*; Lovari 1978; Dendaletche and Saint-Lebe 1991). Small numbers of *C. monedula* shared roost on Islay studied by Still *et al.* (1987).

2. Often less shy than other Corvidae in Britain, sometimes allowing approach to within *c.* 10 m, presumably because less persecuted by man. In Tibet, where not harmed by man, notably tame and nests in occupied buildings (Schäfer 1938). Detailed study of reactions to tourists in south-west Wales (Owen 1985, 1989) found minimum flushing distance averaged 35 ± SD6 m (*n* = 101), whereas minimum distances as low as 9·5–11 m were recorded elsewhere in Britain; when flocking with the shyer *C. monedula*, minimum flushing distance increased. Frequently vocal at all times of year, giving Chwee-ow call most commonly (see 2 in Voice). Lovari (1978) found this call uttered more frequently in flight (71%) than when perched (29%) (*n* = 1207) and given most frequently by flocks. When perched, Chwee-ow call often accompanied by Wing-flirting, wing-tips lifted 4–5 cm (sometimes much more) in quick flicking movement coincident with calling (Fig A). Chwee-ow call often appears to function as

A

contact-call, especially in flight, but combination of this call and Wing-flirting appears to serve wide variety of self-assertive functions, e.g. given before chasing conspecifics from nest area, when feeding birds see others fly over or land nearby, as response

to distant conspecific calling, and when mobbing potential predator. (Holyoak 1972b.) Also given by mildly alarmed birds as
humans approach and by members of pair in sexual contexts
(Owen 1985, 1989, which see for quantitative analyses of 'wing-
flick calling'). Chwee-ow calls accompanied by Wing-flirting and
vigilance behaviour (see Flock Behaviour, below) are commonly
also directed at potential predators. Birds of prey and flying
Ravens *Corvus corax* commonly cause upflight and often elicit
Alarm-calls (Holyoak 1972b: see 3 in Voice) and sometimes
Mobbing-call (see 5 in Voice). *P. pyrrhocorax* may also freeze on
ground when these birds pass nearby (Owen 1985). Rapid
upflight also recorded on sudden approach of *C. corone*, Peregrine *Falco peregrinus*, and less often Herring Gull *Larus argentatus* or Kestrel *F. tinnunculus* (Harrop 1970; Owen 1985).
Various raptors and other potential predators (*C. corax*, *C.
corone*) may be mobbed in flight, especially by flocks, individual
birds sometimes swooping close to predator (Praz and Oggier
1976; Owen 1985; I D Bullock). Other responses to *F. peregrinus*
(known predator of *P. pyrrhocorax*) as follows: Owen (1985)
recorded flock bunching tightly in flight when *F. peregrinus* flew
near; one individual hid low in scrub when *F. peregrinus* passed
nearby (P Bennett); pair hid in rock crevice for 8 min following
unsuccessful attack (Moore 1991). 3 birds were very alarmed
and called noisily when fox *Vulpes vulpes* appeared (Praz and
Oggier 1976). FLOCK BEHAVIOUR. Breeding birds (pairs, and ♂
while ♀ incubating) frequently flock and feed with conspecifics,
both breeders and non-breeders, flying high into the air and up
to several km from nest sites (Holyoak 1972b). Flocks tend to
lack cohesion, with birds joining and leaving in pairs or small
groups. Members of flocks often soar for short periods and
frequently perform skilful aerobatics; commonly, birds make
earthward stoops from high up, straightening out near ground
to flap and glide up again. Flocks feeding in grassland usually
rather loosely dispersed with birds often several metres apart;
closer association within flocks of apparently paired birds is
usually evident. On level feeding areas, average spacing between
individuals (other than pairs or juveniles) tended to be maintained at 5–6 body lengths; on rocky areas 8–9 body lengths.
(Owen 1985.) No evidence of dominance hierarchy in feeding
flocks, but overt aggressive interactions infrequent in Britain.
Segregation of age-classes in communal roost (see Roosting)
implies that age-related hierarchy exists there. Vigilance generally shown by looking about with head raised, its duration
apparently related to degree of threat (Owen 1985). Flight-
intention movements, given by single, paired, or flocking birds,
consist of quick upward flick and slight opening of wing-tips
(Fig B); wings often lifted *c.* 1 cm, but may involve nearly

B

half-raising them; tail usually also moved upward a little
(Holyoak 1972b). See Owen (1985, 1989) for quantitative study
of frequency of Flight-intention wing-flicking and vigilance in
different contexts, especially in relation to disturbance of feeding
birds by tourists. From studies in Sicily, Sorci *et al.* (1971)
suggested that some birds serve as sentries to warn flocks of

possible danger, but little evidence of this in another Italian
study (Lovari 1978). Birds using communal roosts usually arrive
in groups, well before sunset (Islay: Still *et al.* 1987) or just
before nightfall (Italy: Lovari 1978); leave well before dawn in
large flocks that soon disperse into small bands and pairs (Lovari
1978). In Italy, at roost with *P. graculus*, mixed flock usually
made a few flights around roost-site before stooping together
into roosting crevice; pairs of *P. pyrrhocorax* arriving later
entered roost directly (Lovari 1978). At other Italian communal
roosts, birds arrived average 40 min before settling to roost,
perching and flying nearby and giving almost continual noisy
calls, loudest immediately after entering roosting crevice. Evidence from Islay that birds follow each other from communal
roost-sites to productive feeding sites (Monaghan 1989). Communal roost-sites on Islay have good feeding areas nearby that
are used regularly for pre-roost feeding (Bignal and Curtis 1989).
SONG-DISPLAY. No ritualized display described, but see 1 in
Voice for 'song'. ANTAGONISTIC BEHAVIOUR. (1) General. In Britain, apparently less aggressive than *Corvus*: less often seen to
use threat-displays or to chase conspecifics, both in and out of
breeding season (Holyoak 1972b; Owen 1985). One or both of
nesting pair sometimes pursue conspecifics that intrude into nest
area, pursuer(s) often giving loud Chwee-ow calls. At large communal roost, 3rd-year birds significantly more likely to initiate
attacks, but they roosted in densest part of roosting flock; tendency for 1st-year birds to be attacked more than other age-classes;
aggressive interactions also seen (infrequently) in feeding flocks
or in display (see below) (Still *et al.* 1987). Infrequent aggression
in feeding flocks (Holyoak 1972b; Lovari 1978) perhaps related
to small size of most food items, it being pointless to attempt to
displace a conspecific if threatened bird can immediately swallow
disputed item. However in Tibet, Schäfer (1938) found aggression much more prevalent, perhaps because of higher densities;
2 birds fighting fiercely on ground were both found to be ♀♀
when shot. (2) Threat and fighting. Threat-display in feeding
flocks observed only 8 times in study on Isle of Man (Holyoak
1972b): 6 times a bird walked up to another with neck outstretched (Forward-threat posture: Fig C) and other walked

C

away, and twice birds threatened in this way lowered head and
gaped at aggressor, causing it to retreat. This display often
used when displacing another bird at feeding site (Owen 1985;
Goodwin 1986); forager is sometimes actually pushed away
(Owen 1985); see Fig 6c in Sitasuwan and Thaler (1984) for use
of this display by captive birds. Head-up threat posture (Fig D)
regularly used by captive bird threatening Magpies *Pica pica* in
adjacent cage; also reported with captives by Sitasuwan and
Thaler (1984, Fig 6a). Goodwin (1986) noted that this display

D

often leads to serious attack and fighting. Airborne swoops at conspecific away from nest-site frequently observed March, April, July, and September, but not other months (Owen 1985); play behaviour probably involved as well as overt aggression. Fighting also reported in association with apparent heterosexual display (see below). Little submissive posturing reported, merely retreat of attacked birds. Owen (1985) regarded posture in which bird crouches with head and bill high as submissive, but defensive-threat may be involved (compare Head-up threat posture). See Heterosexual Behaviour for 'irrelevant activity' associated with aggressive interactions. HETEROSEXUAL BEHAVIOUR. (1) Pair-bonding behaviour. Little information. Displays that may be related to pair-formation were seen only on 8 of numerous days of observation (late spring, early autumn) by Holyoak (1972*b*). In these, individuals in feeding flocks were seen to hop (not run) quickly towards others, then stop sharply about 0·5 m from them, so that head was lowered abruptly as if they were about to overbalance, tail raised, and wings flirted as in self-assertive calling, though no call given (Bowing-display: Fig E). On several days, up to 3 individuals in feeding flocks

E

gave this display simultaneously, but directed at different birds. Often, bowing bird, or bird it was displaying to, was noisily attacked by other members of flock with Chwee-ow calls, but sometimes little attention paid to it. Then bird that had bowed would step forward and nervously Allopreen the other, with only bill-tip touching other bird's nape or mantle. Twice a bird that had bowed and then Allopreened was seen to regurgitate food and present it to the other, which snatched it 'anxiously' and swallowed it (see also subsection 2, below). This may imply that Bowing-displays are performed mainly or entirely by ♂♂, but no firm evidence. In Tibet, Schäfer (1938) reported very intense display, presumably variant of Bowing-display, from two presumed ♂♂ to presumed ♀. Both ♂♂ had all body feathers fully erected, head lowered with bill perpendicular to ground, wings slightly open and drooped, and tail erected so that at times it almost touched back; displaying thus, they tripped rapidly around ♀ in silence. Similarly, Owen (1985) recorded ♂♂ walking around in vicinity of mate with wings spread widely and shaking, head and bill down, and tail at 45°; variants of this display were seen; not used immediately prior to copulation. Williamson (1959) recorded similar displays in September, but when 2 birds bowed opposite each other they came to blows, and as soon as they became locked together 3–4 others joined in excitedly. Such attacks on displaying birds by other flock members are difficult to account for; the apparently nervous response of ♀ perhaps causes neighbours to mob bird displaying close to her. Various maintenance activities and comfort movements are performed in rather perfunctory or incomplete manner after sexual (and aggressive) displays (Holyoak 1972*b*). These 'irrelevant activities' include bill-wiping (regularly in both sexual and antagonistic contexts) and incomplete preening, feather ruffling, and stretching movements. Performance of bill-wiping, in particular, may be perfunctory as apparent displacement activity, or persistent and apparently stereotyped when it may function in sexual and social display, as may wing-stretching actions of ♂ (Owen 1985). Paired birds frequently Allopreen each other on back of

head, nape, and side of neck, sometimes with breast of one bird touching flank of mate (Holyoak 1972*b*). Paired birds may Allopreen each other simultaneously, reciprocally, or singly; behaviour may occur at intervals during foraging, after Wing-flirting with call (see above), and after Courtship-feeding (Owen 1985). 2-year old nest-helper Allopreened with both parents (Cullen 1989). Aerial evolutions said to play part in, but not confined to, courtship (Witherby *et al.* 1938). In display-flights seen on 7 of 22 days of observation in Valais (mid-April to July, once in October), pair flew down together from nesting crag with very pronounced undulations, wings closed, feet trailing, and many calls (Praz and Oggier 1976). Similarly, Goodwin (1986) saw spectacular flights, probably with display function, by birds in or near breeding areas: presumed ♂♂ (and, less markedly, presumed ♀♀) bounded through the air in series of dives and inverted arcs. ♂ may accompany ♀ very closely and persistently when she is collecting nest-lining (Holyoak 1972*b*), possibly mate-guarding (D T Holyoak). (2) Courtship-feeding. Feeding of ♀ by ♂ (mainly if not entirely on fresh food rather than by regurgitation) is regular away from nest in breeding season, but less frequent at other times, among both breeding and non-breeding birds (Holyoak 1972*b*); it begins before laying and continues well afterwards (Owen 1985). Usually, ♀ walks or flies to ♂ while he is feeding, or turns to face him if she is already near, crouches somewhat with ruffled plumage, and usually flutters part-opened wings in Begging-display very similar to that by newly-fledged juvenile. At higher intensity, ♀ may beg with wings two-thirds to fully spread and quivering, body almost horizontal and bill open. (Owen 1985.) ♀ also uses Begging-display when she is fed at nest on regurgitated food from ♂ (see subsection 4 below). Begging-call described from ♀ by Schifferli and Lang (1940*a*) (see 6 in Voice). Courtship-feeding may lead directly to copulation (see below). Displays similar to those given by ♀♀ begging for food have been recorded in other behavioural contexts and from ♂♂; significance unclear. Thus, Ryves (1948) reported display in which ♂ opened and quivered wings towards ♀ after feeding her. Similar but higher-intensity display with wings widely spread and accompanied by 'series of indescribable notes' was performed by a ♂ while his mate was laying. Goodwin (1986) saw similar display given by bird immediately before it turned and fled from another that was swooping at it. (3) Mating. Copulation (seen 4 times) described by Holyoak (1972*b*) as follows. Each time, pair feeding together on grassland, away from other conspecifics, when ♀ squatted low on ground, with ruffled plumage, partly opened wings, and tail horizontal and quivered slightly or flicked upwards (Soliciting-posture: Fig F); twice

F

this behaviour followed Courtship-feeding, and twice pair were merely feeding close together. ♂ hopped onto back of ♀ and stayed there for 15–30 s with wings fluttering intermittently. After mating they separated and preened briefly before continuing to feed or flying off. Owen (1985) found copulation not usually preceded by display: with feeding pair, ♀ adopted Soliciting-posture facing ♂; ♂ then turned towards ♀ and mounted from rear with occasional silent Wing-flirting; mating took 15–20 s; ♂ then resumed normal posture, dismounted, and both continued feeding. (4) Nest-site selection and behaviour at nest. Roles and behaviour of sexes in nest-site selection inadequately known; Nethersole-Thompson and Nethersole-Thompson

(1943a) reported ♂ and ♀ inspecting various caves and crevices. During incubation, ♂ feeds ♀ (who performs Begging-display), either while she stands over nest or more often after she flies onto rocks part-way towards him when she hears his Chwee-ow calls nearby. At several nests on Isle of Man, ♀ seen to be called off eggs quite regularly by ♂, especially on sunny afternoons; pair would fly up to 1-2 km from nest, either to join temporary flock calling noisily or to feed and preen together. (Holyoak 1972b.) ♀ is not called off nest to feed in other west Palearctic Corvidae. One record of helper at nest feeding incubating ♀ (J M Warnes). ♀ apparently dominates ♂ when at nest (Sitasuwan and Thaler 1984). RELATIONS WITHIN FAMILY GROUP. Eggshells carried away by ♀ (not eaten, though she eats any remnants of membrane adhering to inside of shell or to nestlings) (Schifferli and Lang 1940a). ♀ broods young almost continuously for first 2 weeks, and, using Begging-display, receives food for them from ♂ (García Dory 1983). At nest observed closely by Schifferli and Lang (1940a), ♂ first directly fed nestlings on day 2 and was not directly feeding them as often as ♀ until day 11. Parents feeding older young usually visit nest together, but sometimes singly; parents usually remain near nest for some minutes after feeding young (Matvejev 1955; Owen 1985). When visiting together, usually first one parent (♀ according to Williamson 1959), then the other, enters nest cavity to feed young, although sometimes they go inside together. Helpers may feed young; on Isle of Man, helper was 2-year-old offspring of breeding pair (Bullock et al. 1983a; Warnes 1983; Jennings 1984; Cullen 1989). Both parents remove faecal sacs (except when young very large), but only ♀ cleans and preens young. For first 3 days ♀ ate some faeces and removed others; after this she removed all, sometimes after leaving them temporarily on nest rim. (Schifferli and Lang 1940a.) For first 3 days, ♀ at one nest fed young by inserting bill into throat with single movement; later, food regurgitated deep into throat with repeated shaking movement; ♀ ate spilled food and regurgitated it again. Young beg intensively, both spontaneously and in response to parent's Food-offering call (see 7 in Voice), both in early stages and later when also responding to visual stimuli. (Schifferli and Lang 1940a.) Stay in nest area for several days after fledging, hiding in rock crevices when parents absent (Cowdy 1962). At several nests in North Wales and on Isle of Man, young eventually left nest area 5-9 days after first flight of at least one bird; hid in rock crevices meanwhile, usually close together at first, but more widely 1-2 days before finally leaving the area. When hidden young were widely spaced, often gave 'kwa-ak' calls (see Voice), usually reciprocated by other young out of sight. When parents arrived and gave Chwee-ow call, young scrambled from hiding places towards parent, giving 'kee' food-call and fluttering partly open wings; hid after feeding. (Holyoak 1972b.) Cowdy (1962) found that fledglings began to follow parents 6 days after fledging and were accompanying them by day 8. Owen (1985) described and analysed Begging-displays of juveniles and responses of parents; intense begging by juvenile sometimes resulted in attack or retreat by parent. When they eventually leave nest area, young often taken 1 km or more away, but whole family returns to roost near nest-site for several weeks (see Roosting). Young of 3 broods were accompanied by parents for 4-5 weeks after fledging, although they appear to collect most of their own food after about 3rd week (Holyoak 1972b). Roberts (1985) recorded fledglings staying with parents for 43 days after fledging. Cowdy (1962) noted young beginning to pick up food unearthed by adults on day 6 after leaving nest area. Allopreening of fledged young by both parents recorded (Ryves 1948). ANTI-PREDATOR RESPONSES OF YOUNG. Well-grown nestlings crouch silently in nest in response to disturbance. Fledglings dispersed near nest-site

spend much time in crevices (Cowdy 1962) and seek shelter at first sign of danger (García Dory 1983). PARENTAL ANTI-PREDATOR STRATEGIES. (1) Passive measures. Incubating ♀♀ usually stay on nest until human intruder fairly close, then slip away quietly. (2) Active measures: against birds. Breeding adults commonly pursue conspecifics and other Corvidae that approach nests, although Praz and Oggier (1976) noted C. monedula passing a few metres from nest without provoking any reaction. Osmaston (1925) recorded C. corax taking 'half-fledged young bird' from a nest in spite of repeated but ineffectual attacks from parents. Pair mobbed F. peregrinus energetically when it approached within 100 m of fledglings, causing it to leave (Owen 1985). (3) Active measures: against man. When human observer reaches nest containing well-grown young, parents often perch or fly nearby giving Chwee-ow calls (D T Holyoak).

(Figs by D Nurney: A, B, and F from drawings in Coombs 1978; C-E from drawings in Holyoak 1972b.)           DTH

**Voice.** Conspicuously vocal throughout year. No detailed field studies of calls or their geographical variation. Difficulties with phonetic renderings and with understanding context of calls prevent confident categorization of many of those reported in the literature. Sitasuwan and Thaler (1985) gave sonagrams and brief descriptions of 18 call-types recorded in captivity; these are compared with homologous calls of Alpine Chough P. graculus and hybrids. Several call-types cited by these authors differ only slightly although some are shown to have different meanings; contexts in which certain other calls would be given by wild birds are unclear, and this classification is therefore not followed in provisional scheme used here. For further sonagrams, see Bergmann and Helb (1982). Sitasuwan and Thaler (1985) showed that calls of conspecifics may be copied and used for deception. Reports of sounds that could be rendered as 'k' chuff', 'choff', or 'chuff' appear in the literature (e.g. Witherby et al. 1938) but have not been heard by recent field observers; might derive from mistaken assumptions of origin of the species' English name, which was likely to have been originally pronounced 'chow' not 'chuff' (Williamson 1959; Holyoak 1972b). However, calls of this kind reported from Morocco by Bergmann and Helb (1982) and from northern Iraq by Chapman and McGeoch (1956). For descriptions using musical notation, see Stadler (1927).

CALLS OF ADULTS. (1) Song. Succession of low warbling, chittering, and churring sounds heard on several occasions (Holyoak 1972b), resembling 'songs' described for some other Corvidae. Report of chattering twitter, not unlike song of Starling Sturnus vulgaris (Witherby et al. 1938) may be similar. 'Subsong' of Sitasuwan and Thaler (1985) may also be related. Function obscure, and not known which sex gives these vocalizations. (2) Chwee-ow call. Commonest call. Loud, yelping, drawn-out 'chwee-ow', subject to considerable variation. Also rendered 'kyaa' (Witherby et al. 1938), 'keeaw', 't' cheea' (Goodwin 1986), 'kyeow' (D I M Wallace), and 'kijar' (Bergmann and Helb 1982). Fig I shows 2 variants by same bird, 1st somewhat tonal, 2nd harsher; Fig II shows more piercing

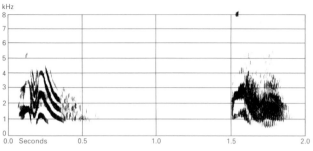

I  V C Lewis  Wales  May 1978

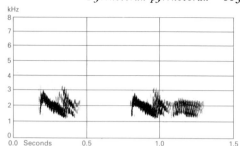

II  P A D Hollom  Iran  April 1977

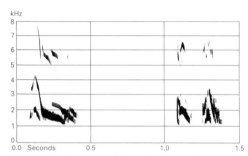

III  E D H Johnson  Pakistan  October 1986

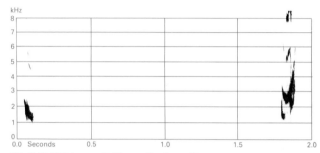

IV  E D H Johnson  Pakistan  October 1986

squeaky variant from pair perched outside nest-hole. Figs III-IV show calls from 2 birds in tumbling flight: Fig III depicts 'KY-aw tya-ka'; in Fig IV calls reduced to clipped sharp 'kaa' (P J Sellar) or 'kya'. Basic call often accompanied by Wing-flirting when function is commonly self-advertising, but also given regularly without wing movements and in flight, when it probably functions as contact-call. Lovari (1978) distinguished contact-call 'kiaa' from excitement-call 'kioo'; young birds gave squeakier versions of both of these. Sitasuwan and Thaler (1985) distinguished variants with apparently different meanings, such as flight-call, expression of excitement, and commands to 'come' and 'go away'. Soft conversational variants noted from pairs sitting quietly together (Goodwin 1986). Call resembling Buzzard *B. buteo* when soaring, and 'dya' or 'dyo' like Jackdaw *Corvus monedula* (Schifferli and Lang 1940a) may also be variants. (3) Alarm-call. Harsh, scolding 'ker ker ker', apparently signalling alarm, heard when raptor or Raven *C. corax* passed close by and when disturbed by human observer suddenly appearing nearby from cover (Holyoak 1972b). According to Lovari (1978), used when predator spotted or mobbed and when bird deeply distressed (e.g. seized by predator or in fight with conspecific); not heard from young less than 6 months old. Probably the same or similar calls, always hard and short, described by Schifferli and Lang (1940a) as several variants of a cry of fear, anger, or both, given by birds being handled, when diving in attack at *B. buteo* and Kestrel *Falco tinnunculus*, and when otherwise angry. Sitasuwan and Thaler (1985) recorded 'Kontaktlaute' and 'Angstschrei' which appear similar. Following calls may be the same or variants: deeper 'kaah' (quite like gull *Larus*

when used by flocks), 'kwuk-uk-uk', and, in alarm, 'keeaw' changing to loud hard 'kwarr' when nest or young in danger; sharp 'quek', short, subdued, tremulous, guttural 'quahrr' (Witherby *et al.* 1938), and shrill 'tchaik-aik' from captive birds shown *C. monedula* in the hand (Goodwin 1986). (4) Screaming 'kreeaa' heard from fighting birds (Goodwin 1986) may be extreme form of call 3 but perhaps more likely distinct. (5) Mobbing-call. Long, loud screech 'chaaaaay', rather like 'alarm screech' of *C. monedula* (presumably call 8a of that species, p. 133) but clearer and higher pitched, was given by 3 captive birds shown head of Eagle Owl *Bubo bubo* (Goodwin 1986). 'Aggressionslaute' of Sitasuwan and Thaler (1985) may be similar. (6) Begging-call of ♀. Whimpering 'veeaygaygaygaygay', similar to food-calls of fledglings (see below) (Schifferli and Lang 1940a). Fig V depicts calls apparently of this type: nasal, squeal-

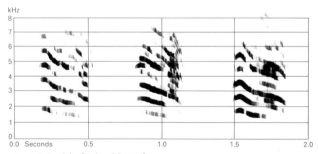

V  J-C Roché  Spain  May 1985

ing, whinnying sounds (of complex broad-band structure incorporating 'mixed harmonics') from bird flying near nest where mate perched. 'Nest-chatter' of Owen (1985), not described in detail, may be the same but ♀ said to have higher pitch and rate of usage, implying ♂ also gives the call; ♀ initiates 'nest-chatter' when ♂ enters breeding area. (7) Food-offering call. A 'croa' or 'graa' given by both sexes feeding young, stimulating them to beg (Schifferli and Lang 1940a). (8) Other calls. Sporadic, brief, low-pitched, harsh clucking sounds, with quality of Pheasant *Phasianus colchicus* (P J Sellar: Fig VI) from pair perched

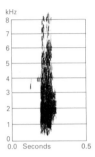

near nest. In another recording, parent perched near nest gives brief, gruff, quiet 'chikut' and 'quit' sounds before feeding well-grown young (P A D Hollom, P J Sellar); Fig VII shows swallowed 'chi-kut'.

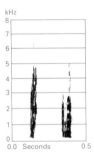

VII   P A D Hollom   Wales   May 1978

CALLS OF YOUNG. Peeping cry heard from before hatching; at first very weak, later louder when food-begging and softly after being fed as apparent signal of contentment (Schifferli and Lang 1940a; Goodwin 1986). On days 8 and 9, young silent, even when begging, but on day 10 began to beg with noise very like young *S. vulgaris*; this gradually changed to food-call (Fig VIII) like call 6 of adult ♀ (Schifferli and Lang 1940a; Goodwin 1986). Pre-

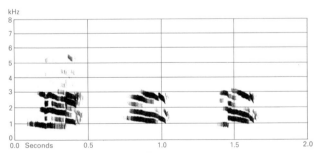

VIII   D Hemingray   Wales   July 1974

sumably the same call of older nestlings and fledglings was rendered by Holyoak (1972b) as high-pitched 'kee kee...'. Recently-fledged young give 'kwa-ak' calls when hiding near nest-site (Holyoak 1972b).                                    DTH

**Breeding.** SEASON. Britain and Ireland: first eggs laid early April, rarely late March, mainly mid- to late April or early May, upland pairs to mid-May; young fledge June; double brood recorded twice (southern Ireland) in May and September in 2 years at same nest (Cabot 1965; Rolfe 1966; Fisher 1969; Holyoak 1972b; Dawson 1975; Bullock *et al.* 1983a, b); eggs laid in Britain June–July probably late 1st or repeat clutches (Goodwin 1986). Alps: laying early to mid-April at 1400 m, mid-April to mid-May at 2300 m (Schifferli and Lang 1940a; Praz and Oggier 1976). Southern France: incubation early May (Mayaud 1934). Picos de Europa (northern Spain): nest-building mid-April; fledged young from June (García Dory 1983). SITE. Crevice in cliff or ledge in cave, shaft, or overhang; sometimes in or on building (Williamson 1959; Bullock *et al.* 1983a); dark but open space, not usually closed hole (Harrop 1970; Holyoak 1972b). Over most of range in rock, but in Hoya de Guadix (southern Spain) in holes in clay cliffs (Soler 1989c); in Ladakh (India) excavated holes 1 m deep in soft sandstone (Osmaston 1925) and has nested successfully in old burrow of rabbit *Oryctolagus* (Bullock *et al.* 1983b). Also in nest-box (Jennings 1984). In Britain and Ireland, most nests on coast; in Ireland, 61% of 135 visible nests were in crevices in cliffs and 39% in caves; 89% of 55 inland nests in natural crags (Bullock *et al.* 1983b); in Wales, 32% of nests inland (Bullock *et al.* 1986). In Ireland only 5% of sites man-made, in Scotland 27% (Bullock *et al.* 1983b; Monaghan *et al.* 1989). Most 10–30 m (1–250) above ground or sea; one nest reached by 60 m tunnel, and 30 m vertical shaft sometimes used as exit (Cabot 1965; Rolfe 1966; Bonham 1970a). Aspect in Ireland and inland Wales north-west, coastal Wales south-west or west, and 4 known sites in Valais (Switzerland) south and west (Praz and Oggier 1976; Bullock *et al.* 1983a). Same site may be used year after year, sometimes beyond possible lifetime of any one pair (Nethersole-Thompson and Nethersole-Thompson 1943a; Bonham 1970a; Goodwin 1986). Nest: bulky, untidy structure of dry twigs, roots, moss, and plant stems; base often solely heather *Calluna* stems, occasionally bound with mud, lined thickly with wool and occasionally other animal hair, man-made material, and thistle down; in absence of suitable materials, one nest in Kerry (Ireland) consisted of sea campion *Silene maritima*; another on Isle of Man consisted solely of lining material; in Ladakh, nest burrows contained lining only and no sticks (Osmaston 1925; Schifferli and Lang 1940a; Williamson 1959; Harrop 1970; Holyoak 1972b; Roberts 1979; García Dory 1983). Dimensions of one nest, Tarasp (Switzerland): outer diameter 35 cm, inner diameter 12 cm, overall height 17 cm, depth of cup 7·5 cm (Schifferli and Lang 1940a). Building:

by both sexes, though ♂ more active at start, ♀ completing nest, including lining, by herself; ♀♀, accompanied by ♂♂, on Calf of Man travel 2 km, including 1 km sea-crossing, to Isle of Man to gather wool for lining; often simply re-lines old nest (Holyoak 1972b); wool may be taken from sheep's back (Walpole-Bond 1905). EGGS. See Plate 57. Sub-elliptical, smooth and glossy; very pale, tinged greenish, creamy, or faint buff, marked overall with small blotches, spots, and streaks of olive-brown and grey; underlying markings, which often predominate, lilac-grey; sometimes veined and streaked like Nightjar *Caprimulgus europaeus* (Walpole-Bond 1905; Harrison 1975). Nominate *pyrrhocorax*: 39·2 × 27·9 mm (33·3–43·2 × 21·1–29·2), $n = 151$; calculated weight 15·7 g. *P. p. erythrorhamphus*: 40·0 × 28·0 mm (36·5–44·1 × 21·1–29·7), $n = 38$; calculated weight 16·2 g. *P. p. barbarus*: 37·3 × 27·5 mm (37·0–37·5 × 27·5), $n = 2$; calculated weight 14·6 g. (Schönwetter 1984.) Clutch: 3–5 (1–6). Of 43 clutches, Islay (Scotland): 1 egg, 2%; 2, 5%; 3, 5%; 4, 9%; 5, 67%; 6, 12%; average 4·7 (Bignal *et al.* 1987). Of 87 clutches, Ireland: 2 eggs, 10·5%; 3, 25%; 4, 46%; 5, 15%; 6, 3·5%; average 3·76. Average in Britain and Ireland 3·88, $n = 236$; Isle of Man 3·39. Inland clutches in Wales tended to be smaller than coastal. (Bullock *et al.* 1983a, b, c.) Over 55% of clutches in Britain and Ireland of 4, except in Scotland where 5 most common (Bullock *et al.* 1983a; Bignal *et al.* 1987). One brood; 1 pair in Cork laid and successfully raised 2 clutches of 4 eggs (Cabot 1965). Eggs laid at intervals of 1–3 days; 30 hrs in nest in Tarasp (Schifferli and Lang 1940a; Holyoak 1972b; Sitasuwan and Thaler 1984). INCUBATION. 17–18(–21) days (Schifferli and Lang 1940a; Fisher 1969; Bonham 1970a; Sorci *et al.* 1971). By ♀ only; usually sits for up to 1 hr followed by break of up to 5–10 min (Mayaud 1933b; Schifferli and Lang 1940a; Holyoak 1972b). Incubation usually starts with first egg, hatching thus usually asynchronous (Walpole-Bond 1905; Cowdy 1962, 1976; Bonham 1970a; Sitasuwan and Thaler 1984). ♀ of pair in Tarasp probably started with 2nd egg, but on day of laying 3rd egg ♀ left nest 11 times for total of c. 5·7 hrs to be fed by ♂ (Schifferli and Lang 1940a). YOUNG. Fed and cared for by both parents; ♀ broods for 2–3 weeks (Holyoak 1972b; Sitasuwan and Thaler 1984). Helpers may feed young (see Social Pattern and Behaviour). For account of brooding, feeding and development of young at one nest see Schifferli and Lang (1940a). FLEDGING TO MATURITY. Fledging period 31–41 days (Schifferli and Lang 1940a; Fisher 1969; Bonham 1970a; Sorci *et al.* 1971; Harrison 1975; García Dory 1983); 38 days in captivity; young may leave nest but remain in nest-cavity for c. 10 days before flying (Sitasuwan and Thaler 1984). First breeding at 2 years on Islay; at 3 years on Bardsey, though evidence suggests at least 4 years more frequent (Roberts 1985). BREEDING SUCCESS. Some pairs do not breed every year (Roberts 1985). Hatching success in Britain and Ireland 2·7 nestlings per pair ($n = 194$); of c. 915 eggs laid in 236 clutches, 61% produced fledged young; success high-

est in Ireland (76%), intermediate in Isle of Man (56%) and coastal Wales (63%), and lowest in inland Wales (50%) and Islay (42%); number of fledged young per pair increased steadily south-west through Britain and Ireland: Islay (1·82, $n = 34$), Isle of Man (1·88, $n = 67$), inland Wales (2·05, $n = 79$), coastal Wales (2·68, $n = 160$), Ireland (2·85, $n = 67$); higher still (3·7, unknown number) in Abruzzo (Italy) (Bullock *et al.* 1983a). In subsequent study on Islay, 2·2 young per pair ($n = 27$) in 1983 and 1·96 ($n = 27$) in 1984 (Monaghan and Thompson 1984; Monaghan and Bignal 1985). In Wales, success lower inland than on coast, possibly due to inexperienced young pairs and higher predation (Bullock *et al.* 1983a). Of 94 nests in Granada (Spain), 8 parasitized by Great Spotted Cuckoo *Clamator glandarius* with 1–4 cuckoo eggs in each; Raven *Corvus corax*, Peregrine *Falco peregrinus*, and fox *Vulpes* took eggs and nestlings (Zuñiga 1989). Asynchronous hatching means that last-hatched nestling often dies through bad weather or poor food supply (Sitasuwan and Thaler 1984). WGM

**Plumages.** (*P. p. erythrorhamphus*). ADULT. Head, body, lesser and median upper wing-coverts, and all under wing-coverts black, glossed purple; tail, flight-feathers, tertials, all greater upper wing-coverts, and bastard wing slightly duller black, gloss more metallic green on outer webs, dull purplish on inner webs, brighter purple or bluish-purple on narrow outer fringe only. *In worn plumage*, abraded feathers duller sooty-black with variable dark grey bloom, especially tail-tips, primary-tips, tertials, and face; some dark grey of feather-bases sometimes visible on breast or belly. NESTLING. Upperparts, upperwing, and thigh covered with dark grey or dark grey-brown down c. 1–2 cm long; remainder naked (Williamson 1939; Schifferli and Lang 1940a). For development, see Schifferli and Lang (1940a). JUVENILE. Rather like adult, but feathering of head, body, and lesser upper wing-coverts markedly shorter and looser, dark grey to black-brown with limited amount of purplish-black on each feather-tip, back, rump, and underparts in particular appearing dark brown-grey and woolly. Tail and wing as adult, more glossy than body and often glossier than worn adult (but see First Adult). FIRST ADULT. As adult, but tail, flight-feathers, tertials, greater and (if any left) median upper wing-coverts, greater upper primary coverts, and bastard wing still juvenile, these slightly less glossy than in fresh-moulted adult; retained feathers fade more rapidly than in adult, becoming dull brown-black; difference in gloss often marked from mid-winter, when tail-tips and primary-tips also more distinctly abraded than in adult. Tip of p10 pointed or narrowly rounded (in adult, broadly rounded; see also Mayaud 1933b); tips of other flight-feathers, tertials, and tail-feathers often also rounded, less truncate than adult. Up to October–November, some scattered juvenile feathers on body retained, especially on neck, belly, and median upper wing-coverts, markedly browner than neighbouring new feathers and with limited amount of glossy black on tip (if any left). Also, brown-grey bases of new body feathers slightly more extensive than in adult, more inclined to show through when plumage worn.

**Bare parts.** ADULT, FIRST ADULT. Iris brown or dark brown. Bill, leg, and foot orange-red, brick-red, bright red, or carmine; bill sometimes pink or dark crimson, leg and foot sometimes purple-red or dark red. Claws black. (Goodwin 1986; ZFMK,

ZMA.) NESTLING. At hatching, bare skin, mouth, and leg flesh-red, bill tinged pink-horn, gape-flanges yellow-white (Schifferli and Lang 1940a). Bill colourless at hatching, changing through whitish-pink and yellowish to pink-yellow or pale red at fledging (Witherby *et al.* 1938). At 6–7 days, bare skin pink-flesh; bill mauve, tip and cutting edges whitish, gape-flanges pale yellow; mouth salmon-pink with white spurs on palate; leg and foot flesh-pink (Williamson 1939). About 1 week before fledging, iris dark brown, bill purplish-brown to greyish horn-black, tip and cutting edges whitish, pale yellow-horn, or salmon-pink, narrow gape-flanges yellow-white, pale yellow, or pale orange-yellow, mouth pink-red or reddish-flesh; leg and foot pink-flesh to orange-pink with grey, purple-brown, or black tinge on scutes at front of tarsus, on upper surface of toes, and on joints; claws black-brown. (Williamson undated; Goodwin 1986; ZMA). From about day 22, leg yellow (♀) or dirty orange (♂) with dark grey scutes (Schifferli and Lang 1940a). At about fledging, iris brown, bill pink-orange with grey tinge on base, narrow gape-flanges yellow-white, leg and foot yellowish flesh-red or flesh-orange. JUVENILE. Iris dark brown. Bill flesh-pink, yellow-brown, dirty orange, or orange. Leg and foot yellow-pink with brown or grey tinge, pale red, orange-pink, or deep reddish-pink. Adult colour obtained during post-juvenile moult. (Hartert 1903–10; Niethammer 1937; Williamson 1939; Schifferli and Lang 1940a; Kumerloeve 1961; Mauersberger *et al.* 1982; Svensson 1992; Goodwin 1986; ZFMK, ZMA.)

**Moults.** ADULT POST-BREEDING. Complete; primaries descendent. In Britain, starts with p1 early June to late July, completed with p10 between late August and early October; average duration of primary moult *c.* 92 days; tail starts from primary moult score *c.* 10–20, completed at *c.* 40–50; secondaries start from score *c.* 20–30, completed with regrowth of p10 (score 50) or slightly later (Holyoak 1974; Ginn and Melville 1983). In Switzerland, primary moult late May or June to late August (Schifferli and Lang 1940a). In Afghanistan, wing, tail, and body in moult mid-June, wing and body late July, and still usually some moult September–October (Paludan 1959). In 2 captive *docilis*, primary moult lasted 174 and 182 days (Winkler *et al.* 1988, which see for sequence and growth rate of individual flight- and tail-feathers). In 45 birds examined May–October, whole geographical range, moult starts with p1 early May to early June; by 10–30 June, primary-scores of 12, 18, 20, 22, 24, 24, and 25 reached in 1-year-olds, 18, 23, 23, and 24 in adults; during 1–20 July, score of 1-year-olds 18, 21, 26, 34, and 36, of adults 16, 17, 18, 20, 23, 24, and 31; during August, scores (ages combined), 34, 37, 38, 38, and 39; in September, 43 and 44; moult completed (score 50) from October, 2 birds from early November still with traces of moult (BMNH, RMNH, SMTD, ZFMK, ZMA, ZMB). Estimated duration of moult at least 150 days (C S Roselaar), in agreement with data of Winkler *et al.* (1988) and contra Holyoak (1974). See also Mayaud (1933b), Dementiev and Gladkov (1954), Keve and Rokitansky (1966), Diesselhorst (1968), and Piechocki and Bolod (1972). POST-JUVENILE. Partial; head, body, and lesser and median upper wing-coverts. Starts shortly after fledging, in Switzerland from early June (Schifferli and Lang 1940a). Fully juvenile birds encountered mid-June to late August, moulting birds late July to late October, birds with moult completed from early September (BMNH, RMNH, SMTD, ZFMK, ZMA, ZMB).

**Measurements.** Nominate *pyrrhocorax*. Britain and Ireland (Vaurie 1954a). Bill (S) to skull.

| | | | | | | |
|---|---|---|---|---|---|---|
| WING | ♂ | 276 | (8) 268–293 | ♀ | 271·5 (5) 266–278 | |
| TAIL | | 133 | (8) 120–141 | | 132 (5) 125–140 | |

| | | | | | |
|---|---|---|---|---|---|
| BILL (S) | 54·7 (8) | 51–59 | 52 | (5) | 50–33 |
| TARSUS | 53 (8) | 49–56 | 50 | (5) | 48–54 |

In 8 ♂♂, wing 257–281, tail 126–145, bill to nostril 41–49, tarsus 55–59; in 10 ♀♀, wing 245–271 (once 293), bill to nostril 39·5–48; in 5 unsexed birds, wing 263–280 (once 297), bill to nostril 45–52 (Witherby *et al.* 1938).

*P. p. erythrorhamphus.* Spain and Alps, all year; skins (BMNH, RMNH, SMTD, ZFMK, ZMA, ZMB). Bill (S) to skull, bill (N) to distal corner of nostril; exposed culmen on average 3·5 less than bill (S). Juvenile wing and tail include retained juvenile wing and tail of 1st adult. Bill includes data of 2nd-calendar-year birds.

| | | | | | |
|---|---|---|---|---|---|
| WING AD | ♂ | 300·3 (10·26; 16) 282–315 | ♀ | 291·2 (2·28; 5) 289–294 | |
| JUV | | 285·6 ( 4·34; 5) 280–292 | | 275·3 (6·61; 8) 265–285 | |
| TAIL AD | | 146·1 ( 6·83; 16) 135–158 | | 139·4 (1·34; 5) 137–141 | |
| JUV | | 141·5 ( 4·46; 6) 135–147 | | 137·9 (9·80; 9) 130–152 | |
| BILL (S) | | 58·4 ( 2·86; 19) 54·9–62·5 | | 53·0 (2·62; 12) 48·2–56·3 | |
| BILL (N) | | 48·0 ( 2·20; 19) 44·5–51·8 | | 43·7 (3·08; 12) 37·2–48·0 | |
| TARSUS | | 56·4 ( 2·00; 17) 53·5–59·5 | | 53·2 (2·78; 13) 48·2–56·0 | |

Sex differences significant, except for adult wing and juvenile tail.

Wing and bill, sexes combined or sex unknown. (1) Bretagne, (2) Alps, (3) Pyrénées; bill is exposed culmen (Mayaud 1933b). (4) Sardinia (*n*=4) and Sicily (*n*=4), combined, 4 of each sex, age unknown; bill to nostril (Londei and Gnisci 1988).

| | | | | |
|---|---|---|---|---|
| WING BILL (1) AD | 285·6 (13·95; 13) 253–299 | — ( — ; 12) | 45–54 | |
| JUV | 266·2 ( 8·30; 11) 257–283 | — ( — ; 7) | 46–51 | |
| (2) AD | 293·8 (10·12; 10) 282–310 | 52·4 (2·62; 9) | 47–55 | |
| JUV | 277·6 (16·46; 7) 245–294 | 52·1 (4·38; 7) | 48–60 | |
| (3) AD | 298·8 (13·82; 3) 283–309 | 55·4 (4·63; 3) | 50–56 | |
| JUV | 285·0 ( 8·37; 4) 273–292 | 53·4 (2·43; 4) | 50–55 | |
| (4) | 286·0 (11·46; 8) 272–300 | 46·2 (2·71; 8) | 41–50 | |

*P. p. barbarus.* Morocco and La Palma (Canary Islands), all year; skins (RMNH, ZFMK, ZMA, ZMB).

| | | | | |
|---|---|---|---|---|
| WING AD | ♂ | 296·9 (6·90; 8) 286–310 | ♀ | 274·5 (3·74; 5) 272–281 |
| TAIL AD | | 144·2 (5·42; 8) 135–152 | | 134·4 (4·06; 5) 128–139 |
| BILL (S) | | 60·9 (2·16; 8) 57·9–62·8 | | 55·5 (1·19; 5) 54·5–57·5 |
| BILL (N) | | 50·4 (1·67; 8) 47·9–52·5 | | 45·4 (1·41; 5) 44·2–47·5 |
| TARSUS | | 55·2 (0·91; 7) 53·5–56·1 | | 52·7 (0·77; 5) 51·5–53·5 |

Sex differences significant. Morocco and Algeria: in 15–17 adult ♂♂, wing 302 (290–310), tail 145 (129–153), bill (S) 63·2 (61–67); in 4 ♀♀, wing 280 (275–284), tail 138 (131–147), bill (S) 57·0 (55–58) (Vaurie 1954a; Rand and Vaurie 1955).

*P. p. docilis.* Crete, Turkey, Syria, Caucasus, and north-west Iran, all year; skins (SMTD, ZFMK, ZMB).

| | | | | |
|---|---|---|---|---|
| WING AD | ♂ | 313·2 (9·96; 12) 299–325 | ♀ | 291·6 ( 4·39; 5) 284–295 |
| JUV | | 306·2 (8·45; 6) 294–314 | | 284·1 (11·64; 8) 271–299 |
| TAIL AD | | 153·2 (8·75; 12) 140–166 | | 145·0 ( 3·29; 6) 141–150 |
| JUV | | 147·5 (3·62; 6) 144–154 | | 139·0 ( 3·56; 8) 134–144 |
| BILL (S) | | 55·6 (1·76; 16) 52·6–59·2 | | 50·6 ( 2·03; 11) 46·9–54·1 |
| TAIL (N) | | 45·3 (2·09; 17) 42·1–49·2 | | 41·0 ( 1·83; 11) 37·3–44·1 |
| TARSUS | | 56·3 (1·21; 18) 54·6–59·5 | | 2·4 ( 1·21; 13) 50·6–54·1 |

Sex differences significant. Southern Yugoslavia: wing, ♂ 312·7 (3) 311–315, ♀♀ 288, 298; tail, ♂ 153·3 (3) 153–154, ♀ 137; tarsus, ♂ 55·3 (3) 53–58, ♀ 51 (Matvejev 1955). Caucasus to Turkmeniya: wing, ♂ 326·1 (21) 290–344, ♀ 305·1 (18) 291–325 (Dementiev and Gladkov 1954). Northern Iran and Turkmeniya: wing, ♂ 319·0 (5·00; 5) 314–327, ♀ 297·8 (5·85; 4) 293–305 (Stresemann 1928b; Vaurie 1955). Iran: wing, ♂ 315 (11) 306–330, ♀ 295 (13) 280–307; tail, ♂ 145 (11) 132–161; bill (S), ♂ 56·8 (11) 52–61 (Vaurie 1954b; Rand and Vaurie 1955). Afghanistan, adult: wing, ♂ 307·2 (4·79; 4) 302–313, ♀ 290·2 (5·12; 6) 282–297; exposed culmen, ♂ 53·8 (0·96; 4) 53–55, ♀ 47·8 (1·60; 6) 46–50 (Paludan 1959).

Adult ♂ only: *centralis*, (1) Tien Shan mountains (*n* = 13), (2) Kashmir and Ladakh (*n* = 11), (3) Altai mountains and Transbaykalia (*n* = 2); *himalayanus*, (4) Nepal and southern Tibet (*n* = 5), (5) Szechwan (*n* = 8), (6) Kansu (*n* = 3); *brachypus*, (7) Shansi and Hopeh (*n* = 2); *baileyi* (8) Ethiopia (*n* = 3) (BMNH, RMNH, SMTD, ZFMK, ZMA, ZMB).

| | WING | TAIL | BILL (S) | TARSUS |
|---|---|---|---|---|
| (1) | 309·8 (293–322) | 163·8 (151–175) | 53·8 (51·0–57·6) | 49·3 (47·3–51·5) |
| (2) | 314·2 (303–323) | 162·9 (154–173) | 56·6 (55·3–57·6) | 52·6 (49·3–55·2) |
| (3) | 321·0 (320–322) | 168·0 (165–171) | 55·8 (55·5–56·0) | 49·6 (49·1–50·1) |
| (4) | 325·6 (317–336) | 159·4 (151–163) | 62·1 (59·1–66·0) | 61·3 (59·1–62·8) |
| (5) | 310·3 (291–328) | 163·4 (153–171) | 59·0 (56·2–61·1) | 61·2 (60·2–62·8) |
| (6) | 311·8 (304–317) | 170·9 (167–173) | 55·8 (54·3–57·6) | 58·0 (56·1–59·5) |
| (7) | 307·5 (304–311) | 165·0 (164–166) | 50·1 (49·0–51·2) | 52·6 (51·0–54·2) |
| (8) | 328·0 (325–332) | 150·0 (149–151) | 66·5 (63·5–70·0) | 59·0 (58·8–59·3) |

*P. p. himalayanus*. Tibet: wing, ♂ 323 (28) 295–357, ♀ 299·5 (26) 280–323 (Vaurie 1972).

*P. p. baileyi*. Simien mountains (Ethiopia): wing, ♂ 318·6 (7) 310–329; tail, ♂ 150·6 (7) 145–155; bill (S), ♂ 60·6 (7) 54–65 (Vaurie 1954a; Rand and Vaurie 1955).

**Weights.** *P. p. erythrorhamphus*. Switzerland: May, adult pair, ♂ 350, ♀ 293; nestlings on day 1, ♂ 11, ♀ 10·6; on day 21, ♂ 262, ♀ 198; on day 42, ♂ 295, ♀ 257; on days 86–114, ♂ 340–360, ♀ 300–320 (Schifferli and Lang 1940a, which see for details). Italy, 230–390 (Sorci *et al.* 1971).

*P. p. docilis*. Crete: ♂ 375 (November); ♀♀ 300 (June), 275 (November) (ZMB). Western Turkmeniya: February, ♂♂ 340, 375; June, unsexed 277 (Dementiev and Gladkov 1954). Afghanistan: June, ♂ 291, ♀ 264; September–October, ♂ 314, ♀ 265·5 (4) 253–283 (Paludan 1959).

*P. p. centralis*. Kazakhstan: ♂ 273·2 (5) 258–290, ♀ 236·3 (3) 219–250 (Korelov *et al.* 1974). Kirgiziya: ♂ 230–260 (*n* = 3), ♀ 227–287 (*n* = 3) (Yanushevich *et al.* 1960). Northern Mongolia: late May and early June, ♂♂ 270, 274; ♀♀ 220, 240; August, ♀ 234 (Piechocki and Bolod 1972). Ladakh (India): April–May, ♂ 331·2 (8·47; 5) 318–341, ♀ 274 (Meinertzhagen 1927).

*P. p. himalayanus*. Sikkim: November, ♂♂ 402, 428; ♀ 358 (Meinertzhagen 1927). Nepal: July–September, ♂ 450, ♀ 327·7 (3) 349–385 (Diesselhorst 1968).

*P. p. brachypus*. Manchuria (northern China): August, ♀ 245 (3) 238–253 (Piechocki 1959).

**Structure.** Wing rather long, broad at base, tip rounded. 10 primaries: p7 longest, p6 0–5 shorter (rarely, p6 slightly longer than p7); in adult (*n* = 15), p8 4–11 shorter, p9 25–44, p5 9–21, p4 42–62, p1 90–120; in juvenile (*n* = 12), p8 0–6 shorter, p9 22–40, p5 16–30, p4 45–74, p1 95–127; difference between p5 and p6 6–20 in adult, 15–28 in juvenile. P10 somewhat reduced, 102–122 shorter than p7 in adult, 95–120 in juvenile; 55–75 longer than longest upper primary covert in adult, 51–67 in juvenile. Outer web of (p5–)p6–p9 deeply emarginated, inner web of (p6–)p7–p10 with distinct notch. Tertials short, longest equal to inner secondaries. Tail rather short, tip slightly rounded; 12 feathers, t6 2–10 shorter than t1. At rest, tail-tip equal to wing-tip or slightly shorter or longer. Bill slightly longer than head; culmen, cutting edges, and gonys gently and evenly decurved towards sharply pointed tip; rather wide and deep at base, culmen and gonys rounded in cross-section; depth of bill at middle of nostril 12·0–13·0 (♂) or 11·5–12·5 (♀) in nominate *pyrrhocorax*, 13·0–14·5 (♂) or 12·0–14·0 (♀) in *barbarus* and *himalayanus*, 12·5–14·0 (♂) or 11·4–12·4 (♀) in *erythrorhamphus* and *docilis*, 11·0–13·0 (♂) or 10·0–12·0 (♀) in *centralis*, 13·0–13·5 (♂) or 11·0–

12·0 (♀) in *brachypus* (see also Vaurie 1954a). Side of both mandibles covered at base by dense and compact layer of short, stiff, bristle-like feathers projecting forward from lore and hiding small rounded nostril; chin with numerous fine hairs. Tarsus and toes strong (relatively heaviest in *erythrorhamphus*, *docilis*, and *himalayanus*); tarsus booted (traces of scutes present only in just-fledged juveniles); middle toe with claw 36·9 (10) 34–41 in *erythrorhamphus* and *barbarus*, outer toe with claw *c.* 78% of middle with claw, inner *c.* 81%, hind *c.* 84%.

**Geographical variation.** Marked. Mainly involves size (as expressed in weight or wing length), relative measurements of tail, bill, tarsus, toes, and bill depth, colour of gloss on body, wing-coverts, flight-feathers, or tail, and (perhaps) wing shape. As retained juvenile wing and tail of 1st adult markedly shorter than those of adult, and 1st adults also have wing differently shaped and with less intense gloss, only adults should be used for study of variation. In general, size smaller in north and in birds of coastal cliffs or inland hills, larger in south and in high-mountain populations; gloss strongly purplish in populations living in humid climates, more greenish in arid regions.

Nominate *pyrrhocorax* from Britain and Ireland is smallest race, average adult wing *c.* 275; bill relatively long, average *c.* 54; gloss on body and wing-coverts purplish, on tail- and flight-feathers bluish-green or slightly purple. *P. p. erythrorhamphus* from Alps and Italy is slightly larger, average adult wing *c.* 292, bill *c.* 55·5; gloss on tail- and flight-feathers (bluish-) green; birds inhabiting Massif Central (France), Pyrénées, and Iberia slightly larger, average adult wing *c.* 299, bill *c.* 58. Birds from Bretagne (western France) intermediate between *erythrorhamphus* and nominate *pyrrhocorax*, but nearer *erythrorhamphus*. *P. p. barbarus* from La Palma (Canary Islands) and north-west Africa similar in size to *erythrorhamphus*, but bill distinctly longer and thicker; gloss on wing-coverts and often on mantle and scapulars more (bluish-)green, less purplish; on La Palma, slightly smaller on average than in Africa (average adult wing La Palma *c.* 292, bill *c.* 58; Morocco and Algeria *c.* 300 and *c.* 60 respectively). *P. p. docilis* from southern Balkans, Greece, and Crete east to Iran, Afghanistan, and southern Turkmeniya greener still and less intensely glossy than *barbarus*; large, but bill relatively short; average adult wing *c.* 304 and bill *c.* 53 in southern Yugoslavia, Greece, and Crete, *c.* 308 and *c.* 54 respectively in mountains of Caucasus, Transcaucasia, and northern Iran; perhaps this race bred formerly in Ural mountains, if any (Johansen 1944; Vaurie 1959). *P. p. docilis* extends east to Afghanistan, grading into *centralis* in western Pamir-Alay ranges. Further east, in central and eastern Asia, situation not yet fully elucidated. In general, relative tail length of these populations slightly longer than further west (tail/wing ratio *c.* 0·53, against *c.* 0·49 in west); tarsus and toes either relatively long and heavy (tarsus/wing ratio *c.* 0·190, similar to nominate *pyrrhocorax*, *erythrorhamphus*, and *barbarus*) or short and slender (tarsus/wing ratio *c.* 0·165); bill to skull either longer or shorter than tarsus. *P. p. himalayanus* from Garhwal, Nepal, and southern Tibet eastwards is large, legs heavy, bill length about equal to tarsus, gloss strongly blue; birds from western Szechwan closely similar but somewhat smaller. Further north, in Kansu, tarsus heavy but bill shorter than tarsus; probably intermediate between *himalayanus* and *brachypus*, but nearer *himalayanus*. *P. p. brachypus*, occurring from Ningsia to Hopeh and probably in neighbouring parts of Inner Mongolia and Manchuria, has tarsus and bill short, but tarsus longer than bill, as bill shorter than in any other race; gloss rather faint, body satin-black, wing and tail slightly purplish. Birds from Kashmir and Ladakh east to Sutlej valley often included in *himalayanus*, as gloss similar, but bill and especially tarsus markedly shorter; measurements closely similar to *centralis* from Tien Shan mountains and here provisionally included in that race, though gloss on wing more bluish, less greenish; tarsus of Tien Shan birds shortest of all races. Wing shape of Tien Shan birds said to differ from other populations (Hartert and Steinbacher 1932–8; Vaurie 1954*a*), but little constant difference found when specimens of same age compared. Only a few birds from Altai mountains and Transbaykalia examined; gloss of these perhaps slightly duller than birds from Tien Shan, but measurements much like these and hence included in *centralis*; bill/tarsus ratio of Altai–Transbaykalia birds 1·11, near that of populations from Tien Shan, Alay range, and Ladakh–Kashmir, different from 0·93–0·94 of *brachypus* and of *himalayanus* from Szechwan. Perhaps *centralis* grades into *brachypus* in southern or eastern Mongolia. Highly isolated *baileyi* from Simien and (perhaps this race: Brown 1967) Bale mountains of Ethiopia is large, but tail rel-atively short; gloss of body, wing, and tail relatively dull and greenish, rather like *docilis*, but duller still and bill longer; bill, tarsus, and amount of gloss rather like *barbarus*, but wing distinctly longer. For possible explanation of present-day distribution, see Guillou (1981).

**Recognition.** Adult easily distinguished from Alpine Chough *Pyrrhocorax graculus* by long decurved red bill, relatively short tail (about equal to wing-tip at rest), broader wing, and more strongly glossed plumage. Just-fledged juvenile with bill and wing not yet full-grown and without red on bill more difficult; best distinguished by extent of feathering on bill base: dense bristles at base of both mandibles equal in extent, covering less than 25% of bill-base (in *P. graculus*, bristles at side of upper mandible cover *c.* 33% of base, but lower mandible bare, almost without bristles). Beware of hybrids between the species (see Social Pattern and Behaviour).    CSR

## *Corvus monedula* Jackdaw

PLATES 7 and 9 (flight)
[between pages 136 and 137]

Du. Kauw    Fr. Choucas des tours    Ge. Dohle
Ru. Галка    Sp. Grajilla    Sw. Kaja

*Corvus Monedula* Linnaeus, 1758

Polytypic. Nominate *monedula* Linnaeus, 1758, Scandinavia, extreme southern Finland, Denmark south to Esbjerg and Haderslev, eastern Germany and Poland east to *c.* 23°E, Czechoslovakia, Austria, Hungary, north-west Rumania including Carpathian mountains, and northern Yugoslavia (central Slovenija to Vojvodina); *spermologus* Vieillot, 1817, western Europe from Netherlands, Rheinland (western Germany), western Switzerland, Italy, and extreme north-west Yugoslavia westwards, from Britain south to Iberia, also Morocco; *soemmerringii* Fischer, 1811, eastern Europe and western Asia, from south-west and central Yugoslavia and southern and eastern Rumania southwards, and from south-east Finland and former USSR eastwards, east to Kashmir and Mongolia; *cirtensis* (Rothschild and Hartert, 1912), north-east Algeria and (formerly) Tunisia.

**Field characters.** 33–34 cm; wing-span 67–74 cm. Less than 75% size of Rook *C. frugilegus*; somewhat smaller and less broad-winged than Chough *Pyrrhocorax pyrrhocorax*. Small, dapper, and bustling crow, with short bill and (on ground) quite high head carriage. Black, with grey rear to head (palest in east) and dull grey underparts from breast to vent. Eye pale in adult. Sexes similar; little seasonal variation. Juvenile separable. 4 races in west Palearctic, of which easternmost most distinctive.

ADULT. Moults: in west of range, mostly June–September (complete); in east, a month later. (1) West European and Moroccan race, *spermologus*. Front of head (face, throat, and cap) black, glossed purple on crown, setting off pale, pearly eye and grey area over rear of head which is sharply defined at edge of crown and mantle, but merges with dusky ear-coverts and throat. Sometimes a trace of paler grey or silvery collar on side of neck. Rest of plumage black, with faintly bluish fringes to back feathers, faint green and blue gloss on wings, and faintly paler greyish cast over fringes and to underpart feathers except thigh, rear belly, and under tail-coverts. With wear, grey area of ♂'s head tends to be paler than ♀'s; from late spring, feather fringes become sufficiently paler to create faintly scalloped pattern on back and body; by early summer, grey area of head (especially ♂'s) becomes more silvery but formation of pale, partial collar still rare, and underparts become duller. Bill and legs black. (2) Scandinavian and central European race, nominate *monedula*. Variation in plumage as *spermologus* but bird in similar state of wear typically paler grey on rear of head and side of neck, with more frequent grey, rarely whitish collar and more uniformly greyish underparts (contrasting with folded wing). (3) Algerian race, *cirtensis*. Resembles *spermologus*, having darker underparts than nominate *monedula*, but never shows even trace of paler collar; cap glossed bluish, not purple. (4) East European and central Asian race, *soemmerringii*. Plumage even more contrasting than nominate *monedula*, with paler silvery rear of head and neck ending in distinct white or cream collar, and paler grey underparts (in south-west of range). On many, more bleached fringes to body feathers create noticeably scalloped or dappled appearance. On some, head cap glossed bluish. Important to note that intergradation of races complex and open to varying interpretation: see Geographical Variation for discussion, particularly of birds with pale collar in western Europe. JUVENILE. All races. Much duller even than ♀, with little-glossed, brown- or sooty-toned plumage in which rear of head duller and less contrasting. Eye dull brown, taking a year to become fully pale.

Unmistakable at ranges where short bill and grey rear of head are visible—or even just compact, bustling form—but sometimes difficult to separate on ground at distance when mixed with frequent companion *C. frugilegus*, while distant flight views may also provoke confusion with choughs. Head-on or from behind, silhouette and flight occasionally suggest small pigeon *Columba*. Flight light and active, with quick, rather jerky beats of backward-directed wings which at distance may appear to 'twinkle'; incorporates gliding patrol, close circling, and marvellously free tumbling (all particularly exhibited by cliff-dwelling birds), and short flips and slow but tight manoeuvres close to ground (typical of flock intent on food search). Given to sudden 'dreads', with flocks (often mixed with *C. frugilegus*) often rising *en masse* ('twinkling' impression then strongest), breaking up, circling, and finally drifting back. Migration flight steady and less jerky. Gait a quite fast, jaunty walk, with occasional bobble of head, roll of shoulders, and forward or sidling jump. Unlike larger crows, small size and relatively short legs produce more ground-hugging, bustling, even shuffling progress over ground. Carries head high when searching for food, but keeps it low (much like Woodpigeon *C. palumbus*) when feeding. Perches freely on any prominence; enters holes at all seasons. Opportunistic and confident, fussing with its fellows, and accompanying calls with various movements in a way that conveys self-importance. Highly sociable, often in hundreds with *C. frugilegus*.

Voice distinctive, with most typical call a short, rather high-pitched but resounding 'tchack' or 'kjack' (sounding like 'chock' at distance); often given in flight and frequently extended into characteristic attractive cackle which at long distance in chorus also has chuckling tone. Other calls include caws and yelps, also a short 'kyow' and extended 'carr-r-r-r' which, in young birds, may suggest *P. pyrrhocorax*.

**Habitat.** Breeds across middle and upper middle latitudes of west Palearctic, in boreal, temperate, steppe, and Mediterranean lowlands, continental and oceanic, up to July isotherm of 12°C (Voous 1960). Tolerates wide ranges of precipitation and settled or unsettled weather, but avoids extremes of heat, ice, and snow. Needs sheltered nesting places, apparently adapting from main reliance on hollow or shady trees to rock crevices (inland or coastal), holes in buildings of various kinds, and even burrows of rabbit *Oryctolagus*. Requirements for enough of these in suffi-

cient proximity to satisfy gregarious instincts probably explains to some extent remarkably patchy and fluctuating distribution, and competition or commensalism with other birds subject to similar demands for breeding sites. For foraging will resort to pastures in preference to croplands, but in common with Rook *Corvus frugilegus* is ready to fly long distances to exploit ample food supplies of varying types (Yapp 1962). Habits of feeding on backs of sheep, donkeys, and other livestock, of assembling on trees bearing crops of acorns *Quercus*, cherries *Prunus*, or other fruits, and of aerial hawking of high-flying insects, illustrate adaptability which takes different forms in different parts of range (Nicholson 1951). Use of ground under woodland canopy for feeding considered exceptional in study near Oxford (England) by Lockie (1955) who found no evidence of moving into woods even in hard weather, though analysis of data from throughout Britain found breeding to occur in over half of surveyed woods, birds occurring in winter in more than 60% of them (Fuller 1982). Much less of an open field bird than *C. frugilegus*, but much less of a woodland bird than Jay *Garrulus glandarius*, and rather attached to mature or old free-standing or grouped trees, avenues, and trees neighbouring other attractions such as cliffs or tall buildings. Explores small enclosed patches of herbage on cliff slopes, under trees, and by roads and railway tracks. Effects of habitat limitations are indicated by patchy distribution in Switzerland along river valleys (highest breeding at 1230 m) either based on human settlements, where breeding is in buildings or old trees, or in cliffs and woods far from habitations, mainly north of Alps (Glutz von Blotzheim 1962). In Russia, also largely in valleys, often breeding in mixed colonies with *C. frugilegus*, pigeons (Columbidae), Kestrel *Falco tinnunculus*, and Red-footed Falcons *F. vespertinus* (Dementiev and Gladkov 1954). Also in wooded steppes; much less dependent on grain than *C. frugilegus* (Goodwin 1986). Regular journeys of several km from nest, and liking for aerial flock manoeuvres up to considerable heights, result in airspace forming unusually important element in habitat. Dependence on cultivation and human refuse, and on wide range of artefacts for breeding, also lead to closer links with people than for other Corvidae, even though remote independent living patterns are also widely maintained. As both tree- and rock-breeder, occupies intermediate place between (e.g.) *C. frugilegus* and choughs *Pyrrhocorax*, which contrastingly remain rigidly conservative in their choice of habitats.

**Distribution.** Recent range expansion in Scandinavia; few other changes reported.

ICELAND. Accidental visitor, but fairly frequent. Pair built nest in Reykjavik 1977, but no eggs laid; only known breeding attempt (Nielsen 1979). FAEROES. Rare visitor, winter and spring (Bloch and Sørensen 1984). FRANCE. Has spread since 1930s, especially in south-east (Yeatman 1976); no significant change recently (RC). NORWAY.

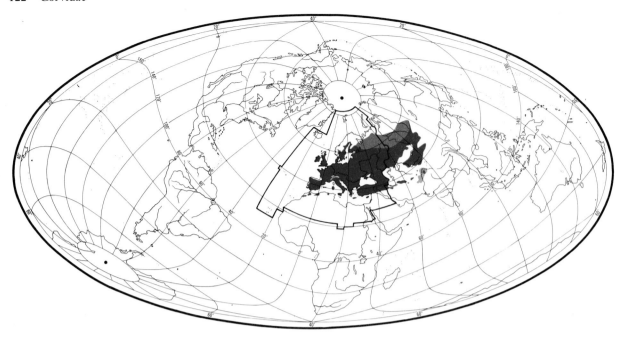

Recent spread to coastal areas in south and west (VR). SWEDEN. Has spread north along Gulf of Bothnia; in 19th century did not breed north of Uppsala (LR). ITALY. Recent range extensions in central and northern areas (PB). MALTA. Formerly common resident; exterminated by human persecution in mid 1950s (Sultana and Gauci 1982). MOROCCO. Locally distributed in mountains, but has recently colonized lowland town of Ouezzane (MT). TUNISIA. Formerly bred in north-west, now extinct (Thomsen and Jacobsen 1979).

Accidental. Iceland, Tunisia, Jordan, Madeira, Azores.

**Population.** Has decreased recently in many parts of Europe.

BRITAIN, IRELAND. More or less stable, *c.* 500 000 pairs (Sharrock 1976). FRANCE. 10 000–100 000 pairs (Yeatman 1976). PORTUGAL. Apparently decreasing in recent years, at least locally (Rufino 1989). BELGIUM. Estimated 21 000 pairs; probably stable (Devillers *et al.* 1988). NETHERLANDS. Estimated 60 000–120 000 pairs (SOVON 1987). GERMANY. 93 000 pairs (G Rheinwald). East Germany: 10 000±4000 pairs (Nicolai 1993); marked decline at beginning of 20th century (SS); decline apparently continuing in north; total population in Rostock, Schwerin, and Neubrandenburg 971 pairs, 1984–5 (Plath 1989). SWEDEN. Estimated 100 000 pairs (Ulfstrand and Högstedt 1976). Marked decrease during 20th century in southern and central Sweden (LR). FINLAND. Marked decrease since 1956 (Hildén 1988). Now stable, or decreasing locally; perhaps 10 000–20 000 pairs (Koskimies 1989). CZECHOSLOVAKIA. Marked decrease thoughout since 1960 (KH). AUSTRIA. Marked decline in some areas due to intensification of forestry and agriculture (H-MB). ITALY.

Increasing in central and northern areas (PB). RUMANIA. Decline in Transylvania in last 60 years (Salmen 1982). LATVIA. Decreased in period 1950–68; now apparently stable (Vīksne 1983). ESTONIA. Estimated 30 000 pairs (Dwenger 1989).

Survival. Britain: mortality in 1st calendar year 38%, in 2nd year 36%, in 3rd–5th years 43% (Holyoak 1971). Finland: annual mortality 35±SE3·2% (Haukioja 1969). Europe: mortality in 1st year of life 45·5% (Busse 1969). Oldest ringed bird 14 years 3 months (Rydzewski 1978).

**Movements.** Resident to migratory, wintering almost entirely within breeding range; birds head mostly west or WSW, so some birds of northern race nominate *monedula* and eastern race *soemmerringii* winter in range of western race *spermologus*. Arrivals and passage mask movements of local birds. Migrates by day in flocks, often in company of Rook *C. frugilegus*. Juveniles migrate more than adults, and over longer distances. (Busse 1969, which see for review; Baltvilks 1970; Yeatman-Berthelot 1991).

Chiefly resident in northern and western Europe, but less so further east, e.g. *c.* 23% migrate from Belgium, up to 70% from Poland (Busse 1963; Lippens and Wille 1972); some French birds (mostly juveniles) leave breeding areas, but probably only small numbers go further than 100 km (Yeatman-Berthelot 1991). In many northern areas, e.g. Finland and St Petersburg region (Russia), birds tend to concentrate near human habitations in winter (Mal'chevski and Pukinski 1983; Haila *et al.* 1986).

In Britain, tends to move more than other Corvidae (Holyoak 1971); widespread in winter, but with western bias; bleaker uplands are vacated, and many birds winter in sheep pastures of downs; others migrate west to milder

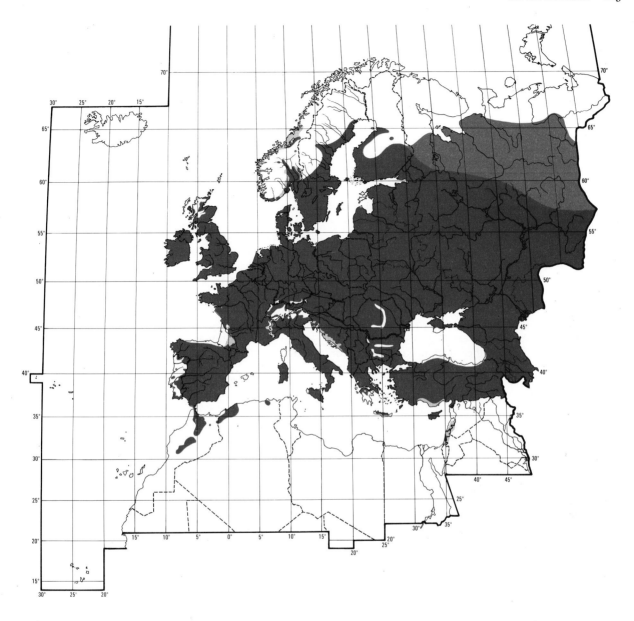

climate of Ireland. Birds ringed Wexford (south-east Ire-land), winter, recovered Wales and northern England (Cumbria and North Yorkshire) in summer. Within Britain and Ireland, 1982-6, 154 recoveries at 0-9 km, 32 at 10-100 km, 6 over 100 km. (Spencer and Hudson 1980, 1981; Mead and Hudson 1983; Lack 1986; Mead and Clark 1987). Large-scale movement reported exceptionally in Cornwall (south-west England), late October 1984, with thousands of birds flying north, west, or south-east; highest total on 23 October, when at least 20 000 flew north (Christophers 1984); in same period, *c*. 4000 invaded Isles of Scilly, then returned to mainland (Rogers 1984*b*). 29 long-distance recoveries show movement to or from Neth-

erlands (12), Denmark (7), Belgium (5), Sweden (3), Norway (1), and Germany (1) (Mead and Clark 1987).

Migrants from northern and western Europe winter at various distances along fairly narrow band extending west or south-west from breeding areas; birds from Low Countries and Denmark reach extreme north of France and eastern Britain, and Scandinavian migrants winter chiefly in Low Countries and Denmark, sometimes reaching northern France and Britain. Some north German birds winter in northern France, and some southern German and Swiss birds winter south of line from Strasbourg to Nantes, reaching extreme south of France. Winter range of birds breeding in eastern Europe spreads more widely;

birds from Baltic states and north-east Poland winter over vast region extending WSW from breeding areas to France; birds from southern Poland, north-west Czechoslovakia, and Ukraine recovered between WSW (as far as France) and south-east (Hungarian plain); 27 widespread recoveries in France of birds ringed eastern Europe from Finland south to Czechoslovakia. (Busse 1969; Yeatman-Berthelot 1991.)

Many recoveries at more than 1000 km, e.g. bird ringed Latvia, January, recovered c. 1000 km ENE in Vologda region (Russia), April, and bird ringed Frankfurt-an-der-Oder (eastern Germany) recovered 1060 km ENE in Belorussiya. 4 Lithuanian-ringed birds recovered 1200-1900 km WSW; bird recovered Belgium had moved c. 1660 km south-west from Finland, and bird ringed Belgium was found 1160 km south-east in Italy; also bird ringed Sweden recovered 1371 km south-west in France. 6-month-old Polish bird recovered c. 2000 km from natal site. Dispersal may result in change of breeding area, e.g. bird breeding in Denmark, April 1932, was 500 km NNE in Sweden in April 1934. (Baltvilks 1970; Lippens and Wille 1972; Dwenger 1989; *Rep. Swedish Bird-ringing*). Data suggest that birds from central European Russia usually migrate up to 300 km, and from further north in Russia up to 700 km (Konstantinov *et al.* 1986). German birds probably migrate up to c. 950 km (Dwenger 1989). Of 84 recoveries of Swedish-ringed birds, 1977-88, 45 at 0-10 km, 23 at 10-50 km, 3 at 50-100 km and 13 at over 100 km (up to 951 km), with movements to Low Countries, Denmark, western Germany, Britain, Finland and Estonia (*Rep. Swedish Bird-ringing*).

Large numbers winter or occur on passage in Belgium; ringing data show movements mostly to or from north-west European countries, but east to Finland and Lithuania and south-east to Czechoslovakia and Italy (Lippens and Wille 1972; Roggeman 1983). In Switzerland, small proportion of local population, chiefly juveniles, departs south-west; widespread winter visitor and regular on passage, with some birds overflying Alps, at least in autumn; bird ringed late May (presumably on breeding grounds) recovered 990 km south-west in north-east Spain, November (Busse 1969; Winkler 1984). In Baden-Württemberg (south-west Germany), where local population predominantly sedentary, 15-20 000 immigrants winter (Hölzinger 1987). Some immigrants reach (mostly north-east) Spain (Muntaner *et al.* 1983), and Italy, where altitudinal movements also occur (Lippens and Wille 1972; Mingozzi *et al.* 1988). Otherwise, little movement reported from western Mediterranean; no recent records on Malta (Sultana and Gauci 1982), and no evidence of passage on Sicily (where many breed) (Iapichino and Massa 1989); c. 10 records on Corsica (Thibault 1983), and only 1 record from Balearics (Bannerman and Bannerman 1983). Occasionally seen crossing Strait of Gibraltar both spring and autumn, so apparently winters on Moroccan side of strait in small numbers (Finlayson 1992). North-west African birds show no evidence of movement (Brosset 1961; Heim de Balsac and Mayaud 1962); exceptionally, several records at Nouadhibou (western Mauritania), 1985-6 (Meininger *et al.* 1990). In eastern Mediterranean, immigrants join residents in Cyprus, Syria, and Israel (Flint and Stewart 1992; Baumgart and Stephan 1987; Shirihai in press), but none reach Libya or Arabia (Bundy 1976; F E Warr), and 2 reports in Egypt were probably misidentified birds (Goodman and Meininger 1989). In Iraq (breeds in north), some birds move south in winter; recorded in Baghdad area chiefly October-December, with fewer January-March (Marchant 1962; Marchant and Macnab 1962); in Iran, resident in north-east and north-west, and common winter visitor to south Caspian lowlands (D A Scott); in Afghanistan, resident in north, uncommon winter visitor elsewhere (S C Madge).

Vacates north of range in western Siberia and northeast Kazakhstan. Many winter in north Caspian area and Ural valley (north-west Kazakhstan), arriving there from north-east; and large numbers occur in winter and on passage, in part with *C. frugilegus*, in foothills of western Tien Shan, arriving from north and probably also from east (western China); not known if local birds there are resident. (Dementiev and Gladkov 1954; Ivanov 1969; Korelov *et al.* 1974.) In Kashmir (north-west India), during post-breeding dispersal some adults ascend above breeding range to at least 3500 m. Winter visitor (mid-October to early March) to Quetta valley in western Pakistan, also south of breeding range in northern Pakistan, occasionally reaching north-west Indian plains. (Ali and Ripley 1972.)

Autumn movement September-November. In Britain, local birds depart September-October, and continental birds arrive on east and south coasts October-November (Lack 1986). Passage begins mid- or late September in northern Denmark and on Polish coast (Møller 1978a; Busse and Halastra 1981), and passage through central Europe chiefly from October (Mildenberger 1984; Winkler 1984; Hölzinger 1987), with main arrival in France from late October (Yeatman-Berthelot 1991). In eastern Germany, relative numbers at winter roosts show that *C. monedula* begins to arrive earlier than *C. frugilegus*, and exodus continues later (Wernicke 1990). In Moscow region, local birds depart 1st half of September, and winter visitors arrive October (Ptushenko and Inozemtsev 1968). Immigrants arrive in Israel mid-October to mid-December, mostly 1st half November (Shirihai in press). In Chokpak pass (western Tien Shan), autumn passage from mid- or late September to early November, with 86% in 2nd half October (peaking later than *C. frugilegus*) (Gavrilov and Gistsov 1985).

Spring movement is early, chiefly February-March, continuing to April or early May. Winter visitors leave Britain mid-February to 3rd week of April (Witherby *et al.* 1938), and in northern Denmark, passage from mid-February, with sharp peak at end of March (Møller

1978*a*). In Switzerland, passage chiefly February–March, with arrival of local birds chiefly March–April (Schifferli *et al.* 1982; Winkler 1984); similarly in Rheinland, arrives March to early April (Mildenberger 1984). Main migration on east Polish coast finishes at end of March (Busse 1976). Most birds leave Syria by beginning of March, slightly later than *C. frugilegus* (Baumgart and Stephan 1987). In Chokpak pass, spring passage from late February to early or mid-May, chiefly 1st half of March (Gavrilov and Gistov 1985).

Displacement westward across Atlantic not infrequent, underlining westerly component to heading. Many records from ocean weather stations in eastern Atlantic, sometimes involving parties of dozens (Mead 1986); also recorded from Atlantic islands (Madeira and Azores) (Bannerman and Bannerman 1965, 1966). In Iceland, up to 1979, 18 birds reported 3 March to 10 May, and 75 (including flocks of 38 and 10) 3 September to 28 December, mostly 10–31 October, with largest influxes 1975 and 1976; unsuccessful breeding attempt recorded in 1977 (Nielsen 1979). Some have reached eastern North America (*c.* 41–47°N) in recent years. In 1982–5, *c.* 12 birds recorded from Pennsylvania north-east to St Pierre-et-Miquelon (off Newfoundland); birds at St Pierre-et-Miquelon stayed for several months; at Lewisburg (Pennsylvania), pair attempted to breed, not known if successful. Flock of 52 at Port Cartier (Quebec), November 1984, probably resulted from man-assisted passage, perhaps in ship's hold. (Hall 1985; Smith 1985; Desbrosse and Etcheberry 1986; Mead 1986.)                                                         DFV

**Food.** Invertebrates, fruits, seeds, carrion, and scraps; sometimes small vertebrates or birds' eggs; food of nestlings predominantly invertebrates. Generally feeds in pairs or small flocks, almost wholly on ground, though will forage seasonally in tree-tops for defoliating caterpillars, beetles (Coleoptera, particularly *Melolontha*), or even acorns *Quercus*, though rarely seen on woodland floor (Gebhardt 1944; Lockie 1955; Weinzierl 1961; Folk 1967*a*; Coombs 1978). Feeds mostly in open areas of short or scattered vegetation in pasture, parks, etc., and a common scavenger at rubbish tips, farmyards, and abattoirs; coastal birds readily forage between tidelines (Richford 1978; Goodwin 1986; Dwenger 1989). Of 606 winter feeding observations, central England, 58% on stubble and pasture; temporary grassland and bare arable fields avoided (Waite 1984*a*, which see for foraging comparisons with other Corvidae); however, will follow plough, etc. (Dwenger 1989), and feed between maize *Zea* plants 30–50 cm high (Strebel 1991); in southern Sweden, in spring, 100% of observations were on grassland and grass heath; pine *Pinus* plantations and tall herbs in study area not visited (Högstedt 1980*a*, which see for comparison with Magpie *Pica pica*). In Switzerland, avoids pasture when vegetation higher than 15–20 cm (Strebel 1991). Often feeds in company with Rook *C. frugilegus*, moving in after

them to scattered dung-pats or disturbed soil (Lockie 1955; Jakubiec 1972*a*; Strebel 1991). Of 672 feeding actions, southern England, 49% picking from surface, 20% turning of clods, stones, etc., 17% surface probing (parting grass with bill), 10% jumping at flying insects, 3% deep-probing (inserting bill in hole in soil then opening), and 2% digging with bill; much more agile than other corvids at jumping up to snatch insects, and also ran quickly between dung-pats catching flies (Diptera) as they scattered; probes far less in soil than *C. frugilegus* since bill too short (Lockie 1956*b*). See also Högstedt (1980*a*) and Waite (1984*a*, *b*) for foraging behaviour and interactions with other Corvidae. Also pulls up grass tufts and shoots to get at root larvae and seeds (Dwenger 1989). Shows agility and adaptability when foraging, acrobatically stretching and hanging head-down to reach caterpillars and fruit in trees, or peanuts in garden holder (King 1976*a*; Jong 1981; Jonkers 1983). Groups of up to a dozen carried stones weighing up to 150 g (over half of body weight) 10–15 cm in bill to get at scraps and invertebrates, and will push aside stones of up to 200 g (Paudtke 1988). Will hawk for flying insects such as ants (Formicidae) high in the air, making rather clumsy dives at them from above (Cornish 1947; Podmore 1948). Particularly in Britain, some birds kleptoparasitize auks (Alcidae) at their colonies, either harrying them into dropping fish brought for young (Mylne 1960; Birkhead 1974*b*; Richford 1978), or actually snatching it from them on ground or ledges (Olsthoorn 1987); in one study, only 7% of 90 attacks on Puffins *Fratercula arctica* successful, and one bird responsible for most successes; birds hid in vegetation and seized fish-carrying *F. arctica* by tail or wing (Corkhill 1973). Will patrol cliffs looking for opportunities, or follow Great Black-backed Gull *Larus marinus* as it harries auks (Birkhead 1974*b*). Observations of 2 birds in December harrying feeding squirrel *Sciurus vulgaris* as if to rob it of food (Whittaker 1990). Takes eggs or young of other birds only rarely, though specialists can have significant local effect: may cause up to 60% of egg loss by Guillemot *Uria aalge* on island of Skokholm (Wales) (Birkhead 1974*b*); in town in Hungary, said to 'live exclusively' on eggs and young of Collared Dove *Streptopelia decaocto*, harrying ♀♀ from nests despite defence by ♂♂ (Bozsko 1977), and has caused considerable egg loss at colonies of Grey Heron *Ardea cinerea* in Britain and Poland (Creutz 1981). Recorded taking eggs of *F. arctica* and Manx Shearwater *Puffinus puffinus* by one luring incubating bird from burrow while another steals egg; also robs unattended eggs of Razorbill *Alca torda* on ledges (Birkhead 1974*b*; Richford 1978). In Switzerland, 2 birds specialized in entering loft to take eggs and young of domestic pigeon *Columba livia* (Zimmermann 1951). Few records of adult songbirds being killed and eaten; dropped on Chaffinch *Fringilla coelebs* on lawn from height of *c.* 2 m and carried it off in bill (Bawtree 1950); Starling *Sturnus vulgaris* killed and mostly consumed (Van Oss 1950). In

England, recorded pulling bats (Chiroptera) from hole in tree and flying off with them (Page 1988). Usually carries food in bill, but record of transporting young thrush *Turdus* in foot, flying some distance pursued by 4 adults (Williams 1946). Seen to pull up peanut- holder by string to get at nuts (Knijff 1977), and has learned to open milk bottles to get at contents (Åbro 1964); group of birds at carpark watched litter bins, individuals disappearing inside if anything thrown in (Wiehe 1988). Perches on domestic animals to remove parasites from hair and wool (Kumerloeve 1969b). On a few occasions seen feeding on fresh dog faeces (King 1985). Holds large food items under one or both feet to pull them apart with bill; picks seeds out of some berries (Birkhead 1974b; Jonkers 1983; Dwenger 1989). Will fly some distance to good food source (Högstedt 1980a), and up to 300 at a time seen in tree-tops 12 km from urban nest-sites feeding on beetles *Melolontha* (Weinzierl 1961). See Wechsler (1988b) for experiments showing that individuals probably do not learn feeding techniques from each other, and Partridge and Green (1987) for experiments which showed that increased specialization of feeding techniques can lead to greater feeding efficiency.

Stores food to a far lesser extent than most other Corvidae; habit mostly observed in captivity (Strauss 1938a; Röell 1978; Goodwin 1986). In field experiment with pieces of bread, never seen to hide any—in contrast to *P. pica* (Henty 1975). One bird seen to bury food in garden, pushing it into soft earth or cavities in rubbish heap (Simmons 1968). Other observations of hiding food inside own nest-box and in tree cavity (Turček and Kelso 1968; Richards 1974). Any food hidden is apparently not covered (Strauss 1938a).

Diet in west Palearctic includes the following. Vertebrates: bats (Chiroptera), voles (Microtinae), mice (Murinae), frogs *Rana*, fish; eggs, young, and adults of many birds, including eggs of White Stork *Ciconia ciconia*, Grey Heron *Ardea cinerea*, gulls (Laridae), auks (Alcidae), and Barn Owl *Tyto alba*; young of gamebirds (Phasianidae) and pigeons (Columbidae); adult songbirds, up to size of Starling *Sturnus vulgaris*; all kinds of vertebrate carrion. Invertebrates: bristle-tails, springtails, etc. (Apterygota), grasshoppers, etc. (Orthoptera: Gryllidae, Tettigoniidae, Gryllotalpidae, Acrididae), earwigs (Dermaptera: Forficulidae), cockroaches and mantises (Dictyoptera), web-spinners (Embioptera), Psocoptera, bugs (Hemiptera: Pentatomidae, Coccoidea), snake flies (Neuroptera: Raphidiidae), scorpion flies (Mecoptera), adult and larval Lepidoptera (Hepialidae, Tortricidae, Pyralidae, Notodontidae, Noctuidae, Lymantriidae, Geometridae), adult and larval flies (Diptera: Tipulidae, Trichoceridae, Bibionidae, Mycetophilidae, Cecidomyiidae, Stratiomyidae, Rhagionidae, Tabanidae, Therevidae, Empididae, Dolichopodidae, Pipunculidae, Syrphidae, Tachinidae, Calliphoridae, Muscidae), Hymenoptera (sawflies Symphyta, Ichneumonidae, ants

Formicidae, wasps Vespidae, bees Apoidea), adult and larval beetles (Coleoptera: Cicindelidae, Carabidae, Hydrophilidae, Histeridae, Silphidae, Staphylinidae, Lucanidae, Geotrupidae, Scarabaeidae, Byrrhidae, Buprestidae, Elateridae, Cantharidae, Dermestidae, Anobiidae, Meloidae, Cleridae, Tenebrionidae, Cerambycidae, Chrysomelidae, Curculionidae), spiders (Araneae: Lycosidae, Thomisidae, Salticidae), harvestmen (Opiliones), ticks (Acari: Ixodidae), woodlice (Isopoda: Ligiidae, Oniscidae, Porcellionidae, Armadillidiidae), shrimps (Decapoda: Natantia), sand-hoppers (Amphipoda: Talitridae), Phyllopoda (Limnadiidae), millipedes (Diplopoda: Julidae), centipedes (Chilopoda: Lithobiidae), earthworms (Lumbricidae), molluscs (Succineidae, Helicidae, Bulimidae, Vitrinidae, Planorbidae). Plants: fruits and seeds of juniper *Juniperus*, beech *Fagus*, oak *Quercus*, chestnut *Castanea*, walnut *Juglans*, apple *Malus*, pear *Pyrus*, plum, cherry, etc., *Prunus*, rowan *Sorbus*, various *Rubus* species, elder *Sambucus*, hazel *Corylus*, fig *Ficus*, mulberry *Morus*, dogwood *Cornus*, barberry *Berberis*, spindle *Euonymus*, olive *Olea*, various *Vaccinium* species, mistletoe *Viscum*, grape *Vitis*, trefoil or clover *Trifolium*, birdsfoot *Orni-*

Table A  Food of adult Jackdaw *Corvus monedula* by stomach analysis. Figures are percentages by dry weight (Hell and Soviš 1958), volume (Folk 1967a), or wet weight (Soler *et al.* 1990), animal and plant material treated separately.

| | Czechoslovakia, winter | Czechoslovakia, all year | S Spain, all year |
|---|---|---|---|
| *Animal material* | | | |
| Vertebrates, eggs | 18·0 | 0·2 | 3·3 |
| Orthoptera | 0 | 0·4 | 18·1 |
| Hemiptera | 0 | 0·1 | 2·5 |
| Lepidoptera | 0 | 2·8 | 4·3 |
| Diptera | 0 | 0·4 | 0·5 |
| Hymenoptera | 0 | 1·0 | 28·6 |
| (ants) | (—) | (1·0) | (26·9) |
| Coleoptera | 2·9 | 94·0 | 19·5 |
| (Scarabaeidae) | (—) | (32·2) | (1·5) |
| (Tenebrionidae) | (—) | (0·6) | (6·5) |
| (Curculionidae) | (—) | (44·3) | (2·6) |
| Millipedes, centipedes | 0 | 1·6 | 5·4 |
| Worms | 0 | 0·2 | 0·1 |
| Snails | 0 | 0·2 | 12·6 |
| Other (inc. carrion) | 79·0 | 0 | 5·1 |
| Total no. of items | 12 | 995 | 9405 |
| *Plant material* | (97·1% of diet) | (70·2% of diet) | (90·1% of diet) |
| Cereal grain | 96·4 | 75·3 | 62·6 |
| Other agricultural | — | — | 30·6 |
| Other seeds | 3·6 | 12·6 | 3·7 |
| Other fruits | — | 6·4 | 0·4 |
| Total sample | 76 stomachs | 277 stomachs | 439 stomachs |
| Source | Hell and Soviš (1958) | Folk (1967a) | Soler *et al.* (1990) |

*thopus*, vetch *Vicia*, sainfoin *Onobrychis*, bean *Phaseolus*, pea *Pisum*, lupin *Lupinus*, chick-pea *Cicer*, potato *Solanum*, buckwheat *Fagopyrum*, dock, etc. *Rumex* and *Polygonum*, buttercup *Ranunculus*, bedstraw *Galium*, poppy *Papaver*, hemp *Cannabis*, mullein *Verbascum*, pondweed *Potamogeton*, orache *Atriplex*, charlock *Sinapis*, spurge *Euphorbia*, turn-sole *Chrozophora*, hemp-nettle *Galeopsis*, storksbill *Erodium*, spurrey *Spergula*, thistle *Cirsium*, knapweed *Centaurea*, sunflower *Helianthus*, wheat *Triticum*, rye *Secale*, barley *Hordeum*, oats *Avena*, millet *Panicum*, maize *Zea*, *Sorghum*, other grasses *Lolium*, *Echinochloa*, *Setaria*, *Digitaria*, horsetail *Equisetum*. (Rey 1907*b*; Newstead 1908; Collinge 1924-7; Madon 1928*b*; Schuster 1930; Campbell 1936; Kluijver 1945; Williams 1946; Paatela 1948; Bawtree 1950; Zimmermann 1951; Kovačević and Danon 1952; Hell and Soviš 1958; Mylne 1960; Turček 1961; Folk 1967*a*; Holyoak 1968; Birkhead 1974*a*, *b*; Richford 1978; Jong 1981; Jonkers 1983; Goodwin 1986; Page 1988; Dwenger 1989; Soler *et al.* 1990.)

Very opportunistic and flexible forager; 2 studies in England gave 72% of diet by volume animal material, $n = 48$ stomachs, all year (Collinge 1924-7), and 84% plant material, $n = 22$, March-June (Campbell 1936). See Table A. Great deal of variation within and between studies of adult diet due to adaptability and opportunism of individual birds and of populations in different habitats. In Czechoslovakia, average weight of plant material per stomach ($n = 277$) was 2·3 times that of animal material in spring and summer, 16 times in autumn, and 19 times in winter (Folk 1967*a*). In England, grain present in most of 222 gizzards all year (in almost 100% during July-February) since in spring and summer taken from animals' troughs and gamebird feeding places; other agricultural crops (peas, beans, root crops) present November-April, fruits and large seeds (plum, elder, apple, acorns, etc.) important in autumn, while weed seeds had highest occurrence in winter; proportions of animal prey in diet changed predictably with seasonal availability, beetles, snails, and worms being taken all year, with most snails in midwinter, and Lepidoptera and Diptera larvae and adults in spring and summer (Holyoak 1968). See also Lockie (1956*b*). In southern Spain, ants were most important part of animal component in diet in 3 separate areas, though barely recorded elsewhere; over the year, average 8·5% of diet by volume animal material, peaking at *c.* 25% April-May, and *c.* 15% in October due to increase in importance of Orthoptera and Hymenoptera; diet in midwinter virtually 100% plant material, maize and chick-peas being important in autumn and winter, though absent in summer, when wheat, barley, and rye were main fractions of plant material (Soler *et al.* 1990, which see for detailed treatment of diet month by month in 3 different areas by weight, volume, frequency, and occurrence). For Netherlands, see Kluijver (1945). In southern England, 41% of 298 items of animal prey 2-6 mm long, 34% 7-12 mm, 24% 13-18 mm, and 1% 19 mm or more (Lockie 1956*b*).

Table B   Food of nestling Jackdaw *Corvus monedula* by collar-sampling. Figures are percentages by volume (Folk 1967*a*; Richford 1978), dry weight (Högstedt 1980*a*; Kamiński 1983), or number (Strebel 1991).

| | W Wales | S Sweden | E Poland days 1-15 | E Poland days 16-35 | Switzerland | Czechoslovakia |
|---|---|---|---|---|---|---|
| Vertebrates, eggs | — | 4·7 | — | — | 0 | 2·1 |
| Hemiptera | 1·6 | 0·2 | 0·3 | 0·7 | 2·5 | 1·3 |
| Lepidoptera | 24·7 | 18·3 | 28·6 | 8·6 | 7·8 | 31·0 |
| (larvae) | (23·3) | (15·4) | (25·9) | (6·8) | (7·0) | (28·1) |
| Diptera | 20·7 | 5·8 | 10·4 | 17·6 | 13·1 | 3·8 |
| (larvae) | (4·6) | 0 | (1·9) | (3·1) | (—) | (0·7) |
| Hymenoptera | 2·0 | 0·2 | 2·1 | 3·0 | 3·3 | 0·3 |
| Coleoptera | 23·2 | 33·4 | 43·0 | 27·6 | 22·3 | 31·6 |
| (adults) | (—) | (33·3) | (31·8) | (23·2) | (—) | (26·9) |
| (Scarabaeidae) | (—) | (0·2) | (7·0) | (1·5) | (6·3) | (—) |
| (Elateridae) | (—) | (2·7) | (19·9) | (1·7) | (8·2) | (—) |
| (Curculionidae) | (—) | (30·2) | (2·2) | (1·2) | (9·7) | (—) |
| Woodlice | 2·6 | 0 | — | — | 1·8 | 0·2 |
| Arachnida | 6·0 | 2·1 | 0·1 | 0·1 | 9·7 | 3·4 |
| Worms | 2·4 | 5·0 | 0·1 | 0·1 | 14·3 | 6·2 |
| Snails | 0 | 0·1 | 0·1 | 0·1 | 2·4 | 0·4 |
| Cereals | 10·8 | 28·8 | 7·4 | 23·5 | (—) | 12·2 |
| Other/scraps | 7·0 | 1·4 | 7·9 | 18·7 | 22·8 | 7·5 |
| Total sample | 357 samples | 6·2 g | 735·8 g, 173 samples | | 800 animal items | 1860 items |
| Source | Richford (1978) | Högstedt (1980*a*) | Kamiński (1983) | | Strebel (1991) | Folk (1967*a*) |

In central England, average energy intake per bird ($n = 60$) on grassland in winter ranged from $3 \cdot 72$ kJ per hr from medium-sized invertebrates above soil surface to $0 \cdot 03$ kJ per hr from small invertebrates beneath surface (Waite 1984a).

For food of nestlings see Table B.

Average animal material content in diet in various studies $c.$ 81% (71-100%). In study in southern England, most important defoliating moths (Lepidoptera) in diet of young, *Tortrix viridana* (Tortricidae) and *Operophtera brumata* (Geometridae), were maximally available (i.e. still larvae, but not too small, and still in tree canopy) early May to mid-June, coinciding with nestling period; 65% by volume of nestling diet ($n = 218$ collar-samples, 7628 items) were such caterpillars, but none found in 577 items from 22 samples from eastern Scotland, where proportionately more beetles, Diptera, and Lepidoptera larvae from grassland were fed to young (Lockie 1955, which see for comparison with *C. frugilegus* and Carrion Crow *C. corone*). In Switzerland, caterpillars accounted for 4% (by number) of animal prey one year, but 27% the following year, and Scarabaeidae 3% and 25%, reflecting abundance; domestic scraps in diet rose during May from 28% to 50% of diet, as grassland invertebrates decreased in availability (Strebel 1991). On Skomer island, most insects brought to nest were longer than 8 mm, though up to 60 larvae less than 5 mm long found in single collar-samples (Richford 1978). In urban study in Zürich, intact beetles 7-10 mm long fed to smallest young; wing-covers and thorax sometimes removed from large *Melolontha*; 100% of diet animal material, 64% of which beetles (40% larvae) (Zimmermann 1951, which see for content of 63 collar-samples, percentage by number). For Switzerland, see also Strebel (1991). Median length of prey item in southern Sweden $7 \cdot 2$ mm, median weight $5 \cdot 0$ mg (96% of items less than 20 mg); in this study, only 71% of diet was animal material (Högstedt 1980a). In Poland (data from 18 nests and 342 food samples), nestlings each consumed average of 10 g of food on day 1 (45 kJ), rising to 133 g on day 24 (600 kJ), then falling to 80 g on day 35 (380 kJ); total intake per nestling over fledging period was 13 500 kJ, 80% of which used for growth (Kamiński 1986).   BH

**Social pattern and behaviour.** Well studied. For pioneering work on free-flying colony of hand-reared birds, see Lorenz (1932, 1952, 1955). Most important recent study (of wild birds in nest-box colony in Groningen, Netherlands) by Röell (1978, 1979). Useful reviews by Coombs (1978), Goodwin (1986), and, especially, Dwenger (1989).

1. Mostly gregarious outside breeding season, although lifelong pair-bond (see Bonds, below) means that pair is basic unit within flocks, e.g. pair continue to visit nest-sites almost throughout the year in parts of range where sedentary. In Groningen, small flocks (1-10 birds) comprised more than 60% of all flocks in most months, and were particularly common April-July (latterly family groups); large flocks (more than 40) less than 10% except in December-January and not seen at all during breeding season;

autumn-winter feeding on grain may favour larger flocks than spring-summer feeding on grassland insects (Röell 1978, which see for references). In Tampere (Finland), Tast and Rassi (1973) found birds dispersed in few, relatively large flocks in autumn; in winter, in many small flocks; in spring in large flocks again. For larger assemblages, and for communal roosting, see Roosting (below), also Flock Behaviour (below). Often associates with Rook *C. frugilegus*, especially for feeding and when migrating or flying to and from roost-sites (Goodwin 1986: see also Roosting, below), e.g. $c.$ 55% of migratory flocks of *C. frugilegus* in Bayern (Germany) included *C. monedula* (Leibl and Melchior 1985). In autumn and winter, north-west England, often associated for feeding with Starling *Sturnus vulgaris*, occasionally Lapwing *Vanellus vanellus* or Common Gull *Larus canus* (Brown 1942). BONDS. Mating system essentially monogamous, exceptionally bigamous (see below) and pair-bond usually life-long and maintained all year (Lorenz 1932; Zimmermann 1951). However, mate changes do occur (Coombs 1978); most detailed study by Röell (1978) as follows. Some pair-bonds known to persist at least 5 years; in 122 pairs lasting 6 months or more, 27 mate-changes of which at least 15 due to death of partner. Even after several unsuccessful breeding seasons, pair-bond not usually dissolved. Thus, if pair-bond lasted more than 6 months, death was almost the only cause of severance. Newly paired juveniles, however, changed mates more often: of 52 pairs formed during 1st year, 22 mate-changes recorded within 6 months; only 5 of these known to be caused by death of mate, although success or failure of breeding attempt did not appear to influence likelihood of divorce. Divorce very costly, incurring loss of both nest-site and rank. ♂♂ and ♀♀ react differently to losing mate: widowed ♂♂ try to retain nest-site but are immediately attacked by neighbours and might be courted by 1 or more unpaired ♀♀; widowed ♀♀ give up nest-site and court ♂♂ (usually paired, but also unpaired site-owners); neither ♂ nor ♀ able to rear young alone. Pair-members share defence of nest-site, nest-building (for exceptions, see Zimmermann 1951, Lebeurier 1955, Groebbels 1960) and feeding young, but only ♀ incubates (and is then courtship-fed by mate: see below) and broods. Single, usually older, ♀ ('follower') not uncommonly associates with pair and sometimes succeeds in breaking it up, replacing original ♀. (Röell 1978; see also Lorenz 1955.) In 7-year study of 20 pairs, 2 cases of bigamy (or at least joint breeding): in 1st case, ♂ with 2 ♀♀ arrived in colony and ousted another pair from nest-box; trio cooperated in building and the 2 ♀♀ laid in adjoining nest-cups; however, one ♀ dominant, laying first, being apparently only recipient of ♂'s food, and ended up with all 6 surviving young (1 from other ♀) in her nest; all 3 adults helped to feed young, also cooperating to defend site throughout winter and re-building the following spring. 2nd case of bigamy similar except that only 1 clutch (apparently dominant ♀'s) hatched; the 2 ♀♀ were closely bonded, Allopreening each other (see Heterosexual Behaviour, below) and courted ♀ often subordinate ♀. (Röell 1979.) Young are fed for $c.$ 4 weeks after fledging, and family bonds (other than pair-bond) severed after week 5 (Dwenger 1978). For apparent adoption of orphaned young see Manson-Bahr (1953). Pair-bonds form in 1st year but pair does not usually breed until 2 years old (Lorenz 1932, 1952), exceptionallly at 1 year (Wontner-Smith 1939; Zimmermann 1951; Bäsecke 1955; Röell 1978). Lorenz (1955) considered sexual immaturity usually prevented 1-year-olds from breeding, but Röell (1978) thought shortage of nest-sites responsible. In study by Peter and Steidel (1990), ♀♀ first bred at 1-4 years, the lower ages coinciding with expanding phase of colony. BREEDING DISPERSION. Typically colonial, but dispersion largely dependent on availability of sites, and solitary pairs not uncommon (Coombs 1978; Röell 1978;

Goodwin 1986). In Wytham Wood (southern England), no preference for solitary or clumped nest-boxes (Heeb 1991). Territory
limited to nest-site and immediate surrounds, e.g. in chimney
nest, defence concentrated on chimney-top. Within colony, often
nest very close together and will occupy contiguous nest-boxes.
(Coombs 1978.) In Israel, often in dense colonies of a few hundred pairs (Shirihai in press). In Bukhara (Uzbekistan) 5 colonies
each of 10–30 nests in 60 km²; distance between nests in any
colony 0·2–2 m, while separate pairs in tree-holes nested 2–800
m apart (Sagitov and Bakaev 1980). In New Forest (southern
England), 2–9 'territories' per km² in unenclosed deciduous
woodland (Glue 1973). In farmland in Britain, average 1–6 pairs
per km² (Sharrock 1976). In open country in Sweden, *c.* 2 pairs
per km² (Ulfstrand and Högstedt 1976). In 3 different regions
(2 on Baltic coast) in eastern Germany, average 3·2–8·4 pairs per
km² over 3 years (Dwenger 1989; Plath 1989). No active nesting
associations, but for case of pair nesting 30 cm from nesting
Magpies *Pica pica*, see Moody (1954). Also recorded nesting
amicably very close to Barn Owl *Tyto alba*, Kestrel *Falco tinnunculus*, and House Sparrow *Passer domesticus* (Dwenger 1989).
Foraging areas of different colonies show considerable overlap,
especially at rich food sources, and no evidence of communally
defended feeding areas; feeding and roosting sites extend sometimes more than 10 km from nest-sites (Röell 1978). After fledging, colony often deserted until autumn (Goodwin 1986). Thus,
in nest-box colony 'residents' (paired birds possessing nest-site)
abandoned sites only July–August (moult), reoccupying them
September; at end of 1st year some new pairs obtained nest-site,
bred, and thus became residents; 'non-residents' (birds of all
ages, paired and non-paired, breeders and non-breeders, but
often young, inexperienced, low-ranking birds) did not defend
nest-sites in winter. Residents typically defended only 1 nest-box
during breeding season, but during winter sometimes defended
more than 1 (up to 5); secondary sites usually neighbouring but
not invariably so, e.g. for 2 months pair defended 2 boxes in 2
colonies over 200 m apart. After death or disappearance of partner, resident either lost interest in nest-site or forfeited it under
pressure from neighbours (see Bonds, above). (Röell 1978.) Site-
fidelity appears to vary, at least partly with type of nest-site
used. Thus in colony, successful breeders used same box year
after year (Röell 1978), but on Skomer Island (Wales) where
nest-sites are burrows (and probably not as limiting a resource
as in Röell's study) about half the birds changed burrow between
years, irrespective of previous breeding success (Richford 1978);
see Wechsler (1989) for similar lack of site-attachment in captivity. Little quantitative information on fidelity to natal colony,
but in study by Peter and Steidel (1990) fidelity increased from
10·8% to 54·4% over 8 years as a result of colony isolation.
ROOSTING. Widely studied. For most of year, both residents and
non-residents (see Breeding Dispersion for definitions) roost
communally at traditional sites. In breeding season, residents
roost mostly in or near nests, while non-residents often use
communal roost, even May–June (Coombs 1978). However,
incidence of communal roosting in breeding season varies (see
Gyllin and Källander 1976 for references). At Groningen colony,
♀♀ roosted in nest during incubation and brooding while mates
continued using communal roost, often switching to vicinity of
nest after hatching (Röell 1978). Communal roost ranges from
several hundred (e.g. Rappe 1965) to several thousand (e.g.
Gramet 1956, Grodziński 1971). Exceptionally large winter
roosts of up to 40 000 in Uppsala district (Sweden) (Lundin
1962) and in Estonia where migrants augment local flocks (Lint
1964, 1971). Often roosts with *C. frugilegus* and sometimes with
Carrion Crow *C. corone* (Borgvall 1952; Coombs 1978; Röell
1978; Rode and Lutz 1991); e.g. in Brno (Czechoslovakia) mixed

roost of 30 000–100 000 contained 10–15% *C. monedula*, rest *C.
frugilegus* (Hubálek 1978b). Other reported associates include
Raven *C. corax* and (once) *P. pica* (Diesing 1984); once with
Oystercatchers *Haematopus ostralegus* in shore roost (Ogilvie and
Ogilvie 1984). Communal roost usually in trees, but sometimes
reedbed or cliff (Goodwin 1986). For roosting (with *C. frugilegus*)
on frozen lake, see Lambert (1965). Especially in northern Europe (notably Sweden, Finland, Baltic states, and Poland), often
roosts in trees or buildings in town centres, and undertakes
lengthy daily commuting flights, maximum 20–30(−35) km,
especially in winter when entire regional population may join
(e.g. Lundin 1962, Lint 1971, Tast and Rassi 1973, Gyllin and
Källander 1976, Gyllin *et al.* 1977, Grodziński 1980). In Örebro
(central Sweden), 4500–5000 roosted densely in clump of 11 tall
trees, always in upper branches but never topmost twigs (Gyllin
and Källander 1976); Örebro study indicated that urban roosting
does not achieve net energy gain for most birds and that other
factors involved (Gyllin *et al.* 1977, which see for wider discussion of function of communal roosting). In mixed roost with
*C. frugilegus*, species usually separate to some extent (Coombs
1978). Most communally roosting *C. monedula* settle in pairs,
partners perching side by side (Groziński 1971; Tast and Rassi
1973). Communal roost-sites often change seasonally, with urban
roosting often favoured (in northern Europe) from autumn to
spring (e.g. Lint 1971) or even earlier, e.g. communal roost of
non-breeders in breeding season near Örebro was joined at end
of breeding season by local family parties before roost shifted
(starting from late June) to central Örebro, this site attracting
entire regional population in winter (Gyllin and Källander 1976).
At Tampere, time gap between stopping feeding and final entry
into roost up to *c.* 90 min throughout year; in morning, all left
simultaneously or over up to *c.* 2 hrs (Tast and Rassi 1973). In
autumn and spring, compared with winter, arrival at roost is
earlier relative to sunset, and departure later relative to sunrise,
both entry and departure taking place in twilight or dark. Bright
weather retards time of entry and advances time of departure,
vice versa in overcast conditions. For details of roosting flock
behaviour, see below. For daily activity rhythm in spring after
nest-building started, see Paatela (1948).

   2. Wary in woodland and open country, but relatively
approachable in urban setting. Readily imprints on man (see
Lorenz 1932, 1952). For possible alarm reaction of flock, see
Flock Behaviour (below); for mobbing of predators (etc.), see
Parental Anti-predator Strategies (below). FLOCK BEHAVIOUR.
Resident birds (see Breeding Dispersion, above, for definition)
generally form flocks (i.e. coalition of pairs) whereas non-
residents do not form stable flocks but instead join flocks of
residents (Röell 1978). When flock is flying or resting, pair-
members keep markedly closer to each other than to other
flock-members (Goodwin 1986). Dominance hierarchy in flocks
first studied by Lorenz (1932, 1952), but his conclusions sub-
stantially modified by findings of Tamm (1977), Lovari (1979),
Wechsler (1988a), and especially Röell (1978). Following
account summarizes main findings of these recent references,
which see also for details of (e.g.) social preferences in relation to
rank, changes in rank-order over time, and other factors affecting
rank. Contra Lorenz (1932, 1952, 1963), no evidence of group
defence, or of any special (policing or cohesive) role of (dominant) α-♂ in flock, or of individuals in flock striving to improve
own rank, or of hierarchy being rigidly fixed. Although dominance hierarchies exist in the wild, at least in small flocks,
rank-order perhaps no more than a convenient description of
observed behaviour in one specific context. Nevertheless the
following rules apply, largely dependent on correct interpretation of rank-relations between pair-members. Only paired,

or very recently widowed ♂♂, possess nest-site and high rank. ♂♂ generally dominant over ♀♀, and ♂ usually dominates his mate throughout year at feeding sites and perch-sites. Rank of paired ♀ depends on presence and rank of mate, i.e. ♀ has about same rank of ♂ only in his presence, thus refining belief of Lorenz (1932, 1952) that pairing confers on ♀ same rank as her mate. Study of flocks in Oxfordshire (southern England) from autumn to spring indicated that low temperatures suppress sexual behaviour and increase aggression. Aggression, apparently representing recrudescence of sexual fighting, also common during warm autumn. (Lockie 1956a.) Diurnal roosting pattern outside breeding season described by Aschoff and Holst (1960), Coombs (1961), Lundin (1962), Riggenbach (1970), Grodziński (1971, 1980), Lint (1964, 1971), Tast and Rassi (1973), and Gyllin and Källander (1976); see these references for details of (e.g.) flock sizes, mixing with *C. frugilegus*, flight speeds, etc. Birds usually well dispersed by day but, well before sunset, start congregating at first of 2–4 staging posts *en route* to roost. Flocks approaching staging posts from different directions funnel along regular routes delineated by landmarks, e.g. rivers, woods, railways (Hubálek 1978b). In autumn and spring, pre-roosting behaviour includes lengthy circling, towering flights, and headlong descents (including zigzag flight) over staging posts (Brown 1942; Lundin 1962). Aerial evolutions sometimes strikingly coordinated but more often not (Witherby *et al.* 1938). Tumbling dives, not uncommonly in company with *C. frugilegus* (see p. 162), especially prevalent in windy weather (Coombs 1960). Following account, from Grodziński (1971, 1980) of entry into roost in late September and early October, is typical of many (e.g. Borgvall 1952, Lundin 1962, Tast and Rassi 1973, Gyllin and Källander 1976). From staging posts, flocks generally joined up to circle noisily over roost-site (mainly using call 2 in Voice), often dispersing, returning, and separating into smaller flocks before finally alighting, first singly then in small groups, finally in masses. Groups then flew noisily and restlessly from tree to tree, 30 min elapsing before total silence, by which time all birds perched densely in canopy. In winter, such manoeuvres can last hours (Gyllin and Källander 1976). In spring and summer, Black Sea coast, dusk roosting flights spectacular as birds poured onto cliff in steep dive; also mass flights (several thousands) from cliffs to freshwater source 4 km away (Frank 1951). Waking behaviour studied at Tampere roost: birds do not apparently all wake together; gradually more became active, performing various activities (e.g. preening, flying short distances, becoming noisier) (Tast and Rassi 1973). Morning exodus from roost reverses pattern of entry: sometimes leisurely but often faster and more direct with less aerial display and fewer stopovers *en route* to feeding grounds (e.g. Lundin 1962, Lint 1971, Tast and Rassi 1973). Aerial display resembling 'dreads' of terns (Sternidae) reported at breeding colony in trees: birds left colony *en masse* in apparent panic, all flying in same direction for *c.* 100 m before returning; no obvious stimulus except once Little Owl *Athene noctua* hunting insects nearby (Podmore 1948). SONG-DISPLAY. Little known about contexts or function of song (see 1 in Voice), but given by either sex when alone, from perch or when flying (Lorenz 1932, 1952; Goodwin 1986). Postures and displays that normally accompany certain calls are also seen when these calls are incorporated in song from perch, e.g. threat-display accompanies call-types 5 and 8a–b (Lorenz 1931, 1952: see Voice). ANTAGONISTIC BEHAVIOUR. (1) General. Dominance relations extensively studied in captive flocks (see Flock Behaviour for nature of hierarchy and for references; see same references, and especially Röell 1978, for influence of rank on outcome of disputes). In the wild, most important studies by Lockie (1956a) and Röell (1978). (2) Threat and fighting. In study of flocks

(Lockie 1956a) disputes occurred both intraspecifically and with *C. frugilegus* (see below for interspecific aggression); threat-display in rivalry over ♀♀ and food as follows (from Lockie 1956a with some additions from Röell 1978 and Katzir 1983). Bill-up posture (Fig A) with sleeked plumage, apparently indicating bird

A

simultaneously appeasing and prepared to hold ground, often performed by birds on outskirts of dense feeding flocks, 'fearing' to enter throng. Röell (1978) found that Bill-up at feeding site did not cause rival to retreat but was sometimes precursor of successful attack. Take-off posture (similar to Bill-up but bird crouched as if for take-off) occasionally seen just before combat when may signal readiness to fly above rival to attack. Bill-down posture (Fig B) frequently used; intensity varies (lowering of

B

bill sometimes barely perceptible); sometimes wings slightly raised and, in similar posture described by Lorenz (1932, 1952), nape and head feathers erected (see Heterosexual Behaviour, below, for related Allopreening as intimidation). In captive study by Katzir (1983), similar posture often accompanied by Facing-away. Bill-down usually maintained by both birds till one retreated or fight followed (Röell 1978). Various intensities of Forward-threat posture occur, with body horizontal and differing mainly in forward thrust of forebody, more extreme forms often being accompanied by bill-jabbing at rival (which nearly always retreats forthwith: Röell 1978) and with various threat-calls (see 5 and 8a–b in Voice). In most intense form (Fig C: 'elaborate full forward'), seen only in disputes over ♀ and

C

nest-site, plumage is ruffled and (according to Lorenz 1932, 1952) tail spread; drawing in Röell (1978) shows wings and tail raised; aggressor displaying thus may run after rival maintaining posture rigidly. Lorenz (1932, 1952) described similar

D

defensive-threat posture (Fig D) in which head and bill lowered, plumage ruffled, and tail spread (apparently on side facing rival if latter is to one side); bird displaying thus usually wins dispute. In flock studied at feeding site, dominance established by outcome of threat-display (as above), supplanting (aggressor flies to and displaces rival from perch-site, usually without physical contact), and fighting. In fight, rivals jump at each other feet to feet and fall to ground with feet interlocked; may lie on their sides pecking at each other, or one may stand over the other. Conspecifics often rush to scene of fight and much excited calling ensues. Loser flies off, seldom pursued by victor. (Röell 1978.) For aerial fighting, see Raevel and Deroo (1980) and Raevel and Roussel (1983). Attackers nearly always win disputes with conspecifics (Lockie 1956a; Röell 1978). During incubation and chick-rearing, residents very successful in supplanting, in front of own nest-box, individuals which would have dominated them elsewhere (Wechsler 1988a). Nevertheless, breeders may suffer intraspecific interference of various kinds from neighbours such as nest-exploration ('peeping in'), fighting, and probably also attacks on nestlings. Interference leads variously to starvation and death (by pecking) of young (but no evidence for direct predation on eggs or nestlings for food) and nest desertion. Apparently only unsuccessful breeding pairs interfered thus with other pairs, leading however to build-up of interference through season and consequently greater threat to late layers. Main motivation for interference seemed to be acquisition of another nest-site (to be claimed on return to colony in autumn); successful pairs, by contrast, returned to their successful sites. (Röell 1978.) Nest-site defended by both pair-members. For aggression over nest-site when one member of pair dies or disappears, see Röell (1978) and Bonds (above). Importance of defending nest-site emphasized by defence throughout the year (in parts of range where sedentary), except for period of post-breeding moult. Pairs recorded flying straight from communal roost to nest-sites and staying there for some time before going off to feed. Residents return to nest-sites periodically during day, and again last thing before returning to roost. Particularly in September–November and February–April, such visits involve intense activity, residents visiting not only own sites but also those defended by others, and disputes are common. (Röell 1978.) Sometimes takes over nests of Stock Dove *Columba oenas* (Glutz von Blotzheim 1962) and roosting holes (to then be used as nest-sites) of Black Woodpecker *Dryocopus martius* (Striegler *et al.* 1982). Expanding colony of *C. monedula* also recorded driving off Choughs *Pyrrhocorax pyrrhocorax* (Glutz von Blotzheim 1962). However, *C. monedula* subordinate in disputes with *C. corone*, *C. frugilegus*, and *P. pica*; often attacked and sometimes even killed by *C. corone* (Röell 1978). *C. frugilegus* drove off *C. monedula* in 93% of 198 encounters in feeding flocks (Lockie 1956a). For experimental demonstration of competition for food between breeding *C. monedula* and *P. pica*, causing reduced breeding success of *P. pica*, see Högstedt (1980b). See Parental Anti-predator Strategies (below) for other reactions to birds. HETEROSEXUAL BEHAVIOUR. (1) General. Throughout year, can be seen in pairs feeding, visiting nest-sites, and roosting; pair-status also detectable by Allopreening, Begging by ♀, and Courtship-feeding by ♂, e.g.

these behaviours typical of visits to nest-site outside breeding season (Röell 1978). (2) Pair-bonding behaviour. Little precise information on how or when pair-formation occurs (Coombs 1978). Pairing seems to involve extended process of familiarization and is achieved with little obvious display. Only after individiual-distance of prospective mates is reduced to minimum do Allopreening, Begging, and Courtship-feeding start. Pair-formation occurs prior to acquisition of nest-site, but because pair-formation generally results in rank promotion for ♀♀ (see Flock Behaviour), ♀ seeking mate sometimes prefers high-ranking paired resident ♂ to low-ranking unpaired non-resident. In practice, however, paired ♀♀ are rarely displaced by such (usually older) 'follower' ♀♀, although bigamy exceptionally results (see Bonds). (Röell 1978, 1979.) Judging by intimate behaviour and close physical contact, young birds probably start pairing from autumn onwards (Morley 1943; Hinde 1947; Coombs 1978; Kaatz 1986). In addition, pursuit-flights, commonly involving 3 birds, occur throughout the year, though mostly in spring, and may be sexual, but precise function unknown; include distinctive glide with wings below horizontal; wing-tips sometimes fluttered during glide (Coombs 1978). In Groningen, newly paired young birds were seen December–June, peaking March–April; re-pairings of adults took place in almost every month but mostly January–March and September–November (Röell 1978). When partner returns to nest, mate may perform Meeting-display (Fig E): wings half-spread and drooped, tail may be rapidly quivered

E

laterally (i.e. outer feathers flicked alternately out and in), and bill pointed down (Coombs 1978; Dwenger 1989). Tail-quivering may also occur without any other posturing (Goodwin 1986), including in autumn (Dwenger 1989). Slight up-and-down tail-quivering also seen (Coombs 1978). Allopreening is mainly on head and neck region. Bird solicits Allopreening by ruffling head feathers and presenting nape to partner, and may also lower bill (Katzir 1983; Goodwin 1986). Within flock of captive juveniles (where Allopreening apparently more prevalent than in the wild), Allopreening by subordinate was brief and hasty, often inducing attack, whereas Allopreening by dominant was deep, long, and sometimes rough, recipient appearing ill-at-ease (Katzir 1983, which see for illustrations and discussion of function). According to Röell (1978), Allopreening mainly by ♀ (consistent with finding by Katzir 1983 that initiators are mostly of low rank), but in study of captive flock, Allopreening (which was almost exclusively restricted to pairs) performed equally by either sex; nor did one sex contribute more than the other to maintaining mutual proximity, i.e. one partner was just as likely to follow as the other (Wechsler 1989). (3) Courtship-feeding. As in Allopreening, almost exclusive to pairs, with ♂ feeding ♀ (Wechsler 1989). ♀ begs from ♂ by crouching, wing-fluttering, and tail-quivering, with bill angled up towards ♂ (see drawing in Blume 1967) and giving Begging-call (4 in Voice). Begging and Courtship-feeding may apparently occur at any time of year (Röell 1978): e.g. in Britain, sightings include probable October record (Rooke 1950), but most commonly February–May. ♂ typically feeds ♀ several times in quick succession. (Tucker

1950.) ♂ continues feeding ♀ when she is incubating and also for 1st week of brooding (Dwenger 1989). (4) Mating. Usually takes place on nest, but also recorded in November flock mixed with *C. frugilegus* (Morley 1943). In Soliciting-posture (Fig F),

F

♀ crouches with wings somewhat drooped and held out from body, head lowered and bill inclined slightly upwards, tail horizontal or slightly raised and often quivered (Coombs 1978). In case of mating recorded during incubation, ♂ mounted ♀ when she was apparently turning eggs; copulation was accompanied by Copulation-call (7 in Voice), not known from which sex; sequence repeated 10 min later (Dwenger 1989). ♂ guards ♀ assiduously around time of laying (mate-guarding) (Röell 1978). (5) Nest-site selection and behaviour at nest. Both members of pair seem to be involved in selecting site. In study of captive flock, nest-site preferences of pair sometimes changed several times before final choice. When pair changed preference, one partner was usually the more active in visiting new site, but mate soon followed suit, so that pair-bond facilitated site attachment (Wechsler 1989). Coombs (1978) described visits to chimney by prospecting pair accompanied by Alloadopting and brief wing-drooping; ♂ examined chimney carefully, lowering himself into it as far as he could while keeping firm grasp with foot on rim. Nest-sites may be chosen in autumn when young birds start pairing (Coombs 1978; Röell 1978). Nest-building may be accompanied (perhaps only in new pairs) by tail-quivering and quiet clucking calls (10c in Voice); other birds, however, make nest without ceremony (Dwenger 1989). For other behaviour at nest see subsections 2–4 of Heterosexual Behaviour. RELATIONS WITHIN FAMILY GROUP. For first few days after hatching, ♂ brings food (in buccal pouch for regurgitation) for ♀ and young while ♀ broods assiduously; brooding continues, with gradually less intensity, up to *c.* 16 days, after which young form 'warmth pyramid' after each feed. During 1st week, ♂ feeds young via ♀, thereafter directly. By 12 days both parents bring food; at this stage only ♀ seen to remove faecal sacs. (Dwenger 1989.) Faeces are not always enclosed in sac, and when not are removed together with some adjacent nest-lining (Goodwin 1986; Dwenger 1989). According to Soler and Soler (1990), eyes of young begin to open on day 8 and are completely open in 90% of nestlings on day 14; however, Dwenger (1989) reported eyes starting to open at 12 days, half open at 14 days, fully open at 17 days. Soft Contact-call (see 1 in Voice) of parent arriving with food stimulates small young to gape (Dwenger 1989), but parents silent when approaching older young which gape as soon as they hear sounds of approach (Coombs 1978). Young fledged at *c.* 35 days but some still in nest at 38–40 days; family does not disperse from nest-area until last of brood leaves nest, even though this may take several days (Dwenger 1989). Nevertheless, some leave nest less developed than others and remain on ground where they suffer high mortality (Soler and Soler 1990). For first few days after fledging, young (not yet able fliers) gather in groups where fed by parents. At night parents remain with young wherever young happen to be. Within 1 week after fledging, young capable of following parents to feeding grounds and communal roost. Around 1 month after fledging, young rarely beg

and family parties gradually break up. (Röell 1978; see also Bonds.) ANTI-PREDATOR RESPONSES OF YOUNG. From day 25 in nest, until fledging, young cease giving food-calls on hearing strange noises (such as human approaching nest) and instead remain silent (Dwenger 1989). PARENTAL ANTI-PREDATOR STRATEGIES. (1) Passive measures. No information. (2) Active measures: general. In study by Lorenz (1932) of reactions to mammalian predators, appeared to differ from other Corvidae in not immediately attacking anything resembling major predator, whether bird or mammal; confirmed by study with stuffed Eagle Owl *Bubo bubo* (Dwenger 1989). Röell (1978) found almost complete lack of alarm-calls and mobbing when potential predators approached nest, and suggested this perhaps due to relative inaccessibility of nests (in boxes) and their lack of crypsis. However, mobbing responses reported from other studies (see below); main mobbing vocalization is Rattle-call (see 8a in Voice). (3) Active measures: against birds. For general dominance relations with other Corvidae, see Antagonistic Behaviour (above). At 2 small colonies, one wild (Zimmermann 1951), one captive (Lorenz 1932, 1952), conspecifics not belonging to colony were collectively mobbed and driven off during breeding season. *P. pica*, habitual robber of colony nests, was similarly mobbed (Lorenz 1932, 1952). Also seen driving off young *C. corax* (Lorenz 1932); in another case, 5–6 *C. monedula* (apparently not breeding nearby) giving alarm-calls surrounded and lunged at *C. corax*, forcing it to fly off, then followed it, calling, for *c.* 200 m (Radford 1991). Aerial attack by 2 birds, with bill-jabbing, also seen on Egyptian Vulture *Neophron percnopterus* (Matt 1983), and Common Tern *Sterna hirundo* also once mobbed (Rolls 1973). When stuffed Sparrowhawk *Accipiter nisus* was put on ground near colony, 2 *C. monedula* arrived and started giving (unspecified) alarm-calls, attracting *c.* 30 others which circled agitatedly over it at heights of 1–4 m; some made dive-attacks but none landed (Dwenger 1989). (4) Active measures: against man. Will mob (with call 8a) intruder at nest or handling fledglings (Goodwin 1986). Parent of chick being ringed flew past intruder and struck blows at his cap (Kalitsch 1943). However, physical contact exceptional and only one similar incident (involving more than 1 attacker) known to Dwenger (1989). (5) Active measures: against other animals. No information.

(Figs A–F by D Quinn from drawings in Coombs 1978.)

EKD

**Voice.** Complex, especially at nest-site in breeding season. Many calls intergrade and this, together with variability in renderings by individual listeners, makes schemes of different authors hard to reconcile. First attempted scheme by Lorenz (1932, 1952). No information on geographical variation. For additional sonagrams see Bergmann and Helb (1982). For oscillograms of main calls, see Lesitsyna and Nikol'ski (1979). Bill-snapping occurs (Lorenz 1952; Bergmann and Helb 1982).

CALLS OF ADULTS. (1) Song. Ceaseless chatter (Daniels and Easton 1985); quiet soliloquy which can last several minutes (Zimmermann 1931), given equally well by both sexes. Structure a medley of call-type units with great variation in loudness and inflection, given in display notable for incorporating, as each call delivered, the posture appropriate to that call. (Lorenz 1932, 1952.) Song given when alone, perched, or flying, and volume and variability sufficient to deceive listener into believing source to be a flock (Goodwin 1986). Although song of hand-reared birds said to include mimicry (Lorenz 1932, 1952), this not known in the wild (Goodwin 1986). (2) Contact-call. Commonest call, given in flight or perched, 'chak' or 'KEak' (J Hall-Craggs) heard in variety of contexts, especially pair-contact, including invitation to fly or share company (Lorenz 1932, 1952), often from bird announcing arrival at nest to mate and young (Dwenger 1989). Fig I

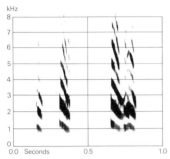

I   W T C Seale   England   June 1985

shows 'ka chack chakchak' from pair sitting at entrance to nest containing well-grown young (W T C Seale). For numerous other renderings and related sounds, indicating great flexibility of basic call, see Goodwin (1986: 'tchak', etc.) and Dwenger (1989: 'kjack', etc.). Sometimes louder and harsher, sometimes softer and more mellow, noisier or more tonal, shorter (e.g. 'tchk': Daniels and Easton 1985) or longer (more disyllabic), vowel sound 'a' or 'o'. In recordings, most calls start tonally but rest of structure varies from tonal to more noisy, e.g. short musical 'kji' or 'kjä' often exchanged by mates in duet (Bergmann and Helb 1982); Fig II shows series (not known if from 1 or 2

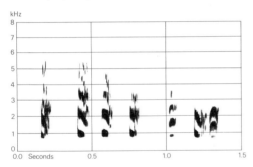

II   P J Sellar   England   July 1982

birds) producing melody of almost entirely tonal 'KEak' notes; Fig III depicts noisier 'kia' or 'kyoio' from flock-members, each call with explosive start due to marked starting transient; emphatic quality perhaps locatory cue facilitating contact between flock-members on the move (J Hall-Craggs). (3) Feeding-call. 'Kiaw' or 'kiaow' like swallowed variant of call 2, used by adults to guide young home or to entice them to join flock, and by ♂ offering food to mate (Lorenz 1932, 1952). Also heard as 'kyow' (Witherby *et al.* 1938) or 'kīa' (J Hall-Craggs). For son-

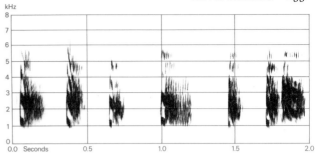

III   S Palmér/Sveriges Radio (1972–80)   Sweden   October 1966

agram see Fig f in Bergmann and Helb (1982). (4) Begging-call. Nagging imploring variant of call 2, rendered roughly 'kyaay' or 'tchaayk', given by ♀ soliciting food from mate (Goodwin 1986). Also described as nasal 'giäää- giäää-giäää' (Zimmermann 1931); drawn-out buzzy 'kjerr' (Bergmann and Helb 1982) perhaps the same. (5) Nesting-call. Rapid series of 'gjück' and 'gjick' sounds given by birds guarding nest-site against rivals (Dwenger 1989). This presumably the call described by Lorenz (1932, 1952) as higher, sharp 'tsick-tsick-tsick' or 'zick, zick, zick' given by ♂ advertising site-ownership to rival ♂♂ and also to attract ♀. (6) Clicking-call. Very quiet 'tut-tut' or 'stut-stut' similar to sound made by human clicking tongue off roof of mouth; given by perched bird (juvenile), sometimes just before sleep and thought to express contentment (Daniels and Easton 1985). Quiet 'tschik-tschik- tschik' from relaxed pair at nest (Dwenger 1989) perhaps the same; see also 10c (below). (7) Copulation-call. Loud, rasping 'rääääh-rääääh-rääääh' heard during copulation at nest; not known from which partner, but also given by juvenile trying to copulate with owner's hand (Dwenger 1989). (8) Alarm- and Warning-calls. (8a) Rattle-call. Loud, repeated, harsh, grating 'kaaarr' (Goodwin 1986: Fig IV) or 'arrrrrrrrrr' (Dwenger

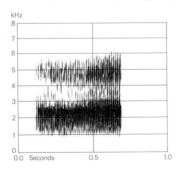

IV   P J Sellar   Scotland   June 1978

1989) heard especially when warning about approach of, and when mobbing, predator. According to Lorenz (1932, 1952), call usually accompanied by craning towards source of disturbance and wing- fluttering, but no such wing movements seen in wild birds by Goodwin (1986). Lower-

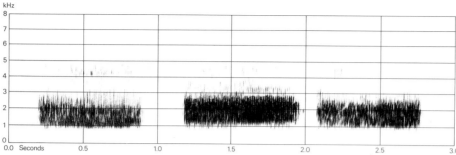

V  V C Lewis  England  May 1969

intensity variant, e.g. in threat between conspecifics, a rasping extended 'kjarr' (Bergmann and Helb 1982); Fig V shows 3 calls from 2 birds, 2nd call from bird nearer microphone (J Hall-Craggs). Also rendered 'kraare' or 'chaairr', given in Forward-threat posture, and often in gliding flight when it is usually preceded and/or followed by contact- calls (Goodwin 1986: see above). (8b) Yip-call. Repeated staccato 'jüp' or 'yup' given in extreme Forward-threat posture in intense disputes over nest-sites (which typically begin with series of call 5). Heard only during breeding season, especially when young present, and attracts conspecifics which subsequently join chorus. (Lorenz 1932, 1952.) (8c) Hiss-call. Hissing puff, given sometimes when warding off rival or excessively importunate mate (Strauss 1938a). Also described (in juvenile more than 160 days old) as quiet throaty coughing or hissing (Daniels and Easton 1985). (9) Distress-call. Shriek (Daniels and Easton 1985); short, abrupt, covering wide frequency range (Morgan and Howse 1973, 1974, which see for sonagrams). (10) Other calls. At least some of the following perhaps not different from calls listed above. (10a) Rolling 'raak', probably excitement-call (Daniels and Easton 1985); drawn-out 'groah' (Bergmann and Helb 1982) perhaps the same. (10b) Gurgling 'gruieet' from ♂ bringing food to ♀ on nest (Dwenger 1989). (10c) Various quiet calls from one or both mates at nest, such as 'gug', 'guh', or 'güh', 'huh' and 'itt', 'ück', and 'gick' (Dwenger 1989).

CALLS OF YOUNG. For development of food-calls and associated behaviour see Korbut (1981, 1982) and Redondo et al. (1988). For detailed ontogeny of repertoire in hand-reared young, see Daniels and Easton (1985). In 1st week, food-call a thin cheeping, changing after 2nd week into piercing screech or squeal which reaches maximum volume and urgency at c. 18 days; thereafter, pitch begins to descend and, in week before fledging, food-calls much less raucous to human ear (Dwenger 1989). 3 other calls emerge around fledging (c. 40 days) but persist only until c. 60 days (Daniels and Easton 1985): firstly a sharp squeaking like child's rubber toy or rusty hinge, apparently to seek attention or reassurance; secondly, high-pitched penetrating 'kwaak', presumably food-contact call described by Goodwin (1986) as 'kwaay' or 'tchaayk' (as in adult call 4), and by Bergmann and Helb (1982) as

drawn-out 'karrr' (c. 1 s) very similar to locatory call of young Rook C. frugilegus but not so harsh or repetitive; thirdly, resonant bell-like 'pwow' or 'kpwow' when addressed by owner while exploring novel objects and apparently signalling 'I hear you but I'm busy'. After 60 days, vocalizations generally quieter and recognizable as full adult calls or at least precursors. (Daniels and Easton 1985.) In Fig VI, for example, structure of juvenile calls

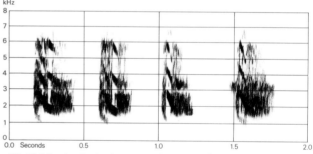

VI  V C Lewis  England  May 1968

quite closely resembles some of adult in Fig III but are generally noisier, even starting with a noise instead of (as in Fig III) brief starting transient (J Hall-Craggs).  EKD

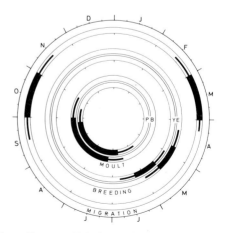

**Breeding.** SEASON. Britain and Ireland: early April to mid-May; average date for laying of 1st egg ranged from

23 April (Ireland, $n=40$) to 28 April (northern England, $n=165$) (Holyoak 1967c, which see for regional variation); average date of 1st egg in central Scotland 23 April in 2 consecutive years, $n=117$ (Chesney 1986); for western Wales, see Richford (1978). Little variation throughout Europe: average laying date in southern Finland 29 April (mid-April to end of May), $n=115$ (Antikainen 1978); average date of 1st egg in southern Spain 29 April, $n=192$ (Soler and Soler 1987, which see for comparison of many studies); peak period in central Europe (see diagram) 2nd half of April (Zimmermann 1951; Folk 1968; Röell 1978; Dwenger 1989; Strebel 1991), in St Petersburg area (north-west Russia) early to mid-May (Eygelis 1958). SITE. Hole or cavity in tree, rock-face, man-made structure (especially chimney, bridge, or similar rather inaccessible place), and also in nest-boxes; very often in old tree-hole of Black Woodpecker *Dryocopus martius*; in some countries (e.g. Britain, Ireland, Belgium, Netherlands) in disused burrow of rabbit *Oryctolagus cuniculus*, and in Finland commonly in nest-box erected for Goldeneye *Bucephala clangula* (Antikainen 1978; Dwenger 1989). Very rarely on tree branches, sometimes in old nest of Rook *C. frugilegus* or Magpie *Pica pica*; such nests often in dense cover of conifer or ivy *Hedera* (Coombs 1978; Röell 1978; Goodwin 1986); no open nests found in survey of 1708 nests in Switzerland and neighbouring France (Riggenbach 1979); 0·7% of 2656 nests in Finland on branches (Antikainen 1978), but 52% of 129 nests in study area in Netherlands were open on conifer branches (Röell 1978). In Poland, 55% of 298 nests in trees or in nest-boxes on trees, 26% on building, and 19% on rocks; in trees, natural holes preferred, then old woodpecker holes, then nest-boxes (Kulczycki 1973). No apparent preference for size of hole or species of tree, but very often in beech *Fagus* and (in Finland) aspen *Populus tremula* since these favoured by *D. martius*; less frequently in oak *Quercus* and lime *Tilia* (Folk 1968; Kulczycki 1973; Antikainen 1978). Average height in tree, Czechoslovakia, 7·1 m (1·2–21·0, $n=310$) and on building 16·0 m (3·4–45·0, $n=229$) (Folk 1968); nests recorded at up to 70 m in towers; often some distance between cavity entrance and nest, e.g. 6 m recorded in chimney (Folk 1968; Antikainen 1978). See Röell (1978) for nest-site choice and interference by Carrion Crow *C. corone*. Nest: very variable in size and structure depending on nature of cavity; in (e.g.) chimney or hollow tree sticks thrown in hole until they lodge, and nests often re-used, so foundation can be huge (3·5 m recorded in St Petersburg area: Eygelis 1958) or suspended just below cavity entrance, or can cover floor of nest-box or similar flat area (Zimmermann 1951; Folk 1968; Röell 1978). Foundation messy accumulation of largish twigs and sticks, often with irregular layer of lumps of mud or dung (in nest-boxes often forming proper bowl 1–3 cm thick) lined with rootlets, rotten wood, stalks, moss, hair (sometimes taken from live animal), feathers, paper, etc. (Zimmermann 1951; Folk 1968; Kulczycki 1973; Antikainen

1978; Soler 1987; Strebel 1991). Cup can be very shallow, or deep inside pile of twigs touching base of cavity; sometimes lining greater in bulk than foundation, which is present only as rough ring of twigs (Kulczycki 1973; Dwenger 1989). Open nests in Netherlands often with some sort of roof (Röell 1978). 31 nests in Poland had average outer diameter 38·5 cm (16–100), inner diameter 12·1 cm (9–23), overall height 24·6 cm (7–100), depth of cup 4·8 cm (3–8); large nests weigh up to 20 kg (Kulczycki 1973; Dwenger 1989). For description of 108 nests in southern Spain, see Soler (1987). Building: by both sexes; either or both birds bring material and both build, ♀ perhaps more on lining; twigs, etc., usually brought singly but sometimes in bundles of small items (Zimmermann 1951; Röell 1978; Dwenger 1989); takes *c.* 20 days (5–27, $n=121$) (Antikainen 1978); 7–68 days recorded at 28 nests in Switzerland (Strebel 1991). Generally collects twigs from ground but sometimes pulls them from trees (Dwenger 1989); if cavity has been used in the past, especially by Feral Pigeon *Columba livia*, ejects old material and droppings before building; can excavate own cavity in rotten wood, soft cliffs, or crumbling mortar (Zimmermann 1951; Kulczycki 1973; Makatsch 1976). May still add to nest when brooding young (Zimmermann 1951; Strebel 1991). EGGS. See Plate 57. Sub-elliptical, sometimes short sub-elliptical, smooth and glossy; pale light blue to greenish-blue with very variable specks and blotches of blackish-brown to light olive, pale grey, or greyish-violet, becoming larger towards broad end; some unmarked (Harrison 1975; Makatsch 1976). Nominate *monedula*: $35·0 \times 24·7$ mm ($30·4–39·4 \times 21·5–25·5$), $n=317$; calculated weight 11·1 g. *C. m. soemmerringii*: $34·8 \times 25·0$ mm ($29·5–39·5 \times 22·6–28·2$), $n=600$; calculated weight 11·3 g. *C. m. spermologus*: $35·0 \times 25·2$ mm ($29·3–40·3 \times 21·0–29·0$), $n=542$; calculated weight 11·5 g. (Schönwetter 1984.) For discussion and comparison of egg size in European studies and dimensions of 814 eggs from southern Spain, see Soler (1988a). Clutch: 4–6 (2–8). Of 293 clutches, western Wales: 2 eggs, 2%; 3, 11%; 4, 33%; 5, 42%; 6, 11%; average $4·4 \pm SE0·1$ (Richford 1978). Average in central Scotland 4·5, $n=117$; average from 6 regions of Britain and Ireland 4·4, $n=646$, with no significant variation (Holyoak 1967c; Chesney 1986). Of 167 clutches, southern Finland: 2 eggs, 4%; 3, 9%; 4, 24%; 5, 35%; 6, 23%; 7, 4%; 8, 1%; average $4·8 \pm SE0·1$ (Antikainen 1978). Little variation in averages throughout Europe: eastern Germany 5·0, $n=204$ (Schmidt 1988); Switzerland 4·7, $n=51$ (Strebel 1991), 4·2, $n=80$ (Zimmermann 1951); Czechoslovakia 4·5, $n=353$ (Folk 1968); St Petersburg area 4·1, $n=18$ (Eygelis 1958); southern Spain 5·2, $n=184$ (Soler 1989b). Eggs laid daily, though sometimes with gap of 2 or even 3 days (Zimmermann 1951; Folk 1968; Richford 1978). One brood; hardly ever a replacement clutch if 1st fails; none recorded in extensive studies in Switzerland, Finland, Netherlands, or eastern Germany (Zimmermann 1951; Antikainen 1978; Röell

1978; Dwenger 1989); 6 replacement clutches of average size over 3 years in Wales (Richford 1978), and 4 recorded in 52 clutches from Rheinland (western Germany) (Mildenberger 1984). INCUBATION. 17-18 (16-20) days; by ♀ only. Starts with 2nd or 3rd egg, rarely with last (Zimmermann 1951; Antikainen 1978; Richford 1978); from laying of last egg to hatching of last young, Czechoslovakia, took 17·2 days (15-20, $n=16$), and from laying of 1st egg to hatching of 1st young 19·1 days (17-22, $n=16$) (Folk 1968); in central Scotland, 17·6 days between laying and hatching of last egg, $n=7$ (Chesney 1986). Hatching over 0·5 days in clutches of 3, 1·5-2 days in larger clutches (Folk 1968; Antikainen 1978; Chesney 1986). YOUNG. Fed and cared for by both parents; fed by ♂ via ♀ until c. 10 days old then fed by both sexes (Zimmermann 1951; Dwenger 1989). Brooded by ♀ for c. 11 days (6-17, $n=28$); ♀ left nest 5 times for a few seconds each time in 16 hrs of observation (Antikainen 1978). For development of young, see Kamiński (1986), Soler and Soler (1990), and Strebel (1991). FLEDGING TO MATURITY. Fledging period in southern Spain 32·4 days (28-36, $n=246$ nestlings), longer in larger broods (Soler 1989a); in southern Finland 33·8 days (28-41, $n=75$) (Antikainen 1978). Young become independent c. 5 weeks after fledging (Dwenger 1989). First breeding usually at 2 years, sometimes at 1 (Zimmermann 1951; Richford 1978; Röell 1978; Dwenger 1989). BREEDING SUCCESS. In southern Spain, of 965 eggs, 70·3% hatched and 26·1% produced fledged young, giving 1·4 fledged young per pair ($n=184$) or 1·7 per successful pair ($n=147$); all eggs lost in 24% of nests (mostly due to predation by Raven C. corax, intraspecific interference, human disturbance, or broodparasitism by Great Spotted Cuckoo Clamator glandarius); 24% of broods failed completely, almost 50% of deaths being by starvation in first 5 days; nestlings also taken by C. corax, and some leave nest before they can fly and are eaten by ground predators, particularly fox Vulpes (Soler 1989b; Soler and Soler 1990). See also Soler (1990), especially for breeding success in relation to nest density. In southern Finland, of 391 eggs, excluding total losses, 62% hatched (28% disappeared) and 39% produced fledged young, giving 1·9 fledged young per pair overall ($n=81$); egg losses due to predation by rat Rattus norvegicus, probably squirrel Sciurus vulgaris, observer disturbance, and possible cannibalism; great majority of nestlings died in first 5 days, and 62% of 37 young which died were last to hatch; extremes of weather and lack of food main causes of nestling death, as well as intraspecific interferences and trampling; success highest in clutches of 4-5, in small colonies, and in tree-holes, especially smaller ones; colony in individual nesting cells of church tower (3·3 fledged young per pair, $n=18$) more successful because of less interference from conspecifics (nest entrance easy to defend) and less observer disturbance (Antikainen 1978, 1981). Breeding success greater in deeper nest-boxes (Röell 1978). In Czechoslovakia (2·0 fledged young per

pair, $n=118$), western Wales (0·7, $n=247$), northern Netherlands (1·1, $n=198$), and central Scotland (1·6, $n=166$), mortality of youngest nestlings high and other C. monedula suspected of taking eggs and young (Folk 1968; Richford 1978; Röell 1978; Chesney 1986). In Netherlands study, eggs and young removed, not eaten, by conspecifics when nest usurped (Röell 1978). For Switzerland, see Strebel (1991). In Wales, quality of nestling diet most important factor in survival of young (Richford 1978). In Scotland, 118 nestling deaths (from 357 eggs) in one year due to being fed grain poisoned by farmers (Chesney 1986). 24% of 46 complete failures in eastern Germany caused by marten Martes (Schmidt 1988). Nests in England robbed by Tawny Owl Strix aluco, weasel Mustela nivalis, and possibly by squirrel S. carolinensis; nestlings sometimes drown in flooded nest-holes (Lockie 1955; Holyoak 1967c). No predation by squirrels noted in 3-year study in southern England in which 86% of 378 nestlings failed to fledge, 90% of all deaths (mostly by starvation) occurring in first 2 weeks after hatching (Heeb 1991). See Gibbons (1987) and Heeb (1991) for discussion of asynchronous hatching, nestling mortality, and brood reduction strategy (i.e. selective feeding of largest members of asynchronous brood to ensure maximum productivity), involving experimental manipulation of clutch size and hatching interval.          BH

Plumages. (C. m. spermologus). ADULT MALE. In fresh plumage (autumn), forehead and crown black with strong purplish-violet gloss, forming small contrasting cap. Bristle-like feathers over nostril dull black with dark grey or brown-grey shafts. Lore to just above eye, feathering at gape, lower cheek, chin, and throat black, deep and velvety on lore, slightly glossed blue on feather-tips just above eye and on lower cheek, chin, and throat; tips of feathers at gape tinged grey, feather-shafts dark grey or brown-grey on chin and throat; black not sharply demarcated from grey of remaining side of head. Hindneck, side of head from eye and rear of cheek backwards, and side of neck silky medium-grey, usually tinged plumbeous-grey and sometimes finely spotted or streaked purple-blue on lower hindneck or entire hindneck, sometimes slightly paler (light grey) on rear of ear-coverts or rear border of side of neck. Upperparts backwards from mantle black, often with faint plumbeous bloom when plumage freshly moulted, feather-tips faintly glossed blue or purple-blue. Underparts backwards from chest and side of breast plumbeous-black, each feather with broad but scarcely paler dark plumbeous-grey fringe, except thigh and under tail-coverts. Tail black, slightly glossed metallic bluish-green or green, bases of outer webs of central feathers often more strongly glossed purple-violet. Wing black; primaries, upper primary coverts, bastard wing, and centres and inner webs of secondaries and tertials glossed dark metallic green, sometimes partly purple-violet on outer webs of inner primaries and greater primary coverts; outer webs of

PLATE 1. Garrulus glandarius Jay (p. 7). Nominate glandarius: 1 ad summer, 2 ad winter, 3 juv. G. g. rufitergum: 4 ad. G. g. fasciatus: 5 ad. G. g. albipectus: 6 ad. G. g. glaszneri: 7 ad. G. g. atricapillus: 8 ad, 9 juv. G. g. krynicki: 10 ad. G. g. brandtii: 11 ad. G. g. cervicalis: 12 ad, 13 juv. G. g. whitakeri: 14 ad. G. g. minor: 15 ad. (DR)

Darren Rees

PLATE 2. *Perisoreus infaustus* Siberian Jay (p. 31). Nominate *infaustus*: **1** ad, **2** juv. *P. i. ruthenus*: **3** ad.  *Cyanopica cyanus cooki* Azure-winged Magpie (p. 42): **4** ad, **5** juv in post-juvenile moult. (DR)

PLATE 3. *Pica pica* Magpie (p. 54). Nominate *pica*: **1** ad, **2** 1st ad, **3** juv. *P. p. melanotos*: **4** ad. *P. p. mauritanica*: **5** ad. (DR)

PLATE 4. *Nucifraga caryocatactes* Nutcracker (p. 76). Nominate *caryocatactes*: **1** ad winter, **2** ad summer, **3** juv. *N. c. macrorhynchos*: **4** ad. (DR)

PLATE 5. *Garrulus glandarius glandarius* Jay (p. 7): **1-2** ad summer. *Perisoreus infaustus infaustus* Siberian Jay (p. 31): **3-4** ad. *Cyanopica cyanus cooki* Azure-winged Magpie (p. 42): **5-6** ad. *Pica pica pica* Magpie (p. 54): **7-8** ad. *Nucifraga caryocatactes caryocatactes* Nutcracker (p. 76): **9-10** ad summer. (DR)

PLATE 6. *Pyrrhocorax graculus graculus* Alpine Chough (p. 95): **1** ad, **2** juv. *Pyrrhocorax pyrrhocorax* Chough (p. 105). Nominate *pyrrhocorax*: **3** ad, **4** 1st ad, **5** juv. *P. p. docilis*: **6** ad. *P. p. barbarus*: **7** ad. (DR)

PLATE 7. *Corvus monedula* Jackdaw (p. 120). *C. m. spermologus*: **1** ad summer, **2** ad winter, **3** juv. Nominate *monedula*: **4** ad winter. *C. m. soemmerringii*: **5** ad winter. *C. m. cirtensis*: **6** ad winter. (DR)

PLATE 8. *Corvus dauuricus* Daurian Jackdaw (p. 140): **1** ad, **2** 1st winter. *Corvus splendens zugmayeri* House Crow (p. 143): **3** ad, **4** juv. (DR)

PLATE 9. *Pyrrhocorax graculus graculus* Alpine Chough (p. 95): **1–2** ad. *Pyrrhocorax pyrrhocorax pyrrhocorax* Chough (p. 105): **3–4** ad. *Corvus monedula spermologus* Jackdaw (p. 120): **5–6** ad summer. *Corvus dauuricus* Daurian Jackdaw p. 140): **7–8** ad. (DR)

PLATE 10. *Corvus ruficollis ruficollis* Brown-necked Raven (p. 197): **1** ad winter, **2** 1st ad summer, **3** juv. *Corvus rhipidurus* Fan-tailed Raven (p. 223): **4** ad, **5** juv. (DR)

PLATE 11. *Corvus corax* Raven (p. 206). Nominate *corax*: **1** ad winter, **2** 1st ad summer, **3** juv. *C. c. varius*: **4** ad. *C. c. laurencii*: **5** ad. *C. c. tingitanus*: **6** ad. (DR)

PLATE 12. *Onychognathus tristramii* Tristram's Grackle (p. 229): **1** ad ♂ summer, **2** ad ♂ winter, **3** ad ♀, **4** juv.  *Sturnus sturninus* Daurian Starling (p. 234): **5** ad ♂, **6** ad ♀, **7** juv. (PJKB)

PLATE 13. *Sturnus roseus* Rose-coloured Starling (p. 269): **1** ad ♂ summer, **2** ad ♂ winter, **3** ad ♀ summer, **4** 1st ad ♂ winter, **5** 1st ad ♀ winter, **6** juv. (PJKB)

Darren Rees

secondaries and tertials and all lesser, median, and greater upper wing-coverts strongly glossed violet-purple, intensity of gloss contrasting sharply with duller remainder of wing and (especially) body, but similar to gloss of cap; gloss of shorter coverts along leading edge of wing more purplish-blue. Under wing-coverts and axillaries dark grey with glossy bluish-black fringes; longest coverts glossy dark grey, like undersurface of flight-feathers. Sexes similar when plumage fresh, but head and neck of ♂ tend to become paler due to more abrasion than in ♀ (Voipio 1968). *In worn plumage*, shows marked influence of bleaching and abrasion. By about March–April, when slightly worn, cap more sharply defined, gloss stronger bluish- or purplish-violet, nasal bristles tinged brown, grey on side of head and neck and along rear border of cap more silvery medium grey, mantle and scapulars more bluish-plumbeous-grey (less sooty); entire underparts virtually uniform dark plumbeous-grey, feather-tips of chin and throat less glossy black, chin and throat contrasting less with remainder of underparts, but shafts slightly paler. From late April or May, feather-fringes of mantle, scapulars, and entire underparts abraded to paler plumbeous-grey or dark silvery-grey (colour and degree of contrast with centre of feather depending on angle of light), appearing scalloped. During June to about July, when heavily worn, black of cap duller; grey of hindneck and of side of head and neck still paler, light silvery-grey, streaked with darker grey; silvery-grey almost uniform along rear border of cap (especially in ♂); dark grey streaks often heavier on ear-coverts; hindneck mottled dull black; silvery-grey only exceptionally forms pale bar across rear side of neck; much dull sooty-brown of feather-bases visible on underparts. NESTLING. Back and (slightly) crown, thigh, and upperwing covered with pale smoke-grey or light grey down *c.* 8–10 mm long; otherwise naked (Witherby *et al.* 1938; Zimmermann 1951; Strebel 1991). JUVENILE. Nasal bristles, cap, lore, ear-coverts, and cheek greyish- or sooty-black, sometimes slightly glossed green, brown-grey of feather-bases showing through, black not sharply demarcated from remainder of head. Remainder of upperparts and side of neck dark sooty-grey, some tips of feathers of mantle and tips of some scapulars and upper tail-coverts black with slight gloss. Underparts entirely dull sooty-black to dull dark grey, chin and upper throat darkest but with some pale grey of feather-bases often visible; plumage short and loose, especially on lower flank, thigh, and vent. Grey to sooty-black of body rapidly turns to brown from shortly after fledging (Bährmann 1968b). Tail and wing as adult, but tertials and tail-feathers narrower, tips more rounded, less truncate; gloss much less intense, mainly dark metallic green, rather dull violet-purple restricted to outer webs of secondaries and to part of outer webs of greater upper wing-coverts; marginal coverts sooty-grey, gloss virtually absent; under wing-coverts dark grey, without black fringes. FIRST ADULT. Like adult, but juvenile tail, flight-feathers, tertials, greater upper secondary and primary coverts, and variable number of median and sometimes lesser coverts retained; tertials and tail-feathers narrower and with more rounded tips than in adult, distinctly browner and more frayed at tip than in adult at same time of year; tips of tail-feathers markedly

less truncate, innermost tertial and (if any) some old median or lesser coverts dark brown, contrasting sharply with new 1st adult scapulars and wing-coverts. New plumage of head and body as in adult, but grey of hindneck and of side of head and neck slightly darker on average (especially on neck), more intensely tinged plumbeous, in ♀ often scarcely paler than remainder of body. As in adult, some variation in tinge of underparts, most birds plumbeous-black with faintly paler dark plumbeous-grey feather-fringes ('*nigriventris*' morph), others medium grey with slight ashy tinge ('*griseiventris*' morph), but latter apparently exceptional in typical *spermologus* from western Europe; exceptionally, underparts of 1st adult virtually uniform deep black ('*nigra*' morph); see Bährmann (1968b).

**Bare parts.** ADULT, FIRST ADULT. Iris pearl-grey, silver-grey, milk-white with blue tinge, or almost pure white. Bill, leg, and foot black. NESTLING, JUVENILE, FIRST ADULT. In 1st week after hatching, bare skin light pink (day 1) to reddish-flesh (day 5–7), gradually becoming blackish on future feather-tracts; bill pink-flesh to dark horn-grey, mouth blood-red or purplish-pink; gape-flanges broad, pale yellow; leg slightly paler yellow-pink than skin. On day 11, skin greyish-purple-flesh. From day 10 to day 18, scutes on front of tarsus and upper surface of toes darken to greyish-black; upper mandible becomes gradually darker from base, starting from day 23, lower mandible from day 30. At fledging, bill, leg, and foot dusky horn-grey. Yellow gape-flanges disappear 1–2 weeks after fledging (at age of 40–60 days). (Heinroth and Heinroth 1924–5; Witherby *et al.* 1938; Zimmermann 1951; Strebel 1991.) Iris grey-blue at fledging; during post-juvenile moult, brown ring develops round pupil, remainder of iris gradually becoming increasingly spotted brown; by autumn and early winter, iris fully brown; later on, gradually paler again; by May of 2nd calendar year, white with prominent brown blotches, these latter gradually disappearing during 2nd year of life (Bährmann 1968; Henderson 1991).

**Moults.** ADULT POST-BREEDING. Complete; primaries descendent. In Britain, primaries start with shedding of p1 mid-May to mid-July, completed with regrowth of p10 mid-August to mid-October; tail moult starts with loss of t1 at primary moult score 10–25, completed with t6 at score 38–50; secondaries start with s1 at primary moult score 10–35, completed from primary score 40 up to slightly later than primaries (over score 50); 1-year-olds start before adults (Holyoak 1974); average start of primary moult 20 June (adult, both sexes), 22 May (1-year-old ♂), or 24 May (1-year-old ♀), duration of primary moult 127 days; secondaries start on average 21 July, duration of moult 94 days; tertials 13 July, duration 69 days; tail 15 July, duration 58 days; in captive birds, p10 shed 143 days after p1, duration of growth of individual primaries *c.* 32 days (p1, p2, p3, p4, p10) to *c.* 45 days (p8, p9); innermost secondary (s6) shed 76 days after s1, duration of growth of each secondary 30–33 days; outer tail-feather (t6) shed 35 days after t1, duration of growth of each feather 37–39 days (Seel 1976, which see for details). In Leicestershire (England), adult moult starts late June, that of 1-year-olds 3 weeks earlier (Henderson 1991). In Netherlands, p1 shed from mid-May in 1-year-old, from mid-June in adult; moult completed mid-September to early October (*n* = 23, ZMA). In France, starts from late June, completed from 1st half of September (Mayaud 1933b). In eastern Germany, 1-year-olds start early June, adults from 2nd half of June; moult completed by late September to mid-October (Bährmann 1937, 1958). In small samples from Scandinavia, Yugoslavia, and Turkey, timing lies within samples above (RMNH, ZFMK, ZMA). In USSR, moult

PLATE 14. *Corvus frugilegus frugilegus* Rook (p. 151): **1** ad summer, **2** ad winter, **3** 1st ad summer, **4** juv.

*Corvus corone* Carrion Crow (p. 172). Nominate *corone*: **5** ad winter, **6** 1st ad summer, **7** juv. *C. c. cornix × corone*: **8–9** ad winter. *C. c. cornix*: **10** ad non-breeding, **11** juv. *C. c. sharpii*: **12** ad winter. *C. c. capellanus*: **13** ad winter, **14** juv.

*Corvus albus* Pied Crow (p. 195): **15** ad. (DR)

July–September (Dementiev and Gladkov 1954). For moult of *soemmerringii* in Mongolia, early July, see Piechocki and Bolod (1972); for Afghanistan, early August, see Paludan (1959). POST-JUVENILE. Partial: head, body, lesser upper wing-coverts (occasionally some juvenile coverts retained), highly variable number of median coverts, rarely some inner greater coverts (tertial coverts), and exceptionally one or a few tertials or tail-feathers. Starts 4–6 weeks after fledging, mainly from July, peaking August–September, duration 9–10 weeks (Mayaud 1933*b*; Bährmann 1958).

**Measurements.** ADULT, FIRST ADULT. *C. m. spermologus*. Britain, France, Belgium, and (mainly) Netherlands, breeding birds, whole year; skins (RMNH, ZMA). Juvenile wing and tail refer to those of 1st adult with retained juvenile flight-feathers and tail. Bill (S) to skull, bill (N) to distal corner of nostril; exposed culmen on average 4·0 mm less than bill (S). Depth is bill depth at middle of nostril.

| | | ♂ | | | ♀ | |
|---|---|---|---|---|---|---|
| WING AD | | 239·2 (5·19; 84) | 228–252 | | 230·7 (4·70; 96) | 220–242 |
| JUV | | 232·0 (4·67; 23) | 224–244 | | 224·4 (3·39; 20) | 217–231 |
| TAIL AD | | 130·1 (4·00; 80) | 123–139 | | 124·3 (3·69; 91) | 118–133 |
| JUV | | 124·2 (4·81; 24) | 116–135 | | 117·6 (2·74; 16) | 114–124 |
| BILL (S) | | 36·3 (1·86; 51) | 34·1–40·1 | | 34·5 (1·08; 52) | 32·5–36·5 |
| BILL (N) | | 23·1 (1·48; 56) | 20·9–25·4 | | 21·9 (0·99; 61) | 20·2–23·7 |
| DEPTH | | 13·9 (0·48; 42) | 13·2–14·8 | | 13·2 (0·44; 45) | 12·4–13·9 |
| TARSUS | | 46·2 (1·26; 53) | 43·8–48·1 | | 44·2 (1·16; 51) | 42·1–46·0 |

Sex differences significant. Adult wing mainly 233–247 (♂) or 222–238 (♀), bill (S) 34–38 (♂) or 33–35 (♀), bill depth 13·4–14·5 (♂) or 12·6–13·4 (♀), tarsus 45–48 (♂) or 43–45 (♀).

Various races. Wing; skins, unless otherwise stated. *C. m. spermologus*. Britain: (1) Leicestershire, adult, live or freshly dead, (2) skins (Henderson 1991, which see for other data). (3) France, adult (Mayaud 1933*b*). Nominate *monedula*. (4) Norway (Fjeldså 1972, which see for other data). (5) Scandinavia, adult (RMNH, ZMA). (6) Scandinavia, adult (Keve-Kleiner 1942). South of former East Germany: (7) adult, (8) 1st adult (Eck 1984); see also Bährmann 1976). (9) Czechoslovakia, freshly dead (Folk 1966, which see for other data; see also Hell and Soviš 1959). (10) Germany and Austria, (11) Carpathian basin, adult (Keve-Kleiner 1942). Intergrades between nominate *monedula* and *soemmerringii*. (12) South-east Poland, south-west Belorussiya, and north-west Ukraine, adult (Dunajewski 1938). Netherlands, winter: (13) adult, (14) 1st adult (ZMA; see also Voous 1960*a*). *C. m. soemmerringii*. (15) Balkans, Turkey, and USSR, adult (RMNH, ZMA). Makhedonija (southern Yugoslavia): (16) adult, (17) 1st adult (Stresemann 1920). (18) Turkey, ages combined (Jordans and Steinbacher 1948; Kumerloeve 1961, 1969*b*, 1970*a*). (19) Asia Minor to Turkmeniya, adult (Keve-Kleiner 1942). (20) Afghanistan to Tien Shan and Kashmir, adult (Lönnberg 1905; Keve-Kleiner 1942). Mongolian Altai, summer: (21) adult, (22) juvenile (Piechocki and Bolod 1972).

| | | ♂ | | | ♀ | |
|---|---|---|---|---|---|---|
| (1) | ♂ | 242 (3·9 ; 74) | 230–251 | ♀ | 231 (6·3 ; 67) | 220–244 |
| (2) | | 240 (3·6 ; 36) | 225–255 | | 231 (4·6 ; 39) | 215–245 |
| (3) | | 243 (6·14; 9) | 233–253 | | 234 (4·01; 13) | 227–238 |
| (4) | | 242·0 (7·3 ; 13) | 227–255 | | 228·3 (7·9 ; 16) | 217–241 |
| (5) | | 237·4 (5·34; 7) | 230–245 | | 231·3 (6·96; 12) | 219–240 |
| (6) | | 233·3 (4·36; 14) | 226–242 | | 224·7 (4·55; 6) | 220–230 |
| (7) | | 239·7 (4·81; 74) | 228–250 | | 230·6 (5·10; 69) | 219–243 |
| (8) | | 231·0 (7·30; 35) | 213–244 | | 223·9 (6·56; 34) | 207–235 |
| (9) | | 237·1 (5·35; 137) | 208–253 | | 228·8 (4·07; 117) | 205–247 |
| (10) | | 233·3 (6·36; 42) | 220–249 | | 222·8 (7·07; 23) | 211–235 |
| (11) | | 237·1 (6·84; 55) | 221–251 | | 227·8 (6·14; 32) | 215–242 |
| (12) | | 238·8 (4·29; 10) | 234–246 | | 229·7 (3·54; 9) | 225–235 |
| (13) | | 240·6 (6·77; 12) | 232–250 | | 229·6 (5·53; 17) | 217–237 |
| (14) | | 235·6 (3·94; 16) | 225–243 | | 227·2 (5·62; 14) | 218–238 |
| (15) | | 237·4 (6·72; 12) | 228–255 | | 229·6 (5·39; 15) | 220–239 |
| (16) | | 236·9 (5·50; 25) | 223–244 | | 229·9 (6·65; 17) | 221–243 |
| (17) | | 232·7 (4·37; 10) | 225–238 | | 223·7 (6·34; 16) | 215–237 |
| (18) | | 235·1 (7·20; 8) | 220–243 | | 221·1 (6·27; 5) | 211–228 |
| (19) | | 231·8 (7·01; 12) | 222–243 | | 226·4 (4·50; 7) | 219–231 |
| (20) | | 239·9 (5·38; 11) | 227–246 | | 236·0 (7·46; 6) | 230–250 |
| (21) | | 241·8 (6·98; 5) | 233–251 | | 229·4 (5·16; 7) | 220–236 |
| (22) | | 235 (— ; 8) | 230–237 | | 232 (— ; 4) | 228–239 |

Wing, sexes combined. *C. m. spermologus*. Spain, adult: 226·4 (8·88; 5) 218–240 (Keve-Kleiner 1942; ZFMK). Morocco, ages combined: 237·7 (9·71; 6) 227–255 (Keve-Kleiner 1942); 234·7 (5·72; 9) 227–245 (Vaurie 1954*b*). *C. m. soemmerringii*. Tien Shan mountains and Kashmir: 241·7 (8·71; 6) 233–255 (Stresemann 1920). Kashmir: 223–252 (*n* = 57) (Meinertzhagen 1926).

Other measurements, nominate *monedula* and *soemmerringii* combined; skins (RMNH, ZMA; samples 5, 13, 14, and 15 from above).

| | | ♂ | | | ♀ | |
|---|---|---|---|---|---|---|
| TAIL AD | | 128·8 (4·07; 27) | 122–138 | ♀ | 123·6 (4·36; 36) | 115–134 |
| JUV | | 124·6 (3·94; 20) | 118–132 | | 118·2 (5·04; 18) | 108–127 |
| BILL (S) | | 36·6 (1·45; 33) | 34·0–39·0 | | 34·7 (1·11; 46) | 32·4–37·2 |
| BILL (N) | | 22·8 (1·15; 40) | 20·6–24·5 | | 21·5 (0·93; 48) | 19·8–23·2 |
| DEPTH | | 13·4 (0·36; 40) | 12·8–14·0 | | 12·7 (0·53; 47) | 11·8–13·4 |
| TARSUS | | 46·1 (1·37; 33) | 44·2–49·0 | | 44·1 (1·32; 47) | 41·2–46·5 |

Sex differences significant.

*C. m. cirtensis*. Algeria, sexes combined, adult; skins (ZFMK).

| | | |
|---|---|---|
| WING | 229·3 (6·22; 6) | 222–236 |
| BILL (N) | 22·8 (1·00; 6) | 21·7–24·5 |
| TAIL | 124·1 (2·97; 6) | 119–127 |
| TARSUS | 44·4 (2·09; 6) | 42·3–46·7 |
| BILL (S) | 35·6 (1·61; 6) | 33·8–37·9 |

Wing: 225–244 (*n* = 23) (Meinertzhagen 1926).

NESTLING, JUVENILE. For growth of various parts of body, see Folk (1967*b*), Andriescu and Andriescu (1972), Lilja (1983), and Soler and Soler (1990).

**Weights.** ADULT, FIRST ADULT, JUVENILE. *C. m. spermologus*. Britain: (1) ages combined, August–December; adult, (2) January–March, (3) April, (4) May–July (Seel 1976, which see for details). (5) Leicestershire (Britain) (Henderson 1991, which see for details). Netherlands: (6) March–June, (7) October–January (ZMA). Nominate *monedula*. (8) South of former East Germany, adult, whole year (Bährmann 1972, which see for monthly variation; see also Eck 1984). (9) Czechoslovakia, whole year (Folk 1966, which see for details). (10) Hungary (Keve-Kleiner 1942). Intermediates between nominate *monedula* and *soemmerringii*. (11) Netherlands, November–February (ZMA). *C. m. soemmerringii*. (12) Greece and Turkey, whole year (Makatsch 1950; Kumerloeve 1969*b*, 1970*a*). (13) Kazakhstan (Korelov *et al.* 1974). Mongolia early July: (14) adult, (15) juvenile (Piechocki and Bolod 1972).

| | | ♂ | | | ♀ | |
|---|---|---|---|---|---|---|
| (1) | ♂ | 256 (16 ; 34) | — | ♀ | 232 (15 ; 42) | — |
| (2) | | 256 (18 ; 50) | — | | 231 (15 ; 54) | — |
| (3) | | 255 (19 ; 80) | — | | 243 (16 ; 90) | — |
| (4) | | 234 (16 ; 262) | — | | 215 (24 ; 212) | — |
| (5) | | 240 (17·2; 71) | 210–259 | | 222 (24·5; 67) | 179–240 |
| (6) | | 239 (18·9; 27) | 185–280 | | 230 (18·8; 22) | 199–250 |
| (7) | | 243 (16·9; 6) | 209–253 | | 245 (12·8; 12) | 209–250 |
| (8) | | 237 (18·4; 59) | 206–280 | | 219 (14·6; 45) | 186–243 |
| (9) | | 233 (20·4; 146) | 174–275 | | 225 (19·2; 116) | 175–271 |
| (10) | | 238 (— ; 27) | 213–259 | | 217 (— ; 16) | 191–247 |
| (11) | | 253 ( 7·8; 5) | 240–260 | | 246 (13·8; 18) | 211–282 |
| (12) | | 240 (17·6; 8) | 210–260 | | 235 (19·3; 4) | 225–264 |

(13)    230 ( — ;   6) 204–246    234 ( — ;   5) 207–260
(14)    216 (16·0;   5) 196–240    193 (10·2;   7) 179–208
(15)    213 ( — ;   8) 196–236    194 ( — ;   4) 180–210

Nominate *monedula*. Norway: ♂ 224–260 (7), ♀ 193–250 (7) (Haftorn 1971). Eastern Czechoslovakia, winter: 218 (23·7; 82) 150–260 (Hell and Soviš 1959).

*C. m. soemmerringii*. Eastern Germany, winter: ♂ 275·3 (3) 269–288 (Eck 1984). Israel: 190–200 (Paz 1987). Afghanistan, early August (mainly juvenile): 202·7 (12·37; 6) 184–217 (Paludan 1959). In Kirgiziya, ♂ up to 300 (Yanushevich *et al.* 1960).

NESTLING. On day 1, 9·9 (1·73; 73) 6–13; peak reached on day 24 (Strebel 1991, which see for growth curves). For growth curves and weights on various dates, see also Zimmermann (1951), Folk (1967*b*), Andriescu and Andriescu (1972), Kamiński and Konarzewski (1984), and Soler and Soler (1990).

**Structure.** Wing rather short, broad at base, tip bluntly pointed. 10 primaries: p8 longest (rarely, slightly shorter than p7), p9 13–19(–27) shorter, p10 73–91, p7 0–3, p6 10–17, p5 30–40, p4 46–55, p1 75–89; p10 (33–)39–50 longer than longest upper primary covert. See also Folk (1966) and Eck (1984). Outer web of p7–p8 emarginated, inner of p8–p9 with notch. Tip of longest tertial reaches to about tip of p1. Tail rather short, tip slightly rounded; 12 feathers, t6 8–14 shorter than t1. Bill rather short, stout; culmen *c.* 75% of head length; wide and deep at base, culmen and gonys gradually curving towards rather sharply pointed tip; tip laterally compressed. Both mandibles with dense layer of bristle-like feathers at base, projecting forward, covering *c.* 40% of base of upper mandible and *c.* 25% of lower; many bristles along upper side of gape. Tarsus and toes fairly strong, distinctly scutellated. Middle toe with claw 34·0 (12) 31–37 in nominate *monedula* and *soemmerringii*, 35·7 (56) 32–40 in *spermologus*; outer and inner toe with claw both *c.* 76% of middle with claw, hind *c.* 86%.

**Geographical variation.** Slight in colour, very slight in size. Variation in size notably small given wide distribution of species; bill of *spermologus* from western Europe slightly heavier and middle toe slightly longer than in other races (see Measurements and Structure); wing of populations of mountains in central Asia slightly larger (but difference too small to separate latter as *ultracollaris* Kleinschmidt, 1919, from eastern race *soemmerringii*. Variation in colour mainly involves depth of grey of rear of head and neck and of underparts, and presence and width of white crescent along side of neck at rear border of grey; also slight variation in colour of gloss on cap. As colour differences also strongly dependent on age, sex, and season, juveniles and 1st adults should not be used in assessment of races, and preferably only series of adults of same sex and in relatively fresh plumage should be compared. In general, body colour darker in north, in mountains, and in more humid climates, and paler in south, in lowlands, and in drier climates; grey of neck darker in more humid climates (Keve 1958*a*; C S Roselaar). Due to abrasion and bleaching, especially from about April onwards, grey of neck gradually becomes paler, more silvery, and mantle, scapulars, and underparts show gradually paler grey fringes; bleaching of grey of neck more pronounced in ♂ than in ♀ (sexual difference increasing) (Voipio 1968), especially in central and eastern races, and bleaching of neck and underparts strongest in ♂♂ from dry and warm climates, especially in populations occurring from Balkans through Turkey and Iran to Turkmeniya, where side of head and neck become silvery-white by June–July, hardly contrasting with white crescent on rear of neck. Nominate *monedula* from south-east Norway, southern

Sweden, and northern and eastern Denmark in general does not show white neck-crescent and thus belongs, with west European *spermologus*, to western group of races, not to eastern *soemmerringii*, which shows white crescent (Hartert 1903–10; Zedlitz 1921; Stresemann 1943*a*; Fjeldså 1972); inclusion of *soemmerringii* into nominate *monedula*, as advocated by (e.g.) many present-day Russian authors (Korelov *et al.* 1974; Stepanyan 1978, 1990) not warranted. Side of head and neck and hindneck of adult nominate *monedula* on average paler grey than in adult *spermologus* of same sex at same time of year; some birds have short, narrow pale grey crescent on side of neck, but usually not a truly white, broad, and contrasting half-collar; grey of side of head and neck of nominate *monedula* paler than fresh underparts (both almost equally dark in *spermologus*); underparts medium ash- or plumbeous-grey, less dark plumbeous-grey than in *spermologus*. Populations of south-west Norway and eastern Sweden variable, former ones in part darker than nominate *monedula*, latter more often with white crescent on neck, apparently because these areas relatively recently colonized by both nominate *monedula* and *spermologus* (in Norway) or *soemmerringii* (in Sweden) (Zedlitz 1921, 1925; Fjeldså 1972); settlers on Åland islands and along southern coast of Finland are nearer nominate *monedula*, those further north and in south-east of Finland nearer *soemmerringii* (Voipio 1969). In Jylland (Denmark), nominate *monedula* grades into *spermologus* on line between about Esbjerg and Haderslev (Voous 1950). Birds occurring in central Europe between *c.* 10° and *c.* 23°E, south to Austria, extreme northern Yugoslavia, and Carpathian basin sometimes separated as *turrium* C L Brehm, 1831 (Kleiner 1939*f*; Keve-Kleiner 1942; Voous 1950; Bährmann 1968*b*); side of head and neck and underparts paler grey than in *spermologus*, but neck without white crescent of *soemmerringii*; included in *spermologus* by Hartert (1903–10) and Vaurie (1954*b*, 1959), but very different from this race (Keve 1958*a*) and best included in nominate *monedula*, to which it is closely similar (Fjeldså 1972; Eck 1984; C S Roselaar); grades over wide area (north-west Germany, German Rheinland, Switzerland, eastern Netherlands to north-east France) into *spermologus* and in a similarly wide zone (eastern Poland, Latvia, western Belorussiya and Ukraine, eastern Rumania) into *soemmerringii* (Dunajewski 1938; Kleiner 1939*f*; Keve-Kleiner 1942). Within *spermologus*, populations from Britain, north-west France, and Morocco darkest, those of southern France and Iberia on average paler on neck and underparts (Keve-Kleiner 1942; Voous 1950), but difference mainly visible in worn plumage (fresh plumages more similar), and recognition of *ibericus* Kleiner, 1939, for Iberian populations and *nigerrimus* Kleiner, 1939, for Moroccan birds not warranted (Hartert and Steinbacher 1932–8; Vaurie 1954*b*, 1959). Italian birds rather pale, like Iberian ones, and occasionally with white crescent on neck (Bährmann 1968*b*), provisionally included in *spermologus* here, following Keve-Kleiner (1942). *C. m. cirtensis* from north-east Algeria and (formerly) Tunisia has underparts uniform dark ash-grey, about intermediate between *spermologus* and nominate *monedula*; grey of side of head and neck nearer to *spermologus*; gloss of cap bluish, more like some *soemmerringii*, not as purplish as most *spermologus*, no pale crescent on neck. *C. m. soemmerringii* differs from other races in presence of distinct white crescent or half-collar on rear of neck; grey of head pale, similar to or even paler than nominate *monedula*, grey of underparts rather pale in west and south of range (similar to nominate *monedula*), darker in east and in mountains in south-east (e.g. Kleinschmidt 1919, Hartert 1921–2, Keve-Kleiner 1942, Piechocki and Bolod 1972); more pronounced bleaching of grey of neck and of fringes on body due to abrasion in southern population is no reason to split off *collaris* Drummond, 1846, in Balkans and Greece (Stresemann

1920), or *pontocaspicus* Kleiner, 1939, in Cyprus, Turkey, Levant, northern Iraq and Iran, and Turkmeniya (Vaurie 1954*b*). Birds with partial or full white crescent regularly occur in western and central Europe in winter (e.g. Voous 1960*a*, Herroelen 1967, Bährmann 1968*b*, Erz 1968, Voipio 1969, Jung 1975, Wille 1983, Eck 1984); these birds apparently mainly nominate *monedula* or populations from intergradation zone between

*soemmerringii* and nominate *monedula* or '*turrium*', with only a few true *soemmerringii* among them (C S Roselaar).

Forms superspecies with Daurian Jackdaw *C. dauuricus*. For further relationships, see Horváth (1977) and Goodwin (1986). For possible distributional history since last glaciation, see Voous (1950) and Fjeldså (1972). CSR

## *Corvus dauuricus* Daurian Jackdaw

PLATES 8 and 9 (flight)
[between pages 136 and 137]

Du. Daurische Kauw    Fr. Choucas de Daourie    Ge. Elsterdohle
Ru. Даурская галка    Sp. Grajilla Daurica    Sw. Klippkaja

*Corvus dauuricus* Pallas, 1776

Monotypic

**Field characters.** 33–34 cm; wing-span 67–74 cm. Size and structure as Jackdaw *C. monedula* but with jowl under chin. Small, dapper, even elegant crow, with general character and behaviour much as *C. monedula*. Typical adult mainly black with whitish collar and underbody, suggesting pattern of 'Hooded Crow' races of Carrion Crow *C. corone*. Juvenile similar but duller, though subsequent 1st-year plumage mainly black. Sexes similar; no seasonal variation.

ADULT. Complete moult, June–October. All-black adults are reported (see Plumages), but normally jet-black (showing purple and green gloss at close range) with striking deep grey-white to cream-white hindneck, neck-sides, breast, and belly; side of head behind eye either all-black or black streaked whitish. Underwing dull black, with silvery tone on undersurface of flight-feathers. Eye dark, no paler than brown (unlike adult *C. monedula*). Bill and legs black. JUVENILE. Much duller than adult, with greyer or fawn cast to collar and underbody and little gloss and much mottling on duller black plumage. Eye no paler than greyish-brown. FIRST-WINTER. Surprisingly, assumes virtually all-dark plumage from August, and thus liable to confusion with *C. monedula*. Mainly black, lacking full gloss of adult but darker than western races of *C. monedula*; often shows greyish-streaked ear-coverts and nape, and broad black bib of adult often apparent due to greyer surround or mottled underparts.

Once size clearly established, pied adult and juvenile unmistakable. Note that some eastern Hooded Crows can, with wear, be as pale as duller individuals of *C. dauuricus*, but show pale back lacking in that species. Dark 1st-year potentially troublesome, closely approaching appearance of juvenile and dull adult *C. monedula spermologus*, but in close view *C. dauuricus* shows diagnostic combination of dark eye, black bib, and only limited pale area on hindneck (not extending up to rear crown). Flight, gait, and beha-

viour said not to differ from *C. monedula*, but (perhaps due to plumage pattern) pied bird looks more portly and less bustling on ground.

Voice said to resemble *C. monedula*, but less cackling and lower pitched, recalling *C. corone*, e.g. 'kaah', not 'kya'.

**Habitat.** Breeds extralimitally as eastern counterpart of Jackdaw *C. monedula* in equivalent east Asian habitats, in hollow trees along river valleys, scattered in fields, or on river islands, and in mountainous Altai region on rock faces with access to meadows for foraging, up to 2000 m. Normally inhabits only open woodlands and clearings, avoiding dense forest and open steppe areas; also lives in open and hilly country and in cultivated areas and pastures with some trees, and in winter feeds on stubble fields. For roosting favours cliffs and old buildings, and even town trees; will build nests in branches of trees, but prefers holes in them where available. Inadequate data available point to no significant ecological distinction from *C. monedula*. (Dementiev and Gladkov 1954; Goodwin 1986; see also Nechaev 1975.)

**Distribution.** Breeds from *c*. 96°E in southern Siberia east to Amurland and Ussuriland, south through Mongolia and Manchuria to northern and western China. Winters south to Russian Turkestan, Korea, Japan, and southern China.

Accidental. Sweden: Umeå, 26 April 1985 (Olsson 1988). Finland: Nykarleby, 1 May 1883, shot (Holmström 1959–62).

**Movements.** Resident to migratory. No information on wintering area or distance of movement of particular individuals or populations. Data suggest adults and juveniles tend to winter in different areas, juveniles moving further south than adults (Nechaev 1975). Migrates in flocks, often with Rook *C. frugilegus*.

Northern breeding areas mostly vacated, though some birds remain at least in mild winters. Thus at Zeya in Amur region (Russia, 53°48′N), *c.* 40 birds successfully wintered in one year (Il'yashenko 1986); some birds remain to winter near habitations in south-west Ussuriland (extreme south-east Russia) (Nechaev 1975) and Mongolia (Kozlova 1933); in Manchuria (north-east China), many winter in neighbourhood of Harbin (*c.* 46°N) in mild years (Piechocki 1959). In China, widespread in north and west in summer, and in most regions except extreme south in winter (Vaurie 1959; Nechaev 1975; Schauensee 1984); extent of movement of Chinese populations not known, but perhaps birds wintering in east originate from colder areas to both north and west. Common in Hopeh province (north-east China) in winter (Wilder and Hubbard 1938) and 'abundant' in lower Yangtse valley (eastern China) (Gee and Moffett 1917); resident in south-east Yunnan (southern China) (La Touche 1923). Vagrant to Taiwan (Schauensee 1984), and 1 record from Hong Kong 30 November 1986 (Tipper 1987). Present in Tibet all year (Schäfer 1938). In Kazakhstan (west of breeding range), irregular in autumn and winter in north-east (Zaysan and Alakol' depressions, and Altai) (Korelov *et al.* 1974). Abundant passage migrant in northern Korea, and common winter visitor to south (Austin 1948; Gore and Won 1971); previously rare in Japan, but since 1953 has wintered regularly in small (sometimes quite large) numbers among *C. frugilegus* flocks in Kyushu (southern Japan), especially in extreme south (Brazil 1991).

Following post-breeding dispersal, birds leave south-west Ussuriland 2nd half of September and October, with isolated groups reported to mid-November when snow already falling; main heading south-west and west, birds migrating in flocks through open country and occurring only very rarely as far east as shore of Sea of Japan in both spring and autumn (Nechaev 1975). At Zeya, latest record 8 October over several years (Il'yashenko 1986). Timing similar in Mongolia; in eastern Hangay (central Mongolia) in one year, at least 2876 birds overflew 2 sites from 27 September to 9 October, with strong passage continuing in October (peak not known) (Mey 1988). In north-east China, earliest records at Beidaihe (Hopeh province) 25 September to 24 October over 4 years, migration continuing to about mid-November; at Peking, earliest records 28 September to 13 October over 17 years. Both adults and juveniles common on passage, but wintering birds at Peking chiefly adults. (Wilder and Hubbard 1924; Hemmingsen and Guildal 1968.)

Spring migration begins early. In Hopeh province, passage from end February or early March, continuing throughout March, with fewer (chiefly 1st-years) in April, occasionally May (La Touche 1920; Hemmingsen and Guildal 1968). Arrives north of wintering grounds in Korea from mid-February with intensive passage until mid-March, but suddenly absent thereafter (Austin 1948). Spring passage at Ulan Bator (northern Mongolia) begins mid-March, often involving flocks of 300–400 individuals (Kozlova 1933). Reaches breeding grounds in Ussuriland from mid- or late February, movement continuing throughout March and early April; adults arrive before 1st-years (Nechaev 1975). Earliest records 8–22 March over 10 years in southern Amur region (Nechaev 1975), 7 March over several years at Zeya (Il'yashenko 1986). Returns to breeding grounds in Chita region (east of Lake Baykal, Russia) in 2nd week of March (Dementiev and Gladkov 1954).                                                   DFV

**Voice.** See Field Characters.

**Plumages.** ADULT. Nasal bristles dull black with narrow grey shaft-streaks. Forehead and crown deep black with purple gloss. Lore and upper cheek deep black, slightly glossed bluish-purple. Dimorphic with regard to remainder of side of head: (1) in dark-headed morph (38% of 26 ♂♂, 56% of 23 ♀♀), side of head behind eye and rear of crown purple-black with some dark grey mixed on feather-tips, ear-coverts blackish with slight grey wash, entire head appearing either blackish with slightly greyer ear-coverts or all-black; (2) in silver-eared morph (62% of 26 ♂♂, 44% of 23 ♀♀), ground-colour of side of head as above, but each feather with narrow contrasting silvery-white or blue-white tip, side of head behind eye (including ear-coverts) and rear of crown appearing closely streaked white on black ground. Contrastingly white collar on rear and side of neck. Upperparts from mantle backward as well as entire tail and wing deep black, glossed metallic purple-blue; however, gloss more purplish or violet-purple on outer webs of secondaries and greater upper wing-coverts, slightly less intense and more greenish-blue on tail, tertials, primaries, and greater upper primary coverts; underwing and axillaries duller black, underside of flight-feathers and greater under wing-coverts tinged silvery-grey. Chin and throat deep black with greenish-blue (chin) or purplish (throat) gloss, forming large patch ending in broad round-ended bib on upper chest, strongly contrasting with white side of chest, breast, belly, and flank; upper flank (below wing) and lower mid-belly partly suffused grey. Thigh, rear vent, and under tail-coverts black with slight purple gloss; tips of under tail-coverts fringed grey. *In fresh plumage* (autumn and winter), white of collar and underparts sometimes suffused greyish, pinkish, or cream (independent of sex). *In worn plumage* (late spring and early summer), nasal bristles browner; cap and upperparts from mantle backwards sometimes slightly duller and more bluish-black; in silver-eared morph, white feather-tips on side of head and on rear of hind-crown partly worn away, appearing speckled grey-white rather than streaked silvery-white, but not as uniformly dark as dark-headed morph; some grey of feather-bases showing through in white of collar and underparts, and white itself sometimes contaminated by soil, tinged partly ochre or isabelline. Also, some individual variation, a few birds (especially at western end of breeding range) showing dark grey spots or streaks on hindneck (sometimes visible only when plumage worn) and more extensive grey suffusion on flank and lower belly, perhaps pointing to gene flow with Jackdaw *C. monedula*. A probable true hybrid with *C. monedula* from Issyk-kul' (Kirgiziya) (ZFMK), had white of body replaced by grey, hindneck partly blackish, and ear-coverts dark grey without conspicuous white spots or streaks; for appearance of other hybrids from Sinkiang (Turfan depression), Tien Shan, Altai, and Kansk and Irkutsk areas, see Nechaev (1975).

An all-black morph ('*neglectus*', similar to 1st adult, below) reported by Svensson (1984*b*) from Korea. A bird from Kansu (China) was intermediate: white replaced by blackish-grey on breast and belly, but rear and side of neck and side of breast white (Eck 1984; SMTD); a bird from Lake Baykal (RMNH) rather similar to this Kansu bird, but ear-coverts mainly medium grey, and iris brown, and this bird perhaps a hybrid with *C. monedula*. No true black-morph adults encountered among birds examined (RMNH, STD, ZFMK, ZMA), nor found by Sushkin (1913), by Kleinschmidt (1935), by Piechocki and Bolod (1972) in large sample from Mongolia, by Nechaev (1975) in USSR, by Goodwin (1976) in many specimens in BMNH, or by Jollie (1985) in many specimens in American Museum of Natural History. In sample of breeding birds from Manchuria, 25% all-black, remainder pied (Meise 1934*a*); in Yunnan (China), 43% of 14 'adults' black, 57% pied (La Touche 1923), but these black birds may have been 1-year-olds; in Mongolia, all adults which were feeding young were pied (Mauersberger *et al.* 1982). JUVENILE. Rather like adult, and thus pied, but body greyer (not as black as 1st adult). Head like adult, but uniform sooty-black, virtually without gloss, side of head slightly mottled grey, throat with broad grey feather-bases and narrow grey feather-tips, appearing heavily mottled, lower edge of dull black on upper chest ill-defined. Collar and lower chest to belly rather variable dirty light grey, light fawn-grey, or greyish-black and mottled with paler grey feather-tips, never fully blackish. Mantle to upper tail-coverts, vent, under tail-coverts, and lesser and median upper wing-coverts sooty-black, faintly glossy on scapulars only. Feathering of body markedly short and loose. Tail, flight-feathers, tertials, and greater wing-coverts as adult, but see 1st adult (below). According to Sushkin (1913), an all-black morph occurs, but none present in small samples examined by La Touche (1923), Piechocki and Bolod (1972), Nechaev (1975), Goodwin (1986), and Jollie (1985), or in RMNH, SMTD, or ZFMK; Kozlova (1933) recorded pied adults feeding pied and black young, but latter may have been just-moulted 1st adult (C S Roselaar). FIRST ADULT. Markedly different from adult and juvenile in being mainly black; however, black morph adults and juveniles occasionally reported (see above), and pied morph 1st adults occasionally occur according to Meise (1934*a*); of 3 pied juveniles in Yunnan, 21–22 August, 1 moulted to black plumage, 2 apparently to pied grey plumage (La Touche 1923). In dark-headed morph (see Adult), plumage similar to dark-headed adult, but white of body replaced by dark sooty-grey or plumbeous-black, slightly paler grey only along rear border of black ear-coverts and along broad black bib on upper chest and sometimes on fringes of feathers of breast and belly (latter then often blotched darker and paler grey). This morph darker than 1st adult of dark races of *C. monedula*, with ear-coverts not grey but black (sometimes faintly streaked grey), neck darker grey with palest part at front (not rear) of neck-side, and throat and upper chest with broad black round-ended bib, often contrasting in colour with greyer side of chest and breast. Silver-eared morph similar to dark-headed morph, but ear-coverts and patch from above eye to nape finely streaked silvery-white or grey-white on black or greyish-black ground (white more prone to wear off than in adult, with heavily worn 1st-summer birds sometimes as black as black-headed morph). In both morphs, black of cap, throat, mantle to upper tail-coverts, tail, and wing slightly duller and less glossy than in adult, gloss more bluish, less purplish. Collar and underparts either black, feathers tipped grey, appearing scaly or mottled ('*griseiventris*'), or black with largely concealed grey or white feather-bases ('*nigriventris*'). In Mongolia, 77% of 31 1st adults were uniform black below, 23% mottled grey and black (Piechocki and Bolod 1972). Tail-feathers narrower than in adult, tip more rounded, less broad and square, more heavily worn than adult at same time of year; tertials and primaries narrowing slightly towards tip, less broad and rounded, in spring and summer in particular browner, less glossy, and more worn than in adult. SECOND ADULT. Like adult, but white of body less pure, more greyish-white (Nechaev 1975).

**Bare parts.** ADULT. Iris dark brown. Bill, leg, and foot black. JUVENILE. Iris grey-brown. Bill, leg, and foot black. FIRST ADULT. Iris medium brown. (Piechocki and Bolod 1972; RMNH, ZFMK.)

**Moults.** ADULT POST-BREEDING. Complete; primaries descendent. In Mongolia, adults not yet moulting late May ($n=4$) or 12 June ($n=8$), but one from 30 June had started (with innermost primaries) and moult advanced (primary score *c.* 31, secondaries, body, and head in full moult) in ♀ from 23 August; in 1-year-olds, 2 from 26 May had replaced black feathers on belly by adult whitish ones, 17 from 12 June had inner 3 primaries new or growing, and ♂ from 6 July had primary score *c.* 17 (Piechocki and Bolod 1972). In Altai (central Asia), moult intense August (Dementiev and Gladkov 1954); p1 shed on 11 June in ♀ from Transbaykalia (RMNH). In China, ♂ from Manchuria had not started on 7 June (SMTD), 2 ♂♂ nearing completion on 4–11 September (Piechocki 1959), single ♀ almost finished on 30 October (Meise 1934*a*); in Szechwan, score 1 in 1-year-old ♀ from 7 June, score 2 in adult ♂ from 14 June, scores (many wing-coverts, tertials, and tail-feathers just shed or growing), score 29 in 1-year-old ♀ from 20 August (SMTD, ZFMK); in Shensi, early November, one almost completed (score 46), 3 completed (score 50) (RMNH, SMTD, ZFMK). See also Nechaev (1975). POST-JUVENILE. Partial: head, body, and lesser and median wing-coverts; rarely some tail-feathers or tertials (Jollie 1985), never flight-feathers or greater coverts. Fully juvenile birds encountered up to late August (ZFMK); in Mongolia, 6 birds from 23 August had replaced pale greyish juvenile feathers by blackish 1st adult (Piechocki and Bolod 1972); in Manchuria, moult nearing completion in early September (Piechocki 1959); moult completed in birds from 20 September, October, and later (RMNH, ZFMK).

**Measurements.** Russia and eastern China south to eastern Sinkiang, Shensi, and Kiangsu, all year; skins (RMNH, SMTD, ZMA). Juvenile wing and tail include retained juvenile wing and tail of 1st adult. Bill (S) to skull, bill (N) to distal corner of nostril; exposed culmen on average 3·9 less than bill (S).

| | ♂ | ♀ |
|---|---|---|
| WING AD | 233·0 (7·35; 14) 220–244 | 223·0 ( 6·88; 12) 211–232 |
| JUV | 230·9 (6·61; 6) 222–238 | 221·7 (10·46; 7) 209–233 |
| TAIL AD | 128·5 (4·14; 14) 123–134 | 119·1 ( 4·15; 12) 114–127 |
| JUV | 123·0 (3·74; 4) 118–127 | 119·0 ( 6·17; 7) 111–128 |
| BILL (S) | 34·8 (1·41; 16) 32·2–37·0 | 32·9 ( 1·63; 12) 30·2–35·4 |
| BILL (N) | 21·4 (1·30; 16) 19·2–23·5 | 20·2 ( 1·18; 10) 19·4–22·1 |
| TARSUS | 45·7 (1·39; 17) 43·1–47·2 | 44·7 ( 1·98; 11) 41·8–47·3 |

Sex differences significant for adult wing, adult tail, and bill.

Wing. Mongolia: (1) adult, (2) juvenile and 1st adult (Piechocki and Bolod 1972); (3) USSR (Dementiev and Gladkov 1954); Mongolia, Manchuria, and Kansu (China), (4) adult, (5) juvenile (Lönnberg 1909; Meise 1934*a*; Stresemann *et al.* 1937; Piechocki 1959); (6) south-east Yunnan (China), mainly winter (La Touche 1923); (7) Szechwan, adult, April–August (SMTD, ZFMK, ZMS); (8) Tibet (Vaurie 1972).

| | ♂ | | ♀ | |
|---|---|---|---|---|
| (1) | 241 ( — ; 12) 234–249 | | 229 ( — ; 10) 225–233 | |
| (2) | 234 ( — ; 20) 210–240 | | 227 ( — ; 16) 220–232 | |
| (3) | 227·2 ( — ; 14) 213–238 | | 221·2 ( — ; 13) 213–231 | |

| (4) | 229·8 (8·10; 8) 219–241 | 229·2 (11·5; 5) 217–246 |
| (5) | 225 ( — ; 1) — | 209·4 (10·8; 5) 195–225 |
| (6) | 232·7 (8·45; 6) 219–241 | 222·3 (4·33; 9) 216–228 |
| (7) | 234·1 (4·65; 8) 227–240 | 225·2 (5·99; 6) 214–230 |
| (8) | 247·4 ( — ; 10) 242–258 | 241·0 ( — ; 7) 232–250 |

**Weights.** Mongolia: (1) adult, late April to August (mainly June); (2) 1st adult (1 year old), late May to early July (mainly early June), (3) juvenile, 23 August (Piechocki and Bolod 1972). (4) Manchuria, July–September (Piechocki 1959). USSR: (5) adult, (6) 1st adult (Nechaev 1975).

| (1) | ♂ 218 (12) 208–230 | ♀ 191 (10) 175–208 |
| (2) | 203 (18) 185–214 | 185 (12) 173–206 |
| (3) | 236 ( 2) 222–250 | 202 ( 4) 182–214 |
| (4) | 211 ( 2) 210–212 | 190 ( 3) 175–210 |
| (5) | 214 (11) 197–235 | 220 ( 2) 205–235 |
| (6) | 207 (11) 194–225 | 199 ( 6) 187–206 |

**Structure.** 10 primaries: p8 longest, p9 12–16 shorter, p7 0–2, p6 6–16, p5 25–40, p4 41–56, p1 75–92; p10 somewhat reduced, 73–85 shorter than p8, 38–48 longer than longest upper primary covert ($n = 10$ adults: RMNH, ZMA). Of 31 adults, 77% had p10 shorter than or equal to p1 (28% in 134 adult *C. monedula*), 19% had p10 between p1 and p2 (66% in *C. monedula*), 3% had p10 longer than p2 (7% in *C. monedula*); little difference between adults of the 2 species in position of p9 relative to p6 (contra Vaurie 1954b), and little difference in wing formula between juveniles of the species (in both, p10 of juveniles is relatively longer than in adult) (Eck 1984, which see for details). Feathering of neck and side of head normal, not long and silky as in *C. monedula*. Depth of bill at middle of nostril 13·0 (6) 12·5–

13·8 in ♂, 12·0 (5) 11·5–12·8 in ♀. Middle toe with claw 35·1 (5) 32–37; outer toe with claw *c.* 74% of middle toe with claw, inner *c.* 71%, hind *c.* 81%. Remainder of structure as in *C. monedula*.

**Geographical variation.** Slight, involving size only. Birds from eastern Himalayas (Yunnan north to Tsinghai, China) slightly larger and perhaps separable as *khamensis* Bianchi, 1906. Wing of Tibetan birds is indeed long (adult ♂ mainly over 245, adult ♀ over 235, elsewhere in geographical range mainly below this) but not of those of Szechwan (see Measurements); situation obscured in winter, when northern birds occur in range of 'khamensis'. Further study needed.

Forms superspecies with Jackdaw *C. monedula*, and sometimes united in single species (e.g. by Hartert and Steinbacher 1932–8, Meise 1934a, Johansen 1944, and Dementiev and Gladkov 1954) because of close similarity in size and structure and because geographical ranges largely complementary. Kept separate here, however, as hybrids in rather narrow overlap zone are infrequent (see Plumages and Nechaev 1975) and difference in adult body and iris colour apparently largely prevents mixed pairings (but see Piechocki *et al.* 1982); also, usual plumage sequence (pied juvenile, black 1st adult, and pied adult) unique among *Corvus*. For probable evolution of both species, see Nechaev (1975).

**Recognition.** Dark birds not always easy to separate from *C. monedula*, especially because of great individual variation in colour among *C. dauuricus*. Even in darkest *C. dauuricus*, iris dark brown and chin to broad bib on upper chest deep black, contrasting slightly with greyer-black of breast and side of chest (in *C. monedula*, iris whitish and chin plumbeous-black merging into dark grey of throat, chest, and breast, with no large contrastingly darker patch). CSR

---

## *Corvus splendens* **House Crow**

PLATES 8 and 15 (flight)
[between pages 136 and 137, and facing p. 280]

Du. Huiskraai    Fr. Corbeau familier    Ge. Glanzkrähe
Ru. Блестящий ворон    Sp. Corneja India    Sw. Huskråka

*Corvus splendens* Vieillot, 1817

Polytypic. *C. s. zugmayeri* Laubmann, 1913, south-east Iran through Pakistan to north-west India (northern Punjab, Jammu, and southern Kashmir); nominate *splendens* Vieillot, 1817, India (except north-west and south-west), Nepal, and Bangladesh, both this race and *zugmayeri* introduced or self-established in many coastal localities in eastern Africa and along Red Sea and Persian Gulf, including Egypt, Israel, Joradn, and Kuwait, populations here often forming variable mixture between both races. Extralimital: *protegatus* Madarász, 1904, south-west India (Kerala), Sri Lanka, and Maldive and Laccadive Islands, grading into nominate *splendens* in north-east Kerala and southern Tamil Nadu; *insolens* Hume, 1874, Burma south to Tenasserim; these races introduced widely in coastal localities of south-east Asia.

**Field characters.** 41–43 cm; wing-span 76–85 cm. 10% smaller than Carrion Crow *C. corone*; has proportionately longer bill with deeper and more curved upper mandible, more domed crown, and (in some attitudes) longer neck and legs. Quite large attenuated crow, lacking bulk of common large European black crows and having distinctive bill and head profile. Plumage suggests hybrid between black and hooded forms of *C. corone*: black on front of head and throat, abruptly grey on nape, neck, and chest, slate on underbody, and black on back, wings, and tail. Sexes similar; no seasonal variation. Juvenile not studied in the field (see Plumages). Racial attribution of west Palearctic birds uncertain (see Geographical Variation).

ADULT. Moults: June–October (complete). Forecrown to behind eye, face, chin, wings, and tail black, with blue gloss particularly on forewing. Rear crown, neck, shoulders, and breast pale to smoky-grey, merging with glossy black back and dull slate underparts. Looks dull and

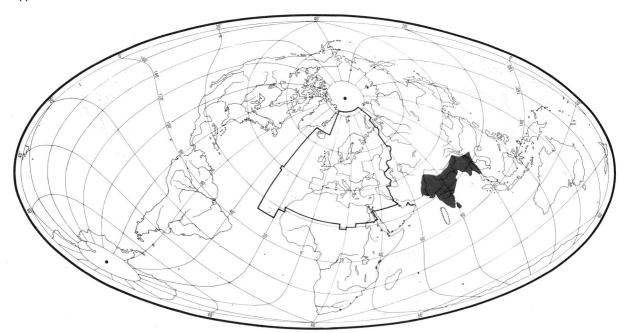

scrawny in worn late summer plumage. Bill and legs black. Note that this description, based on those of Mei-nertzhagen (1954) and Hollom *et al.* (1988), likely to be of both Pakistan race *zugmayeri* and Indian race, nominate *splendens*; some eastern forms are much more uniformly black.

Typical bird unmistakable, but beware confusion with *C. corone sardonius* where ranges overlap in northern Egypt. Plumage also recalls Jackdaw *C. monedula*, but that species over 20% smaller, with short bill and legs. Flight powerful, with silhouette featuring long bill, narrow head, and long, round-cornered tail. Gait as *C. corone* but appears more sprightly due to longer thighs. Thieving scavenger of human habitations, largely shunning natural habitats. Always alert and wary; cowardly when alone.

Commonest call 'kraar'. Voice described by Hollom *et al.* (1988) as quieter and higher pitched than *C. corone*, but can be lower pitched.

**Habitat.** Original range lies in Indian subcontinent, in subtropical and tropical lowlands, also in hills up to some-what more than 2000 m, but does not ascend from Indus valley to highland Baluchistan within that limit, and has established only precarious foothold in Vale of Kashmir. Presence of some trees probably essential; roosts com-munally in mangroves, banyans, coconuts, and in tree plantations, often reached by long high-level flights. Has become intricately enmeshed with human activities in urban and even metropolitan areas, and to lesser extent in small hill-stations where opportunities for easy scavenging exist. Attracted by human feeding activities, including cooking smoke. In Sri Lanka recorded as favouring home-steads and copra drying yards along coastal backwaters.

The longest-established and most complete case of adap-tation by Corvidae from natural to man-made habitats: now nearly always confined to human neighbourhoods, and generally successful in competition with other town scav-engers. (Bates and Lowther 1952; Ali and Ripley 1972; Goodwin 1986.) At Elat (Israel), inhabits low-lying desert areas, mainly coastal near human settlements with scattered patches of trees (Shirihai in press); in Egypt, towns and villages, breeding in trees of parks and gardens (Goodman and Meininger 1989). Can be a serious pest, e.g. in Aden (Yemen) (Jennings 1992).

**Distribution.** In west Palearctic, confined to a few ports and other areas along major shipping routes, where occur-rence due to deliberate introductions by man and self-introductions by ship transportation (Meininger *et al.* 1980; Bijlsma and Meininger 1984). For review of spread, see Lever (1987).

ISRAEL. First recorded 1976, at Elat (Shirihai in press). JORDAN. First recorded 1979, at Aqaba (IJA). EGYPT. Established at Suez in or before 1922, thereafter spreading to other parts of Suez Canal area, Red Sea coast, and probably Sinai (Goodman and Meininger 1989). KUWAIT. First recorded 1972, but birds recorded as hooded form of Carrion Crow *C. corone* in 1957-8 may have been this species. Now recorded annually, chiefly April; bred 1983-4. (Pilcher 1986.)

Accidental. Spain: Gibraltar, 6 March to 5 April 1991, coinciding with return of warships from Arabian Gulf (*Br. Birds* 1992, **85**, 14).

**Population.** ISRAEL. Up to 12 birds at Elat 1976-87. Increased from 1988, with 32 birds February 1990. (Shi-

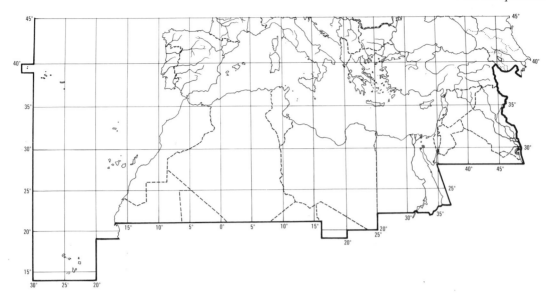

rihai in press). JORDAN. Up to 35 birds in Aqaba town (IJA). EGYPT. Estimated 800–850 birds at Suez 1981 (Bijlsma and Meininger 1984).

**Movements.** Almost entirely sedentary. In India, subject to altitudinal movements in cold northern areas (Ali and Ripley 1972); otherwise, no reports of seasonal movements. Outside natural range (southern Iran to central Thailand) occurs locally in coastal towns and villages, particularly port areas (west to Egypt, south to South Africa), with some dispersal inland, due chiefly to deliberate introductions by man and self-introductions by ship-transportation (Meininger *et al.* 1980; Bijlsma and Meininger 1984; Pilcher 1986; Ash 1988): e.g. birds were brought to Zanzibar from Bombay (India) in 1890s to scavenge on refuse (Vaughan 1930); in November 1950, 4 birds blown onto ship at Colombo (Sri Lanka) flew off inland at Cape Guardafui (Somalia) (Davis 1951); similarly, several reports of ship-borne birds from Sri Lanka reaching Australia, though not established there (Meininger *et al.* 1980); population on Malay peninsula derives from both methods of introduction (Medway and Wells 1976).                                                              DFV

**Food.** Almost all information extralimital. Very dependent on man's rubbish, scraps, offal, and sewage, otherwise any edible invertebrates, small vertebrates, plant material, or carrion. Feeds mostly on ground, but also in trees and on buildings (Bijlsma and Meininger 1984; Goodwin 1986; Berruti and Nichols 1991). Forages mainly at rubbish tips, abattoirs, markets, farms, beaches near fisheries or tourist resorts, etc., travelling up to 16 km from roost to feeding place; carrion taken includes human corpses in India (Jerdon 1877; Ali and Ripley 1972; Tyler 1980; Ryall and Reid 1987; Ash 1988; Feare and Mungroo 1989). Although cautious, will take any opportunity to scavenge or steal food at human habitations, even entering houses to take it from tables, making it a major pest species and health risk over much of its range (Meininger *et al.* 1980; Ryall 1986; Feare and Mungroo 1989; Berruti and Nichols 1991). Will forage only near houses, even when a preferred food (e.g. maize *Zea*) is available in fields nearby (Ali *et al.* 1982). In India, does great damage in fields of maize, *Sorghum*, oats *Avena*, etc., pulling off heads of grain and digging up or pulling out seeds and seedlings (Mason and Maxwell-Lefroy 1912). In Egypt, remains wary of man, often preferring to feed in early morning away from town centres; apparently eats enough then to last for rest of day since little further foraging observed (Bijlsma and Meininger 1984); will also feed at night where there is illumination (Ryall and Reid 1987; Jennings 1988a). Feeds on backs of large domestic animals taking parasites, and around their feet with other birds such as Cattle Egret *Bubulcus ibis*, also robbing these birds of their prey (Ali and Ripley 1972; Gallagher and Woodcock 1980; Tyler 1980; Ryall and Reid 1987). Will hover over water to take aquatic insects, plunging in from *c.* 5 m to seize them with bill (Dodsworth 1911), and can snatch fish from water with bill while in flight (Acharya 1953). Makes 'ungainly' hawking sallies after flying insects, e.g. termites (Isoptera) or locusts (Orthoptera: Acrididae) (Mason and Maxwell-Lefroy 1912; Ali and Ripley 1972; Bijlsma and Meininger 1984). Can hang upside-down on thin branches to reach fruit (Tyler 1980). Kills young sheep, goats, calves and domestic cats, and attacks monkeys and humans (Ryall 1986; Ryall and Reid 1987); repeatedly picked up gerbil *Meriones* in bill, shook it and dropped it again, finally carried it to open space, pecked it until dead, flew with it to post, and pulled it apart while held under foot (Fitzwater 1967). See Sengupta (1969) for prolonged chase of palm squirrel *Funambulus* in tree, and Thiede and Thiede (1974) for description of birds being killed and eaten in

Nepal. Can have severe local effect on other bird populations and seems inordinately aggressive towards great
variety of other species, including swimming waterfowl
(Gallagher 1989*b*; R P Martins); very destructive of colonies of herons (Ardeidae) and weavers *Ploceus* in India,
pulling up hanging nests to get at chicks (Ali and Ripley
1972; Ryall and Reid 1987), and Mombasa Island (Kenya)
'almost devoid' of native species because of its nest-
predation and mobbing of raptors and other large birds,
often marauding in large groups (Ryall 1986; Ryall and
Reid 1987). Robs nests of Feral Pigeon *Columba livia* in
Israel (Shirihai in press). Recorded storing food (crab and
fish fragments) at base of coconut cluster, some of which
fed to young by regurgitation, remainder replaced (Angwin 1977); replaced food plus covering stones and dry
leaves after taking some from cache under railway sleeper
(Biddulph 1954).

Diet includes the following. Vertebrates: bats (Chiroptera, including fruit bats), mice, squirrels, and gerbils
(Rodentia, including gerbil *Meriones*), young sheep and
goats, frogs, lizards (including geckos Gekkonidae and
skinks Scincidae), fish, birds of all kinds including poultry
(eggs, young, and adults). Invertebrates: grasshoppers,
etc. (Orthoptera, including locusts Acrididae), Lepidoptera larvae, termites (Isoptera), beetles (Coleoptera),
Hymenoptera, spiders (Araneae), crabs, etc. (Decapoda),
millipedes (Diplopoda), centipedes (Chilopoda), earthworms (Lumbricidae), Mollusca. Seeds, nuts, fruit, and
nectar of any edible plants; all kinds of carrion and refuse.
(Mason and Maxwell-Lefroy 1912; Fitzwater 1967; Ali
and Ripley 1972; Tyler 1980; Ali *et al.* 1982; Ryall 1986;
Ryall and Reid 1987; Feare and Mungroo 1989; Gallagher
1989*b*; Berruti and Nichols 1991.)

In Bihar, north-east India, 42 adult stomachs contained,
among other material, 5 frogs, 1 lizard, 226 insects, including Orthoptera (Gryllotalpidae, *Schizodactylus*, *Chro-
togonus*, *Brachytrypes*), termites, Lepidoptera (Noctuidae
larvae), ants (Formicidae), beetles (mainly Scarabaeidae
and Tenebrionidae), millipedes, centipedes, and plant
material (Mason and Maxwell-Lefroy 1912). No further
information.                                              BH

**Social pattern and behaviour.** The only significant study in
west Palearctic is by Bijlsma and Meininger (1984) for Suez
(Egypt). Following account therefore draws heavily on studies
in Indian subcontinent (see especially Lamba 1963*a*, 1969, Ali
and Ripley 1972), also Ryall and Reid (1987) for Mombasa
(Kenya).

1. Gregarious throughout year. In Suez, flocks of 5–130, but
within flocks commonly in pairs, also often trios (Bijlsma and
Meininger 1984). At start of breeding season, flocks start breaking up into pairs (Lamba 1963*a*). For further information on
flocks, see Roosting (below). BONDS. Mating system monogamous
although promiscuity not uncommon (Ali and Ripley 1972). In
India, pair-bond maintained all year (see above) and presumably
for life (Ali and Ripley 1972), though in Mombasa pair-bond
thought to last only for breeding season (*c.* 14 weeks) (Ryall and
Reid 1987). Both sexes share in nest-building and tending young

(Ali and Ripley 1972). ♀ incubates, ♂ contributing only occasionally when ♀ absent (Lamba 1963*a*; Ganguli 1975) or perhaps
not at all (Goodwin 1986) (*contra* Ali and Ripley 1972). Young
dependent on parents for some time after fledging (Lamba
1963*a*), but no details. Age of first breeding not known. BREED-
ING DISPERSION. Mainly in colonies but some pairs nest solitarily
near colonies (Sen Gupta 1969). In Suez, described as loosely
colonial (Goodman and Meininger 1989). Sometimes up to 9
nests in one tree (if large), sometimes more than 1 on same
branch. In India, main nest fabric often wire (see Breeding),
and such nests may be enlarged in subsequent years so that
neighbouring nests, initially separate, grow into one another,
forming 'community nest' (Altevogt and Davis 1980). Territory
confined to relatively small nest-area, although size difficult to
define since radius defended varies with intruder; conspecific
neighbours tolerated except when they encroach on nest itself,
but, once eggs and young are in nest, area of *c.* 30 m² around
nest defended against conspecifics which do not belong to colony
(Sen Gupta 1969). Predators elicit attack at even greater distances from nest (see Parental Anti-predator Strategies, below).
Territory occupied and defended during breeding season only,
and serves (usually) for courtship and copulation (Lamba 1969).
ROOSTING. Communal in traditional tree-sites, and in India can
involve thousands of birds. Start arriving before sunset and
disperse again before sunrise (Ali and Ripley 1972; Ryall and
Reid 1987; see Flock Behaviour, below, for details of assembly
and dispersal). In Suez, due to natural wariness (see below),
prefers to feed early in morning at quiet places, loafing more
during day (see also Heterosexual Behaviour, below). In September–November, birds start departing for roost-site (palm
groves up to 6 km away) 16.00 hrs, 25% (of 700 birds travelling
from some distance) arriving 16.00–17.00 hrs, remainder 17.00–
18.00 hrs; including birds which spend day close to roost, total
roost-size 800–850. (Bijlsma and Meininger 1984.)

2. Behaviour typical of commensals of man: bold and confiding
but also wary (Woodcock 1980). Described by Ali and Ripley
(1972) as 'intelligent, inquisitive, and impudently familiar, yet
excessively wary and alert at all times'. Movement on ground
accompanied by constant uneasy wing-flicking (Ali and Ripley
1972). FLOCK BEHAVIOUR. Much given to performing aerobatics
like chough *Pyrrhocorax*, including headlong swoops with wings
closed, twisting, turning, somersaulting, etc. (Ali and Ripley
1972). In Mombasa, small parties display thus around tops of
buildings in late afternoon: call loudly, chasing one another,
sometimes gliding and soaring with wings below horizontal (as in
Jackdaw *C. monedula*) (Bijlsma and Meininger 1984). Approach
roost in steady stream of straggling flocks (Ali and Ripley 1972;
Ryall and Reid 1987) and perform aerial evolutions before settling (Bijlsma and Meininger 1984). On entering roost in Mombasa, clamorous activity occurs as new arrivals vie for perches,
taking up to several hours before all birds settle and noise subsides (though calling sometimes heard during night: see 6 in
Voice); exodus shortly before dawn in parties of 3–20, birds
flying up to 16 km to feed (Ryall and Reid 1987). Birds readily
flock to source of food or anything unusual (e.g. distressed bird);
attracted by calls, circle overhead and join in calling (Dharmakumarsinhji 1955; Ali and Ripley 1972; Ryall and Reid 1987:
see 1 in Voice). For flocking to mob predators, see Parental
Anti-predator Strategies (below). SONG-DISPLAY. None reported.
ANTAGONISTIC BEHAVIOUR. (1) General. Nest-territory defended
by combination of warning calls, threat, pursuit, and attack; one
member of pair constantly on guard (Lamba 1963*a*, 1969). For
constant defence by pair of apparent loafing site (perhaps nest-
site), see Bijlsma and Meininger (1984). Aggressive also over
food (Bijlsma and Meininger 1984: see below) and nest-material

(Sen Gupta 1969), and remarkably hostile to almost all other birds, even causing local extinctions in some places where *C. splendens* has extended its range, e.g. in Mombasa (where *C. splendens* arrived *c.* 1947) supplanted local weavers (Ploceidae), sunbirds (Nectariniidae), and Pied Crow *C. albus* by mid-1960s (Ryall and Reid 1987). (2) Threat and fighting. In flocks, especially when feeding, disputes common, accompanied by much threat but escalating to attack. Protagonists usually perform Forward-threat display (Fig A), usually while calling (see 1 in

A

Voice): body more or less horizontal, neck extended with ruffled throat feathers, wings held close to body or slightly drooped, tail lowered and flicked down with each call; rival threatened thus usually retreats. (Bijlsma and Meininger 1984.) Colony-members collectively attack and expel any strange conspecifics trying to steal nest-material (Sen Gupta 1969). Where well established, tolerates small birds within a few cm of nest, but aggressive towards such intruders after laying (Lamba 1969; Sen Gupta 1969). However, away from colony area, often openly hostile to almost any bird. At Jaisalmer (India), Indian Roller *Coracias benghalensis* sought refuge in tree while 30–40 *C. splendens* 'terrorized' it, poking their heads into foliage; during hour after *C. benghalensis* escaped, *C. splendens* seen to attack Indian Pond Heron *Ardeola grayii*, Teal *Anas crecca* (attacked from above and flushed from water), Pintail *A. acuta*, Coot *Fulica atra*, Red-wattled Plover *Hoplopterus indicus*, Little Stint *Calidris minuta*, Spotted Redshank *Tringa erythropus*, Common Sandpiper *Actitis hypoleucos*, etc. (R P Martins). See also Parental Anti-predator Strategies (below). HETEROSEXUAL BEHAVIOUR. (1) Pair-bonding behaviour. Little detailed information. Courtship described as discreet, including Allopreening and Courtship-feeding (see below). Once pair established, mates keep relatively close together (Lamba 1963*a*). Even outside breeding season, pair often perch huddled together on shady branch by day, taking turns to Alloppreen each other's lowered head for 10 min or more at a time (Ali and Ripley 1972). At loafing site shared by one pair, arrival of mate typically seen to elicit silent Meeting-display (Fig B) from partner: adopted upright posture,

B

flicked wingtips, lowered tail, and ruffled plumage on forehead (Bijlsma and Meininger 1984). Pair-members also seen rubbing bills together (Ryall and Reid 1987). (2) Courtship-feeding. ♂ frequently feeds mate after pairing (Ryall and Reid 1987), especially at nest (Goodwin 1986), but according to Ganguli (1975) even when young have fledged in August. When regurgitating food into own bill for presentation to mate, ♂ gives Feeding-call (see 4 in Voice). ♀ accepts (and presumably first solicits) food with Begging-display: crouches, wing-flirts, and gives Begging-calls (Ali and Ripley 1972: see 3 in Voice). (3) Mating. Takes place in tree, sometimes rooftop or ground (Lamba 1963*a*), very often while ♀ sitting on nest (Ali and Ripley 1972). May be preceded by Alloppreening or Courtship-feeding, but just as often without preliminaries (Lamba 1963*a*). Ali and Ripley (1972) mentioned soliciting call by ♀, but no details. During copulation, ♀ draws in neck and partly spreads wings, while ♂ grasps back of her head with his bill; copulation lasts only a few seconds and may occur at any time of day, most often during nest-building (Lamba 1963*a*). Copulating pair frequently mobbed by conspecifics, and promiscuity probably not uncommon (Ali and Ripley 1972; Ryall and Reid 1987). (4) Nest-site selection and behaviour at nest. Not known which sex selects site but ♀ apparently plays significant role, often being seen prospecting sites with twig in bill, closely followed by presumed mate (with or without twig). Both sexes collect nest-material but only ♀ builds, ♂ passing his material to her. Nest constantly attended, one mate guarding when other absent. Only ♀ sits on nest at night. (Lamba 1963*a*.) See also Bonds for incubation roles. RELATIONS WITHIN FAMILY GROUP. Young not fed, or unable to accept food, until *c.* 1 day old. Eyes said to open at 2–3 days (seems improbably early). Both parents bring food and feed young, one foraging while other guards. (Lamba 1963*a*.) Food often brought in crop but also in bill; young usually begin begging (see Voice) when parent lands at nest (Thiede and Thiede 1974, which see for treatment of bird prey prior to delivery to young). Dead nestlings ejected by parents. After leaving nest, young remain in branches of nest-tree where parents feed them. Later stay close to parents (usually ♀), begging (with 'wing-shuffling') whenever food found. (Lamba 1963*a*.) ANTI-PREDATOR RESPONSES OF YOUNG. No information. PARENTAL ANTI-PREDATOR STRATEGIES. (1) Passive measures. No information. (2) Active measures: against birds. Colony-members cooperate to mob predators fiercely with dive-attacks. Confront raptors within 50–60 m of nest, and chase them off up to 200–300 m. In India, especially hostile toward Koel *Eudynamys scolopacea* (regular brood-parasite), attacking it whenever seen or heard within 100 m of nest. (Lamba 1963*a*, 1969.) In Suez, low-flying raptors, regardless of size, pursued relentlessly for up to 1 km by flocks of 5–30, members taking turns to launch dive-attacks accompanied by Alarm-calls (Bijlsma and Meininger 1984, which see for list of species chased; see also 5 in Voice). Once seen to land on back of White-backed Vulture *Gyps bengalensis* sitting on top of tree, and again after it took off (Rahmani and D'Silva 1986). (3) Active measures: against man and other animals. Man, monkeys, and carnivores regularly subjected to attack if they climb nest-tree or one nearby; attack continues until they retreat 200–300 m from nest (Lamba 1969). Tactics as for raptors, with furious dive-attacks to head and tail, making physical contact; man usually sustains peck-wounds and scratches to face, and reports of cats being blinded and even killed are common. Attacks on snakes recorded. (Ryall and Reid 1987.)

(Figs by D Quinn from drawings in Bijlsma and Meininger 1984.)                                                                      EKD

**Voice.** Extensive repertoire to cater for contingencies such as suspicion, alarm, anger, invitation to copulate, announcement of food-finding, contentment, loss of mate, etc. (Ali and Ripley 1972). However, no descriptions available for some of these; only a few calls described and these mainly from extralimital studies. 'Caw' calls evidently embrace variety of contexts, but few details of corresponding structural variation. Often vociferous during flight (Goodwin 1986).

CALLS OF ADULTS. (1) Contact-alarm call. 'Quah quah' or nasal 'kaau kaau' (Ali and Ripley 1972); 'kwar kwar kwar' or 'waaa waaa waaa' (Hollom et al. 1988). In recordings, calls vary in length and spacing but most sound roughly like 'kraar'. Although said by Hollom et al. (1988) to be quieter and higher pitched than Carrion Crow C. corone, calls depicted in Fig I lower pitched than at least some of C. corone; Fig II shows rapid series of shorter calls (c. 0·3 s). Fig IV depicts variant (perhaps different call) in which each of 4 units begins with typical rasping 'krar-' followed by descending tonal offset, thus 'krarlu'. (W T C Seale.) In recordings, calls given in bouts of 1–5, most often 2 (E K Dunn). The following calls are presumably related: short high-pitched 'vâ-vâ', given in Forward-threat display (Bijlsma and Meininger 1984); guttural 'caw' used by bird that has found food or is in difficulties (Dharmakumarsinjhi 1955), and in either case has effect of attracting conspecifics; at least in case of food-finding, thought to serve specifically as summons to fellow flock-members (Dharmakumarsinjhi 1955; Ali and Ripley 1972; Ryall and Reid 1987). See calls 4–6 (below) for other sorts of 'caw'. (2) Kurr-call. Rather musical 'kurrrrrr' (c. 0·5–1 s long) suggestive of watch-spring running down, given by bird, apparently to itself, while resting on branch (Ali and Ripley 1972). Subdued 'kurr' once heard from adult on nest-rim feeding young (Thiede and Thiede 1974). Perhaps a contentment- or contact-call. (3) Begging-call of ♀. Subdued 'kree-kree-kree' by ♀ displaying to elicit food from mate (Ali and Ripley 1972). (4) Feeding-call of ♂. Short distinctive 'caw' given by ♂ regurgitating food for mate (Ali and Ripley 1972). (5) Alarm-call. Harsh 'crao-crao' given by flock-members mobbing large raptors (Bijlsma and Meininger 1984). 'Krraar' (0·53 s) and protracted 'krrraaar' (c. 1 s), dropping in pitch slightly at end (Fig III) from perched birds disturbed by recordist. (6) Other calls. (6a) Single drawn-out 'caw', unlike anything heard during day (but somewhat resembling call 4) sometimes heard from roosting birds (Ali and Ripley 1972). (6b) A 'kow' with downward inflection, also low-pitched 'kowk' (D I M Wallace).

CALLS OF YOUNG. Food-call of small nestlings a quiet strangled 'schee schee', of older young a strangled 'reckkeck reckkeck', also 'kerckkerck kickkerck' (Thiede and Thiede 1974).                                              EKD

**Breeding.** SEASON. Southern Israel: mid-March to July, eggs laid April to late May (Shirihai in press). Egypt: newly-fledged young recorded mid-August (Goodman and Meininger 1989). Kuwait: nest-building April to mid-May (Pilcher 1986). For Arabia, see Bundy et al. (1989), Gallagher (1989b), which see for nestlings in Yemen end of September, and Richardson (1990); for Kenya, see Ryall (1990); for India, see Lamba (1963a) and Ali and Ripley

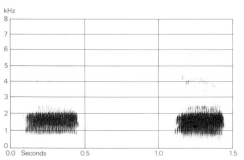

I  M North  Kenya  July 1954

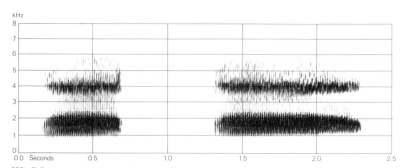

III  B Bertram  India  April 1967

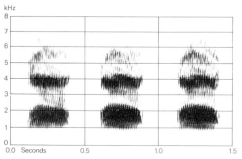

II  B Bertram  India  April 1967

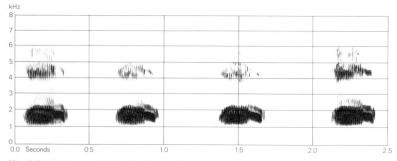

IV  E D H Johnson  Sri Lanka  December 1977

(1972). SITE. Always close to human habitation; in fork near top of tall tree or in outermost branches, often in banyan *Ficus bengalensis* or mango *Mangifera indica* (Lamba 1963*a*; Feare and Mungroo 1989); in Israel, in groves of tamarisk *Tamarix* (Shirihai in press), and in Oman in mangrove swamp (Gallagher and Woodcock 1980). In Kenya, average height of 159 nests 6·8 m, none below 4 m; large trees contained up to 6 nests, though well-separated, and apparently species with dense canopies preferred (Ryall 1990). Has developed remarkable habit, particularly in India, of nesting in busy streets on buildings, street lamps, pylons, etc. (Altevogt and Davis 1980), even recorded on verandah of occupied house on old electricity bracket (Husain 1964). In Kuwait, nested at least 20 m up on cranes although suitable trees nearby (Pilcher 1986). Nest: untidy mass of twigs, often thorny, sticks, plastic, string, assorted pieces of metal, electrical cable, etc.; depression in centre lined with fibres of wood or bark, grass, hair, cloth, and similar soft material, though can be unlined (Lamba 1963*a*; Feare and Mungroo 1989; Ryall 1990; Shirihai in press). In towns, often solely of wire and metal, including items such as spectacle frames, coat-hangers, bicycle pedals, metal sheeting, etc.; such nests may accumulate over years to weigh up to 25 kg and may apparently contain eggs of several pairs, though single nests can be over 8 kg and contain more than 250 m of wire; nests containing much metal last for years, often removing need to build new nest in subsequent years, and may be unlined (Altevogt and Davis 1980). Twig nests in western India had outer diameter 25–30 cm, inner diameter 12–15 cm, depth of depression 7–10 cm (Lamba 1963*a*). Building: both sexes gather material but only ♀ builds; green vegetation may be used in lining, or dry material wetted before use; takes 4–7 days ($n = 14$) (Lamba 1963*a*, which see for details of construction); in Kenya, 14–17 days (Ryall and Reid 1987). Most material from ground or rubbish tips but may break twigs from trees (Ryall 1990). EGGS. See Plate 57. Very variable in shape, size, and colour; generally short oval, some even pyriform; fairly glossy, pale bluish-green with brown or grey speckling, blotches, and streaks (Lamba 1963*a*; Ali and Ripley 1972; Ryall 1990). Nominate *splendens*: 38·0 × 26·5 mm (30·4–44·1 × 23·0–29·1), $n = 225$; calculated weight 13·7 g. *C. s. zugmayeri*: 37·7 × 25·9 mm (31·2–45·0 × 23·5–28·5), $n = 80$; calculated weight 13·4 g. (Schönwetter 1984.) Clutch: 3–5 (2–6) (Lamba 1963*a*). In Kenya, of 15 clutches: 2 of 2 eggs, 5 of 3, 7 of 4, 1 of 5; average 3·9; average of 123 Indian clutches 4·0 (Ryall 1990). In Oman, replacement clutches following loss of 1st clutch laid until midsummer (Gallagher and Woodcock 1980); in Kenya, can be double-brooded (Ryall 1986). Eggs laid at intervals of 1–2 days (Lamba 1963*a*). INCUBATION. Period between laying of last egg and hatching of last nestling 16 days in 15 cases, 17 days in 2 (Lamba 1963*a*). Said to be by both sexes, ♂ relieving ♀ for short periods and only by ♀ at night (Lamba 1963*a*; Ali and Ripley 1972), but only by ♀

throughout according to Goodwin (1986). Probably only by ♀ at one nest in India (Husain 1964). YOUNG. Fed and cared for by both parents (Lamba 1963*a*; Husain 1964); brooded for 15–18 days (Shirihai in press). FLEDGING TO MATURITY. Fledging period 21–28 days (Lamba 1963*a*); in Israel, 27–28 days ($n = 5$) (Shirihai in press). BREEDING SUCCESS. In western India, of 81 eggs, 74% hatched, and 54% produced fledged young (75% in clutches of 3, $n = 12$ eggs, 48% in clutches of 4, $n = 44$, and 56% in clutches of 5, $n = 25$); most nestling deaths caused by starvation in 1st week, and youngest in clutch of 5 usually dies; many broods parasitized by Koel *Eudynamys scolopacea*, whose young are fed alongside own nestlings by *C. splendens* parents (Lamba 1963*a*). BH

**Plumages.** (*C. s. zugmayeri*). ADULT. Forehead to crown, side of head to just behind eye and to shorter ear-coverts, and chin to throat black, glossed purple-blue or green-blue on top of head, blue-green on side of head and on chin to throat; bristles at base of bill dull black. Hindneck to upper mantle, side of neck backwards from longer ear-coverts, side of breast, and chest medium ash-grey, sometimes with slight brown tinge (especially when worn); grey rather sharply defined from black on nape and on side of head, but more gradually merging into black through dark grey on side of throat, lower throat, and on lower mantle, and very gradually passing through dark grey on mid-belly and vent, to greyish-black on flank and side of belly, and to sooty black on thigh and under tail-coverts (on latter, tips of some feathers faintly glossed green). Lower mantle and scapulars black with purple-blue or green-blue gloss, back to upper tail-coverts similar but gloss less intense, rump often partly greyish-black. Tail, primaries, tertials, greater upper primary coverts, and bastard wing black with purple-blue or green-blue gloss, secondaries and remaining upper wing more strongly glossed bluish-purple or violet-purple. Under wing-coverts and axillaries dark grey to greyish-black, glossed green-blue on smaller coverts along leading edge. In fresh plumage, gloss on face, mantle and scapulars, tail, and upper wing stronger, more purplish-blue; grey of neck and chest with slight lavender tinge; *in worn plumage*, gloss greener, but virtually absent from tail and primaries, neck and chest paler ash-grey or light brownish-grey.

NESTLING. No information. JUVENILE. Like adult, but cap and chin to chest hardly glossy, except sometimes for limited purple-blue on feather-tips of forehead and lower throat, mainly dull black, less contrasting with grey of neck and underparts. Colour of neck and underparts less saturated pale grey, duller, feather shorter, more white of feather-bases visible, especially when plumage worn. Lower mantle and scapulars glossed purplish-blue, like adult, but with much dull black of feather-centres visible. Tail-feathers narrower (central pair *c.* 25 mm wide in middle, outermost *c.* 18 mm; in adult, *c.* mm more) tip more rounded (less truncate), duller black, gloss more faintly grey-blue, less purplish. Entire wing less glossy than in adult, especially on primaries, tertials, and greater upper primary coverts, gloss more faintly greenish-blue, but gloss on secondaries and other wing-coverts sometimes as glossy as adult. FIRST ADULT. Like adult, but juvenile flight-feathers, tail (sometimes except central feathers), some or all tertials, and variable number of greater upper wing-coverts (none to all) retained; old feathers less glossy than neighbouring new ones, distinctly browner once plumage worn (difference usually most marked in tertials and greater coverts), not as uniform in colour and gloss as adult; gloss on secondaries and greater coverts (greenish-)blue, only slightly purplish in some lights; tips of primaries and tail-feathers more worn than adult, outer primaries with more pointed tips, tail-feathers narrower. New feathering of head, body, and wing as in adult, but cap, throat, lower mantle, and scapulars glossed greenish-blue or purplish-blue when fresh (depending on light; in adult, purple or bluish-purple); under tail-coverts not glossy or with faint green-blue gloss; gloss on wing-coverts sometimes less intense.

**Bare parts.** ADULT, FIRST ADULT. Iris brown or dark brown. Bill black, mouth brownish-slate or black. Leg and foot black, soles grey. NESTLING. No information. JUVENILE. Iris slaty-white or blue-grey. Bill greyish-black to black; mouth pink to flesh-red. Leg and foot greyish-flesh with bluish-black scutes, rear of tarsus and sides paler pink-flesh. (Ali and Ripley 1972; Goodwin 1986; BMNH, ZMA).

**Moults.** ADULT POST-BREEDING. Complete; primaries descendent. In northern India and Nepal (breeding season mainly April-June), moult starts mid-June to late July, in full moult August-September, and apparently just completed in birds from October; in Sri Lanka (nesting May–August), moult September to late December. Tail and tertials start at primary moult score 6–15 (with shedding of about p3), but body and secondaries not until moult score c. 25–30; at score of c. 35–40, tail, tertials, wing-coverts, and body mainly new, secondaries and head halfway through moult, neck mainly old. POST-JUVENILE. Partial: head, body, lesser and median upper wing-coverts (occasionally, some juvenile outer median coverts or scattered lesser coverts retained), sometimes a few tertials and inner greater coverts, occasionally (in 1 of 10 birds examined), all greater coverts or central tail-feathers. Probably start soon after fledging; in northern India, moult finished in 1 bird from September and 3 from October; of 2 birds from Sri Lanka, November, one fully juvenile, other had partial moult already completed. (BMNH, RMNH, ZFMK, ZMA, ZMB.)

**Measurements.** Adult wing, adult tail, and bill to skull of (1) *zugmayeri*, Pakistan and north-west India, (2) nominate *splendens* and *protegatus*, Nepal, central and southern India, and Sri Lanka; other measurements combined, all year; skins (BMNH, RMNH, ZFMK, ZMA, ZMB). Juvenile wing and tail include juvenile wing and tail of 1st adult. Bill (N) to distal corner of nostril; exposed culmen on average 2·7 less than bill to skull.

| | ♂ | ♀ |
|---|---|---|
| WING (1) | 271·4 (7·10; 14) 263–285 | 261·0 (7·39; 11) 254–270 |
| (2) | 273·0 (5·95; 8) 266–283 | 249·8 (7·76; 5) 242–261 |
| JUV | 251·0 (4·64; 13) 244–258 | 239·8 (5·25; 8) 231–243 |
| TAIL (1) | 166·9 (6·96; 13) 159–180 | 158·3 (8·67; 8) 149–175 |
| (2) | 160·5 (4·74; 8) 153–167 | 148·5 (5·80; 4) 141–155 |
| JUV | 149·2 (8·16; 12) 139–164 | 140·5 (8·70; 8) 129–149 |
| BILL (1) | 50·9 (2·07; 14) 48·0–54·5 | 46·5 (1·63; 7) 43·8–48·3 |
| (2) | 51·2 (1·63; 12) 48·2–54·5 | 47·0 (2·12; 9) 43·6–49·6 |
| BILL (N) | 32·3 (1·71; 22) 29·2–35·0 | 28·9 (1·40;15) 26·9–30·8 |
| TARSUS | 49·1 (2·13; 22) 47·2–52·0 | 46·4 (2·04; 16) 42·2–49·5 |

Sex differences significant.

Mauritius, ages combined, October (Feare and Mungroo 1989).

| | ♂ | | | ♀ | |
|---|---|---|---|---|---|
| WING | 268·2 (8·1; 18) 255–286 | | | 246·6 (8·5; 10) 237–267 |
| CULMEN | 50·9 (3·0; 18)   45–56 | | | 45·8 (1·8; 9)   42–49 |

Wing of *zugmayeri* and nominate *splendens* 253–284 (n=68), of *protegatus* 217–275 (n=5), of *insolens* 230–278 (n=33) Meinertzhagen 1926).

**Weights.** All races combined: adult ♂ 319·8 (28·48; 4) 300–362, 1st adult ♂ 260, ♀ 252–304 (n=5); unsexed 266, 280 (Ali and Ripley 1972; ZMA). Mauritius, ages combined, October: ♂ 316·6 (21·2; 16) 270–371, ♀ 269·7 (13·9; 10) 245–295 (Feare and Mungroo 1989).

**Structure.** Wing rather long, tip bluntly pointed, rather broad at base. 10 primaries: p7 longest, p8 0–6 shorter, p9 19–35, p10 75–110, p6 0–7, p5 16–26, p4 49–54, p3 55–75, p2 65–85, p1 70–90; p10; 40–65 longer than longest upper primary covert. Outer web of (p5-)p6–p9 emarginated, inner web of (p6-)p7–p10 with notch. Tip of longest tertial reaches to tip of p2–p3. Tail rather long, tip rounded; 12 feathers, t6 18–32 shorter than t1 in adult, 12–20 in juvenile. Bill rather long, straight; length of visible part of culmen about equal to head length; broad and deep at base, depth at middle of nostril 16·9 (10) 16·3–17·8 in ♂, 15·8 (7) 15·0–16·5 in ♀; culmen distinctly decurved, cutting edges slightly decurved. Bristles at side and top of upper mandible project forwards, covering c. ⅓ base of bill. Feathering of hind-neck and side of neck dense, soft, and silky; feathers of chin and throat slightly elongated, tips bifurcated. Leg and foot rather short, strong. Middle toe with claw 44·2 (10) 42–46·5; outer and inner toe with claw both c. 72% of middle with claw, hind c. 77%.

**Geographical variation.** Marked, but largely clinal, involving only tone of grey on neck and chest. Grey of nominate *splendens* from much of Indian subcontinent medium ash-grey, fawn-grey, or brown-grey; *zugmayeri* from south-west Iran, Pakistan, and north-west India paler, pale ash-grey or pale smoke-grey when fresh, milk-grey or dirty grey-white when worn. *C. s. protegatus* from Kerala (south-west India) and Sri Lanka has grey distinctly darker, dark ash-grey to slate-grey; in *insolens* from Tenasserim (peninsular Burma), grey replaced by greyish-black or sooty-black, scarcely contrasting with remainder of body (but not glossy). Boundaries between races not sharp, however: e.g. nominate *splendens* from Delhi area averages slightly paler than birds from Bangladesh (but not as pale as *zugmayeri*), and nominate *splendens* grades into *protegatus* in southern India; birds from Rangoon area (Burma) not as black as Tenasserim birds, about intermediate between *protegatus* and typical *insolens*, but nearer latter.

Widely introduced or self-established in coastal areas from East Africa north to Jordan and east to Australia (Meininger *et al.* 1980), and such birds often difficult to assign to a race, either because they are derived from populations intermediate between 2 races, or are a mixture of various races. Most populations in Middle East and East Africa are nominate *splendens*, but populations in Kuwait and Oman are partly *zugmayeri* or intermediates (Pilcher 1986), and *protegatus*-like bird recorded Aden (Yemen) (Meinertzhagen 1954). Birds from Singapore dark, but less so than *insolens*; either derived from *protegatus* or intermediate between *insolens* and nominate *splendens*.   CSR

## *Corvus frugilegus* Rook

DU. Roek    FR. Corbeau freux    GE. Saatkrähe
RU. Грач    SP. Graja    SW. Råka

*Corvus frugilegus* Linnaeus, 1758

Polytypic. Nominate *frugilegus* Linnaeus, 1758, Europe and Asia Minor east to Yenisey river, north-west Altai, and north-west Sinkiang. Extralimital: *pastinator* Gould, 1845, from Yenisey and central Altai east to eastern Siberia, Japan, Korea, and eastern China.

**Field characters.** 44-46 cm; wing-span 81-99 cm. Slightly smaller than Carrion Crow *C. corone*, with more slender bill, proportionately smaller head with steeper forehead, and seemingly deeper body due to loose flank feathers cloaking thighs; in flight, more splayed wing-tips and rounder tail; over 30% larger than Jackdaw *C. monedula*. Quite large and elegant crow, with slender bill, bare and pale face (in adult), and characteristic 'baggy trousers' above legs. Plumage black with heavy gloss. Highly sociable at all times of year. Voice distinctive. Sexes similar; no seasonal variation. Juvenile lacks bare face (thus resembling black races of *C. corone*).

ADULT. Moults: May-October (complete). Length of slender and sharp-pointed black bill exaggerated by contrast with whitish-grey skin round base of bill from lower forehead to rear of gape and upper throat; bare patch gives characteristic aged look. Rest of head black, glossed greenish-purple (most obvious on rear ear-coverts); back and body black, glossed mostly reddish-purple; loose thigh feathers black, glossed greenish; wings black, glossed on upper surface reddish-purple on smaller coverts and outer webs of secondaries but mainly greenish-purple elsewhere; tail black, glossed reddish- and greenish-purple above. At short range in sunlight, iridescence is dazzling and beautiful; at longer range even in dull light, may make black bird appear sheeny and hence strangely pale. JUVENILE. Unlike adult, has black nasal bristles and fully-feathered head, retained until December-April. Loose thigh feathers may be less evident. Rest of plumage duller and less glossy, with brown cast to hindneck and back. Easily confused with black races of *C. corone*, but glossier plumage, steeper forehead, and less curved culmen of more slender bill usually evident.

Separation of adult from black races of *C. corone* not difficult at closer ranges and in good light but may be impossible at distance, particularly in case of single bird when no clues of association or behaviour available. Important also to remember that *C. corone* does regularly form flocks, particularly from late summer into winter, and exploits same food sources as *C. frugilegus* and *C. monedula* at that time. Separation of young *C. frugilegus*, without bare face, tricky (see above). Flight variable: around colony, remarkably agile, allowing birds easy entrance to foliage, frequent indulgence in partly aerial squabbles, and wheeling overhead; less accomplished when moving between feeding areas, with direct and deliberate progress along regular paths achieved by fast-flapping and slightly laborious action with more regular wing-beats and less gliding than in *C. corone*, but birds erratically spaced, producing characteristic straggling flock (roosting flights are even more leisurely and disjunct); on migration, action more purposeful but flocks again break out in wide straggling arcs; in sudden mass 'dread' flights at colony, action more rapid and flock more cohesive, but soon breaks up again after take-off leading to characteristic staggered return. Wing-beat rhythm enhanced in good light by contrast of sheeny upperwing and black underwing, so that action has also flickering aspect not unlike *C. monedula*. Flight silhouette shows noticeably narrow bill and face, well fingered ends to primaries, and rounded tail. Gait mainly a sedate and (probably due to 'baggy trousers') apparently slightly rolling walk; also includes heavy hopping and sidling movements; may hold head high and look tall when searching for food, but when actually probing whole bird sinks lower to ground. Colonial, and highly gregarious all year. Except in early breeding season, tolerant of close neighbours and not excitable (compared to *C. monedula*). Readily accepts close presence of *C. monedula* but usually stays some distance from *C. corone*.

Most calls are associated only with breeding. Commonest a harsh 'kaah', more deliberate and prolonged and less unattractively raucous than similar note of *C. corone*; usually given singly and not in straining triplet so typical of *C. corone*; loudest calls are accompanied with bowing and tail-fanning.

**Habitat.** Breeds only in boreal and temperate middle latitudes of west Palearctic, in both continental and oceanic lowlands, to July isotherm of 12°C, but absent from warmer regions, except in winter. Strong winds, ice, and snow are avoided but rain and mist are tolerated. Range excludes most montane regions. Breeds in England up to 450 m (Yapp 1962), in Scotland usually well below 350 m (Baxter and Rintoul 1953), and in Carpathians not above 600-700 m (Dementiev and Gladkov 1954). Extralimitally, however, in Asia nests up to 2000 m or more. Requires for breeding fairly tall trees, either on edges of forests or woodlands or by preference in clumps, groves, or riverain or other linear forms fronting open grasslands or croplands. Avoids dense woodland, dry,

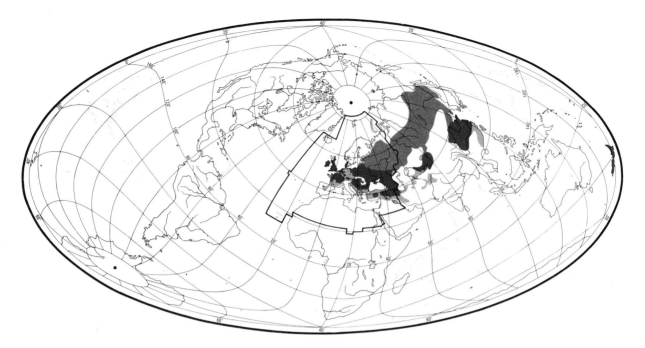

hard, and rocky surfaces, wetlands and other tall dense vegetation, including scrub and thickets, small dissected spaces such as gardens and orchards, and, in most cases, interiors of any but small built-up areas. Breeds in towns and villages only where adjacent countryside is readily accessible. Dependence on agriculture, land improvement for pasture, and conservation of tall trees outside forests have expanded suitable habitats within modern times, favouring types of mixed arboreal and artificially short ground vegetation required. Some modern changes in farming practice appear to have rendered certain habitats either more or less suitable. For feeding, bare ploughland and fallow are often seasonally favoured, and in winter some use is made of woods. Almost all foraging occurs, however, on ground, and in England more than half of it on grassland, especially rich grassland near rivers and streams. (Lockie 1955.) Roads are frequently visited on account of insects, birds, and small mammals killed by traffic. Attitude to human disturbance is pragmatic, commensalism being freely accepted where freedom from persecution can be relied on. Habits of gregarious nesting and roosting involve much flying, sometimes at considerable altitudes, and aerial component of habitat is accordingly significant. For thorough review of complex seasonal habitat changes due to varying agricultural regimes, see Murton (1971).

**Distribution.** Has spread in some northern parts of range, parts of central Europe, and France.

FRANCE. Has spread south in 20th century. In 1930s very rare breeder south of Loire. First breeding colonies near Lyon 1952, Vienne 1964; northern Gironde col-

onized 1974. (Yeatman 1976.) NETHERLANDS. Disappeared almost completely from west after 1950 due to pesticides (see Population); did not recolonize west when population recovered (Directie NMF 1989). SWEDEN. Long-term spread to north. During first part of 19th century bred in only 3 provinces in south. New breeding sites established from 1860s, latest in Jämtland (*c.* 63°N) in 1981. (LR.) FINLAND. Colonization began in 1880s. Large colonies established only in south-west; further north, only scattered pairs or occasional breeding. (OH.) POLAND. Until mid 19th century very local breeder, becoming widespread in 2nd half of 19th century. Has spread again since 1945, colonizing human settlements and even city centres. (LT.) AUSTRIA. Has bred in eastern Steiermark since 1986/7 (H-MB). SWITZERLAND. First bred 1963, subsequently establishing 2 main breeding centres (Winkler 1984).

Accidental. Iceland, Malta, Lebanon, Jordan, Kuwait, Algeria, Azores.

**Population.** Has decreased in parts of central Europe.

BRITAIN, IRELAND. Probably *c.* 1-5 million pairs (Sharrock 1976). General decrease between mid-1950s and mid-1970s followed by slight recovery in some areas; changes probably due to changing land-use (Sage and Whittington 1985). FRANCE. 100 000 to 1 million pairs (Yeatman 1976). Seine-et-Marne (5915 km²): 2533 nests in 56 colonies (estimated 90% of population), 1975 (Jarry 1976). Poitou-Charente region (25 790 km²): 6450 nests, 1977-80 (Bouard 1983). SPAIN. 600-1000 pairs (AN). BELGIUM. 8000 ± 150 pairs. Declining in some areas due to persecution, changing agricultural methods and spread

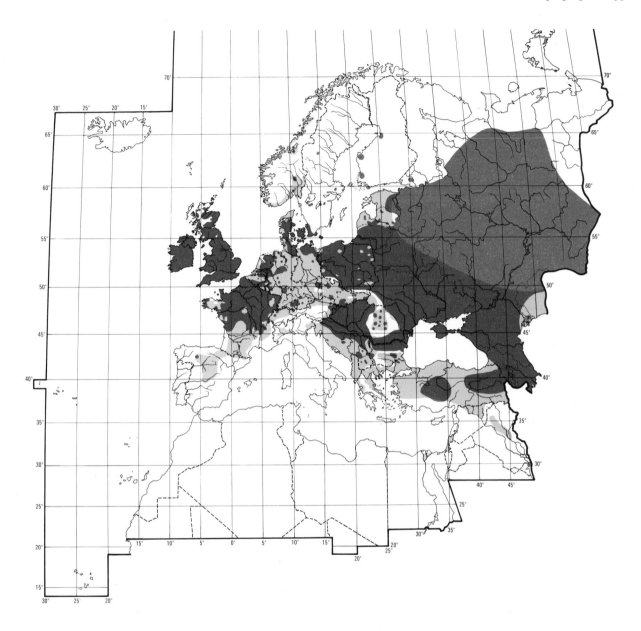

All censuses since 1928 have given populations in range 5000–9000 pairs. (Devillers *et al*. 1988.) LUX-EMBOURG. Estimated 500–600 pairs 1960, 889 pairs 1968, 1757 pairs 1986. Increase associated with urbanization; all colonies now in towns. (Mentgen 1988.) NETHERLANDS. Decrease after 1950, mainly due to pesticides, followed by recovery from 1970s: *c*. 40 000 pairs in 1943; down to *c*. 10 000 in 1970; recovered to *c*. 20 000 in 1980, and *c*. 30 000 in 1985 (SOVON 1987; Directie NMF 1989). GER-MANY. 35 000 pairs (G Rheinwald). Bayern: 2229–2279 pairs 1986 (Franz *et al*. 1987). East Germany: decline from *c*. 100 000 pairs in 1900 to *c*. 8000 pairs in 1941, rising to 29 000 in 1945 then 13 315 in 1960 (Ruge 1988);

12 900 ± 2000 pairs (Nicolai 1993). NORWAY. A little more than 350 pairs, 1986 (Vedum and Tøråsen 1988). SWEDEN. Marked decrease due to pesticides (Malmberg 1971), but has since recovered (LR). Estimated 10 000 pairs (Ulf-strand and Högstedt 1976); increasing, now 10 000–15 000 (LR). FINLAND. Increased until early 20th century, then decreased due to persecution (recently relaxed); *c*. 1000 pairs, probably stable in last few decades (Koskimies 1989). POLAND. Very scarce until mid 19th century; increased in 2nd half of century, then decreased due to persecution; increasing since 1945. Estimate of 211 000 pairs in 1964, probably too low. (Tomiałojć 1990.) CZECHOSLOVAKIA. Bohemia: decline from *c*. 1200 pairs in

1950 to 750 in 1975; *c.* 400 in 1986 (Vondráček 1988). HUNGARY. Decline in number of colonies, colony size, and mean density, especially in east; 254 361 pairs 1980, 118 762 pairs 1984 (Kalotás 1988). AUSTRIA. Increase at least since 1975 (when *c.* 160 pairs); *c.* 250 pairs 1978, now over 500 pairs in 10 colonies (Grüll 1988). SWITZERLAND. About 250 pairs (Winkler 1984). RUSSIA. Declined Yaroslavl' region 1960-85 (Belousov 1986). LATVIA. Estimated 16 000 pairs 1970-3 (Vīksne 1983), then decrease due to human persecution to 7000-9000 pairs 1980-4 (Vīksne 1989). LITHUANIA. 80 000-100 000 pairs (Logminas 1991).

Survival. Britain: mortality in 1st calendar year 59%, in 2nd year 51%, in 3rd-5th years 25% (Holyoak 1971). Europe: mortality in 1st year 54·0% (Busse 1969). Oldest ringed bird 19 years 11 months (Rydzewski 1978).

**Movements.** Resident to migratory, with more birds migrating in cold winters. Winters in Eurasia, within and south of breeding range. Migrates by day in flocks, often following leading-lines such as coastlines and river valleys and frequently accompanied by Jackdaws *C. monedula*. Ringing data have revealed winter quarters of particular populations in unusual detail, and show that mountain ranges act as barriers, thus affecting winter distribution. Adults tend to move less far than juveniles. For review and suggestion that migratory pattern is linked to post-glacial distribution, see Busse (1969).

British and Irish birds almost entirely resident; juveniles may disperse from natal area in 1st winter, but rarely move more than 100 km (Lack 1986). In northern Scotland, however, some upland valleys are abandoned in winter (Munro 1975); in upper Deeside, 1977-8, birds wholly absent from breeding area November-December (distance moved not known); from January they returned, roosting up to 45 km away and making daily visits to colonies; departure and return independent of weather, and probably due to local food shortage (McKilligan 1980; see also Social Pattern and Behaviour). Spanish birds also resident (Ena 1984*b*). Chiefly resident in France, though ringing recoveries show that some birds from north migrate 100-400 km (Giban 1947; Yeatman 1976). Partial migrant in Low Countries and Scandinavia (Busse 1969; Andell *et al.* 1983); in southern Norway, where 2 distinct populations, many from Hedmark and Oppland migrate, but Trøndelag birds probably non-migratory, perhaps due to milder coastal climate (Røskaft 1980*a*). Recently established population at Basel in northern Switzerland is resident (Böhmer 1973), whereas status in Germany varies: thus, Baden-Württemberg population in south-west vacates breeding area entirely (Hölzinger 1987) while proportion migrating from eastern Germany is dependent on severity of weather (Wernicke 1990). Chiefly migratory in Poland and Czechoslovakia (Busse 1963; Hubálek 1980). In USSR, present all year in southern areas, though some southern birds may move south in colder winters; in migratory areas further north, some birds stay irregularly

in certain years (Dementiev and Gladkov 1954). In central region of European Russia, more remain than formerly to winter in breeding areas (Konstantinov *et al.* 1986).

Within Europe, migrants head between west and south, so winter numbers far higher than summer in western Europe, and many migrate also to central and eastern Europe (Busse 1969; Yeatman 1976; Mildenberger 1984; Lack 1986; Hölzinger 1987). Winter range extends south of breeding range in France, with largest concentrations in north; in Seine-et-Marne, breeding population *c.* 10 000, winter population 1975-6 *c.* 500 000 (Yeatman-Berthelot 1991). Migrants enter Spain via western and central Pyrénées, to winter chiefly in Ebro and mid-Duero basins (Ena 1984*b*). In Switzerland (where few breed), winters in low-lying areas in considerable numbers (Schifferli *et al.* 1982); in Bayern (southern Germany), 2229-2279 breeding pairs 1986, but 154 000 wintering birds 1986-7 (Franz *et al.* 1987); in Austria, *c.* 500 breeding pairs, but winter population in Vienna area (where highest concentrations occur) *c.* 100 000 (Grüll 1988). In Greece, abundant winter visitor, especially in north (G I Handrinos). In Italy, now regular winter visitor only to north (previously also to centre and south); very rare on Mediterranean islands (previously more common) (Sultana and Gauci 1982; Flint and Stewart 1992; Thibault 1983; Bogliani 1985; Iapichino and Massa 1989), and exceptional in north-west Africa (Heim de Balsac and Mayaud 1962). Widespread in Israel (Paz 1987); rare and irregular in Egypt, but several recent records of single birds or small groups (Goodman and Meininger 1989). Further east, common and widespread winter visitor to Iraq (Ticehurst *et al.* 1921; Georg and Vielliard 1970), but only 2 records from Kuwait (Bundy and Warr 1980); winters south of breeding range in Iran (Scott *et al.* 1975), and regular in fluctuating numbers in Pakistan and north-west India, chiefly in plains or valleys close to foothills; straggles as far south-east as Ludhiana *c.* 76°E (Ali and Ripley 1972; T J Roberts). Widespread and locally abundant winter visitor to lower areas of Afghanistan, arriving in southern regions usually in late winter (S C Madge).

Ringing recoveries show that range of movement is fairly narrow for most European birds (mainly between west and south-west), much broader for birds from western USSR; distance of movement varies greatly. Data from central European Russia suggest recent tendency to progressively shorten migration route: average distance of movement 1929-38, 2200 km; at end of 1950s, 1900 km; 1971-85, 1400 km; probably accounts for increased numbers wintering in Moscow region (Konstantinov *et al.* 1986), and perhaps also for diminishing numbers in Mediterranean region (see above), though other factors may be involved; see also Busse (1969). Dutch migrants winter in Britain (mostly in east), Scandinavian migrants chiefly in Denmark, some reaching northern Germany and Britain; migrants from north-east France and Belgium move to north-west France. Many Baltic birds winter in eastern

Britain and northern France (north of *c.* 47°N). Central and east European birds (north of Alps and Carpathians, east to Urals) move chiefly WSW: many from Germany, Poland, Czechoslovakia, and western USSR winter in France (mostly in north and central areas), and birds wintering in northern Spain have same origin—recovery at Tudela de Duero 41°35′N 4°35′W of bird ringed as far east as 53°28′N 44°38′E in Penza region (Russia); also Polish-ringed bird recovered 2550 km WSW in south-west Spain. Birds ringed at Belfort gap between Jura and Vosges (where passage concentrated) were recovered mostly in neighbouring areas of eastern France, some in south-west France. Many Polish and western USSR birds winter in Germany and Austria; of birds ringed in winter at Braunschweig (northern Germany), 7 recoveries were 931–2168 km east or ENE, in Finland and USSR; of birds ringed winter in Schleswig-Holstein (northern Germany), 7 recoveries were in Baltic states and adjacent parts of USSR; and from birds ringed in Vienna area, 11 recoveries in European USSR. Wintering birds in Czechoslovakia originate from USSR, chiefly from area between 52–58°N and 37–55°E. From south-west USSR (extending north to Moscow region), movement is in various directions; in general (but with some overlap), birds from west and central Ukraine migrate WSW to plains of northern Italy and Hungary; birds from Moldavia and southern Ukraine, SSW to Bulgaria and southern Rumania; from south-central Ukraine, south-east to West Caspian. (Giban 1947; Busse 1963, 1969; Popov 1978; Ena Alvarez 1979; Grüll 1981; Hubálek 1983; Greve 1983; Mead and Clark 1987; Wernicke 1990; Rode and Lutz 1991; *Rep. Swedish Bird-Ringing*.) Individuals recorded in widely differing areas in different winters, e.g. migrants ringed entering France via Belfort gap recovered in later winters in northern Italy, along middle Danube and middle Rhine (Giban 1947; Busse 1969). Evidence also of winter site-fidelity: 2 birds wintered in Vienna in 3 and 4 consecutive years (Grüll 1981). Long-term study on north-west coast of Black Sea showed that 12% of breeding population remain for winter, and those migrating move south-west via Black Sea coast or Danube valley, to winter chiefly in southern Bulgaria and along Danube, with a few as far north as Rumania; some reach Mediterranean coast in south-east Italy, Greece, and Sea of Marmora (north-west Turkey); a few travel south-east to winter in Dagestan on western shore of Caspian (Ardamatskaya 1968). 2 recoveries January–March show that birds from Astrakhan' (north-west Caspian) winter in Iraq (Georg 1971). Data from Kazakhstan show that many birds winter 1000–2000 km from breeding areas: those breeding in north-west migrate west of Caspian Sea to winter from Gruziya south-east to north-west Iran; those breeding in northern and central areas move south on wide front to winter from northern Iran east to Uzbekistan; those breeding in east (joined by birds from Altai and adjoining areas of Siberia) move south or south-west, turning west on reaching foothills of Tien Shan (with passage concentrated through Chokpak pass) to winter chiefly in Uzbekistan and Tadzhikistan. Birds wintering north of Tien Shan (Alma-Ata and Dzhambul areas) are apparently not of Kazakhstan provenance. (Gistsov *et al.* 1978.) More recently, 5 recoveries in northern Caucasus of birds ringed Tyukalinsk (55°55′N 72°10′E) have indicated wider dispersal for west Siberian birds (Khokhlov and Konstantinov 1983).

Following post-breeding dispersal, autumn departure begins September, with main movement October–November (Busse 1963). Birds leave Moscow region from late September (Ptushenko and Inozemtsev 1968), and earliest record in 10 years was 25 September on Polish coast (Busse and Halastra 1981). In southern Sweden, movement of local birds peaks mid-October (Andell *et al.* 1983). In eastern Germany, winter population arrives from 3rd week of October with rapid increase, and is stable by end of November (Wernicke 1990). In Switzerland, passage usually begins 2nd third of October, with peak 18 October to 1 November, decreasing to end of November (Winkler 1984). Arrivals in Britain and France are late September to November (Giban 1947; Witherby *et al.* 1938), and wintering birds reach Rhône-Alpes (southern France) in last third of October (Lebreton 1977). Post-breeding dispersal reported up to 404 km north or north-east in central Kazakhstan, but at Alma-Ata in south-east birds apparently remain near breeding colonies then (Gistsov *et al.* 1978). Main movement October; at Chokpak pass, 80% passage in 1st third of October (Korelov *et al.* 1974; Gavrilov and Gistsov 1985). Arrives on plains of Iraq very regularly in 3rd week of October, numbers building up to mid-November (Ticehurst *et al.* 1921). At Islamabad (northern Pakistan), earliest record 23 October over 3 years (Mallalieu 1988). Hard-weather movements sometimes reported mid-winter, e.g. in eastern Bayern, 4000 on 16 December 1980 (Leibl and Melchior 1985); in Turkmeniya, large flocks reported flying south-east, end of November to January 1971–5 (Rustamov 1977).

Return movement is early, February–March, exceptionally from January; in USSR, first harbinger of spring (Dementiev and Gladkov 1954). Winter visitors leave Britain mid-February to 3rd week of April (Witherby *et al.* 1938), and Rhône-Alpes by end of March or beginning of April (Lebreton 1977). In Switzerland, passage begins last third of February, peaking 7–11 March, continuing to late March, with stragglers to mid-April (Winkler 1984). At Straubing (south-east Germany), first records last week of January in fine weather, but usually passage begins slowly February to early March, suddenly increasing mid-March and ending early April (Leibl and Melchior 1985). In north-east Germany, winter visitors leave from early March, numbers rapidly declining to end of March (Wernicke 1990). Swedish breeding birds return February–March (Andell *et al.* 1983); arrives on breeding grounds in Moscow region 2nd half of March, in north-west Black Sea area end February to beginning of March (Ardamatskaya 1968; Ptushenko and Inozemtsev 1968). Further east, departure from Iraq begins mid-February, chiefly in March (Ticehurst *et al.* 1921), and most have left India by end of March; at Islamabad, last record 21 March over 3 years (Ali and Ripley 1972; Mallalieu 1988). In central Asia, ringing data show movement sometimes begins as early as end of January, but chiefly February-March (Gistsov *et al.* 1978); in Chokpak pass, 91% of passage 6-20 March (Gavrilov and Gistsov 1985). At Kurghal'dzhino (northern Kazakhstan), average first arrival 16 March over 13 years (Kovshar' 1985).

Spring route is mostly reverse of autumn, but observations in Scandinavia suggest that northern edge of migratory stream is further north; autumn birds follow south Baltic coast, whereas some spring birds cross southern Sweden (Alerstam 1988). Similarly, more conspicuous spring than autumn at Lake Ladoga (St Petersburg region) (Noskov *et al.* 1981). In mid-October 1976, heavy and prolonged storm over Poland blew birds (heading southwest to France and Germany) north of usual route; probably, most wintered in Denmark and Netherlands, some in central Norway; return in spring 1977 was parallel to usual route but further north, resulting in higher numbers off west Baltic coasts and Finland, and breeding at several new localities in southern Sweden (Røskaft 1980*a*). In Switzerland, far more common in autumn, with spectacular numbers sometimes occurring along leading lines (Winkler 1984), but more common in spring at Straubing (south-east Germany) (Leibl and Melchior 1985); perhaps in autumn many birds reach Alpine range before changing to more westerly direction, whereas return movement more direct. This supported by data from Eichstätt (southern Germany): heading chiefly south-west in autumn, but east in spring (Waterhouse 1949).

In Kazakhstan, many 1st-year birds remain on wintering grounds all year; others return only part of the way, e.g. recovery 29 April 440 km south-east of breeding grounds (Gistsov *et al.* 1978). Similarly in study north-west of Black Sea, 1st-years were reported beyond breeding area throughout year, and those that returned to breeding grounds arrived later in spring than adults (Ardamatskaya 1968). At Lake Ladoga, spring passage of 1st-years is later than adults: 1st records of adults mid-March to mid-April, depending on lateness of spring, with main movement mid-April; 2nd wave in mid-May involves chiefly juveniles (Noskov *et al.* 1981).

Analysis of European ringing recoveries shows that juveniles tend to migrate further than adults (Busse 1969); results similar in ringing study on north-west coast of Black Sea (Ardamatskaya 1968), but no age difference in migratory distance detected for birds from Latvia (Baltvilks 1970) or Kazakhstan (Korelov *et al.* 1974).

In eastern Asia, some birds move south. Common winter visitor to southern Korea, some reaching adjacent areas of southern Japan in varying numbers (Austin 1948; Ornithological Society of Japan 1974). In China, winters in southern parts of breeding range and south to Kwangtung and Kwangsi in south-east (Schauensee 1984).

In New Zealand, where introduced, no regular movements reported; isolated rookeries far from normal range result from local introductions, though some perhaps due to vagrancy (Anon 1985). DFV

**Food.** Invertebrates, mainly beetles (Coleoptera) and earthworms (Lumbricidae), plant material (principally cereal grain), small vertebrates, carrion, and scraps of all kinds. Primarily a bird of agricultural landscapes, foraging almost exclusively on ground; only rarely in trees, taking defoliating caterpillars (Lepidoptera) or swarming beetles in spring (Lockie 1955). Forages on both pasture (taking invertebrates) and arable land (taking invertebrates and crops); in spring, feeds on newly-sown cereal or follows plough, etc., for exposed invertebrates, particularly larvae, then moves to pasture, notably where water-table high, when cereal grows to *c.* 15-30 cm (Lockie 1956*b*; Pinowski 1959; Kemper 1964; Feare *et al.* 1974; Bogliani 1985; Hölzinger 1987). In south-west Germany, when air temperature higher than soil temperature at depth of 5 cm prefers pasture to spring-sown cereal, since earthworms come to surface then; also, sown fields dry quickly, so only cereal seeds are then available (Ganzhorn 1986). Food can be short in summer because of necessity of feeding young, dry weather reducing availability of soil invertebrates, and increased energy expenditure caused by moult (Feare *et al.* 1974; Purchas 1980, which see for details of feeding ecology of introduced population in New Zealand). In late summer, favours stubble or maize *Zea* fields after harvest; in autumn, still stubble (especially after burning), ploughed fields, dung-fertilized land, and freshly-sown winter cereals; in addition, takes fruits, nuts, acorns *Quercus*, etc., mostly from ground though at times from trees (Lockie 1956*b*; Kemper 1964; Feare *et al.* 1974; Coombs 1978; Hubálek 1983; Hölzinger 1987; Watson 1989). Autumn is generally period of greatest food abundance (Waite 1985*a*). In winter, mostly returns to grassland, arable being significantly avoided in many places (Waite 1984*a*), though also frequents harvested root-crop fields and sprouting winter cereal (Hubálek 1983; Bogliani 1985; Hölzinger 1987); in severe weather, moves increasingly towards settlements, gathering at farmyards, animal feeding places, rubbish tips, etc.; also occurs in parks and gardens, searching lawns, turning over leaves, and feeding on or standing below bird-tables to catch falling scraps or harass smaller birds; often on shore feeding at low tide (Lockie 1956*b*; Kemper 1964; Grodziński 1976; Coombs 1978; Goodwin 1986). In south-west Germany, rubbish-tips are preferred winter feeding sites, attracting flocks of several thousand, many of which move on to neighbouring fields (Veh 1988). For seasonal variations in foraging strategy and flocking in Ireland, see Macdonald and Whelan (1986). In winter, will fly from urban roost to agricultural land every day to feed; distance of 12 km recorded (Pinowski 1959). Specializes more than other west Palearctic Corvidae in extracting invertebrates from below soil surface; of 569 feeding observations on grassland, southern England, 49% digging with bill and deep-probing (bill pushed into hole of earthworm or Tipulidae larva then opened), 31% surface picking, 10% surface probing (bill inserted in grass tussock then opened), 5% turning of clod, dung, etc., 4% pouncing on and pulling up earthworms, and 2% jumping up to take flying insects; digging occurs when hard ground makes probing difficult; hard blows delivered to walls of hole and dislodged earth tossed aside; often observed scattering cow-pats to get at invertebrates (Lockie 1956b, which see for comparison with other Corvidae). In central England in winter, 36% of 212 observations deep-probing, 18% jabbing into soil, 17% surface picking, 13% turning, 7% digging, 3% surface probing, and 1% pouncing (Waite 1984a). Abundance of preferred food items exploited by feeding in flocks; for disadavantages to juveniles of feeding with adults in large flocks, see Höglund (1985), East (1988), and Henderson (1990). See Waite (1984b) for winter foraging interactions with other Corvidae. In New Zealand, removed areas of turf to average depth of 19 mm (7–37, n = 40) to obtain Scarabaeidae (Coleoptera) larvae; tugged at grass for several seconds than examined exposed soil and underside of turf; no observations of probing (McLennan and MacMillan 1983). Will dig into drills from side to reach potatoes; pulls up, or digs out with bill, sprouting crops to feed on tubers or larvae in roots (Vertse 1943; Feare 1974). In southern England and eastern Scotland, flocks recorded sallying from fences to land with outspread wings to flatten barley crop and so gain access to grain (C J Feare). Flock of c. 40 recorded perching on heads of sunflowers *Helianthus* to take seeds (Plath 1987). For estimates of economic damage caused to agriculture see Eygelis (1961) for USSR, Fog (1963) for Denmark, Pivar (1965) for Yugoslavia, Feare (1974, 1978) for Scotland and southern England, Leever (1982) for Netherlands, and Hölzinger (1987) and Brod (1988) for Germany; for major early studies, see Rörig (1900), Collinge (1924–7), Madon (1928b), and Fisher (1948). Type of crop, farming methods, numbers of invertebrate pest species taken, and many other factors make it impossible to reach any general conclusion on whether *C. frugilegus* is harmful or beneficial to agriculture (Coombs 1978; Hölzinger 1987). Hammers open walnut *Juglans* with tip of bill (Purchas 1975); apparently transports them by making 2 holes for mandible tips to get a grip as otherwise too wide to hold in bill (Wallis 1932); record of walnut being transported 2 km to cache (Turček and Kelso 1968). Large earthworms held under foot and pulled to pieces; acorns, ears of cereal, etc. similarly held and pecked into fragments (Lockie 1956b; Coombs 1978). Much vertebrate food eaten as carrion, but can catch live prey with some skill: dropped on running vole *Microtus* on busy road, picking it up in bill 'with great dexterity' (MacLeod 1987); recorded snatching live fish from water surface in feet and taking it to tree (Grobe 1983). In captivity, all live mice *Mus* introduced into aviary were killed, though some not eaten; did not kill frogs *Rana* or any mammals larger than hamster *Mesocricetus*, and tackled dead mammals only if skin first cut open (Luniak 1977a). One bird killed and removed c. 4 *C. frugilegus* chicks from nest when parents absent (Caldwell 1949); records of dead *C. frugilegus* adults being eaten (Raynor 1948; Hubálek 1983). In Hungary, takes eggs of ground-nesting species, e.g. Skylark *Alauda arvensis*, Kentish Plover *Charadrius alexandrinus*, and terns (Sternidae), or other colonial nesters such as Red-footed Falcon *Falco vespertinus* and herons (Ardeidae) (Kalotás 1986b). Young gamebirds and poultry killed by heavy blows of bill while held under foot (Collinge 1924–7). Observation of bird carrying mussel *Mytilus* into air and dropping it 3 times until shell broke (Priestley 1947); bird with deformed bill ate mussels and slipper limpets *Crepidula* by holding them up after hammering open and extracting contents with tongue, or by inserting tongue between intact shells to hook them out (King and Rolls 1968). In northern Germany, in winter, at times 2 birds together seen on head and back of cow, sometimes pecking in hair presumably searching for parasites (Diesing 1987). Will kleptoparasitize Jackdaws *C. monedula* when feeding in mixed flocks (Höglund 1985); in garden in winter, attacked Starlings *Sturnus vulgaris* until they dropped food (Raevel 1981a); one bird attacked rabbit *Oryctolagus cuniculus* by striking with bill to drive it away from apple (Weiss and Wiehe 1984). When feeding in gardens shows considerable ingenuity and agility in reaching food, such as hanging on suspended bones and fat, or pulling them up by clamping string under foot after each tug, or hanging on string sideways (Richards 1973; Washington 1974). Seen to disappear inside litter bins when scavenging in quiet town centre (King 1976b). Stores food in autumn for later consumption, mainly acorns, walnuts, and pine *Pinus* cones, though earthworms recorded as being stored throughout year—probably for more immediate consumption; many acorns hidden September–December (peak in October), and cached acorns will provide more energy per minute when eaten than grain or invertebrates, although finding and handling times are longer; mostly recovered from stores February–March (Simmons 1970; Waite 1985a, b). As far as known, only European corvid to first dig hole for cache, covered by any nearby loose material after item deposited; apparently prefers grass, but also in bare soil or snow; sometimes just pushes food into grass tussock, e.g. 6 peanuts in 1 tussock, or at other times divided between 2–3 (Simmons 1968, 1970; Andrew 1969; Hubálek 1983). Record of cache in tree bark recovered some days after hiding (Fairhurst 1974). In New Zealand, marked birds spent 30–50% of autumn and winter feeding time hiding, locating, and eating acorns and walnuts; collected from trees or ground and carried singly to open space up to 300 m away where either eaten or cached;

individual birds stored walnuts over 1–2 km² and items were moved 4–5 times if other *C. frugilegus* present; average searching time during recovery 5·8 min per nut (Purchas 1975, 1980); in England, 4·7 min (Waite 1985*a*).

Diet in west Palearctic includes the following. Vertebrates (many as carrion): mole *Talpa*, shrews *Sorex araneus*, *S. minutus*, *Crocidura leucodon*, mice *Micromys minutus*, *Apodemus sylvaticus*, *A. flavicollis*, *Mus musculus*, rat *Rattus norvegicus*, voles *Arvicola terrestris*, *Microtus arvalis*, hamster *Cricetus cricetus*, rabbit *Oryctolagus*, hare *Lepus*, domestic cat *Felis*, lizard *Lacerta agilis*, frog *Rana*, eggs and young of birds, principally ground-nesters such as Anatidae, Phasianidae, domestic fowl, Charadriidae, Sternidae, Alaudidae, etc., or colony-nesters like Ardeidae and *F. vespertinus* and including own species. Invertebrates: springtails (Collembola), dragonflies (Odonata: Libellulidae), grasshoppers, etc. (Orthoptera: Gryllidae, Tettigoniidae, Gryllotalpidae, Acrididae), earwigs (Dermaptera: Forficulidae), cockroaches (Dictyoptera: Blattidae), bugs (Hemiptera: Scutelleridae, Pentatomidae, Coreidae, Reduviidae, Nabiidae, Cicadellidae, Coccoidea), thrips (Thysanoptera), lacewings, etc. (Neuroptera), adult and larval Lepidoptera (Pieridae, Tortricidae, Noctuidae, Lasiocampidae, Geometridae), caddis flies (Trichoptera), adult and larval flies (Diptera: Tipulidae, Bibionidae, Stratiomyidae, Tabanidae, Asilidae, Syrphidae, Oestridae, Tachinidae, Calliphoridae, Muscidae), fleas (Siphonaptera), adult and larval Hymenoptera (Ichneumonidae, ants Formicidae, wasps Vespidae, bees Apoidea), adult and larval beetles (Coleoptera: Cicindelidae, Carabidae, Dytiscidae, Hydrophilidae, Histeridae, Silphidae, Staphylinidae, Lucanidae, Geotrupidae, Scarabaeidae, Byrrhidae, Dryopidae, Elateridae, Cantharidae, Dermestidae, Meloidae, Anthicidae, Tenebrionidae, Nitidulidae, Coccinellidae, Cerambycidae, Chrysomelidae, Curculionidae), spiders (Araneae: Theridiidae, Araneidae, Lycosidae, Linyphiidae, Clubionidae), harvestmen (Opiliones: Phalangiidae), mites (Acari), woodlice (Isopoda: Armadillidiidae), shrimps (Amphipoda: Gammaridae), millipedes (Diplopoda: Julidae), centipedes (Chilopoda: Lithobiidae, Geophilidae), earthworms (Lumbricidae), molluscs (Mollusca: Helicidae, Succineidae, Bulimidae, Planorbidae, Chondrinidae, Valloniidae, Limacidae, Neritidae, Littorinidae, Mytilidae, Tellinidae, Calyptraeidae). Plants: fruits, seeds, and roots of pine *Pinus*, fir *Abies*, spruce *Picea*, oak *Quercus*, hornbeam *Carpinus*, beech *Fagus*, maple *Acer*, horse chestnut *Aesculus*, walnut *Juglans*, mulberry *Morus*, fig *Ficus*, nettle-tree *Celtis*, spindle *Euonymus*, dogwood *Cornus*, barberry *Berberis*, sea buckthorn *Hippophae*, mistletoe *Viscum*, bilberry, etc. *Vaccinium*, crowberry *Empetrum*, elder *Sambucus*, honeysuckle *Lonicera*, snowberry *Symphoricarpos*, *Robinia*, carob *Ceratonia*, locust *Gleditsia*, broom *Cytisus*, vetch *Vicia*, clover, etc. *Trifolium*, thistles *Cirsium*, *Carduus*, ox-tongue *Picris*, knapweed *Centaurea*, dandelion *Tarax-*

*acum*, various *Polygonum* species, goosefoot, etc. *Chenopodium*, buckwheat *Fagopyrum*, orache *Atriplex*, *Amaranthus*, plantain *Plantago*, hemp *Cannabis*, corncockle *Agrostemma*, stitchwort *Stellaria*, mustard *Sinapis*, wild radish *Raphanus*, hemp-nettle *Galeopsis*, buttercup *Ranunculus*, violet *Viola*, rowan *Sorbus*, apple *Malus*, pear *Pyrus*, plum, cherry, etc. *Prunus*, blackberry *Rubus*, hawthorn *Crataegus*, rose *Rosa*, strawberry *Fragaria*, tulip *Tulipa*, Virginia creeper *Parthenocissus*, grape *Vitis*, other cultivated plants such as sunflower *Helianthus*, pea *Pisum*, bean *Phaseolus*, various *Brassica* species, cucumber *Cucumis*, all types of root crops and cereals, especially maize *Zea*, grasses *Echinochloa*, *Setaria*, horsetail *Equisetum*, fungus *Lepista*. (Collinge 1924–7; Schuster 1930; Vertse 1943; Eygelis 1961; Turček 1961; Fog 1963; Kemper 1964; Pivar 1965; King and Rolls 1968; Loxton 1968; Folk and Beklová 1971; Holyoak 1972*a*; Feijen 1976; King 1976*b*; Gromadzka 1980; Hubálek 1983; Ganzhorn 1986; Kalotás 1986*b*; Hölzinger 1987; Jentzsch 1988; Sueur 1990*b*.)

Less of a generalist feeder than some other Corvidae, but wide variation occurs in proportions of animal and plant material taken through the year; on the whole, about equal parts of each, with peak consumption of plant material in winter and of animal prey in spring, when young often fed only invertebrates (Vertse 1943; Lockie 1955, 1956*b*; Kemper 1964; Feare *et al.* 1974; Waite 1984*a*, which see for comparison with other Corvidae). Sampling methods mean that soft-bodied invertebrates, particularly earthworms, may be overlooked in stomachs or pellets, since almost completely digested, so are often underestimated in diet (Lockie 1956*b*, which see for other Corvidae; Fog 1963; Luniak 1977*a*; Veh 1988). For food of adult see Table A. Additional studies: for England and Wales, see Collinge (1924–7), Lockie (1956*b*), and Holyoak (1972*a*); for Germany, Kemper (1964) and Hölzinger (1987); for Denmark, Fog (1963); for Netherlands, Feijen (1976); for Austria, Herrlinger (1966); for Poland, Gromadzka (1980); for Czechoslovakia, Hell and Soviš (1958); for Hungary, Vertse (1943); for Italy, Bogliani (1985) and Fasola *et al.* (1986).

Earthworms more important in adult diet than most studies suggest; accounted for 60–96% by weight of invertebrates taken in winter in England (Lockie 1956*b*; Waite 1981), and present in 91% of 123 stomachs in spring (Holyoak 1968). See also Hölzinger (1987). However, see Luniak (1977*a*) for captive birds' apparent rejection of earthworms and preference for beetles, but Purchas (1980) for similar experiments showing them to be among preferred foods. In general, animal material apparently preferred and cereals only taken in its absence, but modern farming methods mean reduction in invertebrate availability (Kalotás 1986*b*; Hölzinger 1987); according to Kalotás (1986*b*), proportion of invertebrates in diet in Hungary has decreased since study by Vertse (1943) by 15–20%, and in traditionally farmed habitat diet contained 21% more insects than in intensively cultivated study area.

Table A  Food of adult Rook *Corvus frugilegus* from stomach analysis. Figures in italics are percentages by volume, others are percentages by number.

| | Czecho-slovakia, Feb–Jul | | Czecho-slovakia, winter | | S W European USSR, Apr–Jun | Hungary, all year |
|---|---|---|---|---|---|---|
| Beetles | 16·2 | *12·9* | 7·0 | *0·1* | 52·6 | 27·8 |
| (Scarabaeidae) | (2·1) | *(3·5)* | (—) | *(—)* | (45·0) | (1·2) |
| (Curculionidae) | (7·2) | *(6·3)* | (—) | *(—)* | (3·9) | (4·8) |
| (Carabidae) | (2·3) | *(1·4)* | (0·8) | *(0·03)* | (0·7) | (4·6) |
| (Elateridae) | (2·6) | *(1·0)* | (0·8) | *(0·03)* | (0·4) | (1·7) |
| Lepidoptera | 0·5 | *0·9* | 0 | *0* | 18·0 | 4·0 |
| Diptera | 0·3 | *0·1* | 3·1 | *1·8* | 0·6 | 0·9 |
| Orthoptera | 0 | *0* | 0 | *0* | 0·2 | 2·3 |
| Hemiptera | 0·5 | *0·3* | 0 | *0* | 0·1 | — |
| Hymenoptera | 0·7 | *0·1* | 0 | *0* | 1·0 | — |
| Spiders, harvestmen | 0·1 | *0·1* | 0 | *0* | 0·1 | — |
| Earthworms | 0·02 | *0·2* | 0 | *0* | 1·3 | 0·7 |
| Snails | 0·1 | *0·7* | 0 | *0* | 0 | 7·3 |
| Vertebrates, eggs | 0·2 | *4·3* | 7·0 | *4·9* | 0·1 | 6·3 |
| Carrion, scraps | — | *1·5* | 13·3 | *15·9* | — | 4·8 |
| Cereal grain | 62·2 | *54·6* | 46·9 | *69·5* | 13·7 | 15·8 |
| Other crops | 0·1 | *1·0* | 3·1 | *2·2* | 10·7 | 4·2 |
| Other plants | 0·3 | *3·7* | 18·8 | *6·3* | 1·2 | 25·3 |
| Total no. of items | 6930 | | 128 | | 1205 | 4842 |
| Source | Folk and Toušková (1966) | | Folk and Beklová (1971) | | Eygelis (1961) | Kalotás (1986*b*) |

Invertebrates probably taken according to accessibility and seasonal abundance (Holyoak 1972*a*; Gromadzka 1980); availability of earthworms, leatherjackets (Tipulidae larvae), and wireworms (Elateridae larvae) dependent on soil condition and air temperature (Lockie 1956*b*; Holyoak 1972*a*; Ganzhorn 1986). Proportion of animal material in diet begins to decline sharply in late summer or early autumn when cereals, seeds, fruits, etc., increasingly taken; earthworms taken fairly constantly throughout year, with peak in breeding season, while vertebrates probably more frequent in winter (Holyoak 1972*a*; Gromadzka 1980; Waite 1985*b*; Kalotás 1986*b*). In south-west Germany, winter, 90% by weight of 94 pellets was plant material, remainder almost all vertebrates, though invertebrates may be poorly represented in pellets (Hölzinger 1987); *c.* 75% by weight of contents of 131 winter stomachs from Poland was plant material (Gromadzka 1980), as was 90% of 359 stomachs from Czechoslovakia (Hell and Soviš 1958). However, other studies have found 4 times more plant than animal material by weight taken in February–July (Folk and Toušková 1966) or over whole year (Kalotás 1986*b*). Grit in gizzard can often account for 30–50% of weight of contents (Hell and Soviš 1958; Herrlinger 1966; Folk and Beklová 1971). In southern England, 21% of 130 invertebrates taken (excluding earthworms) were 2–6 mm long, 31% 7–12 mm, 31% 13–18 mm, and 17% longer than 19 mm (Lockie 1956*b*).

In central England, in winter, highest average energy intake in 4 microhabitats always obtained from small earthworms, ranging from 0·01 kJ per min from within or beneath dung, to 0·3 from beneath soil surface (Waite 1984*a*, which see for comparison with *C. monedula*, Carrion Crow *C. corone*, and Magpie *Pica pica*); cached acorns provided 9·2 kJ per min, grain 1·7, and invertebrates as a whole 0·8 (Waite 1985*a*). Highest intake in north-east Scotland (*c.* 6 kJ per min) obtained from stubble and newly-sown grain fields (Feare *et al.* 1974). Adult has average daily requirement of *c.* 750 kJ, and, over the year, eats average 10–70 g wet weight of grain and 15–70 g of animal material per day (Gromadzka 1980, which see for details); in north-east Scotland, daily intake varies between *c.* 450 kJ and *c.* 2450 kJ through the year; captive ♀ required 32 g dry weight (*c.* 620 kJ) per day (Feare *et al.* 1974). In New Zealand, 30 earthworms, 50 insects, or 0·5 walnuts required per daylight hour (Purchas 1980). From pellets at winter roost, south-west Germany, each bird estimated to eat average of 10 rodents over 1 winter (Veh 1988).

For food of nestlings, see Table B. In north-east Scotland, fed only invertebrates for first 2 weeks, after which cereal brought, though animal material still dominant (Feare *et al.* 1974). In southern England, proportion of

Table B  Food of nestling Rook *Corvus frugilegus* from stomach analysis (Gerber 1956) and collar-samples (other studies). Figures in italics are percentages by volume, others are percentages by dry weight (Ganzhorn 1986) and by number (other studies).

| | S E England | | South-west Germany | Hungary | S W European USSR |
|---|---|---|---|---|---|
| Vertebrates, eggs | — | — | 11·6 | 0·4 | 0·3 |
| Orthoptera | — | — | — | 14·6 | 0·1 |
| Hemiptera | — | — | — | 1·3 | 0·9 |
| Lepidoptera | — | — | 0·2 | 0·4 | 1·3 |
| Diptera | 12·7 | *5·5* | 12·7 | 0 | 2·2 |
| Beetles | 6·3 | *1·5* | 4·0 | 53·1 | 72·3 |
| (Scarabaeidae) | (—) | *(—)* | (—) | (11·1) | (53·5) |
| (Curculionidae) | (—) | *(—)* | (3·3) | (29·6) | (7·5) |
| (Carabidae) | (—) | *(—)* | (0·1) | (—) | (0·3) |
| (Elateridae) | (—) | *(—)* | (0·1) | (7·5) | (2·0) |
| Spiders, harvestmen | 12·3 | *1·0* | 0·3 | 0 | 0·1 |
| Earthworms | 33·4 | *60·5* | 47·5 | 0 | 8·1 |
| Snails | — | — | 0 | 7·1 | 0·2 |
| Cereals | — | *13·0* | 24·0 | 22·1 | 12·0 |
| Other crops | — | — | 0 | — | — |
| Other plants | — | — | 0 | 0·9 | 2·1 |
| Total sample | 100 samples, 1304 animal items | | 33 samples | 452 items | 1632 items |
| Source | Lockie (1955) | | Ganzhorn (1986) | Gerber (1956) | Eygelis (1961) |

plant material by volume in warm spring at age 1–7 days was 1%, 8–15 days 6%, 16–22 days 22% ($n = 706$ items); in cold spring 16%, 11%, and 37% ($n = 598$). In same 2 years, earthworms 74% and Tipulidae larvae 9% of diet by volume in warm spring, 47% and 0% in cold spring, while Lepidoptera larvae (mainly Noctuidae) rose from 7% to 30%; in some years, Tipulidae larvae too small for adults to bring to nest so earthworms become very important. (Lockie 1955, 1959.) For 103 stomachs in Denmark, see Fog (1963). Energy requirement per nestling per day at 1–5 days *c.* 230 kJ, at 16–20 days *c.* 460 kJ, and at 26–30 days, *c.* 500 kJ; total requirement over nestling period *c.* 13 000 kJ per nestling; each receives *c.* 35 g of cereal and *c.* 1750 g of invertebrates in that time (Gromadzka 1980, which see for details). In north-east Scotland, average of *c.* 2000 kJ per day brought to nest, some of which for ♀ (Feare *et al.* 1974). BH

**Social pattern and behaviour.** Well studied. Important early monographs by Yeates (1934) and Gerber (1956). For more recent studies see Coombs (1960, 1978) for Cornwall (England), and Ena Alvarez (1979) for León province (northern Spain), also Dunnet, Patterson, and colleagues for Ythan valley (north-east Scotland) (see below for details).

1. Gregarious outside breeding season for feeding, roosting, and migration, though pair is the chief constituent unit (see Bonds, below). Foraging mainly in family groups until end of July or early August when these coalesce into flocks. Local dispersal of juveniles and, in parts of range, migration of all age groups, affects flock composition. Although juveniles most often join adult flocks, they sometimes predominate (Coombs 1960, 1978, which see for ratios) or even form independent flocks, often with Jackdaw *C. monedula* (Burkitt 1936; Campbell 1936). In Ythan valley, some evidence of partial segregation of adults and 1st-years in late autumn and early winter (Dunnet *et al.* 1969, which see for details). In overwintering German flocks, average *c.* 45% juveniles (Friedrich 1974, which see for other references). In Bayern (Germany), migrating flocks of up to 100 or several hundred, about half associated with *C. monedula* (Leibl and Melchior 1985). Flock size and dispersion (see Flock Behaviour, below) vary regionally and seasonally with food supply, e.g. in southern Ireland, flocks large in late winter, small at other times (Macdonald and Whelan 1986). In Ythan valley, generally quite large dense flocks near colonies in autumn and winter, more widely dispersed small flocks or singletons in summer, with dramatic reduction in flock size from March or April (for similar seasonal change in Poland, see Pinowski 1959). Birds foraged further from colony when breeding (average *c.* 1·5 km) than in autumn and winter (*c.* 1 km) but furthest immediately after breeding (*c.* 2·2 km). No evidence that each colony has exclusive feeding area; rather, neighbouring colonies overlap. (Patterson *et al.* 1971; Patterson 1975.) In León, foraging radius in breeding season 3 km compared with *c.* 10 km in autumn and winter (Ena Alvarez 1979). In winter, Czechoslovakia, limits of dispersal from roost mostly 20–25 km, exceptionally 45–50 km (Hubálek 1978*b*, 1980). In addition to *C. monedula* (see above), also associates for feeding with Starling *Sturnus vulgaris* and House Sparrow *Passer domesticus*, though interspecific dominance hierarchy prevents free intermixing (Katzir 1981: see Flock Behaviour, below). In Trondheim area (Norway), large mixed flocks of *C. frugilegus*, Carrion Crow *C. corone*, and *C. monedula* forage together (Røskaft 1980*a*). BONDS. Mating system essentially monogamous (for

possible exceptions, see below) and pair-bond maintained throughout year, for several consecutive years, and, at least in established breeders, perhaps life-long (Coombs 1978). However, Røskaft (1980*c*) found that inexperienced birds may sever bond after breeding failure (see also Breeding Dispersion, below, for other consequences of unsuccessful breeding). In colony, small proportion of ♀♀ unpaired and not uncommonly succeed in associating with established pairs (Patterson 1980*b*), producing trios (♂, 2 ♀♀) often interpreted as bigamy though little evidence of ♂ copulating with 'extra' ♀. Extra ♀ sometimes attaches herself to pair's nest, and may attempt to contribute to nest duties, but is usually rebuffed by resident ♀ (Chappell 1946; Dean 1947; Green 1980). In other cases, extra ♀ builds own nest, attached or near to pair's nest, lays, and incubates; both ♀♀ then fed by ♂ (Dunlop 1917; Adams 1948; Lye 1948; Chappell 1949; Green 1982) but both fail to fledge young, and since ♂ alone feeds young until well-developed, extremely unlikely he could ever raise 2 broods simultaneously (Green 1982*a*). In exceptional case, 2 ♀♀ incubated amicably side by side in nest-cup (Dunlop 1917). ♂♂ typically and frequently promiscuous with ♀♀ which are already paired, e.g. paired ♂ seen to attempt to copulate with 4 neighbouring ♀♀ and to join in mating attempts initiated by other ♂♂ (Goodwin 1955*b*). Although extra-pair copulations are typically with incubating (and therefore already fertilized) ♀♀ (e.g. Coombs 1978), likely that both ♂♂ and ♀♀ gain some genetic advantage from promiscuity (Røskaft 1983*a*: see Heterosexual Behaviour, below). Within monogamy, pair share building and defence of nest, but usually only ♀ incubates (♂ feeding her on nest) and broods; ♂ alone brings food for young until nearly halfway through fledging, after which ♀ makes increasing contribution (Røskaft 1981*b*; see also Relations within Family Group, below). Young independent at *c.* 6 weeks after fledging (Røskaft 1985). Age of first breeding typically 2 years (Marshall and Coombs 1957; Patterson and Grace 1984), occasionally 1 (Ena 1984*b*). BREEDING DISPERSION. Typically colonial, with nests densely clustered in tree-tops ('rookery'). Nest density within colony varies and, where large numbers nest in uniform habitat, colony may be hard to define. Thus 'complex' colonies occur where nesting groups are loosely connected (Patterson *et al.* 1971). 'Solitary' nests usually colony-outliers but, according to Dementiev and Gladkov (1954), genuinely solitary nesting replaced colonial habit in Oka Reserve (Russia) after colonies were incessantly molested for several years. Largest colonies recorded in Hungary (maximum 16 000 nests in 1943); other large colonies in Germany (maximum 9000 nests in 1899); in Netherlands, 1944, 9 colonies with more than 1000 nests, 1 with 2240 nests (Feijen 1976); see also below for large Scottish colonies. Exceptionally large concentrations, however, are often complex colonies, e.g. Hatton Castle complex (Grampian, Scotland) held 6085 nests in 1945, 6697 in 1957, and 2669 (in 16 groups) in 1975 (Sage and Vernon 1978). Population declines in west Palearctic (see Population) have been accompanied by widespread fragmentation into smaller colonies, but this trend recently reversed where recovery has occurred. In 1975, reflecting trend to smaller colonies, 68·7% of colonies in England and Wales, a third in Scotland, and just over half in Northern Ireland, contained 25 nests or fewer; only 41 colonies contained more than 500 nests (of which 36 in Scotland, largest 2087 nests); overall, average number of nests per colony ranged from 24·4 in England to 79·1 in Scotland (Sage and Vernon 1978; see also Castle 1977 for Scottish parts of census). Sample census in 1980 showed concentrations into slightly larger colonies since 1975, due to reduction in number of very small colonies (Sage and Whittington 1985). For colony sizes in Poland, see Józefik (1976), Hordowski (1989); for Skåne (southern Sweden),

Malmberg (1971); for Netherlands, Directie NMF (1989); for Luxembourg, Wassenich (1969) and Mentgen (1988); for Germany, Pfeifer and Keil (1956) and various papers in *Beih. Veröff. Nat. Land. Bad.-Württ.* 1988, **53**; for USSR, Dementiev and Gladkov (1954). For pioneering discussion of fluctuations in colony size, including dwindling to extinction and founding of new colonies, see (e.g.) Nicholson and Nicholson (1930), Alexander (1933), Yapp (1951), and Coombs (1961). Studies in Grampian showed that large changes in colony size are caused by variability in number of birds joining colonies; general population increase tends to focus on large colonies which are the most attractive to recruits (Richardson *et al.* 1979; see also below for site-fidelity). In Poland, minimum distance between colonies 2·16 km (Józefik 1976), in Moskva valley (Russia) 3–5 km (Korbut 1985), but density more often expressed as colonies per 100 km² or as nests or pairs per km². In Britain, density varies from less than 10 to more than 50 colonies per 100 km² (Brenchley 1986, which see for comparison of 1944-6 and 1975 surveys). In 1975 survey average nests per km² as follows: Wales 1·9, Scotland 3·2, England 3·9, Northern Ireland 8·0) (Sage and Vernon 1978); highest density in Britain 24 nests per km² in Ythan valley (Patterson *et al.* 1971). In various parts of Europe, typically less than 1 pair per km² (Wassenich 1969; Feijen 1976; Józefik 1976). Territory confined to small area round nest, radius 1–3 m or less (Røskaft 1980b, 1981a); will compress where additional nesting pairs squeeze into cluster (Coombs 1960). Usually several nests per tree, but where only 1 in tree, territory extends to perimeter of tree. When fledglings scatter into trees in colony, defence extends to area occupied by brood (perhaps 80–100 m²), thus overlapping 'extended territories' of other pairs with young out of nests. (Ena Alvarez 1979.) Nests typically 0·5–2 m apart (Røskaft 1981a). However 'multiple' nests, with up to 4 abutting, quite common (Green 1982a), sometimes arising from unpaired ♀♀ building alongside established pairs (see Bonds). Fidelity to colony and nest-site generally high. For assiduous attendance at sites throughout most of year, see Roosting and Colony Attendance (below). 9 pairs studied after failed breeding suggested that old, experienced, and previously successful pairs have strong bond to, and will retain, nest-site; some will, however, move to new site near old one. By contrast, inexperienced birds abandon nest-site after breeding failure and pair-bond may even break down. All pairs which breed successfully return to same site next year. (Røskaft 1980c.) Fidelity to colony site continues even if nests removed (Oskar 1986). Nests destroyed early in breeding season are rapidly replaced, although destruction of colony may also result in wholesale exodus to neighbouring areas (Dyrcz 1966). No quantitative information on natal site-fidelity but significant proportion of young breed in or near natal colony (Richardson *et al.* 1979; I J Patterson); also records of some dispersing hundreds of km (Grodziński 1980). For attendance patterns of young birds at colony indicative of likelihood to recruit there, see Patterson and Grace (1984). No true nesting associations with other species, though in Volga delta and lower Syr Dar'ya, 'mixed colonies' (presumably some segregation) found with herons (Ardeidae), Cormorant *Phalacrocorax carbo*, and Glossy Ibis *Plegadis falcinellus* (Dementiev and Gladkov 1954). Colony near Trondheim included 15 pairs of successfully breeding Fieldfares *Turdus pilaris* (Røskaft 1978, 1980b). Kestrel *Falco tinnunculus* reported nesting in midst of colony without any obvious conflict (Gill 1919). ROOSTING AND COLONY ATTENDANCE. Widely studied. For most of year, members of several colonies typically merge to form communal roosts, often with *C. monedula* (see that species for references). Even in communal roost, members of pair perch close together (Coombs 1960, 1978). Both sexes may roost on prospective nest-site (Nethersole-Thompson and Nethersole-

Thompson 1943a). Once nesting under way in spring, ♀ roosts on nest, ♂ on branch nearby. For roosting of family immediately after fledging, see Relations within Family Group (below). By mid-July in Britain, when family parties form flocks, communal roosting gradually resumed. (Coombs 1960, 1978.) Until the following breeding season, succession of communal sites may be used, numbers and duration of these varying regionally, as following examples show. In Ythan valley, progressively from mid-May to early July, birds from 4 colonies occupied summer roost (c. 1 km from colonies); during September, switched to larger autumn roost 8 km from colonies; site of final winter roost, occupied from October, changed considerably between years; thus, including colony, 4 different sites used in most years. For 19 British colonies, average maximum distance from colony to winter roost 18 km. (Patterson *et al.* 1971.) Pattern in Cornwall differed in that some post-breeders were 'nomadic', roosting near their last feeding place of the day; also, immatures and some failed breeders continued to use winter roost site during breeding season (Coombs 1961, 1978). For similar seasonal patterns, see Aerts and Spaans (1987) for Netherlands and Grodziński (1980) for Poland. In León, pattern distinct from other European populations in that breeding colony serves as roost throughout the year except for mid-July to mid-October (Ena Alvarez 1979). Elsewhere, roosting pattern more influenced by migration, e.g. in Moravia (Czechoslovakia), roost types include one used only for several days during passage (Hubálek 1980). For dispersion of winter roosts and age structure of users in Vienna, see Grüll (1981). Communal roosts may contain up to tens of thousands of birds, e.g. in Grampian (Scotland) 2 roosts (largest in Britain) c. 45 000 and 65 000 (Patterson *et al.* 1971). For review of size of other Scottish roosts, see Munro (1975); for elsewhere in Britain, see Coombs (1961). In Moravia, average combined totals of *C. frugilegus* and *C. monedula* per roost c. 23 000 (200-135 000) of which c. 80% *C. frugilegus*; average area occupied by roost c. 5 ha (0·5-25), average density (both species combined) c. 5000 per ha, and 11 (4-18) per tree (Hubálek 1980; see also Hubálek 1978b for details of roost in Brno). In Hertfordshire (England), winter roost sites typically in different woods (younger and denser) from those used for nesting, and roosting birds sought sheltered south side of wood (Brian and Brian 1948). No such distinctions found in Ythan valley by Patterson *et al.* (1971), but preference for leeward side also found by Swingland (1977) and, in cold north winds, by Hubálek and Horáková (1988). Swingland (1977) also found vertical stratification by age with adults higher (near tree-tops) than younger (subordinate) birds (for juveniles forming separate groups, see also Burkitt 1936). In severe weather, adults descended lower, 'pushing' younger birds further down trees and sometimes even displacing them to another part of wood. However, reasons for stratification not clear, as adults exposed to greater energy loss by being higher up. (Swingland 1977.) Roosting on frozen lake recorded (Lambert 1965; Schlögel 1987). Thermal efficiency presumably accounts for urban setting of winter roosts (trees) in some colder parts of range (see, e.g., Grodziński 1971, 1976, 1980, also references and discussion in *C. monedula*). Diurnal and seasonal rhythm of roosting very similar to *C. monedula* (see that species). For data on *C. frugilegus*, see Dalmon (1932), Philipson (1933), Patterson *et al.* (1971), Feare *et al.* (1974), Stork *et al.* (1976), Swingland (1976), Coombs (1961, 1978), and Hubálek (1978a, b, 1980, 1983). As in *C. monedula*, timing of arrival and departure directly related to time of sunset and sunrise, and to actual light intensity. See Flock Behaviour (below) for assembly at and departure from roost. For daily and seasonal sleep patterns, see Szymczak (1987a, b) and Szymczak *et al.* (1989). In early spring, birds start visiting colony in small numbers in early morning and increasingly through day; however, at this initial stage remain faithful to winter roost site overnight. In March, numbers visiting colony and time spent there increase markedly (arrive earlier in morning, leave later in evening) until winter roost abandoned completely (Burkitt 1935; Busse 1965). Attendance at colony largely discontinued during post-breeding moult but regular visiting resumed in autumn, less so in winter; e.g. in Ythan valley resident breeders visited colonies in all months except August and, even in mid-winter, usually assembled at colony at beginning (dawn) and end (dusk) of feeding activity, and at intervals during day (Patterson *et al.* 1971). For diurnal and seasonal variation in colony attendance in relation to temperature (etc.), see Busse (1962).

2. Generally wary of man in breeding season, but in winter, when weather restricts access to food, allows closer approach (e.g. Grodziński 1976). For flushing distances in colony, see Parental Anti-predator Strategies (below), and for erratic flights as presumed anti-predator strategy, also 'dreads', see Flock Behaviour (below). No other escape reactions recorded except for account of bird apparently shamming death (Moffat 1943). In general alarm, adopts sleeked upright posture. At colony, especially at own nest-site, notably agile, e.g. sliding down twigs towards nest (Ena Alvarez 1979). For somersaulting on wires and hanging upside down, see Wüst (1961), Greenhalgh (1965), and Munro (1977). FLOCK BEHAVIOUR. Widely studied, especially for feeding and roosting. Within flock, closer contact between pair-members is evident (Coombs 1960), but feeding flocks are generally loose (compared with, e.g., Woodpigeon *Columba palumbus*) and mobile. Individual distance in Ythan valley increased from winter to summer, e.g. from *c.* 5 m on grass in January to *c.* 10 m in July. (Patterson *et al.* 1971; Patterson 1975; see also Feare *et al.* 1974.) Similar pattern found in southern Ireland (Macdonald and Whelan 1986). In Leicestershire (England), young birds preferred to be in small flocks, being less efficient foragers in large flocks with higher proportion of adults; this effect not due to inexperience alone (Henderson 1990) and presumably reflects, in part, dominance relationships. Dominance in captive flocks is related to sex (♂♂ dominate ♀♀), fledging weight (Røskaft 1983c), and age (Swingland 1977: see also Roosting). Flocking considered to be adaptation for exploiting localized food sources (Feare *et al.* 1974; Höglund 1985, which see for relationship between flock size and foraging success). For other possible functions of flocking, also for vigilance, dispersion, etc., in relation to food supply, see Waite (1978, 1981). For sexual behaviour in flocks, see Heterosexual Behaviour (below). Study of mixed feeding flocks of *C. frugilegus* and *C. monedula* in Oxfordshire (southern England) from autumn to spring indicated that low temperatures suppress sexual behaviour and increase aggression in both species (Lockie 1956a: see Antagonistic Behaviour, below); predominance of courtship-display in flocks in middle of day (Macdonald and Whelan 1986)) perhaps likewise attributable to higher temperature at that time. Roosting behaviour outside breeding season described by Philipson (1933), Coombs (1960), Grodziński (1971, 1976, 1980), and Hubálek (1978a, b, 1980, 1983); see also references in roosting associate, *C. monedula*. Flocks start leisurely return along regular flightlines to roost in afternoon, punctuated by stopovers (pre-roost assemblies) at which flocks (both resident and migrant, e.g. Burns 1957, Grodziński 1976) gradually join forces. One of the most detailed accounts of final assembly before entering roost (in Ythan valley) as follows. During afternoon and evening, birds (with *C. monedula*) converged from feeding grounds (usually corn stubble) near roost. These enlarging flocks fed actively and noisily at first, this gradually subsiding until most sat in silence, some preening or sleeping, except for fringe birds (more restless than central birds) which attempted to crowd into centre of flock. Fringe birds also initiated entry into roost. Entry slow and noisy, accompanied by erratic flights (interpreted as anti-predator strategy) before descent into roost-trees. Pre-roost assemblies did not occur during breeding season, and only rarely during summer, but began August–September when grain became available; from November to February, both pre- and post-roost assemblies occurred daily. In October, and again late February and March, usually flew directly to colonies from roost in morning, but in midwinter exiting birds (about half of which were immigrants: I J Patterson) headed instead for fields near roost where fed intensively before dispersing further. (Feare *et al.* 1974, which see for weather effects on pre-roost dispersal pattern.) *En route* to roost, from early August to early December but typically September–October (e.g. Savage 1928, Peake 1929), flock (within which pairs evident) often performs characteristic towering, tumbling flights, sometimes known as 'crows' wedding' or 'parliament': birds spiral up to great height, calling, with some chasing, swooping, and tumbling in midst of flock (Wynne-Edwards 1962). Same aerial display also occurs on approach to colony or feeding grounds, especially in windy weather and on warm sunny days (when thermals presumably aid ascent). Flock behaviour at colony also includes sudden outflights or 'dreads': during periods of high attendance, sudden bout of Caw-calls (see 2 in Voice) precedes mass silent departure from all or part of colony; only sometimes traceable to obvious disturbance or alarm. (Coombs 1960, 1978.) SONG-DISPLAY. So-called song (1 in Voice) given by unpaired ♂ from tree-top or other vantage point (e.g. telegraph pole) to attract unpaired ♀♀, mainly in autumn (after moult) and early spring, sometimes for several hours on warm autumn and winter days. Singing accompanied by less mobile variant (Fig A) of Bowing-display (see Antagonistic

A

Behaviour, below) in which legs flexed, tail raised and spread, wings closed but lifted somewhat, head moved forward and back (horizontally); at same time, body may pivot from side to side. May sing thus with full throat-pouch (intending to feed any ♀ attracted). (Coombs 1960, 1978; Richards 1976a.) ANTAGONISTIC BEHAVIOUR. (1) General. For dominance hierarchy, see Flock Behaviour (above). For fighting over food, see Lockie (1956a) and Patterson (1970, 1975). Most nest defence by ♀ (Richards 1976a; Røskaft 1981a). Defence strongest in breeding season, especially during nest-building, with lull from May–June to end of July or early August, marked resurgence September–October and continuing at reduced intensity until spring (Coombs 1960, 1978). (2) Threat and fighting. Following account based on detailed descriptions by Lockie (1956a) and Coombs (1960, 1978) which see for drawings. For associated calls, see especially 2, and 13–17 in Voice. Commonest indication of threat by both sexes in territorial disputes is feather-ruffling, especially on back and flanks. Ruffling accompanied most threat-postures, which are various intensities of Head-down and Forward-threat, e.g. Fig B shows high-intensity Head-down; for apparently related 'elaborate bill-down' (Lockie 1956a) see Mating in Heterosexual Behaviour (below). Bowing-display frequently used as self-

B

advertisement by both sexes before and after territorial disputes, especially during nest-building, and often ends with whole groups of quarrelling birds departing in Pursuit-flight (see Heterosexual Behaviour, below); in Bowing-display (Fig C), bird

C

simultaneously bows deeply and gives Caw-call, with each bow raising spread tail above horizontal and raising wings; intensity of display varies, and at greatest intensity legs flexed. Simultaneous Wing-lifting (folded wings quickly lifted just above back) and Tail-flirting (tail opened slightly, then closed again) may continue between bouts of Bowing-display. Intransigence in ground dispute may elicit Take-off posture (Fig D) in which

D

body angled steeply upwards, head and bill upstretched, all feathers sleeked except forehead, fanned tail raised or horizontal, sometimes (as in Fig D) wings held out and drooped; Take-off sometimes immediately precedes jumping or flying up towards rival and fighting just above ground. Bird initiating any sort of attack usually wins confrontation (Lockie 1956a; Patterson 1975). For appeasement display, sometimes used in disputes, see ♀ soliciting-posture under Mating (below). In feeding flocks, frequency of aggressive encounters much higher in winter than summer, increasing with decreasing temperature, with snow cover, and with closer spacing (Lockie 1956a; Patterson 1970, 1975). Birds (conspecific or other: see Parental Anti-predator Strategies, below) intruding on nest-territory are almost always

chased off, sometimes only after pecking attack (Røskaft 1980b, 1981a; Røskaft and Espmark 1982; Roland 1988). Motives for intrusion on nest-territory notably include stealing nest-material and extra-pair copulations (see Heterosexual Behaviour, below). In interspecific disputes, *C. frugilegus* subordinate to *C. corone* (Coombs 1960, 1978), but exceptionally *C. frugilegus* recorded killing *C. corone* (Morrison 1977). *C. frugilegus* dominant over smaller Corvidae, etc. (Katzir 1981); thus in mixed flocks, successfully supplanted *C. monedula* in 93% of 198 encounters (Lockie 1956a). For apparent competition with *S. vulgaris* over communal roost-site, see Abbott (1931). HETEROSEXUAL BEHAVIOUR. (1) General. Annual cycle extensively studied (Marshall and Coombs 1957; Lincoln *et al.* 1980, which see for associated physiological changes). Sexual cycle begins, and pair-bonds established, in autumn (regular colony attendance), enabling early spring nesting and fledging young before dry summer months when invertebrate food becomes scarce. Courtship-display in colony in autumn probably as intense as in spring. (Coombs 1960, 1978, Lincoln *et al.* 1980.) From June–August, no sexual behaviour and birds appeared not to be paired (but see below), and also reduced sexual activity in coldest months (December–January) (Lincoln *et al.* 1980). For courtship (including copulation) in and around colony in winter, especially in mild weather, see Morley (1943) and Ogilvie (1949). (2) Pair-bonding behaviour. Starts in 1st autumn, and 1-year-olds seen visiting colony from mid-March where show incipient courtship and nest-building activity (Campbell 1936; Goodwin 1955b). When unpaired ♀ is attracted to singing ♂ (see Song-display) pair is formed as soon as she accepts food from him (see subsection 3, below) after which they keep close company (Richards 1976a). Pair-bonding also includes Pursuit-flights, and Allopreening (see below). Pursuit-flights typically take place around breeding colony, especially in autumn and early spring and often after territorial disputes. 2–12 birds may participate, and flight may continue for 20 min, accompanied by much noisy calling (2 in Voice). One or more birds appear to be pursuing but roles may change; ♀ probably more often the pursued. Chasing groups may split up or coalesce. Diagnostic of Pursuit-flight is slow buoyant flight (Fig E) in which wings are stiff, raised further,

E

and upstroke much slower than in normal flight; occasionally, upstroke held in short glide. Pursuit-flights typically end (as they often begin) with bouts of Bowing-display. (Coombs 1960, 1978.) Allopreening occurs between mates in colony in autumn and spring, especially before hatching (Røskaft 1981a); usually confined to head, neck, and mantle; either sex may preen or be preened (Ogilvie 1951; Coombs 1960, 1978) but apparently ♀ initiates Allopreening once pair-bond established (Richards 1976a) and only she Allopreens later in breeding season (Roland 1988). According to Ena Alvarez (1979), ectoparasites are removed by Allopreening. Incubating ♀♀ are usually Allopreened on the nest; in addition, pair sometimes perform Billing (mutual nuzzling of bills). (3) Courtship-feeding. Has critical role in autumn pair-formation (Richards 1976a). ♂ feeds ♀ but ♂ also occasionally begs (Goodwin 1955b). Winter lull occurs (though seen from late December: Firth and Firth 1945), behaviour re-starting in spring among 1-year-olds at colony (Goodwin 1955b) and in breeding pairs from laying until young *c.* 3 weeks old (Røskaft 1981a; 1983b). At colony early in season, ♂'s Bowing-display may elicit Bowing-display from ♀ (Richards 1976a) but more often Begging-display (see below); once nesting starts, ♂'s display wanes and ♀ starts begging as soon as she sees mate approaching (Roland 1988). ♀'s Begging-display (Fig F) is

F

as in juvenile: crouches, wing-flutters in 'loose-at-the-shoulders' fashion, and calls, voice changing as food transferred (Goodwin 1955b; Coombs 1960, 1978; Røskaft and Espmark 1982: see 3–5 and 7–8 in Voice). Once nest built, ♂ regurgitates to ♀ sitting or standing on nest, but during chick-rearing she also leaves nest to be fed (see also Relations within Family Group). For appeasement-display by begging ♀, see Mating (below). After Courtship-feeding, one or both birds bill-wipe, then both usually continue intimate contact for a while (during which they may copulate). (Yeates 1932; Ogilvie 1947; Røskaft 1981a.) In Norway, ♂ fed ♀ on average 1·4 times per hr throughout day (*n* = 26 pairs), at rate directly related to fledging success (Røskaft 1983b). Young ♂♂ fed mates more often (but were absent from nest longer) than older ♂♂ (Røskaft *et al.* 1983, which see for other parameters of reproductive effort in relation to age and sex). (4) Mating. Usually on nest, occasionally on branches nearby; sometimes in prospective nest-site (Nethersole-Thompson and Nethersole-Thompson 1943), or in trees without nests, and, at least in some areas, not uncommonly on ground near colony (Campbell 1936; Coombs 1960, 1978; Røskaft 1983a; Goodwin 1986). Either sex may solicit, but most copulations are promiscuous and forced with no prior display. Soliciting-display of ♂ most often seen early in season, and, although ♀ does not appear cooperative, actual or attempted copulation nearly always follows. Facing ♀, ♂'s display (Fig G) resembles exaggerated Bowing-display but with slower movements, wings lifted (may be held there) and bill down. ♂ intending to copulate may approach ♀ through branches displaying thus (Coombs 1960, 1978). This display the same as 'elaborate bill-down' seen during

G

H

sexual rivalry in winter flocks (Lockie 1956a). If displaying ♀ is sideways on to ♂, he tilts upperside of body towards her (Fig H). ♂ never seen by Goodwin (1955b) to mount own mate without prior display, though this is recorded by Yeates (1932, 1934) and Ena Alvarez (1979). ♀ solicits by crouching with head low and level, wings loose; at higher intensity, wings may be partly spread and tail may be rapidly vibrated both vertically and horizontally (Fig I). As ♂ approaches, ♀ may be still except

I

for nictitating membrane blinking rather slowly. (Coombs 1960, 1978.) When ♀ unreceptive, ♂ may also adopt ♀'s soliciting-posture (thus general appeasement posture) after his own soliciting-display (Goodwin 1955b, 1986). During copulation, which lasts only a few seconds, ♂ wing-flaps vigorously (Roland 1988; for associated calls, see 5 and 12 in Voice.) Decline in copulation between mates after laying is obscured by widespread promiscuity, e.g. in Cornish study 94% of ♀♀ involved in copulation were already incubating and only 17% of attempts were not interfered with by others (Coombs 1960, 1978). Interference typically begins with mates copulating or ♂ (not mate) approaching and attempting to mount incubating ♀. Either case attracts attention of neighbouring ♂♂ who mob and try both to displace copulating ♂, and themselves to copulate with ♀ or simply attack her (peck fiercely); other ♀♀ may leave their nests to repel intruders. (Goodwin 1955b; Marshall and Coombs 1957; Coombs 1960, 1978.) See 7–8 and 12 in Voice for associated calls. Røskaft (1983a) found that ♀♀ were generally receptive to ♂ (not mate) approaching with soliciting-display, but resisted those who did not display; young ♀♀ copulated more with promiscuous ♂♂ than did old ♀♀, while old ♂♂ copulated more often with neighbouring ♀♀ than did young ♂♂ (see Bonds for interpretation of mating strategy). (5) Nest-site selection and behaviour at nest. Nest-sites are chosen, and building occurs, in both autumn and spring; in September, birds seen carrying stones and pine cones as well as sticks, suggesting display significance (Lincoln *et al.* 1980). Both sexes participate in selecting site; in one case, mates chose separate sites before settling for ♂'s; in 2 other cases, ♀'s choice was used (Coombs 1960, 1978). Sometimes pair start to build more than one nest simultaneously, and quite close together, but complete only one (Gramet 1956). For exceptional cases of ♀ building unaided by ♂, see Ogilvie (1947), Goodwin (1955b), and Coombs (1960, 1978). ♀ typically guards nest to prevent neighbours stealing material. Stealing is rife (see Røskaft 1981a for references) especially in morning (peak of building), with raiding parties of up to 10–15 (usually younger) birds on unguarded nests. Most stolen material is dropped and not incorporated into any nest. (Ena Alvarez 1979; Roland 1988.) Foundation sticks are seldom stolen, so nests readily rebuilt (Crook 1921); see Busse (1965) for time taken to rebuild destroyed nests. Nests reaching lining stage are immune from stealing, releasing ♀ to assist mate in collecting lining material (Ena Alvarez 1979); both sexes help to shape nest-lining (Coombs 1960, 1978). ♀ may spend several days on nest before 1st egg laid (Ena Alvarez 1979). For diurnal activity of ♂ and ♀ during incubation and chick-rearing, see Røskaft (1981*b*). During hot weather in León, ♂ sometimes apparently brought ♀ water (or very dilute food) during incubation (Ena Alvarez 1979). RELATIONS WITHIN FAMILY GROUP. When young hatch, some ♀♀ eject eggshells but most seem to eat them (Roland 1988). For 1st half of fledging period (total 32–36 days) all food for ♀ and young brought by ♂, while ♀ broods; during days 11–20, ♀ broods progressively less as she starts foraging for herself and young; from day 21 to fledging, young fed by both parents (Røskaft 1981*b*). Various methods of food distribution, as follows: exceptionally, ♂ feeds only the young in turn (♀ tries to intercept); ♂ feeds young first, then feeds ♀ (♀ begs, but this wanes as young develop); when young older, ♂ gives all food to ♀ who feeds them; ♂ gives ♀ some food and both parents then feed young together; ♂ regurgitates food onto nest-rim for ♀ to distribute between herself and young (Goodwin 1955*b*; Ena Alvarez 1979; Roland 1988). Parents may stimulate inactive young to gape by prodding (Roland 1988). Once eyes open (fully by day 12), young beg from any passing adult, but only from parents after 15 days. After *c.* 4 weeks, begging-posture almost identical to adult ♀'s. (Ena Alvarez 1979; Roland 1988.) Faecal sacs of young 1–2 days old swallowed by ♀ (Crook 1921); when young older, sacs collected and discarded by both sexes, though ♀ apparently does more. ♀ alone removes food debris and fouled nest-material. Eventually young able to defecate over side of nest. (Coombs 1960, 1978; Ena Alvarez 1979; Roland 1988.) Once nestlings feathered, ♀ sometimes Allopreens them. From *c.* 21 days, young exercise in nest and later start exploring branches around nest. Those perched outside were always fed first (thus favouring stronger chicks). Any young falling to ground are ignored and therefore doomed. (Ena Alvarez 1979.) Near fledging, ♀ very agitated, hopping from branch to branch near nest as if enticing young to follow (Crook 1921, which see for photograph). Soon after leaving nest, young start following parents to feeding grounds (Coombs 1960, 1978: see also Bonds). Initially may return to nest to roost at night, or more often to trees near colony (Coombs 1960, 1978), parents roosting nearby. However, families soon resort to communal roost (see Roosting). ANTI-PREDATOR RESPONSES OF YOUNG. In alarm, young in nest fall silent and crouch motionless; near fledging, leave nest and do likewise (Ena Alvarez 1979; Røskaft 1980*b*; Roland 1988). PARENTAL ANTI-PREDATOR STRATEGIES. (1) Passive measures. Disturbed ♀ not an especially tight sitter but this varies between colonies: in main colony in Trondheim study, flushing distance from nest on approach of man only 2–4 m, but 30–50 m in another smaller colony (Røskaft 1980*b*). (2) Active measures: against birds. At Trondheim colony, *C. corone*, *C. monedula*, and Magpie *Pica pica* were tolerated until they entered nest-territory, then were driven off by owners (Røskaft 1980*b*). Raptors mobbed near colony include Buzzard *Buteo buteo*, Peregrine *Falco peregrinus*, *F. tinnunculus*, and Sparrowhawk *Accipiter nisus*; also mobs sundry other large birds, e.g. Grey Heron *Ardea cinerea* (Coombs 1960, 1978). However, at Trondheim colony, *C. frugilegus* fled colony (i.e. no mobbing) when Goshawk *A. gentilis* approached (Røskaft 1980*b*). (3) Active measures: against man and other animals. No active responses to man, nor to red squirrel *Sciurus*

*vulgaris*, cat, dog, or fox intruding in colony (Røskaft 1980*b*; Røskaft and Espmark 1982).

(Figs by D Quinn: from drawings in Coombs 1960, 1978.)                                               EKD

**Voice.** Following scheme and descriptions (including most names of calls) adapted from Røskaft and Espmark (1982, which see for sonagrams); study of 19 pairs in colony at Trondheim (Norway) produced repertoire of over 20 calls, several similar in structure but differentiated according to context. Most calls were from mates in close contact, suggesting special role in strengthening and maintaining pair-bond. For other descriptions of (fewer) calls, see Coombs (1960, 1978), Bergmann and Helb (1982, also for sonagrams), and Goodwin (1986), but difficult to match repertoires these describe with fine distinctions made by Røskaft and Espmark (1982). Few renderings given by Røskaft and Espmark (1982) and anyway difficult and of dubious value in this species where many calls are noisy and raucous; reference to structure, and use of sonagrams, thus essential to appreciate complexity of repertoire. Mimicry rare in the wild, but for alleged cases see Wassenich (1973: barking like dog) and Stephan (1974: bleating like goat), also Hulten (1972) for hand-reared bird mimicking human and making bill-clapping sounds.

CALLS OF ADULTS. (1) Song of ♂. Comprises medley of much or all of call repertoire, but components usually given more softly than in their respective contexts: various soft cawing, gurgling, rattling, and crackling calls, in sum resembling loud Starling *Sturnus vulgaris* (Goodwin 1986). In recording, most units (as in 1st and 3rd of Fig I) are buzzing rasps (W T C Seale). Given when perched,

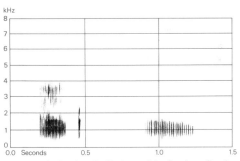

I   S Palmér/Sveriges Radio (1972–80)   Sweden   October 1966

apparently to attract ♀, but no territorial function (Coombs 1960, 1978; Richards 1976*a*). (2) Caw-call. Commonest call, used by both sexes all year for self-advertisement and contact: 'kaah', apparently harsher and 'flatter' than Carrion Crow *C. corone* (Goodwin 1986); deep harsh sonorous 'korr' or 'krah' (Bergmann and Helb 1982). Variable in timbre and pitch, both between individuals and in different contexts, and ♀'s call longer and higher pitched than ♂'s. In Fig II, successive units show progressive shortening, thus 'kaaar kaar kau' (W T C Seale). Heard most often from ♂ perched near nest. Also

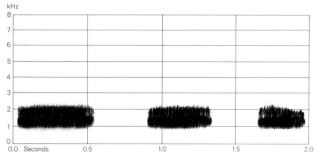

II   J Burton and M Smythe/BBC  England  March 1981

used by ♀ to greet approaching mate, but rarely after onset of incubation (see 3 below). Especially during nest-building, pair sometimes alternate single Caw-calls in duet, typically during Bowing-display; Fig III shows 'kar-kor kar-kor kar-kor kar-kor' in which higher-pitched 'kar' (probably ♀) is rapidly answered by mate's 'kor' (E K Dunn, W T C Seale). Also characteristic of Bowing-display is series of Caw-calls, starting with single short call, followed by 2–6 relatively long ones. (3) Main food-begging calls. Røskaft and Espmark (1982) distinguished 2 main calls: 'hunger call' when ♀ starts Begging-display on approach of mate, changing to 'feeding call' as ♂ starts feeding her. Recordings show that these calls can grade into each other. Fig IV begins with loud long 'kaar' (hunger call) followed by 'gow gowa' (feeding calls) which are shorter and progressively more wavering (W T C Seale). (4) Trembling-calls. 2 variants used by some ♀♀ when fed by mate: either a long continuous call (up to *c.* 2 s), or else regular repetition (at intervals of *c.* 0·1 s) of relatively short units covering wider frequency range. (5) Click-call.

Low frequency clicks (up to *c.* 2 kHz) given either singly or in series (*c.* 0·05 s apart) by ♂ feeding ♀. Click-calls also heard from ♂, less often ♀, during courtship, copulation, and preening, and from both sexes during aggressive encounters. Though they sound mechanical (Goodwin 1986), no indication that clicks are instrumentally produced. (6) Rattle-call. Series of 5–20 staccato sounds, each *c.* 0·5–4 kHz, somewhat resembling rapid tapping of woodpecker (Picidae), occasionally heard from pair in close contact, probably only from ♀. Fig V shows 2 rattles

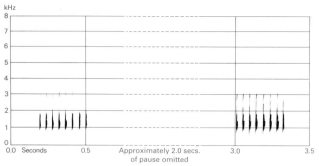

V   A P Radford  England  April 1986

*c.* 2 s apart each sounding like series of hollow clicks (just over 22 per s), strongly resembling drumming of Lesser Spotted Woodpecker *Dendrocopos minor* (W T C Seale). (7) Bass-call. Low-pitched call of variable duration (up to 1 s) sometimes given by ♀ when fed by mate. Prolonged variant (typically 1 s, occasionally up to 2–3 s) probably also from ♀, frequently heard during skirmishes when copulation interrupted by other ♂♂. (8) Quack-call of ♀. Low quacking sound given singly or in series in various

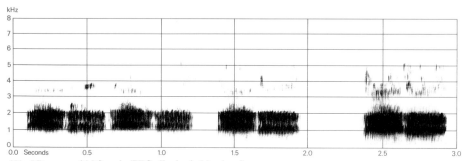

III  J Burton and M Smythe/BBC  England  March 1981

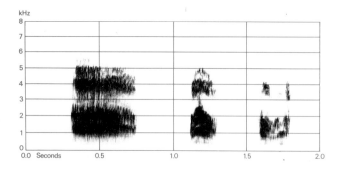

IV   J Burton and M Smythe/BBC  England  March 1981

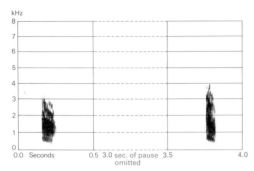

VI S Palmér/Sveriges Radio (1972–80) Sweden October 1966

types of intimate contact with mate, especially when being fed, also when intruding ♂ tries to force copulation. Fig VI shows 2 rather tonal 'ku' calls *c.* 3·5 s apart (W T C Seale); 'kjä'; or 'kju' like Jackdaw *C. monedula* (Bergmann and Helb 1982). (9) Contentment-call. Variable in structure, but usually short low-pitched growling sounds, repeated at short intervals. A contact-call, often exchanged by mates, e.g. just before copulation, during allopreening or billing; also heard from ♀ making shaping movements in nest ('nest-building call'). Similar calls heard when feeding young, especially up to 2 weeks old ('young-feeding call'). Low-pitched croak ('gorrr': W T C Seale) in Fig VII is apparently Contentment-call. (10) Creaking-

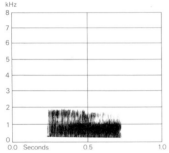

VII J Burton and M Smythe/BBC England March 1981

call. Sound (0·5–2 s long) resembling creaking door, occasionally heard from both sexes in close pair-contact such as courtship and preening. (11) Long tone-call ('long A-call': Røskaft and Espmark 1982). Tonal call with fundamental frequency *c.* 1 kHz, lasting 1–2 s. Occasionally given by ♀ when feeding young. (12) Pip-call. Markedly short call (0·2 s) of narrow frequency range (0·5–1·5 kHz) heard from ♂ in disputes, especially during copulation (see calls 7–8 for calls of ♀ in same context). (13) Snarling-call. Distinctly brief call (*c.* 0·1 s) of wide frequency range (*c.* 1–8 kHz), sometimes given by ♂ threatening conspecifics intruding on nest-territory. (14) Gull-call. Distinguished by marked frequency modulation and resemblance to Kyow-call of Herring Gull *Larus argentatus* (E K Dunn); given sporadically by territorial ♂♂ threatened or attacked by conspecifics. (15) Squalling-call. Reverberating rattling sound, up to *c.* 1 s long, varying in duration and pitch,

perhaps with different levels of excitement. Sonagram in Røskaft and Espmark (1982) suggests some resemblance to 'reception-crowing' of Guillemot *Uria aalge* (E K Dunn). (16) Distress-call ('pain call': Røskaft and Espmark 1982). Harsh noisy rasping sound heard from retreating intruder attacked at nest (Røskaft and Espmark 1982) and presumably also in other contexts when distressed. (17) Alarm-call ('distress call': Røskaft and Espmark 1982). Unusually harsh Caw-calls given in rapid repetition by ♀♀ scared off nest and circling over colony. Call used when mobbing predators (Goodwin 1986) is presumably the same or similar. (18) Other calls. Fig VIII shows

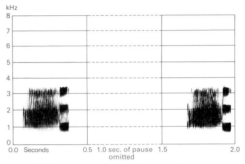

VIII V C Lewis England February 1985

'karlee' call in which 1st part ('kar') rough and rasping like Caw-call, 2nd part tonal and sounding surprisingly high-pitched (W T C Seale); this call not uncommon (P J Sellar), though context not known, and difficult to reconcile with any call in scheme of Røskaft and Espmark (1982).

CALLS OF YOUNG. Food-call of small nestlings a loud squealing chirrup or husky squeak, later developing into piercing 'rra' (Bergmann and Helb 1982; P J Sellar). Food-call *c.* 2 weeks later thus similar to call 2 of adult ♀ but higher pitched, becoming progressively louder and stronger as chick develops (Roland 1988). In recordings, calls of young also include wavering 'wawawawa' (W T C Seale). EKD

Breeding. SEASON. Britain: mean date of 1st egg varied from 20 March in southern Scotland, *n* = 44, to 30 March in northern England, *n* = 119; eggs laid until end of April (Holyoak 1967*c*); mean date in southern England over 6 years, 7–23 March; some nest-building on warm days in February (Owen 1959); possibly 2 nest-building peaks, 1st half of March and mid-April, latter perhaps first-time breeders (Nau 1960); rare records of young in nest in November but no records of successful autumn breeding (Coombs 1978). Central Norway: mean laying date for 2-year-old ♀♀ 18 April (*n* = 13), for older ♀♀ 10 April (*n* = 75); laying can continue until mid-May (Røskaft 1981*b*; Røskaft *et al.* 1983). Northern Germany: see diagram; mean 1st egg date over 4 years, 19 March to 21 April, *n* = 153 (Wittenberg 1988); for Bayern, see Leibl

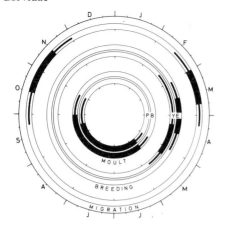

and Melchior (1985). For Poland, see Grodziński (1976). For Croatia, see Pivar (1965). Russia: eggs laid in Moscow region mid-April, in Omsk area, early May; apparently no significant variation with latitude (Dementiev and Gladkov 1954). See Tuajew and Wassiljew (1974) for reedbed nesters in Azerbaydzhan. For New Zealand, see Coleman (1972) and Purchas (1979). SITE. In topmost crown of high tree, exceptionally on horizontal branch or against trunk; trees almost always in rather isolated groups (Kemper 1964; Kulczycki 1973; Coombs 1978). Over 60 species of tree used in Britain and Northern Ireland, 23 of which conifers; 27% of 539 393 nests were in pine *Pinus sylvestris*, 20% in beech *Fagus sylvatica*, 16% in elm *Ulmus*, 12% in sycamore *Acer pseudoplatanus*, 9% in oak *Quercus*, and 8% in ash *Fraxinus excelsior*; no apparent preference except for tall species (Coombs 1978; Sage and Vernon 1978). Of 874 nests in Poland, 26% in various poplar *Populus* species, and 23% in alder *Alnus*, but possibly some preference for sycamore, elm, and lime *Tilia* because of arrangement of branches; average height above ground 19.4 m (7-40), great majority above 14 m (Kulczycki 1973). No preferred tree species in northern Germany, where most nests in beech, which is commonest tree (Kemper 1964). In northern Spain, almost always in poplar at 20-30 m (Ena 1984a). In absence of tall trees in steppe areas of USSR, in low trees and bushes below 4 m (Dementiev and Gladkov 1954). Rarely on church tower, pylon, etc., often because of persecution elsewhere (Benecke 1970; Sage and Vernon 1978); exceptionally on ground (Scott 1959; Kulczycki 1973). In Azerbaydzhan, colony of c. 250 nests 40-150 cm above water in *Phragmites* reedbed (Tuajew and Wassiljew 1974). Nest: fairly regular hemisphere, sometimes slightly flattened; foundation of sticks and large twigs, inside which layer of thin pliable twigs very often of birch *Betula* and willow *Salix*, many with leaves, followed by compact mass of rootlets, moss, etc., mixed with clay to form small cup, which is lined with grass, moss, stalks, feathers, leaves, paper, etc. (Pivar 1965; Kulczycki 1973; Grodziński 1976; Coombs 1978; Leibl and Melchior 1985; Wittenberg 1988). 50 nests in Poland had average outer diameter 42 cm (27-85), inner

diameter 20 cm (16-31), overall height 31 cm (17-60), and depth of cup 12 cm (7-19) (Kulczycki 1973); see also Roland (1988) and Wittenberg (1988). Nests in Azerbaydzhan reedbeds are of bent-over *Phragmites* stems, lined with leaves and fibres as well as stalks of herbs; diameter as normal nests, height and depth of cup slightly less (Tuajew and Wassiljew 1974). Building: by both sexes; probably ♂ brings most material, particularly lining, and ♀ builds more; in study in southern England, many observations of one bird bringing material and building while other apparently guarded against thieving conspecifics (Ogilvie 1951; Coombs 1978; Røskaft 1983b). Lone ♀ capable of building complete nest (Coombs 1960). Most nests probably re-used by pair that built them; twigs mostly broken from trees and some gathered from ground, but in some colonies significant number stolen from neighbouring nests, unclaimed or occupied, or from other colonies; epidemics of thieving and raids for material can suddenly erupt, sometimes directed at single pair; strips of bark can be pulled from trees (Ogilvie 1951; Coombs 1960; Kulczycki 1973; Grodziński 1976; Roland 1988). Completion can take from less than 1 week to c. 4 weeks (Coombs 1960; Grodziński 1976; Roland 1988). See Kulczycki (1973) and Coombs (1978) for details of building technique. EGGS. See Plate 58. Sub-elliptical, smooth and faintly glossy; light blue to dull green with olive-buff to blackish-olive specks, blotches, and hair-streaks, sometimes forming cap at broad end; some unmarked; often large variation within clutch (Harrison 1975; Makatsch 1976). Nominate *frugilegus*: $39.8 \times 29.9$ mm (37.6-47.4 × 25.2-32.0), $n = 1050$; calculated weight 16.0 g (Schönwetter 1984). See Valverde (1953) for dimensions from isolated population in northern Spain. Clutch: 2-6 (1-7); of 292 clutches before April, southern England: 2 eggs, 3%; 3, 12%; 4, 37%; 5, 39%; 6, 9%; 7, 0.3%; average over 6 years 4.2-4.7 (Owen 1959); in England, average before 1 April 4.3 ($n = 151$), thereafter 3.5 ($n = 20$), difference significant (Lockie 1955); for size in various regions of Britain, see Holyoak (1967c). In Norway, average for 2-year-old ♀♀ 2.8, $n = 12$, and for older ♀♀ 3.7, $n = 66$ (Røskaft et al. 1983). In northern Germany, of 153 early clutches: 1 egg, 1%; 2, 15%; 3, 24%; 4, 35%; 5, 24%; 6, 2%; average over 5 years 3.4-4.2 (Wittenberg 1988). Average in Netherlands 3.7, $n = 279$ (Feijen 1976); in northern Spain 3.0, $n = 48$ (Ena 1984b); in central European Russia over 3 years, 4.0-4.2, $n = 537$ (Bykova 1990). Replacement clutch laid after loss of 1st clutch or young brood is smaller, often in same nest (Feijen 1976; Roland 1988; Wittenberg 1988); eggs laid daily (Holyoak 1967c; Makatsch 1976). One brood (Makatsch 1976; Coombs 1978; Goodwin 1986). INCUBATION. 16-18 days (Pivar 1965; Makatsch 1976; Røskaft 1983b); in one nest, in cold spring, 23 days from start of laying to start of hatching (Roland 1988). By ♀ only; see Kane (1960) and Roland (1988) for observations of ♂ sitting on eggs briefly. Probably starts with 2nd or 3rd egg (Lockie

1955; Holyoak 1967c; Makatsch 1976), or even last egg (Roland 1988); hatching asynchronous. Average 97% of daylight hours spent on eggs before 18.00 hrs, and 95% after, $n = 28$ nests (Røskaft 1981b). YOUNG. Fed and cared for by both parents; fed only by ♂ for first 2-4 weeks; brooded only by ♀ (Gerber 1956; Coombs 1978; Røskaft 1983b; Roland 1988). FLEDGING TO MATURITY. Fledging period 30-36 days; usually first flies a few days after leaving nest (Pivar 1965; Coombs 1978; Røskaft 1985; Roland 1988). Fed by parents for *c.* 6 weeks after fledging (Coombs 1978; Røskaft 1985). For development of young, see Ena (1984a) and Roland (1988). Age of first breeding usually 2 years, sometimes 1 (Dementiev and Gladkov 1954; Gerber 1956; Marshall and Coombs 1957; Nau 1960; Ena 1984b). BREEDING SUCCESS. In southern England over 6 years, 85% of 915 eggs hatched and 61% produced fledged young, giving average 2·5-3·7 fledged young per nest per year (omitting total failure of *c.* 5% of nests) (Owen 1959); 90% of young that died in 29 broods in England were last-hatched, and in one year with good food supply (principally earthworms Lumbricidae) broods of 5 produced average 3·5 fledged young, in following poor year 2·8 (Lockie 1955); most losses of eggs and young caused by high winds and starvation, though human persecution is major factor where it occurs (Lockie 1955; Owen 1959; Holyoak 1967c; Feijen 1976; Ena 1984b). Great majority of nestling deaths occur in 1st 10 days (Owen 1959; Røskaft *et al.* 1983). In central Norway over 5 years, 2-year-old ♀♀ produced average 0·6 fledged young per nest, $n = 16$, older ♀♀ 1·9, $n = 79$, but most significant factor in nestling mortality appeared to be age of ♂, less expert foraging of younger ♂♂ leading to increased nestling starvation (Røskaft *et al.* 1983); main predators Carrion Crow *C. corone* and Goshawk *Accipiter gentilis* (Røskaft 1980b, 1983b); see also Røskaft (1985, 1987). In Netherlands, 64% of 346 eggs hatched and 31% produced fledged young, giving 1·2 young per pair overall; eggs and young taken by martens *Martes* and Marsh Harrier *Circus aeruginosus* (Feijen 1976). In northern Germany, 2·3 fledged young produced per successful nest, $n = 70$ (Wittenberg 1988). Only 0·6 young produced per nest in northern Spain, $n = 2938$, where many nests brought down by humans (Ena 1984b). In central European Russia, 1·6 fledged young per nest overall, $n = 537$ (Bykova 1990). Exceptional observation of nestlings being killed and eaten by *C. frugilegus* from neighbouring nest (Caldwell 1949). For New Zealand, see Coleman (1972) and Purchas (1979).

BH

**Plumages.** (nominate *frugilegus*). ADULT. Base of upper mandible, lore, cheek backwards to below eye, chin, and throat bare, skin covered with wart-like tubercles; chin and throat covered with short down-like dark or pale grey feathers in autumn, these gradually wearing off during winter (last to wear is strip at side of throat, where feathers protected in fold which forms when pouch empty). Plumage otherwise entirely black, strongly glossy (more so than in Carrion Crow *C. corone corone*), colour of gloss depending on angle of light: crown and rear of cheek violet-blue to purplish-green, remainder of body strongly violet- or purple-blue on fringes and centres of feathers, more violet-bronze or oily purple-green subterminally. Tail, upper wing-coverts, tertials, and secondaries strongly violet-blue, primaries, greater upper primary coverts, and bastard wing more purple- or greenish-blue. Feathers of body off-white on bases, concealed when bird in normal posture unless plumage heavily worn. *In worn plumage* (about May-August), gloss slightly less intense violet-blue, more purplish-bronze or oily violet-green, especially on wing; tips of primaries and tail-feathers scarcely glossy, deep black. NESTLING. Down short and sparse; on upperparts, upperwing, and thigh only; dark smoke-grey (Witherby 1913; Heinroth and Heinroth 1924-6; Witherby *et al.* 1938). JUVENILE. Face feathered, but feathering scanty below eye, and narrow strip at side of chin bare; skin smooth. Nasal bristles, forehead, crown, and side of head black, sometimes faintly glossed violet-blue on crown and ear-coverts. Upperparts from hindneck to upper tail-coverts as well as side of neck greyish-black, feathers with sooty-black fringes along tips, fringes sometimes glossed violet-blue, especially on lower mantle; feathering of hindneck and rump short, soft, woolly. Chin and upper throat white, black, or variable mixture of black and white; remainder of underparts dark grey, feathers with sooty-black or (occasionally) purplish-black tips, underparts appearing mottled; feathering of underparts loose and soft, especially on vent. Bases of body feathers brown-grey, less pale than adult. Tail as adult, but gloss less strong, more bluish; feathers usually narrower, tips pointed or rounded, less broadly truncate. Wing as adult, glossy black, contrasting with dull greyish body; innermost lesser and median upper wing-coverts with much dark grey (not glossy) of bases visible; gloss of wing less intense than in adult, but fairly strong in first months of life, not becoming dull and brownish-black as rapidly as *C. corone* of same age. FIRST ADULT. As adult, but face feathered, nasal bristles and chin black with silvery-grey shafts (chin sometimes partly white), remainder of face glossy purple-black, like remainder of head and body; gloss on body as adult, but often slightly more dull black of feather-centres visible on underparts, and fringes on belly and vent slightly more bluish, less violet. Lesser and most or all median upper wing-coverts new, strongly glossed violet-blue, as in adult; remainder of wing and all tail still juvenile, slightly less glossy, but contrast in gloss and ground-colour between new and old feathers not as strong as in 1st adult *C. corone*; old secondaries often as glossy violet-blue as new median coverts, but greater upper wing-coverts and (especially) inner tertials duller sooty-black than neighbouring strongly glossy scapulars and median coverts (in adult, contrast less marked); juvenile primaries and tail less glossy than adult, tips more abraded, tail-feathers narrower, less truncate. In Britain, feathers on face gradually lost through moult (not abrasion) from January-March onwards, but down feathers and filoplumes, concealed among normal feathering in autumn, are retained; new feathers growing are degenerate dark grey pins 4-5 mm long with some reduced barbs. Pins appear first on central chin, moult then spreading to base of lower mandible and to lore and area below eye; bristles and feathers on forehead and projecting over nostrils replaced last, up to June. At same time, bare skin of face changes from soft and pink to hard and whitish. Down and new reduced feathering partly worn off during spring. In 1st immature post-breeding moult (at 1 year old), remnants of down and reduced feathers replaced by new down and reduced feathers on chin and throat, but feather papillae on remainder of face show some short knobs at most. No spring moult in 2nd-years and older, these birds attaining reduced feathering and down on chin and throat only in post-breeding moult; this reduced feathering gradually wears off in autumn and winter. (Witherby 1913.) Not all continental birds attain bare face as rapidly as British birds cited above; in birds examined, and according to literature, bare skin sometimes starts to show on chin, below eye, and on cheek from November, but often not until April; skin may become more extensively bare at side of face from January or not until May–July; some birds in March have forehead and side of head bare and chin and throat downy, but nasal and rictal bristles still present, others (especially ♀♀) have bristles and some feathers on throat present up to January–February of 2nd winter; face of most advanced birds sometimes entirely bare (except for traces of down on throat) from May of 1st spring (Chappellier 1932; Bährmann 1966; Dorka 1973; Eck 1984; RMNH, ZMA). Some 1-year-olds have face still well-feathered up to August, and a few retain traces of feathers or bristles until 2 years old (Coombs 1960). Only 18% of over 2600 2nd-winter birds (September–April) examined had face as bare as adult; these advanced birds differed from older ones in smoother skin on face (warts less coarse) and in colour of mouth (see Bare Parts) (Greve 1990).

**Bare parts.** ADULT. Iris brown, hazel-brown, or dark brown. Bill greyish-black or horn-black. Mouth bluish-black. Bare skin of face whitish-grey, light grey, or ash-grey. Leg and foot black. NESTLING. Bare skin brown at hatching, shading to dark violet with velvety sheen later. Gape-flanges broad but thin, pale yellowish-flesh. Mouth pink-flesh. JUVENILE, FIRST ADULT. At fledging, iris light grey-blue; bill, leg, and foot greyish-black; mouth, tongue, and inside of mandibles pink-flesh. Brown iris colour attained in 1st autumn, iris shading from grey through grey-brown to hazel; tip of tongue and terminal part of inside of mandibles black by October. Mouth pink to bright pink-flesh during 1st year of life, becoming dull bluish-grey with some traces of pink in July–August of 2nd calendar year and fully blackish-blue in 3rd. (Heinroth and Heinroth 1924–6; Marshall and Coombs 1957; Coombs 1960; Dorka 1973; Greve and Dornieden-Greve 1982; Greve 1990.)

**Moults.** ADULT POST-BREEDING. Complete; primaries descendent. In Britain, starts with innermost primary (p1) early May to late June (occasionally early July); 1-year-olds start first, followed by older ♂♂, these by older ♀♀ (which perhaps moult more quickly); primary moult completed after c. 107 days with regrowth of p10 mid-August to late October, occasionally from late July (Holyoak 1974a; Ginn and Melville 1983). In captivity, p1 shed 2–3 days after young left nest (Richards 1976b), but in former East Prussia, many adults shed p1–p2 on 23 May when small young in nest (Heinroth and Heinroth 1924–6). Tail moult centrifugal; starts with t1 at primary moult score 15–30 (occasionally, 10–35), shed at same time as p4–p5; completed with regrowth of t6 from primary score 40 or later, occasionally just after completion of primaries. Secondaries start with shedding of s1 at primary score 15–25, completed with regrowth of s6 at same time as completion of outer primaries or slightly later. (Holyoak 1974a.) In another sample from Britain, primary moult started May or early June (occasionally late June), completed September or 1st half of October; adult ♂♂ started on average 19 May, adult ♀♀ 29 May, 1-year-old ♂♂ 13 May, 1-year-old ♀♀ 11 May; moult completed after 154 days. Tail started on average 16 June, completed after 92 days; tertials on 22 June, completed after 88 days; secondaries on 4 July, completed after 105 days. (Seel 1976, which see for many other details.) Small samples examined from elsewhere (Netherlands, Germany, Rumania: RMNH, ZFMK, ZMA, ZMB) fit British pattern.

POST-JUVENILE. Partial: head, body, usually all lesser upper wing-coverts, and highly variable number of median upper wing-coverts (none to almost all); occasionally inner greater coverts (tertial coverts), exceptionally some feathers of bastard wing, some primaries, or a few tail-feathers. Starts as soon as juvenile plumage full-grown, from age of 6–7 weeks (Heinroth and Heinroth 1924–6; RMNH, ZMA); mantle, scapulars, and flank first, soon followed by remainder of body and wing-coverts; head last (Witherby 1913). Fledged full-grown juveniles without moult examined from late May to late July, moulting ones late June to early September, birds with moult completed from early August onwards. For moult of face in spring, see Plumages; for moult of 1-year-olds, see Adult Post-breeding (above).

**Measurements.** Nominate *frugilegus*. Netherlands (mainly), Belgium, and Germany, all year; skins (RMNH, ZMA). Juvenile wing and tail include 1st adult; bill and tarsus of 1st adult similar to adult, combined; bill (S) to skull, bill (F) to feathering on forehead, bill (N) to distal corner of nostril; depth is depth of bill at middle of nostril.

| | | ♂ | | ♀ | |
|---|---|---|---|---|---|
| WING AD | ♂ | 322·9 (6·00; 47) | 311–335 | ♀ 306·6 (5·62; 54) | 297–320 |
| JUV | | 309·9 (8·52; 18) | 298–325 | 295·9 (8·16; 14) | 277–310 |
| TAIL AD | | 176·8 (4·57; 23) | 170–186 | 165·1 (5·30; 27) | 154–176 |
| JUV | | 164·0 (6·72; 14) | 154–176 | 159·2 (7·52; 14) | 147–166 |
| BILL (S) | | 61·7 (2·81; 44) | 57·5–66·5 | 57·6 (1·49; 36) | 55·0–60·0 |
| BILL (F) | | 56·7 (2·33; 21) | 53·0–60·5 | 52·1 (1·71; 29) | 49·5–55·5 |
| BILL (N) | | 38·9 (2·35; 42) | 35·3–43·7 | 36·2 (1·08; 29) | 34·0–38·5 |
| DEPTH | | 18·1 (0·69; 45) | 17·2–19·5 | 16·8 (0·60; 35) | 15·5–17·7 |
| TARSUS | | 56·9 (1·69; 31) | 54·0–59·5 | 54·1 (1·49; 30) | 51·0–56·5 |

Sex differences significant, except juvenile tail.

Bill depth at base, Scotland: ♂ 21·2 (18·5–26), ♀ 19·8 (17–22·5) (Green 1981).

Eastern Rumania, Turkey, and Caucasus area, all year; skins (RMNH, ZFMK, ZMA).

| | | ♂ | | ♀ | |
|---|---|---|---|---|---|
| WING AD | ♂ | 324·3 (2·74; 9) | 322–329 | ♀ 306·2 (6·17; 16) | 298–322 |
| BILL (S) | | 63·6 (1·51; 9) | 61·0–66·1 | 58·2 (2·89; 16) | 54·6–63·0 |
| DEPTH | | 18·1 (0·38; 9) | 17·4–18·5 | 16·8 (0·60; 16) | 15·8–17·7 |

Sex differences significant.

Tien Shan mountains and western Sinkiang (China), summer (ZFMK, ZMA).

| | | ♂ | | ♀ | |
|---|---|---|---|---|---|
| WING AD | ♂ | 319·0 ( — ; 2) | 315–323 | ♀ 302·4 (2·41; 5) | 299–305 |
| BILL (S) | | 63·6 ( — ; 2) | 62·9–64·2 | 57·9 (0·75; 5) | 57·0–58·9 |
| DEPTH | | 17·8 ( — ; 2) | 17·7–17·9 | 16·0 (1·54; 5) | 15·5–16·9 |

Wing; live or freshly dead birds, unless otherwise noted. Nominate *frugilegus*. (1) Scotland, adults (Green 1982b). (2) Braunschweig (Germany), 2nd winter and older (Greve 1990). (3) Lausitz (Germany), adult; skins (Eck 1984; see also Eck 1990). (4) Vienna (Austria), adult (Steiner 1969). (5) USSR (Dementiev and Gladkov 1954). (6) Northern Iran, adult, mainly spring migrants; skins (Stresemann 1928b; Paludan 1940; Schüz 1959; Diesselhorst 1962). C. f. pastinator. (7) Manchuria and Kansu (China), adult; skins (Meise 1934a; Stresemann et al. 1937). (8) Japan, adult; skins (RMNH).

| | | ♂ | | ♀ | |
|---|---|---|---|---|---|
| (1) | ♂ | 321·6 ( 6·7 ; 102) | 305–338 | ♀ 300·8 (7·7 ; 54) | 280–314 |
| (2) | | 320·1 ( — ; 4837) | 300–343 | 302·6 ( — ; 2231) | 285–318 |
| (3) | | 320·3 ( 6·98; 53) | 304–337 | 305·1 (6·56; 82) | 289–319 |
| (4) | | 323·1 ( 7·45; 44) | 307–336 | 309·0 (6·96; 42) | 294–327 |
| (5) | | 319·7 ( — ; 55) | 300–340 | 309·3 ( — ; 43) | 280–340 |
| (6) | | 314·4 ( 7·09; 7) | 304–324 | 303·6 (5·56; 7) | 289–307 |
| (7) | | 314·0 (12·81; 5) | 292–323 | 301·6 (5·56; 7) | 295–308 |
| (8) | | 313·2 (10·94; 6) | 300–331 | 293·0 ( — ; 1) | — |

For data other than wing, see Steiner (1969), Green (1982b), Eck (1984), and Greve (1990); for Britain, see also Witherby (1913) and Green (1981); for Germany, see Bährmann (1960c,

1976), Focke (1966), and Niethammer (1971); for Czechoslovakia, see Hell and Soviš (1959); for Yugoslavia and elsewhere, see Stresemann (1920); for introduced birds in New Zealand, see Niethammer (1971).

**Weights.** Nominate *frugilegus*. ADULT, FIRST ADULT. (1) Scotland, mainly March–May (Green 1981). Netherlands, adult: (2) October–March, (3) April–May (RMNH, ZMA). (4) North-east France, December (Bacmeister and Kleinschmidt 1918). Braunschweig (north-central Germany), ranges approximate, read from figure: (5) 1st winter, (6) 2nd winter, (7) 3rd winter and older (Greve 1990). Lausitz (south-east Germany), October–March: (8) adult, (9) 1st adult (Bährmann 1972, which see for average per month throughout winter; see also Bährmann 1942, 1960c). Niedersachsen (north-west Germany), January, feeding conditions poor: (10) adult, (11) 1st adult (Latzel and Wisniewski 1971). Vienna (Austria), February–March: (12) adult, (13) 1st adult (Steiner 1969). (14) USSR (Dementiev and Gladkov 1954).

| | | |
|---|---|---|
| (1) | ♂ 489 (30·24; 162) 405–560 | ♀ 418 (34·52; 126) 325–525 |
| (2) | 494·4 (61·85; 9) 370–570 | 420·7 (39·91; 11) 365–480 |
| (3) | 482·2 (28·49; 9) 445–540 | 443·3 (44·94; 9) 375–515 |
| (4) | 498·0 (11·68; 5) 488–518 | 462·8 (19·19; 4) 435–477 |
| (5) | 452·0 ( — ; 1874) 385–535 | 397·0 ( — ; 1114) 340–465 |
| (6) | 451·7 ( — ; 1654) 385–570 | 397·5 ( — ; 1010) 345–440 |
| (7) | 495·4 ( — ; 3181) 405–600 | 402·9 ( — ; 1221) 340–425 |
| (8) | 523·1 (38·24; 79) 425–600 | 473·0 (41·18; 97) 386–552 |
| (9) | 482·1 (38·88; 69) 401–585 | 412·1 (40·98; 66) 340–500 |
| (10) | 475 ( — ; 57) 385–520 | 476 ( — ; 49) 415–500 |
| (11) | 449 ( — ; 50) 375–500 | 448 ( — ; 48) 405–495 |
| (12) | 515·4 (42·81; 43) 415–574 | 456·0 (33·72; 40) 415–555 |
| (13) | 484·7 (25·56; 15) 428–515 | 426·0 (58·93; 12) 292–505 |
| (14) | 424 ( — ; 5) 350–490 | 375·7 ( — ; 4) 313–450 |

Averages in Britain: January–April, adult ♂ 489 (92), 1st adult ♂ 446 (25), adult ♀ 432 (207), 1st adult ♀ 401 (29); May–July, adult ♂ 452 (173), 1st adult ♂ 423 (31), adult ♀ 389 (277), 1st adult ♀ 359 (29); August–September, ♂ 476 (22), ♀ 416 (27); October–December, ♂ 501 (14), ♀ 441 (32) (Seel 1976, which see for monthly averages).

In eastern Czechoslovakia, winter: 392·0 (74·4; 452) 235–600 (Hell and Soviš 1959). Northern Iran, February–March: ♂ 456·3 (3) 440–477, ♀ 345 (Schüz 1959). For Britain, see also Feare *et al.* (1974); for Germany, see also Bäsecke (1956), Diesselhorst (1956), and Focke (1966). Lowest weights of exhausted birds, Netherlands: ♂ 335, ♀ 318 (RMNH, ZMA). Exhausted birds in Britain 389 (♂), 282, 282, 362 (♀) (Macdonald 1962, 1963; Ash 1964). Heavy ♀ from Braunschweig, March, 540 (Bäsecke 1956). Dwarf ♀ from Braunschweig 280 (wing 269) (Greve 1990).

*C. f. pastinator*. Mongolia, May: 1st adult ♀ 330 (Piechocki and Bolod 1972).

NESTLING, JUVENILE. For growth, and relationship of fledging weights to dominance hierarchy 1 year later, see Røskaft (1983c).

**Structure.** Wing rather long, broad at base, tip bluntly pointed. 10 primaries: p7 longest, p8 0–10 shorter, p9 18–41, p10 89–125, p6 4–14, p5 31–47, p4 56–81, p3 74–101, p2 93–116, p1 104–130; tip of p10 reaches to (p1–)p2–p3, tip of p9 to p5–p6; p10 50–72 longer than longest upper primary covert. Outer web of p6–p9 emarginated, inner web of p7–p10 with distinct notch. Longest tertial reaches to tip of p1–p2. Tail rather long, tip rounded or bluntly wedge-shaped; 12 feathers, t1 longest, t6 24·0 (30) 18–33 shorter in adult, 16·0 (10) 11–23 in juvenile. Bill long (visible culmen slightly longer than head length), deep and wide at base; culmen gradually and gently decurved, tip of bill sharply pointed (bill less deep in middle than in Carrion Crow *C. corone*, terminal half of culmen less abruptly decurved, tip less blunt). Nostrils rounded. Face largely bare in adult; feathered in juvenile (see Plumages for extent of feathering and change with age); basal 40% of base of upper mandible covered with bristles in juvenile. Feathering of crown, hindneck, and lower flank soft and lax, more so than in *C. corone*. Tarsus and toes rather short, strong; middle toe with claw 46·8 (20) 44–51 in western Europe, 43·1 (4) 41–45 in Tien Shan; outer and inner toe with claw both *c.* 75% of middle with claw, hind *c.* 84%.

**Geographical variation.** *C. f. pastinator* from eastern Asia markedly different, showing feathered lore and chin, only skin at side of bill-base bare; head and neck darker, gloss less bluish, more dull reddish-purple. Slight variation occurs within both nominate *frugilegus* and *pastinator*. Within nominate *frugilegus*, birds in east of range (east from Iran and Turkmeniya to Ladakh and Tien Shan) average smaller, with relatively less deep and more attenuated bill and weaker feet with shorter toes; however, not all birds in east show these characters, and some in west are equally slender-billed (see Measurements), and recognition of slender-billed violet-tinged south Asian race, *tschusii* Hartert, 1903, or slender-billed blue-tinged central Asian race, *ultimus* is not warranted. Birds in north and west of range of *pastinator* sometimes separated as *centralis* (Tugarinov, 1929): said to differ in more bluish gloss and in slightly smaller size (Vaurie 1959), but gloss strongly dependent on abrasion and on age of specimens, and overlap in size large; hence, not recognized either. Situation in Middle East not yet elucidated: birds formerly breeding Palestine said to have head fully feathered (only 2 of 20 breeders with bare face) and gloss less intensely green-black (Tristram 1864, 1884; Stresemann 1928); majority of birds breeding Iran have face feathered (Meinertzhagen 1926, 1930), while migrants in northern Iran have bill markedly slender (Stresemann 1928b; Paludan 1940; Schüz 1959), as have those of north-west India and Sinkiang (western China) (BMNH, ZMA). Breeding areas of nominate *frugilegus* and *pastinator* said by Johansen (1944) to be well separated, with no intermediate birds known, but according to Witherby (1913) some *pastinator* have feathering on chin and throat partly reduced as in nominate *frugilegus*, and face said by Meinertzhagen (1930) to become gradually less bare towards east, or to become gradually fully bare at a later age eastward; Stepanyan (1978) believed intermediates to occur Altai and Sayan mountains.

May form superspecies with Cape Rook *C. capensis* from north-east and southern Afrotropics, which shows some similarities in anatomy and build of bill, but which differs in (e.g.) solitary breeding habits and plumage characters; not considered closely related by Hall and Moreau (1970) and Goodwin (1986).

**Recognition.** Juvenile with feathered face closely similar to *C. corone corone*; apart from differences in colour of gloss and bill shape, differs in relatively short p5: 31–47 shorter than wing-tip and distinctly shorter than p9 (15–31 in *C. corone*, about equal to p9).

CSR

# *Corvus corone* Carrion Crow

Du. Kraai  Fr. Corneille noire  Ge. Aaskrähe
Ru. Черная ворона  Sp. Corneja  Sw. Kråka

*Corvus Corone* Linnaeus, 1758

Polytypic. CORONE GROUP. Nominate *corone* Linnaeus, 1758, from England and Wales south to Iberia and southern slopes of Alps, east to Schleswig-Holstein and Elbe valley in Germany, western Czechoslovakia, and Austria, grading into *cornix* in southern Scotland, southern Denmark, north-west and centre of former East Germany, central Czechoslovakia, border of Austria with Hungary and Yugoslavia, and at southern foot of Alps; *orientalis* Eversmann, 1841, east-central Iran north to Aral Sea and Turgay depression, east in north through southern Kazakhstan to western Altai and in south through southern and eastern Afghanistan to Kashmir, Sinkiang, Mongolia, central and northern China, Korea, and Japan, from western Altai north through Yenisey valley to border of taiga and east to Koryakland and Kamchatka, perhaps also Bol'shoy Balkhan in western Turkmeniya; probably this race vagrant to Turkey. CORNIX GROUP ('Hooded Crow'). *C. c. cornix* Linnaeus, 1758, Ireland, Isle of Man, Scotland, Faeroes, Denmark, and Fenno-Scandia, east to Yenisey valley (there grading into *orientalis*), south in west to eastern part of former East Germany, eastern Czechoslovakia, Hungary, north-west Rumania, northern Yugoslavia, northern Italy, and Corsica, and in east to *c.* 49°N in Russia; *sharpii* Oates, 1889, Sardinia, Sicily, southern mainland Italy, and from coastal and southern Yugoslavia (south from Dalmatia, Hercegovina, Kosovo, and south-east Serbia), eastern Rumania, and Bulgaria south to Greece (including Crete), and from Moldova and Turkey (except south coast and south-east), east through southern Ukraine and northern Kazakhstan to western Altai and through Caucasus area and northern Iran east to central Turkmeniya and northern Khorasan (north-east Iran), grading into *cornix* at *c.* 49°N in Ukraine, European Russia, and Kazakhstan, overlapping and partly hybridizing with *orientalis* in Turkmeniya and Kazakhstan; *pallescens* Madarász, 1904, Cyprus, south-east and south coast of Turkey, Levant, northern Iraq, and Egypt; *capellanus* Sclater, 1876 Iraq and neighbouring south-west Iran, from about Ar Ramadi, Samarra, and Kirkuk south-east to Al Faw and Bushire.

**Field characters.** 45-47 cm; wing-span 93-104 cm (85-110). Slightly larger than Rook *C. frugilegus*, with deeper bill and curved upper mandible, heavier, flatter head, rather longer but less fingered wings, and slightly longer and squarer tail; slightly smaller and more compact than Brown-necked Raven *C. ruficollis* with shorter, squarer tail; 25% smaller than Raven *C. corax*, with markedly smaller bill, no shagginess to head, less full wings, and shorter, squarer tail. Large, quite powerful, heavy-billed crow, either all black (*corone* subspecies-group) or with contrasting grey back and underbody ('Hooded Crow', *cornix* subspecies-group). Lacks any grace or attractive feature, unlike other similar common west Palearctic crows, and does not show loose thigh feathers of *C. frugilegus*. Sexes similar; little seasonal variation but wide range of intergradation between black and hooded birds in Britain, northern Europe, and Mediterranean region. Juvenile separable at close range. 1 black and 4 hooded races in west Palearctic, 3 hooded races easily separable (but distinctions blurred by bleaching in Levant and Iraq).

ADULT. Moults: May-September (complete). (1) West-central and south-west European race, nominate *corone*. All-black, glossed greenish over forehead, greater coverts, and flight-feathers, and bluish- or reddish-purple elsewhere except on breast and belly (thus notably drabber below than *C. frugilegus* and *C. corax*). With wear, plumage dulls noticeably and bird becomes drabbest of all large crows. Bill and legs black. (2) North-west and east

European race, *cornix*. Lower hindneck, back (including scapulars), rump, breast sides, flanks, and rest of underbody and wing-coverts ashy-grey, with dark mesial streaks visible at close range; grey dulls with wear and may show brown tones. Variable necklace of black lanceolate marks at junction of black throat and breast centre. Rest of plumage, and bare parts, as nominate *corone*. (3) North-central Mediterranean, Turkish, and central Eurasian race, *sharpii*. As *cornix* but averages smaller, with rather weak bill and slightly shorter wings. Plumage similar when fresh but paler, even pinkish-toned, when worn and bleached. (4) Iraq race, *capellanus*. Largest and palest hooded form, with pale areas greyish-white even milky, contrasting strongly with rest of plumage. JUVENILE. Black nominate *corone* distinctly scrawnier than adult, with browner, much less glossed plumage. Hooded races have adult pattern but grey plumage distinctly browner overall and mottled above. HYBRIDS. Every intergradation between typical all-black and hooded birds; often isolated black dappling on back and chest.

Black birds need to be carefully distinguished from *C. frugilegus* and *C. corax*, but note differences in structure (see above), flight, and behaviour (see below). Hooded birds unmistakable in Europe and Levant, but racial hybrids with grey restricted to foreparts and just behind head may need care to prevent wishful thoughts of House Crow *C. splendens* or even Pied Crow *C. albus* (see those species). Flight powerful but rather slow, with usually regular and emphatic but sometimes loose, untidy wing-

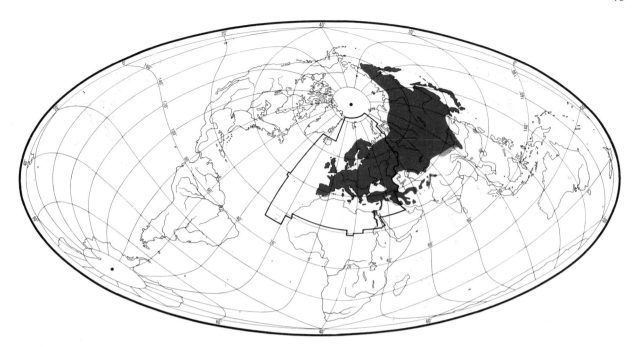

beats producing steady and usually straight progress; capable of laborious hover; rarely soars; action varied much less than *C. frugilegus*, but direct progress on migration over coasts apparently less sustained with odd birds often dropping behind rest of loose group and wheeling. Gait a steady, direct walk, varied by clumsy hop or sidling jump, in more horizontal attitude than *C. frugilegus*. Stance on perch half-upright to almost level (when windy), again usually less upright than *C. frugilegus*. Perceived as predator by many smaller passerines, with sudden appearance tending to flush them. Generally rather solitary, but will flock in autumn and feed close to other crows all year. Wary of man, even in urban areas.

Voice includes croaking gabble approaching song. Most distinctive call an abrupt, deep, rasping croak, 'kaarr', 'krarrr', or 'kawrh', subject to various inflections but rather monotonous; often 3 calls together. Other notes include complaining 'keerk. . .', blocked 'konk' or 'ponk' (suggesting old motor horn), and (when mobbing) short, angry rattle.

**Habitat.** *C. c. cornix* breeds in west Palearctic from sub-arctic and boreal through temperate to Mediterranean, steppe, and desert zones, up to 1000 m in Carpathians and Urals. In Scotland, predominates over nominate *corone* on higher ground, often moorland above *c.* 300 m, and is much more often found nesting on rocks, cliff ledges, and even on banks or islands on ground among heather, ranging up to *c.* 750 m (Baxter and Rintoul 1953); similar sites used in Norway. Apart, however, from these western populations subject to oceanic climates, habitats over greater part of continental range are no less arboreal than those of nominate *corone*, which in Swiss Alps breeds up to much higher levels, even up to 2000 m. In Russia, commonly found in breeding season in forest country, especially forest edges, groves, and river valleys. In winter, often moves near to human habitations, frequently in mixed flocks with Jackdaw *C. monedula*, roosting in tall trees in gardens and parks. (Burton 1974.)

Nominate *corone* in Germany prefers open country with copses and carr woodlands, but tolerates forest edges and rides (Niethammer 1937). In Low Countries, breeds in parks and woodlands intersected by fields or clearings, but also in many polders and wooded dunes, or along lines of trees bordering water channels (IJzendoorn 1950; Lippens and Wille 1972). In Switzerland, prefers open country with windbreaks, copses, carr woodland, or scattered trees, but ascends more thinly into montane forest bordering on grazed pastures, and forages on water-borne refuse from streams (Glutz von Blotzheim 1962). In Britain, habitats are similar, but with marked attraction towards foraging on tidal estuaries, salt-marshes, and coasts. Recently, strong build-up has occurred within even largest towns, although even here arboreal nest-sites are commonly essential. In inner London, large parties will fly from park trees down to muddy shore of tidal Thames as it is uncovered by tide (E M Nicholson). Artefacts, even tall buildings and ships, are freely used as perches. Many studies indicate that given its wide-ranging foraging, omnivorous appetite, and skill in exploiting the most diverse opportunities, habitat constraints are not normally limiting, except when and where they become linked with opportunites for human persecution, now a declining feature over most of range. Although no less

aerial than several congeners this species less commonly uses airspace for playful activities.

**Distribution.** No changes reported, except for expansion of range in northern Russia, Israel, and Egypt.

NORWAY. Nominate *corone* irregular breeder in coastal areas of Rogaland (VR). RUSSIA. In Kola peninsula, confined to forest zone in early 20th century; now common in forest tundra, and some penetration into tundra zone (Mikhaylov and Fil'chagov 1984). ISRAEL. Has expanded range in recent decades, following development of new settlements in 1950s and 1960s (Shirihai in press). JORDAN. Formerly more widespread; now mainly in northern highlands and Jordan valley (I J Andrews). EGYPT. Spread to

northern parts of Nile delta from about 1964 (Goodman and Meininger 1989).

Accidental. Iceland, Malta, Morocco, Tunisia, Libya, Madeira.

**Population.** No marked changes reported, except in southern France, Russia, and Israel.

FAEROES. 500–1000 pairs (Bloch and Sørensen 1984). BRITAIN, IRELAND. Probably about 1 million pairs (Sharrock 1976). Has increased in Ireland since 1924, noticeably so in late 1950s and early 1960s (Hutchinson 1989). FRANCE. 100 000 to 1 million pairs (Yeatman 1976). Has probably increased generally since about 1970; spectacular increase on Mediterranean coast from Camargue to east, where formerly scarce (RC). BELGIUM. Estimated 16 000 pairs (Devillers *et al.* 1988). NETHERLANDS. Estimated 50 000–80 000 pairs (SOVON 1987). GERMANY. 562 000 pairs (G Rheinwald). East Germany: 125 000 ± 60 000 pairs (Nicolai 1993). NORWAY. 0–5 pairs of nominate *corone* (VR). SWEDEN. Estimated 400 000 pairs (Ulfstrand and Högstedt 1976). FINLAND. Has probably increased in recent decades due to diminution of predators (Goshawk *Accipiter gentilis* and Eagle Owl *Bubo bubo*) and increased availability of garbage; perhaps 200 000–500 000 pairs (Koskimies 1989). CZECHOSLOVAKIA. Decrease in last 20 years (KH). HUNGARY. Estimated *c.* 50 000 pairs (Kalotás 1988). RUSSIA. Declined in Yaroslavl' region 1960–85 (Belousov 1986). Marked increase in Vologda region (Lebedev 1986). LITHUANIA. *c.* 70 000 pairs (Logminas 1991). ISRAEL. Increase in recent decades; see Distribution (Shirihai in press).

Survival. Britain and Finland: mortality in 1st calendar year 61%, 2nd year 45%, 3rd–5th years 48% (Holyoak 1971). Finland: annual mortality 47 ± SE2·3% (Haukioja 1969). Europe: mortality in 1st year of life 62·4% (Busse 1969). Oldest ringed bird 19 years (Staav 1983).

**Movements.** Varies from migratory in north of range to sedentary in south and west; many populations partially migratory. Winters almost entirely within breeding range. Migrants move mainly south-west or south. Movements diurnal, often in flocks. For reviews, see Busse (1969), Melde (1984).

Nominate *corone* (breeding western Europe) essentially sedentary. In Britain, almost all ringing recoveries (327 out of 332) up to 1965 were within 30 km of ringing site, with mean distance 7·8 km in northern England and Scotland and 4·2 km in southern England and Wales; most distant record was 115 km from natal site (Holyoak 1971). More recent ringing recoveries within Britain include movements of 135 km for Scottish bird and 91 km for Welsh bird (Mead and Hudson 1985). Movements lack any clear directional trends; short movements may consist only of flights to communal roosts, up to 6–8 km. Most dispersal takes place in winter, with highest proportion of distant recoveries in 1st and 2nd winters. (Holyoak 1971.) In Scotland, some birds move from upland areas to western seaboard (Lack 1986); in southern England, evidence for movement to coast in winter (e.g. in Sussex: Shrubb 1979). No overseas recoveries of British-ringed birds up to 1990 (Mead and Clark 1987, 1991), and no evidence that migrants from other countries reach Britain, though regular sightings on Scottish islands may involve Dutch or Danish birds (Lack 1986). Recoveries (up to 11 years after ringing) of 288 German birds ringed as nestlings show more movement, but little evidence of migration: only 11% recovered more than 25 km from ringing site and only 5% more than 50 km; 5 longest movements (275–950 km) were all west or south-west; long movements mostly involved 1st- and 2nd-winter individuals; birds settled to breed up

to 40 km from natal site (90% within 25 km) (Kalchreuter 1969a, 1970). Extensive local movements occur in some parts of northern Germany; Schleswig-Holstein almost entirely vacated, but these birds perhaps move only short distance further south, as winter arrivals regularly recorded from about end of October in northern Sauerland (Dobbrick 1931; Melde 1984). Chiefly sedentary in France, with some dispersal of juveniles (3 records 100–300 km); recoveries of foreign-ringed birds (23) are from Belgium, Switzerland, and western Germany (Yeatman-Berthelot 1991). In Switzerland, mostly sedentary, but some individuals move considerable distances, reaching south-west France; 5 recoveries of birds ringed as nest-lings were 145–680 km south-west of ringing site; birds from further north occur on passage and in winter (Glutz von Blotzheim 1962; Busse 1969; Winkler 1984). North of breeding range, occurs annually in Sweden (especially in south), chiefly in spring; 575 records 1976–89 (SOF 1990; Tyrberg 1990). Few birds move south of breeding range. Occasionally recorded at Strait of Gibraltar in both seasons (Finlayson 1992), and winters in extreme north of Morocco in small numbers in some years (Pineau and Giraud-Audine 1979). Rare vagrant to Corsica (Thibault 1983).

*C. c. cornix* (breeding northern and eastern Europe) includes essentially sedentary, partly migratory, and almost completely migratory populations. In Ireland and northern Scotland, ringing recoveries show no long-distance movements (Holyoak 1971). Fenno-Scandian populations partially migratory, and more migratory in north than south (Busse 1969). 1st-year birds more migrat-ory than older birds (Haftorn 1971). Movement on fairly narrow front. Migrants from Denmark and Scandinavia head south-west, reaching Low Countries, north-east France, and (in very small numbers) eastern Britain, but with many wintering in intermediate areas. Finnish birds show migratory divide: those from western Finland migrate through south-east Sweden and winter from Low Countries to Sweden; those from south-east Finland migrate south of Baltic and (with populations from Baltic states) reach Germany (chiefly north-east) and northern Poland. (Busse 1969.) Migrants show strong tendency to follow coast in Baltic region (Thomson 1949; Melde 1984). Study of migrating flocks in northern Germany showed that in autumn they mostly flew 10–70 m above ground, in spring at 80–500 m (Rittinghaus 1957). Some breeding areas in Norway almost completely deserted in winter, e.g. Drammen (southern Norway) and woodlands near Trondheim (central Norway) (Haftorn 1971; Slagsvold 1979a). However, recoveries (*n* = 71) of birds ringed as nestlings in western Norway show very few long-distance movements, 97% within 100 km of ringing site (only 14 winter recoveries, however) (Håland 1980). Data from different regions of Norway suggest more ♀♀ and 1st-winter birds move than adult ♂♂, both locally and over long distances (Slagsvold 1979b). Large numbers winter in south-west Norway, presumably due to reluctance to cross sea (Busse 1969). Many birds from Finland and northern Sweden winter in southern Sweden (Andell *et al.* 1983; SOF 1990). Study of marked population in Skåne (southern Sweden) showed that many local birds left area in winter; migrants included immatures and adults, but territory-holding adults remained. Of 12 winter recoveries outside study area, 7 within Skåne and 5 west or south-west in Denmark. (Loman 1985.) In Finland, most local winter recoveries are of adults, although many adults migrate; 1st-winter birds mainly migrate and may remain abroad (especially in southern Sweden) through 1st summer (Holyoak 1971). Birds ringed Kaliningrad region (western Russia), October, were recovered in breeding season in southern Finland, St Petersburg region, Estonia, and Latvia; winter recoveries were along southern coast of Baltic in Poland and Germany with one in Netherlands (Thienemann and Schüz 1931). In displacement experiments, 896 birds on spring passage in Kaliningrad region were ringed and released at 3 main sites 750, 1010, and 1025 km to west and south-west; most of 176 birds recaptured had migrated along traditional north-east bearing but were displaced to west, so that they spent summer in Denmark and Sweden instead of in eastern Baltic states and Russia (Rüppell 1944). Danish birds essentially sedentary; in Kraghede area (northern Denmark), winter population (including immigrants) far larger than summer; resident pairs often remain in territories all year (Møller 1983*b*).

Winter recoveries of Finnish nestlings and juveniles show reduction since *c.* 1940s in proportion migrating and mean distance moved; 46 recoveries 1933–48 included only 7% at less than 8 km, mean distance of the rest 924 km; 121 recoveries 1949–64 included 22% at less than 8 km, mean distance of the rest 745 km (Holyoak 1971). Similarly, widespread reports of decline in numbers wintering in western and central Europe, e.g. in northern Germany (Melde 1984), Poland (Tomiałojć 1976*b*), near Basle (northern Switzerland) (Böhmer 1973), and in France, where winter range has also contracted (Labitte 1955*b*; Sueur 1981; Yeatman-Berthelot 1991). Numbers at western extreme of winter range in eastern and central England have declined progressively over past 60 years, from 'great numbers' on east coast in early 1900s (e.g. Witherby *et al.* 1938), to flocks of hundreds in 1930s, tens in 1950s and 1960s, and scattered individuals in 1980s (Holyoak 1971; Cox 1984; Frost 1986; Mason 1989). Reduction in extent of migration of north European birds probably due chiefly to increasing supplies of winter food at rubbish tips within breeding range, perhaps also to climatic amelioration (Holyoak 1971; D T Holyoak).

In east of range, populations of Russia east to Urals probably at least partly migrate south to spend winter in lower Volga area (Varshavski 1977), others move southwest; birds ringed St Petersburg region in winter recovered north-east to Komi ASSR (up to 1500 km)

(Mal'chevski and Pukinski 1983). Populations of southwest USSR essentially sedentary (Dementiev and Gladkov 1954). No evidence that this race reaches Levant or Middle East; records in Egypt are misidentifications (Goodman and Meininger 1989). Vagrant to Iceland (30 records 1979–88: Pétursson *et al.* 1991), Greenland, Bear Island, Spitsbergen, and Novaya Zemlya (Vaurie 1959).

Autumn movement chiefly October–November. On Kola peninsula in north-west Russia (where since 1950s many birds remain to winter in towns), all birds previously departed by mid-November (Semenov-Tyan-Shanski and Gilyazov 1991). In Uppland (eastern Sweden), 1970–5, arrivals from Finland recorded from end of September, peaking mid-October, and with heavy concentrations on particular days (Johansson and Lundberg 1977). Passage and arrivals in northern Denmark early October to November (Møller 1983), and migrants reach Belgium and northern Germany in 1st half of October (Lippens and Wille 1972; Rutschke 1983), and Rheinland (western Germany), where few winter, in 2nd half of October (Mildenberger 1984). Winter population previously reached eastern Britain 2nd week of October to 3rd week of November (Witherby *et al.* 1938); in 1983–90, recorded (in very small numbers) from 5 October in Suffolk and 28 September in Norfolk (Moore 1984; Seago 1988). In north-east France, average earliest record 17–18 October in 1971–5, but (with decreasing numbers) 26 October in 1976–9 (Sueur 1981).

Spring movement February–April. Wintering birds leave north-east Germany by end of March (Rutschke 1983), and Belgium chiefly in March (Lippens and Wille 1972); winter population previously left Britain mid-March to 3rd week of April (Witherby *et al.* 1938). Passage and departures in northern Denmark late February to early April (Møller 1983), and arrivals in Sweden February–April (SOF 1990).

*C. c. sharpii* and *pallescens* sedentary within Mediterranean region (e.g. Beaman 1978, Goodman and Meininger 1989, Flint and Stewart 1992); individuals occasionally recorded south of breeding range in Israel, south to Elat (Shirihai in press). Vagrant records in Tunisia and Libya (Hollom *et al.* 1988) possibly *sharpii* rather than *cornix*. *C. c. capellanus* (Iraq and south-west Iran) apparently sedentary (Ticehurst *et al.* 1921–2; D A Scott).

Populations of *sharpii* in western Asia vary from migratory to sedentary from north to south of range. Vacates north of range (e.g. Yamal peninsula: Danilov *et al.* 1984), though many birds winter further south in western Siberia (Johansen 1944; Dementiev and Gladkov 1954). Common and widespread on passage in Kazakhstan (typically with south-west heading), and winters throughout but especially in plains of southern Kazakhstan (Korelov *et al.* 1974), and also in adjoining areas of south-central USSR (Rustamov 1958; Ivanov 1969; Abdusalyamov 1973). Widespread in winter in Afghanistan (breeds only in north) (S C Madge), and winter visitor to central and southern plains of Iraq (Ticehurst *et al.* 1921-2; Moore and Boswell 1956). Sedentary in Iran (D A Scott). Rather uncommon but regular winter visitor to northern Pakistan and adjoining regions of north-west India (Ali and Ripley 1972). Autumn movement September–November; departs from Yamal peninsula chiefly September, with main passage through Kazakhstan in October and arrivals in Pakistan November. Spring movement late February to May; leaves Pakistan by end of March, with passage through Kazakhstan late February to April and arrival at extreme north of range April–May. (Ali and Ripley 1972; Korelov *et al.* 1974; Danilov *et al.* 1984.)

*C. c. orientalis* (central and eastern Asia) also migratory to sedentary. Migratory in extreme north of range (northern Siberia), moving south but also west (Johansen 1944). In Tadzhikistan and Turkmeniya, makes only local or altitudinal movements (Rustamov 1958; Abdusalyamov 1973). Recorded as winter visitor in small numbers in north-east Iran (D A Scott). More widespread winter than summer in northern Pakistan and north-west India; population may then include winter visitors (Vaurie 1959; Ali and Ripley 1972). Vagrant 'Carrion Crows' in Turkey probably this race (Beaman 1978). In eastern Russia, present all year in breeding areas in Kamchatka peninsula, tending to concentrate near human habitations (Lobkov 1986); few remain for winter in Ussuriland, where common in summer (Panov 1973a). Winter range extends to southeast China (Schauensee 1984). Only local movements reported from Japan (Ornithological Society of Japan 1974; Brazil 1991). Main passage September–October and March–April (Panov 1973a; Korelov *et al.* 1974). DTH, DFV

**Food.** Principally invertebrates and cereal grain; also small vertebrates, birds' eggs, carrion, and scraps, proportions varying greatly according to local availability. In general, a ground-feeder and scavenger in agricultural landscapes, typically in pasture or rough grassland in spring and summer, arable fields in autumn and winter, when also nearer to towns, farms, woods, etc. Favourite sites include dung-rich pasture, hayfields, fields of cereal after harvest, areas by water (especially seashore), and rubbish tips, often exploiting rich food sources to exclusion of others; commonly follows plough. (Meidell 1943; Lockie 1955; Tenovuo 1963; Tompa 1976; Houston 1977a; Jollet 1984; Studer-Thiersch 1984; Fasola *et al.* 1986). In central England in winter, of 411 observations, 42% surface picking, 33% dung or stone turning, 9% surface probing, 9% pouncing, 5% digging or jabbing, and 3% deep probing (Waite 1984a, which see for comparison with other Corvidae). ♀♀ of 2 pairs took much more food by clod turning and surface picking and less by probing than did ♂♂, perhaps because of shorter bill ($n = 5118$ observations) (Holyoak 1970a). In southern Ireland, on seashore, 42% of 1613 actions were walking and

searching, 20% shell dropping, 17% surface picking, 7% seaweed turning, and 2% probing (Berrow *et al.* 1991, which see for comparison with Jackdaw *C. monedula* and Rook *C. frugilegus*). Will forage in taller vegetation than other west Palearctic Corvidae (Goodwin 1986, which see for review). See also Lockie (1956b), Tenovuo (1963), and Jollet (1984); for contrast between feeding strategies of resident birds and winter migrants in Denmark, see Møller (1983b). When killing small vertebrates or robbing nests often cooperates in pairs or small groups, one bird distracting while another attacks (Tenovuo 1963; Kneis 1977; McKee 1985; Schoof 1988). Commonly forces other birds, including raptors, to drop prey, e.g. Osprey *Pandion haliaetus*, Night Heron *Nycticorax nycticorax*, Common Gull *Larus canus*; also seen to rob surfacing Coot *Fulica atra* of bivalve *Dreissena* (Ouweneel 1970; Tinning and Tinning 1970; Edholm 1979, 1980; Deckert 1980; Loison 1984; Melde 1984; Wiprächtiger 1987). Frequently observed picking fish from water: will hover briefly before seizing them in bill (Jones 1955; Roberts 1955; Winkler and Winkler 1986), or take them, alive or dead, in feet like *P. haliaetus* (Calvert 1988; Tichon 1989). Will even repeatedly alight in, or jump into water deep enough to be completely immersed to take fish or frog *Rana* in bill (Hughes 1976; Coombs 1978), or wade into rock-pool apparently to drive small fish into position to be caught (Dunn 1985, which see for various seashore foraging techniques). Can be very persistent in dropping shells, usually mussels (Mytilidae), onto hard surfaces to break them open; one bird seen to drop mussel 7–8 times from 15–20 m over pebble beach before giving up because wind blew it onto grass (Terne 1978). Uses similar tactic with walnuts *Juglans* (Creutz 1953). In one study, tended to choose largest shells and to drop them from minimum height necessary (Whiteley *et al.* 1990, which see for details). See also Berrow *et al.* (1991). Strikes limpets (Patellidae) from rocks with sharp blow of bill (Greenslade 1979). Many accounts of predation on other birds away from nest, often young ones or in severe weather: one bird seized flying newly-fledged Blue Tit *Parus caeruleus* in bill (Mönke 1975); 2 chased Dunlin *Calidris alpina* in cold weather, one of them knocking it to ground and killing it with blows of bill (Dunn 1990); also in cold conditions, once attacked and killed adult Woodpigeon *Columba palumbus* in flight (Geyer 1985); took young House Martin *Delichon urbica* in bill after pursuit flight (Yapp 1975); Alpine Swift *Apus melba* taken while resting against wall (Hagmann and Dagan 1992); one bird seen to grab young Blackbird *Turdus merula* on ground and carry it off in feet (Bauer 1961); see Coombs (1978), Deckert (1980), and Studer-Thiersch (1984) for further examples. Will also catch insects in flight, e.g. bee *Bombus* captured after twisting pursuit lasting 5 s (Rogers 1982), and observation of small bat (Chiroptera) being taken in air (Arnold 1955). On cliff ledges will seize adult Guillemots *Uria aalge* or Fulmars *Fulmarus glacialis* by wing to drag them off eggs, again often operating in pairs (McKee 1985). Shows considerable ingenuity in exploiting food sources: has learned to peck through milk bottle tops (Bates 1979); will hang upside-down on peanut bags or suspended fat in gardens (Gush 1978); pulls up fishing lines through ice-holes, holding firm under foot (Tenovuo 1963); takes parasites by perching on backs of domestic animals (Kumerloeve 1957); group of birds seen hanging on sunflower *Helianthus* heads feeding on seeds (Strache and Madas 1988), and one bird observed hammering at dead branch like woodpecker (Picinae), removing wood to reach larvae (Piesker 1972). Easily finds birds' nests on ground if able to see people examining them, robbing them later, or going to spot from which sitting bird flushed (Picozzi 1975*b*; Götmark and Åhlund 1984; Salathé 1987). In one study, appeared to remember sites of experimentally placed nests 1 year later (Sonerud and Fjeld 1987). For various aspects of predation on nests of waterfowl or Galliformes, and of songbirds, see Tenovuo (1963) for Finland, Deckert (1980) and Melde (1984) for Germany, Erikstad *et al.* (1982) for Norway, Vercauteren (1984) for Belgium, Elliot (1985) for Scotland, and Gulay (1989) for Ukraine. For role of search-image and learning in foraging behaviour, see Croze (1970) and Frugis *et al.* (1983). Prey carried in bill or in feet (sometimes in 1 foot only), sometimes transferred from foot to bill or vice versa in flight (King 1969; Tinning and Tinning 1970; Edholm 1979; Radermacher 1983). Unable to open carcass of sheep, rabbit, etc., so if this is not done by another predator can only take eyes or tongue (Hewson 1981, 1984). In Scottish study, only *c.* 0·1% of lambs born estimated to have died as result of attacks by *C. corone*, and great majority of those eaten were probably dead or already dying; such attacks occur at time of food abundance so probably not motivated by hunger (Houston 1977*b*, 1978, which see for details). In western Scotland, when presented with opened carcasses, at first fed at them then began to cache pieces of meat of *c.* 10-170 g; of 42 caches, 36% 1-20 m from carcass, 36% 20-40 m, 17% 40-100 m, and 12% more than 100 m; caches were in tufts of grass or moss and covered with dead grass and were made by both sexes; when recovering stored food flew to within 1 m of site, also in snow (Hewson 1981). In southern Norway, birds cached at least 35% of 222 experimentally-placed hens' eggs, only 1% seen to be eaten on spot, others taken to dumps for consumption, not nest; caches were 1-500 m from food source and no eggs cached or eaten less than 140 m from nest; contents carried in gullet to young (Fjeld and Sonerud 1988, which see for details). Seen to cache insects, earthworms, acorns, and scraps (Waite 1985*a*, which see for discussion); see also Waite (1986) for recovery of cache in snow. One bird observed placing pieces of fish under small stone then building pile of stones and fragments of wood around it (Ewins 1986). Will store items singly or several at a time (Goodwin 1955*a*; Hewson 1981). See Strauss (1938*b*) for experiments with captive birds.

Diet in west Palearctic includes the following. Vertebrates: (many as carrion, road casualties, etc.): mole *Talpa*, shrews *Sorex*, *Crocidura*, bat (Chiroptera), hedgehog *Erinaceus*, rabbit *Oryctolagus*, hare *Lepus*, souslik *Spermophilus*, hamster *Cricetus*, mice *Micromys minutus*, *Apodemus sylvaticus*, *Mus musculus*, rat *Rattus*, voles *Microtus agrestis*, *M. arvalis*, *Clethrionomys glareolus*, *Pitymys subterraneus*, weasel *Mustela nivalis*, domestic sheep *Ovis*, roe deer *Capreolus*, lizard *Lacerta*, slow worm *Anguis*, snake (Ophidia), frog *Rana*, shad *Alosa*, trout *Salmo*, bream *Abramis*, bleak *Alburnus*, carp *Carassius*, *Cyprinus*, grass carp *Ctenopharyngodon*, minnow *Phoxinus*, roach *Rutilus*, tench *Tinca*, dace *Leuciscus*, perch *Perca*, eggs, and young and adult birds, chiefly ground-nesters such as Anatidae, Phasianidae, Charadriidae, Alaudidae, etc. but also passerines of all kinds, including own species. Invertebrates: mayflies (Ephemeroptera), damsel flies and dragonflies (Odonata: Coenagriidae, Aeshnidae, Libellulidae), grasshoppers, etc. (Orthoptera: Gryllidae, Tettigoniidae, Gryllotalpidae, Acrididae), earwigs (Dermaptera: Forficulidae), cockroaches (Dictyoptera: Blattidae), bugs (Hemiptera: Scutelleridae, Pentatomidae, Nepidae, Naucoridae, Notonectidae), lacewings (Neuroptera: Chrysopidae), adult and larval Lepidoptera (Cossidae, Arctiidae, Noctuidae), caddis flies (Trichoptera: Phryganeidae, Limnephilidae), adult and larval flies (Diptera: Tipulidae, Tabanidae, Empididae, Syrphidae, Calliphoridae, Muscidae), adult and larval Hymenoptera (Cimbicidae, Mutillidae, Chrysididae, ants Formicidae, wasps Vespidae, bees Apoidea), adult and larval beetles (Coleoptera: Cicindelidae, Carabidae, Dytiscidae, Hydrophilidae, Histeridae, Silphidae, Staphylinidae, Lucanidae, Geotrupidae, Scarabaeidae, Byrrhidae, Elateridae, Cantharidae, Dermestidae, Meloidae, Tenebrionidae, Coccinellidae, Cerambycidae, Chrysomelidae, Curculionidae), spiders (Araneae: Theridiidae, Araneidae, Gnaphosidae, Lycosidae, Clubionidae), harvestmen (Opiliones: Phalangiidae), Crustacea (woodlice Isopoda, crayfish *Astacus*), millipedes (Diplopoda: Glomeridae, Julidae), centipedes (Chilopoda), earthworms (Lumbricidae), Mollusca (Helicidae, Succineidae, Bulimidae, Planorbidae, Acmaeidae, Neritidae, Valvatidae, Viviparidae, Lymnaeidae, Clausiliidae, Physidae, Ancylidae, Patellidae, Mytilidae, Unionidae, Cardiidae, Dreissenidae). Plants: fruits, seeds, and roots of spruce *Picea*, juniper *Juniperus*, oak *Quercus*, beech *Fagus*, chestnut *Castanea*, birch *Betula*, lime *Tilia*, ash *Fraxinus*, walnut *Juglans*, mulberry *Morus*, fig *Ficus*, carob *Ceratonia*, barberry *Berberis*, spindle *Euonymus*, dogwood *Cornus*, sea buckthorn *Hippophae*, grape *Vitis*, Virginia creeper *Parthenocissus*, *Ampelopsis*, olive *Olea*, bilberry, etc. *Vaccinium*, crowberry *Empetrum*, elder *Sambucus*, mistletoe *Viscum*, hemp *Cannabis*, knotgrass, etc. *Polygonum*, buckwheat *Fagopyrum*, orache *Atriplex*, amaranth *Amaranthus*, poppy *Papaver*, charlock *Sinapis*, cabbage, etc. *Brassica*, rowan, etc. *Sorbus*, plum, etc. *Prunus*, apple *Malus*, pear *Pyrus*, strawberry *Fragaria*, hawthorn *Crataegus*, blackberry *Rubus*, rose *Rosa*, *Cotoneaster*, *Amelanchier*, vetch *Vicia*, clover, etc. *Trifolium*, *Hibiscus*, pumpkin *Cucurbita*, watermelon *Citrullus*, ivy *Hedera*, jujube *Zizyphus*, Umbelliferae, paprika *Capsicum*, bedstraw, etc. *Galium*, honeysuckle *Lonicera*, sunflower *Helianthus*, bur marigold *Bidens*, knapweed, etc. *Centaurea*, nipplewort *Lapsana*, pondweed *Potamogeton*, tape-grass *Vallisneria*, grasses (Gramineae), sedge *Carex*, rush *Juncus*, horsetail *Equisetum*, moss (Musci), lichen (Mycophycophyta), and all kinds of agricultural crops, especially cereals (e.g. maize *Zea*, wheat *Triticum*) and root crops (e.g. potato *Solanum*). (Rey 1907*b*; Csiki 1914; Meidell 1943; Arnold 1955; Hell and Soviš 1958; Turček 1961; Glutz von Blotzheim 1962; Tenovuo 1963; Holyoak 1968; Sterbetz 1968; Yom-Tov 1975*a*; Bauer 1976; Houston 1977*a*; Deckert 1980; Kiss and Rékási 1983; Jollet 1984; Melde 1984; Fasola *et al.* 1986; Snow and Snow 1988; Soler and Soler 1991.)

Great variation in diet and seasonal changes depending on habitat, type of agriculture, etc., in addition, soft-bodied invertebrates such as earthworms, snails, and larvae are often underestimated in analysis of stomach or pellet contents (Deckert 1980; Jollet 1984; Studer-Thiersch 1984). In general, higher proportion of plant material (especially cereals and root crops) eaten in autumn and winter (though in areas with little arable agriculture winter diet shows increase in small mammals, carrion, and scraps), and more animal prey (chiefly invertebrates, some eggs, nestlings, and other small vertebrates) taken in spring and summer (Meidell 1943; Hell and Soviš 1958; Tenovuo 1963; Holyoak 1968; Houston 1977*a*; Jollet 1984; Fasola *et al.* 1986). For food of adult, see Table A;

see also Tenovuo (1963) for Finland, Madon (1928*b*) for France, Deckert (1980) for eastern Germany, Csiki (1914) and Sterbetz (1968) for Hungary, Dončev (1958) for Bulgaria, and Popov (1978) for European Russia.

Earthworms recorded in only 1 study in Table A, but (e.g.) present in 10 of 12 stomachs, March–April, in southern England (Lockie 1956*b*, which see for comparison with other Corvidae), in 47% of 86, January–February, in England and Wales (Holyoak 1968, which also see for comparisons), and were commonest animal component by occurrence in major Swiss study, recorded in *c.* 65–100% of 831 stomachs over the year (Studer-Thiersch 1984). In this study 38–48% of diet over the year at lower altitude was animal material ($n = 645$ stomachs), but 59% in Alps ($n = 175$); in non-mountainous areas intensive agriculture means both more cereals and reduction in invertebrate numbers, while at high altitude pasture predominates. Similarly in western Scotland, 62% of winter diet by volume in uplands was carrion and 27% grain from cattle feed; in lowlands, up to 80% grain and only *c.* 5% carrion (Houston 1977*a*). Difference between upland and lowland diets always slight in spring and summer, when chiefly invertebrates in both cases. In central Norway, *c.* 80% of items by number in 50 stomachs over the year were plant material (cereals, potatoes, and berries; some animal components underestimated) (Meidell 1943). In Czechoslovakia, vegetable matter 86% by weight (grit excluded) in 59 winter stomachs (Hell and Soviš 1958), and in southern Spain *c.* 80% by volume (grit excluded) of autumn and winter diet was cereal (Soler and Soler 1991). In central France, 15% of plant material eaten in April was

Table A   Food of adult Carrion Crow *Corvus corone* from pellet analysis (Spain) and stomach analysis (other studies). Figures in italics are percentages by dry weight, others are percentages by number. Studies from Spain and Norway are of animal material only.

| | S Spain | | | | Cent. France, all year | N Italy, all year | Cent. Norway, all year |
|---|---|---|---|---|---|---|---|
| | autumn | | winter | | | | |
| Vertebrates, eggs | 2·5 | *68·6* | 2·6 | *86·6* | 0·5 | 20·1 | 2·5 |
| Orthoptera | 21·1 | *23·9* | 9·7 | *8·4* | 1·4 | 1·3 | 9·2 |
| Hymenoptera | 22·3 | *0·8* | 49·8 | *1·1* | *c.* 0·6 | 1·8 | 10·9 |
| (Formicidae) | (19·8) | (*0·4*) | (46·7) | (*0·9*) | (*c.* 0·6) | (1·3) | (8·9) |
| Beetles | 22·5 | *3·2* | 20·6 | *1·8* | 22·8 | 27·9 | 59·5 |
| (Scarabaeidae) | (0·3) | (*0·03*) | (0·6) | (*0·1*) | (14·6) | (5·5) | (37·4) |
| (Carabidae) | (3·2) | (*0·1*) | (2·5) | (*0·1*) | (1·2) | (8·0) | (12·2) |
| (Curculionidae) | (9·0) | (*0·3*) | (9·5) | (*0·2*) | (—) | (6·8) | (1·0) |
| (Elateridae) | (0) | (*0*) | (0·2) | (*0*) | (4·6) | (3·0) | (1·3) |
| Larvae | 24·5 | *1·6* | 12·5 | *0·4* | 2·7 | 4·3 | 5·6 |
| Spiders, harvestmen | 0 | *0* | 0·1 | *0* | 0·1 | 0·3 | 7·9 |
| Earthworms | — | — | — | — | 5·8 | — | 0 |
| Snails | 1·5 | *0·2* | 1·6 | *0·2* | 0·1 | 9·3 | 0·5 |
| Cereal grain | — | — | — | — | 55·6 | 17·3 | — |
| Other crops | — | — | — | — | 0·6 | 1·0 | — |
| Other plants | — | — | — | — | 2·1 | 8·9 | — |
| Total no. of items | *c.* 1020 animal items | | *c.* 1350 animal items | | *c.* 3130 | 394 | 393 animal items |
| Source | Soler and Soler (1991) | | | | Jollet (1984) | Fasola *et al.* (1986) | Meidell (1943) |

ivy berries (Jollet 1984), yet these apparently never recorded in diet in Britain (Snow and Snow 1988). In southern England, 16% of 67 invertebrates taken (excluding earthworms) were 2-6 mm long, 51% 7-12 mm, 22% 13-18 mm, and 11% longer than 18 mm (Lockie 1956b).

In central England, in winter, average energy intake ranged from 0·001 kJ per min from small invertebrates below soil surface, to 0·38 from large earthworms on surface ($n = 90$ birds); medium-sized invertebrates from dung also important (Waite 1984a, which see for comparison with other Corvidae). In central France, required daily intake varies from 517 kJ in August to 645 kJ in February; required intake in fresh weight ranges from 61 g in June (different from energy minimum because of varying animal/plant ratio over the year) to 83 g in February (Jollet 1984, which see for details). Daily consumption in captivity c. 130 g plant and 25 g animal material (Deckert 1980). See also Møller (1983b) and Melde (1984).

For food of nestlings, see Table B. Diet very like that of adult in spring, and local resources often dominate; in Finland, beetle Cetonia aurata (Scarabaeidae) accounted for over 50% by volume in 71 stomachs (Tenovuo 1963); in western Scotland, 39% by number of animal prey beetles Chrysomelidae ($n = 268$ items from 10 samples), 38% by fresh weight were Chrysomelidae and Geotrupidae (Houston 1977a), and in southern England, where birds fed often in oak woodland, caterpillars comprised 19% of items from 41 samples (Lockie 1955). In Norway, almost half of vertebrate component were frogs, fed mainly to older nestlings; youngest fed mostly beetles

and crane-fly (Tipulidae) adults and larvae; older chicks also given plant material (Meidell 1943). Also in northern Scotland, beetles, crane-fly larvae, and worms were fed more to younger birds, while frequency of occurrence of carrion and small vertebrates in 56 stomachs increased from 22% to 65% with age; barley Hordeum was in up to 80%, mostly older nestlings. Most Carabidae given to young were 5-12 mm long, caterpillars 20-55 mm, crane-fly larvae 10-15 mm, and pieces of meat weighed up to 4 g. (Yom-Tov 1975a.) Holyoak (1970a) found that significantly different prey sizes were fed by ♀ and ♂ parent; average size of items fed by ♀ 14·0 mm ($n = 132$, mostly small insects and spiders), ♂ 22·2 mm ($n = 118$, with more larvae). In western Scotland, lowland birds took wider variety and larger average size of insects for young than those foraging in uplands (Houston 1977a). In Norway, average daily intake of c. 65 g needed when young weighs 100 g, rising to c. 140 g at fledging (Meidell 1943).   BH

Social pattern and behaviour. Well known, with detailed studies of both nominate corone and cornix. Information on these and much sparser data on sharpii, pallescens, and capellanus show no clear differences. Detailed studies by Charles (1972), Picozzi (1975a), and Spray (1978) in Scotland, Tenovuo (1963) in south-west Finland, Loman (1975, 1980d, 1985) in southern Sweden, Slagsvold (1978, 1984, 1985) in Norway, Wittenberg (1968, 1976) and Kalchreuter (1971) in Germany, Tompa (1975), Böhmer (1976a, b), and Richner (1989a, b, 1990) in Switzerland, and Grabovski (1983) in Moscow area. Reviews by Madon (1928b), Coombs (1978), Melde (1984), and Goodwin (1986); popular review by Wilmore (1977). Account includes unpublished data from P J Conder (west Wales and Channel Islands), J R Harpum (south-west England), D T Holyoak (southern England and Isle of Man).

1. Single birds, pairs, and flocks occur at all times of year. Pairs and sometimes single birds hold territories (see Breeding Dispersion, below), in all months in some regions, but principally during breeding season in others. Largest flocks occur at rich food sources or at pre-roost gatherings (see Roosting, below): up to 500 or more, or 800 in English Midlands (Harrison 1982), 900 in roost, southern Rhône (France) (Salathé and Razumovsky 1986). Greatest numbers seen on migration, e.g. crossing Rybachiy (Kaliningrad region, western Russia), where up to 30 000 cornix recorded as continuous stream in 1 day (Thomson 1949). Resident flocks generally largest in winter and smallest during breeding season. Gradual increase in size of flocks during summer in some regions (e.g. southern Sweden: Loman 1985). 1st-winter birds may spend much time in flocks. In southern Germany, flocks exclusively of juveniles form in summer (Kalchreuter 1971), whereas in southern Sweden juveniles joined flocks that also contained older flocking birds and territorial birds. In other studies, juveniles remained on natal territory into 1st winter (see Bonds, below). In southern Sweden, from beginning of 2nd calendar year, increasing proportion of flock-members leave flocks temporarily to 'prospect' for territories and later hold territories (see below) (Loman 1985). However, territory-holding birds often rejoin flocks temporarily. In south-west England, prospecting occurs in spring by single birds, or more usually pairs, often at heights exceeding 100 m (J R Harpum). In Sweden, winter, 1st-years fed in large flocks at places with concentrations of food. Territorial (older) birds also occurred in large flocks but more often in smaller flocks feeding

Table B   Food of nestling Carrion Crow Corvus corone from collar-samples (Scotland) and stomach analysis (other studies). Figures in italics are percentages by fresh weight, others are percentages by number.

| | W Scotland | | Cent. Norway | E Rumania |
|---|---|---|---|---|
| Vertebrates, eggs | 0 | | 8·6 | 9·3 |
| Lepidoptera | 3·0 | 5·4 | 0·6 | 0·6 |
| Diptera | 16·4 | 22 | 18·0 | — |
| Hymenoptera | 1·1 | 1·4 | 2·6 | 1·2 |
| Beetles | 63·8 | 57 | 39·1 | 68·3 |
| (Scarabaeidae) | (2·6) | (9·0) | (1·3) | (0·6) |
| (Carabidae) | (5·2) | (0·8) | (18·6) | (35·1) |
| (Curculionidae) | (0) | (0) | (4·2) | (10·9) |
| (Elateridae) | (6·0) | (2·3) | (1·3) | (2·5) |
| Spiders, Harvestmen | 8·6 | 3·8 | 8·1 | 0 |
| Earthworms | 3·0 | 6·0 | 0 | 0 |
| Snails | 1·1 | 0·1 | 0 | 0·6 |
| Cereal grain | 0 | | 7·4 | 16·8 |
| Other plants | 0 | | 11·2 | 0·6 |
| Total no. of items | 268 items | | 1051 items | 322 items |
| Source | Houston (1977a) | | Meidell (1977) | Kiss and Rékási (1983) |

in areas without obvious food concentrations. 2nd- and 3rd-years without territories showed intermediate flocking behaviour. Some flocking birds, but no territorial individuals, emigrated during winter. (Loman 1985: see Movements.) Winter immigrants to Denmark mainly feed in flocks at concentrations of food, unlike residents (Møller 1983*b*). In Camargue (southern France), where population thinly spread, non-breeding flocks do not occur in breeding season, and flocking seen only outside this period (Salathé and Razumovsky 1986). Distinction in many studies between territory-owners (mainly breeding pairs) and flock-members sometimes accentuated by well-marked spatial segregation (e.g. Charles 1972, Böhmer 1976*b*). In Cheltenham (south-west England), year-round territories stretch across edge and some central areas of town. Flock-birds are based in farmland to south from where they range into town, crossing residents' territories. Urban birds regularly feed in flock area while flock-birds make forays into town. Attempts by adults to establish territories and breed in flock area always fail due to attentions of flock-birds and human interference. Territorial boundaries in urban area fairly stable from year to year but fully established in late autumn when much conflict between territory-holders. (Harpum 1985; J R Harpum: see Antagonistic Behaviour, below.) Territorial defence minimal in summer after breeding, although birds remain on territories in at least some regions (e.g. England: D T Holyoak). In southern Sweden, many territorial birds leave territory in summer and do not reoccupy it until next spring (Loman 1985). In north-west Germany, most birds absent from territories in summer but return later in autumn (Wittenberg 1968). Extent to which territorial birds stay on territory during winter varies regionally. In north-west Germany (Wittenberg 1968), Scotland (Charles 1972; Spray 1978), Switzerland (Tompa 1975), Denmark (Møller 1983*b*), and Hertfordshire (D T Holyoak), territories defended throughout winter; likewise in south-west Germany, but territories temporarily abandoned in heavy snow (Kalchreuter 1971). In southern Sweden, only *c.* 25% of territorial birds were in territory in winter but much variation between pairs, probably related to differences in winter food supply in territories (Loman 1985). Near Trondheim (central Norway), birds leave woodland territories almost entirely during winter when they feed near human settlements (Slagsvold 1978). In north-west Germany, Wittenberg (1968) found territorial limits not strict outside breeding season, owners making occasional excursions outside them. In Scotland, persistent intruders tolerated during winter as '3rd birds' in territories, but evicted from them before following breeding season (Charles 1972) (see Bonds, below). Flocks, pairs, and single birds often feed and roost with Rooks *C. frugilegus* and Jackdaws *C. monedula*. Apparent association with other species such as gulls *Larus* and Starlings *Sturnus vulgaris* may relate to aggregation at common food source, or sometimes to *C. corone* seeking opportunities for stealing food. BONDS. Essentially monogamous. Pair-bond of long duration (Wittenberg 1968; Loman 1985). In Sweden, observation of 16 marked pairs over 34 pair-years showed only 1 pair separating; that pair had established territory but did not breed, and both pair-members may have been 1st-years; in following year, both had different mates, ♂ staying in original territory, ♀ in neighbouring one (Loman 1985). For rapid mate-replacement after death of pair-member, see Breeding Dispersion and Heterosexual Behaviour (below). Strong tendency for established pair to remain together, even when in flock. Loman (1985) noted fewer birds paired in November–December than in other months. Nest-building by both sexes. Incubation by ♀ alone, fed by ♂. Small young brooded by ♀ alone, fed by ♀ with food provided by ♂; larger young fed by both sexes until *c.* 4–5 weeks after fledging (see Relations within Family Group, below). Time

at which juveniles leave natal territory varies regionally. Studies in Scotland (Charles 1972), southern England (J R Harpum, D T Holyoak), and Switzerland (Tompa 1975) show young leaving gradually during 1st winter. In contrast, in southern Germany (Kalchreuter 1971) juveniles leave in late summer, and likewise in southern Sweden (Loman 1985) most juveniles ceased to visit natal territories by September. Scottish study showed apparently intermediate timing of juvenile dispersal, with half the broods independent by September but some others remaining with parents until following breeding season (Picozzi 1975*a*). In north-west Germany, Wittenberg (1968) found juveniles beginning to separate from parents 4–8 weeks after fledging, but they returned for short periods up to beginning of breeding season. Loman (1985) suggested these differences related to relative abundance of food in natal territories and in areas used by flocks. In Sweden, 2 exceptional immatures (both probably ♂♂) remained partly associated with natal territories throughout 1st winter and following breeding season, and were again associated with it the following spring (Loman 1985). In Switzerland, Tompa (1975) observed 3rd bird in 2 out of 36 territories, but ages and origins of extra birds not established. Function of this association uncertain; Loman (1985) thought it unlikely extra birds were nest-helpers. In high-density urban population in Switzerland, however, Richner (1990) recorded helpers at 3 nests (including 2 out of 33 nests studied in detail); 2 of these helpers fed nestlings regularly, 1 of them also fed incubating ♀ and joined in violent territorial disputes against neighbours; in 2 of the cases, helpers were 2-year-old ♂ offspring of pair they helped. In Gloucestershire, philopatry observed in 1st-years which assisted parents in defending their territory from flock incursions as late as mid-March (J R Harpum). Harper (1992) recorded partly albino immature ♂ that remained in territory of breeding pair, despite much aggression from adult ♂ during early April; immature assisted in territorial defence, twice copulated with ♀, and seen to feed ♀ (twice) and nestlings. First breeding reported at 2 years old, but typically at 3 or later (see Breeding); local and individual differences common. Presence of pair in territory not always proof of breeding. Richner (1989*a*) showed body size to be important in acquiring territory. In southern Sweden, some single (mainly ♂) or paired 1st-years were observed prospecting away from flocks (see below) in spring. None of 60 marked 1st-years established territory; 2 out of 30 did so in 2nd year, and 5 out of 13 when *c.* 3 years old. One of 2nd-year territory-holders bred, and 3 of the 5 3rd-year pairs (Loman 1985). In Scotland, Picozzi (1975*a*) recorded 3 birds first holding territories in their 2nd year, 2 in their 3rd year, and 1 in 4th year. Niethammer (1936) found no 1st-years holding territories or attempting to breed. In north-west Germany, no 1st-years paired or established territories. Among 3-year-olds there were 7 unpaired and 5 paired birds, of which only 2 bred. 5 out of 6 4-year-olds bred. (Wittenberg 1976.) In Scotland, Charles (1972) found that young flocking birds established small temporary territories during winter and early spring. Later, during breeding season, these birds were usually back in flocks. This not reported in other studies, but may well be alternative to return of young birds to their natal territories described above. Overall proportions of non-breeders hard to assess because many form scattered flocks and others hold territories. Following examples therefore restricted to detailed studies over large areas. In Sweden, Loman (1985) found *c.* 430 individuals in 20 km² included only 50 breeding pairs, implying *c.* 77% non-breeders. In Scotland, estimates of 50% (*n* = 80) to 73% (*n* = 134) non-breeders in different years (from Picozzi 1975*a*) may be underestimates because they apparently did not take account of territory-holding non-breeders. In Camargue, 16-km² study area held c. 30 birds, of which 12 (40%) non-breeders; latter comprised 3 non-breeding pairs and 6 immature birds (2 pairs + 2 single birds) (Salathé and Razumovsky 1986). Few data available on sex ratios. Larger numbers of ♂♂ than ♀♀ reported in some studies of shot birds (Houston 1974; Picozzi 1975a); attributed by Houston (1974) to better survival of more dominant ♂♂. Extra mortality of ♀♀ shot on nests is another possible explanation. Slagsvold (1979b) gave data on sex (and age) ratios of birds shot at various localities in Norway; sex differences in extent of migration are evident (see Movements). BREEDING DISPERSION. Essentially territorial. Nests sometimes so close together as to form loose colonies where suitable nest-sites are few and concentrated in extensive feeding areas. Exceptional colonial nesting reported in north-east Germany, with 37 pairs in wood of 7·5 ha (Abshagen 1963). In Sweden, with low population densities, nests weakly overdispersed (Loman 1975); at higher densities, more clearly overdispersed (e.g. farmland with plentiful nest trees in southern England: D T Holyoak). Other examples of close nesting include 3 nests in 2 adjacent trees in small London square (England 1970b), 2 successful nests less than 10 m apart in tree in south-west England (Gush 1975), and 6 nests in Scots pines Pinus sylvestris on 65 ha of moor and 5 nests in 16 ha (narrow strip) of woodland in northern Scotland (Rebecca 1985, 1986). 3 nests of sharpii in one tree in Iran (Baker 1932). Size of territory variable. In north-west Germany, with groves of trees widely spaced among open fields, 0·14-0·46 km² (mean 0·25 km²), inter-nest distance 118-510 m (mean 190 m) (Wittenberg 1968); in southern Sweden, in pastures with plentiful trees, mean 0·45 km², mean inter-nest distance 360 m (n = 362 nests over 4 years in 20 km²: Loman 1975). In Scotland, in woods and farmland, mean 0·33-0·44 km² (over 4 years, in 7 km²: Picozzi 1975a). On islets with dense breeding populations in south-west Finland, closest c. 250 m apart (Tenovuo 1963). In Moscow area, Konstantinov et al. (1982) found 32-36 breeding pairs per km² in urban areas, with nests sometimes only 20 m apart; by comparison, only 2-3 pairs per km² in surrounding agricultural areas. In Lenin Hills (Moscow), Grabovski (1983) recorded inter-nest distances of 10-200 m (for 'neighbourhoods' and their behaviour, see Parental Anti-predator Strategies, below). In Rostock area (north-east Germany) (colonial nesting in small wood isolated in farmland), mean inter-nest distance less than 45 m (Abshagen 1963). Charles (1972) demonstrated that experimental addition of tree in large gap between nests led to establishment of new nesting territory between existing ones. Territory may vary markedly in size during year (Kuhk 1931), also from year to year with same breeding pair; over 10 years, territory of single breeding pair in Cheltenham 0·30-0·36 km² (mean 0·32 km²); territory had well-defined 'roof' (15-30 m high) so that it had a finite volume which varied during day depending on activity of owners; during incubation, pair reduced territory to about one third of its full size but roughly doubled maximum roof height. Adjacent territory-holders also maintained a roof; territorial birds flying to countryside from inner urban area flew high, with direct flight, unchallenged over other territories. (J R Harpum.) During breeding season almost all birds can easily be recognized as either territorial (mainly breeders, some non-breeders) or flocking (mainly, if not entirely immatures) (e.g. Harpum 1985, Loman 1985). Territories all-purpose, partly overlapping (Wittenberg 1968) or discrete (Loman 1985), but not necessarily contiguous; in Cheltenham (as in most urban areas), territories may be separated by 'neutral zones' of high-density housing and streets lacking suitable nest-sites (J R Harpum). Territory serves for nesting and most of territory is covered every day in search for food (Wittenberg 1968; Loman 1985; D T Holyoak). Territorial birds rarely (Wittenberg 1968) or occasionally (Loman 1985)

observed outside territory in breeding season. On islets off south-west Finland, breeding birds fly up to several km to feed in seabird colonies (Tenovuo 1963). Overall breeding density: reaches 10-15 pairs per km² in south-west Finland (Tenovuo 1963); 2-3 pairs per km² in 20 km² of mainly pastoral land in southern Sweden (Loman 1985); 2·3-3·0 pairs per km² in 7 km² of woods and farmland in Scotland (Picozzi 1975a); 20 pairs in coastal zone (7·8 km²) of trees in Argyll (Scotland) (Hewson and Leitch 1982); 9 pairs per km² in richest part of 3 km² near Basle (Switzerland) (Böhmer 1976); c. 6 pairs per km² in rural habitat and c. 36 pairs per km² in urban habitat in Switzerland (Richner 1990); 4·9-6·8 nests per km² in 3 areas of south-west Norway (n = 202: Munkejord et al. 1985); 2·0 breeding pairs per km² in Camargue (Salathé and Razumovsky 1986); 6·8 nests per km² on north-west edge of Massif Central (France), mean inter-nest distance 202 m (Jollet 1985); estimated 0-3·2 breeding pairs per km² in large farmland area of northern Italy (Fasola and Brichetti 1983); in Bulgaria, ranges from 0·1 pairs per km² in farmland shelterbelts to 9 pairs per km² in vegetation along river banks in Sofia city (Iankov 1983; B Ivanov); 6-16 pairs over 4 years in 10 km² of marshland in Oka river area, south of Moscow (Melde 1984). Replacements of failed nests typically quite close to original nest-site. In Sweden, 5 replacements were 20-200 m away (mean 66 m) (Loman 1975); in south-west Finland, 6 replacements at 20-100 m (mean 48 m) (Tenovuo 1963). Although territories often held from year to year, seldom re-uses precisely the same nest or site in successive years: in Swedish study, 2 out of 120 nests re-used, and 1 pair re-used same tree-fork when nest of previous year had been destroyed; some nests were several hundred metres from those used in previous year, mode 75 m (Loman 1975). In Scotland, only 3 of 98 nests were re-used the next year; 1 nest, in isolated tree, was re-used 3 times in 4 years; in this population, however, most nests were close to those of previous year in traditional sites (Picozzi 1975a). In Germany, Wittenberg (1968) noted that nests are most often re-used where sites are scarce. In urban territory in Cheltenham, limited number of nest-trees ensured repeated use of trees during 12-year study, but no repetition of actual nest-site (J R Harpum). However, Goodfellow (1977) described 'compound' nest used for 3 successive years, each built on preceding one, and Haverschmidt (1937) and Bäsecke (1950) both recorded use of previous year's nest. No close fidelity of breeding birds to natal site; difficulties in obtaining territory appear to ensure at least local dispersal (e.g. Kalchreuter 1970b, Melde 1984, Loman 1985). In Swedish population at rather low density, territories maintained year after year as long as one of owners survived (Loman 1985). Likewise, in north-west Germany, territories maintained from year to year with little alteration of boundaries (Wittenberg 1968). In contrast, in Scotland, Charles (1972) found single bird could not maintain territory alone after loss of mate. In Sweden when territorial bird dies, mate obtains new partner and keeps territory (Loman 1985); hence most birds in this population probably became territorial by mating with widowed territorial bird (see also Heterosexual Behaviour, below). Similarly, when ♀ of breeding pair shot, ♂ replaced her within 1-2 days (Cooper 1847); shooting repeated twice more in little over a week with same outcome. Other studies (e.g. Patterson 1980a) reported displacement of widowed bird by already paired flock-birds. Charles (1972) found that these replacements took a few hours if ♂ was removed, 2-5(-14) days if ♀ removed. In Cheltenham, death of one member of pair resulted on 3 occasions in take-over of territory by neighbouring owners who enlarged their territories to fill vacated space. Subsequent fighting over new boundaries totally inhibited breeding by at least 1 pair of usurpers on 2 of these occasions. (J R Harpum.) In breeding season, home-ranges of flock birds overlap each other widely, are larger than those of territorial birds, and occur either around concentrated food sources or near to intersection of several territories and far from nests. In some cases, however, nests were close to food source and, at least in one case, territory of nesting pair was regularly used also by flocking birds; this pair was occasionally seen feeding together with flock. (Loman 1985.) Sometimes nests in or close to colonies of *C. frugilegus* (e.g. Coombs 1960, Røskaft 1980*b*). Loman (1985) recorded nest-site remaining in same copse after establishment of *C. frugilegus* colony. Presence of *C. corone* under these circumstances usually tolerated (Bannerman 1953*a*), but pair that nested for several years in same trees were ousted by *C. frugilegus* that had moved in only 2 seasons before (Bond 1946). In contrast, *C. corone* has caused whole colony of *C. frugilegus* to desert (Prendergast and Boys 1983). For interspecific territoriality with Magpie *Pica pica*, see Antagonistic Behaviour (below). ROOSTING. Usually in high trees, or on sheltered cliff ledges, also sometimes on or near ground (see below). Roosts in pairs, singly, or communally, often several hundreds or (*cornix*) thousands together. Near Bergen (Norway), maximum 9100 used communal roost (Håland 1980). Communal roosts usually in woodland or groves, often conifers; commonly in traditional sites used over many years. Normally roosting birds settle high up, but Labitte (1937) observed that in rain or mist birds tended to move further down tree. Communal roosts frequently shared with other *Corvus*, especially *C. frugilegus* and *C. monedula* (e.g. Coombs 1960, Wittenberg 1968, J R Harpum). Loman (1985) has also drawn attention to traditional nature of some roosts and has described roost recruitment areas in which cohesion of feeding and subsequent roosting was seen in flock-birds. Roosting on ground normally occurs in areas devoid of many trees, such as moorlands (e.g. Outer Hebrides, Scotland: Hollom 1971), where both nesting and roosting may occur on or near ground in heather *Calluna*. Roosting also reported in reeds *Phragmites* (Bell 1968) in Sutherland (Scotland) and Nottinghamshire (England). 6-year study in Kent (south-east England) revealed a number of arable fields used as ground roosting sites, surrounding a small wood which was used as one pre-roost gathering site (others were in more distant fields). Roosting birds (up to 230 in one field) formed rough lines, side by side, sometimes on crest of ridge, and near middle of fields. Tended to fly up when disturbed, but dropped back to ground nearby or flew elsewhere. (Stainton 1991.) ♀ roosts on nest for 2–3 nights before laying, and through laying, incubation, and brooding (Coombs 1978). ♂ roosts nearby (Charles 1972) or, if few tall trees, up to 50 m away (J R Harpum). Where territories defended all year (see above), owners typically roost in territory at all times, although occasionally join pre-roost gatherings (Picozzi 1975*a*). In Germany, even in breeding season, a few territory-holders roosted outside territories without suitable roost-sites; outside breeding season, some territory-holders used communal roosts, up to 3 km from territory (Wittenberg 1968). During winter in Cheltenham, most urban territorial pairs flew to communal roost south of town at dusk and returned before dawn; roosted in territories from spring until early autumn with only occasional visits to communal roost in late summer. (J R Harpum.) In Sweden, non-breeding territorial birds usually roosted in territories in breeding season (Loman 1985). In Scottish study, Charles (1972) found much variation in roosting behaviour of territorial birds outside breeding season. In Sweden, during breeding season and also in summer, flocking birds usually roosted within 1 km of daytime feeding area, in smaller communal roosts than those used in winter. 2 radio-tracked individuals frequently changed roost-site. In winter, most used large winter roosts (some of more

than 500 birds), up to several km from daytime activity range. However, some roosted in smaller groups, especially territory-holders. (Loman 1985.) Studies on communal roosting of *C. corone* to test information-centre hypothesis of Ward and Zahavi (1973) indicated only a possibility that the mechanism may operate, and then only at times of food shortage (Loman and Tamm 1980). Richner and Marclay (1991), by providing experimental food source in area containing 5 principal roosts, demonstrated that information-transfer effect was produced by turnover of birds at food patch.

2. Very shy in many areas where persecuted by shooting, flushing at 150 m or more (Bannerman 1953*a*). In urban gardens, takes flight instantly when human appears at window, but outside will permit unhurried approach before flying off at *c*. 30 m (J R Harpum). Often learns to recognize and be especially wary of individual humans when harmed or frightened by them. May become much tamer where not threatened by man and given artificial food, as in London parks, approaching to within 5 m at times although still cautious (D T Holyoak). Aerial evolutions less common than in *C. frugilegus*, *C. monedula*, and *C. corax*, but include soaring in strong winds and diving from a height (Witherby *et al.* 1938); dives may be interrupted by short, spiral evolutions followed by near-vertical dive (J R Harpum). A 'jumping display', in which bird leaps up to 4 m into the air and alights on same spot, occurs in both nominate *corone* (Tebbutt 1949) and *cornix* (Witherby *et al.* 1938) in winter and spring but function not clear. For further details of jumping and aerial displays, see Nethersole-Thompson and Nethersole-Thompson (1940). Mobs predators such as fox *Vulpes* and marten *Martes* (Wendland 1958) and domestic cats (Allison 1975). Humans may be mobbed, especially near nest (see Parental Anti-predator Strategies, below) or if they are carrying anything black and shiny; in one case, young tame *C. corone* perched on shoulder provoked same response (Kramer 1941; Goodwin 1986). Mobs many birds of prey; attacks or mock-attacks delivered in swooping or pursuit-flights (e.g. Melde 1984, Goodwin 1986); see Antagonistic Behaviour (below). On hearing Hawk-alarm call (see 5 in Voice) typically given at appearance of flying Goshawk *Accipiter gentilis*, conspecifics at once fly up, bunch together, and seek to gain height; fly away from source of call unless it is distant, whereupon fly to high perch and keep watch (Löhrl 1950). Slagsvold (1981), however, filmed *C. corone* feeding on ground being caught by *A. gentilis* but with little reaction from rest of flock. Intense mobbing of *A. gentilis* does occur at times (e.g. Wendland 1958, Greening 1992). Hawk-alarm call also heard in response to Eagle Owl *Bubo bubo* (Kramer 1930) and marten (Wendland 1958). Commonest form of 'play' in *C. corone* is dropping inanimate objects in flight, either to be recovered from ground each time (e.g. Hayman 1953) or more usually caught in bill before reaching ground (e.g. Persson 1942, Denny 1950, Stevenson 1950, McKendry 1973), and once seen also transferring objects from bill to foot and back again in flight (King 1969). This behaviour clearly related to habit of dropping mussels *Mytilus* (see Food), although dropping golf balls (Duckworth 1983) is probably an attempt to break hard 'eggs'. Reported to have damaged gold-leaf on roof of Kremlin (Moscow) by sliding down on feet (Boswall 1985*a*). Sliding down slated or tiled roof, culminating in take-off, not uncommon among juveniles (J R Harpum). Upside-down hanging occurs when bird, landing on branch, wire, etc., immediately swings underneath perch where it hangs for short time, looking around, before flying off (e.g. Pettitt and Butt 1949, McIntyre 1953, Michels 1973, Radermacher 1974, Stegemann 1975, J R Harpum). No indication that this is play of kind described for juvenile *C. corax* by Gwinner (1966). In most cases, initial inversion appears an error,

but clasping reflex of foot prevents fall (J R Harpum). Some birds use this technique in raiding nut-feeders (Gush 1978). Various flying activities described as play; some may involve other Corvidae (e.g. mutual aerial chase involving 2 *C. corone* and 3 *C. monedula*: Bates 1979). Supplanting attacks common against Woodpigeon *Columba palumbus* (e.g. Bromley 1947, Geyer 1985) and Rock Dove *C. livia* (e.g. Pearce 1978, Delbove and Fouillet 1986); many observations of both types suggest play (J R Harpum). 'Playful' flight with Marsh Harrier *Circus aeruginosus* described by Laudage and Schroeter (1982). FLOCK BEHAVIOUR. In Sweden, territorial birds (which roosted in territory) observed 4 times to participate in evening gatherings of up to 10 birds, on ground and in trees, in their own or neighbours' territories. Such gatherings not followed by communal roosting, and only seen with certainty in April. (Loman 1985.) 'Social gatherings' like these said to be common in late winter and early spring; sometimes involve advertising display between neighbouring pairs (Goodwin 1986). Goodwin (1986) suggested pair-formation might be involved, but predominance of territorial birds in observations by Loman (1985) casts doubt on this. Noisy gatherings in spring observed several times in Cheltenham along boundaries between 2 territories or at junction between 3, when local territory-holders intercepted small group of prospecting flock-birds in trees. No overt attack, but calling is loud and may persist for 5–20 min (see 2 in Voice), ceasing quite suddenly and participants then dispersing. (J R Harpum.) This behaviour reminiscent of 'Ceremonial Gatherings' of *P. pica* (Birkhead and Clarkson 1985; see that species), although significance not necessarily the same. 'Social gatherings' on Skokholm island (west Wales) and Alderney (Channel Islands), May–September, occurred at all times of day and involved up to 60 birds, many of them paired territory-holders, sometimes including family parties. Individuals often grouped close together on open ground, for no apparent reason, and postured with wing-flicking, fanned tails, and raised head feathers. (P J Conder.) Wing-flicking (flight-intention) signal consists of quick lifting of closed wing-tips *c.* 3–4 cm, often accompanied by slight tail-spreading. Action of considerably larger amplitude than same movement in *C. frugilegus* and helpful in identification at long distance. Besides use as flight-intention signal by alert or nervous bird, wing-flicking also probably serves as threat. During breeding season and in summer, flocking birds spent most of day feeding at 1–2 food concentrations, but some ranged further afield, both singly and in small flocks (Loman 1985). In some areas, flocks (and perhaps also neighbouring territorial birds) regularly intrude into breeding territories of conspecifics, where often responsible for predation of eggs and nestlings (e.g. Tenovuo 1963, Wittenberg 1968, Charles 1972, Yom-Tov 1974b, 1975a), though little evidence of this in detailed study in southern Sweden (Loman 1985). Harper (1992) recorded immature ♂ that robbed nest in company with 3 conspecifics; in following season it helped at nest of same pair and was seen to copulate with ♀ (see Bonds, above). In Moscow, 2 types of winter flock identified, sedentary and wandering. Sedentary groups formed round 1–2 pairs of established nesters and would contain some of their offspring, some neighbours, and some outsiders; fed only in their own area, whereas wandering interlopers attempting to do the same were ranked at bottom of feeding hierarchy (see below). Sedentary flocks also defended their feeding area against incursions by other flocks. (Grabovski 1983.) Individual distance of several body lengths normally maintained between unpaired birds, but this considerably reduced between pair-members (J R Harpum; D T Holyoak). Dominance hierarchy clearly evident in flocks feeding at concentrated sources. Detailed study of winter flock of over 100 in Switzerland, involving provision of artificial food (Rich-

ner 1989b), showed that sex, age, and body size (decreasing order of importance) determined outcome of antagonistic encounters: ♂♂ dominated ♀♀, adults dominated immatures, and (within same age- and sex-class) larger birds dominated smaller ones. During 1 hr or more before joining communal roosts, birds assemble in gradually increasing numbers at one or a few gathering places, usually on open ground, mostly near concentrated food sources, and often traditional; like roosts, often shared with *C. frugilegus* and *C. monedula*. At gathering places birds appear to feed actively before leaving for roost more or less simultaneously (usually over 10 min or less: Loman 1985). Arrived at winter roost in Scotland *c.* 30 min after sunset and either settled immediately or circled and dived at one another for 2–28 min (Picozzi 1975a). In morning, flight pattern approximately reverse of that in evening, with dispersal via same gathering places, although less time spent before reaching feeding sites. Birds using different gathering places in Swedish study fed in different areas. Some birds changed areas during course of winter; such changes commonest with 1st-winter birds, less frequent with 2nd- or 3rd-year flocking birds, and least frequent with territory-holders. (Loman 1985.) Also evidence for changes of roost during winter in Scotland (Picozzi 1975a). Territorial pairs in Cheltenham area, when using communal winter roost, usually leave territories at dusk but return to them before dawn, ahead of any flock-birds (J R Harpum). SONG-DISPLAY. No display recorded accompanying so-called 'song' (see 1 in Voice), but advertising-call (see 2a in Voice) used with display in ritualized manner in advertisement and defence of territory (see Antagonistic Behaviour, below, and 10 in Voice). ANTAGONISTIC BEHAVIOUR. (1) General. For annual cycle of territorial defence, see introduction to part 1 (above). Overt, intraspecific, aggressive behaviour often common and conspicuous, e.g. between flock-members feeding at good food source or between pairs when breeding territories at high density. Under other circumstances, aggression may be rare, such as when paired birds feed on grassland within territories— or infrequent, as between neighbouring pairs when adjacent territories are large, e.g. in spring, territorial birds in urban Cheltenham show almost no aggression towards one another and often join in mutual defence against incursions of flocks or other predators (J R Harpum). For details of dominance hierarchy in winter feeding flocks, see Flock Behaviour (above). ♂♂ apparently dominate ♀♀ in all antagonistic situations, except at nest where ♀♀ often appear dominant (D T Holyoak). Some evidence that in territorial defence each sex mostly threatens or fights intruders of same sex (Bossema and Benus 1985; D T Holyoak). Intra-pair cooperation in territorial defence may involve 'pincer movement' by owners; for further details, see Bossema and Benus (1985). In Sweden, little territorial aggression, but when territorial birds were trapped, neighbours soon encroached on territory; once, when territorial bird slightly wounded in trap it was attacked after release by neighbouring pair and probably killed (Loman 1985). In north-west Germany, at higher densities, territories clearly defended with threat and fighting, intensity of which declined towards borders. Defence most intense during nest-building period. Aggressiveness towards neighbours somewhat less than towards their young and 'foreign' birds, tending to result in some overlap of territories and shared defence of these parts. Outside breeding season, territory defended only occasionally. (Wittenberg 1968.) Interspecific territoriality occurred with *C. frugilegus* and *C. monedula* in Germany (Wittenberg 1968) and also in Cheltenham where *C. frugilegus* and *C. monedula* either did not infringe territorial roof or, if they tried to fly below it, were chased or escorted to border. However, *C. monedula* frequently managed to fly surreptitiously through territory at low level. The only time that *C. frugilegus* was not readily ousted was during competition with resident (territorial) *C. corone* for ripe nuts from local walnut trees *Juglans* (J R Harpum). Territorial relationships with *P. pica* somewhat more complex: both nominate *corone* and *cornix* known to compete with, attack, and even kill *P. pica* (e.g. Baeyens 1981*b*, *c*, Högstedt 1981, Vines 1981, Eden 1987*b*). Where they nest near to each other there is mutual interference with nest-building and predation of nest contents (e.g. Butlin 1959, Love and Summers 1973); material may be stolen from *P. pica* nests (Montier 1977). In Cheltenham area, *P. pica* territories tend to be established along borders between territories of *C. corone*, especially when those contract in size during nesting cycle (J R Harpum). Mosaic pattern of nesting by coexisting *P. pica* and *C. c. cornix* in northern Italy described by Saino and Meriggi (1990). For further discussion of relationships between these species, see Birkhead (1991). In much of western Europe, *C. corone* dominant over other Corvidae with exception of *C. corax* (Bossema *et al.* 1976), although recorded attacking *C. corax* in areas where latter uncommon (Avery 1991). *C. corone* once seen being killed by *C. frugilegus* (Morrison 1977). In Egypt, however, *pallescens* has been displaced from urban areas of Suez by introduced House Crow *C. splendens* (Bijlsma and Meininger 1984). *C. corone* often chases away small raptors by flying just below and slightly behind them (J R Harpum). Interspecific attacks usually related to predation (see Food) or to territorial defence, particularly during nesting period; likewise probably many conflicts with gulls *Larus*. Several accounts of attacks in flight on Swallow *Hirundo rustica* (Hanford 1969; Radford 1970*b*) and waders (Charadrii) (e.g. MacBean 1949, O'Mahony 1977). 'Tail-tweaking' of potentially threatening intruders (bird or mammal) by single bird or pair may be precursor to attack; Grey Heron *Ardea cinerea*, especially, seems to elicit Tail-tweaking (e.g. Birkhead 1972, Cramb 1972). Once, Barn Owl *Tyto alba*, hunting by day, was brought to ground by *C. corone* gripping it in its feet (Dickson 1972). Anxiety-call (see 2d in Voice) commonly given when predator seen. Mobbing-call (see 4 in Voice) given when attacking various flying raptors (e.g. Wendland 1958, Wittenberg 1968, Poppe 1976, Melde 1984, D T Holyoak) up to size of White-tailed Eagle *Haliaeetus albicilla* (Bannerman 1956: but see Wendland 1958); also when attacking birds resembling raptors such as *Ardea cinerea* (Birkhead 1972; Todhunter 1987; D T Holyoak); *B. bubo* is regularly mobbed and thus used as lure during culling of *C. corone* (e.g. Somerkoski 1984). Mobbing-call occasionally also given, but usually less intensely, in apparently semi-playful attacks on conspecifics (Wittenberg 1968; Goodwin 1986). (2) Threat and fighting. Territorial neighbours commonly exchange Advertising-calls (see 2a in Voice), often from prominent perches, in characteristic Cawing-display (Fig A) with neck

A

extended forwards and upwards, head lowered with each call, and neck feathers more or less ruffled. Although these calls are also given by flocking birds in other contexts, little doubt they function in advertisement and defence of territory. However, much territorial advertisement achieved by silent loafing

(Amlaner and Ball 1983) without display, on top of conspicuous perch; corresponds to 'tree-top sitting' of *P. pica* (Møller 1982*a*). Pursuits in flight, often accompanied by Attack-call (see 2c in Voice) are commonest part of higher-intensity aggression over territories. Attacks by territorial pairs are usually effective in driving off intruding conspecifics (Bossema *et al.* 1976). Wing-flicking is common in antagonistic contexts and may signal intention to fly towards threatened bird, as well as general alertness (Charles 1972; Coombs 1978; P J Conder). Dive-attacks on perched birds also common in winter at pre-roost gatherings and in roosts (Picozzi 1975*a*). Threat on ground commonly takes form of quick approach, at a run or with skipping gait. If threatened bird does not retreat, approach may terminate in Forward-threat display (Fig B), aggressor holding its body low

B

and near horizontal, head forward with bill extended (sometimes opened) and plumage ruffled, especially at high intensity. Upright-threat display (Fig C) with upright stance, head raised,

C

but bill pointed down and plumage more or less ruffled, is used less often, apparently in defensive-threat or when birds really close. (Coombs 1978.) In Bristle-head display (Fig D), head

D

feathers raised in threat (Charles 1972; Coombs 1978; P Conder); perhaps low-intensity version of Upright-threat or separate display; Bristle-head display also commonly performed by ♂ of territorial pair when in company with ♀ (see also Heterosexual Behaviour, below), during breeding season or when feeding in family group with ♀ and one or more juveniles in summer and early autumn; in latter case, may be sustained for considerable time especially if neighbours can be heard calling in distance (J R Harpum). For Bowing-display in antagonistic context, see

Heterosexual Behaviour (below). Fighting considerably less common than aggressive posturing or pursuits. Birds occasionally fight breast to breast, but more often a few sharp pecks cause one contestant to retreat. For harassment of and even attacks on albino or part-albino conspecifics in flocks, see (e.g.) ffrench (1991). There is also aggressive reaction to any fluttering of wings by injured or fighting birds (Harrison 1976). Chasing in flight often follows retreat of threatened bird with disputed food item. Birds feeding in flocks often avoid possible aggression by running or sometimes flying from vicinity of conspecifics immediately they obtain food item too large to be swallowed on the spot. Submissive behaviour commonly consists only of retreat. In Submissive-posture, bird crouches with lowered wings and ruffled plumage. Displacement activity (or redirected aggression) in antagonistic contexts often includes hammering perch with bill, pecking at ground, or tearing at objects. (D T Holyoak.) 'Stare down' (Harrison 1965a; Goodwin 1986) in which bird suddenly lowers head and looks at or between its feet occurs in various situations, not all of them antagonistic. C. corone is one of the most frequent performers of image-fighting with mirror-like surfaces (Burton and Burton 1977). In typical example, territorial ♂ attacked same pane of glass (which rapidly became covered with saliva and blood) for a week in May from c. 05.15 hrs to 06.30 hrs each morning, despite attempts to frighten it off; pecking accompanied by Attack-call punctuated by periods of quiet when bird pecked at window ledge, etc., and performed 'stare down'; episode finished when eventually pane of glass was broken; ♀ took no part except to watch from nearby (J R Harpum). For similar attack by single territorial bird, see Simpson (1970b) and Tyler (1971), and by 2 birds see (e.g.) Burnett (1965) and Campbell (1968). There are accounts of mass attacks on several windows in same building by flocks (e.g. Edqvist 1945, Burton 1979) accompanied in some cases by redirected fighting among birds themselves. Reflective surfaces on motor vehicles, including wing mirrors (Trotman 1974: glass eventually broken) and windscreens may also be attacked, with redirected pecking of the car bonnet and sometimes removal of rubber from windscreen wipers (e.g. Rogers 1984a). HETEROSEXUAL BEHAVIOUR. (1) Pair-bonding behaviour. Some pair-bonds formed by young birds in flocks (Wittenberg 1968; Kalchreuter 1971; Charles 1972; Loman 1985). Some ♂♂ may establish territory on their own and then attract a mate (Kalchreuter 1971), but in study by Loman (1985) successful establishment of territories by single ♂ was never observed. In Sweden, pair-formation occurred (apparently on territories) when established birds lost their mates. All territorial birds that lost mates retained their territories in following spring. If loss occurred during breeding season (7 cases) widowed bird sometimes re-mated in same season (4 cases, including one ♀ who lost ♂); in 2 cases, no re-pairing observed until following spring, but may have occurred earlier, unnoticed; no information available in 7th case. At least 2 of the new pairs became permanent. (Loman 1985.) In Scotland, however, Charles (1972) found that single bird could not maintain territory and 'widower' ♂♂ could not therefore form new pair-bond there. Infrequent display-flight probably has sexual significance as well as being part of territorial behaviour. Performer flies with very deliberate wing-beats, wing-tips appearing to pass through wider arc than in normal flight (Goodwin 1986). Melde (1984) describes circular flight in March, of 2 birds 80 m above ground, lasting 10 min. In south-west England, similar circular flight by territorial pairs, without calling, seen at territorial 'roof-height' (30 m) in early spring, which was modified in high winds to triangular flight with downwind heading (100 m long) followed by double 'tack' upwind to regain original position; repeated for 5–15 min. Pairs

of flock birds may also display with circular flight, calling loudly all the time, at considerable height, sometimes above known territories. (J R Harpum.) Bowing-display clearly homologous with that of C. frugilegus (see Fig C of that species) and given mainly, if not entirely, in sexual contexts. Each bow followed by upright head-jerk, closure of nictitating membranes, and spreading of tail. Bowing bird may or may not call. Sometimes both pair-members display thus to each other, at other times with head feathers fully erected, prior to ♀ performing Soliciting-display or begging to be fed. Often 2 hostile ♂♂ stand close together or walk side by side displaying at each other. Particularly at such times body and spread tail may be tilted towards bird at which display directed. (Goodwin 1986.) In similar context, 'pot-bellied posture' may be adopted (Charles 1972): neck and head upright and pushed forward, belly feathers ruffled, wings close to body, and tail tilted down. Bird runs and hops parallel and close to intruder, uttering 'karr' call (Coombs 1978: see 2b in Voice). Bristle-head display (see Antagonistic Behaviour, above) apparently indicates more inhibition against attacking, or more definite sexual motivation. Bird may strut about in this posture, or even maintain it for some time while food-seeking, if sexual or territorial rival in vicinity. In pre-copulatory display, body plumage more fully ruffled, wings somewhat lowered, and tail partly spread. Feathers above eyes sometimes erected to form 'ears' as in C. corax; colour pattern of cornix gives this display particularly striking effect. (Goodwin 1986.) Rival ♂♂ sometimes bow repeatedly, side by side, with 'ears' erected (Goodwin 1986; D T Holyoak). Allopreening of head and nape regular between paired birds (e.g. Summers-Smith 1959) and also occurs between immature siblings (Goodwin 1986). (2) Courtship-feeding. Feeding of ♀ by ♂ apparently not very frequent before laying (Goodwin 1986), although similar feeding is regular during incubation. ♀ begs for food with flapping wings (like juvenile) giving Begging-call (see 8 in Voice). (3) Mating. Usually on or near nest (Haverschmidt 1934; Wittenberg 1968; Melde 1984) or in neighbouring trees (Witherby et al. 1938). Once reported on ground after which ♀ moved forward with tail spread (Gent 1949). Soliciting-display of ♀ used before mating is much like Submissive-posture used in other contexts: bird crouches with partly open and lowered wings and quivering tail (at high intensity wings probably more fully opened and also quivered: Goodwin 1986). Mating lasts 10–15 s and ♂ flaps wings while balancing on ♀'s back. Characteristic loud call during mating (see 6 in Voice) attributed to ♀ by Haverschmidt (1934), to ♂ by Melde (1984). In established pairs, mating may take place without any conspicuous prior display (Wittenberg 1968); Goodwin (1986) suggested that with established pairs mutual understanding and communication by glances and slight posturing may replace full display. After mating, pairs continue with other activities such as feeding, nest-building, or incubation. Copulation occurs only just before and during laying, except for promiscuous matings. Attempts by ♂♂ from neighbouring territories to rape sitting ♀♀ recorded. (Wittenberg 1968.) (4) Nest-site selection and behaviour at nest. Roles of sexes in nest-site selection apparently not described, although in south-west England both sexes seen to carry twigs into several trees for c. 1 week before final site determined (J R Harpum). Nest often approached furtively during building. Incubation by ♀ only, fed on or near nest by ♂; ♀ attracts ♂ initially with Nest-call (see 7 in Voice). She also leaves nest for short periods and may then find food for herself. RELATIONS WITHIN FAMILY GROUP. ♀ broods young, with decreasing intensity, for first 12–15 days, but may start to collect some food for them as early as 8th or 9th day. Small young are fed by ♀ alone on food brought by ♂. Older young are fed and tended by both parents. Food almost always brought in throat pouch (e.g.

Tenovuo 1963). Large items may be carried in bill to near nest, but broken up before being fed to young (Goodwin 1986). Faecal sacs of small young removed from nest by ♀. Nest begins to be soiled on sides after *c*. 3 weeks (Tenovuo 1963). One nest reported as being 'decorated' with fresh oak *Quercus* sprigs (Lieff and Jordan 1950), and continued to be replenished with fresh leaves until August, long after young had left; similar activity recorded by Pounds (1949). Tenovuo (1963) reported that unhatched eggs are eaten, but this seems unlikely in view of experimental results reported by Yom-Tov (1976), unless eggs broken accidentally. Food-calls of young begin on day 5 (see Voice); eyes begin to open on day 8 (Melde 1984). For some days before first flight, young move onto nest-rim and adjoining branches (Loman 1985). Immediately before first flight, much loud calling by young with responses by parents; continues for some time after young have landed in adjacent trees or on ground (J R Harpum). Juveniles spend the few weeks after first flight near nest waiting for parents to feed them (Loman 1985). Regular feeding of juveniles by parents stops *c*. 4 weeks after fledging, but occasional feeding may occur in 5th week or later. After *c*. 4th week, follow parents and mainly feed themselves (Loman 1985), but may try to solicit food discovered by adults when feeding in family groups for some time after this (J R Harpum). Juveniles leave parental territory at different times in different populations from 2nd month after fledging until late in 1st winter (see Bonds). Even within family group, young birds do not necessarily all leave at same time (J R Harpum). ANTI-PREDATOR RESPONSES OF YOUNG. Disturbance nearby causes at least well-grown nestlings to crouch low and silently in nest. Young near fledging clamber away from nest or attempt to fly away from it when human approaches very closely. (D T Holyoak.) PARENTAL ANTI-PREDATOR STRATEGIES. (1) Passive measures. Some incubating or brooding ♀♀ remain crouched low on nest as human or other danger approaches. ♀ may sit very tight on nest when humans intrude on ground, and can only be flushed by knocking or scraping tree trunk (e.g. Campbell and Ferguson-Lees 1972). More often, sitting ♀ leaves nest inconspicuously when human approaches, then watches from distance (Wittenberg 1968). Visits to nest commonly postponed when threat or apparent threat nearby. Wittenberg (1968) recorded that when fledgling fell from its roosting perch on high-tension mast both parents at once flew down and spent whole night on ground with it. (2) Active measures: against birds and other animals. Potential predators near nest are mobbed with noisy Attack-calls, most vigorously when young are large (e.g. Tenovuo 1963). For details of species mobbed, and tactics used, see Antagonistic Behaviour (above). Birds from neighbouring territories and non-breeders commonly join in mobbing (Tenovuo 1963; Wittenberg 1968). Nesting population near Moscow was grouped into 'neighbourhoods'; when danger threatened, birds responded only to their own group alarms and not even to alarms from nearest neighbour to them if it was not part of their neighbourhood group. (Grabovski 1983.) (3) Active measures: against man. Intruders at nests are mobbed, although dive-attacks do not approach very closely. Parent birds are frequently shot at nests with young because of their reduced caution, especially when nestling or fledgling seen to be handled or gives noisy calls. (D T Holyoak.)

(Figs by D Nurney from drawings in Coombs 1978.)

JRH, DTH

**Voice.** Conspicuously vocal throughout the year. According to Kramer (1941), Witherby *et al*. (1938), Melde (1984), and P J Sellar, no clear differences between vocalizations of nominate *corone* and *cornix*, allowing for variability within both, but L Svensson considers nominate *corone* to sound on average harder, more raw, angrier, and with less rolling 'r' than *cornix*, which is slightly softer, more 'open', with generously rolling 'r', nominate *corone* sounding more like Rook *C. frugilegus* than does *cornix*. Geographical variation reported by Goodwin (1986) to occur. Except where specified, sonagrams depict nominate *corone*. Commonest calls of *sharpii* and *capellanus* apparently similar. For additional sonagrams, see Bergmann and Helb (1982). The only instrumental sound reported is bill-snap sometimes used in aggressive interactions (D T Holyoak). Variability of actual calls and difficulties with phonetic renderings cause doubt about number of distinct calls that should be recognized. Considerable uncertainties also arise in attempts to correlate traditional descriptions (Kramer 1941; Wittenberg 1968; Charles 1972; Coombs 1978; Melde 1984; Goodwin 1986) with actual field investigations and sonagrams (Frings *et al*. 1958). Studies by Richards and Thompson (1978) and Thompson (1982) suggest that traditional accounts of vocalizations of *C. corone* (and of American Crow *C. brachyrhynchos*) are incomplete. Thompson (1969*a*, *b*) distinguished 2 major types of vocalization in these corvids, 'structured and unstructured cawing'; see call 10. Calls 1-9 describe traditional scheme of the more obviously different calls and main variants.

CALLS OF ADULTS. (1) Song. Variously described: succession of croaking and bubbling sounds, or croaking song (Witherby *et al*. 1938); soliloquy of very variable calls that seem to represent low-intensity versions of most other calls and, possibly, some vocal mimicry (Goodwin 1986); see also Niethammer (1937), Bacchus (1943), Géroudet (1951*a*), Verheyen (1957*a*), and Bergmann and Helb (1982). Not commonly heard; Goodwin (1986) encountered song only in immatures. 'Nest-song' given by ♀ sitting on nest described as very soft 'crackling' which may continue uninterrupted for several minutes (Bau 1902-3; Zimmermann 1907; Witherby *et al*. 1938). Wittenberg (1968) suspected that this may be same as Nest-call of ♀ (see 7 below). See call 10 for effective song characteristics of structured calling. (2) Advertising-call. Commonest call. Repeated harsh 'kraa' with vibrant or resonant quality but very variable; often delivered as sequence of bouts (see 2a below). Also rendered 'kraah' or 'kraarrr' (Witherby *et al*. 1938), 'karaa', 'aaarr' (Goodwin 1986). Harsher and more vibrant than *C. frugilegus*. Each call generally drops perceptibly in pitch at the end but calls in same sequence are usually identical. 4 principal variants. (2a) Self-assertive call. Loud, harsh 'kraar' (Fig I) delivered in bouts of *c*. 2-6 calls with long pause between bouts (see call 10); given with characteristic head and body movements (see Antagonistic Behaviour). This is commonest variant and most characteristic of Structured Calls (see call 10); described as self-assertive, especially when given by ♂, and commonly given in territorial situations (Goodwin 1986). Also rendered 'kraang' (Charles

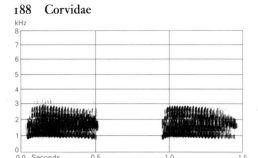

I   V C Lewis   England   May 1969

1972; Coombs 1978); loud and resounding, audible up to
700 m, usually repeated. Variants include 'krong' (like
'honk') and 'kraa' (Coombs 1978); several versions given
by Thompson (1982), as well as higher-pitched, complaining 'keerk keerk keerk' and (Figs II–III) 'motor-

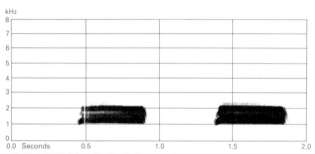

II   P A D Hollom   England   June 1976

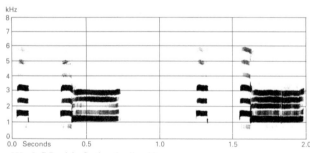

III   J-C Roché   Spain   April 1988

horn' call (Witherby *et al.* 1938); Fig III shows unusually
musical variant rich in harmonics; motor-horn call lacks
any terminal lowering of pitch within each call; 3 birds
calling thus in Cheltenham area (south-west England)
were all ♀♀ (J R Harpum). Loud 'aaah', given *c.* 3 times
followed by a pause, often given at dawn by ♀ of territorial
pair, is somewhat distinct from ♂'s equivalent (Charles
1972; Coombs 1978). (2b) Softer versions of Self-assertive
call often given by ♀ in response to ♂'s Self-assertive call,
but may also be given by both sexes (Goodwin 1986).
Short-range aggressive 'karr' is a single, low, muffled,
guttural call given with bill hardly open, sometimes sounding like a mew and accompanied by head-bobbing; directed
by territory-holder at persistent flock-intruders or at

flock-birds on feeding grounds, and also uttered during
hop and run in 'pot-bellied posture' (Charles 1972;
Coombs 1978) (see Social Pattern and Behaviour). (2c)
Attack-call. Loud, strident 'kar kar' (Fig IV: *cornix*) or

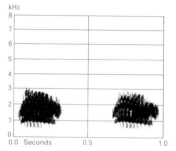

IV   R Margoschis   Scotland   May 1988

'ark ark', usually given 2 at a time in territorial disputes,
particularly by bird flying towards trespassing individual
(Wittenberg 1968; Goodwin 1986). Charles (1972) and
Coombs (1978) label this 'kraah' call. As intruder is
approached, pitch rises and rate of delivery increases.
Similar calls often given when mobbing human intruder
at nest with young, becoming higher pitched with frequent
repetition if young handled (D T Holyoak). Witherby *et
al.* (1938) quoted rapid double or treble, rather high-
pitched screech of anger that may be particularly intense
version of same call. (2d) Anxiety-call. Loud, harsh,
repeated 'kaaaar', like call 2 but harsher, flatter, and more
drawn-out, used as apparent alarm-call when mobbing
ground predators and perched birds of prey. According
to Goodwin (1986), motivated by conflict between fear
and aggression, or fear and curiosity, and also given when
conspecific appears to be in danger, or at sight of remains
of *Corvus*, or what calling bird takes for such. This appears
to be 'distress call', broadcast recordings of which have
been used to disperse *C. corone* from airport runways
(Hardenberg 1965) and roof of Kremlin in Moscow (Boswall 1985). Wittenberg (1968) thought it signalled
defensive-threat. Goodwin (1968) noted similar call given
when several birds attracted to food put out for them,
possibly because individuals were uneasy at presence of
competitors. In both situations, attracts conspecifics
(Goodwin 1986). (3) Rattle-call. Mechanical-sounding,

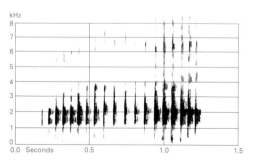

V   P J Sellar   England   October 1984

rattling 'klok klok klok' (Fig V), suggestive of machine-gun fire (e.g. Mayo 1950). Given apparently by ♀ only, in some territorial situations, at social gatherings, and sometimes in response to display or Self-assertive calling of mate (Goodwin 1986). May be uttered with lowered head (similar to Castanet-call of ♀ Jay *Garrulus glandarius*), with head movement, if any, similar to that accompanying Self-assertive call (Goodwin 1986; D T Holyoak; P J Sellar). (4) Mobbing-call. Short, grating rattle, 'krrr'. Given during dive-attack on flying birds of prey (Kramer 1941), and sometimes in apparently playful attacks on conspecifics (Wittenberg 1968; Goodwin 1986) (see Social Pattern and Behaviour). Löhrl (1950) noted it was never uttered when mobbing Goshawk *Accipiter gentilis*, but only when less feared raptors involved. (5) Hawk-alarm call. Series of short, high-pitched, breathless-sounding calls that lack any resonance (Löhrl 1950; Goodwin 1986). Rendered as 'ir' (with short 'i' sound) by Wendland (1958). Apparently given only in great fear, such as in response to flying *A. gentilis* (Kramer 1941; Löhrl 1950) (see Social Pattern and Behaviour). (6) Mating-call. Loud, drawn-out 'rärrärrärr...' or 'rararrarr...' given during copulation (Wittenberg 1968); attributed to ♀ by Haverschmidt (1934), to ♂ (based on close observation) by Melde (1984). (7) Nest-call of ♀. Described by Wittenberg (1968) as short 'rarr' or 'rärr' ('r' sound not audible at distance), with rather nasal tone. Uttered when laying, also at and near nest when ♂ comes with food, from 2 days before laying to more than 2 weeks afterwards; may carry over into call 8. (8) Begging-call of incubating ♀. Loud 'caaa' or 'aaa aaa' (Fig VI: *cornix*), closely similar to food-call of fledged

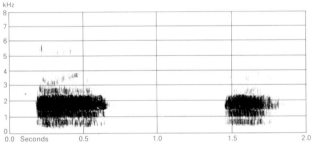

VI  S Palmér/Sveriges Radio (1972–80)  Sweden  May 1969

young (see below). Not clear whether 'high-pitched greeting note' (Witherby *et al.* 1938) is this call or another. (9) Other calls. (9a) Wittenberg (1968) described bleating-call with fluctuating pitch, 'äh', used when returning to nest after territorial strife, or later in breeding season when pair return to find nest empty, through predation or because young have fledged. Same call apparently described by Strauss (1938a) and Géroudet (1951); latter rendered it a nasal-sounding, melancholy 'ou-ain'. (9b) Hand-reared birds may indulge in vocal mimicry, including human speech (Lorenz 1952), but this not reported from wild individuals (Goodwin 1986). (9c) Some calls

(perhaps not different from those listed above) are so soft that they can only be detected when bird is close to a microphone (Bossema 1980). (10) Structured and Unstructured Calling. Summary (above) of traditional view of vocalizations relates most calls to readily identifiable reaction in conspecifics (i.e. unstructured calling). Organized, repetitive sequences of calls (typically call 2, especially 2a) are classed as structured calling (Thompson 1982) and elicit no obvious reaction except for some vocal response in caller's mate or in close territorial neighbours. In study of *C. corone* in Somerset (south-west England), structured calling characterized by repetitive calling in bouts and by discontinuous pattern of emission within sequences. All calls within a sequence were similar but sometimes differed from those in other sequences. In 3 pairs of *C. corone* studied in detail average call duration was 0·4–0·9 s, interval between onset of calls in one bout 1–2 s, and 2–6(–22) calls in a bout. Structured calling was heard in every season and throughout 3-km² study area. 4 calls seemed to be related to individual birds, but individuals capable of giving more than one type of call. (Thompson 1982, which see for details.) Richards and Thompson (1978) had previously found that many calls used in structured form by *C. brachyrhynchos* were also those used in unstructured calling; same is clearly true for *C. corone*. So traditional view that each call in repertoire of *C.corone* has stable meaning is no longer tenable. In study of 8 birds by Bossema (1980), 23 distinguishable categories of structured calling identified. Individuals produced average of 6·5 different structured call-types, some by ♂♂ or ♀♀ alone. Bossema and Benus (1985) found that structured calling was increased markedly in territorial pair by intruding conspecifics, leading in some cases to (structured) duetting between pair-members which stimulated attacks on intruder. Duetting also heard in similar situations in Cheltenham (J R Harpum). Significant conclusion from studies like these is the analogy between structured and unstructured calling on the one hand and song and call in other passerines (Thompson 1978). In this respect, perhaps significant that calls of *C. corone* during dawn chorus take form of structured calling (J R Harpum).

CALLS OF YOUNG. Nestlings begin to give food-calls when 5 days old, commencing with low, weak 'hi(ä)-hi(ä)', monotonously repeated. By day 7 this has changed to 'screaming' 'gschää-gschää-gschää'. By day 9, calls are louder; sharp, insistent 'tschiä-tschi', with very soft 'gschi' at times between feeds. Larger young give increasingly loud food-calls, becoming a more raucous 'rräh-rräh'; occasionally a short 'schrebb' (Melde 1984). Sonagram of food-call ('begging-call') of 9-day-old nestlings by Redondo and Arias de Reyna (1988) shows broad-band, harsh triple call. They found food-call of 3-day old nestlings to be preceded by 'click'-like structure. Such loud calls should attract potential predators and be maladaptive; that this is not the case is discussed by

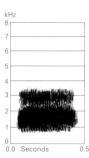

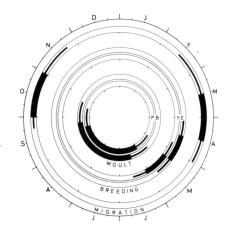

VII  V C Lewis  England  June 1977

Redondo and Arias de Reyna (1988). Food-call of fledged juveniles a loud 'caaa' or 'aaa aaa' (Goodwin 1986: Fig VII).                                      JRH, DTH

**Breeding.** Season. Britain: in southern England, start of laying mid-March to end of May; average date of 1st egg below 160 m ranged from 13 April in Wales ($n=138$) to 27 April in northern Scotland ($n=44$); at $c$. 300–600 m, northern England and southern Scotland, 2 May ($n=5$) (Holyoak 1967$c$, which see for regional variation and comparison with other Corvidae); for north-east Scotland, see also Yom-Tov (1974) and Picozzi (1975$a$). Finland: see diagram; in south, peak laying period mid-April to mid-May; $c$. 2 weeks later in Lapland (Haartman 1969). South-west Germany: eggs laid late March to mid-May, peak early April (Hund and Prinzinger 1981). Russia: eggs laid from end of March in south, end of April in north (Dementiev and Gladkov 1954). Israel: nest-building starts mid-January, eggs laid mid-March to late April (Shirihai in press). Greece: said to lay very late; of 34 clutches, most laid in May (Makatsch 1976). Persian Gulf: full clutches in late February (Goodwin 1986). Site. High in tree at woodland edge, in small stand, or isolated; also on pylon or telephone pole, more rarely on cliff, rock, building, or ground; if no high trees or pylons available (e.g. in north of range) in small tree, bush, or dense low vegetation. Almost always in upper third of highest available local tree, in fork or on branch generally near trunk, or in

topmost twigs of smaller species; in Europe, no fundamental difference between nominate *corone* and *cornix*. (Kulczycki 1973; Coombs 1978; Melde 1984; Goodwin 1986; Korbut 1989$b$). Usually well hidden, and height greater close to houses and where persecuted (Schifferli and Fuchs 1981; Fasola and Brichetti 1983; Jollet 1985). Some indications that in suitable areas conifers slightly preferred to deciduous species, probably because of permanent cover (Tenovuo 1963; Wittenberg 1968; Kulczycki 1973; Korbut 1989$b$). In northern Germany, average height above ground 14·8 m (7–24, $n=119$), and 63% in oak *Quercus* (Wittenberg 1968); in south-west Germany, average height 9·0 m (3–30, $n=362$), 34% in old pear *Pyrus* (Prinzinger and Hund 1981); in northern Norway, 5·9 m (4–7, $n=26$) (Parker 1985). In one study in Denmark, 66% of 46 nests in conifers; in another, 10% of 41 nests (Møller 1981$a$; Henriksen 1989). See Haartman (1969) for Finland, Munkejord *et al.* (1985) for Norway, Kulczycki (1973) for Poland, Picozzi (1975$a$) for Scotland, Jollet (1985) for France, Fasola and Brichetti (1983) for Italy, Korbut (1989$b$) for Russia, Shirihai (in press) for Israel. Nests recorded on ground after nesting trees felled (Bergh *et al.* 1989), near seabird colony whose eggs were food source (Vauk 1980), and as possible response to continual persecution or high nest density (Rebecca 1985, 1986), but also for no apparent reason (Linn 1984). Increasingly on buildings and trees in some cities, e.g. Moscow (Plath 1988$b$) and Milan (Londei and Maffioli 1989). In some areas may prefer pylons to trees since more secure against predators (Heise 1970). In eastern Russia in reedbeds (Dementiev and Gladkov 1954; Melde 1984). Nest: rigid but elastic construction typically in 4 layers: foundation of stout, short twigs mostly snapped off trees and bushes, sometimes with leaves, held together by layer of turf and moss, which is followed by smaller twigs, stalks, roots, and commonly runners of couch grass *Elymus*, then lining of bast, bark strips, grass, wool, feathers, etc., and much soft man-made material (Abshagen 1963; Tenovuo 1963; Wittenberg 1968; Kulczycki 1973, which see for detailed discussion; Jollet 1985). Animal bones and wire often incorporated, sometimes forming whole foundation (Walford 1930; Bannerman 1953$a$; Coombs 1978; Melde 1984). Size very variable depending on position; in Finland, 52–66 nests had average outer diameter 47·1 cm, inner diameter 19·8 cm, overall height 33·0 cm, depth of cup 12·6 cm (Tenovuo 1963); 31 nests in Poland had average outer diameter 29·9 cm (22·5–42·5) (Kulczycki 1973). Building: by both sexes, usually carrying and building own material, but sometimes only ♀ or ♂ carries and/or builds accompanied by mate (Wittenberg 1968; Picozzi 1975$a$; Makatsch 1976; Goodwin 1986). Although nests stable, almost always builds new one each year; in 3 studies only 1–3% of nests re-used, and old material rarely removed (Wittenberg 1968; Picozzi 1975$a$; Melde 1984), though proportion can be much higher where choice of sites limited, e.g. in towns or open fields (Haverschmidt

1937; Abshagen 1963; Coombs 1978). Builds rather casually and time taken varies; in northern Germany, took average 20 days (14-31, $n=9$) from start of building to 1st egg, but only 6-11 days for 9 replacements (Wittenberg 1968); 1st nest can be completed in 4 days (Melde 1984); 4 nests in St Petersburg (Russia) took 7-8 days (Ilyina 1990, which see for many details including energy expenditure). EGGS. See Plate 58. Short sub-elliptical to long oval, smooth and slightly glossy; from light blue to green with very variable speckles, spots, blotches, and scrawls of olive-green to blackish-brown, sometimes very sparse, sometimes obscuring ground colour, often concentrated at broad end; great variation within clutch, and no difference between nominate *corone* and *cornix* (Harrison 1975; Makatsch 1976). Nominate *corone*: 42·9 × 29·9 mm (36·4-51·0 × 26·2-33·6), $n=500$; calculated weight 19·8 g. *C. c. cornix*: 42·3 × 29·7 mm (36·0-51·0 × 25·2-33·0), $n=710$; calculated weight 19·3 g. Mediterranean populations of *sharpii*: 41·6 × 29·1 mm (36·0-50·0 × 27·1-32·0), $n=352$; calculated weight 18·2 g. South-east European *sharpii*: 42·8 × 28·7 mm (37·5-49·3 × 26·5-34·0), $n=140$; calculated weight 18·3 g. *C. c. capellanus*: 43·7 × 29·1 mm (40·6-48·0 × 26·6-31·5), $n=54$; calculated weight 19·1 g. (Schönwetter 1984.) See also Makatsch (1976) and Melde (1984), and also Rofstad and Sandvik (1985) for variation in size within clutch in Norway. Clutch: 3-6 (2-7). In Britain and Ireland, of 216 clutches: 2 eggs, 8%; 3, 23%; 4, 47%; 5, 22%; average of 345 clutches 4·1; local averages varied from 3·6 ($n=32$) on south coast of England to 4·5 ($n=24$) in northern Scotland; some reduction as season progresses and with increasing altitude (Holyoak 1967c). In northern Germany, of 74 clutches: 2 eggs, 7%; 3, 9%; 4, 20%; 5, 51%; 6, 12%; average 4·5 (5·0 in early April to 4·1 early May); average of 32 replacements 3·9 (Wittenberg 1968). In Sweden, average for first-time breeders 3·2 ($n=13$), for experienced birds 4·3 ($n=27$) (Loman 1984). See also Yom-Tov (1974) and Loman (1977) for factors affecting clutch size. Replacements started 5-20 days ($n=23$) after loss of 1st clutch; replacements recorded following loss of young aged up to 14 days; 2nd replacements rare; in one study, 8% of 76 replacements in same nest, mostly on pylon (Wittenberg 1968). Eggs laid daily. One brood (Holyoak 1967c; Makatsch 1976; Melde 1984, which see for review). INCUBATION. 18-19 days (17-20) (Wittenberg 1968; Pettersson 1977; Loman 1980b). By ♀ only; usually starts with 2nd egg (Wittenberg 1968; Holyoak 1970b; Makatsch 1976; Coombs 1978; Goodwin 1986). But see Schifferli (1992) for reliable record of ♂ undertaking 8 stints at one nest. ♀ leaves nest for short periods only and hardly at all towards hatching; in 2 cases where ♂ apparently brought insufficient food, ♀♀ left nest for total of 90 min in 12 hrs (16 breaks of 1-10 min), and 50 min in 14·8 hrs (Wittenberg 1968). YOUNG. Fed and cared for by both parents; ♀ broods for *c.* 70% of daytime for *c.* 1 week, and brooding ceases when young *c.* 2 weeks old (Wittenberg 1968; Coombs 1978; Loman

1980b; Goodwin 1986). See also Tenovuo (1963) and Melde (1984, which see for reports of ♂ brooding), and Schifferli (1992) for observation of ♂ taking 10 brooding stints. FLEDGING TO MATURITY. Average fledging period in northern Norway, 32·2 days (28-38, $n=14$) (Parker 1985); in northern Germany, 33·5 (30-36, $n=10$); young can probably feed themselves on leaving nest; usually independent after 3-5 weeks (Wittenberg 1968). For development of young, see Yom-Tov (1974), Loman (1977), Pettersson (1977), Rofstad (1986), and Richner (1989a). Age of first breeding in most studies 2 years (Kalchreuter 1971; Tompa 1975; Böhmer 1976a; Melde 1984); in dense population in northern Germany, most at 4 years old, some at 3 (Wittenberg 1976). BREEDING SUCCESS. In north-east Scotland, 69% of 186 eggs hatched and 53% produced fledged young; overall success 1·6 fledged young per pair ($n=84$), 2·9 per successful pair ($n=47$); 74% of total failures due to desertion or disappearance of clutch or brood (Picozzi 1975a); in another study, only 30% of 88 eggs resulted in fledged young (1·2 per pair, $n=22$); extra food provided near nest doubled success, since conspecifics will take eggs and young when parents have to leave nest vicinity to forage (Yom-Tov 1974), and adults will abandon young when food in short supply (Houston 1977a). For rest of Britain, see Holyoak (1967c). In northern Norway, where little human interference and few non-breeders (which often rob nests), 3·2 fledged young per pair overall ($n=26$) (Parker 1985); for comparison of various habitats in south-west Norway, where well-hidden nests in gardens most successful, see Munkejord *et al.* (1985). In southern Sweden, 0·7 fledged young per pair produced by first-time breeders ($n=13$), 1·6 by experienced birds ($n=30$) (Loman 1984); see also Pettersson (1977), and for clutch manipulation experiments and breeding success, Loman (1980b, c). In northern Germany over 5 years, 85 pairs had 1·2 fledged young per pair (0·7-1·7); in woodland, 91% of 55 breeding attempts failed but only 51% of 47 on pylons nearby; many losses caused by unpaired conspecifics (Wittenberg 1968, which see for details); in eastern Germany, 1·5 fledged young per pair ($n=25$) (Deckert 1980). For Switzerland, see Tompa (1975), Böhmer (1976a, b), and, for comparison of urban and agricultural habitats, Richner (1989a). Great majority of nestling deaths occur in 1st week and youngest often dies (Tenovuo 1963; Holyoak 1967c; Wittenberg 1968; Loman 1977; Rofstad and Sandvik 1985). Losses through predation/persecution are due mainly to man, conspecific birds, Magpie *Pica pica*, Buzzard *Buteo buteo*, Goshawk *Accipiter gentilis*, probably Eagle Owl *Bubo bubo*, mink *Mustela*, squirrels *Sciurus*, and domestic cats (Holyoak 1967c; Wittenberg 1968; Yom-Tov 1974; Picozzi 1975a; Tompa 1975; Loman 1980b; Shirihai in press). In Mediterranean region, nests often parasitized by Great Spotted Cuckoo *Clamator glandarius*, but apparently little effect on breeding success (Koenig 1920; Shirihai in press). BH

**Plumages.** Nominate *corone*. ADULT. Entirely black, glossed purple-blue in some lights (especially cap, mantle to upper tail-coverts, throat, and much of upperwing), more bluish-green in others (mainly on forehead, primary coverts, and primaries), but gloss less intense than in, e.g. Raven *C. corax* and Rook *C. frugilegus*, and in particular bristles at base of upper mandible, hindneck, side of neck, and belly appear hardly glossy in most lights. Bases of feathers of body light grey. Effect of bleaching and wear limited; in spring and summer, bristles on bill with grey shafts and pale brown tips, hindneck and side of neck sometimes tinged brown, gloss of primaries and tips of tail-feathers less purple-blue, duller bronze-green or largely without gloss, but still deep black. NESTLING. Down short, rather sparse, pale grey, smoke-grey, or whitish; on upperparts only, with traces on vent and thigh. JUVENILE. Head brown-black with slight oily gloss, hindneck to upper tail-coverts as well as lesser and median upper wing-coverts dull sooty-black or brown-black, longer feathers (mainly those of mantle, scapulars, tail-coverts, and median coverts) with rather restricted amount of glossy purplish-black on tip; some light grey of feather-bases often visible (especially on mantle and rump). Entire underparts sooty-black, virtually without gloss, grey of feather-bases showing through or partly visible, barbs of feathers wide apart and loose, feathering appearing woolly, especially on vent. Tail, flight-feathers, tertials, greater wing-coverts, and bastard wing black, as in adult, glossed greenish- or purplish-blue on tail and primaries, more purplish elsewhere, but gloss less intense than in adult, and more subject to more rapid fading, tips of tail-feathers and primaries appearing already greyish-black or brown-black before post-juvenile moult completed. FIRST ADULT. In all races, like adult, but juvenile tail, flight-feathers, tertials, greater upper wing-coverts, greater upper primary coverts, bastard wing, and occasionally some median coverts retained, less glossy and more greyish-black or brown-black than in adult at same time of year, especially greater coverts and bastard wing contrasting markedly with new purplish-black lesser and median upper wing-coverts. Tail-feathers often narrower than in adult, tips more rounded, less bluntly truncated; tips of outer primaries often pointed, not rounded; tips of tail-feathers and primaries markedly browner and more abraded than in adult. In nominate *corone*, innermost tertial distinctly browner and more frayed at tip than neighbouring longer scapulars. See also Bare Parts.

*C. c. cornix*. ADULT. Head, tail, and wing as nominate *corone*. Hindneck backwards to rump and side of neck light ash-grey with faint darker shafts, tips of feathers of upper hindneck washed darker grey, but contrast between black crown and grey hindneck sharp; some birds show some dusky along margins of outer scapulars, but this perhaps due to influence of hybrid-ization with nominate *corone*. Shorter upper tail-coverts ash-grey, middle ones grey with large ill-defined black subterminal blotch, longer ones black with grey base and partial grey fringe. Black of throat extends into black bib on central chest; feathers at side and lower end of bib ash-grey, each with black central streak or blotch, latter wider near central chest, narrower sideward and downward; thigh black, often with some grey suffusion; remainder of underparts light ash-grey with faint dark shafts. Wing-coverts as in nominate *corone*, but some shorter inner upper wing-coverts and many lesser and median under wing-coverts grey; axillaries light ash-grey. Much influence of bleaching and wear on grey of plumage; when fresh, grey slightly bluish or glaucous; when worn, in spring and summer, grey often paler, slightly tinged sandy or brown, some birds showing pink-brown tinge; grey sometimes partly contaminated by soil or grease. Worn primaries and tail-feathers glossed green rather than

purple-blue, as in nominate *corone*, but ground-colour distinctly blacker than in 1st adult. NESTLING. As nominate *corone*. JUVENILE. Head, tail, and wing as in juvenile nominate *corone*, but lesser and median upper and under wing-coverts grey, partly dull black on feather-tips. Body grey, as in adult *cornix*, but grey of upperparts tinged brown or buff, and each feather darker grey or brown-grey on tip and centre, paler grey subterminally and at base, upperparts appearing much less uniform grey than in adult; dull black bib on central chest small, more greyish-black towards side and lower end, bordered at sides and below by grey feathers with restricted amount of dusky on shafts; remainder of underparts light buffish- or brownish-grey, feathering short and woolly. FIRST ADULT. See nominate *corone* (above).

**Bare parts.** ADULT. Iris dark cinnamon to brown-black. Bill black, cutting edges plumbeous-horn to blackish. Inside of mandibles dark grey to black. Leg and foot black, fissures between scutes and soles grey. NESTLING. Bare skin, including bill, leg, and foot, flesh-pink at hatching; gape-flanges narrow, yellowish-flesh or pale yellow. Mouth bright pink. From day 6–8, skin yellowish-green, future feather-tracts dark grey, bill and tarsus brown-violet; from day 11, tarsus grey-brown, iris brown-violet; from *c.* 2 weeks, iris blue-grey, and bill and scutes on leg and foot darken to grey. At fledging, iris grey to bluish-grey-brown, bare skin at gape and throat pink, bill dark plumbeous-grey with blackish culmen ridge, paler blue-grey or whitish cutting edges, and traces of small pale pink gape flanges; mouth bright pink-red; inside of mandibles flesh-pink to white; scutes of leg and foot plumbeous-grey, fissures between scutes, rear of tarsus, and entire soles pink-flesh. During post-juvenile moult, iris brown-grey, and bill, leg, and foot darken to adult colour. FIRST ADULT. Like adult, but iris cinnamon and inside of upper mandible partly grey, partly white or flesh-pink in 1st autumn, dark with pale grey band in spring and autumn of 2nd calendar year. (Heinroth 1924–6; Witherby *et al.* 1938; Dementiev and Gladkov 1954; Kalchreuter 1969*b*; Melde 1984; Svensson 1992; RMNH, ZMA.)

**Moults.** ADULT POST-BREEDING. Complete; primaries descendent. In Britain, starts with shedding of p1 late April to mid-July (mainly May), completed with regrowth of p10 or inner secondaries early August to mid-October (mainly late September and early October); secondaries and tail start from primary moult score 15–25(–30), completed from primary score 40–50 or slightly later than primaries; non-breeding 1-year-olds start *c.* 1–2 weeks before successful breeders (1-year-old ♂ and ♀ start on average 15 May, adult ♂ 23 May, adult ♀ 31 May) (Holyoak 1974; Seel 1976; Ginn and Melville 1983); primary moult lasts *c.* 133 days (Holyoak 1974) or *c.* 172 days (Seel 1976); secondaries start 8 July, sequence s1 to s6, completed after *c.* 125 days; tertials start 6 July, sequence s8–s9–s7–s10–s11, duration *c.* 89 days; tail starts 3 July, sequence centrifugal (t1 to t6), duration 98 days; various tracts of body start on average 14 July to 1 August, duration 74–150 days, but head and neck 22 August, duration 73 days (Seel 1976, which see for details). In Germany, *cornix* starts early May to early June in 1-year-olds (♂ slightly earlier than ♀), from early June in adults (♀ slightly before ♂), completed up to mid-October; t1 starts slightly before s1, body with s1 (Bährmann 1958); nominate *corone* starts with shedding of p1 21 May to 10 June (adult ♂), or from 1 May (1-year-old), p10 shed 1–30 September (adult ♂ and ♀) or between 21 August and 20 September (1-year-old), full-grown after *c.* 1 month; tail starts with t1 and secondaries with s1 mainly in mid-July (adult ♂), in late July (adult ♀), or in early July (1-year-old); tertials (sequence s8–s9–s7–s10–s11) start with s8 *c.* 10 days earlier to 10 days later than s1; body moult mainly mid-July to mid-September; entire moult in population between 10 May and 25 October (Kalchreuter 1969). About 50 moulting adult nominate *corone* from Netherlands and 8 from Spain started primary moult late May to early July (mainly June), completed late September and early October; small samples from Sardinia, Greece, Cyprus, Turkey, and Lake Baykal also fit within this and within British schedule (RMNH, ZFMK, ZMA, ZMB). In Crete and Egypt, moult completed by late August (Meinertzhagen 1926). In USSR, starts from early June (*orientalis*) or July (*cornix*) (Dementiev and Gladkov 1954). In Afghanistan, 3 *orientalis* in moult of wing and tail (not body) on 18 July (Paludan 1959). POST-JUVENILE. Partial: head, neck, and variable number of upper wing-coverts; in *cornix*, lesser coverts (except sometimes for a few longer ones) and a few innermost median coverts in 28% of 25 birds, all lesser and most median in 20%, all lesser and median in 28%, and all these plus a few innermost greater coverts (tertial coverts) in remainder; in nominate *corone* and *sharpii*, median coverts usually all replaced, tertial coverts frequently, and part of remaining greater coverts and some tertials as well as t1 rarely (once in 20 examined) (RMNH, ZMA); in Germany, 38% of 24 birds moulted some tertial coverts (Kalchreuter 1969). Single *capellanus*, Iraq, January, moulted tail, tertials, and some scattered flight-feathers (p5-p6 in both wings, p9 and s1-s3 in one wing), but perhaps aberrant (ZMB). Moult start at age of *c.* 7 weeks (Heinroth and Heinroth 1924-6), sometimes before juvenile feathers full-grown (Bährmann 1958). For sequence of body moult, see Kalchreuter (1969). Fully juvenile birds examined February-March (Iraq) and from early April (Egypt, Iraq) to early September (Netherlands), birds in full moult July to early October, birds with moult completed from August onwards (but probably occur earlier in Egypt and Iraq) (RMNH, ZFMK, ZMA, ZMB).

**Measurements.** ADULT, FIRST ADULT. Nominate *corone*. Netherlands, Belgium, and north-east France, whole year; skins (RMNH, ZMA). Juvenile wing and tail include retained juvenile wing and tail of 1st adult. Bill (S) to skull, bill (N) to distal corner of nostril; depth is depth of bill at middle of nostril; exposed culmen on average *c.* 6·4 less than bill (S).

| | ♂ | ♀ |
|---|---|---|
| WING AD | 330·5 (8·22; 29) 318-340 | 318·4 (7·60; 30) 303-326 |
| JUV | 315·0 (5·00; 17) 306-324 | 302·0 (7·42; 22) 292-312 |
| TAIL AD | 184·5 (6·38; 17) 173-194 | 178·6 (5·69; 18) 170-189 |
| JUV | 177·7 (5·48; 11) 168-185 | 166·5 (5·56; 21) 155-174 |
| BILL (S) | 57·9 (1·52; 20) 55·5-60·5 | 52·3 (1·65; 29) 49·5-56·5 |
| BILL (N) | 37·1 (1·70; 21) 34·5-39·6 | 33·3 (1·18; 29) 31·4-35·5 |
| DEPTH | 19·3 (0·81; 21) 17·9-20·5 | 17·6 (0·54; 28) 17·0-18·5 |
| TARSUS | 61·6 (2·31; 22) 58·0-65·4 | 56·6 (2·17; 31) 52·8-60·0 |

Sex differences significant.

Adult ♂♂ with wing 345 and 353 and adult ♀♀ with wing 332 and 335 excluded from range; dwarf ♀ with wing 268, tail 139, and tarsus 51·0 excluded from table.

Britain and Spain. (1) Hertfordshire (Britain), live or freshly dead (Holyoak 1970, which see for other measurements). (2) South-east Spain, skins (ZFMK).

| | ♂ | ♀ |
|---|---|---|
| (1) WING AD | 329 ( — ; 93) 298-361 | 315 ( — ; 127) 283-387 |
| (2) WING AD | 322·8 (8·50; 4) 314-331 | 311·2 (5·76; 6) 304-320 |
| (2) BILL (S) | 57·0 ( — ; 2) 56·1-57·9 | 53·6 (2·03; 6) 50·9-56·3 |
| (2) DEPTH | 17·8 ( — ; 3) 17·5-18·2 | 16·6 (0·67; 8) 15·7-17·5 |

For bill depth, Germany, see Bährmann (1978*b*).

*C. c. orientalis* (1) USSR, ages combined (Dementiev and Gladkov 1954). (2) Afghanistan through central Asia to Japan (Stresemann *et al.* 1937; Paludan 1959; Piechocki 1959; Piechocki and Bolod 1972; Eck 1984; RMNH, ZMA).

| | ♂ | ♀ |
|---|---|---|
| (1) WING | 348 ( — ; 31) 320-375 | 342 ( — ; 25) 310-370 |
| (2) WING AD | 355·5 (11·06; 11) 345-380 | 333·3 (10·06; 11) 320-349 |
| (2) JUV | 336·2 ( 9·07; 12) 323-346 | 316·7 (11·73; 7) 294-328 |
| (2) BILL (S) | 59·9 ( 2·21; 10) 54·1-64·9 | 54·9 ( 2·46; 6) 50·5-57·3 |
| (2) DEPTH | 19·0 ( 0·94; 10) 17·5-20·0 | 17·4 ( 0·74; 6) 16·7-18·6 |

*C. c. cornix.* Sweden, Belorussiya, north European Russia, Poland, eastern Germany, Faeroes, and Scotland, mainly summer, and Netherlands, mainly winter; skins (RMNH, ZFMK, ZMA, ZMB); adult wing including data from Stresemann (1920) and Bährmann (1950).

| | ♂ | ♀ |
|---|---|---|
| WING AD | 330·2 (8·09; 47) 318-342 | 316·7 (6·44; 37) 304-332 |
| JUV | 321·0 (6·84; 23) 308-335 | 307·5 (8·91; 34) 286-325 |
| TAIL AD | 192·2 (5·55; 16) 182-202 | 181·5 (5·44; 16) 172-191 |
| JUV | 179·3 (6·43; 19) 170-188 | 172·6 (6·92; 33) 159-184 |
| BILL (S) | 58·2 (1·76; 34) 56·0-61·4 | 54·0 (1·70; 33) 50·8-57·2 |
| BILL (N) | 37·4 (1·56; 17) 35·0-39·8 | 34·1 (1·11; 20) 32·0-35·5 |
| DEPTH | 19·3 (0·54; 33) 18·6-20·3 | 17·7 (0·64; 33) 16·7-18·8 |
| TARSUS | 62·4 (2·58; 18) 59·5-68·2 | 58·3 (1·62; 20) 56·0-61·3 |

Sex differences significant.

Wing. Trondheim area (Norway): (1) adult, (2) juvenile (Slagsvold 1980, which see for other measurements, some smaller Norwegian samples, and seasonal variation in size). Lausitz (eastern Germany), breeding season: (3) adult, (4) juvenile (Bährmann 1950; see also Richter 1958, Bährmann 1978*b*, Eck 1990). (5) USSR, ages combined (Dementiev and Gladkov 1954; see also Danilov *et al.* 1984 for Yamal). (6) Adult Hungary, northern Yugoslavia, and northern Italy (RMNH, ZFMK, ZMA, ZMB; Bährmann 1950).

| | ♂ | ♀ |
|---|---|---|
| (1) | 329·8 ( 8·5 ; 83) — | 315·4 (8·3 ; 141) — |
| (2) | 318·3 (11·0 ; 138) | 304·6 (9·7 ; 175) — |
| (3) | 331·0 ( — ; 13) 321-340 | 321·0 ( — ; 19) 307-330 |
| (4) | 312·3 ( — ; 9) 302-326 | 299·0 ( — ; 9) 293-310 |
| (5) | 327·8 ( — ; 70) 292-355 | 315·0 ( — ; 78) 298-340 |
| (6) | 317·7 (10·22; 9) 309-340 | 309·5 (4·10; 9) 305-315 |

For bill depth in Sweden, European Russia, and eastern Germany, see Bährmann (1978*b*). For birds from hybrid zone between nominate *corone* and *cornix* in Scotland, see Picozzi (1975*a*), from similar zone in eastern Germany, Richter (1958). See also Hell and Soviš (1959) for eastern ČSSR. Wing of ♂ in Germany occasionally up to 351 (Kleinschmidt 1938) or 352 (Richter 1958). See Slagsvold (1982*b*) for size variation in winter, Norway.

*C. c. sharpii.* ADULT. Wing, bill to skull, bill depth, and tarsus of (1) eastern Rumania, central and northern Turkey, Caucasus area, southern European Russia, and Kazakhstan, and (2) Sardinia, Sicily, southern Yugoslavia, and Greece, whole year; skins (RMNH, ZFMK, ZMA, ZMB; wing includes data from Stresemann 1920 and Bährmann 1950).

| | ♂ | ♀ |
|---|---|---|
| WING (1) | 325·0 (9·84; 19) 308-345 | 310·6 (8·05; 14) 296-320 |
| (2) | 319·4 (8·39; 21) 297-335 | 304·9 (9·82; 20) 290-324 |
| BILL (1) | 55·5 (1·66; 11) 54·4-57·8 | 51·7 (1·38; 7) 49·5-53·0 |
| (2) | 54·6 (1·64; 13) 52·0-57·0 | 52·1 (3·14; 6) 46·5-54·8 |
| DEPTH (1) | 18·5 (0·44; 10) 17·7-19·2 | 16·6 (0·75; 5) 15·3-17·3 |
| (2) | 18·3 (0·95; 6) 17·2-19·8 | 17·5 (0·37; 5) 16·9-18·0 |
| TARSUS (1) | 60·4 (1·71; 6) 58·1-62·2 | 55·4 (1·84; 6) 53·1-56·7 |
| (2) | 58·9 (2·08; 7) 56·7-62·3 | 56·9 (3·33; 4) 54·0-61·5 |

Sex differences significant, except bill depth and tarsus of (2).

South-east European Russia, ages combined: wing, ♂ 310-335; bill, ♂ 52-58, ♀ 47-54 (Fediuschin 1927). Ages and sexes combined: Balkans, wing 280-333, bill 42-59, depth 16-21 (*n* = 11); southern Russia, wing 314-345, bill 49-59, depth 19-23 (*n* = 47); Sardinia and Corsica, wing 301-329, bill as Balkans (*n* = 18); Crete, wing 313-327, bill 55-61, depth 20-22 (Meinertzhagen 1926). For Turkey, see also Kumerloeve (1961, 1967) and Rokitansky and Schifter (1971). Northern Iran, adult wing: ♂ 324·6 (7·89; 5) 318–337, ♀♀ 302, 312 (Stresemann 1928; Paludan 1940).

*C. c. pallescens*. Cyprus, south-central Turkey, Levant and Egypt, skins (RMNH, ZFMK, ZMB; wing includes data from Stresemann 1920).

| | | | | | |
|---|---|---|---|---|---|
| WING AD | ♂ 313·9 (8·10; 10) 297–327 | ♀ 307·0 (6·86; 5) 299–316 |
| BILL (S) | 54·6 (1·93; 8) 52·0–57·0 | 51·8 (1·77; 4) 50·1–53·8 |
| DEPTH | 17·4 (0·52; 8) 16·8–17·9 | 16·3 (0·95; 4) 15·1–17·0 |

Sex differences significant for bill depth.

Cyprus, ages and sexes combined: wing 285–314, bill 47–53, bill depth 17–20 (n = 7) (Meinertzhagen 1926). Ages combined: Levant, wing, ♂ 292–312 (6), ♀ 278–301 (8); bill, ♂ 51–56, ♀ 45–53, bill depth, ♂ 18–20·5, ♀ 17–19; Egypt, wing, ♂ 286–331 (20), ♀ 286–317 (15); bill ♂ 50–58, ♀ 45–54; bill depth (both sexes) 17·5–20 (Meinertzhagen 1921). Ages and sexes combined: Levant, wing 278–324 (22), Arabia 278–324, Egypt 286–332 (50) (Stresemann 1920; Hartert 1921–2; Meinertzhagen 1926, 1930, 1954; Hartert and Steinbacher 1932–8).

*C. c. capellanus*. Iraq, all year; skins (BMNH, ZFMK, ZMB).

| | | | | | |
|---|---|---|---|---|---|
| WING AD | ♂ 328·1 (5·45; 10) 320–335 | ♀ 322·5 (5·77; 13) 315–334 |
| BILL (S) | 65·3 (2·18; 10) 62·4–69·3 | 61·7 (1·92; 15) 59·4–66·5 |
| DEPTH | 20·5 (0·69; 10) 19·4–21·3 | 19·0 (0·63; 15) 17·8–19·8 |
| TARSUS | 70·8 (1·92; 10) 69·0–75·0 | 67·7 (1·87; 15) 64·0–70·3 |

Sex differences significant except for wing.

Wing 325–350, tail 215–225, tarsus 64–70 (Hartert 1921–2).

JUVENILE. For averages from May of 1st calendar year to June of 2nd calendar year, see Bährmann (1978b). For influence of weather on growth of various parts of body, see Slagsvold (1982b) and Rofstad (1988).

**Weights.** ADULT, FIRST ADULT. Nominate *corone*. Britain: (1) Hertfordshire, November–February (Holyoak 1970); adults England, Wales, and Scotland (perhaps including some *cornix*), (2) November–February, (3) March–May, (4) June–July, (5) August–October (Seel 1976, which see for 1st adults). Netherlands: (6) adult, all year (RMNH, ZMA).

| | | | | |
|---|---|---|---|---|
| (1) | ♂ 561 ( — ; 93) 465–645 | ♀ 523 ( — ; 127) 455–604 |
| (2) | 589 (54·0; 131) — | 517 (36·2; 116) — |
| (3) | 577 (47·8; 181) — | 512 (45·5; 355) — |
| (4) | 516 (64·1; 41) — | 452 (39·6; 34) — |
| (5) | 585 (41·9; 66) — | 488 (38·8; 99) — |
| (6) | 522 (72·1; 23) 418–625 | 490 (55·3; 16) 383–583 |

For hybrids in Scotland, see Picozzi (1975a).

*C. c. orientalis*. Kazakhstan, Mongolia, and Manchuria (China), mainly June–September: ♂ 553·4 (64·4; 8) 460–658, ♀ 494·2 (8) 402–563 (Piechocki 1959; Piechocki and Bolod 1972; Korelov et al. 1974).

*C. c. cornix*. Trondheim area (Norway), August–May: (1) adult, (2) 1st adult (Slagsvold 1980, which see for other Norwegian samples and for graphs of monthly variation). East Germany: pure *cornix*, (3) adult, April–May, (4) adult, whole year, (5) 1st adult, whole year (Bährmann 1972, which see for graphs of monthly variation); birds from hybrid zone (mainly *cornix*), winter, (6) adult, (7) 1st adult (Richter 1958). Netherlands: (8) adult, October–April (RMNH, ZMA). Russia: (9) Yamal peninsula, summer (Danilov et al. 1984).

| | | | |
|---|---|---|---|
| (1) | ♂ 564·6 (38·3 ; 83) — | ♀ 488·2 (34·3; 141) — |
| (2) | 543·7 (43·6 ; 138) — | 472·9 (37·1; 175) — |
| (3) | 535·2 ( 5·53; 24) — | 483·3 ( 4·7; 36) — |
| (4) | 538·9 (39·90; 91) 430–618 | 480·9 (24·2; 83) 410–548 |
| (5) | 528·5 (35·9 ; 40) 440–586 | 450·5 (28·2; 49) 397–500 |
| (6) | 538·7 (40·4 ; 105) 440–670 | 464·5 (36·7; 115) 370–570 |
| (7) | 516·6 (44·2 ; 53) 400–640 | 458·8 (43·9; 67) 370–560 |
| (8) | 559·6 (64·0 ; 18) 471–730 | 476·1 (56·5; 9) 400–550 |

| | | | |
|---|---|---|---|
| (9) | 558 ( — ; 8) 464–595 | 478 ( — ; 6) 435–512 |

Eastern Czechoslovakia, winter: 490·6 (56·5; 75) 360–638 (Hell and Soviš 1959). For graphs of variation throughout year in Norway, see Slagsvold (1982b). Ranges, Sweden: adult, April, ♂ 525–660, ♀ 470–520; May, ♂ 525–615, ♀ 445–515; June–July, ♂ 500–580; August, ♀ 470–475; juvenile, June–July, ♂ 420–570, ♀ 415–425; August–September, ♂ 520–580, ♀ 395–510 (Zedlitz 1926). In USSR, maximum of ♂ 740, of ♀ 670 (Dementiev and Gladkov 1954). For relationship between fat content and body dimensions, see Slagsvold (1982a).

*C. c. sharpii*. Greece: ♀ 500 (Makatsch 1950). Northern Iran, February–March: ♂ 557 (3) 540–575 (Schüz 1959).

*C. c. sharpii* or *pallescens*. Turkey: March, 493; May, ♀ 450; June, ♀ 455; July, ♂ 524; August, juvenile ♂ 436 (Kumerloeve 1961, 1969b, 1970a; Rokitansky and Schifter 1971).

*C. c. capellanus*. Iraq, February: juvenile ♂ 350 (ZFMK). JUVENILE. For growth rates, see Richner et al. (1989).

**Structure.** Wing rather short, broad at base, tip bluntly pointed, 10 primaries: p7 longest, p8 1–8 shorter, p9 29–40 (adult) or 21–38 (juvenile), p10 105–130 (adult) or 95–115 (juvenile), p6 0–9, p5 15–31, p4 52–76, p1 100–126 (10 adult ♂♂), 97–119 (10 adult ♀♀, 10 juvenile ♂♂), or 98–112 (10 juvenile ♀♀), no differences between races. P10 59·1 (30) 47–68 longer than longest upper primary-covert in adult, 58·5 (20) 47–75 in juvenile. Outer web of p5–p9 emarginated, inner web of p6–p10 with notch. Tip of longest tertial reaches tip of p2. Tail rather long, tip rounded; 12 feathers, t6 21·5 (40) 12–31 shorter than t1 in adult, 15·0 (34) 8–25 in juvenile. Bill long, strong; visible culmen c. 105% of length of head; wide and deep at base; basal half of culmen straight, tip strongly decurved (culmen less gradually decurved than in *C. frugilegus*, base and middle deeper, but gape relatively less wide). Bill depth depends on race and sex: see Measurements. Nostril oval, concealed below dense layer of bristle-like feathers projecting forward from base of bill, covering basal half of top and side of upper mandible; numerous fine bristles project obliquely down from lore over gape and at lateral base of lower mandible. Feathers of throat slightly elongated, tip pointed. Leg and foot rather short, but strong. Middle toe with claw 47·0 (20) 44·5–51 in ♂♂ of nominate *corone*, *cornix*, *sharpii*, and *pallescens*, 43·8 (20) 41·5–48 in ♀♀, 53·8 (2) 52·5–55·1 in *capellanus*. Outer toe with claw c. 75% of middle with claw, inner c. 77%, hind c. 86%.

**Geographical variation.** 2 distinct groups. (1) All-black *corone* group, with 2 disjunct races: nominate *corone* in west and *orientalis* in east. (2) Grey-and-black *cornix* group ('Hooded Crow') with a number of races in northern and eastern Europe, from Corsica and Italy eastward, in Middle East, and in northern, western, and central Asia.

Within *corone* group, variation slight, *orientalis* differing from nominate *corone* only in larger size, particularly wing (adult 310–380), tail (195–290), and tarsus (60–68). Within nominate *corone*, birds from Spain average smaller than birds from England and western Europe, especially in bill depth; within *orientalis*, birds from mountains and northern part of distribution on average larger than those elsewhere, and bill slightly longer and more slender in west, thicker and shorter in eastern Siberia and Japan, but differences slight and clinal and no races other than *orientalis* recognized here.

Within *cornix* group, variation slight and clinal, involving size (wing, bill, and tarsus), relative depth of bill, and tone of grey of body; populations become gradually smaller in size and in bill depth towards south and paler towards south and east, but *capellanus* from central and southern Iraq and neighbouring Iran quite distinct, forming pale end of cline in colour but not fitting into cline of size, being much larger than neighbouring populations, especially in bill and tarsus length and size of foot. As bleaching has pronounced influence on depth of grey (birds in arid south and east fading much more rapidly to pale greyish-white or isabelline than birds from more humid north and west), only birds in fresh plumage should be compared for assessment of races, and, as wing length depends strongly on age and sex, only birds of same age and sex should be compared. *C. c. cornix* from northern Europe (west to Faeroes, Scotland, and Ireland) large, close to size of nominate *corone* from western Europe, body light bluish ash-grey; birds from eastern Germany, eastern Czechoslovakia, Hungary, northern Italy, and Corsica have grey on average slightly darker (perhaps due to some introgression of nominate *corone*), and bill more slender (see Measurements), and these sometimes separated as *subcornix* Brehm, 1831; birds of north-west Siberia (east to Yenisey) have grey on average slightly paler, but differences of all these populations slight and all included in *cornix*. *C. c. sharpii* from eastern Rumania, Moldavia, southern Ukraine, and through southern European Russia and southern Transcaspia east to foot of Altai equal in size to *cornix* or slightly smaller, but grey distinctly paler; birds from European part of range larger than those from Asia, especially bill; those from Don basin and lower Volga and Ural rivers sometimes separated as *khozaricus* Fediuschin, 1927. Further south and west, situation more complex; birds from west, central, and northern Turkey, Caucasus area (named *kaucasicus* Gengler, 1919), and northern Iran about similar in size to or slightly smaller than *sharpii*, but grey slightly darker, more or less intermediate between *cornix* and *sharpii*; birds from Sardinia (*sardonius* Kleinschmidt, 1903), Sicily, coastal and southern Yugoslavia, and Bulgaria south to European Turkey, Greece, and Crete slightly smaller than *sharpii*, but grey darker in west, similar to Turkey-Caucasus birds, paler (like *sharpii*) in east; as differences in size and colour rather slight and inconstant, all included in *sharpii*. Birds of central and southern Italy and Sicily are intermediate between *cornix* and 'sardonius' (Keve 1970), but here provisionally included in *sharpii*. *C. c. pallescens* from Cyprus, southern Turkey, Levant, Syria, and (probably this race) northern Iraq is smallest race; grey pale, like typical *sharpii* or even paler; birds from Egypt (named *egipticus* Keve, 1972) on average slightly darker, but small as well, and here included in *pallescens*. *C. c. capellanus* markedly paler than other races, body cream or pale silvery rather than grey when plumage fresh (often with slight vinous-pink tinge), milky-grey or almost white when worn; bill and legs markedly heavy, throat feathers rather long, tail long and rather strongly graduated (matching *orientalis* in this respect); rather similar to Pied Crow *C. albus* in bill and leg size and in pale colour of body, but pattern of greyish-white different (extending to upper tail-coverts and vent; central chest black) and fine dusky shaft-streaks present; sometimes considered separate species, but plumage pattern similar to other races of *cornix* group, proportions rather similar to *orientalis*, and said to intergrade into *sharpii* at foot of Zagros mountains in Iran (Meinertzhagen 1926; Meise 1928; Vaurie 1959).

Geographical boundary between *corone* group and *cornix* group rather sharp, formed by relatively narrow zone in which extensive hybridization occurs. Due to secondary character of hybridization, the groups are sometimes considered separate species; in Asia, present-day overlap perhaps caused by man-made habitat changes, *orientalis* and *cornix/sharpii* perhaps originally well separated. Hybridization zone between *cornix* group and *orientalis* apparently not yet stable (for position of zone, see Meise 1928, Johansen 1944, Vaurie 1954b, and Korelov *et al.* 1974), and, as they overlap locally with only limited hybridization, Russian authors (e.g. Korelov *et al.* 1974, Stepanyan 1978, 1990) consider the groups to be full species; *orientalis* then usually combined with *corone* group into single species *C. corone*, which comprises only 2 races. However, nominate *corone* and *orientalis* are probably less closely related to each other than nominate *corone* is to *cornix* (Eck 1984; C S Roselaar). *C. c. capellanus*, which is in some structural characters nearer to *orientalis* than to *cornix*, may form 4th group in the *C. corone* complex (C S Roselaar), while Collared Crow *C. pectoralis* (synonym *C. torquatus*) from China, which is close in characters to *orientalis* (Kleinschmidt 1922; Meise 1938) may form 5th member. Pending further research, traditional view of single species *C. corone* (comprising *corone*, *orientalis*, *capellanus*, and *cornix/sharpii/pallescens* sub-groups) and a separate *C. pectoralis* is maintained here.

For detailed maps of hybridization zone between nominate *corone* and *cornix*, see especially Meise (1928); also (for more local details), Salomonsen (1930a), Kleinschmidt (1938), Richter (1958), Bährmann (1960a, b, 1978b), Cook (1975), Dybbro (1976), and Picozzi (1976, 1982). For possible history of evolution of subspecies-groups, see Geyr (1920) and Meise (1928). Outside rather narrow zone of true hybridization, a wider border occurs ('introgression zone') in which most birds are typical of one race but some show slight influence of other; also a few birds, especially of *cornix*, summer in heart of distribution of nominate *corone* and form mixed or (rarely) pure pairs there. For highly variable appearance of hybrids, see Richter (1958) and Melde (1984).                                      CSR

---

## *Corvus albus*  Pied Crow

PLATES 14 and 15 (flight)
[facing pages 137 and 280]

Du. Schildraaf      Fr. Corbeau-pie      Ge. Schildrabe
Ru. Пегий ворон      Sp. Corneja pía      Sw. Svartvit kråka

*Corvus albus* P L Statius Müller, 1776

Monotypic

**Field characters.** 45 cm; wing-span 98–110 cm. Large, robust crow, slightly larger than Carrion Crow *C. corone*, with proportionately longer and deeper bill and longer legs. Black, with white chest and collar. Sexes similar; no seasonal variation. Juvenile separable at close range.

ADULT. Head, most of neck, throat, centre of breast,

back, rump, wings, and tail black, glossed blue and purple; rear underbody black, less glossy. Broad white collar across base of neck and round shoulder joins white chest, fore-belly, and flanks. Bill and legs black. JUVENILE. Similar to adult but duller, with less gloss above and none below.

Unmistakable in Afrotropics, but vagrants to west Palearctic must be distinguished from bleached hooded forms of *C. corone* (differing distinctly in greater extension of pale plumage to nape, back, rear underbody, and under wing-coverts). Flight powerful and (unlike *C. corone*) bird often soars; wing shape recalls Raven *C. corax*, but body size and tail shape more like *C. corone*. Gait and behaviour typical of *Corvus*.

Commonest call a deep, guttural croak, 'raark' or 'caw', recalling Rook *C. frugilegus* or even *C. corax*.

**Habitat.** Breeds in tropical low latitudes, usually avoiding arid regions and favouring more or less open country with trees, in which it normally nests; also occurs in forest clearings and in towns and villages, associating freely with man in cultivated or pastoral areas. Feeds largely on ground but is fond of aerial activities, soaring at considerable heights. In Eritrea also occurs in arid uninhabited regions. (Goodwin 1986.) Absent from closed forest, but occurs in all West African vegetation zones (Serle *et al.* 1977). In southern Africa, found at refuse dumps near towns and often seen on roads, feeding on animals and insects run over by cars; often nests in tall exotic trees (Prozesky 1974).

**Distribution.** Resident over almost all of sub-Saharan Africa, Madagascar, and Comores and Aldabra group in western Indian Ocean.

Accidental. Algeria: one reported in extreme south in 1961 (no further details), and another at In Azoua, December 1964 (Dupuy 1969). Libya: Jalo, 24 April 1931 (Ghigi 1932).

**Movements.** Chiefly sedentary, especially in south; in some northern areas, many birds move north towards Sahara in rainy season.

In southern Africa, adults mostly sedentary, though some young birds probably wander some distance before establishing territory; reports of varying numbers suggest there may be local movements; only 1 ringing recovery of more than 20 km: bird ringed as nestling in Transvaal recovered 47 km away after 9 years 4 months (Winterbottom 1975; M P S Irwin, T B Oatley). In East Africa, probably chiefly sedentary, but present only in wet periods in some arid and semi-arid areas, and local movements sometimes result in large concentrations (Britton 1980; Lewis and Pomeroy 1989). Limited movement reported in north-west Central African Republic, where present on plains only October–April, apparently dispersing from mountain breeding areas then (Blancou 1939). Reported

by Lynes (1924) as almost entirely a dry season visitor to western Sudan; common November–May, but birds moved elsewhere for breeding, with only a few stragglers remaining June–September. In Chad, many birds move north into Sahara zone in wet season; thus at Fada (17°14′N) very common July–September, but at Abéché (13°49′N) abundant October–May; also at Fort Lamy (12°10′N) present chiefly from October (Malbrant 1952; Salvan 1967–9).

In West Africa also, northward movement in rainy season regular in some areas. In Mauritania at that time, birds move north almost to Nouakchott (18°09′N) (Gee 1984), and movement also reported Sénégal (Morel and Morel 1990) and Mali (Lamarche 1981). In southern Nigeria, numbers present at Ibadan apparently higher November–April than during wet months, May–October, perhaps due to pre-breeding break-up of flocks rather than to overall movement, since at Lagos data argue against northward movement for breeding (Elgood *et al.* 1973; Elgood 1982); at Malamfatori (north-east Nigeria), occasional in most months, but large flocks reported September–October (Hopson 1964). At Mole (northern Ghana), all records (in considerable numbers) are from 2nd half of dry season, so perhaps birds move further north to breed (Greig-Smith 1977); large-scale influx occurs in dry season at Niamey (north-west Niger) (Giraudoux *et al.* 1988).     DFV

**Voice.** See Field Characters.

**Plumages.** ADULT. Head, neck, and upper chest black, contrastingly bordered by white collar across mantle, side of breast, and lower chest, extending into flank and belly; width of collar variable, depending on posture of bird. Scapulars, back to upper tail-coverts, vent, thigh, under tail-coverts, and all tail and wing including axillaries and under wing-coverts contrastingly black again. All black of head, body, and wing distinctly glossed purple-violet, gloss of flight-feathers and tail slightly duller and more purple. *In worn plumage*, black slightly duller and less glossy, primaries and tail slightly browner (but distinctly less so than in 1st adult at same time of year), white of mantle and underparts sometimes tinged brown, buff, or pink due to contamination by soil or dirt. JUVENILE. Rather like adult; gloss of black parts of body, tail, and wing purple-violet, as in adult, but slightly less intense. Feathers of head short, limited purple on tips, much dull black of centres visible; feathers of throat short, hardly glossy (in adult, elongated and strongly purple). Border between white mantle and belly and black back and vent not sharply defined, feathers at border white with mottled grey tip, some white of feather-bases visible down to lower back. Feathers of vent dull black with restricted purple gloss on tips, markedly short and loose. With wear, black of head and body turns dull brownish-black. Flight- and tail-feathers narrower than in adult, tips of outer primaries often more pointed, those of tail-feathers often more rounded, less broad and blunt than in adult, tail-tip more rounded, outermost pair (t6) 19·7 (11) 15–24 shorter than central pair (in adult, 20–43 shorter). FIRST ADULT. Like adult, but juvenile tail, fledglings, greater upper primary coverts, greater and variable number of median and lesser upper wing-coverts, and tertials retained, duller black and less intensely glossy than in adult, though difference sometimes hard to see;

tertials and old greater and median coverts often distinctly duller black than newer neighbouring 1st adult median or lesser coverts and scapulars; juvenile tail- and flight-feathers narrower than in adult; tail-tip more rounded, less wedge-shaped (see juvenile, above). Rarely, a few tail-feathers new, contrasting in colour and abrasion with juvenile ones.

**Bare parts.** ADULT, FIRST ADULT. Iris brown or dark brown. Bill, tarsus, and toes black, soles and ridges between scutes of leg and foot grey. JUVENILE. Irish blue-grey. Mouth scarlet; narrow white gape-flanges present at fledging; similar to adult after post-juvenile moult.

**Moults.** ADULT POST-BREEDING. Complete; primaries descendent. Timing variable, depending on local breeding season. In birds examined, moult just started (with shedding of p1-p2) in February (Uganda, 2 birds). May (Sudan), June (Ethiopia), or July (Ghana), almost completed (p9-p10 still growing) May (Comoro Islands); none of 10 others examined December–June from Chad and Sudan in moult (RMNH, ZFMK, ZMA). Moult not started in 3 birds from northern Nige, June and early July (Fairon 1975). Moult just started with inner primaries in ♂ collected December in Zaïre (Verheyen 1953). POST-JUVENILE. Partial: head, neck, lesser and variable number of median upper wing-coverts, rarely a few tertial coverts or tail-feathers. Probably starts soon after fledging, as in Raven *C. corax*, but no information.

**Measurements.** ADULT, FIRST ADULT, JUVENILE. Wing and bill to skull from (1) adult, northern Afrotrpics, (2) adult, eastern and southern Africa, (3) adult, Madagascar and Comoro Islands. Other measurements for all localities combined, all year; skins (RMNH, ZFMK, ZMA). Bill (N) to nostril; exposed culmen on average 3·5 shorter than bill to skull.

| | ♂ | ♀ |
|---|---|---|
| WING(1) | 366·0 (8·18; 13) 354–380 | 351·6 (7·46; 12) 341–367 |
| (2) | 368·7 (9·94; 7) 355–382 | 357·0 ( — ; 2) 352–362 |
| (3) | 365·8 (5·49; 5) 361–376 | 346·2 (6·24; 4) 340–353 |
| JUV | 336·7 (5·61; 6) 329–345 | 313·7 (6·50; 6) 305–322 |
| TAIL AD | 188·8 (5·64; 25) 179–197 | 182·0 (4·11; 17) 173–189 |
| JUV | 168·9 (3·26; 6) 163–175 | 160·2 (5·95; 6) 153–170 |

| | ♂ | ♀ |
|---|---|---|
| BILL (1) | 62·7 (2·46; 13) 59·2–66·0 | 58·7 (2·93; 12) 54·5–63·0 |
| (2) | 60·0 (2·52; 7) 57·2–63·5 | 55·7 ( — ; 2) 54·8–56·5 |
| (3) | 59·4 (1·07; 5) 58·0–60·6 | 55·4 (1·54; 4) 53·5–57·1 |
| BILL (N) | 39·3 (1·81; 21) 36·9–43·0 | 36·9 (1·79; 13) 23·8–40·5 |
| TARSUS | 63·3 (1·74; 21) 60·8–67·0 | 61·2 (2·15; 15) 58·5–64·7 |

Sex differences significant, except wing (2).

In Chad, wing 33·7 (8) 320–342 (Salvan 1967–9). NESTLING. For growth, see Mundy and Cook (1977).

**Weights.** Air (northern Niger): June, ♀♀ 407, 445; July, 420 (Fairon 1975). Northern Chad: ♂♂ 465, 560. Namibia: ♂ 550, ♀ 470. (Niethammer 1955.) South Africa: adult 612, juveniles 474, 490, 511 (Skead 1977). Juveniles in 1st week after fledging August–September, northern Nigeria: 504 (17) 400–570 (Mundy and Cook 1977, which see for growth curves under varying conditions).

**Structure.** Wing long, bluntly pointed, rather broad at base. 10 primaries: p7 longest, p8 0–7 shorter, p9 16–30, p10 95–120, p6 5–13, p5 38–55, p4 80–92, p1 135–145; p10 80–95 longer than longest upper primary covert. Outer web of p6–p9 emarginated, inner of (p6–)p7–p10 with distinct notch. Tail rather long, tip wedge-shaped in adult, more rounded in juvenile and 1st adult; 12 feathers, t1 longest, t6 31·7 (22) 20–43 shorter in adult, 19·7 (11) 15–24 in juvenile and 1st adult (no difference between sexes). Bill long and heavy; depth at middle of nostril 21·0 (25) 19·6–22·4 in ♂, 19·5 (17) 17·9–20·7 in ♀. Middle toe with claw 51·8 (13) 47–56 mm; outer, inner, and hind toe with claw 76–78% of middle with claw. Feathers of chin and throat elongated, but less so than in *C. corax*, tips more rounded, less sharply pointed, rarely bifurcated. Remainder of structure as in *C. corax*.

**Geographical variation.** Slight, if any; bill of birds of northern Afrotropics perhaps slightly longer and deeper than elsewhere, but only small samples examined.

Forms superspecies with *C. corax* and Brown-necked Raven *C. ruficollis*; locally hybridizes with *C. ruficollis edithae* where ranges touch in eastern Ethiopia and Somalia (Kleinschmidt 1906; Smith 1957; Blair 1961; Hall and Moreau 1970).      CSR

---

## *Corvus ruficollis* Brown-necked Raven

PLATES 10 and 15 (flight)
[between pages 136 and 137, and facing page 280]

DU. Bruinnekraaf      FR. Corbeau brun      GE. Wüstenrabe
RU. Пустынный ворон      SP. Cuervo desertícola      SW. Ökenkorp

*Corvus ruficollis* Lesson, 1831

Polytypic. Nominate *ruficollis* Lesson, 1831, Cape Verde Islands and North Africa east through Middle East and Arabia to Transcaspia and western Pakistan, south to Sahel zone and Sudan; also Socotra. Extralimital: *edithae* Phillips, 1895, eastern Ethiopia and Somalia south to Kenya.

**Field characters.** 50 cm; wing-span 106–126 cm. 10% smaller than Raven *C. corax*, with proportionately more slender bill and longer head (both lacking depth of *C. corax*), less shaggy throat, longer outer part of wing, less evenly-wedged end to tail, and longer-looking legs. Large crow, with general character most recalling *C. corax* but with slighter build, brownish gloss on rear of head, and bill drooping downwards in flight. Sexes similar; no sea-

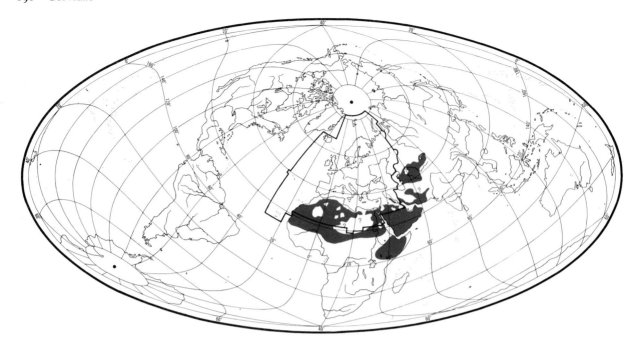

sonal variation except for loss of gloss. Juvenile separable at close range.

ADULT. Moults: June–October (complete). Black, glossed violet to purple on head, neck, and underbody and violet elsewhere. When fresh, sheen on side of head, nape, hindneck, and upper mantle purplish-bronze sufficiently strong to catch light and shine, forming discrete shawl. With wear, bird becomes drab with patches of coppery gloss and even browner shawl which may become indistinct or absent. Rictal bristles, bill, and legs black. JUVENILE. Resembles most worn adults in not showing any bronzy-brown on shawl in the field until 1st autumn moult.

At any distance, or with structure and colour of gloss uncertain, difficult to distinguish from *C. corax* and both species thus subject to frequent confusion in narrow areas of overlap across North African desert boundary and from southern Levant eastwards. Even at closer ranges, brown neck may still not show, but differences in structure (see below) do allow certain identification. Flight much as *C. corax* but less majestic, action being rather lighter in tight spaces and less powerful over long distances; can hang on the wind. Flight silhouette combines proportionately longer, narrower-based, and more pointed wings, rather thinner tail (from which central feathers extend to form slight blob), and drooping bill; said to recall Lammergeier *Gypaetus barbatus* (Hollom *et al.* 1988). On ground, extension of folded wing-tips to near or beyond tip of tail is also helpful, as wing-tips of *C. corax* usually fall well short of tip of tail. Gait a direct, quite free-stepping walk, varied by occasional jumps and sidling movements. Frequently joins other scavengers, wandering widely in search of food source.

Commonest call a harsh 'karr', commonly repeated; varies considerably in pitch and loudness but typically much less deep and croaking than most calls of *C. corax*, more like Carrion Crow *C. corone*.

**Habitat.** Basically in deserts of lower middle latitudes, generally in very warm arid open plain country, but exceptionally resident on islands (Cape Verde), there also inhabiting cultivated land, which is usually avoided (Goodwin 1986). In North Africa, closely linked to *Artemisia* steppe with groups of jujube trees *Zizyphus*, being more attached to desert than any other *Corvus* (Etchécopar and Hüe 1967). Also resorts to date palms *Phoenix dactylifera*, tamarisks *Tamarix*, and large shrubs, and (where available) to rock ledges and artefacts such as clay structures or telegraph poles. Prevailing confinement to lowland deserts and semi-deserts imposes difficulties in finding suitable breeding sites. In northern Afghanistan, birds in post-juvenile moult were met with at around 440 m (Paludan 1959); mountainous areas seem usually to be avoided. Common in March along Red Sea coast of Egypt, sometimes in bands of more than 20 on town rubbish dumps; *c.* 100 roosted in date palms by monastery of St Anthony in arid northern region (P A D Hollom and E M Nicholson). In India, essentially a desert bird, scavenging tamely round encampments and replacing Raven *C. corax* ecologically, but status in upland regions needs clarification (Ali and Ripley 1972).

**Distribution.** No changes reported, except in Israel and Algeria.

ISRAEL. Has spread in recent decades, following devel-

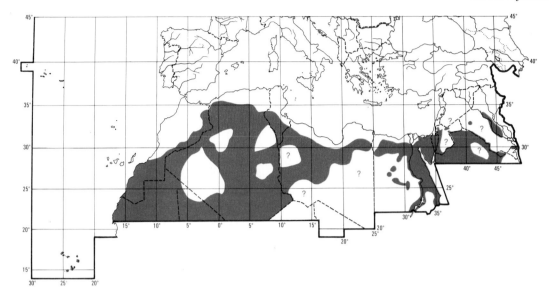

opment of agricultural settlements, army camps, and rubbish dumps (Shirihai in press). ALGERIA. Apparently extending range to north (EDHJ).

**Population.** No information on changes.
CAPE VERDE ISLANDS. Common (CJH).

**Movements.** Essentially sedentary over much of range, but some northern areas in (central Asia) vacated, and some seasonal movements reported elsewhere. No studies of ringed birds.

In Israel, most of adult population resident, but non-breeders (mainly immatures) make local movements to feeding sites and roosts, and sometimes longer seasonal or nomadic movements. Small numbers move beyond desert zone into Mediterranean zone, in autumn and winter and occasionally in summer. Longer-distance movements recorded at Elat, southward mainly July–September and northward February–March. (Paz 1987; Shirihai in press.) Occasional longer movements in Levant and Middle East also evident from records of vagrants in Syria and south-east Turkey (Jakobsen 1986; Hollom *et al.* 1988). In eastern Saudi Arabia, where breeding starts November, recorded wandering outside breeding areas mostly during hottest part of summer, July–August (J Palfery). Resident in northern Yemen (Brooks *et al.* 1987), Iran (D A Scott), and Afghanistan (S C Madge).

Most birds leave Kazakhstan, but some winter in southern parts of Kyzyl-Kum, Muyunkum, and Lake Balkhash area; not known whether these are local birds or immigrants from further north. Leaves Aral Sea region from beginning October, but somewhat earlier further east in Betpak-Dala and northern Lake Balkhash area; returns to south-west in 1st third of March, and to east in last third of March (Korelov *et al.* 1974). Present all year in desert regions of Turkmeniya, but tends to move south from northern areas in severe weather (Rustamov 1954, 1958). At Chokpak pass (foothills of western Tien Shan mountains), occasionally occurs during dispersal, with records in March, April, and October (Gavrilov and Gistsov 1985). In south-east of range, apparently also makes wandering movements, e.g. uncommon winter visitor to Quetta (western Pakistan), arriving November, and recorded straggling to Peshawar valley (northern Pakistan) (Meinertzhagen 1920; Ali and Ripley 1972).

Populations of North Africa and Afrotropics apparently essentially sedentary, with no reports of movements, e.g. from north-west Africa, Cape Verde Islands, Egypt, or Kenya (Heim de Balsac and Mayaud 1962; Goodman and Meininger 1989; Lewis and Pomeroy 1989; C J Hazevoet); local or short-distance movements reported in Mauritania, Mali, and Sudan (Lamarche 1981, 1989; Nikolaus 1987). DTH

**Food.** Ground-dwelling invertebrates, small vertebrates, and carrion; some fruit, grain, and other seeds. Forages generally in open ground, either cultivated or uncultivated, on soil, sand, stones, and short, often grassy, vegetation where sometimes takes contents of birds' nests (Rustamov 1954; Valverde 1957; Bel'skaya 1963; Ilani and Shalmon 1986; Nørrevang and Hartog 1984); often feeds on rubbish-tips, dung-heaps, and animal carcasses (Bannerman 1953*b*; Valverde 1957; Ewins 1979; Walker 1981*a*; Cornwallis and Porter 1982; Phillips 1982; Jakobsen 1986); sometimes feeds on back of camel *Camelus* or donkey *Equus* (Valverde 1957; Heim de Balsac and Mayaud 1962; Brooks 1987; J Palfery); occasionally feeds along shorelines (Bourne 1955; Goodman and Meininger 1989). Forages generally in pairs or flocks (Nørrevang and Hartog 1984); in Yemen, 50–100 in urban areas and on rubbish dumps (Cornwallis and Porter 1982; Phillips 1982; Brooks *et al.* 1987); in Morocco, 30 at swarm of Orthoptera (Brosset 1956). Locates feeding sites (e.g. carcasses) visually in flight, sometimes patrolling large areas of desert (Frazier *et al.* 1984; Paz 1987). Walks on ground, only hopping just before flight (Alexander 1898a; Valverde 1957; Paz 1987); occasionally flies low from one patch to another (Jakobsen 1986). Feeding action in newly turned field has been compared to Rook *C. frugilegus* (Nørrevang and Hartog 1984); also turns stones with bill, and searches in dung for insects (Brosset 1956; Valverde 1957). At fresh carcasses recorded feeding preferentially on eyes; may remain for whole days at carrion, where has been recorded catching large flies (Geyr von Schweppenburg 1918; Koenig 1920; Valverde 1957). When feeding on locusts (Acrididae) on ground, individuals may cooperate, taking turns to act as beaters (Alexander 1898a); also recorded catching locusts *Schistocerca* and termites (Isoptera) in flight with feet and eating them in flight (Valverde 1957; Symens 1990). Perches and hangs head-down on camel and donkey to pick ectoparasites such as tick *Ixodes* from neck, back, sides, legs, and occasionally head; also feeds at cuts and sores (Geyr von Schweppenburg 1918; Koenig 1920; Valverde 1957; Goodwin 1986; Brooks 1987). Catches and kills small vertebrates with bill, ripping larger prey into pieces before eating; hunts small mammals at dusk; in Israel, flock repeatedly flushed and chased hare *Lepus* with low fast flight, attempting to strike prey with feet (Ilani and Shalmon 1986); also recorded chasing hare with eagle (Accipitridae) (Ticehurst and Cheesman 1925); in North Africa, recorded attacking new-born, and occasionally adult, gazelles (Valverde 1957). Takes exhausted migrant birds and plunders nests (Geyr von Schweppenburg 1918); in Arabia, recorded systematically taking eggs of Western Reef Heron *Egretta gularis* (Goodman and Storer 1987). Feeds on dates *Phoenix dactylifera* directly from tree (Geyr von Schweppenburg 1918). To scavenge, will follow humans passing through territory and may tear open bags on pack animals to get at foodstuff (Goodwin 1986; Paz 1987). Recorded hiding and recovering food in captivity (Koenig 1920; Goodwin 1986). Drinks from any available water, but many individuals seem not to drink, presumably obtaining sufficient water from food (Valverde 1957; Marder 1973).

Diet in west Palearctic includes the following. Invertebrates: locusts and crickets (Orthoptera: Acrididae), beetles (Coleoptera), termites (Isoptera), ticks (Acari), snails (Mollusca). Also lizards *Agama*, small snakes, birds and their eggs, small mammals up to size of hare *Lepus*, and stranded fish. Plant material includes seeds of *Lantana* (Verbenaceae) and grasses (Gramineae) including maize *Zea* and rice *Oryza*, and fruits of date palm *Phoenix*. Also takes offal, carrion, and undigested food from animal dung. (Hartert 1915; Koenig 1920; Brosset 1956; Naurois 1969; Goodwin 1986; Paz 1987.) Diet in Kazakhstan and central Asia well known (for which see Dementiev *et al.* 1953,

Rustamov 1954, 1958, Varshavski and Shilov 1958, and Bel'skaya 1987). In North Africa, of 50 pellets taken from roosts, 65% contained fruit, 30% barley, 10% beetles, 5% crickets, 2% snails, 1% lizards, and 1% other seed; of 2 stomachs from birds which had been feeding on rubbish dump, one contained beetles and remains of fish, the other mammal fur and bones (Valverde 1957). On Cape Verde Islands, of 2 stomachs, one contained Orthoptera, the other grains of maize (Naurois 1981).

No information on food of young in west Palearctic. In Turkmeniya, of 225 pellets taken from unknown number of nests, 70·7% contained remains of insects, principally beetles (Carabidae and Tenebrionidae), and Orthoptera; 51·5% contained small mammals, especially gerbils (Gerbillinae), 35·1% Arachnida, predominantly Solifugae; lizards, tortoises, snakes, birds, and carrion all recorded in less than 10% of pellets (Bel'skaya 1963). Also in Turkmeniya, of 99 pellets at 3 nests, 79% contained remains of insects, especially Carabidae and Tenebrionidae, 43% contained Solifugae, 39% small mammals, 12% reptiles, and 10% birds and their eggs (Bel'skaya 1987, which see for comparison of diet with Long-legged Buzzard *Buteo rufinus*). WGM

**Social pattern and behaviour.** Literature scanty, and much of it implies behaviour similar or identical to Raven *C. corax*, but without giving details. Koenig (1905) described breeding in captivity with comments on behaviour; other notes from Egypt (Koenig 1920), Israel (Pitman 1921; Paz 1987; Shirihai in press), Uzbekistan (Lakhanov 1967, 1977), Turkmeniya (Rustamov 1954, 1958, 1984), Kazakhstan (Korelov *et al.* 1974), and in review by Goodwin (1986). Unpublished observations from eastern Saudi Arabia (J Palfery), Cape Verde Islands (C J Hazevoet), western Sahara, and (on *edithae*) northern Kenya (D T Holyoak).

1. At all seasons occurs singly, in pairs, and (where numerous) in flocks. In Israel, adults mainly in territorial pairs all year, while remainder of population, especially immatures, lives in flocks of variable size which tend to congregate at food sources (Shirihai in press). Flocks there may number hundreds at communal roosts (see Roosting, below), with maxima of 950 and 1200 recorded (Paz 1987; Shirihai in press). In Cape Verde Islands, flocks of 200–300 reported, with *c.* 1000 at communal roost (C J Hazevoet). In Kazakhstan, gregarious in autumn and winter, with flocks at carrion, although more often seen singly or in pairs; in mid-August, larger flocks of up to 100, either of unpaired immatures or birds dispersing south-west to winter quarters (Korelov *et al.* 1974). In Saudi Arabia, usually in pairs or small groups, but large flocks noted June–September, typically near water, including records of 500 and 900 (Jennings 1980, 1981a). In eastern Saudi Arabia, flocks of up to 60 occur from late May to end of August, but especially July–August when birds are largely absent from breeding areas, to which they return in late October and early November (J Palfery). In northern Oman, flocks occur at refuse tips from September onwards, but birds disperse into foothills and mountains in late December (Walker 1981a). In western Sahara (southern Morocco, Algeria, northern Niger), in summer and winter alike most sightings were of single birds and pairs, less often groups of 3; flocks of up to *c.* 26 seen repeatedly at a few good food sources (human refuse) but not in open desert. Similarly, in northern Kenya flocks of up to *c.* 33 seen where food plentiful, but elsewhere mainly 2(–4)

together. Largest flock congregated progressively when bread was provided, birds arriving singly, in pairs, or as small groups; within this flock many birds apparently paired, keeping close company and sometimes Allopreening (see Heterosexual Behaviour, below). (D T Holyoak.) Flocking birds regularly mix with Fan-tailed Raven *C. rhipidurus* (Jennings 1981*a*; Paz 1987). At Goulimine (southern Morocco), 70 *C. ruficollis* recorded in mixed flock with 20 *C. corax*, mobbing Short-eared Owl *Asio flammeus* (Smith 1965: see part 2). In northern Niger, individuals and pairs several times seen associating with groups of Pied Crow *C. albus* (D T Holyoak). Several species of vultures (Accipitridae) said to follow *C. ruficollis* to carrion, but competing vultures are harassed or attacked at carcasses (Paz 1987). BONDS. Mating system and nature of pair-bond not studied in detail, although birds commonly in pairs and thus presumably essentially monogamous. In Kazakhstan, adults remain in pairs in winter, although ♂ and ♀ sometimes far apart (Korelov *et al.* 1974). Rustamov (1954) recorded ♂ remaining in area 'searching for mate' for several days after ♀ shot, suggesting close pair-bond. Migrants returning to northern Aral Sea region (Kazakhstan) in March are already paired (Varshavski and Shilov 1958), suggesting long-lasting bond. Nest built by both sexes (Rustamov 1958; Jennings 1987). Incubation entirely by ♀ (Goodwin 1986) or mainly so (Dementiev and Gladkov 1954; Rustamov 1954; Paz 1987). Nestlings fed and tended by both parents (Bel'skaya 1963), possibly also by helpers (Hartert 1915) until after fledging. Duration of post-fledging dependence not known (but see Relations within Family Group, below). Age of first breeding apparently unconfirmed in wild: Rustamov (1958) gave 2nd year, 'as in *C corax*'; Korelov *et al.* (1974) gave at least 2nd year; Koenig (1905) reported first breeding of captive pair in 6th year. Non-breeding adults apparently occur regularly in areas of high density in Israel (Shirihai in press). BREEDING DISPERSION. Territories evidently defended in some areas, but very little information available on size or purpose. In Jordan, 200 nests of *C. ruficollis* and *C. corax* along 320 km of telegraph poles (Hardy 1971), implying territory size smaller than is usual in *C. corax*. Minimum distance between nests in Israel *c.* 2 km (once 1·2 km) and 'average' 3·3 km (Shirihai in press). In Israel, Paz (1987) noted increase in territorial activity at beginning of breeding season (late January and February). In south-west Kyzyl-Kum (Uzbekistan), 5 nests 2-10 km apart, within area of *c.* 50 km². No territory defended at these low densities. Birds fed both in immediate vicinity of nest and up to 3 km from it. (Lakhanov 1967.) In Turkmeniya, Rustamov (1958) recorded nests at least 4-7 km apart, within home-range of several km² used for feeding. In northern Aral Sea region, breeding birds hunted up to 8-12 km from nest (possibly as much as 15 km away at times), although nests sometimes only 3-4 km apart (Varshavski and Shilov 1958). Nests 4-5 km apart in northern Aral Sea region, 3 km in northern Kyzyl-Kum, and 10 km at Taup on northern edge of range (Korelov *et al.* 1974). See Rustamov (1954, 1984) for rates of occurrence at various times of year along transects in Kara-kum (Turkmeniya). Densities in this region, spring: in sand-clay desert typically 0·1-1·0 bird per km², in favourable habitat up to 3 per km 2; in clay desert less than 0·1 per km², near oases 0·1-1·0 per km² (Rustamov 1984). Nest-sites used over several years (Rustamov 1958), but not known whether same pair involved. In Kazakhstan, 'territory' taken up soon after birds return in spring. (Korelov *et al.* 1974.) ROOSTING. In most regions, roosts on rock ledges or in trees. In Israel, commonly also uses overhead wires (Paz 1987; Shirihai in press). Roosting on telegraph pole reported from Turkmeniya (Rustamov 1954). Flocking birds roost communally, and roosts in Israel may contain from tens of birds to over 1000. Largest concentrations occur July-February

and involve mainly immatures which do not hold territories, but some communal roosts remain during breeding season, when non-breeding adults use them. (Paz 1987; Shirihai in press.) In Cape Verde Islands, *c.* 1000 counted at roost (C J Hazevoet).

2. Can be bold and fearless where not persecuted by man, but wary and suspicious if aware of being watched and becomes very shy where shot at (Goodwin 1986). In Israel, territorial birds will accompany military convoys and groups of hikers passing through their territories, to pick up scraps (Paz 1987). In Turkmeniya, shy, wary, and not allowing close approach, both in remote deserts and areas where it frequently encounters man (Rustamov 1954). In eastern Saudi Arabia noted on several occasions to fly or circle over observer sitting or lying on ground, before continuing in flight (J Palfery). Reactions to predators poorly described (see also Parental Anti-predator Strategies, below). In southern Morocco, Smith (1965) described attack by mixed flock of 70 *C. ruficollis* and 20 *C. corax* on *A. flammeus* which was forced to ground and its wings were pulled. Noted attacking Tawny Eagle *Aquila rapax* in Kenya (M E W North). Recorded mobbing or attacking hare *Lepus* in Israel (Ilani and Shalmon 1986). FLOCK BEHAVIOUR. Poorly known. In northern Kenya, birds feeding on bread mostly maintained individual distance of several body lengths, but less in apparently paired individuals (D T Holyoak). Often flies high, and in Israel reported to wheel *c.* 500-600 m above ground in hottest part of day (Paz 1987). Similarly, in eastern Saudi Arabia in spring when young still in nest, adults sometimes noted soaring in middle of day to great heights, such that they were only just visible (J Palfery). In eastern Saudi Arabia, aerial acrobatics noted from small groups and flocks of up to *c.* 40 birds, including chasing, spiralling upwards in thermals and descending (often in pairs) in series of tumbling dives and swoops (J Palfery). Aerial 'play' includes fast spiralling descent from great height (Rustamov 1954). At Tamanrasset (Algeria), winter, observed flying *c.* 8 km from oasis towards roost in mountains *c.* 30 min before sunset (Gaston 1970). In Israel, arrives at communal roosts singly and in flocks (Shirihai in press). In eastern Saudi Arabia, pairs noted emerging from roost-sites at first light, *c.* 10-15 min before sunrise, then perching on rocks and shrubs and calling noisily (J Palfery). Rather quiet at communal roosts in Morocco and Egypt, giving few and mainly subdued calls (P A D Hollom). SONG-DISPLAY. Not reported. ANTAGONISTIC BEHAVIOUR. Threat and submissive displays apparently not described. ♂'s advertising display apparently much like that of *C. corax* (Koenig 1905, 1920). HETEROSEXUAL BEHAVIOUR. No details of pair-bonding behaviour, pair-formation, or sexual displays. Pair reported rolling in 'display flight' in Algeria in February (E D H Johnson). Sexual (pre-copulatory) display of ♂ said to be much like *C. corax*, but no details (Koenig 1905). Allopreening of head and neck of apparent mates (Fig A) seen repeatedly in feeding flock

A

in northern Kenya (D T Holyoak). Courtship-feeding probably occurs: in eastern Saudi Arabia during incubation period, one member of pair crouched beside mate with wings drooped and fluttering, as if begging; at one stage bills touched, but observer too distant to tell if food transferred (J Palfery). Mating and nest-site selection not described. RELATIONS WITHIN FAMILY GROUP. For role of sexes in brood-care, see Bonds (above). Rustamov (1954) recorded ♀ shot at nest containing 4 well-feathered young; on later visit young were independent and flying well, although still accompanied by ♂, suggesting ♂ reared young alone after loss of mate. Development of nestlings in Uzbekistan described by Lakhanov (1967, 1977): eyes begin to open at 8 days, fledge at 37-38 days, but not able to fly properly until 42-45 days. In Israel, occasional families reported to remain together for a few months (Shirihai in press). In Kazakhstan, family reported to stay together long after fledging, young being fed by parents and taught to get their own food (Korelov et al. 1974). In Turkmeniya, Rustamov (1954) also recorded pair accompanying and 'defending' 4 fledglings that could fly well. ANTI-PREDATOR RESPONSES OF YOUNG. At 16-17 days, young crouch in nest, adopt threat-posture and (when large) even attempt to leave nest (Lakhanov 1977). PARENTAL ANTI-PREDATOR STRATEGIES. (1) Passive measures. In Uzbekistan very wary during incubation, flying off nest when man still some distance away (Lakhanov 1967). Similarly, in north-east Turk-meniya adults did not come within 200-250 m of nest while observers present (Dementiev et al. 1956). In eastern Saudi Arabia during incubation, one bird (presumed ♂) would 'stand guard' c. 20 m from nest (J Palfery). (2) Active measures: against birds. In eastern Saudi Arabia, adult seen to fly from nest to mob buzzard Buteo flying c. 1 km away. Egyptian Vultures Neophron percnopterus ignored, even when close to nest area. (J Palfery.) In Israel, Paz (1987) noted that any large bird approaching nest-site is attacked, and migrating raptors such as Buteo or kite Milvus may be killed, but confirmation needed. In Uzbekistan, attempts to distract or drive off predators when young in nest (Lakhanov 1967). (3) Active measures: against man. In eastern Saudi Arabia, whenever man intruded on breeding area '♂' would call a few times, but '♀' remained silent on nest until intruder was seen; she then slipped off nest and joined ♂ nearby. Both birds remained in vicinity, calling from perches and making brief flights, until intruder withdrew. At nest with young, one adult would fly in as soon as man approached nest area, alight on hill in sight of nest, and croak softly while intruder present. Sometimes both adults would fly overhead when man passed by. (J Palfery.)

(Fig by D Nurney from photograph by D T Holyoak.)   DTH

**Voice.** Quite freely used, but not studied in detail. Commonest calls of nominate *ruficollis* apparently lower pitched than those of small Kenyan race *edithae* (D T Holyoak). Most or all calls are less deep and croaking than Raven *C. corax* (e.g. Korelov et al. 1974, Mild 1990), thus suggestive of Carrion Crow *C. corone* or even Rook *C. frugilegus* (Etchécopar and Hüe 1967; Walker 1981a; Goodwin 1986). Reported to have less varied repertoire than *C. corax* (e.g. Hollom et al. 1988) and to be less vocal overall (e.g. Hüe and Etchécopar 1970, P A D Hollom); neither appears true of *edithae* in northern Kenya (D T Holyoak). In northern Niger, voice distinctly deeper than Pied Crow *C. albus*, at least when possible to compare common flight-calls of both species directly (Holyoak and Seddon 1991; D T Holyoak). However, calls recorded from *edithae* described as indistinguishable from *C. albus* (M E W North). Calls considerably deeper and harsher than Fan-tailed Raven *C. rhipidurus*. Scanty information does not allow detailed categorization of vocalizations, which are likely to be as complicated as in closely related *C. corax*.

CALLS OF ADULTS. (1) Karr-call. Harsh 'karr-karr-karr' from adult ♀ when observer near her young (Goodwin 1956). Koenig (1920) described 'korr-korr' and 'kuerk-kuerk' from flying birds. Fig I shows series of sounds given in display (no details); Fig II represents longer, loud, excited calls. Rustamov (1954) distinguished commonest call, a muffled 'kruk kruk kruk' (final 'k' not very clear), from a croaking or cawing 'kaarr' given during attack; both calls used alternately when disturbed. Similarly, Korelov et al. (1974) described 'kruk-kruk' and cawing 'kra-kra-kra'. However, intermediate sounds probably connect these 2 call-types. Adult perching near nest with young gave occasional soft croaks while human intruder present (J Palfery). Main call described for *edithae* as 'karr', not harsh and commonly given in short series in variety of contexts and apparently equivalent to call 2 (and perhaps also call 3) of *C. corax*. Quiet versions resemble calls of *C. frugilegus*. Excited birds gave louder versions which sound considerably harsher and shriller. (D T Holyoak.) From recording of *edithae*, Fig III shows typical low-pitched call, Fig IV a shorter, higher-pitched variant. (2) Knocking-call. In northern Kenya, birds in feeding flock gave short quick series of clucking, clicking, or gobbling sounds, much like shorter version of call 8 of *C. corax* (D T Holyoak). Similar sounds, heard as loose,

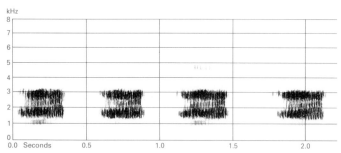

I  E D H Johnson  Algeria  January 1968

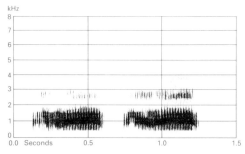

II  C Chappuis  Algeria  October 1971

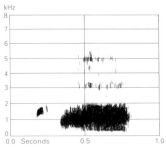

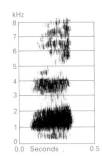

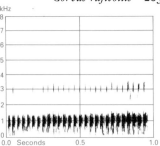

III  M E W North  Kenya  July 1955

IV  M E W North  Kenya  July 1955

V  M E W North  Kenya  July 1955

throaty rattle, recorded in Algeria (E D H Johnson, C Chappuis), e.g. loose, throaty rattle (Fig V). Possibly produced mechanically by tongue (see call 3 in *C. rhipidurus*). (3) Other calls. (3a) Koenig (1920) described variety of guttural, gurgling, and cawing calls from captive birds. Quiet, frog-like gargling or rattle (Fig V), heard as bird stretched its neck and arched its head and bill downwards, may be similar. (3b) Very quiet nasal 'nie-ar' from pair rolling in 'display flight' in Algeria in February (E D H Johnson). (3c) Raucous squealing in 'display' recorded in Algeria in October (C Chappuis). (3d) Meinertzhagen (1954) reported bell-like call. (3e) Low-pitched cawing (Fig VI), apparently similar in structure to call 1 but

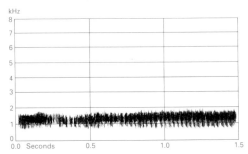

VI  C J Hazevoet  Cape Verde Islands  September 1988

exceptionally long, heard repeatedly from adult perched not far from nest.

CALLS OF YOUNG. No information.                               DTH

**Breeding.** SEASON. Cape Verde Islands: eggs laid mid-November to mid-April, after rains; of 60 clutches, 17% laid mid-November to late December, 67% January–February, 17% March to mid-April (Naurois 1969, 1981). North-west Africa: eggs laid early January to late March; October in southern Mauritania; some pairs fledge young before others lay (Geyr von Schweppenburg 1918; Brosset 1956; Valverde 1957; Heim de Balsac and Mayaud 1962; Goodwin 1986). Egypt: eggs laid late February to early April, young recorded early March to mid-May (Goodwin 1986; Goodman and Meininger 1989). Israel: laying mainly mid-February to early March, some late January; at Elat, hatching recorded late February to early May (Shirihai in press). Arabia: first eggs laid January, main

laying period February, some early March; probably later in mountains of south-west Arabia; young fledge late March and April, some in May (Jennings 1980, 1986b, 1988b; Stagg 1987; J Palfery). Weather in spring known to influence onset of breeding in Cape Verde Islands, Morocco, and Kara-Kum desert (Turkmeniya); tends to breed after rains, and earlier in warmer years (Rustamov 1954; Naurois 1981). SITE. Prefers crown of tree; in Arabia and Egypt, prefers *Acacia*, date palm *Phoenix dactylifera* is last resort (Goodman and Ames 1983; Jennings 1987); of 33 nests in Morocco, 70% in pistachio *Pistacia atlantica* and 15% in jujube *Zizyphus* (Brosset 1956); of 68 nests, Kara-Kum desert, 69% in saxaul *Arthrophytum* and 16% in other trees or bushes (Rustamov 1954). In arid areas of Arabia, sometimes in low bushes; one nest in thornless, leafless *Calligonum*, 80 km from any tree, hidden by lie of land from all directions (Ticehurst and Cheesman 1925; Jennings 1987). Where trees scarce, on cliffs or rocky outcrops, either on open or concealed ledge, or in cracks in cliffs; on Cape Verde Islands, uses coastal cliffs (Ticehurst and Cheesman 1925; Wilmore 1977; Gallagher and Rogers 1978; Naurois 1981; Jennings 1987; J Palfery). In Algeria, in areas with neither trees nor cliffs, nests on ground, amongst tamarisk *Tamarix* shrubs in dunes (Geyr von Schweppenburg 1918). In Arabia, on treeless Red Sea island, one nest recorded in old ground-level nest of Osprey *Pandion haliaetus* (Jennings 1987); in northern Kyzyl-Kum desert (Uzbekistan) nested in old nest of Long-legged Buzzard *Buteo rufinus* in saxaul tree (Stepanyan 1969a). Uses man-made structures and frequently recorded on side-extensions of telegraph poles and power pylons; of 33 nests in Morocco, 15% on telegraph poles (Brosset 1956; Naurois and Bonnaffoux 1969; Hardy 1971; Wallace 1983; Frazier *et al.* 1984; Goodwin 1986; Stagg 1987); in Arabia, often in open-ended barrel on pole used as oil-well marker (Jennings 1980, 1987; J Palfery); in Kara-Kum desert, of 68 nests, 15% on clay buildings (Rustamov 1954; Short and Horne 1981; Goodwin 1986; Jennings 1987). In trees, usually 1·7–7 m from ground (Ticehurst and Cheesman 1925; Rustamov 1954; Stepanyan 1969b; Naurois 1981); in bushes 0·8–1·4 m (Rustamov 1954; Jennings 1980). Nest: bulky structure of sticks similar to that of Raven *C. corax*, but smaller; lined thickly with plant fibre, grass, wool, hair, feathers, paper, etc. (Dementiev *et al.* 1953; Brosset 1956; Valverde 1957; Smith 1965; Wilmore 1977; Naurois 1981; Goodwin 1986). Dimensions of at least 11 nests, Turkmeniya and Uzbekistan: outer diameter 32–60 cm, inner diameter 20–30 cm, overall height 30–39 cm, depth of cup 11–15 cm (Rustamov 1958; Lakhanov 1967; Sopyev 1967); lining may be 15 cm thick (Dementiev *et al.* 1953). Building: by both sexes, sometimes simply relining old nest (Rustamov 1958; Jennings 1987); adds to structure throughout breeding season (Jennings 1988b; J Palfery). Captive pair built nest in 5 days (Koenig 1905). Of 200 *C. ruficollis* and *C. corax* nests destroyed, 70 rebuilt 6 hrs later (Hardy 1971). EGGS. See Plate 58. Sub-elliptical, smooth and glossy; pale blue with spots, small longitudinal streaks or scribbles, and small blotches of olive-buff to olive-brown and blue-grey; markings often pale and sparse (Harrison 1975; Goodwin 1986). Similar to *C. corax* but smaller and narrower (Hartert 1915). Nominate *ruficollis*: 45·3 × 31·7 mm (38·5–52·0 × 28·0–35·4), *n* = 140; calculated weight 23·6 g (Schönwetter 1984). In Morocco and Saudi Arabia, one smaller and better-marked egg sometimes present in clutch (Ticehurst and Cheesman 1925; Brosset 1956). Smaller egg in clutch in Saudi Arabia 32·0 × 23·5 mm (Ticehurst and Cheesman 1925). Clutch: 2–5 (1–7); in North Africa, clutches in semi-desert areas typically 5–6, those in true desert 2–3(–4); in Israel, 4 most frequent (Heim de Balsac and Mayaud 1962; Harrison 1975; Goodwin 1986; Shirihai in press). Of 20 clutches, Morocco: 1 egg, 10%; 2, 5%; 3, 10%; 4, 30%; 5, 25%; 6, 20%; mean 4·15. Clutches larger earlier in season: of 11 laid in March, mean 4·9; whereas of 14 laid in mid- to late April, mean 3·85 (Brosset 1956; Heim de Balsac and Mayaud 1962). Of *c.* 60 clutches, Cape Verde Islands, 2 of 6 eggs, most 4–5, 4 becoming more frequent towards end of season in March–April; clutches of 1–3 eggs probably replacements (Naurois 1969). Probably 1 brood; replacements possible (Bannerman 1953b; Naurois and Bonnaffoux 1969; Harrison 1975; Paz 1987). In Kyzyl-Kum, eggs laid 08.00–10.00 hours, over 5–6 days (Lakhanov 1967, 1977). In captivity, one clutch of 3 eggs laid one per day (Koenig 1905). INCUBATION. 18–23 days (Harrison 1975; Goodwin 1986; Shirihai in press). Mainly by ♀ (Rustamov 1954; Short and Horne 1981; J Palfery). Incubation starts with 1st, 2nd, or 3rd egg of clutch of 4–5, thus young may hatch asynchronously (Rustamov 1954; Lakhanov 1967); in captivity, started after 3rd (final) egg (Koenig 1905). YOUNG. Fed and cared for by both parents (Bel'skaya 1963), and possibly by helpers (Hartert 1915). FLEDGING TO MATURITY. Leave nest at 35–38 days; flying at 42–45 days (Lakhanov 1967; Shirihai in press). Age of first breeding not definitely known; see Social Pattern and Behaviour. BREEDING SUCCESS. In Morocco, of 13 successful nests, 3 young fledged from 2, 4 from 3, 5 from 8; mean 4·46 (Brosset 1956). In Kyzyl-Kum desert, of 33 eggs in 7 clutches, 91% hatched and 78% fledged (Lakhanov 1977). In Morocco, success lower in dry year due to shortage of food; pairs generally could not raise more than 1 young, but 2 pairs near human settlements both raised 3 (Olier 1958).

WGM

**Plumages.** ADULT. Entirely black, strongly glossed purplish-violet on cap, from lower mantle to upper tail-coverts, and on upperwing; hindneck, upper mantle, side of head and neck, chin, and throat less glossy, showing purplish-bronze lustre instead; gloss on remainder of underparts and on flight-feathers and tail purplish-violet, but somewhat less intense than on upperparts and upperwing, especially vent, under wing-coverts, and axillaries rather dull deep black with little gloss. Hidden feather-bases greyish-white or white. Much influence of bleaching and wear; in fresh plumage, more bronze lustre of neck often not contrasting with purplish-black of remainder of head and body (but dependent on light and on attitude of bird); *in worn plumage*, gloss of head and body more bluish-purple, and neck often distinctly brownish-bronze, still depending on light. Gloss on head and body of Raven *C. corax* similar to that of *C. ruficollis* or even stronger and more purplish, but gloss on neck, chin, and throat purplish-violet like remainder of body, not a duller bronze lustre; in worn plumage, head and neck of *C. corax* may become brownish, especially in south of range, but never as pale and bronzy as in *C. ruficollis*. NESTLING. Closely covered with dense and fairly long light brown-grey down; cheek, throat, and belly naked (C J Hazevoet). JUVENILE. Cap, hindneck, and side of head and neck black with faint blue gloss, hindneck distinctly blacker (less brown) than fresh adult; remainder of upperparts and upperwing like adult, but gloss more oil-blue, less intensely purplish-violet. Cheek and entire underparts dull greyish-black, feathers with glossy bluish fringes (brown-grey on belly, lower flank, and vent), longer feathers with blackish-grey of feather-bases visible, giving underparts mottled appearance; lesser and median upper wing-coverts greyish- or sooty-black, remainder of wing as adult, but gloss somewhat less intense, more bluish. Flight-feathers and tail often narrower and with more pointed tip than in adult; feathering of underparts shorter and less dense than in adult, vent and under tail-coverts appearing more fluffy. FIRST ADULT. Like adult, but juvenile flight-feathers, tail, greater upper primary and secondary coverts, and variable number of median and lesser upper wing-coverts retained, less strongly glossy, fading more rapidly to dull sepia-brown due to bleaching and wear; tips of flight-feathers and tail browner and more worn than adult at same time of year. In fresh plumage, gloss on head and body slightly less intense than in adult; gloss on centres of feathers of underparts more restricted, bordered brown along fringes and with more sooty grey of feather-bases visible, underparts appearing scalloped, less uniform than in adult. In worn plumage, head and body browner than in adult, especially neck and underparts. Ageing best done by contrast in colour and abrasion between older browner juvenile upper wing-coverts and tertials and newer blacker 1st-adult wing-coverts and scapulars (in adult, all feathers more or less uniform in colour and grade of wear).

**Bare parts.** ADULT, FIRST ADULT. Iris brown or dark brown. Bill, leg, and foot black, fissures between scutes of leg and foot as well as soles dark grey. Mouth dark grey. NESTLING. Bare skin yellowish-orange. Culmen and bill-tip greyish-black, remainder of bill yellowish-orange; gape-flanges orange-yellow with swollen pale yellow edges. Mouth orange, tongue with some dusky spots. Leg and foot yellowish-flesh-grey, scutes on front of tarsus and on upper surface of toes dark grey. Iris blue-grey once eyes open. JUVENILE. Like adult, but bill with traces of yellow-white flanges in first days after fledging, base of upper mandible and middle of lower grey, base of lower yellowish-flesh; mouth flesh-yellow; iris grey or brown-grey; soles pale grey or whitish. Not known at what age adult colours obtained. (C J Hazevoet; RMNH, ZFMK.)

**Moults.** ADULT POST-BREEDING. Few moulting birds examined. Timing in relation to nesting, and duration of moult, probably close to those of *C. corax*. Primary moult late June or July to October in southern Algeria, completed by August in Iranian Baluchistan (Hartert 1921-2). Single adult from Aïr (Niger), late June, in full moult, wing-coverts and tail largely new (Fairon 1975). Moult just completed in bird from Ennedi (Chad) on 19 September; all plumage new except for growing p9 and old p10 in single ♂ from Sinai, 27 October (ZFMK). In USSR and Iran, moult not started in May, one from late June in heavy moult, one half-way through moult on 10 July, several nearing completion in 2nd half of July, all moult completed September (Dementiev and Gladkov 1954). POST-JUVENILE. Head, body (sometimes excluding some feathers of vent, rump, or tail-coverts), and lesser and median upper wing-coverts (but sometimes retaining a few short rows of lesser coverts and a few to many juvenile median coverts). Starts soon after fledging, sequence as in *C. corax*. Timing depends on fledging date (mainly April-June in north of range, but January-July in south, depending on local rains). In Afghanistan, 4 in full moult late July (Paludan 1959).

**Measurements.** Nominate *ruficollis*. Wing, bill to skull, and tarsus of adults from (1) southern Algeria, northern Niger, and northern Chad, (2) Nile valley of Egypt and northern Sudan, (3) Sinai and Levant; other measurements for all areas combined, whole year; skins (RMNH, ZFMK, ZMA). Juvenile wing and tail include those of 1st adult, and include data from Afghanistan (Paludan 1959); bill (N) to nostril, bill depth at middle of nostril; exposed culmen on average 4·5 less than bill to skull.

| | | | |
|---|---|---|---|
| WING (1) | ♂ 404·2 (18·81; 8) 386-439 | ♀ 380·7 (11·96; 8) 363-397 | |
| (2) | 411·7 (13·15; 7) 396-436 | 396·1 ( 8·16; 5) 388-408 | |
| (3) | 384·0 ( — ; 2) 378-390 | 386·4 ( 8·59; 6) 374-397 | |
| JUV | 385·0 (20·55; 5) 367-416 | 362·0 ( 8·33; 8) 352-374 | |
| TAIL AD | 204·5 (10·86; 17) 186-226 | 197·0 ( 8·32; 20) 181-216 | |
| JUV | 200·4 ( 6·02; 5) 192-206 | 188·6 (10·61; 9) 174-204 | |
| BILL (1) | 71·7 ( 2·40; 8) 68·5-75·0 | 65·6 ( 2·56; 8) 61·7-68·9 | |
| (2) | 68·4 ( 1·00; 7) 66·4-69·5 | 65·2 ( 2·67; 5) 63·0-67·5 | |
| (3) | 66·7 ( — ; 2) 65·9-67·5 | 63·7 ( 2·14; 6) 61·4-66·4 | |
| BILL (N) | 45·0 ( 1·82; 17) 42·0-48·0 | 41·4 ( 2·04; 19) 38·5-45·0 | |
| BILL DEPTH | 22·6 ( 0·93; 16) 21·0-24·0 | 21·1 ( 0·95; 19) 19·5-23·0 | |
| TARSUS (1) | 70·7 ( 2·53; 8) 68·0-74·5 | 66·1 ( 1·74; 8) 63·0-68·4 | |
| (2) | 71·6 ( 2·19; 7) 68·4-73·8 | 69·6 ( 2·67; 5) 67·3-72·5 | |
| (3) | 66·2 ( — ; 2) 63·7-68·8 | 66·0 ( 1·68; 6) 64·0-68·5 | |

Sex differences significant, except for wing (3), juvenile tail, and tarsus (2) and (3).

Wing, bill to feathers of forehead, and tarsus of birds from (1) Cape Verde Islands, (2) Mauritania, Morocco, and Algeria; skins (Naurois 1981, including some data from Dekeyser and Villiers 1950).

| | | | |
|---|---|---|---|
| WING (1) | ♂ 379 (7; 16) 368-392 | ♀ 365 (10 ; 12) 347-378 | |
| (2) | 388 (-; 11) 360-410 | 364 (12 ; 4) 353-380 | |
| BILL (1) | 64·0 (3; 16) 58-70 | 59·5 ( 2·5; 12) 55-61·5 | |
| (2) | 66·3 (-; 11) 59-74 | 60·4 ( 2·2; 4) 58-62·5 | |
| TARSUS (1) | 66·0 (-; 16) 61-70 | 64·5 ( — ; 12) 61·5-66·5 | |
| (2) | 65·9 (-; 11) 58-70 | 63·4 ( 1·9; 4) 62-66 | |

Eastern Egypt and Gebel Elba, sexes combined: wing 376·6 (13·66; 7) 360-393, tail 184·3 (11·21; 7) 165-199 (Goodman and Atta 1987; Goodman 1984). For wing and bill length and depth of many samples, see Meinertzhagen (1921, 1926).

**Weights.** Nominate *ruficollis*. Algeria, January: ♀ 560 (ZFMK). Ennedi (northern Chad), April: ♂ 580 (Niethammer 1955). Aïr (northern Niger): ♀ 500 (Kollmansperger 1959); June, ♂ 595 (Fairon 1975). Israel: 550-830 (Paz 1987). Kazakhstan: ♂♂ 770, 795; ♀ 700 (Korelov *et al.* 1974). USSR: ♀♀ 500, 598, and 647 (Dementiev and Gladkov 1954).

**Structure.** Wing long, with relatively long tip and narrow base compared with *C. corax*. 10 primaries, p7-p8 longest or either one slightly shorter than other, or (occasionally) p8 up to 10 shorter; p9 20-40 shorter than longest (juvenile 15-30), p10 110-135 (juvenile 95-110), p6 10-25 (all ages), p5 60-85 (juvenile 45-70), p4 105-130 (juvenile 75-105), p1 170-200 (juvenile 130-160). Wing differs from *C. corax* mainly in relatively shorter p6, which is 15·0 (4·08; 10) 10-25 shorter than longest primary and 13·0 (4·32; 10) 6-20 shorter than p8 (in 30 *C. corax corax* and *C. c. tingitanus*, p6 shorter than longest primary by $7·8 \pm SD2·81$, range 0-12, and $4·3 \pm SD3·83$ shorter than p8, ranging from 12 longer to 10 shorter), and (in adult) in greater distance from p1 to wing-tip (ratio of this distance to wing length is 0·43-0·48 in adult, 0·40-0·44 in juvenile; in *C. corax*, 0·38-0·42 in adult, 0·36-0·41 in juvenile). Outer web of p6-p9 emarginated, inner web of p6-p10 with distinct notch. 12 tail-feathers, t6 41·5 (15) 32-56 shorter than t1 in adult, 30·5 (6) 24-35 in juvenile. Bill distinctly more slender than in *C. corax* (at least in western populations), ratio of bill length (to skull) to bill depth (at mid-nostril) 3·22 (16) 3·08-3·50 in western Sahara and 2·99 (20) 2·70-3·20 in Nile valley and Levant (in *C. corax corax* on average 2·86, in *C. c. laurencei* on average 2·90, range 2·78-3·03; in *C. c. tingitanus* on average 2·71, range 2·55-2·85); nasal bristles relatively shorter. Feathers of chin and throat relatively shorter (but longer and more pointed than in Pied Crow *C. albus*), tip more often tending to be bifurcated. Middle toe with claw 54·3 (10) 47-61; outer and inner toe with claw both *c.* 77% of middle with claw, hind *c.* 86%. Remainder of structure as in *C. corax*.

**Geographical variation.** Slight within nominate *ruficollis*. In sample examined (mainly RMNH and ZFMK), bill of population west from Libya and Chad tends to be longer and more slender than those from Nile valley eastwards. In birds examined by Meinertzhagen (1921, 1926), those from Egypt (except Sinai) and Khartoum largest but bill often rather slender (in contrast to ZFMK data above), and those of Sinai and Cape Verde Islands smallest, while birds from Socotra combine short wing with relatively large and thick bill. *C. r. edithae* from north-east Africa distinctly smaller than nominate *ruficollis*, wing 310-365 (smaller even than Pied Crow *C. albus*), bill relatively short (50-57), and feather-bases of neck on average whiter; may form separate species within *C. corax* species-group, which, besides *C. ruficollis*, also includes *C. albus*. See Goodwin (1986) for discussion of possible intermediates between nominate *ruficollis* and *edithae*, and comments doubting a report of sympatry of nominate *ruficollis* and *edithae*.

*C. ruficollis* often considered a race of *C. corax* (e.g. Hartert and Steinbacher 1932-8, Dementiev and Gladkov 1954, Ali and Ripley 1972), replacing *C. corax* in arid regions of northern Africa, in Middle East, and in west-central Asia. Some hybridization may occur in east, as some birds there difficult to allocate, but wide overlap exists in (e.g.) Turkmeniya, southern Iran, and Israel, while ranges touch in North Africa, all without clear intergradation. Ecology, voice, and morphology differ, and thus better considered separate species; see Vaurie (1954b), Smith (1965), Vuilleumier (1977), Desfayes and Praz (1978), and Paz (1987).

**Recognition.** Differs from all races of *C. corax* in more slender bill, with depth at middle of nostril less than 24 (♂) or 23 (♀) (in *C. corax*, ♂ over 25, ♀ over 23·5), and in different wing structure (see Structure). In Middle East, differs also from *C. c. corax* and *C. c. laurencei* in shorter bill and more slender foot. In North Africa, shows some overlap with *C. c. tingitanus* in measurements (though not in bill depth), but has wing often shorter and bill and tail longer.                                    CSR

## *Corvus corax* Raven

PLATES 11 and 15 (flight)
[between pages 136 and 137, and facing page 280]

Du. Raaf       Fr. Grand corbeau       Ge. Kolkrabe
Ru. Ворон      Sp. Cuervo      Sw. Korp       N. Am. Common Raven

*Corvus Corax* Linnaeus, 1758

Polytypic. Nominate *corax* Linnaeus, 1758, Europe and Asia from Scandinavia, Britain, and France east to Yenisey basin, grading into *kamtschaticus* between Yenisey and Lena basins and in area round Lake Baykal, south to Pyrénées, mainland Italy, Balkan countries, western Greece, north-east Turkey, Transcaucasia, northern Iran, basin of middle Volga, southern Urals, and Kazakhstan south to *c.* 51°N; grades into *hispanus* (but nearer nominate *corax*) on Sardinia and perhaps Corsica and Sicily; *hispanus* Hartert and Kleinschmidt, 1901, Iberia and Balearic islands; *laurencei* Hume, 1873 (synonym: *subcorax* Severtzov, 1872), eastern Greece (Thrace, Aegean islands, and Crete), Cyprus, Turkey (except north-east), Levant, Iraq, Iran (except north), and from lower Volga basin east through Transcaspia to plains and hills of eastern Kazakhstan (south of *c.* 51°N), Sinkiang (China), Afghanistan, Pakistan, and north-west India; *tingitanus* Irby, 1874, North Africa (Morocco to north-west Egypt); *canariensis* Hartert and Kleinschmidt, 1901, Canary Islands; *varius* Brünnich, 1764, Faeroes and Iceland. Extralimital: *principalis* Ridgway, 1887, northern and eastern North America and Greenland; *kamtschaticus* Dybowski, 1883, eastern Siberia, south to northern and western Mongolia, northern Manchuria, Amurland, and Hokkaido (Japan); *tibetanus* Hodgson, 1849, mountains of central Asia from Tien Shan and Pamir-Alay system to Himalayas and Tibet; *sinuatus* Wagler, 1829, from Rockies (western North America) to Nicaragua.

**Field characters.** 64 cm; wing-span 120–150 cm. As large as Buzzard *Buteo buteo*, differing from all other west Palearctic crows in deep, massive bill, flat head, shaggy throat, long, deep body, long, broad wings, strong legs and feet, and rather long, wedge-ended tail; up to 25% larger than Brown-necked Raven *C. ruficollis* and 30–35% larger than Carrion Crow *C. corone* and Rook *C. frugilegus*. Huge, majestic, and powerful black crow, with superb flight but at times playful nature. Cruciform silhouette when soaring. Sexes similar; no seasonal variation. Juvenile separable. 6 races in west Palearctic, of which North African and east Mediterranean somewhat distinct.

Adult. Moults: mostly July–October (complete). (1) North European and Siberian race, nominate *corax*. Medium-sized. All-black, glossed purple on crown, ear-coverts, and nape, mainly purple-blue on secondaries, wing-coverts, and from back to upper tail-coverts, reddish-purple on tail, reddish- to blue-purple on long, lanceolate feathers of shaggy throat and underbody. In fresh plumage, sheen as marked as *C. frugilegus*, but lost on wings and tail when worn and these can look distinctly brown in early summer. Nasal bristles, bill, and legs black. Bill 25% longer and larger than *C. corone*, with upper mandible noticeably deeper than lower, with high culmen and markedly decurved distal 30%. (2) Faeroes and Iceland race, *varius*. Larger than nominate *corax*, with up to 10% longer wings and 15–20% longer bill. (3) North African race, *tingitanus*. Slightly smaller than nominate *corax*; even higher culmen produces stumpy look to 10% shorter bill. Plumage highly glossed but with much olive, producing 'oily' sheen. Throat feathers short. (4) East Mediterranean race, *laurencei*. Slightly larger than nominate *corax*. When fresh, plumage glossed more steely-blue but when worn becoming distinctly brownish on nape, mantle, and throat, and so suggestive of *C. ruficollis*.

Unmistakable if seen well in north or west of range, with huge, powerful form unique among *Corvus*. Glimpsed or only seen at distance in North Africa and Levant, subject to confusion with *C. ruficollis*; separation best based on comparison of size and structure (not plumage, as both *laurencei* and *C. ruficollis* show brownish nape). Best characters of *C. corax* are (1) deeper and longer bill and head (slighter and often drooped in *C. ruficollis*), (2) larger and broader-based and broader-tipped wings (narrower-based and more pointed in *C. ruficollis*), and (3) fuller, wedge-ended tail (narrower-based and with central feathers just protruding in *C. ruficollis*). Flight powerful and majestic; often performed at great height and includes more gliding (including hanging on the wind) and soaring than any other crow, so inviting confusion with broad-winged raptors. Active flight produced by series of powerful, rather ponderous, noticeably regular but not deep wing-beats, interspersed with glides and occasional falls and rises. Near cliffs, may glide on updraughts for long

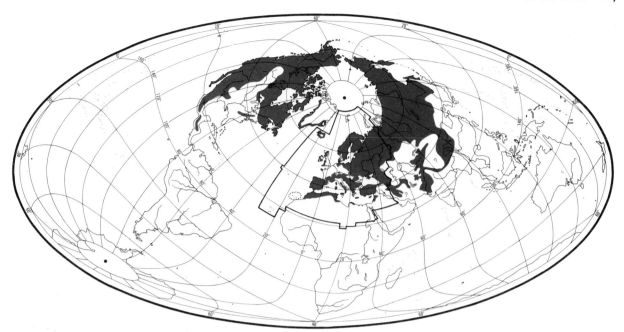

periods, being able to side-slip on the wind for long distance; at all times of year, particularly in early breeding season, may indulge in aerial play, with rolls, tumbles, and dives all performed with evident enjoyment. Rather more agile in air around food sources, using lighter, shallower wing-beats and sweeps, veering into wind to manoeuvre to landing spot. Gait a strong walk, varied by forward leap or sidling hop. Stance on ground noticeably lengthy and horizontal, on branch more upright, but rarely drops tail.

Voice includes 3 distinctive calls, all with strangely percussive quality lacking in smaller *Corvus*: in flight, deep, rather hollow 'pruk pruk'; also in flight, less full and higher-pitched 'toc-toc-toc' or 'kok kok kok'; from nest or in greeting, resonant 'corronk' or 'karronk'.

**Habitat.** So wide-ranging that concept of habitat is hardly applicable. Breeds almost throughout west Palearctic, except for certain densely settled and cultivated regions, from Arctic to tropics, even to July isotherm of 3°C. Ranges from sea-level to *c.* 2400 m in Alps and Altai, and in Tibet exceptionally to 5800 m (Ali and Ripley 1972). Overriding requirements are for nest-site of difficult access, normally on rock-face or tall tree, and wide, largely undisturbed foraging area with tracts of open surface of any kind on which long-range food-gathering, often involving high flights, can be practised. Thus avoids interior of large or dense forests, scrub woodland, thickets, shrubby terrain, wetlands with tall aquatic vegetation, orchards, plantations, field crops, and intensively farmed or grazed lands. Coasts with cliffs, even in windy and chilly climates, often satisfy, especially where they afford respite from human persecution. Ascends to snowline in summer in

Caucasus and will nest in towers of high buildings and in (e.g.) oak *Quercus* close to a house, where not disturbed. Feeds at waste dumps and slaughterhouses as well as on river banks and seashores and open steppes (Dementiev and Gladkov 1954). In Switzerland, prefers steep inaccessible valley walls near open land, usually below treeline and often among boulders or scree (Glutz von Blotzheim 1962); nest-sites are usually well sheltered from rain and snow on rock faces; in summer and autumn forages on open ground, often up to 3000 m in Canton Bern (Lüps *et al.* 1978). In Britain, recolonization of lost ground is accompanied by return to tree-nesting, mainly when cliff or crag sites lacking; few nests above 600 m, and many are along coast; in uplands, prefers areas where sheep carrion plentiful (Sharrock 1976). In 16th century, common in Scottish towns, and bred on rocks of Arthur's Seat (Edinburgh) until 1837; nested up to 600 m (Baxter and Rintoul 1953). Breeds freely on oceanic islands. In North America, ranges over open treeless tundra, on treeless Aleutian Islands, and further south on mountains above *c.* 1000 m, as well as on rocky coasts and tree-clad islands; forages along sea coasts and river banks, and often on human refuse around camps and settlements where not molested; tolerates very hot climates, e.g. Death Valley (Bent 1946), in contrast to habits in west Palearctic where deserts and arid regions are not favoured. Uses airspace freely, for play as well as transit even above 6000 m in Himalayas (Ali and Ripley 1972).

**Distribution.** Recent expansion of range over much of northern and central Europe due to reduction of persecution and active conservation, including reintroduction in Belgium, Netherlands, and Germany.

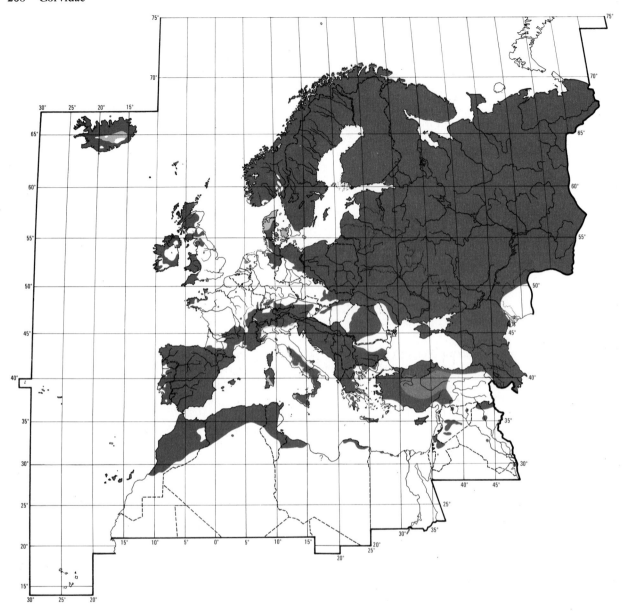

BRITAIN, IRELAND. Formerly widespread, but range much contracted due to persecution in 19th century. Considerable expansion this century. (Sharrock 1976.) FRANCE. Slight expansion of range following reduction of persecution (Yeatman 1976). Further expansion since 1976, mainly in Massif Central and inland areas of Bretagne (RC). PORTUGAL. Range may have been reduced in recent years, but still extensive (Rufino 1989). BELGIUM. Began to disappear as breeding bird from late 19th century; last record 1919 (but 1 isolated breeding record 1948). Reintroduced in east 1973-80 (Delmotte and Delvaux 1981), and bred successfully (Delvaux 1983). NETHERLANDS. Sharp decline in early decades of 20th century; last recorded breeding 1944. Reintroduction star-

ted 1966 and first free-living breeding pairs observed 1976; after a few years of increase population crashed, but recovered 1983-4; 17 pairs 1988, 21 in 1989, 34 in 1990. (Renssen 1988, 1991a.) NORWAY. Recent spread in south; now breeding in nearly all parts (VR). FINLAND. Formerly absent over extensive areas in south; has recently spread (OH). EAST GERMANY. Has spread since 1945 as result of protection, after apparent extinction as breeder (SS); see Glandt (1991). POLAND. Formerly widespread; then range retracted, due to persecution, to mountains in south and area east of Vistula. Recovery from 1930s with recolonization, from north-east, of all areas. (AD, LT.) CZECHO-SLOVAKIA. Expansion from eastern Slovakia from 1944, still continuing (KH). AUSTRIA. Expansion in north-east, and since *c.* 1970 has begun to breed in Wienerwald south-west of Vienna (H-MB). SWITZERLAND. Before 1950 breeding restricted to Alps; has since spread to Jura and lowlands (RW). HUNGARY. Major decline to *c.* 10 pairs in 1970s. Recovered since, probably due to ending of poisoning for pest control; now *c.* 100 pairs. (G Magyar.)

Accidental. Malta.

**Population.** Recent increases over much of northern and central Europe due to reduction of persecution; for major review, see Glandt (1991). Decline reported only in Israel.

FAEROES. 150-350 pairs (Bloch and Sørensen 1984). BRITAIN, IRELAND. Estimated *c.* 5000 pairs (Sharrock 1976). FRANCE. 100-1000 pairs (Yeatman 1976), subsequently increasing (RC). NETHERLANDS. 17 pairs 1988 (Renssen 1988). GERMANY. 6000 pairs (G Rheinwald). East Germany: 3400±900 pairs (Nicolai 1993); has increased since 1945 (SS). DENMARK. About 200 pairs 1985, increasing (TD). SWEDEN. Marked increase began in 1960s (LR). Estimated 10 000 pairs (Ulfstrand and Högstedt 1976). In 1977, population of southern third of Sweden estimated at 4300 pairs. Present total population estimated 15 000 pairs. (LR.) FINLAND. Marked increase in last 20 years, especially in south; estimated *c.* 10 000 pairs (Koskimies 1989; OH). POLAND. Recovery began 1930s; still increasing (AD, LT). CZECHOSLOVAKIA. Marked increase after 1944; still increasing (KH). AUSTRIA. Increasing (H-MB). SWITZERLAND. Increasing since 1950 (RW). ITALY. Increasing in pre-Alpine region (PB). BULGARIA. Marked increase since 1950s (former decline attributed to poisoned bait against wolf *Canis lupus*) (TM). LATVIA. Decreased through persecution in 2nd half of 19th century; recovery began 1940s. Estimated 600-700 pairs 1954; several thousand in 1980. (Vīksne 1983.) ISRAEL. Common up to 1960s, with at least a few hundred pairs. Population then drastically reduced by pesticides and perhaps other causes; recovered somewhat in 1980s, but still only a few pairs. (Shirihai in press). CANARY ISLANDS. Declined, possibly due to changes in agriculture and persecution. Probably now over 100 pairs on each of Fuerteventura, Gran Canaria, La Gomera, and El Hierro; fewer than 100 pairs on Lanzarote, Tenerife, and La Palma. (Nogales 1992.)

Survival. Britain: mortality in 1st calendar year 37%, in 2nd year 52%, in 3rd-5th years 50% (Holyoak 1971). Europe: mortality in 1st year of life 63·5% (Busse 1969). Oldest ringed bird 16 years (Staav 1983).

**Movements.** Populations south of *c.* 60°N essentially sedentary, but some immatures make extensive movements; northern populations mainly sedentary and dispersive, but more prone to make southward movements in winter. For review, see Zink (1981).

Nominate *corax* (most of Europe and northern Asia) sedentary towards south, but immature birds make dispersive movements, apparently chiefly in 1st autumn and winter. In Britain, no ringing recoveries involving movements to or from continental Europe (Mead and Clark 1987, 1991). Of 147 recoveries of birds ringed in Britain, nearly all were from same hill areas, with longest movements 304 km (Anglesey to East Lothian), 192 km, 168 km, and 122 km; one bird moved from Northern Ireland to Scotland. About half of all birds recovered were over 32 km from ringing site by 1st winter. (Holyoak 1971.) Bird ringed as nestling in Wales recovered in Ireland (170 km north-west) in December of its 2nd year (Spencer and Hudson 1978b); also, exceptional recovery in Suffolk (eastern England) of bird ringed 551 km north-west in Antrim (Northern Ireland) (Mead and Clark 1987). Local ringing studies showed median distance of recoveries of birds ringed in northern Wales of 23 km (*n* = 47) (Dare 1986), and movements of similar extent in central Wales (Davis and Davis 1986). In Spain, most distant record is of bird ringed as nestling in Coto Doñana in south, and recovered 45 km south-east in its 3rd year (Zink 1981). In France, occurs more widely in winter than summer and makes some altitudinal movements in montane regions; most distant recovery, in Massif Central, was 50 km west of ringing site (Zink 1981; Yeatman-Berthelot 1991). Recoveries from central Europe likewise show few longer movements, and no clear overall directional trends; those over 100 km are from Swiss Alps (1), Bayerischer Wald in southern Germany (1), Mecklenburg in eastern Germany (4), and western Ukraine (1) (Zink 1981). Recovery in Bulgaria of bird ringed 490 km north-east on north coast of Black Sea (Zink 1981) is thus apparently exceptional. Study of marked birds in Switzerland showed flocking immature birds move around area as large as 10 000 km² (Huber 1991).

In north of range, nominate *corax* more prone to make medium-distance movements. In Fenno-Scandia, longest movements chiefly involve birds from northern areas, and extend south or SSE to 350 km (Norway), 510 km (Sweden), and 380 km (Finland); none recovered overseas. Recoveries following post-fledging dispersal of ringed nestlings showed considerable movement in all directions, but longer movements (over 200 km) were mostly between south-west and south-east (thus showing tendency towards true migration), whereas movements of 100-200 km were chiefly between north-west and north-east. (Zink 1981.) However, only minority in these populations moves far: only 13 of 270 Finnish ringing recoveries show movement of over 200 km (Saurola 1977). In Latvia, most recoveries of juveniles are of 50-200 km (Dementiev and Gladkov 1954). Chiefly resident in USSR; tends to frequent towns in autumn and winter, e.g. in western Siberia and north-west Caucasus (Johansen 1944; Dementiev and Gladkov 1954; Polivanov and Polivanova 1986). Vagrants recorded north of range in Spitzbergen and Novaya Zemlya (Vaurie 1959).

*C. c. varius* (Iceland and Faeroes) resident with some dispersal. In Iceland (205 recoveries), median distance between natal and recovery site 67 km in south-west, 36 km in north-east and 77·5 km in east; longest movement 386 km. Most summer recoveries were inland, but most winter recoveries in coastal regions. (Skarphéthinsson *et al.* 1992.)

*C. c. canariensis* (Canary Islands) and *tingitanus* (North Africa) sedentary except for wandering movements of some young birds (Heim de Balsac and Mayaud 1962; Etchécopar and Hüe 1967).

*C. c. laurencei* (eastern Greece and Crete east to north-west India) sedentary in Levant and Middle East, except for wandering movements of immatures (Hüe and Etchécopar 1970; Flint and Stewart 1992; Shirihai in press); in Turkey, more widely dispersed outside breeding season (Beaman 1978). In Pakistan and north-west India, occurs in varying numbers in non-breeding areas in winter (Ali and Ripley 1972). Vagrant to Kashmir (Vaurie 1959). Himalayan race *tibetanus* makes altitudinal movements in at least some regions. In south-east Kazakhstan, moves down into foothills during autumn and winter, reaching human habitations and even desert steppe, occurring there with Brown-necked Raven *C. ruficollis* (Korelov *et al.* 1974). At Chokpak pass (western Tien Shan, Kazakhstan), occurs regularly on passage in small numbers: in autumn, from end of August to end of October, with main passage 16–25 October; in spring, from 1st week of March to mid-May (Gavrilov and Gistsov 1985). In Nepal, makes local movements to lower altitudes in winter (Inskipp and Inskipp 1985). Reported to descend south of main Himalayan axis to *c.* 3000 m in exceptionally severe winters (Ali and Ripley 1972).

East Asian race *kamtschaticus* apparently dispersive. On Kamchatka peninsula, more widespread winter than summer (Lobkov 1986). Visits Ussuriland in small numbers on passage and in winter, mid-November to mid-March (Panov 1973*a*). Winter visitor to Hopeh (north-east China) and Hokkaido (northern Japan), especially in severe winters (Vaurie 1959; Brazil 1991).

North American races *principalis* and *sinuatus* also essentially sedentary with dispersive movements (Godfrey 1986; Heinrich 1990). In Canada, winters within breeding range north at least to Melville Island and northern Baffin Island, but extends somewhat south of breeding range in Alberta and Saskatchewan (Godfrey 1986). Some evidence for partial migration south from northern Greenland (Bent 1946).                                                               DTH

**Food.** Plant and animal material, taken opportunistically; animal food may be killed with powerful bill, or scavenged as carrion, refuse, etc.; also robs nests and takes invertebrates (especially molluscs on shore); plant material mainly cereals and fruits. Where carrion plentiful, usually takes food mostly by scavenging. (Coombs 1978; Goodwin 1986; Marquiss and Booth 1986; Cugnasse and Riols 1987.) Usually forages on ground away from cover, and commonly on rubbish-tips, near slaughterhouses, on tide-

line, etc., and, where not persecuted, scavenges boldly around dwellings, particularly of nomadic herdsmen; will follow the plough, and frequently on fields where dung has been spread (Goodwin 1986; Sellin 1987; Hauri 1988*a*; Huber 1991). Often flies considerable distance in search of food, singly or in pairs; non-territorial individuals in North America can cover 1800 km² (Heinrich 1988*b*); in Wales, territorial pairs usually forage within 1·5 km of nest (Dare 1986). Searches for live prey and carrion from air or on ground; when actively hunting (e.g.) rabbits *Oryctolagus* and hares *Lepus* flies low over ground; when searching for nests of Rock Dove *Columba livia*, flies repeatedly in front of cliff, and when hunting on ground investigates crevices in rocks, etc., while walking (Lewis 1920; Owen 1950; Marr and Knight 1982; Goodwin 1986). Recorded foot-paddling in rough grass for worms (Ewins 1989). Harries sick and injured animals, even species not normally preyed upon, making darting lunges with bill, often aimed at eyes; also takes advantage of tourist refuse, road casualties, remains of raptor kills, and infestations of defoliating caterpillars, and will hawk for flying ants (Formicidae) (Coombs 1978; Marquiss *et al.* 1978; Hope Jones 1980; Dare 1986; Renssen 1991*b*). In North America, and presumably elsewhere, non-territorial individuals locating large carcass within territory of conspecific call to recruit other non-territorial birds, thereby ovewhelming any opposition from territory owner; in this manner flocks of *c.* 40–100 form rapidly, and where food particularly abundant flocks can reach 150–300(–800) (Mylne 1961; Ewins and Dymond 1984; Sellin 1987; Heinrich 1988*b*). Approaches carcasses with suspicion; in North America, those of unfamiliar species not touched for 25–74 hrs after discovery, familiar ones within 7 hrs (average 3·3 hrs); lands 5–10 m from carcass and walks slowly towards it, stopping frequently to look at it (unlike American Crow *C. brachyrhynchos*, which stops to inspect surroundings); when within 5 m, almost always makes sudden, powerful vertical leaps, assisted by one or more wing-beats, before moving closer and leaping again; finally pecks at carcass (usually at haunches) and takes flight, sometimes not returning for some hours. Then, repeats same jumping actions, often landing on carcass and continuing to jump up and down on it, sometimes for several minutes, until starting to feed. (Heinrich 1990, which see for many details.) Feeding activity on intact carcass of sheep *Ovis* or deer (Cervidae) is concentrated around eyes, mouth, umbilical cord, vent, and groin; often removes eyes intact, then tongue, sometimes tearing hole in floor of mouth from outside; though strong and adept at pulling or tearing, has difficulty penetrating skin unless carcass opened by (e.g.) raptor or carnivore (Hancox 1985; Marquiss and Booth 1986; Heinrich 1988*b*). Very unlikely to kill (e.g.) healthy lamb (Glandt and Jensen 1991, which see for discussion); frequently takes afterbirths, also intestines of shot deer, etc. (Renssen 1991*b*). Often obtains food by robbing other species, e.g. Short-eared Owl *Asio* *flammeus* of voles *Microtus* (Marquiss *et al.* 1978); Golden Eagle *Aquila chrysaetos* of hare *Lepus* (Bille 1980); 2 forced Peregrine *Falco peregrinus* to drop prey, one flying off with it (Latscha 1979); tried to rob vultures *Gyps* as they brought food for young (Guyot *et al.* 1991). In captivity, pair robbed Great Black-backed Gull *Larus marinus* by approaching from different sides, one attacking, the other snatching food (Neale 1899-1900). Pair recorded dropping into roosting flock of Kittiwake *Rissa tridactyla* and killing one adult each (Parmelee and Parmelee 1988). Will pull incubating *R. tridactyla* and *C. livia* from nest by head and bill, taking eggs and killing adults (Marr and Knight 1982; Klicka and Winker 1991); in North America, one bird pulled incubating flicker *Colaptes* out of nest, transferred it to foot, then seized a 2nd from nearby branch and flew off with both (Eells 1980). In flight, carries prey in bill or in feet, and can easily transfer from one to other or eat item held in foot (König 1966; Nitsche 1980; Goodwin 1986). Occasionally attempts to seize birds in flight with feet, usually unsuccessfully; duck *Anas*, *C. livia*, and Ptarmigan *Lagopus mutus* all recorded evading capture, though latter probably taken sometimes (Tinbergen 1953; Schaber 1983; Marquiss and Booth 1986). One clutched Jackdaw *C. monedula* in flight, though it escaped when both fell to ground (Dickson 1969); another recorded carrying *C. monedula* corpse in right foot for 30 m (Owen 1950). Often hides food, particularly when hungry; prefers to cache fat or fatty meat, but also recorded hiding whole eggs, bones, bread, dates, and dung; during breeding season, caches more than at other times and preferentially food suitable for young (Heatherley 1910; Gwinner 1965*b*; Turček and Kelso 1968). Usually caches in holes or under stones (Neale 1899-1900); in snow, digs hole with sweeping motions of bill (Kilham 1989). Bird in Tunisia stored small pieces of bread in throat-pouch before caching them in leeward side of small sand ridges; fragments cached singly by pushing them into sand with head held sideways and mandibles slightly open; covered by raking sand over with side-to-side movements of bill; caches robbed by mate of hoarding bird all contained dung (Simmons 1970). In captivity, bill full of food rammed into soft earth then opened, and cache covered as above or by placing small stone, leaf, etc., on top; for larger pieces, will make depression with bill then place food in it before covering (Gwinner 1965*b*, which see for details).

Diet in west Palearctic includes the following. Vertebrates, many as carrion: mole *Talpa*, shrews (*Sorex, Crocidura*), hedgehog *Erinaceus*, rabbit *Oryctolagus*, hare *Lepus*, squirrel *Sciurus*, lemming *Dicrostonyx*, voles (*Microtus, Arvicola*), mice (*Mus, Apodemus, Micromys*), rat *Rattus*, weasel, etc. *Mustela*, marten *Martes*, otter *Lutra*, fox *Vulpes*, dog *Canis*, cat *Felis*, seal (Phocidae), sheep and mouflon *Ovis*, goat *Capra*, cattle *Bos*, deer (*Cervus, Rangifer*), pig *Sus*, horse *Equus*, human *Homo*, fish (Pisces), frog *Rana*, toad *Bufo*, snakes (Serpentes), lizards

*Lacerta*, birds and their eggs (mostly ground-nesters such as Anatidae, Tetraonidae, Phasianidae, Charadriidae, Laridae, etc., but also passerines of all kinds). Invertebrates: grasshoppers, etc. (Orthoptera: Gryllidae, Gryllotalpidae, Acrididae), earwigs (Dermaptera: Forficulidae), mantises (Dictyoptera: Mantidae), bugs (Hemiptera: Pentatomidae, Cicadidae), larval and pupal Lepidoptera (Tortricidae, Noctuidae), larval Diptera (Calliphoridae), Hymenoptera (ants Formicidae, bees Apoidea), beetles (Coleoptera: Cicindelidae, Carabidae, Silphidae, Geotrupidae, Scarabaeidae, Buprestidae, Tenebrionidae, Cerambycidae, Chrysomelidae, Curculionidae), spiders (Araneae), centipedes (Chilopoda), woodlice (Isopoda), crabs and crayfish (Decapoda: *Hyas, Carcinus, Astacus*), barnacles (Cirripedia: *Balanus*), worms (Annelida: Serpulidae, earthworms Lumbricidae), molluscs (Mollusca: Mytilidae, Ostreidae, Littorinidae, Hydrobiidae, Patellidae, land snails Stylommatophora), sea urchins (Echinoidea: *Echinus*). Plants: seeds, fruits, buds, etc. of juniper *Juniperus*, oak *Quercus*, beech *Fagus*, mulberry *Morus*, fig *Ficus*, barberry *Berberis*, grape *Vitis*, olive *Olea*, service tree *Sorbus*, plum, etc. *Prunus*, cloudberry, etc. *Rubus*, heather *Calluna*, bilberry, etc. *Vaccinium*, crowberry *Empetrum*, potato *Solanum*, cereals (Gramineae, including maize *Zea*, wheat *Triticum*, oats *Avena*, barley *Hordeum*), moss (Musci), seaweed *Corallina*. (Newstead 1908; Bolam 1913; Stewart 1927; Owen 1950; Koshkina and Kishchinski 1958; Turček 1961; Ratcliffe 1962; Marquiss *et al.* 1978; Ewins *et al.* 1986; Goodwin 1986; Marquiss and Booth 1986; Cugnasse and Riols 1987; Amat and Obeso 1989; Huber 1991; Soler and Soler 1991.)

In southern Scotland, of 697 pellets, *c.* 56% contained remains of sheep, 28% eggs, 27% rabbit and hare, 16% voles, 15% insects and spiders, 14% goat, and 12% birds (Marquiss *et al.* 1978); but in Orkney, where different agricultural practice means little sheep carrion, 80% of 945 pellets contained rabbit and hare (principally road casualties), 24% feathers (mainly from larger species, many probably tideline corpses), 20% eggs, 20% shore invertebrates, and 13% remains of large domestic mammals, including sheep (Marquiss and Booth 1986). In Britain and Ireland in general, sheep carrion is important component of diet (Ratcliffe 1962; Newton *et al.* 1982; Ewins *et al.* 1986); e.g. in Mayo (western Ireland), where estimated 22 kg of sheep carrion per km² in some places, comprises main food particularly in spring and early autumn (Watson and O'Hare 1980). See also Holyoak (1968) for comparison with other Corvidae. In southern Spain, autumn and winter, 146 pellets contained 86% by volume plant material and 14% animal; plant material 73% by volume barley, 10% grass seeds, 7% wheat, 2% maize, 2% figs, 2% olives, 1% almonds; animal material 74% by weight remains of birds, 19% small mammals, 5% Orthoptera, 0·5% bettles. Only significant change in animal food with season was in Orthoptera: 12·2% in autumn, only

0·5% in winter. However, commonest prey by number was ants, which accounted for 47·4% of total number of animal items. (Soler and Soler 1991, which see for many details, and comparison with Carrion Crow *C. corone* and Magpie *Pica pica*.) See also Amat and Obeso (1989) for 120 pellets from marshland in south-west Spain. In Massif Central (southern France), 56% (by number) of 1257 items from pellets and observations were plant material (43% cherries *Prunus*), 34% insects (17% grasshoppers), 6% mammals, and 2% birds, but mammals and birds most important by weight (Cugnasse and Riols 1987). In Schleswig-Holstein (northern Germany), of 345 vertebrates identified in 2083 pellets, 43% voles, 13% rats, 12% mice, 12% birds, 7% hare, 6% mole (Looft 1971*b*). Difficult to assess importance of various elements in diet from pellets because remains do not accurately reflect quantities of food taken. In Orkney, seasonal changes appear to reflect availability and abundance of foods, e.g. sheep carrion in early spring lambing season, young rabbits and hares as well as birds' eggs from spring to midsummer, insects from summer to autumn. Calcareous items from seashore, including seaweed *Corallina* (which apparently accounts for significant part of diet), may be selectively taken in breeding season when nestlings need calcium; birds nesting or roosting near heathland or coast took more eggs than those on agricultural land. (Marquiss and Booth 1986.) In Shetland, diet differences between territories reflected local food sources in proportion of items from shore, road casualties, sheep carrion, etc.; in August, refuse from tips present in 70% of pellets (Ewins *et al.* 1986). On tundra of northern Russia, composition of diet depends on rodent cycles; in good years, *c.* 90% of diet rodents, in poor ones *c.* 40%, and much more plant material taken, especially *Vaccinium* berries (Koshkina and Kishchinski 1958). For comparative study in Oregon (USA), see Stiehl and Trautwein (1991).

Diet of young similar to adults'. In captive study, parents fed no meat to nestlings in first 2 days, and preferred food was larvae, flies, and small Orthoptera; larvae crushed first in bill and insect extremities removed. Mice and small chicks pulled into very small pieces on day 2–3, these often kept in throat-pouch before being given to young, and much material cached at this time. Until young *c.* 10 days old, parents very often drank water after filling throat-pouch to soften food. (Gwinner 1965*a*.)   WGM, BH

**Social pattern and behaviour.** Fairly well known, but the only detailed studies of marked wild birds are on *principalis* in Maine (USA) (Heinrich 1988*a*, *b*, *c*, 1990) and non-breeders (nominate *corax*) in Switzerland (Huber 1991). Principal behavioural studies in west Palearctic on captive or semi-captive birds in Germany (Lorenz 1931, 1940; Gwinner and Kneutgen 1962; Gwinner 1964, 1965*a*, *b*, 1966); population studies in central Wales by Davis and Davis (1986); reviews by Coombs (1978, which see for additional drawings of displays) and Goodwin (1986). Rather scanty information on *varius*, *tingitanus*, *canariensis*, and *laurencei* suggests considerable similarities to better known races.

1. Outside breeding season occurs solitarily, in pairs, or in flocks (sometimes large) which commonly include paired birds. Territory-holders commonly remain in territory all year, even in severe weather (e.g. Dare 1986, Davis and Davis 1986), but sometimes join flocks at food in and near to territory (Heinrich 1990). In central Wales, some territory-holders left territories on high ground during daytime, particularly in hard weather, some wandering up to 12 km from territory (Davis and Davis 1986). In Greenland, some breeding pairs leave territory entirely in winter (Salomonsen 1950–1) and same may apply elsewhere in Arctic. Flocks consist of independent immature birds and also non-breeding adults that do not hold territories. Territories often defended outside breeding season; rich food sources within territory such as carcasses are often defended vigorously. (Hurrell 1956; Oggier 1986; Heinrich 1988*a*, *b*, 1990.) Flocks tend to roam and always appear loose and straggling (e.g. Davis and Davis 1986, Heinrich 1990). Largest flocks gather at rich food sources and roosts (see below), but some birds remain in flocks at most if not all times. Flocks of 50–60 birds are frequent in Britain, occasionally several hundred (e.g. Holyoak and Ratcliffe 1968). Ewins and Dymond (1984) reported exceptional flock of 800 feeding at dead whale in Shetland (Scotland) in June; flocks of more than 1000 and *c.* 2000 reported in USA (Heinrich 1990). Study in Switzerland showed non-breeders move around area as large as 10 000 km², mainly in small groups that frequently congregate in larger flocks (Huber 1991). In Maine, flocking non-breeders ranged over at least 1800 km² (Heinrich 1988*b*). In Alaska, radio-tagged birds regularly flew 128 km from communal roost to feeding site (Heinrich 1990). At carcasses, associates with other feeding birds, including other *Corvus*, raptors, and large gulls *Larus*. Evidently dominant over smaller *Corvus* at carcasses, but subordinate to large raptors such as Buzzard *Buteo buteo* and Red Kite *Milvus milvus*. BONDS. Monogamous. Almost certainly pairs for life (e.g. Heinrich 1990), but few detailed studies of marked birds. Pairs remain together throughout year and occupy same territory year after year (e.g. Harlow 1922). Both sexes participate to variable extent in nest-building. Incubation and brooding of small young by ♀ alone, fed at or near nest by ♂. Larger young and fledged juveniles fed by both sexes. Helpers at nest not reported. Juveniles remain on parental territory up to 6 months after fledging, receiving food from parents initially (Goodwin 1986). Age of first breeding not established from marked birds, but generally assumed to be at least 2 years and perhaps older; Davis and Davis (1986) suggested first breeding at 3 years or older. Captive pair held by Neale (1901) part-built a nest in 3rd year and laid for first time in 4th year. Substantial proportions of non-breeders in populations evident in breeding season. In central Wales, Davis and Davis (1986) thought non-breeders perhaps equal in number to breeding population. Marked non-breeders included 1st, 2nd, 3rd, and (once) 5th-year birds. In Valais (Switzerland), of population of *c.* 400, up to 62·5% non-breeders (Oggier 1986). Smaller proportion of non-breeders reported in other studies (e.g. Dare 1986), perhaps due to emigration. BREEDING DISPERSION. Nests solitary within all-purpose territory. Territory much larger than in sympatric *Corvus*, and estimates of mean size include: 2·8 km² (*n* = 99) on Hierro (Canary Islands) (Nogales 1990); 10·4 km² (*n* = 47) in Botosani (Rumania) (Andriescu and Corduneanu 1972); in Britain, 7·39 km² (*n* = 195) in Shetland (Ewins and Dymond 1984), 4·9 km² (*n* = 65) on upland sheepwalks and 11·4 km² (*n* = 14) on farmland in central Wales (Davis and Davis 1986), 17·0 km² (*n* = 67) in Lake District, northern England and 19·1 km² (*n* = 23) in Galloway Hills, south-west Scotland (Ratcliffe 1962), 22·6 km² (*n* = 23) in Orkney, Scotland (Booth 1979); in Germany, 31·3 km² (*n* = 24) in West Mecklenburg, 46·5

km² (*n*=49) in Schleswig (Looft 1971*a*), 53·0 km² (*n*=23) in south-east Holstein (Warnke 1960); in Switzerland, 33·3 km² (*n*=75) in Valais (Oggier 1986), 66·7 km² (*n*=9) in Baselbieter Jura (Böhmer 1974). Regional variation in territory size in Britain evidently related to food supply (Holyoak and Ratcliffe 1968; Newton *et al.* 1982; Ewins and Dymond 1984; Davis and Davis 1986). Not all territories occupied every year: in central Wales, Davis and Davis (1986) found 80–89% occupancy in different years, although slight population decline occurred during that study. In north Wales, Dare (1986) found occupancy averaged 91% over 4-year period. In Iceland, Skarphéthinsson *et al.* (1992) found annual occupancy rates of 59–72% in different regions, but persecution by man thought responsible for low occupancy. Ratcliffe (1962) noted that if territory is deserted it often remains unused for several seasons. In study in central Wales, mean 12% of territorial pairs did not lay eggs, although many of these partly built or repaired nests; some repaired 2–3(–5) nests (Davis and Davis 1986). Similarly, in Iceland *c.* 12% of territorial pairs did not breed (Skarphéthinsson *et al.* 1992). Densities include: 10 pairs per 100 km² (total 490 km²) in Rumania (Andriescu and Corduneanu 1972); 35·6 pairs per 100 km² on Hierro (Nogales 1990); in Britain, 21 pairs per 100 km² on upland sheepwalk (total 315 km²) and 9 pairs per 100 km² on farmland (total 160 km²) in central Wales (Davis and Davis 1986), 6 pairs per 100 km² (total 1142 km²) in Lake District, 5 pairs per 100 km² (total 440 km²) in Galloway Hills (Ratcliffe 1962), 4 pairs per 100 km² (total 523 km²) in Orkney (Booth 1979); in Germany, 3 pairs per 100 km² (total 750 km²) in West Mecklenburg, 2 pairs per 100 km² (total 2280 km²) in Schleswig (Looft 1971), 2 pairs per 100 km² (total 1220 km²) in south-east Holstein (Warnke 1960); in Switzerland, 3 pairs per 100 km² (total 2500 km²) in Valais (Oggier 1986), 1·5 pairs per 100 km² (total 600 km²) in Baselbieter Jura (Böhmer 1974); in Iceland, ranged from *c.* 1·5 pairs per 100 km² in southern lowlands (total 1020 km²) and north-east (total 5200 km²) to 12·4 pairs per 100 km² at Fljótshlíd in south-west (total 105 km²) (Skarphéthinsson *et al.* 1992). In population at high density in central Wales, inter-nest distances 0·6–3·6 km (mean 1·7 km in upland sheepwalk, *n*=65; 2·0 km in farmland *n*=14) (Davis and Davis 1986). 4 nests in 3·2 km of coastal cliffs reported from Anglesey (north Wales), 15 nests in 27 km (with 3 in 1 km) of coast in north Devon and Cornwall (England) (Ratcliffe 1962; Holyoak and Ratcliffe 1968). In Shetland, closest nests 1·0 km (Ewins and Dymond 1984). Exceptional instance of 2 nests 0·35 km apart (Dare 1986). Where density high, territorial behaviour appears to result in rather even spacing of nests (Holyoak and Ratcliffe 1968; Davis and Davis 1986); nests significantly overdispersed in study in north Wales (Dare 1986). Territory contains nest-site(s), and most feeding and all food-caching takes place within it (Oggier 1986; Heinrich 1990). Same nest-site may be used for many years in succession (e.g. Allin 1968), but most territories contain 2 or more sites that are alternated irregularly. In study over many years in north Wales, some pairs used as many as 6–7 sites in territory (Allin 1968). Alternative sites in Welsh study ranged from a few metres to 2·2 km apart; movement to alternative site not related to breeding success (Davis and Davis 1986). Of 11 repeat layings in same season, 5 in same nest and 6 in different nest (Davis and Davis 1986). Studies of marked birds in central Wales (Davis and Davis 1986) and Maine (Heinrich 1990) suggest great majority of juveniles leave natal area in 1st autumn and do not return. Not otherwise known how far first-time breeders settle from natal area, although hardly any birds appear to breed within natal area (Davis and Davis 1986). Old nests of *C. corax* often used by falcons *Falco* such as Peregrine *F. peregrinus*, Gyrfalcon *F. rusticolus* and Kestrel *F. tinnunculus* (e.g.

Nielsen 1986). ROOSTING. Roost-sites typically in trees or on cliff ledges, less often on buildings; large communal roosts recorded on utility pylons in USA (Heinrich 1990). Paired territory-holding adults typically roost at or near nest-sites in territory throughout year (Davis and Davis 1986; Heinrich 1988*b*). Flocking birds roost communally throughout year, sometimes in large numbers. Communal roosts very often develop at sites closely adjacent to rich food source (e.g. Coombs 1946, Bryson 1947, Hurrell 1956, Davis and Davis 1986, Duquet 1986, Heinrich 1990). Some communal roosts are in traditional sites used for many years, others only for a few nights (Heinrich 1990). Roosts of up to 150 birds reported in Cornwall (Coombs 1946), 86 in Devon (Hurrell 1956), *c.* 100 in Switzerland (Hewson 1957), up to 220 in Doubs (France) (Duquet 1986), and over 300 in Denmark (Grünkorn 1991). Communal roosts often associated with other *Corvus*, especially Rook *C. frugilegus* and Jackdaw *C. monedula* (e.g. Coombs 1946, Cadman 1947, Hurrell 1956). Hutson (1945) reported large communal roost in Iraq shared with large numbers of 'Kites' (doubtless Black Kites *Milvus migrans*). For roosting behaviour, see Flock Behaviour (below).

2. From detailed studies of semi-captive birds, Gwinner (1964) found that, in those portions of behaviour in which individual relationships are predominant, displays are often highly variable. Undifferentiated patterns prevail and they may vary in frequency, amplitude, and orientation. In addition, whole chains of displays can be altered, either extended individually into more complex forms, or reduced to basic elements. This lack of rigidity thought to allow communication of more subtle detail of social information to mate or to well-known member of group. Members of pair readily recognize each other individually and transmit modified vocal information directed only at mate, even over long distances. (Gwinner 1964.) Behaviour interpreted as play often reported. Studies on semi-captive birds reveal much more complex play repertoire than reported for any other bird (Gwinner 1966), including hanging upside-down, and sliding down sloping surfaces. Play sequences prone to great individual variation and group-specific play combinations arise by mutual imitation. Some learned play sequences are secondarily employed in other categories of behaviour. Eyes give important indication of mood and intention: in hostile contexts, eyelids tend to be widely open and pupils dilated; in friendly contexts, eyelids slightly closed and pupils somewhat contracted. Captive birds often at once show appropriate responses to friendly, intimidating, or aggressive looks from humans. (Gwinner 1964; Goodwin 1986.) Frequently performs aerobatics, such as soaring on thermals, diving with closed wings, and suddenly rolling over; often in groups, sometimes near roosts but also far from any roost (e.g. Hurrell 1951, 1956, Lockley 1953, Hewson 1957, Heinrich 1990). Calls frequently in flight and at other times. Characteristically shy of man in regions where shot, such as Britain; elsewhere a bold scavenger around human dwellings and encampments (e.g. Goodwin 1986, Heinrich 1990). Apparently owing to large size, relatively immune to predation from hawks, eagles (Accipitridae), and *Falco* (e.g. Williamson and Rausch 1956, Heinrich 1988*b*); consequently pays much less attention to raptors than do smaller *Corvus*. See Heinrich (1990) for comments on interactions in North America with large predators such as wolf *Canis lupus* and their prey, especially caribou *Rangifer tarandus*. FLOCK BEHAVIOUR. Usually maintains individual distance of several body lengths. Paired birds approach each other more closely, but even when paired birds Allopreen (see Heterosexual Behaviour, below) preening bird stands a little away from its partner rather than in body contact (Goodwin 1986). Dominance hierarchies well marked in captive and semi-captive groups (Gwinner 1964; Heinrich 1990) and occur in wild feeding flocks (Heinrich 1990;

Huber 1991). Paired adult territory-holders are generally high-ranking, unpaired immatures in flocks are generally of lower rank (Heinrich 1990; Huber 1991). In Swiss study, Huber (1991) found age-dependent hierarchy at food in winter, with older birds dominant; some marked birds in hierarchy knew each other individually; dominance relations between individuals were consistent and remained unchanged for considerable periods. Flocking birds commonly give Rüh-call (see 4 in Voice) to recruit others to rich food sources, apparently in order to overwhelm defence by resident territory-holders (Heinrich 1988b, c, 1990). Birds roosting communally often arrive at regular gathering place in ones and twos or small groups before flying to roost as flock (e.g. Hutson 1945, Hurrell 1956, Duquet 1986). Small groups commonly arrive at communal roost from different directions (Heinrich 1990; Huber 1991). In Devon, pre-roosting birds regularly assembled on moorland near roost and sometimes indulged in chasing or aerobatics; made 2-3 visits to woodland where they roosted before finally settling down for night. Paired birds apparently roosted close together; varied calls often heard at roost. (Hurrell 1956.) At communal roost in Doubs, pre-roost gatherings in tall coniferous trees 750-1250 m from roost; return to roost began, on average, 50 min before sunset in January (Duquet 1986). SONG-DISPLAY. According to Heinrich (1990), song (see 1 in Voice) usually heard from mid-August until late autumn in Maine. Also in North America, Zirrer (1945) reported several birds singing in chorus in autumn from tops of tall trees, and song may continue for hours at a time (Bent 1946). No associated display known. ANTAGONISTIC BEHAVIOUR. (1) General. Aggressive interactions, especially aerial chases, are frequent at all seasons over food and in disputes over territory (e.g. Heinrich 1990). Actual fighting frequently occurs where flocks concentrated at food sources, and aggressive interactions increase in frequency as numbers crowding to feed increase (Heinrich 1990). In study in central Wales, territorial pairs generally behaved aggressively towards wandering non-breeding individuals within c. 0·5 km of nest-site, especially in breeding season, but they normally ignored conspecifics flying high overhead. A few breeding pairs were unable to defend territory against non-breeding flock and ensuing conflict resulted in nest failure. In one instance, nest destroyed and sticks scattered over wide area. (Davis and Davis 1986.) Hurrell (1956) also recorded non-breeding birds from flock destroying old nest. (2) Threat and fighting. Advertising-display performed by both sexes in both aggressive and sexual contexts. In aggressive contexts, directed at conspecifics of same sex and usually appears to be elicited by combination of aggression and fear. However, may also be given when physically prevented from attacking a conspecific not feared, e.g. by wire-netting barrier. (Gwinner 1964; Goodwin 1986.) ♂'s Advertising-display (Fig A) usually begins with erection of feathers close above each eye so that they look like small ears; other head feathers not much erected, flank feathers sleeked to give trouser-like appearance, and feathers of throat, neck, and breast are erected so as to form a smooth surface together with long raised flank feathers. Plumage of hindneck and mantle also form continuous line; carpal joints held slightly away from body. A copious flow of saliva causes frequent swallowing, emphasizing long shining throat feathers. Bird strides about thus and after some time may give soft 'ko' or 'cho' (see 7 in Voice) or similar calls in time with its steps. From this, ♂ may go into Thick-headed posture (Fig B) in which all head and

B

throat feathers fully erected; similar calls may be given to those in Advertising-display. At high intensity, bird holds itself horizontal and makes forward bowing or retching movements giving Choking-calls (see 7 in Voice); movements appear strained as displaying bird leans or reaches towards its rival; usually stands more upright between series of forward bowing movements and calls. During bowing and calling, tail partly spread and white nictitating membranes often drawn over eyes. ♀'s Advertising-display differs in that 'ears' are shorter and other head feathers are slightly erected so that 'ears' less conspicuous; no calls given. ♀'s Thick-headed display also differs in that bows are lower, leaning towards rival with spread tail and wings held well away from body; usually gives series of squeaking calls or clicking or clappering sounds (see 8 and 12a in Voice) accompanied by upward movements of previously more-or-less down-pointed bill and drawing of nictitating membranes over eyes; tail inclined upwards, and, when calling, head raised above line of back resulting in rather U-shaped outline (♂ at this stage has head, back, and tail in almost straight line). Outright threat-display given in more-intensely aggressive situations, usually after Advertising-display and immediately before actual attack: performer sleeks all plumage except 'ears', stands upright and moves with quick light steps and bounds. Forward-threat display very variable, with bill open and often snapped, and sometimes with wings lifted. Low-intensity version of this display consisting of merely directing bill towards subordinate is usually sufficient to keep it at distance or drive it from food. Frontal-threat seen in pure form only in nest defence: plumage erected, especially on head, and bill opened; at high intensity, wings partly opened and drooped. Defensive-threat display (Fig C) may be more

A

C

intense version of Forward-threat or involve greater fear and thwarting of tendency to flee; differs in that plumage of mantle and back fully erected to give jagged or ruffled profile, and hump-backed posture often adopted; bill held open, and calling (see 6 in Voice) and Bill-snapping occur. Full Defensive-threat display is commonly given by captives that cannot escape from dominant individuals. (Gwinner 1964; Goodwin 1986.) Fights at food are often limited to quick pulling of a wing- or tail-feather, seldom a chase (Heinrich 1990). In more intense fighting, eyes are never harmed; combatants often lie on ground with inter-locked claws, directing pecks mainly at each other's bills and carpal joints. In captivity, weak or injured individuals may be killed by others; apparently no inhibition against inflicting serious injury except to eyes. (Gwinner 1964; Goodwin 1986.) Submission shown by retracting neck and sleeking plumage, so bird tends to look small and thin; bill raised towards that of dominant but held at lower level; turning head away or looking down towards feet (Stare-down) may also signal appeasement. In all appeasing situations, may give juvenile-type calls (see 4 in Voice). Both sexes may beg for food as apparent appeasement behaviour, with same display and calls (see 10 in Voice) as juvenile begging from parents or ♀ soliciting Courtship-feeding (see below). Appeasement to mate or other more-or-less trusted individual may be shown by Solicitation-posture (see Heterosexual Behaviour, below); tame captive birds often direct this display at humans. (Gwinner 1964; Goodwin 1986.) Heinrich (1990) reported immatures in winter occasionally lying on their sides (once on back) in snow in submission to adults. Displacement activity (or redirected aggression) in antagonistic contexts often includes pecking and hammering at perches or other objects, and perfunctory bill-wiping movements. Heterosexual Behaviour. (1) General. Pair-formation apparently takes place in flocks of non-breeding (mainly immature) birds, in which many birds evidently paired (Davis and Davis 1986; Heinrich 1990; Huber 1991). Sexual display often seen in flocks of non-breeders, and observed in captive juveniles and wild 1st-winter birds (Heinrich 1990). However, sexual display must also occur between pair-members at or near nest, where it leads to mating. (2) Pair-bonding behaviour and mating. Sexual display usually develops from ♂'s Advertising-display (see Antagonistic Behaviour, above). Display by ♂ then typically proceeds through intermediate display, in which neck stretched forward, slightly spread tail raised, partly spread wings stretched back, and Choking-call given while nictitating membranes drawn over eyes. This changes into ♂ Pre-copulatory display (Fig D) in which ♂

D

half-spreads wings and holds them drooped but well away from body, tail spread and raised, all body feathers except 'trousers' sleeked, head raised on extended neck with bill either horizontal or inclined down and nictitating membranes drawn over eyes for periods of several seconds. (Gwinner 1964; Goodwin 1986.) ♀ may then invite copulation with Solicitation-display (Fig E) in

E

which she crouches, slightly or markedly opening and extending or drooping wings and shaking or quivering tail (slightly raised at high intensity). Mating may follow (occurs at or near nest); if not, ♂ normally proceeds from Pre-copulatory display to typical Solicitation-display (also used in submissive contexts; see Antagonistic Behaviour, below). (Gwinner 1964; Goodwin 1986.) One ♂ several times reported erecting one median wing-covert of each wing during low-intensity sexual display (Keränen and Soikkeli 1985). Mutual Allopreening of paired birds is regular, but also occurs (at least in captivity) between unpaired birds. Sick captive birds that sit with head feathers ruffled are frequently preened by others including subordinates. Allopreener very commonly lifts long throat (and other) feathers of partner and looks intently at feather-bases as if seeking parasites (Goodwin 1986). Paired birds commonly also touch and grasp each other's bills. (3) Courtship-feeding. ♂ commonly feeds ♀ by regurgitation, at least in spring, and also at or near nest when she is incubating and brooding. ♀ solicits food with Begging-call (see 10 in Voice). Gwinner (1964) implied that some ♀♀ always beg with flapping wings when fed and others do not, though behaviour perhaps varies according to body condition rather than being constant difference between individuals (Goodwin 1986). (4) Nest-site selection. Roles of sexes apparently not described. (5) Behaviour at nest. Both sexes build nest; shares of ♂ and ♀ at some or all stages of building vary individually (Gwinner 1965a; Goodwin 1986; Heinrich 1990). One bird of pair normally does most stick-carrying and building—♀ according to Harlow (1922) and Davis and Davis (1986). Nest may be completed 1–4 weeks before eggs laid. ♀ dominates ♂ at nest. Incubation by ♀ alone; spends most of her time on nest from laying of 1st egg but does not begin true incubation until penultimate or last egg laid. Before incubation begins, ♀ probes beneath eggs and pulls up surrounding material, so that eggs sink into nest-lining. (Gwinner 1965a.) Relations within Family Group. ♀ turns hatching eggs so that bill of young is uppermost and visible to her. She eats eggshell, cleans newly hatched chick, and places its neck so that it rests on an egg. (Gwinner 1965a.) Both sexes feed and tend young. Brooding only or mainly by ♀, although ♂ will stand or crouch over young (Ryves 1948; Gwinner 1965a). Brooding continues until young *c.* 18 days old. In study of captive and semi-captive birds, water was brought to panting young on hot days. (Gwinner 1965a.) Wild birds also reported to bring water to nestlings (Hauri 1956). When young apparently suffering from heat, ♀ may also wet her underparts and brood young thus, or bore hole through nest to provide ventilation from underneath. In cold weather, ♀ may half-bury nestlings in soft nest-lining. (Gwinner 1965a.) Droppings of small young are nearly always vented just after feeding; swallowed by parents (Gwinner 1965a; Heinrich 1990). Unhatched eggs and small dead young quickly disappear from nests, presumably eaten (Davis and Davis 1986). Large nestlings defecate over edge of nest and may regurgitate pellets (Heinrich 1990). Nestlings are Allopreened by both parents, especially by ♀ (Heinrich 1990). Young near to fledging at *c.* 40 days old often move onto branches or ledge near nest (Davis and Davis 1986). 4 young at Welsh nest fledged over 3 days (Goodwin 1986).

Family reported to remain together for up to 6 months after fledging (Goodwin 1986). Davis and Davis (1986) recorded juveniles on territories in Wales until July–August(-September); towards end of period on territory, observed moving about as a small sibling party, feeding and roosting independently of parents; by autumn, wandered away or chased away by adults. In Iceland, most young disperse from territory *c*. 1 month after fledging, when they begin to join non-breeding flocks (Skarphéthinsson *et al.* 1992). In Maine, juveniles follow adults for only *c*. 3 weeks after fledging (Heinrich 1990). ANTI-PREDATOR RESPONSES OF YOUNG. In response to Warning-call of parent (see 11 in Voice), small (and often larger) young typically crouch low in nest and remain silent when human intruder near. Heinrich (1990) found that in this context some broods of large young gave rasping, growling calls (see Voice) whereas others remained silent. Young 30–35 days old leave nest prematurely if human climbs to it, despite being unable to fly properly (Davis and Davis 1986). PARENTAL ANTI-PREDATOR STRATEGIES. (1) Passive measures. While incubating or brooding small young, ♀ may crouch low on nest when human approaches. One parent, probably ♀, usually spends much of its time at or within sight of nest until young have left it (Goodwin 1986). (2) Active measures: against birds. Herring Gulls *Larus argentatus* approaching nest were pursued by ♀, with calls (see 2 in Voice) (Goodwin 1986). Nests also defended against conspecifics (Gwinner 1965*a*; see Antagonistic Behaviour, above). (3) Active measures: against man. Intruders at nests are usually mobbed by parents flying nearby and calling noisily (see 2 in Voice), but in Britain intruders rarely experience very close approach (D T Holyoak). Semi-captive birds made intense attacks on humans climbing to nests (Gwinner 1965*a*). In Oregon, adult reported dropping 7 large pebbles onto human intruder at nest (Janes 1976); not clear, however, that deliberate 'tool-use' involved, rather than mere dropping of objects dislodged or picked up during vigorous pecking and hammering or redirected aggression common under these circumstances (e.g. Heinrich 1990).

(Figs by D Nurney: A from drawings in Coombs 1978 and Heinrich 1990; B–E from drawings and photographs in Gwinner 1964 and drawings in Coombs 1978.)     DTH

**Voice.** Much used, throughout year. Most common calls are low-pitched with gruff, croaking tone, so, despite great variability, these are easily distinguishable from other European *Corvus*—though less easily from some calls of Brown-necked Raven *C. ruficollis* and, especially, Fan-tailed Raven *C. rhipidurus*. Commonest calls of all west Palearctic races appear similar, as do those of North American *principalis* and *sinuatus*, but no detailed comparisons. Principal studies giving sonagrams are of semi-captive nominate *corax* in Germany by Gwinner (1964) and of North American races by Dorn (1972), Brown (1974), Conner (1985), and Heinrich (1988*b*, 1990). For additional sonagrams, see Bergmann and Helb (1982). Despite considerable amount of information available, difficult to provide descriptive list as synthesis of these studies, some of which recognize numerous different categories of calls (over 30 in Dorn 1972, 18 in Conner 1985), while others suggest fewer standard calls and much improvisation (e.g. Goodwin 1986, L Svensson). Difficulty arises because some calls may be used in situations different from those 'normally' eliciting them (Goodwin 1986); also, much indi-

vidual variation, vocal mimicry may be incorporated and combined with innate calls or used instead of them (Gwinner 1964; Goodwin 1986; Heinrich 1990), and perhaps also regional differences. *C. corax* readily learns to imitate human speech in captivity, but wild birds rarely mimic other species (e.g. Gwinner 1964, Heinrich 1990). Bill-snapping often used in aggressive encounters.

CALLS OF ADULTS. (1) Song. So-called 'song' comprises long series of varied, mainly soft sounds, many musical or pleasing in tone. Goodwin (1986) noted that song of young birds may include any or all of innate calls even when no stimuli that normally elicit them appear to be present (see also 3b, below). Not known whether one or both sexes sing. (2) Pruk-call. Commonest call. Short barking 'pruk'; described by Heinrich (1988*b*, 1990) as 'quork', always in sequences of 3–6 evenly-spaced calls, predominant energy at 0·8 kHz (e.g. Figs I–II, both of ♂♂); 'kra' (Gwinner

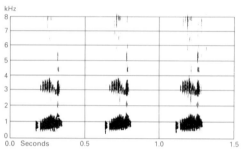

I  P Enggist-Düblin  Switzerland  February 1991

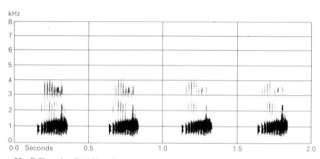

II  P Enggist-Düblin  Switzerland  February 1987

1964). Other renderings include 'krok', 'kruk', 'kro', 'prok', and 'croak'. Varies considerably, between and within individuals (Gwinner 1964; Heinrich 1990). Given in many situations, in flight and perched, but almost always appears to involve real or imagined threat from non-conspecific enemy (Gwinner 1964; Goodwin 1986). Often 3–4 calls in rapid succession in alarm, or, more slowly, as 'conversation' between pair members (L Svensson). According to Gwinner (1964), loud 'kra' also given while defending food. Harsh calls given by parents mobbing human intruder at or near nest are probably similar or same; merge with call 11 (see that call, and Fig VI). Heinrich (1990) thought Pruk-call also functioned in territorial advertisement and defence. Heard from displaying

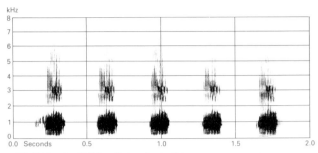

III P Enggist-Düblin Switzerland February 1991

dominant ♂♂ and often appeared to cause departure of conspecifics (Brown 1974). Flight-call described as short 'rapp' or 'krapp' (Fig III) may be variant; used particularly during 'wild flight chases' (Gwinner 1964). (3) Contact-calls. (3a) Gro-call. A 'gro', more uniform and softer in tone than call 2 (Gwinner 1964). Goodwin (1986) rendered it 'a deep, low throaty note' perhaps equivalent to Appeal-call of Jay *Garrulus glandarius*, and, like that species, used in many social contexts. Given only when conspecific already near. Fig IV shows apparently this

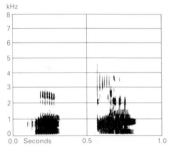

IV P Enggist-Düblin Switzerland February 1991

call, 1 from each of 2 birds calling alternately; pair-members commonly exchange similar quiet calls (duet) during billing and allopreening (P Enggist-Düblin). Slightly different variants are used as contact-call by paired bird offering to feed mate, and by parent offering to feed fledged young. (Gwinner 1964; Goodwin 1986.) Food-call (soft 'korr') given by adults at nest (Heinrich 1990) is apparently a variant, immediately causing still sightless and apparently sleeping young to gape. ♀ with young also gives Gro-call when fed on nest by ♂, even when she does not give any food to young herself (Gwinner 1964). (3b) Soft, tender version of call 3a, 'gru', 'gri', 'gwee', etc.; termed 'Winsellaute' (whimpering or whining) by Gwinner (1964). Appears to have similar social functions to call 3a; given especially by young birds when Allopreened by parents, also by hand-reared birds to humans; has prominent place in song of young birds and one form of song consists entirely of variants of this call. (Gwinner 1964; Goodwin 1986.) (4) Rüh-call. 'Yell' of Heinrich (1988*b*, 1990), described as appearing higher pitched than call 2 and differing in having 2-3 upper

components at 1·8, 2·5, and 3·5 kHz and a fundamental of 0·8 kHz; call given singly or in irregular sequence and apparently corresponds to 'rüh' of Gwinner (1964), 'ky' of Dorn (1972), and 'juvenile kaah' of Brown (1974). Gwinner (1964) noted that loud version of this locatory call is used by hungry fledglings to summon parents, and by incubating ♀ to summon mate. According to Gwinner (1964) and Goodwin (1986), forms of this call are given as nest-call to stimulate mate to join in nest-building, as an appeasement-call in response to actual or implicit threat from dominant individual, and sometimes to dominant individuals in apparently friendly situations, and by ♀ while laying. According to Goodwin (1986), common factors in all these situations are that calling bird feels submissive or dependent and, usually, ill-at-ease. Call sounds much like food-call of juveniles (see below); also given habitually by flocking immature birds, apparently to attract others to good food sources (see Social Pattern and Behaviour) (Heinrich 1988*b*, 1990). (5) Protest-call. Very quiet, rather rattling call with complaining sound, often given by bird being Allopreened or fed rather roughly by another. Indicates low-intensity threat or protest and intergrades with call 3a. (Gwinner 1964; Goodwin 1986.) (6) Defensive-threat call. Loud variable 'kray'; rendered 'krä' by Gwinner (1964). Given by birds that are cornered, defending nest or food, or in similar situation. Apparently strongly dependent on specific 'mood' eliciting it and, unlike other calls, never incorporated in juvenile song. (Gwinner 1964; Goodwin 1986.) (7) Choking-call. In advertising and sexual display (see Social Pattern and Behaviour) ♂ gives choking sound, 'ow', 'row', 'krrooa', or soft 'ko' or 'cho'. ♀ may also call thus in sexual display. (Gwinner 1964; Goodwin 1986.) (8) Repeated knocking, mechanical sound, given primarily by ♀, especially in Advertising-display (see Social Pattern and Behaviour). In Fig V, relatively long sequence comprising 27 units

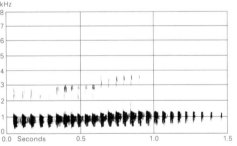

V S Palmér/Sveriges Radio (1972-80) Sweden May 1961-4

showing rallentando and diminuendo. Sonagrams given by Heinrich (1988*b*, 1990) show 10 very brief and evenly spaced units given within *c*. 0·8 s, each with frequency range *c*. 0·2-1·7(-2·0) kHz. This call presumed by Gwinner (1964) to be imitation of Bill-clattering by White Stork *Ciconia ciconia*, but probably not since British and North American birds give similar calls. ♂♂ may learn to give these sounds but they utter them much less often than do ♀♀ (Gwinner 1964). Probably same as 'kwulkulkul' of Brown (1974) and 'rattle' of Conner (1985), although Heinrich (1990) noted that sonagram in Conner (1985) shows rattle at much higher frequency than in Heinrich's recordings. (9) 'Trill', apparently undescribed elsewhere, listed by Heinrich (1988b, 1990): sonagrams show single call, similar in duration to call 2 and resembling it in having most energy at *c*. 0·8 kHz, but differing in having 'harmonics' at 1·9 and 3·1 kHz. Call signals high level of excitement and is possibly an advertising-call of ♂ (Heinrich 1990). (10) Begging-call. Husky 'kra-kra-kra' or 'kreh-kreh-kreh' with pronounced 'r' sound, like fledged young (Gwinner 1964; Goodwin 1986). (11) Warning-call. Oft-repeated 'rrack', given near nest, which causes nestlings to crouch low (Heinrich 1990). Fig VI shows calls transitional between this call and call 2 (D T Holyoak).

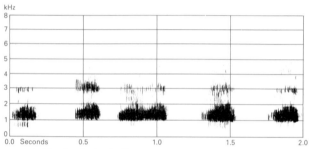

VI  V C Lewis  England  May 1979

(12) Other calls. (12a) In sexual display (see Social Pattern and Behaviour) ♀ gives, especially in 1st year, squeaking or squealing calls, in addition to frequent use of calls 7–8 (Gwinner 1964; Goodwin 1986); however, copied sounds are often incorporated in sexual displays. (12b) Soft cooing sounds given by pair-members close together (Heinrich 1990). Following recordings may belong in this category: sound like motor horn (D T Holyoak: Fig VII), apparently given only by ♀ (P Enggist-Düblin); duck-like quacking (P J Sellar: Fig VIII) given by *tingitanus* (no details of context). (12c) Established pairs may use personal variants of various innate calls, which presumably function to facilitate immediate recognition at a distance as well as, presumably, to enhance pair-bond. When one of pair is lost

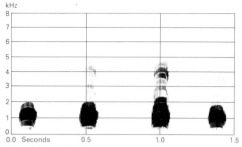

VIII  J-C Roché  Algeria  March 1967

(or removed by human) other often utters particular calls or variants of calls habitually used by lost partner (but not by itself); this has attractive effect on partner and may result in its immediate return. (Gwinner and Kneutgen 1962; Goodwin 1986.) (12d) Disyllabic call, roughly 'teer-do' with marked pitch drop to 2nd syllable (P J Sellar: Fig IX), given by *tingitanus*.

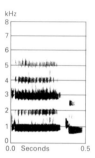

IX  J-C Roché  Algeria  March 1967

CALLS OF YOUNG. Food-call of small young a soft peeping; develops into imploring rasping sound when young well feathered (Heinrich 1990). Large young and fledglings beg with husky 'kra-kra-kra' or 'kreh-kreh-kreh' (Fig X) with pronounced 'r' sound (Goodwin 1986). When human climbs to nest, some broods of well-grown young give long, rasping, growling calls like snarling dogs (Heinrich 1990).                                                DTH

**Breeding.** SEASON. Early. Britain: eggs laid mid-February to mid-April; mean dates for 1st egg ranged from 5 March (*n* = 73) in Wales to 22 March (*n* = 15) in northern Scot-

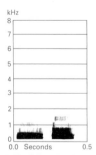

VII  P Enggist-Düblin  Switzerland  March 1991

X  V C Lewis  Wales  June 1979

land (Holyoak 1967*c*); later in Shetland where many 1st clutches laid in 2nd half of April and replacements recorded early May (Ewins and Dymond 1984); fledged young in January at one nest in south-west Scotland exceptional (Mearns and Mearns 1989); see Cullen (1978) for Isle of Man, Booth (1979) for Orkney, and Davis and Davis (1986) for Wales. Northern Finland: eggs laid from late March (Haartman 1969). Schleswig-Holstein (northern Germany): mid-February to end of March, with peak in 1st half of March (Warncke 1960); for various regions in Germany, see Glandt (1991). Switzerland: end of January (rarely) to mid-April, mainly around mid-March (Glutz von Blotzheim 1962; Hauri 1988*a*). Russia: eggs laid 2nd half of March, in far north to mid-May (Dementiev and Gladkov 1954; Ptushenko and Inozemtsev 1968). North Africa: apparently later than Europe; usually from beginning of April, sometimes end of March (Etchécopar and Hüe 1967). SITE. High up on tree, isolated or generally at forest edge, inland or coastal cliff, or man-made structure; where undisturbed, can be much lower, also in tall shrubs or even on ground (Grote 1937; Ratcliffe 1962, which see for discussion; Kulczycki 1973, which see for comparison with other studies). Obviously dependent on availability in landscape; of 277 nests in north-east Germany, 75% in pine *Pinus*, 15% beech *Fagus*, 5% oak *Quercus*, 2% on pylons (Sellin 1991); for 300 nests in Schleswig-Holstein, see Looft (1967). In Poland, 86% of nests in pine, 6% in birch *Betula*, all others also in trees, with average height above ground 21·7 m (13-31, *n* = 50) (Kulczycki 1973), but in Shetland, 96% of 85 nests on cliffs (Ewins and Dymond 1984). In central Wales, of 294 nests, 71% in trees (average 9 m above ground, *n* = 179; conifers preferred), and 29% on cliffs (Davis and Davis 1986). In Mittelland region of Switzerland, has relatively recently started to nest in trees, probably because available rockface sites now occupied (Hauri 1966; Bühler 1968). In Schleswig-Holstein, apparently moving from beech to spreading spruce *Picea* plantations (Looft 1965). In trees, in mid- or upper crown in main fork or near trunk, sometimes on broad branch up to *c*. 2 m from trunk (Kulczycki 1973); on cliffs, generally in upper third, seldom within *c*. 3 m of top or bottom, very often protected by overhang (Ratcliffe 1962; Booth 1979; Ewins and Dymond 1984). For nests on man-made structures, mainly disused, see (e.g.) Kulczycki (1973), Booth (1979), and Cochet and Faure (1987), and for buildings in city-centres, see Ortlieb (1971), Hume (1975), and Hauri (1988*b*). Nest: basically in 4 fairly distinct layers: outer foundation of sticks, twigs, or woody stems up to *c*. 150 cm long and 2·5 cm thick, neatly woven with fresh twigs, which sometimes form distinct rim, then layer of earth, dung, grass, and roots, at times making bowl, but can be completely absent, particularly if ground still snow-covered; this is lined with moss, grass, rootlets, leaves, etc., then finally compacted layer of wool, hair, fur, grass, lichen, stems, etc.; wire and bones not infrequently found as material (Gwinner 1965*a*;

Kulczycki 1973, which see for many details; Coombs 1978; Goodwin 1986). In Poland, 15 nests had average outer diameter 70·2 cm (43-93), inner diameter 27·6 cm (20-37), overall height 44·6 cm (26-65), and depth of cup 11·1 cm (8-16) (Kulczycki 1973). Nests can be 1·5 m high and 1·5 m wide (Nethersole-Thompson 1932; Jonin and Le Demezet 1972). Building: by both sexes, but all possible variations recorded, either ♀ or ♂ doing no, some, or all work at some or all stages, including both doing equal amounts (Gwinner 1965*a*; Coombs 1978; Davis and Davis 1986; Goodwin 1986; Bäumer-März and Schuster 1991). Captive pairs took 7-8 (3-12) days (Gwinner 1965*a*, which see for technique); see also Coombs (1978). Nests frequently re-used, often over many years; each pair generally has *c*. 2-7 alternative sites which are used in irregular rotation (Allin 1968; Davis and Davis 1986; Goodwin 1986). In Berlin area (Germany), 60% of 75 nests over *c*. 10 years were new, 40% refurbished old ones; new nest usually built on top of previous year's rather than (e.g.) old lining replaced (Sömmer 1991); on Isle of Man, 13 of 20 pairs used same nest in successive years (Cullen 1978). In Switzerland, tree nests used year after year more often than those on cliffs: may use old nest of Carrion Crow *C. corone* or raptor (Hauri 1966). EGGS. See Plate 58. Sub-elliptical, smooth and slightly glossy; light blue to blue-green with very variable olive to blackish-brown blotches, spots, scrawls, or hair-streaks usually concentrated towards broad end, though at times completely absent, occasionally with greyish-violet undermarkings; very like egg of *C. corone* but larger (Harrison 1975; Makatsch 1976). Nominate *corax*: 48·8 × 33·8 mm (42·1-68·0 × 29·0-38·5), *n* = 744; calculated weight 28·8 g. *C. c. laurencei*: 50·5 × 34·0 mm (41·9-59·6 × 31·0-36·2), *n* = 100; calculated weight 30·4 g. *C. c. tingitanus*: 47·6 × 32·5 mm (42·4-54·0 × 30·0-35·5), *n* = 220; calculated weight 26·0 g. (Schönwetter 1984.) Clutch: 4-6 (2-7); in central Wales, of 114 complete 1st clutches: 1 egg, 2%; 2, 4%; 3, 13%; 4, 27%; 5, 29%; 6, 24%; 7, 1%; average 4·5; in February 4·8 (*n* = 25), early March 4·7 (*n* = 48), late March 3·7 (*n* = 16), early April 3·4 (*n* = 8). Only 9 replacement clutches recorded in 269 breeding attempts, 5 of them in original nest, average size 3·9; if 1st clutch lost after *c*. 12 days incubation then no repeat laid; 15 days noted in one case between loss of 1st clutch and start of repeat. (Davis and Davis 1986.) In Shetland, only 2 replacements laid following 23 failed clutches (Ewins and Dymond 1984), and repeat clutches hardly recorded in extensive Swiss study (Hauri 1988*a*). 2nd replacements also possible (Makatsch 1976). Average clutch size in Finland 4·7, *n* = 12 (Haartman 1969); North Africa 5·3, *n* = 12 (Etchécopar and Hüe 1967); Britain 4·6, *n* = 139 (Ratcliffe 1962); see Holyoak (1967*c*) for regional variation in Britain, Allin (1968) for north Wales, and Ewins and Dymond (1984) for Shetland. Eggs laid daily, in early morning; sometimes at intervals of 2 days. One brood. (Gwinner 1965*a*; Hauri 1970; Makatsch 1976; Goodwin 1986.) INCUBATION. (18-)

20–21 days, by ♀ only (Gwinner 1965a; Makatsch 1976; Davis and Davis 1986). In one captive study, started with 2nd or 3rd egg (Bäumer-März and Schuster 1991), in another, with last or penultimate (Gwinner 1965a). YOUNG. Fed and cared for by both parents; ♀ broods into 3rd week, intensively for first c. 7–10 days (Gwinner 1965a; Goodwin 1986; Bäumer-März and Schuster 1991). FLEDGING TO MATURITY. Fledging period c. 45 days (35–49), young often leaving nest before able to fly (Makatsch 1976; Davis and Davis 1986; Goodwin 1986). Independent (2–)4–6 months after fledging (Schmidt 1957; Coombs 1978; Goodwin 1986; Bäumer-März and Schuster 1991). Age of first breeding (2–)3 years (Schmidt 1957; Coombs 1978); one ♂ in Orkney very probably first bred at 6 years old (Booth 1986). BREEDING SUCCESS. In central Wales, of 269 pairs that laid eggs, 33% failed to raise fledged young, giving 2·8 young per successful pair ($n = 181$), 1·9 per pair overall; higher nests seemed to be more successful, with 18% above 14 m failing ($n = 34$), 43% below ($n = 145$). Almost all complete failures occur at egg or small nestling stage, and youngest often dies; of 47 failures where cause known, 38% due to egg collectors, 55% other human persecution or disturbance, and 6% weather, nest-collapse, or intraspecific fighting. (Davis and Davis 1986.) For north Wales, see Allin (1968) and Dare (1986); for Isle of Man, Cullen (1978). In Shetland, of 133 breeding attempts, 48% failed completely; of 64 failures, 33% due to human persecution and 22% to interference by Fulmar *Fulmarus glacialis*; on Fetlar, hatching success 57% and fledging success 93% ($n = 13$ clutches), and newly-fledged young very vulnerable to oiling by *F. glacialis* (Ewins and Dymond 1984); see also Robertson (1975a). Frequently evicted from nest by Peregrine *F. peregrinus* (Ratcliffe 1962, which see for review; Dare 1986). Nests robbed by (e.g.) Jackdaw *C. monedula*, Herring Gull *Larus argentatus*, and fox *Vulpes* (Ratcliffe 1962; Holyoak 1967c, which see for review). Success rate in Germany can be low, e.g. 1·2 fledged young per pair overall ($n = 99$) in Schwarzwald in south-west (Eisfeld *et al.* 1991); 1·3 ($n = 110$) in north-east, where almost 50% of natural losses caused by conspecifics (Sellin 1991), though in Berlin area was 2·3 ($n = 91$); martens *Martes* among predators (Sömmer 1991). For Netherlands, see Renssen (1988, 1991b).   WGM, BH

**Plumages.** ADULT. Entirely black, strongly glossed purple, especially on head, neck, upperparts, and chest, more purple-blue when worn and on upper mantle and belly (in part depending on light), more bronzy-purple in some lights on ear-coverts and side of neck. Primaries and tail often slightly duller purple-black than body, but (especially) tertials, secondaries, and lesser and median upper and under wing-coverts strongly purple or purple-blue; greater and primary under wing-coverts and under-surface of flight-feathers and tail glossy brown-black. Feather-bases of head and body light grey, palest (grey-white) on neck and mantle. *In worn plumage* (spring and summer), gloss slightly more bluish (least so on wing-coverts); side of neck tinged bronze; dull black feather-centres of underparts visible, glossy purplish feather-fringes showing as indistinct dark scaling; tips of primaries often scarcely glossy, sooty-black; in some southern populations, hindneck and side of head and neck more strongly bronzy-brown than in north due to stronger bleaching, e.g., south from Yugoslavia (Stresemann 1920), but no difference from northern populations in fresh plumage. Sexes similar, but on average ♀ perhaps slightly less strongly purple than ♂, and with more dull black of feather centres visible. NESTLING. Down mouse-brown, fairly dense but rather short, on upperparts, upperwing, and thigh. JUVENILE. Head including throat sooty-black, feathers with restricted amount of purplish or virtually none at all; neck backwards to rump, vent, and shorter upper and under tail-coverts dull dark grey, feather-tips blackish, bases paler grey; no gloss, except for tips of feathers of lower mantle and scapulars; black feathers with more intense purple or purple-blue feathers appear even before wing full-grown (especially on mantle, back, chest, and belly), giving body mottled appearance, this either late-growing juvenile feathers or early 1st adult. Longer tail-coverts black with slight purple gloss. Wing and tail deep black, primaries, upper primary coverts, and bastard wing glossed purplish-blue, remainder fairly strongly purplish or purplish-blue when just fledged (sometimes more so than adult at same time of year, as these are abraded when young fledge). *In worn plumage*, within a few weeks of fledging, wing and tail considerably duller due to abrasion; much light grey of feather-bases visible, especially on hindneck, rump, and under-parts. See also First Adult (below) and Structure. FIRST ADULT. Like adult, but juvenile flight-feathers, tertials, tail, many wing-coverts (see Moults), and sometimes scattered feathers of body retained, distinctly less glossy and more liable to fading to black-brown than neighbouring 1st adult feathers, duller and browner than adult at same time of year; upper wing-coverts with distinct contrast between newer black and older brown feathers; tertials hardly glossy, much browner, and less broadly rounded at tip than in adult; tips of outer primaries pointed (not narrowly rounded); tail-feathers narrower and with less broadly truncate tips than in adult; tips of primaries and tail-feathers distinctly more abraded than in adult at same time of year and distinctly browner when plumage worn. New 1st adult feathering of head, neck, throat, and upperparts as in adult, but gloss slightly less intense and slightly more bluish on average; underparts down from chest and side of breast decidedly less black than in adult, brownish- or greyish-black, often contrasting with deeper black of lower throat, virtually without gloss, tips of feathers often with paler dull grey or brown fringe, showing as narrow pale scaling, grey of feather-bases partly visible, especially when worn.

**Bare parts.** ADULT, FIRST ADULT. Iris dark brown. Bill black, exceptionally yellow (Löns 1907). Leg and foot black, fissures between scutes and soles grey. Mouth of adult blue-black, of 1st adult flesh, greyish-flesh, or flesh with grey mottling; depending on social status, mouth turns fully black between ages of 8 months (in highly dominant paired birds) and 28 months or more (in subordinate unpaired birds) (Heinrich and Marzluff 1992). In partial albino birds from Faeroes, bill (especially base) and leg (especially joints and rear) partly yellow. NESTLING. At hatching, skin, including bill and foot, flesh-pink. Mouth blood-red or purplish-pink. Gape-flanges rather narrow, pale yellow to yellowish-flesh. At 9–10 days, skin greyish-flesh, gape-flanges and cutting edges pale yellow, bill and scutes of legs and foot dusky grey. At c. 3 weeks, bill blue-grey with black culmen and tip. Eye bluish- or dove-grey once open. JUVENILE. Like adult, but iris blue-grey, changing to dark brown some months after leaving nest; mouth flesh-colour. (Heinroth and Heinroth 1924–6; Witherby *et al.* 1938; RMNH.)

**Moults.** ADULT POST-BREEDING. Complete; primaries descendent. In Britain, moult starts with shedding of p1 mid-April to mid-June, completed with regrowth of p10 early September to late October (Holyoak 1974); starts March–April, completed after *c.* 140 days (Ginn and Melville 1983). Tail moult starts with t1 and secondaries with s1 at primary moult score 15–25 (when p4–p5 shed), completed at same time as p10 (primary score 50), or (inner secondaries, s5–s6) slightly later (Holyoak 1974; Ginn and Melville 1983). In 7 captive breeders followed over several years, Germany, p1 shed 17 March to 22 April (on average, *c.* 2 April), at about time when eggs about to hatch; p2 shed shortly after p1 (on average, *c.* 6 days later), but difference between shedding of subsequent primaries gradually larger, e.g. p8 shed *c.* 20 days after p7; p10 shed 122 (11) 111–133 days after p1, *c.* 18 July to 27 August; tail moult starts with shedding of t1 and secondary moult with s1 shortly before or after shedding of p5, at about time of fledging of young; tertials start with s8 at about shedding of p6, followed by s9; last tail-feather (t6) dropped 60 (11) 45–72 days after 1st (Gwinner 1966). In 4 captive birds, Belgium, p1 shed late March or early April, p10 early August to mid-September; s1 mid-May to mid-June, s6 mid-August to early September; tertials (s7–s11) from early May/mid-June (s8) to late July/August (s11); t1 mid-April to mid-June, t6 late June to late August (Delmotte 1981, which see for details). Moult June–September (Stresemann 1920). Of 34 birds in moult examined (BMNH, RMNH, ZFMK, ZMA), those from North Africa and Ibiza, as well as non-breeding 1-year-olds from western and central Europe, started moult April or early May, completing August and later, other birds from western and central Europe as well as birds from northern Europe, Greenland, and central Asia started early May to early June and completed September–October(–November). In *tibetanus*, primary moult half-way through and body and tail just started April–May; most completed late July and early August (Meinertzhagen 1926). For small samples of moulting birds, see also Stresemann (1928), Dementiev and Gladkov (1954), Paludan (1959), Keve and Rokitansky (1966), and Diesselhorst (1968). POST-JUVENILE. Partial: head, body (except sometimes for some scattered feathers, in particular on vent, outer scapulars, or tail-coverts, especially in birds from Greenland, Iceland, and northern Europe), and variable number of lesser and median upper wing-coverts (in some birds, all lesser and median; in others, only some rows of lesser or median, especially in north of range); occasionally, some tertial coverts or t1 new, mainly in south of range (RMNH, ZFMK, ZMA). Starts at age of *c.* 2 months, completed at *c.* 4 months (Heinroth and Heinroth 1924–6). Fully juvenile birds examined up to early July, birds in moult mid-June to late September, birds with moult completed from early August onwards (from localities as far apart as Ibiza, Turkey, Sweden, and Greenland) (RMNH, ZFMK, ZMA).

**Measurements.** ADULT, FIRST ADULT. Nominate *corax*. Fenno-Scandia, Britain, Netherlands, Germany, and Alps, all year; skins (RMNH, ZFMK, ZMA). Juvenile wing and tail are retained juvenile wing and tail of 1st adult. Bill (S) to skull, bill (N) to distal corner of nostril; exposed culmen on average 7·5 less than bill (S).

| | ♂ | | ♀ | |
|---|---|---|---|---|
| WING AD | 430·4 (11·76; 15) | 407–452 | 412·4 (11·66; 10) | 400–439 |
| JUV | 416·5 ( 7·98; 10) | 403–430 | 401·6 ( 8·96; 9) | 390–417 |
| TAIL AD | 230·0 ( 7·73; 15) | 221–243 | 227·1 ( 9·52; 8) | 213–245 |
| JUV | 231·9 (10·30; 10) | 207–243 | 223·4 ( 5·96; 9) | 214–233 |
| BILL (S) | 78·1 ( 2·97; 24) | 73·0–83·4 | 74·5 ( 3·01; 19) | 70·9–80·3 |
| BILL (N) | 47·5 ( 1·77; 24) | 44·0–50·6 | 45·3 ( 2·23; 19) | 43·2–49·9 |
| TARSUS | 70·7 ( 2·79; 26) | 66·2–75·0 | 68·6 ( 2·12; 20) | 65·0–72·3 |

Sex differences significant, except for tail.

Wing, bill to skull, and tarsus, adults only. (1) *C. c. principalis*, Greenland. (2) *C. c. varius*, Iceland and Faeroes. (3) *C. c. hispanus*, Spain and Balearic Islands. Nominate *corax*, (4) Sardinia, (5) southern Yugoslavia and Greece (wing including data from Stresemann 1920), (6) north-east Turkey and Caucasus. (7) *C. c. laurencei*, Crete and Levant. (8) *C. c. tingitanus*, North Africa. *C. c. canariensis*, (9) Fuerteventura (eastern Canary Islands), (10) Tenerife and Gomera (western Canary Islands). (RMNH, ZFMK, ZMA, ZMB.)

| | ♂ | | ♀ | |
|---|---|---|---|---|
| WING (1) | 457·1 (20·78; 6) | 429–486 | 442·4 ( 9·17; 7) | 434–456 |
| (2) | 436·1 (17·95; 5) | 406–451 | 427·2 ( 6·45; 7) | 420–438 |
| (3) | 424·4 (13·83; 7) | 403–439 | 410·2 ( 1·04; 3) | 409–411 |
| (4) | 425·6 ( 5·22; 5) | 420–434 | 411·9 ( 8·00; 9) | 405–426 |
| (5) | 416·2 (15·08; 6) | 400–438 | 421·2 ( 5·53; 3) | 416–427 |
| (6) | 438·2 ( 8·39; 3) | 428–443 | 417·5 ( — ; 1) | — |
| (7) | 436·2 (10·35; 6) | 423–450 | 428·0 ( — ; 2) | 424–432 |
| (8) | 405·2 (15·98; 6) | 381–423 | 388·8 ( 5·52; 4) | 383–396 |
| (9) | 382·0 ( 6·77; 5) | 370–388 | 383·2 (11·89; 6) | 367–400 |
| (10) | 396·8 (10·67; 8) | 380–411 | 383·2 (10·02; 7) | 368–394 |
| BILL (1) | 90·2 ( 3·11; 5) | 86·5–94·0 | 79·9 ( 2·98; 6) | 76·4–82·9 |
| (2) | 82·8 ( 2·56; 5) | 80·1–85·0 | 78·3 ( 3·29; 7) | 74·9–82·5 |
| (3) | 77·2 ( 1·21; 7) | 75·3–78·5 | 71·5 ( 1·75; 3) | 69·8–73·3 |
| (4) | 78·7 ( 2·19; 5) | 76·5–81·1 | 73·7 ( 1·47; 9) | 71·9–76·0 |
| (5) | 74·7 ( 1·34; 5) | 73·6–76·2 | 71·1 ( — ; 1) | — |
| (6) | 80·9 ( 2·98; 3) | 78·9–84·3 | 74·4 ( — ; 1) | — |
| (7) | 80·4 ( 3·60; 6) | 76·0–85·8 | 77·2 ( — ; 2) | 76·2–78·2 |
| (8) | 68·6 ( 4·05; 6) | 65·2–73·5 | 64·4 ( 0·48; 4) | 63·8–64·9 |
| (9) | 69·8 ( 0·85; 5) | 68·4–70·6 | 67·9 ( 1·09; 5) | 67·2–69·8 |
| (10) | 69·5 ( 2·66; 8) | 65·0–74·2 | 66·0 ( 1·24; 7) | 63·8–67·2 |
| TARSUS (1) | 72·5 ( 3·39; 6) | 69·2–76·0 | 68·8 ( 1·30; 7) | 67·2–71·0 |
| (2) | 71·2 ( 1·24; 5) | 69·5–72·6 | 69·8 ( 1·42; 7) | 68·3–72·6 |
| (3) | 76·3 ( 2·25; 7) | 73·8–80·0 | 72·6 ( 2·25; 3) | 70·0–74·0 |
| (4) | 73·2 ( 1·55; 6) | 71·8–75·7 | 71·4 ( 1·94; 9) | 69·0–75·1 |
| (5) | 70·5 ( 0·75; 5) | 69·8–71·3 | 67·5 ( — ; 1) | — |
| (6) | 70·4 ( 2·74; 3) | 67·4–72·8 | 71·0 ( — ; 1) | — |
| (7) | 73·4 ( 1·29; 6) | 72·0–75·7 | 74·4 ( — ; 2) | 74·0–74·8 |
| (8) | 72·6 ( 1·97; 6) | 69·8–74·5 | 71·1 ( 1·24; 4) | 69·8–72·6 |
| (9) | 67·6 ( 2·37; 5) | 64·0–69·9 | 67·1 ( 2·45; 6) | 64·0–69·7 |
| (10) | 69·0 ( 0·87; 8) | 68·2–70·4 | 65·6 ( 0·84; 7) | 64·7–67·2 |

Ages and sexes combined. Nominate *corax*, (1) Britain, (2) Scandinavia. (3) *C. c. hispanus*, Spain. *C. c. laurencei*, (4) Cyprus, (5) Levant, (6) Iraq, (7) south-east Iran and western Pakistan, (8) southern Afghanistan, (9) south-east Pakistan and north-west India. (10) *C. c. tingitanus*, North Africa (Morocco to north-west Egypt). (11) *C. c. canariensis*, Canary Islands. *C. c. varius*, (12) Iceland, (13) Faeroes. *C. c. principalis*, (14) Greenland and northern Canada, (15) Alaska. *C. c. kamtschaticus*, (16) Komandorskiye Islands (northern Pacific), (17) north-east Siberia and Manchuria (China). (18) *C. c. tibetanus*, mountains of central Asia from Pamirs to eastern Tibet. (Meinertzhagen 1926.)

| | n | wing | bill | bill depth |
|---|---|---|---|---|
| (1) | 17 | 394–445 | 74–84 | 28–32 |
| (2) | 19 | 405–445 | 72–85 | 29–33 |
| (3) | 5 | 410–430 | 67–73 | 27–31 |
| (4) | 8 | 401–434 | 66–74 | 27–30 |
| (5) | 16 | 396–446 | 66–80 | 27–30 |
| (6) | 6 | 419–440 | 70–80 | 27–30 |
| (7) | 13 | 400–450 | 67–77 | 24–29 |
| (8) | 10 | 407–462 | 71–79 | 23–27 |
| (9) | 33 | 389–437 | 66–75 | 23–27 |
| (10) | 38 | 362–430 | 60–69 | 25–28 |
| (11) | 19 | 355–428 | 63–73 | 25–29 |

| (12) | 14 | 417-445 | 73-82 | 28-33 |
|------|----|---------|-------|-------|
| (13) | 6 | 405-434 | 82-87 | 29-31 |
| (14) | 61 | 411-484 | 72-94 | 27-35 |
| (15) | 18 | 418-455 | 68-90 | 24-31 |
| (16) | 22 | 405-460 | 67-89 | 24-30 |
| (17) | 8 | 410-451 | 71-82 | 26-34 |
| (18) | 23 | 455-493 | 75-83 | 27-31 |

Wing. Nominate *corax*. Scandinavia 390-445, once 465 (*n* = 32) (Salomonsen 1935). Poland, live birds, wing: ♂ 423·3 (14·4; 27) 388-442, ♀ 413·8 (11·9; 20) 395-433 (Ruprecht 1990). Balkans: wing up to 475 (Kleinschmidt 1940). USSR: ♂ 441·2 (49) 410-473, ♀ 432·3 (34) 385-460 (Dementiev and Gladkov 1954). European Russia 406-458, West Siberia 413-465, mid-Siberia 415-468 (Johansen 1944).

*C. c. varius*. Faeroes 400-426 (*n* = 9), Iceland 410-445 (*n* = 11) (Salomonsen 1935).

*C. c. kamtschaticus*. USSR: ♂ 455·7 (10) 425-480, ♀ 448·7 (4) 445-455 (Dementiev and Gladkov 1954). East Siberia and Mongolia 423-481 (Johansen 1944). East Siberia, ♂ 492, ♀ 443; Manchuria, Ussuriland, and Vladivostok, ♂ 433·4 (13·45; 7) 411-450, ♀♀ 433, 435 (Meise 1934a; Piechocki 1959).

*C. c. tibetanus*. Tibet: ♂ 486 (13) 462-517, ♀ 467 (12) 432-480 (Vaurie 1972).

*C. c. laurencei*. Afghanistan (perhaps in part intermediate with *tibetanus*): ♂ 461·4 (15·85; 5) 440-483, ♀ 438·2 (7·69; 5) 429-449 (Paludan 1959). USSR, sexes combined: 442 (405-471) (Stepanyan 1978). Whole range, sexes combined: 445 (16) 415-474 (Vaurie 1954b, 1959). Israel: 390-450 (Paz 1987).

*C. c. tingitanus*. North-west Egypt: 359-398 (*n* = 5) (Meinertzhagen 1930). See also Niethammer (1953a) and Vaurie (1954b, 1959).

JUVENILE. For growth in captivity, see Mödlinger (1977).

**Weights.** ADULT, FIRST ADULT. Nominate *corax*. Norway 990-1380 (*n* = 9) (Haftorn 1971). Poland: ♂ 1254·0 (112·6; 10) 1080-1370, ♀ 1147·1 (65·3; 7) 1070-1235 (Ruprecht 1990). USSR: ♂ 1383 (3) 1100-1560, ♀ 1085 (3) 798-1315 (Dementiev and Gladkov 1954). Netherlands, winter: ♂♂ 1205, 1275 (ZMA). North-west Greece, November-December: ♂ 1112·5 (62·92; 4) 1100-1200, ♀ 1150 (Makatsch 1950). Sardinia, January: ♂ 1200 (Demartis 1987). For relationship between measurements and weight, see Terentiev (1970).

*C. c. laurencei*. Israel: 900-1300 (Paz 1987).

*C. c. kamtschaticus*. USSR: 1300 (Dementiev and Gladkov 1954). Chukotskiy peninsula, November: ♂ 1500 (Portenko 1973). Mongolia, May: ♂ 1450, ♀ 1200 (Piechocki and Bolod 1972). Manchuria, September: 1190 (Piechocki 1959).

*C. c. principalis*. Eastern Greenland, juvenile, July: ♂ 1285, ♀ 1195 (ZMA). Northern Alaska: ♂ 1240 (5) 1100-1400, ♀ 1158 (3) 1050-1300 (Irving 1960).

JUVENILE. Average of 5 in captivity: 840 at *c.* 2 weeks, 1034 at *c.* 3, 1050 at *c.* 5 (Mödlinger 1977).

**Structure.** Wing long, bluntly pointed. 10 primaries: p7 usually longest, occasionally p8, rarely p6; in adult, p8 usually 1-15 shorter, p9 20-40, p10 100-150 (mainly 115-135), p6 3-11, p5 48-71, p4 85-116, p3 110-135, p2 135-165; p1 169·0 (15) 154-181 shorter in nominate *corax*, *varius*, and *laurencei*, 194·5 (5) 175-215 in *principalis* and *tibetanus*, 158·1 (5) 149-169 in *tingitanus* and *canariensis*; in juveniles (all races), p8 0-10 shorter, p9 19-35, p10 95-135, p6 4-13, p5 53-60, p4 86-108, p3 105-130, p2 125-142, p1 156·7 (10) 140-170. P10 94·0 (15) 75-110 longer than longest upper primary covert in adults of all races,

95·3 (10) 80-110 in juveniles. Outer web of p6-p9 deeply emarginated, inner web of p7-10 with distinct notch. Tip of longest tertial reaches to tip of p2-p3. Tail long, tip graduated; 12 feathers, t1 longest, t6 50·6 (25) 42-63 shorter in adult nominate *corax*, *varius*, and *laurencei*, 56·6 (5) 49-63 in adult *principalis* and *tibetanus*, 37·2 (10) 34-40 in adult *tingitanus* and *canariensis*, 38·4 (25) 28-50 in juvenile nominate *corax*, *varius*, and *laurencei*, 39·8 (5) 33-52 in juvenile *principalis* and *tibetanus*, 34·5 (5) 30-39 in juvenile *tingitanus* and *canariensis*. Bill markedly long, *c.* 120% of head length; straight and massive, deep and broad at base, strongly compressed laterally at tip; basal half of culmen straight, distal half strongly decurved; tip rather blunt. Depth of bill at middle of nostril 27·5 (28) 24·8-29·2 in ♂ nominate *corax*, 26·2 (27) 23·8-27·7 in ♀; 27·4 (5) 26·1-28·8 and 26·4 (7) 25·5-27·2 in ♂ and ♀ *varius*, 29·2 (5) 28·0-31·0 and 26·1 (7) 24·8-28·0 in ♂ and ♀ *principalis*, 27·1 (7) 26·0-29·0 and 25·1 (3) 24·4-26·0 in ♂ and ♀ *hispanus*, 27·6 (6) 26·4-29·2 and 27·2 (2) 27·0-27·4 in ♂ and ♀ *laurencei*, 25·6 (2) 24·8-26·0 and 24·0 (4) 23·6-24·9 in ♂ and ♀ *tingitanus*, 25·7 (13) 24·8-26·8 and 24·4 (13) 23·3-25·5 in ♂ and ♀ *canariensis* (RMNH, ZFMK, ZMA; see also Meinertzhagen 1930 and Niethammer 1953a). Nostril rounded, covered by dense layer of bristles projecting forward from lateral base of upper mandible and which cover more than half of length of bill; similar but much shorter bristles extend obliquely downwards at lateral base of lower mandible; numerous short stiff bristles project from lateral base of upper mandible over gape. Feathers of chin soft, ending in hairs, feathers of throat narrow and elongated in adult, forming long throat hackles (tips partly broken off and tips ending bifurcated in some *tingitanus*); feathers of side and rear of crown, hindneck, and side of neck long, tips loose, silky; in juvenile, feathers of crown, throat, and neck broad and short, not elongated, barbs of both webs widely spaced, loose, as on remainder of body. Leg and foot short but strong, scutes on front of tarsus and on upper surface of toes distinct; claws strong, short, decurved. Middle toe with claw 54·3 (5) 51-58 in *tingitanus*, 58·6 (5) 57-60 in *canariensis*, 59·8 (30) 55-65 in other races; outer and inner toe with claw both *c.* 75% of middle with claw, hind *c.* 81%.

**Geographical variation.** Slight in intensity of gloss, colour of feather-bases, and length of throat feathers, more pronounced in size (wing, tail, weight), relative tarsus length, and relative length and depth of bill. 3 main groups in size: (1) large *principalis* from northern and eastern North America and Greenland (average wing of adult ♂ *c.* 455, tail *c.* 250) and *tibetanus* from mountains of central Asia (average wing of adult ♂ *c.* 473, tail *c.* 257); (2) small *tingitanus* from North Africa (average wing of adult ♂ *c.* 403, tail *c.* 208) and *canariensis* of Canary Islands (average wing of adult ♂ *c.* 394, tail *c.* 206) (in both these races, wing/tail ratio 1·91-1·94, against 1·82-1·87 in all other races, thus wing relatively longer or tail shorter than others); (3) all other races intermediate (average wing of adult ♂ 421-437, tail 226-239). In both large races, tarsus relatively short (average 71·5 in ♂, range 68-76, 15-16% of wing length). In small races, tarsus short in *canariensis* (68·3 in ♂, range 64-70, 17·3% of wing length), long in *tingitanus* (72·7 in ♂, range 70-75, 18·1% of wing length); in intermediate races, tarsus longer in *hispanus* from Iberia and Balearic Islands and in *laurencei* from east Mediterranean coast and lower Volga east to plains and hills of central Asia (*c.* 74-76 in ♂, range 72-80, 16·9-17·4% of wing), shorter in *varius* from Iceland and Faeroes and nominate *corax* from Scandinavia, Britain, and Pyrénées east to Yenisey basin (*c.* 69-71 in ♂, range 66-75, 15·9-16·6% of wing). In both large races, bill to skull longer in *principalis* (*c.* 87 in ♂, range 86-94); in intermediate races with longer tarsus, bill shorter in *hispanus* (*c.* 77 in ♂, range 75–78, 18·1% of wing), longer in *laurencei* (*c.* 80 in ♂, range 76–86, 18·5% of wing); in intermediate races with shorter tarsus, bill longer in *varius* and in nominate *corax* from Alps and Caucasus (*c.* 82 in ♂, range 79–85, 18·4–19·2% of wing), shorter in remaining populations of nominate *corax* (average *c.* 75–78 in ♂♂ of various populations, range 73–83, 17·8–18·4% of wing); in both smaller races, bill to skull slightly shorter in long-legged *tingitanus*, slightly longer in short-legged *canariensis*, range in ♂♂ of both 65–74 (17·0% of wing in *tingitanus*, *c.* 18·0% in *canariensis*. Bill relatively thick when compared with length in *canariensis*, *tingitanus*, and *hispanus* (depth at nostril 36·8% of length to skull in first 2 races, 35·2% in last), relatively slender in *principalis* and *varius* (depth 32–33% of length), intermediate (*c.* 34·4%) in all others. Races as defined above based mainly on measurements, as differences in colour slight. Gloss of *principalis* similar to nominate *corax*, but feather-bases on body said to be slightly whiter, contrasting more sharply with black tips (Vaurie 1959); however, feather-bases rather variable in colour in nominate *corax*, and some birds similar to *principalis*. In *varius*, feather-bases as in *principalis* (and thus doubtfully different from nominate *corax*), but gloss slightly less intense than in *principalis* and nominate *corax*, more bluish on head, throat, and upperparts, more greenish on remainder of underparts (Salomonsen 1935), but difference not marked in birds examined; on Faeroes, part of population partly white in past, mainly on head, underparts, and bases of flight-feathers and tail; *varius* close to nominate *corax* from Alps and Caucasus in size, and latter tend to show white feather-bases also; maintained here only due to average difference in gloss. *C. c. hispanus* more bluish than nominate *corax*, less intensely glossed purplish. *C. c. laurencei* slightly duller and more oily-bluish than fresh nominate *corax*, often distinctly duller when worn, side of head and neck then tinged coppery-brown (but less so than in *C. ruficollis*), especially in zone from Levant to India, less so in eastern Greece, Turkey, Cyprus, and eastern Iran; throat feathers distinctly shorter, tips sometimes bifurcated (former name '*subcorax*' is synonym of Brown-necked Raven *C. ruficollis*: Hartert 1903–10, 1921–2, Kleinschmidt 1940, Stepanyan 1978, 1990). *C. c. tingitanus* less strongly glossed purple than nominate *corax*, gloss more oily blue-green; throat feathers distinctly shorter, tip often bifurcated; feather-bases grey; feathering on thigh more restricted. *C. c. canariensis* strongly glossed purple, similar to nominate *corax*, but browner in worn plumage, especially in 1st adults of arid eastern Canary Islands (sometimes separated as *jordansi* Niethammer, 1953). *C. c. tibetanus* more strongly glossed purple than nominate *corax*; *hispanus* similar to nominate *corax* when fresh, slightly duller and browner when worn. Birds from Sardinia (sometimes separated as *sardus* Kleinschmidt, 1903) between *hispanus* and nominate *corax* in size, but nearer latter. *C. c. kamtschaticus* from eastern Asia a poor race; large in north-east of range (Kamchatka and probably further north), near *principalis*, and also in northern Mongolia and Transbaykalia, where probably grading into *tibetanus*, but smaller in Manchuria, Amurland, and Ussuriland, hardly differing from nominate *corax*. No information on birds from Pyrénées, northern Spain, Corsica, Sardinia, southern Italy, northern and central Turkey, and northern Kazakhstan; *hispanus* and *laurencei* both probably grade clinally into nominate *corax*.

For relationships, see *C. ruficollis*. CSR

## *Corvus rhipidurus* Fan-tailed Raven

PLATES 10 and 15 (flight)
[between pages 136 and 137, and facing page 280]

Du. Waaierstaartraaf  Fr. Corbeau à queue courte  Ge. Borstenrabe
Ru. Щетинистая ворона  Sp. Cuervo colicorto  Sw. Kortstjärtad korp

*Corvus rhipidurus* Hartert, 1918. Synonym: *C. affinis*.

Polytypic. *C. r. stanleyi* Roselaar, in prep., Sinai, Levant, and Arabia; nominate *rhipidurus* Hartert, 1918, Afrotropics north to Tibesti and Gebel Elba.

**Field characters.** 47 cm; wing-span 102–121 cm. Head and body close in size to those of Brown-necked Raven *C. ruficollis* but with shorter wings and much shorter, stump-like tail. Large, bat-winged black crow, with folded wing-tips extending well past tail and behaviour on ground recalling Rook *C. frugilegus*. Croak recalls gull *Larus*. Sexes similar; some seasonal variation through wear. Juvenile separable.

ADULT. Moults: mainly August–November (complete). When fresh, plumage entirely glossy blue-black, with additional purple and bronze iridescence marked in certain lights. With wear, blue and purple tones change to oily-blue bronze or even copper, and bird may then look browner, though less so than in *C. ruficollis*. Underwing shows contrast between black coverts and somewhat paler, browner underside to flight-feathers. Rictal bristles, bill, and legs black, but grey soles of feet may show against dark plumage in flight. JUVENILE. Duller than adult, with less gloss even in fresh plumage.

Unmistakable. Due to unusual silhouette flight appears less powerful than other large crows, most recalling *C. frugilegus*; much given to soaring and group acrobatics. Gait a walk, varied by sidling jumps and occasional leaps; relative lack of tail makes legs appear long. Has distinctive habit of holding bill open, as if panting. Mounts mammals to search fleeces for food; flocks feeding on ground recall those of *C. frugilegus*.

Commonest call a short, high-pitched, falsetto croak, 'pruk'; often given in series and suggesting *Larus*. Much higher pitched than usual call of Raven *C. corax* and

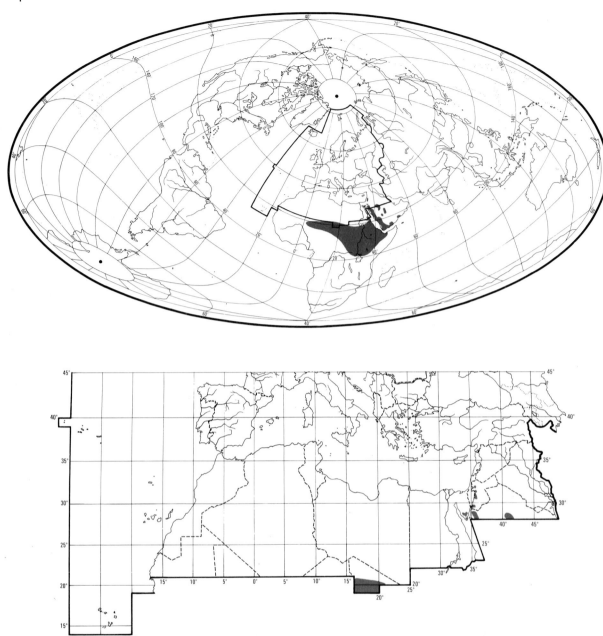

usually shorter and somewhat higher pitched than usual call of *C. ruficollis*.

**Habitat.** Breeds in subtropics and tropics, mainly in semi-arid or arid regions, preferring presence of cliffs or crags suitable for nesting, and creating opportunities for habit of playing and soaring in thermals. Only exceptionally nests in trees. Feeds on ground, often flying to distant feeding sites. (Mountfort 1965; Goodwin 1986.)

**Distribution.** No changes reported.

**Population.** ISRAEL. Estimated 300 pairs; apparently more abundant formerly, perhaps adversely affected by competition with Brown-necked Raven *C. ruficollis* which has spread (Shirihai in press).

**Movements.** Essentially sedentary, with local movements.

In Israel, mainly sedentary but recorded at Elat in south chiefly in winter, and formerly wintered regularly in and near Jerusalem; wanderers (mostly non-breeders) occur singly or in groups along Arava valley (Krabbe 1980;

Shirihai in press). Resident in Saudi Arabia, but apparently some move south from Tuwaiq escarpment September–January (Jennings 1980, 1981a). Birds breeding in Dhofar mountains (southern Oman) wander south to coast and north to Thamarit (*Oman Bird List*). In northern Yemen, large flocks (once totalling *c.* 1000 birds) seen rising on thermals and moving south or west, sometimes with raptors (Phillips 1982); these presumably only short-distance movements.

In Africa, also appears chiefly sedentary. No evidence of seasonal movement on Jebel Marra in northern Sudan (Wilson 1981). In East Africa, where breeds commonly at 400–1500 m, wanders to adjoining plains and highlands, including summit of Mount Nyiru at 2600 m; regular visitor to human habitations (Britton 1980).     DTH, DFV

**Food.** Mainly recorded scavenging near human settlements for offal, scraps, and rubbish; also takes insects and other invertebrates, grain pecked from animal droppings, berries, and fruit (Benson 1946; Jennings 1980; Cornwallis and Porter 1982; Lewis 1989). Drinks frequently (Coe and Collins 1986; Paz 1987). Forages on ground in pairs, or more frequently in flocks at rubbish dumps, wadis, oases, picnic sites, fish-drying areas, and fields of young corn; may travel long distance to find food (Mountfort 1965; Brooks *et al.* 1987; Ash 1988; Jennings *et al.* 1988; Goodman and Meininger 1989; Jennings *et al.* 1991). On ground, feeding actions similar to Rook *C. frugilegus* (Mountfort 1965; Wallace 1984). Usually takes locusts (Acrididae) from ground, though also recorded catching them in the air, and eating them in flight (Gallagher and Woodcock 1980; Symens 1990). Often waits on vantage point near human habitation; when scraps appear, flies down, walks quickly towards food, snatches it up with bill, hops rapidly away, and flies back to perch to eat (Archer and Godman 1961). Frequently perches on backs of goats *Capra* and camels *Camelus*, searching for ticks (Acari) and other ectoparasites; pecks continually at animal's coat, removing tufts of hair; when prey located, throws head back a little to swallow (Lewis 1989). Feeds on ripening dates *Phoenix* from tree, and in Saudi Arabia farmers may put sacks over fruit to reduce losses (Jennings and Salama 1989).     WGM

**Social pattern and behaviour.** Very poorly known, with no detailed studies. Most information from Israel (Shirihai in press) and Saudi Arabia (Jennings 1980, 1981a); review by Goodwin (1986).

1. Occurs singly, in pairs, and in groups (especially at good food sources) all year (e.g. Goodwin 1986, Paz 1987, Hüe and Etchécopar 1970, Shirihai in press). Frequently in large groups (Cornwallis and Porter 1982; Hollom *et al.* 1988). In Israel, largest concentrations (up to 70) July–January. Groups of non-breeders also occur during breeding season. (Shirihai in press; see also Roosting, below.) Reported to flock in late summer in Saudi Arabia where 130 seen together at Riyadh (Jennings 1981a). In northern Yemen, seen in flocks of up to 800 with over

1000 at communal roost (Phillips 1982; Brooks *et al.* 1987). Sometimes flocks with Brown-necked Raven *C. ruficollis* (Jennings 1980, 1981a; Phillips 1982; Goodwin 1986). BONDS. Mating system and nature of pair-bond not known. No information on roles of sexes in breeding. Reports of recently fledged juveniles in family parties in Israel (Shirihai in press) suggest post-breeding dependence of juveniles as in other *Corvus*. Age of first breeding not known. BREEDING DISPERSION. Sometimes colonial, with 2–5 pairs on same cliff ledge and 50–200 m between active nests, but some pairs breed separately in 'more firmly defined territories' (Goodwin 1986; Paz 1987; Shirihai in press). Goodwin (1986) received reports from Israel of nesting pairs holding extensive territories. No other details of size or purpose of territory. ROOSTING. Little known. In Arabia, reported sometimes to roost in palms with *C. ruficollis* (Bates 1936). Breeding birds presumed to roost in rock crevices and on sheltered ledges and occasionally in trees or on buildings, depending on nest-site, but no details. In Israel, gathers in tens or even hundreds (maximum 370 reported) at roosts (Shirihai in press). In northern Yemen, roosts in large groups all year except April–May (Brooks *et al.* 1987); at roost in Sana'a, numbers increased from *c.* 25 on 3 September to over 1000 at end of November (Phillips 1982).

2. Shows very little fear of man where not molested (Hüe and Etchécopar 1970; Goodwin 1986; D T Holyoak). Seems to be much shyer in Middle East than in Ethiopia (Hüe and Etchécopar 1970). Apparent play reported from flying birds in Israel, including catching feathers and papers in the air, and carrying stones or twigs in flight which are dropped and then caught again as they fall (Paz 1987). Reactions to predators not described; seen being mobbed by Long-legged Buzzard *Buteo rufinus* in Jordan (D I M Wallace). FLOCK BEHAVIOUR. Reported to be fond of soaring and playing in thermals, perhaps to even greater extent than other ravens (Goodwin 1986; Paz 1987; Hollom *et al.* 1988). No interspecific aggression seen in mixed flocks (Jennings 1980). SONG-DISPLAY. No details, but see 1 in Voice for Song. ANTAGONISTIC BEHAVIOUR. No information. HETEROSEXUAL BEHAVIOUR. Jennings (1980) noted that in January–April in Saudi Arabia pairs soar almost out of sight, then plummet back to cliffs; pursuit-flights and calling also reported, but unclear whether these were displays. No further information. RELATIONS WITHIN FAMILY GROUP. No details, but see Bonds (above). ANTIPREDATOR RESPONSES OF YOUNG, PARENTAL ANTI-PREDATOR STRATEGIES. No information.     DTH

**Voice.** Poorly known, with no detailed studies. Varied and sometimes conflicting descriptions indicate great variety of calls (Etchécopar and Hüe 1967; Goodwin 1986). This account distinguishes only a few main types, but fuller studies may reveal vocal complexity comparable to that of Raven *C. corax*.

CALLS OF ADULTS. (1) Song. Recording is of medley of call 2, soft clucks, high-pitched squeals, and loud tremolos (Fig I) suggestive of some frogs (Goodwin 1986; M E W North, P J Sellar). (2) Croaking-call. Best-known call, commonly given as series, a falsetto croaking very like White-necked Raven *C. albicollis* of Afrotropics, but perhaps even higher pitched (Goodwin 1986). Described as loud, high-pitched croak, rather like gull *Larus* (Hollom *et al.* 1988); raucous cries and guttural croaks (Archer and Godman 1961). Usually shorter and somewhat higher pitched than call 1 of Brown-necked Raven *C. ruficollis*

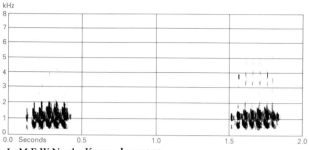

I M E W North Kenya June 1955

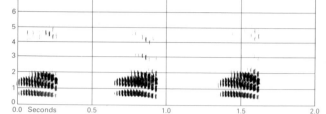

II M E W North Kenya June 1955

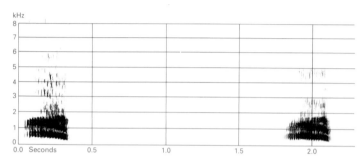

III T C White Kenya January 1983

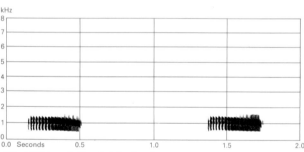

IV Mild (1990) Israel March 1987

(e.g. Etchécopar and Hüe 1967, Jennings 1980, Paz 1987, Mild 1990, J Palfery, D T Holyoak) and much higher pitched than Raven *C. corax* (Hüe and Etchécopar 1970). Recordings show considerable variation in pitch, duration, intensity, and amount of vibrato (Mild 1990; P J Sellar, D T Holyoak): Fig II shows 3 distinct, relatively high-pitched croaks, while Fig III shows 2 quieter, more guttural croaks; in Fig IV, calls are deep and rather hollow-sounding. (3) Knocking-call. 'Clacking of the tongue at two speeds' (Hüe and Etchécopar 1970), presumably comparable to call 8 of *C. corax* and call 2 of *C. ruficollis*. Not clear whether this is mechanical sound produced with tongue or true vocalization. (4) Other calls. (4a) Double 'cuhcuh' given repeatedly in flight by 1–2 birds, in answer to call 2 from accompanying bird (J Palfrey). Repeated low hooting 'hoo hoo' (N Tucker, D J Tombs, P J Sellar) may be similar. (4b) Various short calls with gobbling or bubbling quality (T C White, P J Sellar); may be distinct call or variants of call 2.

CALLS OF YOUNG. No information.                    DTH

**Breeding.** Little known, due to inaccessibility of nest. SEASON. Israel and Jordan: late February to late July; eggs laid early March to late April (Jourdain and Lynes 1936; Mountfort 1965; Goodwin 1986; Paz 1987; Shirihai in press). Arabia: nesting recorded mid-February to May, rarely in December and June; generally later in south (Gallagher and Woodcock 1980; Jennings 1980; Walker 1981*b*; Jennings *et al.* 1990). East Africa: eggs laid mid-April to early June (Archer and Godman 1961; Lewis 1989). SITE. Hole, crevice, or sheltered ledge, usually inaccessible in sheer cliff; rarely on building; in Somalia, 1–2 records of tree nests (Mackworth-Praed and Grant 1960; Archer and Godman 1961; Mountfort 1965; Paz 1987; Shirihai in press). Nest: loosely constructed platform and cup of sticks, twigs, and roots, lined with wool, hair, cloth, freshly plucked twigs, and other soft material (Jourdain and Lynes 1936; Archer and Godman 1961; Harrison 1975; Goodwin 1986). EGGS. See Plate 58. Sub-elliptical to long oval, smooth and glossy; pale greenish-blue, blotched and speckled or faintly streaked with olive-brown and dark brown, and less conspicuous violet-grey undermarkings (Archer and Godman 1961; Harrison 1975). 45·0 × 30·7 mm (41·8–50·4 × 27·6–34·5), $n = 31$; calculated weight 21·8 g (Schönwetter 1984). Clutch: 3–4 (2–6) (Archer and Godman 1961; Etchécopar and Hüe 1967; Goodwin 1986; Shirihai in press). Probably 1 brood (Harrison 1975). INCUBATION. 18–20 days (Shirihai in press). FLEDGING TO MATURITY. Fledging period 35–40 days (Shirihai in press). BREEDING SUCCESS. In Israel, family parties containing 3–5 recently fledged young recorded (Shirihai in press). In Somalia, occasionally parasitized by Great Spotted Cuckoo *Clamator glandarius* (Archer and Godman 1961).                    WGM

**Plumages.** (*C. r. stanleyi*). ADULT. Entirely black, glossed oily-purple or blue-purple all over, except for inner webs of flight-feathers and tail; gloss less strong and less violet than in Raven *C. corax* and Brown-necked Raven *C. ruficollis*, especially on wing. Tips of feathers of chin and (slightly) upper throat grey; feathers of throat slightly elongated, but broad (much shorter and less narrow than in *C. corax* and *C. ruficollis*), tips bifurcated. Feather-bases of hindneck and upper side of neck extensively white, but these usually hidden; bases elsewhere pale grey (in *C. corax* and *C. ruficollis*, bases of feathers of hindneck grey or off-white, less contrasting). *In worn plumage*, gloss even less intense, especially wing and tail hardly glossy, but (in contrast to *C. corax* and *C. ruficollis*), head, neck, wing, and tail still deep black, less inclined to bleach to brown-black or (on neck) brown; grey on chin and throat worn off. NESTLING. No information. JUVENILE. Closely similar to adult when plumage fresh, but gloss of head and body bluish rather than purplish; underparts distinctly duller, amount of gloss on each feather restricted, feathers with faint grey tips. Feathers of chin entirely dull grey, those of throat dull black, both distinctly shorter than in adult, tips rounded (in adult, slightly longer, tip bifurcated, more glossy). Gloss on wing and tail purplish, almost as strong as in adult; tail-feathers narrower than in adult; tip of outermost primary (p10) pointed, tips of tail-feathers narrowly rounded, less blunt. *In worn plumage*, gloss more rapidly lost than in adult, black of wing and tail fading to black-brown when heavily worn (less brown than in juvenile Brown-necked Raven *C. ruficollis*). FIRST ADULT. Like adult, but juvenile flight-feathers, tail, tertials, greater upper primary and secondary coverts, and variable number of median upper wing-coverts retained, occasionally some lesser coverts or scattered feathers of body also; these slightly duller than in adult at same time of year, contrasting slightly with new glossy lesser and median coverts and scapulars. Tail-feathers narrower; tips of inner primaries and p10 more pointed. *In worn plumage*, tips of retained juvenile tail, primaries, and tertials somewhat duller black and more abraded than in worn adult, but difference much less marked than in *C. corax* and *C. ruficollis*.

**Bare parts.** ADULT, FIRST ADULT. Iris brown. Bill black. Leg and foot black, ridges between scutes and soles dark grey. NESTLING, JUVENILE. No information.

**Moults.** ADULT POST-BREEDING. Complete; primaries descendent. No moult in 39 birds from January–May examined from entire geographical range, except for 2 from Aden (13 February and 8 March), which were in last stage of primary moult (score 49); 4 heavily worn 1-year-olds from Dead Sea and Arabia, May, had not started. Single ♂ from central Arabia, 2 June, had just started (score 2); 2 from Ethiopia had score 16 (18 August) and 22 (13 September); October birds had score 30 (central Arabia) and 32 (Ethiopia); single bird from Dead Sea had score 43 on 5 November, another had just completed (score 50) on 12 November; moult completed in 6 birds from December from entire range, except for 1 from southern Sudan, 23 December (score 48). No moult in 3 birds from late June, northern Niger (Fairon 1975). Thus, moult apparently mainly June–July to November–December. Tail starts with shedding of t1 at primary score *c.* 15, moulting centrifugally; body in heavy moult at score 30–43. (C S Roselaar.) POST-JUVENILE. Partial: involves head, body, and many or all lesser and variable number of median upper wing-coverts; some scattered feathers on body occasionally retained, especially on nape, mid-belly, or scapulars. In Sinai, 2 late July birds (both apparently recently fledged) had just started moult, replacing some feathers of mantle and side of breast (ZFMK). Moult completed in birds examined from October and later.

**Measurements.** ADULT, FIRST ADULT. *C. r. stanleyi*. Adult wing, bill to skull, and tarsus of birds from (1) Sinai, Israel, Jordan, and extreme north-west Saudi Arabia (Al Bad), (2) western Arabia from Mecca area south to Aden and Hadhramaut; other measurements combined, skins (BMNH, RMNH, ZFMK, ZMA, ZMB). Juvenile wing and tail are retained juvenile wing and tail of 1st adult. Bill (N) to distal corner of nostril; depth is depth of bill at middle of nostril; exposed culmen on average 2·0 less than bill to skull.

| | | ♂ | | | ♀ | | |
|---|---|---|---|---|---|---|---|
| WING (1) | | 364·3 ( 4·08; 6) | 359–370 | | 356·2 ( 5·74; 4) | 351–363 |
| | (2) | 363·6 ( 5·88; 7) | 355–373 | | 354·4 ( 5·10; 5) | 349–359 |
| JUV | | 347·0 (11·2 ; 4) | 333–357 | | 340·2 (12·7 ; 5) | 328–358 |
| TAIL AD | | 150·5 ( 3·09; 12) | 145–157 | | 148·1 ( 4·61; 8) | 143–155 |
| JUV | | 145·5 ( 3·77; 3) | 141–149 | | 142·0 ( 2·62; 5) | 139–146 |
| BILL (1) | | 54·8 ( 2·17; 8) | 52·6–58·2 | | 52·2 ( 0·91; 7) | 50·9–53·7 |
| | (2) | 55·6 ( 1·62; 7) | 53·6–57·7 | | 51·6 ( 1·15; 6) | 49·9–52·6 |
| BILL (N) | | 34·3 ( 1·79; 15) | 31·7–37·2 | | 32·1 ( 1·60; 11) | 30·0–34·0 |
| DEPTH | | 20·3 ( 0·55; 15) | 19·7–21·3 | | 19·5 ( 0·50; 11) | 18·9–20·5 |
| TARSUS (1) | | 60·7 ( 2·34; 8) | 57·8–63·6 | | 59·4 ( 2·60; 7) | 55·4–62·3 |
| | (2) | 61·7 ( 2·96; 7) | 58·0–65·4 | | 60·0 ( 1·79; 6) | 56·7–61·9 |

Sex differences significant for adult wing, bill, and bill depth.

Nominate *rhipidurus*. Adult wing, bill to skull, and tarsus of birds from (1) Ennedi (Chad), western and northern Sudan (Darfur, Kordofan, Nile valley south to 12°N, and Suakin area), and upper Egypt, (2) Ethiopia and north-west Somalia, and (3) southern Sudan (south of 6°N), Uganda, and Kenya; other measurements combined; skins (BMNH, RMNH, ZFMK, ZMA, ZMB).

| | | ♂ | | | ♀ | | |
|---|---|---|---|---|---|---|---|
| WING (1) | | 400·3 ( 4·62; 3) | 395–403 | | 384·3 (10·0 ; 3) | 373–392 |
| | (2) | 402·1 (12·5 ; 7) | 388–424 | | 397·0 (14·5 ; 4) | 381–415 |
| | (3) | 398·5 ( 9·04; 4) | 393–412 | | 389·0 ( — ; 2) | 384–394 |
| JUV | | 388·9 (12·9 ; 15) | 371–408 | | 365·8 (14·4 ; 13) | 345–381 |
| TAIL AD | | 164·3 ( 5·23; 10) | 155–172 | | 160·4 ( 3·69; 7) | 154–164 |
| JUV | | 156·8 ( 5·87; 15) | 148–168 | | 149·8 ( 5·49; 13) | 141–159 |
| BILL (1) | | 62·4 ( 2·14; 3) | 60·8–64·8 | | 58·1 ( 1·02; 7) | 57·1–59·8 |
| | (2) | 59·8 ( 1·76; 15) | 57·0–62·5 | | 58·1 ( 2·16; 10) | 55·0–61·2 |
| | (3) | 60·2 ( 0·80; 6) | 59·2–61·2 | | 57·6 ( 2·56; 4) | 55·2–60·9 |
| BILL (N) | | 37·2 ( 1·37; 24) | 35·0–40·2 | | 35·6 ( 1·33; 21) | 33·0–37·5 |
| DEPTH | | 22·5 ( 0·84; 24) | 21·4–24·4 | | 21·3 ( 0·90; 20) | 20·0–22·4 |
| TARSUS (1) | | 68·0 ( 1·46; 3) | 66·5–69·4 | | 65·3 ( 2·06; 7) | 63·0–67·3 |
| | (2) | 70·7 ( 2·10; 15) | 67·5–74·0 | | 66·5 ( 2·89; 10) | 62·3–69·8 |
| | (3) | 70·1 ( 2·18; 6) | 67·3–72·8 | | 62·6 ( 4·40; 4) | 58·8–67·5 |

Sex differences significant, except for adult wing and tail, bill (3), and tarsus (1).

See also Meinertzhagen (1926).

**Weights.** *C. r. stanleyi*. Israel: 330–550 (Paz 1987).

Nominate *rhipidurus*. Northern Niger, June: ♂ 705, ♀♀ 565, 610 (Fairon 1975). Northern Chad, April: ♂ 512; ♀♀ 560, 595 (Niethammer 1955; ZFMK).

**Structure.** Wing long, tip elongated, ending in blunt point, arm markedly broad. 10 primaries: p7 longest, p8 1–6 shorter (occasionally, p8 equal to or up to 2 longer than p7), p9 20–35 shorter, p10 95–126 (adult) or 85–95 (juvenile), p6 3–14, p5 35–52 (adult) or 24–45 (juvenile), p4 75–85 (adult) or 60–75 (juvenile), p1 120–142 (adult) or 105–125 (juvenile); in adult Brown-necked Raven *C. ruficollis*, p1 170–200 shorter; ratio of wing-tip (distance from tip of p7 to tip of p1) to wing length 0·32–0·34 in nominate *rhipidurus* at all ages, 0·34–0·38 in *stanleyi*, 0·40–0·48 in *C. ruficollis*. Distal half of outer primaries each markedly narrow, tapering to narrowly rounded (adult) or pointed (juvenile) tip. Outer web of p6–p9 emarginated, inner web of (p6–)p7–p10 with distinct notch. Tail markedly short, wing/

tail ratio 2·36 (0·08; 16) 2·24–2·48 (average 1·97 in *C. ruficollis*), tip rounded; 12 feathers, t6 20·6 (15) 12–28 shorter than t1 in adult, 15·9 (5) 13–19 in juvenile. Bill markedly short and deep (see Measurements), about equal to head length, less wide at base than in Raven *C. corax*; culmen strongly decurved, often with narrow blunt ridge on tip, especially in birds with deeper bills. Unlike other Palearctic Corvidae, upper nasal bristles directed upward, lower forward, forming short fan over nostril and base of upper mandible. Feathers of chin and throat short, broad, not as elongated as in *C. corax* or *C. ruficollis*; tips bifurcated in adult. Tarsus and toes strong; middle toe with claw 48·9 (5) 46–52 in *stanleyi*, 55·1 (6) 50–61 in nominate *rhipidurus*; outer and inner toe with claw both *c.* 78% of middle with claw, hind *c.* 84%.

**Geographical variation.** Marked; involves size only. Birds from Sinai and from Jordan valley south through Saudi Arabia to Yemen and Oman markedly smaller than birds from Afrotropics, warranting recognition of 2 races: nominate *rhipidurus* in Afrotropics (type locality Massawa, northern Ethiopia: see Hartert 1918*a*, *c* for nomenclatural history), *stanleyi* in Arabia, Sinai, Israel, and Jordan (type specimen BMNH 1946.63.10, north-west shore of Dead Sea: Roselaar in prep.). Named in honour of Stanley Cramp OBE, instigator of this Handbook. Races do not differ in colour, but *stanleyi c.* 10% smaller in all measurements; in particular, adult wing, tail, bill to skull, and tarsus hardly overlap in length (see Measurements). No clear variation in size within each race.

Forms species-group with White-necked Raven *C. albicollis* of eastern and southern Africa and Thick-billed Raven *C. crassirostris* of north-east Africa (those 2 together comprising a superspecies). See also Goodwin (1986).                CSR

# Family STURNIDAE starlings

Medium-sized oscine passerines (suborder Passeres). Habitat range wide: grassland, savanna, and steppes, woodland, forest, etc.; some species closely associated with human cultivation and habitation. Mainly arboreal to greater or lesser extent but some (especially in genus *Sturnus*) largely ground-feeders. Many species frugivorous and insectivorous but others more omnivorous, eating variety of animal and vegetable foods—including large insects, small vertebrates, and seeds. Brahminy Starling *S. pagodarum* and (to lesser extent) a few other species specialize in taking nectar, Rose-coloured Starling *S. roseus* and (especially) Wattled Starling *Creatophora cinerea* in feeding on locusts. Family of Old World origin: found mainly in Africa, Eurasia, Indonesia, and western Pacific islands, with main centres of distribution in Afrotropics and southern Asia. A few species commonly kept as cage or aviary birds. Some introduced elsewhere, 2 (Starling *S. vulgaris* and Common Myna *Acridotheres tristis*) widely, including North America and New Zealand (where no native starlings occur) and Australia with only 1 native species. Predominantly sedentary or nomadic but some species migratory. 113 species (Roselaar 1991): 6 in west Palearctic, 5 breeding (including introduced *A. tristis*) and 1 vagrant.

Starlings long accepted as well-defined, homogeneous family of uncertain affinity. Here, following Voous (1977), placed near Corvidae (crows) but also thought at one time or another to be closer to Ploceidae (weavers), Dicruridae (drongos), Oriolidae (orioles), or Icteridae (New World blackbirds and allies); see review by Sibley and Ahlquist 1984. 2 subfamilies: Sturninae (starlings) and Buphaginae (oxpeckers, 2 species); latter confined to Afrotropics and not considered further here. About 24 genera of starlings (variously also called mynas or grackles), including (1) *Aplonis* (shining starlings), 25 species—Indonesia to Pacific, with 1 marginally in Australia; (2) *Onychognathus* (red-winged starlings), 10 species—Afrotropics, with 1 species (Tristram's Grackle *O. tristramii*) in Middle East; (3) *Lamprotornis* (glossy starlings), 15 species—Afrotropics; (4) *Spreo* (superb starlings), 7 species—Afrotropics; (5) *Sturnus* (typical starlings), 16 species—Eurasia to Indonesia (mainly southern and eastern Asia to Indonesia): (6) *Acridotheres* (typical mynas), 7 species—southern Asia and eastern Indonesia; (7) *Basilornis* (crested mynas), 4 species—Indonesia, Philippines. Remaining genera each of only 1–3 species.

Typically rather heavily built with sexes of similar size (♂ slightly larger). Head bristly and almost naked in a few species, some others having wattles, lappets, or bare skin round eye. Bill usually quite stout and pointed, sometimes with hook; often decurved to some extent; characteristically used by ground-feeding species to pry when feeding, i.e. by method known as open-billed probing in which closed mandibles inserted into hole and then pressed open in order to expose and then seize prey; linked with anterior narrowing of skull which enables feeding bird to look forward while prying. Nostrils open and free of feathers but protected by operculum; no nasal or rictal bristles. Tongue of nectar-feeding *S. pagodarum* has brush-like tip. Wing broad and rounded or rather long and pointed; 10 primaries, p10 reduced (length 15–25% of longest) and often lanceolate. Flight strong, fast, and straight in longer-winged species, slower and more deliberate in others. Tail often fairly short and square but long and graduated in some; 12 feathers. Leg and toes sturdy. Tarsus quite long in many species, shorter in others; scutellate, booted at rear. Gait a hop in some species but most walk and run, often with upright stance. Head-scratching by indirect method (see, e.g., Simmons 1957*a*, 1961*a*). Drinking and bathing methods as in most other passerines but, as well as shaking wings when drying plumage after bath, birds often first beat them vigorously. Little information on sunning behaviour except in *S. vulgaris* which performs frequently, especially in late summer, while adopting variety of postures—including lateral and full-spreadeagle; both these postures also reported from *S. roseus* (see Simmons 1986*a* for details). Anting observed in at least 21 species of 9 genera (including 3 *Lamprotornis*, 7 *Sturnus*, and 3 *Acridotheres*), mainly in captivity (Simmons 1957*a*, *b*, 1966): birds collect acid-spraying ants one by one in bill and apply them progressively in growing ball or wad, using direct active method only. *S. vulgaris* commonly seen anting in wild thus, mainly when ants swarming in late summer or autumn; anting (like bathing and sunning) often a social activity in this species. Jungle Myna *A. fuscus* recorded applying millipede (Clunie 1976; discussed by Ehrlich *et al.* 1986).

Voice relatively unspecialized; calls mostly simple, often loud, harsh, and grating, but uttered frequently and often delivered in garrulous manner. Advertising song also typically simple, usually consisting of musical whistles or warblings. Mimicry common, especially from ♂ in courtship song: some mynas (especially Hill Myna *Gracula religiosa*) can be trained to imitate human phrases with surprising accuracy. Most species highly gregarious at times, especially when feeding and roosting. Many loosely colonial when breeding, each pair defending small territory near nest, but some tropical species nest communally (i.e. cooperatively) with helpers and maintain group territory (Feare 1984; Brown 1987). Monogamous mating system probably the general rule but duration of pair-bond variable; evidently sustained and strong in many tropical species but seasonal and less strong in others, e.g. *S.vulgaris* (in which ♂♂ tend to polygamy and often change mates between broods). Although sociable, most starlings maintain clear individual distance and avoid physical contact with other birds of same species (including mate), never clumping, even at roost; allopreening, however,

recorded from a few species (in genera *Leucopsar*, *Onychognathus*, and *Acridotheres*) with persistent pair-bonds (see Harrison 1965*c*). Nest usually a rough but substantial collection of material placed in existing or (in 2 species) excavated hole, but domed nest built in vegetation by a few others, and suspended ploceid-like nest by Long-tailed Starling *Aplonis magna* of New Guinea; usually built by both sexes or ♀ only (but in *S. vulgaris* mainly by ♂). Incubation usually by ♀ only, or by both sexes but with ♀ often taking larger share. Both sexes usually feed young.

Plumage variable: often black or black-and-white, sometimes with patches of rufous, brown, grey, pink, crimson, or yellow; silky and often with glossy, metallic iridescence. A few species have erectile or permanently raised crests. Sexes usually quite similar, with ♀ often duller and (e.g.) lacking bright facial colour of ♂; in a few species, sexes different. Seasonal variation quite marked in a few species, mainly due to changes caused by abrasion or by development of lappets, etc. Juvenile plumage either like duller version of adult or, more usually, quite different, underparts often being white with black streaking. Nestlings naked with rudimentary down, or with plentiful down in some species; no tongue markings but interior of mouth conspicuously coloured (yellow, etc.). Adults have single complete post-breeding moult annually, starting after fledging of last brood of young. Post-juvenile moult partial or complete, starting 1–2 months after fledging.

Relationship to other passerine groups much debated (see above). Egg-white protein data (see summary by Sibley 1970) suggest no affinity with corvids but possibility of some relationship to wood-swallows (Artamidae) especially (as well as to ploceids, drongos, and others). Mockingbirds (Mimidae) not previously considered at all close, but data of Sibley and Ahlquist (1984, 1990) now indicate these to be closest relatives of starlings—close enough, in fact, for both to be placed in same family within superfamily which also includes thrushes (Turdidae) and Old World flycatchers (Muscicapidae) as defined here (see Volumes V and VII).

---

## *Onychognathus tristramii*  Tristram's Grackle

PLATES 12 and 18 (flight)
[between pages 136 and 137, and 280 and 281]

Du. Tristrams Spreeuw    Fr. Rufipenne de Tristram    Ge. Tristramstar
Ru. Тристраммов длиннохвостый скворец    Sp. Estornino irisado    Sw. Sinaiglansstare

*Amydrus Tristramii* Sclater, 1858

Monotypic

**Field characters.** 25 cm (tail 9 cm), wing-span 44–45 cm. Head and body close in size to Starling *Sturnus vulgaris* but wings and especially tail noticeably longer, with silhouette somewhat reminiscent of Ring Ouzel *Turdus torquatus*. Lengthy, strong-billed and strong-legged, starling-like bird, with long, square-cut tail. Plumage looks

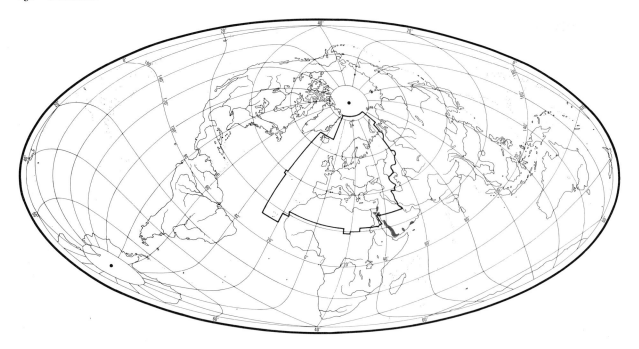

black at distance, with striking pale chestnut patches on primaries. Song melancholy. Flocks dash about. Sexes dissimilar; no seasonal variation. Juvenile separable.

ADULT MALE. Moults: July–September (complete). Glossy blue-violet, with greener tones in iridescence on wing feathers; basal two-thirds of all primaries pale chestnut, forming conspicuous panel on folded wing and bold patch above and below in flight. Eye reddish-brown, giving rather angry look. Bill strong, shaped like that of Golden Oriole *Oriolus oriolus*; black. Legs grey-black. ADULT FEMALE. Less glossy than ♂, especially on head, neck and breast which are ash-grey, streaked purplish-black; wing-patches slightly less contrasting, due to duller primary-tips. JUVENILE. Even less glossy than adult ♀. Head, neck, and breast darker, dull black or sooty-brown like rest of body. Wing-patches enlarged by similarly coloured greater primary coverts. Eye dull brown.

Unmistakable in west Palearctic, with other members of this African genus approaching no nearer than coast of lower Red Sea. Flight fast, action recalling both *Sturnus* and thrush *Turdus*; often performs sweeping, diving, and climbing manoeuvres which, when indulged in by flocks, give impression of fearless and purposeful aerial play. Gait combines bouncing hop and striding walk.

Song a sweet, yet wild series of whistles and fluting notes, with strange melancholy quality. Calls include loud and fluting 'chee-oo-wee' or 'dee-oo-ee-o', recalling *S. vulgaris* and *O. oriolus*.

**Habitat.** Restricted to Arabia and Levant, where basically confined to mountains or other rocky areas, up to 3200 m in Yemen (Cornwallis and Porter 1982), though avoiding barren mountain tops and inhabiting ravines, canyons, and cliffs, down to below sea-level. Roosts and nests in these, using holes or crevices in caves. Needs access to vegetation bearing food, from desert plants such as *Nitraria* to dates *Phoenix*, prickly pears *Opuntia*, and grapes *Vitis* obtained on visits to villages and even towns, especially rubbish dumps. Will also pluck ticks from donkeys, camels, and other grazing animals, wandering to desert margins and even to coastal plains; some evidence of shift to lower ground in winter. (Phillips 1982.) Will fly considerable distances daily to visit food sources.

Association with man at Arad (Israel), 600 m above sea-level, has recently led to urban nesting in holes or crevices in uninhabited buildings 6–18 m above ground; local food supply was formerly exploited only by birds nesting in nearby wadis, some of which continue to fly in from outside, competing with new urban residents (Hofshi *et al.* 1987*a*). Recent increase in numbers and spread of range seem attributable to growing dependence on man, and greater tameness. Raids on plantations, however, sometimes lead to protective shooting. Restricted choice of nest-sites is offset by diurnal mobility.

**Distribution.** Has spread in Israel.

ISRAEL. Up to 1950s, limited distribution in Judaean desert and Dead Sea depression; has since expanded range, following development of agriculture and settlements (Shirihai in press).

**Population.** ISRAEL. Has increased (see Distribution);

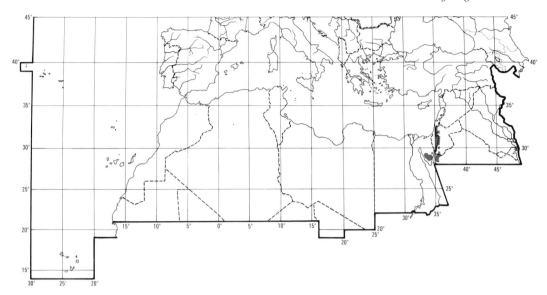

1000-2000 pairs in 1980s (Shirihai in press). JORDAN. 21-40 pairs Petra, 1983 (Wittenberg 1987). EGYPT. Locally common, Sinai (Goodman and Meininger 1989).

**Movements.** Resident and dispersive, with flocks seen out of breeding season well away from breeding range (Bigger 1931; Hofshi *et al.* 1987a; Paz 1987). In Oman, disperses widely into mountains and coastal plains (Gallagher and Rogers 1980) and once reported from Muscat, well out of normal range (Gallagher and Woodcock 1980). In Jordan, south-east movement of 211 birds recorded over Petra, 26-30 April (Wallace 1984); flies considerable distances from breeding or roost site to feed each day (Mild 1990).      CJF

**Food.** Mainly fruit and insects. Feeds in bushes and on ground, sometimes among cattle. Will crack snail shells on anvil like thrush *Turdus* (Jennings *et al.* 1991). Picks ectoparasites from pelts of ibexes *Capra* (though not from adult ♂♂), donkeys, cattle, and camels; with ibexes, concentrates on head and genital area (Gallagher and Woodcock 1980; Walker 1981b; Paz 1987; Yosef and Yosef 1991).

Diet includes the following. Invertebrates: grasshoppers (Orthoptera), butterflies (Lepidoptera), flies (Diptera), bees (Hymenoptera: Apoidea), ticks (Acari), snails (Mollusca). Fruits of desert plants: *Nitraria*, *Ciridia*, *Salvadora*, *Ochradenus*, *Acacia*, *Myoporum*, *Atriplex*, rose hips *Rosa*, olives *Olea*, prickly pear *Opuntia*, figs *Ficus*, grapes *Vitis*, dates *Phoenix*. Takes human food scraps and visits rubbish tips. (Gallagher and Rogers 1980; Walker 1981b; Phillips 1982; Brooks 1987; Hofshi *et al.* 1987a; Paz 1987; Mild 1990.) No detailed studies or specific information on food of young.      CJF

**Social pattern and behaviour.** Little information.

1. Gregarious. Some remain in pairs outside breeding season (see Bonds, below), but usually in flocks of 10-50 (Shirihai in press). Winter flocks of 100-300 recorded, at other times usually small groups of parents and offspring (Paz 1987); *c.* 2 weeks after leaving nest young join flocks of other juveniles (Hofshi 1985). Non-breeding birds flock in summer (Shirihai in press). BONDS. Monogamous, pair-bond retained all year and apparent in flocks. Both sexes participate in searching for nest-site, collecting material, and in building; ♀ incubates alone, both sexes feed young in nest and for *c.* 10 days after fledging (Paz 1987; Hofshi *et al.* 1987a); see also Relations within Family Group (below). BREEDING DISPERSION. Pairs nest both solitarily and in groups (Gallagher and Woodcock 1980), each pair defending small territory, with 10-100 m between nests (Shirihai in press), and foraging far from nest-site; same site used for 1st and 2nd clutches (Hofshi *et al.* 1987a; Paz 1987). At Petra (Jordan), *c.* 3 pairs per km² and 2-5 pairs along *c.* 1300 m (Wittenberg 1987). ROOSTING. Communal, especially outside breeding season. In Israel, some roosts exceed 200 birds (Shirihai in press). In canyons and cliffs of wadis. (Hofshi 1985; Mild 1990.) In urban area, traditional sites used for day roosting, those selected giving protection from wind and rain in winter and from sun in summer (Hofshi 1985). During breeding, ♀ roosts on nest until young completely feathered while ♂ roosts nearby (Hofshi *et al.* 1987a).

2. Orange wing-patches conspicuous in flight. Sometimes tame (Gallagher and Woodcock 1980), but not in Jordan (Wittenberg 1987). When pair-member finds food, usually calls, at which mate flies to join it; single birds forage quietly (Hofshi *et al.* 1987a). FLOCK BEHAVIOUR. Before flying to night roost, forms pre-roost assemblies on elevated sites (Hofshi 1985); pre-roosting flocks perform aerial manoeuvres like Starling *Sturnus vulgaris* (Paz 1987). SONG-DISPLAY. During incubation, ♂ occasionally sings (see 1 in Voice) near nest, if ♀ joins him they fly off together to feed (Hofshi *et al.* 1987a); no further information. ANTAGONISTIC BEHAVIOUR. No postures described, but ♂ guarding young in nest can resort to physical attack against conspecifics and House Sparrow *Passer domesticus* (Hofshi *et al.* 1987a; see also Relations within Family Group and Parental Anti-predator Strategies, below). HETEROSEXUAL BEHAVIOUR. During courtship, ♂ vibrates wings close to body during which wing-patches make 'blurring flash' in sun (Walker 1981b). In Jordan, late April, aerial (sexual)

chasing seen, accompanied by calling (D I M Wallace: see 1 in Voice). Mutual allopreening reported (Paz 1987), but not seen by Hofshi *et al.* (1987*a*). ♂ feeds ♀ with insects during pair-formation, ♂ walking round ♀ with food in bill until she accepts; ♂ also occasionally seen presenting twig to ♀ (Hofshi *et al.* 1987*a*). No further information. RELATIONS WITHIN FAMILY GROUP. ♀ removes eggshells from nest immediately after hatching. During 1st week, ♀ broods nestlings by day (see also Roosting, above). Parents approaching nest call frequently, stimulating young to gape and beg. By 7 days old, young restricted begging response to parental calls only. (Hofshi *et al.* 1987*a*.) Following death of ♂ parent, ♀ continued to feed chicks alone and defended nest against 3 adult ♂ intruders; this included physical attack by ♀, pecking head and body; ♂♂ succeeded in entering nest and killed nestlings (Hofshi *et al.* 1987*b*). ANTI-PREDATOR RESPONSES OF YOUNG. By 7 days old, young in nest freeze on hearing parental warning-calls (Hofshi *et al.* 1987*a*). PARENTAL ANTI-PREDATOR STRATEGIES. In Israel, seen mobbing (with dive-attacks) flying Kestrel *Falco tinnunculus* and both perched and flying Lesser Kestrels *F. naumanni* (*Israel Land Nat.* 1987, **12**, 168). In Saudi Arabia, bird thought to be breeding darted out

from cliff and 'attached itself in anger' for *c.* 3 s to passing Barbary Falcon *F. pelegrinoides* (Jennings 1981*c*). CJF

Voice. Highly vocal, though repertoire not studied in detail. Difference between song and calls not always clear (P J Sellar).

CALLS OF ADULTS. (1) Typical call/song. Variety of loud musical whistles (Figs I–II) quite like human whistles made by blowing while tongue behind top lip or with fingers pressed on top of tongue (E K Dunn), e.g. 'dee-oo-ee-o', 'o-eeou', 'vu-ee-oo', 'sweee-to', 'tsoowheeo', 'wee-o-weee', etc. (Hollom *et al.* 1988; B D Waite). Recording by B D Waite (Israel) indicates whistle types not usually repeated, though size of repertoire not known. Recordings indicate calls may speed up and become higher pitched, presumably with increasing excitement. Recordings contain long whistles from both sexes (♂ in Fig III, ♀ in Fig IV). (2) Subsong. Recording (Figs V–VI) contains

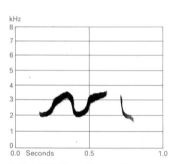

I  P A D Hollom  Yemen  November 1985

II  N Tucker and D J Tombs/BBC  Israel  March 1985

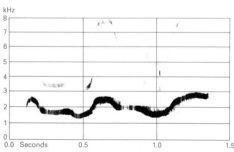

III  Mild (1990)  Israel  April 1989

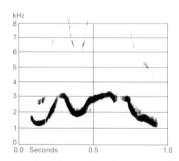

IV  Mild (1990)  Israel  October 1987

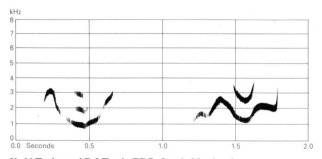

V  N Tucker and D J Tombs/BBC  Israel  March 1985

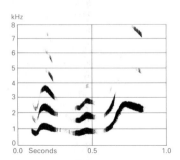

VI  N Tucker and D J Tombs/BBC  Israel  March 1985

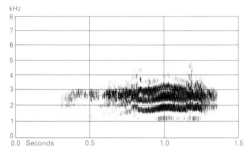

VII Mild (1990) Israel April 1989

rather quiet series of soft conversational whistles, more squeaky than call 1 and sounding like soft leather rubbed on glass (E K Dunn). (3) Less often heard harsh rising 'weeeaagh' (Fig VII: E K Dunn) or 'kraaaah' rather reminiscent of alarm-call of Golden Oriole *Oriolus oriolus* (P J Sellar).

CALLS OF YOUNG. No information. ME

**Breeding.** SEASON. Israel: beginning of March to end of June, first clutches March–April (Shirihai in press). Oman: in Dhofar, nests from January to at least April (Gallagher and Rogers 1980) with fledged young from 20 May (Walker 1981*b*). SITE. Crevices and holes in rocky ravines, on ledges, and in caves, high above ground; in Arad (Israel), has recently begun nesting on roofs and in shutter units of buildings, 6–21 m above ground (Hofshi *et al.* 1987*a*; Paz 1987; Shirihai in press). Nest: of twigs, tamarisk *Tamarix* branches, rarely with green *Acacia* leaves, with lining of softer material including feathers, hair, and paper; nest size and shape adapted to fit cavity but usually a deep plate (Paz 1987); in urban sites, nest a deep bowl, outer diameter 30 cm, cup diameter 12 cm, depth of cup 4 cm (Hofshi *et al.* 1987*a*). Building: by both sexes (Hofshi *et al.* 1987*a*). EGGS. Sub-elliptical, smooth; turquoise-blue, sometimes mottled with sparse light brown markings (Paz 1987) or reddish speckles (Gallagher and Woodcock 1980). $27\cdot2 \times 20\cdot0$ mm ($26\cdot9$–$27\cdot6 \times 19\cdot5$–$21\cdot1$), $n = 5$; calculated weight $5\cdot6$ g (Schönwetter 1984). Clutch: normally 3 but up to 5, 2nd clutches in same nest as 1st (Paz 1987). INCUBATION. About 16 days; by ♀ alone (Inbar 1971; Hofshi *et al.* 1987*a*). YOUNG. Fed and cared for by both parents (Paz 1987). Brooded by ♀. FLEDGING TO MATURITY. Fledging period 28–31 days (Paz 1987; Shirihai in press). Fed by both parents for *c.* 10 days after fledging (Paz 1987); 11–14 days after fledging young fed independently and formed juvenile flocks (Hofshi *et al.* 1987*a*). BREEDING SUCCESS. In Israel, of 23 chicks from 8 broods, 14 fledged; in 1983, mean $2\cdot2 \pm \text{SD}0\cdot2$ nestlings per nest, $1\cdot3 \pm \text{SD}0\cdot5$ fledglings; in 1984, $3\cdot3 \pm \text{SD}0\cdot2$ and $2\cdot3 \pm \text{SD}0\cdot6$ respectively (Hofshi *et al.* 1987*a*). CJF

**Plumages.** ADULT MALE. Apart from primaries, but including greater upper primary coverts, entirely deep black, feather-tips strongly glossed purplish-blue on body and lesser and median upper wing-coverts, more greenish-blue on tail, secondaries, and greater upper primary and secondary coverts, somewhat duller black on vent, under tail-coverts, under wing-coverts, and axillaries; longer under wing-coverts dark grey. Primaries bright tawny-cinnamon; tips black or greyish-black (*c.* 1 cm of tip of innermost primary, 3–4 cm on outermost), outer web of outer visible primary (p9) and strongly reduced p10 greyish-black; tawny-cinnamon contrasts strongly with uniform glossy black of secondaries and of greater upper primary coverts, less so with greyish-black wing-tip. *In worn plumage* (spring and early summer), tawny-cinnamon of primaries bleached to tawny-buff. ADULT FEMALE. Face dark ash-grey to drab-grey with faint narrow black shaft-streaks, almost absent on throat; towards rear of head, shaft-streaks gradually broader and more distinct, hind-crown, rear and side of neck, and lower foreneck rather evenly streaked or spotted deep blue-black on grey ground, this in turn merging into black of body on upper mantle and chest. Upperparts backwards from lower mantle as well as lesser and median upper wing-coverts black, feathers with broad glossy purplish-blue fringes, but fringes slightly narrower than in adult ♂, more dull black of feather-centres visible, and traces of narrow dark drab-grey feather-tips sometimes present on body. Underparts down from chest black, purple-blue tinge rather faint, mainly restricted to chest, flank, and upper belly; feathers of chest with drab-grey or brown-grey fringes (reduced on feather-tips), remainder of underparts with traces of dark drab-grey fringes. Tail and wing as adult ♂, but gloss of black parts slightly less intense; tips of primaries slightly browner, less contrasting with bright tawny-cinnamon; greater upper primary coverts uniform bluish-black. *In worn plumage* (spring and early summer), grey of head and neck partly worn off, head and neck appearing darker and more heavily and sharply streaked; black of body distinctly duller, less glossy; cinnamon of primaries more tawny-yellow or buff, tips of primaries sepia; tail browner. NESTLING. Entirely naked (Paz 1987). JUVENILE. Rather like adult ♀, but head and neck darker and body paler; general colour intermediate between adult ♂ and adult ♀, but not glossy, except for mantle and scapulars when plumage fresh. Unlike adults, greater upper primary coverts tawny-cinnamon (if fresh) or buff (if worn) with poorly defined dull greyish-black tips and bases. Entire head and body virtually uniform dull black (♂) or sooty-brown (♀), sides of feathers of chin, throat, and chest with faint and narrow drab-grey or brown-grey fringes; some glossy black sometimes visible on lower mantle and scapulars; thus, head, neck, chest, and upper mantle not as contrastingly streaked grey and glossy black as in adult ♀, body less dark and glossy. Tail black-brown, hardly glossy, feathers with narrow off-white edges along tip. Upper wing-coverts, secondaries, tertials, and bastard wing dull black, outer fringes of feathers slightly glossed bluish-green if plumage fresh, some outer secondaries sometimes with ill-defined cinnamon spot on outer web; primaries as adult ♀; for primary coverts, see above. FIRST ADULT MALE. Like adult ♂, but juvenile flight-feathers, greater upper primary coverts, and a variable number of tertials, greater (and rarely, median) upper wing-coverts, tail-feathers, and feathers of bastard wing retained, these browner than neighbouring newer 1st adult feathers, less glossy black than in adult; greater upper primary coverts with much cinnamon (if fresh) or buff (if worn) variegation on middle portion (in adult, fully black). Plumage of body on average less glossy than in adult ♂, some more dull black of feather-centres visible, especially on underparts. FIRST ADULT FEMALE. Colour and pattern of head and body about intermediate between adult ♀ and juvenile; part of juvenile feathers of wing and tail retained, as in 1st adult ♂, greater upper primary coverts partly cinnamon or buff (in adult ♀, fully black). Head, neck, and chest on average darker drab- or sooty-grey than adult ♀, shaft-streaks (on rear head) or feather-centres (on neck and chest) duller black, less contrasting (but head and neck not as uniform dark as juvenile); remainder of body rather like adult ♀ (blacker and more glossy than in juvenile), but feathers more sooty-black with narrow drab fringes, less uniform and glossy purplish blue-black, especially on underparts.

**Bare parts.** ADULT, FIRST ADULT. Iris brown, red-brown (♀), scarlet, or dark red (♂). Bill black or greyish-black, cutting edges and base of lower mandible sometimes paler grey or slightly yellowish. Leg and foot black; fissures between scutes, rear of tarsus, and soles often paler, dark grey to greyish-black. (BMNH, ZFMK, ZMB.) NESTLING. No information. JUVENILE. Iris dark brown. Bill dark olive with dark horn culmen. Mouth yellow-flesh. Leg and foot black, soles whitish. (BMNH.)

**Moults.** ADULT POST-BREEDING. Complete; primaries descendent. Only 12 in moult examined: 7 adults and 5 1-year-olds, from entire range. In adults, primary moult scores 13, 20, and 23, tail moult scores 0, 5, and 6 on 2-3 August; primary score 31 and 39, tail score 17 and 29 on 28-30 August; primary score 44 and 48, tail score 30 (moult finished) in singles from 25 September and 23 October. In 1-year-olds, primary score 13 and 21, tail score 0 (not started) on 2 and 13 June; primary score 22 and 23, tail score 8 and 10 on 17 and 27 July; primary score 24 on 15 August. (BMNH, RMNH, ZFMK.) For moult in Afrotropical *Onychognathus*, see Craig (1983, 1988). POST-JUVENILE. Partial, unlike most other Palearctic Sturnidae; involves head, body, lesser upper wing-coverts, many or all median upper wing-coverts, none to all greater coverts, tertials, and feathers of bastard wing, and 1-2 central pairs of tail-feathers (in 30% of 33 examined) or all tail (3%); in 1 of 42 birds examined, some greater upper primary coverts. Fledged but scarcely full-grown juveniles from Sinai, Israel, Jordan, and Saudi Arabia examined from 27 July to 11 September (thus, probably of 2nd broods; those of 1st broods fledge May: Hofshi et al. 1987a); slightly worn juveniles (not yet moulting) examined from 19 July and 16 September; birds in early stages of moult on 1 and 20 September and 2 October, in full moult 16 and 18 September and 1 November, in last stages 26 October and 2 November; birds with moult completed examined from 16-21 September (5), 26 October (1), and November (4). (BMNH, RMNH, SMTD, ZFMK, ZMA, ZMB.)

**Measurements.** Sinai, Israel, Jordan, and extreme north-west Saudi Arabia, all year; skins (BMNH, RMNH, SMTD, ZFMK, ZMA, ZMB). Juvenile wing and tail include retained juvenile wing and tail of 1st adult. Bill (S) to skull, bill (N) to distal corner of nostril; exposed culmen on average 5·4 shorter than bill (S)

| | | ♂ | | ♀ | |
|---|---|---|---|---|---|
| WING | AD | 152·8 (2·70; 8) 149-158 | | 146·7 (3·28; 7) 143-151 | |
| | JUV | 145·4 (3·13; 20) 140-152 | | 140·4 (2·81; 22) 134-145 | |
| TAIL | AD | 108·5 (2·85; 9) 104-114 | | 106·3 (2·71; 6) 103-111 | |
| | JUV | 101·0 (2·06; 9) 98-103 | | 97·5 (3·58; 13) 88-102 | |
| BILL (S) | | 30·1 (1·63; 10) 27·5-32·0 | | 29·2 (1·34; 14) 27·5-31·5 | |
| BILL (N) | | 17·2 (0·78; 10) 15·9-18·1 | | 15·9 (1·06; 13) 14·6-17·2 | |
| TARSUS | | 33·0 (1·25; 11) 31·2-34·8 | | 32·1 (1·06; 15) 30·5-34·0 | |

Sex differences significant for wing and bill (N).

**Weights.** Israel: 98-140 (Paz 1987). Nestlings on day 24, Israel, 102 ± 14 (Hofshi et al 1987a).

**Structure.** Wing rather short and broad, tip bluntly pointed. 10 primaries: p8 longest, p7 and p9 both 0-3 shorter than p8, p6 3-6, p5 12-16, p4 21-26, p1 36-42; p10 reduced, narrow, 78-90 shorter than p8, 3 shorter to 3 longer than tip of longest upper primary covert; not much difference between age-groups. Outer web of (p6-)p7-p8 emarginated; inner of p8-p9 with notch (sometimes faint). Longest tertial reaches to tip of outermost secondary (s1) or p1. Tail rather long, tip square or slightly rounded; 12 feathers, t6 4-10 shorter than t1. Bill rather long, c. 65% head-length; rather deep and wide at base, laterally compressed at tip, culmen gradually decurved towards sharply pointed tip, cutting edges decurved in distal part of bill. Bill less straight than in Starling *Sturnus vulgaris*, upper mandible distinctly less flattened dorso-ventrally, less sharp as seen from the side but less blunt as seen from above. Nostril rounded. Leg and foot strong; middle toe with claw 30·2 (13) 28·5-33 mm; outer and inner toe with claw both c. 71% of middle with claw, hind c. 76%.

**Geographical variation.** None. Birds from Yemen sometimes separated as *hadramauticus* Lorenz and Hellmayr, 1901; originally described as having all-black primary coverts and darker rufous primaries, but these are age-dependent characters, and no differences found in specimens examined.

Forms superspecies with African Red-winged Starling *O. morio* and perhaps Chestnut-winged Starling *O. fulgidus* and Somali Chestnut-winged Starling *O. blythi*, all from Afrotropics.

CSR

---

## *Sturnus sturninus* Daurian Starling

PLATES 12 and 19 (flight)
[between pages 136 and 137, and 280 and 281]

Du. Daurische Spreeuw    Fr. Etourneau de Daourie    Ge. Daurischer Star
Ru. Малый скворец    Sp. Estornino daurico    Sw. Amurstare

*Gracula sturnina* Pallas, 1776. Synonym: *Sturnia sturninus*.

Monotypic

**Field characters.** 18 cm; wing-span 30-33 cm. About 15% smaller than Starling *S. vulgaris*, with proportionately 10% shorter and much stouter bill, 20% shorter tail, and noticeably long toes, giving large-footed appearance. Small and short-billed starling, with remarkably variegated plumage: pale grey head and underbody contrast with dark back, tail, and wings; 2 whitish wing-bars. Sexes dissimilar; no seasonal but much individual variation on rump and upper and under tail-coverts. Juvenile separable.

ADULT MALE. Moults: August–September (complete). Head and underbody (to rear flanks) pale ashy-grey, just darker on head than elsewhere, with variable purple patch on rear of head and fine purple-grey streaks on lower nape above mantle (neither of these features may show at distance or when grey feathers overlay them). Mantle, back, and upper surface of wings appear almost black at distance, with (a) grey to white outer fringes to scapulars, (b) broad pale grey to white tips to median coverts and bastard wing, forming obvious upper wing-bar (perched and in flight), (c) buff, grey, or white fringes and tips to greater coverts, forming 2nd wing-bar (usually obvious when perched), and (d) long white panels on bases of inner webs of primaries and on both webs of secondaries, forming long bar along rear of wing visible in flight only. At close range, mantle and back show dark purple gloss, dark areas of wing mainly green and purple gloss. Rump, vent, and under tail-coverts variable: usually grey-brown, sometimes buff or white, rump occasionally black like back (see Plumages). Underwing basically pale grey, strikingly banded white across larger coverts and translucent bases of flight-feathers; at distance, appears mainly white, with translucent bar emphasizing broad dark trailing edge. Bill greyish-horn to black, with bluish base; no longer than distance from base of culmen to eye, thus shorter even than Rose-coloured Starling *S. roseus*; base stout, with lower mandible tapering noticeably. Iris dark brown; at distance, eye looks like quite large bead on pale head. Legs bluish-grey to earthy-green or ashy-brown. ADULT FEMALE. Sometimes slightly smaller than ♂. Plumage duller, usually greatly so. Head and underbody less immaculately pale grey, suffused overall with buff tone of rump and under tail-coverts. Purple patch on rear of head larger, extending raggedly over nape and occasionally even forward to forecrown; pale rear collar thus restricted. Mantle and back dirty grey-brown, sometimes with faint purple tinge; rump and tail-coverts even more variable than ♂ (see Plumages). Wing pattern as ♂ but ground-colour of upperwing brown-black, with only faint green sheen; wing-bars less broad and duller buffish-white. JUVENILE. Even duller than ♀, with dirty grey head and underbody and pale umber-brown mantle and back; no patch on rear of head, being merely washed purple on rear crown and entire nape and lacking any pale collar. Rump and upper tail-coverts pale buff-white, mottled grey in centre. Wings far less patterned than adult: umber coverts and dull brown flight-feathers marked only by restricted and broken white bar along greater coverts. Note that as in *S. roseus*, may delay or suspend moult until reaching winter quarters.

To inexperienced observer, almost pied plumage might briefly suggest ♂ Snow Bunting *Plectrophenax nivalis*, but once true character established, can only be confused with partially moulted immature *S. roseus* (which lacks white wing-bars) and escaped White-shouldered Starling *S. sinensis* (only one visible broad white band across wing-coverts; obvious white tips to outer tail-feathers; lacks dark back). Flight action varies: over short distance and when flycatching, relatively broad wings and changes in body angle produce loose, rocking progress reminiscent of roller *Coracias*, and steep ascents from ground and curving glides recall bee-eater *Merops*; over long distance, action and progress much reminiscent of *S. vulgaris*. Gait includes purposeful, waddling walk and short, determined, and bounding run, as in *S. vulgaris*. Stance similar to *S. vulgaris*. Scottish vagrant held tail above wing-tips in manner recalling wheatear *Oenanthe*; spent much time flycatching from perch, feeding on ground only in dull weather (Riddiford *et al.* 1989).

Call on flushing a slow, soft, drawn-out 'chirrup', resembling similar call of *S. vulgaris* (Riddiford *et al.* 1989).

**Habitat.** Breeds in east Palearctic in upper middle latitudes in warm continental summer climate, avoiding mountainous regions and favouring river valleys with groves of trees scattered in fields and meadows, or villages and outskirts of lowland towns, where nesting may occur in boxes, cracks in buildings, or rotten posts, substituting for natural choice of tree hollows. Rarely, however, occurs even on borders of large forests, and seems to favour deciduous rather than coniferous tree species. (Dementiev and Gladkov 1954.)

**Distribution.** Breeds across central and eastern Asia: Transbaykalia, Amurland, Ussuriland, Mongolia, Manchuria, northern Korea, and northern China. Winters from southern China south to southern Burma, Malaya, Sumatra, and Java.

Accidental. Britain: adult ♂, Fair Isle (Shetland), 7–28 May 1985 (Riddiford *et al.* 1989). Norway: juvenile shot Lillestrøm (Akershus), 29 September 1985 (Bentz 1987).

**Movements.** Migratory. Limited information.

Winters in large flocks in lowlands of peninsular Malaysia (south to Singapore) and Sumatra; apparently in smaller numbers in Tenasserim (southern Burma), Thailand (chiefly in south, though large concentrations occasionally recorded from south-east, probably passage birds), Cochin Chine and Tonkin (Vietnam), and northern Laos, with records also from southern China (Hume 1874*b*; Vaurie 1959; Tweedie 1960; King *et al.* 1975; Medway and Wells 1976; Lekagul *et al.* 1985; Marle and Voous 1988). Rare visitor to coastal Java (MacKinnon 1988), and 1 record from Borneo (Smythies 1981).

Migrates in flocks west of south, chiefly inland rather than along east coast of China, though regular as far east as Fukien in south-east; scarce passage migrant in Hong

Kong, with several autumn records but only 1 in spring (La Touche 1925-30; Schauensee 1984; Chalmers 1986). Accidental to western Japan, chiefly in spring (Brazil 1991), and has straggled east to Bonin Islands (Vaurie 1959). Occasional records from Nicobar Islands in Bay of Bengal include flock of 70-80 (Hume 1874a; Ali and Ripley 1972). Exceptional record of flock of 17 in Chitral (northern Pakistan) 16 July 1902 (Fulton 1906; Ali and Ripley 1972).

Leaves breeding grounds early, chiefly July in north of range; in middle Amur region (south-east Russia), latest record over 3 years 5 August (Vinter and Sokolov 1983); passage rapid in Hopei province (north-east China), where reported 6 August to 2 September (Hemmingsen 1951); last birds leave Korea in early October (Gore and Won 1971), and passage in Hong Kong 22 September to 12 October (Chalmers 1986). Extreme dates in peninsular Malaysia 14 September to 7 May (Medway and Wells 1976), on Sumatra 9 October to 3 April (Marle and Voous 1988).

In spring, 'extremely abundant' at Tavoy (Tenasserim) in April (Smythies 1986), and not uncommon in lower Yangtse (eastern China) in May (La Touche 1925-30); less conspicuous spring than autumn in Hopei, recorded 13 May to 12 June (Hemmingsen 1951), and arrives in Korea from early May (Gore and Won 1971). Reaches breeding grounds in middle Amur from late April, with single birds and small flocks observed 20-30 April over 3 years; however, main arrival (fairly concentrated) not until mid- or late May, with first records 17-24 May over 3 years in Bureya-Khingan lowlands (Vinter and Sokolov 1983), and similar dates on other breeding grounds in central Asia (Dementiev and Gladkov 1954; Rashkevich 1965).

Despite relatively early date (May) of British record (see Distribution), weather conditions suggest long-distance spring overshooting, rather than short-distance displacement from a European wintering site following autumn vagrancy (Riddiford *et al.* 1989). DFV

**Voice.** See Field Characters.

**Plumages.** ADULT MALE. Top and side of head down to upper mantle and side of chest light ash-grey; purple-black patch of variable size on rear of head, often restricted to nape, sometimes extending forward to mid-crown; border of front and rear of patch not sharply defined, purple-black streaked grey. Lore and narrow ring round eye white. Lower mantle, scapulars, and back black with strong purple-violet gloss, outer scapulars with broad but ill-defined brown-grey, buff, or white outer fringe and tip. Rump and upper tail-coverts either purple-black like back (15% of 46 examined), fully buff-cinnamon to buff-white (20%), or dark grey-brown with broad buff-cinnamon to buff-white feather-fringes (65%). Chin to chest, side of breast, and flank pale ash-grey, sometimes almost white on chin and upper throat; belly, vent, and thigh pale cream or white. Under tail-coverts variable, largely independent of abrasion, bleaching, or locality: in 46 October-February birds, bright light tawny-cinnamon in

14%, paler pink-cinnamon in 60%, isabelline-white in 26%; in small sample from breeding area, paler on average due to bleaching, but still variable. Tail black; central pair of feathers (t1) and outer webs of t2-t5 glossed green; outer web of t6 cinnamon to off-white (colour variable, as under tail-coverts). Flight-feathers, greater upper primary coverts, and greater upper wing-coverts black, strongly glossed green on both webs of coverts and tertials, outer webs and tips of secondaries, and tips of primaries; inner greater coverts and (usually) tertials with small or large triangular grey, buff, or white spot on tip, those of greater coverts sometimes showing as short broken wing-bar; outer web of middle secondaries medium brown-grey, forming patch which increases in width towards outer secondaries; middle portion of outer web of middle primaries (p4-p5) medium brown-grey, increasing to long brown-grey fringe on outermost primaries (p8—p9); basal and middle portion of inner webs of flight-feathers contrastingly white. As on under tail-coverts and outer web of t6, tone of grey-brown and white of flight-feathers highly variable, grey-brown on outer webs sometimes tinged rufous-cinnamon (especially on primaries), white on inner webs sometimes tinged isabelline-pink. Median upper wing-coverts white or light brown-grey, forming distinct bar extending to base of bastard wing and to shorter primary coverts; longer feathers of bastard wing black with brown-grey or white outer fringe. Lesser upper wing-coverts black with strong purple gloss. Under wing-coverts and axillaries white. ADULT FEMALE. Basically like adult ♂, but light grey of head and neck washed brown (sometimes hardly so on forecrown and side of head); purple-black of nape-patch, lesser upper wing-coverts, and lower mantle to back (or upper tail-coverts) replaced by dark brown-grey or dull drab-brown, hardly contrasting with paler brown-grey rear and side of neck; in some birds, nape-patch rather dark, brown-black or plumbeous-black, though virtually without gloss, and these may show dusky blackish spots on lower mantle and scapulars; grey of underparts sometimes slightly suffused buff; tail duller black slightly glossed purple (not green); black of wing duller, much less strongly glossed with purple rather than green; brown-grey or white on median coverts and brown-grey patches on outer secondaries and outer primaries more restricted; white on inner webs of flight-feathers more restricted, less pure, and hardly contrasting; pale spots on tips of tertials more often absent. Much individual variation, as in adult ♂; blackish nape-patch sometimes extends over whole crown; rump and upper tail-coverts either all drab-brown (28% of 32 examined), all tawny-buff to white (28%), or mixed brown and buff (44%); under tail-coverts and outer web of t6 pink-buff (27%), pink-isabelline (30%), or isabelline-white (43%); tips of outer scapulars and of median coverts grey or white, brown-grey patches on outer secondaries and outer primaries tinged cinnamon or pure grey, spots on tips of tertials and inner greater coverts rufous, grey, or off-white. JUVENILE. Upperparts rather like adult ♀, but top of head uniform grey-brown with faint darker grey spot on rear of crown, less pure ash-grey on forehead and side of head; hindneck slightly paler grey-brown, less grey and contrasting than in adult ♀; lower mantle to back less saturated and less dark brown-grey, white on outer webs of outer scapulars more restricted, less sharply defined; rump and upper tail-coverts extensively isabelline-white, variable number of central feathers grey-brown with paler isabelline fringes. Side of head and neck pale grey-brown with some white mottling on lower cheek and side of neck; a faint mottled grey and white eye-ring; supercilium short but fairly distinct (often more so than in adult ♀). Underparts dirty white; chest and side of breast tinged pale grey-brown, marked with poorly defined darker grey-brown shaft-streaks, latter extending onto white of lower throat, breast and side of belly. Tail black, tinged brown-grey towards base, outer web of outermost feather (t6) with ill-defined isabelline-white fringe. Primaries black, p6-p9 with pale rufous-cinnamon or isabelline fringe along outer web (not reaching tip or extreme base); secondaries greyish-black, bases extensively pale rufous-cinnamon to isabelline-white (less sharply defined than in adult ♂). Greater upper primary coverts uniform dull black. Tertials dull black-brown, narrowly fringed white on both inner and outer webs; outer tertials partially glossed green. Greater upper wing-coverts uniform dull black-brown; lesser and median coverts dull drab-brown, median with faint grey tips. See also Structure. Sexes perhaps separable by extent and contrast of isabelline or white on inner webs of flight-feathers, but only a few sexed birds examined. FIRST ADULT. Indistinguishable from adult when no juvenile feathering retained. Not known what proportion of juveniles has complete post-juvenile moult; of 75 birds of all ages examined October–March, 33% retained at least 1 juvenile secondary (on average, 3·6 ± 1·74 secondaries, range 1–6), remaining 67% (adults and 1st adults) had complete moult. Occasionally, also 1-2 juvenile tertials or some wing-coverts retained; in a few retarded birds, also scattered feathers of body or tail. Retained juvenile secondaries markedly different in colour from neighbouring adult ones, tips greyish-black to grey-brown (depending on wear) without gloss, instead of deep black with strong green (♂) or more faintly purple (♀) gloss.

**Bare parts.** ADULT, FIRST ADULT. Iris brown. In winter, bill dusky horn-brown, cutting edges and base of lower mandible whitish- to greenish-blue or pale horn; in summer (from about May to August), bill of ♂ black, of ♀ black with brown base. Leg and foot greenish-horn or grey. (Ali and Ripley 1972; RMNH.) JUVENILE. No information.

**Moults.** ADULT POST-BREEDING. Complete; primaries descendent. Takes place on breeding grounds, apparently completed before start of autumn migration. In central Asia, plumage heavily worn mid-July, but moult not yet started (Dementiev and Gladkov 1954). Moult already complete on arrival in Indonesian winter quarters, October (RMNH, ZMA). See also Vinter and Sokolov (1983). POST-JUVENILE. Complete or almost so. Starts soon after fledging, bird starting autumn migration with moult completed or suspended. In central Asia, single ♂ from 26 September in advanced moult (Dementiev and Gladkov 1954). On arrival in winter quarters, October, unknown proportion of birds have moult already completed (see First Adult in Plumages), others have 1-6 juvenile secondaries retained, as well as (occasionally) a few tail-feathers, wing-coverts, or scattered feathers of body; all these (except secondaries) apparently changed for adult ones later in winter.

**Measurements.** ADULT, FIRST ADULT. Greater Sunda Islands (Indonesia), October–March; skins (RMNH, ZMA). Bill (S) to skull, bill (N) to distal corner of nostril; exposed culmen on average c. 4·4 less than bill (S).

WING      ♂ 108·3 (2·16; 46) 103–112    ♀ 106·8 (2·31; 33) 101–110
TAIL      50·6 (1·72; 18)  48–54         50·7 (1·96; 20)  47–54
BILL (S)  19·1 (0·64; 19) 18·0–20·1      19·2 (0·80; 20) 18·0–20·3
BILL (N)  11·4 (0·54; 18) 10·6–12·3      11·3 (0·54; 20) 10·4–12·3
TARSUS    26·6 (1·01; 19) 25·2–28·4      26·5 (0·80; 20) 25·0–27·6
Sex differences significant for wing.

Wing. Middle Amur river: ♂ 106·3 (49) 103–110, ♀ 104·5 (30) 100–109 (Vinter and Sokolov 1983). Manchuria (China): ♂ 107·2 (2·63; 4) 105–110, ♀ 102·2 (1·26; 4) 101–104 (Meise 1934a).

JUVENILE. Middle Amur river: wing 101·1 (13) 98–104, tail 49·2 (13) 48–51 (Vinter and Sokolov 1983).

**Weights.** Middle Amur river, mid-May to July: adult ♂ 49·0 (3·79; 8) 43·5–55·0, adult ♀ 49·5 (3) 48·5–51·1; nestlings on day 2 5·9 (0·55; 5) 5·3–6·7, day 16 40·2 (4·26; 4) 34·0–43·3 (Vinter and Sokolov 1983, which see for details; see also Rashkevich 1965).

**Structure.** Wing rather long, tip pointed. 10 primaries: in adult, p8-p9 longest or either one 0-1 shorter than other (see also Vinter and Sokolov 1983); p7 3-8 shorter than longest, p6 8-14, p5 13-19, p1 30-37; p10 strongly reduced, narrow, tip pointed, 64-70 shorter than p8-p9, 9-12 shorter than longest upper primary covert; in juvenile, p8 longest, p9 1-3 shorter, p7 3-5, p10 relatively longer, broader, tip more rounded, 52-60 shorter than p8, 6-8 shorter than longest primary covert. Outer web of p7-p8 faintly emarginated, inner web of p8-p9 with faint notch. Longest tertial reaches to about tip of p1. Tail rather short, tip square or central pair of feathers (t1) slightly shorter than others; 12 bluntly pointed feathers. Bill c. 35% of head length, slender, rather broad at base, tip laterally compressed, cutting edges straight, tip of culmen gently decurved towards sharply pointed tip. Leg and foot relatively strong. Middle toe with claw 24·6 (10) 23-27 mm; inner and outer toe with claw both c. 68% of middle with claw, hind c. 77%.

**Geographical variation.** Strong variation (see Plumages) apparently individual, not geographical, but only a few birds from breeding grounds examined. In related White-shouldered Starling *S. sinensis*, geographical variation occurs in white/pink tones, but in winter quarters only (Hall 1953).

Forms superspecies with Red-cheeked Starling *S. philippensis* (synonym *S. violacea*) from Japan, southern Kamchatka, and Kuril Islands, and probably with *S. sinensis* from southern China. ♀ and (especially) juvenile of *S. philippensis* sometimes hard to distinguish from *S. sturninus*, differing mainly in uniform grey-brown scapulars and darker grey flank.          CSR

# *Sturnus sinensis* (Gmelin, 1788)  **Grey-backed Starling**

FR. Etourneau mandarin          GE. Mandarinenstar

A south-east Asian species, breeding in southern China and northern Indo-China and extending south and south-west of breeding range in winter. 2 records at Lågskär (Finland): adult ♀, 27 July–11 September 1975 (Mikkola 1979); 16 May 1990 (*Dutch Birding* 12, 204–9); both may have been escapes (H Jännes).

*Sturnus vulgaris* **Starling**

PLATES 16 and 19 (flight)
[between pages 280 and 281]

Du. Spreeuw    Fr. Etourneau sansonnet    Ge. Star
Ru. Обыкновенный скворец    Sp. Estornino pinto    Sw. Stare    N Am. European Starling

*Sturnus vulgaris* Linnaeus, 1758

Polytypic. Nominate *vulgaris* Linnaeus, 1758, Europe from Iceland, Britain (except Shetland and Outer Hebrides), Ireland, and France east to Perm', Kazan, Ul'yanovsk, and Volgograd, south to Pyrénées, north-east Spain, mainland Italy, western Greece (east to Thessaloniki), western Bulgaria, central Rumania, and Ukraine and European Russia at *c.* 48°N, grading into *poltaratskyi* further east in European Russia and into *tauricus* in Thrace (Greece, east of Nestos river), European Turkey, eastern Bulgaria (east from upper Maritsa and Yantra rivers), eastern Rumania (east of Bucharest and Bîrlad), southern Moldavia, and south-west Ukraine (south of Balta and Kremenchug), east to lower Dnepr; introduced North America and elsewhere; *faroensis* Feilden, 1872, Faeroes; *zetlandicus* Hartert, 1918, Shetland and Outer Hebrides (Scotland); *granti* Hartert, 1903, Azores; *poltaratskyi* Finsch, 1878 (synonym: *menzbieri* Sharpe, 1888), western Siberia west to southern Urals and eastern Bashkiriya (south-east European Russia), east to Lake Baykal, south through northern Kazakhstan to Altai and western Mongolia, grading into *porphyronotus* between southern foot of Altai and northern foot of Dzhungarskiy Alatau and eastern Tien Shan (eastern Lake Balkhash and Alakol' to Markakol and Kara Irtysh basin); *tauricus* Buturlin, 1904, south-east Ukraine (east of Lower Dnepr), Crimea, and shores of Sea of Azov north to Donetsk area and east to Rostov-na-Donu, Krasnodar, and Novorossiysk; also, Asia Minor east to at least central Anatolia; *purpurascens* Gould, 1868, eastern Turkey (east from about Gaziantep in south and Trabzon in north), northern Iraq, and western Transcaucasia (east to Tbilisi and Lake Sevan, west to Sochi and Maykop, grading into *caucasicus* and *tauricus* further north); *caucasicus* Lorenz, 1887, northern Caucasus north-west to Kislovodsk and Stavropol' and north-east to Volga delta, eastern Transcaucasia (east of *purpurascens*), and northern, western, and south-west Iran, grading into nominate *vulgaris* north of Caspian Sea and into *nobilior* in north-east Iran. Extralimital: *nobilior* Hume, 1879, southern Turkmeniya, eastern Iran, and Afghanistan, grading into *porphyronotus* at southern foot of Pamir-Alay mountains; *porphyronotus* Sharpe, 1888, eastern Uzbekistan and Tadzhikistan through Tien Shan to Sinkiang (China), north to Kara Tau and Dzhungarskiy Alatau; *humii* Brooks 1876, western Himalayas; *minor* Hume, 1873, Sind (southern Pakistan).

**Field characters.** 21·5 cm; 37–42 cm. Head and body close in size to Blackbird *Turdus merula* but with rather long, pointed bill and sloping forehead (giving very slender appearance to front of head), distinctly triangular wings, and rather short, square tail; similar in size to Rose-coloured Starling *S. roseus* and Spotless Starling *S. unicolor*. Medium-sized, rakish but full-bodied, bustling passerine, with cheeky mien but marked wariness; epitome of family. Adult looks blackish at any distance but actually intricately patterned and shot with iridescence, so quite beautiful at close range; white-spotted in winter. Juvenile dull brown at any distance, with pale throat. Bill strikingly pale yellow on breeding birds. Flight fast, with silhouette recalling modern fighter aircraft more than any other bird. Gait bustling. Song loud and chattering; includes, as calls, much mimicry. Sexes dissimilar; marked seasonal variation. Juvenile separable. 8 races in west Palearctic, differing little in size, but more markedly in colour of gloss and basic tone of juvenile plumage.

(1) European race, nominate *vulgaris*. Adult Male. Moults: May–October (complete). Fresh autumn bird has whole plumage almost black, with buff tips to feathers of upperparts, white to grey tips to feathers from throat to rear flanks and belly (forming distinct spots) and on vent and under tail-coverts (forming pale chevrons), and pale buff margins and tips to wing-coverts and flight- and tail-

feathers; on larger wing-feathers, pale margins produce strongly linear pattern. Plumage also strongly iridescent, with green gloss on head, reddish-purple on mantle, neck-sides, and breast, bronze on chest, blue-purple on flanks, and green (inside pale margins) on wings. Feathers of throat and breast long and pointed, forming shaggy 'beard' on singing bird. By late March, wear removes buff and grey-white feather-edges and -tips; plumage appears much darker and more glossy, particularly on green crown and bronze and purple underbody. May appear spotless in the field when very worn. Bill proportionately long, slender, and finely pointed, variable in colour: lemon-yellow with grey base in breeding season and grey-brown with yellowish base in autumn and early winter. Legs bright reddish-brown, palest when breeding. Iris dark brown. Adult Female. Body feathers shorter than ♂'s, with broader buff and grey-white tips when fresh (thus appearing more heavily spotted than ♂) and keeping some all year. Less glossy than ♂, especially on greater coverts and secondaries. Bill less clean-toned than ♂'s, with pinkish-white base in breeding season. Iris as in ♂ but usually with pale inner or outer ring. Juvenile. Dark dun-brown above and paler brown below, with plumage features restricted to whitish, slightly streaked chin and upper throat, buff chevrons to upper and under tail-coverts, and quite broad, bright buff margins to larger wing- and tail-feathers. Underwing pale brown-grey. Bill dull grey-brown. FIRST-YEAR. Juvenile plumage entirely replaced from July to November but brown head may be retained until last, giving both sexes distinctive pale-hooded appearance quite unlike dark-headed adult. Pale spotting heavier than in adult. Bill horn-black in winter and eye grey-brown. In 1st spring, both sexes retain fuller pale spotting than adult. (2) Northern Atlantic races, *faroensis* and *zetlandicus*. Noticeably robust, with broader-based bill, slighter longer wings, and strong legs. Adult plumage as nominate *vulgaris*, but ground-colour of juvenile much darker, sooty-brown to sooty-black so that pale black-spotted chin and whitish streaks on belly show more. (3) East European and Russian races, *tauricus*, *purpurascens*, *caucasicus*, and *poltaratskyi*. Similar in size to nominate *vulgaris* but differing in distribution and colour of strongest plumage gloss, most evident on *tauricus* in more reddish-purple breast and bronze underbody, on *purpurascens* in coppery-purple head and throat, on *caucasicus* in more purplish-violet underbody, and on *poltaratskyi* in purplish head and bluer underbody. Vagrancy of extralimital races across Europe given virtually no modern attention; origin of so-called 'black-headed' birds which occur in eastern England not known.

North of range of *S. unicolor*, adult unmistakable at close range but paler juvenile may invite confusion with similarly aged *S. roseus* (see that species). At long distance and in poor light, confusable above all with waxwings *Bombycilla*, but also with small dark thrushes *Turdus* (especially Redwing *T. iliacus*), small *Calidris* waders, Little Auk *Alle alle*, and even hirundines (Hirundinidae). Where range overlaps that of *S. unicolor*, separation subject to considerable pitfalls which have attracted no real study (see that species). Flight swift and usually notably direct, bird even taking beeline from nest to food source; action variable, most commonly over distance combining rapid beats of triangular wings with occasional glides and momentary closed-wing attitudes producing level shooting progress; over short distance and in tight spaces, wheels and sweeps, but still retains shooting element, giving impression of almost too much speed at times and leading to apparently rushed tumbling landing (particularly of flock); take-off when surprised noticeably flapping, leading into sudden escape ascent to nearby high perch or further; when hawking for flying insects (particularly in autumn), flight noticeably different, bird using slower and more graceful wheels and glides and then recalling Swallow *Hirundo rustica*. Flight silhouette distinctive, with rather geometric outline in pointed bill and head, triangular wings and short, sharp-cornered tail, all combining into uncanny recall of V-winged aircraft. Note, however, that appearance of *Sturnus* in flight can be matched by wide range of similar-sized birds (see examples above). Gait most commonly a quick, confident walk with head held up and jerked with each step and clear lifts of feet, instantly developed into run if bird hurrying to steal food and

occasionally interrupted by hop and sideways jump. Behaviour energetic and bustling in flock, with impression of constant alertness and often haste and bad temper. Gregarious, assembling in huge flocks when roosting and combining into feeding parties at any season; quarrelsome, fiercely disputing food sources with fellows and other birds. Normally wary of man.

Song a throaty but lively medley of whistles, gurgles, chirrups, and clicks, having characteristically variable, mostly low-volume output of sound interspersed with regular creaking phrases which pervade whole performance; some birds highly mimetic, introducing calls learnt from wide range of species, e.g. Curlew *Numenius arquata* and Golden Oriole *Oriolus oriolus*. Roosts give out continuous twittering or murmuring chorus. Individual calls include drawn-out, musical 'tsieuw', mostly from ♂, and common, rather grating, querulous 'tcheerr'. Harsh scream when surprised.

**Habitat.** In west Palearctic, breeds from upper to lower middle latitudes marginally to Arctic fringe and thence through boreal and temperate to steppe and Mediterranean climatic zones within July isotherms *c*. 10–30°C (Voous 1960*b*). Mainly in lowlands and uplands but in Swiss Alps breeds regularly up to 800 m and sparsely or locally even to 1500 m (Glutz von Blotzheim 1962); in USSR, in mountains up to 1850 m and extralimitally even higher (Dementiev and Gladkov 1954). During rather brief breeding season must concentrate where suitable holes are available, either naturally in hollow trees, rock or clay crevices, or previously excavated burrows or holes, or artificially in apertures in buildings or other structures, drain pipes, strawstacks, or nest-boxes, which have provided additional opportunities in Silesia since middle of 17th century, and now widely elsewhere (Niethammer 1937). Readiness to fly frequently over considerable distances permits breeding in open forests or near woodland margins, along rocky coastlines and in not too dense human settlements, birds foraging on neighbouring grasslands, field crops, floodlands, vacant sites, and even airfields, refuse tips, and sewage disposal areas. After young fledge, parties move quickly to open country, including grazed hill pastures, upper parts of salt-marshes, heaths, rocky shorelines, and seasonally to orchards and thickets bearing soft fruits or berries. Avoids, however, lower dense cover such as bushes, undergrowth, or tall herbage, but will roost in reedbeds; on ground prefers sparse or low vegetation offering ease of movements and access to food organisms beneath soil surface, though will forage close to thick cover. On buildings, prefers ledges or holes at no great height, except for roosting or where unusually attractive shelter is offered on towers or tall structures. In open country prefers presence of some trees or other raised perches, but avoids bare arid areas and also dense forests (especially conifers), extensive wetlands and precipitous rocky terrain unless it fronts more attractive habitats.

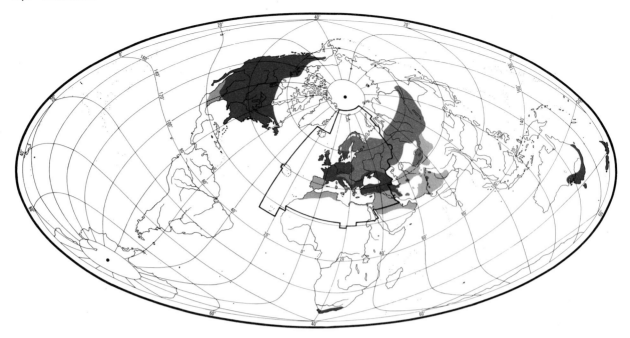

Markedly ready to take advantage of human settlements, artefacts, and managed land, from which it has profited much. Remarkable as ground feeder at home in wide variety of partly closed situations; in parts of range is attracted to water, but elsewhere manages well without it. Social roosting demands secure undisturbed shelter well above ground. Rapid flight at any convenient level enables combination of diverse habitats to be readily exploited.

**Distribution.** Long-term spread in north of range from 19th century to 1950s or 1960s, and more recent southwards spread in France and Italy, reaching northern Spain in 1960. (Beyond Palearctic, introduced and now widespread in North America; also southern Africa, Australia, and New Zealand.)

ICELAND. Long known as winter visitor; breeding first proved 1935. 2 breeding populations established: one in west, and smaller one, now perhaps extinct, in south-east (Thórisson 1981). BRITAIN, IRELAND. Marked range contraction in early 19th century followed by long-term expansion from c. 1830, continuing until northern and eastern areas colonized in 1950s or 1960s (but *zetlandicus* has survived from earlier in Shetland and Outer Hebrides) (Sharrock 1976; Thom 1986; Hutchinson 1989). FRANCE. Formerly absent as breeder in south, but since 1930s has spread slowly south, reaching Hérault in 1947, Toulon in 1974 (Yeatman 1976). First bred Pyrénées Orientales 1967, until 1976 only in mountains areas, in 1977 a few in lowlands (Clergeau 1989). Corsica: occasional breeder; several pairs bred 1932, and breeding may still sometimes occur (Thibault 1983). SPAIN. Not a regular breeder until 1960. In 1960 established 2 colonies in north-east, subsequently spreading and establishing further breeding

areas along Bay of Biscay coast. Spread became slower after contact with Spotless Starling *S. unicolor*, spreading from south (see that species), apparently due to competition. (Ferrer *et al.* 1991.) Breeds occasionally in Balearic Islands (Haffer 1989). SWEDEN. Until mid 19th century bred only in south and centre, and along Gulf of Bothnia. Expansion into interior of northern Sweden began c. 1850, not completed until c. 1950. (LR.) FINLAND. Continual expansion of range up to end of 1960s (OH). ITALY. Recent range expansion in centre and south, and altitudinal extension (to maximum of 2000 m) in Alps (PB). Small breeding population found in Sicily 1979, but had probably been present since at least 1974 (Iapichino and Massa 1989). RUSSIA. Not recorded in Kola peninsula in 19th century; vagrant early 20th century. Now breeds to c. 68°N, but has not reached Murman coast (Mikhaylov and Fil'chagov 1984). AZORES. Common on all islands, especially in coastal areas (GLG). CANARY ISLANDS. Recently colonized Tenerife and Gran Canaria (Emmerson *et al.* 1982).

Accidental. Madeira, Cape Verde Islands, Mauritania.

**Population.** Increase in peripheral parts of range, associated with range expansion (see Distribution), followed by marked recent decrease in several areas, especially northern and central Europe. Decreases in north accompanied by marked decreases in numbers of wintering birds in Britain, France, and Spain (Feare *et al.* 1992), and lower numbers wintering in Israel suggest decreases in breeding populations of central Asia (C J Feare).

ICELAND. Population in west (established 1935) increased to c. 2500 pairs in 1977 (Thórisson 1981). FAEROES. Several tens of thousands of pairs (Bloch and Sør-

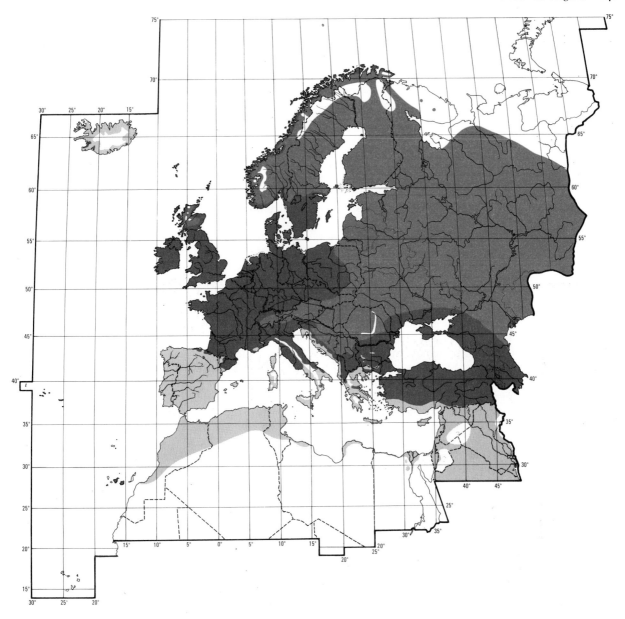

ensen 1984). BRITAIN, IRELAND. Probably 4–7 million pairs; drastic decline early in 19th century followed by increase from *c.* 1830 (Sharrock 1976). Marked decline in Britain during 1960s, continuing in 1980s (Marchant *et al.* 1990). FRANCE. Increase noted 1965–75. Estimated 2·2–6·8 million pairs 1975–85 (Clergeau 1989). BELGIUM. Estimated 430 000 pairs; has increased in last 50 years, probably through extension of suitable habitat and better winter survival (Devillers *et al.* 1988). NETHERLANDS. Estimated 725 000 pairs (Teixeira 1979). GERMANY. Estimated 4 million pairs (Rheinwald 1992). East Germany: 1 500 000 ± 600 000 pairs (Nicolai 1993); has decreased recently (Schneider 1982, 1984; Dathe 1983). NORWAY. Decreasing in many areas, especially in north (VR). SWEDEN. Estimated 2 million pairs; marked decrease in last decade (Svensson 1988; Svensson 1990; LR). FINLAND. Increasing up to end of 1960s, followed by dramatic decrease in late 1970s and early 1980s; reasons not clear, but at least partly due to persecution in winter quarters and pesticides. Recent estimate 100 000–200 000 pairs. (Koskimies 1989.) POLAND. Marked increase from 2nd half of 19th century, associated with colonization of human settlements, followed by decrease in 1970s, at least in some eastern areas, with stabilization in 1980s (LT). CZECHO-SLOVAKIA. Local decreases at least, probably country-wide (KH). ITALY. Recent increase, connected with range extensions (see Distribution) (PB). RUSSIA. Declined in Yaroslavl' region 1960–85 (Belousov 1986). LITHUANIA.

Decrease from *c*. 1950, numbers stabilizing *c*. 1980 (Logminas 1991).

Survival. Britain: mortality in 1st year of life (from 1 August) 48%, in 2nd year 48% (Lack 1946); annual mortality 52·8 ± SE1·0% (Coulson 1960). Finland: annual mortality 46 ± SE4·4% (Haukioja 1969). Czechoslovakia: mortality in 1st year 68·1%, in 9th year, 22·2%, in 10th year 14·2% (Beklová 1972). Oldest ringed bird 20 years 1 month (Dejonghe and Czajkowski 1983).

Movements. Generally migratory in north and east of breeding range, although increasing tendency to remain resident in urban areas; partial migrant or resident in south and west. Young disperse or, in some populations, undertake more extensive directional movements. Direction of autumn migration of adults predominly to south-west, but more southerly in east of range and more westerly in west. Migrant populations winter western and southern Europe, Africa north of Sahara, Egypt, northern Arabia, northern Iran, and plains of northern India. Island races *zetlandicus* and *faroensis* and extralimital *minor* resident. This account refers primarily to nominate *vulgaris*. For review, see Feare (1984) and, for account based on ringing, Fliege (1984). Migrates mainly during day but also at night.

Winter range lies from Britain and Ireland south through western France to Iberia, in all of these areas immigrants outnumbering residents (Spotless Starling *S. unicolor* in Iberia). Irregular winter visitor to Canary Islands. Abundant migrant to southern France, Italy, and countries bordering Mediterranean to Israel and Lebanon. Common migrant and winter visitor on Balearic Islands, Malta, and Cyprus. Common or abundant all across North Africa, common in Iraq, northern Iran, and Arabian Gulf states, but irregular in Oman and rare in Yemen; accidental in Ethiopia and Cape Verde Islands.

Northern and central Europe largely vacated in winter. Recoveries show that birds from north of range travel mainly WSW so that birds from Norway, Sweden, Finland, and northern Poland winter in Britain and north-west France, although there is evidence that birds from different origins within this catchment show some segregation in wintering area (Goodacre 1959). While some birds cross North Sea to reach Britain, there is a major concentrated migration route along southern North Sea coast, through Low Countries. Passage September–November. Early migrants are predominantly Dutch breeders with destinations in Britain; German and Scandinavian birds pass mid-October, while November migrants originate mainly in Poland, Finland, and Russia, with destinations in Belgium and France (Perdeck 1967). Birds from central Poland migrate more south-west to winter in southern France and Iberia, although in 1970s some of these shifted wintering area further north, possibly in response to increase in food availability where maize silage was fed to intensively reared cattle (Gramet

and Dubaille 1983). Birds from southern Poland travel through southern Europe to winter in Tunisia and northern Algeria (Rydzewski 1960; Gromadzki and Kania 1976), where severe damage to olives is reported. Recoveries in southern Europe and North Africa may be over-represented due to hunting and destruction of roosting birds in attempts to protect olive crops. While birds from given breeding area clearly migrate to defined wintering area, individuals do not appear to show great fidelity to particular winter site: birds from Netherlands found wintering up to 1000 km from ringing site in subsequent winter, although mean distance between site of initial capture and capture in subsequent winters decreased from 236 km in 1st winter after ringing to 68 km in 4th winter (Spaans 1977). Failure to recapture winter immigrants at a farm in southern England in years after ringing also suggests lack of fidelity to wintering site (Feare 1984).

Where resident, as in Britain, juveniles disperse after attaining independence from parents. Juveniles of migratory populations undertake longer and more directional movements, as demonstrated in Switzerland (Studer-Thiersch 1969); juveniles in summer travel over 500 km north-west to Low Countries and northern France. From here begin autumn migration, initially mainly to Algeria, but during winter move west to Morocco and Iberia where many Swiss adults winter. Post-breeding movements of Polish juveniles also take them to Low Countries; although earlier than adults, this route nevertheless similar to that taken by adults (Gromadzki and Kania 1976). Young of 2nd broods tend to migrate shorter distance and winter closer to breeding area (Fliege 1984).

Where a summer migrant, among earliest bird species to return to breeding area. In Belgium (where partially migratory), arrival begins with older ♂♂ in February, older ♀♀ some days later and younger ♂♂ and ♀♀ as late as April–May (Verheyen 1969a); in Poland, migration begins February, peaks March (Gromadzki and Kania 1976); in southern Sweden, arrives late February to early March (Karlsson 1983); in southern and western Finland mid-March to mid-April (Korpimäki 1978); around Arkhangel'sk, arrives April (Dementiev and Gladkov 1954); further east, arrives later, in Kazakhstan 2nd half of March to late April (Sema 1978); in north-east of range in Perm' area (58°N) does not arrive until mid-April (Dementiev and Gladkov 1954).                              CJF

Food. Animal and plant material taken at all times of year but animal food predominates in spring and is fed almost exclusively to nestlings. Plant material forms high proportion of diet in autumn and winter (but see Feare and McGinnity 1986 for relative requirement of animal and plant food). Seasonal changes in relative proportions of plant and animal foods paralleled by changes in intestine length: longer when eating plant than when taking animal food (Al-Joborae 1979). Animal food mainly insects and their larvae. Soft fruits taken in summer and autumn, and seeds, including cereals, in autumn and winter. Forages mainly on ground in open areas of short grass (often in association with grazing ungulates) or other short or sparse vegetation, e.g. cereal stubbles, and sometimes follows plough; also feeds in intertidal zone, on sewage treatment beds, refuse tips, farmyards, feeding areas for domestic stock, including troughs and open faces of cereal silage clamps. Sometimes forages in taller vegetation but this common only in summer flocks of juveniles in coastal salt-marsh, upland rough pasture, or moorland. Sometimes arboreal, especially post-breeding flocks of adults and young where caterpillars abundant, and also when eating fruit; in trees shows some agility but less acrobatic than some Asian *Sturnus*. Sometimes forages among fur of ungulates, hawks insects, and drinks nectar (Feare 1984, in press). At farm in southern England, average *c.* 60% of feeding birds in winter recorded in grass fields, *c.* 30% taking cereal component of cattle food (Feare 1981). Much food taken from surface or just below surface of soil or among grass roots. Walks or runs over ground and chases or pecks at items on surface, but specially adapted for probing into substrate (prying or open-bill probing): pushes closed bill into soil, litter, or grass roots, and opens bill to create hole; during bill-opening, eyes rotate forward to give binocular vision. These actions permitted by modifications of skull and musculature: protractor muscles (which open bill) are better developed than in many birds (including extralimital predominantly fruit-eating Sturnidae), so that bill can be opened forcefully; also, front of skull narrow so eyes can rotate forward avoiding necessity of turning head to one side to see into hole (Beecher 1978). While head pointing down in feeding position, eyes can be rapidly rotated backwards to allow all-round vision for detection of predators and movements of conspecifics. Low-light vision also acute which may help detection of prey exposed by probing (Martin 1986). Open-bill probing also used when (e.g.) searching among seaweed or stones at water margins and at sewage treatment beds, parting fur of domestic animals while searching for ecto-parasites, digging cereal seed from germinating crops, and selecting food particles of preferred sizes in troughs of cattle food. Other feeding techniques used include picking insects from branches and foliage of trees, hawking of flying insects (especially flying ants Formicidae), drinking nectar, removal and swallowing whole small fruits, and pecking pieces out of larger fruits. When eating (e.g.) cherries *Prunus*, often takes flesh leaving stone on tree but also removes whole fruit and removes flesh from stone while standing on ground; sometimes swallows entire fruit. Also regularly visits garden bird-tables, less often bird-feeders, when can show acrobatic ability, sometimes hanging upside down. Drinks regularly, especially in winter at farms where cereal forms major component of diet, and then particularly in evening prior to roosting when large flocks visit pools. Recorded kleptoparasitizing Lapwing *Vanellus vanellus* (Källander 1988) and often feeds with

that species, also with Rook *Corvus frugilegus*, Jackdaw *C. monedula*, and other flocking species in grassland.

Diet in west Palearctic includes the following. Vertebrates: nestling and dead adult passerines, lizards *Lacerta*, frogs *Rana*, newts *Triturus*. Invertebrates: may-flies (Ephemeroptera), damsel flies and dragonflies (Odonata: Coenagriidae, Libellulidae), grasshoppers, etc. (Orthoptera: Gryllotalpidae, Tetrigidae, Tettigoniidae, Acrididae), earwigs (Dermaptera: Forficulidae), bugs (Hemiptera: Scutelleridae, Pentatomidae, Lygaeidae, Nabiidae, Miridae, Notonectidae, Coreidae, Cicadidae, Cercopidae, Cicadellidae, Aphidoidea), lacewings, etc. (Neuroptera: Sialidae, Chrysopidae, Raphidiidae), adult and larval butterflies and moths (Lepidoptera: Satyridae, Nymphalidae, Lasiocampidae, Cossidae, Zygaenidae, Yponomeutidae, Hesperiidae, Pyralidae, Tortricidae, Arctiidae, Noctuidae, Lymantriidae, Geometridae), adult and larval caddis flies (Trichoptera: Phryganeidae, Limnephilidae), adult and larval flies (Diptera: Tipulidae, Culicidae, Chironomidae, Bibionidae, Stratiomyidae, Rhagionidae, Tabanidae, Therevidae, Asilidae, Empididae, Dolichopodidae, Syrphidae, Lonchaeidae, Sepsidae, Anthomyzidae, Tachinidae, Calliphoridae, Muscidae), adult and larval Hymenoptera (sawflies Symphyta, Chalcidoidea, Tenthredinidae, Cephidae, Ichneumonidae, Braconidae, ants Formicidae, wasps Vespidae, bees Apoidea), adult and larval beetles (Coleoptera: Carabidae, Dytiscidae, Hydrophilidae, Melyridae, Histeridae, Silphidae, Staphylinidae, Lucanidae, Geotrupidae, Dermestidae, Scarabaeidae, Byrrhidae, Dryopidae, Buprestidae, Elateridae, Dascillidae, Pyrochroidae, Phalacridae, Coccinellidae, Cerambycidae, Chrysomelidae, Curculionidae), spiders (Araneae: Agriopidae, Thomisidae, Lycosidae), harvestmen (Opiliones), woodlice (Isopoda: Asellidae), millipedes (Diplopoda: Glomeridae, Julidae), centipedes (Chilopoda), earthworms (Oligochaeta), molluscs (Gastropoda: Cochlicopidae, Vertiginidae, Bradybaenidae, Helicidae, Zonitidae, Succineidae, Vitrinidae, Lymnaeidae, Planorbidae). Plants: fruits and seeds of yew *Taxus*, acorns *Quercus*, apple *Malus*, pear *Pyrus*, plum, etc. *Prunus*, rowan *Sorbus*, elder *Sambucus*, olive *Olea*, grape *Vitis*, nightshade *Solanum*, bryony *Bryonia*, sea buckthorn *Hippophae*, plantain *Plantago*, pennycress *Thlapsi*, catchfly *Silene*, *Robinia*, wintercress *Barbarea*, vetch *Vicia*, oats *Avena*, barley *Hordeum*, millet *Panicum*, sorghum *Sorghum*, wheat *Triticum*, maize *Zea*, rye *Secale*; leaves of lavender *Lavandula*, flowers of dandelion *Taraxacum*, fungi (*Auricularia*, *Hygrocybe*). Also takes human food waste and garbage. (Kluijver 1933; Szijj 1957; Ray 1965; Gromadzki 1969; Tutman 1969; Radford 1974, 1983, 1985; Warren 1974; Taapken 1976; Yom-Tov 1980b; Raevel 1983; Feare 1984; Flohart 1985; Sueur 1988.)

Opportunistic and adaptable forager, adults eating wide range of plant and animal foods, proportions varying with season and habitat (Feare 1984). In England, where resident all year, 51% by volume of items from 368 stomachs collected throughout year were animal food: in April–June, animal food 65–92% of diet, but in July–October cereals and cultivated fruits predominated (Collinge 1924–7). In Poland, where summer migrant, 85% of 3953 items from 85 stomachs were animal food, with stomachs in March–June containing almost entirely animal remains (Gromadzki 1969). In Czechoslovakia, where also a migrant, 69% of 9917 items from 336 stomachs were animal food, again with almost no plant food in March–May (Havlín and Folk 1965). Over 40% of c. 200 stomachs of wintering birds in Israel contained plant material, including cereal grains, leaves, and olives Olea (Yom-Tov 1980b). Captive birds lost weight when fed only plant food and needed c. 60% animal food (dry weight) for weight maintenance (Feare and McGinnity 1986), but free-living birds appear able to survive periods of snow in Britain by subsisting largely on foods of plant origin (Dunnet 1956; Taitt 1973). In many areas, consumption of cereals and fruit leads to economic loss (Feare et al. 1992). Within wide range of foods taken, usually a few species predominate in diet, often adult and larval beetles larval flies, larval moths, and ants (Table A). When collecting food for young, adults eat smaller items taking larger ones to nest (Tinbergen 1981). Items smaller than 1 mm not taken and most items over 5 mm long; earthworms up to 5 g taken (Dunnet 1955). For detailed treatments of diet, see Kluijver (1933), Havlín and Folk (1964), Gromadzki (1969), and Szijj (1957); for extralimital studies, which

Table A  Food of adult Starling Sturnus vulgaris. Figures in brackets are percentages of birds containing each type of item. Prey types present only in small quantities are not shown.

| | Poland Feb–Sep | Czechoslovakia Mar–Nov |
|---|---|---|
| Animal material | No. of items | No. of items |
| Lepidoptera | 142 | 135 (46) |
| Diptera | 789 | 659 (65) |
| Hymenoptera | 769 | 3022 (65) |
| Formicidae | — | 2817 (29) |
| Coleoptera | 1445 | 2682 (74) |
| Carabidae | 209 | |
| Scarabaeidae | 248 | |
| Spiders | + | 136 (20) |
| Plant material | % of stomachs containing items in different months | No. of items |
| Cereals | 0–60 | 336 (11) |
| Wild seeds | 0–40 | 71 ( 6) |
| Cultivated fruits | 0–70 | 727 (45) |
| Wild fruits | 0–30 | 1898 (10) |
| No. of stomachs | 85 | 336 |
| Source | Gromadzki (1969) | Havlín and Folk (1964) |

Table B  Food of nestling Starling Sturnus vulgaris. Figures are percentages of samples containing each type of item (Poland, urban) or percentages by number of total number of items (other studies). Data from Netherlands and Czechoslovakia obtained from collar-samples.

| | Poland | | Czecho-slovakia | Nether-lands |
|---|---|---|---|---|
| | farm | urban | | |
| Hemiptera | 7·8 | 0 | 0 | + |
| Lepidoptera | 12·9 | 45 | 25·1 | 18·4 |
| (Noctuidae) | (11·3) | (36) | — | (14·5) |
| Diptera | 43·9 | 23 | 8·4 | 39·6 |
| (Tipulidae) | (11·2) | ( 6) | + | (11·4) |
| (Bibionidae) | (26·9) | (17) | (4·9) | |
| (Rhagionidae) | + | — | — | (12·7) |
| Hymenoptera | 1·3 | + | 6·8 | 4·5 |
| Coleoptera | 26·8 | 65 | 51·8 | 25·0 |
| Spiders | 2·1 | 0 | + | + |
| Oligochaete worms | 3·4 | 37 | 4·8 | + |
| Other animal material | 1·8 | + | 0·8 | 12·5 |
| Fruit | — | + | 2·3 | + |
| Food scraps | — | 28 | — | — |
| Total no. of items | 3561 | 1348 | 650 | 17933 |
| Source | Gromadzki (1969) | Gromadzka and Luniak (1978) | Havlín and Folk (1964) | Kluijver (1933) |

show similar dietary constituents, see Coleman (1977) and Russell (1971).

No estimates of energy intake of free-living adults. Intake of captive birds 210–265 kJ per day on animal food, 382 kJ per day on poultry pellets (Taitt 1973).

Food of nestlings almost entirely of animal origin (Table B): typically, small number of invertebrate species predominate, often Diptera (especially larval Tipulidae and adult Bibionidae), larval Lepidoptera (especially Noctuidae), larval and adult Coleoptera, and earthworms (Dunnet 1955; Havlín and Folk 1964; Gromadzki 1969; Gromadzka and Gromadzki 1978; Gromadzka and Luniak 1978; Tinbergen 1981); most food items taken from surface or just below surface of soil, especially grassland. For factors influencing prey selection, see Tinbergen (1981). Where 2nd broods occur, nestling food more diverse and includes aquatic, arboreal, and flying insects and more plant material (Gromadzki 1969). In Scotland and Netherlands, intensive studies (Dunnet 1955; Tinbergen 1981) showed Tipulidae larvae most frequent prey, in Scotland 81% by weight of food brought to nest, but other invertebrate prey also needed for healthy development (Kluijver 1933; Tinbergen 1981); no evidence of change in diet with age of nestlings at 4 colonies in Poland (Gromadzka and Gromadzki 1978), but some studies indicated spiders may be selected for young nestlings (Kluijver 1933; Al-Joborae 1979; Tinbergen 1981). In Netherlands study, average lengths of 4 prey species over 4 years 13·0–36·0 mm, frequency of nest visits by parents increased over first 5–6 days and weight of food brought also increased from 0·25 to 0·6 g (fresh weight) per visit over this period. In more wooded area of Netherlands, increase from 0·55 g (fresh weight) on day 3 to 0·80 g on day 16 for brood of 4; total intake over 19-day nestling period 12 350 kJ (Westerterp 1973). CJF

**Social pattern and behaviour.** Widely studied on account of ready acceptance of nest-boxes, ease of keeping in captivity, and economic damage. Major studies by Hartby (1969), Verheyen (1969a, 1969b, 1980), Merkel (1978, 1980), Evans (1980, 1988), Feare (1984), Pinxten *et al.* (1987, 1989a, 1989b, 1990, 1991), Eens and Pinxten (1990), and Eens *et al.* (1990, 1991). For reviews, see Schneider (1972), Gallacher (1978), and Feare (1984).
  1. Gregarious throughout year. Feeding flocks in fields largest and distances between individuals smallest in winter (Williamson and Gray 1975); see Flock Behaviour (below). In winter in farmland, England, 58% of 417 flocks contained less than 50 birds, 6% over 250 (Feare 1984); flocks smallest in breeding season when commonly feed in pairs, but pairs often coalesce into flocks, usually small but can be hundreds. After breeding, ranges of adults and young may separate (see Movements) when adults remain in small flocks but young can form flocks of hundreds, sometimes thousands, in suitable habitats. Migrates in flocks, birds arriving on Lincolnshire coast after North Sea crossing in flocks of 50–200 (Feare 1984) but on Dutch coast flocks of up to 100 000 recorded (Gruys-Casimir 1965). In winter, flock size increases during day with largest flocks prior to roosting (Clergeau 1990) (see Roosting, below). Especially in winter, forms flocks with other species, often Lapwing *Vanellus vanellus*, Rook *Corvus frugilegus*, Jackdaw *C. monedula*, gulls *Larus*, thrushes *Turdus*, and, in south-west winter range, Spotless Starling *S. unicolor*. Where resident, ♂ may defend nest-site for most or some of winter, taking in nest-material and roosting there (see Roosting), sometimes with ♀. At farm in southern England, showed extreme fidelity to feeding and day-roost sites throughout winter (Feare 1980), while in eastern England site fidelity not demonstrated and range of movement much greater (Summer and Cross 1987); difference between areas probably attributable to different age structures of marked birds (mainly adult in southern England, mainly young in eastern England) and to habitat differences. No site-fidelity between winters (Spaans 1977; Feare 1984). BONDS. Monogamy and polygyny demonstrated in several populations (e.g. Merkel 1978, Feare 1984). In 4-year study in Belgium, 20–60% of ♂♂ succeeded in becoming polygynous, in different years; proportion highest in older ♂♂; polygynous ♂♂ had higher breeding success than monogamous ♂♂ (Pinxten *et al.* 1989a). Polygyny usually successive, occasionally simultaneous; 2(–5) ♀♀ involved (Merkel 1978; Pinxten *et al.* 1989b). Polygynous ♂ defends 2 or more nest-sites and begins acquisition of secondary mate during incubation of 1st clutch; generally assists in feeding only young of primary mate (Merkel 1978; Pinxten *et al.* 1989b) and secondary ♀♀ have lower breeding success than primary ♀♀. Extra-pair copulations occur with already mated ♀♀, usually neighbours (Eens and Pinxten 1990). Intraspecific brood-parasitism occurs widely, with 0–46% of clutches parasitized (Yom-Tov *et al.* 1974; Numerov 1978; Power *et al.* 1981; Karlsson 1983; Feare 1984; Evans 1988; Pinxten *et al.* 1991). In 3-year study at 2 sites in Britain, 16–36% and 11–37% of 1st clutches parasitized (Evans 1988); in 6-year study in Belgium, average 15% of 1st clutches and 2% of intermediate clutches parasitized (Pinxten *et al.* 1991); when laying parasitic eggs, ♀ sometimes removes one

of host eggs (Lombardo *et al.* 1989; Feare 1991; Pinxten *et al.* 1991); parasitic behaviour appears related to disruption of normal laying or to shortage of nest-sites (Feare 1991; Pinxten *et al.* 1991). In resident populations, pairs formed in autumn may remain together over winter and into breeding season but changes do occur; extent not quantified and may be stimulated by attempts to catch adults for identification (C J Feare). In migrant populations, ♂♂ arrive in nesting area first and defend particular sites; ♀♀ do not visit nests until a few weeks later (Kluijver 1933; Verheyen 1980); even in migrants, some pairs may form in autumn and remain together into breeding season (Kluijver 1933). In southern England, some pair-bonds maintained between breeding seasons (C J Feare), but mate changes are frequent between 1st and 2nd clutches (Feare and Burham 1978) and between 1st and replacement clutches (Feare 1984). In Belgium, where partially migratory, pair-bond not recorded to survive between years (M Eens). ♂ defends nest-site and builds most of nest but ♀ contributes to lining; ♀ incubates but ♂ sits on eggs, reducing heat loss, in absence of ♀; ♀ broods for first few days after hatching but both sexes feed young, in at least some instances ♀ taking greater share. Adults other than parents reported visiting nests with young but no good evidence of helping (Flux 1978). Young dependent on parents for food for some days after leaving nest but precise duration and relative roles of parents not established. ♀♀ breed at 1 year old, ♂♂ generally not until 2 years old; sex-ratio biased towards ♂♂, European populations averaging 66% ♂♂ (Coulson 1960); proportion of non-breeders in population not known but in some colonies some ♂♂ unable to procure mates (Kluijver 1933). BREEDING DISPERSION. Usually colonial but sometimes solitary. Distance between nests usually determined by availability of nest-holes, and colonies appear loose, but high degree of synchrony of 1st clutches (see Breeding) suggests social interactions important in colony maintenance. Apparent loose structure of colony renders precise delineation difficult and therefore no good data on colony size, but ranges from a few to probably hundreds of nests. Territory restricted to small area around nest-site, ♂ chasing away others up to *c.* 10 m from nest; most feeding occurs within 200 m of nest but can extend to 500 m (Feare 1984). Within colony, nest spacing determined by hole distribution; nests can be as close as 1 m and large trees may have several nests. Nest density (nests per km²) in parkland 170–809 (Poland: Tomiałojć 1974; Luniak 1977b; Tomiałojć and Profus 1977); 36–625 (Germany: Blümel 1986); beech *Fagus* forest 598 (Poland: Jakubiec 1972); over 200 (England: Williamson 1968); mixed woodland 50–800 (Germany: Blümel 1986); urban areas 6–70 (Poland: Luniak 1977b); 1450 (Germany: Blümel 1986); over larger geographical areas of farmland, woodland, and urban areas in Britain, Poland, and Germany, densities of 7–12 pairs per km² recorded (Feare 1984). Territory mainly for nesting and mate acquisition but courtship and copulation also occur elsewhere (Feare 1984). Replacement nests sometimes in original site, often in nearby site but up to 170 m away (Feare 1991). Some birds nest in same site in consecutive years, occasionally more than 2 years (C J Feare); some change sites between 1st and 2nd clutches within year (Feare and Burham 1978). Low natal site-fidelity recorded in southern England (Feare 1984) and Belgium (Pinxten *et al.* 1989b). In limited areas of overlap, nests in association with *S. unicolor* but breeds earlier (Motis 1985); believed to compete, but form of competition unknown (Ferrer *et al.* 1991). ROOSTING. In breeding season, ♀ roosts in nest-site. Resident birds in winter may also roost in nest-site; migrants also roost in nests after arrival in breeding area but frequently change sites and partners (M Eens). During breeding season small (hundreds) communal roosts comprise breeding ♂♂ and non-breeders

(Feare 1984); in summer, roosts larger with substantial segregation of adults and juveniles (Tahon *et al.* 1978); size of roosts increases to maximum in winter (Eastwood *et al.* 1962) when numbers commonly exceed 100 000 and can reach 1 million. Roost sites include reedbeds, cane fields, plantations (especially young conifers), thickets (especially thorn, e.g. *Crataegus*, *Prunus* etc.), bridges, piers, and buildings in town centres (Feare 1984; Feare *et al.* 1992). Also forms diurnal roosts, usually in exposed trees, overhead wires, television antennae, etc., close to feeding sites, sometimes in more sheltered situations, e.g. evergreen bushes (*Hedera*, *Laurus*); availability of such daytime roost sites an important attribute of good feeding sites (Feare 1984). Normally roosts with more or less regular spacing between individuals on branches or ledges, but especially (though not exclusively) on cold nights birds may huddle together in groups of up to 40 (Peach *et al.* 1987). For pre- and post-roost gatherings see Flock Behaviour (below).

2. Can be tame and approachable in parks and gardens but in rural areas more wary. Flock responds to aerial predators by coalescing tightly and giving repeated Chip-calls (see 8 in Voice); may mob predator, e.g. Sparrowhawk *Accipiter nisus*. Birds frequently give Distress-calls (see 7 in Voice) when caught; in winter, ♀♀ give more Distress-calls than ♂♂ (Inglis *et al.* 1982). FLOCK BEHAVIOUR. Feeds in tight or loose flocks depending on season and feeding site, and to some extent time of day; *c.* 0·25 m between individuals in winter, up to *c.* 2·25 m in summer (Williamson and Gray 1975); flock spacing smaller on stubbles and at cattle feeding areas than in grassland (Feare 1984), and smaller in dusk pre-roost assemblies than earlier in day (C J Feare). Dominance hierarchies established in captive flocks, both feeding (Feare and Inglis 1979; Van der Mueren 1980) and roosting (Feare *et al.* in press), with ♂♂ usually occupying more dominant positions. ♂ dominance may explain prevalence of adult ♂♂ at competitive feeding situations (Feare 1980) and in preferred positions in roosts (Summers *et al.* 1986; Feare *et al.* in press). Flocks usually noisy, but flight often preceded by silence, especially noticeable in leaving day and night roosts; individuals often crouch before flight, especially when warned by behaviour of conspecifics or other birds. On take-off, frequently gives Flight-call (see 3 in Voice). Especially in winter, often performs spectacular manoeuvres over roost site prior to entry, wheeling in cloud- or smoke-like formations, sometimes in flocks of hundreds of thousands from which they stream into roost site. Individuals from smaller flocks that do not indulge in this behaviour tumble erratically into roost site giving Chip-calls. Departure from roost occurs in series of exoduses, usually *c.* 3 min apart (Eastwood *et al.* 1962); in each, birds stream out, often giving Flight-calls on take-off and Chip-calls as they rise from roost. On arrival at feeding sites, often tumble from flocks, giving Chip-calls like birds entering roost from small flocks (C J Feare). SONG-DISPLAY. In breeding season, only ♂ sings (see 1 in Voice)—from perch which may be branch, fence, television antenna, gutter, etc., usually close to nest-site. Adopts characteristic upright posture (Fig A), with bill pointed slightly up,

A

tail pointed down with rump feathers ruffled giving hunch-backed appearance, feathers of lower belly ruffled and long throat feathers puffed out. In more intense song, bill often pointed higher, exaggerating puffed-out throat feathers, but belly feathers more sleeked (Fig B). Towards breeding season, especially

B

March–April in southern England, singing ♂♂ may indulge in Wing-flicking (half-open wings flicked from closed position away from body, often during rattle song-type 3: see Fig B, and 1 in Voice) or Wing-waving (Fig C: rotation of half-open wings

C

around shoulder in fairly slow and exaggerated movement, often during rattle song but sometimes from song perch when not singing but when ♀ flies by) (Feare 1984). Experiments with captive birds suggest main function of Wing-waving is mate attraction (Eens *et al.* 1990). When ♀ attracted to singing ♂, ♂ may fly into nest and sing there. ♂ also sings when close to ♀ during mate guarding (Feare 1984). Song output decreases on mate acquisition (Cuthill and Hindmarsh 1985) and introduction of ♀ into cage of ♂ in experiments led to increase in song output (Eens *et al.* 1990), suggesting major function of song in mate attraction. ♂♂ sing in most months, except for a few weeks during post-breeding moult. Song most intense in spring, with smaller peak of singing in autumn after moult when nest building also occurs; function of autumn song not investigated, but possibly related in resident birds to nest-site tenure (C J Feare); in autumn and winter, both sexes sing in day roosts and both sexes probably sing in night roosts (but see Charman 1965), where singing intensive after arrival at roost and before morning departure and, especially in town-centre roosts, can continue sporadically all night (C J Feare); function of roost song unknown but may play role in defence of perch (Feare *et al.* in press) and in synchronizing morning departure (Feare 1984). ANTAGONISTIC BEHAVIOUR. (1) General. ♂♂ defend area around nest-site in breeding season and to lesser extent in autumn. In breeding season, defence most intense in pre-laying and early laying periods, but intruders (♂ and ♀) tolerated when young in nest (Flux 1978). In day and night roosts, much jockeying for position, and birds in captive flocks fight to defend preferred positions, ability

D

to defend depending on dominance (Feare *et al.* in press). (2) Threat and fighting. Basic threat is to stare at opponent (Fig D, left), during which crest feathers may be raised; more noticeable is Upright Threat-display (Fig D, right) in which body upright, bill above horizontal, and crest raised; at greater intensity graduates into Open-bill Threat-display with bill slightly or widely open and posture may be more erect ('tall posture': Ellis 1966) and with crown feathers raised. If opponent fails to give way, 2 birds may participate in 'fly-up' (Feare 1984; 'dance-fight' of Van der Mueren 1980), flying up and down facing each other (Fig E), sometimes grappling with feet and stabbing with bills

E

and giving loud chattering calls ('attack call': Hartby 1969); this is commonest form of fighting, seen in open fields but more frequently at competitive feeding sites where space limited. Submissive bird adopts more horizontal posture, crouched with feathers sleeked (Fig F); bill-wiping may also indicate submission

F

(Feare 1984). Distress-calls sometimes given by one bird when attacked by another and may have submissive function in this context. ♂ defends small territory around nest and chases intruders. In experiment, introduction of new ♂ into cage of ♂♂ with occupied nest-sites led to occupiers spending more time near nest, occupiers also sang more but not significantly so (Eens *et al.* 1990). HETEROSEXUAL BEHAVIOUR. (1) General. Initiation of first clutches highly synchronous (see Breeding), probably as result of social interaction. In resident populations, nests may be built in autumn and winter and pairs may form, but changes of mate and nest are frequent, and true pair-bonding does not occur until the weeks before laying (M Eens). In migrant populations, ♂♂ arrive first and occupy nest-sites (Verheyen 1969*b*) and pairing can occur within 1 day of arrival of ♀♀ (Eens *et al.* 1990). (2) Pair-bonding behaviour. First stage of mate attraction involves ♂ building bulky foundation to nest, this in a site

actively defended by him; ♂ sings during nest-building period but when ♀♀ in vicinity intensity of song increases and ♂ performs Wing-waving display (see Song-display, above). When ♀ approaches, ♂ often flies into nest cavity and sings inside, may also take in flowers or leaves, stimulating ♀ to enter (Verheyen 1969*b*; Eens *et al.* 1990). Copulation (within-pair and extra-pair) always preceded by song (Eens and Pinxten 1990). ♂'s song stimulates ♀ to solicit. ♂♂ with larger repertoire and those which sing longer bouts have higher breeding success, achieve more copulations, and acquire more mates (Eens *et al.* 1991). In days immediately before and at beginning of egg-laying, ♂ follows ♀ closely, with much 'chackerchacker' calling, resembling Attack-call (see 6 in Voice) (Feare 1984). This mate-guarding chiefly from 4 days before 1st egg to day of 2nd egg, when ♀ followed by ♂ every time she left nest; intensity peaks 3-4 days before laying commences (Pinxten *et al.* 1987); on feeding areas ♂ stays close to ♀, sometimes sings softly (Feare 1984), and chases approaching ♂♂. Allopreening not recorded. (3) Courtship-feeding. Generally does not occur (C J Feare), but ♂ once recorded feeding ♀ on bird table (Chappel 1949). (4) Mating. Often in tree close to nest, also on feeding grounds. ♀ approaches singing ♂ (Eens and Pinxten 1990), moves to his side, and may press her body against his; sometimes pecks at his breast or upper wing-coverts; adopts submissive crouch with tail raised, remains motionless (Verheyen 1968) or shakes tail from side to side (Feare 1984); occasionally gives Copulation-call (see 10 in Voice) (Hartby 1969; M Eens). ♂ mounts and both birds twist tails to achieve cloacal contact; during mounting ♂ often waves wings and holds ♀'s nape feathers with bill (Feare 1984); copulation achieved in a few seconds without repeated cloacal contact. Afterwards, ♂ moves short distance away; both birds often preen and sometimes bill-wipe (C J Feare). Copulation recorded from 10 days before 1st egg to 5 days after, with 61% of 31 copulations within fertile period of ♀, from 5 days before laying to day of 1st egg (Pinxten *et al.* 1987); frequency of copulation peaks 1-2 days before 1st egg (Verheyen 1968) or on day of 1st egg (Pinxten *et al.* 1987); copulation mainly in morning (C J Feare). In extra-pair copulations, ♂ more active in solicitation: approaches ♀ (usually neighbour) when her own mate absent and sings near her; ♂ not furtive in his approach (Eens and Pinxten 1990). (5) Nest-site selection and behaviour at nest. ♂ may select and attempt to defend several neighbouring nest-sites; in migrant populations, older ♂♂ arrive first and select sites; later-arriving younger ♂♂ compete for and may secure some of these (Verheyen 1969, 1980; Pinxten *et al.* 1990). ♂ builds most of nest and shows it to approaching ♀ by flying into cavity and singing from inside; may also carry leaves and flowers into nest. ♀ completes cup, her building activity peaking 4-5 days before laying. ♀ incubates at night and for most of day; ♂ sits on eggs when ♀ absent. Mate often absent at incubation changeover, when no special behaviour observed (C J Feare); when mate present, ♂ occasionally sings quietly in vicinity, apparently stimulating ♀ to leave, and sometimes copulation occurs (this during incubation and outside ♀'s fertile period). Other calls sometimes given by both ♂ and ♀ when approaching nest with mate incubating; ♂ sometimes aggressive towards ♀ at changeover (M Eens and R Pinxten). Eggs often partially covered with leaf, paper, etc., when parent leaves nest in absence of mate (Feare 1984). Approach to nest usually silent, bird arriving in nest tree and fluttering down through branches to nest; when feeding young, approach more direct. Adults leave nest directly, often without Flight-call, though this perhaps more frequent when leaving nest after disturbance (C J Feare). RELATIONS WITHIN FAMILY GROUP. Parents remove eggshells from nest, dropping them on ground nearby. Nestlings brooded during first 5 days, mainly by ♀; both sexes feed young, about equally or with ♀ taking greater share (see Breeding). Both ♂ and ♀ can rear chicks alone after death or desertion of mate, though with reduced success (Feare 1984; Eens *et al.* 1990). Both sexes share nest sanitation: some faecal sacs initially eaten (Kluijver 1933), most removed from nest and dropped a few metres or up to 20 m or more away (Feare 1984). Large chicks may, if nest entrance accessible, raise cloaca and defecate directly out of hole. Chicks 1-2 days old respond to disturbance by raising neck vertically and opening mouth for a few seconds before collapsing (Kessel 1957); usually on 2nd day begin to give thin food-call (see Voice); by day 4, can move around nest-cup using legs and wings and beg strongly (Hartby 1969). Food-calls continue to be used after leaving nest, both in flight and when perched. In week before fledging, contact-calls between adults and young are individually recognizable (Elsacker *et al.* 1986). Eyes open at 6-7 days (Kessel 1957). Fledge at 19-22 days; rarely return to roost in own or another nest after fledging (C J Feare). Duration of family bonds after fledging not known with precision; rapidity with which some pairs re-nest after successful fledging of 1st brood suggests dependence on parents can only last a few days, but can be longer, especially with 2nd-brood young, some of which appear to stay with parents for 2-3 weeks (C J Feare). ANTI-PREDATOR RESPONSES OF YOUNG. Disturbance of nest during first 2 days stimulates begging; after eyes open, disturbance promotes chicks to crouch in nest-cup and remain quiet; at 12 days, begin to attempt to escape when handled (Kessel 1957). From 15-16 days, may leave nest when disturbed; unable to fly at this stage, instead crawling in ground vegetation or freezing there (C J Feare). PARENTAL ANTI-PREDATOR STRATEGIES. (1) Passive measures. When parents leave nest without disturbance, often partially cover eggs with leaf or other nest-material (Feare 1984). Incubating parents can often be caught by hand due to either tight sitting or freezing when predator (man) approaches (C J Feare). (2) Active measures: against birds. Approach of predator (e.g. *A. nisus*) induces birds in colony to stop singing and to give Chip-calls; stimulates parents to mob predator in tight flock. (3) Active measures: against man and other animals. After hatching, parents give Snarl-call (Hartby 1969; see 9 in Voice) when predator approaches nest, intensity increasing up to fledging (Chaiken 1986); distress calls of young stimulate adults to make diving attacks at predator (Chaiken 1992).

(Figs by D Nurney from drawings and photograph in Feare 1984.) CJF

**Voice.** Widely studied. Highly vocal all year (especially ♂) except during moult when virtually silent. Repertoire rich and varied. Most outstanding features of song are its complexity, incorporation of mimicry, and modification of repertoire in adulthood. Other than song, 11 different calls reported (Hartby 1969). Song dialects found in some populations. For additional sonagrams, see Hartby (1968), Feare (1984), Hindmarsh (1984), Adret-Hausberger and Jenkins (1988), Adret-Hausberger (1989), and Eens *et al.* (1989, 1991*a*, *b*, 1992*a*, *b*).

CALLS OF ADULTS. (1) Song of male. Well documented (Hindmarsh 1984; Adret-Hausberger and Jenkins 1988; Adret-Hausberger 1989; Eens *et al.* 1989, 1991*a*, *b*). Full song a lively rambling medley of throaty warbling, chirruping, clicking, and gurgling sounds interspersed with musical whistles and pervaded by peculiar creaky quality (Witherby *et al.* 1938). Uninterrupted sequence of structurally complex sub-phrases ('song-types'), 0.5 to 1.5 s long, progressing from relatively simple pure-sounding whistles to more complex sub-phrases including mimicry and rattles, and ending with high-frequency sub-phrases (Hindmarsh 1984; Adret-Hausberger and Jenkins 1988; Eens *et al.* 1989, 1991*a*, *b*; Böhner *et al.* 1990; Chaiken *et al.* in press). Frequency ranges from less than 1 kHz to more than 10 kHz. Songs or song sentences (or song bouts in terminology of Hindmarsh 1984 and Eens *et al.* 1991*a*) can be up to 60 s long, usually separated by pauses of more than 5 s. Most sub-phrases in a bout are sung 2-3 times before another is introduced. Although majority of song (90% or more: Eens *et al.* 1991*b*) is sung in bouts during breeding season, some whistles and heterospecific imitations are also sung outside song bouts. Each ♂ has unique repertoire of *c.* 20-70 sub-phrases, sung in predictable sequence. All birds' songs conform to common basic pattern, although details of sub-phrase structure differ for each individual. Eens *et al.* (1990, 1991*b*) distinguished 4 categories of sub-phrase. The following analysis of song of 25 ♂♂ is after Eens *et al.* (1991*b*); see also Adret-Hausberger and Jenkins (1988). 55% of songs begin with one or several pure-sounding whistles (e.g. Fig I) (Type 1). Each ♂ has repertoire of 2-11 different whistles, average 6 (see below for details). Unlike sub-phrases from the 3 other categories, whistles sung as part of song bout are mostly not repeated. Introductory whistles are always followed by large number of 'variable' sub-phrases (Type 2). Variable sub-phrases have complex structure: usually contain many different units, many of them covering wide range of frequencies in a short time (e.g. Figs II-IV from bout of variable song by one bird, Figs V-VI from another). Variable sub-phrases very often contain diads, evident in various units of Figs II-VI. Mimicry often

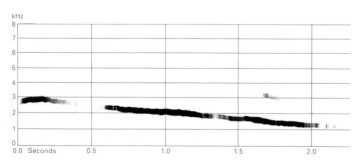

I  G Thielcke/Sveriges Radio (1972-80)  Germany  July 1964

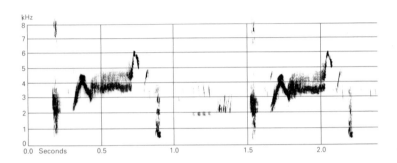

II  G Thielcke/Sveriges Radio (1972–80)
Germany   July 1964

III  G Thielcke/Sveriges Radio (1972–80)
Germany   July 1964

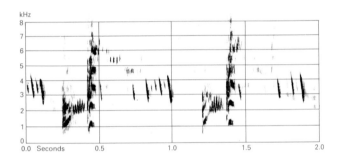

IV  G Thielcke/Sveriges Radio (1972–80)
Germany   July 1964

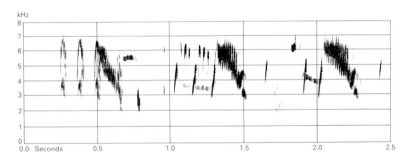

V  D Bower   Wales   May 1976

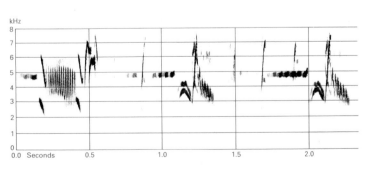

VI  D Bower   Wales   May 1976

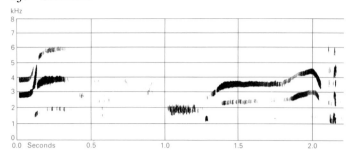

VII  V C Lewis  England  June 1978

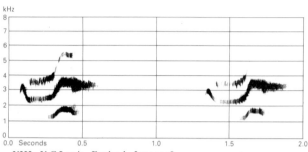

VIII  V C Lewis  England  June 1978

IX  P J Sellar  Sweden  May 1978

incorporated into these sub-phrases, e.g. Fig VII (Lapwing *Vanellus vanellus*), Fig VIII ('kewick' call of Tawny Owl *Strix aluco*, from same section of song as Fig VII), and Fig IX (alarm-chatter of Song Thrush *Turdus philomelos*). Some variable sub-phrases are entirely made up of heterospecific imitations, whereas other imitations are integrated into a sub-phrase with singer's own sounds. Each ♂ has repertoire of 10–35 different variable sub-phrases, average 24. In a song bout, variable sub-phrases pass into rattle sub-phrases (Type 3). Rattle sub-phrases (e.g. Fig

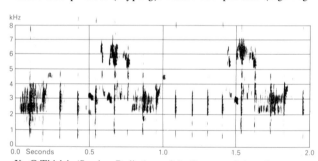

X  G Thielcke/Sveriges Radio (1972–80)  Germany  July 1964

X) are made up of rapid succession of clicks with maximum energy below 4 kHz and sung at a rate of *c.* 15 per s (*c.* 9 per s on sonagram); at same time, several other sounds are made, creating diads. Each ♂ has repertoire of 2–14 different rattle sub-phrases, average 8. Song typically ends with series of high-pitched trills or screams which Eens *et al.* (1991b) called 'high-frequency segments' (high-frequency sub-phrases) (Type 4) with frequency range mainly 6–10 kHz. A ♂ can have maximum of 6 high-frequency sub-phrases, average 3·5. Only 30% of song

bouts end with high-frequency sub-phrases, as song can be interrupted at any stage during a bout. Average length of song bouts varies considerably among ♂♂, 15–35 s. Average length of complete song bouts (i.e. song bouts that end with high-frequency sub-phrases) 21–50 s. The 4 categories of sub-phrase in a song bout are sung with different intensity. Introductory whistles can be sung either quietly or loudly. Variable sub-phrases are sung with lowest intensity, audible only at fairly close range. Rattle sub-phrases are mostly sung louder than variable sub-phrases. High-frequency sub-phrases at end of song bout are loudest. According to Adret-Hausberger and Güttinger (1984) and Adret-Hausberger (1989), whistles can be divided into 8 general classes according to physical structure. 5 among these 8 classes are sung by all or most birds (i.e. 'species-specific themes'). Long descending whistle (Fig I) found in almost all populations studied (Eens *et al.* 1991b; Eens 1992): 'hooid' (Kluijver 1933), 'tsüüü' (Witherby *et al.* 1938), 'seeooo' (Harbard 1989). Hausberger and Guyomarc'h (1981) found considerable local variations in exact form of these 'species-specific themes' (dialects). According to Hausberger and Guyomarc'h (1981) and Adret-Hausberger (1984), relative frequency of occurrence of whistle types varies with season, time of day, and type of activity. An outstanding mimic, producing accurate copies of calls or songs of other bird species, frogs, and mammals (e.g. goat, cat, man) and of mechanical sounds (Witchell 1896; Allard 1939; West *et al.* 1983; Hindmarsh 1984; Eens *et al.* 1991b). 18 mimicked bird species listed by Witchell (1896). Hindmarsh (1984, 1986) found that individual birds have a repertoire of *c.* 15–20 distinct imitations. On average, *c.* 7% of total duration of song bout composed of heterospecific mimicry

(Hindmarsh 1984). Local dialects of mimicked sounds reported by Hindmarsh (1984). In addition to imitations within a song bout, Hindmarsh (1984) found 33% of imitations given as single calls between song bouts, and Eens *et al.* (1991*b*) reported bird singing 11 different imitations outside song bouts. Imitations integrated in song are mostly shorter in duration than imitations sung outside song bouts (M Eens). Some species appear to be preferentially mimicked (e.g. Buzzard *Buteo buteo*, Golden Oriole *Oriolus oriolus*, domestic chicken *Gallus*). See Fig VII–IX for other examples. Heterospecific imitations are probably learned from original models as well as from other *S. vulgaris* (Hindmarsh 1984), but in places where introduced (e.g. North America, New Zealand) European mimicries appear to be absent, indicating that transmission between generations (if any) has not persisted. In aviary study, Eens *et al.* (1992*b*) found that ♂♂ can copy heterospecific imitations from each other. For more details on mimicry of human sounds, see West *et al.* (1983) and West and King (1990). Older ♂♂ have almost twice as large a repertoire as 1st-years (Eens *et al.* 1991*a*, 1992*a*). In laboratory experiments, Böhner *et al.* (1990) showed that ♂♂ aged 11-12 months can learn new sub-phrases, and Eens *et al.* (1992*b*) presented direct evidence that birds go on modifying repertoires in adulthood. Changes in repertoire size from one year to next depend on age of bird. 1st-years increase repertoire size significantly more than older birds. Nevertheless, birds at least 4 years old can still modify repertoire (Eens *et al.* 1992*b*). In contrast to many other temperate zone songbird species, shows remarkably high song output almost all year except for brief period during post-breeding moult (Witherby *et al.* 1938; Böhner *et al.* 1990; M Eens). Most intense singing occurs in spring, with smaller peak in autumn after moult. Function and incidence of autumn song requires further study. Function of song in spring is well-studied. Song activity decreases markedly at pairing (Cuthill and Hindmarsh 1985; Eens 1992), indicating that attraction of ♀♀ is important function. In agreement with this, Eens *et al.* (1990) found in aviary experiment that unpaired ♂♂ increase song output significantly after introduction of a ♀, but not after introduction of a ♂. Mountjoy and Lemon (1991) showed experimentally that ♀ is attracted to playback of song. Aviary experiments indicate that ♀♀ prefer ♂♂ which sing longer song bouts and have larger

repertoires. During fertile period of their ♀♀, song activity of ♂♂ increases again (Kluijver 1933; M Eens). All within-pair copulations are preceded by ♂ singing, suggesting that important function of song after pairing might be to invite and stimulate ♀ to solicit copulation (Hartby 1968; Eens and Pinxten 1990; Eens 1992). Likewise, ♂♂ engaging in extra-pair courtship sing very close to ♀. Song matching between ♂♂ occurs only with whistles, suggesting whistles have primarily intrasexual function (Adret-Hausberger and Jenkins 1988). However, Eens· (1992) showed experimentally that whistles are not sung more frequently in ♂-♂ encounters than in ♂-♀. In autumn and winter, when often in large flocks, birds do periodically sing in day-time roosts. They also sing in night roost after evening arrival and before morning departure; some can be heard singing all night (Feare 1984). Spencer (1966*a*) found roosts relatively quiet only May–June, with high proportion of juveniles or moulting adults. Charman (1965) thought that only ♂♂ sing in roost, but this needs confirmation (Feare 1984). Function of roost singing not known, though volume often striking. According to Feare (1984), roost singing in morning could have a role in synchronizing departure because volume gradually increases until eventually sudden silence precedes exodus of a wave. Song given by juveniles from August (M Eens); full song begins at 6-9 months old, and crystallizes at *c*. 9 months (Chaiken *et al.* in press). (2) Song of ♀. Given occasionally in autumn and winter (Kluijver 1933; Witschi and Miller 1938; Bullough and Carrick 1942; Feare 1948; M Eens), but very little during breeding season (Kluijver 1933; Hartby 1968; Hausberger and Black 1991; M Eens), and function obscure. Testosterone-implanted ♀♀ produced song very close in structure to ♂'s, but without high-frequency sub-phrases (Hausberger and Black 1991). Some ♀♀ in the wild, however, can sing high-frequency sub-phrases, and ♀'s song can also contain heterospecific imitations (M Eens). Repertoire smaller than in most ♂♂ (Hausberger and Black 1991). (3) Contact-alarm calls. (3a) Flight-call. As soon as bird takes flight, frequently utters faint purring 'prurrp' (Hartby 1968; Feare 1984) or drawn-out hoarse 'ärr wrr' (Bergmann and Helb 1982); also when released after being handled. Given with bill closed or nearly so. (3b) A 'querrr' (Fig XI, 3rd unit: P J Sellar) or 'squar' (Bergmann and Helb 1982) given typically in flight, but also when

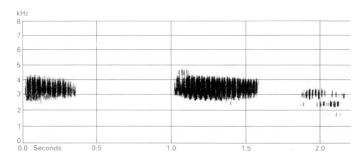

XI V C Lewis England May 1964

perched, by adults with fledged young in summer (C J Feare). (4) Flock-call. Hartby (1968) described harsh sound of rather high intensity, frequently given when post-breeding flocks are formed and during autumn. (5) Threat-call. According to Hartby (1968), this is an aggressive call frequently heard in the wild. Rattling form is used in all kinds of aggressive conflicts. (6) Attack-call. Loud squawks and chattering accompanying fighting (Hartby 1968). According to Feare (1984), incorporates variety of different sounds but main component is harsh rapidly repeated 'chackerchackerchacker'. (7) Distress-call. Repeated, high-pitched, raucous, penetrating scream given when held or cornered by predator (Feare 1984). Sometimes also when one bird grasps another in fight close to nest-site, and by some birds when handled by man (Feare 1984; M Eens). Incidence of this call varies seasonally as well as with age and sex. More juveniles give call than do adults, and perhaps more birds give Distress-calls during moult than at other times. In winter, ♀♀ tend to call much more than ♂♂. (Feare 1984.) Individual variation in acoustic structure. Parents hearing screams of their own young were significantly more likely to make dive-attacks on source (loudspeaker) than parents hearing screams of neighbouring chicks. (Chaiken 1992.) Playback of Distress-calls has been used for dispersing birds (Feare 1984). For role of different signal components within call, see Aubin (1987). (8) Chip-call (mobbing call). Short metallic 'chip' or 'spet' (Fig XII) of wide frequency range (Feare 1984; M Eens) in response to predator, flying, perched, or on ground. If predator near, call may be repeated as rapidly as 7 times per s. (Hartby 1968.) Fig XIII apparently of this type. Call often accompanied by 'concealed' wing-beats (Hartby 1968). Also given by some

birds as they enter roost, by birds leaving roost after most others have gone, and by many birds as they arrive at feeding area from roost first thing in morning. May reflect state of anxiety. (Feare 1984.) Occasionally given before song or between song bouts (M Eens). (9) Snarl-call. Unpleasant noisy sound of rather wide frequency range (Fig XIV: Hartby 1968); low-pitched 'caaar' (Feare 1984). Given by adults close to nest at approach of potential predator. Most often heard in breeding season. Snarling bird approaches predator more closely than when giving Chip-call, although these calls may be alternated. Snarling may promote mobbing from neighbouring birds. (Hartby 1968; Feare 1984.) Nestlings older than 10 days become silent when hearing Snarl-call (Feare 1984; C de Bruyn). Occasionally occurs at other times of year and may even be incorporated into song (Feare 1984). (10) Copulation-call of ♀. Before or during copulation, ♀ may give series of very delicate sounds (Hartby 1968). Precedes only small fraction of copulations (M Eens).

CALLS OF YOUNG. For development of vocalizations in nestlings, see Hartby (1968). For additional sonagrams, see Hartby (1968) and Chaiken (1990). Food-call begins at 1–2 days old; little more than a weak 'tzee' (Feare 1984). As chicks grow, gradually develops into 2 different calls, each given in different situations (Hartby 1968): close food-call a loud, raucous 'cheer cheer cheer' (Fig XI, 1st 2 units; Fig XV) given when adults with food approach chicks, in or out of nest; continuously heard in late April and May wherever there is a colony; distant food-call a quieter and more mellow 'churrr' given in parents' absence. Frequently given by groups of juveniles sitting in bushes and trees awaiting parents' return with food. (Feare 1984.) Distant food-call has wider frequency range,

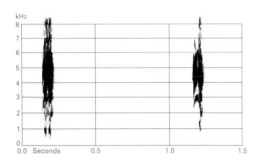

XII  J-C Roché  France  April 1968

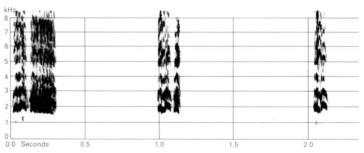

XIII  E D H Johnson  Channel Islands  December 1989

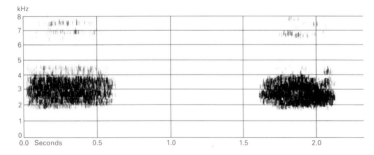

XIV  V C Lewis  England  May 1968

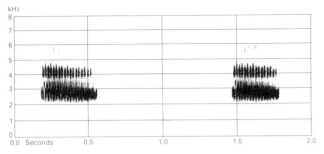

XV P J Sellar Scotland June 1974

of lower overall frequency than close food-call. At 16-18 days old, the 2 calls achieve final form. (Hartby 1968.) Chaiken (1990) described nestlings participating in 3 types of vocal exchange: (1) with each other, while parents away; (2) parents and nestlings exchanging calls during feeding; (3) nestlings exchanging calls with parents approaching from a distance; antiphonal exchanges between parents and young develop shortly before fledging and appear to help family maintain contact thereafter. When handled, nestlings may give Distress-call as described for adult (call 7, above). ME

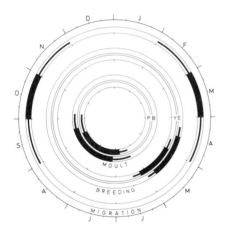

**Breeding.** SEASON. Account refers to nominate *vulgaris*. Breeding often in 3 phases: 1st clutches initiated synchronously throughout colony over period usually of 4-10 days (Verheyen 1980), 2nd clutches less synchronous and started 40-50 days after 1st clutches; between 1st and 2nd clutches, 'intermediate' clutches include replacements of lost 1st clutches, clutches of ♀♀ of polygamous ♂♂ and clutches of late-arriving birds (Pinxten *et al.* 1990). Italy: in one year, 1st clutches initiated 22 March to 9 April, last 2nd clutches initiated 16 June (Brichetti *et al.* 1992). Southern England: over 17 years, 1st clutches initiated 1-24 April, last 2nd clutches initiated 4 June (Feare 1984; C J Feare). Germany: see diagram; in Oberlausitz, 1st clutches initiated 6-23 April, with 2nd clutches in late May (Blümel 1986). Southern Sweden: 1st clutches initiated 26 April-6 May over 7 years, with no 2nd clutches (Karlsson 1983). Northern Finland: 1st clutches initiated 3-10 May (Ojanen *et al.* 1979); 2nd clutches exceptional, laid mid-June (Ojanen and Kylmänen 1984). Russia: at Arkhangel'sk, 1st clutches completed 16-20 May (Dementiev and Gladkov 1954). Occasionally eggs laid autumn and winter (Witherby *et al.* 1938) but clutches rarely completed and rarely hatch (C J Feare). SITE. Hole in tree, cliff, building, pylon, etc.; also in nest-boxes, occasionally in holes in ground where alternatives scarce; uses old holes and will take over current ones of woodpeckers (Picinae) and Sand Martin *Riparia riparia*; often in occupied houses (Feare 1984). Also reported excavating own holes in sand dunes (Stevenson 1866; Summers 1989). No data on tree species selection. In Italy, 72% of nests had entrances facing north and east; in Belgium, 61% of occupied nest-boxes (at least 1 egg) had entrances facing east, south-east, or south (Verheyen 1969a; Brichetti *et al.* 1992). Nests sited 0-15 m or more above ground (Campbell and Ferguson-Lees 1975); higher sites preferred. In experimental investigation of preference for height above ground, 11% of 9 nest-boxes below 2 m were occupied, 42% of 24 at 2-7 m, 56% of 27 above 7 m (Verheyen 1969a). Nest: bulky base of dry grasses and fine twigs, sometimes pine needles, size of this base dependent largely on size of cavity, and in roof spaces can be *c*. 1 m across, 25 cm deep (C J Feare); cup constructed in part of nest cavity remote from entrance, with variable lining of finer materials, including grasses, rootlets, moss, feathers, wool, paper, etc.; lining may be thicker in cooler climates; in later stages of building, fresh leaves or flowers may be incorporated into lining (Feare 1984); these perhaps contribute to mate attraction (Feare 1984) or to defence against nest parasites (Clark and Mason 1988). Cup diameter average 10 cm (9·6-15·3, *n*=11), depth average 9·5 cm (6·8-14·0, *n*=11) (Sema 1978). Building: where migratory, ♂♂ begin building before ♀♀ arrive (Kluijver 1933); base built entirely by ♂; ♀ contributes to lining but fresh green material and flowers brought by ♂, sometimes removed by ♀. In Italy, building commenced 15 March to 2 April and took 4·6 days (2-9) (Brichetti *et al.* 1992); in southern England, some nests built in autumn, and nest-boxes can contain nest-material throughout winter; other nests built rapidly in late March and early April (Feare 1984). In southern England, intermediate and 2nd clutch nests sometimes built on top of early nests, but usually at least some old material cleaned out by ♂ before new nest commences (C J Feare). EGGS. See Plate 59. Sub-elliptical, sometimes ovoid, smooth with some gloss, various shades of pale blue and occasionally white; some eggs, especially in 2nd clutches, with reddish or blackish spots of dried blood from parasite bites on incubating ♀ (Feare and Constantine 1980). Nominate *vulgaris*: 29·7 × 21·2 mm (26·5-34·5 × 20·0-22·5), *n*=1549; calculated weight 7·0 g. *S. v. faroensis*: 31·2 × 22·0 mm (28·1-34·3 × 20·3-23·6), *n*=50; calculated weight 8·0 g. *S. v.*

*zetlandicus*: 31·0 × 21·9 mm (28·1–34·7 × 20·2–24·5), $n = 64$; calculated weight 7·8 g. *S. v. poltaratskyi*: 30·2 × 21·3 mm (29·5–31·4 × 20·2–22·1), $n = 3$; calculated weight 7·2 g. *S. v. tauricus*: 28·9 × 21·6 mm (26·3–30·2 × 20·2–23·1), $n = 19$; calculated weight 7·1 g. *S. v. caucasicus*: 29·8 × 21·5 mm (27·5–30·2 × 19·0–22·6), $n = 25$; calculated weight 7·2 g. (Schönwetter 1984.) Clutch: Italy, 1st clutch 5·4 (4–7, $n = 31$), intermediate 5·4 (3–8, $n = 47$), 2nd 5·2 (3–7, $n = 14$) (Brichetti 1991); southern England, mean clutch sizes, 1st 4·48–4·96 over 7 years, intermediate 3·50–4·71 over 6 years, 2nd 3·87–4·50 over 5 years (Feare 1984); northern Finland, 1st clutch 5·1–5·4 over 9 years (Ojanen *et al.* 1979); given clutch sizes may be influenced by intraspecific nest parasitism (see Social Pattern and Behaviour). Usually one brood in north and east (Dementiev and Gladkov 1954; Ojanen *et al.* 1979), but in many areas, particularly south and west, proportion of pairs regularly but not invariably has 2 broods, 2nd clutch initiated 40–50 days after initiation of 1st; in seasons when 1st clutches are late, 2nd clutches may not be laid (Feare 1984); 2nd clutches locally rare, however, e.g. East Anglia (England) (A Dawson). In most areas lost 1st clutches replaced. INCUBATION. Average 12·2 days (11–15, $n = 84$ clutches), no significant differences between 1st, intermediate, and 2nd clutches (Brichetti *et al.* 1992). Entirely by ♀, who spends all night and estimated 71% of day on eggs of 1st clutch, 82% for intermediate clutch, and 88% for 2nd (Feare 1984); ♂ has poorly developed incubation patch and does not incubate, but does spend up to 29% of day on nest, reducing rate of heat loss from eggs (Feare 1984; H Biebach); shift duration averaged 10·8 min ($n = 41$) for ♀ and 14·1 min ($n = 15$) for ♂ in southern England; ♀ can complete incubation alone but nest then left unattended longer than when ♂ participates (Feare 1984). When left unattended, eggs often covered with leaf (see Social Pattern and Behaviour). Incubation usually begins with penultimate egg (but see Brichetti *et al.* 1992) and hatching asynchronous with last egg hatching up to 24 hrs after others (Feare 1984). YOUNG. ♀ alone broods chicks for first 5–8 days (Kluijver 1933) or ♀ broods at night with ♂ taking over sometimes during day (Westerterp *et al.* 1982); amount of brooding decreases until only night-time brooding by end of 1st week, ceasing by day 10; decrease in daytime more rapid in larger broods and at higher ambient temperatures (Westerterp *et al.* 1982). Fed by both parents about equally (Tinbergen 1981), or ♀ makes more feeding visits (Kluijver 1933), or *c.* 70% of visits with food by ♀ (Feare 1984). ♀ can feed chicks alone, as when ♂ dies or deserts, or with 2nd ♀ of polygamous ♂, but success lower (Merkel 1980; Verheyen 1980). Parents remove faeces from nest, but roles of sexes not established. FLEDGING TO MATURITY. Young remain in nest *c.* 21 days (Westerterp *et al.* 1982; Feare 1984) but will leave earlier (17 days: C J Feare) if disturbed. Stay in vicinity of breeding colony, being fed by parents, for a few days, but duration of this phase not quantified; young of 2nd brood may spend more time with parents, since in southern England begging seen up to *c.* 1 month after fledging of most young in colony (C J Feare). After fledging, young group into flocks of similar age birds. Many ♀♀ breed at 1 year, most ♂♂ do not breed until 2 years old (Coulson 1960; Verheyen 1969b). BREEDING SUCCESS. Hatching success generally high; in studies in England, Scotland, Finland, Poland, Italy, and Kazakhstan, 75–100% of eggs laid survived to hatching in 1st clutches, 42–89% in 2nd clutches (Dunnet 1955; Anderson 1961; Tenovuo and Lemmetyinen 1970; Luniak 1977b; Korpimäki 1978; Gromadzki 1980; Feare 1984; Brichetti *et al.* 1992); many losses of eggs result from observer disturbance, and predation generally not a major source of loss. Later broods generally less successful than earlier ones (see Table A), losses mainly due to starvation, especially of last-hatched chick, with predation and weather playing little part (but see Brichetti *et al.* 1992). Poorer food delivery contributes to lower success of 2nd broods. Postfledging survival up to 9 weeks estimated 39–62%, with highest survival in heavy chicks that fledged early in season (Kremetz *et al.* 1989). CJF

Table A Breeding success of Starling *Sturnus vulgaris* for early and late broods. Figures are percentage of eggs that produced fledged young.

| | Finland | Sweden | Poland | Scotland | England | Netherlands | Italy | Kazakhstan |
|---|---|---|---|---|---|---|---|---|
| Early | 24–44[1] 57–95[2] 71[3] | 75·2 | 62–98 | 79–85[4] 74–84[5] 70[6] | 67–81 | 77 | 14 | 69 |
| Late | | | | 64–78[4] 74–85[5] | 21–45 | 70 | 58 | 50 |
| Source | [1] Tenovuo and Lemmetyinen (1970), [2] Korpimäki (1978), [3] Ojanen *et al.* (1979) | Karlsson (1983) | Gromadzki (1980) | [4] Dunnet (1955), [5] Anderson (1961), [6] Evans (1988) | Feare (1984) | Lack (1948) | Brichetti *et al.* (1992) | Sema (1978) |

**Plumages.** (nominate *vulgaris*). ADULT MALE. In fresh plumage (September–November), head and body black with strong gloss, each feather with contrastingly pale tip, plumage appearing spotted; for colour of gloss, see worn plumage (below). Spots on forehead, crown, and nape rufous-brown to cream-buff, *c.* 1·5–2 mm long and *c.* 1–1·5 mm wide, partly coalescing and thus largely concealing black on forehead; paler towards side, where often forming buff-white supraloral line. Spots gradually larger on upper mantle, rounded-triangular and *c.* 2–3 mm long and wide on lower mantle and scapulars, on outer scapulars extending into narrow brown fringe along outer web; rounded-triangular spots on back and rump buff-brown to grey-buff, extending into narrower fringe of same colour at each side of feather, upper tail-coverts with more even buff-brown or grey-buff fringes *c.* 1·5–2 mm wide. Lore uniform velvety-black, remaining side of head as well as side of neck, chin, and throat with narrow elongate cream or off-white specks of *c.* 1 mm long, 0·5–1 mm wide, slightly larger and browner on lower side of neck, sometimes coalescing on chin; cream or white spots gradually larger on chest, *c.* 2–3 mm long and *c.* 1·5–2 mm wide on belly and flank, partly interrupted by black at shaft, larger still and more cream-buff on rear of flank and thigh, and extending into broad cream-buff fringe on under tail-coverts (fringe *c.* 4–5 mm wide on tip of coverts, *c.* 2 mm at sides). Each tail-feather dark ash-brown with slight bronze lustre, paler on central pair (t1), gradually darkening to grey-black on outer pair (t6); feathers narrowly fringed buff-brown to pale grey-buff (widest at bases of outer webs, faint and narrow on inner webs, except on t1, interrupted by black point on tip of each feather), bordered submarginally by contrasting deep black mark; shafts horn-black; tips of feathers ending in obtuse point; tip of inner web of most outer feathers concave, that of t6 with rather distinct broad grey-buff margin. Flight-feathers, greater upper primary coverts, and bastard wing greyish-black, darkest on outer webs, narrowly fringed buff-brown to pale grey-buff on outer webs and tips of primaries, primary coverts, and bastard wing, more broadly so on secondaries; inner web and tip of p1–p7 with deep black margin, bordered subterminally by ash-brown spot with slight bronze lustre. Tertials dark ash-brown with bronze lustre, outer web and tip fringed buff-brown, bordered submarginally by black mark. Greater upper wing-coverts black, outer web and tip rather narrowly fringed buff-brown, median coverts broadly fringed buff-brown on tip but hardly on outer web, lesser coverts more evenly fringed buff-brown on tip. Under wing-coverts and axillaries dark grey (darkest on shorter and inner feathers), rather narrowly fringed pink-buff or tawny-buff. *When slightly worn* (about November or December to January), spots and fringes of head, body, wing, and tail bleach to pale cream-buff (on upperparts and wing) or white (on underparts and tail), first on tips of feathers; from January–February (western Europe), March (central Europe), or April (eastern Europe) pale spots gradually wear off, throat and chest first, and glossy black of head, body, secondaries, tertials, and upper wing-coverts gradually exposed. Once nesting, only traces of pale spots remain on back to upper tail-coverts, lower flank, vent, under tail-coverts, and upper wing-coverts, and pale fringes of tail and flight-feathers very narrow; shortly before moult (late June or early July) even these largely worn off. Gloss of forehead, crown, chin, and cheek rather dark green, more purplish with oblique light; ear-coverts more strongly green in all lights; rear of crown, side of neck, and elongated feathers of lower throat strongly purple in all lights. Hindneck to upper tail-coverts strongly glossed green, sometimes more blue towards feather-centres, but lower mantle usually partly reddish-purple, occasionally fully purple, and upper tail-coverts sometimes bronzy-green, bluish-green, or purple, espe-

cially when heavily worn; if fully green hindneck to upper tail-coverts is scored 1 and fully purple upperparts scored 5, with 2–4 intermediate stages (2 slightly purple on central mantle, 3 more extensively so, 4 green only on hindneck, outer scapulars, and back), then large sample from western, central, and northern Europe scored 2·6 (99) 1–5, without marked difference between areas or seasons. Elongated feathers of chest, flank, upper belly, and side of belly green, flanks bluish or purplish in oblique light, belly sometimes slightly blue, remainder of underparts slightly green. Elongated feathers of lower throat at border between purple of throat and green of chest 22·3 (80) 19–25 mm (ZMA), 22·9 (104) 20–26·5 mm (Berthold 1964), or 20·5–25 mm (Svensson 1992); feathers of chest 27·8 (101) 26·5–30·5 mm; length of glossy part of throat feathers 12·8 (104) 11–17 mm (Berthold 1964). Narrow black outer borders of secondaries and tertials, broader ones of greater coverts, broad black tips of median coverts, and smaller ones of lesser coverts all strongly glossed green or blue-green, often more strongly blue or purplish-blue on feather-tips; black arcs on tips of tail-feathers, tertials, and primaries often slightly glossed purple. When heavily abraded, from about July, feathers at bill-base sometimes completely worn off. ADULT FEMALE. Like adult ♂, differing mainly in iris colour and (in breeding season) in bill colour (see Bare Parts); spots on head and body in fresh plumage on average broader than in adult ♂, 1–1·5 mm wide on crown, 2·5–4·5 mm on mantle, scapulars, and belly, more often inclined to extend into narrow brown fringe at sides of feather on mantle; faint pale spots on crown often still present when breeding, and spots from mantle to upper tail-coverts and on lower flank, vent, and under tail-coverts on average larger than those of adult ♂ then; gloss as in adult ♂, but sometimes less lustrous, in particular belly mainly dull black; average score of upperparts (see Adult Male) nearer green, 1·9 (59) 1–3(–4); elongated feathers of lower throat shorter than in adult ♂, 19·1 (45) 16–22 mm (ZMA), 17·9 (51) 14·5–20 mm (Berthold 1964), or 15·5–20 mm (Svensson 1992); length of chest-feathers 24·1 (51) 23–26 mm, length of glossy part of throat-feathers 8·2 (51) 6–11 mm (Berthold 1964); tail-feathers more obtusely pointed or slightly rounded on tip. NESTLING. Down fairly long and plentiful, grey-white, restricted to upperparts, sparse or sometimes absent on thigh and belly. For development and growth of feathers, see Hudec and Folk (1961) and Ricklefs (1984). JUVENILE. Upperparts and side of head and neck dark grey-brown, sometimes with narrow buff fringes on mantle and scapulars; lore darker brown or sooty; often with indistinct short paler brown-grey stripes above and below eye. Lower cheek, chin, and throat grey-white, mottled grey-brown on cheek and lower throat and sometimes on centre; side of breast, chest, and flank grey-brown, sometimes with slightly darker feather-tips, showing as faint streaks; belly and vent grey-white with more distinct grey-brown feather-tips or shafts, appearing more clearly mottled or streaked, especially on mid-belly. Thigh and under tail-coverts grey-brown with faintly paler cream-buff fringes. Tail dark grey-brown, like upperparts, faintly edged off-white; flight-feathers, tertials, and all upper wing-coverts dark grey-brown (darkest on tertials, outer greater coverts, and longest feather of bastard wing), all with narrow and indistinct grey-buff fringes, becoming paler grey-white and more distinct when plumage worn; in some birds, fringes more cinnamon, showing as bright patch on closed wing, contrasting with duller brown remainder. Under wing-coverts and axillaries pale brown-grey with darker shafts and indistinct isabelline fringes. Some individual variation in depth of ground-colour and in amount of white on underparts; ground-colour sometimes darker than described above, fuscous-brown or (e.g. in 14% of 70 juveniles from Netherlands examined) sooty-black, grey-white sometimes restricted to small patch on chin and some traces of streaks on mid-belly. Rarely, some gloss on tips of upper tail-coverts (Bährmann 1976). Post-juvenile moult starts a few days after juvenile flight-feathers full-grown, lesser coverts, inner greater coverts, and feathers on side of belly growing first, brown-fringed black coverts and white-tipped black belly feathers contrasting strongly with rest of plumage. When primary- and tail-moult nearing completion, body black with contrasting pale spots (as in adult), but head and neck still juvenile, worn, contrastingly uniform drab-grey. FIRST ADULT MALE. Like adult ♂, but spots on head and body larger, like those of adult ♀, differing from latter mainly in colour of iris or (when nesting) bill; white spots on underparts less divided by long black point along shaft than in adult ♀. Tail-tips more rounded than in adult ♂, pale fringes of feathers less clearly defined than in adult ♂ and ♀ (especially on tips), bordered submarginally by ill-defined dark brown or black-brown edge, latter not extending into sharp black point on tip (see Svensson 1992). Gloss as in adults, colour of mantle to upper tail-coverts scoring 2·0 (56) 1–3(–4) (see Adult Male). Feathers of head, neck, and chest shorter and more gradually tapering to tip than in adult, less drawn out into long point. Length of feathers on throat 18·4 (40) 15·5–21 mm (ZMA), 16·4 (51) 15–18 mm (Berthold 1964), or 16–20 mm (Svensson 1992); length of chest feathers 24·1 (51) 22·5–25·5 mm, length of glossy part of throat feathers 8·0 (51) 6–11 mm (Berthold 1964), thus close to adult ♀. Exceptionally, some juvenile wing-feathers retained. FIRST ADULT FEMALE. Like adult ♀, but spots larger when fresh, 4 mm or more wide on mantle, c. 3–5 mm long and wide on belly, sometimes almost coalescing on lower flank, lower belly, and vent; white of spots on underparts not interrupted by black line, gloss of black less intense; feather-centres of mantle and scapulars sometimes dull black, not glossy, underparts duller, less glossy, in particular belly sometimes dull black without gloss; rarely, tips of primaries without black fringe and grey centre. Colour score of mantle to upper tail-coverts 1·8 (56) 1–3 (see Adult Male), thus greener (less purplish) on average than in other plumages. Length of feathers on centre of lower throat 15·7 (48) 12–18 mm (ZMA), 12·3 (52) 9·5–15 mm (Berthold 1964), or 13–16·5 mm (Svensson 1992); length of feathers of chest 19·1 (51) 17·5–20 mm, length of glossy part of throat feathers 3·5 (52) 1–5·5 mm (Berthold 1964). Tail-tips as in 1st ♂, but as tail of adult ♀ often slightly more rounded and less clearly marked on tip than in adult ♂, ageing of ♀♀ on tail more often problematical than in ♂. For sexing and ageing on feathering, see also Hicks (1934), Bullough (1942a, b), Kessel (1951), Schneider (1957, 1972b), Davis (1960), Delvingt (1961), Folk et al. (1965), Coleman (1973), Klijn (1975), Pyle et al. (1987), and Svensson (1992).

**Bare parts.** ADULT, FIRST ADULT. Iris of ♂ liver-brown to dark brown, of ♀ brown with narrow, contrastingly white, grey-white, cream, pale yellow, yellow-brown, light brown, or deep orange inner or outer ring; occasionally, ring of ♀ scarcely paler brown than remainder of iris, or ♂ with slightly paler brown ring; 1–3% of birds have 'wrong' iris colour for their sex (Schwab and Marsh 1967). Bill black with slight grey, green, or brown tinge in autumn and winter, usually with paler grey-brown to horn-white cutting edges; in non-migratory or partially migratory western and southern populations, bill of adult ♂ gradually turns yellow, starting (from base) November–January in England, December–January in Belgium, and January–February in Netherlands and Germany; c. 2–5 weeks later adult ♀ and 1st adult ♂, c. 5–8 weeks later in 1st adult ♀. In migratory populations, bill turns yellow (January–)February–April, rarely May; bill darkens again (starting from base) at end of nesting cycle before post-breeding

moult starts, sometimes in late May, generally in 2nd half of June and early July. When at least base of upper mandible yellow (mainly April to mid-June), marked difference between sexes in colour of basal part of lower mandible: in ♂, blue-grey, horn-grey, slate, or blue-black, contrasting with yellow remainder of bill; in ♀, pale yellow, pink-white, light pink, or pink-flesh. Leg and foot warm brown, red-brown, or chestnut-brown, light brown during breeding. For details of iris colour and change of bill colour, see Harrison (1928, 1938), Stresemann (1928), Kluyver (1935), Banzhaf (1937a), Witherby et al. (1938), Bullough (1942a, b), Nichols (1945), Kessel (1951), Schneider (1957, 1972a), Davis (1959, 1960), Delvingt (1961), Parks (1962), Thompson and Coutlee (1963), Berthold (1964), Wydoski (1964), Schwab and Marsh (1967), Bährmann (1972, 1976), Coleman (1973), Klijn (1975), and Feare (1984). For hormonal control of bill colour change, see Witschi and Miller (1938). NESTLING. Mouth bright yellow (Witherby et al. 1938) or orange (Feare 1984), gape-flanges pale yellow. JUVENILE. Iris liver-coloured, grey, or grey-brown; at c. 3 weeks after fledging, dark grey-brown, dark brown, or black-brown in ♂, light sepia-brown or light grey with yellow tinge in ♀. Bill horn-black to deep black, cutting edges and base of lower mandible paler brown, grey-brown, yellow-horn, or horn-white, with traces of pale yellow gape-flanges; mouth yellow, some yellow on tongue sometimes present up to November (all dark in adult). Leg and foot grey-brown, red-brown, horn-brown, warm brown, chestnut-brown, or dark brown, rarely blood-red, darkest shortly after fledging (Krätzig 1936; Bullough 1942a; Schneider 1972a; Coleman 1973; Bährmann 1976; Svensson 1992; ZMA.)

**Moults.** ADULT POST-BREEDING. Complete; primaries descendent. In southern England and Wales, starts with shedding of p1 (primary moult score 1) late May to late June, rarely early July, completed with regrowth of p9 (score 45, or 50 if vestigial p10 included) after c. 80 days, mid-August to mid-September, sometimes late September; no difference between sexes; in northern England and Scotland, moult slightly later (Ginn and Melville 1983). For timing of moult in Orkney, Shetland, Faeroes, Iceland, and Norway, see Evans (1986). In Netherlands, moult starts 5–25 June, but some (probably those with 2 broods) 1–15 July; completed mid-August to late September (ZMA). In eastern Germany, starts late June and early July, completed mid-August to end of September (Bährmann 1964; Schneider 1972a). In USSR, starts mid-July in nominate vulgaris from central European Russia, slightly later in north, from late June in some southern races; in purpurascens, central Asia, moult intense July and early August, completed late August or early September (Dementiev and Gladkov 1954). Tail starts at primary moult score 15–25, completed at score 35–45 (ZMA); sequence not always strictly centrifugal (t1–t6), t6 and/or t5 frequently after t2 or t3, or sequence t1–t2–t6–t5–t4–t3 (Bährmann 1964, 1970). Secondary moult usually ascendent, starting with shedding of s1 at primary moult score 23–34, completed with regrowth of s6 at score 45–50 or slightly later (ZMA). For sequence, see also Williamson (1961b) and Bährmann (1964). Primary moult sometimes not strictly descendent, secondary moult not strictly ascendent; moult sometimes temporarily suspended (Evans 1986, which see for details). For hormonal control of moult, see Schleussner et al. (1985) and Schleussner (1990). POST-JUVENILE. Complete, primaries descendent; exceptionally, one or a few secondaries, tertials, wing-coverts, feathers of bastard wing, or p10 retained (Williamson 1961; Scott 1965a; ZMA); moult sometimes temporarily suspended, especially when condition of bird poor, or moult eccentric (Evans 1986). Moult of body easy to follow due to strong colour contrast between grey-brown juvenile feathering and black white-spotted 1st adult plumage. Starts with scattered feathers at side of belly and mantle, as well as lesser coverts, soon followed by shedding of p1. Timing of tail and secondary moult in relation to primary moult as in adult post-breeding or moult of these relatively slightly later; lesser and greater wing-coverts and side of belly moulted at primary moult score 0–30, remainder of coverts and body as well as tertials at primary moult score 10–35(–40); bastard wing, head, and neck start with primary score 33–40, completed at same time as outer primaries, some feathers round or above eye sometimes later. (RMNH, ZMA; 100 moulting birds examined: C S Roselaar.) In southern England and Wales, primary moult starts early June to mid-July, completed at *c.* 90 days late August to early October; moult in northern England and Scotland *c.* 1 week later, from late June or early July on Scottish islands (Ginn and Melville 1983). In Netherlands, starts 10 June to 5 August, completed 5 August to 15 October; birds of 1st broods start mainly 20–30 June, those of 2nd broods 20–30 July (RMNH, ZMA). In eastern Germany, moult starts *c.* 6 weeks after fledging, early July to early September, completed late August to late October (Heinroth and Heinroth 1924–6; Schneider 1972*a*; Bährmann 1964). In Scandinavia, moult of migratory birds from northern Sweden more rapid than in non-migratory birds from coastal Norway further north: in migratory ones, primary moult starts at $63 \cdot 1 \pm 8 \cdot 5$ days old, finishing at $105 \cdot 3 \pm 7 \cdot 4$ days old; in non-migratory ones, from day $67 \cdot 6 \pm 9 \cdot 2$ to day $122 \cdot 4 \pm 10 \cdot 3$ (Lundberg and Eriksson 1984, which see for moult of other feather tracts). In USSR, starts early July to late August, completed early September to October (Dementiev and Gladkov 1954).

**Measurements.** Nominate *vulgaris*. Wing and bill to skull of birds from (1) Scandinavia and Poland east to central European Russia, breeding, (2) England and southern Scotland, breeding, (3) Netherlands, breeding (in 1–3, adult and 1st adult combined); Netherlands, September–April, (4) adult, (5) 1st adult; other measurements combined for these areas, skins (RMNH, ZFMK, ZMA). Bill (N) to distal corner of nostril; exposed culmen on average 4·5 less than bill to skull.

| | | | | |
|---|---|---|---|---|
| WING (1) | ♂ 132·7 (1·73; 8) 131–136 | ♀ 128·8 (3·44; 16) 123–134 |
| (2) | 132·5 (2·23; 13) 129–136 | 129·1 (2·57; 8) 126–134 |
| (3) | 132·1 (2·20; 47) 127–137 | 128·8 (2·55; 33) 124–135 |
| (4) | 134·0 (2·41; 77) 128–141 | 131·3 (2·62; 47) 126–137 |
| (5) | 132·6 (2·21; 42) 128–137 | 129·3 (2·40; 43) 125–134 |
| JUV | 124·4 (2·20; 25) 121–129 | 121·4 (2·68; 21) 117–126 |
| TAIL AD | 64·1 (2·24; 27) 60–68 | 61·8 (2·01; 19) 58–65 |
| 1ST AD | 61·9 (1·40; 15) 59–64 | 61·4 (2·01; 20) 57–65 |
| JUV | 60·4 (1·79; 19) 58–64 | 58·8 (2·20; 14) 55–62 |
| BILL (1) | 29·0 (1·22; 8) 27·6–30·3 | 28·9 (0·76; 16) 27·6–29·9 |
| (2) | 29·8 (1·48; 13) 28·0–32·3 | 28·3 (1·48; 7) 26·2–30·8 |
| (3) | 28·8 (1·12; 42) 26·6–31·2 | 28·1 (1·23; 32) 25·5–30·1 |
| (4) | 29·6 (0·91; 64) 27·8–31·2 | 28·7 (1·04; 36) 27·0–30·5 |
| (5) | 29·2 (0·91; 34) 27·5–30·7 | 28·2 (1·13; 35) 27·0–30·0 |
| BILL (N) | 18·5 (1·03; 45) 16·9–20·6 | 17·7 (0·90; 43) 16·3–19·6 |
| TARSUS | 30·5 (0·81; 58) 28·8–32·2 | 29·7 (0·85; 50) 27·6–31·2 |

Sex differences significant for, except for 1st adult tail and bill (1)–(5). Juvenile wing, tail, bill, and tarsus generally not full-grown until post-juvenile flight-feather moult started; bill and tarsus excluded from table.

Wing. (1) Sweden and Denmark (Svensson 1992). (2) Leipzig area (Germany), breeding (Schneider 1957). Lausitz area (south-east Germany): (3) adult, (4) 1st adult, (5) juvenile (Eck 1985*a*; see also Bährmann 1976, 1978 and Eck 1990). Bodensee (southern Germany), mainly spring and summer: (6) adult, (7) 1st adult (Berthold 1964). (8) Czechoslovakia, March–July (Folk

*et al.* 1965). (9) South-east France, breeding (G Olioso). (10) Southern Yugoslavia, winter (Stresemann 1920). (11) Western Bulgaria, September–March (Doïchev 1973).

| | | |
|---|---|---|
| (1) | ♂ — (—; 147) 126–138 | ♀ — (—; 108) 121–136 |
| (2) | 131·2 (2·51; 60) 123–137 | 127·8 (2·22; 98) 121–133 |
| (3) | 132·2 (2·30; 118) 127–137 | 128·1 (2·52; 83) 120–133 |
| (4) | 131·0 (2·38; 82) 126–138 | 128·1 (2·83; 46) 121–134 |
| (5) | 122·7 (3·35; 23) 116–128 | 120·8 (2·38; 18) 117–126 |
| (6) | 131·1 (2·09; 104) 126–136 | 127·6 (2·22; 51) 122–133 |
| (7) | 128·3 (2·48; 51) 122–135 | 126·4 (2·42; 52) 118–130 |
| (8) | 130·8 (2·66; 60) 124–137 | 128·3 (2·49; 44) 124–133 |
| (9) | 129·0 (2·45; 11) 126–132 | 124·8 (3·39; 16) 118–132 |
| (10) | 131·6 (2·45; 28) 127–136 | 129·8 (2·07; 17) 127–134 |
| (11) | 130·2 (3·07; 146) 121–138 | 127·7 (3·39; 139) 121–136 |

Northern Greece: ♂ 127–131, ♀ 127–132 (Makatsch 1950). For Germany, see also Niethammer (1971); for Netherlands, see Klijn (1975); for USA, see (e.g.) Baldwin *et al.* (1931) and Stegeman (1954).

*S. v. faroensis.* Faeroes, wing: 133–136 (*n* = 12) (Hartert 1921–2); 134–140 (*n* > 100) (Salomonsen 1935); ♂ 136·8 (2·14; 6) 134–140 (ZFMK).

*S. v. zetlandicus.* Wing: 134·0 (2·43; 11) 131–137·5 (Hartert 1921–2); 129–138 (Salomonsen 1935); Shetland, ♂ 130–138 (*n* = 12); South Uist (Outer Hebrides), ♂ 125–136 (*n* = 12) (Witherby *et al.* 1938).

*S. v. granti.* Azores, wing: ♂ 129·2 (1·26; 4) 128–131, ♀ 127·4 (3·68; 4) 123–131·5 (G Le Grand); ♂ 129·8 (1·98; 8) 128–133 (Feare 1984).

*S. v. tauricus.* Central and southern Anatolia (Turkey), April–June, adult and 1st adult combined; skins (RMNH, ZFMK, ZMA). Bill (S) is bill to skull.

| | | |
|---|---|---|
| WING | ♂ 136·9 (2·96; 11) 133–142 | ♀ 133·6 (2·52; 10) 130–138 |
| BILL (S) | 30·0 (0·93; 11) 28·8–31·4 | 30·3 (0·85; 10) 28·9–31·6 |
| BILL (N) | 19·0 (0·62; 11) 18·0–19·7 | 18·4 (0·82; 9) 17·4–19·5 |

Sex differences significant for wing. Sexes combined: tail 64·3 (1·56; 9) 62–66, tarsus 31·2 (0·75; 9) 30·2–32·2 (RMNH, ZMA).

Western and central Asia Minor, wing: ♂ 134·4 (2·58; 6) 132–138, ♀ 131·8 (3·30; 4) 128–135 (Jordans and Steinbacher 1948; Kumerloeve 1961, 1964*a*).

Adult and 1st adult (both sexes) combined: (1) *purpurascens*, western Transcaucasia; (2) *caucasicus*, Caucasus and eastern Transcaucasia; (3) *poltaratskyi*, western Siberia and (in winter) India; (4) *porphyronotus*, Tien Shan (Kirgiziya) and Tarim basin (Sinkiang, China); skins (RMNH, ZMA).

| | | | | |
|---|---|---|---|---|
| WING | (1) 134·5 (2·57; 6) 131–138 | (3) 130·1 (2·16; 22) 126–137 |
| TAIL | 63·3 (3·12; 6) 60–66 | 63·4 (1·45; 21) 61–66 |
| BILL (S) | 21·2 (1·16; 6) 30·1–32·8 | 30·1 (0·93; 19) 28·8–31·8 |
| BILL (N) | 19·7 (0·86; 6) 18·9–21·0 | 18·9 (0·93; 19) 16·9–20·4 |
| TARSUS | 31·4 (0·90; 6) 30·4–32·4 | 30·8 (1·18; 20) 29·2–33·1 |
| WING | (2) 130·9 (2·11; 6) 128–134 | (4) 132·9 (3·17; 14) 128–139 |
| TAIL | 62·2 (2·30; 6) 59–65 | 63·7 (1·82; 15) 61–67 |
| BILL (S) | 30·5 (1·38; 6) 28·2–32·5 | 32·0 (1·39; 15) 29·4–33·7 |
| BILL (N) | 19·3 (0·54; 6) 18·8–20·3 | 20·8 (1·21; 15) 18·8–22·9 |
| TARSUS | 31·4 (0·51; 6) 30·8–32·3 | 31·6 (1·08; 15) 29·9–33·5 |

*S. v. caucasicus.* Wing. Caucasus area: 127–136 (*n* = 68) (Jordans 1923). Talysh area (south-east Transcaucasia): ♂♂ 132, 132, ♀ 128 (Stresemann 1928). Northern Iran: ♂ 126·4 (3·83; 13) 121–134, ♀ 124·3 (234; 6) 121–127 (Stresemann 1928; Schüz 1959; Diesselhorst 1962; Erard and Etchécopar 1970). South-west Iran: ♂ 127·5–129 (*n* = 10) (Hartert and Steinbacher 1932–8).

*S. v. purpurascens.* Wing. Eastern Turkey, breeding: ♂ 134·2 (1·98; 11) 131–138, ♀ 130·4 (3·55; 12) 122–137 (Kumerloeve 1967*a*, 1969*b*); ♂ 133 (12) 130–136 (Erard and Etchécopar 1970).

Northern Iraq, wing: ♂ 135·8 (2·22; 4) 134–139, ♀ 131 (Hartert 1921–2).

*S. v. poltaratskyi.* Wing. Migrants, Afghanistan: ♂ 125·5 (6) 121–130, ♀ 125·4 (8) 122–129 (Paludan 1959). Juvenile, Lake Chany (south-west Siberia): 121·5 (2·16; 11) 118–125 (Havlín and Jurlov 1977).

*S. v. nobilior.* Southern Turkmeniya, wing: ♂ 126·8 (24) 125–132, ♀ 124·8 (11) 122–127 (Dementiev and Gladkov 1954). North-east Iran, May, wing: ♂♂ 134, 135 (Erard and Etchécopar 1970).

*S. v. humii.* Kashmir, wing: ♂ 121·8 (2·49; 5) 120–126, ♀ 117·2 (1·79; 9) 115–120 (P R Holmes and Oxford University Expedition Kashmir 1978 and 1983). India and Pakistan, wing: ♂ 116–121, ♀ 113–119 (Ali and Ripley 1972).

*S. v. minor.* Sind (Pakistan): wing 110–120 (Ali and Ripley 1972).

**Weights.** ADULT, FIRST ADULT. Nominate *vulgaris*. Netherlands: (1) November–February, (2) lighthouse victims, March, (3) April–June, breeding, (4) lighthouse victims, September–October, (5) exhausted birds, all year (ZMA), (6) southern Netherlands, March (Klijn 1975). (7) Britain, found dead in cold weather (Harris 1962). (8) Bodensee (southern Germany), mainly March–June (Berthold 1964). (9) Leipzig (eastern Germany), breeding (Schneider 1957). (10) Lausitz (eastern Germany), February–October (Bährmann 1972; see also Bährmann 1976, see both for monthly data). (11) Helgoland (northern Germany), migrants (Weigold 1926). (12) Western Bulgaria, December–March (Doičev 1973). (13) South-east France, breeding (G Olioso). (14) USA, March (Stegeman 1954).

| | | | | | |
|---|---|---|---|---|---|
| (1) | ♂ | 82·7 (8·43; 14) | 70–94 | ♀ 79·9 (9·17; 7) | 65–91 |
| (2) | | 82·0 (6·85; 14) | 73–94 | 81·2 (7·29; 15) | 72–93 |
| (3) | | 82·7 (7·87; 7) | 74–94 | 78·3 (6·97; 11) | 70–95 |
| (4) | | 79·1 (7·77; 17) | 70–94 | 74·2 (4·29; 14) | 68–83 |
| (5) | | 53·9 (6·47; 9) | 45–61 | 56·2 (5·12; 13) | 50–64 |
| (6) | | 52·0 ( — ; 52) | 36–61 | 53·0 ( — ; 57) | 41–62 |
| (7) | | 67·5 ( — ; 39) | 55–80 | 65·0 ( — ; 35) | 52–78 |
| (8) | | 81·0 (7·68; 128) | 52–108 | 78·3 (9·94; 86) | 58–99 |
| (9) | | 78·0 (4·03; 60) | 70–94 | 80·7 (4·85; 98) | 67–91 |
| (10) | | 80·1 (5·61; 169) | 68–107 | 76·5 (6·80; 122) | 64–101 |
| (11) | | 76·8 ( — ; 8) | 64–82 | 74·2 ( — ; 17) | 60–88 |
| (12) | | 82·3 (5·95; 74) | 66–96 | 77·2 (5·64; 64) | 66–90 |
| (13) | | 79·1 (3·86; 9) | 74–86 | 76·1 (4·39; 8) | 70–81 |
| (14) | | 82·0 (5·22; 240) | 66–98 | 76·2 (5·66; 215) | 60–90 |

Czechoslovakia, all year: 79·3 (6·14; 250) 57–104 (Folk *et al.* 1965, which see for monthly averages). Northern Greece: 78 (11) (Makatsch 1950). Migrants Algeria (*c.* 35·5°N), late September to November: 75·5 (5·1; 51) 65–87 (Bairlein 1988). Sweden, ♂: on arrival 80–91, 1 week later 67–73 (Zedlitz 1926). For weight during various stages of breeding cycle, see Schneider (1957), Bährmann (1972), Westerterp *et al.* (1982), and Ricklefs and Hussell (1984). Maximum of ♂ 112 (Feare 1984), of laying ♀ 105 (Bährmann 1976). For variation in weight and/or fat contents, see Thompson and Coutlee (1963), Blem (1981) (USA), Coleman and Robson (1975) (New Zealand), and Cooper and Underhill (1991) (South Africa). See also Bacmeister and Kleinschmidt (1920), Baldwin *et al.* (1931), Hicks (1934), Baldwin and Kendeigh (1938), Browne and Browne (1956), Dunnet (1956), Schneider (1972b), Taitt (1973), and Ward (1977).

*S. v. zetlandicus* and *faroensis.* For average during moult, see Evans (1986).

*S. v. granti.* Azores: ♂ 78·7 (3) 73–83, ♀ 76·5 (3·32; 4) 74–81 (G Le Grand).

*S. v. poltaratskyi.* Afghanistan, migrants: March, ♂ 74·3 (5)

71·5–78·7, ♀ 76·3 (6) 72·2–80·2; October, ♂ 77, ♀♀ 75, 75 (Paludan 1959).

*S. v. tauricus.* Central Turkey, June: ♂♂ 75, 76; ♀ 78·3 (3) 77–80 (Kumerloeve 1964a).

*S. v. purpurascens.* Eastern Turkey, May to early July: ♂ 87·8 (7·94; 14) 80–105, ♀ 84·8 (8·50; 12) 67–100 (Kumerloeve 1967a, 1969b).

*S. v. caucasicus.* Northern Iran, February: ♂♂ 78, 82 (Schüz 1959).

*S. v. humii.* Kashmir, July–September: ♂ 67·8 (2·86; 5) 64–71, ♀ 61·2 (8·27; 10) 40–71 (P R Holmes and Oxford University Expedition Kashmir 1978 and 1983).

Various races. See also Dementiev and Gladkov (1954), Schüz (1959), Yanushevich *et al.* (1960), Kumerloeve (1967a), Ali and Ripley (1972), Korelov *et al.* (1974), and Piechocki *et al.* (1982).

NESTLING. JUVENILE. For various aspects of growth of nestling, growth curves, fledgling weight in relation to brood size, etc., see Kluijver (1933), Dunnet (1955), Hudec and Folk (1961), Delvingt (1962), Westerterp (1973), Crossner (1977), Ricklefs (1979, 1984), Ricklefs and Peters (1979), Feare (1984), Thompson and Flux (1988), Boyarchuk (1990), and Peskov (1990). In Britain, average peak shortly before fledging in various years 69·6 ± SE1·0 to 79·2 ± SE7·7 (1st broods) or 59·4 ± SE2·4 to 74·7 ± SE1·7 (2nd broods) (Feare 1984). For monthly average June–October, see Bährmann (1976). Fledged juveniles: Netherlands, in moult, ♂ 73·4 (7·11; 16) 60–85, ♀ 70·6 (3·80; 7) 67–78 (ZMA); Öland (Sweden), July, ♂ 71·8 (5·36; 304) 51–87, ♀ 69·6 (5·32; 315) 48–85 (Svensson 1964, which see for influence of time of day and weather); south-west Siberia (*poltaratskyi*), 73·2 (3·49; 11) 66–78 (Havlín and Jurlov 1977); Armeniya (*purpurascens*) 82·5, 83 (Nicht 1961); central Turkey (*tauricus*) 82·7 (3) 81–85 (Kumerloeve 1964a).

**Structure.** Wing rather long, broad at base, tip pointed. 10 primaries: in adult, p9 longest (occasionally 0–1 shorter than p8), p8 0–3, p7 4–10, p6 12–20, p5 19–28, p1 47·5 (12) 42–51 in ♂, 43·5 (9) 40–49 in ♀; in full-grown juvenile, p8–p9 as in adult, p7 6–12 shorter, p6 15–20, p5 22–27, p1 42·5 (10) 39–46. P10 strongly reduced, especially in adult, broader and longer in juvenile; in nominate *vulgaris*, 84·9 (25) 75–92 shorter than wing-tip in adult, 71·6 (25) 68–76 in full-grown juvenile; 13·6 (25) 10–17 shorter than longest upper primary covert in adult, 7·3 (25) 4–11 shorter in juvenile (ZMA); length of p10 of nominate *vulgaris* 10–14 (adult) or 22–27 (juvenile) (Stresemann 1920). Outer web of p7–p8 emarginated; inner web of (p8–)p9 with notch, sometimes faint. Tip of longest tertial reaches to p1–p2. Tail short, tip square, slightly rounded or slightly forked; 12 feathers, tips slightly pointed in adult, more rounded in juvenile and 1st adult. Bill long, straight, about equal to head length, dorso-ventrally flattened; culmen straight, except for slightly decurved tip, gonys virtually straight; both mandibles rather deep at base, shallow at tip, bill ends in rather narrow but blunt tip or in sharp point as seen from above, latter especially in eastern races, but shape apparently depends on abrasion. For rate of growth of bill, see Wydoski (1964). Nostril oval, partly covered by membrane above, bordered by feathering of forehead at rear. Feathering of lore short, scanty, bristle-like. No bristles at gape. Feathers of head, neck, and chest elongated and pointed in adult, slightly also elsewhere on body; less elongated in 1st adult, normal in shape in juvenile. For pterylography, see Moyson (1973). Tarsus and toes moderately long, strong. Middle toe with claw 27·8 (20) 26–29·5; outer toe with claw *c.* 71% of middle with claw, inner *c.* 70%, hind *c.* 81%. Claws short, strong, decurved.

**Geographical variation.** Marked variation in colour of gloss on head, body, upper wing-coverts, and streak on outer web of tertials and secondaries of adult and 1st adult, in colour of under wing-coverts, and in general colour of juvenile. Less marked variation in size (mainly wing length, but *humii* and *minor* of Indian subcontinent markedly smaller in all measurements) and in length of p10. Slight variation in timing of moult, in timing of abrasion of white spots on head and body, and in timing of colour change of bill (timing of events mainly dependent on local breeding season, whether locally resident or migratory, or whether 1 brood or 2). Nominate *vulgaris* of northern, central, and western Europe glossed dark green on cap, chin and upper throat (sometimes purple in oblique light), green on ear-coverts, strongly (bronzy-)purple on nape, side of neck, and upper chest; hindneck to upper tail-coverts glossy green, sometimes more bluish-green on scapulars, rump, and upper tail-coverts, often bronzy-purple on central mantle or on whole mantle as well as upper tail-coverts, rarely also on hindneck and rump, exceptionally entirely bronzy-purple (apparently never in south and east of range); see also Plumages for variation in gloss of upper-parts with age and sex; belly glossed (bluish-)green, flank bluish-green to purple-blue, wing-coverts and fringes of tertials and secondaries bluish-green to purple-blue; under wing-coverts and axillaries dark grey with cream-buff fringes, latter mainly 1–1·5 mm wide in adult. Some individual variation in gloss, but gloss of head and underparts has virtually no link with age or sex; colour of gloss of 1st adult ♀ similar to adult ♂ but glossy patch on each feather less extensive; gloss in autumn partly concealed by brown and white spots and fringes. Gloss of cap, chin, and throat of birds occurring Slovakia to western Greece more often purplish than in birds from further north and west, but still green in some lights and separation of *graecus* Tschusi, 1905, for birds with more purplish head not warranted; gloss of upper wing-coverts of *c.* 15–25% birds of eastern Russia purplish instead of blue, but most birds similar to nominate *vulgaris*, and separation (as *ruthenus* Menzbier, 1891, or *jitkowi* Buturlin, 1904) not warranted. Populations of Atlantic islands largely similar to nominate *vulgaris* in adult plumage, differing mainly in juvenile plumage and in structure. *S. v. faroensis* large, leg and foot strong, bill longer, wider at base (8·5–9 mm; in nominate *vulgaris*, 7–8 mm: Salomonsen 1935); p10 longer and broader, 16–20 mm long in adult (10–15 in adult nominate *vulgaris*, mainly 12–14); juvenile sooty-black with restricted white on chin and belly, white with bold black spots on throat; gloss of adult slightly duller and darker green than in nominate *vulgaris*, pale spots smaller and wearing off more rapidly. *S. v. zetlandicus* from Shetland (except Fair Isle) intermediate in size between *faroensis* and nominate *vulgaris*, bill width at base 8–8·5 mm, colour of adult and juvenile as in *faroensis*, length of p10 as in nominate *vulgaris*. Birds from Outer Hebrides (including St Kilda) often also included in *zetlandicus*, mainly because of dark juvenile; however, measurements similar to nominate *vulgaris*, and dark juveniles not infrequent in range of nominate *vulgaris*, especially in Scotland, occasionally England and Netherlands (see Plumages), and inclusion of birds from Outer Hebrides in nominate *vulgaris* is equally well-founded. *S. v. granti* from Azores similar to nominate *vulgaris*, but slightly smaller, especially leg, and upperparts more often completely purple; p10 narrower, shorter, 8–13 mm long; bill narrower at base, short (Hartert 1903–10); though difference from nominate *vulgaris* not as marked as those in *faroensis*, *granti* and *faroensis* are probably equally old off-shoots of *vulgaris* and both deserve to be separated (Feare 1984).

Variation in remainder of range mainly in colour of gloss; thoroughly analysed by Meinertzhagen (1924), Hartert and Steinbacher (1932–8), Sushkin (1933), and Pateff (1947), and these authors followed here, with help of additional data from Hartert (1918*b*, 1921–2), Jordans (1923, 1935), Stresemann (1935), Meinertzhagen (1953, 1954), Vaurie (1954*c*, 1959), Munteanu (1967), Erard and Etchécopar (1970), Feare (1984), and specimens examined (BMNH, RMNH, ZFMK, ZMA). Races as defined here show little individual variation throughout range, but are connected by broad zones where highly variable populations occur (see list of races at start of this species); many races described from these zones, but none recognized. *S. v. poltaratskyi* from eastern Bashkiriya (south-east European Russia) and central Urals east to Lake Baykal and western Mongolia like nominate *vulgaris*, but head (including ear-coverts) purple, only slightly green in oblique light; nape, side of neck, and upper chest purple or green, depending on light; mantle to upper tail-coverts green, as in many nominate *vulgaris*; mid-belly green but side of belly and flank often more purple-blue than in nominate *vulgaris*; wing-coverts green-blue or (partly) purplish; differs from occasional purple-headed birds of nominate *vulgaris* and from all other races in broad light cinnamon-buff fringes of under wing-coverts and axillaries, mostly over 1·5 mm wide and often up to 5 mm with black on centres strongly reduced. *S. v. tauricus* from Crimea and neighbouring parts of southern Ukraine and western Ciscaucasia as well as western and central Asia Minor markedly different from nominate *vulgaris*: head green, sometimes faintly purple on ear-coverts in oblique light; nape, side of neck, and upper chest reddish-purple (green in some lights), remainder of upperparts reddish- or bronzy-purple, sometimes slightly blue on mantle; belly bronzy-purple, merging into bronze-green on flank; upper wing-coverts bronze-green (sometimes purple-bronze); underwing mainly blackish, fringes white, mainly 0·5–1·5 mm wide; pale spots virtually completely worn off in breeding season, wing longest of all races. *S. v. tauricus* grades into nominate *vulgaris* over wide area, extending from eastern Greece and European Turkey along western and northern shores of Black Sea to Dnepr River (see Pateff 1947 and Munteanu 1967); intermediates sometimes separated as *balcanicus*, but large individual variation; in typical birds, head purple, nape, side of neck, and upper chest green, hindneck to upper tail-coverts green, belly, flank, and upper wing-coverts (bluish-)purple; some birds rather similar to *poltaratskyi*, but underwing dark, fringes 1–1·5 mm wide. *S. v. purpurascens* from eastern Turkey and eastern shore of Black Sea to Tbilisi and Lake Sevan has cap, chin, and upper throat purple (sometimes green in oblique light), ear-coverts green, nape, side of neck, and upper chest green, hindneck to upper tail-coverts green (sometimes blue or purple on outer scapulars, rump, and upper tail-coverts), belly purple, flank bronze-purple, wing-coverts (purple-)bronze; underwing dark, fringes white, mostly *c.* 1 mm wide; wing rather long; birds of eastern Turkey are mainly *purpurascens*, but some are intermediates with *tauricus* (Kumerloeve 1967*a*; Erard and Etchécopar 1970; ZFMK). *S. v. caucasicus* from Volga delta and Stavropol southward through eastern Caucasus and eastern Transcaucasia green on cap, chin, and upper throat (purple in some lights), green on ear-coverts, purple on nape, side of neck, and upper chest, green on remainder of upperparts, bluish- or reddish-purple on belly and flank, bluish-purple on upper wing-coverts, dark on underwing (fringes white, *c.* 1 mm wide); rather like nominate *vulgaris*, but belly, vent, and upperwing more purple; rather like *poltaratskyi*, but head green and underwing darker. *S. v. nobilior* from Afghanistan, south-east Turkmeniya and neighbouring part of Uzbekistan, and eastern Iran similar to *purpurascens*, but ear-coverts purple (not green), belly, flank, and upperwing bright reddish-purple (less bronze), and wing shorter; may grade into *caucasicus* in northern Iran, where birds are similar to *caucasicus* in colour but as small as *nobilior* in size (sometimes separated as *heinrichi* Stresemann, 1928). Situation further south in Iran not elucidated: birds from Neyuriz and Shiraz (southern Iran) (described as *persepolis* Ticehurst, 1928) are perhaps small as well, and plumage rather like *caucasicus* but upperparts more purple and belly, flank, and wing-coverts greener, apparently rather like nominate *vulgaris* and perhaps migrants only. *S. v. porphyronotus* from west-central Asia similar to *tauricus*, though spatially separated; differs only in slightly smaller size and thus doubtfully separable; grades into *poltaratskyi* between Dzhungarskiy Alatau and Altai. Isolated

*humii*, occurring Kashmir to Nepal, bluish- or reddish-purple on cap, chin, and upper throat, green on ear-coverts, purple on nape, side of neck, and upper chest, green on remainder of body and upperwing, but flank and tail-coverts sometimes purple; small in size. *S. v. minor* of southern Pakistan as small as *humii*, but head green, nape, mantle, side of neck, and upper chest purple shading to bronze-green on belly and back, tail-coverts purple.

Forms superspecies with Spotless Starling *S. unicolor*.  CSR

## *Sturnus unicolor* Spotless Starling

PLATES 17 and 19 (flight)
[between pages 280 and 281]

Du. Zwarte Spreeuw  Fr. Etourneau unicolore  Ge. Einfarbstar
Ru. Чёрный скворец  Sp. Estornino negro  Sw. Svartstare

*Sturnus unicolor* Temminck, 1820

Monotypic

**Field characters.** 21–23 cm; wing-span 38–42 cm. Slightly larger than Starling *S. vulgaris* (particularly ♂), though with similar structure; adult has distinctly more elongated feathers on head and throat. Medium-sized starling of similar form and behaviour to *S. vulgaris* but adult differs noticeably in black and evenly glossed, virtually unspotted plumage, while juvenile differs from continental races of *S. vulgaris* in darker, browner plumage. Only 1st-year ♀ has obvious spotting. Sexes similar; little seasonal variation. Juvenile separable.

ADULT MALE. Moults: June–October (complete). Never shows any spots at any time. Black, more evenly and less intensely glossed than *S. vulgaris*: green-purple when fresh, dull purple when worn and then virtually unmarked except for paler purple margins to larger wing-feathers, particularly tertials. Lanceolate feathers of head and throat up to twice as long as those of *S. vulgaris*, making bird look more bearded and also loose-feathered around neck. Bill pale yellow, with bluish-black base in breeding season, becoming dusky or black in winter. Legs pale pink to brownish-flesh, in breeding season some appearing distinctly paler than *S. vulgaris*. ADULT FEMALE. Resembles ♂ but always less glossy and with less elongated feathers on head and neck; when fresh, vent and under tail-coverts finely mottled white, and margins to larger wing-feathers pale gold. Bill-base pinkish-brown to pale rose in breeding season; otherwise as ♂. Eye with pale outer ring. JUVENILE. Drab, dark brown, matching in colour juveniles of *S. vulgaris faroensis* and *S. v. zetlandicus* (races of North Atlantic islands) and showing rather more pronounced off-white supercilium and paler mottled cheek than *S. v. vulgaris* (western Europe). In moult to 1st-winter, has brown-hooded look similar to *S. vulgaris*. Bill dusky. Legs usually dark brown. FIRST-WINTER. Unlike

adult, both sexes show pale whitish tips to face and body feathers (but rarely crown) and on ♀ these strong enough to form small dull arrow-shaped spots; beware resemblance to partly worn *S. vulgaris*. Later progression of plumage and bare-part colours is erratic, but ♀♀ retain some spots until 2nd year (S J Peris).

Mainly sedentary, restricted at all times to Iberia, west Mediterranean islands, and north-west Africa; *S. vulgaris* normally absent from all these regions in summer and early autumn. Separation in winter from *S. vulgaris* little studied, and clearly high risk of overlapping appearance between worn, relatively spotless adult *S. vulgaris* and fresh, relatively spotted immature *S. unicolor*, as both species may moult early or late and show erratic wear. Winter flocks of *S. vulgaris* known to be joined by *S. unicolor* in Iberia and north-west Africa. Best distinctions of full adult *S. unicolor* appear to be (1) colour of gloss (essentially evenly purple in *S. unicolor*, always partly brilliant green in *S. vulgaris*) and (2) lack of many pale feather-margins on wings and tail (particularly in ♂). Separation of immature depends on close observation of (1) size and shape of spots (small and arrow-shaped in *S. unicolor*, small to quite large, and round in *S. vulgaris*) and again (2) wing markings (relatively faint in *S. unicolor*, strong in complex linear pattern in *S. vulgaris*). Differences in leg colour less trustworthy, with much overlap, though extremes are helpful: pale flesh in *S. unicolor*, bright but fully reddish-brown in *S. vulgaris*. Flight, gait, and behaviour much as *S. vulgaris* but said to fly even faster. Wary in winter.

Voice similar to *S. vulgaris*, but song noisier and introductory whistles (e.g. drawn-out 'seeooo') much louder.

**Habitat.** Within restricted warm west Mediterranean range, habitat corresponds closely to that of Starling *S.*

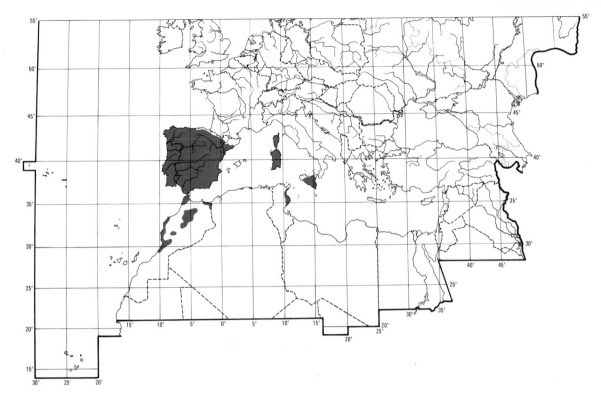

*vulgaris* elsewhere, nesting in buildings (even in inner cities: S J Peris) and in tree-holes. In Spain, prefers open woodland with access to short grass and herbage, and frequently found in association with cattle (see Food); in winter, prefers more open places such as irrigated and cereal fields (Peris 1981; S J Peris); forages extensively over marisma marshlands of Coto Doñana (Mountfort 1958). In Corsica, mainly on littoral plains of west coast, frequenting cultivation but also degraded maquis and out-skirts of certain villages; goes up to *c*. 700 m in interior (Thibault 1983). In north-west Africa, also much attracted to human habitations (Etchécopar and Hüe 1967), but found nesting colonially in holes in large cedars *Cedrus* in Moyen Atlas, making long flights from forest to open plateaux to collect food, at altitudes around 1700-1800 m (Snow 1952); found up to 2500 m (S J Peris). In Tunisia, occurs chiefly near buildings in olive *Olea* plantations, and breeds in old houses, old wells, and ruins (Thomsen and Jacobsen 1979).

**Distribution.** Has spread in northern Spain and estab-lished breeding colonies in southern France.

FRANCE. Has bred near Opoul (Pyrénées Orientales) since at least 1985, and in several villages near Leucate and Sigean (Aude) (RC). SPAIN. Formerly confined to central and south-western parts of Iberia. Began to spread east probably in 1950s, probably due to changing agri-cultural practice, especially irrigation, reaching present limits in 1980s; now sympatric with *S. vulgaris* at or near

north-eastern limit of range, competition with it probably restricting further range expansion. (Motis *et al.* 1983; Ferrer *et al.* 1991.)

Accidental. Greece, Malta, Libya, Madeira.

**Population.** Recent increase in northern Spain, accom-panying spread to east.

SPAIN. Common, increasing. FRANCE. Small numbers in south, where recently spread from Spain. Sparsely dis-tributed in Corsica. ITALY. Sparsely distributed in Sicily. MOROCCO. Locally common to 31°40′S. (No data on status in Algeria and Tunisia.) (S J Peris.)

Survival. Spain: mortality in 1st year 58·3%. Oldest ringed bird 3 years 1 month. (S J Peris.)

**Movements.** Resident, or partial short-distance migrant, subject to nomadic dispersal. Account based largely on Peris (1991), and information from S J Peris.

At least in Sardinia and some parts of Spain, winter range same as breeding (Walter 1965). Flocks of *c*. 400 seen in late October to fly early in morning from Tarifa (southern Spain) to Morocco, and several flocks seen to return from Morocco at dusk (Telleria 1981). Higher density of birds found in southern Spain September-November (Telleria 1981). Of 24 birds ringed as nestlings in Spain, 59% recovered no more than 9 km from nest, 37% 9-99 km, and 4% further than 500 km (Peris 1991). Mean distance moved by adults 32·2 km (0-159, *n* = 19), young birds 41·9 km (0-701 km, *n* = 24). Longest recoveries probably due to individuals flying with wintering flocks of Starling *S. vulgaris*. Autumn and winter recoveries involve longest movements, showing nomadic tendency during these seasons.                                    SJP

**Food.** Principally invertebrates from early spring to summer, seeds and fruits during rest of year; in west-central Spain, *c.* 60% of diet over year animal material. Opportunistic feeder, taking food basically according to abundance; favourite foraging places are improved grassland and pasture, but also feeds in vineyards, olive groves, arable fields (especially stubble), and rubbish tips. Very strong association with cattle all year round, taking plant remains in dung, flushed insects (particularly Orthoptera), and parasites on body and head of animals. In spring, 99·8% ($n = 218$) of feeding observations on ground; in summer, 80% on ground, 15% in trees, 5% in air ($n = 72$); in autumn, 88·5% on ground, 11% in shrubs, 0·5% in air ($n = 87$). Foraging on ground, takes mainly insect larvae and beetles (Coleoptera); in trees and shrubs, primarily cherries *Prunus* and brambles *Rubus*, but also caterpillars; in air, takes (e.g.) Tipulidae in spring, ants (Formicidae) in autumn. (Bernis 1960, 1989*b*; Peris 1981, which see for details.) General strategy and technique very like Starling *S. vulgaris*. In soft soil, commonly uses prying action, pushing closed bill into soil then opening it to enable bird to spot invertebrates; in dry soil, more commonly employs open-bill probing; flocks in winter use 'roller-feeding' movement, often in company with *S. vulgaris*, birds at rear of flock continually moving to front by overflying others (Peris 1981; Bernis 1989*b*, which see for details). In Gibraltar, recorded feeding on nectar of century plant *Agave* (Cortés 1982). In Marbella (southern Spain), nesting birds flew 600–800 m to reach feeding areas in pasture and abandoned olive groves, compared to usually less than 200 m by *S. vulgaris*; possibly because bulk of diet is adult beetles (see Peris 1980*a*, below) which are less secure and more localized food source than soil-dwelling larvae preferred by *S. vulgaris*. This could mean occasional difficulties in finding suitable food for young, and could be part of reason for more restricted range and smaller numbers of *S. unicolor*. (Feare 1986.)

Diet in west Palearctic includes the following. Vertebrates: shrews (Soricidae), mice and rats (Muridae), lizard *Psammodromus*, skink *Chalcides*, frogs (*Rana, Hyla*), toad *Bufo*. Invertebrates: dragonflies (Odonata: Libellulidae), grasshoppers, etc. (Orthoptera: Gryllidae, Tettigoniidae, Gryllotalpidae, Acrididae), earwigs (Dermaptera: Forficulidae), cockroaches (Dictyoptera: Blattidae), bugs (Hemiptera: Pentatomidae, Reduviidae, Aphidoidea), adult, pupal, and larval Lepidoptera (Satyridae, Nymphalidae, Lycaenidae, Tortricidae, Plutellidae, Noctuidae, Lymantriidae, Lasiocampidae, Sphingidae, Geometridae), adult and larval flies (Diptera: Tipulidae, Bibionidae, Syrphidae, Muscidae), adult and larval Hymenoptera (Ichneumonoidea, Cynopoidea, ants Form-

icidae, wasps Vespidae, Sphecidae, bees Apoidea), adult and larval beetles (Coleoptera: Carabidae, Dytiscidae, Hydrophilidae, Histeridae, Silphidae, Staphylinidae, Geotrupidae, Scarabaeidae, Buprestidae, Elateridae, Cantharidae, Lampyridae, Dermestidae, Meloidae, Oedemeridae, Tenebrionidae, Coccinellidae, Cerambycidae, Chrysomelidae, Curculionidae), spiders (Araneae: Lycosidae), harvestmen (Opiliones), mites (Acari), scorpions (Scorpiones), false scorpions (Pseudoscorpiones), Solifugae, woodlice (Isopoda), millipedes (Diplopoda: Julidae), centipedes (Chilopoda: Lithobiidae), earthworms (Lumbricidae), snails (Pulmonata). Plants: seeds, fruits, buds, etc., of oak *Quercus*, elm *Ulmus*, poplar *Populus*, *Robinia*, fig *Ficus*, olive *Olea*, mulberry *Morus*, grape *Vitis*, lentisk *Pistacia*, Loranthaceae, dock *Rumex*, knotgrass *Polygonum*, Caryophyllaceae, Chenopodiaceae, Cruciferae, Rosaceae (including bramble, etc. *Rubus*, strawberry *Fragaria*, cherry, etc. *Prunus*), Thymelaeaceae, Resedaceae, Leguminosae, Umbelliferae, Solanaceae, Globulariaceae, Compositae (including sunflower *Helianthus*), grasses (Gramineae, including maize *Zea* and other cereals), sedges (Cyperaceae). (Gallego and Balcells 1960; Peris 1980*a, b*; Bernis 1989*b*; Pascual 1992.)

In Salamanca area (west-central Spain), 658 stomachs over year, in arable and pasture habitat with *Quercus ilex*, contained 8667 items of animal prey, of which 57% by number adult beetles (20% Scarabaeidae, 10% Curculionidae, 8% Chrysomelidae, 7% Carabidae, 5% Tenebrionidae), 14% Hymenoptera (13·8% ants), and 12% larval beetles, Diptera, and Lepidoptera; remainder Orthoptera, earwigs, millipedes, centipedes, arachnids, and snails. Of 688 items of plant material, seeds of grasses (including cereals) commonest (maximum 88% by number in March, minimum 22% in September), followed by Leguminosae (from 29% in August to 5% in November); peak consumption of other items included 48% *Prunus* in July, 19% acorns *Quercus* in December, 15–38% grapes September–December; figs and olives also favourite foods June–November. Proportion of animal material in diet lowest in January at 18·0% by volume, rising to peak of 98·6% in April; beetles fairly constant over year, falling below 50% by number of animal items only in February; Hymenoptera ranged from 4% in April to 29% in October, and larvae from 3% in August to 31% in April. No difference found in diet of ♂ and ♀. (Peris 1980*a*, which see for detailed treatment.)

Nestlings have more larvae in diet and less adult Lepidoptera and beetles than adult birds; 2nd broods apparently given also plant material such as grapes, cherries, and grasses as well as more adult insects, and only 2nd broods given grit, presumably to deal with these harder items. Collar-samples from nestlings in Salamanca study area (as above) contained 1104 items, mostly beetles (especially Scarabaeidae), larval and pupal Lepidoptera, Orthoptera, Diptera, and earthworms. Large Scarabaeidae fed from day 1, accounting for *c.* 20% of diet by number up to day 5, 45% at days 11–15; 30–60% at this time caterpillars; more variety after young *c.* 2 weeks old, with Diptera and Orthoptera being brought, and also earthworms which are *c.* 6% of diet at fledging, when Lepidoptera larvae and pupae 60%. (Peris 1980*b*, which see for comparison of diets from 2 habitats.) Large insects usually have head and thorax removed before being fed to young; small insects, pupae, and larvae collected together in sticky balls; earthworms longer than *c.* 7 cm are pulled to pieces. Small nestlings also given spiders. (Bernis 1960, 1989*b*; Peris 1980*a*, *b*.) In *Q. pyrenaica* forest in western Spain, 474 collar-samples over 3 years showed main food to be caterpillars and Orthoptera, but not earthworms. Diet varied greatly between years, but other important components were adult and larval Lepidoptera (especially Tortricidae), beetles, millipedes, centipedes, Arachnida, and also reptiles and amphibians including adults. Invertebrates 2–25 mm long, vertebrates 11–131 mm. (Pascual 1992.)                                        SJP

**Social pattern and behaviour.** Major studies in Spain by Feare (1986), Peris (1984*c*, 1991; Peris *et al.* 1991). Information from S J Peris except where otherwise acknowledged.

1. Gregarious all year. Feeding flocks average 90–110 birds in summer and winter, but only 8–12 in April–May. For much larger roosting assemblages see Roosting (below). In immediate post-breeding period, juveniles form into large flocks, subsequently all ages together (Gallego and Balcells 1960; S J Peris). Winter density in Iberia ranged from 0·1–2·8 birds per km in cork oak *Quercus suber* to 10·8 in Pyrenean oak *Q. pyrenaica*, 12·9–16·1 in holm oak *Quercus ilex*, 16·6–17 in ploughed fallow land, 20·2 in pastures, 20–40·5 in orchards, irrigated land and cereals (Arroyo and Tellería 1983; SEO 1985; Motis *et al.* 1987). Readily associates with other species for feeding, notably with Starling *S. vulgaris*, Magpie *P. pica*, Crested Lark *Galerida cristata*, Carrion Crow *Corvus corone*, Stock Dove *Columba oenas*, sparrows *Passer*, and Lapwing *Vanellus vanellus*. BONDS. Little information. Both monogamous and polygamous (Peris 1984*c*). In study of neighbourhood group of 11 ♂♂ in Corsica, 6 monogamous, 4 polygamous (2 ♂♂ had 2 ♀♀, 1 ♂ had 3 ♀♀, exact status of 4th ♂ not known), 1 unpaired; of polygynists, 2 simultaneously bigamous (less than 5 days between laying of ♀♀), rest considered successive (interval 5–11 days) (A Renard). In case of bigamy in Marbella, while ♂'s 1st ♀ incubating, he attracted 2nd ♀ to another nest-hole and copulated with her, while continuing to copulate with 1st ♀ (Feare 1986; see also Heterosexual Behaviour, below). No information on duration of bonds. Normally clutches larger than 7 eggs (frequency of occurrence 19–21% of 1st clutches in April) indicate nest-parasitism ('egg-dumping') or polygamy (Pascual *et al.* 1992; see also Breeding). In Corsica, in 2 years, respectively 24% (*n* = 21) and 33% (*n* = 15) of clutches parasitized; however, these may be underestimates if (e.g.) owner removes dumped eggs (see subsection 3 of Heterosexual Behaviour, below) (A Renard). Both sexes build nest. Studies by Semple (1971) and Feare (1986) suggested only ♀ incubates, but study in west-central Spain indicated that ♂ incubates for at least 20% of the time; *c.* 60–70% of ♂♂ had brood patch (Peris 1984*c*; see also Breeding). In Corsican study, polygamous ♂♂ helped in incubation and returned to 1st ♀ to help feed young (A Renard). Typically both sexes care for young, but ♂♂ less attentive than ♀♀; when ♀

died, ♂ fed the young alone. Young were fed for *c.* 1 week after leaving nest. Age of first breeding 1 year for ♀♀, but many 1st-year ♂♂ fail to breed (in face of competition with older ♂♂). BREEDING DISPERSION. Normally in loose neighbourhood groups but also solitary. In Marbella in area of *c.* 1·5 km², 27 1st clutches (all in buildings) in neighbourhood groups of 8, 7, 3, 3, 2, 2, plus 2 others each relatively isolated; average nearest-neighbour distance 70 m (22–340) (Feare 1986). Much higher density found in central Spain: maximum 22–23 nests in tiled roof of 625 m², average *c.* 6·7 nests per roof, average distance between nests 2 m (1–6) (Peris 1984*c*; S J Peris). In another Spanish study, 180 nest-boxes each 10 m apart were all occupied. Territory confined to area *c.* 20 m around nest (Feare 1986: see Antagonistic Behaviour, below); serves for courtship, nesting and some feeding, but foraging extends to 800 m from nest (Feare 1986; Peris 1984*d*). Other densities in Spain as follows: 3·3 birds per km² in *Q. ilex*, southern Spain (Herrera 1980), 4 birds per km² in *Q. pyrenea*, central Spain (Potti and Tellería 1984); 5 birds per km² in mixed *Quercus* (Torres and Leon 1985); 7 birds per km² in juniper *Juniperus* (Santos *et al.* 1981); in west-central Spain, 2 birds per km² in chestnut *Castanea*-forests, 7 in *Q. pyrenea*, 8–80 in *Q. ilex*, 12 in fallow land with dispersed farmhouses, and 76 in irrigated land with farmhouses; in central Spain, up to 45 nests per ha in open *Q. ilex* woods (Peris 1984*c*, 1991; S J Peris). Forms mixed colonies with *S. vulgaris* in north-east Spain: in small agricultural villages of Catalonia, average 3·5 pairs *S. unicolor* per house (i.e. in tiled roof), compared with 3·1 *S. vulgaris* (Motis 1986). In Sardinia, 895 pairs per km² in small town (Walter and Demartis 1972). In various habitats where *S. unicolor* scarce or absent (such as montane conifers, citrus groves, maquis) population can be increased by putting up nest-boxes. At least ♀♀ show site-fidelity (use same nest-box) from year to year. Competes for sites with House Sparrow *Passer domesticus*. ROOSTING. Mainly communal throughout the year, in rural districts but also in cities. In central Spain in summer roosts typically in bramble *Rubus* bushes and willows *Salix* in river basins, poplar *Populus* plantations or *Q. pyrenea* woods (Pedrocchi 1979; Peris *et al.* 1991). In winter, Morocco, on coastal cliffs (Smith 1965). In winter in Spain, many switch to dovecotes (Peris 1979), and singletons also found roosting in nest-boxes from December to breeding season, using same nest-boxes for this purpose from year to year. Incubating ♀ roosts on nest at night while ♂♂ fly off with neighbouring ♂♂, presumably to roost communally (Feare 1986). Size of communal roosts increases to 1000–20 000 birds from mid-June to late March. During June–July, immatures roost communally with a few adults. From November–March, many Spanish roosts are mixed with *S. vulgaris*, single mixed roosts reaching more than 100 000 birds in many places near the coast or in cities; however, in central Spain, 93–96% of assemblages are *S. unicolor* only, i.e. only 4–7% of roosts are mixed (Peris *et al.* 1991).

2. Wary at all times, but in winter more approachable when associated with human settlements. When approached to *c.* 60–100 m, often stands erect and flies off, landing several hundred metres away. In Spain, mostly ignores raptors like buzzards *Buteo* and kites *Milvus*, but mobs Sparrowhawk *Accipiter nisus*. FLOCK BEHAVIOUR. For flock associations for feeding outside breeding season, see introduction to part 1 (above). In feeding flock, individual distance *c.* 10–20 cm, and birds will progress in same direction in roller-feeding fashion (Peris 1981); if disturbed, all fly up together in compact flock, returning to same place when disturbance over. Aerial evolutions, much like those of *S. vulgaris*, reported before roosting in Morocco and Spain (Smith 1965; Ruthke 1971). SONG-DISPLAY. ♂ sings (see 1 in Voice) mostly from favoured song-post close to nest (5–15 m away) such

as tree-branch, roof, overhead wire, television aerial, but also on feeding grounds when mate-guarding (see Heterosexual Behaviour, below). Song-display as in *S. vulgaris*: ♂ adopts a hunch-backed posture with tail lowered, feathers erected on crown, throat, upper breast, and rump (Fig A); wings are waved at

A

climax of some bouts of song; neighbouring ♂♂ matched one another's whistled calls (Feare 1986). Full song heard late February–June, but also on warm days in autumn (after moult) and winter. Quieter song heard in presence of ♀♀ near nest, e.g. ♂ sang thus on first arrival at nest-site each morning, stimulating ♀ to emerge and solicit copulation (Feare 1986: see Heterosexual Behaviour, below). ANTAGONISTIC BEHAVIOUR. Only some individuals are aggressive in breeding season, and only close to nest (see Breeding Dispersion, above). Little known about threat-display, except that ♂ pushes out breast to render throat feathers conspicuous. Resident ♂ expelled any other ♂, or occasionally a pair, intruding within 20 m of nest, and frequently attacked ♂♂ (apparently strangers rather than familiar neighbours) that approached within 50 m; resident once left his mate to attack a bird more than 100 m away, necessitating flight through another ♂'s territory (Feare 1986). Fights seen at nest between birds of same sex, sometimes resulting in eggs being destroyed (Peris 1984c). For attacks on other species see Parental Anti-predator Strategies (below). Aggression by both sexes towards conspecifics sometimes seen outside breeding season, e.g. ♂♂ displace ♀♀ at feeding grounds (Peris 1984c). HETEROSEXUAL BEHAVIOUR. (1) Pair-bonding behaviour and nest-site selection. No significant differences from *S. vulgaris* (S J Peris). In study by Feare (1986) most ♂♂ apparently already paired, but much of their activity involved inciting ♀ to enter nest-hole (Nest-showing); in late March ♂ seen interrupting bouts of song (see Song-display, above) to break off twigs or (especially after copulation: see Mating, below) green leaves of nearby *Eucalyptus* and take them into hole ('nest decoration'). ♀ followed ♂ into hole and after a few seconds emerged, sometimes carrying piece of nest-material which she dropped. Nest-decoration by ♂ not seen after 31 March. Throughout nest-building and copulation (see below), ♂ remained close to ♀ (mate-guarding): e.g. when she was in nest-hole or perched nearby, ♂ was close-by singing frequently. As soon as ♀ flew off, ♂ followed 1–5 m behind, following every course-change of her zigzag flight, and continued to keep close to her (usually within 2 m) when she reached feeding grounds; there, ♂ interspersed feeding with brief bouts of song. Bigamous ♂ went through complete courtship with 2nd ♀, including Nest-showing, copulation, and mate-guarding. (Feare 1986.) In Corsican study, mate-guarding (judged by frequency with which ♂ followed ♀ out of sight) by pair mainly from 2 days before egg-laying till 3rd day after start of egg-laying (A Renard). (2) Courtship-feeding. None. (3) Mating. Occurs

near nest on the ground, roof, overhead wires, or branches of tree (S J Peris). In study at Marbella, copulation occurred throughout morning period of activity at nest, averaging *c.* 3 copulations per hr, less frequent (typically absent: S J Peris) during middle of day but increasing again in late afternoon (Feare 1986). ♂ solicits by approaching ♀ in full Song-display with Wing-waving; if ♂ stationary, he makes up-down movements of forebody, such that bill sometimes touches perch; on the ground, may run towards ♀ (S J Peris). On most occasions, ♀ approached ♂ and perched close to him, either facing him or away from him. ♂ was sometimes stimulated to mount simply by presence of ♀ alongside, but more often she solicited by sleeking body plumage and with vigorous sideways shaking of tail; also raised her tail, exposing cloaca, at which point ♂ adopts an upright advertising posture and mounts briefly, copulation lasting only a few seconds (Feare 1986; S J Peris). Bigamous ♂ chased off soliciting mate during early incubation and 2 days later began copulating and Nest-showing with 2nd ♀; however, 1st ♀ successfully solicited copulation from him *c.* 1 week after she (i.e. 1st ♀) started incubating (Feare 1986). For duration of mate-guarding (suggesting likely fertile period of ♀) see sub-section 1 of Heterosexual Behaviour (above). Further study needed to establish to what extent promiscuity (additional to polygyny: see Bonds, above) occurs. (4) Behaviour at nest. Nest-material collected mainly by ♂, with ♀ watching from nearby (Feare 1986). In study in Corsica, only one case seen of egg-removal by nest-owner from parasitized nest (see Breeding Dispersion, above) but this may be hard to detect happening. RELATIONS WITHIN FAMILY GROUP. At hatching, young are cleaned and then brooded by ♀. In at least one case, young brooded by apparent ♂ after death of ♀. Both parents feed young and participate in nest-hygiene (swallow faecal sacs for first few days) but ♀♀ more attentive in both these roles. Dead young are removed when small. Nestlings beg upwards for 1st week in response to tactile stimuli; after eyes open at 7–8 days, young orientate bills towards incoming bird (or finger of observer) (Peris 1984c). From 12–13 days, gaping accompanied by wing-shivering. Young stay in general vicinity of breeding area for 2–7 days after fledging, then disperse to form flocks. ANTI-PREDATOR RESPONSES OF YOUNG. Young in nest become immobile on hearing parental alarm-calls. At *c.* 13–14 days give distress-call (see Voice) when handled. When disturbed, some very aggressive from this age, and may leave nest from day 16 (i.e. typically 3 days prematurely). PARENTAL ANTI-PREDATOR STRATEGIES. (1) Passive measures. Some ♀♀ leave nest when intruder approaches to *c.* 5–10 m, but many sit tight during incubation and can be captured by hand on nest. Parents give alarm-calls (see 3 in Voice) away from nest when disturbed, more often when nest contains young. (2) Active measures. Sitting bird sometimes pecks hand of man at nest. Aggressive towards Magpie *Pica pica* and *P. domesticus* encroaching on nest.

(Fig by D Nurney from drawing by S J Peris.)        SJP

**Voice.** Used throughout the year. More information needed on call repertoire (few calls described) which presumably matches diversity of Starling *S. vulgaris*. Descriptions and renderings are by S J Peris except where otherwise acknowledged. For additional sonagrams, see Bergmann and Helb (1982).

CALLS OF ADULTS. (1) Song. (1a) Full song. Resembles Starling *S. vulgaris* in overall structure, duration, and variability, e.g. Fig I. However, compared with *S. vulgaris*, song of *S. unicolor* noisier (Bergmann and Helb 1982) and

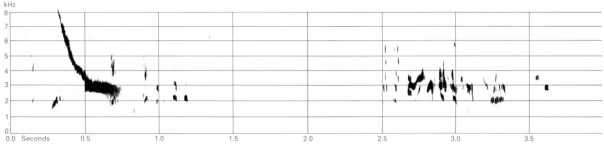

I   C Chappuis   Spain   March 1964

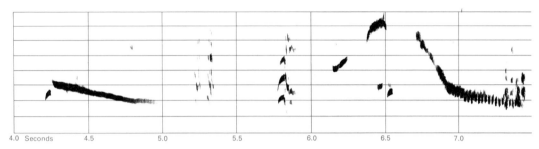

I   *cont.*

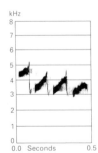

I   *cont.*

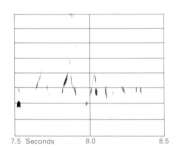

II   J F Burton and D J Tombs/BBC   Spain   May 1982

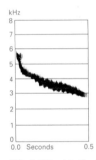

III   J-C Roché   Spain   May 1965

IV   J F Burton and D J Tombs/BBC
Spain   May 1982

introductory whistles much louder (Koenig 1888; Feare 1986). Whistles also given independently as calls, and are among the commonest sounds heard outside breeding season. Whistles in song highly varied (but conform to definite scheme: see below), typically drawn-out, rising or falling (portamento), tonal, e.g. 'seeoooo' (S J Peris), 'suuíh-tju' (L Svensson, H Delin), 'düüiie'; also often harsh, e.g. 'srri', 'srije' (Bergmann and Helb 1982). Based on study in north-east Spain by A Motis: whistled song has same basic structure as in *S. vulgaris*: every ♂ has 4 'species-specific themes' (themes given by all or most singing ♂♂)—inflection theme (high form Fig II, low form Fig III), uniform simple theme, rhythmic theme (Fig IV), and composed theme (terminology of Adret-Hausberger 1983, 1984 for *S. vulgaris*; see Voice of that species for further details); only the 'uniform with harmonics' theme of *S. vulgaris* in central Europe has not been found in *S. unicolor* (or *S. vulgaris*) in Spain (A Motis). Local variations (dialects) in species-specific themes occur, and each ♂ *S. unicolor* also individualistic in possessing other, 'non-species-specific themes' (individually unique themes, or those given only by some singing ♂♂): e.g. sounds ascending in pitch (up to 8 kHz), uniformly pitched sounds, mimicry (see below); average 13·6 themes (8-18, $n = 12$ ♂♂ in different colonies) per ♂; this large repertoire compared with *S. vulgaris* derives from high incidence of non-species-specific themes, related in turn to large size of some breeding colonies of *S. unicolor* in Spain. In area of sympatry with *S. vulgaris* in north-east Spain, intra- and interspecific song-matching, confined to species-specific themes, occurs in mixed colonies; there are cases of precise matching of dialectical variants of species-specific themes by 2 ♂♂ of the different species from same colony. Mimicry reported in North Africa (Etchécopar and Hüe 1967) although not heard in Málaga (southern Spain) (Feare 1986). Elsewhere in Spain, *S. unicolor* an accomplished mimic of many other birds: in central Spain, well known for mimicry of Kestrel *Falco tinnunculus*, Quail *Coturnix coturnix*, Moorhen *Gallinula chloropus*, Little Owl *Athene noctua*, Calandra Lark *Melanocorypha calandra*, Great Tit *Parus major*, Golden Oriole *Oriolus oriolus*, Robin *Erithacus rubecula*, Blackbird *Turdus merula*, Magpie *Pica pica*, and Jackdaw *Corvus monedula*; also reported mimicking Green Woodpecker *Picus viridis* in north-east Spain (A Motis), Short-toed Treecreeper *Certhia brachydactyla*, Buzzard *Buteo buteo*, and protracted bill-clattering of White Stork *Ciconia ciconia* (N J Collar), captive parrot (Psittacidae) (Bernis 1989b) and poultry *Gallus gallus*. Recordings include mimicry of Red Kite *Milvus milvus* (Fig V) and Scops Owl *Otus scops*. Full song, especially

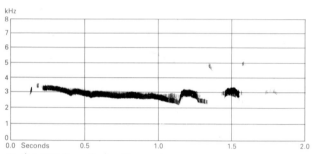

V  P A D Hollom  Spain  May 1976

loud whistles, given by ♂♂ from February–June, also on warm days in autumn and late winter. (1b) Subsong. Given by both sexes outside breeding season. (2) Contact-call. A sound roughly rendered as 'gaa-haa', somewhat like (presumably cackle of) chicken. (3) Threat- and alarm-calls. Usually given when human, cat, or corvid approaches nest. (3a) Threat-call. Rattling 'kökökökörr' or chattering 'tretet...' (Fig VI), similar to Display-flight-call of Great Spotted Cuckoo *Clamator glandarius* (Bergmann and Helb 1982). (3b) Chip-call. A sharp 'fiit...

fiit', similar to Chip-call of *S. vulgaris*. (3c) Snarl-call. A harsh rasp (Fig VII), as in *S. vulgaris*. (4) Excitement-call. A 'geee-geee-geee' heard (e.g.) during roosting. (5) Distress-call. Screaming 'haaa-haaa-haaa' heard (e.g.) during disputes in roost.

CALLS OF YOUNG. Incompletely known. Food-call a 'hiip' from hatching until 4–5 days, later replaced by a more disyllabic sound. Calling increases from 4–10 days, declining abruptly thereafter, and young silent near fledging (at least when observer approaches nest). After day 12, threat-, alarm-, and distress-calls (as in adult) more common and heard increasingly up to departure from nest.

SJP

**Breeding.** Information from S J Peris except where otherwise acknowledged. SEASON. West-central Spain: eggs laid April to early May, 2nd clutches late May to mid-June; fledglings found in one nest in central Spain 10 October (Peris 1984c, d). North-east Spain: c. 2 weeks earlier, though still later than Starling *S. vulgaris* (Motis 1985); see also Gallego and Balcells (1960) and Bernis (1989b). Sicily: eggs laid early April to mid-June (Iapichino and Massa 1989). North Africa: March–April, 2nd clutches probably May–June (Heim de Balsac and Mayaud 1962). SITE. Hole, usually in man-made situation such as under roof-tiles, in wall (often in large towns), agricultural structure, nest-box, etc.; also in tree or rock-face (Bernis 1933, 1945; Gallego and Balcells 1960; Smith 1965; Sudhaus 1969c). In central Spain, 92% in man-made sites (Peris 1984d); in Spain overall, 5% of natural sites in old holes of Green Woodpecker *Picus viridis* and Great Spotted Woodpecker *Dendrocopos major* (Gallego and Balcells 1960; Garzón 1969; Peris 1984d). Principal nest-trees in Spain oaks *Quercus ilex*, *Q. robur*, chestnut *Castanea*, poplar *Populus*, elm *Ulmus*, alder *Alnus*, ash *Fraxinus*, pine *Pinus*; in Sardinia, often in *Eucalyptus*. In trees, average height above ground 3·6 m (1–15). In Marbella (southern Spain), average height on buildings 27 m (10–40, n = 27) (Feare 1986). In Spain, 45–50% of nests orientated east or south-east (Peris 1984d; Motis 1985). Often in old nest of White Stork *Ciconia ciconia*, Bee-eater *Merops apiaster*, Sand Martin *Riparia riparia*, Jackdaw *Corvus monedula*, or sparrow *Passer* (Peris 1984d). For Sardinia, see Walter and Demartis (1972). One nest near Madrid (central

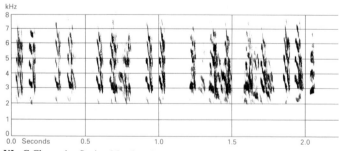

VI  C Chappuis  Spain  March 1964

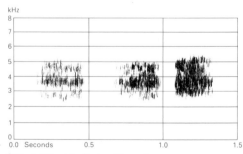

VII  J-C Roché  Spain  May 1965

Spain), on branch of tree. Nest: foundation of twigs, dry grass, and herb and cereal stalks lined thickly with rootlets, grass, leaves, flowers (e.g. Compositae, Boraginaccae), and feathers (apparently mostly black); some nests made of soft material only. In Spain, 56 nests had average outer diameter 20·5 cm (18–22), inner diameter 11·0 cm (9–13), overall height 7–18 cm, depth of cup 4–10 cm. Building: by both sexes; replacement nests in same hole; nests for 2nd brood usually have new foundation; in Marbella, mainly by ♂ at one nest, collecting grass from ground and breaking twigs and green leaves from tree, though ♀ seen once to collect feathers (Feare 1986). EGGS. See Plate 59. Sub-elliptical, smooth, and slightly glossy; pale blue, *c.* 4% in 1st clutches with very small reddish spots, 37% in 2nd clutches. 30·8 × 21·8 mm (28·1–34·2 × 20·5–22·6), *n* = 86; calculated weight 7·7 g (Schönwetter 1984). In central Spain: 30·5 × 21·3 mm (30·1–31·2 × 20·7–21·8), *n* = 414; fresh weight of 209 eggs 7·2 g (5·0–10·0) (Peris 1984*d*). Clutch: 4–5 (2–9). Average 4·1 in central Spain, 4·5 in north-east (Peris 1984*d*; Motis 1985). Significant relationship between clutch size and size of nest-box; 5 commoner in larger boxes, 4 in smaller ones. Usually 44–49 days between 1st and 2nd clutch; average ♀ lays 8·3 eggs per breeding season; extra-large clutches normally laid by 2 ♀♀; intraspecific brood-parasitism occurred in 30–35% of 1st clutches in west-central Spain, in 0–16% of 2nd clutches. (Pascual *et al.* 1992.) In North Africa, of 3 clutches: 2 of 4 eggs, 1 of 3 (Heim de Balsac and Mayaud 1962). INCUBATION. 10·5–11·6 days (10–15); mainly by ♀ (for 81% of time in west-central Spain); ♂♂ have smaller and less vascularized brood-patch; usually starts with 3rd or 4th egg and hatching occurs over 5–10 hrs usually in morning or late evening. At one nest in Marbella, breaks totalled 15% of time (*n* = 479 min); average break 3·2 min (1–8, *n* = 22); ♂ probably did not incubate (Feare 1986). YOUNG. Fed and cared for by both parents; ♀ parent does more feeding than ♂. FLEDGING TO MATURITY. Fledging period 21–22 days (18–25); young fed by adults for 2–7 days after leaving nest. Age of first breeding 1 year for both sexes. BREEDING SUCCESS. In Spain, of 150 eggs laid in natural holes, 71·3% hatched and 50·8% produced fledged young; of 173 eggs in 2nd clutches, 64·8% hatched and 54·4% fledged. Significant difference in success related to clutch size, with 64·5% hatching and 48% fledging from 60 clutches (1st and 2nd broods) of 4–5 eggs, 60% hatching and 39·5% fledging from 12 clutches of 3, 6, and 7 eggs. In nest-boxes, nestling mortality 21·7–57·9%; 5·9 and 6·1 fledged young per breeding pair per year produced in west-central and north-east Spain respectively (Peris 1984*d*; Motis 1985), but very variable according to year and habitat, average 3·5 overall (Pascual 1992). Most losses due to predators, especially rat *Rattus rattus* in roof-tiles, and to intraspecific competition (Peris 1984*c*).

SJP

**Plumages.** ADULT MALE. In fresh plumage (September–January), entirely black. Elongated feathers from crown to back, on side of head and neck, and from chin to breast with slight but variable grey bloom (sometimes absent), fringes along sides of elongated feather-tips sometimes faintly margined grey (especially on mantle, side of neck, and chin to breast), but no pale spots or V-marks on feather-tips; gloss rather faint, oily-purple. Remainder of body generally without grey bloom, unspotted, more glossy; gloss on rump and upper tail-coverts green or purple (depending on light), on flank, belly, and under tail-coverts mainly green. Tail dull black with slight bronze or grey fringe, tips and outer webs of feathers broadly fringed velvety-black (glossed purple in some lights); feather-tips pointed, tips of inner webs of t2–t6 concave. Secondaries, tertials, and inner primaries as tail, showing bronze tinge with contrasting velvety-black fringes along tips; black fringes along outer webs of secondaries and tertials strongly glossed purple or violet-purple; outer webs of outer primaries sometimes with faint and narrow pale edges. All upper wing-coverts deep black, lesser coverts and greater primary coverts strongly glossed green or purple on outer webs (depending on light), median and greater coverts strongly glossed purple or violet-purple. Axillaries and under wing-coverts uniform black, some shorter coverts sometimes with faint grey edges. *In worn plumage* (February–June), grey bloom on head, neck, and fore-part of body worn off, head to mantle, side of head and neck, and chin to chest deep black with oily-purple gloss (slightly green in some lights), remainder of body glossed oil-green (all gloss of body less strong than in Starling *S. vulgaris*), lesser upper wing-coverts and fringes of primary coverts and primaries strongly glossed green, fringes of median and greater coverts, tertials, and secondaries strongly violet-purple. When heavily worn, July–August, head and body less glossy, mainly sooty-black. ADULT FEMALE. In fresh plumage, grey bloom on head, neck, and fore-part of body more pronounced than that present in some adult ♂♂, extending over entire underparts also; narrow grey edges along sides of elongated feathers slightly more distinct, often forming tiny narrow V-marks on feather-tips of mantle and on chin to chest; feathers of lower flank, vent, and thigh as well as tail-coverts narrowly fringed off-white on tip, appearing mottled. Outer webs of flight-feathers, outer tertials, and greater upper primary coverts often narrowly edged buff, edges widest on secondaries, but even there less than 0·5 mm wide. Elongated feathers shorter than in adult ♂, longest on central throat 21–31 mm (in ♂, 24–41: Svensson 1992; S J Peris, RMNH, ZMA). *In worn plumage*, grey bloom and traces of pale spots on body and pale edges on wing worn off, and thus closely similar to adult ♂, but see length of throat feathers (above) and Bare Parts; gloss slightly less intense and general appearance slightly greyish-black, less deep black and glossy than in adult ♂; when heavily worn, body more sooty-grey, with traces of gloss on head and upper wing only. NESTLING. Naked at hatching except for some sparse white or grey down tufts on back and upperwing; on day 7–8, when feather-pins start to appear, down long and rather dense, sooty-grey (Peris 1984*b*, which see for details). JUVENILE. Closely similar to darkest juveniles of *S. vulgaris vulgaris* or about as dark as *S. v. faroensis* or *S. v. zetlandicus*; relatively small patch of isabelline-white on chin and upper throat and rather narrow isabelline streaks on side of throat, lower throat, and mid-breast to vent; compared to most *S. vulgaris*, short off-white supercilium more pronounced and lower cheek more mottled black-brown and isabelline, giving a more distinct facial expression, but much individual variation; fringes of flight-feathers, tertials, tail, and upper wing-coverts narrow, isabelline-white, as in dark *S. vulgaris*. For sexing, see Bare Parts. FIRST ADULT MALE. Like adult ♂, but tail-feathers less pointed (as in 1st adult *S. vulgaris*; for figure, see Hiraldo and Herrera 1974; difference often hard to see once tail worn, spring and summer); length of longest throat-feathers 24-30 mm (in adult, 27-41: Svensson 1992; RMNH, ZMA); length of black along shaft of longest throat-feathers 14·4-23 mm ($n = 32$; in adult 24-38, $n = 28$: Hiraldo and Herrera 1974). In fresh plumage, September or October to January, black of head, neck, and fore-part of body usually tinged grey (but much less so than in adult ♀); tiny pale V-marks on feather-tips of body present, as in adult ♀, but marks often more profuse, extending to head and tail-coverts; traces of pale edges present along flight-feathers. *In worn plumage*, pale marks and edges worn off, plumage uniform glossy black (more so than in adult ♀); as difference in tail-shape difficult to see in spring and length of throat-feathers overlap, separable from adult ♀ by bare parts only. FIRST ADULT FEMALE. Like adult ♀, but tail-feathers less pointed (see 1st adult ♂, above), longest throat feathers 17-23 mm (in adult, 21-31: Svensson 1992; RMNH, ZMA). In fresh plumage, grey bloom on head, neck, and body often extensive, faint purple or green gloss present only on cap, rump, and chest, remainder of body appearing mouse-grey; all feathers (including lesser and median wing-coverts) have pale V-mark on tip, *c.* 1-2 mm long on head, neck, and forepart of body, 2-4 mm on rear and on coverts, full pale buff fringes on tips of tail-coverts; outer webs of tail-feathers, flight-feathers, outer tertials, greater upper primary coverts, feathers of bastard wing, under wing-coverts, and axillaries distinctly fringed pale buff (fringes 0·5-1 mm wide). Thus, much more distinctly marked with pale spots and fringes than in both adult ♀ and 1st adult ♂. In worn plumage, pale spots and edges disappear, but (in contrast to adult ♀ and 1st adult ♂) traces remain on vent, tail-coverts, shorter under wing-coverts, and sometimes elsewhere; throat feathers shorter and plumage greyer, less glossy, and more spotted than in 1st adult ♂, especially on underparts (see also Bare Parts), slightly less glossy and generally more spotted and with less elongated throat-feathers than in adult ♀.

**Bare parts.** ADULT, FIRST ADULT. Iris uniform dark brown (♂ and some ♀♀) or dark brown with pale grey-brown to cream ring (most ♀♀). Bill in autumn dark horn-brown to black-brown, cutting edges paler horn or yellowish; in spring, bill-tip turns yellow, but *c.* 30% of base of upper mandible and *c.* 40% of base of lower turns blue to bluish-black (♂) or pink-flesh to pinkish pale brown (♀); bill turns pale earlier in ♂ than in ♀, earlier in adult than in 1st adult: adult ♂ from late December onwards, adult ♀ from January, but of 1st adults in March 52% of 33 ♀♀ and 11% of 9 ♂♂ still dusky; bill shades to black again during post-breeding moult. Mouth of adult ♀ pale violet below, white above, pink on palate, of 1st adult ♀ yellow below, pale pink grading to orange inwards above, white on palate; ♂ as ♀, but more extensively violet. Tarsus dark brown in 67% of 103 birds in 1st calendar year, light brown in 23%, intermediate in remainder; in 1st half of 2nd calendar year, about equal numbers have tarsus light brown, intermediate, or dark brown; in 130 adults, 55% light brown, 23% intermediate, 22% dark brown, latter perhaps mainly birds in late 2nd or early 3rd calendar year; no variation with season (Peris 1983). However, in adults examined, some had legs puple-flesh to flesh-red in spring (C S Roselaar). NESTLING. At hatching, flesh-coloured, including bill and leg, bill-tip and cutting edges tinged yellow; mouth and gape-flanges yellow, palate bright green. From about day 9-10, iris dark brown; bill turns dark horn-brown, first on tip, spreading gradually towards base, tarsus shades to paler or darker brown. At day 15 (*c.* 5 days before fledging), iris of ♂ dark brown, of ♀ starts to be bicoloured; bill black-brown, except for

paler brown cutting edges, yellowish base of lower mandible, and yellow gape-flanges; mouth orange-yellow, palate pale pink; leg dark brown. JUVENILE. At fledging, iris, bill, and leg as in autumn adult, but up to day 40 traces of yellow flanges at gape and mouth orange below, pale pink above, white on palate. (Hiraldo and Herrera 1974; Peris 1983, 1984*a*; Svensson 1992; Motis 1987; RMNH, ZFMK.)

**Moults.** Based mainly on study by Peris (1988) in Salamanca (west-central Spain); some additional data from specimens originating elsewhere (BMNH, RMNH, ZFMK, ZMA). ADULT POST-BREEDING. Complete; primaries descendent. No difference in timing between sexes or between adults and 1-year-olds. Starts with shedding of innermost primary (p1) mid-June to early July, when chicks of 2nd brood still in nest or just fledged. Primary moult completed with regrowth of p9 mid-September to early October, primary moult lasting *c.* 90 days. Moult of tertials (s7-s9) starts at same time as primaries, s8 first; completed mid-August to late September. Secondaries moult ascendently, starting with s1 mid-July to mid-August, completed with regrowth of s6 mid-September to mid-October. Exceptionally, some flight- or tail-feathers moult in late winter or early spring; occasionally, sequence of flight-feather moult not strictly descendent or ascendent. Moult of tail, wing-coverts, and body at same time as tertials, head somewhat later, like secondaries; tail moult usually centrifugal (t1-t6), occasionally centripetal (t6 to t1). In Morocco, 2 birds had moult completed by 2nd week of September; in Portugal, 1 in last stages of moult November. POST-JUVENILE. Sequence of feather replacement as adult post-breeding; timing highly variable, depending on fledging date; starts 4-6 weeks after fledging (S J Peris). In Salamanca, July-November; moult of birds of 1st broods at about same time as adult post-breeding, those of 2nd brood *c.* 1 month later; p1 shed late June to late August, p9 completed August to late October (16% of August birds examined had not yet started primary moult, 30% in September already completed). Birds in full juvenile plumage recorded early May to late September, birds in 1st adult plumage from early August. No birds examined had retained juvenile feathers into 1st winter.

**Measurements.** ADULT, FIRST ADULT. Whole geographical range, all year; skins (RMNH, ZFMK, ZMA). Bill (S) to skull, bill (N) to distal corner of nostril; exposed culmen on average 4·4 less than bill (S).

| | ♂ | ♀ |
|---|---|---|
| WING | 133·3 (2·51; 20) 129-138 | 131·0 (3·05; 27) 126-137 |
| TAIL | 63·7 (2·29; 12) 59-68 | 62·6 (2·33; 12) 58-67 |
| BILL (S) | 29·6 (1·46; 12) 27·8-31·5 | 28·8 (1·30; 12) 27·0-30·2 |
| BILL (N) | 18·2 (1·02; 12) 17·0-20·0 | 17·3 (0·89; 12) 16·0-18·5 |
| TARSUS | 30·7 (1·21; 12) 29·0-32·6 | 30·7 (1·00; 13) 29·4-32·5 |

Sex differences significant for wing and bill (N). Adult and 1st adult combined above, though wing and tail of 1st adult slightly shorter: wing, adult ♂ 135·1 (2·59; 9) 131-138, 1st adult ♂ 132·2 (1·08; 10) 130-134; tail, adult ♂ 64·4 (2·16; 5) 62-68, 1st adult ♂ 63·2 (2·41; 7) 60-66.

Wing and tail. (1) Spain, mainly (*c.* 75%) south-west, ages combined (Hiraldo and Herrera 1974). Salamanca (west-central Spain): (2) adult, (3) 1st adult (Peris 1989). (4) Sicily and Sardinia, (5) Spain and Portugal (mainly Salamanca), and (6) North Africa (mainly Tunisia), ages combined (RMNH, ZFMK, ZMA).

| | ♂ | ♀ |
|---|---|---|
| WING (1) | 133·8 (3·1 ; 62) 126-142 | 129·9 (3·5 ; 69) 120-138 |
| (2) | 132·8 (3·1 ; 166) 124-143 | 128·4 (3·5 ; 92) 120-138 |
| (3) | 129·5 (5·2 ; 60) 119-136 | 126·8 (4·0 ; 61) 117-132 |
| (4) | 134·1 (2·44; 12) 130-138 | 130·3 (2·15; 8) 128-135 |
| (5) | 132·8 (1·89; 8) 131-136 | 130·2 (2·36; 10) 127-134 |

| | | | | | | |
|---|---|---|---|---|---|---|
| (6) | 134·0 (3·33; | 6) | 129–137 | 131·3 (3·65; | 9) | 126–137 |
| TAIL (1) | 62·7 (2·3 ; | 51) | 56–68 | 60·9 (2·4 ; | 65) | 54–68 |
| (2) | 65·7 (3·1; | 152) | 58–78 | 64·0 (3·9 ; | 91) | 57–77 |
| (3) | 64·7 (4·7 ; | 61) | 54–80 | 63·2 (3·1 ; | 58) | 58–72 |
| (4) | 64·9 (2·71; | 11) | 59–68 | 64·2 (1·78; | 6) | 62–67 |
| (5–6) | 63·0 (1·58; | 5) | 61–65 | 60·9 (1·46; | 6) | 58–63 |

See Hiraldo and Herrera (1974) and Peris (1989) for other measurements and differences between age-classes.

JUVENILE. Wing on average *c.* 10 mm shorter than adult and 1st adult, tail *c.* 5 shorter.

NESTLING. For growth of wing, p9, culmen, and tarsus with age, see Peris (1984*b*) and Motis (1987).

**Weights.** ADULT, FIRST ADULT, JUVENILE. Salamanca (west-central Spain): (1) June–July, (2) August–October, (3) November, (4) December–January, (5) February, (6) March–May (Peris 1984*a*).

| | ADULT ♂ | ADULT ♀ | JUV/1ST AD ♂ | JUV/1ST AD ♀ |
|---|---|---|---|---|
| (1) | 95·4 (4·6; 20) | 85·8 (5·2; 12) | 79·8 (4·9; 12) | 74·6 (6·9; 16) |
| (2) | 90·4 (3·8; 34) | 89·2 (2·3; 17) | 88·2 (4·7; 17) | 85·2 (5·3; 14) |
| (3) | 93·1 (2·8; 20) | 85·7 (5·0; 11) | 89·9 (5·7; 11) | 86·7 (4·5; 27) |
| (4) | 95·8 (5·4; 26) | 90·4 (3·1; 11) | 92·9 (5·0; 11) | 83·8 (3·6; 6 ) |
| (5) | 92·0 (4·5; 16) | 88·4 (3·8; 11) | 86·2 (4·5; 11) | 85·2 (3·3; 9) |
| (6) | 94·1 (5·9; 17) | 90·2 (3·7; 4) | 93·8 (5·4; 5) | 93·8 (6·1; 8) |

South-west Spain, adult and 1st adult: ♂ 90·0 (36) 80–115, ♀ 83·2 (31) 70–100 (Hiraldo and Herrera 1974). Sardinia, March: ♀ 86 (Demartis 1987).

NESTLING. Salamanca: at hatching, 7·1 (1·03; 39) 6–9; peak (day 16) 79·9 (6·73; 39) 57–86; at fledging (day 20) 73·7 (3·73; 39) 68–80 (Peris 1984*b*, which see for details). North-east Spain: at hatching, 7·0 (0·76; 6) 6–8; at day 6–8, 40·5 (9·2; 8) 32–55; at day 12–14, 68·5 (7·3; 29) 54–81; at day 18–20, 78·4 (5·4; 24) 69–88; growth rate about equal to *S. vulgaris*, more rapid than in *S. unicolor* from Salamanca (Motis 1987).

**Structure.** 10 primaries: in adult, p8–p9 longest, p7 2–6 shorter,

p6 10–15, p5 18–23, p1 42·6 (12) 39–46; p10 reduced, 69–88 shorter than p8–p9, 12·0 (10) 10–14 shorter than tip of longest upper primary covert; in juvenile, p8 longest, p7 and p9 2–5 shorter, p10 relatively slightly longer than in adult, 5–7 shorter than longest primary covert. 12 tail-feathers, t3–t4 longest, t1 2–3 shorter, t6 1–2 shorter. Feathers of crown to upper back, of side of head and neck, and of chin to chest distinctly elongated (especially in adult ♂), narrow (in *S. vulgaris*, those of chest only). Throat feathers distinctly longer (Feare 1986). Middle toe with claw 28·2 (10) 26·5–29·5 mm; outer and inner toe with claw both *c.* 68% of middle with claw, hind *c.* 81%. Remainder of structure as in *S. vulgaris*, but bill relatively thicker at base.

**Geographical variation.** Slight, if any. Size apparently declines towards west, but samples small: average wing, sexes combined, Sicily 135·0 (*n* = 4), Tunisia 133·4 (*n* = 11), Sardinia 132·0 (*n* = 16), Madrid and Salamanca (central Spain) 131·3 (14), south-west Spain and southern Portugal 130·7 (*n* = 3), Morocco 129·6 (4).

Forms superspecies with Starling *S. vulgaris*. Sometimes combined into single species, as habits, morphology, moult, and eggs closely similar (Jordans 1970; Eck 1985*a*), but, while races of *S. vulgaris* all grade clinally into each other, *S. vulgaris vulgaris* is abruptly replaced by *S. unicolor*, which differs more in morphology from *S. vulgaris vulgaris* than does any other race of *S. vulgaris*. Following recent spread of both species, breeding ranges now show slight overlap (Lavin Castanedo 1978; Peris 1980; Motis *et al.* 1983; Peris *et al.* 1987; Iapichino and Massa 1989); no hybrids yet known. In captivity, forced mixed pairings of ♀ *S. unicolor* and ♂ *S. vulgaris* either did not breed at all, or laid infertile eggs, or embryos died in egg, or produced young with various abnormalities, none of which survived for more than 18 days; also, much difference in timing of sexual activity, gonads of ♂ *S. vulgaris* already regressing when ♀ *S. unicolor* starts to show interest in nesting; thus, highly likely that the 2 forms are good species (Berthold 1971).                    CSR

---

## *Sturnus roseus*  Rose-coloured Starling

PLATES 13 and 18 (flight)
[between pages 136 and 137, and 280 and 281]

DU. Roze spreeuw    FR. Etourneau roselin    GE. Rosenstar
RU. Розовый скворец    SP. Estornino rosado    SW. Rosenstare

*Turdus roseus* Linnaeus, 1758. Synonym: *Pastor roseus*.

Monotypic

**Field characters.** 21·5 cm; wing-span 37–40 cm. Size as Starling *S. vulgaris* but with shorter bill and more domed crown softening outline of head in subtle but distinctive way. Medium-sized, rather short-billed (in adult) crested starling with pale orange-pink legs. Adult in worn plumage basically blue- to purple-black, with striking pink jacket on back, flanks, and belly; spotted and streaked in fresh plumage. Juvenile pale dun-brown with whitish throat, eye-ring, and belly, prominent brownish-white margins to larger wing-feathers, and striking yellowish base to bill. Sexes similar; some seasonal variation. Juvenile separable.

ADULT MALE BREEDING. As adult male non-breeding (see below), but wear removes tips of feathers so that nearly all dull spots and margins are lost and bird takes on full and immaculate glossy black and rosy-pink contrasts of classic appearance. ADULT MALE NON-BREEDING. Moults: July–October, but later in birds lingering within breeding range (complete). Head and crest (reaching nape), mantle edge, throat, and breast black, glossed reddish-purple but also tipped buff-grey, giving hoary appearance. Wings black, glossed blue-purple and green and margined with buff and whitish on lesser and median coverts; thighs, vent, and under tail-coverts black, glossed blue-purple and tipped buff-grey. Underwing black, mottled white. Dark plumage contrasts with pink pale-brown tipped feathers of mantle, back, rump, chest, flanks, and belly. Bill relatively short and stubby compared to *S. vulgaris*; brownish-pink in winter, pink with black base in spring, and dusky to black in summer and autumn. Legs dull flesh-yellow, becoming bright pink in spring. ADULT FEMALE. Duller than ♂ at all times, with larger pale feather-tips, less gloss, and additional pale buff margins to larger wing feathers; even when worn, pink and black are duller. JUVENILE. Head, breast, flanks, and back warm but pale sandy- or dun-brown; lore and eye-ring paler, contrasting with forehead and dark brown eye. Rump often rather paler, more dun-pink, contrasting with back and tail in flight. Throat, belly, and vent pale brownish-white; under tail-coverts brown with brownish-white margins and tips, forming faint chevrons. Wings and tail dark brown, with conspicuous brownish-white margins or tips to feathers producing quite strong patterning across greater and median coverts, inner secondaries, and tertials; wings generally contrast with rest of plumage, unlike *S. vulgaris*. Underwing brown-grey, looking dark along leading edge and on axillaries, but broad buff-white fringes to longer coverts make rear centre look paler, further emphasizing darker area. Bill brown-horn, with bright yellowish or pale orange base. Legs orange-pink or straw colour. FIRST-YEAR. Resembles adult non-breeding ♀; difficult to sex (see Plumages). Juvenile plumage usually moulted from August in winter quarters, but all or majority of feathers retained as late as November in vagrants to western Europe.

Adult unmistakable. Given reasonable view, so is juvenile, since plumage distinctions from juvenile *S. vulgaris* are further enhanced by rather less frenetic, more placid behaviour. Flight closely resembles *S. vulgaris* but wings look more leaf-like, less triangular in outline and appear to beat more loosely (perhaps due to pale longer coverts and brown inner webs to flight-feathers and their contrast with dark fore under wing-coverts); certainly flocks sweep, circle, and land more slowly, while single *S. roseus* within flock of *S. vulgaris* shows habit of trailing or being 'thrown out' to edge of flock and of landing to feed apart (D I M Wallace). Gait variably described, said to be more sprightly than *S. vulgaris* but that of vagrant juveniles often slow and methodical, at least when feeding in grass or weeds. Gregarious, with social behaviour like *S. vulgaris*; not adapted to urban roosts except where large trees available.

Voice similar to *S. vulgaris*. Calls include frequent, harsh 'tschirr', producing rippling flock chorus, and raucous 'kritsch'. Alarm-call 'quilp', more a yelp than a scream. Noise of feeding flock a rapid, high-pitched chatter, more musical and stronger, less murmuring than *S. vulgaris*.

**Habitat.** In west Palearctic, ranges over lower middle latitudes, mainly in steppe, semi-desert, and Mediterranean lowland zones. Movements often governed by ephemeral localized abundance of gregarious invertebrate food organisms, concentrated in dry, open, often arid spaces, as well as grasslands and stony or rocky terrain. Requires ready access to water, but not dependent on wetlands or sea coasts. Resorts to trees and bushes, usually only in smallish groupings.

Gregarious movements often lead over upland or montane regions. Route over Baluchistan between Indian winter quarters and breeding areas in Asia Minor crosses plateau at *c.* 2000 m, where birds rest in clumps of trees and orchards (Ticehurst 1926–7, E M Nicholson). Arrivals in Turkey in May landed first in an Aleppo pine *P. halepensis*, shifting to spreading crown of mulberry *Morus* 5 m tall; open running water within 100 m was not visited (E M Nicholson). Such arboreal habitats, offering choice of fruits as well as insect food, provide acceptable substitute for treeless tracts infested by locusts (Acrididae) or other plague insects. Availability of natural or artificial rock piles, or cliff faces with plenty of holes or clefts, is normal requirement for nesting; generally below 400 m, although recorded up to 2834 m in Asian USSR. Preference is for foothills and undulating terrain where suitable nest-sites available, which may include artefacts such as fortress walls, railway embankments, salt mines, and spaces between tombstones. (Dementiev and Gladkov 1954.) Generally indifferent to human presence, often occurring in settled areas (Flint *et al.* 1984). Social roosting often occurs in groves, avenues, orchards, or tree plantations, or in thorn scrub or reedbeds, often at some distance from areas serving its normal ground-loving habits.

Aerial mobility, mostly within low or lower middle airspace, promotes flexibility of daily living. In Indian winter quarters, dense masses gather to drink and bathe at puddles, and birds follow cattle (especially on waterlogged ground) in order to catch insects disturbed from grass. Open farmlands, including fields cropped with cereals are frequented, as well as fruit trees. Unlike some of its relatives, however, shows little interest in picking parasites from bodies of grazing animals.

**Distribution.** Erratic, irruptive visitor to central and western Europe, with occasional breeding to west of usual limits.

CZECHOSLOVAKIA. Occasional breeding, recorded in eastern Slovakia 1918 (Schenk 1919). HUNGARY. Small influx May–June 1989 (*Br. Birds* 1990, **83**, 229). ITALY. Breeding recorded 1739 and 1875 (PB). YUGOSLAVIA. Breeds almost every year in Makedonija and southern Montenegro. Last massive invasion 1983. (VV.) Large influx May 1989 (*Br. Birds* 1990, **83**, 16). GREECE. Recorded breeding for first time in northern Makedhonia in 1985, at 2 sites in 1987, and 1 in eastern Makedhonia in 1988; sites not occupied in following years (Limbrunner

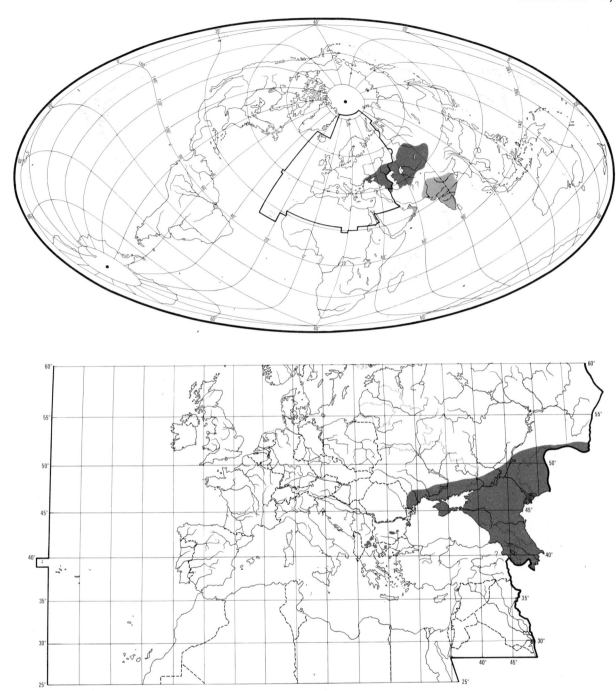

1987; Hölzinger 1992*c*; GIH). Bulgaria. Many old breeding records. Now of almost annual occurrence on coast of Dobrogea, also further south in some years. First recent breeding record near Balchik 1971, at least 2 other breeding records in Dobrogea since; possibly breeds annually. (TM.)

Accidental. Iceland, Faeroes, Britain, Ireland, France, Spain, Germany, Denmark, Norway, Sweden, Finland, Poland, Austria, Hungary, Switzerland, Malta, Albania, Algeria, Tunisia, Libya, Egypt, Lebanon, Jordan, Kuwait.

**Population.** No information on numbers or trends.

Survival. Oldest ringed bird 11 years (Korelov *et al.* 1974).

**Movements.** Migratory, wintering south-east of breeding range in peninsular India and Sri Lanka; migrates in flocks (sometimes huge) by day. Populations from west of range migrate almost directly east before heading south-east into India.

In India, widespread in winter east to Bihar and south throughout peninsula; abundant in Gujarat and Deccan, with fewer further south in Kerala and southern Tamil Nadu. In Sri Lanka, winters mainly in north and in coastal dry zone; sometimes common, at other times rare or absent. Straggles east to Bangladesh and Assam; vagrant or irregular visitor to Andaman Islands. Regular in southern Sind (Pakistan). Also winters in small and irregular numbers in Oman. (Henry 1971; Ali and Ripley 1972; Gallagher and Woodcock 1980; Roberts 1992.)

Autumn migration less well described than more synchronized and spectacular spring return. After young leave nest, adults and young desert colony rapidly (see Breeding) and move together to areas of abundant food, but age-groups soon separate (Schenk 1907; Korelov et al. 1974); during this post-breeding or summer dispersal phase, flocks roam widely (Kovshar' 1966), and may be nomadic (Isakov and Vorobiev 1940; Kazakov 1976). In Uzbekistan, birds (including a ringed individual) flew north-west from breeding colony to vineyards c. 50 km away; also reported remaining in vineyards for 2½ months (Serebrennikov 1931), so they may not resort to nomadism in summer if food remains abundant. Summer dispersal changes almost imperceptibly into directional autumn migration (Ivanov 1969) in which adults leave before juveniles (Hüe and Etchécopar 1970; Korelov et al. 1974). Departure of birds on autumn migration usually noticed as diminution of numbers (Schenk 1934), but mass departures do occur (Serebrennikov 1931; Salikhbaev and Bogdanov 1967). Reported leaving in flocks of 4–6 birds, departure of such flocks lasting at least 2 hrs (Schenk 1934). Birds from east Mediterranean migrate chiefly east, and birds from south-central Asia predominantly south-east; all populations pass through Pakistan on narrow front between foot of Himalayas and Baluchistan, thereafter dispersing more widely over peninsular India (Abdulali 1947).

Spring route is reverse of autumn. Many birds pass through Afghanistan; some continue through northern and western Iran and Iraq (Hüe and Etchécopar 1970; D A Scott); others head for south-central Asia; north-easterly migration recorded in Uzbekistan (Salikhbaev and Bogdanov 1967) and at Alma-Ata (Kazakhstan) (Korelov et al. 1974), birds having migrated to west of Pamirs (Ivanov 1969). In Turkmeniya, migrates mainly north-west along Amu-Dar'ya, Murgab, and Tedzhen rivers, continuing to Caspian or to Atek and Akhal plains (Rustamov 1958). Pre-migratory fat deposited in India in spring (George 1976), but no information on timing. Spring migration chiefly 08.30–11.00 hrs in small parties or flocks varying in size from tens to thousands; millions can pass in

a few days (Serebrennikov 1931; Dementiev and Gladkov 1954; Rustamov 1958; Ali and Ripley 1972).

Summer dispersal starts as soon as most young have fledged. In Uzbekistan, 1929, young fledged mid-June, and adults and young moved to vineyards, remaining there until main departure on 25 August (Serebrennikov 1931); more frequently, main departure from Uzbekistan mid-September, few birds remaining to October (Salikhbaev and Bogdanov 1967). Further north, in Kazakhstan, most leave by late August, with few remaining to mid-September (Korelov et al. 1974). In Hungary and Greece, where breeds only during sporadic influxes, departure July–August, shortly after young leave nest (Schenk 1907, 1934; Hölzinger 1992c). Main passage (sometimes abundant) in southern Caspian area August and early September (Passburg 1959; Feeny et al. 1968; D A Scott). In Afghanistan, movement recorded early July to mid-October, with adults preceding juveniles (Paludan 1959). One of earliest migrants to reach India; passage through Pakistan and north-west India begins early July, peaks late July, and ends chiefly in September; further south in peninsular India, arrival late September to November. Winter occurrence and timing in India probably influenced by flowering of nectar-bearing trees and fruiting of figs *Ficus*, etc. (Abdulali 1947.)

Spring migration begins in March, with large-scale passage through north-west India and Pakistan March–April, and through Afghanistan April–May (Ali and Ripley 1972). In Iran, movement last week of April to 3rd week of May (D A Scott); recorded mostly mid-May in Iraq (Marchant 1963a), and passage records in Syria and Turkey chiefly in May (Beaman 1986; Baumgart and Stephan 1987). One of latest migrants to reach breeding areas, with arrival in Uzbekistan and Turkmenistan mid-April to May (Rustamov 1958; Salikhbaev and Bogdanov 1967), and in Kazakhstan usually mid- or late May, somewhat earlier in south (Kovshar' 1966; Korelov et al. 1974). In Greece, 1980–8, arrived 2nd half of May (Hölzinger 1992c); reached Hungary in early June in 1907 and 1932 invasions (Schenk 1907, 1934). 1 ringing recovery: bird ringed north-east Hungary, June, recovered c. 4800 km south-east in western Pakistan the following April (Ali and Ripley 1972).

In line with irregular colonization (see Social Pattern and Behaviour), spring invasions occur west of normal range in south-east Europe (notably to Yugoslavia and Greece), apparently less frequent and smaller than formerly; similarly, at edge of breeding range, numbers may vary from thousands to few (e.g. in Hungary). At Villafranca in north-east Italy, 6000–7000 pairs bred in exceptional influx in 1875; but in 20th century only vagrant to Italy. (Ferguson-Lees 1971.) At Petrovac (coast of south-east Yugoslavia), several thousand recorded in late May 1989 (Axell 1989).

Further north in Europe, vagrant both seasons (see Distribution, also Alström and Colston 1991). In Britain and Ireland, 160 records before 1958, and 211 in 1958–91, chiefly mid-May to beginning of November; widely distributed, with many in Shetland (Scotland), chiefly May–July, and in Isles of Scilly, chiefly October; 4 records of successful overwintering (Dymond *et al.* 1989; Rogers *et al.* 1991, 1992). In Sweden, recorded mostly May–June (Breife *et al.* 1990). In France, 1900–89, 28 records involving 62 individuals, mainly in west and south; one overwintering record (Dubois and Yésou 1992). Bird wintered on Frégate Island (Seychelles), October 1990 to February 1991 (A Gretton).                          CJF

**Food.** Much information extralimital, especially winter diet. In breeding season mainly insects, especially locusts and grasshoppers (Acrididae), and other swarming Orthoptera; after young fledge, major items are grapes *Vitis* and mulberries *Morus*; takes fruit, nectar, and seeds autumn and winter. Acrididae remain flightless ('hoppers') for *c.* 40–50 days over summer, presenting ideal food resource. During breeding season, most food taken from surface of ground but some Orthoptera caught in the air. When taking Orthoptera from ground, large flocks move in one direction, with birds in front moving (usually running) faster than those behind, birds from rear flying to front in 'roller-feeding' fashion. (Serebrennikov 1931; Korelov *et al.* 1974.) Hoppers are swallowed directly after capture, sometimes first thrown into air; when satiated, captive birds continued to kill Orthoptera by crushing in bill, then dropping them on ground; wings and legs of winged Orthoptera broken off before remainder taken back to young. Food for nestlings brought back in bill, often 6–8 insects at a time but sometimes more, so quantity of food in bill almost as big as bird's head. (Serebrennikov 1931.) At breeding colony, Schenk (1907, 1934) found regurgitated pellets 15–20 × 8–12 mm, containing insect remains; pellets of 34 × 12 mm found in northern Greece (Hölzinger 1992a). Breeding birds fly 10 km or more from colony to feeding areas, sometimes from steppe to forest to forage for 'pests' such as moth *Operophtera brumata* (Dubinin 1953); sometimes accompany domestic animals, though settling on their backs less often than other *Sturnus*; some birds in Bulgaria perched on sheep (Salikhbaev and Bogdanov 1967; Mautsch and Rank 1973). Also feed in trees and bushes. In northern Greece, continual stream of *c.* 9000 birds flew back and forth *c.* 300 m between colony and mulberry trees (Prigann 1992). Aggressive to conspecifics and other birds when taking nectar from flowering trees (Ali and Ripley 1972), and often steals food from conspecifics returning to colony. Feeding of breeding adults inhibited by rain (Serebrennikov 1931).

Diet includes the following. Vertebrates: small lizards (Lacertidae). Invertebrates: grasshoppers, etc. (Orthoptera: Tettigoniidae, Acrididae), mantises (Dictyoptera: Mantidae), termites (Isoptera), bugs (Hemiptera: Pentatomidae, Cicadidae), ant-lions (Neuroptera: Myrmeleontidae), adult and larval Lepidoptera (Noctuidae, Lasiocampidae, Saturniidae, Geometridae), flies (Diptera: Asilidae), Hymenoptera (ants Formicidae, wasps Vespidae, bees Apoidea), beetles (Coleoptera: Carabidae, Dytiscidae, Lucanidae, Scarabaeidae, Buprestidae, Elateridae, Meloidae, Tenebrionidae, Cerambycidae, Chrysomelidae, Curculionidae), spiders (Araneae), harvestmen (Opiliones), Solifugae, woodlice (Isopoda), centipedes (Chilopoda: Scolopendridae), snails (Pulmonata: Helicidae). Plant material: fruit, seeds, leaves, and nectar of date *Phoenix*, fig *Ficus*, mulberry *Morus*, *Streblus*, *Capparis*, silk cotton *Bombax*, *Careya*, grape *Vitis*, Polygonaceae, currant *Ribes*, cherry, apricot, etc. *Prunus*, apple *Malus*, raspberry *Rubus*, *Butea*, *Erythrina*, *Pithecellobium*, *Salvadora*, jujube *Ziziphus*, *Bridelia*, *Lantana*, nightshade *Solanum*, chilli *Capsicum*, honeysuckle *Lonicera*, grasses (Gramineae, including wheat *Triticum*, millet *Pennisetum*, *Sorghum*, *Agrostis*), Liliaceae. (Serebrennikov 1931; Dementiev and Gladkov 1954; Rustamov 1958; Stepanov 1960; Pek and Fedyanina 1961; Kovshar' 1966; Ali and Ripley 1972; Mautsch and Rank 1973; Korelov *et al.* 1974; George 1976; Kekilova 1978; Auezova 1982; Stepanov 1987; Hölzinger 1992a; Prigann 1992.)

Many accounts emphasize importance of Orthoptera in diet of breeding birds (Serebrennikov 1931; Dubinin 1953; Korelov *et al.* 1974; Li *et al.* 1975), and breeding distribution changes annually according to distribution of locusts, etc. (Rustamov 1958; Stepanov 1960); these sources claim *S. roseus* is valuable agent in biological control of locust outbreaks; see also Hölzinger (1992a). Major studies (Table A) show broader diet and other arthropod groups may predominate, e.g. woodlice occurred in 96% of 148 stomachs collected in April (Rustamov 1958). On migration, animal food predominated with 79% of 19 stomachs containing ants, but 52% also contained plant material (Kekilova 1978). After dispersal from breeding colonies, can cause considerable damage by eating grapes and mulberries (Serebrennikov 1931; Rustamov 1958; Korelov *et al.* 1974). In northern Greece, early June, of 85 pellets (see above), 54% contained mulberry seeds, 33% remains of Tettigoniidae, and 13% shells of *Helicella* (Hölzinger 1992a).

Daily intake of each of 2 flightless captive birds (♂, ♀), when released into swarm of locusts was 200 3rd instar, or 150 4th instar, or 120 5th instar. (Serebrennikov 1931.) In feeding experiments, average daily intake per adult 167 adult Orthoptera *Gomphocerus* or 74 *Damalacantha* (Li *et al.* 1975). Extrapolating from captive study, each adult in colony of at least 1500 pairs consumed up to 320 grasshoppers (each 2·5–3·0 g) daily; whole colony therefore consumed *c.* 2·5–3·0 tonnes per day (Rustamov 1958).

Food of nestlings studied in detail only by Auezova (1982), who trapped adults returning with food to colony in south-east Kazakhstan; in 138 samples, 75% of items Orthoptera, mainly Acrididae, 10% of items cicadas, rest made up of woodlice, spiders, mantises, beetles, ant-lions,

Table A Food of adult Rose-coloured Starlings *Sturnus roseus*. A: percentage by number of 1048 items in stomachs of birds caught during and after breeding in Uzbekistan, number of stomachs not given (Serebrennikov 1931). B: percentage of stomachs containing different items in birds killed by hail on 30 April 1956 just after arrival in Turkmeniya, n = 148 (Rustamov 1958). C: percentage of 58 stomachs, Kirgiziya, containing different food items; time and method of collection not given (Pek and Fedyanina 1961).

| | A | B | C |
|---|---|---|---|
| Orthoptera | 66·7 | | |
| Acrididae | (48·9) | <5 | 15·1 |
| Tettigoniidae | (17·8) | | <5 |
| Hemiptera | 6·1 | <5 | 11·3 |
| Coleoptera | 2·9 | 77·7 | 11·3 |
| Tenebrionidae | | 75·0 | 7·6 |
| Curculionidae | | 5·4 | 22·7 |
| Dytiscidae | | 22·9 | |
| Hymenoptera | | | |
| Formicidae | 0·3 | 58·7 | 7·6 |
| Diptera | | | <5 |
| Lepidoptera | 0·1 | <5 | <5 |
| Araneae | 1·0 | <5 | |
| Isopoda | 0·1 | 96·6 | <5 |
| Myriapoda | | <5 | |
| Mollusca | 0·7 | | <5 |
| Plant material | | <5 | <5 |
| Grapes | 3·0 | | |
| Seeds | 19·2 | | |

flies, moths, and fruits. In feeding experiments, nestlings (age not known) each ate average 137 adult *Gomphocerus* or 23 *Damalacantha* daily (Li *et al.* 1975). In northern Greece, newly-fledged young were fed bush crickets *Decticus albifrons* (Tettigoniidae) with wing-span of up to 10 cm, small *Helicella* snails (very abundant around nest-site), and mulberries; of 45 pellets collected in early July (when most young leaving nests), 71% contained remains of bush crickets, 20% mulberries, and 9% cherry stones (Hölzinger 1992a). At another site, large *Ephippiger* bush crickets were brought to nests (Prigann 1992). Shells of hatched eggs, previously removed from nest, sometimes fed to young along with locusts (Serebrennikov 1931). CJF, BH

**Social pattern and behaviour.** Few detailed studies in west Palearctic, due to mainly sporadic occurrence there; most information from Schenk (1929, 1934). Extralimital information (Kazakhstan and central Asia) in Serebrennikov (1931), Rustamov (1958), Salikhbaev and Bogdanov (1967), Ivanov (1969), and Korelov *et al.* (1974), and from winter quarters in Ali and Ripley (1972).

1. Gregarious all year. Little information on size of feeding flocks but tens to hundreds reported in breeding season (Rustamov 1958); in winter, forms small parties or large flocks, sometimes of 'swarm' proportions (Ali and Ripley 1972). Flocks on migration vary in size, tens or often many hundreds or thousands (Serebrennikov 1931; Dementiev and Gladkov 1954; Rustamov 1958). In Kazakhstan, July, flocks of up to 2000 adults and juveniles (Stepanov 1987); for post-breeding flocks of juveniles

alone, see Relations within Family Group (below). During late summer dispersal and autumn migration, forms flocks with other species, including Starling *Sturnus vulgaris* (Hüe and Etchécopar 1970; Paz 1987; see also Roosting, below). BONDS. Generally believed to be monogamous (Dementiev and Gladkov 1954) and suggestions of polygamy (Serebrennikov 1931) possibly due to confusion with activities of unattached ♂♂ (Korelov *et al.* 1974). Some ♀♀ apparently courted by 2–4 ♂♂ and sometimes copulated with 2 ♂♂ in quick succession (Schenk 1934). ♂♂ arrive first in breeding colonies (Ivanov 1969) and pairing and nest-building begin shortly after arrival (Korelov *et al.* 1974), thus seems unlikely that pair-bond survives between years. Dementiev and Gladkov (1954) cited one source which stated birds already paired when colony established. According to Serebrennikov (1931), ♀ builds nest, while Korelov *et al.* (1974) stated that ♂ chooses nest-site and both sexes build. Only ♀ incubates according to Serebrennikov (1931) and Makatsch (1976), but both sexes incubate and feed young according to Salikhbaev and Bogdanov (1967), Ivanov (1969), and Korelov *et al.* (1974). Serebrennikov (1931) indicated ♂♂ feed young as frequently at first but ♀♀ do most feeding of older young; Rustamov (1958) asserted (from observations at one nest) that ♂♂ responsible for most feeding of small young. After fledging, duration of parental feeding and role of sexes not known. In breeding colonies, ♂♂ said to outnumber ♀♀ (Schenk 1934). Reaches 'sexual maturity' at 1 year (Dementiev and Gladkov 1954) but no information on age of first breeding. BREEDING DISPERSION. Colonial. Commonly hundreds, thousands, and tens of thousands, occasionally hundreds of thousands of pairs (Rustamov 1958; Korelov *et al.* 1974), with some colonies extending over large areas, up to *c.* 4 km² (Serebrennikov 1931); in northern Italy, colonies up to 7000 pairs recorded (Moltoni 1969). Dispersion of nests depends on configuration of breeding areas but nests can be contiguous, or up to 0·5 m or more apart (Schenk 1934; Korelov *et al.* 1974); in unusual use of natural holes in willow trees *Salix*, nests much further apart (Schenk 1934); average 15–18 nests per 25 m² (Korelov *et al.* 1974), 4–5 nests per m² (Rustamov 1958), 2 nests per m² (Li *et al.* 1975). Colony in Crimea occupied *c.* 50 years in succession (Dementiev and Gladkov 1954). In many areas, nest-sites not occupied annually and presence depends on availability of superabundant grasshoppers and locusts (Orthoptera) (Serebrennikov 1931; see also Food). ROOSTING. Little information. In Uzbekistan, recorded roosting nocturnally in reedbeds and in steppes; also on slopes, cliffs, trees, and shrubs; loafs during day near drinking and bathing sites (Serebrennikov 1931; Salikhbaev and Bogdanov 1967). According to Serebrennikov (1931), after laying, most ♂♂ left colony temporarily and formed flocks, roosting away from colony at night. In winter roosts, India, in thousands in thorn bushes, trees (including in noisy and illuminated urban roadside avenues), coconut plantations, and reedbeds, communally with parakeets *Psittacula*, crows *Corvus*, mynas *Acridotheres*, sparrows *Passer*, and weavers *Ploceus* (Ali and Ripley 1972).

2. Crest erected in courtship-display and wings part-extended and shaken, displaying contrasting black and pink; during song, throat feathers and crest prominent. For responses to predators, including escape reactions, see Parental Anti-predator Strategies (below). FLOCK BEHAVIOUR. Feeds in dense flocks, especially when among swarms of Orthoptera; flock moves forward in 'roller-feeding' fashion (Serebrennikov 1931; Ali and Ripley 1972; Korelov *et al.* 1974; see also Food). Very noisy while settling in roost (Ali and Ripley 1972; see Roosting, above). SONG-DISPLAY. Territorial song (see 1 in Voice) given by ♂ standing in erect posture, crest raised and throat feathers ruffled (Fig A), usually near nest but sometimes sitting in tree; often

A

chorus of (e.g.) 100 birds singing together in colony (Schenk 1907). Unlike other *Sturnus*, courtship-song given during display on ground (see Heterosexual Behaviour, below). ♀ reported singing at nest after feeding young (Serebrennikov 1931; see 2 in Voice). ANTAGONISTIC BEHAVIOUR. Little information. Serebrennikov (1931) and Prigann (1992) alluded to aggression between ♂♂ during colony establishment, but no details. ♂-♂ fights occur during sexual chases (see Heterosexual Behaviour, below, also for other forms of aggression during courtship). When food decreased near colony, birds forced to fly further into steppe, or will attack, and sometimes rob, ♀♀ returning with food; one record of fight (feathers flying) between 2 ♀♀, i.e. nest-owner attacking bird trying to steal food. (Serebrennikov 1931.) In winter, uses threat-display to defend food sources (flowers with nectar) against other birds: droops, twitches, and shivers wings, spreads tail and flicks it sideways, and accompanies supplanting-attack with angry calls (Ali and Ripley 1972; see 4b in Voice). Same calls given during disputes with sparrows *Passer* (Schenk 1907). HETEROSEXUAL BEHAVIOUR. (1) General. Most accounts indicate brief breeding season with late arrival on breeding grounds, rapid colony establishment, synchronized breeding, and rapid desertion of colony as soon as most young fledged; late eggs and young are deserted by parents at this stage; ♂♂ arrive shortly before ♀♀ but pairing accomplished quickly (probably within 5–6 days) and nest-building and laying begin shortly after (Dementiev and Gladkov 1954; Korelov *et al.* 1974). (2) Pair-bonding behaviour. In courtship, ♂ moves singing round ♀ in crouched posture but with head up, wings and tail continuously vibrated, crest raised, and throat ruffled (Fig B); ♀ quiet at first but then gives loud Soliciting-calls (see

B

7 in Voice), and ♂ and ♀ walk around each other ever faster, ♀ also crouched in horizontal posture and with crest raised; copulation (see subsection 4, below) occurs when ♀ stops and both birds become quiet (Schenk 1907, 1929). This display not reported by other authors and not seen subsequently in other colonies (Schenk 1934), although briefer version in which ♂ circles round ♀ described by Serebrennikov (1931); Ser-

ebrennikov (1931) and Schenk (1934) also described courting ♂♂ giving call resembling chirping of nestlings (see Voice) while flapping wings, bowing, and gaping widely with bill. ♂♂ also chase ♀♀ which, when unreceptive, rush gaping at ♂♂; physical fighting, so intense that grappling ♂♂ could be caught by hand, described by Schenk (1934). (3) Courtship-feeding. Not recorded. (4) Mating. Occurs on ground in colony (Schenk 1934), not recorded elsewhere. Pre-copulatory display (see subsection 2, above) followed by rapid copulation (Schenk 1934; Serebrennikov 1931) and courtship-display may recommence shortly after (Schenk 1934); ♂ initiates courtship, but after copulation ♀ may solicit; ♀ also more likely to commence courtship later in season (Schenk 1907). Serebrennikov (1931) recorded ♂, with ruffled feathers, running off after copulation, both members of pair then gathering nest-material and taking it back to nest. No information on frequency or timing of copulation. (5) Nest-site selection and behaviour at nest. ♂♂ arrive at colony earlier than ♀♀ (Ivanov 1969) and probably select nest-site (Schenk 1934; Korelov *et al.* 1974). Korelov *et al.* (1974) reported ♂ displaying by nest-site (perhaps Nest-showing) giving apparent song. Sometimes ♂ gathers nest-material energetically and in large billfulls, turning around frequently in front of ♀; ♀ often keeps such a ♂ away from nest, ♂ then dropping material or flying off with it (Serebrennikov 1931). RELATIONS WITHIN FAMILY GROUP. Duration of brooding and contribution of each sex not known. For role of sexes in feeding young, see Bonds (above). ♀♀ recorded removing eggshells and faecal sacs, ♂♂ rarely remove latter. Eyes of young open at 3–4 days (Serebrennikov 1931). Serebrennikov (1931) described synchronized mass departure of young, with weakly flying young being fed by parents and swarming in village on trees, roofs, fences, in yards and on roads. Flocks, entirely of young, seen soon after fledging, so dependence on parents presumably short-lived; adults leave area before young (Korelov *et al.* 1974). ANTI-PREDATOR RESPONSES OF YOUNG. Nothing recorded. PARENTAL ANTI-PREDATOR STRATEGIES. (1) Passive measures. Parents may sit so tightly as to be caught by hand (Schenk 1907; Rustamov 1958; Korelov *et al.* 1974). (2) Active measures. Although breeding colonies attract mammalian, bird, and reptilian predators (Salikhbaev and Bogdanov 1967; Korelov *et al.* 1974), responses to predators apparently not well described. At approach of raptor, seeks refuge in bushes, under stone, or in crevice in colony, or if approaching raptor seen, takes off calling and circles above raptor (collective mobbing). Likewise when man near nests, birds fly around him for long time in agitation (Serebrennikov 1931).

(Figs by D Nurney: A from photograph in Hölzinger 1992*c*; B from drawing in Schenk 1929.)                    CJF

**Voice.** No detailed study of repertoire, and following scheme tentative. Probably makes wing noises, as in Starling *S. vulgaris* (Bergmann and Helb 1982, which see also for additional sonagrams and for renderings of calls which are mostly harsh and rarely given singly).

CALLS OF ADULTS. (1) Song of ♂. Has same character as *S. vulgaris* but more harsh and unmusical, a jumble of discordant grating noises mixed with some melodious warbling notes (Witherby *et al.* 1938); long continuous sequence of bubbling, warbling, and whistling phrases, generally less discordant than *S. vulgaris* (Roberts 1992). Less varied and less melodious than *S. vulgaris* (Bergmann and Helb 1982). Recording (Fig I) indicates greatest similarity to the complex section of *S. vulgaris* song (which

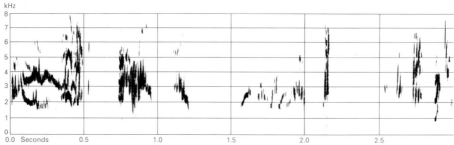

I  B and L Coffey  India  March 1973

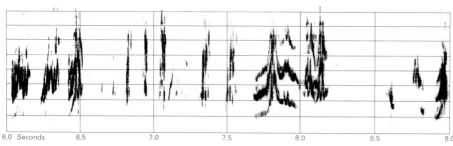

I  *cont.*

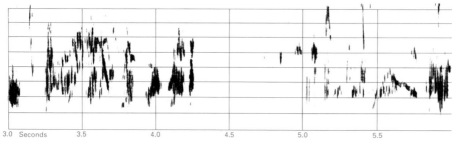

I  *cont.*

follows introductory whistles of that species) but faster (E K Dunn), thus sounding more congested and throttled, a little like Budgerigar *Melopsittacus undulatus*; song sequences shorter than those of *S. vulgaris* (P J Sellar). In contrast to *S. vulgaris* and Spotless Starling *S. unicolor*, no mimicry of other bird species reported (Witherby *et al.* 1938; Dementiev and Gladkov 1954; Roberts 1992). Most frequently repeated sequences of song are short series of rather dry low-pitched squeaks 'kitch kitch kitch kitch'. Flocks at winter roosts keep up continuous excited chattering and bubbling, very like *S. vulgaris*, before settling down. Small feeding groups also indulge continuously in bouts of this chattering and rather discordant grating and squeaking notes. (Roberts 1992.) In colony, one ♂ heard singing continuously for *c.* 30 min (Schenk 1907, which see for attempted rendering of song). (2) Song of ♀. Serebrennikov (1931) mentioned that ♀♀ produce clicking sounds and whistles between feeding trips. (3) Short flight-call, similar to that of *S. vulgaris*, produced when taking off (Dementiev and Gladkov 1954; Roberts 1992).

This is presumably the call variously rendered (for ♂) 'prüi' or 'trüi', or twittering 'pribri', and (for ♀) 'cüij' (transcribed into English as 'tsüiy') (Schenk 1907). (4) Alarm-calls. (4a) Similar to *S. vulgaris* (Roberts 1992) but no details. Recording contains resonant rasp (Fig II) sounding very like Snarl-call of *S. vulgaris* (P J Sellar). (4b) A 'tititiri' of alarm heard from ♂ during disputes (Schenk 1907). Repeated 'chit' calls during disputes over

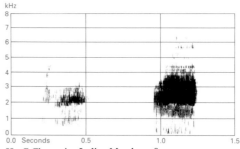

II  C Chappuis  India  March 1978

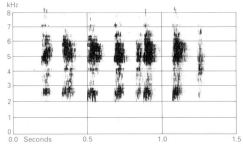

kHz

III C Chappuis India March 1978

feeding sites (Ali and Ripley 1972). (5) Recording contains loud raucous call (Fig III) like Advertising-call of Wryneck *Jynx torquilla* (P J Sellar). Rendered 'dschä... drschä...' (Bergmann and Helb 1982). (6) Feeding-call of ♀. Mentioned by Serebrennikov (1931) who observed antiphonal communication between ♀ and fledged young. (7) Soliciting-call of ♀. Before copulation, ♀ gives special call, not described (Witherby *et al.* 1938). This is presumably call described by Schenk (1907) as series of 4-6 'cilij-' calls (transcribed into English as 'tsiliy'), repetition rate increasing towards end (then 'c' lij' or even 'cij').

CALLS OF YOUNG. Young produce weak bat-like chirp just after hatching (Serebrennikov 1931). ME

**Breeding.** SEASON. Throughout breeding range a late spring migrant. In invasions of Hungary and Greece, nest-building and egg-laying started late May to early June; colonies deserted mid-July to early August (Schenk 1907, 1934; Hölzinger 1992*b*, *c*). In southern Kazakhstan, arrives late April to mid-May and begins nesting within a few days of arrival, while in rest of Kazakhstan nests from mid-May (Korelov *et al.* 1974). Further east, in Sinkiang (China), egg-laying begins early June (Li *et al.* 1975). In some years laying can be later, into July, possibly result of early clutches succumbing to bad weather (Ivanov 1969). SITE. Hole, most commonly among stones, e.g. scree, ideally with stones 10-15 cm diameter (Ivanov 1969); also in crack in rocks or cliff, under railway sleeper, in wall or bridge, under roof, in thatch, in nest-hole of Sand Martin *Riparia riparia*, and sometimes in hole in tree, especially willow *Salix*; nests usually deep among stones but sometimes near surface or even exposed (Serebrennikov 1931; Schenk 1934; Stepanov 1960; Kovshar' 1966; Salikhbaev and Bogdanov 1967; Korelov *et al.* 1974; Hölzinger 1992*b*). Nest: roughly made of thin twigs and grasses, lined with finer grass, often with fresh wormwood *Artemisia* and feathers, usually of *S. roseus* itself (Korelov *et al.* 1974), or of dry stems and leaves of annual herbs, especially giant fennel *Ferula* or grass *Aeluropus* (Rustamov 1958) (see Starling *S. vulgaris* for use of plants with insecticidal properties); in sites used annually, often re-uses nest of previous year (Salikhbaev and Bogdanov 1967); sometimes only a few blades of grass or eggs laid on rock debris in hole (Wilson 1883). Building: by ♀ (Serebrennikov 1931)

or both sexes (Schenk 1907; Ivanov 1969); time taken not known, but most accounts note rapidity of site occupation and nest-building. EGGS. See Plate 59. Pale blue or pale azure without marks and with slight gloss (Dementiev and Gladkov 1954; Rustamov 1958; Li *et al.* 1975). 28·7 × 20·9 mm (25·0-33·0 × 18·5-22·7), *n* = 306; calculated weight 6·6 g (Schönwetter 1984). Clutch: 3-6 (2-10); 8 and above probably laid by 2 ♀♀; in Kazakhstan, average 4·7, *n* = 26 (Stepanov 1960); in Hungary, average 6·3, *n* = 27 clutches (including 13-egg clutch) (Schenk 1934). Normally 1 brood but 2nd broods may occur in years of exceptional abundance of locusts (Acrididae) (Rustamov 1958); replacement clutches probable (Serebrennikov 1931; Korelov *et al.* 1974) but not confirmed from marked birds. INCUBATION. Said to be *c.* 15 days (Korelov *et al.* 1974) but no precise information; by both sexes; may begin before clutch complete, as hatching asynchronous (Korelov *et al.* 1974). YOUNG. Fed by both parents (Serebrennikov 1931; Korelov *et al.* 1974). FLEDGING TO MATURITY. Fledging period *c.* 24 days (Korelov *et al.* 1974); age at first breeding not known, 'sexually mature' at 1 year (Dementiev and Gladkov 1954); independent soon after fledging, forming juvenile flocks. BREEDING SUCCESS. No information. Colonies attract many avian, mammalian, and (possibly) reptilian predators (Serebrennikov 1931; Korelov *et al.* 1974). CJF, BH

**Plumages.** ADULT MALE. In fresh plumage (winter), head (including elongated crest extending from hindcrown), neck, and chest deep black with slight bluish-purple gloss, each feather with 1-2 mm wide fringe along tip, fringe ash-grey on head and chest, brown-grey on hindneck, giving grizzled appearance to black. Longest upper tail-coverts and a few longer outer scapulars black with blue or green gloss, remainder of upperparts light pink, each feather broadly tipped brown-grey; dark fringes largely conceal pink on mantle and back, but pink well-visible on scapulars and rump. Side of breast, lower flank, belly, and vent pale pink, feathers tipped pale brown-grey or ash-grey; dark tips paler and more restricted than on upperparts, mainly confined to side of breast, upper belly, and flank, but even here much pink visible, and vent virtually uniform pale pink. Upper flank, thigh, and under tail-coverts black, feather with pale grey-brown (on flank and thigh) or off-white tips (on coverts) of 1-2 mm wide, up to 4 mm on tip of longest under tail-coverts. Tail black with rather faint oil-green gloss; very narrow edges along feather-tips dirty white, bordered by dull black submarginally. Flight-feathers, tertials, and all upper wing-coverts black, more greyish-sooty on basal and middle portions of inner webs of flight-feathers, glossed green on tertials, secondaries, tips of primaries, and greater, median, and primary coverts, glossed blue on lesser coverts; edges of median and greater upper wing-coverts and tertials glossed blue; tips of lesser and median coverts narrowly fringed off-white, traces of pale edges sometimes on tips of primary coverts, inner greater coverts, or tertials. Under wing-coverts and axillaries dull black, broadly fringed white. *In slightly worn plumage* (March-June), pale fringes and edges of black parts of head, body, upperwing, and tail worn off, as well as dusky fringes from pink of body; head, crest, neck, and chest black with purple gloss, all upper wing black with green gloss, primaries, tail, thigh, and under tail-coverts somewhat duller green-black, remainder of body uniform bright pink, often more reddish-pink on lower belly and vent; traces of pale fringes still visible on longer under tail-coverts, many pale fringes still present on under wing-coverts and axillaries. When heavily worn, July or August to October or November, gloss on black parts duller, especially on flight-feathers, pink paler and appearing buff due to contamination by dirt; some brown of feather-bases sometimes visible, especially at side of breast. ADULT FEMALE. In fresh plumage, like adult ♂, but pale fringes of head, neck, and chest slightly wider and browner, of upper flank, upper and under tail-coverts, under wing-coverts, and axillaries distinctly wider, black parts appearing more heavily mottled pale. Black of flight-feathers, tail, and hindneck often more brown-black or sooty black, less deep black; flight-feathers and tail with distinct but very narrow pale brown edges along outer webs. Pink body-feathers more broadly tipped sandy-brown, visible pink mainly restricted to rump, belly, and vent. Crest-feathers shorter, longest 16-30 mm long (mainly 19-27; in adult ♂, 27-48, mainly 29-40 (Svensson 1992; RMNH, ZMA). *In worn plumage* (spring and summer), closely similar to ♂, but flight-feathers, tail, and (sometimes) hindneck slightly duller sooty black, less glossy; pink of body slightly less saturated, sometimes with traces of brown fringes (especially on mantle); more traces of pale fringes remain on tail-coverts, underwing with more white; most importantly, longest crest feathers shorter, and much dark brown or black of bases of scapulars exposed. NESTLING. Largely naked, but upperparts, thigh, and upperwing covered with sparse light grey down (Serebrennikov 1931); when larger, upperparts dark chocolate-brown with lighter upper tail, becoming pale towards fledging-time (Dementiev and Gladkov 1954). JUVENILE. Forehead and crown dark grey-brown or fuscous, gradually merging into paler grey-brown on hindneck and side of neck, feathers faintly fringed paler brown; no crest. Lore pale brown-grey, contrasting with forehead; side of head down to cheek medium brown-grey with some darker brown mottling, ear-coverts faintly streaked isabelline; eye-ring uniform pale brown-grey. Upperparts backward from mantle light brown-grey, palest and sometimes slightly isabelline-pink on rump, latter contrasting in colour with darker brown-grey lower back and upper tail-coverts (Walsh 1976); tail-coverts fringed paler brown-grey. Chin, belly, and vent greyish- or isabelline-white; chest, breast, flank, and thigh medium brown-grey, chest with contrasting dark brown shaft-streaks (varying in number and extent); under tail-coverts dark grey-brown with broad ill-defined off-white fringes along tips. Tail dark grey-brown, outer webs narrowly fringed pale brown-grey. Flight-feathers, tertials, and all feathers of upperwing dark grey-brown or fuscous-brown, darkest on bastard wing, all feathers fringed pale isabelline to off-white on outer webs and tips. fringes widest on outer webs of secondaries (sometimes showing as pale panel on closed wing) and on tips of greater and median coverts. Under wing-coverts and axillaries brown-grey with broad isabelline-white fringes. FIRST ADULT MALE. Rather variable, both in timing of moult (see Post-juvenile Moult) and in appearance of plumage. In some birds (perhaps retarded ones, irrespective of sex) feathers growing early in post-juvenile moult hardly different from juvenile feathers, those on upperparts dark grey-brown, those on underparts pale grey-brown, in both with restricted amount of cream-pink on base; in vagrants to western Europe, part of body plumage in September-November often like this, easily mistaken for juvenile feathering, advancement of moult becoming better visible once dark under tail-coverts, tertials, or inner primaries visible; in other birds, bases of new feathers more extensively pink, dusky tips paler and more restricted; in both types, new tail- and flight-feathers, tail-coverts, tertials, and upper wing-coverts contrastingly

blacker. Once moult completed, cap (including crest), chest, upper wing-coverts, and tertials deep black, cap and chest faintly glossed purple, coverts and tertials faintly green, all feathers tipped brown-grey; side of head and neck, hindneck, tail-coverts, flight-feathers, greater upper primary coverts, and tail brownish-black, feathers fringed pale brown (narrowest on tail and primaries); chin and throat either black or mottled black and isabelline-white; mantle, scapulars, and back brown-grey, cream-pink of feather-bases scarcely visible, rump isabelline-white, side of breast, lower flank, belly, and vent pinkish-isabelline; under wing-coverts and axillaries dark grey-brown with broad off-white fringes. Crest shorter than in adult ♂, longest feathers 24-34 mm (Svensson 1992; RMNH, ZMA). Occasionally, some juvenile secondaries retained, much browner than blackish neighbouring ones. *In worn plumage*, more similar to adult ♂ (but crest short, black of hindneck tinged brown-grey, pink of mantle and back suffused brown, black of bases of scapulars visible, wing and tail duller brown-black, and more traces of brown or off-white fringes remaining on hindneck, tail-coverts, and upper wing) or (especially) to adult ♀; differs from latter mainly in less black hindneck, more extensive brown wash to mantle and back, and (on average) longer crest, but some probably indistinguishable. FIRST ADULT FEMALE. Variable, as 1st adult ♂, but longest feathers of crest shorter, 13-19 mm (Svensson 1992; RMNH, ZMA); pale tips of black parts of body wider, black appearing mottled pale even in worn plumage; pink on upperparts extensively washed grey-brown, even when worn, only rump uniform cream-pink to isabelline; hindneck dark grey-brown with broad paler brown fringes, chin and throat mottled black and off-white; black of chin and throat mottled white; wing black-brown, slightly glossed bronze on tertials and greater coverts, green on lesser and median coverts; narrow pale edges along secondaries may form pale wing-panel (less conspicuous than in juvenile); some secondaries occasionally still juvenile. SUBSEQUENT PLUMAGES. In 2nd adult ♂ (15-27 months old) pale fringes of head, neck, and chest distinctly wider in fresh plumage than in older ♂♂, head appearing less black; flight-feathers and tail duller black, hardly glossy, secondaries with narrow but distinct pale edges; upper wing-coverts and tertials glossed green (in older birds, partly tinged blue); in worn plumage, wing less glossy than in older birds, more like adult ♀, but crest long and scapulars without black of bases visible. In 2nd adult ♀, fringes in fresh birds wider than in older ♀♀, but many probably indistinguishable, especially when plumage worn.

**Bare parts.** ADULT. Iris bright brown. During November-March, bill pale pink, yellow-flesh, pink-yellow or brownish-pink, culmen and tip slightly darker and browner. Between March and May, base of lower mandible of ♂ turns contrastingly plumbeous-black, first in older birds; remainder of bill pale pink or pink-yellow; from July, entire bill shades to dark horn-grey or plumbeous-black, this all gradually bleaching from base of lower mandible onwards in August-November. In ♀, some birds obtain contrasting black at base of upper mandible, similar to ♂, but others have base dusky horn or grey, less dark than ♂ and without sharp contrast. Mouth plumbeous-black, turning to flesh anteriorly. Leg and foot brownish-yellow with flesh tinge, brightening to flesh or reddish-pink from April-May, gradually browner again from July-October. (Stresemann 1920; Herroelen 1987; RMNH, ZFMK, ZMA, ZMB.) NESTLING. Bare skin, including bill, leg and foot, reddish-flesh; mouth red, gape flanges broad, yellowish-white (Serebrennikov 1931). JUVENILE. At fledging iris dark brown; bill pale pink, brighter pink at base of upper mandible, yellow on basal half of lower mandible; mouth pale yellow, turning to flesh-pink or grey-pink and anteriorly;

leg dull pink. In India, August, bill dark, base of lower mandible yellowish-orange (Berg 1987a). FIRST ADULT. From age of c. 2 months, bill straw-yellow with pink tinge, tip darker horn, leg pale reddish, flesh-pink, or flesh-brown; later in autumn as adult in winter; in May or June, when almost 1-year-old, some obtain dark base of lower mandible, but others do not. (Stresemann 1920; RNMH, ZFMK, ZMA.)

**Moults.** ADULT POST-BREEDING. Complete; in winter quarters. Most start August, on arrival in Indian subcontinent (Berg 1982b); those lingering in north either start there, or delay moult, e.g. one from Afghanistan, 15 October, heavily worn, but no moult (Paludan 1959); none of USSR birds moulting, though heavily worn: (Dementiev and Gladkov 1954). In India, starts July–August, completed by about October; tail starts relatively early, at primary moult score 5–10 (p2 or p3 in moult); secondaries start when p5–p6 in moult, completed shortly after regrowth of outer primaries (primary score 50); secondary moult frequently suspended, with 1–3 old ones retained, these either moulted in March or retained until next post-breeding moult (Naik and Naik 1969; Herroelen 1987). POST-JUVENILE. Complete or almost so; 1–3(–5) juvenile secondaries (especially s6) or (rarely) some other feathers retained until c. 1·5 years old. Birds arrive in Indian winter quarters in juvenile plumage and moult there into 1st adult plumage from August or September, completing December (Naik and Naik 1969) or moult October (Sharpe 1890). Birds lingering in breeding area or straggling further north and west show marked individual variation in timing of moult, probably depending on condition of bird; either start at correct time, occasionally suspending moult when body condition deteriorating, or delay moult markedly; in winter quarters, a few birds markedly late also. Depending on local conditions, juveniles fledge mid-June to early August, mainly mid-July; birds in fully juvenile plumage encountered July (many localities), 21 August (Kashmir), 2–3 September (western Tibet), 16 September (Afghanistan), 27 September (Netherlands, plumage fresh), 5 November (Baluchistan), and 24 November (Netherlands), as well as in c. 12 of 13 birds recorded late August to early November in Britain. Birds with innermost primaries and some body feathers new and some more feathers in moult or with moult suspended occur 18 August (India), early October and 15 and 28 November (Britain), and up to February (Arabia). Inner 3–4 primaries (moult score c. 18–25), tertial coverts, and inner greater coverts new, p5 (p4–p6) and central tail-feathers moulting (tail starting with shedding of t1 from primary score c. 15), and body in early stages of moult in birds from September (several from Afghanistan and Pakistan), 20 October (Netherlands), and 28 October (India); inner 6 primaries and all greater coverts new in one from 22 November (Sri Lanka); inner 7 primaries, 4 pairs of tail-feathers, 2 secondaries, and all wing-coverts and tertials new, with p8–p9 (score 44), t5–t6, and s3–s4 growing and head and chest in full moult in one from 23 December (India); 5 birds in final stages of moult 2–23 January (Sri Lanka), singles just completed 21 February (Sri Lanka) or as late as 7 May (Sri Lanka). (Stresemann 1920; Paludan 1959; Roberts 1982; Holmes et al. 1985; Svensson 1992; RMNH, ZFMK, ZMA.) Thus, primary moult normally starts mid-August to early October, completing January–February, but a few birds start up to February and finish up to May.

**Measurements.** Whole geographical range, all year (but 75%

from west Palearctic May–November); skins (RMNH, ZFMK, ZMA, ZMB). Adult wing and tail are from birds 2 years or older, 1Y is 1-year-olds (including 1st adults with wing and tail moulted), juvenile includes 1st adults with retained juvenile wing and tail. Juvenile wing includes data of Paludan (1940, 1959), Diesselhorst (1962), and Piechocki and Bolod (1972). Bill (S) to skull, bill (N) to distal corner of nostril (adults and 1-year-olds); exposed culmen on average 5·0 less than bill (S).

| | | | | |
|---|---|---|---|---|
| WING AD | ♂ 132·3 (3·07; 43) 127–139 | | ♀ 128·8 (3·35; 15) 125–135 | |
| 1Y | 131·0 (3·55; 7) 126–136 | | 127·7 (4·05; 8) 120–133 | |
| JUV | 127·3 (5·46; 9) 121–139 | | 123·8 (3·19; 6) 120–127 | |
| TAIL AD | 69·5 (2·97; 26) 65–75 | | 68·7 (2·88; 7) 64–74 | |
| 1Y | 68·1 (3·03; 7) 64–73 | | 67·5 (2·65; 6) 65–72 | |
| JUV | 66·1 (2·72; 4) 64–70 | | 63·8 (2·90; 4) 60–67 | |
| BILL (S) | 24·6 (1·22; 35) 22·2–26·5 | | 23·8 (0·92; 16) 22·4–25·6 | |
| BILL (N) | 15·0 (0·69; 33) 14·2–16·3 | | 14·3 (0·93; 15) 13·5–15·8 | |
| TARSUS | 32·1 (1·25; 34) 29·4–34·2 | | 30·9 (1·38; 16) 29·0–32·5 | |

Sex differences significant for wing, bill, and tarsus.

Southern Yugoslavia, wing: ♂ 129·6 (2·71; 42) 125–135 (excluding one each of 121 and 123), ♀ 125·3 (2·38; 15) 122–130. Kirgiziya, wing: ♂ 125–134 (n=25), ♀ 120–132 (n=10) (Yanushevich et al. 1960).

**Weights.** Turkey, Armeniya, Iran, Afghanistan, Mongolia, and some stragglers western Europe; (1) May–June, (2) July–August (Weigold 19..; Paludan 1938, 1959; Schüz 1959; Nicht 1961; Vauk 1968; Kumerloeve 1969b; Piechocki and Bolod 1972; ZFMK, ZMA). (3) Kazakhstan (Korelov et al. 1974). (4) Kirgiziya (Yanushevich et al. 1960).

| | | | | |
|---|---|---|---|---|
| (1) | ♂ 79·6 (5·09; 13) 71–88 | | ♀ 66·9 (8·87; 6) 60–83 | |
| (2) | 81·0 ( — ; 2) 79–83 | | 77·0 (9·64; 3) 70–88 | |
| (3) | 73·9 ( — ; 16) 59–89 | | 71·9 ( — ; 7) 69–77 | |
| (4) | — ( — ; 12) 68–90 | | — ( — ; 3) 67–79 | |

October: ♂♂ 67, 76 (Paludan 1959). Juveniles, July–September: 62·7 (10·83; 6) 53–81 (Paludan 1940, 1959; Piechocki and Bolod 1972; P R Holmes and Oxford University Kashmir Expedition 1983). India: 70–80 (n=7) (Ali and Ripley 1972).

**Structure.** Wing rather short, broad at base, tip pointed; 10 primaries: p9 longest, p8 0–2 shorter, p7 5–8 (adult) or 6–12 (juvenile), p6 11–16 (adult) or 13–19 (juvenile), p5 18–23 (adult) or 19–27 (juvenile), p4 24–30 (adult) or 23–32 (juvenile), p1 40–47 (all ages); p10 narrow, reduced, tip pointed in adult, slightly broader and more rounded at tip in juvenile; p10 75–88 shorter than p9 in adult, 72–86 in juvenile, 13·1 (10) 11–15 shorter than longest upper primary covert in adult, 9·9 (8) 7–12 in juvenile. Outer web of p8 emarginated, inner web of p9 with faint notch. Longest tertials reach to about tip of p1. Tail short, tip square, 12 feathers. Bill deeper at base than in Starling S. vulgaris, appearing stouter, culmen with distinct ridge, less flattened dorso-ventrally, tip of culmen gently decurved, tip laterally compressed, less blunt when seen from above. Feathers of hindcrown narrow, strongly elongated, especially in adult ♂, forming dense crest; no crest in juvenile; no elongated feathers on chest (unlike S. vulgaris and Spotless Starling S. unicolor), and this, together with other colour pattern and different bill-shape, a reason to put Rose-coloured Starling in a separate genus, Pastor. Middle toe with claw 29·4 (12) 27·5–31·5 mm; outer and inner toe with claw both c. 66% of middle with claw, hind c. 77%.

**Geographical variation.** None.

CSR

*Acridotheres tristis* **Common Myna**

Du. Treurmaina    Fr. Martin triste    Ge. Hirtenmaina
Ru. Обыкновенная Майна    Sp. Miná común    Sw. Brun myna

*Paradisea tristis* Linnaeus, 1766

Polytypic. Nominate *tristis* (Linnaeus, 1766), Turkmeniya and southern Kazakhstan, south and east to south-east Iran, Pakistan, and India; *tristoides* (Hodgson, 1836), Nepal and Burma; *melanosternus* Legge, 1879, Sri Lanka, grading into nominate *tristis* in Kerala and southern Tamil Nadu (southern India). Nominate *tristis* or *tristoides* introduced to Caucasus area, Arabia, eastern and southern Africa, south-east Asia, Australia, and many islands in Indian and Pacific Ocean.

**Field characters.** 23 cm; wing-span 33–36·5 cm. Only 10% longer but noticeably bulkier than Starling *Sturnus vulgaris*, with deeper bill, seemingly deeper chest and strong, lanky legs. Yellow bill and facial wattle obvious on black head, warm brown-grey body, brown-black wings with prominent white central panel and white under wing-coverts, white vent, and yellow legs. Sexes similar; no seasonal variation. Juvenile separable.

ADULT. Moults: mainly August (complete). Head (slightly crested at front), throat, and centre of breast glossy black, with short, strong, pointed yellow bill and yellow to almost orange facial wattle from lore, across cheek and on ear-coverts behind eye. Between black plumage and rest of body, a dark grey band most obvious on side of neck and at shoulder. Mantle to rump and chest to flank and central belly deep vinous-grey, contrasting with white vent and under tail-coverts. Wings blackish, with brown edges to inner feathers and smaller coverts and white primary coverts and white base to primaries showing on folded wing as bright patch below greater coverts and extending in flight to form bold, round central panel on both surfaces. Under wing-coverts and axillaries conspicuously white. Tail black, with white tips increasing outwards and forming striking white corners. Vinous-grey thighs usually exposed, making yellow legs look long, almost chicken-like. JUVENILE. Resembles adult but duller, lacking any crest and fully coloured wattle.

Unmistakable in west Palearctic but beware always chance of captive origin. Flight has noticeably flapping action, lacking darting progress of *Sturnus* and reminiscent of Jay *Garrulus glandarius*. Gait mainly a jaunty walk. Extremely self-possessed, being tame and bold around man and exploiting every possible food source. Not often gregarious.

Noisy, at times irritatingly strident. Song repetitive with much mimicry. Calls include liquid 'kiky-kiky-kiky', 'chour chour kok kok kok', and rough notes; 'traaahh' in alarm, somewhat recalling Nutcracker *Nucifraga caryocatactes*.

**Habitat.** Natural breeding range lies within subtropical and tropical lower latitudes of Asia, mainly in lowlands, but up to 3000 m in Himalayas. In Afghanistan, remains in Kabul (1700 m) even in severe winters, although keeping strictly to towns and villages and cultivated fields surrounding them. (Paludan 1959). In highland Baluchistan, however, none were found in summer even in towns (E M Nicholson). Breeds locally in high areas, and in central Asia irregularly into more northerly climates, much influenced by opportunities arising from human modification of habitats. Forages much in the open, from semi-desert to cultivation, even following the plough or herds of cattle; also on pastures, burnt fields, and in scrub jungle and refuse deposits. Roosts in large trees, coconut groves, reedbeds, sugar cane plantations, railway stations, warehouses, and other safe sheltered situations. Nests in any suitable holes, either natural or in buildings, etc. Mobility and gregarious habits confer robust life-style, not dependent on specialized or difficult habitat requirements. (Whistler 1941; Bates and Lowther 1952; Dementiev and Gladkov 1954; Hutson 1954; Ali and Ripley 1972.)

**Distribution and population.** FRANCE. Several (undoubtedly originating from escapes) in Dunkerque harbour since at least 1986; at least 1 pair nested 1988–9 (Hars 1991). RUSSIA. Resident and breeding in small numbers on Black Sea coast in Sochi and Gagra areas (north-west Caucasus) at least from 1978; population probably originates from escaped cagebirds (Mauersberger and Möckel 1987), though may be part of natural range expansion according to Sperl (1992, which see for report of further spread in Turkey west to Trabzon). For major review of distribution changes in central Asia and Kazakhstan, see Sagitov *et al.* (1990).

PLATE 15. *Corvus splendens zugmayeri* House Crow (p. 143): **1-2** ad.
*Corvus frugilegus frugilegus* Rook (p. 151): **3-4** ad winter.
*Corvus corone* Carrion Crow (p. 172). Nominate *corone*: **5-6** ad winter. C. c. *cornix*: **7-8** ad winter.
*Corvus albus* Pied Crow (p. 195): **9-10** ad.
*Corvus ruficollis ruficollis* Brown-necked Raven (p. 197): **11-12** ad winter.
*Corvus corax corax* Raven (p. 206): **13-14** ad winter.
*Corvus rhipidurus* Fan-tailed Raven (p. 223): **15-16** ad. (DR)

Darren Rees

PLATE 16. *Sturnus vulgaris* Starling (p. 238). Nominate *vulgaris*: **1** ad ♂ summer, **2** ad ♀ summer, **3** ad♂ winter, **4** ad ♀ winter, **5** 1st ad summer, **6** 1st ad winter, **7** juv. *S. v. zetlandicus*: **8** juv. *S. v. tauricus*: **9** ad winter. *S. v. purpurascens*: **10** ad winter. (PJKB)

PLATE 17. *Acridotheres tristis tristis* Common Myna (p. 280): **1** ad, **2** juv. *Sturnus unicolor* Spotless Starling (p. 260): **3** ad ♂ winter, **4** ad ♂ summer, **5** juv. (PJKB)

PLATE 18. *Onychognathus tristramii* Tristram's Grackle (p. 229): **1–2** ad ♂ summer.   *Sturnus roseus* Rose-coloured Starling (p. 269): **3–4** ad ♂ summer, **5–6** juv.   *Acridotheres tristis tristis* Common Myna (p. 280): **7–8** ad. (PJKB)

PLATE 19. *Sturnus sturninus* Daurian Starling (p. 234): **1–2** ad ♂.   *Sturnus vulgaris vulgaris* Starling (p. 238): **3** ad ♂ summer, **4–5** ad ♂ winter, **6** juv, **7** flock.   *Sturnus unicolor* Spotless Starling (p. 260): **8–9** ad ♂ summer. (PJKB)

PLATE 20. *Passer domesticus* House Sparrow (p. 289). Nominate *domesticus*: **1** ad ♂ summer, **2** ad ♂ winter, **3** ad ♀, **4** juv. × *italiae*: **5** ad ♂ summer. *P. d. tingitanus*: **6** ad ♂ summer. *P. d. niloticus*: **7** ad ♂ summer, **8** ad ♀ summer. (IL)

PLATE 21. *Passer hispaniolensis* Spanish Sparrow (p. 308). Nominate *hispaniolensis*: **1** ad ♂ summer, **2** ad ♂ winter, **3** ad ♀, **4** juv. *P. h. transcaspicus*: **5** ad ♂ winter, **6** ad ♀. *Passer montanus montanus* Tree Sparrow (p. 336): **7** ad, **8** juv. (IL)

PLATE 22. *Passer moabiticus moabiticus* Dead Sea Sparrow (p. 320): **1** ad ♂ summer, **2** ad ♂ winter, **3** ad ♀, **4** juv. *Passer iagoensis* Iago Sparrow (p. 327): **5** ad ♂ summer, **6** ad ♀, **7** juv. (IL)

PLATE 23. *Passer simplex* Desert Sparrow (p. 331). *P. s. saharae*: **1** ad ♂, **2** ad ♀, **3** juv. Nominate *simplex*: **4** ad ♂, **5** ad ♀. *Passer luteus* Sudan Golden Sparrow (p. 351): **6** ad ♂, **7** ad ♀, **8** juv. (IL)

PLATE 24. *Carpospiza brachydactyla* Pale Rock Sparrow (p. 357): **1** ad winter, **2** ad summer, **3** juv. *Petronia xanthocollis transfuga* Yellow-throated Sparrow (p. 365): **4** ad ♂ summer, **5** ad ♀, **6** juv. (IL)

PLATE 25. *Petronia petronia* Rock Sparrow (p. 371). Nominate *petronia*: **1, 6** ad ♂, **2** ad ♀, **3** juv. *P. p. barbara*: **4** ad ♂. *P. p. puteicola*: **5** ad ♂. *Passer domesticus domesticus* House Sparrow (for comparison): **7** ad ♀. (IL)

PLATE 26. *Passer domesticus domesticus* House Sparrow (p. 289): **1–2** ad ♂ summer, **3** ad ♀. *Passer hispaniolensis hispaniolensis* Spanish Sparrow (p. 308): **4–5** ad ♂ summer. *Passer moabiticus moabiticus* Dead Sea Sparrow (p. 320): **6** ad ♂. *Passer iagoensis iagoensis* Iago Sparrow (p. 327): **7** ad ♂. *Passer simplex saharae* Desert Sparrow (p. 331): **8** ad ♂. (IL)

PLATE 27. *Passer montanus montanus* Tree Sparrow (p. 336): **1** ad. *Passer luteus* Sudan Golden Sparrow (p. 351): **2** ad ♂, **3** ad ♀. *Carpospiza brachydactyla* Pale Rock Sparrow (p. 357): **4** ad winter. *Petronia xanthocollis transfuga* Yellow-throated Sparrow (p. 365): **5** ad ♂. *Petronia petronia petronia* Rock Sparrow (p. 371): **6–7** ad ♂. (IL)

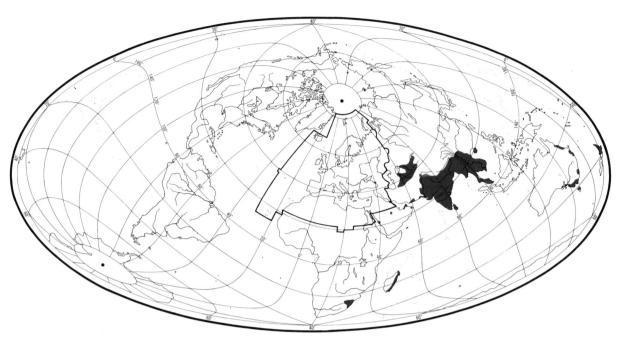

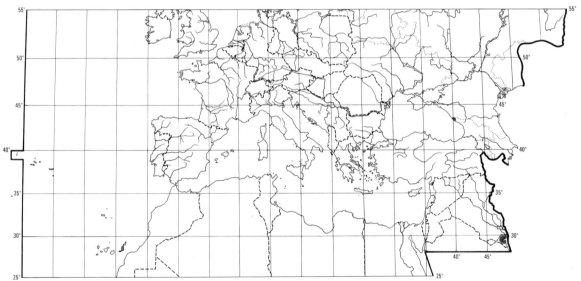

**Movements.** Said apparently to be sedentary in native range in western Asia (Long 1981), and where introduced

in west Palearctic (Mauersberger and Möckel 1987). However, at Chokpak pass (western Tien Shan, Kazakhstan), seasonal movements regular: birds head west throughout autumn, mainly 1st half of September and 6–31 October, averaging 74 birds per year; spring movement eastward, averaging 441 birds per year, chiefly mid-March to 20 April and in 1st third of May (Gavrilov and Gistsov 1985). NK, DFV

**Food.** Omnivore and scavenger. Feeds predominantly on ground, sometimes in trees (e.g. Ali and Ripley 1972,

Macdonald 1973, Kang 1992). Feeding sites include grassy areas, scrub, roads, pavements, and foliage of trees and shrubs; will forage around buildings and livestock or poultry farms; scavenges among rubbish and waste food and on foreshore at low tide; occasionally hawks for insects from perch (Kang 1989). Will follow and feed in tracks of domestic cattle (e.g. Ali and Ripley 1972, Macdonald 1973), Lesser Whistling Teal *Dendrocygna javanica* (Bharucha 1989), the plough (Ali and Ripley 1972), and grass-mowing machines (Kang 1992). During ground foraging, bird walks at steady pace, frequently tilting head to one side to fix gaze on possible food items, and food picked up with both eyes directed forwards at it; frequently makes running dash after fast-moving prey (Counsilman 1971). Large eye orbits with narrow supraorbital isthmus and thin interorbital septum may facilitate forward vision for ground feeding (Dubale and Patel 1975). Able to open bill while probing into ground by means of well-developed protractor muscles which raise upper mandible (see Feare 1984). In western Malaysia, Shukla (1981) observed peaks in ground foraging activity at 08.00–10.00 hrs and 14.00–17.00 hrs. In Singapore, food items of less than 2 cm taken more often than larger items; items of less than 1 cm swallowed with little or no preparation, larger items commonly broken up by stropping against ground (Kang 1989).

Diet is wide, composition varying with location, habitat, and availability of food sources. The following are recorded. Vertebrates: frogs, geckos, lizards, mice; as carrion, snakes, birds, and rats; also reported to prey on eggs and nestlings of seabirds (e.g. Stoner 1923, Byrd 1979, Grant 1982) and some land birds (Edgar 1972, 1974). Invertebrates: springtails (Collembola), dragonflies and damsel flies (Odonata), grasshoppers, etc. (Orthoptera), cockroaches and mantises (Dictyoptera), termites (Isoptera), stoneflies (Plecoptera), earwigs (Dermaptera), bugs (Hemiptera), caddisflies (Trichoptera), adult and larval butterflies and moths (Lepidoptera), adult and larval flies (Diptera), ants, bees, and wasps (Hymenoptera), beetles (Coleoptera), earthworms (Oligochaeta), snails (Gastropoda), crabs (Decapoda), ticks (Acari), spiders (Araneae), centipedes (Chilopoda), millipedes (Diplopoda). Plant food: fruit (e.g. berries, figs *Ficus*, papaya *Carica*, tomato *Lycopersicon*, pear *Pyrus*, apple *Malus*, guava *Psidium*, mangoe *Mangifera*, breadfruit *Artocarpus*, etc.), seeds (e.g. maize *Zea*, peas *Pisum*, beans *Phaseolus*), nectar, and flower parts. Also waste food and plant remains. (Mason and Maxwell-Lefroy 1912; Dementiev and Gladkov 1954; Akhmedov 1957; Wilson 1965; Sengupta 1968, 1976; Ali and Ripley 1972; Feare 1976; Moeed 1976; Shukla 1981; Narang and Lamba 1984; Jalil 1985; Kang 1989.)

♂♂ and ♀♀ appear to have similar diets (Moeed 1976), and composition may vary from month to month (Narang and Lamba 1984). In India, insects are principal food all year, supplemented by fruit and nectar during rainy season (July–September), with scavenging increasing in importance in summer (March–June) and post-monsoon period, when natural sources of food more scarce (Sengupta 1976). At Bhopal (India), May, 43 gizzards contained 32·5% (by number) beetles and larvae, 29·9% crickets (Gryllidae), 27·2% grasshoppers (Acrididae), 10·4% weevils (Curculionidae), 2·0% dates *Phoenix*, 2·0% figs, 1·7% wheat, 1·1% beans, and 2·0% other stone fruit (Moeed 1976). In West Bengal, stomach contents of 632 birds over 9 years comprised (by weight) *c.* 90% animal and 10% plant material: included 19·4% Hemiptera, 18·2% beetles, 15·7% moths and butterflies, 13·3% grasshoppers and crickets, 4·4% Oligochaeta, 4·1% damsel flies and dragonflies, 3·9% Dermaptera, 2·4% Hymenoptera, 2·3% termites; plant material included fruit of neem *Azadirachta indica*, date palm *Phoenix sylvestris*, berries of peepal *Ficus religiosa*, banyan *F. indica*, wild fig *Ficus cunia*, seeds of gram *Cicer arietinum*; during January–March, also fed on nectar from flowers of trees (Sengupta 1976). For further analyses, see Narang and Lamba (1984: India), Shukla (1981: Malaysia), Jalil (1985: Singapore), Wilson (1965: New Zealand).

In western Malaysia, about half the day (6 hrs) spent foraging, giving energy intake of *c.* 208 kJ per bird per day, with energy expenditure of *c.* 201 kJ per bird per day (Shukla 1981).

Nestlings in India (230 stomachs) were fed on soft insects and larvae for first 5–6 days; other invertebrates (e.g. dragonflies, butterflies, beetles, bugs, earthworms) fed from day 6–7, and fruit (e.g. pulp of ripe mango) included from day 10–11 (Sengupta 1976). Nestlings in New Zealand tended to be fed with fewer but larger insects than those eaten by adults (Wilson 1965).    NK

### Social pattern and behaviour.

1. Essentially solitary or in pairs all year, occasionally forming small flocks and feeding aggregations, up to *c.* 100 in Kazakhstan (Lamba 1963*b*; Eddinger 1967; Sengupta 1968, 1976; Ali and Ripley 1972; Korelov *et al.* 1974; Kang 1989; Mahabal and Vaidya 1989). Average flock size during day 1·65 birds (Kang 1989). Will associate loosely with many other bird species while feeding (e.g. Counsilman 1974*a*, Kang 1992). BONDS. Monogamous pair-bond, life-long after a pair has bred successfully, and persisting all year. Pairs for first time within flocks during winter and spring. (Lamba 1963*b*; Eddinger 1967; Wilson 1973.) In captivity, mate-fidelity high in paired birds, even when unpaired birds present (N Kang). Lost mate is replaced within flocks or on territory (Wilson 1973), but bird may remain for some time in vicinity of dead mate before selecting another (N Kang). Both sexes involved in selection of nest-site, nest-building, incubation, feeding of young (for several weeks after leaving nest), and territory guarding, but contributions vary (Lamba 1963*b*; Eddinger 1967; Counsilman 1974*a*; N Kang). First breeding at 1 year old (Eddinger 1967). BREEDING DISPERSION. Territorial (Lamba 1963*b*; Wilson 1965; Eddinger 1967; Sengupta 1968; Counsilman 1974*a*). Pairs nest in close proximity (Eddinger 1967), and Dementiev and Gladkov (1954) reported large nesting colonies, nesting groups of several pairs, and isolated pairs in various parts of central Asia. Breeding

density in Auckland (New Zealand) estimated at *c.* 53 pairs within 2·4 km² (about 22 pairs per km²) (Counsilman 1974*a*). In New Zealand, territory averaged *c.* 9000–14 600 m², used for both breeding and feeding (Wilson 1965; Counsilman 1974*a*); not clear, though, whether all this area is defended. In India, territories averaged 117·04 m² and considered to be used mainly for breeding, with food being obtained elsewhere (Sengupta 1968). In Singapore, home-range averages *c.* 100 000 m², within which breeding and feeding occur, but total range (area bound by furthest sites at which the individual was ever located) *c.* 250 000 km² (Kang 1992). ROOSTING. Typically roosts communally in trees, although dispersed roosting recorded (C J Feare). Wide variety of trees used, generally characterized by good canopy cover (Hails 1985; N Kang, V Yeo). Communal roosts usually formed all year. May be of less than 60 birds in a single tree, or with several thousand birds spread over several adjacent trees (e.g. Counsilman 1974*b*, Grieg-Smith 1982, Mat and Davison 1984, Kang 1989, Mahabal and Vaidya 1989). On Seychelles, tends to roost in small dispersed groups of 2–4 birds, rather than communally in large numbers (C J Feare). In Australia, communal roosting on buildings recorded (J J Counsilman). Radiotelemetry suggests adults tend to be loyal to particular roost (Kang 1992). May (Feare 1976) or may not (Sengupta 1973; Counsilman 1974*b*; Kang 1992) benefit from roosting by increased efficiency in finding food. Birds may fly directly to and from communal roost to feeding areas as far as 3 km away (Sengupta 1973; Mat and Davison 1984). In Singapore, flew average 400 m from communal roost to feeding areas (Kang 1992). During breeding season, number of birds in communal roost may decrease due to incubating and brooding ♀♀ remaining at nest (Counsilman 1974*b*), and radio-tagging indicates that when one member of pair remains at nest other may roost nearby instead of joining communal roost (N Kang). Mat and Davison (1984) recorded increase in numbers even after period when fledglings may be expected to be recruited into roost, so fluctuations in numbers may be due to young birds roosting elsewhere initially. Time at which first birds wake and call apparently related to time of sunrise, with birds waking earlier in breeding season than at other times of year. Break-up of communal roost preceded by long period of calling that continues while birds leave. Time at which birds started to leave, and length of time taken for all birds to leave, were related to sunrise, but birds left later and roost took longer to break up on overcast or rainy mornings than on clear mornings, and departures were earliest during breeding season regardless of weather. (Counsilman 1974*b*; Mahabal and Vaidya 1989.) Birds tended to move away from communal roost directly to territories and feeding areas (Counsilman 1974*b*; Mat and Davison 1984; Kang 1989); departure more rapid and synchronized than evening arrival, with average group sizes larger (Grieg-Smith 1982) (see Flock Behaviour). Prior to evening return, late afternoon pre-roost assemblies are usually formed (Counsilman 1974*b*; C J Feare), on ground or in trees near roost site (Mat and Davison 1984; Mahabal and Vaidya 1989); birds here generally displayed range of behaviour, from feeding and bathing to social interactions and displays. Time at which birds enter roost appears related to time of sunset; latest during breeding season (Counsilman 1974*b*; Mahabal and Vaidya 1989). Birds usually returned to roost from same direction in which they left, indicating that they returned directly from feeding areas (Mat and Davison 1984; Kang 1989). Unlike Starling *Sturnus vulgaris*, no communal flights before entering roost, except for fast twisting approach usually performed by late arrivals (Counsilman 1974*b*). Once in roost, period of movement and social interactions occurs as birds settle, and calling may continue until first few hours of darkness. Calling may be triggered during night by

disturbance, particularly in urban roosts, and heavy rain or strong wind may cause birds to leave roost and move to more sheltered trees (N Kang).

2. Generally approachable, to within 1 m in urban areas. Alarm-call (see 3 in Voice) given on approach of potential predator (e.g. cat, raptor) to nest or fledglings, or on sighting predator away from nest or fledglings (Counsilman 1971). Alarm-calls also given when chased by more aggressive species of bird, e.g. House Crow *Corvus splendens* (Kang 1991). FLOCK BEHAVIOUR. Minimum distance between individuals in flock varies from 25 cm (perched individuals in roost, feeding members of pair) to 50 cm (feeding flock) (Mat and Davison 1984; N Kang). Movements in flocks are coordinated by Flight-calls (see 2 in Voice) given when bird takes off (Counsilman 1971). Apparently no stable dominance hierarchy in the wild, but in captivity almost linear hierarchies may be formed in relation to access to food sources and roost perches, with breeding birds dominant over unpaired individuals (L K Wang, N Kang). SONG-DISPLAY. Sings (see 1 in Voice) throughout year, though more often during breeding season. Used by both sexes in heterosexual context and for territorial defence. (Counsilman 1971.) In captivity, both sexes (more frequently ♂) sing from exposed and elevated perch (tree or building) near or outside nest-hole (N Kang). ANTAGONISTIC BEHAVIOUR. (1) General. Appears to be aggressive when fights occur, but such behaviour relatively infrequent. May or may not involve physical contact and include following elements: jabbing and pecking, ruffling of body feathers with erect posture, sleeking of body feathers with horizontal posture, jumping, and fighting; usually accompanied by calls, especially during courtship and pair-formation. (Kang 1989.) Antagonistic displays may also be seen during disputes over food or roost perches. (2) Threat and fighting. Counsilman (1971, 1977) identified 7 displays as follows. (a) Head-forward display (Fig A) is most common threat

A

display performed between paired and unpaired birds when individual moves into the space or occupies perch of another. Bird faces opponent, with body more or less horizontal, legs bent, and head, neck, and body feathers sleeked but not depressed. Bill held open and may be used to jab, in which case no physical contact made, or peck, which involves physical contact but apparently causes no injury. Forward thrusts of head may be absent or pronounced and repeated, depending on response of recipient. Jabbing and pecking directed at front of opponent's body or feet. Loud single Chip-calls (see 4 in Voice) occasionally given. (b) Fencing-display, consisting of counter-jabbing movements of bill, may develop from 2 birds performing Head-forward display, and posture similar: bird withdraws head, closes bill, and ruffles contour feathers, crown, and nape, creating small crest; lowers bill, but raises and opens it to block jabbing from opponent. As Fencing develops, ruffling spreads from crown to breast, birds stand taller, thrusting and blocking become weaker, and calling becomes louder. Both birds may behave similarly or one may be clearly dominant. Submissive individual crouches, often leaning away, ruffles crown and nape more than opponent, thrusts weakly, and calls loudly. Dominant bird thrusts strongly and frequently. (c) Standing-display (Fig B) occurs between 2 birds during territorial dispute. Both stand fully erect, thrusting

B

D

weakly, with bills held high and close together, calling loudly; contour feathers on upper half of body ruffled; tail slightly spread and wings held away from side of body. (d) Jumping occurs during territorial disputes and sometimes at pre-roost assemblies. 2 or more birds leap *c.* 1 m into the air, attempting to grab each other's head with claws. Single calls may be given. (e) Fighting occurs during territorial disputes when no preliminary threat given. Consists of pecking, particularly at head, and wrestling on ground between periods of Jumping, Standing, and Bowing (see Heterosexual Behaviour, below). Fights are ♂-♂ and ♀-♀. ♀♀ may Stand and Bow instead of fight (Wilson 1973; N Kang). Birds call loudly and almost continuously. (f) Supplanting. Approach of one bird causes fleeing by another before threat or fighting can occur. Most supplanting is by flight and neither bird calls. (g) On surfaces which can be grasped by feet, sub-missive birds may perform Leaning-display: body and neck extended, sleeked, and leaned away from bird approaching from side; face turned away; no calls. HETEROSEXUAL BEHAVIOUR. (1) General. ♂ tends to initiate most displays. Sexual differences occur in some vocalizations (see Kinsey 1972). (2) Pair-bonding behaviour (see Counsilman 1971, 1977). Pairing initiated within flock. In early stages, ♂ and ♀ repeatedly perform prolonged Head-forward and Fencing-displays, though fleeing delayed or absent, and threats and chases gradually decline, to be replaced by pre-copulatory displays, Placing, and Bowing (see below), which become increasingly frequent during and after nesting. In Placing-display (Fig C), mates stand side by side and place open

C

bills together in stylised manner while crouching slightly, necks extended forward, giving song-like and other calls. Often one member of pair does not call but may Bow. Placing usually lasts 5-10 s and may be distinguished from threat displays by response of recipient. (Counsilman 1977.) Bowing performed by both paired and unpaired birds whilst foraging or at communal roost, but most often by pairs within home-range. May be stimulated by activity of mate, e.g. rapid approach, Placing, pecking, Bowing, fighting, or sometimes breast-preening (which Bowing resembles). Crown, neck, and scapulars rapidly ruffled, tail partly spread, and wings slightly drooped; tail then fully spread and neck extended as bill lowered to within a few cm of ground or perch (Fig D); movement highly stylised, lasting 0·4 s. Head

then raised to three-quarters of usual position and bird makes 3-4 (1-10) short bows in which head dropped to level of 1st bow but not fully raised. Each short bow separated by 0·7 s and final raising of head takes *c.* 0·6 s. Neck withdrawn and tail narrowed but contour feathers may remain ruffled for several seconds before being relaxed or Bowing begins again. Many song-like and other calls given during Bowing. Sometimes contour feathers not raised and little or no spreading of the tail, particularly in ♀. (Counsilman 1977.) Placing and Bowing precede copulation, and also occur in many other circumstances at times of probable strain on pair-bond, e.g. during territorial disputes and after long separations during nesting or foraging (Counsilman 1977; N Kang). Allopreening occurs between mates (N Kang), though not seen by Counsilman (1977); distinguishable from light peck-ing (see Counsilman 1977). (3) Courtship-feeding. Not recorded, though presentation of objects such as grass recorded (Eddinger 1967; Sengupta 1968). (4) Mating. Copulation may follow sim-ultaneous Bowing by both mates as described by Counsilman (1977). After Bowing, ♀ sleeks contour feathers, crouches deeply, raises head high, and vibrates tail from side to side. ♂ stops Bowing but plumage stays ruffled as he mounts and balances with partly open wings on back of ♀. ♂ lowers tail (still half-spread) and pushes ♀'s tail aside. Copulation lasts 1-2 s, fol-lowing which birds preen, forage, or rest. May take place on ground or perch, and at various times of day. (5) Nest-site selection and behaviour at nest. ♂ and ♀ both involved in select-ing nest-site, but ♀ may make final selection (Lamba 1963*b*; Eddinger 1967; Sengupta 1968; Counsilman 1974*a*). Both sexes engage in nest-building but in captivity ♀ may spend more time building than ♂ (N Kang). In New Zealand, eggs found in first nests of season after average 7 days from start of building, some taking twice as long. Incubation and brooding also carried out by both sexes (more frequently by ♀: N Kang), but only ♀ incubates and broods at night (no brood-patch in ♂). (Coun-silman 1974*a*.) Changeover at nest usually accompanied by calls (Counsilman 1971): returning adult enters nest and mate usually perches nearby and may preen or Bow; eggs and nestlings fre-quently left unattended while adults remain nearby (N Kang). RELATIONS WITHIN FAMILY GROUP. Nestlings fed by both parents. After fledging, dependent on parents for several weeks, and young perform begging displays. Eventually chased out of territory by parents. (Lamba 1963*b*; Eddinger 1967; Sengupta 1968; Counsilman 1971, 1974*a*; Kang 1989.) ANTI-PREDATOR RESPONSES OF YOUNG. From *c.* 7 days old, respond to disturbance of nest by crouching (Sengupta 1968). Fledged young will also crouch and become silent on hearing Alarm-calls of parents (N Kang). PARENTAL ANTI-PREDATOR STRATEGIES. Give Alarm-calls on approach of potential predator to nest or fledged young, accompanied sometimes by mobbing of predator.

(Figs from Counsilman 1971.)                                    NK

**Voice.** Highly vocal during breeding season, and to lesser extent during rest of year (Dementiev and Gladkov 1954;

Langrand 1990; Roberts 1992). For detailed study of repertoire, see Counsilman (1971). For sonagrams, see Counsilman (1971) and Maclean (1985).

CALLS OF ADULTS. (1) Song. Disjointed, rather noisy, and tuneless, comprising variety of different phrases, each usually repeated quite rapidly several times. Some units are gurgling, others disyllabic and whistling, others rather softer with almost strangled timbre; typical sequence 'tee-uh tee-uh tee-uh krok krok krok cheehtoo cheehtoo cheehtoo'. (Roberts 1992.) According to Langrand (1990), usual song a fluted, resonant, fairly harmonious 'teeyoo teeyoo teeyoo' or 'tweeyoo tweeyoo tweeyoo' delivered rapidly 6–12 times from perch or ground. Maclean (1985) described song as sustained jumble of squawks, whistles, croaks, creaks, and whines, most components repeated 2–4 times, e.g. 'krr krr krr ci ri ri ri krrup chip krrup chirri chirri chirri weeu weeu. . .'. According to Dementiev and Gladkov (1954), normally begins with peculiar croaking. Song longer and more complicated at beginning of spring (Roberts 1992). According to Langrand (1990), mimics perfectly calls of several bird species as well as wide range of sounds heard around human settlements. Said by Roberts (1992), however, generally not to be good mimic, and according to Dementiev and Gladkov (1954) mimicry manifested mainly in captivity. Various piercing, harsh calls also uttered in roost in early morning or evening (Langrand 1990). At roosts, may sing at intervals during night (Sengupta 1973). (2) Flight-call. When flushed or otherwise taking flight, gives rather weak and querulous 'kwerrh' not indicating any great alarm (Roberts 1992). (3) Alarm-call. Loud, drawn-out, grating alarm- or threat-call heard only when mobbing predator (Roberts 1992). Harsh swearing 'kharr' in alarm (Maclean 1985). A 'traaahh' somewhat like Nutcracker *Nucifraga caryocatactes* (Hollom *et al.* 1988). (4) Chip-call. Given by adults with fledglings and between mates which are physically separated, e.g. in communal roost; effective over short to long distances. Primarily descending sounds centred at 2·5 kHz, with harmonics up to 8 kHz or more; less than 0·1s in duration. (Counsilman 1971.)

CALLS OF YOUNG. Food-call of nestlings, a thin tittering squeak, becomes audible 48–72 hrs after hatching. By 2nd week, arrival of parent with food leads to crescendo of rapid twittering. (Roberts 1992.) ME

**Breeding.** SEASON. Central Asia: nests with eggs found May and August, nestlings and fledglings June (Dementiev and Gladkov 1954). India: starts January–September depending on onset of monsoon (Lamba 1963*b*; Ali and Ripley 1972; Rahman and Husain 1988). SITE. Wide variety of holes and crevices (Gibson-Hill 1950; Cairns 1952; Lamba 1963*b*; Eddinger 1967; Sengupta 1968; Ali and Ripley 1972; Macdonald 1973; Counsilman 1974*a*; Medway and Wells 1976; Michel 1986; Smythies 1986; Kang 1990). Natural sites include holes in trunk of trees, gaps in dense vegetation or clusters of fruit, and holes and

crevices in earth banks and cliffs. Other types of hole include those made by other species of birds or animals and wide variety of cavities in man-made structures, even in machinery in use. Height above ground 1·5–25(–40) m, average *c.* 7 m (Counsilman 1974*a*; N Kang). Nest: untidy mass of natural and man-made material such as twigs, leaves, roots, straw, feathers, fur, paper, cloth, string, cigarette ends, etc., with or without shallow cup (Gibson-Hill 1950; Lamba 1963*b*; Eddinger 1967; Sengupta 1968; Ali and Ripley 1972; Macdonald 1973; Counsilman 1974*a*; Medway and Wells 1976; Michel 1986; Smythies 1986; Kang 1991); cup sometimes lined (Dementiev and Gladkov 1954; Kang 1991), though this not reported by Counsilman (1974*a*); size and shape of nest dependent on space in hole, and may be 50 cm or more across (Eddinger 1967; Kang 1991); may be re-used for several years, with new material added (e.g. Lamba 1963*b*, Eddinger 1967, Ali and Ripley 1972, Counsilman 1974*a*). Building: by both sexes (Lamba 1963*b*; Eddinger 1967; Sengupta 1968; Counsilman 1974*a*; Kang 1991); in captivity, ♀ may do more than ♂ (N Kang); for first 2 days, material brought *c.* 8–12 times per hr; initially deposited inside hole near entrance, partly blocking it and perhaps reducing risk of entry by predators (Sengupta 1968); material may be added even after eggs laid (Lamba 1963*b*; Eddinger 1967; Sengupta 1968), making it difficult to determine precise duration of nest-building; estimated 6–8 days in India (20 nests: Sengupta 1968) and 5–12 days in Hawaii (25 nests: Eddinger 1967). EGGS. See Plate 59. Elongated oval or often pear-shaped, with hard glossy texture; pale blue to sky-blue or greenish-blue, unmarked (Lamba 1963*b*). Little geographical variation in average size (Sengupta 1968), e.g. in Turkmeniya and Tadzhikistan, 30·1 × 21·6 mm (28·7–32·4 × 20·4–23·0), $n = 16$, calculated fresh weight 7·7 g (7·2–8·3) (Dementiev and Gladkov 1954); in India, 30·0 × 21·5 mm (27·0–35·0 × 19·0–23·0), $n = 60$ (Lamba 1963*b*), 30·8 × 21·9 mm (27·6–35·0 × 19·2–23·2), $n = 100$ (Baker 1926), 29 × 21 mm (27–31 × 19–23), $n = 45$, calculated fresh weight 8·06 g (7·30–8·80) (Sengupta 1968); in Singapore, 29·5 × 21·3 mm (26·7–32·5 × 20·0–22·4), $n = 28$, calculated fresh weight 7·0 g (6·0–8·1) (N Kang). Clutch: 4–5 (2–6), varying geographically (Gibson-Hill 1950; Dementiev and Gladkov 1954; Lamba 1963*b*; Eddinger 1967; Sengupta 1968; Ali and Ripley 1972; Counsilman 1974*a*; Medway and Wells 1976; Michel 1986; Smythies 1986; Kang 1990); in India, mean 4·4, $n = 20$ (Lamba 1963*b*), mean 3·8, $n = 12$ (Sengupta 1968); New Zealand, mean 3·9, $n = 8$ (Counsilman 1974*a*); Hawaii, mean 3·5, $n = 10$ (Eddinger 1967); Singapore, mean 4·8, $n = 5$ (N Kang). Eggs laid at intervals of *c.* 24 hrs (Dementiev and Gladkov 1954; Lamba 1963*b*; Eddinger 1967; Sengupta 1968; N Kang). Eggs usually laid in morning (Sengupta 1968; Counsilman 1974*a*). No replacements laid after partial loss of clutch (Counsilman 1974*a*), but full clutch may be replaced within a few days (Lamba 1963*b*; N Kang). 1–3 broods per season, varying geographically, with next clutch laid c. 1-2 weeks after previous brood fully independent (Spittle 1950; Lamba 1963b; Eddinger 1967; Sengupta 1968; Counsilman 1974a; Rahman and Husain 1988). INCUBATION. 13-18 days (Lamba 1963b; Eddinger 1967; Sengupta 1968; Ali and Ripley 1972; Counsilman 1974a; Smythies 1986). Starts with laying of 2nd (Sengupta 1968; Counsilman 1974a) or last egg of clutch (Eddinger 1967). Incubation during day is irregular and eggs may be left uncovered for short periods (Eddinger 1967) or during hotter part of day. Both sexes incubate during day (Lamba 1963b; Eddinger 1967; Sengupta 1968; Counsilman 1974a), with average 4·7 visits per hr for ♂ and ♀ combined; only ♀ at night (Sengupta 1968; Counsilman 1974a) and ♀ may also do more than ♂ during day (Dementiev and Gladkov 1954; N Kang). Captive ♀♀ in New Zealand spent c. 30-40% of day incubating, average 9 min per sitting bout; ♂♂ made fewer visits lasting average 4 min (Counsilman 1974a); in India, incubation stints 1-10 min or more (Sengupta 1968). Hatching usually asynchronous, at intervals of 24-48 hrs (Lamba 1963b; Eddinger 1967; Sengupta 1968), although Counsilman (1974a) found that some clutches hatched synchronously. YOUNG. Both parents feed and brood nestlings during day, but night-time brooding performed by ♀ for c. 14-17 days. Nestlings in same brood are not fed equally and do not grow at same rate, earlier-hatched siblings growing faster. (Lamba 1963b; Eddinger 1967; Sengupta 1968; Counsilman 1974a.) FLEDGING TO MATURITY. Fledging period 22-35 days, and young may leave nest before able to fly (Lamba 1963b; Eddinger 1967; Counsilman 1974a; Smythies 1986). Dependent on parents for several weeks. First breeding at 1 year old. BREEDING SUCCESS. Data for hatching success as follows. India: one year, 92% (n = 20 nests) (Lamba 1963b); 97·8% (n = 12) (Sengupta 1968). New Zealand: one year, 48% (n = 8) (Counsilman 1974a). Hawaii: one year, 75% (n = 23) (Eddinger 1967). Data for nesting success (number of fledglings as proportion of number of eggs laid) as follows. India: one year, 61%, average brood size 2·7 (n = 20 nests) (Lamba 1963b); 76%, average brood size 2·8 (n = 12) (Sengupta 1968); two years, for 1st, 2nd, and 3rd broods respectively, 75%, 67%, 33% in one year, and 75%, 57%, 40% in another year (Rahman and Husain 1988). New Zealand: one year, 22% (n = 53) (Counsilman 1974a). Hawaii: one year, c. 14% (n = 18) (Eddinger 1967). Precise causes of nestling mortality and nest failure vary geographically: nestling starvation, shortage of food for adults due to poor weather, poorly sheltered nest-holes, predation, and disturbance. (Lamba 1963b; Eddinger 1967; Sengupta 1968; Counsilman 1974a). Overall nesting success may be low compared to hatching success, and recruitment of young birds into breeding population each year may be less than 1 bird per 2 birds aged 1 year or older (Counsilman 1974a). NK

Plumages. (nominate tristis). ADULT. Forehead and crown sooty-black, glossed oil-green or faintly purplish (depending on light and abrasion), but gloss much less strong than in various species of Sturnus. Side of head black, slightly glossy on ear-coverts, slightly greyish on cheeks. Rear and side of neck dark grey. Mantle and scapulars dark vinous-drab, back and rump paler vinous-drab, slightly tinged pink, upper tail-coverts dark drab-grey. Chin to chest dark grey, gradually merging into dark vinous-grey of side of breast, flank, upper belly, and thigh; lower mid-belly, vent and under tail-coverts white, sometimes with slight cream tinge. Tail black, tinged bronze-brown on central pair (t1); tip of t1 narrowly fringed white, t2-t6 broadly tipped white, amount of white increasing outward (c. 5 mm on inner web of t2, c. 20-25 mm on outer web of t6). Primaries black, bases contrastingly white, amount of white increasing towards inner primaries: p7-p9 black, except for bases, which are largely hidden beneath primary coverts, p1 white except for black tip of c. 2 cm; tips of primaries tinged dark olive-brown. Secondaries and tertials dark olive-brown, slightly glossed bronze in some lights, edges narrowly sooty; all lesser, median, and greater upper wing-coverts bronzy dark olive-brown. All primary upper wing-coverts white; bastard wing black, middle feather with white outer web. Under wing-coverts and axillaries white. Sexes similar. In fresh plumage, feathers of neck and body with faint narrow and ill-defined vinous-pink fringes (more buffy on belly); in worn plumage, feather-tips of mantle and scapulars more vinous-grey, centres darker vinous; chest paler grey, breast and upper flank paler vinous-grey; white on tail-tips partly worn off, especially on t1. NESTLING. Down ash-grey (Sapozhenkov 1962). JUVENILE. Top and side of head dull or sooty-black (less deep than in adult), gradually merging into brown-black of hindneck; tips of feathers on head narrowly fringed brown when plumage fresh; feathers of crown not elongated, less than 10 mm long (in adult, 12-21: Brooke 1976). Mantle, scapulars, tertials, and all upper secondary coverts saturated brown with a slight chestnut tinge, tips of feathers more rufous, especially on coverts, lesser coverts tinged grey; some ill-defined black of bases just visible on greater coverts and tertials, longest tertial with off-white fringe at tip. Back, rump, and upper tail-coverts buff-brown with faint darker bars. Chin, throat, and chest brown-grey, feathers narrowly fringed buff and with blackish centres or shaft-streaks, chest sometimes appearing streaked; brown-grey of chest gradually turns into buffish-grey breast and flank and this into off-white on mid-belly, vent, and under tail-coverts; breast and flank faintly barred rufous-brown. Tail as adult, but black duller and white feather-tips tinged isabelline, less wide, and less sharply defined. In Natal (South Africa), a 1st generation of tail-feathers (brown-black, narrow, tips fawn, length of pale on tip of t6 c. 6 mm) replaced by a 2nd generation (black, tips white with black shafts, length of pale tip of t6 c. 14 mm) c. 2 weeks after fledging (Brooke 1976). Flight-feathers and bastard wing as in adult, but tinged brown, secondaries with partial rufous fringe along tip, primaries with narrow off-white edges, and white of bastard wing and on bases of primaries less sharply defined; middle feather of bastard wing dusky with white patch (in adult, with white outer web); greater upper primary coverts white, as in adult, but tips black, especially on inner webs. FIRST ADULT. Like adult, but white of tail-tips sometimes with all or partly black shaft, and sometimes slight dusky marks on tip of outer 3-4 greater upper primary coverts (Brooke 1976); some juvenile flight-feathers occasionally retained.

Bare parts. ADULT, FIRST ADULT. Iris of nominate tristis brown to brown-red, often with white specks; of melanosternus pink-brown or pink-grey in some, whitish with fine dark specks in others. Bare skin round eye bright yellow to yellow-orange;

eye-lids yellow, edges of eye-lids black. Bill yellow, base of lower mandible blackish-green or brownish-green (paler when in moult). Mouth, including palate, dark slate (in at least some 1st adults, palate dusky pink: Brooke 1976). Leg and foot yellow or orange-yellow, claws horn. NESTLING. No information. JUVENILE. Iris brown. Bare skin round eye yellow. Bill entirely yellow, palate yellow. Leg, foot, and claws yellow. Adult colours obtained during post-juvenile moult. (Sharpe 1890; Hartert and Steinbacher 1932-8; Ali and Ripley 1972; Brooke 1976; RMNH.)

**Moults.** ADULT POST-BREEDING. Complete; primaries descendent. In Tadzhikistan, moult not started by mid-June (Dementiev and Gladkov 1954); in Afghanistan, 2 ♂♂ from 3 July worn, but a few body feathers growing; 2 ♀♀ from 2 August in full moult of wing, tail, and body (Paludan 1959). In India, no clear demarcation between breeding and moulting season, moult in some birds starting when others still have young in nest (Naik and Naik 1969). In introduced population of Réunion, March sample contained birds in fresh and in very worn plumage as well as birds just starting moult (RMNH). POST-JUVENILE. Complete, primaries descendent. Tail moulted *c.* 2 weeks after fledging, and again with all remainder of plumage at end of breeding season (Brooke 1976). In Afghanistan, some body moult in 2 juveniles from early August (Paludan 1959). March sample from Réunion contained just-fledged birds as well as birds in primary moult (moult scores 2–26, *n* = 5); those with scores 25–26 had body, tertials, and lesser and median upper wing-coverts new, tail just starting, and head and neck mainly juvenile (RMNH). In India, some birds breed before post-juvenile moult completed, suspending with some flight-feathers still juvenile.

**Measurements.** Nominate *tristis*. India, and a few from Caucasus, Réunion, Mauritius, Andaman islands, China, and Hawaii; all year, skins (RMNH, ZFMK, ZMA). Bill (S) to skull, bill (N) to distal corner of nostril; exposed culmen on average 7·0 less than bill (S).

| | | | | |
|---|---|---|---|---|
| WING AD | ♂ | 145·5 (4·40; 20) 138–152 | ♀ | 138·8 (3·88; 14) 134–147 |
| JUV | | 133·8 (4·07; 5) 129–140 | | 130·2 (1·89; 3) 128–132 |
| TAIL AD | | 85·3 (3·71; 21) 79–92 | | 80·6 (3·12; 14) 75–85 |
| JUV | | 69·6 (1·65; 4) 68–72 | | 66·7 (2·89; 3) 65–70 |
| BILL (S) | | 26·9 (0·85; 22) 25·1–28·4 | | 27·1 (1·27; 13) 25·8–28·7 |
| BILL (N) | | 15·9 (0·73; 22) 14·8–17·4 | | 15·8 (0·58; 12) 14·9–17·2 |
| TARSUS | | 39·9 (1·66; 24) 37·0–42·2 | | 38·8 (1·61; 17) 36·0–41·6 |

Sex differences significant for adult wing and tail.

India and China (Brooke 1976).

| | | | | |
|---|---|---|---|---|
| WING AD | ♂ | 139·9 (4·0; 9) 134–147 | ♀ | 139·2 (3·4; 6) 133–143 |
| TAIL AD | | 82·8 (5·0; 8) 77–91 | | 82·6 (4·3; 5) 78–88 |
| BILL (S) | | 26·1 (1·1; 9) 24–28 | | 26·0 (1·1; 6) 24–27 |

USSR: wing, ♂ 144·6 (16) 133–149, ♀ 136·4 (21) 131–143 (Dementiev and Gladkov 1954).

*A. t. tristoides.* Natal (South Africa) (Brooke 1976).

| | | | | |
|---|---|---|---|---|
| WING AD | ♂ | 139·8 (3·3; 22) 133–144 | ♀ | 134·1 (3·2; 23) 129–141 |
| JUV | | 130·8 (3·9; 9) 124–135 | | 128·3 (—; 3) 121–133 |
| TAIL AD | | 82·1 (4·7; 18) 74–90 | | 78·6 (3·7; 16) 74–88 |
| JUV | | 66·8 (2·2; 6) 63–69 | | 63·5 (—; 3) 60–67 |
| BILL (S) | | 27·9 (1·6; 20) 25–31 | | 26·8 (1·0; 25) 24–29 |

*A. t. melanosternus.* Sri Lanka, all year; skins (RMNH).

| | | | | |
|---|---|---|---|---|
| WING AD | ♂ | 141·0 (4·25; 10) 136–148 | ♀ | 138·8 (3·47; 10) 135–145 |
| BILL (S) | | 27·2 (0·77; 10) 26·1–28·3 | | 28·0 (1·11; 10) 26·5–29·4 |
| BILL (N) | | 15·7 (0·46; 10) 15·1–16·7 | | 15·7 (0·58; 10) 14·8–16·5 |
| TARSUS | | 39·1 (1·37; 10) 37·2–41·3 | | 39·1 (1·19; 10) 37·2–41·0 |

**Weights.** Nominate *tristis*. Kazakhstan: ♂ 127·5 (7) 123–130·5, ♀ 121·2 (7) 111–143 (Korelov *et al.* 1974). Afghanistan: March, ♂ 126; July–August, adult ♂ 122, 125, juvenile ♂ 114, adult ♀ 105, 116, 125, juvenile ♀ 94 (Paludan 1959). India: 109·8 (17) 82–130 (Ali and Ripley 1972).

*A. c. tristoides* Natal (South Africa): ♂ 123·3 (9·5; 8) 106–134, ♀ 109·5 (7·0; 8) 98–116 (Brooke 1976).

**Structure.** Wing rather short, broad at base, tip rounded. 10 primaries: in adult, p7–p8 longest, p9 2–5 shorter, p6 1–5, p5 6–11, p4 10–18, p1 31–39; in juvenile, p8 longest, p9 1–4, p7 0–3, p6 2–5, p5 7–11, p4 15–20, p1 31–37; p10 reduced, especially in adult, 85–91 shorter than p8 in adult, 75–86 in juvenile, 6·9 (10) 4–10 shorter than longest upper primary covert in adult, 6·1 (7) 3–8 in juvenile. Outer web of (p5–)p6–p8 emarginated, inner of (p6–)p8–p9 with rather faint notch. Tip of longest tertial reaches tip of p1. Tail rather short, tip rounded; 12 feathers, t6 5–13 shorter than t1. Bill rather short, visible part of culmen *c.* 45% of head-length; wide and deep at base, gradually tapering to tip, but distal end of culmen gently decurved; tip laterally compressed. Nostril narrow, covered by feathered operculum above. Patch extending from gape and below eye to large triangle behind eye bare. Feathers of crown narrow and elongated in adult and first adult, forming short crest; length of longest feathers on nape 16·2 (38) 12–21 in ♂, 14·3 (38) 12–18·5 in ♀ (Brooke 1976). Short feathers above nostril directed upwards. Leg and foot strong; middle toe with claw 33·9 (12) 29·5–36; outer and inner toe with claw both *c.* 72% of middle with claw, hind *c.* 82%.

**Geographical variation.** Slight, with only *melanosternus* of Sri Lanka distinct. On Asiatic continent, birds paler and greyer in north-west (Turkmeniya: named *neumanni*, Dementiev, 1957), gradually darker and more vinous towards southern India (nominate *tristis*) and Assam, but difference slight, in part also caused by difference in bleaching, and much individual variation. *A. t. tristoides* of high rainfall areas of Nepal and Burma is also darker, throat and chest darker grey, body more richly coloured (Brooke 1976). *A. t. melanosternus* of Sri Lanka much darker, upperparts and upper wing-coverts purplish-maroon; chin, throat, chest, and thigh darker greyish-black, dark tinge extending to mid-belly, side of belly and flank dark maroon; white at base of primaries much less extensive, hardly reaching beyond tips of greater upper primary coverts, latter with much black on tips (in adult nominate *tristis*, fully white), outer ones in particular sometimes largely black. Birds from extreme southern tip of India intermediate between *melanosternus* and nominate *tristis*. Introduced populations contain pure *tristoides*, pure nominate *tristis*, or intermediates between these, sometimes also apparent intermediates with *melanosternus*. CSR

# Family PASSERIDAE sparrows, rock sparrows, snow finches

Predominantly small, thick-billed oscine passerines (sub-order Passeres). Most found in open, dry or semi-arid country: bush, savanna, and even desert; also (e.g.) in forest, woodland cultivation, high mountain country (*Montifringilla*), and human habitation—some species (most notably House Sparrow *Passer domesticus* and, in Orient, Tree Sparrow *P. montanus*) being highly urbanized and among most successful of all birds. Mostly feed on or near ground, seeds (especially of wild grasses and cultivated grain) forming bulk of diet in many species, but other vegetable food and insects also taken (insects especially for young); bread and other man-made foods predominate in towns and villages. Some species come readily to bird-tables and feeders. Occur naturally in Old World only, with centre of distribution in Afrotropics, but 2 species of *Passer* introduced elsewhere (including New World and Australia). Mainly sedentary, with some minor local or seasonal movement, but some species (mostly in genera *Petronia* and *Montifringilla*) wholly or partially migratory. 34–37 species (C S Roselaar); 11 in west Palearctic, all breeding.

Following Sibley (1970), Voous (1977), Bock and Morony (1978), and Summers-Smith (1988), passerids here classified as a family distinct from Ploceidae (weavers, etc.), Estrildidae (waxbills, etc.), and other weaver-like birds on the one hand and from Fringillidae (finches) and Emberizidae (buntings and allies) on the other. 3 generally accepted genera, each representing major division of family (see Summers-Smith 1988): (1) *Passer* (true sparrows), 21–24 species—Africa and Eurasia; (2) *Petronia* (rock sparrows), 5 species—Africa and Eurasia; (3) *Montifringilla* (snow finches), 7 species—Eurasia (Pyrénées to China and Mongolia, some wintering in India). Pale Rock Sparrow *Carpospiza brachydactyla* of Eurasia also included here following Voous (1977, though retained there in *Petronia*), but its affinities evidently lie elsewhere—probably in Carduelinae (see species account, also Summers-Smith 1988).

Sexes almost of similar size (♀ smaller on average). Bill of seed-eater type: short and conical, tapering to point (for details of structure, see Ziswiler 1965, 1967a). Nostrils large, covered by short bristles and (partly) by feathers of forehead; rictal bristles present but poorly developed. Like Ploceidae and all or most other finch-like birds, passerids bill-flick to uncover food on ground. Wing fairly long to short, broad and bluntly pointed; 10 primaries, p10 considerably reduced (length 10–25% of longest, mainly 10–15% in adult, 20–25% in juvenile). Flight fairly swift and sometimes undulating, with rapidly beating wings. Display-flights of some sort occur in all genera except *Passer*, with elaborate song-flight in *Montifringilla*. Tail short, square-tipped or slightly forked (notched); 12

feathers; often flicked repeatedly down and up. Leg and foot short but stout; tarsus covered with large scutes in front, smooth and sharply ridged at rear. Foot not used in feeding, either to hold food or to scratch for it. Gait basically a hop (*Passer*, *Petronia*) or hop and walk (*Montifringilla*); *Carpospiza* sometimes runs, looking rather like lark (Alaudidae). Head-scratching by indirect method so far as known (see, e.g., Simmons 1957b, 1961a for *Passer*). Sparrows (at least) unusual in that they both bathe and dust frequently, often sociably, though both activities also performed independently, bathing often followed by dusting (e.g. Stainton 1982); bathe in typical 'stand-in' manner or, if water too deep, by 'standing-out' (see Simmons 1985); dusting especially prevalent on sunny days—similar wing actions used as in bathing, with addition of bill-flicking and foot-scraping movements (see Simmons 1954 for *P. domesticus*). Sunning (often by a number of birds together) also common in *Passer*: of both types (feather exposure and heat-basking); postures often simple but versions of both lateral and semi-spreadwing forms recorded—also semi-spreadwing in captive Snow Finch *M. nivalis* (Simmons 1986). Anting not yet reliably documented for any member of family though repeatedly claimed for *P. domesticus*.

Voice generally harsh; calls chirping, grating, wheezing, etc. Song in *Passer* and *Petronia* weak and unspecialized though (unlike that of Estrildidae) has advertising function; more developed in *Montifringilla* and decidedly finch-like in *Carpospiza*. Many species gregarious all year, nesting in loose colonies; others (including all *Montifringilla*) flock outside breeding season only. In *Passer* particularly, large communal roosts often formed outside breeding season. Monogamous mating system seems to be general rule in all 4 genera; detailed study of Rock Sparrow *P. petronia* (Ivanitski 1986) suggested that polygamy may be typical in this species but further investigation required. Pair-bond strong and lifelong in *P. domesticus* and *P. montanus*; no information for most other species, but long-term pair-bond perhaps typical of at least the sedentary species. In *Passer*, several ♂♂ will gather to pursue and display to one ♀; peck at ♀'s cloaca and sometimes attempt to copulate. Though highly gregarious, passerids are not typically close-contact birds, seldom or never clumping, even at roost; in *P. montanus*, however, 2 or more birds usually roost in same hole outside breeding season (same being observed in *P. petronia*), and members of pair often loaf together side-by-side. Allopreening not certainly recorded but may occur rarely in *P. montanus*. Nest built by both sexes: in *Passer*, typically a bulky domed unwoven structure with side entrance (sometimes extended into short tunnel) and lined with feathers, placed in bush, small tree, hole, or nest-box; in *Petronia* and

*Montifringilla*, nest also domed at times but often less elaborate—a shallow, feather-lined pad or cup placed in hole. Incubation usually by both sexes so far as known, but by ♀ only in Dead Sea Sparrow *P. moabiticus*. Young fed by both sexes. In *Passer* at least, old nest sometimes used for roosting; special such nest often constructed by *P. domesticus* at high latitudes in winter.

Plumages typically brown or grey, with dark streaks on upperparts in most species; head of ♂ (in many species) or of both sexes (in others) with contrasting patches or streaks of black, white, chestnut, or tawny and often including black bib. Sexes alike (*Petronia*, *Carpospiza*, *Montifringilla*, and many *Passer*) or markedly different (many other *Passer*). Often some seasonal variation, brown or grey feathers-tips concealing bright colours until abraded. Bill of ♂ typically black only when breeding. Juvenile plumage like adult's (adult ♀'s where sexes differ). Nestling naked in *Passer* and *Petronia* but covered with dense down in *Montifringilla* and *Carpospiza*); gape-flanges yellow, no mouth markings. Single annual adult post-breeding moult; complete, usually starting shortly after fledging of last brood. Complete post-juvenile moult,

starting within 1 month of fledging in the 3 typical passerid genera, but partial in *Carpospiza* (as in most cardueline finches).

Relationship of typical passerids to ploceids and other deep-billed, seed-eating oscines much debated. Both passerids and ploceids often formerly united as 2 subfamilies within Ploceidae but egg-white data (Sibley 1970) and other studies later suggested that, pending further investigation, passerids should be given family rank on own (as here). DNA data (Sibley and Ahlquist 1990), however, suggest that both passerids and ploceids should be again united as subfamilies within an enlarged 'Passeridae' which also includes wagtails and pipits (Motacillidae) and accentors (Prunellidae) as well as Estrildidae. This assemblage found to be distinct from other 9-primaried finch-like birds; latter (including Fringillidae, Parulidae, Thraupidae, Emberizidae, and Icteridae) all placed in enlarged family 'Fringillidae', but both assemblages assigned to same superfamily, together with Alaudidae (larks), Nectariniidae (sunbirds and allies), and some other wholly extralimital birds (see further in Volumes V and VII).

## *Passer domesticus* House Sparrow

PLATES 20 and 26 (flight)
[between pages 280 and 281]

Du. Huismus    Fr. Moineau domestique    Ge. Haussperling
Ru. Домовый воробей    Sp. Gorrión doméstico    Sw. Gråsparv

*Fringilla domestica* Linnaeus, 1758

Polytypic. DOMESTICUS GROUP. Nominate *domesticus* (Linnaeus, 1758), northern Eurasia from Britain and Scandinavia east to Sea of Okhotsk, south to western and northern France, Alps, Hungary, northern Rumania, Ukraine, Crimea, north-east Turkey, western Transcaucasia, northern slope of central and eastern Caucasus, Kazakhstan, Tien Shan, and northern Mongolia; *balearoibericus* Von Jordans, 1923, Mediterranean France, Spain (except north-west), Balearic Islands, Balkans from Yugoslavia and southern and eastern Rumania to Greece, and western and central part of Asia Minor; populations intermediate between this race and nominate *domesticus* occur south-west France and Portugal, and probably elsewhere where ranges meet; *tingitanus* Loche, 1867, Maghreb countries and north-east Libya; *biblicus* Hartert, 1904, Cyprus, and from Levant east to western Iran, north to south-east Turkey (along Syrian border) and (perhaps this race) eastern Turkey and central and eastern Transcaucasia, merging into *indicus* in Negev and Arava valley (Israel and Jordan) and perhaps in northern Saudi Arabia, merging into *hyrcanus* in Talysh area (south-east Transcaucasia), and into *persicus* in western Iran in zone running from Qazvin and Hamadan south through eastern Lorestan; *hyrcanus* Zarudny and Kudashev, 1916, northern Iran, north of Elburz mountains, and neighbouring part of south-west Turkmeniya (extralimital); *persicus* Zarudny and Kudashev, 1916, Iran, south and east of *hyrcanus*, east of *biblicus*, east to south-west Afghanistan and perhaps westernmost Pakistan (extralimital); *niloticus* Nicoll and Bonhote, 1909, Egypt, merging into *biblicus* in northern Sinai and into *rufidorsalis* in Nile valley near Wadi Halfa in extreme south. INDICUS GROUP. *P. d. indicus* Jardine and Selby, 1831, Elat (southern Israel), Arabia (except north-east), southern Afghanistan, and Indian peninsula, south of Himalayas, east to Burma, south to Sri Lanka; *rufidorsalis* C L Brehm, 1855, Nile valley of Sudan (extralimital); *hufufae* Ticehurst and Cheesman, 1924, north-east Arabia from Al Hasa area (Saudi Arabia) to northern Oman (extralimital); *bactrianus* Zarudny and Kudashev, 1916, Transcaspia, from Mangyshlak peninsula east to Lake Zaysan, north in Kazakhstan to *c.* 50°N, south to northern and eastern Afghanistan and north-west Pakistan, overlapping widely with nominate *domesticus* in Kazakhstan, Uzbekistan, and Tien Shan foothills (extralimital); *parkini* Whistler, 1920, Himalayas from Pakistan to Nepal (extralimital).

N.B. Locally forms hybrid populations with Spanish Sparrow *P. hispaniolensis*, of which the following appear to be more or less stabilized: × *maltae* Hartert, 1910, Malta, Sicily, and small islands off Sicily, including Pantelleria; × *italiae*

Vieillot, 1817, southern Switzerland (south of Alps) and northern and central Italy; also Crete (Greece). See Geographical Variation of *P. hispaniolensis*.

**Field characters.** 14-15 cm; wing-span 21-25·5 cm. Slightly larger than Tree Sparrow *P. montanus*, with 10% longer wings but proportionately somewhat shorter tail; averages slightly smaller than Spanish Sparrow *P. hispaniolensis*, most noticeably in Egyptian race *niloticus*. Heavy-billed, rather large-headed, robust passerine, suggesting tubby finch (Fringillidae) but differing in broader wings and square-ended tail; epitome of genus. ♂ boldly patterned: warm brown above, with mainly grey crown and black eye-stripe and bib contrasting with dull white cheeks, dark streaks over back, 2 wing-bars, grey rump, and greyish underparts. ♀ rather featureless: dull brown with indistinct pale supercilium and 2 wing-bars. Flight often fast and direct, with more whirring action than finches. Sexes dissimilar; seasonal variation only by wear in ♂. Juvenile separable at close range. 6 races in west Palearctic, showing gradual intergradation; 2 distinctive western races described here; for rest, see Geographical Variation. Around Mediterranean, populations hybridized with *P. hispaniolensis* occur, particularly in southern Switzerland, Italy, Malta, and Sicily (e.g. Italian Sparrow, × *italiae*, often considered a race of *P. domesticus*); see Geographical Variation of *P. hispaniolensis*.

ADULT MALE. Moults: June-October (complete); loss of feather-fringes produces brightest plumage from April onwards. (1) West, north, and central European race, nominate *domesticus*. Crown and centre of nape grey, tinged buff when fresh; short white streak over and behind eye, bordered narrowly above and at rear below with forward extension of bright chestnut nape, fringed buff when fresh; broad loral streak, line under eye, and large bib black, with last at lower extremities scalloped with whitish fringes when fresh and increasing in size as they wear off; bold cheek patch between bib and nape pale smoky-white, almost grey when fresh. Mantle, scapulars, greater coverts, tertials, and secondaries chestnut and buff, looking dully streaked when fresh but more sharply streaked black on brighter chestnut ground in breeding season; within these areas, distinctive wing pattern created by contrast of uniform chestnut lesser coverts with large, almost white tips to median coverts (forming bold upper wing-bar) and pale buff tips to greater coverts (forming indistinct lower wing-bar) and pale buff edges to tertials. Primaries black-brown, with chestnut outer webs when fresh and basal parts of these contrasting with dark primary coverts. Lower back, rump, and upper tail-coverts dull grey, with brown tinge when fresh but contrasting with mantle and inner wings. Tail black-brown, with buff fringes to all feathers. Underparts from below bib to under tail-coverts dull grey- or buff-white, with darker tone than cheek-patch and dark mottling under tail. Underwing pale buff-grey. Bill short, deep, and heavy, with curved culmen;

black in breeding season, horn-brown or horn-grey at other times. Legs pale pink, grey, or brown. ADULT FEMALE. Lacks ♂'s strong plumage pattern and bright colours, appearing at distance almost uniform dull buff- to olive-brown. At close range, shows subtle but distinctive pattern: quite deep, pale buff supercilium (most obvious behind eye), dull black-brown streaks on mantle and scapulars, buff-white tips to median coverts (forming less obvious upper wing-bar than on ♂), and buff tips to greater coverts and edges to tertials. As in ♂, dark primary coverts contrast with pale greyish-buff basal fringes of primaries. With wear, throat may show ashy central spot, streaks on upper back become stronger, and underparts paler. Bill horn, yellowish at base. JUVENILE. Resembles adult ♀ but has distinctive 'unfinished' look, often showing yellow gape-flanges, brown mottling on crown and rump, and whiter throat and belly. ♂ may show marked ashy-black spot on centre of throat. Bill pale-tipped. Adult plumage assumed September-October, but 1st-summer ♂ shows many whitish fringes on bib. (2) North-west African race, *tingitanus*. ♂ resembles nominate *domesticus* but grey crown streaked black, and cheek-patch and underparts rather cleaner. (3) Hybrids and secondary intergrades. Appearance variable. Mixed parentage with *P. hispaniolensis*, or racial intergradation, most obvious in invasion of crown by chestnut nape, heavier streaks on upperparts, and extension of bib to shoulder and on to fore-flanks. Such birds occur south into Sahara and north to Switzerland (Vaurie 1959); detected even in England (D I M Wallace).

♂ of grey-crowned races unmistakable in good view, but ♂ hybrids and all ♀♀ and juveniles subject to confusion with *P. hispaniolensis* (see that species). Distinction from *P. montanus* not difficult, with clear differences in head pattern, structure, and voice soon learnt (see *P. montanus*). ♀ and juvenile may also suggest ♀ and immature rosefinches *Carpodacus* to inexperienced observer, but lack streaked underparts and distinctive calls of that genus. Confusion with bunting-like Nearctic sparrows (*Ammodramus*, *Passerella*, *Melospiza*, *Zonotrichia*; in Emberizidae) is unlikely, except in case of such a bird hiding head or underpart pattern, but possible with rock sparrows *Carpospiza* and *Petronia* and even immature Lapland Bunting *Calcarius lapponicus* and Greenfinch *Carduelis chloris* (see those species). Flight distinctive, recalling medium-sized finch but differing in more rapid and less fluent wing-beats, making wings appear to whirr unnecessarily (for size of bird) and almost hurtling bird along at times. Track often noticeably direct when on passage or over long distance. Flying flocks usually densely packed, not straggling, with individual members hardly changing position, unlike finches. Stance half-upright to virtually horizontal, with large bill and head and often tail held above low-slung body. When perched, often ruffles body feathers and appears to 'sit down'. Gait a hop, varied by jumps between perches and apparent shuffling when feeding on ground; movements accompanied by flicks of wings and, especially, tail. Behaviour variable; noticeably bustling and bold, even tame around human habitation whose food sources cheekily utilized, but much less so, even shy and wary, when in farmland or wild country. Markedly gregarious, especially at autumn and winter concentrations of food or at roosts. Beware occasional confusing appearance away from habitation with true migration of northern populations.

Noisy; insistent voice exaggerates bustling mien, particularly during long breeding season. Common, loud call, particularly of ♂, 'chee-ip', 'chissip', or 'chirr-eep'; other notes include short 'teu teu' (beware again confusion with *Calcarius lapponicus*), a rattled twitter, and a low, husky 'chreek' when flushed. Chirping notes developed into emphasized, rather than rhythmic, series which serves as song.

**Habitat.** Greatly affected by enormous spread of range within recent historical time, changing breeding habits and diet, and close and flexible association with man; thus liable only to limited interspecific competition, though in parts of range competes with Tree Sparrow *P. montanus* and Spanish Sparrow *P. hispaniolensis* which show similar but much more limited tendencies. In west Palearctic, has spread in recent times and is now established throughout except in Iceland, some small Atlantic islands, the more arid tracts of North Africa, and high Arctic fringe. Avoids closed or dense vegetation, from forests to plantations, large thickets, reedbeds, and some high-density built-up areas, especially where structures are tall and lacking in ledges and vegetation. Except for seasonal foraging in cornfields and on other crops, usually avoids open terrain lacking in shrub, tree, or other cover, and, unlike some congeners, shows little attraction to either fresh water or sea coasts. Wherever constant food supply is assured by human activities shows remarkable indifference to climatic constraints, extending north to 10°C July isotherm fringing tundra, and tolerating extremes of heat, aridity, and moisture. Few places prove too wet, dry, cool, warm, or elevated for it, provided they attract continuous human settlement or activity, especially when accompanied by presence of livestock. Such an element forms the almost essential core of acceptable habitat, around which use may be freely made of wide variety of herbage, shrubs, bushes, trees, rocks, structures, and artefacts of almost any type, although very tall buildings and even tall trees are usually avoided. Subject to that, everyday movements include frequent vertical as well as lateral displacements (e.g. from ground to bushes, trees, or rooftops), but, except for seasonal shifts to (e.g.) harvest fields, birds remain within restricted home-range and rarely fly above lower airspace.

Mainly a ground feeder, and a keen dust-bather, but enjoys perching on sunny flat ledges or rooftops, and conducts much of busy social life in trees and bushes, hedgerows, or climbing plants along walls and rock faces. In view of successful commensalism with man, requires less time for foraging, which tends to yield higher energy inputs than for great majority of small birds. Subject to local interspecific competition for nest-holes.

In regions where other *Passer* compete in towns and villages *P. domesticus* usually wins, exhibiting adaptability almost unmatched in the avian world; successful breeding has even been recorded by birds which became trapped 600 m underground in coal mine, where they survived for some years (Summers-Smith 1985).

Habitat in economically advanced countries rarely includes natural types and consists largely of mosaic of artefacts and humanly modified environments; thus highly varied in composition and constantly changing (see detailed accounts by Nicholson 1951 and Summers-Smith 1963). However, no precise comparative data for extent to which newly created opportunities for food, shelter, and other requirements have been taken up, and the respective timelags in doing so. Data are also deficient concerning instances where habits and habitats are characteristic of a stage prior to adapting to commensalism with man. Extralimitally, central Asian *bactrianus* recorded avoiding proximity to human permanent habitations, while favouring high precipitous river banks, river floodland woods, and groves or thickets, feeding in cereal and other cultivated fields and in orchards, and roosting near water in shrubs and reedbeds. This seems closer to lifestyles of *P. hispaniolensis* and Dead Sea Sparrow *P. moabiticus*, suggesting a more primitive model, matched by evidence from introduced populations in New South Wales (Australia) (Hobbs 1955), uninhabited islands off New Zealand, and isolated cases elsewhere (Summers-Smith 1963). Such cases, however, seem not to occur in Europe.

**Distribution.** Has become established and spread, with man's assistance, on almost all continents and many oceanic islands (see Summers-Smith 1988 for details). In west Palearctic has spread north in 20th century, in wake of human settlements, and has recently expanded range in Israel and Egypt.

ICELAND. Breeding first recorded 1959. Annual breeding in west 1971–80; this population now extinct, but new breeding population established in south-west in 1984. (ÆP.) NORWAY. Recent spread in north (VR). FINLAND. In this century has bred more regularly in northern Lapland, where formerly bred only occasionally (Koskimies 1989). RUSSIA. Has spread north in 20th century; breeding on Murman coast by 1955–7 (Mikhaylov and Fil'chagov 1984). In 19th century spread east from Urals, reaching Pacific coast 1929 (Summers-Smith 1988). ISRAEL. Up to beginning of 20th century bred mainly in Mediterranean areas, with a few colonies in north and

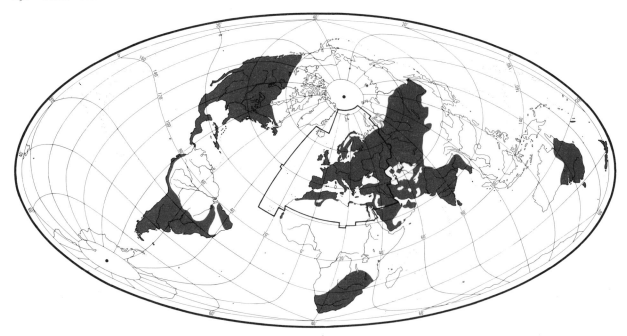

centre. Has since expanded to whole country, following development of new settlements. Population in extreme south (Elat and neighbouring areas) perhaps partly introduced by ships. (Shirihai in press.) EGYPT. Has recently colonized Dakhla and Kharga oases (Goodman *et al.* 1986). WESTERN SAHARA. Recently established at Layoune (MT). CAPE VERDE ISLANDS. Only on São Vicente; introduced (CJH). AZORES. Introduced 1960; now common on all islands, except Flores (colonized 1983) and Corvo and Santa Maria (absent in 1985) (GLG).

**Population.** Some evidence of local increase or decrease, but apparently no major recent changes except those consequent on spread of populations to new areas (see Distribution).

ICELAND. About 7 pairs 1987 (ÆP). BRITAIN, IRELAND. Probably 3·5–7 million pairs (Sharrock 1976). Estimated 5·5–6 million pairs (Lack 1986). FRANCE. Over 1 million pairs (Yeatman 1976). SPAIN. Recently decreased markedly in north, with abandonment of cereal cultivation and loss of nest-sites due to new building techniques (AN). PORTUGAL. May have increased in 1980s (RR). BELGIUM. 2 million pairs 1970 (Lippens and Wille 1972). Estimated 715 000 pairs 1973–7; probably too low, but perhaps some recent decrease (Devillers *et al.* 1988). NETHERLANDS. Estimated 1–2 million pairs (Teixeira 1979). GERMANY. 7·5 million pairs (G Rheinwald). East Germany: 4·5 ± 1·8 million pairs (Nicolai 1993). SWEDEN. Estimated 2·5 million pairs (Ulfstrand and Hogstedt 1976). FINLAND. Increased 1957–74, then decreased (OH). Perhaps 200 000–500 000 pairs (Koskimies 1989). CZECHO-SLOVAKIA. Local decreases in last 15 years (KH). ISRAEL. At least a few million pairs (Shirihai in press). AZORES.

Estimated 50 000–60 000 pairs in whole archipelago 1984 (Summers-Smith 1988).

Survival. Annual mortality calculated as 35–55%, depending on source of data and method of analysis. Post-fledging survival probably higher for ♂♂ than ♀♀. (Dyer *et al.* 1977.) Britain: adult mortality of ♂♂ 47 ± SE4%, ♀♀ 43 ± SE4% (Dobson 1987). Mortality highest in breeding season, *c.* 5% per month April–August, 2–3% per month in rest of year (Summers-Smith 1988). Calculated from February to following February, mortality of 1st-year birds 43%, adults 35% (Gramet 1973). Oldest ringed bird 12 years 11 months (Staav 1983).

**Movements.** Most races sedentary, especially in west of range. Juveniles disperse locally from natal area, but once settled remain within a few km. A small proportion, mainly juveniles, makes more directed migration, mainly to south and south-west but usually limited in extent. Larger-scale movements occur sporadically, mainly involving northern populations. For review, see Summers-Smith (1956*b*). Much migration appears to be diurnal, though casualties at Danish lighthouses suggest at least some nocturnal movement (Krüger 1944).

Nominate *domesticus.* Very small numbers recorded on passage at Falsterbo and Ottenby (Sweden), September–October, but rather irregular with peaks at Falsterbo 1943–4, 1950, 1959, though nil or insignificant numbers 1942, 1949, 1952–8, 1960 (no observations 1945–8, 1951), and peaks at Ottenby 1947–50, 1954, though nil or insignificant numbers 1951–3 and 1955–6. (Rudebeck 1950; Edelstam 1972; Ulfstrand *et al.* 1974.) Little evidence from Denmark to support Swedish observations, though migratory flocks recorded September 1937 and October

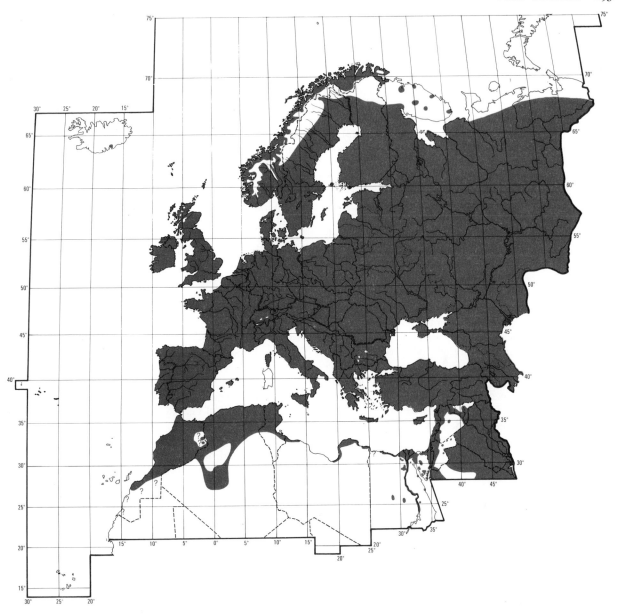

1938 (Krüger 1944); in north-west Germany, passage recorded Helgoland, where breeding population also leaves by September (Gätke 1891), and at Wilhelmshaven in 1949–50 (Bub and Präkelt 1952). Similar erratic movements also recorded for east and south-east coasts of England, though with peaks in only 14 years during 1879–1986. (Nelson 1907; Ticehurst 1909; Ticehurst 1932; Smith 1951, 1953, 1954; Smith and Cornwallis 1953*b*; Lack 1954*b*; Williamson and Spencer 1960; Cohen 1963; Mather 1986.) Only 2 peaks, 1950 and 1952, coincide with Swedish data. On west coast of Britain, small movements across Dee estuary seen at Hilbre Island mid-September to November (Bell 1962; Craggs 1976). Inland, movement noted across North Downs south of Newbury (southern

England) August–December, mainly October, 1962–3 (Lewis 1985). Recoveries of British-ringed birds in Belgium (*c.* 100 km east) and France (*c.* 130 km south) show some movement between Britain and continent (Spencer 1961). Passage occurs on coast of Bretagne (France) from mid-October into November (Guermeur *et al.* 1973), across Bay of Biscay (2 single birds landing on ships October–November), and along Biscay coast of Spain in October (Baker 1926; Snow *et al.* 1955). Inland movement through Swiss Alpine passes late September to early November involved 91% young birds ($n = 35$), with only 29·2% ♂♂ ($n = 144$), but overall numbers small (Jenni and Schaffner 1984). Some × *italiae* from north-west Italy move to Provence and Languedoc (southern France)

(Mayaud 1956); e.g. 2 birds ringed in winter in Camargue (southern France) recovered 300 km north-east in north-west Italy (Hoffmann 1955, 1956). These records, together with immigration of nominate *domesticus* into Corsica (Thibault 1963), confirm general picture of south to south-west movement in winter. Further east, movement seen between Azov and Caspian Seas in October 1942 (Bub 1955). Apart from report of *c.* 2000 flying east from Cyprus in September 1973 (Flint and Stewart 1992), no records from further south. East of Urals, most populations from north of Arctic Circle migrate in winter (Dementiev and Gladkov 1954), though some remain at high latitudes in Kola peninsula (Mikhaylov and Fil'chagov 1984), as they also do in Norway and Canada by coming into buildings (Summers-Smith 1988).

Ringing recoveries indicate movements involve only small proportion of population and small distances. Thus, in Germany, 90% less than 2 km and only 1% over 10 km (*n* = 881) (Preiser 1957); in Britain, 91% less than 2 km and only 3% over 10 km (*n* = 436) (Summers-Smith 1956b); see also Przygodda (1960); juveniles tend to make longer movements than adults (Rademacher 1951). Even birds passing through Col de Bretolet (Switzerland) moved only short distance: only 38% of 47 recoveries over 10 km, maximum 104 km and 174 km (Jenni and Schaffner 1984). Only 18 recoveries reported over 100 km; in addition to those mentioned above, include 2 recoveries of German-ringed birds in Netherlands and France (maximum 545 km) (Hansch 1938; Bub and Präkelt 1952; Hoffmann 1955, 1956; Preiser 1957; Spencer 1961, 1963, 1966b; Lewis 1985). Less evidence of return migration in spring, though a few thousand occur at Leucate (southern France) every year in April (R Cruon).

Two other types of movement are of greater interest. Many populations undertake movement from colony area in late summer to ripening grain fields; this can be up to 2 km, birds remaining in area of grain fields and returning to breeding areas September-October (Summers-Smith 1954b; Fallet 1958a; Cheke 1973). Second is dispersal: occurs October, April-May, and (juveniles wandering after becoming independent) June-August. Difficult to distinguish October and April-May movement from migration discussed above, but seems to involve birds that as result of overcrowding have failed to obtain breeding sites. (Summers-Smith 1956b; Preiser 1957; Craggs 1967, 1976; Cheke 1973; Bezzel 1985.) 3 records of birds on Baltic islands in April and 13 on British offshore islands March-June probably involve dispersal of this type (Krüger 1944; Summers-Smith 1956b). In American natal dispersal study, limited to 5 km, estimated 50-75% of juveniles dispersed 1-5 km, with more ♀♀ dispersing than ♂♂ (52% of 42 ♀♀ ringed, 27% of 70 ♂♂) (Lowther 1979b; Fleischer *et al.* 1984). Dispersal is key factor in recolonization of areas temporarily below breeding capacity (Preiser 1957; Heij and Moeliker 1991, who showed that urban populations in Amsterdam were maintained

by immigration from nearby suburban areas) and rapid colonization of areas to which they have been introduced. Rate of spread from Urals to east coast of Siberia *c.* 40 km per year; colonization of USA after introduction 20-30 km per year; spread in southern Africa from 1950 to 1980, 30-80 km per year (Johnston and Klitz 1977; Summers-Smith 1988, 1990).

Additional evidence of sedentary nature from translocation experiments in north-west Germany: of 143 birds (mainly adults, but could have included some 1st-years) ringed in November-March and transported 3-51 km, only 34% returned, maximum distance 38·5 km; none of 52 1st-years returned (Bub 1962).

Extralimital populations of nominate *domesticus* withdraw from extreme north in winter. Race *bactrianus* breeding in central Asia winters south to 25°N in Pakistan and India, moving south August-September and returning April-May; 6 ringed at Bharatpur in winter recovered in Kazakhstan 1500-2000 km away. Ringing studies in Chokpak pass (western Tien Shan), showed constancy of route and timing of individual birds in spring, with ♂♂ preceding ♀♀ by *c.* 5 days and adults returning 2-8 days earlier than 1st-years. *P. d. parkini* breeding in Himalayas is partial migrant, populations breeding at high altitudes withdrawing south in winter. Other populations show similar movement pattern to nominate *domesticus* in west Palearctic. (Dementiev and Gladkov 1954; Ali and Ripley 1974; Gistsov and Gavrilov 1984.)                     JDS-S

**Food.** Mainly plant material, though nestlings largely fed animal material during first half of nestling period, and some animal material taken by adults immediately prior to and during breeding season. Vegetable food principally seeds, but shoots, buds, and berries taken to lesser extent. Birds living in urban and suburban areas take wide range of household scraps (bread, cooked potatoes, peas, meat, etc.), elsewhere exploit food put out for domestic animals. Forages mainly on low plants or seeds on ground, though commonly perches on ripening cereal heads, frequently breaking stems (see Summers-Smith 1963). Seeds also taken directly from trees (e.g. birch *Betula*: Goodwin 1964). De-husks seeds by pressing them against palate with tongue (Ziswiler 1965, 1979). Elderberries *Sambucus* eaten by squeezing out juice and pulp, dropping skin and seeds, but, with harder fruits like *Pyracantha* and *Cotoneaster*, eats seeds and discards pulp (Snow and Snow 1988). Collects invertebrates by (e.g.) searching leaves and bark of trees (e.g. Dementiev and Gladkov 1954) for slow-moving prey, hovering close to plants and pouncing on prey, catching insects in flight, shaking leaf clusters by grasping with feet and fluttering wings to dislodge insects (Guillory and Deshotels 1981). For review of methods, see Summers-Smith (1963). Examples of opportunism and evolving techniques as follows. Searches spiders' webs for dead insects and car radiators for both dead and live trapped insects; latter first recorded in 1940s (Fitter 1949), but now apparently widespread (e.g. Hobbs 1955, Bankier 1984, Simmons 1984). Hovering at suspended nut-feeders, first specifically recorded 1964–5 (Sharrock 1974); clinging to feeders first reported 1968–70, initially to plastic-net feeders, but soon to wire-mesh feeders, to which they hold on with legs bent, head inclined sideways or down (Chatfield 1970; Spencer and Gush 1973; Jack 1974). As with nut-feeding, exploitation of milk in bottles on doorsteps probably copied from tits *Parus* (Fisher and Hinde 1949). Also recorded hanging onto bottom of tit-bell and feeding on fat (Thorpe 1956). Will catch moths (Lepidoptera) in flight at night, and mayflies (Ephemeroptera) attracted to lights (Malloch 1922), even 80 storeys up Empire State Building, New York (USA) (Brooke 1973). Numerous other unusual techniques reported: clasping food with foot (Tinbergen 1934); softening hard bread by soaking in water (Clarke 1949; Becher 1949); snatching food from bill of Wren *Troglodytes troglodytes* (Boyle 1966), American Robin *Turdus migratorius* (Kalmbach 1940), Mistle Thrush *Turdus viscivorus* (Lloyd-Evans 1948), and Starling *Sturnus vulgaris* (Fitter 1949); pushing aside stones under bird-feeder to exploit hidden seeds (Wade and Rylander 1982); kleptoparasitizing Blue Tits *Parus caeruleus* carrying bread (Summers-Smith 1963), and digger wasps *Sphex* carrying paralysed grasshopper (Acrididae) (Brockmann 1980).

Diet in west Palearctic includes the following. Invertebrates: bristle-tails (Thysanura: Machilidae), springtails (Collembola), mayflies (Ephemeroptera), damsel flies and dragonflies (Odonata: Agriidae, Lestidae, Aeshnidae, Gomphidae, Libellulidae), grasshoppers and crickets (Orthoptera: Gryllidae, Tettigoniidae, Acrididae), earwigs (Dermaptera), cockroaches (Dictyoptera), termites (Isoptera), psocids (Psocoptera: Mesopsocidae), larval and adult bugs (Hemiptera: Pentatomidae, Coreidae, Nabiidae, Cicadidae, Cercopidae, Psyllidae, Aphididae, Coccoidea), thrips (Thysanoptera), larval and adult lacewings, etc. (Neuroptera: Sialidae, Hemerobiidae, Chrysopidae), mainly larval and pupal, also adult butterflies and moths (Lepidoptera: Nymphalidae, Pieridae, Pyralidae, Tortricidae, Yponomeutidae, Coleophoridae, Gracillariidae, Noctuidae, Lymantriidae, Geometridae, Sphingidae), caddis flies (Trichoptera), flies, mainly larvae and pupae, but also newly emerged adults (Diptera: Tipulidae, Trichoceridae, Culicidae, Chironomidae, Bibionidae, Scatopsidae, Cecidomyiidae, Rhagionidae, Tabanidae, Asilidae, Empididae, Dolichopodidae, Phoridae, Syrphidae, Coelopidae, Helomyzidae, Tachinidae, Calliphoridae, Muscidae), fleas (Siphonaptera), larval, pupal, and adult Hymenoptera (sawflies Cephidae, Tenthredinidae, Ichneumonidae, Cynipidae, Mymaridae, ants Formicidae, wasps Sphecidae, bees Apoidea), larval, pupal, and adult beetles (Coleoptera: Carabidae, Hydrophilidae, Silphidae, Staphylinidae, Scarabaeidae, Byrrhidae, Elateridae, Cantharidae, Oedemeridae, Tenebrionidae, Coccinellidae, Cerambycidae, Chrysomelidae, Curculionidae), spiders (Araneae), millipedes (Myriapoda), earthworms (Lumbricidae), molluscs (*Hydrobia*, *Littorina*, Enidae, *Mytilus*). Vertebrates: small frogs *Rana* and mammal carrion, including meat fibres on bones and even dead *P. domesticus*. (*Countryman* 1946 **33**, 257; Hammer 1948; Attlee 1949; Brown 1959; Łącki 1962; Simeonov 1964; Encke 1965; Ion 1971; Grün 1975; Wieloch 1975; Metzmacher 1984; Sánchez-Aguado 1986.) Plant material includes the following (seeds unless stated), mainly seeds of low herbs and grasses, but also tree seeds and to lesser extent buds, flowers, and fruit: poplar *Populus*, birch *Betula*, elm *Ulmus*, plane *Platanus*, mistletoe *Viscum* (berries), hemp *Cannabis*, nettle *Urtica*, Polygonaceae, *Portulaca*, Chenopodiaceae, amaranth *Amaranthus*, Caryophyllaceae, buttercup *Ranunculus*, *Annona*, Papaveraceae, Cruciferae, mignonette *Reseda*, Rosaceae, mulberry *Morus* (fruit), Leguminosae (including flowers), wood-sorrel *Oxalis*, Geraniaceae, flax *Linum*, *Philodendron*, currant *Ribes*, spindle *Euonymus* (berries), Christ's thorn *Paliurus* (berries), *Elaeagnus*, mallow *Malva*, violet *Viola*, *Citrullus*, willowherb *Epilobium*, *Vaccinium*, thrift *Armeria* (buds), bedstraw *Galium*, Boraginaceae, Labiatae, Solanaceae including tomato (fruit) *Lycopersicon*, Scrophulariaceae, plantain *Plantago*, elder *Sambucus* (fruit), vine *Vitis* (grape), Compositae, Gramineae (wild grasses and cereals), reedmace *Typha*, Cyperaceae (Hammer 1948; Fitter and Lousley 1953; Howells 1956; Simeonov 1964; Encke 1965; Grün 1975; Rékási 1976; Dubery 1983; Alonso 1985*b*, 1986*b*; Sánchez-Aguado 1986). Additional plant material includes nectar taken from greengage *Prunus* (Hagger 1961), honey (Soper 1969), and linseed oil from putty (Lay 1970). Regularly attacks flowers (particularly yellow ones) and green leaves (e.g. dandelion *Taraxacum*) in spring, but not certain that these are eaten (Summers-Smith 1963, 1988). Minerals taken include salt from saltlick put out for cattle (Calhoun 1947*b*) and snail shells by ♀♀ forming eggshells (Schifferli 1977). See Barrows (1899) and Kalmbach (1940) for detailed studies of food taken in USA.

For seasonal variation in adult diet, see Table A. When available, clear preference for cereals obtained from ripening grain heads, spilled seed, and animal feed; in USA study ($n = 4848$ stomachs) 77·7% (by volume) of annual intake was cereal of which 60% estimated to be animal feed (Kalmbach 1940). Invertebrates taken during breeding season and in period immediately following. Food varies considerably with local availability, though a few families of plants and invertebrates predominate in each area, e.g. *Polygonum* around Sofia (Simeonov 1964, Table A), grass *Echinochloa* in Spanish study (Alonso 1986*b*); see also Sánchez-Aguado (1986). In Hungary, *Amaranthus* in 15·6% of 235 adult stomachs, *Setaria* in 14·0%, *Polygonum* 10·6%, *Chenopodium* 8·4%, beetles (mainly Curculionidae) 4·1% (Rékási 1976). In USA, beetles 2·0% by volume in 4848 adult stomachs (Kalmbach 1940).

For rate of food consumption, see Table B. Energy

Table A   Seasonal variation in food of adult House Sparrow *Passer domesticus* from analysis of crop and stomach contents; figures are percentages by dry weight.

| | Spain | | Bulgaria, vicinity of Sofia | | | | Bulgaria, industrial area | | |
|---|---|---|---|---|---|---|---|---|---|
| | Feb–Mar | Apr–May | Sep–Nov | Dec–Feb | Mar–May | Jun–Aug | Sep–Feb | Mar–May | Jun–Aug |
| Crop seeds | 82·2 | 96·4 | 38·3 | 63·0 | 35·9 | 11·2 | 62·3 | 73·1 | 51·2 |
| Weed seeds | 13·2 | 0·7 | 58·2 | 37·0 | 55·0 | 40·2 | 37·7 | 25·6 | 28·8 |
| Invertebrates | 4·6 | 2·9 | 3·5 | 0 | 9·1 | 48·6 | 0 | 1·3 | 20·0 |
| Total no. of birds | 16 | 29 | 145 | 106 | 89 | 89 | 82 | 106 | 89 |
| Source | Sánchez-Aguado (1986) | | Simeonov (1964) | | | | | | |

Table B   Estimates of average food consumption (kg dry weight per bird per year) of House Sparrow *Passer domesticus*, based on model using standardized populations (Wiens and Dyer 1977, which see for further details).

| | Finland | Germany | Poland | | | USA |
|---|---|---|---|---|---|---|
| | | | Gdynia | Dziekanów Leśny | Kraków | |
| Invertebrates | 0·48 | 0·55 | 0·13 | 0·16 | 0·96 | 0·09 |
| Weed seeds | 0·80 | 1·11 | 0·60 | 0·73 | 0·50 | 0·61 |
| Cereal grain | 1·87 | 2·37 | 2·82 | 3·10 | 1·94 | 2·17 |
| Totals | 3·15 | 4·03 | 3·55 | 3·99 | 3·40 | 2·87 |

Table C   Food of nestling House Sparrow *Passer domesticus*, by crop analysis (Bulgaria) and collar-sampling (other studies).

| | Bulgaria | | Germany | | Poland, rural | | USA |
|---|---|---|---|---|---|---|---|
| | vicinity of Sofia | industrial area | rural | town | Turew | Barniewice | rural |
| Orthoptera | | | 0 | 0·9 | 19·6 | 1·1 | |
| Hemiptera | | | 5·3 | 71·8 | 0·1 | 0·3 | |
| Lepidoptera | | | 17·6 | 8·0 | 5·5 | 5·3 | |
| Diptera | 96·1 | 63·6 | 36·4 | 11·4 | 5·0 | 18·0 | 90·3 |
| Coleoptera | | | 28·9 | 3·9 | 27·9 | 22·4 | |
| Other invertebrates | | | 0·7 | 3·0 | 26·2 | 33·2 | |
| Plant material | 3·9 | 36·4 | 11 | 1 | 15·7 | 19·7 | 8·6 |
| Grit | — | — | — | — | — | — | 1·1 |
| % by: | weight | | number | | dry weight | | volume |
| No. of birds or samples | 205 | 165 | 529 | 478 | 319 | 177 | 240 |
| Source | Simeonov (1964) | | Encke (1965) | | Wieloch (1975) | | Anderson (1980) |

demands for 14 populations (Europe and USA) 59 000–93 000 kJ per bird per year (Wiens and Dyer 1977, which see for further details).

Food of nestlings predominantly insects (Table C), though in industrial area of Bulgaria over one-third plant material. Considerable variation in type of insects taken, locally and from year to year (Wieloch 1975). Proportion of plant material tends to increase with age of nestlings, though considerable variation between studies. According to Wieloch (1975), small, soft invertebrates, larvae, and adult insects with soft bodies or with hard parts removed are fed for first 5 days; older young fed larger, more chitinous insects. Mean length of prey items fed to nestlings in USA 7·3 mm ($n = 240$), or 10·7 mm ($n = 185$) for 6 most important taxa accounting for 74·1% by volume food (Anderson 1980).   JDS-S

**Social pattern and behaviour.** Extensively studied in Europe (Daanje 1941; Summers-Smith 1954a, b, 1955, 1958, 1963; Fallet 1958a, b; Deckert 1969; Barnard 1979, 1980a, b, c; Barnard and Sibly 1981) and extralimitally (North 1968; Fleischer et al. 1984; Ivanitski 1985b).

1. Gregarious, with dispersion throughout year based mainly on small, loose, but discrete colonies (see Breeding Dispersion, below). In areas of low breeding density and favourable feeding, colonies remain isolated from neighbouring ones except for minor contact, but at higher densities and where food more dispersed (e.g. urban areas) birds from different colonies associate at feeding areas within radius of c. 500 m. Once independent, young form small foraging flocks which later coalesce at suitable feeding places into larger aggregations of young from several neighbouring colonies. These aggregations grow as more young fledge and are joined by adults that have finished breeding. Members of late summer and autumn feeding association roost in feeding area and, for brief period, breeding colony area may be deserted. Once adults have completed annual moult they return to breeding colonies to repossess their old nest-sites. Mates remain together. Young birds remain longer in feeding association, still roosting in feeding area, but gradually disperse locally, attaching themselves to colonies that made up autumn feeding association, some returning to natal colony. Most are back in colony areas by October. Once birds have joined colony, majority remain faithful to it for life, though there may be a 2nd local dispersal at beginning of breeding season by birds that have failed to obtain nest-sites and mates. Late summer and autumn flocks may grow to several thousand birds on edges of towns and prime arable farmland, but size depends on local breeding density, individuals seldom moving more than 1–3 km from colony area. Apart from late summer aggregations, most are generally in flocks of up to c. 30 birds, though outside breeding season may associate with large feeding flocks of Tree Sparrow *P. montanus*, finches (Fringillidae) and buntings (Emberizidae). (Summers-Smith 1954b, 1956b, 1963; Fallet 1958a, b; Heij and Moeliker 1991.) BONDS. Largely based on Summers-Smith (1958, 1963, 1988, which see for further details), supplemented from Daanje (1941) and Deckert (1969); see also Breeding. Essentially monogamous (Daanje 1941; Novotný 1970), with most pairs remaining together for life and normally re-using same nest-site; e.g. colour-ringed pair bred together in same hole for 6 years. Some polygamy occurs: both bigamy—♂ exceptionally maintaining 2 ♀♀ in nearby nests (Summers-Smith 1963) or 2 ♀♀ laying in one nest (Boxberger 1930; Pearse 1940)—and casual polyandry by ♀ soliciting copulation from ♂ other than mate (Summers-Smith 1958); such promiscuity may be quite frequent (see Heterosexual Behaviour, below). Pair-formation takes place at nest-site. Adult ♂♂ return in autumn to nest-site of previous breeding season where joined by mate or acquire replacement mate, either widowed ♀ from nearby nest or one of older young born that year. For infanticide by widowed ♂ or ♀ to facilitate re-pairing, see Heterosexual Behaviour (below). At same time, 1st-year ♂♂ investigate possible nest-sites (see Roosting below), and increasingly adopt sites from late January onwards, with age of first breeding typically 1 year. Both sexes build nest, incubate, brood young, and feed them for up to 2 weeks after fledging; within broods, same fledgling never seen being fed by both parents, suggesting brood-division may occur, but not proved. ♀ recorded raising young after death of mate. Hybridizes extensively with Spanish Sparrow *P. hispaniolensis* in North Africa and sporadically elsewhere in areas of overlap; stabilized hybrid populations (*P. d. italiae*) occur in Italy, Corsica, and Crete (Meise 1936; Johnston 1969a). Hybridization with Tree Sparrow *P. montanus* less common, but regularly reported (Cordero and

Summers-Smith 1993). BREEDING DISPERSION. Mainly in loose colonies of 10–20 pairs, with occasional isolated pairs. Colony typically spread over 0·25–0·5 ha, separated from others by empty spaces of apparently identical habitat. When nesting in holes, spacing determined by their availability, but otherwise nests may be joined together (e.g. in cliff face or tree: Baxter and Rintoul 1953; McGillivray 1980; Ivanitski 1985b), but with nest-entrances no closer than 10–20 cm (Novotný 1970). Colony size apparently not determined by availability of nest-sites: provision of nest-boxes at one study area did not increase number of breeding pairs, though free-standing nests in trees were abandoned in favour of more secure enclosed nests. Young birds surplus to holding capacity of colony disperse to seek vacancies in neighbouring colonies and even maintain non-breeding surplus from which losses in breeding pairs are quickly made up (see above). (Clark 1903; Summers-Smith 1954b, 1963; Schifferli 1978; Anderson 1991; Heij and Moeliker 1991.) Formation of mixed colonies with *P. montanus*, over which *P. domesticus* is dominant, further suggests colony size not determined by availability of nest-sites (Cordero and Rodriguez-Teijeiro 1990). Ultimate factor limiting colony size presumably availability of food: e.g. colony of 5–7 pairs bred on Hilbre Island, Cheshire (England) when tenant present with horse and hens, but died out in 1957 after stock removed, only to become re-established in 1968 when new tenant with stock arrived; population size apparently determined by winter availability of grain put out for animals (Craggs 1967, 1976); see also (e.g.) Baxter and Rintoul (1953). Density depends not only on habitat and latitude, but for semi-colonial species also on size of area studied, small areas tending to give exaggerated values (Table A). Some evidence for reduction in density in 20th century. Thus, in one area of central London (England), fall of 60–70% between 1925–6 and 1948–9; attributed to loss of food through disappearance of horses from streets. More recent decline apparent, though less well quantified, in several areas: in London suburban area, fall of 20–30%

Table A  Breeding densities of House Sparrow *Passer domesticus* in various habitats of western Europe. Densities are given in breeding birds per km²; sample sizes are number of studies averaged. (Based on data from Pinowski and Kendeigh 1977, analysed by Summers-Smith 1988, which see for further details.)

| | All results | | | Results for areas >50 ha | | |
|---|---|---|---|---|---|---|
| | Mean | Range | n | Mean | Range | n |
| Houses with livestock | 3952 | 1714–5428 | 3 | | | |
| Towns: residential areas | 1145 | 10–3020 | 23 | 936 | 730–1080 | 6 |
| Towns: commercial areas | 909 | 268–1670 | 12 | 680 | | 2 |
| Villages | 724 | 14–3810 | 21 | 354 | 14–1292 | 9 |
| Towns: suburban areas | 611 | 44–1884 | 19 | 196 | | 2 |
| Parks in large towns | 394 | 14–3600 | 29 | 115 | | 1 |
| Villages with open fields | 338 | 20–1096 | 18 | | | |
| Old orchard | 160 | | 1 | 160 | | 1 |
| Riparian areas/cemeteries | 124 | 24–360 | 11 | 42 | | 1 |
| Towns: small allotments | 64 | 10–354 | 6 | | | |
| Deciduous forest | 68 | 38–90 | 3 | | | |

between early 1960s and 1970s. (Simms 1962, 1975; Summers-Smith 1988.) One population in 10 ha near Ankara (Turkey) had breeding density of 440 birds per km² (Kiziroğlu *et al.* 1987). ♂♂ normally retain nest-site for life (Summers-Smith 1954*b*, 1958; Deckert 1969; see also Bonds). ROOSTING. For roosting of recently fledged young, see Relations within Family Group (below). Independent young roost communally near foraging area in dense trees or bushes, more rarely in reeds (Holmes and Wright 1969) or long grass (A E Chapman), later being joined there by adults that have finished breeding. In late September, adults return to colonies and roost in old nests, sometimes pair-members together, but more frequently ♀ in nest, ♂ in cover nearby. Young returning to colonies in early autumn form communal roosts there in thick thorn hedges, trees, or bushes; as leaves fall and cover reduced, move to more secure sites, e.g. evergreens, ivy-covered walls, interiors of buildings, particularly if heated (Kalmbach 1940; Jones 1950; Rait Kerr 1950; Woodward 1960*b*). Increasingly thereafter, particularly from early spring, young take over possible nest-holes and use them for roosting. Nest-material may be added in early winter and this has given rise to idea of roost nests (i.e. nests built specifically for roosting in cold winter areas), particularly as these may not be used subsequently for breeding (e.g. Mayes 1926, Kalmbach 1940, Janssen 1983). This, however, could merely be facet of autumn sexual activity and inexperienced young birds using sites unsuitable for nesting, though such insulated sites could clearly be of advantage for roosting. Street lamps in towns are favoured (Lancum 1928; Wynne-Edwards 1927; Fitter 1949). Occasionally roosts on ground, e.g. on floor of coalhouse (Woodward 1960*a*), or even below ground level, birds entering subterranean transformer vault in Missouri (USA) through grating flush with ground (Marti 1974). Communal roosts normally contain up to *c.* 100 birds, though in towns larger numbers occur, e.g. in London up to *c.* 1000, frequently in association with Starling *Sturnus vulgaris*, coming from at least 2·25 km (Cramp *et al.* 1957), 1800–2000 in Berlin in January (Rinnhofer 1965), 3400 birds in Boston (USA) in August (Townsend 1909), 6000 in Lima (Peru) in August (Leck 1973), 29 000 south London in August (Summers-Smith 1963), even 100 000 in Cairo (Egypt) in August, drawn from up to 6·5 km away (Moreau 1931). Roost in Las Ramblas (Barcelona, Spain) estimated to hold 20 000 birds in spring 1925 (Henrici 1927) still in use over 30 years later, despite increased street lighting and traffic (Westernhagen 1956). Communal roosting is a social phenomenon, not determined by environmental factors (Graczyk 1961), though very large town roosts may result from lack of suitable sites. Some communal roosts maintained all year (Cramp *et al.* 1957), presumably involving non-breeding birds.

2. Wary, rather than shy, but if food regularly provided (e.g. in towns) will boldly come down to open-air cafés to feed among tables and even perch on humans to take food (Cramp and Teagle 1952; Bannerman 1953*a*); will even enter buildings (London Natural History Society 1957; Felton 1969). Wariness shown by avoidance of new food supply until birds become accustomed to it (Crawhall 1952; Radford 1966); similarly wary of unfamiliar objects placed near accustomed feeding place (Radford 1971; Galloway 1972). FLOCK BEHAVIOUR. Flocks form on ground only for feeding and bathing, staying close to cover which can be used as refuge when disturbed. Thus, birds feeding on ripening grain spend much time in hedge into which, on alarm, they fly together as flock, trickling back in smaller numbers when danger passed; feeding bouts rarely last more than 1–2 min, birds seldom venturing more than 5–10 m from hedge so that band of standing grain adjacent to hedge may be completely destroyed while rest of crop untouched (Summers-Smith 1963). Feeding flock con-

tains up to 30 or more birds, the larger the flock the further from cover. Vigilance by individual birds decreases with flock size up to 12–16 birds. (Barnard 1979, 1980*a*, *b*, *c*; McVean and Haddlesey 1980; Barnard and Sibly 1981; Caraco and Bayham 1982; Elgar and Catterall 1982; Elcavage and Caraco 1983; Studd *et al.* 1983, which see for further details on feeding and vigilance strategy.) Water- and dust-bathing are also performed in small groups (Summers-Smith 1963), though up to 200 seen dust-bathing at once (Simmons 1954). After such activities, flock remains together preening in nearby tree. Aerial predators greeted by Alarm-call (see 4a in Voice): flock-members dash to cover, where they fall still and silent. Ground enemies are greeted with different Alarm-call (see 4b in Voice), alerting other flock-members which keep enemy in sight, only flying off if it approaches too closely (Daanje 1941; Summers-Smith 1963; Deckert 1969). Feeding or dust-bathing birds threaten rival approaching closer than *c.* 10 cm, but individual distance of only a few cm tolerated by resting birds (J D Summers-Smith). At small local communal roosts, birds arrive 2–3 at a time (McAtee 1950); for larger roosts with larger catchment areas, flocks gather at pre-roost collecting points, where there is Social-singing (see 2c in Voice), giving rise to so-called 'chapels', and move to roost in groups of 10–50 (Moreau 1931; Cramp *et al.* 1957; Konradt 1968). At larger roosts, birds may perform aerial evolutions before finally entering roost (Parry 1948; Holmes and Wright 1969) and noise before settling down is considerable (Henrici 1927; Moreau 1931), more intense than Social-singing, with Threat-calls (see 3 in Voice) mixed in. Even in winter birds arrive at roost up to 1–2 hrs before sunset (Townsend 1909; Parry 1948; McAtee 1950; Konradt 1968; North 1968), suggesting birds have no great difficulty in finding food. SONG-DISPLAY. No true song, but ♂ gives Advertisement-call (see 2a in Voice) at nest-site or from nearby perch during period of autumn sexual recrudescence (September–November) and again in spring, starting late January, to proclaim ownership. Unpaired ♂ switches to Song-display (see 1a in Voice) at approach of possible mate, adopting Advertising-posture with chest thrust forward, showing off black bib, wings held out slightly and lowered and partly rotated towards ♀, showing off white wing-bar, and tail raised and fanned, with grey rump feathers ruffled (Fig A). In this posture ♂ hops round ♀, bowing

A

stiffly up and down; at high intensity, may shiver wings. ♂♂ call regularly at nest in mornings in autumn and increasingly from early spring up to breeding season, unpaired ♂ calling almost incessantly at his nest during day in breeding season (up to 80–90% of daylight hours). (Daanje 1941; Summers-Smith 1963.) ANTAGONISTIC BEHAVIOUR. Defends food, dust-bathing hollow, and nest against rivals. Initially uses Forward-threat display in which head lowered and thrust forwards at rival, wings held out slightly from body, and feathers sleeked. In intense threat, ♀

B

C

new mate quickly from non-breeding surplus (see Breeding Dispersion, above, for references). Activities helping to maintain pair-bond in early spring include cleaning out nest-site and adding new material, but bond not fully established until breeding starts. Away from nest, established pairs clearly recognize each other, tolerating close approach (e.g. feeding side by side) when other birds threatened and chased off. Mates of both sexes lost during breeding season are quickly replaced, remaining bird retaining nest, though occasionally ♂ takes over widowed ♀ at nearby nest and maintains bigamous relationship for one breeding season (see Bonds, above). However, in some cases such a widowed ♀ may destroy eggs or young (infanticide) of neighbouring pair to induce ♂ to desert and obtain his sole attention. Widowed ♂ also sometimes commits infanticide (eggs or young) of neighbouring pair to induce ♂ to desert and leave ♀ to pair with him. This occurs even when non-breeding ♀♀ available, so ♂ perhaps favours experienced ♀ because she will more quickly produce new clutch (Veiga 1990). (3) Courtship-feeding. None reported. (4) Mating. Based mainly on Daanje (1941) and Summers-Smith (1958, 1963). By far the greatest number of observed copulations are at nest-site. ♂ gives Solicitation-display to ♀ (Fig D), similar to Advertising-posture except that he

gapes at opponent (Fig B); ♂ raises tail and rotates wings markedly so that scapulars point forward (Fig C). If rival does not flee, aggressor lunges at it and if, as happens occasionally, other bird remains, fight can develop; in attacks at nest-site, rivals may fall to ground, locked together. Fights of this kind usually only occur between birds of same sex; rarely, one bird is killed (Biedermann-Imhoof 1913). ♂ attacks only unfamiliar intruder at nest-site, not recognized neighbour. In Group-display (see below), ♀ threatens displaying ♂♂ in Forward-threat, biting at those that come too close. Threatens own and other species up to size of Blackbird *Turdus merula* at foraging site. Larger ♂ normally dominant over ♀, but during breeding season ♀ becomes more aggressive and uses Forward-threat towards mate at nest-site (see Heterosexual Behaviour, below) and ♂♂ at feeding sites. (Daanje 1941; Summers-Smith 1954a, 1955, 1963; Amadon 1967; Johnston 1969b; Kalinoski 1975; McGillivray 1980.) HETEROSEXUAL BEHAVIOUR. (1) General. Pair-members frequently together both at nest and away from it when feeding (Summers-Smith 1963). (2) Pair-bonding behaviour and nest-site selection (based mainly on Summers-Smith 1954b, 1955, 1958, 1963; Deckert 1969). Most pair-formation takes place by replacement of lost mate, either on return to nesting colony in autumn or later by ♂ Advertisement-calling (see 2a in Voice) at nest-site. New mate can be widowed ♀ from nearby nest or ♀ hatched earlier in same year, but in approach to and during breeding season widowed ♀ can retain nest and attract ♂ (see below for use of infanticide to induce pair-formation). New pairs of young birds also formed in spring by 1st-year ♂♂ taking up nest-site and attracting mate by Advertisement-calling. Pairs normally form quickly, though the longer an unpaired ♂ has been calling at nest-site the more reluctant he is to accept ♀, showing nest but preventing access for possibly 2–3 days. Once pair established, ♀ roosts with ♂ in nest or even displaces ♂ from it. During breeding, pairs perform Solicitation-display (see Mating, below, for description) on nest-relief or arrival of mate at nest with food for young, though ♀ tends to be dominant and may give Forward-threat rather than solicitation. Solicitation-display becomes abbreviated and is omitted with increasing familiarity of pair. Suitable nest-site appears essential in pair-formation, as ♂ with unsuitable site or site on edge of colony may fail to attract ♀ and continues to give Advertisement-call until June(-July) (Martin 1939). This is not through lack of ♀♀, as ♂ with suitable site who loses mate attracts

D

crouches and shivers wings. ♀ not ready to accept mate performs Forward-threat display and flies off chased by ♂. This attracts neighbouring ♂♂ who follow, and Group-display occurs when ♀ lands in bush or on ground: ♂♂ hop round ♀ in Solicitation-display posture, attempting to peck her cloaca (e.g. Hardy 1932, Harber 1945a); ♀ threatens with Forward-threat any ♂ approaching too closely and may fly off, again pursued by ♂♂, and process repeated; occasionally, forced copulation occurs, but usually unsuccessful (e.g. Cooke 1947), though Møller (1987) saw 3 successful attempts in 69 displays over 2-year study. ♀ ready for mating performs Solicitation-display (Fig E): crouches,

E

shivers wings, and gives soft call (see 2e in Voice). ♂ also then gives soft call (2e in Voice), mounts (sometimes 20 or more times), frequently pecking at ♀'s nape and at ♀'s cloaca before mounting or after initial mount. Copulation seen up to 2 months before 1st egg laid (Summers-Smith 1958; North 1968), but most frequent from 10 days before laying, when there can be 40 copulation bouts per day (Birkhead *et al.* 1987). Occasionally, ♀

solicits copulation from ♂ away from nest (Geyr von Schweppenburg 1942a; Simms 1948); Summers-Smith (1963) observed this twice in 7-year study. Presumably, ♂ involved is not always mate, as DNA-fingerprinting of broods has shown genetic mis-match (Burke and Bruford 1987; Wetton et al. 1987). Such mis-match is not considered result of egg-dumping (though egg-dumping witnessed, Brucker 1985, and inferred by Manwell and Baker 1975 and by Kendra et al. 1988, which see for further details), with one study giving 13·8% of nestlings in 183 broods not sired by ♂ at nest (Wetton and Parkin 1991b), though rate of observed extra-pair copulations (1 in 58) much less than extra-pair fertilizations (73 in 536) (Wetton and Parkin 1991a). Repeated copulation, and cloaca-pecking to induce ♀ to eject sperm, are presumably attempts to ensure paternity. Polyandry by ♀ (see Bonds, above) is presumably insurance against infertile mate. (5) Behaviour at nest. Unpaired 1-year-old establishing ownership of nest-site will begin to add nest-material, but once he has attracted ♀ both share in building until structure complete; pair also share incubation (see Breeding). Especially during early phase of nest-occupation, nest-relief accompanied variously by Solicitation-display (see above) and by Nest-relief call (see 2d in Voice) by arriving ♂ if ♀ present, while ♀ more often gives Threat-call (see 3 in Voice). These displays and calls wane as pair-bond strengthens. Occasionally (especially just after young have hatched), ♂ passes nest-material or food to ♀. (Daanje 1941; Summers-Smith 1955, 1963.) RELATIONS WITHIN FAMILY GROUP. Both sexes brood young. Faecal sacs disposed of by both sexes, swallowed for first 1–2 days, but then carried away (Seel 1960, 1966). Both sexes feed young, which a few days before fledging come to nest-entrance to be fed. Rate of feeding falls off on day young fledge, particularly in ♂ who at this time performs Song-display at nest to stimulate ♀ to begin new brood, though this perhaps also induces young to fledge. Young occasionally return to nest to roost for up to 4 days after fledging, otherwise roost with ♂ until independent. (Marples and Gurr 1943; Summers-Smith 1963; North 1973.) Fledged young fed by both parents up to 14 days, before joining flocks of independent young; up to independence, feeding mainly by ♂, but also by ♀, particularly in case of last brood. (Weaver 1942; Summers-Smith 1963; Deckert 1969.) ANTI-PREDATOR RESPONSES OF YOUNG. Young in nest become silent on hearing Alarm-calls of parents (see in 4a–b in Voice) and 2–3 days before fledging (Sappington 1977). PARENTAL ANTI-PREDATOR STRATEGIES. ♂ giving Threat-call (see 3 in Voice) at nest can attract up to 8 other conspecifics of both sexes from neighbouring nests, all then threatening or mobbing predator (Simmons 1952; Cordero 1991) or other threat, e.g. Mistle Thrush Turdus viscivorus stealing grass from free-standing nest (Summers-Smith 1958). Parents lead fledglings to cover by enticing them with food, but otherwise no particular strategies apart from same Alarm-calls used in flock (see above).

(Figs A–E by R Gillmor from drawings in Summers-Smith 1963.) JDS-S

Voice. Extensive range of calls used at nest and elsewhere. Detailed studies by Daanje (1941), Summers-Smith (1963), Deckert (1969), and R B Warren. Following account based on these except as indicated. Unacknowledged onomatopoeic renderings follow Summers-Smith (1963). For additional sonagrams, see Bergmann and Helb (1982). Other Passer show great similarity in calls used, repertoire of all species comprising 4 basic call-types which vary in structure, intonation, etc., according to situation and species. (a) Chirp-type. A 'chirp' is most

characteristic Passer call, consisting essentially of initial 'ch' ('tch'), middle 'r' or 'rr' (sometimes lacking), and final 'p'; may vary in vowel sounds and be 1–2 syllables ('r' separated from 'p'). (b) Churr-type. A 'churr', initial 'ch' followed by higher-pitched extended chatter or rattle, used in threat. (c) Nasal-type. Series of short, nasal calls with often a predominantly 'u' or 'ew' sound used in alarm. (d) Pleading-type. Quiet sound, predominantly with drawn-out 'ee', used in solicitation between family-members (mate to mate, young to parent).

CALLS OF ADULT. (1) Song. (1a) Song of ♂. Loose sequence of basic chirps and variants (see 1b), given by unpaired ♂ at nest to attract ♀ (Fig I), speeding up and

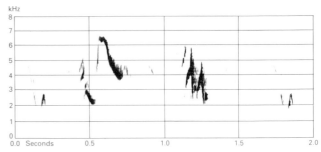

I P J Sellar England April 1992

becoming more excited on her approach; also given by paired ♂ at end of nestling period to induce mate to start another clutch, and by ♂♂ in Group-display. (1b) Subsong. Quiet rambling sequence of chirps, etc., e.g. 'chirrup-tee-chirrup-chirp-chirp' or 'chirri-pip-pip'. Not common; given by ♂ away from nest at any time of day and appears to express contentment. Similar version, mainly from juveniles (aviary-bred and wild) heard up to November, consists of sounds like Greenfinch Carduelis chloris, Linnet C. cannabina, or Canary Serinus canaria (Sick 1957; Wotkyns 1962; J D Summers-Smith) and possibly learned from one of these (Wickler 1982). (2) Other advertisement- and contact-calls. (2a) Basic 'chirp' and variants, e.g. 'chirrup', 'chirrip' (in Fig I, penultimate figure is of this type), 'cheerup', 'chee-up', or 'chillip' (with 1st sub-unit accentuated), or shortened to monosyllabic 'chirp' (Fig II, in which 2 sub-units so close as to be heard as single sound), 'chweep', 'cheep', or

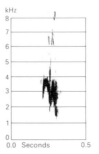

II P A D Hollom Scotland May 1975

'cheerp' ('schilp': Deckert 1969). Uttered at rate of *c.* 1 call per 2 s to 2 calls per 3 s. Used by ♂ at nest-site to proclaim ownership and to attract ♀ (less often by widowed ♀), and, rather more quietly, away from nest at all times of year as flock contact-call, possibly to promote social cohesion. (2b) Flight-call. 'Chirp' (with variants as described above), but more often disyllabic 'churrip', 'churrit', or 'turrip' with deeper 'u' sound in 1st syllable and 2nd syllable accentuated. Used in flight, presumably to maintain contact. Also rendered monosyllabic 'siep' (Daanje 1941), 'jerk' (Deckert 1969), soft 'tch' (Condor 1947), possibly corresponding to higher-pitched and more far-carrying 2nd syllable of disyllabic representations given above. Monosyllabic 'chip' used as flight take-off call; 'siep' (Daanje 1941), 'ssiep' (Deckert 1969). (2c) Social-singing. Mainly quiet chirping sounds, regularly heard outside breeding season, particularly on winter afternoons, a number of birds collecting in tree and calling together. Unlike Social-singing in roost, no threat-calls used and performance sounds conversational. According to Daanje (1941), only ♂♂ take part, but more probably both sexes. Possibly used to maintain social bond of colony. (2d) Nest-relief call. Quiet 'chee', 'tchee', 'dee', or 'pee' ('die die die': Daanje 1941) given by relieving bird at nest, usually accompanied by wing-shivering. Similar to juvenile food-call (see below). Usually given in groups of 2–5 calls. Apparently an appeasement-call used most frequently at beginning of breeding cycle and less often later as mates become more familiar. Most often used by ♂, dominant ♀ normally using call 3. (2e) Copulation-invitation call. Quiet 'quee-quee-quee'; whispered 'iag-iag' (Deckert 1969), 'tee-tee-tee' (Beven 1947). Characterized by drawn-out 'ee' sound, usually given in groups of 2–5 calls, similar to call 2d but even quieter. Used by both sexes as preliminary to copulation. (3) Threat-call. Rattling 'churr-r-r-it-it-it-it' or 'chit-it-it-it-it' or intermediate 'chur-tit-tit-tit'; 'terrettettet' (Daanje 1941), 'terterret' (Deckert 1969), 'chib-ib-ib-ib' (R B Warren). Fig III shows 2 rattles, 1st of which has intro-

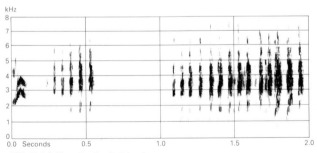

III  P J Sellar  England  March 1977

ductory 'chur-', 2nd without. Given by ♀ arriving at nest during period when she is dominant over mate. Also used against both conspecific and other intruders at feeding sites (Fig III depicts call in this context), and at nest.

Heard frequently from roost as birds settling down, presumably birds disputing perches. (4) Alarm-calls. (4a) Against aerial predator. A 'kruu'. On hearing this, flock-members take cover and fall silent. (4b) Against ground-predator. Nasal, staccato 'quer-quer' (Fig IV

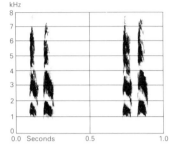

IV  E D H Johnson  South Africa  September 1969

shows 'quer-quer quer-quer') with variants: 'cher-cher', 'quer-it', 'quer-ik', 'ki-quer', and 'ki-quer-tit' ('kew-kew': Daanje 1941) used against (e.g.) cat or human. Neighbouring birds thus alerted and enemy kept in sight, birds flying off if approached too closely. (5) Distress-call. Shrieking 'chree chree' given by bird on being caught by cat or human.

CALLS OF YOUNG. Food-call of nestlings a soft, sibilant, shrill 'sheep-sheep-sheep'; in fledglings, develops into harder, deeper call (Fig V), closer to but noisier than

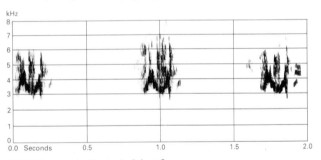

V  R Margoschis  England  July 1989

basic adult chirp. See adult call 1b for juvenile Subsong.  JDS-S

**Breeding.** Extensive study in west Palearctic by Summers-Smith (1963, 1988); see also Seel (1968*a*, *b*, 1969, 1970) and Novotný (1970). SEASON. Start of egg-laying positively correlated with latitude. In Europe normally April–August, but March in Azores (Le Grand 1983) and not until early May in Finland (Alatalo 1975). Considerable variation between years, e.g. in Dziekanów Leśny (Poland), 4 April to 3 May in 14-year study (Pinowska and Pinowski 1977). Regular reports in west Palearctic of successful breeding for all months. Europe: in local populations, laying of 1st eggs of clutches mostly covers period of 93–135 days, but only 23–43 days in

Finland where only 1 clutch. (Summers-Smith 1963, 1988.) Egypt: eggs laid early April to late June (Raw 1921), March–June (Ghabbour 1976). Israel: 1st half February to 2nd half August (Shirihai in press). Cyprus: begins late March (Flint and Stewart 1992). Iraq: beginning of April to mid-August (Kadhim *et al.* 1987; Al-Dabbagh and Jiad 1988). SITE. Usually in hole: in buildings and other man-made structures, e.g. street-lights in 100 out of 868 sites in urban area in Poland (Indykiewicz 1990), parked vehicle or aircraft (Bridgman 1962), tree, earth-bank (including nest-holes of Sand Martin *Riparia riparia*), cliff; also in foundations of occupied and unoccupied large nests (e.g. of Corvidae, birds of prey, White Stork *Ciconia ciconia*), up to 10 nests in one large nest (Dementiev and Gladkov 1954), and free-standing in branches of tree (up to 30 nests in one tree: Ticehurst *et al.* 1921–2), creepers on wall, and telegraph pole. In building, nest may be inside roof-space or even occupied room (Jesse 1902; MacDonald 1960). Enclosed sites preferred (Cink 1976). For 271 nests in rural and built-up areas, see Kulczycki and Mazur-Gierasińska (1968), and Indykiewicz (1990) for 868 nests in urban area. Enclosed nests of swallows and martins (Hirundinidae) frequently appropriated (Summers-Smith and Lewis 1952) and open nests (e.g. Swallow *Hirundo rustica*, Song Thrush *Turdus philomelos*) used as base and domed over (Boyd 1951). Hole occasionally excavated, e.g. between stones in wall (Pitman 1961) or in rotten tree stump (Bourne 1953). (See Summers-Smith 1963, 1968 for further details.) Most nests are 3 m or more above ground; 65·3% 3–7 m, 83·0% 2–8 m, $n=271$ (Kulczycki and Mazur-Gierasińska 1968); 36·8% 3–6 m: Indykiewicz 1990); occasionally at ground level (e.g. in gorse *Ulex*: Baxter and Rintoul 1953) and up to 54·6 m above ground on building (Plath 1983). See Kulczycki and Mazur-Gierasińska (1968), and Indykiewicz (1990), for further details of nest height and location. Nest: free-standing nest is large, domed, roughly globular structure, with entrance at side; loosely woven of dried grass or straw; cup lined with feathers, hair, or other soft material, especially tree bast; in hole, available space normally filled with material, though nest may be reduced to cup if little space (Summers-Smith 1963); free-standing nests had average outer diameter 21·3 cm (17·0–23·5), $n=26$, height 21·9 cm (14·0–31·0), $n=24$, inner diameter 8·9 cm (6·0–12·0), $n=26$ (Kulczycki and Mazur-Gierasińska 1968); in unrestricted cavity, mean outer diameter 40·2 cm, height 14·8 cm, inner diameter 9·6 cm, depth of cup 4·4 cm, $n=55$ (Indykiewicz 1990); weight 117 g (20–500), $n=45$ (Heij 1986). Diameter of nest entrance 6–7 cm (Dementiev and Gladkov 1954). Building: by both sexes; old nest may be removed from hole before fresh material taken in; building may take several weeks, but can be as little as 2 days if nest removed just before ♀ ready to lay (Boyd 1951); both sexes continue to add lining up to hatching, and fresh material added between broods;

most material collected within 20–50 m of nest (Deckert 1969, which see for building technique); see also Summers-Smith (1963, 1988). Eggs. See Plate 59. Sub-elliptical, smooth and only slightly glossy. White or faintly tinted greenish or greyish; very variably marked with spots, speckling, or small blotches of grey, blue-grey, greenish-grey, purplish-grey, black, or purplish-brown; rarely unmarked. (Harrison 1975; Makatsch 1976.) Last egg usually more lightly coloured and pattern more diffuse (Lowther 1988). Nominate *domesticus*: 22·2 × 15·7 mm (18·0–25·0 × 13·4–17·5), $n=1464$; calculated weight 2·89 g. *P. d. tingitanus*: 21·4 × 15·3 mm (20·3–22·5 × 14·7–15·5), $n=33$; calculated weight 2·64 g. *P. d. biblicus*: 21·9 × 15·1 mm (18·9–26·0 × 14·2–16·9), $n=19$; calculated weight 2·64 g. (Schönwetter 1984.) See also Novotný (1970) for data on 1099 eggs from Czechoslovakia. Clutch: 3–5 (2–7). Mode in Britain 4, in continental Europe 5; tends to increase from west to east across Eurasia (Summers-Smith 1988). During breeding season clutch size initially increases and then falls towards end: e.g. in Britain rises from mean 4·0 in April to maximum 4·3 in second half June and then decreases to less than 4. (Summers-Smith 1963, 1988; Seel 1968b; Ivanov 1987.) Eggs laid daily (in Britain, usually before 07.30 hrs). Up to 4 broods per year; in Bulgaria, 18% of 132 pairs laid 4 clutches, but 2–3 more usual; England 2·1, $n=90$; Czechoslavakia 2·6, $n=166$. Interval between broods: mean in England 40·4 ± 3·92 days; usually 2–5 days (but can be up to 10 days) between fledging and laying of next egg, though overlap recorded in USA. (Summers-Smith 1963; Seel 1968a; Mackowicz *et al.* 1970; Novotný 1970; Lowther 1979a.) In Iraq, clutch 2–7, mean 4·8, mode 5; Israel mean 5·1, mode 5 ($n=39$). In Iraq, 1–3 broods, occasionally 4, mean 1·9, $n=67$; Israel 2·3 ($n=39$). Interval between broods in Iraq 42 ± 5 days (33–53). (Kiziroğlu *et al.* 1987; Al-Dabbagh and Jiad 1988; Singer and Yom-Tov 1988.) INCUBATION. Begins with 1st egg, but not significant until 3rd or 4th egg laid; in 50% of 286 cases, began before clutch complete (Novotný 1970). On average occupies 87% of daylight hours, 34% by ♂, 53% by ♀, with eggs seldom uncovered for more than 5 min; average duration of incubation spell 8·9 min by ♂, decreasing markedly 2–3 days before hatching, when ♂ becomes very restless, and 11·0 min by ♀ (1312 hrs observation). Period usually 11–14 days (9–18, mean 12), $n=95$, negatively correlated with mean daily ambient temperature (15·8 days at 16°C, 13·9 days at 24·5°C) (Singer and Yom-Tov 1988). Most eggs hatch early morning, usually within 1 day, but occasionally up to 4 days; over more than 1 day in 28% of 95 clutches (Summers-Smith 1963). See Weaver (1943) for description of hatching process. YOUNG. Brooded up to 6–8 days after hatching (though not capable of thermoregulation until 10·5 days: Seel 1969), with ♀ typically roosting in nest up to fledging. Both sexes feed young, taking almost equal share up to 2–3 days before fledging when ♂ spends more time in Solicitation-display (see Social Pattern and

Behaviour) near nest; testosterone level in ♂ increases at this stage, suggesting that this is to stimulate ♀ to start new brood (Hegner and Wingfield 1986, 1987). Feeding rate peaks early morning, at minimum during 3 midday hours. Young initially fed by regurgitation, with some feeding by regurgitation up to 5 days (Weaver 1942). Widowed birds capable of rearing 9–10-day-old young, i.e. when nearly thermoregulating (Summers-Smith 1963, 1988). See Weaver (1942) and Novotný (1970) for development of nestlings. FLEDGING TO MATURITY. Fledging period: Czechoslovakia 12–16 days (11–18), mean 14·1, *n* = 491 (Novotny 1970); Britain 11–19 days, mean 14·4 (Summers-Smith 1963). Study in Israel showed negative correlation with mean ambient temperature, decreasing from 16·4 days at 17°C to 14·5 days at 25°C (Singer and Yom-Tov 1988). Young begin self-feeding *c.* 7 days after fledging, but continue to be fed by parents up to 10–14 days (Deckert 1969). Age of first breeding usually 1 year, but some apparently breed in year of hatching: ♂♂ with incompletely ossified skulls (ossification complete at 31–32 weeks: Nero 1951) in breeding condition in California (USA) in June (Davis 1953); precocious breeding also reported in Sénégal (M-Y Morel). Young from late broods sometimes do not breed until 2 years old (Selander and Johnston 1967). BREEDING SUCCESS. Hatching success: Britain 71–80% (3 studies), continental Europe 55–95% (17 studies), Israel 70·1%. Fledging success: Britain 45–74%, continental Europe 48–89%. Overall success: Britain 38·4–60·3% (5 studies), continental Europe 37·2–84·6% (18 studies), Israel 52·3%. (Summers-Smith 1988, which see for further details.) Most egg failures result of infertility (11·9 ± 1·6%); flooding also causes loss of eggs and nestlings (Ivanov 1987). Long spells of cold and wet weather cause significant nestling mortality in some years, particularly up to first 5 days of life and of later-hatched nestlings (Mackowicz *et al.* 1970; Seel 1970). In Spain, nest predation rate of 2·8% (*n* = 395 broods) caused mainly by weasel *Mustela nivalis*, with black rat *Rattus rattus*, house mouse *Mus musculus*, and Wryneck *Jynx torquilla* also implicated (Cordero 1991). In another Spanish study, 9–12% of breeding attempts subject to destruction of eggs and young by conspecifics (Veiga 1990). See also Schifferli (1978), for possible removal of eggs by ♂ from neighbouring nest. Young fledglings preyed on by Jay *Garrulus glandarius*, Magpie *Pica pica*, Little Owl *Athene noctua*, cat, and snakes *Malpolon* and *Elaphe* (Cordero 1991). Domestic cat is frequent predator, taking significant numbers June–August, presumably inexperienced juveniles (Churcher and Lawton 1987). Sparrowhawk *Accipiter nisus* locally kills large number of juveniles, particularly near villages and towns (Tinbergen 1946). JDS-S

**Plumages.** (nominate *domesticus*). ADULT MALE. In fresh plumage (autumn), forehead and crown dull olive-brown, some

dull grey of feather centres usually shining through or just visible. Broad stripe backwards from above eye, over upper side of neck, and across nape rich maroon-chestnut, each feather with broad olive-brown or buff tip, latter partly concealing chestnut, especially on side of neck and nape. Lore, stripe below eye, and small patch behind eye black; a small off-white spot above rear corner of eye. Upper cheek and ear-coverts dull olive-green, shorter feathers at border of black below eye often paler grey, upper and longer ear-coverts tipped olive-brown or buff. Lower cheek olive-grey, some white of feather centres partly visible at rear; lower side of neck olive-green with some white shining through. Mantle and scapulars broadly streaked black, each streak bordered rufous-chestnut at side and pale buff-brown at tip. Back, rump, and upper tail-coverts olive-brown, some grey of feather centres shining through or just visible on rump, longer tail-coverts with dark shafts. Chin and throat black, forming rather short and narrow bib, hardly extending to upper chest; feathers of chin narrowly fringed off-white on tips, fringes gradually broader towards lower throat, latter appearing black with broad off-white scalloping. Chest, side of breast, flank, and thigh light ash-grey, feather-tips broadly pale drab-buff, partly or fully concealing grey, some rounded black feather-centres sometimes just visible on upper chest; ash-grey and drab-buff gradually paler towards off-white mid-belly and vent. Under tail-coverts pale drab-grey, fringes pale buff, bordered off-white subterminally. No streaks on underparts, except for dusky shafts of tail-coverts and sometimes on upper belly. Tail dark sepia-brown with black shafts; both webs of central pair (t1) and outer web of t2–t5 narrowly fringed pale buff-brown, tips of feathers and outer web of t6 faintly edged grey. Primaries greyish-black, inner primaries narrowly fringed pale buff-brown on outer web and faintly grey on tip, outer primaries more distinctly fringed pale pink-cinnamon at base and near emarginated part of outer web, but emarginations proper and outer web of p9 without pale fringe. Secondaries dull black with broad rufous-cinnamon fringe along outer web and faint grey-buff fringe along tip. Greater upper primary coverts and bastard wing black, outer webs narrowly fringed rufous-chestnut (except on longest feather of bastard wing). Tertials and greater upper wing-coverts black, outer webs broadly fringed rufous-cinnamon on tertials, more rufous-chestnut on greater coverts, rufous of fringes merging into broad cinnamon-buff or cream-buff fringe along tip and this in turn to dull pink-cinnamon on terminal part of inner web; narrow edges along outer webs pale buff. Median upper wing-coverts white with black bases (latter broadest on inner web), tips washed buff; lesser coverts rich rufous-chestnut. Under wing-coverts and axillaries dirty grey-white, darker grey or dull black bases partly visible, especially along leading edge of wing. Wear and abrasion have marked effect on plumage. By November–February, much dark grey exposed on forehead, crown, and rump, but usually still tinged olive or brown; more chestnut visible behind eye, on upper side of neck, and on nape, pale fringes largely worn off; rear of lower cheek and lower side of neck grey-white; black bib larger, extending into broad rounded patch on chest, but black feathers of lower throat and chest still broadly fringed off-white. *When worn* (about April–June), forehead and crown dark ash-grey, sometimes with faint dark shaft-streaks. Stripe backwards from eye pure maroon-chestnut, extending to upper side of neck and sometimes along nape (in some birds, nape dark grey or mottled chestnut and grey); mantle and scapulars sharply streaked black and rufous-chestnut; back to upper tail-coverts dark grey with olive wash. Small white spot above rear corner of eye distinct; ear-coverts dull grey with olive tinge, sometimes partly mottled white; lower cheek and lower side of neck dirty grey-white. Black bib large, ending broadly rounded on chest, off-white feather-fringes narrow, confined to lower chest; remainder of underparts pale grey, slightly darker grey or olive-grey on flank, grey-white on mid-belly and vent; belly, side of breast, and under tail-coverts with dusky shafts (sometimes faint). Pale fringes of tail-feathers bleached to grey-buff, partly worn off; tips of tertials and greater coverts bleached to pale grey-buff, partly worn off; tips of median coverts pure white, except for faint dark shafts; some dark grey or black of feather-bases sometimes visible on lesser coverts and under wing-coverts. When extremely abraded, July–August, crown speckled dusky grey, much grey of feather-bases visible on nape, mantle, scapulars, and back; ear-coverts, cheek, and side of neck dirty grey-white, black of bib on chest bordered by some black triangular marks, remainder of underparts dirty grey; pale fringes of flight-feathers and tertials worn off, except at bases and near emarginations of primaries. Much individual variation, especially in amount of grey or chestnut on nape and in extent and shape of bib; occasionally, a white supraloral stripe or spot, some dusky streaking on side of breast and flank, or some chestnut mixed in black of bib; grey of cap frequently streaked black; exceptionally, cap entirely chestnut, or grey with chestnut spots, also outside range of Spanish Sparrow *P. hispaniolensis*. For variation, see Stresemann (1928), Calhoun (1947a), Bodenstein (1953), Niethammer (1953, 1969), Piechocki (1954), Hazelwood and Gorton (1955), Löhrl and Böhringer (1957), Johnston (1981), Stephan (1984), and Eck (1985). ADULT FEMALE. In fresh plumage (autumn), forehead and crown dark brown with slight olive tinge, merging into paler drab-brown with darker mottling on nape and upper mantle. Distinct buff-brown or cinnamon-brown supercilium, extending backwards from just above and behind eye, bordered below by dull black stripe over upper ear-coverts. Lore and front part of upper cheek mottled drab-brown and pale grey-buff; remainder of cheek and ear-coverts dark olive-brown, merging into buff-brown on side of neck. Lower mantle and scapulars broadly streaked black and pale rufous-cinnamon; back, rump, and upper tail-coverts drab-brown or dull olive-brown, tail-coverts with dark shafts. Chin and throat variable, either uniform pale greyish-buff, or pale grey-buff with darker grey mottling, latter forming ill-defined mottled grey bib on throat. Chest and upper belly pale greyish-buff, merging into darker olive-brown or buff-brown on side of breast, flank, and thigh and into dirty cream-white on lower belly and vent. Under tail-coverts light drab-brown with broad pale grey-buff or isabelline borders. Underparts unstreaked, except for dark shafts of tail-coverts and (sometimes) on upper belly. Tail as adult ♂. Flight-feathers, tertials, greater upper primary and secondary coverts, and bastard wing as adult ♂, but pale fringes on all feathers dull pink-cinnamon or buff-cinnamon (like those of primaries of ♂), without deeper rufous-cinnamon on secondaries, tertials, and primary coverts or rufous-chestnut on greater coverts; centres of median coverts more extensively dull black, tips narrower, 2–3 mm long, more intensely washed pink-buff (in ♂, broad and whitish); lesser upper wing-coverts dull grey with broad and ill-defined olive-brown fringes. Under wing-coverts and axillaries as adult ♂, but bases of marginal coverts dark grey, less blackish, leading edge of wrist tinged isabelline, less whitish. Effect of wear and bleaching less marked than in adult ♂. *In worn plumage* (April–June), forehead to nape and back to upper tail-coverts dull brown-grey or olive-brown, sometimes with faint dark shafts on cap; supercilium paler buff, but often narrower and extending less far back than in fresh plumage; lore and cheek dirty grey, gradually darker olive-grey towards longer ear-coverts; chin and throat pale isabelline-grey, sometimes with mottled medium grey

throat patch (in those birds which show pronounced grey bib on throat often a rather pale isabelline-white border on chin and side of bib); remainder of underparts dirty light brown-grey, merging into grey-white or cream-white on belly, often with narrow but distinct dark shaft-streaks; pale fringes of tail, flight-feathers, tertials, and along tips of greater coverts bleach to pale grey-buff, sometimes largely worn off; tips of median coverts narrowly buff-white. When extremely abraded, July–August, cinnamon of mantle and scapulars worn off, entire upperparts then dull brown-grey, marked with black-brown streaks on lower mantle and scapulars; side of head and neck and entire underparts dirty pale brown-grey. Exceptionally, bib blackish and lesser coverts with some rufous (Bährmann 1971a; Eck 1985); for birds with intersexual characters, see Mayr (1949) and Harrison (1961). NESTLING. Naked at hatching, but mouse-grey or brownish covering of down soon obtained (Keck 1934). For development and growth, see Weaver (1942), Novotný (1970), O'Connor (1975, 1977a, b, 1978), and Ion and Saracu (1971). JUVENILE. Like washed-out version of adult ♀, differing mainly in showing fresh flight-feathers when those of adult worn, in short loose feathering of body (especially rump, vent, tail-coverts, and under wing-coverts), and in soft, rather narrow tail-feathers. Upperparts as adult ♀, but much bluish-grey of feather-centres visible (especially on nape and rump); centres of feathers of mantle and scapulars black-brown, not as black and sharply defined as in adult ♀, fringes pale olive-brown or buff-brown, scarcely brighter than remainder of upperparts, not as cinnamon as adult ♀. Side of head and neck as in adult ♀, but supercilium and dark stripe through eye short and poorly defined (in some ♂♂, supercilium more distinct, rufous-cinnamon, bordered by sooty black above and below); in ♂, often a small white spot above rear corner of eye. Underparts less uniform than in adult ♀, more mottled due to exposure of grey-white feather-bases; often a poorly defined mottled dark grey malar stripe; in some birds, chest, side of breast, flank, belly, and under tail-coverts with fine dark shaft-streaks, but these vestigial or entirely absent in others. Sexes on average slightly different in colour of underparts: ground-colour of chest, side of breast, and flank usually light drab-grey in ♂, pale drab-brown or greyish-isabelline in ♀, but some overlap; chin and throat of ♂ often with dark grey bib (sharply defined in some birds, gradually merging into mottled drab-grey sides in others), chin and throat of ♀ usually uniform isabelline-white. About 95% of birds can correctly be sexed by colour of chin and throat (see Weaver 1942, Harrison 1961, Cheke 1967, Johnston 1967b, c, Bährmann 1970b, and Cordero 1990). Tail as adult, but ground-colour paler brown-grey; fringes isabelline-buff, poorly defined, tips narrow and frayed. Wing as adult ♀, but ground-colour of flight-feathers, tertials, greater primary and secondary coverts, and bastard wing distinctly browner, especially tertials and greater coverts much less black; fringes narrower (especially those of tertials and greater coverts), more buff, less cinnamon, sharply defined on tips of feathers but not at bases; median coverts dull black with narrow buff or off-white fringe along tip, soon worn off; lesser coverts buff-brown. Body and tail strongly liable to abrasion and bleaching, appearing as worn as adult ♀ at same time of year within a few months, but outer primaries still rather fresh, not as worn as in adult ♀. FIRST ADULT. Like adult, and indistinguishable when post-juvenile moult completed. Last juvenile feathers present during moult are p9–p10 and inner secondaries, these usually still rather fresh (in adult ♀, heavily abraded, especially primaries); for juvenile p10, see also Structure. Fringes of feathers in fresh plumage of ♂ on average broader than in adult ♂, but with much individual variation, some almost resembling ♀, others similar to adult ♂. See Selander and Johnston (1967).

**Bare parts.** ADULT, FIRST ADULT. Iris hazel-brown to dark brown. Bill horn-brown to dark horn-grey, lower mandible paler horn-brown or horn-grey with yellow, horn-white, paler horn-yellow, or brown-flesh base; bill of breeding ♂ black or bluish-black, darkening when gonads become enlarged, in some already from late November, in others not until March, gradually paler again from base onwards just before start of post-breeding moult or during moult (in birds examined, bill becoming paler from between mid-June and late September onwards). For timing of blackening of bill, see also Nichols (1935), Steinbacher (1952), Piechocki (1954), and Löhrl and Böhringer (1957). For hormonal control of bill colour, see Keck (1934), Witschi (1936), Witschi and Woods (1936), Novikov (1938), Lofts *et al.* (1973), Murton and Westwood (1974), Haase (1975), and Dawson (1991). Leg and foot flesh-pink, flesh-brown, light horn-brown, dull flesh-grey, or horn-grey. NESTLING. At hatching, bare skin (including leg) pink-flesh, bill pink; mouth pale flesh to pink-yellow, gape-flanges pale yellow. At *c.* 1 week, bill light flesh-brown, leg and foot pale lilac-grey; mouth yellow. JUVENILE. Iris dark brown. Bill horn-brown or flesh-grey. Leg and foot blue-grey. Adult non-breeding colour obtained during post-juvenile moult. (Heinroth and Heinroth 1924–6; Witherby *et al.* 1938; Bährmann 1972; RMNH, ZMA.)

**Moults.** ADULT POST-BREEDING. Complete; primaries descendent. Exceptionally, some primaries moulted out of sequence (Herroelen 1983); moult occasionally suspended (Casto 1974; Harper 1984). In southern England and Wales, ♂ starts with p1 mid-July to early September (rarely up to late October), primary moult completed with regrowth of p9 (moult score 45) after *c.* 60 days, mid-September to late October or early November; timing in ♀ more variable, some starting late June, some finishing November. Moult in northern England and Scotland similar, ♂ moulting slightly earlier than ♀. (Ginn and Melville 1983.) In eastern Germany, moult recorded early August to mid-October; duration of primary moult *c.* 79 days in individuals (Zeidler 1966, which see for many details). In south-west Spain, starts early July to early August (on average, 24 July), all moult completed after 72 days, primaries after 69 days, early September to mid-October (Alonso 1984a, which see for sequence of moult of various feather tracts). In Afghanistan, just started early or mid-July; by September and early October, mainly new, but outer primaries, a few tail-feathers, and some feathers of body still growing (Paludan 1959). In Kazakhstan, resident nominate *domesticus* starts moult August (occasionally late July or early September), completes mid-September to mid-October (rarely late October); summer visitor *bactrianus* in same area starts mid-July to mid-August (sometimes from late June), completes early September to early October, sometimes from mid-August or up to mid-October (Stephan 1982). For sequence of replacement of feathering, see Zeidler (1966), Bährmann (1967), and Ginn and Melville (1983). For influence of light on moult, see Dawson (1991). See also Dementiev and Gladkov (1954), Haukioja and Reponen (1969), Casto (1974), and Mathew and Naik (1986). POST-JUVENILE. Complete; primaries descendent. Starts at age of *c.* 5 weeks (Heinroth and Heinroth 1924–6) or 4–6 weeks (Summers-Smith 1988), and timing thus as variable as hatching date. In southern England and Wales, moult starts mid-June to early September (occasionally early October); in birds from early broods, completed at *c.* 80 days, in late birds after *c.* 60 days, mainly between early September and early November (Ginn and

Melville 1983). In eastern Germany, early-fledged birds start moult with p1 from 23 May; very late birds not yet started 4 September; moult in early birds slow, 1–2 feathers growing simultaneously in each wing, moult of late birds rapid, 3(–4) feathers growing at same time (Zeidler 1966). In south-west Spain, starts late June to mid-September (on average, 29 July) timing and sequence of moult as in adult post-breeding; late fledglings start moult at younger age than early ones (Alonso 1984a, which see for details).

**Measurements.** Nominate *domesticus* and *balearoibericus*. Adult wing and bill to skull for populations from (1) England, (2) Netherlands, (3) Sweden, (4) Rumania, (5) southern France, (6) Iberia; other measurements for all areas combined (juveniles mainly from Netherlands), all year; skins (RMNH, ZMA). Bill (N) to distal corner of nostril; exposed culmen on average 2·4 less than bill to skull. Adult wing includes data from Lack (1940), Niethammer (1971), Alonso (1985a), and Eck (1985).

| | ♂ | | ♀ | |
|---|---|---|---|---|
| WING (1) | 76·2 (1·87; 199) | 71–81 | 74·3 (1·50; 32) | 72–78 |
| (2) | 80·7 (1·85; 56) | 77–84 | 77·8 (1·44; 27) | 75–80 |
| (3) | 79·4 (1·84; 101) | 75–85 | 76·5 (1·70; 46) | 73–80 |
| (4) | 79·5 (1·16; 16) | 78–82 | 75·5 ( — ; 2) | 75–76 |
| (5) | 81·4 (1·67; 7) | 79–84 | 76·7 (1·04; 3) | 75–78 |
| (6) | 79·4 (1·63; 54) | 75–83 | 75·5 (1·18; 6) | 74–77 |
| JUV | 76·3 (1·75; 15) | 73–79 | 74·8 (2·36; 11) | 71–78 |
| TAIL AD | 56·3 (2·04; 79) | 51–61 | 54·9 (1·81; 33) | 51–59 |
| JUV | 49·9 (2·40; 15) | 46–54 | 48·4 (2·65; 11) | 45–52 |
| BILL (1) | 15·4 (0·51; 14) | 14·5–16·1 | 15·8 (0·35; 4) | 15·5–16·3 |
| (2) | 15·3 (0·59; 69) | 14·0–16·2 | 15·3 (0·70; 36) | 14·2–16·2 |
| (3) | 15·5 (0·63; 5) | 14·9–16·2 | 15·7 (0·48; 5) | 15·5–16·4 |
| (4) | 15·3 (0·76; 16) | 14·2–16·3 | 15·8 ( — ; 2) | 15·5–16·2 |
| (5) | 16·0 (0·47; 7) | 15·2–16·5 | 15·3 (0·40; 3) | 14·8–15·5 |
| (6) | 15·6 (0·57; 12) | 14·6–16·4 | 15·5 (0·87; 6) | 14·5–16·4 |
| BILL (N) | 9·7 (0·45; 94) | 8·9–10·5 | 9·6 (0·52; 45) | 8·8–10·5 |
| TARSUS | 19·9 (0·72; 81) | 18·3–21·2 | 19·6 (0·88; 36) | 18·1–21·2 |

Sex differences significant for wing (2)–(6). Differences in bill lengths above probably due to differences in collecting season, not due to geographical origin: bill longer in summer (when partly feeding on soft insects and bill less liable to abrasion) than in winter (when feeding mainly on hard seeds): see Steinbacher (1952), Davis (1954), Packard (1967b), and Rising (1973) for average bill length in various months throughout year. For differences in measurements between freshly dead birds and those of skins or skeletons, see Bjordal (1983a, 1984).

Wing. (1) Balearic Islands (Mester 1971). (2) South-east France (G Olioso). (3) Stuttgart area (south-west Germany) (Löhrl and Böhringer 1957). (4) Nordrhein-Westfalen (west-central Germany) (Niethammer 1953b). (5) Niedersachsen (north-west Germany) (Nordmeyer *et al.* 1972; Oelke 1973; Scherner 1974; see also Nordmeyer *et al.* 1970). (6) South-west of former East Germany (Niethammer 1969; Eck 1985; see also Piechocki 1954, Grimm 1954, and Geiler 1959). (7) South-east of former East Germany (Eck 1985; see also Bährmann 1968c). (8) Czechoslovakia (Folk and Novotný 1970; Hubálek 1976). (9) Southern Yugoslavia (Stresemann 1920). (10) Southern Greece (Niethammer 1943; ZMA). (11) Central and northern European Russia (Eck 1985). (12) South-east Ukraine (Samchuk 1971). (13) Tyumen' and Tomsk (western Siberia) (Johansen 1944; RMNH, ZMA). (14) Baykal area and Mongolia (Keve 1943; Johansen 1944; Piechocki and Bolod 1972). (15) Northern Turkey (Jordans and Steinbacher 1948; ZMA). (16) Central Anatolia and Taurus (Turkey) (Kumerloeve 1961; Vauk 1973; RMNH, ZMA). (17) Eastern Turkey (Kumerloeve 1967a, 1969a). (18) Azores (G Le Grand).

| | | | | | | | |
|---|---|---|---|---|---|---|---|
| (1) | ♂ | 76·9 (2·08; | 44) | 74–81 | ♀ | 74·9 ( — ; | 21) | 69–78 |
| (2) | | 79·8 (2·39; | 20) | 76–83 | | 76·4 (2·42; | 11) | 72–80 |
| (3) | | 80·3 (1·52; | 246) | 77–86 | | 77·8 (1·51; | 225) | 74–80 |
| (4) | | 77·8 (1·64; | 650) | 70–84 | | 75·1 (1·57; | 626) | 70–80 |
| (5) | | 79·7 ( — ; | 2138) | 70–89 | | 78·9 ( — ; | 1894) | 70–86 |
| (6) | | 79·6 (2·07; | 694) | 73–86 | | 77·2 (1·96; | 581) | 70–83 |
| (7) | | 79·6 (1·87; | 127) | 75–85 | | 77·1 (1·51; | 60) | 73–80 |
| (8) | | 79·6 ( — ; | 835) | 68–88 | | 77·0 ( — ; | 683) | 70–84 |
| (9) | | 78·6 (1·46; | 36) | 76–83 | | 76·0 (3·69; | 17) | 74–79 |
| (10) | | 78·7 (1·07; | 12) | 78–81 | | — ( — ; | ----) | — |
| (11) | | 80·1 (1·91; | 32) | 76–84 | | — ( — ; | ----) | — |
| (12) | | 80·7 (1·44; | 65) | — | | 77·0 (0·92; | 38) | — |
| (13) | | 79·7 ( — ; | 34) | 76–84 | | 77·9 (1·93; | 4) | 76–81 |
| (14) | | 81·2 ( — ; | 19) | 79–84 | | — ( — ; | ----) | — |
| (15) | | 79·3 (2·56; | 5) | 76–82 | | 75·0 ( — ; | 2) | 75–75 |
| (16) | | 79·3 ( — ; | 30) | 77–84 | | 75·9 (1·32; | 6) | 74–78 |
| (17) | | 82·6 (1·52; | 18) | 80–85 | | 79·8 (2·43; | 8) | 76–84 |
| (18) | | 76·6 (2·26; | 110) | — | | 74·4 (1·82; | 87) | — |

Most references cited above give data on other measurements also. England: wing, ♂ 76·2 (150) 72–81 (Hartert 1921–2). Balearic Islands: bill (N), 10·4 (9·5–11·1), ♀ 9·9 (9·4–10·5) (Mester 1971). For Hungary, see Keve (1960); for Yugoslavia, see Urbánek (1959); for USSR, see Terentiev (1966, 1970). For North America, see many references cited at end of Geographical Variation (below). For New Zealand, see Niethammer (1971) and Baker (1980).

*P. d. tingitanus*. Algeria: wing, ♂ 79·3 (1·08; 6) 77–81, ♀ 76·6 (0·75; 4) 76–78; bill to skull, ♂ 16·0 (0·29; 6) 15·6–16·4, ♀ 15·3 (3) 15·1–15·5 (RMNH, ZMA). Western Algeria: wing, ♂ 79·3 (1·56; 19) 77–82, ♀ 76·8 (1·63; 30) 74–79 (Metzmacher 1986*b*, which see for other data).

*P. d. biblicus*. South-east Turkey (Amik Gölü to Ceylanpinar): wing, ♂ 80·6 (1·38; 10) 79–83, ♀ 75 (Kumerloeve 1961, 1963, 1970*a*, *b*). Cyprus: wing, ♂ 80·0 (209) 73–92 (Flint and Stewart 1992). Cyprus and Levant: wing, ♂ 80·2 (1·56; 8) 78–82, ♀ 76·2 (2·08; 5) 73·5–79; bill to skull, ♂ 15·8 (0·39; 6) 15·4–16·5, ♀ 15·5 (3) 15·2–15·8 (Diesselhorst 1962; ZMA). Levant to western Iran: wing, ♂ 81·7 (44) 78–85; culmen, ♂ 14·0 (44) 13·0–15·0 (Vaurie 1949*a*). Iraq (*n*=23): wing, ♂ 79–83·5, ♀ 77·5–82 (Ticehurst *et al.* 1921–2). See also Meinertzhagen (1921).

*P. d. hyrcanus*. Northern Iran: wing, ♂ 77·8 (3·07; 15) (70–) 75–82, ♀ 75·6 (2·35; 12) 72–80 (Paludan 1938, 1940; Schüz 1959; Diesselhorst 1962; see also Stresemann 1928 and Vaurie 1959); wing, ♂ 78·0 (14) 74–81; culmen, ♂ 13·1 (4) 12·0–14·0 (Vaurie 1949*a*).

*P. d. persicus*. South-west Iran: wing, ♂ 77·9 (2·25; 11) 73–81, ♀ 77·6 (0·75; 4) 76–78 (Zarudny 1916; Paludan 1938; Diesselhorst 1962). Iran (except west and Caspian districts) and south-west Afghanistan, ♂: wing 78·7 (97) 73–84, culmen 12·8 (100) 11·0–14·5 (Vaurie 1949*a*).

*P. d. bactrianus*. Western Transcaspia: wing, ♂ 76·5 (1·54; 9) 74–79 (Zarudny 1916). Afghanistan: wing, ♂ 77·5 (34) 75–80, ♀ 74·9 (25) 72–78 (Paludan 1959). Central and northern Afghanistan: wing 77·9 (16) 74–81, culmen 13·3 (16) 13·0–14·0 (Vaurie 1949*a*). See also Gavrilov (1965).

*P. d. parkini*. Kashmir, live birds: wing, ♂ 80·5 (2·09; 38) 76–86, ♀ 76·6 (2·36; 39) 70–81 (P R Holmes and Oxford University Kashmir Exped. 1983); see also Hellmayr (1929), Vaurie (1972), and Ali and Ripley (1974). Kashmir, northern Punjab, and Nepal, ♂; wing, 79·5 (25) 74–85, culmen 13·5 (25) 12·5–14·5 (Vaurie 1949*a*).

*P. d. indicus*. Central and southern India and northern Burma, ♂: wing, 74·4 (31) 70–78, culmen 12·5 (31) 11·5–13·5 (Vaurie 1949*a*; see also Vaurie 1956*f*). Iranian Baluchestan:

wing, ♂♂ 74, 75; ♀ 70·0 (3) 69–71 (Diesselhorst 1962). Elat (Israel): wing, ♂ 76·0 (7) 74–79, ♀ 74·0 (12) 70–77; bill to skull 15·7 (20) 15·0–16·3 (Shirihai in press). See also Saini *et al.* (1992).

*P. d. hufufae*. El Hufuf (northern Saudi Arabia): wing, ♂ 72–77 (9) (Hartert and Steinbacher 1932–8).

*P. d. niloticus*. Egypt: wing, ♂ 75 (20) 72–77 (Vaurie 1956*f*, 1959); ♂ 71–75 (9), ♀ 73 (1) (Meinertzhagen 1921).

*P. d. rufidorsalis*. Sudan: wing, 73·9 (0·74; 5) 73–75, ♀ 74·5; bill to skull, ♂ 13·8 (0·64; 5) 12·8–14·3, 13·9; bill to nostril 8·5 (0·57; 6) 7·7–9·3 (RMNH, ZMB); wing, ♂ 72 (20) 69–75 (Vaurie 1956*f*).

Weights. ADULT, FIRST ADULT. Nominate *domesticus*. (1) Norway (Haftorn 1971). (2) Netherlands, all year (RMNH, ZMA). Germany: (3) Nordrhein-Westfalen, March (Niethammer 1953*b*); (4) Stuttgart area, November–December (Löhrl and Böhringer 1957); (5) Hannover-Wolfsburg-Helmstedt area, January–March (Oelke 1973; Scherner 1974); (6) Leipzig-Halle area, December–February (Grimm 1954; Piechocki 1954; Geiler 1959); (7) Lausitz area, all year (Bährmann 1968*c*, which see for details; also, Bährmann 1972). (8) Czechoslovakia, all year (Folk and Novotný 1970; Havlín and Havlínová 1974; Hubálek 1976). (9) South-east Ukraine, spring to autumn (Samchuk 1971). (10) Lake Chany (south-west Siberia) and Mongolia, July (Piechocki and Bolod 1972; Havlín and Jurlov 1977). *P. d. balearoibericus*. (11) South-east France, all year (G Olioso). (12) Eastern Turkey, May–June (Kumerloeve 1967*a*, 1969*a*). *P. d. tingitanus*. (13) Algeria, November–December (Fairon 1971). *P. d. biblicus*. (14) Southern Turkey and south-west Iran (Paludan 1938; Kumerloeve 1970*a*, *b*). *P. d. hyrcanus*. (15) Northern Iran, February (Schüz 1959). *P. d. persicus*. (16) Southern Iran and Afghanistan, April–June (Paludan 1959; Desfayes and Praz 1978). *P. d. bactrianus*. (17) North-east Iran and Afghanistan, April–October (Paludan 1940, 1959). *P. d. parkini*. (18) Kashmir, August–September (P R Holmes and Oxford University Kashmir Exped. 1983).

| | | | | | | | |
|---|---|---|---|---|---|---|---|
| (1) | ♂ | 31·1 ( — ; | 11) | 28–35 | ♀ | 29·6 ( — ; | 14) | 27–33 |
| (2) | | 30·2 (2·53; | 108) | 24–37 | | 30·2 (2·58; | 33) | 25–35 |
| (3) | | 29·8 (1·42; | 650) | 23–35 | | 29·5 (1·66; 626) | | 24–35 |
| (4) | | 29·5 (1·57; | 738) | 24–36 | | 28·7 (1·52; 672) | | 25–34 |
| (5) | | 31·8 (1·78; | 1308) | 25–38 | | 30·9 (1·93; 951) | | 25–38 |
| (6) | | 32·3 (2·07; | 831) | 26–39 | | 31·3 (2·12; 860) | | 25–39 |
| (7) | | 31·1 ( — ; | 130) | 25–36 | | 30·9 ( — ; | 77) | 26–37 |
| (8) | | 30·4 ( — ; | 915) | 24–38 | | 29·8 ( — ; | 741) | 24–38 |
| (9) | | 30·0 (0·80; | 45) | — | | 29·6 (1·28; | 38) | — |
| (10) | | 29·9 (1·54; | 7) | 28–30 | | 33·4 (6·84; | 3) | 27–41 |
| (11) | | 28·8 (2·04; | 17) | 26–33 | | 28·2 (2·07; | 10) | 15–31 |
| (12) | | 29·1 (1·71; | 18) | 26–32 | | 31·0 (4·08; | 7) | 25–38 |
| (13) | | 24·9 (1·04; | 5) | 23–26 | | 25·4 (1·56; | 4) | 23–27 |
| (14) | | 31·3 (3·25; | 6) | 26–35 | | 28·9 (1·87; | 3) | 27–31 |
| (15) | | 28·8 (1·89; | 4) | 26–30 | | 29·3 (1·75; | 6) | 26–31 |
| (16) | | 23·2 ( — ; | 8) | 22–24 | | 22·0 ( — ; | 4) | 21–23 |
| (17) | | 24·4 ( — ; | 49) | 21–28 | | 24·2 ( — ; | 40) | 21–28 |
| (18) | | 25·3 (1·81; | 36) | 22–30 | | 24·3 (1·41; | 37) | 23–28 |

Nominate *domesticus*. Britain: April–August 27·1 (149) (Seel 1970; for monthly fluctuations, see O'Connor 1972, 1973 and Summers-Smith 1988). Netherlands, exhausted birds, all year: 21·6 (1·54; 10) 19·8–24·5 (ZMA). Central Urals at 57°N, ♂: June 30·9 (1·87; 19), January 33·9 (1·48; 20) (Danilov *et al.* 1969*a*; see also Danilov *et al.* 1969*b*). Kazakhstan: 29·0 (160) 25·5–37·7 (Korelov *et al.* 1974); ♂ 28·3 (32) 25·5–31·7 (Gavrilov 1965). For influence of age on weight, see Löhrl and Böhringer (1957).

For weight at various altitudes in Caucasus, winter, see Baziev (1976).

*P. d. balearoibericus*. Balearic Islands: 27·4 (20) 23–32 (Mester 1971). See also Niethammer (1943), Rokitansky and Schifter (1971), and Vauk (1973).

*P. d. biblicus*. Cyprus: ♂ 28·0 (116) 22–35 (Flint and Stewart 1992).

*P. d. bactrianus*. Kazakhstan: 25·0 (189) 21–29·5 (Korelov *et al.* 1974); ♂ 24·8 (70) 22·5–28·5 (Gavrilov 1965). See also Dol'nik and Gavrilov (1975).

*P. d. indicus*. See Mirza (1973) and Saini *et al.* (1992).

Various introduced races. For North America, see Baldwin and Kendeigh (1938), Packard (1967*a*), Blem (1975), Anderson (1977, 1978), Clench and Leberman (1978), McGillivray (1984), and various references cited at end of Geographical Variation. For South Africa, see Skead (1977).

NESTLING, JUVENILE. Britain: on day 13, 23·8 (1030) (Seel 1970, which see for relation of fledging weight with brood size, time of year, survival, etc.). For various aspects of growth, see O'Connor (1975, 1977*a*, *b*, 1978), Lowther (1979*b*), and Schifferli (1980). Fledged juveniles, Balearic Islands: 24·3 (56) 20·5–28 (Mester 1971).

**Structure.** Wing rather short, broad at base, tip rounded. P8 longest, p7 and p9 0–1 shorter, p6 1–4, p5 5–9, p4 9–14, p3 11–17, p2 13–20; in adult, p1 19·2 (10) 16–23 shorter, in juvenile 16·8 (5) 15–19. P10 strongly reduced, a tiny pin, concealed below reduced outermost greater upper primary covert in adult, less so in juvenile; in adult, 51·2 (10) 45–55 shorter than p8, 6·7 (10) 5–9 shorter than longest upper primary covert; in juvenile, 43·3 (20) 40–47 shorter than p8, 2·9 (20) (0–)2–5 shorter than longest primary coverts (thus, longer than in adult). Outer web of p6–p8 emarginated, inner of p7–p9 with notch (sometimes faint). Tip of longest tertial reaches to tip of p3–p4. Tail rather short, tip square or slightly forked; 12 feathers, t4–t5 longest, t1 and t6 0–5 shorter. Bill strong, about 57% of length of head; depth and width at base both 8·3 (20) 7·7–8·8 in nominate *domesticus*, depth 8·5 (10) 8·1–9·0 and width 7·9 (10) 7·4–8·5 in *biblicus*, depth 8·1 (10) 7·8–8·4 and width 7·8 (10) 7·3–8·2 in *bactrianus* and *parkini*; culmen evenly decurved, ending in sharp tip, cutting edges slightly decurved, gonys slightly convex; bill-tip slightly compressed laterally. Nostril small, rounded, covered by small tuft of feathers projecting from base of bill. A few short bristles project obliquely downward from lateral base of upper mandible. For pterylosis, see Clench (1970, 1973); for wing pterylography and individual lengths of various wing-feathers, see Zeidler (1966); for weight of feathers, see Schifferli (1981). Tarsus and toes rather short, but strong. Middle toe with claw 19·2 (55) 17–21 (no marked difference between races); outer and inner toe with claw both *c.* 67% of middle with claw, hind *c.* 77%.

**Geographical variation.** Marked; involves mainly depth of colour, to lesser extent width of streaking on upperparts, size, and relative bill depth. 2 main subspecies-groups recognized: (1) *domesticus* group in west and north, comprising populations with longer wing and larger bill, and in which ♂ has greyish cheek and ear-coverts, more extensively grey underparts with white mainly restricted to belly, and rufous–chestnut of upperparts rather pale and restricted; (2) *indicus* group in south-east, which is generally smaller and in which ♂ is extensively white on side of head and neck and on underparts and has extensive deep chestnut on upperparts. The groups overlap extensively in breeding season in central Asia, local resident nominate *domesticus* of *domesticus* group already breeding when summer

visitor *bactrianus* of *indicus* group arrives, thus limiting interbreeding, while groups differ also somewhat in ecology, timing of moult, and behaviour, suggesting that they may be candidates for treatment as separate species (see Gavrilov 1965, Gavrilov and Korelov 1968, Radzhabli and Panov 1972, Korelov *et al.* 1974, Yakobi 1979, Stephan and Gavrilov 1980, Stephan 1982, Stepanyan 1983, Ivanitski 1985*b*), but *indicus* of *indicus* group very gradually merges into *biblicus* of *domesticus* group throughout Iran, and occurrence of nominate *domesticus* within range of *bactrianus* is probably too recent to be sure of permanent separation. Thus united in single species here, following Dementiev and Gladkov (1954), Sudilovskaya (1957), and Vaurie (1959).

Nominate *domesticus* from northern Eurasia is a dark race; populations from Britain and Ireland slightly smaller and darker than in northern and central Europe, birds from central and eastern Asia slightly paler and larger, but difference too slight to warrant recognition of *hostilis* Kleinschmidt, 1915, in Britain and of *semiretschiensis* Zarudny and Kudashev, 1916, *sibiricus* Khakhlov, 1928, or *baicalicus* Keve, 1943, in central and eastern Asia (Hartert 1921–2; Stachanow 1931; Vaurie 1956*f*). Populations from Mediterranean France, central and eastern Iberia, Balearic Islands, Balkans from Yugoslavia and southern and eastern Rumania south to Greece, and western and central Asia Minor paler than nominate *domesticus*, about intermediate between nominate *domesticus* and *biblicus* of Levant; cap and rump of ♂ paler, medium ash-grey, fine dark shaft-streaks on cap (if any) more contrasting, rufous fringes of mantle and scapulars slightly paler, more pink-cinnamon when worn, less rufous-chestnut; grey of ear-coverts, upper cheek, lower side of neck, side of breast, and flank paler, light ash-grey, less washed brown or olive; lower cheek and bar at upper side of neck almost pure white, less mottled grey; more frequently, a white supraloral line; medium grey on side of breast and flank more restricted, pale grey along lateral and lower border of bib and on chest, and white of belly more extensive; under tail-coverts grey-white, less isabelline-grey; ♀ paler buff-brown on upperparts than in ♀ nominate *domesticus*, especially on cap, dark stripe along upper border of supercilium more distinct; dark streaks on mantle and scapulars less deep black; side of neck and underparts paler buff, tinged olive-grey on lower flank, extensively off-white on belly and vent. These birds have been separated as *balearoibericus*, being similar in colour to typical birds of that race from Balearic Islands, although latter are generally slightly smaller. *P. d. tingitanus* from North Africa (Morocco to Libya) similar in colour to *balearoibericus*, but ♂ has pronounced black streaking on cap, entire cap sometimes mottled dark grey and black when light ash-grey fringes worn off in abraded plumage. For hybridization with *P. hispaniolensis* in North Africa and for position of × *italiae*, see *P. hispaniolensis*. *P. d. biblicus* from Levant east to western Iran a pale race; cap and rump of ♂ light ash-grey, both marked sandy-buff when fresh (cap then not as grey as in nominate *domesticus*); fringes of feathers of mantle, scapulars, tertials, and greater coverts rufous-cinnamon, paler pink-buff on feather-tips when fresh; ear-coverts and upper cheek pale ash-grey with slight buff suffusion; chest, side of breast, and flank with paler and more restricted pale ash-grey wash than in *balearoibericus*, underparts extensively white; lesser upper wing-coverts rufous-chestnut, not as deep chestnut; cap and rump of ♀ light sandy drab-grey, dark streaks on mantle and scapulars rather narrow, sooty or brownish; underparts pale isabelline-grey; pale fringes of wing paler isabelline-buff, less dull cinnamon-buff. *P. n. niloticus* from Egypt similar to *biblicus* or slightly paler, differing mainly in much smaller size; as large as *indicus* from Elat and Arabia eastwards, but ear-coverts grey (not white) and grey of cap extends usually to nape; as large as *rufidorsalis* from Sudan, but that race has ear-coverts white, nape, mantle, and scapulars extensively rufous-chestnut or maroon, marked with narrower black streaks on lower mantle and scapulars, and rump and upper tail-coverts spotted rufous, not uniform grey-brown. Situation in eastern Turkey and Transcaucasia not clear; birds in south-east Turkey (bordering Syria) and in central and eastern Transcaucasia are *biblicus* (Kumerloeve 1963, 1969a, 1970a; Stepanyan 1978); birds from eastern Turkey (Van area, Erzurum) are near *balearoibericus*, but upperparts slightly darker, nearer nominate *domesticus*, and ear-coverts paler, often silvery-white when plumage worn, size large; perhaps separable as *mayaudi* Kumerloeve, 1969; birds from western Transcaucasia and neighbouring north-east Turkey slightly darker on upperparts than nominate *domesticus* and perhaps separable as *colchicus* Portenko, 1960, but here provisionally included in nominate *domesticus*. Position of *hyrcanus* in northern Iran also unclear: 4 birds examined by Vaurie (1949a, 1950b) considered to be similar to *indicus*, apart from occurrence of blackish streaks on crown and darker chestnut mantle and scapulars, and therefore included in *indicus* group, but larger samples of birds from highlands of Teheran closely similar to nominate *domesticus* in colour and size, and typical *hyrcanus* from northern lowlands like nominate *domesticus* but smaller, much darker than *biblicus* (Stresemann 1928; Paludan 1938; Schüz 1959; Diesselhorst 1962), and *hyrcanus* therefore included in *domesticus* group, following Stepanyan (1978, 1990), though he considered *hyrcanus* to have paler upperparts than population occurring eastern Transcaucasia ('*caucasicus*' Bogdanov, 1879), which is near to *biblicus*. *P. d. persicus* from southern, central, and eastern Iran and western lowlands of Afghanistan intermediate in size and colour between *biblicus* and *indicus*, becoming gradually paler and smaller towards east, there grading into *indicus* (Vaurie 1949a; see also Diesselhorst 1962). *P. d. indicus* from lowlands of Indian peninsula, Arabia (except north-east), and Elat (southern Israel) small, cheeks and ear-coverts of ♂ white (slightly tinged grey at rear of ear-coverts when plumage fresh), crown medium ash-grey, nape, mantle, and scapulars extensively rufous-chestnut, side of breast and flank with restricted light grey, bib extensively bordered white; bill relatively more slender than in *domesticus* group; grades into *biblicus* in northern Saudi Arabia (Meinertzhagen 1954) and in southern Levant (Zedlitz 1912); *hufufae* from north-east Arabia has cap, rump, and upper tail-coverts greyer than in *indicus*, mantle and scapulars paler, less cinnamon, underparts whiter, fringes of greater coverts and tertials paler cinnamon (Hartert and Steinbacher 1932-8). *P. d. bactrianus* from central Afghanistan north to Kazakhstan similar to *indicus*, but larger and even paler below, side of breast and flank with more limited pale grey, mainly white. *P. d. parkini* from Himalayas closely similar in colour and size to *bactrianus*, chestnut of mantle and scapulars darker and richer, extending over nape, and bill and wing slightly larger (Vaurie 1949a, 1959); birds from northernmost Pakistan and Kashmir intermediate between *bactrianus* and *parkini*. For trend of variation in Europe, see Keve (1960) and Johnston (1969c).

Introduced in many places all over the world (see Summers-Smith 1963, 1988). In some areas, founder populations evolved rapidly into divergent forms, making *P. domesticus* an interesting species for study of speed of evolution. For variation and evolution in (especially) North America, see Lack (1940), Calhoun (1947b), Johnston and Selander (1964, 1966, 1971, 1973a, 1973b), Packard (1967a), Selander (1967), Selander and Johnston (1967), Grant (1972), Johnston et al. (1972), Rising (1973), Johnston (1973a, b, 1976), Klitz (1973), Blem (1975), Johnston and Klitz (1977), Hamilton and Johnston (1978), Johnston and Fleischer (1981), Fleischer and Johnston (1984), McGillivray (1984), McGillivray and Johnston (1987), and Parkin (1988); for New Zealand, see Baker (1980); for South Africa, see Crowe et al. (1980); see also Keve (1966c).

Forms superspecies with *P. hispaniolensis*, Somali Sparrow *P. castanopterus*, and Desert Sparrow *P. simplex* (Summers-Smith 1988). CSR

---

*Passer hispaniolensis* **Spanish Sparrow**

PLATES 21 and 26 (flight)
[between pages 280 and 281]

Du. Spaanse Mus    Fr. Moineau espagnol    Ge. Weidensperling
Ru. Черногрудый воробей    Sp. Gorrión Moruno    Sw. Spansk sparv

*Fringilla hispaniolensis* Temminck, 1820. Synonym: *Passer salicicola*.

Polytypic. Nominate *hispaniolensis* (Temminck, 1820), Canary Islands, North Africa from Morocco to north-west Libya, Iberia, Sardinia, Balkan countries, and Greece; introduced Cape Verde Islands and Madeira; grading into next race on Cyprus, and in Turkey, Levant, Caucasus area, and north-west Iran; *transcaspicus* Tschusi, 1903, Iran (except north-west) and Transcaspia east to eastern Kazakhstan and Afghanistan.

N.B. Locally forms hybrid populations with House Sparrow *P. domesticus*, of which the following appear to be more or less stabilized: × *maltae* Hartert, 1910, Malta, Sicily, and small islands off Sicily, including Pantelleria; × *italiae* Vieillot, 1817, southern Switzerland (south of Alps) and northern and central Italy; also Crete (Greece). See Geographical Variation.

**Field characters.** 15 cm; wing-span 23-26 cm. Averages slightly larger than House Sparrow *P. domesticus*, with slightly heavier bill and stronger plumage pattern contributing to bolder form. Rather large handsome sparrow differing distinctly from *P. domesticus* in ♂'s dark chestnut crown, whiter cheeks, and black-splashed chest, flanks, and back, and more streaked appearance, particularly on flanks of ♀ and juvenile. Commonest call distinctive. Sexes dissimilar; seasonal variation only by wear in ♂. Juvenile separable at close range. 2 races occur in west Palearctic. Around Mediterranean, populations hybridized with *P. domesticus* occur, particularly in southern Switzerland, Italy, Malta, and Sicily (e.g. Italian Sparrow, × *italiae*); see Geographical Variation.

(1) West Palearctic race, nominate *hispaniolensis*. ADULT MALE. Moults: July–September (complete); brightest plumage produced by wear. Crown, nape, and sides of neck buff- to grey-brown, wearing to uniform dark chestnut; short white supercilium broken over eye and ending above end of black loral streak and eye-stripe which terminates behind eye; supercilium nevertheless sharp and distinct compared to smaller similar mark of *P. domesticus*. Chin and throat black, forming wide spread of blackish marks over breast and along flanks and lateral tail coverts; with wear, becomes fully black bib and chest, spreading as pronounced black arrowheads along flanks towards base of tail; most arrowheads retain narrow whitish fringes and present highly decorative and unique pattern. Between chestnut nape and black bib, grey-white cheeks (almost white in breeding season) stand out; almost white belly is striking between black-marked flanks. Mantle and scapulars basically grey-buff, wearing to bright buff, but look very dark, with black-splashed scapulars and mantle showing pale edges. Rump dull grey. Wings and tail as *P. domesticus*. Bill horn, becoming black in breeding season. Legs bright brown. ADULT FEMALE. Indistinguishable at distance from *P. domesticus*. Close to, best-marked individuals show somewhat paler, greyer crown and face, more mottled bib, distinct dusky furrows from breast along flanks, whiter belly, and rather sharper streaks on mantle, scapulars, and larger wing-coverts. Tips of median and greater coverts and edges of tertials also cleaner and whiter, forming sharper bars and lines. Bill buff-horn. Legs grey-brown. JUVENILE. Resembles least-marked ♀, and also indistinguishable from juvenile *P. domesticus*. Best aged by gape-flanges and begging behaviour. (2) Hybrids with *P. domesticus*. Full range of intergrades occur. Chestnut or black mottling on crown and heavily black-splashed scapulars are most trustworthy indicator of *P. hispaniolensis* parentage.

♂ easily separated from *P. domesticus*, given clear sight of more vivid head pattern and more extensive black tracts in plumage, but distinguished from hybrid, if at all, only by close inspection. Crucial characters of full-blooded *P. hispaniolensis* are voice (see below) and full black gorget, which joins black-splashed shoulders and back, extends into streaks and arrowheads along entire flanks, and confines pale underparts to belly, vent, and under tail-coverts. ♀ said to be mostly inseparable from *P. domesticus* (Svensson 1992), but many birds, in Jordan at least (resembling eastern race *transcaspicus*), show subtle differences from *P. domesticus* noted above (D I M Wallace). Flight much

as *P. domesticus* but action rather less whirring. General behaviour as *P. domesticus* but still adapted to natural habitat, often breeding in large colonies among tree clumps. Gregarious, in countryside forming flocks just as dense as *P. domesticus*.

Vocal repertoire similar to *P. domesticus*, but calls and notes strung into song are all richer and deeper. Most distinctive is abrupt, contralto 'chup' or 'tcheup'.

Habitat. In contrast to House Sparrow *P. domesticus*, remains confined to narrow lower middle latitudes, largely Mediterranean but extending east into west-central Asia in steppe and semi-desert valleys, sometimes ascending foothills and locally breeding in mountains in Turkmeniya up to 2300 m and in Afghanistan in willow *Salix* scrub up to c. 2300(-2750) m (Paludan 1959). Typically, however, a warm lowland moisture-loving species inhabiting trees, shrubs, thickets, and reedbeds along riversides or irrigation ditches, groves of olives *Olea*, date palms *Phoenix*, *Acacia*, and eucalyptus, and even glades in woods and forests, where nests are often built in foundations of nests of storks *Ciconia* and eagles *Aquila*. In the course of recent evolution it seems that this species tended to diverge from *P. domesticus* partly by becoming adapted to less arid areas and even to moist habitats, and partly by preferring to nest in vegetation and less frequently occupying human cultivation and settlements (Summers-Smith 1963, 1988). Nevertheless, in countries on both sides of Mediterranean, has tended to converge ecologically with *P. domesticus* and has even formed stable hybrid populations, × *italiae* and × *maltae* (see Geographical Variation). In northern Morocco, a colony occupied old nests of House Martins *Delichon urbica* and Red-rumped Swallows *Hirundo daurica* under arches of mosque (Mountfort 1958) in manner of *P. domesticus*, while in southern Morocco a flock was encountered on bare desert on fringe of Sahara (Bannerman and Priestley 1952), and in northern Tunisia breeding occurs mainly in trees in villages and towns, which are later abandoned by a move out into farmland and finally a southward migration to warmer winter quarters. In Tunisia, seems to be a typical steppe species which avoids woodland and mountains and has taken advantage of cultivation of cereals; where *P. domesticus* is locally absent, seasonally occupies the normal urban niche of that species (Bortoli 1973); see also Summers-Smith (1988). In India, winters in large flocks, both in cultivation and semi-desert (Ali and Ripley 1974).

Distribution. Has spread in Balkan region in recent decades. Colonization of Atlantic islands began in early 19th century and not completed until 2nd half of 20th century (Summers-Smith 1988).

FRANCE. Corsica: established in south of island (Thibault 1983). YUGOSLAVIA. Expanding north and northwest, more slowly up coast than inland (Lukač 1988; VV). Since 1950 has spread north to Vojvodina at c. 45°N

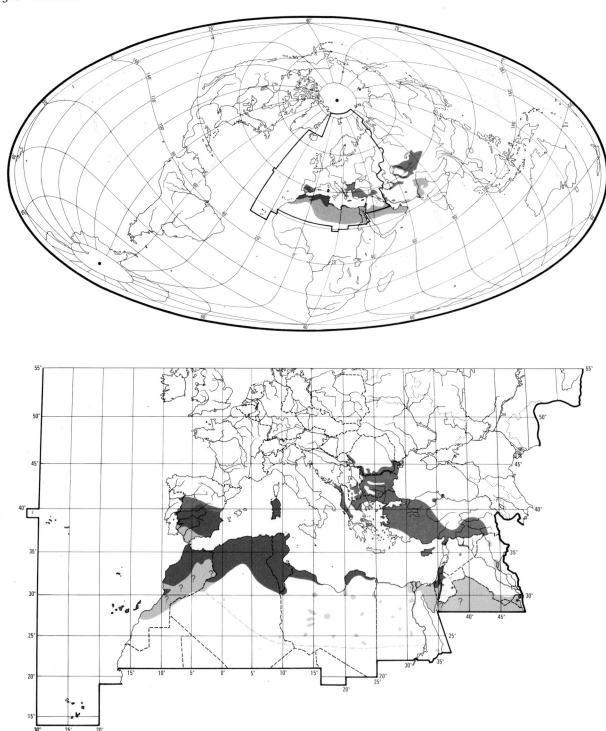

(Summers-Smith 1988). BULGARIA. Spread north along coast to Balchik by 1960, and continuing northward thereafter (TM). RUMANIA. First recorded (in southern Dobrogea) 1964; subsequently spread north to southern edge of Danube delta, Bucharest, and up river Siret to *c.* 46°N

(Tălpeanu and Paspaleva 1973, 1979). USSR. Spread north to Moldavia (*c.* 46°N) since 1950 (Summers-Smith 1988). MADEIRA. Arrived May 1935 after persistent easterly winds and became established (Bannerman and Bannerman 1965). CANARY ISLANDS. Said to have been introduced, but colonization probably a natural westward extension of range. First recorded dates earliest in eastern islands (Lanzarote 1828, Fuerteventura 1830), latest in western islands (La Palma and Gomera 1949, Hierro 1960). (Summers-Smith 1988.) CAPE VERDE ISLANDS. Colonization probably began early in 19th century, not reaching north-westernmost islands until 2nd half of 20th century; not known to have been deliberately introduced. Populations vulnerable to drought; probably became extinct on Brava in droughts of 1940s but recolonized between 1963 and 1982, and may have become extinct on São Nicolau in 1980s. (Summers-Smith 1988; CJH.)

Accidental. Britain.

**Population.** No quantitative data. Has clearly increased in areas of range expansion; few changes reported elsewhere.

PORTUGAL. Apparently decreasing (RR). ISRAEL. At least a few tens of thousands of pairs (Shirihai in press). CAPE VERDE ISLANDS. Island populations have fluctuated markedly, probably due to drought (Summers-Smith 1988); see Distribution.

Survival. Oldest ringed bird 11 years 3 months (Dejonghe and Czajkowski 1983).

**Movements.** Pattern very complex. Some southern populations mainly sedentary, but others partially migratory. Populations in north-west Africa both migratory and nomadic. Eastern populations show more regular migratory behaviour, in some areas moving further north for successive breeding attempts. Winters in Spain, north Africa, Middle East, central Asia, northern Pakistan, and north-west India.

Populations in Cape Verde Islands, Canary Islands, Madeira, and Malta sedentary. Some of population breeding central Iberia moves to south coast of Spain and possibly across to Morocco (Alonso 1986a): one bird ringed in Berkane (Morocco) recovered near Málaga on south coast of Spain (Heim de Balsac and Mayaud 1962), though scarcity of records for Gibraltar (Córtes *et al.* 1980) and lack of observations of passage in Tingitane peninsula of Morocco (Pineau and Giraud-Audine 1979) suggest that any movement across Strait of Gibraltar is small. Winters in very small numbers in Le Rharb (north-west Morocco) (Bachkiroff 1953), but most African birds move south to Western Sahara and northern limits of Sahara south of Atlas mountains, with scattered records as far south as 24°N, e.g. Timimoun and Djanet (Algeria), Fezzan and Libyan Desert (Libya) (Guichard 1955; Heim de Balsac and Mayaud 1962; Etchécopar and Hüe 1964; Summers-Smith and Vernon 1972; Bundy 1976, for which see details of scattered southern Libyan wintering records). Pluckings in colony of Eleonora's Falcon *Falco eleonorae* suggest it is a common passage migrant in south Aegean (Ristow *et al.* 1986). In Middle East, winters in southern Turkey (usually very uncommon, more rarely in quite large num-

bers) and Cyprus (where breeding population is augmented), east to Iraq (common), south to Israel (common) and in small numbers to Sinai and north-west Saudi Arabia; regular (often in large flocks) in Kuwait, but very uncommon elsewhere in Gulf states and south to Oman and Masirah (Allouse 1953; Beaman 1978; Bundy and Warr 1980; Gallagher and Woodcock 1980; Jennings 1981a, b; Martins 1989; Flint and Stewart 1992; Shirihai in press). Also winters commonly in Nile valley in Egypt south to northern Sudan (Cave and Macdonald 1955; Goodman and Meininger 1989). Provenance of wintering birds in Egypt and Sudan not clear. Meinertzhagen (1921) considered them to be nominate *hispaniolensis*; Nicoll (1922) shot a bird of this race in Sudan, but placed a series of others, as well as one from Egypt, in eastern race *transcaspicus*. Vaurie (1959) considered majority *transcaspicus*, but (unlike Goodman and Meininger 1989) attributed all recent material to nominate *hispaniolensis*; support given for European origin by a ringing recovery in Greece; probably both races occur. Further east, pattern is of general displacement south, with some birds remaining on more southerly breeding areas. In Iran, winters commonly on south Caspian coast, but also passes through central plateau to winter in south. Large numbers of *transcaspicus* winter in Turkmeniya, smaller numbers in Tadzhikistan and Kirgiziya, but large numbers also cross Hindu Kush, mostly passing through Afghanistan (though small numbers winter in south-east) and western Tien Shan to winter in northern Pakistan and north-west India, where mixed roosts (with House Sparrow *P. domesticus*) of over 1 million birds recorded, reaching as far south as 25°N. (Ali 1963; Gavrilov 1963; Gistsov and Gavrilov 1984; S C Madge, D A Scott.) 7 ringing recoveries show that birds from Dzhambul district of southern Kazakhstan winter in India (Ali and Ripley 1974).

Studies in Dzhambul and western Tien Shan suggest adults depart after breeding without completing moult, ♂♂ preceding ♀♀, whereas young migrate later after post-juvenile moult (Gavrilov 1963, 1972). In Bulgaria, Baumgart (1980) also found that birds departing August separated into ♂- and ♀-plumaged groups. Similarly Meinertzhagen (1940) observed autumn flocks predominantly of ♂♂ in Sous (southern Morocco). On return in spring, ♂♂ again precede ♀♀. Same routes taken in both seasons, and ringing controls indicate that timing of individual birds is relatively constant from year to year. Migration appears to be diurnal, with birds in flocks of 5–300, keeping to open areas and bushes. (Gavrilov 1962a, 1963, 1972; Gistsov and Gavrilov 1984.) Departure of northern populations of nominate *hispaniolensis* from central Spain begins late September, peaking October (Alonso 1986a). Passage along Atlantic coast of Morocco in October possibly involves some birds from Spain (Smith 1965). Majority of breeding population in northern Tunisia leaves October (Bortoli 1973). Departure of eastern nominate *hispaniolensis* west of Black Sea takes place September–

October, many passing through Cyprus September–November (peaking first half October) (Flint and Stewart 1992) and Levant, though flock on Crete in October (Summers-Smith 1980) perhaps suggests direct passage across Mediterranean to Egypt. Birds depart Caucasus from September (Dementiev and Gladkov 1954) with passage noted Iraq in October (Marchant 1962). Kazakhstan populations of *transcaspicus* begin to move south in September, crossing western Tien Shan September–October (Gistsov and Gavrilov 1984), and Hindu Kush, through North West Frontier Province (Pakistan), early August to November (Whitehead 1909; Ali 1963; Ali and Ripley 1974).

Return migration in spring follows similar pattern with birds leaving winter quarters from mid-March. In Le Rharb (north-west Morocco), breeding birds arrive April, but rear only 1 brood before moving east for another breeding attempt (Bachkiroff 1953). In Tunisia, 1st brood reared in north-east and birds then move west for 2nd brood (Bortoli 1973). Common on passage through Cyprus, though movement very protracted, late March to early May, occasionally late May (Flint and Stewart 1992). Protracted movement confirmed by arrival in Rumania and Vojvodina (Yugoslavia) in April (Baumgart 1980), when large numbers still moving through Rhodes (Jenning 1959), and west along north-east coast of Greece in early May (Raines 1962). According to Baumgart (1980) only 1 brood reared in southern Balkans and then birds move north-east to Danube basin for further broods (see above for similar behaviour in north-west Africa). Begin to leave Iraq at end of March (Marchant and Macnab 1962) and gradually disappear by mid-April (Allouse 1953). Birds leave India mid-March to mid-April with passage through North West Frontier Province from mid-March to mid-May (Whitehead 1909; Ali and Ripley 1974), through western Tien Shan from beginning of April to early July, peaking 26 April to 2 May (Gavrilov 1962a; Gavrilov 1963; Gistsov and Gavrilov 1984; Gavrilov and Gistsov 1985). Arrives in breeding area in Alma-Ata region in 1st half of May, but not until end of May or beginning of June in Chokpak pass region of western Tien Shan (Gistsov and Gavrilov 1984).

In addition to apparent regular migratory behaviour in north-west Africa, extensive ringing of 27 500 juveniles in Morocco showed nomadic pattern of movement with young departing from natal areas in summer and dispersing widely, some ringed at Berkane settling down to breed to the east towards Oran (Algeria), others recovered to west in Le Rharb and Meknes-Fez area and further south-west in Western Sahara (Bourdelle and Giban 1950-1; Bachkiroff 1953; Heim de Balsac and Mayaud 1962). Northward movement of young birds in Bulgaria in August indicates similar dispersive movement (Uhlig 1984). JDS-S

**Food.** Plant material (mainly seeds) and invertebrates.

Plant material taken from low herbs or ground (shed seeds), but cereal seeds taken from ripening heads of grain at 'milk' stage; also buds and fruit from trees (Dementiev and Gladkov 1954; Gavrilov 1962b, 1963). Invertebrates taken mainly from ground, but also by searching leaves of bushes and trees, by fluttering in front of leaves and by catching insects in flight (Ticehurst and Whistler 1938; Baumgart and Stephan 1974; J D Summers-Smith).

Diet in west Palearctic includes the following. Invertebrates: grasshoppers (Orthoptera: Acrididae), earwigs (Dermaptera: Forficulidae), mantises (Dictyoptera), bugs (Hemiptera), larval and adult lacewings (Neuroptera: Ascalaphidae), larval and adult Lepidoptera (Noctuidae), flies (Diptera), Hymenoptera (wasps Sphecoidea, bees Apoidea, ants Formicidae), beetles (Coleoptera: Carabidae, Scarabaeidae, Coccinellidae, Chrysomelidae, Curculionidae), centipedes (Chilopoda: Scolopendridae), spiders (Araneae), and small snails (Gastropoda). (Bachkiroff 1953; Metzmacher 1984.) Plant material mainly seeds, but also green shoots, buds, flowers, and pulpy fruits: hemp *Cannabis*, persicaria *Polygonum*, *Portulaca*, goosefoot *Chenopodium*, *Noaea*, *Amaranthus*, mouse-ear *Cerastium*, spurrey *Spergula*, campion *Silene*, wall-rocket *Diplotaxis*, cress *Lepidium*, mignonette *Reseda*, strawberry *Fragaria*, apple *Malus*, pear *Pyrus*, plum, apricot, and peach *Prunus*, orange *Citrus* (fruit), grape *Vitis* (fruit), fig *Ficus* (fruit), jujube *Ziziphus* (shoots and fruit), pea *Pisum* (including buds), rest-harrow *Ononis*, clover *Trifolium*, chickpea *Cicer*, storksbill *Erodium* (including buds and flowers), cranesbill *Geranium*, *Peganum harmala*, spurge *Euphorbia*, mallow *Malvus*, *Lavatera*, olive *Olea* (fruit), bindweed *Convolvulus*, gromwell *Lithospermum*, dead-nettle *Lamium*, hemp-nettle *Galeopsis*, nightshade, potato *Solanum*, tobacco *Nicotiana*, tomato *Lycopersicon*, sweet pepper *Capsicum*, yellow rattle *Rhinanthus*, cornsalad *Valerianella*, *Artemisia*, sow-thistle *Sonchus*, date *Phoenix*, prickly pear *Opuntia* (fruit), Liliaceae, reedmace *Typha*, Gramineae (mainly seeds, also green shoots and flowers), e.g. wild grasses *Stipa*, *Poa*, *Phalaris*, *Panicum*, *Echinochloa*, *Digitaria*, *Bromus*, *Schismus*, and cultivated cereals, wheat *Triticum*, barley *Hordeum*, oats *Avena*, millet *Panicum*, maize *Zea*, rice *Oryza*. (Bachkiroff 1953; Metzmacher 1984; Alonso 1986b.) Bread and kitchen scraps taken near houses in areas where House Sparrow *P. domesticus* absent (J D Summers-Smith). For additional items in diet of extralimital *transcaspicus* in Kazakhstan, see Dementiev and Gladkov (1954), and especially Gavrilov (1962b).

Plant material predominates in diet of adults, but insects taken throughout year and young fed mainly insects. In Spain, 112 stomachs in winter gave 95·4% by weight vegetable matter, mainly Gramineae seeds (wild grasses 75·1%, cultivated cereals 13·7%), with only 1% invertebrates (Alonso 1986b). In Libya, 55 crops and gizzards (March–April) gave 86% by weight plant material, 14% invertebrates (Mirza *et al.* 1975). In Morocco, of 291 gizzards, November–March, 95·6% contained plant material (43·5% contained cereals) and 26·9% invertebrates; of 259 gizzards, April–October, 97·3% contained plant material (70·8% cereals) and 35·6% invertebrates (Bachkiroff 1953). Similarly in Kazakhstan, of 432 stomachs of extralimital *transcaspicus*, April–August, cultivated cereals in 75·0%, seeds of wild plants in 13·6%, and insects in 20·3% (Gavrilov 1963). Insect food said to be particularly important in spring before appearance of cereal seeds (Dementiev and Gladkov 1954).

In Morocco, Bachkiroff (1953) found one bird typically ate *c.* 20 grains of wheat or barley, or 87 grains (*c.* 2·4 g) of rice, per day. During 170-day stay in Kazakhstan breeding area, 'standardized' population of 100 adults plus offspring consumed estimated 65 kg per km² (1·63 kJ per m²) (20·0% insects, 66·2% cereal grain, 13·8% weed seeds), with young accounting for 22·9% (Wiens and Dyer 1977).

Food of nestlings in Algeria (208 stomachs, April–May) apparently contained more plant material in 2nd half of nestling period than in 1st; up to 6–8 days old, 94 insect items included 64% (by number) grasshoppers and 11% bugs; in remainder of nestling period, of 114 insects, 32% grasshoppers, 21% beetles, 19% ants, and 11% bugs (Metzmacher 1984). In Libya, 2 stomachs of nestlings less than 4 days old contained 75% animal food by weight (Mirza *et al.* 1975). In Kazakhstan, 89·5% of 679 nestling stomachs contained insects, proportion changing little with age and season; stomachs of fledglings also contained high proportion of insects, though this fell as season advanced; 94% on 12 June (*n*=32), 73% on 29 June (*n*=201). On the other hand, proportion of stomachs containing vegetable food varied greatly with the sample, ranging from 8% (*n*=237) in one colony to 82% (*n*=87) in another, no doubt reflecting availability (Gavrilov 1963). JDS-S

**Social pattern and behaviour.** No systematic investigations in west Palearctic. For detailed extralimital studies of *transcaspicus*, see Gavrilov (1962a, 1962a, b, 1963, 1972) and Ivanitski (1984, 1985); compared with House Sparrow *P. domesticus bactrianus*, breeding birds generally less social, with activity confined more to nest and less to surrounding area; *P. hispaniolensis* also has more high-intensity display postures (Ivanitski 1985, which see for further detailed schematic illustrations of display sequences). For comparison with *P. domesticus* of patterns of colony formation in migratory populations, see Ivanitski (1984).

1. Highly gregarious throughout year, more so than *P. domesticus* (Blyth 1867; Meinertzhagen 1954). Independent young form large flocks that roam widely in search of food (Heim de Balsac and Mayaud 1962; Bortoli 1973); may remain separate or be joined later by adults that have finished breeding. If food dispersed, flocks break up into smaller bands for foraging, but re-form at communal night-roost (see Roosting, below). In Spain, autumn, foraging flocks up to 4000, mean 232±SE57 (Alonso 1986b). In migratory populations, adults depart before young, both adults and young migrating in single-sex flocks; similar separation may be maintained in winter quarters (Meinertzhagen 1940; Gavrilov 1962a, b, 1963, 1972; Baumgart 1980). In

Kazakhstan, average flock size migrating in autumn 32 (*n*=130) (Gavrilov 1962a). Behaviour in winter quarters similar to immediate post-breeding period, birds dispersing during day into foraging flocks, size depending on availability of food. In Morocco, numbers collecting at daytime roosts increased in January–February to many thousands preparatory to return migration to breeding area (Bachkiroff 1953). On spring migration, flock on Rhodes (Greece) 50–100 birds (Jenning 1959); in Kazakhstan, 5–300 (Gavrilov 1962b, 1963). In winter, foraging flock associates with other seed-eaters, e.g. Goldfinch *Carduelis carduelis*, Linnet *C. cannabina*, Serin *Serinus serinus*, *P. domesticus*, Rock Sparrow *Petronia petronia* (Nicoll 1912; Steinbacher 1956; Bannerman and Bannerman 1965); in India also with Rose-coloured Starling *Sturnus roseus* (Ali 1963). Flocks foraging in crops move about in 'clouds', settling to feed from time to time (Ali 1963). Bonds. Essentially monogamous (polygamy unknown: Metzmacher 1990), though not known if pair remains together for subsequent breeding attempts that are often in different locations (see also Breeding Dispersion, below). Promiscuity common, but attempts at extra-pair copulation apparently rarely successful (Ivanitski 1985). Hybridizes freely with *P. domesticus* in north-west Africa from eastern Algeria to Tripolitania (Libya); stabilized hybrid populations occur in Italy, Corsica, and Crete, and at isolated North African oases; hybrids occur less frequently in other parts of range, i.e. Cape Verde Islands, elsewhere in Mediterranean basin, central Asia. (Hartert 1913; Stanford 1954; Meise 1936; Johnston 1969; Naurois 1969; Summers-Smith and Vernon 1972; Panov 1989.) For discussion of isolating mechanisms, see Panov and Radzhabli (1972), Baumgart (1980, 1984), and (Panov 1989). 2 hybrids with Tree Sparrow *P. montanus* reported from Malta (Smith and Borg 1976; Sultana and Gauci 1982). Both sexes build nest (Sultana and Gauci 1982), but at times ♂ guards nest when ♀ builds (Baumgart 1980); both sexes incubate and feed young (Metzmacher 1990). Young independent a few days after leaving nest (see Relations within Family Group, below). Age of first breeding not known. Breeding Dispersion. Breeding colonial, with some colonies very large and nests closely packed. Size of colony very dependent on local conditions, ranging from a few pairs to many thousands. In Libya, 2 colonies had 118 pairs in 200 ha, 442 pairs in 70 ha (Mirza *et al.* 1975). In Morocco, one colony had estimated 125 000 nests in 60 ha, with up to 50 (more typically 10–20) nests in one tree (Bachkiroff 1953). In Spain, colony size 2–2500 nests (mean 213, *n*=63), with up to 180 nests per tree (Alonso 1986a). In survey in Algeria in 1983, *c.* 35 310 pairs in 13 colonies bred over *c.* 40 000 km² (0·9 pairs per km²) (Metzmacher 1986d). In Kazakhstan, colonies of 200–800 000 nests, up to 120–130 nests per tree; densely packed with 1 ha of forest plantation holding average 13 000 nests, or strung out linearly over several km; in one area, estimated 1·3 million pairs nested in 1 km² (Gavrilov 1963; Korelov *et al.* 1974). In places, 3–4 nests can be touching (Makatsch 1955). Territory limited to immediate neighbourhood of nest (J D Summers-Smith). Where population sedentary (e.g. Malta), same colony area used year after year and nests may be re-used (Sultana and Gauci 1982); in migratory populations, same colony sites can be used for successive broods in one year (though not recorded if same nest used for subsequent brood) and from year to year, but new sites frequently used and birds may move to new area for successive broods (Bachkiroff 1953; Dementiev and Gladkov 1954; Gavrilov 1962b, 1963; Baumgart 1980; Metzmacher 1986d). *P. domesticus* sometimes a nesting associate. Roosting. Large roosts formed in trees or reedbeds, birds packed closely together (Chambers 1867; Alonso 1986b). In Sardinia, Sicily, Malta, and Atlantic Islands, where *P. domesticus* absent, roosts frequently in shade trees in town squares; one roost in Malta held 15 000-30 000 birds, another in Sardinia estimated to contain tens of thousands in November (Gibb 1951; Steinbacher 1956; Summers-Smith 1979, 1992; Sultana and Gauci 1982). In central Spain, post-breeding roost of mixed *Passer* increased from 20 000 birds in July to maximum 60 000 birds September-October (64% *P. hispaniolensis*), falling to 4000 in November when most had emigrated. In post-breeding daytime roosts formed in willows *Salix*, mean 418 ± SE123 birds, thus larger than local foraging flocks (see above) (Alonso 1986b). Winter roost in India estimated to hold over 1 million *Passer*, including both *P. hispaniolensis* and *P. domesticus* (Ali 1963). Catchment area for large roosts not known, but birds in Malta came from at least 8 km (Gibb 1951), those departing from Sardinia roost crossed high up over 3 km of open water to mainland (Summers-Smith 1992), and birds roosting on Fuerteventura (Canary Islands) traced as far as 10 km (J D Summers-Smith). See Flock Behaviour (below), for details of aerial movements at roosts. Not all birds roost in trees. In Canary Islands, roosts have been reported in wells (Bannerman 1963); in Malta, some birds roost at nest-site (Sultana and Gauci 1982). Massive roosts of *transcaspicus* during spring migration found in field shelter-belts in Kazakhstan (Gavrilov 1962b).

2. Rather shy, foraging flocks not allowing approach of observer closer than 50 m (Guichard 1955; Steinbacher 1956). FLOCK BEHAVIOUR. At large roosts, birds arrive at roosting area in dense flocks from 30 min before to 30 min after sunset (Gibb 1951; Ali 1963) and fly about restlessly performing aerial evolutions for up to 20 min before settling (Adams 1864; Ali 1963). In morning, large flocks form, circling in the air above roost, before streaming away (Summers-Smith 1992). In smaller roosts with only a few thousand birds, no aerial evolutions, and arriving birds either fly directly into roosting trees or first gather in nearby trees, with this procedure reversed for morning departure (J D Summers-Smith). Noise deafening at large roosts when birds settling down in evening and before morning departure (see 2c in Voice) (Chambers 1867; Bannerman 1912; Ali 1963). SONG-DISPLAY. ♂ sings at nest (see 1 in Voice), to attract ♀ and to proclaim ownership. Sings with tail fanned, chest thrust forward showing off black throat, and wings held out, drooped, and violently shivered (Fig A) (J D Summers-Smith). Song-

B

A

display commonest before beginning of egg-laying, particularly in afternoon; declined during incubation, but began again at end of 1st week of nestling period, certain ♂♂ displaying when bringing food to young (Metzmacher 1990). ANTAGONISTIC BEHAVIOUR. During colony formation, ♂ responds to strange ♂ at nest with Forward-threat posture (Fig B), crouching with tail held up and fanned, wings held out, and scapulars thrust forward

(J D Summers-Smith). If rival does not retire, dispute can escalate to fighting with both ♂♂ grappling together, fluttering to ground (Sultana and Gauci 1982). Neighbouring ♂♂ quickly recognize each other (Baumgart 1980), and frequency of aggressive display rapidly decreases so that it is no longer seen by end of incubation period (Metzmacher 1990). ♀ shows similar, though less intense, display against rival ♀ at nest (J D Summers-Smith). ♂ defends nest-site with Advertising-call (see 2a in Voice), which can be given up to 30% of time near nest (Metzmacher 1990). HETEROSEXUAL BEHAVIOUR. (1) Pair-bonding behaviour. Pair-formation occurs mainly during formation of colony. In Morocco, Bachkiroff (1953) saw no pair-formation there and assumed that it occurred in pre-migration daytime assemblies or night-roosts, though in Kazakhstan Ivanitski (1985) considered that it occurs in colony. Initially, unpaired ♂♂ follow ♀♀ moving through colony and direct Song-display at them; ♀♀ either ignore or threaten such birds. ♂♂ then take up and defend suitable site giving Song-display to attract ♀. If unpaired ♀ approaches, ♂ hops round her in Song-display. This period of intense activity and noise, as nest-sites and pairs are being established, lasts for only a few days, though in large colonies the cycle is not closely synchronized and adoption of nest-sites and pair-formation continue in newly established areas as colony grows from its starting point. Pair formed once ♀ accepts site. ♂ now greets arriving ♀ with less intense version of Song-display (see Ivanitski 1985 for variants): wings shivered, but only held out slightly from body, tail raised and fanned. Soon ♀ becomes dominant, spending much time at nest-site, threatening arriving ♂ with mild Threat-call (see 3 in Voice) and associated display (see Antagonistic behaviour above). If ♂ in nest, ♀ wing-shivers on arrival and may attack him. (Gavrilov 1963; Sultana and Gauci 1982; Ivanitski 1985; J D Summers-Smith.) (2) Courtship-feeding. None. (3) Mating. Starts at least 3-5 days after pair-formation (Ivanitski 1985). Copulation (Fig C) occurs close to nest following wing-shivering Solicitation-

C

display by ♀ with calling (J D Summers-Smith: see 2d in Voice), after undescribed noisy display (Metzmacher 1990). ♂ frequently mounts only 1-2 times, though occasionally as many as 20 times in succession (J D Summers-Smith); 1–17 times (average 6·5, *n* = 55) (Ivanitski 1985). If ♀ resists copulation, ♂ pecks at her nape and back; between mountings, ♂ may adopt Song-display posture (Ivanitski 1985). For detailed descriptions of extra-pair copulations and ♂ trying to cuckold neighbour's mate, see Ivanitski (1985). (4) Nest-site selection and behaviour at nest. ♂ initially selects site and advertises it to attract mate. This may be bare site, or rudimentary nest may be started. ♂♂ spend up to 30% of time at nest giving Advertising-call (see 2a in Voice), decreasing from nest-building to end of laying period (Metzmacher 1990). Both sexes co-operate in nest-building once pair formed, though in dense colony where there is much stealing of nest-material, ♂ may remain guarding nest while ♀ builds; ♀ chases away other ♀♀ stealing material. (Baumgart 1980; Sultana and Gauci 1982; Ivanitski 1985; J D Summers-Smith.) ♂ takes major role at beginning, ♀ becoming more involved towards end of building and addition of lining, which is mainly by ♀ (Gavrilov 1963; Metzmacher 1990). Both sexes incubate during day (75% of daylight hours in Algeria: Metzmacher 1990), ♀ taking major part and overnight stint (Gavrilov 1962*b*, 1963). RELATIONS WITHIN FAMILY GROUP. Both sexes brood and feed young (Gavrilov 1962*b*, 1963; J D Summers-Smith), though during latter part of nestling period ♂ again reverts to Song-display (Metzmacher 1990). Young continue to be fed for 4–5 days after they leave nest (Bortoli 1973), though said by Bachkiroff (1953) to start feeding themselves on day after fledging. According to Gavrilov (1963), fledglings return at night to roost in nests in colony or in branches of nesting trees with adults. ANTI-PREDATOR RESPONSES OF YOUNG, PARENTAL ANTI-PREDATOR STRATEGIES. No details, but see 4 in Voice.

(Figs by R Gillmor: A from photograph in Gavrilov 1963; B from photograph in Gavrilov 1962*b*; C from photograph in Di Carlo and Laurenti 1991.) JDS-S

**Voice.** Generally very similar to House Sparrow *P. domesticus*, but advertising-calls typically fuller and louder with strident quality. Information from J D Summers-Smith except where otherwise indicated. For additional sonagrams, see Bergmann and Helb (1982).

CALLS OF ADULT. (1) Song. Strident, rapidly repeated 'cheeli-cheeli-cheeli' (Fig I), more metallic and slightly

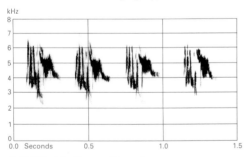

I  J-C Roché  Greece  May 1965

higher pitched than *P. domesticus* (L Svensson); used by ♂ in Song-display at nest to attract ♀ in pair-formation. Audible at range of several hundred metres. (2) Other advertising- and contact-calls. (2a) Basic 'chirp' and variants, e.g. almost disyllabic 'tchweeng' (Fig II, which shows terminal sub-unit not always present, being absent

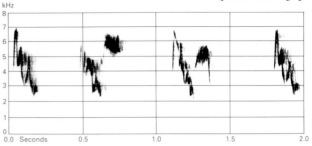

II  P J Sellar  Gran Canaria (Canary Islands)  June 1991

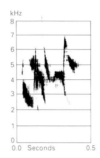

III  C J Hazevoet  Cape Verde Islands  September 1988

from 1st and 4th calls) or multisyllabic 'tchweeligig' (Fig III); given at nest by ♂. According to Metzmacher (1990), ♂♂ 'cheep' during up to 30% of time spent at nest, amount decreasing from nest-building to end of egg-laying period. Also given by ♀ but less frequently. Quieter 'chirp' (Fig IV) used away from nest for social contact. (2b)

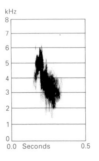

IV  J L Herelle  Spain  June 1990

Flight-call. Rather deeper 'churp' with variants, used as contact-call by members of flock in flight; 'dschwed' (Bergmann and Helb 1982). (2c) Social-singing. Conversational jumble of quiet 'chirps' and variants such as 'cheep', 'chip', 'cheer', 'cheeri', 'chewip', and 'chittup' used in chorus in daytime roost, and at nocturnal roost when collecting prior to arrival, when settling down into roost, or before morning departure. (2d) Solicitation-call. 'Chee-chee-chee-chee' or 'que-que-que-que' used by ♂ (subordinate to ♀: see also call 3) arriving at nest. Also by ♀ inviting copulation. Repeated 'pshie' (Bergmann and Helb 1982). (3) Threat-call. Nasal 'churr-it-it-it' (Fig V) similar to *P. domesticus*, but slightly deeper and rattle shorter. Used by dominant bird arriving at nest when

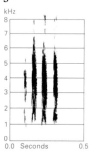

V  J L Herelle  Spain  June 1990

partner already there; immediately after pair-formation, by ♂ towards subordinate ♀, but within a few days ♀ becomes dominant and calls thus to ♂ arriving with nest-material or food for young. Call softens in tone as familiarity develops between pair and is replaced by call 2d. (4) Alarm-call. Nasal 'quer' may be given singly, in 'quer-quer' couplets (Fig VI), or as 'quer-quer-it', rather fuller

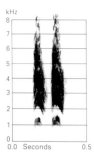

VI  C J Hazevoet  Cape Verde Islands
September 1988

than *P. domesticus*; used by both sexes at nest when enemy near, e.g. human observer. 'Terr' or 'tet-et-et' (Bergmann and Helb 1982).

Calls of Young. No information.  JDS-S

**Breeding. Season.** Laying begins in North Africa in March: Canary Islands mid-March (Bannerman 1963), Libya 1st week March (Mirza 1974), Morocco end of March (Bachkiroff 1953), but not normally until 1st half April in Algeria (Metzmacher 1986d) and mid-April in Tunisia (Bortoli 1973); 2nd clutches started in May in Algeria and Tunisia (Bortoli 1973; Metzmacher 1986d) and breeding continues until July (Metzmacher 1986d; Mirza 1974). Begins early in April in south-west Europe and continues until July or early August (Alonso 1984c; Sultana and Gauci 1982; Iapichino and Massa 1989). Migratory populations do not arrive in Balkans until April and 1st eggs not laid until early May, with breeding continuing until July–August (Baumgart 1980). In northern Caucasus (Russia), nest-building starts mid-April (Kazakov and Lomadze 1984). Unseasonal breeding can occur in sedentary populations, e.g 2 nests with young in Malta October–November (Sultana and Gauci 1982). Progression with season clearly follows spring temperature, but in Cape Verde Islands 2 separate breeding seasons: August–October, following rains, and February–

March (C J Hazevoet). In Kazakhstan, first eggs of *transcaspicus* not laid until 2nd half May or 1st week of June with breeding continuing until late July (Dementiev and Gladkov 1954; Gavrilov 1962b). According to Gavrilov (1963), ♀♀ are ready to begin laying on arrival in breeding area, eggs being laid before completion of nest and some even laid before arrival when still on passage. In all areas, breeding closely synchronized in colony (Bachkiroff 1953; Gavrilov 1963; Alonso 1984c; Metzmacher 1986d); in Kazakhstan, 75–89% of 1st eggs within 5 days (Gavrilov 1963). Site. Very large variety of sites used. Free-standing nests in trees, resting on branch forks, are most common. No particular preference shown for type of tree, pylons and telegraph poles providing suitable tree substitutes; more rarely low bushes (e.g. oleaster *Elaeagnus*), hedges, and reeds *Phragmites*. Nests commonly built into foundations of nests of large birds, including birds of prey, White Stork *Ciconia ciconia*, and crows *Corvus*, at times in company with House Sparrow *P. domesticus*; as many as 20 nests in *C. ciconia* nest in Algeria (Jourdain 1915). Also uses more enclosed sites, such as crowns of palms, cavities in trees, cliffs, or buildings, sometimes taking over old nests of Red-rumped Swallow *Hirundo daurica* and House Martin *Delichon urbica*. Use of man-made structures more common in sedentary populations in areas where *P. domesticus* absent, e.g. Sardinia, Sicily, Malta, Atlantic Islands. Nest height 0·5–30 m above ground. (Jourdain 1936; Bachkiroff 1953; Dementiev and Gladkov 1954; Makatsch 1955; Steinbacher 1956; Mountfort 1958, 1962; Sacarrão and Soares 1975; Sultana and Gauci 1982.) Nest: free-standing, large, untidy, roughly spherical, domed structure with entrance-hole on side. Normally constructed of dry grass or straw and thin twigs, but sometimes fresh green vegetation; leafy sprigs from trees and flower panicles sometimes incorporated. Nests in cavity of similar construction, filling most of available space. Nest-cup a more compact structure of fine grass, plant down, feathers, and animal hair. (Mountfort and Ferguson-Lees 1961; Gavrilov 1963; Sultana and Gauci 1982.) Mountfort and Ferguson-Lees (1961) described nest hanging freely on 25 cm long straw 'rope', recalling nest of weaver *Ploceus*. In Spain, orientation of entrance found to correlate significantly with prevailing wind direction (Rodríguez-Teijeiro and Cordero-Tapia 1983). Outer diameter 15–30 cm, entrance hole 3–4 cm diameter (Dementiev and Gladkov 1954). Average weight of nest 146 g (70–295, $n = 14$) (Gavrilov 1962b, which see for further nest measurements). Building: ♂ begins initial structure, either re-using old nest or new site (Gavrilov 1962b, 1963); ♀ joins in once pair formed, taking increasing role and completing most of lining; material loosely woven round supporting branches and in dense colonies interwoven with neighbouring nests (Gavrilov 1962b, 1963; Baumgart 1980); takes 4–7 days, but frequently, at least in *transcaspicus*, 1st egg laid before nest complete and building continues through laying of clutch and beginning of incubation (Gavrilov 1963; Baumgart 1980). According to Gavrilov (1962b, 1963) when nest made of stiff stems on good supporting fork, complete openwork structure is built in 2 days and then filled in during subsequent days; when suitable support lacking or soft fresh grass used, complete nest is built up from base and finally domed over. EGGS. See Plate 90. Sub-elliptical, smooth and slightly glossy. White or faintly tinted blue or green, marked with specks, spots, or small blotches in various shades of grey, violet-grey, blackish-violet, or purplish, often with darker markings concentrated at large end (Harrison 1975). Nominate *hispaniolensis* 22·2 × 15·2 mm (18·7–25·4 × 13·6–16·8), n = 478; *transcaspicus* 21·8 × 15·3 mm (18·0–25·7 × 13·7–16·8), n = 354 (Schönwetter 1984). × *italiae*: 22·3 × 15·5 mm (19·0–23·7 × 14·1–16·4), n = 55; calculated weight 2·83 g. Weight: Morocco, 2·68 g (1·9–3·3) (Bachkiroff 1953); Kazakhstan, 2·63 ± 0·016, n = 248 (Gavrilov 1963). Clutch: 4–6 (2–8); Cape Verde Islands 3–5 (Naurois 1988); Canary Islands 4–5 (Bannerman 1948); Tunisia 4–5, 1st clutch mean 4·3, 2nd 4·8 (Bortoli 1973); Algeria 1–6, mean 4·6, n = 1751 (Metzmacher 1986d), 3–6, mean 4·79 ± 0·65, mode 5, n = 33 (Metzmacher 1990); Malta 4–6 (Sultana and Gauci 1982); Sicily 4–5 (3–6) (Iapichino and Massa 1989); Bulgaria 2–8, mean 5·9, n = 57 (Makatsch 1955); Spain 1st clutch mean 4·9 ± 0·9, n = 527, 2nd 5·2 ± 0·8, n = 155, 3rd 5·0 ± 0·8 n = 4, overall 5·0 ± 0·9, n = 686 (Alonso 1984c); Kazakhstan, 2–7, mean 4·4, n = 1099 (Gavrilov 1962b, 1963). One egg laid per day. In Tunisia, 2 broods (Bortoli 1973); in Spain 1–3 clutches per nest per year, mean 1·33, n = 486 nests (Alonso 1984c). In Kazakhstan, in different years, 45·4%, 4·6%, and 21·0% of ♀♀ nested twice; if 1st clutch not started until June, no 2nd clutch laid (Gavrilov 1962b, 1963). INCUBATION. In Morocco 11–11·5 days (Bachkiroff 1953), in Tunisia 11 days (Bortoli 1973); in Kazakhstan, typically 12–13 (11–14) days (Gavrilov 1962b, 1963). Incubation by both sexes, but more by ♀. Starts with 3rd egg (Metzmacher 1986d); with 2nd egg in Kazakhstan, leading to asynchronous hatching. YOUNG. Cared for and fed by both parents. FLEDGING TO MATURITY. Fledging period in Tunisia 15 days (Bortoli 1973), in Kazakhstan 11–12 days (Gavrilov 1963). Young fed by adults for 4–5 days after leaving nest (Bortoli 1973). Age of first breeding not known. BREEDING SUCCESS. In Spain, 27·0% of eggs lost during incubation, 4·4% infertile or contained dead embryo, n = 2951; 52·2 ± 1·1% young fledged, n = 2005, overall success 35·5 ± 0·9%, n = 255 nests (Alonso 1984c). In Tunisia, over 3 years, 4·4–13·3% of eggs infertile or non-hatching; overall success 38·6–46·1%, 50% of 4023 clutches producing at least 1 young (Metzmacher 1986d, 1990). In Kazakhstan, 4·9% of eggs infertile, 1·2% contained dead embryo; loss of nestlings increased with clutch size giving 75% success for 2-egg clutches, falling to 55% for 6 eggs, overall success 57·6%; much nestling mortality attributed to asynchronous hatching, later-hatching young failing to survive in competition with older siblings (Gav-

rilov 1962b, 1963). Snake *Elaphe* and Cattle Egrets *Bubulcus ibis* recorded predating nestlings in Algeria (Metzmacher 1990). JDS-S

**Plumages.** (nominate *hispaniolensis*). ADULT MALE. In worn plumage (spring), forehead, crown down to upper ear-coverts, and bar behind ear-coverts deep maroon-chestnut, similar in colour to stripe behind eye of House Sparrow *P. domesticus*, much darker and deeper chestnut than vinous-chestnut of cap of Tree Sparrow *P. montanus*. Narrow white stripe above lore and small white spot above rear corner of eye, both often connected to form a narrow supercilium. Lore, narrow stripe below eye, and short stripe behind eye black; lower ear-coverts, cheeks, and bar at upper side of neck contrastingly white. Hindneck usually maroon-chestnut, like crown, but sometimes black or mottled black-and-chestnut. Mantle and scapulars broadly streaked black, sides of feathers of lower mantle and inner scapulars cream-white, showing as double whitish V-mark on upperparts, outer webs of outer scapulars fringed pale rufous-cinnamon at sides and cream-buff on tips. Back and upper rump black with pale grey-buff feather-fringes, lower rump and upper tail-coverts olive-grey with sandy-buff wash; some individual variation in extent of black on rump. Chin, throat, and chest black, lower chest with traces of off-white fringes on feather-tips; black on side of chest connected with that on side of mantle along lower side of neck, black on chest not appearing as a broad and smoothly round-ended bib as in *P. domesticus*. Lower border of chest marked with triangular black points, these merging into broad black streaks or arrow-marks on side of breast and flank and into narrower black streaks on upper belly; ground-colour of side of breast and flank pale cream-grey, merging into cream-white on belly, vent, and under tail-coverts; longer tail-coverts with black shafts (sometimes obsolete), thigh mixed grey and cream-white. Tail as *P. domesticus*, but pale fringes narrower and paler, pale grey-buff. Wing as *P. domesticus*; fringes of flight-feathers, tertials, and greater coverts rather pale cinnamon- or tawny-yellow, as in *indicus* group of *P. domesticus*, less rufous-cinnamon than in *domesticus* group; lesser coverts maroon-chestnut; underwing as *P. domesticus*. In fresh plumage (autumn and winter, except in Cape Verde populations), all feathers of cap, back, and rump broadly fringed sandy olive-grey, largely concealing chestnut on cap and black on back and rump, some birds (probably 1st adults) then rather resembling ♀ (Königstedt and Robel 1977); pale fringes slightly paler olive-buff on side of crown, forming ill-defined supercilium (white above lore concealed, but white spot above rear corner of eye usually just visible). Pale fringes of mantle and scapulars more cream-buff, less whitish, tips fringed olive-brown, black streaks less predominating. White of cheeks, ear-coverts, and lower side of neck washed pale olive-grey. Upper side of neck buff. Black feathers of chin and throat rather narrowly fringed cream-buff, but black still predominating; those of chest, side of breast, and flank more broadly fringed cream-buff, black concealed except for rounded or triangular dots on chest and side of breast. Fringes of flight-feathers, tertials, and greater coverts slightly warmer pink-cinnamon. Pale fringes of head and body wear off mainly in March-April. In worn plumage, quite different from *P. domesticus*, differing in fully chestnut cap (no grey on midcrown), more pronounced white supercilium, broader black stripes on mantle and scapulars bordered by narrower and paler cream-white fringes (no rufous tinge), black dots on back and upper rump, and in more extensive black chest extending into bold black dots and streaks on lower side of neck, side of breast, and flank; in fresh plumage, surprisingly similar to *P. domesticus* in plumage (see Eck 1977), but chestnut feather-centres on crown and black centres of back, side of breast, and chest usually visible, and mantle and scapulars more contrastingly marked with broader black streaks and paler cream fringes. For exceptional specimens of *P. domesticus* possessing some characters of *P. hispaniolensis* (e.g. chestnut cap or black flank streaks), see *P. domesticus*. ADULT FEMALE. Closely similar to ♀ *P. domesticus* and often indistinguishable. Cap on average slightly greyer, less brown; pale streaks on lower mantle and scapulars paler, paler olive-grey with pale cream-pink tip when fresh, pale greyish cream-white when worn (in *P. d. domesticus*, more cinnamon, even when worn, but in southern populations and races of *P. domesticus* paler pink-cinnamon or pink-buff; in both species, pale fringes liable to wear off in breeding season); ground-colour of supercilium, cheeks, and underparts often paler than in ♀ of *P. domesticus*, but southern populations and races of latter also pale; throat often with broad dark grey spots, forming mottled bib (bib present in some ♀ *P. domesticus* more contiguous, paler grey, forming diffuse rounded or elongate grey patch, but patch in some birds similar to that of *P. hispaniolensis* and frequently no trace of grey at all in either species); chest, side of breast, and flank sometimes with broad diffuse grey streaks (exceptional in *P. domesticus*.) NESTLING, JUVENILE. Indistinguishable from *P. domesticus*. FIRST ADULT. Indistinguishable from adult once last juvenile feathers replaced in complete post-juvenile moult, except perhaps by colour of feathers on 'heel' of ♂ (see Metzmacher 1986b).

Hybrid form × *italiae* combines characters of *P. domesticus* and *P. hispaniolensis*. In adult ♂, cap backwards to upper mantle and upper side of neck rich maroon-chestnut, unlike *P. hispaniolensis*. Lower mantle and scapulars rather narrowly streaked black, fringes of feathers deep rufous-cinnamon at sides when fresh, more tawny towards tips, similar to *P. domesticus*; when worn, pale streaks often partly cream-pink, paler than in most *P. domesticus*, but still broader and more extensively rufous than in *P. hispaniolensis*. Back to upper tail-coverts either uniform sandy olive-grey, as in *P. domesticus*, or with some dull black spots on back, tending towards *P. hispaniolensis*. Narrow white stripe above lore distinct (absent in most *P. domesticus*), but hardly inclined to form complete supercilium, as in many *P. hispaniolensis*. Upper cheek and ear-coverts tinged olive-grey, lower cheek and upper side of neck greyish-buff, rather similar to *P. hispaniolensis* when fresh but generally not as white when worn, paler than in *P. d. domesticus*, but closely similar to Middle East races of *P. domesticus*. Black bib restricted, ending broadly rounded on upper chest, similar to bib of *P. domesticus*, (in *P. hispaniolensis*, forming full black collar when plumage worn), side of breast and flank without black dots and streaks, similar to *P. domesticus*, but in some birds some black dots visible on side of breast in worn plumage, extending along lower side of neck, or some rounded or triangular black dots at lower border of bib on chest. Fringes of tail and wing bright rufous-cinnamon, as in *P. domesticus*. In fresh plumage, bright colours partly concealed below broad light olive-brown feather-fringes on upperparts and below cream-buff on side of head and neck and underparts, as in parent species; for timing of abrasion of pale fringes of ♂, see Bogliani and Brangi (1990). Black of chest and back occasionally partly replaced by chestnut, perhaps more often so than in *P. domesticus* and *P. hispaniolensis*. Adult ♀ of × *italiae* indistinguishable from *P. domesticus* on upperparts, like ♀ of either species on underparts. For other populations which are more or less intermediate, see Geographical Variation.

**Bare parts.** No differences reported from House Sparrow *P. domesticus*. In Mediterranean basin and Asia, bill of adult ♂

turns bluish-black in March or April, becoming gradually paler again during post-breeding moult; in Cape Verde Islands, timing of blackening highly variable, depending on timing of breeding, which depends on local rains.

**Moults.** ADULT POST-BREEDING. Complete; primaries descendent. In south-west Spain, starts with shedding of p1 early July to early August (on average, 24 July), all moult completed late September; in captivity, duration of primary moult in individuals *c.* 66 days, secondaries completed 3 days later; tail starts at primary score 20–25, completed at 40–50; secondaries start at 20–27, completed at 45–50 or later; tertials start at 10–25, completed at 30–40 (Alonso 1984a, which see for other details). In Tunisia, moult July–September (Bortoli 1973). In Kazakhstan, birds leave breeding area mid-June or July, apparently to go to special moulting area, but some in early stages of moult when still near nesting grounds in August (Gavrilov 1962a, 1963); in USSR generally, moult July or early August to September (Dementiev and Gladkov 1954). In birds examined from Mediterranean basin, plumage fresh October, heavily abraded June, none of these in moult, but 3 birds from Cape Verde Islands in heavy moult February, primary moult scores 38, 40, and 41 (RMNH, ZFMK, ZMA). On Cape Verde Islands, peaks in breeding activities late February to May and late August to October (Naurois 1988), moult probably June–August and November–February. On Sardinia, single bird in moult late August (Bezzel 1957); in northern Iran, heavily worn but not yet moulting 24 July, in Afghanistan moult completed October (Paludan 1940, 1959). POST-JUVENILE. Complete; primaries descendent. Timing variable, depending on hatching date; starts *c.* 4 weeks after fledging, but late-fledged birds start relatively earlier than early-hatched ones (Alonso 1984a). In Spain and Balkans, moult starts late June to late August, completed late September or October (Alonso 1984a; RMNH, ZFMK, ZMA). In USSR, moult August–September (Dementiev and Gladkov 1954). For sequence of moult of various feather-tracts, see Alonso (1984a).

**Measurements.** ADULT, FIRST ADULT. Wing and bill to skull of nominate *hispaniolensis* from (1) Cape Verde, Canary, and Madeira islands, (2) North Africa and Spain, (3) Sardinia; (4) *transcaspicus*, Turkey to Kazakhstan; other measurements for all areas combined, whole year; skins (RMNH, ZFMK, ZMA). Bill (N) to distal corner of nostril; exposed culmen averages 2·0 less than bill to skull.

| | | ♂ | | ♀ | |
|---|---|---|---|---|---|
| WING | (1) | 78·1 (1·02; 10) | 76–80 | 75·6 (1·25; 4) | 74–77 |
| | (2) | 78·8 (1·14; 8) | 77–81 | 77·1 (0·89; 5) | 76–78 |
| | (3) | 78·6 (1·14; 11) | 77–81 | 75·1 (1·85; 8) | 73–78 |
| | (4) | 80·4 (1·82; 17) | 78–84 | 78·1 (2·95; 4) | 75–82 |
| TAIL | | 54·7 (2·22; 43) | 51–59 | 53·4 (1·96; 20) | 50–57 |
| BILL | (1) | 15·4 (0·38; 6) | 14·9–16·0 | 15·8 (0·38; 4) | 15·5–16·3 |
| | (2) | 15·6 (0·72; 8) | 14·6–16·3 | 15·9 (0·35; 5) | 15·5–16·3 |
| | (3) | 15·2 (0·56; 11) | 14·5–16·0 | 15·4 (0·64; 8) | 14·5–16·3 |
| | (4) | 15·6 (0·83; 17) | 14·3–17·0 | 15·6 (0·25; 4) | 15·3–15·9 |
| BILL (N) | | 9·8 (0·53; 17) | 9·0–10·8 | 9·7 (0·43; 19) | 9·1–10·5 |
| TARSUS | | 20·4 (0·60; 17) | 19·6–21·5 | 20·3 (0·67; 19) | 19·1–21·5 |

Sex differences significant for wing.

Western Algeria (Metzmacher 1986b, which see for other measurements):

| | | ♂ | | ♀ | |
|---|---|---|---|---|---|
| WING | AD | 79·9 (1·57; 17) | 76–82 | 76·4 (1·83; 19) | 71–79 |
| | 1ST AD | 78·3 (1·29; 12) | 75–80 | 76·7 (1·01; 11) | 75–79 |

Wing data from literature. Nominate *hispaniolensis*: (1) Canary Islands (Naurois 1988), (2) Sardinia (Meise 1936; Bezzel 1957; see also Lo Valvo and Lo Verde 1987); (3) Greece and European Turkey (Makatsch 1950; Kinzelbach and Martens 1965; Rokitansky and Schifter 1971); (4) migrants, Libya (Erard and Lari-gauderie 1972). *P. h. transcaspicus*: (5) Asia Minor (Kumerloeve 1961, 1963, 1964a, 1970b; Rokitansky and Schifter 1971; Vauk 1973); (6) Israel (WIWO, G O Keijl); (7) Iran and Afghanistan (Vaurie 1949a); (8) Afghanistan (Paludan 1959). See also Meise (1936).

| | ♂ | | ♀ | |
|---|---|---|---|---|
| (1) | 76.5 ( 8) | 75.0-77.5 | 75.5(15) | 73-78.5 |
| (2) | 77.8(26) | 74-80 | 74.8( 4) | 74-75.5 |
| (3) | 77.2( 6) | 75-81 | 76.0( 3) | 76-76 |
| (4) | 79.1( 6) | 78.5-81.5 | (—) | 77.5-79.5 |
| (5) | 79.4(45) | 76-83 | 75.2( 9) | 73-76 |
| (6) | 80.9( 7) | 78-85 | 78.2( 4) | 76-80 |
| (7) | 81.1(15) | 78.5-85 | 77.2(15) | 74-81 |
| (8) | 80.7( 8) | 77-85 | 78.6(11) | 76-81 |

Wing of ♂. (9) Cape Verde Islands (Naurois 1988). (10) Canary Islands, (11) Spain, (12) Sardinia (Vaurie 1956f). (13) Canary Islands, (14) Andalucia (Spain), (15) Caceres (Spain) (Alonso 1985a, which see for other measurements and for data of hybrids with *P. domesticus*). (16) Sardinia, (17) Tunisia (Eck 1985). (18) Syria, breeding (Kumerloeve 1969d).

| | | | | | | |
|---|---|---|---|---|---|---|
| (9) | 76.4(—; 16) | 73.5-78 | (14) | 80.2(1.42; 38) | — | |
| (10) | 77.5(—; 10) | 76-80 | (15) | 78.9(1.53; 350) | — | |
| (11) | 79.5(—; 14) | 78-83 | (16) | 77.5(1.55; 47) | 75-81 | |
| (12) | 78.3(—; 23) | 76-81 | (17) | 78.5(1.72; 33) | 74-81 | |
| (13) | 77.1(1.75; 21) | — | (18) | 80.5(—; 3) | 79.5-82 | |

Kizlyar area (Terek valley, north of Caucasus), wing 76.5 (12) 73-79, bill (N) 10.1 (12) 9.4-10.5; Transcaspia, wing 79.1 (6) 77-82, bill (N) 9.3 (6) 9.0-9.7 (Buturlin 1929).

Hybrid form × *italiae*. Northern and central Italy, all year; skins (RMNH, ZMA). Bill (S) is bill to skull.

| | ♂ | | ♀ | |
|---|---|---|---|---|
| WING | 79.4(1.89; 44) | 77-83 | 77.0(1.65; 19) | 74-80 |
| TAIL | 56.2(2.01; 19) | 53-60 | 54.8(1.77; 19) | 52-58 |
| BILL (S) | 15.2(0.64; 44) | 14.0-16.5 | 15.2(0.63; 19) | 14.4-16.2 |
| BILL (N) | 9.4(0.53; 17) | 8.6-10.3 | 9.2(0.51; 19) | 8.5-10.0 |
| TARSUS | 19.5(0.72; 18) | 18.6-20.8 | 19.7(0.68; 19) | 18.9-21.0 |

Sex differences significant for wing and tail.

Wing, exposed culmen, and bill depth of ♂ from (1) northern Italy, (2) central Italy, (3) southern Italy (except Calabria), (4) Calabria, and (5) Sicily and Malta (Lo Valvo and Lo Verde 1987).

| | WING | CULMEN | BILL DEPTH |
|---|---|---|---|
| (1) | 77.1(2.01; 269) | 12.7(0.63; 292) | 8.2(0.39; 272) |
| (2) | 76.6(1.46; 58) | 12.6(0.61; 63) | 8.2(0.31; 59) |
| (3) | 76.4(1.92; 28) | 12.5(0.69; 28) | 8.2(0.29; 27) |
| (4) | 76.3(1.48; 39) | 12.9(0.59; 41) | 8.3(0.27; 39) |
| (5) | 76.1(2.01; 180) | 13.0(0.84; 197) | 8.4(0.33; 167) |

Wing: northern and central Italy, ♂ 78.2 (1.88; 69) 73-82 (Eck 1985); Montecristo Island (off Elba), ♂ 76.7 (3.2; 13) 72-81, ♀ 73.4 (1.6; 14) 71-75 (Baccetti et al. 1981); Corsica, ♂ 78.5 (1.40; 22) 76-82 (Eck 1985); Crete, ♂ 78 (50) 76-82 (Rokitansky 1934).

Hybrid form × *maltae*. Wing: Sicily, ♂ 78.3 (2.01; 11) 76-81 (Steinbacher 1954), 78.4 (1.74; 22) 76-82 (Eck 1985). Malta: wing, ♂ 79.4 (2.04; 5) 77-82, ♀♀ 77, 77.5; bill (S), ♂ 16.3 (0.45; 5) 15.5-16.6, ♀♀ 16.2, 16.4, bill (N) 10.6 (0.47; 7) 9.7-11.2 (ZMA).

**Weights.** Nominate *hispaniolensis*. (1) Sardinia, all year (Demartis 1987). (2) Greece and European Turkey, April-May (Makatsch 1950; Rokitansky and Schifter 1971). (3) Migrants south-west Libya, early April (Erard and Larigauderie 1972). *P. h. transcaspicus*. (4) Asia Minor, May-July (Kumerloeve 1964a, 1970b; Rokitansky and Schifter 1971; Vauk 1973). (5) Israel, March-April (WIWO, G O Keijl). (6) South-west Afghanistan, March-April (Paludan 1959). (7) Kazakhstan (Korelov et al. 1974). (8) India, March-April (Ali and Ripley 1974).

| | ♂ | | ♀ | |
|---|---|---|---|---|
| (1) | 27.2(1.11; 12) | 26-30 | 26.0(2.00; 10) | 22-28 |
| (2) | 27.9(1.02; 4) | 26.6-29.1 | 27.5(0.95; 3) | 26.6-28.5 |
| (3) | 31.1(3.74; 6) | 25.5-36 | 33(—; —) | — |
| (4) | 30.6(—; 28) | 26.6-37 | 30.1(—; 7) | 28-32 |
| (5) | 28.5(1.83; 7) | 27-32 | 28.1(1.11; 4) | 27-29.5 |
| (6) | 27.4(—; 7) | 23-31 | 29.0(—; 7) | 27-31 |
| (7) | 28.7(—; 306) | 22.7-37.5 | 28.5(—; 303) | 22.2-37.5 |
| (8) | 24.9(—; 100) | 20-28 | 23.5(—; 100) | 18-28 |

Spain: ♂ 28.4 (3.08; 561), ♀ 28.1 (3.76; 437), juvenile at fledging average 23 (Alonso 1985c; which see for monthly fluctuations, influence of time of day, etc.).

Cape Verde Islands, June: ♂ 25.3 (3) 23-27, ♀ 23 (J Rabaça). Amik Gölü (central-south Turkey), May: ♂ 23.6 (2.07; 5) 22-27 (Kumerloeve 1961). Syria, June: 26.3 (3) 25-28 (Kumerloeve 1969d). Northern Iran, ♂: early April, 28, 29 (Schüz 1959); late July, 26 (Paludan 1940). Central Afghanistan, October: 33, ♀ 29.5 (4) 29-30 (Paludan 1959).

Hybrid form × *italiae*. Montecristo Island (off Elba, Italy), June-October: ♂ 26.6 (1.9; 13) 23-30, ♀ 25.6 (1.7; 14) 23-28, juvenile 25.2 (1.9; 10) 23-29.5 (Baccetti et al. 1981).

Hybrids Tunisia, May: ♂ 28.0 (1.6; 7) 26.8-29.5, ♀ 26.4 (1.7; 4) 24.5-28.5, juvenile 24.7 (1.3; 8) 22.5-26.5 (G O Keijl).

**Structure.** 10 primaries: p7-p8 longest, p9 0-1 shorter, p6 1-3, p5 5-9, p4 9-12; p1 17.9 (15) 16-20 shorter in nominate *hispaniolensis*, 20.4 (10) 19-22 in *transcaspicus*, 18.2 (10) 17-20 in × *italiae*. P10 reduced; in adult, 45-54 shorter than p7-p8, 7.1 (10) 5-10 shorter than longest upper primary covert. Both depth and width of bill at base 8.6 (30) 8.3-9.1. Middle toe with claw 19.5 (10) 18-20.5; outer and inner toe with claw both *c.* 72% of middle with claw, hind *c.* 82%. Remainder of structure as in *P. domesticus*.

**Geographical variation.** Very slight within *P. hispaniolensis*. In fresh plumage, nominate *hispaniolensis* from Cape Verde and Canary Islands east to Balkans, Greece, and north-west Libya slightly more rufous-pink on fringes of mantle, scapulars, flight-feathers, tertials, and greater upper wing-coverts. *P. h. transcaspicus*, occurring from Iran (except north-west) and Transcaspia eastwards, paler cream-pink; feather-fringes of upperparts of both sexes of nominate *hispaniolensis* more sandy olive-grey, those of *transcaspicus* paler grey-buff; birds from Asia Minor, Levant, Caucasus area, and north-west Iran intermediate. No difference in colour in spring and summer; no constant difference in colour of cap. Birds east from Asia Minor and Levant slightly larger than those from further west (see Measurements). For differences between various populations in western Algeria, see Metzmacher (1986c). Wing-tip of *transcaspicus* longer than in nominate *hispaniolensis* from Sardinia: 22-25 mm from p8 to outermost secondary in *transcaspicus*, 18-22 in nominate *hispaniolensis* (Eck 1977; see also Structure).

Position of Italian Sparrow '*P. italiae*' and other populations of more or less similar appearance is problematical. Form *italiae* sometimes considered to be an early offshoot of *P. domesticus-P. hispaniolensis* species-group, showing certain feather patterns not found in *P. domesticus* or *P. hispaniolensis*, but which are shared with North African form of House Sparrow *tingitanus*; *italiae* then a separate species, which includes *tingitanus* as a race (Stephan 1986). More likely, however, that *italiae* as found in northern and central Italy is stabilized hybrid population between *P. domesticus* and *P. hispaniolensis*: changes in agricultural practice *c.* 3600 years ago in Italy apparently made it possible for *P. domesticus* to invade northern Italy, which was formerly inhabited only by *P. hispaniolensis* (Meise 1936; Johnston 1969a; Baumgart 1984), and both gradually became mixed. Without further major invasions, hybrid populations became stabilized, as are the isolated mixed populations of various Saharan oases and Mediterranean islands. Elsewhere (Iberia, Morocco, western Algeria, Balkans, Greece, and from Turkey and Levant eastward), *P. domesticus* and *P. hispaniolensis* live side-by-side, sometimes more or less separated ecologically or spatially (Panov and Radzhabli 1972; Summers-Smith and Vernon 1972; Baumgart and Stephan 1974; Baumgart 1980, 1984; Metzmacher 1986a; Panov 1989), though here too hybrid populations occur on small scale (Meinertzhagen 1921, 1940; Meise 1934; Steinbacher 1954; Kumerloeve 1961; Gavrilov 1965; Summers-Smith and Vernon 1972; Baumgart and Stephan 1974; Stephan and Gavrilov 1980; Alonso 1984a; Metzmacher 1986a); appearance of these hybrids highly variable (Macke 1965; Baumgart 1984; Alonso 1985a), more so than are stabilized populations of × *italiae* in northern and central Italy. In captivity, crosses appear to produce few offspring (Alonso 1984b).

Throughout mainland Italy and Sicily, populations gradually show more *P. hispaniolensis* characters towards south, with those of western Sicily and Malta almost pure *P. hispaniolensis*; Corsica and 'stepping stones' to Italy (Elba, Montecristo) inhabited by populations similar to × *italiae* from northern and central Italy (Meise 1936; Johnston 1969a, 1972; see also Lo Valvo and Lo Verde 1987 and Iapichino and Massa 1989). Sardinia, which is more isolated (Cheke 1966), has virtually pure *P. hispaniolensis* only. In Greece, invasion of *P. domesticus* came from north (Balkans) and east (Turkey) (Niethammer 1942), but, as these movements are probably older than those into Italy, and less hindered by mountains (C S Roselaar), *P. hispaniolensis* characters are clear in populations of Crete (where virtually inseparable from × *italiae*), less so on Karpathos, and traces only in Pelopónnisos (especially south), Athens area, and on Rhodes (Niethammer 1943; Kinzelbach and Martens 1965; Kinzelbach 1969; Summers-Smith 1980). In Algeria, Tunisia, and north-west Libya, where breaking up of ecological or spatial separation between *P. domesticus* and *P. hispaniolensis* is apparently more recent, populations show extreme individual variation, from pure *P. domesticus* to pure *P. hispaniolensis* and all forms between. Isolated Saharan populations are more stabilized and usually nearer *P. domesticus tingitanus* (which is itself variable, probably due to introgression of *P. hispaniolensis* characters even in its pure form). Those of coastal Tunisia and of north-west Libya are nearer *P. hispaniolensis*, those of north-east Libya near *P. domesticus* (Meise 1936; Summers-Smith and Vernon 1972; Metzmacher 1986a).

When various identifying characters of *P. domesticus* are scored 0 and those of *P. hispaniolensis* 2–6, total score of pure *P. domesticus* theoretically becomes 0(−3), that of *P. hispaniolensis* (15−)17 (Johnston 1969a); scores of these and of hybrid populations then as follows.

*P. domesticus*: (1) north-west Yugoslavia, (2) Spain and Portugal, (3) Balearic Islands, (4) north-east Libya. Hybrids: 'italiae', (5) southern slopes of Alps (northern Italy), (6) northern Italy (Po valley to Pisa and Firenze), (7) Elba, (8) Corsica; 'schiebeli', (9) Crete; 'brutius', (10) central Italy (Rome area to Foggia), (11) southern Italy (Bari to Reggio di Calabria); 'maltae', (12) eastern Sicily, (13) western Sicily, (14) Malta. *P. hispaniolensis*: (15) north-west Libya, (16) Sardinia, (17) Canary Islands. (Johnston 1969a; see also Johnston 1969c.)

| | | | | | |
|---|---|---|---|---|---|
| (1) 0·5 | ( 8) 0–3 | ( 7) 7·6 | (11) 6–9 | (13) 13·5 | ( 29) 10–17 |
| (2) 0·6 | ( 31) 0–3 | ( 8) 7·2 | (55) 0–11 | (14) 12·9 | ( 40) 10–16 |
| (3) 0·3 | ( 44) 0–3 | ( 9) 7·3 | (64) 3–12 | (15) 15·0 | ( 11) 9–17 |
| (4) 4·4 | ( 9) 0–9 | (10) 9·0 | (42) 7–11 | (16) 15·0 | (100) 10–17 |
| (5) 6·3 | ( 85) 3–9 | (11) 10·3 | (63) 8–13 | (17) 15·5 | ( 32) 13–17 |
| (6) 7·5 | (214) 3–11 | (12) 12·2 | (77) 7–16 | | |

For generally smaller samples with scoring system 0–100, see Meise (1936), followed by Niethammer (1943), Steinbacher (1954, 1956), and Lo Valvo and Lo Verde (1987).

If one accepts form *italiae* as hybrid between *P. domesticus* and *P. hispaniolensis*, it is not logical to include it as a race in either *P. domesticus* (as often done, e.g. by Vaurie 1956f, 1959) or *P. hispaniolensis* (e.g. Harrison 1961, Summers-Smith 1988), even though characters and habitat requirements tend towards those of *P. hispaniolensis*; moreover, *italiae* is an older name than *P. hispaniolensis*, and familiar scientific name for Spanish Sparrow would thus have to be changed to *P. italiae hispaniolensis* if these forms are combined. Separating *italiae* as full species simply because it is a morphologically well-recognizable form with little recent inflow from parent species (as advocated by Johnston 1969a) is unacceptable, given its geologically recent origin and hybrid character, stable or not. Also, inconsistent to call morphologically similar birds a separate species in Italy and just a hybrid in eastern Algeria (Metzmacher 1986a; Massa 1989). Here, use of neutral terms 'hybrid form × *italiae*' and 'hybrid form × *maltae*' is favoured, which for convenience are dealt with under *P. hispaniolensis*.

For hybrid zone between × *italiae* and *P. domesticus* in Alps, see Wallis (1887), Meise (1936), Ris (1957), Niethammer (1958), Schöll (1959, 1960), Schweiger (1959), Wettstein (1959), Niethammer and Bauer (1960), Löhrl (1963), Seitz (1964), Klaas (1967), Schifferli and Schifferli (1980), Summers-Smith (1988), and Lockley (1992); hybrids in this area often have black-mottled cap, unlike parents, but rather like *P. domesticus tingitanus* (Hartert 1921–2). CSR

---

## *Passer moabiticus* Dead Sea Sparrow

PLATES 22 and 26 (flight)
[between pages 280 and 281]

Du. Moabmus   Fr. Moineau de la Mer Morte   Ge. Moabsperling
Ru. Месопотамский воробей   Sp. Gorrión del Mar Muerto   Sw. Tamarisksparv

*Passer moabiticus* Tristram, 1864

Polytypic. Nominate *moabiticus* Tristram, 1864, Israel and Jordan; *mesopotamicus* Zarudny, 1904, southern Turkey, northern Syria, Iraq, south-west Iran, and (probably this race) Cyprus. Extralimital: *yatii* Sharpe, 1888, Iranian Seistan and neighbouring south-west Afghanistan.

**Field characters.** 12 cm; wing-span 19–20 cm. Almost 20% smaller than frequent companion Spanish Sparrow *P. hispaniolensis*, with proportionately smaller bill and more compact build. Distinctly trim sparrow, smallest and (♂) most colourful of genus. ♂ has grey head with pale supercilium, streak under eye, and submoustachial stripe turning up round cheek, and black bib; yellow-buff tone to rear supercilium and lower part of moustache is unique within genus. ♀ buffier above and cleaner below than all other *Passer* except Desert Sparrow *P. simplex*; also shows yellow in supercilium and on side of throat; as ♂, lacks bold wing-bars. Sexes dissimilar; some seasonal variation. Juvenile resembles ♀. 2 races in west Palearctic, not separable in the field.

ADULT MALE. Moults: August (complete). Remarkably colourful when plumage worn, in breeding season. Crown, cheek, and nape ash-grey, strikingly decorated by white fore-supercilium, whitish to yellow-buff or pale chestnut rear supercilium, greyish-white patch under lore, and white submoustachial stripe which broadens into bright yellow half-collar, running up behind cheek and below bib. Loral patch dusky; chin and throat black, forming distinct bib of similar size to that of Tree Sparrow *P. montanus*. Mantle, scapulars, and tertials pale chestnut-buff, with black feather-centres forming obvious streaks. Inner wing-coverts bright chestnut, with both median and greater narrowly tipped buff (but not forming obvious bars); rest of wing blackish-brown, with dark primary coverts contrasting with pale patch at base of primaries. Rump and upper tail-coverts ash-grey, latter with blackish streaks on largest feathers. Tail almost black, with pale buff margins to all feathers. Below bib and yellow half-collar, chest banded with grey; flanks and belly greyish-white, flanks softly splashed with buff-grey; under tail-coverts greyish-white, with chestnut centres showing as mottling. Underwing pale cream. When fresh, crown and nape sandy, supercilium dull, bib hoary, and underparts less clean. Bill black-grey when breeding, paler horn with dark ridge at other times. Legs bright greyish-brown. ADULT FEMALE. Lacks decorated head of ♂; crown, cheek, and nape pale brown, supercilium whitish, becoming pale buff behind eye; submoustachial stripe, chin, and throat all faintly buff-white, with indication of yellow below cheek and on rear half-collar. Ground-colours of other plumage lighter and duller than ♂'s but pattern similar. Bill brown-horn. JUVENILE. Resembles ♀.

Unmistakable when small size, ♂'s unique head colours and pattern (lacking pale cheeks) and ♀'s head and upperparts well seen. Not known to overlap with other small *Passer*; streaked back of ♀ and juvenile quickly exclude *P. simplex*. Flight typical of genus, with whirring action and direct track more reminiscent of *P. montanus* than House Sparrow *P. domesticus*. Behaviour much as *P. domesticus* but modified by liking for tall trees or well-grown bushes near water, in which it feeds and nests and from which it descends less often to ground. Flicks wings and cocks tail in excitement. Due to habitat preferences, often associates with *P. hispaniolensis*.

Voice quite distinctive: song more tuneful and more emphatically rhythmic than *P. domesticus*, particularly in colonial chorus: 'tri-rirp tri-rirpe', 'tlir-tlir-tlir', or 'chr-rech-er chr-rech-er'. Commonest call a high-pitched 'trrirp'.

**Habitat.** Patchily distributed in south-west Asia near watercourses or pools in arid usually lowland regions where shrubs such as tamarisk *Tamarix*, thick scrub, or trees afford cover and nest-sites. Breeds near Karak (Jordan) in thicket of tamarisks interspersed with bushes of *Leptadenia spartium*, at *c.* 450 m below sea-level on shore of Dead Sea (Mountfort 1965). In Israel, also in well-vegetated cultivated land (Shirihai in press). In Afghanistan found in stunted badly-cut tamarisk scrub by estuary of lower Farah Rud (Seistan) at *c.* 500 m, in flocks as well as single birds, just before breeding season (Paludan 1959). Attachment to open water and lack of attachment to human settlements seem pronounced. Winters in northern Baluchistan in scrub of *Prosopis*, *Rubus*, tamarisk, etc. (Christison 1941).

**Distribution.** Has spread in Israel, and has probably spread north from Tigris valley in central Iraq to northern Iraq, northern Syria, and Turkey. Range extensions to Cyprus (also first record for Greece) apparently a continuation of westward spread in southern Turkey. (Cramp 1971; Summers-Smith 1988.)

TURKEY. Colony discovered at Birecik (upper Euphrates) 1964, and further colonies 1965; not certain whether this indicates range expansion or increased observer coverage (Summers-Smith 1988). CYPRUS. Has bred on north shore of Akrotiri salt lake since at least 1976; first recorded 1973 (Summers-Smith 1988). SYRIA. Colony discovered 1968 (Kumerloeve 1969a). IRAQ. Has probably spread north along Tigris: known only as far north as Baghdad area in 1915–22, and first found breeding on upper Tigris at Mosul in late 1940s (J D Summers-Smith). ISRAEL. Up to 1930s limited to a few localities in Dead Sea area and southern Jordan valley. Has since spread, due to growth of agricultural settlements and consequent increased availability of water. (Shirihai in press.)

Greece: *c.* 20, Kalithea (Rhodes), 1st week of October 1972 (Andersen 1989).

**Population.** CYPRUS. 35 nests in 3 colonies, 1985 (Summers-Smith 1988).

**Movements.** Poorly known for west Palearctic races, nominate *moabiticus* and *mesopotamicus*; many birds absent from breeding colonies October–March, but this appears to be more of a dispersal into feeding areas in cultivated land than a directed migration (Summers-Smith 1990).

In Cyprus, absent from breeding area in winter and

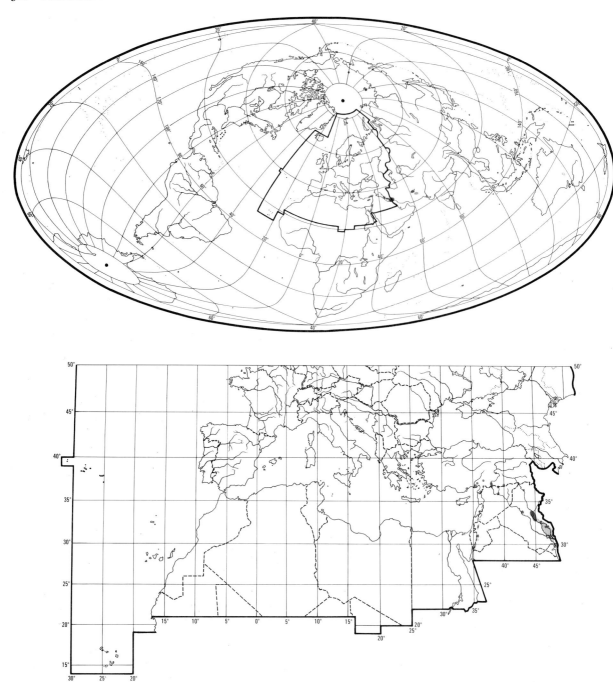

presumed migratory (Flint and Stewart 1983). In Turkey, very local summer visitor (Beaman 1986), with only one published record outside breeding season (Kumerloeve 1969a). Flock of 20 on Rhodes (Greece) October 1972 (Andersen 1989) suggests possible passage south from Turkey. In Israel, only small numbers, if any, remain in breeding areas in winter, most birds wandering erratically in Rift Valley with general southward drift (J Langer, D Yekutiel, Y Yom-Tov); abundant in Elat from mid-

November to end of March, less common from late October and as late as beginning of April (Krabbe 1980); returns to breeding grounds in Arava during March, not until April further north (Paz 1987). 2 winter records from Egypt, both near Nuweiba in southern Sinai: c. 15 winter 1971–2 (Goodman and Meininger 1989); 10 from 30 October to 3 November 1987 (Br. Birds 1991, 84, 11). In central Iraq, some birds remain in breeding areas (Marchant 1962), others moving several hundred km south along

Tigris in winter (Cheesman 1919). First records in Arabia were 70 at Jubail (eastern Saudi Arabia) on 13 November 1991, at least 245 on Bahrain from 19 December 1991 to 13 March 1992, and 4 in Abu Dhabi January–February 1992 (Hirschfeld and Symens 1992).

In extralimital *yatii*, breeding in Seistan on Iran/Afghanistan border, some remain in breeding area (Sarudny and Härms 1912; Paludan 1959; Scott *et al.* 1975), others move south to Baluchistan (Pakistan): common in winter in Chagai in north, and recorded south almost to Makran coast (Sarudny and Härms 1912; Christison 1941). JDS-S

**Food.** Casual observations suggest mainly seeds, especially of grasses. Most food sought on ground (Paz 1987), but seen taking seeds from tamarisk *Tamarix* trees and papyrus *Cyperus papyrus* (J D Summers-Smith), and majority of insect food obtained by searching leaves of bushes and small trees such as tamarisk and willow *Salix* (Cramp 1969) and *Suaeda* (Ticehurst *et al.* 1921). See Lulav (1967) for suggestion that range related to occurrence of beetle *Steraspis squamosa* (Buprestidae) which develops in stem of tamarisk.

Diet in west Palearctic includes unspecified insects, both larvae and adults, and earthworms (Lumbricidae) (Jiad and Bunni 1988). Plant material mainly seeds of grasses (Gramineae), including *Phalaris* and reeds (Mendelssohn 1955; Howells 1956; H Mendelssohn), also papyrus, reedmace *Typha* (Ticehurst *et al.* 1926), tamarisk, and mulberries *Morus* (Kumerloeve 1965*a*).

Young are fed on seeds and insects (Yom-Tov and Ar 1980). JDS-S

**Social pattern and behaviour.** No systematic studies. Unless otherwise attributed, information from J D Summers-Smith.

1. Largely gregarious. Normally in small flocks outside breeding season. Independent young remain together and join to form flocks with other young from breeding colony; adults join flock (typically 50–100 birds) after breeding (Mendelssohn 1955). Flocks persist after breeding season, sometimes associating with House Sparrow *P. domesticus* and Spanish Sparrow *P. hispaniolensis*. BONDS. Monogamous. Duration of pair-bond not known. Nest built by both sexes (see Breeding). ♀ alone incubates and broods and does most feeding of young (Yom-Tov and Ar 1980). BREEDING DISPERSION. Forms small loose colonies, typically of 10–15 pairs (Yom-Tov and Ar 1980), with nests not less than 8 m apart (H Mendelssohn), typically 15–30 m; see Yom-Tov and Ar (1980) for further data; 25 nests in 400 m of tamarisk *Tamarix* by River Jordan (Wallace 1984). In Iraq, nests *c.* 100 m apart with *c.* 100 nests scattered along shore of Tigris (Cheesman 1919), though in same area Marchant (1963*b*) found nests closer together. While ♀ laying, ♂ defends territory of up to 1000 m² but feeds over wider area (Yom-Tov and Ar 1980; H Mendelssohn). Breeding colony maintained from year to year. Densities in Israel and Jordan locally 3000 to *c.* 10 000 nests per km² (Mendelssohn 1955; Lulav 1967; H Mendelssohn, D I M Wallace), but upper figure not necessarily breeding pairs as minority of ♂♂ build 2 nests (not occupied simultaneously) in same tree, 1–2 m apart, and nests last for several seasons (Men-

delssohn 1955); more typically, pair thought to use same nest in season for successive clutches (Yom-Tov and Ar 1980). Sometimes in mixed colonies with *P. hispaniolensis* (Kumerloeve 1969*a*). ROOSTING. Outside breeding season, in trees, bushes (etc.) in small groups, maximum recorded *c.* 5000. During breeding season, ♀ roosts in nest, ♂ nearby (Mendelssohn 1955; H Shirihai).

2. General behaviour elusive, wary, and restless. FLOCK BEHAVIOUR. Flock feeding in open area frequently flies up to cover in trees where birds move about restlessly from branch to branch before trickling back to same feeding area. Approach each other more closely than most *Passer*, both when feeding and when in cover. SONG-DISPLAY. Song (see 1 in Voice) given from beside nest with wings raised and vibrated through limited arc (Harrison 1965*b*: Fig A), rendering conspicuous wing-markings

A

(Mendelssohn 1955) and yellow spot at side of black throat (Kumerloeve 1969*a*). Mendelssohn (1955) referred to social singing by groups of birds, mostly in early morning and evening, presumably similar to 'great din' heard by Kumerloeve (1965*b*) from flock of birds away from breeding area in February; otherwise tends to be rather silent outside breeding season. ANTAGONISTIC BEHAVIOUR. None reported; no aggression even in mixed colonies with *P. hispaniolensis* where their nests only 40–50 cm away (Kumerloeve 1969*a*), and only mild reaction (Threat-calls by ♂: see 3 in Voice) when *P. domesticus* entered nest with eggs (Yom-Tov and Ar 1980). HETEROSEXUAL BEHAVIOUR. Pair-formation begins with ♂ building nest and calling (see Song-display, above) from it or hopping about conspicuously (wing-raising and calling) from branch to branch above it to attract ♀; pair-bond reinforced by reciprocal pecking and passing of nest-material (Mendelssohn 1955; Jiad and Bunni 1965). In presence of ♀, ♂ performs Parade-display (Fig B):

B

wings drooped and shivered, tail cocked, calling throughout; as ♀ hops around ♂, he turns to constantly face her (Y Yom-Tov). Once pair forms, ♀ helps ♂ to line nest. ♀ invites copulation by crouching near nest with wings drooped and shivered, giving soft call (see 2c in Voice). After copulation, both birds nest-build (J D Summers-Smith). Apart from regulating attentiveness according to temperature (see Breeding), at high temperatures incubating ♀ also brings water (evidently in belly plumage, perhaps also in bill) for cooling eggs by direct wetting (Yom-Tov *et al.* 1978; Y Yom-Tov). ♀ sits on eggs at night (Mendelssohn 1955). RELATIONS WITHIN FAMILY GROUP. ♀ appears to undertake most feeding of nestlings, but ♂ enters nest occasionally and may give some assistance. ♂ also helps to feed fledglings (Mendelssohn 1955; Yom-Tov and Ar 1980). Some ♀♀ re-lay while still feeding young of 1st brood (Yom-Tov and Ar 1980). No further information. ANTI-PREDATOR RESPONSES OF YOUNG, PARENTAL ANTI-PREDATOR STRATEGIES. No information.

(Figs by R Gillmor: A from drawing in Mendelssohn 1955; B from photograph in Cramp 1969.) JDS-S

**Voice.** Used mainly in breeding season. Not well studied. Following account based mainly on notes by J D Summers-Smith.

CALLS OF ADULT. (1) Song. Excited, rhythmic, high-pitched 'chilling-chilling-chilling' (Fig I), shriller than

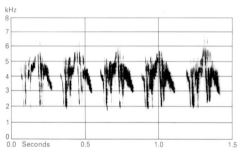

I  Mild (1990) Israel  March 1988

Spanish Sparrow *P. hispaniolensis*. Given by unpaired ♂ in Song-display when ♀ approaches him at nest. Other renderings: clear 'dli-dli-dli' like curtailed song-phrase of Chaffinch *Fringilla coelebs* (Kumerloeve 1965b; Cramp 1969), 'jew-ee' like bunting *Emberiza* repeated 3-4 times (Mountfort 1965). (2) Other advertisement- and contact-calls. (2a) Basic 'chirp'. Regularly repeated 2-syllable 'chip-chew' or 'tcheep-tcheep' (Fig II), 3-syllable 'chip-chip-chew' or 'chip-chip-chizz' given by ♂ at nest, going over into song when ♀ approaches. (2b) Social-singing. Chorus (undescribed) by group of birds, mostly early

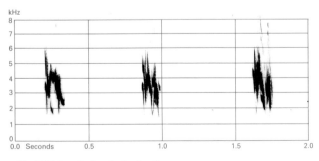

II  Mild (1990) Israel  April 1989

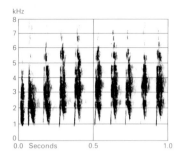

III  Mild (1990) Israel  March 1988

morning (Mendelssohn 1955). (2c) Solicitation-call. Quiet 'tweeng-tweeng' given by ♀ soliciting copulation. (3) Threat-call. 'Chittup', accompanied by tail-flicking, used to express mild alarm, changing to churring 'churr-it-it-it' (Fig III) or 'chit-it-it-it' in more extreme threat; higher pitched than similar call of *P. hispaniolensis*.

CALLS OF YOUNG. No information. JDS-S

**Breeding.** SEASON. Egg-laying in Israel end of March to early July (Yom-Tov and Ar 1980), further north from late April in Cyprus (Flint and Stewart 1983) to late July in Turkey, where young still being fed in nest mid-August (Beaman 1978). Iraq: laying probably April to end of June (Marchant 1963b). SITE. Mostly 1-10 m above ground in branches of trees near water or standing in water (Harrison 1975; Flint and Stewart 1983); occasionally recorded at least 2 km from water, perhaps up to 7 km (Lulav 1967; Cramp 1971), but, since close proximity to water (maximum 100 m) evidently obligatory in Rift Valley of Israel (nests even with full clutches or young deserted if water dries up), confirmation needed that these exceptionally long distances involved successful nests (Yom-Tov and Ar 1980; Y Yom-Tov). Tamarisk *Tamarix*, poplar *Populus*, olive *Olea*, willow *Salix*, *Eucalyptus*, and fruit trees used; bare branches of dead trees preferred in Israeli Rift Valley (Lulav 1967; H Mendelssohn), trees killed by fire used in Cyprus (J D Summers-Smith). Nests also on tall herbs; record of colony of 30 nests with each nest supported by 2-3 thistles bound together (Lulav 1967). Nest: bulky, open globular or cone-shaped structure built of stiff dry twigs, 15-25 cm long, finely interwoven round branches of tree (resembling small nest of Magpie *Pica pica*), lined with thick pad of plant down, seed panicles, fibres, and feathers. Domed, with entrance hole 40 mm wide spiralling down from top to cup. In hot areas near Dead Sea (Israel), nest thicker at top to provide more insulation against sun. Large for size of bird: diameter 20-30 cm, height 25-40 cm, weight 150-1000 g (Mendelssohn 1955); egg chamber 16 cm from top (Lobb 1981). Building: outer structure largely built by ♂, lined by both sexes. Old nest may remain for several seasons, but not re-used from season to season, though may serve as base for new nest (Lulav 1967). EGGS. See Plate 59. Sub-elliptical, smooth and slightly glossy. Ground-colour white or buffish, but often completely obscured by purplish-brown or grey spots and speckling; 1–2 eggs in clutch much lighter with only sparse spotting at large end (Harrison 1975). Nominate *moabiticus*: 18·3 × 13·0 mm (17·1–19·2 × 12·2–13·9), $n = 60$ (Schönwetter 1984). Fresh weight 1·64 g (Yom-Tov *et al.* 1978). Clutch: Israel, 4–5 (1–6), mean 4·0 ($n = 317$), decreasing with season (Yom-Tov and Ar 1980); Iraq, 4–5 (2–7), mean 4·4 (Jiad and Bunni 1988). In Israel, commonly 2(–3) broods; 40–43 days between 1st egg of 1st and 2nd clutches; replacements laid 5–7 days after loss (Yom-Tov and Ar 1980; Paz 1987). Eggs laid daily (Mendelssohn 1955). INCUBATION. In Israel, 9–16 days, mean 12·3 ($n = 185$), negatively correlated with mean ambient temperature: mean 11·6 days at 30°C, 10·3 days at 35°C (Yom-Tov *et al.* 1978; Yom-Tov and Ar 1980); Iraq, 12–16 days, mean 14 days, $n = 6$ (Jiad and Bunni 1988). Almost exclusively by ♀ (Yom-Tov *et al.* 1978), beginning after last or penultimate egg, thereafter 28–52% (mean 39%) of daylight hours; average incubation stint 4·5 min, average break 7·9 min (Jiad and Bunni 1988). Attentiveness to incubation by day varies with ambient temperature: decreases to *c.* 35% up to 35°C (optimum for incubation), then, to prevent over-heating, increases sharply over 35°C (Yom-Tov *et al.* 1978; see also Social Pattern and Behaviour). Eggs hatch over 48 hrs (Yom-Tov and Ar 1980); mean 51·2 hours, $n = 5$ (Jiad and Bunni 1988). YOUNG. Brooded by ♀ alone, fed mainly by ♀ (Yom-Tov and Ar 1980). FLEDGING TO MATURITY. Fledging period 11–13 days (Jiad and Bunni 1988). BREEDING SUCCESS. Hatching success in 2 Israeli colonies 39·8% ($n = 174$ eggs) and 43·0% ($n = 134$); fledging success 28·6% ($n = 109$) and 27·4% ($n = 90$) (Yom-Tov and Ar 1980). Low success attributed to desertion caused variously by intraspecific nest-parasitism (Yom-Tov 1980*a*), interference by other species (House Sparrow *Passer domesticus*, Yellow-vented Bulbul *Pycnonotus xanthopygos*), and predation by snakes (Yom-Tov and Ar 1980, which see for further details).     JDS-S

**Plumages.** (nominate *moabiticus*). ADULT MALE. In fresh plumage (September to about February), forehead, crown, hindneck, and upper mantle sandy-brown, some medium blue-grey on feather-centres just visible from about December onwards. Supercilium distinct; white above lore, pale cinnamon-buff above eye and ear-coverts. Lore dark grey or black, stripe just behind eye over upper ear-coverts black. Ear-coverts and upper side of neck dark grey, feathers tipped brown. Patch just below eye white with some grey mottling, broad stripe across lower cheek off-white, merging into broad lemon-yellow line below ear-coverts, latter turning upwards behind lower ear-coverts. Lower mantle and scapulars rufous-cinnamon with black shaft-streaks *c.* 1–2 mm wide; cinnamon of feather-tips tinged buff once plumage slightly worn. Back to upper tail-coverts buffish-brown, some grey of feather-centres visible; longer tail-coverts dark grey with sandy-buff fringes. Chin and throat black, forming small sharply-defined bib, partly concealed by white feather-fringes of *c.* 1 mm wide. Lower side of neck, chest, side of breast, and flank light greyish-buff, indistinctly streaked light grey and buff at rear of flank; mid-breast, belly, and vent white with slight cream

tinge; under tail-coverts pink-buff with contrasting rufous-cinnamon centres. Thigh contrastingly black. Central pair of tail-feathers (t1) and tertials black with broad pale cinnamon fringes; t2–t6 greyish-black with narrow pale cinnamon to off-white fringes. Flight-feathers, greater upper primary coverts, and bastard wing grey-black (blacker on inner secondaries); primaries and primary coverts narrowly fringed isabelline-pink to off-white along outer webs (widest near base of primaries and just below emarginated parts, forming pale patches on closed wing), outer webs of secondaries more broadly fringed greyish-cinnamon, shading to pale grey on tips. Lesser upper wing-coverts black, longer ones with broad pink-cinnamon or isabelline-white tip, forming wing-bar; median and greater coverts rufous-chestnut with small pink- or isabelline-white tips *c.* 2–3 mm wide, forming 2 other wing-bars. Under wing-coverts and axillaries cream-white, some longer inner coverts tinged cinnamon. *In worn plumage* (April–July), forehead to upper mantle dark ash-grey, nape and hindneck sometimes with traces of brown fringes; feather-tips of lower mantle and scapulars more extensively pale buff due to wear and bleaching; back to upper tail-coverts grey with brown wash; ear-coverts and upper side of neck dark ash-grey, bordered by black eye-stripe above; supercilium paler, white, shading to buff at rear; yellow at end of white stripe over lower cheek bright, sometimes extending upwards behind ear-coverts to meet buff supercilium. Bib uniform black; remainder of underparts off-white, except for light grey wash on chest and buff-grey wash on flank; under tail-coverts off-white, rufous-cinnamon centres more distinctly exposed. Pale fringes of tail and wing (including wing-bars) almost worn off, wing appearing mainly chestnut and black; white wing-bar on lesser coverts often still visible and traces of greyish-cinnamon to off-white fringes still present on tertials and secondaries; patch at base of primaries and frayed edges of primaries pure white. ADULT FEMALE. Closely similar to adult ♀ House Sparrow *P. domesticus*, but smaller, upperparts paler sandy-brown, black streaks on mantle and scapulars narrower; front part of supercilium distinct; lore pale; often some pale yellow at rear of cheek, and underparts paler, unstreaked off-white with buff wash on chest and flank. Upperparts dark sandy-brown (duller and more grey-brown when worn); mantle and scapulars light cinnamon-brown (if plumage fresh) or pale greyish-buff (if worn) with narrow brown-black shaft-streaks. Supercilium distinct, widest above ear-coverts, pale cinnamon-buff (if fresh) or isabelline-white (if worn). Lore and patch below eye off-white with some brown and grey mottling. Cheek grey with slight buff tinge, bordered by buff-brown at rear and by pink-buff line below, latter usually ending in pale lemon-yellow spot; some yellow occasionally near base of bill also. Side of breast, chest, flank, thigh, and under tail-coverts pale greyish-brown, grading into off-white on chin and upper throat and on mid-belly and vent; some light grey sometimes visible on chin and throat, some dark brown-grey on under tail-coverts. Tail and tertials as adult ♂, but centres dark brown-grey to black-brown, less black; flight-feathers, greater primary coverts, and bastard wing slightly browner; upper wing-coverts quite different, lesser uniform brownish-grey, median black with broad pink-buff tips, greater black-brown with broad greyish-cinnamon outer fringe and narrow isabelline tip, latter forming narrow pale wing-bar. *In fresh plumage*, rather similar to adult ♂, but streaks on upperparts, centres of tertials, and t1 browner, less black; black eye-stripe absent, ear-coverts browner, less deep grey; stripe over lower cheek less white, indistinct, not ending in bright yellow patch at rear, or with trace of pale yellow only; black bib absent; chest and flank more buff, no rufous-cinnamon on under tail-coverts; no black and chestnut on wing. *In worn plumage*, buff-brown of entire plumage more greyish, supercilium and pale stripe over lower cheek paler, latter extending upward behind ear-coverts; fringes of tail and wing narrower and whiter, tips of median and greater coverts bleached to off-white, forming narrow wing-bars. NESTLING. Naked (Jiad and Bunni 1988). JUVENILE. Like adult ♀, but upperparts and upper wing-coverts more diluted grey-brown, streaks on mantle and scapulars less sharply defined, pattern on side of head and neck less clear, yellow absent. Ground-colour of tail- and flight-feathers browner. Feathering of body short and loose, particularly rump, vent, and tail-coverts. FIRST ADULT. Generally indistinguishable from adult. In worn plumage, some ♂♂ (perhaps 1st adults) retain many traces of buff on feather-tips on nape and from back to upper tail-coverts (others, perhaps older birds, almost uniform grey here).

**Bare parts.** ADULT, FIRST ADULT. Iris hazel to dark brown. Bill greyish horn-brown with flesh tinge, culmen and tip dark horn, base of lower mandible yellowish or pale flesh-grey; in breeding season, bill of ♂ plumbeous-black: darkens February–April, bleaching again during post-breeding moult. Leg and foot grey-brown with flesh tinge. (BMNH, RMNH, ZFMK.) NESTLING. Bare skin flesh-pink at hatching (Jiad and Bunni 1988). JUVENILE. No information.

**Moults.** ADULT POST-BREEDING. Complete; primaries descendent. Probably starts as soon as water supply in breeding area exhausted: water essential for successful nesting, breeding halted when sources dry out, timing of which is strongly dependent on local circumstances (Yom-Tov *et al.* 1978, Paz 1987). Moult in Israel from August onwards (Mendelssohn 1955; Summers-Smith 1988). POST-JUVENILE. Complete; primaries descendent. In Israel, from August onwards (Mendelssohn 1955; Summers-Smith 1988).

**Measurements.** ADULT, FIRST ADULT. Nominate *moabiticus*. Dead Sea area, all year; skins (BMNH, RMNH, ZFMK, ZMA, ZMB). Bill (S) to skull, bill (N) to distal corner of nostril; exposed culmen on average *c.* 2·4 less than bill (S).

| | | | | | |
|---|---|---|---|---|---|
| WING | ♂ | 62·3 (1·17; 32) | 60–64 | ♀ 59·5 (1·14; 19) | 58–62 |
| TAIL | | 47·3 (1·28; 8) | 45–49 | 45·7 (0·99; 12) | 44–47 |
| BILL (S) | | 11·4 (1·16; 11) | 10·0–13·0 | 11·4 (0·75; 12) | 10·4–12·1 |
| BILL (N) | | 7·4 (0·48; 12) | 6·8–8·2 | 7·5 (0·26; 12) | 7·2–8·0 |
| TARSUS | | 17·2 (0·62; 8) | 16·4–18·0 | 16·6 (0·47; 12) | 16·0–17·4 |

Sex differences significant, except for bill. Bill in summer *c.* 10% longer than in winter.

Dead Sea area: wing, ♂ 59·7 (1·53; 8) 57·6–61·6, ♀ 59·2 (0·66; 4) 58·3–59·9; tail, ♂ 49·4 (8) 47·5–51·8, ♀ 48·9 (8) 48·0–50·2 (Boros and Horvath 1955).

*P. m. mesopotamicus.* Wing of birds from (1) south-central Turkey, May, (2) Iraq and south-west Iran, all year; other measurements combined; skins (BMNH, ZFMK).

| | | | | | |
|---|---|---|---|---|---|
| WING (1) | ♂ | 65·6 (1·01; 10) | 64–67 | ♀ 62·5 ( – ; 1) | – |
| (2) | ♂ | 65·2 (0·98; 11) | 63–67 | ♀ 62·8 (0·52; 6) | 62–64 |
| TAIL | | 49·4 (1·32; 10) | 48–52 | 47·3 (1·38; 7) | 46–49 |
| BILL (S) | | 12·0 (0·31; 10) | 11·6–12·4 | 11·9 (0·52; 7) | 11·3–12·7 |
| BILL (N) | | 7·5 (0·23; 10) | 7·2–7·8 | 7·6 (0·32; 7) | 7·2–8·2 |
| TARSUS | | 17·4 (0·48; 9) | 16·8–18·0 | 17·4 (0·68; 7) | 16·5–18·3 |

Sex differences significant for wing and tail.

South-west Iran: wing, ♂ 63·5–66·5, ♀ 59·5–63·5 (Sarudny 1904).

*P. m. yatii.* South-east Iran, all year; skins (BMNH, RMNH, SMTD, ZFMK, ZMA, ZMB).

| | | | | | |
|---|---|---|---|---|---|
| WING | ♂ | 65·4 (1·32; 17) | 63–67 | ♀ 62·8 (1·25; 6) | 61–64 |
| TAIL | | 49·8 (1·80; 9) | 48–53 | 46·8 (0·84; 5) | 46–48 |
| BILL (S) | | 12·0 (0·50; 14) | 11·8–13·1 | 11·7 (0·15; 6) | 11·5–11·9 |
| BILL (N) | | 7·5 (0·43; 9) | 7·1–8·4 | 7·2 (0·34; 6) | 6·8–7·7 |
| TARSUS | | 17·3 (0·83; 4) | 16·3–18·2 | 17·8 (0·55; 5) | 17·2–18·4 |

Sex differences significant for wing and tail.

Seistan (Iran-Afghanistan border): wing, ♂ 64·2 (1·27; 16) 61·9–66·4, ♀ 61·9 (1·87; 9) 60·1–65·6; tail, ♂ 52·4 (1·93; 16) 48·4–55·1, ♀ 51·0 (2·50; 9) 48·7–55·5 (Boros and Horvath 1955).

JUVENILE. At fledging, wing 76% of adult length, bill 87%, tarsus 88% (Jiad and Bunni 1988).

**Weights.** ADULT, FIRST ADULT. Nominate *moabiticus*. Israel: 11–17 (Paz 1987).

*P. m. mesopotamicus.* Central-south Turkey, May: ♂ 16·3 (2·06; 10) 15–20, ♀ 20 (ZFMK).
NESTLING, JUVENILE. For growth curve, see Yom-Tov and Ar (1980): at day 7, *c.* 5·0 (2·5–7·5); weight constant from day 13; shortly before fledging, at day 13–15, weight *c.* 13 (11·5–15). In Iraq, at fledging, day (11–)13, weight 73% of adult (Jiad and Bunni 1988).

**Structure.** Wing short, broad, tip rounded. 10 primaries: p7–p8 longest, p9 and p6 0–1·5 shorter, p5 2–5, p4 5–8, p1 10–15. P10 reduced, a tiny narrow and pointed feather hidden below greater upper primary coverts; in adult, 37–44 shorter than p7–p8, 3–5 shorter than longest upper primary covert. Outer web of p6–p8 emarginated, inner of (p6–)p7–p9 with faint notch. Tip of longest tertial reaches tip of p2–p3. Tail short, tip square or slightly forked; 12 feathers. Bill as in *P. domesticus*, but distinctly smaller, relatively narrower and less deep at base. Leg and foot short, rather slender. Middle toe with claw 16·1 (5) 15·5–16·5; outer and inner toe with claw both *c.* 69% of middle with claw, hind *c.* 75%. Remainder of structure as in *P. domesticus*.

**Geographical variation.** Rather slight in size, more marked in colour. Nominate *moabiticus* and *mesopamicus* both dark sandy-brown above, whitish below; nominate *moabiticus* from Dead Sea area and Arava valley is small, *mesopotamicus* from south-central Turkey and northern Syria south-east through Iraq to south-west Iran is larger. *P. m. yatii* of Iranian Seistan and neighbouring south-west Afghanistan similar in size to *mesopotamicus*, but upperparts suffused buff in both sexes, underparts yellow: in ♂, distinct pale cinnamon-buff tinge to cap and rump when plumage fresh, grey on cap pure but rump still with olive-green or green-yellow tinge when worn; in ♀, upperparts light buff-grey, less brown-grey than ♀ nominate *moabiticus*; in both sexes, lower throat and belly extensively tinged pale yellow (brighter in ♂), pale fringes of flight-feathers wider, paler pink-buff (less greyish-cinnamon), wing appearing less dark, and under tail-coverts narrowly streaked dusky, without contrasting cinnamon spot. See also Boros and Horvath (1955). Colour of supercilium of ♂ variable, independent of bleaching or wear: yellow, white, white shading to yellow at rear, or white shading to rufous-cinnamon at rear. *P. m. yatii* considered by Boros and Horvath (1955) to be separate species due to difference in colour and size from nominate *moabiticus*, but *mesopotamicus* combines colour of nominate *moabiticus* with size of *yatii*, and difference in colour mainly due to a single character, extensive suffusion of yellow pigment in *yatii* (comparable to difference between races of Cinereous Bunting *Emberiza cinerea*). CSR

# *Passer iagoensis* Iago Sparrow

DU. Kaapverdische Mus    FR. Moineau du Cap-Vert    GE. Rostsperling
RU. Рыжеспинный воробей    SP. Gorrión grande    SW. Brunryggad sparv

*Pyrgita iagoensis* Gould, 1837

Monotypic

**Field characters.** 13 cm; wing-span 17·5–20 cm. About 10% shorter than Spanish Sparrow *P. hispaniolensis*, with noticeably slighter bill, 20% shorter wings, and 15% shorter tail. Rather small, compact sparrow, confined to Cape Verde Islands. Plumage colours and pattern recall House Sparrow *P. domesticus* but ♂ has black crown-centre, rufous supercilium, black rear eye-stripe extending round white cheek, and only narrow black bib. ♀ duller, with strong streaks on scapulars, contrasting pale supercilium, dark eye-stripe, and (on some) dusky bib. Sexes dissimilar; little seasonal variation. Juvenile separable.

ADULT MALE. Moults: probably January–June. Head strongly patterned, with long blaze of rufous-cinnamon on rear supercilium and side of nape; contrasting with this are grey (when fresh) or mostly black (when worn) crown, white supra-loral stripe, and black eye-stripe and rear cheek border; cheeks essentially white, contrasting with eye-stripe and long, rather narrow black bib. Shawl over shoulders dusky, but mantle, scapulars, back, and lesser wing-coverts bright chestnut, with strong black streaks on mantle; rump much duller, essentially brownish-grey. Wings also strongly patterned, with bold white bar across median coverts emphasized by their black bases and grey-centred and rufous-fringed greater coverts; greater coverts also show buff (when fresh) or off-white (when worn) tips, forming less distinct lower wing-bar; flight-feathers greyish-black, fringed buff to off-white except on tertials where margins cinnamon-buff. Underwing pale grey. Tail black, with feathers narrowly edged pale grey to white (when fresh). Underparts white, tinged cream in centre (when fresh) and grey across chest and along flanks. Bill black in breeding season, horn (or even paler on lower mandible) at other times. Legs lead, fading grey or brown. ADULT FEMALE. Suggests faded ♂ with always grey crown, pink-buff or paler supercilium, duller black eye-stripe, and warm brown (not bright rufous) back and lesser wing-coverts. Unlike other west Palearctic ♀ sparrows, at least some show dusky bib. Wings less bright than ♂, with bar across median coverts narrower and duller, cream (when fresh) to white (worn). JUVENILE. Resembles ♀ but sexual differences already clearly apparent in ♂'s black-streaked crown and rufous rear supercilium and ♀'s paler, browner crown and less warm, pale buff to white rear supercilium (see also Plumages). Bill paler than adult, yellow-horn to dull pinkish-grey.

On Cape Verde Islands, confusable only with Spanish Sparrow *P. hispaniolensis*, but that species noticeably larger, with ♂ heavily marked black on chest, back, and flanks. Flight and behaviour typical of genus. Highly gregarious.

Voice apparently much like other *Passer*.

**Habitat.** In west Palearctic only in Cape Verde Islands, which, although tropical and oceanic and having main breeding season at end of the rains, have unusually cool climate, with tendency to increasing dryness. Climatic change appears to be modifying habitats, partly to the advantage of *P. iagoensis* which is typically a bird of open desert, dry scrub, and rock faces, but also occurs in woodland and in cultivated areas and towns. Prefers crevices for nesting, but will also use trees or even sites under stones (Bourne 1955). Listed as occurring in all 9 of the islands' chief habitat divisions, and as being common in desert, scrub, woods, fields, ravines, and oases, and also in towns, plantations, and irrigation. Occurs up to *c.* 1250 m, and will coexist with Spanish Sparrow *P. hispaniolensis* in palm gardens with deep irrigation wells; on coastal sites, occurs on lava flows. (Bannerman and Bannerman 1968; C J Hazevoet.)

**Distribution and population.** CAPE VERDE ISLANDS. Widespread and abundant on all main islands, except Fogo (where probably absent) and Sal (where scarce); also common on Rombos islands and Raso; on Santa Luzia and Branco, absent in some years and present in small numbers in others (CJH). No information on changes.

**Movements.** Apparently sedentary. No confirmed records from Fogo, lying only 15 km from Rombos islands and 17 km from Brava, on both of which it is very common. Small population on Branco and Santa Luzia (where absent in some years) probably augmented by immigration from São Vicente and/or Raso (C J Hazevoet).    JDS-S

**Food.** Limited observations suggest mainly plant material obtained from ground, especially seeds, also young leaves; insects taken by adults during breeding season and fed to young. Insects include Orthoptera (Alexander 1898a; Summers-Smith 1988) and larval Lepidoptera (Summers-Smith 1988). Plant material mainly seeds of small plants, e.g. sparrow-rice *Zygophyllum* (C J Hazevoet), grasses (Bolle 1856; Alexander 1898a), maize *Zea* (Bourne 1955);

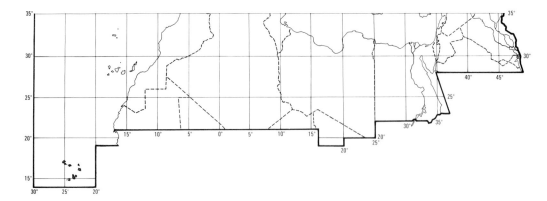

also domestic scraps, mangos *Mangifera*, figs *Ficus*, nectar, and flowers (Nørrevang and Hartog 1984; Hazevoet and Haafkens 1989; Hartog 1990; C J Hazevoet). JDS-S

**Social pattern and behaviour.** No systematic studies. Information from J D Summers-Smith except where otherwise indicated.

1. Generally sociable, forming flocks (sometimes large) outside breeding season, sometimes with Spanish Sparrow *Passer hispaniolensis* (C J Hazevoet), less often with House Sparrow *P. domesticus*. BONDS. Probably monogamous; pairs remain close together away from nest, when both feeding and resting. Both sexes build nest, incubate, and feed young. BREEDING DISPERSION. Pairs breed either alone or in loose aggregations of up to 10–12 pairs, with nests as close as 1 m in holes under roofs, up to 10 nests in one tree (C J Hazevoet). ROOSTING. During breeding season, pairs roost in nest; otherwise nothing recorded.

2. Often very tame, approaching observer to within less than 1 m. Tail-flicks (sharp upward jerk of tail) when alarmed. FLOCK BEHAVIOUR. Individuals stay close together (within a few cm), both in feeding flock and when resting. Flocks tend to stay close to cover, acacias, or bolder-strewn ravines, where they can hide from aerial predator, e.g. Kestrel *Falco tinnunculus*, or shelter from midday sun (Alexander 1898*a*; Naurois 1988). SONG-DISPLAY. ♂ gives Advertisement-call (see 2a in Voice) close to nest-site; switches in unpaired ♂ to Song-display (see 1 in Voice) on approach of ♀. ANTAGONISTIC BEHAVIOUR. Threatens approaching conspecifics at nest or at dust-bathing site with Forward-threat, thrusting forward with wings held out; also attacks *P. hispaniolensis* (C J Hazevoet). HETEROSEXUAL BEHAVIOUR. ♂♂ apparently arrive at breeding sites before ♀♀ (Bourne 1955). Pair-formation apparently occurs at nest-site. On arrival of ♀, ♂ performs Parade-display with excited calling (see Song-display, above), wings drooped (not held out or shivered) and slightly rotated, exposing chestnut rump, with scapular-patches directed forward. Copulation occurs in vicinity of nest. ♀ becomes dominant at this stage. ♂ invites copulation by hopping round ♀ in similar posture, but with wing-feathers and rump ruffled. Unresponsive ♀ pecks at ♂, but when ready invites copulation by crouching. RELATIONS WITHIN FAMILY GROUP. Both sexes feed nestlings, though ♀ takes major role, ♂ at times merely accompanying ♀ but not collecting food. Both parents feed fledglings. Nest-sanitation by ♀. ANTI-PREDATOR RESPONSES OF YOUNG, PARENTAL ANTI-PREDATOR STRATEGIES. No information. JDS-S

**Voice.** Not well studied. Information from J D Summers-Smith except where indicated.

CALLS OF ADULT. (1) Song. Loose series of calls 'cheep chirri chip cheep chirri chip chip'. (2) Other advertisement- and contact-calls. (2a) Basic 'chirp'. 'Cheerp', given in groups of 2–3 calls (3 shown in Fig I),

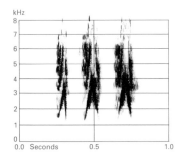

I  C J Hazevoet  Cape Verde Islands  September 1988

lower pitched than call 2a of House Sparrow *P. domesticus*; slurred 'chirrp' (Bourne 1955), given at nest-site. Similar twangy 'cheesp', 'chew-weep', 'chew-leep', or sibilant 'chisk' given by birds away from nest. (2b) Social-singing. Chorus of calls given by group resting together in cover (Bolle 1856); in recording of chorus by C J Hazevoet,

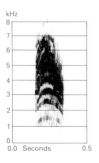

II  C J Hazevoet  Cape Verde Islands  September 1988

calls mainly of 'chirp' type. (3) Threat-call. Soft churring 'chur-it-it-it-it' or twangy 'chur-chur-chur' accompanied by tail-flicking. (4) Alarm-call. Quiet, sibilant single 'chisk' or twangy 'cheesp' (Fig II), expressing mild anxiety; accompanied by bill-wiping.

CALLS OF YOUNG. No information.                                    JDS-S

**Breeding.** SEASON. Triggered by rains in August–September

(Bourne 1955). Nest-building September–October, breeding September–March (Keulemans 1866; Dohrn 1871; Naurois 1969, 1988) or August–MArch (C J Hazevoet). According to Naurois (1969), breeding peaks September–November and February–March, but probably irregular in arid areas, responding to rains (C J Hazevoet). SITE. Mainly in holes: lava cliffs, road cuttings, stone walls, buildings, wells, on ground under boulders (Bolle 1856; Keulemans 1866; Alexander 1898b; Naurois 1969), street lights (J D Summers-Smith), but some free-standing nests in branches of trees (Keulemans 1866; Bourne 1955; C J Hazevoet). Nest: free-standing nest is globular, domed structure with side entrance (Keulemans 1866), c. 15 cm diameter, 20–25 cm long (C J Hazevoet); under boulder, compact open structure of thick plant stems lined with feathers (Alexander 1898b; Naurois 1988). Outer structure coarse dry grass, cup lined with hair and feathers (J D Summers-Smith). Building: by both sexes (J D Summers-Smith). EGGS. See Plate 59. No adequate description: similar to those of House Sparrow *P. domesticus* (Dohrn 1871; Murphy 1924); one clutch described by Alexander (1898b) as uniformly coloured, similar to those of Tree Sparrow *P. montanus*, with one paler egg. 19·0 × 13·0 mm, calculated weight 1·7 g (n = 2) (Schönwetter 1984). Clutch: 3–5 (Alexander 1898b; Bannerman and Bannerman 1968; Naurois 1969). Replacement clutches in November 1–2 (Naurois 1969). INCUBATION. Both sexes incubate, in stints of 5–15 min (J D Summers-Smith). No further information.

                                                                  JDS-S

**Plumages.** ADULT MALE. Forehead, crown, and hindneck dark grey, feathers with black centres, which are concealed when plumage fresh, but exposed when worn, cap then appearing black with grey mottling. Narrow white line from nostril to above front of eye, merging into broad deep rufous-cinnamon supercilium, which extends to side of neck. Lore black; black stripe from just below and behind eye over upper ear-coverts. Lower cheek and remainder of ear-coverts white, ear-coverts washed buff when fresh, longer coverts tipped grey. Hindneck and upper mantle dark grey. Lower mantle, scapulars, back, and lesser upper wing-coverts deep rufous-chestnut, mantle and inner scapulars with black streaks 2–4 mm wide. Rump and upper tail-coverts brown-grey with olive or buff tinge. Long but rather narrow bib on chin and throat black, fringes of feathers white when plumage fresh, partly concealing black; remainder of underparts grey-white or white, slightly tinged cream when plumage fresh; chest, side of breast and upper flank slightly darker grey, lower flank and centres of under tail-coverts brown-grey. Tail greyish-black; both webs of central pair (t1)

and outer webs of t2–t6 narrowly edged pale olive-grey or white, worn off when plumage abraded. Flight-feathers, tertials, greater upper primary coverts, bastard wing, and greater upper wing-coverts greyish-black; outer webs and tips of flight-feathers, primary coverts, and shorter feathers of bastard wing with sharp and narrow pale buff-grey to off-white fringe, those of greater coverts and tertials with broader but less sharply-defined pale cinnamon-buff fringe, bleaching to pale buff or off-white on tip when worn. Median upper wing-coverts white with concealed black base, white forming distinct bar 4–6 mm wide. Under wing-coverts and axillaries pale olive-grey, small coverts along leading edge of wing and under primary coverts mottled dark grey. ADULT FEMALE. Forehead and crown dark drab-grey, feathers with slightly paler olive-grey fringes when plumage fresh; hindneck, side of neck, and upper mantle grey-brown or dark drab-grey; lower mantle to upper tail-coverts drab-brown, less dark and grey than crown, upper tail-coverts slightly greyer, lower mantle and inner scapulars with distinct black streaks 2–3 mm wide. Long isabelline or grey-white supercilium, extending from nostril to side of neck, narrow and sometimes faint above eye, sometimes partly tinged pink-buff when plumage fresh. Lore and stripe through eye over upper ear-coverts dark grey to brown-black. Upper cheek and ear-coverts grey-white, washed buff when plumage fresh. Underparts pale grey-white, almost pure white on chin, throat, mid-belly, and vent, washed brown-grey on flank, thigh, and centres of under tail-coverts; chin and central throat sometimes with indistinct mottled grey bib. Tail dark olive-sepia, narrow edges along outer webs pale olive-grey. Flight-feathers, tertials, bastard wing, and greater primary and secondary coverts as adult ♂, but ground-colour brown instead of black and fringes duller and greyer, less cinnamon. Median upper wing-coverts sooty black, c. 2–3 mm of tips contrastingly pink-cream (if fresh) or white (if worn), forming narrow pale wing-bar; lesser upper wing-coverts brown-grey, feathers partly tipped cinnamon. Under wing-coverts and axillaries as adult ♂. NESTLING. No information. JUVENILE MALE. Rather like a washed-out version of adult ♂. Cap dark brown-grey with blackish streaks; mantle and scapulars buff-brown with rather ill-defined sooty streaks, back and rump rusty-buff. Supercilium off-white in front of eye, cinnamon behind, bordered below by distinct sooty black eye-stripe; upper cheek and ear-coverts cream-white with dark grey mottling, not sharply demarcated from dark eye-stripe. Underparts isabelline-white, tinged grey-buff on side of breast and flank; a fairly distinct sooty bib on chin and central throat, mottled white, not clearly defined at borders. Tail, flight-feathers, greater upper primary coverts, and bastard wing as adult ♂, but ground-colour distinctly browner; tertials and greater upper wing-coverts dark brown, grading to brown-black on tip, bordered by narrow cinnamon-buff to buff-white fringes 1–2 mm wide; median coverts dull black, c. 2 mm of tips white, forming narrow bar; lesser coverts rufous-cinnamon. Feathering of head and body shorter and looser than in adult, especially rump and tail-coverts. JUVENILE FEMALE. Rather like a dilute-coloured adult ♀, and sometimes hard to distinguish from juvenile ♂; feathering shorter and looser than in adult ♀. In contrast to adult ♀, back and rump tinged rufous, less uniform dull drab-brown; often a faint light grey bib; ground-colour of tail- and flight-feathers browner; median coverts dull black, white tips less sharply defined, 1–2 mm long; lesser coverts rufous-brown. Differs from juvenile ♂ in slightly paler and browner cap, which is less streaked black; rear of supercilium pale cream-buff or white, less rufous; dark eye-stripe narrower and less clearly defined, less broad and blackish; bib paler grey and less distinct; lesser coverts bib; lesser coverts paler rufous. FIRST ADULT. Like adult; indistinguishable when post-juvenile moult completed. During moult, outer primaries still fresh, edges smooth (in adult, heavily abraded).

**Bare parts.** ADULT, FIRST ADULT. Iris dark hazel, dark ochre-brown, or dark brown. Bill dark horn or grey-horn, lower mandible paler horn, light grey, or yellow; in breeding season (October–December in birds examined), bill of ♂ black. Leg and foot dark green-grey, horn-grey, or dark horn-brown; in breeding season, those of ♂ dark brown, plumbeous, or grey-black. NESTLING. No information. JUVENILE. Iris brown-grey, hazel, or dark brown. Bill yellow-horn or horn-grey, culmen dark horn to plumbeous-black; base of lower mandible pink-grey. Leg and foot flesh-brown, light brown, grey-brown, or leaden-grey. (Naurois 1988; Summers-Smith 1988; BMNH, ZMA.)

**Moults.** ADULT POST-BREEDING. Complete; primaries descendent. Only 1 of 22 dated adults examined was in active moult; plumage in most birds rather worn by October, heavily worn January; single birds examined with fresh plumage in March, April, October, and November likely to be juveniles which had just completed moult (BMNH, RMNH). Moult starts January, after breeding season August–January (which followed rains in mid-July or August) (Bourne 1955). Single bird nearing end of moult (primary moult score 43) on 6 June is probably adult (bill black; in all other birds examined, bill pale, these probably juveniles). Other data on moult (e.g. Alexander 1898a) probably refer to juveniles. POST-JUVENILE. Complete, primaries descendent. Of 32 dated juveniles examined, 10 in moult. Fresh-plumaged juveniles occur (according to collection examined in BMNH and RMNH) November–June, mostly December–January (7 birds) and April (9); birds in moult had primary moult scores 29 (12 October), 2 (December), 30 (4 March), 19, 26, 35, 38, and 45 (15 April), and 40 (6 June); depending on fledging date, juveniles probably start moult between December and June and complete March–November. Tail starts at about primary moult score 7–18; moult centrifugal, completed with primary score 38 and above; body fairly fresh juvenile when primary moult starts, in full moult at primary score 19–20, mainly new except for hindneck at score 26–30, almost completed at score 38 and above. (C S Roselaar.)

**Measurements.** Adult wing, bill to skull, and tarsus of birds from (1) Santo Antão, (2) São Vicente, (3) Raso, (4) Branco, (5) São Tiago, and (6) Brava; other measurements for all islands combined (data of juvenile mainly from São Vicente); skins (BMNH, RMNH; some additional data of wing and bill to skull from Vaurie 1958b). Bill (N) to distal corner of nostril; exposed culmen on average 2·9 less than bill to skull.

| WING | (1) | ♂ | 64·5 (0·89; | 6) | 63–66 | ♀ | 58·9 (0·22; | 5) | 58–60 |
|---|---|---|---|---|---|---|---|---|---|
| | (2) | | 64·6 (1·30; | 8) | 63–67 | | 60·7 (1·23; | 5) | 58–62 |
| | (3) | | 66·1 (0·87; | 6) | 65–67 | | 61·0 ( — ; | 2) | 60–62 |
| | (4) | | 65·8 (0·84; | 5) | 65–67 | | — ( — ; —) | | — |
| | (5) | | 65·1 (1·24; | 7) | 63–67 | | 59·2 (1·63; | 6) | 57–61 |
| | (6) | | 67·0 ( — ; | 4) | 66–68 | | 62·5 ( — ; | 1) | — |
| JUV | | | 62·8 (1·11; | 14) | 61–65 | | 60·1 (1·49; | 13) | 57–62 |
| TAIL AD | | | 51·4 (2·17; | 15) | 48–55 | | 47·6 (1·22; | 7) | 46–50 |
| JUV | | | 47·2 (1·48; | 9) | 45–49 | | 45·6 (1·58; | 10) | 44–48 |
| BILL | (1) | | 15·3 (0·45; | 6) | 14·8–16·1 | | 13·8 (0·47; | 4) | 13·4–14·4 |
| | (2) | | 15·0 (0·58; | 7) | 14·5–16·0 | | 13·0 ( — ; | 1) | — |
| | (3) | | 15·1 (0·44; | 6) | 14·5–15·8 | | 14·3 ( — ; | 2) | 14·2–14·4 |
| | (4) | | 14·0 (0·35; | 5) | 13·5–14·5 | | — ( — ; —) | | — |
| | (5) | | 15·7 (0·45; | 7) | 15·4–16·4 | | 14·7 (0·26; | 5) | 14·3–15·0 |
| | (6) | | 15·5 ( — ; | 2) | 15·0–16·0 | | 14·1 ( — ; | 1) | — |
| BILL (N) | (1–3) | | 9·9 (0·45; | 16) | 9·3–10·5 | | 8·7 (0·46; | 10) | 7·9–9·3 |
| | (5–6) | | 9·9 (0·44; | 8) | 9·3–10·8 | | 8·8 (0·23; | 3) | 8·5–8·9 |

| TARSUS | (1) | | 19·2 (0·59; | 6) | 18·4–19·7 | | 18·3 (0·58; | 4) | 17·7–19·0 |
|---|---|---|---|---|---|---|---|---|---|
| | (2) | | 19·0 (0·25; | 4) | 18·7–19·3 | | 17·6 ( — ; | 1) | — |
| | (3) | | 19·6 (0·54; | 6) | 9·0–20·5 | | 18·5 ( — ; | 2) | 18·4–18·6 |
| | (5) | | 18·6 (0·47; | 7) | 18·1–19·1 | | 17·8 (0·29; | 5) | 17·5–18·2 |
| | (6) | | 20·1 ( — ; | 2) | 19·6–20·6 | | 18·8 ( — ; | 1) | — |

Sex differences significant.

For more measurements, mainly based on same collections as those above, see Bourne (1955, 1957) and Naurois (1988).

**Weights.** No information.

**Structure.** Wing short, broad at base, tip rounded. 10 primaries: p7–p8 longest, p9 1–2 shorter, p6 0–1, p5 1·5–3; p4 3·5–6, p1 9·5–13. P10 reduced, narrow, pointed, concealed below outer greater upper primary covert; 37–43 shorter than p7–p8 in adult, 34–37 in juvenile; 1–5 shorter than longest upper primary covert in adult, 0–2 shorter in juvenile. Outer web of (p5-)p6–p8 emarginated, inner web of p7–p9 with notch. Tip of longest tertial reaches tip of p2–p3. Tail short, tip square; 12 feathers. Bill short, conical, much as in House Sparrow *P. domesticus*, but relatively smaller and more slender. Tarsus and toes short, but rather strong. Middle toe with claw on most islands 16·6 (10) 15·5–18·0, but 17·4 (5) 17·0–18·5 on Raso; outer and inner toe with claw both *c*. 67% of middle with claw, hind *c*. 77%. Remainder of structure as in *P. domesticus*.

**Geographical variation.** Rather slight; involves depth of general colour (which is, however, influenced by age, abrasion, and bleaching), size (wing, tail, and tarsus length), and relative bill size. 3 races sometimes recognized: *hansmanni* Bolle, 1856, north-west islands (Santo Antão, São Vicente, Santa Luzia, Raso, São Nicolau, *brancoensis* Oustalet, 1883, Branco (in north-west islands, though population not permanent; see Distribution), and nominate *iagoensis* from southern islands (Brava, Rombos islands, and São Tiago); not known what race is on Maio, Boa Vista, and Sal. General colour perhaps paler in north-west islands (except Branco, which has deepest colour of entire species); brighter, more extensively rufous on head, back, and wing and more buffish on underparts in southern islands (Brava and Rombos islands); colour on São Tiago and Boa Vista intermediate (Bourne 1955, 1957; Bannerman and Bannerman 1968). Slightly smaller birds occur Santo Antão, São Vicente, São Nicolau, Boa Vista, and São Tiago, slightly larger birds on Brava, Branco, and Raso; bill relatively small on Branco, relatively large on Brava and São Tiago, intermediate on other islands. See Measurements, also Murphy (1924), Bourne (1955, 1957, 1966), and Vaurie (1958b, 1959). However, all differences are slight and variation is irregular, not constant, and dependent on local conditions on each island; no clear trend (Vaurie 1958b, contra Bourne 1957); also, overlap in characters is large, and thus no races recognized here, following Bannerman (1948) and Vaurie (1958b).

Sometimes united with Great Sparrow *P. motitensis* and its relatives in a single polytypic species (e.g. Hall and Moreau 1970), Iago Sparrow when named *P. motitensis iagoensis* (Clancey 1964; White 1967). Here, all forms considered allospecies of a superspecies, comprising *P. iagoensis*, *P. motitensis* of southern Africa, Kenya Rufous Sparrow *P. rufocinctus* of Kenya, White Nile Rufous Sparrow *P. shelleyi* from south-east Sudan and northern Kenya north-east to southern Ethiopia and north-west Somalia, Socotra Sparrow *P. insularis* from Socotra and Abd-Al-Kuri islands, and Kordofan Rufous Sparrow *P. cordofanicus* from western Sudan; all these are apparently long-isolated relics of a formerly widespread single polytypic species, which are in general successful only where other *Passer* of similar size and habits are absent. Etire superspecies perhaps closest to *P. domesticus* and Desert Sparrow *P. simplex* (Hall and Moreau 1970). For reasons (some doubtful) used to separate *P. iagoensis* from remaining forms, see Someren (1922), Lynes (1926), Grant and Mackworth-Praed (1944), Bannerman (1948), Summers-Smith (1984, 1988), and Bourne (1986); for characters of these forms, see Lynes (1926).                                      CSR

## *Passer simplex*  Desert Sparrow

PLATES 23 and 26 (flight)
[between pages 280 and 281]

Du. Woestijnmus          Fr. Moineau blanc          Ge. Wüstensperling
Ru. Пустынный воробей          Sp. Gorrión sahariano          Sw. Ökensparv

*Fringilla simplex* Lichtenstein, 1823

Polytypic. *P. s. saharae* Erlanger, 1899, Sahara of south-east Morocco, Algeria, southern Tunisia, and central Libya; perhaps this race or nominate *simplex* in Mauritania; nominate *simplex* (Lichtenstein, 1823), southern Sahara from central Mali through Aïr (northern Niger), Tibesti, Borkou, and Ennedi (northern Chad) to central Sudan. Extralimital: *zarudnyi* Pleske, 1896, Turkmeniya and Uzbekistan.

**Field characters.** 13·5 cm; wing-span 22–25 cm. About 10% smaller than House Sparrow *P. domesticus*, with stubbier bill, somewhat longer tail and 10% longer legs contributing to leggy stance. Rather small, relatively large-headed sparrow with pale plumage. ♂ essentially pale grey above and buff-white on rump and body, with bold black bill, short eye-stripe and bib, and pied wings. ♀ remarkably dissimilar, essentially pale sand-buff above, with horn bill and only vestigial wing markings. Juvenile resembles ♀. 1–2 races in west Palearctic; north-west Saharan race, *saharae*, described here.

ADULT MALE. Moults: probably July–August. Crown, back, and lesser wing-coverts silvery-grey. Head marked by black loral stripe, just extending below and behind eye, and white cheeks; striking black bib and chest-centre. Wing shows cream median coverts forming bold upper wing-bar; dark grey to black bases to greater coverts and primary coverts create narrow V-shaped dark band; buffish-white tips to greater coverts and long cream-pink fringes to secondaries and primaries form conspicuous pale panel on folded and extended wing, further emphasized by dark grey-black centres to buff-white-fringed tertials and mainly blackish tips of primaries. Below back, paler greyish-isabelline rump, almost white upper tail-coverts, and pale buff-white fringes to outer tail-feathers form further obvious marks. Tail strikingly dark-centred due to dusky grey on middle of central tail-feathers. Underwing rosy-white. Underparts dull white, often tinged pink to buff from desert soils. Bill black in breeding season, becoming brown- and yellowish-horn for rest of year. Legs light pink- to grey-brown. ADULT FEMALE. Very different from ♂, with characterless appearance recalling other desert passerines such as Bar-tailed Desert Lark *Ammo-manes cincturus* (sympatric in western Sahara), ♀ Sudan Golden Sparrow *P. luteus*, ♀ Trumpeter Finch *Bucanetes githagineus* (sympatric in western Sahara), and ♀ Sinai Rosefinch *Carpodacus synoicus*. Head and upperparts isabelline to pink-buff, with no obvious face markings except paler lore and faint supercilium, and wing pattern much reduced due to much wider margins obscuring dusky marks, these being restricted mainly to primary coverts, tips of flight-feathers, and centres of tertials. In breeding season, bill only rarely blackish-horn on upper mandible, white-horn on lower; paler at other times. JUVENILE. Resembles ♀ but upperparts less buff, underparts uniformly sandy-white; contrast of wing and tail markings even more subdued (see also Plumages). Bill and legs pale flesh.

Unmistakable if adult ♂♂ present, but single ♀ or juvenile could well puzzle observer inexperienced in desert passerines. Flight and behaviour typical of *Passer*, but lives in small groups, becoming commensal in villages. Often localized within sandy areas.

Song much more musical than *P. domesticus*: melodious series of trills which recalls Linnet *Carduelis cannabina* and particularly Greenfinch *C. chloris*. Calls chattering, recalling *P. domesticus* but can sound higher pitched: 'chip-chip' or subdued repeated 'chu'.

**Habitat.** Confined to arid, subtropical, mainly lowland regions, in sandy areas, hollows among dunes, or dry wadis with shrubs, but locally ascends hills and even slopes of mountains. Nesting requirements lead to choice of locations near oases with trees, or sometimes isolated buildings such as forts, while patches of cultivation are also attractive. (Guichard 1955.) Birds in Asia said never to

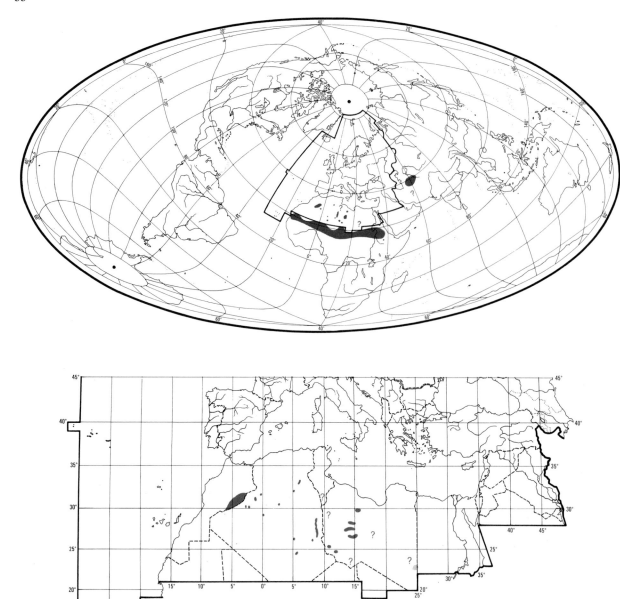

come near human settlements (Dementiev and Gladkov 1954).

**Distribution and population.** No information on changes. In north-west Africa, common in some areas, absent or very rare in others (Heim de Balsac and Mayaud 1962). ALGERIA. Erratic in choice of breeding localities, seldom occupying one locality for more than 2 consecutive years (EDHJ).

Accidental. Egypt: Gebel Uweinat area in extreme south-west (Goodman and Meininger 1989).

**Movements.** Some southward movement in winter (Hollom *et al.* 1988), but erratic appearances (see Berg and Roever 1984; Ash and Nikolaus 1991) suggest this may be more nomadic than directed. Extralimital *zarudnyi* sedentary, with very slight local movements in autumn and winter (Dementiev and Gladkov 1954; Sopyev 1965; Leonovich 1983).                      JDS-S

**Food.** Casual records suggest chiefly plant material with some animal matter in breeding season. Seeds collected on ground (Rothschild and Hartert 1911; Bates 1934) and by flying up and pulling seed heads of larger plants to ground (Densley 1990).

Animal food includes adult, larval, and pupal beetles (Coleoptera) (Bates 1934; Dementiev and Gladkov 1954) and spiders (Araneae) (Guichard 1955; Densley 1990). Stomachs of extralimital *zarudnyi* contained caterpillars (Lepidoptera) and beetles; compared with House Sparrow *P. domesticus* and Saxaul Sparrow *P. ammodendri*, proportionately smaller digestive tract of *P. simplex* (see morphological study of 4 desert *Passer* by Amanova 1977) suggests *P. simplex* has higher proportion of animal food in diet (Rustamow and Sopyew 1990). Plant material mainly seeds: of grasses (Gramineae), e.g. awn grass *Aristida* staple food according to Snigirewski (1928), Densley (1990), and Rustamow and Sopyew (1990), cereal grain (e.g. millet *Panicum*, barley *Hordeum*) (Bates 1934; Guichard 1955), and of desert plants (saxaul *Haloxylon*, saltwort *Salsola*, tragacanth *Astragalus* in Turkmeniya, and *Raetama* flowers) (Heim de Balsac 1929; Rustamow and Sopyew 1990). Probably all water obtained from food (Rustamow and Sopyew 1990).

Nestling diet includes larval Lepidoptera, spiders, and masticated seeds (Densley 1990); grasshoppers and locusts (Orthoptera) and flies (Diptera) (Rustamow and Sopyew 1990).                                                        JDS-S

**Social pattern and behaviour.** Not well studied. Apart from studies of extralimital *zarudnyi* in Turkmeniya by Rustamov (1958), Sopyev (1965), Ponomareva (1983), and Rustamow and Sopyew (1990), most observations have been opportunistic.

1. Sociable outside breeding season with flocks of up to 150–200 recorded in winter in Algeria (Heim de Balsac and Mayaud 1962) and *c.* 80 in Tunisia (Madsen 1990), though mostly of 5–50 (Tristram 1859; Erlanger 1899; Snigirewski 1928; Guichard 1955; Browne 1981; Hirschfeld 1988). In Morocco, mixes with House Sparrow *P. domesticus* for feeding (Destre 1984). BONDS. No information on mating system nor on age of first breeding. Nest-building and incubation by both sexes, but mostly by ♀; both parents feed young, including for some time after leaving nest (Rustamov 1958; Sopyev 1965; Ponomareva 1983). BREEDING DISPERSION. Will form small colonies, e.g. several together in foundation of tree nest of Black Vulture *Aegypius monachus*, Tunisia (Erlanger 1899); colony in cavities in trees, Algeria (Heim de Balsac 1929); 3 nests in crown of palm, 2 within 1 m, Morocco (Densley 1990). Also in isolated pairs, exclusively so in Turkmeniya (Snigirewski 1928); clumped nests in one bush probably old nests of same pair (Sopyev 1965), indicating site-fidelity. Re-use of old nests found by Ponomareva (1983) and Rustamow and Sopyew (1990); not infrequently builds new nest on old one, using some of same material (Ponomareva 1983). No information on territory or density in west Palearctic, but Rustamow and Sopyew (1990) give 2–6 birds per km² in Repetek (Turkmeniya). ROOSTING. In cavity, dense bush, or foliage, e.g. palms, roses (Heim de Balsac 1929), remaining faithful to particular bush (Snigirewski 1928).

2. Rather shy, although Rustamow and Sopyew (1990) described *zarudnyi* as not shy, allowing close approach by man. FLOCK BEHAVIOUR. No information. SONG-DISPLAY. Peak singing activity (see 1 in Voice) by ♂ early in nest-building (Ponomareva 1983). ANTAGONISTIC BEHAVIOUR. No information. HETERO-

SEXUAL BEHAVIOUR. Bundy and Morgan (1969) described ♂♂ hopping after ♀ with outstretched and shivering wings (compare group-display of *P. domesticus*). Near Repetek (Turkmeniya), pair seen copulating (evidently prior to 2nd brood) in shrub. 30 m from nest containing young ready to fledge (Rustamov 1958). ♂ more active than ♀ in search for nest-site but final choice by ♀. Building begins within a few days of site-selection; both sexes build but ♀ does much more than ♂ who may initially confine activity to accompanying ♀ on collecting trips, only gradually contributing to building himself. Initially almost all incubation by ♂, ♀ continuing to build and staying only briefly in nest; within a few days, ♀'s contribution increases. (Ponomareva 1983.) According to Sopyev (1965) most incubation by ♀, but share presumably varies between pairs. In very high temperatures, birds seen sitting by nest entrance (not incubating) with head poking out. Both sexes usually silent when approaching nest, but chirping sounds given by incomer if mate on nest. (Ponomareva 1983.) At one nest, ♀ used regular route, including perch just before final approach (Rustamow and Sopyew 1990). RELATIONS WITHIN FAMILY GROUP. After leaving nest at *c.* 12–14 days, young remain in nearby bushes at least for first few days (Rustamow and Sopyew 1990). No further information.

JDS-S

**Voice.** No systematic study. Calls frequently (Hüe and Etchécopar 1970); ♀ silent (Densley 1990) or rather silent (Rustamow and Sopyew 1990).

CALLS OF ADULT. (1) Song. 7–10 twittering sounds (Fig I) resembling song of Pied Wagtail *Motacilla alba* (Madsen 1990). Fig II shows that conversational calls from group

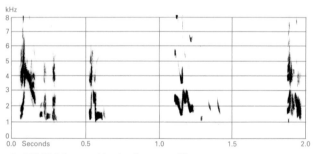

I  E D H Johnson  Algeria  January 1968

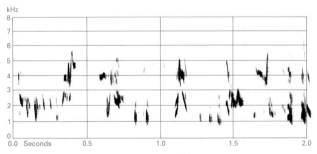

II  E D H Johnson  Algeria  January 1968

of birds at rest in cover (E D H Johnson) are probably also song. (2) Other advertisement- and contact-calls. (2a) Basic 'chirp' (Fig III). Advertisement-call similar to call 2a of House Sparrow *P. domesticus* (Densley 1990); softer

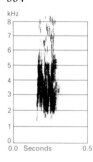

III N Gardner Morocco December 1990

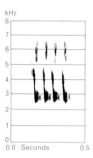

IV E D H Johnson Algeria January 1968

than Spanish Sparrow *P. hispaniolensis* (Bundy and Morgan 1969). (2b) Main flight-call. Quiet twitter (Bundy and Morgan 1969: Fig IV), like delicate version of flight-call of Greenfinch *Carduelis chloris* or mixture of Linnet *C. cannabina* and Chaffinch *Fringilla coelebs* (presumably 2c, below) (Madsen 1990); also recalls Twite *C. flavirostris* (Rustamow and Sopyew 1990). Call likened by Hüe and Etchécopar (1970) to *C. chloris* is presumably the same, likewise melodious trilling designated as song by Densley (1990). (2c) Soft 'jip' heard in flight (Bundy and Morgan 1969). Subdued 'chu' (no other details) mentioned by Berg and Roever (1984) is perhaps the same. (3) Threat-call. Guttural chattering 'chit-it-it' (E D H Johnson).

CALLS OF YOUNG. No information.                  JDS-S

**Breeding.** SEASON. Northern Sahara: end of March to June (Heim de Balsac and Mayaud 1962; Etchécopar and Hüe 1967; Bundy 1976; Densley 1990). Southern Sahara (extralimital nominate *simplex*): September–November (Bates 1934; Heim de Balsac and Mayaud 1962). Turkmeniya (extralimital *zarudnyi*): begins late April or May (Dementiev and Gladkov 1954), 1st brood late April, 2nd end of May to beginning of June (Sopyev 1965; Rustamow and Sopyew 1990). SITE. Variety of holes, 1·5–3 m above ground (Rothschild and Hartert 1911; Densley 1990); 1·5–4 m in Turkmeniya (Sopyev 1965). Varies from tree-cavities (Erlanger 1899; Rothschild and Hartert 1911; Heim de Balsac 1929; Heim de Balsac and Mayaud 1962) to cracks in rocks and human structures (walls, pyramids of stones used as landmarks, wells) (Heim de Balsac 1929; Densley 1990) and among roof rafters (Guichard 1955). According to Heim de Balsac and Mayaud (1962), hole, particularly in tree, is preferred site. Also builds exposed nests closely interwoven with dense branches of trees

(Rothschild and Hartert 1911; Bates 1934; Heim de Balsac 1929), in crowns of palms (Heim de Balsac 1929; Densley 1990); occasionally in foundations of nests of crows (Corvidae) and vultures (Accipitridae) (Erlanger 1899; Heim de Balsac 1929). In Turkmeniya, mainly in trees, including under (rather than inside foundation of) old nests of Brown-necked Raven *Corvus ruficollis* and Golden Eagle *Aquila chrysaetos*, sometimes even under old *P. simplex* nest; habit of building under nests is apparently for shade (Dementiev and Gladkov 1954; Sopyev 1965; Rustamow and Sopyew 1990). Nest: free-standing nest built in 3 parts—loose, untidy outer structure of plant fibre and coarse grass (Densley 1990), inner flask-shaped structure of twigs, grass, and tightly woven plant fibre with entrance on one side sloping up to nest-cup; in Turkmeniya, always faces north or north-east (Dementiev and Gladkov 1954; Rustamow and Sopyew 1990); cup made of fine plant fibre, grass, and feathers (Rothschild and Hartert 1911; Dementiev and Gladkov 1954; Densley 1990). Oval, 13·6–25·0 cm long, entrance hole 3·2–5·0 cm wide (Dementiev and Gladkov 1954; Densley 1990; Rustamow and Sopyew 1990). For further data from Turkmeniya, see Rustamov (1958). Structure of nest in hole not described. Building: by both sexes; nest added to (apparently to improve insulation) up to hatching of 1st egg; old nests re-used (Ponomareva 1983; Rustamow and Sopyew 1990; see Social Pattern and Behaviour). EGGS. See Plate 59. Sub-elliptical, smooth and slightly glossy. White, spotted and blotched with shades of brown and violet-grey (Harrison 1975). *P. s. saharae* 19·1 × 13·9 mm (18·0–21·3 × 13·0–16·3), *n* = 32; calculated weight 1·98 g. Nominate *simplex* 20·0 × 14·6 mm (19·3–20·5 × 14·2–15·0), *n* = 6; calculated weight 2·24 g. (Schönwetter 1984.) See also Rustamow and Sopyew (1990) for 57 eggs from Turkmeniya. Clutch: 2–5 (Etchécopar and Hüe 1967); 4–5, falling to 2–3 in dry years, mean 3·5, *n* = 16 (Heim de Balsac and Mayaud 1962); in Turkmeniya, 3–6 (2–8) (Sopyev 1965). Double-brooded (Heim de Balsac and Mayaud 1962); average 1st clutch 5·5, 2nd 4·4; eggs usually laid daily, occasionally up to 4–5 days apart (Sopyev 1965). INCUBATION. No information for west Palearctic. In Turkmeniya, 12–13 days, starting with 1st egg; young hatch over up to 5 days. By both sexes, mostly by ♀ (see also Social Pattern and Behaviour); average stint by ♀ 8·7 min (1–35), average by ♂ 5·5 min (1–23) (Sopyev 1965). See also Ponomareva (1983) for incubation stints. YOUNG. Fed by both parents (Sopyev 1965). FLEDGING TO MATURITY. Fledging period 12–14 days (Sopyev 1965). BREEDING SUCCESS. In Turkmeniya, 57% of 43 eggs produced fledged young (Sopyev 1965).

JDS-S

**Plumages.** (*saharae*). ADULT MALE. Upperparts from nasal bristles and cap to scapulars and back light ash-grey. Rump pinkish-isabelline, some light grey of feather-centres visible; upper tail-coverts isabelline-white. Lore contrastingly black, extending narrowly below eye into short black triangular patch behind eye, and this in turn sometimes into indistinct dark grey line over upper ear-coverts. Cheek, ear-coverts, and side of neck white. Chin and throat black, feathers with narrow white tips when fresh. Remainder of underparts cream-white, tinged pale grey on side of breast and flank, almost pure white on central belly, vent, and under tail-coverts. Tail greyish-black (more sepia when worn), broad and contrasting pale cream-pink fringe along both webs of central pair of feathers (t1), and narrower whiter fringe along both webs of t2–t6; fringes extend around feather-tips when plumage fresh. Flight-feathers dark grey (sepia when worn); outer webs of outer primaries fringed off-white (except for emarginated parts), outer webs of inner primaries and all secondaries more broadly fringed pink-cinnamon or cream-pink, widest on feather-bases, narrower on inner webs and tips. Tertials dull black, broad outer fringes pink-cinnamon or cream-pink, becoming gradually narrower and paler on tip, narrower inner fringes white. Greater upper wing-coverts pink-cinnamon or cream-pink, basal halves contrastingly black; median coverts cream (bases dark grey, hidden); lesser coverts ash-grey or brown-grey; bastard wing and greater upper primary coverts contrastingly black, narrow fringes along outer webs and tips off-white. Under wing-coverts and axillaries pinkish-white. *In fresh plumage*, tips of feathers of upperparts narrowly washed sandy-buff; when worn, upperparts tinged with some sandy-isabelline, mainly on scapulars; pink-cinnamon or cream-pink of wing bleached to off-white. ADULT FEMALE. Entire upperparts including all upper secondary coverts pale pink-buff, feather-centres often slightly deeper pale vinous-pink, fringes paler sandy-buff; upper tail-coverts pale sandy-buff or isabelline. Side of head and neck pale pink-buff, lore and area just below and behind eye paler isabelline-buff, gradually merging into cream-white chin and throat. Remainder of underparts pale isabelline-buff (if fresh) or isabelline-white (if worn), almost white on mid-belly, vent, and under tail-coverts. Tail sepia-brown; t1 with broad pale pink-buff fringe, t2–t6 more narrowly fringed isabelline-white. Flight-feathers as adult ♂, but pink-cinnamon or cream-pink outer fringes wider; tertials pale pink-cinnamon, basal and middle portions of inner webs with contrasting dark sepia-brown patch, fringes pale buff. Bastard wing black with narrow pink-buff fringes; greater upper primary coverts pink-buff with small but contrasting blackish tips and dark grey centres and inner webs. Underwing as adult ♂. NESTLING. Naked (Sopyev 1965; Densley 1990). For development, see Sopyev (1965) and Rustamow and Sopyew (1990). JUVENILE. Sexes similar. Upperparts like adult ♀, but general colour more isabelline-pink, less saturated pink-buff; underparts entirely isabelline-white, feathers short and loose-webbed. Lore and area round eye sometimes sooty-brown (perhaps in ♂ only: Berg and Roever 1984; Densley 1990). Tail, flight-feathers, and tertials grey-brown or sepia (depending on bleaching), less dark than in adult; pale fringes as in adult, but tertials have ill-defined isabelline-pink fringes, centres without black patch, or (sometimes) a dull ill-defined grey or brown patch. Bastard wing and greater upper primary coverts grey-brown or sepia-brown, pale fringes wider than in adult ♀, almost hiding dark remainder. Lesser, median, and greater upper wing-coverts grey, tips with ill-defined sandy pink fringe, less uniform than in adult ♀; grey at bases of median coverts hidden. See also Densley (1990). FIRST ADULT. As adult, and indistinguishable when last juvenile feather (p10 or s6) shed; p10 longer and with more rounded tip than in adult (see Structure); juvenile s6 much browner than fresh neighbouring feathers and with broader and more ill-defined pale fringe at tip. Some fresh ♂♂ have sandy-buff feather-fringes of upperparts much broader than others, largely concealing grey; these more buff birds probably 1st adults.

**Bare parts.** ADULT. Iris brown. Bill horn-brown, lower mandible pale yellow-horn; in ♂, turns black in spring (in Algerian Sahara, black attained January–February), fading again during post-breeding moult; bill of ♀ occasionally dark, too (Berg and Roever 1984). Leg and foot light horn-pink or pale grey-brown with pink tinge. (Hartert 1903–10; Snigirewski 1928; ZFMK, ZMB.) NESTLING. Bare skin, including bill and leg, pink; gape flanges pale yellow (Rustamow and Sopyew 1990). JUVENILE. Iris hazel or mid-brown. Bill, leg, and foot pale pink-flesh. Gape-flanges pale lemon-yellow. (Densley 1990.)

**Moults.** ADULT POST-BREEDING. Complete; primaries descendent. None of many examined December–February from Algeria in moult, plumage mostly fresh, but a few distinctly worn; a few from April and one from 5 June worn, moult not started. Most birds of small series from central Sudan in moult, but these not dated: primary moult scores 37–38 (6–7 inner primaries new), tail and secondaries in full moult, body new but many feathers still growing on head and neck (ZFMK, ZMB). In Turkmeniya, body new in series from 2nd half of August, but outer tail-feathers and outer primaries still growing (Snigirewski 1928); moult apparently from late July or early August to end of August or early September (Dementiev and Gladkov 1954). POST-JUVENILE. Complete; primaries descendent. Fully juvenile birds with moult not started examined from central Algeria, 2–5 June (*n* = 5); one from southern Algeria, 11 January, had body new, but flight-feathers and tail in moult (primary-score 28, p1–p5 new) (ZFMK, ZMB); 2 ♂♂ from Algerian Sahara, 25 July, in full moult (Berg and Roever 1984). Timing of moult and breeding in southern Sahara probably depends on local rains.

**Measurements.** ADULT. *P. s. saharae*. Algerian Sahara and Fezzan (Libya), December–June; skins (BMNH, RMNH, ZFMK, ZMB). Bill (S) to skull, bill (N) to distal corner of nostril; exposed culmen on average *c.* 2·6 less than bill (S).

| | | | | |
|---|---|---|---|---|
| WING | ♂ | 80·0 (1·65; 37) 76–83 | ♀ | 76·2 (1·86; 29) 73–81 |
| TAIL | | 55·1 (1·42; 19) 53–58 | | 52·2 (1·42; 11) 50–55 |
| BILL (S) | | 12·9 (0·56; 25) 11·8–13·9 | | 13·0 (0·48; 19) 12·0–13·6 |
| BILL (N) | | 7·5 (0·58; 22) 6·6–8·5 | | 7·7 (0·56; 18) 7·0–8·9 |
| TARSUS | | 20·4 (1·00; 19) 19·3–22·2 | | 20·1 (0·60; 11) 19·4–21·0 |

Sex differences significant for wing and tail.

Nominate *simplex*. Sudan: wing, ♂ 76·3 (3) 76–77, ♀♀ 71, 74; bill (S) 12·8 (0·22; 5) 12·5–13·1; bill (N) 7·3 (0·13; 5) 7·1–7·5; tarsus 19·3 (3) 18·8–19·8 (RMNH, SMTD, ZMB). Ennedi (northern Chad), wing: ♂♂ 73, 74 (Niethammer 1955). Wing: ♂ 73–77 (Hartert 1903–10).

*P. s. zarudnyi*. Wing, ♂ 72 (17) 69–74, ♀ 69·9 (9) 68–72 (Dementiev and Gladkov 1954).

JUVENILE. *P. s. saharae*. Algerian Sahara: wing, ♂ 76·2 (3) 75–78, ♀ 75·7 (3) 74–78; tail 50·5 (1·73; 4) 49–53 (ZFMK, ZMB).

**Weights.** *P. s. saharae*. Algeria, December–January: ♂♂ 19, 20; ♀♀ 18, 21 (Niethammer 1955; ZFMK).

Nominate *simplex*. Niger, December: ♂ 20·3 (Fairon 1971). Northern Chad, September: ♂ 19·5 (Niethammer 1957*b*).

*P. s. zarudnyi*. Nestling on day 1 1·3–1·7, at fledging (day 12–14) 17–19 (Rustamow and Sopyew 1990, which see for other details).

**Structure.** 10 primaries: in adult, p8 longest, p9 1–2 shorter, p7 0–1, p6 2–4, p5 7–10, p4 10–14, p1 16–23; in juvenile, p8–p9 longest, p7 1–2 shorter, p6 3–5, p5 8–11, p4 11–15, p1 16–21. P10 reduced, narrow and pointed in adult, slightly wider and longer in juvenile: in adult, 47–56 shorter than p8, 5–8 shorter than longest upper primary covert; in juvenile, 43–48 shorter than p8, 2–4 shorter than longest covert. Outer web of (p5-)p6-p8 emarginated, inner web of (p6-)p7-p9 with notch (sometimes faint). Tip of longest tertial reaches to tip of p2–p5. Tail rather short, tip square; 12 feathers. Bill short and conical, wide and deep at base (depth at base *c*. 7–8 mm), culmen arched. Leg and foot rather short and slender; middle toe with claw 16·6 (5) 16–17·5, outer toe with claw *c*. 74% of middle with claw, inner *c*. 70%, hind *c*. 80%.

**Geographical variation.** Rather slight. *P. s. saharae* of north-west Sahara palest and largest race, ♂ pallid ash-grey on upper-parts (fringed sandy-buff if fresh), ♀ pale pink-buff. Nominate *simplex* of southern Sahara from Mali to Sudan somewhat smaller; ♂ in Sudan slightly darker ash-grey, tinged brown-buff when fresh, underparts tinged pink-buff, less pale cream; ♀ more saturated buff-brown on upperparts, less pinkish, and more pink-buff below; difference slight but distinct, not just due to age of specimens (contra Rothschild and Hartert 1911 and Bannerman 1948). These races perhaps grade into each other: birds from Tanezrouft and Ahaggar (southern Algeria) on average slightly smaller than typical birds from north Algerian Sahara and Tunisia, wing of ♂ in south 79·1 (1·68; 7) 77–81, in north 80·3 (1·67; 13) 77–83, and colour slightly darker grey, but closer to *saharae* than to nominate *simplex* (ZFMK, ZMB); in Aïr (north-ern Niger) somewhat intermediate, but nearer nominate *simplex* (Vaurie 1956*f*). *P. s. zarudnyi* from Transcaspia nearest nominate *simplex* in colour, but even smaller, upperparts purer pale grey (rump sandy-grey), underparts whiter (faintly washed vinous September–November: Snigirewski 1928), black of lore extend-ing both above and below eye to upper ear-coverts (in nominate *simplex*, to just below eye only), and bill more stubby, deeper at base, shorter, and with well-arched culmen; unlike nominate *simplex*, sexes rather similar (Redman 1993).

May form superspecies with House Sparrow *P. domesticus* and Great Sparrow *P. motitensis* (including *iagoensis*) (Hall and Moreau 1970).                                                                CSR

## *Passer montanus* Tree Sparrow

PLATES 21 and 27 (flight)
[between pages 280 and 281]

Du. Ringmus     FR. Moineau friquet     GE. Feldsperling
RU. Полевой воробей     SP. Gorrión molinero     Sw. Pilfink     N. AM. Eurasian Tree Sparrow

*Fringilla montanus* Linnaeus, 1758

Polytypic. Nominate *montanus* (Linnaeus, 1758), Europe and Siberia from Norway, Ireland, and Portugal east to Altai, northern Mongolia, north-west Manchuria, and Sea of Okhotsk, south in west to Spain, Sardinia, Sicily, Malta, and Greece, and in east European Russia and Siberia south to *c*. 40°N, grading into *transcaucasicus* in eastern Bulgaria, eastern Rumania, Crimea, and probably in European Turkey and lower Don basin; *transcaucasicus* Buturlin, 1906, Turkey east from Istanbul, and Caucasus area east to Gorgan in northern Iran and north to valley of lower Volga. Extralimital: *dilutus* Richmond, 1896, Transcaspia east through western Pakistan, Tien Shan, and Zaysan basin to Sinkiang and Gobi desert in northern China; *dybowskii* Domaniewski, 1915, eastern Asia from lower Amur river south to northern Korea, west in Manchuria (northern China) to Great Khingan mountains, grading into *saturatus* on Sakhalin and probably in central Korea; *kansuensis* Stresemann, 1932, Zaidam basin and northern Kansu (China); *tibetanus* Baker, 1925, southern and eastern Tibet; *iubilaeus* Reichenow, 1907, eastern China from Peking area (Hopeh) south to lower Yangtze river, west to Shansi; *obscuratus* Jacobi, 1923, Nepal, Sikkim, north-east India and neighbouring Burma, and in south-west China from Yunnan, Szechwan, and Shensi east to Hupeh; *saturatus* Stejneger, 1885, southern Kuril islands, Japan, southern Korea, and through Ryu Kyu Islands to Taiwan and south-east China (Chekiang to Kwangtung); *malaccensis* Dubois, 1885, south-east Asia from Hainan, Vietnam, and central Burma south to western Indonesia. Various races introduced in North America, Australia, eastern Indonesia, Philippines, and elsewhere.

**Field characters.** 14 cm; wing-span 20–22 cm. Only slightly but distinctly smaller than House Sparrow *P. domesticus* and Spanish Sparrow *P. hispaniolensis*, with proportionately less bulbous bill, trimmer head and body, and narrower and 10% shorter tail. Quite large but always tidy-looking sparrow, with dashing, direct flight even more pronounced than in *P. domesticus*. Sexes alike. Both adult and juvenile have diagnostic combination of black-spotted white cheeks, long white collar emphasizing head, and 2 pale wing-bars. Calls distinctive. 2 races in west Palearctic, differing little; European race, nominate *montanus*, described here. Hybridizes rarely with *P. domesticus* and *P. hispaniolensis*.

ADULT. Moults: June–September (complete), with wear producing little change. Rather short, noticeably round crown and nape uniform dark chestnut; short black streak in front of, just under, and behind eye; rather short, quite narrow black bib; Cheeks and obvious collar (almost meeting on nape in some attitudes) greyish to pure white, contrasting noticeably with bib and crown; conspicuous black patch below and behind eye. Altogether, head pat-tern appears clean and vivid, catching eye much more than in *P. domesticus*; note that obviousness of cheek-patch and collar varies with bird's stance and plumage state (may be hidden on hunched fluffed-up bird). Back and rump basically yellowish-brown, on mantle streaked black and marked rufous. Wings brownish-chestnut, with conspicuous double wing-bar formed by virtually white tips to black median coverts and chestnut-edged, black-centred greater coverts; pale buff bases to primaries contrast with black-brown primary coverts; conspicuous pale margins to black tertials and inner secondaries, at times forming narrow buff panel on folded wing. Tail dark brown, with outer edges of feathers buff. Below bib and collar, breast dusky grey-white and flanks washed dusky brown but broad belly white, adding to bird's clean appearance. Under tail-coverts pale buff-white, with brown centres. Underwing isabelline. Bill black in breeding season but duller and browner with yellowish-horn on lower mandible at other times. Legs pale brown. For distinction of some ♀♀ from ♂, see Plumages. JUVENILE. Less cleanly coloured, and duller than adult. At close range, can be distinguished by dull smoky, even black-tipped crown, less rufous tone on nape, mantle, and wing-coverts, dull, greyish tone to eye-streak and bib, and duller, less white, more buff wing-bars. HYBRIDS. Mixed parentage with *P. domesticus* indicated by invasion of grey into forecrown, vestigial white streaks in front of and behind eye, lack of discrete cheek-patch, and larger bib.

Unmistakable at any range where head pattern visible. With experience, also separable on silhouette and voice (see below). Flight essentially as *P. domesticus* but wing-beats even faster, producing remarkably dashing progress along direct track and greater agility in tight spaces. Escaping flock has habit of towering to *c.* 10 m before flying off. Gait, stance, and wing and tail movements much as *P. domesticus* but with slimmer, daintier form constantly obvious. Behaviour variable; in western Europe, markedly less commensal with man than *P. domesticus*, but in Scandinavia, Russia, and south-east Europe as much a 'house sparrow' as *P. domesticus*.

Voice typical of *Passer* but with experience can be distinguished from *P. domesticus* by (1) more musical and rhythmic repetition of both sharper and fuller notes in song, (2) rather harsh 'teck', 'tett', or 'chet', given up to 4 times when taking flight (and producing characteristic chatter from flock), and (3) different timbres and pitches of calls, all contributing to more incisive, merrier sound.

**Habitat.** In west Palearctic, breeds in middle and (locally) higher latitudes up to July isotherms 12–13°C, mainly in continental but also marginally in oceanic climates, preferring temperate to warm Mediterranean or high boreal regimes, in which House Sparrow *P. domesticus* appears to hold an advantage. Despite scientific name normally a lowland or low upland bird; in Switzerland infrequent above *c.* 700 m (Glutz von Blotzheim 1962), although in northern Caucasus reaches *c.* 1700 m (Dementiev and Gladkov 1954). Within above range, occupies suitable habitats only patchily and with prolonged fluctuations, involving inexplicable colonizations, and desertions of settled areas. In Scotland, these have been particularly

numerous and well documented especially on coasts and islands (Baxter and Rintoul 1953). In Ireland also, after 1960 a remarkable repopulation occurred, especially along coast or in sight of waterbodies, involving mainly nest-sites in ruined buildings of all kinds but also willows *Salix* and other trees; more detailed study needed, but there seems no evidence that any factors in habitats concerned have been responsible. Acceptable habitats, from oceanic to continental, appear to fit series of distinct types: coastal cliffs, especially with ivy *Hedera*; other coastlines, especially with empty or ruined buildings; pollarded willows and other trees with nest-holes along slow-flowing lowland watercourses; quarries and nest-holes of Sand Martin *Riparia riparia*; free-standing trees along roadsides or in groups in parks, cemeteries, or farmland; woodlands, especially where they are small and isolated in open country with well-spaced mature broad-leaved trees; and spacious suburbs especially where nest-boxes provided, e.g. in Berlin (Germany) (Bruch *et al.* 1978). Such urban habitats are not occupied in west of range where they are monopolized by *P. domesticus*. In eastern Asia, the ultimate stage is reached of replacing *P. domesticus*, even in inner cities (Dementiev and Gladkov 1954), while partial replacement occurs in north-west Pakistan (Ali and Ripley 1974).

**Distribution.** Has recently spread in Fenno-Scandia. Range in Britain and Ireland has been subject to marked changes. (Introduced and locally breeding in USA and southern Australia).

FAEROES. Resident breeder until *c.* 1910; since then accidental visitor only (Bloch and Sørensen 1984). BRITAIN. Undergoes major fluctuations. Latest expansion from late 1950s, followed by decline and range contraction since 1976–7. (Sharrock 1976; Summers-Smith 1988.) IRELAND. Long-term decline and range contraction from 19th century to 1950s. No breeding recorded 1959–1960, then recovery from 1961. (Hutchinson 1989.) NORWAY. Scattered; recent spread in parts of south-east, west, and north (VR). SWEDEN. Former breeding population in 19th and early 20th century in Torne Lappmark on Finnish border. Began to breed in Gotland in 1920s, and became common there in 1960s; recently spread north into Medelpad (*c.* 62°30′N). (LR.) FINLAND. Spreading, the 2 formerly separate breeding areas (south-east Finland and Åland) gradually joining (OH). ITALY. Introduced to Sardinia at end of 19th century and now widespread there (Summers-Smith 1988). MALTA. First recorded breeding 1959 (Sultana and Gauci 1982). TURKEY. May breed locally within wintering area (see map) (RPM). MOROCCO. Two ♀♀ with incubation patches at Kenitra, 1985, probably breeding (see map); if so, first record for North Africa. Otherwise known only as accidental visitor. (MT.)

Accidental. Israel, Egypt, Morocco, Algeria.

**Population.** Has increased with recent range extensions

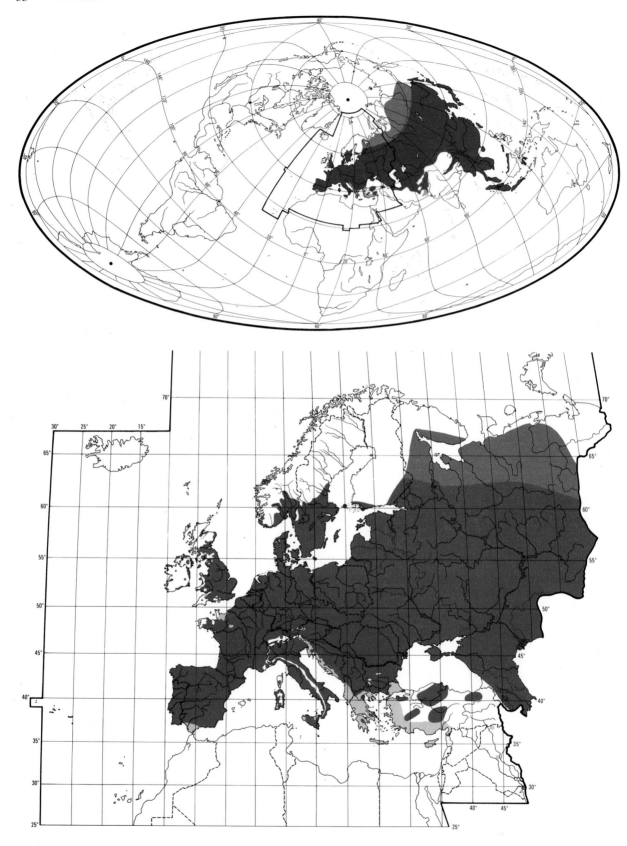

in Fenno-Scandia. Populations in Britain and Ireland subject to major fluctuations. Few recent changes reported elsewhere.

BRITAIN, IRELAND. Major recent changes, with accompanying range expansion and contraction (see Distribution). Estimated 285 000 pairs 1985 (Summers-Smith 1988). FRANCE. 100 000 to 1 million pairs (Yeatman 1976). SPAIN. Decreased in north (AN). BELGIUM. Estimated 210 000 pairs; apparently stable (Devillers *et al.* 1988). NETHERLANDS. Estimated 500 000–750 000 pairs (Teixeira 1979). GERMANY. 2·6 million pairs (G Rheinwald). East Germany: 700 000 ± 350 000 pairs (Nicolai 1993). NORWAY. Varies from very common in some areas to very rare in others (VR). SWEDEN. Estimated 1·5 million pairs (Ulfstrand and Högstedt 1976). FINLAND. Increasing rapidly. Estimated 1000 pairs in early 1970s, *c.* 5000 pairs 1988 (Koskimies 1989). SWITZERLAND. Has apparently decreased since early 1980s, but no exact data (RW). RUSSIA. Population of north-west part of European Russia *c.* 1 million pairs (Noskov 1981).

Survival. Oldest ringed bird 12 years 9 months (Dejonghe and Czajkowski 1983).

**Movements.** Mainly sedentary, especially in west of range, with only small proportion undertaking relatively short-distance migration, mainly to south or south-west. Larger-scale autumn eruptive movements occur from time to time, particularly from more northerly parts of range. Juveniles disperse locally from natal areas before settling down.

Nominate *montanus*. Small numbers of passage migrants recorded annually September–October at Falsterbo (southern Sweden) (Rudebeck 1950; Ulfstrand *et al.* 1974; Roos 1985). Matched by arrivals at Fair Isle (Scotland) (*Fair Isle Bird Obs. Rep.*), and by passage along North Sea coasts, e.g. Denmark October–November (Krüger 1944), Britain from Fife southwards September–November, mainly October (Eagle Clarke 1904; Baxter and Rintoul 1953; Smith and Cornwallis 1953*a*; Lack 1954*b*; Mather 1986). Some may be local coastal movement, but ringing gives evidence of movement between England and adjacent coasts of continental Europe (Netherlands, Belgium, France), lending support to due westward movement from continent suggested by observations on lightships off south-east England (Eagle Clark 1904; Owen 1953). Further south, Lack (1955) noted autumn coastal passage along south-west coast of France into Spain.

Small-scale movements less easy to detect away from North Sea, but passage along Bodensee shore (southern Germany) (Jacoby *et al.* 1970) in October suggests broader-front movement. Also, Jenni and Schaffner (1984) reported movement through Swiss Alpine passes from end of September to beginning of November, with recoveries up to 580 km to south. Movement also recorded at Chignolo d'Isolo (northern Italy) October–November, with maximum in first 10 days of October (Pesenti 1945); 7

birds ringed winter in Camargue (southern France) and recovered 250–550 km ENE in north-west Italy (Hoffmann 1955, 1956) possibly relate to same south-westerly autumn passage. Numbers observed passing across Downs north and south of Newbury (southern England) during September–November 1952–83 (L R Lewis).

Evidence of small scale of migration in west, as distinct from local wandering, given by recoveries of ringed populations. Exceptionally, Verheyen (1957*b*) considered that up to 25% of Belgian population wintered up to 400 km from breeding area, mainly to south-west, but other observers found little evidence of significant movement: of 3322 nestlings ringed Moravia (Czechoslovakia), only 2 recoveries further than 10 km (Balát 1976); none out of 4379 nestlings ringed near Warsaw (Poland) (Pinowski 1965*b*); none out of 1500 adults and young ringed at Wolfsburg (northern Germany) (Scherner 1972*a*); only 8 (all nestlings) out of 9491 ringed at Braunschweig (northern Germany) over 50 km, maximum 292 km (Berndt and Winkel 1987). Same picture emerges from national ringing schemes in west, with only very small number of birds recovered over 50 km. More regular movement further east, but not well recorded in west Palearctic and difficult to form complete picture of proportion of population involved and extent of movement, though some adults included and distances appear to be greater than in west: young bird from St Petersburg region (western Russia) recovered 3 years later in Portugal (3500 km), 1 from Vitebsk region (Belorussiya) later same year in Italy (2500 km). Departure from northerly parts (Kola peninsula, 67°N) and passage through Baltic states from end of September through October. Similar passage immediately to east of Urals—Tyumenskaya region (57°N) September and Chelyabinsk area (55°N) September–October. Movement generally to south or south-west: nestling from Kamyshin (50°N) on Volga recovered in December *c.* 600 km south-west in Rostov region. Passage through Crimea October–November and through Astrakhan' along west Caspian coast to winter quarters in Dagestan in October and beginning of November. (Dementiev and Gladkov 1954; Noskov 1981.) Movement recorded across north-west corner of Black Sea in October, but not known if annual (Korzyukov 1979).

Further evidence of limited southward movement shown by increase in numbers in winter at southern limits of breeding range (e.g. south and west Turkey: Beaman 1978) and erratic reports further south, in north-west Africa (Hollom *et al.* 1988), Mediterranean islands (Mallorca, Pantelleria, Corfu, Crete) (J D Summers-Smith), Cyprus (Flint and Stewart 1992), Israel, Gulf states of Arabia (Hollom *et al.* 1988), and Egypt (Goodman and Meininger 1989).

Numbers involved in movements are subject to considerable fluctuations, forming irregular pattern that suggests eruptions rather than normal annual migration.

Significant movement into Britain occurred in 2nd half of 19th century and 1957–62, with almost complete absence for intervening period and since 1962; 1957–62 peak coincided with one at Falsterbo, but there were also peaks there 1949–52, 1973–75, and 1980–83 that were not reflected in noticeable movement on British coast (Ulfstrand *et al.* 1974; Summers-Smith 1989).

Such migration as takes place mainly involves young birds: of 57 trapped at Col de Bretolet (Switzerland), 97% juvenile (Jenni and Schaffner 1984); 88% of 233 recoveries over 10 km reported by Rademacher (1951) for whole of Germany also juvenile.

Eruptive movements probably responsible for recent southward extensions of range that have occurred in 20th century in Corsica, Sardinia, and Malta, with some over-wintering birds remaining to breed. Recolonization of Ireland after extinction in early 1950s probably arose in same way.

Return movement in North Sea area occurs late March to May. Also some records of small numbers at this time on islands off British coast where no breeding, though not clear whether this is return migration or the more limited spring dispersal undertaken by juveniles (see below) (Summers-Smith 1989). Spring migration more obvious further east: winter quarters in south-west Caspian vacated by April; passage through Gorki (Belorussiya) recorded early April and through Baltic states at end of March and in April; arrival at Lake Ladoga and Petrozavodsk (north of St Petersburg) from end of March to mid-April, but not on Kola peninsula until mid-May (Dementiev and Gladkov 1954).

Most observations indicate autumn movement diurnal: in early morning hours along Baltic coast, but peak not until 2nd half of day in Bulgaria (Noskov 1981). In contrast, passage across Black Sea took place at night (Korzyukov 1979). Spring passage along Baltic also in early daylight hours (Noskov 1981), but on lower Ural (north-west Kazakhstan) both after sunrise and again shortly before sunset (Noskov 1981).

In addition to migratory and eruptive movements discussed above, there is also a more randomly oriented dispersal of juveniles from natal areas before they settle down to join sedentary breeding population. Occurs both during period of autumn sexual display in nesting areas and again immediately prior to breeding season; Pinowski (1965a, b) attributed it to overcrowding in breeding colony. Dispersal mostly small-scale, involving movements of less than 10 km. No doubt such dispersal accounts for widely reported rapid build-up of new breeding colonies in areas where nesting opportunities increased by provision of nest-boxes (see Social Pattern and Behaviour).

Movements of other races seem to be similar. Most individuals appear to be largely sedentary: a few *transcaucasicus* have been recorded in northern Iraq in winter (Ticehurst *et al.* 1921–2); some individuals of *dilutus*, breeding in Turkestan and Afghanistan, winter on Makran coast of Pakistan (Ali and Ripley 1974), and irregular, large-scale, eruptive movements involving *saturatus* have been recorded on east coast of China (Styan 1891; La Touche 1912).                                            JDS-S

**Food.** Plant and animal material, proportions varying with both season and availability. Seeds comprise bulk of plant matter, with fewer buds and berries. Food predominantly sought on ground. Takes seeds from low growing plants, both by flying up and perching, e.g. on cereal stems (particularly at 'milk' stage), or by pulling seed-head to ground in case of weaker plants, and stripping off seed-head. Searches for seeds on ground (e.g. in stubble fields), where birds gather in flocks, often with House Sparrow *P. domesticus*, finches (Fringillidae), and buntings (Emberizidae) and by 'roller feeding', whereby birds at rear of flock continually overfly those at front so that flock progresses in one direction (Deckert 1968b). In winter, not uncommonly associates with crows (Corvidae), taking advantage of their ability to dig in snow; also associates with grazing stock in summer (Noskov 1981). Feeding bouts (8–30 min, average 17) are followed by resting periods in nearby cover (10–26 min, average 15) (Deckert 1962). Food on ground, particularly invertebrates, also sought by removing loose surface soil with flinging movements of bill (Berck 1961–2) (similar action for delving into snow: Noskov 1981) and by turning over leaves or dry grass (Deckert 1962). Bread, cooked potatoes, chicken meal, kitchen scraps, etc. taken on ground around human habitations. Deckert (1962) also described robbing of food from Greenfinches *Carduelis chloris*, particularly dehusked sunflower seeds *Helianthus* which *P. montanus* is unable to open itself. Also takes seeds occasionally from trees, soft fruits, and berries. Reported drinking nectar from greengage *Prunus* (Hagger 1961). Invertebrates collected on ground by searching (small and less mobile animals) and pouncing (larger prey, e.g. grasshoppers Orthoptera, damselflies Odonata) and by turning over surface layers; searches leaves and small twigs of bushes and trees, frequently high in canopy, hovers in front of small twigs (Berck 1961–2), and sallies for flying insects. Bird in Thailand caught moths and other flying insects attracted to floodlight at night (Broun 1971). De-husks hard seeds by pressing seed against palate with tongue, crushing it with rapid to-and-fro movement of lower mandible (Ziswiler 1965); eats soft seeds directly; see Snow and Snow (1988) for method of dealing with berries of elder *Sambucus*. Small animal food is also eaten directly or collected as 'packet' in bill when feeding young. Larger invertebrates held in bill with body hanging down, then killed by beating against branch (etc.) before being eaten or carried back to nest as individual item, wings being removed from flying insects before body eaten; invertebrates never held under foot. Animal food for young frequently obtained from trees close to nest; birds breeding in foundation of nests of Rook *Corvus frugilegus* even forage for scraps of food dropped by host (Murton 1971). Animal prey items mainly 2-7(-30) mm long (Anderson 1978). Takes some mineral matter in addition to grit, including snail shells (Anderson 1978; Szlivka 1983), eggshells (Anderson 1978), and mortar (Kummer 1983; Szlivka 1983), presumably for calcium.

Diet in west Palearctic includes the following. Invertebrates: bristletails (Thysanura), springtails (Collembola), mayflies (Ephemeroptera), damsel flies (Odonata: Lestidae), adult and larval grasshoppers, etc. (Orthoptera: Tettigoniidae, Tetrigidae, Acrididae, Tridactylidae), earwigs (Dermaptera), termites (Isoptera), psocids (Psocoptera), bugs (Hemiptera: Aradidae, Scutelleridae, Coreidae, Pentatomidae, Lygaeidae, Reduviidae, Miridae, Cicadidae, Cicadellidae, Psyllidae, Aphididae, Coccidae), thrips (Thysanoptera), scorpion flies (Mecoptera), lacewings (Neuroptera: Raphidiidae, Chrysopidae), larval, pupal, and some adult butterflies and moths (Lepidoptera: Nymphalidae, Pieridae, Lycaenidae, Zygaenidae, Pyralidae, Tortricidae, Coleophoridae, Noctuidae, Psychidae, Arctiidae, Lymantriidae, Geometridae, Sphingidae), caddis flies (Trichoptera), mainly larval and pupal flies (Diptera: Tipulidae, Culicidae, Chironomidae, Bibionidae, Cecidomyiidae, Stratiomyidae, Rhagionidae, Tabanidae, Therevidae, Asilidae, Empididae, Syrphidae, Sepsidae, Ephydridae, Drosophilidae, Agromyzidae, Chloropidae, Tachinidae, Calliphoridae, Muscidae), Hymenoptera (Pamphilidae, Tenthredinidae, Diprionidae, Ichneumonidae, Braconidae, ants Formicidae, bees Apoidea), larval, pupal, and adult beetles (Coleoptera: Cicindelidae, Carabidae, Hydrophilidae, Silphidae, Staphylinidae, Scarabaeidae, Byrrhidae, Buprestidae, Elateridae, Cantharidae, Tenebrionidae, Nitidulidae, Coccinellidae, Cerambycidae, Chrysomelidae, Curculionidae), spiders (Araneae), mites (Acari), earthworms (Lumbricidae), small snails (Gastropoda). (Hammer 1948; Dementiev and Gladkov 1954; Deckert 1962, 1968b; Dornbusch 1973; Graczyk and Michocki 1975; Grün 1975; Wieloch 1975; Noskov 1981; Szlivka 1983; Sánchez-Aguado 1986; Krištín 1988; Török 1990.) Exceptionally vertebrates, including young frogs (probably as carrion) and discarded lizard tails (Noskov 1981). Plant material mainly seeds of low herbs and grasses: mistletoe *Viscum*, hemp *Cannabis*, nettle *Urtica*, Polygonaceae (e.g. knotgrass *Polygonum*), Chenopodiaceae, Amaranthaceae, Caryophyllaceae, Ranunculaceae, *Portulaca*, fumitory *Fumaria*, poppy *Papaver*, Cruciferae, Rosaceae, Leguminosae, flax *Linum*, cranesbill *Geranium*, Euphorbiaceae, currant *Ribes*, *Elaeagnus*, violet *Viola*, carrot *Daucus*, bedstraw *Galium*, Boraginaceae, Labiatae, Solanaceae, Scrophulariaceae, plantain *Plantago*, elder *Sambucus*, Compositae, *Asparagus*, grasses and cereals (e.g. wheat *Triticum*, barley *Hordeum*, oats *Avena*, millet *Panicum*, maize *Zea*, *Sorghum*, rice *Oryza*), bur-reed *Sparganium*, Cyperaceae; tree seeds include pine *Pinus*, alder *Alnus*, elm *Ulmus*; soft fruits and berries, e.g. *Cotoneaster*, rowan *Sorbus*, elder *Sambucus*, grape *Vitis*, cherry and apricot *Prunus*, mulberry *Morus*; also buds, young shoots, glumes, parts of leaves, and flowers of fruit trees. (Hammer 1948; Dementiev and Gladkov 1954; Turček 1961; Deckert 1968b; Keil 1973; Grün 1975; Wiens and Dyer 1977; Noskov 1981; Szlivka 1983; Sánchez-Aguado 1986.)

Pinowski *et al.* (1973) found that caged birds preferred cereal grains (millet *Panicum*) to weed seeds, yet in field study by Pinowski and Wócjik (1969) cereals hardly featured. Most important factor is probably seed size; seeds of millet (202 seeds per g) appear to be preferred to wheat (*c.* 27 seeds per g) and bristle-grass (600 seeds per g) and the wild *Echinochloa crus-galli* (908 seeds per g). Tables A–B show predominance of plant material with considerable variation in proportion of cereals compared with weed seeds, highest values for cereals occurring in 2nd half of year, suggesting origin more from crops than from fodder. Table B indicates that bulk of weed seeds at any one season and place are provided by only a few plant families (see also Hammer 1948). Invertebrates significant only April–June (Szlivka 1983), in Yugoslavia comprising (by number of items) 73% Coleoptera, 20% Diptera, and 8% Hymenoptera. Keil (1973) divided stomach contents by weight into 46% food and 54% grit.

For rate of food consumption, see Table C. Energy demand for 17 populations ranged from 57 000 to 87 000 kJ per bird per year (Wiens and Dyer 1977, which see for further details).

Table A  Food of adult Tree Sparrow *Passer montanus* by stomach analysis throughout the year. Figures are percentages by number of items (Yugoslavia) or dry weight (other areas).

| | Poland | | | Germany | Netherlands | Rumania | Yugoslavia |
|---|---|---|---|---|---|---|---|
| | Gdynia | Dziekanów Leśny | Kraków | | | | |
| Weed seeds | 94 | 94 | 29 | 28 | 28 | 77 | 63 |
| Cereals | 4 | 5 | 37 | 54 | 58 | 10 | 28 |
| Invertebrates | 2 | 1 | 34 | 18 | 15 | 13 | 9 |

Data from: 'standardized populations' 455 birds

| Source | | Wiens and Dyer (1977) | | | | | Szlivka (1983) |
|---|---|---|---|---|---|---|---|

Table B  Seasonal variation in food of adult Tree Sparrow *Passer montanus* by stomach analysis; data for Spain also include contents of crops. Plant material is all seeds. Figures are percentages by number of items (in italics by dry weight).

| | Germany | Spain | | Yugoslavia | | | |
|---|---|---|---|---|---|---|---|
| | winter | winter | spring | Jan–Mar | Apr–Jun | Jul–Sep | Oct–Dec |
| Invertebrates | 0 | 0·6 | 4·3 | 0 | 44·0 | 5·5 | 4·5 |
| Polygonaceae | | 0·6 | 1·1 | 0 | 0 | 25·5 | 0 |
| Chenopodiaceae | | 32·8 | 23·8 | 0 | 0 | 0 | 0 |
| Amaranthaceae | | 22·8 | 0·2 | 24·0 | 0 | 0 | 0 |
| Caryophyllaceae | | 0 | 1·6 | 0 | 1·5 | 0 | 0 |
| Cruciferae | | 0·3 | 0·8 | 0 | 12·5 | 0 | 34·5 |
| Boraginaceae | 90·4 | 9·1 | 0 | 0 | 0 | 0 | 0 |
| Plantaginaceae | | 0 | 16·4 | 0 | 0 | 0 | 0 |
| Compositae | | 0 | 0·9 | 0 | 6·5 | 0 | 0 |
| Gramineae (exc. cereals) | | 24·2 | 32·0 | 45·0 | 8·0 | 29·0 | 11·0 |
| Other weed seeds | | 3·1 | 3·2 | 10·0 | 1·5 | 4·5 | 0 |
| Cultivated seeds | 9·6 | 6·5 | 15·7 | 21·0 | 26·0 | 35·5 | 50·0 |
| No. of birds | 64 | 37 | 39 | 197 | 100 | 93 | 60 |
| Source | Keil (1973) | Sánchez-Aguado (1986) | | Szlivka (1983) | | | |

Table C  Estimates of average food consumption of Tree Sparrow *Passer montanus* (kg dry weight per bird per year) for European sites at 47°–59°30′N, based on model using standardized populations (Wiens and Dyer 1977, which see for further details).

| | Germany | Netherlands | Poland | | | Rumania |
|---|---|---|---|---|---|---|
| | | | Gdynia | Dziekanów Leśny | Kraków | |
| Invertebrates | 0·48 | 0·47 | 0·04 | 0·04 | 0·90 | 0·43 |
| Weed seeds | 0·77 | 0·88 | 2·36 | 2·56 | 0·78 | 2·47 |
| Cereal grain | 1·48 | 1·83 | 0·10 | 0·13 | 0·99 | 0·32 |
| Totals | 2·73 | 3·18 | 2·50 | 2·73 | 2·67 | 3·22 |

Table D  Food of nestling Tree Sparrow *Passer montanus* as found from collar-samples. Figures are percentages by volume (Poland) and number (Czechoslovakia).

| | Poland, villages | | Poland, pine forest | Czechoslovakia, windbreaks |
|---|---|---|---|---|
| | May | June | July | |
| Orthoptera | 1·3 | 2·1–6·9 | 17·0 | — |
| Hemiptera | 0·9–5·4 | | 10·5 | 12·1 |
| Lepidoptera, mainly larvae | 16·7–25·5 | 2·9–21·4 | 11·0 | 35·5 |
| Diptera | 16·9–33·6 | 0·1–21·5 | 6·8 | 11·8 |
| Hymenoptera | 0–2·0 | | 4·1 | 1·7 |
| Coleoptera | 26·5–31·7 | 49·6–67·5 | 45·4 | 30·1 |
| Arachnida | 2·1–6·9 | | 3·8 | 9·5 |
| Sample size | 133 samples | | 1228 items | 1090 samples |
| Source | Anderson (1984) | | Graczyk and Krištín (1988) | Michocki (1975) |

Food of young predominantly insects, composition depending on age, habitat, and season (Table D). In eastern Germany, in pine *Pinus* woods, of 197 items from collar-samples, 52·8% Lepidoptera (mainly larvae), 21·3% Coleoptera (mainly adult Coccinellidae), 12·2% Diptera, 8·1% Hemiptera, 3·6% Arachnida, 1·0% Hymenoptera, and 1·0% *P. sylvestris* seeds; Lepidoptera proportionately replaced by Coleoptera as season progressed (Dornbusch 1973). Szlivka (1983) reported 82% (by number) invertebrates in diet of nestlings in Yugoslavian agricultural land: 55% Coleoptera, 27% Orthoptera. In Hungary, Török (1990) found 97·5% (by number) animal matter in orchards, mainly Coleoptera (30%), Orthoptera (13·5%), and Diptera (11%); after spraying orchard with insecticide, proportion of Coleoptera increased to 54%. Anderson (1984) found prey length mostly below 20 mm, mode 5 mm, mean 6·5 mm. Török (1990) gave mean lengths ranging from 18·1 mm (Carabidae, n = 59) and 16·4 mm (Orthoptera, n = 203) down to 2·1 mm (Aphididae,

$n = 410$). Initially, nestlings are fed on small aphids and viscera of larger animals; this is followed by soft larvae and, from about day 6, harder insects such as beetles and grasshoppers. While insects are main food, exclusively so for first 4–5 days, plant material may be included later. (Deckert 1962, 1968*b*; Wieloch 1975.) Mansfeld (1950), in cereal area, found over 90% invertebrate food up to day 10, falling to *c.* 60% at day 15–16, whereas in park and orchard breeders it remained over 90% invertebrates throughout; once independent, young feed predominantly on seeds. Rearing one young to fledging requires *c.* 800 kJ (Scherner 1972*a*). JDS-S

**Social pattern and behaviour.** Well studied in Germany (Creutz 1949; Berck 1961–2; Deckert 1962, 1968*b*) and Poland (Pielowski and Pinowski 1962; Pinowski 1965*a*, *b*, 1966, 1967*b*). For major review, including studies in USSR, see Noskov (1981).

1. Largely sociable all year. Outside breeding season, birds from adjacent loose breeding colonies (each covering 1–10 ha) tend to mix and wander over home-range of 10–100 km². After breeding season, adults join flocks of young birds that have already formed at suitable feeding places. Flock-members roost together (see below). Sexual recrudescence occurs in adults after moult at end of September and October, when they return to colony area to roost at night, ♂♂ displaying at nest-sites in mornings (Morning-display: see Song-display, below) before rejoining flock. Communal roosts begin to break up at end of October and 1st-years disperse to colony areas within winter home-range, where some older 1st-year ♂♂ acquire nest-sites and take part in Morning-display, attendance at nests peaking at this time, dying away completely by beginning of December; attendance lower on cold, damp days. Birds continue to roost in colony area, but disperse during the day to form large foraging flocks in fields, at times associating with finches (Fringillidae) and buntings (Emberizidae). (Pinowski 1966.) Large flocks persist to March–April, up to 1000 birds in favourable localities (Pinowski 1967*a*). If there is sufficient snow to cover food plants, flocks break up into smaller groups that scatter to search for food, particularly near human settlements (Pinowski and Pinowska 1985). Variations in sexual cycle occur away from core-area of central Europe, e.g. no autumn display in European USSR where population migrates south in winter (Dementiev and Gladkov 1954); Boyd (1932, 1933, 1934, 1935, 1949) made no mention of autumn display in intensive British study; further south, autumn display extended, e.g. nest-building seen through November in Sardinia (J D Summers-Smith). BONDS. Essentially monogamous with pairs formed for life, though occasionally ♂♂ polygamous, taking over additional widowed ♀♀ in neighbouring nests (Creutz 1949; Deckert 1962, 1968*b*; Weise 1992). One ♂ took over 2 additional nests where ♂♂ had disappeared, subsequently helping in incubation and feeding young at all 3 nests (Deckert 1962). Hybrids with House Sparrow *P. domesticus* are rare, though widely reported (Cordero and Summers-Smith 1993). 2 cases of hybridization with Spanish Sparrow *P. hispaniolensis* reported from Malta (Smith and Borg 1976; Sultana and Gauci 1982). No general agreement on how pair-formation takes place. ♂♂ that have lost mate return to nest-site and can acquire new mate during period of autumn display; new mate may be widowed ♀ or young bird, e.g. 5-month old ♀ paired with widowed ♂ (Creutz 1949). In autumn, ♀♀ that have lost mate abandon nest-site to join widowed ♂♂, but in spring retain the nest-site and quickly obtain new ♂. Creutz (1949) suggested that young birds establish pair-bond in 1st autumn from roosting associations, particularly with birds that roost together in a site that later becomes a nest. Birds that acquire mate or nest breed at 1 year old. Béthune (1961) and Deckert (1962) discounted pairing of young birds in autumn, considering that this does not occur until March–April, when ♂♂ take over nest-sites and begin calling there (see 2a in Voice). Nest-building does not normally begin until pair formed (Deckert 1962). Both sexes take part in building (nest can be ready for eggs some weeks before laying begins) as well as in incubation and brooding. Fledglings begin to feed themselves at 7–9 days after leaving nest (Berck 1961–2) and are fed by parents up to 10–14 days (up to 16 days for last brood), before joining up with other young in field flock. Feeding of fledglings from early broods mostly by ♂, but ♀ may also help while incubating new clutch. Often, each adult feeds preferred young (suggesting brood-division). If either parent dies, other may rear young alone from incubation to fledging. (Deckert 1962, 1968*b*.) BREEDING DISPERSION. Loosely colonial. Provided there is sufficient food, colony size depends on availability of suitable nest-sites. Population size exceeding holding capacity of colony results in dispersal of young birds unable to acquire nest-sites, normally the ones from later broods (Pinowski 1965*a*). New breeding colonies are formed by these dispersing young in areas where new nesting opportunities become available, e.g. by provision of nest-boxes (see below). Up to 4–5 pairs found together (Pfeiffer 1928) in foundation of large nests of other bird species (see Breeding); in latter, nests of *P. montanus* may be touching or within a few cm, but when nesting in individual holes in trees can be up to 1 km apart. Territory limited to immediate vicinity (*c.* 1 m) of nest (Deckert 1962). Density highest in old parks, orchards, and isolated woods of 0·01–1 km² (Noskov 1981), especially when nest-boxes provided: 22 pairs in 0·8 ha (2750 pairs per km²) in a garden, Nottinghamshire, England (N C Moore); population in 35 ha deciduous woodland, Freyburg (eastern Germany), increased to 40 pairs (114 pair per km²) in 6 years where nest-boxes provided, but remained at 2 pairs in 9·85 ha (20 pairs per km²) control area (Schönfeld and Brauer 1972); new population of 65 pairs built up over 8 years in 2·5 ha bird reserve (2510 pairs per km²), Westfalen (Germany) after nest-boxes provided (Gauhl 1984); 1210 pairs per km², Ukraine (Eliseeva 1961). Lower densities in more natural situations: 1100 pairs per km² in orchard, Rumania (Korodi Gál 1958); in Germany, 16–19 pairs in 2·82 ha (620 pairs per km²) of scattered human settlement (Deckert 1962), but only 50 pairs per km² in park in Sachsen (eastern Germany) (Berndt and Frieling 1939), 28·9 pairs per km² in open country in Sweden (Ulfstrand and Högstedt 1976), and average 1 pair per km² of farmland in north-east England (J D Summers-Smith). In Sardinia (where a comparatively recent colonist), urban study area of 10·5 ha had 190 pairs per km² (Walter and Demartis 1972). Pinowski and Kendeigh (1977) gave extensive list of breeding densities, 3–3900 pairs per km²; see also Noskov (1981). For nest-site fidelity, see Breeding. ROOSTING. Independent young form roosting associations in dense-foliaged trees, hedges, or reedbeds in foraging area, later joined there by adults after they finish breeding. Beginning late September, field roosts abandoned for colony areas, adults roosting in old nest-sites (not infrequently, in autumn-built nests) or close by, young birds initially forming roosting associations in nearby trees and bushes. Increasingly, young birds take over holes for roosting, up to 7 recorded in one site (nest-box) (Creutz 1949), and may establish pair-bonds in roosting association (see above). Communal roosting sometimes with tits (Paridae) and Nuthatch *Sitta europaea*; communal roosting under snow also recorded (Noskov 1981). Mates do not always roost together in nest, ♂ sometimes roosting in cover nearby. ♀ usually roosts in nest and always when there

are eggs or young, occasionally tolerating ♂. Away from nest-site, pairs not involved in breeding activity tend to stay together when feeding or bathing and (uniquely in *Passer*) often rest pressed together side-by-side (Fig A).

A

2. Mostly shy and unobtrusive in west Palearctic, though not so in Far East. On seeing distant aerial predator flies immediately to cover (see Flock Behaviour, below), though mobs owls (Strigiformes). (Deckert 1968*b*). FLOCK BEHAVIOUR. Foraging flock stays close to cover, with birds flying into cover every few minutes without obvious reason and then drifting back to feeding area; ♀♀ tend to remain on edge of flock and can continue to feed during resting period (see Food), indicating dominance of ♂♂ (Deckert 1962). Dust-bathing typically flock activity throughout year, birds not more than 5-10 cm apart, and individuals often defending their own pit (Noskov 1981). Aerial predators elicit Warning-calls (see 4 in Voice) and flock immediately flies to cover (Deckert 1968*b*); if other species present, remains silent but responds to their warning-calls (Berck 1961-2). If predator approaches, birds move deeper into cover and freeze silently; not readily driven out (Berck 1961- 2). Ground predators (e.g. cats) are kept in sight but not warned against. Changes feeding location after rest period (see Food) by individuals moving to (e.g.) top of tree, giving Flight-call (see 2b in Voice) and jerking tail; stimulates others to follow and they fly off together (Berck 1961-2). Birds arrive at roosting area in small groups. SONG-DISPLAY. Used by ♂ to attract ♀ for pair-formation (see 1 in Voice, and Heterosexual Behaviour, below). As substitute for song, ♂ also gives Advertisement-calls (see 2a in Voice) from conspicuous perch at nest-site in Enticement-posture (see Pair-bonding behaviour, Fig E, below) during autumn (Morning-display) and again in spring to proclaim ownership and attract a mate. Display also involves addition of nest-material to nest-hole and continues with increasing intensity until start of breeding. (Berck 1961-2.) In September-October and again from January to breeding season, flocks resting in bushes indulge in Social-singing (see 2e in Voice); according to Deckert (1962), has weak sexual motivation but not connected with pair-

B

formation. ANTAGONISTIC BEHAVIOUR. Birds approaching closer than individual distance (*c.* 10 cm: Deckert 1962) during feeding, bathing, or resting are challenged with Forward-threat posture (Fig B); head thrust forward, wings held slightly out; at higher intensity (Fig C), body inclined slightly forward, wings half-

C

raised and twisted to present upper surface to rival. Associated aggressive calls are used (see 3a, b in Voice). If trespasser does not retreat, threat may proceed to attack with birds wing-seizing and fluttering up against each other. ♂ defends nest against rivals by hopping 20-30 cm away with Head-up display (Fig D):

D

plumage sleeked, head held high showing black throat-patch, wings slightly opened and lowered, and tail raised high and quivered. Occasionally escalates to struggle. Pursues rival if it flies off. Similar behaviour used occasionally by ♀, but at lower intensity. ♂ or both pair-members constantly guard nest-hole against threat of occupation by conspecifics or other birds. Not uncommonly displaces tits *Parus* or Pied Flycatcher *Ficedula hypoleuca*, though usually subordinate to *P. domesticus* and *P. hispaniolensis* in competition for sites. (Schönfeld and Brauer 1972; Balát 1974; Löhrl 1978*b*; Noskov 1981.) HETEROSEXUAL BEHAVIOUR. (1) Pair-bonding behaviour and nest-site selection. Berck (1961-2) and Deckert (1962) suggested most pair-formation takes place at nest-hole. ♂ sits in front of nest or hops around in vicinity in Enticement-posture with feathers ruffled and tail slightly drooped (Fig E), giving Advertisement-call. When a ♀ arrives, ♂ advertises nest by flying to it and wing-shivering. If ♀ flies off, ♂ follows in exaggerated 'butterfly' flight, lands beside her, and continues display. If ♀ remains, ♂ enters nest, comes out with material, and with little further ado the pair is formed and ♀ is allowed to go into nest. Replacement-pairing much less conspicuous. ♂♂ also display to ♀♀ away from nest in Group-display (Fig F) (Berck 1961-2; Deckert 1962): ♂ hops

E

F

around ♀ with head up, wings held out and shivered, tail raised and may be spread, feathers ruffled, calling repeatedly ('song': see 1 in Voice). If ♀ flies away, ♂ follows and continues to display when she alights. In most cases only 2 birds involved, but sometimes other ♂♂, up to 3–4, join in (6 out of 42 displays: Deckert 1962). Group-display begins February–March, becoming most frequent in April and during preparations for subsequent broods; possibly serves to stimulate ♀ and appears to occur when she will not accept sexually-ready ♂. ♂ displays to unpaired ♀♀ and to already paired ♀♀ as well as to his own mate. Once formed, bond maintained by close association (see Roosting, above); mates recognize each other by voice (Berck 1961–2). In December–January, tend to associate less during day (Deckert 1962). Moore (1962) described Allopreening, and Boyd (1949) bill-caressing, but these appear to be rare; Deckert's (1962) extensive study did not record Allopreening. (2) Courtship-feeding. Does not occur (Berck 1961–2). (3) Mating. Takes place near nest (within 50 m: Deckert 1962), never on ground (Creutz 1949). ♂ gives Solicitation-display with head up, wings drooped and quivered, and tail raised, giving quiet Copulation-invitation call (see 2g in Voice). Unreceptive ♀ responds with threat and attack. If receptive, ♀ invites with similar Solicitation-display to ♂, crouching slightly with quivering wings, calling softly (see 2h in Voice). Copulation observed at all times of day, starting 6–8 days before egg-laying, but most frequent in morning. In successful copulation, ♂ mounts 5–10 times (Deckert 1962), bill-wiping, bowing, and preening in between. After copulation, both birds preen and add nest-material. Attempted copulation (♀ usually resists), involving young and old birds, also recorded in autumn display (Noskov 1981). (4) Behaviour at nest. Both sexes build. During this period and throughout breeding, ♀ becomes dominant: e.g. if pair-members arrive at nest together with building material or

food, ♀ threatens (see 3a in Voice) and goes in first, sometimes taking material or food from ♂. ♀ does not usually allow ♂ to come into nest to roost. No fixed relief-ceremony (Berck 1961–2); usually exchange takes place immediately on arrival of partner, sometimes with soft appeasement call (see 2f in Voice). During incubation, arriving bird frequently takes in fresh nest-material. RELATIONS WITHIN FAMILY GROUP. Following account based largely on Berck (1961–2) and Deckert (1962); for details of growth and development of young, see Noskov (1981). Both sexes brood nestlings by day, only ♀ at night. On hatching, nestlings brooded for average 75% of daylight hours on day 1, 42% on day 2, and at decreasing rate up to day 9, by which time fully feathered. Faecal sacs eaten by parents on day 1, thereafter carried away and dropped. Eyes open on day 5. Come to nest entrance to be fed from day 10. Both sexes bring food, in first few days ♂ often transferring food to young via ♀. In cases of overlapping broods, ♂ typically cares for 1st brood alone, while ♀ incubates 2nd clutch. Where both parents attend young throughout, rate of feeding, particularly by ♂, falls off on day young fledge; adults very excitable at time of fledging, coming to nest, but not feeding young (Noskov 1981); nestlings frequently leave when adults not present. Initial flight 50–100 m to cover, where they call (see Voice) to attract parents to feed them. Fledglings do not normally return to nest. For first couple of days out of nest, huddle on branches, then follow parents for up to *c.* 2 weeks until independent. ANTI-PREDATOR RESPONSES OF YOUNG. Young become markedly timid at 11 days, crouching in nest (Noskov 1981). No further information. PARENTAL ANTI-PREDATOR STRATEGIES. (1) Passive measures. Adults very wary and slip off nest on hearing disturbance or warning-calls of other species (Löhrl 1978b). Prone to desert if disturbed at nest. (2) Active measures. Young are enticed from danger by exaggerated slow flight and piping calls (Deckert 1962: see 2j in Voice). Adults also give this call if cat approaches when young in nest (Deckert 1962); nestlings at once fall silent. With older nestlings, adults remain nearby giving alarm-call before flying off (Deckert 1962). For alarm- and threat-calls in presence of ground and aerial predators, see 3 in Voice.

(Figs by R Gillmor: A from drawing in Summers-Smith 1988; B and D–F from photographs in Deckert 1962, 1968b; C from drawing in Berck 1961–2 and photograph in Deckert 1968b.)                    JDS-S

**Voice.** No distinctive song as such, but a wide range of calls described, though difficult to say how far these are different onomatopoeic renderings, variants, or genuinely different calls. Comprehensively described by Berck (1961–2) and Deckert (1962, 1968b). Renderings are those of J D Summers-Smith except where otherwise indicated. For additional sonagrams see Bergmann and Helb (1982).

CALLS OF ADULTS. (1) Song. Excited series of 'tschirp' calls (Fig I) by unpaired ♂ when ♀ arrives in territory (Deckert 1962); rapidly alternating high-low series of 'tschilp tschiib' at rate of 65–70 calls per min (Berck 1961–2). Rhythmic 'tschip-tnöke' or rapidly repeated 'schilp schilp', directed by courting ♂ at ♀ away from nest; other ♂♂ can be attracted to join in (Deckert 1962, 1968b). (2) Other advertisement- and contact-calls. (2a) Advertisement-call. Basic 'chirp'. High-pitched, essentially monosyllabic 'chip' or 'chirp' (Fig II), 'dschib' (Berck 1961–2), 'schilp' (Deckert 1962, 1968b), though at times a more clearly disyllabic 'chee-ip' (Fig III) (With-

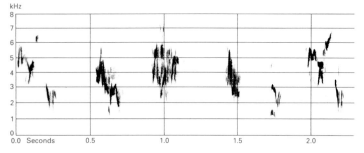

I  W Pedley  England  April 1975

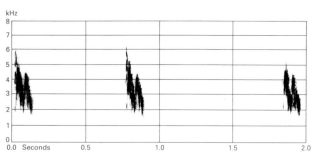

II  W Pedley  England  May 1979

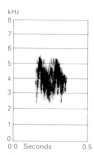

III  P A D Hollom  England  May 1987

erby *et al.* 1938). Used by ♂ to proclaim ownership of nest-site and attract a mate; initially 30–55 calls per min (Berck 1961–2), going over to song when ♀ arrives in territory. Also used by both sexes as pair-contact call (Deckert 1962). (2b) Flight-call. 'Schiep' given by flock-members moving to higher branches in preparation for flying to another location; leads to flight-readiness in flock (Deckert 1962). (2c) Call 2b changes to 'plui' (Berck 1961–2), or 'uik' (Deckert 1962) attracting rest of flock as leading birds fly off. (2d) Flight-contact call. Hoarse 'teck' (Fig IV) (Berck 1961–2). (2e) Social-singing. Rhythmic

same as the quiet 'brrrk' or 'krrrk' given by Berck (1961–2) as pair-contact call, but see also call 2a. (2g) Copulation-invitation call of ♂. Very soft 'dluidluidlui' (Berck 1961–2) or quiet, soft 'wlüg' (Deckert 1962) given by ♂ soliciting copulation. (2h) Copulation-solicitation call of ♀. Quiet 'psiehiehiesissihie' by ♀ soliciting copulation (Deckert 1962). (2i) Quiet, soft 'gäg gäg' or 'schelp schelp' by adults to stimulate young nestlings to feed (Deckert 1962). (2j) Piping 'fuik' to lead young away from danger (Deckert 1962). (3) Threat-calls. (3a) 'Tscherp tscherp', 'gräg gräg', or 'tert tert' (Fig V) by dominant ♀

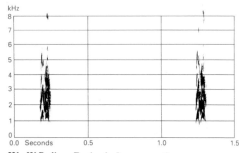

IV  W Pedley  England  January 1978

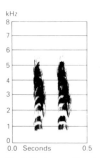

V  S Carlsson  Sweden  July 1984

series of calls, incorporating many of those used in other situations, given by flock-members during rest period in cover, sound ebbing and flowing (Deckert 1962, 1968b). Also rendered as repeated 'chittup-chirrtoowet' (Witherby *et al.* 1938). Included by Berck (1961–2) under general heading 'play song'. (2f) Nest-relief call. Quiet 'psieh-hie' or piping 'hi-huihie' or 'huik' used in appeasement between pair at nest-relief (Deckert 1962); perhaps

threatening ♂ (Deckert 1962). (3b) Rattle-call. 'Tetetätät' series given by both sexes against each other and against rivals at nest (Deckert 1962), probably identical to churring or rattling twitter 'tet-tet' or 'trr', with units given 8 times per s (Berck 1961–2; Fig VI)), used against cats and birds such as crows (Corvidae) and owls (Strigiformes). (4) Warning- or alarm-call. Harsh, loud, single or double 'tait', 'tät', 'däm', 'gäm', or 'däk' on sighting aerial pred-

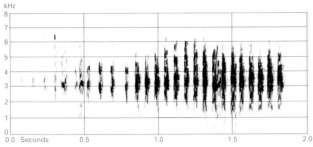

VI  P A D Hollom  England  November 1976

ator; flock-members react by flying into cover (Berck 1961-2; Deckert 1962). Quieter 'krüh-tet' used for distant flying predator (Deckert 1962). (5) Distress-call. Loud, shrieking 'chiu-chiu' (Berck 1961-2) or 'krätsch' (Deckert 1968*b*) given by trapped bird. Nearby birds react by flying off. (6) Other calls. For variety of other calls used in specific situations, see Berck (1961-2) and Deckert (1962, 1968*b*).

CALLS OF YOUNG. Small nestlings solicit food with 'sis-sisib' or 'sib-schilb' (Berck 1961-2); gradually changes to basic 'chilp' ('tschirp': Deckert 1968*b*) as nestlings develop, and used as location-call by fledglings still being fed.                                                                      JDS-S

**Breeding.** SEASON. Egg-laying in Europe normally mid-April to July (beginning of April to mid-August). Spain, end of April to mid-August (Sánchez-Aguado 1984; Cordero and Salaet 1987); Belgium, mid-April to mid-July (Béthune 1961); England, mid-April to end of July (early April to mid-August) (Seel 1964, 1966); Germany, mid-April (exceptionally early April) to July (Eisenhut and Lutz 1936; Creutz 1949; Scherner 1972*a*, *b*; Kaatz and Olberg 1975; Gauhl 1984); Poland, mid-April to end of July (exceptionally early August) (Pinowski 1968); Kursk (south European Russia), begins late April (Eliseeva 1961). For review, see Pinowski and Kendeigh (1977) and Noskov (1981). *P. m. transcaucasicus*: laying from late March in south Caspian (Witherby 1910) to at least May in Armeniya (Lyaister and Sosnin 1942). Start of 1st clutch about 1 week after average temperature reaches 10°C (Seel 1968*a*; Pinowski 1968). SITE. Predominantly in hole: in tree, building, earth bank (including Sand Martin *Riparia riparia* colonies), also in foundation of large nest (e.g. of crows Corvidae, Grey Heron *Ardea cinerea*, birds of prey), and more rarely free-standing in branches of dense conifers and hawthorn *Crataegus*. Holes 22-60 mm diameter, but *c.* 30 mm preferred (Löhrl 1978*b*; Gauhl 1984); also cracks and hollows in trees, built up with nest-material (Löhrl 1978*b*). Of 1073 nests in Britain, 95% in hole, 1·5% in foundation of large nest, 3% free-standing in tree (Seel 1964); of 1253, Yugoslavia, 91% in hole (17% in earth bank), 5% in large nest, 3% free-standing (Szlivka 1983); of 1815, Bulgaria, 94·5% in hole (13·5% in earth bank), 2% in large nest, 3·5% free-standing (Nankinov 1984). 2-5 m (0-24) above ground

(Béthune 1961; Seel 1964), or below ground level in well (Nankinov 1984). Nest: free-standing nest is flattened sphere, with entrance on side leading to nest-cup; in hole, available space normally filled with material, though roof can be omitted; Wasylik and Pinowski (1970) found 48·5% of 321 nests in nest-boxes had no roof; built from plant stems, rootlets, and leaves, lined with moss, wool, hair, and feathers (Creutz 1949; Gauhl 1984); external diameter 12·5 cm, internal diameter 9·3 cm, overall height 8·1 cm, depth of cup 4·8 cm (Dementiev and Gladkov 1954; Wasylik and Pinowski 1970); occasionally made entirely of pine *Pinus strobus* needles (Eisenhut and Lutz 1936; Creutz 1949). Building: by both sexes equally, each visiting nest every 2-6 min (Berck 1961-2; Deckert 1962); normally takes 5 days (Deckert 1962), though usually 1-2 weeks according to Noskov (1981). Nest-site usually kept for life, but occasional pairs alternate use of 2 holes, up to 12 m apart (Berck 1961-2). Material, including fresh green leaves and blossoms, frequently added during incubation (Eisenhut and Lutz 1936; Creutz 1949) and before 2nd or 3rd clutch laid (Wasylik and Pinowski 1970). For building techniques, see Deckert (1962). Some pairs build partial or complete nest in autumn (Löhrl 1978*b*). EGGS. See Plate 59. Sub-elliptical, smooth, slightly glossy; white to pale grey (translucent greenish: Eisenhut and Lutz 1936), heavily marked with spots, small blotches, or speckling, usually dark brown, sometimes purplish or greyish, often heavy enough to obscure ground; markings usually concentrated around broad end; great variation in size, shape, and colour (Harrison 1975; Makatsch 1976). Last egg paler, less marked (Witherby *et al.* 1938). Britain: 19·3 × 14·01 mm (17·5-20·8 × 13·0-15·1), $n=100$ (Witherby *et al.* 1938). Yugoslavia: 17·0-20·0 × 12·5-15·5 mm, $n=5287$ (Szlivka 1983). See Eisenhut and Lutz (1936) and Noskov (1981) for further extensive review of egg dimensions. Average calculated weight of 109 eggs, 2·11 g (Schönwetter 1984). Clutch: typically 2-7 (1-8), 5 most frequent. Multi-brooded, mean clutch size varying through season, with 2nd clutch largest and 3rd smallest: in Spain, 1st 4·92±0·85 ($n=91$), 2nd 5·07±0·74 ($n=171$), 3rd 4·21±0·58 ($n=98$) (Sánchez-Aguado 1984); in Belgium, 1st 4·85 ($n=91$), 2nd 4·96 ($n=77$), 3rd 4·80 ($n=43$) (Béthune 1961); in Czechoslovakia, 1st 4·75 ($n=149$), 2nd 4·92 ($n=87$), 3rd 4·54 ($n=26$) (Balát 1971); in Poland, 4·81±0·97 ($n=515$), 2nd 5·44±1·00 ($n=516$), 3rd 4·77±1·04 ($n=386$) (Pinowski 1968). 13 studies showed 65% (8-97%) lay 2 clutches, 25% (4-66%) lay 3, 3·5% lay 4. Clutch size also varies with habitat: pine forest 5·25±0·77 ($n=85$), deciduous forest 4·63±0·74 ($n=230$) (Balát 1971). Small year-to-year changes, but rarely significant. Slight increase in clutch size and decrease in numbers of broods with latitude, but not significant. For more complete summaries, see Noskov (1981), Gauhl (1984), and Summers-Smith (1988). Interval between clutches 40 days (30-65, $n=61$) with occasional overlapping (Creutz 1949); 38 (31-45, $n=31$) (Béthune

1961). Eggs laid daily (rarely 2-day interval), 06.00–08.00 hrs (Deckert 1962). INCUBATION. 11–14 days (Berck 1961–2; Scherner 1972). By both sexes during day, ♀ only at night. Begins casually with 3rd or 4th egg (Pinowski 1968), but not significant until clutch complete, then on average 50% of daylight hours by ♀, 29% by ♂ (Deckert 1962). Average duration of incubation stints: on day 3 6·5 min, day 8 16·3 min, day 13 37 min (Berck 1961–2). Hatching normally over 1 day, often within 1 hr, occasionally 2–3 days (Deckert 1962). YOUNG. Both sexes feed young almost equally (Berck 1961–2) or ♂ slightly more (Szlivka 1983). Almost all widowed birds, including ♂♂, capable of rearing young (Deckert 1962). FLEDGING TO MATURITY. Fledging period 15–20 days (reports of 12–14 days probably caused by observer disturbance: Summers-Smith 1988), but Nankinov (1984) reported extension by 5 days for nests in wells with more difficult first flight. Young independent 10–14 days after fledging (Deckert 1968b). Age of first breeding 1 year. BREEDING SUCCESS. Hatching success (16 European studies) 58·3–93·5%; fledging success 28·6–73·7% of eggs hatched; overall success 16·7–68·9%. Number of young per pair per year (14 European studies) 1·25–8·8. See Summers-Smith (1988) for details and references, and Noskov (1981) for further data. Losses of complete broods mainly due to bad weather (Pinowski 1968). Partial losses usually involve last-hatched young (Balát 1971), but Seel (1964) found no correlation between brood size and mortality. Predators include Little Owl *Athene noctua*, Wryneck *Jynx torquilla*, Great Spotted Woodpecker *Dendrocopos major*, cat *Felis*, stoat *Mustela erminea*, weasel *Mustela nivalis*, polecat *Mustela putorius*, marten *Martes martes*, red squirrel *Sciurus vulgaris*, dormouse *Eliomys*, and wood mouse *Apodemus* (Boyd 1935; Béthune 1961; Eliseeva 1961; Pinowski 1968; Balát 1971; Cordero and Salaet 1987; Indykiewicz 1988). Pinowski (1967b) also recorded broods destroyed by wasps (Vespidae) and bees *Bombus*. However, most nestling losses caused by cold weather and rain (Pinowski 1968; Mackowicz *et al.* 1970). Young fledglings preyed on by *A. noctua*, Magpie *Pica pica*, Jay *Garrulus glandarius*, cat, and snakes *Malpolon* and *Elaphe* (Cordero 1991). JDS-S

**Plumages.** ADULT MALE. Forehead, crown down to middle of eye, and hindneck dark vinous-chestnut, tips of some feathers narrowly fringed buff when fresh, mainly on hindcrown and hindneck, these fringes faintly edged dusky terminally. Mantle and scapulars boldly streaked black and rufous-cinnamon, latter shading to olive-cinnamon on tip of each feather; inner scapulars almost uniform buff-brown, in some postures contrasting with boldly streaked mantle and outer scapulars. Back to upper tail-coverts uniform buff-brown, longer tail-coverts with darker brown centres. Lore and narrow streak just below and behind eye black, forming dark mask just below chestnut of cap; feathers at gape black. Cheek, ear-coverts, and side of neck white, longer ear-coverts black, forming contrasting dark patch within white; upper ear-coverts washed pale buff and tipped white, forming mottled connection between black patch on side of head and rear of black eye-stripe; tips of white feathers on side of neck dusky

grey or sooty-black, forming narrow dusky bar at rear of white side of neck, running from lower throat to upper side of mantle. Chin and throat black, forming rather narrow round- or square-ended bib. Side of breast, flank, and thigh rufous-brown, slightly paler and more greyish toward rear of flank, merging into light grey or light brown-grey chest; belly and vent cream-white, merging gradually into grey of chest; under tail-coverts grey-brown with broad buff-white fringes. Tail-feathers sepia-brown with narrow buff or pink-cinnamon fringes (widest on central pair, 11). Flight-feathers and tertials sepia-black, outer webs of secondaries and both webs as well as tip of tertials broadly fringed rufous-cinnamon; fringes along outer webs of primaries paler rufous-cinnamon, narrow on middle portion of feather but wide at base and just below emarginated part of outer primaries, forming pale patches in closed wing; bases and middle portions of inner webs of flight-feathers sharply bordered pale isabelline-grey. Greater upper primary coverts and bastard wing sooty black with narrow rufous-cinnamon outer fringe (widest on base of primary coverts). Lesser upper wing-coverts rich rufous-chestnut (brighter than cap, not vinous); median coverts contrastingly black with cream-white tips of *c.* 2–3 mm; greater coverts sooty black with broad deep rufous-cinnamon fringe along outer web and white tip 2–3 mm wide (latter virtually absent on 1 or 2 outer coverts); white tips of median and greater coverts form 2 wing-bars. Under wing-coverts and axillaries pinkish-isabelline with whiter bases, shorter coverts along wrist dark grey with white fringes. Influence of wear and bleaching generally slight; *in worn plumage* (from about January–February onwards), buff fringes on rear of cap generally worn off, cap appearing uniform deep vinous-chestnut; rufous-cinnamon of mantle and outer scapulars and buff-brown of remainder of upperparts tinged grey, less pure; white of side of head and neck purer, less dusky suffusion on feather-tips, black ear-patch either more distinctly surrounded by white or (when heavily worn) fully connected with black of eye-stripe; side of breast, chest, and flank dirty greyish-white, virtually similar in colour to belly and vent, warmer greyish-buff tinge restricted to rear of flank, thigh, and centres of under tail-coverts, faint dusky streaks on breast and side of belly; fringes of tail, flight-feathers, and tertials greyer, less cinnamon, tips of tertials narrowly white when not too heavily abraded; white of wing-bars partly worn off, especially on innermost greater coverts. ADULT FEMALE. Like adult ♂, and often indistinguishable. On average, feathers of crown and hindneck more extensively tipped buff, and buff tips on hindneck sometimes bordered by black subterminally; if series of sexed birds compared, autumn ♂ shows traces of buff on central hind-crown only, ♀ buff on central crown and entire hindcrown; in winter, December–February, ♂ often already uniform vinous, ♀ still with many traces of buff; from March onwards, all ♂♂ uniform, but ♀ still with traces of buff on hindcrown; in June–July, ♀ uniform also, but some show black mottling on hindcrown (unlike any ♂). Amount of buff fringes perhaps depending on age also, 1st adult showing more than older birds, and therefore sexing of known pairs on colour of hindcrown probably not reliable, as 1st adult ♂ of a pair may perhaps show more than his older ♀ partner. Also, cheek-patch of ♀ on average less deep black, patch smaller, more greyish, and less sharply-defined; black on tips of feathers of bib more restricted, more grey showing on bib when plumage heavily worn; worn chest and flank on average more olive than in adult ♂, less pale and greyish. See also Szlivka (1983). NESTLING. Down entirely absent. For development, see Deckert (1962), Noskov (1981), and Sánchez-Aguado (1985). JUVENILE. Like adult, but cap buff-brown or cinnamon-brown, grading at sides into a darker rufous-brown stripe running from nostril over eye to side of upper neck;

feather-tips of hindneck sometimes with dusky brown or sooty marks; mantle to upper tail-coverts pallid buff-brown, less saturated than in adult, much grey of feather-bases visible, black streaks of mantle and scapulars shorter, more sooty; lore, eye-streak, cheek-patch, and bib dark grey rather than black, cheek-patch and bib smaller and less distinctly defined, surrounding white more heavily variegated by grey on feather-tips; in some birds, a short pale rufous-cinnamon supercilium extends behind eye, separating rufous-brown side of crown from dark grey eye-stripe. Underparts light grey with isabelline tinge, merging into off-white of belly; flank less buff-brown than adult, under tail-coverts without dusky sharply-defined central marks. Ground-colour of wing and tail paler than in adult, dark grey (but sooty-black on median coverts), fringes conspicuously paler, buff; tertials narrowly and evenly fringed buff (not broadly cinnamon), median coverts with pale buff tips 1–2 mm wide, greater coverts with traces of white on tips. When a series of sexed birds compared, ear-patch and bib of ♂ darker and more distinct than those of ♀, ear-patch of ♂ a fairly solid sooty-grey spot, of ♀ a dusky smudge restricted to rear of ear-coverts; chin of ♀ light grey, merging to medium grey on throat; of ♂, more uniform dark grey; also, rufous supercilium behind eye on average more distinct in ♀. FIRST ADULT. Like adult; indistinguishable when last juvenile feathers (innermost secondaries, p10) replaced; juvenile innermost secondaries still fresh, narrowly fringed pale buff along outer web and tip (in adult, slightly worn, more broadly fringed cinnamon along outer web only); juvenile p10 reduced, but slightly less so than in adult (see Structure).

**Bare parts.** ADULT, FIRST ADULT. Iris brown to dark brown. Bill horn-black to black, extreme base of lower mandible sometimes brown or yellow; base of lower mandible becomes gradually more extensively pale during post-breeding moult (late June to September), bill in October dark horn-grey with dull yellowish flesh-grey base of lower mandible, darkening to black again November–January, but bill of some birds (both sexes) pale until March. Leg and foot pale brown or light flesh-brown. NESTLING. At hatching, bare skin (including bill and leg) pink-flesh; mouth flesh-red (less pale pink than in House Sparrow *P. domesticus*), sometimes with dark spot at tip of tongue; gape-flanges yellow. JUVENILE. Iris brown or dark brown. Bill horn-brown to black-brown, traces of gape-flanges and base of lower mandible yellow or horn-yellow. Leg and foot flesh-colour, pale greyish flesh-pink, or flesh with leaden tinge. (Richmond 1895*a*; Stresemann 1920; Heinroth and Heinroth 1924–6; Witherby *et al.* 1938; Deckert 1968; RMNH, ZMA.)

**Moults.** ADULT POST-BREEDING. Complete; primaries descendent. In Britain, starts with p1 as soon as breeding completed, late June to early September, mainly 2nd half of July; duration of primary moult *c.* 60–70 days, completed with regrowth of p9 early September to late October. Secondaries start with shedding of s1 at primary moult score *c.* 22, finish with regrowth of p9 (primary score 45) or up to 17 days later; duration of regrowth of s6 *c.* 53 days. (Bibby 1977; Ginn and Melville 1983.) In 20 moulting birds from Netherlands and Italy, primary moult started early July to late August, completed mid-September to early November; of 2 from Turkey, 19 June, one just started, one not (RMNH; ZMA). In Germany, moult August–October, starting when young of last nest fledge (Niethammer 1937; Deckert 1968). In Poland, moult late July to mid-October (Pinowski 1968; Myrcha and Pinowski 1970). In Afghanistan, 17 of 18 *dilutus* (including juveniles) in primary moult 16 September to 16 October, one just completed (Paludan 1959). In Austria, one adult which was feeding young had primary moult suspended, p1 new (Kasparek

1979). For moult of *malaccensis* in Singapore, see Ward and Poh (1968) and Wong (1982). POST-JUVENILE. Complete; primaries descendent. In Britain, moult starts with p1 *c.* 38 days after fledging in 1st broods, *c.* 25 days in 2nd broods; starts late June to mid-September (mainly mid-July to August), completed September to early November, sometimes late August (Bibby 1977; Ginn and Melville 1983); birds hatched late sometimes start within 3 days after fledging; duration of moult *c.* 77 days, but much individual variation; moult completed at age of *c.* 4 months (young of 1st brood), *c.* 3·5 months (2nd brood), or *c.* 2·5 months (3rd brood) (Bibby 1977). In Netherlands (25 in moult examined), primary moult starts early July to mid-September, completed late August to mid-November; tertials, upper wing-coverts, and tail start from primary moult score 3–15, completed at score 25–45; body moult mainly at score 15–40, head and neck 30–45; secondaries from score 20–25, completed with outer primaries or slightly later (RMNH, ZMA). In Germany, moult in early-hatched birds (mid-May) starts at 9–10 weeks old (about early August), completed late September (Heinroth and Heinroth 1924–6); in late-hatched birds, starts at 5–7 weeks; duration 1·5–2 months (Stresemann and Stresemann 1966; Deckert 1968). In Switzerland, moult of 1st broods starts at age of 45 days, of 2nd and 3rd broods at 31–34 days, with moult in latter more rapid (Sutter 1985). In Poland, timing of moult of birds from 2nd brood as in adult post-breeding, but birds of 1st brood start relatively later and moult more slowly, though still completing earlier than birds of 2nd broods (Pinowski 1968; Myrcha and Pinowski 1970).

**Measurements.** Nominate *montanus*. Scandinavia, Germany, Netherlands, Britain, northern Italy, and Rumania, all year; skins (RMNH, ZMA). Bill (S) to skull, bill (N) to distal corner of nostril; exposed culmen on average 2·2 less than bill (S).

| | | | | | |
|---|---|---|---|---|---|
| WING AD | ♂ | 71·3 (1·46; 81) 68–74 | ♀ | 68·6 (1·32; 51) 66–72 | |
| JUV | | 66·0 (1·19; 7) 65–68 | | 65·9 (1·82; 11) 64–69 | |
| TAIL AD | | 52·5 (1·71; 23) 50–55 | | 51·0 (1·46; 18) 48–54 | |
| JUV | | 46·3 (1·55; 7) 45–48 | | 47·0 (1·04; 7) 45–48 | |
| BILL (S) | | 13·3 (0·55; 51) 12·4–14·2 | | 13·1 (0·50; 35) 12·3–14·3 | |
| BILL (N) | | 8·4 (0·32; 25) 7·8–8·9 | | 8·3 (0·31; 20) 7·9–8·9 | |
| TARSUS | | 18·0 (0·40; 18) 17·5–18·8 | | 17·9 (0·42; 15) 17·3–18·7 | |

Sex differences significant for adult wing and tail. In summer, bill *c.* 10% longer than in winter (Clancey 1948*a*).

Wing. Sexed birds. (1) Germany, breeding (Niethammer 1937). Eastern Germany: (2) according to Bährmann (1973; see also Bährmann 1976), (3) according to Eck (1985). (4) Southern Yugoslavia and north-west Greece (Stresemann 1920; Makatsch 1950). (5) Mongolia (Piechocki and Bolod 1972; Piechocki *et al.* 1982).

| | | | | | |
|---|---|---|---|---|---|
| (1) | ♂ | 71·1 ( — ; 25) 67–74 | ♀ | 68·8 ( — ; 9) 66–72 | |
| (2) | | 72·1 (1·92; 93) 67–76 | | 69·6 (1·53; 105) 66–74 | |
| (3) | | 71·5 (1·46; 39) 69–75 | | 69·3 (1·56; 45) 65–72 | |
| (4) | | 71·0 (1·37; 28) 68–73 | | 68·3 (1·55; 13) 66–70 | |
| (5) | | 73·7 (1·30; 12) 72–76 | | 71·2 (1·07; 13) 69–73 | |

Wing. Unsexed birds. South-east France: (6) breeding adult, (7) winter, (8) juvenile, June–July (G Olioso). (9) Eastern Germany (Weise 1992; wing mainly 64–72). (10) Czechoslovakia, adult (Havlín 1976). (11) Lake Chany (south-west Siberia), autumn (Havlín and Jurlov 1977). (12) Malta (Keve 1976*b*).

| | | | | |
|---|---|---|---|---|
| (6) | 69·2 (1·69; 20) 67–72 | (10) | 69·8 (2·77; 477) 60–78 | |
| (7) | 70·8 (1·74; 70) 67–75 | (11) | 68·4 (2·16; 140) 64–75 | |
| (8) | 65·8 (1·95; 31) 59–69 | (12) | 66·9 ( — ; 44) 62–72 | |
| (9) | 67·9 (2·17; 214) 62–78 | | | |

Various races. Wing and bill to nostril, sexes combined. Nominate *montanus*: (1) Britain, Ireland, and Bretagne (western

France), (2) southern France, Iberia, Sardinia, Sicily, southern Italy, and Malta, (3) Sweden, Baltic countries, and northern Russia east to western Siberia, (4) Poland and Germany south through central Europe to northern Italy and Balkan countries, (5) Altai mountains, northern Mongolia, and eastern Siberia east to middle Amur river and Sea of Okhotsk. *P. m. transcaucasicus*: (6) Istanbul east to Transcaucasia and north-west Iran, (7) area north of Caucasus, (8) lower Volga valley. *P. m. dilutus*: (9) Transcaspia through Tien Shan and Lake Zaysan to Gobi desert. *P. m. dybowskii*: (10) Amur- and Ussuriland, eastern Manchuria, and northern Korea. *P. m. kansuensis*: (11) Zaidam basin and northern Kansu. *P. m. tibetanus*: (12) southern and eastern Tibet. *P. m. iubilaeus*: (13) Peking to Shanghai area. *P. m. obscuratus*: (14) Nepal, Sikkim, Szechwan, and southern Shensi. *P. m. saturatus*: (15) Japan, Kuril Islands, Taiwan, and south-east China. *P. m. malaccensis*: (16) Vietnam and Thailand to Java. (Keve and Kohl 1978.)

| | WING | | | BILL (N) | | |
|---|---|---|---|---|---|---|
| (1) | 69·3 | (2·68; 57) | 66-75 | 8·8 | (0·61; 51) | 7-10 |
| (2) | 68·1 | (2·10; 28) | 62-72 | 8·6 | (0·72; 29) | 7-10 |
| (3) | 72·0 | (2·30; 119) | 68-77 | 8·9 | (0·68; 119) | 7-11 |
| (4) | 70·2 | (2·10; 572) | 64-76 | 8·6 | (0·69; 573) | 7-10 |
| (5) | 72·5 | (2·75; 23) | 70-78 | 9·0 | (0·76; 23) | 8-11 |
| (6) | 70·1 | (2·17; 21) | 66-74 | 8·2 | (0·53; 21) | 7-9 |
| (7) | 69·5 | (2·12; 34) | 65-75 | 8·1 | (0·62; 33) | 7-10 |
| (8) | 70·5 | (1·87; 29) | 69-74 | 7·9 | (0·45; 29) | 7-9 |
| (9) | 72·5 | (2·34; 107) | 66-78 | 8·3 | (0·64; 108) | 7-10 |
| (10) | 69·6 | (1·82; 60) | 66-74 | 8·0 | (0·88; 60) | 7-10 |
| (11) | 75·8 | (3·05; 15) | 73-80 | 7·9 | (0·25; 15) | 7-8 |
| (12) | 77·7 | (2·02; 9) | 75-81 | 8·8 | (0·76; 10) | 8-10 |
| (13) | 68·9 | (0·99; 10) | 68-70 | 8·7 | (0·47; 6) | 8-9 |
| (14) | 70·8 | (2·79; 29) | 67-79 | 8·6 | (0·79; 16) | 7-10 |
| (15) | 70·7 | (2·36; 121) | 66-76 | 9·3 | (0·70; 121) | 8-11 |
| (16) | 67·5 | (1·99; 72) | 63-72 | 8·3 | (—; 70) | 7-10 |

Wing and bill to skull, sexes combined. *P. m. transcaucasicus*: (1) northern Iran (Stresemann 1928; Vaurie 1949a; Schüz 1959; Diesselhorst 1962). *P. m. dilutus*: (2) eastern Iran and Afghanistan (Vaurie 1949a; Paludan 1959; Niethammer 1973); (3) south-east Kazakhstan (Vaurie 1949a); (4) Sinkiang (China) (Vaurie 1949a; ZMA). *P. m. saturatus*: (5) Japan, (6) Ryukyu Islands, (7) Taiwan (Vaurie 1956f; RMNH, ZMA). *P. m. malaccensis*: (8) Sumatra and Java (Indonesia) (RMNH, ZMA).

| | WING | | | BILL (S) | | |
|---|---|---|---|---|---|---|
| (1) | 68·1 | (27) | 64-72 | 11·1 | (19) | 10·5-12·0 |
| (2) | 73·6 | (43) | 69-77 | 12·2 | (30) | 11·5-13·0 |
| (3) | 73·3 | (8) | 70-77 | 12·5 | (8) | 12·0-13·5 |
| (4) | 74·6 | (13) | 71-78 | 13·8 | (13) | 12·5-15·1 |
| (5) | 70·0 | (47) | 65-75 | 14·4 | (49) | 13·4-16·0 |
| (6) | 70·2 | (31) | 66-73 | 14·0 | (42) | 12·5-15·5 |
| (7) | 70·0 | (71) | 66-74 | 13·9 | (72) | 12·5-15·0 |
| (8) | 67·4 | (33) | 64-71 | 13·7 | (35) | 12·8-14·5 |

**Weights.** Sexed birds. Nominate *montanus*. (1) Netherlands, adult, whole year (RMNH, ZMA). Eastern Germany: (2) adult, whole year, (3) juvenile, May-August (Bährmann 1972, which see for monthly fluctuations; see also Bährmann 1973). (4) Mongolia, May-July (Piechocki and Bolod 1972; Piechocki et al. 1982). (5) *P. m. dybowskii*: Manchuria (China), July-August (Piechocki 1959). (6) *P. m. transcaucasicus*: north-east Turkey and northern Iran, February-June (Paludan 1940; Schüz 1959; Kumerloeve 1967). *P. m. dilutus*. Afghanistan: (7) April-August, (8) October (Paludan 1959; Niethammer 1973). (9) Kazakhstan (Korelov et al. 1974). *P. m. saturatus*. (10) Taiwan, May-June (RMNH).

| | ♂ | | | ♀ | | |
|---|---|---|---|---|---|---|
| (1) | 22·6 | (2·49; 17) | 19-29 | 20·8 | (2·06; 16) | 18-24 |
| (2) | 23·7 | (1·83; 119) | 19-29 | 23·4 | (1·71; 143) | 19-27 |
| (3) | 20·9 | (—; 24) | 17-24 | 20·8 | (—; 17) | 17-24 |
| (4) | 22·6 | (1·93; 12) | 19-25 | 23·4 | (2·01; 11) | 20-27 |
| (5) | 23·5 | (1·29; 4) | 22-25 | 20·5 | (—; 6) | 20-21 |
| (6) | 22·1 | (1·43; 5) | 20-24 | 21·3 | (0·94; 4) | 20-22 |
| (7) | 22·6 | (1·52; 5) | 21-25 | 21·6 | (3·05; 5) | 18-26 |
| (8) | 23·8 | (—; 9) | 22-26 | 22·4 | (—; 8) | 21-24 |
| (9) | 23·8 | (—; 41) | 17-28 | 23·2 | (—; 35) | 21-27 |
| (10) | 21·1 | (1·27; 16) | 19-23 | 19·7 | (2·42; 3) | 18-23 |

Unsexed birds. Nominate *montanus*. Central Germany: (1) breeding, (2) winter (Scherner 1972). (3) Eastern Germany, mainly spring and summer (Weise 1992). South-east France: (4) breeding, June-July; (5) in moult, late August to November; (6) December-February (G Olioso). (7) Czechoslovakia (Havlín 1976; see also Havlín and Havlínová 1974). (8) Lake Chany (south-west Siberia), August and early September (Havlín and Jurlov 1977). (9) Netherlands, exhausted (RMNH, ZMA).

| | | | | | | | |
|---|---|---|---|---|---|---|---|
| (1) | 24·4 | (—; 46) | 22-27 | (6) | 21·4 | (1·44; 70) | 17-25 |
| (2) | 23·7 | (—; 23) | 19-26 | (7) | 23·0 | (1·61; 447) | 19-28 |
| (3) | 23·1 | (1·32; 213) | 21-27 | (8) | 23·1 | (1·66; 142) | 19-27 |
| (4) | 20·8 | (1·43; 19) | 18-24 | (9) | 17·2 | (0·42; 6) | 17-18 |
| (5) | 20·2 | (1·61; 85) | 17-24 | | | | |

Nominate *montanus* during early post-juvenile moult (primary score 1-24) 22·5 (22), when post-juvenile moult just completed 23·3 (19) (Myrcha and Pinowski 1970, which see for details); see also Bibby (1977). For growth of nestling, see Noskov (1981) and Sánchez-Aguado (1985). For variation of Chinese populations throughout year, see Shaw (1935); for that of *malaccensis* see Ward and Poh (1968).

**Structure.** Wing short, broad at base, tip bluntly pointed. 10 primaries: p7-p8 longest, p9 and p6 0·5-2·5 shorter, p5 3·5-6, p4 7-10, p1 13-15 (juvenile) or 14-18 (adult). P10 reduced, narrow and pointed, hidden below outermost greater upper primary covert; in adult, 43-50 shorter than p7-p8, 6·6 (10) 5-8 shorter than longest upper primary covert; in juvenile, 34-40 shorter than p7-p8, 1·9 (24) 0-4 shorter than longest primary covert. Length of adult p10 6·1 (6) 6-6·5, juvenile p10 11·6 (6) 10·5-13 (Stresemann 1920). Outer web of (p5-)p6-p8 emarginated; inner web of (p6-)p7-p9 with notch. Tip of longest tertial reaches tip of p1-p3 in closed wing. Tail rather short, tip square or slightly forked; 12 feathers, t1 and t6 often slightly shorter than others. Bill short, straight, conical; culmen slightly decurved; relatively less deep and wide at base than bill of House Sparrow *P. domesticus*, appearing less massive; bill longer and more attenuated in summer (late May to July), shorter for rest of year, depending on food (Clancey 1948a). Leg and foot short, but strong. Middle toe with claw 16·8 (15) 15·5-18; outer toe with claw c. 71% of middle with claw, inner c. 69%, hind c. 78%. Remainder of structure as in *P. domesticus*.

**Geographical variation.** Marked, involving size (length of wing, tail, or tarsus), relative length and depth of bill, and depth of ground-colour and width of streaking of upperparts. In general, largest birds in north of range and in plains and tablelands of central Asia, smaller birds in south; bill large and thick on Pacific seaboard from Japan to south-east China, small in Turkey and Caucasus area; pale birds in arid plains of central Asia, dark birds in high-rainfall areas at foot of southern and eastern Himalayas and in south-east Asia. Variation much subject to local influences, forming mosaic pattern without distinct clines, resulting in description of 33 separate races, reduced to 10 by Vaurie (1949a, 1956f, 1959), 22-23 by Keve (1978) and Keve and Kohl (1978), and 10 here (but in part differing from those of Vaurie 1959); survey given here based on Keve (1978), Keve and Kohl (1978), and specimens examined (BMNH, RMNH, ZFMK, ZMA). Size of wing and bill of various races mentioned in Measurements show large apparent overlap, but when sexes considered separately, and tail, bill depth, and tarsus also used in analysis of variation, most races as recognized here are readily separable on size, while races which are similar in size differ considerably in colour. Nominate *montanus* is a dark race of small to intermediate size and with intermediate to fairly strong bill. 5 populations separable within nominate *montanus*, each sometimes considered a separate race, but differences too small to warrant this: (1) *catellatus* Kleinschmidt, 1935, in Britain, Ireland, and Bretagne (France), which is slightly smaller and darker than typical nominate *montanus*; (2) *hispaniae* von Jordans, 1933, in southern France, Iberia, Sardinia, Sicily, southern Italy, and Malta, slightly smaller and slightly paler than typical nominate *montanus*; (3) typical nominate *montanus*, from northern Italy and north-west Greece north to Denmark and Poland, grading into *catellatus* in Netherlands, Belgium, and north-west France and into *transcaucasicus* in eastern Bulgaria, eastern Rumania, and probably European Turkey; (4) *margaretae* Johansen, 1944, from Scandinavia, Baltic, Belorussiya, and northern Ukraine east in Siberia to Yenisey basin, south to *c.* 40°N, probably grading into *transcaucasicus* and *dilutus* further south, which is on average larger and shows slightly longer bill than typical nominate *montanus*; (5) *stegmanni* Dementiev, 1933 (synonyms: *boetticheri*, *dementiewi*) in eastern Siberia, east of Yenisey basin to Sea of Okhotsk, south to Altai, northern Mongolia, Kerulen valley, and middle Amur valley, which is near *margaretae* in size but slightly darker and duller. *P. m. transcaucasicus* (synonyms: *volgensis*, *ciscaucasicus*), from Istanbul and Asia Minor east to Gorgan in northern Iran and north through Caucasus area to lower Volga valley, is about as large as nominate *montanus*, but bill shorter and colours paler, tending rather towards *dilutus*, but nearer nominate *montanus*. *P. m. dilutus* (synonyms *pallidus*, *zaissanensis*, *gobiensis*) from central Asia

(Transcaspia, north-east Iran, and Pakistan east through Kazakhstan, Tien Shan, and Sinkiang to Gobi desert) is large, with large bill (but not when compared with body size), and colours markedly pale and sandy-grey or buff, dark streaks on upperparts narrow. Its range is bordered in south by very large races *kansuensis* (synonym: *pallidissimus*) in Zaidam basin and northern Kansu, with short bill and pale colour, and *tibetanus* (synonym: *maximus*) in southern and eastern Tibet, with long bill and dark plumage. Eastern and southern Asia inhabited by number of small or fairly small races: in north, small-billed and fairly dark *dybowskii* from lower Amur south to northern Korea and west to Great Khingan mountains in Manchuria; further south, from Hopeh to lower Yangtze river and west to Shansi, replaced by *iubilaeus* (synonyms: *tokunagai*, *shansiensis*), which has intermediate bill size and rather pale colour; *obscuratus* (synonym: *hepaticus*) from southern Shensi and Hupeh south and west through northern Burma and Assam along foot of Himalayas to Nepal is among darkest and most saturated races, bill is intermediate; *saturatus* (synonyms: *taivanensis*, *orientalis*, *rikuzenica*, *bokotoensis*, *sititoi*, *manillensis*) from southern Kuril Islands and Japan through Ryukyu Islands and Taiwan to coastal south-west China is dark also, but with markedly heavy bill, especially compared with rather small body; birds from Sakhalin ('*kaibatoi*') intermediate between *saturatus* and *dybowskii*; *malaccensis*, occurring from Hainan, Vietnam, Thailand, and southern Burma to Indonesia, is smallest race; bill intermediate, colour fairly dark. See Keve (1978) for colour characteristics of various races and Keve and Kohl (1978) for many measurements; see also Stegmann (1931*a*), Johansen (1944), Clancey (1948*c*), Keve (1976*b*), and Stepanyan (1990). For morphological, mensural, and/or genetic differentiation of populations introduced outside original range, see Barlow (1973, 1980), Keve (1976*a*), and St Louis and Barlow (1987, 1988, 1991). CSR

---

## *Passer luteus*  Sudan Golden Sparrow

Du. Bruinruggoudmus   Fr. Moineau doré   Ge. Sudangoldsperling
Ru. Жёлтый воробей   Sp. Gorrión aureo   Sw. Guldsparv

PLATES 23 and 27 (flight)
[between pages 280 and 281]

*Fringilla lutea* Lichtenstein, 1823. Synonym: *Auripasser luteus*.

Monotypic

**Field characters.** 13 cm; wing-span 18·5–20 cm. About 15% smaller and slighter than House Sparrow *P. domesticus*, with proportionately smaller bill and head. Small, highly gregarious sparrow, with yellowest plumage of west Palearctic *Passer*. ♂ has yellow head and body and chestnut back; ♀ plain yellow-buff on head and back; ♂ and ♀ share dark wing-coverts with 2 white to buff wing-bars. Flight-call distinctive. Sexes dissimilar. Juvenile separable at close range.

ADULT MALE. Moults: May–August (complete). Whole head, neck, and underparts full canary-yellow; large dark eye obvious on face. Mantle and edges of black-centred tertials bright chestnut. Wings basically black but well patterned with (a) greyish lesser coverts occasionally

showing as pale shoulder-spot, (b) narrow yellow or white tips to median coverts, forming inconspicuous upper wing-bar, (c) broad white tips to greater coverts, forming bold lower wing-bar, and (d) when fresh, rufous-buff fringes to secondaries joining with outer edges of tertials to create almost buff-chestnut wing-panel. Underwing mainly yellowish-white. Rump yellow, contrasting with back, wings, and tail in flight; tail dark grey, with pinkish-buff margins to outer feathers. Bill black when breeding, otherwise dark horn with pink or yellow base. Legs flesh to brown. ADULT FEMALE. Head, neck, and breast pale yellowish-buff, with pale yellow supercilium and throat and browner crown, nape, and cheek-surround visible at close range. Mantle and scapulars buff-brown, with

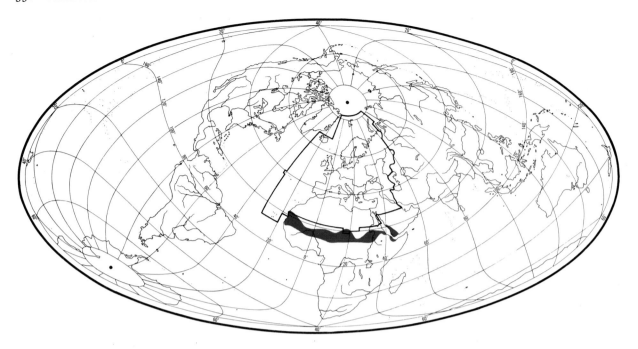

yellowish suffusion and some dull dusky streaks particularly on edges and rear centre. Wing much duller than ♂'s but basic pattern similar; tonal contrasts reduced by blackish-grey rather than fully black ground and buff or only dull white margins and tips to tertials and coverts. Note that tips to median coverts may be bolder than in ♂, creating more distinct upper wing-bar. Rump pale buff; tail as ♂ but less dark-centred. Underparts below breast whitish-buff. Bill always pale horn-grey. JUVENILE. Resembles ♀ but lacks any yellow tones; pale whitish supercilium and underparts distinctive. For 1st-year, see Plumages.

♂ unmistakable. ♀ and immature only confusable with other small sparrows when yellowish head or full strength of wing pattern obscured. Note that in ♀ or immature plumage *P. luteus* can recall Yellow-breasted Bunting *Emberiza aureola* which could conceivably stray into its range. Beware also frequent escapes from captivity. Flight fast and agile, allowing both individual speed and group manoeuvrability; action as reminiscent of small African finches as other sparrows. Escape-flight panicky, birds towering like Tree Sparrow *P. montanus* and making prolonged manoeuvres before resettling. Gait a light, quick hop. Behaviour typical of *Passer*. In Africa, forms vast roving swarms in non-breeding season.

Voice contains individual calls close to classic 'chirp' of *P. domesticus* but in chorus they run together into characteristically more sibilant twitter; when flushed, calls with fast rhythmic 'che-che-che-' (7-8 units), suggesting flight-call of Redpoll *Carduelis flammea*.

**Habitat.** Mainly south of Palearctic in semi-arid tropical lowlands, distribution conforming closely to zone of dry woodland and steppe (Summers-Smith 1988); breeds in largest available trees or bushes such as *Acacia* on margin of Sahel (Guichard 1955) and in Eritrea (Ethiopia) on belt of rocky low hills up to *c.* 100 m or on grassy or arid gravel plains or irrigated deltas. In Sudan, breeds on coastal plains below 300 m in large colonies in dry *Acacia*. Mountains and their foothills and human settlements are avoided. (Smith 1955*b*, 1957.)

**Distribution.** A Sahel species, barely reaching southern fringe of west Palearctic.

MAURITANIA. In west Palearctic area, recorded breeding only at Cansado (20°51′N 17°03′W) near Nouadhibou (RAW). ALGERIA. Occasional breeder, but south of west Palearctic limit; small breeding population found In Guezzam (19°30′N 5°30′E) 1984 (EDHJ). CHAD. Tibesti: recorded breeding Zouar (20°30′N 16°30′E) 1953 (Guichard 1955).

**Movements.** Status in Tibesti (Chad) and Nouadhibou (Mauritania) not known, but populations south of Palearctic are nomadic, following availability of food. Southward movement during breeding season in both Sénégal (Oubron 1976; Ruelle and Semaille 1982) and Niger (Klein 1988) as birds seek new area with sufficient food to rear 2nd brood; ♂♂ move first, leaving ♀♀ to complete rearing of 1st brood. South-west passage of 1·5-2 million birds seen in Sénégal in late September and October between area where one brood had been reared in September and later brood begun in 2nd half of October (Ruelle and Semaille 1982).                    JDS-S

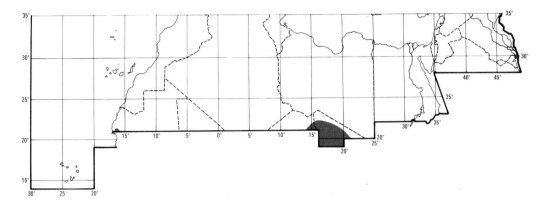

**Food.** All information extralimital (Mauritania, Sénégal, Niger). Diet mainly plant material, chiefly seeds, although nestlings fed largely insects. Seeds collected from seed heads of small plants and from ground, flock 'roller-feeding' with birds at rear overflying those at front (Klein 1988). Seeds dehusked by rubbing before swallowing (Kunkel 1961; Morel and Morel 1978).

The following recorded in diet. Invertebrates: grass-hoppers (Orthoptera: Acrididae), termites (Isoptera), bugs (Hemiptera: Pentatomidae, Cicadidae, Aphididae, Coreidae), larval Lepidoptera, flies (Diptera), ants (Hymenoptera: Formicidae), beetles (Coleoptera: mainly Curculionidae), spiders (Araneae), snails (Gastropoda). Plant material mainly seeds of grasses (Gramineae) e.g. *Panicum*, cultivated grains (millet *Pennisetum*, to lesser extent rice *Oryza* and *Sorghum*); also Molliginaceae and *Salvadora* berries. (Oubron 1967; Bortoli and Bruggers 1976; Bruggers and Bortoli 1976; Morel and Morel 1976, 1980; Klein 1988; L. Bortoli.)

In Sénégal, of 24 stomachs, cultivated grain present in 63%, wild seeds in 58% (Bortoli and Bruggers 1976). According to Morel and Morel (1980), *Panicum* seeds form high proportion of diet. See Klein (1988) for seasonal distribution of plant foods taken in Niger. Hand-reared birds each consumed 2·9 g of grain per day (Klein 1988).

Diet of young in rainy season mainly larval insects. In Sénégal, Morel and Morel (1976) found exclusively insects: Coleoptera 49·8%, Orthoptera 28·8%, Lepidoptera 14·6%, Hemiptera 6·8% (by number of items, $n = 219$). In Niger, Klein (1988) found diet included some grain seeds in 1 out of 3 colonies examined (25% of items in stomachs and crops, plus 7% other seeds, remainder animal material); in 2 other colonies, 1·4% and 3·7% non-cultivated seeds, remainder animal material; in 1st colony, diet for first 4 days 90% animal material, falling to *c.* 50% for older nestlings. In broods raised in dry season seeds supplement insects and by end of fledging period almost exclusively seeds (Bruggers and Bortoli 1976; L Bortoli).

JDS-S

**Social pattern and behaviour.** Not studied in west Palearctic,

but extensive observations made in Mauritania and Sénégal (Oubron 1967; Morel and Morel 1973a, b, 1976, 1978, 1980; Bruggers and Bortoli 1976) and Niger (Klein 1988).

1. Highly gregarious. After breeding, occurs in flocks of 10-100 birds. Frequently associates with Ploceidae (e.g. Buffalo Weaver *Bubalornis albirostris*, *Ploceus*, Black-faced Dioch *Quelea quelea*) and Estrildidae (e.g. Cut-throat *Amadina fasciata*, African Silverbill *Lonchura cantans*). (Bruggers and Bortoli 1976; Klein 1988.) Bonds. Monogamous (Morel and Morel 1973a, b), but not known if pairs remain together for more than 1 brood. Nest-building mainly by ♂ with cup completed by both sexes when pair formed (Morel and Morel 1973a, b; Klein 1988). Both sexes attend nest by day, but, with high ambient temperatures, incubation slight and not continuous during day, by ♀ alone overnight (Oubron 1967; Morel and Morel 1973a; Klein 1988). Both sexes feed young, ♀ taking major role and able to rear young alone if ♂ disappears (Morel and Morel 1973a; Klein 1988). Non-breeding ♀♀ or ♀♀ with failed nests help to feed fledglings, with up to 4 ♀♀ bringing food to one nest (Bruggers and Bortoli 1976). Age of first breeding 1 year (Bruggers and Bortoli 1976). Breeding Dispersion. Forms large colonies in scattered trees: in Sénégal, 1-6 nests per tree—mean 1·45 (Oubron 1967), 1·45 (Morel and Morel 1973a), 1-8 (J D Summers-Smith), in Chad up to 12 (Guichard 1955), in Niger up to 15 (Klein 1988). Colony area very extensive, 6·3 km² recorded (Oubron 1967). Nests can be within a few metres of each other; territory limited to immediate surroundings of nest, birds seeking food in colony area (Morel and Morel 1973a). Breeding densities (nests per km²): in Sénégal, 10 500 (Oubron 1967), 1000-20 000 (Ruelle and Semaille 1982), in dry season 34 000 (Bruggers and Bortoli 1976); in Mali, 10 000 (Bortoli and Bruggers 1976); in Burkino Faso, tens of thousands (Bortoli 1986); in Niger, 12 600-57 000 (Klein 1988). In Sénégal, mean spacing between colonies 85 km (10-159, $n = 7$) (J D Summers-Smith). Some colony sites used traditionally (Klein 1988). Roosting. Roosts gregariously in dense, thorny shrubs and trees and sugarcane fields (Oubron 1967; Bruggers and Bortoli 1976). When food still widely distributed, flocks gather together at night in temporary roosts; with advancing dry season, when food not so evenly dispersed, more permanent, much larger roosts form in favourable feeding areas, usually near water (Bruggers and Bortoli 1976; Klein 1988). In Sénégal, numbers increased from 700-5000 birds following breeding season to 1 million in dry season (Bruggers and Bortoli 1976). Roost can extend over large area: in Sénégal, Oubron (1967) found 400 000 birds occupying 20 ha with 15-400 birds per tree; in Niger, Klein (1988) reported over 1 million birds in 800 ha. For arrival and departure from roost, see Flock Behaviour (below). According to Oubron (1967) day roosts

formed in Sénégal at midday in dry season, but not during rains; if close, site used for night roost can be used by day, otherwise day roost formed near feeding area; however, Klein (1988) found no change in activity during day in Niger.

2. Wary and not easy to approach (Hartert 1921-2). FLOCK BEHAVIOUR. Almost always in flocks, but these lack cohesion, regularly breaking up and reforming (Klein 1988). Birds in feeding flock densely packed on ground. Flock stays close to trees where birds can take refuge from danger or rest between feeding bouts. Very restless, staying only for a few minutes in one place. (Klein 1988.) SONG-DISPLAY. Given by ♂ (see 1 in Voice) on or near partially completed nest with associated wing- and tail-movements (Morel and Morel 1973a). According to Klein (1988), wings held level with back during song, in pauses alternately raised and drooped, showing strong contrast between yellow head and rump and chestnut mantle (Fig A). When ♀ approaches, tempo of calling increases and wings flicked (Fig B, ♂ on left)

A

B

(rather than shivered as is usual in *Passer*) and displaying ♂ hops around ♀ bowing and flying repeatedly up to nest (Nest-showing) (J D Summers-Smith). ANTAGONISTIC BEHAVIOUR. In threat, ♂ takes up horizontal or slightly downward sloping posture with wings held out and drooped, feathers sleeked except for those of crown, bill opened. At higher intensity, wings stretched momentarily up. This display used frequently during formation of breeding colony as nests established; often develops into chasing and even fights with 2 ♂♂ falling to ground (J D Summers-Smith). HETEROSEXUAL BEHAVIOUR. Pair-formation occurs when ♀ attracted by Song-display (see above) of ♂ to partially completed nest (Morel and Morel 1973a). Copulation occurs near nest. ♂ invites by Soliciting-display, crouching with wings raised

and shivered, taking up position below ♀ so that bright plumage well displayed. Unreceptive ♀ pecks at ♂; this attracts other nearby ♂♂ and ♀ flies off followed by ♂♂ in Group-display. If receptive, ♀ adopts similar crouched posture to that of ♂ and calls quietly (see 3 in Voice) with both birds vibrating wings (J D Summers-Smith). ♂ mounts 2-17 times with feathers sleeked and pecks nape of ♀. Bouts of copulation by pair occur up to 10 times per day. (Klein 1988.) ♀ dominant at nest during breeding cycle (J D Summers-Smith). RELATIONS WITHIN FAMILY GROUP. See Bonds (above) for roles of parents in feeding. Negligible brooding of young by both sexes (Klein 1988). Nest sanitation by ♀ (Morel and Morel 1973a). ANTI-PREDATOR RESPONSES OF YOUNG, PARENTAL ANTI-PREDATOR STRATEGIES. No information.

(Figs by R. Gillmor from drawings in Summers-Smith 1988.)
JDS-S

Voice. Very vocal at nest during formation of colony, but otherwise little information.

CALLS OF ADULTS. (1) Song. 5-20 metallic 'ti' sounds in rapid succession (Fig I), series repeated at irregular

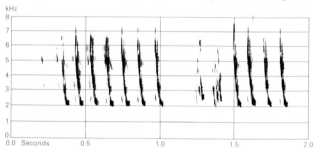

I  G Morel  Sénégal  September 1987

intervals (Klein 1988). Loud, far-carrying 'chitta-chitta-chit-chitta-chitta-churr-chitta-chit' (J D Summers-Smith). (2) Contact-calls. (2a) Flight-call. Twittering like Linnet *Carduelis cannabina* (Witherby 1901); 'che-che-che' with rhythm like Redpoll *C. flammea* (Hollom *et al.* 1988). (2b) Social-singing. Recording (Fig II) contains

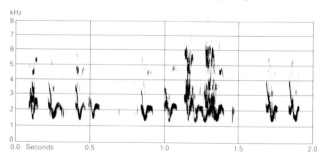

II  C Chappuis  Mali  February 1969

conversational fluty chirping sounds, considerably lower pitched than song, from group of resting birds (C Chappuis); some overlap between birds in Fig II. (2c) Copulation-invitation call. Quiet 'dee' from ♀ to invite copulation (J D Summers-Smith). (2d) Feeding-call. Quiet 'dipp' from ♀ greeting begging young (Kunkel 1961). (3)

Threat-call. Typical *Passer* 'churr' (J D Summers-Smith); probably the same call in captive birds described as long drawn-out rattle (Kunkel 1961).

CALLS OF YOUNG. No information.                      JDS-S

**Breeding.** SEASON. Breeding opportunistic, normally triggered by rains, more rarely in dry season, particularly in irrigated areas (Bruggers and Bortoli 1976; Morel and Morel 1976). Only records for west Palearctic are March in Tibesti (Guichard 1955), mainly March–May (also June–August) in Mauritania (Mahé 1985). See Summers-Smith (1988) for breeding seasons of extralimital populations. SITE. Thorny trees and bushes (*Balanites*, *Acacia*, *Zizyphus*), 1·8–5 m (mean 2·5, *n* = 238) above ground (Morel and Morel 1973a; Bruggers and Bortoli 1976; Ruelle and Semaille 1982); minimum height 1·3 m (Klein 1988). Nest: globular structure wedged between several branches with entrance hole in upper half. Built from 700–1200 thorny twigs, mostly less than 100 mm long, occasionally over 200 mm; more rarely of stiff grass stems; lower half compactly built, upper half rather looser; cup of grass, bark, leaves, flower heads (*Tamarix*, *Suaeda*), feathers; outer diameter 23–31 cm, inner diameter 6 cm, overall height 32–43 cm, diameter of entrance hole 5–6 cm (*n* = 7); mean weight 377 g, minimum 129 g (*n* = 7); old nests occasionally re-used (Oubron 1967; Morel and Morel 1973a; Bruggers and Bortoli 1976; Klein 1988). Building: twigs picked up from ground near nest-tree and frequently stolen from neighbouring nests. Outer structure built (by ♂ only) in 3–10 days, ♂ carrying up to 100 twigs per hr (Morel and Morel 1973a; Klein 1988). Pair continue to add material after eggs laid (Morel and Morel 1973a). EGGS. Sub-elliptical, smooth and slightly glossy; white to light bluish- or greenish-grey, with brown to maroon spots and flecks (Oubron 1967; Klein 1988). Sénégal, 17·5 × 12·9 mm (13·0–19·2 × 12·1–13·9), *n* = 79 (Oubron 1967); Niger, 17·82 × 12·68 mm, *n* = 72 (Klein 1988). Nominate *luteus*: 17·6 × 13·5 mm (17·0–19·0 × 12·7–14·5), *n* = 12, calculated weight 1·70 g (Schönwetter 1984). Clutch: in Niger, 3–5 (mode 4) (Klein 1989); 3·3 ± 0·9, *n* = 51, estimated from spot checks in Sénégal, lower in dry season (Morel and Morel 1976). Eggs laid daily. 2 broods, 2nd clutch laid in different area. INCUBATION. 11 days (Morel and Morel 1973a); 10–12 days (Klein 1989). Incubation only slight and discontinuous, by both sexes, during day; by ♀ alone at night (Morel and Morel 1973a; Klein 1988); because high ambient temperatures effectively start incubation from 1st egg, hatching asynchronous with intervals of 1 day (Klein 1988). YOUNG. Fed by both parents, more frequently by ♀ (Morel and Morel 1973a; Klein 1988). FLEDGING TO MATURITY. Fledging period 14 days (Morel and Morel 1973a), 13–14 days (Klein 1988). BREEDING SUCCESS. Human access to nest contents difficult without destroying nest. In Niger, Klein (1989) sectioned nest so that top could be lifted off and contents examined by mirror: hatching success 82·9%, fledging success

56·5 ± 11·1%, overall 47%; 9·1 ± 2·4% of eggs sterile, 5·1 ± 3·3% failed through adults abandoning nest, 2·9 ± 2·2% taken by predators (rat, snake *Psammophis*); 20·4 ± 8·6% of nestlings died through starvation, 4·7 ± 3·8% taken by predators, 0·8 ± 1·4% abandoned. Morel and Morel (1973a) estimated 87% success in incubation period, 61% during fledging period, overall success 53%; monitor *Varanus* noted as additional nest predator. Not possible to estimate annual production because of insufficient knowledge of number of 2nd broods, but according to Ruelle and Semaille (1982) 2·1 ± 1·5 young reared per nest overall; 3·3 fledglings per occupied nest in 1 year. Fledglings preyed on by Black Kite *Milvus migrans* (Klein 1988), African Harrier-Hawk *Polyboroides radiatus*, Dark Chanting Goshawk *Melierax metabates*, Shikra *Accipiter badius*, Red-headed Falcon *Falco chicquera*, Senegal Coucal *Centropus senegalensis*, and Woodchat Shrike *Lanius senator* (Morel and Morel 1973a). See also Morel and Morel (1976).                      JDS-S

**Plumages.** ADULT MALE. Head, neck, rump, and virtually entire underparts bright yellow, feather-tips on crown, hindneck, and rump washed rufous-grey when plumage fresh, partly concealing yellow; grey fringes wear off last on hindneck and lower rump. Mantle, scapulars, and back contrastingly dark rufous-brown or chestnut, feathers narrowly fringed yellow when plumage fresh, some yellow of feather-bases visible when plumage worn. Upper tail-coverts light grey with slight yellow or buff wash. Thighs grey with yellow wash; under tail-coverts pale yellow or yellow-white with black shaft-streak, but latter largely concealed. Tail dark brownish-grey, darker towards tip; central pair (t1) and outer webs of t2–t5 with light ash-grey bloom, shafts black; fairly broad but indistinct fringe along both webs and tip of t1 and narrower but sharper fringes along other tail-feathers pale pink-cinnamon (if fresh) to off-white (if worn and bleached). Flight-feathers greyish-black; outer webs of primaries narrowly fringed light pink-cinnamon or isabelline (except emarginated parts), tips of inner narrowly white; outer webs of secondaries more broadly fringed darker pink-cinnamon (outer secondaries) or deep rufous-cinnamon (inner ones), tips and inner borders fringed (pink-)white. Tertials deep black with broad and contrasting deep rufous-cinnamon fringe along outer web and tip. Lesser upper wing-coverts dark brown-grey with yellow wash; median and greater coverts as well as greater upper primary coverts and bastard wing contrastingly black, median coverts with narrow yellow or white tips (sometimes showing as pale wing-bar, but frequently inconspicuous, even in fresh plumage, or worn off), greater with broader and more distinct white tips (forming distinct wing-bar), primary coverts and bastard wing with narrow off-white fringes along tips; innermost greater coverts (tertial coverts) black with rufous-cinnamon distal halves and white tips. Under wing-coverts off-white with yellow wash, those along wing-bend black with yellow fringes, axillaries yellow. Not yet established whether adult ♂ has separate non-breeding plumage; a few birds examined with deep rufous-chestnut mantle and scapulars and black wing-coverts (like adult ♂) but with light sandy-grey crown and nape and rather pale yellow face and underparts were either aberrant adult ♂, non-breeding ♂, or advanced 1st adult breeding ♂. ADULT FEMALE. Upperparts entirely buff-brown, palest on hindneck and rump, slightly tinged vinous-pink when plumage fresh; crown, hindneck, rump, and (especially) forehead slightly washed yellow. Lower mantle and inner scapulars with indistinct dark brown or dull black shaft-streaks (more conspicuous when plumage worn). Long and narrow supercilium as well as eye-ring pale yellow, distinct in birds which show little yellow on top of head, more inconspicuous in birds with yellowish-buff forehead and lore; narrow streak through eye brown-grey, merging into pale brown-grey or light vinous-grey side of neck. Cheek light yellowish-grey, chin and throat pale yellow. Side of breast, chest, and flank pale cream-buff when fresh, more pale sandy-grey when worn (then sometimes with slightly darker grey shaft-streaks), merging into off-white belly, vent, and under tail-coverts; when fresh, often a light yellow wash on belly; centres of under tail-coverts pale brown-grey. Tail as adult ♂, but grey slightly browner and tips less blackish. Flight-feathers, greater upper primary coverts, and tertials as adult ♂, but ground-colour dark brown-grey, less blackish (especially on tertials), fringes of inner secondaries and tertials light pink-cinnamon, not rufous, much less sharply-defined. Lesser and greater upper wing-coverts dark grey-brown, like tertials, median coverts darker, sepia or black-brown; lesser coverts tipped buff-brown, medium and greater coverts with broad pale cream-buff to off-white tips (colour depending on bleaching and wear), forming distinct wing-bars (that on median coverts more distinct than in adult ♂); upper wing showing much less black than adult ♂. Under wing-coverts and axillaries grey-white or white, sometimes with slight yellow wash. Some ♀♀ show much yellow on head, neck, rump, and underparts, superficially resembling adult ♀ but yellow paler, less golden, less uniform, mantle and scapulars buff-brown with some dark streaks (in ♂, distinctly darker rufous or chestnut, unstreaked), wing-coverts grey-brown with blackish bases to median coverts and with 2 distinct pale wing-bars (in ♂, black with less distinct wing-bars). NESTLING. No information. JUVENILE. Upperparts like adult ♀, but buff-brown paler and without delicate vinous-pink tinge, more buff on lower scapulars and back to upper tail-coverts; yellow tinge entirely absent; shaft-streaks on mantle and scapulars less distinct. Pattern on side of head as in adult ♀, but no yellow; supercilium rather faint and narrow, pale buff or cream-white, upper ear-coverts pale brown-buff with fine off-white specks, gradually merging into cream-white of cheek. Underparts pale cream-buff, almost white on chin and upper throat and on belly and vent, slightly browner on side of breast, flank, and chest; no yellow. Tail sepia, feathers with poorly defined pale buff fringes. Flight-feathers and bastard wing dark grey-brown or sepia, outer webs narrowly edged pale buff; tertials dark sepia-brown with pale cream-buff fringes of even width (in adult, those on outer web broader than on inner). Lesser upper wing-coverts buff; median coverts sepia with broad pale cream-buff or off-white tip, forming pale wing-bar; greater coverts sepia with slightly less contrasting pale buff tips. Outermost primary (p10) slightly less reduced than in adult (see Structure). FIRST ADULT NON-BREEDING. Like adult ♀, and often hard to distinguish if no juvenile tail- or flight-feathers retained. Develops within a few weeks of fledging. In ♂, plumage similar to adult ♀, but side of head and underparts on average slightly more intensely washed with bright pale yellow; centres of tertials, tail-feathers, and greater upper wing-coverts blacker, less brown-grey, fringes along tips and outer webs of tertials and greater coverts more sharply defined, tinged warmer cinnamon (especially inner greater coverts sometimes rufous-cinnamon); centres of median coverts blacker, tips whiter. First non-breeding ♀ similar to adult ♀, but yellow restricted to faint tinge on face, cap without greenish or greyish tinge, supercilium pale isabelline-buff, less pale yellow and more contrasting than in adult ♀. FIRST ADULT BREEDING MALE. Like adult ♂, but crown slightly tinged green, brown, or (rarely) rufous, contrast between yellow hindneck and rufous-chestnut mantle often less clear; rufous-chestnut mantle and scapulars slightly less deep, feathers more broadly fringed brown-grey or sandy when plumage fresh; upper wing-coverts on average less deep and uniform black, contrasting less with rufous-chestnut scapulars. Forehead, side of head and neck, and underparts either deep yellow, as adult ♂, or (sometimes) less bright and saturated yellow. FIRST ADULT BREEDING FEMALE. As adult ♀.

**Bare parts.** ADULT, FIRST ADULT. Iris pale brown, hazel, or dark brown. Bill pale horn-grey, base tinged flesh-pink or yellow, tip darker horn; in breeding season, bill of ♂ black, upper mandible of at least some ♀♀ dark horn-brown. Leg and foot dark flesh, pale flesh-brown, or light brown. (Bannerman 1948; BMNH, RMNH, ZFMK, ZMA.) NESTLING. No information. JUVENILE. Iris brown. Bill flesh-grey, darker horn-brown on culmen and tip; leg and foot light flesh-brown or lilac-flesh. (BMNH.)

**Moults.** ADULT POST-BREEDING. Complete; primaries descendent. Starts at end of breeding season; nesting highly dependent on local rains (e.g. Morel and Morel 1973a, 1976, Bruggers and Bortoli 1976), so moult season also variable. Moult sometimes suspended if not complete when breeding starts. According to Bannerman (1948), moult in western Sudan early February to early May, in Mali and Niger about May-August, but in birds from Sudan examined (BMNH, ZFMK) moult had started about October and December, in northern Chad June-July; in captivity, moult October-February. POST-JUVENILE. Partial or complete; flight-feather moult interrupted when moult conditions unfavourable. Juvenile plumage of head and body generally replaced by 1st adult non-breeding within 1-2 months of fledging, this followed by partial pre-breeding when 6-11 months old, latter moult often also including those juvenile flight- or tail-feathers not replaced in post-juvenile. Timing highly variable (as in adult post-breeding); moult protracted. See also Paludan (1936).

**Measurements.** ADULT. Whole geographical range (mainly Sénégal and Sudan), all year; skins (RMNH, ZFMK, ZMA). Bill (S) to skull, bill (N) to distal corner of nostril; exposed culmen on average c. 2·0 less than bill (S).

| | ♂ | | | ♀ | | |
|---|---|---|---|---|---|---|
| WING | 64·6 (1·42; 57) | 61-67 | | 62·9 (1·19; 47) | 61-65 | |
| TAIL | 48·5 (1·86; 23) | 45-51 | | 47·9 (2·29; 13) | 45-52 | |
| BILL (S) | 12·5 (0·47; 26) | 11·8-13·7 | | 12·7 (0·35; 13) | 12·1-13·2 | |
| BILL (N) | 7·6 (0·36; 26) | 7·0-8·3 | | 7·6 (0·28; 13) | 7·2-8·1 | |
| TARSUS | 17·2 (0·38; 25) | 16·5-18·0 | | 17·1 (0·42; 13) | 16·4-17·6 | |

Sex differences not significant. Chad: wing 65 (9) 60-72 (Salvan 1967-9).

JUVENILE. Wing 61·4 (0·89; 5) 60-63, tail 43·2 (2·66; 4) 40-46 (RMNH, ZMA).

**Weights.** Northern Niger, June: ♂ 15·2 (3) 14·8-15·7, ♀♀ 14·5, 15·6 (Fairon 1975). Northern Chad: ♂ 14·5-15 (n=3), ♀♀ 11, 11·5 (Niethammer 1955). Chad: 14 (9) 12-16 (Salvan 1967-9). For closely related Arabian Golden Sparrow *P. euchlorus*, see Bowden (1987).

**Structure.** Wing short, broad, tip rounded. 10 primaries: in adult, p8 longest, p7 and p9 0-1 shorter, p6 0·5-2, p5 2-5, p4 5-8, p1 12-15; in juvenile, p7-p8 longest, p6 and p9 0·5-2 shorter, p5 3-6, p4 6-9, p1 13-15; p10 reduced, especially in adult, 39-44 shorter than p8 in adult, 35-38 in juvenile, 4·1 (10) 3-6 shorter than longest upper covert in adult, 2 shorter to 2 longer in juvenile (on average, 0·8 shorter, n=6). Outer web of (p5-)p6–p8 emarginated, inner web of p7–p8(–p9) with notch. Tip of longest tertial reaches tip of p2–p3. Tail short, tip square or slightly forked; 12 feathers. Bill short, thick, base of upper mandible swollen; culmen gradually decurved, ending in sharp tip, often projecting over tip of lower mandible; cutting edges slightly decurved, sometimes with indistinct tooth near base on upper mandible; depth and width at base 6–7 mm. Nostrils small, rounded, partly covered by operculum above, bordered by frontal feathering at rear. Bristles at gape vestigial. Leg and foot short, but rather strong; middle toe with claw 16·1 (20) 14·5–17·5, outer and inner toe with claw both *c.* 68% of middle with claw, hind *c.* 76%.

**Geographical variation.** None.

Forms superspecies with Arabian Golden Sparrow *P. euchlorus.* Latter has head and body entirely yellow, except for some brown suffusion on crown and mantle; somewhat smaller than *P.*

*luteus,* wing 60·6 (27) 57–63 (Bowden 1987); occurs in southern Arabia, also northern Somalia (perhaps as an escape: Meinertzhagen 1954), recently spreading west to Djibouti (Bowden 1987), but not yet touching range of *P. luteus,* which occurs east to south-east Egypt, Sudan, and Eritrea (Hall and Moreau 1970). ♂ Chestnut Sparrow *P. eminibey* (synonym *Sorella eminibey*), occurring from south-east Sudan and south-central Ethiopia to northern Tanzania, has head and body entirely chestnut; sometimes considered to belong to same superspecies (e.g. Hall and Moreau 1970), difference between yellow *P. euchlorus* and chestnut *P. eminibey* bridged by yellow-and-chestnut *P. luteus,* but isolated population of *P. eminibey* occurs in Jebel Marra area (western Sudan), within geographical range of *P. luteus,* without apparent interbreeding, so *P. eminibey* here considered a separate species, forming species-group with superspecies consisting of *P. euchlorus* and *P. luteus.* CSR

---

## *Carpospiza brachydactyla* Pale Rock Sparrow

PLATES 24 and 27 (flight)
[between pages 280 and 281]

Du. Bleke Rotsmus     Fr. Moineau pâle     Ge. Fahlsperling
Ru. Короткопалый воробей     Sp. Gorrión pálido     Sw. Blek stensparv

*Petronia brachydactyla* Bonaparte, 1850

Monotypic

**Field characters.** 13·5–14·5 cm; wing-span 27–30 cm. As long, and as long-winged, as Rock Sparrow *Petronia petronia,* but with slighter, sleeker form most noticeable in smaller (slightly bulbous) bill and slimmer body; slightly smaller and noticeably shorter than Sinai Rosefinch *Carpodacus sinoicus,* particularly in tail length. Fairly slim and rather anonymous sparrow-sized bird of uncertain affinity (see Geographical Variation), recalling ♀ rosefinch *Carpodacus* or odd ♀ bunting *Emberiza* as much as *P. petronia.* Dull pale brown except for double buff wing-bar, pale panel on folded secondaries, and tail with white edges and tip; lacks yellow on breast but shows bright orange-brown legs. Runs. Sexes similar; no seasonal variation. Juvenile separable at close range.

Adult. Moults: at least August–September (complete). Head and upperparts sandy grey-brown, head marked only by indistinct off-white supercilium; underparts basically dusty- or sandy-white, with faint moustachial stripe, dusky-brown malar stripe, and off-white chin and throat. Head and body thus lack distinct features apart from pale horn bill and dark eye. Wings dusky-brown and quite strikingly patterned with (1) off-white tips to median coverts creating quite sharp upper wing-bar, (2) dusty-white fringes and tips to greater coverts forming more diffuse lower wing-bar, (3) dusty-white fringes to tertials, (4) dusty-white outer webs to secondaries overlapping on inner feathers to form quite conspicuous wing-panel, and

(5) creamy tips to dark brown primaries. Tail rather short; blackish-brown with paler edges and all but central feathers with white spot below tips. With wear, head and body plumage becomes paler and wings may lose fully clear marks. Legs bright and translucent-looking, flesh- to brownish-orange. Juvenile. Sandier above and even less coloured below than adult, with only vestigial supercilium, browner ground-colour to flight- and tail-feathers, and fringes of secondaries pink-buff. Bill pale flesh, with dusky culmen; legs greyish-pink.

At first sight, a bird of remarkable anonymity, provoking confusion with ♀ *Carpodacus sinoicus* and even pale ♀ or immature *Emberiza* as much as with *Petronia.* However, has diagnostic combination of rather short but strong and slightly bulbous bill, long, narrow outer wings with pale secondary-panel and double wing-bar, pale edges and white tip to tail (particularly obvious from below), and bright legs. Also differs distinctly from *P. petronia* in slighter form and lack of streaks on head and mantle, from Yellow-throated Sparrow *P. xanthocollis* in somewhat larger size, pale bill, lack of warm brown lesser coverts and boldly white-tipped median coverts, from ♀ Desert Sparrow *Passer simplex* in larger size, greyer appearance, lack of pale, dark-rimmed wings, and from Dead Sea Sparrow *Passer moabiticus* in larger size and lack of mantle streaking. Voice also helpful in diagnosis (see below). Flight action lighter and even more fluent

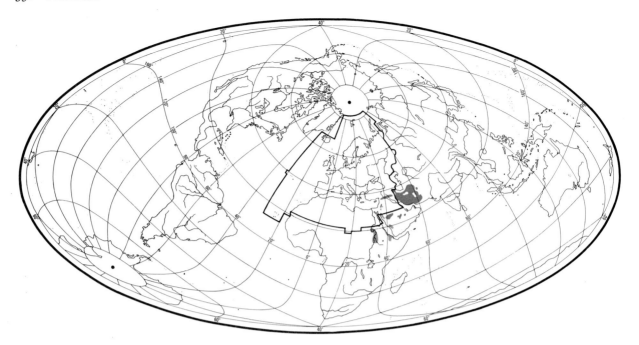

and bounding than *P. petronia*, allowing flocks to swirl like finches; in silhouette, length of wings catches eye more than shortness of tail. Usually runs quite fast on exposed legs, suggesting (with sleek form) lark (Alaudidae) or even pipit *Anthus* (V V Ivanitski); also hops, and shows agility in climbing stems and perching.

Song distinctive: a whistling, then buzzing and persistent 'tss tss tss-tseeeeeeeei' or 'reeze' suggesting wheeze of ♂ Greenfinch *Carduelis chloris*, also Corn Bunting *Miliaria calandra* and cicada. Flight-call suggests distant Bee-eater *Merops apiaster*: soft 'pluip' or soft trill or churr. Also 'twee-ou'.

**Habitat.** In west Palearctic, breeds in low middle latitudes in warm Mediterranean and subtropical zones, largely in arid areas, and often on hillsides and mountains, in Armeniya predominantly at 700–2300 m. Also frequents slopes of barren ravines, overgrown with grass and stunted pistachio *Pistacia* bushes, or areas with scattered bushes and trees which occasionally form sparse thickets or groves. During hot part of day shelters in shade of cliffs, or resorts to water such as springs or streams. During cooler hours descends to ground in search of food, or perches on grass and shrubs such as wild almond, or on walls of abandoned clay dwellings. (Dementiev and Gladkov 1954.) Said by Dementiev and Gladkov (1954) to roost in gullies or rock crevices, which often also serve as nest-sites (see also Social Pattern and Behaviour and Breeding). In Israel, found breeding in June on Mt Hermon at 1100–1800 m on well-drained rocky slopes covered with ground vegetation and dotted with shrubs up to 1–3 m high; nested in shrubs, while neighbouring Rock Sparrow *Petronia petronia* nested in narrow fissures in rock walls (H Mendelssohn). In winter, occurs in more open and cultivated areas further south, especially in cereal crops (Harrison 1982). Winters in Eritrea (Ethiopia) on rocky hills, gravel plains, short grass, and cultivation at less than 300 m (Smith 1957). On migration in Israel occurs on cultivated fields and dry stony and grassy plains, as well as among low scrub and *Acacia* trees in mountains, and in deserts after rainfall (Shirihai in press).

**Distribution.** TURKEY. Apparently a nomadic breeder, occurring within area shown on map; may be abundant in particular localities in one year and absent the next (RPM). LEBANON. Recorded breeding only in Anti-Lebanon range; in flocks in Lebanon range in autumn (Vere Benson 1970). ISRAEL. Breeds Mt Hermon area only (Shirihai in press).

Accidental. Kuwait.

**Population.** ISRAEL. Usually not more than 10 pairs, sometimes 20–30 pairs or none (Shirihai in press).

**Movements.** Short-distance migrant, moving south or west of south to winter in Arabia and north-east Africa. Movements erratic, with great fluctuations in numbers; timing also variable, and dates recorded in scattered observations may not be typical. Passage recorded west to 30°50′E in Egypt, east to Oman, and south on north-east coast of Africa to Djibouti (in spring) (Gallagher and Woodcock 1980; Goodman and Meininger 1989; *Orn. Soc. Middle East Bull.* 1990, **24**, 38). Migration mainly nocturnal, also diurnal (Shirihai in press).

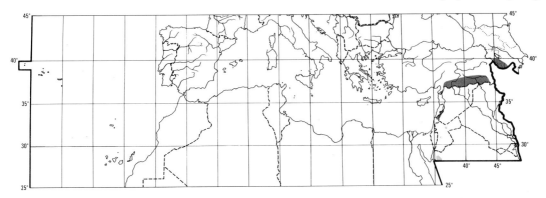

In Saudi Arabia, winters mainly in western highlands, but numbers vary from thousands to few; in Oman and central Arabia, occurs chiefly as passage migrant, with occasional overwintering (Jennings 1981a; Stagg 1985, 1987; Oman Bird List), and most records in Arabian Gulf states are in spring (F E Warr). Only one report (3 birds, 27 December 1965) from southern Yemen (F E Warr), and none from northern Yemen (Brooks *et al.* 1987). Common to abundant in winter in coastal areas of Eritrea (Smith 1955a, 1960); also recorded inland south to 11°N in Danakil desert of Ethiopia, with up to 150 together, and possibly widespread and numerous there (J S Ash). Locally common in Sudan, mostly on coast and in Blue and White Nile region; recorded south to *c.* 12°N, west to 31°E (Nikolaus 1987).

In autumn, leaves breeding grounds July–September. Recorded in Iran to 3rd week of September (D A Scott); Israeli population (breeding Mt Hermon) leaves earlier, mid-July to mid-August (Shirihai in press). Passage in Oman late August to October (*Oman Bird List*), and in Riyadh (central Arabia) August–September in extremely variable numbers (Stagg 1987). Wintering birds arrive south-west Saudi Arabia mid-October (Stagg 1985). In Israel, irregular in very small numbers July–September at Elat, exceptional elsewhere; up to 10 sighted per year (sometimes none), usually in cultivated fields or wadis in mountains; occurs mostly in 2 waves, 17 July to 5 August (probably involving Mt Hermon population) and 28 August to 15 September (presumably birds from further north, e.g. Turkey) (Shirihai in press). In Egypt, where rare and irregular, recent autumn records are late October (Goodman and Meininger 1989). Recorded in Sudan from September (Nikolaus 1987), in Eritrea only from December over several years, with no marked spring or autumn passage; in Danakil desert, 1976–7, noted 29 December to 19 February (Smith 1957, 1960; Ash 1980; J S Ash).

Spring movement begins February and is prolonged; following dates illustrate erratic pattern, especially in Arabia where occasionally breeds. Recorded in Sudan to February (Nikolaus 1987), in Eritrea to early March, though once 9 July, *c.* 100 km inland (Smith 1957; Ash 1980). First Djibouti records were in 1989; *c.* 300 25 February

to 4 March (*Orn. Soc. Middle East Bull.* 1990, **24**, 38). Birds wintering south-west Saudi Arabia leave mid-February (Stagg 1985), and passage reported March–June at Riyadh (Jennings 1986a; Stagg 1987). In 1934 'hundreds or thousands' 9 April on sandy plain inland from Jiddah (western Arabia) were presumed to be about to leave for breeding grounds, though in 1936 still plentiful 31 May further south-east at Bishah (Bates 1936, 1937). In 1986, at breeding site previously unrecorded (perhaps opportunistic following exceptional rain) at *c.* 26°N 39°E in central Arabia, young already hatched in 3rd week of March (Jennings 1986a). Flocks reported late February to mid-April in Arabian Gulf states, with a few records in early to mid-May (F E Warr); in March 1988, exceptional flock of *c.* 300 in United Arab Emirates (Richardson 1989). In Oman, passage February to early April when some may nest (*Oman Bird List*). In Israel, passage from mid-February to end of May (chiefly March and beginning of April), mostly at Elat and along eastern valleys, and almost wholly absent from western Israel; numbers vary greatly, with none or very few in most springs, but sometimes tens or hundreds; in peak spring of 1988 (compare United Arab Emirates, above), at least 1000–2000 between 13 March and 5 April; in influx years, first arrivals at Elat typically by early March, numbers building up and flocks remaining for several weeks then leaving suddenly; present to early or mid-April, occasionally mid-May (Shirihai in press). Common at Azraq (eastern Jordan), where recorded to early May (once July) (Wallace 1982b), and apparently regular along Jabal Hamrin in northern Iraq, especially April (Sage 1960). Returns to Mt Hermon breeding grounds in 2nd half of May (sometimes end of April or beginning of May) (Shirihai in press); 10 April was earliest record in Turkey 1982–6 (Martins 1989). Recorded in Iran from last days of March (D A Scott), and in Turkmeniya from 19 April (Rustamov 1958). Reaches breeding grounds at Yerevan (Armeniya) in 2nd half of May (Adamyan 1965).                          DFV

**Food.** Mainly insects in breeding season, otherwise probably mostly plant material, especially seeds. Forages on ground and on low plants in sandy or rocky places, vineyards, etc.; observation in Jordan of feeding on camel droppings. In winter quarters, feeds in large flocks of up to many thousands in fields of millet *Panicum* or *Sorghum*; these cereal seeds taken on ground, or by hopping up to pull seed from head of grain, or by hanging onto seed head, weighing it down to ground; removes grain completely from one head then moves on to another. Often observed drinking. (Ticehurst *et al.* 1921–2; Bates 1936; Dementiev and Gladkov 1954; Meinertzhagen 1954; Mountfort 1965; Paz 1987; Martins 1989; Shirihai in press.) On breeding grounds, catches insects in flight, sometimes high in the air, in manner of Starling *Sturnus vulgaris*, or picks them from plants while hovering briefly (Adamyan 1965; E N Panov).

In Armeniya, adults recorded in May eating leaves, flowers, and buds of thistle *Cirsium*, sainfoin *Onobrychis*, *Kochia* (Chenopodiaceae), and *Amblyopogon*; 1 stomach at end of June contained 2 cicadas (Cicadidae, Hemiptera) (Adamyan 1965). In north-east Africa, seeds of grasses *Poa*, *Cyperus*, and *Eleusine*, as well as beetles (Coleoptera) recorded in diet (Heuglin 1869–74).

Nestling collar-samples from Armeniya contained 102 insects, *c.* 70% by number cicadas (2–3 species), *c.* 30% larval Lepidoptera, plus a few grasshoppers or crickets (Orthoptera). For first few days after hatching, young fed almost exclusively caterpillars, size of larvae increasing as nestlings grow (one 4-day-old nestling was fed caterpillar 4 cm long), then from about day 5 until fledging diet changes to small cicadas. No plant material recorded in nestling diet in Armeniya, young taking plant food first when independent. (Adamyan 1965.)          BH

**Social pattern and behaviour.** Only detailed study is by Adamyan (1965) in Armeniya (Transcaucasia), and account includes further information from Nakhichevan' (Transcaucasia) supplied by V V Ivanitski and E N Panov; useful general review by Jennings (1986a) for Saudi Arabia.

1. Little information on dispersion in winter, but flocks of *c.* 10 and *c.* 150 recorded Ethiopia 18–19 February (J S Ash); numerous reports of flocks at other times outside breeding season. Flocks typically formed for feeding and watering, and in Israel and Turkmeniya comprise adults and juveniles from late June or early July onwards (Rustamov 1958; Paz 1987; Shirihai in press). In Armeniya, birds still in separate family parties 19 July (Adamyan 1965), but flock of *c.* 250 in eastern Iraq 11 July (Sage 1960), and of *c.* 300 migrants in Saudi Arabia in August (Jennings 1986a); see also Stagg (1987). In Egypt, recent records include late-October flocks of 6–7 and 35 (Goodman and Meininger 1989). In Middle East during spring migration period (February or March to May), in small groups or flocks numbering 50 or up to several hundred or (e.g. when feeding on millet *Panicum*) several thousand (Ticehurst *et al.* 1921–2; Bates 1936; Meinertzhagen 1954; Baumgart and Stephan 1987; Bundy *et al.* 1989; *Orn. Soc. Middle East Bull.* 1990, 24, 38). In Israel during spring influxes, initially solitary or in small flocks which then increase in size and may remain in desert areas for extended period before moving to breeding grounds; exceptionally, flocks of up to 320 birds (Shirihai in press). BONDS. Unlike in Rock Sparrow *Petronia petronia*, monogamous mating system apparently the norm (Ivanitski 1986). Report from Israel of 2 nests

(both with young) very close together in same shrub (H Mendelssohn); perhaps suggests occurrence of mating system other than monogamy, but not clear how many adults in attendance (only one came to feed nestlings at a time), and no information on relationships of birds involved in these breeding attempts (see also Breeding Dispersion, below). Pair-formation evidently takes place on breeding grounds or before arrival there: in Israel, some birds break away from pre-breeding flocks as pairs, ♂♂ singing and performing courtship, but not copulating (Shirihai in press); in Armeniya, first birds to arrive (♂♂ apparently a few days ahead of ♀♀) solitary, later ones already paired. No further information on duration of pair-bond. Only ♀ incubates, but young fed by both sexes, though ♀'s share said to be greater (not clear how sexes distinguished), especially once no longer brooding. Young reported to fledge at 10–12 or 14–16 days (see Relations within Family Group, below); not clear when they become independent, but still fed by parents when able to fly at *c.* 15–18(–25) days. (Adamyan 1965.) In Israel, captive, hand-reared birds began to take food for themselves at 27 days old, suggesting quite long period of dependence (H Mendelssohn). Age of first breeding not known. BREEDING DISPERSION. Solitary and strictly territorial (V V Ivanitski), though territories sometimes more clustered and small neighbourhood groups reported by Dementiev and Gladkov (1954), Rustamov (1958), and Movsesyan and Ayrumyan (1987). On Mt Hermon (Israel), typically in loose groups with (50–)100–150 m between nests, and relatively close to water (Shirihai in press; H Mendelssohn); exceptionally, 2 nests *c.* 20 cm apart in same shrub (H Mendelssohn). In central Saudi Arabia, where common in many places and sometimes even most numerous species present, singing ♂♂ *c.* 50–100 m apart over wide area (Jennings 1986a). Little concrete information on territory size, though extremely large according to E N Panov; in Nakhichevan', however, unpaired ♂ highly mobile within area of only *c.* 0·2 ha (V V Ivanitski). Territory established upon arrival and defended against other conspecific birds, defence intensifying after hatching; nest-material and food gathered within territory, sexes foraging separately and rarely travelling more than *c.* 100–150 m from nest (Adamyan 1965). At Jebel Selma (Saudi Arabia), 30-min census showed 9 ♂♂ along 700 m of wadi bed and plain edge and, 130 km west of Hail, 9 ♂♂ along *c.* 1200 m of open plain; during later visit (19 March) 6 ♂♂ along 800 m of wadi, and eventually 3(–4) nests along 300 m (Jennings 1986a). ROOSTING. Nocturnal; solitary or communal. In Israel, birds rest, and apparently also roost, in shrubs; report from USSR by Dementiev and Gladkov (1954) of roosting in crevices in rocks or earth banks seems questionable (H Mendelssohn); see also discussion of nest-sites in Breeding. Birds forage intensively before coming to roost at dusk; rest in shade of rocks, in gorges, or by water during hottest period of day (Dementiev and Gladkov 1954). In Israel, flocks regularly come to drink morning and afternoon (Shirihai in press).

2. Study of behaviour has shown nothing in common with other Palearctic Passeridae (certainly not with *P. petronia*) and *P. brachydactyla* is apparently a finch (Fringillidae) (E N Panov); see Geographical Variation, also Summers-Smith (1988). Inconspicuous and quiet, spending most of time on ground (Adamyan 1965). In Eritrea, March, large migrant flock feeding on bare fields did not allow close approach (Zedlitz 1911). FLOCK BEHAVIOUR. In Oman, calls (see 2 in Voice) given by typically low-flying flocks; Subsong (see 1b in Voice) noted from flock at roost mid-March (Gallagher 1989a). In United Arab Emirates, mid-August, flock of 12 in thorn tree comprised several families, with some adults flying out and returning with food for their young (Richardson 1989). SONG-DISPLAY. ♂ sings (see 1a in Voice) from ground, bush, or other low perch, sometimes higher

A

C

up on cliff, less often in flight; for upright posture adopted by singing ♂, see Fig A (Dementiev and Gladkov 1954; Moore and Boswell 1957; Kozlova 1975; V V Ivanitski). In Nakhichevan', unpaired ♂ highly mobile in territory, flying from one song-post to another in display-flight (apparently not singing while airborne), characteristic feature of which is that wings slightly retracted and vibrated (much faster wing-beats than normal) or held rigid, bird then gliding, always on straight trajectory (V V Ivanitski). In Armeniya, song noted from 06.00 hrs, ♂ occasionally changing song-post, but not moving far from nest; wanes by evening, ceasing around 20.00 hrs (Adamyan 1965); ♂ sings even at hottest time of day in southern Turkmeniya (Dementiev and Gladkov 1954). In Israel, ♂♂ breaking away from flocks and pairing up before arrival on breeding grounds apparently sing during March and early April (Shirihai in press); song similarly noted from migrants at Azraq (Jordan), April–May (D I M Wallace). Song-period in Arabia covers at least February–May (Gallagher and Woodcock 1980; Jennings 1986a; Gallagher 1989a); intensive singing in southern Turkmeniya in last third of April and 1st half of May (Rustamov 1958). ♂ and ♀ frequently perform well-coordinated antiphonal duet, ♀ responding to ♂'s song with call 3 (or 2: see Voice) (V V Ivanitski, E N Panov). ANTAGONISTIC BEHAVIOUR. In Oman mid-April, many birds seen singing and holding territories, also 2 birds fighting, apparently at start of nesting (Gallagher 1989a). For antagonism between pair-members, see below. No further information. HETEROSEXUAL BEHAVIOUR. (1) General. Near Yerevan (Armeniya), pairs formed within c. 6–8 days of arrival of ♀♀, and pair-formation concluded by end of May (Adamyan 1965). Further south, in Nakhichevan', 2 pairs noted 15 May and ♀ of one pair had started nest-building on following day when pair seen to copulate and unpaired ♂ also present (V V Ivanitski). (2) Pair-bonding behaviour. No detailed study of pair-formation, but apparently characterized by marked antagonism between sexes, this persisting at same level once pair-bond established. ♂ keeps careful watch on ♀'s movements, but tends to avoid coming closer than c. 15–20 m and spends most time singing from perch. Chases, almost always ending in brief fight, take place every c. 15–20 min, ♂ suddenly leaving song-post and flying towards ♀, causing her to take off. (V V Ivanitski.) Chases sometimes initiated by ♀ approaching singing ♂ (Adamyan 1965). (3) Courtship-feeding. Not recorded. (4) Mating. Sometimes follows chase as described above

(Adamyan 1965), or ♀ adopts soliciting-posture when approached by ♂: crouches, with closed tail raised vertically (Fig B, left), keeping it thus during copulation. ♂ then flies to ♀, mounts (only once) and copulates (lasts several seconds). ♂ aggressive also at this point, attempting to peck ♀'s head; also adopts threat-postures (Fig C, left and right) and chases ♀. (V V Ivanitski, E N Panov.) After copulation, ♂ may return to song-post and resume singing. Copulation recorded mid-morning and around midday during nest-building. (Adamyan 1965.) (5) Nest-site selection. Before and after chase and fight, ♀ seen flying from shrub to shrub, apparently searching for nest-site, while ♂ returned (singing) to song-post (V V Ivanitski). Only ♀ builds nest, sometimes accompanied by ♂ when collecting material (Adamyan 1965). ♀ typically vocal and not at all secretive when nest-building; ♂ and ♀ continue to keep their distance, but well-coordinated vocal contact established between them, ♂ singing, ♀ giving call 3 (perhaps also 2: see Voice) from nest and while approaching or leaving it (V V Ivanitski, E N Panov). (6) Behaviour at nest. Laying begins within a day of nest completion. Incubating ♀ gives contact-calls (see 3 in Voice) in response to ♂'s song (as during nest-building, impression is of duet) before leaving nest to feed, etc., sometimes together with ♂. (Adamyan 1965.) RELATIONS WITHIN FAMILY GROUP. Hatching interval generally 2–4 hrs except for last chick which emerges within 24 hrs of penultimate. Eggshells (apart from last egg) removed by ♀ who begins feeding chicks within c. 5–8 hrs of hatching; ♂ and ♀ brood them for c. 5–15 min after each feed, at least during first 4–5 days. ♀ generally silent when visiting nest, ♂ tending to sing; as during incubation, ♀ will call on hearing ♂'s song and then leave. Rapid side-to-side head-turning (probably much as in, e.g., Bullfinch *Pyrrhula pyrrhula*: M G Wilson) is characteristic of small nestlings when food-begging, movement becoming slower with age. Food-begging chick also reported to lift head quickly and as quickly to sink back down; head and neck tremble while gaping (H Mendelssohn). Eyes of young more than half-open by day 7. In Israel, 5 broods fledged at 14–16 days (Shirihai in press), but another report from Israel of young fledging at 11–12 days (H Mendelssohn). In Armeniyan study, young left nest at 10–11 days: moved first onto rim then jumped to ground, calling and moving off to hide in grass or among rocks; able to fly c. 30–40 m at c. 15–18 days. (Adamyan 1965.) ANTI-PREDATOR RESPONSES OF YOUNG. Try to leave (age not stated) if handled (Adamyan 1965). PARENTAL ANTI-PREDATOR STRATEGIES. ♀ sits fairly tightly, sometimes leaving furtively only after direct interference by man; generally more excited when man approaches nest from c. 5–6 days after hatching. In apparent distraction-lure display of disablement type, ♀ runs for a few metres with wings and tail open, then flies up and glides low over ground before landing and hiding. (Adamyan 1965.)

(Figs by R Gillmor from drawings by E N Panov.) MGW

B

**Voice.** Little or nothing in repertoire analogous to vocalizations of Rock Sparrow *Petronia petronia*, and main calls are unlike those typical of Passeridae (E N Panov). Limited information on use of calls outside breeding season (especially in winter), or on which calls are given by which sex. Following scheme provisional.

CALLS OF ADULTS. (1a) Song of ♂. Unlike sparrow *Passer* or *P. petronia*, and more often likened to finch (Fringillidae), bunting (Emberizidae), or insect (Kozlova 1975; E N Panov). Distinctive and persistent, sibilant or wheezing sound strongly recalling wheeze of Greenfinch *Carduelis chloris*; also reminiscent of 'chee-eese' ending of song of Yellowhammer *Emberiza citrinella* and (superficially) like song of Corn Bunting *Miliaria calandra* (Jennings 1986a; Hollom *et al.* 1988). Somewhat grating in quality and monotonous according to Jennings (1986a), though some performances show variation from changes in pitch, duration of units, and rate of delivery (Dementiev and Gladkov 1954). In Transcaucasia, wheezing song given many times in succession, with pauses of 5–20 s between songs (E N Panov). Further descriptions: short or sometimes more drawn-out, muted whistle followed by vibrant sound reminiscent of stridulation of cicada (Cicadidae) (though lower pitched and less piercing), large orthopteran, or *Carduelis* finch (Dementiev and Gladkov 1954; Moore and Boswell 1957; Johnson 1958; Heinzel *et al.* 1972; Gallagher and Woodcock 1980; Paz 1987). Recording contains sibilant, ascending, glissando wheeze (whispered, extended 'z' sound) indeed resembling cicada or faint distant wheeze of *C. chloris* (see that species). Frequency modulation very rapid, persisting throughout wheeze; in recording, *c.* 480 per s; (Fig I upper; Fig I lower shows same call at half-speed playback; both are wide-band sonagrams). Pitch may undulate (as apparent in Fig I and in narrow-band sonagram, Fig II, of another call from same recording), rise throughout wheeze, or rise, fall slightly, and then rise to end. Wheeze starts slowly and gently and (like aerial-predator alarm of many passerines) exceedingly quietly; ends with slight increased ascent in pitch and short diminuendo (Fig II). Amplitude trace (upper part of Fig III) shows gradual crescendo and relatively rapid diminuendo. (J Hall-Craggs.) Sonagram supplied by E N Panov of song given by bird in Azerbaydzhan for over 1 hr comprises rattle of 8 vertical units given simultaneously with 1st half of wheeze, thus splitting it up, roughly 'zt-zt-zt-zt-zt-zt-zt-zt-szweeeeeeeei' (J Hall-Craggs); rendering by Hollom *et al.* (1988), 'tss tss tss-tseeeeeeeei', is presumably also of this variant. Reports of wheeze like *C. chloris* given as call (Witherby 1903; Dementiev and Gladkov 1954) must refer to song. (1b) Subsong. Noted (but not described) for flock at roost in Oman, mid-March (Gallagher 1989a). (2) Loud and rich 'ryurr-ryurr', 'ryuryur ryuryuv' (or 'virvi virvi') given by captive bird originating from Transcaucasia (E N Panov); this (or call 3) is perhaps call reported by Adamyan (1965) to be given by ♀ as contact-call with ♂ just prior to leaving nest for break from incubation. Probably varies somewhat in loudness and timbre as presumably same call is also described as soft trill, purr, or churring sound strongly recalling distant Bee-eater *Merops apiaster* and given in flight (Gallagher and Woodcock 1980; Hollom *et al.* 1988). Full liquid 'pluip' given by flying birds in Jordan also suggests *M. apiaster* (D I M Wallace) and is presumably therefore the same or a related call. (3) Loud, clear, res-

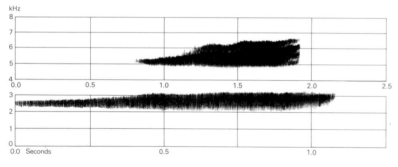

I  P A D Hollom  Iran  April 1971

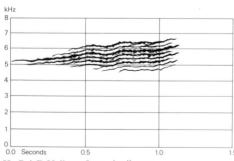

II  P A D Hollom  Iran  April 1971

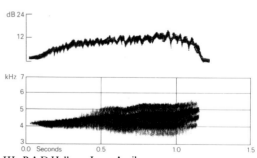

III  P A D Hollom  Iran  April 1971

onant tremolo reminiscent of excited calls of Rock Nuthatch *Sitta neumayer* and Eastern Rock Nuthatch *S. tephronota*; usually of *c.* 10 units, of regular temporal pattern, and with same tonal quality throughout. Given frequently by ♀ when nest-building; not noted from ♂. In response to ♂'s song, nest-building ♀ gives this call (perhaps also call 2), thus creating duet, which apparently continues during incubation and after hatching (see Adamyan 1965). (V V Ivanitski, E N Panov.) (4) Sharp 'twee-ou' (Moore and Boswell 1957; Hollom *et al.* 1988); sound with quality of *Carduelis* finch (Heinzel *et al.* 1972) is presumably the same, as is perhaps also single plaintive note apparently expressing anxiety reported by Jennings (1986*a*); see also Calls of Young (below) for call of ♀ which may similarly belong here. (5) Other calls. (5a) Various chirping sounds recalling *Passer* given (e.g.) while dust-bathing (Tristram 1868; Johnson 1958). (5b) Low twittering sounds (Bates 1936) or constant 'trilling' (Baumgart and Stephan 1987) heard from flocks in Saudi Arabia and Syria; perhaps not distinct from call 5a, and both 5a and 5b may be only alternative descriptions of calls described above.

CALLS OF YOUNG. In Armeniya, cheeping or squeaking food-calls noted from *c.* 4 days old and said by Adamyan (1965) to resemble whistling call of ♀; not clear to which adult call this refers (perhaps call 4).                    MGW

Breeding. SEASON. Armeniya: nest-building early May to early June, eggs laid June, replacement clutches July; fledged young seen mid-June (Grote 1934*a*; Dementiev and Gladkov 1954; Adamyan 1965). Israel: end of May to beginning of July (Shirihai in press). Arabia: nests with full clutch and young recorded mid-March, north-west Saudi Arabia (Jennings 1986*a*); fledglings found one year in August, United Arab Emirates (Richardson 1990); one nest (later abandoned) being built late February, Oman (Walker 1981*b*); see also Gallagher (1989*a*) for probable nesting late April. SITE. In low bush or tree, often thorny, 0·4–1·5 m (0·05–2) above ground, but sometimes touching ground; often near water (Adamyan 1965; Jennings 1986*a*; Paz 1987; Shirihai in press; H Mendelssohn). Old reports of nests in crevices in rocks and buildings (Dementiev and Gladkov 1954; Adamyan 1965) require confirmation since very different from most observations. In Oman, nest built in fork of tree but abandoned before completion (Walker 1981*b*). In Armeniya, mainly in shrub *Atraphaxis spinosa* (Adamyan 1965); in Israel, in hawthorn *Crataegus* and buckthorn *Rhamnus* (H Mendelssohn); in Saudi Arabia, nests recorded in seedling *Acacia*, and in herbs *Echium* and *Forskalia* (Jennings 1986*a*). Nest: open, bulky, untidy hemisphere of generally thorny twigs (e.g. *Acacia*), stalks, roots, leaves, and grass, lined with smooth, felt-like mixture of plant down (especially from thistles e.g. *Cirsium, Echinops, Onopordum*), flowerheads (e.g. *Gnaphalium, Artemesia, Atraphaxis, Alyssum*), grass, soft leaves, bulb scales, and animal hair (Grote 1934*a*; Adamyan 1965; Jen-

nings 1986*a*; Paz 1987; Shirihai in press; H Mendelssohn; E N Panov). Generally resembles nest of Scarlet Rosefinch *Carpodacus erythrinus* or Trumpeter Finch *Bucanetes githagineus* (Grote 1934*a*), though some are small and well-constructed (Paz 1987); rim of nest often contains long twigs (up to 30 cm) and stalks projecting in all directions (H Mendelssohn). 14 nests, Armeniya, had outer diameter 10·1–15·5 cm, inner diameter 5·9–8·1 cm, overall height 6·9–10·0 cm, and depth of cup 4·6–5·8 cm (Adamyan 1965); depth of cup can be as little as 2 cm (Shirihai in press). Building: by ♀ only, sometimes accompanied by ♂ when collecting material; takes 4–5 days (Adamyan 1965, which see for details of technique). EGGS. See Plate 59. Sub-elliptical, smooth and glossy; white (rarely washed pink), with scattered black to reddish-brown spots or commas concentrated at broad end, and greyish undermarkings (Grote 1934*a*; Harrison 1975; Jennings 1986*a*). 21·0 × 14·6 mm (19·0–23·2 × 13·2–15·7), $n = 13$; calculated weight 2·35 g (Schönwetter 1984). 46 eggs from Armeniya had dimensions 17·8–22·5 × 14·1–16·5 mm, weight 1·4–2·9 g (Adamyan 1965). Clutch: 4–5 (3–6); eggs laid daily in early morning, often started within 1 day of nest completion. 1 brood. (Grote 1934*a*; Adamyan 1965; Paz 1987; Shirihai in press.) Replacement of never more than 4 laid in Armeniya within *c.* 10 days of loss of partially-incubated 1st clutch (Adamyan 1965). INCUBATION. 14–16 days, $n = 5$ (Shirihai in press); 13–14 days, by ♀ only, from penultimate egg; stints on eggs 30–45 min, breaks 5–25 min; in adverse weather will sit for up to 2 hrs (Adamyan 1965). YOUNG. Fed and brooded by both sexes (Adamyan 1965). FLEDGING TO MATURITY. Fledging period 10–11 days, though young leave nest before able to fly (Adamyan 1965); 11–12 days (H Mendelssohn), but, also in Israel, Shirihai (in press) gives 14–16 days in 5 nests; young apparently more or less independent at *c.* 20–27 days (Adamyan 1965; H Mendelssohn). BREEDING SUCCESS. In Armeniya, 12 of 14 nests robbed by fox *Vulpes*, reptiles and birds, and only 2 broods fledged (Adamyan 1965). No further information.        BH

Plumages. ADULT. Upperparts dark brown-grey, darkest on forehead, crown, lower mantle, and scapulars, slightly paler and greyer on hindneck, upper mantle, and rump; outer webs of shorter upper tail-coverts bordered white, tail-base showing some white at sides. Supercilium pale isabelline (if plumage fresh) or white (when worn), extending from above eye to over ear-coverts, rather indistinct when plumage heavily worn. Side of head and neck brown-grey, slightly darker and browner on ear-coverts, which may show some rufous on fringes; closely mottled isabelline on lore and cheek, mottling on cheek showing as faint moustachial stripe. Eye-ring indistinct, mottled brown-grey and white, broken by darker brown-grey in front and behind. Faint brown malar streak. Chin, throat, and chest pale isabelline-grey or off-white; side of breast darker brown-grey, remainder of underparts isabelline-white or dirty white, under tail-coverts with broad white tips and (mainly concealed) pale brown centres. Tail black-brown; broad but ill-defined pale brown-grey fringes along both webs of central pair of feathers (t1) and along outer webs of t2-t5; fringe on outer web of t6 inconspicuous, isabelline-white; outer webs of t4-t5 and tips of t2-t6 with narrow white outer edge when plumage fresh; tip of inner webs of (t2-)t3-t6 shows large triangular white spot, $c$. 3 mm long on t3, 8-12 mm on t6, spot on t2 (if any) isabelline-pink. Flight-feathers, tertials, and greater upper primary coverts brown-black, narrowly edged pale greyish-isabelline to off-white along outer web and tip, forming pale panel on secondaries in closed wing. Lesser upper wing-coverts dark grey-brown, longest with narrow and ill-defined pale brown-grey tip; median coverts black-brown with rather contrasting isabelline-white tip (2-3 mm wide); greater coverts dark grey-brown with ill-defined greyish-isabelline-white tip 2-4 mm wide; tips of median and greater coverts form 2 narrow wing-bars, each tinged off-white when plumage slightly worn, but completely absent when heavily abraded. Under wing-coverts and axillaries uniform pale brown-grey. NESTLING. Crown, back, rump, upperwing, and rear of flank closely covered with soft dense down 7-10 mm long, tinged lemon-yellow shortly after hatching, turning to straw-yellow and then to pale straw or white after a few days (Adamyan 1965; H Mendelssohn). For development, see Adamyan (1965). JUVENILE. Like adult, but general colour of upperparts less saturated grey, more sandy-isabelline-grey; underparts dirty white, side of breast, chest, and flank pale isabelline-grey. Pale supercilium indistinct or almost absent; lore whitish. Tail browner than in adult; feathers rather pointed, tips less rounded than in adult. Flight-feathers, tertials, and greater upper primary coverts browner; pale fringes of secondaries warm pink-buff, those of primary coverts and primaries whitish, sharply defined, those of primaries restricted to tip. Lesser and median upper wing-coverts sandy-grey, like upperparts; greater coverts grey-brown, paler than in adult, $c$. 2 mm of tips contrasting isabelline. FIRST ADULT. Like adult, but unknown proportion of birds retain part of juvenile plumage, especially greater upper wing-coverts, these markedly older than neighbouring greater coverts or primaries, fringes along outer webs contrastingly white; some birds retain variable number of juvenile flight-feathers.

**Bare parts.** ADULT, FIRST ADULT. Iris brown or dark brown. Bill pale flesh-horn to dark horn-brown, lower mandible light horn-brown with flesh-pink tinge. Leg and foot light flesh-brown to bright brownish-orange. (Sharpe 1888; Hartert 1903-10; Hollom *et al.* 1988.) NESTLING. At hatching, mouth yellow; gape-flanges narrow, inconspicuous (H Mendelssohn). At 4-7 days, bare skin including bill, leg, and foot pink, bill faintly tinged blue, claws tinged yellow. JUVENILE. Bill pink-flesh with dark grey culmen and tip, leg and foot greyish flesh-pink. (Adamyan 1965.)

**Moults.** ADULT POST-BREEDING. Complete; primaries descendent. Probably starts on breeding grounds, suspended during migration, and completed in winter quarters: in all 29 migrants examined August-September, coastal Sudan, primary moult suspended (inner primaries new, outer old) (Nikolaus and Pearson 1991). Resumed in winter quarters; moult score 45 on 10 January, and 8 others collected 31 December to 16 January had moult completed (BMNH). POST-JUVENILE. Said to be complete (H Mendelssohn), but not so in the few birds examined. Moult starts soon after fledging, but apparently occasionally interrupted during autumn migration: in one undated bird from Ethiopia, just arrived in winter quarters judging from combination of fresh 1st adult and fairly fresh juvenile plumage, moult was suspended with outer 7 primaries, secondaries, tertials, primary coverts, t3-t6, many upper wing-coverts, and many feathers of head and neck still juvenile. Moult resumed in winter quarters;

primary moult scores of 1st-winter birds in central Sudan 45 (1 December), 33, 36, and 41 (10 January), primary moult completed (score 50) in singles from 28 December and 16 January (BMNH). Of 31 1-year-olds, 21 retained only juvenile greater upper primary coverts, 4 retained p1 and all or most primary coverts, 6 retained p1-p2 and primary coverts; occasionally, some outer secondaries (s1-s3) retained (BMNH, ZFMK). In USSR, 4 probable juveniles were moulting body or had moult of body just completed 5-14 July (Dementiev and Gladkov 1954).

**Measurements.** ADULT, FIRST ADULT. Whole geographical range, all year; skins (BMNH, RMNH, ZFMK, ZMB). Wing of 1st adult refers to wing with new outer primaries. Bill (S) to skull, bill (N) to distal corner of nostril; exposed culmen on average $c$. 3·7 less than bill (S).

| | | | | |
|---|---|---|---|---|
| WING AD | ♂ 98·5 (1·71; 13) | 96-102 | ♀ 93·5 (1·58; 13) | 91-96 |
| 1ST AD | 96·3 (1·84; 16) | 93-99 | 92·7 (2·28; 14) | 89-96 |
| TAIL | 50·3 (1·84; 16) | 48-53 | 49·4 (1·68; 18) | 46-53 |
| BILL (S) | 15·4 (0·55; 16) | 14·4-16·5 | 15·0 (0·51; 17) | 14·4-16·9 |
| BILL (N) | 8·8 (0·42; 16) | 8·1-9·5 | 8·6 (0·47; 17) | 7·9-9·4 |
| TARSUS | 19·3 (0·35; 16) | 18·6-19·9 | 19·0 (0·55; 17) | 18·0-19·7 |

Sex differences significant for wing.

Wing and tail of adult ♂, May-June: (1) western Iran, (2) Syria and Lebanon (Vaurie 1949a, which see for other measurements).

| | | | | |
|---|---|---|---|---|
| WING | (1) 96·4 (25) | 90-102 | (2) 96·0 (15) | 93-98·5 |
| TAIL | 46·8 (25) | 42-52 | 48·3 (15) | 46-52 |

Turkey, Armenia, and Iran, wing: ♂ 96·1 (2·51; 5) 93-99, ♀ 91·2 (2·22; 4) 89-94 (Vaurie 1949a; Nicht 1961; Kumerloeve 1970b). Wing, ♂ 95-97, ♀ 90-92 (Hartert 1921-2).
JUVENILE. Wing and tail length about similar to adult, not shorter (unlike *Petronia*).

**Weights.** Ceylanpinar (south-east Turkey), May: ♂ 25 (Kumerloeve 1970b). Armenia, June: ♂♂ 21·4 (3) 21·3-21·5 (Nicht 1961). Israel: 21-25 (Paz 1987).

**Structure.** Wing rather long, broad at base, but narrow at wrist, tip pointed. 10 primaries: p8-p9 longest or either one 0-2 shorter than other; p7 3-8 shorter than longest, p6 10-14, p5 14-19, p4 20-24, p1 32-36. P10 reduced, narrow and pointed, no marked difference in length between adult and juvenile; 60-67 shorter than p8-p9, 8-14 shorter than longest upper primary covert. No emarginations or notches on primaries. Outer secondaries shorter than inner ones; tip of longest tertial reaches to tip of p1-p3. Tail rather short, tip square; 12 feathers, t1 and t6 both 0-3 shorter than t3-t4. Bill short, $c$. 7-8 mm deep and $c$. 7·5-8·5 mm broad at base, conical; culmen and cutting edges decurved; faint tooth in cutting edge on upper mandible, corresponding with faint notch in edge of lower; base of culmen flattened. Tarsus and toes short, strong. Middle toe with claw 17·4 (10) 16·3-18·5; outer and inner toe with claw both $c$. 63% of middle with claw, inner $c$. 66%, hind $c$. 78%.

**Geographical variation.** None. Birds from south-east Iran sometimes separated as *psammochroa* Reichenow, 1916, said to be larger and paler (Reichenow 1916); type specimen (examined, ZMB) is indeed large (wing 98), but similarly large birds occur elsewhere (Vaurie 1949a; BMNH), and pale colour is due to it being a recently fledged juvenile (C S Roselaar).
Often included in *Petronia*, showing tail-spots similar to Rock Sparrow *P. petronia* and general pale colour of (e.g.) Yellow-throated Sparrow *P. xanthocollis* (Vaurie 1949a). However, no yellow on chest, and bill not black in breeding ♂ (Hartert 1903-

10); nest different in location and structure (more like *Fringilla*); eggs quite different, showing sharp spots like *Fringilla*; nestling covered with dense down (almost naked in *Petronia*); voice and behaviour different (H Mendelssohn, E N Panov). Morphology of jaw, tongue apparatus, and skull indicate it to be an early offshoot of finches (Fringillidae), sparrows (Passeridae), or even buntings (Emberizidae) (W J Bock, E N Panov, R L Zusi), and therefore removed from *Petronia* into monotypic genus *Carpospiza* (following Hartert 1903-10, contra Voous 1977) pending further research. CSR

## *Petronia xanthocollis* Yellow-throated Sparrow

PLATES 24 and 27 (flight)
[between pages 280 and 281]

Du. Indische Rotsmus    Fr. Moineau à gorge jaune    Ge. Gelbkehlsperling
Ru. Индийский каменный воробей    Sp. Gorrión pintado    Sw. Gulstrupig stensparv

*Fringilla xanthocollis* Burton, 1838. Synonyms: *Gymnoris flavicollis*, *Petronia xanthosterna*.

Polytypic. *P. x. transfuga* Hartert, 1904, Turkey, eastern Arabia, and from Iraq east to Sind (Pakistan). Extralimital: nominate *xanthocollis* (Burton 1838), east from eastern Afghanistan and northern Pakistan (north and east of *transfuga*).

**Field characters.** 12·5-13 cm; wing-span 23-27 cm. About 10% smaller than Pale Rock Sparrow *Carpospiza brachydactyla* with proportionately longer, finer bill. Rather small, quite slim sparrow, with noticeably long bill. ♂ most colourful west Palearctic member of rather nondescript genus: yellow patch on throat, chestnut patch on forewing, white tips to median coverts, and dark bill. ♀ and juvenile lack yellow and chestnut. Little seasonal variation. Juvenile separable at close range.

ADULT MALE. Moults: July-August (complete). Head, upperparts, and chest rather clean brown- or buff-grey, with black bill (in breeding season), dark grey loral streak, dark eye, faintly paler, more ashy supercilium; underparts almost white, with discrete lemon-yellow spot on lower throat and upper breast (more extensive than in other west Palearctic *Petronia*). Wings boldly patterned, with (a) chestnut lesser coverts, (b) blackish, white-tipped median coverts, creating bold wing-bar (together with chestnut lesser coverts suggesting not dissimilar mark of Brambling *Fringilla montifringilla*), and (c) off-white-tipped greater coverts, forming 2nd thinner wing-bar; in fresh plumage, lacks pale secondary panel of *C. brachydactyla*. Underwing pale grey. Tail dusky-brown, lacking white feather-tips of *P. petronia* and *C. brachydactyla*. Worn bird shows more distinct supercilium, deeper yellow on throat, and brighter wing marks (sometimes including pale panel on secondaries); with extreme wear, may lose wing-bars completely. Bill pinkish-horn to pale brown after breeding season. Legs grey when breeding, otherwise brown or even flesh. ADULT FEMALE. Lacks most of ♂'s distinctive features. Chief character off-white median covert wing-bar, suggesting pattern of ♀ Chaffinch *Fringilla coelebs*. Also shows grey lesser wing-coverts fringed dull rufous,

and inconspicuous pale yellow throat-spot. Bill as non-breeding ♂. JUVENILE. Duller than adult. Lacks face marks and visible throat-spot and has much paler wings with only dull grey or isabelline tips to all coverts, not forming obvious bars. Bill pale flesh; legs pale grey.

♂ unmistakable in good view, but ♀ and juvenile liable to confusion with *C. brachydactyla* and other drab seed-eaters (see *C. brachydactyla*). Within west Palearctic *Petronia* and *Carpospiza*, combination of uniformly coloured tail, pale median covert wing-bar, and dull legs is diagnostic. Also differs in habitat (not normally found in broken and rocky uplands) and voice (see below). Flight even lighter than *C. brachydactyla*, with dipping undulations recalling small pipit *Anthus*. Gait and perching behaviour recall finch as much as *Petronia*.

Song essentially chirruping, like House Sparrow *Passer domesticus*, but softer, sweeter, more rhythmic, and faster: 'chip-chip-chock'. Calls also suggest *P. domesticus* but again softer, more liquid, and tuneful: 'cheep', 'cheelp', 'chilp', or 'chirrup'.

**Habitat.** In west Palearctic, breeds in subtropical lower middle latitudes, mainly in lowlands; in Himalayan foot-hills, however, summers up to 750 m and at other seasons occurs locally up to *c.* 1200 m. In contrast to Rock Sparrow *P. petronia*, favours open dry forest or forest scrub, oases, groves, gardens, and cultivated areas with scattered trees, shrubs, or hedgerows (Harrison 1982). Around Delhi (India), however, avoids gardens and dwellings, preferring thin indigenous woodlands (Hutson 1954). In India, generally inhabits open dry-deciduous forest and thorn jungle, groves, hedges, and trees near villages and cultivation, roosting in thorn thickets and shrubbery and

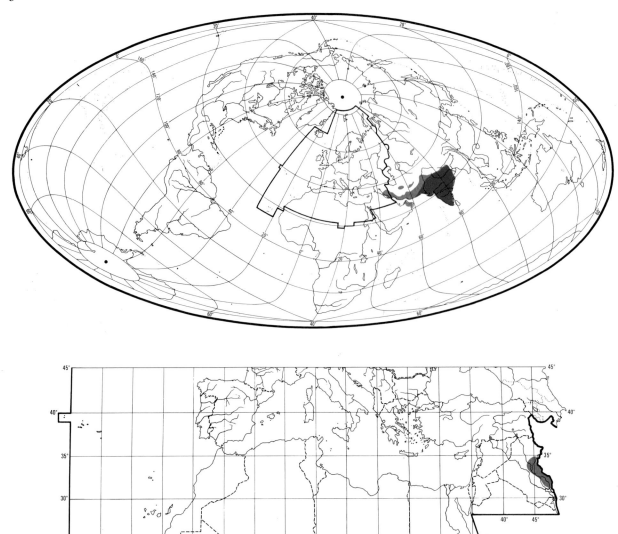

sheltering from midday heat in flocks in leafy trees; will nest in buildings and artefacts as well as in holes in trees (Ali and Ripley 1974).

**Distribution and population.** No information on changes.

Accidental. Israel, Kuwait.

**Movements.** Migratory to resident. Known to winter only in southern Pakistan and India, with occasional winter records from Oman.

Summer visitor to Turkey, Middle East, Iran, northern Pakistan, and some parts of northern India. In Nepal, resident but makes local movements (Inskipp and Inskipp 1985). Common resident and marked local migrant in India, with movements apparently related to rainfall; in some central parts (e.g. Gwalior, Betul), disappears in rainy (post-breeding) season, but common in winter,

appearing irregularly September–November; makes local movements in some southern parts (e.g. Kerala, Deccan plateau); elsewhere in south (e.g. Tamil Nadu, Kodagu), apparently mostly a winter visitor (Ali and Ripley 1974). Only 1 record from Sri Lanka: 'considerable flock' on west coast, October 1876 (Phillips 1978).

Autumn migration begins early. Leaves southern Iraq August–September (Ticehurst *et al.* 1921-2). On Batinah coast of northern Oman, north-west movement of 86 birds in small groups recorded 4 August 1977, with flocks of *c.* 100 and 200 on 25 August 1977 (Walker 1981*a*); on 31 August 1990, flock of *c.* 300 moving east along coast (Eriksen 1990). Some remain in northern Oman until October, however, and passage on Masirah island off east coast 9 September to 6 November (Gallagher and Woodcock 1980; Rogers 1988). In United Arab Emirates, regular until 2nd third of September (Richardson 1990). In Pakistan, not seen at Khanewal *c.* 30°N after end of August over 25 years (Roberts in press), and in Jhang district latest record 10 September in 2 years (Whistler 1922*a*). In southern Pakistan (central and coastal Makran, and lower Sind), resident in small numbers all year (Ticehurst 1926-7; Roberts in press). Present on New Delhi ridge (northern India) until end of September, occasionally October (Gaston 1978).

Spring migration March–April. Breeding population returns to New Delhi ridge by early March (Gaston 1978). In Pakistan, birds apparently travel north via Indus valley; small flocks noted mid-March all along southern Kirthar foothills from Karachi to Karchat (Roberts in press). Reaches central and northern Pakistan late March or early April (Roberts in press), and Iraq in April (Ticehurst *et al.* 1921-2). Regular in Oman and United Arab Emirates from last third of March (Walker 1981*a*; Richardson 1990), and passage 21 March to 17 April on Masirah (Rogers 1988). Present in south-east Turkey by early May (Martins 1989). Vagrant recorded at Elat (extreme south of Israel) 11–12 May 1982 (Paz 1987).     DFV

**Food.** Virtually all information from Pakistan and India. In breeding season, diet insects and plant material, at other times principally plant material, mainly seeds. In breeding season forages more often in trees than sparrows *Passer*; in remainder of year on ground, commonly in large flocks with *Passer*, finches (Fringillidae), and buntings (Emberizidae), under trees or in fields of millet *Panicum* and *Pennisetum* or *Sorghum* taking milky seeds on cereal heads, or on ground in stubble. (Ticehurst 1922; Whistler 1922*a*, 1941; Ganguli 1975; Roberts 1992.) Noted for feeding at flowers on nectar and pollen in summer (for which thin bill well adapted), and at times birds so yellow-stained on head and breast that field identification can be a problem (Ticehurst 1922; Hutson 1954; Ali and Ripley 1974; Ganguli 1975; Roberts 1992). Hawks for insects at dusk, or chases them from trees like flycatcher (Muscicapidae) (MacDonald 1960; Ganguli 1975); sometimes hangs upside-down in trees when foraging (Hutson 1954; Ganguli 1975).    Plant diet includes seeds, nectar, pollen, and berries of mulberry *Morus*, fig *Ficus*, caper *Capparis*, *Salmalia*, coral tree *Erythrina*, *Butea*, *Adhatoda*, *Bassia*, *Panicum*, *Pennisetum*, *Sorghum*, maize *Zea*, oats *Avena*, and other grasses Gramineae (Mason and Maxwell-Lefroy 1912; Ticehurst 1922; Ali and Ripley 1974; Ganguli 1975; Roberts 1992). Of 12 stomachs from India, February–November, 7 contained only insects and 4 only plant material; of 17 insects, 15 were beetles (Coleoptera: 14 Curculionidae, 1 Scarabaeidae), 1 larval Lepidoptera (Geometridae), and 1 ant (Hymenoptera: Formicidae); plant material included pieces of fig, grains of maize and oat, and seeds of herbs and grasses (Mason and Maxwell-Lefroy 1912). 1 stomach in spring in Iraq full of beetles (Ticehurst *et al.* 1921-2).

At nest in Iraq, almost-fledged young were fed large grasshoppers (Orthoptera) (Marchant 1963*b*).     BH

**Social pattern and behaviour.** Information scanty and mostly extralimital.

1. Usually in small, loose-knit flocks, sometimes larger gatherings (for feeding or roosting) outside breeding season, including on migration. On Indian subcontinent, flocks formed after breeding (from May or June) can be of 100 or more; break up in February–March (Baker 1926; Hutson 1954; Ali and Ripley 1974; Ganguli 1975); in Pakistan, usually in pairs or small flocks after arrival, and overall less gregarious than other Passeridae (Roberts 1992). In Oman, small parties recorded during migration, though flocks of 100–300 late August (Walker 1981*a*; Rogers 1988; Eriksen 1990). Sometimes forms mixed feeding and roosting flocks with *Passer* sparrows, finches (Fringillidae), and buntings (Emberizidae) (Whistler 1922*a*; Ali and Ripley 1974; Richardson 1990; Roberts 1992). BONDS. No detailed investigations, but nothing to suggest mating system other than monogamous. In Simla (northern India), pair-formation apparently after arrival on breeding grounds (Jones 1947-8); no further information on pair-bond. Incubation by ♀, but both sexes care for young (Ali and Ripley 1974). Age of first breeding not known. BREEDING DISPERSION. Territorial and often loosely colonial (Ali and Ripley 1974); a dozen or more nests recorded close together in one huge tree (Oates 1890), so defended area presumably small. ROOSTING. Communal and nocturnal. In Delhi (India), flocks seen assembling in 2 small babool trees (Hutson 1954), but not clear if they roosted there. Recorded roosting together with House Sparrow *Passer domesticus* and Black-headed Bunting *Emberiza melanocephala* in thorn thickets and shrubbery (Ali and Ripley 1974). See also Flock Behaviour (below).

2. Shyer than *P. domesticus* (Dharmakumarsinhji 1955). FLOCK BEHAVIOUR. During heat of day, flocks congregate in leafy trees, calling (or singing: see Voice) for hours (Ali and Ripley 1974). SONG-DISPLAY. ♂ sings (see 1 in Voice) persistently during breeding season, often more or less throughout day. Sings from perch near nest when ♀ incubating. (Whistler 1941; Ali and Ripley 1974; Roberts 1992.) In Iraq, song noted late April, also (when tending young) 3 June (Ticehurst *et al.* 1921-2; Sage 1960); in Delhi area, March–May and October (Hutson 1954; Ganguli 1975). ANTAGONISTIC BEHAVIOUR. No information. HETEROSEXUAL BEHAVIOUR. No descriptions of courtship, pair-formation, or copulation. Only ♀ builds nest according to Hutson (1954) and Ali and Ripley (1974), ♂ accompanying her and singing; both sexes build according to Gill (1923) and Ganguli (1975). RELATIONS WITHIN FAMILY GROUP, ANTI-PREDATOR RESPONSES OF YOUNG. No information. PARENTAL ANTI-PREDATOR STRATEGIES. Unlike *P. domesticus* in being wary, not entering nest-hole while man nearby (Dharmakumarsinhji 1955).     MGW

**Voice.** No detailed information on size of repertoire or its use through the year. Song and other calls similar to *Passer* sparrows.

CALLS OF ADULTS. (1) Song of ♂. Series (often long) of not-unattractive chirping sounds (monotonous to some ears), 'chilp-chalp', 'chip-chip-chock', or 'chip' and 'chil-up'; generally likened to *Passer*, but softer, more melodious, higher pitched, and slightly more liquid than House Sparrow *P. domesticus*; also longer, with more definite temporal pattern, faster, and with abrupt end (Witherby 1903; Dharmakumarsinhji 1955; Moore and Boswell 1957; Ali and Ripley 1974; Ganguli 1975; Hollom *et al.* 1988; Roberts 1992). Burst of *c.* 90 s recorded in Pakistan includes sequences such as 'chulp-cheep-chillup-chip-chillup-chip-cheep-chup-chillup-chirrhup-chip' (Roberts 1992). Also rendered methodical 'plit plut put plit plit ploop ploop plit' (R P Martins). Recording from Iran contains rich, fruity chirruping ('cheep', 'chirp', 'chreep', also 'chirrup' or 'chirrip' and attractive 'cheeoo') delivered at first haltingly, in groups of 1–4 units; less hesitant phrase (Fig I) comprises 8 different complex units (1st and 6th show some similarity), all descending in pitch and mostly di- or even trisyllabic (5th most obviously); 7th unit perhaps muted by movement prior to take-off. Seemingly endless song in another recording (Fig II) more varied, with broader frequency range (bird leaping between high and low) and faster delivery. Some units quite shrill, some dry and slightly buzzing, others more delicately tonal. Difference from Fig I perhaps due to geographical or individual variation or to context (song given from perch near nest and thus more overtly territorial, with increased urgency or menace from speed of delivery). The only genuine repetition in available recordings is figure comprising low-pitched and descending high-pitched sound at c. 0·6 s and 3·3 s in Fig II; the same figure occurs (after further examples of different units) also in Fig III, but there precedes example of near-repetition in form of 3 short notes at end. (J Hall-Craggs, M G Wilson.) (2) Main call similarly like *Passer*, but softer and more liquid and tuneful: 'cheep', 'chilp', or 'chirrup' (Jones 1947–8; Ali and Ripley 1974; Hollom *et al.* 1988); loud, resonant 'chee-ah', like miniature Chough *Pyrrhocorax pyrrhocorax* but with *Passer* affinity (L Svensson). (3) Recording (Fig IV) of call perhaps given in contact before take-off or in flight suggests 'kutuk' (J Hall-Craggs) or nasal, dry 'chewet' (M G Wilson). (4) Strained, nasal, drawn-out 'cheeeh cheeeh...' in distress during territorial fights or other close contacts (L Svensson). (5) ♀ said by Gill (1923) to respond to ♂'s 'ceaseless chirrupings' (presumably song

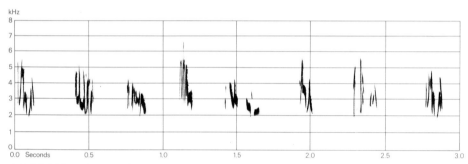

I  P A D Hollom  Iran  April 1972

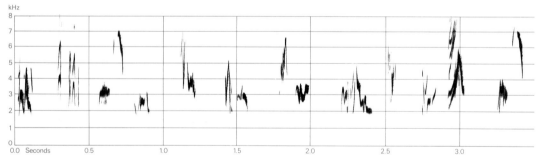

II  J Scharringa  India  February 1986

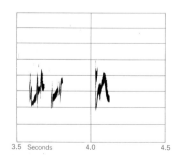

II  *cont.*

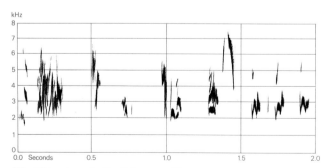

III  J Scharringa  India  February 1986

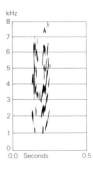

kHz

IV  J Scharringa  India  February 1986

or call 2) with 'slightly modulated' call; not clear whether this differs from call 2.

CALLS OF YOUNG. No information. MGW

**Breeding.** SEASON. Iraq: late April to late July (Tomlinson 1916; Ticehurst *et al.* 1921-2, 1926; Sage 1960; Marchant 1963*b*). Southern Turkey: juvenile being fed by parents recorded early July (Martins 1989). Arabia: April-July, peak May-June (Gallagher and Woodcock 1980; Richardson 1990). For India and Pakistan, see Baker (1934), Ali and Ripley (1974), and Roberts (1992). SITE. Hole or crevice in tree, building, or other man-made structure; may use old hole of woodpecker (Picinae) or barbet (Capitonidae) (Gill 1923), or old, possibly usurped, nest of Red-rumped Swallow *Hirundo daurica* (Ali and Ripley 1974), or of Dead Sea Sparrow *Passer moabiticus* (Iraq: Marchant 1963*b*) and Sind Jungle Sparrow *P. pyrrhonotus* (Pakistan: Roberts 1992), latter sometimes freshly built. In some towns in India and Pakistan commonly inside lamp-post (Ticehurst 1922; Baker 1926). Most often recorded in *Acacia*, tamarisk *Tamarix*, and date palm *Phoenix*, sometimes simply behind bark (Tomlinson 1916; Ticehurst *et al.* 1921-2; Briggs and Osmaston 1928; Richardson 1990; Roberts 1992); also recorded under roofs, inside disused wells, in pipe, and in nest-box (Ali and Ripley 1974; Roberts 1992), and in northern India exceptional nest reported exposed on top of bush like that of House Sparrow *Passer domesticus* (bird obtained for confirmation) (Whitehead 1909). Generally 2-6 m (*c.* 1-10) above ground (Oates 1890; Ticehurst *et al.* 1921-2; Waite 1948; Marchant 1963*b*); up to *c.* 12 nests recorded in large tree (Oates 1890). Nest: shapeless, untidy foundation, often fillng cavity, of dry grass, hair, wool, strips of bark, string, and other man-made material, thickly lined with feathers; cup often only a vague depression in thick pad of material (Oates 1890; Ticehurst *et al.* 1921-2; Gill 1923; Baker 1934). Building: by ♀ only, accompanied by ♂, according to Ali and Ripley (1974), though by both sexes according to Gill (1923) and Ganguli (1975). EGGS. See Plate 59. Sub-elliptical (but great variation from elliptical to pyriform), smooth and very slightly glossy; very variable in colour, even within clutch, ranging from greenish to brownish-white, heavily speckled, blotched, and streaked with all shades of brown, often obscuring ground-colour

(Oates 1890; Gill 1923; Baker 1926; Harrison 1975). *P. x. transfuga*: 18·7 × 13·2 mm (17·4-21·1 × 12·3-14·5), *n* = 50; calculated weight 1·73 g. Nominate *xanthocollis*: 19·0 × 13·9 mm (16·0-21·8 × 12·9-15·0), *n* = 110; calculated weight 1·95 g. (Schönwetter 1984.) For other descriptions and dimensions of *transfuga*, see Ticehurst *et al.* (1921-2), Baker (1934), and Ali and Ripley (1974). Clutch: 3-4 (2-6) (Gill 1923; Waite 1951; Gallagher and Woodcock 1980; Roberts 1992); 3-4 in Iraq (Ticehurst *et al.* 1921-2). In India, replacement clutch laid 7 days after removal of 1st (Oates 1890). 2 broods in Sind (Pakistan), where resident (Roberts 1992), and possibly 2 in central Iraq (Ticehurst *et al.* 1926). INCUBATION. By ♀ only; period unknown (Ali and Ripley 1974). YOUNG. Fed and cared for by both sexes (Ali and Ripley 1974; Ganguli 1975). FLEDGING TO MATURITY. No information. BREEDING SUCCESS. In India, nests robbed by tree pie *Dendrocitta* (Dharmakumarsinhji 1955); no further information.

BH

**Plumages.** (*P. x. transfuga*). ADULT MALE. Entire upperparts uniform brownish drab-grey, sometimes slightly paler and more buffish-drab on rump. Side of head and neck brownish drab-grey, shading to slightly paler drab-grey on cheek and lower side of neck; lore and area round eye sometimes dull grey, a long but faint supercilium more ash-grey but generally inconspicuous. Chin and upper throat greyish-white, sharply divided from drab-grey of cheek. A distinct, sharply defined bright lemon-yellow patch on lower throat, just extending to central upper chest. Side of breast, chest, upper flank, and thigh pale buffish drab-grey, gradually merging into greyish- or creamy-white of remainder of underparts. Tail dark brown-grey, feathers narrowly and inconspicuously edged pale grey or off-white. Flight-feathers, greater upper primary coverts, and bastard wing greyish-black, primaries paler grey-brown towards base; outer webs of primaries faintly edged pale sandy-brown (except on emarginated tips), secondaries more broadly edged pinkish-grey, primary coverts faintly edged brown. Tertials dark brown-grey, blacker on inner webs, broadly but inconspicuously fringed pale drab-grey on outer web. Lesser upper wing-coverts bright rufous-cinnamon, some outer longer coverts drab-grey with cinnamon tips; median coverts greyish-black with contrasting white tips *c.* 3-5 mm long, forming wing-bar; greater coverts brownish drab-grey with indistinct sandy-brown fringes along outer webs and ill-defined greyish- or buffish-white tips forming wing-bar (shorter and less distinct than that on median coverts). Under wing-coverts and axillaries light grey, longer ones partly tipped buff, shorter ones with dark grey bases. In fresh plumage, supercilium, rear of ear-coverts, wing-bars, and fringes of tertials and flight-feathers tinged cinnamon-pink; in worn plumage, ash-grey supercilium sometimes more distinct, throat patch deeper yellow, lesser coverts more chestnut, and wing-bars and fringes of secondaries and tertials whiter (those on secondaries sometimes forming pale panel on closed wing); when heavily worn, just before moult, wing-bars and pale fringes of flight-feathers fully worn off. ADULT FEMALE. As adult ♂, but throat spot ill-defined, paler yellow, and inconspicuous, or just a hint of yellow on whitish lower throat. Lesser upper wing-coverts drab-grey with tawny-cinnamon fringes, less uniform and deep rufous than adult ♂. NESTLING. Naked (Ticehurst 1926). JUVENILE. Like adult, but head paler and less saturated brown-grey; pale supercilium and dark eye-stripe even less distinct or completely absent;

upperparts slightly less saturated, slightly tinged buff when plumage fresh; underparts as adult ♀, but no yellow on lower throat or trace of pale yellow only; flight-feathers greyish-brown, less grey-black; all upper wing-coverts dark grey-brown; lesser coverts without cinnamon or rufous, or tips with a few traces of cinnamon-brown only; median coverts grey-brown instead of blackish, grey-white of tips ill-defined, 2–4 mm wide, forming rather faint wing-bar (absent when plumage worn); greater coverts with ill-defined tips pink-isabelline (if fresh) or off-white (if worn), forming less distinct bar than in adult; tertials with ill-defined dirty white fringe along outer web and tip. FIRST ADULT. Indistinguishable from adult when last juvenile feathers replaced.

**Bare parts.** ADULT, FIRST ADULT. Iris brown. Bill of ♂ (outside breeding season) and ♀ pale horn-brown to dark brown, darkest (bluish-black) on culmen and tip; cutting edges and base and middle portion of lower mandible pinkish-brown or light yellow-horn with flesh tinge; in breeding season, bill of ♂ black, flesh at base of lower mandible (in northern India, birds with black bill appear from January, some not acquiring it until March); bill slowly turns paler during post-breeding moult. Leg and foot brown-flesh, grey-brown, dark purplish-grey, or plumbeous-grey, rear of tarsus and soles flesh-pink; lead-grey when breeding, browner otherwise; claws horn-brown. (Ticehurst 1922; Ali and Ripley 1974; RMNH, ZFMK, ZMA.) NESTLING. No information. JUVENILE. Bill horn-flesh or pale pink; leg and foot pale lead-grey (Ticehurst 1922; Gallagher and Woodcock 1980).

**Moults.** ADULT POST-BREEDING. Complete; primaries descendent. In Iran and Punjab (India), birds heavily worn but not yet moulting May to early July, though in northern Pakistan one heavily worn in January; one ♂ just started moult in July, Punjab, 2 ♀♀ from there in moult August (one, with primary moult score 16, mainly worn but tertials and lesser coverts mainly new; other, with score 34, in full moult of tail and secondaries, wing-coverts, tertials, and body virtually new, head and neck mainly old) (RMNH, ZFMK). No pre-breeding moult. POST-JUVENILE. Complete; primaries descendent. Probably starts soon after fledging, but no moulting bird of known age examined. In western part of range, moult probably July to early September.

**Measurements.** ADULT, FIRST ADULT. *P. x. transfuga.* Iraq and Iran; March–June; summer; skins (BMNH, ZFMK). Bill (S) to skull, bill (N) to nostril; exposed culmen on average 2·7 less than bill (S).

| | ♂ | | ♀ | |
|---|---|---|---|---|
| WING | 84·2 (2·11; 13) | 81–88 | 79·6 (1·39; 6) | 77–82 |
| TAIL | 52·6 (2·43; 13) | 49–56 | 49·0 (2·00; 6) | 47–52 |
| BILL (S) | 15·3 (0·71; 12) | 14·7–16·4 | 15·7 (0·65; 6) | 15·0–16·5 |
| BILL (N) | 9·2 (0·56; 13) | 8·5–10·2 | 9·5 (0·53; 6) | 8·9–10·2 |
| TARSUS | 17·6 (0·42; 13) | 17·0–18·4 | 17·9 (0·37; 6) | 17·5–18·4 |

Sex differences significant for wing and tail.

Nominate *xanthocollis.* Northern Pakistan and India, all year (RMNH, ZFMK, ZMA, ZMB).

| | ♂ | | ♀ | |
|---|---|---|---|---|
| WING | 83·7 (2·23; 11) | 80–87 | 80·0 (3·47; 6) | 75–84 |
| TAIL | 50·2 (2·65; 11) | 46–54 | 47·1 (0·97; 6) | 46–49 |
| BILL (S) | 15·4 (0·55; 11) | 14·6–16·1 | 14·8 (0·68; 5) | 14·0–15·5 |
| BILL (N) | 9·4 (0·51; 11) | 8·7–10·3 | 8·9 (0·62; 5) | 8·1–9·7 |
| TARSUS | 17·1 (0·42; 12) | 16·4–17·8 | 16·6 (0·83; 6) | 15·7–17·5 |

Sex differences

Wing and tail of adult ♂: *transfuga,* (1) western Iran, (2) south-east Iran; nominate *xanthocollis,* (3) northern India, (4) central India, (5) southern India (Vaurie 1949a, which see for other measurements).

| | WING | | | TAIL | | |
|---|---|---|---|---|---|---|
| (1) | 85·1 ( 8) | 83–90 | | 50·5 ( 9) | 47·5–54 | |
| (2) | 84·4 (16) | 80–87·5 | | 51·6 (16) | 47·5–54 | |
| (3) | 82·6 ( 7) | 81–85·5 | | 48·0 ( 7) | 45–51 | |
| (4) | 82·7 ( 6) | 80–85·5 | | 49·0 ( 6) | 47·5–52 | |
| (5) | 84·0 (15) | 81–88 | | 48·2 (15) | 43–51 | |

JUVENILE. Wing on average c. 5 shorter than in adult; tail c. 3 shorter.

**Weights.** *P. x. transfuga.* Southern Afghanistan, late April: ♂♂ 18, 20 (Paludan 1959).

Nominate *xanthocollis.* Southern Nepal and India, February–April: ♂ 17–20 (n = 5), ♀ 18·5 (Diesselhorst 1968; Ali and Ripley 1974). India: ♂ 18 (9) 15–20, ♀ 18 (9) 14–20 (Ali and Ripley 1974).

**Structure.** Wing rather short, broad at base, tip bluntly pointed. 10 primaries: in adult, p8–p9 longest, either one 0–1 shorter than other; p7 0·5–2 shorter than longest, p6 3–8, p5 8–13, p4 13–18, p1 21–29; in juvenile (only 2 examined), other feathers within range of adult. P10 reduced, tiny, hidden below outermost greater upper primary covert; in adult, 49–58 shorter than p8–p9, in juvenile 46–52; in adult, 4–9 shorter than longest upper primary covert, in juvenile 4–7. Outer web of (p6–)p7–p8 emarginated, inner of (p7–)p8–p9 with notch. Tip of longest tertial reaches tip of p1–p2. Tail short, tip square or slightly forked; 12 feathers, t1–t3 1–4 shorter than others. Bill rather long, visible part of culmen c. 50% of head length; depth and width at base both c. 60% of visible culmen length; base of culmen rather flattened; tip of culmen and line of cutting edges slightly decurved; tip sharply pointed. Nostrils small, rounded, directed obliquely up; c. 4 short bristles projecting down at each side of gape. Leg and foot short, slender. Middle toe with claw 15·7 (8) 14–17; outer and inner toe with claw both c. 68% of middle with claw, hind c. 75%.

**Geographical variation.** Slight and clinal. *P. x. transfuga* paler, more brown-grey on upperparts; nominate *xanthocollis* dark drab-grey above, less brown-grey, difference most marked on cap, mantle, and scapulars; rump brown-grey, less buffish-drab; side of head and neck slightly darker and browner; lesser upper wing-coverts chestnut instead of rufous-cinnamon, but much individual variation. Birds from south-west Iran (named *occidentalis* Koelz, 1948) on average perhaps slightly paler and with more slender bill than typical *transfuga* from south-east Iran (Koelz 1948; Vaurie 1949a), but Iraqi birds similar to typical *transfuga* and splitting off further races in Middle East not warranted (Vaurie 1956f). Birds from Afghanistan, Seistan (eastern Iran), Pakistan, Kutch, Rajasthan, and Punjab are intermediate between *transfuga* and nominate *xanthocollis,* those of eastern Afghanistan, northern Pakistan, Punjab, Rajasthan, and Kutch nearer nominate *xanthocollis,* those of Seistan, southern Afghanistan, and southern Pakistan nearer *transfuga* (Vaurie 1949a; Ali and Ripley 1974; BMNH).

Forms superspecies with Yellow-spotted Petronia *P. pyrgita* (Sahel zone from Sénégal to Ethiopia and Somalia, south to northern Tanzania), Bush Petronia *P. dentata* (northern Afrotropics, mainly just south of *P. pyrgita,* from Sénégal to western and northern Ethiopia and south-west Arabia), and Yellow-throated Petronia *P. superciliaris* (southern Africa) (Hall and Moreau 1970); these species, together with *P. xanthocollis,* are rather different from Rock Sparrow *Petronia petronia* in general coloration, structure, and habits, and perhaps better separated from *Petronia* as *Gymnoris. P. pyrgita* sometimes considered a race of *P. xanthocollis,* but differs in larger size, rounder wing-tip

(tip formed by p7–p8, p1 16–20 shorter), relatively longer tail, slightly different position of yellow spot on chest, absence of rufous or chestnut on forewing and of distinct pale wing-bars (wing almost uniform brown-grey), and shorter and thicker bill. *P. xanthocollis* not nearer *P. pyrgita* than to either of the other Afrotropical species.

CSR

## *Petronia petronia* Rock Sparrow

PLATES 25 and 27 (flight)
[between pages 280 and 281]

Du. Rotsmus    Fr. Moineau soulcie    Ge. Steinsperling
Ru. Каменный воробей    Sp. Gorrión chillón    Sw. Stensparv

*Fringilla petronia* Linnaeus, 1766

Polytypic. Nominate *petronia* (Linnaeus, 1766), Canary Islands, Madeira, and Europe (including Mediterranean islands) east to Bulgaria and western Asia Minor; *barbara* Erlanger, 1899, Morocco to Libya; *exigua* (Hellmayr, 1902), Turkey (east from Anatolia), Caucasus area (south from about Stavropol'), northern Iraq, and northern Iran east to Gorgan; *puteicola* Festa, 1894, Levant south from Gâvur Daglari (southern Turkey); *kirhizica* Sushkin, 1925, from lower Volga valley east to Turgay depression and Aral Sea, south in Transcaspia to *c.* 41°N; *intermedia* Hartert, 1901, Iran (south and east of *exigua*) and Transcaspia (south of *kirhizica*) east in west-central Asia to eastern Tien Shan (China) and north-west India. Extralimital: *brevirostris* Taczanowski, 1874, Altai and Mongolia east to western Manchuria, south to northern Szechwan (China).

**Field characters.** 14 cm; wing-span 28–32 cm. Close in size to House Sparrow *Passer domesticus*, but form differs distinctly in 30% longer, deeper, and more conical bill, 25% longer wings, and almost 10% shorter tail. Bulky, long-winged, square-tailed sparrow, with heavy bill on rather large head (apparent size of head due partly to its linear patterning). Differs from other west Palearctic *Petronia* in larger size and strongly striped, streaked, and spotted greyish-brown plumage; diagnostic combination of dark lateral crown stripes and white spots on tail-tip. Calls distinctive. Sexes similar; no seasonal variation. Juvenile separable at close range. 6 races in west Palearctic. European race, nominate *petronia*, described here; eastern forms larger and paler (see Geographical Variation).

ADULT. Moults: July–November (complete). Head well marked with (a) dusky-brown frontal band and wide lateral crown-stripes, isolating pale mottled, tawny to cream centre to crown and nape, (b) obvious broad buffish-white supercilium and lower eye-crescent, (c) dusky ear-coverts, brown enough on upper edge to form dark stripe and further emphasize supercilium, (d) pale off-white sub-moustachial stripe, and (e) paler surround to face, fading into nape. Below bill, chin and upper throat dusky- to buffish-white and lower throat pale yellow, forming fairly distinct spot above breast. Mantle and scapulars dusty-brown, with darker brown-black feather-centres and whitish spotting of feather margins creating both heavily streaked and mottled linear pattern. Back and rump greyish- to buff-brown, dully streaked and spotted dark brown. Wings dark dusky-brown, with (a) almost white tips to black median coverts, forming quite bold upper wing-bar, (b) dusky fringes and almost white tips to terti-als, with brown-black centres creating lines of spots, (c) pale buffish-grey fringes and tips to greater coverts and flight-feathers (but not producing obvious wing-bar or pale wing-panel), and (d) black primary coverts contrasting with pale fringes to primaries. Underwing noticeably pale cream. Tail dark brown, with feathers fringed pale buff and tips conspicuously spotted white; in flight, terminal spots show above as broken line of white patches and below as broad white tip. Underparts below throat dusky to buffish-white, faintly streaked or spotted brown on chest and belly, more markedly on flanks and under tail-coverts; thus looks mottled or pale-spotted from chest to rear belly, dark-streaked or dark-spotted on rear flanks and under tail-coverts. When worn, particularly from February, plumage becomes more contrasted, with pale stripes and spots bleaching almost to white and yellow patch above breast of ♂ noticeably bright, but before moult may lose pale spots and crown-stripe, and underpart streaks may weaken. Large, deep, triangular bill fills face; horn-brown, with paler flesh-brown base to lower mandible. Legs light brownish-yellow to flesh. Eye brown. JUVENILE. Less immaculate than adult, with crown almost wholly dark brown, rear eye-stripe even more distinct, back blotched, and wings and underparts buffier, less patterned. Lacks yellow on throat and has only vestigial tail spots. Bill mainly yellow; legs pink-flesh.

If seen well, quickly separated from other *Petronia* by larger size, proportionately larger bill, strongly striped head, dark-mottled upperparts, and voice (see below). To inexperienced observer, may suggest ♀ or immature of several *Passer*, especially Spanish Sparrow *P. hispaniolensis*, but differs in structure, and obvious white tail-spots of adult are not shared by any *Passer*. Flight much

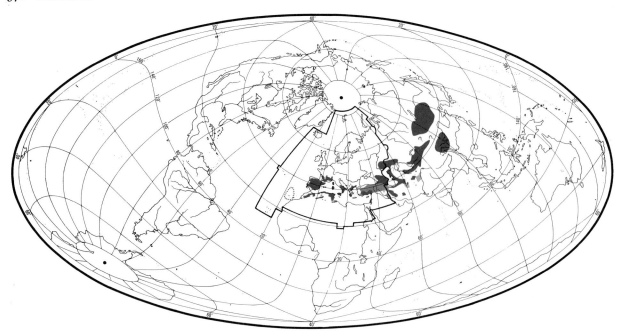

superior to *Passer* in power and speed, with pronounced bursts of fluent wing-beats giving swift, bounding progress and allowing it to make dramatic ascents of cliff faces. Flight silhouette differs from *Passer* in relatively heavier bill and head, much longer outer wings, and shorter, squarer tail, combining to suggest bulky projectile (as does larger Hawfinch *Coccothraustes coccothraustes*). Gait a strong hop; also jumps and shuffles. Generally more active on ground than *Passer*, with alert, rather upright stance often showing thighs above short legs. Restless, often dashing to and from food sources. Always associated with broken ground.

Calls include penetrating, strongly punctuated 'tut' or 'wed', wheezy 'chwee' or 'dliu', and squeaky but sweet 'pey-i' or 'peeuh-ee' recalling Goldfinch *Carduelis carduelis*. Song usually a combination of calls.

**Habitat.** Breeds in middle latitudes of Palearctic in warm temperate Mediterranean, steppe, and desert climates, from sea-level up rocky slopes and hillsides to mountains at 2500 m in Armeniya and to *c.* 4800 m in Himalayas (Dementiev and Gladkov 1954; Voous 1960*b*). Generally frequents rather bare treeless terrain with scanty herbaceous vegetation, ranging from flat desert steppe to rocky slopes or outcrops, screes, stony patches, ravines, cliffs, crags, and clay or earth precipices. In some regions favours less severe environments, such as alpine meadows, grassy or shrubby riversides, vineyards, olive groves, stone walls, ruined castles and other structures on hilltops, and even human settlements, where it may come into competition with House Sparrow *Passer domesticus* or Spanish Sparrow *P. hispaniolensis*. (Harrison 1982; Voous 1960*b*; Dementiev and Gladkov 1954.) Former habitat in south-

ern Germany comprised sunny slopes, crags, and ruins; links with cultivation and some tall vegetation, as well as with various artefacts such as walls, seem to have been common here, but low flat areas were avoided (Niethammer 1937). Largely a ground feeder.

**Distribution.** Range contracted in north in early part of 20th century.

FRANCE. Range contraction in north; disappeared from Alsace and Bourgogne early in 20th century (Yeatman 1976). SPAIN. No certain recent records on Mallorca, and probably extinct there, but still breeds Ibiza and Formentera (J King). GERMANY. Formerly bred in southern Baden, Nassau, Franken, and Thüringen; populations unstable, subject to rapid and wide fluctuations. No certain breeding after 1926, but one pair nest-building in Thüringen 1936. (Niethammer 1937.) AUSTRIA. A few observations from first half of 20th century, mainly from Salzburg Alps near border with Bayern; breeding suspected but no proof. No recent records. (H-MB.) POLAND. Breeding recorded in Sudety mountains, 1897; only an accidental visitor since then (AD, LT). MOROCCO. Known for a long time in Moyen and Haut Atlas; recently discovered in Anti-Atlas and eastern Rif (MT).

Accidental. Britain, Switzerland, Poland, Malta, Cyprus, Azores.

**Population.** FRANCE. 1000–10 000 pairs (Yeatman 1976). GREECE. Probably 1000–10 000 pairs (GIH). ISRAEL. A few thousand pairs (Shirihai in press). JORDAN. At least 80 pairs in Petra area, 1983 (Wittenberg 1987). MADEIRA. Much reduced since appearance of Spanish Sparrow *Passer hispaniolensis* (PAZ, GM).

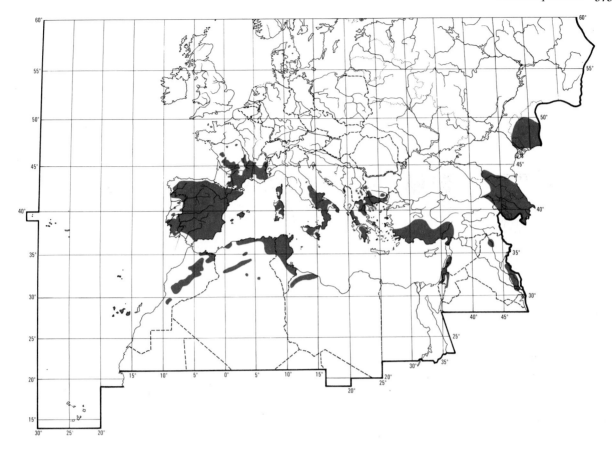

**Movements.** Resident and to some extent dispersive; also altitudinal migrant.

Resident in western Europe (e.g. France, Sicily) and northern Africa, with limited dispersal to cultivated areas in winter (Heim de Balsac and Mayaud 1962; Bundy 1976; Yeatman 1976; Destre 1984; Iapichino and Massa 1989). No seasonal movements reported from Thüringen (eastern Germany), formerly northernmost breeding area in Europe, but now extinct there (Knorre *et al.* 1986). Lack of observations at key migration sites emphasizes sedentary nature: only 1 record on Gibraltar, 7 September 1973 (Cortés *et al.* 1980), and only *c.* 10 on Malta (Sultana and Gauci 1982). One ringing recovery shows marked movement, however: bird ringed Hautes Alpes (south-east France), 9 July 1964, found 150 km south-west in Bouches-du-Rhône, 12 December 1965 (Erard 1968). 9 vagrants reported from Belgium, of which 5 since 1950, chiefly October (Lippens and Wille 1972), 1 from Britain (Norfolk, 14 June 1981: Dymond *et al.* 1989), and 1 from Yugoslavia (Slovenia, 3 May 1987: *Br. Birds* 1988, 81, 338). Resident on Atlantic Islands, though on Madeira makes local visits to outlying islands in winter; 1 record from Azores (Bannerman 1963; Bannerman and Bannerman 1965, 1966).

In Turkey, distribution extends to lower altitudes in winter, birds occurring more widely in west, with fewer records from eastern breeding areas (Beaman 1978). In Syria, some birds apparently move east into desert, e.g. flock reported at Palmyra, February 1981; in very hot weather, flocks of hundreds move to higher altitude to forage (Baumgart and Stephan 1987). In Israel, some populations wholly resident, e.g. eastern Galilee and northern Judean Desert, others are altitudinal migrants: present in Jerusalem hills only from early March to October or mid-November, and in mountainous central Shomron February-November; many birds breeding on Mt Hermon move to lower levels or further afield (apparently eastward), not returning to highest altitudes until April. Winters in large numbers in eastern Israel, south only to north-east Negev; movement beyond Israel is apparently only eastwards towards Jordanian and Syrian mountains. (Shirihai in press.) In Iran, breeding area in extreme north-west apparently entirely vacated in winter; elsewhere, birds disperse locally into adjacent lowlands (D A Scott). Winter visitor to northern Iraq (Allouse 1953). No evidence of longer movements in Middle East; only 2 records (both in November 1982) on Cyprus, and none in Arabia or Egypt (Flint and Stewart 1992; Goodman and Meininger 1989; F E Warr).

Migration more conspicuous (but also primarily altitudinal) further east: marked movement in Ural estuary and eastern shore of Caspian (western Kazakhstan); from south-east Kazakhstan, most birds probably head south to Kirgiziya, Tadzhikistan, and Turkmeniya, with regular passage through Chokpak pass, chiefly in 2nd half of October and 1st half of March (Korelov *et al.* 1974; Gavrilov and Gistsov 1985). In Talasskiy-Alatau mountains (south-east Kazakhstan), numbers diminish from August, though encountered there until mid-October (Kovshar' 1966). In Tadzhikistan, birds probably ascend to sub-alpine and alpine zones after breeding, then descend to valleys after first snowfall, some perhaps moving further; in environs of Samarkand recorded 23 October to 13 March in one year (Sagitov 1962; Abdusalyamov 1977). From Kopet-Dag mountains (Turkmeniya), many descend into northern foothills, sometimes reaching plains (Rustamov 1958). Winters regularly in Gilgit (Kashmir) and northern Pakistan from mid-November to end of March (Ali and Ripley 1974). Probably resident in Afghanistan, with some altitudinal movements, e.g. apparently only a winter visitor to Kandahar (Paludan 1959). Migratory status of Mongolian birds not known; recorded only 7 April to 7 October, but further study required (Piechocki *et al.* 1982; Mey 1988).                              DFV

**Food.** Mostly seeds throughout year, with berries in autumn, and in spring invertebrates, on which young almost exclusively fed, especially caterpillars (Lepidoptera) and grasshoppers, etc. (Orthoptera) (Fenk 1911*a*; Debru 1958; Hollom 1962; Kovshar' 1966; Korelov *et al.* 1974). Forages mainly on ground, running around like pipit *Anthus*, in low herbs and grass, among rocks, in fields of cereal or stubble, etc.; commonly in large flocks, particularly in winter, often with other species, especially finches (Fringillidae) (Debru 1958; Hollom 1962; Korelov *et al.* 1974; Lebreton 1976; Abdusalyamov 1977; Roux 1990). In autumn, in French Alps, often in barberry *Berberis* bushes feeding on fruit (Lebreton 1975, 1976); commonly eats flesh of berries only, and searches dung for seeds (Niethammer 1937). Pursues insects in flight, like Starling *Sturnus vulgaris* (Cullen *et al.* 1952; Lebreton 1976), and recorded hovering at wall to pick off invertebrates (Mayhoff 1911).

Diet in west Palearctic includes the following. Invertebrates: grasshoppers (Orthoptera: Acrididae), adult and larval Lepidoptera, flies (Diptera: Tipulidae), Hymenoptera, beetles (Coleoptera: Chrysomelidae). Plants: seeds, berries, and buds of juniper *Juniperus*, barberry *Berberis*, fig *Ficus*, mulberry *Morus*, cherry *Prunus*, strawberry *Fragaria*, woad *Isatis*, sunflower *Helianthus*, grass and cereals (Gramineae) including maize *Zea*. (Fenk 1911*a*; Mayhoff 1915; Schuster 1930; Niethammer 1937; Cullen *et al.* 1952; Debru 1958; Hollom 1962; Bannerman 1963; Dončev 1981; Roux 1990.)

All detailed studies of diet extralimital, from central Asia. In Turkmeniya, 61 stomachs contained 695 items, of

which 22% (by number) invertebrates, including 11·1% termites (Isoptera), 6·2% beetles (3·7% Carabidae), 1·6% ants (Hymenoptera: Formicidae), 1·0% caterpillars; 78% seeds, including 10·6% gromwell *Lithospermum*, 7·6% viper's bugloss *Echium*, 6·9% rock-rose *Helianthemum*, 5·2% nightshade *Solanum*, 5·0% *Arnebia*, 4·6% milk-vetch, etc., *Astragalus* (Bel'skaya 1974). For 7 stomachs from Uzbekistan, see Salikhbaev and Bogdanov (1967). In Kazakhstan and Tadzhikistan, forages in newly-sown cereals or short grass early in breeding season, taking plant material and invertebrates; in June–July diet almost wholly invertebrates, and in August–September largely seeds and fruits (Kovshar' 1966; Korelov *et al.* 1974; Abdusalyamov 1977). On average, takes larger prey than House Sparrow *P. domesticus* (Lebreton 1976).

In Turkmeniya, mid-April, 158 items from nestling collar-samples were 53% (by number) Lepidoptera (51% caterpillars, 20% hairy), 20% Orthoptera (including Tettigoniidae and Acrididae), 18% beetles (17% adults, including 7% Curculionidae, 1% Tenebrionidae, and 1% larval Scarabaeidae and Elateridae), 5% bugs (Hemiptera), 2% spiders (Araneae), and 1% Diptera; in mid-August, 98% of 80 items (by observation) were Orthoptera, 1% spiders, and 1% millipedes and centipedes (Myriapoda) (Rustamov 1958). Also in Turkmeniya, 263 collar-samples contained 540 items, of which 96% (by number) Orthoptera (95% Acrididae), 3% termites, and remainder caterpillars, beetle larvae, Hemiptera, and cockroaches (Dictyoptera: Blattidae); between days 2 and 11, proportion of larvae in diet fell and adult insects increased (Bel'skaya 1974). Among other items recorded in central Asia were Gryllotalpidae (Orthoptera), Cicadidae (Hemiptera), Mantidae (Dictyoptera), and larval caddis flies (Trichoptera); also berries of hawthorn *Crataegus* and sea buckthorn *Hippophae* (Pek and Fedyanina 1961; Kovshar' 1966; Korelov *et al.* 1974). Young fed by regurgitation and with live insects, of which *c.* 5 can be brought per visit, extremities removed (Mayhoff 1911; Debru 1958; Bel'skaya 1974). Young rarely in nest before June in south-west France, so nestling period coincides with peak abundance of Orthoptera (Debru 1961). Young also fed green buds and 'milky' cereal grains (Fenk 1911*a*).                              BH

**Social pattern and behaviour.** In west Palearctic, studied by Debru (1958, 1961) in southern France and by various authors on former breeding grounds in southern Germany. Major extra-limital study (mainly breeding dispersion and heterosexual behaviour) in southern Turkmeniya by Ivanitski (1985*a*, 1986); account includes additional information from USSR supplied by E N Panov.

1. Typically in compact flocks from late summer to spring (Lindner 1917; E N Panov). Flocks formed for feeding, drinking, and roosting also in breeding season (Kozlova 1930; Salikhbaev and Bogdanov 1967; Bel'skaya 1974). Flocks of up to several hundred widely reported during summer: in Israel, such flocks comprise non-breeders in some years (Shirihai in press; see

Bonds, below); in Syria, June–July, occurs in flocks at higher altitude than normal, evidently to escape heat (Baumgart and Stephan 1987); see also (e.g.) Reiser (1905) for Greece, Lack and Southern (1949) for Canary Islands, Neufeldt (1986) for Altai, and Debru (1958). After fledging, juveniles may form large flocks, later joined by adults (Canary Islands: Bolle 1857), or nomadic family parties may gradually merge into flocks (Kazakhstan: Korelov *et al.* 1974). Autumn flocks reported from (e.g.) Turkmeniya (Rustamov 1958), Mongolia (Mauersberger 1980), and (pre-roost gathering) Israel (Paz 1987). In winter, at least 2000 in coherent flock in Spain (Bernis 1989a); in Syria, hundreds recorded gathering to roost, and birds once reverted to flocking during February cold spell after start of courtship in January (Baumgart and Stephan 1987). Often forms mixed flocks with finches (Fringillidae), buntings *Emberiza*, and other Passeridae, including in Iraq and Altai with Snow Finch *Montifringilla nivalis* (Lindner 1917; Schäfer 1938; Cullen *et al.* 1952; Moore and Boswell 1957; Debru 1958; Smith 1965; Neufeldt 1986; Bernis 1989a). Some reports of large flocks in Germany were perhaps not purely *P. petronia* (see Fenk 1914, also Lindner 1906, 1917). In Mongolia, October, 36 recorded roosting with Choughs *Pyrrhocorax pyrrhocorax* and Tree Sparrows *Passer montanus* (Mey 1988). BONDS. Study in Turkmeniya suggested polygamy typical, though further investigation required: 11 cases of polygyny recorded (♂ paired with 2 ♀♀); figure of 12% polygynous ♂♂ in Badkhyz probably reflects sex-ratio in population and perhaps artificially reduced by study methods (few birds ringed). Almost all well-studied ♂♂ occupied and advertised 2nd potential nest-site after successful pair-formation at 1st; such activity of ♂♂ declined with decrease in number of roving unpaired birds (majority apparently ♀) by mid-April. In 8 of 10 cases of polygyny, 2nd ♀ appeared within a few days of 1st ♀ starting to incubate. While advertising 2nd site, ♂ periodically visits and displays to 1st ♀, but following appearance of 2nd ♀, interacts mainly with her. Pair-bond thus brief, regular contact between ♂ and 1st ♀ lasting only 3–5 days, and increasingly rare thereafter, perhaps assisted by ♀'s aggression towards ♂ which coincides with peak nest-building. ♂ may subsequently attempt to acquire 3rd mate. No aggression recorded between 2 ♀♀ paired with same ♂ regardless of distance between nest-sites, and all 3 birds may associate amicably inside and outside territory. (Ivanitski 1985, 1986.) In Israel in some years, many birds remain in flocks during summer and do not breed, apparently due to water shortage (Shirihai in press). Despite statement in Paz (1987) that incubation mainly by ♀, more likely from mating system (see above) that ♂ takes no part. ♂'s typical role in feeding of young also unclear. Studies in France (Debru 1958), Turkmeniya (Rustamov 1958), and Mongolia (Fiebig and Jander 1987) suggest young fed by both sexes. In southern Germany, all feeding apparently by ♀ (Fenk 1911b); also in one case in Turkmeniya, but ♀ assisted by ♂ Eastern Pied Wheatear *Oenanthe picata* which brought food for nestlings and fledglings, once fed ♀ on nest, and helped to defend territory (Zykova and Panov 1982); ♂ attempting to attract 2nd ♀ when 1st incubating will leave 1st ♀ to rear her young alone (E N Panov); according to Bel'skaya (1974), ♂ helps to feed young, then abandons mate and brood close to fledging in attempt to acquire 2nd mate. One case (in Turkmeniya) of ♂ rearing 3 nestlings alone after ♀ and 3 others killed (Bel'skaya 1974). In Tadzhikistan, young fed for *c.* 7–8 days after fledging (Abdusalyamov 1977). In Transbaykalia (eastern Russia), pair with incomplete clutch occasionally fed fledglings (presumably their own 1st brood) which hung about near nest (Sokolov 1986a). Age of first breeding not known. BREEDING DISPERSION. Solitary or in generally small, loose colonies; rarely, larger colonies spread over wide area (Géroudet

1957). Territorial. In Germany, 2 nests *c.* 60–80 m apart (Salzmann 1909). In Israel, mostly in colonies, with nests *c.* (5–)10–15 m apart (Paz 1987; Shirihai in press). Colonies in Turkmeniya, Kazakhstan, Uzbekistan, and Transbaykalia of up to 20(–30) pairs (Rustamov 1958; Salikhbaev and Bogdanov 1967; Korelov *et al.* 1974; Sokolov 1986a). In (e.g.) large caves with many crevices, nests sometimes only *c.* 10 m apart, but usually much further; hundreds of metres apart when (as often) occupying abandoned nests of Eastern Rock Nuthatch *Sitta tephronota* (E N Panov). Report from Turkmeniya of 5 nests within *c.* 1 m² (Rustamov 1958). On upper Zeravshan river (Tadzhikistan), solitary pairs 200–2000 m apart (Abdusalyamov 1977). Along narrow gorge in Turkmeniya, 5–10 m between song-posts of neighbouring ♂♂; in western Kopet Dag (Turkmeniya), often in colonies of 3–6 ♂♂, each ♂ 8–15 m apart; average distance between neighbours' song-posts in densest colonies $11 \cdot 3 \pm 5 \cdot 2$ m ($n = 15$), sometimes only 2–5 m. Average distance between nests in Turkmeniya (includes only those maximum 40 m apart) $28 \cdot 8 \pm 4 \cdot 4$ m. (Ivanitski 1985, 1986.) Breeding grounds in southern Turkmeniya occupied towards end of February. Territory established and proclaimed by ♂ on arrival, defending core-area of *c.* 2–3 m around potential nest-site; several song-posts usually within *c.* 1–5 m and roost also close by. (Ivanitski 1985, 1986.) In Carcassonne (southern France), core-area said to be 100 cm² (Debru 1961). During early settlement stage, much variation in distance over which ♂ will show aggression. Core-areas of *c.* 30% of ♂♂ in Turkmeniya study sited such that practically no visual or acoustic contact with other ♂♂. After pair-formation, when ♀ nest-building, ♂ shows interest in other ♀♀ inside or outside 1st core-area, and may establish 2nd (in Badkhyz, all ♂♂ do so) average $25 \cdot 6 \pm 8 \cdot 1$ m ($n = 20$) away, though usually returns to 1st core-area to roost; average $4 \cdot 5 \pm 0 \cdot 3$ days ($n = 6$) between start of nest-building and establishment of 2nd core-area. If ♂ attracts 2nd ♀, often then occupies 3rd core-area. ♂ advertises only one core-area at a time, but regularly visits all, and defends large part of newly acquired area; territory size increases markedly once 2nd core-area acquired, and may expand by 100 or more times during season. ♂'s territory contains regular perches (especially song-posts) and sites for roosting and hiding from wind, but most feeding done outside territory. 10 largest territories in Turkmeniya averaged $960 \pm 128$ m²; more or less all this area defended. Boundaries fairly strictly observed, though no true boundary conflicts, with contacts rare even in dense colonies; hence (perhaps) contacts beyond territory limits which are important in early stages of colony establishment (see Flock Behaviour, below). Usually, territory contains all a ♂'s core-areas; less commonly, core-areas separated by neutral ground or by territories of other ♂♂. Average distance between nests of 2 ♀♀ paired with same ♂ in one large territory $26 \cdot 5 \pm 4 \cdot 1$ m; minimum in study by E N Panov *c.* 2·5 m. In 2 cases of polyterritoriality, Turkmeniya, nests of 1st and 2nd ♀ 23 m and 165 m apart. Pair-formation takes place on feeding grounds and in territory; copulation also within territory. (Ivanitski 1985, 1986.) Territory defended by ♂ (E N Panov), though both sexes sometimes involved according to Debru (1961). Recorded using same nest for 1st and 2nd brood, or replacement clutch (Pitman 1921; Debru 1961). Little information on density. In southern Jordan, *c.* 15–20 pairs per km² (Wittenberg 1987). On lower Emba river (Kazakhstan), 0·2–1·3 birds along 10 km (Neruchev and Makarov 1982). In Badkhyz, 800–1000 singing ♂♂ and 1200–1400 nests per km² (Ivanitski 1985); in Kopet Dag, 7 nests along 1 km (Bel'skaya 1974); on steppes of Transbaykalia, 1·5 birds per km² (Sokolov 1986a). Seen successfully competing for nest-sites with *Passer* sparrows (Bundy and Morgan 1969; Piechocki *et al.* 1982) and Starling *Sturnus vulgaris* (Niethammer 1961). Recorded nesting in colony of Bee-eater *Merops apiaster* (e.g. Olioso 1974); see Antagonistic Behaviour (below). In Germany, nest-site used (also as roost) over several years (Fenk 1911*a*); not known whether by same birds. ROOSTING. Solitary or communal, and nocturnal. In hole with narrow entrance in rocks, wall, tree, etc.; sometimes in ivy *Hedera* (Naumann 1900; Géroudet 1957; Debru 1961; Korelov *et al.* 1974; Paz 1987). Communal mainly outside breeding season (but also during pair-formation in March: E N Panov), with several birds in same hole or in holes close together (Naumann 1900). In late summer and autumn (if not cold), flocks may use (e.g.) poplar *Populus* trees until leaves fall, then shift to (e.g.) spruce *Picea* (Naumann 1900). Will also use trees and bushes just to assemble before entering roost-holes (Salzmann 1909; Fenk 1911*a*, 1914). In Turkmeniya, individual ♂♂ used one crevice over long period, usually *c.* 5-8 m from nest-hole, in which ♂ never roosts (Ivanitski 1985, 1986); pair in Bulgaria also did not roost in nest during building (Königstedt *et al.* 1977). In Turkmeniya, early in breeding season, birds tend to come to colony for roosting *c.* 60-90 min before sunset, earlier in good weather, later if cold; first birds return from feeding grounds around 18.00 hrs throughout breeding season. (Ivanitski 1986.) Regularly visits water to drink; in Turkmeniya, throughout day in August (Bel'skaya 1974).

2. Sometimes fairly tame in breeding season, allowing approach to within a few metres; much less so at other times, when single birds and flocks fly off at slightest alarm; in mixed flock, generally the first to fly up (Lindner 1911; Debru 1961). Pair nesting in *M. apiaster* colony did not allow approach within *c.* 70 m (Königstedt *et al.* 1977). Ruffling of crown feathers (more so by ♂) signals excitement, e.g. when man near nest (Fenk 1911*a*; Mayhoff 1911), and simultaneous lowering of head is presumably to display crown-stripe (Baumgart and Stephan 1987). Extended postures to show off yellow throat-patch are typical of *Petronia* in heterosexual or antagonistic display: skin on neck shifts so yellow feathers with black bases are revealed and stretched over protruding part of neck (Abs 1966; Fiebig and Jander 1987). For brief comparison of displays with *Passer* and Père David's Snow Finch *M. davidiana*, see Ivanitski (1986). FLOCK BEHAVIOUR. In Canary Islands, one member of flock spotted observer and took off, others immediately following (Volsøe 1951). In Germany, when *c.* 8 feeding with Yellowhammers *Emberiza citrinella*, each species flushed separately (Lindner 1906, 1917). On feeding grounds in Turkmeniya in spring, ♂♂ often highly excited, adopting various postures and displaying in apparently rather disordered series of interactions. Fly from tree to tree in Quiver-flight (shallow wing-beats, tail fully spread, underpart feathers ruffled); display in Low-hunched posture (Fig A), Head-down posture (Fig B, commonest posture of ♂), less commonly Wings-up posture (Fig C); one bird may hover

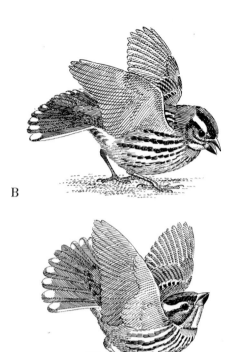

B

C

A

over another (probably ♀) which reacts aggressively by pecking, or may fly up and circle or hover close to first bird, both then separating. When most ♀♀ incubating, displays may erupt simultaneously among flock of 5-7 ♂♂. Early in season, in fine

weather, birds returning to colony to roost typically sing, search for and advertise nest-sites, and fight. Such interactions tend to take place in trees and bushes on slope above colony: sing, chase (♂-♂ and ♂-♀), and adopt Low-hunched and Head-down postures combined with wing-shivering, vibrating of breast feathers, and Quiver-flight; gradually visit territories more frequently and for longer and, as song-intensity declines, increasingly preen, etc., then fly silently in Quiver-flight to roost-hole. Initially, little display following morning emergence from roost, though peak activity (song, etc.) reached by mid-March; even then, all birds fly off to feed around midday. (Ivanitski 1986.) SONG-DISPLAY. ♂ sings or gives long series of Peyee-calls (Fiebig and Jander 1987: see 1-2 in Voice), usually from perch near nest. Sings mainly to advertise nest-hole; intensity of display varies with stage of nesting cycle and proximity of conspecific birds. Before first appearance of ♀, ♂ sings mainly in fairly normal posture (Fig D), tail usually closed and wings only occasionally

D

vibrated at low intensity; yellow throat-patch prominent (E N Panov); bill wide open and throat vibrating (Géroudet 1957). Body typically horizontal, though bird frequently squats or lies down and thus difficult to see (see illustrations in Ivanitski 1986).

More expressive displays occur if single conspecific birds or small flocks fly close by. (Ivanitski 1986; E N Panov.) Sometimes sings in flight, but apparently no ritualized Song-flight (E N Panov), though undulating performance reported from Syria, mid-January (Baumgart and Stephan 1987). In typical sequence, Turkmeniya, ♂ preens for 1–2 min after coming to song-post, sings for 5–8 min, then flies (sometimes in Quiver-flight) to nest-hole; looks inside or enters, and brings out material, tossing it aside or carrying it a few metres in flight; enters hole 3–8 times thus, then returns to perch and sings; flies down periodically to feed for *c.* 2–3 min. In Quiver-flight, ♂ holds Head-down posture (or similar), maintaining this on landing when also hops about and makes bowing movements. Song-bouts generally *c.* 15–30 min (Ivanitski 1986; E N Panov). ♂ frequently leaves territory, sometimes describing arc of radius 100–150 m or circling several times before returning to song-post. Length and frequency of absences increase up to midday and ♂ finally stays on feeding grounds (where may sing and display at any time of day) until evening. In late February in fine weather, much singing prior to roosting; morning singing peaks mid-March when ♂♂ sing intensively up to 14.00–15.00 hrs; by mid-April (♀♀ incubating), ♂♂ sing only up to *c.* 09.00–10.00 hrs before leaving territory. (Ivanitski 1986.) In Saissac (southern France), April, ♂♂ sang for *c.* 1 hr from roofs after dispersing through village on leaving roost (Géroudet 1957). In Kazakhstan, intensive song coincides with start of nest-building. Song from juveniles noted mid-August in Talasskiy Alatau mountains (Kazakhstan). (Kovshar' 1966; Korelov *et al.* 1974.) ANTAGONISTIC BEHAVIOUR. (1) General. Behaviour associated with pair-formation comprises highly vocal chases and fights; much as in House Sparrow *P. domesticus* (Schäfer 1938). (2) Threat and fighting. Threat-posture described from captive birds is similar or identical to Head-down posture (Fig B): plumage ruffled, breast pressed almost to ground, wings raised, and tail raised and fanned, showing white spots; particularly if facing aerial attack, bird will threaten with head pointed up to reveal yellow patch (Fig E),

E

also wing-shiver and give Rattle-calls (see 6 in Voice); if opponent does not yield, pecking attack ensues. If food of captive birds well dispersed, they maintained sleeked plumage and showed no threat. (Abs 1961; Fiebig and Jander 1987.) Aggression of territorial ♂ varies: on return to territory after absence (even of only 10–12 min), generally loath to leave song-post, tolerating conspecific birds if they do not trespass into core-area; if already perched in territory for *c.* 10–15 min (and especially if territory relatively isolated), more likely to attack and give chase, even to conspecific birds flying past several hundred metres away (Ivanitski 1985, 1986). Where ♂♂ advertising sites in dense colony, appearance of unpaired ♀♀ leads to frequent chaotic conflicts (including short fights on ground and in air) involving all pairs in vicinity and newcomers; conflicts also arise on feeding grounds when 2 ♂♂ trying to court same ♀ (Koenig 1888; E N Panov). In Kazakhstan, nest-building adult chased away fledged

juvenile (Korelov *et al.* 1974). Conspecific birds and other nest-competitors such as Swift *Apus apus*, Black Redstart *Phoenicurus ochruros*, and *Passer* are vigorously attacked and chased over several metres, attacker giving Rattle-call and sometimes buffeting with wing (Debru 1961). Rattle-calls also given when pursuing Blackbird *Turdus merula* (Salzmann 1909). In Bulgaria, where took over burrow of *M. apiaster*, that species was also chased (Königstedt *et al.* 1977). In Germany, ♂ defending nest against conspecific intruder gave call 8 and Bill-snapped (see Voice), also (after fight) sang loudly (Salzmann 1911). HETEROSEXUAL BEHAVIOUR. (1) General. In Carcassonne, birds paired up from mid-April; display preceding 2nd brood recorded within *c.* 7–12 days of fledging of 1st brood, though one case of 2nd clutch completed 8 days after 1st brood fledged (Debru 1958, 1961). In Turkmeniya, flocks break up around end of February, but some birds solitary or in pairs in January (Bel'skaya 1974), though pair-formation generally in 2nd half of March and almost all ♀♀ incubating by mid-April (Ivanitski 1986); nest-building in Kazakhstan within *c.* 1–2 days of arrival (Korelov *et al.* 1974). ♂ spends most of breeding season advertising potential nest-sites, trying to attract ♀; pair-formation takes place without interference where nest-sites widely dispersed; simultaneous advertisement of several sites not infrequently leads to polygyny (E N Panov: see Bonds, above). (2) Pair-bonding behaviour and nest-site selection. Early in pair-formation period, ♂ sometimes courts ♀♀ on feeding grounds some distance from colony; 2 ♂♂ may display to same ♀ (E N Panov). Most birds interacting with ♂♂ from mid-March are probably unpaired ♀♀ which are typically highly mobile within colony; hop about mostly in an upright posture with head drawn back or extended and head feathers ruffled; frequently flick out wing-tips, raise tail abruptly, and inspect crevices as if searching for nest-site. Established ♂♂ stimulated by such behaviour are highly active. After pair-formation, ♂ accompanies nest-building ♀, but continues excited, ever ready to court another ♀ within and outside territory, though most such interactions are brief and do not lead to pair-formation. (Ivanitski 1985, 1986.) On feeding grounds, ♂ may land *c.* 1 m from ♀, fan tail and vibrate wings very rapidly, then fly up and hover 15–20 cm above her with tail widely fanned and lowered (E N Panov); see also subsection 4 (below). Other variants of aerial display (also occurring later by nest) include ♂ moving within *c.* 20–30 cm left and right in front of ♀, turning sometimes towards and sometimes away from her, and giving Tsik-call (see 8 in Voice) (E N Panov); see also Salzmann (1911). When ♀ enters territory, ♂ will sing more intensely and adopt Hunched Tail-down posture (Fig F), with breast pressed to

F

ground; also Bill-snaps and may give Tsik-call (Salzmann 1911), makes nodding movements, and shivers wings slightly, this then changing to more intensive flapping (E N Panov). Culminates in most expressive static display (sometimes absent from pair-formation): Wings-up posture (Fig C), held for up to 8 s; lower-intensity variants occur in which body, head, and closed tail horizontal (Ivanitski 1986). ♂ may walk or run (rather than hop) short distance in Head-down posture (Fig B), then fly up and flutter in front of crevice (E N Panov). Often then enters and sings inside for several minutes, this preceded in captive birds by ♀ chasing ♂ (Fiebig and Jander 1987); ♂ hops out if ♀ arrives and flutters back and forth in front of her (E N Panov). Typically emerges from nest-hole in Head-down posture, with markedly ruffled plumage; associated movements include wing-shivering or wing-fluttering, sometimes only of wing nearer ♀ (Ivanitski 1986; Fiebig and Jander 1987). After Nest-showing at one crevice, ♂ creeps in Low-hunched, Head-down, or Wings-up posture (Figs A–C) to another crevice and enters; ♀ looks into or enters 1st, both birds then entering different holes for some minutes, staying at least 2–3 m from each other (♀ presumably eventually chooses hole for nest); ♂ sings and continually displays— bowing, pivoting, sleeking and ruffling head feathers, raising wings, creeping about, or flying in Quiver-flight. Within c. 30–50 min, frequency of Wings-up posture declines, and Head-down posture with wing-shivering predominates. Nest-showing and courtship almost always followed by ♂ attacking ♀, leading to long chase, or sometimes ♀ simply leaves. (Ivanitski 1986.) In French study, pursuits were accompanied by Rattle-calls, and included dives, glides, rapid ascents, and parachute descents (Debru 1958). ♂ always first to return from chase (sometimes in Quiver-flight), and immediately displays in Low-hunched and Head-down postures. Especially in early part of season, ♀ may not return that day, or returns several times and eventually stays (not confirmed by colour-ringing). ♀ may circle and evade ♂'s attacks, also try to enter crevice (often leads to fight), but if persistent enough, gains free access to all of ♂'s territory within 2–3 hrs, and in 3 cases ♀ started nest-building (only sure sign that pair-formation completed) within 2–3 hrs of arrival; each ♀ (and ♂) nevertheless probably interacts with many different ♂♂ (♀♀) before building starts. (Ivanitski 1986.) In Iraq, ♂ seen crouching and wing-shivering with green plant in bill (Moore and Boswell 1957); only ♀ builds, however (E N Panov). ♂ accompanies ♀ if she leaves territory to feed or collect nest-material (up to 20–25 m from nest), and sings from bush in variety of postures (including Low-hunched), also ruffling plumage, fanning tail, wing-shivering, and bowing. If ♀ flies back to territory, ♂ overtakes her and sings near crevice and within 1–3 m of her, in Low-hunched, Head-down, or Wings-up posture; keeps wing-shivering and singing while ♀ enters and cleans out nest-hole; sudden outburst of display (Wings-up, etc.) sometimes followed by attack on ♀. (Ivanitski 1986.) In French study, 2 records of one bird (sex unknown) pecking at upper tail-coverts of other after courtship (Debru 1961). ♂'s advertising displays reach peak just after pair-formation, when ♀ building (E N Panov). During peak nest-building, ♀ aggressive, not allowing ♂ within c. 2–3 m of nest; ♀ will also evade mate and head rapidly for nest (Glayre 1970). ♂ thus increasingly turns attention to 2nd core-area; for details of interactions between ♂ and 2nd ♀, also attempt to attract 3rd ♀, see Ivanitski (1986). (3) Courtship-feeding. In German study, ♀ by nest apparently gave song-like utterance (see 1f in Voice) and moved along branch, calling (not described) and wing-shivering, to meet food-bearing ♂ (Salzmann 1911); ♂ said by Géroudet (1957) also to feed ♀ on nest. No reports in more detailed modern studies. (4) Mating. Quite often follows display as described in subsection 2 (above), and takes place close to nest. ♀ ready for copulation by end of nest-building; tends to sit by nest in a rather upright or more horizontal posture, plumage markedly ruffled; occasionally brings nest-material. At ♂'s approach, ♀ may fly to him and assume Soliciting-posture (Fig G) while wing-shivering; sometimes leads to copulation. ♀ may also call (see 9 in Voice) while soliciting (Lindner 1911). In full sequence of events, ♂ wing-shivers in Hunched Tail-down posture and walks around ♀ who

G

has plumage ruffled; when birds head to head, ♂ flies with tail fanned and lowered and usually hovers over ♀ for up to c. 30 s; then turns 180° to land on her back and copulates; both have wings open and ♂ may beat his and sing excitedly while copulating (Salzmann 1909; Debru 1958). ♂ then flies 2–4 m in Quiver-flight and remounts; or lands, assumes Head-down posture with intensive wing-shivering, and then remounts. Usually mounts 2–3 times in quick succession at intervals of c. 1 hr (E N Panov), average 1·8 ± 0·3 hrs (Ivanitski 1986). In 5 cases, copulation interrupted by ♂ neighbours. (Ivanitski 1986; E N Panov.) ♀ recorded entering nest-hole after copulation, ♂ remaining outside, displaying and calling (Debru 1958), or Bill-snapping long and quietly (Salzmann 1909). (5) Behaviour at nest. ♀ typically silent when approaching (on regular route) and at nest; may call on leaving (Sick 1939). RELATIONS WITHIN FAMILY GROUP. Most food brought by ♂ during first few days when ♀ brooding young (Bel'skaya 1974); see also Bonds (above). Young beg by vibrating wings and gaping (E N Panov); fed initially by regurgitation, later bill-to-bill (Géroudet 1957). In French study, called only when parent arrived. Advanced young come to entrance to be fed, then retreat into hole (Mayhoff 1911; Debru 1958.) Eyes of young fully open at 8 days (Fiebig and Jander 1987, which see for development of comfort behaviour, etc.; see also Bel'skaya 1974 for physical development). Nest kept clean by both adults (Rustamov 1958). Fledging period 16–21 days (see Breeding). ANTI-PREDATOR RESPONSES OF YOUNG. No information. PARENTAL ANTI-PREDATOR STRATEGIES. (1) Passive measures. Adult sits tightly, and can sometimes be seized on nest (Sokolov 1986a). (2) Active measures. Will patter about excitedly, calling and ruffling crown feathers (Piechocki and Bolod 1972). In southern Germany, one of pair called ('djip', 'piw', etc.: see 10e in Voice) and wing-shivered; when perched, also turned side to side and flicked wings and tail (Mayhoff 1911). In Israel, regularly performed distraction-lure display of disablement-type, with 'various noisy antics' (Pitman 1921). Small snakes sometimes attacked when nest threatened (Dimitropoulos 1987; E N Panov).

(Figs by R Gillmor: A–D and F–G from drawings in Ivanitski 1986 and by E N Panov; E from photographs in Abs 1961.)

MGW

Voice. More vocal than *Passer* sparrows (Korelov *et al.* 1974). Repertoire apparently quite large and varied, but many calls poorly known (especially function) and more study needed of which vocalizations may constitute song. Some calls (including song of type 1a) diagnostic, others homologous with and similar to *Passer*. ♂ more vocal than ♀ in breeding season, but sexual differences in use of calls not investigated. Little information on calls given in winter, but at least Peyee-call used then. (Sick 1939; Jonsson 1982; E N Panov.) Instrumental sounds include Bill-snapping given in high-intensity excitement (E N Panov)

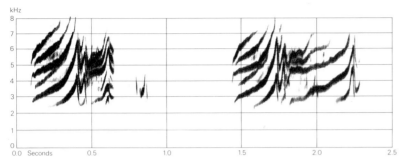

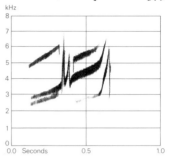

I   C Chappuis   Spain   March 1964

II   R Ranft   Spain   May 1987

and noisy wing- and tail-flicking in courtship (Fiebig and Jander 1987). Apart from pitch differences (see calls 1a and 4) demonstrated by Chappuis (1969, 1976), extent of geographical variation unknown. Account includes descriptions for several races (including nominate *petronia*, *exigua*, *intermedia*, and *brevirostris*). For additional sonagrams, see Abs (1966, which includes comparison with 2 extralimital *Petronia*), Bergmann and Helb (1982), Zykova and Panov (1982), and Fiebig and Jander (1987, which see also for suggestion that high-pitched calls recorded in Mongolia are adaptation to wind noise in open landscape).

CALLS OF ADULTS. (1) Song. Much overlap with Peyee-call (call 2), and many authors make no distinction: e.g. 'vi-viep' or 'viep' as 'attraction call or song' (Jonsson 1982); see also Chappuis (1976) and below. Subdivisions here highly arbitrary. Songs described in 1a–c assumed to be given mainly or exclusively by ♂. (1a) Series of up to *c.* 30-50 units (Lindner 1906), similar to or same as Peyee-call, but often (especially if excited) longer and of notably complex structure often incorporating mixed harmonics relating to 2 fundamentals, implying use of 2 voices (J Hall-Craggs), and giving characteristic shrill, somewhat strained and nasal timbre: 'bäidlid' or 'bäidilid' (Bergmann and Helb 1982). Recording (Fig I) contains 'song-calls' *c.* 1 s long and *c.* 1 s apart. End of unit apparently varies considerably, sometimes sounding abrupt, sometimes more gentle; of up to 5 syllables, extended (as in 2nd unit in Fig I: 'wei-ry-eei-ry-ip') or curtailed (1st unit lacks final syllable). Long calls (perhaps song) in another recording (Fig II) rendered 'eei-ry-eeed', 'd' indicating abrupt termination; bird still employing 2 voices, but harmonics much less complex than in Fig I. (J Hall-Craggs.) Long calls (song-units) from perched bird in Morocco were distinctly lower pitched than those from Spain (as shown in Fig I) and Corsica; difference evident but less pronounced in calls given in flight (Chappuis 1976); see call 4. In study of wild and captive *brevirostris* ♂♂ from Mongolia, units short but otherwise similar to those described above, 1-1·5 s apart, were treated as 'territorial or nest call' (see call 2), while main distinguishing feature of song said to be much faster delivery rate; units vary in length, but all shorter than long complex type shown in Fig I (Fiebig and Jander 1987,

which see for sonagrams). Similar rapid song shown in sonagram of *intermedia* from Turkmeniya (E N Panov). (1b) After death of mate, captive ♂ *brevirostris* gave high-intensity song: rapid, continuous sequence of varied units all above 4 kHz and thus higher pitched than songs described in 1a; similar to varied sequence given by excited ♂ House Sparrow *Passer domesticus* (Fiebig and Jander 1987). In volume, timbre, and pitch, song said to recall Siskin *Carduelis spinus* or sustained twittering of Linnet *C. cannabina* (Lindner 1907; Géroudet 1955, 1957); such descriptions perhaps refer to rapid song of this type; see also Calls of Young (below). (1c) Study in Transcaucasia and Turkmeniya indicated each ♂ has several different song-types (including series of well-spaced long units and fast song-type described in 1a), giving one for a long time before changing. One type, 'tsi-tsi-chirr', like song of Mongolian Trumpeter Finch *Bucanetes mongolicus*; this 'tsiii' or similar typical of excited bird. (Panov 1989; E N Panov; see also Heinroth and Heinroth 1924-6.) (1d) In Syria, song in flight said by Baumgart and Stephan (1987) to be series of 'piah' sounds. (1e) Recordings from France contain great variety of sounds, including quiet song or Subsong (Fig III) in which 'pit-choo' and 'pteruh' form part of lead-in to series of fairly low-pitched husky units (compare 1b) with pronounced but slow frequency modulation (J Hall-Craggs). See also call 1d (below). (1f) Song of ♀. Isolated captive ♀ *brevirostris* from Mongolia gave (from mid-August in one year, in spring of another) quiet song slightly reminiscent of song of Greenfinch *C. chloris* (Fiebig and Jander 1987). Apparent song by ♀ at nest in Germany 'dïdl dïdl-däh dïdl dïdl-däh dïdl dïdl-däh irrrr-irrrr' (Salzmann 1911). (2) Peyee-call. Characteristic, oft-repeated, nasal, rather piercing sound, ascending in pitch, 'pey-i', 'zweh-il', 'bäi', 'bäije', 'süib'; or glissando, falling then rising 'peeyuee' with sweet timbre (L Svensson); given from perch or in flight. Some calls clearly 2-3 syllables, others almost monosyllabic. Given singly or (especially in breeding season) as series, rapidly when excited, and serving to advertise territory or nest. Calls typically louder and longer in breeding season (Schmitt and Stadler 1914; Mayhoff 1915; Lindner 1917; Delamain 1929; Bergmann and Helb 1982; Fiebig and Jander 1987). Considerable

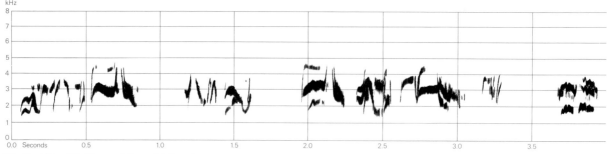

III Roché (1966) France May and July 1965

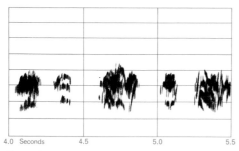

III *cont.*

IV J-C Roché France
May and July 1965

V J-C Roché France
May and July 1965

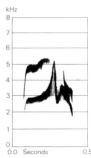

VI R Ranft Spain
May 1987

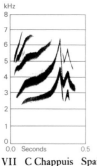

VII C Chappuis Spain
March 1964

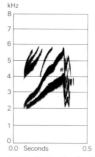

VIII R Ranft Spain
May 1987

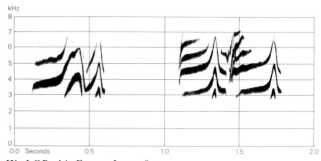

IX J-C Roché France June 1983

variation between individuals (perhaps also according to function), reflected to some extent in transcriptions: 'see-ip', 'schree-ip', 'see-y', plaintive 'wee-eep' or 'weel-eep'; sometimes more sibilant, wheezy 'weez-wee' (Hollom 1959, 1962); 'chwee' (Cullen *et al.* 1952). Often likened to *Carduelis* finch, sometimes to nasal Wayeek-call of Brambling *Fringilla montifringilla* (see renderings in Lindner 1906, 1911, Fenk 1911*b*, and Mayhoff 1915). Recordings suggest 'peyi' or (when loud and close) 'peeyee' (Fig IV), in extended form 'eee-teih' (Fig V), 'eeri-oo' (Fig VI), 'eeery-i' or 'eery-ip' (Fig VII), and 'eeer-yip' with different timbre and abrupt end, resembling *F. montifringilla* (Fig VIII) (J Hall-Craggs, M G Wilson). See also (e.g.) Naumann (1900), Pitman (1921), Meiklejohn (1948), Géroudet (1955, 1957), Moore and Boswell (1957), and Debru (1958). Claim by Lindner (1911) of higher-pitched call in

♀ ('dielit' resembling *C. spinus*) requires confirmation. (3) Antiphonal calling. (3a) Recording (Fig IX) contains 'eeer-y-eip' and 'eeery-ee-ip' whose 2 long ascents strengthen impression of rising pitch. Calls are given antiphonally by 2 birds in close proximity. Sex of birds not known; perhaps 2 ♂♂ advertising nest-holes, though sonagram (albeit of call given while flying to nest) in Zykova and Panov (1982) shows that ♀ does give calls similar to those in Fig IX. (3b) In further example of antiphonal calling by 2 birds initially in close proximity (probably ♂ and ♀ maintaining contact), bird A gives 'eedle-oo' (1st and 3rd units of Fig X), bird B responding with 'dli-yoo' (2nd unit); then close bird A gives 19 'eedle-oo' calls, each alternating (some overlap) with 'pitchoo' or 'dli-yoo' from distant bird B. (J Hall-Craggs.) (4) Calls given (perhaps not exclusively) in flight. Series of 'chi' (Korelov *et al.*

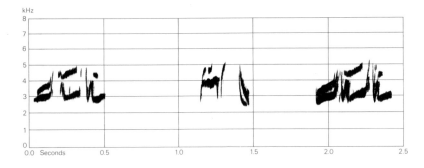

X   Roché (1966)   France   May or July 1965

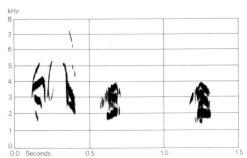

XI   Roché (1966)   France   May or July 1965

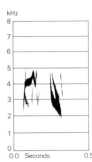

XII   Roché (1966)   France
May or July 1965

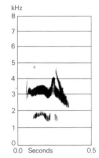

XIII   Roché (1966)   France
May or July 1965

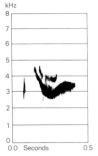

XIV   Roché (1966)   France
May or July 1965

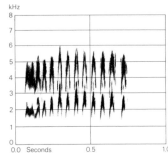

XV   R Ranft   Spain   May 1987

1974) or soft 'düj' sounds (Piechocki *et al.* 1982); sonagram in Fiebig and Jander (1987) shows rapid and more or less continuous sequence of such calls. Single calls given in flight illustrated by Chappuis (1976) for Corsica and (longer, more complex) for Morocco; see also call 2. Call (roughly 'pleeiyoo') in sonagram provided by E N Panov also resembles those in Chappuis (1976), and a presumably similar 'chee-leep', more musical than any calls of *P. domesticus*, was reported as flight-call by Hollom (1962); Géroudet (1957) gave similar transcriptions. Other calls which presumably belong in this category include 'oui' (Jonsson 1982), 'tschli' (Canary Islands) and 'dlü' (Greece) illustrated by Bergmann and Helb (1982). (5) Recordings by J-C Roché contain (in addition to anti-phonal calls and quiet song: see 1d and 3, above) several calls of unknown function, though some perhaps related

to those already described. Fig XI shows 'dl-i-yoo' followed by 2 'uip' units; 'pit-choo' (Fig XII) and 'pee-choo' (Fig XIII) are further descending disyllabic calls, suggesting these vary individually; Fig XIV shows 't-soo-ee'. (J Hall-Craggs.) (6) Rattle-call. Rapid rattling or chattering sound similar to *Passer* given when disturbed or in threat: 'terrettettet trattättät tät' or 'trrr tettettettet' (Naumann 1900; Lindner 1917); 'dedede' or 'rrr' (Bergmann and Helb 1982); 'tee-tr' r' r' r' r' (L Svensson). Varies in volume, speed of delivery, and length (Fenk 1911*b*; Lindner 1917; E N Panov). Recording (Fig XV) contains neat, pronounced buzz preceding 't-k-t-k-t-k-t-k-t'; closely resembles chattering threat-call of *P. domesticus*, and also suggests tremolo purr of Crested Tit *Parus cristatus* (J Hall-Craggs, W T C Seale). Further descriptions perhaps of this or related calls include slightly nasal 'schwe-err' or 'swee-cher-wer' (Hollom 1959), 'schrrye' or 'kriep' (Jonsson 1982), and harsh screeching from birds disputing roost-sites (Salzmann 1909). See also (e.g.) Lindner (1906, 1911, 1917), Schmitt and Stadler (1914), Mayhoff (1915), and Abs (1961). Rattle-call sometimes combined with Peyee-call (Mayhoff 1911) or call 10a (E N Panov). (7) Short 'wäd' apparently used as close-range contact-call between pair-members (Bergmann and Helb 1982; Fiebig and Jander 1987). Further descriptions presumably of this call include muted 'wrä wrä' (Mayhoff 1911), 'wäk' as contact-call or 'wäg wäg' by ♀ in flight, and harsh, strangled 'ra' or 'rä' by ♀ and young (Lindner 1911). (8) Tsik-call. Short, ascending, high-pitched 'tsik' by ♂ displaying in flight to ♀ (E N Panov).

Short, high-pitched 'iss' like Meadow Pipit *Anthus prate-nsis* given by ♂ in display and in fight (Salzmann 1911), also ('issi ssi ss') during copulation (Mayhoff 1915; Lindner 1917) is probably the same or related. (9) Tik-call. Monotonous, relatively high-pitched 'tik' sounds given (like call 8) in high-intensity interactions (E N Panov). A 'zetterittitittittitt' given by ♀ soliciting copulation (Lindner 1911) is perhaps related. (10) Other calls. (10a) Harsh 'zhav' (E N Panov); sonagram shows short, low-pitched, rising and falling unit. (10b) Quiet 'sirrr' or 'zirrr' by ♂ while preening (Lindner 1911). (10c) Sharp 'swit' when flushed (Hollom 1959). (10d) Twittering like song of Swallow *Hirundo rustica* from family parties in Kazakhstan (Korelov *et al.* 1974); probably (at least in part) a medley of calls described above. (10e) Calls apparently expressing anxiety (see also call 6): 'ziwid' and 'tilit' (Lindner 1906, 1917); drawn-out 'djip' and 'ziüp' or 'chirrup' (Fenk 1911*b*; Mayhoff 1911, 1915; Pitman 1921); 'biwiwiwiwi' (from captive ♀), also 'büit', or 'büip bihle bihlewipp bihle', 'byili bili bilibibi', 'piw piji pijü', and 'gwie gweie' (Lindner 1906, 1911, 1917; Mayhoff 1911, 1915). Probably mostly variant renderings of calls described above.

CALLS OF YOUNG. Food-call similar to adult call 2 and presumed to develop into it ('wäi', 'guib', 'pie-iep', etc.), but much higher pitched in small young (Lindner 1906; Fenk 1911*b*; Mayhoff 1911, 1915; Schmitt and Stadler 1914; Heinroth and Heinroth 1924-6); delivery rate faster on hearing parent. Pitch lowers with age, until about same as adult's at *c.* 15 days (see sonagrams in Fiebig and Jander 1987 and Zykova and Panov 1982). Captive young have quiet, twittering song, noted from 34 days (Heinroth and Heinroth 1924-6; Fiebig and Jander 1987); relationship with adult song not clear, but see 1b (above).    MGW

**Breeding.** SEASON. France: end of April to August, peak June–July when 2nd clutches generally laid (Géroudet 1957; Debru 1961). Spain: eggs laid late April to early June (Niethammer 1961; Muntaner *et al.* 1983). North Africa: 1st clutches laid mid-April to early May, 2nd probably late June (Heim de Balsac and Mayaud 1962). Canary Islands: eggs laid mostly in May (Bannerman 1963), but eggs recorded late March and nestlings early August (Martín 1987). Israel: eggs laid end of March to mid-June (Shirihai in press). Germany (formerly): 1st clutches laid May–June, 2nd brood sometimes still in nest at end of August (Fenk 1911*a*, *b*; Mayhoff 1915; Niethammer 1937). SITE. Hole or cavity in rocks, earth bank, tree (especially fruit or nut tree), building or other structure, disused well, etc.; commonly in old, or sometimes usurped hole of other species, particularly bee-eaters *Merops* but also nuthatches *Sitta*, swallows and martins (Hirundinidae), or woodpeckers (Picinae); also in old rodent burrow (Niethammer 1937; Debru 1961; Nicht 1961; Bundy and Morgan 1969; Lebreton 1975, 1976; Königstedt *et al.* 1977; Pforr and Limbrunner 1982; Pannach 1984). On Canary Islands and Madeira, commonly on sea-cliffs (Bannerman

1963; Bannerman and Bannerman 1965). In extralimital Kazakhstan, nests in dense foliage recorded in juniper *Juniperus* trees, as well as nests excavated in thatched roofs (Korelov *et al.* 1974). In Turkmeniya, 65% of nests were in rock crevice, 15% in old nest of Eastern Rock Nuthatch *S. tephronota*, 10% in bee-eater hole, and 10% on building (Bel'skaya 1974). Entrance tunnel of 60 cm recorded (Mayhoff 1915); height above ground varies greatly, depending on suitable sites. Except in hottest areas, site often chosen to maximize amount of direct sunlight (Debru 1958). Nest: largish untidy structure, very like that of House Sparrow *Passer domesticus*; sometimes domed (Harrison 1975; Paz 1987). Foundation principally of grass or straw, occasionally reduced to small pad, lined with feathers, hair, wool, string, cloth, paper, stalks of herbs, rootlets, etc. (Pitman 1921; Niethammer 1937; Debru 1961; Makatsch 1976). Size varies according to cavity; 5 nests in Turkmeniya had average outer diameter 15·4 cm, inner diameter 9·4 cm, overall height 6·5 cm, and depth of cup 5·1 cm (Bel'skaya 1974). Building: by ♀ only (E N Panov). Nest said to be re-used by same individuals over some years, but birds not ringed (Fenk 1911*a*). EGGS. See Plate 59. Sub-elliptical, smooth and glossy; white to brownish-white, with grey or reddish- to blackish-brown speckles and blotches, concentrated at broad end (Niethammer 1937; Harrison 1975). Very similar to *P. domesticus* in colour and variability, but glossier (Hollom 1962; Makatsch 1976). Nominate *petronia*: Europe, 21·5 × 15·7 mm (19·3–23·5 × 14·7–16·9), $n = 110$, calculated weight 2·82 g. Atlantic islands, 21·5 × 15·3 mm (18·3–23·4 × 14·3–15·7), $n = 30$, calculated weight 2·65 g. *P. p. intermedia*: 21·6 × 15·5 mm (17·9–23·4 × 14·5–16·9), $n = 120$; calculated weight 2·75 g. *P. p. puteicola*: 21·8 × 16·2 mm (20·0–24·5 × 15·2–17·5), $n = 20$; calculated weight 3·0 g. (Schönwetter 1984.) For 65 eggs from Catalonia (northeast Spain), see Muntaner *et al.* (1983). Clutch: 4–7(–8). Of 27 clutches, North Africa: 4 eggs, 22%; 5, 26%; 6, 19%; 7, 26%; 8, 7%; average 5·7 (Heim de Balsac and Mayaud 1962; Etchécopar and Hüe 1967). Of 13 clutches, Catalonia: 2 of 4 eggs, 9 of 5, and 2 of 6; average 5·0 (Muntaner *et al.* 1983). Average of 5 clutches in Uzbekistan only 3·6 (Lakhanov 1977). Usually 2 broods, in same nest; perhaps only 1 brood in Israel (Paz 1987; Shirihai in press), and said to have 1 brood in mountains in Greece but 2 in lowlands (Makatsch 1976). One clutch of 5 in south-east France complete 8 days after fledging of 1st brood, and at 2 other nests 40 and 45 days between fledging of broods (Debru 1958, 1961). No significant difference in number of eggs between 1st and 2nd clutches (Niethammer 1937). Replacement clutch laid if 1st lost but not 2nd (Bel'skaya 1974), though Pitman (1921) found 2nd replacement also laid. Incubation: 11–14 days (Debru 1961; Bel'skaya 1974; Makatsch 1976); up to 16 days according to Paz (1987) and Shirihai (in press), but this probably an overestimate, perhaps counted from laying of 1st egg. Probably by ♀ only (Niethammer 1937; Sick 1939;

Makatsch 1976); eggs laid daily, and incubation starts with penultimate (Bel'skaya 1974). YOUNG. Generally fed and cared for by both parents (see Social Pattern and Behaviour). Brooded by ♀ (E N Panov). FLEDGING TO MATURITY. Fledging period given as 16–21 days and probably *c*. 18–19 days (Debru 1961; Bel'skaya 1974; Makatsch 1976; Paz 1987; Shirihai in press). Captive young first picked up food at 20 days and were independent at 30 days (Fiebig and Jander 1987, which see for development of young); also Bel'skaya (1974). BREEDING SUCCESS. In Turkmeniya, of 25 eggs, 72% hatched and 32% produced fledged young; in one area average 4 young fledged per nest (*n* = 4), in another 5·1 (*n* = 8) (Bel'skaya 1974). In Uzbekistan, of 18 eggs, 10 hatched and 8 produced fledged young (Lakhanov 1977). Former German population probably badly affected by nest-site competition with Starling *Sturnus vulgaris* and *P. domesticus* (Niethammer 1937). Nest-site in dry and sunny location preferred, and young can be lost in prolonged wet weather (Debru 1961). In central Asia, small snakes are main nest predator (E N Panov).                                      BH

**Plumages.** (nominate *petronia*). ADULT MALE. Forehead dark fuscous-brown, crown black-brown; central stripe along top of head buff-brown and rather poorly defined on forehead, broadly and contrastingly pale tawny-buff (if fresh) to cream-white (if worn) on crown. Lore mottled grey and pale buff; supercilium broad, well-defined, pale tawny-buff to cream-white, extending from just above and behind eye to above rear of ear-coverts. Narrow eye-ring cream-white. Short stripe from base of upper mandible to below eye pale buff with indistinct grey mottling, continued into black-brown stripe over ear-coverts; ear-coverts sepia-brown, nape and side of neck grey-brown, mottled sepia on central nape. Upperparts from mantle to upper tail-coverts drab-brown; feathers of mantle and scapulars with broad black or brown-black shaft-streaks and greyish-cream spots on sides, showing as intricate pattern of pale spots and short blackish streaks on brown ground; tips of feathers of back and upper tail-coverts sometimes darker drab-brown and with faint pale shaft-streak; rump often tinged ochre (when fresh) or slightly buff (if worn). Chin and throat pale drab-grey, mottled buff-white on centre; patch on border of lower throat and upper chest bright yellow, partly concealed by grey-brown feather-tips when plumage fresh. Ground-colour of chest and side of breast pale drab-grey, of remainder of underparts cream-white, all (except mid-belly, vent, and thigh) with broad but ill-defined dark drab-grey or olive-brown streaks, sharper and darker towards lower flank; under tail-coverts with dark brown subterminal blob, but this largely concealed. Tail-feathers dark drab-brown, shading to black or brown-black on tip; outer web of outer feather (t6) dirty light grey, tipped black; fringes along outer webs of other feathers olive-brown or pale drab-brown; large white blob on tip of inner web of t2–t6, partly bordered black at tip; trace of white on tip of inner web of t1; maximum length of white spot on t6 11·2 (13) 9–13 mm. Flight-feathers black-brown or black, bases tinged drab-grey, tips of secondaries and inner primaries fringed drab-grey, outer webs of secondaries and inner primaries drab-brown, brighter buff-brown towards tips; outer primaries narrowly and sharply fringed yellowish-white (narrowest on middle portion, pale fringes not extending to emarginated parts). Tertials brown-black with drab-brown or olive-brown fringe

along outer web and shallow ill-defined off-white triangular patch at tip; median and greater upper wing-coverts similar, but median coverts lack distinct brown outer fringe, appearing blacker, pale patches on tip showing as broken wing-bar unless heavily worn; greater coverts slightly less black and patches less distinct, especially on outermost coverts, only inner coverts showing broken pale bar. Lesser upper wing-coverts dark drab-brown, tips olive-brown or buffish. Greater upper primary coverts and bastard wing black with narrow olive-brown fringes. Under wing-coverts and axillaries cream-white, shorter coverts with much brown-grey visible at bases. *In worn plumage* (about February–May), pale stripes on head, pale spots on mantle and scapulars, and tips of coverts and tertials bleach to off-white; black streaks on mantle and scapulars and brown streaks on underparts more distinct; yellow patch on upper chest bright, fully exposed, sometimes bordered by black on lower throat. When heavily worn, however (June–July), pale stripe on central crown turns drab-brown (pale feather-tips worn off), trace remaining on hindcrown only; white spots on mantle, scapulars, tertials, wing-coverts, and t1 largely or completely worn off; underparts have streaking less distinct, appearing mottled brown, grey, and dirty white; ground-colour of upperparts and wing duller and greyer, without olive tinge. ADULT FEMALE. Indistinguishable from adult ♂. Yellow spot on upper chest on average paler and smaller, a rounded blob rather than a broader bib, but much overlap: spot large and bright in 22 ♂♂ and 3 ♀♀ examined, small and paler in 4 ♂♂ and 12 ♀♀ (absent or trace only in 4 ♀♀, these probably 1st adult), intermediate (large and pale, small and bright, or intermediate in colour or size) in 6 ♂♂ and 12 ♀♀. Length of white spot on t6 9·7 (12) 7·5–11·5. NESTLING. Naked (Dementiev and Gladkov 1954; Piechocki *et al.* 1982); at *c.* 2 weeks, cap and shoulder downy (Heinroth and Heinroth 1924–6). For development, see Fiebig and Jander (1987). JUVENILE. Cap dark brown; stripes barely visible or absent. Supercilium distinct, pale buff or off-white; lore mottled isabelline-white and buff, upper ear-coverts dark brown, forming distinct dark stripe from behind eye to upper side of neck. Entire upperparts from hindneck backwards buffish grey-brown, tips of feathers of lower mantle and scapulars dark brown or black-brown, forming ill-defined blotches. Underparts pale buff, chest, side of breast, and flank with ill-defined brown spots or short streaks; no yellow on chest. Feathers of head and body shorter than in adult; feather-bases extensively grey, readily visible if plumage worn, cap and underparts in particular appearing mottled. Tail dark brown (less black than in adult); inner web of t2–t6 sometimes with ill-defined off-white or pale isabelline-grey spot on tip, usually faint and frequently absent. Flight-feathers, tertials, and greater upper primary coverts browner than in adult, fringes along outer webs and tips ill-defined brown-grey or buff-brown (sharpest on primaries); p10 longer than in adult (see Structure). Upper wing-coverts buffish grey-brown, tips of longer coverts narrowly edged pale buff, sometimes bordered black-brown subterminally. Under wing-coverts pale greyish-isabelline. FIRST ADULT. Indistinguishable from adult when last juvenile outer tail-feathers or p10 replaced. When in moult, juvenile outer flight- and tail-feathers still fresh (outer feathers of adult distinctly abraded). In some birds (especially 1st-autumn ♀), yellow spot on chest somewhat paler and more concealed than in adult, but much individual variation in size and colour of spot.

**Bare parts.** ADULT, FIRST ADULT. Iris grey-brown, hair-brown, light brown, or dark brown. Upper mandible light or dark horn-brown or greyish-horn with blackish culmen and tip, more yellowish at base in spring; lower mandible yellow with black-brown tip. Leg and foot brownish-yellow, flesh-horn, pale flesh with yellow tinge, or pink-flesh with grey suffusion on toes (Hartert 1903-10; Dittberner and Kage 1991; ZFMK.) NESTLING. Bare skin, including leg and bill, pink-flesh. Mouth flesh-red, gape-flanges yellow. At *c.* 2 weeks, mouth bluish-red; gape-flanges reddish-yellow; bill dusky grey with yellow cutting edges; leg and foot flesh with grey scutes. (Heinroth and Heinroth 1924-6; Debru 1958; Harrison 1975). JUVENILE. Bill horn-yellow with dark grey-horn culmen and tip. Leg and foot pink-flesh with greyish scutes. (BMNH, ZFMK.)

**Moults.** ADULT POST-BREEDING. Complete; primaries descendent. Starts with innermost primary (p1) early July to early August. In September, moult advanced (of 6 birds from whole of west Palearctic range, all were in moult, primary moult score 20-41, average 34); in October, moult completed in 6 birds (score 50), advanced (score 35-45) in 6 others; a few birds still in moult in November. Tail moult centrifugal; t1 starts at primary score 20-30 (RMNH, ZFMK, ZMB). In *barbara*, moult completed mid-October (Meinertzhagen 1940). In northern Iran, 4 *exigua* not yet moulting mid- or late July (Paludan 1940; Diesselhorst 1962). In Afghanistan, *intermedia* just started late July, almost completed September (Paludan 1959). In a few captive *brevirostris*, 1-year-old ♂ started early August, ♀ *c.* 14 days later, completed about 3 October (Fiebig and Jander 1987). In various races of USSR, birds just starting moult encountered 2 July to early August, birds in full moult 1 July to mid-September (mainly from August), birds with moult almost completed 5 July to late September (Dementiev and Gladkov 1954). POST-JUVENILE. Complete; primaries descendent. Sequence as in adult; starts with p1 and scattered body feathers at age of *c.* 6 weeks, thus timing dependent on hatching date (Heinroth and Heinroth 1924-6). Fully juvenile birds examined June-July, moulting ones from mid-July onwards; in 5 birds from September, primary score 29-39, average 34; one from 1 November (score 34) rather retarded in moult, but not known whether juveniles in particular occasionally moult late, as other birds moulting October-November (see Adult, above) had moult too far advanced to be aged (RMNH, ZFMK, ZMB). In USSR, moult in various populations between June and August (Dementiev and Gladkov 1954).

**Measurements.** ADULT. Nominate *petronia*, Atlantic populations. Wing and bill to skull for (1) Canary Islands, (2) Madeira, (3) Portugal and west-central to south-west Spain; other measurements combined; skins (RMNH, SMTD, ZFMK, ZMA, ZMB). Bill (N) to nostril; exposed culmen on average 3·1 mm less than bill to skull.

| | ♂ | | ♀ | |
|---|---|---|---|---|
| WING (1) | 96·5 (1·50; 24) | 94-100 | 93·0 (2·17; 6) | 89-97 |
| (2) | 99·5 (2·45; 6) | 97-103 | 94·0 (1·78; 11) | 92-97 |
| (3) | 96·4 (1·25; 11) | 94-98 | 92·6 (2·29; 4) | 91-96 |
| TAIL | 51·5 (1·19; 11) | 50-54 | 51·1 (1·43; 7) | 49-53 |
| BILL (1) | 16·8 (0·68; 23) | 15·4-17·7 | 17·0 (0·68; 16) | 16·1-17·9 |
| (2) | 17·4 (0·57; 6) | 16·8-18·0 | 17·8 (0·68; 11) | 16·9-18·6 |
| (3) | 16·9 (0·54; 11) | 16·5-17·7 | 17·8 (0·43; 4) | 17·2-18·2 |
| BILL (N) | 10·6 (0·46; 40) | 9·9-11·5 | 10·8 (0·48; 31) | 9·9-11·9 |
| TARSUS | 18·7 (0·77; 11) | 17·4-19·5 | 18·5 (0·89; 7) | 17·5-19·5 |

Sex differences significant for wing and bill (3).

Nominate *petronia*, west Mediterranean populations. (1) Corsica and Sardinia, (2) Mallorca; otherwise as above.

| | ♂ | | ♀ | |
|---|---|---|---|---|
| WING (1) | 95·2 (1·60; 10) | 93-97 | 92·1 (1·22; 7) | 89-93 |
| (2) | 94·7 (1·72; 6) | 92-97 | 92·2 (2·60; 4) | 89-95 |
| TAIL | 50·9 (0·75; 4) | 50-52 | 50·8 ( — ; 2) | 50-52 |
| BILL (1) | 16·8 (0·48; 10) | 16·3-17·5 | 16·6 (0·71; 7) | 15·9-17·5 |

| | ♂ | | ♀ | |
|---|---|---|---|---|
| (2) | 16·1 (0·80; 6) | 15·2-17·2 | 16·5 (0·43; 4) | 16·0-17·0 |
| BILL (N) | 10·5 (0·50; 16) | 9·7-11·2 | 10·6 (0·38; 11) | 10·0-11·1 |
| TARSUS | 18·2 (1·01; 4) | 17·0-19·3 | 18·4 ( — ; 2) | 17·9-19·0 |

Sex differences significant for wing and bill (1). Formentera (Balearics), wing: 87·8 (6) 85-91 (Mester 1971).

Nominate *petronia*, central and eastern populations. (1) Eastern Spain, (2) north-east France and Germany (Sachsen, Thüringen), (3) northern Italy, (4) southern Italy and Sicily, (5) Yugoslavia and Greece; sources as for nominate *petronia*, but German data include skins from Munich Museum and wing of Yugoslavian birds include data from Stresemann (1920).

| | ♂ | | ♀ | |
|---|---|---|---|---|
| WING (1) | 100·8 (1·44; 5) | 99-102 | 95·0 ( — ; 1) | — |
| (2) | 99·4 (1·82; 5) | 99-102 | 97·7 (2·31; 5) | 95-100 |
| (3) | 98·0 (2·65; 8) | 95-102 | 95·9 (1·11; 7) | 94-98 |
| (4) | 96·0 ( — ; 2) | 95-97 | 92·8 ( — ; 2) | 92-94 |
| (5) | 98·4 (1·08; 8) | 97-100 | 94·6 (2·43; 4) | 92-97 |
| TAIL | 53·0 (1·85; 13) | 50-56 | 51·9 (1·18; 12) | 50-54 |
| BILL (1) | 16·9 (0·54; 5) | 16·4-17·5 | 16·6 ( — ; 1) | — |
| (2) | 17·4 (0·94; 5) | 16·3-18·5 | 16·7 (0·53; 4) | 16·3-17·2 |
| (3) | 17·2 (0·63; 8) | 16·3-17·9 | 16·9 (0·66; 7) | 16·2-17·7 |
| (4) | 17·7 ( — ; 2) | 16·9-18·5 | 17·2 ( — ; 2) | 16·9-17·4 |
| (5) | 17·4 (0·52; 4) | 17·1-18·2 | 17·3 ( — ; 2) | 17·2-17·4 |
| BILL (N) | 10·7 (0·45; 24) | 10·0-11·5 | 10·6 (0·33; 16) | 10·0-11·2 |
| TARSUS | 18·8 (0·61; 13) | 17·7-19·5 | 19·0 (0·60; 12) | 18·1-20·0 |

Sex differences significant for wing (1) and (5).

*P. p. barbara.* Morocco to north-west Libya; skins (BMNH, RMNH, SMTD, ZFMK, ZMA, ZMB). Bill (S) to skull.

| | ♂ | | ♀ | |
|---|---|---|---|---|
| WING | 100·6 (1·83; 23) | 98-104 | 97·7 (2·13; 14) | 95-101 |
| TAIL | 55·5 (1·22; 11) | 53-57 | 52·3 (2·73; 7) | 50-55 |
| BILL (S) | 18·1 (0·72; 18) | 17·1-19·3 | 17·6 (0·62; 13) | 17·0-19·0 |
| BILL (N) | 11·5 (0·57; 18) | 10·5-12·4 | 11·3 (0·66; 13) | 10·6-12·3 |
| TARSUS | 19·5 (0·47; 10) | 18·8-20·0 | 19·3 (0·70; 7) | 18·5-20·0 |

Sex differences significant for wing and tail.

*P. p. puteicola.* Levant; skins (BMNH, RMNH, ZFMK, ZMA).

| | ♂ | | ♀ | |
|---|---|---|---|---|
| WING | 101·6 (2·44; 16) | 98-106 | 99·1 (2·40; 17) | 95-103 |
| TAIL | 56·8 (1·82; 9) | 54-59 | 55·0 (1·17; 10) | 53-57 |
| BILL (S) | 18·9 (0·62; 11) | 18·1-20·0 | 18·7 (0·91; 11) | 17·2-19·9 |
| BILL (N) | 12·3 (0·53; 11) | 11·5-13·0 | 12·0 (0·65; 11) | 11·1-13·0 |
| TARSUS | 20·0 (0·83; 9) | 18·9-21·3 | 20·0 (0·82; 9) | 18·4-20·9 |

Sex differences significant for wing and tail.

*P. p. exigua.* Turkey (east from Anatolia), Caucasus, and Transcaucasia; sources as for nominate *petronia*.

| | ♂ | | ♀ | |
|---|---|---|---|---|
| WING | 101·5 (1·83; 18) | 98-106 | 96·5 (1·10; 6) | 95-98 |
| TAIL | 56·1 (1·65; 4) | 54-58 | 52·5 ( — ; 2) | 52-53 |
| BILL (S) | 18·1 (0·72; 18) | 17·3-18·8 | 18·4 (0·98; 6) | 17·4-19·5 |
| BILL (N) | 11·6 (0·79; 10) | 10·5-13·0 | 11·6 (0·77; 6) | 10·9-12·5 |
| TARSUS | 19·2 (0·41; 10) | 18·5-19·7 | 19·3 ( — ; 2) | 19·1-19·5 |

Sex differences significant for wing and tail. North-west Iran: Wing, ♂ 102·8 (3) 100·5-104 (Vaurie 1949a).

*P. p. intermedia:* (1) South-west and north-east Iran, (2) Tien Shan. *P. p. kirhizica:* (3) Ural river mouth and Aral Sea. Other measurements combined. Sources as for nominate *petronia*.

| | ♂ | | ♀ | |
|---|---|---|---|---|
| WING (1) | 101·7 (1·03; 8) | 100-104 | 100·6 (2·33; 5) | 98-103 |
| (2) | 103·5 (1·47; 14) | 101-107 | 99·0 (1·08; 4) | 98-101 |
| (3) | — ( — ; —) | — | 95·5 ( — ; 2) | 94-97 |
| TAIL | 56·3 (1·15; 12) | 54-59 | 55·1 (2·94; 6) | 52-59 |
| BILL (1) | 18·4 (0·66; 8) | 17·6-19·4 | 18·2 (0·48; 5) | 17·5-18·6 |
| (2) | 18·1 (0·71; 13) | 17·5-19·2 | 18·2 (0·39; 4) | 17·7-18·6 |
| (3) | — ( — ; —) | — | 18·3 ( — ; 2) | 18·2-18·4 |
| BILL (N) | 11·5 (0·34; 21) | 11·0-12·2 | 11·5 (0·42; 11) | 10·8-12·1 |
| TARSUS | 19·4 (0·41; 11) | 18·8-20·0 | 19·2 (0·74; 6) | 17·8-19·8 |

Sex differences significant for wing (1).

For data from various populations, see Vaurie (1949a).

JUVENILE. In 9 birds of various races examined, wing and tail both *c*. 5 (1–9) shorter than average adult of race concerned (RMNH, ZFMK, ZMB). In southern Yugoslavia, wing 90·6 (3·13; 5) 88–96, on average 5·2 shorter than unsexed adults (Stresemann 1920). Probable juveniles, Iran: wing 94·5 (6) 93–96 (Vaurie 1956*f*).

**Weights.** Nominate *petronia*. Sicily, December: ♂ 29, ♀ 32 (ZFMK). Greece, May–June: ♂♂ 32, 35 (Makatsch 1950). Western Turkey, July: ♂♂ 28·1, 30 (Rokitansky and Schifter 1971). Formentera (Balearic Islands): 29·4 (6) 26–34 (Mester 1971).

*P. p. exigua*. Southern and eastern Turkey, Armeniya, and northern Iran, May–July: ♂ 33·7 (2·81; 12) 30–39, ♀ 32·1 (2·19; 4) 29·7–35 (Paludan 1940; Nicht 1961; ZFMK).

*P. p. puteicola*. Israel: 26–39 (Paz 1987).

*P. p. intermedia*. Iran and Afghanistan: May–July, ♂ 34·8 (2·03; 4) 33–37, ♀♀ 30, 34·3; September–October, ♂ 34·7 (3) 32–37, ♀ 31·3 (3) 31–32 (Paludan 1938, 1959; Desfayes and Praz 1978).

*P. p. intermedia* or *kirhizica*. Kazakhstan: ♂ 33·9 (18) 28·8–40·0, ♀ 33·9 (13) 30–39 (Korelov *et al*. 1974).

*P. p. brevirostris*. Mongolia, May–July: ♂ 30·6 (2·95; 10) 28–35, ♀ 33·4 (5·81; 5) 27–42 (heaviest ♀ had egg in oviduct) (Piechocki and Bolod 1972).

NESTLING. On day 4, 11·8 (3) 11·5–12, increasing rapidly to *c*. 25 on day 12, then more slowly (Fiebig and Jander 1987, which see for growth curves).

**Structure.** Wing rather long, broad at base, tip bluntly pointed. 10 primaries: in adult, p8 longest, p9 0–1 shorter, p7 1–2, p6 6–8, p5 12–16, p4 18–22, p1 29–34; p10 strongly reduced, narrow, 60–69 shorter than p8, 9·6 (14) 8–12 shorter than longest upper primary covert; in juvenile, p1 22–31 shorter than p8, p10 54–60 shorter; p10 5·7 (10) 3–8 shorter than longest upper primary covert. Length of p10 11·5 (7) 9–13 in adult, 8·1 (7) 7–10 in juvenile (Stresemann 1920). Outer web of (p6–)p7–p8 emarginated, inner web of p8–p9 with faint notch. Longest tertial reaches tip of p2–p3. Tail short, tip square; 12 feathers. Bill strong, markedly broad and deep at base (width and depth at base *c*. 70–80% of length of visible part of culmen); conical, with sharp tip. Base of culmen flattened; culmen and cutting edges straight. Nostril small, rounded; directed obliquely upwards. Bristles at base of upper mandible soft, reduced. Tarsus and toes short but strong. Middle toe with claw 18·6 (30) 17–20; inner and outer toe with claw *c*. 70% of middle with claw, hind *c*. 76%.

**Geographical variation.** Rather slight; involves size (wing, tail, or tarsus length), bill length (relative to other measurements; also varies somewhat with season—longest in summer), and colour (mainly tone of ground-colour, and colour, width, and contrast of dark streaks). Bleaching and abrasion have marked influence, freshly moulted birds tinged tawny or buff, turning to grey after a few months. In some populations, paler and darker birds occur side-by-side (Dementiev and Gladkov 1954). Difference in colour between races often discernible only when series of specimens with same degree of abrasion compared.

Nominate *petronia* a dark race, ground-colour of head and upperparts greyish-brown when plumage fresh, dull cold grey when worn; dark streaks on cap, lower mantle, and scapulars distinct, black or black-brown. Within nominate *petronia*, some variation in both colour and size, but too slight to warrant recognition of *macrorhyncha* Brehm, 1855 (Greece), *madeirensis* Erlanger, 1899 (Madeira), *hellmayri* Arrigoni, 1902 (Sardinia), *idae* Floericke, 1902 (Tenerife), or *balearica* Von Jordans, 1923 (Mallorca). Birds from Canary Islands, Portugal, western Spain, Balearic Islands, Corsica, Sardinia, Sicily, and southern Italy on average smaller than those from Morocco, Madeira, eastern Spain, Pyrénées, Alps, and central Europe to Greece and western Turkey; bill relatively shortest on islands in west Mediterranean; ground-colour of upperparts relatively darker and browner and streaks blacker in birds from Sardinia, Corsica, and (to lesser extent) Canary Islands, Madeira, Portugal, and western Spain (east to Málaga and north to Salamanca), paler and greyer with streaks more black-brown in birds from Balearic Islands and (to lesser extent) Sicily and southern mainland Italy, intermediate in colour in remainder of range. Remaining races of west Palearctic larger than nominate *petronia* (except *kirhizica*) and have bill distinctly longer (especially in *puteicola*) and markedly thick and swollen at base (less so in *exigua*). *P. p. barbara* from Algeria, Tunisia, and Libya paler and greyer than nominate *petronia*; dark stripes on cap dark sepia-brown, less blackish; dark streaks on lower mantle and scapulars slightly narrower, shorter, and more restricted in extent, but as dark as those of nominate *petronia*; ground-colour of upperparts rather pale grey-brown, of underparts more extensively white with less sharply defined streaks. Palest birds occur Libya and Tunisia; gradually darker and bill smaller towards west (Vaurie 1956*f*); birds occurring west of Tlemcen (Algeria) and those of Morocco quite near nominate *petronia* in colour (Hartert and Steinbacher 1932–8), but colour and size nearer *barbara* (BMNH) and included in that race here (C S Roselaar). *P. p. exigua* from Turkey (east from Burdur, Ankara, Ilgaz, and Kastamonu, south to Birecik), Caucasus area, northern Iraq, and northern Iran (east to Gorgan) intermediate between *barbara* and nominate *petronia* in colour; colour and width of marks of upperparts as *barbara*, but ground-colour darker, nearer nominate *petronia*; streaks on underparts more distinct than in *barbara*. *P. p. puteicola* from Levant a well-marked race, large with giant bill; plumage distinctly paler than in previous races, ground-colour of upperparts grey-brown with sandy-buff tinge, less grey than in *barbara*; stripes on cap and streaks on lower mantle and scapulars greyish-olive-brown, contrasting only slightly with remainder of upperparts; underparts extensively whitish; throat spot paler yellow than in previous races. *P. p. intermedia* from south-west Iran east to Turkmeniya and mountains of west-central Asia near *puteicola* in general colour; dark streaks on lower mantle and scapulars narrow and restricted (as *exigua* and *puteicola*), but blackish (as *exigua*), not as pale as in *puteicola*; underparts pale, as *puteicola*, but streaking extensive though ill-defined. *P. p. kirhizica* from lower Volga east to Turgay depression and Aral Sea intermediate in colour between *puteicola* and *intermedia*; smaller in size than both of these. Extralimital eastern race *brevirostris* differs from all western races in less well-marked crown-stripes (distinct at nape only), rather small size, short but thick bill, and dark (though somewhat variable) colour. CSR

## *Montifringilla nivalis* Snow Finch

PLATE 28
[facing page 281]

DU. Sneeuwvink    FR. Niverolle alpine    GE. Schneefink
RU. Снежный вьюрок    SP. Gorrión alpino    SW. Snöfink

*Fringilla nivalis* Linnaeus, 1766

Polytypic. Nominate *nivalis* (Linnaeus, 1766), southern Europe from northern Spain through Corsica, Italy, and Alps to Greece; *leucura* Bonaparte, 1855, southern and eastern Asia Minor; *alpicola* (Pallas, 1831), Caucasus, north-west and northern Iran, Hindu Kush (Afghanistan), and western Pamirs. Extralimital: *gaddi* Zarudny and Loudon, 1904, Zagros mountains (south-west Iran); *tianshanica* Keve, 1943, Alayskiy and Chatkal'skiy mountains through western Tien Shan to Terskey Alatau and Kokshaal-Tau (west of former Soviet Central Asia); *groum-grzimaili* Zarudny and Loudon, 1904, eastern Tien Shan and Bodgo Ula (northern Sinkiang, China), Russian Altai, Tuva area, and western and (perhaps this race) southern Mongolia; *kwenlunensis* Bianchi, 1908, Kun Lun mountains and Astin Tagh (southern Sinkiang) and (perhaps this race) Nan Shan area (north-central China); *adamsi* Adams, 1859, southern and eastern fringe of Tibetan plateau, from Ladakh (India) to eastern Tsinghai (China).

**Field characters.** 17 cm; wing-span 34–38 cm. Size close to Snow Bunting *Plectrophenax nivalis* but recalls particularly sparrow *Passer* in stronger bill, more angular head, slightly longer tail, and somewhat shorter wings. Quite large, montane passerine, mixing sparrow-like structure and behaviour with highly variegated plumage pattern much like *P. nivalis* except for grey head and (in breeding season) solid black bib. Juvenile shows browner head and lacks visible bib. Call distinctive. Sexes dissimilar at close range; some seasonal variation. Juvenile separable. 3 races in west Palearctic separable on head colour; only nominate *nivalis* described here (see also Geographical Variation).

ADULT MALE. Moults: July–October (complete); due to wear, plumage most contrasted in spring. Deep bib black; head to nape bluish-grey with cream forehead, dusky lore, and cheek separated from bib by pale cream malar stripe. Back dark chocolate brown, with dark feather-centres visible at close range; rump and upper tail-coverts black, with buff-grey and white tips creating mottled appearance visible at close range. Folded wings show long white panel bordered black below; on close inspection tertials show black-brown centres and buff fringes, while mainly white primary coverts show as white patch at base of primaries. In flight, wing pattern appears similar to adult ♂ *P. nivalis*, looking wholly white on inner half and black (with white blaze near carpal joint) on outer half: shape of black area dissimilar in close view, however, with that of *P. nivalis* straight-edged on inside and that of *M. nivalis* not so, as primary-bases show increasing area of white inwards. Tail white, with black central feathers and black tips to all but outermost forming inverted T-pattern. Underparts cream-white, with grey cast and brown tips on under tail-coverts. In fresh plumage, head tinged brown-buff, sides of neck tinged grey, bib tipped white, back feathers fringed paler buff, rump feathers more fully tipped buff and white. Bill black, becoming almost orange with black tip in winter. Legs black. ADULT FEMALE. Closely resembles ♂ but plumage less immaculate, with brown tinge to head, smaller and duller black bib, and browner rump, wings, and tail; white on primary coverts more restricted to bases, creating narrow bar rather than patch. Bill brownish-yellow with black tip most obvious in breeding season. JUVENILE. Resembles dull ♀ but crown and nape distinctly brown, mantle, rump, and upper tail-coverts dull brown, and underparts washed buff on breast and flanks. Wings less pied than adult, with blackish mottling on outer webs of secondaries and only a hint of white on primary coverts. Tail pattern also duller, with 4 (not 2) central feathers blackish-brown and all but outermost fringed buff. Similarity with *P. nivalis* most marked at this age.

Comparisons made above with *P. nivalis* mainly necessary to prevent wishful identification as vagrant *M. nivalis*. Given close clear view of grey head and black bib of adult, distinction from *P. nivalis* easy, but at longer range both adult and much browner and duller-winged juvenile may require longer inspection. Flight fast; pied wing pattern produces apparently flickering action as in *P. nivalis* but wing-beats actually rather stiffer, while wings lack loose attitudes so typical of *P. nivalis*. Hops and (apparently) walks, with short quick steps and foreparts held up while tail tends to trail, recalling Chaffinch *Fringilla coelebs* at times. General behaviour on ground recalls both finch and sparrow but does not perch on trees; often stands up, flicking tail upwards. Sociable, usually seen in nomadic parties and occasionally large flocks. Tame, exploiting food sources provided by man.

Song a sparrow-like and monotonous repetition of tri-syllable 'sitticher-sitticher-sitticher...'. Commonest call a penetrating, slightly grating 'tweek' or 'tsweek'; under pressure, develops into longer harsher and wheezy 'tswãã' while taking off. Other calls include short, soft 'pitsch'.

**Habitat.** Breeds in alpine and subalpine elevations of temperate and warm temperate zones across middle latitudes of Palearctic, probably below 10°C isotherm. Breeds in Abruzzi mountains (Italy) to above 2300 m, and in Swiss Alps only exceptionally below 1900 m or above

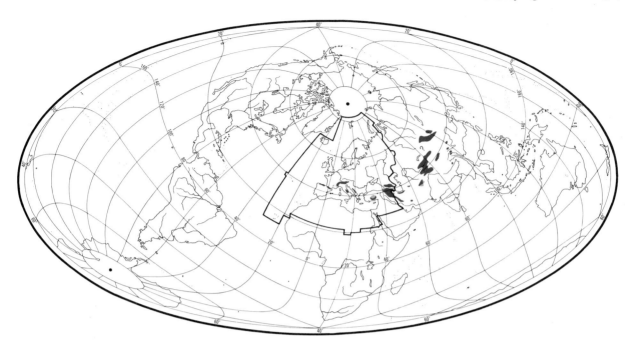

3000 m. In Caucasus, normally at 2750–3160 m, but in Tibet up to 5300 m. In winter commonly remains at similar altitudes, and only occasionally below 1000–1500 m (see Movements). (Dementiev and Gladkov 1954; Voous 1960*b*; Glutz von Blotzheim 1962.)

In Switzerland, inhabits grassy patches above treeline (avoiding even dwarf woody growth), scree slopes, and boulder patches on ridges and peaks or in passes; habitat usually includes suitable choice of crevices either in some kind of rock face (often side by side with Swift *Apus apus* or Alpine Swift *A. melba*) or commonly on buildings such as cowsheds, structures protecting against avalanches, timber stacks, and even busy hotels, whose refuse may contribute significantly to diet (Glutz von Blotzheim 1962; Lüps *et al.* 1978). In USSR, where such artefacts less common at appropriate elevations, habitat consists more largely of boulders, rocks, precipices, screes, and heaps of stones, alternating with patches of alpine meadow watered by rivulets and streams (Dementiev and Gladkov 1954). Occurs in Afghanistan at 3000 m or more at springs or on tussock grass in valleys (Paludan 1959). A ground feeder, tending to avoid snow-covered terrain, although often near it.

Ecologically may be regarded as an alpine counterpart of House Sparrow *Passer domesticus* and has taken similar advantage of man's now widespread exploitation of mountain habitats on which it formerly supported itself unaided in respect both of food and shelter.

**Distribution.** No changes reported, except possibly Bulgaria.

BULGARIA. Perhaps former breeder. Breeding reported Pirin mountains 1962, and possibly central Rila mountains 1964 (Paspaleva 1965), but records rejected by Simeonov (1971) as insufficiently documented (TM). ALBANIA. Perhaps breeds in high mountains in east (EN).

Accidental. Czechoslovakia, Rumania, Malta, Canary Islands.

**Population.** FRANCE. 100–1000 pairs (Yeatman 1976). GERMANY. Estimated 100–300 pairs (Bezzel *et al.* 1980). Probably *c.* 200 pairs (Rheinwald 1992).

**Movements.** Chiefly resident; some birds make altitudinal movements, especially in east of range; in west of range, some birds short-distance migrants. Pattern incompletely known, due in part to lack of observers at high altitude, and low overall numbers; no information on age-class of migrants.

Observations in Switzerland, Turkey, and Kazakhstan show that some birds ascend above breeding grounds in initial post-breeding period, moving down again in autumn (Glutz von Blotzheim 1962; Sutton and Gray 1972; Korelov *et al.* 1974). In Europe, most birds then remain within breeding range; often regarded as wholly resident, with isolated reports far from known range attributed to vagrancy; recently, however, annual records above treeline in outlying mountains indicate that small proportion winters regularly up to at least 300 km from breeding sites; sporadic records further afield are presumably of vagrants. Reported outside breeding areas as early as October, suggesting little dependence on weather conditions; last records in April.

In Switzerland, many winter above 2000 m, some reg-

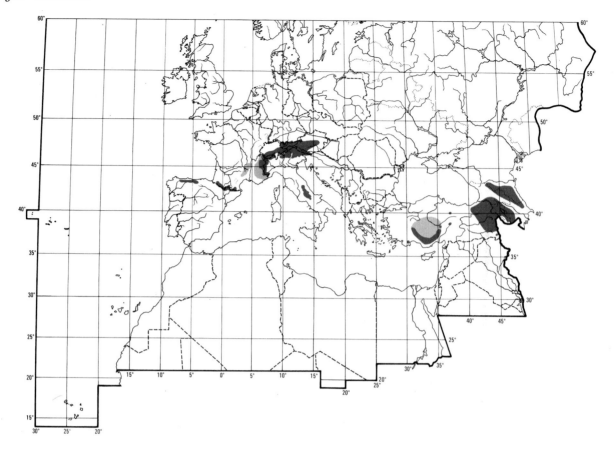

ularly moving lower (to 1500 m) to take advantage of food in tourist areas; reported only occasionally below 800 m (Schifferli *et al.* 1982; Winkler 1984). In area of 19 km² at 1800-4000 m on Kleine Scheidegg in Berner Oberland (Switzerland), where *c.* 50 pairs breed, *c.* 600-800 birds wintered 1981-4, and were widely scattered throughout Berner Oberland (*c.* 650 km²) March-October; some travelled further, with 3 ringed 21-22 February 1984 recovered 275 km ESE in Tirol (Austria) 30 March 1987, and bird ringed 25 April 1982 recovered 437 km south-west in Gard (southern France) 20 October 1982 (Heiniger 1991). Most records of movements are from France, between north-west and south of Alpine breeding area, or north-east from Pyrénées. To west and north-west of Alps, reported from Jura and probably regular as far north as Vosges mountains, with one record at Lac Kir near Dijon at *c.* 230 m (Michelat and Viain 1984; Yeatman-Berthelot 1991). At southern edge of Alps in Haute-Provence (outside breeding area), winters regularly in Mont Ventoux (November-March), Montagne de Lure, Mourre de Chanier, and Cheiron; e.g. 21 January 1982, 3 groups of 90, 40, and 15 birds on Mont Ventoux; more irregular further south, where reported only as singles or very small groups (Cheylan 1973; Besson 1982); has reached coast in Camargue and at Nice (Michelat and Viain 1984). Further

west, Cévennes—equidistant (*c.* 300 km) from Alps and Pyrénées—perhaps also a regular wintering area; groups of birds recorded at scattered sites November 1972 to February 1973 (not a severe winter); provenance not known, but from either breeding area birds would need to leave mountains and traverse river plain (Cheylan 1973). Similarly, origin unknown of wintering or passage birds apparently regular further west in Montagne Noire and at Carcassonne; perhaps birds disperse north-east from Pyrénées, but west or south-west heading recorded twice at Carcassonne (mid- to late October), so Alpine origin also possible (Debru 1960; Cugnasse 1975). Present all year in Pyrénées, and some birds frequent ski resorts; increase in eastern records in winter shows that some birds from central Pyrénées move eastwards, perhaps due to mildness of Mediterranean climate (Muntaner *et al.* 1983). One record at sea-level near Barcelona (north-west Spain) (Soler 1963). In Piemonte and Val d'Aosta (north-west Italy), reported only exceptionally below 1000 m (450 m minimum) (Mingozzi *et al.* 1988); in Belluno (north-east Italy), recorded along Piave river at 330-420 m (Faveri 1989). Movements apparently with character of migration reported in central Italy; further south, on Ustica island (north of Sicily), group of 15 birds 16 October 1976 had presumably travelled 350 km south-west from nearest breeding area, of which 250 km across open sea; this apparently exceptional movement suggested possibility of true migration (Massa 1978). One record from Malta, 28 October 1970 (Sultana and Gauci 1982), 1–2 from Mallorca (Balearic Islands) (Serra 1978), and one 'apparently genuine' record from Tenerife (Canary Islands) *c*. 1841 (Bannerman 1919). In Turkey, descends to lower altitudes in winter, when it also occurs more widely on central plateau (Beaman *et al.* 1975).

In central Asia, present in breeding areas in winter; tends to frequent south-facing slopes; makes irregular altitudinal movements dependent on severity of weather, descending to 1000–2000 m, or exceptionally leaving mountains; also moves to vicinity of human settlements (Ivanov 1969; Korelov *et al.* 1974). In north-west Mongolia, recorded below breeding range from late August or early September (Sushkin 1938). In Afghanistan, early June, and in Tibet, May, large flocks recorded in valleys, presumably forced down from mountains by snowfall (Schäfer 1938; Paludan 1959). DFV

Food. Invertebrates and seeds; in spring and summer insects and spiders, in winter almost wholly seeds, either in wild or provided by man along with scraps. Feeds on ground; in autumn, also on seed-heads of tall herbs, like *Carduelis* finch; often at hotels, tourist resorts (etc.) in winter and sometimes in summer (Münch 1957; Glutz von Blotzheim 1962; Gaston 1968; Wehrle 1989; Heiniger 1991). In Tadzhikistan (extralimital), bulk of adult food is plant material, including buds, leaves, and shoots, except in summer when feeding young (Abdusalyamov 1977). At 2200 m on France/Switzerland border, *c*. 70% of insect food in summer taken from snowfields (Catzeflis 1975). In Berner Oberland (Switzerland) in winter, leaves roosts at up to 3000 m in early morning to glide rapidly down to lower feeding areas (hardly ever below treeline) about 6 km distant, where sunlit slopes have snow-free patches; in severe weather, descends only to *c*. 2100 m, 2–3 km away, to feed on scraps provided by man at mountain-railway stations; returns to roost through afternoon. In spring, preferred areas for foraging for invertebrates are slopes of rusty sedge *Carex ferruginea* and edges of snowfields; many insects wind-borne onto snow where remain immobile on surface; very important food source Tipulidae larvae (Diptera) revealed in June and early July as snow melts; generally too cold in morning for these larvae to be exploited since they exist in water trapped below ice, so other invertebrates taken then; see below for details. (Heiniger 1991.) Also catches insects in flight (Gaston 1968; Korelov *et al.* 1974); feeds in flocks in winter (Wehrle 1989; Heiniger 1991). In Kirgiziya, takes seeds of crops such as wheat *Triticum*, and in Altai mountains feeds at haystacks and in farmyards (Pek and Fedyanina 1961; Neufeldt 1986). Seeds crushed in bill rather than cut open; most efficient method of removing grass-seed husks (Ziswiler 1967*a*). Captive birds selected

seeds for maximum energy gain per unit of time; commercial seeds (especially millet *Setaria* and hemp *Cannabis*) have lower energy content than native Alpine species, though much shorter handling time per g, and artificial supply of seeds important only in conditions of persistent very low temperature and deep snow, and even then taken only in early morning and before going to roost. Otherwise smaller Alpine seeds (1–3 mm), particularly globeflower *Trollius*, are much preferred. (Wehrle 1989.)

Diet in west Palearctic includes the following. Invertebrates: grasshoppers, etc. (Orthoptera), adult and larval Lepidoptera (Lycaenidae), adult and larval flies (Diptera: Tipulidae), beetles (Coleoptera: Carabidae, Scarabaeidae, Curculionidae), Hymenoptera (Tenthredinidae, ants Formicidae), spiders and harvestmen (Arachnida), earthworms (Lumbricidae). Plant material: seeds of Coniferae, lady's mantle *Alchemilla*, campion *Silene*, clover *Trifolium*, mullein *Verbascum*, gentian *Gentiana*, Umbelliferae, thistle *Cirsium*, dandelion *Taraxacum*, sedge *Carex*. (Arrigoni degli Oddi 1929; Münch 1957; Glutz von Blotzheim 1962; Wehrle 1989; Heiniger 1991.) In Kirgiziya and Kazakhstan, feeds on oats *Avena* and wheat *Triticum*; earwigs (Dermaptera) also recorded in diet (Yanushevich *et al.* 1960; Pek and Fedyanina 1961; Kovshar' 1966).

No detailed study of adult diet in west Palearctic. In Berner Oberland, of 14 stomachs of birds found dead in winter, 3 contained seeds of *Alchemilla*, 1 contained seeds of *Silene*, and 1 seeds of *Trifolium* (Wehrle 1989, which see for proportions in diet of 8 species of seeds fed to captive birds). 38 stomachs from Kirgiziya contained bush crickets (Tettigoniidae), Diptera, Hymenoptera (including ants Formicidae), beetles (Carabidae, Scarabaeidae, Tenebrionidae, Chrysomelidae, Curculionidae), and seeds of shrubs, Leguminosae, Polygonaceae, oats, wheat, other grasses, and sedges (Yanushevich *et al.* 1960; Pek and Fedyanina 1961). 10 stomachs from Tadzhikistan contained mainly plant material; in summer and autumn, seeds of *Artemesia*, *Salsola*, *Ceratoides*, and grasses (Abdusalyamov 1977).

When feeding on seeds of Alpine plants in captivity, average daily intake 6·0 g (9961 seeds), with energy content 101·0 kJ; when given commercial food, daily intake also 6·0 g (2290 seeds), 104·2 kJ (21 tests on 3 birds); basic energy requirement for captive bird in winter *c*. 93 kJ per day. Minimum requirement in wild calculated to be 103·6 kJ (6·9 g of seeds) at 17°C; 150·7 kJ (8·9 g) at −10°C, and 168·1 kJ (9·9 g) at −20°C; in mid-winter, if temperature remains below −10°C for more than a few days, more daylight time than is available is needed to collect this quantity of native seeds, hence importance of artificial feeding. (Wehrle 1989.)

Nestling food almost exclusively energy-rich and chitin-low invertebrates, most importantly Tipulidae larvae (22·0 kJ per g dry weight). In Berner Oberland at *c*. 2300 m, 218 collar-samples from nestlings aged 3–9 days contained 64% by dry weight Diptera larvae (60% Tipulidae), 15% Lepidoptera (12% larvae), 7% beetles, 4% Arachnida, and 10% other invertebrates (percentages accurate to 1–3%). Over course of day, Tipulidae larvae in diet rose from 30% in early morning to 80% in afternoon then fell again to 43% in evening, while caterpillars rose from 4% in morning to 30% in evening; adult Lepidoptera taken only in morning. Tipulidae larvae accounted for 70% of diet in June (48% in July), caterpillars 12% (20%), adult Lepidoptera 0% (6%), and earthworms 0% (5%); in 100% snow cover, no Tipulidae and fewer other Diptera larvae taken. Adults very selective in choice of invertebrates gathered for young; supply on snow-free patches was 36% by weight beetles, 21% Arachnida, and 19% Diptera; on snow surface, 34% Diptera, 20% Hymenoptera, 12% Lepidoptera, 12% bugs (Hemiptera) and 10% spiders (Araneae). Observations at 8 nests of nestlings aged 1–20 days in different study area showed 92% of diet by dry weight Diptera (of which 90% Tipulidae), 2% beetles, 2% caterpillars, and 3% seeds. Nests on buildings and pylons ($n = 6$) $c.$ 3 times closer to feeding grounds than those on rock faces ($n = 3$), so adults bringing food to young had flying distances and times of 38·6 km (86 min) and 93·3 km (309 min) per day respectively. Although young in nests on buildings received average 3·8 larvae per feed, and those on cliffs 4·1, intake of former amounted to 22% more per nestling over nestling period, taking account of transport time; this could explain increased nesting on man-made structures. (Heiniger 1991.) Beetles (Carabidae), Lepidoptera (Lycaenidae), and domestic scraps also recorded in nestling diet in Switzerland (Münch 1957). In Altai, grasshoppers (Acrididae) important in diet of young, collected in meadows with Tipulidae (Loskot 1986a; Neufeldt 1986). BH

**Social pattern and behaviour.** No comprehensive studies of nominate *nivalis*; best is that by Lang (1939, 1946a) in Switzerland. Important study of *groum-grzimaili* including comparison with Père David's Snow Finch *M. davidiana*, in Tuva and Altai (south-central Russia) by Ivanitski (1991).

1. Typically gregarious outside breeding season, families joining together in flocks of up to 150 or more from July; some post-breeding flocks of juveniles only (Burg 1922; Lang 1946a; Dementiev and Gladkov 1954; Paccaud 1954; Sutton and Gray 1972; Korelov *et al.* 1974; Brichetti 1983; Wüst 1986). In Alps, flocks largest in autumn (maximum 300 September and November); may be large also in winter (now generally up to 150, formerly many hundreds), e.g. when driven down to lower altitudes by storms, and around buildings where fed by man (Wilson 1887; Lang 1939; Corti 1961; Cheylan 1973; Besson 1982). Winter flocks of 2000–3000 reported in Tibet (Ludlow 1928). In Berner Oberland (Switzerland) at 1800–4000 m, winter population 600–800 birds in 19 km². Flocks of up to $c.$ 20–30 forage up to $c.$ 6 km from roost on sunlit, often snow-free slopes; in bad weather, tend to forage closer to roost (within $c.$ 1–3 km), by railway stations; flocks of 2–10 ascend to $c.$ 3700 m to feed around midday. (Heiniger 1991.) Study by Ivanitski (1991) showed flocks to occur throughout breeding season (see also Flock Behaviour, below). Recorded associating while feeding with finches (Fringillidae), Alpine Accentor *Prunella collaris*, Shore Lark *Ere-*

*mophila alpestris*, and Alpine Chough *Pyrrhocorax graculus* (Géroudet 1957; Kovshar' 1966; Cheylan 1973; Abdusalyamov 1977). BONDS. Mating system monogamous (Ivanitski 1991). At Andermatt (Switzerland), pair-bond and site-fidelity persisted over 2 years (Lang 1946a); no further information on duration of bond. Statement (e.g. in Niethammer 1937, Witherby *et al.* 1938, and repeated subsequently) that both sexes incubate is based on single record of ♂ being flushed from nest by Meiklejohn (1930); in studies by Lang (1939, 1946a), Heiniger (1991), and Ivanitski (1991) only ♀ found to incubate; see also Makatsch (1976). Both sexes feed young (Lang 1939, 1946a), but most during first few days (when ♀ brooding) is done by ♂; share about equal by middle of fledging period (Ivanitski 1991). When 2 parents shot at nest, adult from neighbouring nest fed orphaned brood (Sushkin 1938); not known how long this continued. Young fed for at least 1 week after leaving nest (Potapov 1966); said by Burg (1922) to attain independence within less than 15 days. Study by Ivanitski (1991) found ♀ abandoning family within 3–4 days of fledging, ♂ then caring for young alone. Record from Tirol (Austria) of 2nd-brood nest completed in same hole within 2 days of fledging of 1st brood (Aichhorn 1966); no information on parental roles associated with this. Age of first breeding not known. BREEDING DISPERSION. Solitary or often in small compact or more dispersed neighbourhood groups. In Bayern (southern Germany), almost always in groups of 4–10 pairs (Wüst 1986); similarly at Col de Balme (Switzerland), 2 groups in rocks of 4–5 and 5–6 pairs (Catzeflis 1975), and in Berner Oberland loose groups of 2–6 pairs formed after break-up of winter flocks (Heiniger 1991). Groups of 15–20 pairs reported from central Caucasus (Boehme 1958). In Pamir-Alay mountains (central Asia), nests far apart, and no groups recorded (Potapov 1966; Ivanov 1969); however, in Tadzhikistan, 5 nests in rocky outcrop not exceeding 10 m² (Abdusalyamov 1977). Nests in Tirol minimum 10 m apart on huts, 80 m on cliffs (Aichhorn 1966), while on one hotel complex in Switzerland 4–5 nests annually 10–50 m apart (Lang 1939). Also in Switzerland, of 8 nests in ski-lift supports along $c.$ 1 km between 2050 and 2300 m, closest 50 m, others 100–150 m apart (Pedroli 1967). Nests of solitary pairs in Russian study 100–150 m apart; in more frequently recorded groups (3–10 pairs), 20–50 m, minimum 15 m, between nests. Territory established and defended (only during pair-formation, nest-building, and copulation) by ♂. (Ivanitski 1991.) ♂ known to defend nest (see Antagonistic Behaviour, below), but no information on territory size. Reports from Switzerland (Gordon 1949) and extralimital Asia (Potapov 1966; Loskot 1986a) of birds foraging 300 m or up to $c.$ 1 km or more from nest; average distance for birds nesting on buildings in Berner Oberland 189 m ($n = 6$) and on rocks (up to 300 m above feeding grounds) 622 m (Heiniger 1991); see also Food. Tends to stay in territory for 2nd or replacement clutch, almost always using 1st-brood nest (Aichhorn 1966). In Russian study, no records of 2nd broods by marked birds, and family usually leaves territory immediately after fledging (Ivanitski 1991). Little information on density. In Bayern, large-scale density low, locally quite high; 3–4 pairs along 200 m of rock face over 3 years, $c.$ 5 pairs along 1·5 km, and 4 along 2·5 km (Wüst 1986). In Berner Oberland, winter range of 19 km² holds $c.$ 50 pairs (2·6 pairs per km²), with other $c.$ 500–700 birds spread over at least 650 km² (Heiniger 1991). In Tadzhikistan, at 4000 m, 7 pairs per km² (Abdusalyamov 1977). In Tashanta valley (south-east Altai, Russia), 5 pairs along 6 km, and 30 or more along 5 km, including 12 in 10 ha of scree (120 per km²) (Loskot 1986a). Same nest-site apparently used for more than 1 year, though not known to what extent by same birds (see, however, Bonds, above); one case of ♀ shifting $c.$ 700 m between years (Lang 1946a); see also (e.g.) Potapov (1966).

ROOSTING. Nocturnal and, outside breeding season at least, apparently communal. Generally in well-protected holes in rocks or buildings, where birds, plumage ruffled, may also sit out snowstorm during day (Burg 1922; Dementiev and Gladkov 1954; Abdusalyamov 1977). In Berner Oberland midwinter, birds roosted singly (sometimes in pairs from February or March) in relatively warm, dry, wind-protected cracks and crevices *c.* 10–500 m apart on precipitous rock faces at 2500–3000 m above sea-level. Birds found from ringing to use same roost for quite long period. (Heiniger 1991.) In Turkey, late September, flock of *c.* 60 seen going to roost on sheltered grassy ledges (Mycock 1987); not clear whether birds later entered holes. Also in Turkey, late-summer flock of *c.* 300 recorded roosting 'on cliffs' (Gaston 1968). Nest-hole sometimes used as roost, but birds tend to perch on wall of cavity, though some ♀♀ roost in nest before laying (Aichhorn 1989). In study by Heiniger 1989) birds never roosted in nest-hole in winter, and ♀ only started roosting there after laying 1st egg. Winter flocks come to buildings where fed in early morning and around 16.00 hrs before roosting (Lang 1939). In Berner Oberland, midwinter, birds present at roost 17.00–07.30 hrs; shortly after dawn, glide rapidly down to feeding grounds (Heiniger 1991). In Austria, 19 May (before laying), ♂ emerged from roost at 03.40 hrs (still dark and raining), ♀ from nest 34 min later (Aichhorn 1989).

2. Flicks or jerks tail (e.g. Bruun *et al.* 1986); this (and loose waving of extended wings) said by Whistler (1923) to be characteristic of *Montifringilla* and rosy finches *Leucosticte*. More detailed Russian study showed excited bird to ruffle head feathers while standing fairly erect, and increasingly to wing-flick, sometimes freezing briefly in Wings-high posture (Fig A); also

A

flies more frequently than usual, gliding or performing full display-flight (see Song-display, below). Overall, much less varied repertoire of display postures than *M. davidiana*. (Ivanitski 1991.) Fairly confiding near nest, but shy otherwise (in flocks), though can be very tame and approachable (to within *c.* 1 m) where fed by man, even taking thrown crumbs like sparrows *Passer* (Wilson 1887; Burg 1922; Meinertzhagen 1938; Münch 1957; Schmid 1974; Mauersberger *et al.* 1982). In Switzerland, single bird seen mobbing Peregrine *Falco peregrinus* very high in sky (Gordon 1949). For comparative notes on behaviour of *adamsi* and 4 other *Montifringilla* in Tibet, see Schäfer (1938). FLOCK BEHAVIOUR. Whole flock takes off immediately if one bird gives sharp Szi-call or flies up giving call 3 (Aichhorn 1969: see Voice). While birds may be dispersed for feeding, typically act as compact flock in the air, and may perform aerial evolutions over quite long period, before landing and resuming avid feeding (Burg 1922; Paccaud 1954). Flocks in study by Ivanitski (1991) reported to forage in various habitats including around stock shelters, then forming 'club' on nearby rocks where birds rest, preen, and display (♂♂ only, once incubation has started). In Mongolia, October, flocks of 40–50 came to spring for communal

drinking, bathing, and dust-bathing (Kozlova 1930). SONG-DISPLAY. ♂ sings for territorial advertisement, during Nest-showing and courtship (Aichhorn 1969); in Altai, song given by both sexes prior to copulation (Loskot 1986a); see Voice and Heterosexual Behaviour (below). Rest of account refers to song of ♂, given in flight (see below) or from perch (rock, building, etc.) with wings slightly drooped, head not held very high but black throat-patch revealed through ruffling of feathers, bird either turning frequently or still (Wilson 1887; Meinertzhagen 1938; Lang 1946a; Aichhorn 1966, 1969). Bird performing Song-flight takes off from prominent perch (Ivanov 1969), sometimes on slope (Ludlow 1928), ascends, and, while singing, may then (a) perform steeply undulating flight, with accelerated wing-beats and song during each ascent (Stresemann 1910), (b) glide on motionless raised wings with tail widely fanned, or (c) hang in the wind with wings horizontal (Ludlow 1928; Gordon 1949; Ivanov 1969); overall effect greatly enhanced by prominence of white wing- and tail-patches (Stresemann 1910; Lang 1946a). Flights of up to *c.* 20 s recorded in Russian study; after initial steep ascent to *c.* 5–6 m, bird alternates gliding on raised wings (also descends thus) with bouts of deep, smooth flapping (Ivanitski 1991). Often flies with fluttering action in circle or large figure-of-eight, then recalling Greenfinch *Carduelis chloris* (Meiklejohn 1930; Witherby *et al.* 1938; Aichhorn 1969); sometimes, pair circles together (♂ singing), and ♂ often seen flying while singing towards another pair (Lang 1946a); sings also while descending gently, sometimes in spiralling glide (Ludlow 1928; Witherby *et al.* 1938), and recorded singing while flying from one rocky outcrop to another (Mauersberger *et al.* 1982). In Tibet, Song-flight of *adamsi* includes rapid steep ascent to *c.* 20 m, slow bat-like wing-beats, sudden (and audible) plummeting descent, wing-clapping under body, and somersaulting while descending in fast undulations (Schäfer 1938). Sings from sunrise: often seen to fly immediately to song-post once sun up (Lang 1946a); recorded in Song-flight over nest where ♀ still roosting at 03.34 hrs (Aichhorn 1989). Main song-period in Europe April to May or June, though song noted in Switzerland in one year from 17 March (Lang 1939, 1946a; Di Carlo 1956; Géroudet 1957). In Gruziya, peak song (and courtship) around end of 1st week of May (Dementiev and Gladkov 1954). Birds sing also in autumn (see 1b in Voice), and in other months of year, though not apparently early November to mid-January (Aichhorn 1969). ANTAGONISTIC BEHAVIOUR. (1) General. Aggressive reactions aimed at defence of territory are closely linked to courtship and self-advertising behaviour. Mutual tolerance much increased away from nest, as demonstrated by formation of flocks throughout breeding cycle. (Ivanitski 1991.) (2) Threat and fighting. In early settlement period, ♂♂ display intensively in morning to advertise occupancy of territory, while ♀♀ move about in search of nest-site. Typically brief conflicts occur when ♂♂ follow ♀♀ and thus encroach on others' territories. (Ivanitski 1991.) Threat-call (see 6 in Voice), directed at conspecific and other small birds, given during aerial chase, or on ground in Threat-posture (Fig B): back feathers ruffled, but

B

C

D

crown sleeked, black throat-patch prominent, bill closed; bird faces opponent with head pointed forward and down and moved side to side. In Appeasement-posture (Fig C), wings closed, head (sometimes also tail) slightly raised. Much fighting noted in disputes over roost-, nest-, and feeding sites, also best place for sun- and sand-bathing. (Aichhorn 1969.) Vigorous defence of roost-site also reported by Heiniger (1991). In Swiss study, up to 5 birds recorded foraging together without antagonism, but singing ♂ attacked Wheatear *Oenanthe oenanthe* which approached to within *c*. 4 m, and Rock Thrush *Monticola saxatilis* (Lang 1939, 1946*a*). HETEROSEXUAL BEHAVIOUR. (1) General. In Switzerland, birds recorded visiting nest-sites on fine days in February (Lang 1939); in Berner Oberland, settlement continues up to end of May (Heiniger 1991). Courtship and pair-formation in Apennines (Italy) in April (Di Carlo 1956); peak in Gruziya early May (Dementiev and Gladkov 1954); April–May in Kazakhstan (Korelov *et al.* 1974), but in mountains of Tadzhikistan pairs do not separate off from winter flocks and start courtship until late May or early June (Dementiev and Gladkov 1954; Abdusalyamov 1977). Breeding at 2 sites in Switzerland highly synchronized, all young within each group fledging over 2 days in each of 2 years (Catzeflis 1975). Some resurgence of ♂'s courtship behaviour noted during fledging, though no 2nd broods recorded by Ivanitski (1991). (2) Pair-bonding behaviour. Includes sexual chases (Burg 1922). Completed pair-formation said by Aichhorn (1966) to be indicated by sudden dominance of ♂ at feeding place and roost. In Afghanistan, ♂ seen to make fluttering leaps from rock to rock on one of which ♀ usually perched; in Courtship-posture (recalling *Passer*), ♂ has back hunched, tail fanned, and wings drooped and brushing ground (Meinertzhagen 1938). Presumed mate-guarding indicated by pair keeping close company during nest-building, laying, and (frequent contact-calls: see 2 in Voice) when feeding young (Lang 1939, 1946*a*; Münch 1957; Aichhorn 1989). ♀ may watch while ♂ circles in Song-flight (Lang 1939). (3) Courtship-feeding. ♂ said by Burg (1922) to feed ♀ on nest, but no feeding of ♀ recorded by Aichhorn (1989) or Heiniger (1991). (4) Mating. Often follows ♂'s Song-flight, ♀ soliciting by crouching (much as in Fig C), slightly shivering open wings, and giving Appeasement-call (see 7 in Voice) on ground, post, or wire; call 8 (see Voice) once noted from ♀ during copulation; birds feed quietly afterwards (Lang 1939, 1946*a*; Aichhorn 1969; Abdusalyamov 1977). In Russian study, aerial display by ♂ prior to copulation as follows: flies rapidly to ♀ and circles over her up to 5–7 times (circles *c*. 3 m across, mainly gliding) staying at same height or varying it abruptly; passage through air makes whistling sound. Up to 5 successive mountings noted, ♂ wing-flicking and adopting Wings-high posture (see Fig A) between these. (Ivanitski 1991.) In Altai, early June, pair copulated on ground 4 times in succession (Loskot 1986*a*). Copulation noted after 2 eggs laid (Lang 1946*a*). (5) Nest-site selection. Once territory established, ♂ looks for potential nest-sites and later

(during pair-formation) shows these to ♀, though ♀ also moves about territory searching for suitable site. During nest-showing, ♂ adopts postures as in Fig D left and right, also wing-shivering thereby. (Ivanitski 1991.) In Switzerland, late April, ♂ seen apparently inspecting various holes; finally entered one and sang inside, ♀ (previously passive) following and both remaining inside for some time; ♀ of another pair seen entering previous year's site (Lang 1946*a*). ♂ may pick up plant stem in or near nest-hole, but this part of Nest-showing and all building by ♀ (for details, see Breeding and Aichhorn 1989), though ♂ regularly accompanies her on collecting trips. ♀ recorded tossing out old material (nest-holes can become very clogged), and nest-material sometimes deliberately laid in tunnel and trampled flat. (Lang 1939, 1946*a*; Aichhorn 1989.) (6) Behaviour at nest. Eggs laid within 3–7 days of nest completion (Aichhorn 1989). ♂ often sings near nest, and ♀ will give call 8 (see Voice) in response; ♀ leaves quickly when ♂ approaches and no change-overs recorded. Material may be added to nest throughout incubation. (Lang 1939, 1946*a*; Aichhorn 1989.) RELATIONS WITHIN FAMILY GROUP. No information on brooding. Young fed by ♀ from day 1, and ♂ also recorded coming to nest with food then, though not clear whether he fed young directly (probably did so at least from day 2) or passed food to ♀ for transfer (Lang 1946*a*; see also Aichhorn 1966). Call 8 used to encourage young to gape (Aichhorn 1969). Older nestlings emerge from nest-hole to be fed, then disappear inside again; food-begging (wing-flapping) reminiscent of young Wallcreeper *Tichodroma muraria* as wings flapped apparently through much wider arc (raised almost vertically) than normal in passerine young (Löhrl 1964). Both parents quite often at nest together when feeding young, one sometimes following mate to nest even without food (Münch 1957). One observation (from Switzerland) of nestling faeces being carried from nest and dropped after short flight (Hoffmann 1928). Young exercise wings in nest from 15 days; fledge at *c*. 20–21 days, staying nearby for further 2–3 days, then following parents in response to call 3 (Lang 1939; Aichhorn 1966, 1969: see Voice). ANTI-PREDATOR RESPONSES OF YOUNG. Immediately fall silent on hearing parental alarm (Piechocki and Bolod 1972). Young hide in corners of nest-hole when attempt made to catch them for ringing (Lang 1946*a*). PARENTAL ANTI-PREDATOR STRATEGIES. (1) Passive measures. ♀ typically a tight sitter (Meiklejohn 1930), and remarkably tolerant of disturbance at nest: e.g. may just move aside when hand inserted into nest-hole, soon settling to incubate again, as also after being handled (Lang 1939, 1946*a*); once left nest only after stone removed to enlarge entrance (Meiklejohn 1930). (2) Active measures: against birds. Loud and penetrating Szi-calls (see 2 in Voice) given during dive-attacks on Kestrel *F. tinnunculus* (Aichhorn 1969). (3) Active measures: against man. Birds usually give Alarm-call (see 4 in Voice) when nest threatened (Aichhorn 1969); in Pamirs, marked excitement (Alarm-calls) shown only at nest containing young close to fledging (Potapov 1966); flicking or beating wings (e.g. ♀ while calling: Burg 1922) typical of excited bird (Ivanitski 1991). Excited Szi-calls given by both sexes while nest being inspected (Wilson 1887), and variant of this call (or perhaps call 3: see Voice) noted from both adults in flight (Lang 1946*a*). In

Switzerland, incubating ♀ left nest, flew to ground and there apparently mock-fed (displacement activity or perhaps distraction-lure display), then returned to nest with feather; ♂ leaving another nest gave Alarm-call from nearby roof and was then joined by ♀ who acted like 1st ♀, also returning to nest with material (Meiklejohn 1930). (4) Active measures: against other animals. Alarm-calls given (including for quite large animals such as dog) much as for man (Burg 1922; Aichhorn 1969). When small rodent near nest, calls accelerate with increasing danger; birds may perch on rock facing predator and hover over it while calling (Burg 1922).

(Figs by D Nurney: A and D from drawings in Ivanitski 1991; B–C from drawings in Aichhorn 1969.)   MGW

**Voice.** Detailed study of wild and captive birds by Aichhorn (1969); for additional sonagrams, see also Bergmann and Helb (1982) and Ivanitski (1991, which includes comparison of *groum-grzimaili* with Père David's Snow Finch *M. davidiana*); for descriptions with musical notation, see Stadler (1931). Juvenile song noted from both sexes, adult song apparently only from ♂; report of ♀ and ♂ singing prior to copulation (Loskot 1986a) includes no description of song. No song early November to mid-January; calls 2–3 apparently given throughout year. (Aichhorn 1969.) For brief comparison (of *adamsi*) with 4 other *Montifringilla* in Tibet, see Schäfer (1938). Wing-noise (probably with some signal function) produced during some aerial displays (see Social Pattern and Behaviour).

CALLS OF ADULTS. (1a) Full song of ♂. Phrases of varying length: sometimes only one or a few motifs, or several units and motifs linked together in more continuous series of mainly chirping sounds of varying timbre like sparrow *Passer*; often preceded by 'szi' or 'pink' (see calls 2 and

9), and these, as well as other call-types (e.g. call 3) are incorporated within song (Stadler 1931; Lang 1946a; Géroudet 1957; Aichhorn 1969; Bergmann and Helb 1982; Ivanitski 1991). Much repetition of individual units or motifs apparently typical: e.g. 'sitticher-sitticher' (Witherby *et al.* 1938)—perhaps much as repeated motif in Fig II (see below); 'twi-tju-twi-tju' or 'zü-ti-wi-tju-ti-wi-tju' (Bergmann and Helb 1982); 'pürtzii-pürtzii-pürtzii'; or 'drlätschi-drlätschi-drlätschizi' (Lang 1946a); for further examples of repetition (in *groum-grzimaili*), see Ivanitski (1991). Songs in recording notably buzzy with chirps (many like 'pink' of Chaffinch *Fringilla coelebs*, some isolated), rasp or rattle, and units (probably Szi-call variants) rendered 'schweez' or 'zhwee' and 'zwee'; delivery rate generally slow, but some accelerando and overall effect at times curiously irregular and jerky (M G Wilson); see below. Song-phrase given from perch (Fig I) comprises 23 main units of which 12 (including 1st and last) are small variations of 'pink' of *F. coelebs* (not included in Aichhorn 1969); also 2 descending fast rattles (units 13 and 21), 5 examples of a diad ending in contrary motion (e.g. unit 2), 3 composite (3-part) units (12, 17, 20), and one 2-part unit (15). Song-phrase shown in Fig II is based on same 4 main units as that in Fig I; first parts given while bird still perched, but from beginning of 2nd motif in flight (fades as bird moves away) and then ordered in repeated 4-unit motifs ('sititichurrrwee': M G Wilson); in Fig I, only one such motif (units 10–13). Song given in spiralling descent (Fig III) is based on 6 units comprising high- and low-pitched 'pink', low-pitched noise (double diad, e.g. unit 2), high-pitched noise (unit 8), noisy glissando (unit 9),

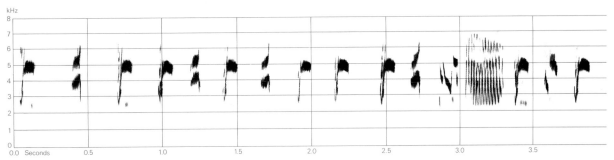

I  J-C Roché  France  June 1963

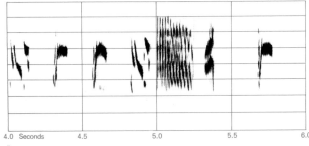

I  *cont.*

and clicking 'tk' (unit 3); song almost metronomic from units 2 to 15, rate slower thereafter. Typically jerky, jolting quality of song (Lang 1946a; Bruun *et al.* 1986) results from contrasting timbre of notes and noises, contrasting loudness levels of adjacent units (see Fig IV with amplitude trace), short units, and longer pauses, giving staccato, syncopated effect. (J Hall-Craggs, W T C Seale.) Complexity of song varies markedly between individuals: one ♂ gave 21 different units in 12·5 s linked together in sub-phrases. Song also varies geographically (dialects) within central Europe. (Aichhorn 1969; Bergmann and

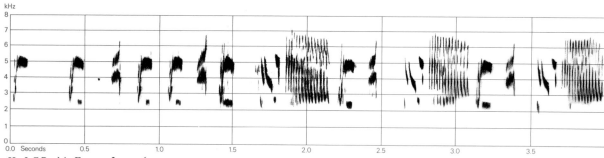

II J-C Roché France June 1963

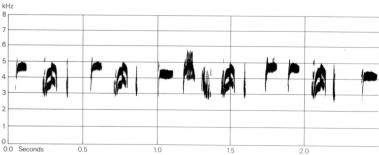

II *cont.*

at all from nominate *nivalis*; song of *groum-grzimaili* in Mongolia a series of sparrow-like sounds—'schiep schiep schtirp schtirp dä dit dit däschip'—given in flight (Mauersberger *et al.* 1982) is evidently much as described for European birds (see above). In Kazakhstan, *tian-shanica* reported to give series of loud Szi-calls, but clear metallic 'chin chin chin' (probably same as 'pink') in song-flight (Korelov *et al.* 1974). (1b) Subsong or 'conversational' song of ♂. Quiet, prolonged, almost warbling Subsong noted in Switzerland in spring (L Svensson);

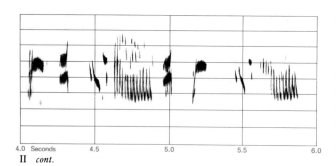

III I Hills Spain June 1976

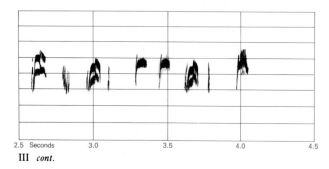

III *cont.*

Helb 1982.) Study of *groum-grzimaili* by Ivanitski (1991) indicated rich repertoire of complex units and motifs (many with tremolo structure), lasting 50–300 ms and separated by fairly constant pauses of 100–200 ms, even in compact sequences typical of high-intensity excitement. Difficult to judge whether this race differs significantly if

subdued song (audible only over a few metres) of exquisite quality reported from Afghanistan by Meinertzhagen (1938) is presumably similar. Study by Aichhorn (1969) found ♂ sometimes (in breeding season and autumn) to give more variable songs than 1a, resembling juvenile song (see Calls of Young, below), though not always as quiet. Not known whether these more variable songs are linked to any particular context(s). (2) Szi-call. Highly variable impure or hoarse 'tseeh', 'tsi', or 'zjih' (Aichhorn 1969; Bergmann and Helb 1982; Bruun *et al.* 1986). Recording (Fig V) of birds disturbed at nest with young apparently contains 2 variants: hoarse, hissing squeal ('szi' or 'sweek') and 'seeoo' which resembles 'zjih' illustrated by Bergmann and Helb (1982), though also perhaps not unlike call 8 (J Hall-Craggs, M G Wilson). In *groum-grzimaili*, timbre of (sometimes disyllabic) 'dzhee' varies from clear to more muted, with noisy, hissing quality (perhaps some overlap with Threat-call described for nominate *nivalis*: see

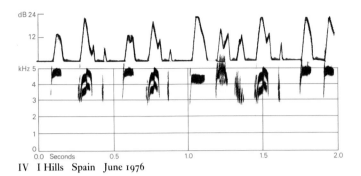

IV I Hills Spain June 1976

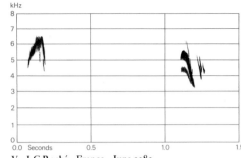

V J-C Roché France June 1983

below); short whistling 'veet' sounds ascending or descending in pitch and given when disturbed are apparently closely related, and intermediates with 'dzhee' occur (Ivanitski 1991, which see for detailed analysis). Other descriptions: harsh 'tsweek' (Witherby *et al.* 1938); 'sts-roui' or 'strui strui' in flight (Biber and Link 1974; Cugnasse 1975); slightly wheezy 'tee teep' (Meiklejohn 1948); see also (e.g.) Lang (1946*a*), Münch (1957), and Corti (1959). Szi-call is given singly or in loose series, in flight (solitary birds or flock) or on ground, and in wide variety of contexts: for contact, when disturbed near or far from nest, during antagonistic interactions at nest, and in encounters on feeding grounds; often precedes, and incorporated in, song (Ivanitski 1991). Anxiety sometimes indicated by delicate and quiet calls, combined with Alarm-call, and likened in melancholy, slightly nasal quality to Siskin *Carduelis spinus* (Stresemann 1910; Stadler 1931; Géroudet 1957); loud, more piercing variants also occur (Aichhorn 1969). Cutting, nasal 'pscheeu' or short 'pschie' given when flying up, sometimes more like 'kaihk' or 'keehk' (Jonsson 1992); presumably this call or mixture of calls 2–3. Further investigation needed to determine whether variety of units illustrated and treated all as 'tsi-call' by Aichhorn (1969) is susceptible to functional and structural subdivision. (3) Quäk-call. Nasal sound similar to, but usually quieter than, Wayeek-call of Brambling *Fringilla montifringilla*; some resemblance also to 'dä' (or 'chay') of Willow Tit *Parus montanus* or harsh fright-call of House Sparrow *Passer domesticus* (Stadler 1931; Aichhorn 1969). Rendered 'quääk' (sometimes drawn-out) (Aichhorn 1969); 'ää', 'jän', 'djää', etc. (Stadler 1931); 'gnièèh èèh' (Géroudet 1957); harsh, wheezy 'tswää' (Witherby *et al.* 1938) probably also in this category. Apparently serves as flight-intention call (Aichhorn 1969), though also given regularly in flight (Wilson 1887). Calls of *groum-grzimaili* similar (in sonagram) to Quäk-calls of nominate *nivalis* are treated by Ivanitski (1991) as variants of call 2 ('dzhee') rather than call 3. (4) Alarm-call. A 'pititit prrt' recalling (especially when preceded by high-pitched 'tsi') Crested Tit *P. cristatus* (Bergmann and Helb 1982). In recording (Fig VI), musical, chipping 2-note 'pit-it' and 'put', also 'tup' (Fig VII), and attractive, delicate, faintly tinny, rippling or purling 'p-t-p-t-p t t t' (Fig

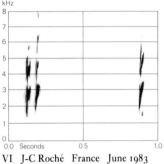

VI J-C Roché France June 1983

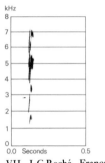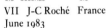

VII J-C Roché France June 1983

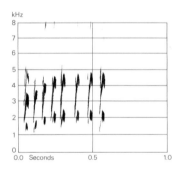

VIII J-C Roché France June 1983

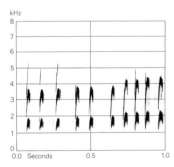

IX I Hills Spain June 1976

VIII), with rapid ascent in pitch over notes 1 to 5; indeed reminiscent of *P. cristatus*, though calls of that species generally faster and higher pitched. In another recording (Fig IX), ripple rendered 'ttt tt p-t-p-t-p'; slower delivery rate and rounder, mellower timbre than in Fig VIII, but no great difference in pitch. (J Hall-Craggs.) Other descriptions: 'pchrrrt' (Jonsson 1992); rapid 'pittittittitt ruititititit', also 'burr-burrr' or similar (Stadler 1931; Géroudet 1957; Aichhorn 1969). Ripple may merge with coarse 'dzhee' (call 2) or hiss (Ivanitski 1991). See also Burg (1922), Hoffmann (1928), Potapov (1966), and Korelov *et al.* (1974). Given when disturbed at nest or in extreme excitement, rarely outside breeding season when normally gives only short, abrupt calls if highly excited, as when frightened or pecked in fight (Aichhorn 1969; Bergmann and Helb 1982). (5) Distress-call. Loud, drawn-out 'krääch' or 'krächz' given (e.g.) in fight and sometimes when handled (Aichhorn 1969). (6) Threat-call. Often loud, drawn-out, high-pitched hissing 'psspss' (Aichhorn 1969; Bergmann and Helb 1982). (7) Appease-ment-call. Rapid series of 'zig' sounds audible within *c.* 10 m and given by both sexes in appeasement (including by ♀ inviting copulation), ♀ generally before laying period, ♂ after (Aichhorn 1969). (8) A 'zion' more or less similar to call 3 given to encourage young to gape (feeding-call), by ♀ on nest in response to ♂'s calls outside, once by ♀ during copulation (Aichhorn 1969). (9) Various calls evidently similar to *F. coelebs* noted by Stadler (1931) and Münch (1957) include quiet, melodious 'jüb' or 'yup', 'pi', 'pink', 'pix', and 'tvic'. Such sounds (notably 'pink') are a regular component of song (see Figs I–III), but no information on their use as separate calls, though short 'pitsch' (probably in this category) apparently given in contact (Niethammer 1937). Sonagram in Bergmann and Helb (1982) shows 'püt' like *F. coelebs* preceding call 2, and call like 'pink' of *F. coelebs* also reported for *adamsi* of Himalayas (Whistler 1923). (10) Other calls. (10a) A 'dü-dü-dü' like Bullfinch *Pyrrhula pyrrhula* given by ♀ flying from nest (Lang 1939); perhaps related is drawn-out 'zjē' or 'zjēb' sometimes given in fast series like singing Pygmy Owl *Glaucidium passerinum* and said by Stadler (1931) to be expansion of call 2. (10b) Quiet, strained 'ejp ejp ejp' given by ♂ approaching ♀ during nest-building (Stresemann 1910). (10c) Ascending 'üüüüü' like twit-tering of Linnet *C. cannabina* or 'ti(x)' recalling Goldfinch *C. carduelis* (Stadler 1931); perhaps related to calls 4 or 7. (10d) Rasping or rattling call given by *groum-grzimaili* when well-grown nestlings threatened; longer, more com-pact and higher pitched than rasps or rattles in song (Ivanitski 1991).

CALLS OF YOUNG. Food-call of nestlings a metallic, hoarse chirping reminiscent of young *P. domesticus* or even young Starlings *Sturnus vulgaris*; very loud after only a few days; rendered 'srieh', 'psi', 'psib', 'pix', etc. (Corti 1939; Lang 1939; Gordon 1949; Aichhorn 1969; Berg-mann and Helb 1982); see also (e.g.) Heinroth and Hein-roth (1924–6), Hoffmann (1928), and Géroudet (1957). Food- and contact-call ('pix-pix') given for some weeks after fledging later develops into adult call 2 (Aichhorn 1969). For further description and sonagrams of nestling and fledgling calls, see Ivanitski (1991). Adult calls 3

(higher pitched in young) and 7 also reported for fledged young (Stadler 1931; Aichhorn 1969). Quiet 'con-versational' juvenile song given by both sexes more vari-able than full song of ♂ (Aichhorn 1969; Bergmann and Helb 1982). MGW

**Breeding.** SEASON. Switzerland: at 2300 m in Berner Oberland, average date of 1st egg over 3 years 15–21 May; nest-building 1st half of May and laying always in 2nd half regardless of temperature, which can be-10°C with 1 m of fresh snow in June (Heiniger 1991); fledging in 10 nests over 2 years at 2400 m always 5–7 July (Catzeflis 1975); nestlings recorded 24 May and 12 August (Glutz von Blotzheim 1962). Italy: young fledged in 1 nest at 2300 m in Abruzzo 25 June (Vaughan 1953). Turkey: young in nests in Ala Daǧlari mountains late June (Gaston 1968); in eastern Anatolia, 2200–2600 m, young in nests and eggs still being laid in mid-June (Kumerloeve 1969c). Central Caucasus: eggs laid mid-June (Boehme 1958). SITE. Crevice or cavity in rock face, sometimes 300 m above ground, boulder scree, earth bank, etc., in darkness at end of entrance passage or fully exposed; on building (even in centre of settlement), pylon, etc., or inside roof, also in nest-box or old mammal burrow (Heinroth and Heinroth 1924–6; Lang 1946a; Dementiev and Gladkov 1954; Glutz von Blotzheim 1962; Kumerloeve 1969c; Aichhorn 1989; Heiniger 1991). Nest: bulky but fairly neat construction with foundation of dry grass with roots, plant stalks, moss, and lichen, lined with fine plant mater-ial, feathers, hair, wool, etc.; very dry materials preferred, to eliminate dampness (Meiklejohn 1930; Makatsch 1976; Aichhorn 1989). 10 nests in Austria had average outer diameter 7·8 cm, inner diameter 7·0 cm, overall height 8·3 cm, and depth of cup 5·0 cm (Aichhorn 1989); 4 in Switzerland had outer diameter 15·6–19·7 cm, inner diameter 8·3–8·6 cm, and depth of cup 4·1–4·8 cm, dimen-sions depending on cavity size (Meiklejohn 1930); another nest in Switzerland had overall height 13·0 cm (Münch 1957). Building: ♀ does all collection of material and con-struction; takes 5–8 (2–13) days (Lang 1946a; Aichhorn 1989; Heiniger 1991). Often built on top of previous year's nest (Lang 1946a; Münch 1957). Material picked up as individual items and bundled in bill; grass pulled from tussocks above snow as well as from snow-free areas (Aich-horn 1989). Nest improved until hatching; most building carried out in early morning and evening (Lang 1946a; Aichhorn 1989; Heiniger 1991). EGGS. See Plate 59. Sub-elliptical, smooth and matt or very faintly glossy; white (Harrison 1975; Makatsch 1976). Nominate *nivalis*: 24·0 × 17·2 mm (21·2–27·0 × 15·2–18·8), *n* = 117; cal-culated weight 3·75 g. *M. n. alpicola*: 22·8 × 15·8 mm (22·4–24·5 × 15·5–16·4), *n* = 6; calculated weight 3·0 g. (Schönwetter 1984.) Clutch: 4–5 (3–6), some disputed records of 7 (Harrison 1975; Makatsch 1976). Of 18 clutches, Austria: 3 of 3 eggs; 13, 4; 2, 5; average 3·9 (Aichhorn 1966). Of 6 clutches, Switzerland: 5 of 4 eggs,

1 of 5; average 4·2 (Glutz von Blotzheim 1962). Eggs laid daily, 1-5 (*n* = 6) days after nest completed (Makatsch 1976; Aichhorn 1989). 2 broods (Aichhorn 1966; Makatsch 1976), possibly overlapping (Glutz von Blotzheim 1962); 9 of 14 ♀♀ had 2 broods, almost all in same nest, which is completed *c.* 2 days after fledging of 1st brood (Aichhorn 1966). INCUBATION. 13-14 days from last egg (Lang 1946a); average at 11 nests 13·1 days (Heiniger 1991). By ♀ only (Lang 1946a; Aichhorn 1966; Makatsch 1976; Heiniger 1991, which see for details); some participation by ♂ according to Meiklejohn (1930), but doubted by other authors. ♀ remains on nest overnight after laying 1st egg, but sits on eggs in daytime only for *c.* 1 hr in early morning then again in late evening, incubation proper beginning after 3rd egg. In first 5 days, takes average of *c.* 30 breaks of 8-10 min per day (*c.* 25% of daylight time), number and duration of breaks increasing with incubation so that in last 2 days only *c.* 50% of daylight time spent on eggs (*c.* 8 hrs per day); when ambient temperature falls below 0°C, stints on eggs of up to 45 min observed, and clutch abandoned if temperature remains at that level for several days as ♀ cannot maintain minimum incubation temperature of 25-27°C, which is lowest known for any passerine. (Heiniger 1991.) Starts with penultimate egg, so last young hatches 1 day after others (Aichhorn 1966). YOUNG. Fed and cared for by both parents (Lang 1946a; Münch 1957; Aichhorn 1966). FLEDGING TO MATURITY. Fledging period 20-21 days (Lang 1939; Aichhorn 1966); for development of young, see Lang (1946a). Captive young fed themselves at 22 days old (Wehrle 1989). BREEDING SUCCESS. No information.

BH

Plumages. (nominate *nivalis*). ADULT MALE. Forehead, crown, nape, and side of head and neck down to upper cheek medium bluish ash-grey, darkest and faintly streaked black on forehead, forecrown, ear-coverts, and upper cheek, slightly paler behind eye (forming faint supercilium) and on side of neck. Lore dark grey with some paler grey mottling; eye-ring white with fine grey specks, distinct, broken by dark grey in front and behind; when fresh (autumn), all grey feathers on top and side of head and neck have grey-brown or buff tips, partly concealing ash-grey, short stripe above lore sometimes uniform isabelline-white; brown tinge gradually disappears in winter, last so on hindcrown, ear-coverts, and side of neck. Mantle, scapulars, back, and rump fuscous-brown, feathers with paler drab- or olive-brown sides; upper tail-coverts black with brown-grey tips, but outer webs of shorter lateral tail-coverts mainly white; when fresh, brown-grey of nape gradually turns into brown on mantle and contrast between brown rump and black tail-coverts not sharp either; when worn, contrast of ash-grey nape and of black tail-coverts with brown remainder of upperparts marked; sepia-brown of mantle and scapulars gradually paler brown during wear, fringes of feathers bleach to grey-brown. Lower cheek pale grey or grey-white, forming rather faint pale stripe, merging into ash-grey of side of neck at rear; chin white or cream-white, sometimes partly mottled black; throat black, feathers with broad white tips when fresh, then partly concealing black; a distinct black bib when worn. Chest and side of breast pale drab-grey,

slightly browner on upper side of breast, merging into ash-grey of side of neck, paler at border of black bib, gradually merging into grey-white on flank and belly and this in turn to white on vent and under tail-coverts; thigh dark brown-grey, tips of longest under tail-coverts dark brown-grey, contrasting with white of vent, tail, and remainder of under tail-coverts. Central pair of tail-feathers (t1) black, outer web narrowly fringed white, tip faintly fringed brown-grey, inner border with faint white edge; t2-t6 white with contrasting black tip 2-10 mm wide (widest on t2, gradually narrower toward t5, virtually or fully absent on t6); concealed bases of t1-t3 with black wedge along shaft. Primaries black, outer webs with narrow white fringes, tips with broader white fringes (faint along outer ones); innermost primary (p1) fully white, p2 with white inner web (except extreme base), p3 black with partly white inner border. Secondaries white, except sometimes for some black at base of outer web of outermost. Tertials sepia-brown, outer webs and tips fringed white and grey-buff. Greater upper primary coverts white, 3-8 mm of tips black (broadest on longest). Bastard wing black. Innermost greater upper wing-coverts (tertial coverts) dark sepia-brown, outer fringed white, remainder of upper wing-coverts entirely white, short lesser coverts partly with narrow dark grey tips and with concealed black bases. ADULT FEMALE. Like adult ♂, but ash-grey of top and side of head and neck replaced by brown-grey, browner than adult ♂ in fresh plumage, less contrasting with mantle, less pure ash-grey when worn, virtually without faint black streaking on face; black of throat more restricted, sometimes absent, forming narrow mottled patch rather than full bib when plumage worn; black of tail and primaries browner, less deep black; greater upper primary coverts either white with 5-12 mm of tip black and with black outer webs to outermost coverts, or black with white bases to innermost, or fully black, except for narrow white fringe; p1 black except for tip and for oblong white mark on inner web; most longer lesser coverts with small grey tips (grey tips not restricted to shorter coverts). See also Bare Parts and Measurements. NESTLING. Down grey-white or white; restricted to 2 fairly long and dense tufts on crown, 1 on back, and some short tufts on upper wing (Ticehurst 1926; Heinroth and Heinroth 1931; Lang 1939, 1946a; Niethammer 1967). JUVENILE. Forehead, crown, and nape uniform brown-grey, gradually merging into isabelline-grey of cheeks and side of neck; eye-ring contrastingly off-white. Upperparts brown-grey, marked with ill-defined dusky streaks on lower mantle and scapulars, less deep and saturated brown than in adult, feathering shorter and softer. Underparts dirty grey-white, tinged grey on throat (black bib absent), tinged isabelline-buff on chest, flank, and belly, dark grey of feather-bases partly visible, feathering soft and loose. Black on tail more extensive than in adult, t1-t2 black, t5-t6 white, t3 and t4 with variable amount of black and white streaks or marks; all feathers tinged tawny-buff along fringes. Primaries and tertials black-brown with buff fringes; secondaries and innermost primary white with variable amount of black on tip and base, partly fringed buff. Greater upper primary coverts black-brown, fringes tinged buff, and inner webs of innermost either mottled white or without white. Shorter lesser upper wing-coverts grey-brown, remaining upper wing-coverts white with buff wash on tip. Too few examined to be certain whether sexes differ. FIRST ADULT. Like adult; probably not separable once post-juvenile moult completed. Individual variation in amount of brown wash on head in fresh plumage or in amount of white and black on greater upper primary coverts perhaps due to age differences, adult ♂ perhaps being purer blue-grey on head and with a smaller number of coverts with smaller black tips than in 1st adult ♂; adult ♀ perhaps with some white variegation at base of inner and middle coverts, 1st adult

♀ without white; those birds which show browner (less whitish) fringes along tips of tail-feathers, outer primaries, and primary coverts perhaps 1st adults.

**Bare parts.** ADULT, FIRST ADULT. Iris hazel, brown, or dark chestnut-brown. In winter (about September to February), bill yellow or orange-yellow (perhaps more orange in ♂, more yellow in ♀), tip of culmen usually black; bill darkens from late February to early April onwards, becoming black in ♂, black-brown with deep straw-yellow basal quarter of upper mandible and basal half of lower mandible in ♀, but in some ♀♀ examined bill horn-black with dark horn-brown lower mandible; bill paler again during post-breeding moult. Leg and foot dark plumbeous-black, black-brown, or black. (BMNH, RMNH, ZFMK, ZMA, ZMB.) NESTLING. Bare skin flesh-pink. Bill light yellow, gape-flanges bright pale yellow, mouth blood-red. Leg and foot dull flesh-grey. (Heinroth and Heinroth 1931; Lang 1939; Niethammer 1967.) JUVENILE. Iris dark grey-brown. Bill pink-yellow or orange-yellow at fledging, small black tip developing 10 days later. Leg and foot dark lead-grey to plumbeous-black. (Lang 1946a; BMNH, RMNH, ZFMK.)

**Moults.** ADULT POST-BREEDING. Complete; primaries descendent. In Switzerland, outer primaries or secondaries frequently still growing October, exceptionally December (Winkler and Winkler 1985). In birds examined (BMNH, RMNH, ZFMK), moult just started in Tien Shan on 27 July (primary score 3) and 6 August (score 2), almost completed in Afghanistan on 9 September (score 34) and in Tadzhikistan on 1 November (score 44), just finished (score 45) 15 October (Mongolia) and later. Moult not started in summer birds from Iran and Afghanistan up to 23 July (Paludan 1959; BMNH); in Afghanistan, moult just started 13–15 July (Niethammer 1973), advanced in ♂ from 13 September (outer primaries and part of body feathering growing); only some feathers of head still growing in ♀ from 6 October (Paludan 1959). In Russian Altai, starts mid-August (Johansen 1944); in Mongolia, ♂ in moult on 23 July (primary score c. 20) (Piechocki and Bolod 1972). ADULT PRE-BREEDING. Perhaps a restricted partial moult of head or throat in March (Witherby et al. 1938), but confirmation required. POST-JUVENILE. Complete; primaries descendent. Starts with p1 at age of 45–50 days (25–32 days after fledging, only 12 days after flight-feathers full-grown) (Lang 1946a; Winkler and Winkler 1985), and timing thus as variable as fledging (which spans June–August: Johansen 1944; BMNH). Duration of primary moult in captivity 93 (4) 84–103 days; secondaries start with s1 c. 28 days after shedding of p1, completed with regrowth of s6 4–21 days after regrowth of p10; middle tertial (s8) shed with p5, immediately followed by t1; tertials moult in sequence s8-s9-s7, tail centrifugal, completed when p9 half-grown (Winkler and Winkler 1985, which see for further details). In birds examined, moult not started June–July; started in birds from 8 August (primary score 26, Switzerland), 9 August (score 5, Tien Shan), and 29 August (score 2, Tien Shan) (BMNH, RMNH). For Afghanistan, see Aichhorn (1970).

**Measurements.** ADULT, FIRST ADULT. Nominate *nivalis*. Southern Europe (mainly Alps), all year; skins (BMNH, RMNH, ZFMK, ZMA, ZMB). Bill (S) to skull, bill (N) to distal corner of nostril; exposed culmen on average 3·5 shorter than bill (S).

| | | | | |
|---|---|---|---|---|
| WING | ♂ | 121·6 (2·46; 32) 117–127 | ♀ | 116·4 (2·00; 19) 113–122 |
| TAIL | | 71·3 (2·47; 12) 68–76 | | 67·0 (2·51; 11) 63–72 |
| BILL (S) | | 16·7 (0·78; 25) 15·4–18·3 | | 17·0 (0·73; 13) 15·8–18·0 |
| BILL (N) | | 10·9 (0·44; 25) 10·0–12·0 | | 11·1 (0·35; 13) 10·6–11·8 |

| | | | | |
|---|---|---|---|---|
| TARSUS | | 23·2 (0·72; 13) 22·2–24·4 | | 23·0 (0·51; 11) 22·3–23·7 |

Sex differences significant for wing and tail.

Bill longer in summer than during rest of year, as summer diet of insects leads to reduced abrasion. Thus, in sample from above, sexes combined: October–March, bill to skull 16·5 (0·75; 18) 15·4–17·6, to nostril 10·9 (0·36; 18) 10·0–11·3; May–July, bill to skull 17·4 (0·66; 9) 16·5–18·3, to nostril 11·2 (0·46; 9) 10·7–12·0. For variation of bill length with time of year in various races, see also Stegmann (1932), Kozlova (1933), and Watson (1961). Wing length increases with age (Lang 1946a).

Switzerland, wing live or freshly dead birds: winter, unsexed 122·4 (2·90; 200) 113–129; all year, ♂ 123·8 (21) 121–128, ♀ 116·7 (15) 114–120 (Lang 1946a). Greece: wing, ♂ 119 (4) 116–120, ♀ 116; culmen, ♂ 13·6 (4) 13·5–14, ♀ 13·5 (Watson 1961).

*M. n. leucura.* Southern and eastern Turkey, May–June; skins (ZFMK; wing and tail include data from Watson 1961).

| | | | | |
|---|---|---|---|---|
| WING | ♂ | 114·6 (2·54; 11) 111–119 | ♀ | 111·3 (3·44; 6) 106–116 |
| TAIL | | 68·5 (1·79; 11) 65–72 | | 66·8 (3·96; 6) 62–73 |
| BILL (S) | | 18·5 (0·27; 6) 18·0–18·7 | | 18·4 ( — ; 2) 18·3–18·5 |
| BILL (N) | | 12·2 (0·44; 6) 11·6–12·6 | | 12·4 ( — ; 2) 12·0–12·8 |
| TARSUS | | 22·1 (0·97; 6) 21·0–23·4 | | 22·0 ( — ; 2) 21·3–22·7 |

Sex differences significant for wing.

Southern Turkey: culmen, ♂ 14·1 (0·42; 5) 13·5–14·5, ♀ 14·1 (0·25; 4) 14–14·5 (Watson 1961).

*M. n. alpicola.* Wing and bill from (1) Caucasus, (2) northern Iran, and (3) Afghanistan, April–July (but Caucasus all year); skins (BMNH, RMNH, ZFMK, ZMA, ZMB).

| | | | | |
|---|---|---|---|---|
| WING (1) | ♂ | 120·5 (1·73; 4) 118–122 | ♀ | 112·1 (1·49; 4) 110–114 |
| (2) | | 121·1 (2·64; 8) 117–125 | | 112·9 (3·28; 4) 110–117 |
| (3) | | 119·5 (2·20; 7) 115–122 | | 113·4 (1·96; 8) 109–115 |
| BILL (S) (1) | | 18·0 (0·68; 4) 17·0–18·6 | | 18·7 (0·64; 4) 17·8–19·3 |
| (2) | | 18·8 (0·47; 8) 18·0–19·5 | | 18·1 (1·28; 4) 16·4–19·2 |
| (3) | | 19·6 (0·42; 6) 19·2–20·3 | | 20·0 (0·68; 8) 19·2–21·2 |
| BILL (N) (1) | | 12·2 (0·73; 5) 11·5–13·3 | | 12·2 (0·77; 4) 11·6–13·2 |
| (2) | | 12·6 (0·38; 8) 12·0–13·1 | | 12·1 (0·52; 4) 11·4–12·6 |
| (3) | | 13·1 (0·40; 7) 12·6–13·6 | | 13·4 (0·60; 8) 12·3–14·4 |

Sex differences significant for wing. Tarsus 22·4 (0·39; 4) 22·0–22·8 (RMNH, ZFMK, ZMA).

Northern Iran: wing, ♂ 118·9 (10) 117–122, ♀ 112·0 (4) 112–116; tail ♂ 68·9 (7) 66–71, ♀ 66·3 (3) 62–71; culmen, ♂ 16·4 (10) 16–17·5, ♀ 15·9 (4) 15·0–16·5 (Vaurie 1949a). Afghanistan: wing, ♂ 118·2 (1·75; 8) 115–121, ♀ 112·5 (2·74; 6) 109–115 (Paludan 1959).

*M. m. gaddi.* South-west Iran, summer: wing, ♂ 119·7 (2·91; 5) 116–124, ♀♀ 109, 115·5; bill (S) 17·7 (3) 16·8–18·2, bill (N) 12·0 (3) 11·5–12·8 (Paludan 1938; BMNH); wing, ♂ 119·3 (19) 116–126, ♀ 114·2 (4) 112–116; tail, ♂ 71·5 (18) 68–76, ♀ 68·5 (5) 65–71; culmen, ♂ 16·2 (18) 15–17·5, ♀ 15·6 (5) 15–16 (Vaurie 1949a).

*M. n. tianshanica.* Alay, Talasskiy Alatau, and western Tien Shan mountains, all year; skins (BMNH, RMNH, ZFMK, ZMA, ZMB).

| | | | | |
|---|---|---|---|---|
| WING | ♂ | 119·9 (2·87; 12) 116–124 | ♀ | 116·3 (0·76; 5) 115–117 |
| BILL (S) | | 18·1 (0·82; 12) 16·8–19·3 | | 17·7 (0·55; 6) 16·9–18·4 |
| BILL (N) | | 11·9 (0·65; 12) 11·0–13·0 | | 11·3 (0·33; 6) 10·8–11·7 |

Sex differences significant for wing. Tarsus 23·1 (0·42; 6) 22·6–23·7 (RMNH, ZFMK, ZMA).

*M. n. groum-grzimaili.* Eastern Tien Shan and Bogda Ula (Sinkiang, China), Altai, and western Mongolia, all year; skins (BMNH, ZMB).

| | | | | |
|---|---|---|---|---|
| WING | ♂ | 118·3 (3·56; 6) 114–124 | ♀ | 113·7 (4·48; 3) 108–117 |
| BILL (S) | | 16·8 (0·51; 6) 16·2–17·5 | | 17·1 (1·62; 3) 15·8–18·0 |

BILL (N)      11·2 (0·30; 6)  10·7–11·5      11·5 (1·05; 3)  10·4–12·5

Sex differences significant for wing. Mongolia: wing, ♂ 114·6 (1·81; 9) 113–118, ♀ 109·8 (2·75; 4) 107–113 (Piechocki and Bolod 1972).

*M. n. kwenlunensis.* Kun Lun to Astin Tagh mountains, southern Sinkiang (China), April–May: wing, ♂ 114·3 (2·91; 5) 110–118, ♀♀ 109·5, 110; bill (S) 17·2 (0·51; 4) 16·5–17·7, bill (N) 10·9 (0·33; 4) 10·5–11·2 (BMNH, ZFMK, ZMA, ZMB). Wing 102–117, bill (N) 10·5–12·5 (Stegmann 1932).

*M. n. adamsi.* Wing, ♂ 109–117, ♀ 106–113; bill (S) 16–17 (Ali and Ripley 1974). Tibet: wing, ♂ 113·9 (88) 106–120, ♀ 109·7 (55) 100–118 (Vaurie 1972).

JUVENILE. Wing and tail on average 8·2 shorter than in adult.

**Weights.** ADULT, FIRST ADULT. Nominate *nivalis.* Switzerland, averages: October–November 36·1 (11), December 44·7 (57), January–March 42·8 (125), April 36·9 (14), breeding season (June–July) 37·5 (16) 35–41·5; sexes similar (Lang 1946a); 39·0 (2·0; 49) in 1st half of November, gradually increasing to peak in 1st half of January, when up to 47·5 (3·8; 77) in severe cold, decreasing to 36·7 (2·1; 56) in 2nd half of April; total range 29–57 (Heiniger 1991, which see for details). Greece: ♂ 38·1 (4) 37–40, ♀ 37 (Watson 1961).

*M. n. leucura.* Southern Turkey, March: ♂ 33·3 (1·64; 5) 31–35·5, ♀ 31·6 (2·10; 4) 28·5–33 (Watson 1961). Eastern Turkey, May–June: ♂ 35·3 (2·48; 6) 33–40, ♀♀ 42, 44 (Kumerloeve 1967a, 1969a; ZFMK).

*M. n. alpicola.* Afghanistan: April, ♂ 32–40 (11), ♀ 35–37 (5) (Niethammer 1967); June, ♂ 32·9 (1·77; 7) 30–35, ♀ 32·3 (1·21; 6) 31–34; September–October, ♂ 34, ♀ 35 (Paludan 1959); July, ♂♂ 32, 36; ♀♀ 34–37 (3) (Niethammer 1973).

*M. n. gaddi.* South-west Iran: May–June, ♂ 35·8 (3) 34·8–36·8, ♀♀ 31, 36·9 (Paludan 1938; Desfayes and Praz 1978).

*M. n. tianshanica.* Kazakhstan: ♂ 34·6 (5) 32·2–36·7, ♀♀ 32·8, 34·9 (Korelov *et al.* 1974). Kirgiziya: ♂ 32·5–43·0 (10), ♀ 31·2–40·0 (8) (Yanushevich *et al.* 1960).

*M. n. groum-grzimaili.* Mongolia, June–July: ♂ 31·4 (1·51; 9) 30–34, ♀ 29·6 (2·51; 5) 28–34 (Piechocki and Bolod 1972).

NESTLING, JUVENILE. For growth, see Lang (1946a).

**Structure.** Wing rather long, broad at base, tip pointed. 10 primaries: p8 longest, p9 0–3 shorter, p7 3–6, p6 12–18, p5 21–27, p1 40–52. P10 strongly reduced, a tiny pin concealed below reduced outermost greater upper primary covert; in adult, 76–88 shorter than p8, 10·0 (15) 8–12 shorter than longest upper primary covert; in juvenile, 64–73 shorter than p8, 5·5 (5) 4–7 shorter than longest primary covert. See Winkler and Winkler (1985) for lengths of flight-feathers in adult and juvenile. Outer web of p7–p8 emarginated, inner web of p8–p9 with notch. Tip of longest tertial reaches to tip of p1–p3. Tail rather short, tip almost square; 12 feathers, t3–t4 longest, t1 and t6 1–3 shorter. Bill rather long, conical; *c.* 50% of length of head, *c.* 8·2 mm deep at base, *c.* 7·9 mm wide; culmen straight, or (when tip slightly elongated in summer) slightly concave; cutting edges straight; gonys straight, often with upward kink at gonydeal angle; tip sharply pointed. Nostril small, rounded; covered by short tuft of feathers projecting from forehead; a strip of short feathers with bristly tips along base of upper mandible, projecting over gape. Tarsus and toes short, fairly strong. Middle toe with claw 20·8 (10) 20–22; outer toe with claw *c.* 72% of middle with claw, inner *c.* 74%, hind *c.* 83%.

**Geographical variation.** Marked. Involves size (wing, tail, and tarsus length, or weight), relative length of bill, colour of head, colour of remainder of upperparts, and amount of black on bases of upper wing-coverts and on bases and outer webs of secondaries. Variation obscured by sexual difference in size and in colour of head, change in bill length with season (see Measurements), marked effect of bleaching and abrasion on colour, and perhaps influence of age on amount of black in wing. In central Asia, variation in amount of black on coverts and secondaries apparently clinal, increasing from Afghanistan and western Pamirs through Tien Shan to Mongolia, and from Tien Shan and Mongolia southwards to southern border of Tibetan plateau. Nominate *nivalis* large, but bill relatively short; head and neck grey, contrasting with deep sepia-brown mantle and scapulars; chest, side of breast, and flank washed grey; upper wing-coverts and secondaries white, except at extreme base. In *leucura* from Asia Minor, wing shorter and bill longer; top of head and neck brown-grey, hardly contrasting with greyish-brown mantle and scapulars, which are lighter than in nominate *nivalis*, underparts whiter; lesser coverts sometimes black on basal half, outermost secondary often with some black visible on base of outer web. Birds from Taurus mountains in southern Turkey sometimes separated as *fahrettini* Watson, 1961, but typical *leucura* from north-east Turkey is similar to Taurus birds (G E Watson, in Kumerloeve 1967a). *M. n. alpicola* from Caucasus, northern Iran, Afghanistan, and western Pamirs has wing longer than in *leucura*; bill long, especially in Afghanistan and western Pamirs (where sometimes separated as *prosvirowi* Zarudny, 1917); cap and nape similar in colour to mantle and scapulars, dark grey-brown or sepia-drab (tinged buff in ♀), slightly darker and browner than in *leucura*; wing as *leucura*. *M. n. gaddi* from Zagros mountains in south-west Iran a poor race; in summer, distinctly paler than *alpicola*, but this mainly due to more intense bleaching, and difference in fresh plumage is slight; otherwise similar to *alpicola*; maintained here, following Vaurie (1949a) and Watson (1961), though not separated from *alpicola* by Mayr (1927), Stegmann (1932), Hartert and Steinbacher (1932–8), Paludan (1938), Vaurie (1956f, 1959), and Desfayes and Praz (1978). *M. n. tianshanica* from western Tien Shan large, but bill shorter than in *alpicola*, longer lesser and median coverts black on basal half, more extensive than in *alpicola*, sometimes not fully concealed; outer web of outermost secondary extensively black, black on bases of other secondaries often just visible, forming wing-bar. Birds from Alayskiy and Talasskiy Alatau mountains are nearer those of Afghanistan and western Pamirs in size, but like birds from western Tien Shan in colour. Birds from eastern Tien Shan, Dzhungaria (China), Mongolia, and neighbouring part of former USSR (Altai, Tuva) separable as *groum-grzimaili*; size intermediate or small, bill short; close to *tianshanica*, but slightly smaller (especially bill), upperparts slightly paler and more sandy-grey; secondaries as in *tianshanica*, median coverts similar or slightly more extensively black. *M. n. kwenlunensis* from southern fringe of Sinkiang (China) has all upperparts fawn or sandy-brown, less grey than in *tianshanica* or *groum-grzimaili*; size small, similar to *adamsi* from southern and eastern fringe of Tibetan plateau; basal 50–75% of upper wing-coverts black, bases of secondaries extensively black, outer webs of outer 1–2 secondaries black (except for tawny fringe), some ♀♀ in particular approaching *adamsi* in amount of black (in some birds from Astin Tagh examined, ♂ labelled *M. n. kwenlunensis*, ♀ labelled *M. adamsi xerophila*). *M. n. adamsi* similar to *kwenlunensis* in size, but upperparts darker brown-grey, lesser and median coverts almost fully black, basal half of greater coverts black, and outer webs of all secondaries black, thus forming end of apparent cline of increasing black on wing, running from *alpicola* through *tianshanica*, *groum-grzimaili*, and *kwenlunensis* to *adamsi*. No birds examined from Nan Shan area, named *xerophila* by Stegmann (1932); described as paler than *adamsi*, upperparts light drab, inner secondaries paler (Stegmann 1932); perhaps intermediate between *kwenlunensis* and *adamsi*.

Forms species-group with Tibetan Snow Finch *M. henrici* of eastern Tibet which overlaps with *M. n. adamsi*; *henrici* some- times considered a race of *M. nivalis* and *adamsi* a separate species (Adams's Snow Finch), but *henrici* rather different from *M. nivalis* (see, e.g., Stegmann 1932) and better separated (Portenko and Vietinghoff-Scheel 1974), while characters of *M. nivalis* seem to merge into those of *adamsi* in central Asia (C S Roselaar).

CSR

# Family PLOCEIDAE weavers and allies

Small to medium-sized, often thick-billed oscine passerines (suborder Passeres). Found in great variety of habitats from semi-arid country (e.g. bush and savanna) to woodland, forest, and cultivation. Mostly feed above ground, seeds forming bulk of diet in many species, but other vegetable food and insects also taken (some species being partly or wholly insectivorous) and even nectar in a few cases. Some species are serious pests of cereal crops, most notably Red-billed Quelea *Quelea quelea*. Of mainly Afrotropical distribution. Mainly sedentary, with some minor local or seasonal movement, but some species migratory. About 124 species (C S Roselaar); 1 introduced species in west Palearctic, Streaked Weaver *Ploceus manyar*.

Following Voous (1977), ploceids here classified as distinct family from Passeridae (sparrows, etc.) and Estrildidae (waxbills, etc.). At least 4 subfamilies—Bubalornithinae (buffalo-weavers, 8 species), Ploceinae (true weavers, 105 species), Plocepasserinae (sparrow-weavers, 8 species), and Sporopipinae (scaly-weavers, 2 species)—but parasitic indigobirds and whydahs (Viduinae, 14 species) also included by many authorities. Only Ploceinae, however, are considered further here; for pioneering review of taxonomy, see Moreau (1960). 9 genera, including: (1) *Euplectes* (bishops and widowbirds), 17 species; (2) *Foudia* (fodies), 7 species; (3) *Malimbus* (malimbes), 11 species; (4) *Ploceus* (typical or arboreal weavers), 64 species; (5) *Quelea* (queleas), 3 species. Remaining genera all monotypic. Majority of weavers found only in Afrotropics (*Ploceus* chiefly in drier parts), but 5 species of *Ploceus* occur in southern Asia (Pakistan to Indonesia) and *Foudia* confined to Madagascar and certain other islands of Indian Ocean (including Seychelles) where one further *Ploceus* also found. A few species introduced elsewhere.

Sexes similar in size. Bill often of typical seed-eater type, short and conical, tapering to point; much variation between species, however, with both thick-billed and thin-billed types occurring and every gradation between. Nostrils small, rounded; a few short bristles present at lateral base of both mandibles. Wing short and rounded or bluntly pointed; 10 primaries. P10 considerably reduced, length 20–39% of longest in *Ploceus* (Moreau 1960). Flight strong in many species. Aerial displays and song-flights (to attract ♀♀) are typical of *Euplectes*; simpler song-flight occurs in some *Ploceus* to entice ♀ to nest (see *P. manyar*), but perched display with waving wings (near, on, or under nest) more typical. Tail short or of moderate length; tip straight or slightly rounded; 12 feathers. Leg and foot rather short but strong; tarsus covered with large scutes in front and with smooth sharp ridge at rear. Foot not used in feeding, either to hold food or to scratch to uncover it. Gait of most species usually a hop so far as known but walk or run reported in *Euplectes*. Head-scratching by indirect method (e.g. Simmons 1957b, 1961a). Bathe in typical stand-in manner so far as known. Dusting behaviour not recorded for majority of species (see Moreau 1967), though once observed in Baya Weaver *P. philippinus* (Ganguli 1975). No information on sunning. Anting recorded, mainly in captivity, for several species of at least 6 genera (including *Euplectes*, *Foudia*, *Ploceus*, *Quelea*); ants applied actively in bill in typical posture (Simmons 1961b, 1963 1966).

Vocalizations variable in character, mainly a wheezy chattering or chirping, often high-pitched and not far-carrying; basic repertoire of *c*. 14 calls (Crook 1969). Song (usually centred on nest-site) often weak and unspecialized, with both courtship and advertising function. Many species gregarious, feeding and roosting in flocks, sometimes (e.g. in *Quelea*) of enormous size. Monogamous mating system in minority of solitary, territorial species (mainly insectivores of forest, savanna, and secondary bush), but majority polygynous and nest colonially to greater or lesser extent (see especially Crook 1964b and, for pioneering review, Friedman 1950). ♂♂ of Jackson's Widowbird *E. jacksoni* unique in forming leks. Although often highly gregarious, ploceids are not typically close-contact birds, seldom clumping, even at roost; allopreening, however, recorded in at least one species, Seychelles Fody *F. sechellarum*), and that species and Speckled-fronted Scaly-weaver *Sporopipes frontalis* are 2 of the few known contact species among Ploceidae (Harrison 1965). Complex nests of Ploceinae (of *Ploceus* especially), and their manner of construction and repair, have attracted large literature (e.g. Friedman 1950, Crook 1964a, Collias and Collias 1964). In most species, typically a woven domed structure suspended from bush or tree and with elongated, downward pointing entrance chamber; built by both sexes in monogamous species, by ♂ alone in polygamous ones. Incubation by both sexes or by ♀ alone. Young fed by both sexes in some species, by ♀ only in others.

Plumages often patterned with red, orange, yellow, black, etc., many having contrasting facial mask; some all-black. Sexes alike (to greater or lesser extent) or ♀ nondescript yellowish, greenish, or greyish, streaked and mottled above but lacking any striking markings—as also ♂♂ of many species in non-breeding plumage. Juvenile plumage nondescript. Nestling with or without down; no mouth markings. Adult post-breeding moult complete, occurring outside breeding season; also a partial pre-breeding moult restricted to body, at least in those species with a non-breeding plumage. Post-juvenile moult complete, starting shortly after fledging; in ♂, leads either to non-breeding plumage or to breeding plumage, depending on time of hatching and/or breeding season.

Ploceidae usually thought to be closest to Passeridae or Estrildidae, either or both of which have sometimes been placed in same family with it (e.g. Ziswiler 1965, 1967*b*, *c*; Sibley 1970; Sibley *et al.* 1974), although trend has been to keep all 3 separate (see above, also Felix 1970). DNA data (Sibley and Ahlquist 1990), however, suggest that the 3 groups may be so closely related as to belong (as subfamilies) in a single family ('Passeridae') which also includes Prunellidae (accentors) and Motacillidae (wagtails and pipits); see Volume V. Same data also indicate that Bubalornithinae, Plocepasserinae, and Sporopipinae all belong within Ploceinae; Viduinae, however, assigned to Estrildidae. See also under Estrildidae and Passeridae.

## *Ploceus cucullatus* (Müller, 1776)  **Village Weaver**

Fr. Tisserin gendarme        Ge. Textor

Widespread resident in almost the whole of sub-Saharan Africa. 7 birds collected on Santiago (Cape Verde Islands) May 1924, with no evidence of breeding; presumably introduced (Bannerman and Bannerman 1968; CJH).

## *Ploceus manyar*  **Streaked Weaver**

PLATE 30
[between pages 472 and 473]

Du. Manyarwever        Fr. Tisserin manyar        Ge. Manyarweber
Ru. Маньярский ткач        Sp. Tejedor listado        Sw. Guldkronad rörvävare

*Fringilla manyar* Horsfield, 1820

Polytypic. *P. m. flaviceps* (Lesson, 1831), Pakistan, India, and Sri Lanka, grading into *peguensis* in eastern Bihar and West Bengal (India); perhaps this race introduced Nile delta in Egypt; *peguensis* Baker, 1925, Bangladesh and north-east India (east of *flaviceps*) to Burma; *williamsoni* Hall, 1957, Thailand to Vietnam; nominate *manyar* (Horsfield, 1820), Java, Bali, and Bawean (Indonesia).

**Field characters.** 15 cm; wing-span 20-22 cm. Size close to House Sparrow *Passer domesticus* but with proportionally even heavier bill and head and shorter wings and tail. Suggests large-billed, chubby sparrow *Passer*. Adult ♂ has distinctive breeding plumage: black head with bright yellow crown, conspicuously streaked back, breast, and flanks, blackish wings, and cream underparts. Non-breeding ♂ and ♀ show long, broad, decurved yellow supercilium and border to ear-coverts. Sexes markedly dissimilar in breeding season; much less so for rest of year. Juvenile separable.

ADULT MALE BREEDING. Moults: June-September (complete); February-April (mostly head and fore-parts).

Long crown bright, slightly greenish yellow, streaked on nape; forms noticeably pale cap to otherwise very dark head. Face, cheeks, and sides of neck black; chin and throat dusky. Below these tracts, strong black streaks extend over pinkish-chestnut back and ochre chest and flanks. Wings dull black; when fresh, fringed pale buff on lesser, median, and innermost greater coverts and tertials (on some, tips of median coverts strong enough to form wing-bar), and narrowly edged yellow-green on flight-feathers. Tail dull black, edged yellow-green at base. Rest of underparts and underwing cream. Bill black. Legs flesh to pale brown. ADULT MALE NON-BREEDING. Resembles breeding ♀ but retains vestiges of yellow cap and dark

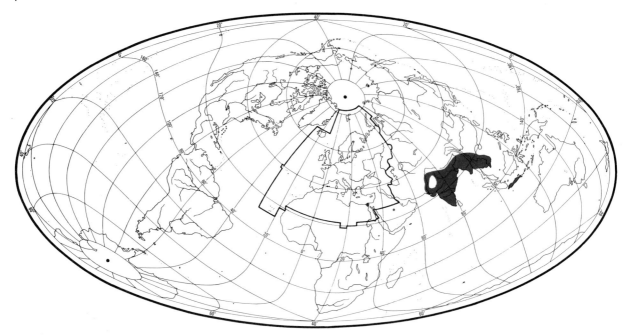

cheeks, while fore-supercilium and face may be brighter yellow. ADULT FEMALE BREEDING. Differs distinctly from breeding ♂ in boldly patterned, mainly yellow head and throat. Long crown, nape, and ear-coverts dusky, finely streaked buff and contrasting with long, broadening lemon-yellow supercilium and contiguous border to ear-coverts. Patch in front of and below eye and ring around it pale yellow; create pale-faced appearance. Incomplete lateral streaks on sides of breast and along flanks narrower than ♂. Bill pale brown-horn, with pinkish or yellowish tinge on lower mandible. ADULT FEMALE NON-BREEDING. Paler, less yellow on head, streaks on underparts even more pointed. JUVENILE. Resembles non-breeding ♀ but plumage pattern more diffuse, with duller wings and tail. Supercilium, border of ear-coverts, and ground-colour of entire underparts isabelline, lacking clarity of ♀; streaks on underparts smudged. Bill yellow-brown, with flesh tone on lower mandible. For 1st-year, see Plumages.

Adult ♂ in breeding plumage unmistakable but non-breeding ♂, ♀, and immature can be confused with similarly-sized *Ploceus* of Africa, which like *P. manyar* may occur as escapes in west Palearctic (see Williams and Arlott 1980, Serle and Morel 1977). Flight, general behaviour, and colonial breeding recall sparrows *Passer*. Like other *Ploceus*, looks front-heavy on perch or ground and in flight.

Flock chorus recalls *Passer domesticus*. Commonest call 'chack', with stony quality suggesting Wheatear *Oenanthe oenanthe*.

**Habitat.** In Nile delta, breeds in reeds *Phragmites australis* up to 3 m high growing in shallow water, in one case in a band *c.* 1 km wide; habitat shared with Clamorous Reed Warbler *Acrocephalus stentoreus* and Fan-tailed Warbler *Cisticola juncidis*; one nest was in reeds in ditch bordering extensive salt-marsh (Meininger and Sørensen 1984). In natural range in India and Burma inhabits flat, swampy, and flooded land and riverbeds, especially reed-mace *Typha* and *Phragmites* standing in water, but is also found in tall grass (Ali 1968; Ali and Ripley 1974).

**Distribution.** EGYPT. Introduced, probably originating from birds escaped from Alexandria Zoo in 1971. First recorded breeding 1978; spread rapidly across Nile delta 1984-5. (Goodman and Meininger 1989.)

**Movements.** Sedentary. Moves about in flocks outside breeding season, though no information for introduced Egyptian population. In India, present in breeding area all year, with no reports of movements other than local ones in search of food (Crook 1963a; Ali and Ripley 1974; V C Ambedkar). Local movements reported from Nepal (Inskipp and Inskipp 1985). In western Pakistan, 2 birds 6·4 km west of Quetta (well outside normal range) 17 July 1913 had probably escaped from bazaar (Ticehurst 1926-7). In Thailand, 10 birds bought at Bangkok market, then ringed and released, were recovered at 16-208 km (average 102 km) in various directions, chiefly between south-west and north (McClure 1974); these movements were presumably stress-related. No other reports of movements.

DFV

**Food.** All information extralimital. Almost wholly gran-ivorous; young fed mainly insects. As well as foraging on ground, often feeds on flower- and seed-heads of grasses (Gramineae), etc., clinging to stems or heads (Roberts

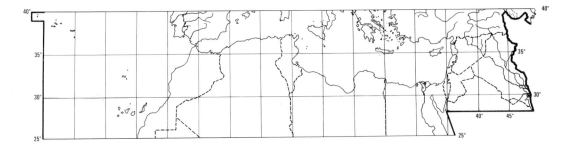

1992). May cause some damage to cereals, but since also consumes seeds of weed grasses and harmful insects need not be regarded as a pest species (Dhindsa and Toor 1990).

Diet includes the following. Invertebrates: grasshoppers (Orthoptera: Acrididae), larval Lepidoptera (Pyralidae, Noctuidae), beetles (Coleoptera: Curculionidae), spiders (Araneae: Araneidae, Lycosidae, Clubionidae, Thomisidae, Salticidae), snails (Gastropoda: Subulinidae, Planorbidae). Plants: seeds of reedmace *Typha* and grasses *Poa, Echinochloa, Saccharum, Phalaris, Cenchrus,* wheat *Triticum,* pearl millet *Pennisetum,* rice *Oryza, Sorghum,* reed *Phragmites.* (Dhindsa 1986; Dhindsa and Toor 1990; Roberts 1992.)

In Punjab (north-west India), feeds on newly-planted rice in spring, and on grains of wheat and *Phalaris* on ground; in rainy season (June–September) takes seeds of pearl millet, *Sorghum,* and *Echinochloa,* and, from September, growing rice. In winter, eats rice grains after harvest, especially in outdoor stores of rice straw after fields ploughed in November, often spending entire day at spread rice straw. At this time of year feeding flocks tend to be unmixed, in contrast to rest of year when often forages with other species. Ceases feeding on standing wheat by mid-April when grains no longer milky. 112 stomachs from all times of year contained 37% by dry weight rice, 28% pearl millet, 14% *Echinochloa,* 7% *Phalaris,* 7% wheat, 5% *Sorghum,* and *c.* 0·6% invertebrates; rice present in 53% of stomachs, insects in 21%. Almost no rice in diet until October then sudden increase to *c.* 90% by weight; in March–June, wheat rose to *c.* 40% of diet, remainder mainly seeds of *Phalaris;* during June–August, *Echinochloa* rose to *c.* 80%, but otherwise pearl millet dominated (up to 80% of diet) and *Sorghum c.* 30% May–November. Average daily consumption in captivity 4·3 g of seeds (*n* = 10 birds over 10 days). (Dhindsa 1986; Dhindsa and Toor 1990, which see for comparison with 2 other *Ploceus;* Roberts 1992.)

Stomachs of 59 nestlings in Punjab contained 61% (by dry weight) invertebrates (51% grasshoppers, 4% beetles, 2% caterpillars, 2% spiders, and 2% snails—see 2nd paragraph for families), 6% plant material (4% wheat, 2% rice), 33% unidentified; spiders, caterpillars, and grasshopper nymphs taken in greater proportions by youngest nestlings, and beetles, snails, and adult grasshoppers by older ones (Dhindsa and Toor 1990). BH

**Social pattern and behaviour.** Little information on introduced population in Egypt, but well studied in India (Crook 1963*a;* Ambedkar 1972; Dhindsa 1986).

1. Sociable all year, e.g. 'enormous' flocks reported in Burma (Smythies 1986). In Java, apparently some all-juvenile flocks formed when 1st broods leave nest, but joined by adults once 2nd broods fledged (Spennemann 1926). In winter, sometimes forms mixed flocks with other *Ploceus* and buntings (Emberizidae) (Oates 1883; Crook 1963*a),* but in northern Thailand, early May, when flocks of Baya Weaver *P. philippinus* also present, the 2 species did not intermingle (Deignan 1945); see also Roosting (below). In Uttar Pradesh (India), ♂♂ leave flocks before ♀♀ to start nest-building, but maintain flocking away from colony (Crook 1963*a).* BONDS. Mating system apparently varies between monogamy and polygamy, with promiscuity by both sexes also recorded, but proportions of each not known. ♂♂ said by Ali and Ripley (1974) to be polygynous like other *Ploceus* and monogamy only occasional. In Punjab (India), *P. manyar* the least polygynous of 3 *Ploceus* studied (Dhindsa 1986). Polygamy reported from Java (Delacour 1947), but Crook (1963*a)* concluded from study by Spennemann (1926) that *P. manyar* is monogamous there. Main Uttar Pradesh study apparently found monogamy the norm: closely observed groups of 3 and 5 nests in one colony all belonged to different ♂♂, and in another colony of 30 nests each ♂ visited only 1 nest (except occasionally to steal material). Unlike Black-breasted Weaver *P. benghalensis* and *P. philippinus* in same area, no ♂♂ recorded building series of nests, and 2 ♂♂ also seen incubating (not recorded in typically polygamous *Ploceus;* see below). In mixed colony with *P. benghalensis, P. manyar* ♂♂ ceased courtship and chasing (see Heterosexual Behaviour, below) long before *P. benghalensis,* though nests of both species were established at same time. (Crook 1963*a.)* Evidence nevertheless of polygyny: 2 Uttar Pradesh ♂♂ each built 2 nests and performed courtship-displays at both structures, while another ♂ built 4 nests in succession and acquired a mate at 3 (4th nest not completed); one ♂ in central India maintained 3 nests at different stages of construction (Crook 1963*a;* Ambedkar 1972). Overall, considerable competition between ♂♂ for ♀♀, and ♂♂ 'vastly in majority' in Sind (Pakistan) (Ticehurst 1922), though not clear whether this reflects true sex-ratio; ♂♂ sometimes rapidly enter another's territory for extra-pair copulations, and neighbouring ♂♂ all seen copulating (apparently successfully) with one particular ♀ during her visits to colony (Crook 1963*a).* In Java, ♂♂ reported to destroy nests not accepted by ♀ and to rebuild in same or adjacent site, one ♂ destroying 3 nests in succession. ♀ apparently chooses (incomplete) nest of a particular ♂, though his

song and display presumably also play a role in mate-choice. (Spennemann 1926.) See also Breeding Dispersion (below). Study in Java found birds normally double-brooded (Spennemann 1926), but no firm evidence of pair-bond persisting for 2nd breeding attempt (though see discussion of parental roles, below), and possibility of polygyny needs to be considered; 2nd broods reported by Dhindsa (1983) separated by a month or more (see Breeding) and persistence of pair-bond in such cases seems even less likely. Abnormally large clutches in Punjab study apparently due to ♀♀ losing nest at critical stage laying in another's (Dhindsa 1983). Apparently hybridizes occasionally with other *Ploceus*, but following factors probably promote reproductive isolation: coloration of breeding ♂♂; courtship sequence; postures, orientation and vocalization during Wing-beating display (most important); site, shape, and material of nest; habitat (see Heterosexual Behaviour, below). In mixed colony with *P. benghalensis*, ♂ recorded mounting soliciting ♀ of that species, and ♀ *P. manyar* (at least 1 already paired and with nest) are occasionally chased and mounted by ♂ *P. benghalensis*, though no case of ♀ *P. manyar* actively selecting ♂ and nest of different *Ploceus*. 2 records of possible hybrids with *P. philippinus* in captivity. ♂ builds nest, but incubation mainly by ♀, though (unusually for *Ploceus*) ♂ frequently does some. (Crook 1963a; Ambedkar 1972; Roberts 1992.) Both sexes feed young (Baker 1934), but ♂ starts only when young *c.* 10 days old according to Ambedkar (1972). At colony in Thatta (Pakistan), young fed by ♀, all ♂♂ occupied in nest-building (Roberts 1992). In Java, when 1st brood fledged, ♂ builds 2nd-brood nest at same site, or may build 2nd nest nearby while ♀ cares for 1st-brood nestlings, then dividing time between displaying on old nest and building new one (Spennemann 1926). No information on age of independence of young. First breeding (for ♂♂ at least) apparently at 2 years old (Dhindsa 1986); see also Breeding Dispersion (below). BREEDING DISPERSION. Territorial and generally colonial; rarely solitary in Burma (Smythies 1986), not uncommonly in India according to Oates (1890). In Egypt, one colony comprised 8 complete nests and 4 under construction (all within a few metres); another comprised 2 complete and 1 incomplete; solitary nest (with flock of 16 birds nearby) also recorded (Meininger and Sørensen 1984; see also Goodman and Meininger 1989). In Punjab, 31 nests in *c.* 100 m² (Whistler 1924); estimated density 94 breeding birds per km² (Dhindsa 1986). Pakistan colonies usually of 12–20 nests, often less than 100 m between colonies (Holmes and Wright 1969; Roberts 1992). In Uttar Pradesh, colonies of 3–30(–60) nests (Crook 1963a; Ambedkar 1972). Colonies in native range can be large where habitat restricted, e.g. one of *c.* 100 nests 20 m across (Oates 1890; Ticehurst 1922; Crook 1963a). Nests in study by Crook (1963a) 2–3 m (1–6) apart; sometimes only *c.* 15 cm in Pakistan (Roberts 1992). Uttar Pradesh colonies all close to 3 other *Ploceus*. In native range, quite often forms colonies mixed with or close to other *Ploceus* (Oates 1890; Crook 1963a; Ambedkar 1972; Dhindsa 1986). Territory apparently restricted to nest and immediate vicinity (Spennemann 1926). Territory established and nest built by ♂ who is able to build 2–3 nests in quick succession. Territory used for courtship and copulation. (Crook 1963a; Ambedkar 1972.) Immature ♂♂ build incomplete nests late in season, usually segregated from colony; such nests apparently not selected by ♀♀, so ♂♂ do not breed (Ambedkar 1972; Dhindsa 1986). ROOSTING. Communal and nocturnal. Study in Punjab found birds roosting with 15 other species throughout year, in sugarcane *Saccharum* and *Typha-Arundo* reedbed (Dhindsa and Toor 1981). In Uttar Pradesh, when ♀♀ incubating or brooding, ♂♂ roost communally away from colony (Ambedkar 1972). Ringing study in Thailand suggested marked

fidelity to reedbed sites (McClure 1974). Arrival at roost in Punjab 32–38 min before sunset January–March, 46–64 min before in other months (see Flock Behaviour, below); during April–June, birds also roosted during hottest part of day in trees with dense foliage (Dhindsa and Toor 1981).

2. Behaviour has much in common with *P. philippinus* and *P. benghalensis* with which *P. manyar* forms species-group (Crook 1963a). In Java, when small hawk (Accipitridae) attacked 3 fighting adults, all birds in colony rapidly flew down into dense thicket, emerging singly *c.* 5 min later (Spennemann 1926). No information on relative shyness towards man at or away from colony. FLOCK BEHAVIOUR. At Punjab roost, birds initially arrive in flocks of 10–15; early arrivals usually take off and circle briefly several times. Give loud chirping chorus just before sunset when all settled in roost. (Dhindsa and Toor 1981.) SONG-DISPLAY. ♂ sings (see 1 in Voice) from nest or nearby, neck upstretched. Change to different song-type, singing in flight low over colony, and Wing-beating display (serving to advertise territory to other ♂♂, and nest to ♀♀) are associated with presence of ♀: for further details, see subsection 2 in Heterosexual Behaviour (below) (Crook 1963a; Ali and Ripley 1974). Song of one ♂ will stimulate whole colony into chorus which is almost continuous in early stages of breeding (Roberts 1992). No information on diurnal rhythm of singing and no details of song-period, though clearly linked to breeding. ANTAGONISTIC BEHAVIOUR. In Uttar Pradesh study, much less trespassing than in *P. benghalensis*; occasional intruders are subjected to supplanting attacks but easily evade them. When flock arrives in colony, each ♂ gives call 4 (see Voice) on going separately to his nest; call perhaps thus has some territorial significance. (Crook 1963a.) HETEROSEXUAL BEHAVIOUR. (1) General. Similar to other *Ploceus* of India. Courtship following first approaches of ♀♀ to territories includes displays near nest and attempted copulation; later, follows sexual chases away from colony. In Uttar Pradesh, 29% of 45 courtship sequences contained marked aggression, 47% ended in copulatory behaviour, 24% ended inconclusively with departure of ♀. ♂ apparently much less aggressive in courtship than *P. benghalensis*, showing strong tendency to behave sexually and to flee from approaching ♀. Courtship attempts frequently frustrated by ♀'s lack of responsiveness rather than by aggression from either sex. (Crook 1963a.) (2) Pair-bonding behaviour and nest-site selection. ♂ chooses site and starts nest in absence of any ♀♀ (Spennemann 1926). Pair-formation takes place when ♀ comes to inspect nest which has reached stage of inverted cup with strap for perching across open bottom. Lower rim of nest sometimes plastered with mud or dung and occasionally adorned with flower petals which may serve to attract ♀. (Baker 1934; Ambedkar 1972; Dhindsa 1986; Roberts 1992.) Mud-plastered 'cock-nests' mentioned by Oates (1890) are not used for laying according to Smythies (1986)—though reference may be simply to nests rejected by ♀♀. ♂ will destroy nests not accepted by ♀ and rebuild in same site or nearby (Spennemann 1926). ♂ will steal material from another's nest (Crook 1963a). Birds of either sex approaching colony give frequent and loud Chirt-calls (see 2 in Voice) over *c.* 100 m, ♂♂ perched at nest responding with dramatic, highly infectious Wing-beating display (Fig A): body upright and facing approaching birds or ♀, bill pointed forward and tail (sometimes slightly fanned) down; wings raised at *c.* 45° above back and beaten vigorously. ♂ gives call 3 or (in courtship to ♀) slightly curtailed song (see Voice). (Crook 1963a.) In Egypt, ♂♂ seen climbing up and down reeds while 'flicking' wings and tail (presumably Wing-beating display) and singing (Meininger and Sørensen 1984); for further descriptions of this and Wing-shivering, see Baker (1934) and below. All ♂♂ may suddenly burst into song and start building if ♀ flies over colony

A

(Ambedkar 1972). ♀♀ entering colony rapidly disperse to territories of displaying ♂♂. ♀ sometimes slips into colony unnoticed and quietly enters a territory whose owner greets her by Wing-beating or (often) attacks and chases her out of colony. ♀ inspecting nest moves nervously, while ♂ flies to roof of nest (Spennemann 1926), or hops about, displaying and singing frequently, especially if ♀ enters nest. ♂ sometimes adopts Stiff-winged posture (Fig B, from drawing of similar posture in *P.*

B

*benghalensis*). (Crook 1963*a*.) In attempt to entice ♀ to nest, ♂ will also fly from reed to reed in or over territory with slow, deliberate wing-beats, legs dangling, and vigorously giving jingling song-type (see 1 in Voice); chases ♀ in same flight if she flies away, but soon returns (Ambedkar 1972; Ali and Ripley 1974). If ♀ leaves colony, ♂ may nevertheless pursue her closely for some distance, and, if she lands, Wing-shivers near her (crouches, vibrating drooped wings and tail, like soliciting ♀); then leads her back to territory, following at once if she diverges; returning birds give Chirt-calls. On arrival at nest, ♂ performs Wing-shivering display on strap. ♂ immediately attempts to mount ♀ if she joins him on strap, ♀ often accepting or may flee pursued by ♂ (or several ♂♂). In 5 cases, ♀'s departure apparently caused by ♂'s aggressive song, but ♀ rarely abandons territory completely. ♂ returning to nest resumes building (Ambedkar 1972) or Wing-shivers. (Crook 1963*a*.) See also subsection 4 (below). In Java, one observation of ♀ helping with nest-building unhindered; otherwise not tolerated by ♂ and driven away; ♀ sometimes brings grass for lining, but does not work it into structure (Spennemann 1928, 1937). (3) Courtship-feeding. Not recorded. (4) Mating. Copulation (attempt) frequently follows return from chase and takes place in vegetation near nest or (10 out of 16 cases) on strap of part-built nest (Ambedkar 1972; Dhindsa 1986). Receptive ♀ will solicit by

crouching, Wing-shivering, and calling (see 5a in Voice); ♂ fluttering about in attempt to mount gives call 5b (see Voice). Extra-pair copulation may occur just as rightful ♂ dismounts. (Crook 1963*a*.) (5) Behaviour at nest. Following pair-formation and ♀'s acceptance of part-built nest, ♂ completes it (usually over *c*. 1 day), ♀ then laying within average 1·36 ± 0·67 days (*n* = 11) (Dhindsa 1983), or *c*. 3–4 days after copulation (Ambedkar 1972). ♂ builds entrance-tube while ♀ laying or incubating, and typically then starts new nests, which may be left unfinished (Ticehurst 1922; Baker 1934; Roberts 1992). See Bonds (above) for successive polygyny associated with multiple nest-building. Ambedkar (1972) recorded little or no work on nest once completed, but other studies (Oates 1890; Roberts 1992) found ♂♂ refurbishing nests or building new ones even after hatching. ♀ frequently leaves nest to feed or bathe, but incubates continuously at night (Ambedkar 1972). RELATIONS WITHIN FAMILY GROUP. After hatching, eggshells dropped through entrance-tube (Spennemann 1926). Young brooded by ♀ for *c*. 1 week; fed by ♀ until ♂ starts at *c*. 10 days (Ambedkar 1972). Nest-sanitation apparently by ♀ alone (Dhindsa 1986), though faecal sacs simply dropped through entrance-tube, and older nestlings defecate out of nest (Spennemann 1926). ANTI-PREDATOR RESPONSES OF YOUNG. No information. PARENTAL ANTI-PREDATOR STRATEGIES. In Uttar Pradesh colony, ♂♂ tried to drive off Jacobin Cuckoo *Clamator jacobinus* perched on nest; did not press home attack and *C. jacobinus* left after *c*. 5 min (Ambedkar 1972). In Java, persistent loud alarm-calls (see 6 in Voice) given when Collared Scops Owl *Otus bakkamoena* in colony and predating young. When nestlings attacked by invading ants *Oecophylla*, highly agitated ♀♀ remained by nests, but did not enter; ♂♂ then detached nests, letting them fall to ground. (Spennemann 1926.)

(Figs by D Nurney from drawings in Crook 1963*a*.)    MGW

**Voice.** Freely used in breeding season; no information for winter. All calls louder, more emphatic than similar calls of Black-breasted Weaver *P. benghalensis* (Crook 1963*a*). Detailed study needed to determine size of repertoire, including what constitutes song: status of some sounds in recordings analysed not clear.

CALLS OF ADULTS. (1) Song of ♂. Attractive series of *c*. 6 high-pitched 'tsi' or similar short whistling notes, culminating in long, drawn-out wheeze; louder and more tuneful than *P. benghalensis*; softer and more musical than otherwise similarly structured song of Baya Weaver *P. philippinus*, also more distinctly rising and falling in pitch and with fewer 'chirr-churr' sounds. Abbreviated song typically directed at ♀ during courtship. (Crook 1963*a*; Ali and Ripley 1974; Roberts 1992.) Transcriptions of song (or part of it) include 'tzrr we tsee tsee tsi tsi tser cheeze we' (Crook 1963*a*), 'see-see-see-see-see-see' followed by 'o-cheeee' or 'o-wheeeez' (Ali and Ripley 1974), and ascending 'tri-tri-tri' preceding lower-pitched, extended 'tee-te-teeh u-u-h dze-ee-h' (Roberts 1992). Recordings contain phrases of this type, but certain other units provisionally treated here as additional song-components. In Fig I, phrase like that above is preceded by 2 subdued, rather plaintive, slightly nasal and diadic 'wheeb' or 'wuh' sounds (disyllabic 'w-wheeb' or 'w-wuh' in another part of same recording); then a diadic 'seeep' followed by delicate tonal 'd-oo pee k-loooo' of remarkable bell-like quality,

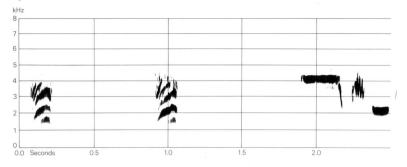

I  C Chappuis  India  March 1978

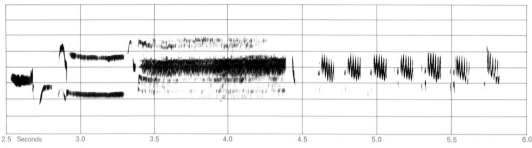

I  *cont.*

and long, thin, strangled, diadic wheezing 'zzzzeeee' reminiscent of wheeze in song of Siskin *Carduelis spinus*; finally, 7 vibrant 'jree' or 'chree' units. In another song (same recording), wheeze is followed by brisk series of 'chiz' units recalling chirping of House Sparrow *Passer domesticus*, while longer tonal series preceding wheeze suggests: 'see ti-PEEE d-oo gee peeee'. (W T C Seale, M G Wilson.) Different song-type given when sighted near nest a pleasing, more spirited, jingling phrase: 'tililileekitee tililileekitee' (Ali and Ripley 1974). Sound of singing birds from colony in Egypt recalled twittering of *P. domesticus* flock (Meininger and Sørensen 1984); perhaps refers to other calls as well as song. Recordings also contain variety of twittering, buzzy chirping, and other sounds sometimes given in compact series which are perhaps song, though some units (e.g. 'srreeu' or 'srroo', 'chrip', and rather delicate, dulcet 'chisrip') equally likely to serve (also) as calls (see calls 3-4, below). Fig II shows sequence of

spirited chirruping or chirping sounds, first 3 (somewhat rasping) and last 5 flanking attractive, delicate, distinctly vibrant unit of liquid timbre and descending pitch: frequency-modulated 'srreeu'. Recording of another sequence (Fig III) suggests 'srr chip chip chip sszip srr

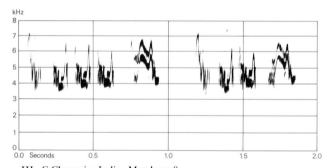

III  C Chappuis  India  March 1978

chip chip sszip', in which 'sszip' a short sucking squeak or wheeze and 'chip' units not unlike flight-call of Redpoll *C. flammea*. (W T C Seale, M G Wilson.) (2) Chirt-call. Birds flying in to colony give loud, rather harsh and short 'chirt chirt chirt' reminiscent of Tuc-call of Wheatear *Oenanthe oenanthe* (Crook 1963a; Meininger and Sørensen 1984). (3) A 'tre tre cherrer cherrer' given repeatedly by ♂ during Wing-beating display (Crook 1963a). (4) Variety of chirring sounds (apparently distinct from calls 2-3) given by ♂, especially when moving to nest after flock arrives in colony (Crook 1963a). (5) Calls associated with copulation. (5a) Thin piping sound given repeatedly by soliciting ♀. (5b) A 'chewe chewe chewe' given by ♂

II  C Chappuis  India  March 1978

during copulation attempts. (Crook 1963*a*.) (6) Persistent loud and plangent calls signalling alarm or distress (Spennemann 1926). (7) Other calls. In Sri Lanka, constant chattering and whistling from flocks (Henry 1971); reference is perhaps only to calls described above.

CALLS OF YOUNG. Food-call twittering, audible considerable distance from colony (Ambedkar 1972; Roberts 1992). MGW

**Breeding.** Information almost wholly extralimital. SEASON. Egypt: eggs at end of 1st week of May, though nests still being built mid-May (Meininger and Sørensen 1984; Goodman and Meininger 1989). India and Pakistan: breeding peaks during monsoon, June–September (Ali and Ripley 1974; Roberts 1992). SITE. In Egypt, in stands of reed *Phragmites australis* in shallow water, nests attached to tops of 1–2 stems 2–3 m above surface (Meininger and Sørensen 1984). In India and Pakistan, also in reeds, etc., always at least 1 m above water, or in trees which may overhang water; 82% of colonies in Punjab (north-west India) above or very close to water (Dhindsa 1986); often in tall cereal crops (Ticehurst 1922; Crook 1963*a*; Dhindsa 1986; Roberts 1992). In northern India, can be very close to houses (Ambedkar 1972). Nest: coarsely-woven retort-shaped brood pouch with short, downward-pointing entrance tube, made of long strips of grass or leaves of various widths; brood pouch often has clumps of mud and dung plastered on outside, and sometimes inside, commonly with embedded flower petals; in Punjab, birds fly some distance to obtain dung, even when there is mud at nest-site (Crook 1963*a*, *b*, which see for comparison with other Ploceinae nests; Dhindsa 1986, which see for description of slightly different nests of nominate *manyar* in Java). Brood pouch height *c.* 20 cm, external diameter *c.* 12 cm; entrance tube 5–10 cm long (sometimes up to 20 cm, exceptionally 70 cm), and *c.* 6 cm in external diameter (Oates 1890; Ali and Ripley 1974; Roberts 1992); nests over water tend to have shorter tubes (Dhindsa 1986). Building: by ♂ only (though see Spennemann 1937 and Social Pattern and Behaviour for observation of ♀ bringing material in Java); 1st stage an inverted basket with strap beneath, hung from a few tied-together leaves, reed-tips, or thin twigs incorporated into structure; nest only completed after ♀ selects mate, brood pouch being finished rapidly in *c.* 1 day, and ♂ often still building entrance tube when eggs already laid. Bird makes strips of reed or palm leaf by grasping leaf near base in bill then flying off towards tip (Crook 1963*a*; Ambedkar 1972; Dhindsa 1986). ♂♂ continue to build new nests while ♀ feeds young, and repair existing nest (Roberts 1992). For details of building technique, see Crook (1963*a*, *b*). EGGS. See Plate 90. Sub-elliptical, white, smooth, and not glossy. *P. m. flaviceps*: 20·4 × 14·4 mm (18·5–21·6 × 13·1–15·1), *n* = 81; calculated weight 2·24 g (Schönwetter 1984); 50 eggs from Punjab had average fresh weight 2·33 g (Dhindsa 1986); see Meininger and Sørensen (1984) for

dimensions of 9 eggs from Egypt, and Crook (1963*a*) and Ambedkar (1972) for descriptions of eggs from northern India. Nominate *manyar*: 20·8 × 14·5 mm (19·5–22·4 × 13·8–15·2), *n* = 36; calculated weight 2·32 g (Schönwetter 1984). Clutch: 2–3 (1–4); some clutches of 4–6 probably have last 1–2 eggs laid by other ♀♀ (Dhindsa 1983, 1986). In Punjab, average size of 165 clutches 2·9, 83% of clutches 2–3 (Dhindsa 1983). In northern India, of 84 clutches: 1 egg, 6%; 2, 17%; 3, 52%; 4, 18%; 5, 6%; 6, 1%; average over 4 years 3·0 (2·4–3·2) (Ambedkar 1972). Eggs laid 1–3 days (average 1·4, *n* = 11) after completion of brood pouch. 2 broods in Punjab in different nests (eggs perhaps laid by different ♀♀); replacement clutch often laid in nest of another pair if 1st lost. (Dhindsa 1983.) In Java, new nest for 2nd or new brood often in same place as 1st, which is bitten loose by ♂ after fledging of young (Spennemann 1926). Eggs laid daily (Dhindsa 1986). INCUBATION. From laying to hatching of last egg takes average 13·2 days, *n* = 12 (Dhindsa 1986); from 1st egg, 14–17 days, *n* = 21 (Ambedkar 1972). By both sexes (Crook 1963*a*; Ambedkar 1972; Dhindsa 1986); ♂ incubates in earliest stage of laying but incubation proper by ♀, in stints of only a few minutes at a time, starting with 2nd egg while ♂ completes construction of entrance tube (Ambedkar 1972). Hatching asynchronous (Dhindsa 1986). YOUNG. Fed and cared for by both sexes; ♂ starts to feed only around day 10, and only ♀ seen to remove faecal sacs; brooded by ♀ for *c.* 1 week (Ambedkar 1972; Dhindsa 1986). FLEDGING TO MATURITY. Fledging period 17·5 days (15–20), *n* = 17 (Ambedkar 1972); 16·5 days, *n* = 24; ♂ first breeds at 2 years old (Dhindsa 1986). BREEDING SUCCESS. In Punjab, of 126 eggs, 43·7% hatched and 23·8% produced fledged young; eggs and young lost to adverse weather (including drought, which kills reeds supporting nests), desertion, human disturbance (mainly harvesting of crops), and predation by snakes (especially in trees, less in reedbeds), lizards, and crows *Corvus*; has lowest clutch-size and lowest breeding success of *Ploceus* in Punjab; all eggs laid parasitically (i.e. in nest other than ♀'s own) failed to hatch because of insufficient incubation time, and only 1 egg of 12 in 3 nests known to be parasitized produced fledged young (Dhindsa 1983, 1986). In northern India, 73% of 70 eggs hatched, and 64% produced fledged young; some nests usurped by tree mouse *Vandeleuria* (Ambedkar 1972). In Java, broods killed by ants *Oecophylla* (Spennemann 1926). BH

**Plumages.** (*P. m. flaviceps*). ADULT MALE BREEDING. In fresh plumage (April–May), forehead and crown bright lemon-yellow, contrasting sharply with greyish-black or black side of head and neck. Hindneck lemon-yellow, mottled or streaked black, especially in centre. Mantle and scapulars streaked with equal amounts of black and pink-cinnamon (centres of feathers black, sides cinnamon), back to upper tail-coverts streaked dull black or greyish-black and paler buff or isabelline-grey (but colour of mantle to upper tail-coverts in part depending on extent of pre-breeding moult: retained worn non-breeding with paler feather-sides than fresh breeding). Chin and throat dark grey, contrasting sharply with ochre or cream-yellow chest, side of breast, and flank, which are marked with pointed black shaft-streaks; streaks on mid-chest sometimes rather inconspicuous when plumage quite fresh, streaks longer and narrower on lower flank. Remainder of underparts pale cream-buff or cream-white, under tail-coverts with faint dark shaft-streaks. Tail greyish-black, both webs of central pair (t1) and outer webs of t2–t5 with broad but ill-defined yellow-green fringe (widest on feather-bases). Flight-feathers, greater upper primary coverts, and bastard wing dull black, outer webs narrowly but contrastingly fringed yellow-green, except for emarginated parts of outer primaries, tips of primary coverts, and longest feather of bastard wing. Tertials and greater upper wing-coverts black, both webs of tertials and innermost greater coverts (tertial coverts) and outer web of other greater coverts contrastingly fringed pale pink-buff or cream-buff; median coverts black, sharply fringed cream-buff (widest on tip); lesser coverts dull black or greyish-black with ill-defined grey-buff fringes. Under wing-coverts and axillaries cream-white with pale buff or yellow suffusion, primary coverts with grey bases. *In worn plumage* (about August–September), pale sides of feathers of upperparts bleached to pale buff or grey-white, partly worn off, black more predominant (less so on rump); black of head and throat greyer, much off-white of feather-bases visible on chin and throat, ground-colour of remainder of underparts bleached to pale buff on chest and off-white on mid-belly and vent, short black streaks more distinct; pale fringes of tail, flight-feathers, tertials, and greater and median upper wing-coverts pale greenish-grey, largely worn off. ADULT FEMALE BREEDING. Cap, hindneck, and side of neck dull black with narrow buff (if fresh) to pale grey or off-white (if worn) streaks. Supercilium lemon-yellow, sharply-defined; narrow above lore, wider above ear-coverts, mottled black above rear of ear-coverts. Lore, ear-coverts, and upper cheek dark grey or brown-grey, almost black on lore and just behind eye, washed olive on central ear-coverts, and with short pale yellow or buff patch from just above gape to below eye; a contrasting bright lemon-yellow bar behind ear-coverts. A narrow pale yellow or white eye-ring. Broad stripe over lower cheek pale yellow or buff-white, bordered below by distinct black malar stripe. Upperparts backwards from mantle as in adult ♂ breeding. Chin and throat uniform pale yellow or white, remainder of underparts as in adult ♂ breeding, but dark streaks on chest, side of breast, and flank slightly narrower and shorter. Tail and wing as in adult ♂ breeding. ADULT MALE NON-BREEDING. Like adult ♀ breeding, but yellow on front part of supercilium, patch below eye, and lower cheek sometimes brighter, black malar stripe sometimes absent, and some yellow sometimes mixed in on cap, some dark grey on side of head, or throat has some dark grey triangular spots. All plumage of head and body equally fresh, sides of feathers of upperparts fringed cinnamon. ADULT FEMALE NON-BREEDING. Like adult ♀ breeding, but dark streaks on chest and side of breast narrower, showing as long points instead of broader arrow-marks; ground-colour of side of head and neck whiter, less yellow; bright lemon-yellow restricted to bar on side of neck; all plumage of upperparts equally fresh, fringes cinnamon. NESTLING. Down fairly long and fairly plentiful, on upperparts, arm, and flank; white (Ticehurst 1926). JUVENILE. Like adult ♀, but streaks on upperparts duller, dark grey, less sharply defined, virtually absent from rump; fringes of feathers of upperparts more diluted cinnamon-buff. Ground-colour of side of head and neck and entire underparts, including supercilium, buffish-isabelline, less yellow and white, dark eye-stripe, stripe on lower cheek, and malar stripe buff-brown, head pattern inconspicuous; streaks on

chest and side of breast slightly broader, less pointed, less clearly defined. Wing and tail as adult, but ground-colour of tail, tertials, and flight-feathers dark grey, less blackish; tips of tail-feathers more sharply pointed, and reduced outermost primary (p10) longer and broader, width (at level of tips of primary coverts) 2–3.5 mm (in adult, 1.5–2.5), 5–8 longer than tip of longest greater upper primary covert ($n = 5$) (in adult, 1–6, average 4.5, $n = 27$). FIRST ADULT NON-BREEDING. Like adult ♀ non-breeding, but part of juvenile feathering sometimes retained, especially some secondaries and outer primaries. On Java (nominate *manyar*), where breeding season long, feathering growing early during post-juvenile moult is similar to adult non-breeding, but that growing later on similar to adult breeding or intermediate between breeding and non-breeding; ♂♂ with such a mixed plumage show mixture of yellow-and-black on cap and black-and-brown on side of head; flight-feather moult of these birds often suspended, with some feathers still relatively fresh juvenile. In northern India and Pakistan, where breeding season is shorter, and birds are probably not sexually active until *c.* 1 year old, moult probably more complete and mixed plumages rare. Birds observed in Egypt in mixed plumage (Meininger and Sørensen 1984) either advanced immatures in mixed breeding and non-breeding plumage, or adults in pre-breeding moult.

**Bare parts.** ADULT, FIRST ADULT. Iris grey-brown, brown, or black-brown. Bill of adult ♂ breeding black or brown-black, in ♂ non-breeding and in ♀ light flesh-horn to dark brown with pink-white or pale yellow lower mandible. Mouth whitish-horn. Leg and foot lilac-pink, pale flesh, reddish-flesh, brown-flesh, pale flesh-brown, or light brown. NESTLING. No information. In *Ploceus* generally, mouth usually scarlet or crimson, gape-flanges pale yellow (Swynnerton 1916). JUVENILE. Iris brown. Bill yellow-brown, lower mandible pale flesh-yellow with brown cutting edges. Leg and foot light brown. (BMNH, RMNH, ZMA.)

**Moults.** ADULT POST-BREEDING. Complete; primaries descendent. In Pakistan and north-west India, birds moult (September–) October–December, after breeding June–September; in non-breeding plumage from about November to February or March. ADULT PRE-BREEDING. Partial; head, neck, and chest, as well as variable amount of feathering on mantle, scapulars, or underparts, sometimes tertials and t1. In Pakistan and north-west India, between February and April. (BMNH, RMNH.) POST-JUVENILE. Starts shortly after fledging; extent variable (see Plumages). Involves at least head, body, lesser and median upper wing-coverts, sometimes tertials, greater coverts, and t1; or all tail and variable number of flight-feathers as well; or moult complete. Moults to non-breeding or to variable mixture of breeding and non-breeding.

**Measurements.** ADULT. *P. m. flaviceps.* Pakistan and north-west India, all year; skins (BMNH). Bill (S) to skull, bill (N) to distal corner of nostril; exposed culmen on average 2.1 less than bill (S).

| | | | | | |
|---|---|---|---|---|---|
| WING | ♂ | 71.2 (1.12; 19) 69–73 | ♀ | 68.2 (1.15; 5) | 67–70 |
| TAIL | | 42.9 (2.20; 9) 39–46 | | 41.9 (1.29; 5) | 40–44 |
| BILL (S) | | 19.1 (0.92; 9) 18.2–20.7 | | 18.3 (0.70; 5) | 17.4–19.0 |
| BILL (N) | | 12.1 (0.42; 9) 11.3–12.6 | | 11.9 (0.14; 5) | 11.7–12.1 |
| TARSUS | | 22.0 (0.64; 9) 21.1–23.0 | | 20.8 (0.41; 5) | 20.3–21.3 |

Sex differences significant, except for tail and bill (N).

Nominate *manyar*. Java, all year; skins (RMNH, ZMA).

| | | | | | |
|---|---|---|---|---|---|
| WING | ♂ | 70.0 (1.35; 21) 67–73 | ♀ | 66.7 (1.08; 10) | 65–69 |
| TAIL | | 43.0 (1.47; 21) 40–46 | | 41.3 (1.23; 10) | 40–43 |
| BILL (S) | | 19.0 (0.38; 20) 18.4–19.9 | | 18.0 (0.48; 10) | 17.4–18.7 |

BILL (N)   12·3 (0·40; 20) 11·8–13·0         11·7 (0·49; 10) 11·0–12·2
TARSUS     21·4 (0·45; 18) 20·8–22·3         21·1 (0·36; 10) 20·7–21·6

Sex differences significant.

JUVENILE. Wing length similar to adult, tail on average c. 1·5 shorter (RMNH).

**Weights.** Thailand: May, ♂♂ 16·4, 18·4 (Melville and Round 1984).

**Structure.** Wing short, broad at base, tip rounded. 10 primaries: p7–p8 longest, p9 1–3 shorter, p6 0–1, p5 1–3, p4 3–6, p3 6–9, p2 8–11, p1 11·0 (25) 7–13; p10 somewhat reduced, narrow, 33·6 (22) 19–37 shorter than p7–p8 (see also Juvenile in Plumages). Outer web of p5–p8 emarginated; inner web of p6–p9 with faint notch. Tip of longest tertial reaches to tip of p3–p4. Tail short, tip slightly rounded; 12 feathers, t6 2–7 shorter than t1. Bill strong, conical; 10·6 (10) 9·9–11·4 mm deep at base, 9·8 (10) 9·0–11·3 wide; tip slightly compressed laterally; culmen gradually decurved towards pointed bill tip, cutting edges slightly decurved, kinked downward at extreme base; gonys straight. Nostril rather small, rounded, bordered by frontal feathering at rear. Bristles at gape strongly reduced. Leg and foot rather short but strong. Middle toe with claw 22·0 (10) 20·5–23 mm; outer toe with claw c. 73% of middle with claw, inner c. 71%, hind c. 84%.

**Geographical variation.** Rather distinct; mainly in colour. Nominate *manyar* from Java differs from *flaviceps* of Indian pen-insula in deeper and more rufous ground-colour and browner tinge of black streaks on body and (in ♂ breeding) of black on head; fringes of mantle and scapulars rufous-cinnamon when fresh (pink-cinnamon in *flaviceps*, but pale buff to off-white in both races when worn), cap of ♂ more golden-yellow; chest, side of breast, upper belly, and flank deeper cinnamon, not as pale cream-yellow, contrasting more with cream-white of mid-belly and vent; amount of dark streaking as in *flaviceps*. *P. m. williamsoni* from south-east Asia intermediate in colour between *flaviceps* and nominate *manyar*, but side of head of ♂ and black streaks tinged brown (like nominate *manyar*), chest and flank pink-cinnamon when fresh, not as rufous as nominate *manyar*, nor as pale as *flaviceps*; *peguensis* of Assam, Bangladesh, and Burma also intermediate in ground-colour, but black streaks heavier, pure black (not as brown as *williamsoni*). See also Hall (1957).

Forms species-group with Baya Weaver *P. philippinus*, Finn's Weaver *P. megarhynchus*, and Black-breasted Weaver *P. benghalensis* (Moreau 1960); all 4 closely similar in measurements and structure (*P. megarhynchus* slightly larger), but geographical ranges overlap widely and thus cannot be united into single superspecies. Adult ♂ breeding *P. philippinus* and *P. megarhynchus* differ mainly from ♂ *P. manyar* in uniform buff chest, side of breast, and flank, without streaking (*P. philippinus* has blackish chin and throat, as in *P. manyar*, *P. megarhynchus* has yellow throat), *P. benghalensis* differs in black neck, upper mantle, and chest. See Crook (1963a) for possible evolutionary history of species-group.                                                    CSR

---

*Quelea quelea* (Linnaeus, 1758)   **Red-billed Quelea**

FR. Travailleur à bec rouge        GE. Blutschnabelweber

Widespread and locally migratory in Afrotropics, often forming immense flocks and breeding in huge colonies. Flock of several hundred recorded on Tenerife (Canary Islands) 23 November 1965, and several dozen still present on 29 November (Hytönen 1972); listed as accidental by Fernandez-Cruz et al. (1985), but the birds were poorly described and escape possibility not considered.

---

# Family ESTRILDIDAE waxbills, grassfinches, mannikins

Tiny to small, often thick-billed oscine passerines (sub-order Passeres); for monograph, see Goodwin (1982). Estrildids typically birds of open country—grassy savanna and thorn scrub—but also found in forest, reedbeds, etc. Feed chiefly on ground or in low vegetation, mostly eating grass seeds but insects also taken, and fruit by some. Of mainly tropical African and Asiatic distribution, with by far largest diversity in Afrotropics, extending to Australasia, where a further but less diverse adaptive radiation has occurred, and to many islands of Pacific Ocean. Some species introduced to areas outside natural ranges (including New World); several also long established as popular cage birds (see also Goodwin 1965b). Largely sedentary though some dry-country species make local movements or are nomadic. 137 species (C S Roselaar); 4 in west Palearctic, 3 introduced breeders and 1 accidental.

Following Steiner (1955), Voous (1977), and Goodwin (1982), estrildids here treated as comprising well-defined, homogeneous family distinct from Passeridae (sparrows), Ploceidae (weavers), Viduinae (indigobirds and whydahs), and Carduelinae (canaries and allies) with which they have at one time or another been associated in pre-DNA classifications (see further, below). 29 genera, falling into 3 main groups (sometimes recognized as tribes), of Afrotropical distribution except as stated: (1) waxbills ('Estrildae') of genera *Lagonosticta* (firefinches, 12 species), *Uraeginthus* (cordon-bleus, 5 species), *Estrilda* (typical waxbills, 17 species), *Amandava* (avadavats, 3 species—also India to Indonesia), *Pytilia* (pytilias, 4 species), *Spermophaga* (bluebills, 3 species), *Pyrenestes* (seed-crackers, 3 species), etc.; (2) grassfinches and allies ('Poephilae') of genera *Emblema* (firetails, 4 species—Australia), *Poephila* (typical grassfinches, 5 species—Australia and Timor), *Erythrura* (parrot-finches, 12 species—Australasia and western Pacific), etc.; (3) mannikins (or munias) and allies ('Lonchurae') of genera *Lonchura* (mannikins, 32 species—some Africa but mainly southern Asia to Australia), *Euodice* (silverbills, 3 species—Africa, Arabia, southern Asia), etc. Monotypic genus *Pholidornis* (Tit-hylia *P. rushiae*, Afrotropics) also sometimes included in Estrildidae but relationships still uncertain.

Sexes generally of similar size. Bill often of typical seed-eater type, deep, short, and well adapted for de-husking seeds (for details of internal structure see Ziswiler 1965); in some species (even seed-eaters), however, quite slender and sharp-pointed. No bristles at base of bill. Wing short and broad with rounded or bluntly pointed tip; 10 primaries, p10 considerably reduced (length 10–30% of longest). Flight variable: e.g. rather fluttering and brief in species frequenting scrub and savanna woodland, strong and undulating in those of open country and reed-beds. Tail usually short and square or rounded, but longer and pointed in some genera (*Lonchura* especially); 12 feathers. As in many other passerines, tail often flicked vertically, laterally, or up-and-down, especially when bird motivated by fear, aggression, etc.; sometimes accompanied by upward flick of closed or partly open wings (see Goodwin 1982 for these and other behavioural characters). Leg and foot short and slender; scutellated in front, booted at sides, sharply ridged at rear. Some waxbills (e.g. *Estrilda*) use feet to hold down grass stems, etc., or to clamp them to perch when feeding or gathering nest-material. Gait a hop in most genera but ground-living quail-finches *Ortygospiza* of Africa also walk and run. Head-scratching by indirect method (see also Simmons 1957b, 1961a). Most species drink by scooping water in normal manner but some Australian dry-country grassfinches, including familiar Zebra Finch *Poephila guttata*, do so by pumping (continuous sucking) (see also Immelmann 1973, Wickler 1961). Bathe in typical stand-in manner, often communally; foliage-bathing also recorded. Sunning common; birds noted to adopt lateral posture but not spreadeagle (see also Simmons 1986). Anting recorded in captive Golden-breasted Waxbill *Amandava subflava*, Red Avadavat *A. amandava*, Melba Finch *Pytilia melba*, and Crimson Finch *Neochmia phaeton* (see also Simmons 1961b, 1966), with one record in wild for Black-bellied Seed-cracker *Pyrenestes ostrinus* (Crook and Allen 1960). Unusually for such small birds, of indirect passive type (ant-exposure); directed at small, acid-squirting ants with

posture most closely resembling that of certain thrushes (Turdidae) and crows (Corvidae); as with Jay *Garrulus glandarius*, ant-application movements also made but without ant in bill (so far as known for certain). Similar responses seen to millipedes (but not to ants) from Red-headed Bluebill *Spermophaga ruficapilla* (Kunkel 1967b) and 2 species of *Uraeginthus* (Goodwin 1971). Dusting not recorded even from species of arid country.

Voice often harsh and discordant. Song loud and musical in some species but mostly rather quiet and intimate without aggressive or territorial significance; primarily sexual in nature. Many species, especially those of open country, often highly gregarious outside breeding season, and some loosely colonial species continue to flock for feeding (etc.) even when nesting. Most, however, breed solitarily. Monogamous mating system the general rule, with strong pair-bond, often maintained throughout year. In most species, mates (and, in some, flock members too) clump together in close body contact when loafing and roosting; allopreening common and widespread but not universal. Nest typically an unwoven domed structure with side entrance (often extended into tube) placed in bush or small tree or in some species in hole (when may lack dome); built by both sexes but with ♂ bringing most of material. Second structure ('cock-nest') often added on top of real nest by many waxbills, some of which also decorate nest with conspicuous and noxious objects. Incubation and care of young by both sexes. Begging behaviour of young distinctive: adopt prone posture with neck twisted through 90–160° so that gape directed upwards. Old nest also sometimes used by adults for roosting, but special nest constructed for this purpose by some species.

Plumage colours variable: many waxbills are brown with contrasting red or blue rump, face, or underparts; grassfinches and allies are quite bright and contrastingly coloured green or brown above with red or black rump or tail with red, black, or blue marks on face, and with white spots, streaks, or bands, or black bars, on underparts; most mannikins and allies more sombre, with contrasting patches of brown, black, and white. Bill often brightly coloured (red, pink, orange, or yellow), but bluish-grey, black, etc., in most species, often with marked waxy sheen. No seasonal variation except in *A. amandava*. Sexes may be alike, rather similar (♀ a little duller than ♂), or markedly different. Juvenile plumage a paler version of adult or quite different. Nestling naked or with varying amounts of down; edge of gape usually with white or black-and-white swellings, some species also having a number of light-reflecting tubercles at base of bill in addition to conspicuous pattern of markings on palate which vary from species to species (for details, see Steiner 1960, Goodwin 1982). Usually single complete adult post-breeding moult annually; starts at end of breeding season but timing highly variable and moult interrupted when conditions suddenly become suitable for breeding again because of local rainfall. Complete post-juvenile moult, usually into adult or near-adult plumage, starts within a few weeks of fledging; completed by time bird 6 months old, though sometimes delayed due to adverse conditions (e.g. drought). Sexual maturity in many estrildids reached at surprisingly early age (e.g. sometimes at 10 weeks in *P. guttata*).

Long accepted that weavers (Ploceidae) are nearest relatives of estrildid finches; this confirmed by egg-white data (Sibley 1970) which also indicate that sparrows (Passeridae)—but not viduines or carduelines—are also related to both estrildids and ploceids but more distantly. DNA data (Sibley and Ahlquist 1990), however, surprisingly suggest that the otherwise ploceid-like viduines (which are nest-parasites on estrildids, even to extent of closely matching appearance of nestlings and juveniles to those of host on a species-to-species basis) are their closest relatives, the 2 groups together forming a subfamily ('Estrildinae') within a greatly enlarged 'Passeridae' which also includes the sparrows and ploceids (see Passeridae and Ploceidae for further details).

## *Lagonosticta senegala*   **Red-billed Firefinch**

PLATE 30
[between pages 472 and 473]

Du. Vuurvinkje    Fr. Amarante du Sénégal    Ge. Senegalamarant
Ru. Обыкновенный амарант    Sp. Bengalí Senegalés    Sw. Amarant

*Fringilla senegala* Linnaeus, 1766

Polytypic. *L. s. rhodopsis* (Heuglin, 1863), Sahel zone from northern Niger to Sudan, western and southern Ethiopia, and north-west Kenya, introduced southern Algeria. Extralimital: nominate *senegala* (Linnaeus, 1766), West Africa south of *rhodopsis*, from Sénégal to Nigeria; *brunneiceps* Sharpe, 1890, highlands of Ethiopia; 5-7 further races in West, East, and southern Africa.

**Field characters.** 9 cm; wing-span 15-16 cm. Slightly smaller and dumpier than Common Waxbill *Estrilda astrild* (more like *Fringilla* finch), with proportionately shorter tail. Diminutive and nervous. ♂ fiery plum-red, with browner back and wings and black tail; ♀ dull brown above, paler below, with contrasting red lore and rump and black tail. Sexes dissimilar; no seasonal variation. Juvenile separable at close range.

ADULT MALE. Moults (Algeria): about May (complete). Head, face, neck, underparts to belly and flank, long rump and upper tail coverts cerise-red, unmarked except for sprinkling of tiny white spots or bars on side of chest (by shoulder); nape and back pale russet-brown, washed with cerise but looking distinctly duller than head; wings russet-brown on coverts but dark brown on flight-feathers, so that folded wing contrasts noticeably with flank and rump; tail black, with red edges to feathers; belly and under tail-coverts pale but dull brown. With wear, red disappears from cap, and spots on back and chest become indistinct. Bill pinkish-red, with black culmen emphasizing sharply pointed shape. Eye red, with narrow yellowish orbital ring. Legs dusky pink-brown. ADULT FEMALE. Head and fore-upperparts dull brown, unmarked except for cerise-red lore and sometimes red wash on face and chin. Underparts pale buff-brown, with small off-white spots or mottling on chest, especially at side, and along upper flank. Orbital ring sometimes white. Otherwise as ♂. JUVENILE. Resembles ♀ but shows distinctive dull grey lore and drabber plumage. Lacks any pale spots on side of chest. FIRST-YEAR. See Plumages.

In Afrotropics, subject to confusion with at least 10 other Estrildidae; within west Palearctic, may be confused with escapes of imported congeners and Red Avadavat *Amandava amandava* (see that species). Flight light and rapid, with fast whirring wing-beats suggesting tiny sparrow *Passer* and allowing sudden manoeuvrability. Hops, but given bird's small size can seem like creeping. Forms small, lively, roving flocks.

Song a short, rather feeble twitter. Commonest call a weak, piping 'tweet tweet' or 'teep teep', often uttered almost incessantly by flocks.

**Habitat.** In Tamanrasset (southern Algeria), occurs near houses, feeding in streets, gardens, and orchards (Gaston 1970). In natural range in Afrotropics, inhabits dry areas with abundant *Acacia*, scrub, or other cover. Has largely colonized cultivated areas, villages, and old towns; avoids close-built modern towns where suitable food gathered on ground is lacking. Readily becomes tame, entering huts and houses. Flies only short distances in lowest airspace. Chiefly in lowlands, but in highland Ethiopia breeds above 1000 m (Goodwin 1982). In East Africa, usually seen feeding on open or bare ground near dwellings; also in scrub thickets and in riverain undergrowth (Williams 1963). In West Africa, abundant and widespread in drier parts of savanna, mainly in towns and villages; less com-

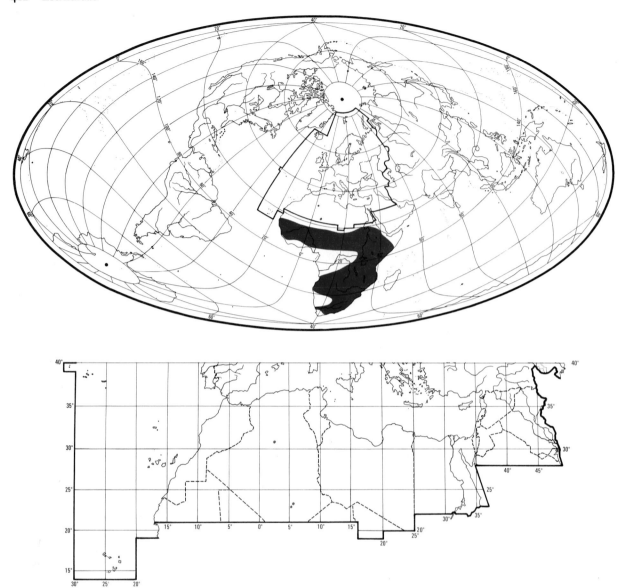

mon in farmland and thickets by streams. Nests usually in grass walls and roofs of huts. (Serle *et al.* 1977.) Habitat similar in southern Africa (Prozesky 1974).

**Distribution.** ALGERIA. Introduced at Tamanrasset *c.* 1940. A further colony since established *c.* 40 km to north-east in atypical rocky habitat. Has occurred, at least since 1972, at El Golea. At In Guezzam, in extreme south, bred in 1954 but disappeared by 1960. (EDHJ.)

Accidental. Morocco.

**Movements.** Chiefly sedentary, but local movements reported from some areas, and these presumably more frequent than currently known.

Strictly sedentary at Richard-Toll (northern Sénégal); of 7759 birds ringed in 10-year study, longest movement recorded *c.* 1100 m; Sénégal river (300–500 m wide) probably acts as barrier to dispersal (Morel 1966). Sedentary or presumed sedentary elsewhere in West Africa also (Elgood *et al.* 1973; Lamarche 1981; Grimes 1987); near Maiduguri in north-east Nigeria, however, present throughout dry season, but when rains begin (late July to August) most birds move north away from advancing rainfront, and those netted at dusk then carried more fat (presumably reserves for migration) than in late dry season; southern edge of desert at *c.* 14°N, *c.* 300 km away, acts as barrier, though no recovery data to give actual distances that individuals move; perhaps lack of seeds, due to germination following rain, causes northward movement, birds returning south to breed when conditions favourable (Jones and Ward 1977; P J Jones). In East Africa, sedentary, with local influxes reported June–September (Lewis and Pomeroy 1989). Mainly sedentary in southern Africa, but evidence of local movement in some areas, e.g. chiefly winter visitor to eastern Orange Free State (Maclean 1985), and sporadic visitor to Barberspan (Transvaal) (Skead and Dean 1977); in 1964, apparent invasion of vagrants into eastern Cape (Skead 1967). In 2-year study in garden of 0·4 ha in Bulawayo (Zimbabwe), birds always present, but ringing revealed steady turnover; thus, small flocks resident for several weeks would depart but reappear 2–3 months later (Harwin 1959). At Lochinvar (southern Zambia), where breeds February–August (mainly March–May), observations and ringing showed that most birds remained all year, but some made movements of several km to sites with permanent water; dispersal not closely synchronized, but mostly in dry season (June–July) and early rains (October–November); no clear direction shown, with some birds moving north to southern edge of Kafue floodplain (which perhaps inhibited dispersal), others east or west; 46 moved 5–14 km, some showing repeated movements, e.g. 3 netted at one site in breeding seasons 1972 and 1973 were 5 km distant in intervening dry season (Payne 1980).

Record of bird ringed Bulawayo September 1953, recovered *c.* 480 km south in Transvaal August 1958, is very doubtful, and perhaps due to mistake with ring number (MacLachlan 1963; Irwin 1981). DFV

**Food.** No information from west Palearctic. Elsewhere, small grass seeds, mainly from ground, and some green plant material; some small insects. Forages in pairs or flocks on open ground and in low vegetation, rarely far from cover; often close to human habitation, where sometimes very tame. Picks most food from ground, or from low-growing seed-heads, though occasionally perches close to elevated seed-head; sometimes seizes and shakes stems or seed-heads to dislodge seeds. Small fragments of green vegetation taken by gripping leaf with bill and making backward jump. Insects picked from ground and vegetation, and possibly whilst searching low in bushes; in captivity, jumps small distances to capture flying termites (Isoptera). In settlements, freely enters buildings and houses, feeding there on spilled cultivated grain (especially millet), fragments of larger grains, breadcrumbs, animal foodstuffs, etc. (Someren 1956; Morel 1969; Irwin 1981; Goodwin 1982; Ginn *et al.* 1989.)

In Sénégal, recorded feeding on grasses *Panicum*, *Setaria*, *Pennisetum*, *Digitaria*, *Dactyloctenium*, *Echinochloa*, *Chloris*, and *Sporobolus* (Morel 1969; Goodwin 1982). In Zambia, crop and stomach contents of 111 individuals, sampled at different dates throughout one year, showed diet dominated by grass seeds 1–2 mm in size; 77% of individuals sampled had fed on *Echinochloa*, 48% *Setaria*, 20% *Urochloa*, 5% *Panicum*, and less than 5% on *Chloris*, *Digitaria*, and *Eleusine*; no great seasonal variation; frequencies closely matched local abundances and therefore may only reflect availability (Payne 1980). In captivity, prefers soaked or germinated millet, though adult diet requires some insect food for successful breeding (Norman 1966; Morel 1969).

In Zambia, diet of small young 90% small whole grass seeds, with a few insects, notably small ground-dwelling termites *Odontotermes* (Payne 1980). In Sénégal, young fed on small seed, broken rice and, very rarely, insects; of 22 insects from crops of 3 young, all were small caterpillars of 4–13 mm (Morel 1969). WGM

**Social pattern and behaviour.** Several studies of captive birds (notably Kunkel 1959, 1967a, which see for comparison with waxbills *Estrilda*; also Harrison 1956, 1962a, c). No information for west Palearctic, and only comprehensive study in the wild by Morel (1969) in Sénégal, from which the present account is drawn except where otherwise acknowledged. For review, see Goodwin (1982). Behaviour in the wild geared to multiple broods over lengthy breeding season (see Breeding). For behavioural relationship with Indigobird *Vidua chalybeata*, see Nicolai (1964), Morel (1969), and Payne (1973).

1. Much less social than most *Estrilda*. Usually in pairs or small parties (typically up to 5 birds) when not breeding, but at any time will congregate at localized food source (Kunkel 1967a; Goodwin 1982). BONDS. Mating system monogamous. Pair-bond close (see Heterosexual Behaviour, below) and generally stable over several breeding attempts (up to 5 per season) if both partners survive, though some divorce occurs. First pairs to nest each season are those which bred the previous season, e.g. 1st pair to breed in 1964–5 successfully raised 2 broods, then ♀ apparently died; during subsequent breeding season, surviving ♂ bred 4 times and changed mate 3 times. Nest-building by ♂ only, but both sexes incubate, brood, and feed the young up to 2–3 weeks after fledging. Family bonds maintained for several weeks after young start self-feeding (Payne 1973). In Sénégal, young hatch between July and following May, and reach sexual maturity at beginning of August following this period; age of first breeding thus varies from 3 months to *c.* 1 year. Also evidence of birds displaying (see Heterosexual Behaviour) and breeding in partly juvenile plumage in Sierra Leone and Zambia (Payne 1976, 1980). BREEDING DISPERSION. Solitary and territorial. Territory limited to nest and its immediate surroundings. No precise details of nest density. Although nests may be quite close to one another, traditional Sénégalese house with cross-sectional area of a few m² rarely accommodates more than 1 nest at same stage of development. Some nests are re-used by same pair (or at least by one of the same partners) or different pair, once or more within season; of 136 nests monitored during one season, 92% used once, 7% twice, and 1% 3 times. Some nests are even re-used over several breeding seasons, and one survived 5 years of use. For detailed case-history of 3 broods raised over 12 months at same site, see Malcolm-Coe (1981). First-time breeders often faithful to natal area or area where they spent juvenile months. ROOSTING. Nocturnal and communal. In thickly foliaged bush or sometimes in building; chosen site varies from day to day, no long-term fidelity (see Flock Behaviour, below, for details of going to roost). ♀ sleeps in nest at night during incubation and first few days of roosting, but nest not otherwise used for roosting by adults or fledged young. At hottest time of day, birds seek shade and will not forage in the open; especially during moult, pair-members often seen loafing side by side in shade of leafy bush.

2. Bold and inquisitive, typically flicking tail jerkily from side to side when excited (Harrison 1956; see also Antagonistic Behaviour). When confronted with large would-be predator (especially man), flies to safe distance or cover; increases alarm-calling (see 5 in Voice), tail-flicks, bill-wipes, displacement-preens, and lightly lifts one or other folded wing. If suddenly alarmed, may also perch motionless, e.g. when pair site-prospecting in house were surprised by *V. chalybeata* (brood-parasite), ♀ hid motionless in dark corner. FLOCK BEHAVIOUR. In Zimbabwe, ringing demonstrated stable flock composition (i.e. same members) over considerable periods (Harwin 1959; see also Movements). Search for roost-site begins well before sunset and is accompanied by much activity, including feeding, drinking, flying from bush to bush, and calling (see 2, 4, and 5 in Voice). Once on suitable roost-perch, bird defends it against newcomers which fly one after another up into bushes. Family groups generally first to settle (from *c.* 1 hr before sunset) while others (especially incubating ♂♂ which have just been relieved on nest) stay feeding on ground until almost dark. Rapid exodus at sunrise to feed. SONG-DISPLAY. Following account mainly from Morel (1969), but including data from Kunkel (1959), Harrison (1962*a*, *c*), and Sullivan (1976). *L. senegala* alone among Estrildidae in having 2 distinct songs. Solitary-song (see 1a in Voice), and its accompanying display, are self-advertisement by ♂ seeking mate. In captivity, Solitary-song is given by ♂ (less often ♀) in absence of mate, e.g. by ♂ when ♀ has died or is sitting out of sight inside nest; ♀ not heard to sing thus when ♂ is in nest. In the wild, Solitary-song typically accompanies Solitary-display (see Heterosexual Behaviour, below); also heard when perched upright, bill uptilted, head plumage ruffled (Payne 1973), or when crouching on ground or branch, sometimes with feather or piece of grass in bill, tail spread and lowered, chest puffed out, head moving up and down, and sometimes sideways; swivels slowly on the spot, singing. Other type of song, Courtship-song (see 1b in Voice) sometimes occurs at end of Feather-display as ♂ bows to ♀ (Kunkel 1959, which see for details) as well as during Solitary-display. ANTAGONISTIC BEHAVIOUR. (1) General. From start of nest-building, both pair-members vigorously defend nest-area against conspecific and other intruders. In captivity, if solitary ♂ confronted by strange ♂, aggression usually follows, but sometimes (more often when stranger is ♀) ♂ performs Greeting-display: stands tensely with head sleeked and withdrawn, tail held to one side and twisted; posture appears aggressive but perhaps facilitates (by exposing critical plumage features) species-recognition ('Recognition Posture': Harrison 1956, 1962*a*). Greeting-display often followed by Allopreening (Kunkel 1967*a*; see below). (2) Threat and fighting. In Upright-threat display, performed on ground or elevated perch, bird stretches neck somewhat upward and directs closed bill at approaching conspecific. This usually induces retreat but if approach persists intruder is chased and pecked on neck and back; aggressor spreads tail and flicks it vigorously sideways and up and down. Tail-movements accompanying alarm-calling and described by Harrison (1956) are presumably the same: spread tail jerked to one side or other with each call, often being momentarily twisted to side (Fig A) or flexed up and down from horizontal (highest point) to vertically down. During normal flock activity, disputes common but short-lived, though sometimes become violent, especially during sexual rivalry, exceptionally even leading to death. In chosen nest-site, resident perches motionless, lightly ruffles plumage on breast and back, lowers tail, slightly lifts wings, points bill forward, and constantly turns head (for surveillance); any intruder in nest-area is expelled. Disputes often arise with neighbours over access routes to nests, established residents chasing newcomers until set-

A

tlement reached. Fierce fighting may result if owners return to discover intruders attempting to usurp nest by adding material to it. HETEROSEXUAL BEHAVIOUR. (1) General. Courtship typically occurs during normal flock activities, and thus not necessarily near eventual nest-site (Morel 1969). In contrast to *Estrilda*, only ♂ *L. senegala* performs Feather-display (Kunkel 1967*a*: see below). Sight of detached feather exerts strong stimulus on ♂ from independence onwards; adult ♂ often tries to steal feather held by juvenile, and if feather drops from bill of one bird it is immediately retrieved by another before it reaches ground. (2) Pair-bonding behaviour. Following account based on Morel (1969) from observations on wild and captive birds; for similar descriptions from captivity, see Kunkel (1959, 1967*a*). Newly independent young show rudiments of courtship-display and many are already paired for breeding by a few months old (see Bonds). Thus at *c.* 1 month, young start catching feathers for displaying to ♀♀ of same age. ♂ seeking mate approaches ♀ and, if she is receptive, performs Solitary-display ('incomplete feather display' of Morel 1969, 'Undirected Display' of Harrison 1956, 1962*a*); derived from, and much simpler than, Feather-display (see below); in contrast to Feather-display, nothing held in bill, and elaborate greeting of Feather-display simplified in that ♂ merely extends neck fully and bows straight forward (not obliquely as in Feather-display), points bill down, and gives Solitary-song. Feather-display, which serves for intimate courtship, differs from Solitary-display not only in use of feather (etc.) and in being more complex, but also in that, at least at close quarters, ♀ approaches displaying ♂ (in contrast to Solitary-displaying ♂ approaching ♀). In Feather-display, unpaired ♂ picks up feather (or piece of grass: hence 'Straw Display' of Harrison 1962*a*) and holds it in tip of bill such that feather points straight out (Harrison 1956: Fig B), and, with head thrown back,

B

chest puffed out, and rest of plumage ruffled, joins group of conspecifics foraging on ground. According to Harrison (1956, 1962*a*), loud wing-whirring of ♂ as he flies towards conspecifics may have signal function, also stiff jerky hops after landing.

When he encounters receptive ♀, ♂ stands at angle to her and cocks spread tail towards her so that it makes acute angle to his own body; then he flexes his legs several times, producing jerky bobbing motion of body, rendering object in bill conspicuous. As he displays, ♂ circles round ♀, gradually closing in on her. After last bob, and accompanied by Courtship-song (not heard earlier in display) ♂ bows deeply in front of ♀ (Fig C) who has

C

remained motionless throughout, and turns his head towards her to present feather just below breast height. ♂ and ♀ then fly off, pair-formation effectively complete. ♀'s response to feather-dropping finale to precopulatory display (see subsection 4, below) was often to fly away, but she often returned almost immediately to preen ♂ (Harrison 1956). Allopreening common during breeding as well as moult, apparently helping to affirm pair-bond. Initiator, with bill forward as if about to peck, approaches mate who responds by ruffling head and back plumage, and may then, with eyes closed, receive lengthy preening of head, neck, breast, and (sometimes) rump; recipient constantly alters posture to give preener best access; finally, preener flies a little way off where partner joins it and reciprocates preening. (3) Courtship-feeding. None. (4) Mating. Takes place on ground or branch, at any time of day from nest-site selection to laying (Kunkel 1967a; Morel 1969). Usually follows ♂'s Feather-display (sometimes repeated several times in succession before mating). Allopreening may also precede mating. In normal sequence, after Feather-display, receptive ♀ solicits by crouching with plumage ruffled, wings slightly drooped, and tail quivering vigorously up and down. ♂ drops feather, approaches mate, and pecks her several times on head and nape (♂ Soliciting-display); with each peck, ♀ lowers forebody further until cloaca is raised higher than head and exposed. ♀ now bends raised tail to left and ♂ mounts, gripping her carpal joints, and copulates with whirring wings. When copulation complete, ♂ again pecks ♀'s back and rump and perches nearby. Finally both may dust-bathe. (Morel 1969.) Like all *Lagonosticta*, *L. senegala* copulates in silence and one copulation (or attempt) never followed immediately by another. If ♂ attempts to mount again after failed attempt, ♀ flies off, likewise if ♂ tries to copulate without first soliciting (pecking head, etc.) No soliciting, however, in extra-pair copulation attempts. (Kunkel 1959, 1967a.) (5) Nest-site selection. ♂ looks for site, slipping through branches giving continual Contact-calls (see 2 in Voice). ♀ accompanies ♂ or stays nearby, e.g. preening and chasing off trespassers. On finding potential site, ♂ gives Nest-call (see 3 in Voice) which summons ♀ to join him for a few moments, then she leaves. This sequence repeated a number of times, at different sites, until final choice made. (6) Behaviour at nest. Once nest-site selected, nest-building begins (at coolest times of day) or else pair re-uses nest from previous breeding attempt (see Breeding Dispersion). During laying, nest-site not guarded continuously. Active building continues during laying, less so thereafter. At nest-relief, incoming bird first perches near nest, quivering tail vertically

and horizontally, continuing to forage, and giving frequent Contact-calls to signal arrival; then makes way to nest, always by same route, and waits until partner leaves. RELATIONS WITHIN FAMILY GROUP. Immediately after hatching, eggshells are eaten by parents. Young brooded for first few days by either parent, only by ♀ at night. Both parents regurgitate food for young, which give food-calls (see Voice) from 1st day (though mostly silent for first few days: Nicolai 1964); feeding is lengthy as adult constantly needs to retrieve food from its crop. Young beg with distinctive side-to-side head-swaying, exposing mouth markings, and with a waving display of tongue; also twist head upside down when parent alongside, and receive food in this posture (Payne 1973: Fig D). No nest-sanitation; faeces accumulate

D

inside nest, rapidly drying out to form granular bedding; even when nest re-used, old faeces not removed. Eyes of young open from day 4 (see Morel 1969 for details of development from hatching to fledging). 2 days before fledging, parents start reducing delivery of food, bringing none on evening prior to fledging. On fledging day, parents commute constantly between ground and nest-entrance with much calling which young reciprocate. Immediately after leaving nest, young are led to refuge where parents continue to feed them (also regurgitating water for them in first few days). Fledged young Allopreen siblings, also sometimes *V. chalybeata* reared with brood (Payne 1973). After young independent (2–3 weeks after fledging), parents reject (by pecking, etc.) entreaties for food (Morel 1969), though parents and offspring continue to associate for some time thereafter (Payne 1973). ANTI-PREDATOR RESPONSES OF YOUNG. Distress-calls (see Voice) of young removed from nest induced parents to perch nearby giving Alarm-calls while siblings remained silent in nest (Payne 1973). Escape-reaction evident from day 3 in nest; when touched, young venture out of nest but soon return. Young recently fledged may initially freeze when threatened, but if danger imminent attempt to flutter away. When flying ability better, flee danger with parents. PARENTAL ANTI-PREDATOR STRATEGIES. ♀ a relatively tight sitter, staying on nest when human intruder first taps nest, only leaving when she comes to entrance hole to check source of disturbance (Someren 1956). Bird leaving nest blocks entrance with feather (Lock 1971), perhaps to conceal contents. Parents defend area around recent fledglings (which stay close together) but flee when threat is from intruders much bigger than themselves.

(Figs by D Nurney: A–C from drawings in Harrison 1956; D from photograph in Payne 1973.)                    EKD

**Voice.** For additional sonagrams see Nicolai (1964) and Payne (1973). ♂ flying to ♀ with feather in bill for display produces loud whirring wing-beats with possible signal function (Harrison 1956, 1962a); see Payne (1973) for sonagram.

CALLS OF ADULTS. (1) Song. (1a) Solitary-song. Short, pleasant, and simple; given by ♂, less often ♀; apparently no significant difference between sexes (Sullivan 1976) or races (Payne 1973). Despite individual variation, not confusable with other Estrildidae, even the most closely related *Lagonosticta* (Nicolai 1964). Starts with 1 (or occasionally more: Payne 1973) introductory Excitement-type units (see 5, below), or slightly quieter variant thereof, followed by 2–6 soft fluting tones (Nicolai 1964) similar to call 2, e.g. loud, clear, melodious 'dwit-ee-dwee' (Harrison 1956, 1962a); in Fig I, introductory unit is 'pik', whole

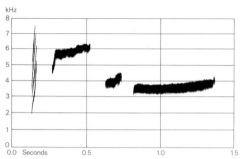

I C Chappuis Mauritania January 1969

phrase 'pik-tseeeee pit-eeeeeeeeeeeeeeee' (J Hall-Craggs). According to Sullivan (1976) main Song-units differ from call 2 in being somewhat longer, more regularly and more closely spaced, and usually louder. Number of units in song varies between individuals, but song typically 1·1–1·2 s long in Sénégal (Morel 1969); in Nigeria, mostly 1·0–1·4 s, units usually longest in song with fewest of them (Payne 1973). Solitary-song given by ♂ seeking mate or by either sex in absence of mate. (1b) Courtship-song of ♂.

Simple sequence of very soft stereotyped clicking sounds, difficult to hear over 1 m away (Morel 1969); 'stip' (Harrison 1956, 1962a), 'zit' (Someren 1956). Sometimes accompanies Solitary-display and bowing finale to Feather-display (Harrison 1956, 1962a). Payne (1973) described 'stip' as a 'call', but Harrison (1962b) suggests it could also be regarded as 'Display Song', likewise Morel (1969); therefore listed as song here. (1c) Subsong. Not previously described, but recording (Fig II) shows apparent Subsong, consisting of rambling amalgamation of sundry calls (which include uncoordinated 2-voice effects, diads, and interference); compared with Solitary-song, Subsong shows wider frequency range and longer improvisatory phrases (J Hall-Craggs). (2) Contact-call. Commonest call of both sexes. Variable (and therefore subject to diverse renderings and interpretation in the literature), but basically soft-toned monosyllabic 'die' (Nicolai 1964), high-pitched quite melodious 'twee' or 'dwee' (Harrison 1956, 1962a), 'uit' (Morel 1969), or soft whistle 'chee' (Sullivan 1976). Clear slurred whistle of varying structure and duration (0·1 s to more than 0·3 s); pitch of each call usually rises smoothly or sometimes only at start or finish (Payne 1973). Fig III ('seee peeet') apparently of this type; perhaps also calls in Fig IV rendered 'erwee zeeek eeee zeeeek', in which 'zeee(e)k' similar to 'peet' (both have strong end-transient) but differs in having buzzy metallic timbre created by very rapid frequency modulation (J Hall-Craggs). (3) Nest-calls. Very quiet sound (audible only at close quarters) repeated rapidly (Morel 1969). Evidently 2 types: those heard by Payne (1973) consisted of 2 units in irregular sequence, 1st a variable whistled note resembling short Contact-call, 2nd an

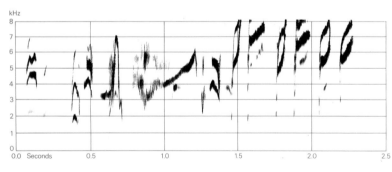

II C Chappuis Sénégal November 1969

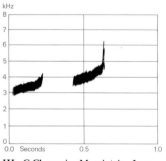

III C Chappuis Mauritania January 1969

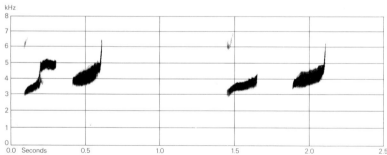

IV C Chappuis Mauritania January 1969

abrupt sound similar to call 5 or perhaps to food-call of small nestlings (see below). However, Nest-call in study by Nicolai (1964) was a quiet whispering 'wis-wiswiswiswiswiswiswiswis', with units of nestling food-call type, given at rate of 8–9 per s. Nest-calls given by ♂ in the few days before laying to attract ♀ to chosen nest-site or nest in course of construction (Harrison 1956; Nicolai 1964; Morel 1969; Payne 1973), also by ♀ inside nest and unable to see mate outside (Morel 1969). (4) Threat-call. Short, high-pitched sound, not far-carrying, and, though apparently distinct from call 2 (Morel 1969), presumed just a variant (Payne 1973). (5) Excitement- or Alarm-call. Single, abrupt, low-pitched nasal 'chuc' (Harrison 1956, 1962*a*), 'zeck' (Nicolai 1964), 'pik' (J Hall-Craggs: Fig V, and 1st unit in Fig I). Repeated at

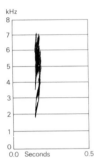

V  C Chappuis  Mauritania  January 1969

varying rate by either sex according to circumstances; often given singly, sometimes in short series, gap between calls generally constant but shortening with increasing excitement (Nicolai 1964; Payne 1973); once heard in rapid rattling fashion from bird mobbing snake (Payne 1973, which see for sonagram). Accompanied by jerky upward and sideways tail-flicking and heard in various contexts such as nest-defence, nest-relief, and any situation arousing alarm (Morel 1969); induces conspecifics to fly away (Sullivan 1976). (6) Distress-calls. (6a) Far-carrying, plaintive, drawn-out, high-pitched call heard from parents returning to nest to find young gone (Morel 1969). (6b) Loud noisy screech occasionally given when handled (Payne 1973; see also below).

CALLS OF YOUNG. Food-calls, first heard from captive young at 5–6 days, are short (less than 50 ms) and more than 0·4 s apart (Payne 1973); described as quiet whispering sounds (Nicolai 1964), probably same as squeaky 'sii sii sii sii' noted by Someren (1956). By day 7–8, call longer (80 ms), consisting of 2 pitch levels. By day 11, call longer still (more than 100 ms), given more rapidly (e.g. 66 in 14 s), and division into 2 pitch bands more distinct. By day 13, yet longer and thus more plaintive ('tet': Nicolai 1964) and alternates with shorter call ('zet': Nicolai 1964) which, in dependent juveniles, becomes indistinguishable from adult Excitement-call and is apparently given more often when disturbed. (Payne 1973.) Typical sequence 'zet-tet zet-tet', gap between 'zet' and

'tet' very short and more or less constant, but gap between each disyllable shortening with increasing excitement (Nicolai 1964). Recording (Fig VI), apparently of juvenile

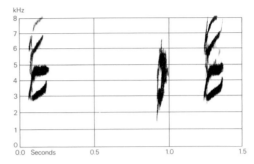

VI  C Chappuis  Mauritania  January 1969

food-begging, rendered 'dwee pwt dwee' (J Hall-Craggs); 'dwee' corresponds to 'tet' and 'pwt' to 'zet'. 'Tet' (juvenile location-call) gradually develops into adult Contact-call. Loud noisy screech of distress (call 6b) heard from day 6, e.g. when young handled. Song heard at 14 weeks in captivity. (Payne 1973.)                                    EKD

**Breeding.** Little known in west Palearctic, but well studied in Sénégal by Morel (1964, 1968, 1969). SEASON. Ahaggar mountains (southern Algeria): eggs recorded mid- to late April and late August (Heim de Balsac and Mayaud 1962). Sénégal: main breeding season August–March (July–May) after rains, during which time may raise 5 broods; earliest breeders tend to be experienced birds (Morel 1969). Zambia: main breeding season March–May during late rainy season and early dry season, though some nest as late as August (Payne 1980). For rest of Africa, see Goodwin (1982). SITE. Well concealed, shaded cavity, depression, or platform among thick vegetation, on or in building, or on ground. Nests in vegetation placed on forked branches in leafy hedges and bushes, in leaf bases of bananas and palms, among tangled roots or in piles of brushwood; those in buildings are in walls, loose thatch, between or among stacked objects, etc.; those on ground are in thick vegetation, recess such as hoofprint, or hole in bank (Bannerman 1953*b*, 1963; Someren 1956; Morel 1969; Payne 1980; Goodwin 1982). Also recorded in disused nest of weaver (Ploceidae) (Someren 1956). Usually no more than 4 m above ground; rarely on ground (Morel 1969). Nest: well-camouflaged structure, mainly of grass stems *c.* 20 cm long, also straw, leaves, paper, etc., thickly lined with feathers and other fine material; depending on site, ranges from closed, domed structure to (less often) open cup; all types rather loosely constructed (material placed together, not woven) (Bannerman 1953*b*; Harrison 1956; Someren 1956; Morel 1969; Payne 1980); mean dimensions of 5 domed nests, Sénégal, height 8·8 cm, width 10·5 cm, entrance diameter 3·1 cm; weight of nests 9–50 g; another nest had 6-cm entrance tube; outer dimensions of cup nests variable, usually *c.* 9 cm high, though cup diameter always *c.* 6 cm (Morel 1969). Building: by ♂, often with ♀ in attendance; very rarely, ♀ assists in collecting lining material; material gathered on ground, mostly (except feathers) within 5–6 m of nest; building behaviour frantic during cooler periods of day; in cup nests, thick base of leaves and long grass stems constructed first, and feather cup shaped by weight of bird and held in place by rings of grass; lining may be added to well into incubation (Harrison 1956; Morel 1969); for building actions, see Kunkel (1959) and Morel (1969). Nests sometimes re-used, though not necessarily by birds that built them; of 136 nests in Sénégal in 1963–4 season, 92% used once, 7% twice, and 1% 3 times in the season; nests sometimes also used over consecutive seasons, once for over 5 years (Morel 1969). EGGS. See Plate 60. Sub-elliptical or oval, with little gloss; white (Priest 1936; Bannerman 1953*b*). Nominate *senegala* and possibly some *rhodopsis*: 13·6 × 10·7 mm (13·0–15·0 × 9·8–11·5), *n* = 288; calculated weight 0·82 g; *L. s. rhodopsis* and *brunneiceps* (north-east Africa): 14·0 × 11·4 mm (13·0–14·7 × 10·5–12·0), *n* = 12; calculated weight 0·95 g (Schönwetter 1984). Clutch: 3–4 (1–6); records of 8–9 eggs probably due to unhatched eggs from previous clutch remaining in nest (Bannerman 1953*b*; Someren 1956; Morel 1969). Of 374 clutches in Sénégal, 74% 3–4 eggs; average 3·4 in 1963–4 season, 3·5 in 1964–5; in latter season, average highest September–October with 3·7–3·8, levelling out at 3·3, falling to 2·9 in April (Morel 1969). Up to 5 successive broods in Sénégal, new clutch started *c.* 2 weeks after independence of young (Morel 1969). INCUBATION. 11–12 days (Goodwin 1982); 13–14 days (Morel 1969). By both parents (Someren 1956; Morel 1969); in Sénégal, ♂ and ♀ share incubation during day; during 20·3 hrs observation, ♂ on nest 56·6% of time, ♀ 43·4%, though varies between pairs; nest rarely left unattended; ♀ incubates at night; temperature changes over season do not seem to affect time spent incubating; 4-egg clutch typically takes 36 hrs to hatch (Morel 1969). YOUNG. Fed (by regurgitation) and cared for by both parents; ♀ broods young at night (Bannerman 1957; Morel 1969). FLEDGING TO MATURITY. 17–19 (14–20) days; fed for *c.* 8 days after fledging by parents, or by ♂ alone. Young remain associated with adults for 2–3 weeks after leaving nest, even when parents start another clutch. (Someren 1956; Brooke 1962; Morel 1969; Goodwin 1982.) Each breeding cycle lasts *c.* 2 months (Morel 1969). Age of first breeding variable, but less than 1 year (see Social Pattern and Behaviour). BREEDING SUCCESS. Data only from Sénégal. Breeding success low, but compensated by large number of broods in single season; in 10-year study, 45% of eggs hatched and 62% of chicks survived to fledging, resulting in overall success of 28% (Morel 1964). Over 2 seasons, number of fledglings raised per pair greatest October–November; average for 1963–4, 2·7 fledglings per pair; for 1964–5, 3·2 (Morel 1969). Regularly parasitized by

Indigobird *Vidua chalybeata*; mortality of *L. senegala* fledglings in parasitized nests 53%, in non-parasitized nests 25% (Morel 1964); average number of fledglings from non-parasitized nests 2·6, from parasitized nests 2·1 *L. senegala* fledglings and 1 *Vidua* fledgling. Harmful effects of parasitism greatest when *Vidua* lays in clutch of 4 or more. However, when *Vidua* parasitizes clutch of 3, egg seems to benefit host by leading to better overall hatching success and fledging success; of 316 broods (all sizes), of which 24% parasitized, fledglings in parasitized broods larger in size and weight than fledglings in non-parasitized broods. (Morel 1967, 1969.)   WGM

**Plumages.** (*L. s. rhodopsis*). ADULT MALE. Forehead and side of head from supercilium downwards deep pink-red. Upperparts from crown to back and including upper wing-coverts dark grey-brown or drab-brown, feather-tips of crown and hindneck extensively washed pink-red (concealing brown to variable extent), mantle and scapulars with pink-red wash. Rump light pink-red, sometimes partly tinged tawny-buff; upper tail-coverts bright vinous-red. Chin to chest deep pink-red, side of chest with fine white spots or short bars, faintly bordered dusky at each side of white. Side of breast, flank, and thigh grey-brown or light drab-brown, feather-tips washed pink-red to variable extent, merging into pink on breast and belly, mid-belly often tinged tawny. Lower belly and vent off-white; under tail-coverts pale isabelline-grey. Tail black, slightly paler brown-grey towards outermost feathers; bases and middle portions of outer webs of feathers extensively tinged bright vinous-red, least so on outermost. Flight-feathers, tertials, greater upper primary coverts, and bastard wing dark drab-brown, outer webs faintly and narrowly fringed paler sandy-brown, inner webs bordered pale isabelline-grey. Under wing-coverts and axillaries cream-white, pure white at base of coverts. *In worn plumage*, pink-red of feather-tips on upperparts abraded, cap appearing dark drab-brown with traces of red fringes (mainly on forecrown and hind-neck), mantle and scapulars almost uniform cinnamon-brown, rump less extensively pink-red; some grey-brown of feather-centres visible in pink-red on side of head and from chin to chest; white spots on side of chest indistinct; belly more extensively washed tawny; tail browner, vinous-pink partly worn off. ADULT FEMALE. Upperparts from forehead to back dark drab-brown, side of head and neck slightly paler drab-brown, entire underparts light grey-brown or sandy-grey, slightly darker on side of breast, chest, and flank, paler on mid-belly, vent, and under tail-coverts. Side of chest with fine white or isabelline spots or short bars; lore bright pink-red, rump pink-red, and upper tail-coverts vinous-red; wing and tail as adult ♂. Rear of body thus similar to ♂, but head and front of body brown, red restricted to lore. *In worn plumage*, upperparts more dark grey-brown, underparts mottled grey-brown and isabelline. NESTLING. Down long, dense, and fluffy, restricted to upperparts, side of breast, and thigh; whitish, light grey, or blackish (probably depending on age) (Bannerman 1949; Morel 1973; Güttinger 1976; Goodwin 1982). JUVENILE. Upperparts light drab-brown; lower rump and upper tail-coverts vinous-red, feathers narrowly tipped drab; side of head and neck, chest, and flank light drab-grey, gradually paler towards pale grey throat, belly, and vent. Wing and tail as adult. Differs from adult ♀ in drab-grey lore (no red) and in generally less saturated colour of body and shorter softer feathers; no spots on side of chest; p10 longer (see Structure) and bill dark (see Bare Parts). FIRST ADULT. As adult, and indistinguishable after juvenile p10 replaced. 1st adult ♂ usually less extensively red than adult ♂; vinous-red restricted to top of head or throat, and restricted red wash on cheek and chest or none at all. Some birds similar to adult ♀, but forehead and crown mottled pink-red.

**Bare parts.** ADULT, FIRST ADULT. Iris dark grey, brown, dark brown, or red-brown. Eyelids yellow in ♂, pale yellow or silvery-white in ♀. Bill pink-red, brown-red, light vinous-red, or bright red, culmen and gonys black-brown (in some ♀♀) or black. Mouth as in nestling. Leg and foot grey, grey-brown, grey-red, brown-red, light brown, dirty flesh, or flesh-brown (Goodwin 1982; ZFMK, ZMA). NESTLING. Bill black. Bare skin yellowish-orange. Mouth white, tip of tongue yellow; 3 black marks on palate, a black U-mark on floor of mouth; 4 little contiguous elevations at each angle of mouth, central 2 purplish-blue, lateral 2 larger and white (Bannerman 1949). JUVENILE. Iris brown. Bill dusky grey to brown-black, base gradually turning vinous-red. Leg and foot reddish-brown or grey-brown. (ZMA.)

**Moults.** ADULT POST-BREEDING. Complete; primaries descendent. In Tamanrasset area (southern Algeria), plumage slightly worn January, distinctly or heavily worn May; of 5 birds from May, 3 not started moulting, one had p1 just shed (primary moult score 2), another in more advanced moult with primary-score 20 (p1-p3 new, p4-p5 growing) but body and tail still old. In northern Niger, July, one bird with plumage fresh, one worn, yet another in last stages of primary moult (Fairon 1975). In northern Nigeria, breeding season protracted, time to moult limited; moult starts asynchronously, mid-December to late February, completed late May or June; moult slow (Fry 1970). In Sénégal, moult February-August; in February-March, mainly p1-p3 in moult, in April p3-p4, in May p5-p6, in June p7-p9, in July p9; moult takes *c.* 3·5 months in each bird (Morel 1973). Further south, timing of moult strongly dependent on local breeding season; in Zambia, moult mainly May-November, in dry season (Payne 1980, which see for many details). POST-JUVENILE. Partial or complete. Starts shortly after fledging, at age of *c.* 6 weeks; head, body, and upper wing-coverts first, in some birds followed by tail- and flight-feathers. Extent probably depends on local breeding season, moult interrupted when conditions suitable for nesting, continuing or starting anew later on. In Tamanrasset area, just-fledged juveniles examined April-May, birds moulting body or with body new but wing and tail still juvenile occur January and May; in one from January, plumage mainly new, but p9-p10, inner secondaries, and some tail-feathers juvenile, moult suspended (ZFMK, ZMA). In Sénégal, recently fledged juveniles without moult occur mainly October-November (some up to February, rarely to June); birds with head and body in moult mainly November-March (some October or up to May, rarely to July); birds with head, body, and inner primaries in moult mainly March-June (some February or July), those with outer primaries in moult mainly July-August (some from April, others to October); thus, birds hatched early (August) have early body moult but complete flight-feather moult not until age of *c.* 270 days, those hatched February-March complete after *c.* 170 days; birds hatched between early August and late December all complete about June (Morel 1973). In Zambia, where rains mainly November-February and breeding mainly (February-)April-July, body moult started shortly after fledging, April-August, but moult of flight-feathers started later, mainly July-August, and was completed after 3·5-4 months in (October-)December. Elsewhere in Afrotropics, moult occasionally arrested, some birds displaying or breeding in partial juvenile plumage. (Payne 1980.)

**Measurements.** ADULT, FIRST ADULT. *L. s. rhodopsis.* Southern Algeria, northern Chad, and western Sudan, all year; skins (RMNH, ZFMK, ZMA). Bill (S) to skull, bill (N) to distal corner of nostril; exposed culmen on average *c.* 1·5 less than bill (S).

| | | | | | |
|---|---|---|---|---|---|
| WING | ♂ | 51·4 (1·51; 21) | 49-54 | ♀ 49·7 (1·89; 11) | 48-54 |
| TAIL | | 37·9 (1·59; 16) | 35-40 | 35·3 (1·91; 11) | 33-38 |
| BILL (S) | | 10·2 (0·55; 16) | 9·3-11·1 | 10·2 (0·60; 12) | 9·5-10·7 |
| BILL (N) | | 6·6 (0·39; 16) | 6·2-7·4 | 6·6 (0·49; 10) | 6·1-7·3 |
| TARSUS | | 13·8 (0·50; 16) | 13·1-14·6 | 13·5 (0·69; 10) | 13·0-14·4 |

Sex differences significant for wing and tail. Wing of birds from southern Algeria from sample above: ♂ 52·2 (0·81; 10) 50·5-54, ♀ 50·8 (1·62; 6) 50-54.

Nominate *senegala.* Sénégal to Nigeria, all year; skins (RMNH, ZMA).

| | | | | | |
|---|---|---|---|---|---|
| WING | ♂ | 49·6 (0·70; 16) | 48-51 | ♀ 49·1 (1·00; 13) | 48-51 |
| TAIL | | 35·6 (1·41; 15) | 33-38 | 34·3 (1·08; 11) | 32-36 |
| BILL (S) | | 10·7 (0·45; 16) | 10·2-11·6 | 10·2 (0·28; 13) | 9·8-10·7 |
| BILL (N) | | 6·7 (0·28; 16) | 6·3-7·3 | 6·6 (0·25; 13) | 6·2-7·0 |
| TARSUS | | 12·6 (0·40; 15) | 12·2-13·3 | 12·6 (0·46; 13) | 11·7-13·2 |

Sex differences significant for tail and bill (S).

Northern Sénégal: wing, ♂ 48·9 (0·96; 52) 48-52, ♀ 49·1 (0·92; 15) 47-50 (Elzen and Wolters 1978).

JUVENILE. Nominate *senegala* and *rhodopsis.* Wing, ♂ 51·4 (2·16; 5) 48-53·5, ♂ 48·7 (3) 48-49; tail, ♂ 37·9 (2·14; 4) 35-40, ♀ 34·5 (3) 33·5-35·5 (ZFMK, ZMA).

**Weights.** *L. s. rhodopsis.* Southern Algeria, January: ♂ 8·8 (0·29; 4) 8·5-9. Northern Chad, February: ♂ 9, ♀ 7. (ZFMK.) Southern Algeria, November: ♂ 8·7 (Fairon 1971). Northern Niger, July: 8·7 (0·84; 12) 7·8-10·6 (Fairon 1975). Chad: 8·5 (57) 6-10 (Salvan 1969).

Nominate *senegala.* Northern Nigeria, October-June: 8·5 (0·3; 37) 7·7-9·9. Northern Ghana, July-September: 8·5 (15) 6·0-11·0. No differences between sexes. (Fry 1970.) See also Greig-Smith and Davidson (1977), Karr (1976), and (for other extralimital races) Britton and Dowsett (1969) and Skead (1974, 1977).

**Structure.** Wing short, broad at base, tip rounded. 10 primaries: p7 longest, p6 and p8 0-1 shorter, p9 1-4, p5 0·5-1·5, p4 2-4, p3 3·5-6·5, p2 5-8, p1 6-9·5. P10 reduced, narrow, pointed; in adult, 29-32 shorter than p7, 2 shorter to 2 longer than longest upper primary covert (on average 0·1 shorter, *n* = 15); in juvenile 25-30 shorter than p7, 2·1 (8) 1-4 longer than longest upper primary covert. Outer web of (p5-)p6-p8 emarginated; inner web of p6-p8 with indistinct notch. Adult p9 narrow at base and middle portion, inner web deeply emarginated *c.* 10 mm from tip, tip only 1·5-2 mm wide; in juvenile, p9 broader, tip less narrow, subterminal constriction more gradual, notches on other feathers almost absent. Tip of longest tertial reaches to tip of p2-p4. Tail rather short, tip square or bluntly rounded; 12 feathers, t6 *c.* 5 shorter than others. Bill short, conical; rather wide and deep at base, ending in sharp point; culmen and cutting edges straight, or slightly decurved towards tip. Nostril small, rounded, covered by short feathering projecting from lore. No nasal bristles. Tarsus and toes slender, short. Middle toe with claw 12·9 (8) 11·5-14 mm; outer toe with claw *c.* 68% of middle with claw, inner *c.* 64%, hind *c.* 76%.

**Geographical variation.** Rather slight. Mainly involves depth of red and brown colours of body and amount of red wash on brown ground-colour of mantle and scapulars or belly. *L. s. rhodopsis* is palest race, in particular birds of southern Algeria paler pink-red on head and underparts and paler brown on remainder of body than birds from northern Chad and Sudan. Nominate *senegala* similar to *rhodopsis* in extent of red, but head and underparts deeper vinous-red, underparts darker brown with vinous-red wash. Other races paler or darker red and brown than nominate *senegala*, but with more extensive brown tinge on upper- or underparts. *L. s. ruberrima* of inland East Africa (Uganda to northern Zambia) darker vinous-red or deep purple-red, belly extensively dusky brown, but upperparts distinctly washed vinous-red. *L. s. rendalli* of East and southern Africa (east and south of *ruberrima*) slightly paler red than *ruberrima*; brown of belly extends to breast, chest more extensively spotted white. For survey of races, see Goodwin (1982). For races in northern Afrotropics, see (e.g.) Wolters (1958, 1962), Elzen and Wolters (1978), and Elzen and König (1983).     CSR

## *Uraeginthus bengalus* (Linnaeus, 1766)  **Red-cheeked Cordon-bleu**

FR. Cordonbleu à joues rouges     GE. Schmetterlingsfink

Breeds in semi-arid habitats in sub-Saharan Africa from Sénégal east to Eritrea and Ethiopia, south to Zambia and Tanzania; generally resident. Occurred in Maadi area of northern Egypt in mid-1960s, but no records since, and presumed not to have become established (Goodman and Meininger 1989). Collected on São Vicente (Cape Verde Islands) 1924, but possibly escapes; no other records (C J Hazevoet).

## *Estrilda astrild*  **Common Waxbill**

PLATE 31
[between pages 472 and 473]

DU. Sint Helenafazantje     FR. Astrild ondulé     GE. Wellenastrild
RU. Волнистый астрильд     SP. Astrilda común     SW. Helenaastrild

*Loxia Astrild* Linnaeus, 1758

Polytypic. *E. a. jagoensis* Alexander, 1898, coastal plain of Benguela and Moçamedes (Angola) and extreme north-west Namibia, introduced Cape Verde Islands, Iberia, São Tomé, Príncipe (Gulf of Guinea), and (race unknown) Brazil. Extralimital: nominate *astrild* (Linnaeus, 1758), southern Namibia and southern Botswana, south to western Cape Province (South Africa), introduced St Helena, New Caledonia, and many islands in Indian Ocean; *kempi* Bates, 1930, Sierra Leone and Liberia; *occidentalis* Jardine and Fraser, 1851, Fernando Poó and from southern Ghana through Cameroun east to central and southern Zaïre; *c.* 12 other races in eastern and southern Africa.

**Field characters.** 11 cm (tail 4 cm); wing-span 12–14 cm. Length similar to Serin *Serinus serinus* but structure distinctly different from finch (Fringillidae): rather small head, short rounded wings, rather slim body, and long graduated tail combine to give slight but lengthy form. Epitome of essentially African genus. Grey-brown above with rufous rump, pale brown to rose below with almost black vent and tail; red bill and red face-patch. Sexes similar; no seasonal variation. Juvenile separable.

ADULT MALE. Moults (Portugal): July–October (complete). Appearance dominated by small, stubby wax-red bill, oval crimson patch from lore to around and behind eye, and almost black vent, under tail-coverts, and graduated tail. Crown to nape brown-grey, mantle, scapulars, inner wings, rump, and upper tail-coverts paler brownish-grey; entire upperparts finely barred blackish-brown, most obviously on rear scapulars and tertials; rump tinged rufous and upper tail-coverts cerise-red. Throat, cheek, and ear-coverts almost white to pale grey, con-trasting with face stripe and fore-edge of mantle but merging with pale rosy to pale brown underparts, finely barred dark brown on flank. Legs dusky-brown to black. ADULT FEMALE. Duller than ♂. Face-patch smaller and tinged orange; bill orange-red. Rump not more than faintly rufous, tail and vent less black. JUVENILE. Duller than ♀, lacking obvious barring, face-mask, and red on underparts. Bill initially brown. FIRST-YEAR. See Plumages.

Beware risk of escape of 2 common cagebirds: Black-rumped Waxbill *E. troglodytes* (with distinctive black upper tail-coverts) and Crimson-rumped Waxbill *E. rhodopyga* (black bill and red fringes to wing-coverts, tertials, and tail). Outside west Palearctic, subject to confusion with *E. troglodytes* in West Africa, *E. rhodopyga* in East Africa, Black-faced Waxbill *E. nigriloris* in Zaïre, and Arabian Waxbill *E. rufibarba* in south-west Arabia (see Geographical Variation). Flight remarkably light, rapid, and direct, with fast whirring wing-beats; has sudden take-off and equally sudden landing. Hops, but often so close to

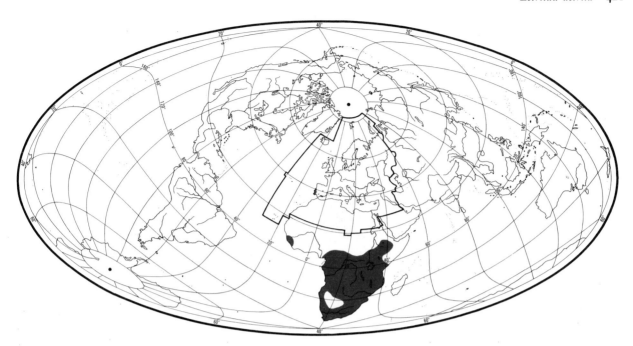

ground that bird appears to shuffle or creep forward; climbs stems. Strongly gregarious, flock members keeping remarkably close together when feeding and in flight.

Song 'tcher-tcher-preee' ('pree' a bubbling sound), rising in pitch. Calls include short but slightly buzzing monosyllable, 'tzep', also 'chip' and 'pit', twittering from flock and becoming more urgent and excited when birds disturbed.

**Habitat.** On Fogo (Cape Verde Islands), occurred up to 1800 m in *Euphorbia* (Naurois 1988). Population on banks of Guadiana (Extremadura, Spain) occupies reedmace *Typha*, reed *Phragmites*, and *Arundo* growth, nesting in willow *Salix*; others in tamarisk *Tamarix* (Guerrero *et al.* 1989). In Portugal, recent spread linked to intensive use of stands of *Arundo donax* and *Phragmites australis* for feeding and roosting (Höller and Teixeira 1983). In Afrotropics, inhabits open country with long grass, marshes, marginal aquatic vegetation such as reedbeds, active and abandoned cultivation, grassy clearings and paths in forest or woodland, gardens, and surroundings of farms and dwellings where cover and seeds from tall plants are available; occurs from sea-level to *c.* 2400 m, though mainly in lowlands; nest-sites may be in tree or creeper up to *c.* 4 m; reedbeds, papyrus swamps, or thick bushes used for roosting (Moreau 1966; Goodwin 1982). In East Africa, often near water (Williams 1963). In southern Africa, found in marshy country with reeds and tall grass, and in grass along streams and rivers (Prozesky 1974). In West Africa, found in forest as well as adjoining savanna, especially in damp places where vegetation is rank; ascends to highland grassland (Serle and Morel 1977).

**Distribution.** Breeding populations in several areas in south all certainly or probably originating from escapes or deliberate introductions.

PORTUGAL. Established in several places since 1967; not certain if range still expanding. Increased probability of escapes, as becoming more popular as cagebird. (RR.) SPAIN. Established in several places; breeding from 1977 (Guerrero *et al.* 1989). AZORES. Flock of *c.* 100 present 1983–4; bred 1984 but subsequently disappeared (GLG). CAPE VERDE ISLANDS. Reported from 6 islands, but now survives only on Santiago, where widespread (CJH).

**Population.** No information on changes.

CAPE VERDE ISLANDS. Locally abundant on Santiago (CJH).

Survival. Oldest ringed bird (Malawi) at least 8 years (Hanmer 1989).

**Movements.** Chiefly sedentary; local movements in drier parts of range. Introduced population in Iberia sedentary, with evidence of local movements; on Atlantic islands, apparently sedentary (Fernandez-Cruz *et al.* 1985).

In southern Africa, regional surveys give evidence of varying pattern of occurrence, presumably due to local movements. In north-west Zambia, regular on Zambezi floodplain during rains, but only occasional there at other times (Britton 1970). Near Cape Town (South Africa), over 9 years, far more common in spring and early summer (September–February) than autumn or winter (March–August), with no May–June records; numbers also vary greatly from year to year (Winterbottom 1962). In Orange Free State, subject to local movement between suitable

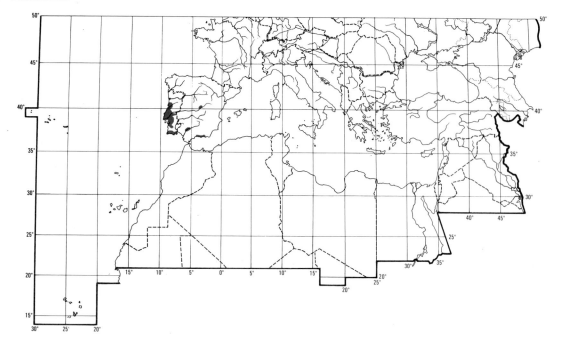

habitats, especially in winter (Earlé and Grobler 1987). Bird ringed Silverton (Transvaal) 27 August 1957 recovered 20 km south-west 28 August 1959 (Elliott and Jarvis 1973), and bird ringed near Durban (Natal) recovered 35 km away (T B Oatley); also bird ringed Cape Province, June, recovered 3 km south, October (Elliott and Jarvis 1970). Nomadic movements reported from East Africa and Gabon (Britton 1980; Brosset and Erard 1986). No evidence of seasonal pattern of occurrence in Nigeria (Elgood et al. 1973; Elgood 1982), and none reported elsewhere in West Africa (e.g. Thiollay 1985, G D Field). Widely-scattered populations (e.g. on St Helena and Seychelles) result from introductions (Goodwin 1982). DFV

**Food.** Grass seeds, taken mainly from flower-heads; rarely some small insects. Forages generally in pairs or flocks, in low herbaceous vegetation, reeds, tall grasses, and on agricultural land; usually in grass-tops, on fallen grass flower stalks, or on ground (Bates 1930; Someren 1956; Immelmann et al. 1965; Ginn et al. 1989). In Portugal, preferred foraging site ditches with reeds Phragmites; also on ground in bean fields (Heinzel and Wolters 1970). Clings to vegetation and holds grass stems and flower-heads under foot when feeding (Goodwin 1982). In Brazil, method of feeding varies with grass species: perches and clings to flower-heads of Guinea grass Panicum maximum and elephant grass Pennisetum purpureum; on Digitaria horizontalis, jumps on flower-heads bringing them to ground (Oren and Smith 1981). Plucks seeds from flower-heads and dehusks them before swallowing (Bates 1930; Oren and Smith 1981). In Africa, occasionally takes flying ants (Formicidae) or termites (Isoptera) in flight (Bannerman 1949, 1953b).

Diet in west Palearctic poorly known. On Cape Verde Islands, recorded feeding on flower-heads of Lantana (Verbenaceae) and sugar cane Saccharum (Bourne 1955). In Portugal, often feeds on seeds of grass Echinochloa crus-galli (Höller and Teixeira 1983). In Africa, seeds of grasses Setaria verticillata, Phalaris minor, and Paspalum are important (Bates 1930; Someren 1956; Siegfried 1968; Goodwin 1982). In Brazil, introduced birds recorded feeding on seeds of cosmopolitan or pan-tropical grasses, especially Panicum maximum, but also Sporobolus indicus and Echinochloa; ignores abundant native South American grasses such as Paspalum repens; feeding recorded on only 1 Neotropical species, sedge Cyperus surinamensis; also fed on amaranth Amaranthus spinosus (Oren and Smith 1981).

Diet of young in west Palearctic unknown. In Africa, parents regurgitate thick white fluid to young (Someren 1956). One captive pair feeding young by regurgitation ate mainly seeds of grasses, some hard-boiled egg yolk, and, after day 5, some small maggots (Davis 1930; Goodwin 1982). WGM

**Social pattern and behaviour.** Much of available information, especially display, derived from captive studies (notably Harrison 1962a, b, Kunkel 1967a, which see for comparison with other Estrilda and Lagonosticta), summarized by Steinbacher and Wolters (1963–5) and Goodwin (1982). Most comprehensive study in the wild (Kenya) by Someren (1956); few data from west Palearctic.

1. Gregarious outside breeding season, usually in small flocks. In Portugal, flocks of up to 300 or more, including immatures (Restall 1975b), but more often 5–50 (Heinzel and Wolters 1970). In south-west Spain, October–November 1987, flocks of up to 40 (Guerrero et al. 1989). On Cape Verde Islands, flocks of up to 50–100 (C J Hazevoet); all birds in flocks in July, breaking up at onset of rains in August, when breeding started (Bourne

1955). Allowing for different breeding seasons, similar dispersion recorded in Africa, e.g. in Kenya and Uganda where flock size large when grass seed nearly ripe (Jackson 1938), and in Brazil (Oren and Smith 1981). In Africa, flocks sometimes include other Estrildidae (Steinbacher and Wolters 1963-5). BONDS. No evidence for other than monogamous mating system. Duration of pair-bond not known. For promiscuity, see Heterosexual Behaviour (below). In captivity has hybridized with at least 5 other *Estrilda* and with African Silverbill *Euodice cantans* (Steinbacher and Wolters 1963-5; Restall 1975b). Both sexes recorded nest-building (Davis 1930), and both typically incubate (usually alternately but, at least in captivity, both pair-members quite often seen incubating simultaneously side-by-side: Steinbacher and Wolters 1963-5); both sexes brood and care for young. Young remain with parents until long after they can fend for themselves (Someren 1956; Goodwin 1982), but no details of age at independence. Age of first breeding not known. BREEDING DISPERSION. Solitary and territorial. Territory limited to nest and its immediate surroundings. Though colonial breeding not known, several pairs may nest in relatively small areas of good food supply (Someren 1956). No data on density in west Palearctic, and very few for elsewhere in range; in Stellenbosch (South Africa), 125 pairs per km² in rural districts, 10 pairs per km² in suburbia (Siegfried 1968). In Africa, nests often parasitized by Pin-tailed Whydah *Vidua macroura* (see Breeding). ROOSTING. Communal; in dense reedbeds (habitually using certain parts of extensive beds), grass, cane, etc., or, if these not available, sometimes in scrub; in reeds, birds perch in rows along strong stems, often until weight collapses stem onto a lower one (Someren 1956). In Cape Verde Islands, typically roosts in strips of cane (Alexander 1898b). In Manaus (Brazil), flocks of 50-200 roost in elephant grass *Pennisetum* at a few selected sites (Oren and Smith 1981). Someren (1956) stated that both pair-members roost overnight on nest during incubation, but in captivity Steinbacher and Wolters (1963-5) found this role confined to ♀. For some days after fledging, parent (usually ♂) leads young back to nest in evening to roost there (Someren 1956); Steinbacher and Wolters (1963-5) also reported young in captivity entering nest on cool days. Few details of activity rhythm. In Kenya, flocks roost at sunset and begin to disperse again just before dawn (Someren 1956); in Brazil, feed mainly 06.00-09.00 and 16.00-18.00 hrs (Oren and Smith 1981). For roosting behaviour, see Flock Behaviour (below).

2. Readily approachable in flocks (which use gardens and garden feeders in Africa: Someren 1958; Siegfried 1968) but 'shy' when nesting (Someren 1958). Like most *Estrilda*, flicks or wags tail from side to side in alarm (Goodwin 1982; Maclean 1985). Birds fly to bush or tree when disturbed (Maclean 1985). FLOCK BEHAVIOUR. Flocks often rise suddenly from cover on approach of observer, and make off noisily together, or more usually in succession of parties, to another patch of cover (Vincent 1949). Roosting behaviour in Kenya described as follows. Flocks converge on roost-site from all directions and, on arrival, fly from place to place with whirring wings and noisy twitter (presumably 2 in Voice). Eventually settle in rows (see Roosting), but a late incomer attempting to squeeze in typically causes others to fly off, circle, and re-settle. Before morning exodus, much activity including calling, tail-wagging, pecking one another, flitting around, and (by some) drinking. Then all suddenly rise and fly to feeding area, dropping into it in small groups until patch full; then feed with much squabbling, calling, and tail-wagging. (Someren 1956.) At other times during day, flock typically silent when crowded on bush or feeding (Jackson 1938). SONG-DISPLAY. Song (see 1 in Voice) given by ♂ in courtship (see Heterosexual Behaviour, below, for details of display) or, at

least in captivity, when separated from conspecifics of either sex, or even from other species with which singer has-formed bond (Harrison 1962b: 'Solitary Song'). ANTAGONISTIC BEHAVIOUR. No details from the wild, although Someren (1956) described nesting pair being powerless to prevent small flock of mannikins *Lonchura* pulling nest to pieces to extract seed heads; apart from hopping about in great agitation, *E. astrild* pair showed no aggression. In captivity, breeding ♂♂ tend to be aggressive, and breeding pair best isolated from conspecifics (Restall 1975b). Overt aggression (in captivity) towards ♂ Black-rumped Waxbill *E. troglodytes* was preceded by a crouched horizontal posture with head extended towards rival (Harrison 1962a). Greeting-display ('Curtseying': Kunkel 1967a) similar to other *Estrilda* and firefinches *Lagonosticta*, occurring during first encounters between birds strange to each other, whether paired or not: quick lowering of body, with turning of head accompanied by soft Contact-calls (see 2 in Voice) given singly by ♀ and in short series by ♂; then slower return to relaxed posture; during display, tail lifted and wagged sideways but with greater amplitude on side facing other bird; greeting also typically interspersed with bill-wiping. Birds not receptive stop displaying quickly, but show no aggression. (Kunkel 1967a.) HETEROSEXUAL BEHAVIOUR. (1) Pair-bonding behaviour. Following account from Kunkel (1967a) and Goodwin (1982) based on captive studies. As described above, ♂♂ perform Greeting-display towards new ♀♀, but once small flock is established, ♂♂ switch to pursuing unpaired ♀♀ and performing Feather-display ('Fluffed Singing': Kunkel 1967a) as follows. ♂ holds feather or piece of nest-material in bill, adopts upright posture (bill pointing upwards) with ruffled plumage on ventral side and flanks, often tilting himself away from ♀ to expose underside colours; tail simultaneously twisted towards ♀, and head turned somewhat (often only very slightly) towards her. In this posture, ♂ jerks stiffly up and down, but without jumping off perch, and starts singing loudly after a few upward movements. ♀ remains silent and normally flees (see subsection 3, below), but if receptive responds with same display. In study by Davis (1930), ♀ began reciprocating display after nest built and lined. Displaying ♂ may pursue ♀ through bush or hop around her on ground ('Circling'; Kunkel 1959). In wild flocks, pursuit and Feather-display are the most frequently seen interactions during breeding season. Establishment of pair-bond indicated by Clumping (♂ and ♀ perch side by side in body contact) and regular Allopreening. Feather-display never directed at mate, but newly formed pairs perform Greeting-display (see above), sometimes for days, although latterly in silence. (2) Courtship-feeding. None. (3) Mating. Copulation between pair-members probably takes place in nest according to Kunkel (1967a), but no details. In captive flocks, paired ♂♂ try to rape only those ♀♀ which are not paired to them but are known to them from previous encounters; such attempted extra-pair copulation is always preceded by ♂'s Feather-display, from which ♀♀ typically flee, so avoiding rape. (Kunkel 1967a.) (4) Nest-site selection and behaviour at nest. In Estrildidae, ♂ chooses site and lures mate with calling and repeated entering (Kunkel 1959); no information specific to *E. astrild*. For incidence and possible function of false (or 'cock') nests, see Breeding. Nest-relief discreet, parents slipping in and out almost unnoticed; at one nest, whenever one of pair approached, it suddenly dived into grass *c.* 5 m from nest and was incubating a few moments later (Granvik 1923). RELATIONS WITHIN FAMILY GROUP. Both parents brood small young and both feed them throughout nestling period. Parents alight just in front of tube-entrance to nest and typically wing-flick, then enter to feed young in main chamber. Often come to nest together; in one such case, ♀ fed brood first, regurgitating and giving a little to each chick and, when she had finished, made room for her mate who distributed food likewise. (Someren 1956.) Thick white fluid regurgitated to young is received with typical estrildine twist of neck to present upturned gape (see Fig C in Red-billed Firefinch *L. senegala*). ♂ also brings small feathers to brooding ♀, presumed 'pair-bond offering' (Someren 1958). Nest not cleaned, and its floor becomes badly soiled as young develop (Chapin 1954). See Bonds and Roosting (above) for post-fledging behaviour. ANTI-PREDATOR RESPONSES OF YOUNG. No information. PARENTAL ANTI-PREDATOR STRATEGIES. Described as fairly tame but always excitable at nest (Someren 1956), suggesting negligible active defence; weak response to raiding flock of *Lonchura* on nest (see Antagonistic Behaviour, above) consistent with this. No further information.                    EKD

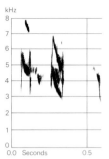

II  C Chappuis  Gabon  November 1972

**Voice.** Rather limited repertoire, although recordings indicate significant variation in contact-calls (see 2, below). Most available information is from captive birds (especially Harrison 1962a, b), and more needed from the wild. No information on geographical variation. For additional sonagrams, see Maclean (1985).

CALLS OF ADULTS. (1) Song of ♂. Short trisyllabic phrase consisting of 2 harsh 'tcher' units followed by throaty bubbling sound rising in pitch, thus 'tcher-tcher-preee' (Harrison 1962a). In southern Africa (nominate *astrild*) repeated 'di-di-di-JEE'; evidently variable, as sonagram of one ♂ shows 'di-di-JEEE di-di-JEEZRREE', in which 1st phrase c. 6·5 s, 2nd c. 9 s (Maclean 1985). Fig I depicts 4 phrases, 1st 'ti-p-t srrreeeet', followed by 3 phrases each rendered 'p-t-srrreeet' (J Hall-Craggs, W T C Seale). Last unit in song described as shrill sibilant call (Davis 1930), reminiscent of Tsree-call of Treecreeper *Certhia familiaris* (P J Sellar). Song similar to Black-rumped Waxbill *E. troglodytes* and, as in that species, has no territorial function; signals that singer is unpaired or, at least in captivity, separated from another individual (which may be of either sex, or even of another species) with which it has formed a bond, hence 'Solitary Song' (Harrison 1962a, b). (2) Contact-calls. As in *E. troglodytes*, 2 contact-calls, both similar to those of that species but less nasal and slightly higher pitched (Harrison 1962a) but no details of differences in function or context. Given frequently and repeatedly, producing noisy twitter from flying flock (e.g. Someren 1956, Steinbacher and Wolters 1963-5). (2a) A 'chip' (Harrison 1962a) or somewhat nasal 'ching' (Maclean 1985). Fig II shows 3 calls. (2b) A 'pit' (Harrison 1962a). (3) Excitement-calls. Abrupt 'tchik' and shrill sharp 'pik', being high-intensity variants of call 2 (Harrison 1962a). Presumably also serve as alarm-calls.

CALLS OF YOUNG. Little known. Food-calls of small young described as twittering (Someren 1956). Call of fledglings a repeated 'chiCHEE' (Maclean 1985).    EKD

**Breeding.** SEASON. Cape Verde Islands: nest-building starts early August, after first rains; first eggs laid late August, with subsequent nest-building from late September for 2nd brood; latest eggs recorded mid-November (Alexander 1898b; Bourne 1955; Naurois 1969, 1988; Hazevoet and Haafkens 1989). Portugal and Spain: breeds February-November (Rufino 1989); some evidence for peak breeding April-July (Bolton 1986; Vowles and Vowles 1987; Campinho *et al.* 1991). In Africa and Brazil, breeding linked closely to rains (Oren and Smith 1981; Fransden 1982; Goodwin 1982). SITE. Cavity in thick vegetation; often low down among grass clumps or in bush, shrub, tree, or creeper (Alexander 1898b; Sclater and Moreau 1933; Bourne 1955; Goodwin 1982). Nests in grass often rest on ground, or less than 50 cm up; elsewhere, usually 2-3 m above ground (Sclater and Moreau 1933; Someren 1956; Immelmann *et al.* 1965; Oren and Smith 1981; Goodwin 1982). Nest: large, untidy, inverted-pear-shaped, domed structure of grass stems, with downward-pointing entrance tube (occasionally absent) from side; often a 'false-nest' on dome (see below)

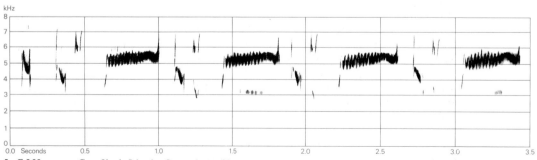

I  C J Hazevoet  Cape Verde Islands  September 1988

(Granvik 1923; Sclater and Moreau 1933; Someren 1956; Immelmann *et al.* 1965; Goodwin 1982; Naurois 1988); main structure of fresh grass stems, with or without flower-heads, criss-crossed, not woven together, to form hollow globe; not woven into surrounding vegetation (Alexander 1898*b*; Bates 1930; Bourne 1955; Someren 1956; Immelmann *et al.* 1965); upper curve of globe continues into entrance tube, which is formed by fine grass stems, heads outward, worked into frame and criss-crossed into loose network (Sclater and Moreau 1933; Vincent 1949; Someren 1956); nests on ground often have cleared area in front of entrance, possibly to aid defence against ants (Formicidae) (Someren 1956); cup lined with feathers, finer grasses, and soft grass flowers (Vincent 1949; Bourne 1955; Someren 1956; Immelmann *et al.* 1965); one nest in Spain built almost entirely from *Piptatherum miliaceum*, and some *Cynodon dactylon*; lined with feathers (Guerrero *et al.* 1989). False-nest is thin, loose-walled, cup-like structure, with or without backward sloping half-dome, on top of true nest and easily removed without damaging it (Granvik 1923; Sclater and Moreau 1933); occasionally lined or ornamented with feathers or grass flowers by ♂, and one pair spent much time re-building and re-lining cup (Someren 1956, 1958; Harrison 1975); frequency of occurrence of false-nest varies over range, e.g. in Cape Verde Islands occurs only sometimes (Goodwin 1982; Naurois 1988); not used for breeding, though occasionally an egg laid in it; once recorded as possible roost site for ♂ (false nest contained droppings during chick-rearing) (Sclater and Moreau 1933; Someren 1956); possibly serves as distraction from real nest (Granvik 1923; Someren 1956); see also Immelmann (1966). Dimensions of one nest, South Africa, length 18 cm, width 9 cm, height 13 cm (Vincent 1949); one nest, Kenya, length 21 cm, width 16 cm, height of false nest dome 6·5 cm (Granvik 1923); entrance tube 7–40 cm long, inner diameter 2·5 cm (Sclater and Moreau 1933; Goodwin 1982). Building: by both sexes (Immelmann *et al.* 1965; Robiller 1979). EGGS. See Plate 60. Sub-elliptical, smooth and non-glossy; white, with faint pink flush when first laid (Harrison 1975). Nominate *astrild*: 12·8 × 10·4 mm (11·0–15·6 × 9·9–11·7), $n = 161$; calculated weight 0·73 g (Schönwetter 1984). *E. a. jagoensis*: 13·4 × 10·6 mm (12·4–14·6 × 10·0–11·0), $n = 22$ (Vincent 1949). Clutch: 4–6 (3–9) (Immelmann *et al.* 1965; Goodwin 1982; Ginn *et al.* 1989). Of 42 clutches in southern Africa, average 4·9 (Maclean 1985); in Cape Verde Islands average possibly lower (Alexander 1898*b*; Bourne 1955; Naurois 1988); no information from Iberia. 2 broods in Cape Verde Islands (Bourne 1955). INCUBATION. 11–12 days (Someren 1956; Harrison 1975; Goodwin 1982). By both sexes, mostly by ♀; in captive pair, ♂ sat most during first few days, ♀ rarely off nest during last few days (Davis 1930; Someren 1956). YOUNG. Fed and cared for by both parents; both sexes brood young when small, and both feed young by regurgitation of thick white fluid (Davis 1930; Someren 1956; Harrison 1975). Nest usually not kept clean and many droppings accumulate (Fransden 1982). FLEDGING TO MATURITY. Fledging period 17–21 days (Someren 1956; Goodwin 1982; Maclean 1985); 22–23 days recorded in captivity (Davis 1930). BREEDING SUCCESS. No information for west Palearctic. Of one successful nest in captivity, 3 young hatched from 4 eggs, and all survived to fledging (Davis 1930). In Africa, parasitized by Pin-tailed Whydah *Vidua macroura*; 40% of nests in Malawi (Belcher 1930; Goodwin 1982). WGM

Plumages. (*E. a. jagoensis*). ADULT MALE. Forehead and crown medium ash-grey, closely marked with fine dark grey bars; central crown and nape slightly tinged brown. Hindneck, mantle, scapulars, back, rump, and upper wing-coverts closely marked with fine pale pink-buff and dark grey bars, pink slightly brighter towards outer and lower scapulars and rump, grading to cerise-red upper tail-coverts, which show fine dusky bars. Distinct crimson patch from lore to ear-coverts. Cheek and chin off-white, throat white with pink wash. Ground-colour of remainder of underparts pale pink or cerise, narrowly and faintly barred dusky grey on chest and side of breast, more distinctly elsewhere; irregular patch on mid-breast and mid-belly uniform cerise (widest on belly); vent and under tail-coverts black. Tail blackish, basal half brown at sides, marked with narrow dusky bars. Flight-feathers sepia; tertials dark brown-grey with indistinct pale brown bars and blackish border. Under wing-coverts, axillaries, and small coverts on leading edge of wing buff-white. ADULT FEMALE. Upperparts like adult ♂, but ground-colour tinged buff, slightly greyish on crown only, dark bars more contrasting than in adult ♂; rump similar to remainder of upperparts, red restricted or almost absent. Red mask tinged orange, extending only a little way behind the eye. Cheek and chin white or pale cream, ground-colour of barred remainder of underparts paler than in ♂, more pink-cream; cerise patch on breast and belly smaller; vent and under tail-coverts dark brown or black-brown. Tail marked with even buff-brown and dusky grey bars; wing as adult ♂. NESTLING. Naked (Harrison 1975). JUVENILE. Upperparts generally greyish-brown, without darker bars or with only traces of them; grey-brown wash on pink-red of rump and upper tail-coverts. No red mask through eye, or a few red feathers only, forming narrow patch on lore. Underparts pale greyish-buff, brighter buff on flank and under tail-coverts; no cerise on breast or belly; centres of under tail-coverts brown. Tail evenly barred brown and dusky grey. (Alexander 1898*b*; ZFMK.) FIRST ADULT NON-BREEDING. Crown and upper wing-coverts grey with well-defined dark bars; remainder of upper-parts brown or grey-brown with some ill-defined and widely-spaced dark bars; back and rump sometimes washed red. Mask through eye irregular, reddish-orange. Underparts brown, some ♂♂ with cerise-red patch on mid-belly; under tail-coverts brown with pale fringes. Sepia juvenile greater primary coverts retained, strongly abraded; sometimes also some flight-feathers, greater upper wing-coverts, or tail-coverts. FIRST ADULT BREEDING (1-year-olds). As 1st adult non-breeding, but red mask well-defined, back and rump usually washed red, breast more extensively pink-red, and under tail-coverts usually tipped light brown; ♂ has mid-belly distinctly washed cerise-red and under tail-coverts mainly black-brown; breast of ♀ deep pink with some cerise-red feathers, under tail-coverts mainly brown or dark brown. SECOND ADULT NON-BREEDING AND BREEDING. Like adult, but longest grey under tail-coverts with dark bars almost as long as outer tail-feathers (in adult, longer than outer feathers); mantle and scapulars tinged brown, less grey; browner greater upper primary coverts and primaries with widely-spaced dark bars, contrasting with greyer remaining upper wing-coverts which have more closely spaced bars (in adult, no contrast in colour or spacing). Rump of ♂ red, back and upper tail-coverts washed red, mid-belly distinctly cerise-red, under tail-coverts uniform brown-black; rump and mid-belly of ♀ deep pink with some cerise-red feathers, sometimes a pink wash on back and upper tail-coverts, under tail-coverts uniform dark brown. (Vowles and Vowles 1987.)

**Bare parts.** ADULT. Iris hazel, brown, red-brown, or red. Bill crimson (♂) or orange-red (♀). Leg and foot brown-grey, dark brown, black-brown, or black. (Naurois 1988; RMNH, ZFMK.) NESTLING. Mouth black, each side of gape marked with distinct bluish-white ring, black gape-flanges of lower mandible with row of bluish-white spots; palate pale flesh with 5 black spots, inside of lower mandible with black crescentic mark, 2 black marks on side of tongue (Swynnerton 1916; Goodwin 1982). JUVENILE, First Adult. Bill brown in juvenile; in 68% of 179 birds, bill turned orange during post-juvenile moult, in remainder still brown then; in spring and summer, all have bill orange. In 2nd adult non-breeding and breeding plumage (2nd winter and summer), bill of ♀ orange, of ♂ orange or red. (Vowles and Vowles 1987.)

**Moults.** ADULT POST-BREEDING. Complete; primaries descendent. In Portugal, moult July–October; starts with scattered feathers of body and upper wing-coverts shortly before shedding of p1, followed by primary coverts and bastard wing; sequence of tail moult either centrifugal or irregular, sometimes incomplete; secondaries start late, outermost (s1) shed when about 4 old primaries left, completed with innermost (s6) after completion of outer primaries. Occasionally partial pre-breeding moult January–February, mainly involving feathers which were not moulted in post-breeding. (Vowles and Vowles 1987.) In Cape Verde Islands, where breeding starts immediately after start of rains, moult starts when rain has stopped and countryside dried up, immediately after end of breeding: when rain fell July–August, breeding followed August–November and moult from about December (Bourne 1955; Bannerman and Bannerman 1968; Naurois 1988; BMNH). POST-JUVENILE. Partial. In Portugal, starts shortly after fledging and involves head, body, and wing-coverts, usually some or all tail-feathers and tertials, but no greater upper primary coverts, and apparently none or only a few flight-feathers; moult continues through January–February as a partial pre-breeding moult, during which remaining juvenile characters are lost (Vowles and Vowles 1987). Of 8 1st adults examined, Afrotropics, all retained juvenile greater primary coverts; 5 retained juvenile flight-feathers, 2 all tail, 2 some tail-feathers and tail-coverts, 1 some outer greater coverts and bastard wing (RMNH, ZFMK).

**Measurements.** E. a. jagoensis. Cape Verde islands and western Angola, all year; skins (BMNH, ZFMK). Bill (S) to skull, bill (N) to distal corner of nostril; exposed culmen on average 1·4 less than bill (S).

| | ♂ | | ♀ | |
|---|---|---|---|---|
| WING | 47·3 (1·00; 11) | 46–49 | 46·2 (1·15; 7) | 44–48 |
| TAIL | 42·8 (2·94; 11) | 39–47 | 42·6 (2·11; 6) | 40–46 |
| BILL (S) | 10·0 (0·62; 11) | 9·4–11·3 | 9·7 (0·49; 7) | 9·1–10·4 |
| BILL (N) | 6·4 (0·35; 11) | 5·8–7·0 | 6·3 (0·25; 7) | 6·0–6·6 |
| TARSUS | 14·4 (0·27; 11) | 14·1–15·1 | 13·9 (0·58; 7) | 13·1–14·5 |

Sex differences significant for tarsus.

On Cape Verde Islands, wing of 3 ♂♂, 1 ♀, and 2 unsexed birds 40–47, tail 41–46; in Benguela (western Angola), wing of

3 ♂♂ and 3 ♀♀ 46–48, tail 42–47 (White 1960; Naurois 1988). Portugal, ♀♀: wing 44, 45·5; tail 41, 46 (Heinzel and Wolters 1970).

**Weights.** E. a. jagoensis. Angola, February: ♂ 9, ♀ 7 (ZFMK).
Nominate astrild. Transvaal (South Africa): 9·1 (0·80; 19) 7·0–10·0 (Skead 1974, 1977; Day 1975). See also Jackson (1989).

**Structure.** Wing short, broad at base, tip rounded. 10 primaries: p7–p8 longest, p6 and p9 0–1·5 shorter, p5 1–2·5, p4 2–4, p3 4–6, p2 5–7·5, p1 7–9. P10 reduced, narrow, hidden below outer greater upper primary coverts; 28–35 shorter than p7–p8, 2 shorter to 2 longer than longest upper primary covert. Outer web of p6–p8 emarginated, inner web of p7–p9 with faint notch. Tip of longest tertial reaches tip of p2–p4 in closed wing. Tail long, tip graduated; 12 feathers, t1 longest, tip pointed, t6 18–26 shorter. Bill short, thick, swollen at base; culmen and cutting edges slightly decurved, tip of upper mandible sharp; bill very similar to bill of Red Avadavat Amandava amandava. Leg and foot short and slender. Middle toe with claw 14·1 (7) 13–16; outer toe with claw c. 70% of middle with claw, inner c. 64%, hind c. 85%.

**Geographical variation.** Marked, but largely clinal. Involves ground-colour of head and body (whiter, browner, or more pinkish), colour and contrast of fine barring, amount of red on rump and upper tail-coverts, colour and extent of red on breast, belly, and vent, and size (as expressed in wing and tail length). As colour also dependent on sex, age, and plumage wear, only birds of same age, sex, and stage of abrasion should be compared. E. a. jagoensis is a pale and grey race: upperparts of adult ♂ have limited brown wash, underparts white on chin and throat and pale grey elsewhere, bars contrasting little, long and rather narrow cerise-red patch on breast and belly; size intermediate; originally described from Cape Verde Islands, where introduced, later found to originate from coastal strip of central and southern Angola (White 1960; Niethammer and Wolters 1966), south to extreme north-west Namibia (Clancey 1989). Nominate astrild from southern Namibia and Botswana to western Cape Province larger, browner, bars slightly darker and heavier, breast and belly with long, narrow crimson patch. West African races occidentalis (Ghana to Zaïre) and kempi (Sierra Leone and Liberia) small, ground-colour of upperparts rather brownish-buff, cheek and throat white, belly with restricted pinkish wash; kempi greyer above, whiter below, and with clearer bars than occidentalis. Elsewhere, larger races in southern Africa and mountains of Ethiopia, smaller races along equator; paler and greyer races in more arid areas, browner ones in moister climates; red more extensive in southern races (markedly so in rubriventris from Gabon, Congo, and south-west Zaïre), less so in races of northern Afrotropics. For survey of the more distinct races, see Goodwin (1982). Birds of population introduced Portugal (more recently spreading to south-west Spain: Guerrero et al. 1989) rather variable in colour, apparently because various races of western Angola were introduced, which now form a mixture; one bird rather grey on upperparts, like jagoensis, but rather extensively red on underparts, pointing to influence of angolensis or rubriventris, another brownish above, close to angolensis (Heinzel and Wolters 1970).

Forms species-group with Black-rumped Waxbill E. troglodytes (Sahel zone from Sénégal to Sudan, mainly north of E. astrild), Crimson-rumped Waxbill E. rhodopyga (Sudan south through interior eastern Africa to Malawi; forms superspecis with E. troglodytes), and Black-faced Waxbill E. nigriloris (very local, near Lake Upemba in southern Zaïre) (Hall and Moreau 1970). Arabian Waxbill *E. rufibarba* of south-west Arabia is also in this group; may be a stabilized hybrid between *E. troglodytes* and *E. rhodopyga* (Elzen and Wolters 1978.) CSR

## *Amandava amandava* **Red Avadavat**

PLATE 31
[between pages 472 and 473]

Du. Tijgervink     Fr. Bengali rouge     Ge. Tigerfink
Ru. Тигровый астрильд     Sp. Bengalí rojo     Sw. Tigerfink

*Fringilla amandava* Linnaeus, 1758

Polytypic. Nominate *amandava* (Linnaeus, 1758), Pakistan, India, and Bangladesh, probably this race introduced in Europe, Arabia, Egypt, Réunion, and elsewhere. Extralimital: *flavidiventris* (Wallace, 1863), Burma and Yunnan (south-west China), also Lesser Sunda Islands from Lombok to Timor; *punicea* (Horsfield, 1831), south-east Asia from Thailand to Vietnam; also Java and Bali, introduced Sumatra and Fiji.

**Field characters.** 9·5 cm; wing-span 13–14·5 cm. Slightly longer than Red-billed Firefinch *Lagonosticta senegala*, but with similar form. Breeding ♂ red with brown wings and black tail; other plumages brown above and grey-buff and dusky-yellow below, with red rump (except juvenile) and black tail; juvenile shows 2 pale wing-bars. Sexes dissimilar; seasonal variation in ♂. Juvenile separable.

ADULT MALE BREEDING. Moults: April–August (complete). Head and most of body crimson, with bright scarlet lower rump and long upper tail-coverts and sooty to black lower belly, vent, and tail. Wings brown-black. At close range in fresh plumage, shows black lore and variable white spots and speckling, most visible on side of breast, flank, tips of larger coverts, and rump. Underwing mainly pale yellow. Wear removes crimson tips to feathers, appearance thus changing to mainly dark brown above and more extensively black below, with bright crimson or scarlet showing only on supercilium, cheek, chin to breast, and parts of rump; white spots and speckles largely disappear. Bill bright wax-red, with black culmen. Legs flesh-brown to pinkish-white. ADULT FEMALE. Markedly different from breeding ♂, but sharing dull scarlet rump, almost black tail, and red bill. Head and upperparts olive-brown, with almost black lore and blackish-brown wings; white speckling on wings usually confined to tips of coverts. Underparts pale buff, whitest on throat, grey-tinged on breast, and strongly dull yellow on centre of underbody. With wear, all body plumage duller, and yellow on underbody more restricted. JUVENILE. Markedly different from ♀, with initially blackish-brown bill, buff rump, prominent pale buff tips to median and greater coverts forming striking wing-bars, pale buff fringes to tertials, and uniformly white centre to underbody. Wing markings and lack of red rump immediately exclude *L. senegala*. FIRST-WINTER MALE. Resembles adult ♀ but wings distinctly blacker on leading coverts, and tertials retaining more white spots. At close range, red patches usually show above lore, on rear face, and on tips of (even greyer) breast feathers.

Given their frequency as cagebirds, *A. amandava*, *L. senegala*, and similar species could escape anywhere. ♂ *A. amandava* easily separated from *L. senegala* by much more uniformly dark red appearance, black vent, and more extensive white spotting on breast, flanks, and wings. ♀ *A. amandava* best distinguished from *L. senegala* by somewhat darker, more olive-toned head and upperparts, blackish (not crimson) lore, and much darker, blackish-brown, faintly speckled wings. Juvenile particularly confusing since observer not recognizing generic character may confuse it with small, wing-barred finches, though none shares its similar frenetic behaviour or voice.

Song a high-pitched, mainly descending, soft, liquid twitter. Calls include high-pitched chirps and squeaks, one recalling commonest call of Penduline Tit *Remiz pendulinus*.

**Habitat.** In Guadiana basin (Extremadura, Spain), introduced birds occupy wetlands dominated by reedmace *Typha* and reed *Phragmites*, as well as meadows and irrigated crops such as lucerne, maize, and tomatoes (Lope *et al.* 1985). In Granada (Spain), found in sugarcane plantations and reedbeds (Anon 1990). In Treviso (Italy), occupies reedbeds and stands of rush *Juncus* and sedge *Carex*, often descending to feed on ground; often favours clay pits, especially those filled with water; survival through cold winters presents severe problems (Mezzavilla and Battistella 1987). In Nile valley (Egypt), lives mainly in reeds (Goodman and Meininger 1989). In natural range in Asia, occurs in tall grass, reedbeds, bushes, and other rank growth, especially by lakes, rivers, and marshes; also sugarcane, grassy clearings in jungle or open woodland, and cultivated areas and gardens; will fly rather high (Goodwin 1982). In India, occurs up to 1800(–2100) m; inhabits dwarf *Ziziphus* or tamarisk *Tamarix* scrub near

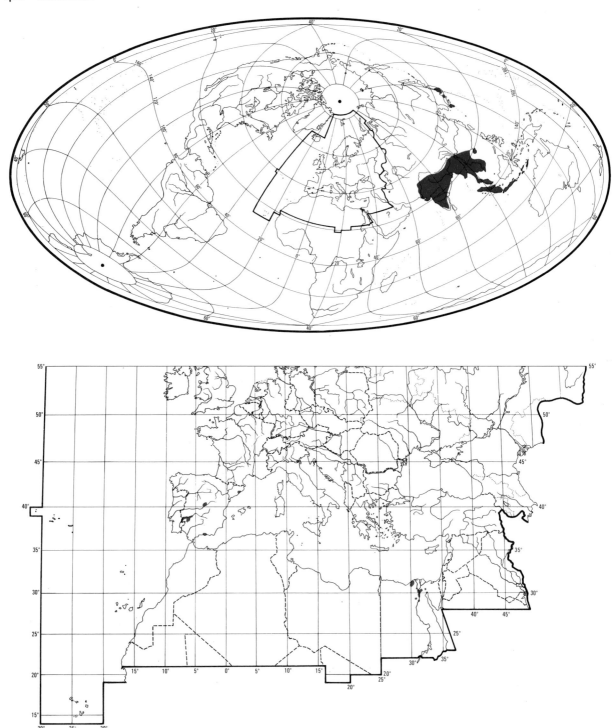

cultivation, and in Assam also villages (Ali and Ripley 1974). Avoids dry and barren plains of the north-west. Chiefly found in well-watered and well-wooded localities; fond of heavy grass jungles and patches of reeds and grass on fringes of pools. (Whistler 1941.)

**Distribution.** Breeding populations established in Spain, Italy, and Egypt, originating from escaped cage-birds.

SPAIN. Seen near Madrid from 1974, apparently breeding. In Extremadura seen since 1978, increasing progressively year by year; major concentration along 110 km of river Guadiana, from south of Badajoz to village of Villanueva de la Serena. (AN.) ITALY. Feral population breeding along river Sile (Treviso) from 1983 (Mezzavilla and Battistella 1987). EGYPT. First recorded 1861; now locally common (Goodman and Meininger 1989).

Accidental. Portugal.

**Population.** SPAIN. No precise data, but increasing; see Distribution. ITALY. About 300 birds, including at least 80-90 breeding pairs, 1983-5; much reduced by exceptionally cold weather in early 1985, subsequently recovering (Mezzavilla and Battistella 1987).

**Movements.** Chiefly sedentary, with local movements.

In study in western Spain, where introduced population breeds along 110 km of Guadiana valley, 15 of 16 recoveries were at site of ringing, and 1 at 12 km; observations confirm sedentary nature (Lope *et al.* 1985).

In Sind (southern Pakistan), present all year, but wanders locally according to availability of seeds (Ticehurst 1922). In Quetta valley (western Pakistan), frequently seen in small flocks, and probably an irregular visitor there, wandering westward from plains (Ticehurst 1926-7). Possibly only a summer visitor to Islamabad (northern Pakistan), where recorded only 31 May to 28 October over 2 years (Mallalieu 1988). In Burma, where *flavidiventris* breeds, nominate *amandava* probably a winter visitor (Smythies 1986). In south-east Asia, apparently wanders a great deal, depending on maturity of rice and other seeds (Delacour 1929). Status uncertain in Hong Kong, where small flocks reported regularly in winter in early 1970s, but scarce since 1978 (Chalmers 1986). DFV

**Food.** Grass seeds, from ground and vegetation; some small insects. Forages generally in pairs or flocks on ground or amongst herbaceous vegetation, particularly tall grasses and reeds, often close to water; in Treviso (northern Italy), ditches are favourite feeding site; in India and Pakistan, also on agricultural land and in forest clearings (Immelmann *et al.* 1965; Evans 1970; Goodwin 1982; Mezzavilla and Battistella 1987; Roberts 1992). On ground, feeds with hopping, or occasionally walking gait, on low-growing or fallen flower-heads, or picks loose seed from bare ground (Immelmann *et al.* 1965; Goodwin 1982; Mezzavilla and Battistella 1987; Roberts 1992). Clings to vegetation to feed on developing, or ripe grass seeds attached to seed-head; wide grasp of foot allows groups of stems to be gripped for support; does not hold food under foot (Harrison 1962*b*; Goodwin 1982). Small insects probably picked from vegetation; in captivity, individuals feeding on flying termites (Isoptera) caught prey in flight and ate them whilst hovering (Baker and Inglis 1930).

Diet in west Palearctic little known. In Guadiana (south-west Spain), germination of stomach and crop contents, of unknown number of birds, revealed presence of viable seeds of *Chenopodium vulgare* and grasses *Setaria*

*verticillata* and *Digitaria sanguinalis*; also recorded feeding at flowers of reedmace *Typha*, rush *Juncus acutus*, maize *Zea*, reed *Phragmites*, and meadow grass *Poa pratensis* (Lope *et al.* 1985). In Treviso, observed on grasses *Phleum*, *Lolium*, *Elymus*, and *Phragmites*, and on *Typha* (Mezzavilla and Battistella 1987). In Germany, free-flying pair fed on half-ripe seeds of grasses *Echinochloa crus-galli* and *Phleum arenarium* (Ammersbach 1960). In captivity, will feed largely on small millet *Panicum* seed, even when other foods available; readily takes unripe grass seed and young shoots of millet and other grasses, and small insects and other invertebrates (Goodwin 1982). In Treviso in winter, feeds on ground and on dead seed-heads; in spring and summer, less on ground and more on vegetation, where takes insects (Mezzavilla and Battistella 1987). In India, takes more insect food when feeding young (Immelmann *et al.* 1965).

Young fed from hatching by regurgitation of whole small seed (Lope *et al.* 1985). Free-flying pair in Germany fed young on half-ripe seeds of grasses *E. crus-galli* and *P. arenarium* (Ammersbach 1960). WGM

**Social pattern and behaviour.** Little information from the wild, at least in west Palearctic, although useful study in Extremadura (Spain) by Lope *et al.* (1984, 1985). Otherwise, data mainly from studies on captive birds, notably Kunkel (1959), Goodwin (1960), Sparks (1963*a*, *b*, 1964, 1965), and Evans (1970). For summary, see Goodwin (1982).

1. Gregarious outside breeding season. In Spain, family parties form small wandering flocks from late November which congregate for roosting (Lope *et al.* 1985; see Roosting, below). In Treviso (Italy), in flocks in late winter and spring (Mezzavilla and Battistella 1987; see Flock Behaviour, below). In India, flocks typically up to 30 birds, often mixed with other Estrildidae (Ali and Ripley 1974). BONDS. No evidence for other than monogamous mating system. Pair-bond probably long-lasting, perhaps life-long as in other Estrildidae (Kunkel 1959), but no hard evidence for wild birds. At least in captive birds bond is strong between mates after breeding season (Sparks 1964; Evans 1970; see Flock Behaviour, below). Both sexes build nest and take roughly equal shares in incubation and feeding of young until after fledging (Goodwin 1960, 1982; Ali and Ripley 1974; Lope *et al.* 1985). Family bonds persist for some time after fledging (Lope *et al.* 1985); young fed by parents for up to 2 weeks after fledging (Immelmann *et al.* 1965); see also Breeding. No information on age of first breeding. BREEDING DISPERSION. Little studied. According to Steinbacher and Wolters (1963-5) several pairs occasionally nest close together but such groupings perhaps, at best, neighbourhood groups. In Extremadura, breeding is not colonial and nests are well separated, closest 18·7 m (Lope *et al.* 1985). This consistent with finding by Sparks (1963*a*) that, unlike other Estrildidae (in which territory confined to immediate nest area), *A. amandava* appears to defend territory stretching 'several yards' in all directions from nest. No data on nest densities. ROOSTING. Large communal roosts formed outside breeding season, typically in reedbeds. For roosting associates in India, see Dhindsa and Toor (1981). Following pattern for Extremadura illustrates main features. Communal roosting begins late January in *Typha* and *Phragmites*, birds settling noisily (rapid high-pitched twittering calls: Roberts 1992) in groups on stems low in cover, often perching very close to one

another (pair-members adjacent), flank to flank (Clumping: see Flock Behaviour, below); communal roosting lasts until end of July–August, after which gradual decline as those birds which have finished moult start defending breeding territories; at end of spring and in summer, roosts are fewer and larger compared with autumn and winter when they are smaller and more dispersed (Lope *et al.* 1985). In January (of one year) 1089 in one roost, 387 in another 11 km away (Lope *et al.* 1984). From day-time flock of 12 captive birds, 8 or more formed roosting Clump (Evans 1970). Incubating ♀ roosts in nest; ♂ sometimes roosts in nest or its entrance, but probably only when suitable sites unavailable elsewhere (Goodwin 1982).

2. Tame and confiding in the wild (Whistler 1941). Flock which fed frenetically at daybreak in cold March weather in Italy was indifferent to presence of man nearby (Mezzavilla and Battistella 1987). Disturbed flock rises *en masse* in tight pack with rapid wing-beats and flies swiftly and directly to nearby trees or to more distant reeds, calling as it goes (Roberts 1992). When excited or disturbed, perched bird performs side-to-side tail movements; as tail is moved sideways, it is often first lowered, then raised again to normal height so that its tip follows U-shaped path; upward movement more rapid and emphatic than downward one (Kunkel 1959; Harrison 1962b); compared with other Estrildidae, however, such tail movements are rare and inconspicuous (Kunkel 1959). FLOCK BEHAVIOUR. See above for flock reaction to disturbance. Following account based mainly on captive birds studied by Sparks (1963a, b, 1964). Flock activities (feeding, resting, bathing, etc.) highly coordinated and synchronized. When not Clumping (see below), resting flock-members keep slightly apart, e.g. minimum 13–17 cm between a landing bird and one nearest, while approach to less than 6–10 cm may elicit low-intensity threat such as Bill-pointing (see Antagonistic Behaviour, below). At regular intervals during day, when not engaged in nesting activities, small groups separate from flock and huddle closely together (Evans 1970), usually on favourite raised perch used as regular rendezvous, but also sometimes on ground and inside nest-box (Sparks 1964). Other than at roost, such Clumping little studied in the wild, although in observations of Italian flock of *c.* 50 feeding at daybreak in March, twice a small group (5 and 4 birds) broke away from flock and formed Clump; in each case another bird soon landed on top of Clump and stayed there for up to 1 min before all rejoined flock (Mezzavilla and Battistella 1987). In study of captive birds (on which all of following is based) daytime Clumps are typically smaller than for roosting (Evans 1970), with 2–4(–9) birds sitting in a row in rather horizontal posture, flank to flank, all facing the same way; rarely, mates sit head to tail; heads of neighbours often averted (Sparks 1965: Fig A); participants may lean or push neighbour so that bird on end of row may be forced to lower wing for support; occasionally one bird leap-frogs

A

another to switch position. Clumping often develops from comfort behaviour after bathing, birds shuffling together as they preen. When a bird is joining Clump, it usually adopts Submissive-posture (ruffles plumage, retracts neck, and avoids unnecessary movement) and does so cautiously as if not to provoke others. (Sparks 1964.) Data on seasonal variation in Clumping partners (Sparks 1964; Evans 1970) as follows. During breeding season, ♂ Clumps only with mate. In study of 2 pairs, mates spent up to *c.* 65% of their time Clumping when not directly engaged in nesting activities. Adoption of non-breeding plumage by ♂♂ after breeding is, through reduction in aggressive signalling, thought to facilitate flock cohesion, while Clumping and Allopreening (see below) apparently increase social bonding between certain flock-members. Although ♂♂ in non-breeding plumage do become less exclusive over Clumping partners, they nevertheless continue to Clump preferentially with mate, and even among unpaired birds the same 2–3 individuals often Clump repeatedly with one another; Clumping relationships thus markedly stable once established, and Clumping partners also show greater coordination of other flock activities than birds which do not Clump. Experiments by Sparks (1964) showed that readiness to Clump was greater when birds were sexually attracted (for criteria, see Heterosexual Behaviour, below) but that temperature had no effect. Clumping birds spend most time crouched in Submissive-posture with eyes partly closed and head feathers slightly ruffled; next most frequent activity is self-preening, followed by Allopreening; in 13 bouts of Clumping by pair totalling 163 min, ♂ self-preened for 10% of time, Allopreened ♀ for 6%; ♀ self-preened 23%, Allopreened ♂ 9% (Sparks 1964). An individual responds to another flock-member joining Clump by pecking it or Allopreening or both. Allopreening is often (not always) preceded by self-preening. Submissive-posture readily invites Allopreening but full Allopreening-invitation-posture (Fig B, left), always adopted while being Allopreened, is intens-

B

ification of Submissive-posture: bird ceases all activity and markedly ruffles feathers on crown, nape, chin, and ear-coverts. Allopreening is mainly on head, occasionally also mantle, upper breast, flanks, and wings. Preener is aided by movements of recipient's head which may be rolled backwards so that bill points vertically up (Fig C), forward (Fig D), or sideways. Allopreening typically directed at nearest neighbour though sometimes a bird leans over neighbour to preen another on outside of row. Invitation to Allopreen sometimes also includes the following: recipient's bill may be opened and closed rapidly at rate of *c.* 10 per s (this action occurs in several contexts, often when birds approach each other); at times combined with almost inaudible call (see 6 in Voice); at other times, soliciting bird buffets the other and nuzzles its head under other's bill. Soliciting does

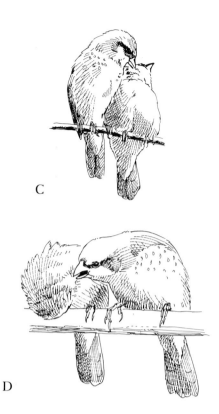

C

D

not guarantee success, however, and bird may solicit at length without being Allopreened. Excessively rough Allopreening may induce plumage-sleeking and flight of recipient. (Sparks 1964, 1965.) SONG-DISPLAY. Following descriptions based mainly on captive birds. Both sexes sing (see 1 in Voice). ♂ sings from favoured perch within nest-territory, evidently to attract ♀ and not to defend territory (Kunkel 1959; Harrison 1962a; Sparks 1963a). Sings typically from top of reed stem, sometimes also in flight (Ali and Ripley 1974). Singing-posture typically upright with plumage somewhat ruffled on underparts and crown (though latter sometimes sleeked when ♂ courts nearby ♀) (Goodwin 1960); ♂ also sings when perched in close contact with ♀ (Goodwin 1960; Harrison 1962b). Fig E shows upright ♂

E

singing to ♀ (in Submissive-posture) on ground. ♂'s display comparable to 'Circling' of *Estrilda* but is static, with ♂ more front-on to ♀ than in *Estrilda*; raises head (not quite so steeply as *Estrilda*), ruffles belly feathers, and sings while turning head arythmically from side to side. Shorter song variant (see 1 in Voice) predominates when ♂ highly aroused, longer variant when less so, but precise functional difference not clear. (Kunkel

1959.) Song (from either sex) also heard during Straw-display and Wings-raised display (see Heterosexual Behaviour, below). ANTAGONISTIC BEHAVIOUR. Account based on data from captive birds. (1) General. ♂ defends nest-territory vigorously against rivals. Harmonious relations are usually established between 2 lone breeding ♂♂, but as soon as ♀ is introduced to aviary they immediately start fighting. ♀♀, and ♂♂ in non-breeding plumage, are attacked by other ♀♀ which ♂♂ are courting (Sparks 1963a). No stable dominance hierarchy in flock (Sparks 1965). (2) Threat and fighting. Based on Sparks (1965). Commonest and mildest form of threat is Bill-pointing at opponent, occasionally developing into stabbing movement in which mandibles are opened and quickly closed again (may gape longer when Bill-pointing at bird flying past); Bill-pointing may also escalate to Bill-snapping. Imminent attack signalled by Head-forward posture in which wings may be partly opened, and plumage sleeked or ruffled depending on degree of attack-flee conflict. Subordinate orients itself horizontally across the perch or away from dominant bird, while well-matched rivals both adopt horizontal Head-forward postures and may (usually rarely and briefly) peck each other. During breeding season, however, ♂♂ may fight longer, and then also launch persistent supplanting attacks in which landing bird may spread tail to expose brightly-coloured upper tail-coverts. Goodwin (1960) observed ♂♂ fiercely attacking other adults which attempted to copulate with their (attackers') fledglings. Appeasement behaviour as follows. After series of supplanting attacks, victim remains motionless and with plumage sleeked after alighting, also stays horizontal with legs bent ready for take-off. If space too restricted to permit escape, adopts submissive-posture, though this rather ineffective at preventing attack. HETEROSEXUAL BEHAVIOUR. (1) General. No information from the wild, although in Spain birds clearly paired by mid-June, ♀ then following ♂ around (Lope *et al.* 1985). ♀♀ play active role in pair-formation (see above for ♀'s song and sexual aggression) and often take initiative by approaching ♂♂. However, only ♂ performs Wings-raised display (see below). Courtship-display resembles that of *Estrilda* and *Lagonosticta*; Straw-display homologous with feather-display of those genera but differs in that both sexes perform Straw-display (though mostly ♂: Sparks 1963a). In captivity, usually displays on ground or on branch lying on ground, occasionally on branch *c.* 1 m up. (Kunkel 1959.) (2) Pair-bonding behaviour. Bird (either sex) attracted to another of opposite sex hops up to it with crown sleeked, and twists its (own) tail (and thus bright upper tail-coverts) towards other bird (Goodwin 1960); performer's iris is contracted and display may be accompanied by song in both sexes ('Sleeked oblique' display: Sparks 1963a). ♂ also performs Wings-raised display (similar to begging of young), perhaps only to ♀♀ other than mate: crouches, raises and flutters wing furthest from ♀, and may extend and flutter wing nearest her; sometimes presses tail to ground and/or twists it towards ♀; usually turns head somewhat towards her and gives Nest-call (see 5 in Voice) or may sing, though song more often comes after or during breaks in display. Significance not clear (Goodwin 1960, which see for further details; see also Davis 1928 for first mention of display), although Kunkel (1959) once observed this display by ♂ *A. amandava* as immediate precursor to copulation with ♀ Black-rumped Waxbill *E. troglodytes*. Both sexes perform Straw-display but ♂ the protagonist in following account by Kunkel (1959): ♂ (Fig F, top) picks up straw (etc.), stands head-on or at angle to ♀, ruffles plumage markedly, bows slowly to one side, raises head again, then bows to other side. At lowest point of bow, turns head rather stiffly to one side, bill pointing down. Between bows, ♂ hops in semicircle in front of ♀, facing her more or less throughout. After 2-3 bows, ♂ starts singing then pauses briefly

F

G

display. When small nestling is approached by parent, it begs in typical estrildid fashion: lowers head to level of feet or lower, and twists neck so as to present gape vertically upwards; older young point bill straight at parent (i.e. not upturned) while fledglings also partly open wings and/or flutter them; then, as parent comes near, raise far wing (Fig G). If partner not on same

perch, other wing may also be partly raised; if parent's approach is frontal, young raises both wings equally high. Alternately, wings may be thrust out obliquely or horizontally. The hungrier the young, the more demonstrative its wing movements; young make no movements if sated or if Clumping with siblings. (Goodwin 1960.) No information on nest hygiene activities by parents, though generally no such behaviour in Estrildidae (Kunkel 1959). During the last days before fledging, young are fed outside but near nest (Lope *et al.* 1985). When young first leave nest, parents greatly excited, flying close overhead and landing alongside. At least in captivity, no brood-division outside nest. (Goodwin 1960.) ANTI-PREDATOR RESPONSES OF YOUNG. Parental Alarm-calls (see 3 in Voice) induce young to freeze with sleeked plumage (Goodwin 1960). PARENTAL ANTI-PREDATOR STRATEGIES. Parents warn fledglings with Alarm-calls; as in other Estrildidae, mostly no active defence (Kunkel 1959), but ♂ of wild pair in Germany seen attacking Wryneck *Jynx torquilla* intruding near nest (Ammersbach 1960).

(Figs by D Nurney: A–D from photographs in Sparks 1965; E from photograph and F from drawing in Kunkel 1959; G from drawing in Goodwin 1960.)                                                    EKD

before bowing 2–3 times again, then singing again, etc. If both birds are on raised perch, straw-carrying bird (♀ in Fig F, middle and bottom) bows only *towards* mate who often ignores display or watches unmoved. (Kunkel 1959.) Of 156 Straw-displays in captive flock of 23 birds, 103 were performed by paired ♂♂; Straw-display thought to maintain pair-bond (Sparks 1963a). (3) Courtship-feeding. None. (4) Mating. Sometimes, in established pairs, ♂ simply hops up to ♀ and mounts (Goodwin 1960, 1982). Otherwise, soliciting ♂ pecks ♀'s nape; if receptive, she quivers tail and ♂ mounts, belly ruffled, usually balancing without using wings, but sometimes beats wings rapidly; if ♂ attempts to mount without nape-pecking, ♀ usually flees; copulation brief (Kunkel 1959). For adults attempting to copulate with fledglings, see Antagonistic Behaviour (above). (5) Nest-site selection and behaviour at nest. No specific information on nest-site selection, but in Estrildidae generally ♂ chooses site and lures mate with calling and repeated entering of site (Kunkel 1959). Bird Nest-calls from actual or potential nest-site (sex not specified, but known to be ♂ in other Estrildidae: Kunkel 1959); same call heard (possibly from incubating ♀) when ♂ enters nest with feather for lining. ♂ performs longer incubation stints during day than ♀. (Goodwin 1960, 1982.) ♂ often adds material to nest when ♀ incubating (Baker 1934). RELATIONS WITHIN FAMILY GROUP. For first few days after hatching, nest attendance (shared equally by both sexes) is high. Eyes of young begin to open at 8 days. From shortly after hatching until after fledging, both parents provide food (regurgitating whole seeds) about equally. (Lope *et al.* 1985.) Unlike most Estrildidae (see Kunkel 1969 for review) young raise wings when begging, as in adult Wings-raised

**Voice.** High-pitched, rather shrill, but sweet calls used in captivity in many contexts (Goodwin 1960, 1982), perhaps indicating wider repertoire than scheme developed here (also derived from captive birds except where stated). Captive-bred birds mimic songs of other Estrildidae (Goodwin 1960, 1982, which see for examples and development of other abnormal songs). Bill-snapping used in threat (Sparks 1965).

CALLS OF ADULTS. (1) Song. Given by both sexes, apparently to attract opposite sex; no territorial function known. Some geographical variation (Goodwin 1982), but no details. Song of ♂ described as very sweet and varied, but rather brief descending sequence of clear liquid notes (Restall 1975a; Goodwin 1982) recalling Willow Warbler

*Phylloscopus trochilus* (Steinbacher and Wolters 1963-5). ♂ recorded singing from top of *Typha* reed in Pakistan had a rather weak soft song comprising rather spaced-out groups of very thin, high-pitched 'tsi' calls (call 2) interspersed with much louder, drawn-out whistles, some rising in pitch at end, some falling slightly; then another series of 'tsi' calls, followed by rather more rapid pretty fluting phrase on descending scale; whole sequence 'tsi-tsi-tsi twe-e-e-e-h tsi-tsi-twe-e-e-h tw-o-o-h tsi-tsi twe-twe-te-dee-slu', repeated 4–5 times (Roberts in press). Fig I (narrow band) depicts recording of allegedly lone ♂ singing from perch: starts with series of relatively rapid, rather subdued liquid fluty notes (although not certain that all these given by same bird) showing overall pitch descent; these notes followed (at *c*. 3·5 s in Fig I) by several types of rather drawn-out clear whistles, in sequence 'swee s-s-sui zeee s(w)eee swee swiwee' (in recording these recur several times in different sequences, suggesting finite

repertoire); to clarify structure, Fig II shows wide-band version of buzzy 'zeee' (4th unit from end in Fig I; 4th unit from start is also of this type) (W T C Seale). Rendering by Roberts (in press) would appear to conform roughly to the first of 2 song-types distinguished by Kunkel (1959) as follows (though Pakistan song apparently lacks its final 'trill'). (1a) Long-song of ♂. Series of soft fluting tones descending overall in pitch, beginning with Contact-call (call 2) and ending in quiet 'trill'. For another description see Steinbacher and Wolters (1963-5). Long-song predominates on days when ♂ not very sexually active (i.e. little display or copulation) (Kunkel 1959). (1b) Short-song of ♂. Comprises introductory Contact-call followed by the first 2–3 whistling notes (descending series) of Long-song; brief pauses separate repetition of this sequence. Predominates on days when ♂ sexually very active (Kunkel 1959). (1c) Song of ♀. Short descending sequence without final 'trill' (Kunkel 1959), this appar-

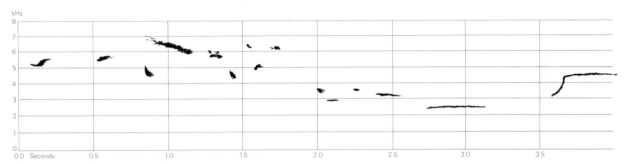

I   T J Roberts   Pakistan   June 1986

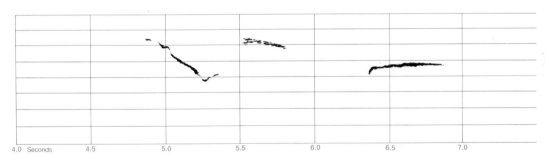

I   *cont.*

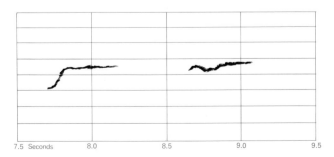

I   *cont.*

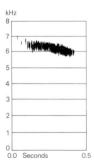

II   T J Roberts   Pakistan   June 1986

ently similar to ♂'s Short-song. Compared with song of ♂, said to be lower pitched (Delacour 1935), thinner, and given mainly in ♂'s absence (Steinbacher and Wolters 1963-5). (2) Contact-call. Short 'psee'; 'pswee' in rapid staccato repetition for long spells by foraging ♂ when separated from ♀ shortly after the 2 birds were first put in aviary together (Harrison 1962a). Also rendered (in song) 'tsi' (Roberts in press), 'zi' (Steinbacher and Wolters 1963-5). (3) Excitement- and Alarm-calls. Rather loud, drawn-out (but still monosyllabic) variation of call 2 given when attacking or threatening evenly-matched rival (Goodwin 1960, 1982); this the 'piercing version' of call 2 listed by Harrison (1962a). Still louder, more emphatic variant given by parents to warn fledglings of apparent danger (Goodwin 1960, 1982). (4) Flight-call. Short, weak 'teet-teet' (Roberts in press). Short 'chic-chic' when flying off after disturbance (Dharmakumarsinhji 1955). (5) Nest-call. Twittering series of rapidly repeated soft but high-pitched 'tee-tee-. . .' or 'teh-teh. . .'; varies in loudness and pitch, apparently with degree of excitement; ♂ gives Nest-call during Wings-raised display; also heard (not known from which sex) from bird sitting in actual or potential nest-site and when ♂ brings feather to nest where ♀ incubating (Goodwin 1960). (6) Other calls. Almost inaudible, high-pitched, rapidly repeated call given by bird soliciting Allopreening (Sparks 1964); not known if additional to repertoire listed above.

CALLS OF YOUNG. Food-call of young, audible at 2 days, a thin chirping (Steinbacher and Wolters 1963-5). No further information. EKD

Breeding. SEASON. Guadiana Basin (south-west Spain): eggs laid from late July, mainly August; latest eggs late November, before onset of autumn rains (Lope *et al.* 1985; Lever 1987). Treviso (northern Italy): main nesting period November-December (Mezzavilla and Battistella 1987). Northern Egypt: at Suez, nest-building recorded early April, fledged young seen May-June; at Lake Qarun, breeding recorded November (Goodman and Meininger 1989); at Giza (where no longer breeding), main nesting period was highly synchronous within colony, beginning August with arrival of birds from unknown wintering area (Nicoll 1909a, b). India and Pakistan: variable; some breed with first rains, but main season later, during monsoon, continuing into dry season (Ticehurst 1922; Goodwin 1982; Roberts 1992). SITE. Well hidden in bush, which is often overgrown with grass or other vegetation, or in reeds, often near, sometimes over, water. Always less than 1 m above ground or water, sometimes touching ground. (Oates 1890; Baker 1934; Whistler 1941; Immelmann *et al.* 1965; Lope *et al.* 1985; Mezzavilla and Battistella 1987; Roberts 1992.) In Guadiana, 15 nests 18-69 cm above ground (average 41·5 cm); recorded in grasses *Bromus* and *Oryzopsis*, reed *Phragmites*, reedmace *Typha*, sedge *Scirpus*, rush *Juncus*, horsetail *Equisetum*, in herbs *Rumex*, *Mentha*, and *Inula* (Lope *et al.* 1985). Nest: hollow sphere

of grass, slightly flattened above and below, often resting on flat platform of grass or twigs; central side entrance; in India and Pakistan, entrance sometimes described as slightly tubular; not elaborately woven, but sometimes strong; from strips of leaf blade up to 80 cm long, flowering stems of various grasses, and occasionally other plant material; each nest contains mix of coarse and fine grass, though one type may predominate; exterior usually coarsest and sometimes held together with spiders' webs; lining usually of finer material, often grass flower-heads, also plant down, feathers, etc. (Oates 1890; Baker 1934; Whistler 1941; Immelmann *et al.* 1965; Ali and Ripley 1974; Goodwin 1982; Lope *et al.* 1985; Roberts 1992); for unknown reason, burnt wood often added to lining, from egg-laying to fledging, mostly around hatching; in 2 nests, 75% concentrated within *c.* 2 cm of entrance (Ammersbach 1960; Goodwin 1982; Lope *et al.* 1985). Nests built mostly of coarser materials are largest; in India, one coarse nest measured *c.* 20·3 cm high, 17·8 cm wide, whereas one particularly small nest, built of fine materials, measured at most *c.* 12·7 × 12·7 cm (Baker 1934); diameter of one entrance hole 3·8 cm (Oates 1890); in Guadiana, average length of nest-platform 22·7 cm ($n = 12$), length of nest (from entrance to back wall) 14·6 cm ($n = 13$), width 12·6 cm, height 10·9 cm, internal length 9·9 cm ($n = 10$), internal width, 8·3 cm, internal height 7·5 cm (Lope *et al.* 1985). Building: by both sexes, taking 2-3 days, though 1 day (by ♂) recorded in captivity; items carried singly or in bundle; materials for main structure collected by ♂, incorporated by ♀; platform constructed first, then walls and nest-cup, roof last; lining collected by both ♂ and ♀; eggs often laid before nest complete and ♂ continues building while ♀ incubates (Davis 1928; Baker 1934; Goodwin 1960, 1982; Vierhaus and Bruch 1963; Whistler 1941; Immelmann *et al.* 1965; Robiller 1979; Lope *et al.* 1985); for details of building technique, see Kunkel (1959). EGGS. See Plate 60. Sub-elliptical to short sub-elliptical, without gloss; white, though may appear pinkish due to translucency of shell (Oates 1890; Whistler 1941; Schönwetter 1984; Roberts 1992). Nominate *amandava*: 14·5 × 11·2 mm (13·0-17·0 × 10·2-12·5), $n = 100$; calculated weight 0·96 g (Schönwetter 1984). Guadiana: 14·8 × 11·3, $n = 30$ (Lope *et al.* 1985). Clutch: in Guadiana 4-7, average 5·3 ($n = 16$) (Lope *et al.* 1985); Treviso 5-6 (Mezzavilla and Battistella 1987); India and Pakistan 5-8 (4-14), though largest clutches possibly by more than one ♀ (Baker 1934; Whistler 1941; Goodwin 1982; Roberts 1992). In Guadiana, clutches started after walls of nest built; one egg laid per day (Lope *et al.* 1985). 2 broods in Pakistan (Roberts 1992); in India, probably 2 broods in mature habitat (Immelmann *et al.* 1965). INCUBATION. In Guadiana, 13-14 days from laying of last egg to hatching of 1st (Lope *et al.* 1985); in India, Pakistan, and captivity 10-12 days (Immelmann *et al.* 1965; Goodwin 1982; Roberts 1992); by both sexes during day, ♀ at night (Goodwin 1982); in 15·5 hrs observation, Guadiana, ♀ contributed

54% of incubation by pair; eggs left uncovered for 14·7% of time, during middle of day (Lope *et al.* 1985). Hatching asynchronous in Guadiana, taking up to 3 days (Lope *et al.* 1985), so incubation presumably starts before last egg laid. YOUNG. Fed (by regurgitation) and cared for by both parents (Goodwin 1982; Roberts 1992). In Guadiana, in 5 hrs of observation during first 4 days, young left unattended for only 20·2% of time; ♀ brooded for 51·3% of remaining time, ♂ for 48·7% (Lope *et al.* 1985, which see for feeding of young). Nest not kept clean and droppings accumulate (Mezzavilla and Battistella 1987). FLEDGING TO MATURITY. Fledging period 17–21 days (Delacour 1935; Ammersbach 1960; Goodwin 1982; Lope *et al.* 1985). In India, young fed by parents for 10–14 days after fledging (Immelmann *et al.* 1965); young of free-flying pair in Germany fledged at 17–18 days, started picking up seed at 21 days, and at 37 days were no longer fed by parents (Ammersbach 1960). Age of first breeding not known. BREEDING SUCCESS. In Guadiana, of 85 eggs laid in 16 clutches during 2 years, 69% hatched and 73% of nestlings fledged giving overall success 51%; 2·69 fledglings per pair; reasons for failures included flooding of nests (Lope *et al.* 1985). WGM

**Plumages.** (nominate *amandava*). ADULT MALE BREEDING. In fresh plumage, upperparts and side of head and neck backwards to upper rump dark crimson; feather-bases dark brown on head, hindneck, side of neck, and mantle, black-brown to dull black from scapulars backwards, bases partly visible within red (especially on side of head, mantle, scapulars, and back, making red look duller). Often some fine contrastingly white specks on back, rump, and outer scapulars, more rarely some tiny white specks on mantle, hindcrown, and rear and side of neck. Lore black. Lower rump and upper tail-coverts bright scarlet, distinctly brighter than crimson of remainder of upperparts; often marked with small white specks, especially on rump. Chin to chest as well as side of breast and flank bright crimson, merging into sooty-brown on breast and upper belly and this in turn into black on lower belly, vent, and under tail-coverts; side of breast and flank with contrasting rounded white spots, sometimes extending as fine white specks to chest, breast, and side of belly. Tail black, (t5-)t6 with white spot or short bar on tip. Flight-feathers, greater upper primary coverts, and bastard wing dark sepia-brown or black-brown; tertials and greater and median upper wing-coverts black or brown-black, each with rounded white spot on tip (sometimes a short bar on tertials); lesser upper wing-coverts brown-black, tips with some red suffusion and sometimes a tiny white speck. Under wing-coverts and axillaries mainly pale yellow, longer coverts cream-white, shorter primary coverts partly grey. When very fresh (feathers still growing), red tinged tawny-gold or coppery-chestnut. *In worn plumage* (at end of breeding season), upperparts dark earth-brown, crimson feather-tips worn off, except sometimes for traces on hindneck, lower scapulars, and back; rump and upper tail-coverts scarlet, mixed with black, white spots largely worn off; side of head dark earth-brown, but red usually still prominent, forming red supercilium and extensive red mottling on cheek, ear-coverts, and side of neck; chin to chest more glossy crimson; breast, belly, and vent more extensively black, but white of feather-bases sometimes partly visible, white spots still distinct only on flank; white spots at tips of outer tail-feathers, tertials,

and upper wing-coverts partly or completely worn off. ADULT FEMALE BREEDING AND NO-BREEDING. Upperparts backwards to upper rump uniform dark earth-brown, tinged olive when fresh, contrasting markedly with crimson lower rump and upper tail-coverts (these slightly duller red than in adult ♂ but similarly marked). Black stripe on lore just over and behind eye, forming small mask, bordered below by pale buff or off-white line from corner of gape to just below eye. Cheek and ear-coverts dark olive-grey, lower side of neck paler olive-grey, gradually merging into dark olive-brown on upperside of neck and side of breast. Chin and upper throat pale buff or off-white, merging into pale greyish-buff on side of throat and on lower throat and chest and into olive-grey on flank; breast, belly, and vent bright buff-yellow, tinged rosy-orange on mid-belly; under tail-coverts buff-yellow to pink-white. Tail as adult ♂, but short white bar on tip of (t4-)t5-t6; outer web of t6 brown-grey. Wing as adult ♂, but white spots on average smaller, tertials and coverts slightly browner (less blackish), and lesser coverts without red tinge. *In worn plumage*, brown of upperparts duller, less olive (upperparts of worn adult ♂ similar, but side of head and underparts quite different); underparts paler, yellow on belly more restricted, chest, breast, and side of belly dirty pale grey-white; some dark grey or black marks often visible on under tail-coverts. ADULT MALE NON-BREEDING. Like adult ♀, but tertials and upper wing-coverts blacker (less brown), white spots larger and more distinct; lore black, usually with red stripe above; upper tail-coverts often more profusely spotted white. Side of head as adult ♀, but often mixed with variable amount of red. Chin, throat, and chest dull olive-grey, some feathers of throat and chest sometimes with traces of red on tips when fresh; distinctly greyer than in adult ♀. NESTLING. Crown and back closely covered with loose brownish down (Güttinger 1976; Goodwin 1982). JUVENILE. Upperparts uniform dark greyish-olive-brown, slightly tinged buff on rump; tips of upper tail-coverts grey-brown or dull tawny-red. Side of head and neck slightly paler greyish-brown, some pale buff speckling on lore and cheek. Chin off-white; side of breast dark greyish-olive-brown, throat, chest, and flank brighter brown-buff, grading into white of mid-belly, vent, and under tail-coverts. Tail black, tips of all feathers and outer webs of outer feathers with buff or white fringe *c.* 1 mm wide. Flight-feathers, tertials, and all upper wing-coverts brownish-black; tertials and greater coverts with contrasting buff fringe along outer web and tip, median coverts with buff tip (thus 2 pale wing-bars). Sexes similar, but ♂ has some red round eye. FIRST ADULT NON-BREEDING AND SUBSEQUENT PLUMAGES. Like adult non-breeding, but juvenile flight-feathers, greater upper primary coverts, bastard wing, and sometimes variable number of greater upper wing-coverts, a few tail-feathers, or tertials retained, browner and more worn than adult at same time of year, primary coverts in particular contrasting in colour and abrasion with neighbouring fresh feathers; ♂ (not ♀) often has scattered red feathers on head, especially in supercilium. 1st adult breeding is similar to adult breeding, but juvenile primary coverts retained; some ♂♂ attain some feathering intermediate in character between non-breeding and breeding. See Moults.

**Bare parts.** ADULT, FIRST ADULT. Iris hazel, brown-orange, light vermilion, or scarlet, independent of sex (bright red recorded in both sexes) but perhaps related to age. Bill bright coral-red or light vermilion, stripe on culmen ridge black or purple-black. Leg and foot light hazel-brown, yellow-buff, brown-flesh, flesh-brown, pink-flesh, or pink-white with slight hazel tinge; claws brown. NESTLING. Mouth pale yellow, gape surrounded by broken ring of tiny slate-coloured spots; also, 3 slate spots on tongue and a few further inside mouth. JUVENILE.

Iris reddish-yellow. Bill black-brown at fledging, bleaching through reddish-brown to red during post-juvenile moult (in Pakistan, turns red about February). (Ticehurst 1922; Steiner 1960; Harrison 1962b; Ali and Ripley 1974; RMNH.)

Moults. ADULT POST-BREEDING. Partial: head, body, lesser and median upper wing-coverts, and sometimes inner greater coverts or a few flight-feathers. Starts at end of breeding season. In Spain (nesting mainly July–October), active moult of body and sometimes flight-feathers November–December; moult suspended mainly January–March (December–April) (Lope et al. 1985). In Indian subcontinent (nesting mainly in 2nd half of year), ♂♂ in breeding plumage August–January (a few July and February), ♂♂ in post-breeding moult occur December–February(–April), ♂♂ in non-breeding plumage January–May (Ticehurst 1922; Vaurie 1949a; Goodwin 1962; Ali and Ripley 1974); only 3 of 15 ♀♀ showed moult September–March (Goodwin 1962), and moult in ♀ perhaps restricted or suppressed. On Java (nesting mainly in 1st half of year, all adults then in breeding plumage), moult to non-breeding plumage August–October; however, post-breeding moult sometimes suppressed, as some birds moult directly from worn breeding plumage into next fresh breeding September–October, and birds in non-breeding plumage are scarce (of 11 birds examined in non-breeding plumage, only 3 were apparently adult) (RMNH, ZMA). ADULT PRE-BREEDING. Complete; primaries descendent. In Spain, mainly April–August; moult of secondaries occasionally regularly ascendent (outermost to innermost, s1 to s6), but usually s6 first, followed by s1 and then ascendent to s5 (Lope et al. 1985). In Indian subcontinent, both sexes moulted flight-feathers and tail from April, soon followed by head and body; many in moult May–July; in last stage of full moult July–September(–October); moult completed July–October (Vaurie 1949a; Goodwin 1962; RMNH). On Java, complete moult August–January; on Sumatra, July–October (RMNH, ZMA). POST-JUVENILE. Extent highly variable. Starts shortly after fledging, at 50–60 days old (Lope et al. 1985), but moult suspended at any stage when conditions unfavourable or when nesting starts. In Spain, body moult October–January (head first), sometimes suspended February–April, followed by moult of body, wing, and tail April–July (Lope et al. 1985). In India and Pakistan, where hatching mainly July–October, most birds have partial post-juvenile moult to non-breeding plumage October–May, followed by almost complete moult to breeding plumage (January–)April–June (Ticehurst 1922; Goodwin 1962). On Greater Sunda Islands (Indonesia), most birds obtain 1st non-breeding plumage July–November (moult excluding flight-feathers and primary coverts), followed by breeding plumage December–March (moult excluding primary coverts and variable number of flight-feathers); a few birds moult directly from juvenile to breeding plumage, or moult into a plumage intermediate between breeding and non-breeding (feathers of underparts extensively buff-white at base, but red on tip and with black subterminal bar); of 20 birds examined from Sumatra in breeding or non-breeding plumage (RMNH, ZMA), 10 without juvenile feathers, these either 1st adults with complete juvenile or 1st pre-breeding moult, or older birds; remaining 10 retained juvenile upper greater primary coverts, 4 of these also all flight-feathers, one p1–p5 and secondaries; of 4 with all flight-feathers juvenile, 3 retained a few juvenile outer greater coverts or scattered juvenile tail-coverts or rump feathers. Occurrence of a non-breeding plumage unusual in Estrildidae; for its function, see Sparks (1963a). As this non-breeding plumage appears to be worn only briefly, mainly during wing moult, can be considered an eclipse plumage; adult ♂♂ in non-breeding

plumage are rare in some populations, perhaps because non-breeding is partly or largely suppressed there (C S Roselaar).

Measurements. ADULT, FIRST ADULT. Nominate amandava. India and Pakistan; skins (RMNH, ZMA). Bill (S) to skull, bill (N) to distal corner of nostril; exposed culmen on average 1·5 shorter than bill (S).

| | ♂ | | ♀ | |
|---|---|---|---|---|
| WING | 49·0 (0·95; 15) | 47–51 | 47·6 (1·28; 12) | 46–50 |
| TAIL | 37·0 (1·74; 15) | 35–39 | 36·3 (2·25; 12) | 33–39 |
| BILL (S) | 11·0 (0·58; 16) | 10·5–12·0 | 10·7 (0·56; 13) | 9·8–11·5 |
| BILL (N) | 6·8 (0·39; 16) | 6·3–7·5 | 6·7 (0·44; 12) | 6·1–7·4 |
| TARSUS | 14·0 (0·43; 14) | 13·2–14·6 | 14·0 (0·46; 11) | 13·3–14·7 |

Sex differences significant for wing.

Spain, live birds (Lope et al. 1985).

| | ♂ | | ♀ | |
|---|---|---|---|---|
| WING | 48·0 (1·14; 100) | — | 47·5 (1·40; 100) | — |
| TAIL | 39·2 (1·38; 100) | — | 38·7 (1·63; 100) | — |

India: wing, ♂ 48·3 (13) 47–50·5, ♀ 47·6 (4) 47–48; tail, ♂ 37·7 (15) 36–41, ♀ 36·5 (4) 35–38; bill, ♂ 10·2 (18) 10–10·5, ♀ 10·0 (10) 9·5–10·5 (Vaurie 1949a).

A. a. punicea. Java and Sumatra (Indonesia), all year; skins (RMNH, ZMA).

| | ♂ | | ♀ | |
|---|---|---|---|---|
| WING | 45·6 (1·12; 52) | 44–48 | 45·6 (0·70; 19) | 44–47 |
| TAIL | 33·4 (1·13; 16) | 32–36 | 33·2 (0·94; 15) | 32–35 |
| BILL (S) | 10·4 (0·34; 16) | 9·8–10·8 | 10·4 (0·53; 15) | 9·5–11·0 |
| BILL (N) | 6·6 (0·32; 16) | 6·1–7·2 | 6·7 (0·30; 15) | 6·3–7·2 |
| TARSUS | 13·4 (0·37; 15) | 12·9–14·0 | 13·8 (0·39; 14) | 13·2–14·4 |

Sex differences not significant.

A. a. flavidiventris. Wing 42–47 (60) (Delacour 1935).

NESTLING, JUVENILE. Juvenile wing and tail both on average 0–1 mm shorter than in adult. For growth curves of nestlings, see Lope et al. (1985).

Weights. Spain, all year: ♂ 10·0 (0·69; 100), ♀ 9·8 (0·81; 100) (Lope et al. 1985, which see for growth curves of nestlings).

Structure. Wing short, broad, tip rounded. 10 primaries: p6–p8 longest, p9 1–4 shorter, p5 0–2, p4 1–3, p3 2–5, p2 4–6, p1 5–8. P10 reduced, 25–30 shorter than p6–p8, 1 longer to 2 shorter than longest upper primary covert. Outer web of (p5–)p6–p8 emarginated, inner of p6–p9 with faint notch. Tip of longest tertial reaches to tip of p4–p6. Tail rather short, tip rounded or graduated; central pair (t1) longest, pointed; t6 10·9 (15) 8–14 shorter. Bill short, conical, rather deep and wide at base (but far less so than in African Silverbill Euodice cantans), tip sharply pointed; culmen and cutting edges slightly decurved, base of culmen rounded in section; tip of bill slightly compressed laterally. Nostrils small, rounded, hidden below frontal feathering. No bristles at base of bill. Leg and foot short and slender. Middle toe with claw 15·0 (10) 14·5–16 mm; outer toe with claw c. 65% of inner with claw, inner c. 61%, hind c. 81%.

Geographical variation. Rather distinct in colour, slight in size. Nominate amandava from Indian subcontinent slightly larger than both other races; general colour of adult ♂ in breeding plumage crimson, belly dark brown to sooty-black, vent and under tail-coverts black; white spots on body c. 1–2 mm across. In punicea from Java and Bali, general colour of ♂ scarlet, belly tawny-brown, vent and under tail-coverts black-brown; white spots 0·5–1 mm across. Birds from south-east Asia (Thailand to Vietnam) as well as Sumatra and Singapore (where introduced) have body colour intermediate between India and Java, but size, belly to under tail-coverts, and diameter of spots similar to birds from Java, and hence included in *punicea*. In *flavidiventris* from Burma and neighbouring south-west China as well as from some Lesser Sunda islands (from Lombok to Timor), colour and size of spots as in *punicea* from Java, but belly and vent golden-orange or buff-yellow, mixed white, and under tail-coverts black with pale yellow mottling. Of 30 adult ♂ *punicea* from western Java, 6 were inseparable in colour from *flavidiventris*, 5 intermediate (belly and vent mixed orange, white, and sooty).

*A. amandava* deliberately or accidentally introduced outside its original range on large scale; disjunct present-day distribution of *flavidiventris* and *punicea* perhaps also due to introduction into Indonesia, or due to parallel mutations in independent populations (Delacour 1935).

For possible relationships, see Delacour (1943), Wolters (1957), Steiner (1960), Harrison (1962*b*), and Güttinger (1976). CSR

---

## *Euodice malabarica* (Linnaeus, 1758)  Indian Silverbill

PLATE 31
[between pages 472 and 473]

FR. Capucin bec de plomb  GE. Malabarfasänchen

A resident Asian species, distributed from southern Arabia east through Indian sub-continent to central Bangladesh, south to Sri Lanka. Has bred in Israel since 1988–9, evidently originating from escapes (Shirihai in press); not clear if yet established.

---

## *Euodice cantans*  African Silverbill

PLATE 31
[between pages 472 and 473]

DU. Zilverbekje    FR. Capucin bec d'argent    GE. Silberschnäbelchen
SP. Monjita pico-de-plata    SW. Svartgumpmunia    N. AM. Warbling Silverbill

*Loxia cantans* Gmelin, 1789. Synonym: *Lonchura malabarica cantans*.

Polytypic. Nominate *cantans* (Gmelin, 1789), Sahel zone from Sénégal to western, central, and southern Sudan; *orientalis* Lorenz and Hellmayr, 1901, southern Arabia, and eastern Sudan through Ethiopia south to northern Tanzania.

**Field characters.** 11 cm; wing-span 15–16 cm. About 20% larger than Common Waxbill *Estrilda astrild*, with proportionately huge, swollen conical bill and narrower, more graduated tail. Medium-sized, strong-billed, beady-eyed waxbill, with sharply pointed tail. Head and upperparts dun-brown, underparts dull white, with black rump and tail. Sexes similar; no seasonal variation. Juvenile separable.

ADULT. Moults: April–May (complete). Head, back, and inner wing-feathers ashy-brown, finely barred but looking dun in strong light; at close range, forehead and face show faintly paler scaling and tertials show dull whitish bars creating distinct vermiculations and white tips. Outer wing-feathers, rump, upper tail-coverts, and tail black, glossed bronze. Entire underparts white; at close range, shows buff suffusion from chin to chest and along flank. With wear, barring less evident and white tips to tertials disappear. Bill proportionately much larger than in *Estrilda*, filling face; pale blue-grey. Orbital ring blue-grey, within buff eye-ring. Legs pink to grey. JUVENILE. Closely resembles adult but separated at close range by

virtually unbarred upperparts and dull brown fringes to outer wing- and tail-feathers.

Combination of bill shape and colour, quite large size, and pale plumage with black rump and tail separate it from all other species except Indian Silverbill *E. malabarica*, not yet established in west Palearctic (diagnostic off-white rump). Black-rumped Waxbill *E. troglodytes* may occur as escape (smaller, with bill red in adult or brown in immature, crimson streak through eye in adult, and whitish fringes and round end to tail). Flight fast and whirring but action rather looser than in *Estrilda*, with short undulations. Gait hopping, also sidling and jumping. Frequently waves, half-spreads, and flicks tail. Sociable but secretive, hiding in tall grass and thorn scrub.

Song a high-pitched 'trill', with falling then rising phrases; whispering quality at distance. Fast 'cheet-cheet-cheet' in flight, recalling Linnet *Carduelis cannabina* but more tinkling; other calls similar to those of *E. malabarica* (see Hollom *et al.* 1988).

**Habitat.** Across Afrotropical and Arabian lowlands, locally up to *c*. 2000 m. Inhabits dry savanna, thornscrub, grassy areas with *Acacia*, cultivated and settled areas, and neighbourhood of water in semi-desert country (Goodwin 1982). Sometimes seen on ground but mostly in scrub, hedges, and dead thorn trees (Butler 1899). In Sudan, commonly breeds in thatched roofs, date palms *Phoenix*, and thorn bushes (Butler 1905); frequent visitor to gardens in Khartoum (Macleay 1960).

**Distribution.** Resident in dry savanna and thornscrub in Afrotropics from Sénégal to Sudan, Ethiopia, and Somalia, south in the east to northern Tanzania, and in southern Arabia.

Accidental. Algeria: one collected Tamanrasset 6 May 1952 (C S Roselaar, ZMA); 2 seen 12 April 1970 at Amsel, 22 km south of Tamanrasset (B P Hall).

**Movements.** Chiefly sedentary, with local movements, but no studies available.

In Ghana, data suggest some birds move south at end of rains (Grimes 1987); movement also reported from Mali (Lamarche 1981), though not from Nigeria (Elgood *et al.* 1973). In Mauretania, regular in south, but only occasional visitor to Nouakchott area in west (Gee 1984). Isolated population at Bilma oasis (Niger, 18°46′N 12°50′E) apparently resulted from birds accompanying camel train (Lavauden 1930). Regularly wanders in flocks in East Africa (Britton 1980). In Sudan, where nominate *cantans* common in centre and west, *orientalis* is mainly a dry season visitor to eastern areas (Nikolaus 1987). In southwest Saudi Arabia, localized breeding resident, but wandering flocks may be encountered throughout the region after breeding (Stagg 1985).                    DFV

**Voice.** See Field Characters.

**Plumages.** (nominate *cantans*). ADULT. In fresh plumage, forehead and crown dark grey-brown, each feather with narrow pale buff-brown or buff fringe and short black shaft-streak, appearing spotted and scaled; lore and narrow ring round eye pale buff; remaining side of head similar to forehead and crown, appearing scaled, but ground-colour more brown and black absent. Hindneck and upper side of neck light vinous-brown with traces of fine dark bars; mantle, scapulars, back, and rump light greyish-vinous-brown with numerous fine dark grey bars. Upper tail-coverts contrastingly black. Chin and upper throat buff, each feather with dark buff-brown centre and pale isabelline-buff fringe, appearing spotted and scaled, gradually merging into darker buff-brown spotted and scaled cheek and into uniform isabelline-buff lower throat. Chest, side of breast, and flank buff or isabelline-buff, rather coarsely barred buff-brown (bars least distinct on central chest); remainder of underparts including thigh and under tail-coverts cream-white or white. Tail black, base and middle portion of central pair (t1) slightly tinged rufous-brown at sides; outer web of t5-t6 faintly greyish and with traces of fine black bars. Flight-feathers, greater upper primary coverts, bastard wing, and outer greater coverts black; remaining upper wing-coverts and tertials greyish vinous-brown with numerous fine grey bars, like mantle and scapulars, out-

ermost of these coverts more dusky grey-brown on tips, tertials slightly paler grey or vinous-grey with more widely spaced bars and narrow white tips. Under wing-coverts and axillaries pale isabelline-buff or cream-white. Sexes similar. *In worn plumage*, pale fringes of head and throat wear off, cap appearing duller grey-brown, side of head buff-brown, chin buff, all with darker feather-centres but without distinct pale scaling; ground-colour of upperparts, upper wing-coverts, and tertials duller brown, less vinous; dark bars on upperparts, side of breast, chest, flank, and upper wing-coverts less conspicuous; white tips of tertials worn off; centre of underparts more extensively dirty white. NESTLING. Naked (Güttinger 1976) or with very sparse white down on mid-back (Ticehurst 1926). JUVENILE. Rather like adult, but ground-colour less saturated yellow-buff; head without paler scaling, feather-centres only slightly darker than fringes; upperparts, upper wing-coverts, tertials, chest, side of breast, and flank without bars, or each feather with single indistinct subterminal bar only, mainly on longer feathers. Upper tail-coverts, tail, flight-feathers, primary and greater coverts, and bastard wing dark brown or black-brown instead of black, tail contrasting slightly less sharply with rump, flight-feathers and greater coverts less sharply with inner wing and forewing; no rufous on t1; tip of outer web of outer tail-feathers white. FIRST ADULT. Indistinguishable from adult once last juvenile feathers replaced. Post-juvenile moult apparently often suspended; of 23 birds examined, 11 retained some juvenile feathers, remaining 12 either 1st adults with complete post-juvenile moult or adults; these 11 all had some or all greater upper primary coverts juvenile; 4 suspended moult with some outer primaries and many secondaries still juvenile, and 2 of these also retained much juvenile feathering on head and neck or part or all juvenile tail; such retained juvenile flight- or tail-feathers and primary coverts are brown, contrasting with neighbouring new black feathers in colour, shape, and abrasion.

**Bare parts.** ADULT, FIRST ADULT, JUVENILE. Iris hazel or brown. Orbital ring blue-grey. Bill light slate-blue or blue-grey, darker on culmen and extreme tip, paler blue-grey at base and on lower mandible. Leg and foot pale vinous-pink, mauve-grey, flesh-pink, or bluish-pink. (ZMA.) NESTLING. Bare skin dark. Gape-flanges wax-yellow, bordered inside by black line, which, together with black horseshoe mark inside both upper and under mandible, forms complete black circle when mouth open (Sharpe 1890; Witherby 1901; Steiner 1960; Glatthaar and Ziswiler 1971; Güttinger 1976; Goodwin 1982).

**Moults.** ADULT POST-BREEDING. Complete; primaries descendent. Several birds in moult examined, but only a few dated. Moult apparently starts immediately at end of breeding season, which is September–April in Sahel Zone (Bannerman 1949; Niethammer 1955; Salvan 1967-9): in 3 from Sudan, late April and early May, 2 had tail old and body either just started or in full moult (both with primary moult score 14), the other had feathering of tail and body growing or new but head and neck mainly old (score 29). One from Ahaggar (southern Algeria), early May, had wing, tail, and tertials rather worn but head and body relatively new, primary score 19. (RMNH, ZMA.) ADULT PRE-BREEDING. Perhaps partial, involving head and body (see Ahaggar bird above), but confirmation needed. POST-JUVENILE. Partial or perhaps complete; at age of *c*. 3 months (Lynes 1924). Timing as in adult, but extent apparently more limited, many birds retaining at least some or all juvenile greater upper primary coverts (see First Adult in Plumages). Moult apparently suspended when environmental conditions unfavourable: 4 birds retained 4-6 juvenile outer primaries and some secondaries without showing active moult; 2 of these birds also had scattered juvenile feathering on head and neck and partly juvenile tail. (RMNH, ZMA.) Full adult plumage attained at age of *c.* 12 months (Lynes 1924).

**Measurements.** ADULT, FIRST ADULT. Nominate *cantans*, Sénégal to Sudan, all year; skins (RMNH, ZMA). Bill (S) to skull, bill (N) to distal corner of nostril; exposed culmen on average 1·2 less than bill (S).

| | | | |
|---|---|---|---|
| WING | ♂ 55·0 (2·05; 10) 52–58 | ♀ 54·0 (1·53; 13) 51–56 |
| TAIL | 43·5 (2·88; 8) 39–46 | 42·4 (2·78; 7) 40–46 |
| BILL (S) | 11·4 (0·25; 10) 11·1–11·8 | 11·4 (0·41; 12) 10·7–11·9 |
| BILL (N) | 7·9 (0·22; 10) 7·6–8·3 | 7·7 (0·19; 12) 7·4–8·0 |
| TARSUS | 13·2 (0·40; 9) 12·8–13·8 | 13·3 (0·43; 13) 12·8–14·0 |

Sex differences not significant.

Wing. Chad: 55 (36) 51–66 (Salvan 1967–9).
JUVENILE. Sexes combined. Sahel zone: wing 53·4 (1·60; 8) 51–56, tail 38·5 (1·71; 9) 36–41 (RMNH, ZMA).

**Weights.** Niger, December: ♀ 11·8 (Fairon 1971). Northern Nigeria, December–June: 10·8 (0·6; 14) 9·8–12·0 (Fry 1970). Chad: 12 (36) 10–14 (Salvan 1967–9). Ennedi (northern Chad), April: ♂ 12, ♀♀ 11·0, 11·5 (Niethammer 1955).

**Structure.** Wing short, broad at base, tip bluntly pointed. 10 primaries, p8 longest, p7 and p9 0–1 shorter, p6 0·5–2·5, p5 2·5–5, p4 5–9, p1 10–14. P10 tiny, hidden below outer upper primary coverts; 32–36 shorter than p8, 0–3 shorter than longest upper primary covert (0–2 in juvenile, 1–3 in adult, but samples small). Outer web of p6–p8 emarginated, inner of p7–p9 with faint notch. Tip of longest tertial reaches tip of p3–p4. Tail rather long, tip graduated; 12 feathers, central pair (t1) longest, tip pointed; t6 12·7 (10) 9–17 shorter in adult, 10·3 (5) 8–12 in juvenile. Bill enormous for such a small bird, proportionately heavier than in Hawfinch *Coccothraustes coccothraustes*; short, base wide, deep, and swollen, tip blunt; culmen and cutting edges slightly decurved; base of culmen broadly flattened. Nostril small and rounded, hidden in frontal feathering; no bristles at base of bill. Leg and foot short and slender. Middle toe with claw 16·1 (5) 15–17; outer toe with claw *c.* 68% of middle with claw, inner *c.* 63%, hind *c.* 77%.

**Geographical variation.** Slight. *E. c. orientalis* from southern Arabia and north-east Africa west to east and south-east Sudan and south to central Tanzania differs from nominate *cantans* from Sénégal to western Sudan in more distinct bars on upperparts, chest, side of breast, and flank, and less buff face, breast, and flank (Ogilvie-Grant 1901; Goodwin 1982; Elzen and König 1983).

Forms superspecies with Indian Silverbill *E. malabarica* from eastern Arabia and Indian subcontinent (Pakistan to Bangladesh, south to Sri Lanka), ranges of the species almost meeting in Oman. Sometimes included in genus *Lonchura* (mannikins of Asia and Australia), but see Güttinger (1970). *E. cantans* and *E. malabarica* sometimes considered to form single species *Lonchura malabarica* (Delacour 1943; Wolters 1957) but *E. malabarica* differs markedly in plumage and voice (Harrison 1964): upperparts uniform light vinous drab-brown, without distinct paler scaling on cap (some black spots on forehead) and without fine dusky bars; contrastingly white lower rump and upper tail-coverts (not black); underparts virtually uniform white, contrasting sharply with side of head, without buff and isabelline scaling on chin and throat and with only faint barring on vinous-grey side of breast and flank; bill dusky, and no bare grey orbital ring; also, size slightly larger, except bill—wing 55·6 (53–59), tail 47·5 (45–51), bill to skull 11·3 (10·4–12·8), to nostril 7·7 (7·0–8·6);, tarsus 14·0 (13·1–15·2) (both sexes combined, *n* = 14).                     CSR

# Family VIREONIDAE vireos

Small to medium-sized oscine passerines (suborder Passeres). Mostly arboreal; feed on insects (mainly by gleaning in outer foliage), also taking fruit (berries) and seeds, especially in autumn and winter. Solitary, sluggish, and rather tame. Except as vagrants, occur only in New World—with more species in Central and South America than in North America; many of more northerly species migratory. 49 species in 3 subfamilies (first 2 confined to Central and South America): Cyclarhinae (pepper-shrikes *Cyclarhis*), 2 species; Vireolaniinae (shrike-vireos *Vireolanius*), 4 species; Vireoninae (vireos), 43 species—28 in genus *Vireo* (typical vireos, mainly North and Central America) and 15 in genus *Hylophilus* (greenlets, mainly South America). Only *Vireo* considered here, with 3 species accidental in west Palearctic.

Sexes almost similar in size (♀ smaller on average). Bill usually rather short; quite thick and slightly hooked. Wing long and pointed in migratory species, shorter and rounder in others; 10 primaries, p10 considerably reduced (vestigial or absent in some). Tail rather short or of medium size; square, rounded, or sometimes notched. Leg rather short; front toes fused at base. No information available on head-scratching, sunning, or anting. Reported to bathe by diving into water from perch as well as in more usual stand-in manner (Slessers 1970).

Plumages usually olive or grey above, white or yellow below, sometimes with contrastingly coloured cap, crown-stripes, eye-stripe, and/or eye-ring; wing with bars or plain. Iris of adults red or white in some species. Sexes alike or nearly so.

DNA evidence (Sibley and Ahlquist 1990) places Vireonidae between shrikes (Laniidae) and a much enlarged Corvidae in same huge superfamily (see Corvidae). Thus, vireos and allies may not be closely related to New World wood-warblers (Parulidae) and other 9-primaried oscines as once widely thought.

# *Vireo flavifrons* Yellow-throated Vireo

PLATE 32
[between pages 472 and 473]

Du. Geelborstvireo    Fr. Viréo à gorge jaune    Ge. Gelbkehlvireo
Ru. Желтогорлый виреон    Sp. Víreo de garganta amarilla    Sw. Gulstrupig vireo

*Vireo flavifrons* Vieillot, 1808

Monotypic

**Field characters.** 12 cm; wing-span 22–24·5 cm. Close in size to Red-eyed Vireo *V. olivaceus* but with heavier, blunter-tipped bill, somewhat shorter tail, and even stronger legs. Plumage highly decorated, with diagnostic combination of yellow spectacle, throat, and breast, and bold double white wing-bar. Sexes similar; little seasonal variation except through wear.

ADULT, FIRST-YEAR. Moults: July–August (complete). Bright green upperparts and bluish-grey rump and upper tail-coverts contrast with basically yellow face and breast, blue-grey inner forewing, black wings and tail, and mainly white rear underbody. Marked with (a) bright sulphur-yellow fore-supercilium and eye-ring, forming striking spectacle, (b) bold white tips to median and greater coverts forming obvious double wing-bar, (c) pale yellow-green fringes to secondaries, and yellow to (distally) white fringes to tertials, creating obvious panel on folded rear wing (brighter and whiter than in immature Icterine Warbler *Hippolais icterina*), and (d) narrow but distinct white edges to tail-feathers. When worn in spring, plumage loses bright green and blue tones, but 1st three distinctive characters remain striking. Bill grey-blue, with dusky culmen and tip. Legs blue-grey to dull black.

Plumage pattern and colours not matched in any similar passerine occurring in west Palearctic (including other *Vireo*) and are only approached in other potentially accidental Nearctic species by adult Pine Warbler *Dendroica pinus* (easily separated by smaller size, fine bill, lack of any bluish plumage, and presence of dusky mottling on side of breast and along flank). Large bill, white spectacle and wing-bars, and bluish-grey rump and inner forewing rule out confusion with yellow-breasted Wood Warbler *Phylloscopus sibilatrix*. Of 5 other Nearctic *Vireo* with white wing-bars, White-eyed Vireo *V. griseus* is most likely to occur, but neither it nor other 4 have yellow throat or bluish-grey on wings and rump. Flight, gait, and behaviour much as *V. olivaceus*; markedly arboreal, however, usually keeping to upper canopy. Undemonstrative and often tame.

Apparently rather silent in autumn.

**Habitat.** Breeds in temperate Nearctic lowland forests and kindred environments, feeding mainly in leafy crowns of tall mature deciduous trees, especially along woodland borders, streams, and roads, and in orchards, avoiding conifers and dense second growth. Has become attached to modern residential areas (Pough 1949), and has grown remarkably trustful of human presence (Forbush and May 1955). In Great Plains is associated with mature, moist deciduous forest, especially on river bottom lands or north-facing slopes; favours open woodlands, foraging within interior foliage, and in this region less often found in wooded residential areas. Studies in Michigan suggested individual birds utilize either high parts of tall trees, totally above ground, or only lower levels including much ground area. (Johnsgard 1979.)

In winter quarters in Venezuela inhabits rain and cloud forest at 800–1800 m, occupying second growth, forest edges, and coffee plantations (Schauensee and Phelps 1978).

**Distribution.** Breeds in North America from southern Manitoba and Minnesota east to Maine, south to eastern Texas, coast of Gulf of Mexico, and central Florida. Winters in Central America from southern Mexico to Colombia, Venezuela, Cuba, and Bahamas.

Accidental. Britain: Kenidjack (Cornwall), 20–27 September 1990 (Birch 1990; Rogers *et al.* 1992).

**Movements.** Migratory.

Autumn migration begins (chiefly late) August, with few remaining in northern states after end of September; more conspicuous than many other migrants, because of September song (Bent 1950; Bull 1974; Janssen 1987). In south, rare after early October in Georgia and Mississippi (Burleigh 1958; Toups and Jackson 1987), and after late October further west in Louisiana and Texas (Lowery 1974; Oberholser 1974). First records mid-September in Central America, with main arrival in October, and present in South America only from November (Monroe 1968; Ridgely and Tudor 1989; Stiles and Skutch 1989). Winter site-fidelity shown in Belize, with one record of bird returning 3 years after ringing (Nickell 1968).

Spring migration begins March, with last reports in Central America in late April (Schauensee and Phelps 1978; Hilty and Brown 1986; Stiles and Skutch 1989). In Gulf of Mexico area, most passage records are from eastern coastal Texas to northern Florida, with only small numbers in southern coastal Texas and southern Florida (Stevenson 1957). Arrives in southern states from mid-March (Imhof 1976; Toups and Jackson 1987), and mid-latitude states (West Virginia, Kentucky) from mid-April (Mengel

1965; Hall 1983). Reaches north of range mostly in 1st half of May (Stewart and Robbins 1958; Speirs 1985; Janssen 1987).

Vagrant north of range only to central Saskatchewan, western Ontario, and Nova Scotia (American Ornithologists' Union 1983). For vagrancy to west Palearctic, see Distribrution. DFV

**Voice.** See Field Characters.

**Plumages.** ADULT. In fresh plumage (autumn), top and side of head backwards to back, as well as side of breast, bright grass-green, purer green (less olive) than in Red-eyed Vireo *V. olivaceus*; sharply demarcated bright sulphur-yellow spectacle on face, formed by yellow line running backwards from nostril and broad yellow eye-ring, narrowly broken in front by black line from gape to eye; similar but sometimes less distinct and narrower dull black or dark grey line extends short distance behind eye. Rump and upper tail-coverts medium bluish-grey, not sharply demarcated from green of back, as grey of feather-centres is exposed on (especially) side of back and feather-tips on rump slightly tinged green. Chin to chest bright sulphur-yellow, sometimes slightly tinged yolk-yellow or orange, much brighter and deeper than yellow on chest of Philadelphia Vireo *V. philadelphicus*. Flank light ash-grey, thigh medium grey; belly to under tail-coverts pure white, contrasting sharply with bright yellow chest and grey flank, almost without yellow tinge (less yellow than *V. olivaceus* or *V. philadelphicus*). Tail black, narrow fringes along both webs of each feather contrastingly white, most conspicuously on outermost feathers. Flight-feathers and tertials black, outer webs and tips of primaries fringed pale grey, outer fringes of secondaries pale yellow-green, those of tertials yellow-green at base, turning to white on tips. Upper primary coverts and bastard wing black with faint dark grey fringes. Lesser upper wing-coverts and tertial coverts medium bluish-grey, tips and outer fringes with green wash; median and greater coverts black with broad and contrasting white tips, forming 2 distinct wing-bars; outer webs of greater coverts narrowly fringed green-grey. Axillaries grey with broad yellow fringes; under wing-coverts white, those of primaries with much dull black of bases exposed. *In worn plumage* (April–July), green of head, neck, and mantle duller and more olive, more like olive-green of fresh *V. olivaceus*; outer scapulars, back, lesser upper wing-coverts, and tertial coverts dull ash-grey, green of feather-tips largely worn off; rump and upper tail-coverts duller grey, less bluish; spectacle slightly less conspicuous, slightly narrower and less deep yellow; chin to chest still deep yellow but slightly less bright, tinged green; flank and side of belly mottled light grey and off-white; white of belly and vent less pure and silky; white fringes of tail-feathers largely worn off, except on outermost feathers; fringes of flight-feathers narrower, those of secondaries and tertials pale green-grey to off-white; white on tips of lesser and median upper wing-coverts somewhat abraded, but wing-bars still distinct. JUVENILE. Upperparts and lesser upper wing-coverts uniform light brownish-grey, slightly tinged olive when plumage fresh; spectacle, chin, throat, and chest pale buff-yellow, grading to deeper yellow on cheek, below eye, and on lower ear-coverts; eye-stripe grey, rather indistinct, scarcely extending behind eye; remainder of underparts greyish-white, tinged ash-grey on flank; wing and tail as adult. Plumage of body shorter, softer, and less dense than adult, especially on underparts. FIRST ADULT. As adult, but juvenile tail, flight-feathers, bastard wing, and part of wing-coverts retained; these generally indis-

tinguishable from those of adult, and ageing usually impossible (Pyle *et al.* 1987; RMNH, ZMA).

**Bare parts.** ADULT, FIRST ADULT. Iris hazel, brown, or dark brown. Upper mandible grey-blue, dusky plumbeous-blue, or plumbeous-black, cutting edges light or dull grey-blue; lower mandible grey-blue or plumbeous-grey with dark tip. Leg and foot light blue-grey, deep leaden-grey, or plumbeous-black. JUVENILE. Iris dark brown. Bill dull flesh-grey with plumbeous-grey culmen and tip. Leg and foot plumbeous-grey. (RMNH, ZMA.)

**Moults.** ADULT POST-BREEDING. Complete; primaries descendent. On or near breeding grounds, July–August; ♀ from Wisconsin, 16 July, had just shed p1–p2. ADULT PRE-BREEDING. Partial; in winter quarters, probably December–February; involves head, body, most upper wing-coverts (not primary coverts and bastard wing), and tertials. POST-JUVENILE. Partial; on breeding grounds, July–August (Dwight 1900; Pyle *et al.* 1987); single ♂ from 1 October in final stages of body moult (Tordoff and Mengel 1956). Involves head, body, and lesser and median upper wing-coverts. FIRST PRE-BREEDING. Timing probably as adult; mainly a continuation of post-juvenile moult, involving those wing-coverts and tertials not replaced in post-juvenile, but extent perhaps sometimes as adult; occasionally some or all juvenile tertials or a few greater coverts retained. Birds from spring with much of plumage new but bastard wing, primaries, and tail slightly worn are probably adult; birds with body slightly worn, tertials and greater coverts either all new, mixed old and new, or all old, and bastard wing, primaries, and tail distinctly abraded are probably 1st adult.

**Measurements.** Eastern USA, May–September, and Guatemala, El Salvador, Costa Rica, and Curaçao, October–March (RMNH, ZMA). Bill (S) to skull, bill (N) to distal corner of nostril; exposed culmen on average *c.* 3·8 less than bill (S).

| | ♂ | | ♀ | |
|---|---|---|---|---|
| WING | 78·3 (1·75; 16) | 76–82 | 76·2 (1·84; 10) | 73–79 |
| TAIL | 49·2 (1·77; 16) | 47–53 | 48·4 (1·68; 10) | 45–51 |
| BILL (S) | 15·8 (0·47; 16) | 15·0–16·6 | 15·4 (0·72; 10) | 14·5–16·5 |
| BILL (N) | 8·5 (0·42; 16) | 7·8–9·0 | 8·3 (0·56; 10) | 7·5–9·1 |
| TARSUS | 19·9 (0·40; 16) | 19·1–20·6 | 19·4 (0·58; 10) | 18·5–20·2 |

Sex differences significant for wing and tarsus.

Birds from west of Allegheny mountains average smaller than those further east. Thus, average wing of ♂: Atlantic coast 78·6 (7), Mississippi valley and Texas 75·3 (6) (Ridgway 1904).

**Weights.** Pennsylvania (USA), May–September: 17·3 (1·65; 10) 15·0–19·2 (Clench and Leberman 1978). Kentucky (USA): April, ♂ 17·0, ♀ 16·6; July, ♂♂ 17·6, 18·6; September, ♂♂ 17·4, 17·9 (Mengel 1965). Georgia and South Carolina (USA), July–August: 16·6 (1·20; 4) 15·1–17·9 (Norris and Johnston 1958). Kansas (USA), early October: ♂ 21·5 (Tordoff and Mengel 1956). Belize: October, ♂ 17·5; November, ♂ 17·2; March, ♀ 16·6 (Russell 1964). El Salvador, October: ♀ 19 (ZMA). Curaçao, March: ♀ 16·1 (ZMA).

**Structure.** 10 primaries: p8 longest, p9 2–4 shorter, p7 0·5–1·5, p6 3–5, p5 9–11, p4 13–16, p1 21–25. P10 strongly reduced, a tiny pin concealed below reduced outermost greater upper primary covert, apparently absent in 1 of 8 birds examined; 47–53 shorter than p8, 6–9 shorter than longest upper primary covert. Outer web of (p6–)p7–p8 emarginated, inner web of (p7–)p8–p9 with notch. Tail relatively short (slightly shorter than in *V. olivaceus* and *V. philadelphicus*), tip square; 12 feathers. Bill as

*V. olivaceus* but slightly thicker in middle, slightly heavier at base, tip blunter. Bristles at base indistinct. Leg and foot slightly heavier than in *V. olivaceus*, tarsus thicker. Middle toe with claw 15·1 (8) 14-16; outer toe with claw *c.* 81% of middle with claw; inner *c.* 73%, hind *c.* 90%. Remainder of structure as *V. olivaceus*.

**Geographical variation.** Very slight, in size only (see Measurements).

Forms species-group with Solitary Vireo *V. solitarius* of west-ern, northern, and north-east North America, Black-capped Vireo *V. atricapillus* of Texas and north-east Mexico, and Hutton's Vireo *V. huttoni* of western North America and of central south-west USA to Mexican plateau (Hamilton 1958); also with White-eyed Vireo *V. griseus* of eastern USA, Grey Vireo *V. vicinior* of south-west USA and neighbouring Mexico, and Carmiol's Vireo *V. carmioli* of mountains of Costa Rica and western Panama, though these not all necessarily closely related (Johnson *et al.* 1988).　CSR

## *Vireo philadelphicus* Philadelphia Vireo

PLATE 32
[between pages 472 and 473]

Du. Philadelphia-vireo　　Fr. Viréo de Philadelphie　　Ge. Philadelphiavireo
Ru. Тонкоклювый виреон　　Sp. Víreo de Filadelfia　　Sw. Kanadavireo

*Vireo philadelphica* Cassin, 1851

Monotypic

**Field characters.** 10-11·5 cm; wing-span 20-22·5 cm. About 15% smaller and noticeably slighter than Red-eyed Vireo *V. olivaceus*, with proportionately shorter bill and more slender legs; slightly longer than Chiffchaff *Phylloscopus collybita* but noticeably more stocky, with short tail. Greenish to dull brown above and yellowish below, with greyish crown, dull white supercilium, and dark brown eye. Sexes similar; no seasonal variation. 1st-winter difficult to separate from adult.

ADULT. Moults: July-August (complete). Crown grey (with olive tinge when fresh in autumn), at some angles showing dark edges; obvious but rather diffuse dull greyish-white supercilium (from behind bill to midway along ear-coverts, most distinct over eye); blackish or dusky eye-stripe, behind eye narrow and extending for only short distance; ear-coverts buff-grey, with diffuse whitish panel under eye emphasizing eye and eye-stripe. Rest of upperparts greyish-olive-green, varying in intensity and becoming olive-brown when worn. Wings dark greyish-brown to almost black, with greyish-olive fringes to feathers. Intensity of wing markings variably described, but (when fresh) paler, whitish tips to at least outer greater coverts form pale shade or bar. Tail dark grey, faintly fringed olive. Underparts dull white, with small green-olive patch on side of breast and dull lemon-yellow wash from lower chin to fore-flanks and lower chest. Eye dark brown. Bill blackish, with blue cutting edges and base. Legs bluish-grey to dull black. FIRST-YEAR. Compared to adult at same time of year, crown more olive and underparts distinctly yellower, even on belly. If retained, juvenile wing-coverts may show more distinct pale bar.

At first glance, could be mistaken for (1) green and yellow Old World warblers, e.g. juvenile Willow Warbler *Phylloscopus trochilus*, (2) one of the less patterned *Vermivora* New World warblers (Parulidae), or (3) other plain *Vireo*. Once short stubby bill clearly seen, confusion with warblers unlikely (much bigger Garden Warbler *Sylvia borin* shows no supercilium, or green or yellow tones in plumage). Only necessary distinction is from other *Vireo*, of which only *V. olivaceus* regularly crosses North Atlantic and is larger, with distinctly longer bill, much stronger head markings including longer, grey, more dark-edged crown and sharper rear supercilium, red eye (in adult), and no obvious yellow tone below except on under tail-coverts. Beware also closely related Warbling Vireo *V. gilvus* which is as likely to cross North Atlantic as *V. philadelphicus* (not dissimilar to dullest *V. philadelphicus* but slightly larger, with even less pronounced supercilium, virtually no eye-stripe, smoky-grey upperparts, and clouded white underparts, only tinged yellow on flank). Flight not as light as Parulidae; suggests small *S. borin* but vagrant observed to hover and fly-catch. Gait a purposeful hop; also clambers. Stance as *V. olivaceus*. Feeds by quite slow search of foliage, working along branches to look under leaves, even hanging upside down, and stretching or lunging for prey. Generally unobtrusive and undemonstrative, often content to feed quietly or rest within foliage for long periods, but allows close approach.

Vagrants silent.

**Habitat.** Breeds in temperate Nearctic lowlands, in often moist broad-leaved woodlands, favouring edges, or second growth in old clearings and burnt areas, and willow *Salix* or alder *Alnus* thickets by streams, ponds, or lakes, feeding both in treetops and lower growth; in mixed woods prefers broad-leaved trees (Pough 1949). Also occupies deserted farms, and occasionally shade trees in villages; in North Dakota, restricted to mature stands of quaking aspen *Populus* with closed canopy (Johnsgard 1979).

**Distribution.** Breeds in North America, from north-east British Columbia, east to central Quebec and Newfoundland, south to North Dakota, northern Michigan, and central Maine. Winters in Central America, from Yucatán peninsula to Panama and northern Colombia.

Accidental. Britain: 1st-winter, Isles of Scilly, 10–13 October 1987 (Rogers *et al.* 1988). Ireland: Galley Head (Co. Cork), 12–17 October 1985 (Brazier *et al.* 1986).

**Movements.** Migratory. Inconspicuous, and song easily confused with Red-eyed Vireo *V. olivaceus*; data thus sparse, and greatly under-represent actual occurrence (Bent 1950; Todd 1963; Cadman *et al.* 1987).

Southward movement begins in August, when migrants appear across southern Canada and northern USA (Sadler and Myres 1976; Sprague and Weir 1984; Janssen 1987). North of 35°N, main movement in September (Palmer 1949; Potter *et al.* 1980; Janssen 1987). Crosses Gulf of Mexico to Central American wintering areas chiefly in October (Monroe 1968; Ridgely 1981; Phillips 1991), with few left in USA by November (Bent 1950; Lowery 1974; Imhof 1976). Migrates chiefly east of Rockies and west of Appalachians (American Ornithologists' Union 1983). Very common in Louisiana in autumn, but uncommon in spring (Lowery 1974); more conspicuous autumn than spring in Alabama (Imhof 1976), but the reverse in Texas (Oberholser 1974). Winter ringing shows site-fidelity in Panama, with returns up to 4 years later (Loftin 1977).

Northward movement begins late March, and by late April most birds have left winter range (Monroe 1968; Stiles and Skutch 1989; Phillips 1991). Crosses Gulf mostly to Texas coast (Stevenson 1957). Some reach southern USA in early April, but few reported owing to scarcity of song then (Oberholser 1974; Imhof 1976; James and Neal 1986). Peak migration in May, when first arrivals detected from Maryland to Quebec and from Colorado to Alberta (Stewart and Robbins 1958; Todd 1963; Bailey and Niedrach 1967; Sadler and Myres 1976). Reaches main breeding areas only in late May or early June (Palmer 1949; Houston and Street 1959; Erskine 1985).

Has strayed beyond accepted range northward to Mackenzie District and Churchill in north-central Canada (Bent 1950; Scotter *et al.* 1985). For vagrancy to west Palearctic, see Distribrution. AJE

**Voice.** See Field Characters.

**Plumages.** ADULT. Forehead and crown dull medium ash-grey, tinged olive-green when fresh, especially on hindcrown; grey cap slightly duller, less bluish, and distinctly shorter than in Red-eyed Vireo *V. olivaceus* (which has blue-grey cap extending to hindneck), not bordered by blackish lines at sides. Supercilium greyish-white, distinct and sharply demarcated from side of upper mandible to above rear of eye, faint and olive-grey further back; lore and short patch behind eye dark grey, forming distinct dark eye-stripe; both supercilium and eye-stripe extend less far behind eye than in *V. olivaceus*. Small ill-defined grey-white patch below eye; lower cheek and ear-coverts olive-grey. Entire upperparts from hindneck backwards greyish-olive, slightly tinged green on hindneck, mantle, and scapulars, but less bright olive-green than in *V. olivaceus*, rump and upper tail-coverts in particular distinctly greyer; grey crown well demarcated from grey-olive hindneck when plumage worn, less so when fresh. Upper chin greyish-white, lower chin, throat, and chest bright pale lemon-yellow, bordered by green-olive on side of throat and on side of breast; remainder of underparts pale greyish-white or white, rather evenly tinged pale yellow when fresh (least so on mid-belly), hardly so when plumage worn. Green-olive patch on side of breast less extensive than in *V. olivaceus*, chin to chest and flank more intensely yellow, belly less white (in *V. olivaceus*, under tail-coverts often distinctly yellower than belly, in *V. philadelphicus* belly and coverts about equally pale yellow-white). Tail dark grey or blackish-grey, outer webs tinged greyish-olive (tail appearing darker and less green than in *V. olivaceus*). Flight-feathers, tertials, upper primary coverts, and bastard wing as in *V. olivaceus*, but fringes along outer webs of feathers narrower, greyer olive, less greenish; lesser, median, and greater upper wing-coverts dark grey with paler olive-grey or ash-grey tips, outer fringes of greater coverts yellow-white or greyish-white, sometimes forming faint narrow pale wing-bar; upper wing-coverts distinctly greyer than scapulars (in *V. olivaceus*, scapulars and coverts both olive-green, but in both species contrast differs when heavily abraded, *V. olivaceus* becoming greyer on coverts, *V. philadelphicus* greyer on scapulars). Under wing-coverts and axillaries as in *V. olivaceus*. JUVENILE. As adult but generally drabber, washed brown; wing-bars more distinct (Pyle *et al.* 1987); similar to *V. olivaceus*, but darker above and distinctly yellow below (Dwight 1897, 1900).

FIRST ADULT. Like adult, but tail-feathers often narrower and more sharply pointed at tip, less broad and round-tipped; crown slightly more extensively tinged olive than in adult at same stage of wear, underparts more extensively and deeply yellow, throat and chest lemon-yellow, belly less white, extensively washed light yellow.

**Bare parts.** ADULT, FIRST ADULT. Iris brown or dark brown. Bill plumbeous-black, darkest on culmen and extreme tip; cutting edges paler blue-grey, base and middle portion of lower mandible plumbeous-blue, tinged pink-blue at extreme base. Leg and foot blue-grey to plumbeous-black. (RMNH.) JUVENILE. Iris deep hazel-brown. Bill pale bistre-brown, tinged flesh on lower mandible. Leg and foot pink-buff. (Dwight 1897.)

**Moults.** ADULT POST-BREEDING. Complete; primaries descendent. On breeding grounds, late July and August. Probably no pre-breeding moult. POST-JUVENILE. Partial; in summer quarters, starting shortly after fledging. Late July and August. (Dwight 1897, 1900; Pyle *et al.* 1987.) Involves head, body, lesser, median, and variable number of greater upper wing-coverts and tertials.

**Measurements.** ADULT, FIRST ADULT. Eastern USA, May–September, and Mexico, winter; skins (BMNH, RMNH, SMTD, ZMA). Bill (S) to skull, bill (N) to distal corner of nostril; exposed culmen on average *c.* 3·6 less than bill (S).

| | | | | | |
|---|---|---|---|---|---|
| WING | ♂ | 68·3 (1·46; 15) | 66–70 | ♀ 67·1 (1·21; 9) | 65–69 |
| TAIL | | 45·9 (1·96; 15) | 43–49 | 43·9 (1·43; 9) | 42–46 |

| | | | |
|---|---|---|---|
| BILL (S) | 13·5 (0·37; 14) 12·8–14·2 | 13·4 (0·51; 9) | 12·7–14·1 |
| BILL (N) | 7·0 (0·25; 15) 6·6–7·3 | 7·0 (0·28; 9) | 6·5–7·4 |
| TARSUS | 17·2 (0·43; 14) 16·6–17·9 | 17·7 (0·56; 9) | 16·7–18·4 |

Sex differences significant for tail.

**Weights.** Pennsylvania (USA): May 12·2 (1·14; 79) 10·3–16·1, September–October 12·1 (167) 9·4–15·4 (Clench and Leberman 1978, which see for details). Illinois (USA), September: 13·4 (9) (Graber and Graber 1962). Kansas (USA), late September and early October: 13·8 (11) 12·0–15·9 (Tordoff and Mengel 1956). Kentucky (USA), September: ♂ 13·6, ♀♀ 10·5, 11·5 (Mengel 1965). Belize, November: ♀ 12·8 (Russell 1964). Oaxaca (Mexico), May: ♀ 12·3 (Binford 1989).

**Structure.** 9 primaries: p7–p8 longest, p9 4–6 shorter, p6 1–2, p5 5–7, p4 8–11, p1 14–17; no trace of a reduced p10 found in the few specimens examined. Outer web of (p5–)p6–p8 emarginated, inner web of p9 with notch (sometimes also a faint one in p6–p8). Bill slightly less deep than in *V. olivaceus*, more flattened dorso-ventrally, tip less compressed laterally, extreme tip of upper mandible less strongly decurved, hook finer; gonys less bulging. Tarsus and toes slender. Middle toe with claw 13·0 (5) 12·5–13·4. Remainder of structure as in *V. olivaceus*.

**Geographical variation.** Slight; in Canada, size increases slightly towards north and west, with smallest birds in central Ontario and largest in Alberta (Barlow and Power 1970).

Forms species-group with Warbling Vireo *V. gilvus* from Canada, USA, and north-west Mexico, and Brown-capped Vireo *V. leucophrys* from mountains of eastern Mexico to Bolivia (Hamilton 1958; Johnson *et al.* 1988). CSR

---

## *Vireo olivaceus* Red-eyed Vireo

PLATE 32
[between pages 472 and 473]

DU. Roodoogvireo  FR. Viréo à oeil rouge  GE. Rotaugenvireo
RU. Красноглазый вирeон  SP. Chivi ojirrojo  SW. Rödögd vireo

*Muscicapa olivacea* Linnaeus, 1766. Synonym: *Vireo virescens*.

Polytypic. OLIVACEUS GROUP. Nominate *olivaceus* (Linnaeus, 1766), northern and eastern North America (except interior of north-west USA), south to central coast of Texas, Gulf coast, and central Florida, vagrant to west Palearctic; *caniviridis* Burleigh, 1960, Washington, Idaho, and northern Oregon (interior of north-west USA) (extralimital). CHIVI GROUP (extralimital). *V. o. chivi* (Vieillot, 1817), central Peru and Bolivia south to northern Argentina; 9 other races in South America north of *chivi*.

**Field characters.** 12·5 cm; wing-span 23–25 cm. Somewhat shorter than Garden Warbler *Sylvia borin*, with 25% longer bill, less rounded head, up to 10% longer wings, and (sometimes when perched) apparently longer legs (due to exposure of thighs and seemingly shorter tail). Rather small but quite robust, weighty, warbler-like bird, with rather long and deep, stubby-ended bill, long head (exaggerated by strong linear pattern through eye), rather peaked crown, full body, pointed wings, and almost square tail. Bold, white, dark-bordered supercilium; eye red in adult, dark brown in 1st-winter. Upperparts mainly dull olive, underparts dull whitish. Sexes similar; no seasonal variation. Juvenile separable.

ADULT. Moults: July–October (partial and variable, suspended during migration). Forehead, crown, and nape uniformly mouse-grey, with bluish cast and narrow dusky to black edge; long supercilium off-white, emphasized by dark edge to crown and shorter but still distinct dusky eye-stripe. Red eye further emphasized by off-white lower crescent. Lower lore and ear-coverts dusky olive; side of neck buff-olive. Back and rump olive, with green cast when fresh but browner when worn. Wings dusky olive, with faint patterning on greater coverts and tertials due to paler fringes, yellowish-green when fresh but dull buff when worn; similarly coloured fringes to secondaries can create dull wing-panel. Underwing pale sulphur-yellow to off-white. Tail dusky olive, with outer feathers fringed as wing-feathers. Underparts dull white, with buff- to yellowish-olive wash on side of chest (extending towards centre on some) and quite well down flanks; under tail-coverts distinctly suffused pale yellow. Bill black with grey to horn base to lower mandible. Legs and feet rather heavy; bluish- to lead-grey. JUVENILE. Much juvenile plumage can be retained by trans-Atlantic vagrants in autumn. Noticeably drabber than adult. Crown dull brown, lacking bluish tone and all but vestigial dusky edge; supercilium dull white, suffused olive at rear. Eye dark brown. Lower face, side of neck, and chest clouded, with chest sometimes appearing softly mottled. FIRST-AUTUMN. Freshly moulted bird similar to fresh adult (see Plumages).

In brief glimpse, can be confused with smaller, plain Old World warblers (Sylviidae), but, when seen well, combination of strong, rather stubby-ended bill, grey crown, long whitish supercilium, and red eye of adult soon distinguish *V. olivaceus* from rather smaller, plain congeners occurring in west Palearctic. Juvenile more confusing but distinctive call always helpful (see below).

Flight lacks lightness of *Phylloscopus* warblers or Nearctic Parulidae, recalling more that of *Sylvia*, *Hippolais*, and even *Acrocephalus*; capable of remarkable acceleration and momentum, dashing through cover and across open spaces. Stance rather level, often carrying tail at similar angle to back, with wing-points slightly drooped; sustained both on perch and when moving. Gait hopping but varied in rhythm so that bird moves both slowly (feeding sluggishly from foliage) and quickly (snatching at insect prey). Generally secretive, usually staying in cover for long periods and even sitting still; occasionally excitable, interacting with food competitors and then giving burst of calls.

Commonest call (in excitement or alarm) a nasal 'quee' or 'chway', with distinctly complaining timbre.

**Habitat.** Breeds in boreal, temperate, and sub-tropical zones of Nearctic, where before forest clearance it was held to be the most abundant bird species, as it still is wherever stands of trees exist within its range; inhabits mainly canopy, although nesting lower (Pough 1949). Since forest clearance, has adapted to successor habitats on farms, and even in cities if trees are present (Forbush and May 1955). In Rocky Mountains, found up to *c.* 2000 m (Niedrach and Rockwell 1959). On Great Plains to the east, inhabits deciduous forests, especially those with semi-open canopies in both upland and river-bottom forests, where there are tall trees for singing perches (Johnsgard 1979). In Canada, occupies open mixed woods and shrubbery along streams and lakesides; also second growth on burnt or logged areas and parklands (Godfrey 1979). In tropical and subtropical winter quarters, occurs in mangroves, gardens, plantations, rain forest, and second growth up to 1650 m, foraging from lower layers to treetops (Schauensee and Phelps 1978).

**Distribution.** Breeds over much of North America (except south-west), and South America south to northern Argentina. North American populations migratory, wintering in South America; South American apparently resident except southernmost, which winters in northern tropical areas.

Accidental. Iceland, Britain, Ireland, France, Netherlands, Germany, Malta, Morocco.

**Movements.** Migratory to resident, all birds wintering in northern South America (American Ornithologists' Union 1983).

Migrants begin to leave breeding areas before mid-August (Squires 1976; Sprague and Weir 1984; Cannings *et al.* 1987), and earliest birds appear in Central America and West Indies before end of August (Bond 1985; Phillips 1991). By early September most have left north-west of range (Munro and Cowan 1947; Burleigh 1972); main movement September through northern and mid-latitude states (Stewart and Robbins 1958; Fawks and Petersen 1961; Mengel 1965; Janssen 1987), and by end of October

nearly all are south of USA (Oberholser 1974; Kale and Maehr 1990). Migrates on broad front, from Rockies to Atlantic, and from eastern Mexico to Bahamas and Cuba (Bent 1950; Peterson and Chalif 1973; Brudenell-Bruce 1975; Bond 1985). Regular in very small numbers west to California (Garrett and Dunn 1981), and common as far east as Bermuda (Wingate 1973).

Northward migration starts early, with movement through Central America (Land 1970; Schauensee and Phelps 1978) and from Texas to Georgia (Oberholser 1974; Stoddard 1978) during March, but movement then pauses while winter retreats (Bent 1950). Perhaps only birds that breed in southern states move in March. In mid-April, a few advance as far as Minnesota and Maryland in some years (Stewart and Robbins 1958; Janssen 1987), but peak movement through northern states is 2–3 weeks later. Main passage into north-east breeding areas early to mid-May (Bailey 1955; Tufts 1986), and arrives late May or early June in western Canada where it is one of latest spring migrants (Houston and Street 1959; Erskine 1985; Cannings *et al.* 1987). Most birds cross Gulf between Louisiana and Alabama; thereafter, migration is on broad front, but fewer in California than autumn, and rare in Bermuda (Bent 1950; Stevenson 1957; Wingate 1973; Garrett and Dunn 1981).

Vagrant to Middleton Island and Anchorage (southern Alaska), and to Greenland (Salomonsen 1950–1; American Ornithologists' Union 1983). Rare autumn vagrant to west Palearctic, appearing with some regularity in Britain and Ireland, where 65 records up to 1990, especially south-west England and southern Ireland, mainly late September to 3rd week of October; exceptional influxes of 13 in 1985 and 11 in 1988 (Rogers *et al.* 1986*b*, 1988, 1989, 1991).                                                    AJE, PRC

**Voice.** See Field Characters.

**Plumages.** (nominate *olivaceus*). ADULT. In fresh plumage, forehead, crown, and hindneck medium bluish-grey, sometimes slightly washed green on forehead; bordered at sides by distinct sooty-black line, often greyish-black or dark grey and less distinct at side of forehead, gradually merging into grey of side of hindneck above rear of ear-coverts. Long, distinct, greyish-white or off-white supercilium, wider between nostril and front of eye, narrower above front half of eye, wider again above ear-coverts, from there becoming increasingly tinged olive-grey towards rear. Distinct eye-stripe formed by dark grey lore and long triangular dark grey patch from behind eye and over upper ear-coverts. Small off-white patch just below eye; lower cheek, ear-coverts, and side of neck light greyish-olive-green. Entire upperparts from mantle backward olive-green, sometimes not sharply demarcated from blue-grey of hindneck. Side of breast and upper flank light olive-green or yellowish-olive; chin to chest, side of belly, and lower flank white with grey wash (due mainly to grey of feather-bases showing through white feather-tips, giving rather mottled appearance), lightest on chin and throat, slightly suffused yellow or yellow-green, especially on chest and flank; mid-belly and vent white with variable light yellow wash; under tail-coverts pale lemon-yellow to yellow-white. Thigh dark grey, mixed with pale yellow. Tail dark grey, both webs of central pair of feathers (t1) and outer webs of t2–t5 olive-green. Flight-feathers, greater upper primary coverts, and bastard wing dark grey to blackish-grey, outer webs of secondaries olive-green, outer webs of other feathers with narrow olive-green fringe (absent on emarginated parts of primaries). Upper wing-coverts and tertials olive-green, dark grey of inner webs (greater coverts, tertials) or of bases (lesser and median coverts) largely concealed. Under wing-coverts and axillaries white with variable pale yellow wash, under primary coverts grey with pale yellow or white mottling. *In worn plumage* (May–July), grey of top of head slightly duller, less bluish; remainder of upperparts more greyish-olive, less bright olive-green; supercilium whiter; lower cheek to side of neck, side of breast, and upper flank duller olive-grey or brown-grey; underparts greyish-white, belly, vent, and under tail-coverts pure white, all with very little or no yellow or with hint of yellow on under tail-coverts only. JUVENILE. Entire upperparts including upper wing-coverts uniform grey-brown or buff-brown, greater and median coverts narrowly fringed and tipped pale buff-yellow or off-white. Supercilium buff-white or white, rather less sharply defined than in adult; dark eye-stripe fairly distinct. Underparts from cheek and chin downwards white, washed buff on side of breast, slightly pale buff or pale yellow elsewhere; feathering of underparts rather short and loose. Tail and flight-feathers as adult. FIRST ADULT. In autumn, plumage fresh; then closely similar to fresh spring adult, but tertials perhaps on average rather more extensively olive-green (less grey) and tail-feathers narrow with distinctly pointed tips (less broad and rounded than in spring adult). However, rather easy to separate from autumn adult: juvenile wing and tail of 1st adult still new, showing smooth tips, and primaries greyish-black with narrow off-white edge at tip; in adult, flight-feathers and tail old, though generally not heavily abraded (some birds have a number of inner primaries and central tail-feathers new or moulting, unlike any 1st adult), tips of feathers slightly frayed, tips of primaries dark brownish-grey without white edge. In spring, inseparable from adult, but occasional birds retaining odd old secondary are perhaps 1st adult.

**Bare parts.** ADULT. Iris bright red (rarely chestnut: Parkes 1988a). Bill slate-blue, slate-black, or black, cutting edges grey-pink or grey, base of lower mandible light blue-grey, in some birds with pink-flesh tinge. Leg and foot bright slate-blue, dull blue-grey, dark grey, or plumbeous-black; soles dull yellow-grey or blue-grey. JUVENILE, FIRST ADULT. Iris brown, chestnut-brown or grey-brown (see Wood and Wood 1972 for variation), turning red in winter quarters, but slight brown tinge sometimes retained through spring (Pyle *et al.* 1987); bill, leg, and foot as adult, but traces of flesh-pink gape-flanges present in early autumn.

**Moults.** ADULT POST-BREEDING. Partial (perhaps occasionally complete); on breeding grounds, July–October. Involves head, body, and variable number of wing-coverts, tertials, tail-feathers, and flight-feathers; moult suspended during autumn migration (Pyle *et al.* 1987), but 8 of 26 migrants examined by Cannell *et al.* (1983) were in active flight-feather moult and some others were in light body moult; one from September was in heavy body and tail moult (Parkes 1988a). In 10 autumn migrants examined, head and body new, tertials generally new (in 2 birds, 1 tertial old); upper wing-coverts new (in 7 birds), new apart from some greater coverts (1 bird), or mainly old (2 birds); tail mainly old, but t1 new in 3 birds, t1–t2 in 1 bird; flight-feathers and greater upper primary coverts old, but 2 innermost new in 1 bird; in 2 birds, innermost secondary (s6) much older than others, probably

*c.* 15 months old (if bird in 2nd calendar year) or *c.* 18 months (if bird adult). Moult completed after arrival in winter quarters; primary moult descendent. ADULT PRE-BREEDING. Extent unknown; moult of flight-feathers, tail, and wing-coverts in winter quarters probably a continuation of adult post-breeding, unless part of feathering moulted twice; head, body, and many wing-coverts and tertials probably replaced again February–April, just before spring migration. For occasional retention of 1 secondary, see above. POST-JUVENILE. Partial; on breeding grounds, July–September; suspended during autumn migration. In autumn migrants examined, head and body new (occasionally, a few scattered juvenile feathers retained), most upper wing-coverts new, but up to 5 outer greater coverts retained in 5 out of 10 birds, all retained in another; 1–3 tertials new in 4 out of 11 birds, tail and flight-feathers still juvenile in all. FIRST PRE-BREEDING. In winter quarters; apparently complete, as all plumage new during spring migration.

**Measurements.** Nominate *olivaceus*. Eastern North America, May–September, and migrants through West Indies, September–November; skins (RMNH, ZMA). Bill (S) to skull, bill (N) to distal corner of nostril; exposed culmen on average 4·7 less than bill (S).

| | | | | | |
|---|---|---|---|---|---|
| WING AD | ♂ | 82·0 (1·05; 26) 80–84 | ♀ | 78·6 (1·06; 18) 77–81 |
| TAIL AD | | 52·1 (1·73; 19) 49–54 | | 49·1 (1·50; 18) 46–52 |
| BILL (S) | | 17·8 (0·53; 18) 17·0–18·7 | | 16·9 (0·48; 18) 16·3–17·8 |
| BILL (N) | | 9·5 (0·44; 18) 8·7–10·2 | | 8·9 (0·40; 18) 8·2–9·7 |
| TARSUS | | 18·6 (0·50; 18) 17·8–19·3 | | 18·3 (0·48; 18) 17·4–19·0 |

Sex differences significant, except for tarsus.

Tail up to 58 in birds measured by Eck (1985a).

**Weights.** Pennsylvania (USA) (Clench and Leberman 1978, which see for sex and age differences).

| | | | | |
|---|---|---|---|---|
| MAY | 16·4 (664)12·6–22·7 | | AUG | 16·9 (533)11·5–23·1 |
| JUN | 16·8 (158)13·7–21·5 | | SEP | 17·4 (506)12·0–25·7 |
| JUL | 16·6 (176)13·9–18·6 | | OCT | 19·0 ( 18)17·0–23·7 |

Averages Illinois (USA), September: adult ♂ 19·2 (*n*=7), adult ♀ 19·1 (*n*=6), 1st adult ♂ 19·6 (*n*=8), 1st adult ♀ 18·8 (*n*=9) (Graber and Graber 1962). Eastern USA: April–July, ♂ 16·9 (0·51; 5) 16·3–17·5, ♀ 17·2 (3) 15·6–18·0; August–September, 17·4 (*n*=15) (Baldwin and Kendeigh 1938; Mengel 1965), 17·3 (50) 15·2–21·3 (Grant 1968). Georgia and South Carolina (USA), July–August: ♂ 17·7 (2·78; 4) 14·3–20·8, ♀ 17·6 (0·52; 5) 16·7–17·9 (Norris and Johnston 1958). Kansas, 1st adult, late September and October: ♂ 21·2 (2·60; 38), ♀ 19·3 (2·16; 23) (Tordoff and Mengel 1965). Louisiana (USA), spring: 17·3 (1·82; 148) (Rogers and Odum 1966). East Ship Island (off Mississippi, USA), spring: 15·5 (1·6; 1691) (Kuenzi *et al.* 1991, which see for weight gains during stay).

Migrants stranded on ships in Caribbean, October: ♂ 17·7 (1·28; 13) 16·7–19·8, ♀ 17·6 (1·80; 10) 14·8–20·0 (Roselaar 1976; ZMA). Belize, October–November: ♂ 16·9 (3) 16·6–17·2, ♀ 13·4 (Russell 1964). Panama, autumn: 16·8 (2·00; 203) 11·4–21·6, once 11·2 (Rogers and Odum 1966; see also Rogers 1965). Curaçao, autumn: exhausted ♂♂ 11, 12 (ZMA). Mexico and Martinique, migrants May: ♂ 14·5, unsexed 14·5 (Schreiber and Schreiber 1984; Binford 1989). Britain, October: 1st adult 16·2 (Grant 1968). Netherlands, October: ♀ 24·5 (ZMA).

For fat-free weights, see Connell *et al.* (1960), Odum *et al.* (1961), and Roberts and Odum (1966).

**Structure.** Wing rather short, broad at base, tip bluntly pointed. 10 primaries: p8 longest, p9 and p6 2·5–5 shorter, p7 0–2, p5 7–11, p4 12–15, p1 18·5–24. P10 strongly reduced; a tiny spike-like feather hidden below reduced outer upper primary covert, 56–61 shorter than p8 in adult, 49–56 in juvenile; 9–12 shorter than longest upper primary covert in adult, 3–7 in juvenile. Outer web of (p6–)p7–p8 emarginated, inner web of (p7–)p8–p9 with faint notch. Tip of longest tertial reaches tip of p2–p3. Tail short, tip square; 12 feathers. Bill rather short, visible length of culmen c. 50% of head length; straight, thick at base and in middle, tip of culmen sharply hooked; cutting edges straight, but small notch in upper mandible just behind hook; gonys bulging to tip. Nostril small, oval. Some short stiff hairs mixed between feathering at base of upper mandible and at side of upper chin; 4 bristles project down at each side of upper mandible over gape. Leg and foot rather short, slender. Middle toe with claw 14·0 (10) 13–15; outer toe with claw 76% of middle with claw, inner c. 72%, hind c. 89%.

**Geographical variation.** Slight in North America; *caniviridis* from interior north-west USA differs from nominate *olivaceus* in paler and more olive-tinged cap with less conspicuous black line

at side, paler and greyer olive-green on remainder of upperparts, and clearer white underparts (Burleigh 1960); validity sometimes doubted (e.g. Monroe 1968). For variation through Canada, see Barlow and Power (1970). South American races (*chivi* group) occasionally considered a separate species, but closely related (Johnson and Zink 1985): mostly smaller, often brighter, and with more rounded wing; for description of races, see Hellmayr (1935); for Measurements, see Hamilton (1958). Yellow-green Vireo *V. flavoviridis*, occurring from Mexico to central Costa Rica and Isla Coiba (see Peters 1931), is sometimes included in *V. olivaceus*, geographical range being complementary and colour rather similar, but biochemically well-isolated from both *olivaceus* group and *chivi* group (Johnson and Zink 1985; Johnson *et al.* 1988).

Forms superspecies with *V. flavoviridis* (see above), Black-whiskered Vireo *V. altiloquus* and Yucatan Vireo *V. magister* from Caribbean region, and Noronha Vireo *V. gracilirostris* of Fernando de Noronha island off north-east Brazil. CSR

# Family FRINGILLIDAE finches

Small to medium-small, stout-billed, 9-primaried oscine passerines (suborder Passeres). Principally seed-eaters, found in many habitats in both Old World and New, including savanna, steppe, scrub, and even desert, but mostly in woodland, forest, parkland, cultivation, and the like. Bill strong and typically conical, but shows much adaptive variation; in all species, however, structurally designed internally for shelling seeds (with aid of tongue and strong jaw muscles). Unlike buntings (Emberizidae), finches much prefer dicotyledonous seeds, slicing them open using sharp edges of lower mandible (see Ziswiler 1965 for trials on captive birds and many other details; also Newton 1967a).

Fringillidae now usually classified in 2 subfamilies: Fringillinae (chaffinches) and Carduelinae (typical finches). Drepanididae (Hawaiian honeycreepers) also included by some authors; not considered further here. Relationships of 9-primaried, hard-billed, and often finch-like birds to one another is much debated (e.g. review by Sibley 1970), and Emberizidae (buntings, etc.) also often included in Fringillidae by earlier authors (e.g. Witherby 1941). Following Voous (1977), west Palearctic 9-primaried passerines here all treated as separate families: Fringillidae, Parulidae (New World wood-warblers), Thraupidae (tangers), Emberizidae, and Icteridae (New World blackbirds, etc.). DNA data, however (Sibley and Ahlquist 1990), indicate all these groups are close enough to form single family, comprising 2 subfamilies: 'Fringillinae' (including Hawaiian honeycreepers) and 'Emberizinae' (including Emberizidae as defined here plus parulids, tangers, and icterids).

# Subfamily FRINGILLINAE chaffinches

Fairly small finches, inhabiting forest, woodland, parkland, and farmland. Pick seeds from ground and insects from foliage or, to lesser extent, by aerial fly-catching; insects and other small invertebrates figure prominently in summer diet, forming sole food of nestlings (unlike in cardueline finches). Food-piracy upon more specialized seed-eating cardueline finches reported. 3 closely related species in single genus *Fringilla*. All found in west

Palearctic, 2 (Chaffinch *F. coelebs* and Brambling *F. montifringilla*) being well distributed in Europe and (especially in case of *F. montifringilla*) extending into Asia. Blue Chaffinch *F. teydea* restricted to pine forests at high altitude on some of Canary Islands, where *F. coelebs* also occurs (in different habitat).

Sexes differ somewhat in size, ♂ slightly larger than ♀. Bill relatively long; pointed with straight culmen. Nostrils sited near base; open, oval in shape, and largely hidden by feathers; a few rictal and nasal bristles present. Food brought to young in bill (not regurgitated as in cardueline finches). Wing fairly long, bluntly pointed; 9 primaries (10th minute and hidden). Flight typically undulating. Tail fairly long, slightly forked; 12 feathers. Leg rather long; tarsus scutellate, booted at side, forming sharp ridge at rear. Foot not used for holding or uncovering food, or for clinging to vegetation when feeding. Gait both a walk and a hop. Head-scratching by indirect method (see Simmons 1957b for *F. coelebs*). Bathe in typical stand-in manner. Use both main methods of sunning (see Simmons 1986). Anting by *F. coelebs* recorded a few times in wild and by *F. montifringilla* in captivity; of active (direct) type only (see Simmons 1957a, 1960).

Lacks twittering calls of most cardueline finches. Song of ♂ of typical advertising type: relatively simple but loud, clear, and more stereotyped than longer, more generalized song of cardueline finches (Newton 1972), though in *F. coelebs* with marked individual and regional variation; much reduced in island forms (including *F. teydea*). Although often gregarious outside breeding season, feeding in flocks and roosting communally, fringilline finches typically solitary and territorial when nesting, some birds maintaining contact with territory in winter where possible. Monogamous mating system with strong pair-bond, usually of seasonal duration only but same birds often pair up again in successive years. Even when roosting, avoid physical contact with conspecifics, never allopreening so far as known. Nest a neat cup; built by ♀ only. Incubation by ♀ only (sometimes fed by ♂ in case of *F. montifringilla* only). Young fed by both sexes.

Sexes markedly dissimilar in appearance; see species accounts for details of plumages (including characteristic wing markings), moult, etc.

## *Fringilla coelebs* Chaffinch

PLATES 29 and 38 (flight)
[between pages 472 and 473]

Du. Vink   Fr. Pinson des arbres   Ge. Buchfink
Ru. Зяблик   Sp. Pinzón común   Sw. Bofink   N. Am. Common Chaffinch

*Fringilla coelebs* Linnaeus, 1758

Polytypic. COELEBS GROUP. Nominate *coelebs* Linnaeus, 1758, from Scandinavia, Netherlands, and France east to central Siberia, south to Pyrénées, mainland Italy, Balkan countries, Greece (including Crete), European Turkey, and *c.* 48°N in Ukraine and European Russia; *solomkoi* Menzbier and Sushkin, 1913, Crimea, eastern shore of Sea of Azov, north-east shore of Black Sea east to about Sukhumi, and perhaps northern slope of Caucasus; *caucasica* Serebrowski, 1925, Caucasus area (except south-west slope of Caucasus east to Sukhumi and perhaps not on northern slope), Transcaucasia, Iranian Azerbaijan, and Asia Minor; *alexandrovi* Zarudny, 1916, northern Iran from Gilan to eastern Mazanderan, in winter in Middle East; *transcaspia* Zarudny, 1916, north-east Iran and neighbouring part of south-west Turkmeniya, in winter in Middle East; *syriaca* Harrison, 1945, Levant and Cyprus, probably grading into *caucasica* in southern Turkey; *gengleri* Kleinschmidt, 1909, Britain and Ireland; *balearica* Von Jordans, 1923, Portugal, Spain, and Balearic islands; *tyrrhenica* Schiebel, 1910, Corsica; *sarda* Rapine, 1925, Sardinia. SPODIOGENYS GROUP. *F. c. africana* Levaillant, 1850, Morocco, Algeria, and north-west Tunisia; also (perhaps this race) Cyrenaica (north-east Libya); *spodiogenys* Bonaparte, 1841, Tunisia (except north-west) and north-west Libya. CANARIENSIS GROUP. *F. c. canariensis* Vieillot, 1817 (synonym: *tintillon* Webb, Berthelot, and Moquin-Tandon, 1841), Tenerife, Gran Canaria, and Gomera (central Canary Islands); *palmae* Tristram, 1889, La Palma (western Canary Islands); *ombriosa* Hartert, 1913, Hierro (western Canary Islands); *maderensis* Sharpe, 1888, Madeira; *moreletti* Pucheran, 1859, Azores.

**Field characters.** 14·5 cm; wing-span 24·5–28·5 cm. Similar in size to House Sparrow *Passer domesticus* but with smaller, sharper bill, smaller, more peaked head (especially in ♂), and particularly longer wings and tail, giving classic form of *Fringilla*. Noticeably long, medium-sized, and rather elegant passerine, epitomising its family but actually sharing its general character with only 2 congeners (and long buntings *Emberiza*). At all ages and in both sexes, plumage pattern dominated by contrasting white panels on leading wing-coverts, buff-white wing-bar on secondaries and inner primaries, and striking white outer tail-feathers which show well on perched bird and twinkle on flying one. Eurasian ♂ shows more colours than any other west Palearctic finch, with blue-grey crown, pale red to pink face and body, and mainly black ground to white wing and tail markings. ♀ and juvenile much more sober, with dusky olive-brown head and upperparts and dusky white underparts showing little pattern; wings and tail duller than ♂'s but show similarly striking marks. Song and several calls distinctive. Sexes dissimilar; some seasonal variation. Juvenile separable. 17 races in west Palearctic, 7 isolated on Atlantic and Mediterranean islands and 10 occupying continental Europe and north-west Africa. Races form 3 distinct groups of which ♂♂ separable. Only 2 European races described here; see also Geographical Variation.

(1) North European and Russian race, nominate *coelebs*. ADULT MALE. Moults: June–October (complete). Full contrast of plumage colours comes from wear of non-breeding plumage. Forehead black, giving distinctive dusky-faced look; crown, nape, neck shawl, and upper border of mantle pure, rather pale grey-blue; mantle dull chestnut-brown, appearing maroon in some lights and with greenish tinge on scapulars; lower back and rump quite bright yellowish-brown, contrasting with mantle and tail; upper tail-coverts blue-grey. Wings intricately patterned and coloured: upper lesser coverts blue-black (usually obscured): lower lesser coverts and median coverts pure white, forming deep and quite long blaze from shoulder to mid-scapulars; greater coverts black with pale yellow to white tips forming narrow but long wing-bar across mid-wing to tertial-bases; flight-feathers brown-black, with narrow white basal marks on inner primaries extending wing-bar along lower edge of black primary coverts; pale fringes most evident on inner secondaries (forming faint greenish-yellow panel) and tertials (forming narrow but obvious buff lines). Underwing white. Tail slate-grey on central pair, black on rest with quite broad white edges, often as eyecatching as shoulder blaze. Lore, patch around eye, and cheek pure rich vinous-pink; throat and upper breast similarly coloured but from lower breast paler pink, becoming pinkish-white by belly and cream-white under tail. In fresh autumn plumage, immaculate contrasts of breeding plumage obscured by buff and green tips: forehead appears dark grey rather than black, crown shows brown wash, mantle looks dark brown, upper tail-coverts become brownish-grey, and pink of lower face and underbody darkens and dulls; on wings, bar across tips of greater coverts becomes deeper and more yellowish-white and fringes of tertials are wider and brighter greenish-buff. Bill bright lead-blue, with dusky tip and paler base to lower mandible. Legs pale grey to brown. ADULT FEMALE. Lacks colourful head and body plumage of ♂, but wing and tail markings identical in pattern and almost as vivid. Head dusky-grey, darkest on crown but purer on centre of nape and sides of neck creating faint shawl; softly patterned by diffuse, pale grey-buff stripe in front of and behind eye (hardly forming true supercilium), similarly coloured lower eye-crescent, and paler, greyish border to rear cheeks. Mantle yellowish- or olive-brown; lower back and rump dull yellowish-green; upper tail-coverts dull yellowish-brown. Underparts basically greyish-white, cleanest on upper throat, belly and under tail but sullied olive- to brown-grey from lower throat to chest and over most of flanks. Bill pale horn with dusky tip. When plumage fresh, appearance changes much less than ♂'s: head

and upperparts become browner in tone, underparts more sullied brownish-grey, and wing markings take on buff tinge. JUVENILE. Resembles non-breeding ♀ but shows distinctive greyish-white nuchal spot, even browner mantle, dull brownish-green rump, and paler underparts. At close range, ♂ shows warm buff cheeks, chestnut tinge to mantle, buff chest and fully bright wing and tail markings of adult ♂. Both sexes assume adult plumage July–September. (2) British and Irish race, *gengleri*. Resembles nominate *coelebs* but, particularly in worn breeding plumage, ♂ darker on underparts, with distinct brown wash over face and underbody; looks dirty compared to immigrants of nominate *coelebs*. ♀ also duskier above and below than nominate *coelebs*. Differences difficult to ascertain on individual but often quite evident when flock can be seen or compared with immigrants.

Unmistakable at close range; no other finch or bunting of similar form has such bold white marks on wing and tail. Brambling *F. montifringilla* shows similar wing pattern but has long white rump, all-black and more forked tail, and much whiter underbody. Flight light and free, with action and silhouette more reminiscent of pipit *Anthus* and bunting *Emberiza* than smaller, shorter-tailed finches. In cover, flight quite fast and level with rapid wing-beats and much spreading of tail when slowing or turning; over longer distances (particularly on migration), more leisurely and soon developing undulations typical of family, with bursts of wing-beats alternated with wing closures and creating both forward and upward shoots and falls within airspace which produce characteristic rhythm in progress of individual bird and flock. With birds high overhead, undulations obscured but opening and shutting of wings still evident. Escape-flight fairly level and usually directed into nearest leaf cover (from hedge to tree canopy height), rarely to more distant ground cover. Flocks quite shy, may indulge in repeated escape behaviour when only slightly disturbed. All flight actions subject to degree of dancing illusion, caused by flickering display of white marks on wing and tail. Also capable of gliding and hovering. On migration, flies in small parties or large loose flocks, less bunched than smaller migrant finches and usually maintaining constant heading into wind. Gait a hop, varied by distinctive tripping walk, accompanied by short quick steps and just visible nod of head. Stance on perch rather upright with tail often drooped and in singing ♂ head held up; on ground much less so, with tail length creating illusion of long body line. General behaviour lacks distinctive movements, bird content to feed intently and only rarely betraying anxiety by slight flick of tail. Highly gregarious after breeding, forming large flocks which sometimes divide into sexual or racial groups and regularly attract company of other finches, buntings, and sparrows. Becomes tame around human habitation but remains wild in natural habitats.

Song a short accelerating and descending series of cheerful, quite musical notes, usually ending in trisyllabic

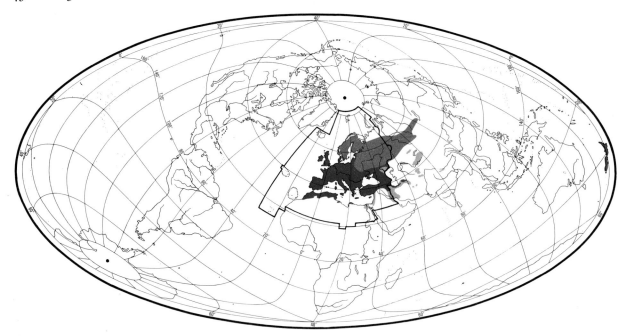

flourish: 'chip-chip-chip-tell-tell-cherry-erry-erry-tissi-cheweeo' (Garstang 1923) or 'chink-chink-chink-tee-tee-tee-terree-erree-erree-chissee-CHU-EE-OO'; acceleration of notes aptly compared to rhythm of bowler's run-up (in cricket). Subsong a low and rambling warble. Calls vary by sex and season and in intensity; commonest all year an abrupt, rather metallic 'chink' or 'chwink', recalling Great Tit *Parus major* and a low 'chup' or 'tsŭp', often repeated, in flight (especially from parties and particularly from migrants); loudest in spring from ♂ a clear loud, far-carrying 'wheet', at times suggesting Redstart *Phoenicurus phoenicurus*, and a penetrating 'chit', 'chuit', or 'tsit'; quietest from ♀ in interactions with ♂ a thin 'tzit-tzit-tzit' and an even softer chirrup, rapidly repeated. Song and calls of Atlantic islands races differ distinctly (see Voice).

**Habitat.** Breeds in west Palearctic in temperate wooded areas, from Mediterranean and marginally steppe zones up to boreal, and in places to edge of tundra; between July isotherms 12–30°C, usually occurs within shifting climatic boundary, beyond which Brambling *F. montifringilla* replaces it to the north (Voous 1960b). Basically arboreal, and in breeding season occupies deciduous, mixed, and coniferous woods and forests, at densities varying greatly according to their species composition. In Fenno-Scandia, density highest in deciduous woods, followed by spruce *Picea* and pine *Pinus*. A Netherlands study showed consistent preference for occupancy of mixed deciduous wood over a neighbouring pine wood, which tended to be settled by 1st-years unable to win territories in the mixed wood (Newton 1972). Studies in Britain showed that, apart from wood of beech *Fagus* and

hornbeam *Carpinus*, occupancy of woodlands was virtually restricted to breeding season, during which it extended to almost all tree species in varying degrees, and to contrastingly structured stands, from upland scrub birch *Betula* to woods of mature pedunculate oak *Quercus robur* with thick understorey of hazel *Corylus avellana*, and to beechwoods on chalk with no undergrowth. Found to colonize young conifer plantations *c*. 7–10 years old. In modern times, parklands, gardens, and farm hedgerows have also been extensively settled, even in suburbs and locally interiors of large cities. Wayside trees, orchards, olive *Olea* groves, and cemeteries are among similar habitats chosen (Yapp 1962). Stages of growth of trees and bushes, their spacing, and access for foraging are significant. In winter, open areas of farmland up to some distance from tree cover are frequented where they offer sufficient food. In mountains, ascends generally to tree-line; in Caucasus, to 2200–2500 m (Dementiev and Gladkov 1954).

**Distribution.** No changes reported, except for occasional recent breeding in Iceland and Faeroes. (Introduced in New Zealand.)

ICELAND. Formerly fairly frequent accidental visitor; first bred 1986, also 1987–8 (Magnússon 1986; Hafsteinsson and Björnsson 1988; AP). FAEROES. Bred 1972 and 1981 (Bloch and Sørensen 1984). ITALY. Colonized Aeolian Islands in 1980s (Iapichino and Massa 1989). MALTA. Occasional breeder (Sultana and Gauci 1982); bred successfully 1984 (Galea 1987). ISRAEL. No evidence of breeding since reported by Tristram (1884) as breeding in lower parts of Mt Hermon (Shirihai in press).

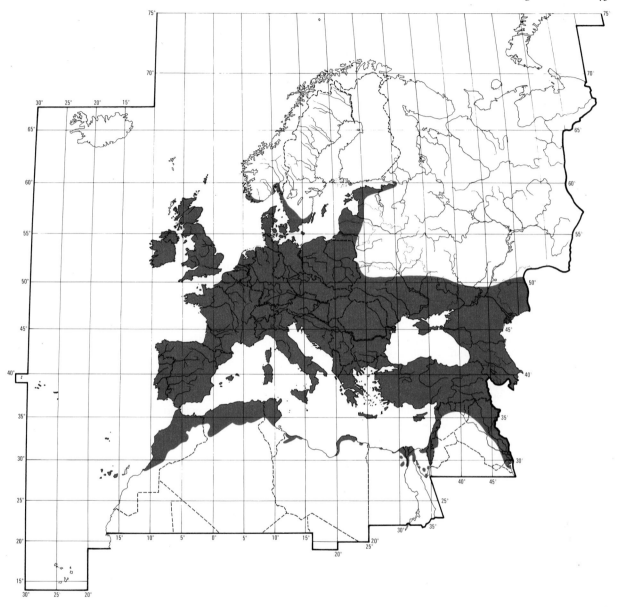

**Population.** No major changes reported.

BRITAIN, IRELAND. About 7 million pairs (Sharrock 1976). Slow increase in Britain from early 1960s. FRANCE. Over 1 million pairs (Yeatman 1976). BELGIUM. Estimated 250 000 pairs, apparently stable in most areas (Devillers *et al.* 1988). NETHERLANDS. Estimated 250 000-400 000 pairs (SOVON 1987). GERMANY. Estimated 10·9 million pairs (Rheinwald 1992). East Germany: 2·5 ± 0·9 million pairs (Nicolai 1993). SWEDEN. Estimated 11 million pairs (Ulfstrand and Högstedt 1976). FINLAND. Has increased this century, perhaps due to fragmentation of forests; estimated 10-20 million pairs (Koskimies 1989). USSR. Declined in Yaroslavl' region 1960-85 (Belousov 1986).

**Survival.** Britain: annual mortality of adult ♂♂ 37 ± SE3%, adult ♀♀ 35 ± SE3% (Dobson 1987). Finland: annual mortality 43 ± SE5·8% (Haukioja 1969). Czechoslovakia: for age structure and mortality, see Pikula (1989). Russia: annual adult mortality significantly lower (42·3%) in birds hatched in late breeding seasons than in early seasons (53·3%), and lowest (37·3%) in birds hatched from 1st broods in late seasons; mortality of adults from broods of 4 lower (43·5%) than those from broods of 5 (51·2%) and 6 (71·4%) (Paevski 1981); annual adult mortality, Kaliningrad region, 47% (95% confidence limits 44-51%), fluctuating widely from year to year; no difference between ♂♂ and ♀♀ (Bardin 1990). Birds ringed in northern and central Europe and recovered in Iberia: annual mortality of adults 48% (Asensio and Carrascal 1990).

Oldest ringed bird 14 years (Barriety 1965).

**Movements.** Sedentary to migratory, wintering chiefly within breeding range in Europe, but further south in Asia. In Europe, migrates south-west on fairly narrow front, with western populations wintering furthest west, and more eastern populations progressively further east. Winter visitors greatly augment populations of western and southern Europe (including Britain), and regularly reach North Africa. Migrates by day in flocks, most actively in morning hours, sometimes in company with Brambling *F. montifringilla*. In many areas the most numerous visible autumn migrant, e.g. at Pape in Latvia (Vīksne 1983), Skåne in southern Sweden (Alerstam and Ulfstrand 1972), Kaliningrad coast, western Russia (Dol'nik 1982), Bodensee area, southern Germany (Jacoby *et al.* 1970). In south-east Denmark, outnumbers all other day migrants both spring and autumn (Christensen and Rosenberg 1964). Well studied. For metabolism and many aspects of migratory behaviour (including experimental studies), see Dol'nik (1982) and references cited therein. For ringing data, see Paevski (1971) for Kaliningrad region, Verheyen (1960) for Belgium, and Osthaus and Schloss (1975) and Krägenow (1986) for Germany.

Scandinavian birds migrate through Denmark, north-west Germany, and Low Countries (especially in west) to winter chiefly in Britain, also in Low Countries, western France, and western Iberia. Birds from Finland, Baltic states, Poland, and north-west Russia (east to Komi) migrate through eastern Low Countries, Germany, and Switzerland to winter chiefly in south-west France and western Iberia; many Finnish and Russian birds follow south Baltic coast as leading line, but some cross southern Sweden. (Perdeck 1970; Paevski 1971; Asensio 1985*a*; Loneux 1988; SOF 1990.) Southern German and Swiss migrants winter in southern France and eastern Iberia (including Balearics) and south-east European migrants in northern and central Italy (Blondel 1969; Savigni and Massa 1983; Asensio 1985*a*). Some northern birds also winter in Switzerland and southern Germany (Schifferli *et al.* 1982; Wüst 1986). Birds from western Czechoslovakia winter in southern France, with a few reaching Iberia, and birds from eastern Czechoslovakia winter in northern Italy (Hájek and Bašová 1963).

British birds (*gengleri*) very sedentary; 90% move no further than 5 km from natal site, and the rest (almost entirely 1st-year birds) less than 50 km (Newton 1972). Some local Scottish breeding areas vacated, however; at Blairbeg wood (Inverness-shire), none present November to late February (Swann 1988). Immigrants winter mainly in southern and central Britain (south of 54°N) and Ireland, and are chiefly of Scandinavian origin; data show heavy passage via Low Countries; 1322 recoveries to or from Britain up to 1986 are from Low Countries (62%), Scandinavia and Denmark (25%), Germany (9%), Finland (2%), France (2%), with a few from Baltic states, North Sea, Faeroes, Poland, and Portugal (Perdeck 1970; Mead and Clark 1987).

French birds mostly sedentary; a few juveniles move west or south-west in autumn, and very occasionally cross Pyrénées to Spain. Birds from Low Countries and Scandinavia east to Russia and Czechoslovakia winter in France. (Blondel 1969; Perdeck 1970; Yeatman-Berthelot 1991.) At Lac de Grand-Lieu (Loire Atlantique, western France), 8 ringing recoveries show long-distance movements to or from Sweden (1), Finland (2), Kaliningrad coast (1), St Petersburg region (1), and Belgium (3) (Marion and Marion 1975). Winter population in French Mediterranean region includes birds ringed both in breeding season and on passage in northern Italy, Switzerland, Germany (south of 52°N), Czechoslovakia, southern Finland, Baltic states and adjoining regions of western Russia, also on passage in Austria and Poland (Blondel 1969). In cold spells, huge flocks assemble in Rhône delta (Blondel and Isenmann 1981).

In countries further east and north, proportion of migrants increases. In Sweden, overwinters in small numbers in south, and occasionally north to extreme north (SOF 1990). In Norway, overwinters especially on coast, occasionally north to Trondheim fjord (Rendahl 1968). In southern Finland, small flocks commonly seen near human habitations in early winter, but numbers usually decrease, probably mostly due to mortality (Haila *et al.* 1986). Winters only in very small numbers in St Petersburg region (Mal'chevski and Pukinski 1983); irregular in Moscow region (Ptushenko and Inozemtsev 1968). In Belgium, *c.* 60% sedentary; local juveniles recorded up to 880 km away in France (Lippens and Wille 1972). In Switzerland, southern Germany, and Czechoslovakia, only a small proportion of local birds winters (Hájek and Bašová 1963; Schifferli *et al.* 1982; Wüst 1986). In Spain (including Balearics), recoveries are of birds ringed both in breeding and non-breeding seasons in all countries further north-east from Scandinavia and Low Countries east to north-west Russia, Poland, Czechoslovakia, and Italy (no breeding season recoveries from Belgium or Italy); mid-winter recoveries are mostly in coastal areas and in southern third of Spain. Reaches Spain both east and west of Pyrénées. (Asensio 1985*a*.) In Italy, 530 recoveries 1939–79, from Czechoslovakia (31·5%), Switzerland (17·3%), Hungary (14·3%), USSR (13·6%), France (7·9%), Germany (7·4%), Yugoslavia (4·2%), and small numbers from Finland, Poland, and Low Countries. Most recoveries are north of 42°N, especially in Toscana and Umbria. Birds from Czechoslovakia and Baltic states tend to arrive via Alpine and Po areas, and Hungarian birds via northern Adriatic. (Savigni and Massa 1983.) Data from Sicily and Malta show similar pattern (Sultana and Gauci 1982; Iapichino and Massa 1989).

In Greece, Turkey, and Cyprus, winter population far larger than summer (Beaman 1978; Flint and Stewart 1992; G I Handrinos). Common winter visitor to Syria (Baumgart and Stephan 1987) and northern highlands of Jordan (I J Andrews), and abundant throughout Israel, but especially in northern and central parts (Shirihai in press). Common in winter in plains of Iraq (Allouse 1953), but in Arabia recorded only from Gulf states (mostly Kuwait), where scarce and irregular (Bundy and Warr 1980; F E Warr). European birds reach north-west Africa in considerable numbers (many crossing Strait of Gibraltar), and African birds disperse slightly south of breeding range (Heim de Balsac and Mayaud 1962; Finlayson 1992). Recovery in Morocco of bird ringed in Kaliningrad region, 3240 km north-east (Paevski 1971). Single record as far south as Mauritania, ♂ 24–25 October 1988 (Meininger *et al.* 1990). Migrants also reach Libya; sometimes common in coastal Tripolitania, but scarce in Cyrenaica (Bundy 1976), and few reach northern Egypt (Goodman and Meininger 1989). Bird ringed Cairo (Egypt) recovered at *c.* 57°30′N 82°E in western Siberia (Mackintosh 1941). Atlantic islands races apparently mainly sedentary, but some reports of altitudinal movements (Bannerman 1963; Bannerman and Bannerman 1965).

Migratory route to Britain involves exceptional changes of direction for Scandinavian birds, and especially for Norwegian birds, many of which head east of south initially. Detailed observations from 18 localities showed that mean standard direction changes from SSW in Scandinavia to south-west by south in Denmark and northern Germany, WSW in Netherlands, and west by north at Cap Gris-Nez in north-east France (Perdeck 1970). Some birds cross North Sea directly from Scandinavia, but most cross from Low Countries or extreme north-east France. Abundance on Netherlands coast results from narrowing of inland broad-front migration. With easterly winds in Netherlands, birds tend to fly high and cross 180 km WSW to Norfolk (eastern England), but with westerly winds they fly low and continue along coast to Cap Gris-Nez or beyond, and approach Britain from south-east or even from south (e.g. arriving at Portland Bill, Dorset) (Newton 1972). Some then head for Ireland, e.g. WNW from north Devon (Lack 1957). Reaches Iceland annually in varying numbers; in 1979–90, 2–95 reported (301 in all) (Pétursson *et al.* 1992). Bird ringed Belgium, November 1964, recovered Iceland, December 1966 (Lippens and Wille 1972). In spring, more birds than in autumn cross North Sea on broad front, especially between Norfolk and Netherlands (Newton 1972).

Autumn migration August–December, chiefly September–November. Movement peaks in last two-thirds of September in northern Sweden, and last third of September and 1st third of October in central and southern Sweden; passage far more intense at Falsterbo in south-west than Ottenby in south-east, most birds evidently migrating inland (Rendahl 1968). In Low Countries, pas-

sage peaks October (Perdeck 1970; Van Hecke and Verstuyft 1972; Schols 1987), and arrivals in Britain September–November, chiefly October (e.g. Witherby *et al.* 1938, Taylor *et al.* 1981, Taylor 1987). Abundant in east Baltic region, e.g. at Ladoga and Chudskoe lakes and in Gulf of Riga (Rootsmäe and Veroman 1967; Mal'chevski and Pukinski 1983); occurs in 2 peaks at Lake Ladoga, 5–14 September and 29 September to 5 October (Noskov *et al.* 1981). In Kaliningrad region, local birds leave 20 August to 20 September, and passage of more northerly birds begins *c.* 20 September (Paevski 1971). On Polish coast also, local birds begin migration before onset of passage from further north (Kania 1981). In Switzerland, passage begins mid-September and peaks 3–12 October, decreasing to end of October or 1st third of November (Winkler 1984); most local birds leave from end of September (Schifferli *et al.* 1982). Immigrants reach Camargue (southern France) and Yugoslavia at beginning of October (Cvitanić 1980; Blondel and Isenmann 1981), and Spain mostly from 2nd third of October (Asensio 1985a). At Bosphorus (western Turkey) in one year, eastward passage recorded 24 September to 8 November, peaking 6–13 October (Porter 1983). In south of Mediterranean region, passage continues to early December (Iapichino and Massa 1989; Finlayson 1992), and even to January (Shirihai in press).

Spring migration begins early and is prolonged, February–May, chiefly March to mid-April in central Europe. In Strait of Gibraltar, passage earlier than other finches, from start of February, peaking 1st half of March and continuing to early April; in recent years, small flocks of *africana* also recorded (Finlayson 1992). In Sicily, Cyprus, and Israel, passage mid-February to early or mid-April (Iapichino and Massa 1989; Flint and Stewart 1992; Shirihai in press). In Switzerland (less abundant than autumn), begins last third of February, peaks mid-March and ends 2nd third of April (Winkler 1984). In Britain, movement (less conspicuous than autumn) mid-March to mid-May (Witherby *et al.* 1938). In Denmark, spring passage almost entirely in south-east, chiefly last week of March to end of April (Christensen and Rosenberg 1964). Passage and arrivals in southern and central Sweden early March to late April, median 28 March, and in northern Sweden mid-March to late May, median 12 April (Rendahl 1968); arrivals in northern Finland mostly 16 April to 19 May (Mikkonen 1985b). Movement in southern Baltic area early April to early or mid-May. In Kaliningrad region, arrival of local birds tends to begin later than passage of birds bound further north, and continues to end of passage period (Paevski 1971; Paevski and Vinogradova 1974). On southern shore of Finnish Gulf (where abundant in spring, but only small numbers in autumn), passage most intense in morning, but continuing with equal intensity all day on peak days (Gaginskaya 1969). Local birds return to St Petersburg region mostly in 2nd half of April (Mal'chevski and Pukinski 1983).

In southern Sweden in spring, retromigration sometimes conspicuous; e.g. in Västergötland, many birds arriving in early April, with thaw, retreated south-west in headwinds when blizzard followed (Alerstam 1990). For retromigration in Sweden in autumn, in company with *F. montifringilla*, see that species.

In east of range also, migratory in north (e.g. Volga-Kama region and Siberia), progressively less so further south (Dementiev and Gladkov 1954; Ravkin 1973; Popov 1978). Winters south to Pakistan. In Kazakhstan (breeds in north), common on passage, especially in Ural valley and Volga-Ural area; wintering widespread, but uncommon and irregular, and more numerous in south (Korelov *et al.* 1974). Also winters in Tadzhikistan (Abdusalyamov 1977) and Uzbekistan (Tret'yakov 1978). In Iran (breeds in north), common winter visitor throughout north and west, straggling to south-east (D A Scott), and fairly widespread visitor to low-lying areas and valleys of Afghanistan (S C Madge). In Pakistan (eastern extreme of winter range), irregular and sometimes common in northern Baluchistan, sporadic further north (Roberts 1992). Autumn movement mainly September–November, and spring movement end of February or early March to late April or early May (Dementiev and Gladkov 1954; Gavrilov *et al.* 1968; Korelov *et al.* 1974). In Iran, winter visitors present 3rd week of October to end of March or 1st week of April (D A Scott).

In autumn, ♀♀ migrate before ♂♂, and juveniles before adults. In Kaliningrad, autumn passage of ♀♀ from further north is 7–10 days earlier than ♂♂ (Paevski 1971); juveniles predominate at start and end of passage (Dol'nik 1982). In Switzerland, first passage birds juvenile ♀♀, followed in turn by adult ♀♀, juvenile ♂♂, and adult ♂♂, at intervals of 3–4 days; passage of juveniles more prolonged, beginning before and ending after adults (Schifferli 1963). Similarly in Belgium, juveniles begin passage earlier than adults (Van Hecke and Verstuyft 1972), and ♀♀ before ♂♂; proportion of ♂♂ increases during passage; in September, 42% ♂♂ and 58% ♀♀, but in November 66% ♂♂ and 34% ♀♀ (Verheyen 1960). In Israel, more than 80% of November arrivals ♀♀, but by mid-January ♂♂ have increased, though ♀♀ still predominate, especially in south (Shirihai in press). In spring, ♂♂ migrate before ♀♀, and arrive earlier on breeding grounds (e.g. Van Hecke and Verstuyft 1972). Also, older birds migrate before younger birds. On Kaliningrad coast, ♂♂ arrive 10 days earlier than ♀♀, and older birds 5 days earlier than younger birds. (Dol'nik 1982.)

♀♀ more migratory than ♂♂, at least in some areas, tending to move further, and differential wintering occurs. Most birds remaining to winter in Sweden are ♂♂ (hence 'coelebs' = bachelor), and ♂♂ predominate in winter in Low Countries and Britain, but ♀♀ predominate in Ireland (Lippens and Wille 1972; Newton 1972). In central Europe in winter, proportion of ♂♂ to ♀♀ *c.* 3:1 from Grenoble (south-east France) north-east to Poland, but further south in France proportion of ♀♀ increases, and they predominate in Rhône delta. In Switzerland (*c.* 75% ♂♂), proportion varies from one region to another, with higher proportion of ♂♂ in Alps than at lower levels. (Schifferli 1963; Marfurt 1971.) Similarly in St Petersburg region and Czechoslovakia, ♂♂ predominate (Hájek and Bašová 1963; Mal'chevski and Pukinski 1983). In Dalmatia (Yugoslavia), influxes in cold spells are chiefly ♂♂ (Cvitanić 1980), suggesting they remained further north initially. In Cyprus, ♀♀ often predominate (Flint and Stewart 1992), also in Israel (Zaharoni 1991; Shirihai in press). Ringing data from Kaliningrad coast show that ♀♀ migrate further than ♂♂ on average; 221 ♂♂ moved 380–2890 km (mean 1564 km), and 214 ♀♀ moved 350–2900 km (mean 1743 km) (Dol'nik 1982). Among birds departing from or on passage through Switzerland, no difference detected in distance or direction travelled by ♂♂ and ♀♀ (Schifferli 1963); and no difference found in main direction (224°) in south-east Netherlands (Schols 1987). Only limited evidence from Spanish ringing recoveries that ♀♀ migrate further than ♂♂; in northern Spain, 50 ♂♂ and 42 ♀♀, but in south 41 ♂♂ and 61 ♀♀ (Asensio 1985a).

Migration typically in succession of waves. Study on Kaliningrad coast found that waves occur on very similar dates each year (usually 6 from mid-September to mid-October), and apparently relate to fat levels; birds feeding at stopovers become simultaneously ready for ongoing flight; no evidence of link with geographical area of origin, sex- or age-groups, or weather. Waves continue for 1–7 days (usually 3), and are followed by pauses of 1–8 days (usually 3). On 1st day, migration of fat birds begins at sunrise and continues for 4 hrs, with a further 2 hrs in evening; highest numbers usually on 2nd day, when many less fat and lean birds participate, apparently drawn on by intensity of movement. Retrap data show that 58% of individuals migrate within same 10-day period in different years. (Dol'nik and Blyumenthal 1967; Dol'nik 1982.) Similarly on Polish coast, succession of 6 waves identified. Evidence that birds moving on first waves migrate to furthest wintering grounds, and those on last waves to nearest wintering grounds; those on middle waves migrate to all wintering grounds, and are most numerous group at medium distances. (Kania 1981.)

Some evidence of winter site-fidelity. In Britain, several ringed migrants have returned to same area in successive winters (Newton 1972); in Jezre'el valley (northern Israel), ringed birds retrapped after several years (Zaharoni 1991). Also, individuals migrate different distances in different years. Birds found wintering in Britain in one year recorded at some point along migration route or near breeding areas in later years (Newton 1972). In Spain, 10 recoveries in winter of birds ringed further north (as far as Low Countries) in previous winters (Asensio 1985a).

In southern Sweden, autumn, migration recorded by radar showed little correspondence in time and space with visible migration. Birds flying at very low altitude compensated for lateral wind forces and tended to follow topographical leading lines; birds flying at considerable altitude became strongly deflected by crosswinds. (Alerstam and Ulfstrand 1972.) For influence of wind, see also Deelder (1952), Rootsmäe and Veroman (1967), Vleugel (1974), and Alerstam (1990).                                    DFV

**Food.** Mainly seeds and other plant material; in breeding season mainly invertebrates. In Oxford (southern England), had widest-ranging and most varied diet of all Fringillidae in major study. (Newton 1967a.) Seeds (principally Gramineae, Cruciferae, Fagaceae, Polygonaceae, Chenopodiaceae) taken generally on ground, notably freshly-turned soil, not direct from plant, except in shrubs and trees; feeds with rapid pecking action unsuited to removing seeds from herbs or grasses, flower or seed-head simply being knocked away. In Oxford, *c.* 80% of foraging over year was on ground; in another study in southern England, 63%. (Beven 1964; Newton 1967a, 1972.) See also Kear (1962) and, for Schleswig-Holstein (northern Germany), Eber (1956). Feeds most often in trees in spring and summer when taking invertebrates (especially defoliating caterpillars), and more on ground during rest of year; commonly forages in large flocks in open country outside breeding season (Eber 1956; Glutz von Blotzheim 1962; Newton 1967a; Dol'nik 1982; Krägenow 1986). Most feeding done in simple standing or perching position, if on herb stem then often to pick small invertebrates (rather than seeds) from head. Sometimes clings to bark like treecreeper *Certhia* searching for invertebrates, and fairly commonly makes flycatching sallies to take insects in air; also hovers briefly when feeding on small caterpillars hanging on silk from leaves. At perch, removes insect wings and legs with typical de-husking action of bill; also snatches falling conifer seeds in flight, but seldom takes them from hanging cone (Eber 1956; Newton 1967a; Dol'nik 1982.) In mainly coniferous plantation, northern England, March–October, foraged preferentially in sycamore *Acer pseudoplatanus* and larch *Larix*, and significantly avoided hemlock *Tsuga*, beech *Fagus*, birch *Betula*, and rowan *Sorbus* (Peck 1989). Hardly ever seen to use feet when handling food, though has been observed doing so with pine *Pinus* seeds while removing kernel (Marler 1956a), and seemingly never learns to do so in captivity when together with other Fringillidae that continually use feet (Newton 1967a). At bird-tables, will hover beside suspended fat or peanuts, or slide down string, but never seen to feed hanging upside-down (Kear 1962); seems to specialize in robbing other birds, even snatching seeds from their bills, or standing on ground to collect dropped items (Simmons 1986b; Sykes 1986). One bird repeatedly dropped pine seed onto hard surface until it split open (Petretti 1979). Makes good use of superabundance of invertebrates, e.g. 300–400 birds ceased ground-feeding to perch in trees and hawk after swarming Diptera until dark (Plath 1984); many fed together on trunks of spruce *Picea* on plague of bark

beetles (Scolytidae) (Fuchs 1984). Will drink sap on tree trunks (Glutz von Blotzheim 1962). Birds observed entering water, putting heads under surface, to pick up caddis fly (Trichoptera) larvae, removing casings in bill like seed husks (Källander 1982). On migration through Helgoland, fed extensively on leaves of cabbage although plenty of wild seeds available (Drost 1940a). In captivity, especially preferred seeds of hemp *Cannabis* and sunflower *Helianthus*, followed by grasses, rape *Brassica napus*, and linseed *Linum*. To de-husk seed, took from average 2·4 s for rape to 41·4 s for sunflower; to gather 1 g of kernel, 4·7 min for grass *Phalaris* to 25·4 min for linseed (Kear 1962, which see for many details, including seed-handling learning in young birds). For details of handling, de-husking, etc., in relation to bill shape, and comparison with other seed-eaters, see also Eber (1956), Ziswiler (1965), and Newton (1967a, 1972). Small or long seeds crushed and swallowed, not de-husked (Newton 1967a; Dol'nik 1982).

Diet in west Palearctic includes the following. Invertebrates: springtails (Collembola), mayflies (Ephemeroptera: Ephemeridae), dragonflies (Odonata: Libellulidae), stoneflies (Plecoptera: Nemouridae), grasshoppers (Orthoptera: Acrididae), earwigs (Dermaptera: Forficulidae), cockroaches (Dictyoptera: Blattidae), bugs (Hemiptera: Pentatomidae, Acanthosomidae, Nabiidae, Cicadellidae, Psyllidae, Aphidoidea, Coccoidea), lacewings, etc. (Neuroptera: Chrysopidae, Hemerobiidae, Raphidiidae), scorpion flies (Mecoptera: Panorpidae), Lepidoptera (Nymphalidae, Lycaenidae, Pieridae, Tortricidae, Pyralidae, Gracillariidae, Arctiidae, Noctuidae, Geometridae), caddis flies (Trichoptera: Glossosomatidae), flies (Diptera: Tipulidae, Culicidae, Chironomidae, Bibionidae, Rhagionidae, Tabanidae, Therevidae, Empididae, Asilidae, Syrphidae, Tachinidae, Muscidae), Hymenoptera (Pamphilidae, Tenthredinidae, Ichneumonidae, Braconidae, ants Formicidae), beetles (Coleoptera: Carabidae, Staphylinidae, Scarabaeidae, Helodidae, Elateridae, Cantharidae, Cerambycidae, Chrysomelidae, Curculionidae, Scolytidae), spiders (Araneae), harvestmen (Opiliones), millipedes (Diplopoda), centipedes (Chilopoda), earthworms (Lumbricidae), snails (Pulmonata). Plants: seeds, flowers, buds, etc., of yew *Taxus*, cypress *Chamaecyparis*, juniper *Juniperus*, *Thuja*, larch *Larix*, spruce *Picea*, pine *Pinus*, fir *Abies*, Douglas fir *Pseudotsuga*, willow *Salix*, poplar *Populus*, birch *Betula*, alder *Alnus*, hornbeam *Carpinus*, beech *Fagus*, chestnut *Castanea*, oak *Quercus*, elm *Ulmus*, *Robinia*, *Zelkova*, nettle-tree *Celtis*, mulberry *Morus*, tulip-tree *Liliodendron*, sycamore, etc. *Acer*, lime *Tilia*, ash *Fraxinus*, olive *Olea*, mistletoe *Viscum*, nettle *Urtica*, knotgrass *Polygonum*, dock *Rumex*, goosefoot *Chenopodium*, beet *Beta*, orache *Atriplex*, glasswort *Salicornia*, amaranth *Amaranthus*, chickweed *Stellaria*, mouse-ear *Cerastium*, corn spurrey *Spergula*, buttercup *Ranunculus*, *Corydalis*, Cruciferae, meadowsweet *Filipendula*, rose *Rosa*, bramble *Rubus*, strawberry *Fragaria*, cinquefoil *Potentilla*, herb bennet

*Geum*, apple *Malus*, pear *Pyrus*, cherry *Prunus*, rowan *Sorbus*, hawthorn *Crataegus*, *Cotoneaster*, *Amelanchier*, clover *Trifolium*, *Oxalis*, cranesbill *Geranium*, spurge *Euphorbia*, currant *Ribes*, sea buckthorn *Hippophae*, violet *Viola*, willowherb *Epilobium*, evening primrose *Oenothera*, dwarf alpenrose *Rhodothamnus*, bilberry, etc., *Vaccinium*, crowberry *Empetrum*, grape *Vitis*, primrose *Primula*, bedstraw *Galium*, thrift *Armeria*, comfrey *Symphytum*, forget-me-not *Myosotis*, Labiatae, plantain *Plantago*, nightshade *Solanum*, figwort *Scrophularia*, elder *Sambucus*, snowberry *Symphoricarpos*, teasel *Dipsacus*, Compositae, sedge *Carex*, rush *Juncus*, various grasses and cereals (Gramineae). (Eber 1956; Tuřcek 1961; Prokofieva 1963*b*; Newton 1967*a*; Tutman 1969; Dol'nik 1982; Sabel 1983; Fuchs 1984; Guitián Rivera 1985; Krägenow 1986.) In Oxford, July–April, 75 stomachs contained 925 items, of which *c.* 46% by number cereal grains, 10% seeds of goosefoot, 7% chickweed, 6% Cruciferae, 3% grasses, 2% beech, 2% mouse-ear, 2% *Rumex*, 5% beetles, 2% Lepidoptera, 2% aphids, and 2% Diptera; seeds ranged in weight from 0·1 mg (*Artemisia*) to 230 mg (*Fagus*) (Newton 1967*a*). Also in England, all year, 128 stomachs contained 56% by volume weed seeds, 23% insects, 9% cereal grains, 5% buds, 4% fruit pulp, 2% earthworms, and 1% spiders (Collinge 1924–7). In St Petersburg area (north-west Russia), April–September, 631 items of animal prey were 50·7% by number beetles (46% Curculionidae), 14·6% Hymenoptera (7% ants), 12·5% Lepidoptera (12% caterpillars), 7·8% spiders, 6·2% caddis flies, 3·6% Hemiptera, and 2·4% Diptera (Prokofieva 1963*b*, which see for monthly figures). In Schleswig-Holstein, 49% of 3944 feeding observations on or under plants (all year) were on cereals, 28% beech, 5% Compositae (especially *Artemisia*), 4% Cruciferae, 4% Polygonaceae (especially *Rumex*), 3% grasses, 3% birch, 3% alder (Eber 1956). For north-west Spain, see Guitián Rivera (1985). In Kaliningrad region (western Russia), animal material 90% of diet in breeding season, 3% during rest of year (Dol'nik 1982); in Moscow area, June–July, 85% animal prey (Krägenow 1986); in Schleswig-Holstein, 9% of 4836 observations over year were of birds feeding on invertebrates; 70% May–July (*n*=331), varying in remainder of year from 0·5% in October (*n*=1032) to 11% in April (*n*=381) (Eber 1956). See Newton (1967*a*) for further plant:animal ratios in diet from literature and, in Oxford, for possible difference between winter diets of resident and immigrant birds. Plant material taken throughout year more or less according to availability, though probably feeds much less on seeds of Compositae than most other Fringillidae; in spring, eats half-ripe seeds, shoots, etc., of Chenopodiaceae, Polygonaceae, and Compositae, as well as buds and catkins of deciduous and coniferous trees and newly-sown grain; in summer, seeds of wide variety of herbs as well as cereals from beneath standing crop; in autumn, very fond of grain picked up in stubble fields and Cruciferae seeds, and

occasionally extracts seeds from Rosaceae fruits; in winter takes seeds of trees, especially beech, grain in fields and around farmyards, and frequents parks, gardens, etc. (Eber 1956; Newton 1967*a*; Dol'nik 1982).

In captivity, consumed average 3·9 g per day (*n*=12 birds) of kernels, i.e. de-husked weight (Kear 1962). Maximum daily intake in wild in one study 7 g fresh weight (138 kJ). Basal metabolic rate of caged birds over year 32·2–41·6 kJ per day, standard rate 64·1–88·4 kJ. (Dol'nik 1982, which see for many details.)

Diet of young almost wholly invertebrates: leaf-dwelling insects, especially aphids, and caterpillars; Lepidoptera and Diptera caught in air, plus beetles and their larvae, earwigs, spiders, earthworm cocoons, snails, etc., from ground (Eber 1956; Eggermont 1956; Newton 1967*a*; Fellenberg 1988). In St Petersburg area, of 1574 items from 311 collar-samples, 20·6% (by number) beetles (*c.* 7% Curculionidae, 5% Cantharidae, 4% Elateridae), 14·0% Diptera (including 8% Tipulidae, 4% larvae), 11·6% Lepidoptera, virtually all larvae (including 5% Geometridae), 9·1% Hymenoptera (*c.* 8% Tenthredinidae larvae), 7·2% Hemiptera, 7·2% caddis flies (*c.* 4% larvae), 6·9% spiders, 6·2% mayflies, 4·0% stoneflies, 1·4% snails, and 9·5% plant material (Prokofieva 1963*b*). In Kaliningrad region, *c.* 56% of diet by dry weight adult Diptera (52·0% Chironomidae, 1·8% Asilidae, 1·6% Tipulidae), 23·7% spiders and harvestmen, 6·9% adult caddis flies, 5·0% adult beetles, 4·8% larval Tenthredinidae, 3·6% caterpillars (Dol'nik 1982). In England, 32 stomachs contained 98% by volume aphids, Geometridae caterpillars, and beetles (Collinge 1924–7). At one nest, parents broke up caterpillars until young *c.* 11 days old, then fed them whole (Barrett 1947); in St Petersburg region, diet of young 3–4 days old contained 16% spiders (*n*=25 invertebrates), at more than 4 days old 7·4% (*n*=1518) (Prokofieva 1963*a*). Each nestling in wild receives average 0·1 g of food during day 1 (2·4 kJ), 4·2 g (81·9 kJ) on day 9, and 3·7 g on day 12. Basal metabolic rate of caged nestlings *c.* 4 kJ per day on day 1 rising to *c.* 33 kJ on day 14. (Dol'nik 1982, which see for many details.)          IN, BH

**Social pattern and behaviour.** Well studied. Following account based largely on study in Cambridge (England) by Marler (1956*a*). For reviews see Newton (1972) and Krägenow (1986).

1. Mainly gregarious outside breeding season, forming flocks for feeding and migration (see also Roosting, below). Flock composition and timing of flock-formation vary regionally according to local migratory status (see Movements). Account by Marler (1956*a*) typifies changes in dispersion where resident population (*gengleri*) is augmented by winter immigration: from July to mid-September, flocks mainly juveniles (start flocking at 40–50 days old), thereafter joined by local adults finishing moult, and in late October by large influx of nominate *coelebs* (which typically form larger and more cohesive flocks). In groups of *gengleri*, sexes roughly equal but in nominate *coelebs* one or other sex often predominates (representing differential migration: see Movements), e.g. ♀♀ in Ireland, ♂♂ in southern England, Nether-

lands, Fenno-Scandia, Germany, and Switzerland (Deelder 1949; Marler 1956*a*; Schifferli *et al.* 1982). In late winter, *gengleri* disperses to territories in fine weather while nominate *coelebs* remains in substantial flocks until March–April. Flock size at Cambridge rose sharply at end of October from *c*. 2–5 to *c*. 30 or more (maximum *c*. 100–200) for rest of winter. At Falsterbo (southern Sweden), autumn and spring, migrant flocks typically 5–30 birds but hundreds or thousands assembled for resting and feeding, typically mixed with Bramblings *F. montifringilla* (Lindström and Alerstam 1986). In northern Finland (as elsewhere in north of range), ♂♂ arrive in spring *c*. 2 weeks before ♀♀, usually singly or in small flocks (mean 5 birds, *n* = 479 flocks, excluding singletons); flocks associated with *F. montifringilla* quite common although within assemblages each species tended to aggregate with conspecifics (Mikkonen 1984). Flocking tendency weaker in *F. coelebs* than in *F. montifringilla* (see that species). *F. coelebs* also commonly associates with other Fringillidae (Marler 1956*a*), and Great Tits *Parus major* also seen joining flocks of *F. coelebs* (Hinde 1952). Little information on other races of *F. coelebs* but, in Azores, *moreletti* forms post-breeding flocks which apparently separate by sex (Marler and Boatman 1951). BONDS. Mating system essentially monogamous. Exceptionally, successive bigamy by ♂ arises as follows: if ♂ dies or is driven from part of territory by mate's aggression (see Antagonistic Behaviour, Heterosexual Behaviour, below) vacated area is annexed by neighbouring ♂ who thus acquires 2 mates, each occupying separate parts of now expanded territory; several such cases recorded, although still rare compared with monogamy. (Marler 1956*a*, *b*.) In 3-year study, Sheffield (England), 3 out of 76 breeding attempts involved such bigamy; ♂ never seen to feed young of secondary ♀ (B C Sheldon). Lovaty (1985) describes unusual case of successive bigamy arising from 2nd ♀ settling in ♂'s territory after 1st ♀ had started incubating; ♂ apparently helped to feed (non-overlapping) broods of both ♀♀. Monogamous ♂♂ readily promiscuous (Marler 1956*a*; Lovaty 1985; B C Sheldon) and some evidence that ♀♀ also protagonists in promiscuity (see subsection 4 of Heterosexual Behaviour, below). At least in resident *gengleri*, pair-bond probably often persists from year to year: pair-members commonly seen together in or near territory in winter (see also Roosting, below), though no obvious pair-bond when in flock (Marler 1956*a*). One marked pair stayed together 3 years (bred for first 2 years) until both accidentally killed (Hendy 1939). In south-west Finland, however, of 4 pairs ringed and surviving 2 years, only 1 pair bred both seasons (Bergman 1953). ♀ sometimes divorced ♂ after nest predated (Hanski *et al.* 1992). For evidence of floating population of unpaired ♂♂, see Bergman (1956) and Hanski *et al.* (1992), for unpaired ♀♀ see Sæther and Fonstad (1981). For occasional hybridization with *F. montifringilla*, see that species. Role of sexes in nest-duties as follows. ♀ alone builds nest, incubates, broods, and does most feeding of young until fledging (Barrett 1947; Marler 1956*a*); ♂ did not start feeding nestlings until 3rd–8th day and then accounted for only 15% of visits (to 2 nests) (Marler 1956*a*). In Sheffield study, ♂ contributed 33·7 % (11·0–57·1) of feeds at 11 nests and his visits were shorter than ♀'s (suggesting he brought less food per visit) (B C Sheldon). In Finnish study, some ♂♂ (perhaps preparing for 2nd brood) played no part in feeding fledglings and resumed singing instead (Hanski *et al.* 1992). At nest containing pair's 2nd brood, *c*. 40% of visits by ♂; ♀ continued to do most feeding of fledged 1st brood while building 2nd nest, and roles continued thus until she started incubating, whereupon ♂ took over feeding 1st brood unaided; same role-division in transition from 2nd to 3rd brood (Barrett 1947); brood-division observed by Peitzmeier (1942), Marler (1956*a*) and Hanski *et al.* (1992). Parents continue to

feed young for *c*. 3 weeks after fledging (Barrett 1947). Age of first breeding 1 year (Marler 1956*a*). BREEDING DISPERSION. Solitary and territorial. Territory size varies with habitat and stage of breeding (see below), but in any case limits hard to define. Thus Marler (1956*a*) and Newton (1972) considered that most feeding done inside territory demarcated by border song-posts. Bergmann *et al.* (1982) showed that some feeding occurred in undefended area outside song-territory (total area used therefore home-range). However, in southern Finland visual- and radio-tracking showed that ♂♂ spent considerable time feeding (including collecting food for young) outside song-territories, making home-range 4–8 times larger than song-territory; resident ♂♂ also spent time in territories of other ♂♂, apparently in search of extra-pair copulations (Hanski and Heila 1988; Hanski *et al.* 1992; see also Heterosexual Behaviour, below). In alder *Alnus* woodland in central Norway, song-territory also considered of only minor importance as source of food (Sæther and Fonstad 1981). Perhaps song-territory and feeding-territory only coincide in densely occupied optimal habitat. Song-territories are smaller in deciduous than coniferous woods (Newton 1972): in Cambridge, average of 17 territories 0·7 ha (0·1–1·2), smallest in optimal habitat (open deciduous with abundant secondary growth) (Marler 1956*a*, *b*). In northern Sweden, territory size in mixed pine *Pinus* and deciduous woodland *c*. 5 ha (calculated from Udvardy 1956). At start of breeding season, most ♂♂ (including experienced breeders returning to former territory) tend to extend territory until checked by neighbour-pressure; territory initially large, shrinking as parts forfeited to subsequent settlers (Bergman 1953, 1956; Marler 1956*a*; Mikkonen 1985*a*). For further details of territory establishment, see Antagonistic Behaviour (below). Apart from feeding (discussed above) territory serves for courtship and nesting (Marler 1956*a*). Contra Merikallio (1951) which suggested interspecific territoriality between *F. coelebs* and *F. montifringilla*, studies indicate marked and amicable territorial overlap between them (Udvardy 1956; Mikkonen 1985*a*); for dominance relations, see *F. montifringilla*. Reed (1982) found territorial overlap between *F. coelebs* and *P. major* on the Scottish mainland but mutual exclusion on adjacent islands (where foraging behaviour of the two species converges). In favoured habitats, density of *F. coelebs* fairly stable from year to year but in suboptimal habitat may fluctuate greatly (Glas 1960). Outside woodland, density depends on availability of large trees as feeding sites and song-posts. In well-wooded hedgerows, up to 3–4 pairs per km² (Newton 1972; Schifferli *et al.* 1982). Usually in a field mosaic, territories cluster where hedges join each other or woodland (Newton 1972). The same 'edge effect' explains higher density (up to 142 pairs per km²) in small Finnish parks (2 ha or less) compared with those exceeding 10 ha (maximum 39 pairs per km²) (Suhonen and Jokimäki 1988). In some places, density up to 300 pairs per km² (Sharrock 1976; Paevski 1978) or more (e.g. Öland, Sweden: Fritz 1989), but generally less than half of this. Highest densities vary regionally from deciduous or mixed woodland to coniferous, e.g. in Netherlands 76 ♂♂ per km² in mixed wood, 35 in pine (Glas 1960). In Fenno-Scandia also, generally more numerous in deciduous (49–145 pairs per km²) than coniferous forest (12–102) (Newton 1972, which see for references; Ulfstrand and Högstedt 1976). In Białowieża (Poland) up to 157 pairs per km² in deciduous forest, 100 in coniferous (Tomiałojć *et al.* 1984). In Britain, favours deciduous and mixed woodland: average 58 pairs per km² for all woodland (Sharrock 1976), 21 on farmland (Marchant *et al.* 1990); however, Simms (1971) recorded highest densities in native Scots pine *P. sylvestris* (e.g. 631 birds per km² in Abernethy, Scotland: Hill *et al.* 1990), pine plantations and, in Ireland, young mixed plantations; in Scotland, also dominant in alder and co-dominant with Willow Warbler *Phylloscopus trochilus* in birch *Betula* (Simms 1971; Marchant *et al.* 1990). In 9-year study of developing oak *Quercus robur* forests in Bourgogne (France), up to 56 pairs per km² (Ferry and Frochot 1970). In Switzerland, density generally higher in conifers at all altitudes than in pure deciduous forest, e.g. maximum 110 pairs per km² in mixed oak–hornbeam *Carpinus*, 190 in subalpine spruce *Picea*; highest recorded 46 pairs in 7–8 ha of larch *Larix*; markedly variable in gardens and parks, only 60 pairs per km² in orchards (Glutz von Blotzheim 1962; see also Schifferli *et al.* 1982). Densities in Mediterranean generally lower: in Italy, 6·5 pairs per km² in Lombardy (Cambi and Micheli 1986), 37 territories per km² in Tuscany (Lambertini 1981). 77 pairs per km² in Provence (southern France), 64 in Corsica (Blondel 1979). In north-west Spain, 99·5 birds per km² in *Pinus radiata* (Carrascal 1987). In Morais (northern Portugal), 19·6 territories per km² in cork oak *Q. suber* (Mead 1975). In Morocco, 37·5 pairs per km² in maquis (Thévenot 1982). In Tenerife (Canary Islands), 9 birds per km² in *P. radiata* (Carrascal 1987). Site-fidelity (return to former breeding territory) appears to vary regionally: strong for ♂♂ in England (Marler 1956a, b) and for both sexes in northern Finland, but even stronger in southern Finland (Bergman 1953), suggesting site-fidelity may increase from north to south (Mikkonen 1983, which see for other references to strong site-fidelity in *F. coelebs*). More information needed on regional variation in natal site-fidelity, but in Kaliningrad (western Russia) over 90% of 1-year-olds (surviving from 2938 nestlings ringed) first bred within 1 km of natal site (Sokolov 1986b). Typically builds new nest for successive clutches within season, but one case of same pair using same nest twice in season (Tracy 1927). ROOSTING. Outside breeding season, migrants typically form large communal roosts, whereas residents (in Britain) usually roost alone or sometimes in pairs in thick cover (usually evergreen or thorny) in their territories (see start of part 1, above) (Marler 1956a; Newton 1972). Communal roosts traditional (used night after night and often year after year) and typically in clumps of evergreens (including conifer plantations), bramble *Rubus* or other thorny scrub, also oaks and beeches *Fagus* retaining dead leaves; may contain hundreds or thousands of birds, in Britain usually 50–100, but regularly up to 2000, often mixed with *F. montifringilla* and other Fringillidae (Newton 1972; I Newton). In Israel, communal roosts in citrus groves and avocado orchards (Shirihai in press). In France, communal roosting occurs from end of October to February (Labitte 1937). Vleugel (1941) describes seasonal and diurnal variation in arrival and departure times for mixed Dutch roost (with small numbers of *F. montifringilla*) used for 2 winters, December–March; birds foraged up to 2 km from roost-site. Few details for non-migrants, but one observation of pair regularly roosting together at end of December (Morley 1943). For diurnal activity pattern in midsummer, northern Sweden (24 hrs daylight), see Armstrong (1954).

2. Relatively confiding and, in urban areas where commensal with man, will tolerate close approach. Mobbing reactions widely studied: with crest raised and neck somewhat extended (plumage otherwise rather sleeked) moves restlessly (bowing, pivoting, tail-flicking, etc.) around object of alarm, alternately approaching and retreating; intense display, e.g. to perched hawk *Accipiter* or owl (Strigiformes) includes dive-attacks, Chink-calls, and, in breeding season, Rain-calls (Hinde 1954b; Marler 1956a; Korbut 1989: see 2 and 5 in Voice). For habituation (etc.) to mobbing response, see Hinde (1954b, 1960). For attacks on would-be predators, and distraction-display when apparently no nest or young involved, see Parental Anti-predator Strategies (below).

A

Escapes from flying *Accipiter* by direct or zig-zag flight to cover; on alighting, adopts Freezing-posture (Fig A): sleeks feathers and conceals wing-bars; eyes bulge and scan upwards. If already in cover, adopts same posture, often accompanied by Chink- and Seee-calls (see 4 in Voice). If *Accipiter* alights nearby, eye-fixation by *F. coelebs* precedes escape-flight or mobbing response. Freezing-posture usually elicited only by aerial predators in winter, but in breeding season (especially after young out of nest) also by ground predators and sudden appearance of any large (otherwise innocuous) bird. For response of flock to attack by *Accipiter*, see Flock Behaviour (below). Lindström and Nilsson (1988) recorded case of possible 'fright-moulting' as confusion-strategy by *F. coelebs* when separated by Sparrowhawk *A. nisus* from flock and pursued at length. FLOCK BEHAVIOUR. Described as follows for winter flocks in Cambridge (Marler 1956a, which see for other references by same author). Alighting singleton typically gives Chink-calls, reciprocation of which encourages flock-formation. Individual distance *c.* 20 cm between ♂♂, 5–10 cm between ♀♀ and between ♂ and ♀. Small groups commonest early morning, gradually merging to maximise flock size from *c.* 2–3 hrs after sunrise until *c.* 1 hr before sunset when departure for roost begins. Some movements of feeding flock synchronized whereby Flight-call (see 3 in Voice) of flock-member initiates mass exodus to another often quite distant feeding site; other movements are of silent, piecemeal, drifting type. As winter proceeds, flock integration increases, decreasing again January–February as local birds resort to territories (see start of part 1, above). Behaviour of feeding flocks on migration at Falsterbo subjected to attacks (rate of which increased with flock size) by *A. nisus* as follows: feeding starts typically when small nucleus group in exposed tree-top, calling loudly, lands on ground near cover; this attracts others and, as flock enlarges, foraging extends to more exposed areas. Attack by *A. nisus* scatters flock into smaller groups which flee to tree-tops. Apparently spontaneous upflights also common. In autumn, flocks dense and movements highly synchronized, in spring looser and less coordinated. (Lindström 1989.) Following account of roosting behaviour based on combined descriptions of Ash (1949), Nelson (1950), Marler (1956a), and Newton (1972). Flock-members leave feeding grounds singly and in small groups which, on reaching roost-site, circle high overhead for some time with hesitant bounding flight before suddenly plummeting *en masse*, silent but for audible wing noise, on to topmost twigs. First settlers often restless, giving Chink-calls, taking off and re-settling before dropping down to lower perches. Competition for favoured perches sometimes considerable, birds milling around inside roost with distinctive slow whirring flight in somewhat upright posture. Vleugel (1941) recorded 2 cases of birds starting spring migration directly on departure from roost. SONG-DISPLAY. Performed by ♂ (see 1 in Voice) to demarcate territory and attract ♀; typically delivered from high conspicuous perch in upright posture with head back, plumage relaxed, wing-bars partly concealed (Marler 1956a). Bergman (1953) separates this 'territorial' song given in morning and evening peaks (see below) from 'courtship' song heard mainly at other times and distinguished from territorial song essentially by delivery, being given from lower perch in forward posture, tail lowered, head sleeked, flanks somewhat ruffled. Singer changes perch regularly (i.e. patrols territory); may move between song-perches in normal flight, but often in undulating Moth-flight in which wings are fully extended and beat rapidly through shallow arc; head and tail are depressed and typically song is resumed just before alighting (England 1945b; Marler 1956a). Subsong (see 1c in Voice) heard mainly from birds foraging on the ground within territory, or from dense cover, from young birds establishing territory, and from old birds visiting territories in winter; also exceptionally from ♀♀ (Marler 1956a). For references to singing by ♀♀, see Marjakangas (1981). For 'battle song' see Antagonistic Behaviour (below) and for 'congested song' see sub-section 4 of Heterosexual Behaviour (below). In adults (resident *gengleri*), song develops (initially from incomplete phrases) after little or no Subsong during warm spells in winter (any month) or spring; in 1st-year ♂♂, transition from Subsong to song is longer. In most pairs, song almost ceases for 1–2 weeks after pair-formation and then gradually returns, but never regains intensity of unpaired ♂♂. (Marler 1956a, b.) Song-rate declines to very low level in 6 days before mate lays 1st egg (i.e. during her peak fertile period) (B C Sheldon). Unpaired ♂♂ sing every 7–15 s for most of day, i.e. *c.* 360 times per hr (Newton 1972). Study in Bavarian Alps (Germany) typifies diurnal and seasonal variation across much of range: song occurred throughout day (minimum afternoon) from sunrise to sunset, increasing from 14 hrs in March to 18 in May–June (song began early February and continued until beginning of August, but song-period mainly March to mid-July, average 146 days). In March, April, and June, morning peak exceeded evening peak; 2 seasonal peaks, 1st before 1st brood, 2nd (June) before peak of 2nd or replacement broods. (Bezzel 1988.) For other studies of diurnal song-activity see Falconer (1941) and Bergmann *et al.* (1982). For influence of temperature on morning onset of song, see Haartman (1952) and Scheer (1952). Spring migrants start singing before reaching breeding grounds; song heard on 27% of days on migration through Helgoland (Moritz 1982). ANTAGONISTIC BEHAVIOUR. (1) General. ♂ dominant to ♀ in winter flocks, but vice versa during nesting (Hinde 1955–6; Marler 1956a; see also Heterosexual Behaviour, below). Description of territory occupation based collectively on Bergman (1953), Marler (1956a), and Mikkonen (1985a): 1st-year ♂♂ begin prospecting territories in early spring, often encountering opposition from established adults, but on finding suitable site begin singing (see Song-display) or giving Chink-calls. Gradually adopt favoured song-posts which define territory and become focus of disputes with rivals. During mild spells in winter, older ♂♂ often defend former territory spasmodically with Chink-calls; one ♂ gave more than 100 calls within a few minutes of first entering territory (Mikkonen 1985a). ♂'s arrival on territory precedes ♀'s. Once paired (but not until), ♀ quickly becomes co-owner, attacking other ♀♀ which trespass, especially near nest; both sexes continue to defend territory with similar intensity throughout nesting period (Marler 1956a) although ♂ apparently defends sharper boundary than ♀ (Mikkonen 1985a). (2) Threat and fighting. Unless otherwise acknowledged, following account based on Marler (1956a) for captive aviary flock. For other descriptions of postures and displays, see Hinde (1953, 1955). Rivalry occurs in winter feeding flocks, in breeding season, and between juveniles during incipient courtship in late summer and autumn. Disputes include Supplanting-attacks in which aggressor usually exposes wing-bars, sleeks plumage (especially on head), and may Bill-snap; on approach of aggressor, victim sleeks plumage, turns away, crouches, raises wings somewhat to expose wing-bars, and usually flees. Less readiness to flee (by either contestant) may be signalled by Head-forward display: bird crouches head-on to rival, retracts

head and points bill (sometimes Bill-snapping) at rival; with increasing intensity, exposes wing-bars and sleeks plumage; ultimately raises wings at shoulder, exceptionally flirts them (quickly raised and lowered), and may Bill-thrust at opponent (Fig B, upper). Appeasement is signalled by a rounded Fluffed-posture (Fig B, lower); ♀ thus inhibits aggression from mate in pair-

B

formation. In breeding season, territory-owner flies at intruder (usually ♂) in Flirting-flight (rapid series of dashes with whirring wings, audible at distance, wing-bars highly conspicuous) followed by Supplanting-attack or, if rival holds ground, Head-forward display with Bill-snapping etc. Established ♂♂ engage in boundary-disputes which may repeatedly pass to and fro across boundary; chased bird may sing in flight (Bergman 1953). Rivals in boundary dispute commonly indicate attack-flee conflict with variations of Head-up display (Fig C): head and neck sleeked and

C

extended, bill *c.* 30° above horizontal, breast ruffled, wing-bars exposed; performer pivots constantly and jerkily, tail-flicking and often wing-flirting slightly (above back) with each turn of body. Head-up display of intruder or known rival *within* non territory is less aggressive and sometimes includes crest-raising. Chink-calls are given by territory-owner in boundary-disputes, while song may also accompany mild intention to attack (battle song: Bergman 1953). Rivals (usually ♂♂, exceptionally ♀♀) occasionally fight, tumbling to ground and rolling around locked together; for illustration of ♂ fighting another which intruded on territory to attempt extra-pair copulation, see Birkhead and Møller (1992); for ♂ pecking another to death, see Anon (1976). Interspecific fighting rare in breeding season (see Barrett 1947 for examples, also Parental Anti-predator Strategies, below). For interspecific territoriality, see Breeding Dispersion (above). HETEROSEXUAL BEHAVIOUR. (1) General. In migrant populations ♂♂ arrive on breeding grounds and occupy territories ahead of ♀♀ (Bergman 1953; Dementiev and Gladkov 1954). 1-year-old ♂♂ pair *c.* 2 weeks after establishing territory, usually later than older ♂♂ (Marler 1956a, b). Dominance of ♀ over mate in nest-area a significant factor in pair-bond relations. (2)

Pair-bonding behaviour. Unless otherwise acknowledged, following account based on Marler (1956a). Various postures and displays also described by Coombs (1945), Beven (1946), Hinde (1953, 1955-6) and Dol'nik (1982), and repertoire summarised by Newton (1972). ♂ is usually joined by ♀ a few days after he develops full song (see Song-display). ♀♀ prospecting for mates (typically 1-4 weeks after ♂ occupies territory) move from tree to tree giving frequent Flight-calls which may induce courtship in ♂; ♀ approaches ♂ tentatively (tail-flicking, sleeking plumage, some wing-raising) and, as ♂ nears her, she adopts stationary Fluffed-posture (see above). ♀ soon begins moving away from ♂ (often inducing ♂ to try to 'lead' her), exploring territory, during which she may examine nest-sites (see sub-section 5, below). Territorial ♂ attracted by ♀ sleeks plumage to highlight wing-bars, then sings and takes off with Moth-flight (see Song-display) to fly at lower level than ♀ so that his upper side is visible to her. On landing (beneath and beyond ♀), ♂ performs Crouching-lopsided display (Fig D): with head sleeked, turns

E

D

sideways to ♀, crouches rather stiff and motionless, and glances at her over his shoulder; exposes wing-bars, enhanced at higher intensity by slight lifting of closed wings. At this point, lopsided element comes in, ♂ tilting body sideways and at same time raising still-closed wing nearest ♀, exposing red flank and belly to her. Wing nearest ♀ is kept raised for several seconds while ♂ periodically glances at her. Then he relaxes and performs Moth-flight to another perch to repeat display. If he turns (relative to ♀), other wing is raised in display. Song now ceases, and ♂ begins to use Kseep-call (see 6a in Voice) not previously heard, and initiates another Moth-flight beneath her and towards core of territory. If she follows, ♂ leads her around territory, giving Kseep-calls whenever she moves. If she stays still, he returns and displays anew, always avoiding close contact and, if she is persistently unresponsive, he often attacks or chases her and occasionally attempts copulation (see below). ♂'s general strategy is thus to contain her within territory by combination of vocalization and display. Marler (1956a) distinguished between early (in season) first meetings between old unpaired ♂ and ♀ (♂'s aggression strong but sexual responses not fully developed) and late first meeting (copulatory urge ascendant in ♂). Only the latter markedly different from typical pattern (above) and therefore described here: ♂ occasionally has to wait in territory 3-4 weeks before ♀ arrives, whereupon first response to ♀ is usually to replace song immediately with 'tchirp' call (see 5 in Voice). ♂ then sleeks plumage and may momentarily adopt Upright-lopsided display (Fig E), more erect (and submissive) than Crouching-lopsided; approaches ♀ in short Moth-flights, body becoming more erect and legs extended as he does so. As ♂ nears ♀ he still perches side-on to her but switches from left to right with increasing frequency, each time raising wing nearest her. If ♀ keeps still (as is usual), ♂'s approach ceases. Occasionally he performs Moth-flight to fork in tree and, perched there, makes circling movements (see sub-section 5, below) giving occasional

'tchirp' calls. If she follows, he flees, and this may develop into Supplanting-attacks by ♂. If ♀ sleeks plumage, crouches, then flees on ♂'s approach, rapid zig-zag chase often ensues, ♂ in pursuit (rapid version of Moth-flight) with 'tchirp' calls or sometimes low-intensity song. Often after short spell of contact with ♀, ♂ breaks away to give combination of Seee- and Rain-calls (see 4 and 5 in Voice). In early courtship, sexual responses alternate with aggression. ♂ initially dominant to ♀ but gradually dominance reverses, ♀ remaining dominant till end of breeding season; switch in dominance coincides with transition in ♂ from Crouching-lopsided to Upright-lopsided display, also with increasing incidence of sexual chases, Rain-calls and Seee-calls. However, Crouching-lopsided is not abandoned entirely but just excluded from intense courtship, e.g. ♂ adopts Crouching-lopsided on settling after (dominant) ♀ has made Supplanting-attack on him. During pre-nesting phase, pair spend most of time on territory, maintaining close contact with Chink-calls and ♂'s Rain-calls. (3) Courtship-feeding. None reported. ♂ never seen to feed ♀ on or off nest (Marler 1956a); exceptional cases of ♀ on nest soliciting food from mate by waving head or body from side to side like juvenile (see below); such begging usually unsuccessful (Hendy 1943; Barrett 1947), though occasionally she may intercept food intended for young (Newton 1972). (4) Mating. Fertile period probably starts 6-10 days before laying; onset of Mate-guarding indicated by decline in distance between pair-members from c. 6 days (minimum 2 days) before onset of laying (B C Sheldon). As start of nest-building approaches, ♂ increasingly tries to copulate with mate who, at first, typically attacks him or precipitates sexual chase. Attempts may be solicited by either sex, though only 9% by ♂. ♂'s attempts only succeed if ♀ performs Soliciting-display which is confined to period between completion of nest and start of incubation (Marler 1956a). Following description of ♀'s Soliciting-display (Fig F) after Marler

F

(1956a) which see for references to numerous other descriptions: wings first slightly raised and extended, then (with increasing intensity) shivered and half-raised, also tail raised and somewhat spread, and breast pushed out. Climax of ♀'s display accompanied by crescendo of Seep-calls (see 6b in Voice) heard as frenzied squeaking, audible 80 m away, attracting other ♂♂ (B C Sheldon); also tail raised higher, head sleeked and drawn backwards and upwards. If Soliciting-display fails to attract ♂, ♀ performs Moth-flight to another perch where she displays anew. Exceptionally, ♂ solicits with ♀-display, but typically confronts ♀ with intense Upright-lopsided display, dancing to and fro before fluttering up and mounting. Sometimes, when ♀ solicits, ♂ flies straight on to her back giving low-intensity ('congested') song as he approaches. ♀'s Seep-calls and ♂'s song continue during copulation, ceasing as ♂ dismounts. Shortly after, ♀ may resume soliciting, and further display by ♂ and copulation follow. After one or more copulations, ♂ flies off to perch some distance away, giving Seee- and Rain-calls. In Sheffield study, pairs copulated frequently, peaking at up to 6 times per hr, 3–5 days before 1st egg; pair may copulate 180 times for 1 clutch, this together with Mate-guarding (see above) providing paternity-protection. However, once ♀ starts incubating, ♂ often furtively visits neighbouring territories to seek extra-pair copulations. Intrusions on other territories peaked 2 days before 1st egg if ♀ in encroached territory, and were commonest before and after fertile period of intruder's own mate. Some evidence that ♀ shares initiative in seeking extra-pair copulations by evading mate during her peak fertile period. Of 253 copulations seen, 7·9% were extra-pair copulations. 14·8% of 27 young (and 50% of 8 broods) contained extra-pair paternity-young. (B C Sheldon.) After nest-desertion, building may begin again within 1–2 days and mating cycle is repeated (Marler 1956a; see Breeding for multiple broods). (5) Nest-site selection. By ♀. Moving around at various levels in tree, ♀ alights on branch and hops to a fork; if fork is in deep crutch she examines it briefly with turning and scraping movements (Marler 1956a; Pokrovskaya 1968), then leaves, often giving excited Chink-calls as she alights to prospect another site. ♂ may occasionally visit sites and display mildly during pair-formation (see above), but plays no part in final selection (Marler 1956a). (6) Behaviour at nest. Nest built exclusively by ♀ although ♂ typically (though not invariably: see Barrett 1947) attends mate assiduously and often follows her on collecting trips (Marler 1956a). For calls associated with nest-building, see Barrett (1947). Only ♀ incubates; usually gives Chink-calls after leaving nest and again on return. Courtship wanes at this time and there is resurgence of singing by ♂. (Marler 1956a.) RELATIONS WITHIN FAMILY GROUP. Immediately after eggs hatched, ♀ ate eggshells (Mascher 1952). ♀ may begin feeding young within 1 hr of hatching but initially spends more time brooding. Brooding gradually declines to nought on c. day 6, but recurs at any age during heavy rain. ♀ does most feeding of nestlings (see Bonds), small share by ♂ perhaps associated with ♀'s dominance in nest-area. ♂ approaching with food often sings or gives Chink-calls above nest. ♀ eats faecal sacs for first few days, thereafter ♀ (less often ♂) carries them away (in Moth-flight: Barrett 1947). Newly-hatched young beg silently, later raise body and give 'cheep' calls (see Voice). Exercising begins c. 4 days before fledging. On fledging, young at first fly clumsily and are led by parents into dense cover; show no gregarious urge, dispersing through territory for a few days before wandering further afield. Parents communicate with young by Flight-calls. Fledglings beg with 'chirrup' calls (see Voice) delivered whilst crouching, crest raised, and rocking body from side to side, sometimes accentuated by wobbling head movement (which may develop just before leaving nest). (Marler 1956

a, b.) ANTI-PREDATOR RESPONSES OF YOUNG. If alarmed, nestlings crouch, also scatter from nest from 11 days (I Newton). Nestlings and fledged young fall silent on hearing the following parental calls: Rain-call, Seee-call, possibly also Chink-call (Marler 1956a). PARENTAL ANTI-PREDATOR STRATEGIES. (1) Passive measures. On hearing ♂'s Seee-call, ♀ on nest stays motionless with plumage sleeked (Marler 1956a). ♂ seldom visits nest till after hatching (I Newton). (2) Active measures: against birds. ♀♀ may attack other species, especially tits *Parus* and Dunnock *Prunella modularis* intruding near nest-site (Marler 1956a, b). After eggs hatched, ♂ drove off Tree Sparrows *Passer montanus* whenever they approached to within c. 2 m of nest; ♂ usually made sudden darts or slow stalking approach but once applied wing-cuffing (Barrett 1947). Attacks on Jay *Garrulus glandarius* and stuffed Cuckoo *Cuculus canorus* also reported (Marler 1956a, which see for references). (3) Active measures: against man. When flushed from nest containing eggs or young, ♀ flies off silently with weak fluttering flight; if nest is relatively high ♀, 'parachutes' towards ground; if nest low, she flies very low over ground (B C Sheldon). ♂ occasionally performs mobile distraction-lure display of disablement type (injury-feigning), usually to protect young (Freitag 1978). In England, May (no nest or young found), ♂ approached to c. 3 m of observers, lay on ground, tail vertically upwards, body tilted away from observers and partly supported by open wing. On closer approach, hobbled and flopped away, wings spread; moved thus for c. 20 m, collapsing whenever observers halted, finally allowing approach to within 1 m before flying off. (Harwood 1959.) 2 similar accounts for ♂♂ in Germany, both described as 'broken wing' type, suggest that distraction-display extends beyond breeding season: Freitag (1978) recounts incident of 10 April in which no nest or young present and ♂ was apparently protecting mate who crouched nearby. 2nd example on 26 September when no ♀ or young seen (Metz 1981).

(Figs by D Nurney: A, C, and F from photographs in Marler 1956a; B and D from photographs in Marler 1956a and drawings in Newton 1972; E from drawing in Newton 1972.)     EKD

**Voice.** Originally studied in detail by Marler (1956a, b), but more recent work is extensive. Pioneering study by Thorpe (1958) on song learning. Song varies between individuals, with considerable local sharing but without clear dialect pattern (Slater and Ince 1979), except for some island races. Call 5 (Rain-call) shows mosaic dialect distribution (e.g. Haartman and Numers 1992). For comparison with Blue Chaffinch *F. teydea*, see that species.

CALLS OF ADULTS. (1) Song. (1a) Full song of ♂. Short phrase of 1·5–3·0 s (Marler 1956a, b). ♂♂ have 1–6 song-types (Slater 1981) with fixed sequence of unit types; includes 'trill' of 2–4 differing segments within each of which an identical unit is repeated several times, followed by terminal flourish (occasionally given twice, rarely 3 times: W T C Seale) of unrepeated units which are longer and often of broader frequency. Segments within the 'trill' may also be separated by unrepeated transitional units. (Marler 1956a; Slater and Ince 1979.) Later segments in song tend to be lower pitched than earlier ones (Thorpe 1958; Slater *et al.* 1984). Figs I–IV show 4 different but typical songs given by same ♂ during relatively short bout of singing; Figs V–VI show 2 songs by another ♂, and Figs VII–VIII one song each from further 2 ♂♂. Subdued

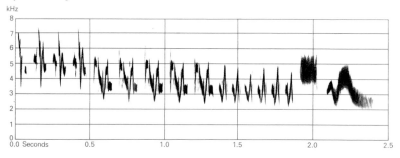

I  P J B Slater  England  May 1981

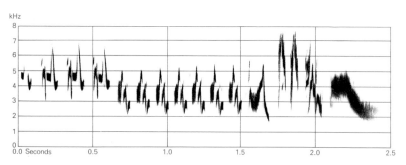

II  P J B Slater  England  May 1981

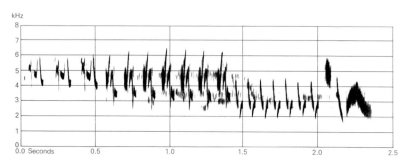

III  P J B Slater  England  May 1981

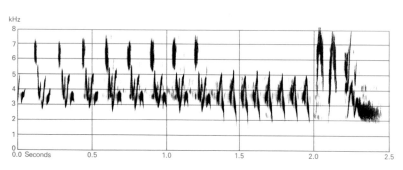

IV  P J B Slater  England  May 1981

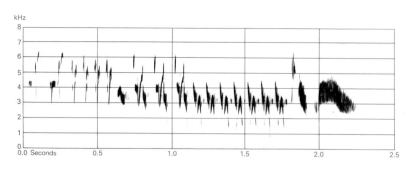

V  P Jenkins  England  May 1972

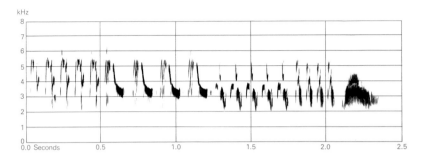

VI   P Jenkins  England  May 1972

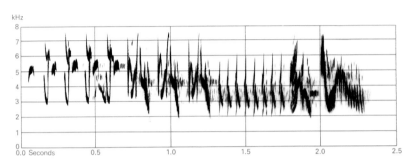

VII   P J B Slater  England  April 1982

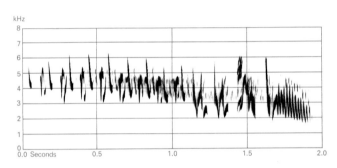

VIII   P J B Slater  England  April 1982

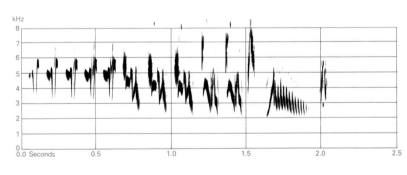

IX   W T C Seale  England  May 1985

initial units omitted in all sonagrams shown here except Figs V–VI and IX. ♂♂ often sing at rate of *c*. 6 songs per min, and those with repertoire of more than one song-type sing series of one before switching to another, often singing each type in turn before returning to 1st (Hinde 1958*a*; Hansen 1981*a*; Slater 1983). Song is learnt during 1st year of life (Thorpe 1954, 1958; Slater and Ince 1982), although ♂♂ can sometimes change repertoire at later age (Nürnberger *et al.* 1989; Goodfellow and Slater 1990). Neighbouring birds frequently share song-types, but many birds also sing songs unique to themselves (Slater and Ince 1979). While mix of song-types varies between localities there are not usually overall differences such that local dialects can easily be described (Slater *et al.* 1984), and structure of song does not differ between *gengleri* and nominate *coelebs* or over their range. However, in wide areas of Europe, though only rarely in Britain (W T C Seale), song may be followed by a 'kit' unit, similar to that of Great Spotted Woodpecker *Dendrocopos major* (Witherby *et al.* 1938; Thielcke 1969: Fig IX, in which final

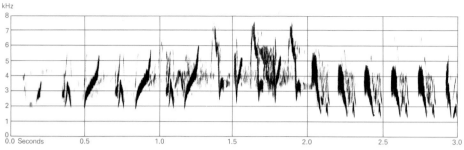

X   P J Sellar   Gran Canaria (Canary Islands)   May 1991

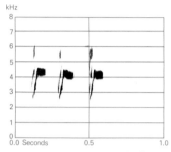

XI   P A D Hollom   Wales   April 1975

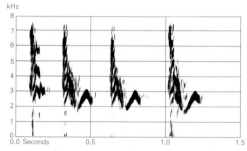

XII   P J B Slater   Tenerife (Canary Islands)   May 1984

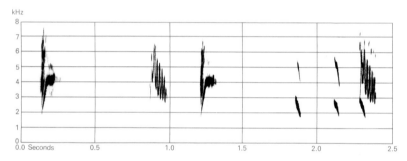

XIII   S Carlsson   Sweden   July 1983

unit is 'kit'). For song in Corsica and Balearic Islands, see Chappuis (1976). In Azores, *moreletti* has 'trill' which is sometimes not clearly split into separate sections, with flourish shorter or absent (Marler and Boatman 1951). In Canary Islands, *canariensis* tends to have larger song repertoire, with longer units but without clear terminal flourish (Slater and Sellar 1986: Fig X). (1b) Song of ♀. Occasionally heard (e.g. Hendy 1941, England 1945a). Simpler than song of ♂; noted as series of identical units, lacking terminal flourish, not unlike song of Lesser White-throat *Sylvia curruca* (England 1945a). (1c) Subsong. Low-pitched, quiet, rather amorphous chirping, warbling, and rattling heard from ♂ mainly early in season before full song developed (Marler 1956a, b). Rapid continuous quiet song ('congested song') also given by ♂ just before and during copulation. (2) Chink-call. Distinctive 'chink' (Fig XI, also units 1 and 3 in Fig XIII), similar to that of Great Tit *Parus major*. Most frequent call, given by both sexes during breeding season in varied situations: mild alarm, mobbing, territory establishment, separation from mate (Marler 1956 a, b; Poulsen 1958). Intensity of call varies with context (Thorpe 1958). Exact form is learnt, though similar across range, except in Azores where equivalent is 'gai' (Knecht and Scheer 1968) and in Canary Islands where 'chee-choo chee-choo-choo'; see Fig XII for version rendered 'chee-chooi-chooi-chooi' produced in alarm at cat. (3) Flight-call. A 'tupe' or 'tsüp' (Witherby *et al.* 1938) given by both sexes in flight and when about to take off, also in communication with fledglings (Marler 1956a, b). Recording suggests softer 'whib' (P J Sellar: units 4–5 in Fig XIII). Units not run together when repeated (as those of Greenfinch *Carduelis chloris* often are: see call 2 of that species) (Witherby *et al.* 1938). (4) Seee-call. High-pitched thin 'seee' which, as in many passerines, is produced in strong alarm, especially in response to overhead hawk *Accipiter*, also to dog and man when young out of nest; similar call heard (in series with call 5) from ♂ during breaks in intense courtship and after copulation; also given by injured birds (Marler 1956a, b; Marler and Hamilton 1966). For sonagram of homologous call in *F. teydea*, see Fig IV in that species. (5) Rain-call. Regionally variable call given in many different contexts

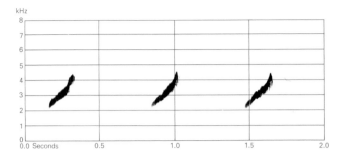

XIV  R Margoschis  England  April 1977

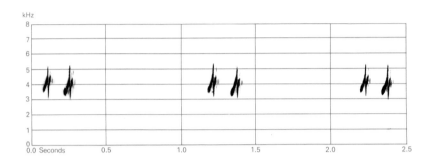

XV  R Ranft  England  June 1989

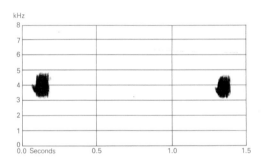

XVI  P A D Hollom  England  May 1984

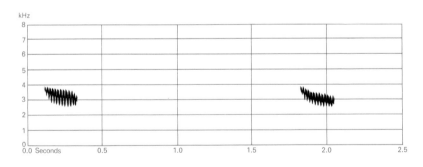

XVII  P J Sellar  England  June 1983

by breeding ♂, sometimes when alarmed but often apparently spontaneously. Occurs in long series. May take form often described as 'huit' (Fig XIV) but also, in England, as 'hreet' or 'breeze' (Marler 1956a, b). Another variant, frequently heard in Scotland, is 'limmik' (Fig XV), occasionally 'limmikik' (P J Sellar) or, in Denmark, 'hrit hrit' or 'it it it' (Poulsen 1958). Further variant, Germany, rendered 'rülsch' (Baptista 1990); English recording suggests 'chizz' (P J Sellar: Fig XVI). Variant common in south-east England is descending 'eerb' with vibrant quality (P J Sellar: Fig XVII); 'tchirp' (Marler 1956a, b) is perhaps the same. In much of Finland, call is 'hüitt' but given as 'rrüp' in southern part of south-west archipelago, with sharp boundary between these dialect areas (Haartman and Numers 1992). Foraging ♂♂ sometimes utter Rain- and Chink-calls alternately for long periods (P J Sellar). Fig XVIII depicts version given by *canariensis* in Gran Canaria, rendered 'zees chee chee' (P J Sellar). In

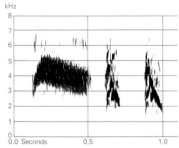

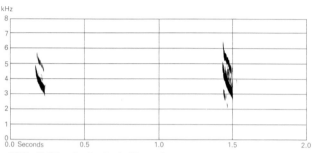

XVIII  P J Sellar  Gran Canaria (Canary Islands)  May 1991

XIX  V C Lewis  England  May 1962

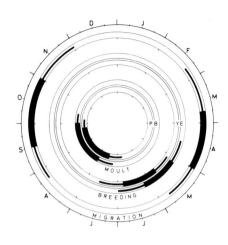

Swedish recording, one bird alternates 'chink' with 'chirr' (similar to juvenile food-call) while 2nd gives 3 'whib' flight-calls (Fig XIII: W T C Seale). (6) Courtship-calls. (6a) Kseep-call. A 'kseep' given by ♂ early in season when still likely to attack ♀. (6b) Seep-call. A 'seep' given by ♀ when soliciting copulation (Marler 1956a, b; Poulsen 1958); see Fig XIX for variant suggesting explosive 'tsioo' (C Chappuis, P J Sellar). (7) Aggressive-call. Low buzzing sound given by ♂ during fights (Marler 1956a, b), and by both sexes when attacking conspecific or other intruders near nest (Poulsen 1958). (8) Other calls. Short, thin, very high-pitched 'see', much shorter than call 4, only audible at close range, sometimes given by ♀ in response to call 5 of ♂ while feeding and on take-off from feeding area (P J Sellar).

CALLS OF YOUNG. Newly hatched young give short 'peep' or 'cheep', increasing in volume and duration, as well as structural complexity, with age. From 1 week old, young develop longer food-calls, of fluctuating frequency and c. 0·5 s duration, together with shorter more penetrating 'cheep' (Wilkinson 1980). Fledglings beg with 'chirrup' or 'chirr' (W T C Seale), which is also used socially and in mobbing contexts before adult calls develop. In alarm, young birds produce 'tew' call for several weeks after fledging, and this call is also occasionally heard from adults (Marler 1956a, b). For further development of adult repertoire by young, see Marler (1956a).                                        PJBS, PJS

Breeding. SEASON. Start and pattern of laying related to spring temperature, becoming later from south-west to north-east in Europe (Newton 1964a; Svensson 1978; Dol'nik 1982). Britain: most clutches started late April to mid-June, range mid-March to mid-July (Newton 1964a, which see for regional variation). Sweden: see diagram. Northern Scandinavia: median date for start of laying (calculated from collections) 29 May (Svensson 1978). Kaliningrad region (western Russia): eggs laid early May to early July, peak late May to early June (Dol'nik 1982); see also Nankinov (1978). Germany: eggs laid beginning of April to mid-July; major studies by Schreiber (1987) in Niedersachsen, Krägenow (1986) in Sachsen-Anhalt, Eifler (1990) in Sachsen, and Glück (1983) in Baden-Württemberg. SITE. In fork of tree or bush, on branch or on several thin twigs. In fruit trees, south-west Germany, 7% of 28 nests 1-2 m above ground, 11% 2-3 m, 28% 3-4 m, 25% 4-5 m, 14% 5-6 m, and 15% over 6 m; average 4·4 m; nest-tree highest of 5 Fringillidae studied, though chosen according to abundance; of the 5 species, average thickness of nest-carrying branch greatest at 6·2 cm; 14% of nests against trunk, 32% up to 2 m away, 46% 2-4 m, and 8% more than 4 m; site itself not especially well-hidden and receives much sunlight, but nest well camouflaged (Glück 1983). In Sachsen, 59% of 276 nests in deciduous trees or shrubs, and elder Sambucus in hedges and villages slightly preferred, though generally nest-tree species in line with abundance; most nests 1·8-2·5 m above ground (1-16, n=81) (Eifler 1990). In Poland, 71% of 132 nests in deciduous species at average 3·9 m; no species preferred, but nest needs strong support (Tomek and Waligóra 1976, which see for tree species). Average height in Canary Islands 5·1 m (2·5-9, n=19) (Martín 1987). Sometimes on buildings and recorded on ground (Haartman 1969; Makatsch 1976); for other unusual sites, see Krägenow (1986); nests at over 20 m reported (Marler 1956a). Nest: compact and neat, with firm walls and deep cup, clad with lichen and moss thus looking green or greyish; pliable and yielding to touch; outer layer of lichen, moss, bark, and fibres bound with spider silk, then grass and stalks lined with rootlets, hair, and feathers (Dobben 1949; Haartman 1969; Makatsch 1976; Tomek and Waligóra 1976). In Poland, of 72–76 nests, average outer diameter 9·0 cm, inner diameter 5·3 cm, overall height 7·1 cm, and depth of cup 4·0 cm (Tomek and Waligóra 1976). See also Newton (1972), Nankinov (1978), and Krägenow (1986). Building: by ♀ only, usually accompanied by ♂ (Dobben 1949; Marler 1956a; Krägenow 1986, which see for reports of ♂ building; Schreiber 1987); takes 7 days (3–18, *n*=11) from start of building to laying of 1st egg; old nest often demolished for material (Marler 1956a, which see for details); 1st nest takes 7–14 days, 2nd 3–7 days (Svensson 1978); see also Barrett (1947). EGGS. See Plate 60. Sub-elliptical, smooth and slightly glossy; very variable, pale bluish-green to reddish-grey with purple-brown blotches (at times edged with pink), scrawls, and hair-streaks, concentrated at broad end (Harrison 1975; Makatsch 1976). See Bunyard (1932) for wide variety in collection. Nominate *coelebs*: 19·3 × 14·6 mm (16·6–22·8 × 13·2–15·8), *n*=1264; calculated weight 2·16 g. *F. c. gengleri*: 19·9 × 14·7 mm (17·1–22·9 × 13·0–16·2), *n*=121; calculated weight 2·24 g. *F. c. africana*: 20·7 × 15·2 mm (18·5–24·5 × 13·9–16·3), *n*=101; calculated weight 2·61 g. *F. c. maderensis*: 22·1 × 15·5 mm (20·0–24·5 × 14·8–16·7), *n*=40; calculated weight 2·78 g. (Schönwetter 1984.) See also Svensson (1978) for discussion of size variation. Clutch: 4–5 (3–6). In Britain, average size 4·3 (*n*=3182), 4·4 until late May then rapidly falls to 3·5 by late June (Newton 1964a); see also Snow and Mayer-Gross (1967). In Finland, of 318 clutches: 2 eggs, 1%; 3, 3%; 4, 25%; 5, 65%; 6, 5%; 7, 0·3%; 8, 0·3%; average 4·7 (Haartman 1969); for Scandinavia, see also Svensson (1978). Average in Czechoslovakia 4·64; 4·52 in April (*n*=51), 4·73 in May (*n*=92), 4·52 in June (*n*=19) (Pikula and Folk 1970); in Kaliningrad region, 4·82 in May (*n*=88), 3·43 in early July (*n*=23), average 4·64 (*n*=955) (Dol'nik 1982). For Germany, see Krägenow (1986) and Eifler (1990). Average in north-west Africa 3·9, *n*=157 (Martín 1987). Eggs laid daily in early morning, average 2·6 days (0·3–7, *n*=210) after nest completion, interval decreasing as season progresses (Dol'nik 1982). Up to 3 replacement clutches recorded following clutch losses (Newton 1964a), and new nest can be started 1 day after loss (Marler 1956a). 2 broods apparently unusual (Marler 1956a; Newton 1964a; Haartman 1969; Krägenow 1986); in Kaliningrad region ♀ 3% of pairs double-brooded (Dol'nik 1982). 2nd brood always in new nest (Schreiber 1987); at one site, nest of 3rd clutch ready 2 days after 2nd brood flew (Barrett 1947). INCUBATION. Average 12·6 days (10–16, *n*=200) (Newton 1964a); by ♀ only, probably from penultimate egg; 2 ♀♀ had average break of 8 min (5–13, *n*=18) and average stint of 38 min (12–71, *n*=19) (Marler 1956a). In Kaliningrad region, during daylight, on eggs average 2 hrs per day after 2nd egg laid, 5 hrs after 3rd, and 13 hrs after 4th (Dol'nik 1982). YOUNG. Fed and cared for by both parents, though more by ♀; ♂♂ at 2 nests made 15% of 104 feeding visits and rarely before young 3–8 days old; brooded by ♀ for *c.*

6 days (Marler 1956a). At one nest brooded for *c.* 11 days; ♂ brought food on 38% of 293 visits (Barrett 1947). See also Dol'nik (1982) and Krägenow (1986). FLEDGING TO MATURITY. Average fledging period 13·9 days (11–18, *n*=200) (Newton 1964a); Haartman (1969) gave 12·7 days (10–15, *n*=15) in Finland. Fed by parents for *c.* 3 weeks after fledging, though young able to partly feed themselves after 2 weeks (Barrett 1947; Marler 1956a). For development of young, see Marler (1956a), Dol'nik (1982), and Krägenow (1986). Age of first breeding 1 year (Marler 1956a; Paevski 1985). BREEDING SUCCESS. In Britain, of 10 967 eggs, 59% hatched and 41% produced fledged young; excluding total failures, 88% of 7021 eggs produced fledged young. Percentage of nests from which at least 1 young flew, 1950–60, varied between 18% (*n*=161) and 60% (*n*=207); within years, proportion of successful clutches increased from 38% of those started in 1st half of April to *c.* 50% in 2nd half of May, attributable to reduced predation as cover thickened. Most failures due to predation, especially by man, followed by desertion (generally caused by death of 1 parent), and bad weather; most losses caused by Corvidae, grey squirrel *Sciurus carolinensis*, and cats, and many doubtless by weasel and stoat *Mustela*. 2·6–4·5 fledged young per successful nest depending on month and clutch size; clutches of 5–6 most productive. (Newton 1964a, which see for many details.) In southern England, successful nests varied from 11% (*n*=133) in woodland to 29% (*n*=1127) in 'rural habitat' (Snow and Mayer-Gross 1967, which see for discussion of methods and comparison with Newton 1964a). Early clutches similarly vulnerable in southern Sweden, where 90% of those in 1st half of May taken by Jay *Garrulus glandarius* and red squirrel *S. vulgaris* (Svensson 1978). In Kaliningrad region, 63% of 3859 eggs hatched and 52% produced fledged young; average 3·1 young flew from successful nests (Dol'nik 1982, which see for many details). Main predators in St Petersburg area (north-west Russia) Corvidae and red squirrel (Nankinov 1978). In Sachsen, 3·7 fledged young per successful nest (*n*=20), 2·4 overall (*n*=31) (Eifler 1990); in Sachsen-Anhalt, in 47 cases where cause of complete failure known, 32% due to Corvidae and other aerial predators, 26% man, 17% red squirrel, 17% weather, and 4% cats (Krägenow 1986, which see for many records of birds of prey as predators). For Niedersachsen, see Schreiber (1987). Success on islands off southern Finland improved following removal of predators (Bergman 1956). In contrast to cardueline finches, nestlings fed mainly on caterpillars, so mortality is high if they are in poor supply when young in nest, e.g. in cold spring or late in season (Newton 1964a). IN, BH

**Plumages.** ADULT MALE. In fresh plumage (autumn), forehead mottled black and buff-brown; crown, hindneck, and side of neck medium blue-grey, each feather fringed drab-brown or warm brown on tip, partly concealing blue-grey, especially on side of crown, nape and side of neck. Lower mantle and scapulars dark umber-brown, feathers fringed with slightly paler olive-brown or buff-brown, largely concealed centres brighter cinnamon-rufous. Back, rump, and shorter upper tail-coverts bright yellowish olive-green; longest upper tail-coverts blue-grey with green fringe along sides and buff-brown on tip. Lore finely mottled grey, buff, and vinous; side of head from just above eye and upper ear-coverts down to lower cheek deep vinous-cinnamon, each feather faintly tipped pale drab-brown or buff, partly concealing cinnamon, in particular above eye, on rear of ear-coverts, and on cheek, border between blue-grey top of head and side of neck and vinous-cinnamon remainder ill-defined. Chin, throat, chest, and side of breast deep pinkish vinous-cinnamon, feathers narrowly fringed buff on tip; flank, upper belly, and side of belly gradually paler, vinous-pink, merging into cream-white or pink-white of mid-belly, vent, and under tail-coverts. Central pair of tail-feathers (t1) dark grey with black shaft-streak, often widening into black blob on tip; narrow fringe along outer web green, narrow border along inner web and tip white; t2–t6 black, sometimes with faint green outer edge and faint white edge along tip; inner web of t4 sometimes with small white blob or wedge near tip, inner web of t5 with broad and contrasting white wedge 25–30 mm long on tip, inner web of t6 with white wedge c. 40 mm long, occupying entire distal half of inner web (except blob at shaft near tip) and extending to middle portion of outer web. Flight-feathers black, tinged grey on inner web; basal one-third of secondaries white, but this concealed by greater coverts on upperwing; basal one-quarter of inner and central primaries white, just visible as narrow bar along tips of upper primary coverts; terminal half of outer web of secondaries and of inner 3–4 primaries fringed pale yellow, outer webs of other primaries more fully edged pale yellow, but narrow edge along outer web of p9 and along emarginated parts of p5–p9 white; basal and middle portions of inner webs of flight-feathers with long white wedge, visible on underwing only. Tertials black, broad fringe along base of outer web pale yellow, grading through yellow-buff or green-buff into drab-brown or brown-grey fringe along tip. Bastard wing, greater upper primary coverts, and greater upper wing-coverts deep black; outer webs of primary coverts narrowly fringed green, greater upper wing-coverts broadly tipped pale yellow or white, tips c. 2–3 mm wide on outermost coverts, c. 6–8 mm on inner, forming distinct wing-bar; innermost (tertial coverts) black with broad drab-brown fringe. Median and longer lesser upper wing-coverts white, washed pale yellow, forming conspicuous white shoulder; shorter lesser slate-blue or blackish-grey, partly fringed white. Under wing-coverts and axillaries white, coverts along leading edge of wing dotted black and white. *In worn plumage* (spring), buff and brown fringes of head and body worn off, forehead with contrasting velvet-black band, crown to upper mantle and side of neck uniform medium blue-grey, lower mantle and inner scapulars umber-brown to tawny-brown, forming contrasting 'saddle', outer scapulars blue-grey with some brown or green on tips; back to shorter upper tail-coverts bright green, longer coverts blue-grey; side of head rufous-cinnamon with vinous cast, strongly contrasting with blue-grey of crown and nape; chin to chest and side of breast deep vinous-cinnamon, merging into pink-vinous of belly and flank. By about May, pale fringes of t1 and tertials bleached to grey, pale yellow fringes of flight-feathers and of upper wing-coverts bleached to whitish. When heavily abraded (late June and July), cap more ash-grey, saddle broken into brown spots interspersed with grey and green, green of rump duller, mixed grey, pale fringes of t1 and tertials worn off. ADULT FEMALE. In fresh plumage (autumn), upperparts from forehead to scapulars as well as longer upper tail-coverts dull drab-brown or brown-grey, feathers with faintly brighter olive-

grey borders on mid-crown, lower mantle, scapulars, and tail-coverts, slightly darker and more uniform dark olive-brown or black-brown on broad stripe on each side of crown and (in particular) on side of nape, stripes on nape isolating paler green-grey central patch. Upper back drab-brown, grading to bright green or olive-grey on lower back, rump, and shorter upper tail-coverts. Side of head and neck light drab-grey or dull ash-grey with slight olive tinge, palest at base of bill, mottled cream-buff above and below eye (forming broken eye-ring), faintly streaked off-white on shorter ear-coverts; ear-coverts often browner than greyish rear of supercilium and side of neck. Chin to upper belly, side of belly, and flank pale drab-grey with slight olive or buff wash; white of feather-bases on chin and throat partly visible, white on lower throat frequently washed pink, centres of feathers of chest and upper belly occasionally with slight vinous-pink suffusion. Mid-belly, vent, thigh, and under tail-coverts cream-white, not sharply demarcated from drab-grey of remainder of underparts. Tail as adult ♂, but t1 tinged olive-brown, less grey, without black along shaft, black of t2–t6 slightly greyer, less deep, white wedges slightly less sharply contrasting; white on inner web of usually t4 absent (Niethammer 1962), wedge on t5 15–25 mm long, wedge on t6 30–40 mm long. Wing as in adult ♂, but flight-feathers and primary coverts slightly greyer, especially basal parts of outer webs less deep black, pale yellow fringes along distal half of outer webs of secondaries and inner primaries less sharply contrasting, outer webs of primary coverts fringed dull olive (in adult ♂, either fringed bright green or uniform deep black); black of bastard wing and greater upper wing-coverts slightly less deep, more sooty, outer web fringed dull olive (uniform black in adult ♂), tips of greater coverts pale yellow or whitish, as in adult ♂, but slightly less sharply defined; median coverts pale yellow or white, forming broad wing-bar; lesser coverts mainly dark plumbeous (♀ showing less white at shoulder than in adult ♂); shorter under wing-coverts mottled dull black or dark grey and white. *In worn plumage* (spring), forehead to back duller drab-brown, faintly fringed grey rather than green, cap and stripe at each side of nape sepia-brown, grey patch on central nape more distinct, underparts paler and more uniform olive-drab-grey, white of mid-belly and vent even less contrasting, pink suffusion on lower throat, chest, or upper belly (if present) more pronounced; pale fringes of tail, tertials, and along outer webs of primary coverts and greater coverts greyish, partly worn off from about May; wing-bars whiter, sometimes slightly narrower, sometimes with some black visible at bases of median coverts; narrow fringes of flight-feathers grey-white, when heavily abraded. NESTLING. Down long, plentiful, pale smoke-grey, on upperparts, upperwing, thigh, and vent. (Witherby *et al.* 1938; Krägenow 1986.) JUVENILE MALE. Rather like adult ♀, but feathering of head and body short and soft; tail, tertials, and flight-feathers new at the time that those of adult ♀ are worn or moulting. Top of head and neck drab-grey with dark grey-brown stripe at each side, centre of nape with ill-defined pale grey patch. Lower mantle and scapulars diluted tawny-brown with green tinge, rump buffish- or greenish-yellow. Underparts entirely off-white, tips of feathers of throat and side of breast to upper belly and flank washed greyish- or cream-buff. Tail as adult, but feathers narrower, tips more pointed (less broadly rounded); t1 paler grey, partly tinged olive, without black on tip, fringes along outer web and tip greenish-brown, less sharp, narrow, and green than in adult; black on t2–t6 tinged brown, less deep than in adult ♂, less contrasting with white on outer feathers; no white on t4, except sometimes for white fringe on tip. Lesser upper wing-coverts dark grey, longer ones with broad white tips; median coverts mainly white; greater sooty-grey with green outer fringe and white tip (latter narrower than in adult, 1-2 mm on outer coverts, 5-7 mm on inner). Greater upper primary coverts sooty grey with narrow olive fringe along outer web and tip (in adult, deep black with narrow green fringe along outer web of outermost only); remainder of wing as in adult. JUVENILE FEMALE. Like juvenile ♂, but lower mantle and scapulars brown-grey or olive, without tawny tinge; tips of feathers of underparts paler, more creamy, less buff; 11 grey-brown instead of grey; longer lesser upper wing-coverts more narrowly fringed white (0-2 mm rather than 2-4 mm). FIRST ADULT MALE. Like adult ♂, but juvenile tail, flight-feathers, tertials, greater upper primary coverts, bastard wing, and sometimes a few outer greater coverts retained; tail-feathers more pointed and less deep black than in adult, primary coverts, bastard wing, and (if any) outer greater coverts greyer than neighbouring new greater coverts, which are deep black; from about April, 11, tertials, primary coverts, bastard wing, and tips of primaries distinctly browner and more worn than in adult at same time of year. In some birds, head and body as in adult; in others, grey of cap more broadly fringed brown when plumage fresh, traces of brown remaining when worn, lower mantle and inner scapulars partly tinged green, and vinous of underparts less deep and intense. Ageing in southern populations often more difficult, as many birds apparently replace tail and tertials in post-juvenile moult. FIRST ADULT FEMALE. Like adult ♀, but part of juvenile feathering retained, as in 1st adult ♂. Ageing more difficult than in 1st adult ♂, especially when no juvenile outer greater coverts are retained; juvenile primaries not much different in colour from new 1st adult greater coverts, but different shape of tail-tips and (in spring) more worn juvenile tertials, primary coverts, and primaries often useful.

**Bare parts.** ADULT, FIRST ADULT. Iris brown or dark brown. Bill of ♂ in spring and summer light slate-blue or lead-blue with small black tip and cutting edges; during rest of year and in ♀ dull grey, horn-grey, greyish-horn, or whitish-horn with slight flesh, lilac, mauve, or reddish tinge, palest at base of lower mandible, darker horn-grey or steel-grey on culmen and tip. Leg and foot dark pearl-grey, pale grey, brown-grey, blue-grey, or dark grey with slight reddish, flesh, or mauve tinge. (Hartert 1903-10; RMNH, ZMA.) NESTLING. Bare skin pink-flesh. Mouth rose-red or carmine, orange on palate, grey along inside border of mandibles. Gape-flanges yellow-white, cream-white, or whitish. (Heinroth and Heinroth 1924-6; Witherby *et al.* 1938.) JUVENILE. Iris hazel or brown. Bill light flesh-horn, flesh-grey, or brown with flesh tinge, paler at base of lower mandible. Leg and foot flesh-grey or lilac-grey. (RMNH.)

**Moults.** ADULT POST-BREEDING. Complete; primaries descendent. In Britain, starts with shedding of p1 between mid-June and late July, rarely late May or early August; completed with regrowth of p9-p10 after *c.* 70 days late August to early October (Newton 1968; Ginn and Melville 1983). In Finland, starts early to late July, completed after 72-73 days early or mid-September; tail mainly starts at primary moult score 8-20, completed at primary score 37-44; secondaries mainly start from primary score 17-30, completed at same time as outer primaries (score 45) or a few days later (Haukioja 1971*a*). In southern Sweden, inner secondaries still growing in singles from 29 September and 3 October; sample of 22 in moult from Netherlands fell within British data (RMNH, ZMA). In Yugoslavia, moult July-September, rarely from late June, occasionally still traces of moult October (Stresemann 1920). In European Russia, timing and duration of moult depend on latitude: in Kareliya (*c.* 62·5°N), average flight-feather moult *c.* 5 July to *c.* 1 October; at Lake

Ladoga (*c.* 61°N), average *c.* 23 July to *c.* 15 October; in St Petersburg area (*c.* 60°N), starts 20 June to 31 July, completed 20 August to 31 October, on average *c.* 15 July to *c.* 20 September; on Kurische Nehrung (eastern Baltic, *c.* 55°N), starts *c.* 20 June to *c.* 20 July, completed September to early October, on average *c.* 27 June to *c.* 23 September; in Crimea (*c.* 45°N), moult on average *c.* 10 July to *c.* 25 October, some starting late June (Dol'nik and Blyumenthal 1967; Dol'nik and Gavrilov 1974, 1980; Noskov 1975; Noskov *et al.* 1975; Gavrilov 1979; Rymkevich 1990). In western Siberia, moult from July (Johansen 1944). In northern Iran, moult just started late July to mid-September, almost completed from 10 September onwards (Vaurie 1949*b*). On Sardinia, single ♂ almost completed 23 August (Bezzel 1957). In North Africa, moult completed mid- or late October (Ticehurst and Whistler 1938; Meinertzhagen 1940). On Azores, 2 ♂♂ in advanced moult late September, others had completed in 2nd half of October (Chavigny and Mayaud 1932). For influence of day length on timing and duration of moult, see Dol'nik and Gavrilov (1974, 1980), Dol'nik (1975), and Noskov (1977). No pre-breeding moult. POST-JUVENILE. Partial: head, body, lesser, median, and many or all greater coverts (rarely innermost only; occasionally, 1-4 juvenile outer ones retained), sometimes 1-3 tertials, rarely central pair of tail-feathers (t1). Starts at age of *c.* 5 weeks (Heinroth and Heinroth 1924-6) or 20-30 days after complete juvenile plumage attained; duration of moult *c.* 45 days at 60°N in Russia, *c.* 50 days in Ukraine and Crimea; depending on hatching date, moult starts early June to early September, completed late July to mid-October, before autumn migration (Dol'nik and Blyumenthal 1967; Dol'nik and Gavrilov 1979; Dol'nik 1980; ZMA).

**Measurements.** ADULT, FIRST ADULT. Nominate *coelebs*. Wing, bill to skull, and bill depth at feathering of base of breeding birds from (1) Fenno-Scandia, (2) France, Germany, Denmark, Czechoslovakia, northern Yugoslavia, and Poland, and (3) Netherlands; other measurements combined, skins (RMNH, ZMA: C S Roselaar). Bill (N) to nostril; exposed culmen on average 2·7 less than bill to skull. Tail, bill (N), and tarsus include data of wintering birds in western and central Europe (A J van Loon).

| | | ♂ | | | ♀ | | |
|---|---|---|---|---|---|---|---|
| WING | (1) | 89·6 (2·22; 17) | 86-93 | | 83·3 (0·97; 9) | 82-85 | |
| | (2) | 89·6 (2·90; 20) | 86-95 | | 85·2 (1·35; 5) | 83-86 | |
| | (3) | 89·4 (2·19; 35) | 86-93 | | 83·4 (1·04; 14) | 81-85 | |
| TAIL | | 64·5 (2·10; 133) | 60-70 | | 59·8 (2·31; 67) | 56-64 | |
| BILL | (1) | 14·8 (0·51; 17) | 14·2-15·6 | | 14·6 (0·90; 9) | 14·3-14·9 | |
| | (2) | 15·1 (0·77; 15) | 14·1-16·3 | | 15·0 ( — ; 2) | 14·5-15·5 | |
| | (3) | 15·2 (0·41; 33) | 14·4-15·9 | | 14·3 (0·71; 14) | 13·5-15·2 | |
| BILL | (N) | 9·7 (0·43; 127) | 8·8-10·8 | | 9·3 (0·50; 54) | 8·6-10·2 | |
| DEPTH | (1) | 7·2 (0·36; 13) | 6·8-7·8 | | 7·0 (0·13; 7) | 6·9-7·2 | |
| | (2) | 7·4 (0·22; 13) | 7·0-7·8 | | 7·0 ( — ; 2) | 6·8-7·2 | |
| | (3) | 7·5 (0·31; 20) | 7·1-8·0 | | 7·2 (0·35; 14) | 6·7-7·7 | |
| TARSUS | | 18·0 (0·60; 125) | 16·5-19·5 | | 17·6 (0·57; 48) | 16·5-18·8 | |

Sex differences significant, except for bill (1)-(2) and bill depth (1)-(2). Ages combined, though retained juvenile wing and tail of 1st adult distinctly shorter than in older birds. Thus, in ♂♂ from samples above, wing of adult 90·9 (1·64; 40) 87·5-95, 1st adult 87·5 (1·24; 23) 85-90 (C S Roselaar), or, in larger sample including wintering birds: wing, adult ♂ 90·4 (1·97; 56) 86-95, 1st adult ♂ 88·0 (2·05; 83) 84-93, adult ♀ 83·9 (1·19; 17) 82-86, 1st adult ♀ 83·0 (1·77; 51) 78-88; tail, adult ♂ 65·8 (1·80; 43) 62-69, 1st adult ♂ 63·8 (1·99; 69) 60-70, adult ♀ 59·4 (1·71; 16) 56-62, 1st adult ♀ 59·8 (1·92; 43) 56-64 (A J van Loon).

Wing; live birds unless stated. (1) Sweden, skins (Bährmann and Eck 1975). (2) Tomsk (western Siberia), skins (Johansen 1944). (3) Bonn area (Germany) (Niethammer 1962; see also Niethammer 1971). (4) Eastern Germany, April, freshly dead

(Haensel 1967; see also Bährmann and Eck 1975, Bährmann 1976, Krägenow 1986, Eck 1985a, 1990). (5) Czechoslovakia (Pikula 1973; see also Mlíkovský 1982). (6) Poland, on spring migration (Busse 1976). (7) South-east France, winter (G Olioso). (8) Southern Yugoslavia, winter, skins (Stresemann 1920).

| | ♂ | | ♀ | |
|---|---|---|---|---|
| (1) | 89·5 (—; 33) | 86–93 | — (—; —) | — |
| (2) | 89·8 (2·14; 16) | 87–94 | — (—; —) | — |
| (3) | 89·2 (2·50; 103) | 83–95 | 83·8 (1·89; 39) | 78–87 |
| (4) | 88·0 (2·52; 391) | 81–95 | 81·6 (1·98; 94) | 77–86 |
| (5) | 88·3 (2·7; 384) | 80–99 | 83·3 (3·1; 228) | 80–89 |
| (6) | 88·5 (2·42; 1899) | — | 82·6 (2·03; 2189) | — |
| (7) | 90·3 (2·10; 88) | 86–96 | 83·7 (1·59; 46) | 80–86 |
| (8) | 89·2 (1·74; 33) | 86–92 | 83·5 (1·70; 14) | 81–87 |

Wing, bill to skull, and bill depth at base of (1) nominate *coelebs*, Crete (Stresemann and Schiebel 1925; Niethammer 1942; BMNH: A J van Loon); (2) *caucasica*, northern Turkey, (3) *syriaca*, Cyprus (RMNH, ZMA: C S Roselaar); (4) *transcaspia*, north-east Iran (BMNH: A J van Loon); other measurements combined, skins (A J van Loon).

| | | ♂ | | ♀ | |
|---|---|---|---|---|---|
| WING | (1) | 88·0 (2·16; 7) | 83–91 | 82·8 (—; 4) | 80–87 |
| | (2) | 89·3 (1·21; 6) | 88–91 | 83·6 (1·24; 5) | 81–85 |
| | (3) | 90·3 (2·18; 9) | 87–93 | 85·0 (—; 3) | 84–86 |
| | (4) | 89·5 (—; 3) | 88–92 | 85·0 (0·71; 5) | 84–86 |
| TAIL | | 65·8 (2·39; 19) | 61–70 | 61·3 (2·10; 14) | 58–64 |
| BILL | (1) | 15·6 (0·61; 5) | 14·9–16·4 | — (—; —) | — |
| | (2) | 15·7 (0·41; 6) | 15·2–16·1 | 15·2 (0·24; 4) | 14·9–15·5 |
| | (3) | 15·5 (0·61; 8) | 14·8–16·3 | 15·2 (—; 3) | 15·0–15·4 |
| BILL (N) | | 10·2 (0·45; 15) | 9·3–10·9 | 9·7 (0·58; 7) | 8·6–10·2 |
| DEPTH | (2) | 7·8 (0·38; 6) | 7·5–8·4 | 7·0 (0·21; 4) | 6·8–7·2 |
| | (3) | 7·5 (0·21; 8) | 7·3–7·8 | 7·3 (—; 3) | 7·2–7·4 |
| TARSUS | | 18·4 (0·65; 16) | 17·4–19·3 | 17·9 (0·79; 7) | 17·0–18·9 |

Sex differences significant, except for bill (3).

*F. c. caucasica.* Turkey, mainly May–August: wing, ♂ 87·8 (2·48; 11) 83–91, ♀ 81·0 (2·92; 6) 77–86 (Kumerloeve 1961, 1964a, 1967a, 1969b; Rokitansky and Schifter 1971).

*F. c. syriaca.* Troödos mountains (Cyprus): wing, ♂ 88·1 (16) 82–92, ♀ 83·1 (39) 77–89 (Flint and Stewart 1992).

*F. c. transcaspia.* Iran: culmen, ♂ 14·1 (6) 13·5–14·5, ♀ 13·2 (7) 12·5–14·0 (Vaurie 1949b).

*F. c. alexandrovi.* Northern Iran: wing, ♂ 86·1 (2·30; 12) 80–89, ♀ 81·0 (1·26; 6) 80–83 (Stresemann 1928; Paludan 1940; Schüz 1959); culmen, ♂ 14·4 (7) 13·5–15·5, ♀ 14·0 (3) 13·5–14·5 (Vaurie 1949b).

*F. c. solomkoi.* Crimea and south-west Caucasus, ♂ (n=6): wing, 86–94, bill (N) 10·2–11·0, depth of bill at base 8·5–9·0 (Menzbier and Suschkin 1913). Crimea, summer: wing, adult ♂ 95 (12) 93–100, juvenile and 1st adult ♂ 92 (92) 87–95, ♀ 88·7 (80) 82–91 (Noskov *et al.* 1975).

Wing, bill to skull and to nostril, and depth of bill at base of (1) *gengleri*, Britain and Ireland, (2) *balearica*, Portugal and Spain, (3) *tyrrhenica*, Corsica, and (4) *sarda*, Sardinia; other measurements combined, breeding birds only; skins (RMNH, ZMA: C S Roselaar).

| | | ♂ | | ♀ | |
|---|---|---|---|---|---|
| WING | (1) | 85·8 (2·26; 39) | 82–90 | 80·6 (1·87; 13) | 77–84 |
| | (2) | 86·6 (2·28; 17) | 83–91 | 81·9 (1·37; 7) | 79–84 |
| | (3) | 86·7 (2·02; 3) | 84–89 | — (—; —) | — |
| | (4) | 86·7 (1·43; 7) | 84–89 | — (—; —) | — |
| TAIL | | 63·7 (2·10; 68) | 60–69 | 59·0 (1·67; 20) | 57–64 |
| BILL (S) | (1) | 14·8 (0·61; 28) | 13·8–16·2 | 15·0 (0·46; 6) | 14·3–15·5 |
| | (2) | 14·7 (0·37; 17) | 14·1–15·4 | 14·6 (0·49; 7) | 14·0–15·2 |
| | (3) | 15·7 (0·29; 3) | 15·4–16·0 | — (—; —) | — |
| | (4) | 15·4 (0·48; 7) | 14·8–16·2 | — (—; —) | — |
| BILL (N) | (1) | 9·8 (0·45; 28) | 9·0–10·5 | 9·6 (0·50; 12) | 8·8–10·3 |
| | (2) | 9·4 (0·35; 17) | 8·9–10·0 | 9·3 (0·23; 7) | 9·0–9·6 |
| | (3) | 10·9 (0·40; 3) | 10·5–11·3 | — (—; —) | — |
| | (4) | 10·5 (0·36; 7) | 10·0–10·9 | — (—; —) | — |
| DEPTH | (1) | 7·0 (0·23; 26) | 6·6–7·4 | 6·9 (0·25; 6) | 6·6–7·2 |
| | (2) | 7·0 (0·21; 17) | 6·7–7·4 | 7·0 (0·33; 6) | 6·6–7·3 |
| | (3) | 8·0 (0·22; 3) | 7·8–8·3 | — (—; —) | — |
| | (4) | 8·1 (0·22; 7) | 7·8–8·4 | — (—; —) | — |
| TARSUS | | 17·9 (0·66; 51) | 16·3–19·1 | 17·6 (0·38; 19) | 16·9–18·4 |

Sex differences significant for wing and tail.

As in other races, retained juvenile wing and tail of 1st adult shorter than in older birds. Thus, wing of combined samples from above: adult ♂ 87·2 (1·43; 36) 84–91, 1st adult ♂ 84·4 (1·37; 29) 81–87, adult ♀ 82·7 (0·93; 6) 81–84, 1st adult ♀ 80·3 (1·27; 11) 78–82. ♂♂ of short-winged races from above readily separable on wing length (in skins) from nominate *coelebs* and other long-winged races, especially when age taken into account: adults of long-winged races have wing mainly 89 or more, short-winged adults 88·5 or less; long-winged 1st adults have wing 86 or more, short-winged 1st adults 85·5 or less; 10% of 128 birds wrongly classified by this.

*F. c. gengleri.* Ireland: wing, ♂ 84·4 (2·30; 5) 82–87 (Marler 1949). South-west Scotland, breeding ♂: wing 83–91, tail 61·5–72, bill (S) 15–17, tarsus 18–20 (n=18) (Harrison 1937). Wing: England, breeding ♂ 84·9 (2·18; 38) 80–89; Scotland, ♂ 85 (2·4; 70) 80–90, ♀ 80·0 (2·4; 10) 76–83 (Niethammer 1971).

Wing and bill to skull of (1) *spodiogenys*, Tunisia, and (2) *africana*, Algeria and Morocco; other measurements combined; skins (BMNH, RMNH, ZMA: A J van Loon).

| | | ♂ | | ♀ | |
|---|---|---|---|---|---|
| WING | (1) | 89·1 (2·24; 18) | 86–92 | 83·7 (2·32; 10) | 80–87 |
| | (2) | 90·5 (3·19; 17) | 84–97 | 85·3 (2·05; 16) | 83–90 |
| TAIL | | 68·9 (2·81; 35) | 62–76 | 63·6 (2·67; 26) | 59–69 |
| BILL | (1) | 15·5 (0·76; 17) | 14·5–17·0 | 15·6 (0·80; 10) | 14·0–16·9 |
| | (2) | 15·9 (0·50; 14) | 14·8–16·6 | 15·7 (0·42; 13) | 14·9–16·5 |
| BILL (N) | | 10·1 (0·36; 34) | 9·5–11·0 | 10·2 (0·45; 25) | 9·4–11·1 |
| TARSUS | | 19·0 (0·63; 34) | 18·0–20·6 | 18·9 (0·59; 26) | 17·8–20·2 |

Sex differences significant for wing and tail.

Cyrenaica (Libya): wing, ♂ 88–92, ♀ 83–85; bill, ♂ 16–18, ♀ 15–16 (Stanford 1954); bill ♂ 17·4 (1·49; 8) 16·5–18·0 (in Tunisia, 14–16, n=21) (Vaurie 1956a). Atlas Saharien (Algeria): wing, ♂ 89–98·5 (Ticehurst and Whistler 1938). Bill depth at base: 8·1 (0·29; 12) 7·7–8·7 (RMNH, ZMA). See also Grant (1979).

*F. c. canariensis.* Tenerife and a few from Gran Canaria and Gomera (Canary Islands), all year; skins (BMNH, RMNH, ZMA: A J van Loon).

| | ♂ | | ♀ | |
|---|---|---|---|---|
| WING | 84·8 (1·88; 26) | 81–90 | 79·5 (2·19; 13) | 76–83 |
| TAIL | 70·8 (1·87; 25) | 67–75 | 65·7 (1·76; 13) | 63–69 |
| BILL (S) | 17·2 (0·58; 17) | 16·2–18·0 | 16·6 (0·63; 14) | 15·7–17·7 |
| BILL (N) | 10·8 (0·68; 16) | 9·4–12·0 | 10·7 (0·45; 13) | 10·0–11·5 |
| TARSUS | 21·5 (0·62; 18) | 20·4–22·5 | 21·2 (0·34; 13) | 20·7–21·8 |

Sex differences significant for wing and tail. For large samples of this and other races of Atlantic islands, see also Eck (1975) and Grant (1979). Bill depth 8·1 (0·40; 10) 7·5–8·6 (RMNH, ZMA).

*F. c. palmae.* La Palma (Canary Islands), November–April; skins (BMNH, RMNH, ZMA: A J van Loon).

| | ♂ | | ♀ | |
|---|---|---|---|---|
| WING | 86·6 (1·93; 22) | 83–90 | 81·7 (1·47; 7) | 80–84 |
| TAIL | 73·7 (1·67; 22) | 70–76 | 68·2 (2·10; 7) | 64–70 |
| BILL (S) | 17·8 (0·57; 12) | 16·9–18·8 | 17·1 (0·45; 6) | 16·5–17·6 |
| BILL (N) | 11·9 (0·40; 14) | 11·2–12·7 | 11·5 (0·33; 7) | 11·0–12·1 |
| TARSUS | 21·6 (0·66; 13) | 20·3–22·6 | 21·0 (0·73; 7) | 20·2–22·1 |

Sex differences significant for wing and tail. Bill depth 8·1 (0·36; 7) 7·7–8·6 (RMNH, ZMA).

*F. c. ombriosa* Hierro (Canary Islands), November–April; skins of ♂: wing 87·4 (1·25; 4) 86–89, tail 73·5 (2·08; 4) 71–76, bill (S) 16·8 (0·66; 4) 16·1–17·6, bill (N) 11·3 (0·14; 4) 11·1–11·5, tarsus 21·7 (0·37; 4) 21·3–22·2 (BMNH, RMNH: A J van Loon). Wing: ♂ 88 (7) 85–89 (Vaurie 1959).

*F. c. maderensis.* Madeira, all year; skins (BMNH, RMNH, ZMA: A J van Loon).

| | ♂ | ♀ |
|---|---|---|
| WING | 84·1 (1·91; 19) 79–86 | 78·0 (1·40; 16) 76–82 |
| TAIL | 71·2 (1·88; 19) 68–76 | 65·5 (1·69; 16) 62–68 |
| BILL (S) | 17·5 (0·75; 14) 16·3–18·9 | 16·8 (0·49; 14) 16·2–17·7 |
| BILL (N) | 11·0 (0·39; 17) 10·4–11·8 | 10·7 (0·40; 15) 10·0–11·7 |
| TARSUS | 21·7 (0·40; 18) 21·0–22·6 | 21·0 (0·50; 16) 20·0–21·9 |

Sex differences significant, except for bill (N). Bill depth 7·2 (0·19; 10) 7·0–7·5 (RMNH, ZMA).

*F. c. moreletti.* São Miguel (Azores), March–June; skins (BMNH, RMNH: A J van Loon).

| | ♂ | ♀ |
|---|---|---|
| WING | 83·9 (1·87; 14) 80–87 | 78·0 (1·82; 10) 75–81 |
| TAIL | 68·0 (1·88; 14) 64–71 | 62·9 (1·40; 10) 61–65 |
| BILL (S) | 19·6 (0·47; 13) 18·7–20·2 | 17·7 (0·74; 10) 16·3–19·1 |
| BILL (N) | 12·2 (0·47; 14) 11·4–13·0 | 11·7 (0·05; 10) 10·7–12·9 |
| TARSUS | 21·1 (0·45; 14) 20·5–22·1 | 20·9 (0·61; 10) 20·2–21·6 |

Sex differences significant, except for tarsus. Live birds: (1) São Miguel, Pico, São Jorge, and Faial, (2) Flores; bill is exposed culmen (G Le Grand).

| | | ♂ | ♀ |
|---|---|---|---|
| WING | (1) | 83·6 (2·06; 234) 77–88 | 76·6 (1·86; 160) 71–82 |
| | (2) | 85·9 (1·90; 27) 80–89 | 77·3 (3·50; 24) 64–83 |
| BILL | (1) | 16·1 (0·92; 239) 10·6–18·0 | 15·3 (0·82; 161) 12·8–17·0 |
| | (2) | 17·1 (1·61; 17) 15·6–18·5 | 16·2 (0·43; 24) 15·5–17·5 |

Pico and Faial, skins (Chavigny and Mayaud 1932).

| | | ♂ | ♀ |
|---|---|---|---|
| WING AD | ♂ | 85·8 (12) 83–89 | 77·1 (7) 73–80 |
| 1ST AD | | 82·9 (18) 79–85 | 75·6 (9) 72–79 |
| TAIL AD | | 68·3 (13) 64–72 | 58·8 (7) 56–64 |
| 1ST AD | | 63·2 (19) 60–68 | 56·8 (9) 54–59 |

See also Eck (1975) and (in particular) Grant (1979).

**Weights.** ADULT, FIRST ADULT. Nominate *coelebs*. (1) Norway, spring and autumn (Haftorn 1971). (2) Falsterbo (southern Sweden), September (Scott 1965b). (3) Kazakhstan, winter (Korelov *et al.* 1974). (4) Poland, on spring migration (Busse 1976). (5) Helgoland (Germany), migrants (Weigold 1926). (6) Eastern Germany, April (Haensel 1967; see also Eck 1985a). (7) Breeders Bonn area, Germany (Niethammer 1962; see also Niethammer 1971). Czechoslovakia, all year: (8) Havlín and Havlínová 1974, which see for monthly averages), (9) (Pikula 1973). Netherlands: (10) September–November, (11) December–February, (12) March–April, (13) May–August (RMNH, ZMA). (14) France (Chavigny and Mayaud 1932). (15) South-east France, winter (G Olioso). *F. c. caucasica.* (16) Turkey, mainly May–July (Kumerloeve 1961, 1964a, 1967a, 1969b; Rokitansky and Schifter 1971). *F. c. syriaca* and nominate *coelebs*. (17) Cyprus (Flint and Stewart 1992). *F. c. alexandrovi.* (18) Northern Iran, February–July (Paludan 1940; Schüz 1959). *F. c. africana.* (19) Haut Atlas (Morocco), September (Grant 1979). *F. c. canariensis.* (20) Tenerife (Canary Islands), February and August (Grant 1979). *F. c. palmae.* (21) La Palma (Canary Islands), August (Grant 1979). *F. c. moreletti.* Azores: (22) Pico and Faial (Chavigny and Mayaud 1932), (23) São Miguel, August (Grant 1979), (24) Pico, São Jorge, Faial, and Flores, May (G Le Grand).

| | ♂ | | | | ♀ | | | |
|---|---|---|---|---|---|---|---|---|
| (1) | 23·7 (— ; | 41) | 20–28 | | 22·1 (— ; | 13) | 20–24·4 |
| (2) | 23·2 (— ; | 3) | — | | 21·5 (1·07; | 17) | — |
| (3) | 22·7 (— ; | 27) | 18·5–26·4 | | 20·6 (— ; | 5) | 19·9–21·3 |
| (4) | 21·2 (1·62; | 1238) | | | 18·5 (1·42; | 1477) | |
| (5) | 21·8 (— ; | 10) | 19–24 | | 20·2 (— ; | 15) | 15–23 |
| (6) | 25·4 (— ; | 379) | 19·8–31·2 | | 23·0 (— ; | 75) | 18·0–26·6 |
| (7) | 22·7 (1·38; | 103) | 19–26 | | 21·7 (1·42; | 39) | 19–25 |
| (8) | 23·9 (1·59; | 153) | 20–30 | | 22·6 (1·64; | 87) | 18–27 |
| (9) | 24·0 (1·7; | 289) | 19·8–30·0 | | 23·2 (2·9; | 173) | 20·0–40·0 |
| (10) | 22·1 (2·88; | 48) | 15·4–27·0 | | 18·9 (2·67; | 29) | 15·6–25·0 |
| (11) | 24·6 (3·72; | 19) | 16·2–30·0 | | 20·7 (3·07; | 9) | 16·5–24·0 |
| (12) | 21·7 (3·30; | 25) | 16·5–29·0 | | 18·5 (2·87; | 31) | 15·0–23·7 |
| (13) | 23·1 (2·14; | 8) | 20·0–26·0 | | 19·7 (2·84; | 6) | 14·7–22·7 |
| (14) | 23·6 (1·20; | 10) | 22·0–25·8 | | 20·3 (1·47; | 6) | 18·3–21·6 |
| (15) | 23·1 (1·80; | 87) | 19–27 | | 21·0 (1·21; | 45) | 18–24·5 |
| (16) | 22·8 (2·12; | 8) | 20–26 | | 20·1 (— ; | 2) | 20–20·2 |
| (17) | 22·6 (— ; | 35) | 20–26 | | 21·1 (— ; | 193) | 16–26 |
| (18) | 22·8 (1·02; | 4) | 21·9–24 | | 23·7 (2·73; | 4) | 19·8–26 |
| (19) | 23·4 (0·95; | 14) | — | | 22·0 (2·08; | 8) | — |
| (20) | 26·1 (1·55; | 4) | — | | 23·2 (1·43; | 6) | — |
| (21) | 27·5 (1·50; | 20) | — | | 25·7 (— ; | 19) | — |
| (22) | 24·6 (2·47; | 6) | 22·1–29 | | 23·0 (1·54; | 7) | 21·1–25·4 |
| (23) | 24·2 (1·99; | 86) | — | | 21·5 (1·74; | 88) | — |
| (24) | 26·2 (1·59; | 22) | — | | 23·8 (1·68; | 13) | — |

Nominate *coelebs*. Sweden, ♂: on arrival, March and early April, 26–29; later in April and in May 22–26 (Zedlitz 1926). Lake Chany (south-west Siberia), August–September: ♂♂ 22·1, 23·1 (Havlín and Jurlov 1977). Autumn migrants eastern Baltic: 22·0 (*n* = 5352) (Dol'nik and Blyumenthal 1967). Migrants Algeria (35·5°N), October–November: 20·7 (0·7; 9) 19·8–21·7 (Bairlein 1988). For premigration fattening, see Dol'nik and Blyumenthal (1967) and Dol'nik (1975).

*F. c. gengleri.* Exhausted birds: Skomer (Wales), January, 12·4 (1·38; 4) 11–14 (Harris 1962). See also MacDonald (1962, 1963) and Ash (1964). For New Zealand, see Robertson *et al.* (1983).

*F. c. sarda.* Sardinia, March: ♀ 21 (Demartis 1987).

*F. c. spodiogenys.* Tunisia, May: ♂ 25·8, ♀ 26·5 (G O Keijl).

*F. c. maderensis.* Madeira: ♂ 23 (Grant 1979).

**Structure.** Wing rather long, broad at base, tip rounded, 10 primaries: in nominate *coelebs*, *caucasica*, and *syriaca* p7 longest, p8 0–0·5(–1) shorter, p9 1–5, p6 0–1(–2), p5 4–9, p4 11–18, p3 15–22, p2 18–25; p1 24·7 (10) 23–27 in ♂ nominate *coelebs*, 21·0 (10) 19–23 in ♀, 21·3 (20) 17–25 in both sexes of *caucasica* and *syriaca*; *spodiogenys* group as nominate *coelebs*, but p5 3–8 shorter than p7, p4 10–17, p3 14–20, p2 16–22, p1 20·6 (10) 17–24; in *gengleri*, *balearica*, and *tyrrhenica*, p7 longest, p8 0–1 shorter, p9 2–6, p6 0–3, p5 4–8, p4 12–16, p1 20·3 (36) 17–24; in *sarda*, p7 longest, p6 and p8 0–1 shorter, p9 3–6, p5 3–7, p4 8–13, p1 18·5 (7) 15–21; in *canariensis* group, p6 longest, p7 0–1 shorter, p8 1–3, p9 3–8, p5 1–5, p4 6–11, p3 9–15, p2 11–17, p1 15·2 (12) 13–18. Tip of p9 falls between p4 and p5 in *canariensis* group, between p5 and p6 in *spodiogenys* and nominate *coelebs* groups; tip of p5 to tip of p9 4·5 (67) (1–)2·5–8 in continental Europe and Balearic Islands, 4·3 (23) 3–6·5 (Marle and Hens 1938) or 2·3 (20) 0·5–4 (ZMA) in Britain, 2·9 (7) 1–5 on Cyprus (ZMA), but only 1·7 (30) 0–3·5 on Sardinia (Marle and Hens 1938; Bezzel 1957; ZMA) and 2·0 (10) 0·5–4·5 in northern Turkey (ZMA). Wing-tip to outermost secondary (s1) of ♂ 25·6 (33) 24–28 in Sweden, 25·0 (100) 22–29 in eastern Germany, 23·4 (21) 20–26 in Morocco, and 18·2 (51) 16–21 in races of Atlantic islands (Bährmann and Eck 1975; Eck 1975). For sexual difference in wing-shape, see Mlíkovský (1982). P10 strongly reduced, a tiny pin, concealed by reduced outermost upper primary covert which is 1–4 mm longer; p10 in nominate *coelebs* 58–67 shorter than p7, 9–13 mm shorter than longest upper primary covert. Outer web of (p5–)p6–p8 emarginated; in nominate *coelebs*, p5 often without emargination, but sometimes a distinct one; in Britain and Mediterranean basin generally distinct on p5, in races of Atlantic islands exceptionally also on p4. Inner web of (p6–)p7–p9 with (generally faint) notch. Tip of longest tertial reaches to tip of p2-p3 when plumage fresh. Tail rather long, tip slightly forked; 12 feathers, t5 longest, t1 4·6 (10) 3-7 shorter than t5, t6 0-2 shorter. Bill rather short, less than half head length, swollen at base; culmen slightly decurved, more so at tip, often ending in fine hook; cutting edges virtually straight, apart from faint 'teeth' at base of upper mandible; tip of bill slightly compressed laterally. Nostril small, oval, covered by small tuft of feathers projecting from base of upper mandible; some short but distinct bristles over nostril and lateral base of upper mandible. For depth of bill at base, see Measurements; width of bill at base about similar to depth of bill at base. Tarsus and toes rather short, slender. Middle toe with claw 17·3 (15) 16·5-18·0; outer toe with claw *c.* 74% of middle with claw, inner *c.* 68%, hind *c.* 79%; for toes and claws, see also Pikula (1973).

**Geographical variation.** Marked and complex. 3 distinct groups of races, differing especially in pattern of head and colour of upperparts of ♂: *coelebs* group in western Eurasia and Middle East, *spodiogenys* group in North Africa, and *canariensis* group on Atlantic islands. Each group probably better considered a separate species, forming a superspecies or species-group together with Brambling *F. montifringilla* and Blue Chaffinch *F. teydea* (Eck 1975). Some characters link these 5 taxa closely, e.g. uniform top and side of head and neck with contrasting pale nape-patch of ♂♂ in *spodiogenys* group are shared with *F. montifringilla*, and *teydea* closer in morphological characters to *canariensis* group than latter is to *coelebs* group; genetic difference between *canariensis* and *coelebs* groups is large, but that between *spodiogenys* and *coelebs* group less so (Baker *et al.* 1990*a*, *b*). Probably best to give all 5 same rank as a full species within species-group, but traditional view of Hartert (1921-2) followed here, in which *spodiogenys*, *canariensis*, and *coelebs* groups are considered to belong to single species. For possible evolutionary history on Atlantic islands, see Grant (1979, 1980).

♂♂ of *coelebs* group characterized by vinous, rufous, or cinnamon ear-coverts and cheeks, contrasting with black lore and with blue-grey crown and side of neck, similar in colour to or slightly darker than underparts; also, lower mantle and inner scapulars various shades of brown, back to upper tail-coverts green. In comparisons given below, only birds from May-June used, as colour strongly affected by wear. Within *coelebs* group, 2 subgroups: one with shorter wing and more rounded wing-tip with races in Britain, Ireland, Iberia, Balearic Islands, Corsica, Sardinia, and northern Iran, and another with longer and more pointed wing, occurring from central and northern Europe east to western Siberia, Middle East, Balkans, and Italy. Within longer-winged group, 3 distinct races: nominate *coelebs* in Fenno-Scandia and north European Russia (mantle of ♂ umber- or tawny-brown, rump bright green, side of head deep rufous-cinnamon with vinous cast, chin to chest vinous-cinnamon, belly pink-vinous), *syriaca* from Cyprus and Levant (mantle of ♂ paler tawny- or orangey-brown, rump yellowish-green, side of head deep vinous-cinnamon, underparts uniform pale pink-mauve or vinous-pink apart from white mid-belly to under tail-coverts; upperparts of ♀ paler drab-brown than in ♀ nominate *coelebs*, stripes of cap paler brown, underparts with restricted light ash-grey tinge, less dark and extensive olive-drab), and *solomkoi* of Crimea and south-west slope of Caucasus mountains (like nominate *coelebs*, but larger, bill markedly heavier, feather-tips of mantle of ♂ with rather restricted dull umber-brown, partly mixed green, rump duller and more restricted olive-green, underparts slightly paler vinous-pink); *solomkoi* probably also on northern slopes of Caucasus (Dementiev and Ptushenko 1939). Other races of long-winged subgroup less well-defined: *caucasica* from southern slope of Caucasus (except western end), Tran-

scaucasia, north-west Iran, and eastern and northern Turkey intermediate in bill size and colour of underparts between *solomkoi* and *syriaca*, upperparts nearer to *solomkoi*. Birds from Balkans and Greece (including Crete) sometimes included in *balearica* (e.g. Harrison 1934, Niethammer 1943), in *caucasica* (e.g. Matvejev and Vasić 1973), or, on Crete, separated as *schiebeli* Stresemann, 1925; upperparts, underparts, and wing length rather like *caucasica* or intermediate between nominate *coelebs* and *caucasica*, bill rather long, like *caucasica*, but slender at base, like nominate *coelebs*; here included in nominate *coelebs*. Birds from central Europe (France through Netherlands, Germany, and Switzerland to Poland, Czechoslovakia, and Hungary) sometimes separated as *hortensis* C L Brehm, 1831: chin to chest of ♂ sometimes more tawny-cinnamon than rich pink-vinous of ♂ from typical nominate *coelebs* from Fenno-Scandia, belly more vinous-cinnamon, less pink-vinous, but many indistinguishable from Fenno-Scandian birds (e.g. 69% of 45 breeders from Netherlands), contra Harrison (1947*a*, *b*) and Bährmann and Eck (1975); also, a few from Fenno-Scandia are as cinnamon below as birds from central Europe; bill length and depth of birds from central Europe on average slightly larger than in Fenno-Scandia, but marked individual variation everywhere and bill length varies with time of year (slightly longer in summer). Birds from western Siberia (sometimes separated as *wolfgangi* Teplouchov, 1921) average slightly larger and paler than nominate *coelebs* from Fenno-Scandia, but much overlap in size and colour (Johansen 1944). *F. c. transcaspia* of north-east Iran (east of *alexandrovi*) and south-west Turkmeniya is rather similar to *alexandrovi* (see below), but larger and paler (Vaurie 1949*b*). Of short-winged races of *coelebs* group, *gengleri* from Britain and Ireland is distinct (Hens 1931; Hens and Marle 1933; Harrison 1934, 1947*a*, *b*; Clancey 1946); similar to nominate *coelebs* on upperparts, or mantle of ♂ very slightly darker brown, more mixed with green; side of head of ♂ rufous-cinnamon or ochre-brown, less tinged vinous than ♂ nominate *coelebs*, underparts distinctly darker and more uniform in most birds, rufous-cinnamon or tawny-cinnamon, sometimes paler and tinged vinous-cinnamon on belly, but no vinous on throat; some slight variation between various populations (see, e.g., Clancey 1938, 1940, 1943, 1945, 1946; Marle 1949; Meinertzhagen 1953), but too slight to warrant recognition of *scotica* Harrison, 1937 (south-west Scotland) or *hibernicus* Van Marle, 1949 (south-west Ireland) (C S Roselaar). For birds introduced to South Africa, see Wattel (1971); for New Zealand, see Niethammer (1971) and Baker *et al.* (1990*a*). *F. c. balearica* from Portugal, Spain, and Balearic Islands similar in size to *gengleri*; upperparts and side of head of ♂ as in nominate *coelebs*, chin to chest pale vinous-cinnamon or pink-vinous, belly vinous-pink, mid-belly extensively white; underparts paler than nominate *coelebs*, but rather similar to *caucasica*; difference from

PLATE 29. *Fringilla coelebs* Chaffinch (p. 448). Nominate *coelebs*: **1** ad ♂ summer, **2** ad ♂ winter, **3** ad ♀ summer, **4** ad ♀ winter, **5** juv. *F. c. gengleri*: **6** ad ♂ summer. *F. c. solomkoi*: **7** ad ♂ summer. *F. c. africana*: **8** ad ♂ summer, **9** ad ♀ summer. *F. c. maderensis*: **10** ad ♂ summer. *F. c. canariensis*: **11** ad ♂ summer. *F. c. palmae*: **12** ad ♂ summer. *F. c. moreletti*: **13** ad ♂ summer.

*Fringilla teydea* Blue Chaffinch (p. 447). Nominate *teydea*: **14** ad ♂ summer, **15** ad ♀ summer, **16** juv. *F. t. polatzeki*: **17** ad ♂ summer.

*Fringilla montifringilla* Brambling (p. 479): **18** ad ♂ summer, **19** ad ♀ summer, **20** ad ♂ winter, **21** ad ♀ winter, **22** juv. (CR)

PLATE 30. *Ploceus manyar flaviceps* Streaked Weaver (p. 401): **1–2** ad ♂ breeding, **3** ad ♂ non-breeding, **4** ad ♀ breeding, **5** juv. *Lagonosticta senegala* Red-billed Firefinch (p. 411): **6–7** ad ♂, **8–9** ad ♀, **10** juv. (CR)

PLATE 31. *Estrilda astrild jagoensis* Common Waxbill (p. 420): **1–2** ad ♂, **3** ad ♀, **4** juv. *Amandava amandava amandava* Red Avadavat (p. 427): **5** ad ♂, **6–7** ad ♀, **8** juv. *Euodice malabarica* Indian Silverbill (p. 437): **9** ad. *Euodice cantans cantans* African Silverbill (p. 437): **10–11** ad, **12** juv. (CR)

PLATE 32. *Vireo flavifrons* Yellow-throated Vireo (p. 440): **1** ad. *Vireo philadelphicus* Philadelphia Vireo (p. 442): **2** ad dark, **3** ad light. *Vireo olivaceus* Red-eyed Vireo (p. 444): **4** ad autumn, **5** 1st ad autumn. (CR)

PLATE 33. *Serinus pusillus* Red-fronted Serin (p. 499): **1** ad ♂ summer, **2** juv. *Serinus serinus* Serin (p. 508): **3** ad ♂ summer, **4** ad ♂ winter, **5** ad ♀, **6** juv. *Serinus syriacus* Syrian Serin (p. 521): **7** ad ♂ summer, **8** ad ♀ summer, **9** juv. (CR)

PLATE 34. *Serinus canaria* Canary (p. 528): **1** ad ♂ summer, **2** ad ♀ summer, **3** juv, **4** hybrid with Goldfinch *Carduelis carduelis*, **5** captive-bred variant. *Serinus citrinella* Citril Finch (p. 536). Nominate *citrinella*: **6** ad ♂ summer, **7** ad ♀ summer, **8** 1st ad, **9** juv. *S. c. corsicana*: **10** ad ♂ summer. (CR)

PLATE 35. *Carduelis chloris* Greenfinch (p. 548). Nominate *chloris*: **1** ad ♂ summer, **2** ad ♂ winter, **3** ad ♀ summer, **4** juv ♂, **5** juv ♀. *C. c. aurantiiventris*: **6** ad ♂ summer. *C. c. chlorotica*: **7** ad ♂ summer, **8** ad ♂ winter. *C. c. bilkevitchi*: **9** ad ♂ summer. (CR)

PLATE 36. *Carduelis carduelis* Goldfinch (p. 568). Nominate *carduelis*: **1** ad ♂ summer, **2** ad ♂ winter, **3** ad ♀ summer, **4** 1st ad ♂ summer, **5** juv. *C. c. britannica*: **6** ad ♂ summer. *C. c. parva*: **7** ad ♂ summer. *C. c. niediecki*: **8** ad ♂ summer. *C. c. loudoni*: **9** ad ♂ summer.(CR)

PLATE 37. *Carduelis spinus* Siskin (p. 587): **1** ad ♂ summer, **2** ad ♂ winter, **3** ad ♀ summer, **4** 1st ad ♀ winter, **5** juv. (CR)

PLATE 38. *Fringilla coelebs coelebs* Chaffinch (p. 448): **1** ad ♂ summer, **2** ad ♀ summer. *Fringilla teydea teydea* Blue Chaffinch (p. 474): **3** ad ♂ summer. *Fringilla montifringilla* Brambling (p. 479): **4** ad ♂ summer, **5** ad ♀ winter. *Serinus pusillus* Red-fronted Serin (p. 499): **6** ad ♂ summer. *Serinus serinus* Serin (p. 508): **7** ad ♂ summer, **8** juv. *Serinus syriacus* Syrian Serin (p. 521): **9** ad ♂ summer. (CR)

PLATE 39. *Serinus canaria* Canary (p. 528): **1** ad ♂ summer. *Serinus citrinella citrinella* Citril Finch (p. 536): **2** ad ♂ summer. *Carduelis chloris chloris* Greenfinch (p. 548): **3** ad ♂ summer, **4** ad ♀ summer. *Carduelis carduelis carduelis* Goldfinch (p. 568): **5-6** ad ♂ summer. *Carduelis spinus* Siskin (p. 587): **7** ad ♀ summer. (CR)

PLATE 40. *Carduelis cannabina* Linnet (p. 604). Nominate *cannabina*: **1** ad ♂ summer, **2** ad ♂ winter, **3** ad ♀, **4** juv. *C. c. autochthona*: **5** ad ♂ winter. *C. c. bella*: **6** ad ♂ summer. *C. c. guentheri*: **7** ad ♂ summer. *C. c. harterti*: **8** ad ♂ summer. (CR)

PLATE 41. *Carduelis flavirostris* Twite (p. 625). Nominate *flavirostris*: **1** ad ♂ summer, **2** ad ♂ winter, **3** ad ♀ summer, **4** juv. *C. f. pipilans*: **5** ad ♂ winter. *C. f. brevirostris*: **6** ad ♂ winter. (CR)

nominate *coelebs* in colour often slight (e.g. Marle and Hens 1938, Jordans 1950, Niethammer 1957), but clear difference in size (see Measurements). *F. c. sarda* from Sardinia and *tyrrhenica* from Corsica both short-winged, like *gengleri* and *balearica*, wing-tip of (especially) *sarda* strongly rounded, mantle rather dull brown, often tinged green, rump dull olive-green; bill of both heavy at base, width and depth at base *c*. 8 mm in ♂ (*c*. 7 in *gengleri* and *iberiae*); underparts of ♂ *tyrrhenica* largely uniform deep pink-vinous, contrasting with rufous-cinnamon of side of head, underparts of *sarda* mainly tawny-cinnamon like side of head, tinged vinous on belly; underparts of ♂ *tyrrhenica* rather like *caucasica*, but vinous deeper; underparts of *sarda* rather like *gengleri*, but more ochre, less rufous. *F. c. alexandrovi* from southern shore of Caspian Sea in northern Iran also short-winged; mantle rather dull brown, rump dull olive-green, side of head and underparts vinous-brown.

In *spodiogenys* group of North Africa, crown and nape of ♂ medium blue-grey (in *africana*, occurring Morocco to north-west Tunisia) or light blue-grey (in *spodiogenys* of eastern and central Tunisia); nape with white spot; band on forehead as well as lore contrastingly black; side of head down to upper cheek and side of neck bluish ash-grey with contrasting white eye-ring (latter broken by black in front and behind); mantle and inner scapulars bright green, forming contrasting saddle, back and rump green, mixed green and blue-grey, or (occasionally) almost fully blue-grey, outer scapulars and upper tail-coverts blue-grey; lower cheek and chin down to upper belly and side of belly pale vinous-pink (*africana*) or pink-white (*spodiogenys*), side of breast and flank with restricted ash-grey, mid-belly to under tail-coverts white; tail more extensively white than in *coelebs* group, t5-t6 white except black terminal third of outer web, black shaft, and extreme base, t4 black with white wedge 25-45 mm long on inner web, t3 often with white spot or wedge up to 30 mm long, t2 sometimes with small white subterminal spot; wing as *coelebs* group, but basal half of outer web of longest tertial and basal and middle portion of inner 2-3 secondaries extensively white, and white tips of greater coverts longer, innermost sometimes all-white. ♀♀ of *spodiogenys* group markedly pale, light drab-grey on upperparts and side of head and neck, paler ash-grey on middle of crown and nape, on rear of supercilium, and on side of neck; tinged green on mantle and scapulars, more extensively bright green on rump; underparts off-white, tinged grey on side of breast and flank; t6 extensively white, t4-t5 with white wedge, t3 sometimes with white spot; bases of inner 2 secondaries white; darker birds rather like pale ones of *syriaca* of *coelebs* group. Much geographical variation within *spodiogenys* group, colour gradually darker from eastern Tunisia to north-west Morocco, size larger in Atlas Saharien of Algeria (wing on average *c*. 3 mm longer than elsewhere), smaller in Tanger and Er Rif area of northern Morocco (wing on average *c*. 3 mm shorter than elsewhere: see Grant 1979 for details), but only medium-sized pale birds separated as *spodiogenys* and all others united in *africana*,

following Meinertzhagen (1940) and Vaurie (1956a, 1959), though birds from Atlas Saharien in eastern Algeria scarcely darker than *spodiogenys* from Tunisia. Birds from Cyrenaica (northern Libya) have longer bill than those from remainder of North Africa, colour near to *africana*, but underparts tinged yellow-buff instead of white, and white nape-patch large (Stanford 1954); here included in *africana*, following Vaurie (1959), but perhaps separable.

In ♂ *canariensis* of *canariensis* group, occurring Gran Canaria, Tenerife, and Gomera (Canary Islands), upperparts largely deep slate-blue, black bar on forehead scarcely contrasting, but rump contrastingly bright green; in fresh plumage, feather-tips of upperparts broadly olive-brown, largely concealing plumbeous (especially in 1st adult). Pattern on side of head differs strongly from *coelebs* and *spodiogenys* groups: lore, broad ring round eye, and front part of cheek ochre or orangey-buff, contrasting with dark blue-grey remainder of side of head and neck. Chin to upper belly and side of belly ochre, grading into cream-white or white mid-belly, vent, and under tail-coverts; belly tinged vinous; side of breast and lower flank rather contrastingly grey. Tail with less white than *coelebs* group, white wedge on inner web of t6 20-40 mm long, on t5 5-35 mm, on t4 short or absent; white on wing less extensive also, white on bases of flight-feathers and on tips of greater coverts more restricted, less white on tips of lesser coverts, forewing with broad white band formed by median coverts rather than largely white 'shoulder'; fringes of flight-feathers and tertials deeper and broader greenish-yellow. ♀ dark olive-brown on upperparts with bright green rump when fresh, dark brown-grey with green tinge when worn; almost without darker streaks on cap and nape; side of head, neck, throat, and breast light olive-brown or brown-grey, tinged ochre or buff on lore, eye-ring, and forecheek, washed ochre or buff on chin, throat, and (sometimes) mid-chest to belly; white on tail and wing and fringes of wing as in ♂. *F. c. palmae* of La Palma (western Canary Islands) similar to *canariensis*, but rump of ♂ dark blue-grey, not green; belly and upper flank white, contrasting sharply with ochre chest; larger, especially bill; ♀ less green on rump than ♀ *canariensis*, belly more extensively pure white. *F. c. ombriosa* of Hierro (western Canary Islands) like *palmae*, but rump with traces of green and underparts slightly less extensively white, somewhat tending towards *canariensis*. For variation within Canary Islands, see also Lack and Southern (1949), Grant (1976, 1979), and Baker *et al.* (1990b). *F. c. maderensis* from Madeira like *canariensis*, but nape and top of head of ♂ dark blue-grey, less deep plumbeous than in races of Canary Islands, black band on forehead contrasting more sharply; lower mantle, scapulars, back, and rump bright green, outer scapulars and upper tail-coverts blue-grey with green wash; face and chin to throat ochre or warm buff, as in *canariensis* or slightly paler, merging into pale vinous-pink of upper belly and side of belly; side of breast and flank rather more extensively grey; size as in *canariensis* but bill slightly longer, more slender at base; tail and wing as *canariensis*. ♀ *maderensis* like ♀ *canariensis*, but cap and nape sometimes with more distinct sepia stripes and mantle and scapulars greener if fresh. *F. c. moreletti* of Azores like *maderensis*, but bill markedly longer and heavier at base (near bill of Blue Chaffinch *F. teydea*), upperparts more extensively green, chest browner, belly less vinous; amount of white in tail markedly reduced, largely replaced by light grey (as in *F. teydea*); for variation between islands, see Hartert and Ogilvie-Grant (1905) and Grant (1979). CSR

PLATE 42. *Carduelis flammea* Redpoll (p. 639). Nominate *flammea*: **1** ad ♂ summer, **2** ad ♂ winter, **3** ad ♀ summer, **4** ad ♀ winter, **5** 1st ad ♂ winter, **6** juv. *C. f. cabaret*: **7** ad ♂ summer, **8** ad ♂ winter, **9** ad ♀ summer, **10** 1st ad ♂ winter, **11** juv. *C. f. islandica*: **12** 1st ad ♂ winter. *C. f. rostrata*: **13** 1st ad ♂ winter.

*Carduelis hornemanni* Arctic Redpoll (p. 661). *C. h. exilipes*: **14** ad ♂ summer, **15** ad ♂ winter, **16** ad ♀ summer, **17** 1st ad ♂ winter, **18** juv. Nominate *hornemanni*: **19** ad ♂ winter, **20** 1st ad ♂ winter. (CR)

## *Fringilla teydea* **Blue Chaffinch**

PLATES 29 and 38 (flight)
[between pages 472 and 473]

Du. Blauwe Vink    Fr. Pinson bleu    Ge. Teydefink
Ru. Голубой зяблик    Sp. Pinzón azul    Sw. Kanariebofink

*Fringilla teydea* Webb, Berthelot, and Moquin-Tandon, 1841

Polytypic. Nominate *teydea* Webb, Berthelot, and Moquin-Tandon, 1841, Tenerife (Canary Islands); *polatzeki* Hartert, 1905, Gran Canaria (Canary Islands).

**Field characters.** 16·5 cm; wing-span 26·5–31·5 cm. Up to 10% larger than Chaffinch *F. coelebs*, with similar silhouette but noticeably heavier and more robust, with distinctly longer bill. Plumage closely resembles *F. coelebs* but relatively more uniform: ♂ bluish-grey, uniform at distance but showing dull wing marks echoing those of *F. coelebs* at close range. ♀ and juvenile dusky-olive, with brighter wing-bars. Sexes dissimilar; little seasonal variation. Juvenile separable. 2 races, ♂♂ separable.

(1) Tenerife race, nominate *teydea*. ADULT MALE. Moults: January–June (complete); wear intensifies blue plumage. Head, back, rump, upper tail-coverts, and underparts to rear flanks and belly dusky blue-grey, wearing to rather bright leaden-blue, these areas marked only by narrow black band over base of bill, white upper and lower eye-crescents, and paler, whitish chin and throat. Wings black, far less strikingly marked than *F. coelebs* but showing, at close range, broad bluish-white tips to median coverts (forming upper wing-bar), pale bluish-grey fringes and tips to greater coverts (forming less distinct lower wing-bar), and pale bluish-grey fringes to secondaries and tertials, on tertials wide enough to show as paler lines; tiny white spot on base of primaries near primary coverts may just be visible. Markings difficult to distinguish at any distance, and flickering pattern of *F. coelebs* thus not apparent in flight. Tail dull black, with broad grey-blue edges on central pair of feathers and narrow grey-blue or dull white ones on others; thus lacks bold pattern of *F. coelebs*. Rear underparts greyish-white, becoming fully white under tail. Bill light blue with black tip in breeding season; duller, more horn-grey at other times. Legs grey-brown. ADULT FEMALE. Plumage pattern as ♂ but ground-colour greyish-olive-brown above and brownish-ashy-grey below, becoming greyish-white on belly and under tail. With wear, becomes greyer above. Pale buff eye-ring. Tips to median and greater coverts whiter than ♂'s, forming stronger wing-bars. Bill horn-grey. JUVENILE. Resembles ♀ but browner above and buffier below, with even stronger wing-bars. (2) Gran Canarian race, *polatzeki*. Up to 10% smaller than nominate *teydea*. ♂ distinctly duller, being no brighter than ashy-grey but with broader, almost white tips to both median and greater coverts which form more obvious double wing-bar, and deeper black frontal band over base of bill. ♀ paler below than nominate *teydea*, especially on belly.

Unmistakable, since even noticeably blue-backed Canarian races of *F. coelebs* have full white wing-marks, pink breast (in ♂), and fully white belly. Flight much as *F. coelebs* but larger size evident in rather slower or more gentle undulations. Gait and general behaviour as *F. coelebs*; specific habits include persistent hawking of and searching of bark for insects, long periods of silent, motionless perching, and less gregarious behaviour, with parties not exceeding 10 birds. Far more restricted in habitat preference than *F. coelebs*, rarely leaving montane pine forest.

Song resembles *F. coelebs*, but 1st part frequently shorter and simpler, and terminal flourish often repeated. Commonest call 'whit-chooee' which may have either clear or chirping quality.

**Habitat.** Restricted to small forest areas on Canary Islands in subtropical eastern Atlantic, mainly above 1200 m, and as high as 1830 m in the most favourable areas on southern slopes. Virtually confined to forests of pine *Pinus* of varying types, lowest with tangle of undergrowth of tree-heath *Erica*, holly *Ilex*, and other shrubs, while at higher levels undergrowth is lacking, forest floor being strewn with pine needles on which much time is spent foraging for seeds and insects. Forages almost exclusively in pine forests at c. 1000–2000 m on Tenerife (slightly lower on Gran Canaria), occasionally descending to cultivated areas or even gardens at 500–700 m in adverse weather, and at other times in high-altitude scrubland. (Bannerman 1963; Collar and Stuart 1985; Martín 1987.) For details of feeding habitat, see Collar and Stuart (1985). During heat of day sits quietly on pine branches. Need for water and liking for bathing may lead to daily local journey of some distance. Does not fly high or far. Not even displaced by winter snowfalls. (Bannerman and Bannerman 1965.)

**Distribution and population.** Restricted to Tenerife and Gran Canaria in Canary Islands. Tenerife (nominate *teydea*): distribution dependent on pinewoods, formerly in continuous belt round the island, now reduced to isolated stands; reafforestation has resulted in slow but steady recovery locally. Gran Canaria (*polatzeki*): in clear decline and in need of protection, though in 1984 still present, but local, in Pinar de Tamadaba and Pinar de Pajonales. (Martín *et al.* 1984; Collar and Stuart 1985.)

**Movements.** Sedentary, making only very local movements in forests of pine *Pinus canariensis*. Ringed juveniles retrapped up to 3 km from nest area (Martín *et al.* 1984). Occurs at high altitude even at times of deep snow cover, though probably makes limited altitudinal movements then. Recorded occasionally above treeline, and exceptionally at some distance from usual range: 2 birds at Teror (580 m) in north of Gran Canaria, May 1910, well away from pinewoods, and vagrant (race undetermined) in north-west Lanzarote, 18 October 1967. (Bannerman 1963; Trotter 1970.)                                    DFV

**Food.** Seeds (principally Canarian pine *Pinus canariensis*) and invertebrates (Koenig 1890; Thanner 1910; Etchécopar and Hüe 1957; Bannerman 1963). Seems to feed mainly in places where tall mature trees are closely spaced and understorey fairly low, average height 1·6 m (Carrascal 1987, which see for comparison with other species). Spends much time foraging on ground taking mainly pine seeds, but also seeds of broom *Spartocytisus supranubius*, forget-me-not *Myosotis*, and chickweed *Stellaria media*; picks up pine seeds loose, or extracts them from open cones, and dehusks them; invertebrates, especially beetles (Coleoptera), also taken on ground (Godman 1872; Meade-Waldo 1889*b*; Etchécopar and Hüe 1957; Martín *et al.* 1984; Carrascal 1987; Bergmann *et al.* 1988). Perhaps spends more time on forest floor in morning, in trees in afternoon, then on ground again in evening (Koenig 1890). In trees, searches crevices in pine bark for invertebrates, especially moths (Lepidoptera); starts at top of tree and examines each branch from base to tip before flying down to next one; also hangs from open cones to remove seeds (Meade-Waldo 1889*b*; Bannerman 1963; Pérez Padrón 1981; Carrascal 1987). Commonly hawks after Lepidoptera; said to prefer Lepidoptera or grasshoppers (Acrididae) to seeds in June (Meade-Waldo 1889*b*; Bannerman 1963; Martín *et al.* 1984). Often moves through trees in small groups, and in winter sometimes in mixed flocks with Chaffinch *F. coelebs* (Meade-Waldo 1889*a*; Lack and Southern 1949). Said to break up pine cones with bill like crossbill *Loxia*, and large, pointed, slightly laterally compressed bill apparently adapted both for extracting seeds from cones, rather like Goldfinch *Carduelis carduelis*, and for crushing them (Godman 1872; Volsøe 1951). However, see Meade-Waldo (1889*b*), who says bird is incapable of breaking up cones.

Diet includes the following. Insects: grasshoppers, etc. (Orthoptera: Tettigoniidae, Acrididae), ant-lions (Neuroptera: Myrmeleontidae), adult and larval Lepidoptera (Noctuidae, Lymantriidae, Sphingidae), beetles (Coleoptera: Heteroceridae, Tenebrionidae, Curculionidae). Plants: seeds of pine *Pinus*, escobon (broom) *Chamaecytisus proliferus*, retama (broom) *Spartocytisus supranubius*, bean-trefoil *Adenocarpus viscosus*, bramble *Rubus*, forget-me-not *Myosotis*, chickweed *Stellaria*, and perhaps fruit of myrtle *Myrica faya*. (Godman 1872; Meade-Waldo

1889*b*; Koenig 1890; Thanner 1903, 1910; Bannerman 1963; Pérez Padrón 1981; Martín *et al.* 1984; Collar and Stuart 1985; Naurois 1986.)

Young fed mainly adult and larval insects (many caterpillars being taken from broom *Chamaecytisus* understorey), and pine seeds softened in crop of adult (Thanner 1903; Collar and Stuart 1985; Naurois 1986). Insects include bush crickets (Tettigoniidae), ant-lions, beetles (Tenebrionidae, Curculionidae), and Lepidoptera (Sphingidae, Lymantriidae, Noctuidae) (Martín *et al.* 1984; Collar and Stuart 1985).                                         BH

**Social pattern and behaviour.** Little known, and that mostly from basic observations around 1900, but valuable recent study of nominate *teydea* by Martín *et al.* (1984). For reviews, see Bannerman (1963), Collar and Stuart (1985), Ruelle (1987), and Bergmann *et al.* (1988). Behaviour of nominate *teydea* and *polatzeki* considered identical (Polatzek 1909; Thanner 1910), but no detailed studies.

1. Relatively gregarious outside breeding season, forming small (up to 10 birds), loose, roving flocks, especially of younger birds, which break up at start of breeding season; flocks sometimes associate with other species, especially Chaffinch *F. coelebs* (Meade-Waldo 1889*a*; Koenig 1890; Lack and Southern 1949; Bannerman 1963). Dispersion in pairs seems to begin in February (see Heterosexual Behaviour, below). BONDS. No evidence for other than monogamous mating system. No information on duration of pair-bond, but observation of presumed pairs within flocks in winter (see Heterosexual Behaviour, below) suggests that, as in *F. coelebs*, pair-bond lasts for much of year and perhaps from year to year. Ruelle (1987) reports a hybrid resulting from captive breeding of ♂ *F. teydea* and ♀ *F. coelebs*, but none known from the wild. ♀ alone builds nest, incubates, and almost exclusively feeds nestlings (Martín *et al.* 1984) (see similar, if less extreme, provisioning bias of ♀ in *F. coelebs*); however, other sources, perhaps quoting Thanner (1903), suggest that both parents feed young; provisioning role of sexes not known after fledging. No information on age at independence but young remain together till late autumn (Thanner 1903), this being interpreted by Collar and Stuart (1985) and by Ruelle (1987) to mean that the 'family' stays together till late autumn. Age of first breeding 1 year, though 1-year-olds breed much later in season than older birds (Thanner 1903). BREEDING DISPERSION. Solitary and territorial. Martín *et al.* (1984) estimated territory size at *c.* 3 ha (in pinewood *Pinus canariensis* with understorey of *Chamaecytisus*) by putting ♂ in cage at variable distances from nest, trapping territory-owner when he approached caged bird, and summing area enclosed by successful trappings. In late March–early April, Tenerife, 67 birds per km² in *P. radiata*, 25 in *P. canariensis* (Carrascal 1987). For usurping nest of Canary *Serinus canaria* see Antagonistic Behaviour (below). ROOSTING. No information. Observers frequently report extended periods of daytime (especially midday) loafing, birds perching quietly in trees for long spells between feeding bouts (Thanner 1910; Bannerman 1912; Bergmann *et al.* 1988; Butler 1989). Fond of bathing (Bannerman 1963; Bergmann *et al.* 1988).

2. Very confiding, allowing close approach (Koenig 1890; Bergmann *et al.* 1988). Meade-Waldo (1889*b*) described pair entering tent to pick up seed, and to catch butterfly released from fingers. Runs rapidly on ground, sometimes raising crown feathers when excited or alarmed (Hüe and Etchécopar 1958; Bergmann *et al.* 1988). For alarm-call in response to overhead

hawk *Accipiter* (etc.) see 3 in Voice. FLOCK BEHAVIOUR. Little known, but see start of part 1 (above) for winter flocking. Drinking evidently an important part of feeding regime (Bergmann *et al.* 1988) and Koenig (1890) reports flocks coming to water to drink. SONG-DISPLAY. Given by ♂ from elevated perch. Sometimes heard on fine days in late winter (e.g. subdued but complete phrases, 22 February, Tenerife) but song not full before April-May, at which time serves to demarcate territory (Bergmann *et al.* 1988). Heard frequently near nest up to and including incubation, declining in nestling period, but not ceasing completely (Thanner 1903; Bannerman 1963). Given mostly in morning (Hemmingsen 1958). In good habitat, not unusual to hear more than one singer in same area (Ruelle 1987). ANTAGONISTIC BEHAVIOUR. No information on threat-display, but experiment to determine size of territory (see Breeding Dispersion) demonstrates active territorial defence by ♂ against encroaching rivals. One report of 2 ♂♂ fighting noisily (Bannerman 1963). Some observations suggest that, as in *F. coelebs*, ♀ may be dominant to ♂ during nesting: firstly, ♀ usually chased off ♂ accompanying her as she collected nest-material (Meade-Waldo 1889b); secondly, ♀ does almost all feeding of nestlings (see Bonds), suggesting that (as Marler 1956a found for *F. coelebs*) ♂ may be hesitant to enter nest-area. Martín (1985) reported possibly exceptional case in which nest of *S. canaria* was taken over (and its clutch presumed eaten) by *F. teydea* which laid her own eggs in, and built up, the nest. For possible factors ameliorating competition with sympatric *F. coelebs* see Martín *et al.* (1984). HETEROSEXUAL BEHAVIOUR. Very little known. Most sources suggest that pair-formation occurs March-May (Koenig 1890; Thanner 1903; Martín *et al.* 1984), but Bergmann *et al.* (1988) observed birds in pairs from end of February on Tenerife, while parties on Gran Canaria, early February, apparently consisted mainly of pairs (Bannerman 1911). According to Meade-Waldo (1889b) ♂ apparently chooses nest-site (this presumably the source for the same view quoted by Bannerman 1963 and Bergmann *et al.* 1988), but no evidence offered. ♀ builds nest alone; ♂ attends but plays no active role. Detailed sequence of nest-building described by Hemmingsen (1958): on Gran Canaria, ♀ seen building for much of day (15 May) with visits every 30 min

during midday heat, shortening to 2-10 min at 17.00 hrs; by this time ♀ accompanied on almost every nest-visit by ♂ who, however, carried no material but perched 2-3 m below nest while she incorporated material; ♂ sometimes became 'impatient' and flew to nest. After bout of building, pair flew off together, giving 2-part call (presumably 2 in Voice). Nest takes *c.* 3 weeks to complete (Bergmann *et al.* 1988). ♀ incubates assiduously, occasionally leaving to drink (Thanner 1903). According to Thanner (1903) (presumably the source quoted by Volsøe 1951, Bannerman 1963, and Bergmann *et al.* 1988), ♂ feeds ♀ on nest, but this must be doubted in absence of supporting evidence (also no courtship-feeding in *F. coelebs*). No further information. RELATIONS WITHIN FAMILY GROUP. Nestlings fed mainly or exclusively by ♀ (see Bonds), but no details. Older young beg like *F. coelebs* with sideways movements of head and frequent ruffling of head feathers, also with horizontal and vertical movements of tail (Thanner 1903: see also Voice). No further information.

EKD

**Voice.** Song not unlike Chaffinch *F. coelebs* but with simpler 'trill' and longer flourish (Slater and Sellar 1986). Most frequent call is 'whit-chooee' (call 2), which appears equivalent to 'chink' of *F. coelebs*, but varies considerably between localities. Study by Bergmann *et al.* (1988) was on Tenerife. Descriptions of calls given below are by P J B Slater and P J Sellar except as otherwise acknowledged.

CALLS OF ADULTS. (1) Song. As in *F. coelebs*, song consists of 'trill' followed by flourish (Fig I). However, 'trill' in *F. teydea* typically only of 1-2 segments and flourish often repeated, sometimes several times (can also be repeated in *F. coelebs*; see that species). Figs I-II show 2 similar songs from different sites on Tenerife, differing mainly in length of 'trill' (Fig II shows only part of 'trill'). Each ♂ usually has only one song-type (Slater and Sellar 1986; Bergmann *et al.* 1988). While song does vary between birds and between localities, constituent units

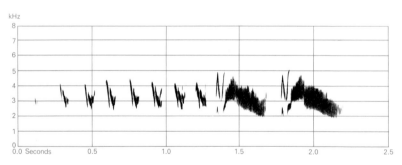

I  P J B Slater  Tenerife (Canary Islands) May 1984

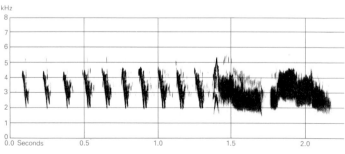

II  P J B Slater  Tenerife (Canary Islands)  May 1984

are less variable than in *F. coelebs*. 'Trill' units are usually very short and rapidly descending in pitch ('tik tik tik. . .'), while units in the flourish are long, and rise and fall in pitch ('whee-oo whee-oo'). (2) Whit-chooee call. Given by both sexes during breeding season in situations of mild alarm, and probably equivalent to call 2 of *F. coelebs*. Fig

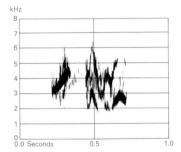

III P J Sellar Gran Canaria (Canary Islands) May 1991

III depicts call of ♀ on Gran Canaria: 1st unit ('whit') short and simple, with clear tonal quality; 2nd ('chooee') more complex and variable, often nasal and harsh with overtones (Bergmann *et al.* 1988), and frequently repeated. Exact form of units varies between birds and between localities (P J B Slater). Described as 'dju-delii' by Bergmann *et al.* (1988). On Tenerife, October, ♀ gave fruity, sparrow-like 'ip-truip' or 'truip'; ♂ a clear liquid 'p-lup p-lip', last unit higher pitched; repeated persistently (D J Brooks). (3) Seee-call. Thin, high-pitched 'seee' (Fig IV) which, as in *F. coelebs*, is produced in strong alarm, especially in response to overhead hawk *Accipiter*. (4) Rain-call. Rarely heard but possibly equivalent to Rain-call (call

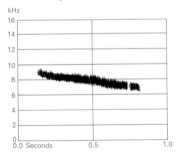

IV P J Sellar Tenerife (Canary Islands) May 1989

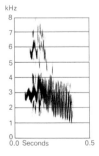

V P J Sellar Tenerife (Canary Islands) May 1989

5) of *F. coelebs*. Fig V depicts call rendered 'teeurb' given by ♂ from perch after chasing another ♂ (P J Sellar). (5) Flight-call. Short 'pit-pit' (Bergmann *et al.* 1988). (6) Short 'pie' of unknown significance also described by Bergmann *et al.* (1988).

CALLS OF YOUNG. Food-call said to be like that of Hawfinch *Coccothraustes coccothraustes* (Thanner 1903).

PJBS, PJS

**Breeding.** SEASON. Tenerife: late breeder, with snow in early spring not uncommon on breeding grounds at *c.* 1000–2000 m; nest-building starts end of May, eggs laid 2nd half of June, in south-facing areas possibly May (Thanner 1910; Bannerman 1963; Martín 1987); eggs recorded 8 June to 25 August (Volsøe 1951), some even later (Thanner 1903); nest-building observed July (Koenig 1890). Gran Canaria: perhaps a few weeks earlier because of lower altitude (*c.* 700–1200 m) (Bannerman 1912, 1963; Collar and Stuart 1985); nest-building observed at one site mid-May, slightly in advance of average starting time on Tenerife (Hemmingsen 1958). SITE. Almost always in pine *Pinus canariensis*, generally *c.* 10 m (1·5–20) above ground at end of thin branch, sometimes against trunk (Bannerman 1963; Martín 1987; Bergmann *et al.* 1988). Unusual nest recorded 6–7 m above ground against trunk of laurel *Laurus* at altitude of 700 m (Naurois 1986); see Volsøe (1951) and Bannerman (1963) for other sites. See also Martín *et al.* (1984) for eggs laid in nest of Canary *Serinus canaria*. Nest: foundation of twigs of pine, tree heath *Erica arborea*, or broom *Chamaecytisus*, herb stalks, pine needles, moss, lichen, and spiders' webs (last 2 materials giving many nests white appearance), lined with grass, plant down, hair, feathers, etc. (Meade-Waldo 1889*b*; Hemmingsen 1958; Bannerman 1963; Moreno 1988); one nest had outer diameter 12·4 cm, inner diameter 6·5 cm, overall height 6·5 cm, and depth of cup 4·7 cm (Koenig 1890). Building: by ♀ only; takes *c.* 3 weeks (Meade-Waldo 1889*b*; Thanner 1903; Hemmingsen 1958; Bannerman 1963; Martín *et al.* 1984). EGGS. See Plate 60. Sub-elliptical to long oval, smooth and slightly glossy; light blue or greenish-blue with chestnut or purplish speckles and streaks at broad end; faint undermarkings of violet or brownish-purple blotches. Larger and more brightly coloured than those of Chaffinch *F. coelebs*, and with little variation. (Bannerman 1963; Harrison 1975; Schönwetter 1984; Moreno 1988.) Nominate *teydea*: 23·8 × 16·5 mm (22·5–26·0 × 15·6–17·6), *n* = 46; calculated weight 3·37 g (Schönwetter 1984). Clutch: 2; apparently complete clutches of 1 sometimes recorded (Meade-Waldo 1889*b*; Thanner 1903; Bannerman 1963; Collar and Stuart 1985; Martín 1987). 4–6 days recorded between eggs; replacement laid if 1st clutch lost (Thanner 1903). Probably 1 brood, though record of ♀ building new nest on day young fledged (Martín 1987). Said to have clutch of 3–4 on Gran Canaria (Hemmingsen 1958), but this likely to be wrong. Eggs of *F. teydea* and *S. canaria* recorded in same nest (Martín 1985). INCUBATION. 13-14 days, starting usually with 2nd egg, perhaps sometimes with 1st; by ♀ only (Meade-Waldo 1889b; Thanner 1903; Volsøe 1951; Martín 1987); 15-16 days according to Moreno (1988), but perhaps counted from laying of 1st egg. YOUNG. Fed and cared for by both parents (Thanner 1903; Bannerman 1963). FLEDGING TO MATURITY. Fledging period 17-18 days (Moreno 1988). Captive young first took own food at 10 weeks old. ♂♂ still in immature plumage recorded breeding at 1 year old, and such birds possibly account for late breeders in July-August. (Thanner 1903). BREEDING SUCCESS. Nests robbed by Sparrowhawk *Accipiter nisus* (Meade-Waldo 1889b; Martín *et al.* 1984), and possibly by Kestrel *Falco tinnunculus* (Hemmingsen 1958). Of 36 eggs, 3 failed to hatch (Meade-Waldo 1889b). No further information.                                                              BH

**Plumages.** (nominate *teydea*). ADULT MALE. Entire upperparts dark slate-blue, rather glistening in some lights when worn, slightly duller grey at base of upper mandible, slightly greyer blue on rump and upper tail-coverts; forehead not as black as ♂ of Tenerife race of Chaffinch *F. coelebs tintillon*, crown to mantle not as dark slaty blue-black, scapulars to rump not washed green. Lore and patch below eye medium grey, mottled darker grey, merging into dark slate-blue of remainder of side of head and side of neck (side of head less glistening and bluish than cap); short stripes on upper and lower eyelid white, forming broadly interrupted eye-ring. Chin to chest and side of breast as well as lower flank medium slate-blue; remainder of flank, upper belly, and side of belly pale blue-grey, merging into white of mid-belly, thigh, vent, and under tail-coverts; some grey or black of feather bases sometimes showing through on belly, thigh, or under tail-coverts. Central pair of tail-feathers slate-blue, variegated black along shaft; t2-t5 black with slate-blue fringe along outer web, t6 sooty-grey with narrow off-white outer edge and highly inconspicuous slightly paler grey wedge on tip of inner web (white in all races of *F. coelebs*); tips of inner webs of (t3-)t4-t6 narrowly fringed white when plumage fresh. Flight-feathers, greater upper primary coverts, and bastard wing black, slightly paler grey-black on tips of primaries and along borders of inner webs; outer webs of secondaries, primary coverts, and shorter feathers of bastard wing contrastingly fringed slate-blue, primaries with sharp and narrow pale blue-grey fringes (faint, brown-grey, on p9 and along emarginated parts of p5-p8); p5 with small white spot at base of outer web, often concealed by primary coverts. Tertials black with broad slate-blue outer border. Lesser upper wing-coverts dark slate-blue; median coverts black with slate-blue fringe along sides and contrasting pale blue-grey tips 3-5 mm wide, forming wing-bar; greater coverts black with slate-blue outer fringes merging into paler blue-grey fringe along tip, latter forming somewhat less clearly-defined wing-bar than that across median coverts. Under wing-coverts and axillaries medium grey with paler grey fringes; grey darker towards leading edge of wing. Influence of bleaching and wear limited; in fresh plumage, some traces of dark olive-brown fringes sometimes present on upperparts and chest; in worn plumage, blue of upperparts more glossy, grey fringes of tail, tertials, flight-feathers, and tips of median and greater coverts bleached and partly worn off. ADULT FEMALE. Upperparts dark olive-brown, some dark ash-grey of feather-bases partly visible, especially on forehead, side of crown, and rump. Side of head and neck dark ash-grey, slightly washed dark olive-brown, especially on ear-coverts and side of

neck; lore and patch below eye mottled dark and pale grey; short off-white streaks on upper and lower eyelid form broadly interrupted eye-ring. Chin pale grey or off-white, merging into medium ash-grey throat, chest, side of breast, and lower flank; grey partly washed olive-brown, especially on lower flank; remainder of flank, upper belly, and side of belly pale ash-grey with slight olive wash, merging into white of mid-belly, vent, and under tail-coverts. Tail as adult ♂, but t1 more olive-grey and grey of fringes along outer webs of t2-t5 less blue, less contrasting. Wing as adult ♂, but black replaced by sepia-black, and slate-blue fringes replaced by dark olive-brown, palest along outer webs of primaries; tips of median and greater coverts pale grey or dull white, forming 2 rather contrasting wing-bars; grey of underwing tinged olive-brown. *In worn plumage* (spring), olive-brown feather-tips of upperparts and throat to chest and side of breast partly worn off, dark ash-grey more fully exposed (much duller, browner tinged, less slaty, and less uniform than adult ♂), tips of median and greater coverts and fringes of outer primaries (except for emarginated parts) bleached to off-white, partly worn off. NESTLING. Down thick and plentiful, black (Harrison 1975; Bergmann *et al.* 1988). JUVENILE. Rather like adult ♀, but upperparts dull brown, less olive, darkest on tips of feathers, paler grey-brown on centres; sides of head and neck dull buffish-grey, lore and narrow broken eye-ring paler, mottled cream-grey and brown-grey; chin to chest and side of breast as well as lower flank light grey-buff with pale grey feather bases, remainder of underparts off-white with cream or pale buff suffusion on feather-tips. Best distinguished from adult ♀ by shorter and looser character of feathering. Upper wing-coverts and tertials brown-black, tips of greater and median coverts contrastingly pale cream or off-white (forming 2 wing-bars, each 2-4 mm wide, more sharply contrasting than in adult ♀), innermost greater coverts and tertials with ill-defined buffish borders along outer web or tip. FIRST ADULT MALE. In fresh plumage, closely similar to adult ♀ in fresh plumage: upperparts dark olive-brown, side of head and neck mainly brown-grey (except for paler lore and broken off-white eye-ring), chin to chest and side of breast medium olive-brown; however, feather centres on upperparts slate-grey (more bluish than in adult ♀), on side of head and neck and on chin to chest and side of breast light blue-grey (less ashy than adult ♀); this slate- or blue-grey largely concealed when plumage quite fresh, but soon exposed with wear of feather-tips, bird becoming gradually more bluish (unlike adult ♀), but never as uniform and glistening slate-blue as adult ♂, some brown tinge remaining. Also, juvenile tail-feathers retained, tips pointed, less truncate or broadly rounded than in adult; some outer greater coverts and sometimes a few median coverts still juvenile, browner and with narrower whiter tips than neighbouring black new ones, which have broader blue-grey tips. FIRST ADULT FEMALE. Like adult ♀, but upperparts uniform dark olive-brown, virtually without ash-grey on feather bases; side of head and neck and chin to chest and side of breast paler olive-brown, with pale grey mottling on lore, just below eye, and chin, without ashy tinge of adult ♀. Tail, flight-feathers, and primary coverts still juvenile, brown-black instead of brown, fringes more olive, less grey, tips of tail-feathers pointed; variable number of median and greater upper wing-coverts juvenile, dark olive-brown with narrow buff-white tips contrasting with newer greyish-black inner coverts (if any) which have broader grey-white tips.

**Bare parts.** ADULT, FIRST ADULT. Iris red-brown or brown. Bill of ♂ in breeding season light blue with black tip and cutting edges, in rest of year duller bluish-horn or grey-horn; bill of ♀ dark horn-grey with paler horn lower mandible. Leg and foot dark flesh-brown, grey-brown with pink tinge, slate with pink-brown tinge, or dark horn-grey; darkest, slate-grey, in breeding ♂. NESTLING. No information. JUVENILE. Bill completely dusky. (Bannerman 1912, 1963; Martín *et al.* 1984; BMNH, RMNH, ZFMK.)

**Moults.** ADULT POST-BREEDING. Complete; primaries descendent. Information limited. Of *c.* 70 *polatzeki* of all ages examined, January–May, none in moult (C S Roselaar). Of 70 nominate *teydea* of all ages, most birds were January–June, and only 11 adults from probable moult season July–December; of these 11, 2 from July worn but not yet moulting, single ♀ from November in moult (primary moult score 37: p1–p6, p7–p8 growing, p9–p10 old; part of feathering of head and neck growing, remainder new), as were 1 of 4 ♂♂ (score 44) and 2 of 4 ♀♀ (score 39, 44) from December (outer primaries growing, remainder of plumage new). On Tenerife, 3 ♂♂ in wing- or tail-moult on 20 September; tail-moult centrifugal (A J van Loon). POST-JUVENILE. Partial: head, body, lesser and variable number of median and greater upper wing-coverts; juvenile under tail-coverts apparently sometimes retained. Only a few July–December birds available; moult probably starts soon after fledging; a few July–August birds examined were fully juvenile or just starting moult; single ♀ from October had body new but head, neck, and under tail-coverts in moult (mainly old). Some birds from November–February had moult apparently suspended; juvenile median and greater coverts, tertials, tail, under tail-coverts, and flight-feathers retained, though others from this period had median coverts, innermost greater coverts, and under tail-coverts new. (BMNH, RMNH, ZFMK, ZMA.)

**Measurements.** ADULT, FIRST ADULT. Nominate *teydea*. Tenerife, whole year; skins (BMNH, RMNH, ZFMK, ZMA: A J van Loon, C S Roselaar). Bill (S) to skull, bill (N) to distal corner of nostril; exposed culmen on average 2·4 less than bill (S).

| | | | | |
|---|---|---|---|---|
| WING | ♂ | 101·8 (2·40; 35) 96–105 | ♀ | 93·6 (2·34; 24) 88–98 |
| TAIL | | 81·8 (2·26; 35) 77–86 | | 74·4 (1·88; 24) 71–78 |
| BILL (S) | | 19·5 (0·53; 33) 18·5–20·5 | | 19·4 (1·02; 23) 17·8–20·9 |
| BILL (N) | | 13·7 (0·44; 34) 12·6–14·7 | | 13·4 (0·51; 21) 12·6–14·6 |
| TARSUS | | 22·8 (0·60; 35) 21·6–24·5 | | 22·6 (0·88; 23) 20·3–24·3 |

Sex differences significant for wing and tail.

Wing: ♂ 101·9 (2·46; 23) 98–104 (once 109), ♀ 92·7 (1·93; 24) 89–97 (Martín *et al.* 1984); ♂ 103 (10) 100–107 (Vaurie 1959); ♂ 101·2 (96–107), ♀ 91·7 (80–97) (total *n* = 122) (Bannerman 1912).

*F. t. polatzeki.* Gran Canaria, all year; skins (BMNH, RMNH, ZFMK, ZMA: A J van Loon, C S Roselaar).

| | | | | |
|---|---|---|---|---|
| WING | ♂ | 96·1 (1·54; 15) 93–98 | ♀ | 89·3 (1·71; 13) 87–93 |
| TAIL | | 76·2 (1·07; 7) 75–78 | | 71·5 (2·92; 5) 69–74 |
| BILL (S) | | 18·5 (0·56; 12) 17·7–19·4 | | 18·2 (0·70; 12) 17·3–19·3 |
| BILL (N) | | 12·7 (0·38; 14) 12·0–13·3 | | 12·7 (0·21; 13) 12·4–13·1 |
| TARSUS | | 22·0 (0·40; 6) 21·5–22·7 | | 21·6 (0·42; 5) 21·0–22·0 |

Sex differences significant for wing and tail.

Wing: ♂ 94 (90–97), ♀ 87 (85–97) (total *n* = 76) (Bannerman 1912). For both races, see also Vaurie (1959) and Bährmann (1976).

**Weights.** Nominate *teydea*. Tenerife: ♂♂ 30·9 (2), ♀ 29·2 (Grant 1979).

**Structure.** 10 primaries: p6–p7 longest or either one 0–1 shorter than other; p8 0–2 shorter than longest, p9 4–9, p5 2–6, p4 10–14, p3 15–19, p2 17–23, p1 19–25; tip of p9 between p4 and p5. Outer web of p5–p8 emarginated, inner web of p4–p9 with slight notch. Bill distinctly larger than in *F. coelebs*, *c.* 11–12·5 mm deep and 10–11 mm wide at base, culmen and gonys gradually tapering to sharply pointed tip, distal half of culmen less decurved than in *F. coelebs*. Middle toe with claw 20·6 (8) 19·5–21·5 in nominate *teydea*, 19·5 (4) 19–20 in *polatzeki*; outer toe with claw *c.* 74% of middle with claw, inner *c.* 69%, hind *c.* 76%. Feathering of body markedly long and dense. Remainder of structure as in *F. coelebs*.

**Geographical variation.** Slight in colour, more marked in size. *F. t. polatzeki* of Gran Canaria smaller than nominate *teydea* from Tenerife in all measurements (see Measurements); ♂ *polatzeki* differs from ♂ of nominate *teydea* mainly in presence of velvet-black band on forehead (in adult), slightly duller and more restricted slate-grey on chin to chest, slightly paler greyish-white tips of median and greater coverts, and narrower slate-blue fringes of upper wing-coverts and tertials, upperwing appearing blacker; ♀ *polatzeki* paler and less extensively grey (adult) or brown-grey (1st adult) on chin to chest, belly more extensively off-white.

For relationships, see *F. coelebs*. CSR

---

## *Fringilla montifringilla* **Brambling**

DU. Keep    FR. Pinson du Nord    GE. Bergfink
RU. Вьюрок    SP. Pinzón real    SW. Bergfink

*Fringilla Montifringilla* Linnaeus, 1758

Monotypic

PLATES 29 and 38 (flight)
[between pages 472 and 473]

**Field characters.** 14 cm; wing-span 25–26 cm. Similar in body and wing size to Chaffinch *F. coelebs* but stubbier bill, seemingly larger head, and slightly shorter, more distinctly forked tail produce less attenuated silhouette on ground and in flight. Medium-sized, elegant finch, with general character and behaviour of *F. coelebs* and rather similar basic plumage pattern but different, less varied colours in ♂. Both sexes show diagnostic combination of long, oval white rump and almost completely black tail. ♂ distinguished in breeding season by glossy black head and mantle, bordered by orange blaze from breast across shoulder and below back; in winter by black-speckled face and crown and black-splashed mantle. ♀ and juvenile distinguished by mottled dark brown head, with broad buff supercilium and grey sides to neck. Sexes dissimilar; marked seasonal variation in ♂. Juvenile separable from ♂.

ADULT MALE. Moults: July–September (complete); immaculate breeding plumage produced by wear. Seemingly large head and deep face, mantle, lateral rump and upper tail-coverts, and tail black, showing at close range blue gloss on head and mantle and white fringes on side of rump. Centre of lower back and rump pure white, forming long oval patch. Scapulars yellow-buff, combining with orange of fore-underparts and leading wing-coverts to form warm blaze from chest through shoulder and along edge of back. Wings black, boldly marked with (a) bright yellow-buff lesser coverts, (b) pale buff median-coverts, with almost white tips, (c) almost white tips to greater coverts, forming conspicuous wing-bar below blaze, (d) white bases to 6 innermost primaries forming bright patch by primary coverts and extending wing-bar onto outer half of wing, and (e) variable pale buff edges to tertials and inner secondaries, usually restricted by wear to outer edges. Underwing bright lemon-yellow on axillaries, white on coverts. Ground of underparts pure white but throat and particularly breast, chest, and fore-flanks pale orange-buff, while middle and rear flanks boldly and roundly spotted black. Tail lacks visible white on sides. When plumage fresh, dark areas extensively covered by greyish-buff fringes and tips: appearance ♀-like except for blackish face and strong black centres to crown and lateral nape feathers, and deeper orange-buff on fore-underparts, leading wing-coverts, and scapulars, with difference in tone striking and allowing, with blackish-patched head, easy sexual differentiation. Bill blue-black when breeding, orange-yellow with black tip at other times. Legs brownish-flesh, darker when breeding. ADULT FEMALE. When fresh, head pattern complex, with (a) brown-black crown with buff feather-fringes dividing into blackish lateral rear crown- and nape-stripes which border obvious whitish-grey central patch on lower nape and end above large whitish-grey patch on sides of neck behind ear-coverts, (b) quite obvious brownish-buff supercilium, clearly separating lateral crown- and nape-stripes from (c) greyish-brown lore and ear-coverts, and (d) pale whitish-buff chin and throat. Mantle and scapulars brown-black, with wide buff fringe isolating dark feather-centres into quite distinct lines of spots. Lower back, rump, and tail as ♂, but tail duller black. Wings dull black, as ♂ but with tri-coloured marks duller and differing particularly in lack of uniformly-coloured lesser wing-coverts, these showing brown-black centres and less white, more buff, tone to tips of greater coverts; these combine with pale inner primary-bases into wing-bar as in ♂. Chest and flanks pale orange-buff, looking washed out compared to ♂ and flanks less distinctly dark-spotted. When worn, buff feather-tips

less prominent and brown-black crown and back thus more uniform. Bill dull grey when breeding; straw-yellow with dusky tip at other times. JUVENILE. Resembles fresh ♀ but distinguished at close range by (a) darker black-brown median and greater coverts which have only buffish-white tips and thus look mottled or barred rather than pale-patched as in adult and (b) yellowish cast to rump and belly. For differences between 1st-year ♂ and ♀, see Plumages.

Unmistakable when seen well but similar wing pattern may invite confusion with distant *F. coelebs*. Goldfinch *Carduelis carduelis* and Bullfinch *Pyrrhula pyrrhula* both show white rump and black tail but rest of plumage pattern and colours differ distinctly (see those species). Flight recalls *F. coelebs* but action differs in slightly quicker wing-beats, producing more bouncy progress, and apparently (perhaps due to lack of white in it) less spreading of tail in manoeuvres; flight silhouette less attenuated than *F. coelebs*. Migrant flocks move in similar manner to *F. coelebs* but tend to be more compact. Gait and behaviour not noted as different to *F. coelebs* but feeding flocks pack even closer, indulging in swirling manoeuvres and escaping mainly to trees. Feeds mainly on ground, especially on beechmast, but autumn migrants are at least as persistent as *F. coelebs* in searching for larvae in tree canopies. Gregarious, highly so in occasional mass immigrations and winter assemblies.

Song a grating, monotonous, repeated 'dwee' or 'dzweea', strongly recalling Greenfinch *Carduelis chloris* and sometimes followed by harsh rattle. Most audible call a rasped, rather tinny 'tsweep', or wheezing, nasal 'teh-ehp', given with increased volume in alarm. Other calls include rather subdued 'tchuc' or 'jek', often repeated rapidly in flight, especially by migrants.

**Habitat.** Breeds across boreal and subarctic zones of west Palearctic between 10° and 18–19°C July isotherms (Voous 1960b). Owing to northerly range and arboreal requirements does not extend much up mountains, but is common on uplands in the more open birch *Betula* woods, and in mixed forests of birch and conifers. Sometimes ranges beyond into lower growth of juniper *Juniperus*, willow *Salix*, or alder *Alnus*. Tall and dense stands in forest appear to be less favoured than open growth with clearings. Also found in riverine belts of willows. (Bannerman 1953a; Dementiev and Gladkov 1954; Voous 1960b). Change after breeding season from insectivorous to largely seed diet involves shift to ground feeding, sometimes in stackyards and on farmland, but attachment to woodland remains strong, especially in case of beech *Fagus* where crop of fallen mast is ample (Bannerman 1953a). Almost unmatched among Fringillidae for scale and frequency of movements and corresponding use of lower airspace in characteristically dense linear formations. Switzerland is often the goal of these mass movements leading to settlement from late December until mid-March

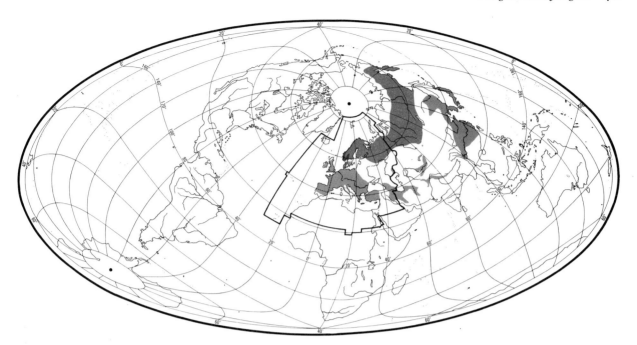

in woodland roosts serving a foraging radius to some 20 km, mainly in beechwoods rich in mast. When these sometimes become covered in snow the mast can be got out from under it up to a depth of *c.* 15 cm, but beyond that foraging has to be switched elsewhere. (Lüps *et al.* 1978). Except in severe weather, remarkably independent of all kinds of human settlements, artefacts, and food resources.

**Distribution.** Signs of recent spread in both north and south, but long-term changes uncertain, tending to be obscured by marked annual fluctuations in breeding range. In Norway and Sweden, breeding numbers vary from year to year according to abundance of defoliating geometrid caterpillar *Epirrita* (*Oporinia*) *autumnata* (Hogstad 1985; Lindström 1987).

ICELAND. First recorded as accidental in 1930; since then recorded more commonly, and breeding proved 1974, though had probably bred earlier. Breeding now recorded nearly every year. (ÆP.) FAEROES. Bred 1967 and 1972, and probably 1982 (Bloch and Sørensen 1984). BRITAIN. First recorded breeding in northern Scotland 1920, and bred also (e.g.) 1979 and 1982. (Sharrock 1976; Thom 1986.) NETHERLANDS. 1–5 pairs bred annually, 1973–7 (Teixeira 1979). 50–150 ♂ territories annually, but ♀♀ rare, and seldom any young fledged (SOVON 1987). WEST GERMANY. Small breeding population (continuous with Netherlands population), perhaps descended from released decoy birds; estimated *c.* 5 pairs (Rheinwald 1982, 1992). DENMARK. Total of *c.* 10 breeding records (TD). SWEDEN. Probable south-west spread since *c.* 1970 but obscured by changes from one year to another. Occasional breeding records south to *c.* 57°N. (LR.) FINLAND.

Marked annual fluctuations in breeding range (OH). CZECHOSLOVAKIA. Bred in extreme north-west (Erzgebirge) 1928 (KH). AUSTRIA. At least 2 breeding records confirmed, Carinthian Alps, 1952 and 1988; a few more records of summering in Alps (H-MB). ITALY. 2 confirmed breeding records in period 1975–81, and several possible (Brichetti 1982). YUGOSLAVIA. Almost regularly observed in Slovenia (Julian Alps) during breeding season; possibly breeds (Grošelj 1983; Šere 1986). LATVIA. At least 2 breeding records, 1892 and 1960 (Vīksne 1983); also 2 singing ♂♂, 1980–4 (Vīksne 1989). LITHUANIA. 2 breeding records, 1978 and 1983 (Logminas 1991).

Accidental. Lebanon, Jordan, Kuwait, Canary Islands.

**Population.** Marked annual fluctuations associated with changes in breeding distribution, and data inadequate to separate one from the other (see Distribution).

SWEDEN. Estimated 2·5 million pairs (Ulfstrand and Högstedt 1976). FINLAND. Perhaps 2–5 million pairs; marked annual fluctuations (Koskimies 1989; OH).

Oldest ringed bird 14 years 9 months (Staav 1983).

**Movements.** All populations migratory, wintering almost entirely south of breeding range. European birds head between west and south, chiefly south-west. Extent of movement is strongly dependent on food availability (chiefly seed of beech *Fagus*); local numbers wintering fluctuate greatly, and concentrations of millions of birds occur, especially in south-central Europe. Ringing data give evidence of winter site-fidelity, but also of individuals wintering in widely differing areas in different years. ♂♂ predominate in areas closer to breeding range. Migration

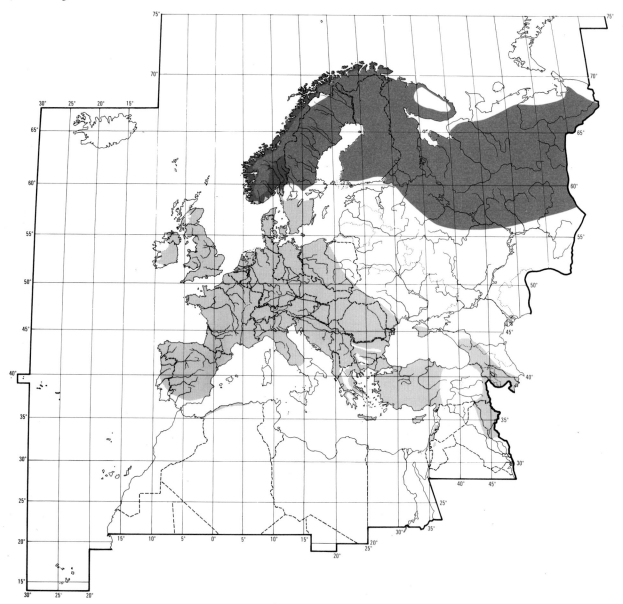

mostly diurnal (chiefly in morning), especially inland, but nocturnal migration observed on coasts and at sea (Jenni 1985). Chief studies (primarily for Switzerland) by Jenni (1982, 1985) and Jenni and Neuschulz (1985). For migratory restlessness in caged birds, see Lundberg (1981).

In Europe, wintering area lies immediately south and west of breeding area, extending south chiefly to southern France and northern Italy. In Britain, wintering numbers vary widely, perhaps between c. 50 000 and c. 2 million birds; patchily distributed over much of Britain in areas where beech *Fagus* occurs, but markedly fewer in Ireland; readily undertakes hard-weather movements west and south (Lack 1986). Observations and ringing recoveries show that some birds reach Britain via coast of Denmark and Low Countries, but others cross North Sea directly (Newton 1972). In Sweden, where breeds south to c. 60°N, winters regularly in varying numbers north to c. 65°N (SOF 1990). Probably winters annually in southern Finland (breeds further north), but in greatly varying numbers, and especially in good rowan *Sorbus* years; wintering birds appear in December (long after passage has ended), having perhaps moved gradually south (Eriksson 1970b; Haila *et al.* 1986). In France, widely distributed in plains and at mid-altitude; highest numbers in east (Lorraine, Franche-Comté), south-east (Rhône-Alpes), south-west (Pyrénées), and Ile de France in north (Yeatman-Berthelot 1991). Many winter in central Europe, with mass influxes in different regions in different years, e.g., 1958-9 and 1964-5 in north-central Europe, 1970-71 in Netherlands, Belgium, and Luxembourg, 1965-6 and 1977-8 in Switzerland (where mass influx first recorded 1413) (Jenni 1985; Jenni and Neuschulz 1985). Usually irregular in Spain, occasionally in large numbers west to Extremadura (Lope *et al.* 1983). Regular in very small numbers on passage at Strait of Gibraltar (Finlayson 1992), and in winter in north-west Africa, chiefly in extreme north of Morocco, with occasional records further south, e.g. in Moyen Atlas and Haut Atlas, once as far south as Beni-Abbès (30°05'N 2°35'W) in Algerian Sahara (Dupuy and Johnson 1967; Fornairon 1977; Pineau and Giraud-Audine 1979; Thévenot *et al.* 1982). 1st records for Mauritania 9 January and 5 April 1988 at Nouadhibou on north-west coast (Meininger *et al.* 1990). Very scarce but almost annual autumn migrant and winter visitor to Malta (Sultana and Gauci 1982). In Turkey, widely dispersed and usually uncommon passage migrant, wintering in western two-thirds (Beaman 1986); scarce on Cyprus (Flint and Stewart 1992). Infrequent winter visitor to northern Iraq (Moore and Boswell 1957); in Iran, common in south Caspian lowlands and fairly common in Tehran area; rather scarce elsewhere (D A Scott). In north-west Caucasus, sometimes numerous, sometimes rare (Polivanov and Polivanova 1986). Very scarce and irregular in Arabia (F E Warr). In Israel, scarce or uncommon in most years, sometimes fairly common (Shirihai in press). Rare winter visitor to Egypt, mostly to Nile delta (not recorded south of Cairo), also at St Katherine in southern Sinai, and south to Safaga (26°44'N) on Red Sea coast (Goodman and Meininger 1989).

Two major components to migratory pattern, perhaps involving different parts of population: (1) regular movement south-west towards southern Europe, and (2) tendency to remain as far north-east as possible, depending on food availability, and resulting in mass concentrations and influxes (Jenni 1982, 1987).

(1) Ringing recoveries indicate general south-west direction of movement. Most birds wintering in Britain and Belgium are from Scandinavia (Verheyen 1954; Lack 1986; Mead and Clark 1987). In France, Norwegian birds winter in north-west, Swedish birds in centre and south, Finnish and west Russian birds in east or south-east (Yeatman-Berthelot 1991). Swiss winter population is from Finland and north-west Russia (Jenni 1982). Most recoveries in same season of birds ringed autumn in Germany and Belgium are south-west of ringing site (Verheyen 1954; Hilprecht 1965). Swiss autumn passage birds winter chiefly on southern edge of Alps (northern Italy and southern France), with some reaching Spain and central Italy (Jenni 1982); of birds ringed winter in Steiermark (eastern Austria), 11 of 14 foreign recoveries in later winters were in north-east Italy (Haar 1975). Norwegian birds recovered chiefly in north-west Europe (including one in south-east Iceland, 1130 km north-west, 9 days after ringing), also east to eastern Germany and south to Spain and Portugal

(Holgersen 1982; Runde 1984); recoveries of Finnish birds average markedly further east (Eriksson 1970). Swedish birds ringed 1972-88 were found west to France, south to Spain and Italy, east to Finland and Kaliningrad region (north-west Russia) (*Rep. Swedish Bird-Ringing*). Many recoveries show movements between western Europe (including Britain) and northern Russia, north to Murmansk region and east beyond Urals to Tyumen' region (Spencer 1972; Dobrynina 1981). In Baltic region, route apparently differs according to season; in autumn, few on southern shore of Gulf of Finland, but large numbers further south at Lake Chudskoe, and on Kaliningrad coast; in spring, only small numbers on Kaliningrad coast, but extremely common on southern shore of Gulf of Finland (Paevski 1968; Gaginskaya 1969). Birds ringed on autumn passage on Kaliningrad coast show 2 headings: west and south-west through northern and central Germany and Belgium, and SSW through Switzerland and northern Italy; many from both headings reach France, others winter in intermediate areas; no correlation found between direction of movement and early or late timing of passage. Birds from north of Kaliningrad (Baltic states and St Petersburg and Murmansk regions) perhaps take mostly SSW heading. (Paevski 1968; Dobrynina 1981.) Evidence of fidelity to winter or passage site also points to regular pattern of movement; in Steiermark, 66 retraps in later winters (up to 6 years after ringing) (Haar 1975); in Baden-Württemberg (south-west Germany), 15 winter retraps up to 6 years after ringing (Schlenker 1976); in Belgium, autumn or winter, 21 retraps in later years within 12 km of ringing site (Verheyen 1954). Data from Steiermark also show site-fidelity in same winter; individuals trapped over 96, 74 and 72 days (Haar 1975).

(2) Studies indicate tendency to remain as far north-east as possible. Birds migrate across central Europe on broad front, stopping when they reach good beechmast area; if this area sufficiently extensive, very large numbers build up; later, if beechmast becomes inaccessible due to deep snow cover, mass onward movement occurs, resulting in heavy concentrations progressively further south (especially in southern Germany and Switzerland), but also some dispersal; on edge of beechmast area, therefore, where winters relatively mild (e.g. Britain, Greece, Yugoslavia), birds tend to be associated with hard winters. Analysis of data for central Europe (southern Sweden south to Switzerland), 1900-83, has revealed strong link between mass concentrations of birds and extent of beechmast crop and snow cover; concentrations further west (central France) or east (south-east Europe) show no clear relation with those in central Europe. (Jenni 1987.) So influxes depend on feeding conditions in main winter quarters; no evidence (unlike in 'invasion' species) of dependence on density on breeding grounds. Also, these influxes involve many adults, whereas invasions chiefly involve young birds. In Switzerland, influxes sometimes involve several million birds in good beechmast years;

no evidence of any difference in origin between birds wintering in normal and influx years. (Jenni 1982.) Mass influxes into Switzerland occur later than normal passage, which is itself reduced in mass influx years as some birds stop off at good beechmast areas further north (Jenni 1985). In 1977, when autumn passage had ended, influx began early November throughout Switzerland, and large roosts were established until 2nd half of January when heavy snowfall began; birds then moved on and formed new roosts to north and (mostly) west, but these areas also soon abandoned, and large numbers probably left Switzerland; in mid-February, many birds were in snow-free area on south foot of Jura. In 1982, large flocks observed in Switzerland in November, but extensive roosts built up mostly from early or mid-December, coinciding with immense concentrations in southern Germany; heavy snowfall in eastern Switzerland in February caused many birds to move to western Switzerland (Jenni and Neuschulz 1985). For other mass influxes into Switzerland and southern Germany, see (e.g.) Sutter (1948), Schifferli (1953), and Zinnenlauf (1967); for eastern Germany, see Ruthenberg (1968). For annual concentrations in Pyrénées, see Alberny et al. (1965). Hard-weather movements in large numbers may occur as late as April: in Bergisches Land (west-central Germany), 4 April 1984, snowfall much lighter than in surrounding regions, and movement north-west, then south-east, of at least 20 000 birds observed at 09.00-10.00 hrs (Radermacher 1984).

Extreme changes of winter quarters involving Swiss-ringed birds are associated with influx years; 51 recoveries in later winters were west, north, or east of Switzerland, as far as Ireland, Rumania, Greece, and Gruziya (Jenni 1982). Apparently, birds are caught up in mass movements and drawn away from normal heading. Birds ringed Belgium autumn or winter recovered in later winters in Italy; those recovered in south-west France and north-east Spain had probably also travelled by more eastern route (Switzerland, or eastern France) than in year of ringing (Verheyen 1954). Birds ringed winter in Britain recovered in later years as far south-east as Italy and Yugoslavia (Lack 1986). Several birds ringed winter in central Europe recovered October-May in later years in south of former USSR, from Volgograd (48°40′N 44°25′E) south to Gruziya and Azerbaydzhan (Dobrynina 1981).

Ringing data (mainly from Switzerland) show that ♂♂ winter further north-east than ♀♀; proportion of ♂♂ decreases from north to south in winter range, and adults probably winter further north than juveniles (Jenni 1982, 1985). In normal years, flocks on passage or wintering in Switzerland have well-balanced sex ratio, but in mass influx years (involving birds which have attempted to winter further north) ♂♂ predominate. ♂♂ also tend to winter further east than ♀♀: of 42 birds ringed autumn in Switzerland and recovered further south in same winter, those east of 7°E (in Italy) predominantly ♂♂ (74%, n = 19), those west of 7°E (France, Spain) only 22% ♂♂ (n = 23).

(Jenni 1982.) Winter population in Israel comprises only about 30% ♂♂, and in more southern areas percentage even smaller (Shirihai in press). Limited data suggest ♂♂ more likely to winter in Fenno-Scandia than ♀♀ (Eriksson 1970), and in Kaliningrad region, throughout year, percentage of ♂♂ trapped (52-55%) regularly higher than ♀♀ (Paevski 1968).

In autumn, short-distance reversed migration regularly occurs in extreme south of Sweden; after reaching coast (where no food available), some birds, in company with Chaffinch F. coelebs, return inland 20-40 km north-east to feed chiefly in fields of rape Brassica napus; reverse migrants weigh less than normal migrants, and include more juveniles than adults; this thought to be adaptive response of birds with small fat reserves, on reaching barrier of sea crossing (Lindström and Alerstam 1986).

Autumn passage in north-central Europe (Poland, northern Germany, Denmark, southern Sweden) begins mostly in 2nd half of September, and continues to early or mid-November; peaks 1st half of October in Finland, Poland, and eastern Germany, from mid-October further west. Some evidence that overall passage in north-central Europe in 2 waves, peaking in 1st and 2nd half of October. (Eriksson 1970; Jenni 1985.) Trapping data on Kaliningrad coast suggest ♀♀ begin autumn migration earlier on average than ♂♂ (Paevski 1968). In Switzerland, passage begins end of September and peaks mid-October, decreasing sharply and ending mostly by end of October; in mass influx years, 2nd arrival from November (Jenni 1985). Usually arrives in Britain October or November, but may be inconspicuous at first and may move on elsewhere in Europe (Lack 1986). Reaches Basses-Pyrénées in south-west France from mid-October, chiefly November (Alberny et al. 1965), and Mediterranean region October-November (Sultana and Gauci 1982; Goodman and Meininger 1989; Finlayson 1992; Flint and Stewart 1992). Present in winter quarters chiefly November-February (e.g. Verheyen 1954, Yeatman-Berthelot 1991).

Spring migration February-May. ♂♂ leave winter quarters earlier than ♀♀ and arrive earlier on breeding grounds (Zinnenlauf 1967; Paevski 1968; Mikkonen 1985; Shirihai in press). Leaves Britain mostly early March to April (Newton 1972). Begins to leave Pyrénées in 2nd half of February; most birds depart in early March, with stragglers to about mid-April (Alberny et al. 1965). In Belgium, passage and departures from late February to early April (Verheyen 1954). Main movement March in Switzerland, with only small numbers after mid-April (Winkler 1984), and passage in Berlin area from last third of February, mostly mid- or late March to mid-April (Witt 1988). In Brandenburg (eastern Germany), passage peaks in last third of March (Rutschke 1983). On Kaliningrad coast, main passage 10-30 April, and in St Petersburg region mid-April to mid-May (Noskov et al. 1981; Mal'chevski and Pukinski 1983). Movement mostly end of March to beginning of May in Sweden (SOF 1990).

Reaches Budal in central Norway when snow-cover 30–50%, usually in last 10 days of May (Hogstad 1982). At Juorkuna in northern Finland, 1967–73, arrivals mostly 3–25 May; movement later than in *F. coelebs*, and less drawn-out (Mikkonen 1985b). On Kola peninsula (north-west Russia), over 44 years, average earliest record 8 May (20 April to 22 May), and movement continues to end of May or early June; arrival more simultaneous than in many other migrants (Semenov-Tyan-Shanski and Gilyazov 1991). In some years when spring is late, some birds curtail their migration and breed further south than usual, notably in Fenno-Scandia but also recorded from Baltic region and Denmark (Newton 1972).

In Asia, winters in Tadzhikistan and southern Kazakhstan (irregularly north to western Siberia) (Johansen 1944; Korelov *et al.* 1974; Abdusalyamov 1977), in Afghanistan (Paludan 1959), and in northern Pakistan and north-west India, occasionally east to Nepal (Ali and Ripley 1974; Inskipp and Inskipp 1985). Also occurs throughout much of China (Schauensee 1984), southern Korea (Austin 1948), and Japan (mostly Honshu southwards) (Brazil 1991). In north of range (e.g. Yamal peninsula and Kamchatka), last autumn records in September (Danilov *et al.* 1984; Morozov 1984; Lobkov 1986). Occurs throughout Kazakhstan on passage, mostly October–November; in west, migrates chiefly along Ural valley, but also across Aral-Caspian steppe (Korelov *et al.* 1974); at Chokpak pass (western Tien Shan), migration peaks 16–20 October (Gavrilov and Gistsov 1985). Present in India and Pakistan November to March or April (Ali and Ripley 1974). In Mongolia, passage recorded in both east and west (Piechocki and Bolod 1972), and common on migration in Ussuriland (south-east Russia) (Panov 1973a). Reaches Japanese winter quarters late September to November, via both Hokkaido and Sea of Japan (Brazil 1991). Spring movement late March to April in north-east China (where very common) and Japan (Williams 1986; Brazil 1991), early or mid-March to April in Kazakhstan (Korelov *et al.* 1974). In Kamchatka, over 5 years, average earliest arrival 14 May (5–22 May) (Lobkov 1986); reaches Yamal peninsula usually in last week of May, later if snow persists (Danilov *et al.* 1984). Rare migrant in western and central Aleutian Islands, occasional elsewhere in Alaska (Kessel and Gibson 1978). Also casual visitor to British Columbia, with other scattered records in North America, some perhaps involving escaped cage birds (American Ornithologists' Union 1983; Godfrey 1986).     DFV

**Food.** Seeds, berries, and (in summer) invertebrates, especially Lepidoptera larvae and beetles (Coleoptera); in winter quarters specializes in beechnuts *Fagus sylvatica*. On breeding grounds feeds mainly in trees, but at other times mostly on ground, commonly in flocks. (Bannerman 1953a; Schifferli 1953; Newton 1972; Mikkonen 1984.) In central Norway, late April to mid-May, 73% of 136 foraging birds observed were in trees (97% of 288 observations mid-May

to late June), 23% (1%) on ground, 3% (2%) in air, and 1% (0·3%) in bushes; average foraging height 8·2 m, and 82% of observations in outer third of branch (Sæther 1982); see also Hogstad (1988). In study in western Norway, foraged at average 11·9 m above ground (Angell-Jacobsen 1980). Similarly, in northern Finland during breeding season, 84% of foraging was in trees, 10% on ground, and 6% in air; mostly in upper half of tree towards outer foliage, and of 138 actions, 44% at base of leaf or needle, 43% on twig, 10% on thin branch, and 3% on branch (Virkkala 1988, which see for comparison with other species, especially Willow Warbler *Phylloscopus trochilus*). 3 foraging techniques on breeding grounds; gleaning (taking prey from ground or vegetation when standing), hawking (taking prey from vegetation in flight), and flycatching (aerial feeding); 90% of 3768 observations in central Norway gleaning (Hogstad 1988); in northern Norway, recorded picking insects from conifer trunk by mixture of flutter-hovering, sallies, and gleaning, like tit *Parus* or flycatcher (Muscicapidae) (E K Dunn). In mixed forest, western Norway, fed almost wholly in pine *Pinus* (Angell-Jacobsen 1980). In usual feeding site, high up in outer parts of branches, overlaps little with other species; concentrates on seasonally abundant resources (Sæther 1982; Virkkala 1988). In winter and on migration feeds on agricultural land, especially cereal fields, and in beech woods (Olsson 1954; Nilsson 1979; Nøhr 1984; Jenni 1987). In northern Germany, winter, 56% of 2655 feeding observations were in woodland, 26% in fields, 14% in hedges, 2% in farmyards, and 2% in parks, gardens, etc.; 53% of birds were eating beechnuts, 38% seeds of grasses (including cereals), and 9% other seeds and fruits (Eber 1956). In deep snow moves into human settlements, where feeding can be intensely competitive, with dozens of birds fighting on bird-tables (Bannerman 1953a; Rinnhofer 1968; Schäpper 1986; Jenni 1987). Takes beechnut on ground by pushing aside leaf litter with bill or by digging tunnels 30–40 cm deep through soft snow on slopes, using bill and 'swimming' action of wings (Lanz-Wälchli 1953; Schifferli 1953; Berg-Schlosser 1978; Jenni and Neuschulz 1985). According to Nardin and Nardin (1985), also uses feet to dig in snow, though such observations disputed, e.g. by Berg-Schlosser (1978). Sometimes eats beechnuts on trees, at times pulling them from cups while hovering; seeds of birch *Betula* also taken *in situ* (Newton 1967a; Jenni and Neuschulz 1985). In Switzerland, seeds of spruce *Picea* important in winter diet at higher altitudes (Schifferli 1953). Feeding method on ground generally rapid pecking, leaves (etc.) being flicked aside by closed bill, but has great difficulty in picking seeds from flexible herbs (though still more successfully than Chaffinch *F. coelebs*), and captive birds never learned how to remove pappus of Compositae seeds on ground, although they had seen carduline finches do it; feet never used in feeding (Newton 1967a). Can be inventive and aggressive at bird-tables or in other novel feeding situations, and various such activities recorded: catching falling peanuts in the air and snatching them from Greenfinch *Carduelis chloris*· (Brodie 1985); hovering beside hanging net full of nuts, striking blows, then flying down to pick up fallen fragments, but later apparently learning from watching *C. chloris* to hang onto net and wire container (Hughes 1972; Chittenden 1973); following thrushes *Turdus* in trees to take seeds from apples they had hacked open (Löhrl 1982). Eats seeds and flesh of rowan *Sorbus* berries, but only seeds of (e.g.) guelder rose *Viburnum* (Eber 1956; Jenni and Neuschulz 1985). Recorded taking aphids (Aphidoidea) from buds with wiping motion of bill rather than typical picking action of *F. coelebs* (Diesselhorst 1971a). In Switzerland in winter, because of danger from aerial predators, e.g. Peregrine *Falco peregrinus* and particularly Sparrowhawk *Accipiter nisus*, fed on exposed ground for average of only *c.* 10 s at a time, carrying beechnuts (etc.) into trees for consumption, but very cold weather and heavy snow can result in much longer foraging periods, and reduction in fear of predators and man, birds even entering houses in search of food and occurring in such numbers near built-up areas that thousands of birds can be killed on roads and traffic forced to stop (Osieck 1973; Schäpper 1986). See also Lindström (1990) for effect of *A. nisus* pressure on feeding behaviour of birds migrating through southern Sweden. Average time taken by ♂ to open beechnut and eat all fragments 51 s ($n = 10$), but only 25 s ($n = 8$) to eat sprouting nut in February; by ♀, 93 s ($n = 8$) and 33 s ($n = 9$) (Schäpper 1986). Captive birds took average 9 s to open beechnut (time to open other seeds ranged from 3 s for *Brassica rapa* to 16 s for sunflower *Helianthus*); in choice experiments, 80% of seeds taken were beech, 11% Canary grass *Phalaris*, 9% hemp *Cannabis*, and 1% *B. rapa* (Ziswiler 1965, which see for bill morphology and seed-handling in comparison with other species). See also Eber (1956) and Newton (1967a) for similar studies.

Diet in west Palearctic includes the following. Invertebrates: springtails (Collembola), stoneflies (Plecoptera), earwigs (Dermaptera), bugs (Hemiptera: Cicadellidae, Psylloidea, Aphidoidea), adult and larval Lepidoptera (Geometridae), caddis flies (Trichoptera), adult and larval flies (Diptera: Tipulidae, Culicidae, Phoridae), Hymenoptera (Tenthredinidae, Ichneumonidae, ants Formicidae, wasps Vespidae, bees Apoidea), adult and larval beetles (Coleoptera: Carabidae, Hydrophilidae, Staphylinidae, Scarabaeidae, Cantharidae, Elateridae, Cerambycidae, Chrysomelidae, Curculionidae), spiders (Araneae), harvestmen (Opiliones), mites (Acari), pseudoscorpions (Pseudoscorpiones), centipedes (Chilopoda), woodlice (Isopoda), snails (Pulmonata). Plants: seeds, fruits, etc., of spruce *Picea*, pine *Pinus*, larch *Larix*, yew *Taxus*, juniper *Juniperus*, poplar *Populus*, birch *Betula*, alder *Alnus*, hornbeam *Carpinus*, beech *Fagus*, oak *Quercus*, maple and sycamore *Acer*, lime *Tilia*, rowan, etc. *Sorbus*, blackthorn, etc. *Prunus*, hawthorn *Crataegus*, lady's mantle *Alchemilla*,

tormentil *Potentilla*, ash *Fraxinus*, olive *Olea*, privet *Ligustrum*, elder *Sambucus*, guelder rose *Viburnum*, dogwood *Cornus*, *Robinia*, crowberry *Empetrum*, bilberry, etc. *Vaccinium*, knotgrass *Polygonum*, sorrel, etc. *Rumex*, goosefoot *Chenopodium*, orache *Atriplex*, chickweed, etc. *Stellaria*, spurrey *Spergula*, buttercup *Ranunculus*, globeflower *Trollius*, rocket, etc. *Sisymbrium*, rape, etc. *Brassica*, charlock, etc. *Sinapis*, radish *Raphanus*, shepherd's purse *Capsella*, trefoil *Lotus*, wood-sorrel *Oxalis*, flax *Linum*, violet *Viola*, chickweed wintergreen *Trientalis*, rock-rose *Helianthemum*, evening primrose *Oenothera*, dead-nettle *Lamium*, hemp-nettle *Galeopsis*, cow-wheat *Melampyrum*, plantain *Plantago*, cudweed *Gnaphalium*, wormwood, etc. *Artemesia*, sunflower *Helianthus*, wood-rush *Luzula*, sedges *Carex*, grasses (Gramineae), wheat *Triticum*, maize *Zea*; all kinds of seeds and nuts at feeding places. (Guéniat 1948; Schifferli 1953; Dementiev and Gladkov 1954; Gothe 1954; Olsson 1954; Eber 1956; Creutz 1961; Turček 1961; Newton 1967a, 1972; Ptushenko and Inozemtsev 1968; Tutman 1969; Eriksson 1970b; Jenni and Neuschulz 1985; Hogstad 1988; Snow and Snow 1988; Semenov-Tyan-Shanski and Gilyazov 1991.)

In central Norway, mid-May to mid-June, 2626 seeds from 58 stomachs comprised 62·7% (by number) *Vaccinium* (almost all bilberry *V. myrtillus*), 17·1% grasses and sedges, 8·2% crowberry, 4·2% tormentil, 2·3% *Ranunculus*, 1·7% chickweed, 1·7% other herbs, and 2·2% unidentified; from mid-June to mid-July, 2344 invertebrates from 23 stomachs comprised 55·8% (by number) beetles (51·1% adults; 33·4% Curculionidae, 14·3% Chrysomelidae, 1·6% Staphylinidae), 21·1% larval and pupal Lepidoptera (Geometridae), 12·0% Diptera (7·5% adults), 9·8% Hymenoptera (9·0% adults; 9·3% Tenthredinidae), and 1·7% spiders. (Hogstad 1988.) In Switzerland, winter, 201 invertebrates in 133 stomachs comprised 17·9% (by number) spiders, 16·5% beetles (12·5% adults; 9·0% Curculionidae, 0·5% Staphylinidae, 0·5% Chrysomelidae), 16·4% earwigs, 14·4% springtails, 11·9% Diptera (1·5% adults), 2·5% pseudoscorpions, 2·0% Hemiptera, 1·5% centipedes, 1·0% harvestmen, 1·0% mites; remainder unidentified; only 69% of stomachs contained invertebrates and proportion by weight compared to beechnuts in stomachs extremely small; proportions of invertebrates in diet corresponded to supply on beech wood floor (Jenni and Neuschulz 1985). For Kola peninsula (north-west Russia), see Novikov (1952) and Semenov-Tyan-Shanski and Gilyazov (1991). In central Norway, almost all foraging on ground until early June; late June–early July *c.* 90% of feeding in trees, then in late July increase again in ground feeding. Takes almost exclusively seeds on ground but also beetles, spiders, and Tipulidae larvae (Diptera), then sudden rise in invertebrate food (mainly leaf-eating caterpillars and beetles) to 100% of diet around mid-June, this change complete within *c.* 10 days; in cold spells may return to foraging on woodland floor and proportion of invertebrates can fall again. (Hogstad 1988.) See also Angell-Jacobsen (1980) for western Norway. See Mikkonen (1984) for change in feeding habitat in Finland between arrival and breeding. Like *F. coelebs*, eats larger seeds than Carduelinae (Ziswiler 1965); beechnuts eaten in fragments 4–12 mm long (Guéniat 1948). Average length of invertebrate prey items (excluding larvae) 6·4 mm; average size of beetle in adult stomach 5·9 mm ($n = 923$), larger than average of those available in trees, 4·4 mm ($n = 1209$); early in season, *Epirrita* caterpillars present in large numbers in canopy but only taken later when 16–19 mm long (Hogstad 1988).

In southern Sweden, birds feeding on rape seeds had average energy intake 87·9 kJ per hr (21 seeds per min) ($n = 38$); feeding on beechnuts in trees 0·1 nuts per min ($n = 21$), and on ground 0·4 per min ($n = 30$) giving average intake per bird feeding on beechnuts of 41·3 kJ per hr. Upper limit of processable food *c*. 163 kJ per day, so limit reached after *c*. 2 hrs feeding in rape fields and 4 hrs in beechwood; both these food sources particularly energy-rich. (Lindström 1990.) See also Nilsson (1979); for daily consumption in Switzerland, see Guéniat (1948) and Jenni and Jenni-Eiermann (1987). Winter birds often described as very fat (Granvik 1916; Guéniat 1948). For metabolism of captive birds, see Pohl (1971*a*).

In central Norway, 23 nestling stomachs contained 773 items, of which 58·1% (by number) larval Geometridae, 29·6% adult beetles (19·5% Curculionidae, 9·7% Chrysomelidae), 10·7% Tenthredinidae (8·4% adults, 2·3% larvae), 1·3% adult Diptera, and 0·3% snails. Proportion of caterpillars in nestling diet fell from 88% by number during days 1–2 to 20% on day 13, and beetles rose from 12% to 70%. Average size of prey items (excluding larvae, almost all of which 20–25 mm long) 7·4 mm. Of 753 invertebrates sampled in surrounding trees, 18% were Geometridae caterpillars, 12% beetles, and less than 1% Tenthredinidae; relatively large and slow-moving insects taken preferentially. (Hogstad 1977, 1988.) Tipulidae larvae (Diptera) can also be important in diet of young (Newton 1972). For Moscow region (Russia), see Ptushenko and Inozemtsev (1968).                                    BH

**Social pattern and behaviour.** Overall, less well-studied than Chaffinch *F. coelebs*, and better known in winter (especially roosting and flocking) than in breeding season, best study of which (including comparison with *F. coelebs*) is summarized in Mikkonen (1985*b*).

1. Generally gregarious outside breeding season, including during migration, and sometimes forms enormous flocks for feeding, while those for roosting may be of many millions. In Switzerland, flocks on autumn migration during 1965 irruption averaged 70–100 birds, several such flocks often passing per minute (Zinnenlauf 1967). Particularly large concentrations occur when beechmast *Fagus* crop good over large area (e.g. Switzerland and southern Germany); in 1950–1 winter, *c*. 100 million present in or passed through Switzerland according to Schifferli (1953), though different method suggested 4–10 million more likely total (Jenni and Neuschulz 1985, which see for discussion of census methods); for mass concentrations in Europe

1900–83 and their dependence upon beechmast and effect of snow cover, see Jenni (1987). Examples of feeding flocks in Switzerland include 200 000–300 000 in *c*. 4 ha of forest (5–8 birds per m²) and once a flock covering *c*. 1 km² (Schifferli 1953); in eastern Germany, tens of thousands in *c*. 3 ha of forest (Ruthenberg 1968). Roosts sometimes contain up to *c*. 20 million birds (Jenni 1987); in Pyrénées-Atlantiques (France), *c*. 15 million together with *c*. 120 000–150 000 Starlings *Sturnus vulgaris* (Alberny *et al.* 1965); also in Basses-Pyrénées, *c*. 410 000 (Hémery and Pascaud 1981). For details, see Roosting (below). For winter dispersion according to sex, and for winter site-fidelity, see Movements. Flocking during spring migration (April–May) studied in northern Finland. Tendency to flock more pronounced than in *F. coelebs*, though some migrate singly: *c*. 95% of birds were in flocks and *c*. 81% of all flocks pure *F. montifringilla*, others mixed with *F. coelebs*. Average size of pure *F. montifringilla* flocks 12·9 ± SE1·2 ($n = 412$), of mixed *F. montifringilla/F. coelebs* (former predominant) 36·6 ± SE7·0 ($n = 95$). Flock size greatest at peak of migration (largest *c*. 750), though large flocks occasionally recorded late May. In bad weather, local breeders may abandon territories and feed again in flocks for several days. (Mikkonen 1981*a*, *b*, 1984.) In Switzerland, autumn passage migrants are mixed with *F. coelebs* and feed on fields with other Fringillidae; during irruptions pure *F. montifringilla* flocks follow later and feed in beechwoods (Zinnenlauf 1967). Flocks mixed with *F. coelebs* are especially well integrated (Newton 1972), but also commonly associate with (e.g.) buntings (Emberizidae), sparrows *Passer*, and thrushes *Turdus* (England 1951; Richters 1952; Schifferli 1953; Verheyen 1954; Rinnhofer 1968; Marquardt 1975). BONDS. Mating system monogamous (Fonstad 1984) and, unlike in *F. coelebs*, of ♀-defence rather than resource-defence type (Mikkonen 1985*a*). In Hedmark (Norway), 2 ♀♀ recorded simultaneously, each with 3–4 eggs, in double nest (Mehlum 1978); no information on ♂(♂) paired with these birds. Of floating population of 40 non-breeders at Ammarnäs (Swedish Lappland), 75% ♂ (Cederholm *et al.* 1974); in central Finland, 10–30% of territorial ♂♂ (including occasionally some early settlers) unpaired (Mikkonen 1985*a*). ♂ paired with ♀ *F. coelebs* in western Norway had 4 hybrid young which died (Folkestad 1967); further reports of (apparent) hybrids from Finland (Vickholm *et al.* 1981), Switzerland (Jenni 1985), Italy (Moltoni 1956), Croatia (Rucner 1973*a*), western Russia (Paevski 1970*b*), and (captive birds) Cvitanić (1986). ♂ associated with *F. coelebs* pair in Scotland helped to feed young (apparently pure *F. coelebs*) and all 3 birds showed alarm when nest threatened (Mullins 1984*c*). Pair-formation often takes place in flocks on breeding grounds (see Heterosexual Behaviour, below). Little information on duration of pair-bond, but in one case (Yamal peninsula, northern Russia) maintained when nest predated (Danilov *et al.* 1984); not known to persist between years, and unlikely to do so in view of low incidence of site-fidelity (see Breeding Dispersion, below). Role of sexes in nest-duties as follows: only ♀ builds nest and incubates, but both sexes (probably some variation) feed young (Witherby *et al.* 1938; Hogstad 1977, 1982). ♀ recorded rearing young alone (Stašaitis 1982; Danilov *et al.* 1984). Age of first breeding 1 year (Paevski 1985); not known when young become independent. BREEDING DISPERSION. Territorial and sometimes solitary, but more commonly in loose neighbourhood groups, with report of several ♂♂ defending same nest (Montell 1917; Udvardy 1956; Mikkonen 1985*a*). In central Finland, usually 3–5, maximum 8, territories in group (Mikkonen 1985*a*). At Ler (Norway), where birds in colony of Fieldfare *Turdus pilaris* (see below for details), nests clumped, 2 successful ones only 19 m apart (Slagsvold 1979*c*). On southern edge of range in Gor'kiy region (east European Russia), c. 500 m between one nest and nearest other pair (Anikin 1963). Settlement and territory establishment more synchronous than in *F. coelebs*: primarily correlated with snowfall and temperature. Unlike in earlier-arriving *F. coelebs*, birds separating from migrant flocks tend to disperse in loose groups (or singly), group of singing ♂♂ eventually remaining in one site and ♀♀ also joining or staying near such 'display flocks'. First ♂♂ tend to settle around good food-supply or (see Slagsvold 1979c) in *T. pilaris* colony when forest virtually free of snow. (Mikkonen 1984, 1985a.) In Budal (central Norway), after arrival during last third of May, territories established within 1 week (Hogstad 1982). ♂♂ starting to sing regularly display to ♀♀ and defend small, unstable song-territories in groups of 2–8, tens of metres apart. Such groups attract newcomers, which also sing and court ♀♀, to settle nearby. (Mikkonen 1985a.) On small island (southern Finland), lone, unpaired ♂ defended song-territory of c. 700 m², but shifted frequently (up to c. 100 m) over 25-day period (Bergman 1952). Study in northern Finland found part (sometimes whole) of singing group changing, especially during cold weather in early spring. Annually, in breeding season, c. 20–40% of territorial ♂♂ left study area. Late-arriving pairs settle at vacant sites or weakly defend part of another's territory. If birds pair up in flock, they arrive at nesting location together and ♀ chooses nest-site, usually within ♂'s song-territory or nearby, ♂ then defending area around ♀ and nest, though nest and vicinity also defended by ♀. (Mikkonen 1985a.) On Yamal peninsula, when ♂ experimentally removed, ♀ defended part of territory alone (Danilov *et al.* 1984). Definite nesting-territory established (and simultaneously expanded) only after nest-site chosen, usually just before or after pairing, and maintained only while nest in use (Mikkonen 1985a). At Ammarnäs, breeding ♂♂ found to wander far from territory when ♀ incubating (Cederholm *et al.* 1974). In Murmansk region (north-west Russia), family stayed within c. 8 ha for c. 1 week after fledging (Gavrilo 1986). On Yamal peninsula, limits of song-territories unstable, but ♂♂ sang in discrete, non-overlapping territories during nest-building and later (Danilov *et al.* 1984). Territory limits in central Finland initially diffuse, but stabilized (in some cases) within c. 3 weeks; may change if territory-owners die or disappear. ♂'s song-posts tend to be near boundaries or in core area near nest. Feeding and watering places often located outside territory. (Mikkonen 1985a.) Little concrete information on size of nesting territory. ♀ in Scotland generally fed within c. 30–50 m of nest, ♂ over wider area (Bucknall 1983). For ♂ associating with nesting ♀ *F. coelebs*, see Hornbuckle (1984). In Budal, when nest abandoned after 1st egg laid, 2nd nest built c. 30 m away (Fonstad 1981). Density high where food supply temporarily abundant (Mikkonen 1985a). Typically the most numerous species (often together with Willow Warbler *Phylloscopus trochilus*) in most forest types of Fenno-Scandia, notably in birch *Betula*, sometimes also spruce *Picea* (e.g. Moksnes 1971, Røv 1975, Virkkala 1989). At Ammarnäs in subalpine birch forest, over 9 years, 22–107 territories per km² (Enemar and Sjöstrand 1970; Cederholm *et al.* 1974); density positively correlated with that of *Epirrita* larvae (Lepidoptera), with up to 125 ♂♂ per km² at peak (Lindström 1987). See also (e.g.) Enemar (1964), Ytreberg (1972), Ulfstrand and Högstedt (1976), and Järvinen and Pietiäinen (1982). At Budal, average over 3 years 39 territories per km² (35–45) in mixed forest, 41 (32–52) in heath birch forest; average 40·3 (25–52) over 1966–76 (Hogstad 1969, 1988). High densities also recorded in some coniferous or mixed forest habitats: in Forradal (Norway), 100 territories per km² in spruce forest of c. 20 ha, up to 65 in mixed spruce and birch (Enemar 1964; Enemar *et al.* 1965; Moksnes 1971); see also Moksnes (1972). Locally 80–100 pairs per km² also in mixed riverine forest

(predominantly spruce and larch *Larix*) of Yamal peninsula (Danilov *et al.* 1984), while on Kola peninsula (north-west Russia), 21–29 pairs per km² in riverine spruce-birch forest (Semenov-Tyan-Shanski and Gilyazov 1991). At Utajärvi (64°57′N in central Finland), 1967–72, 18–24 pairs per km² in 68 ha of mostly pine *Pinus sylvestris* forest, or 20·2–27·7 calculated from 48 ha of forest (Mikkonen 1983). At Sodankylä (Finland, 68°N), *F. montifringilla* the most numerous species in all forest types, maximum 50·2 pairs per km² in birch; fivefold increase in density compared with 1950s; significantly higher densities recorded in virgin than in managed forest (for details, see Virkkala 1987, 1988, 1989). In isolated willow *Salix* forest, Bol'shezemel'skaya tundra (north-east European Russia), 10–16 pairs per km² (Morozov 1987). See also Shutov (1990) for 8 years' data from study in northern Urals. Commonly nests in same woods as *F. coelebs* and territorial overlap up to 90% (or even complete). For relationship with *F. coelebs*, see Antagonistic Behaviour (below). Interspecific feeding competition or territoriality reported with *P. trochilus* (Hogstad 1975; Røv 1975; Angell-Jacobsen 1980), but reduced territorial overlap more likely due to habitat preferences (Fonstad 1984; Mikkonen 1985a). Sometimes nests (like *F. coelebs*) close to *T. pilaris*, late arrival of *F. montifringilla* allowing it to select successful *T. pilaris* colonies: one colony in Norway occupying c. 12 ha contained 21 *F. montifringilla* territories out of 44 found, only 4 out of 27 *F. coelebs* territories (Slagsvold 1979c). Study in central Finland suggested fidelity to breeding site low or nil: of 68 birds ringed while breeding in study area, none found there in subsequent years. Settlement dependent on conditions in new area each year: spring weather, local food supply, also social establishment of nesting territory. (Mikkonen 1983, 1985a.) The few recoveries indicate some birds at least are nomadic, breeding at sites up to 600 km apart in different years (Lindström 1987). ROOSTING. In winter, nocturnal and typically communal, huge roosts (sometimes shared with other species) containing many thousands or millions of birds (Jenni 1987). Generally in trees, with dense conifers favoured (e.g. Granvik 1916, Guéniat 1948): in Switzerland, mainly in firs *Abies alba* 5–10 m tall (Mühlethaler 1952); in another study, perched at 5–15 m, less commonly 2–5 m (Lanz-Wälchli 1953). Birds sometimes assemble first in tall trees, then move to dense undergrowth (including *Rhododendron*) for roosting (e.g. Vleugel 1941, England 1951, Baur 1981, Norman *et al.* 1981). In south-east England, evidence suggested individuals regularly use same perch (England 1951). Emergency sites used particularly in bad weather include fruit trees and house roofs, also on ground in grass or small walled enclosure (Saxby 1874; Bolam 1912; Mühlethaler 1952; Zinnenlauf 1967). Sleeps in crouched posture, head tucked under wing (Mühlethaler 1952; Nardin and Nardin 1985). Very large roosts often in hollow or valley (typically on slope c. 100 m above main valley) and well protected against cold air and wind (Guéniat 1948; Mühlethaler 1952; Lanz-Wälchli 1953; Jenni 1991, which see for discussion of microclimate and energetic advantages). Predation apparently less than in small roosts. Large roosts rarely in best feeding areas, direction of flight and location of feeding grounds being strongly dependent on weather and snow cover (Jenni 1980, 1986). Birds in larger roosts (1–6 million) tend to be heavier and fitter; smaller roosts (10 000–30 000) are formed when snow cover hampers foraging, and birds then lighter (Jenni and Jenni-Eiermann 1987, which see for variation with sex and age). In Switzerland, early-arriving birds use well-scattered small roosts, in some cases for only one or a few nights, others for months; snow and ice may cause birds to abandon roost temporarily. Roosts grow rapidly from December, but numbers fall again as snow cover increases, until all birds shift to snow-free areas.

(Jenni and Neuschulz 1985.) In another Swiss study, birds used several temporary roosts during week before leaving winter quarters (Lanz-Wälchli 1953). Timing of arrival at and departure from roost clearly linked to sunset and sunrise (Vleugel 1941). Birds returned to one Swiss roost at *c.* 16.15 hrs in early January, 17.00–17.15 hrs in February, *c.* 17.30 hrs in late March (Guéniat 1948). Arrival times also depend on distance to feeding grounds, and in early April birds returned later when overcast than when sunny (Mühlethaler 1952). ♀♀ tend to change roost more than ♂♂ in snow (Schäpper 1986; see Food). Birds may travel up to *c.* 50 km from roost to feed; catchment area for one Swiss roost *c.* 300 km² (Guéniat 1948; Mühlethaler 1952; Schifferli 1953; Norman *et al.* 1981). For movement to and from roost, see Flock Behaviour (below). From end of February up to departure, birds in Switzerland recorded roosting in trees also during day (Géroudet 1952). On breeding grounds at *c.* 67°55′N on Kola peninsula, birds roost for *c.* 3½ hrs per night late May to late July (Novikov 1949); at Kilpisjärvi (69°N, Finland) during nestling phase, resting period of adults around midnight and chiefly controlled by light factor (Peiponen 1970); at Budal (62°42′N), 22.00–02.00 hrs (Fonstad 1981). At Ammarnäs (65°58′N), June–July (over 2 years), peaks of activity (indicated by maximum number of birds trapped) in early morning and evening (Frederiksson *et al.* 1973). For diurnal activity rhythm in captive birds, see Pohl (1971*b*) and Daan (1972).

2. Rather more wary than *F. coelebs* (Bannerman 1953*a*; feeding flocks observed by Guéniat (1948) extremely shy, but may also allow approach to *c.* 2–5 m, though more wary when increasingly excited closer to departure from wintering grounds (Lanz-Wälchli 1953*a*). Extreme tameness sometimes reported can be due to hunger (e.g. Schäpper 1986), but not always (Bannerman 1953*a*; Zimmermann 1987). Captive birds showing tendency to flee often moved wings and tail synchronously up and down with Flight-call (see 3 in Voice) (Hinde 1955–6). Flocking Behaviour. A wintering flock feeding on stubble was once seen to make sudden mass departure for roost (Alberny *et al.* 1965). Birds in southern Sweden seen to assemble first in deciduous wood, calling, flying far off and returning, numbers gradually increasing before moving to roost proper (Granvik 1916). Flight to roost usually direct, though short stopovers recorded: birds may form flocks of varying size, or one or several long columns each of which may contain millions. Attacks by various raptors frequent: in one Swiss study, birds in main column made sharp turns and dives when attacked from above by 3–4 Sparrowhawks *Accipiter nisus*. (Granvik 1916; Guéniat 1948.) For attempts to calculate numbers in such columns, see (e.g.) Mühlethaler (1952). Upon arrival at roost, may congregate in trees other than those used for roosting; call frequently and loudly (probably mainly Wayeek-call: see 2 in Voice), but abruptly silent when alarmed (Mühlethaler 1952). Prior to roosting and up to sunset, flocks typically perform aerial evolutions over forest, twisting and wheeling in unison (Norman *et al.* 1981); seen to move in 2 huge carousels, inner birds in different direction from outer, also flocks of 50–500(–1000) separating off from main flock, all members moving in perfect coordination, even simultaneously defecating over water (Guéniat 1948). Often then return to trees where earlier congregated (Mühlethaler 1952). At roost, fall silent for *c.* 1–2 s at any suspicious sound; fly up noisily if raptor approaches, dispersing rapidly then reassembling in different trees (Guéniat 1948), or zigzag around trees in tightly packed flock (Baur 1981); may first fly down to ground when disturbed, then hesitantly up to trees again (Granvik 1916). In northern England final move to roosting sites seen to involve flocks moving rapidly through trees, many then dropping directly into bushes for roosting (Norman *et al.* 1981); see also England (1951).

Roost-trees may be occupied from top to bottom; birds typically tightly packed on perches, sometimes touching (Granvik 1916; Guéniat 1948; Mühlethaler 1952). Swiss study found much competition over perches at dusk, and large, heavy ♂♂ generally in centre of roost, young ♀♀ on periphery (Jenni 1986). Call loudly as last birds settle, but cease when dark (Granvik 1916; Guéniat 1948; Baur 1981), though other studies have shown that birds never completely quiet (call when alarmed) and, particularly in upper part of roost, movement continues after those lower down are mostly asleep (Nardin and Nardin 1985). For further description of behaviour at roost, see Zinnenlauf (1967). Before leaving roost in morning, birds give loud chorus of twittering (see Voice); suddenly fall silent, then (sometimes after flying about) depart in small or several large flocks or continuous stream; generally more dispersed in rain or snow. Sometimes unite to form huge snaking column (e.g. taking *c.* 45 min to pass and comprising many millions). First birds to depart from one Swiss roost were from lowest perches, last from top. Usually fly direct to feeding grounds, though may rest briefly en route. (Granvik 1916; Guéniat 1948; Mühlethaler 1952; Alberny *et al.* 1965; Baur 1981.) Autumn flocks (mixed with *F. coelebs*) feeding on fields in southern Sweden flew into cover when attacked by *A. nisus* or for no obvious reason, re-emerging and progressing in leap-frogging movement also reported in beechwoods (Lindström 1990; also Mikkonen 1984, Nardin and Nardin 1985). With increase in flock size, birds often feed more in open. When flock attacked by raptor, birds may scatter and then re-assemble in trees, or large flock may fly up intact. Loud and intensive alarm-calls (presumably 2 in Voice) apparently given only after take-off, when predator nearby, but birds often discover danger *c.* 50–100 m away. (Lindström and Alerstam 1986; Lindström 1989, 1990.) Birds sometimes show little concern once raptor has killed and is still plucking prey, e.g. dozens perched within *c.* 50 cm of Goshawk *A. gentilis* (Mühlethaler 1952; also Schäpper 1986). Flocks on ground and in flight typically dense and highly synchronized in autumn, looser, less coordinated in spring. In beech forest (unlike in fields) seldom inactive for long periods. In trees and on ground below regularly attempt to steal unshelled beechnuts from one another, but little aggression recorded in fields. (Lindström and Alerstam 1986; Lindström 1989, 1990.) Individual distance often reduced virtually to nil when feeding in fields (Mikkonen 1984). Typically when birds hungriest (following long flight from roost), conflicts arise over food even where beechmast abundant; birds with beechnut tend to face away from opponents and are sometimes forced to edge of flock, but (where predation pressure low) return to centre after eating beechnut (see also Antagonistic Behaviour, below). Where threat from predators is high, birds may descend briefly from trees to feed (usually only *c.* 10 s) before flying back to perch for a few minutes, etc. Birds on edge of flock warn of danger, and all move off if *A. nisus* arrives and stays. Alarm-calls of other species elicit silence, birds then taking appropriate evasive action. (Schäpper 1986.) Feeding flocks in central Finland changed site several times during day, birds leaving site in groups or long chains (Mikkonen 1985*a*). In spring, migrants fly high on sunny mornings, often lower and in denser flocks when cloudy and later in day. Well-integrated flocks most common during peak migration, with cohesion presumably aided by calls 2–3 (see Voice) and particular plumage features. Large migrant flocks settling in trees and bushes may split into groups, with birds resting, preening or singing (Wheezing-song or chorus of Subsong and other calls: see Voice), also frequently chasing and fighting, all in highly synchronized fashion. Several ♂♂ often sing in loose group from one or a few trees (tend to perch over 50 cm apart, 20–50 cm otherwise when perched), song and display attracting ♀♀ and leading to courtship. (Mikkonen 1984, 1985a.) See also Song-display, Antagonistic Behaviour, and Heterosexual Behaviour (below). SONG-DISPLAY. ♂ gives Wheezing-song (see 1a in Voice) from tree-top or other well-exposed perch (Roseveare 1951; Bergman 1952; Mikkonen 1985a). Song-posts used are high up in early song-period, later also lower down, as usually with Subsong (see 1b in Voice) during courtship. Of 62 records of singing by one ♂ in central Finland, most song-posts used were at 6-14 m, higher than those of *F. coelebs* in overlapping territory. (Mikkonen 1985a.) Perches of one ♂ in northern England 3-5 m (1-20) (Hornbuckle 1984), of autumn flock in Switzerland *c.* 4-10 m (Lanz-Wälchli 1953). Song also noted from bird on ground (Bergman 1952). Groups of singing ♂♂ typical of migrant flocks (Mikkonen 1984). Each unit of Wheezing-song given with bill wide open and head thrown back (further than in *F. coelebs*) so that bill pointed up at *c.* 60° or almost vertical (Fig A). Wings

A

drooped (more so than in otherwise similar posture of *F. coelebs*) and white rump exposed; wings sometimes also flicked outwards, and tail partly spread. (Ruthke 1939b; Bergman 1952; Hornbuckle 1984.) Like *F. coelebs*, will make small jumps between song-bouts and may turn 180° (Bergman 1952); also sometimes makes low and silent flights of *c.* 5-10 m, gliding or with slow, flicking wing-beats (Bergman 1952; Hornbuckle 1984). North of Arctic Circle, singing begins around midnight or shortly after, later as season advances (e.g. Brown 1963). Thus at 67°55′N on Kola peninsula begins 00.30-00.55 hrs late May to early June, 00.25 hrs mid- to late June, 02.00-03.20 hrs late July; in late May-June, finishes 19.10-22.25 hrs; most songs (over 1280 per hr) around 15.00 hrs (Novikov 1949). In various studies in continuous daylight, song noted for most of day, activity pattern varying with latitude, season, and local circumstances: e.g. on Kola peninsula, 12 June, peaks at *c.* 03.00 hrs and (lower) *c.* 19.00 hrs, rest period *c.* 12.00-18.00 hrs (Novikov 1949). At 68°38′N, Finland, late June to early July, one ♂ had peaks after midnight and *c.* 16.00 hrs, with trough *c.* 19.00-23.00 hrs (Palmgren 1935). At Abisko (68°21′N, Sweden), late June, none singing *c.* 22.00-24.00 hrs (Armstrong 1954). At Budal (62°45′N, Norway) late May to early July, song noted mainly 02.00-22.00 hrs, average 0·6 ± SD1·4 songs per min (Fonstad 1981). Sings from late winter or spring before departure or en route for breeding grounds (Witherby *et al.* 1938; Moritz 1982; Witt 1988), also in autumn flocks (e.g. Witherby *et al.* 1938, Lanz-Wälchli 1953, Semenov-Tyan-Shanski and Gilyazov 1991). In eastern England, lone ♂ sang late May to early July (May 1951). Following break-up of winter flocks in southern Finland, scattered ♂♂ sang from mid-March (Eriksson 1970b); song-period late April to mid-July for birds unusually staying to breed there (Bergman 1952). At 64°57′N in central Finland, song given

in late May during laying (Mikkonen 1985a); similarly, at Budal, peak of song in 2 years early June, coinciding with egg-laying (Fonstad 1981). On Kola peninsula, birds sing from arrival in early May to end of July, peaking during first half of June (Semenov-Tyan-Shanski and Gilyazov 1991). On Bol'shezemel'skaya tundra over 4 years, song-period early June to late July (Morozov 1987). ANTAGONISTIC BEHAVIOUR. In flocks just arrived at feeding sites from roost, hungry birds frequently involved in brief conflicts over beechnuts, but no obvious dominance of one sex over the other (Schäpper 1986), though Hinde (1955-6) found captive ♂ to dominate ♀ in winter. In Threat-posture, bird crouches and faces opponent, raising wings, sometimes moving them rapidly up and forward; may also gape, Bill-snap, and give Threat-call (Hinde 1955-6: see 5 in Voice). Bird under attack tries to keep back to opponent and, if further pressed, will flutter, hop, or run to edge of flock where several form ring or semicircle, facing outward (Schäpper 1986). On perch, may sink down, head withdrawn, plumage markedly ruffled, or crouch in a horizontal posture, feathers sleeked, prior to fleeing (Hinde 1955-6). Little evidence of antagonism between conspecific birds on breeding grounds, but on Yamal peninsula during early settlement period, fights and chases noted following encroachment on song-territories (Danilov *et al.* 1984). Sometimes behaves very aggressively towards *F. coelebs*, ♂ (dominant over ♂ *F. coelebs*) especially during peak courtship, ♀ especially near nest, though the 2 species forage (see Food) and sing at different heights (Mikkonen 1984, 1985a); see also Järvinen and Väisänen (1979) and Sæther (1982). In Finland, unpaired ♂ dominated *F. coelebs* on feeding grounds and in own territory, usually quickly chasing that species away, and particularly attacking ♂♂ singing from low perch (Bergman 1952); in England, ♂ seen locked together in aerial tussle with ♂ *F. coelebs*, birds almost falling to ground (May 1951); see also Hornbuckle (1984). In northern Sweden, where scarce *F. coelebs* population overlaps with *F. montifringilla*, no interspecific aggression over feeding recorded where *F. montifringilla* already breeding when *F. coelebs* arrived unusually late (Udvardy 1956). On Kola peninsula, aggression recorded near nest and on feeding grounds towards (e.g.) White Wagtail *Motacilla alba* and Pine Grosbeak *Pinicola enucleator*, and larger species such as Siberian Jay *Perisoreus infaustus*, Hawk Owl *Surnia ulula*, and *A. gentilis* (Semenov-Tyan-Shanksi and Gilyazov 1991). Where *F. montifringilla* coexists with *P. trochilus*, the 2 species will forage close together amicably; in all chases observed, *P. trochilus* the pursuer (Fonstad 1984). HETEROSEXUAL BEHAVIOUR. (1) General. ♂♂ apparently leave winter quarters first (e.g. Mühlethaler 1952, Zinnenlauf 1967), and migrate ahead of ♀♀ through western Russia (Paevski 1968). In central Finland, first ♀♀ arrived only when *c.* ⅓ of migration period over (Mikkonen 1981a); some ♀♀ present in early flocks reaching Yamal peninsula (Danilov *et al.* 1984). In spring, pair-formation often rapid (unlike in *F. coelebs*), taking place in flock (mobile display groups) or in song-territory, before nesting territory established; first pair formed within *c.* 4 days of arrival of first ♀♀. (Mikkonen 1984, 1985a.) At Budal, most birds are paired within a few days of establishment (within a week of arrival) of territories (Hogstad 1982). Captive birds performed courtship May-August, also (rarely) in autumn following moult (Natorp 1940). For comparison of courtship behaviour with *F. coelebs* and other Fringillidae, see Hinde (1955-6). (2) Pair-bonding behaviour. In eastern England, early June, ♂ seen flying with shallow, quivering wing-beats towards ♀ *F. coelebs* (May 1951). Captive ♂ often attacked and chased ♀ early in season (Hinde 1955-6). ♂ displaying to ♀ at close quarters (e.g. within *c.* 30-50 cm) adopts Courtship-posture (Fig B): slightly crouched and tends to be side-on to ♀; may turn or

B

sway from left to right (Hinde 1955-6; Roworth 1983; Mikkonen 1985*a*), and one displaying ♂ circled stationary ♀ *F. coelebs* (Roworth 1983). Wings drooped, that closer to ♀ more so, with shoulder slightly raised and carpal joint held away from body. Whole colour pattern of one side displayed to ♀, with attention drawn to bright orange-buff scapulars, white wing-bars, conspicuously ruffled white rump, and upper surface of tail which is slightly spread and raised; head (feathers ruffled) extended forward and often down. Displaying ♂ gives Wheezing-song or Subsong. (Natorp 1940; Roworth 1983; Hornbuckle 1984; Mikkonen 1985*a*.) When displaying to *F. coelebs*, one ♂ stretched head and neck forward, drooped wings, and raised tail at each 'call' (presumably song-unit: see 1a in Voice), and each display sequence completed with vigorous bill-wiping (Roworth 1983). Courtship-posture equivalent to Lop-sided display in *F. coelebs* (Hinde 1955-6). (3) Courtship-feeding. None, as in *F. coelebs*, according to Newton (1972). In Scotland however, ♂ visited nest 13 times in *c.* 6 hrs, feeding incubating ♀ 8 times; ♀ also left frequently to feed herself (Bucknall 1983). See also Mating (below). In Iceland, ♂ also occasionally fed incubating ♀ (Bárdharson 1986). (4) Mating. In Scotland, ♂ mounted ♀ after chase and fed her during copulation; birds afterwards went their separate ways (Bucknall 1983). Captive ♂ seen to approach ♀ (perhaps as prelude to copulation) in an upright posture resembling Upright-lopsided posture of ♂ *F. coelebs*; also performed series of fluttering dives near mate (Hinde 1955-6). (5) Nest-site selection. By ♀, ♂ playing no part and (like *F. coelebs*) having no nest-calling display (Hinde 1955-6; Mikkonen 1985*a*). (6) Behaviour at nest. In Iceland, ♂ sang from perch near nest when ♀ incubating (Bárdharson 1986). Scottish ♀ left nest quickly and (as on return) silently (Buckland and Knox 1980). RELATIONS WITHIN FAMILY GROUP. In Iceland, nestlings brooded by ♀ when small; fed by both parents, who also both disposed of faecal sacs and sometimes arrived at nest together (Bárdharson 1986). ♂ sometimes passes food to ♀ for transfer to young (Newton 1972: Fig C). Young leave nest at 10-14 days, earlier if

C

disturbed (Hogstad 1977; Danilov *et al.* 1984; Gavrilo 1986). One study found fledglings to disperse on ground within *c.* 25 m of nest, moving little for 1-3 days and having no contact with one another for *c.* 10 days; then follow parents, brood sometimes

reuniting at this stage (Gavrilo 1986). ANTI-PREDATOR RESPONSES OF YOUNG. Left nest at 9 days when disturbed (Danilov *et al.* 1984). Single nestling about to fledge pecked at observer (Erdmann 1972). PARENTAL ANTI-PREDATOR STRATEGIES. (1) Passive measures. ♀ a tight sitter (Bannerman 1953*a*). (2) Active measures: against birds. Birds nesting in *T. pilaris* colony more aggressive than that species towards stuffed Cuckoo *Cuculus canorus*; punctured own eggs or abandoned nest (Moksnes and Røskaft 1988). Pair seen vigorously attacking *P. infaustus*, driving them away from nest (Vorobiev 1963); for attacks on this and other species, see also Antagonistic Behaviour (above). (3) Active measures: against man (no information for other animals). Alarm at nest normally indicated by Wayeek-call from ♀ and call 4a (see Voice) from (often more excited) ♂ (Ruthke 1939*b*; Bourne and Nelder 1951; Armstrong 1954); Wheezing-song also given (not clear whether by both sexes) when disturbed (Danilov *et al.* 1984). In northern Germany, ♂ guarding single nestling close to fledging attacked observer several times and finally stayed *c.* 2 m away with plumage ruffled and tail spread, giving Wayeek-call (Erdmann 1972).

(Figs by D Nurney: A from drawing in Jonsson 1979; B based on descriptions in the literature; C from photograph in Newton 1972.)                                                    MGW

**Voice.** Song markedly different from Chaffinch *F. coelebs*, though *F. montifringilla* occasionally mimics song and calls of that species and voice of rarely occurring hybrids may further cause confusion (see call 1c). Nasal Wayeek-call and Flight-call given freely throughout year (Witherby *et al.* 1938; Natorp 1940). Deafening chorus of 'twittering' (see call 1b) from enormous roosting flocks (see Social Pattern and Behaviour) audible over several km, and considerable wing-noise (no evidence of signal function) created by such flocks leaving roost (Granvik 1916; Guéniat 1948). Bill-snapping occurs in antagonistic context (Hinde 1955-6). For additional sonagrams, see Marler (1957), Thorpe (1958), Bergmann and Helb (1982), and Fonstad (1984).

CALLS OF ADULTS. (1) Song. (1a) Wheezing-song of ♂. Drawn-out, somewhat dreary or melancholy wheezing or bleating sound with initial crescendo after quiet start: 'vzhee' (Danilov *et al.* 1984), 'dshweeh' (Jonsson 1979), twanging, drawled 'dree-e-e' (Bannerman 1953*a*), or 'rrrrhee' (Bruun *et al.* 1986). Like wheeze of Greenfinch *Carduelis chloris*, but evenly pitched (not falling); given in persistent and monotonous series. Each unit *c.* 1 s, separated by pauses of 2-5 s; 20-32 units per min in one study. (Hortling 1938; Ruthke 1939*b*; May 1951; Roseveare 1951; Bergman 1952; Bannerman 1953*a*; Creutz 1961; Bergmann and Helb 1982; Bruun *et al.* 1986.) 1-2 harsh chirping sounds (Buckland and Knox 1980) sometimes interspersed between wheezing units, but long bouts of wheezing without chirps are common (P J Sellar). Recording (Fig I) contains wheeze preceded by chirping 'trrt' closely resembling a Rain-call of *F. coelebs*. Wheezes in other recordings from Sweden and Finland show moderate individual differences in pitch, band-width, and duration (see also Witherby *et al.* 1938 and Natorp 1940, including suggestion that wheezes vary geographically).

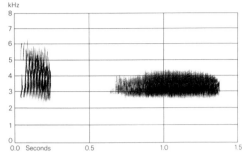

I   P J Sellar   Sweden   May 1971

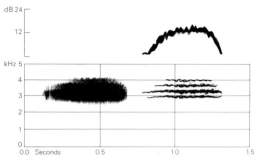

II   J-C Roché   Finland   June 1964

Wheeze shown in Fig II (wide- and narrow-band analysis, with amplitude trace above) divides and concentrates sound energy between 4 discrete narrow bands sounding simultaneously at half-tone to whole-tone intervals between 2·9 and 3·9 kHz (see narrow-band analysis in Fig II). Amplitude trace is good example of gradual crescendo followed by short diminuendo to end. (J Hall-Craggs.) Chirps also likened to food-call of recently fledged thrush *Turdus* (Jonsson 1979) and sometimes given in isolation (Witherby *et al.* 1938); see call 4c. Quiet twittering between wheezing units (Natorp 1940) probably same as chirping, though 'tjüpp' mentioned for this context by Bergman (1952) is perhaps more likely call 3. In Norway, neighbouring ♂♂ often sang alternating units, one giving wheeze, other responding with rasping chirps (E K Dunn). Recording by S Carlsson (Sweden) of large spring flock giving Wheezing-song suggests electric and mechanical buzzing or sawing; reminiscent also of loud chorus of cicadas (Cicadidae) (M G Wilson). On Yamal peninsula (northern Russia), when ♂ experimentally removed, typical Wheezing-song given by ♀ defending territory (Danilov *et al.* 1984). Wheezing-song reported by Danilov *et al.* (1984) to serve as alarm-call may refer to the frequent calls intermediate between wheezes and call 2 noted by Hinde (1955-6). (1b) Subsong. Various harsh chirping and squeaky units given at faster rate than Wheezing-song. In fragment illustrated by Bergmann and Helb (1982), 'zip chii tschrr'; similar 'tschrr tschrie tschra tschirrr tschrii' reported by Hortling (1938); 'ksiip' and 'grrr' given by ♂ during courtship (Mikkonen 1985a), harsh rattle grading into 'chirrl' like *F. coelebs*, and 'song' comprising series

of warbling notes with various calls incorporated (Hinde 1955-6) are probably also Subsong. Twittering (song) noted from autumn flocks (e.g. Hortling 1938, Lanz-Wälchli 1953) perhaps Subsong, or medley of calls, notably call 2. Witherby *et al.* (1938) treated Wheezing-song as 'spring-call of ♂', and 'true song' (given only at beginning of breeding season) said to be sweet and melodious, with several flute-like notes, somewhat recalling Redwing *T. iliacus*; probably refers to subsong, as nothing similar reported by other authors—though Wheezing-song said by Natorp (1940) sometimes to contain attractive fluting component. (1c) Songs recalling *F. coelebs* (including mimicry) and song given by possible hybrids. In recording (Fig III), song of high-pitched, rather metallic 'see' units

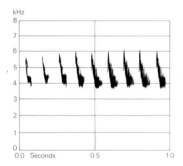

III   S Palmér/Sveriges Radio (1972-80)   Sweden   June 1964

suggests *F. coelebs* (Palmér and Boswall 1981) and may be mimicry. Another recording (Fig IV) contains truncated copy of *F. coelebs* song, comprising (in terminology of

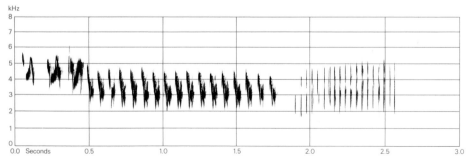

IV   C Watson   Scotland   June 1987

Thorpe 1958) sub-phrase 1a and greatly extended sub-phrase 1b, but omitting sub-phrase 2 and 3a–b; this followed by remarkable copy of rattle from song of Wren *Troglodytes troglodytes*. Same bird also precedes song with good imitation of *F. coelebs* 'huid' (Fig V), perhaps in place of chirping 'trrt'. (J Hall-Craggs.) Song of bird

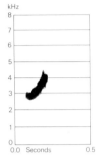

V   C Watson   Scotland   June 1987

thought to be hybrid with *F. coelebs* resembled that species, but less vigorous (Vickholm *et al.* 1982). (2) Wayeek-call. Characteristically nasal, hoarse, incisive and strangled, rather metallic sound—'wayeek' or 'wayk' (Fig VI). Further descriptions: 'teh-ehp' (Bruun *et al.* 1986);

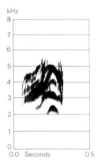

VI   J-C Roché   Finland   June 1964

'eahp' or 'chway', also harsh 'aehp' (Jonsson 1979); 'dschäe' or 'quäig' (Bergmann and Helb 1982); 'tsweek' or 'tswee-ik' (Witherby *et al.* 1938); 'zeeeuw' or 'zeeuw-ee', and 'chew' given in alarm (E K Dunn) probably also belong here. Given in flight or when perched, serving as contact-call in migrating flocks, also between pair-members on breeding grounds, and (especially loud and harsh) as main alarm-call (Hortling 1938; Witherby *et al.* 1938; Ruthke 1939*b*; Natorp 1940; Bergmann and Helb 1982). Complex call with wide frequency range given when mobbing owls (Strigiformes) and illustrated by Marler (1957) is probably also a Wayeek-call or variant. (3) Flight-call. Short, hard, slightly hoarse 'yeck', 'tjek', or 'chucc' given typically in series, rapid especially when taking off, and serving for close contact; sometimes conversationally quiet and unhurried. Some similarity to Flight-call of *F. coelebs*, but usually readily distinguishable through being more nasal, harder, and lower pitched. (Witherby *et al.* 1938; Natorp 1940; Jonsson 1979; Bergmann and Helb 1982;

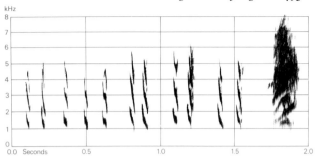

VII   S Carlsson   Sweden   September 1983

Bruun *et al.* 1986.) Recording (Fig VII) contains series of Flight-calls followed by nasal 'wayk' (call 2). (4) Alarm-calls (see also call 2). (4a) Quite loud, high-pitched, hard, ringing series of 'slitt' (Bruun *et al.* 1986) or penetrating, far-carrying 'zlip' sounds (Natorp 1940). Further descriptions: shrill, rather piping 'tsip' (Witherby *et al.* 1938); loud 'styp', like alarm-call of Tree Pipit *Anthus trivialis*, given up to 30 times per min by ♂ in northern England (Hornbuckle 1984); see also Hortling (1938). Recording (Fig VIII) similarly suggests alarmed *A. trivialis* (M G

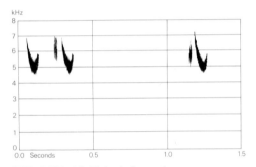

VIII   J-C Roché   Finland   June 1964

Wilson). Perhaps given mainly (even exclusively) by ♂ (Natorp 1940; Hornbuckle 1984). (4b) Calls described here probably at least related to 4a, and some perhaps only alternative renderings (see Bourne and Nelder 1951). High-pitched 'pee-eep pee-eep' given by ♀ disturbed at nest (Witherby *et al.* 1938); similar series of 'ruI' sounds noted from ♀ by Hortling (1929), 'choo-ee', also 'chee' (perhaps from both sexes) by E K Dunn, 'tsi' (Rosenberg 1953), and 'dsih' from ♂♂ in Lapland said by Ruthke (1939*b*) to resemble Yellow Wagtail *Motacilla flava*. (4c) Regular churr or rattle given by ♀ when disturbed (Witherby *et al.* 1938); milder version of rasping chirp in ♂'s song (see 1a) given by ♀ (E K Dunn) is probably the same. (4d) Rather hoarse, low-pitched 'chick-chick-chick' closely resembling call 3, but apparently serving as alarm-call of ♀ (Witherby *et al.* 1938). (5) Threat-call. Harsh 'zicker icker icker' and/or rattle reported by Hinde (1955–

6). (6) A 'chink' resembling 'pink' of *F. coelebs*, probably given by both sexes (E K Dunn).

CALLS OF YOUNG. Harsh 'dshr' and 'fd fd' reported for fledged young (Hortling 1938); hard 'chirr' (Witherby *et al.* 1938) presumably same as first of these.      MGW

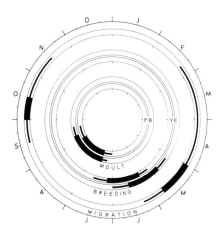

**Breeding.** SEASON. Finland: see diagram; eggs laid May in south, end of May to beginning of July in north (Haartman 1969); median laying date in central Finland over 6 years 27 May, *n*=6 (Mikkonen 1981, which see for comparison with Chaffinch *F. coelebs*). Norway: laying starts 2nd half of May in south, July(-August) in north (Haftorn 1971; Slagsvold 1979*c*, which see for comparison with *F. coelebs*; Hogstad 1982; Sæther 1982). Nest-building observed in Scandinavia alongside newly-fledged young, and season can vary considerably from year to year (Bourne and Nelder 1951; Bannerman 1953). On Kola peninsula (north-west Russia), nests built late May to end of June (Semenov-Tyan-Shanski and Gilyazov 1991). Scotland: eggs in one nest hatched early June (Bucknall 1983), in another year eggs found beginning of July (Buckland and Knox 1980). SITE. High in tree, often against trunk of conifer or in fork of deciduous tree (Hortling 1938; Haftorn 1971); 58% of 130 nests in Finland were in birch *Betula*, 26% in spruce *Picea*, 8% in pine *Pinus*, and 4% in alder *Alnus*; whether in deciduous tree or conifer depends on composition of forest. In Finland, average height above ground 3·3 m (1·0-10·0, *n*=59) in north, 6·0 m (1·5-15·0, *n*=58) in south where trees taller. (Haartman 1969.) Of 19 nests on Kola peninsula, 16 in fir *Abies* (9 against trunk, 7 on branch within 1 m of trunk), 2 in fork of birch, and 1 in pine; average height of 14 nests, 4·1 m (Semenov-Tyan-Shanski and Gilyazov 1991). Exceptionally in rocky scrub, even on ground (Makatsch 1976). Nests can be fairly close together, and often within colonies of Fieldfare *Turdus pilaris* (Bannerman 1953; Slagsvold 1979*c*). Nest: similar to that of *F. coelebs* but larger and more loosely built; outer structure of moss, lichen, grass, heather, strips of birch or juniper *Juniperus* bark, and cobwebs, lined with feathers (often many), moss, plant

down, soft grass, hair, fur, and sometimes paper, string, etc. (Hortling 1938; Haartman 1969; Buckland and Knox 1980; Bucknall 1983; Semenov-Tyan-Shanski and Gilyazov 1991). Outer diameter 10·5-15·0 cm, inner diameter 5·2-7·0 cm, overall height 7·0-12·0 cm, depth of cup 3·0-6·0 cm (Haftorn 1971; Bucknall 1983); see also Hortling (1938) and Semenov-Tyan-Shanski and Gilyazov (1991) for dimensions, and Tomek and Waligóra (1976) for comparison with other *Fringilla* nests. Building: by ♀ only (Haartman 1969; Haftorn 1971; Hogstad 1982). EGGS. See Plate 60. Sub-elliptical, smooth and glossy; very like those of *F. coelebs*, but greener, and with same wide variation; from clear light blue to dark olive-brown, with sparse to dense pink to rusty spots and blotches, and sometimes fine hair-streaks (Bannerman 1953; Harrison 1975; Makatsch 1976). 19·4 × 14·5 mm (18·1-22·2 × 13·5-15·7), *n*=663; calculated weight 2·14 g (Schönwetter 1984). For 155 eggs from Norway, see Haftorn (1971). Clutch: 5-7 (3-8); rare clutches of 9(-10) probably laid by 2 ♀♀. Of 53 clutches, Finland: 3 eggs, 2%; 4, 8%; 5, 26%; 6, 36%; 7, 26%; 8, 2%; average 5·8 (Haartman 1969). Average of 12 clutches in Norway 6·9 (Slagsvold 1979*c*), though in another study only 5·6, *n*=28 (Hogstad 1982); see also Slagsvold (1985) for clutch manipulation experiments; average of 14 clutches from Kola peninsula 5·4, and of 20 from Yamal peninsula (extralimital northern Russia) 5·5 (Danilov *et al.* 1984; Semenov-Tyan-Shanski and Gilyazov 1991). Eggs laid daily; replacement laid in new nest if 1st clutch lost (Hortling 1938; Haftorn 1971; Danilov *et al.* 1984). One brood in Scandinavia according to most authors, though Newton (1972) thought July clutches in south perhaps 2nd broods; 2 broods normal in Kareliya (north-west European Russia) (Belopol'ski 1962). INCUBATION. Last egg laid takes 11·5-12 days, *n*=15 (Haftorn 1971); by ♀ only (Hortling 1938; Haartman 1969; Haftorn 1971), beginning with last egg (Bannerman 1953). At nest in northern Scotland, ♀ left nest 21 times in 6 hrs, average break 4·7 min (maximum 15 min) and average stint on eggs 13·2 min (73 min) (Bucknall 1983); at another nest in same region, average break over 1·6 hrs of observation 5·3 min (4-7), average stint 15·3 min (12-18) (Buckland and Knox 1980). Eggs usually hatch within 1·5 days, but can be spread over several days (Haftorn 1971). YOUNG. Fed and cared for by both parents (Haartman 1969; Haftorn 1971; Hogstad 1982), though ♀ sometimes does much more than ♂ (Hortling 1938). FLEDGING TO MATURITY. Fledging period 13-14 days, *n*=20 (Hogstad 1977, which see for development of young); Haartman (1969) gave 12 days as typical, and apparently as little as 10-11 days recorded (Hortling 1938; Haftorn 1971; Danilov *et al.* 1984). Age of first breeding 1 year (Paevski 1985). BREEDING SUCCESS. In Norway, in year of low insect abundance, half of population left study area in 2nd half of June and no fledglings recorded; other insect feeders appeared to be more successful, and *F. montifringilla* perhaps poorly adapted to variable food supply (Hogstad 1982). Of 40 eggs, 90%

hatched (Slagsvold 1985); success apparently greater for nests within colonies of *T. pilaris* (Slagsvold 1979c). On Yamal peninsula, 59% of 109 eggs hatched and 57% produced fledged young (Danilov *et al.* 1984). Nests robbed by Carrion Crow *Corvus corone*, Siberian Jay *Perisoreus infaustus*, and vole *Clethrionomys*; often parasitized by Cuckoo *Cuculus canorus* (Bannerman 1953; Danilov *et al.* 1984). Of 12 nests on Kola peninsula, 3 destroyed, 1 abandoned, and 1 brood lost because of heavy rain (Semenov-Tyan-Shanski and Gilyazov 1991). BH

**Plumages.** ADULT MALE. In fresh plumage (autumn), forehead, crown, and each side of nape light greyish-buff, round-ended bluish-black centre of each feather just visible; central nape pale greyish-buff, some white and grey of feather-centres showing through. Mantle, scapulars, and back as crown, but often some more black visible and fringes sometimes warmer cinnamon-buff or dull cinnamon. Rump and shorter upper tail-coverts contrastingly white, slightly tinged cream when plumage quite fresh; lateral feathers black, partly fringed white or pale buff-grey, longer upper tail-coverts black with extensive light grey bases and fringes, partly washed buff on tips. Side of head down to lower cheek light greyish-buff or greyish-cinnamon, side of neck paler buff-grey, feathers with blue-black centres partly visible, side of head and neck appearing mottled, but black on lore, supercilium, and at rear of ear-coverts largely concealed. Chin to breast and upper flank tawny-cinnamon to deep cinnamon-rufous, feather-tips indistinctly fringed pale grey or buff, chin sometimes paler tawny-yellow; lower flank cinnamon-buff, some white of feather-bases showing through, each feather marked with rounded black dot subterminally. Thigh mixed black and cinnamon; belly and vent white, under tail-coverts white with cream or pale cinnamon wash. Tail black, central pair (t1) broadly fringed grey or grey-buff on tip; outer webs and tips of t2-t5 narrowly and inconspicuously edged grey-white; basal half of outer web of t6 white, bordered by oblong white patch along shaft on middle of inner web, extending into grey wedge towards tip, visible on undersurface of t6 only. Flight-feathers greyish-black, merging into deep black on basal two-thirds of outer web; outer web of p9 faintly edged white, outer web of p6-p8 fringed pale yellow in middle and light grey on emarginated part, terminal half of outer web of p1-p5 and of outer secondaries fringed pale yellow, terminal half of outer web of inner secondaries yellow-buff to dull cinnamon; extreme base of outer web of secondaries and p1-p6 white, partly tinged yellow when plumage quite fresh, white on bases of secondaries contiguous with cinnamon to buff-white tips of greater coverts, white on bases of primaries forming contrasting bar *c.* 2-6 mm wide along tips of primary coverts; long white wedge along basal border of inner web of flight-feathers, visible from below only. Tertials deep black with broad rufous-cinnamon or tawny fringe along outer web. Greater upper primary coverts and bastard wing deep black; greater upper wing-coverts deep black with contrasting tawny-buff to rufous-cinnamon tips 6-10 mm long on inner ones, narrower and paler buff-and-white on outer ones; lesser and median under wing-coverts bright tawny-orange, yellow of feather-centres showing through on lesser coverts, paler buff or whitish on tips of median coverts. Shorter under wing-coverts and bases of axillaries bright yellow, tips of axillaries, longer coverts, and under primary coverts white. Much influence of abrasion; by mid-winter and early spring, top and side of head and neck as well as mantle and scapulars glossy black with narrow buff scalloping (latter virtually absent on face, wider on rear of

supercilium, side of neck, and scapulars), central nape mottled black, buff, and white; in summer, June-July, when more heavily worn, top and side of head and neck and upperparts down to back uniform bluish-black (head still mottled brown in *c.* 20% of birds: Hogstad and Röskaft 1986), nape mottled white and grey, rump contrastingly white (some dull grey of feather-bases sometimes partly visible), chin to upper flank paler, more tawny, lower flank and under tail-coverts off-white, flanks more distinctly spotted black, tail without pale fringes, except at base of outer web of t6, pale yellow of fringes of flight-feathers bleached to whitish, partly worn off, fringes of tertials and tips of greater coverts pale cinnamon, bleached to white on extreme tips, medium coverts whiter, lesser coverts more yellow-orange. ADULT FEMALE. In fresh plumage (autumn), forehead and crown olive-grey to dull brown, each feather with limited amount of dull black on centre, largely concealed (black not as deep and glossy as in ♂, less broad, less contrasting with tips of feathers); feathers at side of crown (from above eye backwards) and along rear of crown more extensively and deeper black, sometimes slightly glossy, forming black border to crown. Lore and feathering round eye light brown-grey or grey-buff, cheek and ear-coverts darker brown-grey with faint paler shaft-streaks, sometimes black-brown at lower border of lower cheek, forming faint and short dusky malar stripe. Broad supercilium (backwards from eye), rear of cheeks, side of neck, and nape light ash-grey, feather-tips extensively washed buff-brown, partly concealing grey (especially on supercilium and cheek); each side of nape with contrasting deep black stripe, isolating extensive ash-grey patch on central nape. Mantle and scapulars dark grey-brown or black-brown, each feather with ill-defined paler brown-grey fringe; outer scapulars often blacker and with broader and more sharply demarcated dull cinnamon fringe. Back and side of rump dull black, mottled grey and off-white; centre of rump contrastingly white. Shorter upper tail-coverts black with narrow brown-grey fringes, longer tail-coverts dark grey-brown with broad pale grey-buff fringes. Chin to chest, side of breast, and upper flank tawny-cinnamon to cinnamon-rufous, often almost as bright as in adult ♂, but partly concealed by pale cream-buff feather-fringes, some white of feather-bases showing through on chin and throat, extending less far down to breast, and less sharply-defined from brownish ash-grey of cheek and white of belly. Flank tawny-buff, mixed with white and grey, lower flank with dark grey dots. Belly and vent cream-white; under tail-coverts white with pink-cinnamon or cream suffusion. T1 dark grey, remainder of tail greyish-black; narrow fringes along outer webs of t1-t5 pale yellow, pattern on t6 as adult ♂. Flight-feathers, greater upper primary coverts, and bastard wing greyish-black, basal outer webs of flight-feathers, greater coverts, and bastard wing not as deep black as adult ♂, but pattern and colour of pale fringes along tips of flight-feathers and of white bars along bases similar. Tertials and greater upper wing-coverts dull black, extent of pale fringes of tertials and of tips of greater coverts as in adult ♂, but colour duller rufous-cinnamon, merging into buff-and-white on tips of outer greater coverts. Lesser and median coverts rather different from those of adult ♂; mainly black, sometimes slightly glossy, median with contrasting rufous-cinnamon (outermost) or buff-white (innermost) tips *c.* 3-6 mm wide, lesser with narrower dull cinnamon fringes. Under wing-coverts and axillaries as adult ♂, but marginal coverts dark grey rather than black, and yellow of shorter coverts and axillaries sometimes paler. As in ♂, marked effect of bleaching and wear. By mid-winter, dull black of feather-centres of cap, mantle, and scapulars more exposed, but fringes still much broader and duller brown than in adult ♂ in same stage of wear; ash-grey of central nape, side of neck, and cheek more exposed, black stripe at each side of crown and nape and along rear of crown more distinct; tips of median and greater coverts whiter. When heavily abraded (June–July), cap dull black with fine grey scalloping, nape extensively paler ash-grey with pronounced black stripe at each side, side of head and neck ash-grey, palest round eye, darker and tinged brown on supercilium and ear-coverts, short black malar stripe often distinct; chin and upper throat dirty white, lower throat, chest, and side of breast with rather restricted amount of pale tawny-cinnamon, remainder of underparts dirty white with cream-buff tinge on upper flank and under tail-coverts and dark grey dots on lower flanks; fringes of tail, flight-feathers and tertials bleached to white, partly worn off; pale tips of median and greater coverts narrower, whiter, distinct. NESTLING. Down long and dense, white, restricted to tufts on crown, back, rump, and upper wing, more scanty on thigh and vent. JUVENILE. Head and body rather like adult ♀ in fresh plumage, differing mainly from adult ♀ by showing fresh plumage when ♀ heavily abraded; centres of feathers of cap, mantle, and scapulars duller than in adult ♀, dark grey-brown, less contrasting in colour with more olive-brown fringes; central nape and side of head and neck light brown-grey, palest round eye; rump tinged pale yellow; chin to chest buff-brown, less bright tawny-cinnamon, belly tinged yellow. Tail as adult ♀, but tips of feathers ending in obtuse point at shaft, less rounded (see Svensson 1992); t1 of ♀ dark grey, of ♂ grey with variable amount of black on centre. Wing as adult ♀ but fringes of flight-feathers slightly less sharply demarcated, primary coverts and bastard wing with faint pale fringe (widest in juvenile ♀); greater upper wing-coverts dull black or greyish-black with buff or white tips 2–4 mm wide, median and lesser coverts dark grey-brown or black-brown with buff-white or pale grey-buff tips, paler and wider on median coverts. FIRST ADULT MALE. Like adult ♂, but juvenile tail, flight-feathers, tertials, greater upper primary coverts, bastard wing, and variable number of outer greater upper wing-coverts retained. Tips of tail-feathers more pointed than in adult at same time of year, t1 often more extensively grey; black of bases of flight-feathers, primary coverts, bastard wing, and old outer greater coverts slightly more greyish-black, contrasting with deep black new inner greater coverts, less uniform deep black than in adult; sharp contrast in colour, width, and degree of abrasion between pale tips of new 1st adult and old juvenile outer greater coverts (in adult, tips of outermost also whiter, narrower, and more worn than those of innermost, but characters change more gradually, not abruptly). Head, body, and lesser and median upper wing-coverts as in adult ♂, but bases of median and longer lesser coverts sometimes more extensively black, partly visible among cinnamon-brown of lesser coverts (latter less rusty-orange and yellow than in adult ♂), lesser coverts often all or partly marked with black specks or streaks, less uniform than in most adults; buff fringes of cap, mantle, and scapulars on average broader than in adult at same time of year, black duller, traces of brown fringes remaining in summer (see Hogstad and Røskaft 1986 for variation); black on centres of feathers of mantle ends in blunt point, less rounded than in adult (see Svensson 1992), chin sometimes whitish. FIRST ADULT FEMALE. Like adult ♀, but part of juvenile plumage retained, as in 1st adult ♂. No clear difference from adult ♀ in colour of tail, flight-feathers, primary coverts, or bastard wing, but difference in shape of tips of tail-feathers and contrast in colour and wear between new inner and unmoulted outer greater coverts often visible; also, pale tips of median coverts sometimes narrower, 2–3 mm rather than 2–6 mm, throat, chest, and side of breast paler, buff-cinnamon, less rufous, and tips of lesser and median coverts sometimes duller cinnamon-buff.

**Bare parts.** ADULT, FIRST ADULT. Iris brown. Bill of ♂ light blue in summer, tip and distal part of cutting edges black; bright orange-yellow with black tip during rest of year; bill of ♀ dull blue-grey with black tip in summer, base variably tinged pink-flesh or yellow, ochre-yellow with grey-black tip during rest of year. Leg and foot flesh-pink, dull flesh-horn, brown-flesh, greyish-flesh, or dark flesh-brown, darkest in breeding ♂; upper surface of toes tinged light grey or dark plumbeous-grey. NESTLING. Mouth blood-red, gape-flanges yellow-white. JUVENILE. Bill, leg, and foot flesh with grey tinge, culmen darker leaden-grey or horn-grey. (RMNH, ZMA.)

**Moults.** ADULT POST-BREEDING. Complete; primaries descendent. In northern Sweden (c. 66°N), starts with p1 mid- to late July, completed with regrowth of p9–p10 mid-August to early September; estimated duration of primary moult 46 (♀) to 48 days (♂); secondary moult starts with shedding of s1 at primary moult score 14–27, completed with regrowth of s6 a few days after last primaries; tertials (sequence s8-s9-s7) mainly start at primary moult score 5–15, completed at 28–38 (Ottosson and Haas 1991). In northern Norway (70°N), moult well under way early August, completed late August to mid-September (Evans 1971). In Finland, moult mid-July to early September; ♂ starts c. 2 weeks before ♀; duration of primary moult c. 55–58 days (Haukioja 1971c; Lehikoinen and Niemelä 1977). In north-west Russia, moult starts late June to early August, on average c. 20 July; moult completed 20 August to late September, occasionally October, mainly late August or early September (Rymkevich 1990). In western Russia, moult on average 1 July to 20 August (Dol'nik and Gavrilov 1974). In western Siberia, starts late July to mid-August (Johansen 1944). POST-JUVENILE. Partial: head, body, lesser and median upper wing-coverts, and variable number of greater coverts (outer 1–6 usually retained, sometimes all). Moult at same time as adult post-breeding, generally completed before start of autumn migration.

**Measurements.** ADULT, FIRST ADULT. Scandinavia, mainly summer, and western and central Europe, mainly September–April; skins (RMNH, ZMA: A J van Loon). Bill (S) to skull, bill (N) to distal corner of nostril; exposed culmen on average 3·7 shorter than bill (S).

| | ♂ | | ♀ | |
|---|---|---|---|---|
| WING AD | 93·2 (1·74; 22) | 91–97 | 87·3 (1·86; 17) | 84–91 |
| 1ST AD | 92·8 (1·82; 48) | 89–97 | 86·4 (1·92; 42) | 81–90 |
| TAIL AD | 66·3 (2·30; 22) | 63–71 | 62·0 (2·11; 17) | 57–66 |
| 1ST AD | 65·4 (1·84; 47) | 62–70 | 60·7 (0·94; 40) | 56–65 |
| BILL (S) | 15·6 (0·75; 25) | 14·5–17·1 | 15·4 (0·41; 22) | 14·8–16·5 |
| BILL (N) | 9·6 (0·43; 70) | 8·7–10·5 | 9·3 (0·43; 59) | 8·4–10·2 |
| TARSUS | 18·8 (0·57; 65) | 17·1–19·7 | 18·6 (0·53; 58) | 16·9–19·4 |

Sex differences significant for wing, tail, and bill (N).

Wing. Live birds except as noted. Western Norway, autumn: (1) adult, (2) 1st adult (Albu 1983). (3) Central Norway, May–July; (4) Finland, May–June (Hogstad 1985, which see for variation between years and in different areas of Scandinavia). Eastern Germany: October–March, (5) adult, (6) 1st adult (Creutz 1961); (7) April, ages combined (Haensel 1967). Switzerland, winter: (8) adult, (9) 1st adult (Mühlethaler 1952). Netherlands, autumn: (10) adult, (11) 1st adult (A J van Loon). (12) South-east France, winter (G Olioso). (13) South-east Yugoslavia, winter; skins (Stresemann 1920). (14) Iran and Afghanistan, winter; skins (Vaurie 1949b). (15) Russian Lapland and Yamal peninsula, summer (Danilov et al. 1984; Semenov-Tyan-Shanski and Gilyazov 1991). (16) Japan, mainly adult, winter; skins (ZMA).

| | | | | | |
|---|---|---|---|---|---|
| (1) | ♂ | 95·1 (1·92; 364) | — | ♀ 89·2 (1·85; 110) | — |
| (2) | | 93·9 (1·94; 1445) | — | 89·0 (1·82; 950) | — |

| (3) | 90·3 (1·91; | 91) | 85–96 | 86·6 (1·72; | 49) | 82–91 |
|---|---|---|---|---|---|---|
| (4) | 91·1 (2·2 ; | 18) | — | 86·2 (1·1 ; | 7) | — |
| (5) | 93·3 ( — ; | 126) | 89–98 | 87·4 ( — ; | 59) | 82–91 |
| (6) | 91·7 ( — ; | 91) | 87–98 | 86·7 ( — ; | 73) | 84–91 |
| (7) | 90·3 (2·67; | 23) | 85–97 | 84·6 (2·18; | 24) | 80–88 |
| (8) | 93·8 ( — ; | 59) | 89–98 | 88·3 ( — ; | 26) | 85–95 |
| (9) | 92·8 ( — ; | 62) | 90–97 | 87·8 ( — ; | 43) | 85–91 |
| (10) | 94·3 (1·83; | 38) | 90–98 | 89·0 (1·71; | 26) | 85–92 |
| (11) | 94·0 (2·19; | 156) | 88–100 | 88·9 (1·73; | 156) | 85–93 |
| (12) | 93·4 (1·84; | 34) | 88–97 | 88·2 (1·66; | 45) | 85–93 |
| (13) | 91·8 (2·54; | 17) | 86–98 | 86·2 (2·12; | 23) | 82–90 |
| (14) | 92·2 ( — ; | 26) | 88–97 | 87·2 ( — ; | 21) | 82–91 |
| (15) | 90·3 ( — ; | 69) | 86–97 | 85·0 ( — ; | 23) | 82–88 |
| (16) | 93·7 (2·10; | 14) | 90–97 | 87·8 (1·77; | 9) | 85–91 |

Wing of breeding ♂, central Norway: adult 91·8 (1·4; 71) 89–95, 1st adult 89·3 (1·3; 34) 85–92 (Hogstad and Røskaft 1986).

Bill, Iran and Afghanistan, 14·0 (46) 13·0–15·0 (Vaurie 1949b); in Japan, bill to skull 16·0 (0·67; 22) 15·0–17·2, to nostril 10·1 (0·43; 21) 9·5–10·8 (ZMA).

**Weights.** ADULT, FIRST ADULT. Netherlands: (1) August–October, (2) November–December, (3) January–February, (4) March–May (A J van Loon, E R Osieck, ZMA, RMNH). (5) South-east France, winter (G Olioso). (6) Autumn migrants, Col de Bretolet (south-west Switzerland); (7) northern Switzerland, February (Jenni and Jenni-Eiermann 1987, which see for fat contents and for influence of age, temperature, roost size, and snowfall on weight; see also Mühlethaler 1952). (8) South-west Germany, January–February (Hoehl 1939). (9) Eastern Germany, April (Haensel 1967; see also Eck 1985a). (10) Czechoslovakia, winter (Havlín and Havlínová 1974; see also Havlín 1957). (11) Russian Lapland and Yamal Peninsula, summer (Danilov et al. 1984; Semenov-Tyan-Shanski and Gilyazov 1991). (12) Kazakhstan, winter (Korelov et al. 1974). Exhausted birds: (13) eastern Germany, winter (Creutz 1961); (14) Netherlands, all year (ZMA; see also MacDonald 1962 and Ash 1964).

| (1) | ♂ | 23·3 (1·80; 383) 18·5–28·5 | ♀ | 21·8 (1·86; 369) 17·0–27·9 |
|---|---|---|---|---|
| (2) | | 25·5 (2·53; 8) 22·0–30·0 | | 24·8 (3·68; 9) 18·9–30·0 |
| (3) | | 28·3 (3·26; 32) 22·9–34·9 | | 25·7 (1·84; 35) 22·6–29·8 |
| (4) | | 22·6 (3·07; 12) 18·5–28·5 | | 23·6 (3·89; 5) 17·0–27·0 |
| (5) | | 25·5 (1·49; 34) 23·0–28·5 | | 24·1 (1·63; 45) 17·5–27·0 |
| (6) | | 22·7 (1·4; 390) — | | 21·2 (1·3; 392) — |
| (7) | | 28·5 (2·4 ; 507) — | | 25·9 (2·2 ; 269) — |

| (8) | 26·2 ( — ; 137) 18·9–31·2 | 23·5 ( — ; 95) 18·7–28·0 |
|---|---|---|
| (9) | 25·4 (1·32; 23) 23·0–27·0 | 24·1 (1·68; 24) 21·6–27·2 |
| (10) | 25·6 (1·95; 82) 22·0–31·0 | 24·6 (2·19; 45) 21·3–33·5 |
| (11) | 22·6 ( — ; 67) 19·3–25·9 | 22·7 ( — ; 23) 17·9–26·5 |
| (12) | 24·7 ( — ; 13) 22·0–31·0 | 23·4 ( — ; 6) 18·3–27·0 |
| (13) | 20·4 (1·63; 6) 17·9–22·3 | 19·9 (1·56; 5) 18·6–22·1 |
| (14) | 16·8 ( — ; 5) 15·3–18·0 | 15·6 ( — ; 4) 15·0–16·0 |

Averages, eastern Germany: October, ♂ 25·1 (36), ♀ 24·1 (20); January, ♂ 26·6 (54), ♀ 24·9 (40); February, ♂ 26·7 (234), ♀ 24·2 (138); March, ♂ 26·3 (63), ♀ 24·6 (46) (Creutz 1961, which see for age differences). Central Norway: late May (arrival at breeding area), ♂ 23·1 (14), ♀ 22·6 (9); June, ♂ 22·7 (0·71; 70), ♀ c. 23·8; July, c. 22·0–22·5 (Hogstad 1982, which see for graph and for daily variation; Hogstad 1985). See also Weigold (1926). For graph of weight during moult, see Ottosson and Haas (1991): weight at start of moult c. 18·5–24·5; near end c. 21–27. For pre-migration fattening, see Lofts and Marshall (1960); for weight under various light conditions, see Pohl (1971a).

NESTLING, JUVENILE. For growth curve, see Hogstad (1977).

**Structure.** Wing rather long, broad at base, tip bluntly pointed. 10 primaries: p8 usually longest, p7 0–1 shorter, but sometimes p7 up to 1 longer than p8; p9 0·5–2·5 shorter than longest, p6 (1·5–)3–4, p5 10–15, p4 15–20, p3 18–24, p2 21–26, p1 24–28; p10 strongly reduced, 57–67 shorter than reduced outermost upper primary covert. Outer web of p6–p8 emarginated, inner web of p7–p9 with notch (sometimes faint). Tip of longest tertial reaches to tip of p1–p3. Tail forked, t1 8·7 (20) 6–12 shorter than t5–t6. Bill rather long, strong, conical; 7·9 (15) 7·5–8·5 deep at feathering near base, 7·4 (15) 7·0–8·0 wide; culmen and gonys straight, but tip of culmen slightly decurved, ending in fine tip. Middle toe with claw 17·8 (15) 17–19; outer toe with claw c. 74% of middle with claw, inner c. 69%, hind c. 78%. Remainder of structure as in Chaffinch F. coelebs.

**Geographical variation.** Slight or none. Colours gradually brighter towards east, ♂♂ from eastern Asia more intensely black and deeper rusty-orange than those of Europe (Johansen 1944); birds wintering Japan therefore sometimes separated as subcuneolata Kleinschmidt, 1909, but difference in birds examined too slight to warrant recognition of a separate eastern race.

For relationships, see F. coelebs.    CSR

# Subfamily CARDUELINAE typical finches

Small to medium-sized, seed-eating finches inhabiting wide variety of habitats (see Fringillidae) and specializing in extraction and breaking of edible seeds; these supplemented by buds in a few cases but insects (etc.) taken to much lesser extent than by fringilline finches, young being fed (by regurgitation) on seeds or mixture of seeds and invertebrates, never invertebrates exclusively. Found in both Old and New Worlds but mainly in Palearctic; only 35 species strictly confined to Afrotropics and Arabia (mainly in genus Serinus), 25 to Americas (mainly Carduelis), and 1 each to Indonesia and Philippines (see below). Some successfully introduced elsewhere outside normal range (e.g. Australia and New Zealand); several (especially Goldfinch Carduelis carduelis) kept as cage-birds, and one (Canary Serinus canaria) long domesticated. Sedentary, partially migratory, or fully migratory; some species irruptive or nomadic. About 126 species (C S Roselaar) of which 26 breed in west Palearctic and 2 accidental (1 from North America).

About 24 genera, including (e.g.) (1) Serinus (serins and canaries), 37 species—Eurasia and Africa with one species (Mountain Serin S. estherae) confined to Indonesia; (2)

*Carduelis* (Goldfinch and allies), 30 species—Eurasia, North America, and South America (*c.* 15 species); (3) *Loxia* (crossbills), 4 species—all Palearctic but 2 also in North America and 1 in Philippines; (4) *Carpodacus* (rosefinches), 20 species—Eurasia (mainly central Asia) with a few in North America; (5) *Rhodopechys* (Crimson-winged Finch *R. sanguinea*), monotypic—Morocco, Middle East, and central Asia; (6) *Rhodospiza* (Desert Finch *R. obsoleta*), monotypic—Middle East and Asia; (7) *Bucanetes* (trumpeter finches), 2 species—Spain and North Africa east to central Asia; (8) *Pinicola* (pine grosbeaks), 2 species Eurasia and North America; (9) *Pyrrhula* (bullfinches), 6 species—Eurasia, with one (White-cheeked Bullfinch *P. leucogenys*) confined to Philippines; (10) *Eophona* (Asiatic grosbeaks), 2 species—eastern Asia; (11) *Coccothraustes* (Hawfinch *C. coccothraustes*), monotypic—Eurasia and North Africa; (12) *Hesperiphona* (New World grosbeaks), 2 species—North and Central America. These genera fall informally into at least 4 subgroups: (A) typical cardueline finches (genera 1–4); (B) desert finches (genera 5–7, but these sometimes united in single genus *Rhodopechys*; (C) bullfinches, etc. (genera 8–9); (D) grosbeaks (genera 10–12).

Sexes almost the same size (♀ usually slightly smaller). Bill mainly of typical seed-eater type but very variable, even within genus; deepest (and as wide as high) in *Coccothraustes*, finest in some *Carduelis* (e.g. *C. carduelis*); uniquely crossed at tip (for extracting pine seeds from cone) in *Loxia*. Both sexes of some species in groups B and C develop gular storage pouches in breeding season. Nostrils partly covered by feathers; rictal and nasal bristles present but not well developed. Wing usually rather long and bluntly pointed; 9 primaries (p10 minute and hidden). Flight fast and undulating, typically with periodic closure of wings; more erratic and flitting, with light, dancing action, in some smaller species. ♂♂ of several genera perform song-flights: butterfly-like with slow-beating wings in all or most group A species; spectacularly undulating in most group B species (genera *Rhodopechys* and *Bucanetes*, but not *Rhodospiza*). Tail short or of moderate length; square, rounded, or (usually) forked to lesser or greater extent; 12 feathers. Leg short and strong; tarsus scutellate, booted, and ridged as in Fringillinae. Some species of *Carduelis* (at least) make simple use of foot in feeding, to hold down object (e.g. plant stem) and/or clamp it to perch (see Newton 1972); smaller, lighter species (e.g. Redpoll *C. flammea*) or larger species with relatively short, thick legs (e.g. Crossbill *L. curvirostra*) adept at hanging from vegetation to feed, sometimes upside-down (like tits *Parus*). In *Loxia* also, bill used to clamber about in trees in manner recalling parrots (Psittacidae). Gait on ground typically a hop, sometimes a shuffling walk. Head-scratching by indirect method (see Simmons 1957*b*, 1961*a* for *Serinus*, *Carduelis*, *Pyrrhula*, and *Eophona*). Bathing usually by stand-in method so far as known (e.g. Sabel 1963 for Citril Finch *S. citrinella*); information sparse,

however, and Scottish Crossbill *L. scotica* once recorded dive-bathing repeatedly from perch (Christie 1927). Sunning, with use of both main postures, reported from several species in the wild (see Kennedy 1969; Simmons 1986). No records of anting in the wild, apparently, but active (direct) anting reported from 3 species in group D, (*C. coccothraustes*, Evening Grosbeak *H. vespertina*, and Yellow-billed Grosbeak *E. migratoria*) in captivity (see Simmons 1957*a*, 1961*b*; Whitaker 1957); some smaller species of genera *Serinus* and *Carduelis* did not ant in captivity even when given the opportunity (Poulsen 1956).

Many species highly vocal especially when flocking, with a number of noisy often twittering calls. Song less specialized than in fringilline finches, being less stereotyped, longer, and quieter but sometimes highly melodious (Newton 1972); used for courtship (etc.) as well as advertisement by most species. However, in (e.g.) group C (in which both sexes sing) and group-D, song soft, undemonstrative, and individually variable. Majority of species highly gregarious, flocking outside breeding season and in case of group A and group B even during it (especially for feeding, and in group B for drinking); group C and group D species less gregarious forming small parties in autumn and winter only (if then). Most species typically territorial when nesting; territories usually small and grouped in loose colonies; group D carduelines sometimes solitary, however, and group C always solitary and non-territorial. Monogamous mating system general, usually with seasonal pair-bond; bond particularly strong and sustained throughout year in group C. Although sociable, all species (even in groups) are essentially non-contact birds, though roosting *L. curvirostra* reported huddling together (Khan 1986); never (e.g.) allopreen so far as known, but billing and courtship-feeding common. Nest typically an open, compact cup, usually hidden in tree or bush but distinctive 2-layered nest of twigs and roots in groups C and D; built by ♀ (with ♂ in close attendance) in most groups, but foundation platform laid by ♂ in group D. Incubation by ♀ only (fed by ♂). Young fed by ♂ until brooding period (by ♀) over, then by both sexes.

Except in a few species (e.g. *C. carduelis*), sexes differ, with ♂ the brighter bird to greater or lesser extent. Plumages highly variable: often brown, grey, or olive on upperparts, usually with dusky streaks, and sometimes with contrasting white, yellow, or red rump; usually white or yellow on underparts, frequently with streaks; many species have contrastingly coloured black, red, yellow, or white patches or stripe on cap, side of head, or throat; tail frequently with contrasting white spots, wing with contrastingly white or yellow bar; fringes of flight- and tail-feathers often contrastingly coloured white, yellow, pink, or red; adult ♂ often largely red in *Loxia*, *Carpodacus*, *Pinicola*, and some related genera (but this colour generally not attained until 2 years old). Juvenile usually brown or grey above and white or yellow below, marked with dusky streaks; except for wing and tail, lacks contrastingly coloured patches of adult. Nestling covered with down, usually plentifully; no spots on tongue. Adults have single complete post-breeding moult annually, starting after fledging of last brood of young. After moult, many species have cryptically coloured feather-tips on head and body and typically attain brightly coloured breeding plumage by abrasion and not by pre-breeding moult. Post-juvenile moult usually incomplete, restricted to body and to wing-coverts, but complete or nearly complete in species or populations breeding in tropics and subtropics; starts within 1-2 months after fledging.

## *Serinus pusillus* **Red-fronted Serin**

PLATES 33 and 38 (flight)
[between pages 472 and 473]

Du. Roodvoorhoofdkanarie  Fr. Serin à front d'or  Ge. Rotstirngirlitz
Ru. Корольковый вьюрок  Sp. Verdecillo carinegro  Sw. Rödpannad gulhämpling

*Passer pusillus* Pallas, 1811

Monotypic

**Field characters.** 12 cm; wing-span 21–23 cm. Slightly larger than Serin *S. serinus* but with similar stubby form on ground and in flight. Distinctive serin, with sooty head and breast, fiery orange-red forecrown, and heavily streaked back and flanks; rump orange in centre. Juvenile has rufous-buff face, cheeks, and throat. Sexes similar; some seasonal variation. Juvenile separable.

ADULT MALE. Moults: July–October (complete). Forehead and forecrown to above eye fiery, even luminous orange-red to scarlet, setting off otherwise sooty-brown, almost black head, nape, throat, and breast, finely mottled yellowish-grey when fresh. From nape, heavy black-brown streaks form lines on yellow ground of mantle and scapulars; from breast, similar streaks form lines on pale yellow to greyish-white ground of lower chest and along flanks. Wings basically black but intricately patterned with whitish to orange-buff fringes and tips to median and greater coverts, tertials, and flight-feathers; effect is to create pale double wing-bar, pale edges to tertials and patches of more distinctly orange-yellow on overlapped greater coverts and along bases of secondaries and primaries. Underwing mostly pale yellow. Centre of rump yellow to orange, contrasting less noticeably with rest of upperparts than in other *Serinus*. Tail almost black, with pale yellow-orange edges obvious on bases of outer feathers. In flight, appears very swarthy at front. Bill and legs dusky brown to black. ADULT FEMALE. Resembles ♂ but with smaller red patch on forehead, greyer cast to black plumage, and fully streaked rump. JUVENILE. Lacks adult's near-black foreparts and red forecrown, having instead sooty-brown face, cheeks, and throat suffused pale orange-yellow to rufous, particularly on lore and below eye. Upperparts, including rump, streaked brown-black and tawny-buff; underparts tawny-buff when fresh and indistinctly streaked. Wing duller than in adult, with pale markings rufous (not yellow) in tone. In moult to adult plumage, head and breast noticeably mottled black.

Within *Serinus*, unmistakable at all ages, but beware confusion with (1) vagrant Redpoll *C. flammea* which also has usually red but occasionally paler orange forehead in adult plumage and can appear quite swarthy and heavily streaked at distance, and (2) Turkish race of Twite *C. flavirostris* whose black-splashed chest and flanks recall young *C. pusillus* in 1st autumn transitional plumage. Flight bouncing, apparently as *S. serinus*. Gait a hop. Frequently feeds on ground, mixing freely with other finches.

Song a pleasant and powerful outpouring of liquid notes, with full timbre recalling Goldfinch *Carduelis carduelis* and lacking sibilant, sizzling quality of *S. serinus*. Call a rippled 'drillt-drillt', recalling *S. serinus*.

**Habitat.** Breeds in south-east of west Palearctic, in middle and upper tree belts of mountains, subalpine meadows, and in wide and narrow ravines along rivers; in Caucasus, at 600–3000 m. Sings from upper branches of low birches *Betula* or pines *Pinus*, or rock ledges, but is often on ground or on stony or rocky terrain, nesting in rock crevices but occasionally in lower branches of juniper *Juniperus*, rose *Rosa*, or other shrub. Occurs in *Rhododendron* zone and among juniper, descending in winter to valleys of lower mountain zone, and in snowy conditions even to foothill plains and town orchards, but rarely travels far. Very trusting, visiting courtyards and streets of mountain villages, and feeding in vegetable gardens (Dementiev and Gladkov 1954; Flint *et al.* 1984). In central and western Afghanistan, found breeding at *c.* 2300–3300 m, in scrub within and above coniferous forest, and among scattered junipers on sparsely vegetated mountain slopes (Paludan 1959). In Himalayas, breeds at 2400–4300 m, in summer

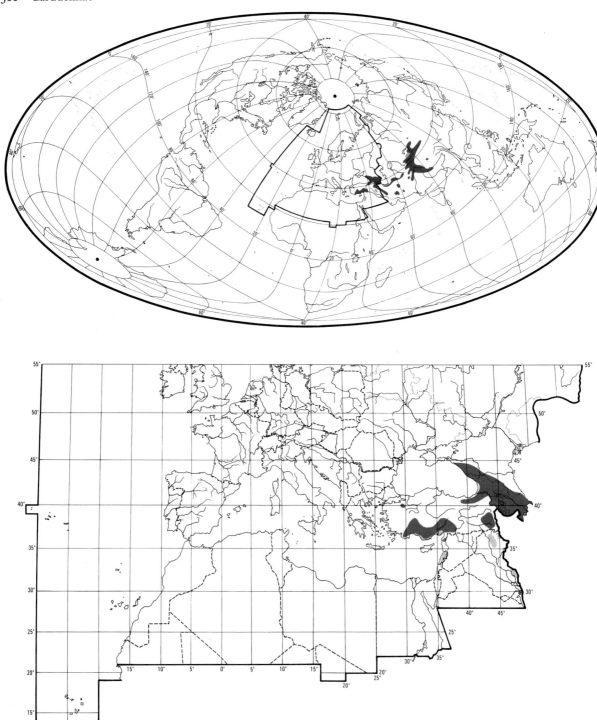

frequenting dwarf junipers at and above treeline, rocky hillsides with stunted bushes, shingle screes, and birches and willows near cultivation. Winters at c. (750–)1500–3300 m, frequenting open hillsides and stony ground with bushes and coarse herbage. Feeds mostly on ground; may be seen drinking and bathing at all times of day at any available water. (Ali and Ripley 1974.)

**Distribution and population.** No information on changes.

LEBANON. Apparently not proved to breed, but occurs in high parts of Lebanon range in autumn and winter (Vere Benson 1970). IRAQ. Probably breeds in north (Čtyroký 1987).

Accidental. Austria, Greece (Chios), Cyprus, Egypt.

**Movements.** Mainly altitudinal migrant.

Chiefly resident in Turkey, dispersing to lower altitudes in winter (mostly November–March), and more widespread then in southern coastlands (Beaman *et al.* 1975); recently recorded in autumn and winter as far west as Chios island off western Turkey (*Br. Birds* 1988, 81, 121; G I Handrinos); probably disperses more widely in eastern Turkey in winter than mapped here, but data lacking (R P Martins). Reaches Cyprus only exceptionally (Flint and Stewart 1992), but some Turkish birds move south inland, apparently especially in cold years, to winter locally in Syria, Lebanon, and northern and central Israel, chiefly above 800 m. Numbers in Israel vary greatly: rare or very rare in most years, but sometimes minor influxes occur, with small numbers reported from various areas; at main site, Baal Hazor (31°59′N 35°17′E), 15 reported 1984–5, 82 1986–7; singles appear irregularly in Jerusalem, but groups of up to 20 in 1966–7; exceptional as far south as southern Israeli deserts and Egypt. (Goodman and Meininger 1989; Shirihai in press.) Arrives in Israel from end of October to end of December, majority reaching winter sites December or beginning of January; movement inconspicuous, with occasional sightings of individuals or groups of 2–4 from scattered localities; most are 1st-year birds. Spring movement through Israel (also inconspicuous) from early February to mid-March, chiefly February. (Shirihai in press.)

Mainly resident in Iran, with small numbers descending to Caspian lowlands in winter (D A Scott). Apparently irregular winter visitor to highlands of northern Iraq (Ticehurst *et al.* 1921–2; Allouse 1953; Marchant and Macnab 1962; Čtyroký 1987).

In north-west Himalayas, locally common below breeding range November–March (Whistler 1924; Ali and Ripley 1974); at Quetta (western Pakistan), flocks descend to plains in severe weather (Meinertzhagen 1920).

In Caucasus, abundant after snowfall at lower altitudes in Severo-Osetinskaya ASSR and Kabardino-Balkarskaya ASSR (Dementiev and Gladkov 1954). In mountains of central Asia (Pamir-Alay, Tien Shan, Talasskiy Alatau), rare in most breeding areas in winter; birds descend to lower mountain levels, foothills and adjacent plains, usually moving not more than 50–70 km. Records and observations of passage show that these altitudinal movements have no directional bias; in Kazakhstan, fairly regular in towns and settlements (e.g. Alma-Ata, Zaysan) in foothills north-west of breeding areas, and flocks sometimes reach Chu and Ili valleys; bird ringed 14 January 1977 at Alma-Ata, 43°19′N 76°55′E, recovered 1 March 1978 as far south as Kyzyl-Kiya, 40°15′N 72°06′E. (Kovshar' 1966;

Korelov *et al.* 1974; Abdusalyamov 1977; Dobrynina 1982.) Numbers involved and extent of movement depend on severity of winter; at Bol'shoe Almatinskoe lake in Tien Shan (within breeding range), 1971–5, reported fairly frequently only in mild winter of 1974–5, and rarely in other years when common at low altitude in Alma-Ata (Kovshar' 1979); at wintering grounds on lower Vakhsh (Tadzhikistan), few in 1940–1 though very numerous 1939–40 (Ivanov 1969). In Shakhdara valley (western Pamirs), where breeds in large numbers, only upper reaches are vacated in winter, though in extreme weather birds descend to lowest third of main valley; most such wintering birds are 1st-years (Stepanyan 1969b). Following post-breeding dispersal (when some birds ascend above breeding range), autumn migration begins September and main movement from mid-October, with few remaining in breeding areas by early November. Return movement gradual, beginning at first sign of spring, sometimes from 2nd half of February but mostly in March, with main arrival on breeding grounds in April when last birds still in winter quarters. In 1959, birds appeared in breeding area in Kirgizskiy mountains as early as February, though renewed cold weather caused them to retreat again until 2nd week of March. In Alma-Ata, latest records in 4 years 6–14 March. (Kuznetsov 1962; Kovshar' 1966; Ivanov 1969; Korelov *et al.* 1974; Abdusalyamov 1977.)    DFV

**Food.** Almost all information extralimital. Seeds, fruits, and other plant material; sometimes small insects. Feeds on ground, in herbs, and in trees. Recorded in juniper *Juniperus*, birch *Betula*, alder *Alnus*, and willow *Salix* (Meiklejohn 1948; Hollom 1955; Sutton and Gray 1972; Korelov *et al.* 1974; Desfayes and Praz 1978; Roberts 1992; Shirihai in press). Will forage gregariously all year (see Social Pattern and Behaviour), and described as very restless and animated forager (Géroudet 1963; Ali and Ripley 1974; Korelov *et al.* 1974; Roberts 1992). In southern Turkey, of 74 birds watched in measured area, 51 stayed for less than 1 min and none for more than 5 min, while some Serins *S. serinus* remained for 20 min (Sutton and Gray 1972). To get at seed-heads, will jump and flutter up from ground to snatch seeds from low herbs and grasses (Ali and Ripley 1974), or cling to stalk to pick at head (Sutton and Gray 1972), or stand on stem to bend it over and feed on head held under foot (Meiklejohn 1948). Very agile in trees; behaviour reminiscent of Siskin *Carduelis spinus*, at times hanging upside-down at ends of birch twigs to reach catkins (Korelov *et al.* 1974; Desfayes and Praz 1978). Picks seeds from snow or when softened in thawed patches (Korelov *et al.* 1974; Kovshar' 1979). Will fly some distance to sources of salt and minerals, some places being visited daily May–August (Kovshar' 1966; Korelov *et al.* 1974).

Following recorded in diet. Invertebrates: aphids (Hemiptera: Aphidoidea). Seeds, fruits, and green parts of conifers (Coniferae, including spruce *Picea*), birch *Betula*, alder *Alnus*, mulberry *Morus*, dock *Rumex*, chickweed *Stellaria*, barberry *Berberis*, kale *Crambe*, shepherd's purse *Capsella*, rocket *Sisymbrium*, *Prunus*, lady's mantle *Alchemilla*, rest-harrow *Ononis*, St John's wort *Hypericum*, *Plectranthus* and *Coridothymus* (Labiatae), elder *Sambucus*, thistles *Carduus* and *Cirsium*, fleabane *Inula*, wormwood *Artemisia*, dandelion *Taraxacum*, sawwort *Saussurea*, salsify *Tragopogon*, viper's grass *Scorzonera*, *Cousinia*, grasses *Poa*, *Dactylis*, *Alopecurus*. (Schüz 1959; Géroudet 1963; Veger 1968; Ali and Ripley 1974; Korelov *et al.* 1974; Kovshar' 1979; Mycock 1987; Fergmann 1988; Roberts 1992.)

Feeds principally on seeds of Cruciferae, Compositae, and grasses (Gramineae) (Kovshar' 1979; Elzen 1983; Roberts 1992).

In Kazakhstan, June–July, main food is unripe seeds or closed flower-heads of dock, salsify, viper's grass, and dandelion; in August, seeds of dock, shepherd's purse, lady's mantle, thistles and grasses; in January, 1 stomach contained Cruciferae seeds (Kovshar' 1966, 1979; Korelov *et al.* 1974). In summer, 4 of 25 stomachs contained insects (Korelov *et al.* 1974). In captivity, particularly fond of half-ripe seeds of chickweed, Cruciferae, Compositae, and grasses, as well as shoots and other green plant material (Bielfeld 1981); see also Senger (1977), Elzen (1983), Dathe (1986), and Magnusson (1988).

Diet of young mostly seeds, but probably some insects (Roberts 1992). Important component in Kazakhstan is unripe seeds of dandelion (Kovshar' 1979). In captivity, parent fed young seeds of Cruciferae and Compositae and also aphids (Veger 1968; Winkler 1979b; Fey 1982). BH

**Social pattern and behaviour.** No detailed study; best is that by Kovshar' (1979) in Tien Shan mountains (central Asia).

1. Sociable all year, but largest flocks outside breeding season (Dementiev and Gladkov 1954; Ivanov 1969). In Talasskiy Alatau mountains (Kazakhstan), following March arrival, typically in small flocks (up to 10) during April, flocking continuing during summer; feeding flocks can be of several tens, otherwise smaller during day, larger and mixed with other species (see below) by evening (Kovshar' 1966, 1979). Further reports of small flocks during May in Turkey (Hollom 1955) and Iran (Meiklejohn 1948), large ones on alpine meadows of Zeravshan basin (central Asia) in June (Carruthers 1910). Summer flocks in Kirgiziya mainly of ♂♂ (Yanushevich *et al.* 1960). Flocking more pronounced after fledging, birds then also moving to lower altitudes; occasionally up to 200–300 together in former USSR, and flocks of several thousand noted by late October or early November in Tadzhikistan. Winter flocks (sometimes of juveniles only) reported from various localities in former USSR. (Dementiev and Gladkov 1954; Kovshar' 1966, 1979; Stepanyan 1969b; Korelov *et al.* 1974; Abdusalyamov 1977; Reber 1988.) For winter counts of birds per km² or along 1-km stretches, see (e.g.) Vtorov (1967, 1972), Abdusalyamov (1977), Shukurov (1986). In Israel, average flock size 11; up to 82 in stormy conditions; juveniles apparently have greater tendency to flock than adults (Fergman 1988). Often associates for feeding with Serin *S. serinus* and other Fringillidae (Dementiev and Gladkov 1954, Kovshar' 1966; Fergman 1988; Hollom *et al.* 1988; Roberts 1992); in Israel, less

commonly also with Corn Bunting *Miliaria calandra* (Fergman 1988). Bonds. Nothing to suggest mating system other than monogamous. Pair-formation evidently takes place in flocks, as in many Fringillidae (E N Panov), and pair-bond strong (Dementiev and Gladkov 1954; Korelov *et al.* 1974). Incubation by ♀ alone (Dementiev and Gladkov 1954; Reber 1988), but young fed by both sexes (Abdusalyamov 1977; also Fey 1982 for captive birds), mainly by ♂ after leaving nest (Dementiev and Gladkov 1954); study of captive birds suggested brood divided between parents (Winkler 1992). Parental post-fledging care of at least 1 week suggested by Korelov *et al.* (1974) and Abdusalyamov (1977), but ♂ seen feeding brood 16 days out of nest (Kovshar' 1979); captive young fledged at *c.* 2 weeks and independent within *c.* 5 weeks, though self-feeding to some extent before this (Fey 1982). Of 5 captive birds (3 ♂♂, 2 ♀♀), 1 ♀ first bred at 1 year (Fey 1982); both members of another captive breeding pair were also 1-year-olds (Winkler 1992). Breeding Dispersion. In Pakistan, often 'semi-colonial' (Roberts 1992); suggests existence of neighbourhood groups, as does report from central Caucasus, where nests 'close together'—e.g. 2 *c.* 4–5 m apart and several more pairs in area (Boehme 1958); in Kazakhstan, pairs separated by *c.* 50–150 m (Korelov *et al.* 1974). Known to defend nest, but little information on establishment or size of territory (see, however, Antagonistic Behaviour, below); most feeding apparently done far from nest (Ivanov 1969), and thus outside territory, e.g. up to several hundred metres away in Talasskiy Alatau (Kovshar' 1979). Limited information on density exclusively from mountains of central Asia: in Tien Shan, 10 birds per km² (Vtorov 1967), or (data for 2 years in each of 5 areas) 8–400 birds per km² (Shukurov 1986); in Zeravshanskiy range (Tadzhikistan), 40–50 pairs per km² in juniper *Juniperus* forest (Abdusalyamov 1977, which see for further counts, not necessarily all true nesting densities, in various localities and habitats). Long-term site-fidelity in Kazakhstan confirmed by ringing: e.g. 1st-year ringed October and retrapped in August after 5 years (Dobrynina 1982); 2 records of juveniles retrapped at or near ringing site in following year, but juveniles generally less site-faithful than adults (Kovshar' 1979). Roosting. Nocturnal and (at least for part of year) communal; in trees or shrubs, e.g. in Tadzhikistan, in tall, dense juniper shrubs (Abdusalyamov 1977), or, in winter, in deciduous trees which have retained some leaves (Stepanyan 1969b). In Pakistan, recorded roosting on ground, amongst rocks and boulders (T J Roberts). In Turkestanskiy range (Tadzhikistan), early May, active from first light (*c.* 05.30 hrs), feeding in flocks up to *c.* 11.00–12.00 hrs, then flying (sometimes long distance) to water (Abdusalyamov 1977); flocks and pairs also regularly visit clay banks and buildings to take salt (Korelov *et al.* 1974; Kovshar' 1979). Birds roost in juniper shrubs, or seek shade under rocks (etc.), during hottest part of day; after evening feeding bout (*c.* 17.00–18.00 hrs to dusk), settle in shrubs for night (Abdusalyamov 1977).

2. Some reports suggest little fear of man, birds coming into mountain villages or perching on tents (Carruthers 1910; Shnitnikov 1949; Dementiev and Gladkov 1954). In contrast, April flocks in Kashmir shy and restless (Whistler 1922b). Flock Behaviour. Flying flocks typically dense and compact, performing synchronous, sometimes complex evolutions (Salikhbaev and Bogdanov 1967; Korelov *et al.* 1974); 2 large flocks in Kashmir, April, loose and straggling, birds breaking away then rejoining others when disturbed (Whistler 1922b). Birds give frequent Ripple-calls (see 2 in Voice) both in flight and when feeding on ground; typically restless; will perch (including when disturbed) on bare tops of trees or shrubs, flock then giving rapid chorus of Ripple-calls (Hollom 1955; Ali and Ripley 1974; Roberts 1992; T J Roberts). Feeding flocks of *c.* 30 in Iran, May, always broke

up and flew off in pairs when disturbed (Meiklejohn 1948). March flocks in USSR fairly compact (minimum individual distance 20-30 cm); both ♂-♂ conflicts and courtship occur, displays being accompanied by chorus of song and calls (E N Panov). See Song-display, Antagonistic Behaviour, and Heterosexual Behaviour (below). Early-May flocks of 50-60 in Tadzhikistan noisy when settling to roost (Abdusalyamov 1977). SONG-DISPLAY. Sings (see 1 in Voice) from perch (rock, tree, etc.) or in flight (Dementiev and Gladkov 1954; Boehme 1958; Kovshar' 1979); in Kashmir, late April, in undergrowth according to Whistler (1922b). Characteristic is communal singing and display (apparently linked to pair-formation) of up to 4 ♂♂ from perch or on ground when ♀(♀) nearby (similar when lone ♂ displaying to ♀): each ♂ sings, droops and partly spreads wings, fans and slightly raises tail, and prominently ruffles red crown (Fig A); may crouch, also turn body or head to left and right in

A

front of another ♂, or birds circle one another while singing. Usually ends with whole flock taking off suddenly. (Dementiev and Gladkov 1954; Boehme 1958; Kovshar' 1979; Roberts 1992; E N Panov.) Singing in flight also frequent: individual ♂ may take off singing after display on ground or perch as described, hover over ♀ for *c.* 15-30 s, then descend with rapid wing-beats and land next to her (Boehme 1958; Korelov *et al.* 1974), or glide down onto rock, ruffling crown feathers on landing (Kovshar' 1966; Sutton and Gray 1972). In longer Song-flights, bird flies (apparently in straight line) with wings and tail fully spread, wings beaten steadily to *c.* 15° above and below horizontal, sometimes more shallow and quivering; may ascend slightly during flight, then drop down again, in undulating pattern (E N Panov), or Song-flight may appear slow-motion and wavering, bird breaking into it from normal flight (Hollom 1955). Not clear whether report (by Meiklejohn 1948) of ♂ singing frequently 'while jumping up and down in the air' refers to brief Song-flight or leaps from perch; no other similar descriptions. In Tadzhikistan, flocks of 6-8 fly to trees after feeding bout and start to sing around midday; some ♂♂ sing during day-loafing (Abdusalyamov 1977). Singing and associated displays continue for some weeks before nesting (Roberts 1992). In mountains of central Asia, song-period end of March to early August; marked variation in activity of individual ♂♂, but much song noted during nest-building (early June) in Talasskiy Alatau; song in July-August no less vigorous than in May-June (Kovshar' 1966, 1979); may cease with hatching (Abdusalyamov 1977). Captive ♂♂ sang from mid-January, full song from early February (Winkler 1979b). ANTAGONISTIC BEHAVIOUR. Highly sociable and rarely aggressive towards conspecific birds, even near nest; one record of ♂ attacking another that was singing within *c.* 30 m of first's nest; other species (e.g. tits *Parus*, warblers *Phylloscopus*) are chased out of nest-tree (Kovshar' 1979). Captive juvenile subordinate to adult for feeding and roosting faced older bird with plumage ruffled

when it sang nearby (Winkler 1979b). Frequent song-duels noted between 2 ♂♂ of captive pairs (Winkler 1992). Displays and postures of antagonistic and heterosexual significance (no clear distinction) include opening and shivering wings (to varying extent), also holding tail horizontal or often raised, and sometimes quite widely fanned; bird may sway from side to side in posture shown in Figs A and (also moving sideways on perch) B, while nodding movements performed in posture shown in Fig C (E N Panov); in Turkey, early May, bird displaying thus was

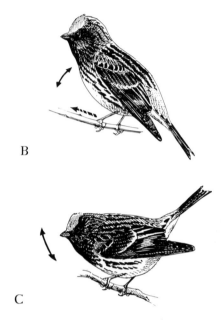

B

C

chased away by another (Hollom 1955). HETEROSEXUAL BEHAVIOUR. (1) General. In Caucasus, pair-formation from early May (Dementiev and Gladkov 1954); in Tien Shan, most birds in pairs by end of April or May (Kovshar' 1966, 1979); see also Flock Behaviour (above) for report from Iran in May of birds keeping together as pairs in flocks. (2) Pair-bonding behaviour. For further details of displays and posture, in some cases probably associated with pair-formation, see Song-display and Antagonistic Behaviour (above). Pairs may leave flock after pair-formation, though 2-3 pairs sometimes remain together (E N Panov). Captive ♀ chased ♂ when she left nest, ♂ displaying by drooping wings and flicking tail (Winkler 1979b). (3) Courtship-feeding and mating. ♂ sometimes feeds ♀ before copulation, ♀ first giving Begging-call (see 5 in Voice) and wing-shivering, then crouching (Korelov *et al.* 1974); ♂ mounts and copulates, after which both fly off (Dementiev and Gladkov 1954). In Talasskiy Alatau mountains, copulation noted 9-10 June (Kovshar' 1966). Captive ♀ soliciting intensively caused ♂ to hide, but further pressure from ♀ (temporary reversal of dominance roles) eventually elicited symbolic Courtship-feeding (no food seen to be transferred). Incipient copulatory behaviour (one soliciting, the other hovering over it) noted in captive birds at 30 days old. (Winkler 1979b.) ♂ also regularly feeds ♀ on nest: in Tien Shan, one ♂ fed mate 4 times in 4 hrs during which ♀ left nest twice (each time for 3 min) (Kovshar' 1979); see also Fey (1982) for captive birds. (4) Nest-site selection and behaviour at nest. Site apparently chosen by both pair-members, birds keeping close together during period before building and flying from one shrub or tree to another. Nest built by ♀, occasionally with some assistance by ♂ (Kovshar' 1979; Reber 1988); captive ♂♂

seen flying about with nest-material (Fey 1982), displaying with drooped wings, also calling (not described) from new site, evidently attempting to encourage ♀ (feeding young) to start 2nd nest (Winkler 1979b). Usually, ♂ stays within c. 10–30 m of nest-building ♀, singing or silently watching her; may accompany her, especially on longer excursions for material (up to 300 m or more). Up to 3 days between nest-completion and laying, or laying may even start before nest completed. (Kovshar' 1979.) ♂ and ♀ similarly keep close company when incubating ♀ leaves nest to feed or drink (Abdusalyamov 1977). RELATIONS WITHIN FAMILY GROUP. ♀ broods young intensively for first few days (Kovshar' 1979); at nest in Tien Shan, decreasingly up to day 11 (Kydyraliev 1972). In initial period after hatching, ♂ brings food, regurgitating it into ♀'s sublingual pouch, ♀ then distributing it to young (Kydyraliev 1972; Kovshar' 1979); see also Fey (1982) for captive birds. Captive ♀ rearing young alone (♂ removed) carried away nestling faeces up to 11–12 days, but left them on nest-rim thereafter (Winkler 1979b). Young fledge at c. 14–17 days (Dementiev and Gladkov 1954; Kydyraliev 1972; Korelov et al. 1974; Kovshar' 1979). ANTI-PREDATOR RESPONSES OF YOUNG. One brood left nest prematurely (at c. 12 days) when observer approached (Kuznetsov 1962). PARENTAL ANTI-PREDATOR STRATEGIES. ♀ a tight sitter (Korelov et al. 1974), especially close to hatching (Kuznetsov 1962). On Zeravshan river, attacked man fearlessly (no further details) when eggs were being removed (Carruthers 1910).

(Figs by D Nurney from drawings by E N Panov.)     MGW

**Voice.** Much in common with Serin *S. serinus* and certain *Carduelis* finches. Freely used, at least in breeding season; no information for winter.

CALLS OF ADULTS. (1) Song of ♂. Attractive and melodious, rapid and continuous, high-pitched bubbling twitter, with frequently interspersed ripples or (at times slightly hoarse), 'trills' of varying length (some similar to call 2), and hoarse 'kveeh' sounds; powerful outpouring delivery (Whistler 1922b; Meiklejohn 1948; Dementiev and Gladkov 1954; Ali and Ripley 1974; Kovshar' 1979; Hollom et al. 1988). Parts of song rendered 'tsi-tsu-teeh-tsi-tit-tit-tit-tit', also some discrete, but rapidly delivered, short and lower-pitched whistles, 't-ri-ri-ri-ri-ri-ri-ri' (Roberts 1992). Recalls song of Linnet *Carduelis cannabina* (though that species typically has more buzzy units: W T C Seale) and Goldfinch *C. carduelis*; also reminiscent of *S. serinus*, but lacks its sibilant, sizzling quality (Hollom 1955; Svensson 1984a; Mild 1987; Hollom et al. 1988);

*contra* Sutton and Gray (1972), lower pitched than *S. serinus* (W T C Seale); less variable, softer in timbre, and with more rapid and prolonged 'trills' than *C. carduelis* (Roberts in press). Recording from Iran contains tittering, rippling, and twittering sounds (at times suggesting song of Swallow *Hirundo rustica*), also short purrs; several units have nasal quality and song is especially rich in diads (best example in Fig I is longest unit); in series of brief units not all of which are diadic, apparently a tendency for series to start with diads. Rate of delivery of song changes considerably from relatively slow at start of recording (Fig I) to extremely rapid nearer end (Fig II), very rapidly

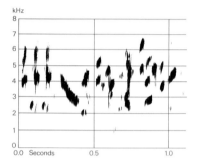

II  P A D Hollom  Iran  April 1972

delivered sections having attractive liquid quality. (W T C Seale, M G Wilson.) Song given in flight and (more developed) from perch (Hollom 1955). Short, rather weak song given by captive ♀ (Fey 1982). (2) Ripple-call. Short, or sometimes longer, bubbling tremolo or high-pitched, soft, rippling twitter, resembling equivalent call of *S. serinus* (see call 2 of that species), but more delicate, weaker, and softer, with 'singing' and slightly nasal or metallic sound (Meiklejohn 1948; Dementiev and Gladkov 1954; Svensson 1984a; Bruun et al. 1986; Mild 1987; Fergman 1988; Roberts in press; W T C Seale). Ripple-call given in flight and on ground or perch; serves for contact, perhaps also signals excitement or alarm. Recording contains liquid tremolos, longer calls in particular descending in pitch (see Hollom 1955); notes (of which first few are diadic) shown in Fig III form delightful silvery cascade (W T C Seale). Ripple-call difficult to

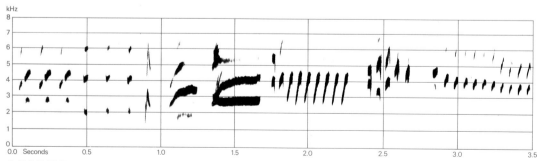

I  P A D Hollom  Iran  April 1972

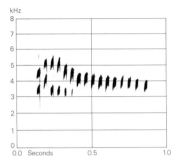

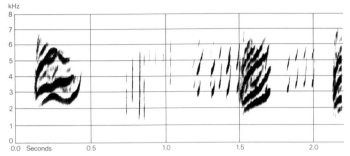

III  Mild (1987)  Kirgiziya  May 1987

IV  Mild (1987)  Kirgiziya  May 1987

transcribe, very roughly rendered 'prrrrrree' (W T C Seale, M G Wilson) or soft, tinkling 'firrrrrrrrrr' (Alström and Colston 1991). For further descriptions, see (e.g.) Schüz (1959), Ali and Ripley (1974), Jonsson (1982), and Hollom *et al.* (1988). In another recording (Fig IV; 2 birds calling), apart from nasal sounds (see call 3) and wing-noises following 1st of these, other sounds illustrated are typical ripple units and presumed shorter variant of same (W T C Seale, M G Wilson). (3) A 'dshUee' or 'djuee', reminiscent of equivalent call of Greenfinch *C. chloris* or *C. carduelis* (Svensson 1984*a*; Mild 1987), but most closely resembling alarm-call of *S. serinus* (call 6b of that species), and may well have similar function. Recording (1st unit in Fig IV) suggests 'chewee' with nasal timbre. At slow-speed playback, 3 tones (triad) are readily audible, and astonishing complexity of call revealed in narrow-band, scale-magnified sonagram made at half-speed playback (Fig V). Fig IV contains 2 further

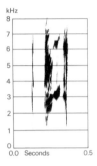

VI  Mild (1987)  Kirgiziya  May 1987

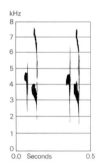

VII  Mild (1987)  Kirgiziya  May 1987

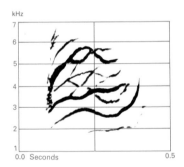

V  L Svensson  Kirgiziya  June 1983

roughly transcribed as 'peen' (E N Panov); function not known.

CALLS OF YOUNG. Food-calls (not described) of captive young barely audible when first noted at 10-11 days (Winkler 1979*b*). No further information on food- or contact-calls. Song of captive birds given from 24 days less clear and pure-sounding than adult's (Winkler 1979*b*).                                MGW

examples of nasal calls (diadic sound suggesting 'jwee', differing in structure from 'chewee' described above, but probably only individual variations, sonagram showing clearly that more than one individual is calling). (W T C Seale.) (4) Recording contains brief nasal twittering 't-chewrit' (Fig VI), recalling *C. carduelis*, and 2 'jerit' sounds (Fig VII) (W T C Seale); relationship with other calls not clear. (5) Begging-call of ♀. Squeaking or cheep-ing sound (Korelov *et al.* 1974; Winkler 1979*b*). (6) Call somewhat resembling 'pink' of Chaffinch *Fringilla coelebs*,

**Breeding.** SEASON. Turkey: April–July; eggs laid mid-April at *c.* 1500 m (Danford 1877–8; Makatsch 1976). Caucasus: eggs laid 2nd half of May, 2nd clutch in July (Dementiev and Gladkov 1954; Boehme 1958). Iran: nest-building recorded mid-May at 3000 m (Desfayes and Praz 1978). For Kazakhstan, see Kovshar' (1979); for Pakistan, see Roberts (1992). SITE. Low in dense bush or tree, generally growing at top or on ledge of inaccessible cliff, or high up in conifer, though still well-protected above by foliage; also in rock crevice, hole in scree, etc. (Dementiev and Gladkov 1954; Boehme 1958; Ali and Ripley 1974; Korelov *et al.* 1974; Kovshar' 1979; Roberts 1992). Report of nests on open ledges disputed by Kovshar' (1979). Very often in juniper *Juniperus*, in tall as well as in stunted or creeping forms; also commonly in rose *Rosa* or bramble *Rubus* bushes (Carruthers 1910; Hüe and Etchécopar 1970; Ali and Ripley 1974; Korelov *et al.* 1974; Kovshar' 1979; Roberts 1992). In one area of Talasskiy Alatau mountains (Kazakhstan), all of 40 nests in spruce *Picea* at 1·7–19 m above ground; 54% at less than 4 m, 28% 4–10 m, and 18% 10–19 m; most nests on branches 0·2–3 m from trunk, some by trunk near top of tree (Kovshar' 1979). Nest: neat and compact, appearing large and thick-walled for size of bird; foundation of dry grass, bark strips, stalks, moss, lichen, and sometimes twigs, lined thickly with plant down, feathers, etc., spiders' webs often incorporated (Dementiev and Gladkov 1954; Himmer 1967; Veger 1968; Hüe and Etchécopar 1970; Korelov *et al.* 1974; Kovshar' 1979; Roberts 1992); nests recorded with foundation only of juniper bark (Carruthers 1910); 6 nests, Kazakhstan, had outer diameter 78–103 mm, inner diameter 42–52 mm, overall height 44–70 mm, depth of cup 25–44 mm, and weight 6·4–12·4 g (Korelov *et al.* 1974; Kovshar' 1979). Building: by ♀ only, accompanied by ♂; rarely, ♂ helps with bringing of material or even with building (Nicolai 1960; Korelov *et al.* 1974; Kovshar' 1979); captive ♂ also recorded carrying material (Winkler 1979*b*; Fey 1982). One nest in the wild took 13 days to build (Kovshar' 1979, which see for details), but in captivity 2–7 days (Winkler 1979*b*; Fey 1982). EGGS. See Plate 60. Short sub-elliptical, smooth and faintly glossy; bluish-white, sparsely flecked, with pink or reddish-brown to purple-black scrawls, speckles, and blotches, mostly at broad end; sometimes unmarked (Hüe and Etchécopar 1970; Harrison 1975; Makatsch 1976). 17·2 × 12·6 mm (15·2–19·0 × 11·5–14·0), *n* = 70; calculated weight 1·43 g (Schönwetter 1984). For eggs from Kazakhstan, see Korelov *et al.* (1974) and Kovshar' (1979). Clutch: 3–5. In Talasskiy Alatau, of 22 clutches: 3 eggs, 5%; 4, 50%; 5, 45%; average 4·4 (Kovshar' 1979). In Himalayas, of 9 clutches: 3 of 3 eggs, 5 of 4, 1 of 5; average 3·8 (Hüe and Etchécopar 1970). Clutches of 5 rare in Pakistan (Roberts 1992). Laying can begin before completion of nest, or up to 3 days after (Kovshar' 1979); 5 days recorded in captivity (Winkler 1979*b*); eggs laid daily (Korelov *et al.* 1974; Fey 1982). Often 2 broods (Dementiev and Gladkov 1954; Boehme 1958; Scheifler 1968; Veger 1968), though no proof of 2nd brood in Kazakhstan, according to Kovshar' (1979). INCUBATION. 11–16 days, depending on how calculated (Dementiev and Gladkov 1954; Korelov *et al.* 1974; Kovshar' 1979); *c.* 13 days in captivity, starting with 3rd egg (of 5) in one case (Winkler 1979*b*; Fey 1982). In the wild, said to start after 1st egg (Kovshar' 1979, which see for incubation stints). By ♀ only (Veger 1968; Korelov *et al.* 1974; Kovshar' 1979; Fey 1982). YOUNG. Fed and cared for by both parents; faeces left on nest-rim after

nestlings *c.* 11 days old (Winkler 1979*b*; Fey 1982). For brooding, see Kovshar' (1979). FLEDGING TO MATURITY. Fledging period 14–16 days (Kovshar' 1979; Winkler 1979*b*; Fey 1982). Captive young first picked up seeds at *c.* 3 weeks old and still fed by parent at 5 weeks (Winkler 1979*b*; Fey 1982), but independent at 3–4 weeks in the wild according to Dementiev and Gladkov (1954) and Kovshar' (1979). BREEDING SUCCESS. In Talasskiy Alatau, average brood 3·7 (*n* = 5), 84% of average clutch; 4% of 67 eggs infertile. High rate of nest failure: 60% of 25 nests with eggs produced no young; 48% predated, mainly by Magpie *Pica pica* and squirrels (Sciuridae), at least 8% of broods lost due to weather. (Kovshar' 1979.)　　BH

**Plumages.** ADULT MALE. In fresh plumage (autumn), forehead and forecrown glistening flame-orange to scarlet-red, forming contrasting rounded or squarish patch *c.* 9–12 mm long. Nasal bristles dark grey-brown or black-brown, remainder of head as well as all neck and chin to chest deep sooty-black, each feather narrowly fringed pale yellow or grey, partly concealing black on hindneck, side of neck, and chest. Mantle, scapulars, and back black, feathers broadly fringed bright yellow on mantle and inner scapulars, paler yellow or yellow-white on outer scapulars and back. Centre of rump uniform flame-orange, orange-yellow, or golden-yellow for 10–15 mm, sides black with golden-yellow and white feather-fringes. Upper tail-coverts black, broadly fringed yellow on sides and white on tips. Side of breast, flank, upper belly, and side of belly pale yellow, each feather with broad black central spot or streak; yellow ground-colour paler towards rear of body, yellow-white or off-white, dark streaks greyer and less sharply-defined towards rear, mid-belly and vent yellowish-white with faint grey or yellow streaking. Under tail-coverts orange-yellow or bright yellow, often paler on lateral coverts and on tips of longest. Tail black, tip of each feather contrastingly fringed white, fringes broader and grading to orange-yellow or bright yellow towards bases, except on outermost feather (t6). Flight-feathers greyish-black (darkest on tip of outer webs, paler grey along borders of inner webs); primaries with contrastingly white fringe on tip and bright yellow or orange-yellow fringe along outer web; outer web of secondaries broadly bordered yellow at base (forming pale bar across inner wing), tawny or pink-cinnamon at terminal third, border black on middle portion. Tertials black with broad golden-buff or pink-cinnamon outer border, grading into white fringe on tip. Greater upper primary coverts and bastard wing black with narrow yellow fringe along outer web and narrow white fringe along tip, except for longest feather of bastard wing. Greater upper wing-coverts black, broadly bordered orange-yellow or golden-yellow along terminal half of outer web, merging into white fringe along tip; lesser and median coverts orange-yellow or golden-yellow, bases with some (mainly concealed) black, tips narrowly fringed white when plumage quite fresh. Under wing-coverts and axillaries yellow, longer coverts pale grey with broad white fringes. Bleaching and wear have marked effect. *In worn plumage* (about April–May), pale tips of black feathers of head, neck, and chest worn off, these fully sooty or greyish black with strongly contrasting scarlet to orange patch on forehead and fore-crown; yellow to off-white fringes of mantle, scapulars, back, and upper tail-coverts partly worn off, narrower, broad black streaking predominant, not as evenly streaked black-and-yellow as in fresh plumage; yellow on underparts partly worn off also, black of chest extending into broad and conspicuous black streaks along side of breast, flank, and side of belly. When heavily worn (June–July), yellow or whitish fringes of mantle, scapulars, upper tail-coverts, side of breast, tips of primaries, tertials, and greater upper wing-coverts virtually completely worn off, rump and lesser and median upper wing-coverts heavily spotted black, fringes along terminal part of outer web of secondaries bleached to white and partly worn off, bird appearing much blacker than in fresh plumage. ADULT FEMALE. Like adult ♂, but flame-orange to scarlet patch on forehead on average smaller, forming band across forehead of *c.* 6–8 mm wide; rump distinctly less bright, yellow with grey or black streaks, not as extensively and uniform orange to golden yellow on centre as in ♂; black of head, neck, and chest slightly greyer, pale tips of feathers broader, hindneck, side of neck, and lower chest less dark in fresh plumage than in adult ♂; fringes along bases of tail-feathers, primaries, and upper wing-coverts bright yellow to golden-yellow, not as orange-yellow as in some ♂♂; under tail-coverts yellow or yellow-white. In worn plumage, when much yellow of feather-fringes worn off, as extensively black as adult ♂, but patch on forehead on average narrower; black of head, neck, and chest greyer, less sooty; yellow of rump and under tail-coverts paler and more restricted. NESTLING. Down short, fairly plentiful, on upperparts, upper wing, thigh, and vent; pale grey (Ticehurst 1926). JUVENILE. Nasal bristles pale grey-brown or buff. Forehead and forecrown sooty-brown, feathers narrowly fringed grey-brown; remainder or crown, hindneck, and side of head and neck warm rufous-brown, sometimes faintly streaked or spotted grey. Lore and patch below eye pale buff with grey and white mottling. Entire upperparts backward from mantle closely streaked dull black and tawny-buff; dark streaks most pronounced on lower mantle and scapulars, least on rump. Lower cheek, chin, and upper throat buff-yellow or pale sulphur-yellow, finely speckled dusky grey, merging into rufous-brown of lower throat, side of breast and chest; some grey of feather-bases shining through, especially on chest, some faint short dark grey streaks sometimes visible on side of breast and (very faintly) chest. Flank, belly, vent, and under tail-coverts tawny-buff (if plumage fresh) to cream-white (if worn), upper belly, flank, and under tail-coverts marked with ill-defined dark grey streaks. Tail as adult, but fringes along tips of feathers tinged pale pink-cinnamon and fringes along bases pale yellow or yellow, less orangey than in some adult ♂♂. Flight-feathers and tertials as adult, but fringes along tips of primaries and tertials and along terminal halves of outer webs of tertials pale vinous-cinnamon, not as white as in adult. Lesser, median, and greater upper wing-coverts sooty-black, outer webs of greater coverts and broad tips to all coverts vinous-cinnamon or pink-cinnamon (in adult, yellow); extreme tips of greater coverts bleach to white when plumage worn. Sexes almost identical, but head of ♂ more uniform brown than in ♀, less streaked with black; throat blacker in ♂, browner in ♀. FIRST ADULT NON-BREEDING. Body and lesser and median upper wing-coverts like adult. Head and neck variable, either still juvenile, largely rufous-brown, or mixture of juvenile (especially on part of crown and ear-coverts, here conspicuously rufous-brown, and on part of throat, buff or yellow) and 1st non-breeding plumage, which is similar to adult breeding, but without red or orange-red on forehead and fore-crown: new feathering dull black or black-brown with narrow buff-brown feather-fringes. Flight-feathers, tail, tertials, greater upper primary coverts, and variable number of greater upper wing-coverts still juvenile; tips of tail-feathers and primary coverts often more pointed than in adult, more worn at same time of year; tertials and part of greater upper wing-coverts with pink-cinnamon fringes, contrasting with brighter orange-yellow fringes of new coverts, if any (in adult, fringes of tertials and greater coverts more extensively golden-

yellow, grading to vinous-pink or white on tips); however, pink-cinnamon of fringes at all ages often bleached to white by mid-winter, especially on tertials. A few red feathers appear among juvenile or 1st adult non-breeding plumage of forehead from mid-October to March onwards. FIRST ADULT BREEDING. Like adult breeding, and sometimes hardly distinguishable. Some birds show fully red forehead of adult breeding by November, others not until April; some rufous-brown juvenile feathering on part of crown, cheek, or ear-coverts sometimes present up to late March. Tail, flight-feathers, and greater upper primary coverts still juvenile, more worn and more pointed at tip than in adult at same time of year; some tertials and outer greater coverts often still juvenile, fringes of tertials bleached to white, those of greater coverts pale cinnamon to off-white without or with limited amount of golden-yellow.

Bare parts. ADULT, FIRST ADULT. Iris dark brown. Bill very dark brown to black, base of lower mandible paler brown, gape whitish (Hartert 1903–10; BMNH, ZMA.) NESTLING. Skin pale flesh with red cast; mouth and tongue pale pink, tinged violet on palate. Gape-flanges and cutting edges of bill yellow-white, bordered by narrow black line inside. Bill pale grey. (Neufeldt 1970.) JUVENILE. Iris dark brown. Bill dark brown. Leg and foot black. (Sharpe 1988; BMNH.)

Moults. ADULT POST-BREEDING. Complete; primaries descendent. In USSR, starts from mid-July (Dementiev and Gladkov 1954); in Iran and Afghanistan, also from mid-July, completed by or before end of October (Vaurie 1949b). In Kopet-Dag (western Turkmeniya), single ♀ had just started on 6 July (primary moult score 6) (RMNH). In northern Iran, pair in full moult on 26 August (Diesselhorst 1962). Moult not started in Afghanistan up to 25 July (Paludan 1959) or in 6 birds from western Himalayas on 15–17 July (ZMA). POST-JUVENILE. Partial: head, body, lesser, median, and variable number of greater upper wing-coverts, and sometimes a few tertials or t1 (Ali and Ripley 1974; RMNH, ZMA). Starts late July in USSR (Dementiev and Gladkov 1954), completed late September to early October in Iran and Afghanistan (Vaurie 1949b; Paludan 1959). Birds in full juvenile plumage examined throughout late June to late October; in September–October, body and lesser and median upper wing-coverts generally like adult breeding, but head either juvenile or in mixed juvenile and 1st adult non-breeding plumage (latter like adult, but red of forecrown and forehead replaced by sooty). Some red feathers appear on forehead from mid-October, but usually December–January; some still without red on cap until late March; occasionally birds show full red cap by November, but many others not until February–March. (BMNH, RMNH, ZMA.)

Measurements. ADULT, FIRST ADULT. Wing and bill to skull of (1) Asia Minor and northern Iraq, (2) Caucasus, (3) Elburz mountains (northern Iran) and Kopet-Dag and Bol'shoy Balkhan mountains (western Turkmeniya), (4) Afghanistan to Tien Shan mountains, (5) western Himalayas; other measurements combined, whole year; skins (BMNH, RMNH, ZMA: A J van Loon, C S Roselaar). Bill (S) to skull, bill (N) to distal corner of nostril; exposed culmen on average 3·3 shorter than bill to skull.

| | | | | | | |
|---|---|---|---|---|---|---|
| WING (1) | ♂ | 74·2 (1·54; 12) | 72–77 | ♀ | 72·2 (2·27; 8) | 68–75 |
| (2) | | 77·0 ( — ; 2) | 76–78 | | 72·7 (1·37; 9) | 70–75 |
| (3) | | 75·2 (1·04; 7) | 74–77 | | 72·6 (1·74; 6) | 70–75 |
| (4) | | 77·8 (1·18; 13) | 74–80 | | 76·0 (2·18; 5) | 73–78 |
| (5) | | 75·1 (0·86; 6) | 74–76 | | 73·6 (1·78; 5) | 71–76 |
| TAIL | | 55·7 (2·98; 40) | 50–61 | | 53·2 (2·43; 23) | 49–57 |

| BILL (1) | 10·1 (0·65; 9) 9·5–11·0 | 10·3 (0·39; 7) 9·8–10·9 |
|---|---|---|
| (2) | 10·9 ( – ; 2) 10·8–11·0 | 10·8 (0·52; 9) 9·9–11·6 |
| (3) | 10·3 (0·44; 6) 9·8–10·9 | 10·5 (0·14; 6) 10·3–10·7 |
| (4) | 10·6 (0·38; 11) 9·9–11·3 | 10·6 (0·53; 5) 9·8–11·2 |
| (5) | 10·4 (0·70; 6) 9·6–11·1 | 10·3 (0·41; 5) 9·7–10·5 |
| BILL (N) | 5·9 (0·22; 37) 5·4–6·4 | 6·0 (0·22; 22) 5·5–6·3 |
| TARSUS | 14·3 (0·40; 38) 13·4–14·9 | 14·2 (0·61; 19) 13·2–15·3 |

Sex differences significant for wing (1)–(4) and tail. For wing, tail, and bill of other samples from same areas, see Vaurie (1949*b*).

Wing. Turkey: ♂ 75·3 (3) 74–77 (Kumerloeve 1964*a*, 1967*a*, 1969*b*). Northern Iran: ♂ 72·0 (1·08; 4) 71–73·5, ♀ 67 (Schüz 1959; Diesselhorst 1962). Eastern Afghanistan: ♂ 75·0 (1·15; 4) 74–76, ♀ 70·7 (3) 70–71 (Paludan 1959; Nogge 1973). Western Himalayas: ♂ 77·6 (15) 74–78, ♀ 74·8 (10) 73–78 (Vaurie 1972). Kazakhstan: ♂ 73–79 (22), ♀ 70–77 (19) (Korelov *et al.* 1974).

Weights. Asia Minor, ♂: May 11, June 11, November 13 (Kumerloeve 1964*a*, 1967*a*, 1969*b*). Iran: March, ♂♂ 12, 13; May–June, ♂ 11·7 (3) 11–12, ♀ 12 (Schüz 1959; Desfayes and Praz 1978). Eastern Afghanistan, May–September: ♂ 9·9 (0·38; 7) 9–10, ♀♀ 9, 10 (Paludan 1959; Niethammer 1973). USSR: ♂ 11·4 (4) 10·5–12·7, ♀ 10·5 (Dementiev and Gladkov 1954). Kashmir (India), late August to mid-September: adult ♂ 11·7 (1·25; 6) 10·0–12·9, adult ♀ 12·5 (0·69; 4) 11·5–13·1; sex unknown, all ages, 10·6 (0·77; 9) 9·6–12·0 (P R Holmes and Oxford University Kashmir Expedition 1983). Kazakhstan: ♂ 11·6 (10) 10–13·5, ♀ 11·5 (10) 9·5–13·5 (Korelov *et al.* 1974).

Structure. Wing rather long, broad at base, tip bluntly pointed. 10 primaries: p7–p8 longest or one of them slightly shorter than other; p9 1–3 shorter than longest, p6 0–2, p5 4–8, p4 10–13, p3 12–16, p2 15–19, p1 16–22; p10 strongly reduced, a tiny pin hidden below reduced outermost greater upper primary covert, 50–56 shorter than p7–p8, 6–10 shorter than longest upper primary covert. Outer web of p6–p8 emarginated, inner of (p6–)p7–

p9 with faint notch. Tip of longest tertial reaches tip of p1 or to secondary tips. Tail rather short, tip forked; 12 feathers, t1 7–13 shorter than t6. Bill short, length of exposed culmen *c.* 40% of length of head; conical, *c.* 5·2–6·0 deep and *c.* 5·0–5·8 wide at base; culmen and cutting edges slightly decurved, gonys straight. Nostril small, rounded, covered by tuft of short feathers projecting from forehead over base of upper mandible. Base of bill without bristles. Feathering of body long and lax, especially in winter. Tarsus and toes short, slender; claws rather long but fine, decurved. Middle toe with claw 14·6 (8) 13·5–16; outer toe with claw *c.* 67% of middle with claw, inner *c.* 65%, hind *c.* 79%.

Geographical variation. Very slight, and study hindered by pronounced effect of bleaching and abrasion. In fresh plumage, birds from Caucasus slightly darker on upperparts than those from central Asia, mainly because feather-fringes are narrower and duller, less pale yellow and grey-white; birds from Iran slightly paler on average; side of head and throat more sooty in birds from Afghanistan and Iran, less deep black than in Caucasus or Tien Shan area (central Asia). In worn plumage, Tien Shan birds slightly deeper black on side of head and throat and on upperpart streaks, less sooty than birds from Caucasus, Iran, Afghanistan, and western Himalayas. Variation in tone and extent of red on forehead individual or sexual, not geographical; in all populations, some more golden-orange, others more scarlet. (Hellmayr 1929; Vaurie 1949*b*; BMNH, RMNH.) Single fresh-plumaged ♂ from northern Iran had fringes of flight-feathers dark orange-yellow and those of lesser upper wing-coverts bright red-brown, not as yellow as in fresh birds from Caucasus and Tien Shan, but this perhaps individual rather than geographical variation (Diesselhorst 1962); similar birds occur elsewhere, especially among adult ♂♂ (BMNH, RMNH). No races recognized, following Meinertzhagen (1927), Hellmayr (1929), and Vaurie (1949*b*, 1959).                                                 CSR

*Serinus serinus* **Serin**

PLATES 33 and 38 (flight)
[between pages 472 and 473]

Du. Europese Kanarie      Fr. Serin cini      Ge. Girlitz
Ru. Канареечный вьюрок      Sp. Serín      Sw. Gulhämpling

*Fringilla Serinus* Linnaeus, 1766

Monotypic

Field characters. 11·5 cm; wingspan 20–23 cm. 15% smaller and noticeably more compact than Linnet *Carduelis cannabina*; slightly smaller than other west Palearctic serins *Serinus* and Siskin *C. spinus* but with proportionately long wings and deeply forked tail. Diminutive, stubby-billed, rather compact finch, epitome of genus. Adult has rather green, streaked upperparts with bright yellow rump; ♂ brilliantly yellow on forehead, face, throat, and breast; ♀ only dull yellow on face. Juvenile lacks yellow rump and is more heavily streaked below than either adult. Flight light, noticeably bouncing. Voice

distinctive. Sexes dissimilar; no seasonal variation. Juvenile separable.

ADULT MALE. Moults: July–November (complete). At distance, appears bright yellow on face, fore-underbody, and on small area of rump, and streaked green on rear crown, cheeks, and back, with duller wings and tail. Pattern somewhat recalls ♂ *C. spinus* but lacks black crown and bold yellow wing and tail markings; strongly invites confusion with escaped dark derivatives of Canary *S. canarius* and with Syrian Serin *S. syriacus* in Levant (see below and those species). Important therefore to close range on any serin and with ♂ *S. serinus* to look particularly for: (1) bill shape (see below); (2) bright lemon-yellow head (especially when worn) marked with (a) grey-green sides of crown and nape, streaked black-brown on nape, (b) dusky olive ear-coverts, upper edge emphasizing pale almost white lore and yellow eye-ring and supercilium, rear and lower borders (with nape and edge of mantle) creating broad yellow surround from rear supercilium round ear-coverts to cheek, and (c) paler centre of cheek, forming yellow patch when worn (head pattern shared only by fully worn ♂ *S. canarius*); (3) pure lemon- or golden-yellow throat, breast, and rump, brightest when worn (lacks golden tone of *S. canarius*, looks brighter on rump than any congener except *S. syriacus*); (4) yellowish-brown back, broadly streaked black-brown (looks more yellowish than any congener except Citril Finch *S. citrinella*); (5) black-brown wings, with yellowish-green fringes and tips to median coverts (not forming bar), yellowish-buff tips to greater coverts (forming quite distinct bar particularly when worn, but lacking contrast of black band across secondaries shown by *C. spinus* and *S. syriacus*), and pale brownish-grey fringes and almost white tips to tertials forming pale lines when worn (wing pattern shared only by *S. canarius*); (6) yellowish-buff to yellow-white breast-sides and flanks, streaked black-brown, most contrastingly so when worn (includes longest streaks of west Palearctic *Serinus*); (7) black-brown tail with yellowish-green fringes to outer feathers, but these never obvious (lacks width and paleness of fringes of *S. syriacus* and *S. citrinella*). Bill dusky horn, paler at base of lower mandible; noticeably short, broad based, and swollen, both mandibles rounded in outline. Eye almost black, looking beady. Legs brown. ADULT FEMALE. Basic pattern as ♂ but much duller, looking less green and more brown above, more streaked and less yellow on head and breast, and duller, less pure white, on lower body. Obvious yellow tones restricted to short supercilium, area around eye and lower ear-coverts, chin and throat, and rump, but these features all paler, less eye-catching, and on some birds hardly visible in the field. Confusion with ♀ and juvenile *C. spinus* and other *Serinus* even more likely than in ♂; certain distinction requires close examination of bill shape (noticeably stubby in *S. serinus*), wing pattern (narrow wing-bars), and breast markings (uniformly heavy streaks, extending evenly along flanks). JUVENILE. Resembles ♀ but noticeably duller, with autumn vagrants retaining this plumage, even appearing washed out. Ground-colour of plumage brownish-buff, with only faint yellowish or whitish tone below and almost complete overlay of dark brown streaks which partly obscure slightly paler rump. Wing markings initially indistinct, due to dull buff fringes and tips. Confusion with juvenile *C. spinus* and other *Serinus* all too easy but, as with ♀, bill shape and full extent of streaks on underparts are trustworthy characters.

Below montane range of *S. citrinella* and north and west

of Turkey, *S. serinus* overlaps widely only with *C. spinus* (see that species, and above); within and south of Turkey, overlaps only with distinctive Red-fronted Serin *S. pusillus* but approaches similar *S. syriacus* (see that species, and above). Anywhere in west Palearctic, further risk of confusion with escaped (wild caught or domestic) *S. canarius* and other (wild caught) African *Serinus*. Of latter, Yellow-fronted Canary (Green Singing Finch) *S. mozambicus* most likely to appear but importantly it differs distinctly from *S. serinus* in heavier bill, much stronger, linear head pattern including dark eye-stripe and moustache, even greener back, wholly yellow rump and underparts lacking any streaks, and white-tipped tail. *S. serinus* thus no easy target for inexperienced observer and worth long study when first found; important to remember that it has smallest size, heaviest and sharpest streaks, and stubbiest bill of west Palearctic *Serinus*. Flight noticeably light and seemingly fast, with alternation of rapid wing-beats and closed wings producing almost dancing or skipping progress and lacking such regular undulations as those of larger finches; most recalls Redpoll *C. flammea cabaret*. Song-flight flitting with almost bat-like wing-beats and glides; strongly recalls diminutive Greenfinch *C. chloris*. Spends much time on ground, with rather fast hopping or seemingly creeping gait and low carriage when feeding but remarkably upright posture in excitement or alarm; escape-flight is into bushes or trees. Sociable at all seasons, but flocks typically small.

Song a distinctive sibilant jingle of rushed units, reminiscent of distant Corn Bunting *Miliaria calandra* but less discordant at close range; noticeably accelerated at times, then sounding excited, even hysterical. Flight-call a distinctive rippled 'tirrilillit', varied into shorter, rushed 'trillet' or more pronounced 'tittertee'. In alarm, 'tsooeet', suggesting *S. canarius* but with tinny quality.

**Habitat.** Confined to west Palearctic, originally in Mediterranean zone, spreading north in 19th century into temperate drier and warmer regions of central Europe, and continuing in 20th century to fringe of boreal and steppe zones, and sparsely towards oceanic margins, not yet beyond July isotherm of 17°C (Voous 1960*b*). Vulnerable to cold wet weather and unable to cope with more northerly winters, or with higher altitudes except in south of range, where ascends to subalpine zone (Harrison 1982). In Transcarpathia, ascends from lower foothills to timberline at 1300 m (Dementiev and Gladkov 1954). In Switzerland, occurs commonly up to 800 m, and often higher where conditions favourable, especially in sunny, open, dry areas close to human settlements, breeding exceptionally up to nearly 2000 m (Glutz von Blotzheim 1962). Geographical spread has been accompanied by shift of habitat from mainly forest edges and clearings or scattered clumps and rows of trees on hills and mountain slopes to parkland, cemeteries, orchards, vineyards, garden suburbs, avenues, and other well-mixed, sunny, dry situations offering nest-

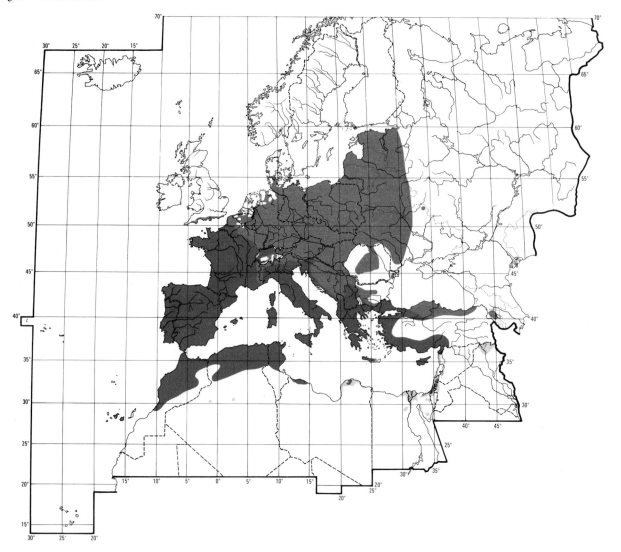

sites, often in conifers, and song-perches, often on posts or cables. Fond of flying in or just above lower airspace, and highly tolerant of human presence.

Among trees and shrubs seems most at home in conifers, including various exotic forms. Shows marked preference for mosaic patterns of vegetation of diverse heights, spacing, and composition, avoiding dense or uniform examples, or large blocks. Prefers areas with good margin of resources: struggling at limits of subsistence or climatic endurance is foreign to its nature, though not slow to take advantage of new opportunities. In northern Germany, 28% of 529 feeding observations were in parks and gardens, 46% on wasteland, and 26% in areas of cultivation; no records in this study of foraging in woodland or farmyards (Eber 1956, which see for comparison with other Fringillidae and buntings Emberizidae). In spring, more in parks and gardens, moving to wasteland, etc. in late summer; in autumn, agricultural land and vegetable gardens frequented, and in winter railway embankments, waste ground, and similar areas of ruderal vegetation; not seen to eat cereals (Quépat 1875; Eber 1956; Sabel 1983).

**Distribution.** Major northward spread began in 19th century; continued in 20th century (Mayr 1926), but now apparently stopped. Only minor range changes reported in south (Italy, Israel).

BRITAIN. Accidental until comparatively recently. Increase in records after 1960, and first recorded breeding 1967. Breeding still sporadic. (Sharrock 1976; Spencer *et al.* 1989.) FRANCE. Spread north in 2nd half of 19th century, continuing into 20th century. Paris area regularly occupied from *c.* 1906, Nord in 1950s. Spread west more slowly, reaching Finistère 1970s. (Yeatman 1976.) DENMARK. Sporadic breeding records 1948–65; then regular breeder for some years. Subsequently decreased markedly, and now probably not a regular breeder.

(UGS.) SWEDEN. Only 3 records up to 1939. First breeding records in 1940s; no obvious further spread in last decade. (LR.) FINLAND. Only 4 records 1921–57, 9 in 1960s, 77 in 1970s; reduction in 1980s to 3–10 per year. Only 7 records of confirmed (1967 and 1976) or probable breeding. Population not yet permanently established. (OH.) EAST GERMANY. Has bred since beginning of 20th century; initially in south, now up to Baltic coast (SS). Began to winter in Mecklenburg in 1940s (Klafs and Stübs 1987), very small numbers also in Brandenburg (Rutschke 1983). POLAND. Breeding first recorded 1853; now breeds over whole country, but still patchily (LT, AD). CZECHOSLOVAKIA. Colonized 1840–60; no recent changes (KH). USSR. Latvia: first recorded 1935; first breeding record 1938 (Vīksne 1983). Lithuania: first breeding record 1957 (Logminas 1991). St Petersburg region: northward spread, reaching south coast of Gulf of Finland 1976 (Noskov and Shamov 1983). ITALY. Colonized Aeolian and Egadi islands in early 1980s (Iapichino and Massa 1989). MALTA. Occasional breeder (Sultana and Gauci 1982). ISRAEL. Breeding first recorded 1977 on coastal plain; has since spread in coastal areas and western Negev (Shirihai in press). SYRIA. Probably breeds in cultivated country near Damascus, but confirmation needed (Baumgart and Stephan 1987). JORDAN. Breeding first recorded 1983, Petra (Wittenberg 1987). EGYPT. Probably local breeder in Nile delta (Goodman and Meininger 1989). CANARY ISLANDS. Recent colonizer, perhaps originating from escapes (Moreno 1988).

Accidental. Ireland, Norway, Madeira.

**Population.** Has increased with northward range expansion, but some evidence of recent decline.

FRANCE. 100 000 to 1 million pairs (Yeatman 1976). BELGIUM. Estimated 2500 pairs in 1970s; decline since 1979, but similar declines have occurred before (Devillers *et al.* 1988; Van der Elst 1990). NETHERLANDS. 450–550 pairs in 1978–9; not more than 100–150 in 1983–5 (SOVON 1987). GERMANY. 563 000 pairs (G Rheinwald). East Germany: 100 000 ± 50 000 pairs (Nicolai 1993). DENMARK. Declined after peak in late 1970s, and now probably does not breed regularly (UGS); see Distribution. SWEDEN. Estimated 10 pairs (Ulfstrand and Högstedt 1976); now 25–30 pairs (LR). FINLAND. Probably 0–5 pairs annually in 1980–7 (Koskimies 1989). USSR. Latvia and St Petersburg region: recent increases (Noskov and Shamov 1983; Vīksne 1983). ITALY. Has increased recently in Sicily (Iapichino and Massa 1989). ISRAEL. Recent increase; perhaps a few thousand pairs in late 1980s (Shirihai in press).

Survival. Spain: annual mortality 40 ± SE8% (Senar and Copete 1990*b*). Oldest ringed bird 8 years 7 months (Rydzewski 1978).

**Movements.** Sedentary to migratory, wintering within and south of breeding range.

Most birds vacate northern parts of range, but winter records show that small numbers remain, at least in some years. In centre and south of range, amount of movement masked by passage and arrivals from further north, but observations and ringing data show that even in Mediterranean countries a considerable number are migrants, contrary to earlier assumptions. Status in various areas as follows. In Mecklenburg (north-east Germany), wintering birds previously rare, now not uncommon, individuals remaining for several weeks: bird ringed 8 January 1967, retrapped 26 March 1967 (Klafs and Stübs 1987). In Luxembourg, occasional winter records in recent years of local birds or immigrants (Melchior *et al.* 1987). Winters in small numbers in Rheinland (western Germany) (Mildenberger 1984) and low-lying areas of Switzerland (Schifferli *et al.* 1982). Birds from northern France chiefly migratory, but wintering occurs locally, e.g. now regular in increasing numbers in Fontainebleau area; highest winter numbers usually in west and south, but more widely dispersed in milder winters; data from Spain show that considerable number from southern France also migrate (Asensio 1985*a*; Siblet 1988; Yeatman-Berthelot 1991). In Iberia, highest numbers of winter visitors are in Mediterranean coastal provinces, especially in northern half; ringing data show some local birds sedentary, others migratory (Asensio 1985*a*). In Italy, at least some breeding areas are mostly vacated: in Piemonte and Val d'Aosta in north, winters (in reduced numbers) in mildest localities only (Mingozzi *et al.* 1988); in Firenze area, very common in breeding season but only occasional midwinter (Dinetti and Ascani 1990). Many winter in Sicily; local birds not wholly sedentary, as some areas are deserted after breeding season (Iapichino and Massa 1989). In Malta (where breeds rarely), present in winter, but numbers vary (Sultana and Gauci 1982); locally common in winter in Greece (Bauer *et al.* 1969). In Turkey, winters mainly in south and west; status of local birds varies, e.g. resident at Köyceğiz-Dalyan in south-west, but Kizilcahamam in north is vacated (Beaman 1978; Bariş *et al.* 1984; Kiliç and Kasparek 1989). Common winter visitor to low ground on Cyprus; some local birds make altitudinal movements, with few remaining above 1000 m on Troodos mountains in winter (Flint and Stewart 1992). In west of range, winters only slightly south of breeding areas, but in east of range movement extends further. In north-west Africa, recorded south to Béni-Abbès 30°11′N on edge of Algerian Sahara (Dupuy 1966, 1969); local birds apparently mostly sedentary, but those breeding at Oukaimeden (Haut Atlas, Morocco) make altitudinal movements (Barreau *et al.* 1987). In Libya, resident in north-west (coastal Tripolitania) and winter visitor to north-east (Cyrenaica); recorded south to Murzuq *c.* 26°N in west and to Sarir 27°40′N in east (Bundy 1976). Regular winter visitor to Syria and Lebanon (Vere Benson 1970; Baumgart and Stephan 1987); in Jordan, common in northern highlands, with fewer further south (I J Andrews), and abundant in all Mediterranean and semi-desert regions of central and northern Israel south into northern Negev; few reach Elat in south, but regular in Sinai, and small flocks recorded annually in Nile delta and valley (Egypt); southernmost record Aswan 24°03′N, 22 March 1986 (Krabbe 1980; Goodman and Meininger 1989; Shirihai in press). Apparently a winter visitor in small numbers to northern Iraq (Goodwin 1955b; Moore and Boswell 1956), and vagrant to Iran (Vaurie 1959; Scott et al. 1975).

Main autumn heading south-west for west European birds and south for east European birds (reverse in spring), with many recoveries over 1500 km. German birds head south-west to south, but those moving south or east of south are apparently chiefly from eastern Germany: birds ringed in breeding season in Dresden area c. 51°N 13°30′E have been recovered in both Spain and Italy (Meyer and Schloss 1968), suggesting migratory divide in this region which may reflect route of colonization. Entry into Spain chiefly via eastern Pyrénées, though marked passage also in north and north-west Spain. In Spain, 102 recoveries of foreign-ringed birds up to 1984: from western Germany 36 (of which 19 ringed in breeding season), France 33 (21), eastern Germany 18 (9), Switzerland 10 (1), and Belgium 5 (1); recoveries mostly in winter, with none as far south as Strait of Gibraltar area, suggesting that most birds are winter visitors rather than passage migrants. Iberian birds, however, show marked tendency to head towards extreme south, and many winter near Strait of Gibraltar; heavy passage across Strait probably involves chiefly Iberian birds, and this supported by wing formulae; in one autumn, 6th most common visible migrant, with c. 15000 recorded; 1 recovery in Morocco of Spanish-ringed bird. (Snow et al. 1955; Tellería 1981; Asensio 1984, 1985a; Finlayson and Cortés 1987.) Ringing data from Italy show movement to and from Czechoslovakia, Yugoslavia, southern Poland, Austria, and Germany; from Genova area in north-west Italy, recoveries also west or south-west in south-east France (1) and Balearic Islands (1) (Bendini and Spina 1990). In Sicily, 6 recoveries December–March of birds ringed April–August in Czechoslovakia (4), Yugoslavia (1), and central Italy (1) (Iapichino and Massa 1989). In Malta, 9 recoveries December–March of birds ringed July–October in Czechoslovakia (5), Yugoslavia (3), and Hungary (1) (Maltese Ringing Reports, Il-Merill). 3 birds ringed on Kaliningrad coast (western Russia) recovered 1600 km south-west in south-east France (September), 1120 km SSW in Yugoslavia (October), and 1380 km SSW in Italy (April) (Paevski 1971). Birds cross Mediterranean regularly, especially at Strait of Gibraltar, but also from Malta and Crete (Phillips and Round 1975; Cortés et al. 1980; Sultana and Gauci 1982); only 1 definite record of passage south from Cyprus (Flint and Stewart 1992). Movement through northern and central Israel on broad front, with highest concentrations in west (Shirihai in press).

Autumn movement (August-)September–November,

chiefly October. On Polish coast, 1961–70, sporadic observations 17 August to 22 October (Busse and Halastra 1981), and on Kaliningrad coast passage peaks mid-October (Odinzowa 1967). In Luxembourg and Rheinland, movement September to mid-October (Mildenberger 1984; Melchior et al. 1987). At Col de Bretolet (western Switzerland), passage from end of September or beginning of October, peaking 8–17 October, continuing to end of October, with stragglers throughout November (Winkler 1984). Few immigrants reach Spain before October (Asensio 1985a), and highest numbers cross Strait of Gibraltar 2nd week of October to 1st week of November (Cortés et al. 1980). At Settat 33°N 7°37′W in north-west Morocco, passage chiefly November to early December (Thouy 1976). Firenze area (Italy) vacated early October (Dinetti and Ascani 1990), and present in Malta mostly from mid-October (Sultana and Gauci 1982); first arrivals in Cyprus usually late October (Flint and Stewart 1992). In Israel, movement early October to mid-December, chiefly end of October to end of November (Shirihai in press); recorded in Egypt from mid-September (Goodman and Meininger 1989), and small flocks present in Cyrenaica from October (Bundy 1976).

Spring movement February–May, chiefly March–April. At Settat, passage from February to end of April or early May (Thouy 1976); at Strait of Gibraltar from end of February or March, chiefly end March to 1st 3 weeks of April (Pineau and Giraud-Audine 1979; Cortés et al. 1980). Present in Malta to early April (Sultana and Gauci 1982); in Sicily, some evidence of spring passage late February to late March, with a few into early May (Iapichino and Massa 1989); on Crete, in one-year, regular from at least mid-March to early April, with stragglers later (Phillips and Round 1975). Present in Egypt chiefly to mid-March (Goodman and Meininger 1989), and winter populations leave Israel and Syria mainly in March (Baumgart and Stephan 1987; Shirihai in press), Cyprus chiefly late March to early April (Flint and Stewart 1992). Earliest record 7 March at Kizilcahamam (Turkey) (Bariş et al. 1984). Returns to breeding grounds in Firenze area sometimes end of February, otherwise well into March (Dinetti and Ascani 1990). In Switzerland, passage occasionally from mid-February, usually from 2nd third of March, peaking 1st third of April and continuing to end of April, occasionally to beginning of May (Winkler 1984). Reaches breeding grounds in Rheinland from mid-March (Mildenberger 1984), and Luxembourg in April (Melchior et al. 1987). In study in southern Belgium, 1986–9, most passage April, but timing varied considerably from year to year, with range of approximately a month (c. 21 March to 21 April) for earliest arrivals; fidelity to passage-site shown by bird ringed 15 April 1989 retrapped 20 October 1989 (Rion 1990). Similarly in Mecklenburg earliest records vary greatly from year to year, averaging 4 April (5 March to 28 April) (Klafs and Stübs 1987). At extreme north of range, reported in Kaliningrad region chiefly in

April (Odinzowa 1967); on southern shore of Gulf of Finland, individuals and small groups of 2-4 birds recorded at end of April and beginning of May annually since 1960 (Noskov and Shamov 1983).

In Britain, recorded in all months, with distinct peaks October–November and especially April–May (Dymond *et al.* 1989). Long-term expansion of range across Europe shows marked north-east tendency (towards St Petersburg region); expansion into Scandinavia slower (Olsson 1971), and still only sporadic in Britain, suggesting reluctance to cross open water (Spencer *et al.* 1988*b*, 1989), though data above show that some cross Mediterranean regularly.

DFV

**Food.** Seeds and other plant material; occasionally small invertebrates. Forages principally on herbs and on ground; tree-foraging probably mainly in spring (Mayr 1926; Eber 1956; Glutz von Blotzheim 1962; Olsson 1971). Especially in winter, forages in large flocks, often with other seed-eaters (see Social Pattern and Behaviour). Feeds energetically and with agility like Linnet *Carduelis cannabina* (in tall herbs) or Siskin *C. spinus* (in trees); extracts ripening seeds from heads of Compositae by carefully pulling down bracts surrounding inflorescence one by one, and removing petals or pappus using bill and tongue; pulls buds and catkins to pieces (Eber 1956). Will use feet to hold items on ground while feeding (Newton 1972).

Diet in west Palearctic includes the following. Invertebrates: aphids (Hemiptera: Aphidoidea), larval Lepidoptera (Coleophoridae), spiders (Araneae). Seeds, buds, flowers, etc., of spruce *Picea*, *Thuja*, larch *Larix*, birch *Betula*, alder *Alnus*, elm *Ulmus*, mulberry *Morus*, nettle *Urtica*, hemp *Cannabis*, dock *Rumex*, knotgrass *Polygonum*, buckwheat *Fagopyrum*, orache *Atriplex*, amaranth *Amaranthus*, chickweed *Stellaria*, sandwort *Minuartia*, poppy *Papaver*, Cruciferae, rose *Rosa*, willow herb *Epilobium*, evening primrose *Oenothera*, pimpernel *Anagallis*, forget-me-not *Myosotis*, lavender *Lavandula*, plantain *Plantago*, Compositae, grass *Poa*, millet *Setaria*. (Quépat 1875; Mayr 1926; Tutman 1950; Eber 1956; Nicolai 1960; Turček 1961; Glutz von Blotzheim 1962; Olsson 1971; Tucker 1980*b*; Elzen 1983; Sabel 1983.)

In northern Germany, of 491 feeding observations, 20% were on mugwort *Artemisia*, 15% shepherd's purse *Capsella*, 15% birch, 10% elm, 6% knotgrass, 5% hedge mustard *Sisymbrium*, 5% pennycress *Thlaspi*, 4% dock, 4% *Brassica*, 3% charlock *Sinapis*, 3% dandelion *Taraxacum*, 2% *Poa*, 2% chickweed, 2% evening primrose, 1% alder, 1% rose. In spring, when green parts of plant important, favourite food apparently buds of elm, birch catkins, and dandelion heads; in summer seeds of Cruciferae; similarly in autumn, especially *Brassica*, and also seeds of knotgrass. Diet in general mostly composed of small, round, oil-rich seeds up to 3 mm long. Of 523 observations of seed-eating, 31% Cruciferae and 25%

Compositae. (Eber 1956.) In France, feeds on various cultivated Cruciferae in gardens in autumn, but seems to be especially fond of seeds of lettuce *Lactuca* and *Capsella* (Quépat 1875; Sabel 1983). In winter, very often recorded on *Artemisia vulgaris* (Olivier 1949; Eber 1956; Kroymann 1965; Rinnhofer 1969). *Thuja* seeds reported to be favourite autumn food in Lithuania (Ivanauskas 1961). Birds in captivity preferred either seeds of *Brassica* or thistle *Carduus*, followed by millet, Canary grass *Phalaris*, and poppy (Ziswiler 1965, which see for details of handling, eating, and bill structure; Sabel 1983).

Young fed seeds, softened in crop of adult, e.g. Cruciferae, thistle, *Poa* (Dupont 1944; Nicolai 1960; Makatsch 1976; Tucker 1980*b*; Sabel 1983). In captivity sometimes fed aphids by parents (Meyer-Deepen 1954), but usually fledges successfully on diet of seeds alone (Sabel 1983, which see for details).

BH

**Social pattern and behaviour.** Useful early observations by Quépat (1875), and general reviews by Géroudet (1957) and (some aspects) Olsson (1971), but important details of pair-formation and displays still lacking.

1. Gregarious outside breeding season, forming small flocks for feeding and migration, usually fewer than 100 birds (Géroudet 1957) but not uncommonly more on migration and in favoured wintering sites. In Berlin (Germany), June–September, feeding flocks typically 8-14 birds (Herrmann 1977), in Bonn up to 40 together in winter (Linke 1975). On spring migration, St Petersburg region (Russia), seen singly and in groups of 2-4 (Noskov and Shamov 1983). On autumn migration, north-west Spain, feeding flocks probably of several thousand occur (Snow *et al.* 1955). In Israel, autumn migration mainly in flocks of tens or more; in spring, hundreds assemble at staging posts; in main wintering areas, flocks sometimes of thousands gather for feeding, drinking, and roosting (Shirihai in press). Freely associates with other Fringillidae for feeding (for species involved, see, e.g., Géroudet 1957, Rinnhofer 1969, Linke 1975), though *S. serinus* usually in minority in mixed flocks (Géroudet 1957). For references to association with Red-fronted Serin *S. pusillus*, see that species; in mixed flocks with *S. pusillus* in Israel, winter, *S. serinus* usually predominated (Fergman 1988). Association with Citril Finch *S. citrinella* on Corsica reported (Armitage 1937). See also Flock Behaviour (below). In Strait of Gibraltar area, *S. serinus* apparently widely scattered at low densities in winter: in maquis, *c.* 8 birds per km² (Finlayson 1992); 2 per km² in matorral, 6·3 in pasture (Arroyo and Tellería 1984). BONDS. No evidence for other than monogamous mating system. No information on duration and fidelity of pair-bond. In Hungary, observations on marked birds suggested ♂ who lost mate replaced her in 3 days (Barta 1976). Nest-building mainly or entirely by ♀; only ♀ incubates and broods (Quépat 1875; Gnielka 1978). Both sexes feed young. Broods may overlap: ♀ seen starting to build new when 1st brood (close to fledging) still in old nest (Gnielka 1978). After 1st brood fledges, ♂ presumably feeds them while ♀ incubates new clutch (Tucker 1980*b*). Steinfatt (1942) recorded parents occasionally feeding (but more often ignoring) fledged 1st brood in territory while also attending 2 8-day-old young in nest. Young independent *c.* 9 days after leaving nest (Géroudet 1957). Age of first breeding 1 year (C S Roselaar). BREEDING DISPERSION. In optimal conditions, dispersed in neighbourhood groups, otherwise solitary (Witherby *et al.* 1938; Herrmann 1977; Paz 1987). Territory size variable,

typically 1 ha in gardens and parks (Géroudet 1957). In Bernberg (eastern Germany) average *c.* 80 m between territory centres; 3 nests separated by 33, 54, and 65 m, and along avenue of false acacia *Robinia* 11 nests in *c.* 1100 m, with 8–330 m between nests (Herrmann 1977). In Israel, breeding groups each contain a few tens of pairs, minimum 5 m between nests (Shirihai in press). Territory serves for courtship, feeding (in one territory, foraging flights rarely exceeded 100 m from nest: Steinfatt 1942) and nesting; however Song-flights (see below) overlap territory boundaries (Herrmann 1977). In Halle (eastern Germany), singing ♂♂ occupied 50% of territories by mid-April and all settled by first third of May; first nests were found in early-occupied territories (Gnielka 1978). Density varies markedly, often highest in cemeteries, e.g. in western Ukraine 3–4 pairs in 1 ha (Marisova *et al.* 1990). In Switzerland, density generally high (100–200 pairs or more per km²) in gardens and parks around Geneva and Lyss, but often less, and these high densities rarely achieved in orchards and forest (Glutz von Blotzheim 1962); 110–170 pairs per km² over 26 years in cemetery in Lausanne, but in built-up area (340 ha) in Zurich, 5–15 pairs per km² in 1978; mountain villages often only have isolated pairs (Schifferli *et al.* 1982). In Germany, 76–133 pairs per km² in cemetery in Bonn, 1967–75 (Linke 1975); in Halle, 89 pairs per km² in cemetery, 1964–77; 28 in riverine woodland, 1959–77 (Gnielka 1978). In Bulgaria, 6–9 pairs per km² in predominantly silver fir *Abies alba* (Simeonov 1975; Petrov 1988) to 108 in predominantly juniper *Juniperus oxycedrus* (Petrov 1982). In Italy, 11 territories per km² in mixed coniferous–deciduous woodland in Tuscany (Lambertini 1981), *c.* 500 pairs per km² in suburbs of Palermo (Sicily) (Iapichino and Massa 1989). In Morais (northern Portugal) 3·6 territories per km² (Mead 1975). In Spain, 6·4 birds per km² in *Pinus radiata*, Basque region (Carrascal 1987); in Sagunto (Valencia), 290–296 pairs per km² in 2 years (Gil-Delgado 1981). In Morocco, average 125 pairs per km² in forest (all types), 18·2 in maquis (all types), maximum (for any habitat) 36 in semi-arid *Thuja* scrub (Thévenot 1982). No information on site-fidelity from year to year. In Halle, once young fledged, nests too damaged to be re-used but several cases of old nest being partly dismantled to build new one; 3 broods per season in many territories, including successive nests in same tree (Gnielka 1978). ROOSTING. Little information. ♀ said to roost with young in nest at night (Quépat 1875) though young alone at one nest studied (Steinfatt 1942). Roosts communally outside breeding season, e.g. in Israel, large roosts in woodland, birds dispersing daily a few km to feeding sites (Paz 1987; Shirihai in press).

2. Confiding and bold in breeding season, allowing approach of man to within a few metres (Géroudet 1957; Barta 1976). When habituated to observer at one nest, regularly came to within 1 m (Steinfatt 1942). Shyer outside breeding season, but in Erzgebirge (eastern Germany) birds in mixed flocks would allow approach to *c.* 20 m; when Kestrel *Falco tinnunculus* flew in low, whole flock scattered, then re-settled when danger passed (Rinnhofer 1969: see also Flock Behaviour, below). If flying bird threatened by raptor, darts sideways and seeks safety of cover in zigzag flight (Quépat 1875). During mild disturbance, may give short fragments of song (Bergmann and Helb 1982: see 1 in Voice). In study of birds handled in Spain, resident more often gave alarm-calls (6b in Voice) than passage birds (Senar *et al.* 1986). FLOCK BEHAVIOUR. Feeding flocks restless and highly mobile. Flock-members periodically gather in tree where they preen briskly, call (2–3 in Voice) and sing in chorus before returning to feed (Géroudet 1957). Flock fairly tight-knit, but association with other Fringillidae typically loose, e.g. *S. serinus* and Redpolls *Acanthis flammea* feeding on low plants and on ground almost always separated when flock took off (Rinnhofer

1969). SONG-DISPLAY. ♂ sings typically from tall exposed tree, overhead wire, etc. (e.g. Géroudet 1957), sometimes from much lower perch, even from ground (Cawkell 1949); also often leads directly to relatively brief (but often reported) Song-flight over and beyond territory (Herrmann 1977). Following account mainly from Quépat (1875), Géroudet (1957), Tucker (1980*b*), Bergmann and Helb (1982), and Harris *et al.* (1989). ♂ starts singing in Advertising-posture (Fig A: ♂ left, ♀ right): upright

A

with wings somewhat drooped, tail often half-cocked, throat ruffled, and body trembling with effort, head stretched up and turning from side to side. Newton (1972) described ♂ singing (perhaps to ♀) while hopping from foot to foot. Soon he launches almost vertically into the air, ruffles plumage, spreads wings and tail, and, with slow deep wing-beats (sometimes interspersed with glides) follows flitting erratic course *c.* 10–20 m above ground like Song-flight of Greenfinch *Carduelis chloris*; path includes wide arcs and performer often throws himself from side to side in rolling movement; descent is in circles (Newton 1972), with slow parachute drop at end (Etchécopar and Hüe 1967) to land on perch from which flight began (Witherby *et al.* 1938), often beside ♀ (Newton 1972). There, continues singing in Advertising-posture or, after short pause, starts another Song-flight. ♂ typically displays alone, although Herrmann (1977) occasionally saw 2nd ♂ in Song-flight behind 1st—and perched birds not uncommonly singing 15–20 m apart without hostility (more often 50–80 m). In southern France, ♀♀ also claimed to have performed Song-flights (Flegg 1974). ♂♂ sing virtually all year; outside breeding season, heard especially from mid-winter onwards. In Provence (France) song almost as common in winter as summer (Quépat 1875), and in Malta often sings persistently in winter (Sultana and Gauci 1982). Further north also (e.g. Germany), song regular through winter from September, mainly on sunny days (Lütgens 1955; Kumerloeve 1974; Linke 1975; Zucchi 1975); in winter in Erzgebirge, several ♂♂ in small area sang alternately, not synchronously as others have reported (Rinnhofer 1969, which see for references). On Helgoland (Germany), song heard on 54% of 20 days of spring passage (Moritz 1982). However, main song-period April–August (Géroudet 1957); in Cyprus February–August (Flint and Stewart 1992). In south-west England song from paired ♂ heard from beginning of May to mid-August (Tucker 1980*b*). ♂'s song-activity declines within each nesting attempt, i.e. through incubation and chick-rearing (Olsson 1971; Tucker 1980*b*), so resurgence of song indicates next clutch imminent, e.g. in study by Steinfatt (1942) ♂ re-commenced vigorous Song-display when brood *c.* 5 days from fledging. While ♀ nest-building, ♂ sang almost continuously, almost all day, although, after incubation presumed to have started, song was mainly confined to late afternoon

(Tucker 1980*b*). ANTAGONISTIC BEHAVIOUR. Song-display of ♂ (see above) serves to demarcate territory and defend it against intruders (Géroudet 1957). Open hostility rare, or at least rarely seen. In Berlin study, the only serious dispute between 2 ♂♂ occurred in late May: 2 ♂♂ singing from perches *c.* 40 m apart; one flew towards and landed near other which then approached intruder, attacked and chased him in short vigorous pursuit-flight in which feathers were pulled out (Herrmann 1977). For aggressive element in courtship, see Heterosexual Behaviour (below). During incubation, ♀ seen driving off importunate 1st-brood fledgling in territory (Gnielka 1978). At bird-table in Hessen (Germany), 3 birds repeatedly drove off *C. chloris* (Zucchi 1975). For Submissive-display common to all *Serinus*, see Canary *S. canaria*. HETEROSEXUAL BEHAVIOUR. (1) General. In Halle, 1st eggs laid *c.* 23 days after territory occupied by ♂, and 4–9 days after start of nest-building (Gnielka 1978). (2) Pair-bonding behaviour. Presumably starts in flocks, as in other Fringillidae, but no information. Best description of intimate courtship is by Quépat (1875): on landing after Song-flight (see above), ♂ perches, adopts Advertising-posture, quivers drooped wings, bows quickly, ruffles plumage, stretches neck, raises and fans tail, and struts in front of ♀ who remains passive; ♂ suddenly pounces on ♀ and chases her repeatedly. Newton (1972) described chasing ♀ and crouching with wings out, plumage ruffled on head and throat. (3) Courtship-feeding. Occurs late in incubation, ♀ sitting continuously between ♂'s visits (Quépat 1875). Courtship-feeding sequence in one pair as follows. ♂ announced arrival with brief 'chirr' calls (see 2 in Voice) *c.* 20–30 m away, and ♀ at once answered with intensive Begging-calls (5 in Voice) until he reached nest; sometimes she left nest for *c.* 1–2 min to defecate but returned at once to beg by sinking deep into nest while raising head and bill and wing-shivering rapidly; ♂ regurgitated food in small portions over 15–20 s and fed her regularly thus at intervals of *c.* 1 hr. (Olsson 1971.) ♂ continues feeding ♀ during brooding (see Relations within Family Group, below). (4) Mating. The only description concerns pair copulating when young in nest near fledging, presumably to fertilize next clutch; copulated 3 times in same evening (19.55–20.50 hrs) when both came to nest to feed young. On 1st occasion, ♀ at nest (not feeding young) called quite frequently and excitedly (no further details of call) and flew to tree *c.* 15 m away where ♂ suddenly mounted her while she crouched with wings spread and head turned up towards ♂; one bird (presumed ♂) gave quiet delicate calls, and copulation lasted *c.* 3 s (Steinfatt 1942). (5) Nest-site selection and behaviour at nest. ♂ accompanies ♀ constantly when she inspects prospective nest-site (Géroudet 1957), but not known which bird makes final choice. ♀ visited nest-site (in course of construction) cautiously, never entering or leaving tree directly (Tucker 1980*b*). Géroudet (1957) found ♂ to accompany ♀ during building as closely as during nest-site selection, but in other studies he was absent (Gnielka 1978; Tucker 1980*b*). RELATIONS WITHIN FAMILY GROUP. At Swedish nest watched from hatching, ♀ continued brooding young for many days as assiduously as she had incubated; ♂ regurgitated all food to ♀ and, as soon as he left, she in turn delivered it (presumably from sub-lingual pouch) to young (Olsson 1971). No doubt ♀ also eats some of ♂'s food; in cold weather, Gnielka (1978) saw ♀ brooding young almost continuously; ♂ fed her on nest, occasionally also feeding any chick which emerged from beneath her. When young older, ♂ and ♀ both forage in territory for them; parents seen foraging together and always returning together to feed young; each feed lasted 15–30 s, young stretching high and gaping (Steinfatt 1942). At first ♀ eats faecal sacs, but after 7–9 days young defecate on nest-rim and it becomes heavily encrusted, droppings also littering ground below nest. During last days before fledging, young flutter and jump on nest-rim. (Olsson 1971.) Will explore branches around nest but return to it overnight (Quépat 1875). Young may fledge over 1–2 days and disperse into nest-tree where, despite persistent food-calls, difficult to locate (Gnielka 1978); at nest in Devon, fledglings at times up to 250 m apart (Tucker 1980*b*). ANTI-PREDATOR RESPONSES OF YOUNG. From *c.* 10 days, young become erect at slightest noise, listening attentively; crouch in nest-cup as soon as they hear ♀'s alarm-call (see 6 in Voice). PARENTAL ANTI-PREDATOR STRATEGIES. (1) Passive measures. ♂ gives alarm-call as soon as he sees raptor or cat approaching nest; ♀ responds by crouching motionless in nest, and covering any young (Quépat 1875). ♀ sits tightly, often not leaving nest until human within 0·5 m; no cases of desertion from observer checking nest (Gnielka 1978). (2) Active measures. Distraction-lure display ('injury-feigning') by ♀ observed in Germany (Witherby *et al.* 1938; Gnielka 1978), but no details. In one case, young had just hatched; in another, young *c.* 2 days old were being examined with mirror on stick (Gnielka 1978).

(Fig by D Nurney from drawing in SOVON 1987.)  EKD

**Voice.** Vocal all year, notably in continuing song outside breeding season (see Social Pattern and Behaviour). For additional sonagrams, see Bergmann and Helb (1982). Analysis of recordings (below) and their renderings by W T C Seale.

CALLS OF ADULTS. (1) Song. Fast jumble of high-pitched sizzling, tinkling (etc.) sounds given by ♂ from perch or in song-flight (see below for possible song by ♀). Persistent succession of relatively long phrases of harsh jingling, chirping, twittering sounds, somewhat like splinters of glass being rubbed together (Jonsson 1982; Bruun *et al.* 1986). Reminiscent of Corn Bunting *Miliaria calandra*; utterly different from rich warbling of Canary *S. canaria* (Holman and Madge 1982). Also recalls Siskin *Carduelis spinus* (Olsson 1971) but much higher pitched (W T C Seale); see Holman and Madge (1982) for comparison with *C. spinus* and other *Serinus*, also Nicolai (1957) for comparison with other *Serinus*. Much faster and higher pitched than Syrian Serin *S. syriacus* (K Mild, E K Dunn). For comparison with Red-fronted Serin *S. pusillus*, see that species. Part of phrase of *S. serinus* shown in Fig I (half speed) is typical in consisting partly of rapid series of different unit-types, partly of rapid reiterations of same type (i.e. segments); some composite units are repeated in the same, or modified, form at different places in the phrase. Further structure derives from recurrence of sections within the phrase: whole phrase includes repetition of at least 2 types of section; part depicted in Fig I shows repetition of one section which first appears at *c.* 0·19–0·93 s and again at 1·88–2·65 s. Short fragments of song given when mildly alarmed; rate of delivery so rapid as to make individual units scarcely distinguishable to human ear (Bergmann and Helb 1982), so no renderings attempted here—but see Quépat (1875) and Olsson (1971). Song often preceded by call 6b (Quépat 1875; Géroudet 1957). Apparently rare case of mimicry of song of Wren *T. troglodytes* in Denmark, May (Hansen 1975, which see for sonagram). For alleged ♀♀ in full song in

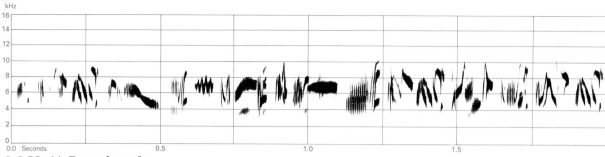

I J-C Roché France June 1982

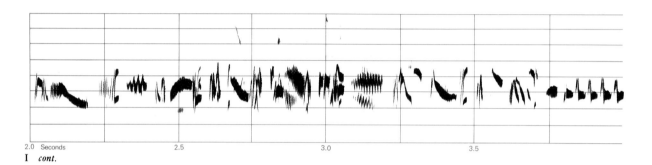

I cont.

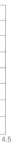

I cont.

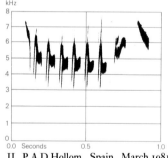

II P A D Hollom Spain March 1981

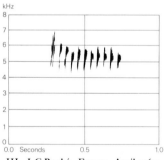

III J-C Roché France April 1967

southern France, see Flegg (1974). (2) Ripple-call. Rather dry 'trillilit' (Holman and Madge 1982) given by both sexes as contact-call (Tucker 1980b), typically (though not always) in flight. Also rendered 'tirrilillit', with many variants, e.g. 'tittertee' (Witherby *et al.* 1938); high-pitched, metallic, twittering 'zr-r-litt', not unlike timbre of juvenile Pied Wagtail *Motacilla alba* (Bruun *et al.* 1986). Fig II shows quite penetrating, descending 'ch-ik-ik-ik-ik-ik-ik' (followed by 'si see') with ringing metallic quality reminiscent of *M. alba*. Sometimes no more than 'tirr' (Jonsson 1982). Calls described by Bergmann and Helb (1982) apparently also short variants: high-pitched tremolo 'tri' or more drawn-out 'girr' which sometimes has detached introductory unit, thus 'psitirr'. Call in Fig III apparently of extended 'girr' type: rather musical rattling 't-r-r-r-r-r-r-ri' with timbre reminiscent of Tic-calls of Robin *Erithacus rubecula*. 'Chirr' of ♂ approaching nest with food for ♀, and longer sequences heard by Olsson

(1971), perhaps also belong here: 'ti-ti-ti-chirr' like Blue Tit *Parus caeruleus*, or sometimes 'chiii chi-chi-chi-chirrr', 'chirr' in both cases sharp and distinctive. Ripple-calls are usually preceded by call 6b (Quépat 1875; Taylor 1989), but perhaps only when flushed. (3) Rattle-call. Hard 'chit-chit-chit' recalling *T. troglodytes* (Peterson *et al.* 1983). Fig IV shows dry rattling 'ti-ti-ti-ti-ti-ti-ti-ti-ti-ti-', slightly reminiscent of alarm-rattle of *T. troglodytes*, and very similar to, but higher pitched than, rattles in Fig IV of *S. syriacus*. Function not known. (4) Chirp-calls. Chirruping sounds like sparrow *Passer* noted by Witherby *et al.* (1938) perhaps include weak short 'chuip' and 'chitt-itt' exchanged by ♂ and ♀ (Olsson 1971). Some such calls may be abbreviated variants of calls 2–3. (5) Begging-call of ♀. High-pitched penetrating 'zizizizi...' given to solicit food from ♂ (Gnielka 1978); intense rapid 'chi-chi-chi-chi...' from ♀ on nest answering 'chirr' calls (see 2, above) of approaching mate (Olsson 1971).

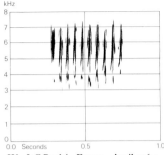

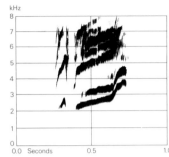

IV  J-C Roché  France  April 1967

V  J-C Roché  France  April 1967

VI  E D H Johnson  Corsica  May 1989

Begging-call apparently derived from food-call of young (see below). (6) Excitement-, alarm-, and distress-calls. (6a) High, piercing, repeated 'tsiiii' directed at dog (Olsson 1971). High-pitched nasal 'schwid' in lower-intensity excitement (Bergmann and Helb 1982) is perhaps the same. (6b) Common alarm-call, mostly disyllabic with rising pitch, but diversity of descriptions indicates real variation, e.g. rather tinny 'tsooeet' like *S. canaria* (Witherby *et al.* 1938); soft call with upward inflection, rather like 'wheet' of Chaffinch *F. coelebs*, usually followed by call 2 (Taylor 1980); 'djui' like *C. chloris* (Jonsson 1982); drawn-out, slightly shrill 'tvuih' (Géroudet 1957); broad-band 'dschäi' given in more intense alarm than call 6a (Bergmann and Helb 1982). Fig V shows nasal 'hu-hooee' similar to homologous calls of other *Serinus* and some other Fringillidae. The following are apparently related but noisier: broad-band 'grii' given by handled birds but less often than characteristic 'tuit' (Senar *et al.* 1986, which see for differences in rate of calling between resident and passage birds). (7) Other calls. Calls heard from flock feeding on spring migration in Corsica (E D H Johnson) included rather nasal, ringing, diadic calls (Fig VI), notes showing rapid frequency modulation, overall reminiscent of very small tin trumpet; thus similar to units in 'chant' of Citril Finch *S. citrinella corsicana* (see Fig V of that species). Calls depicted in Fig VII are of 2 types, both

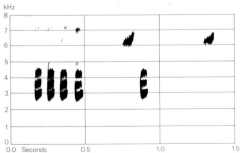

VII  E D H Johnson  Corsica  May 1989

rapidly modulated: high pitched 'zi' and lower pitched, slightly nasal, diadic 'di'; whole sequence 'di-di-di-di zi di zi'.

CALLS OF YOUNG. Food-call of small young 'ci ci ci ci ci', weak yet audible at 'IO-I2 paces' (Quépat 1875). Older nestlings and fledglings give regularly repeated, very high-pitched 'tsi' (Bergmann and Helb 1982), also rendered variously as 'zeee', 'zeeez', 'ziz', e.g. sequence when excitedly following parent 'zeeez-zeeez ziz-ziz-zeee', like contact-calls of Spotted Flycatcher *Muscicapa striata* (Tucker 1980a, b). Young also heard to give fairly high-pitched 'prrrp', most often disyllabic and thus said to be indistinguishable from disyllabic variants of adult call 2 (Tucker 1980a). In recording by J-C Roché, these calls (roughly 'di-du') all disyllabic but do not sound like any adult call; regular monotonous delivery alone enough to distinguish them as young.                    EKD

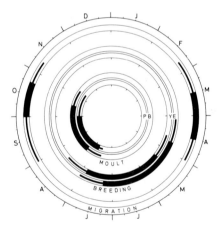

**Breeding.** SEASON. Estonia: nest-building 2nd half of May (Kumari 1958). Germany: see diagram; median date for start of 1st clutch in conifers, eastern Germany, 9 May ($n=177$), $c.$ 1 week later in broad-leaved trees ($n=55$); 2nd and possibly 3rd clutch laid June to end of July (Gnielka 1978); for south-west Germany, see Glück (1983). France: eggs laid early May in north-east, 2nd clutch end of June to early July (Quépat 1875). England: one nest built early May; fledged young recorded late June and probable 2nd brood fledged end of August (Tucker 1980b). Spain: peak laying period mid-March to mid-April, 2nd clutch June to early August (Gil-Delgado

1981; Gil-Delgado and Gómez 1988). North Africa: eggs laid from February (Etchécopar and Hüe 1967). SITE. Generally in conifer rather than broad-leaved tree, though also in bush; also commonly in fruit trees (Quépat 1875; Mayr 1926; Glutz von Blotzheim 1962; Makatsch 1976; Mildenberger 1984). In Mediterranean area, often in citrus groves (Gil-Delgado 1981; Shirihai in press). In large cemetery in eastern Germany, 53% of 387 nests in spruce *Picea* (much preferred in relation to abundance), 9% lime *Tilia*, 7% yew *Taxus*, 5% pine *Pinus*, and 5% larch *Larix*; birch *Betula* and oak *Quercus* avoided. In broad-leaved woodland, 54% of 78 nests in elm *Ulmus* (preferred) and 26% elder *Sambucus*; oak avoided. Preferred position in outermost twigs, followed by top of conifer, then on branch against trunk; average height above ground 5·7 m (1·5–13·5, n=387), elsewhere, nests recorded up to 18 m. (Gnielka 1978.) See also Jung (1955). In area of scattered orchards, south-west Germany, 56% of 43 nests in apple *Malus* and 33% in pear *Pyrus*; generally twice as far from trunk as from periphery, average height 3 m, maximum 6 m (Glück 1983). Average height of 131 nests, Czechoslovakia, 2·9 m (Hudec 1983). For France, see Quépat (1875); for north-east Spain, see Muntaner *et al.* (1983). Nest: small and compact, of fine twigs, stalks, sometimes strips of bark, roots, grass, moss, or lichen, lined neatly and thickly with rootlets, hair, feathers, plant down, etc. (Harrison 1975; Makatsch 1976; Gnielka 1978); outer diameter 8·5–10·0 cm, inner diameter 4·5–7·0 cm, overall height 4·5–5·0 cm, depth of cup 3·0–4·0 cm; weight of one nest 15·9 g (Köhler 1943; Dupont 1944; Géroudet 1957); for detailed descriptions from various regions of France, see Quépat (1875). Building: takes 4–11 days from start of building to laying of 1st egg (Gnielka 1978; Gil-Delgado 1981). Generally by ♀ only, ♂ sometimes bringing material and rarely assisting with lining (Quépat 1875), though other authors report no participation by ♂ (Makatsch 1976; Gnielka 1978; Tucker 1980b); ♂ in captivity recorded building nest alone in 5 days (Meyer-Deepen 1954). Material almost all from ground, though bark perhaps pulled from birch trees (Gnielka 1978). EGGS. See Plate 60. Sub-elliptical, smooth and slightly glossy; pale bluish-white, sometimes greenish-white, sparsely spotted and streaked rusty and purplish, mostly at broad end, sometimes forming circle (Harrison 1975; Makatsch 1976). 16·1 × 12·0 mm (14·4–17·8 × 11·0–13·3), n=337; calculated weight 1·21 g (Schönwetter 1984). For 53 eggs from north-east Spain, see Muntaner *et al.* (1983); for change in egg size over season in eastern Spain, see Gil-Delgado and Gómez (1988). Clutch: 3–4 (2–5). In eastern Germany, of 222 clutches: 2 eggs, 0·5%; 3, 8%; 4, 82%; 5, 9%; average 4·0; little change over season (Gnielka 1978). In eastern Spain, of 185 clutches: 2 eggs 5%; 3, 18%; 4, 77%; 5, 0·5%; average 3·7 (Gil-Delgado 1981; Muntaner *et al.* 1983). Average in western Germany 3·5, n=75 (Mildenberger 1984); average in Czechoslovakia 4·0, n=67, including 1 clutch of 6 (Hudec 1983); average

in south-east France 3·9, n=23 (G Olioso). According to Quépat (1875), clutches of 5–6 not uncommon in southern France. 1–2 broods in Mediterranean area, perhaps up to 3 in central Europe (Steinfatt 1942; Glutz von Blotzheim 1962; Gnielka 1978; Gil-Delgado 1981; Shirihai in press), possibly because of greater abundance of food in late summer in north (Mayr 1926). In Rheinland (western Germany), said to have 2 broods in climatically favourable areas, otherwise 1 (Mildenberger 1984). Records of 2nd clutch started 4 days after fledging of 1st brood (Glutz von Blotzheim 1962), and of ♀ building new nest with young still in 1st (Gnielka 1978). In Rheinland, 3rd clutch laid following death of 2nd brood (Mildenberger 1984). Eggs laid daily (Meyer-Deepen 1954; Makatsch 1976; Gnielka 1978). INCUBATION. By ♀ only (Harrison 1975; Makatsch 1976; Gnielka 1978); from laying of last egg to hatching of last young, average 12·6 days, n=8 (Gnielka 1978), 12·8 days, n=11 (Gil-Delgado 1981). Starts with penultimate or last egg (Quépat 1875; Meyer-Deepen 1954; Géroudet 1957; Gnielka 1978). YOUNG. Fed and cared for by both parents; faeces left on nest-rim after young 7–9 days old (Steinfatt 1942; Olsson 1971; Makatsch 1976). ♀ brooded *c.* 8-day-old young continuously for some days in light rain (Gnielka 1978). FLEDGING TO MATURITY. Average fledging period 15·2 days (13–18, n=14) (Gnielka 1978), or 14 days (12–16, n=15) (Gil-Delgado 1981, which see for development of young). See also Rohner (1980). Independent *c.* 9 days after fledging (Dupont 1944; Géroudet 1957). BREEDING SUCCESS. In eastern Germany, of 587 eggs, 90·3% hatched; of 450 nestlings, 92% fledged; *c.* 5·5 young fledged per pair per season in optimum cemetery habitat (wide variety of tree species), *c.* 4·0 in broad-leaved woodland (Gnielka 1978). In eastern Spain, of 290 eggs, 67% hatched and 33% produced fledged young; of 83 eggs in March, 59% hatched and 43% produced fledged young; of 47 in April, 34% and 6%; of 25 in May–July, 76% and 0%; 37% of losses caused by predation, 20% agricultural activity, 20% infertility and nest damage, 13% weather, and 10% abandoned; 44% of nestling losses due to starvation, remainder predation. (Gil-Delgado 1981; Gil-Delgado and Gómez 1988.) In Switzerland, nests destroyed by Corvidae, especially Magpie *Pica pica*, cats, and wind (Géroudet 1957; Glutz von Blotzheim 1962). For review of productivity, see Senar and Copete (1990b). BH

**Plumages.** ADULT MALE. In fresh plumage (autumn and early winter), forehead and band across hindneck bright yellow, partly obscured by sandy-grey and black mottled feather-tips on forehead and by olive mottling on hindneck; forehead not sharply defined. Crown light sandy-brown, marked with short and ill-defined dull black streaks, some yellow or green of feather-bases sometimes shining through. Mantle and scapulars light drab-brown or grey-brown, sometimes with tawny tinge, heavily streaked black, sides of feathers often partly yellow or light green on base, latter mainly concealed. Back closely streaked black and green-yellow; rump contrastingly uniform yellow, partly tinged green on feather-tips at side. Upper tail-coverts grey with olive border, black shaft-streak, and sandy-brown fringe. Lore and front part of supercilium dark grey with some yellow-green mottling. Supercilium (backwards from above eye), short stripe just below eye, and bar across side of neck bright yellow, but partly mottled olive and not sharply defined. Cheeks and ear-coverts with indistinct dusky grey streaks and some green suffusion; some yellow of feather-bases partly visible, especially in lower cheek, forming ill-defined stripe from base of lower mandible to lower side of neck, bordered by indistinct dusky green or grey moustachial and malar stripe above and below. Chin, throat, chest, and mid-belly bright yellow or golden-yellow, feathers narrowly tipped pale pink-grey, giving light grey cast to (especially) chest; side of breast and flank tawny- or sandy-buff, merging into yellow on side of belly, closely marked with dull black or dark grey streaks. Vent and under tail-coverts white with slight yellow or buff suffusion, under tail-coverts with concealed dark grey arrow-marks on centres. Tail greyish-black, fringes along outer webs of central 5 pairs (t1–t5) yellow or green-yellow (widest near bases), faint fringes along tips and inner webs of t1–t5 and along both webs of t6 pale pink-buff or pale grey. Flight-feathers, greater upper primary coverts, and bastard wing greyish-black; narrow fringes along p1–p8 (except for emarginated part of p6–p8), primary coverts, and shorter feathers of bastard wing pale yellow or green-yellow, those along secondaries slightly broader and more green or buff-green; fringes along outer webs of secondaries narrower towards base, but without contrasting black band across base, unlike Syrian Serin *S. syriacus* and Red-fronted Serin *S. pusillus*. Tips of flight-feathers narrowly fringed pale grey. Tertials dull black, outer webs broadly fringed green-buff or sandy-buff, more greyish-buff towards tips. Lesser upper wing-coverts yellow-green with dull black centres; median coverts dull black with pale yellow or green-yellow tip *c.* 2 mm wide; greater coverts dull black with green-yellow outer fringe and pale yellow or buff-yellow tip *c.* 3–4 mm long, hardly extending to inner web. Under wing-coverts and axillaries pale grey, almost white on tips of longest feathers, mottled green-yellow and grey along leading edge of wing. *In worn plumage* (about April–July), forehead bright yellow with some dusky mottling along base of bill, merging into green-yellow crown, latter with narrow dusky streaks; side of crown and (sometimes) hindcrown olive-grey with similar streaks; forehead distinctly less sharply defined than that of *S. syriacus*. Hindneck and side of neck with well-defined bright yellow semi-collar, which shows faint olive streaks or spots; pattern on side of head more clearly defined than in fresh plumage, side of head grey or olive-grey with fairly distinct yellow supercilium, short yellow stripe below eye, and yellow spot or short streak on lower cheek. Mantle and scapulars less sharply streaked black, brown of feather-tips worn off, yellow-green of lateral feather-bases more exposed; rump green-yellow, strongly contrasting. Chin to mid-belly uniform bright yellow, sometimes tinged green, merging through yellow-white into white of flank, belly, vent, and under tail-coverts; side of breast and flank marked with contrasting black streaks. Tail and flight-feathers browner, pale fringes along tips worn off, yellow fringes along primaries narrow; more black of bases of median and greater coverts exposed, tips green-yellow, more sharply contrasting than in fresh plumage. ADULT FEMALE. In fresh plumage, upperparts browner than in adult ♂, hindneck, mantle, and scapulars virtually without yellow, purer green-yellow restricted to rather narrow and ill-defined patch on rump; sides of feathers of forehead with some yellow (in ♂, entire basal half yellow). Side of head and neck more evenly streaked and mottled green and brown, greyer on lore, slightly yellower on supercilium and just below eye, without distinct stripes and

marks of adult ♂. Chin to breast and flank pale yellow with slight cream-buff wash, all streaked with dusky grey or black, least so on chin and throat (in ♂, extensively deep green-yellow, only slightly tinged grey or pale buff on chest, streaks mainly restricted to side of breast and flank). Tail, flight-feathers, primary coverts, and bastard wing as adult ♂, but ground-colour slightly greyer, less greyish-black, and fringes along outer webs slightly narrower, paler yellow, especially those of outer primaries pale and faint. Black of greater and median upper wing-coverts greyer, tips slightly narrower pale buff-yellow, less broad and contrasting than adult ♂. *In worn plumage* (spring), some yellow present on forehead, supercilium, and across hindneck and side of neck, but mottled olive and grey, less bright and uniform than adult ♂; mantle and scapulars more olive-grey, less contrastingly streaked yellow and black; rump paler yellow, more intensively mottled black and olive on borders, amount of pure yellow restricted; dark marks on side of head and neck as adult ♂, but less distinctly defined, yellow less pure and bright, mottled and suffused olive; underparts with yellow paler and more restricted than in ♂, usually finely streaked dusky on chin and throat and with broader black streaks across chest to upper belly, but a few birds of either sex intermediate, yellow intermediate in tinge and extent, chin and throat unstreaked, chest to upper belly with ill-defined faint grey streaks. NESTLING. Down fairly long but rather scanty, pale grey; mainly on upperparts and upperwing, traces of down on belly and lower flank (Witherby *et al.* 1938; ZMA). For growth and ageing of nestling, see Rohner (1980). JUVENILE. Entire upperparts including rump closely marked with dark grey and pale yellow-buff streaks; dark streaks less pronounced on hindneck and rump, pale streaks on latter more yellow-white, dark streaks more sharply dull black on lower mantle and scapulars. Side of hindneck yellow-buff, finely speckled dark grey; faint and spotted supercilium from above eye backwards slightly more yellow or yellow-white, upper and rear border of ear-coverts more buff-brown. Underparts entirely diluted pale yellow, almost yellow-white on lower flank and vent, tinged cream-buff on thigh and under tail-coverts, entirely marked with narrow dark grey streaks which widen somewhat near tip of each feather. Tail, flight-feathers, tertials, greater upper primary coverts, and bastard wing as in adult, but tip of most tail-feathers and outer web and tip of tertials with broader and less well-defined cream-buff or yellow-buff fringe (in adult, fringe slightly narrower and more sharply defined pale grey or off-white); lesser and median upper wing-coverts dull black with sharply defined even buff or rusty fringe along tip, greater coverts with even buff or rusty fringe along outer web and tip (in adult, centres of coverts deeper black, tips of lesser and median coverts more broadly green-yellow, greater coverts with yellow-green fringe along outer web, widening on tip, and extending into a narrow and faint pale fringe on tip of inner web). FIRST ADULT. Like adult, and indistinguishable when post-juvenile moult complete, but 2nd-calendar-year birds with less complete moult often hard to distinguish also, as retained juvenile feathers too worn to establish juvenile character. In autumn and winter, most birds from northern part of breeding range retain juvenile tail, tertials, flight-feathers, primary coverts, bastard wing, and outer greater coverts, and especially contrast in colour, pattern, and wear between new inner and juvenile outer greater coverts facilitates ageing (for colour and pattern, see above, Rohner 1981, or Svensson 1992); in south, many birds replace all greater coverts and tertials in post-juvenile moult, as well as a variable number of central tail-feathers; these to be aged by contrast in shape, colour, and abrasion between central and outer tail-feathers (in adult, all approximately similar); however, by early summer feathers of all age-groups heavily worn. Some ♀♀ (especially in south of breeding range) have ground-colour of underparts extensively white in spring, with some pale yellow (if any) restricted to chest; these perhaps 1-year-olds, but 1st adult character of these birds hard to ascertain, either due to heavy abrasion or perhaps due to complete post-juvenile moult. For sexing and ageing, see also Rohner (1981).

**Bare parts.** ADULT, FIRST ADULT. Iris dark brown or black-brown. Bill dark horn-brown, dark horn-grey, or steel-grey (darkest in breeding ♂), slightly paler flesh-grey or pale grey at base of lower mandible. Leg and foot flesh-grey, dull flesh-brown, or purplish-brown, toes darker grey. (RMNH, ZMA.) NESTLING. Mouth bright pink, gape-flanges pink, edges of mandibles greenish-yellow (Witherby *et al.* 1938). JUVENILE. Iris brown. Bill greyish-horn, dark horn-grey, or horn-brown with slight flesh tinge, base of lower mandible paler flesh-grey. Leg and foot flesh-grey. (Heinroth and Heinroth 1924–6; RMNH.)

**Moults.** ADULT POST-BREEDING. Complete; primaries descendent. Late July to mid-October, occasionally up to early November (Meinertzhagen 1940; Ginn and Melville 1983; RMNH, ZMA). No information on duration in individuals. No pre-breeding moult. POST-JUVENILE. Partial, but in south of range apparently sometimes complete (Ginn and Melville 1983). Starts shortly after fledging, and timing of moult thus highly variable: just-fledged juveniles examined from late April (southern Portugal) to August (Netherlands) (RMNH, ZMA), and sometimes occur March (Stanford 1954) to September (Mayr 1926; Gnielka 1978) or, rarely, mid-October (Rohner 1981). Birds in complete 1st adult plumage examined from July onwards (Belgium), and undoubtedly occur earlier in south of range. Moult includes head, neck, lesser and median upper wing-coverts, and variable number (none to all) of greater coverts (in central Europe, often 1–7 outer retained, on average 3·7, n = 12), occasionally central pair of tail-feathers (rarely, several pairs) or a few or all tertials (RMNH, ZMA). 15% of 75 1st adults from Switzerland apparently replaced all greater coverts (Rohner 1981). See also First Adult in Plumages.

**Measurements.** ADULT, FIRST ADULT. Wing and bill of birds from (1) Belgium, Netherlands, and western Germany, April–November; (2) Portugal, Spain, Algeria, and Italy, March–July; (3) Turkey and Cyprus, summer, and northern Iraq, winter; other measurements combined, skins (BMNH, RMNH, ZMA: A J van Loon). Bill (S) to skull, bill (N) to distal corner of nostril; exposed culmen on average 2·9 less than bill (S).

| | | ♂ | | ♀ | |
|---|---|---|---|---|---|
| WING | (1) | 72·3 (1·94; 15) | 69–76 | 68·5 (1·64; 12) | 66–72 |
| | (2) | 71·6 (2·24; 22) | 68–76 | 67·5 (1·26; 6) | 66–69 |
| | (3) | 74·4 (1·46; 10) | 73–78 | 70·5 ( — ; 2) | 70–71 |
| TAIL | | 50·6 (2·23; 45) | 46–55 | 48·1 (1·84; 19) | 46–53 |
| BILL (S) | (1) | 10·3 (0·42; 17) | 9·9–11·0 | 10·3 (0·44; 14) | 9·7–11·0 |
| | (2) | 10·3 (0·26; 24) | 9·8–10·6 | 10·5 (0·61; 7) | 9·9–11·3 |
| | (3) | 10·5 (0·15; 11) | 10·3–10·8 | 11·1 ( — ; 2) | 11·0–11·2 |
| BILL (N) | (1) | 5·7 (0·30; 17) | 5·4–6·3 | 5·8 (0·24; 14) | 5·4–6·2 |
| | (2) | 5·8 (0·25; 24) | 5·5–6·3 | 5·7 (0·19; 7) | 5·5–6·1 |
| | (3) | 5·9 (0·17; 11) | 5·6–6·2 | 6·0 ( — ; 2) | 5·9–6·2 |
| TARSUS | | 13·6 (0·48; 38) | 12·8–14·4 | 13·4 (0·52; 13) | 12·6–14·3 |

Sex differences significant for wing and tail. Ages combined, though retained juvenile wing and tail of 1st adult on average slightly shorter than in older birds; thus, in ♂ from samples above: wing, adult 73·0 (1·80; 12) 71–76, 1st adult 72·0 (2·77; 16) 68–75(–78); tail, adult 51·7 (2·02; 12) 48–55, 1st adult 50·1 (2·70; 14) 46–54.

Wing. (1) Germany, breeding (Niethammer 1937). Zürich area (Switzerland), April–November, live birds: (2) adult, (3) 1st adult. Migrants Col de Bretolet (Swiss Alps), September–November, live birds: (4) adult, (5) 1st adult. (Rohner 1981.) Spain, live birds: (6) breeders Segovia area, (7) migrants Tarifa (Cadiz) (Asensio 1984). (8) Southern Yugoslavia to Turkey; skins (Stresemann 1920; Makatsch 1950; Kumerloeve 1964a; Rokitansky and Schifter 1971).

| | | ♂ | | ♀ | |
|---|---|---|---|---|---|
| (1) | ♂ | 72·0 ( —; 18) | 69–75 | ♀ 68·5 ( —; 4) | 65–73 |
| (2) | | 72·8 (1·7; 47) | 69–76 | 70·1 (1·9; 55) | 66–75 |
| (3) | | 71·7 (1·6; 85) | 68–76 | 68·9 (1·7; 57) | 66–73 |
| (4) | | 74·8 (1·4; 36) | 71–78 | 71·2 (1·1; 29) | 69–73 |
| (5) | | 74·3 (1·4; 42) | 71–77 | 71·1 (1·0; 26) | 70–73 |
| (6) | | 71·2 (3·2; 30) | — | 69·1 (3·1; 43) | — |
| (7) | | 70·7 (1·5; 5) | — | 69·5 (1·6; 10) | — |
| (8) | | 73·6 (1·8; 4) | 71–75 | 69·0 ( —; 2) | 69–69 |

South-east France, sexes and ages combined, live birds: 72·2 (1·68; 19) 70–75 (G Olioso). Tuscany (Italy), ages combined, skins: ♂ 72·8 (1·68; 15) 70·5–76 (Eck 1985). Elba: ♂ 69–71 (7) (Trettau 1964). Tunisia, sexes and ages combined, live birds, May: 70·6 (1·73; 29) 66–74 (G O Keyl).

**Weights.** ADULT, FIRST ADULT. Switzerland: Zürich area, April–November, 11·8 (0·8; 203); migrants Col de Bretolet (Alps), September–November 11·5 (495) (Rohner 1981, which see for variation throughout day and monthly averages). South-east France, summer: adult ♂ 11·6 (0·48; 4) 11–12, adult ♀ 12·3 (0·27; 5) 12–12·5 (G Olioso). Czechoslovakia: April–October, 12·5 (1·08; 48) 10·9–16·8; average April–June 11·6 (15), July 12·4 (5), August–October 13·1 (28) (Havlín and Havlínová 1974). Tunisia, late May: adult 11·3 (0·78; 15) 10·5–13·0 (G O Keyl). Entire geographical range, all year: ♂ 10·4 (2·05; 8) 8–14, ♀♀ 11, 11·5 (Mountfort 1935; Makatsch 1950; Kumerloeve 1964a; Rokitansky and Schifter 1971; Eck 1985; Eyckerman *et al.* 1992; RMNH, ZMA).

NESTLING, JUVENILE. For growth curves of nestling, see Rohner (1980). Juvenile, Switzerland, July–August: 11·1 (24) (Rohner 1980; see Rohner 1981 for monthly averages). South-east France: juvenile 11·2 (0·54; 10) 10–12·5 (G Olioso). Tunisia, late May: juvenile 10·7 (0·3; 14) 10·5–11·1 (G O Keyl).

**Structure.** Wing rather short, broad at base, tip bluntly pointed. 10 primaries: p7–p8 longest, p9 and p6 0·5–2 shorter, p5 6–8, p4 10–13, p3 12–16, p2 14–19, p1 16–20. P10 strongly reduced, 45–54 shorter than p7–p8, 6–9 shorter than longest upper primary covert. Outer web of p6–p8 emarginated, inner web of p7–p8 (p6–p9) with notch (sometimes faint). Tip of longest tertial reaches to tip of p1–p2. Tail rather short, forked; t1 7–11 shorter than t6. Bill very short, conical, swollen at base; 5·9 (50) 5·5–6·3 deep at base, 5·7 (50) 5·4–6·2 wide; culmen and cutting edges slightly decurved, gonys virtually straight; bill relatively deeper and more swollen at base than in *S. pusillus* and *S. syriacus*, shorter, tip blunter; rather like bill of Canary *S. canaria* in shape, but relatively shorter and blunter; bill of Citril Finch *S. citrinella* markedly finer, longer, more sharply pointed, and more compressed laterally on tip. Tarsus and toes short and slender. Middle toe with claw 14·3 (10) 13·5–15; outer and inner toe with claw both *c.* 66% of middle with claw, hind *c.* 79%. Remainder of structure as in *S. pusillus*.

**Geographical variation.** Slight, if any. Birds from south-east of range average slightly larger (average wing 73·7, bill to skull 10·6, tarsus 13·9), those of south-west slightly smaller (average wing 70·7, bill 10·3, tarsus 13·4). Birds in north of range, which was colonized only from early 19th century (Mayr 1926; Olsson

1969), are similar in size to those from south-west, though bill perhaps slightly less heavy at base: average depth and width at base *c*. 6·0 in Iberia (*n* = 19), *c*. 5·8 in Turkey and Cyprus (*n* = 9), *c*. 5·7 in Belgium and Netherlands (*n* = 19) (ZMA). See also Measurements. Notwithstanding relatively recent occurrence in north, populations here sometimes separated as *germanicus* Laubmann, 1913, (e.g. by Hartert 1903-10, Stresemann 1920, Mayr 1926, Vaurie 1949*b*, Kumerloeve 1961), differing from populations south of Alps in having underparts greener-yellow, less pure yellow. However, much variation in colour in north, greener tinge of some specimens caused mainly by contamination of yellow underparts with soot and industrial dust, greenish birds becoming yellow again after cleaning (Mayr 1926), though apparently not in all samples (Eck 1985). Central Europe perhaps invaded from south-west by greener birds from west Mediterranean basin and from south-east by more yellow birds from east Mediterranean, accounting for apparent local differences (Mayr 1926; Schnurre 1959), but samples from Belgium, Netherlands, Portugal, Spain, North Africa, Italy, and Turkey all closely similar in colour at same time of year (BMNH, RMNH, ZMA).

For relationships, see *S. canaria*. CSR

## *Serinus syriacus*  Syrian Serin

PLATES 33 and 38 (flight)
[between pages 472 and 473]

Du. Syrische Kanarie    Fr. Serin syriaque    Ge. Zederngirlitz
Ru. Сирийский канареечный вьюрок    Sp. Verdecillo sirio    Sw. Gulhämpling

*Serinus syriacus* Bonaparte, 1850

Monotypic

**Field characters.** 12·5 cm; wing-span 21·5-24 cm. Nearly 10% larger than Serin *S. serinus*, with less compact form most evident in relatively longer tail; matches Canary *S. canarius* in size except for smaller bill. Small (not tiny), rather long-tailed serin, with rather subdued pale greyish-olive-yellow plumage which lacks strong streaks except on mantle. Breeding ♂ has remarkably open foreface, with almost orange forehead, pale yellow eye-ring, and conspicuous pale yellow greater coverts and edges to inner flight-feathers and tail. ♀ less colourful, greyer above and paler on face and below with indistinct streaks on rear flanks. Sexes dissimilar; some seasonal variation in ♂, due to wear. Juvenile separable.

ADULT MALE. Moults: July-September (complete). Forehead orange-yellow, wearing to almost orange in spring and combining with broad pure yellow ring round dark, beady eye and deep yellow throat to give characteristic open foreface; rest of head pale greyish-olive with yellowish wash and only faint grey streaks on rear crown and nape. Mantle and scapulars bright olive-yellow, streaked black or grey; rump contrastingly bright yellow. Wings noticeably pale, with grey-black ground obscured by (a) conspicuous golden-yellow lesser and median coverts, forming bold bright panel by shoulder, (b) long, similarly coloured tips to greater coverts, creating broad bright wing-bar and emphasized by (c) black bar across bases of inner secondaries, (d) conspicuous yellow fringes and tips to tertials and secondaries, showing as quite distinct lines on tertials and diffuse panel on secondaries. Tail black but with all feathers broadly margined yellow and white, thus appearing pale-sided or pale-based in some attitudes. From breast to under tail-coverts, underparts olive- or greyish-yellow, wearing paler and more yellow, and unmarked except for greyish patches on flanks. Bill grey when breeding, browner otherwise. Legs dark horn to flesh-brown. ADULT FEMALE. Resembles ♂, but plumage more clouded overall, lacking orange forehead and bright wing markings; streaks on upperparts broader. JUVENILE. Quite different from ♀, being sandy- to rufous-brown overall and well streaked on back, while crown and ear-coverts show striations, and underparts mottled dusky. Within rather amorphous pattern, only noticeable features are pale whitish eye-ring and bar on greater coverts. Yellow in plumage restricted to fringes of flight-feathers and tail until autumn moult when it appears also on face and rump.

Not difficult to distinguish from *S. serinus*, due to larger size, more uniform, much less streaked plumage, and different voice (see below), but beware confusion with potential vagrant Citril Finch *S. citrinella* which though greyer and less bright yellow on face and body shows similar wing and tail pattern (see that species). Flight light, similar to *S. serinus*; note that in flight, plumage somewhat cryptic making bird hard to follow. Gait a hop, at times a creeping shuffle. Like all serins, moves freely between treetops and ground. Sociable, forming parties and even large flocks.

Song a soft assortment of 'trilling', purring, and plaintive notes, recalling Goldfinch *Carduelis carduelis* and Linnet *C. cannabina*, and lacking vehemence of *S. serinus*; beware confusion with Red-fronted Serin *S. pusillus*. Flight-call a rather dry 'tirrrh' or 'tsirr', less ringing and musical than *S. serinus*. Other calls include thin, high-pitched, nasal 'shkeep', and a low, soft twitter.

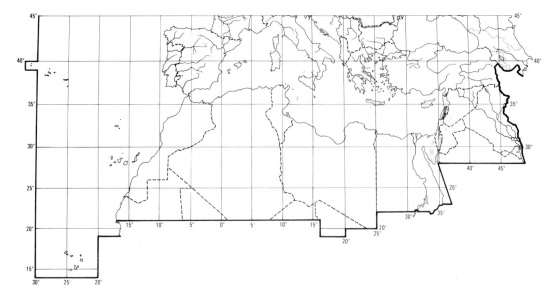

**Habitat.** Breeds in restricted part of east Mediterranean sector of west Palearctic in dry warm sunny climate on high upland slopes and ridges, often rocky, carrying sparse open woodland or clumps of low bushes. Access to drinking water is essential. Early nests are at 900-1500 m, soon after snow melts, but 2nd broods in July are reared at *c.* 1750 m. Sometimes breeding occurs in fruit orchards. (Paz 1987; Shirihai in press). In 19th century, found breeding near cedars *Cedrus* of Lebanon (Dresser 1871-81). Found in breeding season in grassy glades among junipers *Juniperus* in gorge in central Jordan (Mountfort 1965).

In winter, migrates to lower ground, especially desert areas with some trees and water sources, or well-vegetated wadis or cultivated land. Regularly occurs on coast at Elat (Israel), wintering mainly in wadis surrounding the town. In Sinai, migrants on passage frequent wadis of high mountain region; hundreds recorded in fruit gardens there in March. Necessity for migration arises from commitment to breeding at relatively high altitude where habitat seasonally untenable. Resulting shift, however, is much more than mere altitudinal displacement.

**Distribution and population.** No information on changes.

SYRIA. Recently confirmed breeding west of Damascus (Baumgart and Kasparek 1992). ISRAEL. 180 pairs in 1982-3 (Shirihai in press).

**Movements.** Resident, and altitudinal and short-distance migrant. Movements erratic and poorly known; more conspicuous in spring than autumn. Migration mainly nocturnal (Shirihai in press).

Apparently mainly resident in Syria and Lebanon (Hollom *et al.* 1988), though Israeli breeding area (Mt Hermon) vacated in autumn (Shirihai in press). Some birds descend to lower levels for winter, e.g. to villages on edge of snowline or to Beqa'a plain in Lebanon (Tristram 1884; Meiklejohn 1950; Kumerloeve 1962b). Others move further, heading between north-east and south; some winter in Tigris valley in extreme north of Iraq (Ticehurst *et al.* 1926; Göwert 1978), some in eastern and southern Syria, and Jordan (Shirihai in press); others head south. In Israel, now rare or locally uncommon on passage or in winter (Shirihai in press), though formerly locally common: at Yafo on coast, regular winter visitor in thousands (almost exclusively ♂♂) over many years (Aharoni 1942). Most records are at Elat in extreme south, where numbers fluctuate: in 1984 120 autumn passage migrants recorded and flock of 45 wintered, but in 1985 only 9 recorded (Krabbe 1980; Shirihai in press). Apparently irregular winter visitor to Sinai (Egypt): at St Katherine, 28°31'N, common in March 1970, and recorded also in December 1978 and March 1979; a few at El Arish 31°09'N March 1986. Elsewhere in Egypt, singles recorded at Helwan 29°51'N 31°20'E December 1919 and at Bahig 30°56'N 29°35'E November 1969; further south, at Tuna el Gebel 27°46'N 30°44'E singing ♂ and 'apparent' ♀ recorded February 1979 (Short and Horne 1981; Goodman and Meininger 1989).

In autumn, birds gradually leave Mt Hermon August to mid-November, majority descending to lower levels to form pre-migratory concentrations mid-September to mid-October; most passage records in central and southern Israel are November. At Elat, 1984, peak of 34 on 26 November; in 1985, peak of 9 on 15 November. (Shirihai in press.) Recorded in northern Iraq from October (Göwert 1978).

Spring migration early February to mid-April, mainly in March. In southern Israel, peak 20 February to 15 March, with movement chiefly along eastern valleys;

at Elat, 1985, 350 between 4 February and 6 April, with maximum of 94 on 3 March. Returns to Mt Hermon breeding grounds 2nd half March and April. (Shirihai in press.)  DFV

**Food.** Preferred food in captivity sprouting or half-ripe seeds of Canary grass *Phalaris canariensis* (Elzen 1983). For diet in captivity, see also Göwert (1978) and Bielfeld (1981). No further information.  BH

**Social pattern and behaviour.** Little known, except for dispersion during and outside breeding season.

1. Gregarious outside breeding season. Flock of up to 45 winters at Elat; elsewhere in Israel winters in smaller flocks or singly. In spring, migrates in larger groups than in autumn (see below) and up to 30 seen together in desert areas at clumps of trees or water source; at Elat, small flocks stay up to 20 days in spring, and, in years of good passage there, flocks exceed 50 birds. From July (end of breeding season) in Mt Hermon region, birds gather into flocks of a few tens up to 300 or more (see also Roosting, below). In autumn, migrates singly or in groups of 3–8 birds. (Shirihai in press.) Also in Mt Hermon region, April, loose flocks of 5–6 recorded, often in association with a few Linnets *Carduelis cannabina* (P A D Hollom). BONDS. No information, other than that birds pair at least for breeding season. BREEDING DISPERSION. In Israel, scattered, with a few tens or hundreds of metres between pairs; however, neighbourhood groups in some places, with less than 10 m between nests (Shirihai in press). In Lebanon, 3–4 pairs in isolated clump (*c.* 2·4 ha) of old cedars *Cedrus* (Meinertzhagen 1935). In Jordan, at least 18–21 pairs per km² in Barra forest (juniper *Juniperus* and oak *Quercus*) (D I M Wallace). ROOSTING. No information for breeding season. At Mt Hermon, post-breeding flocks assembled in large roosts (see above for flock sizes); flocks also regularly seen drinking, mainly morning and afternoon (Shirihai in press).

2. Shy in winter (Meinertzhagen 1930). In Barra forest, juniper and oak serve as escape cover (D I M Wallace). FLOCK BEHAVIOUR. See Roosting (above). No further information. SONG-DISPLAY. Usually given from exposed perch in tree or bush (Mild 1990); e.g. in Lebanon, sings typically from topmost branch of cedar (Meinertzhagen 1935), while in Israel, April-May, heard singing from bushes up to *c.* 1·5 m high on otherwise bare rocky hillside, also from posts (P A D Hollom). Unlike Serin *S. serinus*, apparently no special Song-flight, though one bird sang in normal flight as it approached tree-top where it settled to continue singing (Hollom 1959). No information on duration of song-period, in particular whether singing continues outside breeding season (as in *S. serinus*). Isolated observations include ♂ singing, Egyptian Nile, 25 February (Short and Horne 1981), another Jordan 24 March (I J Andrews). ANTAGONISTIC BEHAVIOUR. No information on interactions with conspecifics. On Egyptian Nile, February, Goldfinch *C. carduelis* landing on song-perch of ♂ *S. syriacus* was briefly chased before ♂ returned to same perch and resumed singing (Short and Horne 1981). HETEROSEXUAL BEHAVIOUR. Practically nothing known, but see Song-display which evidently functions, at least partly, for mate-attraction. Captive birds seen feeding each other (Göwert 1978), but no details, and precise relevance, if any, to courtship-feeding not known. RELATIONS WITHIN FAMILY GROUP, ANTI-PREDATOR RESPONSES OF YOUNG. No information. PARENTAL ANTI-PREDATOR STRATEGIES. In several nests found in Lebanon, incubating bird (no details of sex) sat very tight (Hollom 1959).  EKD

**Voice.** Commonest calls well known but no information on those used in intimate pair-contact or in alarm. Also no information on calls of young. Seasonal use of song also requires further study. All analysis of recordings and their renderings by W T C Seale.

CALLS OF ADULTS. (1) Song of ♂. Long phrase of trilling, chirping, harsh twittering, and high-pitched jingling sounds (Mild 1990); includes 'siou' sounds and purring cardueline 'trrrr', and generally not loud (Hollom *et al.* 1988). Recordings show great diversity in rate of delivery and unit type, so that while particular phrases recall other *Serinus*, overall variability precludes consistently close resemblance to any one *Serinus*. Considerably lower pitched than Serin *S. serinus* (W T C Seale) and also lacks sizzling sounds so typical of that species (Hollom 1959). Compared with *S. serinus*, also generally slower and more drawn out, with more varied themes and tempo (Mild 1990). However, delivery rate of vigorous song is relatively fast; contains many tremolos fluctuating widely in rate and timbre (from liquid to nasal or buzzy and thus variously recalling Greenfinch *Carduelis chloris* and Linnet *C. cannabina*: W T C Seale). Figs I–II depict same bird: Fig I contains good selection of buzzy components, tremolos, and more drawn-out units, etc.; Fig II consists of 2 tremolos, given at different rates, followed by some rather extended nasal units; markedly similar overall to Red-fronted Serin *Serin pusillus*. Song of another bird (Figs III–IV) generally more measured, containing rather nasal sounds, in this and in other parts of same recording resembling *C. chloris*; Fig III also starts with 'tiu-lee', which recurs several times in recording and is probably a call: at *c.* 2·0 s in Fig III is an overlap of 2 birds; Fig IV shows rattle-calls (see 2, below) followed by brief burst of song in flight (part buzzy, part more tonal, overall slightly like hirundine song); however, song-display of *S. syriacus* typically from perch, and no ritualized song-flight known. (2) Ripple- and Rattle-calls. Recordings show variety of calls ranging from liquid (rippling) to drier (rattling). Not known to what extent these are related functionally (overall probably contact-calls) but grouped here for convenience. Middle call in Fig V represents liquid ripple ('tchi-r-r'), distinctive for its relatively slow delivery; in Fig VI, sub-units very similar to 'r' sub-units of 'tchi-r-r' but more rapidly delivered to produce an emphatic dry churring rattle. A 2nd group of calls shares sub-units which form inverted V shape in sonagrams, different from those in Figs V–VI; some of these are associated with song, e.g. 1st complex unit in Fig I, penultimate unit in Fig III; slower than these is 't-r-r-r' (Fig VII); 1st and 3rd complex units of Fig IV, heard as rippling rattles (respectively 'pi-ti' and 'pi-ti-ti-ti') begin with similar type of sub-unit as Fig VII. Figs V–VII hard to reconcile with descriptions in literature, but Fig VII perhaps the same as 'flight-call' (given both in flight and when perched) rendered as 'tearrrh', 'tirrrh', or 'tsirrr', drier, less ringing, and less musical than *S. serinus* (Hollom 1959; Hollom

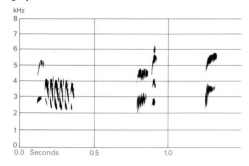

I   Mild (1990)   Israel   April 1989

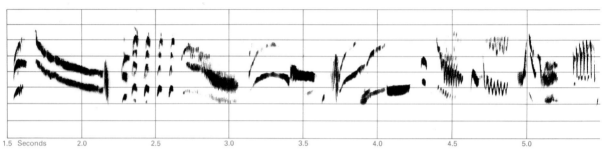

I   *cont.*

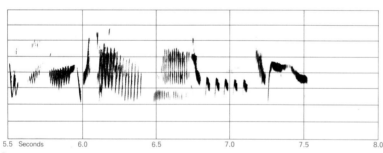

I   *cont.*

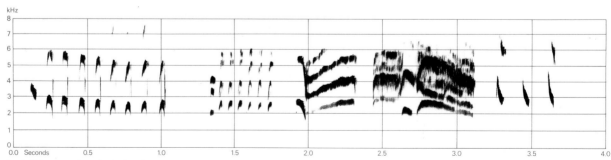

II   Mild (1990)   Israel   April 1989

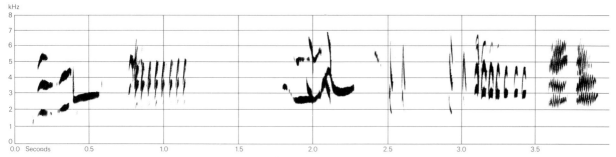

III   P A D Hollom   Israel   April 1980

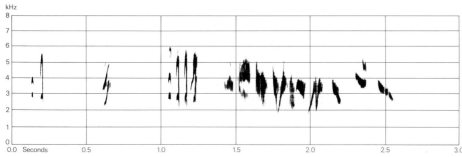

III   *cont.*

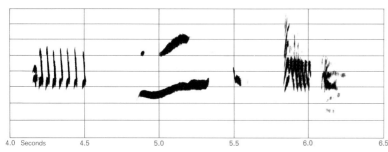

IV   P A D Hollom   Israel   April 1980

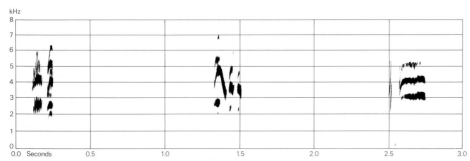

V   Mild (1990)   Israel   April 1989

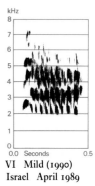

VI   Mild (1990)
Israel   April 1989

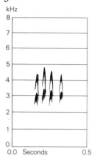

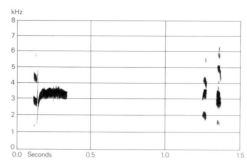

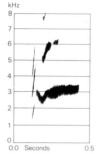

VII  P A D Hollom
Israel  April 1980

VIII  P A D Hollom  Israel  April 1980

IX  P A D Hollom  Israel  April 1980

*et al.* 1988). (3) Twittering-calls. Quiet 'tree-dar-dee' or 'tree-der-doo', not carrying far; given in flight or from perch (Hollom 1959; Hollom *et al.* 1988). (4) Thin nasal 'shkeep' (Hollom 1959; Hollom *et al.* 1988); nasal 'pe-tchee' (Mild 1990). Perhaps related are a ringing 'pwee' (Fig VIII, 1st unit), 'tueee' (Fig IX), and 't-paer' (Fig V, last complex unit). (5) Other calls. Recordings include the following sounds which may possibly be regular calls: the 2 brief units ('chi-chu') which end Fig VIII resemble Flight-calls of Citril Finch *S. citrinella* (see Fig VI of that species); 2nd component of 1st call in Fig V is similar, and preceded by a rather grating sound, the 2 together comprising 'cht-ik'.

CALLS OF YOUNG. Nothing known.                                    EKD

**Breeding.** SEASON. Israel: nest-building at 900–1500 m starts with snow-melt mid-April to May, and eggs laid May–June; most pairs ascend later to *c.* 1750 m July–August for 2nd brood (Shirihai in press). Syria and Lebanon: April–June, but very variable, nest-building recorded early May at 1500–2000 m (Aharoni 1931; Meinertzhagen 1935; Hollom 1959; Hüe and Etchécopar 1970). SITE. Nests recorded 1–2 m above ground in oak *Quercus*, maple *Acer* bush, cedar *Cedrus*, juniper *Juniperus*, hawthorn *Crataegus*, and almond *Prunus dulcis*; can be rather conspicuous (Tristram 1884; Aharoni 1931; Meinertzhagen 1935; Hollom 1959). Possibly prefers cedar (Hüe and Etchécopar 1970; Bielfeld 1981). Nest: rather like that of Goldfinch *Carduelis carduelis* though less neat; cup shallow (Tristram 1868, 1884). Building: no information. EGGS. See Plate 60. Sub-elliptical to oval, smooth and glossy; very pale blue, sparsely speckled reddish- or purplish-brown mostly at broad end; sometimes only spotted or scrawled in circle at broad end and rest of egg unmarked (Aharoni 1931; Hüe and Etchécopar 1970; Harrison 1975). 17·2 × 12·6 mm (16·1–18·2 × 12·0–13·0), *n* = 25; calculated weight 1·45 g (Schönwetter 1984); see also Hüe and Etchécopar (1970). Clutch: 4 (3–5, *n* = 7) in Israel (Shirihai in press); normally 4, maximum 5 (Aharoni 1931; Hüe and Etchécopar 1970). Often 2 broods, sometimes 3 (Paz 1987; Shirihai in press). INCUBATION. 12–14 days, *n* = 7 (Shirihai in press). Probably by ♀ only. YOUNG. No information. FLEDGING TO MATURITY. Fledging period

14–16 days, *n* = 7 (Shirihai in press). BREEDING SUCCESS. No information.                                    BH

**Plumages.** ADULT MALE. Forehead golden-yellow or orange-yellow, forming bright and contrasting patch. Nasal bristles, lore, and short contrasting stripes just above and below eye pale yellow or bright sulphur-yellow, lore partly spotted grey, spots forming narrow dark stripe before eye. Remainder of top and side of head and neck light brown-grey, each feather with darker grey central streak, most distinct on crown; grey-brown of sides and tips of feathers partly suffused green-yellow or golden-yellow, showing as yellow wash, especially on crown, rear of cheek, and side of neck, on latter often forming contrasting golden-yellow bar, which sometimes extends across hindneck. Mantle and inner scapulars light brown-grey with ill-defined dark grey streaks and some yellow-green suffusion; outer scapulars extensively yellow-green, feathers tipped light brown-grey. Back bright yellow with sharply-defined deep black streaks, rump uniform bright yellow. Shorter upper tail-coverts light grey with yellow suffusion and faint darker grey central streak, longer coverts green-yellow with more distinct dark grey central streak. Chin and throat rufous-yellow, golden-yellow, or green-yellow, merging into light green-grey of lower cheek and into light brown-grey of side of neck. Chest and side of breast light brown-grey or sandy-grey, partly suffused golden-yellow or green-yellow, especially on middle of chest, merging into bright yellow of flank and belly and this in turn to cream-white on thigh, vent, and under tail-coverts; tips of feathers of belly and upper flank partly cream-grey when plumage fresh, lower flank marked with some narrow black streaks, though general appearance of underparts otherwise unstreaked. Tail black, inner web and tip of each feather broadly fringed white, outer web fringed bright yellow (widest on bases of central feathers); inner web of outermost feather (t6) completely white except for black streak along shaft, outer web pale grey. Flight-feathers, greater upper primary coverts, and bastard wing greyish-black or black, outer webs narrowly and contrastingly fringed bright yellow (except p9, emarginated parts of outer primaries, and longest feather of bastard wing), tips and inner webs with rather ill-defined pale grey or off-white fringes; basal outer webs of secondaries black, forming contrasting short dark bar across inner wing. Lesser, median, and greater upper wing-coverts yellow-green or golden-green (far more uniform than in other west Palearctic *Serinus*), broken by black bar along base of greater coverts; tips of greater coverts faintly fringed off-white. Under wing-coverts and axillaries yellow, sometimes golden-yellow along leading edge of wing; bases of shorter coverts grey, tips of longest coverts grey-white. *In fresh plumage* (autumn), feathers on forehead slightly suffused sandy-grey, feathers from crown to mantle and of side of head and neck more broadly fringed sandy-grey, less green-yellow or yellow-green of centres visible, golden-yellow on neck concealed, but short yellow streaks above and below eye still conspicuous; feathers of rump tipped off-white; feathers of chin to upper flank and upper belly broadly tipped pale sandy- or pinkish-grey, partly concealing yellow. *In worn plumage* (late spring and early summer), pale fringes of tail, flight-feathers, and tertials partly or largely worn off, some black of bases of median upper wing-coverts partly visible. ADULT FEMALE. Like adult ♂, but forehead paler and less contrastingly yellow, sometimes with grey mid-line, usually less uniform golden or orange than in adult ♂; ground-colour from crown backwards to mantle and scapulars more sandy-grey, less suffused yellow-green, especially on crown and outer scapulars, dark streaks on centres of feathers slightly broader; yellow of rump slightly paler, more greenish-yellow; yellow on lore and just above and below eye paler and less contrasting than in adult ♂, remainder of side of head and neck more mottled sandy and grey, less tinged golden, but usually with mottled grey-and-yellow collar across hindneck to side of neck; chin pale yellow, merging into light grey on lower throat, side of breast, chest, and upper flank, which show less and paler golden or green-yellow suffusion than in ♂; remainder of underparts mainly cream-white with limited amount of pure yellow on upper belly and side of belly and some yellow tinge on mid-belly; side of breast and flank with some fine dusky streaks. Tail, flight-feathers, and tertials as in adult ♂, but fringes sometimes paler yellow or light green-yellow; upper wing-coverts as in adult ♂, but green duller, more olive, less golden. NESTLING. No information. JUVENILE. Rather different from adult: head, body, and upper wing-coverts extensively tinged sandy-buff gradually merging into buff-and-white mottled mid-belly, vent, and under tail-coverts. No contrasting patch on forehead (faint yellow suffusion at most), no yellow on lore or above and below eye, side of head and neck uniform sandy-buff. Lower mantle and scapulars usually marked with rather ill-defined dark brown-grey streaks. Chin tinged yellow; belly sometimes with pale yellow wash. Tail as adult, but fringes along tips tinged pink-cinnamon or greyish-buff, not as white as adult; flight-feathers and primary-coverts as adult, but green-yellow fringes along outer webs sometimes tinged buff or rufous, less yellow; fringes along tertials greyish- or pink-cinnamon, not yellow; lesser, median, and greater upper wing-coverts extensively dark grey with broad buff-cinnamon outer webs and tips (in adult, more uniform golden-green). Rather similar to juvenile Red-fronted Serin *S. pusillus*, but cap and side of head more buffish, less sooty-brown; hindneck, mantle, scapulars, and rump more uniform buff with narrower and less contrasting dark grey streaks (heavily streaked black in *S. pusillus*); side of breast and flank buff or tawny, sometimes with faint traces of grey streaks (in *S. pusillus*, more cinnamon, usually with distinct black streaks). FIRST ADULT MALE. Like adult ♂, but tail, flight-feathers, greater upper primary coverts, and sometimes some outer greater upper wing-coverts still juvenile; difference of these from those of adult often hard to see, especially once plumage worn, as more cinnamon or greyish-buff fringes bleach rapidly to off-white; tips of tail-feathers more pointed than those of adult, relatively more worn than those of adult at same time of year. Central pair of tail-feathers (t1) sometimes new, contrasting in colour, shape, and degree of abrasion with other feathers (all uniform in shape and colour in adult); juvenile outer greater coverts (if any) distinctly more extensively dusky in centre and with cinnamon or grey-buff

fringe along outer web, not as extensively golden-green as neighbouring feathers. FIRST ADULT FEMALE. Like adult ♀, but part of juvenile feathering retained; characters of this as in 1st adult ♂.

**Bare parts.** ADULT, FIRST ADULT. Iris brown. Bill dark horn-brown or dark horn-grey, paler horn at base of mandible; during breeding season, bill of ♂ dark steel-grey. Leg and foot dark flesh-brown or horn-brown, tinged grey in ♂ in breeding season. (BMNH, ZFMK, ZMA.) NESTLING. No information. JUVENILE. No information, but (judged from skins), similar to adult but bill paler horn or grey with flesh or yellow tinge at base and leg and foot flesh-brown (BMNH).

**Moults.** ADULT POST-BREEDING. Complete; primaries descendent. Moult not started in 5 birds from Mt Hermon in June, but ♀ from 26 July in full moult, primary moult score 11 (p1 new, p2–p4 growing), part of tail and many of feathers of body growing; moult completed in bird from October (BMNH, ZFMK). No further information. POST-JUVENILE. Partial: head, body, lesser, median, and some or all greater upper wing-coverts, variable number of tertials, sometimes t1. Probably starts soon after fledging and completed before autumn dispersion, but no details available; 5 birds from June–July examined, Mt Hermon and Lebanon, were in fresh juvenile plumage, another from 2 August worn but not moulting (BMNH).

**Measurements.** ADULT, FIRST ADULT. Lebanon, south-west Syria, Israel, and Sinai (Egypt), mainly February–July; skins (BMNH, RMNH, ZFMK, ZMA: A J van Loon, C S Roselaar). Bill (S) to skull, bill (N) to distal corner of nostril; exposed culmen on average 2·9 less than bill (S).

| | ♂ | | ♀ | |
|---|---|---|---|---|
| WING | 77·8 (1·35; 17) | 75–80 | 74·6 (1·74; 11) | 71–77 |
| TAIL | 55·2 (1·78; 13) | 52–57 | 53·4 (1·52; 8) | 52–56 |
| BILL (S) | 10·6 (0·44; 11) | 9·8–11·3 | 10·7 (0·73; 8) | 9·5–11·5 |
| BILL (N) | 6·3 (0·19; 12) | 5·9–6·5 | 6·2 (0·30; 8) | 5·7–6·5 |
| TARSUS | 14·8 (0·48; 13) | 14·0–15·9 | 14·4 (0·63; 9) | 13·7–15·5 |

Sex differences significant for wing.

**Weights.** Mainly Mt Hermon, ages combined: March, ♂ 13; June, ♂ 11·4 (1·14; 5) 10–13; July, ♂ 12, ♀♀ 11, 11; August, ♂ 13; October, ♀ 12 (BMNH, ZFMK, H Hovel). Israel: 10–14 (Paz 1987).

**Structure.** 10 primaries: p7–p8 longest, p9 0–1 shorter, p6 0–2, p5 6–9, p4 12–14, p3 14–17, p2 16–20, p1 18–22; p10 reduced, 50–55 shorter than p7–p8, 6–10 shorter than longest upper primary covert. Outer web of p6–p8 emarginated; inner web of p7–p9 with notch. Tip of longest tertial reaches to tip of p1–p2. Tail rather short, forked; 12 feathers, t1 8–12 shorter than t6. Bill short, conical; *c.* 5·8–6·4 deep at base, *c.* 5·6–6·6 wide. Middle toe with claw 15·5 (5) 14·5–16·7; outer toe with claw *c.* 66% of middle with claw, inner *c.* 64%, hind *c.* 78%. Remainder of structure as in *S. pusillus*.

**Geographical variation.** None.

Sometimes considered a race of Serin *Serinus serinus* or Canary *S. canaria* (e.g. Hartert and Steinbacher 1932–8), but quite distinct from these, and in structure and plumage characters more like *S. pusillus*, though differing especially in strong reduction of black pigment.                                                                  CSR

## *Serinus canaria* Canary

Du. Kanarie     Fr. Serin des Canaries     Ge. Kanarengirlitz
Ru. Канарейка     Sp. Canario     Sw. Kanariefågel     N. Am. Common Canary

*Fringilla Canaria* Linnaeus, 1758

Monotypic

**Field characters.** 12·5 cm; wing-span 20–23 cm. About 10% larger than Serin *Serinus serinus*, with noticeably less stubby bill, proportionately shorter wings (and primary extension), and more attenuated rear body and tail. Small (not tiny), rather long-tailed finch; ancestor of larger domestic canary but resembling only its 'mule' variant, being far from wholly yellow and looking less green, more grey than *S. serinus*; brightest wing markings of west Palearctic *Serinus*. Song vigorous and musical. Sexes dissimilar; no seasonal variation. Juvenile separable.

ADULT MALE. Moults: post-breeding (complete). Face greenish-yellow to golden-yellow, offset by blackish streaks on pale dusky olive crown and nape and similarly coloured lore and surround to ear-coverts; thus emphasizes yellow of forehead, diffuse supercilium, and surround and centre to cheek; pattern recalls *S. serinus* but rather less contrasted. Throat, chest, and lower flanks bright greenish-yellow, flanks ashy on upper edges and streaked grey on rear flanks; bird appears very yellow head-on. Mantle, scapulars, and upper tail-coverts ashy, broadly streaked black-brown. Wings ashy-black, with yellow tips to median and greater coverts forming wing-bars and pale grey fringes to tertials and innermost secondaries also forming quite obvious lines. Underwing mainly yellow. Rump uniform olive-yellow, far less bright than *S. serinus*, but contrasting with back and upper tail-coverts. Tail ashy-black, fringed dull yellowish-white on outer webs. Belly to under tail-coverts white, latter tinged yellow. Bill horn. Legs pinkish-brown. ADULT FEMALE. Distinctly duller than ♂, with less extensive and less bright yellow on face and foreparts of body. Wing markings whiter and narrower than ♂'s. Rump dull yellow and streaked, far less eye-catching than ♂'s. JUVENILE. Much duller than adult, even ♀; lacks green or yellow tones, having instead buffish-grey head and upperparts overlaid with sooty streaks, and cream underparts, softly and narrowly marked sooty along sides of chest and body. Dark eye emphasized by cream stripes above and below, not strong enough to form obvious supercilium or cheek-bar. Wing and tail markings cream or almost white; wing-bars narrow.

Unmistakable on Azores and western Canary Islands, where other *Serinus* not known. As an escaped bird elsewhere in west Palearctic (in wild form or as one of variety of domestic types) liable to confusion with all congeners except Red-fronted Serin *S. pusillus*. Short extension of folded primaries beyond tertials is always helpful. Flight fast, action resembling *S. serinus* but larger size and espe-

cially longer tail create silhouette recalling also Linnet *C. cannabina* and Redpoll *C. flammea*. Escape-flight as all above species. Song-flight includes fluttering ascent, butterfly-like circles, and parachuting or gliding descent to prominent perch, last phase recalling Tree Pipit *Anthus trivialis*. Gait a hop, giving bustling progress. Highly gregarious, forming flocks after breeding.

Song loud, sustained, and remarkably musical (like domestic bird but of greater variety and extended pitch range), lacking sibilant splutter of *S. serinus*. Calls include loud, clear 'tsooeet' or 'sweet' and pleasant twittering from flock.

**Habitat.** Resident on several west Palearctic Atlantic Islands, at all altitudes from sea-level to 760 m or more in Madeira, to *c.* 1100 m in Azores, and even above 1500 m in Canary Islands. Sometimes in stands of pines *Cryptomeria*, *Eucalyptus*, or in laurel forest and thickets of tamarisk *Tamarix*, but more usually in open countryside with small trees, gardens, vineyards, orchards, and even on sand-dunes. Attracted, especially at nesting time, to banana trees bearing green clusters, camellias, and orange trees, and to shrubs such as heath *Erica scoparia* and broom *Cytisus*, as well as to cultivation of tomatoes and other crops, and to hedges. More rarely extends to arid areas and to pure tree-heath *E. arborea*. Vigorously aerial, especially in display, and has been tracked crossing wide bay in Azores to roost in pines *Pinus*. Perches on highest treetops. While making full use of human activities and (e.g.) nesting on a garden pergola, shows little sign of seeking human company. Contrasts with other Fringillidae inhabiting the same islands in being highly adaptable and able to succeed over almost the entire range of available habitats. (Bannerman 1963; Bannerman and Bannerman 1965, 1966.)

**Distribution and population.** Breeds Madeira, Canary Islands, and all islands of Azores. No information on changes.

**Movements.** Resident, with local movements. Tends to wander in flocks outside breeding season. Extent of inter-island movement apparently varies. In Canaries, not recorded from eastern islands. In Madeiran group, part of population leaves main island of Madeira in autumn; apparently rare in summer but common in winter on Porto Santo (further north-east); birds arrive occasionally last

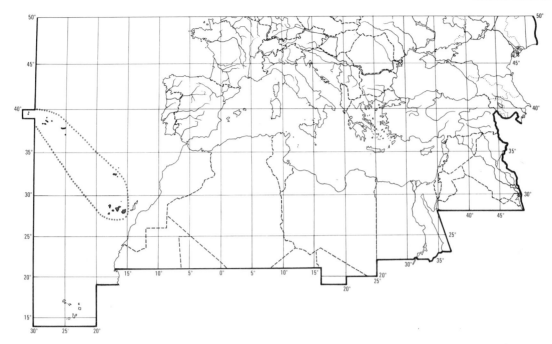

week of August, but chiefly September–October, departing February–March. Inter-island movements also reported from Azores. (Volsøe 1951; Bannerman 1963; Bannerman and Bannerman 1965; G Le Grand.) Small breeding population on Bermuda presumably results from introduction, probably from Azores at some time between 1870 and 1930 (Wingate 1958, 1973).      DFV

**Food.** Seeds and other plant material, occasionally small insects; little-studied in the wild.

Forages mainly on ground (Bolle 1858a; Bannerman and Bannerman 1966; Moreno 1988). Mixed flocks, especially with Goldfinch *Carduelis carduelis* and Linnet *C. cannabina*, can cause damage to crops, e.g. flax *Linum* in Azores (Godman 1870) and cotton *Gossypium* in Canary Islands, though most birds seen in cotton fields by Ennion and Ennion (1962) were feeding on weeds.

Diet includes the following. Invertebrates: aphids (Hemiptera: Aphidoidea). Plant material: seeds, fruits, and green parts of conifers (Coniferae), birch *Betula*, fig *Ficus*, mulberry *Morus*, amaranth *Amaranthus*, chickweed *Stellaria*, *Brassica*, charlock *Sinapis*, cotton *Gossypium*, flax *Linum*, mercury *Mercurialis*, *Eriobotrya* (Rosaceae), Labiatae, dandelion *Taraxacum*, sow-thistle *Sonchus*, lettuce *Lactuca*, purple lettuce *Prenanthes*, mugwort *Artemisia*, ox-tongue *Picris*, ragwort, etc. *Senecio*, agave *Agave americana*, grasses including Canary grass *Phalaris*. (Bolle 1858a; Godman 1870; Hartwig 1896; Ennion and Ennion 1962; Bannerman and Bannerman 1966; Elzen 1983; Sabel 1983; Bergmann 1991.)

Eats primarily seeds of grasses (Gramineae), Cruciferae, Compositae, and similar ruderal or cultivated families;

ripening seed-pods of charlock by far favourite food followed by buds and shoots of chickweed and Compositae seeds (Ennion and Ennion 1962; Bannerman and Bannerman 1966; Elzen 1983; Sabel 1983; Radtke 1986). Also very fond of seeds and flesh of fig *Ficus* (Bolle 1858a; Bannerman and Bannerman 1966; Bielfeld 1980). On El Hierro (Canary Islands), small flocks seen to pick at base of *Agave* flowers to get at nectar (Bergmann 1991). Seems to take aphids when possible (Hartwig 1886; Bielfeld 1981). On Tenerife, stomachs of birds in spring full of green Cruciferae seeds (Bolle 1858a). No quantitative information on diet in the wild. In captivity, very readily takes leaves and stems of spinach *Spinacia*, lettuce, and especially *Stellaria media* (Bielfeld 1980), also seeds of Canary grass, which is important part of diet in the wild (Sabel 1983, which see for details of captive feeding); see also Elzen (1983) and Radtke (1986).

Young generally reared on seeds alone, though sometimes fed small insects (Newton 1967a; Bielfeld 1980).      BH

**Social pattern and behaviour.** No comprehensive studies of wild birds, especially in breeding season. The most detailed descriptions of display are based on domesticated birds (Hinde 1955–6; Nicolai 1960; Tsuneki 1962, which see for references to earlier papers by same author) and such findings cited below should be regarded with caution until corroborated from wild birds.

1. Gregarious all year. During breeding season flocks persist for both feeding and roosting (Meade-Waldo 1893; Ennion and Ennion 1962); according to Volsøe (1951) summer flocks seem to consist of sexually mature ♂♂. Outside breeding season, roving flocks typically large and associated variously with Rock Sparrow

*Petronia petronia*, Spanish Sparrow *Passer hispaniolensis*, Fringillidae, and Corn Bunting *Miliaria calandra* (Ennion and Ennion 1962; Morphy 1965; Bannerman and Bannerman 1965, 1966; G Le Grand); in Madeira, April, mixed flocks of over 200 comprised *c.* 30% Goldfinches *C. carduelis*, 10% Linnets *C. cannabina*, and 10% *S. canaria* (Berthold and Berthold 1987). On Canary Islands flock sizes declined by end of February (Ennion and Ennion 1962). BONDS. No evidence for other than monogamous mating system. In domesticated pairs, mate-fidelity the norm between years (Tsuneki 1962). For hybridization in captivity with various Fringillidae, see Nicolai (1960), Brichovsky (1968), Güttinger *et al.* (1978), and Güttinger and Claus (1982). Nest-building and incubation mainly by ♀ (Bolle 1858*a*; Nicolai 1960; Ennion and Ennion 1962); Bannerman (1963) reported ♂♂ sitting on nests, and incubation by ♂ recorded in domesticated pairs (Hinde 1955-6). Both sexes care for young for unspecified time after fledging (Bolle 1858*a*). No information on age of first breeding. BREEDING DISPERSION. Little information. In neighbourhood groups (Bergmann and Helb 1982), each pair defending not very large territory (Bolle 1858*a*). Few data on breeding density. In Tenerife, 7 birds per km² overall; 5 per km² in pine *Pinus canariensis*, 24 in *P. radiata* (Carrascal 1987). ROOSTING. Communal all year. During breeding season on Tenerife, apparent roosting flights of *C. cannabina* and *S. canaria* (mixed flocks) seen heading towards forests high on southern slopes (Ennion and Ennion 1962). In Azores, post-breeding flocks of 10-20 *S. canaria* headed in steady stream for dense pine forest until dusk (Bannerman and Bannerman 1966).

2. In Madeira, very confiding in breeding season, entering gardens and towns (Bolle 1858*a*). FLOCK BEHAVIOUR. In Madeira, flocks seek shelter from storms in urban trees (see above), then (when storm past) disperse with loud twittering chorus; flock-members call constantly in flight (Bolle 1858*a*; see 2 in Voice). For social organization of domesticated flocks, see Tsuneki (1960). SONG-DISPLAY. ♂ sings (see 1 in Voice) from usually high perch throughout song-period (see below) but also performs Song-flights in spring. Song from perch given with head raised and turning from side to side, bill only slightly open, throat feathers markedly ruffled, and wing-tips slightly lowered (Bolle 1858*a*; Hinde 1955-6; Bergmann and Helb 1982). For song used in threat between rivals, see Antagonistic Behaviour (below). Song-flight the same as in Serin *S. serinus*: ♂ begins by ruffling plumage until he appears twice normal size, then (still ruffled) flies up from perch and, with slow bat-like wing-beats flies from tree to tree or in circle back to starting point (Hartert and Ogilvie-Grant 1905; Nicolai 1960). Final descent vertical or gliding with outstretched quivering wings (much like Tree Pipit *Anthus trivialis*), to alight on highest branch of tree, still singing (Godman 1872; Bannerman 1963). ♂ continues singing intermittently while mate incubating (Bolle 1858*a*). Song heard almost all year but peaks in early part of breeding season. In Tenerife, full song at beginning of January (start of observations) but little by end of February and beginning of March (Ennion and Ennion 1962); however, continues mid-March to early April (Lack and Southern 1949), late June to late August (Cullen *et al.* 1952). In Madeira, sings more or less all year (Bolle 1858*a*), including December-January (Buxton 1960). In Azores, Bannerman and Bannerman (1966) occasionally heard low-intensity song September-October. For song by ♀, see 1c in Voice. ANTAGONISTIC BEHAVIOUR. (l) General. In studies of captive domesticated flocks, ♂♂ usually dominant over ♀♀, though ♀♀ dominant (including towards mate) during height of breeding activity (Shoemaker 1939; Nicolai 1960; Tsuneki 1960). (2) Threat and fighting. Following description based mainly on Nicolai (1960); for similar postures, see Hinde (1955-6). In Threat-display (Fig

A

A), all plumage sleeked except thighs; ♂ aggressor approaches and moves round rival with springing hops and exaggerated up-down tail-flicking; if rival is ♂ he does likewise and both birds hop around each other, usually on ground, tail-flicking. If rival holds ground, excitement increases, aggressor (Fig B, right) stands more upright, slightly droops wings, and, with ruffled throat, begins to sing loudly and directly at rival (Fig B, left: ♂

B

in a defensive posture). If rival still does not yield, fight ensues: both birds sing and circle each other tightly with small pattering steps, each trying to fly onto other's back. Paired or nest-building or incubating ♀♀ generally attack any ♂ singing at them; for response of unpaired bird, see Heterosexual Behaviour (below). Interactions between ♂ and ♀ (mates) described by Hinde (1955-6): in intense Forward-threat display, sleeked ♂ crouches facing mate, wings raised to level of back or above, bill open (sometimes singing or calling); ♂ may then peck ♀ or attack her more forcibly; for dominance relations between mates throughout the year, see Hinde (1955-6); see also Heterosexual Behaviour (below). In Submissive-display, common to all *Serinus* and most other Fringillidae, bird raises plumage, droops wings slightly, and makes very slow hesitant movements as if feigning sickness, thus suppressing attack and allowing own escape (Nicolai 1960). HETEROSEXUAL BEHAVIOUR. (1) Pair-bonding behaviour. ♂ attracts ♀ by singing from perch and in display-flight (see above); also much chasing at this time (Hartert and Ogilvie-Grant 1905). Following account of intimate courtship based mainly on Nicolai (1960). In common with all carduline finches, ♂ performs display derived from Courtship-feeding: ♀ perches near ♂ who approaches making slow opening movements of bill. If ♀ receptive, she follows suit, then ♂ takes hold of her bill (Bill-flirting). As soon as pair-bond forms, pair start nest-building (see subsection 4, below) and ♀ becomes dominant. In domesticated birds, ♂ seen hopping around ♀ in low-intensity aggressive display carrying piece of grass or feather in bill (Hinde 1955-6). (2) Courtship-feeding. Described only from domesticated birds.

♀ begs in posture closely resembling Soliciting-display (see sub-section 3, below), shivering wings and sometimes giving Begging-calls (see 3 in Voice); plumage usually more ruffled than ♂ who is often sleeked and stands upright with head higher than hers. As ♂ starts to regurgitate, ♀ opens bill and turns head to one side to receive food. Early in breeding season, feeding often incomplete and bills merely touch (Bill-flirting); at start of Courtship-feeding phase, display is vigorous, but display and food-passing often perfunctory later. Complete Courtship-feeding not usually frequent until some time after ♀ starts nest-building, and shortly before copulation. In one pair, ♂ often incubated and begged like ♀ when she returned, ♀ then feeding him on nest. (3) Mating. In Soliciting-display, ♀ crouches, plumage often slightly ruffled, with breast lowered, tail raised, and wings drooped, somewhat extended, and shivered. ♀ solicits on approach of ♂ who either edges towards her (showing signs of aggression) along perch before mounting, or flies directly onto her back from afar. Exceptionally, ♂ solicits (as described for ♀) by ♀'s side with Begging-call or, in one instance, song. Just before mounting, both birds may give Begging-call. While copulating, ♂ often lowers his bill to upturned bill of partner. After dismounting, ♂ and ♀ usually adopt sleeked wings-raised posture then fly off before relaxing. (Hinde 1955-6, which see for variations in sequence of precopulatory display and discussion of aggressive component.) Dominance of ♀ with onset of building results in very little copulation taking place around nest-building time or during laying, ♀ defending herself vigorously against attempted mating; however, dominance relations can change rapidly at this stage, and if (e.g.) ♀ continually solicits copulation, ♂ may attack her viciously (Nicolai 1960). (4) Nest-site selection and behaviour at nest. In domesticated birds both members of pair start visiting nest-sites (coincident with incipient Courtship-feeding) (Hinde 1955-6) and some ♂♂ repeatedly enter and leave nest-site as if attracting mate's attention to it (Tsuneki 1962). Domesticated ♂♂ collect nest-material (Hinde 1955-6: see Pair-bonding behaviour) and may present it to ♀, but only ♀ builds (Hinde 1955-6; Nicolai 1960). Incubating ♀ regularly gives Begging-call to summon ♂ and elicit food from him (Bolle 1858a; Nicolai 1960). RELATIONS WITHIN FAMILY GROUP. Practically nothing known from the wild, although young cared for by both parents (mainly ♂) for some time after fledging (Bolle 1858a). Both parents feed young by regurgitation; during brooding, ♂ partly gives food to ♀ for transfer to young. In domesticated birds, both parents swallow faecal sacs when young are small, later (when young deposit faeces in nest-cup) carry them away; finally, well-grown young eject faeces onto nest-rim and parents leave them to accumulate. (Tsuneki 1962.) ANTI-PREDATOR RESPONSES OF YOUNG, PARENTAL ANTI-PREDATOR STRATEGIES. No information.

(Figs by D Nurney from drawings in Nicolai 1960.) EKD

**Voice.** Freely used throughout the year. More study needed of wild birds. Comprehensive studies of domesticated birds by Nicolai (1960) and Mulligan and Olsen (1969); from these extensive repertoires (much larger than known for wild birds) a limited number of key calls has been included here where information for wild birds is lacking, but homologous calls of wild birds may well differ from these in pitch, timbre, etc. (see 1, below). For additional sonagrams see Güttinger (1981, 1985: wild and domesticated birds) and Bergmann and Helb (1982: wild birds, Canary Islands). No information on geographical variation, though song perhaps differs between the various island populations of Canary Islands (Bolle 1858a). All analyses (below) of recordings, and their renderings, by J Hall-Craggs.

CALLS OF ADULTS. (1) Song. (1a) Full song of ♂. Rich, varied, and musical. Consists of phrases or nearly continuous delivery of variously chirping, twittering, trilling, wheezing and piping sounds. Some organization into segments (defined here as series of uniform sounds, the 'tours' of the domesticated bird) differing in duration, tempo, and timbre. Thus long song-phrase (Fig I) of Tenerife bird begins—typically for *S. canaria*—with a sub-phrase containing rather whining, ascending and descending portamento notes and an ascending glissando leading to the first two (of 17) segments; these usually alternate with contrasting sub-phrases, but at times with so little as a simple motif (as at *c*. 3·2 s) or even a single unit (as at *c*. 4·4 s) separating them. Two segments may be adjacent (e.g. *c*. 1·7–3·0 s) and many may form a long chain of uninterrupted but discernibly different segments (e.g. ten, from *c*. 14·3 s to end of song-phrase). Whole phrase (*c*. 23·5 s) comprises 13·8 s of segments (mean 0·8 s per segment, $n = 17$) and 9·7 s non-segments. Another phrase (17·7 s) by same bird contains fewer (6) segments totalling 6·3 s (mean 1·1 s) and 11·4 s of non-segments. Another recording from Tenerife includes phrase with 4 segments (total 2·3 s, mean 0·58 s) and 2·6 s of non-segments; song of this bird includes (not uncommonly for *S. canaria*) diads (2 shown in Fig II, each with simultaneous descending and quieter ascending tones). However, in all recordings described above, segments may be unusually frequent and long for wild birds. Analyses by Güttinger (1981, 1985) (including 3 Tenerife recordings) showed that song of wild birds is strikingly lacking in long segments compared with domesticated birds: incidence of single units (i.e. not in segments) was 80–96% in wild birds not stimulated by playback, and segments longer than 0·4 s were rare, whereas most segments in domesticated birds exceed 1 s. Wild birds also differ from domesticated birds in lacking very low-pitched units (2–8 kHz, compared with 1–4 kHz for those domesticated breeds selected for lower frequency range of human pitch-discrimination) and in having much greater complexity and larger repertoire of unit-types (more than 400), notably including rather harsh, strangled-sounding, drawn-out units. (Güttinger 1981, 1985; Bergmann and Helb 1982.) On Canary Islands, some songs long and persistent (presumed unpaired ♂♂), others short (Löhrl 1978a). Song given from perch or in display-flight, but not known if delivery (etc.) varies accordingly. (1b) Subsong. Apparently heard occasionally in Azores, September–October (Bannerman and Bannerman 1966). December recording from Tenerife includes *c*. 17 s of Subsong or quiet twittering between song-phrases. (1c) Song of ♀. No reports from wild birds but sometimes given by domesticated ♀♀ (Gerber 1963; Pesch and Güttinger 1985). (2) Contact-calls. (2a) Ripple-call. Loud twittering 'tjüdididi' typically given as flight-call (Bergmann and

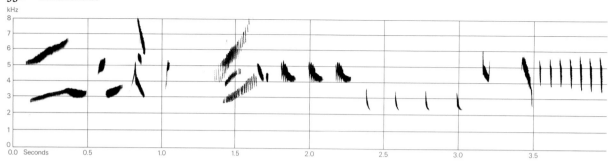

I   P J Sellar   Tenerife (Canary Islands)   May 1984

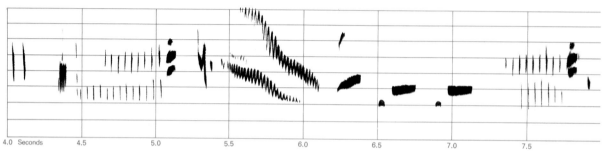

I *cont.*

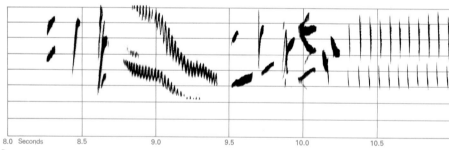

I *cont.*

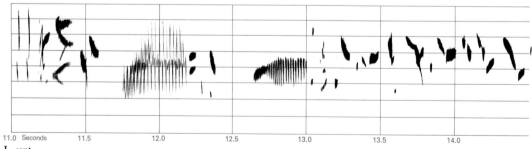

I *cont.*

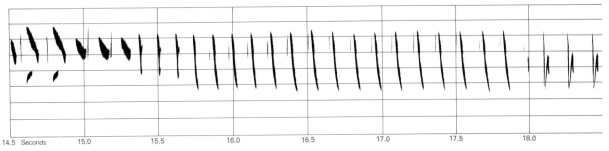

14.5 Seconds    15.0    15.5    16.0    16.5    17.0    17.5    18.0

I *cont.*

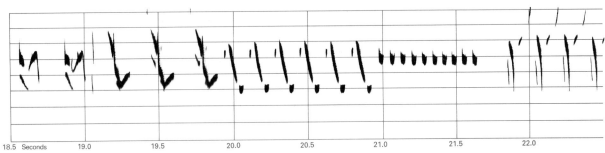

18.5 Seconds    19.0    19.5    20.0    20.5    21.0    21.5    22.0

I *cont.*

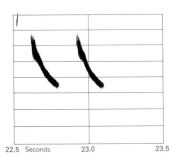

22.5 Seconds    23.0    23.5

I *cont.*

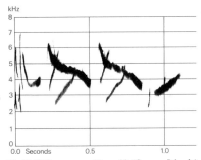

II  H-H Bergmann  Tenerife (Canary Islands)  December 1982

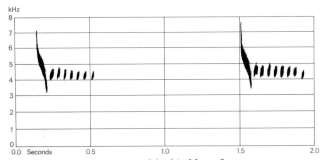

III  P J Sellar  Tenerife (Canary Islands)  May 1984

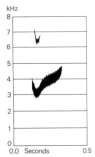

IV  H-H Bergmann  Tenerife (Canary Islands)
December 1982

Helb 1982). Same or related calls probably include 'chiuri-i-it' (Moreno 1988), liquid 'tweetetee', 'deedero', and rolling bell-like 'tseer' (Fig III) not unlike ending of song of Blue Tit *Parus caeruleus* (P J Sellar); also heard as 'tsee' followed by gently and rapidly descending ripple (J Hall-Craggs). Recordings suggest individual variation occurs. Calls presumably of this kind, usually referred to as trills or tremolos, are exchanged by domesticated mates during courtship and nest-building, and when separated from each other (Hinde 1955-6; Nicolai 1960; Mulligan and Olsen 1969). (2b) Metallic 'dit-dit' similar to Citril Finch *S. citrinella* (Knecht 1960). (2c) High-pitched 'chi-íit' ('Advertisement call': Moreno 1988). (3) Begging-call. In domesticated birds, series of rapid 'pee' sounds given by pair-members (especially ♀) to solicit courtship-feeding or copulation (Mulligan and Olsen 1969). Also rendered as thin high-pitched 'zi zi zi' (Hinde 1955-6) and 'didi-dididididi' (Nicolai 1960). (4) Anxiety- and alarm-calls. (4a) Drawn-out rising then falling 'psiäh' or 'iäh' similar to other Fringillidae (Bergmann and Helb 1982). (4b) Clear ascending whistle 'twee' (P J Sellar: Fig IV).

CALLS OF YOUNG. In domesticated birds, food-call of fledged young a disyllabic complaining 'zi-tü zi-tü-tü zi-tü' similar to *S. citrinella* (Nicolai 1960). Domesticated birds begin singing (initially Subsong) at *c.* 1 month old (Poulsen 1959). EKD

**Breeding.** SEASON. Canary Islands: nest-building from late January, eggs laid late January to July (peak April); young recorded end of January (Ennion and Ennion 1962; Martín 1987; Moreno 1988). Average date depends on altitude, though in same area birds can have young in nest while others laying (possibly 2nd clutches) (Koenig 1890; Bannerman 1963; Löhrl 1978a). Madeira: March–June, mostly mid-April to end of May (Bannerman and Bannerman 1965). Azores: eggs laid end of March to July, peak May–June (Bannerman and Bannerman 1966; G Le Grand); see also Chavigny and Mayaud (1932). SITE. In tree or bush in woodland or hedge, commonly evergreen or species coming into leaf early; usually well-hidden on fork or at end of branch, or in top of small tree (Bolle 1858a; Bannerman 1963; Reber 1986; Moreno 1988; G Le Grand). On Tenerife (Canary Islands), 55% of 94 nests in tree-heath *Erica arborea*, 16% pine *Pinus*, 6% almond *Prunus*, 4% laurel *Laurus*, and 4% *Picconia* (Oleaceae); average height above ground 3·1 m (0·9–6·0) (Martín 1987). Nests recorded inside bunches of growing bananas (Bannerman 1963). In Azores, often in myrtle *Myrica* hedges and tamarisk *Tamarix* trees (Bannerman and Bannerman 1966). See also Koenig (1890) for Canary Islands. Nest: small, compact, often deep cup; foundation of twigs, stalks, rootlets, grass, moss, or lichen, lined with much plant down, also hair, feathers, or soft leaves (Bannerman 1963; Bannerman and Bannerman 1965; Moreno 1988; G Le Grand). According to Bolle (1858a), some nests entirely of plant down, and on Azores, some only of

sheep's wool (Bannerman and Bannerman 1966). On Tenerife, 17 nests had average outer diameter 10·0 cm, inner diameter 5·4 cm, overall height 5·6 cm, and depth of cup 2·8 cm (Koenig 1890); on São Miguel (Azores), 10 nests had outer diameter 8·6 cm, inner diameter 5·3 cm, overall height 6·8 cm and depth of cup 4·0 cm (G Le Grand). Building: by ♀ only, accompanied by ♂ (Nicolai 1960; Radtke 1986). In captivity only by ♀, ♂ occasionally carrying material (Hinde 1958b; Hinde and Steel 1972, which see for many details of captive birds; Bielfeld 1980). See also White and Hinde (1968) for details of building routine. EGGS. See Plate 60. Sub-elliptical, smooth, not glossy; pale light blue or bluish-green, darker than those of Serin *S. serinus* and more heavily marked with violet, red, or rust spots and blotches, rather concentrated at broad end (Bannerman 1963; Harrison 1975); see also Moreno (1988). 17·4 × 13·4 mm (15·6–19·1 × 12·0–14·0), $n = 101$; calculated weight 1·54 g (Schönwetter 1984). For 64 eggs from Azores, see Chavigny and Mayaud (1932). Clutch: 3–4(–5). In Azores, of 22 clutches: 3 eggs, 32%; 4, 68%; average 3·7 (Chavigny and Mayaud 1932; Bannerman and Bannerman 1966); average on Tenerife 3·9, $n = 28$ (Martín 1987). Never more than 5 according to Bolle (1858a). Eggs laid daily (Bielfeld 1980; Speicher 1989). See Hinde (1959) for clutch size in captivity. In Canary Islands 2 broods, probably 3 (Bielfeld 1980; Dathe 1986; Martín 1987; Moreno 1988); in Funchal area of Madeira, probably 2–3 (Hartwig 1886; Bannerman and Bannerman 1965) or even more (Bolle 1858a); 2 recorded in Azores (G Le Grand). INCUBATION. 13–14 days; by ♀ only (Bolle 1858a; Harrison 1975; Reber 1986; Moreno 1988). Starts with last or penultimate egg (Bielfeld 1980). YOUNG. Fed and cared for by both parents (Bolle 1858). FLEDGING TO MATURITY. Fledging period 15–17 days (14–21) (Harrison 1975; Speicher 1989). Domesticated young feed themselves at 3–4 weeks, fully independent at *c.* 5 weeks (Bielfeld 1980). BREEDING SUCCESS. No information on wild birds. For captive studies, see Hinde (1959). BH

**Plumages.** ADULT MALE. In fresh plumage (autumn), forehead and short supercilium yellow-green, gradually merging into light grey or pale drab-grey of crown and nape; forehead often partly speckled olive or grey, crown and nape narrowly streaked dark grey or sooty, bases of feathers of crown and nape with varying amount of yellow-green suffusion, which is sometimes largely concealed, sometimes prominent, and which often forms ill-defined yellowish band across lower nape. Mantle and scapulars pale drab-grey, each feather with broad sooty streak on centre and slight pale greyish-yellow suffusion on side. Back variegated drab-grey, yellow, and sooty; rump greenish-yellow, feathers slightly brighter yellow on bases. Upper tail-coverts like scapulars; rump rather sharply contrasting with both back and upper tail-coverts. Lore grey, mixed with some pale yellow; short stripe behind eye olive-grey or dark grey. Eye-ring narrow, pale yellow, broken by grey in front and behind. Patch below eye yellow-green, bordered below by ill-defined olive-grey or dark grey line, which gradually turns into olive-green of shorter ear-coverts; longer ear-coverts, rear of supercilium, and side of neck light grey, neck often with ill-defined yellow-green bar. Rear of upper cheek yellow-green, bordered below by olive-green or olive-grey moustachial stripe. Chin and throat green-yellow; side of breast light grey or pale drab-grey; chest green-yellow, tips of feathers with varying amount of olive or grey suffusion. Upper flank and belly bright yellow, merging into pale buff-white on lower flank and into yellow-white or pure white on vent and under tail-coverts; flank and under tail-coverts with dark grey or sooty streaks, darkest and more distinct on lower flank. Tail black, tip and inner web of feathers narrowly fringed off-white, outer web narrowly fringed pale yellow or yellow-white (widest on central pair and towards bases). Flight-feathers, greater upper primary coverts, and bastard wing greyish-black, outer webs narrowly fringed pale green-yellow (except on p9, on emarginated parts of outer primaries, and on longest feather of bastard wing), tips narrowly fringed white, inner webs faintly bordered pale grey; fringes much duller and less contrastingly yellowish and basal outer webs of secondaries much less contrastingly black than in Red-fronted Serin *S. pusillus* and Syrian Serin *S. syriacus*. Tertials dull black, outer web and tip with broad but ill-defined pale grey border, black of outer web sometimes with slight olive-green suffusion. Lesser and median upper wing-coverts yellow-green, median and longer lesser with black of centres visible; greater coverts dull black with broad yellow-green or green-yellow outer border and tip, latter faintly fringed pale grey. Under wing-coverts and axillaries pale yellow, tips of longest feathers extensively white; shorter coverts along leading edge of wing mottled black and bright yellow. *In worn plumage* (about March–June), yellow-green tinge on cap more pronounced, yellow-green of forehead not contrasting with crown (unlike *S. pusillus* and *S. syriacus*); yellow-green bar across hindneck and side of neck more pronounced; black streaks on mantle and scapulars more pronounced; rump brighter yellow, with less green suffusion; yellow of underparts brighter, chest yellowish-olive, flank more distinctly streaked black; fringes of tail, flight-feathers, and tertials paler, off-white, partly worn off; yellow-green of tips of upper wing-coverts partly worn off, more dull black of bases exposed. ADULT FEMALE. Like adult ♂, but yellow-green on top and side of head and neck restricted, on top confined to forehead (where finely speckled black, unlike ♂) or completely absent, on side of head confined to short yellow-green supercilium, eye-ring, and short stripe below eye; remainder of cap and hindneck light grey or pale drab-grey, as in adult ♂, but dark streaks on average blacker and more sharply-defined; side of head and neck otherwise grey, faintly streaked darker grey on ear-coverts, sometimes slightly tinged green on centre of ear-coverts or rear of cheek; no clear moustachial stripe; yellow of rump duller, greener, and with more extensive dusky streaking than in ♂; yellow of underparts paler and more restricted than in adult ♂, mainly confined to chin as well as upper belly and side belly, throat, chest, and upper flank either completely light ash-grey or grey with some yellow-green suffusion. Wing and tail as adult ♂, but pale yellow of fringes largely replaced by white, and greater and median upper wing-coverts with more restricted and less bright pale green or grey-green tips. NESTLING. Down rather short and sparse, smoky blue-grey (Harrison 1975; ZMA). For development, see Ziswiler (1959). JUVENILE. Upperparts and side of head and neck like adult ♀, but without green or yellow tinge (except sometimes on rump), entirely buffish-grey with rather ill-defined dusky grey streaking; yellow stripes on side of head replaced by ill-defined cream-buff to off-white lines above and below eye. Side of breast, chest, and flank marked with rather narrow and ill-defined dark grey or sepia streaks, ground-colour of underparts either cream-white to off-white or pale yellow (merging into off-white on mid-belly, vent, and under

tail-coverts); tone of underparts perhaps dependent on sex (more yellow in ♂, whiter in ♀), but only a few juveniles of known sex examined. Wing and tail as in adult, but fringes along outer webs paler than in adult, tips of greater primary coverts and tail-feathers more pointed than in adult; tips of lesser, median, and greater upper wing-coverts rather narrowly pink-cream (when fresh) to off-white (when worn), less broad and yellow than in adult. FIRST ADULT. Like adult, sometimes indistinguishable, but some birds show new t1 with rounded tip, contrasting with pointed and worn juvenile outer tail-feathers (in adult, all feathers about equal in shape and wear); some retain juvenile outer greater coverts, with narrow and worn light pinkish-grey to off-white tips, contrasting with newer broadly green-yellow tipped inner coverts; especially in spring, tips of outer tail-feathers and greater upper primary coverts more pointed and more heavily abraded than in adult at same time of year.

**Bare parts.** ADULT, FIRST ADULT. Iris brown. Bill light or dark horn-grey or horn-brown, sometimes with pale flesh or pink tinge at base of lower mandible. Leg and foot pink-brown, horn-brown, or dark brown. (Hartert 1903–10; Bannerman 1912, 1963; RMNH.) NESTLING. Mouth orange-red, gape-flanges cream-yellow (Harrison 1975). JUVENILE. Iris brown to black-brown. Bill light horn-brown with flesh tinge, palest on lower mandible. Leg and foot flesh-brown, light horn-brown, or black-brown. (Sharpe 1888; RMNH.)

**Moults.** ADULT POST-BREEDING. Complete, primaries descendent. Information limited; none of 34 birds examined November–May in flight-feather moult (BMNH, RMNH, ZMA); moult probably starts shortly after breeding season, which is from February or March to June or July. POST-JUVENILE. Partial: head, body, lesser and median upper wing-coverts, many or all greater upper wing-coverts, frequently t1, and occasionally one or a few tertials. Probably starts soon after fledging, completed within a few months: fully juvenile birds examined July, birds with moult completed September–October (BMNH, RMNH, ZMA).

**Measurements.** ADULT, FIRST ADULT. Canary Islands (Tenerife, Gran Canaria, Hierro) and Madeira, mainly November–May; skins (BMNH, RMNH, ZMA: A J van Loon). Bill (S) to skull, bill (N) to distal corner of nostril; exposed culmen on average 3·3 less than bill (S).

| | ♂ | | ♀ | |
|---|---|---|---|---|
| WING | 72·8 (1·39; 23) | 71–76 | 70·3 (2·00; 14) | 66–74 |
| TAIL | 59·0 (1·73; 23) | 56–62 | 57·4 (2·69; 14) | 53–62 |
| BILL (S) | 11·9 (0·57; 22) | 11·0–13·1 | 11·8 (0·42; 14) | 11·0–12·5 |
| BILL (N) | 7·0 (0·36; 23) | 6·4–8·0 | 6·9 (0·40; 13) | 6·4–7·8 |
| TARSUS | 16·8 (0·50; 22) | 16·2–18·0 | 16·9 (0·50; 12) | 16·1–17·8 |

Sex differences significant for wing and tail.

Azores (São Miguel, São Jorge, and Pico), all year; live birds (G Le Grand). Bill is exposed culmen, tarsus and culmen probably measured slightly differently from above.

| | ♂ | | ♀ | |
|---|---|---|---|---|
| WING | 72·6 (1·88; 99) | 68–77 | 70·0 (2·36; 68) | 65–73 |
| TAIL | 58·5 (2·07; 96) | 53–64 | 56·5 (2·29; 68) | 52–62 |
| BILL | 9·9 (0·55; 100) | 8·6–11·5 | 9·8 (0·53; 68) | 8·9–11·0 |
| TARSUS | 17·5 (0·94; 99) | 14·7–19·5 | 17·5 (1·36; 68) | 14·5–19·0 |

**Weights.** ADULT, FIRST ADULT. Azores, all year: ♂ 15·2 (1·28; 101), ♀ 15·3 (1·26; 59), latter including egg-laying ♀♀ of up to 18–20 (G Le Grand). For growth of nestling, see Zimka (1968).

**Structure.** Wing rather short, broad at base, tip fairly rounded. 10 primaries: p7–p8 longest, p9 and p6 (0–)0·5–1·5 shorter, p5

4–7, p4 8–12, p3 11–15, p2 12–17, p1 14–20; p10 reduced, 46–51 shorter than p7–p8, 4–8 shorter than longest upper primary covert. Outer web of (p5–)p6–p8 emarginated, inner of p6–p9 with notch. Tip of longest tertial reaches to tip of p3. Tail rather long, tip forked; 12 feathers, t1 8–12 shorter than t6. Bill as in *S. pusillus*, but slightly thicker and more swollen at base; depth at base 6·3–7·1, width 5·9–6·7, culmen and gonys slightly more curved. Middle toe with claw 16·6 (5) 16–17; outer toe with claw *c.* 68% of middle with claw, inner *c.* 71%, hind *c.* 86%. Remainder of structure as in *S. pusillus*.

**Geographical variation.** None.

Sometimes combined with Serin *S. serinus* into polytypic species which is sometimes thought also to include *S. syriacus* (e.g. Hartert 1903–10, 1921–2; Stresemann 1920; Hartert and Steinbacher 1932–8; Vaurie 1949*b*). However, *S. canaria* differs markedly from both in relatively shorter and rounder wing, longer tail (and hence, high tail/wing ratio: Vaurie 1956*a*, Eck 1985), larger and heavier tarsus and foot, song, and behaviour, and therefore kept separate here, following Stresemann (1943), Wolters (1952), Vaurie (1956*a*, 1959), and Nicolai (1960). Closest relative probably an Afrotropical species, perhaps Yellow-crowned Canary *S. canicollis* of eastern and southern Africa (Vaurie 1956*a*; Nicolai 1960; Hall and Moreau 1970).   CSR

## *Serinus citrinella*  Citril Finch

PLATES 34 and 39 (flight)
[between pages 472 and 473]

Du. Citroenkanarie     Fr. Venturon montagnard     Ge. Zitronengirlitz
Ru. Лимонный вьюрок     Sp. Verderón serrano     Sw. Citronsiska

*Fringilla Citrinella* Pallas, 1764

Polytypic. Nominate *citrinella* (Pallas, 1764), continental Europe; *corsicana* (Koenig, 1899), Corsica, Sardinia, Elba, and off-lying islands.

**Field characters.** 12 cm; wing-span 22·5–24·5 cm. Slightly larger and less compact than Serin *S. serinus*, with sharply tapered bill and proportionately longer wings; 20% smaller than Greenfinch *Carduelis chloris*. Small, elegant finch, form and behaviour recalling small *Carduelis* as much as any *Serinus*; plumage less contrasted and less streaked than any other west Palearctic *Serinus*. Mainland race basically grey-green, yellower on foreface, lower body, and rump; wings blackish with 2 greenish-yellow wing-bars. Flight-call distinctive. Sexes rather dissimilar; little seasonal variation. Juvenile separable. 2 races in west Palearctic, ♂♂ easily separated.

(1) Mainland race, nominate *citrinella*. ADULT MALE. Moults: July–August (complete). Forecrown, diffuse ring round eye, and cheeks golden-green; bluish-grey on rear crown, nape, ear-coverts, and (especially) sides of neck. Mantle and scapulars grey to green, showing at close range very faint dark streaks. Lower back and rump greenish-yellow; upper tail-coverts similarly coloured but with dusky centres to longest feathers; thus has longest pale area above tail of genus in the region, but far from brightest. Wings have distinctly black ground-colour softly marked with yellowish-green lesser and median coverts, grey tips and yellowish-green distal halves to greater coverts (forming obvious bar widening inwards), and yellowish-green fringes to flight-feathers, brightest on tertials and primary-tips. Underwing yellowish-white. Tail black, fringed as wings. From chin to under tail-coverts, underpart colour grades from grey-washed yellowish-green (breast), to green-tinged grey (flanks), to greenish-yellow (belly and under tail-coverts); underbody thus has

least yellow and most clouded underbody of any west Palearctic *Serinus* except Red-fronted Serin *S. pusillus*. With wear, mantle streaking becomes more obvious but yellow tones do not brighten. Bill horn, with dark grey tip and culmen. Legs light horn. ADULT FEMALE. As ♂, but distinctly duller in direct comparison, in particular with less yellow-green on crown, browner nape, neck, and mantle (also has mantle more streaked), and brownish-grey throat. JUVENILE. Head and body differ distinctly from adult, being basically buff-brown above and brownish-grey to yellowish- and whitish-buff below, with quite strong brown-black streaks on crown and mantle, on paler buff rump, and on all underparts except belly and vent. Wings and tail patterned as adult but all pale marks distinctly buff to cream. FIRST-YEAR MALE. Retains brownish, more streaked mantle and less green, more buff edges on wings and tail. (2) Corsican and Sardinian race, *corsicana*. ADULT MALE. Plumage more contrasting than in nominate *citrinella*, with diagnostic darker rusty-brown mantle with more obvious streaks, and paler, more clearly yellow and white underparts. For ♀, see Geographical Variation.

Unmistakable within normal montane habitats in Europe but vagrant elsewhere could be confused with any west Palearctic *Serinus* except *S. pusillus*. Appearance converges closely with Syrian Serin *S. syriacus* but that species slightly larger and paler, with much brighter yellow foreface in both sexes, fully yellow rump, prominent yellow fringes to tail-feathers, and white lower body. Distinction from *S. serinus* and from escaped adult Canary *S. canarius* is simple, as both show distinct streaks, but separation of juveniles needs care (see Plumages of all

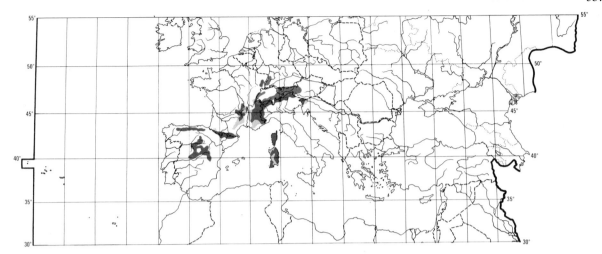

species). Beware, too, wishful confusion with ♂ *C. chloris*. Flight light and fast, rapidly developing into dancing action of *C. carduelis* and Siskin *C. spinus*; song-flight also as those species. Escape-flight is to bushes or trees. Gait a hop. Sociable, usually seen in small compact parties which feed on ground near heath patches or outlying conifers.

Song a fast, liquid twitter, recalling both *C. spinus* and *C. carduelis* and most frequently uttered in flight. Distinctive flight-call a ringing 'didididid'; also a strikingly metallic and nasal, 'tweek', 'creak', or 'chik'. Calls from perched bird include melancholy but quite musical piping 'tsüü' or 'pjee-u', and thin 'tsee'.

**Habitat.** Largely restricted to montane south-west Palearctic, in cool alpine climates, upwards from 700 m in Schwarzwald (Germany) (Niethammer 1937) and from 1000 m in Switzerland up to treeline at *c*. 1300-1500 m, and after breeding to 3300 m or more (Glutz von Blotzheim 1962). Heavy winter snowfalls usually render mainland breeding habitats untenable then, and emigrants to southern France and elsewhere near the Mediterranean often feed in birches *Betula* and alders *Alnus*; ground feeding also common.

In breeding areas, prefers open light marginal woodland composed at least partially of spruce *Picea*, often bordering on alpine meadows or clumps of spruce scattered on open terrain, and frequently having alpine huts as further attraction (Lüps *et al*. 1978). While far from embracing man-made environments to extent of (e.g.) Serin *S. serinus*, not reluctant to take advantage of their overlap with its natural habitat. Firmly tied to trees, but appears never to accept dense closed stands of forest. (Glutz von Blotzheim 1962.)

Favourite foraging sites are open spaces with nearby trees, especially trampled grass, ski-runs, mown meadows, forest clearings, etc., and also roads, scree, roofs of alpine huts, wasteland, or gardens (Glutz von Blotzheim 1962;

Sabel 1965; Mau 1980; Dorka 1986; Marzocchi 1990).

In Corsica and Sardinia, resident race inhabits dry scrub-covered rocky ground, breeding from sea-level to far above present treeline, at least to 1650 m, nesting in low tree-heath *Erica arborea* or *Genista* bushes, although sometimes occupying areas of scattered conifers (Armitage 1937; Thibault 1983).

**Distribution.** No information on range changes.

Accidental. Britain, Netherlands, Poland, Czechoslovakia, Algeria.

**Population.** GERMANY. 870 pairs (G Rheinwald). Evidence for recent changes, but overall trends unclear. Probable decline in Bayern 1967-87, based on numbers trapped (Bezzel and Brandl 1988), and in southern Schwarzwald since 1960 (Hölzinger 1987), but considerable recent increase in northern Schwarzwald, and conspicuous increase in number of records for Thüringer Wald since *c*. 1970 (Sabel 1983). Perhaps some temporary benefit from acid rain damage leading to forest clearance and new plantations, and from increase in ski slopes creating wide rides suitable for food plants (Sabel 1983; Hölzinger 1987).

Survival. Spain: average annual mortality 46 ± SE 15·3% (Borras and Senar 1988). Oldest ringed bird 5 years 8 months (Bezzel and Brandl 1988).

**Movements.** Short-distance and vertical migrant, wintering at middle altitudes, chiefly above 1000 m, though heavy snowfall causes birds to move lower temporarily; sedentary in some southern areas.

Most birds leave Alps in winter; in Switzerland, winters regularly (in varying numbers) only in Valais canton in south-west, less regularly on other south-facing slopes of Jura and Alps; numbers difficult to estimate as birds tend to move from place to place; more remain in milder winters, but probably never more than a quarter or third of population. In Garmisch-Partenkirchen (German Alps), observed in all months, but only irregularly November–February. Regularly remains in breeding areas of French/Italian maritime Alps; many French and Swiss birds winter in limited area in mountains of southern France—Cévennes, south-east Massif Central, and western edge of Alps from Vercors to Monts de Vaucluse; in absence of ringing data, not known whether some French birds move further afield from this area; western Pyrénées, however, almost entirely vacated, with very few winter records even at low altitude. (Hauri 1957; Crousaz and Lebreton 1963; Märki 1976; Boutet and Petit 1987; Bezzel and Brandl 1988.) Apparently chiefly sedentary in Trentino and Aosta areas of northern Italy; very scarce and irregular winter visitor to Lombardia, and exceptional south of Po basin (Brichetti 1973; Märki 1976). In Spain, chiefly sedentary in Asturias in north-west (Noval 1986); in northern Catalonia in north-east, most breeding-sites are abandoned in winter, birds dispersing to lower levels and sometimes to coast (Muntaner et al. 1983). In Corsica and Sardinia, more widespread winter than summer, with many birds moving to lower levels and to coasts (Newton 1972; Marzocchi 1990).

Recoveries of birds ringed in breeding season or on passage in Switzerland (mostly at Col de Bretolet in west or adjacent pass of Col de la Golèze, France) are chiefly in Mont Ventoux area (extreme south-west edge of Alps) and Cévennes (Märki 1976), and bird from Garmisch-Partenkirchen also recovered there, 600 km south-west of ringing site (Bezzel and Brandl 1988); this concentration of recoveries not due to hunting activities, so apparently reflects true situation; most are below 850 m, but tend to coincide with very cold or snowy spells when birds are temporarily forced down from higher ground. Some Swiss birds move further: bird recovered October near Barcelona (north-east Spain) was ringed June at Col de Bretolet, 625 km north-east, where presumably it was breeding, and bird ringed on passage at Col de la Golèze was recovered December just north of eastern Pyrénées; in this area others (presumed passage migrants) also observed in autumn, and numbers markedly higher in winter than summer in Sierra Caborella and Sierra Demanda in northern Spain; so birds wintering in Spain perhaps include considerable proportion of immigrants. (Märki 1976.) Occasional records in Balearics (Parrack 1973; Bannerman and Bannerman 1983) also imply longer movements. In Valle d'Aosta (north-west Italy), many birds winter on south-facing slopes at 800–1800 m, sometimes lower; most are presumably local, but occasional concentrations, e.g. 200 on 25 October, probably involve birds from across Alps (Bocca and Maffei 1984); scarce but regular passage reported both seasons near Biella c. 45°30′N 8°E (Mingozzi et al. 1988). 2 recoveries November–December in Italy of birds ringed at Col de la Golèze August–September: Vercelli (c. 46°N 8°30′E) and apparently as far south as Spoleto (42°44′N 12°44′E) (Moltoni 1973).

Col de Bretolet apparently especially important as migration route, with seasonal totals of 2000–3000, and few reports of passage elsewhere. Autumn movement begins there in last third of September, peaking 13–17 October and ending 1st third of November; local birds leave end of September or beginning of October following long period of dispersal; most passage is in the morning. (Crousaz and Lebreton 1963; Winkler 1984.) In Monts du Minervois and Corbières in southern France (south-west of main wintering area), observed from c. 10 October (Märki 1976).

Spring passage at Col de Bretolet begins early March, peaking end of March to mid-April; end of passage varies, as late snowfall may penetrate lower parts of breeding area until mid-May (Winkler 1984), though latest birds in plains and at low altitude are usually in late April (Praz and Oggier 1973). Earliest reports on breeding grounds vary markedly: in Vallée de l'Orbe (western Switzerland), 16 February to 3 April over 3 years (Glayre and Magnenat 1984); in Garmisch-Partenkirchen, 10 February to 13 April over 26 years (Bezzel and Brandl 1988).    DFV

Food. Small to medium-sized seeds (possibly more grass Gramineae seeds than most other Fringillidae), and sometimes green material from wide variety of plants (Sabel 1963, 1965, 1983; Mau 1982; Dorka 1986). Seeds of spruce Picea and pine Pinus important at times (Newton 1972; Borras and Senar 1991). Probably takes suitable food as available, remaining in same place for long periods if food abundant; important criterion apparently that seeds should be reachable without bird having to hang on plant, otherwise feeds on ground (Märki 1976; Mau 1980, which see for many details of techniques). In central Spain in winter, 79% of 29 feeding observations on ground and 21% on branches 8–18 m above ground (Carrascal 1988). See also Praz and Oggier (1973) and Märki (1976) for winter foraging. Seems to prefer to open or probe into seed-heads while standing on ground, fence, etc., though bill too short to probe deeply like Goldfinch Carduelis carduelis (Sabel 1963; Mau 1980), but when perched on herb or tall grass allows stem to bend to ground and holds seed-head under foot while picking out seeds (Lang 1948). Will hang from seed-head to extract seeds, though rarely completely upside-down, and in general appears less agile than (e.g.) Redpoll C. flammea or Siskin C. spinus, also when removing seeds from conifer cones in trees or feeding on catkins (Bieri 1945; Mau 1980; Bezzel and Brandl 1988). Newly-captive bird took average 26 s to pick up, de-husk, and swallow seed of Guizotia (Compositae), but only 1·8 s 3 days later, when times for other seeds ranged from 1·6 s for Brassica to 4 s for hemp Cannabis. In captivity, preferred Guizotia seeds, followed by thistle Cirsium or Carduus, flax Linum, and chicory Cichorium (Ziswiler 1965, which see for many details of seed-handling technique and comparison with other seed-eaters). See also Elzen (1983) and Sabel (1983) for food in captivity. Group of *c*. 20 seen clinging to stone bridge taking grit or minerals (Frost 1985).

Diet includes the following. Invertebrates: aphids (Hemiptera: Aphidoidea) and larval Lepidoptera. Plant material: seeds, pollen, and buds of spruce *Picea*, pine *Pinus*, willow *Salix*, birch *Betula*, alder *Alnus*, dock *Rumex*, orache *Atriplex*, goosefoot *Chenopodium*, amaranth *Amaranthus*, chickweed *Stellaria*, buttercup *Ranunculus*, whitlow-grass *Erophila*, tormentil *Potentilla*, lady's mantle *Alchemilla*, mountain avens *Dryas*, burnet *Sanguisorba*, clover *Trifolium*, melilot *Melilotus*, forget-me-not *Myosotis*, wood sage *Teucrium*, lavender *Lavandula*, thyme *Thymus*, nightshade *Solanum*, plantain *Plantago*, scabious *Knautia*, Compositae, rush *Juncus*, sedge *Carex*, grasses *Poa*, *Dactylis*, *Glyceria*, *Agrostis*, *Alopecurus*, *Anthoxanthum*, *Phalaris*, *Briza*, *Phleum*, *Calamogrostis*; also spore-capsules of mosses (Musci). (Glutz von Blotzheim 1962; Sabel 1965, 1983; Praz and Oggier 1973; Märki 1976; Fouarge 1980; Mau 1980; Génard and Lescourret 1986; Bezzel and Brandl 1988; Marzocchi 1990.)

No quantitative studies of diet in the wild. In Switzerland in winter, 88% of 116 observations of feeding on herbs were on *Teucrium*, 9% on *Rumex*, and 2% on *Lavandula* (Märki 1976). See Mau (1980) for detailed observations of feeding in Schwarzwald (south-west Germany).

Food of young known mainly from captivity; can be mostly invertebrates or entirely plant material (Sabel 1983). Nestlings given pulp of regurgitated seeds as eaten by adults (Glutz von Blotzheim 1962). In one study, up to 20% of items fed by parents were aphids or pupae of ants (Hymenoptera: Formicidae) (Sabel 1963, which see for preferences). Young in aviary with growing native plants were fed half-ripe seeds and seed-capsules of chickweed *Stellaria*, shepherd's purse *Capsella*, and grass *Poa*, and aphids (Mau 1982). In northern Italy, young were fed mostly seeds of dandelion *Taraxacum* (Maestri *et al.* 1989). In Alps, 1st broods can be reared mostly on spruce seeds in years of good crop (Newton 1972).        BH

**Social pattern and behaviour.** No major studies, and less information overall for *corsicana* (which differs vocally: see Voice) than for nominate *citrinella*.

1. Gregarious outside breeding season, with winter flocks of 20 up to several hundred recorded in Switzerland and France (e.g. Bourrillon 1961, Crousaz and Lebreton 1963, Praz and Oggier 1973, Märki 1976); up to 20 in Corsica (Guillou 1964). In Sistema Central mountains (Spain), 13 birds per km² in normal (non-snowstorm) conditions (Carrascal 1988). Small flocks and pairs reported from Col de Bretolet (Switzerland) during spring migration (Crousaz and Lebreton 1963), and in Corsica during May (Armitage 1937). Flocks of 5 or more occur rarely in Werdenfelser Land (Bayern, southern Germany) May–July (ie. throughout breeding season). Flocks (within which pairs keep together) are larger (20–100) when bad weather forces birds to abandon breeding grounds temporarily and to descend to lower altitudes; ♂♂ apparently in majority (see Bonds, below). (Brandt 1960; Wüst 1986; Brandl and Bezzel 1988.) Young of 1st broods

in Switzerland flock from late June or July, and are then joined by adults and other complete families and become nomadic; larger flocks sometimes unstable, families separating off; in Jura, flocks of 5–20 in August, 10(–50) on migration (Bieri 1945; Glutz von Blotzheim 1962; Crousaz and Lebreton 1963). Similar pattern in Schwarzwald (southern Germany), though flocks of up to 150 in August (Mau 1980; Dorka 1986); up to 35 in Sardinia (Bezzel 1957). Average flock size over 20 years 2·6–7·1 in Werdenfelser Land (Bezzel and Brandl 1988). In Valais (Switzerland), generally in single-species flocks, but tends to mix more with other Fringillidae and buntings (Emberizidae) when feeding on plains (Praz and Oggier 1973). Association with other Fringillidae (including, in Corsica, with Serin *S. serinus*) reported by Armitage (1937), Hauri (1957), Brandt (1960), and Mau (1980); in Switzerland, additionally with Water Pipit *Anthus spinoletta* and Black Redstart *Phoenicurus ochruros* (Corti 1952). BONDS. No evidence of other than monogamous mating system, though need for detailed investigation suggested by apparently biased sex-ratio revealed by ringing in Bayern during bad weather in spring: of 404 adults trapped 1967–85, 33·7% ♀♀, 66·3 ♂♂ (Brandl and Bezzel 1988; see also Brandt 1960). Captive birds immediately associated with Yellow-crowned Canary *S. canicollis* on being introduced into aviary, and has hybridized in captivity with Canary *S. canaria* (Nicolai 1957, 1960). Apparent hybrid with Greenfinch *Carduelis chloris* reported in Sardinia (Arrigoni degli Oddi 1931). Birds apparently pair up in flocks before reaching breeding grounds or soon after arrival there (e.g. Guillou 1964, Praz and Oggier 1973, Bezzel and Brandl 1988). Courtship-feeding (see part 2) recorded in captive *corsicana* pair in October perhaps suggests autumn engagement (Mau 1982), though no evidence of this for wild birds, and no further information on duration of pair-bond. Nest-building and incubation by ♀ alone, but both sexes feed young (Lang 1948; Géroudet 1957; Maestri *et al.* 1989). In Bayern, post-fledging care in one case 14 days (Bezzel and Brandl 1988); captive bird leaving nest at 18 days made first attempts at self-feeding at 22 days and was no longer fed by parents from 37 days (Sabel 1963). Age of first breeding 1 year in captive *corsicana* (Mau 1982). BREEDING DISPERSION. Often in small neighbourhood groups ('loose colonies'); of (1–) 2–3 pairs in Italian Alps (Brichetti 1986); maximum 6 pairs at 1250–1700 m in Werdenfelser Land (Bezzel and Brandl 1988); 3–5(–10) pairs in Schwarzwald, with apparently suitable habitat between groups unoccupied (Dorka 1986). Nests sometimes only a few metres apart in Switzerland (Schifferli *et al.* 1982); map showing 2 loose groups in Sondrio (northern Italy) indicates inter-nest distances of less than 50 m up to 200 m or more (Maestri *et al.* 1989). Captive birds reported to defend territory (see Antagonistic Behaviour, below), but little information on its size or configuration in the wild: Song-flights of ♂♂ (see part 2) overlap and sometimes extend over several hundred metres (Schifferli *et al.* 1982); in Pyrénées, ♂ sang near nest (frequently in flight), but also (mainly from perch) up to 200–500 m away where ♀ collecting nest-material (Fouarge 1980). In Corsica, food for brood close to fledging collected *c*. 1600 m below nest (Armitage 1937). In Sondrio, after 1st nest predated (young 2 days old), replacement built *c*. 2 m away (Maestri *et al.* 1989). Counts of singing ♂♂ tend to produce higher figures than actual breeding densities, which should be based on counts of nests (Schifferli *et al.* 1982). Density data (none based on nest counts) as follows. In Corsica, 3·6–52·6 pairs per km² in various habitats, highest in low maquis (Blondel 1979). In Basque country (northern Spain), 3·5 birds per km² in pine *Pinus radiata* plantations (Carrascal 1987). Variation between different forest types in French Alps: 13 pairs per km² in *Pinus sylvestris*, 21 in dry spruce *Picea*, 18–30 in larch *Larix* (Lebreton 1977). In Sondrio, 17 territories per km² (Maestri *et al.* 1989). Average summer population in Werdenfelser Land (Bayern) 0·52-0·83 pairs per km² over 1440 km² (Bezzel and Lechner 1978; Bezzel and Brandl 1988). ROOSTING. No information on nocturnal roosting. During hottest part of day in late summer in Valais (Switzerland) at 1800-2300 m, birds rested (sometimes for hours) well inside foliage of solitary larches, there preening or sun-bathing with plumage ruffled; in rain, sought shelter in roofs or on sills of alpine huts; feeding flocks seen to shift from dark to sunlit slope around sunset (Bieri 1945).

2. Not especially shy (Corti 1949), though considerable variation noted in France in winter (Crousaz and Lebreton 1963). Will allow cautious approach to within *c.* 15-20 m in open terrain, when feeding or gathering nest-material to within *c.* 2-5 m (Corti 1935, 1952; Armitage 1937; Brandt 1960; Mau 1980). Usually flies into tree when disturbed by man; reacted similarly when Carrion Crow *Corvus corone* flew over (Mau 1980). In excitement, whips tail up and down like *S. canaria* (Nicolai 1957); on ground, movement a measured raising and lowering of tail according to Géroudet (1957). FLOCK BEHAVIOUR. Migrant flocks in Switzerland generally looser, movements in flight less synchronized than *Carduelis* finches; flock often stops if some are trapped, more and more then entering nets (Crousaz and Lebreton 1963). In Corsica in May, flocks flew and called (see 3 in Voice); when flushed with *S. serinus*, only *S. citrinella* called (Armitage 1937). In Switzerland mid-October, flock sang in chorus from tree (Hauri 1957). For aggressive interactions in flocks, see Antagonistic Behaviour (below). SONG-DISPLAY. ♂ sings (see 1 in Voice) from top of tree or shrub; adopts rather erect posture (similar to *S. serinus*), carpal joints held slightly away from body, wing-tips drooped; bill open widest when giving most-strangled sounds (Sabel 1963); may turn to left and right (Brandt 1960). Song given sometimes in normal flight, but Song-flight proper resembles *S. serinus* or *C. chloris*: bird circles, tail spread, beating wings slowly, giving bat-like impression (Ferguson-Lees 1956; Fouarge 1980; Bergmann and Helb 1982); may sing while flying with shallow wing-beats from tree to tree or around tree (Géroudet 1957). In alpine range, song noted in most months of year; irregular in winter, mainly when mild, much as other Fringillidae (Praz and Oggier 1973), also in autumn, following moult (Bezzel and Brandl 1988; also Sabel 1963 for captive birds). Main song-period of nominate *citrinella* late February or March to June or July, with peak during nest-building and incubation; song also noted (including juveniles) in August (Jouard 1930; Corti 1949, 1952; Géroudet 1957; Mau 1980; Wüst 1986; Bezzel and Brandl 1988). In Corsica, song also from late February (Marzocchi 1990); noted in May (Armitage 1937). ANTAGONISTIC BEHAVIOUR. In flocks disputes over food or perch signalled by opening of wings, calling (see 5 in Voice), and chases (Mau 1980); chases involving at a given moment 2 birds from flock resting in tree during day also reported by Bieri (1945). In captivity, especially aggressive towards conspecific birds, ♂ marking territory by singing, defending it with fights (Sabel 1963). HETEROSEXUAL BEHAVIOUR. (1) Pair-bonding behaviour. In 2 captive pairs, sexual chases (less vigorous and persistent than in *S. serinus*) recorded over *c.* 2-4 days around mid-May, ♂ singing from perch between chases; bond thereafter close (probably following 1st copulation), with repeated Courtship-feeding (see below). Wild ♂♂ seen to flutter down from Song-flight and land close to ♀ calling on ground. (Sabel 1963.) (2) Courtship-feeding. ♂ feeds ♀ by regurgitation on nest, occasionally also away from it; visits in one study every 60-90 min. ♀ gives Begging-call (see 7 in Voice), including before ♂ arrives at nest, opens bill and wing-shivers; ♂, with bill wide open and neck feathers slightly ruffled, makes average

21 (6-36, *n* = 24) feeding movements per bout. (Lang 1948; Sabel 1963; Mau 1982; Maestri *et al.* 1989). (3) Mating. One observation, Switzerland: ♀ of pair with nestlings was perched on wire and calling (apparently reacting to disturbance) when ♂ suddenly arrived, mounted, and copulated, ♀ remaining afterwards in copulation posture with tail raised, wings raised and quivering (Lang 1948). (4) Nest-site selection and behaviour at nest. Observations on captive birds (both nominate *citrinella* and *corsicana*) suggested nest-site chosen by ♂; seen to fly about and eventually go to potential site with or without nest-material, deposit material, shuffle about as if shaping nest, and call (see 6 in Voice), encouraging ♀ to approach. Nest-building by ♀ alone, though ♂ (generally with markedly ruffled plumage) always accompanies her, sometimes singing from perch nearby while she builds. (Sabel 1963; Mau 1982; also Fouarge 1980.) In Swiss study, ♀ left nest during incubation apparently mainly to defecate, perhaps also to drink; always landed at same place on rim before settling (as did ♂ when visiting nest); wings spread in rain (Lang 1948). RELATIONS WITHIN FAMILY GROUP. In Switzerland, hatching at 2 nests over 3-4 days; brief brooding stints noted at 10-12 days, though young brooded at night almost up to fledging (Lang 1948); captive ♀ ceased brooding altogether at 12 days (Sabel 1963). Adults approaching nest give Flight-calls (see 3 in Voice). Young fed by regurgitation. ♂ increases frequency of visits after hatching, usually transferring food to young via ♀, though will feed young direct, whether or not ♀ present. Young gape initially when ♀ gets up, later do so when adult lands on nest; parent seen to touch back of unresponsive chick to stimulate gaping. (Lang 1948.) Nest-hygiene apparently mainly by ♀ for whom nestling faeces are perhaps important food; faecal sacs swallowed (up to *c.* day 12 in captive birds, occasionally carried away); young later deposit faeces on rim of nest, from where only fresh pellets are removed (Lang 1948; Sabel 1963); rim of nest can become encrusted with faeces (Armitage 1937). Eyes of captive *corsicana* fully open at 8 days (Mau 1982). Young fledge at 15-18 days (see Breeding). Close to fledging, may make short excursions onto nearby branches, returning to nest when adult arrives (Lang 1948). ANTI-PREDATOR RESPONSES OF YOUNG. Captive bird 12 days old crouched in nest, clinging tight to bottom (Sabel 1963). PARENTAL ANTI-PREDATOR STRATEGIES. In Corsica, adults called excitedly (presumably alarm-call: see 4 in Voice) from perch near nest, which ♀ nevertheless still visited when man nearby; agitation much increased when young out of nest (Armitage 1937). ♀ left nest when dog by nest-tree (Lang 1948).     MGW

**Voice.** Sings mainly during breeding season, also irregularly in autumn and winter (Géroudet 1957); little information on use of other calls through the year or on sexual differences. Considerable differences between nominate *citrinella* and *corsicana* (Chappuis 1976); available recordings confirm this with regard to song, but call repertoires of both races (especially *corsicana*) have yet to be investigated in detail, and following scheme therefore provisional. For additional sonagrams, see Bergmann and Helb (1982); for study with musical notation, see Stadler (1926).

CALLS OF ADULTS. (1a) Song of ♂ nominate *citrinella*. Most often likened, in twittering and tinkling or rapid, at times strained, babbling quality to certain *Carduelis* finches and Serin *S. serinus* (see below). Often a short introduction of one to several well-spaced units (apparently mostly calls or derivatives: e.g. 'wi' sounds, nasal ascending 'zeî' or 'gnin') followed by series of characteristically varied units or mostly short motifs: includes buzzes and rattles, trills, 'pi' or 'pink' sounds, and frequent Flight-calls or similar (see Figs I–II and call 3, below); delivery typically brisk (Stadler 1926; Witherby *et al.* 1938; Géroudet 1957; Chappuis 1976). Song almost continuous, with repetition of particular motifs (Stadler 1926; Bergmann and Helb 1982), or in relatively short phrases separated by pauses of varying length (Witherby *et al.* 1938; Jonsson 1982). Pitch and timbre of notes vary considerably, and song at times shows undulating pattern (overall ascent and descent in pitch: see Figs I–II). Recordings from France and Switzerland contain songs showing fair degree of consistency between them in both form and content; duration *c.* 3–7 s; comprise mostly units of wide-ranging portamento type, also many buzzes and rattles. Songs in Figs I–II well defined by quiet step-wise

ascent through introductory units to 1st figure: rattle in Fig I and, in Fig II, segment strongly resembling part of song of Chaffinch *Fringilla coelebs* (similarity to *F. coelebs* also noted by Stadler 1926). Initial ascent balanced in Figs I–II by similar but portamento descent ('cadence') at the close. (J Hall-Craggs.) Sequence of units frequently changes and each individual thus has rich repertoire of song-types, giving (e.g.) 20 successive songs each different in structure or one type many times in succession with little variation (Stadler 1926). Song lacks sizzling quality of higher-pitched *S. serinus* song, softer and more melodious, containing fewer strained sounds; much less monotonous, and quieter (though volume varies), but otherwise similar (Stadler 1926; Corti 1949, 1959; Géroudet 1957; Sabel 1963), especially in autumn, when more closely resembles autumn song of *S. serinus* (Sabel 1963). Especially similar to Red-fronted Serin *S. pusillus* and Syrian Serin *S. syriacus* (L Svensson) and to Yellow-crowned

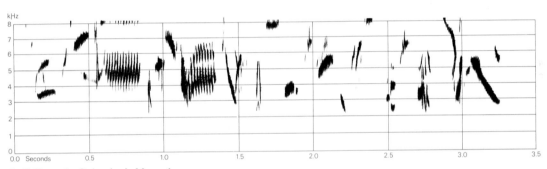

I  C Chappuis  Switzerland  May 1963

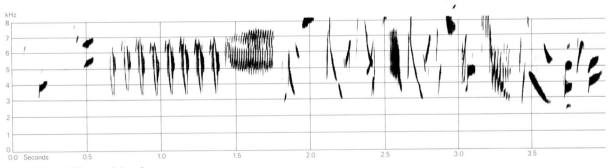

II  J-C Roché  France  July 1987

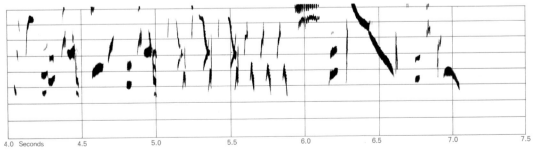

II  *cont.*

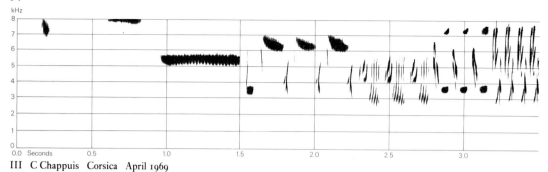

III   C Chappuis   Corsica   April 1969

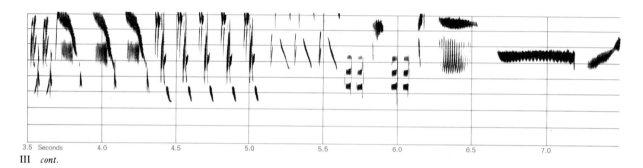

III   *cont.*

III   *cont.*

Canary *S. canicollis* of southern Africa (Nicolai 1957). Some resemblance also to Goldfinch *C. carduelis*, but harsher and less sweet overall (Stadler 1926; Brandt 1960). Nasal twittering further recalls Siskin *C. spinus*, but softer, sweeter, more musical than that species (Ferguson-Lees 1956; Géroudet 1957). (1b) Song of *corsicana*. As noted by Chappuis (1976), easily distinguished from nominate *citrinella* because of its segmented structure, with form closely akin to song of Wren *Troglodytes troglodytes*; to lesser extent than that species, song also contains figures comprising heterogeneous units. Shares with nominate *citrinella* predominance of (mostly descending) portamento-type units, and buzzes, buzzing tones, and rattles (fewer in *corsicana*); further feature of *corsicana* is strong frequency and amplitude modulation to whole figures in song. In Fig III, quiet high note ('rii') followed by even higher steady tone (2nd), then longer, lower-pitched, loud and steady, amplitude-modulated tone (3rd). These 2nd and

3rd units (occurring otherwise only in 'chant': see 2, below) are reintroduced near end of Fig III (before coda of 3 short segments), shorter and higher tone (2nd) now lower and combined with rattle to form extraordinary diad. 2nd segment of song (Fig III) comprises gentle knocking sounds recalling song of Thrush Nightingale *Luscinia luscinia*. Recordings also contain songs starting with rattling buzz, or (as in Fig IV) series of Flight-calls ('yu' and 'hui': see call 3, below). Resemblance to notes given in 'chant' occurs only in coda of Fig IV: final tone, its preceding grace note (noisy in 1st half), and note like Flight-call (compare introductory section of Fig IV) before it. (J Hall-Craggs.) (2) Slow song ('chant') of *corsicana*. Analysis of recordings by C Chappuis has revealed existence in this race of a 'chant' (function unknown) comprising series of mostly long and very steady tones (6 types differing in pitch and timbre) of remarkable beauty delivered at very slow rate (*c.* 51 per min). In Fig V, shorter 1st unit bell-like; 2nd and 4th (same as 2nd but embellished with brief bell-like grace note) have diad and harmonics producing timbre of miniature high-pitched trumpet; 3rd and 5th are buzzing or purring owing to frequency and amplitude modulation (in another recording, frequency modulation is pronounced in all notes of 'chant'); flute-like 7th unit preceded (unusually closely) by repeat of 1st bell-like tone of series; penultimate note preceded by hollow and wooden-sounding 'kuk' grace note; final unit in Fig V a noise, roughly transcribable as 'ri'. No chant described for nominate *citrinella*, though descriptions of calls in some sources perhaps refer to notes of this kind. (J Hall-Craggs.) (3) Flight-call. In nominate *citrinella*, nasal,

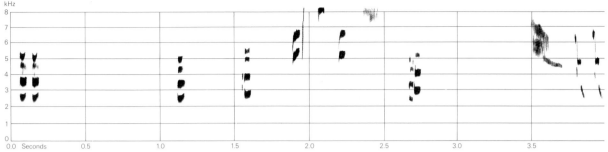

IV J-C Roché Corsica June 1966

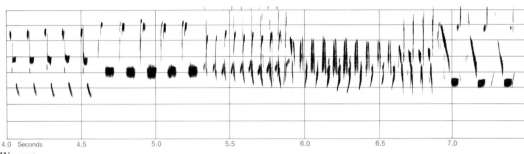

IV *cont.*

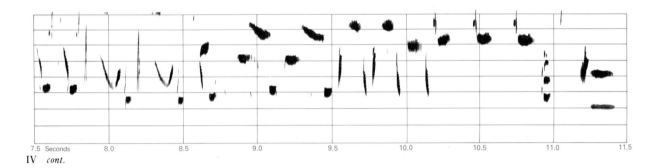

7.5 Seconds

IV *cont.*

creaky and metallic 'di', 'dit', 'tweck', '(t)chet', 'tweek', 'chiht', etc.; sometimes given in rapid series—'tweek-eek-eek', 'didididdid', or attractive, ringing 'pirriti' (Witherby *et al.* 1938; Géroudet 1957; Brandt 1960; Newton 1972; Bergmann and Helb 1982; Jonsson 1982; Bruun *et al.* 1986). Serves as contact-call in flight and at take-off, also frequently incorporated in song (Stadler 1926; see Figs I–II). Depending on context, varies in length and volume (Sabel 1963), also pitch pattern (Stadler 1926). According to Nicolai (1957), 'di di di di' used in close contact similar to homologous calls of Canary *S. canaria* and *S. canicollis* and descends in pitch (no evidence of this in recordings), while variant 'DIE di di DIE di di' said to function as contact-call at greater distance. Captive birds apparently calming down after excitement gave 'dudüd—düdüdüdüd' while flying between perches (Sabel 1963). Recording of calls given by migrating birds (Fig VI) suggests delicate nasal bleats—'hui yu hui yu yu'.

Only description for *corsicana* refers to metallic whinnying twitter, allegedly similar to call of *C. spinus* (Armitage 1937). Recording (Fig VII) contains series of 'de' or 'yu' sounds (compare introduction to Fig IV); similar to nominate *citrinella*, but both Fig VI and sonagram in Bergmann and Helb (1982) from another Swiss recording show ascending 'hui' calls perhaps characteristic of that race and not given by *corsicana* (J Hall-Craggs, M G Wilson). As in nominate *citrinella*, calls regularly incorporated in song (see Fig III–IV). (4) Contact-alarm calls. Renderings in many published sources are supported by little or no information on context or probable function and highly arbitrary sub-division adopted here is based primarily on real or apparent pitch differences. Figs and apparently all descriptions are of calls given by nominate *citrinella*; see below, however, for suggestion of racial difference. (4a) Clear, plaintive, apparently descending call given when disturbed; musical, piping 'tsüü' or 'tsi-ew' (Witherby *et*

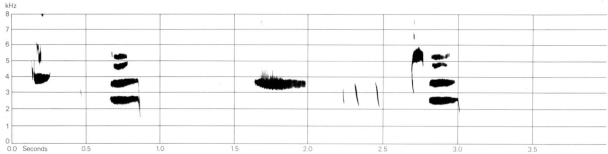

V   C Chappuis   Corsica   April 1969

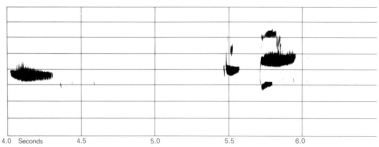

V   *cont.*

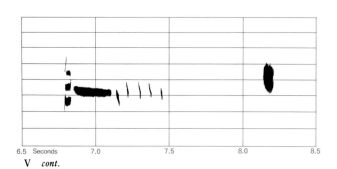

V   *cont.*

VI   C Chappuis   Switzerland   May 1963

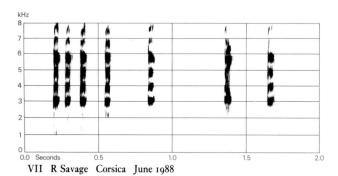

VII   R Savage   Corsica   June 1988

notes of song in Fig I), and 'kees' (1st unit in Fig IX). The following descriptions are probably of this call-type: thin 'tsee' (Witherby *et al.* 1938); slightly nasal 'si', 'sIi', and 'wisst' like call of Water Pipit *Anthus spinoletta* (Stadler 1926); 'bzih' (Géroudet 1957); 'djied' from captive ♀ (Sabel 1963); 'pweet' (Bruun *et al.* 1986). (4c) Disyllabic calls ascending in pitch: 1st unit in Fig X (perhaps from ♀) a squeaky 'hui' with strong fundamental similar to 'hwee' in Fig VIII; 2nd (perhaps from ♂) relatively low-pitched buzzy 'zui'; 2nd unit in Fig IX suggests 'ki-tee' (J Hall-Craggs) and same call rendered 'pchi' by Bergmann and Helb (1982). Further renderings probably of this call-type: 'djöid' and 'zuieg' given by captive breeding ♀ (Sabel 1963); 'ptuee' (Bruun *et al.* 1986); 'touic', and (apparent overlap with 4b) 'pluî pli psih' (Géroudet 1957). Presumably referring to one of calls 4a–c, Chappuis (1976) reported short and strongly frequency-modulated call (given from perch), with complex harmonics for nominate

*al.* 1938; Ferguson-Lees 1956); 'zi-ä' (Bergmann and Helb 1982); 'tiie' recalling *C. spinus* (Jonsson 1982) presumably also belongs here, though named 'attraction call'. 1st unit in Fig VIII a 'zeee-zii' given by alarmed ♂. (4b) Monosyllabic calls ascending in pitch include 'hwee' (2nd unit in Fig VIII, given by ♀ in alarm; similar to introductory

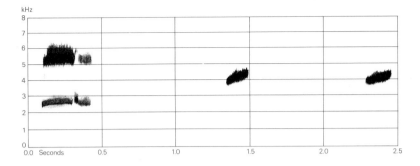

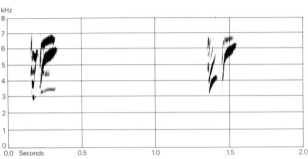

IX  C Chappuis  Switzerland  May 1963

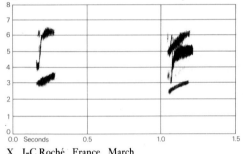

X  J-C Roché  France  March

*citrinella*, and longer, barely frequency-modulated call, sometimes with single harmonic, for *corsicana*. (5) Quiet, harsh 'ch' reminiscent of *C. carduelis* given during fights by nominate *citrinella* (Stadler 1926). 'Wutzetern' (harsh chattering or rattling) mentioned for same context by Mau (1980) is presumably the same or related. (6) Calls apparently associated with Nest-showing (described from captive birds). A 'zizizizi' given by ♂ nominate *citrinella* (Sabel 1963); 'tück tück' given by ♂ *corsicana* to summon ♀ (Mau 1982). (7) Begging-call of ♀. Chirping sounds noted from captive *corsicana* (Mau 1982). (8) Other calls (all apparently given by nominate *citrinella*). Further transcriptions (at least some perhaps only alternative renderings of calls described above) include soft, melodious 'ü' (Stadler 1926), 'chwick' or 'twick' (Witherby *et al.* 1938); 'churr-rr-rah' with stress on final syllable (Witherby *et al.* 1938), also apparently similar low 'gurr' (Géroudet 1957); 'zizèzizizèzi', as part of twittering chorus from flock (Géroudet 1957).

CALLS OF YOUNG. Initially faint food-calls develop with age from thin, high-pitched 'sri', 'sie', 'tsip', or 'ziet' (6-10 days) (occasionally 'deek' from *corsicana*) to mostly double calls 'zigi', 'zitied', 'zitähd', 'zit-zit', or 'zäh-zäh' ('zueg' from *corsicana*) towards end of nestling phase and after fledging; equivalent of adult call 3 at 32 days rendered 'dädädädäd' (Armitage 1937; Lang 1948; Corti 1952; Sabel 1963, 1965). Quiet twittering song of captive nominate *citrinella* noted at 32 days (Sabel 1963); captive *corsicana* gave song, less varied than adult's, from October (Mau 1982).                                                      MGW

Breeding. SEASON. Southern Germany: at 1200-1600 m, nest-building end of April to June, fledged young observed end of June to early September; below 1000 m, fledged young recorded late May (Wüst 1986; Bezzel and Brandl 1988). Switzerland: eggs generally laid from 2nd half of April or early May, 2nd clutches June and 1st half of July; in exceptional years, eggs laid late February and fledged young reported at end of March (Glutz von Blotzheim 1962). Northern Italy: at 2500-3000 m, nest-building from beginning of April to end of May and eggs laid late April to mid-June (Maestri *et al.* 1989). Northeast Spain: at 500-1600 m, eggs laid early April to June (Borras and Senar 1991). Corsica: nest-building mid-March to late May (Payn 1926; Armitage 1937; Guillou 1964; Marzocchi 1990). SITE. On mainland, almost always in conifer, usually close to trunk in upper part of tree and protected by dense twigs above, though sometimes out on branch (Glutz von Blotzheim 1962; Kaniss 1970; Bezzel and Brandl 1988; Maestri *et al.* 1989). In Switzerland, 63% of 43 nests in spruce *Picea* and fir *Abies*, 19% larch *Larix*, 16% pine *Pinus*, and 2% beech *Fagus*; generally 3-10 m (1·5-15·2) above ground (Glutz von Blotzheim 1962). In Bayern (southern Germany), almost always in spruce; average height 9·5 m (6·8-20, *n*=9) (Kaniss 1970; Wüst 1986), though in northern Italy average height of 17 nests was 2·9 m, and of nest-trees (all pines *P. mungo* and *P. uncinata*) 3·6 m (Maestri *et al.* 1989). In some valleys in Italian alps without conifers, will nest in broad-leaved trees, e.g. alder *Alnus viridis* (Mingozzi *et al.* 1988). Nest-sites on Corsica very different; majority in tree-heath

*Erica arborea c.* 1 m (*c.* 0·3–2) above ground, or in juniper *Juniperus phoenicea* (Payn 1926; Armitage 1937; Marzocchi 1990). Nest: foundation of dry stalks, grass, roots, lichen, and spiders' webs, smoothly lined with hair, wool, feathers, rootlets, paper, etc. (Sabel 1965; Kaniss 1970; Maestri *et al.* 1989); on Corsica, flimsy and shallow, mostly of fine grass lined with hair, feathers, moss, or plant down (Payn 1926; Armitage 1937). In Bayern, 8 nests had outer diameter 8·5–9·3 cm, inner diameter 4·6–5·0 cm, average overall height 5·0 cm, and depth of cup 2·5 cm (Kaniss 1970); for 10 nests from northern Italy, see Maestri *et al.* (1989). Building: by ♀ only, accompanied by ♂ (Nicolai 1960; Fouarge 1980; Wüst 1986); in captivity, ♂ seen to carry some material but never build; takes 2–3 days, but 2nd nest can be completed in a few hours (Sabel 1963; Kirner 1964; Mau 1982). EGGS. See Plate 60. Sub-elliptical, smooth and glossy; pale blue, sparsely marked towards broad end with small dark rust spots and scrawls and reddish-violet undermarkings (Harrison 1975; Makatsch 1976). Nominate *citrinella*: 16·7 × 12·7 mm (15·2–18·5 × 11·7–14·1), *n* = 64; calculated weight 1·41 g. *S. c. corsicana*: 15·9 × 12·1 mm (15·4–16·7 × 11·9–12·3), *n* = 3; calculated weight 1·22 g. (Schönwetter 1984.) For 53 eggs from northern Italy, see Maestri *et al.* (1989). Clutch: (3–)4–5 in nominate *citrinella*. In northern Italy, of 12 clutches, 5 of 4 eggs and 7 of 5; average 4·6 (Maestri *et al.* 1989). In Switzerland, of 11 clutches, 1 of 3 eggs, 6 of 4, and 4 of 5; average 4·3 (Glutz von Blotzheim 1962). On Corsica 3–4(–5) (Armitage 1937). Eggs laid daily (Sabel 1963; Mau 1982). Commonly 2 broods, though perhaps only at lower altitudes in southern Germany (Bezzel and Brandl 1988); captive ♀ started 2nd nest when young *c.* 12 days old (Kirner 1964). Replacement clutch laid in new nest *c.* 9 days after 1st clutch predated (Maestri *et al.* 1989). INCUBATION. 13–14 days; by ♀ only (Sabel 1963; Kirner 1964; Mau 1982; Maestri *et al.* 1989). Starts properly with 2nd or penultimate egg (Sabel 1963; Mau 1982). At one nest in the wild, ♀ took 5 breaks (of up to 7 min) in 6 hrs (Lang 1948). YOUNG. Fed and cared for by both parents; ♂ feeds young directly only towards end of nestling period; faecal sacs removed, mostly by ♀, until oldest nestling *c.* 12 days old then left on nest-rim (Lang 1948, which see for feeding routine; Sabel 1963; Mau 1982). Captive ♀♀ brooded young without leaving nest for 8–12 days (Sabel 1963; Mau 1982). FLEDGING TO MATURITY. Fledging period 16–17 (15–18) days (Glutz von Blotzheim 1962; Sabel 1963; Mau 1982; Maestri *et al.* 1989). Young in captivity picked up seeds when 22 days old, but still fed by parents until *c.* day 30–37 (Sabel 1963, which see for development of young; Kirner 1964; Mau 1982; Bezzel and Brandl 1988). BREEDING SUCCESS. In northern Italy, 45% of 55 eggs produced fledged young; average 2·8 fledged young per successful nest (*n* = 9), and 1·8 per nest where eggs laid (*n* = 14) (Maestri *et al.* 1989). Severe weather reduces success either through smaller clutch size or nestling mortality (Brandl and Bezzel 1988). BH

**Plumages.** (nominate *citrinella*). ADULT MALE. Forehead and forecrown yellow-green, sometimes slightly purer yellow at border of upper mandible, partly mottled grey on crown. Length of patch on forehead *c.* 11·5 (21) 9–14 mm (Brandl and Bezzel 1989). Rear of crown, hindneck, and upper mantle light ash-grey, sometimes some yellow-green of feather-centres visible. Lower mantle and scapulars bright pale green with faint grey shaft-streaks, on back merging into bright greenish-yellow of rump and upper tail-coverts; longest upper tail-coverts partly tinged olive-grey. Short supercilium, short stripe below eye, and front part of cheeks bright greenish-yellow; lore mottled light grey and pale yellow; a short grey patch just in front of and behind eye. Remainder of side of head and all side of neck light ash-grey. Chin and mid-throat yellow (rarely green), patch 3–5 mm wide (Brandl and Bezzel 1989); chest green-yellow; side of throat and side of breast light ash-grey. Belly and under tail-coverts bright yellow, yellow of side of belly merging into yellow-green and light ash-grey on flank; vent yellow-white or fully white; thigh mixed light grey-white; longer under tail-coverts with concealed dark grey centres. Tail black; outer webs fringed light yellow (widest on central pair, t1), fringes along tips of tail-feathers and faint edge along outer web of t6 pale grey or off-white. Flight-feathers, greater upper primary coverts, and bastard wing black; outer webs of flight-feathers (except of p9 and of emarginated parts of p6–p8), primary coverts, and shorter feathers of bastard wing narrowly but distinctly fringed pale yellow, basal half of visible part of outer web of secondaries contrastingly black, forming black bar across inner wing; tips and basal inner borders of flight-feathers and outer web of p9 fringed pale grey or off-white. Tertials black, outer web broadly fringed yellow-green, merging into light grey on tip. Lesser, median, and greater upper wing-coverts bright yellow-green or golden-green, black of centres and inner webs of greater coverts partly visible. Under wing-coverts and axillaries white with pale yellow wash on tips; shorter coverts bright green-yellow. *In worn plumage* (about April–July), some yellow-green partly visible among grey of hindcrown, hindneck, and ear-coverts; mantle and scapulars duller green with more extensive ill-defined dusky grey streaks; grey of side of throat, side of breast, and flank partly worn off, underparts more extensively yellow; yellow fringes of tail, flight-feathers, and tertials somewhat bleached and partly worn off, black on bases of secondaries more conspicuous; green-yellow tips of greater coverts partly worn off, more black of centres exposed; often some black of bases visible on median and longer lesser upper wing-coverts. ADULT FEMALE. Like adult ♂, but yellow-green on forehead more restricted, length of patch *c.* 8 (11) 6–10 mm (Brandl and Bezzel 1989); grey from crown to upper mantle duller, less pure ash-grey, slightly tinged brown, without yellow-green on feather centres. Mantle and scapulars more olive or brown-olive with slightly duller grey centres, fringes partly washed green, not as bright pale green as adult ♂ (distinctly more olive and much less streaked than cinnamon-brown of *corsicana*). Yellow of supercilium, streak below eye, and front of cheek greener and more restricted, less contrasting with dull light grey or light brown-grey of remaining side of neck. Patch on chin and upper throat green (occasionally, grey), 1–4 mm wide (Brandl and Bezzel 1989), usually contrasting with light ash-grey of lower throat. Flank more extensively ash-grey or brown-grey, sometimes with traces of slightly darker grey streaks, chest and breast less saturated pure yellow, chest and side of belly partly washed grey. Under tail-coverts white with variable amount of pale yellow wash. Wing and tail as in adult ♂, but black on bases of median and greater upper wing-coverts often more extensive. NESTLING. Down fairly long and plentiful, brown-black, on upperparts, upper wing, lower flank, and (very short) vent (Witherby *et al.* 1938). JUVENILE. Upperparts buff-brown, closely marked with rather ill-defined sooty or sepia streaks; ground-colour of hindneck and rump paler, more buff-brown. Side of head and neck and chin to throat and side of breast brown-grey, marked with ill-defined dark brown or sepia shaft-streaks (faint on lore and side of neck). Chest and flank buff with ill-defined brown streaks, more or less suffused with pale yellow. Belly, vent, and under tail-coverts buff- or cream-white, sometimes faintly tinged yellow, coverts with dark sepia centres. Tail, flight-feathers, tertials, and greater upper primary coverts as adult, but pale outer fringes often greyish-green rather than pale yellow, soon bleaching to off-white; upper wing-coverts sooty-grey with rather narrow and sharply-defined buff to cream-white fringe along tip (in adult, more extensively green-yellow). FIRST ADULT. Like adult, but juvenile tail (frequently except t1), flight-feathers, greater upper primary coverts, often tertials, and usually outer greater upper wing-coverts retained, these more worn than in adult at same time of year, tips of tail-feathers and primary coverts more pointed than those of adult, less broadly rounded. In birds with extensive post-juvenile moult, all greater coverts, tertials, and t1 new; in these, bright green-yellow outer web of tertials contrasts with paler green and more worn outer web of neighbouring secondaries, white fringe along smoothly rounded tip of t1 contrasts with worn and pointed tips of t2–t6, which have pale tip worn off. In birds with less extensive moult, tertials and all tail-feathers old, not contrasting with neighbouring feathers, but outer greater coverts usually juvenile, showing conspicuously narrower, paler, more sharply-defined, and more worn fringes than broadly green-yellow tipped neighbouring coverts. Head and body of 1st adult ♂ as in adult ♂ but length of patch on forehead *c.* 10·5 (15) 8–12 mm (Brandl and Bezzel 1989); hindneck sometimes slightly duller and browner grey; mantle and scapulars slightly duller and more streaked dusky on average; patch on chin and upper throat green or grey, 1–4 mm wide (Brandl and Bezzel 1989); yellow of underparts slightly greener, more washed with ash-grey on side, more mottled white on belly. Head and body of 1st adult ♀ as in adult ♀, but length of patch on forehead *c.* 6 (19) 4–9 mm (Brandl and Bezzel 1989); mantle and scapulars slightly more brown-olive, less tinged green; patch on chin and upper throat usually grey, only occasionally green (Brandl and Bezzel 1989); yellow of underparts paler, throat to chest and side of belly more extensively washed ash-grey or light brown-grey.

**Bare parts.** ADULT, FIRST ADULT. Iris brown. Bill flesh-horn with dark grey culmen and tip, darkest in adult ♂ breeding. Leg and foot flesh-horn or light horn-brown. (Hartert 1903–10; BMNH, ZFMK.) NESTLING. Mouth pale flesh or light flesh-pink; gape-flanges ivory-white or creamy-white with pink spot at base (Witherby *et al.* 1938; Harrison 1975). JUVENILE. No information.

**Moults.** ADULT POST-BREEDING. Complete; primaries descendent. In birds examined, moult just started with scattered feathers of body or p1 in part of a series from Pyrénées, July (BMNH); no moult yet in 2 others from Pyrénées, 21 July (ZFMK). Moult advanced in ♂ from Corsica, 15 August: primary moult score 29 (p1–p4 new, p9–10 old), t1–t3 new, t4–t6 growing or absent, moult of secondaries just started, upper wing-coverts and tertials new or growing, feathers of head and body old or growing (ZMA); primary-score *c.* 32 in ♂ from Sardinia, 17 August (Bezzel 1957). POST-JUVENILE. Partial: head, body, lesser, median and some inner to all greater upper wing-coverts, sometimes tertials and t1(–t3). In advanced birds (especially ♂♂), all greater coverts, tertials, and t1 new; in more retarded birds (especially

♀♀), 1–8 greater coverts, all tertials, and all tail still juvenile. In southern Germany, 2–7 outer greater coverts retained; 59% of ♂♂ retain 2–3 coverts, 33% of ♀♀ (Brandl and Bezzel 1989). Starts shortly after fledging, and timing thus highly variable, as birds in Alps fledge between late May and early September (Bezzel and Brandl 1988) and in north-east Spain perhaps from April (Borras and Senar 1991). Moult completed in birds examined from October and later (BMNH, ZFMK); moult just started in 2 ♂♂ from Sardinia, 15 August (Bezzel 1957).

**Measurements.** ADULT, FIRST ADULT. Nominate *citrinella*. Wing and bill of birds from (1) northern Spain and Pyrénées, (2) Alps, from eastern France to Switzerland and northern Italy; other measurements combined, whole year; skins (BMNH, RMNH, ZFMK, ZMA: A J van Loon, C S Roselaar). Bill (S) to skull, bill (N) to distal corner of nostril; exposed culmen on average 3·6 less than bill (S).

| | | | | | | |
|---|---|---|---|---|---|---|
| WING (1) | ♂ | 78·2 (1·90; 12) | 75–82 | ♀ | 76·2 (1·36; 8) | 75–79 |
| (2) | | 78·7 (1·65; 18) | 76–83 | | 77·1 (1·28; 6) | 75–79 |
| TAIL | | 54·4 (1·93; 28) | 51–58 | | 53·8 (2·04; 14) | 50–58 |
| BILL (S) (1) | | 11·9 (0·42; 11) | 11·1–12·4 | | 11·8 (0·48; 8) | 11·1–12·5 |
| (2) | | 11·8 (0·43; 17) | 10·9–12·5 | | 11·6 (0·40; 6) | 10·9–12·0 |
| BILL (N) (1) | | 7·3 (0·31; 12) | 6·9–7·8 | | 7·2 (0·32; 8) | 6·9–7·9 |
| (2) | | 7·4 (0·30; 15) | 7·1–8·2 | | 7·3 (0·21; 6) | 7·0–7·6 |
| TARSUS | | 14·4 (0·70; 26) | 13·4–16·0 | | 14·2 (0·67; 14) | 13·1–15·0 |

Sex differences significant for wing.

Wing. Bayern (southern Germany): (1) adult, (2) 1st adult (Brandl and Bezzel 1989). (3) Adult, Jura (north-west Switzerland) (Märki and Biber 1975).

| | | | | | | |
|---|---|---|---|---|---|---|
| (1) | ♂ | 79·0 (1·51; 266) | 72·5–83 | ♀ | 77·1 (1·52; 135) | 73–82 |
| (2) | | 77·6 (1·55; 231) | 74–81·5 | | 75·9 (1·37; 211) | 71–79·5 |
| (3) | | 79·0 (1·85; 77) | 76–84 | | 76·6 (1·91; 61) | 72–80 |

*S. c. corsicana.* Wing and bill of birds from (1) Elba, (2) Corsica, (3) Sardinia; other measurements for Corsica and Sardinia only, combined, all year; skins (BMNH, RMNH, SMTD, ZFMK, ZMA: A J van Loon, C S Roselaar).

| | | | | | | |
|---|---|---|---|---|---|---|
| WING (1) | ♂ | 74·8 (0·99; 7) | 73–77 | ♀ | 72·4 (1·64; 5) | 70–74 |
| (2) | | 73·5 (1·42; 17) | 70–76 | | 70·9 (1·97; 13) | 68–74 |
| (3) | | 72·0 (1·77; 10) | 70–75 | | 70·8 (1·41; 8) | 68–73 |
| TAIL | | 53·1 (1·67; 21) | 50–57 | | 50·7 (1·38; 15) | 48–53 |
| BILL (S) (1) | | 11·7 (0·63; 7) | 10·9–12·8 | | 11·9 (0·37; 5) | 11·5–12·4 |
| (2) | | 11·7 (0·33; 17) | 11·2–12·5 | | 11·7 (0·35; 13) | 11·2–12·2 |
| (3) | | 11·5 (0·32; 9) | 11·1–12·1 | | 10·9 (0·48; 7) | 10·3–11·3 |
| BILL (N) (1) | | 7·4 (0·34; 7) | 7·1–8·1 | | 7·3 (0·11; 5) | 7·2–7·4 |
| (2) | | 7·4 (0·26; 17) | 6·9–7·8 | | 7·3 (0·26; 13) | 6·9–7·7 |
| (3) | | 7·2 (0·23; 9) | 6·9–7·5 | | 6·7 (0·44; 7) | 6·1–7·2 |
| TARSUS | | 14·7 (0·32; 19) | 14·1–15·3 | | 14·7 (0·46; 14) | 13·9–15·7 |

Sex differences significant for wing and tail.

**Weights.** Nominate *citrinella*. Switzerland: 1st half of September 12·4 (28) (Jura) or 12·2 (59) (migrants, Col de Bretolet); early October 12·4 (17) (Jura) or 12·9 (118) (Col de Bretolet) (Märki and Biber 1975, which see for graphs and influence of time of day. For correlation of weight with wing length, see Brandl and Bezzel (1989); for relation with age and trapping method, see Borras and Senar (1986). Eastern Pyrénées, July: ♂♂ 12·5, 14 (ZFMK).

*S. c. corsicana.* Sardinia, August: ♂ 11·5 (Bezzel 1957).

**Structure.** Wing rather short, broad at base, tip bluntly pointed. 10 primaries: p8 longest, p9 0·5–1·5 shorter, p7 0–0·5, p6 1–2·5, p5 7–9, p4 11–15, p3 13–18, p2 15–21, p1 17–24 (19–24 in nominate *citrinella*, 17–21 in *corsicana*). P10 strongly reduced; 51–55 (nominate *citrinella*) or 47–51 (*corsicana*) shorter than p8,

5-10 shorter than longest upper primary covert. Outer web of p6-p8 emarginated, inner web of p7-p9 with faint notch. Longest tertial reaches approximately to tip of p1. Tail forked; t1 6-12 shorter than t6. Bill distinctly longer, more slender, more sharply pointed, and more compressed laterally on tip than in other west Palearctic *Serinus*, closely similar to bill of Siskin *Carduelis spinus*; width and depth at base on average 6·0 in nominate *citrinella*, 5·5 in *corsicana*. Middle toe with claw 15·5 (5) 14-17 in nominate *citrinella*, 14·8 (5) 14-16 in *corsicana*; outer and inner toe with claw both *c.* 67% of middle with claw, hind *c.* 81%. Remainder of structure as in Red-fronted Serin *S. pusillus*.

**Geographical variation.** 2 well-marked races, sometimes considered separate species. Isolated *corsicana* from Sardinia, Corsica, Capraia, and Elba differs from nominate *citrinella* from southern continental Europe in colour, size (mainly wing and tail: see Measurements), habitat (nominate *citrinella* confined to subalpine areas, but see Borras and Senar 1991; *corsicana* also in lowlands), and voice (Chappuis 1976; Martin 1980). United here in single polytypic species, pending further research. Yellow of face of ♂ *corsicana* brighter yellow (less green) than in ♂ nominate *citrinella*; yellow-green of forecrown on average more restricted; mantle and scapulars conspicuously different,

cinnamon-brown with pronounced dark grey streaks (in nominate *citrinella*, almost uniform yellow-green); rump duller yellow-grey (not bright green-yellow); black stripe on lore distinct (not faint and not mottled grey and yellow); chest purer yellow (less green); under tail-coverts mainly yellow-white or white (not mainly yellow). ♀ *corsicana* differs from ♂ in having brown-grey crown and hindneck, streaked darker brown, grey lore, and paler and more restricted yellow on underparts (in nominate *citrinella*, ♀ differs from ♂ mainly in grey instead of yellow chin and throat; hindcrown, nape, mantle, and scapulars of ♀ somewhat browner than in ♂, belly less intensely yellow, but difference not as marked as in *corsicana*). No clear geographical variation within nominate *citrinella*. Within *corsicana*, birds from Elba largest, those of Sardinia smallest (see Measurements), but no clear difference in colour (contra Trettau 1964). Birds recorded nesting Mallorca (Henrici 1927) said to be nominate *citrinella* by Vaurie (1959), *corsicana* by Mauersberger (1960).

Sometimes included in *Carduelis* because of bill shape (e.g. Hartert 1903-10, Witherby *et al.* 1938, Mauersberger 1960), but in many characters a true member of *Serinus* (Heinroth and Heinroth 1924-6; Wolters 1952; Nicolai 1957, 1960; Ackermann 1967). Yellow-crowned Canary *S. canicollis* from eastern and southern Africa is perhaps nearest relative (Nicolai 1957, 1960).                                                                    CSR

## *Carduelis chloris*  Greenfinch

PLATES 35 and 39 (flight)
[between pages 472 and 473]

Du. Groenling        Fr. Verdier d'Europe        Ge. Grünling
Ru. Зеленушка        Sp. Verderón        Sw. Grönfink

*Loxia Chloris* Linnaeus, 1758. Synonym: *Chloris chloris*.

Polytypic. Nominate *chloris* (Linnaeus, 1758), northern Europe from Belgium, Netherlands, and Scandinavia east to Urals, south through Rhein valley to northern slope of Alps, Hungary, and Ukraine and European Russia at *c.* 48°N; perhaps also northern Scotland; *harrisoni* (Clancey, 1940), southern Scotland (at least south of line from Aberdeen to Greenock), England, Wales, and Ireland; *aurantiiventris* (Cabanis, 1851), France (south from Loire valley and *c.* 47°N in east, grading into nominate *chloris* further north), central and eastern Spain, Balearic Islands, northern Tunisia, Sicily, mainland Italy, coastal Yugoslavia, Albania, and western Greece; *vanmarli* (Voous, 1951), north-west Spain, Portugal, and north-west Morocco (north of Atlas); *voousi* Roselaar, 1993, Atlas mountains of Morocco and Algeria; *madaraszi* (Tschusi, 1911), Corsica and Sardinia; *muehlei* Parrot, 1905, eastern Yugoslavia (Serbia and Makedonija, grading into nominate *chloris* in Vojvodina and Slavonia and into *aurantiiventris* in Bosnia-Hercegovina), Rumania (grading into nominate *chloris* in extreme west and probably in north), Moldavia, Bulgaria, Greece, Crete, Cyprus, and western Asia Minor; *chlorotica* (Bonaparte, 1850), Levant and (probably this race) northern Egypt, grading into *muehlei* in south-central Turkey; *bilkevitchi* (Zarudny, 1911), Caucasus, Transcaucasia, northern Iran, and south-west Turkmeniya east to Ashkhabad, grading into nominate *chloris* in northern Caucasus and perhaps Crimea and probably into *muehlei* in northern Turkey. Extralimital: *turkestanicus* (Zarudny, 1907), central Asia from Kugitangtau, Samarkand, and Kzyl-Orda east through Alay, Karatau, and Kirgiz ranges to 75°E in Kirgiziya and to central Tadzhikistan.

**Field characters.** 15 cm; wing-span 24·5-27·5 cm. Close in size to House Sparrow *Passer domesticus*, with strong bill and deep head and body also recalling that species but tail distinctly forked; 25% larger than Siskin *C. spinus* and serins *Serinus*. Medium-sized, robust, plump and noticeably short-tailed finch, with stout conical bill; form most compact of west Palearctic finches except for even

larger Hawfinch *Coccothraustes coccothraustes*. ♂ olive-green and yellow, looking bright only in sunlight; ♀ dull olive-brown and yellowish-buff, faintly streaked on back; juvenile dirty buff-brown and pale buff, fully but not sharply streaked; all show striking yellow patches on primaries and side of tail, shining on ♂, duller on ♀, and dullest on juvenile. Large bill pale flesh. Voice distinctive. Sexes dissimilar; little seasonal variation. Juvenile separable. 9 races in west Palearctic, ♂♂ of some separable on size and on intensity of colour; see Geographical Variation for those not described here.

(1) North European and Russian race, nominate *chloris*. ADULT MALE. Moults: July–November (complete). Brightening of plumage entirely due to wear. Crown, nape, mantle, and back greyish-brown, tinged green and looking olive; lower back and rump yellowish-green, looking pale and bright but not contrasting vividly with rest of upperparts as in *Serinus*; upper tail-coverts grey-green. Sides of head appear uniform with crown and back at any distance but show subtle pattern at close range: (a) dusky lores, emphasizing pale flesh bill, (b) faint yellowish eye-ring and line above eye, not forming obvious supercilium but emphasizing dark brown eye, (c) dusky, almost olive ear-coverts, (d) faint dusky-brown malar stripe, and (e) greyish-yellow cheeks, throat, and surround to ear-coverts. Breast greyish yellow-green merging with brownish-grey flanks and bright greenish-yellow underparts which becomes whiter under tail. Wings strongly patterned, with (a) basically olive ground to coverts but with noticeably mauve-grey ends to inner feathers, (b) dusky secondaries and tertials, with dark grey-brown centres emphasizing broad grey fringes on tertials and subtle contrast between olive-brown bases and grey fringes on distal halves of secondaries, (c) dark olive primary coverts, contrasting with (d) brilliant greenish-yellow basal fringes to primaries which form long blaze on lower edge of folded wing and bold patch when extended in flight and contrast with (e) brown-black primary-ends which are margined grey. Underwing brilliant yellow; loose feathers occasionally show as bright line above primary coverts. Tail brown-black, with 4 outer pairs brilliant yellow and creating bright, 'flashing' basal patches. With wear, upperparts become greener, underparts brighten to clear yellowish-green, and grey areas of wings intensify in tone. Legs pale flesh. ADULT FEMALE. Resembles ♂, but in comparison noticeably duller overall, with indistinct but quite broad streaks from crown to mantle and often mottled appearance on fore-underparts; lacks dusky lores and full yellow tints of ♂, particularly on lower underparts which are no brighter than dull yellowish-white. With wear, tips of inner greater coverts and fringes of secondaries become pale greyish-white. JUVENILE. Resembles ♀ but even duller and quite heavily and much more distinctly streaked above and below; lacks obvious green and yellow tones but shows distinctive warm brown tinge over wing-coverts and on edges of secondaries. Face more patterned, with paler moustachial stripe and chin than ♀ emphasizing mottled ear-coverts and dusky malar stripe. (2) South European and west Mediterranean *aurantiiventris*. Brighter green above and yellower below in both sexes, ♂ with almost golden tone on lower body. (3) Levant race, *chlorotica*. With wear, becomes yellower above and on face than any other race. (4) Caucasus race,

*bilkevitchi*. Noticeably greyer at all seasons than other races.

When size evident, adult unmistakable. All other yellow-green finches of west Palearctic smaller and much less uniformly patterned. When calling or in flight, juvenile not difficult to identify but its more strongly streaked plumage does invite confusion with ♀ and immature rosefinches *Carpodacus*, juvenile crossbills *Loxia*, and ♀ and immature sparrows *Passer* and *Petronia*, but none of these show yellow on primaries and sides of tail. Flight quite powerful, with long wings beaten in bursts which soon produce marked undulations, longer in wave than those of Chaffinch *Fringilla coelebs*; shows little agility and much flapping when feeding in trees, etc. Flight silhouette compact and noticeably short-tailed; bird looks heavy on the wing, unlike other finches except *C. coccothraustes*. Escape-flight usually short, to tree canopy or hedge-top where bird perches prominently. Song-flight remarkably bat-like, with circular progress sustained in spite of seemingly too slow and erratic beats of splayed wings. Gait a hop. Stance usually rather upright on perch but more crouched and level when feeding on ground. Rather undemonstrative except in sexual interactions but dominates smaller finches in food competition.

Song assembled mainly from calls, run together in powerful twittering 'trill' given from perch or in flight; used in association with persistent, loud, nasal territorial wheeze, 'tsweee', 'djeeeesh', or 'dzhwee', strongly recalling similar call of Brambling *F. montifringilla*. Flight-calls include repeated 'chichichichichit' (softer and fuller-toned than Linnet *C. cannabina*) and rather quiet couplets 'chúp chúp', 'cheu cheu', or 'teu teu'. Alarm-call a longer, slightly lilted 'tsooeet', recalling Canary *Serinus canaria*.

**Habitat.** Breeds almost throughout Europe, to south of Arctic Circle or of July isotherm of 14°C in boreal, temperate, Mediterranean, and steppe zones, extending also to North Africa and western Asia (Voous 1960b). Attached to tall densely leafed trees and to diet of seeds accessible under appropriate trees, on bushes, or on crop, weed, and other plants in fields. Has expanded from natural woodland edge, scrub, streambanks, and groups of trees on grassland to tall hedgerows, lines of planted trees, orchards, conifer plantations, cemeteries, churchyards, parks, large gardens, and other situations where tall trees, sunny aspects, and ready access to ground or other sources of seeds, fruits, and insect food are present together in breeding season. At other seasons, may utilize areas away from trees, on farm fields, salt-marshes, shingle banks, and other open sites. (Voous 1960b; Fuller 1982.) Restless nature and liking for flying about above lowest airspace over some distance form complement to pronounced selectivity in breeding and foraging habitat, resulting in somewhat local but flexible distribution pattern. Mainly a lowland species, in Switzerland normally below 900 m, although found more sparsely up to c. 1400 m, especially

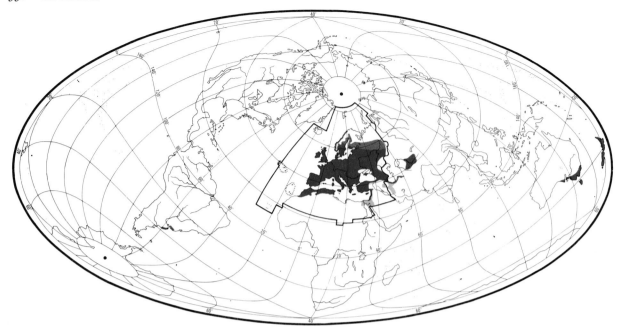

in conifers, where, however, density is much lower than in lowland parks, etc. (Glutz von Blotzheim 1962). In USSR, respects similar altitude limits, preferring to keep to margins of coniferous and hardwood forests, cultivated areas, and floodland scrub, avoiding extensive forests and dense thickets (Dementiev and Gladkov 1954). Prefers thickets and plantings of hemp and sunflower (Flint *et al.* 1984).

Belongs to most loosely territorial Carduelinae, and among these is characterized by links with habitat extremes of tall mature trees and areas of dense seed crops on fields and salt-marsh. Has recently been well suited by new patterns of human land use, including low-density urban developments (Newton 1972).

**Distribution.** Has spread north in Fenno-Scandia, and has expanded range in Israel and Egypt. (Introduced in New Zealand, south-east Australia, and very locally in South America.)

NORWAY. Has spread in south; also northwards, especially in 1980s (VR). SWEDEN. Range expansion since 2nd half of 19th century, which has continued slowly and is not yet complete in interior of northern Sweden (LR). FINLAND. Recent northward spread; probably influenced by intensified winter feeding (OH), also planting of rose *Rosa* bushes and abandonment of arable land (Koskimies 1989). ISRAEL. Has spread in recent decades, following development of agricultural settlements (Shirihai in press). EGYPT. Formerly possible breeder in Sinai. Has recently colonized Nile delta; first recorded breeding 1985 (Goodman and Meininger 1989). MADEIRA. Apparently formerly accidental, probably not breeding. First proved to breed in 1968 (Zino 1969). AZORES. Introduced from Portugal *c.*

1890; now resident breeder São Miguel, Terceira, and Faial (GLG).

Accidental. Faeroes.

**Population.** Has increased in Fenno-Scandia, and in other areas of range expansion (see Distribution).

BRITAIN, IRELAND. Probably 1–2 million pairs (Sharrock 1976). Britain: *c.* 800 000 pairs (Hudson and Marchant 1984). FRANCE. More than 1 million pairs; possibly some recent decline, due to changing agricultural practice (Yeatman 1976). BELGIUM. Estimated 70 000 pairs; apparently stable (Devillers *et al.* 1988). NETHERLANDS. 40 000–80 000 pairs (SOVON 1987). GERMANY. Estimated 2·9 million pairs (Rheinwald 1992). East Germany: 750 000 ± 350 000 pairs (Nicolai 1993). SWEDEN. Estimated 200 000 pairs (Ulfstrand and Högstedt 1976). Marked increase in last 50 years (LR). FINLAND. Marked increase in recent decades (see Distribution); perhaps 100 000–200 000 pairs (Koskimies 1989).

Survival. Finland: average annual mortality 43 ± 8·3% (Haukioja 1969). Czechoslovakia: for age structure and mortality, see Pikula (1989). Oldest ringed bird 12 years 7 months (Rydzewski 1978).

**Movements.** Partially migratory in most of range; some southern populations apparently resident and dispersive. Diurnal migrant. Birds head chiefly south-west to winter almost entirely within breeding range, with concentrations in Mediterranean region. Western populations winter furthest west, and more eastern populations progressively further east.

British birds (*harrisoni*) partially migratory; distribution in winter tends to be more concentrated in lowland and

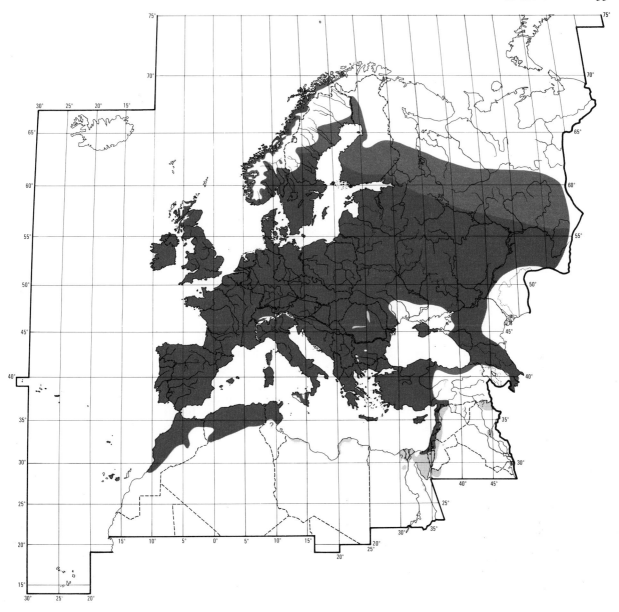

coastal areas than in summer (Lack 1986). Analysis of data for southern and central Britain shows that proportion of recoveries decreases progressively with distance from ringing site: of recoveries over 20 km in 1970-9, *c.* 50% at 20-70 km, *c.* 50% of remainder at 70-120 km, and so on; pattern more typical of dispersal than of true migration. Many more birds from central and eastern England move over 20 km than those from south-west. October–March recoveries chiefly south and west of where ringed, and April–September recoveries north and east. 8 recoveries at or near ringing site of birds which had been controlled at 50-360 km give evidence of return movement. (Boddy and Sellers 1983.) Recoveries at over 25 km of birds ringed in winter in Devon (south-west England) are

chiefly in south-east England, with a few from northern England; many records of individuals retrapped in successive winters (Gush 1980). Ringing data in Oxfordshire (southern England) show that some local birds are sedentary and others winter further south and west; some birds from further north and east winter in Oxfordshire, and others pass through to winter further south and west (Holmes 1982; Brucker *et al.* 1992). Most overseas recoveries are from coastal areas of northern France and Belgium, with decreasing numbers further inland (up to 1100 km in Spain). Small movements also between Britain and Ireland. (Boddy and Sellers 1983; Lack 1986.) British birds have apparently become more mobile in recent years; this is consistent with reduction in rural winter feeding sites following mechanisation. In 1969–75, 63% of recoveries were within 10 km of ringing site and only 6% moved more than 100 km; but in 1977–86, 49% were within 10 km and 11% at over 100 km. (Marchant *et al.* 1990.) Some birds from north-west Europe pass through or winter in Britain. Up to 1986, movements to or from Britain (including local birds) were from France (122), Belgium (45), Netherlands (13), Norway (6), Germany (4), and Spain (1) (Mead and Clark 1987). More recent recoveries include several Norwegian birds, some midwinter (e.g. Mead and Clark 1990, 1991).

Birds heading south-west between Scandinavia, Low Countries, and Iberia migrate on fairly narrow front (as in Linnet *C. cannabina*), *c.* 325 km wide; front extending south-west from Finland and Baltic states is apparently wider, *c.* 500 km (Verheyen 1955*b*; Yeatman-Berthelot 1991). In Sweden, widespread in winter, though some northern areas are vacated (SOF 1990); many birds winter at or near same site in successive years (Rendahl 1958). Foreign recoveries in winter (December–February) of birds ringed Sweden 1972–89 are from Germany, mostly north of 52°N (48), Denmark (23), Belgium (10), France (10), Netherlands (3), Spain (3, with longest-distance recovery at 2269 km), and Poland (2) (*Rep. Swedish Bird-ringing*). Norwegian birds winter north along coast to *c.* 66°N (Haftorn 1971); some reach Britain in winter (see above). In Finland, some birds migrate, but winter population increased 8-fold in 1970–85, due to artificial feeding and milder winters (Koskimies 1989); some evidence of movement in midwinter, but less pronounced than in other cardueline finches (Haila *et al.* 1986). In Belgium, *c.* 24% of local population remains for winter, with some birds moving towards coast (Lippens and Wille 1972); 14 of 15 long-distance recoveries of birds ringed as nestlings were between south and WSW (chiefly south-west) in France (10) and Spain (4), and one was 475 km north-east at Lübeck (Germany) (Verheyen 1955*b*). More common in Belgium winter than summer, and large numbers occur on passage. 317 recoveries in Belgium of birds ringed abroad were chiefly from Germany (34%), Netherlands (30%), and France (19%), with others from Denmark (8%), Britain (3%), Sweden (3%), Spain (2%), USSR up to 2375 km (1%), and 1 from Finland. Recoveries further south-west of birds ringed on passage through Belgium are chiefly from France, also from Iberia, with 2 reported from Morocco, up to 1915 km (Lippens and Wille 1972). Data from Rheinland (western Germany) and Netherlands show similar pattern, with recovery area of birds ringed in breeding season or on passage extending south-west through France, some individuals reaching southern Iberia (Mildenberger 1984; Speek and Speek 1984). In France, local birds mainly sedentary and dispersive; those migrating are mostly from eastern areas. Many immigrants winter; passage or wintering birds from Scandinavia and Low Countries occur chiefly in western and central areas, birds from Finland, Baltic states, Germany, and Swit-

zerland in east and south-east; single individuals from Poland and Czechoslovakia also recorded. (Yeatman-Berthelot 1991.) Recoveries in Camargue (southern France) are mostly of Swiss-ringed birds (passage or breeding); birds ringed on passage or breeding in Germany and Belgium also occur, and birds ringed on passage in north-west Italy, Yugoslavia, and Czechoslovakia (Blondel 1969). 2 birds ringed Sweden in winter were recovered in Massif-Central (southern France) in subsequent winter, showing that individuals winter at widely differing distances from breeding areas in different years (see also Spain, below) (Yeatman-Berthelot 1991). In Switzerland, one of most numerous Fringillidae in winter, when resident birds joined by immigrants chiefly from Germany and Czechoslovakia; Swiss migrants winter in southern France, Spain, and northern Italy (Schifferli *et al.* 1982; Winkler 1984). Also numerous on passage and as winter visitor to southern Germany (Wüst 1986). Long-distance recoveries further south of birds ringed Germany, 1947–67, are west to south-west Spain and east to Yugoslavia (Zink 1969). Birds ringed Kaliningrad coast (western Russia) recorded south-west to Yugoslavia and northern Italy (with one WSW to Belgium), and north-east to Finland and Latvia (Paevski 1971). Birds from Hungary, Czechoslovakia, and USSR occur in Bulgaria (Dontschev 1986), and birds from central and eastern Europe (Germany and Switzerland east to Poland and Czechoslovakia) in Italy (Schifferli *et al.* 1982; Bendini and Spina 1990). Local birds make hard-weather movements in northern Italy (Mingozzi *et al.* 1988). In Malta, common in autumn, with smaller numbers in winter; recoveries show movements chiefly to or from Italy, Yugoslavia, and Czechoslovakia (Sultana and Gauci 1982, 1985*b*, 1988).

Many immigrants winter in Spain; widespread recoveries (with highest numbers in north, north-east, and extreme south) are from Belgium, 77 (including 29 ringed in breeding season), France 63 (33), western Germany 11 (7), Netherlands 6 (2), Sweden 5 (1), Switzerland 5 (1), eastern Germany 3 (1), Britain 1 (1), Denmark 1, Norway 1. 9 birds ringed in winter further north in Europe were recovered in later years in Spain. Similar numbers of ♂♂ and ♀♀ occur, and no evidence that ♂♂ occupy different areas from ♀♀. Recent studies show that Spanish birds (Mediterranean race *aurantiiventris*) are also partially migratory; 7 recoveries over 50 km further south of birds ringed breeding season; 2 were on southern coast in October, perhaps bound for Africa. Wing formulae suggest that most birds on passage through southern Spain to Africa are of Iberian origin rather than from further north; this supported by timing of migration (see below). (Asensio 1986*b*.)

In north-west Africa, where many immigrants winter, no evidence that local birds move south of breeding area (Heim de Balsac and Mayaud 1962; Pineau and Giraud-Audine 1979; Destre 1984); in Tunisia, coastal movements probably involve local birds; no observation of passage at Cap Bon (Thomsen and Jacobsen 1979). In western Mauritania, a few recent records of birds of unknown origin (Meininger *et al.* 1990). Occurs in Libya only in winter (Bundy 1976). In Egypt (where breeding birds probably Levant race *chlorotica*), winter visitors include *aurantiiventris* (Goodman and Meininger 1989). Common winter visitor and passage migrant in Cyprus (also recently colonized) (Flint and Stewart 1992). In Israel, local birds (*chlorotica*) chiefly resident and dispersive; many *chlorotica* from further north pass through and winter; *muehlei* winters in varying numbers, and a few nominate *chloris* reach northern areas. Recovery November of bird ringed Ryazan' (54°37′N 39°43′E, Russia) shows that some wintering birds are from distant breeding grounds. Most migration is in western Israel, especially coastal regions. (Shirihai in press.) Uncommon winter visitor to Tabuk in north-west Saudi Arabia (Stagg and Walker 1982); not recorded elsewhere in Arabia (F E Warr). Present throughout year in Crimea (southern Ukraine), though some seasonal movements reported (Kostin 1983). Mostly migratory in western Ukraine (Strautman 1963); few winter in Moscow region, but fairly common there on passage (Ptushenko and Inozemtsev 1968). Further east, some birds winter in lower Ural valley, south of breeding area (Levin and Gubin 1985). *C. c. turkestanicus* common in winter immediately south of breeding range in northern Iran, and fairly common in west (D A Scott); sometimes reported from Iraq (Marchant 1962; Marchant and Macnab 1962). Isolated population of *turkestanicus* in central Asia makes short-distance and altitudinal movements; regular on passage through Chokpak pass (Ivanov 1969; Korelov *et al.* 1974).

Autumn migration begins late; main movement October–November in most of range. In Britain, usually begins October or November, and may continue to January or even February (Lack 1986). At Ottenby (south-east Sweden), passage mid-September to November, peaking mid- to late October (Enquist and Pettersson 1986), and in Belgium (September-)October–November (Verheyen 1955*b*; Lippens and Wille 1972). Peaks mid-October in Switzerland (Winkler 1984), and recorded from early October in Malta (Sultana and Gauci 1982). Arrives in Spain from October, peaking 1st third of November; passage through Strait of Gibraltar, however (probably Iberian birds, see above), begins earlier, from late September, with maximum mid-October and decrease in November (Asensio 1986*b*). Arrives from late October or early November in Cyprus (Flint and Stewart 1992). In Israel, main passage in 2 waves, 2nd half of October (mainly *chlorotica*) and November (Shirihai in press).

Spring migration chiefly March–April. In Strait of Gibraltar, recorded early February to early May, mostly 1st half of March (Finlayson 1992). Recorded in Libya to March (Bundy 1976), in Malta to mid-April (Sultana and Gauci 1982), and passage and departure from Cyprus March–April (Flint and Stewart 1992). In Israel, main passage mid-February to end March (Shirihai in press). Passage through central Europe March–April (e.g. Verheyen 1955*b*, Mildenberger 1984, Klafs and Stübs 1987), and in southern Sweden March to early May (Roos 1984). Northward movement in Britain also March–April (Boddy and Sellers 1983).                                                      DFV

**Food.** Fairly large (often hard) seeds, mainly of Cruciferae, Polygonaceae, Rosaceae, Compositae, and cereals, also of many trees and shrubs; a few invertebrates taken in breeding season and also fed to young. Eats a wider range of seeds than probably any other Carduelinae in west Palearctic; in Oxford (southern England), takes seeds and other parts of all 30 commonest plants in agricultural landscape. (Eber 1956; Newton 1967*a*; Blümel 1983*a*.) Readily takes to introduced shrubs in gardens (Snow and Snow 1988). In Schleswig-Holstein (northern Germany), 44% of 2203 foraging observations were in open agricultural areas, 23% on waste ground, 16% in hedges, 14% in parks and gardens, and 3% farmyards (Eber 1956). In Oxford, 52% of food obtained from vegetation, 48% from ground; in eastern England, 44% of 191 feeding observations on ground, 22% at less than 0·45 m above ground, 9% 0·45–1 m, 13% 1–1·7 m, and 11% at more than 1·7 m (Kear 1962; Newton 1967*a*). See also Beven (1964). Perches in shrubs and trees to feed but more rarely on herbs, though will stand on stem to bend seed-head or flower onto ground; usually picks seeds from ground, jumps up to pull head down, or sometimes bites through stem so head falls; only occasionally clamps food under foot (Newton 1967*a*; Blümel 1983*a*). Will feed upside-down at feeder or slide down string (Kear 1962, which see for many details of feeding position, morphology, etc., and comparison with other seed-eaters); reported hovering to pluck fruit of yew *Taxus* from tree (Snow and Snow 1988). Generally eats only seeds of fruits and not flesh; feeding on yew, pulls whole fruit from tree and mandibulates it to extract seed, similarly with drupelets of bramble, but fruits of whitebeam *Sorbus*, *Cotoneaster*, and rose hips *Rosa* left on plant and seeds removed one at a time (Beven 1964; Kalden 1983; Snow and Snow 1988). Takes seeds ejected by thrushes *Turdus* in faeces or pellets (Lack 1986). Probably the only west Palearctic Carduelinae to crack open seed pods and capsules rather than waiting until they are open or stripping one side off; flicks loose soil aside with bill to expose seeds (Newton 1967*a*). For many details of handling and de-husking procedure in relation to bill shape, and comparison with other seed-eaters, see Eber (1956), Kear (1962), Ziswiler (1965), and Newton (1967*a*, 1972). Over past few decades, rapid spread through Britain and Ireland of habit of eating unripe fruit of spurge laurel *Daphne mezereum*; bites through still-soft stone to reach kernel, and bush will be stripped of berries in a few days (Pettersson 1959; Snow and Snow 1988). Habit now well-documented in Germany

(Kalden 1983). Also very fond of rose hips, and observed feeding for 30 min at a stretch from one perch (Beven 1964; Johannsen 1974; Snow and Snow 1988). Often feeds gregariously, e.g. in 68% of 386 records in trees and shrubs in southern England, and in Oxford hundreds accumulated over 4 days in July at area with many charlock *Sinapis* plants, remaining for 10 days; many adults present must have flown some km to this source, though, since seeds are collected in gullet for young, long foraging flights are normal (Eber 1956; Newton 1967a; Snow and Snow 1988). In Czechoslovakia, *c.* 90% of damage in sunflower *Helianthus* plantations said to be caused by *C. chloris* (Havlín 1988). Small hard nuts of hornbeam *Carpinus* very readily eaten, though hardly by other birds; in wooded hills, western Germany, 90% of 888 feeding observations, October–March, were in or under hornbeam (Radermacher 1977, 1983). Captive birds greatly preferred seeds of hemp *Cannabis* and sunflower, followed by *Guizotia* (Compositae) and thistle *Carduus*; took average 2·8–5·9 s to de-husk seed depending on species (Kear 1962; Ziswiler 1965); dealt with average 3·7 seeds per min when feeding on rose hips (Snow and Snow 1988); preferred sunflower seeds to nuts at bird-table (Sawle 1988). See Hake and Ekman (1988) for experiments in search efficiency and flock size.

Diet in west Palearctic includes the following. Invertebrates: bugs (Hemiptera: Psyllidae, Aphidoidea), Lepidoptera (Tortricidae, Yponomeutidae), flies (Diptera: Muscidae), Hymenoptera (ants Formicidae), beetles (Coleoptera: Nitidulidae, Coccinellidae, Curculionidae), spiders (Araneae). Plants: seeds, flowers, buds, leaves of yew *Taxus*, cypress *Chamaecyparis*, juniper *Juniperus*, *Thuja*, redwood *Sequoia*, larch *Larix*, spruce *Picea*, pine *Pinus*, willow *Salix*, birch *Betula*, alder *Alnus*, hornbeam *Carpinus*, hop hornbeam *Ostrya*, beech *Fagus*, elm *Ulmus*, *Zelkova*, *Robinia*, mulberry *Morus*, tulip-tree *Liriodendron*, tree of heaven *Ailanthus*, sycamore, etc. *Acer*, plane *Platanus*, lime *Tilia*, ash *Fraxinus*, dogwood *Cornus*, barberry *Berberis*, lilac *Syringa*, mistletoe *Viscum*, grape *Vitis*, hemp *Cannabis*, hop *Humulus*, knotgrass, etc. *Polygonum*, sorrel, etc. *Rumex*, buckwheat *Fagopyrum*, goosefoot, etc. *Chenopodium*, beet *Beta*, orache *Atriplex*, glasswort *Salicornia*, seablite *Corispermum*, amaranth *Amaranthus*, chickweed, etc. *Stellaria*, mouse-ear *Cerastium*, corn spurrey *Spergula*, pink *Dianthus*, buttercup, etc. *Ranunculus*, wild cabbage, etc. *Brassica*, garlic mustard *Alliaria*, charlock *Sinapis*, radish *Raphanus*, rocket, etc. *Sisymbrium*, sea rocket *Cakile*, sea kale *Crambe*, common whitlow-grass *Erophila*, pennycress *Thlaspi*, shepherd's purse *Capsella*, rose *Rosa*, bramble, etc. *Rubus*, strawberry *Fragaria*, burnet *Sanguisorba*, agrimony *Agrimonia*, apple *Malus*, pear *Pyrus*, cherry, etc. *Prunus*, whitebeam, etc. *Sorbus*, hawthorn *Crataegus*, *Cotoneaster*, *Amelanchier*, firethorn *Pyracantha*, sainfoin *Onobrychis*, vetch *Vicia*, clover, etc. *Trifolium*, broom *Cytisus*, flax *Linum*, cranesbill *Geranium*, dog's mercury *Mercurialis*, spurge *Euphorbia*, sea buck-thorn *Hippophae*, spurge laurel *Daphne*, violet, etc. *Viola*, primrose, etc. *Primula*, evening primrose *Oenothera*, thrift *Armeria*, bedstraw *Galium*, comfrey *Symphytum*, houndstongue *Cynoglossum*, borage *Borago*, viper's bugloss *Echium*, hemp-nettle *Galeopsis*, self-heal *Prunella*, plantain *Plantago*, honeysuckle *Lonicera*, guelder rose *Viburnum*, snowberry *Symphoricarpos*, field scabious *Knautia*, devilsbit scabious *Succisa*, golden-rod *Solidago*, mugwort, etc. *Artemisia*, sunflower *Helianthus*, bur marigold *Bidens*, tansy, etc. *Tanacetum*, corn marigold *Chrysanthemum*, coltsfoot *Tussilago*, groundsel, etc. *Senecio*, burdock *Arctium*, thistles *Carduus*, *Cirsium*, knapweed, etc. *Centaurea*, goatsbeard *Tragopogon*, viper's grass *Scorzonera*, chicory *Cichorium*, sow-thistle *Sonchus*, lettuce *Lactuca*, dandelion *Taraxacum*, hawksbeard *Crepis*, catsear *Hypochoeris*, hawkweed *Hieracium*, *Crocus*, rush *Juncus*, sedge *Carex*, reed *Phragmites* and various other grasses and cereals (Gramineae). (Collinge 1924–7; Schuster 1930; Witherby *et al.* 1938; Kovačević and Danon 1952, 1959; Eber 1956; Mal'chevski 1959; Berndt 1960; Turček 1961; Glutz von Blotzheim 1962; Newton 1967a; Ptushenko and Inozemtsev 1968; Kiss *et al.* 1978; Blümel 1983a; Mal'chevski and Pukinski 1983; Sabel 1983; Havlín 1988; Snow and Snow 1988.)

In Oxford, *c.* 23% of *c.* 16 000 feeding observations were on burdock, 13% Cruciferae (especially charlock), 11% cereals, 7% groundsel, 5% *Polygonum*, 4% dandelion, 4% elm, 4% cultivated *Brassica*, 3% bramble, 3% dog's mercury, 3% chickweed, 2% rose hips, 2% thistles (Newton 1967a). In England, of 1294 items from 42 stomachs, 62% by number seeds of chickweed, 18% wheat *Triticum*, 9% *Rumex*, 7% mouse-ear, 3% Cruciferae (Collinge 1924–7). In gardens, etc., in southern England, of 386 records of feeding on native and exotic fruits, 55% were on yew, 23% rose hips, 14% whitebeam, 3% *Cotoneaster*, and 2% spurge laurel (Snow and Snow 1988). In agricultural area of Schleswig-Holstein, *c.* 24% of 1886 feeding observations were on beet, 10% cereals, 10% rose hips, 8% mugwort, 5% snowberry, 5% cultivated *Brassica*, 5% burdock, 4% elm, 4% bramble, 3% shepherd's purse, 3% charlock, 3% sorrel, and 3% dandelion (Eber 1956). Weight of seeds taken ranges from 0·1 mg (*Artemisia*) to 230 mg (*Fagus*), but large ones clearly preferred, overlapping in size to large extent with those taken by Hawfinch *Coccothraustes coccothraustes*, and illustrated by large proportion of cereal grains in diet compared with wild grasses. In Oxford, 8% of seeds taken were less than 0·5 mg, 9% 0·5–1·0, 27% 1·0–10, 54% 10–100, and 2% greater than 100 mg. (Newton 1967a, 1972.) The 2 major studies in Schleswig-Holstein and Oxford show broadly same changes in diet over year. In spring and early summer, takes half-ripe seeds of elm, chickweed, dog's mercury, dandelion, groundsel, freshly-sown cereals and beet, and in Schleswig-Holstein in June particularly fond of large, hard, ripening seed-heads of Boraginaceae, which are only eaten by adults; in late summer in Oxford, seeds of various Cruciferae and cereals formed *c.* 80% of diet; in autumn, charlock, rose hips, *Rumex*, burdock, and (especially) cereals important; in Oxford in winter, seeds of burdock main food, also bramble and persicaria *Polygonum*, while in Schleswig-Holstein snowberry is favourite where available, though main item is seeds of beet, but burdock and bramble also important; in both areas, cereals will always be taken, either freshly-sown, under standing crop, or in stubble. Depending on local resources, moves to agricultural land, gardens, woods, or coast in cold weather, and there is constant movement of groups between food sources. In one severe winter in Oxford, 97% by volume of contents of 223 gullets were peanuts from garden feeders. Over whole year, invertebrates only *c.* 1% of diet in Oxford, both by gullet examination and observation; during September–March, only invertebrates recorded were beetle larvae picked up with thistle and burdock seeds; aphids and caterpillars taken in breeding season (maximum 6% of 224 observations in May) mostly fed to young. (Eber 1956; Newton 1967a, 1972.) In Britain, has arrived at garden feeders earlier each winter since *c.* 1975; due to shortage of natural food, or perhaps a gradual learning process (Lack 1986). In captivity, average consumption of de-husked kernels 6·5 g per day (Kear 1962).

Diet of nestlings same as adults in breeding season, though probably slightly more invertebrates. Plant and animal material generally brought by parent in gullet though invertebrates sometimes carried in bill. (Eber 1956; Berndt 1960; Doerbeck 1963; Newton 1967a.) In Oxford, 1004 gullets contained 24% by volume seeds of Cruciferae, 21% cereal grains, 16% dandelion, 11% elm, 11% dog's mercury, 7% groundsel, and 6% invertebrates (almost all aphids and caterpillars). Maximum proportion of invertebrates 11% May–June (*n* = 22 broods), 3% July–August (*n* = 12); invertebrates disappeared from nestling diet around days 7–11 and late broods probably fed only plant material. (Newton 1967a.) In England, of 28 stomachs, 68% contained seeds and invertebrates, 32% seeds only; invertebrates included caterpillars, adult Diptera, and spiders (Collinge 1924–7). At one nest, eastern Germany, gullets of young contained 100% chickweed seeds on day 5, chickweed and dandelion on day 6, 50% pine seeds on day 7, 50% pine seeds and 50% young spruce needles on day 9, 100% pine seeds on day 10, and 50% pine seeds and 25% aphids on day 11, when each nestling also received 2–3 4-mm caterpillars. At other nests, young fed Tortricidae and Yponomeutidae caterpillars, aphids, Psyllidae, Curculionidae, and seeds and leaf buds of *Rumex*, *Plantago*, and broom. (Blümel 1983a.) In St Petersburg and Voronezh regions (European Russia), 84% of 128 collar-samples contained only plant material, mostly elm seeds and some of grass *Echinochloa*; animal prey, possibly picked up with seeds, included small beetles Coccinellidae and Curculionidae, and caterpillars (Mal'chevski 1959). See also Mal'chevski and Pukinski (1983). In south Wales,

peak month for visiting peanut feeders was June, perhaps because peanuts are valuable food for young (Cowie and Hinsley 1983). BH

**Social pattern and behaviour.** Most aspects well studied, notably behaviour in captive study (Hinde 1954a). Complex mating system only recently elucidated (Eley 1991). For general reviews see Newton (1972) and Blümel (1983a).

1. Gregarious outside breeding season from early autumn to spring, forming feeding flocks which are largest (up to several thousands in Britain) and most compact in late winter when food is well dispersed and highly localised (Dickinson and Dobinson 1969; Newton 1972). In Morocco, small flocks of not more than 12 birds found December–March (Destre 1984). Flocks in main part of range are commonly associated with other finches (Fringillidae), sparrows (Passeridae) and buntings (Emberizidae) (e.g. Witherby *et al.* 1938, Ptushenko and Inozemtsev 1968, Paz 1987). *C. chloris* also gregarious for roosting (see below) and migration; e.g. in Israel (mainly *chlorotica*) migrates in flocks of up to a few hundreds (Shirihai in press). BONDS. Following account based on 3-year study by Eley (1991) in Sussex (southern England): mating system mainly monogamous but significant degree of polygyny also occurs. Of 769 nests found at any stage where parents were known (i.e. ringed), 24% involved polygamy, nearly always polygyny. 55–62% of ringed ♂♂ and 65–79% of ♀♀ were only ever monogamous; 6–12% of ♂♂ and 10–27% of ♀♀ were only ever polygamous. Most polygynous ♂♂ were bigamous (74%, *n* = 34) but 21% nested simultaneously with 3 ♀♀, and there were single cases of adult ♂♂ with respectively 4 and 5 ♀♀ nesting simultaneously. In terms of number of mates taken, adult ♂♂ were not more polygynous than 1-year-olds, but tended to be polygynous for greater part of the season than 1-year-olds. Most monogamous ♂♂ only paired with 1 ♀ throughout season (i.e. for successive broods), though 1 ♂ paired with 3 ♀♀. Mate-swapping did not, therefore, appear to be adaptive for ♂♂, and its occasional occurrence may have been forced by ♀ leaving. In each of 2 years there were also single cases of cooperative polyandry involving different individuals: in 1988 both ♂♂ Courtship-fed the ♀ (see Heterosexual Behaviour, below) and fed the young; similarly in 1990 both ♂♂ copulated with the ♀ and fed the young. In each of 2 years there were also single, though unconfirmed, cases of polygynandry, each involving 2 ♂♂ and 2 ♀♀; no individual was involved in both years; these cases possibly arose simply from an unpaired ♂ courting both ♀♀ of a bigamous ♂ since ♂♂ will sometimes feed fledged young dependent on the ♀ they are courting; in 12 such cases where ♂ was unpaired, 4 ♂♂ later bred with ♀ whose young they had fed. Only a minority of non-breeders were seen feeding fledged young in this way, and most such supernumerary birds were ♀♀. Apparent non-breeders were mostly ♂♂, making up 28–34% (over 3 years) compared with 2–10% for ♀♀. In addition to variable mating system, promiscuity was common. (Eley 1991: see Heterosexual Behaviour, below.) Only ♀ builds nest and incubates; ♂ feeds incubating and brooding ♀ (see also Relations within Family Group, below). Young fed mostly by ♂ for 1st week after hatching, by both parents equally thereafter. (Ptushenko and Inozemtsev 1968; Levin and Gubin 1985; Gnielka 1986.) No information on parental roles of polygamous ♂♂. Mal'chevski and Pukinski (1983) report cases of overlapping broods in which ♂ simultaneously feeds 1st brood (fledglings) and ♀ incubating 2nd clutch; in one case ♂ fed young that fledged in July (no date) right up to hatching of 2nd brood (5 August). Age at independence less than *c.* 14 days, probably longer for last brood (Blümel 1983a). Age of first breeding 1

year, though in Sussex study 1-year-old ♂♂ were less likely to have a known nest than older ♂♂; 1-year-olds tended to pair with 1-year-olds, and older birds with older birds (Eley 1991). Hybrids reported with various *Carduelis*, mostly in captivity, but with Goldfinch *C. carduelis* in the wild; also in captivity with Chaffinch *Fringilla coelebs*, Bullfinch *Pyrrhula pyrrhula*, and Canary *Serinus canaria* (Gray 1958). BREEDING DISPERSION. Solitary or in neighbourhood groups of 4–6 pairs (Newton 1972). No territory as such, with defence confined to immediate nest-area; owner may travel considerable distance from nest: some marked ♂♂ collected food for young and mated promiscuously (see below) more than 3 km from nest (D G C Harper). In Israel (where 'dozens' of nests occur in quite small groves of orange *Citrus*), minimum distance between nests *c.* 10 m (Shirihai in press); in Ural valley (Kazakhstan), 2 pairs nested amicably 1·5 m apart (Levin and Gubin 1985). In Westfalen (Germany), *c.* 30 nests in a row of chestnut trees (species unknown), trees *c.* 10 m apart; there were nests in every tree, or every other tree (Kobus 1967). Bigamous ♂♂ can have widely separated nests (C C Eley). Dispersion in neighbourhood groups results in evaluation of density being highly dependent on size of area sampled, very small areas often producing very high densities and vice versa (see Table 6 in Blümel 1983*a*). In Switzerland, locally more than 30 pairs in 10 ha of gardens and parks (Glutz von Blotzheim 1962); highest recorded 23 pairs in 6·3 ha (365 pairs per km²) in cemetery in Lausanne; in woodland habitat, density at most 10–20 pairs per km² (Schifferli *et al.* 1982). In birch *Betula*-oak *Quercus* wood in north-east Germany maximum density, 15 'territories' per km², occurred in open more varied patch of juniper *Juniperus* and pine *Pinus* (Sellin 1988). In Mecklenburg (eastern Germany), highest in cemeteries: average 119 pairs per km² (60–212) (Klafs and Stübs 1987, which see for densities in other habitats). In Halle (eastern Germany) where average 3 pairs per km², highest density again in cemeteries, e.g. 229 pairs per km²; other densities (pairs per km²) in descending order as follows: 50 (40–90) in gardens, 10–40 in isolated copses in river valleys, 2·5 in heathlands, 1·4 in pine forest (Gnielka 1986, which see for further examples). Other densities include: in Britain in 1972, average 6 pairs per km² in farmland, 9·2 in woodland; seldom more than 20 pairs per km², maximum 34 (Sharrock 1976); in Poland, less than 1 pair per km² in farmland (Górski 1988), up to 6 pairs per km² in ash *Fraxinus*-alder *Alnus* woodland (Tomiałojć *et al.* 1984); in deciduous woodland in Finland and Öland (Sweden), respectively 3 and 11 pairs per km² (Palmgren 1930: Fritz 1989); for density in coastal meadows, Öland, see Ålind (1991); in Tuscany (Italy), 11 'territories' per km² in mixed coniferous-deciduous woodland (Lambertini 1981); in Morocco, density varies from 5 pairs per km² in maquis and sub-humid woodland to 15 in semi-arid *Thuja* scrub (Thévenot 1982). As in other Carduelinae, size and location of neighbourhood groups may change over time, e.g. same pair caught twice in same season at nests 1 km apart (Newton 1972), but breeding site may also be traditional, e.g. in Lucerne (Switzerland) 1 (of *c.* 15) nests in street was used and refurbished for 5 successive years; some nests were apparently dismantled and re-built nearby (Schwab 1969). Gnielka (1986) reports 3 cases of robbed nests being partially dismantled. Ludwig (1984) describes exceptional case of pair nesting 1·3m from nest of Dunnock *Prunella modularis* and helping to incubate and rear *P. modularis* young as well as their own; ♂ *C. chloris* even fed incubating *P. modularis*. ROOSTING. Communal outside breeding season in hedges, plantations, and shrubberies, with preference for evergreens like box *Buxus*, laurel *Laurus*, and *Rhododendron* (Witherby *et al.* 1938; Dickinson and Dobinson 1969), also dense ivy *Hedera* on building (Kiis 1985, 1986). In Germany, frequently in conifers (especially

denser younger trees, but also in mature forest), quite often in small stands of deciduous trees and reedbeds (Stiefel 1976). Report of 20–30 birds roosting in reedbed in Dorset (England) in November (Cawkell 1951); for use of reedbeds elsewhere, see Stahlbaum (1957) and Dorsch (1970). Main roost-sites (of whatever sort) are used night after night and, if not disturbed, year after year, though other sites are used only occasionally (see below). Roost may contain hundreds or thousands of birds, in southern England usually 50–300, but regularly up to 2000 (Newton 1972). In general, roosts become fewer and larger as winter progresses. Roost in Århus (Denmark) largest on cold windy nights (Kiis 1985). Size also varies in short term because, as proven by ringing, some birds use network of roosts in area and choose different sites (perhaps nearest to particular food supply) on different nights; thus roost at Bagley (Oxfordshire, England) usually held less than 300 birds but on some nights up to 1200; most recoveries within radius of 8 km; at another roost in the area, never more than 50 birds on any night, but 700 ringed there in less than 3 years, indicating substantial turnover. Age ratios at Bagley varied markedly from year to year; in one year only 15% adults, which indicated influx of 1st-year birds from outside area. Most birds join communal roost in late summer after finishing moult; spring communal roosting overlaps with start of breeding, with over 50 birds recorded at Bagley roost up to end of April each year. (Dickinson and Dobinson 1969.) Activity rhythm studied in detail at roost of 4–462 birds (mean 163) in ivy at Århus, late November to mid-March: flocks started arriving *c.* 1 hr before sunset; the first birds entered roost on average 7 min before sunset, the last 12 min after. At low light intensities birds settled earlier and departed later (in morning) than expected, and vice versa at high light intensities. Arrival was later relative to sunset on shorter winter days. The more birds roosting, the earlier was morning departure. Presence of patrolling Sparrowhawks *Accipiter nisus* also influenced departure times. (Kiis 1985, 1986.) See Flock Behaviour (below) for behaviour associated with arrival and departure. Communal roosts often include sparrows or Yellowhammers *Emberiza citrinella* (Witherby *et al.* 1938). Reedbed roost in Neuruppin (eastern Germany), November, contained 50–100 *C. chloris* and a small number of Starlings *Sturnus vulgaris* (Stahlbaum 1957).

2. Relatively approachable in urban areas, less so in woodland (etc.). For escape reactions by flocks threatened by *A. nisus* see below. FLOCK BEHAVIOUR. Size and dispersion of winter flocks vary with food supply. After a small group has found a good feeding site, flock may increase rapidly within hours or days until food exhausted. Access of flock-members to food is influenced by dominance: thus at feeding tray a few (dominant) birds get most food while others wait for a time in trees nearby and, if not soon successful in gaining access, move away. (Newton 1972.) ♂♂ usually dominate ♀♀ in flock (including at roost: Kiis and Møller 1986). Experimental study of costs and benefits demonstrated that foraging in flock increased speed of locating food (and therefore reduced risk of starvation) but also reduced intake per individual (Hake and Ekman 1988). Roosting behaviour well studied. Allowing for local variation, assembly pattern as follows. Birds arrive singly or in small flocks and form pre-roost assembly, typically in tree-tops near roost-site. Assembly accompanied by much calling, switching of perches, and periodic mass wheeling upflights over roost. Entry into roost is sudden, silent, and typically *en masse* (sometimes in groups). (Stahlbaum 1957; Dickinson and Dobinson 1969; Newton 1972; Stiefel 1976.) At Århus roost, when assembling birds were disturbed by *A. nisus*, they rose high into the air and coalesced into one dense flock; pre-roosting activities lasted *c.* 1 hr, during which birds perched in trees progressively nearer roost, and circling flocks passed lower

and more frequently over roost. In morning, birds dispersed in flocks either directly from roost or (less commonly, and more often late leavers) from nearby post-roost assembly (same trees as for pre-roost assembly). (Kiis 1985, 1986; see first of these for roost as possible information-centre for food-finding.) SONG-DISPLAY. Given by ♂ from song-post in exposed tree-top or in Song-flight (typically launched from song-post), sometimes also in normal flight. When given from perch, body rather upright, head raised, wings slightly lowered; wings and tail vibrated in time with 'chi' segment of song but not with 'tsweee' (Hinde 1954a: see 1a in Voice). In Song-flight, wing-beats deep and slow ('bat flight'), body rolling from side to side, path weaving erratically over and among tree-tops above breeding area, bird singing and calling the while (Newton 1972; Blümel 1983a); course of one Song-flight estimated to be 50m long (Blümel 1983a). For aberrant vertical Song-flight see Jepson (1987). ♂'s song is often associated with Sleeked Wings-raised display (see Heterosexual Behaviour, below) and is often matched against other ♂♂ (Song-duels). ♀ also sings (see 1b in Voice) but less often and less intensely than ♂. (Hinde 1954a.) For report of ♀ singing at nest see Gerber (1963). In spring, ♂♂ leave flock for part of day to sing in breeding area (Lebeurier and Rapine 1937). In Oberlausitz (eastern Germany) song typically starts in January and ends mid-August (Blümel 1983b); in Halle sometimes starts even at end of December; first Song-flight there 8–18 February over 5 years (in some years with bad weather not till March) (Gnielka 1986). In Germany, period of high-intensity singing and Song-flights continues till beginning of July, declining rapidly towards end of July; song begins after sunrise, i.e. relatively late compared with many other species (Blümel 1983a, which see for references). Song-period in Britain mainly mid-March to late July, but also heard less often mid-January to late August, exceptionally October–November (Witherby *et al.* 1938); for similar song-period in St Petersburg region (western Russia) see Mal'chevski (1959). In Tafilalt (Morocco), where probably breeds, song heard January–May (Destre 1984). Song also heard on spring migration through Helgoland (north-west Germany) (Moritz 1982). ANTAGONISTIC BEHAVIOUR. (1) General. In captive study, ♂♂ appeared to be dominant over ♀♀ in February–March, but dominance reversed some time between then and mid-April so that ♂♂ usually (except in certain contexts) subordinate to their mates. ♀ manifested dominance by Head-forward display (see below) and Supplanting-attacks, and once her dominance was well established ♂ usually avoided her when she approached. (Hinde 1954a.) (2) Threat and fighting. Aggressive display comprises Supplanting-attacks, Head-forward display, and, in appeasement, Fluffed-posture (Hinde 1954a, on which following descriptions based). In Head-forward display bird sleeks plumage and crouches slightly, usually facing rival; often gapes. According to intensity, wings are lifted slightly or raised to level of back, rarely above back (but, if so, wings are waved up and down with upper surface often directed slightly forwards). Head-forward may be combined in varying degrees with Supplanting-attacks. During breeding, but not at other times, rivals may perform Head-up display: performer sleeks plumage, usually holds wings just out from body, stretches neck upwards, and jerks bill up and down again; Head-up often accompanied by Threat-call (see 5 in Voice) and leads sometimes to attack, sometimes to escape-flight. In Fluffed-posture (Fig A), plumage ruffled, especially forebody; subordinate displays thus in winter as appeasement to inhibit attack. If posture fails to appease, performer sleeks plumage and flees. For aggressive elements in sexual display, see Heterosexual Behaviour (below). Little information on aggression towards other species, but ♀ reported driving off Chiffchaff *Phylloscopus collybita* trying to

A

steal material from nest (Gnielka 1986). HETEROSEXUAL BEHAVIOUR. (1) Pair-bonding behaviour. Account based on Hinde (1954a). Useful earlier descriptions of some displays by Bacchus (1941) and Barnes (1941). ♂ courts ♀ by hopping or flying continually after her (sexual chase) performing Sleeked Wings-raised display (Fig B) when perched: sleeks plumage and raises

B

wings to level of back. Tail may be raised or lowered and sometimes slightly spread. Display alternates with ♂ threatening ♀ with Head-forward display (which Sleeked Wings-raised display closely resembles: see above) and gaping, sometimes also Supplanting-attacks. Sometimes while hopping after ♀, ♂ raises wings high and partly spreads them. In less aggressive form of display (Figs C–D), neck is elongated upwards, bill uptilted and

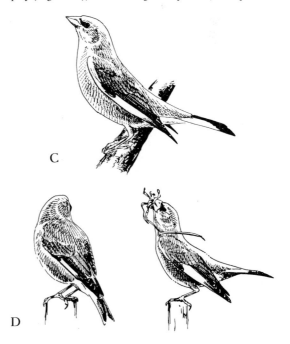

C

D

closed, tail raised a little, and wing-tips slightly lowered, exposing rump. Sleeked Wings-raised display (of whatever intensity) sometimes accompanied by pivoting body about vertical axis (similar to Pivoting-display of *C. carduelis*), usually with 'pee-oo' (see 6a in Voice). ♂ displays to ♀ in bouts (which start suddenly but finish gradually) during chase lasting up to 20 min; ♂ usually silent at start of chase but soon starts giving Contact-calls (see 2 in Voice), wings and tail being vibrated up and down with each call, later on other calls and even full song. 'Tsweee', often associated with song (see 1a in Voice), seems to attract ♀. In display (especially less aggressive form), ♂ often picks up nest-material (Fig D); see also subsection 5, below. Displaying ♂ also regularly adopts intense Fluffed-posture (Fig A), effecting large rounded appearance; like all ♂'s displays, this increasingly common up to period of successful copulation (see Mating, below) and suggests conflict between tendency to flee from and behave sexually towards mate. Hinde (1954a) also recorded ♂ in courtship period performing Butterfly-flight (slow deep wing-beats but this not Song-display) and more rarely Moth-flight (rapid shallow wing-beats). Displays by ♀ include Sleeked Wings-raised display (similar to ♂ but much less common, except at low intensity), Fluffed-posture, and Butterfly-flight. Mutual display includes Bill-flirting (see subsection 3, below). (3) Courtship-feeding. Mates perch side by side and engage bills for ♂ to feed ♀ by regurgitation (Fig E). Usually both lower

E

wings somewhat and ♀ may wing-shiver and give rapid Begging-calls (see 4 in Voice). Most attempts at Courtship-feeding are incomplete, the birds merely inclining bills towards each other (Bill-flirting), sometimes touching, before one flies off or both relax. Courtship-feeding starts in some pairs during nest-building. (Hinde 1954a.) ♂ feeds ♀ on nest during incubation and for a few days after hatching (Ptushenko and Inozemtsev 1968). One report of Courtship-feeding in tree near nest 1 week before young fledged (Hepworth 1946). (4) Mating. Often in dense cover, either in tree canopy ($n = 65$) or on ground ($n = 28$); other locations included peanut-feeder, telegraph wire, and bushes (Eley 1991). For sites of mating in aviary study, see Hinde (1954a, on which rest of account based unless otherwise acknowledged). Either sex may initiate, but more often ♂ who solicits by performing Hovering-display just above ♀, his legs lowered, and giving Soliciting-call (see 3 in Voice); receptive ♀ responds by crouching and ♂ mounts with beating wings; success unlikely if ♀ does not respond thus. ♀ initiates mating by crouching alongside ♂ in mild Fluffed-posture, wings drooped and lightly shivered; raises tail and gives Soliciting-call, attracting ♂ to hover and mount. Copulation rarely lasts more than 2 s (Eley 1991). After successful copulation, ♂ almost always flies suddenly and rapidly from ♀ and gives Sweeet-call (see 6b in Voice). Unsuccessful attempts occurred in bouts of 1–3 or more, which recurred as often as every 15 minutes. ♂ often adopted intense Fluffed-posture between bouts. See Hinde (1954a) for various sequences following solicitation (by either sex) with various outcomes (usually aggression by one or other partner) and for com-

parison with same in *Fringilla coelebs*. Attempted copulation may occur at least as late as day 2nd egg laid. In Sussex study, other conspecific ♂♂ interrupted 17% ($n = 117$) of copulations. 26% of copulations ($n = 69$) between identified partners were extra-pair copulations, and these were more likely to be interrupted than copulation between mates. ♂♂ seen to perform extra-pair copulations up to more than 3 km from own nest. (Eley 1991.) (5) Nest-site selection and behaviour at nest. Based mainly on Hinde (1954a). Some days or even weeks before start of nest-building both pair-members frequently visit potential sites, but without bringing nest material. Usually perch on or near site, or turn round a little on it. After transport of material has begun, and sometimes before, ♀ may perform elaborate mock-building movements on site (e.g. squatting, turning, pressing, weaving); for exceptional case of ♂ doing likewise (crouching and bill-sweeping) see Radford (1970a). Soon both sexes begin to show 'searching' behaviour, hopping on ground as if looking for material but not collecting any. This leads to 'unoriented' transport of material which often continues for weeks before (but stopping with) nest-building; see Fig D for use of such material in display. Most or all building by ♀; role of ♂ varies from some building, to carrying material and not building, to not carrying any material except sometimes for display (Hinde 1954a). According to Levin and Gubin (1985) and Gnielka (1986) all building by ♀, ♂ simply accompanying her as she collects material; one observation on 26 September of ♂ (♀ following behind) carrying feather to potential site (old nest of Song Thrush *Turdus philomelos*) (Gnielka 1986). Warmbier (1973) reported ♀ stealing material from nest of Serin *Serinus serinus* in course of construction. In New Zealand, completed nests reported remaining empty for up to 3 weeks at start of breeding season (Robertson 1985). ♂ feeds incubating ♀ on nest (Ptushenko and Inozemtsev 1968). RELATIONS WITHIN FAMILY GROUP. ♀ seen carrying eggshells from nest (Hepworth 1946) or eating them (Gnielka 1986). ♀ broods young for first 5–6 days during which ♂ does most feeding (mate and young); thereafter parents share feeding (Ptushenko and Inozemtsev 1968; Levin and Gubin 1985). Variations on this pattern include one case where both parents brooded for 1st week, changing over at feeding visits (Spillner 1973, which see for details of regurgitation of food for young). At another nest (1 week before fledging) all feeding by ♀; occasionally she regurgitated food on to nest-rim and fed it piecemeal to young, then often stayed to brood for a few minutes (Hepworth 1946). Both parents participate in nest-sanitation, at least latterly when both are feeding young (e.g. Spillner 1973). According to Gnielka (1986) nest-sanitation till young *c.* 8 days old, or till 2–3 days before fledging (Levin and Gubin 1985) after which young defaecate over side of nest (Hepworth 1946). During period of parental hygiene, faeces are initially eaten, presumably by ♀ (e.g. parent seen swallowing faeces of 5-day-old chick: Gnielka 1986), thereafter carried away (Spillner 1973). Eyes of young open at 4–5 days; just before fledging, young tend to sit around rim of nest or on nearby branches (Mal'chevski 1959). ANTI-PREDATOR RESPONSES OF YOUNG. Young crouch in nest on hearing calls of approaching parent (Spillner 1973). Seriously disturbed young will quit nest 3–4 days before normal fledging age (Mal'chevski 1959; Levin and Gubin 1985). PARENTAL ANTI-PREDATOR STRATEGIES. (1) Passive measures. ♀ mostly a tight sitter during incubation, often able to be touched without flushing (Kobus 1967; Purrmann 1973), sometimes flying off quietly (Levin and Gubin 1985). Only leaves reluctantly, then calls ('mournful twitter') from tree or ground nearby (Pitman 1921). (2) Active measures: against man. Various responses reported though these presumably highlight the more demonstrative pairs. While incubating ♀ sat tight (see above), ♂ very agitated, making repeated dive-attacks on observers; by time of hatching, however, pair had habituated to intrusion and allowed close approach with impunity (Purrmann 1973). ♀ may perform aerial distraction-display, apparently of disablement type: several reports of flying close to ground with slow shallow wing-beats (Mal'chevski 1959: St Petersburg region); ♀ flew from place to place with constant agitated trills (Levin and Gubin 1985: Kazakhstan). Also one report from England of 'injury-feigning from eggs' (no other details) (*Br. Birds* 1952, **45**, 75). No further information.

(Figs by D Nurney: A from photograph in Hinde 1954*a*; B and C from photographs in Hinde 1954*a* and drawings in Newton 1972; D from photograph in Blümel 1983*a*; E from drawing in Newton 1972.)                                                    EKD

**Voice.** Fullest treatment concerns captive birds studied by Hinde (1954*a*), on which following scheme based. For additional sonagrams, see Bergmann and Helb (1982), and papers by Güttinger (details below).

CALLS OF ADULTS. (1) Song. (1a) Song of ♂. Full song consists of groups of pleasant rolling tremolos, of which any particular bird has several variants differing subtly in pitch, timbre etc.; tremolos punctuated by more slowly delivered repetition of tonal and more noisy units, also by single longer and rather nasal 'chewlee' and the familiar lazy buzzing wheeze. Fig I depicts song containing several of these components: starts with tremolo (incomplete in sonagram) of 'yp' units, then 'chewlee', followed by 3 tremolos composed of respectively 'yip', 'lip' and 'yp' units, finally an extended wheeze 'tzeeEEee'. Song in Fig II (same individual as Fig I) more liquid, with greater variety of tremolo segments, and lacks 'chewlee' and wheeze; more liquid timbre derives notably from 2nd and penultimate stepped (2-part) segments each comprising

firstly 'jwee-' then lower-pitched 'joo-' sounds, also 3rd and 4th segments sounding like 'chiririr...'. (W T C Seale.) For other descriptions of full song see Witherby *et al.* (1938), Bergmann and Helb (1982, including variant when singer disturbed) and Blümel (1983*a*). Song not uncommonly much simpler than full song: e.g. song shown in Fig III an alternation of firstly a pure descending whistle 'teeeeooo' and secondly a somewhat nasal (diadic) ascending 'joowee' (W T C Seale). Also, the well-known nasal wheeze (e.g. 'tzeeEEee' in Fig I) is not uncommonly given repeatedly on its own; also described as loud drawn-out 'djeeeesh' recalling song of Brambling *Fringilla montifringilla* (Bruun *et al.* 1986); from any individual, wheeze can be quite variable, e.g. pitch sometimes even, at other times falling (Güttinger 1974); for additional structural analysis see Wilkinson (1975); wheeze shorter on Canary Islands (P J Sellar). Song given from perch or in Song-flight, sometimes also in normal flight (Bergmann and Helb 1982). Most detailed analyses by Güttinger (1974, 1977, which see for geographical variation, locally in Germany, and between Germany, France, Spain, Denmark, Britain, and New Zealand), Güttinger *et al.* (1978, which see for individual variation, also for comparison with song of Canary *Serinus canaria* and captive-bred hybrids with that species), and Güttinger (1978, which see for comparison with closely related *Carduelis*). Summarising from these sources, main features of song as follows: No phrase as such, rather song organized into segments ('tours') of different units, and these may follow one another for minutes without interruptions longer than 0·8 s. Segment length rather constant at 0·6–0·8 s (0·2–

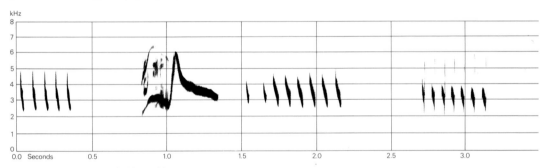

I   W T C Seale   England   May 1986

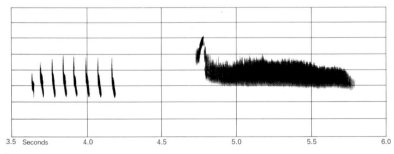

I   *cont.*

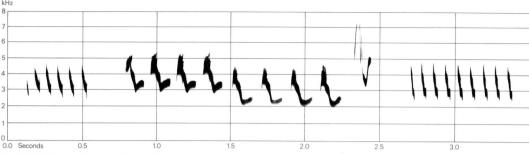

II  W T C Seale  England  May 1986

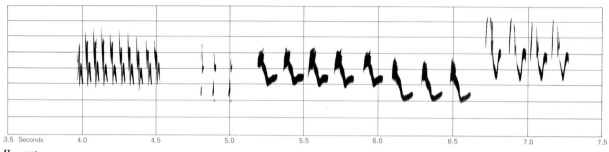

II  *cont.*

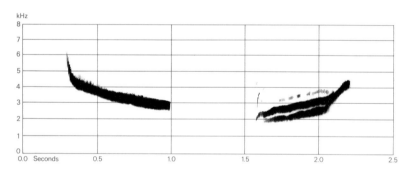

III  P J Sellar  England  July 1971

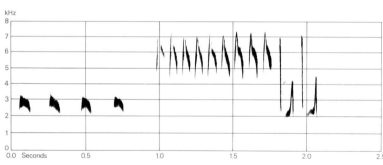

IV  P A D Hollom  England  April 1976

1·2), and sequence of segments not random (see Güttinger 1977 for details). Nasal wheeze relatively common, occurring in association with 10% of segments. Song regularly includes mimicry (or sounds that happen to resemble other species), notably 'excitation' call of Nuthatch *Sitta europaea* (e.g. last 2 complex 'dwip' units in Fig IV: W T C Seale), less commonly calls of Pied Wagtail *Motacilla alba*,

Blue Tit *Parus caeruleus*, Bullfinch *Pyrrhula pyrrhula*, and Blackbird *Turdus merula* (alarm-call); see also Blümel (1983*a*); 'chief-chief-chief-chief' like song of Marsh Tit *Parus palustris* may be variation on more normal 'yip-yip-yip-yip' rather than actually mimicry (L Svensson). (1*b*) Song of ♀. Rare. As in ♂, but less intense (Hinde 1954*a*). Song of one ♀, given for some time without inter-

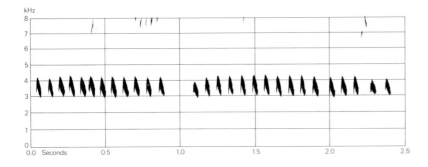

V  V C Lewis  England  June 1965

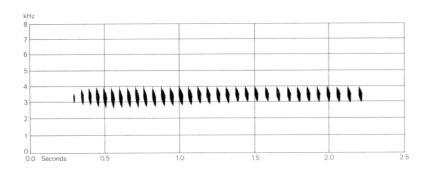

VI  P A D Hollom  England  April 1982

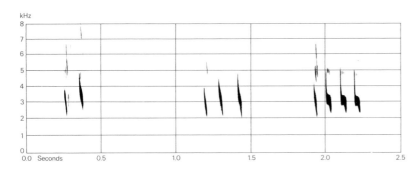

VII  P J Sellar  England  October 1983

ruption whilst waiting to feed nestlings, described as series of rising and falling notes with whimpering quality, interspersed with 'normal' tremolos and a muted wheezing (Gerber 1963). (2) Contact-calls. Commonest call a rapid twittering 'chichichichichit' (Witherby *et al.* 1938), also 'chill ill ill ill' (Hinde 1954*a*), 'ti-ti-ti...' (W T C Seale: Fig V). Slightly liquid tremolo (Fig VI, in which units similar to Fig V but without a marked starting transient: W T C Seale), given by incubating ♀ to ♂, is probably this call but perhaps call 4. Also given as isolated units, thus repeated monosyllabic 'chüp' (recalls Crossbill *Loxia curvirostra* but less hard and less explosive), 'cheu' or 'teu' and variants, commonly in flight but also when perched (Witherby *et al.* 1938). Fig VII shows 'typ' or 'jib' calls (similar to, but higher pitched than, Chaffinch *F. coelebs*) in 2 series of respectively 2 and 3 units, followed by 'jib jub jub jub jub' (W T C Seale). Other renderings ('flight calls') include rapid rolling 'djururUT', or simply short 'djup' more emphatic than *F. coelebs* (Bruun *et al.* 1986);

hard ringing tremolo, e.g. rapid 'gigigi' (Bergmann and Helb 1982). Apart from communication in flight (etc.), also exchanged between mates in closer contact, e.g. heard from ♂ in sexual chases, also when hopping about near nest being built by mate (Hinde 1954*a*). (3) Soliciting-call. Tremolo similar to call 2 but rapid, purer, more soft and gentle, given by either sex to invite copulation (Hinde 1954*a*). (4) Begging-call of ♀. Rapid 'tsit it it' given with wing-shivering, etc., when inviting ♂ to courtship-feed (Hinde 1954*a*). Very like food-call of young (see below), and often heard also when prospecting for nest-site (Güttinger 1978, which see for sonagram). (5) Threat-call. Coarse 'tsk tsk tsk' associated with Head-up display in disputes among breeding birds, leading sometimes to attack or escape-flight; this call, similar to but less harsh than homologous call in Goldfinch *C. carduelis* (see call 5 of that species), is apparently also heard in sexual chases (Hinde 1954*a*). However, 'tsk' rendering does not convey rattling quality, and call is better described as repeated

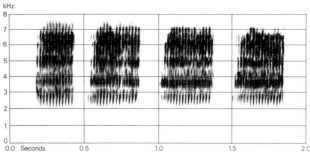

VIII  P A D Hollom  England  June 1979

sharp rattle 'tsrr' (Bergmann and Helb 1982). In recording (Fig VIII) of aggression between 2 ♂♂ in presence of ♀, rapid whirring rattle 'cherrrrr' (W T C Seale) very like juvenile Starling *Sturnus vulgaris* (P J Sellar). (6) Excitement- and alarm-calls. (6a) Alarm-call. Slightly hoarse 'diUwee' or 'dshUee' like *S. canaria* (Bruun *et al.* 1986), 'diu' or 'dschwüid' (Bergmann and Helb 1982). 2 examples illustrated: Fig IX shows one ('chiwheee') of several variants given in bout of calling by anxious ♀;

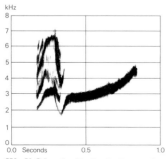

IX  V C Lewis  England  June 1965

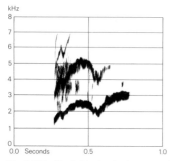

X  A P Radford  England  August 1968

Fig X shows slightly hoarse, subdued, strongly disyllabic 'jouroRI' (W T C Seale) by bird alarming at stuffed Kestrel *Falco tinnunculus*. Similar calls heard in response to Little Owl *Athene noctua*. (W T C Seale.) A 'pee-oo' which usually accompanies part of ♂'s courtship (pivoting in Sleeked Wings-raised display) (Hinde 1954*a*) may be related. In recordings, Alarm-call gives stronger impression of sweet than harsh timbre, and though typically

ascending, sometimes descends in pitch. (6b) Sweeet-call. High-pitched whistling 'sweee(t)' typically given on detecting aerial predator and indicating strong escape tendency, also heard from ♂ flying rapidly from ♀ after copulation (Hinde 1954*a*). (7) Distress-call. Shrill screeching sound when handled (Bergmann and Helb 1982). (8) Other calls. Recording of feeding birds includes descending vibrant 'dzrreeoo' (W T C Seale: Fig XI) not

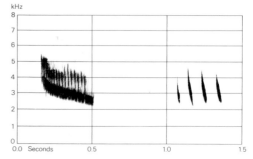

XI  P J Sellar  England  October 1983

easily reconciled with other calls and sonagrams listed above; in same context, also a very quiet low-pitched 'joooee', rather like low-pitched version of 2nd unit in Fig III (W T C Seale).

CALLS OF YOUNG. Food-call of young up to fledging a prolonged sibilant sound, almost a 'trill' (Witherby *et al.* 1938); in recording a shrill whirring squeak (W T C Seale) sounding like 'srrr...' (E K Dunn). Fledged young initially give a rather shrill insistent variant of call 2 (Witherby *et al.* 1938) rendered an abrupt emphatic 'cheeu' (W T C Seale: Fig XII), also described as rounded pleasant 'djil' or 'dju' (Blümel 1983*a*), or 'djüj' (Bergmann and

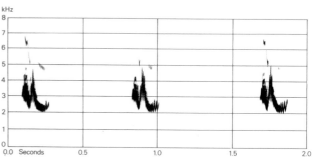

XII  P J Sellar  England  September 1973

Helb 1982). This the most typical call but older fledglings also give call resembling tremolo-variant of adult call 2, e.g. in recording a clear, slightly liquid 'si-si-si-si...' (W T C Seale). Thus, rather harsh muffled timbre (derived from rapid frequency modulation) of food-call disappears as fledgling develops, leading both to adult call 2 and (in ♀) call 4 (Güttinger 1978, which see for sonagrams).

EKD

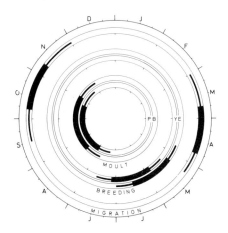

**Breeding.** SEASON. Britain: clutches complete late April to mid-August, with peak mid-May and smaller peak in 2nd half of June (Monk 1954, which see for regional differences). Sweden: see diagram. Southern Finland: eggs laid late April to early July, in town sometimes in March; eggs laid after May are probably 2nd broods (Haartman 1969). Germany: eggs laid mid-March to late July, peak in May; *c.* 10 days earlier in town than in country (Blümel 1983b; Gnielka 1986); in south-west, population in orchards started laying end of April, perhaps because of absence of evergreens (see Site, below) (Glück 1983, which see for comparison with other Fringillidae). In Switzerland, unfledged young recorded into September (Glutz von Blotzheim 1962). Spain: late March to early August, with peaks in May and early July (Gil-Delgado and Catalá 1989). SITE. Against trunk or in strong fork of dense bush, small tree (often in hedge), or creeper; conifers or other evergreens slightly preferred, especially early in season (Monk 1954; Glutz von Blotzheim 1962; Blümel 1983b; Gnielka 1986). In Oberlausitz (eastern Germany), 23% of 229 nests in spruce *Picea*, 10% hornbeam *Carpinus*, 5% *Thuja*, 4% lime *Tilia*, 4% vine *Vitis*; height above ground 0·5-20 m (Blümel 1983b); in another study in eastern Germany, average height 4·5 m (0·9-16·5, *n* = 1047) (Gnielka 1986). In fruit trees, south-west Germany, 48% of 46 nests less than 1 m from trunk on branches of average 3·5 cm diameter (Glück 1983). Many nests recorded inside old nests of other species (Adolph 1943; Lawton 1959; Wagner 1981; Blümel 1983a), and in flower-boxes on balconies, open nest-boxes, or on covered bird-tables (Glutz von Blotzheim 1962; Krüger 1973; Spillner 1973; Blümel 1983a); up to 3 pairs nested over 5 years in station roof 5 m above platform (Nilsson 1971). Nest: stout, robust structure with foundation of dry twigs, grass, moss, and lichen lined with fine grasses, rootlets, plant down, hair, feathers, or man-made material (Harrison 1975; Blümel 1983a); outer diameter 12 cm, inner diameter 5 cm, overall height 7·5 cm, depth of cup 4 cm (Makatsch 1976). 5 dry nests weighed average 20·6 g (15-27) (Blümel 1983a, which see for details of structure and materials).

Building: by ♀ only, accompanied by ♂ (Harrison 1975; Makatsch 1976; Gnielka 1986); rarely, ♂ helps; takes *c.* 8-12 days; material gathered from ground nearby and dead twigs sometimes snapped from trees; old nests sometimes improved and re-used or material used to build new nest; occasionally no re-building at all (Lawton 1959; Schwab 1969; Blümel 1983a); same nest exceptionally used for 2nd brood (Glutz von Blotzheim 1962). EGGS. See Plate 60. Sub-elliptical, smooth and slightly glossy; greyish-white to bluish-white, or beige, sparsely spotted and blotched (rarely scrawled) reddish, purplish, or blackish, concentrated at broad end occasionally forming ring; some with pink or violet undermarkings, others hardly marked (Harrison 1975; Makatsch 1976). Nominate *chloris*: 20·0 × 14·6 mm (18·0-24·1 × 12·2-16·2), *n* = 1124; calculated weight 2·17 g. *C. c. harrisoni*: 20·6 × 14·8 mm (18·6-24·1 × 12·2-16·1), *n* = 100. *C. c. aurantiiventris*: 20·5 × 14·8 mm (18·5-23·5 × 12·5-16·0), *n* = 113; calculated weight 2·31 g. *C. c. chlorotica*: 19·4 × 14·3 mm (16·8-22·4 × 13·1-15·5), *n* = 128; calculated weight 2·10 g. *C. c. turkestanicus*: 20·9 × 15·1 mm (20·1-22·8 × 14·2-16·0), *n* = 103; calculated weight 2·50 g. (Witherby *et al.* 1938; Schönwetter 1984.) Clutch: 4-6 (2-7). In Britain, of 593 clutches: 2 eggs, 0·7%; 3, 2·9%; 4, 20·9%; 5, 63·6%; 6, 12·0%; average 4·83; no suggestion of change in clutch size with latitude, but increases to maximum in early June then declines (Monk 1954); see also Snow and Mayer-Gross (1967) and Eley (1991). In eastern Germany, 3·7 in March, 5·3 May, and 4·2 July; average 4·75 (*n* = 1020) (Gnielka 1986); see also Blümel (1983a). In eastern Spain, 4·5 March-April (*n* = 49), 4·7 May (*n* = 36), 4·1 June (*n* = 19), and 3·9 July (*n* = 15) (Gil-Delgado and Catalá 1989). Average in Finland 5·3 (*n* = 37); later clutches do not seem to be smaller (Haartman 1969); in Switzerland, 5·12 (*n* = 110), maximum 5·22 in July (*n* = 22) (Glutz von Blotzheim 1962). For influence of mating system and ♀'s age on clutch size, see Eley (1991). Eggs laid daily, generally 1-2 days after nest completion (Monk 1954; Makatsch 1976; Blümel 1983a), but up to 3 weeks recorded (Newton 1972). 2 broods apparently usual in Europe, ♂ probably caring for 1st if ♀ starts building before they leave nest (Glutz von Blotzheim 1962; Harrison 1975; Makatsch 1976; Blümel 1983a), though apparently only small proportion of pairs in Finland double-brooded (Haartman 1969), and may not be regularly double-brooded in Britain (Monk 1954). Record of 3 broods successfully reared by ringed pair: 1st brood fledged *c.* 1 June, eggs laid in 2nd nest from *c.* 29 May, young leaving nest 1-4 July, and 3rd clutch laid from *c.* 14 July (Creutz 1962). INCUBATION. Average 12·9 days (11-15, *n* = 121) (Monk 1954); same average found for 53 clutches by Gnielka (1986). By ♀ only, after last egg (Harrison 1975; Makatsch 1976; Blümel 1986a); may also start before last egg according to Glutz von Blotzheim (1962). ♀ rarely leaves nest and only for very short breaks (Blümel 1986a). YOUNG. Fed and cared for by both parents; faeces removed only up to day 8–9 (Spillner 1973; Makatsch 1976; Blümel 1983a; Gnielka 1986). FLEDGING TO MATURITY. Fledging period 14·4 days (14·1–14·7, $n = 160$) (Monk 1954); in German study, 15·8 days (14–18, $n = 31$) (Gnielka 1986). Young of 1st brood fed for less than c. 14 days after leaving nest, those of last brood probably for a good deal longer since observations of dependent young made in late September (Blümel 1983a, which see for development of young). Age of first breeding 1 year (Eley 1991). BREEDING SUCCESS. In Britain, 68·6% of 1589 eggs in 300 clutches hatched and 52·9% produced fledged young, or 2·55 per breeding attempt; 36% of 418 clutches failed completely; most losses due to human predation, others caused by cats, rodents, Corvidae, gulls *Larus*, and weather (Monk 1954). Only 37% of 625 rural nests in southern England produced fledged young (89% of eggs in successful clutches), and 35% of 118 suburban and urban nests (90%), so success unaffected by habitat, though increases as season progresses (Snow and Mayer-Gross 1967). In eastern Germany, 61·5% of 1199 eggs in 257 clutches hatched and 47·3% produced fledged young, or 2·21 per breeding attempt (4·2 per successful clutch, $n = 136$) (Blümel 1983a). Another study showed that of 687 losses, 40% were due to loss or predation at egg stage, 19% desertion of eggs, 36% predation of young, 5% weather, and 1% death of adult; predators probably included Corvidae, squirrel *Sciurus vulgaris* and mouse *Apodemus flavicollis* (Gnielka 1986). In Moscow region (Russia), 38% of nests successful when nest density high, 46% when low ($n = 60$); Carrion Crow *Corvus corone* important predator (Shurupov 1986). In orange plantation, eastern Spain, losses increased through season: in March–April, 51% of 223 eggs produced fledged young (10% of nestlings dying of starvation); in May, 49% of 171 eggs (18%); in June, 26% of 79 eggs (28%); in July, 3% of 58 eggs (47%); this probably caused by cutting of herb layer below trees, reducing seed supply (Gil-Delgado and Catalá 1989). BH

Plumages. (nominate *chloris*). ADULT MALE. In fresh plumage (autumn), forehead and supercilium green, feather-tips narrowly fringed pale grey, merging into green-grey at rear. Feather-tips of crown, nape, mantle, and scapulars ash-grey, centres extensively tinged green, latter tinge shining through grey, often partially visible, especially when plumage slightly worn (from about November); feathers of mantle and scapulars slightly tinged brownish-olive subterminally. Back, rump, and upper tail-coverts bright green, feather-tips fringed ash-grey, faintly on rump, more distinct on tail-coverts. Lore and ring round eye dull grey with green tinge; remaining side of head and neck light ash-grey with some green shining through, like crown and nape, but grey slightly paler; grey almost uniform on surround of ear-coverts and on ill-defined malar stripe, green more predominant at base of bill, just below eye, on centre of ear-coverts, on stripe over lower cheek, and on bar behind ear-coverts, but contrast and extent of grey and green dependent on abrasion. Chin yellowish-green, merging into green on throat, feather-tips narrowly fringed light grey; this in turn merging into darker green of side of breast, chest, and flank, feathers of which show

slightly broader ash-grey tips, largely concealing green on side of breast and upper flank; lower flank and thigh tinged brown-olive. Upper belly and side of belly paler green, faintly tinged grey, merging into bright pale yellow of mid-belly, into white of vent, and into pale yellow on under tail-coverts. Central 2 pairs of tail-feathers (t1–t2) and c. 10–20 mm of tips of others black, fringes (including those along tip) ash-grey, not ill-defined; basal two-thirds of outer web of t2 and basal and middle portion of t3–t6 bright yellow, not sharply defined from black on tip, yellow of each feather broken by black on shaft. Flight-feathers greyish-black, darkest (deep black) on distal ends of primaries, tips of all flight-feathers and outer webs of secondaries with rather broad and contrasting ash-grey fringe, basal borders of outer webs of outer secondaries tinged green, basal and middle portions of inner webs of flight-feathers bordered light grey; basal and middle portions of outer webs of primaries contrastingly bright yellow, yellow reaching to shaft at base of each feather, but not extending onto emarginated parts of p6–p8. Greater upper primary coverts black, outer web broadly fringed green, tinged yellow towards base, merging into ash-grey on tip; bastard wing yellow, inner web of longest feather green-grey, tip of outer web sometimes slightly green, shaft black; marginal primary coverts yellow. Tertials ash-grey, extensively tinged black subterminally. Greater upper wing-coverts ash-grey, basal centres tinged green on outer web, dark grey on inner web; lesser and median coverts bright green. Under wing-coverts and axillaries yellow, longest tipped whitish or pale green-grey. *In worn plumage* (from about April), grey feather-tips of head and body worn off, green fully exposed, often slightly glossy from June onwards; forehead and supercilium yellowish-green merging into pure green of crown to mantle and scapulars; back, rump, and shorter upper tail-coverts often paler and more yellowish-green; side of head and neck green, except for dark grey lore and ring round eye and more olive-green surround of ear-coverts and faint malar stripe; underparts green, merging into green-yellow of chin and into bright yellow of belly (latter sometimes contaminated by dirt, tinged olive or grey), lower flank tinged olive-grey; grey tips of tail-feathers and primaries and grey fringes and tips of tertials partly or fully worn off, black more exposed. ADULT FEMALE. Forehead to mantle and scapulars as well as longer upper tail-coverts brown-grey, each feather with ill-defined darker grey-brown or olive-brown streak, sides of feathers tinged green towards base; cap and nape less uniform ash-grey than in adult ♂, mantle and scapulars browner, more streaky, less washed olive or green; rump and shorter upper tail-coverts green, paler and slightly yellowish on shorter coverts, as in adult ♂. Side of head and neck brown-grey, lore dark grey, brown-grey darker on ill-defined malar stripe, slightly paler and faintly tinged green on ill-defined supercilium, on short stripe below eye, and on centre of cheeks. Side of breast and lower flank brown-grey or olive-brown, chest and upper flank paler brown-olive, flank with faint darker brown and paler olive-green streaks, tips of feathers of chest and flank washed ash-grey when plumage fresh. Brown of chest and flank grades into green on upper belly and side of belly, this in turn to pale green-yellow on mid-belly; feather-tips of belly more extensively pale ash-grey or off-white than in adult ♂, less uniform yellow; vent and thigh more extensively white than in adult ♂; under tail-coverts pale yellow or white. Tail occasionally as in adult ♂, with base and middle portion of t3–t6 yellow (except for black shaft), but t1–t2 dark grey, less black; more usually inner webs extensively grey-black, with or without long yellow wedge on base and middle portion of inner webs, yellow on basal and middle portion of outer webs not reaching to shafts (in study in England, 6·5% of 214 ♀♀ had ♂ tail pattern, 1·3% of 297 ♂♂ had ♀ tail pattern:

Eley 1991). Flight-feathers as adult ♂, but ground-colour of distal ends slightly greyish-black, less deep black, pale grey fringe at tip still contrasting; yellow fringes of outer primaries as in adult ♂ or slightly narrower, but those on inner primaries not reaching shaft at base; see also Svensson (1992). Greater upper primary coverts slightly more greyish-black, less deep black, green of outer fringes similar, but grey fringes of tips sometimes less extensive. Tertials and greater upper wing-coverts either as in adult ♂, or grey slightly tinged brown, less pure ash-grey, and black on inner webs of tertials sometimes less deep and contrasting; lesser and median coverts green with olive-brown wash on tips and dark grey centres, less uniform and dark green than adult ♂. *In worn plumage* (from about April), forehead to mantle and scapulars as well as side of head and neck and lesser and median upper wing-coverts tinged grey-brown or grey, mottled olive, less uniform green than in adult ♂, purer yellow-green restricted to supercilium and some on cheek and side of neck; chin, throat, and belly diluted pale yellow, washed pale ash-grey; side of breast, chest, and flank mottled olive and ash-grey, lower flank, vent, and under tail-coverts extensively pale brown-grey and off-white. NESTLING. Down grey-white, fairly long and plentiful in tufts on crown, back, and upperwing, short and scanty on thigh and vent. For development, see Blümel (1976). JUVENILE. Upperparts buffish olive-brown, slightly mottled buff-yellow on forehead and mantle; lower mantle and scapulars with short and ill-defined dark grey or sooty streaks, similar streaks sometimes on cap (especially in ♀); rump slightly paler and buffier than remainder of upperparts, sometimes with faint dusky mottling (especially in ♀). Side of head and neck olive-brown (♂) or brown-grey (♀), tinged yellow (♂) or pale buff-grey with slight yellow wash (♀), especially at base of bill, on lore, and round eye. Ground-colour of underparts diluted pale yellow to off-white with faint yellow wash on belly (especially ♀ often whitish, but much variation), chest, side of breast, flank, and thigh washed brown-olive or brown-grey, chest, belly, and flank marked with dusky streaks (varying in width and contrast, ranging from sharply-defined olive-brown to faintly brown-olive, independent of sex). Tail as adult, but feathers pointed, less broad and rounded, tips of feathers less washed grey; basal two-thirds of tail yellow in ♂ (as in adult ♂ and some adult ♀), but yellow in ♀ restricted to yellow fringe along basal two-thirds of outer web and ill-defined border along inner web, yellow not reaching shaft of outer feathers (in contrast to ♂); rarely, yellow of ♀ as extensive as in ♂. Flight-feathers as adult, yellow on outer webs of primaries of ♂ bright and extensive, as in adult ♂, those of ♀ generally narrower and often slightly paler, less broad than in some adult ♀♀. Tertials as in adult ♂, tips and outer webs tinged brown, not as grey as in adult ♂; tips more rounded and frayed than in adult, less squarish and smoothly bordered. Greater upper primary coverts as adult, but narrow grey fringe restricted to extreme tip, not extending over distal end of outer web, green of fringe on outer web more extensive. Upper wing-coverts brown-olive with ill-defined grey centres and warmer pale tawny-brown tips (in adult ♂, greater coverts extensively washed grey on tips, but those of adult ♀ more brown-grey, only slightly less tawny than in juvenile); outer webs of bastard wing of juvenile ♂ extensively yellow, a yellow fringe only in juvenile ♀, leading edge of wing of ♀ much less yellow than in ♂. FIRST ADULT MALE. Like adult ♂, but variable number of juvenile feathers retained; in north of range in particular, juvenile outer greater coverts, some or all tail-feathers, sometimes tertials and usually bastard wing, and always flight-feathers and primary coverts retained, tips of juvenile tail-feathers more pointed and worn than neighbouring fresh feathers (if any), clear contrast in colour, shape, and degree of abrasion

between inner and outer greater coverts. In birds with extensive post-juvenile moult (especially in south of range, but see Moults), all greater coverts and tail as in adult; often, variable number of flight-feathers new (especially central or outer primaries, these contrasting in colour and abrasion with old ones, unlike adult); some birds retain part of secondaries and primary coverts only. Head and body as in adult ♂, but fringes of feathers of upperparts often slightly tinged brown in fresh plumage, appearing faintly streaked (but less so than in adult ♀); green and yellow of underparts slightly less deep, grey feather-tips of fresh plumage slightly broader, less soon worn off. FIRST ADULT FEMALE. Like adult ♀ but variable amount of juvenile feathering retained, as in 1st adult ♂. Head and body sometimes as in adult ♀, but usually much less extensively green and yellow; in latter birds, cap to back as well as upper tail-coverts dull brown-grey with ill-defined darker streaking, contrasting with olive-green rump, side of head and neck brown-grey or olive-grey with indistinct purer olive-green or buff-grey supercilium, face, and patch round eye, underparts pale brown-grey with ill-defined darker grey streaking, throat off-white (sometimes faintly streaked dusky), upper belly usually with slight green-yellow wash, belly to under tail-coverts extensively grey-white or off-white.

**Bare parts.** ADULT, FIRST ADULT. Iris brown. Bill pale pink, pale greyish-flesh, or whitish-flesh with small slate-black tip. Leg and foot pink-flesh or purplish-flesh, tinged blue-grey on toes and joints. NESTLING. Bare skin pink-flesh. Mouth bright red, grading to rose-pink at borders; gape-flanges yellow-white. JUVENILE. At fledging, bill, leg, and foot greyish-flesh. (Heinroth and Heinroth 1924–6; Witherby *et al.* 1938; RMNH, ZMA).

**Moults.** ADULT POST-BREEDING. Complete; primaries descendent. In Britain, starts with p1 mid-July to mid-August (occasionally, from late June or up to early September); completed with p9–p10 after *c.* 85 days, between early October and early November, occasionally from late September or up to late November; on average, moult 4 August to 28 October (Newton 1968; Ginn and Melville 1983). Moult duration 13–15 weeks; in captivity, 1st secondary (s1) shed with loss of p5–p7, gradually followed by other secondaries; last one (s6) shed when p9 growing, completed up to 19 days later than p9; body starts 3 (6) 1–5 weeks after p1, completed at same time as innermost secondary (Newton 1967d). In Finland, duration of primary moult 75–85 days (Lehikoinen and Niemelä 1977). Primary moult exceptionally arrested (e.g. in ♀ from Belgium, 20 August: Herroelen 1980). In Algeria, ♂ in full moult mid-July (ZMA); in Morocco, moult in 4 birds almost completed 13 October, completed in single from 1 November (Meinertzhagen 1940). In northern Iran, 4 of 6 birds 20–27 July had just started moult (Stresemann 1928; Vaurie 1949b). Small samples of moulting birds examined from Spain, Rumania, and Greece fell largely within sample from Britain, but one from southern Greece had moult almost completed on 21 August already (RMNH). In USSR, duration of moult 95·3 ± 1·2 days (Noskov and Smirnov 1979). For influence of light intensity on moult, see Oordt and Damsté (1939) and Damsté (1947). POST-JUVENILE. Partial; occasionally complete, especially in south of range. When not complete, involves head, body, lesser and median upper wing-coverts, highly variable number of greater coverts and tertials, frequently some to all tail-feathers (usually central pair only), and occasionally bastard wing and central primaries. Starts shortly after fledging, and timing of moult thus as variable as fledging date; in Berlin, birds hatched (March–)April start moult July at 4 months old (Westphal 1976), later-hatched birds may start at 6 weeks (Heinroth and Heinroth 1924–6). In Britain, moult late

August to mid-October; tertial coverts (innermost greater coverts, gc7–gc9) almost always replaced, but 49% of 788 birds replaced all coverts (53% of 471 ♂♂, 44% of 317 ♀♀), 15% retained 1 covert, 13% 2, 6% 3, 4% 4, 11% retained 5–7 coverts; 1% of all birds may have complete moult (Ginn and Melville 1983). In Berlin (Germany), c. 50% of 77 birds examined November–December had moult restricted to head, body, some tertials, lesser and median coverts, and variable number of inner and central greater coverts (usually, 1–7 old greater coverts retained); c. 30 birds also replaced part of tail (mainly t1, t2, and t6, sometimes all tail) and usually all greater coverts and many or all tertials; 13% from sample of 133 birds replaced 1–10 primaries (but only exceptionally secondaries and apparently never all greater upper primary coverts), primary moult starting descendently from p1 (in 4 birds), p4 (in 3), p5 (in 7), or p6 (in 2) in August or early September; this moult called 'sectoral primary moult' (Westphal 1976). Such partial moult of primaries also recorded Switzerland (Winkler and Jenni 1987) and Balearic Islands (Mester and Prünte 1982). Elsewhere in eastern Germany, 43·8% of 361 birds retained 1–7 greater coverts (Blümel 1976). In Portugal, 10–20% of birds perhaps moult completely (Ginn and Melville 1983). In Finland, 15% of 140 birds replaced all juvenile tail, 23% replaced 1–2(–5) pairs of tail-feathers, remainder replaced none (Lehikoinen and Laaksonen 1977). In north-west Russia, moult starts early July to early September, completed late September to early November (Rymkevich 1990). In northern Iran, 3 of 10 juveniles from 11–27 July had started body moult (Stresemann 1928; Vaurie 1949b).

**Measurements.** ADULT, FIRST ADULT. Nominate *chloris*. Netherlands, all year; skins (RMNH, ZMA: A J van Loon). Bill (S) to skull, bill (N) to distal corner of nostril; exposed culmen on average 2·4 less than bill (S). First adult wing and tail refer to retained juvenile wing and tail of 1st adult.

| | | | | | | |
|---|---|---|---|---|---|---|
| WING AD | ♂ | 88·3 (2·51; 22) | 83–92 | ♀ | 85·8 (2·65; 13) | 81–90 |
| 1ST AD | | 87·4 (1·76; 35) | 83–91 | | 85·3 (1·68; 32) | 82–88 |
| TAIL AD | | 57·2 (2·15; 21) | 53–61 | | 55·6 (2·85; 13) | 51–60 |
| 1ST AD | | 56·9 (1·61; 34) | 53–60 | | 55·9 (1·42; 32) | 54–59 |
| BILL (S) | | 16·7 (0·61; 24) | 15·5–17·5 | | 16·6 (0·78; 17) | 15·2–17·5 |
| BILL (N) | | 10·2 (0·54; 40) | 9·0–11·5 | | 10·1 (0·48; 35) | 9·3–10·9 |
| TARSUS | | 17·7 (0·58; 35) | 16·4–18·5 | | 17·5 (1·57; 34) | 16·6–18·9 |

Sex differences significant for wing and tail.

*C. c. harrisoni.* England and Scotland, April–September; skins (ZMA: A J van Loon).

| | | | | | | |
|---|---|---|---|---|---|---|
| WING | ♂ | 87·5 (1·79; 13) | 84–91 | ♀ | 84·4 (0·99; 8) | 83–86 |
| TAIL | | 57·0 (1·73; 13) | 55–60 | | 55·6 (1·19; 5) | 54–57 |
| BILL (S) | | 16·9 (0·61; 13) | 16·1–18·0 | | 16·6 (0·43; 7) | 16·0–17·1 |
| BILL (N) | | 10·3 (0·46; 13) | 9·5–10·9 | | 10·1 (0·38; 6) | 9·4–10·5 |
| TARSUS | | 17·6 (0·49; 13) | 17·0–18·5 | | 17·8 (0·47; 6) | 17·0–18·2 |

Sex differences significant for wing.

*C. c. voousi.* Atlas Saharien (Algeria), March–July; skins (ZMA: C S Roselaar).

| | | | | | | |
|---|---|---|---|---|---|---|
| WING | ♂ | 88·8 (2·08; 9) | 85–92 | ♀ | 85·0 (2·30; 6) | 82–88 |
| TAIL | | 56·6 (2·39; 9) | 53–59 | | 53·5 (1·57; 6) | 51–56 |
| BILL (S) | | 17·9 (1·05; 9) | 16·7–19·5 | | 18·0 (0·89; 5) | 17·3–19·2 |
| BILL (N) | | 11·2 (0·61; 9) | 10·5–12·0 | | 10·7 (0·33; 6) | 10·3–11·1 |
| TARSUS | | 18·0 (0·61; 9) | 17·1–19·1 | | 17·8 (0·46; 6) | 17·1–18·5 |

Sex differences significant for wing and tail.

*C. c. vanmarli.* Portugal, April–May; skins (RMNH, ZMA: A J van Loon, C S Roselaar).

| | | | | | | |
|---|---|---|---|---|---|---|
| WING | ♂ | 82·8 (1·15; 15) | 80–85 | ♀ | 81·0 (1·35; 4) | 80–83 |
| TAIL | | 52·8 (1·73; 14) | 50–56 | | 50·6 (0·75; 4) | 50–52 |
| BILL (S) | | 16·2 (0·55; 13) | 15·5–17·1 | | 16·5 (0·13; 4) | 16·4–16·7 |
| BILL (N) | | 9·9 (0·45; 14) | 8·9–10·7 | | 10·0 (0·26; 4) | 9·7–10·3 |

TARSUS 16·7 (0·34; 13) 16·1–17·2     17·0 (0·39; 4) 16·7–17·6
Sex differences significant for wing and tail.

*C. c. chlorotica.* Levant and Suez, all year; skins (BMNH, ZMA: A J van Loon).

| | | | | | | |
|---|---|---|---|---|---|---|
| WING | ♂ | 83·4 (2·07; 18) | 80–87 | ♀ | 81·1 (1·88; 8) | 78–84 |
| TAIL | | 54·5 (1·44; 18) | 52–58 | | 51·8 (1·60; 8) | 49–55 |
| BILL (S) | | 15·6 (0·74; 11) | 14·7–16·8 | | 15·8 (0·40; 8) | 15·2–16·5 |
| BILL (N) | | 9·8 (0·45; 11) | 9·1–10·5 | | 9·7 (0·30; 8) | 9·3–10·1 |
| TARSUS | | 16·2 (0·58; 11) | 15·3–17·1 | | 17·0 (0·63; 8) | 15·8–17·1 |

Sex differences significant for wing and tail.

*C. c. bilkevitchi.* Crimea, Caucasus, Transcaucasia, northern Iraq, and northern Iran, all year (BMNH, ZMA: A J van Loon).

| | | | | | | |
|---|---|---|---|---|---|---|
| WING | ♂ | 87·7 (2·38; 15) | 84–92 | ♀ | 84·2 (0·34; 8) | 81–87 |
| TAIL | | 55·9 (1·37; 12) | 54–58 | | 53·7 (1·89; 6) | 51–56 |
| BILL (S) | | 16·4 (0·70; 10) | 15·4–18·4 | | 16·2 (0·80; 6) | 15·5–17·3 |
| BILL (N) | | 10·8 (0·38; 10) | 9·8–11·1 | | 10·2 (0·52; 6) | 9·6–10·8 |
| TARSUS | | 17·5 (0·49; 10) | 16·7–18·2 | | 17·3 (0·54; 7) | 16·5–18·1 |

Sex differences significant for wing and tail. Wing: 88·1 (11) 85–91 (Zarudny 1911); ♂ 86–90 (4), ♀ 83 (Hartert and Steinbacher 1932–8).

*C. c. turkestanicus.* Wing: ♂ 92·1 (2·59; 12) 88–96, ♀ 86·0 (0·73; 5) 85–87 (Sarudny 1907).

Wing. Data mainly from Voous (1951) (♂ only) and Beretzk *et al.* (1969), latter corrected for different method of measuring; skins, unless otherwise noted. Nominate *chloris*. (1) European Russia (Beretzk *et al.* 1969). (2) Norway and Sweden (Voous 1951; Beretzk *et al.* 1969). Berlin (Germany), whole year, live birds: (3) adult, (4) 1st adult (Westphal 1981; see also Prill 1975, Blümel 1976, and Eck 1985a, 1990). (5) Western Germany (Niethammer and Wolters 1969; see also Niethammer 1971). Belgium, winter, live birds: (6) adult, (7) 1st adult (Robert 1977). (8) Schiermonnikoog (Netherlands), live birds (A J van Loon; see also Voous 1951). Basel (Switzerland), winter, live birds: (9) adult, (10) 1st adult (Sutter 1946, which see for increase of wing length with age). (11) Hungary (Beretzk *et al.* 1969). *C. c. harrisoni.* (12) Britain (Voous 1951; Niethammer and Wolters 1969). (13) Sussex (England), all year, live birds (Eley 1991). *C. c. aurantiiventris.* (14) Western France and Pyrénées (Voous 1951). (15) South-east France, breeding, live birds (G Olioso). (16) Spain (Voous 1951). (17) Live migrants Tarifa, southern Spain (Asensio 1984). (18) Balearic Islands, live birds (Mester 1971). (19) Italy, western Yugoslavia, and Albania (Voous 1951; Beretzk *et al.* 1969). (20) Western Slovenija (Geister 1974). *C. c. vanmarli.* (21) Portugal and northern Morocco (Voous 1951). *C. c. voousi.* (22) High Atlas, Morocco (Voous 1951). *C. c. madaraszi.* (23) Corsica and Sardinia (Voous 1951). *C. c. muehlei.* (24) Crete (Niethammer 1942). (25) Eastern Yuogoslavia, Rumania, Bulgaria (Beretzk *et al.* 1969). (26) Makedonija, southern Yugoslavia (Stresemann 1920). (27) Peloponnisos, Attika, and Karpathos, southern Greece (Niethammer 1943; Kinzelbach and Martens 1965; RMNH). (28) Turkey (Jordans and Steinbacher 1948; Kumerloeve 1961; Rokitansky and Schifter 1971; RMNH).

| | | | | | | | |
|---|---|---|---|---|---|---|---|
| (1) | ♂ | 89·6 (2·17; 17) | 85–92 | ♀ | 86·0 (3·48; 6) | 82–92 |
| (2) | | 89·2 (1·83; 58) | 86–94 | | 86·9 (1·61; 13) | 85–90 |
| (3) | | 90·9 (1·87; 111) | 86–96 | | 88·2 (1·76; 100) | 83–94 |
| (4) | | 90·0 (1·79; 290) | 85–96 | | 87·6 (1·68; 278) | 83–93 |
| (5) | | 87·6 (1·88; 35) | 84–92 | | 85·2 (—; 11) | 83–87 |
| (6) | | 90·9 (2·26; 122) | 84–96 | | 88·2 (2·17; 128) | 82–93 |
| (7) | | 89·2 (2·17; 143) | 83–95 | | 87·1 (1·86; 108) | 82–92 |
| (8) | | 89·8 (1·97; 28) | 85–94 | | 87·5 (2·10; 11) | 86–93 |
| (9) | | 90·0 (1·92; 132) | 86–95 | | 87·1 (1·73; 95) | 83–92 |
| (10) | | 88·3 (1·48; 222) | 84–92 | | 86·0 (1·56; 155) | 82–90 |
| (11) | | 89·3 (1·70; 111) | 86–93 | | 86·2 (2·04; 88) | 81–92 |

(12)  87·1 (1·89; 109) 83-92    84·7 (1·69; 38) 81-88
(13)  90·0 ( — ; 297) 83-98    87·0 ( — ; 214) 81-91
(14)  86·4 (1·56; 11) 84-89    — ( — ; — ) —
(15)  88·1 (1·83; 20) 86-93    86·0 (1·93; 27) 83-90
(16)  83·3 (1·86; 15) 81-88    — ( — ; — ) —
(17)  84·2 (3·19; 12) —    82·5 (2·29; 16) —
(18)  87·3 ( — ; 8) 84-90    83·8 ( — ; 8) 81-88
(19)  87·2 (2·26; 46) 82-91    84·0 (2·03; 8) 82-87
(20)  88·7 (0·89; 117) 86-96    86·5 (0·57; 92) 83-91
(21)  83·4 (1·05; 22) 81-85    — ( — ; — ) —
(22)  89·8 (1·30; 4) 88-91    — ( — ; — ) —
(23)  84·7 (1·73; 19) 81-88    — ( — ; — ) —
(24)  — ( — ; 10) 79-84    79·0 ( — ; 2) 79-79
(25)  88·2 (2·00; 34) 83-91    86·5 (2·56; 15) 83-92
(26)  85·9 (1·66; 13) 83-88    84·3 (2·81; 7) 81-89
(27)  83·2 (1·50; 4) 82-85    — ( — ; — ) —
(28)  86·8 (1·91; 7) 84-90    83·8 (2·36; 4) 82-87

Wing of live birds on average 2·0 mm longer than in skins from same area (ZMA). *C. c. aurantiiventris*. Mallorca (Balearic Islands), wing: ♂ 83-88 (*n* = 30), ♀ 81-85 (*n* = 8) (Jordans 1924). Introduced birds Azores (probably *vanmarli*: Voous 1951), wing: ♂ 83·8 (2·28; 13) 81-88 (Voous 1951), ♀ 81·4 (2·20; 8) 77-84 (G Le Grand).

**Weights.** ADULT, FIRST ADULT. Nominate *chloris*. Sussex (England): (1) summer, (2) winter (Eley 1991). Netherlands: (3) August-October, (4) November-February, (5) April-July (A J van Loon; ZMA, RMNH). (6) Belgium, November-April (Robert 1977). Berlin (Germany): (7) July-September, (8) October-November, (9) December-February, (10) March (Westphal 1981). (11) Czechoslovakia, June-December (Havlín and Havlínová 1974). *C. c. aurantiiventris*. (12) South-east France, breeding (G Olioso). *C. c. muehlei*. (13) Greece and Turkey, April-July (Niethammer 1943; Makatsch 1950; Rokitansky and Schifter 1971). *C. c. turkestanicus*. (14) Kazakhstan (Korelov *et al.* 1974).

(1)  ♂  27·2 ( — ; 147) 24·8-30·4  ♀  28·1 ( — ; 77) 24·8-33·5
(2)  28·5 ( — ; 147) 25·7-35·6  28·3 ( — ; 92) 25·1-32·1
(3)  26·8 (2·07; 32) 20·3-30·4  26·0 (2·90; 51) 19·0-30·5
(4)  26·3 (4·45; 20) 18·7-34·5  27·5 (4·03; 10) 23·2-35·2
(5)  24·3 (2·28; 12) 19·0-27·2  25·9 (1·88; 6) 23·7-29·0
(6)  29·8 (2·48; 305) 25·0-35·0  29·2 (2·73; 236) 22·5-37·5
(7)  27·4 (1·8 ; 39) 23·5-32·5  26·9 (1·7 ; 43) 20·0-30·0
(8)  29·4 (1·5 ; 90) 26·5-33·0  28·8 (1·8 ; 89) 25·0-32·5
(9)  29·7 (1·7 ; 108) 23·5-32·0  29·5 (1·9 ; 129) 24·0-35·0
(10)  27·9 (2·5 ; 27) 24·5-35·0  28·7 (2·0 ; 23) 25·0-32·0
(11)  30·4 (2·82; 339) 24·0-38·5  29·6 (2·31; 195) 21·7-36·5
(12)  25·8 (1·94; 20) 24·0-30·0  26·1 (2·46; 26) 21·0-31·0
(13)  26·1 (2·11; 6) 22·5-28·0  26·3 (3·21; 3) 24-30
(14)  29·0 ( — ; 11) 26·1-31·5  31·7 ( — ; 7) 28·8-35·1

Nominate *chloris*. Norway, September-April: 29·6 (14) 26·5-34·4 (Haftorn 1971). Helgoland (Germany): ♀ 26 (12) 25-30; frost victims 21·7 (3) 19-24 (Weigold 1926). Paris area (France): ♂ 27·6 (5) 26·4-28·7, ♀ 25·7 (Mountfort 1935). For weight variation throughout year, see Miles (1968), Prill (1974), and Westphal (1981).

*C. c. harrisoni*. Skokholm (Wales): 26·7 (7) 25·5-28·4 (Browne and Browne 1956). Exhausted birds, southern England: 19·3 (7) 16·2-22 (Harris 1962; MacDonald 1962; Ash 1964). See also Lloyd-Evans and Nau (1965) and Newton (1972).

*C. c. aurantiiventris*. Balearic Islands: 24·2 (17) 22·5-26·1 (Mester 1971). Tunisia: May, ♂ 25 (G O Keyl). Egypt: November, ♂ 21·8 (Goodman and Watson 1983).

*C. c. vanmarli*. Azores: June-August, ♀ 22·5 (1·79; 6) 20-24 (G Le Grand).

*C. c. madaraszi*. Sardinia: February-March, ♂♂ 20, 26 (Demartis 1987).

*C. c. chlorotica*. Israel: 20-26 (Paz 1987); March, ♂ 23 (G O Keyl).

NESTLING, JUVENILE. For growth curve, see Blümel (1976).

**Structure.** Wing rather short, broad at base, tip rounded. 10 primaries: p8 longest, p7 and p9 0-0·5(-1·5) shorter (one of each sometimes up to 0·5 longer than p8), p6 2-6, p5 9-14, p4 13-19, p3 17-23, p2 20-26, p1 27·5 (10) 25-30 shorter in Sweden, 25·4 (10) 23-28 in Netherlands, 25·9 (11) 24-29 in Britain, 24·9 (16) 23-27 in Portugal, 25·8 (14) 22-28 in Algeria, 23·0 (5) 22-25 in Tunisia; p10 strongly reduced, a tiny pin, concealed below reduced outermost greater upper primary covert, 53-63 shorter than p8, 7-11 shorter than longest upper primary covert. For wing formula, see also Prill (1975) and Robert (1977). Outer web of p6-p8 emarginated, inner web of p7-p9 with slight notch. Tip of longest tertial reaches to about tip of p2-p4. Tail short, tip forked; 12 feathers, t1 5-10 shorter than t6, inner webs cut off at sharp angle. Bill short, heavy, conical, tapering to sharp tip; depth at feathering of base 10·5 (10) 9·7-11·0 and width at lateral base of lower mandible 9·3 (10) 8·9-9·6 in nominate *chloris* from Sweden, depth 10·2 (31) 9·5-10·9 and width 9·0 (34) 8·5-9·6 in nominate *chloris* from Netherlands, depth 10·3 (20) 9·7-10·8 and width 9·2 (20) 8·7-9·7 in *harrisoni*, depth 10·8 (9) 10·3-11·4 and width 9·2 (10) 8·6-9·8 in *muehlei*, depth 10·2 (17) 9·5-11·1 and width 9·2 (20) 8·6-9·7 in *vanmarli*, depth 11·8 (16) 11·0-13·1 and width 10·2 (16) 9·7-10·8 in *voousi*, depth 10·8 (5) 10·5-11·2 and width 9·8 (5) 9·4-10·3 in *aurantiiventris* from Tunisia. Culmen and cutting edges straight or slightly decurved towards sharply pointed tip. Nostril small, rounded, covered by short tuft of feathers projecting from forehead; some reduced stiff feathers and short bristles along lateral base of upper mantle. Tarsus and toes rather short and slender. Middle toe with claw 18·0 (15) 16·5-20·5; outer toe with claw *c.* 72% of middle with claw, inner *c.* 70%, hind *c.* 76%.

**Geographical variation.** Marked. Involves size (length of wing, tail, or tarsus), relative length, depth, and width of bill, colour and contrast of forehead, colour and intensity of green on upperparts, throat, and chest, and colour and intensity of yellow of belly to under tail-coverts, tail-base, and fringes of flight-feathers. Variation in size largely clinal, with larger birds in Fenno-Scandia and northern Russia, intermediate ones in Britain, France, mainland Italy, Balkans, Turkey, and Caucasus area, smaller ones in Iberia, north-west Morocco, north-east Tunisia, Balearic Islands, Corsica, Sardinia, Sicily, southern Greece, Cyprus, and Levant. Isolated *turkestanicus* in central Asia is largest race; *voousi* from Atlas mountains in Morocco and in mountains of northern Algeria markedly larger than birds of north-west Morocco and north-east Tunisia, about as large as birds from northern Europe. Bill relatively slightly heavier than elsewhere in Iberia, Balearic Islands, north-west Morocco, and (especially) in *voousi* (see Measurements and Structure). Variation in colour more difficult to assess: depends greatly on age and sex of bird, and strongly influenced by bleaching and wear; when adult ♂♂ in equally worn plumage are compared, variation in tone and intensity of colours more marked than implied by (e.g.) Vaurie (1956a, 1959), especially in Mediterranean basin. Adult ♂ of nominate *chloris* from northern and central Europe and of *harrisoni* from Britain and Ireland both brownish olive-green on upperparts, tinged grey on cap when fresh, fringed brown on remainder of upperparts then; when worn, olive-green of mantle and scapulars still distinctly tinged brown, unlike other races, olive-green of crown contrasting with green-yellow (nominate *chloris*) or yellow-green (*harrisoni*) forehead and with yellow-green (nominate *chloris*) or pale olive-green (*harrisoni*) rump; belly of nominate *chloris* lemon-yellow, throat green-yellow, both contrasting with olive-green chest and flank; in *harrisoni*, underparts markedly uniform olive-green, yellow of belly tinged green, side of breast and flank darker grey-brown than in nominate *chloris*; *harrisoni* perhaps darkest in southern Scotland and northern England, gradually merging into nominate *chloris* further south and replaced by nominate *chloris* in northern Scotland (Voous 1951; Vaurie 1956a, 1959), but no cline found by Niethammer and Wolters (1969), who considered all birds from southern Scotland, England, and Ireland to be *harrisoni*, while birds in northern Scotland, mainly collected at lighthouses, are probably migrating Scandinavian birds (P A Clancey). See also Clancey (1940, 1943, 1946, 1948d). In worn ♂ of *muehlei* from central and eastern Balkans (from Serbia, Makedonija, and Greece eastwards) as well as in western Asia Minor and south-west Ukraine, upperparts olive-green, less brown than in nominate *chloris*, fringes of feathers broadly grey when plumage fresh, not olive-brown; forehead rather contrasting green-yellow, rump yellow-green, throat and belly rather deep golden-yellow, hardly contrasting with yellow-green chest, but side of breast and flank rather dark and grey; in general, *muehlei* nearer to nominate *chloris* than to *aurantiiventris*, but a valid race (Laubmann 1912a; Stresemann 1920; Harrison and Pateff 1933; Niethammer 1943; Beretzk *et al.* 1969). In *aurantiiventris* from central and southern France, Spain (except probably northwest), Balearic Islands, mainland Italy, Sicily, and north-east Tunisia, upperparts of ♂ brighter yellow-green than in *muehlei*, hardly contrasting in colour with slightly yellower forehead and rump; when fresh, fringes of feathers of upperparts narrowly grey; underparts like *muehlei*, golden-yellow on throat and belly, but chest and flank often rather extensively olive-green, flank paler than in *muehlei*; populations from coastal Yugoslavia, Albania, and western Greece are rather variable, some nearer *muehlei*, others nearer *aurantiiventris*, but most inseparable from latter; see Beretzk *et al.* (1969). Adult ♂ of *vanmarli* has upperparts olive-green, duller and less yellow than *aurantiiventris*, forehead

rather contrasting, green-yellow, rump hardly contrasting, yellow-green; underparts extensively yellowish olive-green, with restricted bright yellow patch on belly and slightly darker olive flank. *C. c. voousi* rather like *aurantiiventris*, but all colours paler, less saturated, upperparts diluted olive-green, contrasting with green-yellow forehead, but hardly with pale green rump; yellow of belly rather pale, sulphur-yellow, remaining underparts pale yellowish-green, flank with restricted amount of olive. Adult ♂ of *chlorotica* bright yellow-green on upperparts, hardly contrasting with paler green-yellow forehead and rump; throat and belly golden-yellow, chest and flank extensively yellowish-green; fringes of flight-feathers paler yellow than in *aurantiiventris*. *C. c. madaraszi* from Corsica and Sardinia is dark and extensively green; like *vanmarli*, but forehead dark yellowish-green in worn plumage, less yellow; less brown than in nominate *chloris*. Situation on islands in east Mediterranean not clear; according to Beretzk *et al.* (1969), population on Crete dark and small, like *madaraszi*, birds from Karpathos like *chlorotica*, and one bird from Cyprus similar to *muehlei*, but according to Niethammer (1942) birds from Karpathos and Crete similar, both slightly yellower than *muehlei* from Peloponnisos, though not as yellow as *aurantiiventris*; here included in *muehlei*. *C. c. bilkevitchi* from eastern Turkey, Caucasus area, northern Iran, and south-west Turkmeniya rather like *chlorotica*, but greyer, less yellow; ♂ paler and greyer on upperparts than ♂ nominate *chloris*, paler yellow on underparts and fringes of flight-feathers. *C. c. turkestanicus* from central Asia like *bilkevitchi* and *chlorotica*, but larger and slightly paler and greyer. Variation in ♀ more or less similar to that of ♂; upperparts of nominate *chloris* and *harrisoni* various shades of brown, remaining races grey-brown, palest in *chlorotica* and *turkestanicus*; upperparts of ♀ *voousi* diluted brownish ash-grey, underparts extensively ash-grey with limited green-yellow on belly and (sometimes) throat.

Forms species-group or superspecies with Oriental Greenfinch *C. sinica* from eastern Asia (south to China), Yunnan Greenfinch *C. ambigua* from south-west China, and Blackheaded Greenfinch *C. spinoides* from Himalayas and southern Indochina. See Güttinger (1978) for comparison of voice of these species with that of Goldfinch *C. carduelis*.     CSR

## *Carduelis carduelis*  Goldfinch

PLATES 36 and 39 (flight)
[between pages 472 and 473]

DU. Putter     FR. Chardonneret élégant     GE. Stieglitz
RU. Щегол     SP. Jilguero     SW. Steglits     N. AM. European Goldfinch

*Fringilla Carduelis* Linnaeus, 1758

Polytypic. CARDUELIS GROUP. *C. c. britannica* (Hartert, 1903), Britain, Ireland, western and north-west France, and coastal zone of Belgium and Netherlands; nominate *carduelis* (Linnaeus, 1758), from Denmark, eastern Netherlands and Belgium, and central France east through Europe to central Urals, south to mainland Italy, northern Yugoslavia, north-west Rumania, northern Ukraine, and central European Russia, grading into *britannica* in west-central France, into *parva* in southern France, into *balcanica* in northern Yugoslavia and western Rumania, and into *volgensis* in European Russia; *parva* Tschusi, 1901, southern France (Mediterranean coast and Pyrénées), Iberia, Balearic Islands, Azores, Madeira, Canary Islands, and North Africa from Morocco to Cyrenaica (Libya): *tschusii* Arrigoni, 1902, Corsica, Sardinia, Elba, and Sicily; *balcanica* Sachtleben, 1919, Yugoslavia (except north) and southern Rumania to Crete and European Turkey; *niedecki* Reichenow, 1907, eastern Sporades, Rhodes, and Karpathos (Greece), Asia Minor (except east), northern Iraq, Transcaucasia north to southern slope of Caucasus, Zagros mountains (south-west Iran) south to Shiraz, Cyprus, Levant, and Egypt; *loudoni* Zarudny, 1906 (synonym: *brevirostris* Zarudny, 1899, preoccupied), northern Iran from Azarbaijan to Gorgan and (perhaps this race) eastern Turkey; *colchicus* Kudashev, 1915, south-west slope of

Caucasus, east to above Gagra, northern slope of Caucasus to *c.* 46°N, and Crimea; *volgensis* Buturlin, 1906, southern Ukraine, north-east Rumania (at least in winter), and south-east European Russia, south to plains north of Caucasus where probably grading into *colchicus*, grading into *frigoris* between Ural river and 70°E; *frigoris* Wolters, 1953 (synonym: *major* Taczanowski, 1879, preoccupied), western and central Siberia east and south to Yenisey, western Altai, Zaysan basin, and northern Kazakhstan, in winter to south-east European Russia, Transcaspia, and perhaps Middle East. CANICEPS GROUP (extralimital). *C. c. ultima* Koelz, 1949, south-east Fars and Kerman, southern Iran; *paropanisi* Kollibay, 1910, north-east Iran (east of Gorgan) and northern Afghanistan east through Turkmeniya, Uzbekistan, and Tien Shan to Dzhungarskiy Alatau and Tarbagatay in south-east Kazakhstan, hybridizing with *loudoni* in Gorgan area; *subalata* (Gloger, 1833), south-central Siberia from central and southern Altai through western Mongolia and Tuva to Angara and Lake Baykal, hybridizing with *major* in Zaysan basin, Altai, Kuznetsk Alatau, and Yenisey valley; *caniceps* Vigors, 1831, western Pakistan north to Himalayas, east to central Nepal.

**Field characters.** 12 cm; wing-span 21–25·5 cm. Size close to Linnet *C. cannabina* but with 10% shorter tail; slightly larger and longer winged than Siskin *C. spinus*. Small, delicate, beautifully marked finch, with noticeably pointed bill and light dancing flight. At all ages, displays diagnostic shining, golden-yellow panel along centre of black wing. Adult has unique head pattern of seemingly vertical bands of red–white–black and has tawny-brown back contrasting with wings, bold whitish rump, and black tail. Juvenile brown and streaked, with pale 'unfinished' head, but shows diagnostic wing-panel. Voice distinctive. Sexes closely similar; little seasonal variation. Juvenile separable. 10 races in west Palearctic, varying little in size and body plumage tones. Only west European races described here (see also Geographical Variation).

(1) North and central European race, nominate *carduelis*. ADULT. Moults: July–October (complete). Brightening of plumage entirely due to wear. Front of head (to eye) deep, brilliant crimson, with black lores and border to base of bill; edge of crown behind eye, rear half of ear-coverts, and surround to crimson bib white; top of rear crown, upper nape, and narrowing band down edge of ear-coverts black; bold white spot below upper nape (and between upper sections of bands). Mantle, back, breast-band, and flanks pale rufous to sandy-brown. Rump buffish-white, upper tail-coverts white, mottled black, both forming pale area contrasting with back and black tail, tipped white on all feathers. Wings noticeably black, with brown tips on leading coverts rarely visible and white spots or tips on all flight-feathers far less eye-catching than bright, almost gold outer webs on greater coverts, bases of secondaries, and all but outermost primaries which form 'flashing' panel across extended wing. Under-wing yellowish-white. With wear, brown plumage becomes paler, white areas even purer, and white tips to flight- and tail-feathers decrease in size or are lost. At close range, ♀ may be distinguished by broad grey-brown tips to lesser coverts. Bill pink-white, with fine dark tip, legs pale flesh. JUVENILE. Wholly unlike adult in head and body plumage which is greyish-buff (except for paler face and belly) and softly spotted and streaked dull brown. Wings and tail as adult but flight- and tail-feather tips buff, not white. Bill greyish-horn. (2) British and Channel Islands race, *bri-*

*tannica*. Head pattern slightly clouded, due to less glossy crimson and greyer white bands; mantle, breast-band, and flanks darker, more tawny than sandy-rufous in tone. (3) Atlantic Islands and south-west European race, *parva*. Up to 10% smaller than nominate *carduelis*, with weaker bill and usually greyer back.

Adult unmistakable but juvenile liable to confusion with *C. spinus* and *Serinus*, best separated by length and width of panel on black wings, black tail, amorphous head, and voice (see below). Flight light with noticeable acceleration and rise after burst of wing-beats and then glide or descent with wings closed up beside body, producing dancing, even skipping progress over short distances and more bounding undulations over long tracks than in larger finches. Escape-flight flitting and rapid, usually short and ending in descent to perch on tree or tall plant (noticeably bunched in case of flocks). Gait a hop, varied by clambers and jump in tit-like manner. Stance as *C. chloris*. Sociable but rarely associates with other finches.

Song a characteristic and lively liquid tinkling, also containing purrs, rattles, and rather nasal notes, uttered with pleasant vehemence; based on calls but much elaborated, recalling Canary *Serinus canaria* and *C. cannabina*. Commonest call a trickling, liquid, pleasant trisyllable, 'tswitt-witt-witt. . .' or 'quilp-ilp-ilp'. Other calls include grating 'geez'.

**Habitat.** Breeds over west Palearctic north to July isotherms of 17°C in boreal, temperate, Mediterranean, and steppe zones, both Atlantic and continental (Voous 1960*b*). Predominantly in lowlands, but in Switzerland breeds generally up to 1000 m, quite commonly to 1300 m, and occurs in late summer and autumn up to 2400 m, resorting to alpine meadows and areas near chalets (Glutz von Blotzheim 1962). In breeding season, shows preference for orchards, cemeteries, parks, gardens, avenues, and tree nurseries, often in or near human settlements, and especially where patches of tall weeds and other concentrated food sources are present. Also favours streamside or fen woodlands, open or fringe woodlands and heathlands, and commons with well-grown hawthorn *Crataegus*, gorse *Ulex*, and other scrub or thicket species. Overlaps considerably with Greenfinch *C. chloris* but does

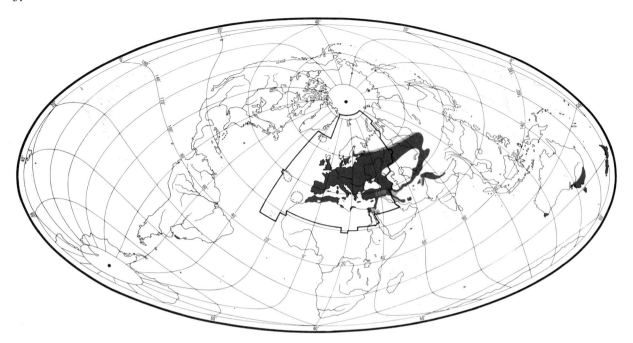

not share its attachment to tall trees and is not attracted to nesting in bushes.

Outside breeding season, reliance on tall Compositae such as thistles *Carduus*, dandelions *Taraxacum*, ragwort *Senecio*, and burdocks *Arctium* dictates movements to rough grasslands, vacant sites, overgrown rubbish dumps, etc., although much use is made of woodlands in winter; alder *Alnus* and pine *Pinus* are also favoured food sources (Newton 1972). Even more widely mobile than *C. chloris*, but tends to fly in lowest airspace and to move in brief stages, inconspicuously pursuing the business in hand.

In USSR, predominantly in hardwood and mixed open forest, avoiding pure coniferous forest but favouring oak stands, well-lit groves, and thinly wooded floodland forests alternating with pastures, meadows, and clearings. Avoids extensive and damp forests, and open spaces amid arboreal vegetation are indispensable. Increasingly favours cultivated arboreal areas, often linked with human settlements, including gardens, parks, and avenues bordered by irrigation ditches. In Caucasus, ascends mountains along river valleys and ravines towards treeline, reaching 2000 m. After breeding season, flocks at stands of tall weeds, also sunflower *Helianthus* crops and hemp *Cannabis*, permitting feeding despite snowfall. (Dementiev and Gladkov 1954; Flint *et al.* 1984.)

**Distribution.** Has spread in Ireland, parts of Fenno-Scandia, Israel, and Egypt. (Introduced in New Zealand, southern Australia, Bermuda, and very locally in South America.)

IRELAND. Has expanded range in last 50 years (Hutchinson 1989). NORWAY. Has recently expanded range to south-west and north (VR). FINLAND. First bred in 1860s, and expanded range up to 1950s (see Population) (Koskimies 1989; OH). ISRAEL. Has expanded range in recent decades, following development of settlements and agriculture (Shirihai in press). EGYPT. Formerly apparently confined to Cairo area. Has recently spread to many other areas; evidence of breeding in Western Desert oases from 1981. (Goodman and Meininger 1989.) AZORES. Introduced at end of 19th century; now breeds on all islands except Corvo (GLG). CAPE VERDE ISLANDS. Introduced to Santiago; bred 1963–5, but not seen since (CJH).

**Population.** Has increased in Britain, Ireland, and Belgium, due to decrease in bird-catching. In Fenno-Scandia, increase followed by recent decrease. In extreme south (Israel, Egypt) has increased recently, with range expansion (see Distribution).

BRITAIN. Decreased in 19th century due to bird-catching, with recovery after 1881 (Sharrock 1976). About 300 000 pairs (Hudson and Marchant 1984). Declined in 1980s, after reaching high population levels in 1970s (Marchant *et al.* 1990). IRELAND. Increase in last 50 years due to reduction in bird-catching (Hutchinson 1989). FRANCE. More than 1 million pairs (Yeatman 1976). BELGIUM. Estimated 1500 pairs 1961–8, 6700 pairs 1973–7. Increase probably due to laws controlling bird trade, introduced in 1970s. (Devillers *et al.* 1988.) NETHERLANDS. 4000–7000 pairs. Population increasing in west (*britannica*, virtually absent before *c.* 1950); decreasing in east (nominate *carduelis*). (SOVON 1987.) GERMANY. 931 000 pairs (G Rheinwald). East Germany: 190 000 ± 100 000 pairs (Nicolai 1993). NORWAY. Estimated 200–400 pairs

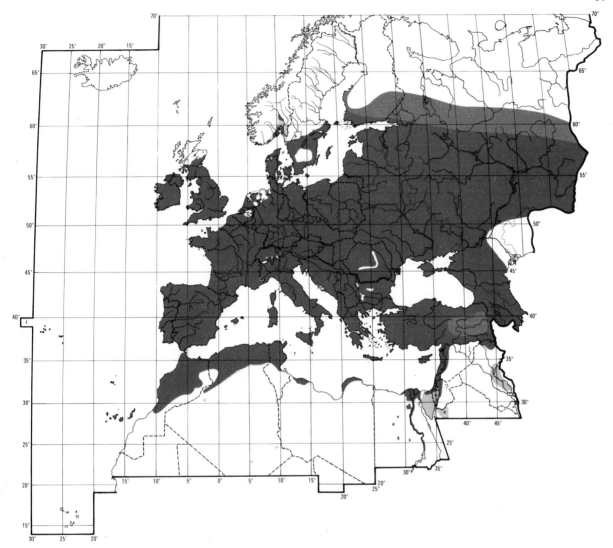

Thought to be less than 10 000 pairs (Ulf-strand and Högstedt 1976); some decrease in last decade (LR). FINLAND. After increase with range expansion (see Distribution), perhaps due to climatic amelioration and increase in weedy areas following urbanization and indus-trialization, population has more than halved since *c.* 1960 for unknown reasons. Now rare breeder, with some hun-dreds of pairs. (Koskimies 1989; OH). CZECHOSLOVAKIA. Slow increase in last 30 years (KH). ISRAEL. At least a few hundred thousand pairs in 1980s (Shirihai in press). JORDAN. 11–20 pairs in Petra area 1983 (Wittenberg 1987).

Survival. Britain: annual mortality of adult ♂♂ 54 ± 10%, adult ♀♀ 60 ± 13% (Dobson 1987). Czecho-slovakia: for age structure and mortality, see Pikula (1989). Birds ringed in northern and central Europe and recovered in Iberia: annual adult mortality 65% (Asensio and Carrascal 1990). Oldest ringed bird 8 years 5 months (Rydzewski 1978).

**Movements.** Partially migratory, wintering almost entirely within breeding range, with concentrations in Mediterranean region. Areas of higher ground are vacated. Migrates by day in flocks. Western European migrants head mostly south-west or SSW on narrow front; east European birds range more widely, between west and south-east. Some Mediterranean populations perhaps sedentary. For migratory activity in caged birds, see Glück (1982).

In Britain (*britannica*), some birds remain to winter but most depart, apparently more ♀♀ then ♂♂; ringing data show that individuals migrate in some years but remain in Britain in other years. Hard-weather southward movement within Britain also occurs, e.g. January–February 1983. Passage on east coast of England may include birds of continental race nominate *Carduelis* in some years; recov-ery at Balsham (Cambridge), May 1976, of bird ringed Schönholz (Nordrhein-Westfalen, western Germany),

August 1975, indicates that at least individuals of that race may reach Britain. As in Linnet *C. cannabina*, initial heading of migrants is south-east to Belgium (by shortest sea-crossing), then SSW to western France and Iberia; recoveries lie in narrow band, extending less far east in Iberia than *C. cannabina*; winter recoveries are from any points along route, suggesting that birds vary greatly in distance they migrate. Passage observed on south and south-east coasts of Ireland; ringing data show that some birds reach Ireland from Britain; in absence of long-distance recoveries, not known whether Irish birds migrate. British ringing recoveries up to 1986 show movement to or from Spain (273), France (172), Belgium (75), Portugal (7), Netherlands (6), Morocco (2), Germany (1), and Malta (1). (Newton 1972; Spencer and Hudson 1978a; Lack 1986, 1988; Mead and Clark 1987; Hutchinson 1989.) Third recovery from Morocco, c. January 1988, was furthest south, at Casablanca (30°39′N, 7°35′W) (Mead and Clark 1989).

Many west European migrants winter in Iberia; 96% of 753 recoveries are from France (34%), Britain (23%), Germany (23%), Switzerland (10%), and Belgium (6%); remaining 4% from Czechoslovakia, Austria, Yugoslavia, northern Italy, Netherlands, and Russia. Winter distribution in Spain more widespread and less coastal than in Chaffinch *Fringilla coelebs*, Serin *Serinus serinus*, and Siskin *C. spinus*; abundant in Strait of Gibraltar area. Proportion of ♀♀ greater in southern than in northern Spain, suggesting ♀♀ tend to go further. Ongoing movement into Morocco is more pronounced than in other finches; most recoveries of *C. carduelis* in Spain are from passage periods, and only 23·5% in December–February, whereas more than 50% in *F. coelebs* and *S. serinus*. Local birds partially migratory (previously thought sedentary); 10 recoveries show movement between south-west and south-east towards southern Spain, probably en route for Africa; also, wing formulae indicate that passage across Strait of Gibraltar includes considerable number of Iberian birds. (Asensio 1984, 1986a.)

In France, many local birds depart, and many immigrants winter, especially in south (Yeatman-Berthelot 1991). Abundant on passage to Spain through Aquitaine in south-west (Boutet and Petit 1987). Recoveries of local or passage birds ringed at Lac de Grand-Lieu (Loire Atlantique, western France) are chiefly south or SSW between 2–6°W in Spain (Marion and Marion 1975). In Belgium, about a third of population resident; many passage birds occur, especially in east, with highest numbers from Germany, considerable numbers also from Britain and Netherlands, and smaller numbers from further north-east (including 8 recoveries of birds ringed in Kaliningrad region, western Russia) (Paevski 1971; Lippens and Wille 1972). Most Swiss birds depart, wintering in southern France and Spain (Schifferli *et al.* 1982). In Germany also, partially migratory; winters regularly in plains, and birds breeding in mountains make altitudinal

movements (Mildenberger 1984). Ringing data from southern Germany show movement on narrow front south-west through southern France to south-east half of Spain (Glück 1982). In north-west Italy, birds occur above breeding areas at 2200–2500 m during post-breeding dispersal, and winter chiefly below 1200 m (Mingozzi *et al.* 1988). Present throughout year also in northern Europe (Fenno-Scandia, Poland, Baltic states), though some birds depart (Viksne 1983; Haila *et al.* 1986; SOF 1990; Tomiałojć 1990); only small numbers remain in St Petersburg region (Mal'chevski and Pukinski 1983). In Moscow region, a part of local population departs, and is replaced by immigrants (Ptushenko and Inozemtsev 1968). Many birds from eastern Europe winter in Yugoslavia (especially Dalmatian coast) and italy. Ringing data from Yugoslavia show movements to or from most east European countries north to Kaliningrad region, also south-east to Bulgaria and Cyprus, south to Malta and Tunisia, and west to Spain. Winter visitors to Yugoslavia include more ♀♀ than ♂♂. (Grittner 1941; Cvitanić 1980; Yugoslav Ringing Reports.) Ringing study in Hungary has shown that most birds are sedentary, but some migrate, with headings ranging between south-west (to northern Italy) and south-east (as far as Chios island off western Turkey) (Schmidt 1960a).

In north-west Africa, occurs south of breeding range in winter; not known whether movements involve local birds as well as immigrants; reported south to Beni Abbès, c. 30°N in Algeria; in Tafilalt (south-east Morocco), numbers higher winter than summer (Heim de Balsac and Mayaud 1962; Destre 1984). At Settat (33°N 7°37′W on west Moroccan coast), where breeds, numbers increase markedly at passage periods (Thouy 1976). New influxes into Morocco occur midwinter in severe weather (Thévenot *et al.* 1981). Fairly common on passage and in winter in Malta (Sultana and Gauci 1982), and common winter visitor to Sicily (where also breeds) (Iapichino and Massa 1989). Winter visitors swell numbers in Turkey, Crimea, Israel, and north-west Saudi Arabia (Dementiev and Gladkov 1954; Beaman 1978; Stagg and Walker 1982; Shirihai in press), and apparently some also reach Egypt (Goodman and Meininger 1989). Scarce winter visitor to Gulf States of Arabia (Bundy and Warr 1980; F E Warr). In Cyprus, many immigrants winter; local birds breeding at high levels usually descend below c. 1200 m; birds breeding at low levels mostly sedentary; recovery in Egypt of bird ringed as juvenile in Cyprus indicates that some birds move long distances (Flint and Stewart 1992). Locally common in winter in Iraq (Allouse 1953), and common winter visitor to northern and western Iran (where also breeds) (D A Scott).

Passage periods prolonged. Autumn migration August–December, chiefly September–November. Main departure from Britain mid-September to late October (Newton 1972). In northern Denmark, movement from early August to late November, peaking in late October (Møller 1978a).

On Polish coast, peaks late September and early October (Busse and Halastra 1981). At Col de Bretolet (western Switzerland), passage early or mid-September to mid-November, peaking 28 September to 7 October, earlier than other finches (Winkler 1984). Department of juveniles from Baden-Württemberg (south-west Germany) begins early August, chiefly in September; population study over 4 years showed that young from early broods leave within 6 weeks of fledging, but last broods present until October; latest records (adults or juveniles) 22 October, though passage of more northerly birds continues later (Glück 1982). Arrives in Spain from October, peaking end of October or beginning of November; French and Swiss birds arrive *c.* 10 days before German, Belgium, and British birds (Asensio 1986*a*). At Strait of Gibraltar, main passage mid-October to mid-November; small numbers continue to move south in December, and some birds move north in January; movement thus virtually continuous (Finlayson 1992). At Settat, passage mid-September to November (Thouy 1976). Arrives on Dalmation cost from 2nd week of October (Cvitanić 1980). At Bospthorus (north-west Turkey), in one year, passage eastward recorded 14 August to 31 October, mostly 2nd half of October (Porter 1983). In Israel, winter visitors arrive October to mid-December, chiefly November; passage mostly along coastal strip (Shirihai in press).

Spring migration February–May. Arrival times vary markedly from place to place. Leaves south of range chiefly from mid-February, e.g. western Morocco (Thouy 1976), Sicily (Iapichino and Massa 1989), and Cyprus (Flint and Stewart 1992). Main passage in Gibraltar area March to early April (Finlayson 1992). In Switzerland, passage early or mid-March to early May, peaking mid April (Winkler 1984). Departures and passage on Dalmatian coast chiefly mid-March (Cvitanić 1980). In Baden-Württemberg, 1972–8, average earliest arrival on breeding grounds 20 April (5 April to 1 May) (Glück 1982). Returns to north-east Germany and Scandinavia March–April (Klafs and Stübs 1987; SOF 1990); in northern Denmark, passage peaks at end of March (Møller 1978*a*); in St Petersburg region, timing variable, with movement sometimes chiefly March, at other times April (Mal'chevksi and Pukinski 1983). Main passage in Belgium April–May (Lippens and Wille 1972), and birds return to Britain chiefly mid-April to early May (Newton 1972; Taylor *et al.* 1981).

In east of range also (*caniceps* group), partially migratory, but limited information; birds disperse in flocks, some remaining in breeding areas all year, even in north or range, others migrating (apparently mostly short-distance) or making altitudinal movements (Dementiev and Gladkov 1954). Passage chiefly September–November and March–May. In Kazakhstan (breeds only in north), winters chiefly in south-east, mid-October to March; uncommon or rare in most other areas (Korelov *et al.* 1974). In Tadzhikistan, birds descend to lower levels from end of August or early September, and in some years move right down to valleys in October or November, remaining there until February (Sagitov 1962). In northern Caucasus, numbers augmented by winter visitors chiefly mid-November to early March. Some birds of eastern race *subulata* move west; recorded in winter from Voronezh and Kama regions. (Dementiev and Gladkov 1954.) In Pakistan, local birds disperse more widely in winter, and visitors arrive from further north, occurring mostly in lower hills (Roberts 1992). Subject to altitudinal movements in India (Ali and Ripley 1974). DFV

**Food.** Small seeds, mainly Compositae; in breeding season, also small numbers of invertebrates. Prefers seeds in milky, half-ripe state, so changes food plants constantly over year, and continually on the move from one patch of suitable species to another, which can be several km away, sometimes following the same route every day. (Newton 1967*a*; Glück 1980, 1985.) For behaviour associated with foraging, see Social Pattern and Behaviour. Generally takes seeds directly from flower or seed-head on plant, mostly on herbs, rarely grasses (Gramineae), in wasteland, open countryside, copses, etc., less often in parks or gardens; in winter regularly in trees, principally alder *Alnus* and pine *Pinus*; only exceptionally feeds on berries. In breeding season, forages in pairs or groups of up to *c.* 30, at other times of year often in larger flocks (Eber 1956; Newton 1967*a*; Glück 1980, which see for many details of group foraging, distance of feeding sites from nests, etc.; Sueur 1990*a*). In southern England, 81% of 108 feeding observations on herbs, 8% on grass, 5% on ground, 5% in shrubs, and 0 in trees or in air (Beven 1964). In eastern England, 26% of 135 feeding actions were on ground, 51% on plants less then 0·5 m above ground, 16% 0·5–1 m, and 7% more than 1 m (Kear 1962). Among most animated and agile Fringillidae, making constant and skilful use of feet both to maintain balance when reaching for seeds and to hold seed-head (etc.) while feeding (Sokolowski 1962; Newton 1967*a*; Glück 1980). In south-west Germany, 42% of actions perching or leaning forward, 23% clinging to sloping stem, 7% to vertical stem, 7% hanging sideways, 7% standing on ground, 6% head downwards, and 6% upside-down. Often lands at base of herb stem then moves up until flower-head bends over and can then be held under foot; on grass, several stems can be pulled together with bill then grasped in feet for support; in trees, frequently upside-down exploring cones or searching for insects in outermost leaves. 3 main feeding methods, illustrating function of long, pointed tweezer-like bill: picks seeds from exposed positions on seed-heads; inserts bill into half-hidden positions and extracts seeds after first opening bill to expose them; pierces tough involucral bracts of closed head to remove ripening seeds, an action performed otherwise only by Siskin *C. spinus* with similarly pointed bill. Milky seeds do not have to be de-husked. (Newton 1967*a*; Glück 1980, 1986.) For details of extraction and handling of seeds in relation to bill and leg structure, and for comparison with other seed-eaters, see Eber (1956), Kear (1962), Ziswiler (1965), and Newton (1967a, 1972). Slightly longer bill of ♂ allows it to reach seeds of teasel *Dipsacus* between spines, which ♀ has first to bend aside (thus ♀ feeds on this plant much less often) (Newton 1967a). When seeds of herbs unavailable, takes those in open pine cones, nipping off wings before cracking in bill (Volsøe 1949; Jahnke 1955). Occasionally eats leaves, (e.g. of cultivated *Brassica* or chickweed *Stellaria*), catkins of birch *Betula*, and various parts of flowers of apple *Malus*, pear *Pyrus*, or cherry, etc. *Prunus* (Eber 1956; Newton 1967a; Pfirter 1975; Glück 1980). Aphids (Aphidoidea) can be taken by turning over leaf, securing under foot, and picking from underside; caterpillars picked up in bill-tip; one bird made sallies from perch for 30 min to catch flies (Diptera) in the air (Glück 1980).

Diet in west Palearctic includes the following. Invertebrates: bugs (Hemiptera: Aphidoidea, Adelgidae), Lepidoptera (mainly larvae: Tortricidae, Coleophoridae, Amatidae, Noctuidae), adult and larval flies (Diptera), larval Hymenoptera, beetles (Coleoptera: Tenebrionidae, Curculionidae). Seeds, buds, flowers, and fruits of cypress *Chamaecyparis*, *Thuja*, larch *Larix*, fir *Abies*, spruce *Picea*, hemlock *Tsuga*, pine *Pinus*, poplar *Populus*, birch *Betula*, alder *Alnus*, elm *Ulmus*, plane *Platanus*, olive *Olea*, nettle *Urtica*, knotgrass *Polygonum*, dock *Rumex*, goosefoot *Chenopodium*, orache *Atriplex*, seablite *Corispermum*, chickweed *Stellaria*, mouse-ear *Cerastium*, buttercup *Ranunculus*, poppy *Papaver*, Cruciferae, meadowsweet *Filipendula*, burnet *Sanguisorba*, apple *Malus*, cherry, etc. *Prunus*, pear *Pyrus*, rowan *Sorbus*, melilot *Melilotus*, clover *Trifolium*, willowherb *Epilobium*, evening primrose *Oenothera*, viper's bugloss *Echium*, alkanet *Anchusa*, forget-me-not *Myosotis*, self-heal *Prunella*, snapdragon *Antirrhinum*, plantain *Plantago*, field scabious *Knautia*, small scabious *Scabiosa*, devilsbit scabious *Succisa*, teasel *Dipsacus*, golden-rod *Solidago*, cocklebur *Xanthium*, sea aster *Aster*, elecampane *Inula*, sunflower *Helianthus*, butterbur *Petasites*, yarrow *Achillea*, mugwort *Artemisia*, bur marigold *Bidens*, corn marigold *Chrysanthemum*, coltsfoot *Tussilago*, groundsel *Senecio*, burdock *Arctium*, thistles *Carlina*, *Cirsium*, *Cnicus*, *Onopordum*, *Carduus*, knapweed *Centaurea*, goatsbeard *Tragopogon*, viper's grass *Scorzonera*, chicory *Cichorium*, lettuce *Lactuca*, nipplewort *Lapsana*, sow-thistle *Sonchus*, wall lettuce *Mycelis*, *Chondrilla*, dandelion *Taraxacum*, catsear *Hypochoeris*, hawkbit *Leontodon*, hawksbeard *Crepis*, hawkweed *Hieracium*, ox-tongue *Picris*, grasses *Poa*, *Dactylis*, *Agrostis*, *Phleum*. (Collinge 1924–7; Witherby *et al.* 1938; Eber 1956; Turček 1961; Glutz von Blotzheim 1962; Sokołowski 1962; Newton 1967a; Tutman 1969; Glück 1980; Sabel 1983; Sueur 1990a.) In New Zealand, spiders (Araneae) recorded in diet (Lane 1984).

In Oxford (southern England), recorded feeding frequently on invertebrates in 2 years when arrival in April

was before first seeds became available (Newton 1967a). Different populations seem to have different 'food traditions', e.g. groundsel not recorded in diet in northern Germany by Eber (1956) but in Oxford in April was 49% of 514 observations (Newton 1967a), and 37% of 38 during April and June in Somme (northern France) (Sueur 1990a) (although ignored in Oxford if preferred dandelion, thistles, or, above all, knapweed present); mugwort was hardly noted in Oxford, and accounted for only 1% of 267 September records in Somme, yet was 11% of diet in northern Germany (n = 547 observations). In Australia, 58% of 33 food plants eaten by introduced population were, in turn, introduced Compositae (Middleton 1970). Preference experiments in captivity generally use hard (not milky) seeds, so probably do not reflect choice of food in the wild (Glück 1980); for such experiments, or reviews, see (e.g.) Kear (1962), Ziswiler (1965), Hoppe (1976), and Glück (1980). In captivity, took 1·2–2(–8) s to open seed after 3 days of habituation; hemp *Cannabis* took 4 times longer than any other seed used (Ziswiler 1965); in the wild, average 1·1 s (n = 30) to consume dandelion seed, and in ideal circumstances can eat up to 98 seeds per min (Glück 1985, 1986).

In south-west Germany, over 2 breeding seasons, 25% of diet (by observation) seeds of thistle *Cirsium oleraceum*, 20% dandelion, 14% buds and flowers of apple, 8% scabious *Knautia*, 6% knapweed, 5% dock, 4% hawkbit, 3% coltsfoot, 3% grass *Dactylis*, 2% meadowsweet, 4% other Compositae, 2% aphids, which can account for almost 25% of diet in some weeks of June–July (Glück 1980). For northern Germany, see Eber (1956). In Oxford, over 3 years, 30% of 7648 feeding observations on thistles, 17% groundsel, 14% burdock, 11% teasel, 7% dandelion, 3% ragwort *Senecio*, 3% pine, 3% alder, 2% chickweed, 2% sow-thistle, 2% meadowsweet, 2% aphids (Newton 1967a). In Somme, 40% of 991 observations, all year, were on thistle *Cirsium arvense*, 28% alder, 12% *C. oleraceum*, 5% goosefoot, 2% *Carduus*, 2% dandelion (Sueur 1990a). In England, 27% (by number) of 1226 items in 54 stomachs were insects, mainly caterpillars and aphids (Collinge 1924–7). In Australia, insects never more than 8% by volume of diet over year, though can be 20% in peak month, mostly aphids and other Hemiptera, and caterpillars (Middleton 1970). Early in year commonly in trees, mainly alder, birch, and pine, feeding on seeds and catkins (Eber 1956; Newton 1967a; Glück 1980); 99% of 123 observations in Somme, January–March, were on alder (Sueur 1990a); in spring and early summer, diet almost wholly Compositae, e.g. 87% in Somme (April–June, n = 38) and 80% in Oxford (April–June, n = 1934), principally dandelion (up to 75% in late May, south-west Germany), hawksbeard, goatsbeard, groundsel, and knapweed; often in woods at this time in Oxford taking elm seeds and invertebrates (16% of diet in April aphids, n = 514 observations), and in fruit-growing area in south-west Germany c. 50% of food can be buds and flowers of fruit trees; from July into autumn, thistles become very important (especially *Cirsium oleraceum*), comprising 80-90% of diet in all studies; in late autumn and winter, seeds of teasel and burdock often eaten, and, in Somme, goosefoot and ragwort before December, when 98% of 141 observations were again in alder (Eber 1956; Newton 1967a; Glück 1980; Sueur 1990a). In Poland, basic diet in spring apparently cocklebur Xanthium (Sokołowski 1962).

Captive birds consumed average 2·2-3·9 g fresh weight per day (Kear 1962); requirement in wild ranges from 4·2 to 27·6 g depending on seed energy content, e.g. grass *Dactylis*, teasel, thistle *Cirsium vulgare*, and burdock contain most energy per g, *C. oleraceum*, coltsfoot, buttercup, and sow-thistle least; daily energy requirement *c.* 52 kJ, thus hypothetical number of seeds needed per day would range from 384 (burdock, gathered in 7 min) to 43 662 (coltsfoot, gathered in 13·5 hrs). If only 'inefficient' seeds available, may not be enough daylight to achieve necessary intake, so birds forced to move to winter quarters with different plants or more daylight; e.g. during October in southern Germany, daylight decreases from 11·5 to 10·0 hrs. (Glück 1980, 1985, which see for many details.) See Korodi-Gál (1965) for additional data on energetics. See Glück (1986, 1987) for increased intake when feeding in flocks because of reduction in individual vigilance.

Food of young little studied. In Poland, fed some insects for first 10 days, thereafter only seeds; insects recorded were caterpillars (Noctuidae, Amatidae) and beetles (Tenebrionidae) (Sokołowski 1962). In Oxford, early broods and those near woodland were fed more insects than later ones, and all fed only seeds towards end of nestling period (Newton 1972). Diet presumably same as adult at this time; most efficient food plants available in spring, and very likely fed to young in south-west Germany, were goatsbeard, knapweed, grass *Dactylis*, and dock; seeds brought in crop, insects in bill (Glück 1980, which see for distances flown from nest to different foraging patches according to energy content of food plants).                                                            BH

**Social pattern and behaviour.** Most aspects well studied. Pioneering study by Conder (1948) in Eichstätt (Bayern, Germany) and descriptions of displays refined by study on captive birds (Hinde 1955-6, which see for comparison with other Fringillidae, also Newton 1972). Important recent studies, especially on flock behaviour, in Baden-Württemberg (south-west Germany) by Glück (1980, 1982, 1985; see below for other publications by same author).

1. Typically gregarious outside breeding season from late summer to spring, also often small flocks of off-duty birds and non-breeders in breeding season (Conder 1948; Dementiev and Gladkov 1954). In Europe and European USSR, family parties begin to gather into loose nomadic flocks in mid-August, and as autumn approaches these in turn coalesce into more compact flocks of up to hundreds and sometimes even thousands (Dementiev and Gladkov 1954; Géroudet 1957; Glück 1980, 1982; Lack 1986). In Baden-Württemberg, early autumn flocks (maximum 80-100 birds) comprised groups of independent juveniles and

family parties; flocks and roosts larger wherever food source substantial (see Roosting and Flock Behaviour, below); in winter and early spring, singletons also occur, usually young ♂♂, or flock-members reconnoitring feeding sites (Glück 1980, 1982). For similar dispersion in European USSR, see Dementiev and Gladkov (1954). Key to understanding flock size on daily and seasonal basis is distribution of food supply: in autumn when food is plentiful, flocks exceeding 100 birds not uncommon but as winter progresses and food becomes scarcer, flocks mostly less than 10 birds. In Britain, ♂♂ predominate in resident winter flocks, indicating that more ♀♀ than ♂♂ emigrate. (Lack 1986.) Never migrates singly, but typically in flocks of 10-60 (3-100) (Glück 1982). Influx of wintering birds produces very high local densities in Spain, e.g. *C. carduelis* the most abundant wintering species (average 649 birds per km²) in pastures on the northern side of Strait of Gibraltar, with lower, but still significant numbers in matorral (mean 70 birds per km²) (Arroyo and Tellería 1984). Because of its specialized winter foods, *C. carduelis* seldom mixes with other finches (Fringillidae) in winter (Géroudet 1957; Lack 1986) though singletons readily associate with Siskins *C. spinus* and Linnets *C. cannabina* (Géroudet 1957). BONDS. No evidence for other than monogamous mating system. In New Zealand unusual case of nest in which both sexes (exceptionally: see below) incubated; from hatching, 3rd bird participated amicably in feeding the young, all 3 making simultaneous visits to nest (Lane 1984). No information on duration of pair-bond. In Baden-Württemberg, 5-10% of ♂♂ unpaired in breeding season (Glück 1980). *C. carduelis* sometimes hybridizes with Greenfinch *C. chloris* in the wild; also recorded hybridizing with various other *Carduelis*, Canary *Serinus canaria*, Serin *S. serinus*, and Bullfinch *Pyrrhula pyrrhula* in captivity (Gray 1958, which see for references). ♀ alone builds nest, incubates and broods. ♂ feeds ♀ on nest and provides most food for mate and young while ♀ brooding, both sexes participating thereafter (Conder 1948; Spillner 1975). Young fed by parents for at least 10 days after fledging (Sokołowski 1962; Newton 1972; Spillner 1975). Age of first breeding 1 year (Glück 1982). BREEDING DISPERSION. Mostly in neighbourhood groups of 2-3 pairs (Newton 1972): singles, or groups of 2-9 pairs with strong social contact (Glück 1980, 1985). Territory relatively small, serving for mating and nesting, but foraging almost entirely outside (Conder 1948), up to 800 m from nest but mostly less than 400 m, mean 154 m (Glück 1980); territory maintained only for duration of nesting period. During courtship, territory *c.* 250 m², contracting to *c.* 10 m² after eggs laid. Nests are typically close, in same or adjoining trees; e.g. in Eichstätt, 2 nests 5-7 m apart, and also a group of 3 nests in *c.* 125 m². (Conder 1948.) Walpole-Bond (1938) reported 3 pairs nesting in same tree; in Israel more than 5 (niediecki) in tree (Shirihai in press). The need to locate nests for successive and replacement broods near feeding areas, and the rapidly changing dispersion of food sources through the breeding season, means that particular nest-sites are used only once per season (Glück 1980). Few details, but in pair which attempted 3 broods, 3rd nest 25 m from 2nd (Spillner 1975). Density in given area varies greatly from year to year (Schifferli *et al.* 1982) and in given year from place to place, e.g. 6 pairs or more may breed in 1 ha while apparently similar habitat may hold none (Géroudet 1957). Density estimates therefore highly dependent on size of area sampled. In Britain, 5-10 pairs per km² in several rural, farmland, and coastal-scrub sites, average 2·2 pairs per km² for all farmland plots (Sharrock 1976). In Mecklenburg (eastern Germany) local differences can be considerable: 10-60 pairs per km² in parks and cemeteries, 2·5-23 in riverine trees and scrub, 21-77 in copses and fields, 10-38 in villages; 1·2-25 pairs per km of tree-lined country roads (Klafs and Stübs 1987). In Bulgaria, density ranges from 0·1–0·2 pairs per km² in agricultural shelterbelts (B Ivanov) to 60 in habitat dominated by lime trees *Tilia* (Simeonov and Petrov 1977); in Sofia city, maximum 40 pairs per km² in 'street vegetation' (Iankov 1983). In mainly cork oak *Quercus suber*, Morais (northern Portugal), 3·6 'territories' per km² (Mead 1975). In maquis, Morocco, 30·8 pairs per km² (Thévenot 1982). No regular association with other species, but for record of nesting close to Mistle Thrush *Turdus viscivorus*, apparently for protection, see Raevel (1981*b*). ROOSTING. No information for breeding season except that brood-members initially roost together from fledging up until 1–2 weeks after independence (Glück 1982). Outside breeding season, *C. carduelis* roosts communally, site often containing hundreds of birds in continental Europe, typically much smaller and more temporary in Britain (Newton 1972). Typical site is dense inner branches of trees and shrubs (Dementiev and Gladkov 1954), notably oaks *Quercus* and beeches *Fagus* which have retained dry leaves (Witherby *et al.* 1938); in Leipzig region (Germany) in willow *Salix* thickets in autumn and spring (Dorsch 1970); in Israel uses maquis and plantations (Paz 1987). Winter roost of up to 360 birds in trees at Billingham (northern England) (Bell 1978). In inner London (England), at least from late December to early April 1964–5, over 100 birds regularly roosted at c. 10–15 m in roadside plane *Platanus* trees spaced over a few hundred metres; roosting complex included several trees (which were *c.* twice as high as roosting height), and preference shifted during the period so that 1–2 trees were favoured for a spell, then others nearby (Ruttledge 1965). Sometimes uses mixed roosts: in Bedfordshire (southern England) up to 70 birds regularly joined *C. chloris*, *C. cannabina*, and Chaffinches *Fringilla coelebs* in elder *Sambucus* and hawthorn *Crataegus* scrub in old quarry (Newton 1972). In Eichstätt, September, *c.* 50 roosted in lime trees with *C. chloris*, *F. coelebs*, House Sparrows *Passer domesticus*, and Tree Sparrows *P. montanus* (Conder 1948). For flocks assembling to roost see Flock Behaviour (below). For locomotor activity rhythms of captive birds see Glück (1978). Migrate in first half of day, feeding after 14.00 hrs (Glück 1982).

2. Little information on flushing distances but, in feeding flocks, juveniles much less nervous (flush at 20–30 m) than adults (*c.* 50 m) (Glück 1982). See Glück (1987) for adjustment of vigilance and foraging behaviour to experimental exposure to overflying Merlin *Falco columbarius*. See also below for relationship of vigilance to flock size. In captive study, Hinde (1955–6) found little difference between behaviour of ♂♂ and ♀♀. FLOCK BEHAVIOUR. Flock of *C. carduelis* is a highly organized mobile unit, with marked development of communication signals, for finding and exploiting patchy and shifting food sources. Small winter feeding parties may travel many kms in the course of a day, and often re-visit the same sites on successive days (Newton 1972). Flocks leave roost early morning and feed assiduously till mid-morning, then move to higher trees to rest, during which (from January) ♂♂ spend 2–3 hrs quietly singing (see Song-display, below); resume feeding *c.* 14.00–15.00 hrs, and move to roost in evening (Dementiev and Gladkov 1954). Foregoing diurnal pattern assumes single flock but, typically, large roosting flock breaks up by day into smaller groups (5–30 birds) which reassemble 15.00–16.00 hrs into one flock, then feed before roosting (Glück 1982). In breeding season, after waking up, members of nesting groups visit each other briefly and sing; often bathe late morning, sometimes also late afternoon, then feed; diet of milky seeds associated with regular bill-wiping after every feeding bout (Glück 1980). If only one food source available, with water nearby, flock stays near it all day; while feeding, and during pauses to preen (etc.), exchange of Flock-calls (see 2b in Voice) is typical; however, small groups or singletons continually break away to prospect for other food sources up to 1–2 km away. In flight, flock-members (whether foraging or migrating) communicate with Contact-call. (Glück 1982: see 2a in Voice.) Party seeking new food source flies out low over ground, calling the while, and if it finds unoccupied feeding site, stays up in trees for minutes to make sure site is safe before flying down; if site already occupied by conspecifics or other Fringillidae, hesitancy still occurs but is shorter. During feeding, in which flock moves in a rolling wave, some birds continue to stay vigilant in trees. (Glück 1980.) The larger the flock the less time each individual spends on vigilance and so the higher its feeding rate; increase in energy intake is considerable up to flock size of 8 birds, only slight thereafter (Glück 1986). The following behaviours are interpreted by Glück (1980) as means of communicating location of feeding sites to fellow members of nesting group (see Breeding Dispersion, above): (1) Several times during day, group-members assemble at a specific place ('group meeting') after which all may fly off together to feeding site; (2) ♂ group-members may pause *en route* to known feeding site to sing briefly in tree-tops, or fly directly to site giving Contact-calls; (3) Group-members overflying feeding site (in breeding season or on migration) may be called down by Attraction-calls (see 2c in Voice); (4) Singing by unpaired ♂♂ (primarily to attract ♀♀) in good feeding sites doubtless attracts other birds to feed. For aggression in feeding area, within and between nesting groups, see Antagonistic Behaviour (below). SONG-DISPLAY. Mainly by ♂ (for mate attraction and territorial defence; see Antagonistic Behaviour, below), occasionally also by ♀ (Géroudet 1957); see 1 in Voice. In captive study, ♀♀ sang almost as often as ♂♂ (Hinde 1955–6). ♂ sings mostly from elevated perch; wings usually slightly drooped, and pivoting movements (see Antagonistic Behaviour, below) may accompany Excitement-call (see 3 in Voice) and other calls when these are included in song (Hinde 1955–6). ♂ also has a Song-flight: flies hesitantly with slow deep wing-beats (Butterfly flight), plumage ruffled, and tail spread, sometimes interspersed with gliding for 1–2s (Conder 1948; Newton 1972); similar to Song-flight of *C. chloris* but not so erratic, less common, and perhaps serves wider function (see Antagonistic Behaviour, below). ♂ sings inside and outside territory, usually when accompanying mate; Song-display by ♂ most intense when ♀ nest-building (including when escorting her on trips to collect material) and laying; also sings during incubation. (Conder 1948.) In western Europe, song reported throughout the year but mainly mid-March to mid-July, less often late January to late August, not uncommonly October (Witherby *et al.* 1938; Géroudet 1957). Song heard in winter quarters near Rome (Italy) (Alexander 1917), and on spring migration (stopover in late March, Denmark) (Volsøe 1949). ANTAGONISTIC BEHAVIOUR. (1) General. Markedly aggressive early in territory occupation (Newton 1972). Food relatively scarce at this time and therefore, unlike at other times, discrete feeding sites quite strongly defended, e.g. one pair regularly defended pine *Pinus* tree against other pairs (Glück 1982). ♀ typically the more aggressive sex during nest-building but roles reversed after onset of incubation; ♀ sometimes came off nest to assist mate (Conder 1948). At one nest, ♂ returned every 5–30 min and chased off (with Threat-calls: Conder 1948; see 5 in Voice) conspecifics and other small passerines intruding closer than 4 m (Sokołowski 1962). Relations between members of nesting group (see Breeding Dispersion, above) amicable unless individual distance (see below) or nest-area infringed; greater hostility shown to trespass, e.g. into feeding area, by strange conspecifics (Glück 1980, which see for details). (2) Threat and fighting. ♂ defends territory by making himself conspicuous, sitting on tops of bushes and occasionally singing, but adopting no special postures; this alone enough to

deter most rivals, but if one approaches too closely, owner flies nearer, sings loudly and, failing that, attacks intruder (Newton 1972). When resting, feeding, or drinking, both sexes respond to conspecifics or other small passerines infringing individual distance (10–15 cm) by slight opening and lowering of wings, Threat-calls, and sometimes attacking by leaping up in extended posture, bill pointed at intruder; latter may retreat in silence or with alarm-call (Zablotskaya 1975: see 6b in Voice). These wing-lifting movements in resisting supplanting-attack perhaps not different from 'Sleeked wings-raised posture' mentioned by Hinde (1955–6). Commonest form of threat in close encounter is Pivoting-display (Fig A) in which bird crouches and swings from

A

side to side; depending on degree of flexure of legs, body may be more or less upright; wings usually lowered and tail spread, exposing respective markings, and red 'blaze' expanded on face (upturned red face of incubating/brooding ♀ also intimidatory); Pivoting-display usually accompanied by Excitement-call, occasionally also Threat-call (Conder 1948; Hinde 1955–6; Newton 1972). The more excited the bird the more exaggerated the display; thus with decreasing intensity, Pivoting arc diminishes from *c.* 90° (maximum) to 10°, or to only one side from midline; instead of Pivoting body, bird sometimes jerks tail to one side only or fans it with no sideways jerk. Owner and intruder often Pivoted in turn and apparently tried to perch higher than each other. (Conder 1948.) At least in captive study, Pivoting-display much more common between mates (see Heterosexual Behaviour, below) than between winter flock-members; sometimes one bird will Pivot after supplanting or being supplanted by its mate, or after chase or fight (Hinde 1955–6, which see for incidence of Pivoting-display in different contexts). In disputes, submissiveness may be signalled by Fluffed-posture in which plumage is ruffled all over, as in *C. chloris* (see that species). However, *C. carduelis*, in contrast to other Fringillidae, also ruffles plumage extensively in attacks and sexual chases (Hinde 1955–6; see also Heterosexual Behaviour, below), e.g. Fluffed-posture sometimes adopted by dominants in winter flock. This presumably the 'Threat-posture' described by Conder (1948). If intruder does not yield to initial threat-displays, owner performs expelling-flight or attacks. (Conder 1948.) Expelling-flight may escalate to fierce fight in which combatants drop to ground, aiming pecks, clawing at each other, and giving Threat-calls (Conder 1948; Newton 1972). After dispute, territory-owner returns to perch and sings, or performs Song-flight over territory (Newton 1972). Heterosexual Behaviour. (1) General. In Baden-Württemberg, first birds back in breeding area (early to mid-April) are highly secretive, moving in pairs (♂-♀) or, more rarely, trios; pairs in which both partners are older than 1 year generally arrive earlier than 1-year-old pairs (Glück 1982). In England, while some have already taken up territories, others are still in flocks (e.g. Bell 1978). (2) Pair-bonding behaviour. Arrival of birds already paired on breeding grounds (see above) indicates pair-formation occurs in flock (Conder 1948; Géroudet 1957). Exact sequence of courtship not clear but includes the following elements, com-

bining descriptions of Conder (1948) and Hinde (1955–6): Pivoting-display (as in Antagonistic Behaviour, above) by either or both partners; when both display they perch, often on same twig 15 cm apart, and each Pivots at different speed, i.e. movements not synchronized. Pivoting-display sometimes precedes or follows Courtship-feeding and copulation (see below). Vigorous sexual chases (described mainly from captive birds), often accompanied by Threat-calls, in which pursuer, of either sex, often has plumage markedly ruffled ('Fluffed chase': Hinde 1955–6). Pursuer may even attack partner but often makes no real effort to catch up; frequently uses Moth-flight over short distances (less than *c.* 6 m: Conder 1948) in which wing-beats rapid and shallow, and tail raised; short glides and Butterfly-flight also seen in this context. Several sexual chases may follow in quite rapid succession, after which ♂ (if pursuer) sleeks plumage, stops chasing, and begins to Pivot and sing. Sexual chasing especially frequent just before copulation (♀ then usually pursues). Between chases, and at other times in pre-copulatory period, pursuer (either sex) often perches in low-intensity Soliciting-posture (Fig B, left): crouches with plumage ruffled, wings

B

drooped (sometimes shivered), and tail raised. (3) Courtship-feeding. Various intention movements anticipate (i.e. occur a few days prior to) Courtship-feeding, and are described below in chronological order. Unless otherwise acknowledged, account combines similar observations of Conder (1948) and Hinde (1955–6). From time to time, one member of pair turns head towards other and opens and closes bill rapidly, which may cause mate to approach closely and do likewise (Bill-flirting). Sometimes bills meet (Bill-touching or 'kissing') repeatedly in rapid series (see Conder 1948 for details) and in greater excitement ♀ may turn her head so that open bill fits into his. These preliminaries sometimes lead to Courtship-feeding which (unlike Bill-flirting and Bill-touching) ♀ invites with Soliciting-display with wing-shivering, bill upturned (Fig B, left) and Begging-calls (see 4 in Voice). This may result only in bills touching without food transfer ('mock-feeding': Conder 1948). Sometimes Courtship-feeding takes place with little ceremony, suggesting pair-bond well established. In captive study, Courtship-feeding seemed more frequent in pairs where dominance was fairly evenly balanced with little aggression. ♂ continues feeding ♀ during incubation and brooding (see below); sitting ♀ shows great excitement (wing-shivers, Begging-calls: Fig C) as ♂ arrives at nest giving Excitement-calls. In study by Spillner (1975) of pair with 2nd clutch, incubating ♀ was fed every 20 min; during 11.00–14.00 hrs, depending on weather, gaps between feeds were much longer, and on some days ♂ did not feed ♀ at all; in such cases ♀ left nest when ♂ arrived and both went off to feed

C

together. (4) Mating. Once nest framework built, copulation occurs mostly on nest (Dementiev and Gladkov 1954). Rest of account based on Conder (1948) and Hinde (1955–6). Copulation may be initiated by either sex performing Moth-flight (see above) which seems to be aerial extension of Soliciting-display. ♀ invites copulation with Begging-calls and intense Soliciting-display, much as for Courtship-feeding (above). ♂ flies (often in rather upright posture) to ♀, hovers above her, then lands alongside, again rather upright, wings quivering, before mounting; may also land directly on her back. After successful copulation, ♂ often briefly adopts ♀'s Soliciting-posture; ♀ usually ignores this (flies off), may threaten ♂, or exceptionally mount him. In Eichstätt, copulation usually occurred after ♀ had been nest-building; during week before eggs laid, pair copulated at least once per day, and ♂ attempted copulation after 1st, 2nd, and 3rd eggs laid; resurgence in copulation *c.* 3 days before young fledged. (5) Nest-site selection. In Eichstätt, site apparently selected very early in, or even just before, territory established. Pair prospected over wide area, passing through territories of other pairs. Not known which sex chooses site; one after the other, ♂ and ♀ entered possible sites, squatting and turning, sometimes with Begging-calls. (Conder 1948.) According to Nethersole-Thompson and Nethersole-Thompson (1943*b*), site chosen after concentrated 'brooding' in chosen crotch by ♀ who calls softly; this seen fully 1 week before building began. (6) Behaviour at nest. ♂ does not usually assist in building but occasionally brings insignificant amounts of material, with excited wing-flicking, when he escorts ♀ to nest. Usually drops material, sometimes gives it to ♀ for adding to nest, exceptionally adds it himself. Material included some robbed from conspecifics and *F. coelebs.* (Conder 1948.) Late nests may include material taken from earlier nests (Paz 1987). See above for Courtship-feeding of incubating ♀. RELATIONS WITHIN FAMILY GROUP. ♀ broods assiduously for 7–9 days, in cold weather sometimes longer (Sokołowski 1962; Spillner 1975), up to 13th day (Conder 1948). When young are small, ♂ gives all food to ♀ (who begs for it, as in Courtship-feeding) for self-consumption and for transfer to young (Conder 1948; Spillner 1975). ♀ fed by ♂ up to 38 times per day, storing food in crop for distribution to young (Glück 1988). ♂ remained at nest long enough to deliver food and, as soon as he left, ♀ fed young; ♂ did not feed young directly until young 7 days old (Spillner 1975), although ♂ also recorded feeding young directly at 3 days when both adults came to nest together (Conder 1948). Eyes of young begin to open at 5 days, fully open at 7 days (Sokołowski 1962). When young are small, faecal sacs are swallowed by ♀ but from *c.* 1 week young begin defaecating onto (and over) nest rim from where parents carry some (nearest nest cup) away for disposal (Spillner 1975; Glück 1988). As the young's digestive efficiency improves, ♀ eats

ever fewer faeces, stopping completely at *c.* 10–13, days by which time young can assimilate over 90% of energy in food (Glück 1988). According to Blair and Tucker (1941) and Conder (1948), both parents eat faecal sacs. Young fledge from 13 days and are difficult to flush even on 12th day (nest-tenacity associated with relatively high siting of nests in trees); on leaving nest they can instantly fly vertically upwards into tree-tops, and in another few days can follow parents (Sokołowski 1962; Newton 1972). When young were fledging, parents occasionally flew *c.* 6–7 m from nest-tree and returned in Butterfly-flight, perhaps indicative of general excitement. On 3 occasions, young seen Pivoting (see Antagonistic Behaviour, above) just before and after leaving nest, and when about to be fed, but usual begging-display of juveniles is like Soliciting-display of adult ♀; after leaving nest, young sometimes seen landing on back of arriving parent. (Conder 1948.) For overlapping broods see Breeding. ANTI-PREDATOR RESPONSES OF YOUNG. No information. PARENTAL ANTI-PREDATOR STRATEGIES. (1) Passive measures. Eggs (of *niediecki*) are sometimes hidden at bottom of nest beneath thick layer of lining. ♀ a tight sitter, not flushing until intruder at nest. (Pitman 1921.) Incubating ♀ once seen crouching low in nest for 2–3 s as Buzzard *Buteo buteo* flew over (Conder 1948). (2) Active measures: against birds. No information other than that in Antagonistic Behaviour (above) for repelling small birds from nest-area. (3) Active measures: against man. Only one report of injury-feigning, but no details of display: in New Zealand, adult feigned injury most intensively when disturbed during hatching, less so when brooding young up to 4 days old, after which response disappeared (Campbell 1972). (4) Active measures: against other animals. During nest-building in Eichstätt, disturbance by cats elicited alarm-calls (see 6b in Voice) and display resembling Pivoting-display, but outer tail feathers twisted down—on right side when body swung to left (Conder 1948) and presumably vice versa. According to Hinde (1955–6) this display different from Pivoting-display (see also Hinde 1954).

(Figs by D Nurney: A from drawing in Newton 1972; B from drawing in Conder 1948; C from photograph in Spillner 1975.)

EKD

**Voice.** Diverse repertoire used throughout the year. 15 calls of nominate *carduelis* (Moscow region) and 16 of *caniceps* group (Tadzhikistan and Kazakhstan) listed by Zablotskaya (1975), but not all transliterations and sonagrams easily reconciled with other published sources (for other sonagrams see Güttinger 1978, Glück 1980, 1982, Bergmann and Helb 1982), so only a selection from Zablotskaya (1975) included in following scheme which, in excluding a number of possibly minor variants, is also more conservative than that of Zablotskaya (1975). In general, frequency range of *caniceps* group is wider, and fundamental frequency mostly higher, than nominate *carduelis* (Zablotskaya 1975). Following account owes much to descriptions from study by Conder (1948) in Eichstätt (Bayern, Germany).

CALLS OF ADULTS. (1) Song. (1a) Song of ♂. Rather tinny, tinkling, jingling which, when heard close by, can be fairly incisive; short snatches confusable with Linnet *C. cannabina* but less nasal and twangy (W T C Seale). Pleasing liquid twittering elaboration of call-type units with variations, recalling Canary *Serinus canaria* (Witherby *et al.* 1938). Also resembles Siskin *C. spinus* but

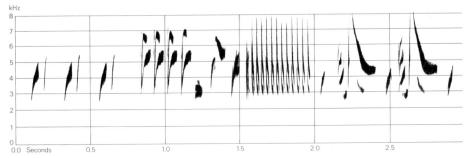

I  W T C Seale  England  June 1992

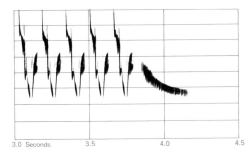

I  *cont.*

distinctive for inclusion of main Contact-call (see 2a, below) as well as attractive mewing sounds and rasps recalling Sand Martin *Riparia riparia* (presumably call 5) (Bruun *et al.* 1986). Mostly in phrases (separated by short pauses), sometimes almost continuous; phrases delivered at rapid tempo (including units inseparable to human ear: Güttinger 1978), often beginning with rapid sequence of diagnostic Excitement-type calls (see 3, below), followed by various 'trills' and, towards end, typically harsh drawn-out nasal units such as 'ziär' similar to those given by *C. spinus* (Bergmann and Helb 1982). Detailed analysis of *caniceps* showed that song typically cardueline in consisting of segments separated by barely audible pauses (less than 0·1 s); longer pauses (more than 1 s) between phrases; phrase length highly variable, comprising 2–20 segments, but mostly 5–6; restricted number of unit-types begin song (see above), e.g. in study of *caniceps* 77% of songs began with particular unit-type (Güttinger 1978).

Fig I (from solitary bird) typical of *britannica* song and, like *caniceps*, exemplifies marked segmental structure; Fig II (from bird in small flock) less clearly segmented, and also notable for inclusion of 7 consecutive Threat-type calls (see 5, below), followed by longer buzzy unit which occurs in other recordings and may be diagnostic (W T C Seale). In different races, song variation limited to a few predictable characteristics, primarily rate of frequency change of units, whereas spacing of segments and phrases remains rather constant. (Güttinger 1978, which see for comparative song organization of nominate *carduelis*, *caniceps*, and various species of greenfinch *Carduelis*.) Compared with nominate *carduelis*, *subulata* less clear and ringing, less loud and more melodious (Johansen 1944; Dementiev and Gladkov 1954). On basis of one recording (Fig III), song of *niediecki* (Cyprus) less rigidly structured than nominate *carduelis* and *britannica*, showing less segmentation and less unit-repetition (W T C Seale), but more information needed. For structure of song of captive *C. carduelis* × *S. canaria* hybrids, see Güttinger and Clauss (1982). (1b) Song of ♀. At least in captive birds, song tended to be less vigorous and less sustained than that of ♂ (Hinde 1955-6). (1c) Subsong. Recording (Fig IV), not known if from adult or juvenile (song emanated from small flock), is subdued chirruping and chattering, undulating markedly between higher and lower pitches, sometimes reminiscent of Swallow *Hirundo rustica* or *R. riparia*; includes various high squeaky components, periodically also units reminiscent of subdued Chink-call of Chaffinch *Fringilla coelebs* (W T C Seale). (2) Contact-

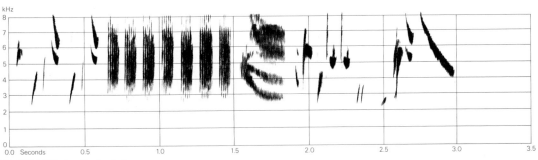

II  W T C Seale  England  June 1992

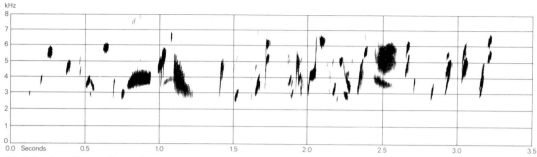

III  J Gordon  Cyprus  June 1984

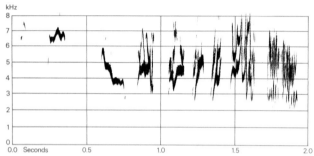

IV  W T C Seale  England  August 1989

'didudide', 'didud', etc., and at lower intensity a quiet 'dud' (Bergmann and Helb 1982). Delivery varies: given slowly and quietly with conversational quality in intimate pair-contact, but at other times more urgently, apparently signalling varying intention to fly (Conder 1948). Described as highly variable multi-unit call, with regional differences in pitch pattern (Güttinger 1978, which see for sonagrams). (2b) Flock-call. A 'chlü' given while feeding in flock, and during pauses to preen, etc. (Glück 1982, which see for sonagram). Call depicted in Fig VI rendered

calls. (2a) Main call. Typically 2, less often 3 or more clear-sounding units of varying pitch, given by both sexes when perched or in flight, delivered in rapid succession; highly diverse, and different renderings represent genuine variation within and between individuals, although pair-members tend to give same call-types (Bergmann and Helb 1982). In Fig V (representing single bird), all calls from 6th unit onwards (but excluding 5th from end) are of this type, individual units sounding like 'wit' and 'pit' with distinctive 'popping' timbre; last few units in this series fade as bird, hitherto perched, flies off; remaining sounds in Fig V are call 2c (W T C Seale). Other renderings of Main call include 'teetut', 'teetü', 'titee', 'tutti', with emphasis on either unit; commonest 3-unit type 'tütitee' and 'tütitü' (Conder 1948); sharp high-pitched 'tickeLIT' (Bruun *et al.* 1986); 'didelit' or 'stiglitt' (hence German name Stieglitz for *C. carduelis*) (Glück 1982); 'dudidelit',

VI  W T C Seale  England  June 1992

'weeju' (W T C Seale). May indicate mild alarm (see 6b, below). (2c) Attraction-call. Call of 2–4 units, e.g. 'hidit', 'zidid' (Bergmann and Helb 1982), or 'zididitt', given by birds already feeding, and has effect of attracting others (directed especially at fellow members of nesting group) (Glück 1980, 1982, which see for sonagrams). In Fig V,

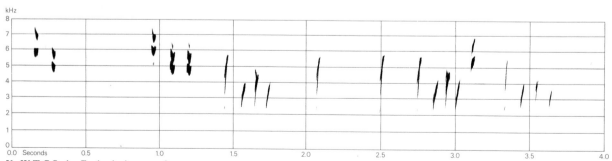

V  W T C Seale  England  August 1989

first 5 units ('jit jut, jit jut jut', with incisive timbre) this call, also apparently 5th unit ('jit') from end; other units in Fig V are call 2a (W T C Seale). (3) Excitement-call. Mainly disyllabic calls which accompany Pivoting-display (see Social Pattern and Behaviour), especially in territorial defence; more varied than call 2a and differs in showing definite slur between syllables, e.g. 'wee ee', 'weeyü', 'waya', 'wairhi', 'wühee', 'tüwit', and 'tsiwa' (Conder 1948). According to Hinde (1955–6), probably the most common is 'tuleep', with sometimes 3rd syllable added, e.g. 'tu-wee-oo'. In Fig VII, 2nd (complex) unit apparently of

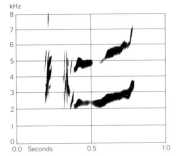

VIII   R Margoschis   England   July 1979

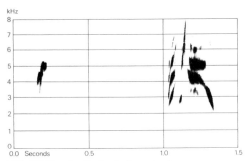

VII   V C Lewis   England   August 1967

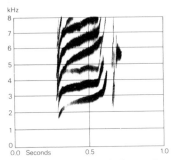

IX   V C Lewis   England   June 1965

this type, rendered 'petchitoo', 1st unit rendered 'tik' (W T C Seale). According to Zablotskaya (1975), call basically 'pikiu' but subject to complex frequency changes. (4) Begging-call. High-pitched, fairly rapid 'tee tee tee' given mainly by ♀ performing Soliciting-display to invite courtship-feeding or copulation from ♂ (Conder 1948; Hinde 1955–6); varies from 'ti' through 'tee' to 'tü' and repeated at rate of *c*. 2 calls per s; used by ♀, only in presence of ♂ and in territory, also infrequently by ♂ (Conder 1948). Described by Zablotskaya (1975) as high-pitched whistling 'tsi-tsi-tsi'. (5) Threat-call. Harsh drawn-out 'tzzz' given in disputes (often chase or fight) (for similar, but less harsh-sounding, call of Greenfinch *C. chloris*, see call 5 of that species) with conspecifics and other small passerines, often accompanied by the more common types of calls 2a and 3 (Conder 1948). Also rendered hard rattling 'tschrr' or 'trr', sometimes in series (Bergmann and Helb 1982), e.g. series of 7 incorporated in song (Fig II). The 'zicker' aggressive-call, very occasionally given in Pivoting-display (Hinde 1955–6), presumably the same. (6) Alarm-calls. (6a) Emphatic staccato 'titt wittit' heard occasionally when alarmed at nest (Conder 1948). (6b) Nasal rising 'wäi' or similar (e.g. 'delüi', 'düii'), shorter than homologous call in some other Fringillidae, given when disturbed by ground- or aerial predator (Bergmann and Helb 1982). Recordings show great variability, e.g. rising 'pit-dooeey' (Fig VIII) from ♀ near nest; rather nasal, diadic 'doooee-chk' ascending steeply in pitch (Fig IX) (W T C Seale). Other renderings (or variants) of presumably this call include: monosyllabic

'whü', similar to Bullfinch *Pyrrhula pyrrhula*, when cat near nest (Conder 1948); grating 'geez' comparable to 'tsweee' of *C. chloris* but shorter and much coarser (Witherby *et al.* 1938); 'vyaa' (not harsh) given variously by subordinate in dispute, when threatened by other passerines, at approach of human intruder (causes conspecifics to fly off), and similar call heard from feeding birds when Kestrel *Falco tinnunculus* nearby (Zablotskaya 1975). (6c) Recording from which Fig VIII made also includes 'hwee duseeyu' (Fig X); in 2nd (complex) unit,

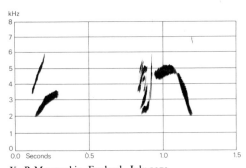

X   R Margoschis   England   July 1979

final pitch descent very noticeable (W T C Seale); 'duseeyu' occurs in this recording in several different contexts and presumably represents discrete call.

CALLS OF YOUNG. Food-calls of well-grown nestlings a squeaky 'ee ee', with 2-unit adult Contact-calls developing around fledging (1st unit usually strong, 2nd a weak 'ter' or 'to'); harder sounds also now develop, e.g. 'tzee', 'stee', 'cheeter', 'peeter', 'zittu', also 3-unit calls which are usually harsh with accented 1st unit (Conder 1948). Other renderings of fledged young include frequent 'di-wetwet...' (Bergmann and Helb 1982); 'tsi-vit tsi-vit pit' in which both 'tsi' and 'vit' rise in pitch; delivery rate speeds up in excitement (Zablotskaya 1975). In Fig XI,

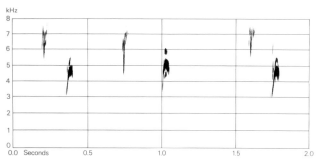

XI  W T C Seale  England  August 1989

'ti pin ti pin ti pin' in which 'pin' has metallic popping timbre, bearing some resemblance to Chink-call of *F. coelebs*, perhaps precursor of adult call 2a (W T C Seale). For experimental study of song-learning by juveniles see Cvitanić and Tolić (1988).                    EKD

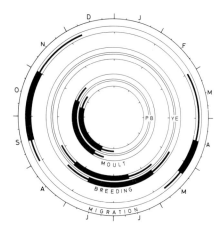

**Breeding.** SEASON. Britain: see diagram; eggs laid mid-May to early August; in southern England exceptionally in April, in hot summers into September (Newton 1972). Finland: eggs laid 1st half of May to early July (Haartman 1969). Poland: 1st clutch laid late April to June, 2nd clutch June to end of July; young sometimes in nest late September (Sokołowski 1962). Germany: in south-west, peak laying end of May to mid-June; nest-building starts when trees in leaf (Glück 1984); one pair in warm weather started building 11 August, young left nest mid-September (Ullrich 1986). North Africa: eggs laid mostly April–May (Etchécopar and Hüe 1967). Azores: eggs laid May–June

(G Le Grand). SITE. Well hidden in inaccessible outermost twigs of tree, and cover seems more important than support (Conder 1948; Makatsch 1976; Glück 1984). In Poland, 2% of 146 nests less than 2 m above ground, 18% 2–4 m, 29% 4–6 m, 34% 6–8 m, 14% 8–10 m, 3% above 10 m; average 6·3 m (Sokołowski 1962). Average height in fruit trees, south-west Germany, 4·3 m ($n = 238$); of 190 nests, 6% against trunk, 14% up to 1 m away, 18% 1–2 m, 20% 2–3 m, 17% 3–4 m, 14% 4–5 m, 11% more than 5 m, and average distance from periphery of crown 0·8 m; average diameter of nest-carrying branch 1·7 cm ($n = 238$); tree species used according to abundance (Glück 1983). In Rheinland (western Germany), 40% of 94 nests in fruit trees, 54% in other deciduous species, 6% in conifers (Mildenberger 1984); in Polish study, 17% of 146 nests in lime *Tilia*, 13% horse chestnut *Aesculus*, 12% ash *Fraxinus*, 10% apple *Malus*, 8% plum *Prunus*, 7% *Robinia*; 19 other species all less than 5% (Sokołowski 1962). Although nests vulnerable to wind, no particular orientation preferred in extensive study by Glück (1983), though see Conder (1948), where majority of 17 nests in Bayern, southern Germany, were sheltered from prevailing wind. Nest: very neat and compact cup of moss, roots, grass, and spider silk, which sometimes binds foundation to twigs, thickly lined with plant down, wool, hair, and occasionally feathers; sometimes 'decorated' on outside with aromatic flowers (Pitman 1921; Conder 1948; Newton 1972; Harrison 1975; Makatsch 1976; Glück 1984). In southern England, birds apparently learned from each other how to untie labels from fruit trees to use as material (Muir 1959). 3 nests in Azores had average outer diameter 7·6 cm, inner diameter 4·3 cm, overall height 4·8 cm and depth of cup 3·5 cm (G Le Grand). Building: by ♀ only, accompanied by ♂ (Sokołowski 1962; Spillner 1975; Makatsch 1976; Glück 1984); ♂ may very rarely carry material (Conder 1948). Takes c. 7 days (4–14), with 3 main bouts of activity during day: 07.00–09.00 hrs, 12.00–13.00 hrs, and 16.00–18.00 hrs (Glück 1984). May add to nest for some time after 1st egg laid (Conder 1948, which see for details). EGGS. See Plate 60. Sub-elliptical, smooth and slightly glossy; very pale bluish-white, sparsely spotted or scrawled reddish or purplish-brown, concentrated at broad end sometimes in faint ring; occasionally reddish-brown blotches or greyish-violet undermarkings (Harrison 1975; Makatsch 1976). Nominate *carduelis*: 17·3 × 13·0 mm (15·5–19·1 × 12·0–14·4), $n = 264$; calculated weight 1·53 g. *C. c. britannica*: 17·0 × 12·8 mm (15·5–19·0 × 12·2–13·6), $n = 100$; calculated weight 1·48 g. *C. c. tschusii*: 16·8 × 12·8 mm (15·8–17·8 × 12·0–13·0), $n = 42$; calculated weight 1·45 g. *C. c. parva*: 16·6 × 12·4 mm (14·0–17·2 × 12·0–13·0), $n = 18$; calculated weight 1·34 g. *C. c. niediecki*: 17·1 × 13·0 mm (15·7–19·0 × 12·3–13·7), $n = 68$; calculated weight 1·50 g. (Schönwetter 1984.) Clutch: 4–6 (3–7). In south-west Germany, of 146 clutches: 2 eggs, 1·4%; 3, 2·7%; 4, 20·5%; 5, 68·5%; 6, 6·8%; average 4·77 (in May 4·88, $n = 68$; June 4·89, $n = 45$;

July 4·45, $n = 29$; August 3·3, $n = 3$). All marked pairs had 2 broods, 12% had 3; overlapping noted in 2 cases: ♀ ceased brooding after 4 days to build new nest, another started laying when young 10 days old. New nest built for replacement clutches until about beginning of July following loss of eggs or small young. (Glück 1984.) Average clutch in Rheinland 4·92, $n = 61$ (Mildenberger 1984). In Britain perhaps more usually 5–6; occasionally 3 broods (Witherby *et al.* 1938; Bannerman 1953*a*). Average in North Africa 4·2, $n = 17$ (Etchécopar and Hüe 1967). Eggs laid daily in early morning on nest completion or 1–2 days after, though see Building (above) (Conder 1948; Sokołowski 1962; Glück 1984). INCUBATION. By ♀ only; average 12·1 days (9–14, $n = 41$) from laying of last egg to hatching of last young (Glück 1984). In another study, 11·3–12 days ($n = 3$); in 203 hrs of observation, eggs covered for average 87–96% of time (from 24% for 1st egg to 96% for 5th); most stints 30–80 min (1 of 202 min), most breaks 1–3 min (Conder 1948). Almost always started with 3rd egg and hatching over 2(–4) days (Glück 1984), but can also be within 27 hrs (Conder 1948). See also Sokołowski (1962). YOUNG. Fed and cared for by both parents (Harrison 1975; Makatsch 1976). ♀ broods young for 7–9 days continuously, at one nest for 20% of time on day 12 (Conder 1948, which see for stints and feeding of young; Sokołowski 1962; Spillner 1975). FLEDGING TO MATURITY. Fledging period 14·7 days (13–18, $n = 40$) (Glück 1984); 19 days at one nest, according to Hui and Hui (1974). Fed by parents for *c.* 7–10 or more days after leaving nest (Sokołowski 1962; Harrison 1975; Spillner 1975). BREEDING SUCCESS. In south-west Germany, 23% of 204 nests where eggs laid successful, 57% suffered loss at egg stage, and 21% at hatching or nestling stage; 187 young flew from 47 successful nests (4·0 fledged young per successful nest, 0·9 per nest overall); 75% of clutches and broods taken by predators, including Magpie *Pica pica* and Kestrel *Falco tinnunculus*; in poor food year, youngest chick often dies (Glück 1984). In Moscow area (Russia), 38% of nests successful when nest density high, 59% when low ($n = 77$ nests); Carrion Crow *Corvus corone* important predator (Shurupov 1986). Nests frequently lost to high winds since built at ends of branches (Conder 1948). BH

Plumages. (nominate *carduelis*). ADULT MALE. In fresh plumage (autumn), nasal bristles, narrow strip of feathering along base of bill, lore, and feathering at lower border of eye black; remaining face glossy crimson-red, forming contrasting fiery patch, extending backwards to forecrown, upper rear border of eye, forecheek, and over *c.* 6–12 mm of chin. Centre of crown and hindcrown glossy deep black, extending into bar across upper side of neck. Side of head (between red of face and black of side of neck) contrastingly cream-white, extending into band across throat; white often partly tinged drab-buff, especially on rear of ear-coverts and cheeks. Narrow bar at rear of black of hindcrown buff-white, sometimes inconspicuous. Nape, lower side of neck, mantle, scapulars, and back saturated cinnamon-brown, feather-tips slightly paler cinnamon and feather-centres slightly vinous

when plumage quite fresh. Rump and upper tail-coverts white, feather-tips suffused tawny-brown, partly concealing white. Side of breast and flank saturated buff-brown, extending into contrasting patch on upper side of belly and often into a narrow buff-brown or buff-white bar across upper chest; remainder of underparts pure white, often with some lemon-yellow at border of patch on upper side of belly and with some tawny-buff or cream-buff suffusion on tips of under tail-coverts. Tail deep black, central pair (t1) with contrasting white tip 3–4 mm long, t2–t3 (–t5) with narrower white tip, partly broken by black; terminal fringe of white suffused buff when plumage quite fresh; inner web of (t2–)t5–t6 with large square-cut white patch 13–18 mm long, terminal 6–8 mm of inner web black; patch on t5 slightly shorter than on t6, that on t4 (if present) mostly 5–10 mm long, rounded. Flight-feathers, greater upper primary coverts, and bastard wing deep black; basal and middle portion of outer web of flight-feathers contrastingly yellow (on primaries, extending 20–25 mm beyond tips of primary coverts, on secondaries, *c.* 15 mm beyond greater coverts), forming broad bar across upper wing; outer web of p9 black, tips of secondaries and of inner and central primaries with contrasting broad-triangular white spot, reduced to faint speck on tips of outer primaries; basal and middle portion of inner web with contrasting off-white wedge along border, visible on underwing only. Tertials black, often with slight metallic-blue glossy (like black of outer webs of secondaries), tip with large cream-buff to buff-white rounded-triangular spot, soon fading to white. Greater upper wing-coverts glossy black with bright yellow tips, tips narrow and soon fading to white on outermost coverts (gc1–gc2), black at base reduced and yellow *c.* 10–15 mm long on central coverts (gc4–gc5), tips narrow again and partly suffused pink-buff on tertial coverts (gc7–gc9). Lesser and median coverts deep black, sometimes with narrow brown fringes. Axillaries and under wing-coverts white, axillaries and some inner coverts often partly suffused pale yellow, marginal coverts spotted black and white. *In worn plumage* (spring), red of face brighter, more glossy, some black of feather-bases often partly visible; off-white bar on nape more extensive, forming pale patch; mantle, scapulars, and back slightly greyer, less tinged cinnamon, more drab-brown; rump and upper tail-coverts more extensively white, some black on centres of longest coverts often visible; white on rear side of head cream-white, grey-white, or pure white, less tinged buff; white of underparts less pure, slightly tinged grey; white tips of tail-feathers, primaries, and tertials reduced to narrow fringe or fully worn off. When heavily abraded (about July–August), red of face tinged pink or orange, mixed black; black of crown mixed with some grey of feather-bases; broad bar along nape and lower side of neck mixed dirty white and pale brown-grey, white of rump, upper tail-coverts, and underparts partly worn off, less extensive, tinged or mixed grey. ADULT FEMALE. Like adult ♂, but nasal hairs light brown-grey to blackish-grey (deep black in adult ♂); red of face reaching slightly less far backwards, hardly extending to forecrown, rear corner of eye, or upper throat, extending 4–8 mm on chin; on chin often round-ended below, less squarish than ♂; red of face more orange-red or flame-scarlet, with more black or dark grey of feather-bases visible among red, especially on chin. Black of crown less deep and glossy, sometimes mottled grey-buff or drab, especially at rear; rear side of head more heavily washed drab-buff, less extensively white; buff-brown of side of body on average more extensive, more often with full band across chest, patches on each side of upper belly almost meeting, usually without yellow at border or sub-terminally on feathers; white patches on inner web of t5–t6 on average smaller, rarely present on t4; lesser and sometimes inner median coverts dark brown or dark grey with broad grey-brown or buff-brown fringes and black subterminal bar, less extensively black than in adult ♂, especially innermost. See also Müller (1982). Effect of wear as in adult ♂, red of face less deep and more heavily mottled grey in worn plumage than in adult ♂. NESTLING. Down fairly long, dark grey, restricted to dense tufts on crown, upper wing, back, and thigh, with some scanty down on rump and vent (Witherby *et al.* 1938; Harrison 1975; RMNH). JUVENILE. Ground-colour of upperparts and side of head and neck from nasal bristles to back light drab-grey; paler, more cream-buff on nape, supercilium, and indistinct stripe below eye, slightly warmer buff-brown on fringes of feathers of scapulars; each feather with rather ill-defined dark grey-brown or sooty spot on tip, head and neck appearing finely speckled, mantle and scapulars more coarsely mottled, sometimes an indistinct dark brown-grey malar stripe. Rump and upper tail-coverts white, partly washed cream-buff, mottled grey, longer coverts with black centres. Ground-colour of underparts pale grey-buff, purer cream-white on belly, dirty off-white when worn; chin and throat speckled dark grey, chest, flank, and often belly more coarsely marked with sooty blotches or short streaks. Tail as adult, but feathers narrower, tips sharply pointed (less rounded than in adult), white mark on inner web of t5–t6 often an oblong rounded patch, smaller and less square than in adult, absent on t4, small and occasionally absent on t5; white tips of tail more intensely washed buff than in fresh adult, marked with dusky speck on tip. Flight-feathers, primary coverts, bastard wing, and tertials as adult, but pale tips of flight-feathers and tertials larger and buffier than in adult, forming short fringe along tip of p6–p8. Lesser and median upper wing-coverts dark grey, sooty, or black (perhaps greyer in ♀, blacker in ♂, but sample examined small), tips contrastingly buff, those on median coverts forming narrow wing-bar; greater coverts sooty or deep black, tips of inner and central ones contrastingly pale cinnamon-buff for 5–10 mm, gradually narrower and whiter to outer coverts, outermost covert dull black with *c.* 3 mm broad buff-white or off-white fringe along tip and inner web (in adult, outermost deep black, tip with yellow triangular spot). When post-juvenile moult advanced, but head (including nasal bristles) still juvenile, sexing often possible on new lesser coverts (mainly black in ♂, brown in ♀). FIRST ADULT. Like adult, but part of juvenile feathering retained; in north of range, outer greater coverts and frequently some tertials still juvenile, contrasting in colour, pattern, and degree of abrasion with new neighbouring feathers; tail either juvenile, or with some feathers newer and more rounded at tip than in others (all uniform in shape in adult); birds with all greater coverts, tertials, and tail new are often difficult to age; in south of range, some primaries sometimes newer than others (unlike adult), but difference hard to see in spring when all plumage abraded. Sex differences as in adult, nasal bristles of ♂ sooty-grey to deep black, of ♀ light grey or whitish; red of head and colour of lesser coverts as in adult.

**Bare parts.** ADULT, FIRST ADULT. Iris dark hazel to black-brown. Bill horn-white, cream-white, or pink-white, extreme tip of culmen and gonys dark horn-grey to black, somewhat varying in extent, sometimes flesh-pink or pale flesh-grey subterminally; in summer, bill sometimes contaminated with dark sticky substance. Leg and foot pale flesh or light flesh-horn. NESTLING. Tongue and floor of mouth crimson, paler on angles of tongue; roof of mouth dark lilac or purplish; gape-flanges cream-white. JUVEN-ILE. At fledging, bill light yellowish- or pinkish-horn with black tip, gape with traces of pale cream-yellow flanges; leg and foot pale greyish flesh-horn to flesh-pink. (Heinroth and Heinroth 1924–6; Witherby *et al.* 1938; Harrison 1975; RMNH, ZMA.)

**Moults.** ADULT POST-BREEDING. Complete; primaries descendent. In Britain, starts with shedding of p1 mid-July to mid-August, occasionally from early July, rarely from late August; primary moult completed with regrowth of p9 after *c.* 70–72 days, mid-September to late October, rarely early September or early November, primary moult on average 29 July to 17 October; tail moult starts with shedding of p4–p5, secondaries with p5–p6; moult of secondaries, head, and neck finished a few days after primaries (Newton 1968; Ginn and Melville 1983; BMNH, RMNH, ZMA). In north-west Russia, starts early July to early August, peaking *c.* 20 August; completed October (Rymkevich 1990). In southern Yugoslavia, no moult up to 18 August; single bird in moult 23 August, 2 nearing completion 13 October (Stresemann 1920). On Corsica, primary moult scores 2, 8, and 19 on 15–18 August (ZMA); on Sardinia, single ♀ in moult 21 August (Bezzel 1957). In North Africa, moult almost finished or just completed in last days of October (Ticehurst and Whistler 1938; Meinertzhagen 1940). No moult in 6 birds from northern Iran, 13–29 July (Stresemann 1928; Paludan 1940). For timing and sequence of moult in birds introduced to Australia, see Middleton (1969). POST-JUVENILE. Partial (in north of range) or sometimes almost or entirely complete (in south). In northern and central Europe, involves head, body, and upper wing-coverts except 1–4 outer greater coverts; moult August–September (Svensson 1992). In north-west Russia, starts early August to mid-September, completed from October onwards (Rymkevich 1990); in USSR generally, largely juvenile head and neck sometimes retained to mid-October (rarely late November) (Dementiev and Gladkov 1954). In Netherlands, moult August–October, sometimes November; 84% of 38 birds retained 1–3 outer greater coverts (remainder 0), most replaced all tertials, 24% replaced t1 (a few t2 and t6 also); on Corsica and Sardinia, only 2 of 10 birds retained 1 outer greater covert, none retained tertials, and only 4 retained scattered tail-feathers (mainly t4–t5); in Iberia, 6 of 10 adults retained juvenile outermost greater covert, 5 (t2–) t3–6, none tertials (BMNH, RMNH, ZMA). On Balearic Islands, 67% of 117 birds moulted 3·14 (1·96; 79) 1–8 primaries, starting descendently with p1 (in 8 birds), p2 (in 2), p3 (in 8), p4 (in 23), p5 (in 20), or p6 (in 18), August–September; 99% of 79 birds with this 'sectorial' primary moult had p6 new after completion of moult, 77% p5, 52% p4, 34% p7, 23% p3, 13% p2, 10% p1, and 6% p8; also frequently s1–s2(–s3) or s5–s6 replaced, usually most tertials, often t1, t2, and t3 (frequently all tail), and occasionally bastard wing (Mester and Prünte 1982). Similar sectorial primary moult at least occasionally recorded Switzerland (Winkler and Jenni 1984) and Greece (Kasparek 1981); of other juveniles recorded in flight-feather moult, not clear whether moult is complete or is a more restricted sectorial primary moult, e.g. in birds from Yugoslavia, Italy, Sardinia, Sicily, Greece, Spain, Portugal, and Britain (Stresemann 1920; Bezzel 1957; Newton 1972; Kasparek 1981; Ginn and Melville 1983; Fraticelli and Gustin 1987). In birds with sectorial primary moult, many or all greater upper primary coverts and at least central secondaries retained.

**Measurements.** ADULT, FIRST ADULT. Nominate *carduelis*. Eastern Netherlands and Germany, all year; skins (RMNH, ZMA). Bill (S) to skull, bill (N) to distal corner of nostril; exposed culmen on average 3·2 less than bill (S).

| | ♂ | | ♀ | |
|---|---|---|---|---|
| WING | 81·0 (1·71; 21) | 78–84 | 78·4 (2·04; 19) | 75–82 |
| TAIL | 49·9 (1·49; 14) | 47–53 | 46·2 (2·37; 10) | 43–50 |
| BILL (S) | 16·3 (0·86; 17) | 15·2–17·5 | 15·3 (0·74; 12) | 14·2–16·5 |
| BILL (N) | 11·2 (0·44; 18) | 10·5–11·9 | 10·5 (0·55; 12) | 9·5–11·5 |
| TARSUS | 14·8 (0·38; 20) | 14·0–15·4 | 14·7 (0·25; 16) | 14·1–15·6 |

Sex differences significant, except for tarsus.

*C. c. britannica.* Britain, all year; skins (BMNH, RMNH, ZMA: A J van Loon).

| | | | | | |
|---|---|---|---|---|---|
| WING | ♂ | 78·8 (1·92; 21) 76–82 | ♀ | 75·9 (1·36; 13) 74–78 | |
| TAIL | | 46·8 (1·64; 18) 44–50 | | 46·2 (0·91; 13) 44–48 | |
| BILL (S) | | 15·8 (0·62; 19) 14·7–17·2 | | 15·1 (0·53; 13) 14·0–16·2 | |
| BILL (N) | | 11·1 (0·57; 19) 9·8–12·0 | | 10·5 (0·50; 13) 10·1–11·6 | |
| TARSUS | | 15·0 (0·37; 16) 14·3–15·5 | | 14·3 (0·30; 13) 13·6–14·8 | |

*C. c. parva.* Wing and bill to skull of (1) Iberia, (2) North Africa, (3) Madeira and Canary Islands; other measurements combined, all year; skins (RMNH, ZMA: A J van Loon).

| | | | | | |
|---|---|---|---|---|---|
| WING (1) | ♂ | 76·3 (2·23; 9) 74–78 | ♀ | 73·5 (1·98; 11) 69–76 | |
| (2) | | 76·5 (1·46; 13) 74–79 | | 72·8 (0·76; 3) 71–74 | |
| (3) | | 75·5 (1·64; 5) 74–78 | | 71·0 (3·50; 4) 67–75 | |
| TAIL | | 45·5 (1·55; 25) 41–49 | | 43·9 (2·38; 17) 39–48 | |
| BILL (1) | | 15·7 (0·35; 10) 15·0–16·1 | | 14·8 (0·46; 13) 14·2–15·6 | |
| (2) | | 16·2 (0·91; 12) 14·7–17·3 | | 15·1 (0·93; 3) 14·1–15·9 | |
| (3) | | 15·9 (0·78; 5) 15·1–16·9 | | 14·5 (0·31; 4) 14·1–14·8 | |
| BILL (N) | | 11·4 (0·73; 27) 10·2–13·4 | | 10·4 (0·51; 20) 9·4–11·3 | |
| TARSUS | | 14·2 (0·50; 29) 13·0–15·9 | | 13·9 (0·36; 18) 13·2–14·4 | |

Sex differences significant, except tail.

*C. c. tschusii.* Corsica and Sardinia, all year; skins (BMNH, RMNH, ZMA: A J van Loon).

| | | | | | |
|---|---|---|---|---|---|
| WING | ♂ | 77·6 (1·99; 27) 75–81 | ♀ | 74·0 (2·21; 20) 70–77 | |
| TAIL | | 46·2 (1·68; 26) 43–50 | | 45·8 (1·69; 20) 42–48 | |
| BILL (S) | | 15·5 (0·54; 15) 14·2–16·5 | | 15·0 (0·80; 10) 13·7–16·0 | |
| BILL (N) | | 10·8 (0·43; 25) 9·9–11·8 | | 10·2 (0·49; 20) 9·5–10·9 | |
| TARSUS | | 14·4 (0·47; 27) 13·1–15·2 | | 14·3 (0·40; 21) 13·7–15·1 | |

Sex differences significant for wing and bill.

*C. c. balcanica.* Western and southern Yugoslavia, Albania, Bulgaria, and mainland Greece, mainly April–August; skins (BMNH, RMNH, ZMA: A J van Loon).

| | | | | | |
|---|---|---|---|---|---|
| WING | ♂ | 78·9 (1·92; 16) 76–83 | ♀ | 76·4 (2·79; 11) 73–81 | |
| TAIL | | 46·7 (1·15; 15) 45–49 | | 45·8 (2·46; 9) 43–49 | |
| BILL (S) | | 15·7 (0·58; 16) 14·3–16·5 | | 14·9 (0·68; 11) 13·5–16·0 | |
| BILL (N) | | 11·1 (0·43; 16) 10·1–11·9 | | 10·2 (0·75; 11) 8·4–11·2 | |
| TARSUS | | 14·6 (0·63; 16) 13·3–15·3 | | 14·5 (0·54; 10) 13·4–15·2 | |

Sex differences significant for wing and bill.

*C. c. niediecki.* Asia Minor, Cyprus, and northern Iraq, all year; skins (BMNH, RMNH, ZMA: A J van Loon).

| | | | | | |
|---|---|---|---|---|---|
| WING | ♂ | 79·6 (1·99; 21) 77–84 | ♀ | 77·4 (2·27; 19) 75–80 | |
| TAIL | | 47·8 (1·48; 17) 45–51 | | 46·8 (1·42; 18) 44–50 | |
| BILL (S) | | 16·1 (1·03; 9) 14·8–17·4 | | 15·3 (0·72; 17) 14·3–16·5 | |
| BILL (N) | | 11·5 (0·73; 21) 10·0–12·7 | | 11·0 (0·60; 17) 9·9–12·2 | |
| TARSUS | | 14·8 (0·55; 15) 13·7–15·4 | | 14·2 (0·42; 16) 13·5–15·0 | |

Sex differences significant for wing, bill (S), and tarsus.

*C. c. loudoni.* South-east Azerbaydzhan and northern Iran, January–March; skins (BMNH, ZMA: A J van Loon).

| | | | | | |
|---|---|---|---|---|---|
| WING | ♂ | 80·9 (2·57; 7) 78–84 | ♀ | 77·8 (1·86; 4) 75–80 | |
| TAIL | | 48·6 (1·78; 7) 46–52 | | 47·3 (1·48; 4) 45–49 | |
| BILL (S) | | 15·6 (0·63; 8) 14·3–16·3 | | 15·0 (0·98; 3) 14·2–16·1 | |
| BILL (N) | | 11·6 (0·49; 7) 10·6–12·1 | | 10·8 (0·36; 4) 10·4–11·2 | |
| TARSUS | | 14·9 (0·10; 3) 14·8–15·0 | | 14·8 (— ; 2) 14·7–14·8 | |

*C. c. frigoris.* South-west Siberia and Tien Shan, all year; skins (BMNH, ZMA: A J van Loon).

| | | | | | |
|---|---|---|---|---|---|
| WING | ♂ | 87·8 (1·74; 8) 85–91 | ♀ | 87·5 (— ; 2) 87–88 | |
| TAIL | | 53·3 (2·61; 8) 48–56 | | 52·5 (— ; 2) 51–54 | |
| BILL (S) | | 17·7 (0·82; 7) 16·7–18·9 | | 16·6 (— ; 2) 16·4–16·8 | |
| BILL (N) | | 12·7 (0·43; 7) 12·0–13·3 | | 12·3 (— ; 2) 12·0–12·6 | |
| TARSUS | | 14·9 (0·56; 7) 14·3–16·0 | | 14·7 (— ; 2) 14·1–15·3 | |

*C. c. volgensis.* Ulyanovsk (south-east European Russia), Volga Steppe, and north-east Rumania: wing, ♂ 83·0 (1·52; 9) 80·5–85·5, ♀ 80·3 (1·23; 6) 78·3–82; bill to skull, ♂ 16·9 (0·54; 9) 16·5–18·3, ♀ 15·4 (0·24; 6) 15·2–15·7 (Buturlin 1906; ZMA).

*C. c. paropanisi.* Turkmeniya to Tunisia, all year; skins (BMNH, ZMA: A J van Loon).

| | | | | | |
|---|---|---|---|---|---|
| WING | ♂ | 84·2 (2·16; 12) 80–87 | ♀ | 81·5 (2·32; 10) 78–85 | |
| TAIL | | 51·1 (2·24; 7) 49–55 | | 49·8 (2·30; 8) 48–55 | |
| BILL (S) | | 18·1 (1·01; 12) 16·7–19·5 | | 18·1 (0·74; 9) 17·0–19·0 | |
| BILL (N) | | 13·1 (0·90; 12) 11·6–14·7 | | 13·0 (0·72; 9) 11·9–14·0 | |
| TARSUS | | 14·7 (0·29; 6) 14·2–15·0 | | 14·4 (0·28; 8) 14·0–14·8 | |

Sex differences not significant.

Wing; skins, unless stated. Nominate *carduelis*. (1) Scandinavia and eastern Germany to Baltic states (Niethammer 1937; Vaurie 1956a; Eck 1985a). (2) North-east Poland and westernmost Russia (former East Prussia), breeding (Tischler 1931). (3) Rhein area, western Germany (Tischler 1931; Niethammer 1937; see also Sachtleben 1918 and Niethammer 1971). South-east France, live birds: (4) breeding, (5) winter (G Olioso). *C. c. volgensis.* (6) South-east European Russia (Vaurie 1956a). *C. c. britannica.* (7) Britain (Niethammer 1971). *C. c. parva.* (8) Balearic Islands, live breeders (Mester 1971). (9) Live migrants Tarifa, southern Spain (Asensio 1984). *C. c. tschusii.* (10) Sardinia (Vaurie 1956a; Bezzel 1957). *C. c. balcanica.* (11) Rumania (Munteanu 1966). (12) Southern Yugoslavia, Greece, and European Turkey (Stresemann 1920; Niethammer 1943; Rokitansky and Schifter 1971; see also Munteanu 1966). *C. c. niediecki.* (13) Asia Minor (Jordans and Steinbacher 1948; Kumerloeve 1961; Rokitansky and Schifter 1971). (14) Cyprus, summer, live birds (Flint and Stewart 1992). (15) Zagros mountains, south-west Iran (Paludan 1938; Vaurie 1949b). *C. c. loudoni.* (16) Iranian Azarbaijan (Vaurie 1949b). (17) Northern Iran (Stresemann 1928; Paludan 1940; Schüz 1959). *C. c. frigoris.* (18) Central Siberia (Dementiev and Gladkov 1954; Vaurie 1956a). *C. c. ultima.* (19) Kirman and Fars, southern Iran (Vaurie 1949b; see also Koelz 1949). *C. c. paropanisi.* (20) North-east Iran and Afghanistan (Paludan 1940, 1959; Vaurie 1949b). (21) Turkmeniya to Tadzhikistan, (22) Tien Shan (Dementiev and Gladkov 1954). *C. c. subulata.* (23) South Siberia and Mongolia (Hartert 1903–10; Dementiev and Gladkov 1954; Vaurie 1959; Piechocki and Bolod 1972). *C. c. caniceps.* (24) Kashmir and Punjab, India (Vaurie 1949b; ZMA).

| | | | | | |
|---|---|---|---|---|---|
| (1) | ♂ | 80·5 (— ; 60) 77–84 | ♀ | 76·8 (— ; 18) 74–80 | |
| (2) | | 81·0 (1·54; 31) 78–85 | | 77·0 (1·25; 15) 74–79 | |
| (3) | | 77·9 (— ; 22) 75–80 | | 75·6 (— ; 10) 74–79 | |
| (4) | | 79·8 (1·48; 5) 78–82 | | 76·9 (1·35; 7) 75–79 | |
| (5) | | 80·2 (1·83; 108) 75–84 | | 77·9 (1·81; 61) 74–83 | |
| (6) | | 85·0 (1·79; 6) 83–88 | | — (— ; —) — | |
| (7) | | 77·8 (1·47; 24) 75–80 | | 75·0 (1·8; 6) 73–78 | |
| (8) | | 75·8 (2·06; 26) 72–78 | | 72·4 (— ; 4) 70–75 | |
| (9) | | 76·8 (2·18; 178) — | | 74·6 (1·85; 269) — | |
| (10) | | 76·2 (1·22; 14) 74–78 | | 73·0 (— ; 2) 73–73 | |
| (11) | | 81·5 (— ; 18) 77–84 | | 77·3 (— ; 3) 76–78 | |
| (12) | | 78·6 (1·59; 50) 75–83 | | 75·5 (1·74; 20) 73–79 | |
| (13) | | 79·6 (1·43; 7) 77–81 | | 74·9 (2·25; 4) 73–78 | |
| (14) | | 78·3 (— ; 63) 73–81 | | 76·5 (— ; 39) 72–80 | |
| (15) | | 80·4 (— ; 30) 77–84 | | 77·4 (— ; 24) 75–80 | |
| (16) | | 81·7 (— ; 9) 80–85 | | 79·1 (— ; 8) 76–83 | |
| (17) | | 78·6 (2·57; 7) 74–81 | | 76·0 (1·41; 4) 75–78 | |
| (18) | | 86·1 (1·95; 7) 84–89 | | 80·5 (— ; 20) 76–86 | |
| (19) | | 83·9 1·02; 6) 83–85 | | 81·7 (— ; 3) 80–85 | |
| (20) | | 82·1 (1·63; 45) 79–86 | | 79·9 (— ; 26) 76–81 | |
| (21) | | 80·4 (— ; 20) 78–84 | | 78·8 (— ; 9) 76–81 | |
| (22) | | 81·4 (— ; 22) 78–84 | | 79·3 (— ; 19) 74–83 | |
| (23) | | 84·8 (— ; 23) 81–88 | | 82·1 (2·31; 8) 79–86 | |
| (24) | | 81·1 (1·51; 10) 78–84 | | 80·0 (— ; 2) 80–80 | |

Wing of birds breeding Caucasus: ♂ 78–81 (*n* = 5), ♀ 75–78 (*n* = 3) (Hofer 1935). Sachtleben (1918), Jordans (1924), Ticehurst and Whistler (1928), Hartert and Steinbacher (1932–8),

Hofer (1935), Meinertzhagen (1940), Niethammer (1942), Johansen (1944), Koelz (1949), Vaurie (1949b, 1956a, 1959), Makatsch (1950), Stanford (1954), Trettau and Wolters (1967), and Eck (1985a).

**Weights.** ADULT, FIRST ADULT. Nominate *carduelis*. (1) Eastern Netherlands and Germany, all year (Eck 1985a; RMNH, ZMA; see also Niethammer 1937). (2) Czechoslovakia, May–November (Havlín and Havlínová 1974). South-east France: (3) breeding, (4) winter (G Olioso). *C. c. parva*. (5) Balearic Islands, summer (Mester 1971). (6) Azores, April–September (G Le Grand). (7) Tunisia, May (G O Keijl). *C. c. balcanica*. (8) Greece and European Turkey, May–July (Niethammer 1943; Rokitansky and Schifter 1971; see also Makatsch 1950). *C. c. niediecki.* (9) Asia Minor, Israel, and south-west Iran, April–October (Paludan 1938; Schüz 1959; Kumerloeve 1961; Rokitansky and Schifter 1971; G O Keijl; see also Eyckerman *et al.* 1992). (10) Cyprus, summer (Flint and Stewart 1992). *C. c. loudoni.* (11) North-east Turkey and Iran, all year (Paludan 1940; Schüz 1959; Kumerloeve 1967a). *C. c. frigoris.* (12) Kazakhstan (Korelov *et al.* 1974). *C. c. paropanisi.* (13) North-east Iran and Afghanistan (Paludan 1940, 1959). *C. c. paropanisi* and/or *subulata.* (14) Kazakhstan (Korelov *et al.* 1974). *C. c. subulata.* (15) Mongolia, July (Piechocki and Bolod 1972).

| | | | |
|---|---|---|---|
| (1) | ♂ 16·4 (2·00; 13) 13·0–19·5 | ♀ | 14·8 (0·86; 11) 13·1–16·0 |
| (2) | 16·9 (1·30; 33) 14·0–20·0 | | 16·6 (0·96; 16) 15·0–19·0 |
| (3) | 16·0 (1·16; 5) 14·0–17·0 | | 14·9 (0·69; 7) 14·0–15·8 |
| (4) | 16·5 (1·11; 113) 13·5–19·6 | | 15·5 (1·16; 62) 13·6–19·0 |
| (5) | 13·7 ( — ; 25) 11·7–15·7 | | 12·7 ( — ; 21) 10·8–15·0 |
| (6) | 14·4 (0·79; 6) 13·5–15·5 | | 13·7 (0·89; 8) 12·5–14·7 |
| (7) | 14·6 (1·1; 6) 13·5–16·0 | | 14·8 (0·8; 5) 14·0–15·8 |
| (8) | 15·7 (1·16; 7) 14·5–18·0 | | 16·7 (1·76; 3) 15·0–18·5 |
| (9) | 16·4 (2·02; 6) 14–20 | | 14·9 (1·15; 3) 13·8–16·1 |
| (10) | 15·1 ( — ; 59) 13·0–21·5 | | 14·8 ( — ; 35) 13·0–19·2 |
| (11) | 17·8 (2·05; 5) 16–21 | | 17·0 ( — ; 1) — |
| (12) | 24·9 ( — ; 8) 21·4–30·0 | | 22·8 (0·91; 3) 21·8–23·5 |
| (13) | 16·9 (1·05; 12) 15·7–19·0 | | 17·6 (1·14; 5) 16–19 |
| (14) | 19·6 ( — ; 30) 15·5–23·4 | | 18·6 ( — ; 18) 15·6–23·1 |
| (15) | 19·0 (1·00; 5) 18–20 | | 20·3 (3·21; 3) 18–24 |

Nominate *carduelis*. Netherlands, exhausted birds: 11·1 (0·86; 11) 9·5–12·1 (ZMA; for *britannica*, see Ash 1964).

*C. c. ultima.* Southern Iran, May–June: ♂ 20 (Desfayes and Praz 1978).

*C. c. caniceps.* Kashmir, August–September: 17·0 (1·12; 13) 15·5–19·1 (P R Holmes and Oxford University Kashmir Exped. 1983).

NESTLING, JUVENILE. Juveniles in summer or early autumn: south-east France (nominate *carduelis*) 15·5 (1·26; 43) 13·5–18 (G Olioso); Balearic islands (*parva*) 12·9 (1·16; 122) 10·1–15·4 (Mester 1971); Tunisia (*parva*) 13·9 (0·5; 11) 13–14·5 (G O Keijl).

**Structure.** Wing rather long, broad at base, tip pointed. 10 primaries: p8 longest, p9 equal or (occasionally) up to 1 shorter; p7 0–2 shorter than p8, p6 2–5, p5 7–13, p4 12–18, p3 14–21, p2 18–24; p1 21–25 in *parva*, 22–27 in nominate *carduelis*. P10 strongly reduced, a tiny pin concealed below reduced outermost greater upper primary covert, which is 0–3 longer; p10 49–58 shorter than p8, 7–11 shorter than longest upper primary covert. Outer web of p6–p8 emarginated (rarely, p5 faintly also or emargination on p6 faint); inner web of p7–p9 with notch (often faint). Tip of longest tertial reaches to tip of p3–p4 when fresh; sometimes slightly longer than p4. Tail rather short, tip with shallow fork; 12 feathers, t5–t6 longest, t1 5·2 (10) 3–7 shorter.

Bill markedly longer than in other species of *Carduelis* (Newton 1967b, 1972), slightly longer than half of head length; depth at edge of feathering on base 6·9 (35) 6·5–7·4 mm in nominate *carduelis*, *britannica*, and *caniceps*, width at base of rami of lower mandible 6·1 (35) 5·7–6·6 mm; in other races, bill sometimes heavier or more slender at base, e.g. depth in *parva* 6·8 (30) 6·2–7·4, width 5·9 (32) 5·4–6·6, depth in *tschusii* 6·4 (15) 5·8–7·0, width 5·6 (15) 5·2–6·0, depth in *balcanica*, *niediecki*, and *loudoni* 7·0 (25) 6·6–7·7, width 6·2 (25) 5·6–6·6, depth in *volgensis* 7·5 (6) 7·3–7·9, width 6·4 (6) 6·0–6·6, depth in *frigoris* and *subulata* 8·4 (5) 7·9–9·0, width 6·9 (5) 6·6–7·3, depth in *paropanisi* 8·0 (6) 7·6–8·3, width 6·6 (6) 6·3–7·0. Culmen straight at base, slightly decurved towards sharply pointed tip, gonys slightly concave, cutting edges virtually straight, but with faint tooth near base on upper mandible. Bill laterally compressed at tip. Nostril small, rounded, covered by short stiff feathering, extending along lateral base of both mandibles. Leg and foot rather short, slender. Middle toe with claw 16·2 (10) 15–17·5 mm; outer toe with claw *c.* 72% of middle with claw, inner *c.* 71%, hind *c.* 88%.

**Geographical variation.** Slight in west Palearctic, involving mainly wing length and general colour of upperparts; marked in Asia. 2 distinct groups: (1) *carduelis* group in western Palearctic and east to northern and western Iran and through western Siberia to western Altai and Yenisey valley, with black cap, nape, and bar at side of neck and white triangular spots on tips of tertials and flight-feathers,; (2) *caniceps* group in Asia, in southern Iran and from north-east Iran to Himalayas and through west-central Asia to western Mongolia and Lake Baykal, characterized by drab-grey head and neck (apart from red face, black lore, short white stripes above and below eye, and white border along red of cheek and throat), often with faint dark streaks on crown; also, rump and shorter upper tail-coverts extensively (cream-)white, tips of flight-feathers black, without white spots, tertials with long white stripe on outer web (forming white streak on inner wing), outermost tail-feather (t6) extensively white on base (but white spot on t5 often more reduced than in *carduelis* group, or sometimes absent), and longer bill. These groups sometimes considered separate species (e.g. Stepanyan 1983, 1990), but *loudoni* of *carduelis* group grades into *paropanisi* of *caniceps* group in narrow zone at Gorgan, Iran (Vaurie 1949b; Schüz 1959), and *frigoris* of *carduelis* group merges into *subulata* of *caniceps* group over much wider zone in central Asia; also, *subulata*, though grey-headed, shares some *frigoris* characters in part of geographical range (Johansen 1944). For some vocal differences between groups, see Zablotskaya (1975) and Voice. Within *carduelis* group, only *frigoris* distinct, being much larger and paler, and with heavier bill than west Palearctic races; variation in west slight, largely clinal (Vaurie 1956a), and probably too many races recognized, but traditional view of Hartert and Steinbacher (1932–8), Hofer (1935), Vaurie (1949b, 1956a, 1959), and Stepanyan (1978, 1990) more or less followed here, though not always supported by specimens examined (BMNH, RMNH, ZMA). Note that colour of upperparts, considered to be important for identification of races, is highly influenced by bleaching and wear, grading from cinnamon-brown when fresh to dull brown-grey when worn in all races. Nominate *carduelis* from southern Scandinavia, eastern Netherlands and Belgium, and central France east to central Urals and south to mainland Italy, northernmost Yugoslavia, western Rumania, and northern Ukraine is rather large, upperparts and patches at side of belly saturated cinnamon-brown, ear-coverts with limited cinnamon (almost pure white when worn). *C. c. britannica* from Britain, Ireland, western and north-west France, and coastal

Netherlands and Belgium slightly smaller, upperparts and belly-patches slightly duller brown when fresh, faintly tinged olive, less cinnamon, ear-coverts more extensively brown (greyish-cream-buff when worn), belly less pure white when worn than in nominate *carduelis*, but differs only marginally from nominate *carduelis* in both colour and size. *C. c. parva* from Mediterranean France, Pyrénées, Iberia, Balearic Islands, North Africa, and Atlantic islands more distinctly demarcated from nominate *carduelis*, mainly due to small size, in particular wing and tail (see Measurements); colour and bill size within geographical range as delimited here not entirely uniform, birds from North Africa (named *africana* Hartert, 1903) and Balearic Islands (named *propeparva* Von Jordans, 1923) nearer to nominate *carduelis* in colour, those of western Spain and Portugal (named *weigoldi* Reichenow, 1913) slightly browner, near *britannica*, those of Atlantic islands slightly paler cinnamon-brown, but difference in colour and bill size too slight to warrant recognition of races other than *parva*; in large series examined by Jordans (1924), no constant variation in colour. *C. c. tschusii* from Corsica, Sardinia, Elba (Trettau and Wolters 1967), and (apparently this race) Sicily (Hofer 1935; see also Eck 1985*a*) is like *parva* in size, but bill slightly shorter and finer at base, and upperparts darker and duller earth-brown than in other races, less cinnamon when fresh, less grey when worn, ear-coverts rather extensively brown, belly-patches olive-brown, less cinnamon than in other races. *C. c. niediecki* from Asia Minor (except east), Levant, Egypt, Cyprus, northern Iraq, south-west Iran, Rhodes, Karpathos, and Transcaucasia similar in size to nominate *carduelis*; in fresh plumage, upperparts and belly-patches slightly paler drab-brown than in nominate *carduelis*, less cinnamon, paler and greyer than nominate *carduelis* when worn, especially in Egypt, Levant, and south-west Iran. *C. c. balcanica* from Balkans and Greece (from Yugoslavia and southern Rumania south to Crete) and *colchicus* from northern and western slopes of Caucasus west to Crimea are poor races, similar in size to *niediecki* and nominate *carduelis*; between them in colour, though nearer to *niediecki*. For distribution of *balcanica* and nominate *carduelis* in Balkans, see Munteanu (1966) and Matvejev and Vasić (1973). *C. c. loudoni* from extreme south-east Azerbaydzhan, north-west and northern Iran, and (probably this race: Kumerloeve 1967*a*) eastern Turkey like *niediecki*, but drab-brown colour deeper, more umber-brown, more extensive on belly-patches, darker and browner than in *britannica*. Only limited number of specimens from Caucasus area, northern Iran, northern Iraq, and Asia Minor examined (BMNH, RMNH, ZMA), and delimitation of races here not definitive; compare also views of Buturlin (1906), Sarudny (1906), Meinertzhagen (1924), Stresemann (1928), Hartert and Steinbacher (1932-8), Hofer (1935), Vaurie (1949*b*, 1956*a*, 1959), Dementiev and Gladkov (1954), Kumerloeve (1961, 1967*a*), and Stepanyan (1990). *C. c. volgensis* from southern Ukraine, south-east European Russia (between 46-54°N), and (at least in winter) north-east Rumania slightly larger than nominate *carduelis*, bill longer, thicker at base; upperparts and belly-patches cinnamon-brown in fresh plumage, as in nominate *carduelis*, but colour paler and less saturated, pale patch on nape larger, and feather-bases of back extensively white. *C. c. frigoris* of west and central Siberia slightly paler than *volgensis*, but much larger in all measurements (see Measurements and Structure). Variation in *caniceps* group largely clinal (Johansen 1944; Vaurie 1949*b*, 1956*a*); *subulata* from Tarbagatay to Lake Baykal and western Mongolia is palest and largest race of *caniceps* group (bill to skull 18-20 mm); upperparts light drab-grey, tinged cinnamon in north-west of range, belly-patches light ash-grey, small. *C. c. paropanisi* from north-east Iran through Tien Shan to Dzhungarsky Alatau slightly smaller, slightly darker drab-grey, belly-patches larger, and more distinct grey band across upper chest; isolated *ultima* from southern Iran rather similar in colour and size, but bill longer (Koelz 1949; Vaurie 1949*b*). *C. c. caniceps* of western and central Himalayas smaller than *paropanisi*, size as nominate *carduelis* but bill slightly longer, extent of grey as in *paropanisi* but distinctly darker drab-grey on both upperparts and underparts.     CSR

# *Carduelis spinus*  Siskin

PLATES 37 and 39 (flight)
[between pages 472 and 472]

Du. Sijs     Fr. Tarin des aulnes     GE. Erlenzeisig
Ru. Чиж      Sp. Lúgano     Sw. Grönsiska

*Fringilla Spinus* Linnaeus, 1758. Synonyms: *Spinus spinus*, *Acanthis spinus*.

Monotypic

**Field characters.** 12 cm; wing-span 20-23 cm. Slightly smaller than Goldfinch *C. carduelis*, with noticeably shorter tail and more compact form; marginally larger than Redpoll *C. flammea cabaret* but with shorter tail and stouter build. Small, quite tubby but elegant finch with short but noticeably forked tail, sharing with *C. flammea* arboreal, tit-like behaviour and (with serins *Serinus*) yellow-green and streaked plumages. At all ages, shows fine bill (like *C. carduelis* and unlike all serins except Citril Finch *S. citrinella*) and striking combination of yellow band on wing and yellow basal patches on tail (unlike any confusion species). ♂ strikingly yellow and green, with diagnostic black crown, much black in wings and tail, emphasizing yellow marks, and greenish-yellow rump; ♀ more greenish, distinctly streaked above and below, especially from sides of breast to rear flanks. Juvenile buff-brown above, even more heavily streaked above and below; lacks pale rump but shows characteristic wing and tail pattern. Flight light and dancing, more direct than *C. flammea*. Voice distinctive. Sexes dissimilar; little seasonal variation. Juvenile separable.

ADULT MALE. Moults: June-October (complete).

Crown, chin, and centre of upper throat usually black with grey feather-tips when fresh; some birds have these tracts less black, even dusky; crown bordered by bright lemon-yellow supercilium. Sides of head, throat, breast, chest, and flanks bright yellow, with greyish-green cast to rear ear-coverts, sides of breast, and upper flanks, and sharp, black streaks along flanks. Nape and mantle yellowish-green, faintly streaked black, contrasting with uniform greenish- or lemon-yellow rump. Wings intricately patterned with green, black, and brilliant yellow: lesser and median coverts as mantle, greater coverts dusky black, tipped pale yellowish-green; tertials with almost black centres but greyish-white tips and dull yellow fringes; secondaries black-brown, with basal quarter of outer webs bright yellow, combining with greater-covert tips to form prominent pale wing-panel; primaries black-brown, with inner 5 marked basally with bright yellow, continuing prominent wing-panel onto outer wing. On folded wing, yellow wing-panel lacks backward extension of yellow so obvious on primaries of *C. carduelis* but nevertheless striking. Underwing greenish-yellow. Cleft tail black, stemming from grey-green coverts and brilliantly edged yellow along basal two-thirds of all feathers, forming patches almost as bold as wing-panels. Belly and under tail-coverts white, latter streaked black. With wear, black on head becomes more intense and rest of plumage loses grey tones, becoming vividly green and yellow. Bill deep at base but tapers acutely to rather long and fine point; bright yellowish-horn. Eye dark brown. Legs dark brown. Adult Female. Lacks full colours and strong head markings of ♂. Upperparts greenish-grey, streaked black-brown overall, even on yellower rump. Head buffish- to greenish-brown, subtly marked with cleaner yellowish supercilium and surround to ear-coverts. Lores, chin, and belly silvery-white; breast, deep fore-flanks, and narrower rear flanks yellower and greyer and sharply streaked black; under tail-coverts white, streaked black on outer feathers. Wings and tail as ♂ but yellow areas neither as extensive nor as bright, with tips of greater coverts and tertials often almost white. Bill darker brown-horn than in ♂. Juvenile. Resembles ♀ but much duller and browner, with heavier streaks above and below. Head pattern distinctive, with even whiter ground-colour and more spots than streaks on face; ear-coverts pale brown. Wings and tail as adult ♀ but contrasts reduced by merely buff tips to greater coverts and tertials. Bill as adult ♀. First-year. ♂ more streaked above and wing-panel less bright than adult; ♀ browner on head and mantle.

♂ unmistakable. ♀ and juvenile less distinctive, but fine bill distinguishes them from Serin *S. serinus*, while streaked underparts separate them from *S. citrinella* (see accounts of these and other *Serinus* for further comparisons). At distance, juvenile may also be difficult to isolate from frequent companion *C. flammea*, but at close ranges difference in wing pattern, fuller extent of streaks on underparts, and voice soon allow distinction. Flight light and quick, with bursts of wing-beats lifting bird into sudden ascent and soon developing rapidly dipping undulations in speedy direct progress. Song-flight of ♂ circling, with spread tail and exaggerated wing-beats. Escape-flight fast, to dense canopy of tree or thicket. Hops on ground, but movements in foliage and twigs remarkably adept and tit-like. Undemonstrative in winter, feeding quietly on tree seeds. Quite tame but often difficult to find, even when calling (due to somewhat ventriloquial nature of voice).

Song a sweet twitter, lively and quite varied, often starting with 'dluee' and ending with long, nasal wheeze. Shrill, clear 'tsüü', given frequently in flight, is often first signal of presence.

**Habitat.** In west Palearctic, breeds in both lowland and mountain forest, coniferous or mixed, mainly in boreal and temperate zones, north to July isotherm of 13°C (Voous 1960*b*). Mainly occupies spruce *Picea* but also fir *Abies* and pine *Pinus*, especially where these are well-grown and well-spaced, and sometimes mixed with broad-leaved trees. Streamside locations are often preferred, especially outside breeding season where much foraging is in alders *Alnus* and birches *Betula* along watercourses. Has in recent years begun nesting more widely and frequently in fresh areas in England, apparently due to afforestation with conifers and to use of planted introduced conifers in parks and gardens, but in Switzerland in formerly neglected native stands, largely in montane regions at *c*. 1200–1800 m, but not infrequently in lowlands, with marked annual fluctuations. Prefers spruce but will also occupy other conifers such as larch *Larix* (Glutz von Blotzheim 1962).

Principally arboreal, but comes down to ground to take (e.g.) fallen seed (see Food). In USSR in winter, during wanderings in steppe regions, occurs only in orchards and parks of towns and villages and in floodland groves on river banks (Dementiev and Gladkov 1954). Less of a forest edge species and less aerial than (e.g.) Goldfinch *C. carduelis*, and slower in adapting to fresh opportunities in man-made habitats, but not inadaptable, given time. Great gap between west Palearctic and far eastern range despite presence of apparently suitable intermediate habitats indicates inexplicable limitations. (Newton 1972.)

**Distribution.** Has expanded range in Britain, Ireland, and other parts of western Europe, following spread of conifer plantations. Breeding range subject to local expansion and contraction, especially on southern fringe, due to marked annual population fluctuations (see Population).

Faeroes. Regular visitor in last decade; first recorded breeding 1985, Tórshavn (DB). Britain. As breeding bird, probably confined to Scots pine *Pinus sylvestris* of Scottish highlands until mid 19th century. Has spread since, with spread of conifer plantations; major expansion since 1950. (Sharrock 1976; Thom 1986.) Ireland. Has expanded range this century due to afforestation (Hutch-

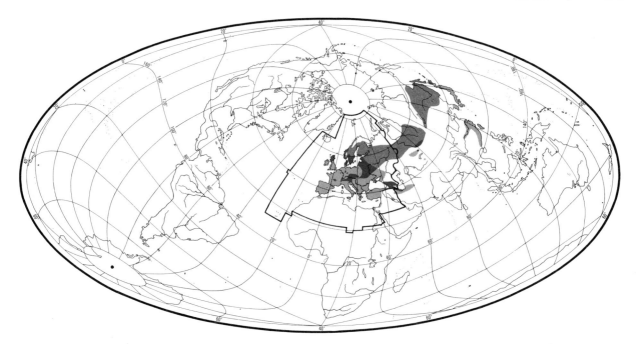

inson 1989). SPAIN. Small numbers breed in mountain woodlands, usually following winters when many have immigrated (AN). PORTUGAL. First recorded probably breeding (birds with incubation patch) 1984, Trás-os-Montes (Dos Santos *et al.* 1985). BELGIUM. Formerly an irregular breeder. Apparently a regular breeder since at least 1960, but range contracted after 1977. (Devillers *et al.* 1988.) NORWAY. Has recently spread north (VR). POLAND. Breeds irregularly in areas other than those mapped (AD, LT). ITALY. Bred Sardinia 1949 (Moltoni 1950). Breeding proved on Sicily (Mt Etna) 1984–5 (Iapichino and Massa 1989). YUGOSLAVIA. Breeds regularly only in north-west (Slovenia); irregularly elsewhere, in some years even in small lowland conifer plantations (VV). TURKEY. Probably more widespread breeder in Black Sea coastlands than data indicate (RDM).

Accidental. Iceland, Madeira, Canary Islands, Azores, Jordan, Kuwait.

**Population.** Has increased in parts of western Europe, following spread of conifer plantations. Marked fluctuations in numbers are characteristic of northern Europe.

BRITAIN, IRELAND. Major increase since 1950. Tentatively estimated at well over 20 000 pairs, possibly 40 000 by 1972. (Sharrock 1976.) FRANCE. 1000–10 000 pairs (Yeatman 1976). BELGIUM. Estimated 760 pairs in 1975; decrease after 1977 (Devillers *et al.* 1988). NETHERLANDS. Estimated 300–700 pairs 1979–83; increasing (SOVON 1987). GERMANY. Estimated 23 000 pairs (Rheinwald 1992). East Germany: 10 000 ± 6000 pairs (Nicolai 1993). SWEDEN. Estimated 1 million pairs (Ulfstrand and Högstedt 1976). FINLAND. Marked annual fluctuations. No

clear long-term trends, but has probably increased due to modern forestry's preference for spruce *Picea abies*; 1–2 million pairs. (Koskimies 1989; OH.) POLAND. Marked fluctuations; very scarce in central provinces (Tomiałojć 1990). CZECHOSLOVAKIA. Marked fluctuations (KH).

Survival. Czechoslovakia: for age structure and mortality, see Pikula (1989). Birds ringed in northern and central Europe and recovered in Iberia: average annual mortality of adults 61% (Asensio and Carrascal 1990). Oldest ringed bird 10 years 11 months (Rydzewski 1978).

**Movements.** Mostly migratory in northern breeding areas, but some southern populations may be resident. Many birds winter in different areas in different years but some are faithful to same area—even exactly the same site. Most are nomadic during winter, but minority becomes resident at same site for several months. Numbers migrating vary greatly from year to year, and more distant movements are recorded when large numbers of birds involved (eruption years). Availability of seed crops on favoured trees (alder *Alnus*, birch *Betula*) seems to be major determinant of strength of movement away from breeding area.

Many birds stay in or close to breeding areas in Britain, Ireland, and southern Scandinavia when plenty of natural food available (Stolt and Mascher 1971; Lack 1986; Thom 1986; Hutchinson 1989), or delay movement until midwinter (or later) in southern Finland (Haila *et al.* 1986). Further east, fluctuating numbers remain to winter in Baltic states and in Russia, e.g. St Petersburg, Moscow, and Volga-Kama areas (Ptushenko and Inozemtsev 1968; Gaginskaya 1969; Popov 1978; Mal'chevski and Pukinski 1983). Throughout Europe south to Mediterranean, win-

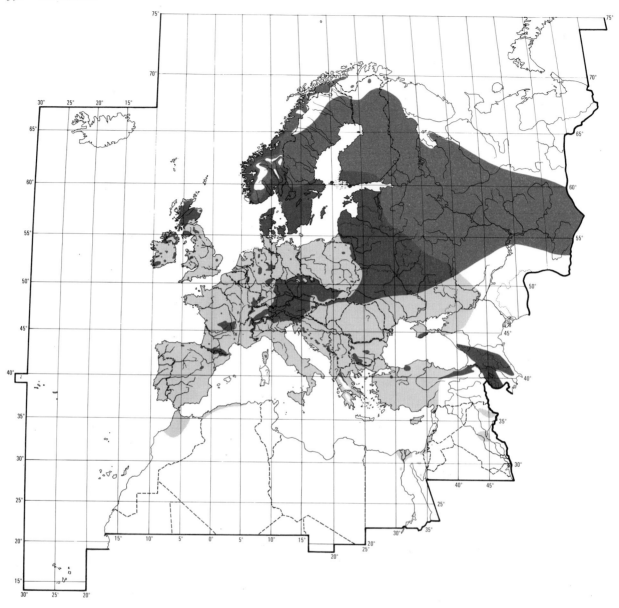

tering birds may be found anywhere with suitable seed crops in irruption years, but are sparse (or do not penetrate far south) in other years. Southern limits are along north coast of Africa, and in Sinai, Israel, Iraq, northern Iran (particularly in south Caspian), and southern Kazakhstan (Vaurie 1959; Etchécopar and Hüe 1967; Korelov et al. 1974; Shirihai in press; D A Scott). Small numbers also reach Arabia (Kuwait, Bahrain, United Arab Emirates, central Saudi Arabia and southern Oman: F E Warr). Isolated eastern population exhibits similar irregularity of migration, and winters in southern Mongolia, Korea, China south to Kwangtung, Hong Kong, Taiwan, Ryukyu Islands, and Japan (Vaurie 1959; Schauensee 1984; Chalmers 1986, Chalmers et al. 1991).

Routes in Europe are complex (see ringing studies, below), with many birds using different wintering areas in successive years, or moving long distances in winter during same year. Mead and Clark (1993) summarized all recoveries involving Britain and Ireland to end of 1991. Total of 512 was as follows: Belgium (198), Netherlands (69), Germany (59), Norway (43), Sweden (33), Baltic states (30), France (24), Finland (11), Denmark (8), Italy (8), Spain (7), USSR (5), Poland (4), Portugal (4), Algeria (3), Austria (2), Czechoslovakia (2), Switzerland (1), and Morocco (1). Longest-distance recoveries for British-ringed birds: west-east (south-west England to Rostov, Russia) 3140 km; north-south (Scotland to Algeria) 2450 km (C J Mead). Many other smaller studies show similar widespread movements, but 2 involve large numbers of records: Spain (399 records: Asensio 1985*b*) and Netherlands (1116 records mapped: Speek and Speek 1984). Previously thought that no birds returned to winter in same distant area in successive years (Newton 1972), but various recent studies show this does happen, e.g. in south-east England (Martin 1990), with 21 retraps from 845 birds in Surrey in later winters; 2 of these birds were controlled in Lithuania, and 2 winters later were re-trapped in same garden where originally ringed (Lack 1986). In eastern Asia, bird ringed in winter on Honshu (Japan) found in summer in north of Sakhalin island (McClure 1974).

Since 1960s, regular feeding in gardens has been observed (see Food), first seen with birds coming to take peanuts in south-east England (Spencer and Gush 1973). This habit, now common in many areas, develops after much of existing natural food in an area is exhausted. Garden feeding noted particularly in March and early April when birds fattening for spring migration (Durham and Sellers 1984; Cooper 1987; Martin 1990). At British sites, recoveries prove that birds involved are from Scottish, Fenno-Scandian, and Russian breeding areas. Senar *et al.* (1992) demonstrated that, for 14 wintering sites in southern England and around Barcelona (north-east Spain), small proportion of birds were resident at feeding stations, but majority were transients. Transients were often caught only once at a site, and moved up to 40 km per day. Residents were caught on 70% or more of trapping days at site over several months; consistently heavier than transients and fed through the day rather than mainly in early morning.

Autumn migration in northern areas may start as early as August, but generally peaks late September or October. Timing and strength vary from year to year, and significant numbers may move as late as December or even January, when most migration stations not manned. At Falsterbo (southern Sweden), visible migration records 1949–88 show peak autumn days (with 8500–9000 birds) as 15 August 1988, 26 September 1974, 1 October 1949, and 1 October 1975; in 1958, peak day was last day (17 November) for the year's observation, with 16% of autumn total (Ulfstrand *et al.* 1974; Roos 1991). Edelstam (1972), analysing visible migration data from Ottenby (south-east Sweden), showed comparable fluctuations in timing over 1947–56; annual totals varied from 66 in 1956 to 6541 in 1953. Further south, on Netherlands coast, annual catches at Wassenaar in 1963–77 varied from 1 (1976) to 300 (1964) (Bezemer 1979). Towards southern edge of wintering range, in Spain (Asensio 1985*b*) and Gibraltar (Finlayson 1992), birds appear in October but peak arrival (passage) November. Small numbers regular, but massive irruptions in 1949–50, 1959–60, 1961–2, 1965–6, 1966–7, 1972–3, and 1988–9. Eastwards along Mediterranean, arrivals mostly mid- or late October to late December, e.g. Malta, Cyprus, Israel (Sultana and Gauci 1982; Flint and

Stewart 1992; Shirihai in press). Totals fluctuate from year to year. On southern edge of wintering range in Arabia, arrivals November to mid-December (F E Warr).

Some autumn recoveries demonstrate rapid movement: St Petersburg to Spain and Latvia to central England (160 km per day), and Lithuania to Belgium (113 km per day) (Lippens 1968; Harrison 1982; Mal'chevski and Pukinski 1983).

Spring departure from regular southern wintering areas is weak and may start in early February, but in most areas continues to mid-April (Sultana and Gauci 1982; Asensio 1985*b*; Finlayson 1992; Flint and Stewart 1992; Shirihai in press). Passage in Switzerland from end of February to mid- or late April (Winkler 1984); peaks in April in northern Denmark and Falsterbo (Møller 1978*a*; Andell *et al.* 1983), and in 2nd week of May at Ottenby (Enquist and Pettersson 1986). On south side of Gulf of Finland (where passage very conspicuous) and in St Petersburg area, waves of migrants pass east or north-east until early or mid-May (Gaginskaya 1969; Mal'chevski and Pukinski 1983). Movement out of south-east England sometimes finishes before end of March, but in most years continues to mid-April, with few if any birds remaining by last week of April (Durham and Sellers 1984; Cooper 1987; Martin 1990). CJM

**Food.** Seeds, especially of conifers (Coniferae, particularly spruce *Picea*), alder *Alnus*, birch *Betula*, and herbs; some invertebrates in breeding season. Of herbs, favours Compositae, Polygonaceae, and Corylaceae, and tends to avoid (e.g.) Cruciferae, Rosaceae, and grasses (Gramineae). (Eber 1956; Glutz von Blotzheim 1962; Newton 1967*a*; MacDonald 1968; Stolt and Mascher 1971.) Very dependent on spruce (or pine *Pinus*) on breeding grounds; for fluctuations in numbers in relation to conifer seed crop, or to birch and alder crop in wintering areas, see Svärdson (1957), Hogstad (1967), Eriksson (1970*b*), Stolt and Mascher (1971), and Shaw (1990). In Schleswig-Holstein, northern Germany, 45% of 1406 foraging observations over the year were in parks and gardens, 39% in damp woodland, and 15% on waste ground (Eber 1956). Outside breeding season, often in mixed flocks with other seed-eaters, especially Redpoll *C. flammea* (MacDonald 1968; Buckland *et al.* 1990). Feeds principally in trees, moving to tall herbs or ground when cones empty and seed has dropped; many invertebrates apparently taken from ground and at water's edge (Pokrovskaya 1956, 1963); also has habit, mainly in winter, of feeding on seeds washed up on shoreline, where favourite alder trees are common (Newton 1967*a*; MacDonald 1968; Kington 1973; Buckland *et al.* 1990). In Switzerland, recorded on water-lilies *Nymphaea* 5 m from shore picking pieces from leaf edges (Hess 1975). Invertebrates also picked from leaves and needles of trees, herbs, bracken *Pteridium*, etc., and insects can be caught in flight (Diesselhorst and Popp 1963; Haapanen 1966; MacDonald

1968; Fellenberg 1988*b*). Moves restlessly through trees examining cones, acrobatically clinging to cones and twigs, often starting at crown and moving towards base; hangs and perches on stems and seed-heads of herbs with equal agility, spending less time on ground than in plants (Eber 1956; Szabó and Györy 1962; Newton 1972; Lack 1986); 34% of feeding positions in captivity clasping vertical stems, 27% bent stems, 18% upside-down, and 9% normal perching; foot/bill coordination highly developed, pulling objects in with bill and clamping under foot and also placing detached items under foot (Newton 1967*a*). At garden feeder, usually hangs on side rather than upside-down (Spencer and Gush 1973). Tweezer-like bill used to extract or pry out (bill inserted then opened) seeds (etc.) from tight spaces (e.g. cones, closed seed-heads, buds, catkins), very like Goldfinch *C. carduelis*, though unable to probe as deeply; pierces flower-bases of Compositae or strips off involucral bracts to get at ripening seeds, often while perched on neighbouring stem (Eber 1956; Newton 1972). Very dexterous in removing seeds from just-opening pine cones, inserting bill at all angles to grasp wings which are then nipped off before seed cracked (Volsøe 1949); cannot open closed cones so has to wait until they are naturally open, or can take seeds from those which have already been worked by crossbills *Loxia* (Turček 1961; MacDonald 1968; Davis 1977). In recent years (in southern England since early 1960s and Ireland since 1980s) frequents bird-tables and garden feeders in winter much more regularly than in past, particularly after alder seed crop exhausted; perhaps first entered gardens to feed on seeds of ornamental cypress *Cupressus* or *Chamaecyparis* (Spencer and Gush 1973, which see for chronicle of habit; Davis 1977; Lack 1986; Hutchinson 1989). Another habit apparently acquired recently is eating of beechmast *Fagus*, first recorded in central England in 1950s (Newton 1967*a*, which see for handling method). In Spain, winter, 30–40 birds regularly picked up *Eucalyptus* seeds from road (Ingram 1965). In captivity, seed handling and consumption time ranged from 1·4 s (*Brassica*) to 10 s (hemp *Cannabis*); Compositae seeds of various species much preferred (Ziswiler 1965, which see for many details of bill morphology and de-husking behaviour). See also Eber (1956) and Newton (1967*a*). At bird-tables, preferred hemp, linseed *Linum*, and Compositae seeds (Pannach 1990). In captive tests, distinguished between aborted and edible pine seeds by briefly 'weighing' in bill (Senar 1983).

Diet in west Palearctic includes the following. Invertebrates: mayflies (Ephemeroptera: Siphlonuridae, Baetidae), damsel flies (Odonata: Agriidae, Lestidae), bugs (Hemiptera: Pentatomidae, Aphidoidea, Adelgidae), Lepidoptera (Nymphalidae, Tortricidae, Coleophoridae, Gracillariidae, Noctuidae, Geometridae), flies (Diptera: Mycetophilidae), Hymenoptera (Tenthredinidae, Diprionidae, ants Formicidae), beetles (Coleoptera: Carabidae, Scarabaeidae, Elateridae, Coccinellidae, Curculionidae, Scolytidae), spiders (Araneae: Argyronetidae, Araneidae), earthworms (Lumbricidae), molluscs (Mollusca: Succineidae, Sphaeriidae). Plants: seeds, buds, fruits (etc.) of cypress *Chamaecyparis*, juniper *Juniperus*, *Thuja*, redwood *Sequoia*, larch *Larix*, spruce *Picea*, pine *Pinus*, hemlock *Tsuga*, Douglas fir *Pseudotsuga*, willow *Salix*, poplar, etc. *Populus*, birch *Betula*, alder *Alnus*, beech *Fagus*, elm *Ulmus*, maple *Acer*, plane *Platanus*, *Eucalyptus*, lilac *Syringa*, olive *Olea*, hemp *Cannabis*, hop *Humulus*, nettle *Urtica*, dock *Rumex*, goosefoot *Chenopodium*, seablite *Corispermum*, chickweed *Stellaria*, water-lily *Nymphaea*, poppy *Papaver*, gold of pleasure *Camelina*, charlock *Sinapis*, shepherd's purse *Capsella*, hoary alison *Berteroa*, wintergreen *Pyrola*, meadowsweet *Filipendula*, apricot, etc. *Prunus*, rowan *Sorbus*, pea *Pisum*, *Acacia*, willowherb *Epilobium*, evening primrose *Oenothera*, bilberry *Vaccinium*, plantain *Plantago*, golden-rod *Solidago*, cone flower *Rudbeckia*, mugwort *Artemisia*, groundsel *Senecio*, burdock *Arctium*, thistles *Carduus*, *Cirsium*, knapweed *Centaurea*, chicory *Cichorium*, sow-thistle *Sonchus*, lettuce *Lactuca*, *Chondrilla*, dandelion *Taraxacum*, hawkbit *Leontodon*, cat's-ear *Hypochoeris*, hawkweed *Hieracium*, lily-of-the-valley *Convallaria*, grasses (Gramineae), fungus. (Schuster 1930; Dementiev and Gladkov 1954; Pokrovskaya 1956, 1963; Turček 1961; Béress and Molnár 1964; Newton 1967*a*, 1972; MacDonald 1968; Tutman 1969; Hess 1975; Sabel 1983; Hirschi 1986; Pannach 1990.)

In Schleswig-Holstein, of 1283 feeding observations over the year, 46·5% on seeds of alder, 35·4% birch, 10·2% mugwort, 2·3% dock, 2·1% elm, 1·8% dandelion, 1·8% thistles; 2·9% of 1406 observations were of birds feeding on invertebrates (Eber 1956). In northern Rumania, winter, of 48 observations, 16 were on *Thuja*, 7 spruce, 7 alder, 6 birch, and 2 chicory; also Compositae, Chenopodiaceae, and other trees (Béress and Molnár 1964). In eastern Germany, 18 of 41 winter observations on alder, 9 birch, 3 larch, and 2 pine; also seeds of beech, poplar, spruce, Compositae, and Chenopodiaceae (Pannach 1990). In spring, feeds almost exclusively on conifer seeds when cones open, also on buds and shoots; in addition, takes milky Compositae heads and insects (Glutz von Blotzheim 1962; MacDonald 1968; Stolt and Mascher 1971); if arrival on breeding grounds is early, has to feed on insects, many obtained from larch buds (Newton 1967*a*). In Finland, spruce seed has dropped by May and pine only later that month, so brief food shortage can result, and invertebrates and Compositae thus become important (Haapanen 1966). In Schleswig-Holstein, also eats birch catkins and elm flower buds; in April, 31% of 52 feeding observations were insects, May 50% of 14, and July 9% of 114 (Eber 1956). In summer, eats much wider variety of herb seeds, mainly Compositae and Polygonaceae (since conifer seed has fallen), and perhaps fewer invertebrates; also birch seeds (Eber 1956; MacDonald 1968; Stolt and Mascher 1971; Shaw 1990); in autumn, moves to alder and birch trees for seeds, but also to waste ground in places, feeding on thistles, etc.; often on ground in late autumn and winter when alder seeds have dropped, and alder remains main food over winter; in extreme south of wintering range feeds on *Acacia* seeds (Eber 1956; MacDonald 1968; Stolt and Mascher 1971; Newton 1972; Davis 1977); also noted in autumn in groups feeding extensively on aphids taken from leaves of fruit trees (Fellenberg 1988b).

Young fed conifer seeds and invertebrates (Glutz von Blotzheim 1962; Diesselhorst and Popp 1963; Newton 1967a); aphids can be important and perhaps also elm seeds in places (Witherby *et al.* 1938; Diesselhorst and Popp 1963). In northern Europe, 1st brood possibly reared more on spruce seeds and 2nd more on pine seeds and insects such as aphids and caterpillars (Newton 1972). Invertebrates essential in diet of nestlings in captivity (Röder 1991). In St Petersburg region (western Russia), of 382 items of animal prey in 92 collar-samples, 28·3% by number Hymenoptera (20·7% ants, 5·8% larval Tenthredinidae), 24·1% beetles (12·3% larval Elateridae, 5·0% Carabidae), 18·9% larval Lepidoptera (11·8% Nymphalidae, 7·1% Noctuidae), 10·0% larval mayflies, 7·6% Hemiptera, 5·5% spiders, 1·3% larval damsel flies; of total 488 items, 21·7% plant material, mainly oat *Avena* seeds, old *Vaccinium* berries, strawberries, and fragments of fungus (Pokrovskaya 1956). See also Pokrovskaya (1963) and Mal'chevski and Pukinski (1983); in latter study, also St Petersburg region, invertebrates said to be much overestimated in earlier work.                                    BH

**Social pattern and behaviour.** No comprehensive study, but most aspects fairly well known. For review and comparison with other Fringillidae, see Newton (1972).

1. Gregarious outside breeding season. Winter flocks of up to several hundred, sometimes several thousand (e.g. Ptushenko and Inozemtsev 1968, Stolt and Mascher 1971, Leonovich 1976); roosting flocks of 10-20(-50) (Newton 1972; see also Roosting, below). Study in Belgium showed birds tend to limit winter home-range once chosen, and some records of winter site-fidelity between years (Verheyen 1956); see Movements. At 14 study sites in Spain and Britain (where excess food available), wintering population found to comprise residents (occupying small home-range and usually moving less than 3 km per day) and transients (majority, staying one or a few days); both continued to arrive throughout winter; high turnover of transients apparently due to high level of aggression between them at feeders (Senar and Metcalfe 1988; Senar *et al.* 1992); see also Flock Behaviour and Antagonistic Behaviour (below). In Sistema Central mountains (Spain), 9 birds per km² recorded in pine *Pinus* forest in normal (non-snowstorm) conditions (Carrascal 1988). Feeding flocks sometimes associate with other Fringillidae, notably Redpoll *C. flammea* and Goldfinch *C. carduelis*; in France in larch *Larix* forest, frequently with mixed tit *Parus* flocks (Laurent 1986). On Gulf of Finland (western Russia), flocks of 3-50 in early April sometimes migrated with Chaffinch *Fringilla coelebs*; later in April, most birds in pairs, though some flocks of 20-25 (Gaginskaya 1969). In Bükk mountains (Hungary), small flocks noted in early May comprised pairs, and some birds already breeding (Szabó and Györy 1962); certain

tendency to gregariousness continues throughout breeding season, ♂♂ feeding together when ♀♀ incubating (e.g. Witherby *et al.* 1938, Nethersole-Thompson and Watson 1974). Flocks, sometimes predominantly or exclusively juveniles, reported in post-breeding period (e.g. Witherby *et al.* 1938, Ptushenko and Inozemtsev 1968, Paevski 1970a); at Memmingen (southern Germany), juvenile flocks stayed at least 40 days in breeding area (Diesselhorst and Popp 1963). BONDS. Mating system not investigated, though little to suggest other than monogamy. Study in eastern Germany found ♂♂ (most paired, some perhaps nonbreeders) to move about singing and displaying during incubation (Weber 1959); not known, however, whether this sometimes leads to polygamy or promiscuity. Pair-formation evidently takes place in flocks, with (e.g.) Courtship-feeding (see Heterosexual Behaviour, below) reported in Spain in March, and in Scotland most birds paired when settle to breed (Géroudet 1957; Ptushenko and Inozemtsev 1968; Nethersole-Thompson and Watson 1974; Senar and Copete 1990a). Occasionally hybridizes with other Fringillidae, e.g. *C. flammea* (Lorenz 1890) or Linnet *C. cannabina* (Lönnberg 1918). In captive birds, Allofeeding (see also Flock Behaviour and Antagonistic Behaviour, below) frequently recorded between 2 ♂♂ November-June (peak February-June, and especially in pre-hatching period; always between birds with clear hierarchical relationship, subordinate feeding dominant ♂; perhaps aids flock integration by reducing aggression) (Senar 1984b). Role of sexes in nest-duties as follows: study in Germany found only ♀ to incubate (Diesselhorst and Popp 1963), but occasional participation by ♂ reported by Dost (1957) for captive pair, and Ptushenko and Inozemtsev (1968) for wild birds. Both sexes feed young (e.g. Witherby *et al.* 1938). In case of late breeding in Norway, ♀ built 2nd nest while both apparently still fed 1st-brood fledglings (Bringeland 1964); captive ♂ cared for young from 7 days old when ♀ started new nest (Jourdain 1930). In another study of captive birds, 1st-brood juvenile helped parents to feed 2nd brood from 10 days up to fledging at 16 days (Senar 1984a). Observations in Scotland suggested young tended by both parents for several weeks after fledging (Nethersole-Thompson and Watson 1974). Age of first breeding 1 year (Paevski 1985). BREEDING DISPERSION. Territorial, but not conspicuously so (see below); usually in loose neighbourhood groups of 2-3 or up to *c.* 15 or more pairs, especially in areas of heavy cone crops (Newton 1972; Sharrock 1976). In Cairngorm region (Scotland), often in groups of up to 6 pairs, with (e.g.) 2 pairs *c.* 20 m apart; in higher forests, well dispersed, and large groups exceptional (Nethersole-Thompson and Watson 1974). In spruce *Picea* wood at Memmingen, nests clumped, and area apparently equally suitable for nesting remained unoccupied; in group of 10 pairs on woodland edge nests 14-17 m apart; by glade in wood (50-60 m from edge) nests of 2nd group (*c.* 15-17 pairs) separated by *c.* 20-25 m (Diesselhorst and Popp 1963). On tributary of Ob' river (just east of Urals), nests in groups of 4-20 birds usually 25-30 m apart, in some cases perhaps only 3-4 m (Lykhvar' 1983). Limited information suggests typical of Carduelinae in that apparently small nesting-territory (exact size not known) established by ♂ only after pair-formation and selection of nest-site, and courtship (etc.) takes place both within and outside territory. (Witherby *et al.* 1938; Newton 1972.) Birds forage over wide area, probably mainly outside territory; at Memmingen, just after hatching, within *c.* 100 m of nest (Diesselhorst and Popp 1963). Study in Scotland found low level of territory defence (by ♂, occasionally chasing other ♂♂ out of particular trees and fighting) during nest-building, then rare after 1st egg laid until ♀ incubating; ♂ often absent and far from territory when foraging (Nethersole-Thompson and Watson 1974). In St Petersburg region (western Russia), fledglings said by Pokrovskaya (1963) to remain for c. 2 weeks in nesting-territory, with adults foraging both there and outside. Breeding density in pine *Pinus* forest, Abernethy (Scotland), 1·99±SE0·34 birds per ha (Hill *et al.* 1990). In Fenno-Scandia: 0·05 in virgin pine forest and 1·5 pairs per km² in uniform forest in northern Finland (Virkkala 1987); average 4·0-4·6 territories per km² over 3 years in 3 plots of coastal subalpine birch *Betula* forest, south-west Norway (Ytreberg 1972); 3-11 pairs per km² in various habitats on Åland Islands (south-west Finland) (Palmgren 1930; Haila *et al.* 1979); overall average in Sweden 5 pairs per km² (Ulfstrand and Högstedt 1976); in Finland, 14-24 pairs per km² in pine stands, up to 45 in spruce *Picea*, with 4-8 pairs per km² in managed, 24 in virgin spruce forest (Haapanen 1965). At Memmingen, one group of 10 pairs along 200 m, another of 15-17 pairs along 300-400 m (Diesselhorst and Popp 1963). In Białowieża forest (eastern Poland), 1-6 pairs per km², highest in pine and bilberry *Vaccinium* (Tomiałojć *et al.* 1984). Also in Bulgaria, high density (60 pairs per km²) in forest predominantly of *Pinus nigra* (Simeonov 1975). Average on tributary of Ob' river 200-300 pairs per km² (Lykhvar' 1983). Site-fidelity demonstrated by ringing study in south-west Scotland where 8·0% of ♂♂ (n=247) and 9·0% of ♀♀ (n=144) returned, including 9 over more than 1 season (1 ♂ over 4, 2 ♀♀ over 3 consecutive seasons); 1 juvenile out of 104 returned (Shaw 1990). ROOSTING. Nocturnal and, at least outside breeding season, communal. ♂ of captive pair sometimes roosted on nest-rim (Dost 1957). Sites used include tall conifer clumps, thorn scrub, or hedgerows, less commonly low alders *Alnus* or reedbeds (Naumann 1900; Witherby *et al.* 1938; Newton 1972). In Spain, November, during irruption, flock of 7-8 roosted at c. 7-8 m in poplars *Populus* during 2 consecutive nights; indicative of roosting in small groups despite high numbers in irruption year (Copete 1990). Winter roosts in Britain usually temporary and of similar size to feeding flocks (Newton 1972). At 60°N in Finland, mid-winter, birds roost for 17-17·5 hrs per day (Sulkava 1969); at 60°30'N, mid-June, c. 20.55-02.15 hrs (Paatela 1938). For diurnal activity rhythm in captive birds, see Pohl (1971b) and Daan (1972); for further experimental studies on bioenergetics of roosting, see Davydov (1976) and Saarela *et al.* (1988).

2. Can be rather secretive when breeding (Diesselhorst and Popp 1963), but not especially shy, more curious (Szabó and Gyory 1962); very tame, almost indifferent to man when feeding in gardens (Spencer and Gush 1973), or elsewhere (e.g. Hirschi 1986). Subject to sudden panic flights, however (see below). In low-intensity alarm (sight of unfamiliar object), captive bird turned head towards object, slightly opened wings and lowered tail, also gave hoarse 'pee' (see 4d in Voice). On sighting or hearing predator, crouches in horizontal posture, sleeks plumage and gives 'vyaa' quietly 1-2 times (see 4h in Voice), head pointed at danger; may freeze in same posture for up to c. 20 min. Loud 'peee' (see 4g in Voice) given by captive in long series (mixed with 'peelee': see 4b in Voice) with bill wide open when family of Jays *Garrulus glandarius* near cage; stretched up, legs fully extended, crown ruffled, tail slightly spread, turning body left and right. Bird may interrupt feeding and give long series of 'tyulee' mixed with 'tyuee' (see 4a in Voice) and (♂♂ only) 'peelee', with head extended up and moved side to side while remaining concealed; 'tyulee' and other similar calls (see Voice) are frequently given by lone bird in flight (see below). (Zablotskaya 1978b.) According to Nicolai (1957c), in high-intensity sexual or antagonistic excitement, tends to move widely fanned tail horizontally, unlike *Serinus* which whip tail up and down; however, *C. spinus* giving call 4f (see Voice) reported by Zablotskaya (1978b) to move tail up and down in time with call.

FLOCK BEHAVIOUR. Winter flocks usually silent when feeding; typically erupt in dense swarm, making short circular flight and giving Twitter-calls (Bruun *et al.* 1986: see 3 in Voice); if alarmed, birds may first fly close to ground before making off at greater height (Naumann 1900); erratic manoeuvres performed by compact flock in longer flights (Géroudet 1957c). In Austria, early January, birds apparently taking minerals from house wall made 'panic flights' to nearby trees; occasional brief scuffles occurred when birds attempted to displace others already on wall (Flaxman 1983). Allofeeding, sometimes involving 2 birds of same sex, also noted in winter flock; accompanied by call 4f or Twitter-call (Zablotskaya 1978b); see Antagonistic Behaviour (below). Closer to spring, activities in flock include courtship (associated with pair-formation), singing (often given quietly by ♂ perched above ♀, though chorus of song and various calls can be quite loud), Song-flights (see below), fights, and sexual or ♂-♂ chases accompanied by calls (see 5b in Voice) (Witherby *et al.* 1938; Neuschwander 1973; Nethersole-Thompson and Watson 1974; Zablotskaya 1978b). In southern England, March-April, birds arriving in garden from roost first assembled in trees (where sang and preened), then dispersed to feeding areas, usually in small parties (Spencer and Gush 1973). Several ♂♂ feeding together in same tree in summer may call 'peek' (see 6a in Voice) and 'peelee'. Bird seeking contact with flock will circle area where flock feeding over several days and call ('tyulee', 'tyuee', 'peelee'); feeding flock will respond with 'tyulee' on hearing this call from passing flock and, given by one bird, call may signal flock's departure. (Zablotskaya 1978b.) For studies showing transient birds attracted by calls of residents, and presumed advantages accruing to small flocks of residents if transients join them, see Senar and Metcalfe (1988) and Senar *et al.* (1992). SONG-DISPLAY. ♂ gives Advertising-song (see 1a in Voice) from exposed perch on top of tree or concealed in canopy, also in flight (see below). Several ♂♂ often sing close together, both on spring migration and when breeding, and song unlikely to have territorial function (Volsøe 1949; Diesselhorst and Popp 1963); for other song-types, associated respectively with antagonistic and heterosexual interactions, see also 1a in Voice. On perch, adopts erect posture, with legs extended and plumage sleeked, turning body or head from side to side. (Newton 1972; Zablotskaya 1978b.) Song sometimes preceded by 'tyulee' (Zablotskaya 1978b). No detailed information on frequency of Song-flights, but performed only occasionally according to Bergmann and Helb (1982). Occur chiefly at start of breeding season (during courtship and incubation), especially in sunny weather, several ♂♂ sometimes in the air together, so that their flight-paths cross and recross. May ascend (repeatedly) quite high, taking off in fluttering flight from perch where had previously sung. Bird has plumage ruffled and tail widely spread; flies in loops and circles at tree-top height or higher and over or close to nest-site. Rapid, shallow wing-beats reported by Ptushenko and Inozemtsev (1968), but other authors refer to hesitant, fluttering butterfly- or bat-like flight, with exaggerated deep wing-beats, so that wings appear almost to meet over back. Sings throughout performance, which resembles that of *C. carduelis*, and also on perch between flights. (Naumann 1900; Holstein 1934; Witherby *et al.* 1938; Géroudet 1957c; Szabó and Gyory 1962; Ptushenko and Inozemtsev 1968; Newton 1972; Nethersole-Thompson and Watson 1974.) When 3-4 birds perform Song-flight together, may all suddenly stop singing and descend rapidly to trees or fly off through forest, calling 'tleeu' (Nethersole-Thompson and Watson 1974: see 4c in Voice). ♀ also sings (see 1b in Voice), but not known to perform Song-flight (Diesselhorst and Popp 1963), though ♂ sometimes sings in flight above flying ♀; also during courtship when perched (see Heterosexual Behaviour, below). In Tipperary (Ireland), breeding birds sang daily to *c.* 14.00 hrs, then silent (Carroll 1916). Sings more or less all year, except during moult, with peak March–April (Newton 1972). Regular song in winter reported from (e.g.) central Italy (Alexander 1917) and Netherlands (Bezemer 1979). Study in southern Britain showed irregular or only subsong (see 1b in Voice) early to mid-February and November to early December, fairly frequent full song February to end of April (Witherby *et al.* 1938). Sings also during spring migration (e.g. Volsøe 1949 for Denmark). At Memmingen, breeding ♂♂ sang vigorously during nest-building, then suddenly ceased (Diesselhorst and Popp 1963). On breeding grounds further north in St Petersburg region, song-period mid-April to early August, then in autumn to end of October (Mal'chevski 1959); similar in Finland (Hildén 1972). ANTAGONISTIC BEHAVIOUR. (1) General. Study of captive birds in winter found interactions mainly concerned (a) members of future pairs, (b) several ♂♂ presumably competing for one ♀, or (c) high-ranking individuals; probably serve to establish hierarchy, and perhaps also have pairing function. ♂♂ found to dominate ♀♀ but no straight linear hierarchy, and age, size, and kinship not correlated with dominance (Senar 1985). Dominants and subordinates typically used threat rather than attack to defend resource (perch, food, water); subordinates in possession significantly increased success rate in encounters with dominants (Senar *et al.* 1989). In Spain in winter, residents dominant over transients, though transient ♂♂ nevertheless initiated most interactions with both sexes, and showed more aggression among themselves than towards residents; ♂♂ directed more aggression at ♀♀ than vice versa (Senar *et al.* 1990). Study of birds feeding on nuts in garden found marked aggression, especially towards conspecific birds; one bird dominant, others waiting for opportunity to supplant it (Spencer and Gush 1973). (2) Threat and fighting. Individual-distance disputes occur while resting, and supplanting attacks among feeding birds. At lower intensity, bird may give threat-call (see 5a in Voice) with bill closed while turning towards opponent and crouching slightly. At higher intensity, variant of threat-call given in full Threat-posture: legs

bent, body horizontal, plumage sleeked or (♂♂) breast and flanks ruffled; wings bent at carpal joint, raised level with back and quite often vibrated; head (level with back) slightly drawn in; bill opened wide (threat-gaping) and directed at opponent. For movements associated with increasing intolerance leading to adoption of full Threat-posture, see Fig A and Senar (1983). At peak intensity (notably in breeding-season disputes between ♂♂), one bird may give further threat-calls or change to battle-song (see 5a–b, also 1a in Voice), then attack, attempting to seize opponent's tail or wing. Attacker (or pursuer if chase develops) often flies with ruffled breast and flanks (not typical of Carduelinae, but see *C. carduelis* for similar performance). When threatened or attacked, weaker bird may give 'vyaa'. (Zablotskaya 1978b.) Fierce fights sometimes develop, protagonists becoming interlocked, but bird trespassing during owner's absence usually flees immediately when territory-owner returns and dive-attacks it (Nethersole and Watson 1974). Allofeeding between 2 ♂♂ frequently recorded among captive birds. Initial aggression (mutual bill-pecking) gradually led to less violent Billing, etc. True passing of food (Fig B) occurred closer to breeding season

B

and always between birds with clear hierarchical relationship, from subordinate to dominant ♂. Feeding bird adopted upright posture, with plumage mostly sleeked, and regurgitated food into other's bill (as in Courtship-feeding; see Heterosexual Behaviour, below); recipient had head drawn in, plumage ruffled, and sometimes quivered wings slightly; gentle pecking (by either bird) at other's bill, also equally gentle taking of food reminiscent of Courtship-feeding. (Senar 1984b.) Recorded stealing material from nest of *F. coelebs* (Ellison 1910), but pair only flew about excitedly when their nest similarly plundered by that species (Hamilton-Hunter 1909). HETEROSEXUAL BEHAVIOUR. (1) General. Courtship and pair-formation evidently take place in flocks during late winter or early spring, in some cases while still on wintering grounds or during spring migration. At *c.* 1800 m in Switzerland, albeit in exceptionally mild winter, had settled to breed by early March (Geyr von Schweppenburg 1930). At Memmingen, birds were nest-building within a few days of arrival, this activity and laying being highly synchronous within nesting groups (Diesselhorst and Popp 1963). In Carduelinae generally, pair-formation prolonged, ♂ approaching ♀ in flock, and courtship involves Bill-flirting (see below) and strong aggressive elements (Newton 1972). (2) Pair-bonding behaviour. Sexual chases occur, both birds sometimes ascending to some height, ♂ fanning tail, rapidly beating wings and singing above ♀; alternatively, birds move through branches or zigzag between trees; either bird may call (see 5b in Voice) (Witherby *et al.* 1938; Nethersole-Thompson and Watson 1974; Zablotskaya 1978b). In

A

courtship on perch (including in flock), ♂ facing ♀ ruffles black crown and yellow rump and breast, droops and quivers wings, and spreads and slightly raises tail (in European Carduelinae, tail-spreading most marked in *C. spinus*, also *C. carduelis* and *C. chloris*). ♂ also gives quiet courtship-song (see 1a in Voice) and various calls: 'tyu-pee' and 'tyueelee', occasionally also 'peek' (see 6a–c in Voice). If ♀ flies to ground, ♂ follows (mate-guarding, as also later during building) and stands close to her, calling still, also sometimes raising wings at *c.* 45°. (Witherby *et al.* 1938; Newton 1972; Panov 1973*a*; Nethersole-Thompson and Watson 1974; Zablotskaya 1978*b*.) Prominent in courtship, as in many Carduelinae, is Bill-flirting ceremony usually initiated by ♀: approaches ♂ and, slightly crouched, points bill at him, rapidly opening and closing it and calling (see 6d in Voice); ♂ (silent or sometimes giving same call) stretches up slightly and inserts bill into ♀'s from above. Ceremony (lasting up to 2–3 min) performed several times per day. ♀ not ready will give high-intensity threat-call (see 5a in Voice) and may attack persistent ♂, both then flying up and ♀ continuing to call in flight. (Zablotskaya 1978*b*.) (3) Courtship-feeding. ♂ feeds ♀ regularly on nest or sometimes close to it, including during building (Ptushenko and Inozemtsev 1968); in Scottish study, usually every 30–120 min, ♀ also leaving nest several times per day to feed herself (Nethersole-Thompson and Watson 1974). ♂ approaching nest may give Twitter-call or call 4f (see Voice), ♀ calling similarly or with insistent 'tew' (presumably same as 'teeyu': see 4c in Voice) (Mundinger 1970; Nethersole-Thompson and Watson 1974; Zablotskaya 1978*b*). At Memmingen, ♂ came to nest silently, perched within *c.* 1 m, then made sudden final approach and fed ♀, who afterwards hopped out of nest to preen close by before returning; ♂ would sit motionless on perch for up to *c.* 2 hrs before flying off silently (Diesselhorst and Popp 1963). Reciprocal feeding (Allofeeding) noted in captive pair (Dost 1957). (4) Mating. Takes place on perch, sometimes close to nest, during building (at least 12 days before 1st egg laid) and laying (Ptushenko and Inozemtsev 1968; Nethersole-Thompson and Watson 1974). In Carduelinae generally, preliminary to copulation is short 'moth flights' (rapid, shallow wing-beats) by ♂ usually in cover when ♀ present (Newton 1972). In *C. spinus*, normally no elaborate ritual, ♂ chasing ♀ briefly along branch, then mounting when she crouches with wings drooped and tail raised; or may fly across glade (perhaps in 'moth flight') and land fluttering on ♀'s back, ♀ flying off if not ready and then chased by ♂ (Nethersole-Thompson and Watson 1974). (5) Nest-site selection. By ♀, generally in presence of ♂ (who sings quietly): several potential sites examined and tested, ♀ creeping in and out of crotches, crouching and rotating in them usually briefly, but may remain in one motionless and silent for up to *c.* 1 hr (Nethersole-Thompson and Nethersole-Thompson 1943*b*; Nethersole-Thompson and Watson 1974). Nest-building also by ♀, usually accompanied by ♂ who is reported occasionally to pick up material and drop it again quickly, or even to bring it to nest (Geyr von Schweppenburg 1930; Witherby *et al.* 1938; Ptushenko and Inozemtsev 1968; Nethersole-Thompson and Watson 1974). ♀ recorded attacking mate when he landed close to her during building (Szabó and Győry 1962). (6) Behaviour at nest. ♀ may start laying before nest completed (Ptushenko and Inozemtsev 1968). May fly direct to nest, giving loud Twitter-calls (Ellison 1910), or first land on branch some distance away and creep towards nest; ♂ may direct quiet song at incubating ♀ from nearby perch (Carroll 1916). RELATIONS WITHIN FAMILY GROUP. Captive ♀ apparently helped chick to hatch, enlarging hole and consuming eggshell fragments (Senar 1989). Shells may also be carried away. Young brooded much of day and all night for first 4–5 days, when ♂ brings most food, passing it to ♀ for transfer (by regurgitation) to young; ♀ some-

times also stands up, allowing ♂ to feed small young direct. (Witherby *et al.* 1938; Nethersole-Thompson and Watson 1974.) Fledge at *c.* 13–17 days (Newton 1972). First threat noted among captive birds at 18–22 days, little threat among fledglings (Senar 1985). ANTI-PREDATOR RESPONSES OF YOUNG. No information. PARENTAL ANTI-PREDATOR STRATEGIES. (1) Passive measures. ♀ usually sits tightly, allowing approach to within *c.* 1–2 m (Szabó and Győry 1962; Ivanchev 1987), though often leaves soon after observer starts to climb tree (Nethersole-Thompson and Watson 1974). (2) Active measures: against man (no information for other predators). Include warning-calls from ♂, alarm-calls and excited flitting about near nest or circling over it, by ♀ or both birds (Szabó and Győry 1962; Pokrovskaya 1963; Nethersole-Thompson and Watson 1974).

(Figs by D Nurney: A from drawing in Senar 1983; B from photograph in Senar 1984*b*.) MGW

**Voice.** Several calls (and song) given all year, most familiar being attractive tonal calls such as 'dluee' or similar (see 4, below). Detailed study of wild and captive birds in Russia revealed 3 song-types and large repertoire of calls, most of which given by both sexes (4b, 6a–c, and 8 by ♂ alone) (Zablotskaya 1978*b*); scheme modified in following account. In experimental study of vocal responses in captive birds, 9 calls distinguished, 6 of which used for contact (see Oehler 1977); for study of dynamics of acoustic communication, see Oehler (1978). For additional sonagrams, see Bergmann and Helb (1982); for description allied with musical notation, see Stadler (1956). Voice of Pine Siskin *C. pinus* of North America similar (Reinsch 1967); capable of mimicking *C. spinus* calls (Mundinger 1970, 1979).

CALLS OF ADULTS. (1) Song of ♂. 3 types distinguished (but not described in detail) by Zablotskaya (1978*b*): advertising-song, given mainly to attract ♀ (descriptions in other sources presumably refer to this type; see below); battle-song, associated with high-intensity aggression and given before disputes between ♂♂; courtship-song. Various calls regularly incorporated in song. (1a) Advertising-song of ♂. Rapid, lively, persistent, and quite sweet twittering chatter, at times recalling Serin *Serinus serinus* (Géroudet 1957*c*) or Sedge Warbler *Acrocephalus schoenobaenus* (Nethersole-Thompson and Watson 1974); more or less continuous or sometimes in phrases. Often preceded by species-specific 'dluee' or 'diu-li' which, like several other calls, is also incorporated in song. Song contains repetitions of units or motifs in regular temporal pattern and numerous easily recognized imitations of other birds. Towards end, typically a feeble, drawn-out and nasal wheeze lasting *c.* 1 s. (Witherby *et al.* 1938; Stadler 1956; Bergmann and Helb 1982; Bruun *et al.* 1986.) Most birds include mimicry in song, and a few are expert, including imitations of 5–10 different species in repertoire. Varied, mimetic song of some birds in Scotland contained copies of 'song trill' of Crested Tit *Parus cristatus* and calls of Scottish Crossbill *Loxia scotica* (Nethersole-Thompson and Watson 1974). In eastern Germany, mid-April, bird mimicked calls of Nuthatch *Sitta europaea*, House Spar-

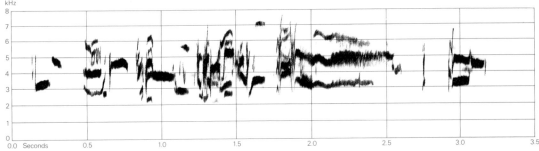

I  V C Lewis  England  March 1969

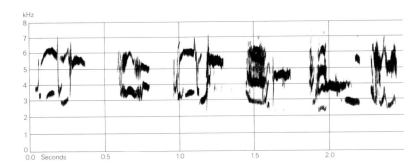

II  V C Lewis  England  March 1969

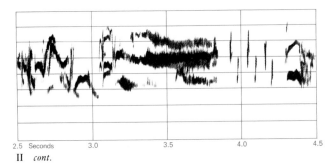

II  *cont.*

row *Passer domesticus*, Crossbill *L. curvirostra*, and Linnet *C. cannabina*; captive birds imitated Chaffinch *Fringilla coelebs* (song), Song Thrush *Turdus philomelos*, Great Tit *P. major*, and Starling *Sturnus vulgaris* (Kneis 1987; see also Helm 1894, Koch 1914). Recording (Fig I) contains short, musical phrase culminating in long wheeze of harmonic structure and flanked by 'diu-li' calls (see 4a, below). In another recording (Fig II), song incorporates several calls; starts with 'dee-u-li' (given again, less smoothly, as note 3: see call 6c, below); twittering units preceding wheeze recall song of Swallow *Hirundo rustica* and are perhaps mimicry; other notes (including wheeze) much as in Fig I, but attractive short coda comprising 't-k-t-k-t-k' and mixture of tone and noise slightly reminiscent of song-flourish of *F. coelebs*. Another part of same recording (Fig III) contains 'tut' (Twitter-call), then 'tiu-li' preceding longer series of Twitter-calls (or variants), then 4 flanked by tonal 't-loo yee' and final 'eeoo-yeeu'. (J Hall-Craggs, M G Wilson.) (1b) Song of ♀. Mentioned but not described by Diesselhorst and Popp (1963). Captive ♀ gave much quieter, weaker song than ♂'s while ♂ still alive, but became ever louder when ♂ died (Gerber 1963). Weak warbling song often noted in winter (Witherby *et al.* 1938); this perhaps given by both sexes. (2) Recording contains well-spaced calls (or motifs) leading in to song: in Fig IV, whistling sounds suggesting 'tsioooee tuilee tiuyear'; in Fig V, 'chileep' (see call 6a, below), 't-ch-chooi keiu te-sioooee sup-sup tik tik tik';

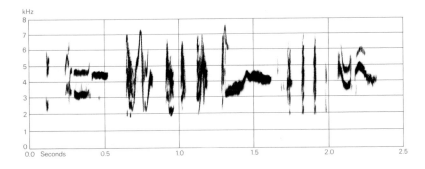

III  V C Lewis  England  March 1969

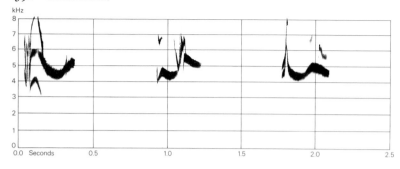

IV   S Elliott   England   May 1990

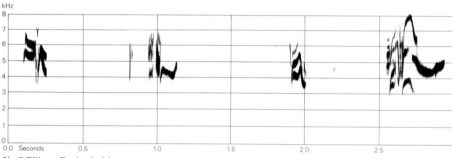

V   S Elliott   England   May 1990

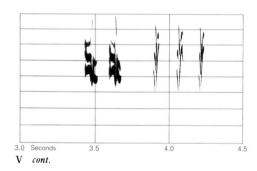

V   *cont.*

VI   S Elliott   England   February 1988

perhaps similar to slow song-like utterance noted in other *Carduelis* (see, e.g., *C. cannabina*) (J Hall-Craggs). (3) Twitter-call. Short, dry 'tet', 'tut', 'ket', or 'chek', often developing into longer twittering tremolo (3-8 or more units) with rapid crescendo: 'tetetet', 'tetterett', 'tcht-chtcht', or 'kette KETT'; series of 'tut' sounds shown in Fig VI, these and related sounds incorporated in song (see Fig III); like buzzing twitter when given by flock (Bergmann and Helb 1982; Bruun *et al.* 1986); descriptions in Zablotskaya (1978*b*) clearly relate to this call, but only 1 out of 3 sonagrams resembles Fig VI and that in Bergmann and Helb (1982). Another study found more variation in Twitter-calls, typical 'tet' units being combined with tonal components; call also shown to vary individually, each bird being capable of copying mate's call (see Twite *C. flavirostris*) (Mundinger 1970). In Fig V, last 5 units (see call 2, above) perhaps represent 2 further types of Twitter-calls. Twitter-call given by both sexes when feeding or resting, hopping about in tree or

as signal for take-off on short or longer flight ('flight-call' of some authors), thus apparently serving as short-range contact-call; also noted during courtship-feeding (usually given by ♂), or allofeeding in captive birds (Zablotskaya 1978*b*). Descriptions of other calls mainly follow Zablotskaya (1978*b*), but transcriptions (transliterations from Russian) are very approximate, and poor quality of sonagrams further increases difficulty in making comparison with other descriptions in this account. Following scheme thus necessarily provisional. (4) Contact-alarm calls. Calls in this category characteristically slightly melancholy, loud whistling sounds, some clear-toned, others with harsher, nasal component; much variation within and between individuals, and subdivision adopted here not to be regarded as rigid. Calls most commonly given in flight are 4c (descending pitch) and 4e (ascending), mixed apparently at random (L Svensson); 4a-b and 4d also given in flight. (4a) A 'tyulee' or 'tyuee' (slightly more complex structure, with noisy component at end) serving as contact-call (including close contact between pair-members), given from perch or in flight. Further renderings: 'DLU-ee' (Bruun *et al.* 1986); 'tüli' or 'uli', and (perhaps also) monosyllabic 'ih' (Bergmann and Helb 1982). Recordings suggest 'diu-li' (1st and last units of song-phrase in Fig I) or 'tiulee' (2nd unit in Fig IV). (4b) Whistling 'peelee' given by ♂, including in flight when may alternate with 4a; otherwise noted when several ♂♂ feeding together in tree, and also incorporated in ♂'s courtship-song; like call 6a (see below), apparently indicates ♂'s readiness to pair. (4c) Soft, whistling 'teeyu' descending in pitch and serving as contact-call between pair-members, e.g. when feeding close together, and at take-off (causing mate to call and act similarly). Calls rendered 'DLEE-u' (Bruun *et al.* 1986), and 'tsüü' (Witherby *et al.* 1938) or 'tsii' (Bergmann and Helb 1982) perhaps also belong here. For reports of sexual difference (longer call in ♂), see Witherby *et al.* (1938) and Géroudet (1957*c*). (4d) Quiet 'pee' descending in pitch; noisy at start, tonal at end, and with very variable amplitude modulation. Given by both sexes, in series or in pairs; in flight, on landing, and while feeding, apparently serving for contact. Alleged hoarse variant (relationship not obvious from sonagrams in Zablotskaya 1978*b*) said to express slight alarm. Recordings contain descending calls perhaps of this type: 'tsiiuu' (Fig VII), and, in Fig VIII,

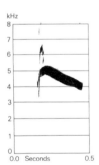

VII  S Carlsson  Sweden  May 1983

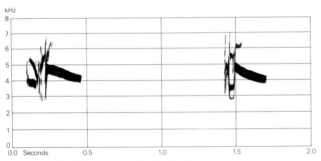

VIII  P A D Hollom  Scotland  April 1983

'ontsiiuu' (with diadic, nasal onset) followed by 'tluliu', much as 'tlüli' illustrated by Bergmann and Helb (1982). (4e) Recordings contain ascending portamento calls typical of *Carduelis* and suggesting 'tsoooee': bleating and nasal in Fig IX and (diad more prominent) Fig X; smoother,

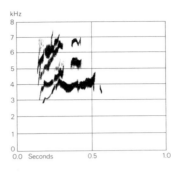

IX  S Elliott  England February 1988

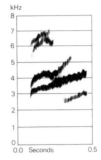

X  S Carlsson  Sweden  May 1982

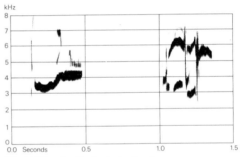

XI  V C Lewis  England  April 1969

portamento ascent shown in 1st unit of Fig XI, 2nd unit suggesting 'dee-u-li' (compare Fig XIII) (J Hall-Craggs, M G Wilson). Further descriptions probably of this call-type: shrill creaky 'tsooeet' (Witherby *et al.* 1938); extended nasal 'tsouïht' or 'kchêi' (Géroudet 1957*c*). Presumably contact- or (more likely by analogy with other *Carduelis*) alarm-call. (4f) Clear 'cheech-veen' comprising noisy sub-unit followed by longer, noisy sub-unit on end of which 3rd component superimposed. Given by both sexes, singly or 2-3 in succession. Serves for contact between pair-members (given also during Courtship-feeding: see call 6e, below) and members of flock. According to Zablotskaya (1978*b*), same as 'flight-call' (see call 3, above) illustrated by Mundinger (1970); however, this not supported by comparison of sonagrams, and 'cheech-veen' probably same as call 4e (above) or related. (4g) Loud 'peee', said by Zablotskaya (1978*b*) to comprise 4 overlapping components, 2nd carrying main energy, but highly variable (in amplitude modulation) even within given individual; noted from resting or preening birds, also lone flying bird, in pair-contact, and when alarmed (serves as mobbing-call), loudness increasing with excitement. (4h) A 'vyaa' given singly or in series (calls and series longer with increasing excitement) by both sexes to signal alarm, also apparently fright or appeasement when threatened or attacked. (4i) Distress-call. Loud harsh sound, somewhat reminiscent of 'kee' in call 8 (see below), but less hoarse. (5) Calls associated with antagonistic interactions. (5a) Series of 2–12 grating and hissing sounds lasting *c*. 1·5 s, and (at higher intensity), slightly higher-pitched, much louder variant comprising series (up to *c*. 2 s) of longer noisy units with whistling quality given in interactions with conspecific or other small birds; variant occurs mainly in breeding-season disputes between ♂♂, when often changes to call 4b or battle-song. (5b) Shrill squealing sound given by ♂ attacking or chasing another, also by ♂ or ♀ in sexual chases. (6) Calls given (mainly) during pairing, courtship, etc. (6a) Series of 3–5 'peek' sounds given by ♂ only, often preceding Advertising-song or between such songs, also when accompanying feeding or nest-building ♀; apparently indicates ♂'s readiness to pair. Call perhaps of this type shown in Fig V (1st unit) almost monosyllabic 'chileep' reminiscent of call of *P. domesticus* (M G Wilson). (6b) A 'tyu-pee' given (usually in series of 3 calls) by ♂ during courtship, including as part of courtship-song. (6c) A 'tyueelee' of rather complex structure and with variation (in frequency and amplitude modulation), including within given individual; like 6b, serves as courtship-call of ♂. Trisyllabic calls in recordings (perhaps of this general type) include 'dontleeuu' (Fig XII), and 'dee-u-li' (Fig XIII), also 1st and 3rd units in Fig II

(song). Apparently trisyllabic call illustrated by Oehler (1977) reported to be that most frequently used for long-range contact. (6d) Weak squeaking sounds given mainly by ♀ during Bill-flirting ceremony. (6e) Whistling 'fee' rising, then rapidly descending in pitch (perhaps related to 'pee': call 4d, above); one variant short and evenly pitched, another (much the rarest) higher-pitched and longer 'fee' with little pitch descent; 'fee' sounds given by either sex during courtship-feeding, also (by ♂) in courtship-song. (6f) Quiet chirping sound given by ♀ during copulation (Dost 1957). (6g) Post-copulatory call reported (without details) by Zablotskaya (1978*b*). (7) Quiet 'pyaa' noted as antiphonal duet in captive birds of same or different sex: 2–6 calls given by one bird, then by the other, or alternating calls by each with regular pause between calls; such duetting (perhaps expresses well-being) may last up to 30–40 min. (8) A 'te-te-kee' (less commonly 'te-kee') in which 'kee' extended and hoarse; given all year, with bill wide open and throat ruffled, by captive ♂♂ on waking, and during ensuing comfort behaviour; also incorporated in song. (Zablotskaya 1978*b*.)

CALLS OF YOUNG. Food-calls barely audible at 4 days, loud and clear at 10–12 days (Mal'chevski 1959). In recording by E and M Nobles (Wales), food-calls of fledglings are rapid series of high-pitched ringing 'pee' sounds with silvery quality. Juvenile song (not described) given by captive ♂♂ before moult (Dost 1957).　　　MGW

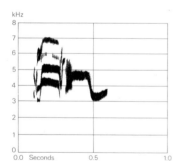

XII P A D Hollom Scotland April 1983

XIII V C Lewis England March 1969

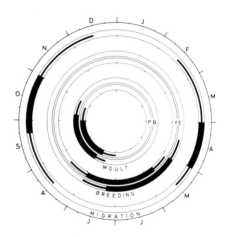

**Breeding.** SEASON. Southern Scotland: laying starts usually in April, but can be early to mid-March in years of good spruce *Picea* cone crop, mid-May in poor years (Shaw 1990). Ireland: eggs can be laid in March; mostly from early April (Ellison 1910; Witherby *et al.* 1938). Finland: see diagram; eggs laid late April probably to beginning of August, with peak early May to early June; earlier in good spruce years (Haapanen 1966; Haartman 1969; Hildén 1972). Southern Norway: nest with eggs 19 March (Berger 1968). Bayern (southern Germany): start of laying early April (Diesselhorst and Popp 1963). Switzerland: can lay late February in good spruce years (Glutz von Blotzheim 1962). Russia: laying starts St Petersburg region late April to May; nest with young late August (Mal'chevski and Pukinski 1983). For review, see Newton (1972). SITE. Generally inaccessible, at considerable height and in outer hanging twigs of conifer, usually spruce (Glutz von Blotzheim 1962; Harrison 1975; Makatsch 1976); also recorded against trunk (Carroll 1916; Pokrovskaya 1963, which see for many details). In Finland, of 45 nests, 69% in spruce, 27% pine *Pinus*; average height above ground 9·5 m (1·5–20·0, n = 28), though probably biased towards lower nests (Haartman 1969). Average height of 9 nests in Bayern 3·1 m (2·6–4·0) (Diesselhorst and Popp 1963). In Ireland, prefers spruce even where pine dominant (Carroll 1916), then, after those species, larch *Larix* and exotic conifers; 20 of 28 nests higher than 10 m (Hamilton-Hunter 1909; Ellison 1910); often in larch in northern Scotland (Nethersole-Thompson and Watson 1974). In years of high population density will nest in birch *Betula* (Newton 1972), and habit apparently spreading in Scotland (Bates and Bates 1989). One nest in north-east Scotland in hanging flowerpot 1·6 m above ground on porch of occupied cottage, although suitable pines nearby (Billett 1989). Nest: small hemispherical construction of (mostly) conifer twigs, heather, grass, moss, bark fibres, and spider's web lined with hair, fur, rootlets, plant down, and sometimes feathers, often with external camouflaging of moss and lichen; occasionally woven into hanging twigs (Dementiev and Gladkov 1954; Diesselhorst and Popp 1963; Haartman 1969; Nethersole-Thompson and Watson 1974). 2 nests in Hungary had outer diameter 7 and 9 cm, inner diameter 3·5 and 4 cm, overall height 4·5 and 9 cm, and depth of cup 3 and 3·5 cm (Szabó and Györy 1962). Building: by ♀ only, accompanied by ♂ (Witherby *et al.* 1938; Diesselhorst and Popp 1963; Haartman 1969; Nethersole-Thompson and Watson 1974); ♀ breaks off little birch and larch twigs or gathers twigs and grass from below trees, plucks moss and lichen from trees, and makes excursions to woodland edge (up to 500 m from nest recorded) for hair, plant down, etc.; can take less then 5 days, or 3–4 weeks if weather poor (Szabó and Györy 1962; Diesselhorst and Popp 1963; Nethersole-Thompson and Watson 1974). EGGS. See Plate 60. Sub-elliptical, smooth, and slightly glossy; rather variable in colour, size, and shape; pale bluish-white or blue sparsely marked with purplish- to blackish-brown spots and scrawls, mostly at broad end; also rusty blotches and reddish-violet undermarkings (Dementiev and Gladkov 1954; Harrison 1975; Makatsch 1976). 16·4 × 12·2 mm (14·7–20·2 × 10·5–13·5), n = 281; calculated weight 1·29 g (Schönwetter 1984); see Witherby *et al.* (1938) for 100 eggs from Britain. Clutch: 3–5(–6). In south-east Ireland, of 42 clutches: 3 eggs, 10%; 4, 45%; 5, 45%; average 4·36 (Hamilton-Hunter 1909; Carroll 1916); for discussion of clutch size in Ireland, see Ellison (1910). 46 clutches from Cairngorms area (northern Scotland) had average 4·04

(Nethersole-Thompson and Watson 1974). In Finland, of 55 clutches (not all complete): 1 egg, 9%; 2, 9%; 3, 7%; 4, 18%; 5, 53%; 6, 4%; probably 2 broods (Haartman 1969). Average in Bayern 4·6 (n = 9); eggs laid daily 4–5 days after start of nest-building (Diesselhorst and Popp 1963); average in Switzerland 4·3 (n = 7); probably 2 broods (Glutz von Blotzheim 1962). 2 broods in north-east Scotland (Buckland *et al.* 1990); in other places apparently only some pairs double-brooded, e.g. Tipperary (Ireland) (Carroll 1916) and USSR (Dementiev and Gladkov 1954). 2 broods normal in captivity, sometimes 3, even if all young survive (Dost 1957); 2nd captive brood started after fledging of 1st (Röder 1991), but in wild in Norway 2nd nest built while pair still feeding brood in nest, and contained 4 eggs 7 days later (Bringeland 1964). INCUBATION. 12–13 days (10–14); by ♀ only (Dementiev and Gladkov 1954; Diesselhorst and Popp 1963; Newton 1972; Nethersole-Thompson and Watson 1974); 13 days in captivity (Röder 1991). May start with penultimate or antepenultimate egg (Witherby *et al.* 1938; Glutz von Blotzheim 1962; Nethersole-Thompson and Watson 1974; Harrison 1975), but also with last (Nethersole-Thompson and Watson 1974); in captivity, all 4 eggs hatched on same day (Röder 1991). ♀ takes only very short breaks (Diesselhorst and Popp 1963; Röder 1991); leaves eggs for 3–6(–10) min 3–4 times per day, and for up to 20 min in evening (Nethersole-Thompson and Watson 1974). YOUNG. Fed and cared for by both parents. Brooded by ♀ only, continuously for first 4–5 days, at night for up to 10 days. (Dementiev and Gladkov 1954; Diesselhorst and Popp 1963; Nethersole-Thompson and Watson 1974.) In captivity, fed for first 2–3 days by ♀ only, with food brought by ♂, then fed directly by both parents (Röder 1991). FLEDGING TO MATURITY. Fledging period 13–15 (–17) days (Glutz von Blotzheim 1962; Newton 1972; Nethersole-Thompson and Watson 1974; Mal'chevski and Pukinski 1983); 14 days in captivity (Röder 1991). Age of first breeding 1 year (Paevski 1985). BREEDING SUCCESS. High in years of good spruce crop, while in poor years many birds do not breed, or breed late resulting in lower productivity (Diesselhorst and Popp 1963; Shaw 1990). In Bayern, of 41 eggs followed, all produced fledged young (Diesselhorst and Popp 1963). In Tipperary, some broods drowned in thunderstorm while parents collected food (Carroll 1916). In Cairngorms, main predator probably squirrel *Sciurus vulgaris* and perhaps Sparrowhawk *Accipiter nisus* (Nethersole-Thompson and Watson 1974).

BH

**Plumages.** ADULT MALE. In fresh plumage (autumn), forehead and crown black, each feather narrowly fringed grey-white, widest on rear of crown, black forming neatly rounded cap. Hindneck to scapulars and back bright yellowish-green, slightly more green-yellow on inner web of feathers, faintly mottled grey on hindneck, usually with narrow but distinct dull black shaft-streaks on lower mantle, scapulars, and back; tips of feathers bordered pale ash-grey when plumage quite fresh. Rump bright sulphur-yellow, feather-tips partly washed olive, feather-centres sometimes with black streak. Upper tail-coverts olive-green with duller olive centres and light ash-grey tips. Lore and nasal bristles dusky grey or sooty, sometimes mottled grey or with narrow green-yellow supraloral stripe; narrow eye-ring dusky grey with yellow mottling. Broad supercilium backwards from above eye green-yellow, mottled olive-grey at rear. Short black stripe behind eye. Ear-coverts olive-green, feathers tipped ash-grey, especially at rear. Cheek, side of throat, and bar at side of neck green-yellow, mottled olive-green, cheek not sharply demarcated from ear-coverts. Often a small black patch on chin, sometimes square and sharply demarcated, sometimes indistinct, fringed yellow; occurrence independent of age (Reinsch 1960). Throat, chest, upper belly, and side of belly yellow, tip of each feather tinged green and narrowly fringed grey-white, faintly streaked black on side of belly. Side of breast olive-green with faint darker olive mottling; flank white with some yellow and ash-grey suffusion and broad distinct dull black streaks. Mid-belly and vent white, sometimes with slight yellow wash; under tail-coverts yellow with white tips and black central marks. Central pair of tail-feathers (t1) black with narrow green-yellow outer fringe and narrow pale grey or yellow-grey fringe along inner web and tip; t2-t6 with bright yellow outer web, paler yellow inner web, and contrasting black tip $c$. 6-14 mm long; black of tip narrowly fringed yellow on outer web, grey along tip; outer web of t6 black. Flight-feathers black; primaries with narrow green-yellow outer fringe (except for emarginated tips) and narrow off-white fringe along tip, bases of outer webs of inner primaries bright yellow, forming short bar just beyond tips of greater upper primary coverts, visible basal one-third of outer webs of secondaries bright yellow, middle portion deep black, terminal portion black with broad green-yellow fringe; bases of inner webs of flight-feathers with yellow wedge. Tertials black, terminal part of outer web broadly fringed pale yellow, grading to white on tip. Greater upper primary coverts and bastard wing black, coverts finely fringed green, shorter feathers of bastard wing narrowly fringed green-yellow. Greater upper wing-coverts deep black, tips contrastingly yellow, tinged green on outer and terminal border, tips $c$. 4 mm long on outer coverts, $c$. 6-8 mm on central and inner ones. Median and lesser upper wing-coverts yellow-green, black of feather-bases partly visible. Axillaries and under wing-coverts pale yellow, longest coverts pale grey, marginal coverts spotted yellow and grey. *In worn plumage* (spring), grey feather-tips on upperparts worn off, cap uniform black, hindneck to back more uniform yellow-green with more distinct black streaks, rump yellow with some dusky grey and olive mottling; chin-patch (if any) uniform sooty, remainder of underparts yellow grading to rather extensive dull white on belly and flank; yellow of underparts often contaminated, tinged grey or olive; pale fringes of primaries and tertials and tips of greater coverts partly bleached to grey-white. ADULT FEMALE. Rather like adult ♂, but cap and chin without black patch and yellow on tail-base and tips of greater coverts more restricted. Upperparts from forehead to scapulars and back greenish-grey, marked with ill-defined dark grey spots on crown and mantle, streaked dull black on scapulars and back (streaks broader than in adult ♂, less black and less sharply contrasting); bar across hindneck and stripes over middle of mantle slightly more yellowish-green. Rump yellow, distinctly streaked black (not as uniform and contrasting with back as in adult ♂); upper tail-coverts as in adult ♂. Supercilium yellow, narrowest above eye, mottled olive-grey, especially above lore and at rear. Lore and narrow eye-ring grey with fine yellow specks. Cheeks yellow with grey spots and streaks, merging into dark olive-green ear-coverts. Side of neck ash-grey, indistinctly mottled yellow and darker grey. Chin

off-white with slight yellow suffusion; throat, chest, and upper belly off-white with yellow suffusion on feather-tips, faintly streaked dark grey on sides. Side of breast olive-grey, flank white with some yellow suffusion, both broadly marked with dull black streaks (broader than in adult ♂), most contrasting on lower flank; mid-belly, vent, and under tail-coverts white, tail-coverts streaked black. Tail as adult ♂, but black of tips extends broadly along shafts, yellow restricted to borders along basal and middle portions of both webs; outer web of t6 black. Wing as adult ♂, but yellow tips of black greater upper wing-coverts $c$. 2 mm long on outer coverts and $c$. 5 on inner, shading to white on tip (less broad and less saturated yellow than in adult ♂); deep black of bases of median coverts and dull black bases of lesser coverts more extensive, readily visible. *In worn plumage* (spring), upperparts more greyish-green, more distinctly streaked dusky on cap, mantle, and scapulars, contrastingly streaked yellow and black on rump; underparts largely off-white, less tinged yellow, marked with more contrasting fine dusky streaks on chest and broad ones on side of belly, flank, and under tail-coverts. NESTLING. Down rather long; smoky blue-grey (Harrison 1975). JUVENILE. Upperparts buff-brown, streaked black-brown; ground-colour of hind-neck and rump paler, buffish-grey, streaks slightly narrower. Side of head buff-brown with dusky brown mottling; supercilium (extending backwards from above eye) and bar along side of neck pale yellow or yellow-white with fine dusky grey mottling. Underparts white, slightly tinged yellow on throat, chest, and belly when just fledged, side of throat, side of breast, chest, flank, and side of belly streaked black-brown. Upper wing-coverts and tertials as adult, but tinged dark grey, less deep black, and pale tips of coverts and fringes of tertials narrow, 1-2 mm wide, white with pale yellow or pink-buff tinge, soon bleaching to white. For tertials, see also Cooper and Burton (1988); for tail, see 1st adult (below). FIRST ADULT MALE. Like adult ♂, but juvenile flight-feathers, tail, tertials, greater upper primary coverts, and some to many outer greater upper wing-coverts retained; tips of tail-feathers more pointed than in adult, less rounded, black on tips less sharply defined, more extensive, often extending into broad point along shaft on middle portion, usually only bases of t2-t5 fully yellow, but much individual variation (Bacmeister and Kleinschmidt 1920) (in adult, bases and middle portions extensively yellow, black on tips restricted, more square-cut); a distinct contrast between juvenile outer coverts (dark grey with narrow white tips) and newer 1st adult inner coverts (black with broad yellow tips). Head and body as adult ♂, but light grey fringes of upperparts slightly wider when plumage fresh, black of cap slightly less exposed, less sharply demarcated from hindneck, hindneck to back slightly more grey, less yellow-green, more similar to adult ♀ but dusky shaft-streaks narrower; rump less pure yellow and less contrasting than in adult ♂, mottled olive and grey on yellow ground; lore, cheek, and ear-coverts dusky olive-grey with yellow mottling, lore less black than in adult, dark line behind eye less distinct; yellow on underparts sometimes slightly paler and less extensive, streaks on flank and side of belly slightly broader and more extensive, like those of adult ♀. FIRST ADULT FEMALE. Like adult ♀, but part of juvenile feathering retained, as in 1st adult ♂. Basal and middle portions of juvenile tail-feathers on average more narrowly fringed yellow (but much individual variation), tips distinctly pointed; tertials and contrast between old and new greater coverts as in 1st adult ♂. Rump scarcely paler than remainder of upperparts, ground-colour less extensively yellow than in adult ♀ or 1st adult ♂, closely streaked green and dusky olive; supercilium, upper chin, and side of neck less yellow than in adult ♀, heavily mottled grey, less contrasting with remaining side of head; ground-colour of underparts white, hardly yellow, more heavily streaked black on side of breast, flank, and side of belly, finely on chest. For sexing and ageing, see also Hájek (1974), Cooper and Burton (1988), and Svensson (1992).

**Bare parts.** ADULT, FIRST ADULT. Iris dark brown. Bill steel-grey with slight flesh tinge; culmen and gonys black, cutting edges paler greyish-yellow or-flesh. Tarsus and toes dull flesh-brown, horn-brown, or dark brown, darkest on joints and toes. (Hartert 1903-10; RMNH, ZMA). NESTLING. Mouth bright red or crimson, gape-flanges yellow (Heinroth and Heinroth 1924-6; Harrison 1975). JUVENILE. Iris dark brown. Bill pinkish yellow-horn or horn-brown with flesh tinge at base. Leg and foot flesh-pink with grey tinge. (Heinroth and Heinroth 1924-6; ZMA.)

**Moults.** ADULT POST-BREEDING. Complete; primaries descendent. Starts late June to early August (Ginn and Melville 1983). Duration of primary moult *c.* 70 days in 6 captive birds (Newton 1968*a*). In north-west Russia, starts 10 July to 10(-20) August, completed 20 September to late October (Rymkevich 1990). In northern Iran, 2 ♂♂ in heavy moult on 24 July (Vaurie 1949*b*). Moult completed in 6 birds from early September in Manchuria (China) (Piechocki 1959). In Spain, some moult on mantle, chest, or elsewhere on body in a few November-February birds; more frequently (27% of 60 birds), some moult in March (Senar 1988). POST-JUVENILE. Partial: head, body, lesser, median, and variable number of inner greater coverts (rarely all 9), and occasionally 1-2 tertials (Ginn and Melville 1983). In Britain, birds from late winter retained average 3·07 ♀±1·15 outer greater coverts (*n* = 107), range 1-6, mainly 2-3 (Sellers 1986). No tertials replaced in sample of *c.* 450 birds (Cooper and Burton 1988). Moult July-September (Witherby *et al.* 1938; Svensson 1992; ZMA). In north-west Russia, starts mainly 20 July to late August, completed early September to November (Rymkevich 1990, which see for details). Occasionally, some outer primaries replaced (Winkler and Jenni 1987).

**Measurements.** Europe and North Africa, all year; skins (RMNH, ZMA: A J van Loon). Juvenile wing and tail refer to retained juvenile wing and tail of 1st adult. Bill (S) to skull, bill (N) to distal corner of nostril; exposed culmen on average 3·5 less than bill (S).

| | ♂ | | ♀ | |
|---|---|---|---|---|
| WING AD | 73·7 (1·31; 22) | 71-76 | 71·9 (1·82; 5) | 69-74 |
| JUV | 72·5 (1·66; 30) | 69-76 | 70·5 (1·32; 26) | 67-73 |
| TAIL AD | 47·0 (1·53; 22) | 45-50 | 46·2 (2·39; 5) | 43-49 |
| JUV | 46·1 (1·57; 30) | 43-49 | 45·0 (1·20; 26) | 43-47 |
| BILL (S) | 13·0 (0·45; 32) | 12·1-13·8 | 12·9 (0·53; 17) | 11·9-13·7 |
| BILL (N) | 8·5 (0·38; 47) | 7·8-9·3 | 8·4 (0·39; 31) | 7·6-9·1 |
| TARSUS | 13·7 (0·47; 40) | 12·8-14·6 | 13·8 (0·51; 23) | 12·8-14·6 |

Sex differences significant for wing and juvenile tail. Age differences significant in ♂.

Japan, all year; skins (ZMA: A J van Loon). Ages combined (mainly 1st adults).

| | ♂ | | ♀ | |
|---|---|---|---|---|
| WING | 72·1 (1·47; 10) | 70-75 | 71·0 (1·69; 12) | 68-74 |
| TAIL | 45·2 (1·41; 9) | 43-48 | 43·8 (1·36; 11) | 40-46 |
| BILL (S) | 13·4 (0·34; 8) | 12·8-13·8 | 13·3 (0·47; 12) | 12·5-14·0 |
| BILL (N) | 8·9 (0·28; 11) | 8·2-9·2 | 8·9 (0·45; 12) | 8·3-9·7 |

Sex differences significant for tail.

Wing. Live birds, except as noted. Gloucestershire (England), winter: (1) adult, (2) 1st adult (Sellers 1986, which see for details). Schiermonnikoog (Netherlands), autumn: (3) adult, (4) 1st adult (A J van Loon). Nordrhein-Westfalen (Germany), winter: (5) adult, (6) 1st adult (Abs 1964). (7) Taunus mountains (Germany), winter (Eck 1990). (8) Eastern Germany and neighbouring part of Czechia; skins (Eck 1985). Czechia, autumn: (9) adult, (10) 1st adult (Hájek 1969, 1974). Northern Poland, spring migrants: (11) adult, (12) 1st adult (Busse 1976).

| | ♂ | | ♀ | |
|---|---|---|---|---|
| (1) | 73·6 (1·00; 25) | 71-76 | 71·6 (1·52; 35) | 68-74 |
| (2) | 73·0 (1·31; 59) | 70-77 | 70·7 (1·20; 37) | 68-73 |
| (3) | 74·6 (2·05; 29) | 69-79 | 72·9 (2·08; 20) | 69-76 |
| (4) | 73·5 (1·74; 103) | 69-79 | 71·7 (1·58; 126) | 67-75 |
| (5) | 75·8 (1·6 ; 10) | 74-78 | 72·1 (1·9 ; 7) | 69-74 |
| (6) | 73·6 (1·6 ; 41) | 70-77 | 71·4 (1·8 ; 38) | 68-76 |
| (7) | 74·1 (1·57; 750) | — | 71·9 (1·61; 784) | — |
| (8) | 72·7 (1·47; 49) | 69-75 | 70·9 (1·32; 38) | 68-74 |
| (9) | 73·1 (1·34; 28) | 70-75 | 70·9 (2·12; 21) | 68-75 |
| (10) | 71·6 (2·01; 102) | 68-76 | 70·2 (1·36; 85) | 67-73 |
| (11) | 72·8 (2·00; 135) | — | 71·4 (1·81; 101) | — |
| (12) | 72·6 (1·75; 236) | — | 70·5 (1·72; 220) | — |

**Weights.** ADULT, FIRST ADULT. Gloucestershire (England): (1) February and 1st half of March, (2) 2nd half of March and April (Sellers 1986, which see for influence of time of day and other details). Schiermonnikoog (Netherlands): (3) September, 1st adult, (4) October, 1st adult, (5) October, adult (A J van Loon). (6) Netherlands, August-April (RMNH, ZMA). (7) Nordrhein-Westfalen (Germany), October-February (Abs 1964, which see for details). (8) South-east Germany, October-April (Eck 1985). Czechia: (9) October (Hájek 1969, 1974); (10) October-January (Havlín and Havlínova 1974). (11) Poland, spring migrants (Busse 1976).

| | ♂ | | ♀ | |
|---|---|---|---|---|
| (1) | 12·9 (0·83; 58) | 11·0-15·0 | 12·7 (0·96; 63) | 10·5-14·8 |
| (2) | 14·2 (1·33; 77) | 12·1-18·2 | 13·4 (1·09; 68) | 10·8-16·3 |
| (3) | 12·8 (1·33; 35) | 10·6-15·7 | 12·3 (1·18; 34) | 10·5-16·3 |
| (4) | 13·8 (1·46; 128) | 10·7-18·5 | 13·2 (1·29; 143) | 10·3-18·0 |
| (5) | 13·4 (1·60; 23) | 11·4-16·8 | 12·7 (1·41; 18) | 10·5-16·1 |
| (6) | 12·8 (1·48; 13) | 10·1-15·0 | 13·2 (1·61; 8) | 11·0-16·0 |
| (7) | 13·0 (1·0; 51) | 11·0-15·5 | 12·7 (1·45; 11) | 11·5-15·0 |
| (8) | 12·6 ( — ; 15) | 10·2-13·8 | 12·6 ( — ; 7) | 12·0-14·9 |
| (9) | 14·0 (1·36; 294) | 11·0-18·0 | 13·4 (1·18; 258) | 10·0-17·0 |
| (10) | 13·5 (1·2; 13) | 12·5-16·0 | 12·7 (0·9 ; 19) | 11·0-15·0 |
| (11) | 12·4 (1·19; 342) | — | 12·4 (1·18; 323) | — |

In spring, Britain: ♀ with fully vascularized brood-patch 13·5 (0·84; 18), ♀ without visible brood-patch 12·3 (1·21; 14) (Shaw 1990, which see for details). For monthly averages of various populations throughout winter in Sussex (England), see Davis (1976); for increase in weight prior to spring migration, see Cooper (1985). Exhausted birds, Netherlands: ♂ 9·1 (0·61; 7), ♀ 9·0 (1) (ZMA). See also Bacmeister and Kleinschmidt (1920), Piechocki (1959), Haftorn (1971), Newton (1972), and Chesney (1987).

**Structure.** Wing moderately long, broad at base, tip bluntly pointed. 10 primaries: p8 longest, p9 0-1(-3) shorter, p7 0-1, p6 4-6, p5 9-13, p4 14-18, p3 17-20, p2 19-24, p1 21-26; p10 strongly reduced, 47-54 shorter than reduced outermost upper primary covert. Outer web of (p6-)p7-p8 emarginated, inner web of p8 (p7-p9) with faint notch. Tip of longest tertial reaches tip of p1-p3. Tail rather short, tip forked; 12 feathers, t1 7·0 (10) 5-10 shorter than t6. Bill rather short, conical; 6·7 (15) 6·2-7·3 deep at base, 5·8 (15) 5·3-6·3 wide (see also Sellers 1988); culmen slightly decurved, mainly at tip, cutting edges straight or faintly decurved, gonys straight or slightly concave; tip sharply pointed, laterally compressed. Nostril small rounded, covered by short tuft of bristly feathers. Tarsus and toes rather short, slender; middle toe with claw 13·1 (10) 12·4-14·0; outer toe with claw *c.* 74% of middle with claw, inner *c.* 70%, hind *c.* 90%.

**Geographical variation.** Despite large gap in distribution in central Asia, populations of western Eurasia and of eastern Asia do not differ in plumage and scarcely in size (see Measurements); no variation within western Eurasia (Hartert and Steinbacher 1932-8); colonization of western Siberia possibly rather recent (Johansen 1944).

May form superspecies with Pine Siskin *C. pinus* of North America (Mayr and Short 1970); closely related (Vaurie 1959), and in captivity clearly react to each other (Reinsch 1967), but *C. pinus* also rather closely related to a group formed by Dark-backed Greenfinch *C. psaltria*, American Goldfinch *C. tristis*, and Lawrence's Goldfinch *C. lawrencei*, all from North America (Marten and Johnston 1986).

CSR

## *Carduelis pinus* (Wilson, 1810)  **Pine Siskin**

Fr. Tarin des pins    Ge. Fichtenzeisig

A North American species, breeding mainly in coniferous forest from Alaska and Labrador south to southern Mexico, wintering throughout breeding range and in non-breeding areas in southern USA and Mexico. One occurred among large fall of migrants on eastbound ship in North Atlantic 8-11 October 1962, staying on board from *c.* 65°W to 30°W (Durand 1963).

## *Carduelis cannabina*  **Linnet**

PLATES 40 and 44 (flight)
[between pages 472 and 473, and 640 and 641]

Du. Kneu      Fr. Linotte mélodieuse      Ge. Bluthänfling
Ru. Коноплянка      Sp. Pardillo común      Sw. Hämpling

*Fringilla cannabina* Linnaeus, 1758. Synonym: *Acanthis cannabina*.

Polytypic. Nominate *cannabina* (Linnaeus, 1758), Europe and western Siberia, west to Ireland, Wales, and England, south to Pyrénées, northern Italy, northern Yugoslavia, Rumania, Ukraine (except Crimea), and northern Kazakhstan; *autochthona* (Clancey, 1946), Scotland; *guentheri* Wolters, 1953 (synonym: *nana* Tschusi, 1901), Madeira; *meadewaldoi* (Hartert, 1901), western Canary Islands; *harterti* (Bannerman, 1913), eastern Canary Islands; *mediterranea* (Tschusi, 1903), Iberia, Balearic Islands, Corsica, Sardinia, Sicily, southern Italy, coastal and southern Yugoslavia, Albania, Bulgaria, and Greece; race unknown, north-west Africa from Morocco to Tunisia and probably northern Libya; *bella* (C L Brehm, 1845), Crimea, Turkey, and Levant east through Caucasus area and northern Iraq to Iran and south-west Turkmeniya; also, central Asia from Afghanistan to south-west Altai and Bogdo Ula mountains (Sinkiang, China).

**Field characters.** 13·5 cm; wing-span 21-25·5 cm. Similar in bulk to Twite *C. flavirostris* but with shorter tail and thus looking slightly more compact; distinctly larger and longer than smallest Redpoll *C. flammea cabaret* but matched in size and structure by larger northern *C. f. flammea* and *C. f. rostrata*; juvenile noticeably larger than all juvenile serins *Serinus*. Small, quite compact, sociable but nervous finch; epitome of smaller, streaked members of *Carduelis*. Except in ♂, rather featureless plumage essentially brown above and pale buff-brown below, marked mainly by white edges to primaries and tail-feathers, quite heavy streaks on back and underparts, and paler eye-crescents and throat markings. ♂ in worn plumage becomes colourful, with pale crimson forecrown on grey head, crimson chest and flanks, almost chestnut mantle, scapulars, and wing coverts (lacking obvious streaks), and white fringes to upper tail-coverts, as well as white wing- and tail-edges. ♀ greyer headed than juvenile. Bill at all times grey or dull horn, never straw like *C. flavirostris* in winter. Flight somewhat sparrow-like. Voice distinctive. 7 races in west Palearctic, 3 isolated on Atlantic islands and 1 from Turkey eastwards separable (see Geographical Variation). Only European and west Siberian race, nominate *cannabina*, described here.

ADULT MALE. Moults: June-October (complete). Wear has marked effect on appearance. Crown and nape buff, streaked dark brown; hindneck, sides of neck, and ear-coverts rather uniform grey-buff; short crescents above and below eye and patch on lower cheeks cream-buff, throat pale buff. Mantle and scapulars chestnut, appearing softly streaked due to indistinct blackish streaks and buff margins; back and rump paler chestnut, almost unmarked. Wings lack obvious bars: wing-coverts as mantle, tertials blackish-brown, fringed buff; secondaries as tertials, primary coverts almost black, contrasting with bright white fringes to inner primaries which fade out before dark tips; wing thus shows striking white-lined panel below secondaries when folded, and spread of white lines over inner primaries in flight. Underwing white. Tail almost black, inner pair margined buff, all others brightly edged white so that tail appears white-sided. Throat shows broken dark malar stripe or patch; breast buff, with some pink showing through; flanks buff, softly streaked brown; belly and under tail-coverts white with pale buff tinge, under tail-coverts streaked. With wear, bird becomes brighter and more simply and strongly patterned. By April, centre of crown pale crimson; rest of head cleaner grey, with marks round eye and chin and throat almost white, throat less streaked; mantle, scapulars, and wing-coverts uniform bright chestnut; breast crimson, varying in extent but always complete band under throat; flanks uniform or dappled buff; rest of underparts cleaner white. Gives bird vivid prettiness quite unlike its relative drabness in winter. By July, some ♂♂ become even brighter, with pearl-grey head, white throat, and deep crimson breast. Bill dark greyish-brown horn to greyish-black. Eye virtually black. Legs dark brownish-flesh to horn-black. ADULT FEMALE. Resembles adult ♂ in winter but never shows any crimson or pink tones, while upperparts umber-brown and more heavily streaked and underparts fully streaked over breast and flanks. Head less grey but white wing-panel and tail-sides present. May show faintly paler wing-bars on median and especially greater coverts. Appearance changes little with wear, but head may appear cleaner and back more uniform. JUVENILE. Closely resembles ♀ but has uniformly buff ground to most of plumage and narrower streaks. Chin and throat unmarked buff-white. Wing-panel and dull wing-bars often indistinct but white tail-sides already well developed.

Adult ♂ in worn breeding plumage has diagnostic combination of grey head, white panel in wing, and white sides to tail. Non-breeding ♂ in fresh plumage, ♀, and juvenile liable to confusion particularly with *C. flavirostris* and also *C. flammea*, Siskin *C. spinus*, and ♀ and juvenile serins *Serinus*. Thus important to study ♀ *C. cannabina* fully before attempting identification of other small streaked finches; voice (see below) as useful as rather unremarkable plumage. Flight light and rapid with easy take-off; progress wavering in short flights, but if longer shows short, steep undulations produced by bursts of wing-beats followed by wing-closure. Escape-flight is to hedges and low bushes if close but may also go off on long high circuits in open country. Over food sources, frequent flights by flocks are less undulating, more direct, recalling tiny sparrows *Passer*. Gait a hop, varied by jumping and clambering; less acrobatic than *C. spinus* and *C. flammea*. Stance low and horizontal on ground, more upright on perch, almost erect in singing ♂. Flicks wings and tail. Highly gregarious, forming large roaming flocks after breeding. Keeps own company but will mix with other ground-feeding finches. Fidgety and difficult to approach except in breeding season.

Song a cheerful, musical, quite varied warble-cum-twitter, interspersed with scratchy and twanging notes which suggest plucking of loose-stringed instrument; carries only short distance except in chorus. Flight-calls sound much less metallic than *C. flammea* and much less nasal than *C. flavirostris*: fast, soft, twittering, rather musical 'djit-djit djit'; dry but lilted 'tihtihtihtihtit', quite loud in chorus at close range but fading quickly at distance; slightly nasal 'terrettt' or 'tett' from migrants. Alarm-call a more plaintive 'tsooeet'.

Habitat. In west Palearctic, breeds almost throughout European boreal, temperate, Mediterranean, and steppe climatic zones, extending to coastal North Africa and south-west Asia, in both continental and, to lesser extent, Atlantic climates (Voous 1960b). Mainly a lowland bird, but also widespread in suitable hilly regions, and in Swiss Alps nests up to 2200–2300 m on dry moors, alpine meadows, or scree where dwarf conifers or rough vegetation flourish (Glutz von Blotzheim 1962; Lüps *et al.* 1978). Generally avoids dense tall forests and woods except those of open type with clearings or glades and ready access to stands of seed-bearing ruderal or similar plants. Occupies fen woodlands but prefers scrub and heath vegetation with dry sunny aspect, farmland with hedges or low trees, vineyards, orchards, maquis, fields left uncultivated, young plantations, and untended forest edges. These afford both nest-sites in low vegetation such as shrubs and bushes, and ready access to foodplants and ground foraging areas. Although indirectly much dependent on human land use, normally avoids towns, suburbs, gardens, and villages, although resorting freely to farmyards. Exceptionally in parts of Fenno-Scandia (after abandonment of farmlands following changes in agricultural practices) has, since 1950, colonized parks, gardens, industrial wasteland, railway yards, and other habitats in small towns and villages (Newton 1972).

Flies freely for considerable distances, not only in lower but in middle airspace, and spends much time on ground as well as perching on bushes and artefacts such as walls and wires, usually quite low. Outside breeding season will shift to more open habitats, such as salt-marshes, shingle banks, and sand-dunes, as well as farmlands (Fuller 1982). Mobility enables use of dissimilar habitats.

Eastern race *bella* apparently differs markedly in preferring dry rocky mountain steppe with bushes of various prickly shrubs, and shrub thickets on mountain slopes up to 2000 m, favouring human neighbourhoods and avoiding cultivated areas and plains (Dementiev and Gladkov 1954). Prefers forest edges, roadside plantings, and thickets on mountain steppe in USSR, but found in inhabited areas (Flint *et al.* 1984). Wintering in Indian subcontinent occurs in open country and cultivation, as well as on stony slopes at base of hills (Ali and Ripley 1974).

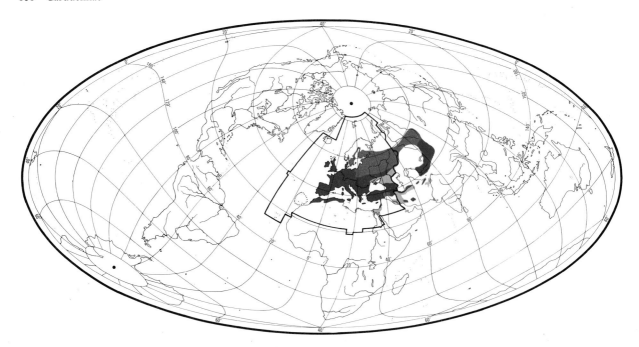

**Distribution.** No major changes except in north, where range limit affected by varying severity of winters.

BELGIUM. Local until *c.* 1930, since when has expanded over whole country (Devillers *et al.* 1988). NORWAY. Has disappeared from Nordland (decline from 1955; few observations in 1970s, none in 1980s); some spread in south-east and west (VR). FINLAND. Marked changes. In 19th century, fairly common in south and centre; at beginning of 20th century nearly disappeared from most parts of former range. Slow recovery since late 1940s. During mild winters of early 1970s expanded range to north, but contraction followed hard winters of 1980s. (Tast 1968; OH.)

Accidental. Faeroes, Kuwait.

**Population.** Has declined in Britain, Germany, and probably elsewhere due to intensive agriculture, with control of weeds; but recovery in Finland attributed to improved foraging opportunities following habitat changes.

BRITAIN, IRELAND. Estimated 800 000 to 1·6 million pairs 1968–72 (Sharrock 1976). In Britain, steep decline since 1977 or earlier due to modern farming practices, especially chemical control of weeds; estimated 700 000–800 000 pairs 1982 (Hudson and Marchant 1984). FRANCE. More than 1 million pairs (Yeatman 1976). BELGIUM. Estimated 150 000 pairs; perhaps stable (Devillers *et al.* 1988). NETHERLANDS. Estimated 60 000–130 000 pairs (SOVON 1987). GERMANY. Estimated 669 000 pairs (Rheinwald 1992). East Germany: 180 000 ± 90 000 pairs (Nicolai 1993). Considerable recent decline in Schleswig-Holstein and elsewhere, due to changing agricultural practice (Busche 1991). SWEDEN. Estimated 200 000 pairs (Ulfstrand and Högstedt 1976); now probably less, gen-erally agreed to be decreasing (LR). FINLAND. Drastic decrease at beginning of 20th century. Recovery since late 1940s due to new opportunities for foraging in abandoned fields and urbanized areas; about 50 000–100 000 pairs (Koskimies 1989). MALTA. A few pairs (Sultana and Gauci 1982). ISRAEL. A few thousand pairs in 1980s (Shirihai in press).

Survival. Britain: average annual mortality of adult ♂♂ 61 ± SE7%, adult ♀♀ 53 ± SE7% (Dobson 1987). Birds ringed in northern and central Europe and recovered in Iberia: average annual mortality of adults 66% (Asensio and Carrascal 1990). Oldest ringed bird 8 years 11 months (Rydzewski 1978).

**Movements.** Partially migratory, most birds moving south-west or SSW on narrow front to winter within and slightly south of breeding range, with concentrations in Mediterranean region. Areas of higher ground are vacated. Diurnal migrant. Atlantic island races sedentary, and probably also some populations of Mediterranean race *mediterranea*, e.g. Malta, Sicily (Bannerman 1963; Bannerman and Bannerman 1965; Sultana and Gauci 1982; Iapichino and Massa 1989).

In Britain, some remain to winter but others depart; proportion probably varies from year to year, and ringing data show that individuals migrate in some years and not in others. Initial heading is south-east to Belgium (by shortest sea-crossing), then SSW to western France and Spain (mostly as far west as 4°W in north of Spain and 6°30′W in south); almost all recoveries lie in band *c.* 500 km wide, and winter recoveries are from many points along route, suggesting birds vary greatly in distance they

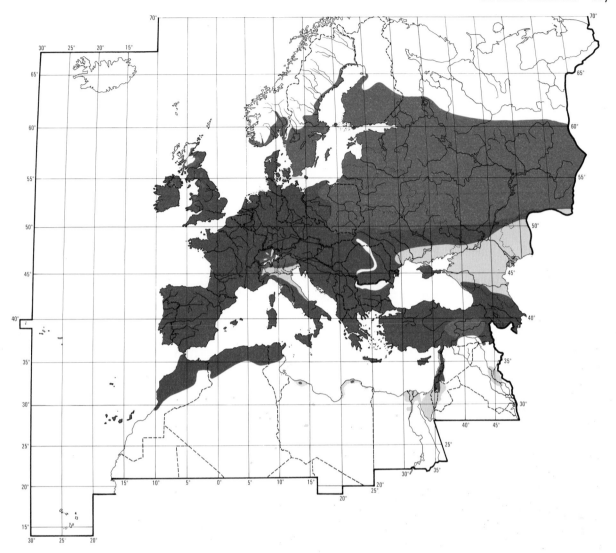

migrate. Relatively few birds winter in Ireland, and many depart from south-east, presumably for France and Spain, but no ringing data. Movement also observed westward from Britain towards Ireland. In hard winters, some residents move south within Britain, not known if by same route as migrants. Numbers slightly higher than average on coasts of north-east England and south-east Scotland, and perhaps include some Scandinavian birds. British ringing data up to 1986 show movement to and from France (487), Spain (184), Belgium (39), Netherlands (10), Portugal (3), Italy (2), Norway (2), Morocco (2, south to 32°20′N), Denmark (1), and Germany (1). (Newton 1972; Spencer and Hudson 1982; Lack 1986; Mead and Clark 1987; Hutchinson 1989.)

In France, most breeding birds depart but many immigrants winter, with highest concentrations on Atlantic coast and abundant, prolonged passage through Aquitaine in south-west; birds wintering in French Mediterranean area

are chiefly from Germany (south of 52°N), Czechoslovakia, and Switzerland (Blondel 1969; Boutet and Petit 1987; Yeatman-Berthelot 1991). Ringing study in southern Bretagne (north-west France) confirmed that summer and winter populations differ, with local birds moving south on narrow front along Atlantic coast and through northern Spain, to winter chiefly in southern Spain, especially Strait of Gibraltar area, with 1 recovery from Morocco (Mahéo 1969); data suggest more French than German birds reach Morocco (Asensio 1984). In Belgium, *c.* 7% of breeding population resident; recoveries of passage and local birds are chiefly in west-central and south-west France, or in eastern and southern Spain, almost entirely within band 370 km wide running from Sweden to southern Spain (Verheyen 1955a; Lippens and Wille 1972; Castelli 1988). Analysis of long-term ringing data up to 1965 shows that 20–30% of German birds are resident, and older birds tend to migrate less far than younger ones;

heading ranges between west and south-east, but west German birds migrate chiefly south-west or SSW; recoveries (up to 2240 km) lie west to 4°23′W in Spain, east to 14°44′E in Sicily, with 2 in north-west Africa; many are in Balearic Islands, indicating regular use of long sea-crossing. A few Danish and Swedish birds winter in northern Germany, and some north German birds in southern Germany; most east German birds winter in Italy. (Retz 1966a, b, 1968.) In Rheinland (western Germany), large gatherings (including birds descending from higher ground and immigrants) winter in valleys and plains (Mildenberger 1984). Winters locally in small numbers in Switzerland (Schifferli et al. 1982). Almost all Swedish birds migrate, though individuals remain every year in south and occasionally north to Värmland, c. 60°N; recoveries (up to 2210 km) are mostly from Belgium, France, and Spain, south to 38°28′N and west to c. 3°W (SOF 1990; Rep. Swedish Bird-ringing). In Finland, usually scarce in winter, but large flocks recorded in some years (Haila et al. 1986). From Kaliningrad coast (western Russia), where regular on passage in very small numbers, 5 recoveries south-west or SSW to northern Italy (4) and south-east France (1) (Paevski 1971). Winters regularly in Poland, especially in west (Tomiałojć 1976a). In USSR, varies from migratory in north to resident in south (Dementiev and Gladkov 1954).

Common on passage and in winter throughout Mediterranean region. Widespread in Spain, with heavy concentrations in south; foreign recoveries are of birds ringed Britain east to Finland, Denmark, and Germany; wing formulae show that abundant passage across Strait of Gibraltar includes considerable number of Iberian birds (Asensio 1984, 1986c; Finlayson and Cortés 1987). Maltese recoveries chiefly involve eastern Europe: Hungary (11), Italy (10, mostly east of 13°E), Czechoslovakia (9), Yugoslavia (6), Austria (3), Lithuania (2), Switzerland (2), Poland (2) (Il-Merill, Maltese Ringing Reports). Italian records show similar pattern, and include recoveries from Germany and 1 from Tunisia (Moltoni 1973; Bendini and Spina 1990). No ringing data from east Mediterranean, but major influxes occur, and winter population much higher than summer on Dalmatian coast (Yugoslavia), in Greece, western and central Turkey, and Cyprus (Bauer et al. 1969; Beaman 1978; Cvitanić 1980; Flint and Stewart 1992). Abundant and widespread in Israel (breeds in north and centre) (Shirihai in press), and regular in Iraq (Moore and Boswell 1956; Marchant and McNab 1962). In Iran, far more widespread winter than summer in north and west, and occasional on Gulf coast (D A Scott). In Arabia, occasionally recorded in east (F E Warr).

Arrivals mask movements of north-west African populations; either local birds or immigrants penetrate Algerian Sahara at least in some years (Heim de Balsac and Mayaud 1962; Dupuy 1969), and many migrate through western Morocco (Thouy 1976). Accidental in Mauritania, and 1 record as far south as northern Sénégal (Lam-

arche 1988; Morel and Morel 1990). In western Libya, recorded south to Sebha 27°N, and common in winter in coastal zone (where possibly breeds); also winters commonly on Cyrenaican coast (eastern Libya), and small flocks noted throughout winter in cultivated areas along north Egyptian coast, in Nile delta and valley, and Sinai (Bundy 1976; Goodman and Meininger 1989). Apparently vagrant to Sudan, though in one winter not uncommon in Nile valley at 21°55′N; also exceptional record 5 km south of Khartoum (Mathiasson 1972; Nikolaus 1987).

Passage periods prolonged. Autumn migration August–November, peaking mid-September to mid-October in northern Europe, and October in south. In Britain and France, movement noted from late August (Newton 1972; Lack 1986; Yeatman-Berthelot 1991); in southern Bretagne, local birds depart from mid-September, and movement south is rapid, with none retrapped within 400 km after 10 October; passage birds abundant to 10 October, continuing to 25 October (Mahéo 1969). In Sweden, passage much heavier at Falsterbo in extreme south-west than at Ottenby in south-east, usually peaking early October (Edelstam 1972; Ulfstrand et al. 1974). In Rheinland, passage peaks from end of September to mid-October (Mildenberger 1984). At Col de Bretolet (western Switzerland), passage from end of September or beginning of October, peaking 8–17 October and continuing to November (Winkler 1984), and main arrival on Dalmatian coast 2nd week of October (Cvitanić 1980). Crosses Strait of Gibraltar earlier than other Fringillidae, beginning 2nd half of September, with mid-October peak sometimes extending into November (Finlayson 1992). Passage later in central and east Mediterranean, mostly mid-October to November in Cyprus, Sicily, and Malta (Flint and Stewart 1992; Sultana and Gauci 1982; Iapichino and Massa 1989); in Israel, broad-front passage mid-October to mid-December, chiefly November (Shirihai in press).

Spring migration begins early, with movement from February widely reported in both south and centre of winter range. In Strait of Gibraltar area, begins early February and peaks 2nd half of March, with last birds late April (Finlayson 1992), and passage February to late March in Sicily (Iapichino and Massa 1989). Departing flocks very common on northern capes of Cyprus mid-February to April (Flint and Stewart 1992), and latest records April in Egypt and north-east Libya (Bundy 1976; Goodman and Meininger 1989). Passage at Col de Bretolet from early March to early May, chiefly last week of March to 20 April (Winkler 1984). At Riederalp (Switzerland), birds arrived when study area 12% snow-free: 15 April in 1985 and 1987, 3 May in 1986 (Frey 1989a). Wintering birds leave southern Bretagne in 2nd half of February, and breeding birds arrive from late March, with peak 10–15 April and stragglers to early May (Mahéo 1969). Arrives in Britain mid-March to early May, chiefly in April (Newton 1972; Lack 1986). Reaches Sweden end of March to April; at Uppsala (central Sweden), mean earliest record 6 April over 15 years (Lundberg and Edholm 1982; SOF 1990). Arrives in St Petersburg region (north-west Russia) in March in early springs, but usually in April, with passage continuing throughout April–May (Mal'chevski and Pukinski 1983).

Further east, in Kazakhstan, most birds migrate, but winters regularly in small numbers in south and west (Korelov *et al.* 1974). Many birds descend to foothills and valleys from mountains of south-central Asia, or move further afield (Ivanov 1969; Abdusalyamov 1977). Some reach Pakistan: irregular and uncommon south to *c.* 32°N in north, and in Quetta area in west (Ali and Ripley 1974). In autumn, main movement September–October; at Chokpak pass (western Tien Shan), passage peaks 11–31 October; migrates early in spring, from 1st half of March; main passage through Chokpak pass completed by 25 March (Kovshar' 1966; Korelov *et al.* 1974; Gavrilov and Gistsov 1985). At Novosibirsk (55°N), 1st records 9–17 April over 3 years (Kazantsev 1967).          DFV

**Food.** Small to medium-sized seeds; probably takes fewer invertebrates than any other west Palearctic finch apart from crossbills *Loxia* or Twite *C. flavirostris*, although importance in nestling diet unclear. In Schleswig-Holstein, 53% of 2844 foraging observations over year were on agricultural land, 44% waste ground, 2% parks and gardens, 1% hedges (Eber 1956). In eastern England, 36% of observations were in bushes less than 1 m above ground, 32% on ground, 29% on herbs, 3% higher branches; of 139 feeding actions, 52% while standing on ground, 44% normal perching, and 4% in flight; did not hang upside-down while feeding (Kear 1962). Similarly in captivity, 54% of postures perching, 30% standing, 15% clinging to bent stems; only very briefly on vertical stems and not upside-down; intermediate in agility between Greenfinch *C. chloris* and (e.g.) Goldfinch *C. carduelis* or Siskin *C. spinus*; uses foot to hold down (e.g.) pappus while seed picked off, but seemingly not to hold seed-heads (etc.) pulled in with bill. Unable to pierce involucral bracts to get at ripening Compositae seeds so strips them downwards. (Newton 1967a.) Skilfully opens Cruciferae pods without seeds falling out (Eber 1956). Seen feeding briefly on floating vegetation, sometimes hovering (Keymer 1975). In winter, forms large mixed flocks with other seed-eaters in open country, feeding much more on ground than in summer, though tends to remain together in groups within such flocks (Lack 1986; Deunert 1989); also forages in groups in breeding season when food abundant; new patches of food-plants quickly discovered this way, and predators more readily detected, an important consideration when feeding in vegetation close to ground. Food for young and preferred seeds collected up to 1–1·2 km from nest and transported in large gullet (Deunert 1989; Frey 1989b). For many details of seed handling, de-husking, bill morphology, etc., see Eber (1956), Kear (1962), Ziswiler (1965), and Newton (1967a). 15% of

seeds eaten were less than 0·5 mg, 31% 0·5–1·0 mg, 52% 1–10 mg, 2% greater than 10 mg; size intermediate between those taken by *C. chloris* and Redpoll *C. flammea* (Newton 1967a, 1972). In captivity took smallest seeds along with *C. carduelis* (Kear 1962).

Diet in west Palearctic includes the following. Invertebrates: mayflies (Ephemeroptera: Siphlonuridae, Baetidae), damsel flies (Odonata: Agriidae), grasshoppers (Orthoptera: Acrididae), bugs (Hemiptera: Coreidae, Aphidoidea), scorpion flies (Mecoptera: Panorpidae), Lepidoptera (Tortricidae, Plutellidae, Geometridae), caddis flies (Trichoptera: Limnephilidae), flies (Diptera: Tabanidae, Empididae, Muscidae), Hymenoptera (Cimbicidae, Empididae), beetles (Coleoptera: Carabidae, Scarabaeidae, Elateridae, Bruchidae, Curculionidae), spiders (Araneae: Araneidae), earthworms (Lumbricidae), snails (Pulmonata: Arionidae). Plants: seeds, fruits, etc., of conifers (Coniferae), poplar *Populus*, birch *Betula*, alder *Alnus*, elm *Ulmus*, dogwood *Cornus*, buckthorn *Rhamnus*, guelder rose *Viburnum*, privet *Ligustrum*, hemp *Cannabis*, nettle *Urtica*, knotgrass, etc. *Polygonum*, sorrel, etc. *Rumex*, goosefoot, etc. *Chenopodium*, orache *Atriplex*, beet *Beta*, glasswort *Salicornia*, seablite *Corispermum*, chickweed, etc. *Stellaria*, mouse-ear *Cerastium*, corn spurrey *Spergula*, sand spurrey *Spergularia*, buttercup, etc. *Ranunculus*, poppy *Papaver*, rape, etc. *Brassica*, hoary cress *Cardaria*, rocket, etc. *Sisymbrium*, wall rocket *Diplotaxis*, charlock *Sinapis*, radish *Raphanus*, garlic mustard *Alliaria*, sea kale *Crambe*, common whitlow-grass *Erophila*, scurvy-grass *Cochlearia*, pennycress *Thlaspi*, gold of pleasure *Camelina*, shepherd's purse *Capsella*, meadowsweet *Filipendula*, strawberry *Fragaria*, cinquefoil *Potentilla*, avens *Geum*, burnet *Sanguisorba*, rowan *Sorbus*, hawthorn *Crataegus*, trefoil *Trifolium*, *Oxalis*, flax *Linum*, purple loosestrife *Lythrum*, willowherb *Epilobium*, evening primrose *Oenothera*, bilberry, etc. *Vaccinium*, primrose, etc. *Primula*, alkanet *Anchusa*, forget-me-not *Myosotis*, speedwell *Veronica*, eyebright *Euphrasia*, thyme *Thymus*, dead-nettle *Lamium*, self-heal *Prunella*, plantain *Plantago*, scabious *Knautia*, *Succisa*, *Scabiosa*, bindweed *Convolvulus*, goldenrod *Solidago*, daisy *Bellis*, mugwort, etc. *Artemisia*, bur marigold *Bidens*, chamomile *Anthemis*, scentless mayweed *Matricaria*, tansy, etc. *Tanacetum*, coltsfoot *Tussilago*, butterbur *Petasites*, groundsel *Senecio*, burdock *Arctium*, thistles *Carduus*, *Cirsium*, knapweed *Centaurea*, goatsbeard *Tragopogon*, chicory *Cichorium*, sow-thistle *Sonchus*, lettuce *Lactuca*, *Chondrilla*, dandelion *Taraxacum*, hawkbit *Leontodon*, hawksbeard *Crepis*, catsear *Hypochoeris*, hawkweed *Hieracium*, grasses and cereals (Gramineae). (Collinge 1924–7; Eber 1956; Pokrovskaya 1956; Glutz von Blotzheim 1962; Newton 1967a; Dornbusch 1981; Sabel 1983; Deunert 1989; Frey 1989b.)

Particularly dependent on weeds of open country and waste ground (especially Polygonaceae, Cruciferae, Caryophyllaceae, and Compositae), so habits determined to large extent by agricultural practices. (Eber 1956; Newton

1967a; Frey 1989b.) In Swiss Alps, preferred dandelion *Taraxacum* and mouse-ear *Cerastium* seeds; avoided coltsfoot *Tussilago*, probably because seeds too small for efficient feeding (Frey 1989b). Diet in Oxford area (southern England) reflected plant abundance (Newton 1967a). In captive tests, preferred seeds of flax *Linum*, *Brassica*, hemp *Cannabis*, thistle *Carduus*, and lastly grasses (Gramineae) (Kear 1962; Ziswiler 1965). In Oxford, took seeds of 25 of 30 commonest weeds in agricultural area, second only to Greenfinch *C. chloris* in range (Newton 1967a), and same breadth of diet recorded in Schleswig-Holstein (northern Germany) (Eber 1956), although here cereal grains were important component, while none recorded in diet of Oxford birds though cereals common in study area. Milky seeds preferred, ripe ones often avoided; only rarely eats buds or seeds of fleshy fruits (Newton 1967a; Frey 1989b). In Oxford, of *c.* 12 500 feeding observations, 33% by number on seeds of Cruciferae (especially charlock), 11% *Polygonum*, 11% goosefoot, 9% chickweed, 5% thistles, 5% dandelion, 3% *Rumex*, 3% rocket, and 3% mugwort. 405 gullets, September–April, contained 50% by volume seeds of Cruciferae (mainly charlock), 24% *Polygonum*, 14% goosefoot, 2% thistles, 1% meadowsweet, and 1% mugwort; only 1 out of 38 adult stomachs obtained in breeding season contained any animal material: 2 small beetle larvae probably picked up with seeds. (Newton 1967a.) In England, 39 stomachs contained 857 items, of which 42% by number seeds of charlock, 22% *Senecio*, 18% *Rumex*, 7% chickweed, 3% self-heal, 3% knotgrass, and 2% adult and larval Lepidoptera (Collinge 1924–7). In Schleswig-Holstein, of 2677 observations, 19% by number *Brassica*, 17% cereals, 13% mugwort, 10% *Polygonum*, 6% *Rumex*, 5% charlock, 4% rocket, 4% mouse-ear (Eber 1956). In spring, main food is milky seeds of chickweed, dandelion, and, to some extent, grasses such as *Poa* or *Echinochloa*; in summer, probably takes same seeds as fed to young, though on average smaller, e.g. unripe seeds of *Rumex*, *Polygonum*, and *Cruciferae*; in Bretagne (western France), very fond of colza *Brassica oleifera* at this time of year; in autumn, feeds on thistles, and in Schleswig-Holstein study area on rape and cereals, and in winter on goosefoot or any available Cruciferae. Often at coastal vegetation and on salt-marshes in winter, also in ploughed fields and, increasingly, in birch trees; even in severe weather very seldom comes to bird-tables, etc., and in snow depends on finding tall plants such as grasses, goosefoot, or Cruciferae. (Eber 1956; Mahéo 1964; Newton 1967a.) At *c.* 2000 m in south-west Switzerland, before seeds appear in early June, takes those of previous year on ground as snow melts; fresh seeds taken generally according to abundance (though see above) and seed stage. In early June, diet 100% (by observation and calculation) dandelion seeds, proportion gradually decreasing over summer; late July, 36% buttercup, 24% hawksbeard, and 22% avens; early August, 40% hawkbit, 24% buttercup, and 22% mouse-ear; mid-August, 90%

shepherd's purse; late August, 66% *Polygonum*, 24% shepherd's purse, and in early September, 57% shepherd's purse, 35% *Polygonum* ($n = 101 165$ in total). (Frey 1989b.) Captive birds took average 2·6–5·2 g fresh weight of seed kernels per day (Kear 1962).

Apparently some disagreement about extent of invertebrate component in nestling diet; modern studies have found diet almost entirely seeds, with little trace of animal material, but according to Collinge (1924–7) and Pokrovskaya (1956), diet largely insects. These discrepancies could be caused by sampling methods. In Oxford, of 3979 samples, 17% by volume seeds of Cruciferae (mainly charlock), 16% dandelion, 11% chickweed, 11% thistles, 8% elm, 7% catsear, 6% sorrel. Insects found in only 2 of 62 broods: in one, 15% of diet for first 9 days aphids and caterpillars; in other, a few small beetle larvae from seed heads. (Newton 1967a.) Also in England, however, 30 stomachs contained larval Diptera, caterpillars, aphids, beetles, and spiders, and apparently no seeds (Collinge 1924–7). In central Germany, 6 collar-samples contained half-ripe seeds of dandelion, shepherd's purse and other Cruciferae, chickweed, corn spurrey, and *Rumex*; invertebrates found were aphids, caterpillars, and 1 bug (Hemiptera), all probably collected incidentally with seeds (Dornbusch 1981). No invertebrates found in another German study, in which early broods were fed almost only coltsfoot seeds, later ones *Brassica* and grass seeds (Handtke and Witsack 1972). Hachfeld (1979) found 1st broods given mostly half-ripe grass seeds, and 2nd broods also only seeds, though more varied. Deunert (1989) found 90% of diet unripe seeds. In major study in Swiss Alps, nestling diet basically as adults', with no insects found: mostly seeds of dandelion early in season, avens, buttercup, and chamomile later on (Frey 1989b). In St Petersburg region (western Russia), 607 collar-samples contained 1304 items, of which 25·5% by number beetles (11% Elateridae, 10% Chrysomelidae), 20·7% Diptera (19% Muscidae), 12·1% larval mayflies, 10·7% Lepidoptera (10% Geometridae), 5·0% spiders, 2·9% Hymenoptera (2% Braconidae), 1·8% larval caddis flies, 1·6% Hemiptera; 14·3% plant material (old berries of *Vaccinium*, seeds of *Oxalis*, alder catkins, young nettle leaves, and some cereal grains). Plant material accounted for 50% of food at one nest in early August, remainder of samples from mid-May and mid-June. (Pokrovskaya 1956.) Also in St Petersburg region, Mal'chevski and Pukinski (1983) considered invertebrates much overestimated in earlier studies and nestling diet to be in fact mostly seeds. Nestlings at 8 days old received 0·4–0·6 ml (0·2–1·9) dry volume per hr (Dornbusch 1981). Young probably given larger seeds on average than in adult diet, e.g. more elm seeds, which weigh 50 mg (Newton 1967a); but see Handtke and Witsack (1972) above, where some broods fed mostly very small seeds of coltsfoot. BH

**Social pattern and behaviour.** Most detailed study is that by Frey (1989*a*, *b*) at Riederalp (Switzerland). For review and comparison with other Fringillidae, see Newton (1972).

1. Typically gregarious outside breeding season, forming flocks (sometimes of hundreds or even thousands) for feeding, roosting, and migration (e.g. Thienemann 1910, Aplin 1911, Steinfatt 1937*b*, Witherby *et al.* 1938, Ptushenko and Inozemtsev 1968, Mahéo 1969). Roosts hold 50-300(-1500) birds or more. Near Oxford (southern England) in winter, numbers peaked at *c.* 1000 in September, falling to 400-500 by October, and to less than 100 on first day of snow (Newton 1967*a*, 1972). At Riederalp, first to arrive on breeding grounds in spring ascend in small or larger flocks, following retreating snow (Frey 1989*a*). Study in eastern Germany found flocks re-forming in bad weather after ♂♂ had split away at start of breeding season (Handtke and Witsack 1972). Still sociable when breeding, several pairs sometimes associating during nest-building and when foraging for young; ♂♂ (predominantly) flock during incubation, and birds also congregate to roost (Braun 1902; Meineke 1979). In Bretagne (France), summer flocks forage within 5-10 km radius, but stick to same roost (Mahéo 1969). As in many Fringillidae, families join together in flocks after fledging, though juvenile-only flocks (1st broods) also reported (e.g. Géroudet 1957, Ptushenko and Inozemtsev 1968, Mahéo 1969). Said by Witherby *et al.* (1938) to be more exclusive than some Fringillidae, but often associates (for feeding and roosting) with other Fringillidae, buntings (Emberizidae), sparrows *Passer*, etc. (e.g. Glutz von Blotzheim 1962, Kovshar' 1966); in Belgium, October, flock of *c.* 30 associated closely with *c.* 20 Stonechats *Saxicola torquata* (Liedekerke 1970). BONDS. Mating system apparently mainly monogamous, though occasional occurrence of polygyny (2 ♀♀ paired with same ♂) suggested by reports of nests apparently of 2 different ♀♀ less than 1 m and *c.* 25 m apart respectively, both nests in each case apparently attended by 1 ♂; also 2 rudimentary nests sited within *c.* 10 m of active nest and all 3 evidently associated with same ♂ (Lehmann 1952, 1962). Known to hybridize occasionally with other Carduelinae, including Twite *C. flavirostris*, Redpoll *C. flammea*, Greenfinch *C. chloris*, and Siskin *C. spinus* (e.g. Witherby 1915, Lönnberg 1918, Sevesi 1939, Panov 1989). In Moscow region (Russia), pairs formed upon arrival (Ptushenko and Inozemtsev 1968). Typical of Carduelinae in pairing up in flocks; pair-bond strong (Newton 1972); according to Handtke and Witsack (1972), ♂♂ leave flocks before start of pair-formation. Nest-building by ♀ (e.g. Steinfatt 1937*b*, Witherby *et al.* 1938), but see subsection 5 in Heterosexual Behaviour (below) for suggestion of 'cock-nests' built by ♂. Incubation also by ♀ alone (Steinfatt 1937*b*; Dementiev and Gladkov 1954), though ♂ reported sometimes to take over for short stints (Witherby *et al.* 1938; Ptushenko and Inozemtsev 1968). Young fed by both sexes, ♂ (as typical in Carduelinae) spending more time on parental duties than in defence of territory (see Breeding Dispersion, below), and bringing all food for first few days after hatching (Newton 1967*a*, 1972). If one bird dies when young still small, mate unable to rear them alone (Frey 1989*a*). Fledglings normally fed for *c.* 2 weeks (Géroudet 1957; Newton 1972). Several reports of overlapping broods, ♀ starting new nest and sometimes laying while 1st brood still in nest, ♂ then taking over care of fledglings within *c.* 2-7 days of them leaving nest (Oakes 1953; Ptushenko and Inozemtsev 1968; Mahéo 1969; Tast 1970; Frey 1989*a*). Age of first breeding 1 year (C S Roselaar, from specimens). BREEDING DISPERSION. Territorial, though defended area apparently small (see below), and loose neighbourhood groups of 2-12 or up to several dozen or more pairs often occur, with some nests only a few metres apart; many pairs solitary (Witherby *et al.* 1938; Driver 1957; Glutz von Blotzheim 1962; Strokov 1962; Ptushenko and Inoz-

emtsev 1968; Tast 1968, 1970; Newton 1972). At Oxford, groups of *C. cannabina* were the largest among 9 Carduelinae studied. Groups presumably formed through social attraction, but food-supply also important for dispersion: in Oxford study, only 1-2 pairs present early in season when food scarce, but at least 10 pairs moved in to nest later when thistles (Compositae) began to seed (Newton 1972). In central Valais (Switzerland), groups formed where few potential nest-sites in rocky steppe habitat. At Riederalp, where no such shortage, birds evenly dispersed over *c.* 34 ha (out of *c.* 73-ha study area), with average distance to nearest simultaneously active nest 91 m ($n = 43$), 92 m ($n = 45$), and 180 m ($n = 30$) in 3 years. Only 5% of 121 nests 20-30 m apart, and attempt to build within 6 m of established nest thwarted by 1st ♂'s aggression, this being evident even when new pair tried to settle up to 59 m away. (Frey 1989*a*, *b*.) In Bretagne, when density low, dispersion more random and nests further apart; more evenly spaced with increasing density, and territoriality more pronounced, though notable concentration when area of suitable habitat much reduced (see below). At start of breeding season, *c.* 50-60 m between pairs, and new pairs allowed to settle closer (within 13·5 m) only later when territoriality waned; average between nests over 7 years 14 m ($n = 49$) to 34 m ($n = 15$). (Eybert 1980, 1985.) Examples of distances between nests in groups: at Tampere (Finland), 2 *c.* 10 m apart, 3rd *c.* 50 m away; 3 within *c.* 20 m (Tast 1970); in Moscow region, sometimes only 3-10 m, once 10 nests within 13-15 m (Ptushenko and Inozemtsev 1968); in garden in Germany with 8-10 pairs, some nests only *c.* 20 m apart, and over 100 nests recorded within radius of *c.* 1000 paces (Naumann 1900); at Halberstadt (eastern Germany), where exceptionally high density (see below), nests sometimes very close but usually separated by several trees (Handtke and Witsack 1972). For distance between nests perhaps of polygynous ♂, see Bonds (above). In Moscow region, 3 territories maximum *c.* 500-600 m apart (Zablotskaya 1982). In Pennines (northern England), where breeding closer to *C. flavirostris*, 2 nests *c.* 45 m apart (Orford 1973). Territory variously reported to comprise only nest (Handtke and Witsack 1972), nest-shrub (Géroudet 1957; Hiss 1979), area of *c.* 220-250 m² (Zablotskaya 1982), or *c.* 700 m², with many nests having small spruces *Picea* used by ♂ as song-posts and look-outs within *c.* 15-20 m (Frey 1989*b*, which see for discussion and further references). Where 2 Riederalp pairs *c.* 46 m apart, each always made detour around other's territory (Frey 1989*b*). Halberstadt study found territorial fights rare and several ♂♂ often singing close together (Handtke and Witsack 1972); see Antagonistic Behaviour (below). Territory of Carduelinae typically established after nest-site selected and maintained only while nest in use (Newton 1972). Defended (in *C. cannabina*) by both sexes, ♂ usually attacking other ♂♂ (occasionally ♀♀), ♀ displacing intruders of either sex (Frey 1989*b*), though ♀♀ at Halberstadt showed no reaction to dummy ♂♂ set up near nest (Handtke and Witsack 1972). At least some courtship (including pair-formation) takes place in flocks away from territory (no information for copulation). As typically in group-nesting Carduelinae whose young are fed primarily seeds, feeding mostly done communally at some distance from nest (Newton 1967*a*, 1972; Tast 1968; Mal'chevski and Pukinski 1983); see Food for ability to transport large amounts. At Riederalp, before seeds ripe, some foraging done within *c.* 67 m of nest; with increasing availability of preferred seeds, up to 1190 m away (Frey 1989*b*). At Oxford, birds mostly nest in gorse *Ulex* at beginning of breeding season, often then shifting some distance (like other Carduelinae) to establish new territory, usually in deciduous shrubs (Newton 1972). At Tampere, sometimes move *c.* 500 m from wood-pile to tree or shrub, though may continue using same feeding area

(Tast 1968). At Riederalp, replacement nests averaged 84 m ($n = 4$) from 1st; one 2nd-brood nest c. 110 m away. When 2nd nest built c. 121 m from 1st, adults flew back to defend old territory (or 1st brood in bush, though young not defended on feeding grounds) (Frey 1989b). In Moscow, same site used for 2 consecutive broods by same pair (Strokov 1962); perhaps also at Halberstadt (Handtke and Witsack 1972). Highest density recorded at Halberstadt, where exceptional concentration of up to 59 pairs in 0·6 ha of dense spruce plantation (9833 pairs per km²) including 14 pairs in 700 m² (20 000 pairs per km²) (Handtke and Witsack 1972). At Paimpont (Bretagne) over 5 years, average 18-69 pairs in c. 2 ha (900-3450 pairs per km²); great increase in density following dry spring and habitat destruction caused by fire (Eybert 1980). Also in Bretagne, 420 pairs per km² recorded on island, 600 on moors (Mahéo 1969). At Herzberg (Harz mountains), average 17 pairs in 2·2 ha (773 per km²) (Meineke 1979). In Moscow, 12-30 nests in 4 ha (300-750 per km²); 0·9-13(-20) nests along 1-km stretches of road (Strokov 1962). Average for woodland and farmland census plots in Britain c. 7 pairs per km²; 75 pairs per km² recorded in coastal scrub (mainly sea buckthorn *Hippophae*), and 40 nests in less than 2 ha of gorse (c. 2000 per km²) (Newton 1972; Sharrock 1976); see also (e.g.) Benson and Williamson (1972). In Switzerland, 50-70 pairs per km² in cemetery (Glutz von Blotzheim 1962) and 40-60 over 3 years at Riederalp (Frey 1989a); for further data, see Schifferli et al. (1982) and Glutz von Blotzheim (1987). In northern Germany (mainly Schleswig-Holstein), c. 2000 pairs in 1405 km² (1·4 per km²); maximum 10 pairs per km² near human habitations, 0·2-6·7 in other habitats, though 24 per km² in hilly area with hedges (Busche 1991). In Bourgogne (France), 6-41 pairs per km² in oak *Quercus robur* shelterwood (Ferry and Frochot 1970). Various forest types in Bulgaria hold 4-36 pairs per km², that dominated by juniper *Juniperus* favoured (Simeonov 1975; Petrov 1982, 1988). In Morocco, 6 pairs per km² in maquis (Thévenot 1982). On Öland (Sweden), average 4·9 territories per km² in 7 coastal meadows totalling 5·9 km² (Ålind 1991). Sometimes associates for breeding with other Carduelinae: in Pennines, small numbers with *C. flavirostris* (Orford 1973); at Tampere, solitary pairs usually close to *C. chloris* (Tast 1970). Ringing studies in Belgium and Bretagne indicated strong fidelity to natal site for up to 4 years (Verheyen 1955a; Mahéo 1969). For nest-sites used over more than 1 year (not known if by same birds), see Lehmann (1962). No site-fidelity of breeding adult between years recorded in Riederalp study (Frey 1989b), and no sightings in subsequent years of 417 birds ringed as juveniles at Halberstadt (Handtke and Witsack 1972). ROOSTING. Nocturnal and solitary or (chiefly outside breeding season) communal, assembling (sometimes with other species) mainly at traditional sites such as coniferous or deciduous trees (e.g. young oaks, or beeches *Fagus* which retain leaves throughout winter), dense shrubs, hedges, reedbeds, or often among coarse grass and herbage on ground; less commonly, gables of thatched roofs (Naumann 1900; Witherby et al. 1938; Kruseman 1942; Boase 1948; Newton 1967a, 1972; Dorsch 1970; Stiefel 1976). At Braunschweig (Germany), roost on island in lake used over several years from late March to end of April and end of June to end of September, peaking August to mid-September with 800-1000 birds (Bäsecke 1942). At Riederalp, sites used before nesting were often close to feeding grounds and shared with *C. flammea*. Roosts broke up as ♂♂ became increasingly aggressive and pairs formed, birds then roosting in pairs or (once ♀ incubating) singly near nest or in forest; main roost was in stand of alders *Alnus*, used for over 40 years, apparently by whole population of breeding area studied, from late June (fledging of 1st broods) into September. (Frey 1989a, b.) Pattern in Harz mountains similar,

though ♂♂ apparently continued using communal roost (c. 500 m from breeding area) during incubation. Birds returned from roost c. 30 min before sunrise in April, at least 1 hr before in May. (Meineke 1979). In Kazakhstan, regularly visits particular sites to drink and to eat salt (Korelov et al. 1974).

2. Fairly tame when breeding, at times almost indifferent to man (Géroudet 1957; Strokov 1962; Mal'chevski and Pukinski 1983). Alarm-calls (see 5g in Voice) given for man or dog close to feeding flock, birds sometimes then escaping into tree or shrubs, as when threatened by raptor (Naumann 1900; Creutz 1967; Zablotskaya 1982). Also when excited (e.g. mobbing predator), may give call 5f (see Voice), ruffle crown, slowly raise and lower tail, and look about, calls encouraging other birds to join in; may then fly off, giving call 5e (see Voice), as typically when flushed, etc. (Zablotskaya 1982). FLOCK BEHAVIOUR. Usually flies to roost singly or in small flocks and, like many other Fringillidae, birds typically assemble in trees or on wires, giving contact-calls (see 5a in Voice) and singing, as also when circling (for up to c. 1 hr) repeatedly over roost and dropping into bushes (Kruseman 1942; Dorsch 1970; Newton 1972; Stiefel 1976; Meineke 1979; Zablotskaya 1982). At Braunschweig, vigorous song given by ♂♂ also from shrubs where roosting; birds completely silent c. 30 min after sunset. When disturbed (probably by Tawny Owl *Strix aluco*), all birds circled in tight-knit flock, gradually returning to shrubs via taller trees. (Bäsecke 1942.) In Tadzhikistan, birds dropped suddenly into reedbed from circling flight; before morning exodus, called and sang in twilight (Ivanov 1969). May rest in shrubs around midday, giving contact-calls, as when feeding (Zablotskaya 1982). Mixed feeding flocks usually move as unit on ground, but once airborne *C. cannabina* tends to separate off and may circle some distance away, while others head for cover (Newton 1972); see also *C. flavirostris*. Lone bird or small flock flying past usually joins bird(s) giving call 5a, while larger flocks tend to continue flight and call similarly, this sometimes encouraging original caller(s) to join them; lone straggler giving 'tuwee' (see 3b in Voice) may join or be joined by passing flock (Zablotskaya 1982). In spring, snatches of song and calls are typically given by actively migrating birds (e.g. Thienemann 1910). SONG-DISPLAY. As in other Carduelinae, song of little territorial significance and given almost throughout year, in early breeding season (also in late-summer and autumn flocks), often by several ♂♂ together. ♂ gives Advertising-song (see 1a in Voice) within a few days of return to breeding grounds from elevated perch (occasionally on ground) and in flight. (Witherby et al. 1938; Stadler 1956; Newton 1967a; Zablotskaya 1982.) While singing, or at least giving preliminary to full song (see 6a in Voice), bird will turn head rapidly to left and right or sway back and forth (Zablotskaya 1982); also ruffles forecrown, and may droop wings and spread tail, repeatedly raising and rapidly vibrating both. In early period, interrupts song to preen or to fly down to feed (Witherby et al. 1938; Ptushenko and Inozemtsev 1968; Zablotskaya 1982). Sometimes sings in normal flight, but true Song-flight also occurs, though apparently less frequent than singing from perch, and perhaps performed only by some ♂♂ (Géroudet 1957; Bergmann and Helb 1982). ♂ variously reported to take off from perch while singing and to ascend c. 10 m, then to sing while flying in erratic circles with rapid wing-beats, tail widely spread (once ascended to top of tall tree thus and there joined by ♀); slow flapping flight recorded, also gliding, sometimes with wings held below horizontal like displaying Redshank *Tringa totanus* (includes bird flying a few metres behind one gliding on flat wings); descent fluttering or gliding in spiral, bird returning to same perch, or landing by ♀ or other conspecific birds (Rendell 1947; Boase 1948; Fitter 1948; Fitter and Richardson 1951; Ptushenko and Inozemtsev

1968; Newton 1972). After territory established, ♂ reported to fly round boundary, and to sing briefly (see 1b in Voice) from each song-post, then close to nest (1b or full Advertising-song). ♂ accompanying mate while she is quietly feeding, inspecting potential nest-site, or nest-building gives Advertising-song with head raised, crown ruffled, wings drooped, and tail raised, breast also slightly ruffled; may also patter with feet and sway from side to side. Sings when disturbed and approaching to feed young (see 1c–d in Voice). (Zablotskaya 1982; also Kovshar' 1966, Hachfeld 1979.) ♀ known to sing occasionally or to give series of calls or motifs (perhaps song: see 2 in Voice), sometimes in duet with ♂. Captive ♀ sang only when deprived of ♂ (Gerber 1955). Main song-period in Britain late March to mid-July, also mid- to late September, though fairly frequent January–March, mid- or late July to early September, and through October; exceptional or only Subsong early or mid-November (Witherby *et al.* 1938; Boase 1948). Similar pattern in other parts of European range (e.g. Géroudet 1957, Mal'chevski 1959, Tast 1970, Mal'chevski and Pukinski 1983), with peak singing activity during nest-building and incubation (Hachfeld 1979). Song outside breeding season full and of fine quality, or Subsong (see 1e in Voice), often associated with mild sunny weather and given by flocks (e.g. at roost) (Stadler 1956; Géroudet 1957; Newton 1972). ANTAGONISTIC BEHAVIOUR. (1) General. Typical of Carduelinae (and unlike Chaffinch *Fringilla coelebs*) in being not markedly aggressive, and having short, inelaborate territorial behaviour (Braun 1902; Newton 1972); higher-intensity antagonism linked to defence of ♀ rather than of small nesting territory (Zablotskaya 1982). (2) Threat and fighting. Threat-call (see 4 in Voice) given in individual-distance disputes on feeding grounds or while resting, including attempt to land on branch amidst flock. Attacker may also adopt Threat-posture: head slightly drawn into shoulders and open bill pointed at opponent; wings raised at *c.* 60° and, at higher intensity, slightly vibrated; plumage slightly ruffled. Attacked bird sometimes calls (see 5e in Voice) and usually flees. (Zablotskaya 1982.) During pair-formation especially, ♂ defends mate within *c.* 50 cm on feeding grounds (where he never attacks other ♀♀); both ♂♂ may adopt Threat-posture and give Threat-call, then flying up (repeatedly) *c.* 1 m breast to breast, fighting with bill and feet, until rival chased away (Zablotskaya 1982; Frey 1989b). ♂ marks nesting territory by conspicuous perching, with ruffled breast (Frey 1989b). During establishment of territory and when patrolling boundary (see Song-display, above), may give call 5f (see Voice), sometimes incorporating it into song, in response to song of another ♂ (Zablotskaya 1982). Both sexes defend territory by giving Threat-call and flying or leaping towards intruder in displacing attack, this occasionally leading to fight, though intruder usually flees and may be chased. Other species driven away in Swiss study included Black Redstart *Phoenicurus ochruros* and Citril Finch *Serinus citrinella*. (Zablotskaya 1982; Frey 1989b.) HETEROSEXUAL BEHAVIOUR. (1) General. In southern Bretagne, birds settle to breed almost immediately after arrival, peak 10–15 April (Mahéo 1969). Riederalp study found timing of flocks' first visits to study area, pair-formation, and nest-site selection (not nest-building) clearly linked to thaw; most marked synchrony in breeding cycle at beginning of season (Frey 1989b). At Halberstadt, most ♀♀ started laying 1st clutches within 9–18 days (Handtke and Witsack 1972). (2) Pair-bonding behaviour. Few detailed observations of courtship, but ♂ known to display most conspicuous plumage features in presence of ♀ by spreading tail and drooping wings (white patches) and ruffling breast and crown, while hopping in rather upright posture towards or round ♀ and singing loudly; singing bird will also sway from side to side (Witherby *et al.* 1938; Poulsen 1954; Newton 1972; Frey

1989a). For calls associated with courtship, see 6 in Voice. Following typical of Carduelinae (including *C. cannabina*) studied by Newton (1972): wing-raising in threat (early stages); ♂ attempting to Bill-touch with several ♀♀, bond probably developing with one that reciprocates, eventually leading to Courtship-feeding (see below). ♂ will also chase ♀ through branches or in flight: in southern England, early May, when ♀ joined singing ♂ on wire, birds flew off low, ♂ flying above and behind ♀ in peculiarly erratic flight with tail lowered, hovered briefly at intervals, beating wings rapidly, before continuing chase, both birds finally disappearing into bushes (Summers-Smith 1951); apparent song fragments given by ♂ hovering over or near ♀ (Boase 1948). (3) Courtship-feeding. Takes place away from nest, including (as typically in Carduelinae) in flock, and on or near nest during incubation. ♂ provides most food for incubating ♀, though she also leaves nest to feed herself; ♂ feeds mate by regurgitation, ♀ calling (not described) and wing-shivering (Harber 1945a; *Br. Birds* 1945, **38**, 333; Géroudet 1957; Newton 1972; Frey 1989a). At Riederalp, ♂ fed ♀ on average every 12 min during 130-min period; of 11 observations, 9 when ♀ in laying phase; Courtship-feeding recorded only between pair-members and never associated with copulation (Frey 1989a). (4) Mating. ♀ solicits by wing-shivering, and pair copulate on perch (not known whether close to nest), sometimes several times in succession (Naumann 1900; Witherby *et al.* 1938). (5) Nest-site selection. Suggested by Lehmann (1962) that ♂ may build several 'cock nests' (nest-starts) one of which selected by ♀. Other studies indicate both sexes may search, though ♀ the more active and always makes final choice, ♂ usually remaining close to mate and singing or calling (Nethersole-Thompson and Nethersole-Thompson 1943b; Ptushenko and Inozemtsev 1968; Frey 1989a). Nest also built by ♀, though ♂ accompanies her, singing frequently. At Riederalp, ♀ collected material within average 81 m of nest ($n = 99$), and completed nest within average 4·6 days ($n = 22$). ♂ occasionally seen carrying material and making jerky movements of head when ♀ present. (Veroman 1978; Frey 1989a.) (6) Behaviour at nest. In Swiss study, often several days between nest-completion and laying. When cold, ♀ took more frequent and regular breaks from incubation; during 5–15 min, preened, was fed by ♂ (see above), and made short flights. (Frey 1989a.) Stands in nest to ventilate eggs on warm days (Hachfeld 1979). ♂ reported to take over incubation rarely (Géroudet 1957); no information on change-over. Otherwise, ♂ usually perches close to nest when ♀ incubating, and escorts her back to nest. At Halberstadt, usually approached nest from above (compare Parental Anti-predator Strategies, below). (Steinfatt 1937b; Handtke and Witsack 1972.) RELATIONS WITHIN FAMILY GROUP. Young brooded by ♀ for first few days, in extreme case (bad weather) for up to *c.* 13 days; at night at least to 11 days (Steinfatt 1937b; Handtke and Witsack 1972; Frey 1989a). When nestlings still small, ♂ brings food, passing it to ♀ by regurgitation and ♀ then feeding young by same method; both parents collect food for older nestlings (Steinfatt 1937b; Frey 1989a). Young 5–7 days old beg by gaping with neck extended, giving quiet food-calls; several young fed per visit (Hachfeld 1979). Faecal sacs of nestlings usually swallowed during first few days, later carried away (Naumann 1900; Hachfeld 1979). Young will leave nest at 9–10 days if disturbed, but normal fledging period longer (up to 17 days recorded: see Breeding). After fledging, disperse in bushes near nest and remain there several days, calling quietly and being fed by parents; eventually follow parents to feeding grounds (Tast 1970; Frey 1989a). ANTI-PREDATOR RESPONSES OF YOUNG. Usually remain still and silent on hearing parental alarm (Zablotskaya 1982). At 9 days, recorded crouching low in nest and gaping (perhaps in threat); may give fright-call and attempt to escape through vegetation (Hachfeld 1979). PARENTAL ANTI-PREDATOR STRATEGIES. (1) Passive measures. Sensitive to disturbance during laying and at start of incubation (Meineke 1979), but behaviour of incubating ♀ varies later: some sit tightly (can even be touched), others leave when man several metres away (Sermet 1967; Reinsch 1977; Mal'chevski and Pukinski 1983). In study by Handtke and Witsack (1972), ♀ always moved down to escape. (2) Active measures (all information relates to man as potential predator). When nest or fledglings threatened, birds may show excitement by ruffling crown and moving tail, ♂ sometimes singing (see 1c in Voice) and one or both adults giving variety of alarm-calls (see 5e–g and 5i in Voice), sometimes from cover or prior to fleeing (see Zablotskaya 1982 and Voice for details). Occasionally, ♀ flushed from nest drops down and performs distraction-lure display of disablement type, slowly creeping away and 'falling on one wing' (Mal'chevski and Pukinski 1983); see also Hachfeld (1979). At Halberstadt, one record of 3 ♂♂ and 1 ♀ gathering when fledgling on ground threatened; called from nearby trees; 1 ♂ flew to ground and there performed apparent distraction-threat display, ruffling breast, raising tail, and shivering slightly open wings (Handtke and Witsack 1972).                                                    MGW

**Voice.** Major study of wild and captive birds in Moscow region (Russia) by Zablotskaya (1982) found rich and complex repertoire of 4 song-types and many calls, most of which are given all year. For detailed description of song using musical notation and comparison with other *Carduelis*, see Stadler (1956); for additional sonagrams, see Bergmann and Helb (1982). Apparently capable of mimicry, but no evidence that this is regular in song of wild birds (Poulsen 1954; Stadler 1956; Kneutgen 1969).

CALLS OF ADULTS. (1) Song. (1a) Advertising-song of ♂.

Pleasing, musical and varied, rapid, lively twittering, incorporating many short units (some delicate), also 'trills' or tremolos (some with metallic timbre), attractive and melodious drawn-out and flute-like whistles, and so called 'crowing' (see call 6b, below) or twanging sounds; usually not very far-carrying. Typically begins with series of calls or modified call-units, 'gigigi' given in long accelerando sequence, sometimes rising and falling in pitch; certain other calls may precede or be included in song. (Witherby *et al.* 1938; Poulsen 1954; Stadler 1956; Géroudet 1957; Bergmann and Helb 1982; Zablotskaya 1982.) Each ♂ has large repertoire of song units, and constantly changing sequence produces great variety of different songs; repetition of units or motifs within a song (segments) are another typical feature (Boase 1948; Stadler 1956; Zablotskaya 1982). Individual ♂♂ show some preference for particular units or motifs which are frequently given (Poulsen 1954); contra Poulsen (1954), however, possibility of genuine regional and individual variation (in both song and calls) should not be discounted. Song given in discrete phrases lasting up to *c.* 5 s, or may continue for *c.* 10 s, typically with abrupt pause midway (Stadler 1956), or *c.* 17 s (Zablotskaya 1982). Often gives truncated song, typically containing whistles (Poulsen 1954; Stadler 1956). Short 'trills' and thin, soft whistles (e.g. 'peeuu', 'trrrü', 'tukeeyü') combined with Twitter-calls (see 3a, below) and given in flight (Bruun *et al.* 1986) refers to snatches of song; see also other song-types (below). Recording (Fig I) contains delicate song with Twitter-calls (including at start), many buzzy units ascending or descending in pitch (most with frequency and/or amp-

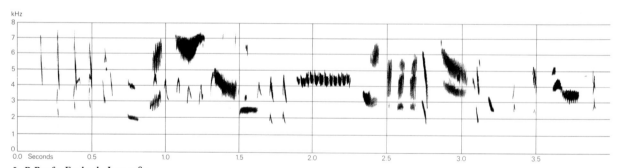

I  R Ranft  England  June 1987

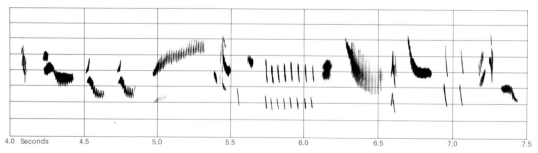

I *cont.*

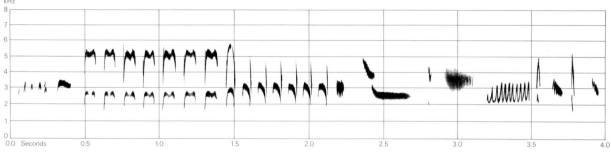

II P A D Hollom England May 1983

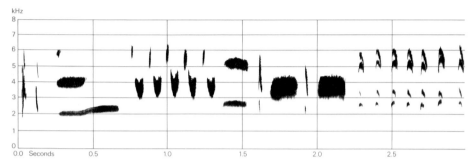

III P J Sellar Sweden May 1978

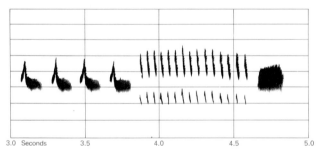

III *cont.*

litude modulation), pronounced tonal tremolo, and short rattle segment. In another recording (Fig II), faint introduction followed by 2 segments (7 'ku' and 7 'ki' notes) which are strongly reminiscent of subdued piping of Oystercatcher *Haematopus ostralegus*, then 'te-ooor t-chee tsurrrr-tk-ze-tk-ze'. Songs in recording from Sweden characterized by segmentation and buzzy sounds: in Fig III, 2 quiet introductory Twitter-calls followed by long tone (see Fig XII) and 5 segments, 1st, 2nd, and 4th of which are buzzy, 5th a rattle; in Fig IV, only 2 segments and no real tones, with whirring sound after introduction, and section from *c.* 2·3s to end comprising all differently paced and pitched buzzes. (J Hall-Craggs.) Following additional song-types reported by Zablotskaya (1982), but no strict division between these and 1a. (1b) Short song (0·62–2·43s), quieter than type 1a, and generally of narrower frequency range; at least 3–4 variants per ♂. Given by ♂ when patrolling territory boundary and directed only at

close neighbours. (1c) Song comprising rapid series of 'ti' sounds (presumably much as 'gi' in 1a) and variety of notes delivered in more flowing manner than 1a and lasting 3·5–6s; given by ♂ once and loudly for any danger near nest. (1d) Short (*c.* 1·5 s) quiet song given by ♂ approaching nest to feed young; 3 best-studied ♂♂ gave only 2 variants, only one or both in succession. (Zablotskaya 1982.) (1e) Subsong. In November recording by V C Lewis, presumed ♂ gives quiet twittering and tinkling, also delicate and some rather strained whistles; some components of song 1a, but delivery more continuous. (2) Series of well-spaced delicate tinkling and cheerful twittering calls or motifs given by both sexes, sometimes as antiphonal duet (e.g. Jonsson 1978, 1992); perhaps type of song. Recordings of ♀ calling thus near nest contain variety of sounds: e.g. 'chi' preceding frequently uttered, mournful, whirring 'pt-rreee' (Fig V); 'che chee te-yooo kit-up chp tyk-chee tien' (Fig VI), 'tzi-tiu' (Fig VII), and 'pt tieeeoo' (Fig VIII). Recording of slower, less varied utterance from ♂ (mate of ♀ above), given also near nest after fledging, contains rather harsh descending 'zweeeah' (Fig IX), and in Fig X 'che tk-ooo' in which 'ooo' rather plaintive like Bullfinch *Pyrrhula pyrrhula*, then 'zri-tiu' from ♀ in background and, finally, rippling, descending twitter from ♂. In recording of pair in antiphonal duet (some overlap), 'pee-che-yurr' attributed to ♂ and 'pee-yer' to ♀ predominate; Fig XI shows these calls (♂ then ♀) without overlap, and Fig XII shows 'chirrup' (probably ♀), then whining penetrating 'mew' from ♂, with overlapping 'pee-yer' from ♀. In Fig XIII, snarling sound from ♂ (♀ responding with 'pee-yer') is perhaps same as

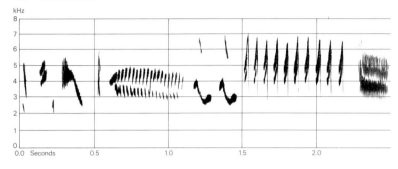

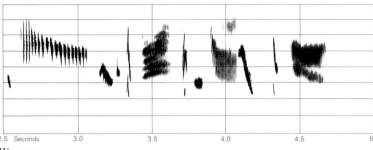

IV  P J Sellar  Sweden  May 1978

IV  *cont.*

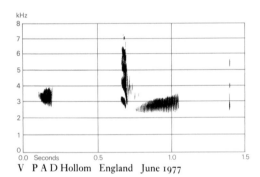

V  P A D Hollom  England  June 1977

M G Wilson). Other probably related sounds are 'tuwee' given especially by isolated bird (Zablotskaya 1982), and 'düje' or 'glü' (Bergmann and Helb 1982). (3c) Descending nasal 'djäe' given for ground predators (Bergmann and Helb 1982). (3d) Recording of ♀ approaching nest contains 'bzwee' followed by 'piu' (Fig XX), also Twitter-call ('p-t') and short descending 'kit' (J Hall-Craggs). (4) Threat-call. Hard, noisy, rattling 'dsrr', longer in increasing excitement (Bergmann and Helb 1982; Zablotskaya 1982). (5) Probably some overlap between the following calls and with calls 6–7 (reported by Zablotskaya 1982), also with those listed above, but poor quality of sonagrams and confusing transcriptions make comparison difficult. (5a) High-pitched whistling 'trill', 'fu-ti fu-ti fi-tee', given by bird alone or in flock, also between pair-members (but call 5b commoner when breeding). (5b) Whistling 'piu' descending markedly in pitch (in 2–3 steps) and serving as main contact-call between pair-members in variety of contexts (sometimes apparently with warning function). (5c) Short rattle or tremolo (up to 13 units), 'rre', given by ♂ approaching to feed young (perhaps same as 'zreeeeee' given in duet: see 2, above). (5d) Strident 3-part whistling 'feeeu' given by ♂ to summon and lead away fledglings; equivalent call of ♀ shorter. Perhaps related to calls shown in Fig XI (see 2, above). (5e) Short whistling 'fi' given before short flight and (variant) expressing fright (given during disputes, escape, when flushed, etc.); closely related 'fiu' (slightly lower pitched and descending more) associated primarily with flock coordination (when feeding, before roosting, etc.), and sometimes included in song. (5f) A 'ti ti fiu' given when disturbed, mobbing

or related to 'crowing' described below (see call 6b). Other motifs in duet include nasal 'te-keehn' (Fig XIV), nasal and squeaky 'hi hi hi hi' (Fig XV), and 'zreeeeee' (Fig XVI). (J Hall-Craggs.) (3) Contact- and alarm-calls. (3a) Twitter-call. Slightly metallic twittering, e.g. 'kekeke-keke', 'gegege', 'djek-djek', 'tett-tett-terrett', 'knetETT', 'chichichichit' (Witherby *et al.* 1938; Newton 1972; Jonsson 1978; Bergmann and Helb 1982; Bruun *et al.* 1986). Given singly, commonly 2 in quick succession, or longer series to integrate flock or family in flight ('flight-call' of many authors) or on ground (Zablotskaya 1982). For comparison with Twite *C. flavirostris*, see that species. Recording contains 'chi chi t-t-tp p-t-pt' (Fig XVII) and 'chit-ip kit-ip kit-ip terreeeeeee' (Fig XVIII) (J Hall-Craggs). (3b) Calls given when disturbed include 'tsooeet' (Witherby *et al.* 1938) and presumably same or related 't' chew-ee' (Boase 1948); recording (Fig XIX) suggests plaintive, mewing, slightly strained 'hoooi' (J Hall-Craggs,

VI  P A D Hollom  England  June 1977

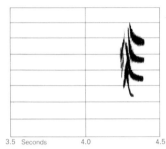

VI  *cont.*

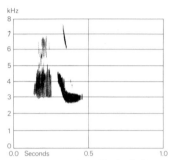

VII  P A D Hollom  England  June 1977

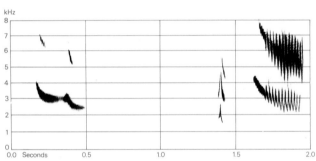

VIII  P A D Hollom  England  June 1977

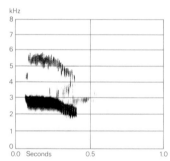

IX  P A D Hollom  England  June 1977

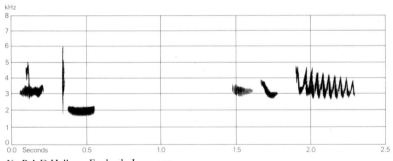

X  P A D Hollom  England  June 1977

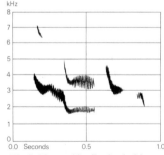

XI  R Margoschis  England  May 1990

predator and in various contexts associated with advertisement or defence of territory (incorporated also in song). Probably related is series of 'ti' sounds followed by descending 'fi' (10–12 units in all) noted from ♂ when eggs or young threatened. (5g) Shrill, noisy 'tri' given singly or in short series when alarmed (perhaps only for ground predators), at or away from nest. (5h) All-clear call. Call 5g or similar preceded by short series of 'ti' sounds. (5i) Short descending nasal whistle, 'coo'; used by both sexes when nest threatened. Despite marked difference in transcriptions, closely resembles 'zia' illustrated by Bergmann and Helb (1982). This or similar ('cuv') sometimes com-

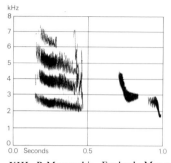

XII R Margoschis England May 1990

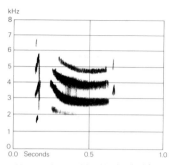

XIII R Margoschis England May 1990

XIV R Margoschis England May 1990

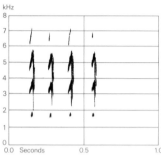

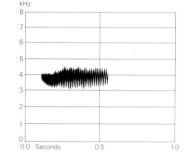

XV R Margoschis
England May 1990

XVI R Margoschis
England May 1990

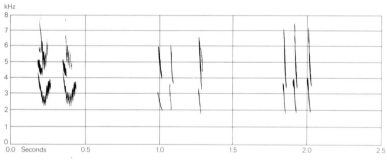

XVII R Ranft England June 1987

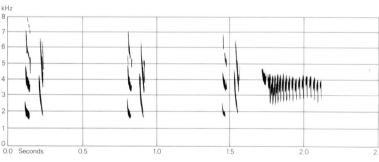

XVIII R Ranft England June 1987

bined with call 5g ('tri') as warning-call of ♀, or (noted from ♀ feeding fledglings) with more evenly pitched whistling 'tuvi'. (5j) Whistling 'tswi' with slightly nasal timbre (some resemblance to 'tuvi') given by ♀ leading young away from danger. (6) Advertising- and courtship-calls of ♂. (6a) Noisy metallic-sounding 'kvik kvik kvik' given usually before song 1a and sometimes elicited by song of another ♂ or calls of another ♂ Carduelinae. Perhaps no more than introduction to song described under 1a

(above). (6b) Advertising-call. Descriptions to some extent contradictory; further study needed. Hoarse and harsh 'crowing' sound descending in pitch—'gae' (Voigt 1933; Poulsen 1954; Stadler 1956); frequently incorporated (often as short segment: Poulsen 1954) in songs 1a–b, and also given as separate call from song-post and during courtship (Zablotskaya 1982, which see for sonagram of call rendered 'cryav' and similar to Fig XIII, though evenly pitched, not descending); according to Bechstein

XIX  V C Lewis  England  May 1964

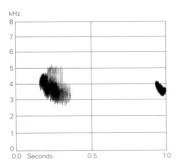

XX  V C Lewis
England  July 1965

(1853), frequent crowing component of song is clear, sonorous sound. (6c) Further calls given by ♂ during courtship include 'tukewee' and 'tu-trya'. (7) Calls mainly associated with feeding, resting, and comfort behaviour. (7a) Whistling 'tiwu' of descending pitch, and similar but more evenly pitched 'tiru', sometimes given in mixed sequence by feeding birds; also similar to 'tiru' though shorter, is a quiet, descending 'itu'. These whistling calls bear some similarity to those in 5e (above), but provisionally separated here. (7b) Quiet, low-pitched, slightly descending 'cuv' often given in long series, most frequently by pair in antiphonal duet (see call 2, above), with regular pauses between calls of each bird, for up to c. 30-40 min; also noted (presumably as feeding-call: see also call 5i) from adult at nest with young. Longer and more evenly pitched but otherwise similar 'tuvi' given (sometimes in long sequence) in same contexts. (7c) Quiet 'chuv' and similar, but lower-pitched, soft 'piv' given rarely while feeding, resting, and preening. (Zablotskaya 1982.)

CALLS OF YOUNG. Food-call of nestlings at 7 days a quiet 'ssst-ssst'; shrill sound given when frightened at 9 days (Hachfeld 1979). Other descriptions (apparently all of fledglings) refer to persistent chirruping 'tsoo-eet' (more grating than adult call 3b) or 'tjui' (Witherby *et al.* 1938; Bergmann and Helb 1982; Harris *et al.* 1989). Russian study described food- and contact-calls developing from 'chi-chuv' to 'chuv-chuv' and 'tveet' (raspy-screeching quality when being fed) within a few days of fledging. Adult call 3a given from c. 20 days. (Zablotskaya 1976a, 1982.) Quiet twittering song given by isolated, hand-reared birds when fully feathered; subdued twittering

mixed with fragments of adult song 1a by wild juveniles trapped in August (Poulsen 1954).     MGW

**Breeding.** SEASON. Southern England: see diagram; eggs laid mid-April to early August (Newton 1972). Southern Finland: start of laying late April, peak period beginning of May to end of June (Tast 1970); see Haartman (1972) for late nesting. Bretagne (western France): peak period for 1st clutch late April to early May; for 2nd clutch, 1st week of June (Eybert 1980); see also Mahéo (1969); for northern France, see Godin *et al.* (1977); for Bourgogne (eastern France), see Brochot and Petitot (1964). Rheinland (western Germany): 50% of 530 clutches over 40 years started by 2 June (Mildenberger and Schulze-Hagen 1973); for other parts of Germany, see Handtke and Witsack (1972), Hachfeld (1979), Meineke (1979), Glück (1983), Deunert (1989), and Gassmann (1989-90). Southwest Switzerland: at 2000 m, eggs laid mainly late May and early June; latest clutches 2nd half of July (Frey 1989a). Russia: in St Petersburg region (north-west Russia), eggs laid April to August with peak in June (Mal'chevski and Pukinski 1983); for Moscow area, see Strokov (1962). North Africa: laying from late March, mostly April (Heim de Balsac and Mayaud 1962). SITE. Very low in dense, often thorny tree, bush, scrub, or hedge, or on ground; frequently in young conifer plantation. Evergreens commonly preferred early in season for 1st broods, later nests more often in deciduous trees and bushes when cover thick, though conifers especially can be preferred all season (Tast 1968; Mildenberger and Schulze-Hagen 1973; Godin *et al.* 1977; Deunert 1989; Frey 1989a). In central England, April, 21% of 291 nests in gorse *Ulex*, 1% in deciduous shrubs; in July, of 132 nests, 6% and 23% respectively (Newton 1972). At Halberstadt (eastern Germany), in spruce *Picea* plantation, 59% of 252 nests on side twig against trunk, 26% between 2 hanging branch-tips, and 15% on stump; 90% less than 2 m above ground (Handtke and Witsack 1972). See also Hachfeld (1979) for 92 nests in Niedersachsen (north-west Germany). In Bretagne, nests of 172 pairs averaged 0·4

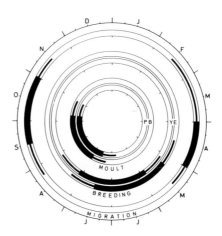

m above ground (Eybert 1980). In south-west Germany, nest-tree or bush significantly lowest of all Fringillidae studied at 1·5 m ($n = 9$), average nest-height 0·9 m, and diameter of nest-carrying branch smallest at 1·0 cm (Glück 1983, which see for comparison with other species). In some places, at times very close to human (even industrial) activity, nesting in cavities in concrete or wood-piles, even where suitable trees available (Tast 1970; Deunert 1981, 1989; Menzel 1983). In northern England, often in clumps of rush *Juncus* even when more usual gorse, hawthorn *Crataegus*, etc., is nearby; at coast in tall salt-marsh herbs (Orford 1959) and marram grass *Ammophila* (Witherby *et al.* 1938). Nest: foundation of small twigs, roots, stalks, and moss lined with hair, wool, plant down, sometimes feathers, paper, etc. (Handtke and Witsack 1972; Hachfeld 1979; Deunert 1989; Frey 1989a). In northern France, 122 nests had average outer diameter 9·7 cm, inner diameter 6·0, overall height 8·5, and depth of cup 3·5 cm (Godin *et al.* 1977); 23 nests from Bourgogne had outer diameter 11·2 cm (Brochot and Petitot 1964). For many details of material, see Hachfeld (1979). Building: by ♀ only, accompanied by ♂ (Handtke and Witsack 1972; Deunert 1989; Frey 1989a); 22 nests in Switzerland took average 4·6 days, almost all building in morning (Frey 1989a, which see for many details); in Niedersachsen, 11 nests took average 7·2 days (4–9) (Hachfeld 1979); can be completed in 2 days; material gathered close to nest and old nests can be demolished for re-use of material; complete old nest only rarely re-used (Handtke and Witsack 1972). EGGS. See Plate 60. Sub-elliptical, smooth, and non-glossy, very variable in shape and colour; pale to whitish-blue, sometimes light grey, with distinct spots, speckles, and streaks of pink, purple, or purplish-brown, also larger blotches of various shades of red to blackish-purple; markings mostly at broad end (Harrison 1975; Makatsch 1976). Nominate *cannabina*: 18·0 × 13·3 mm (15·0–20·8 × 11·9–14·5), $n = 1007$; calculated weight 1·66 g. *C. c. bella*: 18·4 × 13·6 mm (17·2–19·5 × 12·6–14·5), $n = 29$; calculated weight 1·78 g. *C. c. guentheri*: 17·1 × 12·9 mm (16·5–17·8 × 12·7–13·0), $n = 67$; calculated weight 1·50 g. (Schönwetter 1984.) Clutch: 4–6 (3–7). In Oberlausitz (eastern Germany), of 63 completed clutches: 3 eggs, 5%; 4, 27%; 5, 58%; 6, 10%; average 4·4 (April 3·8, $n = 7$; May 5·0, $n = 27$; June 5·0, $n = 20$; July 4·3, $n = 6$) (Deunert 1989). At Halberstadt, 1st clutch 4·8, $n = 128$, 2nd or replacement 5·1, $n = 96$ (Handtke and Witsack 1972). In southern England, average in April 4·6, May 4·8, June 5·0 (more than 50 clutches in each sample) (Snow and Mayer-Gross 1967); similar pattern in Rheinland (Mildenberger and Schulze-Hagen 1973), and in Bourgogne (Brochot and Petitot 1964). Average in Finland 4·8, $n = 31$ (Haartman 1969); in Swiss Alps 4·7, $n = 64$ (Frey 1989a, which see for comparisons with other studies); in European Russia 4·6, $n = 94$ (Mal'chevski and Pukinski 1983; Shurupov 1986), and in North Africa 4·6, $n = 73$ (Heim de Balsac and Mayaud 1962). Eggs laid daily

in early morning; in Alps, 1st egg laid average 4 days after nest completion in May, 1 day after in June–July (Handtke and Witsack 1972; Frey 1989a). 2 broods, 3 in favourable conditions; in Rheinland, *c.* 20% of *c.* 500 pairs probably had 3 broods in good years (Mildenberger and Schulze-Hagen 1973); at one nest in Oberlausitz, 2nd brood started 10 June, 3rd on 1 August; total of 14 young fledged (Deunert 1981); for intervals between laying and fledging of broods, see Deunert (1989). In England, one pair started 2nd nest when young in 1st 4 days old (Newton 1972). See also Frey (1989a). INCUBATION. 11–13 days (9–15) (Tast 1970; Handtke and Witsack 1972; Meineke 1979); in Alps, took average 12·8 days at 10 nests from laying of last egg to hatching of last young, 12–14 days for marked eggs (Frey 1989a). By ♀ only (Handtke and Witsack 1972; Hachfeld 1979; Frey 1989a); in Alps, started on average with egg 2·9 ($n = 43$), but wide variation, from 1st to last; probably earlier than at lower altitude, where usually starts with penultimate or last egg. ♀ on eggs for 85% of daylight time, *c.* 93% towards hatching; when ambient temperature falls, breaks apparently lengthen and become more frequent, as ♀ probably risks becoming too cold by remaining on nest (Frey 1989a.) YOUNG. Fed and cared for by both parents (Handtke and Witsack 1972; Makatsch 1976; Frey 1989a); when young small, ♂ feeds from crop via ♀, later both feed directly; ♀ broods until fledging in Alps if weather poor (Frey 1989a). At lower altitude, only brooded intensively for first *c.* 7 days (Handtke and Witsack 1972; Hachfeld 1979). ♂ feeds alone if ♀ builds new nest before young fledge (Frey 1989a). FLEDGING TO MATURITY. Fledging period 10–17 days (Tast 1970; Handtke and Witsack 1972); average over 3 years 11·3–13·2 days, $n = 52$ (Handtke and Witsack 1972); in Alps, 14·6 days, $n = 12$ (Frey 1989a). For development of young, see Rheinwald (1973), Hachfeld (1979), and Frey (1989a). Young independent *c.* 14 days after leaving nest (Haartman 1969). Age of first breeding 1 year (C S Roselaar). BREEDING SUCCESS. In southern England, 58% of 1227 rural, and 65% of 178 suburban or urban nests failed completely (no young flew); failure rate dropped over season from *c.* 75% to *c.* 45%; excluding total failures, 85% of eggs laid produced fledged young, rate approximately same for all habitats (Snow and Mayer-Gross 1967). See also Newton (1972). In northern France, of 736 eggs in 154 nests, 64% hatched and 57% produced fledged young, giving 4·4 per successful nest ($n = 95$), 2·7 overall. Most egg loss due to humans, and 21% failed to hatch; 50% of nestlings lost died in nest, and severe drought in one year probably responsible. (Godin *et al.* 1977.) Similar rates of success recorded in Finland (Tast 1970), at Halberstadt (Handtke and Witsack 1972), Niedersachsen (Hachfeld 1979; Meineke 1979), Rheinland (Mildenberger and Schulze-Hagen 1973), and Moscow region (Shurupov 1986). In Swiss Alps, 38% of 353 eggs over 3 years produced fledged young (in one year only 17%), giving 4·4 per successful nest ($n = 31$), 1·6 overall ($n = 85$); of 178 eggs lost, 53% failed to hatch, and 75% of 57 nestling deaths probably occurred during hatching; fledging success fairly constant, so hatching success determines failure rate; *C. cannabina* here at altitudinal limit of range where low temperatures can interfere with egg development and cause hormonal disturbance in adults (Frey 1989*a*, which see for many details and comparisons). Cool, wet years can have severe effect, apparently because of increased predator activity rather than direct reduction of brood viability (Handtke and Witsack 1972). Higher density in nesting groups probably leads to more predation and reduced success (Eybert 1980; Shurupov 1986). Low nest-site especially vulnerable to predators, which in above studies were mainly crows (Corvidae), weasels, martens, etc. (Mustelidae), fox *Vulpes*, rodents, cats, man, and, in places, Red-backed Shrike *Lanius collurio*. See also Orford (1973) for poor success rate of nests in colonies of Twite *C. flavirostris* on moorland, northern England; in contrast to *C. flavirostris*, in this marginal habitat many nests lost to severe weather and predators.          BH

**Plumages.** (nominate *cannabina*). ADULT MALE. In fresh plumage (autumn), forehead and forecrown closely streaked dark brown-grey and olive-brown or buff, each feather-centre with dull dark red spot, fully concealed; streaks gradually less contrastingly brown-grey and olive-brown on hindcrown; nape to upper mantle olive-brown with faint paler buff-olive streaking. Lower mantle, scapulars, and back deep rufous-brown or chestnut-brown, feathers with ill-defined buff or pale cinnamon fringes at sides. Upper rump rufous-brown, merging to olive-buff on shorter upper tail-coverts, sometimes marked with broad but ill-defined black streaks on centre, some white of feather-bases often partly visible or shining through on side of rump; longer upper tail-coverts black with broad grey-buff to buff-white fringes. Side of head and neck dull brown-grey with olive tinge; bristles at base of bill, short but contrasting supercilium, and patch below eye cinnamon-buff, forming contrasting patch round eye, broken by dark brown stripe on lore and short brown stripe behind eye; ill-defined buff patch at rear of cheek; side of neck faintly mottled olive-buff. Chin and central throat pale buff, heavily marked dull black, forming streaked patch; stripe at side of throat uniform buff, bar across lower throat buff with olive-brown wash at lower border and dusky grey triangular marks. Chest and upper belly pale buff or cream-white, each feather with broad dull rose-red spot on centre, partly concealed by cream feather-fringes, generally well-visible; side of breast, flank, and side of belly buff-brown or cinnamon with paler buff feather-tips and dull black or dark olive-brown shaft-streaks. Mid-belly, vent, and under tail-coverts white with pink-buff or cream tinge, longest under tail-coverts with black central mark. Tail black; central pair (t1) narrowly fringed cinnamon, soon bleaching to buff-grey; t2–t5 with contrastingly white fringe *c*. 1 mm wide along outer web, pale brown to off-white fringe *c*. 1 mm wide along tip, and contrastingly white border along inner web; t6 with narrow off-white outer edge and broad white border *c*. 4 mm wide along inner web. Primaries black, basal and middle portions of outer webs with contrastingly white fringe (widest on innermost, faint along entire length of p9), basal inner webs bordered light grey, tips of inner primaries with rather contrasting brown-white fringe; inner webs and tips of secondaries dull black, bases of outer webs dark grey or brown-grey merging into dark rufous-brown on middle portions of outer webs, tips

contrastingly fringed pale grey-buff or off-white. Greater upper primary coverts and bastard wing black, outer web of coverts narrowly fringed grey, shorter feathers of bastard wing fringed grey-brown. Tertials and greater upper wing-coverts deep rufous-brown or chestnut-brown, outer webs and tips narrowly fringed pale cinnamon-buff, on tip soon bleaching to pale buff-grey; concealed inner webs of outer coverts and longer tertials dark grey or dull black, shading to greyish-cinnamon on tip. Lesser and median upper wing-coverts chestnut-brown to rufous-chestnut with narrow cinnamon-buff fringe on tips. Under wing-coverts and axillaries white, longer coverts light grey, marginal coverts partly dotted black. *Bleaching and wear* have marked effect. By mid-winter, buff of head and body bleached to off-white, streaking on crown, lower mantle, scapulars and flank, patch round eye, and spot on cheek more contrasting, red on forehead partly visible but still dull, red on chest and upper belly more exposed, appearing scalloped by white feather-fringes about February–March, lower belly to under tail-coverts whiter. By April–May, top and side of head and neck dark grey, forecrown with contrasting carmine patch, hindcrown closely streaked black, feathering on face, short stripes above and below eye, and patch on cheek off-white; mantle, scapulars, and back chestnut-brown, sometimes with narrow black shaft-streaks or traces of grey-buff feather-fringes; lower rump and upper tail-coverts contrastingly streaked black and white, tinged buff to variable extent, black streaks sometimes partly reduced; chin and throat off-white with contrasting dark streaks or spots on centre and lower border; chest and side of upper belly uniform glossy red (colour varying from dark vinous-red through bright carmine and rosy-red to orange-red, depending partly on abrasion); side of breast, flank, and lower side of belly tawny-cinnamon or tawny-buff; mid-belly to under tail-coverts dirty white; pale tips of tail- and flight-feathers and on upper wing-coverts worn off, white outer fringes of tail and primaries strongly contrasting. When extremely abraded, July–August, red on cap and chest partly worn off, blue-grey of feather-bases partly visible on head, neck, and underparts, mantle, scapulars, and upper wing-coverts bleached to tawny-brown, less uniform dark and saturated than in spring. ADULT FEMALE. In fresh plumage (autumn), upperparts and side of head and neck rather as in adult ♂, but cap more heavily streaked black and buff, forecrown without concealed red feather-centres; hindneck streaked pale grey-buff and sooty, less uniform olive; mantle, scapulars, and back more evenly streaked dull black, olive-brown, warm brown, and buff-brown, without rufous or chestnut tinge on feather-centres. Buff marks on side of head and pale buff to off-white chin and throat with dull black dots or streaks on centre and at lower border as in adult ♂. Chest, side of breast, flank, upper belly, and side of belly warm buff or cinnamon-buff, each feather with distinct black-brown or sooty streak on centre, without trace of broad red spot, streaks on side of breast and flank more distinct than in adult ♂. Mid-belly to under tail-coverts as adult ♂. Tail, primaries, primary coverts, and bastard wing as adult ♂, white fringes of tail and primaries not obviously narrower, except at base of inner primaries (see Svensson 1992); secondaries often less tinged grey on bases of outer webs and less broadly fringed with duller brown or buff-brown on middle portion of outer web. Tertials and upper wing-coverts as adult ♂, but ground-colour deep olive- or sepia-brown, less chestnut. *In worn plumage* (spring and summer), no red in plumage; top and side of head and neck dark brown-grey with darker streaking on cap and dirty white feathering at base of bill, round eye (broken in front and behind), and at centre of cheek; mantle, scapulars, back, upper wing-coverts, and tertials brown with cinnamon cast, partly marked with narrow dusky shaft-streaks, tips of feathers partly worn to buff-grey; ground-colour of underparts dirty white, washed tawny on flank and side of belly, heavily marked with dusky streaks, except for side of throat and mid-belly to under tail-coverts. NESTLING. Down fairly long and plentiful, restricted to tufts on head, back, upper wing, and thigh, with traces on belly; smoke-grey. For development, see Heinroth and Heinroth (1924–6) and Rheinwald (1973). JUVEN-ILE. Upperparts light buff-brown (when fresh) to pale grey-buff (when worn), evenly streaked sooty grey on cap, mantle, and scapulars, closely similar to adult ♀, but ground-colour of mantle and scapulars paler, more buffy, and streaks blacker, more sharply defined; feathering of rump short and fluffy, buff or white with dusky grey marks. Side of head and neck finely mottled and streaked brown and pale grey-buff, not as dark as adult ♀, and without distinct pale pattern, except for ill-defined buff lore and supercilium. Underparts pale cream-buff to off-white, streaked dusky grey on chest, side of breast, flank, and side of belly (streaks shorter and narrower than in adult ♀), chin and throat off-white, without black marks on centre or with faint grey specks only. Tail as adult, but each feather tapering to pointed tip, less broad and rounded; black pattern on centre of each ends in sharp point, not rounded. Flight-feathers as adult, but white fringes of primaries sometimes slightly narrower, especially in ♀; basal outer webs of ♂ tend to show more grey than ♀ and brown middle portions tend to be broader and warmer tinge, as in adult ♂, but much overlap; narrow fringes of primary coverts more brown-grey, less pure grey than in adult, usually extending into clear narrow fringe along tip (unlike adult); tertials and upper wing-coverts as adult ♀, but broad and ill-defined tips cinnamon-buff, partly bleaching to off-white, centres of coverts with black shaft-streaks; no clear difference from adult in extent and pattern of black on outer greater coverts. FIRST ADULT MALE. 2 types, more advanced birds (perhaps early-hatched) mostly like adult ♂, with red on forecrown and chest (partly concealed in autumn), more retarded birds more heavily streaked on body than adult ♂, with feather-centres on forecrown and chest grey-brown instead of red, sometimes restricted in extent and then hardly differing from ♀. Retarded birds comprise 32% of sample of 34 1st adult ♂♂ from Netherlands examined (ZMA); adult ♂♂ exceptionally brown instead of red in wild, commonly in captivity. In both types of 1st adult, at least most juvenile secondaries and greater upper primary coverts retained; in north of geographical range, varying number of juvenile outer greater coverts retained (showing distinct difference in colour and abrasion from new neighbouring ones), as well as all juvenile flight-feathers and many or all tail-feathers and tertials (if some tertials or t1 new, these contrasting in shape and abrasion with old ones, unlike adult); in south of range, greater coverts, tertials, and sometimes all tail new, as well as variable number of flight-feathers (mainly central primaries, sometimes all flight-feathers except central secondaries), showing mutual contrast in abrasion, unlike adult. Head and body of advanced type as in adult ♀, but red on chest sometimes less bright and not reaching as far down on belly, sometimes more concealed when plumage fresh, especially on forecrown. Fresh head and body of retarded type as in adult ♀, mantle and scapulars streaked dull black, hardly tinged chestnut; dusky streaks on mid-chest and upper belly narrower and less elongate than in adult ♀, showing as dusky triangular spots; feathers of at least lower chest with broad round-ended brown-grey or pink-brown mark, unlike ♀; in worn plumage, brown feather-centres of forecrown and chest exposed, glossy orange-brown or golden-brown. FIRST ADULT FEMALE. Like adult ♀, but part of juvenile feathering replaced, as in 1st adult ♂, amount varying with latitude; characters and contrast of juvenile plumage as in 1st adult ♂;

like these, ageing often difficult once plumage heavily abraded in spring. White fringes of primaries and tail sometimes largely worn off, especially on inner tail-feathers and outer primaries.

**Bare parts.** ADULT, FIRST ADULT. Iris dark brown to black-brown. Bill dark horn-grey to greyish-black, often tinged paler grey, green, olive, flesh-grey, or grey-white at base and on lower mandible. Leg and foot dark brown-flesh to horn-black with slight purple or flesh tinge. NESTLING. Mouth bright rose-red, edges of upper mandible tinged light blue-grey to whitish inside; gape-flanges pink-yellow to pale yellow. Bare skin, as well as bill and leg, flesh-pink at hatching, bill and leg gradually darkening to grey. JUVENILE. At fledging, bill, leg, and foot grey with pale yellow, pink, or ivory-white tinge, darkening to adult colour during post-juvenile moult. (Heinroth and Heinroth 1924–6; Witherby *et al.* 1938; BMNH, RMNH; A J van Loon.)

**Moults.** ADULT POST-BREEDING. Complete; primaries descendent. In Britain, starts with shedding of p1 early July to mid-August, occasionally from mid-June or up to late August; primaries completed with regrowth of p9–p10 after *c.* 70 days between early September and late October, sometimes from late August; on average, primary moult 27 July to 2 October (Newton 1968; Ginn and Melville 1983). Primary moult occasionally arrested (Herroelen 1980). In Netherlands, starts with p1 mid-July to late August, completed from late August onwards (RMNH, ZMA, A J van Loon). In north-west Russia, starts late June to late July, all birds in full moult August, completed before mid-October (Rymkevich 1990). In Yugoslavia, moult late August to mid-October (Stresemann 1920). In Iran, starts late July or early August, completed early October (Stresemann 1928; Vaurie 1949b). POST-JUVENILE. Partial; extent depending on latitude. In Netherlands and Germany, head, body, lesser, median, and usually (in 93% of 30 birds) a number of greater coverts, with average 2·8 (28) 0–6 juvenile outer coverts retained; in 10% of 30 autumn birds, also all tertials; in 13% 1–2 tertials; in 30% of 30 birds also t1, in 3% t1–t3, in 7% all tail; starts mid-July to late September, completed late August to early November (RMNH, ZMA). Of 25 1st adults from Canary Islands and Mediterranean basin, moult often more extensive, and apparently occasionally complete (Ginn and Melville 1983), but latter birds indistinguishable from adult. In sample of 25 birds examined from south of range, all birds retained juvenile greater upper primary coverts (in 4 birds, a few new); in 9 birds 1–3(–6) outer greater coverts old (in remaining 16 birds, all new); in 4 tail old, in 8 tail partly new (usually t1, t2, and t6), in others all new; in 12 all flight-feathers old, in 3 all primaries new but at least central secondaries old; in 10 a series of central or outer primaries new, series most often including p6 (in all 10 birds), p7 (in 9), p8 (in 7), p5 and p9 (in 5), rarely p4 (in 2) or p3 (in 1) (RMNH, ZMA: C S Roselaar). For similar moult on Balearic Islands, see Mester and Prünte (1982). Depending on hatching date, moult in south of range June–October (BMNH, RMNH, ZMA).

**Measurements.** ADULT, FIRST ADULT. Nominate *cannabina*. Netherlands, April–November; skins (RMNH, ZMA). Bill (S) to skull, bill (N) to distal corner of nostril; exposed culmen on average 3·4 less than bill (S).

| | ♂ | | ♀ | |
|---|---|---|---|---|
| WING | 80·8 (1·59; 75) | 78–85 | 78·7 (1·75; 42) | 76–82 |
| TAIL | 52·6 (2·01; 30) | 49–56 | 52·6 (2·00; 25) | 49–57 |
| BILL (S) | 13·0 (0·55; 41) | 11·8–14·0 | 13·0 (0·61; 12) | 11·9–13·8 |
| BILL (N) | 8·2 (0·37; 19) | 7·5–8·8 | 7·9 (0·40; 12) | 7·4–8·5 |
| TARSUS | 16·1 (0·52; 27) | 14·9–16·9 | 15·8 (0·63; 23) | 14·6–16·9 |

Sex differences significant for wing. Retained juvenile wing of 1st adult shorter than in older birds; thus, wing, adult ♂ 83·3

(1·04; 7) 82–85, 1st adult ♂ 80·8 (1·46; 36) 78–84 (ZMA). Bill in summer slightly longer than in rest of year: bill (S), ♂, May–August 13·2 (0·60; 16) 12·3–14·0, September–April 12·9 (0·49; 25) 11·8–13·6 (ZMA).

*C. c. guentheri*. Madeira, all year; skins (BMNH, ZMA: A J van Loon).

| | ♂ | | | ♀ | | |
|---|---|---|---|---|---|---|
| WING | 75·0 (1·52; 16) | 72–78 | | 73·4 (1·78; 10) | 70–76 | |
| TAIL | 49·8 (1·54; 16) | 47–52 | | 49·2 (1·65; 10) | 48–53 | |
| BILL (S) | 12·6 (0·33; 11) | 11·9–13·1 | | 12·3 (0·41; 8) | 11·7–13·0 | |
| BILL (N) | 7·9 (0·37; 11) | 7·2–8·6 | | 7·6 (0·28; 8) | 7·3–8·1 | |
| TARSUS | 14·9 (0·36; 12) | 14·4–15·6 | | 14·7 (0·46; 10) | 14·0–15·4 | |

Sex differences significant for wing.

*C. c. meadewaldoi*. Tenerife, Gran Canaria, and Gomera (Canary Islands), all year; skins (BMNH, RMNH, ZMA: A J van Loon).

| | ♂ | | | ♀ | | |
|---|---|---|---|---|---|---|
| WING | 77·4 (1·84; 18) | 74–80 | | 75·0 (1·29; 14) | 73–78 | |
| TAIL | 50·6 (1·47; 16) | 48–54 | | 49·7 (1·58; 14) | 47–53 | |
| BILL (S) | 12·8 (0·50; 15) | 11·8–13·4 | | 12·6 (0·40; 12) | 12·3–13·4 | |
| BILL (N) | 8·3 (0·36; 16) | 7·5–8·5 | | 8·1 (0·27; 12) | 7·8–8·7 | |
| TARSUS | 15·6 (0·47; 17) | 14·7–16·2 | | 15·6 (0·58; 13) | 14·4–16·5 | |

Sex differences significant for wing.

*C. c. harterti*. Fuerteventura and Lanzarote (Canary Islands), all year; skins (BMNH, RMNH, ZMA: A J van Loon).

| | ♂ | | | ♀ | | |
|---|---|---|---|---|---|---|
| WING | 75·4 (1·36; 22) | 73–78 | | 73·6 (1·36; 19) | 70–76 | |
| TAIL | 49·3 (1·32; 21) | 47–52 | | 48·8 (1·85; 18) | 45–52 | |
| BILL (S) | 12·4 (0·37; 12) | 11·7–12·9 | | 11·9 (0·44; 14) | 11·1–12·6 | |
| BILL (N) | 8·1 (0·34; 13) | 7·6–8·7 | | 7·8 (0·42; 15) | 6·7–8·4 | |
| TARSUS | 15·2 (0·48; 14) | 14·6–16·5 | | 15·0 (0·51; 15) | 14·2–15·8 | |

Sex differences significant for wing and bill.

Wing and bill of (1) *mediterranea*, southern Europe (Portugal to western Yugoslavia), (2) race unknown, Morocco to Tunisia; other measurements combined, April–July; skins (RMNH, ZMA).

| | ♂ | | | ♀ | | |
|---|---|---|---|---|---|---|
| WING (1) | 76·8 (1·55; 7) | 74–79 | | 74·7 (1·60; 5) | 73–77 | |
| (2) | 76·5 (1·76; 7) | 74–79 | | 76·5 (—; 2) | 76–77 | |
| TAIL | 50·6 (1·62; 10) | 48–53 | | 49·5 (1·64; 5) | 47–52 | |
| BILL (S) (1) | 12·8 (0·38; 7) | 12·3–13·3 | | 12·4 (1·56; 3) | 11·9–13·0 | |
| (2) | 13·2 (0·40; 7) | 12·7–13·9 | | 12·0 (—; 1) | — | |
| BILL (N) (1) | 8·1 (0·40; 7) | 7·5–8·6 | | 7·9 (0·55; 3) | 7·4–8·5 | |
| (2) | 8·4 (0·36; 7) | 7·9–9·0 | | 8·0 (—; 1) | — | |
| TARSUS | 15·7 (0·28; 10) | 15·2–16·1 | | 15·4 (0·36; 5) | 15·1–16·0 | |

Sex differences not significant, but samples small.

Wing and bill of *bella*: (1) Cyprus, Levant, Turkey, Caucasus, and Crimea, (2) Tien Shan and Tarbagatay (central Asia); March–July, skins (RMNH, ZMA).

| | ♂ | | | ♀ | | |
|---|---|---|---|---|---|---|
| WING (1) | 81·0 (1·58; 5) | 79–84 | | 78·7 (1·04; 3) | 77–80 | |
| (2) | 82·0 (3·11; 4) | 78–85 | | 82·0 (—; 1) | — | |
| BILL (S) (1) | 13·3 (0·21; 5) | 13·0–13·5 | | 13·3 (0·50; 3) | 12·8–13·8 | |
| (2) | 13·0 (0·50; 4) | 12·5–13·5 | | 12·9 (—; 1) | — | |
| BILL (N) (1) | 8·6 (0·26; 5) | 8·3–9·0 | | 8·5 (0·72; 3) | 7·7–9·0 | |
| (2) | 8·7 (0·52; 4) | 8·2–9·2 | | 8·3 (—; 1) | — | |

Wing, sexes combined. Nominate *cannabina*: 78–84(–86) (n = c. 90). *C. c. bella*: central Asia (79–)81–87 (n = 31), Asia Minor and Caucasus 78–83 (n = 7). *C. c. mediterranea*: Dalmatia 74–80 (n = 10), southern Italy 76–80 (n = 8), Corsica and Sardinia 75–80 (n = 15), Balearic Islands 73–79 (n = 35) (Jordans 1924).

Wing. Nominate *cannabina*. (1) Pskov (western Russia) and Sweden (Vaurie 1956a). Germany: (2) live birds, ages combined (Creutz 1967), (3) skins (Niethammer 1937; Eck 1985a). Race unknown, but probably including some wintering nominate *cannabina*. (4) Southern Yugoslavia and north-west Greece, mainly winter (Stresemann 1920; Makatsch 1950). (5) Florence (Italy) (Eck 1985a). (6) Iberia and north-west Africa (Vaurie 1956a). (7) Migrants, Tarifa (southern Spain) (Asensio 1984). *C. c. mediterranea*. (8) Balearic Islands (Stresemann 1920). (9) Crete and Karpathos (Greece) (Niethammer 1942; Kinzelbach and Martens 1965). *C. c. bella*. (10) Turkey, summer (Stresemann 1920; Jordans and Steinbacher 1948; Kumerloeve 1964a, 1967a, 1969a; Rokitansky and Schifter 1971; Vauk 1973). (11) Cyprus, live breeders (Flint and Stewart 1992). (12) Levant (Vaurie 1956a). (13) Caucasus area and northern Iran (Stresemann 1928; Paludan 1940; Vaurie 1949b; Nicht 1961). (14) South-west and southern Iran (Paludan 1938; Vaurie 1949b). (15) South-east Kazakhstan and Kirgiziya (Vaurie 1956a).

| | ♂ | | | ♀ | | |
|---|---|---|---|---|---|---|
| (1) | 82·0 (1·84; 14) | 79–86 | | — (—; —) | — | |
| (2) | 82·3 (—; 41) | 79–86 | | 80·2 (—; 24) | 78–84 | |
| (3) | 80·7 (—; 62) | 77–86 | | 79·2 (—; 10) | 77–82 | |
| (4) | 80·3 (1·53; 45) | 77–83 | | 77·4 (1·54; 18) | 75–80 | |
| (5) | 79·6 (1·73; 11) | 77–82 | | — (—; —) | — | |
| (6) | 79·3 (1·67; 21) | 75–82 | | — (—; —) | — | |
| (7) | 79·6 (1·64; 44) | — | | 76·8 (1·71; 52) | — | |
| (8) | 76·1 (1·92; 11) | 74–79 | | 75·0 (—; 1) | — | |
| (9) | 76·7 (1·89; 4) | 74–78 | | — (—; —) | — | |
| (10) | 80·5 (1·53; 22) | 78–83 | | 78·2 (1·34; 12) | 76–80 | |
| (11) | 81·0 (—; 12) | 77–85 | | 79·9 (—; 24) | 76–85 | |
| (12) | 81·1 (1·45; 9) | 79–83 | | — (—; —) | — | |
| (13) | 81·9 (—; 26) | 79–85 | | 79·2 (0·98; 6) | 78–80 | |
| (14) | 82·2 (—; 40) | 79–86 | | 81·0 (—; 1) | — | |
| (15) | 83·4 (1·75; 11) | 80–86 | | — (—; —) | — | |

Karpathos (Greece), live birds, autumn: 75·0 (55) 71–79 (Kinzelbach and Martens 1965).

NESTLING. For growth, see Rheinwald (1973) and Frey (1989a).

**Weights.** ADULT, FIRST ADULT. Nominate *cannabina*. Netherlands: (1) April–May, (2) June–August, (3) September–November (A J van Loon; ZMA). (4) South-east Germany, March (Creutz 1967). (5) Exhausted birds (Weigold 1926; Harris 1962; ZMA). *C. c. bella*. (6) Turkey and Armeniya, mainly May–July (Nicht 1961; Kumerloeve 1964a, 1967a, 1969a, 1970a; Rokitansky and Schifter 1971; Vauk 1973). (7) Cyprus, summer (Flint and Stewart 1992). (8) Iran, May–July (Paludan 1938, 1940; Desfayes and Praz 1978). (9) Kazakhstan (Korelov *et al.* 1974).

| | ♂ | | | ♀ | | |
|---|---|---|---|---|---|---|
| (1) | 19·4 (1·49; 15) | 17·5–23·0 | | 18·5 (1·79; 22) | 15·4–20·5 | |
| (2) | 17·8 (0·77; 12) | 16·8–19·0 | | 19·7 (1·87; 3) | 18·2–21·8 | |
| (3) | 20·0 (2·14; 9) | 15·9–22·2 | | 17·3 (2·52; 3) | 15·0–20·0 | |
| (4) | 21·2 (—; 41) | 18·6–25·7 | | 19·4 (—; 24) | 15·8–21·7 | |
| (5) | 13·9 (0·75; 8) | 13–15 | | 13·4 (1·62; 6) | 11·5–16·0 | |
| (6) | 19·8 (1·38; 11) | 18–22 | | 19·2 (1·36; 3) | 17·5–21 | |
| (7) | 16·9 (—; 21) | 13·8–22·2 | | 16·7 (—; 24) | 14·5–18·5 | |
| (8) | 17·6 (0·99; 4) | 16·8–18·9 | | 18·0 (1·88; 4) | 16·3–20·6 | |
| (9) | 19·2 (—; 25) | 17–24·8 | | 19·5 (—; 21) | 16·1–22·7 | |

Probable nominate *cannabina*. North-west Greece, winter: ♂ 19·8 (2·22; 4) 17–22, ♀ 16 (Makatsch 1950). Sénégal, March: ♂ 24 (Jarry and Larigauderie 1974).

Probable *mediterranea*. Sardinia, February: ♀ 14 (Demartis 1987).

NESTLING. For growth curves, see Rheinwald (1973) and Frey (1989a).

**Structure.** Wing rather short, broad at base, tip bluntly pointed. 10 primaries: p8 longest, p9 0–0·5(–1) shorter, p7 0·5–1·5, p6 3–6, p5 9–13, p4 15–19, p3 18–22, p2 21–26; in nominate *cannabina* and *bella*, p1 26·1 (35) 23–29 shorter, in races of Canary Islands and western Mediterranean area p1 23·6 (25) 21–26 shorter; p10 strongly reduced, a tiny pin concealed below outermost upper primary covert, 53–59 shorter than p8, 8–11 shorter than longest upper primary covert. Outer web of p6–p8 emarginated; inner web of (p7–)p8–p9 with notch. Tip of longest tertial reaches to tip of p3 when fresh. Tail rather short, tip forked; 12 feathers, t1 6·6 (21) 4–10 shorter than t1. Bill strong, conical; depth at feathering of base 7·1 (35) 6·7–7·4 in nominate *cannabina*, width of lower mandible at base 6·4 (35) 6·0–6·8; average of depth and width in *mediterranea* 7·0 and 6·4 respectively (n = 10), in Canary Islands races 7·1 and 6·5 (n = 20), ranges as in nominate *cannabina*; in north-west Africa, average depth 7·6, width 6·8 (n = 9), in *bella* from Middle East depth 7·5, width 6·9 (n = 8), range of depth in these populations 7·2–8·0, width 6·5–7·4. Culmen straight at base, slightly decurved at tip; cutting edges slightly decurved, with faint 'tooth' at base; gonys straight or slightly concave; tip of bill sharp, slightly compressed laterally. For structure of red feathers of ♂, see Görnitz (1927). Tarsus and toes rather short, slender. Middle toe with claw 17·1 (10) 15·5–18·0; outer toe with claw *c.* 68% of middle with claw, inner *c.* 70%, hind *c.* 83%. Remainder of structure as in Goldfinch *C. carduelis*.

**Geographical variation.** Rather slight; involves size (length of wing, tail, or tarsus), relative length, depth, and width of bill, and depth of colour, especially grey of hindneck of ♂ and brown of mantle and scapulars. Colours strongly affected by bleaching and wear, especially in southern populations; variation also confused by numerous nominate *cannabina* wintering in breeding range of Mediterranean and Middle East races. 4 groups separable on size and bill depth: (1) nominate *cannabina* (northern Eurasia from Ireland to western Siberia, south to Pyrénées, northern Italy, northern Balkans, and Ukraine) and *autochthona* (Scotland) with rather long wing (average of ♂ *c.* 80–82, of ♀ *c.* 78–80; see also Measurements), and rather slender bill, depth *c.* 7·0–7·2 (see Structure); (2) *bella* (Turkey, Cyprus, Levant, and Crimea through Caucasus area and Iran to south-west Turkmeniya) with wing length as nominate *cannabina* or slightly longer but bill thicker at base, depth *c.* 7·5–7·6; (3) Atlantic island races (*guentheri* on Madeira, *meadewaldoi* on western Canary Islands east to Gran Canaria, *harterti* on eastern Canary Islands), and *mediterranea* (Iberia, southern Italy, Dalmatia, Albania, Greece, and many islands in Mediterranean from Balearics to Crete and Karpathos in Greece) with rather short wing (average of ♂ *c.* 75–77, of ♀ *c.* 73–75) and rather slender bill (as nominate *cannabina*); (4) populations of North Africa, as yet unnamed (occurring Morocco and mountains of Algeria and neighbouring Tunisia, perhaps also in lowlands of Algeria and Tunisia and in northern Libya) with short wing (as in 3rd group) and thick bill (as in 2nd). Multivariate character analysis of wing, bill, and tail length, bill depth, and bill width clearly separates these groups. Groups 3 and 4 also often separable from groups 1 and 2 by shorter wing-tip (p1–p8 mainly 25 mm or more in groups 1 and 2, mainly 24·5 mm or less in groups 3 and 4: see Structure) and often in extent of post-juvenile moult (see Moults).

2 distinct groups separable on colour, corresponding with grouping on bill depth in samples above: (1) darker, more slender-billed nominate *cannabina*, *mediterranea*, and Atlantic races; (2) paler, thicker-billed North African population and *bella*. In *bella*, top and side of head and neck of ♂ distinctly paler than in nominate *cannabina*, mantle, scapulars, and back cinnamon-brown or hazel (less saturated chestnut-brown than in nominate *cannabina*), back and upper rump sometimes with some

rosy-red feather-centres; lower rump and upper tail-coverts white, black streaks narrower or almost absent; red of underparts paler, more rosy-pink, extending less far down, only feather-fringes brighter fiery-red when plumage worn, less uniform carmine than in nominate *cannabina* (extent and colour of pink somewhat comparable with that of Arctic Redpoll *C. hornemanni*, that of nominate *cannabina* with Redpoll *C. flammea*); red patch on forecrown smaller and paler; flank rather extensively tawny-cinnamon; head of ♀ paler and greyer than ♀ nominate *cannabina*, upperparts more buff-brown when fresh (less warm brown), more brown-grey when worn, dark streaks on underparts less blackish, slightly narrower. Some variation within *bella*, some birds averaging paler and these considered separate races (*taurica* Kudashev, 1916, in Crimea, *persica* Kudashev, 1916, in Iran, *fringillirostris* Bonaparte and Schlegel, 1850, in central Asia), others slightly darker (e.g. northern Turkey and Caucasus), but difference slight, overlap large, and none recognized here. *C. c. bella* probably grades into nominate *cannabina* in eastern Balkans and on islands off western Turkey. Population from North Africa closely similar in colour and bill depth to *bella*, but wing shorter; lower rump of ♂ more intensely spotted dusky olive or black, flank paler and less extensively tawny-buff. *C. c. mediterranea* slightly paler than nominate *cannabina*, but much less so than *bella* or North African birds, differing mainly from nominate *cannabina* in shorter wing; dark streaks on head of ♂ slightly more contrasting, central nape with pale cream-buff streaks tending to form pale spot, mantle and scapulars saturated rufous-brown, slightly less chestnut than in nominate *cannabina*, flank dark rufous-cinnamon; some variation between populations, those of Portugal slightly darker than others, those of Balearic Islands slightly paler. *C. c. mediterranea* probably grades into nominate *cannabina* over wide area, small birds of southern Iberia, Balearic Islands, Sardinia, Sicily, southern Italy, southern Dalmatia, western Greece, and Crete forming end of cline of decreasing size. Scottish race *autochthona* on average darker grey on hindneck in ♂ than ♂ nominate *cannabina*, mantle and scapulars rather dull dark brown, rather heavily streaked; probably grades clinally into nominate *cannabina* towards south. *C. c. meadewaldoi* from western Canary Islands inseparable from *mediterranea* on size and scarcely separable on colour; mantle and scapulars of ♂ rather dark rufous-brown, slightly darker than nominate *cannabina*, but paler than *guentheri*, grey of hindneck slightly darker and more uniform, flanks more extensively rufous-cinnamon on average than in *mediterranea* or nominate *cannabina*; ♀ close to ♀ *mediterranea*. *C. c. harterti* from eastern Canary Islands pale; mantle, scapulars, and flank of ♂ rufous-cinnamon (deeper cinnamon and more saturated than in North Africa, cinnamon on flank deeper and more extensive); hindneck medium grey with slight buff tinge; streaking on underparts of ♀ narrower and more restricted than in other races. *C. c. guentheri* from Madeira, Porto Santo, and Desertas darkest race; mantle and scapulars of ♂ dark rufous-brown, extending to rump; flank extensively deep rufous-cinnamon; as in other Atlantic races, red of forecrown and chest bright ruby-red, not as dark scarlet or carmine as nominate *cannabina*; underparts of ♀ markedly heavier and more extensively streaked than ♀♀ of other races. For geographical variation, see also Jordans (1924), Hartert and Steinbacher (1932–8), and Vaurie (1949b, 1956a, 1959).

Forms superspecies with Yemen Linnet *C. yemenensis* from south-west Arabia and perhaps with Warsangli Linnet *C. johannis* from north-east Somalia (Hall and Moreau 1970); for *C. yemenensis*, see Bowden and Brooks (1987). Forms species-group with Twite *C. flavirostris*.                    CSR

# *Carduelis flavirostris* Twite

Du. Frater    Fr. Linotte à bec jaune    Ge. Berghänfling
Ru. Горная чечетка    Sp. Pardillo piquigualdo    Sw. Vinter hämpling

*Fringilla flavirostris* Linnaeus, 1758

Polytypic. Nominate *flavirostris* (Linnaeus, 1758), Norway, adjacent part of Sweden, and Kola peninsula (north-west Russia); *pipilans* (Latham, 1787), Britain and Ireland, except Outer Hebrides; *bensonorum* (Meinertzhagen, 1934), Outer Hebrides (Scotland); *brevirostris* (Moore, 1856), eastern Turkey and Caucasus to northern Iran and perhaps south-west Iran; *kirghizorum* (Sushkin, 1925), Volga-Ural plains east through northern Kazakhstan (47-51°N) to Irtysh and Ayaguz. Extralimital: *korejevi* (Zarudny and Härms, 1914), eastern Kazakhstan from Kalbinskiy mountains and southern Altai through Tarbagatay to central Tien Shan north of Issyk-Kul'; *altaica* (Sushkin, 1925), central and eastern Altai (east from Chibit) to Tuva area (north to southern slope of western Sayan), south to Goviraltay; *pamirensis* (Zarudny and Härms, 1925), eastern Zeravshan and Alay ranges through Darvaz, Pamir, and Afghanistan to Gilgit and Baltistan (Pakistan); *montanella* (Hume, 1873), fringes of Tarim basin from Kokshaal-Tau and Terskey Alatau (Tien Shan), through Terek Tau and Sarykol (eastern Pamirs) to Karakorum, Kun Lun, and Astin Tagh ranges; *miniakensis* (Jacobi, 1923), Nan Shan mountains and western Kansu through eastern Tsinghai to western Szechwan; *rufostrigata* (Walton, 1905), Himalayas from Ladakh to south-east Tibet.

**Field characters.** 14 cm; wing-span 22-24 cm. Close in size to Linnet *C. cannabina* but with over 10% longer tail and stubbier bill creating subtly different outline. Small, robust, but somewhat attenuated finch; counterpart of *C. cannabina* in mountains of south-west Asia and coastal periphery and hills of north-west Europe. Plumage pattern strongly recalls ♀ or juvenile *C. cannabina* but ground of upperparts more tawny in tone and of face, throat, chest, and flanks distinctly warmer buff, becoming even orange around eye and under bill, while belly clean white. Shares white wing-panel and tail-sides with *C. cannabina* but pale buff tips to median and greater coverts create more noticeable wing-bars. Bill grey in breeding adult but otherwise straw-yellow, recalling not *C. cannabina* but Redpoll *C. flammea*. ♂ has long pink rump, recalling *C. flammea*. Flight silhouette longer than *C. cannabina*. One call diagnostic. 5 races in west Palearctic, 1 distinctive in adult ♂.

ADULT MALE. (1) North-west European race, nominate *flavirostris*. Moults: July-October (complete). Ground-colour of head, upperparts, wing-coverts, and underparts (except upper and under tail-coverts) buff, with distinctive yellowish, even orange tone on face and throat; black or black-brown marks create spotted crown and well streaked ear-coverts, back, breast, chest, and flanks, while warm buff tips to median coverts and pale greyish-buff tips to greater coverts form indistinct upper and fairly distinct lower wing-bar. Long rump pink, when fresh tipped buff and softly marked brown. Flight-feathers brown-black, with broad buff-brown fringes to tertials and conspicuous white edges to inner primaries, last forming panel like that of *C. cannabina*. Underwing buff-white. Tail brown-black, with 2 central pairs margined buff and rest edged white; white not as prominent as in *C. cannabina*. Belly, vent, and under tail-coverts almost pure white, with just a few

dark streaks on under tail-coverts; chest and flanks contrast strongly with belly and vent, unlike *C. cannabina*; combination of white vent and pink rump makes bird appear noticeably pale around base of tail. With wear, upperparts lose buff margins and become noticeably darker, nape takes on grey hue, rump becomes clean pink, and streaks on underparts strengthen. Bill stubby but sharp; bright straw October-March; pale grey in breeding season, sometimes with yellow tinge, always with dusky tip. Beady eye appears black against yellow-buff eye-ring. Legs dark brown to black. (2) Turkish race, *brevirostris*. Adult ♂ distinctive, with paler pink rump and black-dappled sides to breast. ADULT FEMALE. As ♂, but rump uniform with mantle, rarely tinged pink, and wing-panel slightly less vivid. JUVENILE. Resembles ♀ but crown and nape slightly duller, back more rufous, and face and throat paler buff with distinctive streaking, even on chin. Bill becomes straw-colour from August. Legs pale flesh, remaining paler than adult through 1st winter.

Inseparable from *C. cannabina* at distance, but when closer shows distinctive pink rump (♂), warm buff face and throat (adult), more distinct wing-bars (particularly lower), and 2-toned underparts. Flight, gait, and behaviour much as *C. cannabina* but flight silhouette differs in longer tail. Escape-flight often long. Associated with coastal barrens and open moorland but also at home in barren mountains. Adapts willingly to presence of bushes and small trees. Gregarious, forming large flocks.

Song like *C. cannabina* but less musical, with more metallic, jangling, and twanging units, thus sounding more resonant, even percussive and chattering; phrasing rather slow and disjointed, distinctively incoherent; most commonly used notes 'tee-leu teedl-eu', often repeated in simpler version of song. Twittering with harder timbre than *C. cannabina* and contact-alarm call distinctly more penetrating than *C. cannabina*, with harsh nasal and twanging quality: 'cheewk', 'twoo-eek', 'naa-eet', 'twa-it', or 'tchooik'.

**Habitat.** Occupies tundra, boreal, and marginally temperate zones, extending north to about July isotherm of 10°C (Voous 1960*b*). In contrast to Linnet *C. cannabina*, occupies terrain more or less free of trees and shrubs or bushy growth, in cool, windy, and often rainy climates without much sun or warmth, often on stony, rocky, or hilly ground, including sea–cliffs and inshore islands. In Britain and Ireland, largely a lowland bird, favouring heather moors, hill farms, and upland pastures, but not mountains or precipitous areas (Sharrock 1976). In Scandinavia, breeds at high altitude on fjelds, and on barren slopes near crags or precipices, moving later to newly mown fields, and in coastal regions to gardens (Bannerman 1953*a*).

In winter, many shift to coastal lowlands, including salt-marshes. Other passerines share taste for these, but only *C. flavirostris* uses seaward fringe, feeding on seeds of *Salicornia* and *Aster*; sand-dunes, shingle banks, and cliffs are also well used (Fuller 1982). In Poland, has recently begun to make use of large beds of introduced goldenrod *Solidago*, while wartime bombing led to use of weeds on sites of destroyed buildings, and to roosting on high buildings in some German cities (Newton 1972).

Among eastern races, *brevirostris* breeds in alpine and subalpine meadows, thickets of creeping Caucasian rhododendron, and upper stands of dwarf birches *Betula* and pines *Pinus* at 2500-3000 m, descending in winter to mountain slopes and valleys, and even into foothills,

resorting there to vegetable gardens (Dementiev and Gladkov 1954). Also found in USSR on open and rocky steppes (Flint *et al.* 1984).

Sociable and aerial, and has shown capacity to take advantage of fresh ecological opportunities where not inhibited by competition.

**Distribution.** No evidence of change, except for recent range contraction in Britain and Ireland.

FAEROES. Bred Nólsoy 1938-48; otherwise accidental (Bloch and Sørensen 1984). BRITAIN, IRELAND. Considerable range contraction in recent years, perhaps due to amelioration of climate (Sharrock 1976; Thom 1986; Hutchinson 1989). SWEDEN. Fewer than 10 breeding records, but probably annual breeder (LR). FINLAND. Only one known breeding record, east of Lake Kilpisjärvi 1974, but a few pairs may breed more regularly in extreme north-west (OH). IRAQ. May breed. Flock of *c.* 30 birds seen end of April 1957 in north-east mountains (Boswell and Naylor 1957).

Accidental. Spain, Switzerland, Austria.

**Population.** Has declined recently in Britain and Ireland.

BRITAIN, IRELAND. Estimated 20 000-40 000 pairs 1968-72 (Sharrock 1976); has recently decreased (Thom 1986; Hutchinson 1989). NETHERLANDS. Wintering population estimated 3000-6000, 1978-83 (SOVON 1987). SWEDEN. Probably not more than 100 pairs (LR). FINLAND. Perhaps *c.* 5 pairs (Koskimies 1989). Wintering population has apparently declined in recent years (OH).

Survival. Oldest ringed bird 6 years 1 month (Hickling 1983).

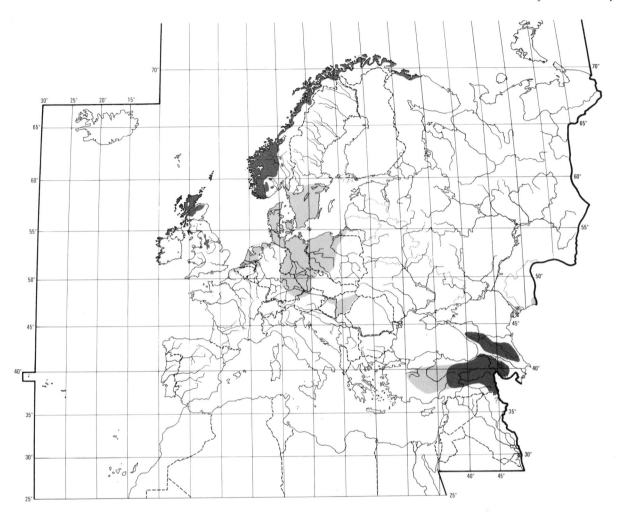

**Movements.** Sedentary to migratory; most Asian races make altitudinal movements. South-eastward direction of migration supports theory (Voous 1960) of central Asian origin, European birds having become isolated following last glaciation.

BRITISH AND IRISH RACES, *pipilans* and *bensonorum*, winter chiefly along coasts, with fewer records inland. In Scotland, upland areas are almost entirely vacated; few data from ringing, but bird ringed Killimster in extreme north-west, December, was at Inverasdale 165 km WSW in May. Birds breeding on Scottish coasts and islands are more sedentary, but some make regular local movements between islands; on Fair Isle, formerly wintered in thousands, but now mainly a summer visitor, departing late August to November and returning late February to early May. From west coast, some birds move south-west to northern Ireland: bird ringed Isle of Skye, August, recovered Donegal, 255 km south-west, February, and bird ringed Copeland Island (Down), April, recovered Mull of Kintyre, 73 km NNW, May. (Newton 1972; Spen-

cer and Hudson 1982; Mead and Hudson 1984; Lack 1986.) Irish birds apparently sedentary, with some local dispersal to estuaries and salt-marshes (Hutchinson 1989).

Birds breeding in southern Pennines (north-central England) move south-east to winter chiefly on east coast from Lincolnshire to northern Kent; a few remain inland, e.g. on arable farmland in Cambridgeshire fens, and at Chasewater (Staffordshire) where recorded regularly since 1948. Some birds cross North Sea, mainly to Low Countries, with recoveries to or from Netherlands (9), Belgium (8), France (3), and even Vicenza in north-east Italy (1), *c.* 1300 km south-east of ringing site; data show that individuals sometimes winter on different sides of North Sea in successive years. Exodus from Pennines begins late August, birds reaching east coast chiefly September–November; return movement from late January, with main departure in 1st half of March and few remaining at coastal sites thereafter, though many settle at breeding sites only in May or even June. (Spencer 1969; Hume 1983; Lack 1986; Mead and Clark 1987; Davies 1988.)

NORTH EUROPEAN RACE, nominate *flavirostris* (breeding Norway and adjacent areas of extreme north of Sweden, Finland, and Russia) chiefly migratory, wintering in northern and eastern Europe; numbers wintering in east have increased in recent decades. Many winter around cities, roosting on buildings especially in Germany, e.g. Berlin (Fiuczynski 1961), Chemnitz (Rinnhofer 1972), and Hannover (Moll 1986). Overwinters in varying numbers north usually to *c.* 64°N in Norway, and *c.* 61°N in Sweden (Haftorn 1971; Bentz and Génsbøl 1988; SOF 1990). Some birds winter along western and southern coasts of Finland, recently *c.* 100-400 per year; these mostly visitors from Scandinavia, as Finnish breeding population very small (Koskimies 1989). Observations suggest some Scandinavian birds reach eastern Britain, though no evidence from ringing; arrivals reported on Scottish east coast in autumn, with parties up to 50 on Isle of May both seasons; in eastern England (where *pipilans* winters regularly), numbers often increase late December to January (Lack 1986; Thom 1986). Highest numbers winter on mudflats with glasswort *Salicornia* in coastal areas of North Sea (Low Countries to Denmark), and many also stop off there to feed (Bub 1985; Bub and Pannach 1988). Winter quarters extend south-west only to northern France; in Low Countries and France, wintering grounds overlap with *pipilans* (see above). Regular on northern French coast and adjacent inland areas west to Cotentin (*c.* 1°30′W), irregular further west and south, where noted mostly in cold spells; numbers vary greatly, with sometimes 1000-2000 on coast of Picardie in north-east, but in January 1981 less than 300 throughout French winter quarters (Yeatman-Berthelot 1991). South-eastwards, winter range extends to Yugoslavia and Bulgaria, with numbers progressively decreasing. In Germany, most common in north and fewer further south (Mildenberger 1984; Wüst 1986; Klafs and Stübs 1987); only 2 records from Switzerland (Winkler 1984). In Poland, winter population has apparently increased since 1900; numbers highest in north-west, with fewer east and south, though locally numerous, e.g. in Wrocław and Warsaw areas; many records (passage or winter) in Vistula valley (Tomiałojć 1967). In central Czechoslovakia, first recorded 1952 and now regular and common on passage, especially along tributaries of Danube, some birds remaining to winter (Kaňuščák 1979). Occasionally reported from Austria, notably at Neusiedlersee in north-east (Bauer and Rokitansky 1951; Ganso 1960). In Hungary, most records are from Danube eastwards; sporadic further west, with only 1 record (flock of 400-500) at Lake Balaton over 30 years (Beretzk and Keve 1971). In Vojvodina (north-east Yugoslavia), considerable numbers winter in some years, none in other years (Antal *et al.* 1971). Very occasionally reported from Italy, mostly from north-east (Brichetti 1976). In Rumania, recent records (including on Black Sea coast) suggest regular wintering in small numbers (Dijksen 1976). In Bulgaria, now frequently observed in western mountains, and occasionally elsewhere (T Michev).

Ringing in Norway (mostly south-west) up to 1980 produced 100 foreign recoveries: Germany (54), Netherlands (20), Belgium (18), Denmark (3), Czechoslovakia (2), Sweden (1), France (1), and western Ukraine (1); 4 were 1500-1900 km from ringing site (Bernhoft-Osa 1965; Haftorn 1971; Holgersen 1982). Ringing has revealed marked fidelity to wintering or stopover sites, with *c.* 3000 retraps at site of ringing or within 20 km, and several retraps in successive years, up to 7 years after ringing. In eastern Germany, 5·6% of 26 378 birds retrapped within 20 km of ringing site (5·2% at same site); in western Germany, 2·9% of 46 049 birds (2·2% at same site). Retraps fewer in Netherlands (1·4% of 7551 birds) and Belgium (0·4% of 2441 birds within 30 km). Site-fidelity also recorded in Scandinavia, Poland, and Czechoslovakia. At Wilhelmshaven on west German coast, 19 November 1958, 15 birds ringed 1-4 years earlier were retrapped in one small area. (Bub and Pannach 1988.) Data show that ♀♀ tend to migrate earlier than ♂♂ in autumn, but ♂♂ earlier in spring (Bub 1986).

Main heading south initially to north-west Germany, then south-east across continental Europe, often via broad river valleys, e.g. Weser, Vistula, Danube (Tomiałojć 1967; Beretzk and Keve 1971; Bub and Vries 1973; Bub and Pannach 1988). Very common on passage in Denmark (Møller 1978a; Meltofte and Fjeldså 1989) and north-west Germany (Vries 1982; Bub and Pannach 1988). Regular but sparse passage throughout Sweden; many birds depart from south-west, with average 2622 per year reported 1973-90 at Falsterbo; numbers much lower at Ottenby in south-east (Edelstam 1972; Ulfstrand *et al.* 1974; SOF 1990; Roos 1991). On central Polish coast, 1972-9, reported each year except 1974, with 62 records of 716 individuals (both seasons) (Górski 1982). Rare in Baltic states (Tomiałojć 1967; Vīksne 1983). In St Petersburg region (north-west Russia), fairly regular on passage west of Lake Ladoga and sometimes winters, with several tens recorded each year, but exceptional in east of region (Mal'chevski and Pukinski 1983); these records perhaps involve birds breeding on Kola peninsula (north-west Russia).

Autumn migration gradual. Passage in southern Norway from last third of August, but peak not until 1st third of October (Bernhoft-Osa 1965); in northern Denmark, average earliest record over 6 years 15 September (28 August to 23 September), with marked peak mid-October (Møller 1978a); few reach north-west German coast before 2nd third of October (Bub and Vries 1973). At Falsterbo, passage October-November, peaking end of October (Roos 1991). Main movement across continental Europe late October to November, continuing into December (Tomiałojć 1967; Beretzk and Keve 1971; Kaňuščák 1979; Mildenberger 1984; Klafs and Stübs 1987). Recorded 22 October to 10 April over 8 years in Hannover (Moll 1986).

Spring movement begins early, from mid- or late February; most birds leave Czechoslovakia then, and central Europe by end of March (Kaňuščák 1979; Rutschke 1983; Mildenberger 1984; Knorre *et al.* 1986). Passage on Polish coast mid-March to early April (Górski 1982). Most birds have left north-west Germany by mid-April (Bub and Pannach 1988), and passage in northern Denmark from late February, peaking early April, with stragglers in May (Møller 1978a).

OTHER RACES. Middle Eastern *brevirostris* makes chiefly altitudinal movements. Birds in eastern Turkey descend to lower levels and extend west across central plateau (Beaman 1986); in Caucasus, some descend to foothills but many remain on mountain slopes even in coldest months, foraging behind flocks of sheep which churn up snow (Dementiev and Gladkov 1954); in Iran, common winter visitor to plains of Azarbaijan in north, and occasional on southern slopes of Elburz mountains (D A Scott). Races in northern India and Tibet also altitudinal migrants; *montanella* descends to *c.* 1500 m, but *rufostrigata* not recorded below *c.* 3000 m (Schäfer 1938; Ali and Ripley 1974). From Pamir-Alay (*pamirensis*), some birds move south to Badakhshan, others west along Alay valley to Tadzhikistan (Abdusalyamov 1973). Most birds breeding in northern and central Kazakhstan (*kirghizorum* and *korejevi*) move south of breeding range; common in winter at mouth of Ural, also in great river floodlands and southern foothills (e.g. in lower Syr-Dar'ya and Talasskiy Alatau); some continue south beyond Kazakhstan. Autumn movement (far more conspicuous than spring) mid- or late September to November; spring departure early, from mid-February, with most birds returning to breeding grounds by end of March (Dolgushin 1968; Korelov *et al.* 1974). Mongolian race *altaica* apparently chiefly resident, but some birds disperse as far as central Gobi (Dementiev and Gladkov 1954; Vaurie 1959).               DFV

**Food.** Small seeds; perhaps a few invertebrates in breeding season. Forages on ground or on low herbs, sometimes in trees; in breeding season, in open areas of pasture and cultivation, by roadsides, at tideline, and by fresh water, feeding mainly on seeds of Compositae, Polygonaceae, and Caryophyllaceae; in winter, in fields, waste ground, allotments, by rivers, etc., and very commonly on coastal salt-marshes, mostly on Chenopodiaceae and Compositae. (Tomiałojć 1967; Orford 1973; Marler and Mundinger 1975; Bub and Hinsche 1982; Davies 1988; Deunert 1989.) In northern England, birds breeding on heather *Calluna* moorland fly 1–1·5 km downhill to feed on favoured weeds in pastures, though incubating ♀♀ will feed close to nest; very important feeding ground here is burned *Molinia* grassland (Orford 1973, which see for many details). In winter, occasionally forages in stands of birch *Betula*, mostly on ground (Tomiałojć 1967, which see for comparison with Linnet *C. cannabina* and Redpoll *C. flammea*; Bub 1969; Smart 1978; Mildenberger 1984). See also Bub (1989). In eastern England, outside breeding season, forages far out on salt-marshes in flocks of up to *c.* 2500 with little competition from other species, which tend to feed closer to land; at high tide moves into fields but quickly returns when water retreats (Davis 1988). Feeds in small-

ish groups in breeding season (Marler and Mundinger 1975), but at other times in large mixed flocks of several thousand with other seed-eaters, notably *C. cannabina* (Schmidt 1960b; Newton 1967a; Davies 1988; Deunert 1989, which see for other species in flocks). Feeding habits very similar to *C. cannabina*, though generally found closer to human habitation, especially in winter (Newton 1972; Deunert 1989); perhaps takes seeds more from plants than from ground (Newton 1967a). Perches on tall herbs to feed (Rinnhofer 1972), and also stands on stem to bend it over, feeding on seed-head on ground, holding it under one foot (Newton 1967a; Abdusalyamov 1977), though in another study not seen to use feet at all when feeding (Kear 1962). When feeding in birch trees, will forage in agile fashion like *C. flammea* or Siskin *C. spinus* (Bub and Hinsche 1982). At tideline, forages in seaweed (Bub 1969), or even eats seaweed itself, perhaps because preferred foods scarce (Jardine 1992a); at water's edge in reservoir, fed on washed-up seeds and other plant debris, mainly alder *Alnus*, birch, and duckweed *Lemna* (Sell 1984). Pulls up cultivated *Brassica* seed-leaves (Saxby 1874), and recorded eating fresh tips of juniper *Juniperus* twigs (Bub 1969). Even when feeding in cereals or at sheaves, is usually searching for smaller weed seeds (Saxby 1874; Dolgushin 1968). In central Europe, frequently associated with stands of tall, introduced herbs goldenrod *Solidago* and evening primrose *Oenothera*, particularly in snow when heads still protrude, and winter distribution possibly determined to some extent by spread of these plants (Tomiałojć 1967; Bub 1969; Beretzk and Keve 1971, which see for discussion and comparison with other studies). Apparently fond of *Artemisia campestris* but rejects *A. vulgaris* (Tomiałojć 1967; Rinnhofer 1972). On tideline, ♂♂ fed at average 33·5 pecks per min, ♀♀ at 45·2 (D C Jardine).

Diet in west Palearctic includes the following. Invertebrates: flies (Diptera: Cecidomyiidae), beetles (Coleoptera: Curculionidae). Plants: seeds, leaves, buds, etc. of juniper *Juniperus*, birch *Betula*, alder *Alnus*, oak *Quercus*, hemp *Cannabis*, nettle *Urtica*, sorrel, etc. *Rumex*, knotgrass, etc. *Polygonum*, goosefoot, etc. *Chenopodium*, sea purslane *Halimione*, orache *Atriplex*, glasswort *Salicornia*, seablite *Suaeda*, saltwort *Salsola*, *Camphorosma*, amaranth *Amaranthus*, chickweed, etc. *Stellaria*, mouse-ear *Cerastium*, corn spurrey *Spergula*, sea spurrey, etc. *Spergularia*, buttercup, etc. *Ranunculus*, poppy *Papaver*, greater celandine *Chelidonium*, cabbage, etc. *Brassica*, rocket, etc. *Sisymbrium*, wall rocket *Diplotaxis*, sea rocket *Cakile*, charlock *Sinapis*, radish *Raphanus*, pepperwort *Lepidium*, pennycress *Thlaspi*, shepherd's purse *Capsella*, wintercress *Barbarea*, cinquefoil, etc. *Potentilla*, bramble *Rubus*, mignonette *Reseda*, broom *Cytisus*, clover *Trifolium*, flax *Linum*, St John's-wort *Hypericum*, loosestrife *Lythrum*, willowherb *Epilobium*, evening primrose *Oenothera*, wild carrot *Daucus*, heather *Calluna*, crowberry *Empetrum*, gromwell *Lithospermum*, self-heal *Prunella*, woundwort *Stachys*, snapdragon *Antirrhinum*, sea lavender *Limonium*, thrift *Armeria*, vervain *Verbena*, plantain *Plantago*, goldenrod *Solidago*, daisy *Bellis*, sea aster, etc. *Aster*, yarrow, etc. *Achillea*, mugwort, etc. *Artemisia*, chamomile *Anthemis*, tansy, etc. *Tanacetum*, coltsfoot *Tussilago*, groundsel, etc. *Senecio*, burdock *Arctium*, thistles *Carduus*, *Cirsium*, knapweed, etc. *Centaurea*, chicory *Cichorium*, *Chondrilla*, dandelion *Taraxacum*, hawkbit *Leontodon*, catsear *Hypochoeris*, hawkweed *Hieracium*, Rannoch-rush *Scheuchzeria*, duckweed *Lemna*, rush *Juncus*, sedges (Cyperaceae), grasses (Gramineae, including reed *Phragmites*), seaweed (Phaeophyta). (Witherby *et al.* 1938; Tomiałojć 1967; Bub 1969; Newton 1972; Marler and Mundinger 1975; Rékási and Sterbetz 1975; Sabel 1983; Sell 1984; Jardine 1992.) For extensive list of plants in diet, see Bub and Hinsche (1982).

In Hungary, winter, of 875 seeds in 25 stomachs, 43% by number amaranth, 23% goosefoot, etc., 19% orache, 5% vervain, 3% *Camphorosma*, 2% gromwell; only 2 insects found, both beetles (Rékási and Sterbetz 1975). In Norway, preferred food in breeding season seeds of dandelion, but these available in unripe state for only *c.* 3 weeks, so then changes to *Rumex* and chickweed (Bub 1969; Marler and Mundinger 1975). Favourite foods on north European wintering grounds are seeds of goosefoot, glasswort, sea purslane, charlock, rocket, radish, wintercress, sea aster, and *Artemisia*, as well as evening primrose and goldenrod (see above) (Witherby *et al.* 1938; Tomiałojć 1967; Bub 1969). On salt-marshes of eastern England, feeds almost only on glasswort and sea aster, taking seeds directly from sea aster, otherwise from mud; also very fond of seablite and orache; seeds of glasswort and sea aster available about September to mid-January, so February–March could be period of food shortage (Davies 1988, which see for details). See also Bub and Hinsche (1982). No, or very few incidentally taken insects found in stomachs in extralimital south-central Asia and Kazakhstan (Potapov 1966; Dolgushin 1968); one stomach in Tadzhikistan contained insects and caterpillars (Abdusalyamov 1977).

Young in northern England fed on regurgitated seeds obtained from pastures below heather moor breeding area (Orford 1973). In captivity, young given only seeds from crop (Gentz 1971). In south-central Asia, young fed mainly plant food, and the few invertebrates in diet are probably picked up with seeds (Potapov 1966; Abdusalyamov 1977), but see Bannerman (1953a), where diet of young said to be largely or wholly insects. BH

**Social pattern and behaviour.** Fullest study in Scotland and Ireland (*pipilans*) and Norway (nominate *flavirostris*) by Marler and Mundinger (1975). For review and comparison with other Fringillidae, see Newton (1972).

1. Outside breeding season typically in flocks (up to hundreds or thousands) for feeding, roosting, and migration. In Pennines (northern England), a few nomadic flocks of 40–60 reported in winter (Taylor 1935; Orford 1973). Larger numbers occur on coast: e.g. around Wash (eastern England), 3600–17 000 birds

in 420 km² (8·6–425 per km²), flocks of several hundred or up to 2500, breaking up in latter part of winter (Davies 1988). In Germany, foraging flocks typically smaller and scattered, roosting assemblages larger (Ringleben 1981); on Schleswig-Holstein coast, of 100 reports of flocks, 39% 1–10, 44% 11–50, once 2000–3000 in mid-November (Schmidt 1960b); also 1000 (with *c.* 30 Shore Larks *Eremophila alpestris* and 10 Snow Buntings *Plectrophenax nivalis*; see below) in area of *Salicornia c.* 15 000–50 000 m² (Gloe 1982); inland in eastern Germany (Brandenburg), predominantly in flocks of 1–100 October–March (up to 2000 recorded), maxima during 2 passage periods but normally migrates in small flocks, though once *c.* 2000 passing over Berlin (Bruch and Löschau 1960; Fiuczynski 1961; Dittberner and Dittberner 1971); for Germany, see also Gentz (1971). Of 264 flocks in Poland, 67% of 1–10 birds; of 3202 birds, 32% in flocks of 6–20 (Tomiałojć 1967); see also Górski (1982). For sex- and age-classes in Germany in winter, see Bub (1987). For winter site-fidelity, see Movements. Roosts quite often of 50–200, sometimes several hundred and maxima (when cold spell forces birds to abandon some sites) several thousand (Newton 1972; Stiefel 1976; Ringleben 1981; Bub and Pannach 1991). Flocks also reported during breeding season, birds congregating to feed, like other Fringillidae (Potapov 1966; Orford 1973; Mauersberger *et al.* 1982; Neufeldt 1986). After fledging, flocks sometimes of juveniles only (e.g. Dolgushin 1968, Orford 1973); on Skye (Scotland), some solitary juveniles also seen foraging by roads (Marler and Mundinger 1975). Associates outside breeding season with other Fringillidae, buntings (Emberizidae), sparrows *Passer*, and larks (Alaudidae, particularly *E. alpestris*); association usually loose and temporary, though often close and persistent with Linnet *C. cannabina* (Schmidt 1960b; Fiuczynski 1961; Beretzk and Keve 1971; Dittberner and Dittberner 1971; Rinnhofer 1972; Tauchnitz 1972; Creutz 1988); in Poland, more often in single-species flocks than mixed (Tomiałojć 1967). See also Flock Behaviour (below). BONDS. Mating system apparently mainly monogamous, but occasional occurrence of promiscuity or polygyny suggested by observations in Norway: ♂♂ (whether or not paired to another ♀) attracted to sexually receptive ♀♀ and, though no extra-pair copulations recorded, 2 ♂♂ (one certainly paired) each seen to visit nesting areas of other pairs and to follow them during copulation period (Marler and Mundinger 1975). Pair-formation takes place in spring flocks (e.g. Orford 1973). Pair-bond apparently strong, extending through and perhaps beyond one season and pair-members typically keep close company: e.g. on Fair Isle (Shetland), ♂ and ♀ of pair often trapped together, in one case 4 times, 7 June to 11 October; another ♀ trapped with one ♂ 11 August, then with another 25 June after 2 years (Williamson 1955); see also below. Nest-building and incubation by ♀, but young fed by both sexes (Witherby *et al.* 1938; Gentz 1971); one case on Fair Isle of ♀ apparently feeding young alone (Littlejohn 1952). In Norway, overlapping broods recorded in 2–3 pairs; in one case, both pair-members cared for 1st-brood fledglings while nest-building and laying. Pair-bond persisted in these cases, also in another when birds re-laid following predation. (Marler and Mundinger 1975.) Young independent within *c.* 15 days of fledging (Witherby *et al.* 1938). For suggestion of strong bond between brood-siblings when one held temporarily captive, see Mylne (1957); 2 brood-siblings ringed Staffordshire (England) in June recovered still together in Wash in February 2 years later (Orford 1973). Age of first breeding not known. BREEDING DISPERSION. Territorial; widely dispersed and solitary where scarce or in extensive areas of suitable habitat, or often in neighbourhood groups, in which nests sometimes quite close together (Butterfield 1906; Witherby *et al.* 1938), e.g. 10–15 m apart in Kazakhstan (Dolgushin 1968). Groups of 4–6 pairs reported in Britain (Newton 1972), 5–7 in south-east Altai (Loskot 1986a), and 12–15 in Kazakhstan (Dementiev and Gladkov 1954). At Selva (Norway), nests *c.* 500 m apart along *c.* 3 km of road, with 'nesting areas' of 4 pairs separated by average 470 m (415–575) (Marler and Mundinger 1975). Where particularly numerous in Pamirs, pairs 100–150 m apart (Potapov 1966). Study in Norway by Marler and Mundinger (1975) suggested ♂'s aggression serves primarily to defend mate (mobile territory) against close approach by other ♂♂ in or away from area surrounding nest. Following pair-formation (apparently in flocks away from nesting area) and once nest-site is selected and becomes focal point of pair's activity (♀ initiating nest-building, then laying and incubating, and pair also copulating near nest), ♂ more obviously defends nesting territory (intruders are most often rival ♂♂ trying to approach owner's mate). Each bird found to engage in significant number of interactions with other members of population. Of 99 such interactions involving at least one member of 4 well-studied pairs, 62% within *c.* 100 m of nest were aggressive, only 15% (*n* = 20) of those further away. Little or no advertising (including conspicuous singing) by ♂ in absence of other conspecific birds; for sentinel behaviour, see Antagonistic Behaviour (below). Adjacent territories not contiguous, being separated rather by unoccupied areas *c.* 340–425 m across in which various birds, mostly those from adjacent areas, forage and collect nest-material without antagonism (at least between neighbouring pairs). Some feeding done within territory or, as typically in group-nesting Carduelinae (Newton 1972), up to 1 km or more away (Marler and Mundinger 1975); in northern England, within *c.* 400–1700 m of nest. (Nuttall 1972; Orford 1973.) No boundary disputes recorded between neighbours, and territory limits difficult to define, though resident ♂♂ rarely showed aggression further than 75 m from nest (suggests defended area *c.* 1·8 ha, though much smaller presumably with reduction in inter-nest distance: see Newton 1972). (Marler and Mundinger 1975.) Apparently stays in territory for new nesting attempts if undisturbed: in Norway, 2 pairs re-nested within *c.* 15–30 m of their 1st nest, but when brood of another pair predated, birds shifted *c.* 415 m for fresh attempt (Marler and Mundinger 1975). Little information on breeding density. On moorland in Orkney, 13 pairs in 1098 ha (1·18 per km²) (Lea and Bourne 1975). In Staffordshire (England), 9·4 pairs per km²; of 25 2-km squares, 12 occupied, overall 1·9 pairs per km², and this density applied to data from 2-km squares in 2 areas of northern England gives estimated 5·0 and 7·8 pairs per km² (Davies 1988). Island of Rundʌy (*c.* 12 km²) in Norway held 5–7 pairs (0·4–0·6 per km²) (Schmidt 1960b). On heather moors in western Norway, estimated densities from line-transect counts 0·7–4·3 pairs per km² (Munkejord 1987, which see for caveat regarding method). Near Kobi (Gruziya), 200–300 singing ♂♂ per km² (Boehme 1958). Recorded nesting close to Redpoll *C. flammea* in habitat (plantations) typical of that species (Orford 1973). ROOSTING. Nocturnal. Almost all information relates to period outside breeding season, when typically communal and sometimes mixed with other species, including other Fringillidae (Newton 1972), though only one case (association with Greenfinch *C. chloris*) reported by Bub and Pannach (1991). In Tadzhikistan, when ♀ incubating, ♂ roosts alone nearby (Abdusalyamov 1977). Wide range of sites reported for communal roosting: dense ground vegetation, including 3–5 m apart in small hollows (also used in wind) among dune grasses (Schmidt 1960b), dispersed in heather (Ericaceae) or bracken *Pteridium* (Orford 1973), stubble fields or unharvested flax *Linum* (Dittberner and Dittberner 1971), reedbeds over dry or wet ground (Beretzk and Keve 1971; Stiefel 1976; Bub and Pannach 1991), tussocks of cotton-grass *Eriophorum* in ponds

(Orford 1973), shrubs such as privet *Ligustrum*, willow *Salix*, and gorse *Ulex*, and bramble *Rubus* (Witherby *et al.* 1938; Newton 1972; Piechocki *et al.* 1982; Bub and Pannach 1991), coniferous trees (Dornbusch 1972; Stiefel 1976), which are also used as shelter in bad weather (Dementiev and Gladkov 1954), boulder scree (Orford 1973), self-made holes in straw stacks (Saxby 1874), in sea cave (Ireland), cracks and crevices in rocks or walls (Potapov 1966; Abdusalyamov 1977; Bub 1987; Bub and Pannach 1991), and reported mainly from Germany (at least since 1950s) on buildings; for review see Bub and Pannach (1991). In reedbed near Leipzig (Germany), birds perched on lower third of stems over *c.* 5–10 cm of water (Dorsch 1970; see also Bub and Pannach 1991). On buildings, birds roost on ledges, etc., also in snow-holes (excavated by birds) on thatched roofs (Lange 1960; Stiefel 1976; Ringleben 1981; Bub and Pannach 1991). Fairly isolated, in twos, or sometimes several quite close together, e.g. 5–6 in crack 60–65 cm long (Schmidt 1964; Moll 1986); apparently mostly maintain individual distance, however (see Flock Behaviour, below, for disputes), which is facilitated in some cases by structure of building (see photographs in Stiefel 1976 and Bub and Pannach 1991). Sites near top of building (under roof overhang) first to be occupied (Moll 1986) and most hotly contested (Hilprecht 1964). In Hannover (Germany), 3 buildings used regularly over 15 years (Moll 1986); as shown by ringing, in some cases by same birds over several winters (Bub and Pannach 1991; see also Tomiałojć 1976b, 1988 for Poland). For further reports of roosting on buildings, see (e.g.) Dien (1965), Pfeifer (1966), and references cited by Bub and Pannach (1991). Near Leipzig, February to mid-March, birds arrived at reedbed roost 25–40 min before sunset, last settling 0–15 min before (Dorsch 1970). In Hannover, where roosting on building, assembled 17.45–18.28 hrs 20 March and quiet within *c.* 30 min of arrival of last flock; exodus next morning 05.40–06.34 hrs (Moll 1986). Active long after street lights come on (Lange 1960), and birds apparently prefer most brightly lit side of building (Bub and Pannach 1991). During day, will rest on wires, also at other sites much as used for nocturnal roosting (Dittberner and Dittberner 1971). In Kazakhstan in summer, birds regularly visit water to drink and bathe morning and evening (Dolgushin 1968). For diurnal activity rhythm in Tadzhikistan, see Abdusalyamov (1977).

2. In Kazakhstan, often close to man, and typically tame and confiding (Dolgushin 1968). Studies in Germany found approachability to vary considerably: single birds or small flocks may allow approach to within a few metres (e.g. where fed on window sills or where natural food restricted), larger flocks only to 30–50 m (Müller 1953; Hammer 1958; Meise 1958; Dittberner and Dittberner 1971; Rinnhofer 1972). FLOCK BEHAVIOUR. Prior to roosting, birds usually first assemble, often closely packed, in trees or on nearby buildings, giving variety of Flight-, contact-, and alarm-calls (see 2–3 in Voice) and song. May circle first over roost much as other Fringillidae, and reported to continue calling and singing from roost-sites on building, though silent immediately after entering reedbed. Calls and song also precede morning departure from roost. Passing Kestrel *Falco tinnunculus* caused birds to fly up en masse from trees and (rapid escape) from roost. (Dorsch 1970; Stiefel 1976; Ringleben 1981; Moll 1986.) Roost sites on buildings defended against an attempted displacing attack once occupied (Lange 1960; Hilprecht 1964). In eastern Germany, associated closely for feeding with *C. cannabina* (sometimes even in feather contact). Mixed flock flushed by man flew up giving Flight-calls (see 2 in Voice), *C. flavirostris* almost always slightly behind *C. cannabina*; all birds soon settled together within *c.* 10–20(–150) m. (Rinnhofer 1972.) Brief fights noted in mixed feeding flock, though no antagonism

where birds feeding on specially strewn seed (Kreibig 1957). Study in eastern England found flocks typically highly vocal when commuting between salt-marsh and freshwater for drinking and bathing (Davies 1988). Birds resting in (often close-knit) flocks in late winter or spring call and sing vigorously, several ♂♂ sometimes performing Song-flight (see below) simultaneously (Schäfer 1938; Piechocki et al. 1982); fall silent when threatened, and further calls then signal take-off (Saxby 1874). SONG-DISPLAY. ♂ sings (see 1 in Voice) from rock, post, tree, etc., or in flight; on perch, in 'normal' posture, but crown and nape ruffled and also flicks tail periodically (Witherby et al. 1938; Marler and Mundinger 1975). In Song-flight (performed by birds in flocks, or over immediate nesting area once breeding), ♂ circles several times, sometimes with erratic manoeuvres; alternately flaps and glides, singing during gliding phase in which wings widely spread and lowered, tail also slightly lowered; resumes flapping after song, then descends like Tree Pipit Anthus trivialis (Schäfer 1938; Marler and Mundinger 1975; Piechocki et al. 1982). In study by Marler and Mundinger (1975), 80% of flight-songs (see 1a in Voice) elicited by rival ♂♂. Song (in flocks) noted in various contexts during winter, also autumn and spring (Schmidt 1960b; Tomiałojć 1967; Dolgushin 1968; Haftorn 1971); see also Flock Behaviour (above). In Britain, song exceptional or only subsong (not described) early to mid-March and early to mid-August, irregular but fairly frequent to mid-April after which regular to end of July (Witherby et al. 1938). ANTAGONISTIC BEHAVIOUR. (1) General. ♂ generally most aggressive when ♀ active and sexually receptive (i.e. during nest-building, laying, and copulation), and pair may be followed around by another ♂ (92% of 36 such cases were in period of ♀'s sexual and early-breeding activity) (Marler and Mundinger 1975). In Tadzhikistan, fights between ♂♂ recorded in late May during pair-formation (Abdusalyamov 1977). (2) Threat and fighting. In Norway, when no intruder visible, ♂ remained relatively inconspicuous, silent and watchful on look-out, rump patch ruffled and prominent; flew from there to tackle intruders. When ♀ on or near nest, ♂ reacted to presence of other ♂♂ by performing Song-flights and, where necessary, launching supplanting attacks, fighting, and chasing; also flew up to 'escort' for short distance strange conspecific birds overflying territory. (Marler and Mundinger 1975.) In defence of mate or territory, also during individual-distance disputes (when resting) or over food, as response to threat or attack (including by other Fringillidae), ♂ will give Threat-call or 'tveet' (see 4 and 3b in Voice): sometimes initially 3–5 units of Threat-call given while simultaneously turning head and slightly inclining body towards opponent without launching attack, then longer sequence in Threat-posture (crouched, with head forward), bird also Threat-gaping before attack. Threatened bird ruffles feathers and calls 'tri' (see 3c in Voice) once or twice, staying put or moving slightly; gives longer and rapid series of calls and flies off if attack more serious. (Zablotskaya 1978a.) Study by Marler and Mundinger (1975) found some intrusions to elicit only limited or no aggression, e.g. strange juveniles allowed to forage within c. 10–20 m of nests. In Shetland, ♂ defended nest against inquisitive young Wheatears Oenanthe oenanthe, chasing them c. 25–30 m (Littlejohn 1952). HETEROSEXUAL BEHAVIOUR. (1) General. Several reports (USSR, Mongolia) of pair-formation taking place in flocks in March–April, sometimes at lower altitudes before moving up to breed (Dementiev and Gladkov 1954; Boehme 1958; Dolgushin 1968; Piechocki et al. 1982); in Altai, not before early June (Sushkin 1938). Pair-formation typically prolonged in Carduelinae (Newton 1972). (2) Pair-bonding behaviour. When ♀ on ground or rock, ♂ (sometimes 2–3 ♂♂ together) displays in the air above her, beating wings rapidly or gliding, and moving in circles or zigzags (Dementiev and Gladkov 1954); presumably much as in Song-flight, though not clear whether ♂ sings. ♂ then hops about in front of or around ♀, singing vigorously (both song-types described by Marler and Mundinger 1975: see 1a–b in Voice); fans tail (revealing white), droops wings (so pink rump displayed), and continually opens and closes primaries in shuffling movement; ♂ may also crouch or tilt body to one side (Witherby et al. 1938; Dementiev and Gladkov 1954; Newton 1972; Marler and Mundinger 1975). Courtship in one study recorded only during presumed attempts at pair-formation, never between ♂ and ♀ of established pair (Marler and Mundinger 1975). Other elements of courtship (described for montanella) include both birds facing each other, each rapidly opening and closing bill (presumably Bill-flirting typical of Carduelinae, perhaps with element of threat: see Newton 1972), and giving long series of 'tsi' and occasional quiet 'chuv' Flight-calls (see 5a and 2 in Voice, and sub-section 3, below); ♂ also seen gently pecking at ♀'s neck on ground or perch, and sometimes gives 3–9 clicking 'chuv' sounds (see 5c in Voice), flicking tail up at each call, while moving away from ♀, thereby also extending and slightly raising body, and sleeking plumage (Zablotskaya 1978a). Especially during nest-building and ovulation, ♂ guards ♀ closely away from territory (Marler and Mundinger 1975). (3) Courtship-feeding. ♂ feeds ♀ (by regurgitation) on nest or away from it, when may approach ♀ frontally. Sometimes follows courtship as described above. ♀ gives begging-call (see 3b in Voice) on sighting mate and immediately before being fed; also wing-shivers and has head and tail raised; on nest, crouches low and points open bill at ♂. (Potapov 1966; Gentz 1971; Marler and Mundinger 1975.) Study of montanella found ♂ sometimes (also during courtship involving bill movements: see above) gives 'veet' calls (see 5b in Voice) before transferring food (Zablotskaya 1978a). (4) Mating. Following information from study in Norway. Takes place almost always during laying (e.g. 12 out of 16 times in one pair), on bare branch or rock within c. 20 m of nest. ♂ solicits briefly (much as when food-begging: see above), mounting directly or hovering before landing on ♀'s back. Post-copulatory display described from 2 ♂♂: next to ♀ or while still mounted, raised head and tail, and ruffled underparts, with quivering movement evident; perhaps sang in one case. (Marler and Mundinger 1975.) (5) Nest-site selection. ♂ accompanies ♀, guarding her and singing nearby while she examines various sites and (presumably) makes final choice; ♂ also stays close to ♀ when she is building (Nethersole-Thompson and Nethersole-Thompson 1943b; Gentz 1971). (6) Behaviour at nest. No information. RELATIONS WITHIN FAMILY GROUP. No information on hatching or duration of brooding. In early days, ♂ regurgitates food into ♀'s bill for transfer to young (Gentz 1971). Captive young quiet between feeds; called while being fed (Marler and Mundinger 1975). At nest in Shetland, most feeds lasted 30–60 s. ♀ removes nestling faeces deposited on nest-rim overnight; at c. 6 days, c. 25% of faeces swallowed by ♀, rest carried well away; by 11 days, none swallowed (Littlejohn 1952). Young fledge at 11–17 days (Dementiev and Gladkov 1954; Newton 1972). ANTI-PREDATOR RESPONSES OF YOUNG. No information. PARENTAL ANTI-PREDATOR STRATEGIES. ♀ a tight sitter, leaving only in serious disturbance; after flushing, may flit about and call ('twittering': probably Flight-call), ♂ sometimes behaving similarly (Armitage 1927; Gentz 1971). ♂ will also express agitation by singing, and either bird by various calls (Marler and Mundinger 1975; Zablotskaya 1978a); see 3c–e in Voice.                                                                                MGW

Voice. Important study of nominate flavirostris (Norway) and pipilans (Ireland and Scotland) found 7 basic call categories, probably representing entire (small) repertoire; all calls given by ♂, all except song by ♀ (Marler and Mundinger 1975), though singing by ♀ reported (without details) by Beretzk and Keve (1971) and Haftorn (1971). Study of captive *montanella* from Kirgiziya by Zablotskaya (1978a) distinguished 12 calls (8 of which given throughout year, 4 only in breeding season) and 3 song-types (scheme modified in following account). For additional sonagrams, see these 2 sources, also Bergmann and Helb (1982). For description of song with musical notation and comparison with other *Carduelis*, including Linnet *C. cannabina*, see Stadler (1956).

CALLS OF ADULTS. (1a) Advertising-song of ♂. Some similarity to *C. cannabina* in general type, but not preceded by 'gigigi' typical of that species and overall much less musical and delicately twittering, more metallic and jangling with nasal timbre, harsher chatter, and pronounced resonant twangy quality; often includes buzzy sounds

(showing frequency and amplitude modulation) and hard rattles; units often smoothly linked but almost all sound strangled and closely resemble or are identical to calls (see below); some quite musical, however, and song may comprise only these, e.g. 'teet-sweet teedle-eu twee-teedl-ee teedl-eu'. Song given rapidly and at uniform volume, in phrases of varying length, or continuous. (Witherby *et al.* 1938; Stadler 1956; Marler and Mundinger 1975; Bergmann and Helb 1982; W T C Seale.) For another rendering of part of song, see (e.g.) Haftorn (1971). In study by Marler and Mundinger (1975), Advertising-song termed 'flight-song' (though given both in flight and from perch), usually preceded by Flight-calls (see 2, below) which, together with calls 3–4, often also conclude it. Recording (Fig I) contains continuous song comprising many calls (see 2 and 3a–c, below): 'tzeeip' (e.g. 1st unit and last 3 units), Flight-calls (e.g. segment following 1st 'tzeeip'), 'wheet' (following these Flight-

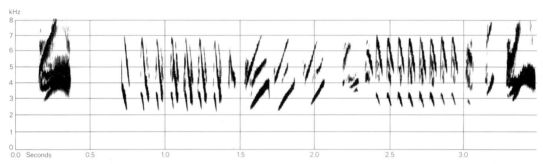

I  P J Sellar  Scotland  May 1974

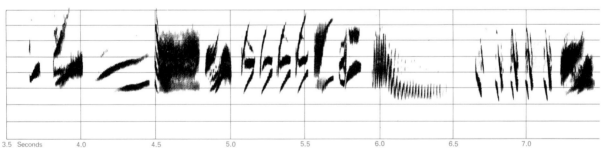

I  *cont.*

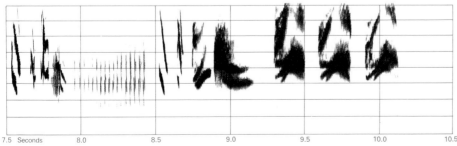

I  *cont.*

calls), and (perhaps) 'tooee' call (at *c.* 4·3 s and preceding prominent buzz near middle of song) (W T C Seale). (1b) Long, progressive sequence (some repetition, etc.) of complex units or motifs (including various calls) given at slow, measured rate, usually continuously for several minutes, but somewhat disjointed, with frequent pauses

(Witherby *et al.* 1938; Marler and Mundinger 1975). Treated as 'rambling song' by Marler and Mundinger (1975), and 'interrupted song' by Zablotskaya (1978*a*); 'gapped song' not unlike 'song-calls' of Rock Sparrow *Petronia petronia* (J Hall-Craggs). Advertising-song sometimes mixed with this utterance, which is given when

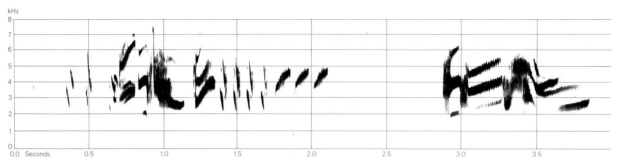

II   P A D Hollom   Scotland   May 1975

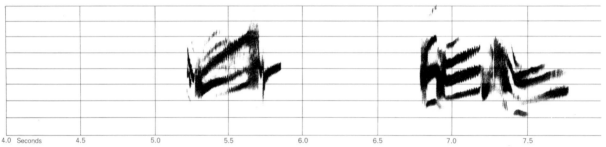

II   *cont.*

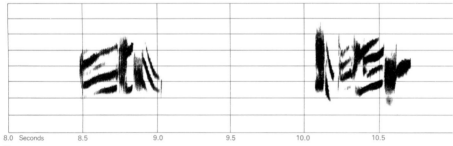

II   *cont.*

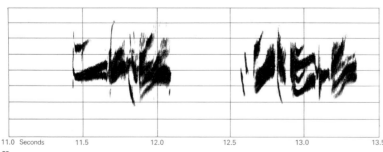

II   *cont.*

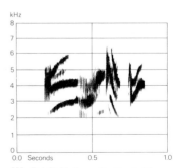

III   P A D Hollom   Scotland   May 1975

disturbed, during mobbing, and in various antagonistic contexts (Marler and Mundinger 1975). In recording (Fig II), 1st phrase (apparently a short song-type 1a) is followed by well-spaced series of complex stereotyped motifs (or very short phrases) often given in progressive sequence, each characteristically buzzy, nasal, and rather twangy, and some components obviously derived from calls (notably 2 and 3a, or stylized versions of these). Excluding 1st phrase and following pause, average length of motifs 0·72 s (0·55–1·0, $n = 11$; last 4 motifs not shown in Fig II), and of pauses 0·86 s (0·48–1·34, $n = 10$). In another recording, 6 such motifs are of 2 basic types, suggesting 't-w-ee-zo-wh-ee' for that shown in Fig III and 't-wee-soo-z-r-r-r-r' for Fig IV; not clear whether motifs

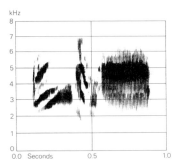

IV  P A D Hollom  Scotland  May 1975

('song-calls') are given by one bird or by 2 alternately (though such antiphonal calling is known for other Carduelinae: see, e.g., *C. cannabina*). (J Hall-Craggs, W T C Seale.) (1c) Courtship-song of ♂ mentioned as further category by Zablotskaya (1978a) but not described in detail. (2) Flight-call. Bouncy chattering or twittering sound: 'tup-up-up', 'chut chululutt', 'tjep-ep-ep', 'jek jek', etc. (Marler and Mundinger 1975; Jonsson 1979; Bergmann and Helb 1982; Bruun *et al.* 1986). Recording (Fig V) suggests 'typ-yp-yp' (W T C Seale); for Flight-calls

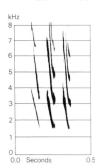

V  W T C Seale  Scotland  July 1991

incorporated in song, see Fig I. Closest to equivalent calls of Redpoll *C. flammea* and Brambling *Fringilla montifringilla* (Bruun *et al.* 1986); perceptibly harder and more metallic than *C. cannabina* (Witherby *et al.* 1938). For further renderings, see (e.g.) Müller (1953) and Pfeifer (1966). Often 3 units in succession, but sometimes gives

longer series; speed of delivery also varies, with fastest series typically given in flight. Commonest call, serving for flock integration; given by both sexes throughout year, when feeding or resting, but mainly at take-off and in flight. (Marler and Mundinger 1975; Zablotskaya 1978a.) Differences noted between individuals and between regions within west European range of nominate *flavirostris* and *pipilans*; also, apparently more marked, between these populations and *montanella* whose calls ('chuv-chuv-chuv') said by Zablotskaya (1978a) to have longer, lower-pitched component than west European birds. Calls of pair-members often share same distinctive pattern (presumably facilitates contact), and, like some other *Carduelis*, birds apparently able to modify calls by learning throughout life. (Marler and Mundinger 1975.) (3) Contact- and alarm-calls. (3a) Distinctive, hoarse, nasal, loud, and rather twangy rasping sound ascending in pitch and suggesting 'TWEit', 'tzeeip' (e.g. 1st unit in Fig I: W T C Seale), 'tsooeek', 'tchooik', etc. (Witherby *et al.* 1938; Marler and Mundinger 1975; Bergmann and Helb 1982; Bruun *et al.* 1986; Harris *et al.* 1989). For further descriptions, see (e.g.) Müller (1953), Pfeifer (1966), Rinnhofer (1972), and Jonsson (1979, 1992). Onset often faint, but rapid crescendo leading to loud and harsh final vibrato. Audible up to *c.* 400 m away and given, singly or in series, as main contact-call, or when feeding or resting, by pair when nest-building, or by bird when others fly over. (Marler and Mundinger 1975; also Zablotskaya 1978a.) Calls of 3 west European populations strikingly similar (Marler and Mundinger 1975), that of *montanella* (also given when disturbed, then being combined with 3c and 3e) apparently a longer and more complex 'tyu-veeu' (Zablotskaya 1978a). (3b) A 'wheet', similar to call 3a, but ascending less steeply and lacking its final harsh vibrato; no differences noted between *pipilans* and nominate *flavirostris* (Marler and Mundinger 1975); calls of this type ('wheet' or 'weip') are shown in Fig I after first series of Flight-calls (W T C Seale). Other renderings, presumably of this call (from context or from comparison of sonagrams), include 'jib jib jib' (Gentz 1971) and 'wäo' (Bergmann and Helb 1982). Analogous call of *montanella* a quiet 'tveet', longer and apparently more complex (Zablotskaya 1978a). Given by both sexes for contact, in series of 6–8 units as begging- and soliciting-call of ♀, and by ♂ in various antagonistic interactions (Marler and Mundinger 1975; Zablotskaya 1978a). (3c) Ascending 'tooee' given by both sexes when disturbed; perhaps uncommon or easily overlooked. Variation noted in small sample of recordings, and some calls apparently intermediate between 'tooee' and 'tew' (call 3d); series combining these 2 calls also noted. (Marler and Mundinger 1975.) In Fig I, perhaps similar sound precedes buzz near middle of song (W T C Seale). Analogous call expressing alarm or fright (notably by subordinate in dispute) in *montanella* is 'tri' with noisy component (as in 'tooee' from Norway illustrated by Marler and Mundinger

1975); given in accelerating series of up to 13 calls with increased excitement (Zablotskaya 1978a). (3d) Gradually ascending, short 'tew' with several overtones and short final descent in pitch; calls similar in nominate *flavirostris* and *pipilans*, but vary within a population, some calls being close to 3c. Given by both sexes when disturbed at nest, during fights, after sudden noise, or when handled. (Marler and Mundinger 1975.) (3e) Harsh, nasal 't' veeyu' reported for *montanella*; disyllabic, harmonics overlaid with noise at start of 2nd syllable. Given when disturbed and when mobbing predator, alternating when already alarmed with call 3a; with increasing danger, up to 20 times in succession and sometimes combined with 'tri' (call 3c); loud and frequent when attempt made to catch bird. (Zablotskaya 1978a.) Sonograms in Zablotskaya (1978a) show similarities between this call, and both longer 3a and shorter call 3c; relationship also likely (but difficult to establish by comparison of descriptions and sonograms) with calls 3c-d of west European birds. (3f) All-clear call of *montanella* 'ti-tri': noisy 'ti' preceding call 3c (Zablotskaya 1978a); see *C. cannabina* for similar call (5h) in that species. (4) Threat-call. Described only for *montanella* and showing some resemblance (in sonogram) to call 2: squeaky, sibilant 'chuv' given in rapid short series (3-5 units) or, with increasing urgency, in slightly lower-pitched sequences of up to 17 units, which may accelerate until individual units more or less merge (Zablotskaya 1978a); quiet 'chuv' unit combined with call 5c in courtship probably also belong here. (5) Courtship-calls of *montanella*. (5a) Quiet squeaking 'tsi' sounds given by both sexes in long sequence. (5b) Delicate piping 'veet-veet-veet' given by ♂ in series of 7-20 units, including during courtship-feeding. (5c) Clicking 'chuv' sounds given by ♂; similar to call 2 in rate of frequency change and tonal quality, but differing in frequency range and energy distribution. (6) Calls noted from feeding and resting *montanella*. (6a) Disyllabic 'tyuryu' of complex (apparently diadic) structure. (6b) Short, abrupt and noisy 'fi'. (Zablotskaya 1978a.) Short 'chee' given when feeding on ground (Jonsson 1979) is perhaps the same. (7) Other calls. (7a) An 'arr' (Müller 1953); context not known, but perhaps related to call 4. (7b) Anxious 'jief' from flock at take-off (Müller 1953); perhaps only alternative rendering of call 3a.

CALLS OF YOUNG. Nestling food-calls high pitched, soft, and rarely heard except during actual feeds. Apparently vary between broods. Closer to fledging, call resembles adult call 3a; and, like adult calls 2 and 3d appearing in repertoire at same time, highly variable within given individual. Acoustically isolated brood of 4 developed similar flight-calls but quite different from parents'. (Marler and Mundinger 1975.)                    MGW

**Breeding.** SEASON. Britain: rather late breeder; in northern Scotland, eggs laid mid-May to mid-August, with peak around mid-June (Williamson 1965; Buckland *et al.* 1990);

in northern England, eggs laid from end of April, mostly late May and June (Taylor 1922, 1935; Orford 1973; Davies 1988). Norway: eggs laid from beginning of April to August; peak in central Norway mid-May to mid-June (Haftorn 1971; Marler and Mundinger 1975). Caucasus: eggs laid from about mid-May (Dementiev and Gladkov 1954). SITE. On or very close to ground in heather *Calluna*, bilberry *Vaccinium*, bracken *Pteridium*, grass tussocks, cotton-grass *Eriophorum*, rush *Juncus*, etc.; often under rock or in crevice and sometimes in dry-stone wall; also on cliff ledge with or without vegetation, and on young conifer in plantation (Taylor 1922; Orford 1973; Marler and Mundinger 1975; D C Jardine). In parts of central Asia, in stone heaps, piles of brushwood, and tall herbs (Gentz 1971); recorded in mammal holes (Saxby 1874; Hüe and Etchécopar 1970). In northern England, 90% of 128 nests in heather, 10% in bracken (Orford 1973, which see for heights above ground of 92 nests in heather). In Britain, of 566 nests, 68% less than 0·3 m above ground, 18% 0·3-0·9 m, 14% greater than 0·9 m; in Scotland, apparently far more frequently in gorse *Ulex* than in England (D C Jardine). In Orkney, in tall *Fuchsia* bushes, and recorded 3 m up in thorn hedge (Lack 1942-3; Balfour 1968). Nest: compact, well-built structure with thick, woven walls and deep cup; foundation of small twigs of heather, etc., roots, stalks, fronds of bracken, grass, moss, etc., lined thickly with felted mass of wool, hair, and sometimes feathers (Gentz 1971; Orford 1973); see also Saxby (1874) and Sloan-Chesser (1937). One nest had outer diameter 9·5 cm, inner diameter 4·0 cm, overall height 5·0 cm, depth of cup 3·0 cm (Makatsch 1976). Building: by ♀ only, accompanied by ♂ (Taylor 1935; Gentz 1971); material gathered in immediate vicinity, except in rocky upland areas (Gentz 1971; Orford 1973). At one nest took 8 days (Saxby 1874, which see for details of building). EGGS. See Plate 60. Sub-elliptical, smooth, slightly or non-glossy; very like Linnet *C. cannabina* but more often with scrawled pattern; pale to darkish-blue with varied specks, spots, small blotches, and scrawls of rust-red to purplish-brown towards broad end, and pink-violet undermarkings (Harrison 1975; Makatsch 1976). Nominate *flavirostris*: $17·4 \times 13·2$ mm ($16·5$-$19·3 \times 12·3$-$13·5$), $n = 17$; calculated weight 1·59 g. *C. f. pipilans*: $17·2 \times 12·9$ mm ($15·3$-$20·0 \times 11·8$-$14·0$), $n = 324$; calculated weight 1·50 g. *C. f. brevirostris*: $16·7 \times 12·4$ mm ($15·0$-$18·0 \times 11·8$-$12·5$), sample size more than 9; calculated weight 1·35 g. (Schönwetter 1984.) Clutch: 4-6 (3-7). In northern England, of 23 clutches: 5 eggs, 26%; 6, 61%; 7, 13%; average 5·87; clutches of 3-4 commoner early in season (Taylor 1935). Average in Scotland 5·41, $n = 22$; in England, 5·53, $n = 256$ (D C Jardine). In central Norway, average 5·8, $n = 6$; all of 4 pairs followed had 2 broods, at least 2 of them starting new nest before 1st brood fledged; of 3 2nd clutches, 2 were of 7 eggs (Marler and Mundinger 1975). Probably 2 broods in Scotland (Buckland *et al.* 1990), and perhaps sometimes in England

(Taylor 1922), but said to be probably 1 brood on Fair Isle (Williamson 1965). Apparently 1 brood in Caucasus but 2 in eastern USSR (Dementiev and Gladkov 1954). See also Gentz (1971). Eggs laid daily (Taylor 1935; Dementiev and Gladkov 1954; Makatsch 1976); at one nest, first egg laid 2 days after nest completed (Saxby 1874). INCUBATION. Takes 12–13 days (11–14); by ♀ only (Taylor 1935; Dementiev and Gladkov 1954; Gentz 1971; Haftorn 1971). Probably starts with 3rd or 4th egg, and young can hatch over 3–4 days (Taylor 1935; Williamson 1965; Harrison 1975); in captivity, started with last egg and ♀ sat on eggs all day at times (Gentz 1971). YOUNG. Fed and cared for by both parents, though at some nests apparently only ♀ feeds, or ♂ via ♀; brooded by ♀ (Williamson 1965; Gentz 1971; Watson 1972). FLEDGING TO MATURITY. Fledging period 11–12 days (10–15) (Taylor 1935; Newton 1972); 10–13 days at 5 nests in Norway (Haftorn 1971). Young fed for *c.* 2 weeks after fledging (Taylor 1935; Marler and Mundinger 1975). For development of young, see Williamson (1965). BREEDING SUCCESS. In Scotland, average brood size at fledging 4·6 ($n=13$ successful nests), 85% of average clutch size; in England 4·89 ($n=157$), 88% of clutch size (D C Jardine). In northern England, nests alongside those of *C. cannabina* are more successful than Scottish nests (above), apparently because of being lower down and better hidden, so less likely to suffer from predation and weather (Orford 1973).                                    BH

**Plumages.** (nominate *flavirostris*). ADULT MALE. In fresh plumage (autumn), upperparts and side of neck closely streaked dull black or black-brown on tawny-brown ground; brown on forehead rather restricted, black predominating; black on nape paler and more diluted, ground-colour of nape and side of neck more buff, forming ill-defined paler shawl. Lower rump and shorter upper tail-coverts rather dull rosy-pink, feather-tips often partly bordered grey or buff, but pink hardly concealed; (60–)80–100% of surface pink (Bub 1978, which see for variation in colour). Longer upper tail-coverts brown with tawny fringes and black shafts. Nasal bristles buff; broad supercilium and short stripe below eye warm buff-cinnamon, separated by dark grey stripe on lore and short stripe behind eye, not sharply defined at rear. Upper cheek and ear-coverts dull black, finely streaked buff-cinnamon (especially on centre of ear-coverts); lower cheek buff-cinnamon, sometimes with indistinct black-mottled moustachial stripe. Chin and throat warm cinnamon-buff, chin with dark grey mottling to distinct black streaks in *c.* 24% of 50 birds examined. Chest, side of breast, and flank buff, marked with short dull black streaks or elongate triangular spots, often narrower on upper chest or mid-chest, browner and less clearly defined on lower flank. Belly, vent, and under tail-coverts white, sometimes with cream tinge, rather sharply contrasting with buff of chest and flank; concealed centres of longest under tail-coverts dull black. Tail greyish-black or black, central pair (t1) tinged brown-grey and with rather ill-defined dull cinnamon (when fresh) to pale grey (when worn) fringes; t2–t6 with sharply defined white fringe along outer web and broad grey-white border along inner web, fringe widest towards feather-bases, faint on t6. Flight-feathers, greater upper primary coverts, and bastard wing black, outer web of p1–p6(–p7) with broad and strongly

contrasting white fringe, forming white panel on closed wing, fringe of outer web of (p7–)p8–p9 pale brown-grey, faint; terminal half of outer web of secondaries with tawny-buff or grey-brown fringe, widest on middle portion of feather; outer webs and tips of primary coverts and of shorter feathers of bastard wing narrowly fringed pale grey; basal inner border of flight-feathers fringed pale grey. Tertials black, tinged brown-grey on outer web and at base, outer web and tip fringed cinnamon-brown, soon bleaching to brown-grey on outer web and grey-white on extreme tip. Greater upper wing-coverts dark brown-grey or fuscous on innermost, gradually blacker towards outermost; fringes along outer webs tinged brown on inner coverts, fringed cinnamon on outer coverts; tips broadly fringed cinnamon-buff, soon bleaching to pale buff or buff-white, especially on outer coverts. Lesser and median upper wing-coverts black-brown or fuscous-grey with ill-defined warmer brown fringes. Under wing-coverts light grey, axillaries and longer inner coverts white. *In worn plumage* (spring), ground-colour of upperparts paler greyish-buff, more cream-grey on hindneck, black more prominent; pink of rump slightly brighter, more rosy-red on feather-tips, some white of feather-centres sometimes partly visible; ground-colour of side of head and neck paler grey-buff, warmer buff restricted to patch round eye (broken in front and behind) and rear of supercilium; chin and throat dirty pale buff, merging into buff-white of chest and flank and this in turn to dirty white of belly; dark streaks or spots on chest, side of breast, and flank more distinct; pale fringes of t1 and tertials worn off, those of tertials often largely so, but white fringes of t2–t5 and of p1–p6 still present, often more contrasting with blacker tail and wing; tips of greater coverts bleached to white, forming bar 2–3 mm wide. Rarely, chest with concealed rose-red tinge (Bub 1976b). ADULT FEMALE. Like adult ♂, but rump and upper tail-coverts streaked black and tawny- or buff-brown, like remainder of upperparts, only occasionally (13% of 54 birds examined) with a few traces of pink on feather-tips (Bub 1978). In contrast to some ♂♂, apparently never black streaks on chin. NESTLING. Down fairly long, plentiful; pale buff-grey. Restricted to dense tufts on crown, back, upperwing, thigh, and (scanty) on rump and belly. JUVENILE. Like adult ♀, but dark streaks of upperparts dull, sooty, broader and less sharply defined, fringes narrower, more buff-brown (when fresh) or pale buff-grey (when worn). Ground-colour of side of head and neck and of underparts cream-buff (when fresh) or dirty buff-white (when worn), warm cinnamon tinge absent; feathering shorter and looser than in adult. Tail as adult, but feather-tips often bluntly pointed, less rounded than in adult. Wing as adult, no constant difference in pattern or colour of greater coverts; greater upper primary coverts narrowly fringed buff when fresh, less pure light grey than in fresh adult, but fringe in both soon fading to off-white. FIRST ADULT. Like adult, ♂ with pink rump, ♀ without (according to Bub 1978, *c.* 59% of 121 1st adult ♀♀ examined showed a few traces of pink, but this not apparent in skins examined; see also Svensson 1992). Juvenile tail, flight-feathers, tertials, and greater primary and secondary coverts retained, on average more worn than those of adult at same time of year; tips of tail-feathers pointed, pale fringes worn off by about November (in adult, tips rounded, off-white fringes along tips present to about March); occasionally, t1 new, contrasting in shape and wear with t2–t6 (in adult, all feathers more or less uniform). Ageing often difficult on breeding grounds, when heavily worn.

**Bare parts.** ADULT. Iris dark brown. Bill in summer light or dark horn-grey, sometimes tinged yellow, green, olive, or horn-white, culmen ridge darker horn-grey or black-brown; between August and early October, changes through grey-green, olive-green, and yellow-green to bright yellow, usually with dark brown tip on culmen; shades to horn and grey again from March–April. Leg and foot brown-black or black. NESTLING. Mouth flesh, darker purplish at back; gape-flanges ivory or cream-white. JUVENILE, FIRST ADULT. At fledging, iris dark brown, bill pale or dark horn-grey with yellow or pink tinge, changing to pale yellow-white in August and to yellow with small dusky tip (like adult) in September. Leg and foot flesh-pink, shading to brown in August–September; in winter, purplish-brown or dark brown, less black than in adult. (Witherby *et al.* 1938, Cornwallis and Smith 1964; Bub 1977; BMNH, RMNH, ZMA.)

**Moults.** ADULT POST-BREEDING. Complete; primaries descendent. In Britain, starts with p1 in July (median date 20 July), but occasionally starts late June or up to mid-August; completed with re-growth of p9 after 70–75 days mid-September to mid-October, on average *c.* 25 September (Newton 1968, 1972; Ginn and Melville 1983). In western Norway, single birds on 15 and 20 September had primary moult scores 34 and 43 (ZMA). In Afghanistan, 20 July to 25 September, 15 birds either heavily worn or in full moult; in Kashmir, 13 adults collected 13 September to 6 October all in moult (Vaurie 1949*b*). In Mongolia, single bird had just started 30 June, 2 others still without moult (Piechocki and Bolod 1972). In Pamir-Alay ranges, just starting 18 August; in Altai, about mid-August (Dementiev and Gladkov 1954). POST-JUVENILE. Partial: head, body, lesser, median, and variable number of inner greater coverts, and occasionally 1–2 tertials. Starts shortly after fledging, and timing thus highly variable, fledging occurring June–August (Johansen 1944). In Britain, moult July–September (Ginn and Melville 1983). In western Norway, moult just started in single bird from 14 September (ZMA). In Afghanistan, 14 birds collected 8 August to 25 September either fully juvenile or in moult; in Kashmir, 5 birds collected 13 September to 6 October in moult (Vaurie 1949*b*).

**Measurements.** ADULT, FIRST ADULT. Nominate *flavirostris*. Scandinavia, April–October, and Netherlands, winter; skins (BMNH, RMNH, ZMA: A J van Loon, C S Roselaar). Bill (S) to skull, bill (N) to distal corner of nostril; exposed culmen on average 2·8 less than bill (S).

| | | | |
|---|---|---|---|
| WING | ♂ | 78·0 (1·07; 44) 76–80 | ♀ 76·1 (1·74; 27) 73–79 |
| TAIL | | 59·0 (1·67; 22) 56–62 | 57·9 (2·32; 13) 54–62 |
| BILL (S) | | 11·4 (0·31; 23) 10·8–11·9 | 11·4 (0·46; 14) 10·6–11·9 |
| BILL (N) | | 6·8 (0·30; 23) 6·3–7·5 | 6·9 (0·27; 13) 6·4–7·3 |
| TARSUS | | 15·7 (0·56; 18) 14·8–16·5 | 15·8 (0·38; 9) 15·0–16·3 |

Sex differences significant for wing.

Wing. (1) Norway, June–July, live birds. North-west Germany, autumn and winter, live birds: (2) adult, (3) 1st adult (average and *n* from other sample than range). (Bub 1976*a*). (4) Germany (Eck 1985*a*). (5) Slovakia, autumn (Kaňuščák 1979; see also Kaňuščák and Šnajdar 1972).

| | | | | | |
|---|---|---|---|---|---|
| (1) | ♂ | 78·4 ( 12) 76–80 | ♀ | 76·5 ( 2) 76–77 |
| (2) | | 79·4 ( 67) 76–82 | | 76·9 (101) 72–80 |
| (3) | | 78·5 (520) 74–83 | | 76·4 (521) 70–80 |
| (4) | | 78·1 ( 15) 77–79 | | 76·5 ( 10) 75–78 |
| (5) | | 76·7 (154) 73–82 | | 74·3 (152) 70–79 |

See also Beretzk and Keve (1971).

Britain. Wing and bill of (1) *bensonorum*, Outer Hebrides (Scotland), (2) *pipilans*, Shetland, Orkney, Highland area (Scotland), and northern England; other measurements combined, March–September; skins (BMNH, RMNH, ZMA: A J van Loon).

| | | | | |
|---|---|---|---|---|
| WING (1) | ♂ | 76·2 (1·62; 10) 75–79 | ♀ 73·4 (1·97; 7) 70–76 |
| (2) | | 76·4 (1·72; 11) 74–79 | 74·3 (2·62; 12) 69–79 |

| | | | |
|---|---|---|---|
| TAIL | 57·9 (2·34; 20) 54–63 | 55·9 (1·98; 18) 52–61 |
| BILL (S) (1) | 11·6 (0·24; 8) 11·3–11·8 | 11·4 (0·34; 7) 10·7–11·9 |
| (2) | 11·6 (0·45; 11) 10·7–12·4 | 11·6 (0·53; 13) 10·8–12·4 |
| BILL (N) (1) | 7·0 (0·29; 9) 6·6–7·6 | 6·8 (0·21; 7) 6·5–7·1 |
| (2) | 6·9 (0·39; 10) 6·4–7·5 | 6·8 (0·33; 11) 6·1–7·2 |
| TARSUS | 15·9 (0·66; 19) 14·9–17·2 | 15·7 (0·65; 16) 14·7–16·8 |

Sex differences significant for wing and tail. Live birds Oronsay (Argyll, Scotland), wing: ♂ 79·0 (1·85; 27) 76–81, ♀ 77·0 (1·78; 34) 73–81 (D C Jardine).

*C. f. brevirostris.* Eastern Turkey and Caucasus area to northwest Iran, all year; skins (BMNH, RMNH: A J van Loon).

| | | | |
|---|---|---|---|
| WING | ♂ 76·2 (1·25; 11) 74–79 | ♀ 74·0 (1·44; 12) 72–76 |
| TAIL | 59·6 (1·68; 11) 57–63 | 58·2 (1·47; 12) 56–61 |
| BILL (S) | 12·0 (0·37; 9) 11·4–12·8 | 11·9 (0·29; 11) 11·5–12·1 |
| BILL (N) | 7·4 (0·27; 10) 6·9–7·9 | 7·2 (0·40; 11) 6·4–7·7 |
| TARSUS | 15·8 (0·51; 10) 15·0–16·6 | 15·6 (0·63; 10) 14·7–16·8 |

Sex differences significant for wing and tail. Eastern Turkey, wing: ♂ 74·9 (1·24; 5) 74–77 (Kumerloeve 1967*a*, 1969*a*). North-east Turkey: wing of 8 ♂♂ and 3 ♀♀ 74–77 (Jordans and Steinbacher 1948). Caucasus and Transcaucasia: wing, ♂ 75·4 (15) 73–80·5, ♀ 71·7 (6) 71–72·5 (Dementiev and Gladkov 1954). Northern and south-west Iran: wing, ♂ 75–80 (Sarudny and Härms 1914).

*C. f. kirghizorum.* Wing: ♂ 74·3 (14) 72–77·5, ♀ 73·2 (10) 72–75 (Dementiev and Gladkov 1954).

*C. f. korejevi* and *pamirensis.* Wing: ♂ 70–76, ♀ 68–75 (Sarudny and Härms 1914).

*C. f. pamirensis.* Afghanistan, wing: ♂ 73·5 (2·38; 4) 71–76, ♀ 71·7 (3) 71–73 (Paludan 1959); range of 8 ♂♂ and 7 ♀♀ 71–77 (Meinertzhagen 1938).

*C. f. altaica.* Mongolia, wing: ♂ 78·3 (2·59; 13) 73–82, ♀ 77·0 (2·00; 6) 74–80 (Piechocki and Bolod 1972; Mauersberger *et al.* 1982; Piechocki *et al.* 1982).

*C. f. miniakensis.* Wing: ♂ 77 (5) 75–78 (Vaurie 1959).

*C. f. rufostrigata.* Wing: Ladakh (India), ♂ 80–81 (5), ♀ 73–81 (5) (Meinertzhagen 1927); Himalayas, ♂ 80·5 (5) 77–85 (Vaurie 1959); Tibet, ♂ up to 87, ♀ up to 82 (Vaurie 1972).

**Weights.** ADULT, FIRST ADULT. Nominate *flavirostris*. (1) Norway, September (Haftorn 1971). (2) Germany, October–April (Bub 1976*a*, which see for many details). (3) Netherlands and Germany, October–March (Eck 1985*a*; ZMA). (4) Slovakia, November (Kaňuščák 1979). *C. f. pipilans*. (5) Oronsay (Argyll, Scotland), April (D C Jardine). *C. f. brevirostris*. (6) Eastern Turkey, May, June, and November (Kumerloeve 1967*a*, 1969*a*). *C. f. kirghizorum* and/or *korejevi*. (7) Kazakhstan (Korelov *et al.* 1974). *C. f. altaica*. Mongolia, (8) May–August, (9) February (Piechocki and Bolod 1972; Piechocki *et al.* 1982). *C. f. pamirensis*. (10) Afghanistan, June and October (Paludan 1959).

| | | | |
|---|---|---|---|
| (1) | ♂ | 15·7 ( — ; 6) 14·8–16·6 | ♀ — ( — ; 3) 14·5–15·7 |
| (2) | | 16·4 ( — ; 1066) 11–21 | 16·0 ( — ; 1187) 11–23 |
| (3) | | 15·5 (0·88; 6) 14·0–16·5 | 15·7 (0·55; 10) 14·5–16·2 |
| (4) | | 16·6 ( — ; 154) 14·5–19·0 | 15·9 ( — ; 152) 13·5–19·0 |
| (5) | | 16·2 (0·85; 32) 14·4–18·6 | 16·0 (0·87; 38) 14·4–18·2 |
| (6) | | 14·8 (0·45; 5) 14–15 | — ( — ; — ) |
| (7) | | 15·3 ( — ; 24) 13·2–17·7 | 14·4 ( — ; 8) 13·5–17·0 |
| (8) | | 15·0 (1·50; 9) 12·0–17·0 | 14·0 ( — ; 3) 12·0–15·0 |
| (9) | | 15·7 ( — ; 3) 15·0–17·0 | 15·5 ( — ; 3) 15·0–16·5 |
| (10) | | 13·2 (0·50; 4) 13–14 | 13·0 ( — ; 3) 12–14 |

Exhausted birds, Netherlands: 12·0 (1·43; 5) 10·0–13·2 (ZMA). See also Weigold (1926) and Newton (1972).

**Structure.** Wing rather long, broad at base, tip bluntly pointed. 10 primaries: p8 longest, p9 and p7 0–1 shorter, p6 2–3·5, p5 8–

11, p4 13–16, p3 15–19, p2 18–22, p1 20–24; p10 strongly reduced, 49–58 shorter than p8, 7–11 shorter than longest upper primary covert. Outer web of p6–p8 emarginated, inner web of (p7–)p8–p9 with faint notch. Tip of longest tertial reaches tip of about p2. Tail rather long, tip forked; 12 feathers, t1 9·9 (15) 7–14 shorter than t6. Bill short, conical; depth and width at base each 6·0 (20) 5·5–6·5 in nominate *flavirostris* and *pipilans*, depth 6·0 (12) 5·6–6·5 and width 5·8 (12) 5·3–6·1 in *brevirostris*, *kirghizorum*, *korejevi*, and *montanella*, depth 6·7 (5) 6·3–7·0 and width 6·5 (5) 6·2–6·8 in *rufostrigata*. Culmen straight, except for slightly decurved tip, cutting edges and gonys straight; tip slightly compressed laterally. For length of individual flight- and tail-feathers, see Busching (1988). Tarsus and toes rather long and slender. Middle toe with claw 16·1 (10) 15–17·5; outer toe with claw *c.* 67% of middle with claw, inner *c.*71%, hind *c.* 84%. Remainder of structure as in Goldfinch *C. carduelis*.

**Geographical variation.** Slight in western populations, mainly involving general colour; more pronounced in Caucasus area and central Asia, involving colour and size. *C. f. pipilans* from Britain and Ireland (except Outer Hebrides) slightly darker in fresh plumage than nominate *flavirostris* from Scandinavia, ground-colour of upperparts more tawny-brown, less buff-brown, hind-neck and side of neck similar in colour to remainder of upperparts (in nominate *flavirostris*, usually paler than remainder, more buff or greyish-buff), streaks on mantle on average slightly broader; chin to chest and flank deeper cinnamon-buff, hardly contrasting with cream-buff belly (in nominate *flavirostris*, paler, more buff, contrasting more with white belly), streaks on chest and side of breast on average slightly more marked and extending further down on lower flank; tertials and upper wing-coverts more black-brown, less grey-brown, tips of greater coverts cinnamon-buff, less pale buff or whitish; when worn (spring), colour rather similar to that of fresh nominate *flavirostris*, but worn nominate *flavirostris* in spring has ground-colour of upperparts more pale buff-grey, of underparts more buff-white. *C. f. bensonorum* from Outer Hebrides more heavily marked with deeper black streaks than *pipilans*, ground-colour of upperparts duller brown. See also Clancey (1954) and Vaurie (1956a). Differences between single individuals of nominate *flavirostris*, *pipilans*, and *bensonorum* often not marked, but races readily separable when series of skins compared. Ground-colour of most races of south-east Europe and Asia markedly paler, contrasting distinctly with dark streaking; pink on rump generally paler, sometimes partly replaced by white; voice also markedly different (Zablotskaya 1978a). Colour, width, and extent of dark streaking on head and body of *brevirostris* from eastern Turkey and Caucasus area to Iran similar to that in nominate *flavirostris*, but ground-colour of upperparts contrastingly pale tawny-cinnamon (when fresh)

to grey-buff (when worn); ground-colour of side of head and of chin to chest buff, of remainder of underparts cream-buff, black streaks heavy, sharply contrasting; tail-feathers more broadly fringed buff to white; upper wing-coverts and tertials brown-grey, greater coverts and tertials broadly tipped cream-white; terminal half of outer web of secondaries broadly fringed buff-grey, primaries more broadly fringed white. *C. f. korejevi* from eastern Kazakhstan paler than *brevirostris*, streaks of upperparts more sepia, less black-brown, but sharply defined, ground-colour pale sandy-buff, warmer cinnamon on outer scapulars; upper wing-coverts and tertials cinnamon-brown or grey-brown, tips and fringes of tail, greater coverts, tertials, and flight-feathers broader than in *brevirostris*, inner webs of secondaries with white border. *C. f. kirghizorum* from Volga–Ural steppe east to Ayaguz in eastern Kazakhstan close to *korejevi*, but streaks on upperparts on average narrower, those on chest narrower and reaching less far down; rather a poor race, sometimes included in *korejevi* (e.g. Stepanyan 1978, 1990), but maintained here, following Sushkin (1925), Johansen (1944), Dementiev and Gladkov (1954), and Korelov *et al.* (1974). *C. f. pamirensis* from Pamir-Alay ranges to Afghanistan and northern Pakistan also close to *korejevi*, but slightly darker, streaks on upperparts, chest, side of breast, and flank slightly blacker, reaching further up to throat and down on flank; ground-colour of upperparts more cinnamon, of side of head and chin to chest warmer cinnamon-buff; wing-coverts and tertials darker brown, pale fringes and tips narrower; approximately intermediate between *korejevi* and *brevirostris* (Sarudny and Härms 1914; BMNH). *C. f. altaica* from central Altai to Tuva and western Mongolia large and dark; streaks on head and body broad and black, but rather poorly defined; ground-colour of upperparts mealy cinnamon-brown. *C. f. montanella* from mountains surrounding Tarim basin is palest race; like *kirghizorum*, but streaks of upperparts narrower, paler grey-brown, less sharply defined, ground-colour extensively sandy pink-buff; chin and throat pale sandy-buff merging into cream-white of belly, streaks of chest brown-grey, restricted and poorly defined, sometimes almost absent; in winter, occurs widely in Tien Shan, Tarim basin, and northern Pakistan. *C. f. miniakensis* from central China as dark as *brevirostris*, but ground-colour of upperparts more rufous-brown, dark streaks more diffuse, but less broad than in *altaica*; smaller than *altaica*. *C. f. rufostrigata* from Himalayas is darkest race of central Asia; streaking as in nominate *flavirostris*, but ground-colour of upperparts more rufous-tawny, of underparts extensively cinnamon; bill heavier than in other races.

For relationships, see Linnet *C. cannabina*.

**Recognition.** For difference of juvenile from juvenile *C. cannabina*, see Svensson (1992).                                          CSR

---

*Carduelis flammea* **Redpoll**

PLATES 42 and 44 (flight)
[facing page 473, and between pages 640 and 641]

Du. Barmsijs        Fr. Sizeron flammé        Ge. Birkenzeisig
Ru. Обыкновенная чечетка        Sp. Pardillo sizerín        Sw. Gråsiska        N. Am. Common Redpoll

*Fringilla flammea* Linnaeus, 1758. Synonym: *Acanthis linaria*.

Polytypic. Nominate *flammea* (Linnaeus, 1758), northern Eurasia from Norway and Baltic countries east to Kamchatka and Anadyrland; also, northern North America from Alaska east to Labrador and Newfoundland; *cabaret* (P L Statius Müller), Britain and Alps, spreading recently along North Sea coast from north-west France to Denmark, southern

Sweden, and south-west Norway, and in hills of central Europe from central France, eastern Belgium, and Vosges east to Czechoslovakia, Poland, and (isolated) probably northern Rumania and western Bulgaria; *rostrata* (Coues, 1862), Iceland, southern Greenland, and Baffin Island.

**Field characters.** 11·5–14·5 cm; wing-span 20–25 cm. Size varies noticeably: British and montane European race *cabaret* ('Lesser Redpoll') slightly smaller and more finely built than Linnet *C. cannabina*, with wing-span under 22·5 cm; Holarctic race, nominate *flammea* ('Mealy Redpoll'), is as large as *C. cannabina*, with wing-span over 21 cm (but smaller than Greenland race of Arctic Redpoll *C. hornemanni hornemanni*); Icelandic, Baffin Island, and Greenland race *rostrata* ('Greenland Redpoll') marginally larger than *C. cannabina* (and *C. hornemanni exilipes* of Holarctic tundra); ♀♀ average nearly 5% smaller than ♂♂. Small, attractive finch, with form somewhat like tit *Parus* and mainly arboreal behaviour in *cabaret* and nominate *flammea* (akin to Siskin *C. spinus*), but more robust, with form more like sparrow *Passer* and more terrestrial behaviour in *rostrata* (akin to Twite *C. flavirostris*). At distance, recalls *C. cannabina* and *C. flavirostris* but at close range adult shows diagnostic combination of red forecrown, blackish frontal band and lores, black chin, double wing-bars, and pale tips to tertials. Between races, upperpart tone varies from most tawny in *cabaret* to darkest brown in *rostrata* and greyest in nominate *flammea*, with pale rump, particularly in nominate *flammea* inviting confusion with *C. hornemanni*. Juvenile less easy to separate from ♀♀ and young of *C. cannabina* and *C. flavirostris*, but lores and chin greyish-black, while streaks on underparts much sharper and darker. In *cabaret*, specific identification quickly completed by sharply pointed bill, rather delicate form, great agility when feeding, dancing flight, and muttered flight-call; in nominate *flammea* and *rostrata*, structure more robust but flight action and call still distinctive. Sexes dissimilar, particularly in breeding season; marked seasonal variation in ♂. Juvenile separable. 3 races in west Palearctic, but variation and intergradation are complex (see Geographical Variation); some hybridization with *C. hornemanni* may also occur (see *C. hornemanni*). Following account concentrates on nominate *flammea*.

(1) Holarctic race, nominate *flammea*. ADULT MALE. Moults: July–October (complete). Intermingling of black-brown and buff-white to almost white lines on upperparts, wings, breast, and flanks makes nominate *flammea* a noticeably grey bird, overall tone also enhancing its size. On some worn birds, rump can appear almost unstreaked and white, and such individuals easily confused with *C. hornemanni*. On some fresh-plumaged birds, rump noticeably streaked, thus having white only in lines or patches; easily confused with *rostrata*. Nominate *flammea* generally the most variable race, with quite wide extremes of plumage tones and marks visible in any flock. Brightening of pink plumage and loss of initial tawny tones in breeding season due entirely to wear. Forecrown

bright, glossy crimson-red, rear crown and nape tawny-brown, noticeably streaked brown-black. Forehead buff; narrow band over bill, lores, and chin-patch black; pale greyish-buff supercilium and surround to ear-coverts contrasts with crown and nape and with dusky-speckled buff-brown ear-coverts. Mantle and scapulars tawny-brown, broadly lined brown-black, particularly on lower feathers; in centre, ground-colour almost white, increasing contrast of 1–2 pairs of lines; back as centre of mantle; rump buffish-rose, noticeably less streaked than back and upper tail-coverts. Cheeks, lower throat, and sides of breast to fore-flanks rose-pink with buff or white feather-tips; flanks buff, strongly marked with lines of dark brown streaks, usually 2 distinct and 1 faint; lower body white, streaked brown on under tail-coverts. Wings black-brown, quite boldly marked by buff-white tips to median and greater coverts (forming both upper and striking lower wing-bar) and similarly coloured fringes to tertials, distal halves of secondaries, and outer webs and (particularly) tips of primaries; creates intricate pattern at close range, featuring in particular dark bases to secondaries (contrasting with lower wing-bar as in Goldcrest *R. regulus*) but lacking striking pale blaze along folded primaries shown by *C. cannabina* and *C. flavirostris*. Underwing buffish-white. Tail black-brown, feathers fringed whitish-buff on outer webs and looking pale-edged side on. With wear, upperparts lose tawny tone and contrast there of dark streaks and more greyish-white feather-fringes increases noticeably, giving characteristic grey or frosty look; underparts and rump become noticeably more pink and brighter; wings and tail lose narrower fringes and width of wing-bars and tertial-marks decreases though with whitening may look just as striking. Note that extent of pink plumage varies individually and may even be replaced by orange to yellow, while colour of forecrown may also be orange. Rather small, sharply pointed, triangular bill straw-yellow, with dusky tip and culmen ridge. Legs dark brown. ADULT FEMALE. Forecrown red, orange,

PLATE 43. *Loxia leucoptera bifasciata* Two-barred Crossbill (p. 672): **1** ad ♂ summer, **2** ad ♂ winter, **3** ad ♀ summer, **4** 1st ad ♂ winter, **5** juv.

*Loxia curvirostra* Crossbill (p. 686). Nominate *curvirostra*: **6** ad ♂ summer, **7** ad ♂ winter, **8** ad ♀ summer, **9–10** 1st ad ♂ winter, **11** juv ♂, **12** juv ♀. *L. c. corsicana*: **13** ad ♂ summer, **14** ad ♀ summer. *L. c. balearica*: **15** ad ♂ summer, **16** ad ♀ summer. *L. c. poliogyna*: **17** ad ♂ summer. *L. c. guillemardi*: **18** ad ♂ summer.

*Loxia scotica* Scottish Crossbill (p. 707): **19** ad ♂ summer.

*Loxia pytyopsittacus* Parrot Crossbill (p. 717): **20** ad ♂ summer, **21** ad ♀ summer. (CR)

PLATE 44. *Carduelis cannabina cannabina* Linnet (p. 604): **1** ad ♂ summer, **2** ad ♀, **3** juv. *Carduelis flavirostris flavirostris* Twite (p. 625): **4** ad ♂ winter, **5** ad ♀ winter. *Carduelis flammea* Redpoll (p. 639). Nominate *flammea*: **6** ad ♂ winter, **7** 1st ad ♂ winter. *C. f. cabaret*: **8** ad ♂ winter, **9** 1st ad ♂ winter. (CR)

PLATE 45. *Carduelis hornemanni exilipes* Arctic Redpoll (p. 661): **1** ad ♂ winter, **2** 1st ad ♂ winter. *Loxia leucoptera bifasciata* Two-barred Crossbill (p. 672): **3** ad ♂ summer, **4** ad ♀ summer. *Loxia curvirostra curvirostra* Crossbill (p. 686): **5–6** ad ♂ summer, **7** ad ♀ summer. *Loxia pytyopsittacus* Parrot Crossbill (p. 717): **8** ad ♂ summer. (CR)

PLATE 46. *Rhodopechys sanguinea* Crimson-winged Finch (p. 729). Nominate *sanguinea*: **1** ad ♂ summer, **2** ad ♂ winter, **3** ad ♀, **4** juv. *R. s. aliena*: **5** ad ♂ winter. *Rhodospiza obsoleta* Desert Finch (p. 739): **6** ad ♂ summer, **7** ad ♀, **8** juv. (HB)

PLATE 47. *Bucanetes mongolicus* Mongolian Trumpeter Finch (p. 748): **1** ad ♂ summer, **2** ad ♂ winter, **3** ad ♀, **4** juv. *Bucanetes githagineus* Trumpeter Finch (p. 754). *B. g. crassirostris*: **5** ad ♂ summer, **6** ad ♂ winter, **7** ad ♀, **8** juv. *B. g. amantum*: **9** ad ♂ summer. (HB)

PLATE 48. *Carpodacus erythrinus erythrinus* Scarlet Rosefinch (p. 764): **1** ad ♂ summer, **2** ad ♂ winter, **3** ad ♀, **4** juv. *Carpodacus synoicus synoicus* Sinai Rosefinch (p. 783): **5** ad ♂ summer, **6** ad ♂ winter, **7** ad ♀, **8** juv. (HB)

PLATE 49. *Carpodacus roseus* Pallas's Rosefinch (p. 789): **1** ad ♂ summer, **2** ad ♀, **3** 1st ad ♀ autumn. *Carpodacus rubicilla rubicilla* Great Rosefinch (p. 792): **4** ad ♂ summer, **5** ad ♂ winter, **6** ad ♀, **7** juv. (HB)

PLATE 50. *Rhodopechys sanguinea sanguinea* Crimson-winged Finch (p. 729): **1** ad ♂ summer, **2** ad ♀. *Rhodospiza obsoleta* Desert Finch (p. 739): **3** ad ♂ summer, **4** ad ♀. *Bucanetes mongolicus* Mongolian Trumpeter Finch (p. 748): **5** ad ♂ summer, **6** ad ♀. *Bucanetes githagineus crassirostris* Trumpeter Finch (p. 754): **7** ad ♂ summer, **8** ad ♀. *Carpodacus erythrinus erythrinus* Scarlet Rosefinch (p. 764): **9** ad ♂ summer, **10** ad ♀. (HB)

PLATE 51. *Carpodacus synoicus synoicus* Sinai Rosefinch (p. 783): **1** ad ♂ summer, **2** ad ♀. *Carpodacus roseus* Pallas's Rosefinch (p. 789): **3** ad ♂ summer, **4** ad ♀. *Carpodacus rubicilla rubicilla* Great Rosefinch (p. 792): **5** ad ♂ summer, **6** ad ♀. *Pinicola enucleator enucleator* Pine Grosbeak (p. 802): **7** ad ♂, **8** ad ♀. (HB)

PLATE 52. *Pinicola enucleator enucleator* Pine Grosbeak (p. 802): **1** ad ♂ summer, **2** ad ♂ winter, **3** ad ♀ summer, **4** ad ♀ winter, **5** 1st ad ♂ winter, **6** juv. (HB)

PLATE 53. *Pyrrhula pyrrhula* Bullfinch (p. 815). Nominate *pyrrhula*: **1** ad ♂ summer, **2** ad ♀ summer. *P. p. pileata*: **3** ad ♂. *Coccothraustes coccothraustes coccothraustes* Hawfinch (p. 832): **4** ad ♂, **5** ad ♀. *Hesperiphona vespertina* Evening Grosbeak (p. 847): **6** ad ♂, **7** ad ♀. (HB)

PLATE 54. *Uragus sibiricus* Long-tailed Rosefinch (p. 814): **1** ad ♂ summer, **2** ad ♀, **3** 1st ad ♀ winter. *Pyrrhula pyrrhula pyrrhula* Bullfinch (p. 815): **4** ad ♂, **5** ad ♀ summer, **6** ad ♀ winter, **7** juv. (HB)

PLATE 55. *Pyrrhula pyrrhula* Bullfinch (p. 815). *P. p. europoea*: **1** ad ♂, **2** ad ♀ summer. *P. p. pileata*: **3** ad ♂, **4** ad ♀ summer. *P. p. iberiae*: **5** ad ♂, **6** ad ♀ summer. *P. p. murina*: **7** ad ♂, **8** ad ♀. (HB)

PLATE 56. *Coccothraustes coccothraustes* Hawfinch (p. 832). Nominate *coccothraustes*: **1** ad ♂, **2** ad ♀, **3** juv. *C. c. nigricans*: **4** ad ♂. *C. c. buvryi*: **5** ad ♂. *Hesperiphona vespertina* Evening Grosbeak (p. 847): **6** ad ♂, **7** ad ♀, **8** 1st ad ♂ winter. (HB)

yellow, or even brown. Rarely shows any pink, though a few have hint around face and on buffish-white rump. Black chin sometimes larger than in ♂; throat and upper breast warm buff, streaked dark brown and joining with flanks; fringes and tips to wing-feathers paler than ♂. JUVENILE. Resembles ♀ but distinguished at close range by lack of crimson forecrown, indistinct dusky chin, and heavier streaks on ear-coverts, throat, and breast. Rump streaked black-brown; wing-marks noticeably pale whitish-buff. FIRST-YEAR MALE. Resembles adult ♀ but most have traces of pink on cheeks and rump. (2) Icelandic, Baffin Island, and Greenland race, *rostrata*. More distinctive in appearance than other races. Noticeably larger, with strong, almost sparrow-like bill (with curved culmen) and postures, and distinctly dark plumage mixing features of both *flammea* and *cabaret*. At close range, ground-colour of upperparts strongly tawny and marked with broad, heavy black-brown streaks in regular linear pattern (lacking pale centre to back of nominate *flammea*); ground-colour of face, breast, and flanks warmer buff and intensely marked on side from breast to rear flanks with 3 almost black streaks ('zebra stripes'). Grey on rump confined to short lines above long grey-brown and black streaked area above and on upper tail-coverts; less obvious than on nominate *flammea*; strong greyish-white wing markings as striking as on *flammea*. Adult ♂ less widely drenched pink, this colour being concentrated on face and upper breast. For differences of some Icelandic birds, see Geographical Variation. (3) British and montane European race, *cabaret*. Smallest and most tawny-brown race, sharing heavy streaks of *rostrata* but these, by lack of contrast with pale ground-colour, actually less visible than in nominate *flammea*. ♀ and juvenile show strong yellowish suffusion on throat and sides of breast. At close range, forecrown noticeably dark, crimson, lacking glossy flash of other races; rump rather dull and well streaked and wing markings often no brighter than buff. General appearance warmest and least patterned of all races; worn ♂♂ become heavily drenched with almost red-pink on face, breast, chest, upper rump, and lower back. Note that, at least in north Scottish populations, birds closely resembling *rostrata* are also present, and that apparent intergrades with nominate *flammea* occur within winter flocks of *cabaret*.

When distinctive flight-call heard, can be confused only with *C. hornemanni*, but at distance or in poor view silent bird may need to be distinguished from *C. cannabina*, *C. flavirostris*, *C. spinus* (frequent close companion on migration and in winter), and juvenile serins *Serinus* (see above and those species). Chief problem is difficulty of racial determination; often insurmountable due to extent of taxonomic variation and hybridization. Thus important to make racial claims in full context of either breeding distribution or pronounced irruptions (to which all northern races are particularly prone) and associated vagrancies of sympatric species; latter helpful in case of *rostrata*.

Flight noticeably dancing, lighter and more steeply undulating or bounding than *C. cannabina*; in smaller races, recalls *Serinus*. Song-flight looping and circling, with wings beaten more slowly than normal and outstretched in glides. Escape-flight variable, usually to nearest cover, or in tundra at first downwind and then circling back to food source. Gait a hop, varied by jumps; capable of marked agility among twigs and foliage, moving and clinging (even upside down) like tit, and moving quickly over and through low plants. Stance on perch variable; on ground, low and hunched in smaller races, but often erect with tail held up in at least *rostrata*. Shy when breeding, when most keep within tree canopy and more often heard than seen, but gregarious and often quite tame on migration and in winter, when tight-packed flocks—often in company with *C. spinus* and rarely *C. hornemanni*—are relatively easy to observe in seed-bearing trees and plants. Behaviour and muttered calls convey impression of constant business. As aid to identification in Britain, most significant timings of movements are (a) post-breeding dispersal of *cabaret* from early September, (b) east coast passage or occasional marked irruptions of nominate *flammea* in October–November, and (c) northern arrivals of vagrant *rostrata* from mid-September to early November.

Commonest call, in flight or perched, a chattering, twittering, or, at distance, muttering, staccato 'chuch-uch-uch-uch' or variant with metallic echo which gives ready differentiation from *C. cannabina* and *C. flavirostris*; chorus of flocks with buzzing timbre. Song combines common call with rattle, 'chuch-uch-uch-errrrrr'. Other common call when perched, a plaintive 'dsooee'.

**Habitat.** Varies widely across west Palearctic, being associated with climates ranging from arctic to relatively warm (Voous 1960b). In arctic zones, where Arctic Redpoll *C. hornemanni* competes, will occupy treeless tundra, preferably with shrubby growth of creeping osier *Salix* and dwarf birch *Betula* as well as stunted forests and open taiga (Dementiev and Gladkov 1954; Flint *et al.* 1984). In southern Greenland, avoids foggy coastal zones, preferring warmer dry summer climate of interior below *c.* 200 m, especially sheltered places protected from wind, and slopes or hollows with luxuriant growth of willow *Salix*, birch *Betula*, juniper *Juniperus*, alder *Alnus*, and rowan *Sorbus*. Climatic amelioration has led to its replacement of *C. hornemanni* in certain areas, confirming that many arctic habitats, although occupied, are marginal, and that taller and more extensive stands of woody vegetation are favoured. (Salomonsen 1950–1.) In Norway also, interior birchwoods are preferred to coastal regions, but birds breed from sea-level to *c.* 1000 m, and from dry to marshy ground, in birch and willow bushes but also in open birchwoods and even in stands of pine *Pinus*; shows liking for vicinity of colonies of Fieldfares *Turdus pilaris*, which may help to guard against predators (Bannerman 1953a).

Disjunct population in Britain has spread and increased

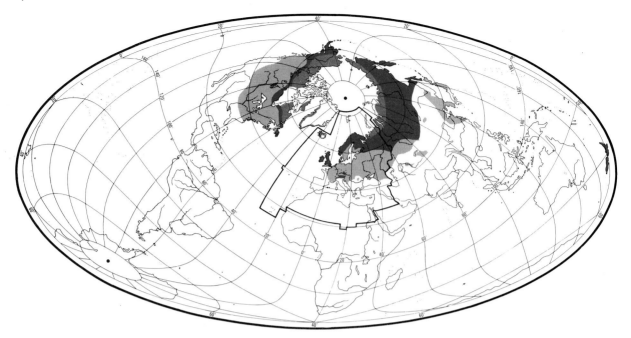

remarkably, if erratically, during 20th century. Mainly distributed in open scrub woodland, often on hillsides and heaths and in field hedgerows, gardens, alder carrs along streams, and increasingly in young conifer plantations, though large and dense forest stands are still avoided. (Sharrock 1976.) It has been suggested that some regional preferences may be shown for apple *Malus* and pear *Pyrus* trees in East Anglia, tall oaks *Quercus* in the west, and tall willows in Berkshire (Bannerman 1953*a*), but any such preferences, perhaps favoured by social breeding habits, may well prove ephemeral.

In Switzerland, once recorded at summit of Matterhorn (4477 m), but breeds principally in subalpine conifer woods, mainly above *c.* 1400 m, in sunny situations with trees at least 3 m tall (especially larch *Larix*) adjoining alpine meadows and pastures. Sometimes occurs by alpine huts and stock shelters. (Glutz von Blotzheim 1962.)

An 'edge-species', evidently responsive to certain climatic influences and showing pronounced fluctuations and adaptations in habitat. Has taken advantage of some recent changes in landscape, e.g. in colonization of dunes in Netherlands and Denmark following spread of birch and alder since cessation of grazing. (Newton 1972.) Restless and aerial, flying freely over distances, mainly in lower airspace.

**Distribution.** C. *f. cabaret* (until 1920 almost confined to Britain, Ireland, and Alps) has spread, apparently from Britain, to coastal areas of north-west Europe and inland to parts of central Europe. Expansion began in 2nd half of 19th century, stagnated in 1st half of 20th century, then resumed with new areas colonized especially in 1950s and

1970s. Main cause of expansion probably large-scale habitat changes due to planting of conifers for forestry, shelter-belts, and amenity. (Ernst 1988.) (Introduced in New Zealand.)

FAEROES. Possibly bred Tórshavn 1960; otherwise scarce visitor (Bloch and Sørensen 1984). BRITAIN, IRELAND. Spread in lowland Britain 1900–10, followed by range contraction; then spread again from 1950, aided by afforestation (Sharrock 1976). Some evidence for westward spread in Ireland in 20th century (Hutchinson 1989). FRANCE. First recorded breeding in Franche-Comté 1983 (Duquet 1984); spreading (Duquet and Pépin 1987). Range expansion to Fribourgois plateau; apparently breeding (Beaud and Savary 1987). BELGIUM. 2 separate populations, one in Ardennes, the other coastal. Coastal population discovered in 1975; since 1980 has spread irregularly, in small numbers. Ardennes population larger (see Population); first recorded breeding 1974. (Devillers *et al.* 1988.) NETHERLANDS. First bred (Wadden islands) 1942, spread starting from 1960s; 80% still breed mainly in coastal area of west and north, remainder inland in sandy country (SOVON 1987). DENMARK. Has recently spread to all islands (TD). NORWAY. Has spread recently in lowland coastal areas south and west of previously known range (VR). SWEDEN. Small population (of *cabaret*) in south-west first recorded in 1970s (Götmark 1981). FINLAND. Very marked annual fluctuations, in both range and numbers (see Population). Normal range is mapped here, but in peak years breeds well to south. (OH.) GERMANY. Probably breeding all along Mecklenburg coast (Müller 1983). CZECHOSLOVAKIA. First recorded breeding 1952; subsequently spread, and still doing so (KH). YUGO-

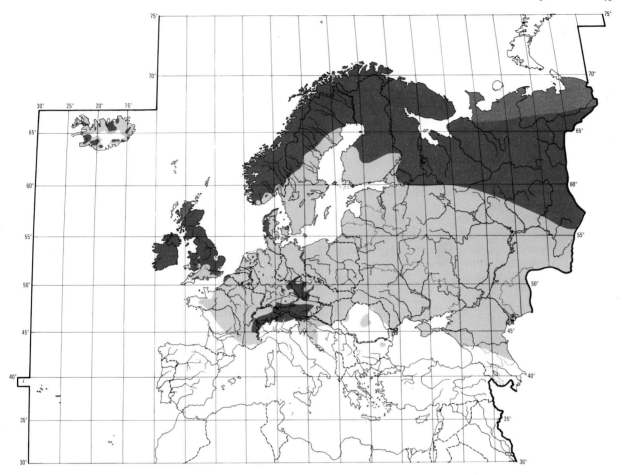

SLAVIA. *C. f. cabaret* breeds in restricted area in north-west, and possibly further south-east (VV). LATVIA. Breeding proved only once, in 1931 (Vīksne 1983).

Accidental. Spain, Albania, Malta, Turkey, Cyprus, Morocco.

**Population.** Has increased in parts of western and central Europe, following range expansion (see Distribution). Northern populations fluctuate markedly, according to abundance of birch *Betula* and spruce *Picea* seed crops.

BRITAIN, IRELAND. Rough estimate 300 000–600 000 pairs 1968–72 (Sharrock 1976). Britain: declining from high levels in 1970s; estimated 140 000–150 000 pairs (Hudson and Marchant 1984; Marchant *et al.* 1990). FRANCE. 1000–10 000 pairs (Yeatman 1976). BELGIUM. Ardennes population (see Distribution) estimated at 250 pairs in 1975, then decreased (Devillers *et al.* 1988). NETHERLANDS. Estimated 600–1000 pairs (SOVON 1987). GERMANY. Estimated 7400 pairs (Rheinwald 1992). East Germany: 1800 ± 800 pairs (Nicolai 1993). DENMARK. Increasing; now very common in western Jutland (TD). SWEDEN. Numbers fluctuate; population estimated at *c.* 1 million pairs in peak years. Small population in south-west

(see Distribution) estimated at 21–26 pairs in 1978 (Götmark 1981). FINLAND. Very marked annual fluctuations; peak years in north coincide with rich birch *Betula* seed crops, and in south with rich spruce *Picea* seed crops. Perhaps 500 000 to 1 million pairs, but several million pairs in peak years. (Koskimies 1989; OH.)

Survival. Britain: annual mortality of adult ♂♂ 59 ± 11%, adult ♀♀ 56 ± 14% (Dobson 1987). Oldest ringed bird 8 years (Rydzewski 1978).

**Movements.** Short-distance migrant in western Europe, but longer-distance from more northern and eastern breeding areas. Sometimes eruptive. Direction of movement chiefly south-east; distance varies from year to year, hence winter range also varies. Avoids long sea-crossings where possible (except *rostrata*).

British populations of *cabaret* often winter within Britain but, in years when food scarce, also further south and east in Netherlands, Belgium, France, western Germany, and very occasionally Iberia. Isolated populations breeding central European Alps are largely resident, moving only to lower altitudes in winter (Schifferli *et al.* 1982; Ernst 1983*a, b*). Fenno-Scandian populations of nominate *flammea* winter chiefly in European Russia, but variable numbers remain in Fenno-Scandia, more in years when seed crop of birch *Betula* is large (Eriksson 1970*a*). *C. f. rostrata* migrate from southern Greenland to Iceland, and a few reach Britain and Ireland (chiefly north-west Scotland) each year. Icelandic population of *rostrata* probably resident. In North America, nominate *flammea* breeding in Alaska and Canada move south in winter into northern half of USA; eruptive movements extending as far south as 30°N are seen in some winters; these occur synchronously across whole continent. (Bock and Lepthien 1976.) For Asia, see below.

Northern British birds move south to south-east in autumn, with no particular routes or concentrations detected. Proportion crossing southern North Sea to Low Countries is high when seed crop of birch in southern England is low and vice versa (Evans 1966, 1969*a*). At least a few individuals, however, return to same wintering site in successive years irrespective of changes in size of birch crop, as proved by ringing in western Germany (Mohr 1967). But some birds ringed in Britain in one winter have been found in Belgium in subsequent winters. Individuals deposit varying amounts of fat (up to 10% of total body mass) before migrating southward from northern England. Migrant flocks are composed of up to 100 birds of all ages and both sexes. Evidence for use of same route in successive years extremely limited. (Evans 1966, 1969*a*.)

Most Fenno-Scandian birds move 1000–1500 km south-east to ESE in autumn, whether or not an eruption takes place, although 2 ringed birds were found exceptionally far east after 1965 eruption, 2900 and 3410 km. Most birds winter in central Russia, where temperatures are, on average, lower than in Fenno-Scandia but where snow generally much less deep and food thus accessible. Rates of movement, from retraps of ringed birds, reach 86 km per day. (Peiponen 1967; Eriksson 1970*a*.) A small number of ringed birds from Sweden and Norway have also been recovered south and SSW in winter (Eriksson 1970*a*), and a few reach eastern Britain in most autumns. A large invasion occurred in 1910 (Evans 1911). Although most ringed Fenno-Scandian birds have been recovered in Russia in winter, they may leave by different routes in different years (through southern Sweden in 1960, 1962, through southern Finland in 1961, 1964, 1965). This may be related to changes in breeding range between years. No definite proof of winter site-fidelity and 3 recoveries indicate wintering in Finland or Sweden in one year and Russia in a later year. This would accord with the marked winter-to-winter fluctuations in numbers staying in southern Finland. (Peiponen 1957, 1967; Eriksson 1970*a*.) Similarly, none of 1879 birds ringed in south-east Germany during invasion in late autumn 1972 (presumably a southward eruption from Fenno-Scandia) were retrapped there in later winters (Ernst 1983*a*). Direction in spring appears to be reverse of autumn. Greenland birds migrate

south-east in autumn to reach Iceland. Large influxes occur into north-west Britain in some years, e.g. 1955 and 1959 (Williamson 1956, 1963*a*). Interpreted as eruptions but may also result in part from displacement of migrants from intended track to Iceland by strong north-west winds (Williamson 1961*a*). No comprehensive studies published on directions of generally southward autumn migration of North American populations of nominate *flammea*, but recoveries (probably relating to both *C. flammea* and Arctic Redpoll *C. hornemanni*) show individuals moving south-east from Alaska to south-east Canada and north-east USA; also birds wintering at long distances (often more than 2000 km) east or west of winter sites of previous years (Troy 1983).

British birds studied in Northumberland (northern England) leave breeding grounds there in mid-September and migrate southward later in month; pass through central England in late September and early October and reach southern England from early October onwards. These timings apparently not influenced by numbers migrating in any year. (Evans 1966, 1969*a*.) In years of large eruptions, departures across North Sea may continue until early November (Boddy 1984). Most recoveries of ringed British birds in Belgium have been in October, fewer in November (Evans 1966). Arrives at wintering site in Rhein-Main-Nahe area (western Germany) from mid-October (Mohr 1967).

Most Fenno-Scandian birds pass through south-west Finland in early to mid-October each year, but start of passage there varies considerably from year to year, and was as early as mid-September in 1965, a year of major eruption. Passage through bird observatories on islands in Gulf of Bothnia decreases markedly in mid-November. (Eriksson 1970*a*.) Birds wintering southern Finland in years of good birch seed crop become much scarcer January–February, possibly as result of further emigration (Peiponen 1967) but perhaps through mortality (Eriksson 1970*a*). Further south, birds reached Chemnitz (south-east Germany) in large numbers in November 1972, a year of major irruption; many passed through by late December, by which time 2 birds ringed in November had been recovered at Trieste (north-east Italy) and nearby in Yugoslavia (Ernst 1983*a*).

Greenland birds leave breeding areas end of August to early October (Salomonsen 1950–1), earlier than Fenno-Scandian birds, and main irruptions into north-west Britain occur September. In North America, movement mostly September–October (Bent 1968).

In spring, *cabaret* leave Rhein-Main-Nahe area by mid-April (Mohr 1967) and pass through Kent (south-east England) in April (Newton 1967*c*) to reach breeding areas in Northumberland by late April or early May (Evans 1966). Nominate *flammea* return from wintering areas to southern Finland in early April. In most years they continue migration to northern Fenno-Scandia to reach fjeld zone and begin breeding mid- to late May. In some years, however, migration ceases in spruce *Picea* zone of southern Finland and 1st broods are reared there, fledging early June, after which further northward movement of adults takes place to traditional breeding areas, which are reached in mid-June. (Peiponen 1957; Hildén 1969a.) For mass retromigration on Kanin peninsula (north European Russia), June–July, see Leonovich (1976).

*C. f. rostrata* return to Greenland from 2nd half of April to beginning of June (Salomonsen 1950-1). In North America, movement mid-March to May (Bent 1968). Birds pass through interior of Alaska in series of waves (Kessel and Springer 1966). In 1966, 1st main wave passed Fairbanks in last days of March and start of April, and 2nd wave in mid-April. Birds arriving towards end of April often stay to breed. (West *et al.* 1968.)

In Asia, migrates in highly fluctuating numbers (Dementiev and Gladkov 1954). In Kamchatka, north-east Russia (where breeds), sometimes common in winter, at other times rare (Lobkov 1986). On Yamal peninsula, where both *C. flammea* and *C. hornemanni* breed, small numbers of one or both species remain in south all winter (Danilov *et al.* 1984). In Ussuriland (extreme south-east Russia), occurred in considerable numbers on passage and in winter 1959 and 1962, but virtually absent 1960-1 (Panov 1973a). In Kazakhstan, winters mostly in north in mild winters, with many also in south in severe winters (Korelov *et al.* 1974). Rare further south (Tadzhikistan) (Abdusalyamov 1977). Many birds reach Mongolia, wintering chiefly in Hentiy, but also further south in central and southern Gobi (Piechocki and Bolod 1972). Birds breeding on northern Sakhalin island move south to southern Sakhalin and Kuril Islands (Gizenko 1955). In China, recorded mostly from north-east, but occasionally south to Fukien and west to eastern Kansu (Schauensee 1984). Autumn survey 1986-90 at Beidaihe (north-east China) indicated erratic occurrence (Williams *et al.* 1992). In Japan, winters in greatly varying numbers in Hokkaido and northern Honshu, also further south in some years (Brazil 1991).

In autumn, begins to leave breeding grounds in 2nd half of August, movement continuing to November. On Yamal peninsula and in Taymyr region, most migrants leave by mid-September (Krechmar 1966; Danilov *et al.* 1984). In Kamchatka, movement from late August continuing to early November (Lobkov 1986). Reaches Kazakhstan from end of September, main movement October in north and November in south (Korelov *et al.* 1974). Present Mongolia from mid-October (Piechocki and Bolod 1972). At Beidaihe over 5 years, passage recorded 7 October to 18 November (Williams *et al.* 1992). Spring migration begins March. In Kazakhstan, passage early March to mid-April (Korelov *et al.* 1974), and most birds leave Mongolia in April (Piechocki and Bolod 1972). In southern taiga of central Siberia, leaves by end of April in early springs, and beginning of May in late springs (Reymers 1966). In Ussuriland, passage recorded March (Panov

1973a). Arrives Kamchatka early April to end of May (Lobkov 1986); on Yamal peninsula, main arrival late May and early June (Danilov *et al.* 1984).     PRE, DFV

**Food.** Very small seeds, especially birch *Betula*; invertebrates in breeding season. Forages principally in trees, but moves to ground when seed in trees exhausted or has fallen; on ground, also eats seeds of herbs, particularly Compositae, Onagraceae, and Chenopodiaceae. (Newton 1967a; Davis 1977; Lack 1986; Thies 1990.) Generally in trees in summer, shifting to ground increasingly throughout autumn and winter as birch crop diminishes, though may remain in birch or alder *Alnus* all winter; on ground, searches both below trees for fallen seed and in all kinds of open country, waste ground, etc.; in some places specializes in seeds of spruce *Picea* or larch *Larix*; in tundra, forages in low vegetation all year (Eriksson 1970a; Neub 1973; Ernst 1986; Schmitz 1987; Thies 1990). Not uncommonly feeds very close to human habitation, especially outside breeding season (Pfeifer 1974; Ernst 1986). In southern England, 73% of 55 feeding observations over year were in trees (90% of these in birch, 10% in willow *Salix*, neither of which among 6 commonest trees and shrubs in study area), 25% in herbs, and 2% in shrubs (Beven 1964); in Oxford area (southern England), 95% of foraging in vegetation, 5% on ground (Newton 1967a). One of the most acrobatic European Fringillidae, characteristically clinging to cones and catkins in outermost twigs (Kear 1962; Newton 1967a; Thies 1990); in captivity, 42% of feeding positions were clinging to vertical stems, 31% to bent stems, 12% hanging upside-down, 11% normal perching, and 3% standing on ground; bill and foot coordination was highest of finches studied (with Siskin *C. spinus* and Goldfinch *C. carduelis*); particularly adept at pulling in birch catkins with bill then clamping them under foot (Newton 1967a), though in another study thought to be not as expert as *C. carduelis* (Kear 1962). For relationship between population size, irruptive breeding, invasions, etc., and availability of tree seeds, see (e.g.) Svärdson (1957), Evans (1966), Enemar and Nyström (1981), Pulliainen and Peiponen (1981), Götmark (1982), and Schmitz (1987). Recent habit noted of eating leaf and flower buds of trees if usual foods unavailable, both in England (Davis 1977) and among introduced birds in New Zealand (Stenhouse 1962a). In Schleswig-Holstein (northern Germany), autumn, fed on large numbers of larvae of leaf-mining moth *Coleophora laricella* during infestation of larches, and appears to specialize on these caterpillars, opening their casings neatly and quickly as if de-husking seeds (Thies 1990). Recorded fluttering to pick items from seed-covered water if unable to alight (England 1974). In extralimital Alaska (USA), will feed on seeds on ground in holes under snow if all vegetation covered (Cade 1953); see also Pohl (1989) for various winter foraging techniques in this region. In Finland, reported eating pieces of wood, as well as mortar, ash, etc., presumably for minerals, in particular calcium (Pulliainen *et al.* 1978; Sundin 1988). For composition of mixed winter feeding flocks in Germany, see Neub (1973) and Thies (1990). In Oxford area, 80% of seeds eaten weighed under 0·5 mg, 18% 0·5–1·0 mg, and 2% over 1·0 mg; ranged in weight from 0·05 mg (willowherb *Epilobium*) to 5 mg (thistles *Carduus*, etc.), average seed weight lowest of 4 Fringillidae studied. Bill shape similar to Linnet *C. cannabina* and Greenfinch *C. chloris*, but very narrow at tip, presumably to get at birch seeds. (Newton 1967a.) In captive trials, range of average times to de-husk seed after 3 days was 1·1 s for *Brassica* to 2·5 s for grass *Phalaris*; exceptionally, took 12 s to open seed of hemp *Cannabis*; order of preference was linseed *Linum*, Compositae, hemp, *Brassica*; linseed much preferred, and grasses (Gramineae) rejected (Ziswiler 1965, which see for bill morphology, handling and consumption of seeds, and comparison with other species). See also Kear (1962) and Newton (1967a).

Diet in west Palearctic includes the following. Invertebrates: springtails (Collembola), dragonflies and damsel flies (Odonata), bugs (Hemiptera: Aphidoidea, Adelgidae), adult and larval Lepidoptera (Tortricidae, Coleophoridae, Geometridae), stoneflies (Plecoptera), booklice (Psocoptera), fleas (Siphonaptera), flies (Diptera: Chironomidae, Culicidae), Hymenoptera (ants Formicidae), beetles (Coleoptera: Curculionidae), spiders (Araneae), mites (Acari), snails (Pulmonata). Plant material: seeds, buds, etc., of juniper *Juniperus*, larch *Larix*, spruce *Picea*, pine *Pinus*, willow *Salix*, poplar *Populus*, birch *Betula*, alder *Alnus*, beech *Fagus*, ash *Fraxinus*, privet *Ligustrum*, lilac *Syringa*, dogwood *Cornus*, buckthorn *Rhamnus*, hop *Humulus*, nettle *Urtica*, mistletoe *Viscum*, knotgrass *Polygonum*, dock *Rumex*, goosefoot *Chenopodium*, orache *Atriplex*, chickweed *Stellaria*, mouse-ear *Cerastium*, buttercup *Ranunculus*, treacle mustard *Erysimum*, charlock *Sinapis*, radish *Raphanus*, shepherd's purse *Capsella*, meadowsweet *Filipendula*, cinquefoil *Potentilla*, burnet *Sanguisorba*, lady's mantle *Alchemilla*, rose *Rosa*, bramble *Rubus*, rowan *Sorbus*, hawthorn *Crataegus*, apple *Malus*, pear *Pyrus*, currant *Ribes*, Virginia creeper *Parthenocissus*, clover *Trifolium*, St John's wort *Hypericum*, crowberry *Empetrum*, cranberry *Vaccinium*, cow parsley *Anthriscus*, carroway *Carum*, willowherb *Epilobium*, evening primrose *Oenothera*, viper's bugloss *Echium*, elder *Sambucus*, guelder rose *Viburnum*, goldenrod *Solidago*, daisy *Bellis*, mugwort *Artemisia*, bur marigold *Bidens*, chamomile *Anthemis*, mayweed *Matricaria*, tansy *Tanacetum*, groundsel *Senecio*, thistles *Carduus*, *Cirsium*, knapweed *Centaurea*, dandelion *Taraxacum*, hawkbit *Leontodon*, hawksbeard *Crepis*, hawkweed *Hieracium*, yarrow *Achillea*, rush *Juncus*, grasses (Gramineae), sedges (Cyperaceae). (Turček 1961; Peiponen 1962; Newton 1967a; Gentz 1970; Antikainen *et al.* 1980; Enemar and Nyström 1981; Pulliainen and Peiponen 1981; Sabel 1983; Ernst 1986; Nyström and Nyström 1987; Thies 1990.)

In Chemnitz area (eastern Germany), mostly October–March, 2705 feeding observations included 32% on seeds of birch, 18% tansy, 13% nettle, 9% orache, 7% alder, 6% mugwort, 4% viper's bugloss, 3% *Rumex*, and 3% goosefoot (Ernst 1986, which see for monthly variation). Apparently prefers tansy, orache, nettle, and mugwort, moving into birch and alder trees only when seeds of these exhausted (Ernst 1990). In south-west Germany, winter, of 174 feeding observations, 21% seeds of birch, 12% tansy, 11% mugwort, 8% nettle, 7% goldenrod, 6% alder, 5% beech, 2% thistles, and 2% fruit tree buds; remainder other weeds (Neub 1973). In Iceland, of 10 stomachs in breeding season, 5 contained 5–26 *Cidaria* caterpillars (Geometridae), 3 contained seeds, and 1 contained 25 aphids; winter birds took only seeds of shepherd's purse, rowan, and birch (Timmermann 1938–49). In central Finland, winter, 15 stomachs held 90% seeds, mostly spruce, plus a few insects and spiders (Antikainen *et al.* 1980); for northern Finland, see Peiponen (1962). In Swedish Lapland, if crop of birch seed poor, switches to insects as substitute, but if continuing cold weather keeps invertebrate activity low then food shortage can be severe (Nyström and Nyström 1991). In southern and central England in spring, main foods are invertebrates (50–60% by number of observations, mostly Hemiptera and Lepidoptera larvae taken from larch, sycamore *Acer*, ash, and oak *Quercus*), seeds and ovules of willow (30–40%), and birch buds and seeds of chickweed, dandelion, etc. (c. 10%); in summer, birch, *Rumex*, and grasses; later, very fond of meadowsweet and willowherb seeds, as well as various Chenopodiaceae, Cruciferae, and Compositae; in autumn into winter, still feeds on meadowsweet and willowherb in addition to tree seeds, mainly birch (up to 90% of winter diet) but also alder; when birch crop poor, prefers willowherb, meadowsweet, mugwort, tansy, goosefoot, and alder (Newton 1967a). Willowherb and meadowsweet also common food in northern England August–September (Evans 1966). In Chemnitz area in spring, fond of parts of flowers of various trees, seeds of larch, pine, willow, and herbs, as well as aphids, then later in year main foods are seeds of orache, nettle, mugwort, and especially tansy, which is taken through winter where available, in addition to birch, alder, and spruce seeds (Ernst 1986).

In Alaska, needs to forage for c. 8·5 hrs per day at −22°C to achieve necessary energy intake of c. 108 kJ (Pohl 1989). Gross daily intake at lowest tolerable temperature in captivity was 134 kJ, and minimum weight of food 7·3 g; unhusked birch seeds contain 23 kJ per g, a much higher energy content than most other seeds, enabling *C. flammea* in the wild to survive on this diet at c. −54°C (Brooks 1968, which see for many details). See also Evans (1969a).

In Swedish Lapland, 53 nestling gullet-samples contained 5973 items, of which 79·9% by number seeds (76·8% birch), 13·1% Diptera (4·8% larvae and pupae), 2·9% springtails, and 1·4% beetles (1·1% larvae and pupae); also Hemiptera, Odonata larvae, caterpillars, booklice, stoneflies, ants, spiders, mites, and snails; in year of low birch crop, 24·9% of 1043 items were birch seeds. Invertebrates *c.* 25% of diet in 1st week, *c.* 9% in remainder of nestling period, but large variation between samples (Enemar and Nyström 1981). However, using 'animal equivalents' based on energy content and size of prey, diet of nestlings up to 8 days old in 46 nests (which included some of Arctic Redpoll *C. hornemanni*, but authors say diet identical) in same area averaged 68% (range 44–97%, *n* = 2197 seeds and 4550 animal equivalents), i.e. invertebrates predominated, in contrast to findings based solely on number of items (Nyström and Nyström 1991); see also Nyström and Nyström (1987). In northern Finland, 14 nestling stomachs contained 61% by volume insects and 39% seeds, mostly birch; diet in June *c.* 85% seeds, in July *c.* 80% insects, though considerable variation (Peiponen 1962). Similarly, according to Svärdson (1957), 1st brood can be reared mostly on seeds, 2nd on insects. In Greenland, ♀ caught flies while brooding and fed them to young (Nicholson 1930). BH

**Social pattern and behaviour.** No comprehensive study, but most aspects well known. Account includes data on *cabaret*, nominate *flammea* (Palearctic and Nearctic populations), and *rostrata* from Greenland and Iceland; some data from Iceland may relate to pale birds, actually Arctic Redpoll *C. hornemanni* (see Geographical Variation).

1. Typically gregarious outside breeding season, and some flocking also when breeding (see below). Especially large flocks occur on migration, particularly during irruptions (see Movements): e.g. up to 4000–6000 in southern Finland, October (Peiponen 1957), and in Moscow region in some winters 500–1000 birds per km², then flocks of 200–6000 during peak spring migration (Ptushenko and Inozemtsev 1968). Roosting flock of *c.* 1000 observed in western Germany (Thies 1990). In northern Germany during 1972–3 irruption, many flocks of hundreds (once 2000 resting on migration), also smaller ones and single birds; in Schleswig-Holstein, average flock size 59 in west, 70 in east (Weber 1954; Busche *et al.* 1975; Müller 1977); in Berlin, *c.* 65% of flocks are of up to 20, but maximum 380 (November), flocks breaking up in December, building up again (January–)February, then declining again by early April (Elvers *et al.* 1974); in Bayern (southern Germany), 80% of flocks 2–50 (Altrichter 1974). For Denmark during same influx, see Braae (1975); for Netherlands during a later one, see Lensink *et al.* (1989). In North America in March, tens of thousands reported migrating in continuous stream (Bent 1968); see also West *et al.* (1968). At Karigasniemi (northern Finland) following June arrival, noted in pairs and occasionally flocks of 4–6 (birds of a nesting group: see Breeding Dispersion, below); flocks of up to 50 occur from July (Hildén 1969a). At Kilpisjärvi (northern Finland), flocks of up to 30 adults and juveniles noted mid breeding season presumably originated from breeding grounds further south (Peiponen 1957); see Breeding Dispersion (below). Mid-summer flocks of 120–400 reported from Finland by Antikainen *et al.* (1980). In north-east Russia, following cold spring, apparently no pair-formation, birds remaining in flocks of 10–50 up to mid-July (Sorokin 1977). In Switzerland, mid-August flock composed mainly of ♀♀ and juveniles (Stingelin 1935), and up to 400 juveniles recorded together in northern England after breeding (Evans 1969a). Frequently reported to associate for feeding with

other Fringillidae, especially Siskin *C. spinus*, also with tits *Parus*, though in many cases probably only chance encounters (Stingelin 1935; Kelm 1936; Neub 1973; Altrichter 1974; Elvers *et al.* 1974; Busche *et al.* 1975) and, in one French study (of migrants), association loose and not persistent even with *C. spinus* (Erard 1966). German studies found *c.* 80% of flocks (Thies 1990) or 80·6% of 160 observations (Altrichter 1974) were of *C. flammea* only. BONDS. Mating system not studied in detail, but some observations indicate deviations from monogamy. In Canada, apparent case of sequential polyandry: 2 ringed adults at one nest were confirmed by DNA fingerprinting as parents of nestlings present; ♀ of this pair built 2nd nest *c.* 200 m away and was accompanied, eventually also fed, by an unringed ♂; not known which ♂ fertilized ♀'s 2nd clutch; polyandry perhaps more frequent than indicated by this single record (Seutin *et al.* 1991); see below. In Switzerland, frequent records of 2 ♀♀ associating closely; once seen billing and attempting to copulate (Stingelin 1935); see Flock Behaviour (below). Supernumerary birds reported from Finland (Hildén 1969a), and flocks apparently of non-breeders (mainly ♂♂) in north-east Russia, though breeding ♂♂ forage some distance from nest (Kishchinski 1980); see also helpers at nest and Breeding Dispersion (below). Pairs formed in flocks (Zablotskaya 1981), and bond maintained through season, sometimes for several breeding attempts: e.g. on Yamal peninsula (northern Russia), 2 pairs each attempted 2nd brood after successful 1st, 3rd pair made 3 nesting attempts (abandoned 2nd nest) (Alekseeva 1986); in Alaska, bond also persisted through 3 attempts (Troy and Shields 1979); captive pair reared 3 (overlapping) broods successfully and attempted 4th (Ludewig 1989). Pair-bond not known to persist beyond one season however; in German study, breeding ♀ ringed in May paired with different ♂ *c.* 3 km away in following year (Ernst 1990). For alleged hybridization with *C. hornemanni*, see that species. Mixed nominate *flammea* × *cabaret* pair reported in Netherlands (Dijksen 1989). Hybridization with Linnet *C. cannabina* recorded for wild and captive birds (Blok and Spaans 1962). In southern Germany, ♂ nominate *flammea* (perhaps escape) associated closely for *c.* 2 weeks with Greenfinch *C. chloris* (Diesselhorst 1971a). Suggested by Dice (1918) that ♂ plays no part in parental duties apart from fertilizing eggs, but role of sexes reported in other studies as follows. All incubation (also brooding) by ♀ (e.g. Grinnell 1943, Hildén 1969a), though some (presumably brief) stints by ♂ according to Zablotskaya (1981), and in Iceland ♂ said by Hantzsch (1905) to incubate regularly. Young fed by both sexes, though probably some variation in share taken by ♂: captive ♂ brought food for first 3 days, but most feeding by ♀ after *c.* 1 week (Timmermann 1938). On Yamal peninsula, unpaired ♂ recorded as helper at each of 2 nests: one brought food over 2 days up to fledging, the other made several visits on one day, neither encountering hostility from nest-owners (Danilov *et al.* 1984). In apparently not uncommon overlapping broods, ♀ may continue to share feeding of 1st-brood young (also brooding them) while building new nest and re-laying, ♂ then assuming sole responsibility for 1st brood and also feeding mate on 2nd nest (Timmermann 1938; Alekseeva 1986). Both parents care for young after fledging according to Evans (1966), but other studies found fledglings tended only by ♀ (Noë 1983) or (more often) by ♂, and for up to *c.* 16–17 days after leaving nest (Timmermann 1938; Alekseeva 1986; Seutin *et al.* 1991). Age of first breeding 1 year (Blok and Spaans 1962; Troy and Shields 1979). BREEDING DISPERSION. Solitary or (often) in neighbourhood groups; territorial (Hildén 1969a), but not rigorously so (Salomonsen 1950–1; Wynne-Edwards 1952), and defended area anyway small (see below). Data from Finland: at Kilpisjärvi, along 5 km of road, 4 nests 30–40 m apart and 5th

300 m away, also 2 groups each of 3 nests 150 m apart; group of 9 nests along 300 m, closest separated by 30 m (Peiponen 1962); at Karigasniemi, nests also 30-40 m apart, once 3 nests (not used simultaneously) separated by 11 and 20 m; at Häme, 30-50 m apart (Hildén 1969a); groups at Lahti small, largest 8 (Peiponen 1967). Nesting groups often of 5-10 pairs; in eastern Germany, up to c. 50 pairs on raised bog (presumably concentrated in several small groups, not evenly dispersed); nearest-neighbour distances 15-16 m (Ernst and Thoss 1977; Ernst 1988). In Greenland, some single pairs or groups of 6-12(-30) pairs (Nicholson 1930). For further data, see Danilov et al. (1984) for Yamal peninsula, Grinnell (1943) for Manitoba (Canada), and Maestri et al. (1989) for Lombardia (Italy). According to Zablotskaya (1981), territory only 50-80 m² (radius of 4-5 m round nest). Territory perhaps established (by ♂) only after pair-formation and nest-site selection as typically in Carduelinae, so courtship takes place outside territory, though C. flammea recorded copulating in nest (see Heterosexual Behaviour, below). ♂ has no fixed area for song and display (Enemar 1963), and birds typically forage up to several km from nest (e.g. Salomonsen 1950-1, Peiponen 1962, Duquet 1984). No territory defence recorded in Greenland, groups of 1-4 ♂♂ tolerated near pairs (Salomonsen 1950-1). In Alaska, 2nd and 3rd nests of one pair within c. 7·5 m of 1st (Troy and Shields 1979). On Yamal peninsula, 2nd-brood nest c. 14 m from successful 1st; 3rd nest of another pair c. 100 m from abandoned 2nd (Alekseeva 1986). See also discussion of site-fidelity (below). Difficult to obtain reliable density values from territory mapping because of pronounced tendency to wander far when breeding, lack of display and territorial behaviour, and dispersion in loose groups (Enemar 1963; Ytreberg 1972). Figures distorted in many cases owing to concentrations in small area: e.g. in Finland, 8 nests in 0·25 ha (3200 per km²) at Lahti, 7 nests in 1 ha (700 per km²) at Pihlajavesi (Peiponen 1967; Hildén 1969a); on Yamal, 33 pairs in 4·5 ha (733 per km²) (Danilov et al. 1984). In Fenno-Scandia (nominate flammea), considerable fluctuations depending on food supply: e.g. at Ammarnäs (northern Sweden), average over 19 years (from study plots, mapping and line transects) 26·2 territories per km² (2-90), peak years coinciding with heavy birch Betula seed crop (Enemar and Nyström 1981); see also Peiponen (1962), Enemar (1969), and Nyström and Nyström (1987). In boglands, northern Finland, 103 pairs per km² (Hildén 1969a), while line transects in various parts of Finland over 3 years showed 2-410 pairs per 100-km stretch (Peiponen 1957); see also Antikainen et al. (1980). Study in northern Finland by Virkkala (1987, 1989) found slightly higher (but markedly fluctuating) densities in managed and fragmented coniferous forest (pine Pinus and spruce Picea) than in virgin and uniform forest of similar composition. Examples from range of cabaret as follows. On Anglesey (north-west Wales), 7 pairs per km² in young conifer plantation (Insley and Wood 1973); up to 50 or more pairs per km² on heaths and in upland birch forests of Britain (Sharrock 1976). In mixed larch Larix forest, Switzerland, maximum 100-400 pairs per km² (Glutz von Blotzheim 1962). 2-23 pairs per km² recorded in various habitats in French Alps, up to 43 in damp birch forest of French Jura (Lebreton et al. 1976; Duquet 1984). In Czechoslovakia, 3-73 pairs per km² in conifer stands on boglands, up to 300-400 in spruce-birch forest; similarly, in eastern Germany, density ranges from 6 pairs per km² in cemetery to maximum 167 in mainly pine Pinus mugo forest, and 135-189 in birch bogland (Ernst 1990). Often nests within colonies of Fieldfare Turdus pilaris (Hildén 1969a; Pulliainen 1982; Danilov et al. 1984; Ernst 1988, 1990). Ammarnäs study suggested birds either nomadic or site-tenacious (Enemar and Nyström 1981). However, no proof of adult nomadism through ringing, and no

evidence in recent years of birds breeding twice in same year in 2 widely separated areas (Götmark 1982), such as reported by Peiponen (1957) for Finland; see also Hildén (1969a), Seutin et al. (1991), and Movements. Evidence of site-fidelity (including natal) in several studies: of 547 birds ringed in spring at Wicken Fen (eastern England), 52 (9·5%) returned in subsequent springs (Moss 1979); in Czechoslovakia, same bird returned after 1, 7, and 9 years (Hudec 1983); see also Ernst (1990) for eastern Germany. Recorded using old nest relined (Wynne-Edwards 1952); perhaps at best remains of old nest serving as foundation for new one (Hildén 1969a), but no proof in such cases of same birds returning to their previous year's site. ROOSTING. Nocturnal and often communal. In northern Norway, late summer, fed and roosted communally in low birch scrub (Harris et al. 1965; Evans et al. 1967); in northern Germany, used stands of spruce (Thies 1990). Recorded roosting in self-made snow-burrows in Finland: birds dropped directly into snow from tree and immediately started burrowing, making descending furrow to more solid snow if not able to go deep enough straight away; then made tunnel average 37 cm (27-40) long ending in roost-chamber 6-11 cm below snow surface; tunnel partly plugged by snow pushed behind; shallower burrows not containing faeces presumably used for short stays, birds not sleeping there. For departure, breaks snow roof of chamber and flies up directly. Of 3 birds, 2 roosted within c. 1 m of each other, 3rd within c. 2 m. (Siivonen 1963; Sulkava 1969.) For experimental study of temperature in snow-refuges, see Korhonen (1981). At Kilpisjärvi (69°N, Finland), in continuous daylight (21 May to 22 July), ♀ incubated almost uninterruptedly and ♂ fed her; rest period for early-summer broods mainly late, (ending c. 01.00 hrs) than for late-July broods (rest period around midnight); chiefly controlled by light factor (Peiponen 1970). At 67°44′N in Finland, ♀ always on nest c. 22.00-02.00 hrs, incubation stints averaging 443 min (382-524, n=9); average rest 5·5 hrs (4·3-6·6), mainly before midnight (Pulliainen 1979a). For further details of diurnal rhythm on northern breeding grounds, see (e.g.) Palmgren (1935) and Franz (1949). For experimental study of circadian rhythms, see Pohl (1972, 1974, 1980). Flocks recorded moving about in moonlight before dawn (Bernhoft-Osa 1978), and experimental work by Brooks (1968) further confirmed ability (also, slightly more so, in C. hornemanni) to continue feeding in arctic environment at very low light intensities. In northern Norway, August, birds left roost c. 04.00 hrs and started feeding, returning to roost c. 20.00 hrs (Evans et al. 1967).

2. Typically fairly confiding and approachable on breeding grounds and when feeding outside breeding season (Walcher 1918; Weber 1954; Dick 1973; Pfeifer 1974), though breeding birds in Greenland shy and wary (Salomonsen 1950-1). Low-intensity excitement expressed with short Rattle-calls (see 3 in Voice), bird sometimes also ruffling crown (see also Song-display, below). A 'pyaa' (see 2f in Voice) signals alarm for man or raptor (see also Antagonistic Behaviour, below). Prior to mobbing predator, may first observe silently from a distance, then give Dsooee-calls, longer ones when closer, then abbreviated 'pyu' (with open bill) at very close range (see 2b in Voice), calls attracting up to 15-20 other birds; Dsooee-calls given again with decreasing excitement if predator stays put. (Zablotskaya 1981.) In comparative study of mobbing by Fringillidae, C. flammea found to have marked lateral body movements, moderate (mostly downward) tail movements, and moderate or marked wing-flicking (Hinde 1954b). FLOCK BEHAVIOUR. Frequent Chatter-calls (see 2a in Voice) assist flock cohesion, being given at take-off (when even large flocks still compact) and during migration (Weber 1954; Thies 1990). Lone feeding birds, swaying from

side to side and sometimes ruffling crown, give Dsooee-calls vigorously on hearing passing flock; latter (especially if small) responds with same call and usually lands by solitary bird; Dsooee-call may also signal take-off or return to ground where flock previously feeding (Zablotskaya 1981). 2 flocks of *c.* 200–300 in Austria visited wall to peck at mortar (presumably for grit or salt) morning and evening at regular times through winter, always first landing in tree; members of each flock generally kept separate; birds quiet, also when feeding in birches which they worked over systematically, with some leap-frogging (Walcher 1918). In Iceland in May, disputes and chases noted in flocks re-formed after pair-formation (Hantzsch 1905). Studies of wild and captive birds showed strict linear hierarchy, ♂♂ dominating ♀♀, typical of flocks outside breeding season (dominance roles reversed in breeding pairs). Rattle-calls given in threat in large post-breeding flocks probably reflect increase in disputes to settle hierarchy. Lowest-ranking bird (♀) in captive flock was almost always in submissive-posture (head drawn in, belly and flanks ruffled: Fig A); least dominant ♂ attacked ♀ after he was beaten

A

by another ♂; Allofeeding (see *C. spinus*) observed between 2 ♀♀. (Dilger 1960; Zablotskaya 1976*b*.) SONG-DISPLAY. Song comprising series of well-spaced units or motifs (see 1a in Voice) given by ♂ without visual display when feeding, preening, or flying between perches; apparently indicative of low-intensity sexual motivation. Short Rattle-calls (perhaps kind of song) given for up to 20–25 min from perch (tree-top, exposed branch, etc.) while turning left and right, looking about, and preening. Higher-intensity Advertising-song (see 1b in Voice) serves to attract ♀ and has no territorial significance, being given before territory occupied. Series of Advertising-songs often followed by Dsooee-calls, bird then ascending for Song-flight, in which often joined by up to 8 other birds. (Ernst and Thoss 1977; Zablotskaya 1981.) Sings while flying on undulating course in loops and circles, beating wings slowly and hesitantly, or wing-beats shallow and quivering; may also glide at *c.* 5–6 m over trees (Witherby *et al.* 1938; Dementiev and Gladkov 1954; Peiponen 1962; Zablotskaya 1981; Röttler 1985*b*), and also descends in glide to rock or tree, ♀ meanwhile calling occasionally from (often concealed) perch (Salomonsen 1950–1). Song-flights in Iceland said by Hantzsch (1905) to comprise brief ascent and descent. Few Song-flights once pairs formed, ♂ then usually giving initially Advertising-song and Courtship-song (see 1c in Voice) or intermediates, though more Courtship-song (perhaps some territorial significance as well) closer to nest-building (Zablotskaya 1981). Captive birds apparently sang to discourage encroachment on territory (Dilger 1960). Main song-period in Britain mid-March to mid-August, and song fairly frequent from then to early October; song exceptional, or Subsong only (see 1d in Voice), late December and January (Witherby *et al.* 1938). In Bayern (southern Germany), vigorous song noted up to 1st third of July, ceasing from about mid-month (Kraus and Gauckler 1970); further east in southern Germany, 2 peaks for Song-flights mid-May and end of July (Ernst 1990). See also Röttler (1985*b*) for Hagen (north-west Germany). Birds breeding further

south than usual in Finland sang and displayed to end of May, while at Karigasniemi (regular breeding area in north) much song late June and early July (Hildén 1969*a*). ANTAGONISTIC BEHAVIOUR. Compiled mainly from descriptions by Dilger (1960) and Zablotskaya (1981). Frequent chases noted at Karigasniemi were probably 2 ♂♂ competing for same ♀ (Hildén 1969*a*). In threat-postures associated with (e.g.) individual-distance disputes, also in defence of ♀ during courtship or nest-building (e.g. Timmermann 1938, Hildén 1969*a*), plumage sleeked to varying degree, crown ruffled when escape tendency strong. In Defensive Threat-posture, all plumage ruffled, neck drawn in, and head pointed at opponent while gaping (Fig B). In low-intensity Forward Threat-posture (Fig C) may just point closed

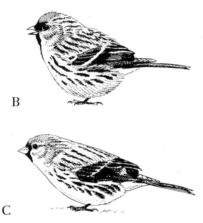

B

C

bill at opponent or, with increasing excitement, make pecking movements at adversary; indications of increasing aggression and imminent attack include quick, perfunctory, sometimes repeated Chin-lifting which exposes black chin (Fig D) (similar movement

D

noted in, e.g., Hawfinch *Coccothraustes coccothraustes*); more pronounced sleeking of plumage and Threat-gaping while facing up to opponent (within *c.* 10 cm), body slightly raised, legs extended, also raising still-closed wings to varying degree (extreme posture shown in Fig E) and giving Rattle-calls; may then leap toward opponent in pecking attack. (Dilger 1960; Zablotskaya 1981.) Subordinate in dispute may stay and assume Submissive-posture, or give alarm-call (see 2f in Voice) and fly off. In breeding season, both protagonists give Rattle-calls, but (unlike in

E

disputes at other times) may adopt sleeked-upright posture (body almost vertical) side-on to each other, one eventually attacking, attempting to leap onto other's back and peck at throat or ventral area; both give 'chak' (see 4 in Voice) during fight, but weaker bird may give alarm-calls before fleeing, as also in disputes with other species, e.g. when chasing away Song Thrush *T. philomelos* from nest. ♀, giving 'chak' calls, seen to fly towards other intruding ♀♀ in displacing attack; no fights, intruders always yielding. (Zablotskaya 1981.) On Yamal peninsula, ♀ on 1st-brood nest tolerant of strange conspecific birds nearby, but very aggressive in defence of 2nd-brood nest, launching attacks, including on ♂ dummy (Alekseeva 1986). HETEROSEXUAL BEHAVIOUR. (1) Pair-bonding behaviour. Sexual chases occur, one or both birds giving Rattle-calls. In courtship, ♂ recorded hopping back and forth between branches, along branch, or on ground towards ♀, then round her, apparently trying to stay side-on. ♂ gives Courtship-song and calls (see 5a and 7b in Voice); legs extended, body slightly erect, head often held up slightly, plumage slightly ruffled; wings drooped, extended horizontally, or raised high over back and slowly fluttered; tail fanned and raised (sometimes steeply); ♂ will also take off suddenly, make short flight and return to ♀. (Witherby *et al.* 1938; Dilger 1960; Zablotskaya 1981.) Bill-flirting ceremony, initiated by either sex and accompanied by series of click calls (see 5b in Voice), much as in other Carduelinae (see, e.g., *C. spinus*); for full description of performance in *C. flammea*, see Dilger (1960) and Zablotskaya (1981); performed during late pair-formation, and may continue for up to *c.* 30 min with pauses; at peak excitement, ♂ may give Chatter-calls; unreceptive ♀ (dominant over ♂ once paired) will give Rattle-calls in threat. (Zablotskaya 1981.) (2) Courtship-feeding. Regular, usually taking place on or close to nest (Windsor 1935; Peiponen 1962). Approaching ♂ gives Rattle- and Chatter-calls, ♀ variously responding with similar calls, 'pyulee' (see 2c in Voice), and, in Begging-posture (head thrown back, bill open, wings shivered), quiet 'tseet' calls (see 5c in Voice); ♂ about to feed ♀ also calls 'pyulee' and first brings up food from crop (Fig F) before feeding mate (Windsor 1935; Peiponen 1962;

F

Gentz 1970; Zablotskaya 1981). Develops out of Bill-flirting in last stage of pair-formation (Dilger 1960, which see for details; Zablotskaya 1981). On Yamal peninsula, ♂ fed ♀ every 40–50 min (Danilov *et al.* 1984); in northern Finland, about once per hour, frequency per day increasing slightly during incubation (Pulliainen 1979a). (3) Mating. Typically takes place during nest-building when ♂ guarding mate (Ernst 1990); sometimes preceded by chase, during which ♀ gives alarm-calls, as also noted after unsuccessful copulation attempt. ♀ solicits by giving quiet but rapid 'tseet' calls while crouching (in nest or on perch) with head and tail raised, fluttering slightly spread wings. (Timmermann 1938; Zablotskaya 1981.) Captive ♂ hovered with rapid, shallow wing-beats over perched ♀ and, if ♀ assumed soliciting-posture, immediately landed on her back and attempted copulation (lasting *c.* 3s); copulation followed in ♀ by immediate preening, etc., while ♂ usually flew off and sang (Dilger 1960); see also Timmermann (1938). ♀ not ready to copulate recorded attacking ♂ while give 'chak' and 'chee' calls (see 4 and 5d in Voice) (Zablotskaya 1981). (4) Nest-site selection. Both sexes search, though ♀ apparently more active and seen to try out various potential sites (Nethersole-Thompson and Nethersole-Thompson 1943b). ♂, sometimes ruffling crown in excitement, may just guard mate and sing, as also later during nest-building and when ♀ incubating (Timmermann 1938; Bille 1978; Zablotskaya 1981). Nest built by ♀ (Stingelin 1935; Timmermann 1938; Hildén 1969a); ♂ occasionally reported to pick up material, but to drop it again immediately (Peiponen 1962); according to Hantzsch (1905), both sexes build, though ♂ mostly brings material. Several records of birds using material from old or recently deserted nest to build new one (Blathwayt 1903; Glayre 1979; Duquet 1984). (5) Behaviour at nest. Often starts laying before nest fully lined (Hildén 1969a), and recorded bringing feathers even during incubation and nestling phase (Pulliainen 1979a, which see for detailed observations at nest in Finnish Lapland). Rapid series of 'tseet' calls given by ♀ before laying and for 15–20 min after. On hearing Dsooee-calls from ♀ off nest, ♂ usually goes to nest and may then cover eggs (presumably briefly). While ♀ incubating and ♂ nearby, birds may perform antiphonal or overlapping duet of 'pyulee' calls (see 2c in Voice). (Zablotskaya 1981.) RELATIONS WITHIN FAMILY GROUP. Asynchronous hatching the norm (Wynne-Edwards 1952). At nest in Finland, ♀ ate eggshells (Pulliainen 1979a). Nestlings variously reported as brooded by ♀ for first 3 days, then only at night from *c.* 1 week (Timmermann 1938 for captive birds), in northern Finland for 6·2 days, with decreasing intensity and most intensively around midnight (Pulliainen 1979a), and little daytime brooding at 5–9 days in study by Hildén (1969a). On Yamal peninsula, ♀ rarely and only briefly absent from nest for first 9–10 days (Alekseeva 1986). Eyes of chicks open at 3–5 days (Peiponen 1962; Bent 1968; see these references for details of physical development). Young fed bill-to-bill or by regurgitation (e.g. Grinnell 1943). In Finnish study, during days 1–7, ♂ fed ♀ and she then fed young; on days 5–7, when ♀ and young begged (actions much as during Courtship-feeding), ♂ fed ♀ or young direct; ♀ returning to nest fed young immediately before or after warming them, rarely (during days 8–10) flew off again after feed; when ♂ came to nest during ♀'s absence (days 8–14), fed nestlings direct (Pulliainen 1979a). In Greenland, ♀ often snapped at insects from nest and fed them direct to young (Nicholson 1930). Study by Ernst and Thoss (1977) found nestlings to gape only on hearing quiet Feeding-call (see 6 in Voice) of parent. ♀ recorded rummaging in nest, presumably for parasites (e.g. Gentz 1970, Ernst and Thoss 1977). Nest-hygiene by both sexes, adult sometimes stimulating defecation by pecking at chick's cloaca, and faeces swallowed or carried away after each feed; during last few days, faeces deposited on nest-rim and not removed from there or below nest (Grinnell 1943; Bent 1968; Hildén 1969a; Ernst and Thoss 1977; Glutz von Blotzheim 1987), though this not invariable (e.g. Pulliainen 1979a). Young may leave nest at 9–10 days if disturbed (Peiponen 1962). Normally fledge at 11–14 days (see Breeding). Brood of 5 in Netherlands fledged over 3 days (Blok and Spaans 1962). Period tends to be 2–3 days shorter in latitudes with continuous daylight (Grote 1943*b*). Able to fly short distance and follow parents at 12–13 days (Pulliainen 1979*a*); Dsooee-calls given by ♀ for contact with fledglings (Zablotskaya 1981). Captive juvenile still not self-feeding at 16 days (Timmermann 1938). ANTI-PREDATOR RESPONSES OF YOUNG. No information for nestlings. Fledged young sometimes permit close approach (Dick 1973), even allowing themselves to be handled (Richard 1928). PARENTAL ANTI-PREDATOR STRATEGIES. (1) Passive measures. ♀ sits tightly and can even be pushed off nest, when may simply hop onto rim or nearby twig, returning at earliest opportunity (Hildén 1969*a*); similar reports by (e.g.) Hortling and Baker (1932), Timmermann (1938), Grinnell (1943), and Danilov *et al.* (1984). (2) Active measures: against birds. Young Siberian Jay *Perisoreus infaustus* mobbed by pair for *c.* 25 min until it flew away (Zablotskaya 1976*c*). (3) Active measures: against man. If flushed from nest (e.g. observer brushes against tree), drops at angle and flutters short distance close to ground (Hildén 1969*a*); perhaps rudimentary distraction-lure display. Also reported to fly 2–3 m into cover, then to make fast, low escape-flight (Grote 1943). Frequent Dsooee-calls given by ♀ perched conspicuously and with ruffled crown when man at or near nest or fledglings, while ♂ may sing in flight (Grinnell 1943; Temple Lang and Devillers 1973); 'tseet' calls also given by ♀ in this context (Zablotskaya 1981). (4) Active measures: against other animals. Persistent Dsooee-calls given by ♀ alarmed by cat (Rettig 1985).

(Figs by D Nurney: A–E from drawings in Dilger 1960; F from photographs in Gentz 1970 and Ernst and Thoss 1977.) MGW

**Voice.** Study of wild and captive nominate *flammea* showed 3 song-types and variety of calls, many of which (including best-known and most frequently uttered Chatter-call and Dsooee-call: see below) are given all year (Zablotskaya 1981, which see for sonagrams). Study in Belgium (of wild and captive birds) found nominate *flammea* much more vocal (frequent Chatter-calls) than Arctic Redpoll *C. hornemanni*, and certain calls also to differ between the 2 species (see *C. hornemanni*); geographical variation in calls further noted between nominate *flammea* and *cabaret* (Herremans 1989, which see for sonagrams). Voice of *rostrata* (Greenland) similar to *cabaret* (Nicholson 1930). For additional sonagrams, see also Bergmann and Helb (1982). For study of *cabaret* using musical notation, see Stadler (1926).

CALLS OF ADULTS. (1) Song of ♂. Sometimes difficult to distinguish between song and calls (Stadler 1926). Song-types in scheme of Zablotskaya (1981) comprise only 16 unit-types, variation coming from different combinations within this limited repertoire; of 16 unit-types, 9 given only in song, others are calls (including 2a–c and 3: see below). (1a) Single units (8 out of 16 in repertoire, including 2a–c and 3) or short segments characteristically separated by long pauses. Much variation in order of units or motifs, each ♂ having at least 8 variants, some being shared by birds in vocal contact. Very loosely constructed phrases in this song-type may last *c.* 6–17 s, with bouts of up to several minutes. Typically given by ♂ on breeding grounds from arrival up to pair-formation and apparently indicative of low sexual motivation. (Zablotskaya 1981.) Description in Bergmann and Helb (1982) similar, though only 3 unit-types mentioned: Chatter- and Rattle-calls occasionally combined with longer nasal units ascending ('dsooee' type) or descending in pitch and lasting 0·5–1 s; see also Borror (1961). Recording (Fig I) contains song evidently of this type given by ♂ *cabaret*: segments of

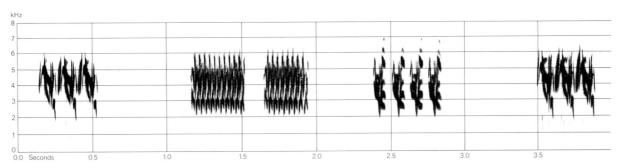

I W Pedley England July 1975

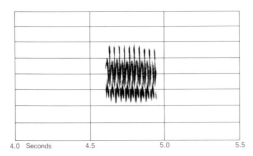

I *cont.*

Chatter-calls ('schyp' and 'chip'), also 'tirrr' rattles, separated by relatively long pauses (W T C Seale). For many authors, song (given in flight) is simple combination of Chatter- and dry Rattle-calls (or variants): e.g. 'chuch-uch-uch-errrrrr' (Witherby *et al.* 1938), 'chut chut chut serrrrrr' (Bruun *et al.* 1986), 'tsche-tsche-tschrrr', sometimes only 'tschrrr', given by ♂ mainly in breeding season (Ernst 1990). Combination of Chatter- and Rattle-calls shown in Figs II–IV: in Fig II (*cabaret*), 2-part segment of 3 'chup' and 3 'chip' units followed by vibrant whirring 'tirrrrrrrr', (after pause) 4 'chyp' units preceding another

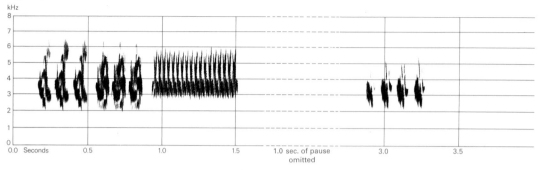

II  J-C Roché  France  May 1973

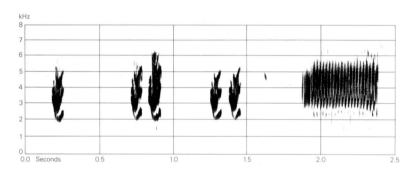

II  *cont.*

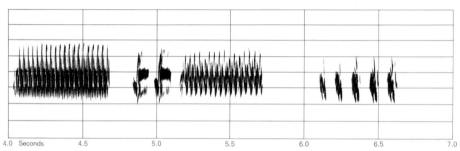

III  J Gordon  Norway  July 1976

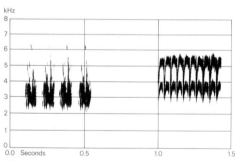

IV  P A D Hollom  Iceland  May 1991

rattle, then 2 'chee' units, rattle, and segment of 5 'chyp' units; in Fig III (nominate *flammea*), 5 'chi' units followed by whirring 'trrrrrrreee' with buzzing onset; in Fig IV (Iceland), 4 'chip' units of rather heavy, solid timbre precede tinkling 'tiriririririr' (W T C Seale). In Russian study, Rattle-call ('tirr') said to serve as advertising-call;

when given for long periods by territorial ♂ from perch, perhaps equally classifiable as 'perched song' (see also call 3, below); 'che-che-tirr' (combination described above) categorized as 'flight-call' rather than song (Zablotskaya 1981), and by Molau (1985) as contact-call, given all year by both sexes, including by young birds. (1b) Advertising-song ('normal song') of ♂ more varied than 1a (14 unit-types used out of 16) and typically given with much shorter pauses between units; variation derived from changing sequence of units. Lasts up to *c*. 29 s. Considerable variation between individuals in richness of repertoire; several variants shared by several ♂♂. Advertising-song given by ♂ around pair-formation. (Zablotskaya 1981.) More varied songs, presumably of this type, given by ♂ *cabaret*, shown in Fig V (1st segment is end of previous phrase; many units apparently derived from Chatter-calls, 'dsooee' near end similar to Fig X) and in Fig VI. In Fig VI, 1st and 10th units suggest 'dsooee',

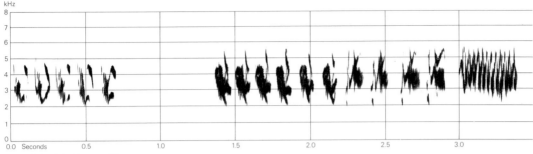

V  J-C Roché  France  May 1973

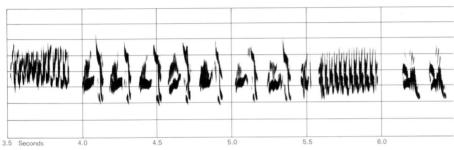

V  *cont.*

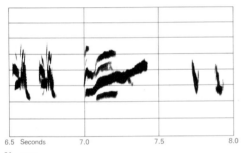

V  *cont.*

with wavering pitch ascent, others 'chids' (2nd and 3rd) and 'chip' or 'chup' (2 segments before 2nd 'dsooee'); same units given later in song, and additionally a notably high-pitched 'ti ti' preceding 'chids chids' (W T C Seale). For brief description of song of *rostrata* (Greenland), see Nicholson (1930). (1c) Courtship-song of ♂. Similar to 1b, but quieter, more delicate and overall more continuous; very varied (containing all 16 unit-types) and with fast delivery, especially when highly excited; lasts *c.* 4–33 s (Zablotskaya 1981). (1d) Subsong. Reported, without detailed description, for *cabaret* by Witherby *et al.* (1938). (2) Contact- and alarm-calls. (2a) Chatter-call. Quite far-carrying, noisy chattering, somewhat metallic 'che', 'dsched', 'chut', etc.; often 2–3 calls in succession (2–8), sometimes long series in which length of pause varies (Zablotskaya 1981; Bergmann and Helb 1982; Bruun *et al.* 1986). Commonest call, given while feeding or resting (quieter, single 'che' typically when preening, etc.), in

response to other conspecific birds flying past, at take-off, and in flight (in synchrony with wing-beats), apparently helping to integrate flock (Zablotskaya 1981). Chatter-call varies geographically (not clear, however, whether differences revealed in sonagrams are consistent or easily discernible in the field): in nominate *flammea*, fast, staccato, rather low-pitched 'che' sounds; call of *cabaret* said by Herremans (1989) to be less staccato, higher-pitched 'tji' series with narrower frequency range, also rendered as metallic twittering 'chuch-uch-uch-uch' (Witherby *et al.* 1938). However, no difference between nominate *flammea* and *cabaret* heard by L Svensson in Sweden. For Chatter-call of nominate *flammea* (much as illustrated by Herremans 1989), see Fig III. Rapid series of 'chi' units given by *cabaret* shown in Fig VII (for Chatter-calls or variants in song, see also Figs I–II, V–VI); for Icelandic birds, see Figs IV and VIII; no recordings of *rostrata* from Greenland available for analysis, but reported to be flatter, louder, and harsher than nominate *flammea*, at least in winter (Witherby *et al.* 1938; D I M Wallace). A 'pee-chuv-chuv' given when feeding (Zablotskaya 1981) apparently consists of Chatter-calls combined with call 2e. (2b) Dsooee-call. Somewhat plaintive and nasal 'dsooee', 'pyuee', 'tsooeet', 'wüid', 'wäid', 'bäi', or clear whistled 'pweet', of complex diadic structure, ascending rapidly in pitch and given singly or in series (Stadler 1926; Witherby *et al.* 1938; Zablotskaya 1981; Bergmann and Helb 1982; Herremans 1989). Given from perch, sometimes by lone bird in long series between feeding and preening (perhaps in such cases a kind of song, as suggested by Stadler 1926). Expresses alarm, being given when mobbing predator

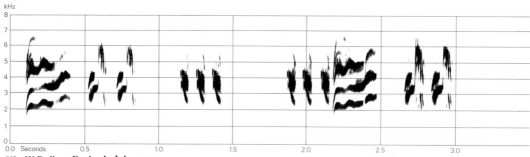

VI   W Pedley   England   July 1975

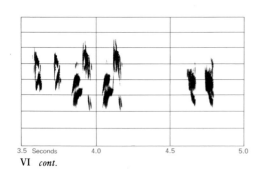

VI   *cont.*

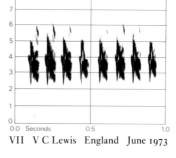

VII   V C Lewis   England   June 1973

VIII   P A D Hollom   Iceland   May 1991

IX   J Gordon   Norway   July 1976

('pyuee' or 'pyu-eee', faster with increasing excitement and at peak, sometimes just 'pyu') (Zablotskaya 1981). Calls of nominate *flammea*, *cabaret*, and Icelandic birds essentially similar, though *cabaret* distinguishable through more S-shaped main component according to Herremans

(1989); for *rostrata* (Greenland), see Nicholson (1930). Recordings suggest 'dzooee' and more disyllabic 'dzoo-EE' for nominate *flammea* (Fig IX), 'dsooee' (Icelandic birds: 4th unit in Fig VIII); for *cabaret*, see sonagrams of song (Figs V–VI) and Fig X which shows 3 alarm-calls ('dsooi') given by ♀ and which differ in structure from previous examples (W T C Seale). Further study needed to determine extent of variation within and between races. (2c) Whistling 'pyulee' ascending in pitch, but with undulating pattern; overall, similar to call 2b ('pyuee'), including in structural complexity. Used in pair-contact, sometimes in long antiphonal or simultaneous duet (Zablotskaya 1981). (2d) Series of clear 'tin' ('teen') sounds given from spring, initially by captive ♂ nominate *flammea* in association with song and display to ♀ who then also called thus; apparently serves for pair-contact in breeding season (Herremans 1989). (2e) Evenly-pitched 'pee' signalling take-off for short flight or given when hopping from branch to branch; ♀ usually gives 2 calls in succession

X   V C Lewis   England   June 1973

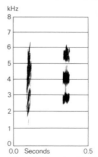

XI   V C Lewis   England   June 1973

when leaving nest (Zablotskaya 1981). (2f) Strangled, dull, noisy 'pyaa' given when disturbed, sometimes in long series, calls becoming longer, louder and faster with increasing excitement (Zablotskaya 1981). (2g) Recording of anxious ♀ *cabaret* contains (in addition to 'dsooi' calls: Fig X) abrupt and rather percussive 'chk' and 'tek' (W T C Seale: Fig XI) presumably also expressing alarm. (2h) Distress-call. Loud, long, harsh sound given by both sexes if trapped or handled (Zablotskaya 1981). (3) Rattle-call. Dry, buzzing or purring rattle: 'serrrrrr', 'tirr', 'tschrr', etc. Variously reported to serve as advertising-call or (usually longer) threat-call (Zablotskaya 1981), excitement-call (Schmidt 1962), or, especially when combined with Chatter- and certain other calls, as song (see 1a, above). Some variation evident in fine structure of Rattle-call (see Figs I–V), but not known whether (e.g.) consistent differences exist between races; *rostrata* (Greenland) described as 'thicker', more slurred than *cabaret* (Nicholson 1930). (4) Chak-call. Series of 'chak' sounds signalling aggression, louder and faster with increasing excitement; given by ♂♂ during territorial disputes in breeding season (Zablotskaya 1981). Presumably same call rendered harsh 'cheh' (Dilger 1960). (5) Calls given mainly in heterosexual contexts. (5a) Courtship-call of ♂. A 'tpyuee' ascending in pitch, sometimes steeply, and with crescendo towards end (Zablotskaya 1981); sonagrams suggest close relationship with calls 2b–c; see also 7b. (5b) Repeated, short, quiet clicking sounds with wide frequency range, given typically in long series with regular temporal pattern during Bill-flirting ceremony (Zablotskaya 1981). (5c) Quiet delicate 'tseet' descending in pitch and given by ♂ often in long series (calls sometimes delivered in pairs) during nest-building (in nest), laying, and early in incubation, including as begging- and soliciting-call (Zablotskaya 1981); similar high-pitched, quiet 'zitt zitt zitt' reported for captive *cabaret* (Gentz 1970), though also rendered 'tié-tié-tié' (Richard 1928). (5d) Long series of 'chee' sounds (units sometimes given in pairs) similar to 5c, but higher pitched and with squeaky quality; given by ♀ in much the same contexts as 5c; stronger aggressive element (Zablotskaya 1981). A 'jib jib jib', louder than 5c (Gentz 1970) probably also belongs here, and descending portamento 'tiu' of Icelandic birds

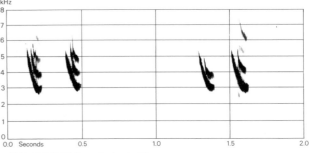

XII   P A D Hollom   Iceland   May 1991

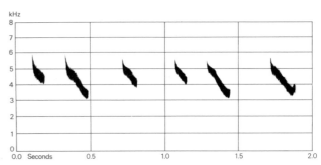

XIII   V C Lewis   England   June 1973

(Fig XII) is perhaps related. (6) Feeding-call. Recording (Fig XIII) contains irregular alternation of descending portamento whistles with penetrating and ringing timbre ('tsi' and 'tseeu') given by ♀ *cabaret* to encourage young to gape (W T C Seale). (7) Other calls. (7a) A 'geez' like Goldfinch *C. carduelis* (see call 5 of that species) given (by *cabaret*) in fights (Witherby *et al.* 1938); perhaps same as call 2f or related. (7b) Quiet hissing sound noted from ♂ *cabaret* during courtship (Witherby *et al.* 1938).

CALLS OF YOUNG. In Iceland, first audible food-calls noted from *c.* 1 week (Timmermann 1938). From 1–2 days before and for *c.* 2 weeks after fledging, contact-call of nominate *flammea* a series of noisy, low-pitched 'che' sounds (each unit of 3 sub-units); despite transcription, unlike adult call 2a or other adult calls (Zablotskaya 1981). Recording contains 'chzip' calls, higher pitched and of different structure than those shown in Zablotskaya (1981). A 'pyuee' (some similarity to adult calls 2c and 5a) given on hearing adult approaching with food or on losing contact with brood-siblings (Zablotskaya 1981). Fledgling *cabaret* gave short 'errr' and 'drd' (probably equivalent of 'che' in nominate *flammea*: see above) for first few days, only soft 'die' ('dee') once fully fledged (Dick 1973). MGW

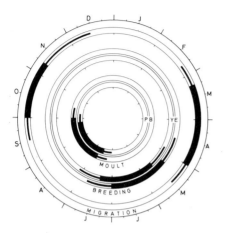

**Breeding.** SEASON. Britain and Ireland: eggs laid from 2nd half of May or early June, occasionally April (Witherby *et al.* 1938; Evans 1966). Iceland: laying 1st half of June, from mid-May in mild springs (Timmermann 1938). Finland: see diagram; in north, eggs laid mid-May to late July, possibly beginning of August, with peak around 1st half of June; laying period varies between a few days and 2 months depending on weather (Hildén 1969; Pulliainen and Peiponen 1981); in central latitudes, from mid-April, with peak 1st half of May; early eggs can be laid with thick snow on ground and night temperature of-20°C (Antikainen *et al.* 1980). In Scandinavia, eggs sometimes laid in March in years of good spruce *Picea* crop (Svärdson 1957). Russia: in far north, eggs laid from mid-June, mostly July (Dementiev and Gladkov 1954). Germany: in Sachsen, laying from mid-April to beginning of August, with peak period in May (Ernst 1990). Switzerland: April-August, peak period late May and early June, mid-June at 1500 m (Glutz von Blotzheim 1962, 1987). Northern Italy: eggs laid late April to late July at 2000 m (Maestri *et al.* 1989). SITE. In shrub or tree, at varying heights depending on habitat and species; occasionally on ground. In northern Finland, 57% of 310 nests in juniper *Juniperus* (average 0·7 m above ground), 25% spruce (4·3 m), 14% birch *Betula* (2·7 m), and 4% pine *Pinus* (3·6 m); 62% against trunk, 18% less than 0·5 m away, 10% 0·5-1·0 m, and 10% more than 1 m (Pulliainen and Peiponen 1981). In southern Finland, 50% in spruce, 22% in pine, and 10% in juniper (*n*=119); as season progresses, builds increasingly in birch and less in juniper (Hildén 1969). In Sachsen (eastern Germany), 25% of 134 nests in spruce, 21% pine, 11% birch, 5% apple *Malus*, 5% elder *Sambu-*

*cus*, 4% willow *Salix*; average height 4·2 m (*n*=128); sometimes only 0·5 m above ground very close to human activity, other nests far out in high crown like those of Siskin *C. spinus*; often builds in colonies of Fieldfare *Turdus pilaris* (Ernst 1990). In Switzerland, 75% of nests in larch *Larix*, at up to 18 m above ground, *n*=69 (Glutz von Blotzheim 1962). For review of central European studies (48 tree and shrub species recorded), see Ernst (1988). Nest: foundation of twigs of spruce, juniper, birch, Ericaceae, stalks, leaf stems, etc., with inner layer, often densely packed, of roots, grass, bark, moss, flower-heads, leaves, etc., thickly lined with plant down, hair, wool, and feathers, whole lining often very white in appearance (Timmermann 1938; Hildén 1969; Glutz von Blotzheim 1987; Ernst 1990). In Sachsen, 30 nests had average outer diameter 9·1 cm (6·5-12·0), inner diameter 4·9 cm (3·7-6·0), overall height 6·2 cm (5·0-8·0), depth of cup 3·9 cm (2·9-5·3) (Ernst 1990). Weights of 5 nests ranged from 8·8 g to 23·0 g (Glutz von Blotzheim 1987). Building: by ♀ only, accompanied by ♂ (Hildén 1969; Glutz von Blotzheim 1987; Ernst 1990); takes *c.* 1·5-3 days (Peiponen 1962; Hildén 1969); in one study, 4·5-7 days between start of building and 1st egg (Ernst 1990); material frequently added right up to nestling stage (Hildén 1969; Pulliainen 1979a); often uses material from old or abandoned nests (Glayre 1979; Duquet 1984). EGGS. See Plate 60. Sub-elliptical, smooth, and slightly, or non-glossy; bluish-white to pale blue-green with variable rust-red blotches, violet-pink undermarkings, and purplish-brown specks and scrawls towards broad end, often forming ring (Harrison 1975; Makatsch 1976). Nominate *flammea*: 16·9 × 12·6 mm (14·8-19·8 × 11·5-13·7), *n*=540; calculated weight 1·40 g. *C. f. cabaret*: 15·8 × 12·1 mm (14·1-17·8 × 10·0-13·2), *n*=139; calculated weight 1·21 g. *C. f. rostrata* (Greenland): 18·0 × 13·0 mm (16·0-20·0 × 12·3-13·4), *n*=27; calculated weight 1·57 g. *C. f. rostrata* (Iceland): 17·7 × 12·6 mm (17·0-18·4 × 12·0-13·0), *n*=9; calculated weight 1·46 g. (Schönwetter 1984.) Clutch: 4-6 (2-7). In northern Finland, of 94 clutches: 2 eggs, 1%; 3, 1%; 4, 16%; 5, 62%; 6, 18%; 7, 2%; average 5·0 (Pulliainen and Peiponen 1981); average in another study 5·2 (*n*=137); 5·7 (*n*=25) in 1st third of June, 5·0 (*n*=32) in last third, and 4·3 (*n*=11) in mid-July (Hildén 1969). In Swedish Lapland, 5·5 (*n*=101) in years of peak population, 4·9 (*n*=43) in non-peak years, difference significant (Enemar and Nyström 1981). Average in central Finland 4·8, *n*=38 (Antikainen *et al.* 1980); in Sachsen, 4·5, *n*=33 (Ernst 1990); in northern Italy, 4·1, *n*=14 (Maestri *et al.* 1989). Generally 2 broods, though in some places apparently only in years of good food supply or only by some pairs (Glutz von Blotzheim 1962; Peiponen 1962; Evans 1966; Dick 1973; Ernst 1990). In captive study in Iceland, new nest started for 2nd brood when young of 1st 7 days old (Timmermann 1938). In Alaska (USA), ringed pair had 3 clutches: 2nd nest finished 1-6 days after 1st clutch of 5 lost; 2nd clutch of 5 abandoned; 3rd clutch started 3-4 days later and all 5 eggs hatched (Troy and Shields 1979). See Movements for northward movement in Finland between 1st and 2nd broods. INCUBATION. 10–12 days (9–14), by ♀ only (Timmermann 1938; Peiponen 1962; Pulliainen 1979a; Ernst 1990); average at 20 nests in northern Finland 10·7 days (Pulliainen and Peiponen 1981). Probably begins with 3rd egg, though some perhaps from 1st or 2nd (Glutz von Blotzheim 1962; Peiponen 1962; Pulliainen 1979a; Ernst 1990), and hatching generally over 2–3 days (Hildén 1969; Pulliainen and Peiponen 1981). ♀ sits very tightly; sat for 96·6% of time at one nest, taking breaks of average 3·3 min (0·3–44, *n* = 142) (Pulliainen 1979a). YOUNG. Fed and cared for by both parents; fed by ♂ via ♀ for first *c.* 3–6 days (Dick 1973; Pulliainen 1979a); brooded by ♀ only, continuously for first 6–9 days (Timmermann 1938; Hildén 1969; Pulliainen 1979a; Ernst 1990). FLEDGING TO MATURITY. Fledging period 9–14 days; in northern Finland, average 11·5 days at 14 nests (Pulliainen and Peiponen 1981); may leave nest before able to fly (Hildén 1969; Ernst 1990). Captive birds first picked up seeds at 17–21 days old, fully independent at *c.* 26 days (Timmermann 1938). Age of first breeding 1 year (Evans 1966); 1-year-old ♀ in Alaska laid 3 clutches, last of which produced nestlings (Troy and Shields 1979). BREEDING SUCCESS. In Sachsen, 57% of 53 breeding attempts where eggs laid mid-April to mid-May failed completely, giving 3·7 fledged young per successful nest (*n* = 23) and 1·6 overall; 36% of 36 later attempts failed, giving 4·2 (*n* = 23) and 2·7 fledged young; main predators were crows (Corvidae), especially Magpie *Pica pica*, squirrel *Sciurus vulgaris*, and cats (Ernst 1990). In Lombardia (northern Italy), of 57 eggs, 68% hatched and 47% produced fledged young; 2·5 young produced per successful nest (*n* = 11), 1·9 overall (*n* = 14) (Maestri *et al.* 1989). In northern Finland, 43% of 70 clutches contained unfertilized eggs, perhaps caused by high population density, and only *c.* 25% of eggs resulted in fledged young; Corvidae chief predators, including Siberian Jay *Perisoreus infaustus*; probably nests in *T. pilaris* colonies to escape predation (Hildén 1969). Similarly in Swedish Lapland, many eggs failed to hatch in years of peak population, although average brood size in all years fairly constant compared to clutch (see above): 4·5 (*n* = 69) in peak years, 4·6 (*n* = 52) in others (Enemar and Nyström 1981). In dense stand of young spruce in central Finland, up to 80% of eggs produced fledged young (Antikainen *et al.* 1980). See also Pulliainen and Peiponen (1981); for effect of sudden cold weather in Swedish Lapland, see Nyström and Nyström (1991). For review, see Ernst (1988). BH

Plumages. (nominate *flammea*). ADULT MALE. In fresh plumage (autumn), feathers at base of culmen dull black with narrow grey-white or pink fringes, latter occasionally concealing black; towards sides, fringes broader, dark of bases fully concealed, forming grey-white supraloral stripe, extending into long grey-white or pure white supercilium. Forecrown glossy crimson-red, forming contrasting patch 11·5 (189) 9–14 mm long (Molau

1985); some dark grey of feather-bases usually visible between crimson; exceptionally, forecrown yellowish- or orange-red (commonly so in captivity). Hindcrown and nape evenly streaked dusky grey and buff-brown, side of neck and sometimes part of crown or nape streaked paler buff or buff-white and dusky grey. Mantle and scapulars tawny-brown, streaked dull black, feather-fringes slightly paler and streaks on centres slightly sharper and deeper black on outer mantle and inner scapulars, forming 'braces'. Back and upper tail-coverts pale buff or buff-white, centre of each feather with broad dark grey-brown mark, fringes of lower back and shorter tail-coverts often partly suffused rosy-pink; rump pale buff or white with rosy-pink suffusion, marked with dark grey-brown or dull black streaks, latter most conspicuous on centre, more diffuse and sometimes inconspicuous towards sides of rump. Lore dull black, forming distinct stripe; narrow eye-ring white or pink-white, broadly interrupted in front and behind. Short and ill-defined stripe below eye mottled grey-white and dark brown-grey; upper cheek and ear-coverts dark brown-grey with faint pale buff or off-white streaks, cheek and lower border of ear-coverts sometimes partly tinged rosy-pink or rufous-brown. Front part of lower cheek and side of throat rosy-pink, partly concealed by white feather-fringes, rear off-white or pale buff with some grey mottling, sometimes slightly tinged pink or tawny. Chin and upper throat dull black, forming contrasting squarish patch; lower throat, chest, and often upper belly and side of belly rosy-pink, partly concealed below broad off-white feather-fringes. Side of breast tawny-buff, merging into pale buff or buff-white on flank, marked with long dark brown-grey streaks; remainder of underparts greyish-white or off-white, sometimes with faint cream-pink suffusion; centres of under tail-coverts with dark brown-grey marks, those on longest coverts 3·6 (223) 1–7 mm (mainly 4) wide (Molau 1985). Tail greyish-black, both webs of central pair (t1) and outer webs and tips of others narrowly but distinctly fringed grey-white. Flight-feathers, greater upper primary coverts, and bastard wing greyish-black, outer webs and tips narrowly but distinctly fringed pale grey (except on emarginated parts of p6–p8 and on longest feather of bastard wing); terminal half of outer web of secondaries often more broadly fringed pale grey or pink-grey. Tertials dull black, middle portion of outer web fringed buff, tip fringed off-white. Greater upper wing-coverts black, tip contrastingly tawny-buff, 2–5 mm wide (widest on outer web; soon bleaching to white), outer webs with faint grey fringe (paler on outermost covert); median coverts black with tawny-buff tip 1–2 mm wide; lesser coverts dark grey-brown with ill-defined paler brown-grey fringes. Axillaries and under wing-coverts grey-white, dusky grey feather-bases partly visible. Marked influence of bleaching and abrasion. *When moderately worn* (spring), buff and tawny feather-fringes of upperparts bleached to light brown-grey, dark streaks on centres of feathers more contrasting, ground-colour of fringes of mantle and rump virtually white; ground-colour of side of neck, side of breast, and flank buff-white, dark streaks more sharply contrasting; white feather-fringes of rosy-pink on cheek and underparts largely worn off, rosy-pink more exposed, often more vinous-pink; tips of median and greater coverts white. When heavily worn (summer), much dull black of feather-bases visible in crimson of crown, much dark grey on remainder of upperparts; pale fringes of upperparts largely worn off, upperparts mainly black-brown with narrow grey-white streaks, latter most distinct on rump; pale supercilium and side of neck mottled grey, less conspicuous; dark grey feather-bases of underparts partly visible, belly dirty white; pale fringes of tail and flight-feathers and white tips of greater and median coverts largely worn off. ADULT FEMALE. Like adult ♂, but rosy-pink restricted to spots on front of lower cheek and slight wash on chest, shorter upper tail-coverts, or upper side of belly, or pink entirely absent; lower cheek and side of throat grey-white, washed buff when fresh; dull black patch on chin and upper throat on average larger, more rounded, and less sharply defined than in adult ♂; chest buff (when fresh) to grey-white (when worn), marked with short dusky streaks on side, sometimes with short fine streaks on centre. Otherwise as adult ♂; length of crimson patch on crown 9·9 (101) 7–13 mm, tinged orange or yellow in c. 35% of birds, dark streak on centre of longest under tail-coverts 3·8 (108) 2–6 mm wide (Molau 1985); back, rump, and side of body on average more heavily streaked. For sexing, see also Brooks (1973). NESTLING. Down fairly long and plentiful, medium grey; restricted to tufts on crown, back, upper wing, thigh, and (very short) on rump and belly. For development, see Peiponen (1962). JUVENILE. Rather like adult, but crimson on crown absent and no rosy-pink else-where on body. Streaks on upperparts (including entire cap) blacker and more sharply defined than in adult, brown feather-fringes on crown, mantle, and scapulars duller and much nar-rower; forehead, supercilium, nape, and side of neck closely mottled dull black and pale buff or off-white; mid-mantle, back, and rump densely streaked black and buff-white; upper tail-coverts tawny-buff with narrow and short black shaft-streak. Side of head black-brown with fine buff and grey speckling; a narrow mottled grey-buff eye-ring. Chin with small ill-defined dark grey patch; lower cheek, side of throat, lower throat, and chest buff-white or grey-white with rounded dark grey spots. Remainder of underparts as adult ♀, tinged buff on upper belly and flank when fresh, dirty white when worn; streaks shorter, narrower, and blacker; vent and under tail-coverts short, woolly, dirty white; longest coverts with short black central mark. Tail as adult, but tips pointed, less rounded than in adult, without clear grey-white fringe along tip; fringes along sides of feathers often browner, less grey. Wing as adult, but fringes along outer webs of flight-feathers and primary coverts browner, less pure grey; tips of median and greater coverts and fringes along tips of tertials tawny-buff at a time when those of adults are bleached to white and heavily abraded. FIRST ADULT MALE. Like adult ♀, but juvenile tail, flight-feathers, and greater upper primary cov-erts retained, tips of outer tail-feathers generally more sharply pointed and less clearly fringed pale grey than in adult (see Svensson 1992 for shape); sometimes a clear difference in colour and degree of abrasion between moulted 1st adult inner greater coverts and more worn retained juvenile outer coverts. More often a rosy-pink wash on lower cheek, shorter upper tail-coverts, or chest than in adult ♀; exceptionally, rosy-pink as extensive as in adult ♂. Length of crimson crown-patch 10·6 (106) 5–14 mm, tinged orange or yellow in c. 20% of ♂♂; width of dusky mark on longest under tail-coverts 4 (207) 2–7 mm (Molau 1985). FIRST ADULT FEMALE. Like adult ♀, but part of juvenile feath-ering retained, as in 1st adult ♂. Only exceptionally some rosy-pink, mainly on lower cheek. Length of crown-patch 9·4 (98) 7–12 mm, tinged orange or yellow in c. 36% of birds; streak on under tail-coverts 3·9 (98) 1–7 mm wide (Molau 1985).

**Bare parts.** ADULT, FIRST ADULT. Iris dark brown. In summer, upper mandible dark horn to plumbeous-black, lower mandible straw-yellow or greyish-yellow, tinged light horn on tip, some-times partly flesh at base, cutting edges pale yellow; during rest of year, bill orange-yellow with plumbeous culmen and black streak on tip of culmen and on gonys. Leg and foot dark brown, sometimes with pink tinge, soles brown-flesh or brown-grey. (A J van Loon; RMNH, ZMA.) NESTLING. Bare skin dark. Mouth red with 2 small pale spots on palate; gape-flanges yellow with red at extreme base (Witherby et al. 1938). JUVENILE. At fledging,

bill lime-green, leg and foot pink; bill-tip, leg, and foot darkening to adult colour and bill-base to orange-yellow shortly before or at start of post-juvenile moult (Wynne-Edwards 1952; Evans 1966; ZMA).

**Moults.** ADULT POST-BREEDING. Complete; primaries descend-ent. In northern England, *cabaret* starts when young inde-pendent, sexes at same time or ♂ slightly before ♀; moult starts with shedding of p1, early to late August (on average, 6 August), completed mid-September to early October (on average, 25 Sep-tember); duration of primary moult 51 (43–56) days in ♂, 54 (48–56) in ♀; secondaries start with s1 at primary moult score 20–27, completed with regrowth of s6 at same time as p9 (score 45) or slightly later; sequence of tertials s8–s7–s9, starting at primary score 0–15, completed at 25–35 (Evans 1966, which see for many details). In southern England, moult of ♂ starts mainly late July to mid-August, average c. 4 August, ♀ mainly August, average 11 August; completed after c. 54 (53–60) days early September to early October (occasionally, late August to early October); within populations, some overlap between breeding and start of moult (Boddy 1983). In eastern Finnmark (Norway, c. 70°N), p1 shed on average 24 (♀) or 28 (♂) July, primary moult com-pleted on average 14 (♂) or 16 (♀) September (Evans et al. 1967, which see for details); many birds in moult in northern Finnmark 24 July to 15 August (Harris et al. 1965). In northern Sweden (c. 68°N), moult of advanced birds c. 2 weeks ahead of those from eastern Finnmark (Pepper and Kennedy 1970). In Finland, moult starts on average 20 July (Lehikoinen and Niemelä). Single nominate *flammea* in Belgium had primary moult score 6 on 5 July (DeBrun 1988). In Alaskan birds, moult early July to early September, duration in population 61 days (Brooks 1968). In captivity, moult of *cabaret* (n = 6) started on average 18 July, range 10–31 July, in nominate *flammea* (n = 2) 25–31 July; dur-ation of primary moult 52·3 (5·32; 6) 44–58 and 53–55 days, respectively; duration of secondary moult 38·5 (4·59; 6) 32–44 and 36–45 days; duration of entire moult 58·7 (8·57; 6) 44–58 and 53–65 days (Newton 1969, which see for details on sequence of moult and duration of growth of many individual feathers). POST-JUVENILE. Partial: head, body, and lesser and median upper wing-coverts; in early-fledged birds, also variable number of inner greater coverts, rarely 1–2 tertials, exceptionally some tail-feathers (Boddy 1981). For sequence of replacement of feathers, see Evans (1966) and Boddy (1983). Starts shortly after fledging, and timing thus variable; moult recorded mid-July to late October, mainly August-September (Witherby et al. 1938; Boddy 1981; Ginn and Melville 1983). In southern England, starts mid-August to September, occasionally from late July, but sometimes many birds not started by mid-September; moult completed from early September onwards (Boddy 1983). In northern England, juvenile body plumage generally lost by late August (Evans 1966). In northern Sweden, some had moult almost completed 18 August, others not started by 25 August (Pepper and Kennedy 1970). In north-west Russia, starts early July to early August, mainly c. 20 July, a few up to early Sep-tember; completed late September to late October, mainly c. 10 October (Rymkevich 1990). In Greenland, some *rostrata* in full moult August, others had finished September (RMNH, ZMA); on Baffin Island (Canada), in varying stage of moult on 11 August (Wynne-Edwards 1952).

**Measurements.** ADULT, FIRST ADULT. Nominate *flammea*. Wing and tail of (1) Fenno-Scandia, summer, (2) Netherlands, winter; other measurements combined, skins (RMNH, ZMA: A J van Loon, C S Roselaar). Bill (S) to skull, bill (N) to distal corner of nostril; exposed culmen on average 4·5 less than bill

(S). Juvenile wing and tail are those of 1st adult. *Holboellii* form (see Geographical Variation) excluded; birds here defined as belonging to *holboellii* if bill (S) 13·0 (♀) or 13·5 (♂) and over, bill (N) 8·7 and over.

| | ♂ | | ♀ | |
|---|---|---|---|---|
| WING (1) | 75·2 (1·53; 14) | 72-78 | 73·4 (1·02; 7) | 71-75 |
| WING AD (2) | 76·0 (1·35; 27) | 73-78 | 73·9 (1·03; 4) | 73-75 |
| JUV (2) | 75·1 (1·48; 34) | 72-78 | 73·8 (1·45; 21) | 71-76 |
| TAIL AD (2) | 54·3 (1·59; 22) | 51-57 | 53·6 (1·93; 4) | 51-56 |
| JUV (2) | 55·2 (1·49; 28) | 52-58 | 53·4 (1·36; 17) | 51-56 |
| BILL (S) | 12·4 (0·56; 60) | 11·4-13·4 | 12·2 (0·54; 25) | 11·3-13·0 |
| BILL (N) | 7·6 (0·38; 73) | 6·8-8·5 | 7·3 (0·44; 30) | 6·7-8·5 |
| TARSUS | 14·9 (0·49; 33) | 13·8-15·7 | 14·7 (0·51; 17) | 13·8-15·4 |

Sex differences significant for wing.

Birds of *holboellii* form from samples above, ages combined; skins (ZMA).

| | ♂ | | ♀ | |
|---|---|---|---|---|
| WING | 77·8 (2·54; 8) | 75-83 | 74·7 (2·08; 5) | 72-78 |
| TAIL | 56·4 (1·92; 7) | 54-59 | 53·4 (1·14; 5) | 51-55 |
| BILL (S) | 14·7 (0·86; 8) | 13·8-16·5 | 13·5 (0·73; 5) | 13·0-14·8 |
| BILL (N) | 9·3 (0·83; 8) | 8·8-11·2 | 9·2 (0·42; 5) | 8·9-9·9 |

Sex differences significant, except for bill (N).

*C. f. cabaret.* Netherlands, May-October; skins (RMNH, ZMA: A J van Loon, C S Roselaar).

| | ♂ | | ♀ | |
|---|---|---|---|---|
| WING | 70·9 (1·56; 27) | 68-74 | 69·2 (1·92; 32) | 66-73 |
| TAIL | 54·3 (2·34; 14) | 51-58 | 53·4 (2·26; 16) | 50-57 |
| BILL (S) | 11·7 (0·39; 22) | 11·1-12·2 | 11·6 (0·32; 36) | 10·9-12·1 |
| BILL (N) | 7·1 (0·41; 21) | 6·5-7·6 | 7·0 (0·35; 24) | 6·6-7·7 |
| TARSUS | 14·0 (0·49; 13) | 13·3-14·8 | 14·1 (0·62; 15) | 13·3-15·1 |

Sex differences significant for wing.

Larger samples of (1) nominate *flammea*, (2) *cabaret* (both mainly live birds, Belgium), (3) *rostrata* from Greenland (breeding birds only; skins), and (4) *rostrata* from Iceland (dark 'islandica', breeding and non-breeding birds combined; skins) (Herremans 1990a). Bill (F) is culmen to feathering, depth is depth of bill at feathering of base.

| | ♂ | | ♀ | |
|---|---|---|---|---|
| (1) WING | 77·6 (2·18; 222) | 70-83 | 75·2 (1·75; 162) | 70-80 |
| TAIL | 55·2 (2·04; 222) | 49-61 | 54·2 (1·89; 162) | 49-60 |
| BILL (F) | 9·1 (0·59; 222) | 7·9-10·4 | 8·8 (0·56; 162) | 7·5-9·9 |
| DEPTH | 5·9 (0·25; 222) | 5·3-6·8 | 5·7 (0·24; 162) | 5·2-6·5 |
| (2) WING | 71·1 (1·92; 66) | 65-75 | 69·4 (1·58; 34) | 67-75 |
| TAIL | 52·3 (1·59; 66) | 48-56 | 51·5 (1·56; 34) | 47-56 |
| BILL (F) | 9·0 (0·46; 66) | 7·7-10·2 | 8·8 (0·36; 34) | 8·1-9·5 |
| DEPTH | 6·0 (0·23; 66) | 5·5-6·5 | 5·8 (0·26; 34) | 5·2-6·5 |
| (3) WING | 80·3 (1·96; 58) | 74-86 | 78·8 (2·62; 25) | 75-85 |
| TAIL | 59·2 (2·74; 58) | 53-67 | 59·0 (2·75; 25) | 54-66 |
| BILL (F) | 9·5 (0·55; 58) | 8·3-11·0 | 9·3 (0·57; 25) | 7·8-10·2 |
| DEPTH | 6·7 (0·26; 58) | 6·3-7·5 | 6·6 (0·25; 25) | 6·3-7·3 |
| (4) WING | 79·5 (2·30; 36) | 75-84 | 76·5 (2·22; 31) | 72-81 |
| TAIL | 59·6 (2·55; 36) | 55-66 | 58·2 (2·30; 31) | 54-63 |
| BILL (F) | 8·9 (0·53; 36) | 8·0-9·9 | 8·7 (0·61; 31) | 6·6-10·0 |
| DEPTH | 6·6 (0·26; 36) | 5·9-7·1 | 6·4 (0·30; 31) | 5·7-7·0 |

Wing. Live birds except as noted. Nominate *flammea*. (1) Northern Norway, summer (wing worn, unflattened) (Evans *et al.* 1967). (2) Russian Lapland (Semenov-Tyan-Shanski and Gilyazov 1991). Torne Lappmark (Sweden), summer: (3) adult, (4) 1st adult (Molau 1985). (5) Västerbotten (Sweden) (Lindström *et al.* 1984). (6) Poland, spring migrants (Busse 1976). (7) Eastern Germany; skins (Eck 1985a). Belgium, winter: (8) adult, (9) 1st adult (Herremans 1973). *C. f. cabaret.* Britain: (10) northern England (Evans 1966, which see for influence of age), (11) skins (Niethammer 1971), (12) southern England (Boddy 1981); see also Prato and Prato (1978). (13) Alps, skins (Niethammer 1971). (14) Erzgebirge (south-east Germany) (Eck 1985a).

| | ♂ | | ♀ | |
|---|---|---|---|---|
| (1) | 73·4 (2·35; 25) | 69-80 | 71·8 (1·74; 44) | 68-76 |
| (2) | 75 (—; 13) | 73-78 | 74 (—; 11) | 72-78 |
| (3) | 74·1 (1·69; 244) | 70-78 | 71·4 (1·46; 112) | 68-75 |
| (4) | 72·7 (1·69; 235) | 69-77 | 70·7 (1·55; 100) | 65-75 |
| (5) | 75·2 (1·54; 27) | 73-79 | 73·5 (1·79; 18) | 70-77 |
| (6) | 74·4 (1·83; 18) | — | 73·3 (2·08; 14) | — |
| (7) | 74·8 (2·39; 16) | 71-80 | 73·7 (—; 12) | 71-74 |
| (8) | 75·3 (—; 4) | 71-79 | 73·1 (—; 31) | 71-77 |
| (9) | 73·9 (—; 58) | 70-79 | 72·0 (—; 24) | 69-75 |
| (10) | 68·4 (1·77; 110) | — | 67·3 (1·74; 116) | — |
| (11) | 70·2 (1·36; 21) | 67-73 | 67·1 (1·61; 13) | 65-72 |
| (12) | 70·8 (1·54; 387) | 67-74 | 68·4 (1·67; 266) | 64-73 |
| (13) | 70·9 (—; 5) | 70-74 | — (—; 3) | 67-71 |
| (14) | 72·4 (1·77; 39) | — | 69·7 (1·61; 58) | — |

Nominate *flammea*. Northern Sweden, juvenile: wing 72·0 (2·20; 42) 68-77 (Pepper and Kennedy 1970). For measurements of *holboellii* form, see Junge (1942) and Herremans (1990a).

*C. f. cabaret.* Wing. Halland (southern Sweden): 71·0 (2·44; 24) 66-74 (Lindström *et al.* 1984). South-west Norway: 71·1 (1·93; 64) 67-75 (Grimsby and Røer 1992a). Austria: ♂ 72·2 (1·44; 12) (Eck 1985a). Eastern Germany: 67-76 (*n*=44) (Ernst and Thoss 1977).

*C. f. rostrata.* Scotland, autumn: wing 79 (28) 73-82 (Williamson 1956, which see for other measurements). Massachusetts (USA), winter: 78·5 (32) 72-84 (Wetherbee 1937). Greenland: wing, ♂ 79·9 (1·51; 23) 77-83, ♀ 78·7 (2·65; 12) 75-82 (Salomonsen 1928); tarsus 16·0 (0·74; 13) 14·9-17·4 (A J van Loon). Greenland, wing 75-82, bill 9·0-10·8; Iceland, wing 75-81, bill 8·0-9·2; bill depth in both 6·4-7·6 (Timmermann 1938-49). Bill to nostril: Greenland 7·8 (0·44; 30) 7·0-9·0, Iceland 7·2 (0·42; 31) 6·5-8·2; only 10 birds from Ireland over 7·4 and only 4 from Greenland below 7·5, thus only 23% of 61 birds wrongly classified by bill length (Vaurie 1957).

Various races. See Vaurie (1956a, 1959) and Svensson (1984b, 1992). For mixed sample of nominate *flammea*, *holboellii*, and *C. hornemanni exilipes*, see Danilov *et al.* (1984); see also Terentiev (1970).

**Weights.** ADULT, FIRST ADULT. Nominate *flammea*. Northern Norway (including some *C. hornemanni exilipes*): (1) late July to early September, (2) August-September (Evans *et al.* 1967). (3) Russian Lapland, summer (Semenov-Tyan-Shanski and Gilyazov 1991). (4) Yamal peninsula (north-west Siberia) (including some *C. h. exilipes*) (Danilov *et al.* 1984). (5) Northern Alaska and north-west Canada, May-July (Irving 1960). (6) Eastern Germany (Eck 1985a). Belgium, autumn: (7) typical *flammea*, (8) *holboellii* form (Herremans 1990a). *C. f. cabaret.* Northern England, August-September: (9) moulting adult, (10) 1st adult (Evans 1966, which see for details). (11) Netherlands, mainly autumn (RMNH, ZMA).

| | ♂ | | ♀ | |
|---|---|---|---|---|
| (1) | 14·3 (—; 9) | 13·0-16·5 | 14·4 (—; 15) | 12·5-18·0 |
| (2) | 14·1 (1·08; 48) | 12-18 | 13·5 (1·01; 74) | 11-17 |
| (3) | 13·0 (—; 13) | 10·2-16·1 | 13·1 (—; 11) | 9·9-14·0 |
| (4) | 14·4 (—; 136) | 11·9-17·6 | 12·7 (—; 203) | 10·9-15·3 |
| (5) | 12·9 (—; 21) | 10·2-15·0 | 13·1 (—; 9) | 10·1-14·2 |
| (6) | 13·4 (1·33; 15) | 11·0-16·0 | 12·8 (0·82; 12) | 11·5-14·0 |
| (7) | 13·3 (0·96; 181) | 11·0-16·0 | 12·5 (0·80; 141) | 10·3-14·5 |
| (8) | 14·9 (1·05; 18) | 13·0-16·8 | 13·5 (0·88; 15) | 12·0-15·3 |
| (9) | 11·8 (0·72; 46) | — | 11·2 (0·48; 29) | — |
| (10) | 11·9 (0·65; 41) | — | 11·6 (0·63; 40) | — |
| (11) | 10·9 (1·08; 11) | 9·0-12·3 | 10·6 (0·91; 10) | 8·8-12·0 |

Nominate *flammea*. Southern Norway, October-January: 13·8 (13) 12·5-15 (Haftorn 1971). Helgoland (Germany): 14·4 (19) 12-17·5, exhausted bird 9·5 (Weigold 1926). For graph of 2-week average throughout year in captivity, see Brooks (1968);

for weight during moult, see Newton (1969). See also Wetherbee (1937), Clench and Leberman (1978), and Newton (1972).

*C. f. cabaret*. Central and southern England: September 11·6 (1208) 9–14·5, October 11·9 (1343) 9–16·5, November 11·9 (389) 9·5–16·5, December 12·3 (73) 10–16·5, January 12·6 (145) 11–14·5, 1st half of February 11·7 (104) 10–14·5 (Boddy 1984, which see for graph of weight throughout year). For weight during moult, see Newton (1969) and Boddy (1984). Exhausted birds: 7·8, 8·4 (Evans 1969a, which see for fat contents and other details). See also Newton (1972).

*C. f. rostrata*. Scotland, autumn: 16·3 (37) 10·7–21·7; shortly after arrival 14·8 (2·39; 6) 10·7–17·1 (Williamson 1956). Iceland: 12·5–17 (12) (Weigold 1926). Massachusetts (USA): winter 18·1 (32) 15·6–20·0 (Wetherbee 1937); March–April 17·8 (32) 13·8–22·8 (Shaub 1950).

NESTLING, JUVENILE. Nominate *flammea*. For growth curve of nestling, see Peiponen (1962): at day 0 *c.* 1·7; at day 10, peak of *c.* 13; at fledging (day 12–13) *c.* 12. Juvenile, northern Norway, July–September: 14·1 (38) 11·5–17·5 (Harris *et al.* 1965); 13·5 (1·17; 253) (Evans *et al.* 1967, which see for details). Juvenile, northern Sweden, August: 12·1 (0·92; 48) (Pepper and Kennedy 1970). *C. f. cabaret*. For graph of average weight England, see Boddy (1984).

**Structure.** Wing moderately long, broad at base, tip bluntly pointed. 10 primaries: p8 longest, p9 0·5–2 shorter, p7 0–1·5, p6 2–5, p5 8–14, p4 12–18, p3 15–20, p2 16–24, p1 18–26 (21–26 in *rostrata*, 20–25 in nominate *flammea*, 18–22 in *cabaret*); p10 strongly reduced, 44–60 shorter than p8, 6–10 shorter than longest upper primary covert. Outer web of p6–p8 emarginated, inner web of p7–p9 with faint notch. Tip of longest tertial reaches tip of p1. Tail rather short, tip forked; 12 feathers, t1 9 (20) 5–12 shorter than t5–t6 in nominate *flammea* and *cabaret*, 11 (7) 9–12 in *rostrata*. Bill short, conical; average depth × width at base 6·1 × 5·9 mm in nominate *flammea*, 7·0 × 6·5 in *holboellii* form, 6·1 × 5·7 in *cabaret*, and 7·0 × 6·5 in *rostrata* (*n* = 8–10 in each); for depth, see also Measurements; culmen very slightly decurved, mainly at tip, cutting edges and gonys almost straight; tip laterally compressed (somewhat less so in *rostrata*), sharply pointed, especially in *holboellii* form of nominate *flammea*. Entire base of bill covered by short bristle-like feathers, concealing small oval nostril. Plumage dense, soft, and lax. For length of individual feathers of wing and tail of nominate *flammea* and *cabaret*, see Busching (1988); for weight of feathering of body and wing, see Brooks (1968) and Evans (1969a). Tarsus and toes rather short and slender; claws long and slender, decurved, but varying greatly in length with wear. Middle toe with claw 12·6 (10) 11·5–14·0 in nominate *flammea*, 12·6 (8) 11·5–13·5 in *cabaret*, 14·0 (6) 12·9–14·7 in *rostrata*; outer toe with claw *c.* 73% of middle with claw, inner *c.* 76%, hind *c.* 98%.

**Geographical variation.** Marked; involves size, relative depth and width of bill, and general colour. *C. f. cabaret* from Britain and north-west France to southern Sweden and south-west Norway as well as (isolated) in Alps and other mountain areas of central Europe is smallest race, wing mainly under 73 mm, tail mainly under 54 mm, but bill not much different from nominate *flammea*; streaks on upperparts and sides of body slightly broader, blacker, and more sharply defined, especially on nape to upper mantle (in nominate *flammea*, nape to upper mantle often paler buff and less well streaked than lower mantle and scapulars), feather-fringes of upperparts darker tawny- or rusty-brown, and rump heavily streaked dull black and buff, usually without white showing through (in nominate *flammea*, streaked

white, pale buff, and dark grey); side of head, chest, side of breast, and flank conspicuously tinged tawny- or buff-brown, heavily streaked dull black on side of breast and flank, contrasting with cream belly (in nominate *flammea*, pale cream-buff, less heavily marked with dark grey streaks, not contrasting with whiter belly; in adult ♂, which shows rosy-pink throat to upper belly, difference between these races less marked, though flank of *cabaret* usually more distinctly tawny and more heavily streaked dull black than in nominate *flammea*); fringes of tail and tertials and tips of greater and median upper wing-coverts cinnamon-buff, less pale buff (when fresh) or off-white (when worn) than in nominate *flammea*; rosy-pink of adult ♂ on average slightly deeper and extending further down on belly. In worn plumage, upperparts of *cabaret* dull brown-grey with broad black streaks, darker than in nominate *flammea*, less streaked grey; rump broadly streaked black and grey-white; ground-colour of underparts bleached to dirty white, as in nominate *flammea*, but streaks on side of breast and flank broader and blacker. Breeding range of *cabaret* is expanding to north and east (see Ernst 1988); will perhaps meet nominate *flammea* in south-west Sweden or south-west Norway, where *cabaret* now established (e.g. Lindström *et al.* 1984, Grimsby and Røer 1992a) and where nominate *flammea* breeds in some years (e.g. Götmark 1982). Birds from western Scotland and Ireland sometimes separated from *cabaret* of Alps as *disruptis* Clancey, 1953; ground-colour of upperparts and side of body richer tawny-brown than in birds from Alps, streaks purer black; birds from shores of North Sea intermediate in colour (Clancey 1953; Meinertzhagen 1953); most birds in small series examined from Alps and Scotland were similar, and only a few of Alps were slightly paler buff-brown (mainly on hindneck), and *disruptis* therefore not recognized here, following Vaurie (1959). For sexing and ageing of *cabaret*, see Evans (1966), Jones *et al.* (1975), Prato and Prato (1978), Boddy (1978, 1981), and Herremans (1990a); for means of separation from nominate *flammea*, and for possibility that intermediates occur, see Lindström *et al.* (1984), Herremans (1977, 1989), and Grimsby and Røer (1992a, b).

*C. f. rostrata* from southern Greenland (north to *c.* 70–71°N: Salomonsen 1931b) and northern Baffin Island distinctly larger than nominate *flammea*, wing mainly over 79 mm in ♂, over 77 mm in ♀, tail mainly over 57 mm; bill deeper and wider at base, culmen and gonys clearly convex (not as straight as in nominate *flammea*); darker than nominate *flammea*, colour near *cabaret*, with width, extent, and colour of dark streaks as in *cabaret* and ground-colour brown, but ground-colour paler buff-brown, less rusty or tawny, rump buff and white with broad black streaks; contrast between buff chest and flank and cream-white belly as in *cabaret*; fringes of tail-feathers and tertials and tips of greater and median coverts as in nominate *flammea*, buff when fresh, off-white when worn; rosy-pink of underparts of adult ♂ averages paler and less extensive than in nominate *flammea*. *C. f. rostrata* may grade into North American population of nominate *flammea* in north-east Labrador and southern Baffin Island (Salomonsen 1928); intermediates have been named *fuscescens* Coues, 1862, but breeding ranges of *rostrata* and nominate *flammea* appear separated in eastern Canada, and *fuscescens* not recognizable (Godfrey 1986), though intermediates sometimes reported in winter quarters (Wetherbee 1937, but see Shaub 1950). Dark birds occurring on Iceland (named *islandica* Hantzsch, 1904) very similar to *rostrata* in colour and size; wing and bill on average slightly shorter, bill slightly less bulging, and upperparts (especially rump) sometimes slightly paler; Greenland and Iceland birds often separable on bill length (Vaurie 1957), but analysis by Herremans (1990a) indicated difference too slight to warrant recognition of *islandica*. Iceland also inhabited by pale birds, for which no name available; these rather variable in size and colour, with colour and bill size on average closer to *C. hornemanni exilipes* and wing- and tail-length rather near *rostrata*; population of dark and pale birds on Iceland sometimes considered to be a hybrid swarm (Salomonsen 1950-1), most likely between *rostrata* and *C. h. exilipes*, but as pale birds differ consistently in measurements from dark birds (especially in tail and bill length) this is not likely, unless incomplete introgression and assortative mating occurs. Iceland perhaps originally inhabited by pale birds only, which may have been different enough from other races of *C. hornemanni* to deserve recognition as separate race, but which is now rare (Bird 1935, 1936; Timmermann 1938-49), having been replaced by recently invading *rostrata*, with which some introgression occurs (Herremans 1990a). Further study of Iceland situation needed.

Long-billed form '*holboellii*' occurs side-by-side with nominate *flammea* and thus sometimes considered to be a different race (Salomonsen 1928; Hellmayr 1938) or even a different species (Hortling and Baker 1932); best considered a long-billed variant of nominate *flammea* with bill length to skull at least 13·0 (♀) or 13·5 (♂) and bill to nostril at least 8·7 (see Measurements), or bill to feathers at least 11 mm (Knox 1988b) or at least 10·5 in ♂ and 10·0 in ♀, with maximum in 82 birds of 13·5 (Svensson

1984b) or maximum in 54 of 12·9 (Herremans 1990a); though wing, tail, and bill depth average slightly greater than in nominate *flammea* and rosy-pink of adult ♂ slightly deeper, other plumages and call notes are closely similar and complete intergradation in measurements occurs (Salomonsen 1928; Vaurie 1956; Troy 1984; Molau 1985; Herremans 1987, 1990a; Knox 1988b); also, interbreeding between the forms has been noted (Svensson 1984b). In various west Eurasian samples, c. 6-9% of nominate *flammea* are of *holboellii* form (Molau 1985; Herremans 1987; Knox 1988b); in breeding area, *holboellii* form may occur mainly along northern edge of breeding range of nominate *flammea* (Vaurie 1956a); in various years on Yamal peninsula (north-west Siberia), 9-37·5% of samples were *holboellii* form (Danilov *et al.* 1984). This form more common in north-east Siberia than in Europe or western Siberia (Johansen 1944), with *c.* 12% *holboellii* in Anadyrland and 65% in winter sample from Altai ($n = 51$) (Grote 1943b). Proportion of *holboellii* recorded depends, however, on the exact bill length measurement used as a definition of the form.

For birds introduced New Zealand, see Westerskov (1953), Stenhouse (1962b), Niethammer (1971), and Fennell *et al.* (1985).

For relationship with *C. hornemanni*, see that species.  CSR

---

## *Carduelis hornemanni*  Arctic Redpoll

PLATES 42 and 45 (flight)
[facing page 473, and between pages 640 and 641]

Du. Witstuitbarmsijs  Fr. Sizerin blanchâtre  Ge. Polarbirkenzeisig
Ru. Тундряная чечетка  Sp. Pardillo ártico  Sw. Snösiska  N. Am. Hoary Redpoll

*Linota Hornemanni* Holboell, 1843

Polytypic. Nominate *hornemanni* (Holboell, 1843), northern Greenland, south to Scoresbysund (70°N) in east and to about Upernavik (73°N) in west, and on Ellesmere, Bylot, and northern Baffin Islands (Canada), occasionally to north-west Europe in winter; *exilipes* (Coues, 1862), northern Eurasia and northern North America (south of nominate *hornemanni*); race breeding Iceland not known (see Geographical Variation).

**Field characters.** 13-15 cm; wing-span 21-27·5 cm. Size varies considerably: main Holarctic race *exilipes* similar to north Eurasian and North American race of Redpoll *C. flammea flammea* but often looking plumper, particularly around neck; Baffin Island and Greenland race, nominate *hornemanni* ('Hornemann's Redpoll'), is up to 10% larger, exceeding *C. flammea rostrata* of Greenland and overlapping with Twite *C. flavirostris*. Beware fact that ♀ nominate *hornemanni* overlaps with ♂ *exilipes*. Small but long, ghostly finch, with similar behaviour and flight to northern races of *C. flammea*, but rather loose plumage has much greyer and whiter tone. Rump of adult and 1st-winter ♂ white and usually unmarked. Plumage pattern like *C. flammea* but buff tones restricted to head, pale supercilium more marked, ear-coverts paler, wing marks white and more contrasting, rump white, and underparts white and far less streaked. Juvenile and some 1st-winter ♀♀ much closer in appearance to *C. flammea* showing streaked rump but paler underparts. Bill stubby, par-

ticularly in *exilipes*, and partly hidden by profuse feathering at base of bill. Sexes dissimilar; some seasonal variation in ♂. Juvenile separable. 2 races occur in west Palearctic, but note that *exilipes* perhaps hybridizes to limited extent with *C. f. flammea* in northern Fenno-Scandia where *c.* 2% of birds intermediate (Svensson 1992), while nominate *hornemanni* perhaps hybridizes with *C. f. rostrata* in Greenland (see Geographical Variation).

(1) Main Holarctic race, *exilipes*. ADULT MALE. Moults: July-September (complete). Brightening of plumage entirely due to wear. Differs in fresh plumage from *C. f. flammea* in (a) more distinct, virtually white supercilium, (b) paler ear-coverts, more buff than brown, (c) broader white lines on centre of mantle and upper back, (d) long white rump, tinged pink but lacking more than a few faint streaks, (e) fully white tips and fringes to wing and tail-feathers, creating bold double wing-bar and sharp margins to tertials, (f) smaller, duller chin-patch, far less pink throat and breast, and essentially white underbody,

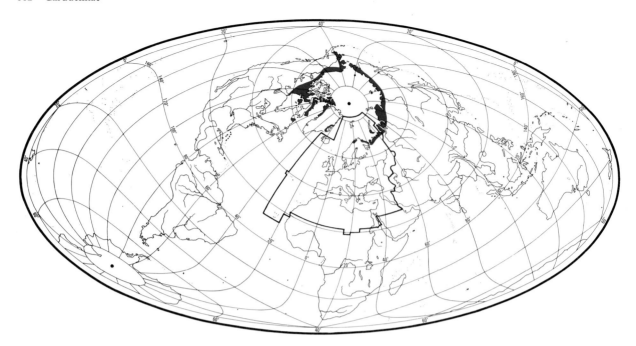

and (g) far fewer black-brown streaks on sides of breast and flanks, forming 1–2 lines at most. With wear, loss of pale grey-buff and white fringes to rear crown, nape, and mantle feathers reduces bird's paleness above, wing marks become narrower, and streaks may appear in centre of rump. Otherwise, retains differences noted above. Bare parts as *C. flammea*. ADULT FEMALE. Resembles ♂ more closely than *C. flammea* but lacks any pink tones, throat and breast being pale cream-buff and this tone being evident also on hindneck, mantle, and on tips of wing-coverts; rump and flanks slightly more heavily streaked. JUVENILE. Much darker than adult, with streaked rump, recalling paler individuals of *C. f. flammea* but ground-colour of rump and underparts still markedly pale, cream-buff wearing to grey-white, while wing-bars when fresh are broader. (2) Greenland and north-east Canadian race, nominate *hornemanni*. Large size emphasized by even paler appearance and bulkier bill than *exilipes*. In all plumages, most ghostly of all forms of *C. flammea* and *C. hornemanni*; ♂ can appear white-faced. Well-marked adult ♂'s plumage differs from *exilipes* in (a) paler face and sides to head, with deeper and purer white supercilium, leading into almost white surround to paler ear-coverts, (b) paler mantle and upper back, sometimes apparently lined black and white, (c) even longer pure white rump, with less streaked upper tail-coverts, (d) even wider fringes and tips to wing-feathers, with bar across greater coverts noticeably bold, and (e) virtually unstreaked underparts (or with isolated striations not forming lines). Adult ♀ resembles ♂ but shows buff tinge on face, breast, mantle, and wing markings, and a few more marks on flanks. Juvenile not studied in the field.

Adult unmistakable when pure white rump seen but certain identification of *C. hornemanni* in juvenile and 1st-winter plumages dogged by its convergence, however caused, with *C. flammea*, particularly of sympatric *C. f. flammea*. Since some birds daunt museum taxonomists, they must defeat field observer; yet with extensive experience of *C. flammea*, many claims of *C. hornemanni* can be sustained given acute observation of characters noted above, particularly short bill, longer and denser frontal feathering (hiding bill and flattening face), thick-necked appearance, generally paler ground-colour to plumage, and little-streaked underparts. Flight, gait, and behaviour as *C. flammea* (certainly in *exilipes*), but with rather looser wing-beats (perhaps illusion due to white plumage) and greater love of ground feeding in nominate *hornemanni*.

Song similar to *C. flammea*. Flight-call may be slower and more spaced out than in *C. flammea*. Other calls include plaintive 'dyeeeu'.

**Habitat.** Distinguishable from that of Redpoll *C. flammea* only by being confined to more northerly Arctic latitudes, where, however, the need for some kind of dwarf willow *Salix* or other shrub growth remains indispensable in breeding season. In Scoresbysund region (eastern Greenland) breeding has been found commonly at heads of fjords at altitudes of 270–400 m but not in outer lower areas (Salomonsen 1950–1). Distribution seems governed by climatic rather than directly by habitat features, and breeding extends to areas with July temperatures as low as 3°C (Voous 1960*b*).

In winter, main population remains in or near breeding latitudes, coping with night temperatures down to *c.*-60°C

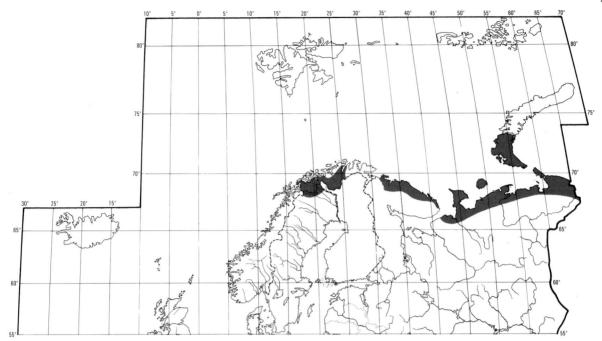

in central Alaska, and foraging for as long as possible in low light available. This degree of hardiness is all the more surprising since closely related forms of *C. flammea* inhabiting temperate climates quite commonly move south in winter, even to Mediterranean regions. Also remarkable in view of fact that even in winter food of arboreal origin is essential. While *C. flammea* will at extremes of range manage with equally stunted shrubs, *C. hornemanni* appears adapted to nothing better; otherwise its environmental habits seem virtually indistinguishable (Newton 1972). In tundra regions of Russia, found in rock-slides, tundra scrub and cut-over taiga, overlapping into stunted forests and northern margins of taiga, but remaining within nesting range all year (Flint *et al.* 1984).

**Distribution.** No evidence of changes. Maps show breeding areas only, where largely resident; for other wintering areas, see Movements.

NORWAY. Breeding distribution not well known (VR). FINLAND. In most years, a small percentage of Finnish redpoll population is *C. hornemanni* (OH); confined to birch *Betula* and willow *Salix* zone of northern Lapland (Koskimies 1989).

Accidental. Faeroes, Britain, Netherlands, France, Belgium, Germany, Poland, Czechoslovakia, Austria, Hungary, Yugoslavia.

**Population.** No information on trends.

SWEDEN. Estimate of 1000 pairs (Ulfstrand and Högstedt 1976) supported by recent censuses in parts of northern Sweden (LR).

**Movements.** Relatively short-distance, partial migrant from circumpolar breeding areas, usually to lower latitudes in winter, most birds probably moving south of Arctic circle.

Populations of *exilipes* breeding across Eurasian Arctic zone, as far west as Lapland and northern Fenno-Scandia, winter over much wider latitudinal range, as result of southward movement by at least part of population (for Asia, see below). Birds from European Arctic zone reach as far south and west as southern Baltic (and sometimes beyond) in small numbers. Numbers recorded in Denmark accompanying Redpoll *C. flammea* were higher in 2 invasion years of that species, 1965–6 and 1972–3 (Braae 1975). In Britain, 109 records 1958–85 (chiefly autumn), almost entirely confined to east coast from Shetland to Kent, thus presumably chiefly from northern Eurasia, though both nominate *hornemanni* and *exilipes* recorded (Dymond *et al.* 1989). Marked influxes 1990–1 (Rogers *et al.* 1991, 1992). In Netherlands, 20 birds recorded 1980–90 (Berg *et al.* 1992); in western Germany, only 2 records 1977–86 (Bundesdeutscher Seltenheitenausschuss 1989); in France, 8 records in 20th century, all since 1966 (Dubois and Yésou 1992). In Poland, January 1989, largest-ever influx reported inland in north-west, with flocks of up to *c.* 100 in company with *C. flammea* (*Br. Birds* 1990, **83**, 16).

Populations of *exilipes* breeding in North America (chiefly in west Canadian Arctic and Alaska) winter regularly as far south as southern Canada and across northernmost states of USA. Considerable numbers, however, winter as far north as Arctic circle in Alaska, e.g. abundant at Fairbanks (interior Alaska at 65°N) with *C. flammea*

(Brooks 1968), and numerous throughout year at Kot-zebue Sound (Bent 1968).

Some nominate *hornemanni* remain on north Greenland breeding grounds in winter, vacating only northernmost areas (Salomonsen 1950-1); others move further south. Some, together with birds from Ellesmere and Baffin Islands, reach Southampton Island and Ungava (northern Quebec) at mouth of Hudson's Bay and overwinter there with *exilipes*; others move still further south, exceptionally as far as Great Lakes region of Canada/USA (Snyder 1957).

Movement generally southward, but few details. Troy (1983) mapped redpoll recoveries in North America without distinguishing species, but summer recovery on Alaskan coast of bird ringed in winter in south-east Canada was most probably *exilipes* (rather than *C. f. flammea*).

Data on timing of migration are sparse; passage often reported in company with *C. flammea* throughout much of range. In autumn, movement in Scandinavia prolonged, September–December (Haftorn 1971; SOF 1990), and recorded chiefly from October in north-west Europe (Dymond *et al.* 1989; Berg *et al.* 1992). North American birds reach wintering grounds at Fairbanks late September and early October (Kessel and Springer 1966), and south of winter range (northern USA) from late October (Janssen 1987). Greenland birds begin moving south as early as August (Salomonsen 1950-1); on Southampton Island, recorded from 21 September in one year (Sutton 1932).

In spring, movement in Sweden mostly March (SOF 1990). Last records in northern USA in April (Janssen 1987), and movement in Alaska March–May (Gabrielson and Lincoln 1959). In one year, birds left Fairbanks late March, and more passed through, *en route* to breeding areas, in mid-April (West *et al.* 1968). Passage on Greenland coast from mid-April (Salomonsen 1950-1).

Returns to breeding grounds earlier than *C. flammea*; on Baffin Island, arrived 2 weeks earlier in one year (Wynne-Edwards 1952). In northern Fenno-Scandia, arrives 'at least several days' before *C. flammea* (Knox 1988*b*), presumably at end of April.

In Asia, winter range extends less far south than that of *C. flammea*. In Siberia, reaches Tomsk in considerable numbers, but scarce further south, with only small numbers, e.g. in Altai and Zaysan depression (Johansen 1944). Very rare in Kazakhstan (Korelov *et al.* 1974). Reaches northern Mongolia (Kozlova 1933), at least in some years, and also recorded (rarely) in northern China (Schauensee 1984; Cheng 1987). Apparently fairly regular on Sakhalin Island. Vagrant to Japan. (Brazil 1991.) Autumn movement begins 2nd half of August; most migrants have left Taymyr region and Yamal peninsula by mid-September (Krechmar 1966; Danilov *et al.* 1984). Departure from Chukotskiy peninsula gradual, with stragglers to mid-October (Portenko 1973). Present in northern Mongolia mid-October to beginning of April (Kozlova

1933). In spring, arrives Kamchatka peninsula April–May (Lobkov 1986), and on Chukotskiy and Yamal peninsulas mostly late May to early June (Portenko 1973; Danilov *et al.* 1984). PRE, DFV

**Food.** Almost all information extralimital. Diet comprises small seeds, particularly birch *Betula*, alder *Alnus*, willow *Salix*, various herbs, and grasses; some small invertebrates in summer. Diet probably very similar to that of Redpoll *C. flammea*, and majority of feeding observations involve mixed groups of both species (Sutton 1932; Gabrielson and Lincoln 1959; Bent 1968; Portenko 1973; Kishchinski *et al.* 1983), but differences in structure of foot, tail, and bill indicate possible divergence of diet and foraging technique between the species (C S Roselaar). Forages in trees like *C. flammea* (Bent 1968); in snow, above all in winter, when one of very few passerines to remain in Arctic, foraging restricted to scrub, tall herbs and catkins above snow, seeds on surface, snow-free patches at coasts or on windy slopes, roadsides, and rubbish tips, which are also source of insects in breeding season (Dementiev and Gladkov 1954; Freuchen and Salomonsen 1959; White and West 1977; Kishchinski *et al.* 1983; Pohl 1989). In breeding areas, most commonly feeds in dwarf birch and willow, taking buds, catkins, and insects; also apparently fond of seeds of cottongrass *Eriophorum* in midsummer (Gabrielson and Lincoln 1959; Bent 1968; Alsop 1973).

Diet includes the following. Invertebrates: bugs (Hemiptera: Aphidoidea), larval Lepidoptera (Geometridae), flies (Diptera: Simuliidae), adult and larval beetles (Coleoptera: Chrysomelidae), spiders (Araneae). Plant material: seeds, buds, catkins, etc., of willow *Salix*, birch *Betula*, alder *Alnus*, knotgrass *Polygonum*, amaranth *Amaranthus*, crowberry *Empetrum*, grasses, including cereals (Gramineae), sedges (Cyperaceae), rushes (Juncaceae). (Salomonsen 1950-1; Wynne-Edwards 1952; Dementiev and Gladkov 1954; Gabrielson and Lincoln 1959; Bent 1968; Portenko 1973; Kishchinski *et al.* 1983). See also White and West (1977) for diet of this species and *C. flammea* together in Alaska.

Few details of adult diet. In Canada, of 11 stomachs, 6 full of birch and alder seeds, 5 from further south contained seeds of knotgrass, amaranth, grasses, and sedges (Gabrielson and Lincoln 1959). In north-east Russia, crop of one bird held *c.* 500-600 small seeds (Portenko 1973); another in Baffin Island (Canada) filled with grass and sedge seeds plus 2 spiders (Wynne-Edwards 1952). In Alaska (data do not distinguish between this species and *C. flammea* in mixed flocks), seeds of spruce *Picea*, willow, dock *Rumex*, and willowherb *Epilobium*, accounted for *c.* 30% of diet in spring and summer, insects, especially Diptera and caterpillars, for *c.* 20-25%; in late summer, willow catkins important; in March-April, foraged by roadsides for seeds of goosefoot *Chenopodium* and columbine *Aquilegia* (White and West 1977, which see for details of time spent foraging).

In captive experiments, gross energy intake over 7 hrs of daylight *c*. 157 kJ per day at −30°C, *c*. 75 kJ at 0°C, *c*. 54 kJ at 30°C; intake of seeds at lower limit of temperature tolerance calculated to be 8·2 g, but actual weight of food eaten in wild, especially high-energy birch seeds, would enable birds to survive temperature of *c*. −67°C. More efficient than *C. flammea* in extremes of temperature, and also forages at lower light intensities; additional adaptation to severe conditions, common to both species, is crop-like oesophageal pocket where up to 2 g of seeds can be stored. (Brooks 1968; Pohl 1989.) See also White and West (1977).

In Swedish Lapland, nestlings in one nest received 48 birch seeds, 6 Diptera, and 1 Geometridae caterpillar; no different from *C. flammea* samples (Nyström and Nyström 1987); see also *C. flammea*. In Baffin Island and Alaska, young given seeds of willow and cottongrass as well as adult and larval insects (Wynne-Edwards 1952; Bent 1968). Young can probably be raised on seeds alone according to Pohl (1989). BH

**Social pattern and behaviour.** Less well known than Redpoll *C. flammea*, but studies at Ammarnäs (Swedish Lapland) and in Greenland found no significant differences from that species (Salomonsen 1950–1; Nyström and Nyström 1987, 1991).

1. Gregarious more or less all year, though especially outside breeding season. In Kamchatka (eastern Russia), in winter, recorded singly, in 'pairs' (see Bonds, below) or in flocks of up to *c*. 40 (Lobkov 1986). In Greenland, flocks formed after breeding sometimes large (Dupont 1939); during migration and in winter, rarely more than 10–20 (Salomonsen 1950–1). Post-breeding flocks in north-east Russia comprised adults and juveniles (Kishchinski 1980), sometimes only juveniles (Portenko 1939). On Kola peninsula (north-west European Russia), flocks of 4–10, occasionally 30–60, noted in October and March–May (Semenov-Tyan-Shanski and Gilyazov 1991). Several reports of breeding-season flocks: e.g. in north-east Russia, some single birds and pairs, but flocks (probably mostly non-breeders: see Bonds, below) of 5–40 recorded throughout summer in Kanchalan valley and on montane tundra (Kishchinski *et al.* 1983); however, breeding ♂♂ sometimes form small flocks for feeding (Bent 1968 for North America); see also Alsop (1973) and Portenko (1939, 1973). Often associates for feeding with *C. flammea* in winter and during breeding season (West *et al.* 1968; Veprintsev and Zablotskaya 1982); in Poland during exceptional influx, *c*. 100 recorded together with *c*. 1000 *C. flammea* in January (*Br. Birds* 1990, 83, 16). In Koryak mountains (north of Kamchatka), birds also associated loosely for feeding with *Phylloscopus* warblers (Kishchinski 1980). BONDS. Mating system not studied, but probably mainly monogamous. Deviations from monogamy reported in Arctic Canada: at one nest, certainly 2 ♂♂ fed brooding ♀ and 3rd ♂ came to nest, but did not feed her; antagonism noted between 2 ♂♂ (see Antagonistic Behaviour, below), but ♀ showed same excited response whichever ♂ arrived; 2 ♂♂ seen to feed ♀ at 2nd nest (Alsop 1973). On Kola peninsula, ♂♂ found to move about during breeding season and some probably do not breed (Mikhaylov 1984); in north-east Russia, perhaps only part of population breeding in June, others not until July (Kishchinski *et al.* 1983). Little information on pair-bond or its duration; no hard evidence for report by Lobkov (1986) of some 'pairs' in Kamchatka in winter; in both Arctic Canada and north-east Russia, however, birds

already paired on arrival (Sutton 1932; Portenko 1973). All incubation and brooding by ♀ (Bent 1968; Portenko 1973; Tomkovich and Sorokin 1983; Nyström and Nyström 1987). In Ammarnäs study, ♀ mainly provisioned by ♂ in early period after hatching, but sometimes foraged near nest then (Nyström and Nyström 1987), and some food collected presumably fed to nestlings. In Alaska, young fed by both sexes, ♂'s share gradually increasing until equal to ♀'s by day 10 (Bent 1968). Role of sexes in care of fledglings not clarified, but 2 broods in Canada each accompanied by ♂ only (Seutin *et al.* 1991). Age of first breeding not known. Major review by Knox (1988*b*) of allegedly frequent hybridization between *C. hornemanni* and *C. flammea* found no clearly documented cases of successful interbreeding anywhere within vast area of sympatry, and direct evidence for hybridization largely based on assumed identity of presumed hybrid offspring rather than on observations of mixed pairings. 'Intermediates' probably part of normal range of plumage variation within the 2 species (see also Svensson 1992). Strong positive assortative mating among *C. hornemanni* and *C. flammea* frequently observed in both Palearctic and Nearctic. In some cases, species separated by habitat (e.g. Molau 1985 for northern Sweden), though such separation not found at Ammarnäs by Nyström and Nyström (1987)—or *C. hornemanni* arrives and settles to breed earlier than *C. flammea* (Wynne-Edwards 1952; Watson 1957*b*). One observation (in Greenland) of ♂ *C. flammea* copulating with ♀ *C. hornemanni* (Salomonsen 1950–1), but reports of 'mixed pairs' need to be regarded with caution, as both species often in flocks during breeding season, display within these, nest in groups apparently with low degree of territoriality (see Breeding Dispersion, below), and hence often difficult to decide what constitutes a pair. (Knox 1988*b*, which see for details and other references; see also Kishchinski 1980, Kishchinski *et al.* 1983, Danilov *et al.* 1984, Svensson 1992, and Geographical Variation). BREEDING DISPERSION. Like *C. flammea*, typically in neighbourhood groups, often tight-knit (e.g. several nests in small clump of bushes) (Wynne-Edwards 1952; Bent 1968). Territorial disputes reported in north-east Russia by Veprintsev and Zablotskaya (1982), and *C. hornemanni* may defend small area around nest like some other Carduelinae, but considered probably non-territorial by Wynne-Edwards (1952); on Kola peninsula, no ♂♂ observed singing on 'nesting territory' (Mikhaylov 1984), and in Alaska no singing from perch near nest from mid to late nesting period, and only 1 antagonistic chase seen (Bent 1968); see also Antagonistic Behaviour (below). Breeding birds also forage far from nest, sometimes in widely overlapping area with conspecific birds and *C. flammea* (Portenko 1973; Knox 1988*b*). Information on density only from north-east Russia (in view of tendency to flock and to wander far during breeding season, some density values should be treated with caution): in Kamchatka, in forest of birch *Betula ermani* with understorey of prostrate pine *Pinus pumila*, 22–50 pairs per km² (some *C. flammea*, but predominantly *C. hornemanni*), and on hummocky tundra 10 pairs per km² (Lobkov 1986); near Uelen (Chukotskiy peninsula), on various types of tundra over 3 years, 2–98 birds per km² (Tomkovich and Sorokin 1983); on Wrangel Island, 0·2–10 pairs per km² of tundra (Dorogoy 1982), 1–2 or 20–50 birds per km² in dense willow *Salix* scrub of river valleys; on Geral'd Island (off Wrangel Island), 100 or more birds per km² in optimal habitat (Stishov *et al.* 1991). Sometimes breeds close to *C. flammea* (e.g. in Greenland, one mixed nesting group with 20 times as many *C. flammea*, and in Canada nests of the 2 species *c*. 50 m apart), but no interbreeding (Taverner and Sutton 1934; Salomonsen 1950–1; Kishchinski *et al.* 1983). On Baffin Island (Canada), same nest allegedly used over several years (Wynne-Edwards 1952), but no proof that same birds involved. ROOST-

ING. Little information on roosting habits; perhaps uses snow-burrows in winter (see Sulkava 1969 and *C. flammea*). At 66°50′N in Arctic Canada, active for 22 hrs in each 24-hr period of continuous daylight when feeding young (Alsop 1973); at 69°23′N in Alaska, early July, ♀ roosted at nest with young for *c.* 3-4 hrs around midnight (Karplus 1952). For experimental study of adaptations to arctic environment, see Brooks (1968) and *C. flammea*.

2. Breeding birds in north-east Russia confiding (Portenko 1973). On hearing Dsooee-call (see 2b in Voice), other conspecific birds may immediately fall silent or approach and express alarm with same call (Veprintsev and Zablotskaya 1982). FLOCK BEHAVIOUR. Bird separated from flock will give Dsooee-calls until flock-members call similarly and join it, or it locates and joins them; Chatter-call alone or combined with Rattle-call (see 2a and 3 in Voice) given by birds in flying flock similarly elicits Dsooee-calls from lone feeding bird, flock usually then flying down to join it (Veprintsev and Zablotskaya 1982). Flocks of 2-4 singing ♂♂ noted Kola peninsula 19-25 June, occasionally up to early August; in June, flocks (*c.* 20-50 m apart) rested in trees every *c.* 100-200 m, then flew on singing (Mikhaylov 1984). SONG-DISPLAY. ♂ sings (see 1a in Voice) from perch (low tree or shrub) or in flight; often precedes song with Dsooee-calls (or incorporates these in song), this tending to elicit same call or vigorous singing in other ♂♂ (Veprintsev and Zablotskaya 1982). For Song-flight, ascends from song-post while continuing to sing (Veprintsev and Zablotskaya 1982). Observations in north-east Russia, mid-June, found birds to ascend gradually and somewhat jerkily, in some cases higher than pipits *Anthus*; quite often flew long distance in Song-flight, but then returned to starting point, singing (or at least giving Chatter-calls) almost uninterruptedly, then Dsooee-calls on landing (Portenko 1973). On Kola peninsula, song noted from early June (by which time area up to 80% snow-free) to early August (Mikhaylov 1984); during June and on 7 July in north-east Russia (Portenko 1973). For song given when close to ♀, see Heterosexual Behaviour (below). ANTAGONISTIC BEHAVIOUR. During individual-distance disputes while feeding or resting, also (apparently rare) territorial disputes when breeding, dominant bird gives Rattle-call in threat, subordinate then fleeing, or adopting submissive-posture (head drawn in, flanks and belly ruffled) and giving Dsooee-calls (Veprintsev and Zablotskaya 1982). HETEROSEXUAL BEHAVIOUR. (1) Pair-bonding behaviour. Especially when close to ♀ during nest-site selection, nest-building, or incubation, ♂ gives excited Courtship-song (see 1b in Voice) (Veprintsev and Zablotskaya 1982). For calls associated with courtship, see 4 in Voice. (2) Courtship-feeding. ♂ feeds ♀ by regurgitation, off nest before start of incubation, on it when ♀ incubating; ♂ approaching nest and while actually feeding mate gives 'chiv' calls (see 4c in Voice), ♀ responding with begging-calls (see 4b in Voice) and wing-shivering (Portenko 1973; Veprintsev and Zablotskaya 1982; Nyström and Nyström 1987). ♀ of Ammarnäs pair also foraged for herself briefly near nest, apparently mainly when ♂ delayed (Nyström and Nyström 1987). (3) Mating. On Baffin Island, late May, birds copulated on boulder (Wynne-Edwards 1952); in north-east Russia, repeatedly in nest (under construction) or nearby (Tomkovich and Sorokin 1983). In Arctic Canada, early July, ♂ seen to break off feeding and fly to ♀ who gave rapid 'twittering calls' (probably 4b in Voice) and fluttered wings; ♂ hovered in front of mate, calling frequently, then turned gradually so as to be above ♀'s back; landed on ♀, beating wings to maintain balance and copulated, then took off; 2nd copulation attempted, ♂ then flying *c.* 5 m to resume feeding (Alsop 1973). (4) Nest-site selection and behaviour at nest. Nest-site chosen by ♀, who also builds nest, accompanied by ♂ (Bent 1968; Porte-

nko 1973; Veprintsev and Zablotskaya 1982). No further information. RELATIONS WITHIN FAMILY GROUP. Eggshells eaten by ♀ after hatching, which (like fledging) is typically asynchronous. In Alaska, all brooding by ♀; declined from 85% on day 1 to 30% by day 4, 27·5% by day 7, and 1% by day 10 (Wynne-Edwards 1952; Bent 1968). In Canada, ♂ gave loud Chatter-calls while approaching nest, landed *c.* 5-10 m away, then moved through cover to nest; ♀ gave frequent begging-calls while rapidly fluttering wings and, after receiving food from ♂, fed young (also by regurgitation) (Alsop 1973). Nestling faeces reported by Bent (1968) to be swallowed; probably carried away in some cases. Fledging period of 11-12 days recorded Greenland (Salomonsen 1950-1); typically (12-)14 days Alaska (Bent 1968); 9 days at 2 nests in Arctic Canada (Alsop 1973). ANTI-PREDATOR RESPONSES OF YOUNG. Nestlings generally silent (e.g. at 7 days); at *c.* 11 days, leapt into grass and hid during serious disturbance (Portenko 1973). PARENTAL ANTI-PREDATOR STRATEGIES. (1) Passive measures. ♀ typically sits tightly (on eggs and at least small chicks), leaving only for serious interference (for details, see Portenko 1973). (2) Active measures. Study in north-east Russia showed ♂ to approach immediately when nest threatened and usually to fly from bush to bush while calling (sometimes chased ♀ and tried to copulate); both sexes recorded close to nest when observer nearby (Portenko 1973); Dsooee-calls (or longer variant: see 2b in Voice) typically given by both sexes for any danger, such calls often attracting neighbours or small flocks, including *C. flammea* (Veprintsev and Zablotskaya 1982; Kishchinski *et al.* 1983): e.g. on Kola peninsula, 4 ♂♂ sang excitedly by nest from which ♀ had flushed, moving only when she had calmed down and settled on nest (Mikhaylov 1984). ♀♀ leaving nest variously recorded performing apparent distraction-lure display of disablement-type ('impeded flight'), perching conspicuously, or flying down to ground where they fed (perhaps mock-fed) or drank (Portenko 1973).                    MGW

**Voice.** In study of wild and captive birds, *exilipes* silent after moult and overall much less vocal than Redpoll *C. flammea* (Herremans 1989, which see for sonagrams). Distinct, though sometimes subtle differences reported to exist between calls of *exilipes* and *C. f. flammea*. Vocal isolation apparently well developed: captive bird responded vigorously on hearing or seeing another *C. hornemanni*, but showed little or no reaction to flock of *C. f. flammea*. (Knox 1988b; Herremans 1989.) Some alleged differences not confirmed by analysis of recordings, or at best probably too subtle to be of use for field identification. North American studies (Wynne-Edwards 1952; Borror 1961) found no significant differences between the species in song or calls. Limited number of recordings of nominate *hornemanni* (Greenland) available for analysis. Major study (including sonagrams) of *exilipes* in north-east Russia recognized 2 song-types and 9 calls of which 5 given by both sexes all year, 4 (2 by ♂ only, 1 by ♀ only, 1 by both sexes) only in breeding season; further investigation required to determine size of juvenile repertoire (Veprintsev and Zablotskaya 1982).

CALLS OF ADULTS. (1) Song. (1a) Advertising-song of ♂. Various units or motifs (some in form of short segments) given with pauses of 0·1-0·5 s in phrases lasting average 7-8 s; sometimes several such phrases strung together in sequences lasting up to *c.* 30 min. Most frequently used

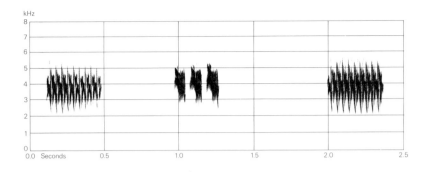

I J-C Roché Finland June 1964

units are calls ('pyuee', 'peee', 'che', 'chuv', 'tirrr'; see below) or derivatives, few apparently given only in song. Considerable variation (within and between individuals) arises from grouping of units to form motifs, and sequence of motifs in song phrase: each ♂ has several different song-types. Advertising-song apparently serves mainly to attract ♀; given around time of pair-formation. (Veprintsev and Zablotskaya 1982.) Simple song-type, combining only Chatter- and Rattle-calls given in flight (as reported for *C. flammea*) also exists: see Fig I and call 3 (below). (1b) Courtship-song of ♂. Similar to 1a, but shorter (*c.* 4-6 s) and quieter, delicately rippling phrase containing more tremolo-type motifs (segments) and some with marked changes in frequency. Given close to ♀ during courtship, nest-building, and incubation. (Veprintsev and Zablotskaya 1982.) (2) Contact- and alarm-calls. (2a) Chatter-call. Short, noisy 'che', 'chuv', or similar, typically given in series (usually 2-3 units), often with regular temporal pattern. Chatter-call given while feeding, resting, preen-

ing or in flight, serving for contact and synchronization of activities between pair-members or within family or flock. Variation noted in fine structure of Chatter-call, probably indicating call susceptible to change throughout bird's life, as in certain other Carduelinae (see, e.g., Twite *C. flavirostris*). (Veprintsev and Zablotskaya 1982.) Recordings of *exilipes* from Finland contain Chatter-calls whose units vary in structure and timbre, but all suggest rendering 'chi' or 'chip': see Figs II-III, also middle 3 units of song (Fig I). Chatter-calls of *exilipes* from Canada (Fig IV) very similar to Fig III; compare also Fig a (groups of 2-3 'tschip' sounds) in Bergmann and Helb (1982) and slightly differently-structured calls in Herremans (1989) (see below). Fig V shows 2 chatter-calls of nominate *hor-*

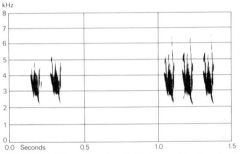

IV W W H Gunn/Sveriges Radio (1972-80) Canada June 1961

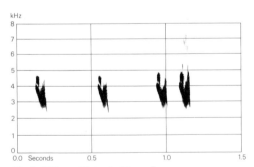

II A G Knox Finland May 1984

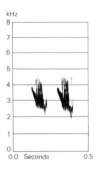

V P J Sellar Greenland August 1985

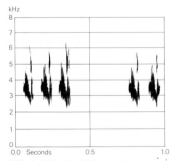

III J-C Roché Finland June 1964

*nemanni*: slightly harsh 'djib djib' bearing some resemblance to *exilipes* (Fig I) and to *C. flammea* from Iceland (see Figs IV and VIII of that species). (W T C Seale, M G Wilson.) Calls of nominate *hornemanni* reported to be huskier and rougher than *C. f. rostrata* (Sutton 1932).

Further comparison between *exilipes* and *C. flammea* as follows: in northern Sweden, *exilipes* said by Molau (1985) to be coarser than *C. f. flammea*, more like House Sparrow *Passer domesticus*; some *exilipes* in Britain reported as louder, slower, higher pitched, lacking tinny quality of *C. f. flammea* (Lansdown *et al.* 1991). Study by Herremans (1989) found Chatter-calls of *exilipes* disyllabic or nearly so, higher pitched than *C. f. flammea*, with slower rate of delivery (see below), narrower frequency range, and lower pitched than *C. f. cabaret*; rather hoarse, less clear-sounding series of 'djeet' sounds; solitary birds gave composite calls such as 'tjee-DJEET'. (Herremans 1989.) In recordings analysed, no appreciable difference between fastest delivery rates in the 2 species (W T C Seale). (2b–d) Diadic and nasal calls with marked overall ascent in pitch and, in some cases, brief final pitch descent. Sonagrams in Veprintsev and Zablotskaya (1982) clearly show pitch descent in some calls and suggest marked degree of overlap between calls 2b–d. Probably in most cases difficult if not impossible to distinguish from *C. flammea* in the field. (2b) Dsooee-call. A 'dsooee' or 'pyuee' mostly for contact between pair-members, parent and young, and flock members (given also when isolated from flock); also signals excitement or alarm, call then typically longer and higher pitched, more like 'pyueelee', but changing back to 'pyuee' as excitement wanes; given singly or in series, with considerable variation in length of pauses between calls (Veprintsev and Zablotskaya 1982). Recordings of *exilipes* contain at most only slight final descent: rather sibilant 'dsooee' (Figs VI and narrow-band Fig VII), and nasal 'dzooee' (Figs VIII–IX). Like those in Veprintsev and Zablotskaya (1982), differ structurally (not composed of several short sub-units) from complex, hesitating, and hoarse 'pwljeeu' (stressed, descending end syllable) illustrated by Herremans (1989). Recording of nominate *hornemanni* (Fig X) contains rather subdued, plaintive, nasal, mewing 'pzooee' and 'pzooelee' (perhaps closer to call 2c: see below), apparently with slight descent at or near end of one component of diad: calls sound different from *exilipes*, but not clear how significant this difference might be (W T C Seale, M G Wilson). (2c) Diadic 'pyulee' ascending in pitch; given in pair-contact, most often as antiphonal duet, with average 1·3 s between calls of ♂ and ♀ (Veprintsev and Zablotskaya 1982). (2d) Alarm-call. A 'peee' similar to call 2b, but with harsher, noisy quality and typically abrupt start; longer and louder with increasing agitation. Given in response to aerial or ground predator or other danger (sometimes elicited by alarm-calls of other species); serves as mobbing-call and, especially in long series typical of this context, intermediate forms between 2b and 2d occur. (Veprintsev and Zablotskaya 1982.) From recording in Canada, call given when disturbed (same as Fig IX) also rendered 'tschräit' (Bergmann and Helb 1982) and closer to call of *C. flammea* than to alarm-call of *C. hornemanni* in Veprintsev and Zablotskaya (1982). According to Herremans (1989), alarm-calls of *C. hornemanni* and *C. f. flammea* do not differ. (3) Rattle-call. A 'tirr' of tremolo structure comprising short (highly loc-

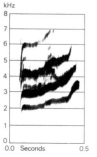

VI   A G Knox   Finland
May 1984

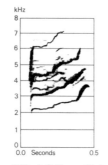

VII   A G Knox   Finland
May 1984

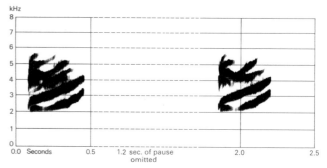

VIII   J-C Roché   Finland   June 1964

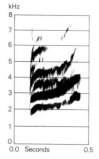

IX   W W H Gunn/Sveriges Radio (1972–80)   Canada   June 1961

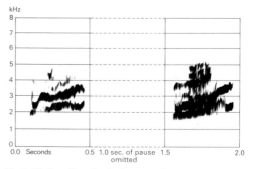

X   P J Sellar   Greenland   August 1985

atable), noisy units. Serves as threat-call (tremolo longer, and slightly wider frequency spectrum, with increasing excitement), given by both sexes in individual-distance and other disputes. Usually short calls of slightly narrower frequency range, sounding higher pitched and looser than threat-call, are given in other contexts: outside breeding season, mainly in flight, especially when about to land; in breeding season, mainly by ♂ from conspicuous perch (perhaps then functions as advertising-call, or 'courtship trill' after pair-formation; see Vorobiev 1963); Rattle-call sometimes combined with Chatter-calls and given by lone bird or flying flock. (Veprintsev and Zablotskaya 1982.) Such a combination often treated as 'flight song' in both *C. flammea* and *C. hornemanni*: in Fig I, vibrant Rattle-calls show subtle structural differences from (albeit highly variable) Rattle-call of *C. flammea* (see Figs I–V of that species), but unlikely to be easily distinguishable in the field. (4) Calls associated with courtship and other heterosexual interactions. (4a) Courtship-call of ♂. Rather quiet, delicate 'tpyuee' with somewhat lilting quality; comprises quiet chirp or twitter ('t') and main diadic part of call; energy concentrated in higher part of spectrum than in species' other calls (Veprintsev and Zablotskaya 1982). Similar or identical to call 5a of *C. flammea* (see that species). (4b) Rapid series of 'ti' ('tee') sounds, longer when excited, given by both sexes during breeding, in connection with nest-building, feeding, etc. (L Svensson); by ♀ as begging-call (stimulating ♂ to feed her), also as soliciting-call (given immediately before copulation) or to accompany various nesting activities (Veprintsev and Zablotskaya 1982). Presumably this call given by captive ♀ associating with pair or with ♂ of *C. flammea* rendered 'tin tin tin tin', and apparently adopted from ♂ *C. flammea* (see Herremans 1989). (4c) Series of 2–12 'chiv' sounds given by ♂ approaching to feed ♀ and while actually feeding her, then sometimes overlapping with call 4b from ♀ (4b rendered 'chiv chiv chiv' by Portenko 1973). Loud, rather strident call, similar to 'chuv' variant of call 2a (see above), but with wider spectrum and energy more evenly spread: series louder and longer with increasing excitement. (Veprintsev and Zablotskaya 1982.)

CALLS OF YOUNG. Recording of nominate *hornemanni* apparently contains mixture of adult Chatter-calls and food-calls of fledglings: 'chi' and 'chup' sounds and (perhaps from young) 'dzib', 'zib', and 'zzib' with buzzy timbre, also nasal 'dzar' sounding evenly pitched (W T C Seale) or curious tuneless 'urr' (P J Sellar).      MGW

**Breeding.** SEASON. Northern Sweden: eggs laid throughout June, peak probably around mid-June (Swanberg 1951*a*; Makatsch 1976; Nyström and Nyström 1987, 1991, which see for suggestion that breeding season and habits indistinguishable from Redpoll *C. flammea*). Kola peninsula (north-west Russia): eggs laid from beginning of June to early July (Mikhaylov 1984). Eastern Siberia: about early June to mid-July (Portenko 1973; Kishchinski

*et al.* 1983; Stishov *et al.* 1991). Northern Canada: about late May to late June (Wynne-Edwards 1952; Alsop 1973; Harrison 1978*a*). North-west Greenland: from end of May to end of June, mostly beginning of June (Salomonsen 1950–1). SITE. Dwarf tree or shrub, almost always willow *Salix*, poplar *Populus*, birch *Betula*, or alder *Alnus*; in crotch or on branch or twigs generally close to trunk, almost always less than *c.* 2 m above ground, though usually in upper part of shrub (Sutton 1932; Bent 1968; Portenko 1973; Kishchinski *et al.* 1983). 5 nests in Alaska averaged 71 cm (30·5–99·0) above ground (Walkinshaw 1948). In northern Sweden, one nest recorded 5 m up in birch in dense forest (Nyström and Nyström 1987). Also in rock crevice, scree, grass tussock, etc., but almost always concealed by vegetation (Salomonsen 1950–1; Wynne-Edwards 1952; Portenko 1973); nests found in driftwood on Wrangel Island (Stishov *et al.* 1991). Not uncommonly on buildings and other man-made structures (Portenko 1973; Stishov *et al.* 1991; Vaughan 1992). Nest: robust structure with foundation of small twigs, bark, stems, roots, grass, catkins, etc., warmly lined with hair, fur, plant down, and many feathers, especially white ones of grouse *Lagopus*; lining sometimes only of feathers, and often higher than outer wall (Wynne-Edwards 1952; Gabrielson and Lincoln 1959; Portenko 1973; Kishchinski *et al.* 1983). In Alaska, 5 nests had average outer diameter 10·4 cm, inner diameter 4·9 cm, overall height 7·8 cm, depth of cup 3·7 cm (Walkinshaw 1948). Building: probably by ♀ only, accompanied by ♂; takes *c.* 4 days (Bent 1968; Portenko 1973; Mikhaylov 1984). EGGS. See Plate 60. Very like *C. flammea*, but perhaps larger and paler with heavier markings more often forming ring at broad end (Witherby *et al.* 1938; Walkinshaw 1948; Makatsch 1976). *C. h. exilipes*: 17·1 × 12·8 mm (15·8–18·5 × 12·1–13·4), $n = 51$; calculated weight 1·44 g. Nominate *hornemanni*: 18·2 × 13·0 mm (16·7–20·7 × 12·4–14·0), $n = 14$. (Makatsch 1976.) See also Bent (1968). Clutch: 4–5 (3–7). In northern Sweden, of 8 clutches (some perhaps incomplete): 3 of 4 eggs, 4 of 5, 1 of 6; average 4·8 (Swanberg 1951*a*). On Baffin Island (Canada), of 4 clutches: 2 of 5 eggs and 2 of 6; average 5·5 (Wynne-Edwards 1952), and average of 20 North American clutches 4·7 (Bent 1968). Average in eastern Siberia 4·3; possibly 2 broods since 2 distinct laying peaks about mid-June and mid-July (Kishchinski *et al.* 1983). Eggs laid daily (Portenko 1973; Makatsch 1976). At one nest in Alaska, 4 days between nest completion and laying of 1st egg (Gabrielson and Lincoln 1959). See Nyström and Nyström (1991) for data on mixed population with *C. flammea*. INCUBATION. 11–12 days; by ♀ only (Salomonsen 1950–1; Wynne-Edwards 1952; Alsop 1973; Portenko 1973). Starts with 2nd or 3rd egg according to Bent (1968), with last egg according to Makatsch (1976). YOUNG. Fed and cared for by both parents, and brooded by ♀ for *c.* 10 days (Alsop 1973; Harrison 1978*a*). FLEDGING TO MATURITY. Fledging period 10–12 days (9–14) (Salomonsen 1950–1; Wynne-

Edwards 1952; Bent 1968; Alsop 1973). BREEDING
SUCCESS. On Baffin Island, all of 20 eggs hatched and 18
fledged; some *C. flammea* possibly included (Wynne-
Edwards 1952). In eastern Siberia, large numbers of nests
destroyed by dogs or high water (Dementiev and Gladkov
1954; Portenko 1973). See Nyström and Nyström (1991)
for effect of poor birch seed crop and cold weather on
success of mixed *C. hornemanni* and *C. flammea*
population.                                             BH

**Plumages.** (*C. h. exilipes*). Traditionally considered to be separable from rather similar Redpoll *C. flammea* by paler ground-
colour of upperparts, unstreaked rump, and less streaked flank,
but since investigations by Molau (1985) and Knox (1988b) now
known that pronounced sex and age variations occur, with ♀♀
and 1st adults being darker and more streaky than ♂♂ and
older birds, 1st adult ♀♀ in particular being often fairly heavily
streaked on rump and flank, approaching adult ♂ *C. flammea* in
these respects, though ground-colour of upperparts paler; these
streaky birds formerly sometimes considered to be hybrids
between the species (Salomonsen 1928, 1950-1; Payn 1947;
Houston 1963; Harris *et al.* 1965; Troy 1985), but now known
to belong to normal variation of *C. hornemanni*, with true hybridization uncommon, if it occurs at all (Svensson 1992). ADULT
MALE. Rather like adult ♂ *C. flammea flammea*, but ground-
colour of hindcrown, nape, mantle, scapulars, and back pale
cream-grey, slightly tinged buff on side of mantle and scapulars,
almost white on centre of mantle and back (in *C. f. flammea*,
ground-colour tawny-brown or buff-brown); extent and width
of dark streaking from hindcrown to back about as in *C. f.
flammea*, but streaks paler, dark brown-grey, not contrastingly
dull black or black-brown. Rump and shorter upper tail-coverts
white, some dull black streaks on feather-bases sometimes showing through on centre of rump, shorter upper coverts usually
delicately tinged pale pink (not as extensive and deep rose-pink
as in *C. f. flammea*), longer upper tail-coverts dark brown-grey
with broad cream or off-white fringes, sometimes faintly tinged
pink. Length of red patch on forecrown 10·4 (61) 7-14 mm; in
11% of 61 birds, red tinged orange or yellow; 2 cm of rump
unstreaked (rarely 1 or 3 cm; in 4% of birds, rump white with
some dusky streaking) (Molau 1985). Feathering at base of culmen dull black with off-white fringes; supercilium, eye-ring, and
side of neck cream-white or off-white, supercilium narrowest
above front of eye, rear of supercilium and side of neck marked
with ill-defined dark grey specks or streaks, sometimes inconspicuous; eye-ring broken in front and behind. Lore dull black;
upper cheek and ear-coverts pale cream-buff and dark brown-
grey. Side of head and neck distinctly paler than in *C. f. flammea*,
not buff-brown, supercilium more distinct. Rather small patch
on chin dull black or sooty grey; lower cheek and side of throat
cream-white or off-white, sometimes faintly spotted pink at lateral
base of lower mandible (in *C. f. flammea*, black of bib extensively
bordered deep rose-pink at sides). Lower throat and chest delicate pale pink, partly concealed below white feather-tips (pink
less deep than in *C. f. flammea*, not extending to belly). Remainder of underparts off-white, marked with some short and narrow
dark brown-grey streaks on sides of breast and upper flank;
longest under tail-coverts either uniform white or with narrow
concealed dusky shaft-streak, 0·5 (70) 0-2 mm wide (Molau
1985). Face and underparts sometimes stained yellow by pollen
(Alsop 1973). Tail, flight-feathers, upper primary coverts, and
bastard wing as in *C. f. flammea*, but ground-colour greyer-black,
less deep black, and fringes along outer webs off-white, not

wider but more contrasting than brown fringes of *C. f. flammea*;
ground-colour of tertials and upper wing-coverts greyer black
also, fringes along outer webs of tertials pale cream-buff to off-
white (in *C. f. flammea*, fringes brown, soon bleached to white
on tips); tips of median and greater coverts cream, soon bleaching
to white, broader and less brown than those of *C. f. flammea*,
forming more contrasting wing-bars; on greater coverts, tips *c.*
3-5 mm long (in *C. f. flammea*, *c.* 2-3 mm), on median coverts *c.*
2-4 mm (in *C. f. flammea*, 1-2 mm). Under wing-coverts and
axillaries white. *In fresh plumage* (autumn), upperparts (excluding rump) and side of head and neck tinged pale cream-buff;
*in worn plumage* (June-July), ground-colour of head and body
grey-white, contrastingly streaked dusky from hindcrown to
back, streaks sometimes extending over centre of rump; worn *C.
f. flammea* often also rather white on mid-mantle, back, and
rump, but marked here and on remainder of upperparts with
sharply defined black streaks. ADULT FEMALE. Like adult ♂, but
ground-colour of upperparts (except rump) and side of head and
neck slightly warmer buff when plumage fresh (up to about
November), more pale grey (less grey-white) when worn, but
always paler than upperparts of ♀ of *C. f. flammea* at same stage
of wear; streaks on upperparts slightly darker and more sharply
defined than in adult ♂, especially on nape, extending to upper
rump; usually no pink on upper tail-coverts, throat, or chest;
supercilium pale buff and ear-coverts warm buff when fresh
(rather like those of *C. f. flammea* when worn), grey-white with
cream-buff wash on ear-coverts when worn. Feathering at base
of culmen and patch on chin as in adult ♂. Underparts cream-
white or off-white, washed cream-buff on chest, side of breast,
and upper flank when plumage quite fresh; side of breast and
upper flank marked with long dark grey streaks 1(-2) mm wide,
a few sometimes extending to lower flank (streaks less deep black
and generally less broad and extensive than those of ♀ *C. f.
flammea*). Tail and wing as adult ♂, but white tips of median
and greater coverts slightly narrower, 3-4 mm on greater, 2-3
mm on median, more liable to wear off when plumage heavily
abraded (in ♀ *C. f. flammea*, *c.* 2 mm buff when fresh, less than
1 mm white when worn). Red patch on crown 8·7 (39) 6-12 mm
long, tinged orange or yellow in 63% of birds; 65% of *c.* 40
birds examined had rump 1-2 cm unstreaked white, remainder
had 2 cm white with some dusky streaking (exceptionally 1 or 3
cm white with streaking), 15% had some pink tinge; width of
dusky streak on longest under tail-coverts 1·22 (41) 0-3 mm wide
(if over 1 mm, often diffuse, not sharply defined) (Molau 1985).
NESTLING. No information. JUVENILE. Like juvenile of *C. f.
flammea*, but ground-colour of upperparts and side of head and
neck distinctly paler, cream-buff when fresh, grey-white when
worn; fringes along outer webs of tail, flight-feathers, and tertials
cream or off-white, less brown, tips of greater and median coverts
broader and paler (as in adult ♀ of *C. h. exilipes*). FIRST ADULT
MALE. Like ♀, distinguishable only by retained juvenile tail-
feathers, which have distinctly pointed tips (rounded in adult);
by November-January, tips of tail, primaries, and primary coverts slightly frayed, pale fringes of tertials partly worn off (in
adult, tips still smoothly edged). Length of red crown-patch 9·6
(45) 6-13 mm, tinged orange or yellow in 27% of birds; rump
uniform white for 1-2(-3) cm in 63% of birds, remainder had
1-2 cm white with dusky streaking; pink flush to shorter upper
tail-coverts in 33% of 62 birds; width of dusky streak on longest
under tail-coverts 1·66 (56) 0-5 mm, mainly 1-2 mm, often diffuse
when over 2 mm (Molau 1985). FIRST ADULT FEMALE. Like
adult ♂, but juvenile tail, flight-feathers, primary coverts, and
tertials retained, as 1st adult ♂. Rump and lower flank more
often marked with some dark grey streaks than in adult ♀ and
1st adult ♂. Red patch on forecrown 8·1 (42) 5-10 mm long,

tinged orange or yellow in 48% of birds (more often than in 1st adult ♀ *C. f. flammea*); rump white for 1–2 cm, uniform in 51%, streaked in 49% of *c.* 40 birds examined, never with pink flush; width of streak on longest under tail-coverts 2·12 (44) 0–5 mm, mainly 2 mm, often diffuse when broader (Molau 1985). For identification, see also Recognition (below) and Wetherbee (1937), Molau (1985), Olsen (1987*b*), Knox (1988*b*), Herremans (1990*a*), Lansdown *et al.* (1991), and Nikander and Jännes (1992).

**Bare parts.** Similar to *C. flammea*.

**Moults.** ADULT POST-BREEDING. Complete; primaries descendent. In Russia, in intense moult by end of August (Dementiev and Gladkov 1954); in arctic North America, moult July–September (Pyle *et al.* 1987). In Alaska, starts from early July, duration *c.* 40 days (Brooks 1968). POST-JUVENILE. Partial: head, body, and lesser and median upper wing-coverts; starts shortly after fledging, moult recorded July–September (Pyle *et al.* 1987; BMNH).

**Measurements.** ADULT, FIRST ADULT. *C. h. exilipes*. Northern Europe (Tromsö to Pechora river), April–July, and a few from south of breeding range, winter; skins (BMNH, RMNH, ZMA). Bill (S) to skull, bill (N) to distal corner of nostril; exposed culmen on average 3·7 less than bill (S).

| | | | | | | |
|---|---|---|---|---|---|---|
| WING | ♂ | 77·1 (1·16; 17) | 74–80 | ♀ | 73·0 (1·21; 26) | 70–76 |
| TAIL | | 58·2 (2·27; 17) | 55–62 | | 56·2 (2·26; 26) | 52–62 |
| BILL (S) | | 11·6 (0·49; 14) | 10·9–12·3 | | 11·4 (0·38; 20) | 10·8–12·1 |
| BILL (N) | | 6·9 (0·34; 16) | 6·4–7·4 | | 6·5 (0·30; 24) | 6·1–7·2 |
| TARSUS | | 15·0 (0·42; 14) | 14·3–15·6 | | 14·4 (0·45; 15) | 13·7–15·1 |

Sex differences significant, except for bill.

Whole geographical range. (1) *C. h. exilipes*, (2) 'islandica' (pale and intermediate birds), (3) nominate *hornemanni*; bill (F) from feathering, depth is depth of bill at feathering near base (Herremans 1990*a*).

| | | | | | | |
|---|---|---|---|---|---|---|
| (1) WING | ♂ | 76·3 (1·72; 47) | 72–81 | ♀ | 73·6 (1·99; 37) | 70–78 |
| TAIL | | 58·1 (1·67; 47) | 54–62 | | 57·4 (1·60; 37) | 53–61 |
| BILL (F) | | 7·9 (0·45; 47) | 6·8–8·8 | | 7·7 (0·47; 37) | 6·3–8·5 |
| DEPTH | | 6·0 (0·19; 47) | 5·6–6·4 | | 5·9 (0·22; 37) | 5·3–6·3 |
| (2) WING | | 78·5 (1·96; 61) | 72–85 | | 77·5 (2·31; 28) | 74–82 |
| TAIL | | 60·6 (2·47; 61) | 54–67 | | 60·8 (2·67; 28) | 54–66 |
| BILL (F) | | 8·4 (0·43; 61) | 7·4–9·3 | | 8·3 (0·62; 28) | 7·7–9·1 |
| DEPTH | | 6·4 (0·25; 61) | 5·9–7·0 | | 6·3 (0·08; 28) | 5·7–6·8 |
| (3) WING | | 85·5 (2·43; 47) | 81–92 | | 82·7 (2·13; 42) | 79–89 |
| TAIL | | 65·5 (2·44; 47) | 59–70 | | 64·4 (2·69; 42) | 58–70 |
| BILL (F) | | 8·9 (0·58; 47) | 7·8–10·6 | | 8·7 (0·52; 47) | 7·7–10·0 |
| DEPTH | | 7·1 (0·24; 47) | 6·3–7·5 | | 6·9 (0·28; 42) | 6·1–7·8 |

*C. h. exilipes*. Wing, northern Sweden: (1) adult, (2) 1st adult (Molau 1985).

| | | | | | | |
|---|---|---|---|---|---|---|
| (1) | ♂ | 75·6 (1·71; 72) | 72–78 | ♀ | 72·4 (1·56; 41) | 70–76 |
| (2) | | 74·2 (1·83; 65) | 70–79 | | 71·8 (1·57; 45) | 68–75 |

Northern Europe: wing, ♂ 71–79 (*n*=72), ♀ 69–77 (*n*=42); bill 7·0–9·6 (*n*=142) (Svensson 1992).

Nominate *hornemanni*. Wing, ♂ 80–88 (*n*=20), ♀ 79–85 (*n*=20); bill 8·3–10·8 (*n*=40) (Svensson 1992). Bill (S) 13·2 (0·57; 6) 12·8–14·0, bill (N) 8·1 (0·61; 6) 7·3–8·7, tarsus 17·1 (0·58; 5) 16·4–17·7 (RMNH). See also Bird (1935) and Knox (1988*b*).

**Weights.** ADULT, FIRST ADULT. *C. h. exilipes*. (1) North-central Alaska, February–July; (2) north-west Yukon (Canada), mainly May–June (Irving 1960).

| | | | | |
|---|---|---|---|---|
| (1) | ♂ | 12·7 ( — ; 30) 10·7–16·1 | ♀ | 12·8 ( — ; 24) 10·4–14·8 |

(2) 13·6 (0·58; 11) 12·6–14·6     12·7 (0·75; 3) 12·0–13·5

Netherlands, winter: ♂♂ 12·1, 13·7 (ZMA). Mongolia, February: ♂ 13, unsexed 12 (Piechocki and Bolod 1972). Massachusetts (USA), winter: 12·3 (6) 10·7–14·2 (Wetherbee 1937). In captivity, average weight *c.* 1 g more than *C. f. flammea*, except May–June when similar (Brooks 1968, which see for graph of weight throughout year and for relationship between weight and ambient temperature).

**Structure.** 10 primaries: p8 longest, p9 (0–)1–2 shorter, p7 0–1, p6 2–3·5, p5 8–12, p4 13–16, p3 16–18, p2 17–21, p1 19–25 (*exilipes*); p10 strongly reduced, 51–56 shorter than p8, 8–10 shorter than longest upper primary covert. Outer web of p6–p8 emarginated. Tip of longest tertial reaches to secondary tips. Tail rather long, on average *c.* 0·77% of wing length; tip forked, t1 9·5 (10) 7–11 shorter than t5–t6. Bill short; smaller and more conical than in *C. f. flammea*; 6·2 (20) 5·6–6·7 deep at feathering of base in *exilipes*, 6·0 (20) 5·4–6·5 wide, 7·3 (6) 6·7–7·6 deep in nominate *hornemanni*, 6·8 (6) 6·5–7·1 wide (see also Measurements); depth on average 0·94% of length of bill (to nostril), width on average 0·90%; bill appears distinctly shorter than in *C. f. flammea*, projecting less beyond denser and longer feathering at base of bill; culmen almost straight, tip of bill less attenuated, less compressed laterally. For pink pigment of chest, see Troy and Brush (1983). Feathering long and dense, even more so than in *C. flammea*; body feathering *c.* 12% heavier than that of *C. flammea* (Brooks 1968, which see for details). For length of individual tail- and flight-feathers, see Busching (1988). Tarsus short and slender; toes short, but strong, shorter than in *C. f. flammea*. Middle toe with claw 11·8 (10) 10·5–14; outer and inner toe with claw both *c.* 76% of middle with claw, hind toe with claw equal in length to middle with claw; claws long and strong, decurved. Remainder of structure as in *C. flammea*.

**Geographical variation.** Rather marked, involving size and general colour. *C. h. exilipes* from entire Holarctic except Iceland, Greenland, and arctic islands of eastern Canada rather small: wing mainly less than 80, tail less than 60, bill depth less than 6·3. Nominate *hornemanni* from northern Greenland and neighbouring part of arctic Canada distinctly larger (wing mainly more than 80, tail more than 60, bill depth more than 6·3), bill heavier, deeper and wider at base (see Measurements and Structure); plumage on average whiter, less warm cream-buff on upperparts and side of head and neck when fresh, purer and more extensively white when worn; streaking less extensive; white fringes of tail-feathers and tertials and tips of median and greater coverts broader; supercilium broader, forehead and forecheek often whiter, especially in ♂, appearing white-faced. Pale birds on Iceland probably form separable unnamed population of *C. hornemanni* (for which name 'islandica' not available); see Geographical Variation of *C. flammea* and Herremans (1990*a*).

Relationships of *C. flammea* complex and much debated. Various opinions include: (1) *C. hornemanni* is a colour morph of *C. flammea* and cannot be recognized (nominate *hornemanni* would then be included in *C. f. rostrata*, *exilipes* in *C. f. flammea*) (F Salomonsen, in Timmermann 1938–49); (2) *hornemanni* and *exilipes* are both races of *C. flammea*, occurring north of main range of *C. f. flammea* and *C. f. rostrata*, inhabiting different habitats, *exilipes* and *C. f. flammea* forming broad zone of intermediates where they meet (Salomonsen 1928; Payn 1947; Williamson 1961*c*; Dementiev and Gladkov 1954; Houston 1963; Harris *et al.* 1965; Troy 1985; Danilov *et al.* 1984); (3) *C. hornemanni* and *C. flammea* are different species, overlapping widely in breeding range and habitat, with little or no hybridization (e.g Wynne-Edwards 1952); (4) *C. flammea* complex should be split into 3-5 different species (e.g. Todd 1963). Considering *C. hornemanni* to be colour morph of *C. flammea* not valid, as differences exist in measurements and structure as well as colour, unlike colour morphs reported in other bird species; also, reduced interbreeding between light and dark birds does not support this opinion. Neither is *C. hornemanni* a race of *C. flammea*: breeding ranges overlap widely, and, though favoured habitats of the forms may differ locally (e.g. Lundevall 1952, Molau 1985), they breed side by side in many places (Johansen 1944; Wynne-Edwards 1952; Harris *et al.* 1965; Watson 1957*b*; Nyström and Nyström 1987, 1991), with no or at most only occasional interbreeding (Molau 1985; Knox 1988*b*; Svensson 1992), a situation which is beyond traditional subspecies concept. As overlap is more than marginal, *C. hornemanni* and *C. flammea* do not form superspecies, contra Mayr and Short (1970) and Voous (1977). Mixed pairings largely avoided because *C. hornemanni* tends to arrive earlier in breeding area, being already paired or nesting when groups of *C. flammea* arrive (Wynne-Edwards 1952; Molau 1985; Knox 1988*b*). Difference in bill, tail, and foot structure between the species probably means that each exploits different food source with use of different foraging technique (C S Roselaar). Thus, view of Hartert (1903-10), Johansen (1944), Wynne-Edwards (1952), Vaurie (1956*a*, 1959), Howell *et al.* (1968), and Stepanyan (1978, 1990)

followed here, considering *C. hornemanni* and *C. flammea* different species. However, not certain if nominate *hornemanni* and *exilipes* are closer to each other than either is to *C. flammea*; both may have arisen independently from a *C. flammea* ancestor, and whole complex may thus comprise 3 species or (if pale Icelandic birds, *C. f. rostrata*, and/or *C. f. cabaret* also separated) even more (Todd 1963; Molau 1985; Herremans 1990*a*). For genetic difference between *exilipes* and *C. f. flammea*, see Marten and Johnston (1986). For possible evolutionary history of *C. flammea* complex, see Salomonsen (1928, 1972), Knox (1988*b*), and Herremans (1990*a*).

Recognition. For identification in the hand on plumage characters, see Plumages and references cited therein. Identification in the hand on measurements often possible, but depends on correct measuring procedures (see Herremans 1990*a*). Exposed culmen of ♂ under 9·0, ♀ under 8·8, culmen/wing ratio under 8·0, and culmen/bill depth ratio under 1·4 in most *exilipes*, more than these values in *C. f. flammea* (Molau 1985, which see for graphs). In formula 12 × tail + 120 × bill depth- 6 × wing- 89 × bill to feathers, value of over 200 points to *exilipes*, below 200 to *C. f. flammea* (Herremans 1990*a*, which see for separation of other races of both species). CSR

## *Loxia leucoptera* Two-barred Crossbill

PLATES 43 and 45 (flight)
[between pages 640 and 641]

Du. Witbandkruisbek    Fr. Bec-croisé bifascié    Ge. Bindenkreuzschnabel
Ru. Белокрылый клест    Sp. Piquiterto franjeado    Sw. Bändelkorsnäbb

N. Am. White-winged Crossbill

*Loxia leucoptera* Gmelin, 1789

Polytypic. *L. l. bifasciata* (C L Brehm, 1827), Eurasia. Extralimital: nominate *leucoptera* Gmelin, 1789, northern North America; *megaplaga* Riley, 1916, mountains of Hispaniola.

Field characters. 15 cm; wing-span 26-29 cm. Slighter than Crossbill *L. curvirostra*, with relatively weaker bill, smaller head, and longer tail creating rather slim form. Rather small, elegant crossbill, with somewhat more *Fringilla*-like outline compared to *L. curvirostra* and bold white wing-panels recalling Chaffinch *Fringilla coelebs*. Adult ♂ has pink-red head, mantle, and underbody (lacking orange tone of other *Loxia*) and black scapulars, wings, and tail. ♀ also looks bright, with more streaked head and body than *L. curvirostra* and much brighter, purer yellow rump. Juvenile shows diagnostic wing-marks. Voice distinctive. Sexes similar; some seasonal variation. Juvenile separable.

ADULT MALE. Moults: mainly September-December (complete). Head, back, rump, and underparts to end of flanks and belly mainly brilliant cerise-pink, more uniform in tone than *L. curvirostra* but at close range showing (a) dusky lores, cheeks, and ear-coverts, (b) dusky-grey cast on mantle and round shoulder, (c) blackish marks on scap-

ulars, (d) dusky-black upper flanks, and (e) black upper tail-coverts tipped pink-white. In flight, pink appears brightest on rump. Wings basically black, strongly patterned with (a) white forewing-panel, formed by pinkish to white tips to lesser coverts and virtually white median coverts, (b) broad white wing-bar, formed by pinkish-white distal halves to inner feathers and broad tips to rest (and thus narrowing towards primary coverts), (c) striking white tips and pink-white fringes to tertials and innermost secondaries, and (d) duller, apparently pale grey tips and fringes to other flight-feathers. Underwing off-white. Tail black, with white fringes. Belly, vent, and under tail-coverts white, under tail-coverts streaked brown-black. With wear, pink plumage becomes almost carmine and all pale tips and fringes reduce in size or are lost. Bird thus looks even more immaculate, with virtually black scapulars making wings seem larger and wing markings reduced to narrower panel and bar on coverts and isolated white tips on tertials. Bill noticeably more slender than in *L.*

*curvirostra*, looking less crossed due to elongation of less curved points of mandibles; dark horn above, pale horn below. Eye dark brown, seemingly set nearer bill than in *L. curvirostra*. Legs blackish-brown. ADULT FEMALE. Head, back, and underparts to end of flanks and belly basically greenish-brown, with quite heavy brown-black streaks emphasized by cleaner fringes than in *L. curvirostra*. Lower back, rump, and all but longest upper tail-coverts quite bright greenish-yellow, with faint dusky marks; area noticeably paler than dull yellowish-olive rump of ♀ *L. curvirostra*. Rear underparts dusky-white, sharply spotted brown-black on under tail-coverts. Longest upper tail-coverts, wings, and tail as ♂, but ground-colour less fully black, while white tips and fringes tinged yellow, not pink. JUVENILE. Resembles ♀ but ground-colour of head and body noticeably buff-brown, with sharper streaks above than in *L. curvirostra* and whiter centre to breast and underbody, with as many spots as streaks. Wings dusky-black, with forewing panel almost as large as adult but wing-bar distinctly narrower; both marks more discrete than exceptional diffuse whitish marks shown by a few *L. curvirostra*. FIRST-YEAR. ♂ has mixture of adult and juvenile colours, showing pale crimson, brown, and yellow fringes and dark feather-centres on head and body, these largest on back which appears noticeably dark. White wing markings retain yellow (not pink) cast.

Unmistakable if full extent and shape of white wing markings seen well. Commonest calls also distinctive (see below). For fuller details of rare *L. curvirostra* with wing-bars, see that species. Flight lighter than *L. curvirostra*, with slightly more pointed wings and proportionately longer tail combining with smaller head and slimmer body to give less robust, more *Fringilla*-like flight silhouette. Escape-flight direct to tree canopy. Gait a hop, varied by jump and clamber in foliage; movements as acrobatic as *L. curvirostra* and noticeably less clumsy. Bolder than *L. curvirostra*. Less numerous than *L. curvirostra*, wandering in small parties or as single birds (latter often joining flocks of congeners).

Song better-developed and more musical than *L. curvirostra*: long, rich, varied tremolos and trills recalling domesticated Canary *Serinus canaria*, Redpoll *Carduelis flammea*, and Siskin *Carduelis spinus*. Calls (perched and in flight) often disyllabic like *L. curvirostra* but distinctly higher pitched, drier, and less metallic. 3 main types: 'kip-kip' or 'tyip tyip'; 'chut-chut' suggesting *C. flammea*; remarkable sound like toy trumpet.

**Habitat.** After last glaciation, immense spread of coniferous forests across northern Palearctic presented an evolutionary challenge, owing to firm lock-up of great nutrient resources within billions of hard fir cones in crowns of spruces *Picea*, pines *Pinus*, and larches *Larix*. These resources, moreover, are in many cases retained on trees for much longer than most other fruits, and from time to time fail over large areas. To exploit them called for exceptional physical adaptations, and unusual flexibility in use of habitat. Challenge was successfully taken up by a group of cardueline finches evolving strong crossed mandibles backed by powerful musculature, including the legs, and by adaptability to different problems posed by the cones of various tree species, leading to evolution of distinct specialized forms of crossbill: Parrot Crossbill *L. pytyopsittacus* and Scottish Crossbill *L. scotica* associated mainly with pine, Crossbill *L. curvirostra* throughout its main boreal strongholds with spruce, and *L. leucoptera* depending mainly on larch. (Newton 1972.)

While all *Loxia* are constrained by dependence on conifers to conform to northerly distribution pattern, *L. leucoptera* seems additionally inhibited by climatic factors, occupying the most northerly regions between July isotherms 13–20°C (Voous 1960*b*). Extraordinary exception, however, has been establishment of small isolated population in Neotropics on Caribbean island of Hispaniola in stands of pine. Can subsist on most conifers, and when necessary on berries and buds of other trees, but larch is main food tree, and it may occur in mature larch plantations; also favours firs *Abies* (Harrison 1982). In Alaska, seen feeding in small flocks high in tops of spruces and firs. Will, however, sometimes feed on ground, or visit sites with salt, for which it has marked liking. (Gabrielson and Lincoln 1959.) Although given to large-scale wanderings, on occasion has failed to settle in many areas where its food trees are abundant, or to establish colonies in more southern montane regions of Eurasia. While living commonly in dense coniferous forest, some accounts suggest that it has a taste for forest edges and open growth, or even detached groves, e.g. in Canada (Godfrey 1979). Success may have been won by adopting a softer and less thoroughgoing approach than other *Loxia* to toughest aspects of the conifer challenge.

**Distribution.** West Palearctic is at or beyond western limit of regular breeding range. Breeding in Scandinavia and most of Finland is irregular, usually after invasions (see Movements), and cannot be precisely mapped. No changes established, but breeding in Scandinavia has perhaps become more frequent in last 100 years (Fischer *et al.* 1992). Maps show breeding areas only, when largely resident; for other wintering areas, see Movements.

GERMANY. Bred in Berlin 1991; first record for central Europe (Fischer *et al.* 1992). NORWAY. Occasionally breeds, after invasions. First documented attempt (unsuccessful) 1987 near Bergen (Larsen and Tombre 1988). SWEDEN. Occasional breeder (LR). Several breeding records in north after major influx 1985–7 (Fischer *et al.* 1992). FINLAND. Probably breeds in most years, mainly Lapland and eastern Finland; but totally absent in years with poor crop of spruce *Picea* seeds. 'Regular' breeding area cannot be mapped with any exactness. (OH.) USSR. Apparently does not breed regularly in western part of European Russia (Dementiev and Gladkov 1954).

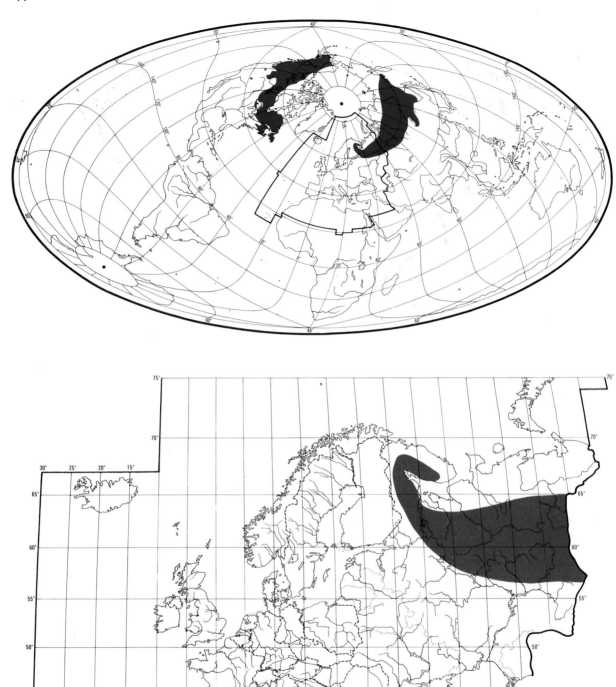

Accidental. Britain, Ireland, France, Belgium, Netherlands, Germany, Poland, Switzerland, Austria, Hungary, Czechoslovakia, Yugoslavia, Bulgaria.

**Population.** Very variable, but little information.

Finland. Perhaps *c*. 500 pairs, but many more in peak years (Koskimies 1989).

**Movements.** Resident and dispersive; also eruptive.

In most years makes only limited movements in response to local food shortages. In occasional years, as in Crossbill *L. curvirostra*, makes eruptive movements, associated with shortage of preferred food (seed of larch *Larix*) and high population density. Apparently involves chiefly juveniles, at least in some years (Lack 1954*a*). In Europe (*bifasciata*), eruptions often coincide with *L. curvirostra* (in 36 of 47 *L. leucoptera* eruptions recorded 1800-1965), and extend more widely than in Parrot Crossbill *L. pytyopsittacus*.

Birds move west or south-west, regularly reaching Finland and Sweden and occasionally various parts of eastern, central, and western Europe, depending on provenance and intensity of movement. (Newton 1972; Larsen and Tombre 1989.) In exceptional 1889 invasion, reported south to Switzerland, northern Italy, and Hungary, west to Britain and Netherlands; many recorded in Czechoslovakia and immediately east at Uzhgorod (Ukraine); only record in Rheinland (western Germany) up to 1980s was in 1889 (Benson 1890; Niethammer 1937; Strautman 1963; Mildenberger 1984). In more typical 1956 invasion, many observations in Finland (chiefly Helsinki area), with small numbers continuing south-west to Sweden, Norway, Denmark, Germany, and Belgium; in Sweden, most reports in south, with highest numbers January-February (Markgren and Lundberg 1959; Jonasson 1960; Lippens and Wille 1972).

Invasions usually reach Scandinavia and Britain in July, continuing chiefly to September or October (Dymond *et al.* 1989; SOF 1990); recorded in central Europe later, mostly from September (e.g. Belgium, Austria) (Lippens and Wille 1972; Brader 1989). Return movement February or March to early June (Haftorn 1971; Møller 1978*a*). Reported in Britain until early April (once early June) (Dymond *et al.* 1989; Rogers *et al.* 1992); in Netherlands, 2 long-staying individuals present until 24 January and 1 February (Berg *et al.* 1992).

In Sweden, occurs every year in strongly varying numbers, and increasingly since 1976; recent major invasions (600-800 birds recorded) in 1979 and in 3 consecutive years (probably for first time since records began in 1786) 1985, 1986, and 1987; largest ever in 1990 (more than 2300 birds), but only *c.* 170 in 1991. In 1979, movement chiefly in eastern Sweden (Gästrikland southward), but in 1985-7 noticeable first in northern Sweden; in 1990, less marked in north but unusually high numbers in south-east (Öland and Gotland). Sometimes breeds in spring following invasion (e.g. 1957, 1987). (Markgren and Lundberg 1959; Tyrberg 1987, 1988, 1991*b*, 1992; SOF 1990.) Only 1 Swedish ringing recovery up to 1989; juvenile ringed August 1966 on Gotska Sandön island (58°22′N 19°15′E) recovered November 1966 in southern Finland (60°10′N 24°53′E) (*Rep. Swedish Bird-ringing*). Recent major invasions in Finland in same years as Sweden; in 1985 most conspicuous at Tauvo in north-west (e.g. Hildén and Nikander 1986, 1988, 1991). In Norway, invasions tend to occur every 7 years: 1952 (probably), 1959, 1966, 1979, and 1986; none in 1972, but some reached Britain then, and perhaps birds moved quickly through Norway due to poor cone-crop; few in 1985, despite Swedish influx. Some evidence that Norwegian invasions linked to food supply further east; in European Russia and Urals, good cone-crops of *Larix* occur every 7 years; in Kareliya (north-west

Russia), cone-crop good in 1978 but poor in 1979 (no data available for other years), so 1979 eruption may have been triggered by insufficient food following year of successful breeding. Perhaps birds reaching Norway originate only as far east as Urals, as cone-crop cycle differs further east. Earliest birds usually reported from Trøndelag (63°-64°N), then movement spreads rather quickly southward along west coast; some birds continue west to Britain, others follow coast to Oslofjord area (south-east Norway), where most records not until October. (Larsen and Tombre 1989.)

In Britain, at least 110 birds recorded up to 1991, usually in company with *L. curvirostra*. Arrivals mainly July-September; October-February birds (exceptionally later) may have arrived earlier and remained to winter. Records chiefly in east but also in west, with 1 in Wales and 4 in Ireland, and may have included Nearctic race nominate *leucoptera*. Distribution varies; in 1966, 5 birds recorded from Cheshire (western England) south to Dorset and Surrey; in 1972, 6 birds of which 5 from Yorkshire (north-east England) north to Shetland (Scotland) and 1 in Devon. In 1987, unprecedented influx confined to north Scottish islands, with at least 25 on Shetland; all records in August, almost entirely juveniles; no simultaneous movement of *L. curvirostra*. In 1990 (when invasions of *L. curvirostra* and *L. pytyopsittacus* also occurred), at least 18 birds reported from Shetland south to Kent and south-west to Devon, of which 12 on north Scottish islands. (Smith *et al.* 1967, 1973; Dymond *et al.* 1989; Rogers *et al.* 1990, 1991, 1992.)

Rare in western Germany, with only 4 records in 1977-89, but 5 in 1990, from Helgoland south to Baden-Württemberg (Bundesdeutscher Seltenheitenausschuss 1992); on Helgoland, only 6 records 1900-66 (Vauk 1972). In Netherlands, *c.* 12 birds in 1990-1; otherwise, only 7 records since 1889 invasion (Berg *et al.* 1992). Also in 1990-1, recorded from France, Belgium, Hungary, and Belorussiya (*Br. Birds* 1991, **84**, 235; 1992, **85**, 14; P J Dubois). In France, only 5 records in 20th century, furthest south-west at Carcans Maubuisson in Gironde, November 1986 (Dubois and Yésou 1992; P J Dubois). In Belgium, 25 records 1938-66; distribution suggests 2 lines of arrival, from north and east (Lippens and Wille 1972). In Mecklenburg (north-east Germany), 11 records 1962-83 (Klafs and Stübs 1987). See Tomiałojć (1990) for Poland, Wüst (1986) for southern Germany, Brader (1989) for north-west Austria.

In east of range (Siberia), movements similar to west, but no details (Dementiev and Gladkov 1954). In western Siberia, disperses south, e.g. to Tomsk and Omsk, usually occurring in flocks of *L. curvirostra* (Johansen 1944). Only 1 record as far south as Kazakhstan (Korelov *et al.* 1974). On Yamal peninsula (north-west Siberia), where breeds irregularly, present throughout winter 1977-8, though not recorded in other winters (Danilov *et al.* 1984).

In North America also (nominate *leucoptera*), irruptive

when cone-crop fails, and sometimes occurs well south of breeding range (as far as Nevada, New Mexico, Texas, and Florida). In Cascade mountains (north-west USA), irruptions typically begin late July, flocks spreading south or towards coast during autumn, usually diminishing by mid-October. (Gordon *et al.* 1989.) In Wisconsin (north-central USA), where immigrants usually arrive mid- to late September, major movement 1989 began mid-July, but chiefly from August; still widespread and numerous January–February, with fewer reports later (Lange 1990; Robinson 1990); exceptional numbers also reached neighbouring state of Minnesota, mostly from mid-October (Janssen 1990). For movements in north-east USA and adjacent Canada according to food availability, see Benkman (1987a). Sometimes wanders north of range, exceptionally to extreme north of Alaska and Baffin Island (Bent 1968). One record in Russia, in Chukotskiy peninsula in extreme north-east (Tomkovich and Morozov 1982). West Indian race *megaplaga* (breeding Hispaniola) resident, but apparently occasionally disperses some distance; several seen in Jamaica December 1970 to April 1971 were presumably this race (American Ornithologists' Union 1957, 1983; Bond 1985). DFV

**Food.** Conifer seeds, principally of larch *Larix* and spruce *Picea*; some invertebrates in breeding season. Larch is main food in some parts of range (e.g. eastern Russia: Dementiev and Gladkov 1954; Danilov *et al.* 1984), spruce in others (e.g. Fenno-Scandia, Murmansk region of north-west Russia: Svärdson 1957; Kokhanov and Gaev 1970; Pulliainen 1971, 1972); differences presumably due to distribution and abundance of these trees. Foraging methods identical to Crossbill *L. curvirostra* and Parrot Crossbill *L. pytyopsittacus* when all 3 species watched together in Murmansk region and Finnish Lapland (Kokhanov and Gaev 1970; Pulliainen 1972); for details, see *L. curvirostra*. Probably more readily hangs on cones to extract seeds, taking them less often to perch than other *Loxia* (Bannerman 1953a; Bent 1968; Kokhanov and Gaev 1970). For details of handling, bill morphology, efficiency, etc., see Kokhanov and Gaev (1970) and Benkman (1987b, 1988d, 1989a). Thinner bill more suitable for extracting seeds from between thin and relatively short cone scales, e.g. larch and spruce (Newton 1967a; Benkman 1987b). Nesting pair in cemetery in Berlin (Germany) often foraged in broad-leaved trees, e.g. fed on seeds of birch *Betula*, hanging head-down, although conifers available (Fischer *et al.* 1992). In eastern Siberia, fed in alders *Alnus* with Redpoll *Carduelis flammea*, and also picked up seeds from ground and rubbish-tips near houses (Mezhennyi 1979; Morozov 1984), and in Finland fed on spruce seeds lying on branches (Pulliainen 1972). In North America, away from conifers, recorded feeding in alder and birch, taking seeds on sunflower *Helianthus* plants and on herbs in fields and on roadsides, eating various berries and fruits, as well as insects and larvae, sometimes taken

from behind bark; also seen feeding on snails (Gastropoda) on tideline (Bent 1968). In central and eastern USA, in winter, has acquired habit of extracting seeds from fruit of sweetgum *Liquidambar* tree, using ability to hang upside-down and make powerful use of bill; will also hang from neighbouring twig to pull fruit in with foot; may have learned technique from American Goldfinch *Carduelis tristis* (George 1968). Caught insects in flight in captivity (Massoth 1989, which see for captive diet). Ate decayed wood from wall of hut, presumably for minerals (Sundin 1988); many accounts of fondness for salt, etc. (Bent 1968). In captive trials, seed-husking time ranged from average 0·9 s per seed for spruce ($n=459$) to 8·4 s for one species of pine ($n=337$); more efficient than *L. curvirostra* when foraging in spruce and extracting seeds from open cones, less efficient in pines and in handling closed cones (Benkman 1987b, which see for many details).

Diet in west Palearctic includes the following. Invertebrates: bugs (Hemiptera: Miridae, Aphidoidea), Lepidoptera (larval Geometridae), Hymenoptera (larval Tenthredinidae), spiders (Araneae). Plants: seeds, buds, shoots, etc. of larch *Larix*, spruce *Picea*, pine *Pinus*, birch *Betula*, alder *Alnus*, knotgrass, etc. *Polygonum*, thrift *Armeria*, rowan *Sorbus*, crowberry *Empetrum*, grasses (Gramineae). (Dementiev and Gladkov 1954; Pulliainen 1972; Ree 1977; Pulliainen and Tuomainen 1978; Harvey 1990; Fischer *et al.* 1992.) See also Kokhanov and Gaev (1970) for food taken by 3 *Loxia* species.

In Finnish Lapland, July–August, 8 stomachs contained 300 items, of which 90% by number (89% by volume) new spruce seeds, 6% (9%) old seeds, 2% by number Hemiptera (1·3% Miridae), 1% (3%) Tenthredinidae larvae, 1% spiders (Pulliainen 1972, which see for comparison with 2 other *Loxia* species). 5 stomachs from same area, April–May, contained only spruce seeds (Pulliainen 1971). During irruption in Scandinavia and Germany, July–March, 79% of 203 feeding observations were on rowan, 9% alder, 6% larch, 3% spruce, 1% pine, and 1% birch; rowan preferred to spruce seeds, which were untouched when rowan available, but larch most preferred where common (Jonasson 1960). But in Finnish Lapland, rich rowan berry crop ignored in favour of spruce in March–April; in July, birds fed on new spruce cones, otherwise picked seeds from old cones on ground (Pulliainen 1972). For importance of rowan, including in captivity, see also Bannerman (1953a) and Massoth (1989). In north-east USA and neighbouring Canada, feeds on seeds of *Larix* and *Picea glauca* from summer to late autumn or winter, then switches to *P. mariana* in order to maximize intake rate, broadly preferring this species until summer; different conifers peak in seed maturity and availability at different times of year, and in October–November an intake of *c.* 0·4 mg of seed per s seems to be minimum rate triggering move to find more profitable trees (Benkman 1987a, which see for many details and comparison with *L. curvirostra*). Stomach of bird on Pribilof

Islands (Alaska) in August was full of Calliphoridae (Diptera) and seeds of *Brassica* (Bent 1968).

In Murmansk region, consumed average 1540 spruce seeds (4·3 g dry weight) per day during winter and spring, taking average 128 min to do this, or 12 seeds per min, excluding search time; slowest rate of 3 *Loxia* species (Kokhanov and Gaev 1970). In North American studies, *c.* 0·6–1·4 mg per s (*c.* 2·2–5·0 g per hr) consumed in spruce; in larch, *c.* 1·0 mg per s; intake required for survival in January *c.* 0·2 mg, and actual intake *c.* 0·5 mg (Benkman 1987*a*, *b*, 1990, which see for details). In flock, average consumption per bird 1·3 times greater than for single bird (Benkman 1988*b*). 4 birds detached 59 spruce cones in 30 min, removing only 4–5 seeds from each (Bent 1968).

Young fed regurgitated conifer seeds. In Murmansk region, average daily intake of nestling was 220 spruce seeds (0·6 g dry weight) at age 1–12 days, and 1760 (4·9 g) at 13–23 days, i.e. 65 300 seeds (183 g) per brood over whole nestling period (Kokhanov and Gaev 1970). In captivity, parents gave young also some aphids and insect larvae (Massoth 1989). BH

**Social pattern and behaviour.** Information on west Palearctic *bifasciata* limited and account includes data on nominate *leucoptera* of North America, notably from review by Benkman (in press); *megaplaga* of Hispaniola not studied.

1. Occurs in flocks all year, though pairs and single birds also recorded. In Chita region (south-east of Lake Baykal, eastern Russia), most numerous small passerine wintering in area; flocks of 3–15, and 0·6 birds per km transect (Leontiev 1965). Probably because main breeding area lies in larch *Larix* forests of eastern Russia, few reach as far west as Scandinavia (and rest of northern Europe) during irruptions (see Movements) and large flocks consequently rare there (Markgren and Lundberg 1959): 20–40 (Ottow 1912; Jonasson 1960; Pulliainen and Saari 1976); larger flocks recorded include *c.* 75 in February, up to 100 (most juveniles) and *c.* 200 (all juveniles) in autumn (Lack 1954*a*; Bernhoft-Osa 1959; Larsen and Tombre 1988), with peak *c.* 300–400 (Larsen and Tombre 1989). Pattern evidently similar in North America: flocks up to *c.* 350, but usually smaller (Bent 1968; Robinson 1990); over 3 years, 1–150 (*n* = 1092); in non-breeding populations, average when not feeding 5·2 ± SE0·83 increasing to 10·2 ± SE0·88 in foraging flocks; among breeding populations, correspondingly 2·5 ± SE0·15 and 3·6 ± SE0·29, apparently single birds (breeders or non-breeders) joining foragers in breeding season (Benkman in press). In Quebec (Canada), late July, 'vast congregation' apparently exclusively of ♂♂ dispersed in small flocks and larger gatherings along *c.* 15–20 km in spruce *Picea* forest; presumed build-up prior to migration (Wynne-Edwards 1962); see also Song-display (below). During invasion in south-east Manitoba (Canada), October–April, average 10·4 ± SD9·3, *n* = 118 (Sealy *et al.* 1980); see also Janssen (1990). Like other *Loxia*, often forages in flocks when breeding (Benkman 1988*b*): sometimes only ♂♂, but later joined by ♀♀ and eventually fledged young to form larger gatherings (Mezhennyi 1979 for eastern Russia; Messineo 1985 for North America). For flock sizes in northern Russia east of Urals, spring to end of July, see Danilov *et al.* (1984) and Morozov (1984); for post-breeding densities in coastal taiga, Okhotsk (eastern Russia), see Kuzyakin and Vtorov (1963). Frequently associates for feeding (etc.) with

other Fringillidae, notably other *Loxia* and Pine Grosbeak *Pinicola enucleator*, also with waxwings *Bombycilla* (Markgren and Lundberg 1959; Leontiev 1965; Bent 1968; Leinonen 1978). North American study suggested *L. leucoptera* typically no longer associates with Crossbill *L. curvirostra* after leaving tree where previously feeding; also, more likely to be joined by (e.g.) Pine Siskin *Carduelis pinus* and Redpoll *C. flammea* than the reverse (Benkman in press). BONDS. Mating system not studied, but probably mainly monogamous, though ♂♂ generally outnumber ♀♀ in *Loxia* (Newton 1972), and North American studies of *L. leucoptera* found ♂:♀ ratio of 1·19:1, this perhaps leading to occasional occurrence of serial polyandry (Smith 1978; Benkman 1989*b* and in press). Duration of pair-bond not known, but some ♂♂ and ♀♀ associate during non-breeding period as if paired, and some apparently paired when they arrive on breeding grounds, others perhaps pairing just before nesting (Benkman in press). Casual observations in northern Utah (USA), August–May, suggested all adults of both sexes paired, but population of unpaired 1st-year ♂♂ also exists, such birds being probably physiologically capable of breeding (Bond 1938; Smith 1978); no further information on age of first breeding. Hybridization recorded in captivity with *L. curvirostra* (Nethersole-Thompson 1975). Recorded feeding *C. pinus* in mixed flock (Salt 1984); see also *L. curvirostra*. Nest-building and incubation by ♀ alone (Bent 1968; Newton 1972); see also Heterosexual Behaviour (below). Young fed by both sexes (Bent 1968; Fischer *et al.* 1992). When cone crop allows high intake rates (see Food) and immediate 2nd nesting attempt follows 1st, ♂ usually cares for 1st-brood fledglings; if pair single-brooded, both adults tend fledged young, and some records of brood tended by ♀ alone (Benkman 1989*b* and in press). Not known if brood sometimes divided between parents (as reported for *L. curvirostra*). Not clear how long family remains intact, though foraging efficiency of young probably continues to increase for well over 60 days (Benkman in press). Period of dependence reported to be *c.* 4–6 weeks in *Loxia* generally (Kokhanov and Gaev 1970; Newton 1972); see also Massoth (1989) for breeding of captive *L. leucoptera* pair. BREEDING DISPERSION. Tends to nest in neighbourhood groups; one group in New Brunswick (Canada) of *c.* 12 pairs (Bent 1968; Messineo 1985; Benkman 1988*b*). Before nesting, up to 10 or more ♂♂ may sing simultaneously within less then 15 m radius. Paired (perhaps also unpaired) ♂ sings from same area (radius *c.* 50 m) daily in early stages of nesting. Typical of *Loxia* in defending only ♀, and small area around nest before and during laying. (Smith 1978; Benkman 1988*b* and in press.) In New York State (USA), larger area (*c.* 0·4–0·8 ha) around tree used by ♂ as song-post and perhaps for nesting (Messineo 1985), though unlikely that all this defended; also in New York, territory reported to be less defined once active nesting has started (Crumb 1985). In North America, colour-ringed birds seen to fly regularly up to *c.* 1 km between nesting and feeding area; winter home-range probably much larger, though may be concentrated temporarily, e.g. in given grove of trees (Benkman in press). No further information on size or use of territory. Data on breeding density all from extralimital Russia: on Yamal peninsula, sample plot of *c.* 14 ha containing much dense spruce *Picea* held 2–3 pairs (14–21 pairs per km²) (Danilov *et al.* 1984); in Siberia east to Lena river, 5–20 birds per km² recorded in taiga of varying composition, dense riverine forest apparently preferred (Ravkin 1984; Vartapetov 1984; Shmelev and Brunov 1986); see also Rogacheva (1988) which includes some densities relating to migration periods. On Vitim plateau (north-east of Lake Baykal), 0·9–2 birds per km² in larch taiga (Izmaylov and Borovitskaya 1967). Nomadic habits mean birds unlikely to return regularly to breeding site or regular winter home-range (Benkman in press). ROOSTING. In North America, birds roost at night communally in dense foliage of conifers (up to 6 birds in same tree) (Benkman in press). In Norway, late March, c. 35 seen resting high up in dense larches in late afternoon (Larsen and Tombre 1989); not known if they spent night there.

2. Like other *Loxia*, usually tame and easily approached when feeding (e.g. Tschusi zu Schmidhoffen 1890, Ottow 1912 for birds outside normal breeding range). In Maine (USA), recorded coming very close to humans, even landing on one, also foraging close to and inside forest cabin (Bent 1968). In New York, in post-fledging period, much warier, flying off at slightest noise or movement, and young in particular reluctant to descend to ground to drink or take grit (Messineo 1985). Generally evades predators by remaining still and concealed among foliage or by flying up and above raptors. ♂♂ are more colourful, perform more elaborate displays (though no detailed description), and have more complex song than *L. curvirostra* (see Voice); foraging differences linked with dispersion of cones mean intraspecific conflicts also more frequent in *L. leucoptera*. (Benkman in press). FLOCK BEHAVIOUR. Birds feeding in flocks do not defend feeding territory, probably because such defence would reduce intake rate (Benkman 1988b). Flying birds will abruptly change course and join feeding flock on hearing 'sharp' call. ♂-♂ and ♂-♀ chases frequent, especially before nesting, though physical contact rare. Members of feeding flocks often displace, chase, or grapple with one another. During 393 min observation in September, no aggression in flocks of 2–3 birds; in larger flocks, c. 1 aggressive interaction per 10 min, increasing slightly with flock size. In one study, interactions most intense when foraging on black spruce *Picea mariana*, probably because of its small cone-bearing crown and scarcity of other food. Captive birds maintain individual distance (including for roosting); occasionally closer for foraging in the wild depending to large extent on dispersion of cones. Dominance hierarchies form among captive birds; in the wild, ♂♂ tend to dominate ♀♀, older individuals dominate 1st-years. (Benkman in press.) At Lake Ontario (New York), October, all birds in flock of c. 25 recorded bill-wiping methodically for c. 15 min, probably to rid themselves of resin from cones (Sunderlin 1978). On Magdalen Islands (Canada), breeding ♀ apparently foraged in flock, returned with other birds, dropped down to nest, and called attracting some members of flock to join her briefly (Bond 1938). SONG-DISPLAY. ♂ sings (see 1 in Voice) from perch (usually tree-top) (e.g. Godfrey 1986, Fischer *et al.* 1992), though not normally near nest (Benkman in press). Recorded frequently twitching tail and ruffling crown while singing (Townsend 1906). Also sings in flight, sometimes beginning on one perch and continuing while flying to another (Borror 1961), or performs Song-flight lasting several minutes, slowly flapping wings while flying in wide circles and singing continuously, sometimes several ♂♂ circling at once over perched ♀♀, then landing on tree (Townsend 1906; Bent 1968; Godfrey 1986). In Quebec, ♂♂ sang excitedly and almost incessantly in chorus, mostly in flight, slow wing-beats being most conspicuous in descent phase (Wynne-Edwards 1962). Paired ♂ will sing while flying towards pairs that fly over his song-perch; flies slowly behind for 10–15 m, then returns to mate (Benkman in press). ♀ also sings (see 1c in Voice), but no information on contexts; only ♂ does so regularly (Bent 1968). In Sweden and Finland, song noted June–October (Elmberg 1992). In northern Russia (Urals to Taymyr peninsula), March to early June, occasionally mid-July (Portenko 1937; Krechmar 1966; Danilov *et al.* 1984); further east, in Yakutiya, vigorous song noted in fog and temperature of −57°C late January, and also mid-August (Vorobiev 1963). In North America, song given mostly just prior

to nesting (Benkman in press). Song-period in northern Utah (USA) August–May, adult ♂♂ ceasing once paired and nesting, but 1st-years continuing (Smith 1978); see also (e.g.) Janssen (1990) and Robinson (1990). ANTAGONISTIC BEHAVIOUR. Little information. Threat-posture similar to *L. curvirostra* (see that species), bird often giving harsh Chatter-calls (see 2b in Voice), also flicking wings and tail; sometimes leads to chase and fights with pecking and grappling; appeasement much as *C. flammea* (see that species) (Benkman in press). Apparent territorial chases noted Yamal peninsula early June (Danilov *et al.* 1984). In New York, January–February, pairs and small flocks responded aggressively (no details) to playback of both conspecific and *L. curvirostra* calls (Messineo 1985). HETEROSEXUAL BEHAVIOUR. (1) General. On Kola peninsula, pairs noted from end of February in one year (Kokhanov and Gaev 1970). In Olekminsk (southern Yakutiya), birds recorded just settling to breed in 2nd half of May when *L. curvirostra* has already reared young (Mezhennyi 1979). (2) Pair-bonding behaviour. Once paired (and especially during laying), ♂ often perched alertly above ♀, guarding her while she forages, takes grit or drinks (Mezhennyi 1979; Benkman in press). In Alaska (USA), late April, ♂♂ performed Song-flights over ♀♀ and sexual chases also took place (Bent 1968). ♀ flies rapidly through trees, apparently to evade ♂♂, or slowly flaps wings as if inviting pursuit (Gabrielson and Lincoln 1959; Benkman in press). Captive pair highly vocal (song from ♂), also chased frequently, over c. 2 weeks in February (Massoth 1989). (3) Courtship-feeding. Noted at Olekminsk in trees before nesting at end of May (Mezhennyi 1979). In North America, recorded initially billing (touching and nibbling), ♂ later feeding ♀ (on or away from nest) by regurgitation; billing and Courtship-feeding seen New York in February (Messineo 1985; Benkman in press). ♂ may give loud 'piping' calls (presumably Trumpet-calls: see 3 in Voice) during approach and for several minutes from nearby perch before flying to nest to feed ♀; often performed Song-flight after feeding mate (Tufts 1906). (4) Mating. Presumed mounting attempt noted in Ontario (Canada) late December, ♂ ascending with rapid wing-beats almost vertically over unreceptive ♀ (Bent 1968). Copulation (no details) recorded New York in February (Messineo 1985). (5) Nest-site selection. Nest-site apparently chosen by ♀ (Mezhennyi 1979). Only ♀ of Norwegian pair seen carrying nest-material (Larsen and Tombre 1988). Building by ♀ in North America also, though accompanied by ♂ who sometimes carries material to nest and helps with building (Palmer 1949; Benkman in press). In Hispaniola, both members of *megaplaga* pair seen collecting material, though ♀ did more, ♂ feeding frequently but staying close to ♀, who tossed out some material added to nest by her mate (Kepler *et al.* 1975). (6) Behaviour at nest. ♀ may leave nest briefly every c. 2–3 hrs (Benkman in press). No further information. RELATIONS WITHIN FAMILY GROUP. In North America, all brooding by ♀, probably almost continuously for first c. 5 days (Benkman in press). Young fed by both sexes by regurgitation. In Berlin (eastern Germany), ♂ seen carrying away nestling faeces (Fischer *et al.* 1992). For general information on 3 *Loxia* species on Kola peninsula, at least some referring specifically to *L. leucoptera* (e.g. feeding rates), see Kokhanov and Gaev (1970); fledging period for 3 *Loxia* in same study 22–24 days (see also Breeding). Can fly quite well within c. 1 week of leaving nest, then follow parents closely (Benkman in press). Fledglings beg by fluttering wings and giving food-calls; still fed by regurgitation (Crumb 1985; Benkman 1989b; Groth 1992b). ANTI-PREDATOR RESPONSES OF YOUNG. May leave nest prematurely (e.g. Leinonen 1978). Fledglings remain hidden until able to fly well (Messineo 1985). PARENTAL ANTI-PREDATOR STRATEGIES. (1) Passive measures. ♀ a tight sitter and can even be handled, though (like *L. curvirostra*) will peck at hand (Kokhanov and Gaev 1970). After flushing, ♀ recorded flying to join ♂ singing nearby (Bent 1968), or returning within 'a few feet' and calling while nest photographed, calls briefly attracting other conspecific birds (Bond 1938). (2) Active measures: against birds. In Berlin, when Jay *Garrulus glandarius* or Carrion Crow *Corvus corone* came within *c.* 50–100 m of nest, alarmed ♂ gave series of Trumpet-calls, often also turning body side to side and flicking one wing (Fischer *et al.* 1992). In North America, Gray Jays *Perisoreus canadensis* approaching nest also typically elicit alarm-calls (Chatter- or Trumpet-calls: Benkman in press); in Colorado (USA), ♂ called and attacked 2 *P. canadensis* threatening fledgling on ground (Groth 1992b). (3) Active measures: against man and other animals. Apparently much as for birds. In Alaska, close to hatching, both adults called while flying from tree to tree (Bent 1968); alarm-calls also given for squirrels *Tamiasciurus* (Benkman in press).                    MGW

**Voice.** Distinctly different from other *Loxia*. Song varies to some extent geographically; noted in most months in different parts of range, and calls also given all year, especially in flight. Silent or nearly so when feeding unmolested, but, like other *Loxia*, typically calls suddenly before take-off; apart from frequent calls, sometimes also sings during longer flights. (Bent 1968; Benkman in press; see Social Pattern and Behaviour.) Members of captive nominate *leucoptera* pair exhibited call matching (of all call-types) apparently by imitation (Mundinger 1979), much as reported for Twite *Carduelis flavirostris* (see that species). For additional sonagrams, see (for *bifasciata*) Fischer *et al.* (1992) and (nominate *leucoptera*) Mundinger (1979). Song of *megaplaga* not described, but calls said to differ from nominate *leucoptera* (C W Benkman).

CALLS OF ADULTS. (1a) Song of ♂ *bifasciata*. Less well

known than nominate *leucoptera* from which said by Elmberg (1992) to differ markedly. Further study clearly needed, however, as other descriptions, and certainly recording analysed, suggest that the 2 races have in fact much in common. Recording from Finland contains loud song comprising long series of different segments mostly delivered at very rapid rate (some variation); also some more well-separated units (calls). Segments include rattles, trills, and buzzy units; some harsh, others more musical. Basic structure (segments of rattling and other units) and consequent similarity especially to Redpoll *C. flammea* and domesticated Canary *Serinus canaria* are shared with nominate *leucoptera* (see 1b, below), though *bifasciata* apparently exhibits greater variety, at times suggesting other species (see below), though not known whether mimicry involved. (J Hall-Craggs, M G Wilson.) Breeding ♂ in Germany gave (apart from calls) 'trills' like domesticated *S. canaria* and other sounds suggesting both Siskin *C. spinus* and *C. flammea* (Fischer *et al.* 1992); similarity to *C. spinus* (not obvious in recording) noted earlier by Bruun *et al.* (1986). In Fenno-Scandia, comparatively long, rich, and varied, comprising mainly clear, metallic notes and slurred whistles, but also chattering sounds and wheezes; calls 2a–c (see below) frequently interspersed. Song said thus to resemble Crossbill *L. curvirostra*. (Elmberg 1992.) Song in recording (Fig I) comprises slow introductory segment of 5 unusual and arresting 'T-ing' sounds (loud, rattling 'tik' or almost 'tok' preceding high tone) followed, after pause of *c.* 1 s, by long musical trill of 31 notes, each a tonal diad alternating with very brief, high-pitched notes; trill segment runs directly into 5 'pink' units (similar

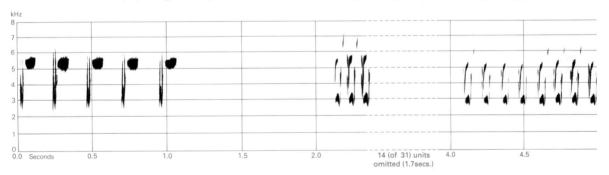

I   H and M Hanhela   Finland   March 1979

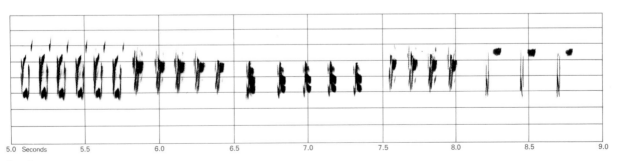

I *cont.*

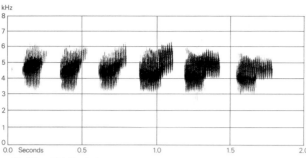

II  H and M Hanhela  Finland  March 1979

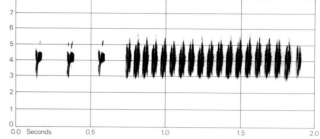

III  H and M Hanhela  Finland  March 1979

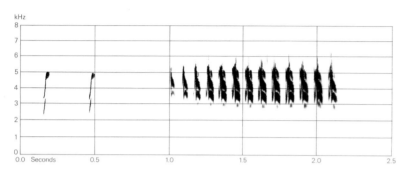

IV  H and M Hanhela  Finland  March 1979

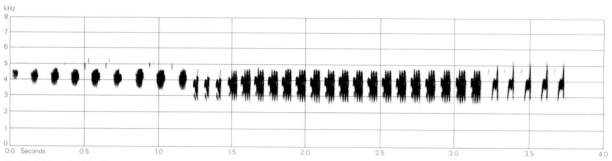

V  Roché (1990)  Canada

to 'pink' of Chaffinch *Fringilla coelebs* and Blackbird *Turdus merula*), then slow, rather noisy segment of 5 'ch' sounds, one of 4 'pink' units, and (faded-out) coda of 3 'T-ing' notes. Fig II shows segment of 6 buzzy 'zizz' units suggesting part of song of Yellowhammer *Emberiza citrinella*. In Fig III, 3 'pink' calls precede rattle segment, and in Fig IV 'ch' units similar to those in antepenultimate segment in Fig I are given as faster rattle and are preceded by 2 'wit' calls (see 2a, below). (J Hall-Craggs, M G Wilson.) (1b) Song of ♂ nominate *leucoptera*. Better-described than *bifasciata*, but also essentially a long series of loud tremolo-type segments (some with buzzy effect) at varying pitches and delivery rates. Pitch generally steady within a segment, but may rise within longer ones. As in *bifasciata*, song delivered with great vigour and urgency, though this varies according to unit-type. Buzzing segments (tremolos) recall Rattle-call of *C. flammea*, others reminiscent of trills in song of Greenfinch *C. chloris*. (Borror 1961; Bergmann and Helb 1982; Elmberg 1992;

M G Wilson from recordings.) Songs last 4–6(–15) s and contain (1–)3–13 segments. Each segment comprises up to 35 units, delivered at rate of 10–16 (3–37) per s. Each bird has repertoire of 15–20 segment-types, these being variously combined to make songs; variation noted between individuals. (Borror 1961; see also Bergmann and Helb 1982.) More varied and attractive song, apparently continuing (presumably with at least brief pauses) for several minutes, described by other authors: frequent changes in volume, and, apart from rattle or trill segments, song contains notably loud, emphatic, ascending 'whee whee whee', chirping sounds, and quiet, sweet warbling; at times very similar to domesticated *S. canaria* (see Townsend 1906, Dale 1924, Bent 1968, and Godfrey 1986, e.g. for renderings of song). Song may carry several hundred metres (Bent 1968). Recording from Canada contains songs showing great technical virtuosity with high delivery rates (one segment running into next) and merging into diads from simple onsets; less interesting than *bifasciata*

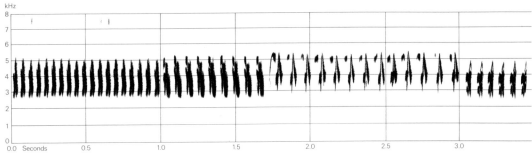

VI Roché (1990) Canada

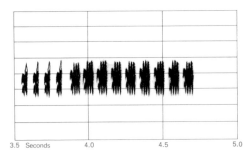

3.5 Seconds    4.0    4.5    5.0

VI *cont.*

overall, being more strongly reminiscent of domesticated *S. canaria*, though higher pitched (J Hall-Craggs). In Fig V, 9 'ziz' units with overall crescendo (recall song of *E. citrinella* and 'zrik' calls of Arctic Warbler *Phylloscopus borealis*) followed by 3 'ch' which then merge with segment of 18 'che' units, song becoming suddenly louder at this point; final segment comprises 5 tonal 'pip' units (sounding more like 'pip-ip-ip-ip-ip'). Song in Fig VI starts with high-speed tremolo of 19 units, broadening out at 20th to 9 similar but longer-duration units, then 13 higher-pitched ringing sounds with initial high tonal onsets; final 2 segments 'ch' and 'che' units as in Fig IV. (J Hall-Craggs, M G Wilson.) (1c) Song of ♀. Noted (apparently irregular, but much as ♂'s song) for nominate *leucoptera* (Bent 1968). Recording of captive ♀ *bifasciata* (caged in forest) contains probable subsong, most notes being short and quiet, though louder calls (2b-c) also given. In Fig VII, 'wit' calls (see 2a, below) are followed by remarkable, high-pitched purr; in Fig VIII, 3 'pink'-like calls (only tonal component making them readily distinguishable from 'wit'), then 4 'ch' calls (some resemblance to Tep-calls of other *Loxia*) becoming 'chit' at calls 5-6. (J Hall-Craggs, M G Wilson.) (2) Contact- and alarm-calls. Generally muffled or rather weak, lacking metallic ringing quality of other *Loxia* calls (Kokhanov and Gaev 1970; Elmberg 1991a). Calls 2a-c given by both sexes in flight and when perched. (2a) In *bifasciata*, high-pitched, liquid, bouncing 'glip glip', 'plitt plitt', quiet 'kip-kip', etc.; reminiscent of

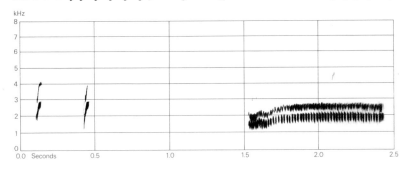

VII  M Schubert  captive (Germany)  May 1978

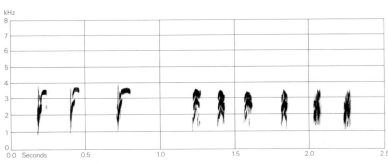

VIII  M Schubert  captive (Germany)  May 1978

advertising-call of Quail *Coturnix coturnix* or 'pwit' of Ortolan Bunting *E. hortulana*, rarely closer to Dyip-call of *L. curvirostra*, though usually weaker, thinner, softer, and more liquid (Bruun *et al.* 1986; Delin and Svensson 1988; Elmberg 1991*a*; Jonsson 1992). In Germany, call (rendered 'gebb') given in series as contact-call mainly by ♂ of breeding pair (Fischer *et al.* 1992). See also Tschusi zu Schmidhoffen (1890). Calls suggesting 'wit wit' indeed reminiscent of *C. coturnix* shown (in each case as 1st 2 units) in Figs IV and VII; 'pink' calls (1st 3 units of Fig VIII) apparently closely related (J Hall-Craggs). Apparently no equivalent call described for nominate *leucoptera*. (2b) Chatter-call. Dry, quite flat, somewhat harsh chattering sounds (often 2-3 together) strongly recalling Chatter-call of *C. flammea* ('chet', 'chek', 'chuch', etc.), but with *Loxia* quality, though not likely to be confused with Dyip-call of *L. curvirostra*, being softer and higher pitched (Bruun *et al.* 1986; Elmberg 1991*a, b*; Olsen 1991*a*; Svensson 1991; Jonsson 1992). Series of Chatter-

calls (suggesting 'chit') given by captive birds shown in Fig IX (presumed ♂) and Fig X (♀). Another recording (Fig XI) contains series of obviously closely related, easily locatable 'chik' or 'pik' sounds similar to pre-roost calls of *T. merula* but less metallic. (J Hall-Craggs.) Chatter-calls similar to *C. flammea* ('chif-chif', 'chut', 'cheet', 'tchet', etc.) given in flight, and apparently sometimes expressing alarm and threat, also reported for nominate *leucoptera* (Witherby *et al.* 1938; Bent 1968; Godfrey 1986; Gordon *et al.* 1989; Benkman in press). 'Barking' calls, probably of this type, reported by Groth (1992*b*). Sonagrams in Mundinger (1979) show bursts of 3-4 Chatter-calls; also a presumably related, higher-pitched rattle, probably much like rattles in song. Liquid 'trrrrrrr' apparently serving as warning-call (Bent 1968) likely to be also this type. (3) Trumpet-call. In *bifasciata*, distinctive nasal sound resembling toy (tin) trumpet or subdued song-note of Trumpeter Finch *Bucanetes githagineus*. Given in flight and when perched, about to take off or land, also serving as alarm-call, and may be incorporated in song; nothing similar reported for other *Loxia* (Bruun *et al.* 1986; Elmberg 1991*a, b*; Olsen 1991*a*; Svensson 1991). Further descriptions: series of 3-5 calls given as warning-call by ♂ of pair in Germany 'wäd-wäd-wäd-wied-wied' (slight pitch ascent; compare discussion of recordings below) or soft, slightly hoarse 'gää-gää-gää-gää' (Fischer *et al.* 1992); far-carrying 'meep' (Holman and Kemp 1991); 'bib-bib-bib-bib' (Kokhanov and Gaev 1970). Trumpet-calls in recordings have penetrating, tinny timbre. Calls often given in pairs of notes which are not fixed: Fig XII shows 4 different call-types in 1st 2 pairs, while notes in 3rd pair

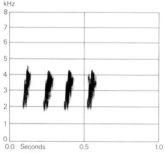

IX   C Chappuis   September 1991

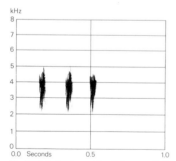

X   M Schubert   captive (Germany)   May 1978

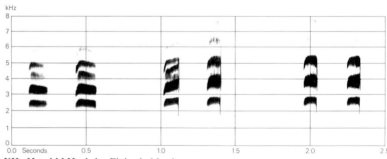

XII   H and M Hanhela   Finland   March 1979

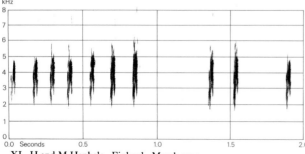

XI   H and M Hanhela   Finland   March 1979

uniform. Stepped tonal ascent from 1st to 2nd note of pair also occurs; in some cases, pitch descends very gradually stepwise over series of calls. Conspicuous feature of some notes is strong (mainly end) transients, especially clear in Fig XIII where calls also exhibit differing loudness levels; in Trumpet-calls of captive ♀ (Fig XIV) starting and end transients very weak or non-existent. (J Hall-Craggs.) Presumed equivalent call of nominate *leucoptera* rendered 'kank' (Mundinger 1979, which see for sonagram). Other descriptions presumably referring to this call include nasal, querulous 'cheit' (Gordon *et al.* 1989), 'pit' from ♀

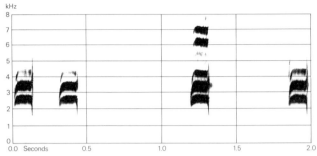

XIII C Chappuis Belgium December 1990

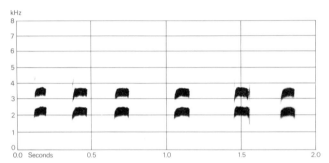

XIV M Schubert captive (Germany) May 1978

disturbed at nest (Bent 1968), and sweet, whistled 'twee' (though this perhaps same as call 3b: see below) or 'peet' from flying flocks (Witherby *et al.* 1938; Bent 1968). Perhaps some geographical variation. (4) Other calls. All reported only for nominate *leucoptera*. (4a) A 'keck' apparently unlike any call of *bifasciata*. (4b) A 'tu-tuee' ascending in pitch and with complex harmonic structure, apparently resembling call of several west Palearctic Carduelinae. (4c) Short trill comprising 4–5 hook-shaped units. (Mundinger 1979, which see for sonagrams.)

CALLS OF YOUNG. No information for nestlings. Food-call of nominate *leucoptera* fledglings 'chit chit' (Groth 1992b, which see for sonagram suggesting calls typically paired). MGW

Breeding. SEASON. Murmansk region (north-west Russia): eggs laid February to mid-May in years of good seed crop of spruce *Picea* and pine *Pinus*, with peak February–March; in years of poor larch *Larix* or spruce crop, delayed until mid-June to end of August when pine seeds available (Kokhanov and Gaev 1970). Finland: February–August, peak probably normally around June (Haartman 1969; Leinonen 1978; Pulliainen and Tuomainen 1978). North America: eggs recorded at least January–August; only nests when seed intake sufficient for both egg-formation and raising of brood (Bent 1968; Benkman 1990 and in press). SITE. In conifer (usually spruce), from low down to over 20 m above ground, almost always against trunk (Bent 1968; Hakala and Nyholm 1973; Pulliainen and Tuomainen 1978). In Murmansk region, usually in lower two-thirds of tree; 3 nests averaged 3·2 m (2–6)

above ground (Kokhanov and Gaev 1970). In Berlin (Germany), nest recorded at end of branch of false cypress *Chamaecyparis* (Fischer *et al.* 1992); one nest in Canada in maple *Acer* (Dale 1924). Nest: foundation of dead conifer twigs, stalks, grass stems, lichen, bark, etc., lined with fine grass, rootlets, stems, and moss, then inside layer of hair, fur, rootlets, moss, feathers, and plant down (Bent 1968; Kokhanov and Gaev 1970; Fischer *et al.* 1992); 3 nests in Finland had average outer diameter 10·5 cm (10·0–11·5), inner diameter 6·3 cm (6·0–7·0), overall height 7·5 cm (5·5–9·0), and depth of cup 2·8 cm (2·0–3·5) (Pulliainen and Tuomainen 1978, which see for details of structure). Building: by ♀ only, accompanied by ♂; takes 7–10 days (Kokhanov and Gaev 1970; Benkman in press). EGGS. See Plate 61. Sub-elliptical, smooth and slightly glossy; very like those of Crossbill *L. curvirostra*; pale whitish-blue to whitish-green, sparsely marked with specks, spots, and scrawls of purple or purplish-black, usually at broad end; sometimes undermarkings of pink or violet (Bent 1968; Harrison 1975). *L.l. bifasciata*: 21·3 × 15·1 mm (20·3–23·5 × 14·2–16·5), *n* = 55; calculated weight 2·54 g. Nominate *leucoptera*: 21·2 × 15·1 mm (18·5–23·5 × 13·5–17·0), *n* = 28; calculated weight 2·53 g. (Schönwetter 1984.) See also Kokhanov and Gaev (1970) for 10 eggs from Murmansk region. Clutch: 4 (3–5); in Murmansk region, 3 clutches were each of 4 (Kokhanov and Gaev 1970); 5 clutches in collection were 1 of 3 eggs, 3 of 4, 1 of 5; average 4·0 (Makatsch 1976). Eggs laid daily; probably 2 broods in good seed-crop years (Kokhanov and Gaev 1970; Makatsch 1976); in North America, 1–2 or more broods depending on spruce crop (Benkman in press); 2 broods in captivity (Massoth 1989). INCUBATION. 14–15 days; by ♀ only (Kokhanov and Gaev 1970; Massoth 1989). Possibly starts with 1st egg; hatching over 3–4 days (Kokhanov and Gaev 1970; Danilov *et al.* 1984); at one Finnish nest, 3 eggs hatched over 2 days (Pulliainen and Tuomainen 1978). In captivity, apparently started with 3rd egg (Massoth 1989). YOUNG. Fed and cared for by both parents and brooded by ♀ for *c.* 10–12 days (Kokhanov and Gaev 1970; Massoth 1989; Fischer *et al.* 1992). In northern USA, by ♂ only after ♀ starts building 2nd nest (Benkman 1989b). For development of young, see Pulliainen and Tuomainen (1978). FLEDGING TO MATURITY. Fledging period 22–24 days (Kokhanov and Gaev 1970); in 2 captive broods, 16 and 18 days (Massoth 1989). Only independent *c.* 4–6 weeks after fledging, when mandibles properly crossed, although every stage of development is extended (Kokhanov and Gaev 1970; Newton 1972). BREEDING SUCCESS. Little information; in Murmansk region, of 12 eggs, only 4 hatched and none fledged, and at one other nest all 4 young fledged; predators included Siberian Jay *Perisoreus infaustus* and squirrels *Sciurus*; some human persecution (Kokhanov and Gaev 1970). BH

Plumages. (*L. l. bifasciata*). ADULT MALE. Head and body

almost entirely rosy- or ruby-red, less scarlet than in adult ♂ Crossbill *L. curvirostra*, not as orange-red as 1st adult ♂ *L. curvirostra*, more similar to colour of ♂ Pine Grosbeak *Pinicola enucleator*, but brighter red, less dull and vinous; outer scapulars contrastingly dull black or brown-black, some dull black of feather-bases sometimes visible on forehead, hindneck, or inner scapulars; rump paler, bright pink-red; upper tail-coverts black with broad white fringes, latter partly tinged red. Feathering at lateral base of bill, lore, and stripe behind eye over upper ear-coverts dull black, stripe sometimes reduced to black spot behind eye and black patch at upper rear of ear-coverts. Patch below eye as well as upper chin and side of chin mottled off-white and grey. Lower mid-belly and vent pink-white or cream-white, under tail-coverts white, sometimes partly tinged pink, centres with dark grey mark (largely concealed). Tail dull black, inner webs and tips narrowly and sharply fringed white, outer webs narrowly fringed pink-buff or pink-white. Flight-feathers, greater upper primary coverts, and bastard wing dull black, tips narrowly but distinctly edged white (more clearly so than in fresh-plumaged *L. curvirostra*, but liable to abrasion in both species), outer webs narrowly fringed pink-buff or pink-white (palest on outer primaries). Tertials dull black, tip of longest with white fringe *c.* 2–4 mm wide (black of centre extending into point on centre), both others with tip entirely white for 5–9 mm. Greater upper wing-coverts dull black with broad and contrasting white tips, latter *c.* 4–6 mm long on outer web of outermost coverts, *c.* 9–13 mm on innermost coverts; median coverts similar, but bar formed by contrastingly white tips more even in width, *c.* 6–9 mm on all coverts. Lesser upper wing-coverts black, with pink-red or rosy fringes. Under wing-coverts and axillaries off-white, longest and marginal coverts with dark grey centres. *In worn plumage* (breeding season), red of plumage becomes more glossy and more orange, but still more rosy than in *L. curvirostra*; some black of feather-bases partly visible on underparts, some white of feather-bases shining through on rump and underparts (bases less grey than in *L. curvirostra*, rump and underparts appearing brighter and more fiery rose-red); white edges along tips of tail- and flight-feathers and white tips of tertials and wing-coverts partly worn off, but effect of abrasion generally negligible in adult, not as marked as in 1st adult. For distinction from *rubrifasciata* variant of *L. curvirostra* (with pink-white tips to coverts and tertials), see Recognition (below). ADULT FEMALE. Head and body as adult ♀ *L. curvirostra*, but dusky feather-centres from forehead to back and on side of head and neck slightly more elongate, black (on cap) or dark grey (on remainder), appearing spotted dusky when fresh and more streaky when worn; green-yellow feather-fringes slightly paler and more yellow, less green, sometimes partly washed light grey when fresh, giving hoary cast to (especially) forehead, ear-coverts, hindneck, and side of neck when plumage fresh. Upper tail-coverts dark grey with broad and often rather contrasting green-yellow fringes. Rump and underparts distinctly more yellow, rump in particular bright lemon-yellow, appearing pale and bright mainly because ground-colour whiter (in *L. curvirostra*, much more grey shows through); in contrast to *L. curvirostra*, many elongated dusky spots on chin and throat and more distinct dark olive-grey streaks on centre of feathers of chest, side of breast, and flank. Tail, flight-feathers, greater upper primary coverts, and bastard wing as adult ♂, but outer fringes pale green-yellow (on tail and secondaries) to yellow-white (on outer primaries), without pink; fringes of tail-feathers and along tips of flight-feathers and primary coverts far more distinct than in ♀ *L. curvirostra*. Remainder of upperwing greyish-black, lesser coverts fringed green-yellow, median and greater coverts as well as tertials broadly tipped yellow-white or pure white; width of

white on coverts scarcely less than in ♂, that on tertials often slightly less, 5–10 mm on shorter tertials, 1·5–3 mm on longest, more invaded by black along shaft. Under wing-coverts and axillaries grey with white tips. NESTLING. Down dark smoke-grey (Kokhanov and Gaev 1970). JUVENILE. Head and body as in juvenile *L. curvirostra*, upperparts and side of head and neck closely streaked dark brown-grey and off-white, tinged olive on lower mantle and scapulars; rump pale yellow, marked with ill-defined dusky streaks; upper tail-coverts black-brown, darker and more contrastingly fringed buff-white than in *L. curvirostra*; underparts off-white, streaked black-brown on chest, side of breast, flank, and centres of under tail-coverts, narrower and greyer on throat and belly. Tail and flight-feathers as in adult ♀, pale fringes slightly more distinct than in juvenile *L. curvirostra*, especially on tail. Tertials black, tips with broad yellow- or buff-white fringe, partly broken by black extending into point along shaft; fringe *c.* 3–7 mm long in ♂, 1–5 mm in ♀. Greater and median coverts with contrasting white tips, but these not as square-cut at base as in adult; width of white wing-bars of juvenile ♂ near those of adult, those of juvenile ♀ *c.* 3–8 mm wide only. When worn, white tips of tertials and outer greater coverts sometimes worn off, especially in ♀. FIRST ADULT. Highly variable, as in 1st adult *L. curvirostra*. Advanced birds similar to adult, but juvenile flight-feathers, tail, usually all tertials, and often at least outer greater upper wing-coverts retained. In more retarded birds, many juvenile wing-coverts retained, as well as scattered juvenile feathering on body (especially on neck and belly); in ♂, plumage mixture of red (like adult ♂), green-yellow (like adult ♀), and juvenile. Head and body of advanced 1st adult ♂ orange- or rosy-red, less bright and less deep rosy than in adult ♂; tertials and feather-bases on body browner, less blackish, bases more exposed, especially on hindneck, scapulars, and side of head and breast; rump bright light pink-red. Head and body of 1st adult ♀ greyer than in adult ♀, less extensively tinged yellow-green or buff-green rather than extensively green-yellow; feather-centres paler, greyer, less contrasting; rump patch narrower, less pure bright lemon-yellow. In 1st adults with inner greater coverts or a few tertials new, some contrast in abrasion and in extent and pattern of white between these and retained juvenile coverts (see Svensson 1992); when coverts and tertials all juvenile, white tips partly wear off, especially white on tertials and outer greater coverts of ♀ sometimes completely lost. Also, white fringes along tips of tail- and flight-feathers liable to wear off; often no clear white fringe along inner web of t6, in contrast to adult. Shape of tail-feathers not reliable for ageing. For difference from *rubrifasciata* variant of *L. curvirostra*, see Recognition (below).

**Bare parts.** ADULT, FIRST ADULT, JUVENILE. Iris brown or dark brown. Bill dark horn-grey to greyish-black, darkest on culmen and tip of upper mandible; cutting edges paler horn-grey or buff-grey. Leg and foot dark plumbeous-horn to greyish-black. (BMNH, ZMA.) NESTLING. Mouth bright purplish-red (Harrison 1975).

**Moults.** ADULT POST-BREEDING. Complete; primaries descendent. Timing erratic, depending on local breeding season; June–November (Pyle *et al.* 1987); mainly September–November (Svensson 1992). In birds examined, plumage worn in 2 from July (but moult just completed in single bird from captivity then); body strongly abraded but primary moult just started in one from 14 September (primary moult score *c.* 8) and in one from 28 September (primary score 24; inner 4 primaries and 2 central pairs of tail-feathers new, outer 4 primaries and 3 pairs of tail-feathers old); plumage fresh in 2 from October–November and in 3 birds from December (BMNH, RMNH, ZFMK, ZMA). In Murmansk area (Russia), 2 birds not yet in moult July, but 5 from October all in moult; completed in 4 birds from November (Kokhanov and Gaev 1970). Occasionally, limited pre-breeding moult, March–April (Pyle *et al.* 1987). POST-JUVENILE. Partial: head, body, and lesser as well as some to all median and greater upper wing-coverts; occasionally, a few tertials. Starts shortly after fledging, and timing thus as variable as breeding season. Moult June–November (Pyle *et al.* 1987). In north-west Russia, less than *c.* 15% of 35 birds examined September in moult, but 40–100% of samples from October–November (total *n* = 45) (Rymkevich 1990). In birds examined, plumage of some still largely juvenile in September, in others mainly 1st adult by then (BMNH, RMNH, ZFMK, ZMA).

**Measurements.** *L. l. bifasciata.* Northern Europe, all year; skins (BMNH, RMNH, ZMA). Bill (S) to skull, bill (N) to distal corner of nostril; exposed culmen on average 4·0 less than bill (S). Depth is depth of closed bill at base, width is width of lower mandible at lateral base of rami.

| | | | | | |
|---|---|---|---|---|---|
| WING | ♂ | 93·6 (5·53; 23) | 88–96 | ♀ 90·4 (1·85; 18) | 87–94 |
| TAIL | | 61·1 (2·01; 14) | 57–65 | 58·5 (2·04; 10) | 56–62 |
| BILL (S) | | 20·9 (1·05; 21) | 19·4–22·4 | 20·5 (1·13; 18) | 19·3–22·4 |
| BILL (N) | | 15·0 (0·80; 13) | 13·9–16·0 | 14·0 (0·88; 10) | 13·3–15·5 |
| DEPTH | | 10·3 (0·44; 18) | 9·7–11·0 | 10·1 (0·38; 14) | 9·4–10·7 |
| WIDTH | | 9·4 (0·50; 10) | 8·7–10·2 | 8·9 (0·45; 7) | 8·3–9·5 |
| TARSUS | | 16·4 (0·32; 10) | 15·8–16·8 | 16·2 (0·41; 9) | 15·8–16·8 |

Sex differences significant for wing, tail, and bill (N). Ages combined above, though juvenile wing and tail average 1·2–1·6 shorter than in adult; thus, in ♂, wing of adult 94·5 (1·55; 13) 91–96, of juvenile and 1st adult 92·9 (2·58; 9) 88–96.

Northern Eurasia; skins (Eck 1981).

| | | | | | |
|---|---|---|---|---|---|
| WING | ♂ | 92·6 (2·28; 21) | 88–97 | ♀ 88·9 (2·43; 15) | 84–92 |
| TAIL | | 62·8 (1·96; 20) | 60–66 | 61·1 (1·67; 15) | 58–65 |

Excluded from above sample are 3 ♂♂ with wing:tail ratio similar to *L. curvirostra*: wing 96, 96, 99, tail respectively 60, 61, 61 (Eck 1981). Range of wing length in total sample of *c.* 100 birds: ♂ 87–99, ♀ 85–94 (Niethammer 1937; Witherby *et al.* 1938; Kokhanov and Gaev 1970; Danilov *et al.* 1984; Svensson 1992).

Nominate *leucoptera.* Northern North America, winter; skins (RMNH, ZMA); exposed culmen on average 3·4 less than bill (S).

| | | | | | |
|---|---|---|---|---|---|
| WING | ♂ | 90·1 (1·68; 8) | 88–93 | ♀ 85·9 (1·28; 11) | 83–88 |
| TAIL | | 59·0 (2·15; 8) | 55–62 | 57·0 (2·50; 11) | 53–61 |
| BILL (S) | | 19·6 (1·26; 8) | 18·0–21·2 | 18·6 (0·60; 10) | 17·4–19·4 |
| BILL (N) | | 14·3 (0·84; 8) | 13·5–15·4 | 13·4 (0·76; 10) | 12·3–14·4 |
| DEPTH | | 8·4 (0·29; 8) | 7·8–8·8 | 8·2 (0·41; 10) | 7·6–9·0 |
| WIDTH | | 7·9 (0·21; 5) | 7·5–8·2 | 7·5 (0·32; 7) | 7·0–8·0 |
| TARSUS | | 15·8 (0·89; 4) | 15·5–16·2 | 15·9 (0·51; 7) | 15·1–16·5 |

Sex differences significant for wing, bill, and bill width.

**Weights.** *L. l. bifasciata.* (1) Russian Lapland (Semenov-Tyan-Shanski and Gilyazov 1991). (2) Murmansk area (Russia) (Kokhanov and Gaev 1970). (3) Yamal peninsula (Russia) (Danilov *et al.* 1984).

| | | | | | |
|---|---|---|---|---|---|
| (1) | ♂ | 34·0 (13) | 30–40 | ♀ 31·6 (5) | 29–34 |
| (2) | | 34·8 ( 8) | 29·7–37·9 | 30·7 (4) | 29·5–31·5 |
| (3) | | 30·4 (14) | 26·3–33·6 | 28·9 (6) | 24·9–35·3 |

Norway: December, ♀ 35; September–October, juveniles 28·7, 30·5 (Haftorn 1971). Eastern Germany, November: ♂ 35 (Wieczorek 1975).

Nominate *leucoptera.* North-central Alaska, July: ♂ 21 (Irving 1960).

**Structure.** Wing rather long, broad at base, tip bluntly pointed. 10 primaries: p8 longest, p9 0·5–4 shorter, p7 0·5–2, p6 3–6, p5 12–17, p4 18–22, p3 23–26, p2 26–30, p1 30–35 in *bifasciata*, 26–30 in nominate *leucoptera*; p10 reduced, 60–67 shorter than p8, 8–12 shorter than longest primary covert. Outer web of p6–p8 emarginated, inner web of p7–p9 with notch. Tip of longest tertial reaches tip of p1–p2. Tail of average length, wing:tail ratio 1·53 (42) 1·41–1·63, mainly (in 33 birds) 1·48–1·58; tip forked; 12 feathers, t1 7·1 (20) 4–10 shorter than longest (t5 or t6). Bill as *L. curvirostra*, but less deep and wide at base (especially in nominate *leucoptera*; see Measurements); tips of both mandibles more gradually curved, more elongated, finer. Upper mandible crosses to right in *c.* 25% of birds (in *L. curvirostra*, *c.* 50%): in 33% of 69 birds (Ticehurst 1910a), 27·8% of 784 birds (Benkman 1988c), or 27·4% of 322 birds (James *et al.* 1987). Middle toe with claw 16·0 (5) 15–16·4; outer toe with claw *c.* 75% of middle with claw, inner *c.* 69%, hind *c.* 90%.

**Geographical variation.** Rather slight; involves size (length of wing, tail, tarsus, or bill), relative depth of bill at base, and colour of feather-centres on head, body, and colour of wing and tail. Wing:tail ratio of all races similar (Eck 1981). No variation within Eurasia. Nominate *leucoptera* from North America smaller than *bifasciata* (see Measurements), bill much less deep and wide at base; feather-centres on head and body more extensively black, more exposed on lore, border of ear-coverts, mantle, scapulars, upper tail-coverts, and flank, these appearing blacker; ground-colour of upperwing, tail, and tertials blacker; pale fringes of tail and flight-feathers narrower or almost absent; white on tips of greater coverts and tertials narrower, 2–10 mm on greater coverts, 1–4 mm on tertials; red of head and body of ♂ slightly deeper, more rosy-crimson. *L. l. megaplaga*, restricted to Hispaniola (see, e.g., Kepler *et al.* 1975), has bill much heavier than in nominate *leucoptera* and slightly thicker than in *bifasciata*, but wing and tail shorter than in *bifasciata*, near nominate *leucoptera*.

**Recognition.** Due to great individual variation in bleaching and wear of plumage even within a sample on a single date, differences in colour of head and body often not reliable for distinction from *L. curvirostra*. White wing-bars and tertial-tips distinct and broad in adults of both sexes and usually in juvenile and 1st adult ♂ (tertial-tips sometimes largely worn off), unlike *L. curvirostra*, but bars and tertial-tips of juvenile and 1st adult ♀ of *L. leucoptera* narrower than in adult and ♂, closely similar in width to bars and tertial-tips of rare *rubrifasciata* variant of *L. curvirostra*, which shows contrasting pink-red, yellow-green, or white tips on median and greater coverts, usually also on tertials, and occasionally on longer lesser coverts. In 2 adults of *rubrifasciata* variant examined, tips of median coverts (pink-)white for 3–4 mm, greater coverts for 3–6 mm (more extensive on outer web than on inner), tertials for 1–4 mm (largely interrupted by black at shaft); extent thus close to that in juvenile and 1st adult ♀ *L. leucoptera bifasciata*. In juvenile *rubrifasciata* (and in 1st adults with retained juvenile wing), white tips restricted to rather narrow but sharply contrasting fringes, 0·5–3 mm wide on median and greater coverts as well as on tips of tertials (mainly 1·5 mm, *n* = 7), largely interrupted by black at shaft (not as even in width as shown in Svensson 1992), extent matched by only a few worn 1st adult ♀ *L. leucoptera*. Lower wing-bar even in width on all coverts (except when inner greater coverts are new and outer juvenile); in *L. leucoptera*, bar gradually wider towards inner coverts. Also, wing, bill depth, bill width, and (especially) tarsus of *L. leucoptera bifasciata* shorter than in *L. curvirostra*, and wing:tail ratio different: 1·53 (42) 1·41–1·63 in *L. leucoptera*,

1·70 (108) 1·61–1·80 in *L. curvirostra*. See also Eck (1981) and Svensson (1992). *Rubrifasciata* variant rare, e.g. only 4 among almost 15 000 *L. curvirostra* examined in north-east Germany (Klafs and Stübs 1987).

CSR

## *Loxia curvirostra* Crossbill

PLATES 43 and 45 (flight)
[between pages 640 and 641]

Du. Kruisbek  Fr. Bec-croisé des sapins  Ge. Fichtenkreuzschnabel
Ru. Клест-еловик  Sp. Piquituertocomún  Sw. Mindre Korsnäbb  N. Am. Red Crossbill

*Loxia Curvirostra* Linnaeus, 1758

Polytypic. Nominate *curvirostra* Linnaeus, 1758, Eurasia from Britain to Sea of Okhotsk, south to Cantabrian mountains and Pyrénées (northern Spain), Abruzzi mountains (central Italy), northern Balkan countries, and European Russia and Siberia at *c.* 55°N, grading into *balearica* in central and southern Spain, into *poliogyna* in southern Italy and Sicily, into *guillemardi* in southern Balkan countries and Greece, into *altaiensis* along northern fringe of Altai, Sayan mountains, and mountains south of Lake Baykal, and into *japonica* from Lake Baykal to Sea of Japan and perhaps Sakhalin and Kuril Islands; *balearica* Homeyer, 1862, Balearic Islands; *corsicana* Tschusi, 1912, Corsica; *poliogyna* Whitaker, 1898, North Africa; *guillemardi* Madarász, 1903, Crimea, Caucasus area, Turkey, and Cyprus. Extralimital: *tianschanica* Laubmann, 1927, Tien Shan, Pamirs, Karakorams, and (probably this race) Dzhungarskiy Alatau and Tarbagatay; *altaiensis* Sushkin, 1925, Altai, Tuva area, and Mongolia; *japonica* Ridgway, 1885, Japan; *himalayensis* Blyth, 1845, Himalayas through eastern Tibet to Kansu (China); *meridionalis* Robinson and Kloss, 1919, southern Annam (Vietnam); *luzoniensis* Ogilvie-Grant, 1894, Luzon (Philippines); *c.* 7 further races in North and Central America.

**Field characters.** 16·5 cm; wing-span 27–30·5 cm. Up to 15% larger than Greenfinch *Carduelis chloris* but with similar compact form; intermediate in size between Two-barred Crossbill *L. leucoptera* and Scottish Crossbill *L. scotica*; proportionately smaller-billed and smaller-headed than larger and heavier Parrot Crossbill *L. pytyopsittacus*. Beware sexual differences and racial variation in bulk; beware also adaptation of bill size to seed sources, south-east European race *guillemardi* approaching bulk of *L. pytyopsittacus*. Large, powerful, somewhat clumsy and noisy finch, with heavy, crossed bill and sharply forked tail; epitome of genus. ♂ basically orange to red, with dusky wings and tail; ♀ and immatures grey or olive, juveniles heavily streaked. Plumage shows few features except for paler rump and vent in adult. Flight strong and bounding, accompanied by distinctive bursts of loud disyllabic calls. Feeds and moves somewhat like small parrot. Sexes dissimilar; no seasonal variation. Juvenile separable. 5 races in west Palearctic, of which 3 isolated on islands or in restricted forest habitat (see Geographical Variation). Separation of races away from breeding locality impractical.

ADULT MALE. Moults: in Europe, July–November (complete). By end of 2nd year, head and body basically rosy-red with brightest pink tone on long rump and parts of face and flanks. Vividness of rump enhanced by contrast of dusky mottling on back and dusky-black wings and tail. At close range, variable grey or dusky wash and patches show on crown, ear-coverts, and breast and create soft mottling, sometimes producing paler cheeks and super-cilium. From mid-belly to under tail-coverts, ground-colour becomes almost white and (due to acrobatic feeding) dark grey spots on under tail-coverts can be visible. Paler fringes of wing and tail-feathers show rosy to grey tone but rarely catch eye. Quite exceptionally, a few have nar-row white tips on median and greater coverts and tertials; unlike *L. leucoptera*, these marks extend into even nar-rower and fading white fringes on greater coverts and tertials (see Plumages and Recognition of *L. leucocephala*, and Svensson 1992). When worn, head and body become even redder with noticeable gloss in sunlight, and wings and tail look virtually uniform dusky-black. Bill laterally compressed and massive, appearing thick at base, with elongated, much curved, sharply pointed and fully crossed tips to mandibles (note particularly that lower mandible curves upwards sufficiently to expose tip above decurved culmen, unlike *L. pytyopsittacus*); horn-brown (unlike *L. pytyopsittacus*, paler horn cutting edges usually incon-spicuous). Legs grey to brown. ADULT FEMALE. Head, back, and underparts basically greenish-grey but appear-ing olive or brown in some lights, relieved most by long bright yellow-green rump and dusky wings and tail. At close range, shows distinctly yellowish mottling on crown, paler supercilium and dark cheeks (on some), dark streaks on mantle and scapulars, dull grey-green fringes to wing-and tail-feathers, mottling or soft streaking on underbody, and dusky spots on whitish vent and under tail-coverts. Rarely, has whitish wing markings as can occur in ♂.

When worn, much brighter and greener, with bronzy-yellow tone on rump and belly and head and throat showing more spotting and streaking than mottling. JUVENILE. Distinctive, with typically much browner, or more olive ground-colour than ♀ and copious blackish streaks over head and entire body. Rump dull, with isabelline ground-colour obscured by streaks. On some, streaking much reduced with result that bird has puzzling pale-headed and pale-rumped appearance; on others, streaking much increased with result that bird appears noticeably blackish on head and body (see also Plumages). Beware shortness and often lack of visibly crossed mandibles in bill of juvenile, creating potential confusion with ♀♀ and juvenile rosefinches *Carpodacus*. FIRST-YEAR MALE. Highly variable, but most have plumage resembling brighter ♀♀ or are even yellower, with orange but rarely any pure red tones until 1st complete moult. See Plumages for details. FIRST-YEAR FEMALE. Resembles adult but retains dull, worn wing and tail.

Unmistakable as a *Loxia* if heard or seen well; crossed bill and parrot-like form and behaviour distinctive. *L. curvirostra* constitutes, however, minefield for field observer, due to (1) rare type (1 in 1000) with thin white wing-bar suggesting *L. leucoptera* and (2) racial variation in bill size and general bulk prompting confusion with *L. scotica* and *L. pytyopsittacus*. 1st confusion well understood (see above and *L. leucoptera*) but 2nd largely ignored, with no tangible evidence on incidence of racial overlap and little or no practicality in field study. Difficult not to assume that, since periodically numerous populations of nominate *curvirostra* regularly erupt west and south (as far as Greenland and northern Morocco), mixing of races (and of *L. curvirostra* with vagrant *L. pytyopsittacus*) could occur anywhere within range of the species and particularly in western Europe. Accordingly, separation of *L. curvirostra* from its 2 larger congeners highly problematic, with only bill shape (not just size), food preference, and voice affording trustworthy clues even to expert observers, and subspecific identification a hopeless quest. Flight fast and powerful, with bursts of strong wing-beats soon producing fast bounds and undulations over distance; usually flies high above trees in breeding habitat but may drop nearer to ground-level on migration. Flight silhouette bulky, with noticeably oval head and body, leaf-shaped wings, and rather short, cleft tail. Escape-flight markedly ascendant to thick canopy above disturbance or out of sight, bird then returning slowly. Gait on ground a clumsy hop, in trees more adept with hop, jump, sidling, and clambering all used in remarkably parrot-like fashion; even uses bill as grasping aid. Able to hang upside down when feeding and uses foot to hold cones while extracting seeds. Often comes to water, giving excellent opportunity for close study. Gregarious, forming urgent, bustling parties and flocks throughout year. Erupting birds make sudden appearances over all habitats but seek out conifers for rest and food.

Song variably described: loud but hesitant, even staccato phrase, interspersed by call, 'cheeree-cheree-cheuf-glipp-glipp-glipp-cheree...'; sweet, musical warble, somewhat like *C. chloris*, 'chip-chip-chip-jee-jee-jee-jee', with first notes trilled but last loud and creaking and again interspersed with 3-4 calls; high-pitched twittering, interspersed with 'tiwee-tiwee' notes and more nasal sounds, altogether thinner, faster, and more ethereal than *L. pytyopsittacus*; curious soft, at times almost inaudible phrases recalling Bullfinch *Pyrrhula pyrrhula* or Starling *Sturnus vulgaris*. Song accompanied by wing-shaking and flapping. Commonest call in flight a disyllabic, explosive 'chiip chiip', 'chip chip', 'jip jip', or 'glipp glipp', emphasized in alarm. When feeding, a quieter 'chük chük'.

**Habitat.** In west Palearctic occurs mainly in boreal and subarctic coniferous forests, but also well represented, often by distinct races, in temperate and Mediterranean insular and montane areas. Extralimitally in Asia ascends to 4500 m, extending to tropical pinewoods. (Voous 1960b; Harrison 1982). Ecologically divided largely into pine-dwelling populations in more southerly parts of range and spruce-based populations in more northerly, latter being more commonly subject to eruptive movements, during which they may remain to breed, temporarily or for longer, in hitherto unoccupied areas, especially where mature conifer plantations or shelterbelts have recently been developed (Newton 1972).

Apparently equally at home in deep dense forest and on edges or in open or detached stands, and readily occupies suitable mature conifers even where they grow in built-up areas, not excluding small towns, showing disregard of neighbouring human activities, and sometimes using artefacts such as overhead cables or drinking from rooftop water tanks. Resorts to ground only infrequently, being specialized in exploiting conifer seeds before these fall from tree; will switch to non-coniferous diet and accompanying habitat changes only when coniferous supplies fail, often forcing long-distance journeys to fresh habitat. Even where sedentary, given to fairly extensive flights, sometimes above lower airspace, apparently serving through frequent calling to make contact with conspecifics which have discovered other food sources.

Frequently visits water, and this must influence choice of habitat in areas where such sources may be widely scattered. Fond also of perching on topmost point of tall trees, especially as song-posts, but shows no apparent preference for upper as against lower levels within crowns, sometimes perching quite near ground (Bannerman 1953a). Unusual in that habitat is normally in stands of single species of tree, commonly spruce *Picea*, pine *Pinus*, or larch *Larix*, rather than mixed forest. Whether topography is lowland, level, hilly, or mountainous appears of little consequence (Dementiev and Gladkov 1954). In Switzerland, breeds less frequently in lowlands than at montane and subalpine levels, preferring spruce to pine

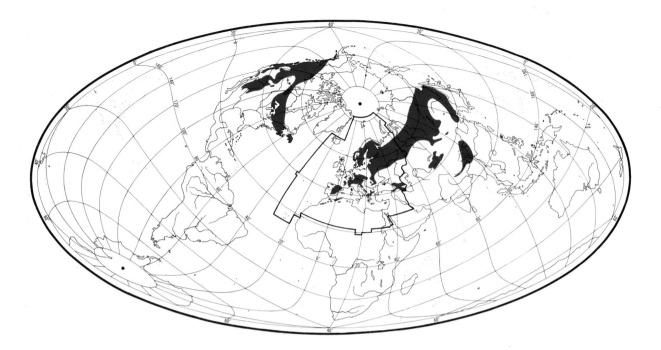

and larch, and favouring light open forest, but not forest edges or plantings in parks and gardens (Glutz von Blotzheim 1962).

In many parts of Britain, modern conifer plantings, windbreaks or clumps of conifers have provided much expanded habitat opportunities which are being somewhat erratically taken by this species, although not often by Scottish Crossbill *L. scotica* (Sharrock 1976).

Mostly found in northern or elevated habitats, but can breed successfully in any season even in coldest areas where trees can grow. Climate, therefore, seems to bear no direct part in habitat choice. Unique adaptations to harvesting conifer seeds also give it virtual immunity from competition, while its uninhibited mobility enables it to take advantage of new habitat opportunities, though extent to which these are taken is uneven.

**Distribution.** Breeding sporadic in much of west and south of west Palearctic, following irruptions. Maps show breeding areas only, where largely resident; for other wintering areas, see Movements.

BRITAIN. Breeding sporadic, following irruptions, except in a few areas where well established, especially East Anglia, New Forest, and Kielder Forest (Sharrock 1976). IRELAND. Scarce and irregular breeder, following irruptions (Hutchinson 1989). FRANCE. Occasionally breeds in Orléanais, Vendée, and Riviera, following irruptions (Yeatman 1976). SPAIN. Irregular and scarce breeder in north, nearly always following invasions (AN). PORTUGAL. Breeding range probably not permanent (RR). BELGIUM. Breeding population established after 1988-9

irruption (Devillers *et al.* 1988). NETHERLANDS. Occasional breeder until mid 1970s, then regular (SOVON 1987). NORWAY. Breeding distribution not fully known (confusion with Parrot Crossbill *L. pytyopsittacus*); probably breeds in all areas with spruce *Picea* woodland, predominating over *L. pytyopsittacus* in south-east (VR). ITALY. Sicily: breeding first confirmed 1981 (Iapichino and Massa 1989). YUGOSLAVIA. Regular breeder in native pine *Pinus* forest; occasional breeder in planted pine woods (VV). GREECE. Crete: first recorded breeding 1984 (Massa 1984). TURKEY. Probably more widespread breeder in forested areas of south coastlands than data indicate (RPM). ISRAEL. Occasionally breeds after irruptions, e.g. 1974, 1982-4 (Shirihai in press).

Accidental. Iceland, Malta.

**Population.** Fluctuates widely, depending on conifer seed crop.

BRITAIN, IRELAND. Estimated *c.* 5000 pairs 1968-72 (Sharrock 1976). Britain: fluctuates between irruptions; probably fewer than 1000 birds in low years, several thousand in good years (Marchant *et al.* 1990). FRANCE. 1000-10 000 pairs (Yeatman 1976). BELGIUM. No estimates, but low in 1979-81, high 1984-5 (Devillers *et al.* 1988). NETHERLANDS. 350-400 pairs 1973-7; at least 400-500 pairs in non-invasion years 1978-83; 700-1000 pairs in 1980 and 4000-5500 pairs 1984, both after irruption in previous year (SOVON 1987). GERMANY. Estimated 16 000 pairs (Rheinwald 1992). East Germany: 5000±2500 pairs (Nicolai 1993). SWEDEN. Marked fluctuations (LR); estimated 500 000 pairs in peak years (Ulfstrand and Högstedt

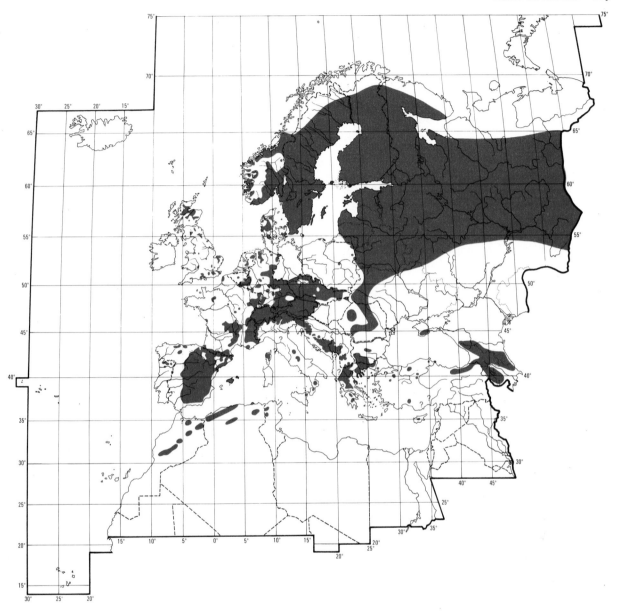

1976). FINLAND. Marked fluctuations (Hildén 1988). Long-term changes poorly known; rough estimate 100 000–200 000 pairs (Koskimies 1989).

Survival. Oldest ringed bird 7 years 1 month (Staav 1983).

**Movements.** Resident and dispersive, also irruptive.

In most years, birds disperse short distances in mid-summer to find new feeding areas, moving in flocks in various directions but remaining within regular range. Local numbers may therefore fluctuate greatly from year to year, dependent on varying state of conifer seed-crops, especially spruce *Picea*; timing of movement coincides with formation of new *Picea* cones. In irruption years (mostly

involving nominate *curvirostra*), birds move much further (up to 4000 km), mainly in one direction; such movements vary considerably in extent and duration, and tend to begin earlier and end later than in normal years. Irruptions probably result from high population levels coinciding with poor or moderate seed harvests; early departures suggest that crowding may sometimes alone stimulate movement, but further study required. (Newton 1972, which see for discussion.)

Earliest recorded invasion into Britain 1251, also 1593; in 1900s 16 major invasions into western Europe up to 1963, involving rapid spread over many areas. In some years migrants originate chiefly in Fenno-Scandia (heading between west and south), in other years chiefly in north-west Russia (heading between west and south-west). (Newton 1972.) Invasions sometimes extend south to Spain (e.g. 1930, 1963, 1972, 1983) (Muntaner *et al.* 1983; Schloss 1984; Senar and Borras 1985) or Malta, where highest numbers recorded July 1909 (over 200) and August–October 1930 (over 550) (Sultana and Gauci 1982). On Madeira (Atlantic islands) recorded December 1930 (1) and July 1943 (2) (Bannerman and Bannerman 1965). Several irruptions have reached Iceland, e.g. 1909, 1953, 1990 (G Pétursson), and 4 records in Greenland 1928–53 were probably also all nominate race (Hagen 1956). In north-west Africa (where *poliogyna* makes only local movements), presumed migrants reported 1855 (1), 1909 (1), and several in north-east Tunisia August 1930 to March 1931 (Heim de Balsac and Mayaud 1962). South European populations show little movement; in Bavarian Alps (southern Germany) birds apparently leave alpine woods entirely for short periods (Bezzel 1972), and some birds on passage at Col de Bretolet (western Switzerland) are probably local (Davis 1964); western Pyrenean birds only disperse locally (Boutet and Petit 1987; Génard and Lescourret 1987). Mediterranean races *balearica*, *corsica*, and *guillemardi* resident (Bannerman and Bannerman 1983; Thibault 1983; Flint and Stewart 1992).

Birds frequently stay to breed in invasion areas, reinforcing local populations or colonising new sites; these settlements usually temporary, but occasionally permanent, e.g. colony in East Anglia (eastern England) dates from 1909 invasion. Most birds move on after breeding; at least some return in direction of origin, and some young birds raised in invasion areas also head north-east, but passage inconspicuous then. So birds apparently make only one annual movement, as in non-invasion years; this supported by ringing data. (Weber 1971–2; Newton 1972; Schloss 1984.) Of birds ringed Col de Bretolet in 1959 and 1963 invasions, 16 of 17 recoveries within year of ringing were further south, in north-west Italy, France and Iberia (reaching west and south-east coasts), and all 4 recoveries in later years were in north-west Russia (Newton 1972). Birds ringed at Lago di Garda (northern Italy) in 1930 recovered mostly in southern France, west to Atlantic coast (Dejonghe 1984). Recovery of birds ringed at Serrahn (north-east Germany), 1962–8, shows headings in 3 directions: to Austrian and Italian Alps and adjacent north-west Yugoslavia, to western Switzerland and southern France, and to Low Countries (Weber 1971–2). Birds ringed elsewhere in Germany show similar pattern; foreign recoveries of German-ringed birds (including Serrahn) are from northern Italy (72), Austria (26), western Czechoslovakia (15), France, north to 47°45′N (11), Low Countries (10), Iberia (7, up to 2340 km south-west), Yugoslavia (3), Poland (2), Greece (1), Switzerland (1), Estonia (1), Sweden (1), and Finland (1); also 20 (including at least 1 bird raised in invasion area) in USSR, mostly at 56–62°N (up to 62°06′N and 62°21′E), with 16 at 2000 km or over (up to 3750 km) (Schloss 1984). Above data suggest that almost all birds on passage through Germany and Switzerland originate from USSR rather than Fenno-Scandia. Some Swedish-ringed birds, however, do pass through Germany; no recent records, but 21 recoveries in 1962–6 were from USSR (8), Italy (7), Austria (2), Germany (2) and Spain (2) (*Rep. Swedish Bird-Ringing*); perhaps some birds ringed Sweden had in fact previously emigrated there from USSR. 13 recoveries of birds ringed Kaliningrad region (western Russia) were from Italy (8), southeast France (2), Austria (2) and Poland (1) (Paevski 1971), and 10 recoveries of birds ringed Poland 1955–62 from Italy (5), southern France (3) and Spain (2) (Szczepski 1970, 1975). Only 4 recoveries involving Britain, showing movement to or from Norway (2), Italy (1) and eastern France (1); bird (presumably of Scandinavian provenance) ringed Fair Isle (Scotland) 6 September 1966 was back in Norway two weeks later; and bird ringed Fair Isle 6 July 1953 was recovered north-west Italy 25 August 1953, suggesting correction for displacement (Thomson and Spencer 1954; Spencer 1967; Mead and Clark 1987, 1991).

Invasions 1962 (summer and autumn) and 1963 (summer) well documented. In 1962, many headed west through Estonia late June to late July, suggesting this was major channel for summer movement; Fenno-Scandia was probably main source of autumn movement, with brief influxes reported end of July to August in southern Sweden, and more prolonged passage there and in southwest Finland in September. Both movements reached Netherlands and France, but only summer movement reached Switzerland. For eastern Germany, see below. In Britain, summer influx chiefly late June to mid-August; most early records were in southern England (Suffolk and Kent west to Devon), suggesting arrival via continental Europe, with Scottish records mainly from mid-July; autumn influx began mid-September, peaking October. Invasion also conspicuous from July in Ireland. (Williamson 1963*b*.) Many birds remained in Britain, with midwinter reports from 62 areas, chiefly fewer than 10 birds, but up to 500 present December–March in Gloucestershire (western England); diminishing records in March suggest some birds moved out of wintering areas then; breeding established or strongly suspected in 21 areas (17 counties), mostly March–May (Davis 1964).

In 1963, invasion into north-west Europe (peaking later than 1962) probably came mainly from Scandinavia, and invasion into south-west Europe mainly from central Europe; movements converged in France. Passage in southern Sweden reported from late June, chiefly late July to August, with daily totals up to 1410 birds at Falsterbo; in Low Countries and north-west Germany also, reported from June, with numbers peaking in August. In Britain (where numbers probably higher than 1962), first records of probable immigrants in 2nd week of July in northern Norfolk and Shetland, but main invasion not until August; reported along most of east coast in 1st week of August, with new arrivals and westward spread throughout month, reaching *c.* 96 areas; last new arrivals mid-September, with stragglers later in Scottish islands; few reached Ireland. In France, invasion from both north-east (presumably chiefly Scandinavian birds) and east (presumably chiefly central European birds); reported in north-east from end of June and further south from July; in north-east, many remained to feed in coniferous forests, partly checking progress south-west; probably both branches of movement reached Atlantic coast, however, where passage reported both northward and southward. In central Europe, marked movement June–July in eastern Germany (see below) and Czechoslovakia. At Col de Bretolet, heavy passage 2nd and 3rd week of July (probably exodus of Alpine breeders) and again in mid-August. Several records in north-east Spain south-west to Madrid. (Davis 1964; Erard 1964.)

In north-east Germany, birds dispersing from central European mountains sometimes occur in autumn in non-irruption years. Study in Serrahn 1962–8 showed passage in 2 phases in irruption years, mid-May to September (chiefly mid-June to late July) and mid-September to end of November; birds head chiefly south-west (between WSW and south-east). Many breed in invasion area January–February or even earlier. Spring return passage north or ENE observed from mid-March, of varying strength dependent on previous summer or autumn invasion, with juveniles accompanying adults; continues to May or even June, and sometimes overlaps mid-June with birds heading south-west in new invasion (e.g. in 1967). In 1962, summer passage 21 June to mid-September in 3 waves, with adults predominating June–July and juveniles from late July; continued strongly throughout August, in contrast to all previous years. Autumn passage peaked early October, conspicuous throughout October, lessening in November and ceasing at onset of severe weather in early December. In 1963, summer passage began 19 May and continued chiefly to end of July, with adults predominating at first and main arrival of juveniles from mid-June; from 10 August to mid-September only adults recorded; autumn passage weak. (Weber 1971–2.) 1985 irruption mostly involved extreme north of Europe. In early June, influx (probably from further east) reported in northern Finland. Widespread records in Orkney and Shetland (northern Scotland) from mid-June, chiefly to late July or early August. Further south in Britain, vanguard arrived 3rd week of June, and in late June to July small numbers reported at east coast observatories, and unusually large numbers widespread inland; some flocks were presumably local, however, following excellent breeding season. On Faeroes, *c.* 125 recorded 25 June to 10 July. (Taylor 1985, 1986; Booth *et al.* 1986; Dalziel *et al.* 1986.) Major influx into Iceland began 3rd week of June with sharp peak July to early August; 345 records, chiefly in east, south-west and north, with reports continuing mostly to early December (Pétursson and Ólafsson 1988; Pétursson *et al.* 1991).

1990 irruption intensive and far-reaching, and more prolonged than in 1985. Strong movement throughout Sweden July–August, with more than 1000 daily 16–20 August at Falsterbo in south; movement conspicuous also in Estonia and Latvia, but only weak to moderate in Finland, despite high breeding density (Hildén and Nikander 1991; *Br. Birds* 1991, **84**, 11). In Netherlands, reported from June, with peaks late July to early August, and 2nd half September to early October (Lensink and Hustings 1991). In France, invasion probably largest since 1968, with records from June to at least October. Flocks of 10–20 widespread in Hungary. (*Br. Birds* 1991, **84**, 235.) Movement began markedly early in Switzerland, from mid-May, birds becoming widespread both in mountain forest and at low altitude, and remaining numerous throughout winter, at least in west (Géroudet 1991). Also recorded in northern Germany from May, with numbers quickly building up (Weber 1990*a, b*). In Malta, up to 50 in early July; in Gibraltar (where only 4 previous records) 78 birds, 7 September to 13 November. On Faeroes, reported from 4 June, with numbers yet higher than 1985. (*Br. Birds* 1991, **84**, 11, 235.) Exceptional numbers (at least 800) also reached Iceland, with first reports both in south-east and north-west in 1st week of June, and main arrival June–July; small numbers remained throughout autumn and into 1991; records widespread, with most in south-east and south-west, but many also in north and west; one found on ship *c.* 114 km west of Iceland (G Pétursson). In Britain, invasion conspicuous especially in Scotland and northern England; hundreds arrived in Shetland and Orkney from late May and quickly moved on, heading south and west. Recorded in southern England from 2nd week of June, both on coast and inland, reaching exceptional numbers locally, e.g. in Thetford (Suffolk) from July. Arrivals in Scotland continued mostly to September (but recorded daily on Fair Isle 29 May to 25 October); survey suggested at least 500 000 present in Scotland during winter, possibly 1–5 million; southward movement apparently peaked November, with widespread records of large flocks, e.g. from Shropshire, Suffolk, and Derbyshire; in Kielder forest (Northumberland) *c.* 40 000 birds (including local population) mid-November to December, far higher than ever previously recorded. (Nightingale and Allsopp 1990; Harvey 1991; Jardine 1991, 1992*b*; Allsopp and Nightingale; Piotrowski 1991.)

Status similar in east of range, but few details. In Turkey (where breeds locally) occurs more widely during irruptive movements (Beaman *et al.* 1975). In Cyprus (where *guillemardi* resident), some records September–December may involve immigrants (Flint and Stewart 1992). Irruptions sometimes reach Israel, presumably from south-west Asia, e.g. several 10s of birds present from December 1971 to July 1974, and again from January 1981 to February 1986; both invasions resulted in breeding (Shirihai in press). Recorded very rarely in Iran and Afghanistan (S C Madge, D A Scott). In India (*himalayensis*), very erratic; recorded at 2700–4000 m at all seasons, exceptionally descending to 1500 m in winter (Ali and Ripley 1974). Siberian birds occur irregularly south of range (Griscom 1937). In Japan (*japonica*), winter visitor in some years to Hokkaido and northern Honshu, with occasional reports of breeding (Brazil 1991).

North American races also erratic and irregular in movements; often absent from parts of breeding area for several years, and frequently recorded outside normal range. In northern Maine, breeds abundantly in some years, but absent in other years. Newfoundland birds apparently migrate west and south-west in some years (perhaps via St Lawrence valley), reaching Illinois, Wisconsin, and Ontario, and occasionally move south along Atlantic seaboard to north-east states. (Griscom 1937.) Of birds ringed USA up to 1983, 5 recoveries over 200 km, of which furthest 2284 km from Massachusetts to Minnesota, and 1654 km from North Dakota to British Columbia (Payne 1987).                                          DFV

**Food.** Conifer seeds, generally spruce *Picea*, but in some parts of range (e.g. England and Mediterranean region) mostly pine *Pinus* (Mould 1974; Nethersole-Thompson 1975; Voous 1978; Massa 1987); also readily feeds on larch *Larix* (Newton 1967a; Fellenberg 1986). Very agile and acrobatic forager, easily fluttering from twig to twig, sidling along branches, hanging from cones, and clambering around, often using bill as help like parrot (Psittacidae); either works at cones, usually riper ones, extracting seeds *in situ* while hanging on cone, or snips them off (sometimes taking whole sections of twig) to carry to perch, often in fork, where held under foot and seeds removed. Often flies to perch carrying cone as heavy as bird itself; quite able to hold loose cone against underside of branch and extract seeds while upside-down; legs and toes are adapted for grasping and securing cones, as bill is for extracting seeds. (Robertson 1954; Newton 1967a; Mould 1974; Schubert 1977; Pfennig 1986.) If cone held in one foot, top points slightly to one side; orientation of cone, and foot used, depends on direction in which mandibles are crossed; if gripped in both feet, held alongside perch. Inserts bill-tips between scales of cone from side, moves lower mandible (which is more angled than upper) sideways, flat against top scale, causing tip of upper mandible to push bottom scale downwards, then scoops out seed with tongue; on thin-scaled cones, upper mandible can be used to hook seeds out. If seed is still fast, can open scales further by inserting closed bill and turning. (Newton 1967a; Benkman 1987b; see both these sources for details.) See Benkman (1988a) for experiments in foraging efficiency when crossed mandible tips removed (able to extract seeds only from open cones), and Benkman (1988d) for comparison with other Carduelinae when feeding on seeds other than conifers. For preferences in such seeds, see Ziswiler (1965). For methods on larch cones, see Delaveleye (1964), Pflumm (1978), and Pfennig (1986, which also see for treatment of beechnuts *Fagus*). Uses similar

technique to prise off bark when searching for invertebrates or cutting through apple *Malus* to reach seeds; picks up seeds on ground using tongue and upper mandible; detaches small apples like cones, and splits berries in base of bill; seen in captivity to collect aphids (Aphidoidea) by running leaf through bill to gather them together. In captivity, 46% of feeding positions were clinging to vertical stems, 28% hanging upside-down, 12% on bent stems, 12% normal perching, 2% stretching forward; in the wild, 99% of all observations on vegetation, almost all trees. (Newton 1967a.) In central Spain in winter, 96% of 115 feeding observations on cones, 2% branches, 2% on ground when clear of snow; most activity was above c. 16 m (Carrascal 1988). When foraging in larch, birds detached cones every 2–4 min; carried only 4% of cones to perch in flight (n = 96); most cones dealt with *in situ* were worked in less than 18 s, those detached in 18–48 s (Pflumm 1978). In another study, maximum c. 15 min spent on one cone; only 20–30% of seeds per cone eaten (Pfennig 1986). In Murmansk region (north-west Russia), took average 48% of seeds from each spruce cone, n = 25 (Kokhanov and Gaev 1970). When feeding on closed green pine cones in southern England in late summer, only removed seeds from top third of cone at first, but after 6 days opened 80–90% of scales (Mould 1974). In North America, feeding efficiency increased in flocks because good patches found earlier (Benkman 1988e). Away from conifers, readily feeds in broad-leaved trees, taking buds as well as fruits and insects, particularly caterpillars and aphids (Glutz von Blotzheim 1962; Newton 1972; Pfützner 1988); took rolled-up leaves of elm *Ulmus* to perch to get at aphids (Pfennig 1988). On migration, frequently feeds in atypical places, e.g. on herbs, on ground, even in seaweed, etc.; for accounts see (e.g.) Barraud (1956), Davis (1964), Erard (1964), Fellenberg (1986), Pfützner (1988). Birds pulled off closed heads of dandelions *Taraxacum* and took them to perch to extract seeds (Bub *et al.* 1959); fed for some time on grains in pellets of gulls (Laridae), although seed-bearing herbs nearby (Arnold and Ellis 1957); at garden peanut feeder, apparently preferred to pick fallen nuts from ground, finding it difficult to pull them through mesh (Khan 1987); in Orkney, during eruption, one bird flew to roof carrying slug (Pulmonata) which it held under foot and pulled to pieces; no suitable plants available in vicinity (Booth and Reynolds 1992). Often reported eating mortar, putty, ash, and similar nutrient-rich substances, and very frequently drinking (Nethersole-Thompson 1975; Fellenberg and Pfennig 1986; Sundin 1988).

Diet in west Palearctic includes the following. Invertebrates: bugs (Hemiptera: Lygaeidae, Aphididae, Adelgidae), Lepidoptera (Tortricidae larvae), flies (Diptera: Empididae), Hymenoptera (Pamphilidae larvae, Tenthredinidae larvae, Cynipidae), spiders (Araneae: Clubionidae), slugs (Pulmonata). Plant material: seeds, buds, shoots, etc., of juniper *Juniperus*, *Thuja*, fir *Abies*, larch

*Larix*, spruce *Picea*, Douglas fir *Pseudotsuga*, pine *Pinus*, poplar *Populus*, walnut *Juglans*, birch *Betula*, alder *Alnus*, hornbeam *Carpinus*, hazel *Corylus*, beech *Fagus*, oak *Quercus*, elm *Ulmus*, mulberry *Morus*, maple, etc. *Acer*, lime *Tilia*, ash *Fraxinus*, *Robinia*, sea buckthorn *Hippophae*, lilac *Syringa*, mistletoe *Viscum*, sorrel, etc. *Rumex*, orache *Atriplex*, yellow-horned poppy *Glaucium*, campion *Silene*, cabbage, etc. *Brassica*, apple *Malus*, plum, etc. *Prunus*, hawthorn *Crataegus*, rowan, etc. *Sorbus*, bramble *Rubus*, pea *Pisum*, buckthorn *Rhamnus*, evening primrose *Oenothera*, ivy *Hedera*, hogweed *Heracleum*, alexanders *Smyrnium*, heather *Calluna*, bilberry *Vaccinium*, crowberry *Empetrum*, thrift *Armeria*, tomato *Solanum*, plantain *Plantago*, elder *Sambucus*, guelder rose *Viburnum*, honeysuckle *Lonicera*, teasel *Dipsacus*, daisy *Bellis*, ragwort *Senecio*, burdock *Arctium*, thistles *Carduus*, *Cirsium*, *Onopordum*, knapweed, etc. *Centaurea*, goatsbeard *Tragopogon*, sow-thistle *Sonchus*, dandelion *Taraxacum*, hawkbit *Leontodon*, sunflower *Helianthus*, sedges (Cyperaceae, including cottongrass *Eriophorum*), grasses (Gramineae, including oats *Avena*, wheat *Triticum*). (Witherby *et al.* 1938; Baxter and Rintoul 1953; Westerfrölke 1958; Bub *et al.* 1959; Smith 1959; Turček 1961; Glutz von Blotzheim 1962; Upton 1962; Davis 1964; Erard 1964; Vauk 1964; Newton 1967a; Pulliainen 1972; Schubert 1977; Sabel 1983; Fellenberg 1986; Schmidt 1991; Seago 1991; Booth and Reynolds 1992.)

In Finnish Lapland, July–August, 806 items in 16 stomachs were 77% by number (60·3% by volume) new spruce seeds, 11% (11·9%) old seeds, 5·8% by number Hemiptera (5·7% Aphididae), 4·7% (27·8%) Hymenoptera (3·8% by number Tenthredinidae larvae), 0·4% by number Diptera, 0·4% by number spiders. In spring and summer, fed in mixed flocks with other *Loxia* on spruce cones in trees and on ground, and in July fed also on new cones. (Pulliainen 1972.) In Kareliya (north-west Russia), contents of total of 25 crops and stomachs as follows: in January, 100% spruce seeds; February, 97% spruce, 2·8% pine, 0·2% spruce flower buds; May–June, 100% pine seeds plus some Cynipidae cocoons; July, 34% Tortricidae larvae, remainder spruce and grass seeds; August–September, 100% spruce seeds. In years of good spruce and pine crop, takes seeds of both equally. (Neufeldt 1961.) In Finnish Lapland, however, pine seeds taken in early spring and spruce in late spring, probably because pine cones in year of study were 2 years old and so beginning to open (Pulliainen 1971, 1974). Similarly, apparent move by migrant birds in Britain from pine in 1st half of year to larch in 2nd half could be due to many pine cones being unripe in any one year while larch seeds are always available in autumn (Davis 1964), although birds will tackle hard green cones in England then (Southern 1945; Mould 1974). In Oxfordshire (southern England), August–April, migrants fed almost exclusively on larch (commonest conifer in study area), though number of observations in pine increased in spring (Newton 1967a).

Spruce drops seed around April–May, and new cones not formed until about late June, so opening pine cones during April to July are important food source (Newton 1970, which see for discussion of eruptions in relation to food supply). For general review, see Nethersole-Thompson (1975), and for North America, see Benkman (1987a).

In Murmansk region, winter and spring, average daily intake 2100 spruce seeds (5·9 g dry weight); consumed in average 117 min, or *c.* 18 seeds per min, excluding search time (Kokhanov and Gaev 1970). Captive birds from Pyrénées ate average 925 *Pinus uncinata* seeds per day in winter, 3000 in July, and 1550 in August; daily intake in wild, winter and spring, *c.* 5·0–9·5 g (Génard and Lescourret 1987, which see for differences between actual and theoretical consumption). In western Germany, March, took *c.* 4000–5000 larch seeds per day, ♀ average 26·4 per min, ♂ 22·8 (Pflumm 1978, 1984). For North American studies, see Benkman (1987b, 1990).

Young probably reared entirely on regurgitated conifer seeds, only those in nest at appropriate season perhaps receiving some small insects (Robertson 1954; Newton 1972; Nethersole-Thompson 1975). In Colorado (USA), mash fed to young just after hatching was of darker colour than later on, suggesting invertebrate component (Bailey *et al.* 1953). In Murmansk region, average daily intake at 1–12 days old *c.* 300 spruce seeds (*c.* 0·8 g dry weight), at 13–23 days *c.* 2400 seeds (*c.* 6·7 g); over nestling period, brood receives *c.* 90 000 seeds (252 g) (Kokhanov and Gaev 1970). In Netherlands, birds that remained to breed late in season, in mid-April, suffered severe food shortage because conifer seeds had already fallen, and attempted to feed young with buds, decayed wood, and bark (Bijlsma *et al.* 1988). BH

**Social pattern and behaviour.** Most aspects well known from studies of nominate *curvirostra*: in East Anglia (eastern England) by Robertson (1954), also reviews and comparison with other *Loxia* on Kola peninsula (north-west Russia) by Kokhanov and Gaev (1970), and in Britain by Nethersole-Thompson (1975). Isolated races of west Palearctic little known. Account includes data on North American birds, notably from useful study of *benti* in Colorado (USA) by Bailey *et al.* (1953).

1. Gregarious all year, though less so during breeding season when birds (breeders and non-breeders) nevertheless congregate to feed (Witherby *et al.* 1938; Göttgens 1989). Breeding-season flocks in Colorado typically loose-knit and constantly fluctuating; comprise (variously) ♂♂ whose mates incubating, pairs feeding nestlings, or birds tending fledglings (Bailey *et al.* 1953). After fledging, family parties amalgamate into flocks (e.g. up to 75 in southern Germany: Schubert 1977) which rove about in breeding area; juvenile-only flocks of 12–20 reported in Colorado (Bailey *et al.* 1953). Depending on food supply, part of or whole population may leave breeding area, sometimes travelling great distances in large flocks. (Newton 1970; Schubert 1977; Mal'chevski and Pukinski 1983; Lack 1986.) During irruptions into Europe, flocks sometimes of several hundred birds, but often (perhaps typically) less than 100 (e.g. Bub and Kumerloeve 1954, Smith 1959, Williamson 1963b, Davis 1964). In Cyprus, where resident, tens or sometimes 100–200 recorded June–September;

at lower altitude, up to 8, mainly December–March (Flint and Stewart 1992). Typically in small flocks most of year in Tien Shan (Kovshar' 1979). In North America, over 4 years, 50% of 148 flocks 1–4, median 3 (Benkman 1988e). In Kielder forest (northern England), large flocks found to comprise 'feeding units' of 2–10 birds (Elliott 1991). Associates for feeding (etc.) with other *Loxia*, sometimes with other Fringillidae and sparrows *Passer* (Barraud 1956; Nethersole-Thompson 1975; Frost 1985; Khan 1986; Benkman 1987a); in larch *Larix* forest of Alpes-Maritimes (France), winter, birds joined mixed flocks of tits *Parus*, but highly gregarious intraspecifically within these, and association closest otherwise with Siskin *Carduelis spinus* (Laurent 1986). In North America, when more than 1 size-class (race) of *L. curvirostra* appears in a given area, each apparently flocks separately (Benkman 1987a). BONDS. Mating system apparently essentially monogamous, though no definite study; no definite proof of bigamy, but observations (Netherlands and Switzerland) of ♂ sometimes associating with 2nd ♀ away from and also briefly at nest (Bauwens *et al.* 1976; Maurizio 1978). Hybridization reported with other *Loxia* (see accounts of other species); once with Greenfinch *C. chloris* in captivity (Newton 1972); in South Dakota (USA), one record of apparent hybrid (presumed wild) with Pine Siskin *C. pinus* (Tallman and Zusi 1984). In East Anglia, courtship and pair-formation take place in autumn for early breeders, more often late winter or early spring (Robertson 1954); noted also in Finland October–November (Excell *et al.* 1974); in Tien Shan, birds pair up from *c.* 2 weeks before nesting (Kovshar' 1979); may occur when nomadic flocks make brief stopover in area without breeding (Nothdurft *et al.* 1988). Little information on length of pair-bond. Probably severed at fledging at least in some single-brooded pairs, but apparently persists (not confirmed by ringing) when more than 1 breeding attempt (e.g. Nethersole-Thompson 1975 for southern Scotland). In Colorado, incubating ♀ apparently not visited by ♂, but later associated with ♂ and 2 fledglings; presumed to relate to double-brooded pair rather than ♀ losing mate and re-pairing (Bailey *et al.* 1953). No evidence for bond being maintained beyond 1 breeding season, and this presumably unlikely to occur, at least in highly nomadic populations. Under optimal conditions, given population may breed continuously for *c.* 9 months, but if cone crop poor, most or all do not breed (Newton 1970; Nethersole-Thompson 1975; Berthold and Gwinner 1978). Role of sexes in nest-duties as follows: ♀ builds nest (see part 2) and incubates (Naumann 1900; Witherby *et al.* 1938), though ♂ perhaps occasionally covers eggs briefly (see Nolte 1930 and other *Loxia* accounts). Nestlings fed by both sexes; initially, most food brought by ♂, ♀ then sharing feeding from *c.* 4–7 days (Robertson 1954; Ptushenko and Inozemtsev 1968; Bauwens *et al.* 1976), or (in Tien Shan) *c.* 10 days (Kovshar' 1979), or only over last few days (Christensen 1957). In northern Germany, ♀ apparently cared for young alone (Nolte 1930); not clear whether they fledged successfully. Post-fledging care protracted, presumably at least in part because young able to feed themselves effectively only from *c.* 45 days (Ternovski 1954), though all stages of development are extended (Newton 1972). According to Kokhanov and Gaev (1970), fledglings (also of Parrot Crossbill *L. pytyopsittacus* and Two-barred Crossbill *L. leucoptera*) are fed and cared for by both parents for *c.* 10–12 days near nest (after fledging at *c.* 22–24 days), then for further *c.* 33–35 days during dispersal, when brood typically divided between parents; brood-division also reported in East Anglia (Robertson 1954) and Mallorca (Munn 1931), while almost all 'families' in southern Germany comprised only ♂ and young (Schubert 1977), suggesting ♀♀ incubating 2nd clutches (A G Knox). In Colorado, fledglings tended for average 18–20 days, once fed at 33 days

(Bailey *et al.* 1953). Captive birds fully independent at *c.* 2 months (Mal'chevski and Pukinski 1983). Wild young recorded begging for food from parents (sometimes successfully) even when well able to feed themselves (Robertson 1954); in Wyoming (USA), food-begging (perhaps not always from own parents) noted at 1 year or even older (Peabody 1907). When 2nd breeding attempt made, ♂ takes over care of 1st-brood fledglings. In Scotland, 3 fledglings accompanied ♂ to 2nd nest where ♀ incubating; one juvenile came frequently and was fed by ♀ (see Relations within Family Group, below), another later fed 2nd-brood nestlings, though was attacked by ♂ (Nethersole-Thompson 1975). Exact age of first breeding not known, but probably physiologically capable of breeding within months or even weeks of birth, as adaptation to irregular occurrence of favourable breeding conditions (Berthold and Gwinner 1972); see also McCabe and McCabe (1933). In Alpes-Maritimes, July–August, both members of pair with young in nest were in juvenile plumage (Besson 1968). BREEDING DISPERSION. Solitary, or often in loose neighbourhood groups of 2–3 (Haftorn 1971; Newton 1972) or 3–12 pairs (Nethersole-Thompson 1975). Territorial, but apparently not markedly so (e.g. Pfennig 1986); see below. In St Petersburg region (north-west Russia), dispersion uneven as very much dependent on spruce *Picea* cone crop, also because birds tend to settle close to those already established, leading to locally high density; further relevant factors include relative abundance of other food types and breeding success in previous year (Mal'chevski and Pukinski 1983). Reports from various parts of range indicate nests (in some cases pairs or singing ♂♂) generally 50–100 m apart (Bailey *et al.* 1953; Christensen 1957; Nethersole-Thompson 1975; Schubert 1977; Kovshar' 1979; Mal'chevski and Pukinski 1983; Maestri *et al.* 1989); in southern Scotland, 2 nests separated by *c.* 100 m, but *c.* 300 m normal (Nethersole-Thompson 1975); see below for suggestion of nests less than 50 m apart. Pair-formation follows courtship which takes place within and outside apparently small territory finally established only after nest-site selected; feeding done mostly in flock outside territory (Newton 1972). Difficult to gauge degree of territoriality and exact size of territory; perhaps some regional variation, but reports anyway suggest often tolerant of conspecifics once nest established. Report from California (USA) of other singing ♂♂ tolerated within *c.* 20 m of one pair's nest (Payne 1972), and in Tien Shan *c.* 40–50 m (Kovshar' 1979). In East Anglia, ♀ recorded building in tree next to one holding nest with young; no friction, but few encounters (Nethersole-Thompson 1975). In Colorado, one observation of 3 ♂♂ fighting suggested territory extending *c.* 50–60 m from nest; 7 out of 16 territories had tree *c.* 60 m from nest used by ♂ as song-post and regular perch (Bailey *et al.* 1953). ♂ frequently sings otherwise from nest-tree or close by (e.g. Gilroy 1922, Payne 1972, McNair 1988). Several reports of birds (small flocks) being tolerated in territory, even feeding in nest-tree (e.g. Bailey *et al.* 1953, Nethersole-Thompson 1975, Maurizio 1978, Kovshar' 1979). In southern England, 2 ♀♀ (both nest-building nearby in adjacent trees) seen to be fed simultaneously by their respective mates when perched close together, no aggression resulting (Gosnell 1947). In Denmark, re-laid within *c.* 50–100 m of 1st nest (Christensen 1957); in southern England, after gale damage, birds re-nested close by in next tree (Gosnell 1932). In larger forests, shows preference for particular parts, this leading to local concentrations; density and population size also decisively influenced by size of cone crop, and hence fluctuate markedly from place to place and between years (Nothdurft *et al.* 1988). At Kuopio (southern Finland), number of pairs per 120-km transect over period 1927–37 was closely correlated with variation in spruce crop (Reinikainen 1937). As nests can be hard to find, easier to base census on counts of singing ♂♂ or behaviour indicative of breeding, and to complete census during 1st part of breeding cycle, including nest-building (Göttgens 1989). In Jämtland (Sweden), exceptional concentration of 10–12 nests in young spruce plantation of *c.* 1 ha (1000–1200 per km²) (Jonsson 1949). Locally up to 400 pairs per km² in same habitat in St Petersburg region (Mal'chevski and Pukinski 1983); in Harz (Germany), 3–8 (Nothdurft *et al.* 1988; Oelke 1992). Following large influx 1974–5, *c.* 100 pairs per km² in spruce-larch forests of southern Belgium (Collette and Fouarge 1978). In Netherlands, after 1983 influx, 0·1 pairs per km² in coastal belt to 4·9 (maximum 20·7) in Veluwe where greatest area of coniferous forest (Bijlsma *et al.* 1988). Densities (pairs per km²) in (predominantly) pine *Pinus* forest include 4 in Vosges du Nord, France (Muller 1987), 6–29 in Bulgaria (Simeonov 1975), 34 in Italian Alps (Maestri *et al.* 1989), 14–40 birds in Pyrénées in spring over 5 years (Génard and Lescourret 1987), 13–20, or 48 (based on sample plot) and probably 40–60 over 8–10 km², in mixed pine-spruce forest of northern Germany (Nothdurft *et al.* 1988; Göttgens 1989). In Finland, locally 5–20; favours open and broken forest and forest edges near bogs and clearings (Koskimies 1989). ROOSTING. Nocturnal and communal; usually in tall, dense conifers (Naumann 1900); 5–10(–50 or more) birds may use temporary roosts wherever feeding at the time (Ptushenko and Inozemtsev 1968; Newton 1972). In Devon (south-west England), October, flock of *c.* 300 roosted in tight clump of spruce, well concealed in upper canopy *c.* 5 km from pond used for drinking and bathing before flying to roost (Khan 1986); see Flock Behaviour (below). Fledglings may initially return to nest to roost (Bailey *et al.* 1953; Ptushenko and Inozemtsev 1968). For diurnal rhythm of migrants in eastern Germany, see Weber (1971–2). For roosting of captive North American birds, see Tordoff (1954).

2. Birds presumed to be from remote uninhabited regions are tame during invasions, permitting approach to *c.* 2–3 m, or in spring (return passage) *c.* 6–7 m; others shyer (Weber 1971–2), seldom allowing man closer than *c.* 25–50 m (Génsbøl 1964); see also *L. pytyopsittacus*. Bird typically unusually restless and excited prior to eruption (Newton 1972). Agitated captive bird flicked tail while giving Dyip-calls (Braun 1989: see 2a in Voice). Captive birds in North America remained motionless and gave 'tuck' calls for raptors, including those flying very high (Tordoff 1954). FLOCK BEHAVIOUR. In Kielder forest, ♂♂ in feeding flocks generally higher than ♀♀, and adult ♂ calling or singing quietly on highest perch apparently acts as sentinel (Elliott 1991). Birds often silent when feeding or give quiet Dyip-calls (e.g. Delaveleye 1964, Pfennig 1986); sometimes long series of Tep-calls (Pfennig 1986: see 2b in Voice), Subsong or similar, or true Social-singing (see 1a–b in Voice), which is highly infectious, attracting other conspecific birds, perhaps also *C. spinus* (Elliott 1991; also Bos *et al.* 1945, Pfennig 1986). Prior to take-off, some birds typically start giving more frequent and louder Dyip-calls, ever more joining in, until flock finally leaves, calling vociferously (Bailey *et al.* 1953; Pfennig 1986); may return to pick up stragglers (Delaveleye 1964). Calls, also song fragments, frequently given in flight and when landing (e.g. Grätz and Grätz 1985). Flocks in spring ('ceremonial gatherings') often noisy, with much calling and movement associated with courtship (including Courtship-feeding) and pair-formation, also threat (e.g. strange ♂ attempts to approach paired ♀) and chases; may continue thus for *c.* 30 min or more, flock then tending to break up as pairs leave (Gosnell 1932; Witherby *et al.* 1938; Robertson 1954). In Kielder forest, flocks reacted to attacks by Sparrowhawk *Accipiter nisus* and Merlin *Falco columbarius* by taking off and circling while giving harsh Dyip-calls of alarm (Elliott

1991). In Finland, flock mobbing perched Pygmy Owl *Glaucidium passerinum* similarly circled it and called anxiously (probably mixture of Dyip- and Tep-calls), dispersing when owl left (Nyström 1925). Birds at Devon roost called in flight and noisily while jostling for perches in topmost branches of trees near roost to which they gradually flew across in twos and threes and huddled together for night (Khan 1986). See also Tordoff (1954) for captive birds which were strongly imitative, e.g. for bathing and sunning. SONG-DISPLAY. Loud song-phrases and full song (see 1a in Voice) apparently given by ♂ only (Elliott 1991; A G Knox); ♀ also sings, but perhaps only Subsong or similar (see 1b–c in Voice), though both sexes participate in Social-singing (Bos *et al.* 1945; Elliott 1991) and ♀'s contribution to that not clarified. At least some reports of singing by ♀ probably refer to not uncommon green-plumaged ♂♂ (A G Knox). ♂ sings from top of tall tree, sometimes lower down (e.g. Gilroy 1922, Witherby *et al.* 1938), usually close to nest or ♀ and, especially if from low perch, normally quietly (Bailey *et al.* 1953; Mal'chevski and Pukinski 1983; Göttgens 1989). Frequently turns side to side (Naumann 1900), and sometimes shakes or flaps wings vigorously like singing Starling *Sturnus vulgaris* (Gilroy 1922). Often also sings in flight (Newton 1972; Mal'chevski and Pukinski 1983): in Kielder forest, Song-flights performed by pairs, ♀ (apparently not singing, but see below) taking off first from forest edge (where nesting) followed by ♂, birds then flying in wide circle and returning to take-off point; ♂, beating wings slowly (particularly close to landing) gave mellower Dyip-calls and loud song (Elliott 1991); descent fluttering and gliding (Laurent and Mouillard 1939). In Colorado, ♂'s song more continuous and complex when given in flight (Bailey *et al.* 1953). In Kostroma (Russia), mid-February, ♀ seen to sing while flying between 2 trees (Bubnov 1956); see also Heterosexual Behaviour (below). Will sing while actively migrating (Schubert 1977). Sings at any time of day; maximum rate in Tien Shan study 168 songs in 15 min (Kovshar' 1979); occasionally noted at night (Witherby *et al.* 1938). In Kielder forest, full song often given by paired ♂ between Song-flights, and some time after dawn (Elliott 1991). Song noted in all months of year, but peak associated with pair-formation and breeding—often autumn through winter to spring: e.g. St Petersburg region (Mal'chevski and Pukinski 1983), Finland (Suormala 1938; Nethersole-Thompson 1975), Netherlands (Bos *et al.* 1945), Belgium, where also generally linked to periods of high numbers (Schmitz 1989), Germany (Schubert 1977; Pfennig 1986), though in Harz most regular May–July (Nothdurft *et al.* 1988), Pyrénées (Génard and Lescourret 1987), and Cyprus (Flint and Stewart 1992). May decline with start of building (northern Germany: Göttgens 1989), or (in Tien Shan) incubation, though some ♂♂ sing while tending fledglings (Kovshar' 1979). ANTAGONISTIC BEHAVIOUR. (1) General. Occasional minor disputes with threat (see below) occur in feeding flocks outside breeding season, but serious fights rare then; more overt territorial behaviour associated rather with pairing and start of nesting (Munn 1931; Weber 1971–2; Schubert 1977). (2) Threat and fighting. In Kielder forest, other conspecific birds (especially ♂♂) attracted into tree where ♂♂ giving loud song; rarely led to aggression, but apparently dominant ♂ would give 'petulant' song-phrase and flutter up briefly as flock passed (Elliott 1991). Threat, fights (probably less common), and chases (all apparently involving only ♂♂) reported in association with boundary disputes (Bailey *et al.* 1953; Pfennig 1986), conflicts among captive birds over roost-site (Tordoff 1954), and intrusions during nest-site selection (Robertson 1954) and nest-building; in German study, however, pair tolerated ♀♀ and juveniles at building stage (Schubert 1977). ♂ gave excited Dyip-calls while (and before) chasing ♂ of intruding pair (Schubert 1977); similar calls noted from ♀ of different pair when her mate chased away another ♂ (Robertson 1954). In close confrontation, opponents face each other and threat-gape like Greenfinch *C. chloris*, and sometimes sway body or head side to side (Schubert 1977). In full Threat-posture (Fig A), captive

A

*benti* advanced with head lowered and neck extended while gaping; sometimes also gave high-pitched buzzing call, this usually causing subordinate to retreat, though pecking occasionally ensued (Tordoff 1954). Protagonists sometimes fly at each other, flutter up briefly while pecking, perhaps also (captive birds) striking with feet and wings so feathers fly, then descend locked together and separate on or just above ground (Bailey *et al.* 1953; Tordoff 1954; Schubert 1977). In northern Germany, Crested Tit *Parus cristatus* coming close to nest under construction was chased by ♂ (Göttgens 1989). HETEROSEXUAL BEHAVIOUR. (1) General. Breeding (and hence pair-formation, etc.) largely dependent on food supply and timing therefore varies (Naumann 1900); see Bonds (above) and Breeding. In *Loxia* generally, pair-formation prolonged, ♂ approaching ♀ in flock; aggressive elements strong, some displays mutual; courtship includes 'bill-scissoring' (Hinde 1955–6; Newton 1972). (2) Pair-bonding behaviour. In captive birds, ♂ quite often threatened ♀ in pre-breeding period, sometimes alternating this with feather-ruffling, also (when highly excited) drooping wings to expose rump (slightly ruffled), and often giving 'tup' calls (see 2a in Voice); sometimes hopped after ♀, flanks often ruffled, or swayed body while calling 'whee-oo' like Goldfinch *C. carduelis* (perhaps song) (Hinde 1955–6). In Tien Shan, from *c.* 2 weeks before nesting, ♂ seen to allow ♀ to extract seeds from cone on which he was feeding (Kovshar' 1979). Other descriptions are mainly of sexual chases, occurring when ♀ arrived after ♂ had been singing and giving 'zock' calls (see 2b in Voice), also flying restlessly between trees (Naumann 1900); starting when ♀ in Song-flight, pair flying twice round tree then away (Bubnov 1956), or taking birds through branches, often close to ground, and leading to Courtship-feeding (Bailey *et al.* 1953). In German study, ♂ twice closed on ♀ in flight, almost touching her; calling noted from both, brief song from ♂ (Röttler 1985*a*). (3) Courtship-feeding. Takes place sometimes in flock in spring, then regularly during nest-site selection, building, and incubation, on or close to nest (Gosnell 1932; Bailey *et al.* 1953; Maurizio 1978, which see for details). ♂ may call ♀ off nest to be fed (e.g. Ussher 1889), or ♀ summons mate to nest (Christensen 1957). Captive ♂ typically rather upright in approach, bending over ♀ who crouched and extended neck towards ♂, bill open; ♀ also turned head, shivered wings, usually raised tail and gave Begging-calls (see 5 in Voice); ♀ fed by regurgitation, ♂ starting to regurgitate before approaching ♀ or only after she has started to beg; either may ruffle plumage, and bill-touching and bill-scissoring also noted (Hinde 1955–6). On Kola peninsula, April, average (for 3 *Loxia*) 7 feeding visits by ♂ per day (Kokhanov and Gaev 1970). (4) Mating. Usually takes place during building, on nest or nearby branch (Hollom 1940; Bauwens *et al.* 1976; Kovshar' 1979; Göttgens 1989). In German study, ♂ in feeding flock started

giving Dyip-calls, flew to ♀ and body-swayed with wings slightly open, calling frequently, before mounting ♀ and copulating (Schubert 1977). ♂ seen to make hovering approach to (unresponsive) ♀; ♀ calls (see 6 in Voice) and adopts posture much as in other Fringillidae (Hinde 1955–6). One pair copulated twice within *c.* 30 min (Schubert 1977); in another study, 3 times within 8 hrs, at various times of day (Kovshar' 1979). (5) Nest-site selection. Both sexes involved, though ♀ apparently the more active, and probably makes final choice; closely examines potential sites, breaks off obstructive twigs, and tests crotches by crouching in them; pair sometimes creep about branches, perhaps searching together; ♂ may sing quietly near ♀ or loudly from tree-top (Bailey *et al.* 1953; Nethersole-Thompson 1975; Kovshar' 1979). Nest built by ♀, ♂ occasionally taking small part (e.g. Witherby *et al.* 1938, Bailey *et al.* 1953, Nethersole-Thompson 1975, Bauwens *et al.* 1976). ♀ collects material from nest-tree or up to several hundred metres away, and is then accompanied by ♂ who sometimes sings or calls quietly, also from nearby perch while ♀ building; ♀ may give quiet Dyip-calls (Gosnell 1932; Suormala 1938; Robertson 1954; Melchior 1975; Schubert 1977; Maurizio 1978). In Austria, ♀ reported to have built 6 nests in succession, using material from 1st (where disturbed) to build 2nd (Hanf 1887). Nest built over 3–5 days (Kovshar' 1979). (6) Behaviour at nest. Eggs laid straight after nest-completion (Kovshar' 1979), or within 2–5 days (Bailey *et al.* 1953; Mal'chevski and Pukinski 1983). ♂ often sings loudly from perch near nest when ♀ incubating (Nethersole-Thompson 1975), but contact-calls given at nest are quiet (Nothdurft 1972). ♂ accompanies ♀ back to nest after break, calling loudly, but ceasing once she has settled (Maurizio 1978); if ♀ forced to leave nest (e.g. when disturbed), ♂ will chase her back, then often perches nearby and sings (Christensen 1957). ♀ usually leaves nest to defecate; one report from Russia of her not doing so, faeces accumulating underneath it (Kokhanov and Gaev 1970). RELATIONS WITHIN FAMILY GROUP. Asynchronous hatching typical (Newton 1972): e.g. over *c.* 3 days (Ptushenko and Inozemtsev 1968). ♀ rarely leaves nest for first few days (Robertson 1954); brooding may continue during day for *c.* 10 days (Kovshar' 1979), ceasing even at night, despite cold, from *c.* 12 days (Mal'chevski and Pukinski 1983). If temperature low, young sometimes torpid when adult returns after long absence and have to be brooded before being fed (Ternovski 1954). Eyes of nestlings start to open from *c.* 5–8 days (Nolte 1930; Robertson 1954; Newton 1972); for typically slow physical development of young *Loxia*, see Kokhanov and Gaev (1970). In early stages, ♂ brings food and feeds ♀ who then regurgitates food for young (Witherby *et al.* 1938); within *c.* 2–3 days, ♀ may move aside for ♂ to feed nestlings directly (Robertson 1954). Young beg by moving head up and down, giving food-calls (Robertson 1954); also flap wings, being stimulated to beg by calls of adult, and later (from *c.* 13 days) when parent landed on nest (Maurizio 1978). Juvenile of earlier brood recorded begging at nest where ♀ feeding nestlings (Nethersole-Thompson 1975: see Fig B). In

B

Colorado study, both parents kept nest clean, swallowing nest-ling faeces or dropping them some distance from nest (Bailey *et al.* 1953). ♀ regularly loosens nest-lining (etc.), and European studies (e.g. Robertson 1954, Ptushenko and Inozemtsev 1968) suggested only ♀ removes faeces; those of small young contain some undigested seeds and may supplement ♀'s diet (Ternovski 1954). In later fledging phase, young defecate onto or over nest-rim and faeces no longer removed (Christensen 1957; Ptushenko and Inozemtsev 1968). Usually fledge at *c.* 20-25 days (16-28) (see Breeding); according to report in Witherby *et al.* (1938), still unable to fly at 24 days. When begging, fledgling flaps wings, spreads tail and calls (Delamain 1912). Fed regularly bill to bill (i.e. with seeds directly after extracting them from cone) rather than by regurgitation (Robertson 1954). ANTI-PREDATOR RESPONSES OF YOUNG. Remained silent in nest when squirrels *Sciurus* nearby (Maurizio 1978); fledglings similarly silent and motionless for any danger (Bailey *et al.* 1953). Left nest pre-maturely when it was touched (Nyström 1925). PARENTAL ANTI-PREDATOR STRATEGIES. (1) Passive measures. Generally rather quiet and secretive at nest, calling (loudly) only when *c.* 30-40 m away (e.g. Mal'chevski 1959, Nothdurft 1972, Maurizio 1978). ♀ typically a tight sitter, sometimes allowing herself to be touched (e.g. Ussher 1889, Bailey *et al.* 1953). When flushed, may drop down, then emerge on other side of cover (Naumann 1900); quick to return to nest (Excell *et al.* 1974). Especially when nesting near House Sparrow *P. domesticus*, which sometimes interferes with *L. curvirostra* nests, ♀ tends to stay on nest from 1st egg (Nethersole-Thompson 1975). (2) Active measures: against birds. Jay *Garrulus glandarius* near nest elicits alarm-calls (high-pitched screeching, presumably variant of Dyip-call), these sometimes attracting other ♂♂ who join in mobbing of predator (Christensen 1957); similarly for Magpie *Pica pica* near fledglings (frequent 'gip': Dyip- or Tep-calls) (Delamain 1912). (3) Active measures: against man. Includes, when nest or fledglings threatened, excited Tep-calls and song (perhaps also expressing alarm) from ♂ while ♀ on nest, ♀ or both birds (once small flock) coming close to intruder and calling (e.g. Nyström 1925, Bailey *et al.* 1953, Nethersole-Thompson 1975, Schubert 1977, Elliott 1991). In Wyoming (USA), ♀ recorded flitting about, tail-flicking and calling when man climbing up to nest (Peabody 1907). ♂ defending fledglings landed on man's arm (Gilroy 1922). ♀ on nest quite often pecks if attempt made to touch her (Ussher 1889; Kokhanov and Gaev 1970). (4) Active measures: against other animals. Squirrels (serious predator) and (e.g.) dogs are mobbed, birds giving frenzied alarm-calls (Ternovski 1954; Kovshar' 1979).

(Figs by D Nurney: A from drawing in Tordoff 1954; B from photograph in Nethersole-Thompson 1975.) MGW

**Voice.** Large and complex repertoire yet to be analysed and documented in full; further study also needed of dis-tinctions from other *Loxia* (see those species). Con-siderable variation (including tonal quality) within certain call categories, notably call 2a, which varies according to context and apparently also geographically (see below). A vocal species, calling frequently in flight and when perched (birds silent or typically give quieter calls when feeding). Generally quieter after pairing according to Suormala (1938), but family parties highly vocal (Berg-mann and Helb 1982). Calls 2a-b and some song (probably mainly Subsong or similar: see 1a-b, below) given by both sexes more or less all year, louder song-phrases only by ♂♂

and mainly before and during breeding season. (Elliott 1991.) Some (perhaps many) reports of singing by ♀ prob-ably relate to green-plumaged ♂♂ (A G Knox). Not known whether bill-snapping occurs such as reported for Scottish Crossbill *L. scotica*. Heavy, whirring wing-noise typically given by *Loxia* at take-off (Bergmann and Helb 1982; Elliott 1991), but no evidence of signal function. For additional sonagrams, see Nethersole-Thompson (1975) and Bergmann and Helb (1982); for description with musical notation, see Hoffmann (1925). All detailed information for west Palearctic relates to nominate *cur-virostra*, and study needed of isolated races *balearica*, *cor-sicana*, *poliogyna*, and *guillemardi*. Geographical variation in calls reported from North America: see (e.g.) Peabody (1907), Tordoff (1954), Benkman (1987a), and (especially) study in Appalachian mountains where consistent differ-ences found in all calls of 2 populations (Groth 1988). Calls of Nearctic birds not treated here in detail, but some are briefly described in Social Pattern and Behaviour.

CALLS OF ADULTS. (1a) Song. Highly variable and com-plex, sometimes difficult to discern where calls develop into song (Robertson 1954; Hinde 1955-6; Melchior 1975). Pleasing and attractive overall (Gilroy 1922). One component (given by both sexes, e.g. when feeding) is quiet, continuous, sweet warbling or twittering (see also Subsong, below). Loud notes or short phrases, carrying up to several hundred metres (Campbell 1973; Mal'chevski and Pukinski 1983), are typically given by ♂, separately, or combined with warbling in full songs (noted only from paired ♂♂) lasting up to several minutes, though usually less. (Naumann 1900; Witherby *et al.* 1938; Campbell 1973; Elliott 1991.) Song of this type, mixed with various calls, given for up to *c.* 1 hr by flock (Social-singing); reminiscent of chorus of Starling *Sturnus vulgaris*, Red-wing *Turdus iliacus*, or various Fringillidae (Gilroy 1922; Laurent and Mouillard 1939; Bos *et al.* 1945). Recording (Fig I) contains almost continuous twittering 'tyi ti ti ti. . .' overlaid with tones, buzzes, and purrs given by feeding flock; in Fig II, basically variants of Dyip-call (see below), first 3 units suggesting 'kitik', last 4 'kit kit p-teik p-teik'. Song-phrases often comprise 2 sub-phrases (with repe-tition prominent, especially in 2nd sub-phrase); other typ-ical features are contrast in pitch and volume and hence in timbre of notes and noise-types; sometimes also con-siderable difference in duration of 2 sub-phrases and units. (J Hall-Craggs, M G Wilson.) Great variety of song-phrases within this general pattern (A G Knox). Relatively unmusical, rather scratchy sub-phrase may be followed by another of clear, ringing, and musical notes often based on 'dyip' or similar (Elliott 1991, which see for renderings). Examples of song-phrases in recordings: explosive phrase comprising short introduction and 5 loud 'tcheet' sounds given by ♂ guarding mate in response to overflying con-specific birds (Fig III); sub-phrase of 2 'tru-t-chi', then 3 descending 'dreeia' motifs (Fig IV). (S Elliott, J Hall-Craggs). Resemblance to songs of other species quite often

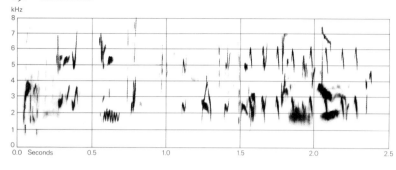

I  S Elliott  England  February–March 1990

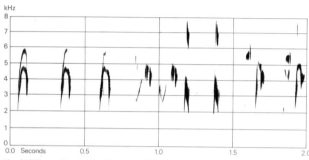

II  S Elliott  England  February–March 1990

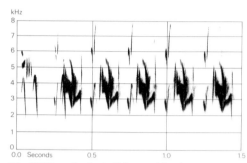

III  S Elliott  England  February–March 1990

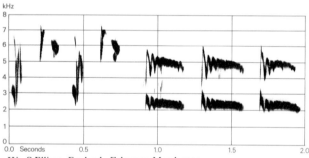

IV  S Elliott  England  February–March 1990

reported (perhaps in some cases mimicry), e.g. Woodlark *Lullula arborea*, Great Tit *Parus major*, and various Fringillidae; tremolo series of 'chip' calls followed by creaking 'jee' sounds often likened to Greenfinch *Carduelis chloris*; for descriptions of such phrases and other renderings, see Witherby *et al.* (1938), Robertson (1954), Hinde (1955–6), Schubert (1977), and Bergmann and Helb (1982). Recording (Fig V) contains song closely resembling mel-

lifluous song of *L. arborea*, with descending pitch and decreasing loudness; each sub-phrase begins with 'kit' and tones in song alternate with locatory 'clik' sounds. In Fig VI, brief introduction followed by sub-phrase strongly recalling *P. major*, upper notes being based on 'dyip' calls, and in Fig VII last 5 units suggest 'pink' and are perhaps copy of *P. major* or Chaffinch *Fringilla coelebs*. First sub-phrase in Fig VIII comprises 4 'p-t' units, 2nd a single motif suggesting 'cheedlee'; in another version of this song-type (Fig IX), 4 'dyip' calls precede 2 'chee-dl-eee' motifs; similar songs allegedly of *L. scotica* (see Nethersole-Thompson 1975) are more likely *L. curvirostra* (A G Knox); see also Parrot Crossbill *L. pytyopsittacus* for similar phrases. (J Hall-Craggs, M G Wilson.) Harder sounds reported in song include 'tup', 'turrup' (Hinde 1955–6), and 'tuk' like pre-roost call of Blackbird *T. merula* (perhaps mimicry) given by ♂ singing near sunset (Barraud 1956); rattles such as hard 'trt trt' (Bergmann and Helb 1982), including at end of song (Barraud 1956; Melchior 1975). For further descriptions, see (e.g.) Nys-

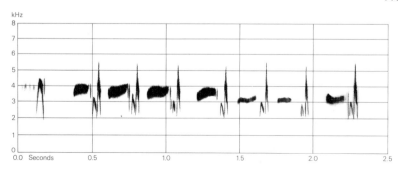

V  S Elliott  England  February–March 1990

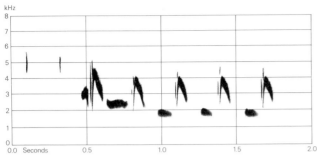

VI   S Elliott   England   February–March 1990

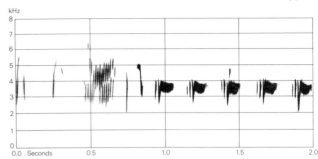

VII   S Elliott   England   February–March 1990

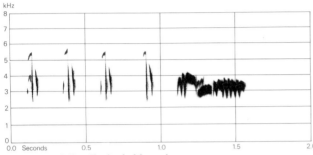

VIII   P J Sellar   England   May 1963

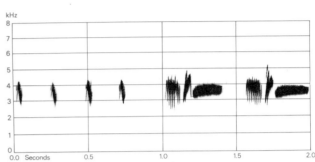

IX   J-C Roché   France   February 1968

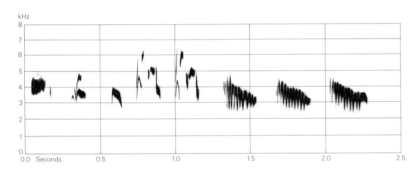

X   S Elliott   England   February–March 1990

tröm (1925), Christiansen (1935), Bos *et al.* (1945), Mal'chevski and Pukinski (1983), and Wüst (1986); for songs of various Nearctic races, see Borror (1961) and Bent (1968). (1b) Subsong. Subdued warbling, twittering, or murmuring like song of distant Linnet *C. cannabina*, intermingled with calls, sometimes 'trill' rising and falling in pitch (Witherby *et al.* 1938; Robertson 1954). Thus apparently much as described above (see 1a), though slightly louder, and includes occasional loud ringing calls or motifs such as 'treea treea', 'ktipktip', and 'tiktikoodee', but lacks boldness and clarity of song-phrases (Elliott 1991; S Elliott). Recording (Fig X) contains moderately loud Subsong with very loud, reiterated 'zeeerrr' calls (J Hall-Craggs, M G Wilson). (1c) Song of ♀. Quiet twittering or delicate 'trilling', less expressive than ♂'s song and apparently lacking its loud phrases (Naumann 1900; Bubnov 1956; Mal'chevski and Pukinski 1983; S Elliott). (2) Contact- and alarm-calls. (2a) Dyip-call. Loud, clear, incisive, metallic, and moderately high-pitched 'dyip',

'chip', 'jip', 'plipp', 'glipp', or 'klip' (Gilroy 1922; Nethersole-Thompson 1975; Bergmann and Helb 1982; Elmberg 1991; Svensson 1991), with wide range of variation in timbre, volume, etc., noted within flock and between different locations (Elliott 1991; A G Knox). Similar to rapid calls of *C. chloris*, but fuller and harsher, with peculiar almost 'explosive' emphasis (Naumann 1900; Witherby *et al.* 1938; Elmberg 1991). Given by both sexes singly or in series of varying length (calls often paired: A G Knox), widely and frequently in flight and when perched, serving for contact and also expressing excitement or alarm (especially for aerial predator, when typically harsh); intention to take off signalled by rapid series in crescendo, often after quiet bout of feeding (Naumann 1900; Elliott 1991). Recording (Fig XI) contains series of Dyip-calls; more tonal 'dyee' sounds lacking transients shown in Fig XII. In Fig XIII, quiet 'chi chi' followed by (and partly overlapping with) 2 quite penetrating tones, the whole perhaps representing warning from 1–2 birds for Spar-

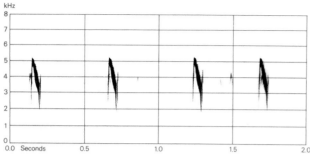

XI  S Elliott  England  February–March 1990

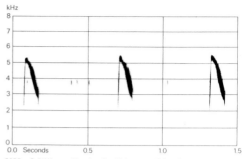

XII  S Elliott  England  February–March 1990

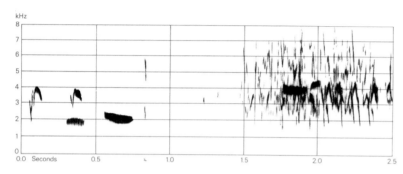

XIII  S Elliott  England  February–March 1990

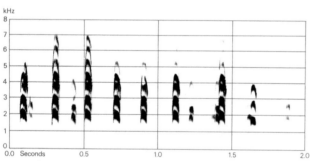

XIV  S Elliott  England  February–March 1990

rowhawk *Accipiter nisus*; after pause of 0·6 s, following attack by *A. nisus*, alarm expressed in overlapping commotion of Dyip-calls from several birds. (S Elliott, J Hall-Craggs.) Other renderings of alarm-calls ('ki', 'kip', 'köp', etc.) (Nyström 1925; Melchior 1975) probably combination of calls 2a and 2b. Further descriptions presumably at least to some extent reflect marked variation within call 2a and/or similarly relate to both 2a and 2b. Of 1430 birds released after ringing in German study, 54·8% called 'gip', 45·2% higher-pitched 'pitt' (Weber 1971–2). For (contradictory) reports of sexual difference in pitch of Dyip-call, see Witherby *et al.* (1938) and Robertson (1954). Quiet calls probably only variants of Dyip-call used for close contact include twittering 'peep-peep', softer and higher pitched than equivalent call of *L. scotica* (see that species), and 'tup' also noted from ♂ displaying to ♀ (Hinde 1955–6; Nethersole-Thompson 1975); see also Naumann (1900), Nÿstrom (1925), Bos *et al.* (1945), Frost Larsen and Aagaard Andersen (1965), Melchior (1975),

Bauwens *et al.* (1976), and Møller (1981*b*). (2b) Tep-call. Characteristically hard (harder than 2a), rich, wooden-sounding 'tep', 'tek', 'gip', 'kik', or rather low-pitched 'chiik-chiik' or 'chük-chük' given by both sexes, sometimes in long series; expresses varying degrees of anxiety (can sound harsh in alarm) and excitement, given (e.g.) in antagonistic interactions. Closely similar ('tooping') calls given by *L. scotica* and *L. pytyopsittacus*. (Nethersole-Thompson 1975; Bergmann and Helb 1982; Bruun *et al.* 1986; Elliott 1991; Jonsson 1992; A G Knox.) Recording reveals some variation: 'tep', inclining to 'tip' and 'typ', at rate of 5 calls per s in Fig XIV, audibly shorter 'jip' or 'jib' given at same rate in Fig XV (J Hall-Craggs, M G Wilson). Also rendered (e.g.) 'göp', 'köp', 'giöp', 'kiöp', and 'tjöpp-tjöpp' (Nyström 1925; Poulsen 1949; Göthel 1969; Melchior 1975); presumably also in this category (or 2c: see below) are low-pitched 'zock zock' (Naumann 1900) and deep, ringing 'choop' (Bradshaw 1991). (2c) A 'tjük' illustrated but not ascribed to context by Bergmann and Helb (1982) is clearly distinct from calls 2a–b; 'tyook' given (perhaps by ♀) in spring flock (Witherby *et al.* 1938) is perhaps the same. (2d) Distinct, rather high-pitched 'dick' sounds following 'tjük' series (calls 2c) shown in Fig d of Bergmann and Helb (1982); clearly differs from calls 2a–c, but no details of context. (3) Petulant 'gaarr' given during disputes in flock, apparently serving as threat-call (Elliott 1991). (4) Quiet, low-pitched 'trrrrr' noted from ♀ perched close to ♂ (Elliott 1991); 'tierr-türr-türr' (Bos *et al.* 1945) perhaps same or related. (5) Begging-calls of ♀. Commonest call in this context resembles 'chit-oo' of young (A G Knox); see below. A

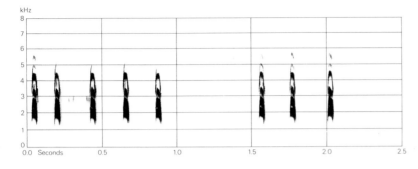

XV  S Elliott  England  February–March 1990

'zi zi zi zi zid' noted from captive bird (Hinde 1955–6). Sonagram of call given during courtship-feeding in Nethersole-Thompson (1975) lower pitched and shorter, but otherwise similar to conversational 'tup' (see call 2a, above) (J Hall-Craggs). (6) Soliciting-call of ♀. Quiet 'lil ill ill ill' (Hinde 1955–6). (7) Other calls. Descriptions in (e.g.) Hoffmann (1925), Reboussin (1931), Delaveleye (1964), and Bauwens *et al.* (1976) probably in most cases related to calls in above scheme.

CALLS OF YOUNG. Food-call of nestlings twittering (Naumann 1900), or sibilant chirp, developing into 'chic' (Witherby *et al.* 1938). Fledglings typically give variable series of 'pit', 'chit', 'sit', or 'see' sounds combined with slightly lower-pitched 'chu', 'cher', or 'too', showing some tendency to give the 2 types in pairs, e.g. 'chit-oo chitoo chit chit' or chattering 'sit-seecher sit-seecher sit-sit-seecher' (Witherby *et al.* 1938; Robertson 1953, 1954; Groth 1992, which see for sonagram of North American birds); see also (e.g.) Naumann (1900), Delamain (1912), Mal'chevski (1959), and Melchior (1975). Recording (Fig XVI) suggests 'pit-chu chu chu pit-chu' (J Hall-Craggs);

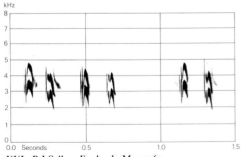

XVI  P J Sellar  England  May 1963

compare other *Loxia*. Soft 'kviik' (perhaps of distress) given when handled (Nyström 1925). No descriptions of juvenile song from west Palearctic; for North American birds, see (e.g.) Peabody (1907) and Bailey *et al.* (1953).

MGW

**Breeding.** SEASON. Stimulated to breed by abundance of food whenever it may occur (Newton 1972); for detailed discussion, see Benkman (1990). Britain and Ireland: in

southern Scotland, in spruce *Picea*, eggs laid August–April; in eastern England, in pine *Pinus*, December–June, peak February–April; in Ireland, *c.* 1 month later (Nethersole-Thompson 1975). Murmansk region (north-west Russia): in years of good spruce and pine seed crop, eggs laid February to mid-May; when only pine crop good, end of May to August, sometimes September (Kokhanov and Gaev 1970). Finland: eggs found mid-January to mid-May, rarely late summer, peak mid-March to late April; newly-fledged young seen November (Haartman 1969; Excell *et al.* 1974). For Denmark, see Christensen (1957). South-west Germany: most clutches laid by mid-March; some pairs nest-building early May when others had fledged young (Schubert 1977). For review of breeding season related to spruce seed abundance in north-central Germany, see Nothdurft *et al.* (1988). For Netherlands, see Bijlsma *et al.* (1988). Switzerland: recorded breeding in every month with apparent exception of September, peak December–May (Glutz von Blotzheim 1962; Maurizio 1978). Pyrénées: end of February to July, peak mid-April to mid-May (Génard and Lescourret 1987). Morocco: eggs laid November–June (Brosset 1957*b*). For Mallorca, see Munn (1931). For Cyprus, see Flint and Stewart (1992). SITE. High in conifer, usually standing isolated or at woodland edge, generally close to top of tree, covered from above by overhanging twigs; of 121 nests in Finland, 54% in pine and 46% in spruce; in spruce, almost always against trunk, in pine out on branch; average height above ground 7·1 m (0·8–18·0, *n*=82) (Haartman 1969; Excell *et al.* 1974). Average height in Czechoslovakia 25·1 m (11–35, *n*=7) (Hudec 1983). In northern Italy, at 2000 m, average height of nest 3·0 m, and of nest-tree 3·7 m, *n* = 10 (Maestri *et al.* 1989). For 100 nests in Britain and Ireland, see Nethersole-Thompson (1975). In (e.g.) Switzerland, also in larch *Larix* and fir *Abies* (Glutz von Blotzheim 1962). Nest: foundation of dead conifer twigs (length 27·5 cm recorded), strips of deciduous bark, moss, lichen, etc., lined with dry grass, decayed wood, plant down, hair, wool, and sometimes feathers (Robertson 1954; Kokhanov and Gaev 1970; Nethersole-Thompson 1975; Schubert 1977); in northern Italy, 6 nests had average outer diameter 12·7 cm (10·5–15·0), inner diameter 6·7 cm (5·8–8·0), overall height 7·3 cm (6·5–8·5), depth of cup 3·9 cm (3·5–4·5) (Maestri *et al.*

1989). Building: by ♀ only, usually accompanied by ♂; twigs, lichen, and moss gathered in trees, and grass (etc.) collected on ground and bundled in bill; takes *c.* 5 days (Kokhanov and Gaev 1970; Schubert 1977; Maurizio 1978); see Nethersole-Thompson (1975) for occasional reports of ♂ participating. EGGS. See Plate 61. Sub-elliptical, smooth and slightly glossy; creamy to bluish-white, very sparsely marked with dark purplish specks, spots, and short scrawls, concentrated at broad end, and violet-grey undermarkings (Harrison 1975; Makatsch 1976). Nominate *curvirostra*: 21·9 × 15·9 mm (19·4–24·4 × 14·4–18·0), *n* = 506; calculated weight 2·95 g. *L. c. balearica*: 20·9 × 15·9 mm (20·2–21·5 × 15·4–16·5), *n* = 15; calculated weight 2·75 g. *L. c. poliogyna*: 21·7 × 15·6 mm (20·0–24·0 × 14·6–16·9), *n* = 61; calculated weight 2·76 g. (Schönwetter 1984.) Clutch: 3–4 (2–5). In England, of 262 clutches: 2 eggs, 0·4%; 3, 32·4%; 4, 63·4%; 5, 3·8%; average 3·7 (Nethersole-Thompson 1975, which see for averages per month). In Denmark, of 31 clutches: 3 eggs, 39%; 4, 56%; 5, 5%; average 3·7 (Christensen 1957). Average in Netherlands also 3·7, *n* = 86 (Bijlsma *et al.* 1988), in Murmansk region 4·0, *n* = 8 (Kokhanov and Gaev 1970), and in Algeria 3·3, *n* = 78 (Heim de Balsac and Mayaud 1962). Eggs laid daily, usually immediately after nest completed (Glutz von Blotzheim 1962; Schubert 1977); in study in Colorado (USA), interval 4–5 days (Bailey *et al.* 1953). Replacement clutch laid 7–8 days after loss of 1st clutch in one study (Christensen 1957). Probably 2 broods over most of range in years of adequate food supply; records of 2 (possibly 3) broods in southern Scotland (Nethersole-Thompson 1975); in southern England, ♀ built 2nd nest 2–3 weeks after fledging of 1st brood but laid no eggs (Hollom 1940). INCUBATION. 14–15 (13–16) days, by ♀ only; often starts with 1st egg (Glutz von Blotzheim 1962; Kokhanov and Gaev 1970; Newton 1972), but also with 2nd, 3rd, or last (Bailey *et al.* 1953; Christensen 1957; Nethersole-Thompson 1975; Schubert 1977). ♀ at one nest took only 6 breaks of 1–13 min over 15 hrs (Newton 1972); at another, break of 28 min recorded towards end of incubation period (Maurizio 1978). YOUNG. Fed and cared for by both parents; brooded by ♀ for *c.* 7–12 days (Kokhanov and Gaev 1970; Makatsch 1976); ♀ often takes long breaks, e.g. young 3–5 days old left for 45 min in rain (Maurizio 1978); see also Robertson (1954) and Newton (1972, where recorded leaving young in −30°C). Care of 1st and 2nd broods can overlap (Newton 1972). FLEDGING TO MATURITY. Fledging period 20–25 (16–28) days (Christensen 1957; Kokhanov and Gaev 1970; Nethersole-Thompson 1975; Maurizio 1978); apparently 11–12 days at one nest in Algeria (Le Du 1935). Still fed by parents 3–6 weeks after fledging, probably until mandibles fully crossed, although every stage of development is extended (Bailey *et al.* 1957; Kokhanov and Gaev 1970; Newton 1972). Can breed in autumn a few months after fledging, though seemingly not in spring when hatched in autumn (Newton 1972); see also Berthold

and Gwinner (1972). BREEDING SUCCESS. In Netherlands in 1984, only 18% of 156 pairs raised fledged young, but in 1975–1980 79% of 52, giving respectively 2·6 and 3·2 fledged young per successful pair; poor success in 1984 due to delay in start of breeding until April, when warm weather caused conifer seed to fall resulting in food shortage and abandonment of eggs and young (Bijlsma *et al.* 1988). In northern Italy, average 3·3 per successful pair, *n* = 6, or 2·5 per pair overall, *n* = 8 (Maestri *et al.* 1989). In Murmansk region, 72% of 32 nestlings in 8 nests fledged, giving 2·9 fledged young per nest; Siberian Jay *Perisoreus infaustus* and squirrels *Sciurus* among predators (Kokhanov and Gaev 1970). In Colorado, 76% of 33 eggs hatched and 33% produced fledged young, resulting in only 1·0 per pair overall, *n* = 11; high winds had little effect but some nestlings froze because ♀ did not brood at night (Bailey *et al.* 1957). In southern Scotland, only 38% of eggs produced flying young; in eastern England, main predators various Corvidae and squirrels, but House Sparrows *Passer domesticus* pull nests apart, presumably for material, and even evict occupants; eggs not infrequently destroyed by frost (Robertson 1954; Nethersole-Thompson 1975). For Finland, see Excell *et al.* (1974). In south-west Germany, early nests apparently more successful than late ones because predators, especially squirrels and Jay *Garrulus glandarius*, less active then (Schubert 1977). Asynchronous hatching can mean last-hatched young often dies (Newton 1972). BH

**Plumages.** ADULT (nominate *curvirostra*). ADULT MALE. In fresh plumage (winter and spring), upperparts rosy-scarlet, appearing rather dark because dark brown-grey feather-centres show through or are partly visible; these centres darker, sepia or fuscous on outer scapulars; crown sometimes with small paler pink subterminal specks; feather-centres of rump and shorter upper tail-coverts paler, grey, more concealed, rump appearing brighter fiery-scarlet; longer upper tail-coverts brown-black with narrow rosy-red or pink-brown fringes. Nasal bristles, lore, patch round eye, and short streak behind eye grey-brown to dusky-grey, sharply demarcated from cap, less so from cheek and ear-coverts, bristles sometimes light isabelline-grey. Cheek and ear-coverts dark brown-grey, narrowly streaked or spotted rosy-scarlet, feathers below eye and shorter ear-coverts with paler grey shaft. Side of neck medium brown-grey with rosy-scarlet suffusion on feather-tips. Underparts rosy-scarlet, less dark than on cap, mantle, and scapulars, feather-centres paler, medium brown-grey, showing through less, but usually partly visible on chin, throat, and side of breast; mid-belly, vent, thigh, and under tail-coverts medium or light brown-grey or ash-grey, sometimes partly suffused rosy, under tail-coverts with dark grey centres (widest on longest coverts) and pale grey or off-white fringes. Tail greyish-black, outer webs narrowly fringed rosy-brown or olive-brown, inner webs faintly edged pale grey. Flight-feathers, tertials, bastard wing, and all upper wing-coverts dark fuscous-brown to brown-black, darkest on outer webs and tips of flight-feathers, greyer at basal inner borders of latter; outer webs of flight-feathers narrowly edged rosy-brown, soon bleaching to grey-brown on outer primaries, sometimes edged green on inner primaries; narrow fringes along outer webs of greater coverts and along tips of median coverts and broader fringes along lesser coverts rosy-brown, but these generally narrow and inconspicuous (more so than in 1st adult ♂), and virtually absent from primary coverts and tips of greater coverts. Under wing-coverts and axillaries dark grey-brown, paler on longer feathers, axillaries and tips of shorter coverts suffused rosy-brown. Frequently, some to (rarely) many yellow-green feathers mixed among scarlet of head and body, these mainly acquired by birds which start post-breeding moult early; 4–5% of ♂♂ entirely yellow (Weber 1971–2). *In worn plumage* (July–September), red of body more glossy, purer scarlet, but some brown-grey of feather-centres visible on cap and underparts and much sepia or fuscous on hindneck, lower mantle, scapulars, cheek, and ear-coverts; rosy-brown fringes of tail and primaries bleached to off-white, partly or (sometimes) fully worn off, those of tertials and most upper wing-coverts still brown but largely worn off. ADULT FEMALE. Nasal bristles pale isabelline-grey or ash-grey. Forehead dark ash-grey, merging into yellow-green with contrasting dark grey feather-centres on crown, this in turn to dark grey or brown-grey with slight green suffusion on nape and upper mantle; side of nape, lower nape, and upper mantle sometimes with broad but ill-defined pale greyish-isabelline streaks. Lower mantle, scapulars, and back marked with broad ill-defined dark grey and green streaks, feathers sometimes with sooty grey subterminal spot. Rump and shorter upper tail-coverts bright yellow-green, some duller grey or olive of feather-bases sometimes showing through; longer upper tail-coverts dark grey or olive-grey with ill-defined dull green fringes. Side of head and neck dark ash-grey or brown-grey, darkest round eye, on stripe behind eye, and on longer ear-coverts; sometimes a faintly paler supraloral stripe or short supercilium; mottled paler grey on lore, narrowly and faintly streaked paler grey or off-white below eye and on ear-coverts; cheek and ear-coverts sometimes suffused green, side of neck sometimes with broad but ill-defined isabelline-grey streaks. Chin and throat medium or pale ash-grey, feathers sometimes with small dusky subterminal specks or streaks. Chest, side of breast, flank, upper belly, and side of belly green or yellow-green, ash-grey of feather-centres sometimes partly visible, in particular on side of breast; mid-belly, vent, thigh, and under tail-coverts pale grey to off-white, sometimes with slight yellow-green suffusion; under tail-coverts with broad dusky grey centres and off-white fringes. Occasionally some red, ochre, or pink feathering on body (Phillips 1977). Tail and entire wing blackish-grey, slightly less dark than in adult ♂; tail-feathers with green outer fringe and faint white inner edge; primaries with narrow pale grey-green outer fringe; outer web of secondaries, tertials, primary coverts, and greater upper wing-coverts with faint ill-defined dark olive-grey fringe, scarcely paler than remainder of feather, tips of lesser and median coverts with ill-defined dull green fringes. Under wing-coverts and axillaries ash-grey to off-white, spotted dusky grey on centres of coverts; axillaries and shorter coverts sometimes suffused yellow-green. *In worn plumage* (July–September), green of head and body brighter, more bronzy-green or bronzy-yellow, brightest (sometimes almost golden-yellow) on rump and belly, cap and throat sometimes more clearly marked with dusky spots or streaks, ground-colour of body more brown-grey, less ash-grey; green fringes of tail and wing partly bleached to pale grey-green, partly worn off. NESTLING. Down very dark grey or almost black, restricted to dense tufts on upperparts, upperwing, and thigh, shorter on vent (Nolte 1930; Witherby *et al.* 1938). JUVENILE. Cap, hindneck, side of neck, and upper mantle closely marked with broad and contrasting black-brown and off-white streaks. Lower mantle, scapulars, and back less contrastingly streaked or spotted black-brown and olive-green, darkest on outer scapulars. Rump and shorter upper tail-coverts yellow-white or isabelline,

rather narrowly but sharply streaked black-brown, streaks sometimes partly bordered olive; longer upper tail-coverts black-brown with dull olive-grey fringes. Nasal bristles and ground-colour of side of head pale isabelling-grey to off-white; front part of lore, supercilium, eye-ring, patch below eye, and bar along rear of ear-coverts faintly speckled or streaked brown-grey, mainly pale; patch in front of and behind eye, ear-coverts, and cheek contrastingly darker, mainly dark brown-grey, finely streaked off-white on cheek and shorter ear-coverts only. Entire underparts off-white, tinged yellow-cream or isabelline when fresh, closely marked with contrasting black-brown streaks, widest on under tail-coverts, reduced in extent on vent. Much individual variation in width of dark streaks on head and body; in some birds, streaks short and narrow, virtually absent from rump, throat, and vent, these birds showing largely white head, neck, and rump, except for dark stripe through eye, short dark malar stripe, and sparse dark spotting elsewhere; in others, dark streaks broad, almost coalescent on top and side of head and neck, looking black-headed with heavy dark streaks on rump and underparts. No difference between sexes in density of streaking, contra Ticehurst (1915). In all birds, tail as in adult ♀ but ground-colour slightly greyer, less sooty, and feathers generally rather sharply pointed, less rounded; however, intermediate tail shape occurs in birds of all ages, and ageing by tail sometimes difficult, especially when worn; ageing by tail shape not useful according to Phillips (1977), but see Svensson (1992). Entire wing as adult ♂, but tertials more tapering to rounded tip, less broad and truncate than in adult, especially innermost; tips of tertials and median and greater upper wing-coverts narrowly fringed grey-green or yellowish, distinctly paler than remainder of feathers (in adult, tips uniform), soon bleaching to off-white on tips of greater upper wing-coverts and tertials, forming narrow pale wing-bar (absent in adult); however, much variation in width of tips: in some, clear green-white fringe of up to 2 mm wide, in others a trace of less than 0·5 mm only, even when fresh, soon wearing off. For rare *rubrifasciata* variant with broader pale tips to tertials and coverts (occurring at any age), see Recognition of Two-barred Crossbill *L. leucoptera*; also, Göthel (1969), Mauersberger (1976), Berthold and Gwinner (1978), Scherner (1979), Berg and Blankert (1980), and Berthold and Schlenker (1982). FIRST ADULT MALE. Head and body highly variable, depending on timing of moult (see Weber 1971-2). In general, birds fledged in late winter or early spring replace streaked juvenile feathering of head and body slowly and gradually by more uniform yellow-green (like adult ♀) or golden-yellow, sometimes with orange tinge but usually without red; new wing-coverts and (if any) tertials of these birds as in adult ♀; appearance completely yellowish except for dark eye-stripe and (partly) juvenile tail and wing by May–August; no red plumage acquired until 1st complete moult when *c.* 1·5 years old. Later-fledged birds grow yellow-green or golden-yellow feathers on head and body and ♀-type coverts in early stage of moult, rosy-red or orange-scarlet feathers and ♂-type coverts later on (from about June–September), resulting in highly variable appearance with piebald mixture of streaked, green-yellow, and red feathering on head and body. Birds starting post-juvenile moult as late as June–September moult directly from streaked juvenile into red adult plumage, and these have head, body, and new wing-coverts similar to adult from about November onwards, though scarlet of head and body often somewhat paler and more rosy; more brown-grey of feather-centres visible, especially on hindneck and upper mantle. In all birds, juvenile tail, flight-feathers, greater upper primary coverts, often outer greater coverts and tertials, and sometimes outer median coverts retained, fringes of tail and flight-feathers green (contrasting strongly in colour with new feathers in red birds, unlike adults), outer greater coverts and tertials with at least traces of off-white fringes along tips (unlike ♀). Shortly before 1st complete post-breeding moult, when 1–1·5 years old, colour of head and body of red birds brighter orange-scarlet, of others golden-yellow with orange tinge or mixture of yellow and scarlet; juvenile tertials, greater coverts, primary coverts, tail, and flight-feathers heavily worn, and juvenile characters hard to establish; some birds which started moult early have new yellow mixed within older red of body, like some adults at same time of year, and at same time birds of next generation occur with new flight-feathers and green, yellow, red, or mixed body colours, adding to confusion, especially during late summer and early autumn invasions, which contain mainly birds of up to 1·5 years old and relatively few older birds. FIRST ADULT FEMALE. Like adult ♀, but juvenile tail, flight-feathers, greater primary coverts, and variable number of greater upper wing-coverts and tertials retained, as in 1st adult ♂; tips of tail-feathers more sharply pointed than in adult ♀, tertials narrower and less truncate, greater coverts with pale fringe along tip; when older than c. 9 months, tips of primaries and primary coverts more heavily worn than in adult at same time of year. Some birds (both sexes, proportion varying between years) replace a number of inner, central, or outer primaries during post-juvenile moult, sometimes all, but only exceptionally some primary coverts or secondaries, these birds showing contrast in colour of fringes and abrasion between feathers (a few adults have similar contrast). SECOND ADULT. Birds with a row of 4–6 old secondaries (more worn and with greyer fringes than neighbouring feathers) are probably 3rd calendar year (as in Lapwing *Vanellus vanellus* and Golden Plover *Pluvialis apricaria*); older birds frequently retain some old flight-feathers too, but these generally more scattered throughout wing and less worn.

**Bare parts.** ADULT, FIRST ADULT. Iris brown. Bill horn-brown, dark horn-grey, or dull lead-grey, darkest on culmen and tips, cutting edges yellow-horn, paler grey-horn, or light brown, widest on distal half, but often inconspicuous, though this depends on light and on position of bill. Leg and foot grey-brown, horn-brown, or burnt umber-brown with flesh or purple tinge. (BMNH, RMNH, ZMA.) NESTLING. Bare skin almost black. Mouth, including tongue, deep red; gape-flanges yellow-pink; bill flesh-red. JUVENILE. At fledging, bill light horn with yellow- or horn-white cutting edges and tip; mouth still deep red, but flanges paler, inconspicuous. Leg and foot dull grey-flesh. (Ticehurst 1910b; Nolte 1930; Heinroth and Heinroth 1931). For development, see Nolte (1930).

**Moults.** ADULT POST-BREEDING. Complete or largely so; primaries descendent; occasionally 1–2(–5) old inner primaries retained, more frequently a variable number of secondaries unmoulted, e.g. c. 50% of sample of 31 birds from Belgium retained 1–6 secondaries, and 60% of 10 *poliogyna* from North Africa retained 3·2 (1–6) secondaries (Herremans 1982, 1988; BMNH, ZMA). In central Europe, gonads enlarged, birds breeding, and no moult occurring late December to mid-February, gonads reduced and birds generally in moult September–November; gonads intermediate and either breeding or moult may occur mid-February–August (Berthold and Gwinner 1978), but not simultaneously (Tordoff and Dawson 1965); however, occasionally in moult during incubation or when young in nest (Hartert 1903–10). In central and northern Europe, starts with p1 and scattered feathers of body mid-July to late August, birds arriving in bad condition during invasion sometimes not until late September; completed with outer primaries late Sep-

tember to late November (based on data of c. 55 moulting birds: Ticehurst 1915; Williamson 1957; Phillips 1977; Herremans 1988; BMNH, RMNH, ZMA), but sometimes largely complete by 24 September (Herremans 1982). In eastern Germany, moult late July or August to September(–November); duration of moult c. 75 days (Weber 1971–2) or c. 85 days (Ginn and Melville 1983). In north-west Russia, starts late June to late September, mainly August, completed from September onwards (Rymkevich 1990). In Montana (USA), starts 1st half of August (Kemper 1959). In ♂, feathers grown before August are yellow (body) or yellow-fringed (tail, flight-feathers, wing-coverts), those from August onwards red or red-fringed; difference depends on diet, red pigment due either to more insect food being eaten from August onwards (Weber 1971–2) or to availability of fresh seeds with different composition in new pine cones (Ginn and Melville 1983). In captivity, no red obtained without special diet, unless cages very large (Völker 1957). Some birds start moult immediately after mid-winter breeding season, suspending moult with 2–6 inner primaries and scattered number of body feathers new when nesting again April–May, resuming moult from August onwards (Herremans 1988; BMNH, RMNH, ZMA); this better considered an early post-breeding moult rather than separate partial pre-breeding, contra Tordoff (1952). In populations of Mediterranean basin, where newly fledged young occur October–June rather than March–August as further north, skins examined indicate moult starts late April to early June (occasionally up to late July), completed mid-August to early October (based on c. 40 birds in moult: Massa 1987; BMNH, RMNH, ZMA). POST-JUVENILE. Partial, but extent highly variable. Some birds retain part of juvenile feathering of head and body in 1st year of life and even breed in this plumage; others retain part of juvenile secondaries and primary coverts only. In most birds, involves head, neck, lesser and median upper wing-coverts, and variable number of greater coverts and tertials. Starts shortly after fledging, chest, side of breast, scapulars, mantle, throat, and lesser coverts first. In birds of mid-winter nesting season, moult starts (January–)February at slow rate; as in adult ♂, new feathers of 1st adult ♂ acquired before August mainly yellow, early-hatched birds acquiring fully yellow head and body from May onwards, later birds growing red feathers amidst yellow from June–August onwards, acquiring chequered plumage; birds fledged May–June replace juvenile feathering mainly directly by red, acquiring fully red head and body from September onwards. (Weber 1971–2; RMNH, ZMA.) Occurrence of partial flight-feather moult apparently highly variable between years; in general, only rarely recorded (Phillips 1977; Herremans 1982; Winkler and Jenni 1987); in sample of c. 70 birds from 1979–80 invasion in Belgium, none had flight-feathers new, c. 4% t1, c. 12% 1–2(–3) tertials, c. 75% retained 4–9 juvenile outer greater coverts, a few retained outer median coverts (Herremans 1982); in sample of c. 170 birds from 1983–4 invasions, c. 15% had some or all primaries new, these moulted descendently from p1 (in 8 birds), p4 (in 4), p5 (in 4), or p6 (in 7) outward, less often from other feathers, moult frequently suspended before p9 reached (thus, e.g., only p6–p8 new); also, 20% replaced some tail-feathers (4% all tail), 57% 1–3 tertials, and most birds 3–9 greater coverts (Herremans 1988, which see for details). In sample of 34 birds from various invasions in Netherlands, 85% had 2·6 (1–3) tertials old, 100% had 5·4 (1–9) greater coverts old, 26% also some median coverts; in 6%, t1 new (RMNH, ZMA). FIRST POST-BREEDING. Like adult post-breeding, starting at c. 1–1·5 years old.

**Measurements.** ADULT, FIRST ADULT. Nominate *curvirostra*. Northern, western, and central Europe, all year; skins (RMNH,

ZMA). Juvenile wing and tail include retained juvenile wing and tail of 1st adult. Bill (S) to skull, bill (N) to distal corner of nostril; exposed culmen on average 3·9 less than bill (S). Depth is depth of closed bill at feathering, width is width of lower mandible at basal feathering of mid-rami.

| | | | | |
|---|---|---|---|---|
| WING AD | ♂ 99·5 (1·41; 20) | 97–102 | ♀ 96·0 (1·60; 17) | 94–99 |
| JUV | 97·9 (2·35; 67) | 93–104 | 94·9 (1·69; 37) | 91–98 |
| TAIL AD | 58·4 (1·75; 11) | 57–62 | 57·2 (1·63; 14) | 55–60 |
| JUV | 57·6 (1·59; 25) | 54–60 | 56·4 (1·43; 16) | 53–59 |
| BILL (S) | 23·0 (1·04; 42) | 21·0–24·5 | 22·4 (0·85; 28) | 21·0–24·0 |
| BILL (N) | 15·9 (0·68; 42) | 14·5–17·0 | 15·7 (0·71; 28) | 14·5–17·0 |
| DEPTH | 11·2 (0·38; 40) | 10·5–11·8 | 11·0 (0·33; 27) | 10·5–11·7 |
| WIDTH | 10·6 (0·43; 41) | 9·9–11·3 | 10·6 (0·44; 28) | 9·7–11·3 |
| TARSUS | 18·0 (0·55; 37) | 17·1–19·2 | 18·0 (0·58; 26) | 17·2–19·0 |

Sex differences significant for wing, juvenile tail, and bill (S).

Wing. (1) South-west Scotland, live local birds (Marquiss 1980, which see for other measurements. Northumberland (England), live birds: (2) adult, (3) 1st adult (D C Jardine; one ♂ 92, one ♀ 107, excluded from range). Belgium, live birds of 1983–4 invasions: (4) adult, (5) 1st adult (Herremans 1988, which see for other measurements; see Herremans 1982 for 1979–80 invasion). (6) Live invasion birds Helgoland (Germany), ages combined (Drost 1930; see also Vauk 1964). (7) Eastern Bayern (Germany), breeding (Niethammer 1937).

| | | | | |
|---|---|---|---|---|
| (1) | ♂ 96·9 (2·5 ; 16) | 93–101 | ♀ 95·2 (2·1 ; 16) | 92–99 |
| (2) | 100·7 (2·96; 15) | 98–106 | 98·4 (1·83; 12) | 95–102 |
| (3) | 99·9 (2·43; 50) | 94–104 | 97·4 (2·81; 43) | 89–101 |
| (4) | 99·9 (2·14; 16) | 95–104 | 97·1 (1·35; 15) | 95–100 |
| (5) | 99·0 (1·99; 72) | 94–105 | 96·4 (1·78; 99) | 91–101 |
| (6) | 98·7 (— ; 70) | 94–105 | 94·8 (— ; 34) | 91–98 |
| (7) | 95·6 (— ; 13) | 91–99 | 94·0 (— ; 11) | 92–97 |

See also Dementiev and Gladkov (1954), Hagen (1956), Davis (1964), Haftorn (1971), Weber (1971–2), Knox (1976), Eck (1981), Massa (1987), Semenov-Tyan-Shanski and Gilyazov (1991), and Svensson (1992).

Eastern Pyrénées (local breeders), June–July, and northern Portugal (probably breeding nearby), October–December; skins (BMNH).

| | | | | |
|---|---|---|---|---|
| WING | ♂ 97·7 (2·74; 9) | 94–102 | ♀ 96·9 (1·88; 6) | 94–98 |
| BILL (S) | 23·5 (0·97; 9) | 22·1–25·2 | 23·2 (0·86; 6) | 21·9–24·5 |
| BILL (N) | 16·3 (0·79; 9) | 15·3–17·4 | 16·2 (0·66; 6) | 15·8–17·5 |
| DEPTH | 11·2 (0·39; 9) | 10·7–11·6 | 11·0 (0·37; 6) | 10·5–11·5 |
| WIDTH | 10·7 (0·34; 9) | 10·2–11·3 | 10·8 (0·51; 5) | 10·4–11·4 |

Sex differences significant for wing.

Intermediates between nominate *curvirostra* and *balearica*, south-east Spain (Cuenca to Granada), June–October; skins (BMNH).

| | | | | |
|---|---|---|---|---|
| WING | ♂ 97·8 (2·34; 8) | 95–101 | ♀ 95·8 (1·50; 4) | 94–98 |
| BILL (S) | 22·3 (1·21; 8) | 21·2–24·2 | 21·4 (— ; 2) | 21·0–21·8 |
| BILL (N) | 14·8 (0·57; 8) | 14·1–15·6 | 14·0 (0·21; 3) | 13·8–14·2 |
| DEPTH | 10·8 (0·13; 8) | 10·5–10·9 | 10·9 (0·12; 3) | 10·8–11·0 |
| WIDTH | 10·3 (0·34; 8) | 10·2–10·5 | 10·8 (— ; 2) | 10·4–11·2 |

Sex differences not significant.

Italy: (1) north (nominate *curvirostra*); (2) Calabria (south), (3) Sicily (both intermediate between nominate *curvirostra* and *poliogyna*) (Massa 1987). Bill is exposed culmen.

| | | | | |
|---|---|---|---|---|
| WING (1) | ♂ 96·3 (2·28; 25) | 90–98 | ♀ 92·1 (3·07; 18) | 86–97 |
| (2) | 96·7 (2·23; 19) | 92–100 | 94·4 (1·61; 4) | 92–96 |
| (3) | 95·8 (2·00; 8) | 92–97 | 92·1 (1·50; 4) | 90–93 |
| BILL (1) | 19·7 (0·89; 25) | 18·3–21·0 | 19·1 (1·00; 18) | 17·6–20·8 |
| (2) | 19·5 (1·20; 19) | 17·6–22·6 | 19·9 (0·89; 4) | 18·8–21·0 |
| (3) | 19·8 (1·10; 8) | 18·5–21·0 | 20·0 (0·80; 4) | 19·2–20·4 |
| DEPTH (1) | 11·5 (0·37; 25) | 10·7–12·1 | 11·3 (0·64; 18) | 10·0–12·2 |
| (2) | 11·6 (0·61; 19) | 10·5–12·6 | 11·2 (0·51; 4) | 10·7–11·9 |

(3) 11·7 (0·60; 8) 11·2–12·1 11·2 (0·50; 4) 10·7–11·9

*L. c. balearica.* Mallorca (Balearic Islands), January–July: skins (BMNH).

| | | | | |
|---|---|---|---|---|
| WING | ♂ 94·4 (2·33; 12) | 91–98 | ♀ 92·1 (1·47; 9) | 89–94 |
| BILL (S) | 21·7 (0·92; 12) | 20·8–23·2 | 21·1 (0·58; 8) | 20·5–22·2 |
| BILL (N) | 14·8 (0·63; 12) | 14·1–16·0 | 14·3 (0·72; 8) | 13·2–15·1 |
| DEPTH | 11·2 (0·38; 12) | 10·6–11·7 | 10·7 (0·42; 8) | 10·2–11·4 |
| WIDTH | 10·6 (0·27; 12) | 10·3–11·1 | 10·2 (0·32; 9) | 9·7–10·7 |

Sex differences significant for wing, bill depth, and bill width. See also Massa (1987).

*L. c. corsicana.* Corsica, November–April, skins: wing, ♂ 99·8 (0·87; 4) 99–101, ♀ 95·2 (0·32; 4) 94–97; 3 ♂♂ and 3 ♀♀ combined, bill (S) 23·1 (1·39; 6) 21·8–24·9, bill (N) 16·1 (0·68; 6) 15·1–17·1, depth 11·5 (0·34; 6) 11·0–12·0, width 11·0 (0·32; 6) 10·5–11·4 (BMNH, ZMA). See also Massa (1987).

*L. c. poliogyna.* Atlas Saharien (Algeria), all year; skins (BMNH, RMNH, ZMA).

| | | | | |
|---|---|---|---|---|
| WING | ♂ 99·1 (2·38; 12) | 95–103 | ♀ 96·2 (2·30; 16) | 93–101 |
| TAIL | 58·2 (2·33; 12) | 54–62 | 54·6 (1·58; 13) | 51–57 |
| BILL (S) | 23·0 (1·38; 10) | 21·3–25·2 | 22·9 (1·12; 16) | 21·2–25·1 |
| BILL (N) | 15·8 (0·87; 9) | 14·7–17·0 | 15·5 (0·57; 16) | 14·9–17·0 |
| DEPTH | 11·6 (0·53; 10) | 11·0–12·5 | 11·1 (0·35; 14) | 10·6–11·9 |
| WIDTH | 10·6 (0·31; 11) | 10·2–11·3 | 10·6 (0·45; 15) | 10·0–11·5 |
| TARSUS | 18·6 (0·32; 12) | 18·0–19·1 | 18·5 (0·55; 14) | 17·8–19·4 |

Sex differences significant for wing, tail, and bill depth. See also Ticehurst and Whistler (1938), Knox (1976), and Massa (1987).

*L. c. guillemardi.* Cyprus, April–September; skins (BMNH, ZMA).

| | | | | |
|---|---|---|---|---|
| WING | ♂ 99·0 (2·03; 12) | 95–102 | ♀ 96·5 (2·65; 6) | 92–99 |
| BILL (S) | 24·1 (1·13; 11) | 22·6–26·3 | 23·7 (1·43; 5) | 22·3–25·5 |
| BILL (N) | 16·4 (0·82; 11) | 15·4–17·8 | 16·6 (1·01; 5) | 15·4–17·8 |
| DEPTH | 11·8 (0·21; 10) | 11·5–12·2 | 11·9 (0·42; 5) | 11·4–12·5 |
| WIDTH | 11·2 (0·41; 12) | 10·6–11·9 | 11·3 (0·27; 5) | 10·9–11·6 |

Sex differences not significant.

Turkey, wing: ♂ 101·5 (3·77; 5) 95–104, ♀ 94; juvenile, ♂ 93–100 (4), ♀ 94–96 (3) (Jordans and Steinbacher 1948; Kumerloeve 1967a; Rokitansky and Schifter 1971). Caucasus, ♂ (*n* = 10): wing 92–100, tail 62–64, bill (N) 15·4–18·0, bill depth 12·0–12·4 (Buturlin 1907). Crimea, ♂ (*n* = 11): wing 98·0 (92·7–102·2), bill (N) 16·5 (15·3–17·5), bill depth 11·8 (10·5–13·0) (Dementiev and Gladkov 1954).

Wing, bill to nostril, and bill depth of ♂ of (1) *tianschanica*, Tien Shan, (2) *altaiensis*, Altai, (3) *japonica*, Transbaykalia and Japan, and (4) *himalayensis*, Himalayas to central China (Stresemann *et al.* 1937; Dementiev and Gladkov 1954; Vaurie 1956b, 1959; RMNH, ZMA).

| | WING | BILL (N) | DEPTH |
|---|---|---|---|
| (1) | 95·5 (44) 89–99 | 14·9 (30)13·7–16·0 | 9·7 (30) 9·1–10·3 |
| (2) | — (16) 91–98 | 15·4 (16)15·0–17·0 | 9·5 (16) 9·2–9·6 |
| (3) | 95·3 (26) 89–101 | 15·4 (17)13·9–16·4 | 10·0 (17) 9·2–11·0 |
| (4) | 87·9 (33) 85–92 | — (—)12·5–15·5 | — (—) 8·5–9·5 |

Weights. Nominate *curvirostra*. Ages combined. (1) South-west Scotland, December–February (Marquiss 1980). (2) Northumberland (England), March–May (D C Jardine). (3) Norway, September–December (Haftorn 1971; see also Hagen 1942). (4) North-west European Russia and Yamal (north-west Siberia) (Kokhanov and Gaev 1970; Danilov *et al.* 1984; Semenov-Tyan-Shanski and Gilyazov 1991). (5) Netherlands, July–March (RMNH, ZMA). Belgium: (6) September–February 1979–80, (7) July–September 1983–4 (Herremans 1982, 1988; see also Arnhem and Van Ammel 1964). Helgoland: (8) 1927–9 invasions (Drost 1930), (9) 1962–3 invasions, July–September (Vauk 1964).

| | | | | | | | | |
|---|---|---|---|---|---|---|---|---|
| (1) | ♂ | 40·7 (2·7 ; 10) | 37–45 | ♀ | 40·5 (3·4 ; 11) | 37–46 | | |
| (2) | | 41·2 (2·34; 52) | 37–46 | | 42·2 (2·80; 45) | 37–48 | | |
| (3) | | 41·3 ( — ; 33) | 36–46 | | 39·5 ( — ; 23) | 36–47 | | |
| (4) | | 42·4 ( — ; 46) | 32–49 | | 41·7 ( — ; 30) | 36–47 | | |
| (5) | | 39·4 (5·64; 10) | 32–51 | | 38·9 (4·69; 7) | 31–45 | | |
| (6) | | 41·2 (2·92; 39) | 35–48 | | 40·3 (2·84; 38) | 34–46 | | |
| (7) | | 40·8 (2·69; 90) | 34–53 | | 39·6 (2·61; 113) | 35–49 | | |
| (8) | | 36·6 ( — ; 36) | 30–44 | | 33·0 ( — ; 22) | 26–38 | | |
| (9) | | 37·9 ( — ; 66) | 29–45 | | 36·1 ( — ; 93) | 29–51 | | |

Southern Germany, all year: 38·2 (216) 29–46; average of ♂ 39·0 (106), ♀ 37·9 (66), juvenile 36·8 (44) (Prinzinger and Hund 1975). Eastern Germany, average in various years: adult, ♂ 38·2–41·9, ♀ 36·8–39·6; 1st adult, ♂ 35·8–40·8, ♀ 34·4–37·3 (Weber 1971–2). Exhausted ♂♂, Netherlands: 29·6 (3) 28–32 (ZMA).

*L. c. poliogyna.* Tunisia, May: ♀ 37·7 (G O Keijl).

*L. c. guillemardi.* Turkey: July, ♂ 49·1 (Rokitansky and Schifter 1971); November, ♂ 44·2 (3·77; 4) 39–48 (Kumerloeve 1967a).

*L. c. tianschanica.* Kazakhstan: ♂ 36·3 (28) 31·0–42·4, ♀ 36·3 (10) 31·0–40·0 (Korelov *et al.* 1974). Kirgiziya: ♂ 34·0–35·5 (*n* = 5), ♀ 31·5 (Yanushevich *et al.* 1960).

Nominate *curvirostra* or *altaiensis.* Mongolia, June–July: ♂♂ 36, 41; ♀ 34·2 (4·60; 5) 30–42 (Piechocki and Bolod 1972).

*L. c. himalayensis.* Central China: ♂ 24·7–28·3 (*n* = 5), unsexed 22·5, 27 (Weigold 1926; Stresemann *et al.* 1937). India, November: ♂ 26·5, ♀ 25·5 (Ali and Ripley 1974).

**Structure.** Wing rather long, broad at base, tip pointed. 10 primaries: p8 longest, p9 0–2·5 shorter, p7 1–4, p6 5–10, p5 14–22, p4 21–29, p3 23–34, p2 27–37, p1 31–42; p10 strongly reduced, a tiny pin hidden below reduced outermost greater upper primary covert, 61–72 shorter than p8, 9–14 shorter than longest upper primary covert; no marked difference between races, e.g. p1 35·4 (15) 31–39(–42) shorter than p8 in nominate *curvirostra*, 33·2 (10) 31–37 in *poliogyna*. Outer web of (p6–)p7–p8 emarginated, inner of p8–p9 with notch (sometimes faint). Tip of longest tertial reaches tip of p2–p3. Tail rather short, tip forked; 12 feathers, t1 7·1 (40) 4–11 shorter than t6 (nominate *curvirostra*, all ages; perhaps slightly deeper in ♂ than in ♀: Hansen and Oelke 1976) or 7·6 (10) 5–10 (*poliogyna*). Bill massive, long, laterally compressed; tips elongated, crossed, much curved (upper obliquely downward, lower obliquely upward), ending in sharp point. Approximately equal numbers of birds have upper mandible curved to left (lower to right) and to right (lower to left): in various samples, upper mandible curves to left in 49·3% of 3823 birds (Drost 1930; Weber 1971–2; ZMA), in 48·0% of 999 birds (James *et al.* 1987), or in 48·0% of 3647 birds (Groth 1992a); see also Vauk (1964) and Herremans (1988). Young have upper mandible elongated into point by time of fledging (*c.* 3·5 weeks old), both mandibles crossed at 4–6 weeks (Heinroth and Heinroth 1931; Nolte 1930). For function of bill during feeding, see Böker (1922), Robbins (1932), Benkman (1987b, 1988a), and James *et al.* (1987). Base of upper mandible covered by short bristle-like feathers projecting forwards from forehead and lore, concealing small rounded nostril; bristles at gape virtually absent. Tarsus and toes short but strong. Middle toe with claw 21·6 (20) 19·5–22 in nominate *curvirostra*, 21·0 (6) 20–22·5 in *poliogyna*, 21·9 (5) 20·5–22·5 in *tianschanica*; outer and inner toe with claw both *c.* 72% of middle with claw, hind *c.* 87%.

**Geographical variation.** Rather slight and mainly clinal in west Palearctic, more marked in central and eastern Asia and

North and Central America. Involves depth of grey ground-colour, colour and extent of red (♂) or green (♀) on feather-tips of head and body, size (wing, tail, and tarsus length, or weight), and relative length, depth, and width of bill. Sometimes proportion of red versus yellow-green ♂♂ in populations used to characterize race (e.g. Hartert 1903–10, Vaurie 1959, Massa 1987), but adult ♂♂ (3rd calendar year and older) of all populations are red, varying proportion mainly reflecting relative number of yellow-green 1st adult ♂♂ in race (which may depend partly on timing of moult); in small-sized races, body moult more rapid than in larger ones, 1st adults in small races usually almost fully red or yellow, less often chequered than in larger nominate *curvirostra*. Colour of red in ♂ and green of ♀ strongly dependent on abrasion, becoming more glossy deep scarlet in ♂, bronzy- or orange-green in ♀ (but in some ♀♀, green of feather-tips largely worn off, plumage becoming mainly dull grey).

In Europe, grey ground-colour of body becomes gradually paler south from Cantabrian mountains, Pyrénées, and central Italy, bright colour of feather-tips paler and more reduced in extent, cline ending in *poliogyna* of North Africa. Ground-colour hiof *poliogyna* pale ash-grey, less dark and saturated than in nominate *curvirostra* from central and northern Europe eastward, feather-tips of head and body of adult ♂ pink-red, much grey of feather-bases visible; rump uniform rosy-pink; belly rosy-red with some white spots or streaks on feather-centres; lore, ear-coverts, tail, and wing distinctly paler brown-grey; fringes of tail- and flight-feathers paler pink or salmon; ♀ has green restricted or virtually absent, liable to wear off, head and body mainly grey, without clearly darker spots or streaks on crown, mantle, and scapulars (in contrast to nominate *curvirostra*), rump pale grey-green, ear-coverts scarcely darker grey than remaining side of head (but small dark patch in front of eye); underparts grey with green wash on chest and flank; bill slightly shorter and deeper at base than in nominate *curvirostra*, culmen more strongly curved, but much overlap. Birds from central and southern Spain as well as Balearic Islands slightly paler and greyer than nominate *curvirostra*, intermediate between *poliogyna* and nominate *curvirostra*; green wash on body of ♀ restricted, as in *poliogyna*; *balearica* from Balearic Islands slightly smaller than both nominate *curvirostra* and *poliogyna*, bill shorter and less deep; birds from central and southern Spain (sometimes separated as *hispana* Hartert, 1904, but not recognized here) like *balearica* but slightly larger and red of ♂ and green of ♀ slightly darker and more extensive, more or less intermediate between nominate *curvirostra* and *balearica*. Birds from Calabria (southern Italy) and Sicily intermediate between nominate *curvirostra* and *poliogyna* (Massa 1987), bill longer than in *poliogyna* but similar in depth, wing longer and bill longer and heavier than in *balearica*. *L. c. corsicana* from Corsica has ground-colour of head and body slightly darker than in nominate *curvirostra*, adult ♂ slightly darker scarlet-red, hardly distinguishable from ♂ nominate *curvirostra*, but green of ♀ rather restricted, head and body mainly dark grey (less pale than *balearica*); dark streaks of juvenile broader; bill slightly longer, deeper, and broader than in nominate *curvirostra*, but much overlap. *L. c. guillemardi* from Turkey, Crimea, Caucasus area, and Cyprus on average slightly larger than nominate *curvirostra* (see Measurements), bill longer, thicker, and broader at base, even more so than *corsicana*, like bill of Scottish Crossbill *L. scotica*, but slightly longer though slightly less deep and wide; grey ground-colour of head and body slightly darker than in nominate *curvirostra*, ♂ with more diluted darker red, ♀ darker grey with green rather restricted (extent as in *poliogyna* and *corsicana*; grey darker than in *poliogyna*, cap to mantle more heavily spotted than in *corsicana*); several races described (*caucasica* Buturlin, 1907, Caucasus; *mariae* Dementiev, 1932,

Crimea; *vasvarii* Keve, 1943, north-west Turkey), but all closely similar in colour and size (and little different from *corsicana*). Not known what race breeds Crete (Massa 1984), mainland Greece, and southern Balkan countries: single ♂ from northern Greece deep red (Kumerloeve 1964a), like *guillemardi*; birds of southern Yugoslavia included in *mariae* (Matvejev and Vasić 1973), but *mariae* here considered inseparable from *guillemardi*; probably nominate *curvirostra* grades into *guillemardi* throughout region.

Races of central and eastern Asia smaller than nominate *curvirostra*, particularly *himalayensis* (Himalayas and eastern Tibet north to Kansu) and *luzoniensis* (Luzon, Philippines). Bill markedly heavy in *meridionalis* (southern Annam, Vietnam), fairly short and slender in *japonica* (Transbaykalia to Japan), *tianschanica* (Tien Shan through Pamirs to Karakoram: Wolters 1968), and *altaiensis* (Altai, Tuva, and Mongolia; these 3 listed in order of decreasing bill depth), or markedly fine (*himalayensis*

and *luzoniensis*). Colours darkest in *himalayensis* (♂ brown-red, ♀ sooty-grey) and *altaiensis* (♂ dark red, ♀ dark grey with olive tinge), slightly paler (near nominate *curvirostra*) in *japonica* and *luzoniensis*, paler still in *tianschanica* and *meridionalis*; intermediates between *altaiensis* and nominate *curvirostra* (occurring just north of Altai to Lake Baykal) sometimes separated as *ermaki* Kozlova, 1930.

For North American races, see Griscom (1937) and Payne (1987). Locally, some races appear to breed side-by-side, and thus more species perhaps involved, comparable with situation among *Loxia* in Scotland and northern Europe. See Kemper (1959), Phillips (1974), and Groth (1988).

Forms species-group with Scottish Crossbill *L. scotica* and Parrot Crossbill *L. pytyopsittacus*. For possible evolutionary history, see Murray (1978), Eck (1981), and Tyrberg (1991a).

**Recognition.** See Two-barred Crossbill *L. leucoptera*.     CSR

---

## *Loxia scotica*  Scottish Crossbill

PLATE 43
[facing page 640]

Du. Schotse Kruisbek     Fr. Bec-croisé d'Ecosse     Ge. Schottischer Kreuzschnabel
Ru. Шотландский клест     Sp. Piquituerto escocés     Sw. Skotsk korsnäbb

*Loxia curvirostra scotica* Hartert, 1904

Monotypic

**Field characters.** 16·5 cm; wing-span 27·5–31·5 cm. Intermediate in size between Crossbill *L. curvirostra* and Parrot Crossbill *L. pytyopsittacus*, with size and structure of bill and head closer to latter. Resident crossbill of relict forests of Scots pine *Pinus sylvestris* in north-central Scotland, formerly regarded as race of one or other congeners. Of differences from *L. curvirostra*, only head/bill structure perceptible in the field. Sexes dissimilar; some seasonal and marked individual variation, especially in ♂♂. Juvenile separable.

All plumages as *L. curvirostra*. Head larger than in *L. curvirostra*; bill deeper (especially upper mandible) and hence blunter; see Plumages. Slightly greater body bulk, with wings and tail averaging 5% longer, not discernible in the field.

*L. curvirostra* known to have bred within breeding range of *L. scotica* (A G Knox) and widespread occurrence there is considered possible (Thom 1986), so assumption that all crossbills in Scotland from Perthshire northwards are *L. scotica* is no longer valid. No observed difference from *L. curvirostra* in flight, gait, or general behaviour, but *L. scotica* forms only small parties (up to 20 birds) after breeding, not erupting widely as *L. curvirostra*. Strong food preference for seed of *Pinus sylvestris*; *L. curvirostra* usually prefers seeds of spruce *Picea*. Regrettably, the few identification clues that fit *L. scotica* also point to *L. pytyopsittacus*, no longer a rare vagrant to

Britain; thus, field identification of *L. scotica* is virtually impractical away from pines of breeding habitat, and may be unsafe even within them.

Voice apparently intermediate between *L. curvirostra* and *L. pytyopsittacus*, but (in direct comparison) 'chup' generally distinguishable from equivalent 'chip' or 'dyip' of *L. curvirostra*.

**Habitat.** Restricted within western boreal zone of west Palearctic to forests and smaller stands of Scots pine *Pinus sylvestris* in north-east of Scottish Highlands, breeding in level lowlands and on gentle or steep slopes, up to sunny hilltops where trees are scattered. Mature plantations are no less attractive than primeval relict forest, and while old trees are preferred their height and size matter little. Breeding may occur in interior of well-spaced woodland, as well as in openings or clearings and on edges. Nests sometimes located in stunted pines on small islands or hillocks, and in forest bogs, also in clumps of pines undergrown with tall rank heather *Calluna*. Requires easy access to water, in water-holes, peat runnels, or other sources. Habitat choice is discriminating, even within favourite woods where open sunny parts are preferred to closed canopies, even in primitive forests. (Nethersole-Thompson 1975.) Almost entirely arboreal, resorting to all levels of tree crowns. Flies freely over distances and at some height. In eastern Scottish Highlands, currently sole

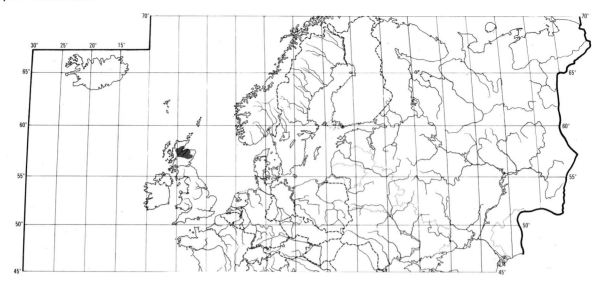

breeding *Loxia* on upper Deeside above Aboyne, but no more than 6 km eastward *L. curvirostra* replaces it in areas where plantations of spruce *Picea* and larch *Larix* become frequent. (Buckland *et al.* 1990.)

**Distribution.** Limits uncertain in some peripheral areas, due to difficulty of distinguishing from Crossbill *L. curvirostra* (Nethersole-Thompson 1975; Lack 1986), but no evidence of major changes. Distribution in valley of Dee and tributaries probably unaltered since *c.* 1800 (Knox 1990*b*).

**Population.** Probably 300–400 pairs in 1980s (Thom 1986). Estimate of *c.* 1500 adults in early 1970s (Nethersole-Thompson 1975) may include some Crossbills *L. curvirostra*.

**Movements.** Resident and dispersive.

In most years, birds disperse after breeding to seek better food supplies, settling in adjacent or fairly close woods or plantations in Scottish highlands. When large populations build up after several years of successful breeding, birds sometimes move further in general exodus, e.g. in summer of 1936 many birds in upper Strathspey dispersed in large flocks, and few remained there to breed in spring 1937. Not known if any birds join flocks of immigrant Crossbills *L. curvirostra* which periodically pass through breeding woods, and not recorded south of 56°21′N. (Nethersole-Thompson 1975; Catley and Hursthouse 1985; Knox 1990*c*.)                    DFV

**Food.** Conifer seeds, primarily of Scots pine *Pinus sylvestris*; some small invertebrates in breeding season. Feeding methods as other *Loxia*. Moves acrobatically in groups through pines, usually high up in end branches, apparently selecting best cones; seeds extracted *in situ* or cone twisted off and taken to perch for handling, often a fork closer to trunk, where cone is steadied with feet on one part of fork while tail is braced against other part for increased leverage. For details of seed extraction technique, see Crossbill *L. curvirostra*. Feeding perch often used repeatedly, even (rarely) for cone brought from other trees. Cone rotated during feeding and seeds generally removed only from central portion. Some feeding trees visited regularly because their exposed position means cones ripen earlier in increased sunlight. (Nethersole-Thompson 1975, which see for details; Knox 1987*b*.) When ♀ forming eggs or incubating, needs increased protein intake, which is obtained from invertebrates, nutrient-rich ♂ pine cones and bud tissue, and seemingly also fragments of bone and flesh from carrion; searches branches and pulls off flakes of bark and pieces of lichen to get at invertebrates, and snaps at flying insects while on nest (Nethersole-Thompson 1975; Bates and Whitaker 1980; Thompson and Nethersole-Thompson 1984). Forages for invertebrates on forest floor, and once in old drey of squirrel *Sciurus*; often reported eating mortar, putty, and similar mineral-rich substances, and visits drinking places frequently (Nethersole-Thompson 1975; Bartlett 1976). When pine seed scarce or unavailable, feeds on larch *Larix*, spruce *Picea*, fir *Abies*, Douglas fir *Pseudotsuga*, and presumably other introduced conifers, as well as on conifer blossom, buds, and shoots. Before these exotic conifers were planted, probably turned to seeds of rowan *Sorbus*, birch *Betula*, and herbs. Starts to feed exclusively on pine around March when breeding season approaches peak. (Nethersole-Thompson 1975.) In early February, recorded foraging in beech *Fagus*, but food not noted (Bartlett 1976). In good seed year, one ♂ dealt with 6 cones in average 2·4 min (1·1–3·8) per cone, extracting average 16 seeds (8–35) from each, or *c.* 7 seeds per min. ♀ fed on ♂ pine cones at rate of *c.* 4 billfuls per min, and took 8 billfuls of bud tissue in 3 min (Nethersole-Thompson 1975; Thompson and Nethersole-Thompson 1984).

Young can be raised on seeds alone, though invertebrates taken when available (Knox 1987b). Only regurgitated pine seeds seen fed to young in extensive study by Nethersole-Thompson (1975), but virtually impossible to tell by observation if animal material present. ♂ gathers seeds intensively in rich feeding trees up to *c*. 800 m from nest, bringing enough per feed to supply ♀ and young for some minutes. Diet of young doubtless similar to *L. curvirostra* and Parrot Crossbill *L. pytyopsittacus*. At one nest, nestlings given seed from faeces by sibling, and it is possible that faeces could be used as food reserve in poor weather (Nethersole-Thompson and Whitaker 1984).                                                      BH

**Social pattern and behaviour.** Some aspects better studied than for Crossbill *L. curvirostra*, but apparently no significant differences from that species. Following account based almost entirely on comprehensive investigation in north-east Scotland (also comparison with *L. curvirostra* and Parrot Crossbill *L. pytyopsittacus*) by Nethersole-Thompson (1975), which incorporates data from earlier study by Ross (1948). Almost certain that observations in both these studies refer at least in part to *L. curvirostra* rather than exclusively to *L. scotica* (A G Knox).

1. Generally in flocks outside breeding season, i.e. summer through to mid- or late winter (Knox 1987b). Winter flocks often large in good seed years; prior to breeding, large 'mating' flocks split into smaller ones composed of pairs (Nethersole-Thompson 1975); see Flock Behaviour (below). In breeding season, ♂♂ often solitary or in twos for foraging, but flocks also occur then commonly (Nethersole-Thompson 1975), including many paired breeders (Knox 1987b); see Bonds (below). After fledging, families often remain near nest; sometimes move (unknown distances) while young perfect flying skills (Knox 1987b). Flocks forming at this time comprise several families or part-families (see Bonds, below), perhaps also non-breeders or failed breeders (Nethersole-Thompson 1975). Sometimes forms mixed flocks with *L. curvirostra* (Knox 1990c), and may associate for feeding (presumably only loosely) with other Fringillidae and Coal Tit *Parus ater* (Nethersole-Thompson 1975). BONDS. Mating system probably normally monogamous, breeding pattern discouraging polygamy which would impose too great a strain on ♂ as provider. Some indications of occasional deviations from monogamy nevertheless. In early pairing phase (when ♀ still dominant and aggressive), 2–3 different ♂♂ may feed and attempt to mount same ♀ within a morning. One record of ♂ associating with and feeding 2 ♀♀, but not known whether either or both attempted to breed; 2 further records of ♂♂ each apparently paired polygynously with 2 ♀♀, laying not simultaneous, and 2nd ♀ in each case unsuccessful (in one case, eggs infertile); once also 2 nests (both with eggs) in same tree, but no further details of owners. (Nethersole-Thompson 1975.) Similar observations in earlier study include apparently bigamous ♂ regularly coming to one nest with both ♀♀ and, after 1st had settled, escorting 2nd to her nest *c*. 300 m away (Ross 1948). Sex-ratio apparently skewed towards ♂♂, and flocks predominantly of ♂♂ (though some ♀♀ and non-breeding pairs) occur at most population levels; some such birds (♂♂) establish flexible territories, sing and fight, and pair up later in season. 2 ♂♂ sometimes form varyingly close association: if unpaired, almost like homosexual pair (though no

copulation recorded); in case of breeding birds, 2 ♂♂ may forage together regularly for their incubating mates, but each normally aggressive towards other near own nest. Hybridization recorded with *L. curvirostra* in captivity, but none among wild birds (Nethersole-Thompson 1975), despite sympatric breeding in at least 9 years out of 13; however, mixed pairs with *L. curvirostra* difficult to detect, as members of *L. scotica* pairs often differ noticeably in bill structure (Knox 1990b). Birds start to pair up in flocks as winter progresses, apparently then dispersing to look for nest-site (sometimes a few pairs together) (Nethersole-Thompson 1975; Knox 1987b). Duration of pair-bond not clear (no ringing): perhaps persists for replacement clutches (possibility of 2nd broods not proven: see Breeding), but unlikely (as in *Loxia* generally) to do so in longer term. Nest built by ♀, ♂ accompanying her but rarely making significant contribution to building; ♀ also incubates (rare records of ♂ covering eggs briefly). Both sexes feed young; nestling recorded feeding brood-sibling (see Relations within Family Group, below, and Food). After fledging, whole family may spend 10–14 days in or close to nesting territory, brood then often divided between parents, each caring for 1–2 fledglings. Young dependent for at least 2 months after leaving nest and during this period apparently sometimes fed by adults (perhaps failed breeders) other than own parent. (Nethersole-Thompson 1975; Knox 1987b.) Age of first breeding not known exactly, but 'immature' birds quite often build nest in which, however, they do not lay, or only after delay, and failure rate high (Bates and Whitaker 1980). BREEDING DISPERSION. Territorial and solitary or, especially in years of high numbers and good seed crop (not invariably even then), in loose neighbourhood groups of 2–6 pairs; such clusters typically formed when pairs in small mobile flock simultaneously settle in same small wood or particular part of larger one. In larger woods, nests usually 75–150 m (50–300) apart; 50–400 m recorded in groups; for 2 nests in same tree, see Bonds (above). Birds more dispersed when numbers low, with solitary (more markedly territorial) pairs separated by 800–1600 m. (Nethersole-Thompson 1975.) Territorial structure poorly understood (Knox 1987b). Early territorial behaviour apparently primarily defence of ♀, but in some years (of low population level), or in case of ♂♂ not pairing up in flock, orthodox song-territory of *c*. 500-m radius may be established and later used for nesting. Other ♂♂ may leave flock only temporarily, establishing song-territory, but then abandoning it and rejoining flock; once 5 such territories within *c*. 100 m. Territory normally established before nest-site chosen and used for courtship and copulation, also for some feeding in early period (a few days), but birds often feed well away from nest after start of laying, though ♀ (especially if ♂ slow to bring food) sometimes feeds close to nest during incubation, ♂ also after hatching. During incubation, some ♂♂ defend rudimentary feeding territory (tree or group of trees) against all but regular foraging companion (see Bonds, above). Some territories have pool where ♀ drinks, and almost all contain tree slightly taller than nest-tree and often adjacent to it used by ♂ both as staging post and favourite perch (Knox 1987b). In one case, territorial boundary *c*. 50 m from nest-tree, but area defended (mainly by ♂, though ♀ will attack other ♀♀) and aggression apparently reduced with advance of season; flocks feed with impunity in occupied territory during ♂'s absence, but at least nest-tree is normally defended against such birds and regular foraging companion up to nestling phase and even after fledging, though another pair occasionally allowed to build near nest containing young. ♂ will sing in response to brief Song-flight (see part 2) by another, but rarely defends air-space over territory otherwise. Defence breaks down for predators (see Parental Anti-predator Strategies, below). (Nethersole-Thompson

1975.) In 2 cases of apparent polygyny, nests of 2 ♀♀ 200–300 m apart (Nethersole-Thompson 1975); in study by Ross (1948), apparently bigamous ♂'s favourite song-post was midway between 2 nests and in sight of both; not clear how much ground defended in such cases. Replacement nests close to 1st or up to c. 200 m away (Nethersole-Thompson 1975). Maximum density recorded by Nethersole-Thompson (1975) 6 pairs in 26 acres (c. 57 per km²); in Abernethy forest, density (from point counts) $1.08 \pm SE0.45$ birds per ha (Hill et al. 1990). Factors controlling distribution not well understood; much variation, including between years (Knox 1987b). No ringing to investigate possible site-fidelity, but probably unlikely, and only 1 case of perhaps same isolated pair nesting in same small wood over 4 years (Nethersole-Thompson 1975). ROOSTING. No information for period when not breeding, but probably much as in other Loxia (see L. curvirostra). During incubation (or from building), some ♂♂ may roost outside territory, perhaps in tree where they regularly forage, and visit ♀ only from early morning; others spend night in or near nest-tree, accumulated faeces suggesting use of regular perch. 2 observations indicated ♀ roosts on or near nest before laying, and perhaps also covers incomplete clutch at night. Fledged young may return initially to nest to roost. (Nethersole-Thompson 1975.)

2. Anxious or excited bird giving Tep-call (see 2b in Voice) often flicks wings and tail. Flying Sparrowhawk Accipiter nisus or Merlin Falco columbarius elicit nervous reaction and repeat 'tak' calls (see 3 in Voice). (Nethersole-Thompson 1975.) FLOCK BEHAVIOUR. Flock silent or birds give quiet 'chuck' calls (presumably Dyip-calls: see 2a in Voice) when feeding (Ross 1948). Disputes and fights (see Antagonistic Behaviour, below) frequent in flocks in autumn. In late winter and early spring, birds often excitable and restless, flying about and giving loud Tep-calls, also tending to land on toips of trees for typically noisy gatherings in which pair-formation takes place: alternate between long tranquil spells (interspersed with communal singing and Song-flights; some evidence of hierarchy with dominant ♂♂ singing from high perches) and bouts of chattering (see 3 in Voice), supplanting attacks for best song-perches, frequent fights, elements of courtship (early generally incomplete courtship-feeding or bill-scissoring, attempted copulation, and sexual chases). Departure by a pair often causes display to break up, but flock may soon land again on same or nearby tree. (Ross 1948; Nethersole-Thompson 1975.) See also Song-display, Antagonistic Behaviour, and Heterosexual Behaviour (below). In smaller flocks composed mainly of pairs splitting away from 'ceremonial gathering', birds of either sex will break off twigs and carry them for short distance before dropping them. Unpaired ♂♂ in small flock break away and temporarily defend tree with fights and song, before regrouping. Non-breeders in nomadic flock often intrude into occupied territory, sometimes through associating with owners and flying back with them after foraging; normally chased away, at least from vicinity of nest. (Nethersole-Thompson 1975.) SONG-DISPLAY. ♂ sings (see 1a in Voice) from perch (usually tree-top) or in flight. Song serves for territory establishment and advertisement (given whether or not ♀ present) and, if ♂ does not pair up in flock, to attract ♀. May stimulate others to sing. Lone ♂ will sing from one or more trees. Loud song by perched bird typical of period before territory finally established. Rarely sings for long or loudly near nest, and overall less than some L. curvirostra. (Nethersole-Thompson 1975.) Flock may give chorus of song, especially in sunny, calm, and frosty weather in late winter or early spring; loud phrases of individual ♂♂, in flight, just about to land or when perched (occasionally also of ♀♀, when perched above ♂♂ feeding below them), stand out against background babble which includes vari-

ous calls and Subsong (see 1b–c in Voice). ♂ on perch may turn head or sway body side to side, and intermittently flutter up and down briefly while singing loudly. (Ross 1948; Nethersole-Thompson 1975.) Recorded flapping wings while singing, much as L. curvirostra. Bird may sing in Butterfly-flight: with tail spread and mantle slightly ruffled, beats wings in slow, deep flicking motion (e.g. when overflying another's territory, occasionally before alighting in own), or glides with wings horizontal or slightly raised in arc from tree to tree; wing-beats sometimes rapid and shallow (Moth-flight). Sometimes circles or describes figure-of-eight. (Witherby et al. 1938; Nethersole-Thompson 1975.) For further details, see Ross (1948). For song in other contexts, see Antagonistic Behaviour and Heterosexual Behaviour (below). Sings in any month, but most frequently and vigorously in spring; will resume within 1–2 days of nest-loss. Noted at night late December. (Ross 1948; Nethersole-Thompson 1975.) ANTAGONISTIC BEHAVIOUR. Various forms of threat (vocal and visual displays), displacing attacks, fights, and chases occur in flocks and later when breeding (defence of territory). Both sexes involved, e.g. all 4 birds of 2 pairs (♂-♂, ♀-♀, or either sometimes attacks opposite sex), though ♀ of pair passive in some cases, and recorded remaining on nest, wing-shivering and 'chittering' (presumably Begging-call: see 5b in Voice) while ♂ fought. Trespassers near top of tree more likely to be attacked than birds below them which are perhaps of lower status. Interactions include adoption of Threat-posture with Threat-gaping (as in L. curvirostra: see Fig A of that species), ♀ generally posturing less dramatically; also wing- and tail-flicking, Song-duels (see 1b in Voice), exchange of noisy calls ('tak' and 'tep', which sometimes inhibit fight), and Bill-snapping from ♂♂ (see Voice). Loud and strident song often precedes and sometimes accompanies attack which may be in Butterfly- or Moth-flight. Opponents may engage in aerial tussle (song becoming almost screeching), fluttering, pecking, and grappling with feet. Sometimes sings loudly after evicting intruder, rarely in flight. Both sexes recorded attacking other Fringillidae, including Chaffinch Fringilla coelebs (usually tolerated, but ♂ on tree-top sometimes jostled), Siskin Carduelis spinus, and Bullfinch Pyrrhula pyrrhula. (Nethersole-Thompson 1975.) See also Ross (1948). HETEROSEXUAL BEHAVIOUR. (1) General. Display by ♂♂ recorded all months, with or without ensuing copulation (Ross 1948). Especially in years when numbers high, pairing behaviour complex (Nethersole-Thompson 1975). (2) Pair-bonding behaviour. Initial advances by ♂ (e.g. attempt to touch bills with ♀ or to feed her, including in flock) may cause ♀ to assume Threat-posture, ♂ then showing readiness to flee by ruffling plumage. Other interactions reported: ♂ singing while fluttering about tree-tops; sexual chases (♀ giving excited Dyip-calls: see 2a in Voice), sometimes after ♀ has snapped off twig. At close quarters, birds may face up, wings held slightly away from sides, bills open (perhaps threat), touch, or even link bills, swaying side to side or back and forth (in some cases while bills caught together), this often preceding copulation (see below). (Ross 1948; Nethersole-Thompson 1975.) ♂ guarding mate while she is feeding typically gives Subsong from tree-top. Displaying to already-paired ♀, perched ♂ seen to sing with tail spread and wings flapping, then to approach her in Butterfly-flight. (Nethersole-Thompson 1975.) (3) Courtship-feeding. Starts while birds still in flock (Witherby et al. 1938); during laying, takes place on or off nest, usually on nest once ♀ incubating firmly. In early stages (when feeding attempts sometimes unsuccessful), ♂ may approach ♀ by sidling or hopping along branch, or simply reaches down to ♀ below him; various postures (crouched or more erect) adopted by both birds; bill-touching and bill-scissoring also occur; ♀ otherwise wing-shivers and opens bill into which ♂ inserts his (as later on nest). ♂ gives Dyip-calls while approaching nest; ♀, recognizing these, becomes alert, stretching up in nest, flicking wings above back and calling (squeaks: see 5b in Voice). ♂ may land within 50-100 m and give Tep-calls while wing- and tail-flicking; calling becomes quieter, then stops altogether with closer approach to nest. ♀, head back and bill open, continues to call while wing-shivering; ♂ then calls (see 5a in Voice) and bends down to feed mate (Fig A); ♂

A

regurgitates food 6-24 times per visit. Overall, amount of calling and calls used vary (see 5 in Voice). ♂ may call or sing quietly before flying off, ♀ often continuing to give Begging-calls after his departure. Average 7-10 feeding visits by ♂ per day, at intervals of 30-150 min, but ♀ in some cases also leaves nest to feed herself (etc.) nearby. (Nethersole-Thompson 1975.) See also Ross (1948). (4) Mating. Birds copulate on branch often during building and early part of laying, usually near nest; may follow Courtship-feeding. Early (often unsuccessful) attempts occur in flock. Preliminaries include, variously, ♀ flying to ♂ (singing quietly) and soliciting (crouches, raises tail, wing-shivers); ♂ approaching in Butterfly- or Moth-flight, adopting upright-posture, giving rather harsh song, as also during ensuing chases which are sometimes frenetic, birds weaving through branches and spiralling round trunks, ♂ then overtaking ♀ and copulating (perhaps sometimes rape). ♂ may hover first over ♀, then mounts, tail lowered and fanned, and beats wings (Fig B); copulation

B

often silent, or both birds call (see 6 in Voice); ♀ occasionally turns head and bill-touches with ♂. Preening by ♀, and bill-wiping by both sexes, recorded after copulation. (Nethersole-Thompson 1975; also Ross 1948.) (5) Nest-site selection.

Sometimes by pair together after both have prospected and examined potential sites. ♀ apparently the more active in most cases: may sit (♂ also, though less than ♀) in crotch intermittently for up to 7 days before building there, and seldom accepts choice made by ♂. (Nethersole-Thompson 1975; also Ross 1948.) Building by ♀, ♂ often accompanying her (close mate-guarding), but rarely assisting: out of 30 pairs, ♂ seen carrying material to nest (in which eggs eventually laid) in only 4 (Nethersole-Thompson 1975, which see for many details, including variation between pairs); see also Ross (1948) and Knox (1987b). Generally quiet during building, though some song given by ♂ and Twittering-calls (see 4 in Voice) by ♀, who may further show excitement by bill-wiping, false-preening, ruffling feathers, and making slow pecking movements. Material gathered up to *c.* 300 m from nest, ♀ often eventually establishing set route. (Nethersole-Thompson 1975.) (6) Behaviour at nest. Nest sometimes remains empty for *c.* 1 week before start of laying, and a few feathers may be added to nest containing 1st egg (Nethersole-Thompson 1975). During incubation, ♀ absent from nest for only short periods (Knox 1987b). ♀ recorded rummaging in bottom of nest, presumably for parasites and/or to ventilate eggs, also carrying away her own faeces (usually defecates onto side of nest). Birds generally quiet and furtive at nest and in flights to and from it; ♂ may give quiet burst of song to call ♀ off eggs. When feeding young, ♂ and ♀ usually fly to nest one just behind the other, often below tree-top level, and move away discreetly afterwards. (Nethersole-Thompson 1975; also Ross 1948.) RELATIONS WITHIN FAMILY GROUP. Eggs hatch over 1-3½ days; eggshells eaten by ♀ at or away from nest, sometimes discarded away from nest. Young brooded by ♀ day and night for *c.* 5 days; at night and for brief stints during day for further *c.* 5 days, exceptionally at night up to fledging. Almost all food brought by ♂ for first few days and passed (by regurgitation) to ♀ for transfer to young, or ♀ sometimes stands up, allowing ♂ to feed young direct, though she may beg from nest-rim (Fig C). Increasingly, both pair-members

C

forage together; during visits (typically *c.* 15-30 min), ♂ may wait while ♀ broods and feeds young, or both come to nest together and feed young from opposite sides; at 2 weeks, chicks reared up and shivered wings when begging, but no calls heard (see, however, Voice for piercing calls otherwise given at this age). Nest kept clean by both sexes, though mainly by ♀; nestling faeces (deposited on nest-rim by older chicks) swallowed or carried away and dropped, but regular nest-hygiene generally abandoned after *c.* 2 weeks. (Nethersole-Thompson 1975.) Nestling recorded begging from and being fed by sibling with seeds extracted from faeces on nest-rim (Nethersole-Thompson and Whitaker 1984). Incipient aggression among brood-siblings in study by Ross (1948) included dabbing at one another in nest and (just able to fly) pushing one another off branches. Fledge at *c.* 3 weeks (see Breeding). May initially (over several days) just clamber along branches and flutter about or make short flights, then return to nest, as do some for roosting even after longer excursions. Calls from parents (see 2d in Voice) apparently encourage young to fly; adult may also fly at fledglings to induce them to move to another tree. Young otherwise generally remain quiet on concealed perch, calling only when about to be fed. (Nethersole-Thompson 1975.) ANTI-PREDATOR RESPONSES OF YOUNG. Fledglings will move to centre of tree and freeze (A G Knox). PARENTAL ANTI-PREDATOR STRATEGIES. (1) Passive measures. On hearing (e.g.) calls of crows (Corvidae) or Black-headed Gull *Larus ridibundus*, ♀ seen to adopt posture as shown in Fig D, also to move bill up and down. As in other *Loxia*, ♀

D

sits tightly, even allowing herself to be touched. (Nethersole-Thompson 1975, which see for reaction of ♀ at 11 nests.) (2) Active measures: general. Anxious calls from ♂ (for details, see Ross 1948) cause mate to crouch lower in nest, ♀ assuming more active defence role when obvious her nest is target. Sometimes 2 pairs unite in defence of 1 nest. (Ross 1948; Nethersole-Thompson 1975.) (3) Active measures: against man. ♀ will peck at fingers when handled, also flutter and land on hand after finally forced off nest. 2 ♀♀ regularly dropped from nest almost to ground before flying away in steep ascent; perhaps distraction-lure display used originally for squirrel *Sciurus*. Other reactions include 'tak' calls (see 3 in Voice) and agitated wing- and tail-flicking by both sexes. (Nethersole-Thompson 1975.) Birds sometimes fly about calling noisily (A G Knox). (4) Active measures: against other animals. For squirrel (main predator) much as for man: calling, fluttering or flicking wings, etc. (see Ross 1948, Nethersole-Thompson 1975 and Fig E).

E

(Figs by D Nurney: A from photograph in Newton 1972; B and E from drawings, and C–D from photographs, in Nethersole-Thompson 1975.)          MGW

**Voice.** Only major study, including sonagrams and comparison with Crossbill *L. curvirostra* and Parrot Crossbill *L. pytyopsittacus*, is by Nethersole-Thompson (1975) which, however, includes some misidentified material (A G Knox). Additional descriptions of song and calls by Ross (1948) perhaps similarly refer, at least in part, to *L. curvirostra*. Song and at least calls 2a–b given all year. Bill-snapping noted in antagonistic interactions (Nethersole-Thompson 1975). For wing-noise at take-off, see *L. curvirostra*. Main calls of *L. scotica* closely resemble those of *L. curvirostra* (for which geographical variation yet to be investigated in depth) and much overlap anyway through variation in pitch and timbre. Similarity of certain calls to *L. pytyopsittacus* also noted (A G Knox). Some calls of *L. scotica* and *L. curvirostra* diagnostic according to Knox (1990a); see 2a–b for possible distinctions. Much variation in song of both and apparently no consistent differences between the species.

CALLS OF ADULTS. (1a) Song of ♂. Units and short motifs variously combined in brief phrases usually comprising 2 sub-phrases. Closely resembles *L. curvirostra*, sharing with that species repetition of units (segments) and characteristic contrast between 1st and 2nd sub-phrase (in duration of each or of units, timbre and loudness of units). Recording (Fig I) contains 3 motifs (only 2nd and 3rd

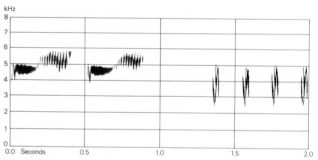

I  A G Knox  Scotland  April 1982

shown) suggesting 'tee-weez' and reminiscent in sawing quality of Great Tit *Parus major*, 4 rather hard 'chik' sounds following in 2nd sub-phrase. More complex song with no repetition in 2nd sub-phrase shown in Fig II: 'tk-whee' preceding descending 3-note motif given 3 times (light tinkling notes like very rapid song of *P. major*), leading into 2nd sub-phrase (or coda)—'tiup rrreee pri-ooo'. In Fig III, buzzy rattling 'schweerr schweerr' precedes quiet melody with no repetition, as 2nd sub-phrase much like quiet warbling of Blackbird *Turdus merula*. (J Hall-Craggs, M G Wilson.) Song-phrases given when advertising territory, to attract ♀, and in various other contexts: e.g. by ♂ perched close to incubating ♀, and in disputes (when especially loud and strident, rising almost to screech). One common song-type (of 'chip' and 'gee' segments) somewhat jerky and unmelodious, resembling song of Greenfinch *Carduelis chloris*. Social-singing also noted from birds in flocks of varying size perched openly

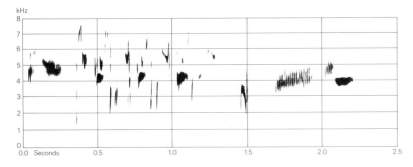

II  W Sinclair  Scotland  March 1987

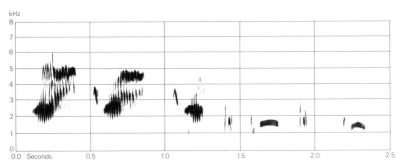

III  W Sinclair  Scotland  March 1987

or in cover; chorus of sweet notes mixed with other harsher and more strident sounds (suggesting that *L. scotica*, like *L. curvirostra*, sometimes gives longer songs, combining louder notes or phrases with more continuous twittering or similar). (Nethersole-Thompson 1975, which see for details, including transcriptions of song-units; see also Ross 1948 and *L. curvirostra*.) (1b) Subsong. Quiet and melodious warbling or bell-like phrases mixed with sub-dued calls ('chip' and 'toop') given by both sexes, often with bill closed. Renderings include 'tip-tip-tip-toohee-toohee-tip-tip-too-hee-quik-quik', 'sip-sip-sip-whee-whee-hoo-hoo-hoo-hoo-ha', and one recalled song of Woodlark *Lullula arborea* (see Fig V of *L. curvirostra*); series of quiet 'chak' calls, with abrupt change to loud 'toop' calls (expressing aggression, e.g. in song-duels), also treated as Subsong. (Nethersole-Thompson 1975.) (1c) Song of ♀. Loud phrase (see 1a, above) noted exceptionally from ♀ in temporary reversal of sexual roles (Nethersole-Thompson 1975). Otherwise, song of ♀ is of Subsong type (see above); for another song-like utterance of ♀, see call 4 (below). (2) Contact- and alarm-calls. (2a) Dyip-call. Penetrating and metallic 'dyip' or 'chip', often loud, though both volume and timbre vary with context. Given by both sexes, perched or in flight, sometimes in long series, but most often in pairs of calls (A G Knox). Used for contact, including by ♀ to summon ♂ to nest for feeding, or by ♂ to call mate off nest. Typically louder and more urgent when alarmed or otherwise excited, e.g. prior to take-off (by pair or flock). (Nethersole-Thompson 1975.) Can be difficult to separate from *L. curvirostra*, but usually distinguishable when direct comparison possible, being generally louder and fuller, and more rounded

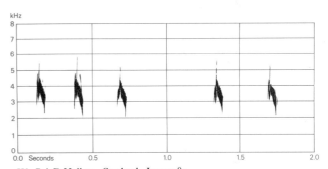

IV  P A D Hollom  Scotland  June 1985

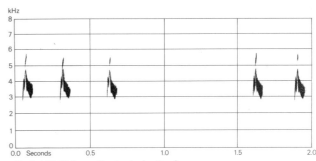

V  P A D Hollom  Scotland  April 1983

(Nethersole-Thompson 1975); short, compact 'chup', overall much less variable than Dyip-call of *L. curvirostra* (Knox 1990*a*; A G Knox). Recordings (Figs IV–V) contain series of rich, full 'dyip' or 'dyeep' sounds, some perhaps more piping than *L. curvirostra*; well-spaced units shown in Figs IV–V and (more so) faster series (typically

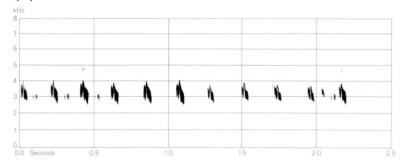

VI  P A D Hollom  Scotland  April 1983

given prior to take-off and once airborne) in Fig VI are also lower pitched and show structural differences from Dyip-calls of *L. curvirostra* (compare Figs XI–XII of that species). Important to note, however, that Fig 7 in Nethersole-Thompson (1975) shows *L. scotica* 'chip' call more like 'quip' of ♂ feeding young (lower-pitched units in Fig VII); also, Fig 7a in same work shows 'chip' call of

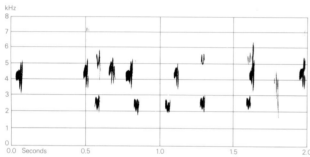

VII  D McGinn  Scotland

*L. curvirostra* unlike Figs XI–XII in that account. (J Hall-Craggs, M G Wilson.) More conversational 'wik' sounds given, as almost continuous subdued twittering, while moving about trees and feeding, also prior to take-off (Nethersole-Thompson 1975); presumably only quiet variant of Dyip-call. (2b) Tep-call. Very similar to equivalent calls of *L. curvirostra* and *L. pytyopsittacus* (Knox 1990a; A G Knox): compare Fig VIII which shows calls reminiscent of distant Jackdaws *Corvus monedula* ('chek' or 'tep' with slightly nasal, twangy quality), Figs XIV–XV of *L.*

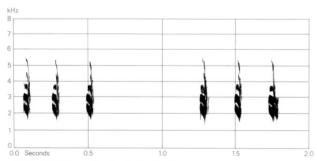

VIII  A G Knox  Scotland  April 1982

*curvirostra*, and Fig XII of *L. pytyopsittacus* (M G Wilson). According to Nethersole-Thompson (1975), call differs between sexes (low-pitched 'toop' given by ♂, explosive, more metallic 'zoop' by ♀), and may be distinguishable from *L. curvirostra* through being lower pitched. In small sample of recordings, heavy tonal section in robust Tep-call of *L. curvirostra* 3–5 kHz, in apparently more delicate call of *L. scotica* nearer 2·8–4·2 kHz; difference perhaps not consistent however, and anyway probably too subtle to be useful in diagnosis (J Hall-Craggs, M G Wilson). Tep-call expresses varying degrees of anxiety or excitement, given (e.g.) by ♂ before flying to nest to feed ♀, also when in flock, apparently associated with wide variety of emotional situations (Nethersole-Thompson 1975; A G Knox). (2c) Nasal 'chow-chow' or 'teow-teow' (once a single 'chick') given by ♀ disturbed at nest with young; less frequent than other alarm-calls, perhaps indicating less urgent excitement. (2d) Birds tending young just prior to fledging give series of 'chit' sounds, sometimes combined with typical juvenile 'chitoo'. (2e) Twangy 'tewng-tewng' (otherwise a juvenile call: see below) noted from adults in flock not containing fledglings. (3) Calls associated with antagonistic interactions include excited chattering 'tak' sounds (apparently also expressing alarm), also harsh 'zip' and loud, low-pitched 'chok-chok' delivered at slow rate and recalling rattle of *T. merula*; calls of ♀ said by Nethersole-Thompson (1975) to be separable from ♂'s, but no details. (4) Twittering-call of ♀. Long series (up to 7–8 min) of 'sip' calls given while building and just before laying; with increasing excitement may develop into song-like utterance—e.g. 'sip-sip-whee-sip-sip-whee-sip-sip'. (Nethersole-Thompson 1975.) (5) Calls associated with Courtship-feeding. Typical fledgling calls ('chitoo': see 2d, and Calls of Young, below) frequently given in this context (A G Knox). (5a) Conversational 'chack' and a few dull, wooden, non-resonant 'chuck' sounds given by ♂ before feeding ♀ (Nethersole-Thompson 1975). Quiet 'quilp', and series of 'quip' sounds (see Fig VII and Calls of Young, below) given by ♂ feeding young are perhaps same as 'chuck' or related (J Hall-Craggs, M G Wilson). (5b) Begging-calls of ♀. If ♂ slow to feed mate, she gives 'chik' or 'chuk' sounds singly or in slow deliberate series (Nethersole-Thompson 1975). Begging-calls in half-speed sonagram (Fig IX:

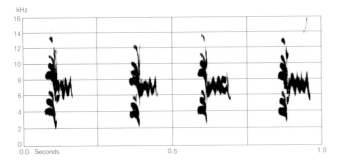

IX   D McGinn   Scotland

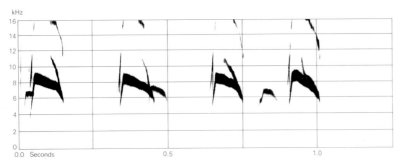

X   D McGinn   Scotland

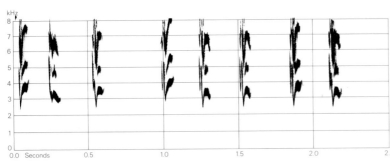

XI   A G Knox   Scotland   April 1983

'chip' or 'chit') are perhaps same. Other call mentioned by Nethersole-Thompson (1975) distinctly different: series of shrill, mouse-like squeaking 'chee' or 'sip' sounds; presumably at least closely related to call 4 (see above). (6) Calls associated with copulation. Metallic twittering given by both sexes, sometimes additionally soft 'whee-whee' from ♀ before ♂ mounts (Nethersole-Thompson 1975); transcriptions suggest close relationship with calls 4 and 5b (see above).

CALLS OF YOUNG. Food-call of small nestlings shrill and squeaking, changing to trilling 'itic-itic' when older and, just before fledging, to insistent rasping 'chit chit chitoo' similar to other *Loxia* (Nethersole-Thompson 1975). Twangy 'tewng-tewng' also occasionally given by fledged young (Nethersole-Thompson 1975). Recording (Fig X: half-speed playback) of very young nestlings contains immensely piercing, descending 'tzee' calls. In recording of nestlings at *c.* 14 days (Fig VII), calls are piercing 'zip' sounds showing some structural similarity to 'chip' or 'chit' of adult ♀ (see 5b, above); below these in Fig VII are presumed feeding-calls of ♂—easily distinguished smoother, rounder 'quip' sounds. Recording (Fig XI) of fledged young contains series of calls suggesting 'chi tup tup chi tup tup chi tup', 'chi' calls ascending in pitch, 'tup' descending or cut short. (J Hall-Craggs.)   MGW

**Breeding.** SEASON. North-east Scotland: eggs recorded in all months February–June; of 169 clutches, 11% completed in February, 46% March, 37% April; in general, synchronized with ripening of cones of Scots pine *Pinus sylvestris* so that young can feed themselves before cones are empty of seed (Nethersole-Thompson 1975; Knox 1987*b*). SITE. Almost always in old pine, just in from woodland edge or clearing, high in fork at centre of crown or near end of spreading branch (Knox 1987*b*). Of 123 nests, 4% less than 6 m above ground, 37% 6–12 m, 54% 12–18 m, 6% 18–24 m; average 12·4 m. Of 197, 59% well out on branch, 34% in or close to crown, 8% against or close to trunk; a few nests recorded in spruce *Picea*, larch *Larix*, and Douglas fir *Pseudotsuga*. (Nethersole-Thompson 1975.) Nest: bulky structure, with foundation of twigs of pine, larch, or birch *Betula*, heather, moss, and grass, lined with lichen, fine grass, fragments of pine bark, dead leaves, fur, hair, and a few feathers (Nethersole-Thompson 1975; Knox 1987*b*). Nests of wool and moss recorded, though not laid in; probably built by pairs in immature plumage (Bates and Whitaker 1980). Building: by ♀ only, accompanied by ♂ (Nethersole-Thompson 1975, which see for reports of ♂ helping; Knox 1987*b*); takes 9–12 days, even longer early in season, 4–5 days for replacement; material gathered close to nest, though lining can be brought from up to 300 m away; seems to prefer fresh twigs bitten from tree (Nethersole-Thompson 1975, which see for building stints). EGGS. See Plate 61. Sub-elliptical, smooth and slightly glossy; creamy to bluish- or greenish-white, with sparse reddish to blackish blotches, specks, and sometimes short scrawls, concentrated at broad end; can be flushed pink (Witherby *et al.* 1938; Nethersole-Thompson 1975). 21·6 × 15·9 mm (18·6–24·0 × 14·6–17·3), n = 100; calculated weight 2·85 g (Witherby *et al.* 1938; Schönwetter 1984). Clutch: 3–4 (2–6). Of 174 clutches: 2 eggs, 1%; 3, 30%; 4, 60%; 5, 7%; 6, 1%; average 3·78. Average in February 3·90 (n = 19), March 3·78 (n = 80), April 3·9 (n = 66), May 3·14 (n = 8). 18 ♀♀ had 1st clutch of 3·61 and replacement of 4·1 (n = 22); up to 3 replacement clutches recorded. More than 7 days can elapse between completion of nest and laying of 1st egg, and 8 days (7–11) between loss of clutch and replacement; eggs laid daily in morning. (Nethersole-Thompson 1975.) Possibly 2 broods if food supply adequate (Knox 1987*b*); one ♀ built 2nd nest while feeding fledged young but laid no eggs (Nethersole-Thompson 1975). INCUBATION. At 6 nests, average 13·2 days (12·5–14·5); by ♀ only; probably starts with 2nd or 3rd egg; sits very tightly, taking breaks of 1–17 min 3–4 times per day (Nethersole-Thompson 1975, which see for records of ♂ incubating briefly). YOUNG. Fed and cared for by both parents (Nethersole-Thompson 1975; Knox 1987*b*); ♀ broods day and night for *c.* 5 days, with breaks of up to 20(–120) min; nightly and brief daytime stints continue until young *c.* 10 days old (Nethersole-Thompson 1975). Nestling recorded being fed by sibling (Nethersole-Thompson and Whitaker 1984). FLEDGING TO MATURITY. Fledging period *c.* 21 days (17–25); abundance of seeds may be factor in determining length of period, and some young apparently capable of flight remain in nest; still fed by parents for at least 8 weeks after fledging until mandibles fully crossed, although every stage of development is extended (Newton 1972; Nethersole-Thompson 1975; Knox 1987*b*). For age of first breeding, see Social Pattern and Behaviour. BREEDING SUCCESS. Of 152 eggs in 41 nests, 75·6% hatched and 46·6% produced fledged young, giving 1·7 per nest overall (human disturbance and persecution not included); squirrel *Sciurus vulgaris* most serious predator of eggs and young, though probably has little effect on numbers; eggs hardly ever lost to cold weather (Nethersole-Thompson 1975, which see for other possible predators). Egg-laying delayed and failure rate

high in pairs breeding in immature plumage (Bates and Whitaker 1980).                                                                     BH

**Plumages.** Identical to Crossbill *L. curvirostra curvirostra*, differing only in size, especially bill depth (see Measurements and Structure).

**Bare parts.** As *L. c. curvirostra*.

**Moults.** As *L. c. curvirostra*. 4–5 old secondaries retained in 3 adults examined, and these birds perhaps in 2nd–3rd year of life, as in Parrot Crossbill *L. pytyopsittacus*; moult not started in single worn adults from 12 July and 1 September, but one from 16 July had just started (primary moult score 16) (BMNH). Of 15 1st adults examined, August–March, 12 had median upper wing-coverts new, 9 also had tertial coverts (inner greater coverts) new, and 4 had tertials new, thus moult on average perhaps more extensive than in *L. pytyopsittacus*, but samples small (BMNH, RMNH, ZMA).

**Measurements.** Scotland, ages combined; all year, skins (BMNH, RMNH, ZMA). Bill (S) to skull, bill (N) to distal corner of nostril; exposed culmen on average 3·9 less than bill (S). Depth is approximate depth at base of closed bill; width is width of lower mandible at point where sides of rami meet feathering in middle.

| | | | |
|---|---|---|---|
| WING | ♂ 100·2 (2·18; 32) 97–105 | ♀ 98·1 (2·19; 29) 92–103 |
| TAIL | 58·6 (1·80; 15) 56–62 | 57·0 (1·49; 13) 54–60 |
| BILL (S) | 22·8 (0·94; 15) 21·3–24·9 | 22·3 (1·02; 13) 20·9–24·1 |
| BILL (N) | 15·6 (0·90; 15) 14·3–17·1 | 15·2 (0·66; 13) 14·4–16·4 |
| DEPTH | 12·5 (0·40; 18) 11·8–13·3 | 12·3 (0·41; 24) 11·7–13·2 |
| WIDTH | 11·5 (0·31; 25) 11·0–12·2 | 11·5 (0·36; 29) 10·8–12·2 |
| TARSUS | 18·3 (0·42; 15) 17·6–19·0 | 18·2 (0·89; 13) 16·8–19·1 |

Sex differences not significant.

Scotland, ages combined; skins (Knox 1976). Bill is exposed culmen.

| | | | |
|---|---|---|---|
| WING | ♂ 99·4 (2·4; 39) 95–104 | ♀ 96·6 (2·5; 21) 90–100 |
| BILL | 18·9 (0·6; 39) 17·5–20·0 | 18·8 (0·6; 21) 18·0–20·0 |
| DEPTH | 11·6 (0·6; 20) 11·0–13·0 | 11·5 (0·4; 10) 11·0–12·0 |
| WIDTH | 12·0 (0·4; 35) 11·0–13·0 | 11·6 (0·5; 21) 10·5–12·5 |

**Weights.** All year, mainly spring: adult ♂ 44·6 (6) 42–49; adult ♀ 41·8 (5) 36·5–46; unsexed juveniles 40·2 (5) 38·5–43·6 (A G Knox).

**Structure.** 10 primaries: p8 longest, p9 0–2 shorter, p7 0·5–2, p6 5–8, p5 15–18, p4 21–15, p1 32–37; p10 reduced, 64–71 shorter than p8, 10–14 shorter than longest primary covert. 12 tail-feathers, t1 4–10 shorter than t6. Bill as in *L. c. curvirostra*, but distinctly deeper and (in particular) wider at base, though less so than in *L. pytyopsittacus*; culmen and gonys more strongly curved, tip appearing blunter, tips of mandibles less projecting. See Knox (1976) and Measurements (above).

**Geographical variation.** None.

Formerly usually considered a race of *L. curvirostra* (Hartert 1903–10; Witherby *et al.* 1938; Vaurie 1956*b*, 1959; Voous 1960*b*) or as a race of *L. pytyopsittacus* (Hartert and Steinbacher 1932–8; Niethammer 1937; Dementiev and Gladkov 1954), measurements being more or less intermediate between these. However, lengths of wing, tail, bill, and tarsus are virtually identical to those of *L. c. curvirostra* and very different from those of *L. pytyopsittacus*;

only degree of curvature of culmen, bill depth, and bill width are about halfway between *L. c. curvirostra* and *L. pytyopsittacus* (see Measurements of those species; also Knox 1976), favouring position of *scotica* as a race of *L. curvirostra* which has bill adapted to feeding on cones of Scots pine *Pinus sylvestris*, rather than those of spruce *Picea* as in *L. c. curvirostra*. However, behaviour keeps nominate *curvirostra* and *scotica* reproductively isolated in Scotland (Nethersole-Thompson 1975) and though ranges overlap at present and probably in past, hybridization

appears to be non-existent (Knox 1990b, c). *L. scotica* has thus reached species status, parallel to situation in continental northern Europe where pine specialist *L. pytyopsittacus* overlaps widely with spruce generalist *L. c. curvirostra* without interbreeding. See Knox (1976, 1990b) and Voous (1978) for discussion of status.

**Recognition.** Not always separable, even in the hand, from *L. c. curvirostra*; see Knox (1990a).                              CSR

---

*Loxia pytyopsittacus* **Parrot Crossbill**

PLATES 43 and 45 (flight)
[between pages 640 and 641]

DU. Grote Kruisbek          FR. Bec-croisé perroquet      GE. Kiefernkreuzschnabel
RU. Клест-сосновик          SP. Piquituerto lorito        SW. Större Korsnäbb

*Loxia Pytyopsittacus* Borkhausen, 1793

Monotypic

**Field characters.** 17·5 cm; wing-span 30·5–33 cm. ♂ noticeably larger and bulkier than Crossbill *L. curvirostra*, with striking parrot-like bill (mandibles less crossed than in *L. curvirostra*), little or no forehead, flat crown on large head, and thick neck, pot-belly, and short-tailed appearance; some ♀♀ and juveniles 5% smaller, with less deep bill and more obvious forehead but still thick neck; size overlaps with Scottish Crossbill *L. scotica*. Plumage similar to *L. curvirostra* but adult duller on wings. Flight as *L. curvirostra* but ♂ looks front-heavy. Voice said to include distinctive calls. Sexes dissimilar; no visible seasonal variation. Juvenile separable.

ADULT MALE. Moults: June–November (complete). Plumage as adult *L. curvirostra* but tones less pure, usually duller red overall and duskier on nape, mantle, and particularly wings, which look less black and contrast less with body (though wing of 1st- and 2nd-year *L. curvirostra* has similar contrast to that in *L. pytyopsittacus*). Bill shape diagnostic: compared to *L. curvirostra*, noticeably deeper and relative to head shorter and broader-based, with little or no step between base of culmen and forehead, and shorter curved mandible tips which only just cross; at close range, shows sharply and evenly decurved arch of upper mandible and noticeable S-curve along cutting edge of lower mandible formed by blunt tooth and shorter, much recurved tip (which hardly projects over culmen). Bill dusky horn with dark culmen; distinctive bright bluegrey to ivory cutting edges to both mandibles create paler panel along centre of closed bill than on most *L. curvirostra* and further enhance its depth. Note that separation from *L. scotica* has hardly been studied. ADULT FEMALE. Plumage resembles *L. curvirostra* even more closely than does ♂, but also shows greyer, less black wings. Bill structure similar to ♂, but averages 10% smaller and can be 40%

smaller. JUVENILE. Resembles *L. curvirostra* in plumage. Until about halfway through post-juvenile moult (some time between May and September), bill smaller than adult's; may even look uncrossed, showing only tiny point of upper mandible.

Typical large, bull-necked, parrot-billed ♂ distinctive, but smaller individuals of either sex at any age converge with *L. scotica* and even with biggest *L. curvirostra*, so separation requires close observation of bill structure (see above) and hearing of apparently distinctive calls (see below). Flight and behaviour much as *L. scotica* and *L. curvirostra*, but shape in flight is most compact, heaviest (particularly at front), and most powerful of genus. Perched and (particularly) on ground, stance less upright than *L. curvirostra* and silhouette not only heavy at front but also pot-bellied and short- and thin-tailed. Erupts far less regularly and in smaller numbers than *L. curvirostra* but mixes freely with that species.

Voice generally slightly deeper and louder than *L. curvirostra*, but individual calls may be indistinguishable given different degrees of emphasis and apparent regional variation. Song generally slower, better enunciated, and deeper. Calls considered specific to *L. pytyopsittacus*: deep 'kop kop', 'choop choop', or 'chok', recalling ♂ Blackbird *Turdus merula* (given by feeding bird); very hard 'cherk cherk', given in alarm and deeper than 'chük-chük' of *L. curvirostra*; 'püt püt püt', louder and more melodious than *L. curvirostra*. Beware higher pitch of juvenile's calls.

**Habitat.** In boreal north-west Palearctic, between July isotherms c. 14–18°C (Voous 1960b). Distinguished from Crossbill *L. curvirostra* by almost total specialization to pine *Pinus* forests, usually mature and open, but also on marshy land as well as more usual dry or mountain terrain.

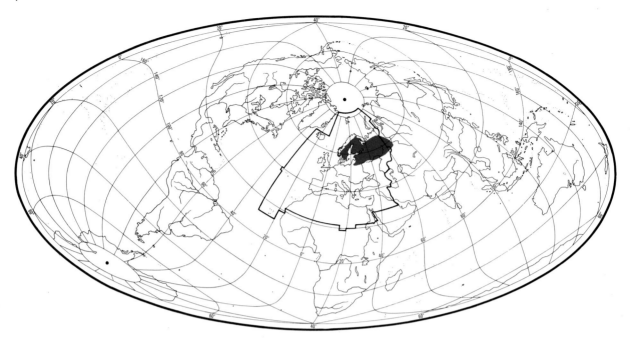

Sometimes in mixture of pine and other conifers. Despite successful adaptation to pines, has not spread to many pine forests occupied by *L. curvirostra* outside boreal zone, or even to some extensive pine forests unoccupied by *Loxia* within it (Newton 1972; Harrison 1982). In Russia, associated with tall dry forest (Dementiev and Gladkov 1954; Flint *et al.* 1984). Various Scandinavian accounts quoted by Bannerman (1953a) stressed habitual occurrence in mixed forest, where it is even said to prefer seeds of spruce *Picea* to those of pine when both available; larch *Larix* and rowan *Sorbus* also attacked in mixed forest.

**Distribution.** Not well known, due to confusion with more widespread Crossbill *L. curvirostra*.

BRITAIN. 1–2 pairs bred at different sites in northern and eastern England in 3 successive years 1983–5 (Spencer *et al.* 1988a). NETHERLANDS. Bred successfully at 2 sites in 1983, following invasion in 1982 (SOVON 1988). DENMARK. Occasional breeder Jylland; last recorded 1960 (TD). NORWAY. Breeding distribution not well known (confusion with *L. curvirostra*); much commoner than *L. curvirostra* in west, and also occurs further north (VR). POLAND. Few breeding records in 20th century, and 19th century records hard to evaluate. Confirmed breeding on Baltic coast 1962, and strongly suspected later. (AD, LT.) LATVIA. No confirmed breeding 1980–4 (Vīksne 1989).

Accidental. Iceland, France, Czechoslovakia, Austria, Italy, Yugoslavia.

**Population.** Fluctuates, but apparently not as widely as Crossbill *L. curvirostra*.

SWEDEN. Estimated *c.* 50 000 pairs in peak years (Ulf-strand and Högstedt 1976). FINLAND. Never as abundant as *L. curvirostra*; seems to fluctuate less. Rough estimate 20 000–50 000 pairs. (Koskimies 1989.)

**Movements.** Resident and dispersive; also eruptive.

In most years, makes only limited movements in response to local food shortage, but occasionally makes eruptive movements, often in same year as Crossbill *L. curvirostra* (in 22 of 27 *L. pytyopsittacus* eruptions recorded 1800–1965), but less extensive and reaching north-west Europe later in autumn. Main food source, pine *Pinus*, often flowers poorly in same year as spruce *Picea* (main food of *L. curvirostra*) and larch *Larix* (main food of Two-barred Crossbill *L. leucoptera*), but pine cones then take nearly 2 years to mature, larch and spruce cones less than a year; so same-year exodus not always attributable to simultaneous food shortage, and perhaps pre-eruption excitement spreads from one species to another. Pine-cone crops fluctuate less than those of spruce, and eruptions of *L. pytyopsittacus* are less frequent than those of *L. curvirostra*. *L. pytyopsittacus* probably sometimes overlooked, owing to similarity with *L. curvirostra*. (Newton 1972; Nethersole-Thompson 1975; Catley and Hursthouse 1985.) Very few ringing data.

Migrating birds head chiefly south-west, reaching Denmark in most years, and extending further south and west in years of eruption. Recent major eruptions 1962 (coinciding with *L. curvirostra*), 1982, and 1990 (coinciding with both *L. curvirostra* and *L. leucoptera*). Movement reached Belgium and Britain in 1962, but was more widespread in 1982, with records west to Britain and east to Mecklenburg (north-east Germany) (Lippens and Wille

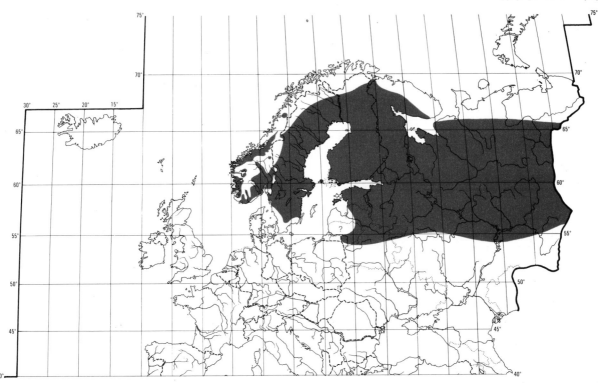

1972; Catley and Hursthouse 1985; Klafs and Stübs 1987); southernmost record (June 1983, presumably also from 1982 invasion) at Doubs (*c.* 47°50′N, eastern France) (Dubois and Yésou 1992). Evidence from Sweden suggests 1982 eruption had northern Scandinavian origin; also, pine-cone crop failed almost entirely in Norway (Catley and Hursthouse 1985). Irruption in 1990 was also widespread, reaching Britain, Belgium, eastern France, and south-west Germany (Rogers *et al.* 1991; Bundesdeutscher Seltenheitenausschuss 1992; *Br. Birds* 1991, **84**, 235; 1992, **85**, 15; P J Dubois). Birds sometimes remain to breed in invasion areas, e.g. in Norfolk (probably also Suffolk), eastern England, 1984 and 1985 (Dymond *et al.* 1989), Veluwe area (central Netherlands) 1983 and 1984, and at several sites in Denmark 1983 (Catley and Hursthouse 1985).

Autumn movement late; first records outside breeding areas mid- or late September, with main passage October–November (Møller 1979; Hirschfeld *et al.* 1983; Catley and Hursthouse 1985). In spring, birds leave wintering areas mostly mid-February to March (Catley and Hursthouse 1985; Schekkerman 1986*a*). In Denmark, movement February–May, with most records in March; highest numbers in Skagen in north-east (Møller 1978*a*). Norfolk birds returned to Holkham breeding area, probably from Wolferton *c.* 27 km south-west, late January to February (Seago 1986, 1987).

In Britain, former status of Scottish Crossbill *L. scotica* as race of *L. pytyopsittacus* has confused older reports.

Before 1958, 13 acceptable records involving 18 individuals; more recently, irruptions in 1962, 1982, and 1990, and occasional reports in other years (Catley and Hursthouse 1985; Rogers *et al.* 1991). Not reported from Ireland (Hutchinson 1989; O'Sullivan and Smiddy 1991). Invasions show clear-cut pattern, with initial arrivals on east coast north to Scottish islands (few on mainland Scotland) from mid- or late September but chiefly in October, much later than *L. curvirostra*, and often coinciding with arrival of other Scandinavian winter visitors and Asiatic vagrants; inland records are later, showing sporadic wintering of small flocks; once established, such flocks very stable November–January, but less so February–March as birds begin to depart. In 1962-3, arrival was concentrated in north Scottish islands, with 61 of 85 reports on Fair Isle; also seen in eastern England at Spurn (October) and Lincoln (wintering flock from January, still present late May), and in Surrey in south-east (May). In 1982-3, minimum total 104 individuals, but comparison with 1962 not possible, owing to great increase in observers and identification skills; arrivals reported on east coast from Norfolk to north Scottish islands; wintering flocks present in Pennines (chiefly South Yorkshire and Derbyshire) from 30 October; most dispersed by mid-February, but at least 11 present 23 February to 7 May at Langsett (South Yorkshire). (Catley and Hursthouse 1985.) In 1990-91, invasion later than usual, not peaking until last 3 weeks of November, with further increase in early January; recorded from Shetland south to Kent, but chiefly from

Durham to Lincolnshire; in all, over 200 reports by end of March (Nightingale and Allsopp 1991; Rogers *et al.* 1991, 1992).

In Skåne (southern Sweden), 1982 movement exceptional, with highest numbers in October; 2199 *L. pytyopsittacus*, 214 *L. curvirostra*, 1957 undetermined (Hirschfeld *et al.* 1984 1983). In Denmark, recorded almost every year, usually in small numbers; distribution varies, with most reports sometimes in west, sometimes in east; 0-125 birds in most years 1960-1 to 1975-6, but 208-1227 in 5 invasion years; 2 invasions (1966, 1972) were same year as *L. curvirostra*, others (1963, 1968, 1975) were independent of other species. Highest total in 1975-6, though no evidence of invasion further south or west, but only 120 birds in 1962-3, when conspicuous in Britain. 58% ♂♂ in irruption years, 47·8% in other years. (Møller 1979.) High numbers also recorded in Denmark 1982 and 1990 (*Br. Birds* 1983, **76**, 276; 1991, **84**, 235). In western Germany, 1977-90, 13 reports involving *c.* 70-80 birds, all in 1982, 1983, 1985, and 1990, mostly in north-west, but south to Lörrach (47°37′N); probably underrecorded in 1982 and 1990 (Bundesdeutscher Seltenheitenausschuss 1989, 1991, 1992). In Netherlands, only 11 records 1900-81, but *c.* 230 in 1982-3, mostly in coastal forests of Noordholland; rare in 1984-9, but probably *c.* 1415 in 1990-91, reaching further south and more inland than in 1982-3 (Berg *et al.* 1991, 1992; B de Bruin). Rare records in France are chiefly in east; 5 individuals in 20th century, in October 1962, June 1983, May 1986 (furthest west, at *c.* 2°E), January 1991, February 1991; southernmost record was at Crest, 44°44′N, in 1896 (Dubois and Yésou 1992; P J Dubois). At Serrahn (north-east Germany), 1962-71, 1-13 birds in most years, but 35 in 1966 (Klafs and Stübs 1987); following 1990 invasion, up to 58 birds (all adults) recorded on Rügen island (north-east Germany) mid-February to beginning of April 1991, with peak at end of February (Dittberner and Dittberner 1992). On Polish coast, 1960-67, 1 bird caught in 1960 and 8 in 1962 (Tomiałojć 1976). On Kola peninsula (north-west Russia), where breeds, invasions 1940-41, 1958-9, 1973-4, 1977-8 and 1987 (Semenov-Tyan-Shanski and Gilyazov 1991), thus not in same years as further south-west. Occasionally recorded south-east of range in western Siberia: at Tomsk (85°E), 1911, and at Novosibirsk (83°E), 1935 (Johansen 1944). DFV

**Food.** Conifer seeds, mainly pine *Pinus*, especially *P. sylvestris*, but also spruce *Picea*; some invertebrates in breeding season. Foraging method identical to that of Crossbill *L. curvirostra* and Two-barred Crossbill *L. leucoptera* when all 3 species watched together in Murmansk region of north-west Russia (Kokhanov and Gaev 1970) and Finnish Lapland (Pulliainen 1972); for details, see *L. curvirostra*. In central England, wintering flock fed mostly on pine, sometimes on larch *Larix*, but not on spruce; typically agile and acrobatic when feeding, biting off cones

and carrying them to branch, sometimes in flight, to extract seeds; green, closed cones dealt with as easily as ripe ones (Catley and Hursthouse 1985); deals with closed cones more efficiently than *L. curvirostra* (Nethersole-Thompson 1975). Also recorded transporting detached cones to tree-stump for treatment; distance between point of removal and perch 1-5 m (Dittberner and Dittberner 1992). In western Norway, average 0·8 mature pine seeds left in cone (*n* = 71 cones) and immature cones avoided as unprofitable; distribution of mature seeds in cones unpredictable, so birds empty each one (Tombre-Steen 1991*a*, which see for details). Similarly in central England, cones were emptied systematically, in contrast to haphazard method of *L. curvirostra* (Catley and Hursthouse 1985), though according to other observers extracts only 10-20 seeds per cone (Olsson 1960; Kokhanov and Gaev 1970), or only those at top of cone (Hildén 1974*a*). In eastern England, exhausted migrants fed on ground and on seeds of thistle *Cirsium*; on Fair Isle, seed of thrift *Armeria* was principal food, as well as thistles and oats *Avena*, picked from stooks and stubble (Davis 1963; Catley and Hursthouse 1985). In Finnish Lapland, ate pieces of decayed wood from house walls, probably for minerals (Pulliainen *et al.* 1978). Often seen drinking fresh and salt water (Land and Lewis 1986; Dittberner and Dittberner 1992). Recorded turning over horse dung for seeds (Bannerman 1953*a*).

Diet in west Palearctic includes the following. Invertebrates: bugs (Hemiptera: Coccoidea), Hymenoptera (Tenthredinidae larvae, Diprionidae larvae, Ichneumonidae pupae), beetles (Coleoptera: Curculionidae). Plants: seeds, buds, shoots, etc., of larch *Larix*, pine *Pinus*, spruce *Picea*, poplar *Populus*, alder *Alnus*, thrift *Armeria*, crowberry *Empetrum*, bilberry *Vaccinium*, rowan *Sorbus*, thistle *Cirsium*, oats *Avena*. (Bannerman 1953*a*; Dementiev and Gladkov 1954; Turček 1961; Olsson 1964; Pulliainen 1972; Nethersole-Thompson 1975; Catley and Hursthouse 1985.) See Kokhanov and Gaev (1970) for list of items taken by 3 *Loxia* species in Murmansk region.

In Finnish Lapland, July-August, 489 items in 15 stomachs were 71·6% by number (61·2% by volume) new spruce seeds, 25·4% (30·4%) old seeds, 2·9% (8·4%) Tenthredinidae larvae, 0·2% by number Ichneumonidae pupae. Mixed flocks of all 3 *Loxia* species fed on spruce cones in trees and on ground in spring and summer, moving to new cones in July. (Pulliainen 1972.) In south-west Sweden, fed on spruce during incubation but changed to pine as soon as eggs hatched (Olsson 1964). Birds in July gorged themselves on scale insects (Coccoidea) (Bannerman 1953*a*).

In Murmansk region, winter and spring, average daily intake was 2450 spruce seeds (6·9 g dry weight) consumed in average 116 min, or 21 seeds per min, excluding search time; slightly faster than *L. leucoptera* and *L. curvirostra*; dealt with pine cone in average 3·4 min. (Kokhanov and Gaev 1970.) Birds can work at 1 cone for up to 30 min (Bannerman 1953*a*). Feeding rates for pine seeds in other studies were from 18 per min (Tombre-Steen 1991*a*, which see for details) to 40 per min (Olsson 1960).

Only regurgitated conifer seeds recorded in diet of young, though invertebrates probably taken occasionally. Breeding season timed so that hatching coincides with opening of pine cones (Catley and Hursthouse 1985). In Murmansk region, average daily intake at 1–12 days old *c*. 350 spruce seeds (*c*. 1·0 g dry weight), at 13–23 days *c*. 2800 seeds (*c*. 7·8 g); over nestling period, brood receives *c*. 105 000 seeds (*c*. 294 g) (Kokhanov and Gaev 1970). In south-west Sweden, brood received average *c*. 2 g of pine seeds per feed (350 seeds, or 3500 per day), giving 70 000–85 000 per brood over nestling period (Olsson 1960, 1964). BH

**Social pattern and behaviour.** No comprehensive study. Some aspects investigated by Olsson (1960, 1964) in Östergötland (Sweden); further useful data in comparative study of 3 *Loxia* on Kola peninsula (north-west Russia) by Kokhanov and Gaev (1970), in review by Nethersole-Thompson (1975), and in description of exceptional breeding in Norfolk (eastern England) by Davidson (1985) and Land and Lewis (1986).

1. Dispersion through the year much as in other *Loxia*. Largest flocks (single species, or often mixed with Crossbill *L. curvirostra*) occur during invasions (see Movements): in Denmark, September–June, mean flock size peaked at 15·0 ± 28·2 (*n* = 46) in December and 13·0 ± 26·4 (*n* = 38) in March, with low 1·6 ± 1·3 (*n* = 18) in May; flocks up to 570 in 1982–3 irruption (Salomonsen 1948; Møller 1979, 1981*b*; Catley and Hursthouse 1985); for Netherlands in 1982–3, see Schekkerman (1986*a*). On Fair Isle (Shetland), during 1962 irruption, flocks of 20–33 (Davis 1963); on Rügen island (eastern Germany), 1990–1, single birds or up to 31 together (Dittberner and Dittberner 1992). Like other *Loxia*, breeding ♂♂ form small parties when foraging for their incubating mates (Kokhanov and Gaev 1970; Hildén 1974*a*; A G Knox); in Norfolk, 2 non-breeding ♂♂ regularly fed with paired breeding ♂ (Davidson 1985); see below. BONDS. Mating system not studied in detail, but nothing to suggest other than monogamy; see Scottish Crossbill *L. scotica*. Hybridization recorded with *L. curvirostra* in captivity (Gray 1958). Pair-formation evidently takes place in flocks as in other *Loxia* (see Heterosexual Behaviour, below). In Norfolk, when ♂ one of ♂ disappeared (probably killed), ♀ re-paired with one of 2 non-breeding ♂♂, which had regularly associated with original pair, until her young fledged (see below); occasionally accepted food from new mate, but rejected his copulation attempts (Davidson 1985). Only ♀ builds nest and incubates; fed by ♂ (Dementiev and Gladkov 1954; Olsson 1964); in Norfolk, ♂ recorded covering eggs briefly during ♀'s absence (Davidson 1985; Land and Lewis 1986). Young fed by both sexes in nest and (like other *Loxia*) for *c*. 4–6 weeks after fledging (Land and Lewis 1986; Semenov-Tyan-Shanski and Gilyazov 1991), but apparently some variation. In Östergötland, at one nest, ♂'s share of feeding increased (and none by ♀) from day 16; at another nest, both fed young up to day 18 and perhaps to fledging; at 3rd nest, ♀ cared for young alone after ♂ disappeared on day 16 (Olsson 1964). In Norfolk, ♂ brought all food for first *c*. 6–7 days, then ♀'s share rapidly increased until equal with ♂'s by time he vanished; ♀ then almost doubled feeding rate to meet nestlings' demands up to fledging, her new mate apparently not assisting (Davidson 1985). Also in Norfolk, in following year, when ♀ built new nest and re-laid, 1st-brood fledglings tended mainly by ♂, but ♂ still

brought more food than ♀ to 2nd nest for first 7–10 days, ♀ then associating increasingly with ♂ and 1st-brood young; parents and both broods perhaps re-united after 2nd brood fledged; first attempts at self-feeding by young at *c*. 4 weeks after fledging and independent *c*. 2 weeks later (Land and Lewis 1986). Of 5 birds (none ringed) seen at nest with 2 young in Norway, 2 presumed to be parents; status of others not clear (Tombre-Steen 1991*b*); however, independent young of *Loxia* continue to follow parents about (helper at nest recorded in *L. curvirostra*), and other conspecific birds (including non-breeders) frequently associate closely with breeders (see Relations within Family Group, below, and other *Loxia* accounts). Age of first breeding 1 year (C S Roselaar, from specimens); report by Catley and Hursthouse (1985) suggested one or both (successive) mates of Norfolk ♀ were 'immature'. BREEDING DISPERSION. Only information is from Norway, where apparently typically in loose neighbourhood groups: e.g. 7 pairs within 'small radius' east of Kristiansand (Valeur 1946); nests generally 100–200 m apart (Haftorn 1971; Spjøtvoll 1972). Territory 'small' according to Naumann (1900), though reported as 200 m across in 3 *Loxia* studied by Kokhanov and Gaev (1970). In Norfolk, nest and its close vicinity defended by ♂ against 2 non-breeding ♂♂ (Davidson 1985). Much courtship presumably takes place in flock (as in other *Loxia*), but copulation normally in territory (see Heterosexual Behaviour, below). Most feeding probably done outside territory: e.g. in Östergötland and Finland, ♂ collected food for ♀ up to 1 km from nest (Olsson 1964; Nethersole-Thompson 1975). Norfolk ♂ recorded defending 2 favourite feeding trees; pair and 1st-brood fledglings fed up to *c*. 300 m away after 2nd brood started in nest separated from 1st by *c*. 60 m (Land and Lewis 1986). As in *L. curvirostra*, breeding density tends to be higher where good cone crop (Catley and Hursthouse 1985). In Finland, nests mainly in mature pine *Pinus* forest, locally up to 5 pairs per km² (Koskimies 1989); in uniform forest in north, 1·4–2·5 pairs per km²; in fragmented forest, 1·8 (Virkkala 1987); on Åland islands, 0·06 pairs per km² (Haila *et al.* 1979); at Kirkkonummi, 12–27 birds along 16 km in January–February of best out of 11 years, and of 8 nests found, 3 were in area of 2·5 ha (120 per km²); even in peak years, much scarcer than *L. curvirostra*, the 2 species not normally breeding in same forest in same year according to Hildén (1974*a*). On Kola peninsula, maximum in mixed forest in June 0·5–2·0 pairs per km² (Semenov-Tyan-Shanski and Gilyazov 1991). On Hel peninsula (northern Poland), out of 200–300 birds in *c*. 4 km², *c*. 70% this species, the rest *L. curvirostra* (Gotzman and Wisiński 1965). ROOSTING. Probably much as in *L. curvirostra* (see Naumann 1900 and that species). In Finland, March–April (after breeding), ♂♂ recorded roosting on quite open bare branches, ♀♀ not visible, so perhaps in dense foliage (A G Knox). On Kola peninsula, 31 May, ♀ left nest 05.37 hrs, making final visit for nocturnal brooding at 18.47 hrs; on following day, 04.50 and 16.34 hrs respectively (Kokhanov and Gaev 1970, which see for details of activity rhythm).

2. Typically tame and approachable when feeding (e.g. Naumann 1900, Salomonsen 1930*b*, Bannerman 1953*a*); more confiding than *L. curvirostra*, and recorded being trapped with hands for ringing (Génsbøl 1964; Møller 1981*b*); always warier when on ground (for drinking or grit) (A G Knox), e.g. not permitting approach within *c*. 20 m (Dittberner and Dittberner 1992). When alert or otherwise excited, alternates between sleeking and ruffling of head feathers (Catley and Hursthouse 1985; Svensson 1991). FLOCK BEHAVIOUR. Flocks, particularly large ones, are noisy, calling (see 1a in Voice) in flight and when landing and starting to feed; variety of calls, some accompanying disputes (see, e.g., 3d in Voice), noted from feeding birds (Bannerman 1953*a*; Catley and Hursthouse 1985; Dittberner and

Dittberner 1992). On Rügen, while flock feeding, some birds on high perch apparently acted as sentinels, loud 'göp' calls (see 2a-c in Voice) from one such bird signalling alarm (for raptors such as Buzzard *Buteo buteo* and Sparrowhawk *Accipiter nisus*) and causing whole flock to take off and sometimes to fly about restlessly (Dittberner and Dittberner 1992). In Norway, flock recorded mobbing (no details) fledgling Great Grey Owl *Strix nebulosa* (Bannerman 1953a). Will also mob Tengmalm's Owl *Aegolius funereus* (Finland: A G Knox). SONG-DISPLAY. ♂ sings from perch (usually tree-top) or in flight; ♀ also reported to sing (e.g. perched birds, Rügen: Dittberner and Dittberner 1992), but few details (see 1 in Voice), and overall probably very rare (A G Knox). May use any one of several perches in territory, including close to nest where, however, Finnish study (see Nethersole-Thompson 1975) found birds tended to give only quiet Subsong (see 1b in Voice). Song given with bill closed (Dittberner and Dittberner 1992) or (certainly for loud song) open (A G Knox). Noted from coastal migrants during irruption in Britain, not connected with territory (Catley and Hursthouse 1985); on Rügen, also given by birds in flock between feeding and comfort behaviour, occasionally up to 5 ♂♂ together, and individual ♂♂ in bouts of up to 5-6 min (Dittberner and Dittberner 1992). Song-flights much as described for other *Loxia* (see, especially, *L. scotica*): recorded ascending and fluttering about, singing continuously, before returning to take-off point (Dementiev and Gladkov 1954). Excited ♂ (e.g. advertising territory) may perform butterfly-like Song-flight in arc from tree to tree: has tail spread and beats wings at varying speed and depth, also glides, while singing loudly and persistently (Naumann 1900; Bergmann and Helb 1982; Davidson 1985). Song noted virtually all year, including in very low winter temperatures (Naumann 1900). On Kola peninsula, marked increase in singing activity with pair-formation, and song vigorous mid- to late January; singing occurs even in polar night (Nethersole-Thompson 1975; Semenov-Tyan-Shanski and Gilyazov 1991). In Britain, during 1982-3 (breeding attempted), song and Subsong given by perched birds October-April (Catley and Hursthouse 1985). In West Jylland (Denmark), ♂ sang early July, during nest-building (Mortensen and Birkholm-Clausen 1963). At Kirkkonummi (Finland), in one year, little song noted from ♂♂ before or during breeding, except for quiet Subsong given near nest (Hildén 1974a). At Kuusamo (central Finland), after breeding, birds sang at least into May (A G Knox). ANTAGONISTIC BEHAVIOUR. Threat and fighting occur among feeding birds (Catley and Hursthouse 1985; A G Knox); disputes between ♂♂ on Kola peninsula, mid-February, were associated with pair-formation (Semenov-Tyan-Shanski and Gilyazov 1991). Captive ♀ threatening Hawfinch *Coccothraustes coccothraustes* adopted threat-posture similar to *L. curvirostra* (see Fig A of that species); after forcing it to move to another perch, continued to threaten bill-snapping *C. coccothraustes* in flight (Scherner 1969). HETEROSEXUAL BEHAVIOUR. (1) General. On Kola peninsula, pair-formation recorded in some years in October, in others January-February, single pairs or sometimes several together splitting away from main flock (Kokhanov and Gaev 1970; Nethersole-Thompson 1975; Semenov-Tyan-Shanski and Gilyazov 1991). In Sweden, breeding synchronized over given area (Olsson 1964). (2) Pair-bonding behaviour. Only report refers to ♀ approaching ♂, and subsequent chasing ending in copulation, after which pair maintain close contact (Naumann 1900). (3) Courtship-feeding. ♂ feeds ♀ regularly during building and incubation, usually on or close to nest. At one nest, Kola peninsula, mid-April, 7 visits to nest by ♂ 07.53-18.02 hrs; decline towards end of day (Kokhanov and Gaev 1970); feeds every 2-2½ hrs in Finland and Sweden (Olsson 1964; Nethersole-Thompson 1975); in Norway, 11 feeds

averaged 85 min (42-150) apart (Spjøtvoll 1972); in Norfolk, increased in regularity towards end of incubation; 7 feeds 41-113 min apart (Davidson 1985). ♀ clearly able to recognize mate's calls ('gyp': see 2a in Voice) at some distance, whether he comes to nest alone or in flock. In Swedish study, ♀ left nest as ♂ approached, defecated, flew down to drink, then back to nest to await his arrival. There, ♀ crouches and gradually raises head until bill vertical (Fig A); also gives 'gyp' calls and wing-shivers,

A

eyes half-closed; ♂ then regurgitates food and feeds ♀. (Olsson 1960, 1964.) (4) Mating. Takes place on nest or nearby branch during building (Naumann 1900). See also subsection 2 (above). (5) Nest-site selection. By ♀, assisted by ♂, in 3 *Loxia* studied by Kokhanov and Gaev (1970). ♀ also builds nest and often accompanied by ♂ who sometimes sings (Naumann 1900; Olsson 1964; Spjøtvoll 1972). (6) Behaviour at nest. Most ♀♀ incubate from 1st egg, but incomplete clutch sometimes uncovered, even in frost (Valeur 1946; Kokhanov and Gaev 1970). ♀ generally leaves nest infrequently and briefly (see Breeding), to defecate and drink; In Norfolk, ♂ once settled on eggs when mate absent, but was displaced by ♀ when she returned, and then fed her (Davidson 1985). When not foraging, ♂, usually perched in tree-top near nest, may call (see 3a in Voice); in response to begging-calls ('gyp') from ♀, ♂ flies off to collect food (Olsson 1964). RELATIONS WITHIN FAMILY GROUP. On Kola peninsula, young (of 3 *Loxia*) hatching in winter months March-April were brooded by ♀ with few interruptions for *c.* 10-12 days; more likely to be left uncovered at earlier stage later in year (Kokhanov and Gaev 1970). Chicks show remarkable ability to survive in very low temperatures (Valeur 1946; Nethersole-Thompson 1975); presumably aided by diet of fat-rich seeds (Tombre-Steen 1991b). In Swedish study, 6-day-old nestlings left for up to 90 min at 2°C sometimes torpid when ♀ returned and needed brooding for several minutes before being fed; apparently thermoregulate by day 9 (Olsson 1964). At one Norfolk nest, no brooding during day after *c.* 1 week (Land and Lewis 1986). Eyes of nestlings open from *c.* 7-9 days (Olsson 1964); for further details of physical development, see Olsson (1964) and Kokhanov and Gaev (1970). In early stages, when ♀ brooding, all food brought by ♂ and regurgitated for ♀ who then feeds nestlings; this pattern still followed at nests in Sweden for a time even when both parents returned to nest together; ♀ moved bill to left and right when feeding young, but fair distribution not ensured as sometimes no food regurgitated; recorded 'mock-feeding' (no food passed) when none left or ♂ long absent (Olsson 1964); for ♀ regurgitating seeds and feeding young, see Fig B. First direct feeding of young by ♂ (alongside transfer as described) recorded at 8-10 days (Olsson 1964; Land and Lewis 1986) or 4-6 days (Kokhanov and Gaev 1970); see these references, also Spjøtvoll (1972), for further details of feeding visits.

B

At 12–14 days, chicks evidently recognize parents' calls and start to call to them at *c.* 30–50 m, ignoring calls of strangers (Kokhanov and Gaev 1970). Nest kept clean by both sexes; nestling faeces swallowed initially by ♀, ♂ assisting once he starts to feed young direct; chicks later begin to defecate onto nest-rim where faeces accumulate, as parents much less assiduous in removing them especially from about day 15 (Olsson 1960, 1964; Kokhanov and Gaev 1970; Spjøtvoll 1972); various displacement movements by adult noted when young failed to defecate after feed (Olsson 1960). Close to fledging, young were typically fed on nest-rim or on nearby branch (Spjøtvoll 1972). Leave at *c.* 21–23 days (19–25) (see Breeding); one Norfolk brood over 2 days (Davidson 1985). Excited calls of adults perhaps acted as stimulus, young leaving immediately after parents visited nest; 1st-brood fledglings followed ♂ about, continuing to beg from him even while he fed 2nd-brood nestlings (Land and Lewis 1986); see *L. curvirostra*. ANTI-PREDATOR RESPONSES OF YOUNG. No information. PARENTAL ANTI-PREDATOR STRATEGIES. ♀ typically sits tightly once clutch complete (Valeur 1946). In Norway, even allowed herself to be touched and, after leaving, soon returned (observer still in nest-tree); ♂ would also approach nest-tree during disturbance (Spjøtvoll 1972); see also Frendin (1943). ♀ less likely to peck than other *Loxia* studied by Kokhanov and Gaev (1970). More anxious after hatching; flew about nervously when young handled (Spjøtvoll 1972). Alarm-calls (presumably most likely of type 2b: see Voice) given whenever Jay *Garrulus glandarius* near nest or young (Olsson 1964; Nethersole-Thompson 1975).

(Figs by D Nurney: A from photograph in Olsson 1964; B from photograph in Nethersole-Thompson 1975.)   MGW

**Voice.** Claims that it is generally quieter than Crossbill *L. curvirostra* when feeding (e.g. Bannerman 1953*a*, Møller 1981*b*, Catley and Hursthouse 1985) are unsubstantiated (A G Knox). As in *L. curvirostra* (see that species), pitch, volume, and timbre of calls vary considerably, and some are confusingly similar, though *L. pytyopsittacus* reported to be generally slightly louder and lower pitched, with in some cases distinctive timbre (Møller 1981*b*; Knox 1990*c*; Olsen 1991*b*; Svensson 1991). Can also sound similar to Scottish Crossbill *L. scotica* (see that species). For additional sonagrams, see Bergmann and Helb (1982) and Schekkerman (1986*a*).

CALLS OF ADULTS. (1a) Song of ♂. Mostly short phrases, each often comprising segments of repeated units or motifs. Sounds reported are hoarse and loud, some quieter,

attractive, and fluting, and variously modified contact- or flight-calls are frequently incorporated. Some songs similar to *L. curvirostra* and *L. scotica*, but overall lower pitched, and can be slower, with notes more clearly enunciated. (Naumann 1900; Bergmann and Helb 1982; Harris *et al.* 1989.) Large repertoire of unit-types. Shares with *L. curvirostra* and *L. scotica* generally rather uniform song-phrase and sub-phrase duration (unless in long singing bout: A G Knox), also contrast in quality of sounds (timbre of notes and kinds of noises), intensity, and unit duration (J Hall-Craggs). Selection of song-phrases from small sample of recordings analysed as follows. Recording of bird singing in flight (Fig I) contains short phrase comprising loud, audibly 2-part 'dyip' calls, the whole sounding like a slow, squeakily musical rattle. In Fig II, isolated 'cheek' call (probably a contact-call such as quite often given before song) precedes short, delicate and quite musical song of 2 3-note motifs recalling Great Tit *Parus major* but faster (compare *L. curvirostra* and *L. scotica*).

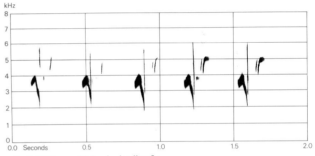

I   A G Knox   Finland   April 1984

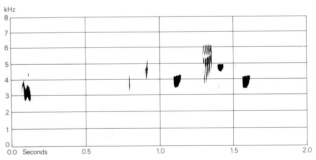

II   A G Knox   Finland   April 1984

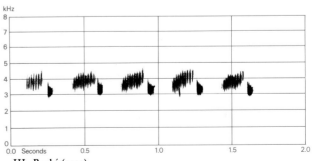

III   Roché (1990)

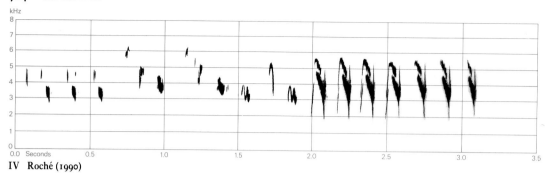

IV  Roché (1990)

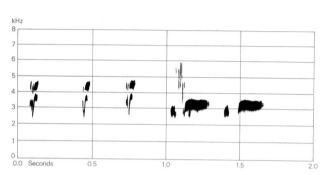

V  S Palmér/Sveriges Radio (1972-80)  Sweden  March 1963

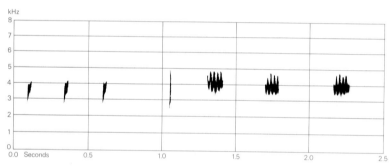

VI  J Lindblad  Sweden  March 1958

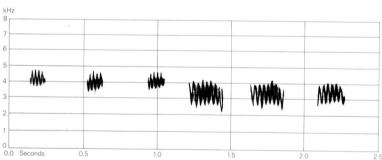

VII  J Lindblad  Sweden  March 1958

In another song-phrase (Fig III), motifs each of 2 units contrasting in tonal quality and duration and delivered at 3 per s suggest 'zrree-tu' and also recall *P. major*. In another recording (Fig IV), 3 quiet sub-phrases, (1) 't-de t-de t-de' (2) 'p-te-che p-te-che p-te-che' (3) 'tp kit tp', are followed by (4) long segment of 7 'chip' or 'dyip' units. Recording from Sweden (Fig V) contains striking song-phrase: 5 (only 3 shown) short, fairly quiet, scratching 'zik' notes in accelerando series, then 2 'ch-weeng' motifs

of piercing tonal quality. Other renderings of introductory units include 'tlip' (Bergmann and Helb 1982) and 'chit', and of final motifs 'tcho-ee' (Harris *et al.* 1989), 'chweng' (Palmér and Boswall 1981), and 'cheeLER-cheeLER' (Bruun *et al.* 1986); see also Dittberner and Dittberner (1992) and Jonsson (1992). Slow song-phrase shown in Fig VI comprises segment of 'tik' units (like gentle tapping on glass) followed by 3 'chwee' sounds with prominent frequency modulation; in Fig VII, 3 'chwee' units precede

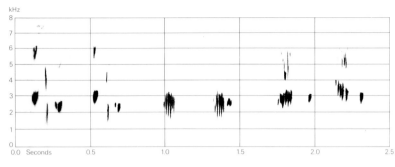

VIII P S Hansen Denmark February 1976

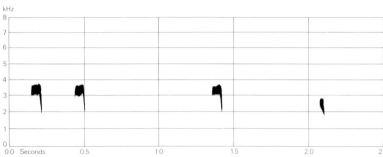

IX A G Knox Finland April 1984

extremely loud, lower-pitched, but also strongly frequency-modulated 'chweeer', rendered in other studies (Naumann 1900; Dittberner and Dittberner 1992) 'errr' and 'ih-ih-tschirh'. (J Hall-Craggs, M G Wilson.) (1b) Subsong. Quiet chattering or twittering comprising flight-calls and other sounds (Bergmann and Helb 1982); like a more rambling, toned down, broken version of full song (1a), mostly subdued, at times becoming slightly louder (A G Knox). Recording of presumed Subsong contains 4 main sound-types: tones, clicks, churring rattles, and tonal rattles; in Fig VIII, 1st (repeated) motif comprises 2 tones separated by click (suggests hurried song of *P. major*), followed by 2 'tirrr' rattles and 2 tonal rattles ('cree') each followed by quiet tone (J Hall-Craggs). (1c) Song of ♀. Reported by Naumann (1900) to be quieter and less persistent than ♂'s song. Captive bird gave song much as described for ♀ *L. curvirostra* (Scherner 1969). See other *Loxia* accounts for caveat regarding reports of ♀ singing perhaps relating to green-plumaged ♂♂. (2) Contact- and alarm-calls. Considerable variation (reflected in transcriptions) and not clear how many different calls exist (some liable to be confused with *L. curvirostra* and *L. scotica*). Comments in some sources which include comparison with *L. curvirostra* (see, e.g., Catley and Hursthouse 1985) probably refer to more than one call variant of *L. pytyopsittacus*. (2a) Contact- and flight-calls. As in *L. curvirostra*, these vary geographically, with only one dominant type usually found in given area, though some overlap between regions; detailed investigation needed (Knox 1992; A G Knox). Calls (often 2 in quick succession) in this category are given quietly when feeding, loudly and more penetratingly when excited or alarmed, prior to take-off and in flight, also after landing. Tran-

scriptions include 'tyoop', 'choop', 'tüpp', 'pyk', 'küp', and soft 'göp'. Differs from hard, clipped call of *L. curvirostra*, being more melodious, lower pitched, and less incisive and far-carrying. (Salomonsen 1930b; Millington and Harrap 1981; Catley and Hursthouse 1985; Harris *et al.* 1989; Olsen 1991b; Svensson 1991; Dittberner and Dittberner 1992; Jonsson 1992.) One call-type in recording (Fig IX; for last unit, see call 3c, below) suggests 'peep' (A G Knox), 'keep' (P J Sellar), or loud piping 'dweep', less ringing than *L. curvirostra* and probably easiest of calls reviewed here to distinguish from *L. curvirostra* (M G Wilson). Another recording (Fig X) con-

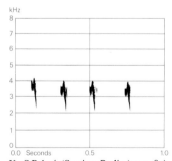

X S Palmér/Sveriges Radio (1972-80) Sweden March 1963

tains series of 4 'chip' calls given in flight; also illustrated by Bergmann and Helb (1982) where rendered 'djip'. Recording of apparently anxious bird (Fig XI) suggests squeaky, scratchy, somewhat strained 'dyip' (J Hall-Craggs, M G Wilson); see also Bergmann and Helb (1982) for this and an obviously related but differently structured call there rendered 'kik'. A 'kip-kip' like emphatic repeated

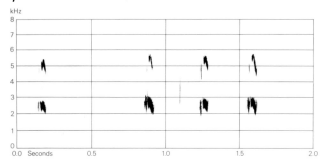

XI   P S Hansen   Denmark   February 1976

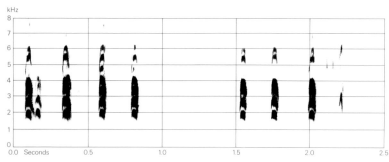

XII   A G Knox   Finland   April 1984

call of woodpecker (midway between Great Spotted Woodpecker *Dendrocopos major* and Three-toed Woodpecker *Picoides tridactylus* (Elmberg 1991); perhaps same as 'kik' or alternative rendering of call shown in Fig IX. (2b) Tep-call. Similar to *L. curvirostra* and *L. scotica* (see those accounts), and like them expresses varying levels of excitement or anxiety (up to full alarm). Recording (Fig XII) suggests 'tip', 'chet', or 'chek', slightly recalling distant Jackdaw *Corvus monedula* (M G Wilson). Typical alarm-call (elicited by, e.g., imitation of Pygmy Owl *Glaucidium passerinum*): hard and low-pitched, loud, harsh 'cherk-cherk' (Bruun *et al.* 1986; Jonsson 1992). (2c) Calls described here provisionally kept separate, but some may be same as or at least closely related to call 2b. Both 2b and 2c are given mainly when perched. Deep 'chok' resembling 'tuk' of Blackbird *Turdus merula* noted from birds feeding calmly (Bruun *et al.* 1986), but obviously same call (series of muffled 'tak' sounds recalling *T. merula* and *C. monedula*, given by both sexes) considered by Scherner (1969) to be same as stuttering 'tsu-tsu-tsu-tsu'

of alarm reported by Olsson (1964); see also Davidson (1985). Recording (Fig XIII) contains series of 'ktiup' calls like slowly-uttered first few units of *T. merula* alarm-rattle, and therefore presumably of type described above (J Hall-Craggs). Further renderings of calls probably in this category include 'quop', 'gop', 'kop', 'göp', and 'quap' (Naumann 1900; Poulsen 1949; Davis 1963; Catley and Hursthouse 1985). (3) Other calls. Probably some overlap with calls described above, but extent not clear. (3a) Weak 'sve-sve-sve-sve' descending in pitch noted only from ♂ close to nest and presumed to serve as contact-call (Olsson 1964). (3b) Quiet 'tip' audible only at close range given by ♂ apparently in association with courtship (Naumann 1900); probably only quiet version of call 2a (above). (3c) Soft 'jüb' recalling Flight-call of Chaffinch *Fringilla coelebs*, associated with resting and low-activity phases (Dittberner and Dittberner 1992; presumably same or related is soft 'jup' shown as last unit of Fig IX. Perhaps a subdued version of call 2b (A G Knox). (3d) Quiet 'chrrih' given during occasional disputes while detaching

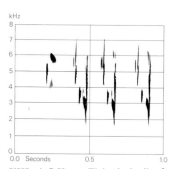

XIII   A G Knox   Finland   April 1984

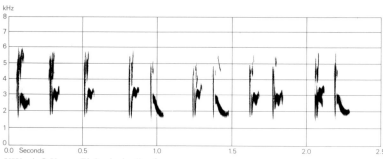

XIV   A G Knox   Finland   April 1984

cone and at take-off (Dittberner and Dittberner 1992); quavering trill recalling Dunlin *Calidris alpina* (Salomonsen 1930b; Bannerman 1953a) is perhaps the same or related.

CALLS OF YOUNG. Trilling sounds noted from nestlings (Olsson 1964; Davidson 1985). Close to fledging (Davidson 1985) and after leaving nest, young give clear, ringing, rather hard 'tii-tu-tit-tit' (Witherby *et al.* 1938; Bergmann and Helb 1982) or 'chit-er chit-er chit-er chit-chit chit chit-er' (Olsson 1964); see also Berndt and Frantzen (1987). One brood in eastern England ceased giving calls of this type ('chitoo') *c.* 5-6 weeks after fledging (Land and Lewis 1986). Recording (Fig XIV) contains sounds superficially like *L. curvirostra* (see that species), but much noisier with very strong (highly locatory) starting transients: 'tink tink tink ti-choo ti-choo tink tink ti-choo' (J Hall-Craggs). Young reported to give adult contact-call ('gyp-gyp': see 2, above) from *c.* 16 days (Olsson 1964). MGW

**Breeding.** SEASON. Finland: eggs laid from beginning of February to late June, sometimes into August, peak mid-March to mid-May; influenced strongly by availability of pine *Pinus* seeds (Haartman 1969, 1973; Hildén 1974a). Scandinavia: of 44 clutches, 5% laid February, 30% March, 52% April, 14% May (Nethersole-Thompson 1975); in south, season mainly March and 1st half of April, pine cones opening when young hatch; record of eggs around beginning of September (Olsson 1964; Haftorn 1971; Tombre-Steen 1991b). Murmansk region (northwest Russia): February to mid-May in years of good pine and spruce *Picea* crop, mid-May to August or September when only pine abundant (Kokhanov and Gaev 1970). England: see Davidson (1985) and Land and Lewis (1986). SITE. High in conifer at woodland edge, by clearing, track, etc., very rarely in dense forest; in spruce close to trunk, in pine in fork among dense twigs, usually a few metres from trunk; can be 20 m above ground, higher nests tending to be closer to trunk (Olsson 1964; Gotzman and Wisiński 1965; Harrison 1975). In Murmansk region, 88% of 24 nests in spruce, 12% in pine; 75% against trunk, 21% on side branch, 4% in fork of trunk; average height 7·0 m (2-15, *n*=18) (Kokhanov and Gaev 1970). For Finland, see Haartman (1969); for Norway, see Spjøtvoll (1972). Nest: foundation of dry conifer twigs, bark (often from deciduous trees), dead leaves, moss, lichen, etc., lined with dead grass, plant down and fibres, hair, sometimes feathers; near man, occasionally fragments of rope, etc.; very like that of Crossbill *L. curvirostra*, perhaps larger, more robust (Olsson 1964; Haartman 1969; Haftorn 1971). In Murmansk region, 6 nests had average outer diameter 14·1 cm (12·0-16·3), inner diameter 6·4 (6·0-6·6), overall height 7·0 (5·8-9·1), and depth of cup 3·8 cm (3·6-4·0) (Kokhanov and Gaev 1970). Building: by ♀ only, usually accompanied by ♂; almost always in morning; takes 7-12 days (Olsson 1964; Kokhanov and Gaev 1970; Haftorn

1971; Davidson 1985); much material gathered from trees, since ground often snow-covered (Olsson 1964; Haartman 1969); in Poland, bottom of nest said to be hard bowl cemented by saliva of ♂ (Gotzman and Wisiński 1965). ♂ recorded partially completing nest alone (Davidson 1985). EGGS. See Plate 61. Sub-elliptical, smooth, and slightly glossy; very like *L. curvirostra*, slightly larger with bolder markings; yellowish-white to pale blue-green with rust to purplish-brown spots, small blotches, and sometimes scrawls, mostly at broad end (Witherby *et al.* 1938; Harrison 1975; Makatsch 1976). 23·1 × 16·6 mm (19·8-26·3 × 15·0-18·0), *n*=236; calculated weight 3·26 g (Schönwetter 1984). Clutch: 3-4 (2-5). In Finland, of 27 clutches (some perhaps incomplete): 2 eggs, 7%; 3, 30%; 4, 48%; 5, 15%; average 3·7 (Haartman 1969). Of 17 clutches in collection: 4 of 3 eggs, 12 of 4, 1 of 5; average 3·8 (Makatsch 1976); see also Nethersole-Thompson (1975). In Murmansk region, average 4·0, *n*=10 (Kokhanov and Gaev 1970). Eggs laid daily (Kokhanov and Gaev 1970; Makatsch 1976); pair in Norfolk (eastern England) had 2 broods, ♀ starting 2nd nest about same time as 1st brood fledged (Land and Lewis 1986); occasionally 2 broods elsewhere (Catley and Hursthouse 1985). INCUBATION. 14-16(-17) days; by ♀ only (Kokhanov and Gaev 1970; Haftorn 1971); ♂ recorded covering eggs briefly (Davidson 1985; Land and Lewis 1986). Probably usually starts with 1st egg (Bannerman 1953a; Olsson 1960; Kokhanov and Gaev 1970; Davidson 1985), but recorded starting with last (Spjøtvoll 1972). ♀ generally sits very tightly; at one nest in Murmansk region took breaks of less than 1 min (Kokhanov and Gaev 1970), though in southern Sweden, while one ♀ did not leave eggs for more than 10 min at a time, another took breaks of up to 45 min despite cold weather (Olsson 1964); in Norfolk, ♀ had average stint on eggs of 110 min (26-183), and breaks of 2-14 min (Davidson 1985). YOUNG. Fed and cared for by both parents (Olsson 1964; Kokhanov and Gaev 1970; Spjøtvoll 1972); brooded by ♀ for *c.* 7-12 days (Land and Lewis 1986; Kokhanov and Gaev 1970), though 6-day-old young can survive being left uncovered in very cold weather for up to 90 min at a time (Olsson 1964); 12-day-old nestlings left for 4 hrs in one nest in low temperature (Land and Lewis 1986). FLEDGING TO MATURITY. Fledging period 21-23 days (19-25); young fed by parents for *c.* 4-6 weeks after fledging, perhaps until mandibles fully crossed, although every stage of development is extended (Olsson 1964; Kokhanov and Gaev 1970; Newton 1972; Land and Lewis 1986). For age of first breeding, see Social Pattern and Behaviour. BREEDING SUCCESS. In Murmansk region, 15 of 24 nestlings fledged, giving 2·5 fledged young per nest, *n*=6 (Kokhanov and Gaev 1970). In Norfolk, only 1 of brood of 4 survived (Davidson 1985). BH

**Plumages.** ADULT MALE. Very similar to adult ♂ Crossbill *L. curvirostra*: when fresh, upperparts dull crimson, more brown-red on hindneck and outer scapulars, black on longest upper tail-coverts; lore and broad stripe through eye to upper ear-coverts dark grey-brown, mottled buff below eye, increasingly spotted red towards lower ear-coverts; lower cheek, side of neck, and entire underparts dark rosy-red, apart from dark olive side of breast, light grey vent, and dull white black-marked under tail-coverts; however, tail, tertials, and entire upper wing on average less blackish than in adult *L. curvirostra*, tinged dark olive-brown, greater number of lesser and medium upper wing-coverts with narrow rufous-brown fringes, secondaries and greater coverts more clearly marked with narrow olive-brown to dull pink fringes (adult ♂ of *L. curvirostra* hardly shows paler fringes along longer lesser, median, and greater coverts, as well as secondaries, but some 1st adults have such fringes on part of feathering); *in worn plumage*, head, mantle, rump, and underparts bright scarlet-red with duller red-brown mantle, dark grey stripe through eye, and grey of feather-bases visible on hindneck and mid-underparts, similar to worn adult ♂ *L. curvirostra*, except for browner (less blackish) wing. Some individual variation in colour of outer fringes of flight-feathers, probably in relation to age: birds with all fringes dull pink or brown-pink (except usually for greenish ones along inner primaries) probably older ones, those with fringes of secondaries partly or mostly olive-green perhaps in 3rd year of life (green-fringed secondaries older than pink-fringed ones, but difference in age sometimes hard to see), those with fringes along some clearly abraded secondaries pale green-grey probably in 2nd year of life (these secondaries probably retained from juvenile plumage). Irrespective of age, some birds have a few olive feathers mixed in red of plumage, especially on mantle, scapulars, and chest, apparently because feathers growing early during adult post-breeding moult are less colourful than those growing later. ADULT FEMALE. Completely similar to adult ♀ of *L. curvirostra*: upperwing on average slightly more olive-grey, less shiny greyish-black, but some individual variation; as in *L. curvirostra*, some birds greener (rump, chest, and flank even yellowish-green when fresh), others duller grey, especially when worn. NESTLING. Down rather sparse at hatching, but dense and fairly long at age of *c.* 1 week, dark lilac-brown, covering top of head, mid-back, upperwing, and thigh (Olsson 1960, which see for development). On day 1, same as *L. curvirostra* (Kokhanov and Gaev 1970). JUVENILE. Only 6 examined; all these completely similar to juvenile *L. curvirostra*, closely streaked dusky grey and pale green to off-white from forehead to upper mantle and on side of head and neck, spotted dusky grey on olive-green of scapulars and back, streaked dark olive and pale yellow on rump and upper tail-coverts, pale green-yellow to off-white on underparts, where closely marked with olive shaft-streaks 1–2 mm wide. In just-fledged birds with wing not yet full-grown, bill almost as heavy at base as in adult (much thicker than in *L. curvirostra*), but tips not yet crossed; in other fresh juveniles (with full-grown wing) bill-tips crossed and shape as in adult, except perhaps for slightly less strong gonys bulge. FIRST ADULT. ♀ as adult ♀, ♂ highly variable, but both with juvenile tail, flight-feathers, tertials, and greater as well as usually median coverts retained; all fringes of tail- and flight-feathers green (unlike adult ♂), tips of median and greater coverts pale buff-yellow or yellow-green (less olive than in adult ♀), but latter rapidly worn off; 1st adult ♂ variable mixture of olive-green and reddish-orange on body as in *L. curvirostra*, sometimes mixed with some streaked juvenile feathers (especially on neck and underparts), sometimes largely reddish-orange, but none as crimson or scarlet as adult ♂; in heavily worn plumage, when *c.* 1 year old, head, rump, and underparts scarlet-orange with much dull grey of feather-bases visible. Separable from *L. curvirostra* by size and bill-shape only (see Measurements and Structure; also Catley and Hursthouse 1985).

**Bare parts.** ADULT, FIRST ADULT, JUVENILE. Similar to Crossbill *Loxia curvirostra*; cutting edges of both mandibles sometimes pale blue-grey to ivory-white, widest on terminal half, but these not always readily visible (depending on light and position of bill), and some *L. curvirostra* have equally pale cutting edges; when present, often more visible than in *L. curvirostra* due to more concave terminal half of cutting edges, which are less fully concealed when bill closed. NESTLING. Mouth light red-violet, gape-flanges bright yellow; bill red-violet, from day 4 developing conspicuous yellow swelling on each side of base of lower mandible (Olsson 1960).

**Moults.** ADULT POST-BREEDING. Complete or almost so; primaries descendent. In north-west Russia, starts mid-June to mid-July (peak late June), completed mid-September to November (peak early October); many or all birds examined late June to late July ($n = 14$) and mid- and late September ($n = 10$) were in moult, but only 10–20% of 110 birds examined October (Rymkevich 1990). Moult in Murmansk area mainly August–October, but one of 6 birds in July and 1 of 8 in November were also in moult (Kokhanov and Gaev 1970). In birds examined, plumage worn June to early August (2 red ♂♂ from August had a few new yellow-green feathers growing on body), fresh in another bird from August and in birds from October–April (RMNH, ZMA). Moult completed in ♂ from 13 September, Lapland; moult likely to be variable, depending on breeding season (Svensson 1992). Of 11 birds older than 1 year, 5 had secondary moult complete (♂♂ among these were deep scarlet), remainder retained 2–5 old secondaries, these feathers differing from new ones especially in colour of outer fringe, and (to lesser extent) in degree of abrasion; body of ♂♂ with partly old secondaries was more orange-red than ♂♂ which had complete moult, and at least some of former were in 2nd year, old secondaries being heavily worn, fringed grey, and apparently juvenile, but others had old secondaries scarcely more worn than new ones, fringes green, these birds possibly in 3rd year. Of 6 birds, old s2 and s6 retained in 2, s3 in 4, s4 in 5, and s5 in all 6. POST-JUVENILE. Partial: head, body, and lesser upper wing-coverts; in contrast to *L. curvirostra*, only occasionally includes median coverts (in 5 of 12 birds) and some inner greater coverts (tertial coverts in 5 of 12 birds, 1–3 inner greater coverts in 2 of 12); no tertials replaced in this sample. Birds examined included juveniles dated late April to early August, birds just starting moult late April to early September, and birds with moult completed from August onwards, but moult perhaps occurs earlier as well. In north-west Russia, none of 67 birds examined late June and July in moult; starts between August and early September (many or all of 47 birds examined September to early October in moult); completed from mid-October, sometimes from late September (Rymkevich 1990).

**Measurements.** ADULT, FIRST ADULT. Fenno-Scandia, all year, and Scotland, Netherlands, and Germany, mainly August–April; skins (RMNH, ZMA). Wing and tail of 1st adult are retained juvenile wing and tail; bill (S) to skull, bill (N) to distal corner of nostril, exposed culmen on average 3·9 less than bill (S). Depth is approximate depth of closed bill where feathering reaches base, width is width at base of lower mandible in middle of rami.

| | | | | |
|---|---|---|---|---|
| WING AD | ♂ | 105·5 (2·25; 7) 103–109 | ♀ | 103·5 (1·32; 3) 102–105 |
| 1ST AD | | 104·9 (1·75; 15) 102–108 | | 102·2 (1·52; 15) 99–105 |
| TAIL AD | | 64·7 (1·55; 7) 62–67 | | 60·7 (2·47; 3) 59–64 |
| 1ST AD | | 62·0 (2·39; 15) 58–66 | | 60·3 (2·23; 15) 57–64 |
| BILL (S) | | 24·2 (0·73; 21) 23·2–25·6 | | 23·7 (0·96; 17) 22·6–25·6 |
| BILL (N) | | 16·4 (0·62; 22) 15·5–17·8 | | 16·2 (0·69; 17) 15·2–17·4 |
| DEPTH | | 14·2 (0·30; 21) 13·7–14·6 | | 13·8 (0·37; 17) 13·0–14·3 |

| WIDTH | 12·6 (0·37; 20) 12·0–13·2 | 12·7 (0·37; 17) 12·0–13·3 |
| TARSUS | 19·2 (0·57; 20) 18·3–20·3 | 19·6 (0·67; 17) 18·3–20·5 |

Sex differences significant for wing and bill depth.

Another sample of skins from whole geographical range, ages combined (Knox 1976):

| WING AD | ♂ 104·7 (2·6; 29) 100–109 | ♀ 101·6 (1·1; 14) 100–103 |
| CULMEN | 20·3 (0·6; 29) 18·5–22·0 | 19·7 (0·4; 14) 18·5–20·5 |
| DEPTH | 13·4 (0·5; 11) 12·5–14·5 | 13·0 (0·4; 13) 12·5–13·5 |
| WIDTH | 12·3 (0·6; 28) 11·5–14·5 | 13·1 (0·3; 13) 12·5–13·5 |

Wing. (1) Fenno-Scandia (Svensson 1992). (2) Belgium (Symens 1991). (3) Russian Lapland (Semenov-Tyan-Shanski and Gilyazov 1991). (4) Murmansk area (Russia) (Kokhanov and Gaev 1970). (5) USSR (Dementiev and Gladkov 1954).

| (1) | ♂ — ( — ; 58) 99–110 | ♀ — ( — ; 57) 95–107 |
| (2) | 107·2 (1·75; 8) 103–108 | 103·8 (2·06; 4) 101–106 |
| (3) | 103 ( — ; 13) 99–107 | 99 ( — ; 8) 88–106 |
| (4) | 102·6 ( — ; 18) 97–108 | 102·0 ( — ; 11) 98–108 |
| (5) | 103·5 ( — ; 37) 98–111 | 99·3 ( — ; 17) 93–103 |

Bill depth: 13·1–15·0 (115) (Svensson 1992); ♂ 13·6 (0·75; 8) 12·3–14·9, ♀ 12·9 (0·25; 4) 12·5–13·0 (Symens 1991); ♂ 14·6 (18) 13·8–15·6, ♀ 14·4 (10) 13·4–15·0 (Kokhanov and Gaev 1970).

JUVENILE. Bill still not full-grown until about halfway through post-juvenile moult.

**Weights.** ADULT, FIRST ADULT. (1) Netherlands and Belgium, November–March (Symens 1991; K Terpstra, ZMA). (2) Murmansk area (Russia) (Kokhanov and Gaev 1970). (3) Russian Lapland (Semenov-Tyan-Shanski and Gilyazov 1991). (4) USSR (Dementiev and Gladkov 1954).

| (1) | ♂ 52·8 (2·18; 12) 49–56 | ♀ 50·3 (1·97; 6) 48–52 |
| (2) | 55·0 ( — ; 20) 48–69 | 53·8 ( — ; 15) 49–61 |
| (3) | 54·6 ( — ; 13) 51–59 | 52·7 ( — ; 8) 49–58 |
| (4) | 52·9 ( — ; 19) 47–58 | 53·6 ( — ; 11) 44–58 |

Fair Isle (Shetland), October: adult 37·5–54·7, 1st adult and juvenile 32·4–41·8 (Davis 1963). Norway: September–December,

51·2–53 (4), ♀ 49–51·2 (3); July, ♂♂ 49, 54; April, ♀ 67 (Haftorn 1971). Germany: 47·5–56·8 (7) (Niethammer 1937).

NESTLING. For growth, see Olsson (1960, 1964); growth curve in Olsson (1960).

JUVENILE. Norway: July 46·5; August 50·2 (7) 48–52 (Haftorn 1971).

**Structure.** 10 primaries: p8 longest, p9 0–1(–3) shorter, p7 0·5–2, p6 5–8, p5 16–19, p4 22–27, p3 26–33, p2 30–37, p1 34–42; p10 strongly reduced, 66–76 shorter than p8, 11–16 shorter than longest upper primary covert, 2 shorter to 2 longer than reduced outermost upper primary coverts. Tail slightly forked, t1 4–11 shorter than t6. Bill markedly deeper and broader at base than in *L. curvirostra* (see Measurements), culmen more strongly decurved, base of culmen running almost straight into line of flattened cap as seen from side; cutting edges of both mandibles parallel at base, but that of lower mandible with blunt tooth in middle and with distal half concave, ending in relatively short and thick tip, which is shorter and more strongly curved upward than in *L. curvirostra*; gonys with broad and strong bulge; tip of lower not or hardly projecting beyond line of culmen in closed bill. Depth of upper mandible at base 9–10 mm; depth of visible part of lower mandible at base 4·9–5·6 mm (on average, 5·2, $n = 38$) (Hartert 1903–10; RMNH, ZMA). Middle toe with claw 21·8 (10) 20·5–23·5 mm; outer toe with claw *c.* 74% of middle with claw, inner *c.* 72%, hind *c.* 90%. Remainder of structure as in *L. curvirostra*.

**Geographical variation.** None.

In past, race *estiae* Piiper and Härms, 1922, sometimes recognized for population of Estonia (e.g. by Hartert and Steinbacher 1932–8). Said to have more elongated bill with less strongly curved culmen and gonys, but birds with similar bill shape occur occasionally elsewhere in species' range and bill rather variable in shape everywhere; *estiae* therefore not recognized, following Vaurie (1956b). For possible evolutionary history of thick-billed *Loxia*, see Tyrberg (1991a).

CSR

---

## *Rhodopechys sanguinea*  Crimson-winged Finch

PLATES 46 and 50 (flight)
[between pages 640 and 641]

DU. Grote Woestijnvink    FR. Roselin à ailes roses    GE. Rotflügelgimpel
RU. Краснокрылый чечевичник    SP. Camachuelo ensangrentado    SW. Bergsökenfink

*Fringilla sanguinea* Gould, 1838

Polytypic. Nominate *sanguinea* (Gould, 1838), Levant, Turkey, and Caucasus area east to Tien Shan and Tarbagatay mountains; *aliena* Whitaker, 1897, north-west Africa.

**Field characters.** 15 cm; wing-span 30–33·5 cm. 5–10% larger than Rock Sparrow *Petronia petronia* but with rather similar bill, wing, and tail structure; 20% larger than Trumpeter Finch *Bucanetes githagineus*, with proportionately larger bill and head and shorter tail. Quite large, heavy-billed, robust, ground-haunting finch, with bounding flight and calls recalling Woodlark *Lullula arborea*. Restricted to rocky mountainsides and summits above 1500 m in breeding season. At any distance, appears nondescript dark brown but at close range displays strikingly pink, dark-rimmed wings, dark crown, and intricate pattern of face markings and body spotting. ♂ shows pink basal patches and white tips to tail. Sexes somewhat dissimilar; no seasonal variation. Juvenile separable. 2 races in west Palearctic, geographically isolated and easily separated.

(1) West Asian race, nominate *sanguinea*. ADULT MALE. Moults: July–September (complete). Crown dull black,

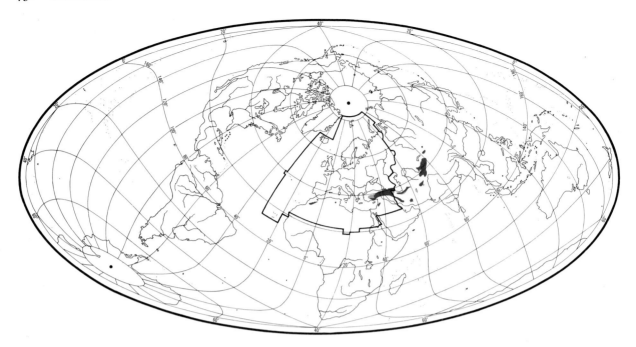

with small, faint buff tips; lore and foreface carmine-red; rear supercilium and surround to ear-coverts greyish-buff; ear-coverts warm brown, speckled with black on cheeks. Centre of nape and mantle buff- to rufous-brown, with dark centres and pale margins creating lines of streaks. Long rump warm buff, on lower part mottled rose-pink. Wings colourful: lesser and median coverts brown, streaked black; greater coverts basically black, with rufous and pale rose margins and tips, bases of secondaries and primaries black with broad carmine-pink fringes forming, with edges of coverts, pink panel along folded wing and across outer two-thirds of extended wing and framed by black, rosy-buff-fringed tertials and pink-fringed and white-tipped black ends to primaries and secondaries. Under wing-coverts mainly white. Chin and throat rufous-buff, spotted black on throat; side of upper breast warm buff, also strongly spotted black. Band between breast and flanks and lower chest pale pink, forming with pink strip across middle of breast distinct transverse divide, extending as pink patch onto belly. Flanks yellowish-buff, foreparts strongly spotted black but rear parts less so. Vent and under tail-coverts warm white. Tail colourful: feathers black-centred, pink-fringed, and white-tipped; appears pink-sided at base. By spring, wear increases contrast of black cap, face pattern (particularly supercilium), rose tone on rump, and red on face; becomes less pink, more black-spotted below, and pink on wing and tail becomes redder. Bill greyish-horn, becoming dull warm yellow in breeding season. Legs brown. ADULT FEMALE. Generally duller than ♂, with broader buff tips to crown, reducing capped look, less full pink and red tones on wings and virtually none on tail, and whiter, less

spotted underparts. Bill always greyish-horn. JUVENILE. Head and underparts warm buff, becoming cream from belly to under tail-coverts; upperparts warm brown, paler on rump. Lacks complexity of adult's head and underpart markings but wings and tail already show adult pattern (see also Plumages). (2) North-west African race, *aliena*. Markedly paler and duller in all plumages than nominate *sanguinea*, with fully ashy-grey nape, rose-white throat, wing-feathers only narrowly fringed rose, rump uniformly brown, and tail without obvious white. On at least some Moroccan birds, supercilium and surround to ear-coverts pale buff to cream.

Unmistakable, with diagnostic combination of large size, dark cap, pale pink centres to wings, white under wing-coverts, and spotted or streaked underparts. Flight fast and powerful, with long wings obvious in silhouette; over distance, action produces weighty, deep undulations; recalls *P. petronia* as much as other arid-country finches. Gait a powerful hop. Stance noticeably level, with heavy bill and head held up. Sociable, forming flocks after breeding. Restless, roaming far in search for food and water.

Song a short, melodious phrase, often given in flight. Flight-call distinctive, a soft 'chee-rup', 'tureep', 'turcep', or 'dy-lit-di-lyt', somewhat recalling *L. arborea*. Other calls include twitter, recalling Linnet *Carduelis cannabina*, and disyllabic 'chee-chee' suggesting House Sparrow *Passer domesticus*.

**Habitat.** Largely complementary to Trumpeter Finch *Bucanetes githagineus*, being situated somewhat further north, in warm temperate zone, and ranging over higher altitudes: 1700-3200 m in Tadzhikistan (Dementiev and

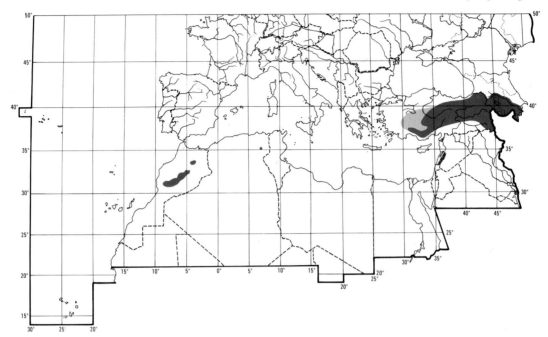

Gladkov 1954), and 2800 m upwards in Atlas of north-west Africa (Etchécopar and Hüe 1967). In USSR and Turkey, found on bare eroded mountains, virtually devoid of vegetation, or on bare pebbly slopes or in tree-shrub mountain zone with stands of juniper *Juniperus*. Occurs both on rocky and clay slopes, and on sand. In Afghanistan, found at *c.* 2700 m on more fertile patches on lower slopes fringing very desolate mountains, and feeding among flowering *Salvia* (Paludan 1959). Generally found in sparse arid scrub on stony slopes and ridges, in scrub and juniper zone above trees, on bare stony slopes with sparse herbage, and on almost bare, dry eroded clay hills, wintering at lower altitudes on bare areas and arable cultivation (Harrison 1982). In Israel, found on rocky slopes with sparse vegetation, nesting above 1900 m primarily in low bushes but also among rocks and on cliff ledges (Paz 1987). Mainly a ground bird, but also on rocks; more rarely in bushes or trees (Flint *et al.* 1984).

**Distribution.** No changes reported, except for probable northward expansion of range from Little Caucasus to Great Caucasus.

USSR. Records in Terek valley (Caucasus) represent probable natural range expansion (Mauersberger and Möckel 1987). ALGERIA. Recorded from Aurès mountains in 1840-2; then not again until rediscovered in July 1970 (Schoenenberger 1972). ISRAEL. Breeds only in Mt Hermon area (Shirihai in press). SYRIA. Presumed to breed in Antilebanon mountains, but confirmation needed (Baumgart and Stephan 1987).

Accidental. Iraq.

**Population.** No information on trends.

ISRAEL. About 30 pairs in 1970s and 1980s (Shirihai in press).

**Movements.** Short-distance migrant in north of range (south-central USSR), altitudinal migrant elsewhere.

Moroccan race *aliena* descends below breeding range in Haut and Moyen Atlas, e.g. recorded at Azrou, *c.* 2300 m, November, and at Taddert, 2000-2600 m, February-March (Heim de Balsac and Mayaud 1962). Present at Oukaïmeden at *c.* 2600 m all year, at least in some years (Thévenot *et al.* 1982). Also winters at much lower levels; recorded at Taza (550 m, eastern Morocco), and in semi-arid area of Oued Dadès, just south of Haut Atlas (Brosset 1961; Smith 1965; Thévenot *et al.* 1981).

Nominate *sanguinea*. In Turkey (breeds in east), descends to lower altitude in winter, and occasionally wanders to western Turkey (Beaman 1978). In Israel, breeds on Mount Hermon in north and not reported elsewhere at any season; during post-breeding dispersal, some birds descend to 1600 m, but most ascend above breeding range; soon all depart, presumably moving east to Syria; only a few remain irregularly in September and October, occasionally into December. Returns to breeding sites in 2nd half March and April. (Shirihai in press.) In Iran, descends southern slopes of Elburz mountains in winter, and also occurs then on plains of Azarbaijan in north-west (D A Scott); some lowland winter records in Afghanistan (S C Madge). Birds breeding in southern Kazakhstan depart south, and recorded there only exceptionally in winter. Movements are gradual: in autumn, mostly September-October; in spring, birds appear in foothills in March, at middle levels in mid-April, and alpine regions in May. (Korelov *et al.* 1974.) In Chokpak pass (Kazakhstan), regular but not common on passage, and more frequent autumn than spring; in autumn, average earliest record over 7 years 22 August (7 August to 4 September), with most records 21 August to 5 September, and last birds end October or beginning November; in spring, average earliest record over 9 years 4 March (24 February to 11 March), with most records 16 March to 15 April, and last birds mid-May (Gavrilov and Gistsov 1985). ♂♂ arrive in breeding areas before ♀♀ (Kovshar' 1966). Winters regularly in Uzbekistan (Korelov *et al.* 1974) and Tadzhikistan, where birds make considerable altitudinal movements, descending to foothills and valleys, mainly in 2nd half September and October; migrants returning in spring are markedly thinner than birds that have remained locally, suggesting they have moved some distance. Some winter at high levels, e.g. in Agalykskiy mountains (where far more common winter than summer); such birds leave only when deep snow cover persists, returning as soon as it melts. (Sagitov 1962; Abdusalyamov 1977.)    DFV

**Food.** Much information extralimital, from central Asia. Diet comprises seeds of low vegetation; a few invertebrates in breeding season. Forages mostly on bare rocky ground, scree, snowfields, etc., or in and around desert-type tussocky herbs and shrubs, in manner described as slow and heavy, taking seeds of limited range of plants, principally Chenopodiaceae, Boraginaceae, Cruciferae, and Compositae; also in overgrown gardens, sown fields, and in settlements, particularly in winter, where flocks feed (e.g.) on spilled seed of cultivated plants. (Brosset 1957a; Hollom 1959; Kovshar' 1966; Beaudoin 1976; Salikhbaev and Bogdanov 1967; Mauersberger and Möckel 1987.) In Morocco, during breeding season, flocks, apparently all ♂♂, fly from nesting area to lower altitude in early morning to feed and collect seeds for ♀♀ and young, which are stored in large throat-pouches that develop at this time of year; when full, these can be about same size as bird's head; flocks depart abruptly later in day to return to nests; these birds seen to feed mostly on seeds of *Alyssum granatense* (Cruciferae) (Brosset 1957a; Roux 1990). In Kazakhstan and central Asia, seems particularly fond of various species of houndstongue *Cynoglossum* (Boraginaceae) (Sagitov 1962; Kovshar' 1966; Ivanov 1969). When eating seed-heads while standing on ground, either bites through stem or pulls seed-head down; otherwise stands on stem, bending it over, and extracts seeds from head; also noted hanging head-down, and recorded digging up Liliaceae bulbs (Kovshar' 1966; Korelov *et al.* 1974). In Turkey, fed on blossom buds of alkanet *Anchusa* (Lehmann and Mertens 1969). Often seen eating salt (Kovshar' 1966). For food in captivity, see Glück and Massoth (1985).

Diet includes the following. Invertebrates: larval Lepidoptera, larval Diptera, beetles (Coleoptera). Plants: seeds, buds, fruits, etc., of honeysuckle *Lonicera*, knotgrass *Polygonum*, goosefoot *Chenopodium*, *Bassia*, sandwort *Arenaria*, Ranunculaceae, woad *Isatis*, shepherd's purse *Capsella*, pennycress *Thlaspi*, alison, etc. *Alyssum*, whitlow-grass *Draba*, stonecrop *Sedum*, cinquefoil, etc. *Potentilla*, sainfoin *Onobrychis*, goat's-thorn *Astragalus*, *Schrenkia*, bedstraw *Galium*, houndstongue *Cynoglossum*, alkanet *Anchusa*, forget-me-not *Myosotis*, *Rindera*, thistles *Carduus*, etc., goatsbeard *Tragopogon*, *Carthamus*, viper's-grass, etc. *Scorzonera*, *Cousinia*, leek, etc. *Allium*, grasses (Gramineae), sedges (Cyperaceae). (Meinertzhagen 1940; Sagitov 1962; Kovshar' 1966; Kowschar 1966; Salikhbaev and Bogdanov 1967; Lehmann and Mertens 1969; Ivashchenko and Kovshar' 1972; Korelov *et al.* 1974; Beaudoin 1976; Heinze and Krott 1979; Glück and Massoth 1985; Roux 1990.)

In Kirgiziya, 10 stomachs contained seeds of Polygonaceae, Ranunculaceae, Leguminosae, Compositae, and other plant fragments; 2 contained beetles and Diptera larvae (Pek and Fedyanina 1961). 35 stomachs in western Tien Shan (Kazakhstan) contained no animal material, but mainly seeds of goatsbeard as well as of Boraginaceae (including *Cynoglossum*) and *Rindera* (Kovshar' 1966). In Morocco, gullets of 12 birds held only seeds of *Alyssum granatense* (Brosset 1957a). In Tadzhikistan, *Cynoglossum* seeds found in stomachs in July, while winter stomachs contained Leguminosae, Cruciferae, and grasses (Sagitov 1962); birds arriving on breeding grounds in spring feed in newly-sown fields (Abdusalyamov 1977). In Caucasus, one ♂ very full of grass seeds (Zhuravlev and Afonin 1982), and in Uzbekistan, summer, stomachs held Polygonaceae seeds (Salikhbaev and Bogdanov 1967).

Diet of young almost wholly regurgitated seeds, usually brought considerable distance (e.g. 2–3 km recorded in Turkey) in throat-pouches of adults, which hold enough to feed whole brood in one trip (Kowschar 1966; Lehmann and Mertens 1969; Gubin 1979). In western Tien Shan, 17 crop-samples from young aged 1–7 days contained only de-husked seeds; at first only of goatsbeard, then also honeysuckle plus 2–3 other species, mainly Boraginaceae including *Cynoglossum*; 25 samples all contained some seeds of goatsbeard (Kovshar' 1966; Kowschar 1966). In Turkey, nestlings given pulp composed of unripe seeds of wild oat *Avena*, and green shoots and buds of herbs, including alkanet; brood 2–3 days old fed for 5–8 min per meal (Lehmann and Mertens 1969). In Morocco, young probably fed seeds of *Alyssum granatense* brought from lower altitude (Brosset 1957a); adults seen to give young caterpillars (Heinze and Krott 1979). For food in captivity, see Glück and Massoth (1985).    BH

**Social pattern and behaviour.** No detailed studies. Account includes data on *aliena* in Haut Atlas (Morocco) and nominate *sanguinea* further east in range (notably Turkey, Kazakhstan, and central Asia).

1. Typically gregarious, including to some extent when breeding (Heim de Balsac and Mayaud 1962; Korelov *et al.* 1974; E

N Panov). Winter flocks of up to 25–30 reported in Tadzhikistan (Sagitov 1962; Abdusalyamov 1977). Arrives Mt Hermon (Israel) in flocks of 5–10 (Shirihai in press); in Kazakhstan (Kovshar' 1966) and Tadzhikistan (Abdusalyamov 1977), during arrival period, recorded in pairs, small flocks (of ♂♂ only where these arrive before ♀♀), or, at peak, flocks of up to 100. Small breeding-season flocks (often of ♂♂ only, presumably when ♀♀ incubating) widely reported (Brosset 1957a; Olier 1959; Kovshar' 1966; Lehmann and Mertens 1969; Roux 1990). After fledging, family parties may initially roam about in breeding area (e.g. Shirihai in press), then amalgamate into flocks for feeding, etc.: up to 100 reported in Turkey late August to early September (Gaston 1968), also in Tadzhikistan prior to dispersal (Abdusalyamov 1977). In Kazakhstan, from late summer into autumn (also during pre-breeding period), mainly smaller flocks occur during day, larger (up to several hundred) for roosting (Kovshar' 1966; Korelov *et al.* 1974; Gubin 1979). In Haut Atlas, associates in winter and through to May with other Fringillidae, including Trumpeter Finch *Bucanetes githagineus*, and Rock Sparrow *Petronia petronia* (Roux 1990); in Kazakhstan and central Asia at various times, with buntings (Emberizidae), other Fringillidae, and Shore Lark *Eremophila alpestris* (Kovshar' 1966; Salikhbaev and Bogdanov 1967; Ivanov 1969; Abdusalyamov 1977). BONDS. Nothing to suggest other than monogamous mating system. Pair-formation takes place in flocks (e.g. Roux 1990). In Talasskiy Alatau mountains (Kazakhstan), ♂♂ reported to arrive before ♀♀, birds then pairing on breeding grounds (Kovshar' 1966; Gubin 1979); in Tadzhikistan, old birds perhaps arrive already paired (Abdusalyamov 1977); see also Heterosexual Behaviour (below). Duration of pair-bond not known, though perhaps extends to post-fledging care (Gubin 1979); see below. Only ♀ incubates (Lehmann and Mertens 1969; Ivashchenko and Kovshar' 1972), but young fed by both sexes (Olier 1959), mainly by ♂ during first few days, until share about equal from day 5 (Kovshar' 1966; Gubin 1979). In Israel, parent-young bond reported to persist for a few weeks after fledging (Shirihai in press); in Kazakhstan, young said to be tended for at least 1 week, but still fed when able to fly well (Kovshar' 1966; Korelov *et al.* 1974). Captive young independent *c.* 3 weeks after fledging (Glück and Massoth 1985). In Morocco, fledglings seen to beg mainly from ♀ (Roux 1990); ♂'s role at this stage not clear (see also Song-display, below). Age of first breeding not known. BREEDING DISPERSION. Territorial, but apparently not markedly so. Solitary or sometimes clumped: e.g. on Mt Hermon, groups of a few pairs, nests separated by a few tens of metres (Shirihai in press); 'colony' also reported from Haut Atlas by Roux (1990). In Talasskiy Alatau, 2 nests (under construction) 50 m apart (Gubin 1979), and in another study 200 m (Ivashchenko and Kovshar' 1972). In Tadzhikistan, where density fairly high (upper Zeravshan), pairs separated by 800–900 m; where scarcer (Mogoltau and south slope of Gissarskiy mountains) 2–4 km (Abdusalyamov 1977). Study in Turkey by Lehmann and Mertens (1969) found no marked territorial behaviour; intruders not always attacked. In Talasskiy Alatau, territory normally defended by ♂ during building and incubation. However, when 2 ♀♀ accompanied by mates were collecting nest-material at same site (within 10 m of one pair's nest), this caused no fights, even when one pair landed on other's nest-site, though the 2 ♂♂ cooperatively drove off other ♂♂ (Gubin 1979). At least when feeding young, adults may forage far from nest: in Talasskiy Alatau, well down mountain, in order to collect particular seeds (Kovshar' 1966); in Turkey, travelled up to 2–3 km (Lehmann and Mertens 1969). In Ala Dag mountains (Turkey), ♂ seen singing and apparently holding territory in late summer (Gaston 1968); significance not clear. No further information on size or

use of territory and none on breeding density. ROOSTING. Only reports (Talasskiy Alatau and Tadzhikistan) refer to communal roosting in late summer in cliffs (not clear whether in crevices) near top of mountain ridge (Kovshar' 1966) and on south-facing slopes (Abdusalyamov 1977). In Haut Atlas in July, flocks of ♂♂ arrive early on feeding grounds; seek out shade during hottest part of day (Roux 1990). For diurnal rhythm of flock activity (including rest periods, visits to water, feeding grounds, and places to eat salt), see Kovshar' (1966), Korelov *et al.* (1974), and Abdusalyamov (1977).

2. Reports on relative tameness vary. In Morocco, generally tame and approachable, both when breeding and outside breeding season (Heim de Balsac 1948; Brosset 1957a; Smith 1965); in autumn, flew to wires as temporary refuge, to rocks high up on hillside in more serious disturbance (Meinertzhagen 1940). In Lebanon, 2 pairs permitted approach within *c.* 30 m (Hollom 1959); other birds extremely shy, flying rapidly downhill when disturbed (Meinertzhagen 1935). Winter flocks similarly shy and unapproachable in central Asian mountains (Ivanov 1969); wary at nest (Kowschar 1966); see also Parental Anti-predator Strategies (below). For calls given when flushed, in flight, etc., see 2 in Voice. Ruffles crown in excitement (E N Panov). FLOCK BEHAVIOUR. In southern Turkey, flocks in evening highly sociable, and no disputes recorded (Lehmann and Mertens 1969). In flock comprising 3–4 pairs regularly visiting Moroccan village apparently to take grit, ♂♂ sang from rocks for *c.* 1 hr daily (Heim de Balsac 1948). SONG-DISPLAY. ♂ sings (see 1 in Voice) from rock, tussock, or bush, also in flight (Lehmann and Mertens 1969; Abdusalyamov 1977), including (in Haut Atlas) during frequent chases (Brosset 1957a). In Talasskiy Alatau, typical display involving 2–3 ♂♂ noted before 09.00–10.00 hrs June to early July as follows: one ♂ arrives and sings from rock or cliff-top, 1–2 others then land close by and sing, but 1st ♂ (wings drooped) attacks and chases them short distance before returning to same spot; other birds circle a few times before returning, so that whole performance repeated, and so on over 1–2 hrs (Gubin 1979). ♂ sometimes performs brief undulating Song-flight, starting from and returning to perch or nearby (Lehmann and Mertens 1969). Performance at times spectacular (Fig A), bird

A

circling high and apparently singing in rhythm with deep undulations, alternating fluttering ascents and gliding descents (Mild 1990; see also 1 in Voice); similar to Song-flight of *B. githagineus* (E N Panov). Sings frequently in Haut Atlas in June (Heim de Balsac and Mayaud 1962); in Taurus (Turkey), song noted from ♂ 19 July while ♀ feeding fledglings (Beaudoin 1976); in Ala Dag, late August and early September (Gaston 1968). Birds singing in southern Iran, mid-May, showed no territorial activity and were probably still on passage (Desfayes and Praz 1978). Song-period in Tadzhikistan May-July (Abdusalyamov 1977). ANTAGONISTIC BEHAVIOUR. In Tadzhikistan, frequent fights between ♂♂ recorded 2nd half of May (Abdusalyamov 1977).

Sometimes aggressive towards other species, e.g. chasing away small passerines from feeding grounds; captive bird hostile toward Brambling *Fringilla montifringilla* and Red-headed Bunting *Emberiza bruniceps* (Kovshar' 1966). HETEROSEXUAL BEHAVIOUR. (1) General. In Haut Atlas, pairing during 2nd half of May, but birds still mainly in flocks then and few signs of sexual activity, though more marked by June when most in pairs (Roux 1990). Pair-formation in Zeravshan mid-May (Ivanov 1969); in Talasskiy Alatau, within 10-14 days of arrival in late February or March (Gubin 1979). (2) Pair-bonding behaviour. Sexual chase noted in Lebanon in May, birds flying erratically low over ground (Hollom 1959). ♂-♀ disputes (perhaps connected with pairing) in Haut Atlas in June (Roux 1990). (3) Courtship-feeding. ♂ feeds ♀ on nest during incubation (Lehmann and Mertens 1969; Gubin 1979). (4) Mating. No information. (5) Nest-site selection. Apparently by ♀, though ♂ usually accompanies her and sings (Heim de Balsac 1948; Gubin 1979); ♀ also builds nest, collecting material within *c.* 5-100 m in one study (Gubin 1979). (6) Behaviour at nest. ♀ lays within a day of nest completion; during incubation, ♂ sings and guards territory (Gubin 1979). Studies in Morocco and Turkey found birds wary and secretive at nest, usually landing *c.* 2-3 m or more away and running in through vegetation (Olier 1959; Lehmann and Mertens 1969). RELATIONS WITHIN FAMILY GROUP. During first few days after hatching, young brooded by ♀ most of day (breaks of 10-15 min, increasing to *c.* 30 min by day 4-5). Brooding continues for up to 10 days, ♀ then starting to forage with ♂. (Kowschar 1966.) In another study, brooding totalled 6 hrs out of 16 on day 3 (stints at 2-3 days 72-140 min morning and evening, otherwise during day 6-30 min), 2 hrs 35 min on day 6, but chicks no longer brooded during day at 12 days, though ♀ still broods them at night (Gubin 1979). Eyes of young open to slit by day 5 (Kovshar' 1966, which see for further details of physical development). Food brought by adults in sublingual pouch (see Food). In early stages, provided by ♂ and transferred to young via ♀. Adult may perch on stone near nest, making 15-20 regurgitation movements before moving to nest to feed young (sticks bill into open bill of nestling and moves head forward 5-6 times); several nestlings or whole brood fed each visit. Young highly vocal by day 11. (Kowschar 1966; Lehmann and Mertens 1969; Glück and Massoth 1985.) Nest kept clean by both parents, swallowing chicks' faeces after each feed while young small (at this stage, faeces contain some undigested seeds), but faeces may start to accumulate on edge of nest from *c.* 7-8 days, and then either swallowed or carried away up to 50-100 m, though not all eventually removed (Kovshar' 1966; Lehmann and Mertens 1969; Gubin 1979). Young leave nest at 10-17 days (see Breeding), in most cases apparently still unable to fly. Adult may attempt to lure young from nest by calling (Lehmann and Mertens 1969). Fledglings seek out hidden perch, calling loudly when adult approaches (Kovshar' 1966); beg by flapping or shivering wings while gaping widely (Lehmann and Mertens 1969); in Morocco, early July, chased after and then placed themselves in front of parents to beg (Roux 1990). ANTI-PREDATOR RESPONSES OF YOUNG. No information. PARENTAL ANTI-PREDATOR STRATEGIES. ♀ sits tightly on eggs or small young, allowing very close approach, such that sometimes even caught on nest. In Turkey, pair (♂ especially) came within a few metres of observers when nestlings threatened. (Kovshar' 1966; Lehmann and Mertens 1969; Ivashchenko and Kovshar' 1972; Gubin 1979.) For calls, some of which apparently given when disturbed, see 2 in Voice. No further information.

(Fig by D Nurney from drawing in Mild 1990.)      MGW

**Voice.** Freely used, at least during breeding season (Meinertzhagen 1940; Sutton and Gray 1972). Further study needed to determine exact size of repertoire, function of certain calls, geographical variation, and possible sexual differences (song reported only for ♂ by Abdusalyamov 1977, but see 1 below, for report of duet by pair).

CALLS OF ADULTS. (1) Song of ♂. Clear, melodious, but at times somewhat wheezy or grating short phrase, often given many times in quite rapid succession without much variety (Hollom 1959; Hollom *et al.* 1988; Mild 1990). In Turkey, song-units (motifs) rendered 'tchwili' and 'tchwilichip', vaguely reminiscent of sparrow *Passer* calls (Lehmann and Mertens 1969); in Morocco, 'tirlui tirlui titurlui...' recalling Woodlark *Lullula arborea* (see call 2a, below), but faster and more emphatic (Brosset 1957a); full song-phrase, a rippling 'turdel-edel-weep-ou' noted from flying bird in Lebanon (Hollom 1959; L Svensson). Recording (Fig I) contains sprightly and jigging phrase

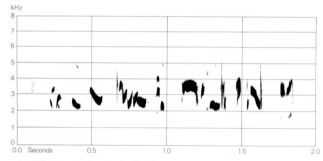

I  Mild (1990)  Israel  April 1989

with compound temporal pattern (each beat divided by 3): 'di did-dle-de did-dle-de did-dle-de diu'; initial 'di' weak (as are 'dle-de'), 'did' and final 'diu' strong. Jaunty pattern recalls song of Whitethroat *Sylvia communis* (P J Sellar). Song-phrase shown in Fig II louder and with added introduction and coda ('diu-dle-de'). Recording of bird singing in flight contains as prime constituent irregularly recurring phrase of 4 units (or figures) interrupted by interspersed, shorter sub-phrase of 1-3 units. Fig III shows main 4-unit phrase followed after pause of 2 s by 2-unit sub-phrase. Sub-phrase shown in 2nd part of Fig IV (1st unit is same as 4th of main phrase) comprises same 2 units as in Fig III (2nd part), and additional 3rd unit not repeated in this recording. In Fig V, single-unit sub-phrase (1st of main phrase), then again main 4-unit phrase, which in Fig VI is extended by addition of new figure. Some variation in duration of pauses between units (figures), this being only vaguely correlated with overall phrase length of 0·85-1·02 s; temporal pattern of song given in flight reported by Mild (1990) to match flight undulations, but this relationship not clear from analysis of recording. (J Hall-Craggs.) Recording of bird singing from perch in Turkey contains phrase closely similar to that shown in Figs III and V-VI, but differs in adding final quite loud, ringing

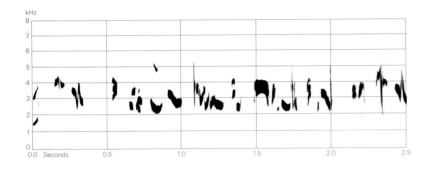

II   Mild (1990)   Israel   April 1989

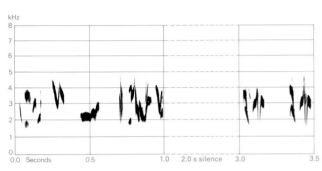

III   Mild (1990)   Israel   April 1989

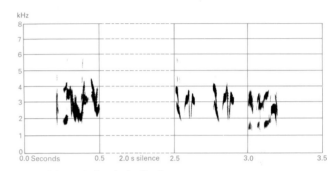

IV   Mild (1990)   Israel   April 1989

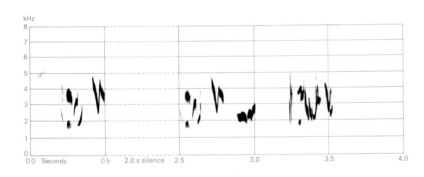

V   Mild (1990)   Israel   April 1989

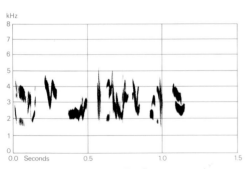

VI   Mild (1990) Israel   April 1989

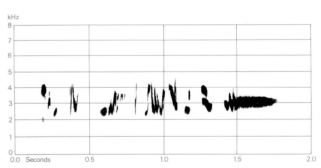

VII   L Svensson   Turkey   June 1989

and slightly shrill whistling 'freee' or 'whreee': 'too-di hui choo-dle oo-dle freee' (Fig VII); in other phrases in same recording 'freee' followed by attached coda of 1-4 motifs. Same bird once gives attractive, rippling or tinkling trem-

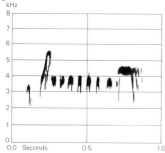

VIII  L Svensson  Turkey  June 1989

olo (Fig VIII) whose relationship with other song components not clear. (J Hall-Craggs, M G Wilson.) Recording from Morocco suggests delicate, delightfully tinkling warble in short, hurried phrase of 3 sections: 'b-doo-lu b-dee b-doo b-doo-lu' (Fig IX); this structure,

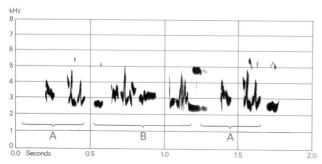

IX  J-C Roché  Morocco  May 1967

with 3rd section being recapitulation of 1st, and middle section differing, is apparently rare in bird song. Overlapping song from 2 birds (duet by ♂ and ♀ of pair according to J-C Roché) in same recording is not organized as duet, though structure shown in Fig IX, and other reduced or extended forms, would certainly lend themselves to organized duetting. (J Hall-Craggs.) (2) Contact- and alarm-calls. Multiplicity of transcriptions and, in many cases, lack of contextual detail or suggested function make it difficult to determine how many different calls exist; possibility of geographical variation not to be discounted. Following subdivision highly arbitrary; almost certainly much overlap, at least between calls 2a–c, which are given mainly in flight. (2a) Rich, musical, softly fluting whistles often reported as reminiscent of calls of *L. arborea*: 'dü-leet dü-leet', 'tureep tureep', or similar (Hollom *et al.* 1988; Jonsson 1992); series of 'tuili', 'tyui', or 'tilui' (Korelov *et al.* 1974; Abdusalyamov 1977; E N Panov); 'tlweep' when flushed and flying short distance (Hollom 1959), also loud 'wheet' and 'klee' (Smith 1965). Apparent extension of this call-type (in some cases apparently song-fragments): 'tirlui tirlui titurlui' given when flying up after allowing close approach (Brosset 1957a); 'tchu-che-ly' from flying bird (Jonsson 1982); 'tchili-dili'

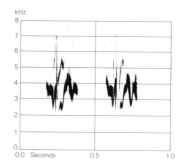

X  J-C Roché  Morocco  May 1967

or 'tili-dulu' carrying only *c.* 15 m and given by ♂♂ running about near nest (Lehmann and Mertens 1969); musical, rather quiet 'whee-tell-ee(r)' also from ♂ on ground (Hollom 1959). (2b) Recording (Fig X) contains calls suggesting rich, fruity, slightly shrill chirruping of sparrow *Passer*: 'tschilip' (J Hall-Craggs, M G Wilson). Calls obviously of this type, descending in pitch (Norton 1958) reported as follows: 'tchilip' or clear, frequently uttered 'tschilpip' (Lehmann and Mertens 1969; Heinze and Krott 1979); 'chee-rup' (Hollom *et al.* 1988); 'tswee-er' (Smith 1965). (2c) Nasal 'shee-chee' or 'chee-chee' like *Passer* (Hollom *et al.* 1988; Mild 1990). Further descriptions of calls probably in this category: short, hoarse, somewhat constrained 'chshi' said by E N Panov to be characteristic; loud 'chee-wee' given in flight and apparently expressing anxiety (Hollom 1959); sparrow-like chirps also reported by Smith (1965) and presumably belong here or in 2b. (2d) Repeated plaintive 'kouïc' given on ground at approach of man apparently as alarm-call (Brosset 1957a); provisionally kept separate here, but may be same as 2a or related. (2e) Recording from Morocco contains calls reminiscent of Pied Wagtail *Motacilla alba* or Grey Wagtail *M. cinerea*: 'tewissick tewissick' (J Hall-Craggs: Fig XI); perhaps related to one of sparrow-like

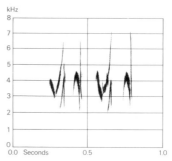

XI  J-C Roché  Morocco  May 1967

calls described above (2b–c). (2f) Fairly weak, plaintive and prolonged 'tuut' given by perched ♀♀ in Morocco (Brosset 1957a). (2g) Harsh 'tchu-chu' given when flushed (Mild 1990); perhaps same or related are hard, but fairly quiet 'tuk' sounds recalling Twitter-call of Linnet *Car-*

*duelis cannabina* in recording of Moroccan flock by E D H Johnson.

CALLS OF YOUNG. A 'tcheep' or 'tchiweep' noted from nestlings in Turkey (Lehmann and Mertens 1969). In Kazakhstan, fledglings gave clear, loud chirruping sounds recalling young sparrows (Kovshar' 1966).    MGW

**Breeding.** SEASON. Morocco: at 2200-3300 m, young recorded in nest mid-May and late June, and fledged young early July (Olier 1959; Heinze and Krott 1979; Roux 1990). Israel: at *c.* 2000 m, breeding begins when snows melt, and eggs laid from end of May or beginning of June (Shirihai in press). Southern Turkey: at 1100 m, laying starts from beginning of April to beginning of May (Lehmann and Mertens 1969); at 2100 m, eggs laid 3rd week in June at one site (Beaudoin 1976). Kazakhstan (extralimital): season extended, probably depending on orientation and altitude, with eggs in nests from about mid-May to July (Kovshar' 1966; Korelov *et al.* 1974; Gubin 1979). SITE. In stony places with little vegetation; on ground under overhanging rock, grass tussock, thorny cushion-type scrub, or in crevice between boulders; also in low bush and on cliff ledge (Olier 1959; Stepanyan 1969b; Ivashchenko and Kovshar' 1972; Paz 1987). Frequently on steep scree slope (Kowschar 1966), and in Turkey inside and outside old volcanic craters under slab of lava, protected against rain and sun (Lehmann and Mertens 1969). In Tadzhikistan, one nest 1·3 m above ground in wall of derelict house (Abdusalyamov 1977); in Turkey, also recorded on vineyard slopes (Beaman *et al.* 1975). Nest: neat and loosely constructed, with foundation principally of tough dry grasses, including cereals, herb stalks, and roots lined with fine grass and plant fibres; rarely animal hair (Kowschar 1966; Lehmann and Mertens 1969; Ivashchenko and Kovshar' 1972; Abdusalyamov 1977). In Morocco, one nest entirely of roots of scrubby plants (Olier 1959). 5 nests in western Tien Shan (Kazakhstan) had average outer diameter 15·0 × 12·7 cm, inner diameter 8·2 × 7·7 cm, depth of cup 5·4 cm (Gubin 1979); see also Ivashchenko and Kovshar' (1972). Building: ♀ scrapes out depression in soil and constructs nest (Lehmann and Mertens 1969; Gubin 1979). In western Tien Shan, ♀ seen collecting material within *c.* 5-100 m of nest, apparently bringing wet stems and blades of previous year's grass, though plenty of dry material available; took 3-4 days (Gubin 1979, which see for details). EGGS. Sub-elliptical, smooth, and slightly glossy; light sky-blue with small violet-brown spots concentrated at broad end (Lehmann and Mertens 1969; Harrison 1975). Nominate *sanguinea*: 21·6 × 16·4 mm (19·5-23·4 × 15·2-17·4), n = 34; calculated weight 3·05 g. *R. s. aliena*: 23·8 × 16·8 mm, n = 1; calculated weight 3·51 g. (Schönwetter 1984.) In western Tien Shan, fresh weight 3·1 g, n = 10 (Gubin 1979). Clutch: 4-5. In Turkey, 3 clutches all of 5 (Lehmann and Mertens 1969), and 6 clutches all of 5 in western Tien Shan (Gubin 1979); clutches of 4 recorded

in Kazakhstan (Kovshar' 1966) and in Israel (Shirihai in press). Apparently 2 broods in Turkey, and captive ♀ laid 3 clutches of 5, including 1 replacement, 3rd started 4-5 days after young of 2nd clutch fledged (Glück and Massoth 1985). Perhaps only 1 brood in Kazakhstan (Kovshar' 1966) and Israel (Shirihai in press). INCUBATION. (12-) 13-15 days, by ♀ only (Lehmann and Mertens 1969; Ivashchenko and Kovshar' 1972; Abdusalyamov 1977; Shirihai in press); from 3rd or 4th egg (Kovshar' 1966; Gubin 1979). In captivity took 12-13 days, started with 3rd egg (Glück and Massoth 1985). YOUNG. Fed and cared for by both parents; ♀ broods continuously for *c.* 10 days (Kowschar 1966; Lehmann and Mertens 1969). In captivity, fed only by ♀ until *c.* day 7 (Glück and Massoth 1985). FLEDGING TO MATURITY. Fledging period 13-15 (10-17) days; often leaves nest before able to fly (Lehmann and Mertens 1969; Gubin 1979; Glück and Massoth 1985; Shirihai in press). Captive young fed by parents for *c.* 3 weeks after fledging (Glück and Massoth 1985). BREEDING SUCCESS. No information. Breeding season perhaps long due to replacement clutches caused by frequent losses (Kovshar' 1966).    BH

**Plumages.** (nominate *sanguinea*). ADULT MALE. In fresh plumage (September-November), forehead and crown black, each feather narrowly fringed cinnamon-buff; black patch narrowing towards rear, central hindneck mottled dull black and sandy-grey. Mantle, scapulars, and back cinnamon-buff, fringes slightly greyer once frayed, centres of feathers of lower mantle and of scapulars rufous-brown with black shaft-streak, partly concealed. Rump and upper tail-coverts cinnamon-buff, feather-tips broadly paler buff, centres of feathers of lower rump and of upper tail-coverts rosy-pink, but these largely concealed. Supercilium, ear-coverts, cheeks, and side of neck cinnamon-buff, streaked dark brown on fore-cheek and ear-coverts; narrow strip of feathers along lateral base of both mandibles, lore, and short stripe just above and below eye bright pink-red. Chin, throat, side of breast, side of belly, and flank cinnamon-rufous, each feather usually with small contrasting black subterminal spot (except chin), on flank with dusky shaft-streaks; feather-tips on chin with cream spot (chin and upper throat occasionally all-cream), those on side of body with broad cream or sandy-grey fringe. Chest pink-white, forming broad pale crescent between cinnamon-rufous of throat and that of side of breast, connected to light rose-pink stripe over mid-breast, latter widening to large and contrasting light rose-pink or pink-white patch on belly. Rear of flank, vent, thigh, and under tail-coverts white with pink or cream tinge. Tail black, feathers fringed white bordered by some brown subterminally, white merging into rosy-pink on bases of outer webs; tip of inner web of central pair (t1) with shallow white notch, inner web of t2-t4 with larger white patch on tip (up to 5-8 mm long on t4), and basal halves of inner web with increasing amount of white hidden below tail-coverts; t5 white, except for black subterminal spot or smudge on both webs; t6 white, except usually for black shaft. Flight-feathers, greater upper primary coverts, and bastard wing black, outer webs broadly and contrastingly fringed bright rosy-pink (except for outer web of p9, emarginated parts of p6-p8, and longest feather of bastard wing), tips of flight-feathers contrastingly fringed white, concealed basal one-third of flight-feathers white; at rest, aspect of wing rosy-pink, apart from black longest feather of bastard wing, some black on tips of longest primary coverts, and black tips with white crescents of primaries. Greater upper wing-coverts and tertials black or black-brown, outer webs brown-grey or cinnamon shading to pink along margin. Lesser and median upper wing-coverts rufous-brown with narrow black shaft-streak and broad paler cinnamon-buff or sandy fringe. Under wing-coverts and axillaries white, suffused rosy-pink on axillaries, spotted rosy-pink on marginal coverts. Marked influence of bleaching and wear. *In fairly worn plumage* (winter), mainly as in fresh plumage but fringes along tips of feathers of mantle, scapulars, tertials, and upper wing-coverts bleached and abraded to pale sandy-grey or buff; rosy-pink of outer edges of flight-feathers brighter rosy-red. In spring, when more heavily worn, cap uniform black, supercilium and bar across hindneck and side of neck cream-white with variable amount of cinnamon suffusion, in particular broad rear part of supercilium contrasting markedly with black of crown and rufous-brown ear-coverts; mantle and scapulars to upper rump rusty-brown, streaked black on (especially) mantle; lower rump and upper tail-coverts more conspicuously washed rosy-pink or rosy-red; red at base of bill, on lore, and round eye more conspicuous; some contrast between black-streaked rufous-brown ear-coverts and uniform paler rufous-cinnamon lower cheek and lower side of neck; chin paler; rufous-cinnamon of chest, side of breast, and flank more distinctly spotted black, more sharply contrasting with whiter (less pink) crescent on chest and with mid-breast and belly; white on tips of tail- and flight-feathers partly or fully worn off; pink fringes of tail, flight-feathers, and of greater upper primary and secondary coverts contrastingly bordered bright ruby-red. ADULT FEMALE. Like adult ♂, but black of cap duller, tinged grey or brown, feathers more broadly fringed dull greyish-buff, not forming a sharply-defined patch; centres of feathers of mantle to rump as well as lesser and median upper wing-coverts duller dark grey-brown, less rusty, less clearly marked with black along shafts, aspect duller and more uniform grey-brown when fresh, dull earth-brown when worn; lower rump and upper tail-coverts gradually paler grey-brown, feathers tipped isabelline, virtually without pink suffusion; lore grey, eye-ring cinnamon-buff, no contrasting red spectacle, except for some faint pink-red spots around eye; remainder of side of head and neck rather uniform warm buff, less clearly streaked dusky on ear-coverts than adult ♂, rear of supercilium less contrastingly pale when worn; chin buff-white, bordered by faint spotted grey malar stripe at side (absent in ♂); ground-colour of throat, side of breast, and flank tawny-buff, distinctly less rufous than in adult ♂, that of crescent on chest, mid-breast, and belly cream-white, virtually without pink, less sharply defined from tawny-buff; chest and side of breast without contrasting black marks, flank with faint grey shaft-streaks only. Tail as adult, but white outer webs almost without pink at bases, t5 white with full broad black subterminal bar 30–40 mm long, and t6 white with broad black or dark grey blotches on tips, no uniform white; flight-feathers, primary coverts, and greater coverts as adult ♂, but pink fringes along outer webs narrower, *c.* 1 rather than *c.* 2 mm wide in middle of feather, extensively white towards base on primaries and primary coverts, especially when worn, only middle portions of primaries and primary coverts bordered pink-red when worn. NESTLING. Down sparse, white, on upperparts and upperwing only (Harrison 1975). JUVENILE. Upperparts entirely rusty- or sandy-brown, paler on lower rump and shorter tail-coverts. Side of head and neck, chin to chest, and flank cinnamon-buff, belly, vent, and under tail-coverts cream-white; feathers of vent and tail-coverts short and loose. For tail, flight-feathers, primary coverts, and bastard wing, see 1st adult (below). Upper wing-coverts and tertials grey-brown or sandy-brown, tips of coverts

and outer borders of tertials with broad but ill-defined grey-buff or tawny-buff fringes. FIRST ADULT MALE. Like adult ♂, but tail, flight-feathers, tertials, greater upper primary coverts, all or part of greater upper wing-coverts, and bastard wing still juvenile. Ground-colour of tail, flight-feathers, bastard wing, and inner webs of tertials and primary coverts slightly browner, less black, especially in spring; tip of inner web of tail-feathers more sloping, less square; tips of tail, flight-feathers, tertials, and most greater coverts more worn than in adult at same time of year, white of tail- and primary-tips worn off by March–April (in adult, present up to about June), tips of outer greater coverts contrasting in wear with those of median coverts and (if new) inner greater; tail more extensively black than in adult, t5 with broad black subterminal bar, t6 with grey or black subterminal spot on one or both webs (like adult ♀). FIRST ADULT FEMALE. Like adult ♀, but part of plumage still juvenile, as in 1st adult ♂; characters as in 1st adult ♂, but tail-feathers often more distinctly pointed, terminal half of t5 black except for white patch on tip of inner web, t6 white with partly dark grey outer web and broad dark grey border along inner web; outer fringes of tail- and flight-feathers, primary coverts, greater coverts, and shorter feathers of bastard wing mainly pale buff or off-white, much less pink than adult ♀ and 1st adult ♂, hardly red when worn, pink restricted to middle portion of primaries. Black of cap more restricted than in adult ♀.

Bare parts. ADULT, FIRST ADULT. Iris brown and dark brown. Bill of ♀ yellow to brown-yellow with blackish tip, more greyish-horn with yellow-horn base to lower mandible in winter and in ♀. Leg pale or dark brown, foot dark brown to blackish, soles sometimes yellowish (Whitaker 1898; Hartert 1903-10; Ali and Ripley 1974; BMNH.) NESTLING. Bare skin pink (Harrison 1975); no further information. JUVENILE. No information.

Moults. ADULT POST-BREEDING. Complete; primaries descendent. Heavily worn but moult not started in 5 ♂♂ and 1 laying ♀ from Afghanistan, 18–27 July; single ♂ from 29 August in full moult (Vaurie 1949b). Body and tail new, but head, secondaries, and outer primaries in moult in single ♂ from Kirgiziya, 14 August, primary moult score 35; body and tertials in heavy moult, tail just started, head and secondaries old in single ♂ with primary moult score 22 on 23 August, Turkmeniya (ZFMK). In last stages of moult in mid-August (Dementiev and Gladkov 1954). In eastern Afghanistan, 9 September, single ♂ not started, 2 ♂♂ half-way through moult, single ♀ had completed (Niethammer 1973). POST-JUVENILE. Partial: head, body, lesser and median upper wing-coverts, sometimes inner greater upper wing-coverts or central pair of tail-feathers (t1). Starts shortly after fledging, and hence timing variable. In Tadzhikistan, August, some still fully juvenile, others in 1st adult plumage (Johansen 1944); singles in moult 7 August, almost completed 28 August (Dementiev and Gladkov 1954); completed in birds examined October and later (Vaurie 1949b; ZFMK). In Morocco, some moult of head and body in October, moult completed in November (Meinertzhagen 1940; BMNH).

Measurements. ADULT, FIRST ADULT. Nominate *sanguinea*. Mainly Turkey, Lebanon, Israel, Syria, Armenia, and northern Iran, a few from central Asia (Turkmeniya to eastern Kazakhstan and northern Sinkiang, China), all year; skins (BMNH, RMNH, ZFMK, ZMA). Bill (S) to skull, bill (N) to distal corner of nostril; exposed culmen on average 4·3 less than bill (S).

| | | |
|---|---|---|
| WING | ♂ 107·3 (2·60; 33) 103–112 | ♀ 102·6 (2·01; 15) 100–106 |
| TAIL | 58·6 (1·73; 14) 56–62 | 55·5 (1·66; 15) 52–59 |
| BILL (S) | 16·4 (0·67; 14) 15·6–17·5 | 16·3 (0·75; 15) 15·6–17·4 |

| | | | | | |
|---|---|---|---|---|---|
| BILL (N) | 9·9 (0·54; 14) | 9·0–10·8 | | 9·7 (0·16; 15) | 9·4–10·0 |
| TARSUS | 20·5 (0·61; 13) | 19·5–21·3 | | 19·7 (0·48; 15) | 19·2–20·9 |

Sex differences significant for wing and tail.

Wing and tail of ♂. (1) Turkey, Lebanon, Israel, and Syria (RMNH, ZFMK, ZMA). (2) Lebanon, (3) Azarbaijan (northwest Iran), (4) Lorestan (south-west Iran), (5) Afghanistan (Vaurie 1949*b*; Erard and Etchécopar 1973). (6) Turkmeniya to eastern Kazakhstan and northern Sinkiang (RMNH, ZFMK, ZMA).

| | WING | | | TAIL | |
|---|---|---|---|---|---|
| (1) | 108·3 (1·64; 9) | 106–112 | | 58·0 (1·58; 9) | 56–61 |
| (2) | 106·0 ( — ; 5) | 104–109 | | 56·0 ( — ; 5) | 55–57 |
| (3) | 105·9 ( — ; 7) | 105–109 | | 59·1 ( — ; 6) | 56–63 |
| (4) | 106·8 ( — ; 15) | 105–110 | | 56·2 ( — ; 15) | 54–60 |
| (5) | 105·4 ( — ; 11) | 100–110 | | 56·3 ( — ; 9) | 54–59 |
| (6) | 109·9 (2·04; 5) | 106–112 | | 59·6 (1·64; 5) | 57–62 |

*R. s. aliena*. Wing of at least 7 ♂♂ and 3 ♀♀: 102 (♀)–111 (♂) (Meinertzhagen 1940).

**Weights.** Nominate *sanguinea*. Turkey, May and early June: ♂ 39·6 (1·95; 5) 37–42 (Kumerloeve 1969*a*, ZFMK). Israel: 30–33 (Paz 1987). Armeniya, early June: ♂ 36·2 (Nicht 1961). Kazakhstan: ♂ 37·5 (46) 32·7–47·7, ♀ 38·8 (8) 32·4–45·7 (Korelov *et al.* 1974). ♂♂ 33, 44; ♀♀ 32, 33 (Desfayes 1969).

**Structure.** Wing long, broad at base, tip pointed. 10 primaries: p8 longest, p9 0–2 shorter, p7 0·5–3·5, p6 8–11, p5 13–20, p4 22–25, p3 27–30, p2 31–34, p1 32–39; p10 reduced, a minute pin, 67–79 shorter than p8, 9–14 shorter than longest upper primary covert. Outer web of (p5-)p6–p8 emarginated, inner web of (p7-)p8–p9 with notch. Tip of longest tertial reaches to tip of p1 or to tips of secondaries when plumage fresh, short of secondaries when worn. Tail short, tip square or slightly forked; 12 feathers, t1 1–5 shorter than t6. Bill short, thick, conical; depth at feathering of base *c.* 11·4–12·0, width *c.* 9·4–10·4; tip of culmen slightly decurved, ending in small hook; cutting edges of upper mandible with faint tooth in middle and with kink at base. Nostril small, rounded, covered by short bristle-like feath-ers projecting from lore; many short bristles projecting obliquely downward along base of upper mandible. Leg and foot short, but strong. Middle toe with claw 19·9 (10) 18–22; outer toe with claw *c.* 68% of middle with claw, inner *c.* 69%, hind *c.* 77%.

**Geographical variation.** Fairly strong; involves colour only. Isolated *aliena* from north-west Africa rather different in colour from nominate *sanguinea* of Turkey and Levant eastward: upper-parts of ♂ as ♂ nominate *sanguinea*, but black on cap more restricted, hindcrown extensively dark ash-grey; mantle, scapulars, back, and side of head dark grey-brown, tinged vinous when fresh, no rufous; black marks on centres of mantle feathers and scapulars indistinct; rump to upper tail-coverts drab brown, sometimes tinged vinous or rufous, without pink on feather-tips; face with less red, none on front of supercilium; ear-coverts, chest, and side of breast without black spots; chin and throat grey-white with rosy tinge, contrasting with tawny-brown collar across lower cheek to upper chest and side of breast, collar marked with pale pink spots (bleaching to white or wearing off); tail with limited pink at base, more extensively black on outer feathers; pink or red fringes along flight-feathers narrower. ♀ *aliena* like ♂ but cap dull earth-brown, less black than in ♀ nominate *sanguinea*; upperparts greyer than nominate *sanguinea*, rump less pink, underparts without pink suffusion on chin, throat, and belly, and much contrast between cream-white chin and throat and tawny cheek and chest; red on face absent or traces only. No marked variation within nominate *sanguinea*, colours of birds at same stage of wear similar throughout geographical range (Vaurie 1949*b*); no variation in size. Geographical variation in amount of white in tail, as found by Vaurie (1949*b*), not supported by specimens examined, difference being related to age, not locality (C S Roselaar).

Probably related to Red-browed Rosefinch *Callacanthis burtoni* of Himalayas, and both of these to *Acanthis* group of *Carduelis* (e.g. Linnet *C. cannabina*), not to trumpeter finches *Bucanetes*, though latter occasionally included in *Rhodopechys* (Desfayes 1969).                                                                        CSR

---

## *Rhodospiza obsoleta*  Desert Finch

Du. Vale Woestijnvink          Fr. Roselin de Lichtenstein          Ge. Weissflügelgimpel
Ru. Буланый вьюрок          Sp. Camachuelo desertícola          Sw. Ökenfink

*Fringilla obsoleta* Lichtenstein, 1823. Synonym: *Rhodopechys obsoleta*.

Monotypic

**Field characters.** 14·5 cm; wing-span 25–27·5 cm. Similar in size and in shape of bill and head to Greenfinch *Carduelis chloris*; slightly larger than Pale Rock Sparrow *Carpospiza brachydactyla*, with proportionately shorter wings but longer, forked tail. Quite large, dumpy finch, with dark stubby bill, well-forked tail, and diagnostic purring call; perches often in trees. Head and body uniform pale buff, contrasting with dark-rimmed, pink and white wings, and dark, white-edged tail. ♂ has black loral stripe and (when breeding) bill. ♀ and juvenile duller than ♂. Sexes dissimilar; little seasonal variation. Juvenile separable.

ADULT MALE. Moults: July–October (complete). Loral stripe and feathers surrounding bill black; bill black when breeding, paler (to yellow-horn) in winter. Head, upperparts, and underparts mostly uniform pale buff, tinged sandy and pink when fresh, grey when worn; upper tail-coverts light chestnut; flanks dappled chestnut-buff; belly and under tail-coverts virtually white. By contrast, wings and tail multi-coloured and strongly patterned: lesser and

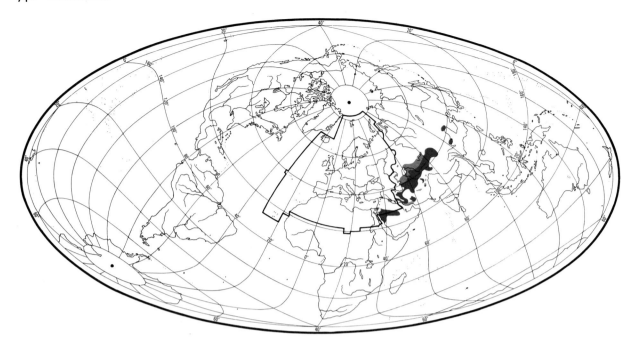

median coverts carmine, greater coverts rosy-pink; secondaries broadly fringed rose; primaries fringed white; together these areas form striking pale panel along folded wing, or over most of extended wing, which contrast with darker rim formed by black patches on bastard wing and primary coverts, black primaries with strikingly wide white tips and edges, and black tertials with bold white fringes. Underwing white; in flight, shows translucent patch from below. Forked tail black, with full white margins to all feathers making sides appear wholly white. Legs dark flesh to greyish-horn. ADULT FEMALE. Bill of some birds almost black but in others dull brown to yellow-horn; lore no darker than brown. Body duller than ♂, lacking warm rump, dappled flanks, and sandy and pink wash. Wings and tail less vividly coloured than ♂, with dark areas of wing and tail only dull black or dark brown. JUVENILE. Much duller than ♀, with brown tone to body, greater coverts, and tertials. Bill straw-colour, becoming pale horn. For 1st-years, see Plumages.

Unmistakable if seen well, but in brief glimpse juvenile might suggest (smaller) Mongolian Trumpeter Finch *Bucanetes mongolicus*, Trumpeter Finch *M. githagineus*, or *Carpospiza brachydactyla* (see those species). Voice diagnostic (see below). Flight light and undulating; silhouette recalls small *Carduelis* finch. Gait a hop. Perches upright in trees, dropping to ground to feed in manner of *C. chloris*; note that *Bucanetes* finches do not normally use such perches. Forms small flocks after breeding which are nomadic in winter.

Song melodious and pleasing, with variety of slow trills; recalls *C. chloris* and Linnet *C. cannabina*. Calls include: quiet whistled 'feenk-feenk' or 'pink pink'; distinctive soft purring 'prrryv', 'prrrt', or 'turr' from breeding adults; similar 'r-r-r-r-r-ee' falling then rising in pitch; sharp 'shreep' in flight; nasal 'hear' in contact.

**Habitat.** In west Palearctic, mainly in lowlands like Trumpeter Finch *Bucanetes githagineus* but is less a desert bird and also ascends mountainous valleys. Occurs where some open tree or shrub growth present, in plantations, orchards, rows of trees, oases, areas of irrigation, and arid places with sparse herbage and scattered trees or shrub thickets, feeding in weedy or fallow cultivated areas. Like *B. githagineus*, needs access to water. Mainly in lowland arid and semi-arid areas but will ascend foothills and valleys to *c.* 2000 m (Harrison 1984). Feeds mostly on ground, but perches freely on bushes, trees, railings, and telegraph wires (Ali and Ripley 1974).

In USSR, found in forest plantings near shores of bodies of water, and in gardens and parks; also alongside irrigation ditches, but usually avoids dense tree plantations. Favours dry clay plains, not too rich in herbaceous vegetation, with isolated shrubs or thickets, especially beside streams or irrigation channels. Movements after breeding season, often to lower ground, apparently involve little change of habitat (Dementiev and Gladkov 1954; Flint *et al.* 1984). In Afghanistan, Paludan (1959) found pairs living in hotel garden below 1000 m, and ♂ singing in tree beside river at *c.* 560 m. In Baluchistan (Pakistan), it is a garden bird, nesting in pollarded vines *Vitis*, roses *Rosa*, almonds, apricots *Prunus*, and also in roadside trees (Ticehurst 1926-7).

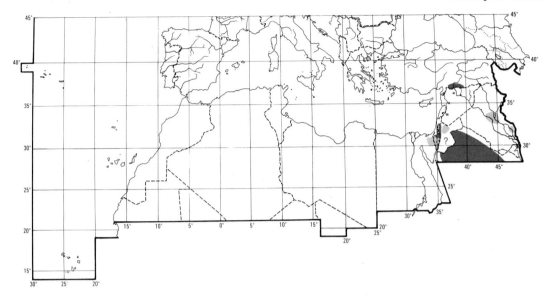

**Distribution.** No information on range changes, except in Israel.

ISRAEL. Up to 1950s an irregular winter visitor, sometimes in large numbers, some remaining to breed after large irruptions. Establishment of permanent breeding population began in late 1950s, following development of agricultural settlements which provided trees and constant water supply, necessary for breeding. (Shirihai in press.)

Accidental. Egypt (north-east Sinai).

**Population.** ISRAEL. A few thousand pairs in 1980s (Shirihai in press).

**Movements.** Chiefly sedentary, with small-scale movements, mainly within breeding range.

In Israel, occurred mainly as irregular winter visitor until 1950s, in very fluctuating numbers; now locally common breeding bird. Part of adult population and apparently some juveniles resident, though in winter most populations tend to flock and wander locally; others move further, with decrease in some breeding areas, and higher numbers in winter than summer in southern and eastern Israel (but not recorded south-west of Sinai: Goodman and Meininger 1989). Post-breeding dispersal and local wandering noticeable from end June, but longer-distance movement (inconspicuous) not until September–October; arrives at Elat (where occurs mainly as winterer) from mid-October, chiefly November, and departs mainly in March. (Shirihai in press.) Birds in north-west Saudi Arabia show no evidence of more than local movement, e.g. not reported from Gulf states, Oman, or Yemen (F E Warr). In Iran, mainly sedentary; wanders beyond breeding area in winter, notably to south-west coast (D A Scott). Apparently occurs only as winter visitor to Iraq (Baghdad area) (Moore and Boswell 1957; Marchant 1962). In Pak-

istan, resident in Baluchistan, with some evidence of local movements: at Quetta, numbers augmented late March by arrival of summer visitors; irregular winter visitor in north (Chitral and North-West Frontier Province) (Meinertzhagen 1920; Ali and Ripley 1974; Roberts 1992).

Status similar in Uzbekistan, Turkmeniya, and southern Kazakhstan. In Kazakhstan, movement mostly inconspicuous, but noticeable in Chokpak pass (chiefly autumn) and Chuya valley (April); birds migrate through Chokpak pass from end of August to beginning of November (chiefly October), and mid-March to early May (in 2 waves) (Korelov *et al.* 1974; Gavrilov and Gistsov 1985). In upper Chuya valley (Kirgiziya), local birds leave breeding grounds at end of August; highest numbers (involving passage birds) occur early September, with few by end of September; spring arrivals recorded from early February, with main passage 11–20 February (Umrikhina 1969). At Badkhyz (south-east Turkmeniya), wholly sedentary in milder winters; in harsh winters, some birds remain in breeding area, but most depart, returning as soon as conditions ameliorate; in unstable weather, they may leave and reappear several times between November and March (Sukhinin 1959). In south-west Kyzyl-Kum (western Uzbekistan), migrates chiefly 08.00–11.00 hrs and 16.00–18.00 hrs in autumn, and 06.00–11.00 hrs and 16.00–19.00 hrs in spring (Tret'yakov 1978). Present all year in western Sinkiang (western China) (Kozlova 1975).          DFV

**Food.** Almost all information from central Asia. Diet seeds and other parts of plants; a few insects in breeding season. Feeds mostly on ground, picking up seeds of desert plants in dry stony places, also in fields, orchards, etc.; sometimes in shrubs or trees taking buds and shoots (Moore and Boswell 1957; Rustamov 1958; Pek and Fedyanina 1961; Yosef 1991; Roberts 1992). In USSR, will feed in village streets on (e.g.) spilled sunflower *Helianthus* seeds with sparrows *Passer* (Umrikhina 1969; Korelov *et al.* 1974). Seems particularly fond of seeds of camel's thorn *Alhagi* and elm *Ulmus* (Rustamov 1958; Sukhinin 1959; Salikhbaev and Bogdanov 1967; Korelov *et al.* 1974). Very often in small flocks at water (Rustamov 1958; Salikhbaev and Bogdanov 1967; Sopyev 1967; Korelov *et al.* 1974). For food in captivity, see Bernasek (1985), Klimanis (1987), and Schäfer (1991).

Diet includes the following. Invertebrates: grasshoppers, etc. (Orthoptera). Plants: seeds, buds, shoots, etc. of elm *Ulmus*, Polygonaceae, Chenopodiaceae (saltwort *Salsola*), Ranunculaceae, Cruciferae (wintercress *Barbarea*, *Euclidium*, *Malcolmia*), Leguminosae (camel's thorn *Alhagi*, *Halimodendron*), Boraginaceae (stickseed *Lappula*), Compositae (sunflower *Helianthus*), grasses (Gramineae, including *Arista* and wheat *Triticum*). (Pitman 1921; Rustamov 1958; Pek and Fedyanina 1961; Salikhbaev and Bogdanov 1967; Umrikhina 1969; Sopyev 1979.)

In Uzbekistan, of 15 crops and stomachs, 11 contained only seeds of wild plants, 2 contained small seeds (probably Cruciferae), and 2 contained seeds and insects (Salikhbaev and Bogdanov 1967). In Kazakhstan, 11 stomachs contained mainly saltwort seeds in May; in other areas, main food seeds of elm or *Halimodendron* (Korelov *et al.* 1974). In Kirgiziya, 6 stomachs held only seeds of Polygonaceae, Ranunculaceae, Leguminosae, and grasses including wheat (Pek and Fedyanina 1961); from April to August, fed also on elm, *Euclidium*, wintercress, stickseed, and Cruciferae in gardens, causing some damage (Umrikhina 1969). In Turkmenistan, early April, 5 stomachs contained only elm seeds, average dry weight per stomach 150-250 mg; in January-February fed on elm buds (Sukhinin 1959).

Young given mostly seeds, occasionally insects. In Uzbekistan, 2 nestling stomachs held only insects (Salikhbaev and Bogdanov 1967), while in another study, almost all crops examined plus 2 collar-samples contained *Malcolmia* seeds (Sopyev 1967, 1979). Tree buds and seeds of grass *Arista* also recorded in diet of young (Rustamov 1958; Roberts 1992). In Israel, ♂♂ fed nestlings milky fluid mixed with seeds and probably green shoots (Yosef 1991). For diet of young in captivity, see Klimanis (1987) and Schäfer (1991). BH

**Social pattern and behaviour.** No comprehensive study. Much of the following is based on extralimital material from Kazakhstan and central Asia.

1. Gregarious all year (see also Flock Behaviour, below), but especially outside breeding season. At end of breeding season, family parties congregate into flocks for feeding, drinking, roosting (see Roosting, below), and migration (Rustamov 1958; Sukhinin 1959; Umrikhina 1970; Korelov *et al.* 1974). In Israel, nomadic flocks of tens or hundreds form after breeding, diminishing September-October due to emigration; elsewhere local populations augmented by large influx (Shirihai in press). In

Baluchistan (Pakistan), small flocks seen summer and winter; rarely forms as large flocks as sympatric Red-fronted Serin *Serinus pusillus* or Goldfinch *Carduelis carduelis* (Roberts 1992). In central Asia, flocks usually small, e.g. in Turkmeniya 3-10' but sometimes 100 or more; winter flocks break up later in cold springs (Rustamov 1958). Post-breeding flocks quite often associated with other species, e.g. sparrows *Passer* and finches (Fringillidae) (Sukhinin 1959; Stepanyan 1970; Umrikhina 1970). BONDS. No evidence for other than monogamous mating system. Pair-bond long-lasting; in summer usually encountered feeding in pairs even when not actively nesting (Roberts 1992). Nest built mainly or exclusively by ♀, ♂ at most carrying some material (Rustamov 1958; Kovshar' 1966; Korelov *et al.* 1974) or taking no part other than remaining in close attendance (Umrikhina 1970; Ali and Ripley 1974; Ponomareva 1974; E N Panov). Perhaps real variation in role of sexes in incubation: by both sexes according to Rustamov (1958), probably more by ♀ (Sopyev 1967). However, only ♀ found incubating by Korelov *et al.* (1974) and ♂♂ shot at this time had no brood patches; incubation exclusively by ♀ also found by Sagitov and Bakaev (1980), Ponomareva (1981), Yosef (1991), and (in captivity) Klimanis (1987). Both sexes care for young (Sopyev 1967; Umrikhina 1970; E N Panov) but most feeding by ♂, at least up to fledging (see Relations within Family Group, below). ♂ seen caring for 1st brood while ♀ building nest for 2nd (Korelov *et al.* 1974); similar report by Ticehurst (1926-7) but sex roles not specified. Such overlap presumably explains observation of nest-building ♀ accompanied by adult ♂ and 'full-grown juvenile' (Paevski *et al.* 1990). Young fed for 14-16 days after leaving nest (Umrikhina 1970). Age of first breeding not known. BREEDING DISPERSION. Solitary or in small neighbourhood groups or colonies. In Israel in loose colonies of usually a few tens of pairs, 1·5-4 m (minimum 1 m) between nests (Paz 1987; Shirihai in press); in Beersheba (north-west Negev), 1958, large isolated colony of 100-150 nests extending over *c.* 1·5 km (Hovel 1960); in Sede Boqer, breeding area comprised stands (each 10-45 m²) of trees, each stand occupied by discrete nesting group (Yosef 1991). At least where grouped, territories of neighbouring pairs overlap widely and defence confined to immediate nest-area (E N Panov; see Antagonistic Behaviour, below). Study by Ponomareva (1981) in Uzbekistan indicates that availability of nesting habitat influences nest dispersion: at Repetek (south-east Kara-Kum) no nesting groups, and pairs at least 500 m apart, but in Kyzyl-Kum shortage of cover resulted in nests being concentrated in saxaul *Arthrophytum* plantations, e.g. 16 pairs in *c.* 0·35 ha of saxaul near houses; nests on average 5-6 m apart (minimum 1·5 m, i.e. 2 nests in same shrub). In Badkhyz (south-east Turkmeniya) density higher in poplar *Populus* groves than tamarisks *Tamarix*; in one area 6 pairs per km², and 3 pairs along 70-80 km of gorge; also 4 nests in 1500 m² (nest-trees 3-8 m apart); solitary nests reported elsewhere in Turkmeniya (Rustamov 1958; Sukhinin 1959). In Kazakhstan, usually well dispersed but sometimes more concentrated; near Nikolaevka, pairs 1-1·5 km apart; in Ili valley 200-500 m apart (Korelov *et al.* 1974). In Israel, number of breeding pairs in particular area varies quite markedly from year to year depending on changes in crops and water supply as well as natural nomadic fluctuations (Shirihai in press). However, same colony sites sometimes used in successive years, e.g. Beersheba colony (see above) reoccupied in 1959 (Hovel 1960). In Uzbekistan, optimal sites, including individual bushes (especially dense saxaul, well protected from wind, sun, and predators) used for many years in succession (Ponomareva 1981). ROOSTING. Mainly shrubs and trees serve for communal roosting and daytime loafing; e.g. eucalyptus groves in Israel (Pitman 1921), conifers at Persepolis (Iran) (Erard and Etchécopar 1970). In

Uzbekistan, also recorded roosting in crevices in steep faces (Salikhbaev and Bogdanov 1967). In Israeli study, dependent fledglings usually roosted together in nest-tree (Yosef 1991). *R. obsoleta* active morning and evening in summer, all day in winter (Salikhbaev and Bogdanov 1967). See Flock Behaviour (below) for roosting behaviour.

2. Wary in open country but tame and confiding near human habitation; e.g. foraged around tents in Israel, June (Pitman 1921), and in Kazakhstan often seen feeding on roads and pavements with sparrows (Korelov *et al.* 1974). No further information. FLOCK BEHAVIOUR. In Israel, winter, huge noisy flocks assembled in *Zizyphus*, *Mimosa*, and other thorny shrubs before retiring to roost in nearby eucalyptus groves; entry to roost protracted, with much preliminary 'fuss' and circling, then alighting in and leaving several parts of grove before finally settling in for the night (Pitman 1921). In Iran, birds dispersed from roosts in the morning over radius of a few hundred metres, most often in groups of 7–8 birds, sometimes up to 15–20, also in pairs (Erard and Etchécopar 1970). During breeding season, frequent flights by small flocks to water for drinking; e.g. in Kazakhstan, May, flocks of 2–6, almost all ♂♂ (fewer singletons) seen circling over lake, settling in reeds, then looking around before flying down to water; rapid departure after drinking (Korelov *et al.* 1974). For song in flocks see below. SONG-DISPLAY. ♂ sings from exposed elevated perch while turning from side to side, including slight nodding, swinging of head (Fig A). In

started almost immediately, indicating that pair-formation occurs in flock; pairs difficult to discern in initially large flocks but evident after these broke down into groups of 4–10 birds; average 16·8 days between formation of these smaller flocks and start of laying (Yosef 1991). Study by Pitman (1921) in Israel also consistent with pair-formation in flocks: pairing seemed complete at end of April and birds stayed in pairs till end of May, after which all apparently left, presumably to breed elsewhere. Likewise in Kazakhstan, soon after arrival on breeding grounds, flocks break up into pairs which seek territories (Korelov *et al.* 1974). In south-east Turkmeniya, first pairs noted mid-February, and pair-formation complete early April (Sukhinin 1959); pair-formation slightly later further north (Ashkhabad) (Rustamov 1958); observed at end of April, Kirgiziya (Umrikhina 1970) and early May, Uzbekistan (Salikhbaev and Bogdanov 1967). (2) Courtship-feeding. ♂ feeds ♀ from before nest-building through incubation to brooding (see Relations within Family Group, below). In Israel, Courtship-feeding observed from early March when flocks first broke up (see above); during 2nd half of May, pairs that had successfully fledged young resumed Courtship-feeding prior to starting 2nd clutch in 2nd week of June (Yosef 1991). During collection of nest-material, ♂ feeds (by regurgitation) ♀ frequently near nest; on ♂'s approach, ♀ performs Begging-display: wing-shivers and, during transfer of food, rapidly beats half-open wings (Fig B); during Courtship-feeding, ♂

A

B

mild weather during January, or more often early February, Turkmeniya, song heard from birds still in flocks (Rustamov 1958; Sukhinin 1959). After arrival on breeding grounds, ♂♂ sing from trees where nests will eventually be built (Umrikhina 1970). ♂ also sings when closely attending ♀ as she collects nest material and builds (Erard and Etchécopar 1970; Ponomareva 1974), also near nest-site during incubation (Hovel 1960). In India, song heard from beginning of March (Whitehead 1911; Ali and Ripley 1974); in central Asia from January (see above), but most song in February (Sukhinin 1959). Little information on end of song-period; in eastern Turkey heard in July (P S Hansen). ANTAGONISTIC BEHAVIOUR. Few details. Sukhinin (1959) noted that even where nests were close together there was no aggression between pairs. However, owners resist (sometimes by slight Threat-gaping) intrusion by neighbours into immediate nest-area (E N Panov). Yosef (1991) likewise found ♂♂ vigilant in immediate vicinity of nest, keeping away conspecifics during incubation and after hatching. HETEROSEXUAL BEHAVIOUR. (1) Pair-bonding behaviour. Little known about display sequence, but presumably involves ♂'s Song-display (see above). In Sede Boqer, newly arrived flocks, early March, were of mixed sex, and Courtship-feeding (see below)

gives Feeding-calls (E N Panov: see 3 in Voice). When ♀ incubating, ♂ usually lands within 10–15 m of nest, calls, and ♀ flies to him and begs (Korelov *et al.* 1974). (3) Mating. No details of display. Seen at nest, during building and especially after nest completed (Rustamov 1958). In Israel, for 1st clutches, starts 10 days before laying; initially *c.* 2 copulations daily per pair, increasing to *c.* 19 around start of laying, declining to zero 6–8 days thereafter (Yosef 1991). (4) Nest-site selection and behaviour at nest. Site selected over *c.* 2 weeks (Umrikhina 1970) but no information on roles of sexes. ♂ attends ♀ closely (presumably mate-guarding) during nest-building and collection of nest material (e.g. Erard and Etchécopar 1970, Ponomareva 1974); see also Song-display (above). During hottest time of day, incubating bird (see Bonds, above) sits rather high in nest and frequently flutters wings (presumably for ventilation, but perhaps sometimes Begging-display); off-duty bird seen to perch on nest, lower head, and give water to mate (Sopyev 1967). In captive study, ♂ seen driving ♀ back to nest during incubation (Klimanis 1987). RELATIONS WITHIN FAMILY GROUP. Hatching may take 3 days (Korelov *et al.* 1974). Eyes of young open at 2–3 days (Rustamov 1958; Sagitov and Bakaev 1980). While ♀ broods and shades young (see below) ♂ provides all food for her and brood (Korelov *et al.* 1974). ♀ always present and attentive at nest but never seen to feed young (Yosef 1991), though not clear if this implies ♂ alone provides food up to fledging (other sources suggest at least partial provisioning role by ♀: see Bonds, above). ♂ calls (see 2a in Voice) when approaching nest containing young (E N Panov) and on arrival regurgitates food to ♀ for self-consumption and transfer to young (Sagitov and Bakaev 1980). One parent (often ♀) almost constantly at nest shading young at hottest time of day (Sopyev 1967). Every 10-15 min, ♀ moves onto nest-rim and ventilates nest (presumably as in incubation: see above); ♂ also shades young but only occasionally and briefly (Korelov *et al.* 1974; Sagitov and Bakaev 1980). 2 records of parents giving water to young (Sopyev 1967), although this must be hard to distinguish from the typical milky fluid (mixed with seeds, etc.) regurgitated to young up to fledging. Both parents eat faecal sacs (E N Panov). After fledging, parents continue to feed young for *c.* 2 weeks (see Bonds, above) during which family roosts together (see Roosting, above). ANTI-PREDATOR RESPONSES OF YOUNG. No information. PARENTAL ANTI-PREDATOR STRATEGIES. (1) Passive measures. ♀ a tight sitter and may almost be touched before slipping off nest (Hovel 1960; Paz 1987). In Kazakhstan, near human habitation, ♀♀ typically sit tight on nestlings until tree is tapped, and one sitting ♀ allowed herself to be handled; however in saxaul scrub, sitting ♀♀ much more wary, flushing when man *c.* 10 m away, and returning only after 20-40 min (Korelov *et al.* 1974). (2) Active measures. In Israel, ♀♀ reported simulating injury (no details) when forced to leave nest, especially containing small young (Paz 1987). No further information.

(Figs by D Nurney from drawings by E N Panov.)     EKD

**Voice.** Following account includes extralimital sources, and more study needed in west Palearctic. No information on geographical variation.

CALLS OF ADULTS. Song of ♂. Quiet, introspective suc-cession of mainly call-type units, notably tremolos, nasal, buzzing, and occasional harsh units, overall with a chattering or chortling quality reminiscent of Budgerigar *Melopsittacus undulatus*. Some songs distinctly phrased (see recording from Iran, below), others not. Thus, in recordings from Israel, rambling improvisatory song obscures phrasing, rather a disconnected medley of highly variable ripple sounds (tremolos: 'trrr', 'prrrp', etc.) and some much louder harsh sounds and shrieks: Fig I begins with 3 ripples followed respectively by scraping sounds and rather scratchy chirps, 2 more ripples, and finally an extended harsh noise (see also 3, below); in this recording song lapses for a time into very quiet Subsong in which units are interspersed with clicks (apparently vocal) giving impression that bird is marking time before resuming song. Song from a different bird more buzzing, including ascending glissandos (some of which are diads): Fig II shows a chattering followed by long buzzing 'zwooeee'. By comparison with Israeli recording, song from Iran a much faster more lively melody with well-defined recurring phrases: Figs III-IV illustrate 2 phrases, difficult to render (especially Fig III) but Fig IV roughly 'prrrrp prp-zink tzank urrrrrrrr'. (J Hall-Craggs.) Evidently song of this type in Iran also described as a little melody consisting of repeated 'prruii prruii' (see call 2a), rolled sounds, and grating nasal tremolos; song distinctive, though recalled Goldfinch *Carduelis carduelis*, Linnet *C. cannabina*, and especially Greenfinch *C. chloris* with its terminal drawn-out units (Erard and Etchécopar 1970); this ending, rendered 'dzhee', also a call (E N Panov: see 3, below). In eastern Turkey, short 2-part song heard as

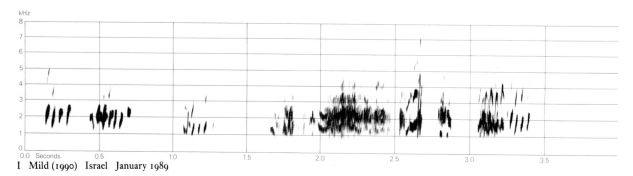

I  Mild (1990)  Israel  January 1989

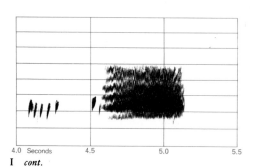

I  *cont.*

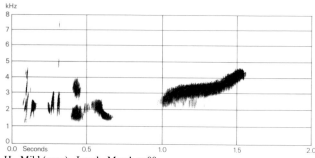

II  Mild (1990)  Israel  March 1988

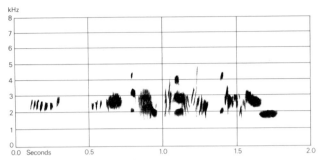

III  P A D Hollom  Iran  May 1977

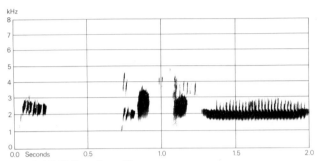

IV  P A D Hollom  Iran  May 1977

Paludan 1959) quite unlike thrush *Turdus*; began with staccato croaking sounds, some high pitched, others low and almost like frog, followed by drawn-out 'toy trumpet-like' squeaky buzzing 'beeze' or 'zweee-e-ah' sounds, interspersed with higher-pitched squeaky whistles, each rising and falling (e.g. 'quee-ah'); later also some soft twittering (Roberts 1992). (2) Contact-calls. (2a) Main call a quiet, soft, but melodious purring 'prrrrrrl', given when perched or in flight; slightly reminiscent of Bee-eater *Merops apiaster*; varies in length, and often interspersed with call 3 (Mild 1990). Call is a tremolo, less variable than those in song: Fig VI shows 'trrrrrr' followed by a louder, lower-pitched 'churrrr' (J Hall-Craggs). Other descriptions as follows: distinctive purring 'prrut prrut' (Pitman 1921: Israel); attractive 'trrryu trrryu-rrryu trrryu' (in which 'yu' sometimes replaced by 'ee') (Korelov *et al.* 1974); muffled throaty 'trree' given by ♂ approaching nest to feed young (E N Panov); 'trreeüt' or 'trrük' rather like Skylark *Alauda arvensis* (Jonsson 1992); 'prruii' component of song (Erard and Etchécopar 1970), though this mistakenly aligned by Ali and Ripley (1974) with call 2b. In recording by P A D Hollom (Iran) brief dry 'trr' sounds presumably represent curtailed variant. Fig VII shows 2 short 'prip' sounds (each preceded by quiet grace-note: J Hall-Craggs) described as flight calls. (2b) Interrogative 'pink pink pink' by ♂ in breeding season (Williams and Williams 1929); also a cheery 'prink prink' widely heard in breeding season (Christison 1940). In Fig V, 3 such calls are apparently part of song. In recordings by P A D Hollom (Iran) this call given in series interspersed with, respectively, calls 2a and 3. (3) Feeding-call. Short,

uniformly pitched 'djiggi', like *C. cannabina*, followed by quiet extended 'dyyyyy' in which pitch initially falls then rises again (P S Hansen). Song recorded in Pakistan described as more reminiscent of Trumpeter Finch *Bucanetes githagineus* than any cardueline, and (contra

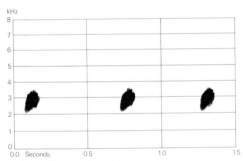

V  P A D Hollom  Iran  May 1977

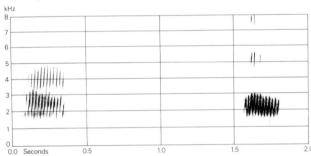

VI  P A D Hollom  Iran  May 1977

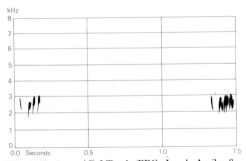

VII  N Tucker and D J Tombs/BBC  Israel  April 1985

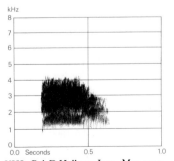

VIII  P A D Hollom  Iran  May 1977

slightly nasal 'sshroe' (Mild 1990). Wheezy 'dzhee-dzhee-dzhee' or 'dze-dze-dze' given by ♂ courtship-feeding ♀ (E N Panov); 'veeip' recalling harsh flight-call of Twite *Carduelis flavirostris* (Jonsson 1992). In recordings, rather like harsh call of distant Jay *Garrulus glandarius* (P J Sellar: Fig VIII). Occurs in combination with calls 2a-b (though less common than 2a: Mild 1990), also in song (see last unit in Fig I).

CALLS OF YOUNG. No definite information, but quite long delicate 'zirrrr' heard when young being fed (Klimanis 1987) is perhaps food-call. EKD

**Breeding.** SEASON. Israel: eggs laid late March to mid-April; 2nd clutches May–June, probably mostly 2nd half of June (Yosef 1991; Shirihai in press). Southern Turkey: young recorded in nest mid-May (Martins 1989). Iran: nest-building recorded early May (Erard and Etchécopar 1970). Baluchistan (Pakistan): eggs laid from late March, 2nd clutches mid-June to July (Ticehurst 1926-7; Williams and Williams 1929). Kazakhstan and central Asia: 2 peak laying periods, late March to early May depending on weather, then June (Rustamov 1958; Sukhinin 1959; Korelov *et al.* 1974; Ponomareva 1981); for review, see Bel'skaya (1989). SITE. In horizontal or vertical fork of shrub or tree generally 1–5 m above ground, frequently in cultivated species in orchard, garden, etc., e.g. various *Prunus* species, vine *Vitis*, or pistachio *Pistacia*; often very conspicuous, though can be well-hidden (Salikhbaev and Bogdanov 1967; Erard and Etchécopar 1970; Korelov *et al.* 1974; Ponomareva 1981; Roberts 1992). In Israel, 2–4 m up in planted *Eucalyptus*, *Acacia*, or pine *Pinus* (Yosef 1991). In Turkmeniya, of 116 nests, 36% in poplar *Populus*, 13% elm *Ulmus*, 11% mulberry *Morus*, 11% tamarisk *Tamarix*, 10% *Prunus*; average height 1·5 m (0·4–3·5, *n* = 86) (Sukhinin 1959). In desert areas, many nests in saxaul *Arthrophytum* shrubs (Rustamov 1958; Sopyev 1967), and in Kirgiziya often in elm or oak *Quercus*, some at 25 m (Umrikhina 1970). Nest recorded in Pakistan in tall herb (Umbelliferae) (Roberts 1992). Nest: foundation of twigs and coarse herb stalks, lined with thick felt-like layer of plant down and other soft plant material, especially of cotton and poplar, sometimes fur, hair, cloth, etc.; some nests only of soft material (Ticehurst 1926-7; Rustamov 1958; Sukhinin 1959; Yosef 1991). Shape said to be often oval (Umrikhina 1970; Yosef 1991); 12 nests in Turkmeniya had average outer diameter 10·6 × 8·5 cm, inner diameter 6·3 × 6·0, overall height 7·4, depth of cup 1·2 cm (Sukhinin 1959); Rustamov (1958) gave depth of cup 4·0–5·0 cm; see also Yosef (1991) for nests in Israel, where 2nd nests smaller in outer dimensions than 1st nests. Building: apparently by both sexes, though probably mainly or wholly by ♀; ♂ reported to help with foundation by Korelov *et al.* (1974) and Yosef (1991); both sexes build, roles not specified, according to Rustamov (1958), Sukhinin (1959), Kovshar' (1966), Salikhbaev and Bogdanov (1967), and Sopyev (1967), while Umrikhina (1970),

Ponomareva (1974), and Roberts (1992) recorded building by ♀ only, accompanied by ♂. Takes *c.* 5–7 days (Rustamov 1958; Umrikhina 1970). Removes foundation material from old nests for 2nd nests (Yosef 1991). EGGS. See Plate 61. Sub-elliptical, smooth and slightly glossy; white to pale greenish-blue, with small purplish-black specks and very fine hair-streaks at broad end (Harrison 1975; Schönwetter 1984; Yosef 1991). Considerable variation in size and shape (Williamson and Williamson 1929; Schönwetter 1984). 19·0 × 14·3 mm (17·4–22·2 × 13·0–15·1), *n* = 120; calculated weight 2·05 g (Schönwetter 1984); average fresh weight of 21 eggs, Israel, was 2·17 g (Yosef 1991). Clutch: 4–6 (3–7). In Israel, of 51 clutches: 4 eggs, 33%; 5, 66%; average 4·7; average 1st clutch 4·8 (*n* = 27), 2nd 4·5 (*n* = 24). Of 16 1st clutch replacements: 3 eggs, 44%; 4, 56%; average 3·6. (Yosef 1991.) In south-east Turkmeniya, of 43 full clutches: 3 eggs, 16%; 4, 9%; 5, 23%; 6, 49%; 7, 2%; average 5·1 (Sukhinin 1959). ♀ can lay up to 4 clutches, including 2 replacements (Bel'skaya 1989). Eggs laid daily; generally 2 broods (Umrikhina 1970; Korelov *et al.* 1974; Yosef 1991; Roberts 1992). ♀ can build 2nd nest while young still in 1st (Ticehurst 1926-7). INCUBATION. 12–15 days (Rustamov 1958; Sukhinin 1959; Korelov *et al.* 1974; Ponomareva 1981); in Israel, average 13·8 days (*n* = 22), by ♀ only, starting with 2nd egg (Yosef 1991). By ♀ only also according to Korelov *et al.* (1974, where apparently from 1st egg), Sagitov and Bakaev (1980), and Ponomareva (1981), but many reports of participation by ♂, e.g. Rustamov (1958), Salikhbaev and Bogdanov (1967), Sopyev (1967, where said to start with 4th egg). At one nest, Uzbekistan, average stint on eggs was 15·1 min (1–116) and there were 39 changeovers between sexes in 14 hrs of observation; at another, 45 min (6–107) and 16 changeovers in 13 hrs (Sopyev 1967). YOUNG. Fed and cared for by both parents; ♀ broods intensively (Sopyev 1967; Umrikhina 1970; Korelov *et al.* 1974; Roberts 1992). At all nests watched in Israeli study, ♀♀ seen to remove faecal sacs but not to feed young (Yosef 1991). In captive studies, ♂ fed young only from *c.* day 5 (Klimanis 1987; Schäfer 1991). FLEDGING TO MATURITY. Fledging period 13–14 days (12–16) (Umrikhina 1970; Korelov *et al.* 1974; Yosef 1991; Shirihai in press); said to be *c.* 21 days in captive study (Schäfer 1991). Young independent 14–16 days after fledging (Umrikhina 1970; Korelov *et al.* 1974; Yosef 1991). In captivity, fed themselves *c.* 11 days after leaving nest (Klimanis 1987). BREEDING SUCCESS. In Israel, of 295 eggs in 67 nests, 60% hatched, 24% lost to predation, weather, and human disturbance, and 16% infertile (Yosef 1991). In Turkmeniya, 18·5% of 119 eggs infertile; only 24% of 81 nests produced flying young, 54% of clutches lost at egg stage, 22% at nestling stage; predators included Magpie *Pica pica*, Little Owl *Athene noctua*, wild cat *Felis silvestris*, snakes *Coluber*, and ants (Formicidae) (Rustamov 1958; Sukhinin 1959). In Uzbekistan, average 2·6 fledged young per nest, 50% of clutch size; many eggs infertile, predation high, and strong winds destroyed nests (Ponomareva 1981).

BH, MK

**Plumages.** ADULT MALE. In fresh plumage (September–February), upperparts from forehead to rump delicate sandy-buff with slight but distinct tawny wash on top of head; rump and upper tail-coverts bright rufous-cinnamon, feathers narrowly tipped sandy-buff, not concealing rufous. Narrow strip of short bristle-like feathers along sides of both mandibles and on upper chin velvety-black; stripe on lore black, bordered above and below by some tawny-rufous. Cheeks and ear-coverts tawny-buff, merging into sandy-buff of side of neck. Chin to chest, side of breast, and upper flank pale sandy-buff or cream-buff with slight pink tinge, palest on chin, merging into buff-white or cream-white of mid-belly, vent, and under tail-coverts; lower flanks brighter tawny-rufous or rufous-cinnamon. Tail black, feathers with contrasting cream-white fringes along both webs and tip; fringes narrower on outer web of t6 and inner web of t4–t6, wider and covering entire outer web on t3–t5. Flight-feathers black, outer webs and tips contrastingly fringed white; fringe narrow along outer edge of p9, on tips of p7–p9, and on emarginated parts of p6–p8, but covering almost entire outer web on remainder of primaries; basal 75% of outer webs of secondaries contrastingly rosy-pink. Broad borders along basal and middle portion of inner web of primaries and basal 70% of inner webs of secondaries white, visible from above only when wing fully spread, but resulting in largely white undersurface of flight-feathers from below. Tertials black with broad and contrasting white fringe, tinged cream-buff when plumage quite fresh. Outer webs of greater upper primary coverts rosy-pink, inner webs and tips contrastingly black; bastard wing black, shorter feathers with pink outer fringe. Lesser upper wing-coverts sandy-buff; median warmer tawny-cinnamon with rosy-red fringe; greater contrastingly paler rosy-pink with contrasting (but mainly concealed) black centres and inner webs, pink-white fringe along tip, and some off-white or sandy-grey on extreme base. Under wing-coverts and axillaries white, some feathers partly suffused pink. *In worn plumage* (about April–July), upperparts including crown duller drab-grey, less buff, forehead distinctly brighter rufous-tawny, lower rump and upper tail-coverts more contrastingly uniform rufous; side of head and neck, chin, and throat paler, more cream-buff, black of lore more contrasting; chest, side of breast, and upper flank pink-buff, grading into rufous on lower flank, more contrasting with purer and more extensive white of belly and vent. Wing and tail as in fresh plumage, but white of tertial-fringes purer, and white fringes on tips of outer tail-feathers, outer primaries, and tertials narrower or worn off; outer edges of pink outer webs of secondaries, primary coverts, and greater coverts sometimes tinged rosy-red (but generally less so than on wing of worn Crimson-winged Finch *Rhodopechys sanguinea*). In very fresh plumage, during and just after moult, pink on wing sometimes tinged grey or vinous. ADULT FEMALE. Like adult ♂, but upperparts and side of neck entirely sandy-drab-grey, less buffish than in ♂, without tawny tinge on forehead and crown and without contrasting rufous-cinnamon on lower rump and tail-coverts. Feathering at base of bill cream-buff, lore tawny-buff or brown, remaining side of head light sandy-buff; no black along base of bill and on lore. Underparts as adult ♂, but chest, side of breast, and flank slightly less pink when worn, more pale grey-buff, white of belly hardly contrasting; lower flank without contrasting rufous-cinnamon. Tail as adult ♂, but fringes less pure white and centres more greyish- or brownish-black, that on t1 grey-brown merging into black-brown on tip. Flight-feathers, primary coverts, and bastard

wing as adult ♂, pink and white of fringes of feathers equally extensive, but black of tips and inner webs of feathers duller, more greyish- or brownish-black; tertials quite different, dark grey-brown or drab-brown with rather narrow and ill-defined pale grey-buff or off-white fringe (in ♂, black with broad and contrasting white fringe); lesser and median upper wing-coverts sandy drab-grey, like upperparts, median not as tawny as in adult ♂ and without red or rufous along fringe; greater coverts drab-grey, shading to dull black on tip of inner web, bordered pink or greyish-pink along outer web (fringe narrower and much less contrasting than in ♂), narrowly off-white (on outer coverts) or tawny (on inner) along tips. Under wing-coverts and axillaries white with cream tinge, rarely with some slight pink suffusion. *In worn plumage*, upperparts duller drab-grey, lore more tawny, pale fringes along tips of tail-feathers, primaries, and tertials largely worn off, these latter less contrastingly patterned than in adult ♂. NESTLING. Down greyish-white (Sopyev 1967) or white, on head and upperwing only, developing on day 2 (Rustamov 1958). For development, see Sagitov and Bakaev (1980) and Yosef (1991). JUVENILE. Upperparts, side of head and neck, and lesser and median upper wing-coverts pale sandy-brown. Underparts dirty white, tinged sandy-brown on throat to chest and on flank. Greater upper wing-coverts and tertials brown, tertials washed pale sandy on both webs, greater coverts with pale sandy outer fringe and tip. FIRST ADULT MALE. Like adult ♂, but juvenile flight-feathers, tail, tertials, greater upper primary coverts, bastard wing, and (sometimes) a few outer greater upper wing-coverts retained. Flight-feathers and primary coverts as adult, but tail, tertials, bastard wing, and (if any juvenile) outer greater coverts different; white fringes of tertials, outer greater coverts, and along tips of most tail-feathers partly or fully washed cream-buff (cream-white or fully white in adult), tips of feathers of bastard wing distinctly pointed, not rounded. When plumage worn, fringes of tail and tertials bleached to white or partly worn off, but shape of bastard wing usually still distinct. FIRST ADULT FEMALE. Like adult ♀, but part of juvenile feathering retained, as in 1st adult ♂. No marked differences in tertials or tail (tips of feathers of 1st adult more worn at same time of year than in adult), but tips of feathers of bastard wing more pointed than in adult, as in 1st adult ♂. Also, rose-pink on secondaries, primary coverts, and greater coverts on average paler and more restricted than in adult ♀.

**Bare parts.** ADULT, FIRST ADULT. Iris hazel-brown, pale brown, or brown. Bill deep black in breeding ♂, greyish-black to yellow-horn in winter; bill generally brown-horn in ♀, but 5 of 9 ♀♀ examined March–May had bill almost as deep black as ♂. Leg and foot dark flesh, flesh-brown, horn brown, or dull flesh-grey (RMNH, ZFMK, ZMA). NESTLING. Naked skin pink at hatching (Rustamov 1958). JUVENILE. Iris brown. Bill yellow-horn or straw-colour, sometimes with pink tinge at base, changing to pale brown when older. Leg and foot flesh-brown. (Hollom *et al.* 1988; BMNH.)

**Moults.** ADULT POST-BREEDING. Complete; primaries descendent. Single ♀ from Turkmeniya, 17 September, mainly in fresh plumage, but some feathers of head and outer primaries still growing (ZFMK); 2 ♂♂ and 4 ♀♀ from Afghanistan in last stages of moult on 14 October (Vaurie 1949b). In Afghanistan, a nesting pair (♀ laying), 23–25 July, had some body feathers and tertials in moult (Paludan 1959). POST-JUVENILE. Partial: head, body, lesser, median, and none to all greater upper wing-coverts. Starts within a few weeks of fledging, completed in birds examined from October onwards. Moult occasionally complete (H Mendelssohn).

**Measurements.** ADULT, FIRST ADULT. Whole geographical range, mainly February–May; skins (RMNH, ZFMK, ZMA). Bill (S) to skull, bill (N) to distal corner of nostril; exposed culmen on average 4·0 less than bill.

| | | | | | |
|---|---|---|---|---|---|
| WING AD | ♂ | 88·1 (1·81; 18) | 85–92 | ♀ 85·8 (1·93; 12) | 83–89 |
| TAIL | | 59·9 (2·64; 18) | 55–64 | 58·1 (2·48; 12) | 55–62 |
| BILL (S) | | 14·7 (0·61; 17) | 14·1–15·7 | 14·6 (0·60; 11) | 13·8–15·4 |
| BILL (N) | | 8·9 (0·38; 18) | 8·2–9·5 | 9·0 (0·32; 12) | 8·3–9·4 |
| TARSUS | | 17·3 (0·51; 18) | 16·4–18·1 | 17·1 (0·56; 12) | 16·2–17·9 |

Sex differences significant for wing.

Iran and Afghanistan: wing, ♂ 89·1 (6) 87·5–92, ♀ 84·8 (14) 83–87; tail, ♂ 58·0 (6) 56–60, ♀ 55·7 (12) 54–60 (Vaurie 1949*b*). Sede Boqer (Israel), breeding: wing, ♂ 86·3 (2·3; 16) (Yosef 1991). Syria: wing, ♀ 81 (Čtyroký 1987). USSR: wing, ♂ 85·7 (38) 80·5–91·5, ♀ 82·1 (19) 78·5–85·5 (Dementiev and Gladkov 1954). Iran: wing, ♂ 86·2 (5) 85–87 (Erard and Etchécopar 1970).

**Weights.** ADULT, FIRST ADULT. Sede Boqer (Israel), March–May: ♂ 23·8 (1·3; 16), ♀ 26·4 (6·6; 7) (Yosef 1991). Kazakhstan: ♂ 23·8 (10) 22–26, ♀ 23·8 (6) 17·5–28 (Korelov *et al.* 1974). Afghanistan, July: ♂ 25, ♀ (laying) 26, unsexed 24 (Paludan 1959). NESTLING. At day 1, 1·7 (0·2; 22) (Yosef 1991). See also Rustamov (1958) and Vorobiev (1980).

**Structure.** Wing rather long, broad at base, tip bluntly pointed.

10 primaries: p8 longest, p7 and p9 0–1·5 shorter, p6 4–6, p5 11–15, p4 16–20, p3 19–25, p2 22–28, p1 25–31; p10 strongly reduced, 1–3 shorter than reduced outermost greater upper primary covert under which it is concealed, 55–66 shorter than p8, 9–14 shorter than longest primary covert. Outer web of (p6–)p7–p8 emarginated, inner web of p8 (p7–p9) with notch (sometimes faint). Tip of longest tertial reaches tip of p1–p3. Tail rather short, tip forked; 12 feathers, t1 7–12 shorter than t6. Bill short, strong, conical; depth at feathering 9·7–10·8, width at base 7·8–8·6; closely similar to bill of Greenfinch *Carduelis chloris*, but slightly shorter, slightly narrower at base, and with base of upper mandible more swollen at side. Short and dense tufts of bristle-like feathers cover nostril and lateral base of upper mandible. As in trumpeter finches *Bucanetes* and rosy finches *Leucosticte*, a small gular pouch (Bock and Morony 1978). Tarsus and toes short, fairly slender. Middle toe with claw 16·7 (10) 15·7–18·0; outer toe with claw *c.* 70% of middle with claw, inner *c.* 67%, hind *c.* 73%.

**Geographical variation.** No differences in colour between series of specimens at same stage of wear (Vaurie 1949*b*), but some slight variation in size, too small to warrant recognition of races: wing of ♂ 86·7 (4) 85–89 in Levant and Turkey, 89·3 (7) 86–92 in Tarim basin (western China), intermediate in Transcaspia (Kazakhstan to Turkmeniya): 87·9 (7) 86–91 (C S Roselaar).

CSR

---

## *Bucanetes mongolicus* Mongolian Trumpeter Finch

PLATES 47 and 50 (flight)
[between pages 640 and 641]

DU. Mongoolse Woestijnvink    FR. Roselin de Mongolie    GE. Mongolengimpel
RU. Монгольский снегирь    SP. Pirrula mongólica    SW. Mongolisk ökentrumpetare

*Carpodacus mongolicus* Swinhoe, 1870. Synonym: *Rhodopechys mongolicus*.

Monotypic

**Field characters.** 13 cm; wing-span 25·5–27·5 cm. Slightly larger than Trumpeter Finch *B. githagineus*; 10% smaller than Desert Finch *Rhodospiza obsoleta*, with proportionately longer wings. Medium-sized, stubby-billed, thick-headed, long-winged but rather short-tailed finch. Looks rather ghostly at distance but shows 2 strikingly pale panels across greater coverts and secondaries at close range. Unlike *B. githagineus*, head sandy, and black-centred tail broadly fringed pink or off-white at all seasons; bill never red. Sexes dissimilar; marked seasonal variation, especially in ♂. Juvenile separable at close range.

ADULT MALE. Moults: July–October (complete). Head and upperparts pale dun- to sandy-brown, only softly marked with faintly darker feather-centres on crown and mantle but, particularly when worn, showing short pale rosy-buff fore-supercilium, narrow whitish eye-ring, and pale rosy rump. Wings patched pale, with (a) dusky-brown median coverts tipped dull buff, (b) white-based greater coverts with rose fringes and dull black and pink tips

forming bold upper wing-panel, (c) almost white secondaries with dusky tips forming long, bold lower wing-panel (paler than all other desert finches), and these framed by (d) almost black tips to bastard wing and primary coverts, (e) black-centred, dun-fringed tertials, and (f) dark brown ends of primaries with dun tips. Wing pattern most striking when plumage worn, dominating rest of plumage at close range and giving ghostly look to flying bird. Tail black, with broad almost white fringes. Underparts dull white to cream-buff, with rose-pink flush strongest on throat. Bill yellowish-horn; legs brownish-yellow. ADULT FEMALE. Duller than ♂, with less boldly patterned wings, less rosy tint (especially on ill-defined rump), and buffier flanks. JUVENILE. Resembles ♀ but lacks all rosy tints, with tawny wash to breast and flanks. Wings noticeably duller, with duller, sandier margins and tips to wing-coverts forming double wing-bar (not pale panels as in adult).

Closely related to *B. githagineus*, but much more

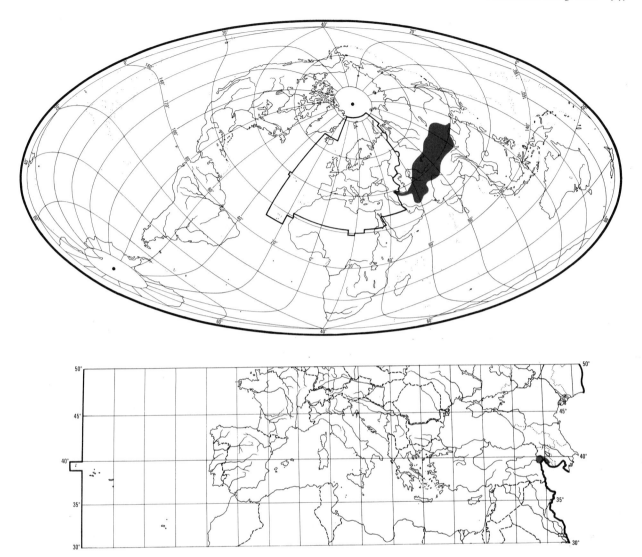

strongly patterned wing and tail immediately exclude that species. Desert Finch *Rhodospiza obsoleta* has forked tail and Crimson-winged Finch *Rhodopechys sanguinea* has dark cap and dappled underparts; both are larger. Flight similar to *B. githagineus* but action more powerful, allowing greater speed. Gait includes swift run. Stance as *B. githagineus*; usually terrestrial but observed to perch on plants near water (D I M Wallace). Sociable and tame.

Song a slow, melodious, but rather mournful phrase interspersed with pleasing chirping notes. Calls quiet but emphatic: 'dju-vüd', 't-yuk t-yuk', and 'tuck tuck tuck', lacking buzzing quality of *B. githagineus*.

**Habitat.** Breeds in dry sunny mountainous regions of south-central Palearctic, breeding at altitudes of 1000–4000 m or even higher. A ground bird, avoiding trees and even shrub growth when these are present in habitat. Favours steep and broken terrain with precipices and hollows, often including stony or clayey as well as rocky slopes and sparse grassy patches of desert or semi-desert type. (Dementiev and Gladkov 1954.) Tends to occur at higher altitude than Trumpeter Finch *B. githagineus* (see Bulatova 1972, Barthel *et al.* 1992). In Afghanistan, found commonly in a valley bottom at 2900 m with small springs and low growth of scattered *Scirpus*; also found in narrow stony side-valley and near desolate mountain slopes adjoining a few cultivated fields (Paludan 1959).

Wintering birds in India occur usually in flocks at *c.* 1500–3000 m; fly regularly in morning and evening to drink at desert springs (Ali and Ripley 1974).

**Distribution.** Early records (1911–15) from Armeniya and adjacent eastern Turkey (Barthel *et al.* 1992, which see for further details) overlooked in most subsequent literature, and not generally recognized as occurring in west Palearctic until 1969, when found breeding at Aza

(Nakhichevan', south of Caucasus), where sympatric with Trumpeter Finch *B. githagineus* (Panov and Bulatova 1972; Panov 1989, which see for detailed comparative study in area of sympatry). In 1989 re-discovered in eastern Turkey near Doğubayazit (where pair seen feeding fledged juveniles in 1990), and in 1992 near Iğdir (Barthel *et al.* 1992; *Birding World* 1992, **5**, 252, 294, 335).

**Population.** TURKEY. One count of 25 birds (including juveniles) in 1992 near Doğubayazit (*Birding World* 1992, **5**, 335).

**Movements.** Some birds short-distance migrants, others make altitudinal movements; some remain in breeding areas all year, making only local movements. Limited information.

In Nakhichevan' (south of Caucasus), recorded from 8 April in one year, though perhaps present earlier (Panov and Bulatova 1972). 3 winter records from Bahrain (Bundy and Warr 1980). Most birds depart from Kazakhstan, apparently wintering further south; in Chokpak pass (western Tien Shan), fairly common on passage in autumn (October), though rare in spring. In eastern Kazakhstan, some evidence of temporary colonization, as in Trumpeter Finch *B. githagineus*; bred at Mointy (north of Lake Balkhash) in 1958, but not in 1959, and irregular breeding also noted in Dzhungarskiy Alatau mountains. (Korelov *et al.* 1974; Gavrilov and Gistsov 1985.) In Tadzhikistan (breeds from *c.* 1200-4700 m), birds vacate Pamir-Alay mountains, but fairly common in winter at lower levels; some roam in flocks within breeding area, others make limited altitudinal movements. Usually leaves Pamir-Alay by end of September; in spring, earliest birds recorded at beginning of May, but most arrive late June or early July (Ivanov 1969; Abdusalyamov 1977). In Shakhdara valley (southern Tadzhikistan), wintering birds observed in 1966-7, but not in 1965-6, and spring migration in 1966 began mid-May (Stepanyan 1969*b*). In north-west India and (mostly northern) Pakistan, fairly common winter visitor October-May to Ladakh, Gilgit, Baltistan, Chitral, and Quetta, 1500-3000 m, and presumed to breed at higher levels (Ali and Ripley 1974); also present at Ladakh in summer in some years—recorded there early summer 1982, and early August 1987 (Harrop 1988). Vagrant to northern Nepal, where reported in severe winter 1982, also in June 1976 (Inskipp and Inskipp 1985). Breeding grounds in western Mongolia are vacated; regular observations indicate birds return exceptionally from late April, usually in May (Kozlova 1930; Piechocki *et al.* 1982). Winter records from Mongolia, e.g. in central Gobi desert, show that some birds remain near breeding areas (Kozlova 1975). Movements of birds breeding in western China not known; winter population in northern China from Nangsia east to northern Hopeh (where none breed) (Schauensee 1984) presumably involves at least some Mongolian birds.                    DFV

**Food.** Most information extralimital. Diet seeds of grasses (Gramineae) and low herbs. Forages almost wholly on ground on rocky slopes and mountainsides in sparse semi-desert type vegetation, very rarely on small shrubs or herbs. (Rustamov 1958; Hüe and Etchécopar 1970; Piechocki and Bolod 1972; Ali and Ripley 1974; Barthel *et al.* 1992; Roberts 1992.) Also feeds on, and at edges of, alpine and subalpine meadows (Potapov 1966; Neufeldt 1986). Foraging behaviour described as similar to Twite *Carduelis flavirostris* (Ali and Ripley 1974; Abdusalyamov 1977). Occasionally lands on stems of herbs, bending them over to get at seed-head (Korelov *et al.* 1974). Will pick up grain from hay put down for domestic animals and extract seeds from dung (Neufeldt 1986). Commonly recorded at water and also at salt (Korelov *et al.* 1974; Kozlova 1975; Mauersberger *et al.* 1982; Neufeldt 1986). For diet in captivity, see Hachfeld (1984).

Diet includes seeds and other parts of Chenopodiaceae (*Agriophyllum gobicum*, *Krascheninnikovia ceratoides*, saltwort *Salsola*), Compositae (wormwood *Artemisia maritima*), grasses (Gramineae, including *Bromus*, *Stipa*, *Elymus*), and sedge *Carex* (Korelov *et al.* 1974; Kozlova 1975; Abdusalyamov 1977; Panov 1989; Barthel *et al.* 1992; Roberts 1992).

In spring, feeds on milky seeds and green parts of plants (Abdusalyamov 1977); in Altai (southern Russia), in July, fed in *Stipa* grass, spending some time de-husking seeds (Neufeldt 1986).

Young apparently fed wholly on unripe seeds brought in large amounts in adult throat-pouches; ♂ forages far from nest, filling pouches on each trip, ♀ brings smaller amounts from vicinity of nest (Potapov 1966; Neufeldt 1986; Panov 1989).                    BH

**Social pattern and behaviour.** Not well studied in west Palearctic, which it only just reaches; main studies in Nakhichevan' (Transcaucasia) by Panov and Bulatova (1972) and Panov (1989), including comparisons with sympatric Trumpeter Finch *B. githagineus*.

1. In flocks outside breeding season, and these generally larger than in Trumpeter Finch *B. githagineus*. In Armeniya, flocks usually of 4-20, sometimes associated with Spanish Sparrow *Passer hispaniolensis* and Rock Sparrow *Petronia petronia* (Panov 1989). In Kazakhstan, flocks mainly 15-30(-50), 100 or more exceptional (Korelov *et al.* 1974); in Mongolia, up to 120 (Mauersberger *et al.* 1982). In Pakistan, flocks form immediately after breeding; become quite large in winter, and smaller flocks persist until June (Roberts 1992). BONDS. Apparently monogamous mating system, but not studied in detail. Young fed and cared for by both parents; brooded by ♀ (Potapov 1966; Abdusalyamov 1977; Roberts 1992). BREEDING DISPERSION. Semi-colonial, breeding in small neighbourhood groups, nests sometimes only 30-100 m apart; no obvious territoriality (Panov 1989). ROOSTING. Flocks in Armeniya usually roost on rock wall or other steep face where many cavities. Strict observance of individual distance (*c.* 1 m) in choosing roost-site; trespass leads to many conflicts (Panov 1989). In Tadzhikistan, roost among rocks or under perennial herbs (Abdusalyamov 1977).

2. Marked differences in displays and other aspects of beha-

viour suggest that relationship to *B. githagineus* is not close (Panov 1989, which see for detailed comparison). FLOCK BEHAVIOUR. Very silent in winter flocks; birds fly in flocks and keep close together, only occasionally giving flight-calls (Roberts 1992). Flocks visit water sources, to drink, twice daily (09.00 and 16.00–17.00 hrs) (Kozlova 1930). Pair-formation (see below) takes place in flocks, and flocks continue into breeding season (mainly ♂♂ during incubation), both for feeding and for visits to water sources (Abdusalyamov 1977; Neufeldt 1986). SONG-DISPLAY. Sings on ground; also regularly in flight. Song begins in spring, while still in flocks, several birds often singing together (Kozlova 1930; Dementiev and Gladkov 1954; Korelov *et al.* 1974; Panov 1989). In Mongolia, ♂♂ gave full loud song from last third of May, in feeding flocks; later, ♂♂ sang only among rocks, presumably near nest (Kozlova 1975). During pair-formation (see below), ♂ accompanies ♀ and sings a great deal (Korelov *et al.* 1974). Sometimes sings at night, in complete darkness (Roberts 1992). ANTAGONISTIC BEHAVIOUR. No antagonism observed on feeding grounds; but conflicts common at roost-sites (see Roosting, above). Aggressive displays involve ruffling of head and body feathers, repeated abrupt movements of closed wings (wing-flicking), and deep bowing of whole body, bill frequently touching ground (Fig A: Panov 1989). HET-

EROSEXUAL BEHAVIOUR. Pair-formation apparently takes place in flocks, on feeding grounds, or during stopovers on migration to breeding areas (Korelov *et al.* 1974; Panov 1989). In Courtship-display (Fig B), ♂ leans forward, with head feathers ruffled and

body plumage sleeked; remains in this posture for Bill-touching ceremony, in which ♂ and ♀ stand facing same way and touch bill-tips (Fig C). In single mating observed, Armeniya, ♂ adopted

courtship posture, moved towards ♀, approaching her from front, then flew up, hovered, turned, and flew in above ♀ from behind. When mounted, maintained balance with open wings and during copulation pressed bill against ♀'s bill. Soliciting ♀ squatted on ground with bill and tail raised. (Panov and Bulatova 1972.) Mating observed Mongolia was preceded by Courtship-feeding, ♂ feeding ♀ from crop; no other details (Piechocki *et al.* 1982). RELATIONS WITHIN FAMILY GROUP. No detailed study. Both parents feed young. ANTI-PREDATOR RESPONSES OF YOUNG, PARENTAL ANTI-PREDATOR STRATEGIES. No information.

(Figs by D Nurney from drawings in Panov and Bulatova 1972.)          DWS

**Voice.** Based mainly on studies in Transcaucasia by Panov and Bulatova (1972) and Panov (1989), which see for sonagrams. A vocal species, calling in flight, when feeding, for contact with conspecifics flying past at some distance, etc. (Panov and Bulatova 1972).

CALLS OF ADULTS. (1) Song of ♂. In Transcaucasia, 2 distinct types. (1a) Series of varied, musical units, *c.* 0·1–0·3 s in length, each unit either rising or falling in pitch (or first rising, then falling) and with marked harmonic structure; intervals between units much shorter than units themselves. This song-type very similar to whistling song of Scarlet Rosefinch *Carpodacus erythrinus*. (1b) Series of more spaced-out, less musical units of varied and more complex structure, of which characteristic component at or near end is sonorous 'growling' rattle. (Panov 1989.) Based on Panov's recordings, described as including rising and falling *Carpodacus*-like whistles, followed by shorter chirps, 'towit-toowhit-tu-tu-churrrh' or 'whi-whi-churrh', this description apparently combining components of both song-types (Roberts 1992). (2) Other calls. In flight, may give series of calls, shorter in length but of similar harmonic structure to units of song-type 1b (Panov 1989). Calls given by birds in flight or in feeding flocks variously described as slightly nasal 'vzheen' and clear, somewhat melancholy 'vee-tyu' (Potapov 1966); melancholy 'piu' (Korelov *et al.* 1974); quite loud, short, attractive fluting sounds (Kozlova 1975). The following are probably renderings of same call: clear, subdued 'witjwitj' (Mauersberger 1982); 'dju-vüd' alternating with 'djudju-vü' (Paludan 1959); twittering 'tyuk-tyuk-. . .' suggesting flight-call of Chaffinch *Fringilla coelebs* (Dementiev and Gladkov 1954); very similar to 3-syllable 'ke-ke-ke' of Trumpeter Finch *B. githagineus* (Roberts 1992). See also Barthel *et al.* (1992).

CALLS OF YOUNG. No information.          DWS

**Breeding.** Almost all information extralimital. SEASON. Nakhichevan' (south of Caucasus): eggs laid around 2nd half of April; fledged young recorded mid-May, and season very extended (Panov and Bulatova 1972; Panov 1989; see these sources for comparison with Trumpeter Finch *B. githagineus*). Eastern Turkey: adults seen feeding young out of nest mid-July (Barthel *et al.* 1992). Northern Iran: nestlings apparently not before early May (Panov 1989); nest-building recorded June (Hüe and Etchécopar 1970).

Tadzhikistan (south-central Asia): above 3000 m, eggs laid 2nd half of July (Potapov 1966); at lower altitude, 2nd half of May and 1st half of June (Ivanov 1969; Abdusalyamov 1977). Mongolia: eggs laid early June and July (Piechocki and Bolod 1972; Piechocki *et al.* 1982). For Kazakhstan, see Korelov *et al.* (1974); for Altai (southern Russia), see Neufeldt (1986). SITE. On ground under bush, grass tussock, rock, etc., or in niche or crevice in rock face, between boulders in scree, or in clay wall of building or inside ruin; entrance tunnel of up to 40 cm recorded; some ground nests almost open (Paludan 1959; Salikhbaev and Bogdanov 1967; Hüe and Etchécopar 1970; Kozlova 1975; Neufeldt 1986). Nest: loose, bulky, flattish foundation of small twigs, rough stalks, stems, and leaves with inside layer of rootlets, grass, leaves, etc., lined with wool, hair, and sometimes plant material (Dementiev and Gladkov 1954; Piechocki and Bolod 1972; Piechocki *et al.* 1982; Barthel *et al.* 1992); 3 nests in Tadzhikistan and Kazakhstan had outer diameter 11·3–13·5 cm, inner diameter 6·0–7·0, overall height 3·7–6·0, depth of cup 2·1–4·5 cm (Potapov 1966; Korelov *et al.* 1974); nests can have 'path' of plant stems leading to them if cavity deep enough (Korelov *et al.* 1974). Building: by ♀ only, accompanied by ♂; one ♀ seen gathering material 60 m from nest; reports of ♀ excavating scrape in soil require confirmation (Potapov 1966; Korelov *et al.* 1974; Kozlova 1975; Hachfeld 1984). EGGS. Sub-elliptical, smooth and slightly glossy; pale blue or slightly greenish-blue, sparsely marked with small brownish-black specks, rarely hairstreaks, at broad end (Harrison 1975; Schönwetter 1984). 19·2 × 14·2 mm (17·8–22·0 × 13·8–15·0), *n* = 56; calculated weight 2·04 g (Schönwetter 1984). Average fresh weight of 6 eggs, Mongolia, 1·91 g (Piechocki and Bolod 1972). Clutch: 4–6 (3–8) (Piechocki and Bolod 1972; Korelov *et al.* 1974; Piechocki *et al.* 1982). 2 broods in north-east Iran (Panov 1989). INCUBATION. No certain information on period. ♂♂ have brood-patch so possibly by both sexes (Kozlova 1975). YOUNG. Fed and cared for by both parents (Abdusalyamov 1977; Roberts 1992). ♀ brooded until at least day 5 (Potapov 1966). FLEDGING TO MATURITY. Fledging period in captivity 18 days; young independent 12–17 days after fledging (Hachfeld 1984). BREEDING SUCCESS. In Tadzhikistan, nests destroyed by weasel *Mustela* and fox *Vulpes* (Abdusalyamov 1977). No further information.

BH

**Plumages.** ADULT MALE. In fairly worn plumage (February–April), upperparts light drab-brown, feathers with slightly darker grey-brown centres and narrow dark brown shaft-streaks (especially on cap, mantle, outer scapulars, and upper tail-coverts), fringes of feathers slightly paler, light sandy-grey; forehead and back partly tinged pink, rump contrastingly bright rosy-pink. Side of head and neck light drab-brown, paler and greyer on lore and behind ear-coverts; supercilium rather broadly rosy-pink, bordered by silvery-grey along side of cap. Lower cheek, throat, chest, side of belly, and flank rosy-pink, brightest on throat, partly concealed below traces of cream-white feather-

fringes on chest, side of belly, and flank, extensively tipped tawny-buff on lower flank. Chin off-white; side of breast light drab-brown; mid-belly, vent, thigh, and under tail-coverts white with slight cream tinge. Tail black or brown-black (greyer and washed isabelline on tip of outer web of outermost pair), outer webs with contrasting cream-white or white fringe *c*. 1 mm wide, tips and inner webs with equally broad but less contrasting buff to grey-white fringe. Flight-feathers, greater upper primary coverts, and bastard wing black (paler greyish-black on basal and middle portions of flight-feathers, which are bordered pale buff-grey or isabelline), all except longest feather of bastard wing with broad rosy-pink fringe along outer web; terminal one-third of primaries (except for emarginated parts) and tips of all flight-feathers contrastingly fringed isabelline or white, rosy-pink of inner secondaries bordered white submarginally. Tertials black, broadly bordered cream-white or white on outer webs, less contrastingly fringed cream-buff on tip and inner web; outer fringe of white border of longer tertials tinged pink. Greater upper wing-coverts black or brown-black, outer webs fringed rosy-pink bordered by silvery-grey or white submarginally and at base, tips indistinctly fringed pink or cream-buff; median coverts black-brown with broad but ill-defined greyish-buff tips, which are partly tinged pink; lesser coverts dark drab-grey, tips tinged rosy-red or rufous. Under wing-coverts and axillaries silvery-white with pink tinge on bases, longer coverts light isabelline-grey. *In fresh plumage* (October–November), upperparts and side of head and neck more uniform drab-grey with slight vinous tinge; pink of rump partly concealed by broad light drab-grey feather-tips; pink feathers of supercilium and throat fringed drab-grey on tips, pink hardly visible and not contrasting with remainder of head; pink on underparts partly concealed by broad pale isabelline-grey feather-fringes; pink of flight-feathers, primary coverts, and greater coverts slightly duller, tinged vinous-grey. *In worn plumage* (about May–August), pink on narrow stripe along forehead, supercilium, rump and underparts slightly brighter and more rosy, more conspicuously fringed silvery-grey on side of cap, variegated by off-white of feather-bases on underparts; fringes of flight-feathers, primary coverts, and greater coverts more rosy-carmine, partly worn and thus narrower, more black of centres of greater coverts visible, these appearing streaked red-and-black with extensive white on bases; white outer borders of tail and tertials and off-white tips of secondaries and (particularly) primaries partly worn off. Rather similar to adult ♂ of Trumpeter Finch *B. githagineus*, but forehead of *B. mongolicus* less extensively pink, centre of crown not ash-grey, and remainder of upperparts not washed pink, except for contrastingly pink rump; feather-centres of crown, mantle, and scapulars darker, browner, and with dusky shaft-streak, less uniform drab-grey than in *B. githagineus*; side of head browner, with rather contrasting pink supercilium and pink lower cheek (in *B. githagineus*, pink confined to strip along side of bill, merging into ash-grey above eye and on ear-coverts); underparts distinctly less extensively pink, pink tinge not reaching chin, nor belly to under tail-coverts; fringes of tail-feathers white (not pink), fringes along tips of primaries and along outer webs and tips of secondaries and (especially) tertials extensively white, less uniform pink; greater coverts blacker with more contrasting white bases and white-and-pink outer borders, median coverts dusky with grey-and-pink tip, no coverts as featureless drab-and-pink as in *B. githagineus*; in fresh plumage, difference sometimes difficult to see, but characters of tail-feathers and greater coverts still distinct; in heavily worn plumage, white patch at base of tertials and inner secondaries and white bar across bases of greater coverts diagnostic. ADULT FEMALE. Closely similar to adult ♂, but pink of rump paler, more restricted, mixed brown-grey; pink supercilium short and narrow, inconspicuous, not bordered by grey along cap; pink of throat, chest, side of belly, and flank paler, more restricted, sometimes hardly extending further down than to chest; pink fringes of flight-feathers and greater coverts on average slightly narrow, white on tertials and greater coverts slightly less extensive, primary coverts and underwing often without pink. Darker, browner, and less uniform on upperparts and side of head and neck than adult ♀ of *B. githagineus*, belly to under tail-coverts whiter (less pale pink-grey), more contrasting with pink of chest and upper side of belly; outer fringes of tail-feathers white, not mainly pink, centres of tertials and greater and median upper wing-coverts blacker, outer webs of tertials and secondaries extensively white, not pink; bases of greater coverts whiter. NESTLING. Down dense, plentiful, white; dense tufts on cap and back, rudimentary tufts on upperwing, rump, vent, thigh, tibia, and belly (Potapov 1966; Neufeldt 1986). JUVENILE. Rather like adult ♀, but upperparts paler and less saturated sandy-brown; underparts without pink; chest and side of body suffused tawny; upper wing-coverts sandy-brown with paler sandy-buff tips. For tail and flight-feathers, see First Adult, below. Rather like juvenile *B. githagineus*, but cap, mantle, and scapulars with indistinct dusky streaks, and head and upperparts more sandy-brown, less dull orange-brown. FIRST ADULT MALE. Like adult ♂, but tail, flight-feathers, greater upper primary coverts, and bastard wing still juvenile, distinctly browner than adult at same time of year, less black, especially in spring; fringes along outer webs of primaries narrower, *c.* 0·5 mm wide (in adult, *c.* 1 mm), partly white (not all uniform pink); fringes of secondaries white, outer ones less extensively pink; primary coverts sepia-brown with pale grey fringes (in adult, grey-black, with less contrasting mixed pink-and-grey fringes); fresh pink-fringed outer greater coverts contrast with abraded brown primary coverts (in adult, no contrast between greater coverts and primary coverts, fringes of all coverts pink and grey), tips of tail-feathers and primary coverts more pointed, less broadly rounded than in adult, distinctly frayed from March onwards (in adult, not until about May-June). Also, pink of body less bright and less extensive, not always reaching side of belly or lower flank. FIRST ADULT FEMALE. Like adult ♀, but part of juvenile feathering retained, as in 1st adult ♂, browner and more worn than in adult ♀; fringes of secondaries, primary coverts, and bastard wing without pink, greater coverts with pink more restricted; contrast between primary coverts and greater coverts as in 1st adult ♂. Supercilium, lower cheek, chest, and lower flank virtually without pink, belly extensively cream-white.

**Bare parts.** ADULT, FIRST ADULT. Iris brown or dark brown. Upper mandible pale yellow-white, yellow-brown, or dull yellow with brown or yellow-brown culmen, lower mandible pale yellow to yellow-white; unlike ♂ *B. githagineus*, bill not orange-red in summer. Leg pale brown, brown-yellow, yellow-flesh, or orange, independent of sex; toes darker yellow-brown to dark brown. (Richmond 1895*b*; Ali and Ripley 1974; Hollom *et al.* 1988; BMNH, RMNH, ZMA.) NESTLING. Skin orange-pink, mouth and tongue violet-red, gape-flanges white-yellow (Neufeldt 1986). JUVENILE. No information.

**Moults.** ADULT POST-BREEDING. Complete; primaries descendent. Single ♂ from 27 July, Afghanistan, had just started, unlike 3 other ♂♂ and single ♀ in same area 21-27 July; plumage fresh and moult completed in birds from late October in Afghanistan and Kansu (China) and from early November in Iran (Vaurie 1949*b*). 26 birds from Afghanistan, 20 September to 14 October, all in various stages of moult; body feathers new or growing

(Paludan 1959). Moult not started in single ♂ from 5 July, Pakistan (RMNH), or in birds up to 18 July, Mongolia (Piechocki and Bolod 1972). No moult in July birds from Tarim basin (China), moult probably starting early August (Dementiev and Gladkov 1954). POST-JUVENILE. Partial: head, body, lesser, median, and greater upper wing-coverts. Probably starts soon after fledging, and timing thus variable, as fledged birds occur late June to late August (Johansen 1944; Dementiev and Gladkov 1954). Moult not started in bird from 17 August (Dementiev and Gladkov 1954); completed in birds examined from October (BMNH, RMNH).

**Measurements.** ADULT, FIRST ADULT. Wing and bill from (1) western Turkmeniya, and north-east Iran, (2) central Asia from Tien Shan to Kashmir; other measurements combined, all year; skins (RMNH, ZMA). Bill (S) to skull, bill (N) to distal corner of nostril; exposed culmen on average 3·5 less than bill (S).

| | | | | | |
|---|---|---|---|---|---|
| WING (1) | ♂ | 89·0 (1·87; 7) | 87-92 | ♀ 86·3 ( — ; 3) | 85-88 |
| (2) | | 92·5 (1·90; 12) | 89-96 | 88·4 (1·28; 8) | 86-90 |
| TAIL AD | | 54·1 (1·93; 17) | 51-58 | 51·8 (1·47; 11) | 49-54 |
| BILL (S) (1) | | 12·1 (0·21; 7) | 11·9-12·5 | 12·0 ( — ; 3) | 11·8-12·4 |
| (2) | | 12·8 (0·54; 11) | 12·1-13·5 | 12·5 (0·24; 8) | 12·0-12·8 |
| BILL (N) (1) | | 6·8 (0·18; 7) | 6·6-7·6 | 6·7 ( — ; 3) | 6·5-6·8 |
| (2) | | 7·1 (0·31; 12) | 6·7-7·5 | 7·1 (0·22; 8) | 6·7-7·4 |
| TARSUS | | 17·0 (0·45; 17) | 16·6-18·1 | 17·2 (0·51; 11) | 16·5-18·1 |

Wing. (1) Eastern Iran and (mainly) Afghanistan (Vaurie 1949*b*; Paludan 1959). (2) USSR (Dementiev and Gladkov 1954). (3) Mongolia (Piechocki and Bolod 1972; Piechocki *et al.* 1982). (4) Tibet (Vaurie 1972). (5) Kansu (China) (Stresemann *et al.* 1937).

| | | | | | |
|---|---|---|---|---|---|
| (1) | ♂ | 91·9 ( — ; 22) | 88·5-96 | ♀ 88·5 (2·88; 6) | 86-94 |
| (2) | | 89·8 ( — ; 25) | 83·5-92·5 | 86·6 ( — ; 19) | 84-91 |
| (3) | | 92·3 (2·52; 15) | 89-98 | 88·0 (2·00; 7) | 85-90 |
| (4) | | 92·0 ( — ; 22) | 87-96 | 88·7 ( — ; 11) | 85-92 |
| (5) | | 90·3 ( — ; 10) | 87-92 | 89·4 ( — ; 8) | 87-93 |

**Weights.** ADULT, FIRST ADULT. (1) Afghanistan, September-October (Paludan 1959). (2) Mongolia, mid-May to mid-July (Piechocki and Bolod 1972; Piechocki *et al.* 1982). (3) Kazakhstan (Korelov *et al.* 1974).

| | | | | | |
|---|---|---|---|---|---|
| (1) | ♂ | 20·9 ( — ; 16) | 18-24 | ♀ 20·6 ( — ; 10) | 19-23 |
| (2) | | 21·4 (1·30; 15) | 19-24 | 22·4 (2·64; 7) | 19-26 |
| (3) | | 21·0 ( — ; 5) | 19-22 | 21·0 ( — ; 4) | 18-26 |

Afghanistan, June: ♂ 20·8 (4) 20-21, ♀ 19 (Paludan 1959). Kansu (China), October-November: ♂♂ 20·3, 22·1; ♀♀ 18·3, 20·4 (Stresemann *et al.* 1937).

**Structure.** Wing rather long, broad at base, tip pointed. 10 primaries: p8 longest, p9 0-1 shorter, p7 1-3, p6 8-10, p5 13-16, p4 18-21, p3 22-25, p2 25-28, p1 28-32; p10 strongly reduced, 61-66 shorter than p8, 9-12 shorter than longest upper primary covert. Outer web of p7-p8 emarginated (sometimes faint); inner web of p8-p9 sometimes with faint notch. Tip of longest tertial reaches to tip of p1-p2. Tail rather short, tip slightly forked; 12 feathers, t1 4-8 shorter than t6. Bill short, conical; *c.* 7·5-8·0 deep at feathering on base, 6·8-7·8 wide; culmen and cutting edges slightly decurved, gonys convex; bill heavy, but less deep and swollen at base than in *B. githagineus*. Middle toe with claw 16·2 (5) 15·5-17 mm; outer and inner toe with claw both *c.* 68% of middle with claw, hind *c.* 76%. Remainder of structure as in *B. githagineus*.

**Geographical variation.** Slight. No differences in colour between birds from Iran, Afghanistan, and Kansu (China) in fresh plumage (Vaurie 1949*b*). However, in birds examined, those from Tarbagatay and Tien Shan larger than those from western Turkmeniya (Ashkhabad area), and neighbouring north-east Iran (see Measurements), and western birds paler than eastern ones in comparable stage of plumage wear, upperparts colder pale drab-grey, less saturated drab-brown; pink of face and underparts paler; side of head, side of neck, and side of breast distinctly paler, light grey, less drab-brown; western birds perhaps deserve recognition as separate race, but no specimens from type locality of *mongolicus* (eastern China) examined, and birds from Afghanistan, western Pakistan, Ladakh, and south-west Sinkiang tend to be intermediate between western and Tien Shan birds in bill length, though wing and colour near Tien Shan birds (C S Roselaar).

Sometimes considered a race of *B. githagineus* (e.g. Hartert 1903-10, Dementiev and Gladkov 1954), but both forms breed sympatrically without hybridization in Nakhichevan' (Tran-scaucasia), northern and eastern Iran, southern Turkmeniya, central Uzbekistan, northern Afghanistan, and perhaps Armenia and eastern Turkey (Vaurie 1949*b*; Dementiev and Gladkov 1954; Erard and Etchécopar 1970; Panov and Bulatova 1972; Portenko and Stübs 1976; Stepanyan 1983). In overlap area, *B. mongolicus* tends to inhabit higher altitudes with more rocky environment (Vaurie 1949*b*); behaviourally well-separated, *B. mongolicus* rather nearer in some features to Scarlet Rosefinch *Carpodacus erythrinus* than to *B. githagineus* (Panov and Bulatova 1972). Bill of eastern *B. githagineus crassirostris* thicker in overlap area than that of non-overlapping races of *B. githagineus*, bill of *B. mongolicus* in overlap area smaller than elsewhere, pointing to character displacement.

**Recognition.** For differences from *B. githagineus*, see Plumages (above).                                                                                 CSR

## *Bucanetes githagineus*  Trumpeter Finch

PLATES 47 and 50 (flight)
[between pages 640 and 641]

Du. Woestijnvink          Fr. Roselin githagine          Ge. Wüstengimpel
Ru. Пустынный снегирь          Sp. Camachuelo trompetero          Sw. Ökentrumpetare

*Fringilla githaginea* Lichtenstein, 1823. Synonym: *Rhodopechys githagineus*.

Polytypic. *B. g. amantum* (Hartert, 1903), Canary Islands; *zedlitzi* (Neumann, 1907), North Africa from Mauritania through Maghreb and Libya to western and northern Egypt (in north, to desert north-east of lower Nile); also southern Spain and (perhaps this race) Aïr (northern Niger), Tibesti, Borkou, and Ennedi (northern Chad); nominate *githagineus* (Lichtenstein, 1823), central and southern part of Nile valley and south-eastern desert of Egypt, south to north-central and north-east Sudan; *crassirostris* (Blyth, 1847), western Arabia, Sinai, Levant, and south-central Turkey, east through Bahrain, Iran, and Transcaucasia to Uzbekistan, Afghanistan, and western Pakistan.

**Field characters.** 12·5 cm; wing-span 25-28 cm. About 10% larger than Serin *Serinus serinus*, with similar dumpy form; slightly smaller than Mongolian Trumpeter Finch *B. mongolicus*, with proportionately shorter wings. Rather small, stocky finch with bulbous bill, deep head, and rather short tail. Plumage rather uniform dusky- or sandy-pink, with orange- to wax-red bill (breeding ♂), large dark eye, darker flight- and tail-feathers, slightly paler rump, and orange-flesh legs. Song distinctive. Sexes dissimilar; some seasonal variation. Juvenile separable. 4 races in west Palearctic, but differences complex (see Geographical Variation).

ADULT MALE BREEDING. Moults: June–August (complete). In breeding season, bright, waxy, pink- to orange-red bill catches eye. Plumage rather uniform, basically pearl-grey on head and buff-grey elsewhere. At close range, shows narrow whitish eye-ring and carmine-pink wash on face, rump, tips of wing-feathers, and (particularly) underparts; dark-centred wing-coverts and blackish primaries and tail-feathers all have quite sharp pink and pale grey margins. Underwing pale grey. After moult, bill yellowish-horn and plumage much greyer and even less marked. Legs bright flesh-brown or flesh-orange.

ADULT FEMALE BREEDING. Plumage duller than ♂, with ground sandier and browner in tone but usually showing faint rosy tints. Bill brown- to grey-horn, only faintly tinged yellow or pink. JUVENILE. Resembles ♀ but shows characteristic yellowish-buff lores, ear-coverts, and wing-feather margins, retained into 1st winter.

Remarkably featureless, requiring close approach for plumage details to be seen. Red bill of breeding ♂ diagnostic; uniformity of plumage virtually so, but shared in Levant by sympatric ♀ Sinai Rosefinch *Carpodacus synoicus*, which is 10% larger and longer-tailed, with ginger face and different voice. Beware also risk of confusion with *B. mongolicus* (slightly larger, whitish wing-panel, different call), Desert Finch *Rhodospiza obsoleta* (15% larger, with usually dark bill, rosy-white wing-panel, black tertial-centres, primary-tips, and tail, and different call), and Pale Rock Sparrow *Carpospiza brachydactyla* (10% larger, with pale double wing-bar, white-tipped outer tail-feathers, and different call). Flight light and bounding, even skipping; rarely over long distance when breeding or in winter quarters. Spends most of time on ground; gait a hop, varied by shuffle and creep. Stance variable, noticeably low when feeding but often very upright when

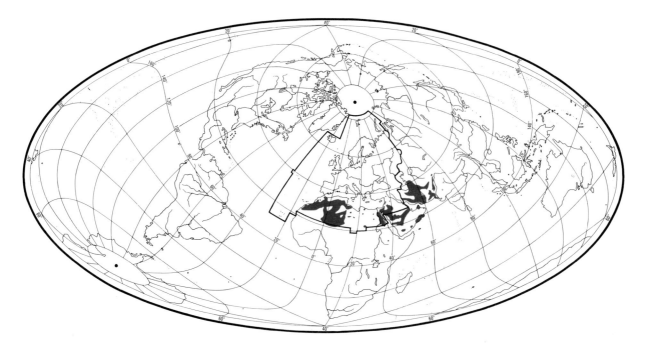

alert or alarmed. Sociable, forming flocks in winter. Vagrants appear along stony coasts and on sea isles.

Song an extended, distinctly nasal buzz or subdued ring, 'cheeeee', recalling toy trumpet; varies from quite loud and clear to harsh and prolonged. Calls variably described: short conversational 'chee' or 'chit' from feeding birds, perhaps same as cackling 'kek-kek kek-keheck' or 'kä kä kä'; in alarm a loud 'schak schak'.

**Habitat.** Patchily distributed across warm arid mainly lowland or hilly subtropical regions from western Sahara across North Africa and Middle East. Concentrates in deserts, semi-deserts, and steppes with minimum of vegetation and much stony or gravelly surface, preferably fronted by rocky crags or vertical exposures with plenty of crevices and sparse growth of small bushes and grasses. Lives mainly on ground but requires daily access to water and is ready to fly some distance to it, especially towards evening. Tolerates very high daytime temperatures. Sometimes enters deep wells with walls, and occasionally nests in holes or gutter-pipes on houses, although normally far from human habitation except for remote guard-posts, where its confiding disposition leads to it readily making itself a home. (Bannerman 1963; Paz 1987.) Found at desert wells in every oasis, and bred in 19th century among old tombs at Thebes. Favours stony, treeless, and hot localities, especially gravelly areas, but avoids sand. Attracted to small patches of cultivation fringing skirts of mountains (Dresser 1871–81). Around Azraq (Jordan) is one of the few species inhabiting inhospitable basalt tracts with silty patches, wadis, and scarps (Nelson 1973). Extralimitally found by Paludan (1959) in Afghanistan in desolate hilly country with stony slopes, apparently at 1500–1900 m. In Chitral (Pakistan), recorded up to 3000 m; inhabits bare hills and stony semi-desert. Perches on rocks or stones; not bushes. (Ali and Ripley 1974.) In USSR, inhabits desert sections of steppes and mountain slopes with pebbly ground overgrown with sparse herbaceous vegetation. Also found on rocky fields devoid of woody vegetation and on surfaces of moraines. Requires availability of fresh or slightly saline water. Flies low over ground. (Dementiev and Gladkov 1954.)

**Distribution.** Has recently spread in Spain, Morocco, and Israel.

Morocco. Has recently spread north, to Moyen Atlas, Massif de Khatouate (south-east of Casablanca), and Mediterranean coast (MT). Turkey. Recorded breeding at one site in 1977; not certain if permanent breeder (RPM). Israel. Spread in 1970s and 1980s, following development of agricultural settlements and army camps providing constant water (Shirihai press).

Accidental. Denmark, Sweden, Austria, Cyprus, Kuwait, Cape Verde Islands.

**Population.** Israel. Rough estimate a few thousand pairs, but large annual variation (Shirihai in press).

**Movements.** Resident and dispersive or nomadic. Need for daily water-supply leads to erratic movements, also to temporary colonization and frequent small-scale changes of range.

In north-west Africa (*zedlitzi*), erratic pattern of movement leads to local abundance in some years and absence

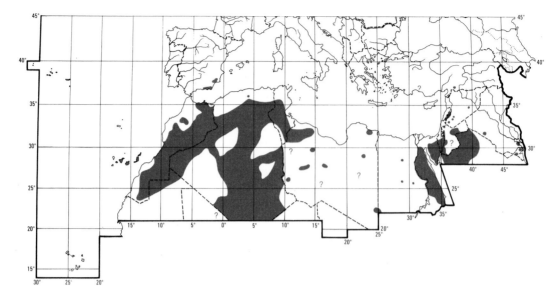

in others, both in summer and winter (Heim de Balsac and Mayaud 1962; Dupuy 1969); e.g. in Morocco, 1957, when weather severe at high altitude, many birds wintered exceptionally on Debdou-Tlemcen ridge in north-east, and remained to breed; area entirely abandoned later, with none present in 1958 (Brosset 1961). Has extended Moroccan range north and west in recent years (Thévenot *et al.* 1982; *Br. Birds* 1985, **78**, 644–5), and several records north of usual breeding area in Tunisia at end of 1960s and beginning of 1970s (Thomsen and Jacobsen 1979). This northward movement is reflected in increasing records in southern Spain, with many wintering in Almería in 1969, and first recorded breeding in 1971 (Cano 1968; Cano and König 1971); also 1st records elsewhere in western Europe (presumably chiefly this race); most such records mid-May to late July, coinciding with post-breeding dispersal (Wallace *et al.* 1977; Noeske and Hoffmann 1988). Usually very scarce in Malta, but flocks of up to 50 in late June and early July 1977 (Sultana and Gauci 1982), and recorded from Mallorca (Balearic Islands) in spring 1971 (Wallace *et al.* 1977). 6 records in Britain, 1971–87, all May–September, mostly in east and extending from Kent in south-east to Orkney (Scotland) (Dymond *et al.* 1989). Also 1 record each from Alderney, Channel Islands (1973), western France (1974), Denmark (1982) and Sylt Island, north-west Germany (1987) (Wallace *et al.* 1977; Boertmann *et al.* 1986; Noeske and Hoffmann 1988; *Br. Birds* 1981, **74**, 263). In Sweden, 2nd record (1971) falls into same pattern and probably *zedlitzi*, 1st record (1966) perhaps north-east African race, nominate *githagineus* (specimen taken) (Westin 1973; Breife *et al.* 1990). In Nile Valley (nominate *githagineus*), flocks frequently visit cultivated areas outside breeding season (Goodman and Meininger 1989).

Middle East race, *crassirostris*. Status uncertain in Tur-key; recorded from southern coastlands and east, and probably rare summer visitor; occurrence perhaps due more to nomadic than seasonal movements (Martins 1989). Occasionally reaches Cyprus, with 6 records March–April and one in December (Flint and Stewart 1992); also 4 records from Greek islands (including Crete) (G I Handrinos). Basically resident in Israel, where breeds in most highland and upland deserts in south and central areas; post-breeding flocks make local movements; those breeding in desert plateaus move down to lower and more sheltered localities, e.g. many leave eastern Negev September–November and winter in Wadi Arava, returning to Negev in March. Breeding grounds on Mount Hermon and in eastern Shomron vacated, birds departing September–October and returning April–May. (Shirihai in press.) In Arabia, widespread but rather uncommon breeding resident in most rocky areas, common on Tuwaiq escarpment (central Arabia); forms flocks and disperses widely after breeding, occurring randomly in winter; at Taif (western Arabia), common November–February, but absent in breeding season (Jennings 1981a; Stagg 1987; F E Warr). In eastern Saudi Arabia, breeds quite commonly in rocky areas in some years, but almost completely absent in other years; also mostly absent June–September (following breeding), but winter foraging flocks occur later, especially November, remaining through winter (J Palfery). Winter records in Afghanistan indicate dispersal from foothills to southern plains (S C Madge). Mainly resident in Iran, where breeds widely; occasionally reaches southern coastal lowlands in winter (D A Scott); also recorded winter from Iraq (Vaurie 1959). In Nakhichevan' (south of Caucasus), in one year, present on breeding grounds from 28 March (Panov and Bulatova 1972). In Pakistan and north-west India, more widespread in winter, reaching Makran coast, Rajasthan, and Punjab, and fairly common then in low hills of Sind (Ali and Ripley 1974).

On Canary Islands (*amantum*), wanders over plains in small flocks outside breeding season; no evidence of other than local movements (Bannerman 1963). DFV

**Food.** Much information extralimital. Diet seeds and other parts of grasses (Gramineae) and low herbs; also a few insects. Forages almost wholly on ground, generally in rocky areas with scattered semi-desert vegetation; flits, creeps, and runs mouse-like around stones and shrubs searching for seeds. (Dementiev and Gladkov 1954; Bannerman 1963; Kozlova 1975; Wallace *et al.* 1977; Löhrl 1987*a*; Panov 1989; Roberts 1992.) On Canary Islands, also feeds in places of abandoned cultivation (Martín 1987, which see for probable food plants there). In Egypt, has been recorded in flocks of thousands in cereal fields (Koenig 1926), and commonly feeds by roadsides on spilled grain (Goodman and Meininger 1989). Vagrants in England and Germany avoided continuous vegetation, even grass, picking up seeds among saltmarsh plants or on stony paths and trampled areas in company with (e.g.) finches *Carduelis* (Wallace *et al.* 1977; James 1986; Noeske and Hoffmann 1988). Digs fairly deeply into soil to find seeds (Hachfeld 1983). Perches on stems of grass to bend them over to get at seed-heads (Polozov 1990). Captive birds severed blades and stems of grass at base and nibbled them in bill near tip to break them into fragments (Harrison 1978*b*); crushes hard Chenopodiaceae seeds in bill (Roberts 1992). Often at water, at times entering wells (Guichard 1955; Valverde 1957; Bannerman 1963; Gaston 1970; Roberts 1992). On Canary Islands, picked up birdseed in hotel garden (Löhrl 1987*a*), and extracted grains of maize *Zea* from camel dung (D F Owen).

Diet includes the following. Insects: nymphs of grasshoppers, etc. (Orthoptera). Plants: seeds, seedlings, buds, etc., of dock *Rumex*, glasswort *Salicornia*, *Schouwia* (Cruciferae), Leguminosae, clary, etc., *Salvia*, *Nicotiana* (Solanaceae), mugwort *Artemisia*, grasses (Gramineae, including *Bromus*, *Stipa*, cereals). (Bolle 1858*b*; Koenig 1926; Guichard 1955; Valverde 1957; Roche 1958; Hachfeld 1983; Löhrl 1987*a*; Paz 1987; Noeske and Hoffmann 1988; Bergier 1989; Goodman and Meininger 1989; Polozov 1990; Roberts 1992.)

In Morocco, May–June, 15 stomachs contained only seeds, mostly *Salvia aegyptiaca* (Valverde 1957). In Algeria, November, 2 stomachs also held only seeds (Fairon 1972); others contained seeds of *Rumex vesicarius*, one of commonest local plants (Roche 1958). In Kirgiziya, spring to autumn, 25 stomachs contained only seeds of Polygonaceae, Leguminosae, and grasses (Zlotin 1968); in another study, 3 stomachs held seeds (including Polygonaceae) and insects (Pek and Fedyanina 1961). On Fuerteventura, main food was fallen seeds of *Nicotiana glauca*, not extracted from flowers as commonly done by other species (Löhrl 1987*a*). Especially fond of grass *Bromus* in Pakistan (Roberts 1992). Captive birds

appeared to prefer smaller, more oily seeds and rejected large ones; also showed no interest in live insects, or chewed on larvae for a long time before giving pieces to young (Koenig 1926; Harrison 1978*b*). See also Glück and Massoth (1985) for food in captivity.

Nestlings at one nest in Spain fed green seeds (Rodríguez 1972). In Kirgiziya, gullets of adults with young in nest and stomachs of fledglings contained only seeds (Zlotin 1968). Young fed pulp of seeds from large throat-pouches of adults (Hachfeld 1983), and seeds often collected far from nest (Panov 1989). BH

**Social pattern and behaviour.** No detailed studies. Main study, in Nakhichevan' (Transcaucasia), by Panov and Bulatova (1972) and Panov (1989), including comparisons with sympatric Mongolian Trumpeter Finch *B. mongolicus*.

1. More or less gregarious outside breeding season; flock sizes variable, but usually not large. In Nakhichevan', no true flocks formed; birds singly or in twos, occasionally loose groups of 3–5 (Panov 1989). In Canary Islands, feeding flocks of 10–20 rather common, even in middle of breeding season, when composed mainly of ♂♂ (Volsøe 1951); in January, usually in groups of 5–10 (Löhrl 1987*a*). In Saudi Arabia, feeding flocks of up to 100 in winter (Jennings 1980); up to 20, breaking up in early February (J Palfery). In Pakistan, winter flocks of up to *c.* 12 (Roberts 1992). BONDS. Apparently monogamous mating system, but further study needed, as 2 ♂♂ observed associating with nesting ♀ over 3-day period (small nestlings), both giving alarm-calls when nest disturbed by observer though only one ♂ seen to go to nest (Panov and Bulatova 1972). Young fed and cared for by both parents (Dementiev and Gladkov 1954); brooded by ♀ only (Makatsch 1976). BREEDING DISPERSION. Varies from completely isolated pairs to groups of pairs nesting only short distances apart (but local concentrations of nests less characteristic than in *B. mongolicus*); no obvious territoriality (Panov 1989). ROOSTING. Communal; on ground, among rocks. In central Sahara, February, birds flew to roost from 20 min before to 4 min after sunset, in parties of up to 15 (Gaston 1970). Active very early in morning; recorded flying around more than 100 min before sunrise, at light intensities of less than 1 lux (Hald-Mortensen 1970). By day, may roost or loaf communally in shade of rocks or overhanging cliffs, apparently regularly. 86 birds loafing in shade of rocks from mid-morning onwards, Saudi Arabia (J Palfery); more than 500 on 4 July in shade of overhanging cliff and *c.* 300 in same place 22 August, Libya, thought to represent population of area not less than 30 km² (Willcox and Willcox 1978).

2. Gregarious (but less so than *B. mongolicus*) and highly nomadic. Never perches in trees. On ground, squats on tarsi and difficult to see; when feeding, runs over ground in mouse-like fashion. Regularly visits water to drink. Tame and confiding, allowing close human approach. (Roberts 1992.) FLOCK BEHAVIOUR. Not studied in detail, true flocks apparently not occurring in Nakhichevan' where most closely studied (Panov 1989; see above). In Fuerteventura (Canary Islands), small groups of 5–10 found associating with other species (Linnet *Carduelis cannabina* and Spanish Sparrow *Passer hispaniolensis*); when alarmed, did not fly with them but always remained on ground (Löhrl 1987*a*). Captive bird scared by intruding human crouched, with legs widely splayed, breast tilted forward so as to almost touch ground, and bill lowered so that tip touched ground; remained for a little time motionless in this position (Harrison 1978*b*). SONG-DISPLAY. ♂ sings on ground; also gives modified form of song in display-flight (see Heterosexual Behaviour, below; also

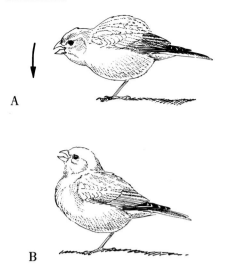

A

B

which song given (see Song-display, above). In Bill-touching ceremony, ♂ and ♀ stand facing same way and touch bill-tips (see Fig C in *B. mongolicus*). (Panov and Bulatova 1972.) In captivity, ♂ displayed to ♀ by picking up and holding leaves or stems in bill; with movements becoming more excited and jerky, ♂ usually gathered 4–5 fragments; stood very upright with plumage sleeked, except for ruffled lower belly and flank feathers, and crown feathers erected into blunt crest; wing-tips lowered, exposing pink rump and basal area of partly spread tail. In this posture made rapid hops from side to side, facing ♀, and uttering loud calls (see 2f in Voice). ♀ either unresponsive or ruffled plumage and threatened with open bill. Infrequently, and later in breeding cycle, ♂ also displayed to ♀ in crouched posture, with feathers of head and neck sleeked, body plumage ruffled, wings drooped, and tail raised a little; moved towards ♀ with rapid, frantic repetition of song which sometimes degenerated into frenzied, prolonged buzzing. (Harrison 1978*b*.) (3) Courtship-feeding. Not recorded. (4) Mating. Sequence of events rigid. ♂ in pre-copulatory posture (Fig D) moves toward

D

♀ (hopping or walking), approaches her from front, flies up, hovers, turns, and flies in above ♀ from behind. Mounts, maintaining balance with open wings and pressing bill against ♀'s. (Panov and Bulatova 1972.) RELATIONS WITHIN FAMILY GROUP. No detailed study. Both parents feed young. At nest with small young attended by 2 ♂♂ (see Bonds, above), one ♂ passed food to ♀, who fed young; ♂ also carried away eggshell (Polozov 1990). Soon after breeding season, juveniles band together in flocks (Roberts 1992). ANTI-PREDATOR RESPONSES OF YOUNG. No information. PARENTAL ANTI-PREDATOR STRATEGIES. ♀ with small young left nest only when hand extended to within 0·3 m of her (Polozov 1990).

(Figs by D Nurney from drawings in Panov and Bulatova 1972.) DWS

Voice). Of 4 song-types recognized (see Voice), Nakhichevan', 2 usually given in characteristic postures (Figs A–B), first abrupt notes accompanied by bowing/nodding movements of head (Fig A), or head jerked back at each note (Fig B). (Panov and Bulatova 1972.) In Pakistan, ♂ typically crouches near top of hill or hillock and sings continuously for 2–3 min (Roberts 1992). Song-period apparently same as breeding season, but not reported in detail: February–March, Saudi Arabia (Jennings 1980); from late March, Pakistan (Roberts 1992). ANTAGONISTIC BEHAVIOUR. Extreme mutual tolerance is characteristic; individual distance not great. Mutual alertness rather than aggression shown when individuals approach closely, but sometimes brief conflicts between ♀♀ competing for nest-material. ♂♂ unusually tolerant of one another; encounters may be marked by display (see below), only occasionally leading to aerial chase. ♀♀ use some display postures, but unlike ♂♂, quite often attack. Practically no territorial behaviour up to start of incubation, pair moving about widely in general environs of nest. In aggressive posture (Fig C), bird ruffles head and body feathers and makes repeated

C

abrupt movements of closed wings; performs bowing movements and, at high level of motivation, mock-feeding (pecking movements at ground). (Panov and Bulatova 1972; Panov 1989.) HETEROSEXUAL BEHAVIOUR. (1) General. Pair-formation apparently takes place in small groups, immediately prior to breeding season. (2) Pair-bonding behaviour. Display-flight is slightly ritualized form of normal flight: ♂ flies fast in wide circles, with erratic changes of direction; regular alternation of active flight (4–5 vigorous wing-beats with gentle ascent) and gliding (losing height slightly); gives slightly modified call or (less commonly) song. Courtship-displays on ground also include postures in

Voice. Remarkable for nasal/metallic sounds, comparable with sound made by child's tin or plastic trumpet. Calls freely, on ground and in the air. Account includes unpublished notes from eastern Saudi Arabia by J Palfery. Repertoire requires further study, and following scheme provisional.

CALLS OF ADULTS. (1) Song of ♂. Variously described, and may vary geographically, but descriptions agree in giving drawn-out, nasal, trumpet-like unit as component, sometimes associated with shorter units. In Canary Islands, described as simply a short, loud note, like child's trumpet (Volsøe 1951); or as drawn-out, metallic-nasal units, 'ewääd', connected by 1–3 short, nasal 'wik' units (Bergmann and Helb 1982, which see for sonagram). In

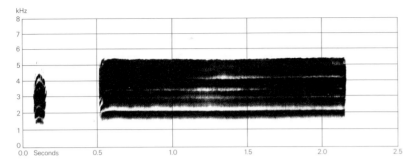

I   J-C Roché   Morocco   March 1966

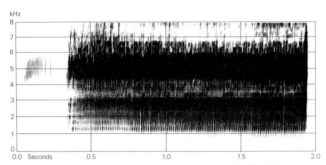

II   C Chappuis/Sveriges Radio (1972–80)   Morocco   April 1966

III   J-C Roché   Morocco   March 1967

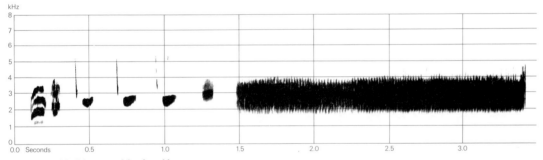

IV   J-C Roché   Morocco   March 1966

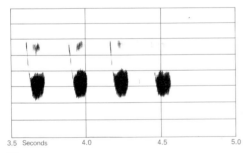

IV   *cont.*

Morocco, may consist of single, drawn-out 'paaaaaa'; also drawn-out unit preceded either by much shorter unit (Fig I), subdued pulse-train (Fig II), or shorter syllable, 'dze-ziaar' (Fig III); also more complex song comprising a particularly drawn-out unit of *c.* 2·0 s, preceded by clear piping notes, clicks, and brief unit with buzzy timbre, and followed by series of buzzy units, thus 't-yu t-yu t-yu zi dzzzaaaaar choo choo choo choo' (Fig IV: W T C Seale). In Arava valley (Israel), 2 song-types described: drawn-out nasal buzz, and, less often heard, nasal buzz preceded by a few high-pitched, metallic clicking sounds and whistles (Fig V, *crassirostris*: Mild 1990). 5 fainter 'pyee' units in middle of Fig V, apparently from a more distant bird, suggest possibility of antiphonal singing (W T C Seale). In Saudi Arabia, 2 kinds of song: one consisted of 3 sections, beginning with 4 calls (apparently contact-call or versions of it), followed by warbled phrase, and finishing with wheeze; other noted simply as 'zeeeer zeee-er'. (J Palfery.) In Transcaucasia, 4 song-types distinguished, showing very different harmonic structure of nasal units, 2 consisting only of drawn-out nasal unit, other 2 of shorter nasal unit preceded by abrupt clicking sounds (Panov and Bulatova 1972; Panov 1989, which see for sonagrams). (2) Other calls. Various descriptions, listed below, undoubtedly include much overlap: principally of buzzing units (similar in quality to main component of

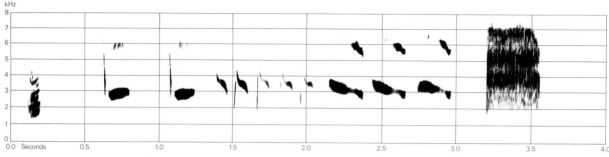

V  Mild (1990)  Israel  March 1988

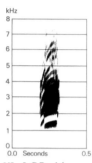

VI  J-C Roché
Morocco  March 1966

VII  J-C Roché
Morocco  March 1966

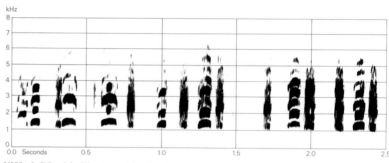

VIII  J-C Roché  Morocco  March 1966

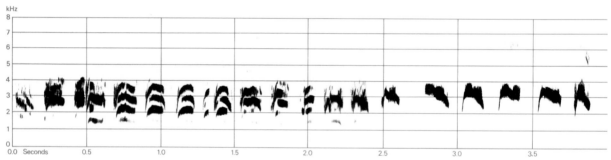

IX  J-C Roché  Morocco  March 1966

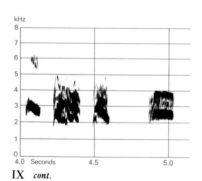

IX  cont.

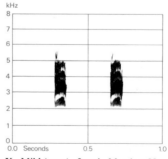

X  Mild (1990)  Israel  March 1988

song) and shorter units rendered 'chik', 'kek', 'tset', etc., without apparent buzzing or nasal quality, though timbre of very brief units difficult for human to perceive. (2a) Call given in flight (Canary Islands) a short, unmistakable 'töd', with harmonics spanning wide frequency range. (2b) Call given at nest, when disturbed, a drawn-out (*c.* 1·5 s), metallic-nasal 'wäääd', or 'dsääd' with low-volume initial sound. (Bergmann and Helb 1982, which see for sonagram;

this however very similar to Fig II, provisionally regarded here as song.) (2c) Common calls in Morocco are single (Fig VI) or pairs of units (Fig VII) of slightly nasal quality; cackling (Fig VIII) and squealing (Fig IX) calls also recorded, from birds in flocks. (2d) Calls in Arava valley, Israel, described as basically monotonous, varying in pitch, timbre, and strength, but all with nasal buzzing quality; those given on ground generally quieter than flight-calls; occasionally, 2 repeated notes, 'tchu-tchu' (Fig X, *crassirostris*) (Mild 1990). (2e) In eastern Saudi Arabia 4 calls distinguished. (i) Commonest call, given by both sexes, in flight and at rest, a short, low-pitched 'dzit', also noted as 'zuk', 'zik', 'week', etc.; also heard from foraging flocks; apparently a contact-call. (ii) Buzzing, nasal 'bzee', falling in pitch; given by ♂ (perhaps also by ♀), sometimes in combination with (i). (iii) More braying 'bzeeow' or 'bzeeaaw', noted only with certainty from ♂. (iv) Sparrow-like 'chuWIT chuWIT', given both in flight and on ground; noted in April when fledged young were about and may have been given only by juveniles. (J Palfery.) (2f) ♂ displaying to ♀ (in captivity) with leaves or stems in bill (see Social Pattern and Behaviour, above) gave rapidly repeated 'dwick', higher-pitched than other calls (Harrison 1978*b*).

CALLS OF YOUNG. No information (but see above).

DWS

**Breeding.** SEASON. Southern Spain: eggs laid about early May; nest-building recorded mid-May (Cano and König 1971; Rodríguez 1972). Canary Islands: eggs laid 1st half of March (January to mid-May) (Bannerman 1963; Martín 1987; Braunberger 1990). North-west Africa: eggs laid February–June, mostly April (Valverde 1957; Heim de Balsac and Mayaud 1962; Dorka *et al.* 1970). Israel: 1st clutch March–May, 2nd May–June (Shirihai in press). Jordan: fledged young seen late April (Wittenberg 1987); nest-building early May (Wallace 1984). Nakhichevan' (south of Caucasus): eggs laid from late April; 2nd clutches from mid-May (Panov and Bulatova 1972). SITE. Depression on ground under rock, shrub, tussock, etc., in cleft between stones, or in cavity in rock face or wall of house, generally 3–6 m above ground; also inside derelict building (Williams and Williams 1929; Bannerman 1963; Makatsch 1976; Martín 1987; Panov 1989); in Jordan, nest recorded 13 m above ground (D I M Wallace). Nest: untidy foundation of small twigs, stalks, roots, rough grass, etc., neatly lined with dry grass, plant down, wool, hair, and (rarely) feathers; sometimes lined with grass only (Dementiev and Gladkov 1954; Bannerman 1963; Rodríguez 1972; Makatsch 1976); outer diameter 11–14 cm, inner diameter 6–7 cm, overall height 3–4.5 cm, depth of cup 2–3.5 cm (Dementiev and Gladkov 1954; Makatsch 1976); cavity entrance may be 'paved' with small stones (Paz 1987). Building: by ♀ only, accompanied by ♂ (Geyr von Schweppenburg 1937; Panov and Bulatova 1972; Hachfeld 1983; J Palfery). Captive ♂ formed rough nest on floor of aviary, ♀ taking no part (Harrison 1978*b*). ♀ in captivity took 3–4 days (Hachfeld 1983). Material gathered from up to 300 m and more from nest (Panov 1989). EGGS. See Plate 61. Sub-elliptical to short oval, smooth and slightly glossy; pale blue, sparsely marked with rusty to purplish-black spots and speckles, generally at broad end; sometimes small reddish-violet under-markings (Bannerman 1963; Harrison 1975; Makatsch 1976). Nominate *githagineus*: 19.2 × 14.2 mm, n = 29; calculated weight 2.04 g (Schönwetter 1984); 3 eggs from Egypt all 18.0 × 13.0 mm (Koenig 1926). *B. g. amantum*: 19.2 × 14.2 mm (16.9–21.0 × 12.9–15.0), n = 88+; calculated weight 2.04 g. *B. g. zedlitzi*: 19.8 × 14.3 mm (18.0–21.6 × 13.7–15.3), n = 127; calculated weight 2.12 g. *B. g. crassirostris*: 18.4 × 14.7 mm (18.2–19.5 × 13.5–15.1), n = 14, calculated weight 2.10 g. (Schönwetter 1984.) See also Bergier (1989). Clutch: 4–6. In north-west Africa, of 106 clutches: 4 eggs, 11%; 5, 64%; 6, 25%; average 5.13 (Heim de Balsac and Mayaud 1962). On Canary Islands, apparently rarely more than 5 (Bannerman 1963). Eggs laid daily (Makatsch 1976; Hachfeld 1983). Some pairs in Nakhichevan' and Israel have 2 broods (Panov and Bulatova 1972; Shirihai in press), and probably in Morocco (Valverde 1957). In captivity, ♀ started replacement nest 8 days after 1st clutch removed, laying clutch *c.* 4 weeks after 1st, and 2nd replacement 2 weeks after removal of 1st replacement (Harrison 1978*b*). INCUBATION. 13–14 days, n = 6 (Shirihai in press); 11–12 days according to Glück and Massoth (1985). By ♀ only, starting with last egg (Makatsch 1976; Hachfeld 1983), though ♂ has sizable brood-patch, suggesting some participation (Dementiev and Gladkov 1954; Panov 1989). At one nest in Saudi Arabia, stints on eggs of 15–20 min and breaks of 10 min noted (J Palfery). YOUNG. Fed and cared for by both parents (Dementiev and Gladkov 1954; Rodríguez 1972, which see for routine at one nest; Roberts 1992); brooded intensively by ♀ (Makatsch 1976; Hachfeld 1983). FLEDGING TO MATURITY. Fledging period 13–14 days, n = 6 (Shirihai in press); 12–13 days according to Glück and Massoth (1985), young leaving nest before able to fly. For development of young, see Rodríguez (1972). Independent *c.* 11 days after fledging (Harrison 1975). BREEDING SUCCESS. No information.

BH

**Plumages.** (*B. g. zedlitzi*). ADULT MALE. In fresh plumage (September–February) forehead and crown pale ash-grey, feathers on forehead inconspicuously tinged pink, grey on crown partly concealed by broad pale drab-brown or sandy-brown feather-tips, widest on hindcrown. Hindneck, mantle, scapulars, and back light drab-brown, feathers of lower mantle and scapulars with faint dull rosy-pink tinge. Rump dull rosy-pink with drab-brown or ash tinge, pink brighter than on remainder of upperparts but rump generally rather inconspicuous. Upper tail-coverts light drab-grey with ash-grey borders and faint pink fringe. Lore pale buff with slight pink tinge, eye-ring conspicuously cream-white; side of head along bill-base pale ash-grey with slight pink tinge, gradually more intensely suffused with light sandy-brown towards rear, side of neck and side of breast virtually uniform drab-brown or sandy-brown. Underparts entirely pale ash-grey, each feather with pale rosy-pink suffusion on tip, ash-grey predominant on chin and throat, rosy-pink predominant on belly, vent, lower flank, and under tail-coverts. Tail black, central 5 pairs of feathers (t1–t5) with broad rosy-pink fringe along outer web, outer web of t6 and inner webs of all feathers narrowly edged white. Flight-feathers, greater upper primary coverts, and shorter feathers of bastard wing black, outer webs broadly fringed dull rosy-pink, pink partly tinged ash-grey subterminally and merging into white fringe along feather-tip; pink absent along outer web of p9 and on emarginated parts of p7–p8; longest feather of bastard wing fully black. Greater upper wing-coverts and tertials drab-brown with broad but faint pink outer border, which is partly washed ash-grey subterminally; lesser and median coverts drab-brown with dull pink suffusion on tips. Under wing-coverts and axillaries pale grey with pink suffusion. When plumage very fresh (feathers still growing or just completed, about July–August), all pink of upperparts, wing, and tail dark vinous-grey, of underparts dull vinous-pink. Wear has marked effect, all rosy-pink becoming gradually brighter in spring, first on feather-tips: *when fairly worn* (about March–May), feathers of forehead, along lateral base of bill, and on chin tipped rosy-pink or rosy-red; crown, ear-coverts, and cheek purer pale ash-grey, brown feather-tips worn off, cheek sometimes with pink tinge; mantle, scapulars, and rump brown with more distinct rosy tinge, rump and upper tail-coverts bright rosy-pink, feather-tips more rosy-red, some ash-grey of feather-bases visible; underparts more clearly rose-pink, but still with much ash-grey of feather-bases on chin, throat, and side of chest; pink outer fringes of tail- and flight-feathers, greater upper primary and secondary coverts, and tertials brighter rosy-pink, often worn to bright rosy-red along edges, submarginal ash-grey on greater coverts and tertials more contrasting (but feather-centres not as pale as those of Mongolian Trumpeter Finch *B. mongolicus*); when extremely abraded (late June and July), most pink worn off, plumage predominantly sandy-grey and brown, but traces of fiery pink-red or orange-red fringes on rump, flight-feathers, greater coverts, and from belly to under tail-coverts. ADULT FEMALE. Rather like adult ♂, but forehead, crown, and side of head drab-brown in fresh plumage, similar to remainder of upperparts, without pink and ash-grey, except sometimes for some faint pink at base of bill; eye-ring buff-white, ear-coverts often darker sandy-brown than remaining side of head (in adult ♂, much paler, ash-grey); underparts as adult ♂, but chest and flank to under tail-coverts slightly washed sandy-brown, each feather with narrow but usually fairly distinct grey-brown shaft-streak; sometimes some dark mottling on side of throat, forming faint malar stripe. Wing and tail as adult ♂, but pink fringes along outer webs of feathers distinctly narrower, browner ground-colour of flight- and tail-feathers better visible when wing and tail not spread. *When worn* (March–May), face, lore, and eye-ring pale buff, either without rosy or with faint rosy-pink suffusion above lore and eye and on forepart of lower cheek; remainder of top and side of head light sandy-brown, without ash-grey or with restricted amount of ash-grey on crown only (especially ear-coverts less pale and grey), rump and upper tail-coverts mixed drab-brown and rosy-pink, without ash-grey, rosy-pink sometimes less extensive than in adult ♂; entire underparts sandy-pink (less grey-and-pink), usually with narrow dusky shaft-streaks; worn tail- and flight-feathers with more black-brown or sepia of centres exposed; when heavily worn, edges of pinkish feathers above lore and of rump, belly, secondaries, and greater coverts become partly rosy-red or fiery orange-red. NESTLING. Down fairly long, dense, covering entire upperparts, upperwing, and thigh; greyish-white (Rodríguez 1972; Harrison

1975). For development, see Rodríguez (1972). JUVENILE. Entire upperparts, side of head and neck, and lesser and median upper wing-coverts uniform pale tawny- or buff-brown, some dull light grey of feather-bases sometimes visible on hindneck and rump; tinge slightly paler sandy-buff on rump and upper tail-coverts, fringes of wing-coverts more rufous-tawny, chest, side of breast and flank light tawny-buff, gradually paler towards buff or isabelline-white chin, vent, and under tail-coverts; some grey of feather-bases visible on belly. Tail dark brown-grey, shading to sepia-black on tip; central pair of feathers (t1) narrow but sharply fringed pale tawny- or pink-buff on both webs and tip, t2–t6 on outer web and tip. Flight-feathers and greater upper primary coverts brownish-black, fringed tawny- or pink-buff on outer web, pale buff-grey on tip, widest on outer web of secondaries. Tertials and greater upper wing-coverts dark brown-grey with broad but ill-defined tawny-buff borders and tips. Axillaries and under wing-coverts cream-white. FIRST ADULT MALE. Like adult ♂, indistinguishable when moult complete, but easy to identify when only part of flight-feathers, primary coverts, or tail moulted: these new feathers contrast markedly in colour, abrasion, and shape with neighbouring retained juvenile feathers (in adult, all feathers equal in colour and wear); tips of retained juvenile tail-feathers sharply pointed, less rounded than in adult, ground-colour browner, fringes along outer webs buff, less pink (except at extreme base); tips of juvenile flight-feathers bordered brown-grey (in adult, more clearly fringed off-white); primary coverts grey-brown, outer web and tip narrowly fringed pink-buff or buff-white, less extensively pink. New feathers of head and body as in adult ♂, but crown and ear-coverts on average more extensively tinged brown, less pale ash-grey, underparts sometimes slightly tinged brown, under tail-coverts and (occasionally) flank with dusky shaft-streaks; head and body of dullest 1st adult ♂♂ approach brightest adult ♀♀ in colour and streaking, though latter in general with less ash-grey on crown and ear-coverts and with more creamy (less pinkish) mid-belly. FIRST ADULT FEMALE. Like adult ♀, but part of juvenile feathering retained in an unknown proportion of birds, extent and characters as in 1st adult ♂♂. Easy to age when part of tail- or flight-feathers juvenile and part new, but sometimes only with difficulty when all juvenile flight-feathers and primary coverts retained, especially when worn. Crown without traces of ash-grey, even when worn; rosy-pink or red on rump, face, and under tail-coverts restricted or absent.

**Bare parts.** ADULT, FIRST ADULT. Iris cinnamon, light brown or dark brown. Bill of breeding ♂ pink-red, orange-red, or coral-red, during rest of year and in ♀ duller pale brown to grey-horn, often with salmon-pink, flesh, dull yellow, or orange-yellow tinge; lower mandible of ♀ sometimes more orange during breeding. Leg and foot pale brown, yellow-brown, flesh-brown, brownish-flesh, or flesh-grey, in breeding ♂ sometimes brighter pink-flesh, orange-flesh, or pink-red. (Hartert 1903–10; Bannerman 1912; Meinertzhagen 1940; Desfayes and Praz 1978; Hollom *et al.* 1988.) NESTLING. At hatching, mouth orange, gape-flanges yellow-white; later on, mouth crimson, flanges pale yellow (Harrison 1975). Bill pale orange or light yellow-horn, gradually darker with age, leg pale orange (Bannerman 1963; Harrison 1975). JUVENILE. Bill greenish- or brownish-horn. Leg and foot pink-buff (Meinertzhagen 1940; Bannerman 1963).

**Moults.** ADULT POST-BREEDING. Complete; primaries descendent. Starts shortly after breeding, and timing thus rather variable. In southern Algeria, starts with inner primaries or some feathers of body from early or late June (RMNH, ZFMK, ZMA); in Aïr (northern Niger), 2 birds in full moult in 2nd half of June, another had moult of wing and tail almost ready, and 3 birds from early July also nearing completion (Fairon 1975). In northern Algeria, moult starts mid- or late July, with primary moult scores of 38, 43, and 46 reached on 8 September, when remainder of feathering almost completely new (ZFMK, ZMA). In central Arabia, 31 July, single ♂ heavily worn, single ♀ halfway through moult (Meinertzhagen 1949). 10 birds from northern Iran all in last stages of moult on 29-30 August (Vaurie 1949*b*). In Transcaspia, some birds moult from May, others not until late June (Dementiev and Gladkov 1954). POST-JUVENILE. Partial, involving at least head, body, and lesser, median, and greater upper wing-coverts, but sometimes complete or virtually so, especially in south of breeding range. Timing highly variable, depending on hatching date; generally starts shortly after fledging. Birds in juvenile plumage occur February (southern Algeria) to July (Aïr, Algeria, Afghanistan). In southern Algeria, birds started complete moult with body and inner primaries in May (score 14 on 27 May); 2 from northern Algeria had score 4 and 13 on 18 July (ZMA); in Aïr, single bird started complete moult early July (Fairon 1975). Other birds in Algeria and Tunisia start with middle primaries (from p4–p8 outwards), retaining juvenile inner primaries; some of these may suspend primary moult later on, when conditions unfavourable for moult or when breeding starts, these birds retaining some juvenile inner and outer primaries, with 2–5 central ones new; such birds often moult all tertials and greater upper wing-coverts as well, and also usually tail and sometimes s6; remaining secondaries and most primary coverts still juvenile (RMNH, ZFMK, ZMA). In other birds (from whole geographical range), moult includes head, body, lesser, median, and many or all greater upper wing-coverts, and variable number of tertials, but no flight-feathers (ZFMK, ZMA), e.g. single ♂ from Iran in body moult on 27 September (Vaurie 1949*b*).

**Measurements.** *B. g. zedlitzi.* Algeria and Tunisia, all year; skins (RMNH, ZMA). Adult wing includes that of 1st adult with new outer primaries, juvenile wing that of 1st adult without. Bill (S) to skull, bill (N) to distal corner of nostril; exposed culmen on average 2·9 less than bill (S).

| | | | | | |
|---|---|---|---|---|---|
| WING AD | ♂ | 89·5 (2·01; 11) | 87–93 | ♀ 86·7 (1·94; 6) | 84–89 |
| JUV | | 86·0 (1·26; 7) | 84–88 | 85·4 (2·25; 4) | 82–87 |
| TAIL AD | | 50·9 (1·80; 11) | 48–53 | 50·7 (2·08; 5) | 48–53 |
| JUV | | 48·2 (2·26; 7) | 46–52 | 49·4 (1·52; 5) | 47–52 |
| BILL (S) | | 12·8 (0·60; 17) | 11·8–13·6 | 12·7 (0·52; 10) | 11·8–13·3 |
| BILL (N) | | 7·4 (0·36; 18) | 7·0–8·2 | 7·4 (0·26; 10) | 7·0–7·7 |
| TARSUS | | 18·1 (0·61; 18) | 17·1–19·0 | 18·1 (0·61; 10) | 17·3–19·1 |

Sex differences significant for adult wing. No marked difference between northern and southern birds; e.g. wing, adult ♂, Atlas Saharien (northern Algeria) and Tunisia 89·4 (2·51; 7) 87–93, Ahaggar (southern Algeria) 89·5 (0·41; 4) 89–90.

Adult and 1st adult ♂: (1) Biskra area (northern Algeria), (2) Aïr (northern Niger) (Vaurie 1956*b*).

| | | | | |
|---|---|---|---|---|
| WING (1) | 89·1 (1·51; 11) | 86–91 | (2) 85·6 (1·42; 9) | 84–88 |
| TAIL | 51·2 (2·09; 11) | 47–54 | 47·1 (2·26; 9) | 43–50 |
| TARSUS | 18·1 (1·03; 11) | 17–20 | 16·9 (0·65; 9) | 16–18 |

(1) *B. g. amantum*, Canary Islands, and (2) nominate *githagineus*, Egypt and northern Sudan, adult and 1st adult ♂ only; skins (RMNH, ZFMK, ZMA; wing includes data from Vaurie 1956*b*).

| | | | | |
|---|---|---|---|---|
| WING (1) | 84·8 (2·18; 21) | 82–88 | (2) 84·5 (2·07; 22) | 80–89 |
| TAIL | 50·8 (4·33; 6) | 47–54 | 48·8 (1·15; 5) | 47–50 |
| BILL (S) | 11·8 (0·71; 6) | 11·1–12·9 | 12·6 (0·45; 5) | 12·0–13·1 |
| BILL (N) | 7·3 (0·36; 6) | 6·7–7·7 | 7·3 (0·39; 5) | 6·8–7·8 |
| TARSUS | 18·0 (0·64; 6) | 17·2–18·8 | 17·3 (0·55; 4) | 16·5–18·0 |

Wing of nominate *githagineus* in southern Egypt and Sudan 78–84(–86) (Neumann 1907; Meinertzhagen 1930; RMNH).

*B. s. crassirostris.* Adult and 1st adult ♂: wing 88·0 (2·18; 18) 84–92 (Vaurie 1956*b*; RMNH, ZMA); tail 49·4 (1·15; 4) 48–51, bill to skull 12·2 (0·61; 4) 11·8–12·8, bill to nostril 7·6 (0·23; 4) 7·3–7·8, tarsus 18·0 (0·59; 4) 17·3–18·7 (RMNH, ZMA). Wing: Iran and Afghanistan, ♂ 88·0 (6) 84–90, ♀ 85·5 (4) 84–87 (Vaurie 1949*b*); Lebanon, May, ♂ 81·5 (Kumerloeve 1952*b*).

**Weights.** *B. g. zedlitzi.* Algeria, November–January: ♂♂ 17·0, 19·7 (Fairon 1971); ♂ 21–25 (8), ♀ 19–22 (3) (Niethammer 1955); Aïr (northern Niger): February, ♂♂ 17, 18 (Niethammer 1955); June–July, ♂♂ 16·5, 18·1, 21·3, ♀♀ 16, 19·3, unsexed 19·8 (Fairon 1975).

*B. g. crassirostris.* Israel: 16–22 (Paz 1987). Southern Iran, May–June: ♂ 20·7 (3) 20–21, juvenile 18 (Desfayes and Praz 1978). Afghanistan, early July: juvenile ♂ 19 (Paludan 1959).

Race unknown. Sweden, June: ♀ 13·7 (Westin 1973).

**Structure.** Wing rather long, fairly broad at base, tip bluntly pointed. 10 primaries: in *zedlitzi* and *crassirostris*, p8 longest, p9 0–1 shorter, p7 0·5–3, p6 6–8, p5 11–16, p4 16–20, p3 21–24, p2 24–28, p1 28–32; in *amantum* and nominate *githagineus*, p3 17–22 shorter, p2 20–26, p1 22–29. P10 reduced, a tiny pin hidden below reduced outermost greater upper primary covert, 52–62 shorter than p8, 8–13 shorter than longest upper primary covert. Outer web of p7–p8 emarginated, inner web of (p8–)p9 with faint notch. Tip of longest tertial reaches to tip of p2–p3 if not abraded. Tail rather short, tip slightly forked; 12 feathers, t1 3–10 shorter than t6. Bill markedly short and thick, length almost equal to bill depth; culmen and cutting edges decurved towards blunt tip; lower mandible deep at base, with pronounced gonydeal angle; depth where feathering reaches bill in 12 *zedlitzi* 9·2 (8·5–10·0) mm, width 7·6 (7·2–8·2); in 5 birds of nominate *githagineus*, depth 9·0 (8·5–9·5), width 7·6 (7·0–8·0); in 5 *crassirostris*, depth 9·6 (9·2–9·9), width 7·8 (7·5–8·1); in 5 *amantum*, depth 8·8 (8·5–9·0), width 8·3 (8·0–8·7); bill of *amantum* shorter, wider, and more swollen laterally than in other races. Nostril small, rounded; covered by short tuft of bristle-like feathers projecting from base of bill; a tuft of short feathers and many fine bristles project laterally down gape. Tarsus and toes short, but strong. Middle toe with claw 16·7 (20) 15·5–18 mm; outer and inner toe with claw both *c.* 68% of middle with claw, hind *c.* 77%.

**Geographical variation.** Marked. Involves depth of ground-colour of (especially) upperparts (paler or darker drab-grey to drab-brown), extent and depth of rose-red on head, body, and wing, size (wing, tail, or tarsus length), and depth and width of bill. 2 groups separable on size: (1) large northern birds (average wing of ♂ *c.* 88·0–89·5) from south-east Spain and central Morocco through Algeria (including Ahaggar in south), Tunisia, and Libya to northern Egypt (*zedlitzi*), and from Sinai and Levant east to west-central Asia (*crassirostris*); (2) smaller southern birds (average wing of ♂ *c.* 85) in central and eastern Canary Islands (*amantum*), Aïr (northern Niger), and northern Chad (Tibesti, Borkou area, and northern Ennedi) (race unnamed), and southern Egypt and neighbouring north-central and eastern Sudan (nominate *githagineus*); not known what race inhabits western Sahara from Mauritania to south-west Morocco (probably *zedlitzi*) and oases in western Egyptian desert (but those of Gebel Uweinat in extreme south-west are *zedlitzi*: Goodman and Meininger 1989). Larger northern birds and smaller southern ones probably grade into each other in central Egypt. Within northern group, variation in colour strongly clinal, ground-colour of hindneck, mantle, and scapulars of *zedlitzi* paler drab-grey, extensively tinged rosy-pink, *crassirostris* slightly darker and browner, almost without pink tinge; pink fringes along outer webs of tail- and flight-feathers on average narrower in *crassirostris*, chin and throat more extensively rosy-pink or rosy-red; boundary between *zedlitzi* and *crassirostris* here arbitrarily drawn at Gulf of Suez, but even in Tunisia birds average slightly greyer than those from Algeria (Vaurie 1986b). In southern group, populations more heterogeneous. In both sexes of *amantum* from Canary Islands, ground-colour of hindneck, mantle, and scapulars darker drab-brown than in *zedlitzi*, and plumage of ♂ more extensively tinged with deep pink-red, ruby-red, or orange-red (colour depending on abrasion); bill shorter but wider and more swollen at base than in other races. Birds from

Aïr rather like *zedlitzi* (apart from size), but ♂ extensively suffused with darker rose-pink on rump and underparts, ♀ more cream-yellow on belly, and fringes of scapulars, greater coverts, and flight-feathers darker, broader, and brighter reddish-pink, rather as in *amantum* (Niethammer 1955; Vaurie 1956b; ZFMK). Nominate *githagineus* has ground-colour of upperparts rather dark grey-brown or earth-brown, darker than in *zedlitzi*, but paler than in *amantum*; amount of rosy-pink on body and wing of ♂ restricted, as in *crassirostris*, with none on hindneck, mantle, and scapulars; ground-colour slightly darker than in *crassirostris*, less drab-grey; smaller in size (Neumann 1907; RMNH, ZFMK).

May form species-group with Mongolian Trumpeter Finch *B. mongolicus*, and both sometimes included in *Rhodopechys*, but *R. sanguinea* probably not closely related (Desfayes 1969); see *B. mongolicus* and *R. sanguinea*.                                  CSR

## *Carpodacus erythrinus*  Scarlet Rosefinch

PLATES 48 and 50 (flight)
[between pages 640 and 641]

Du. Roodmus       FR. Roselin cramoisi       GE. Karmingimpel
RU. Обыкновенная чечевица       SP. Camachuelo carminoso       SW. Rosenfink       N. AM. Common Rosefinch

*Loxia erythrina* Pallas, 1770

Polytypic. Nominate *erythrinus* (Pallas, 1770), Europe, east in Siberia to Lena basin, south to Barnaul and Krasnoyarsk areas, grading into *ferghanensis* in Altai and Sayan mountains and north-west Mongolia, and into *grebnitskii* in north-east Mongolia, Transbaykalia, and eastern Siberia; *kubanensis* Laubmann, 1915, northern Turkey, Caucasus area, and through northern Iran to Khorasan (north-east Iran) and to Bol'shoy Balkhan and Kopet-Dag in western Turkmeniya. Extralimital: *ferghanensis* (Kozlova, 1939), mountains of central Asia, from Afghanistan through Pamir-Alay and Tien Shan to Dzhungarskiy Alatau and Tarbagatay, south-west to western Kun Lun and western Himalayas east to Lahul and Rupshu, there grading into *roseatus*; *roseatus* (Blyth, 1842), central and eastern Himalayas and neighbouring Tibet, in north-east to Kansu and southern Shensi (south-west China); *grebnitskii* Stejneger, 1885, Kamchatka and shores of Sea of Okhotsk south to Amurland and Manchuria (north-east China).

**Field characters.** 14·5–15 cm; wing-span 24–26·5 cm. Close in size to House Sparrow *Passer domesticus* but with relatively smaller, more swollen bill and longer cleft tail; slightly smaller than Pallas's Rosefinch *C. roseus* and Sinai Rosefinch *C. sinoicus*. Medium-sized, bulbous-billed, quite stocky but long finch, with all plumages except adult ♂ reminiscent of ♀ *P. domesticus* and Corn Bunting *Miliaria calandra*. Adult ♂ drenched scarlet on head, rump, and fore-underparts; ♀ and immature dull olive-brown, softly streaked above and below. All ♀♀ and immatures show quite marked double wing-bar. Epitome of Holarctic genus but beware confusion with vagrant or escaped congeners. Voice distinctive. Sexes dissimilar; some seasonal variation. Juvenile like ♀. 2 races in west Palearctic, ♂♂ separable (see Geographical Variation); only north-east European race, nominate *erythrinus*, described here.

ADULT MALE. Moults: September–November (complete). Head, throat, breast, and dappling on upper flanks scarlet, with browner cast from lore, through eye, and on rear ear-coverts (and narrow greyish tips or speckles on

crown, lore, and below eye showing at close range). Mantle, scapulars, and back warm (slightly olive-)brown, washed scarlet. Rump pale scarlet. Wings dark brown, margined pink; tips of median coverts pale enough to create upper wing-bar but those of greater coverts less distinct than edges of tertials; ♂ thus lacks classic double wing-bar associated with *C. erythrinus* in other plumages (and shown by *C. roseus*). Tail dark brown, fringed pink. Lower underparts pale buff, tinged pink and faintly streaked on lower flanks but almost clear on under tail-coverts. With wear, scarlet areas become more brilliant, back browner and pale tips to coverts even less distinct. Bill quite long but looks stubby, due to marked curve to upper mandible producing bulbous appearance; grey with horn base. Eye dark brown. Legs dull flesh to dusky brown. ADULT FEMALE. Head, back, rump, wings, and tail brown, tinged yellowish- to olive-buff except on rump and edges of flight-feathers where overtone distinctly olive. On head and back (and faintly on rump), darker brown centres to feathers form soft lines of spots or streaks and face may show buff wash; wings show pale buff tips to median and greater coverts, forming conspicuous double wing-bar, and pale buff margins to tertials. Underparts pale buff to buffish-white, subtly patterned with whitish moustachial mark, dotted, dusky malar stripes, whitish chin, soft spots and streaks on central throat, breast, and flanks, and almost white vent and under tail-coverts. With wear, bird loses olive tones, becoming browner, even bleached dun above, greyer on rump, and more whitish below; tips to median and greater coverts become almost white and even more conspicuous. Dark eye obvious on pale face. Base of bill may be pinkish-horn. JUVENILE. Resembles adult ♀ but plumage darker, with more fully brown ground to upperparts and larger, more intense marks, extending onto rump above. Double wing-bar and tertial margins sometimes noticeably paler than in adult, being yellowish-buff to off-white and contrasting more vividly with rest of wing. Bill buff-horn. Dark eye obvious, with pale buff eye-ring. Legs may be quite bright brown.

Traditionally regarded as the only rosefinch of Europe, confusion being most likely with escaped Purple Finch *C. purpureus* and House Finch *C. mexicanus* of North America. *C. purpureus* noticeably larger, with distinctive call—dull metallic 'tick' or 'pink'; ♂ noticeably more vinaceous than *C. erythrinus* (but not purple), ♀ and juvenile very heavily streaked on face and below. *C. mexicanus* similar in size to *C. erythrinus*, call a distinctive, harsh, nasal 'che-urr', suggesting *P. domesticus*; ♂ bright red only on face, breast, and rump and heavily streaked on flanks, and ♀ and juvenile more continuously streaked below. Given potential vagrancy of *C. roseus* and other east Asian congeners, these species should be considered. Most serious problem appears to be confusion of young ♂ *C. erythrinus* with ♀ *C. roseus* (see that species). Voice important (see below). Flight slightly heavy, recalling both sparrow *Passer* and Chaffinch *Fringilla coelebs*; over distance, soon becomes undulating. Escape-flight usually to dense bush or similar cover. Gait a hop. Clambers along, even hanging from foliage while feeding. Stance rather upright, with head seemingly sunk into shoulders so that bird can look bull-necked and dumpy, yet quite long-tailed, on perch. Undemonstrative, and sometimes inactive for quite long periods. Sociable, forming large flocks where common, but most westward vagrants occur singly.

Song a short phrase of loud, clear piping notes, suggesting distant Golden Oriole *Oriolus oriolus* but less fluted though still pure and mellow: 'tiū-fiū-fi-tiū', 'WEEje-wüWEEja', or 'pleased to see you'; twittering also recorded. Usual call a soft, not far-carrying 'teu-eik', 'twee-eek', or fresh, pure 'ueet' suggesting Canary *Serinus canaria*, *F. coelebs*, and Twite *Carduelis flavirostris*. Other calls: quiet 'sip' or 'plip', 'peut', and 'pik' or 'peck', all shorter than usual call; alarm-call 'JAY-ee', almost like Greenfinch *Carduelis chloris*.

Habitat. Extends into west Palearctic mainly in temperate continental climatic zone and in lowlands, but with dis-

junct population in foothills and mountains of south-east of region. Lowland form now extending much further west. In USSR, this form avoids desert zones and extends only moderately into forest steppe. Favoured habitats are thickets near forest edges, forest clearings, and patches of regrowth (coniferous or broadleaf), groups of shrubs or isolated trees in humid meadows or river valleys, thickets of osier *Salix* or bird-cherry *Prunus*, and sometimes orchards, graveyards, or thorn hedges (Dementiev and Gladkov 1954; Flint *et al.* 1984). Further west in Europe, drier sites on farmland in briar and scrub patches or gardens are not favoured, choice of breeding places falling on fairly low thickets of alder *Alnus*, poplar *Populus*, or willow *Salix*, with a few taller trees as song-posts (Newton 1972). In different parts of range, markedly different habitats are preferred, from moist (even swampy) bushy or scrub types in western lowlands to low open or marginal forests and drier fields with shrub or thicket patches in more eastern plains. Extends into taiga or forest steppe and then, in sharp contrast, to more elevated foothill forests, shrub thickets of azalea or juniper *Juniperus*, mountain river valleys, alpine meadows, and even cultivated land with trees and shrubs, up to 2500–3700 m (Dementiev and Gladkov 1954).

Extralimitally in Afghanistan, bred in scrub along river, singing from willows or sometimes bare rocks, or isolated conifers in forest glades at 2600–3000 m (Paludan 1959). At similar elevations in uplands of Kashmir, local concentrations found breeding in valley bottoms, especially among dwarf willows but also in bracken *Pteridium* (Bates and Lowther 1952). In India, favours willow and tamarisk *Tamarix*, as well as bushes on edge of cultivation, orchards, slopes with junipers, briar, and rose bushes, and other stony areas. Breeds in Himalayas sometimes down to 2000 m but commonly up to 4000 m or higher. Winters normally below 1500 m in open forest, scrub jungle, or scrub and bushes in cultivated areas. (Ali and Ripley 1974.)

An adaptable and successful species across wide ranges of altitude and climate, but avoids dense forest, deserts, or any open land devoid of some kind of woody vegetation. Favours vicinity of water, but does not commonly occupy coastal habitats or interior of wetlands, and shows some preference for sunny aspects. Appears to spend as little time as possible on ground, and while not always avoiding human settlements shows no sign of seeking them out. Quite mobile but appears less aerial than some of its relatives.

Distribution. Has undergone major expansion to west, in 2 stages. Began in early 19th century, followed by regression in late 19th century. 2nd stage began in 1930s and still in progress; this stage is also taking place in east of range, but fewer details available. (Bozhko 1980.) Recent spread in west of range attributed to long, warm autumns, favouring westward dispersal of 1st-year birds

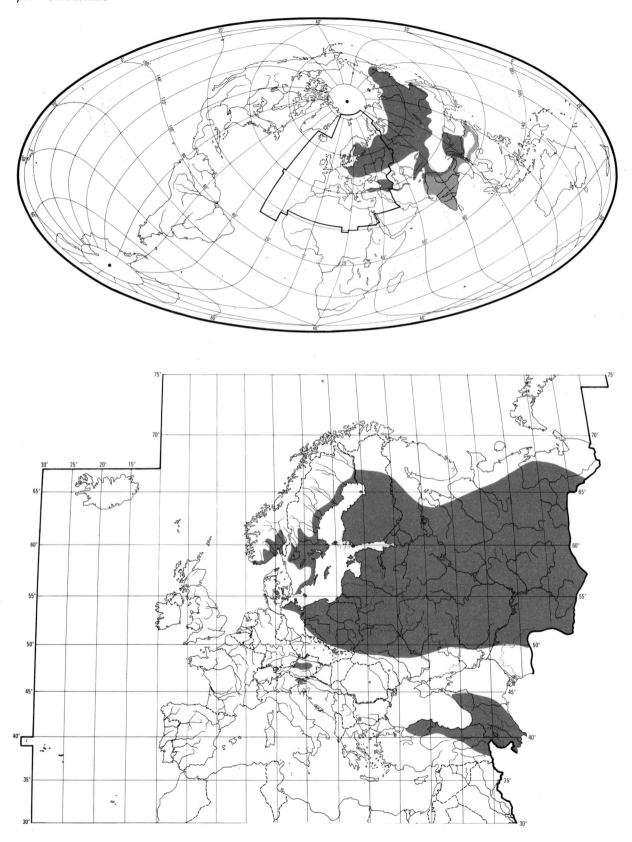

which apparently act as pioneers (Jung 1983); also, in Finland, to increased breeding success in habitats created by man (Koskimies 1989).

BRITAIN. Breeding first recorded Highland (Scotland) 1982 (Mullins 1984*b*). Probable breeding at up to 5 localities (mostly Scotland) 1983-90, and confirmed in Shetland 1990 (Spencer *et al.* 1991, 1993). Major influx 1992 (see Movements) followed by first breeding in England: 3-5 pairs Humberside, 2-3 pairs Suffolk (Lassey and Wallace 1992; Nightingale and Allsopp 1992). FRANCE. First recorded breeding, Doubs, 1985 (Pochelon 1992). NETHERLANDS. First bred 1987 (CSR). WEST GERMANY. First recorded breeding Schleswig-Holstein 1982, Bayern 1983 (perhaps also 1981-2), Niedersachsen 1985 (perhaps also 1983) (Hill 1986), Westfalen 1986 (Fellenberg 1988), Helgoland 1987 (Barth and Moritz 1988). DENMARK. First recorded 1943; next record 1952, thereafter regular in spring and breeding first recorded 1972 (Bozhko 1980). Now regular, but in fewer than 10 localities (UGS). NORWAY. First bred 1970. Spread rapidly in 1970s and 1980s, reaching Rogaland in late 1980s; singing ♂♂ recorded north to Lofoten Islands. (VR.) SWEDEN. One record in 19th century, none 1900-30. Colonization began 1930s. (Risberg and Risberg 1975.) FINLAND. Sparse breeder in south-east until early 20th century. Major expansion since mid-1940s. (Koskimies 1989.) EAST GERMANY. Regular breeder since 1968; rapid colonization of coastal belt followed by slower spread inland. First bred in lower Oder valley 1970s. (Bozhko 1980.) POLAND. Early in 19th century scarce breeder in east, south-east, and south-west Silesia. Later retracted range to extreme east; then since *c.* 1900 spread west along Baltic coast. Now breeds over whole country. (AD, LT.) CZECHOSLOVAKIA. First recorded breeding 1959; continuing expansion thereafter (KH). AUSTRIA. First recorded breeding 1973; isolated singing ♂♂ reported much earlier (Bozhko 1980; H-MB). HUNGARY. Records increasing, some indicating breeding, but no proof yet (G Magyar). SWITZERLAND. Annual summer visitor in very small numbers since 1979. First (unsuccessful) breeding attempt recorded 1983 (Kälin 1983; Winkler 1984); first successful breeding 1989 (Schmid 1991). YUGOSLAVIA. First recorded breeding 1978; subsequently spreading (VV). USSR. Expansion to north and east since 1930s; in 1966 reached White Sea coast. Now breeds throughout Karelia to Murmansk area; observed north to Arkhangel'sk, and breeding recorded in south of Kanin peninsula. (Bozhko 1980.) Crimea: first bred 1991 (Tsvelykh 1993).

Accidental. Iceland, Faeroes, Ireland, Malta, Greece, Cyprus, Morocco, Jordan, Kuwait.

**Population.** Has recently increased very markedly in all areas affected by westward and northward range expansion (see Distribution), but few figures available; at edge of range, prevalence of pioneering singing ♂♂ makes assessment of breeding population difficult.

BRITAIN. See Distribution. NETHERLANDS. 1 pair bred 1987, 15 pairs 1989, more than 20 pairs 1991 (*Br. Birds* 1992, **85**, 443-63), *c.* 45 pairs 1992 plus other territorial ♂♂ (*Dutch Birding* 1992, **14**, 157). GERMANY. Estimated 1000 pairs (Rheinwald 1992). East Germany: 280 ± 80 pairs (Nicolai 1993). DENMARK. Estimated 40-50 pairs (UGS). SWEDEN. After colonization, population increased about threefold every 5th year. Surveys in 1974 indicated 1409 singing ♂♂. (Risberg and Risberg 1975.) FINLAND. Has increased at least 30-fold since mid-1940s; perhaps 200 000-500 000 pairs (Koskimies 1989). Numbers may now have levelled off (OH). CZECHOSLOVAKIA. Bohemia: increasing; *c.* 30-50 pairs 1973-7, *c.* 270-400 pairs 1985-9 (Šťastný and Bejček 1991). USSR. Crimea: 9 territorial ♂♂ 1992 (Tsvelykh 1993).

Survival. Finland: average annual mortality of ♂♂ 21 ± SD7%, ♀♀ 27 ± SD10%. Oldest ringed bird at least 9 years. (Stjernberg 1979.)

**Movements.** Migratory. Most populations long-distance migrants; south-east populations in part altitudinal migrants. Migration nocturnal and diurnal. For review, and discussion of movements of European birds, see Bozhko (1980) and Jung (1983).

Winters south of breeding range from Pakistan east to eastern China, south to southern India (not Sri Lanka), northern Burma, Thailand (mostly in west), and northern Laos. In Pakistan, winters only in small numbers; in India, most abundant in central and western areas; winters up to *c.* 1500 m in Himalayan foothills. Precise winter range of different races not known; nominate *erythrinus* winters east to Nepal, and presumably *kubanensis* and *ferghanensis* also winter in west of range; *roseatus* winters from India east to south-east Asia and southern China, and *grebnitskii* in south-east China. (Whistler 1924; Peters 1968; Ali and Ripley 1974; King *et al.* 1975; Schauensee 1984; Roberts 1992.) *E. e. kubanensis* also winters in very small numbers in eastern Israel (annual since 1981), and occasionally in Oman and Sinai (Egypt); rare and irregular in passage periods further west in Egypt (Goodman and Meininger 1989; Oman Bird List; Shirihai in press). For records suggesting possibility of birds seeking new winter quarters in southern Europe, see Józefik (1960), Bozhko (1980), and Jung (1983).

Western populations (nominate *erythrinus*) migrate east or south-east in autumn (central European birds perhaps initially heading north-east towards St Petersburg and Moscow regions) (Bozhko 1980), passing north of Caspian Sea (common and widespread in Volga-Ural area: Gavrilov *et al.* 1968). Further east and south-east, lower numbers in autumn than spring in various areas suggest many birds overfly. Inconspicuous in north-central Kazakhstan (Korelov *et al.* 1974), so probably most birds head south-east from Volga-Ural area across central Asian deserts rather than continuing east. Marked passage southward in north-east Kazakhstan (Korelov *et al.* 1974) presumably involves chiefly Siberian breeders. In autumn survey, recorded at various desert sites from Aral Sea east to Lake Balkhash, but especially at Baygora (south of Lake Balkhash) (Yablonkevich *et al.* 1985a). In south-west Kyzyl-Kum (Uzbekistan), far fewer autumn than spring; most birds (both seasons) fly at 10-50 m above ground, others at 50-100 m or above (Tret'yakov 1978). In study at Ala Archa canyon (at *c.* 2100-2300 m in Kirgizskiy Alatau, western Tien Shan), where *ferghanensis* breeds, passage of nominate *erythrinus* was apparently at great height, and only grounded birds were seen (Iovchenko 1986). Populations (including *ferghanensis*) converge in mountains of south-central Asia (Tien Shan, Pamir Alay), and high concentrations occur in valleys. In Chokpak pass (western Tien Shan), considerably fewer in autumn than spring; migrates chiefly along slopes of Talasskiy Alatau (Gavrilov and Gistsov 1985). In Gissar valley (Tadzhikistan), large numbers swarm through parks and gardens of towns and settlements (Ivanov 1969). One of most numerous migrants in Alay valley (north-west Pamir Alay), last valley for building up fat reserves before mountain crossing; in 1981 study, stopovers up to 7 days recorded; birds showed pre-migratory restlessness in evening (Vinogradova *et al.* 1985). East of Lake Issyk-Kul' (northern Tien Shan), 1981-2, stopovers up to 10 days recorded (Ostashchenko *et al.* 1985). *E. e. kubanensis* migrates east (apparently mostly skirting Iraq to north) through north-east half of Iran and Afghanistan (D A Scott, S C Madge). Mass migration continues through Pakistan and north-west India. In Pakistan, birds move through North West Frontier Province and Punjab in vast numbers, but few reach Sind (Roberts 1992). At some sites, e.g. Lahore, Islamabad, and Kangra valley, absent or in small numbers in autumn, though common in spring (Whistler 1924; Mallalieu 1988).

The few long-distance recoveries (5) exemplify northwest to south-east movement, from Finland to northern Kazakhstan, Norway and Finland to Uzbekistan, Kaliningrad coast (western Russia) to east Turkmeniya, and Bharatpur (northern India) to Ulyanovsk region (*c.* 4000 km, Russia) (Ali and Ripley 1974; Tret'yakov 1978; Bozkho 1980). Also 2 birds ringed south-east of Lake Ladoga (western Russia) recovered north-west in Finland (Mal'chevski and Pukinski 1983).

Autumn migration starts early, 2nd half of July and beginning of August, throughout boreal area of distribution; southern birds leave later (Bozkho 1980). Adults migrate before juveniles (Peiponen 1974; Gavrilov and Gistsov 1985; Vinogradova *et al.* 1985). Netting over 5 years on southern Finnish coast showed peak movement in 2nd half of July and early August, with proportion of juveniles higher in August (Peiponen 1974). In St Petersburg and Moscow regions, adults disappear mostly in late July and early August, juveniles and some ♀♀ remain longer, with last records early September (Ptushenko and Inozemtsev 1968; Noskov *et al.* 1981). In Volga-Ural area,

passage from end of July or beginning of August to mid- or late September (Gavrilov *et al.* 1968). In north-east Altai (southern Siberia), 75% of local population leaves by end of July; passage in tundra continues to end of August, and in lower hills chiefly to mid-September (Ravkin 1973). At various desert sites in central Asia, of 419 birds caught 12 August to 29 September 1982, 90% in August (Yablonkevich *et al.* 1985a). In Ala-Archa canyon (western Tien Shan), movement from last third of July; most local adults (*ferghanensis*) depart 1-20 August (coinciding with peak passage of nominate *erythrinus*), and most local juveniles 10-30 August; in 1980, when rowan *Sorbus* crop good, many juveniles still present in 1st half of September, but not in 1983 when crop poor (Iovchenko 1986). In southern Kazakhstan, recorded only in small numbers after mid-September, with stragglers to early November (Korelov *et al.* 1974). Leaves Caucasus (*kubanensis*) at end of August (Polivanov and Polivanova 1986). Arrives in Pakistan and India August-October (Ali and Ripley 1974).

Spring migration late and rapid, April-May(-June). Leaves India and Pakistan mostly April to early May (Ali and Ripley 1974); at Islamabad (northern Pakistan), recorded 2 April to 17 May over 3 years (Mallalieu 1988). Returns to Caucasus and Turkey from early May (Polivanov and Polivanova 1986; Martins 1989), and first-ever record in Iraq (at Mosul in north) was 7 May 1968 (Georg 1969). Birds breeding in central Asian mountains return from mid- or late April, but main arrival at high levels (e.g. Ladakh and Issyk-kul' depression), where spring arrives later, not until last third of May or June (Osmaston 1927; Bozkho 1980). Peak passage 1st third of May in Gissar valley (Ivanov 1969), and 1-20 May in central Asian deserts (Yablonkevich *et al.* 1985b). The most numerous migrant in south-west Kyzyl-Kum, passage chiefly at 06.00-08.00 hrs and 17.00-19.00 hrs (Tret'yakov 1978). Main movement in north-east Altai mid-May to early June (Ravkin 1973); in western Siberia, earliest record over many years 23 May at Tomsk, 27 May at Tobol'sk (Johansen 1944). Passage in Volga-Ural area almost entirely in May (Gavrilov *et al.* 1968), and reaches European Russia in 2nd third of May (Bozkho 1980). In St Petersburg region, earliest records averaged 18-19 May in 1950s-1970s, several days earlier than at start of 20th century (Bozkho 1980). Similarly at Jõgeva (Estonia), average earliest arrival 21 May 1936-69, but 19 May in 1960s; also in 1960s, average earliest 18-19 May in southern Finland, but 27 May *c.* 500 km further west at same latitude in Sweden. Observations in southern Finland show arrival from east and SSE (Peiponen 1974). Main passage to Sweden apparently via Åland islands (southwest Finland) (Risberg 1970). Arrives in Austria last third of May and 1st third of June (Mazzucco 1974).

In Europe, range of nominate *erythrinus* has expanded (see Distribution) in 2 directions, westward towards Britain and north-west Germany, and south-west towards

Austria and Yugoslavia (in Germany, most records are in north or south-east, and apparently few reach France); coasts of northern Europe, and rivers and mountain valleys of central Europe, act as leading lines (Stjernberg 1985; Hill 1986). Not known whether records furthest south-east (western Black Sea coast) involve nominate *erythrinus* or *kubanensis* (Jung 1983). In Britain (see below), increase in records has involved primarily 1st-year birds in autumn, apparently pioneers dispersing west; this probably linked with rapid increase in numbers breeding in Fenno-Scandia. In north-west Germany, however, spring records have increased, while autumn records have remained fairly stable; this probably linked with smaller increase in north-east Germany leading to slower expansion (Jung 1983); in Helgoland (north-west Germany), 1972–87, annual in increasing numbers, 65·5% May–June, 34·5% July–October (Barth and Moritz 1988). In south-east Europe and Switzerland also, most records in spring (Jung 1983; Mosimann 1988).

In Britain and Ireland, 200–300 records before 1958 (only 1 in spring), 882 in 1958–85; all between April and November (1 in February), with 14% in spring (mostly mid-May to early June) and 86% in autumn (mostly September). 62% of total in Shetland (northern Scotland), elsewhere records well scattered in spring (none in Ireland), and mainly on east coast and in south-west in autumn (suggesting south-west drift after arrival). (Dymond *et al.* 1989.) In 1992, unprecedented spring influx into Britain, with at least 80 birds mid-May to June; highest numbers in north Scottish islands, but also widely scattered on east coast, and reaching south-east Ireland and Guernsey (Channel Islands) (Lassey and Wallace 1992; Nightingale and Allsopp 1992; M L Long); for resultant breeding, see Distribution.

In Iceland, 29 records up to 1989, including both spring and autumn (Pétursson *et al.* 1992). In France, only *c.* 20 records since 1900, of which 13 in 1986–9, chiefly in October at Ouessant in north-west (Dubois and Yésou 1992).

Autumn migration of *grebnitskii* is later than in nominate *erythrinus*. Leaves breeding areas mid-August to early September, e.g. Mongolia (Mauersberger 1980), Kamchatka, north-east Russia (Lobkov 1986), central Siberia (Reymers 1966). Rare in southern Ussuriland both seasons (Panov 1973a), so presumably passage is further inland; also rare in Korea and Japan (Gore and Won 1971; Brazil 1991). Passage in north-east China mostly September (Hemmingsen 1951). Recorded in Hong Kong 28 October to 24 April, mostly late December to early March (Chalmers 1986). Spring passage in north-east China May to early June (Hemmingsen and Guildal 1968; Williams 1986). Passage and arrivals in Mongolia mid-May to early June (Mauersberger *et al.* 1982; Piechocki *et al.* 1982), and reaches central and east Siberia in last third of May (Bozkho 1980). In Kamchatka, average earliest record over 8 years 28 May (Lobkov 1986).    DFV

**Food.** Seeds, buds, and most other parts of plants; some invertebrates. Forages in grass and herbs, on arable land, in bushes, and in trees up to crown. (Peiponen 1974; Stjernberg 1979; Bozhko 1980.) In southern Finland, spring, of 392 feeding observations, 44% on ground (32% eating newly-sown seeds), though ground-feeding declines as crops grow; of 402 min of feeding observations, spring and early summer, 73% on ground, 27% in trees taking catkins and leaf and flower buds; moves into trees and bushes in June, then in late July often in grass and herbs, including reeds *Phragmites* where probably takes invertebrates (Stjernberg 1979). In Lahti area of southern Finland, where less arable land than in study area above, 90% of foraging in May and early June done in bushes (Peiponen 1974). Removes outer scales from bud, eating only soft nuclei; nibbles pieces from leaves and takes fresh conifer needles; when feeding on parts of flowers, either eats them on the spot or flies off with them to perch; larger seeds de-husked, and grass or cereal grains generally preferred unripe and milky; seeds of berries eaten and pulp usually discarded; skin sometimes left hanging *in situ* following extraction of seeds; will hang head-down when feeding on (e.g.) catkins. When feeding on ground, pulls seed-heads down while standing on ground, or snips through stem; also perches on stem to bend it over, or reaches over from neighbouring twig, etc.; makes hole in side of Compositae heads and extracts seeds in bundle, biting off pappi. (Czikeli and Busch 1974; Peiponen 1974; Stjernberg 1979; Bozhko 1980; Mauersberger 1980; Mauersberger *et al.* 1982; Brown 1985; Barth and Moritz 1988.) Takes 3–4 s to consume elm *Ulmus* seeds in trees, and *c.* 7 s to deal with caterpillar; de-husks seeds in 1–3 s (Stjernberg 1979). Eats *c.* 15 buds in 5–10 min (Peiponen 1974). For discussion of bill-shape in relation to diet, and comparison with (e.g.) Bullfinch *Pyrrhula pyrrhula* and Pine Grosbeak *Pinicola enucleator*, see Ziswiler (1965), Peiponen (1974), and Stjernberg (1979).

Diet in west Palearctic includes the following. Invertebrates: dragonflies and damsel flies (Odonata), bugs (Hemiptera: Cercopidae, Psyllidae, Aphidoidea, Coccoidea), adult and larval Lepidoptera (Tortricidae, Yponomeutidae, Geometridae), caddis flies (Trichoptera), flies (Diptera: Syrphidae, Muscidae), beetles (Coleoptera: Elateridae, Lathridiidae, Chrysomelidae, Curculionidae), spiders (Araneae), mites (Acari). Plants: seeds, buds, leaves, shoots (etc.) of juniper *Juniperus*, larch *Larix*, spruce *Picea*, pine *Pinus*, willow *Salix*, aspen, etc. *Populus*, birch *Betula*, alder *Alnus*, oak *Quercus*, elm *Ulmus*, maple, etc. *Acer*, lime *Tilia*, lilac *Syringa*, mulberry *Morus*, hemp *Cannabis*, knotgrass *Polygonum*, dock *Rumex*, buckwheat *Fagopyrum*, chickweed *Stellaria*, pearlwort *Sagina*, campion *Silene*, buttercup, etc. *Ranunculus*, radish *Raphanus*, currant *Ribes*, bramble, etc. *Rubus*, apple *Malus*, pear *Pyrus*, cherry, etc. *Prunus*, rowan *Sorbus*, *Amelanchier*, meadowsweet *Filipendula*, lucerne *Medicago*, vetchling *Lathyrus*, cranesbill *Geranium*, buckthorn *Rhamnus*, alder buckthorn *Frangula*, spurge laurel *Daphne*, willowherb *Epilobium*, bilberry *Vaccinium*, cow parsley *Anthriscus*, comfrey *Symphytum*, speedwell *Veronica*, plantain *Plantago*, elder *Sambucus*, guelder rose *Viburnum*, snowberry *Symphoricarpos*, honeysuckle *Lonicera*, various Compositae, wood-rush *Luzula*, sedge *Carex*, grasses (Gramineae, including cereals). (Haas 1939; Turček 1961; Ptushenko and Inozemtsev 1968; Müller 1970; Risberg 1970; Bozkho 1974, 1980; Czikeli and Busch 1974; Peiponen 1974; Stjernberg 1979.)

In southern Finland, of 381 feeding observations, late May and early June, 42% on cereal grains, 20% alder buds, 17% willow seeds and catkins, 8% cherry buds, 5% rowan buds, 4% *Ribes* buds, remainder other buds and seeds; some aphids also taken (Stjernberg 1979, which see for many details); in Lahti area, of 305 items by observation involving *c.* 400 birds, 28% (by number) alder buds, 21% grass seeds, 10% willow buds and catkins, 8% *Populus* buds and catkins, 8% rowan buds, 8% aphids, 5% dandelion *Taraxacum* seeds, 4% cherry buds, 2% nectar of elder flowers, remainder buds and seeds. In May, 75% of diet buds, 13% seeds, 10% catkins, 2% insects; 1st half of June, 36% buds, 40% seeds, 23% insects; 2nd half of June, 7%, 79%, 14%; July, 0%, 98%, 2%; insects almost all aphids; allowing for abundance, prefers alder leaf buds (47% of diet, 17% of trees) and avoids birch; in June, seeds chiefly dandelion and grasses *Melica* and *Milium*; these grasses comprise 87% of seeds eaten in July, very high compared to their abundance; insects may be under-represented because data collected by observation. Wild birds in captivity preferred catkins of *Populus* and willow, leaf buds of alder, grass seeds, and hemp seeds. (Peiponen 1974.) For diet in captivity, see also Giebing (1992). In Russia, spring arrival coincides with peak availability of buds and flowers on trees and bushes; takes ripening seeds of elm and willow also, though in north flowers and buds are main food as seeds emerge later; in general, diet according to abundance; commonly feeds on ripening cereal for milky grain (Bozkho 1980). In Moscow region, 5 summer stomachs contained 2 Curculionidae, 1 Odonata, 5 other insects, plus 24 weeds (8 Cruciferae, 16 others); in autumn, diet included fruits of juniper, alder buckthorn, cherry, honeysuckle, and guelder rose (Ptushenko and Inozemtsev 1968). In Uzbekistan and Kazakhstan, can cause considerable damage in cherry orchards, and elm, willow, and *Populus* trees can be stripped bare; in autumn, mulberries and plums taken in large quantities (Salikhbaev and Bogdanov 1967; Korelov *et al.* 1974). In European Russia in autumn, eats honeysuckle, snowberries, juniper, cherries, alder buckthorn; can also damage cereal crops at this time, though more by breaking stems than eating grain. Numerous observations of birds consuming salt and salty minerals, probably because of dietary deficiency; mortar from walls, urine-soaked earth around stables, etc., and flocks recorded at fish-salting works. (Bozkho 1980.) Diet in winter almost wholly plant

material, similar to that eaten at other times, e.g. buds, mulberries, cereal flowers and grain (Roberts 1992).

Young fed on seeds and invertebrates; delivered by adults as pulp from gullet (not gular pouch), collected up to 2 km from nest. In southern Finland, of 172 nestling gullets, 18% (examined through skin of gullet) contained caterpillars (13% Tortricidae, 3% Yponomeutidae, 2% Geometridae), 3% beetles, 25% unidentified insects. Of 30 collar-samples, 12 contained Hemiptera (aphids and Psyllidae), beetles (Chrysomelidae, Lathridiidae), and mites. 44% of nestlings contained invertebrates (36% mid-June, 61% early July) and 97% plant material. Of 195 gullets, 56% (by examination through skin) contained Ranunculaceae seeds (5% at 1–3 days old, 36% at 4–6 days, 15% at 7–9 days), 50% Compositae (2%, 31%, 17%), and 24% elm (3%, 9%, 12%); 42% held larvae (4%, 27%, 11%); by volume, plant material far more important than invertebrates, particularly elm, buttercup, dandelion, and sow-thistle *Sonchus*. (Stjernberg 1979.) In St Petersburg parks, 61 collar-samples from birds 2–12 days old contained 118 invertebrates, of which 70% (by number) Hemiptera (almost all Cercopidae), 14% larval and pupal Lepidoptera (11% Tortricidae), 6% larval beetles (5% Curculionidae), 3% larval Diptera, and 3% spiders; 64% of samples contained grass *Setaria* seeds, 28% *Rumex*, 23% grass *Glyceria*, plus other grasses, Polygonaceae, Caryophyllaceae, Cruciferae, and Compositae. Likely that most invertebrates picked up accidentally with plant material, though some feeds almost wholly animal prey; no great change in diet between days 2 and 12. (Bozkho 1974, 1980.) In Moscow region, nestlings received unripe seeds of chickweed, Ranunculaceae, Leguminosae (including lucerne and vetchling), Umbelliferae, comfrey, goatsbeard *Tragopogon*, sedges, and grasses; also (in small numbers) Hemiptera (Psyllidae and aphids), adult and larval Lepidoptera, adult and larval beetles (including Curculionidae), and spiders (Ptushenko and Inozemtsev 1968). Regurgitated pulp in 1 delivery enough for 15 feeds (Steinfatt 1937a). For diet in captivity, see Giebing (1992). BH

**Social pattern and behaviour.** Well known. Major study of nominate *erythrinus* at Kristinestad (western Finland) by Stjernberg (1979); some aspects investigated by Björklund (1989a, b, 1990a) at Rättvik (central Sweden); review by Bozkho (1980) includes original data for St Petersburg (north-west Russia). *C. e. ferghanensis* studied in Tien Shan mountains by Kovshar' (1979) and Iovchenko (1986).

1. Large flocks reported in winter (Witherby *et al.* 1938); e.g. up to 150 in Andhra Pradesh (India) (Price 1979). In spring, migrates singly (especially red-plumaged ♂♂) or in mostly small flocks (e.g. Peiponen 1974, Stjernberg 1985), though up to 50 recorded (Sushkin 1938; Dementiev and Gladkov 1954) and *c.* 200 in Mongolia, where small flocks often uniform, i.e. comprising only red-plumaged ♂♂ or brown-plumaged birds (Mauersberger 1980). Arriving first on breeding grounds, ♂♂ form small flocks for feeding and are joined later by ♀♀ (Stjernberg 1979; Levin and Gubin 1985); birds sometimes flock

when foraging in breeding season (Korelov *et al.* 1974). Family parties and flocks (up to 20) occur in post-breeding period and during autumn migration (Boehme 1958; Kovshar' 1966; Peiponen 1974; Bozhko 1980). Sometimes associates loosely with other species (e.g. other Fringillidae) when feeding (Zimmermann 1913). BONDS. Generally monogamous (Stjernberg 1979; Björklund 1989a), but some polygyny (e.g. Dementiev and Gladkov 1954) and, with reports of additional ♂♂ associating with pairs (see below), mating system perhaps more complex than previously assumed (Müller and Wernicke 1990). Suspected case of polygyny (2 nests *c.* 460 m apart) at Kristinestad (Stjernberg 1979), another in eastern Germany (Rost 1992), and one definite in Swedish study (Björklund 1990a). For 2 nests exceptionally close together, see Breeding Dispersion (below). Polygyny perhaps facilitated by tendency of ♂ to wander from nest during incubation (perhaps in order to attract another ♀) and to invade others' territories, though such visits commonest during nestling period, and more likely to involve unpaired immature ♂♂: see below) (Stjernberg 1979; Björklund 1990a). At Rättvik, paired ♀♀ courted by other ♂♂, but no extra-pair copulation recorded (see Heterosexual Behaviour, below, for mate-guarding), and monogamy apparently the rule; after moving away during incubation, all ♂♂ later visited their nests and helped to feed young (Björklund 1990a). Sex-ratio may be skewed towards ♂♂, especially before breeding, and excess of 1-year-old ♂♂ typical especially of peripheral populations (Bozhko 1980). At Rättvik, 1♂: 0·71 ♀♀; perhaps due to predation on incubating ♀♀ (Björklund 1989a). Several reports of trios, additional (usually 1-year) ♂ associating with pair and visiting their nest (e.g. Sick 1938, Bozhko 1980). May be driven away by red-plumaged territory-owner (Thienemann 1903), but such aggression not the rule; many supernumerary ♂♂ move away and attempt to pair up (Bozhko 1980). In St Petersburg, at each of 2 nests, 2-3 additional ♂♂ were trapped and, at each of a further 3 nests, 3-4, such birds helping to limited extent with feeding of young (Nankinov 1974). At Anklam (eastern Germany), 1-year-old ♂♂ replaced (no details) red ♂♂ in 3 out of 5 territories and one (status unknown) seen to feed young (Müller and Wernicke 1990); see also Breeding Dispersion (below). In Talasskiy Alatau (Tien Shan), when pair feeding fledglings, ♀ at times accompanied by red ♂ rather than her own brown mate (Kovshar' 1979); not clear whether re-pairing took place. No reports of hybridization, but bird wintering in south-west England associated and even attempted to copulate with ♀ Chaffinch *Fringilla coelebs* (Stevens 1945). ♂♂ tend to reach breeding grounds before ♀♀ arrive (Bozhko 1980); see also Heterosexual Behaviour (below). Swedish study found marked competition between ♂♂ for ♀♀ (see above for sex-ratio), but mate choice essentially neutral with regard to ♂ characteristics, and not important for reproductive success; all 105 ♂♂ recorded over 4 years became paired and no unpaired ♀♀ observed (Björklund 1990a). Study in Ural valley (north-west Kazakhstan) found pair-bond to persist for replacement after nesting failure, but not between years: of 8 marked pairs, 4 ♂♂ and 1 ♀ returned, but all re-paired (Levin and Gubin 1985), as did both members of pair returning to Kariniemi, Finland (Peiponen 1974). Similarly, at Kristinestad, 36 cases in which returning birds took new mates, but 6 (14%) where bond maintained at least 2 years (see Breeding Dispersion, below, for site-fidelity). One pair nested together in 4 successive years: in year 1 and 2, bred only after 1st clutch deserted; in 3rd year, ♀ re-paired and laid 1st clutch, even helping new ♂ to drive off her former mate, but then deserted nest and re-paired with 1st partner with whom also bred in following (4th) year. (Stjernberg 1979.) Role of sexes in nest-duties as follows. Nest-building

and incubation by ♀ alone (Steinfatt 1937a; Risberg 1970). In northern Poland, ♂ once recorded on 4 young and 1 egg (Haas 1939), and brown-plumaged birds singing when flushed from nest presumed by Lüttschwager (1926) and Kuusisto (1927) to be further evidence for incubation by ♂; see, however, Voice. Young fed by both sexes, but reports vary regarding share taken by each sex: most food brought by ♂ early on, but at equal rate by both sexes from 6-7 days (Kovshar' 1979; Stjernberg 1979); on Usedom (northern Germany), ♂ still brought all food at *c.* 7 days (Kraatz 1979); in contrast, ♂ observed by Steinfatt (1937a) did very little feeding during first days, but up to later. For helpers at nest, see above. Fledglings are tended by both parents. At Kristinestad, mainly by ♀, typically staying longer with brood than ♂ (Stjernberg 1979). In Kirgizskiy Alatau (Tien Shan), however, no confirmation of reports that late fledglings tended by ♀ alone after departure of ♂. In all cases where laying started after 15 July, nestlings and fledglings were fed by both sexes, and broods with ♂ and ♀ in attendance were recorded after mid-August (Iovchenko 1986). In Moscow region (Russia), young feed themselves to some extent from *c.* 6-7 days after fledging, but are fed by parents for further *c.* 10-15 days (Ptushenko and Inozemtsev 1968). Tien Shan studies suggested young independent at 18 days (Kovshar' 1979) or 24-27 days old (Iovchenko 1986). Some ♂♂ breed at 1 year (e.g. Thienemann 1903, Sick 1938); still not determined at what age first breed (Stjernberg 1979). At Kristinestad, *c.* 10% of all breeding ♂♂ were yearlings and *c.* ⅓ of surviving ♂♂ first bred at 1 year (Stjernberg 1979). Similarly, at Rättvik, small floating population (*c.* 10% per year) of 1-year-olds; of 2-4 in this age-class advertising in study area, all paired up successfully; in 3 cases, bred locally, rest left with mate (Björklund 1989a, b, 1990a); see Breeding Dispersion (below). Tien Shan study suggested breeding by 1-year-old ♂♂ dependent on number of ♂♂ 2 years or older, sex-ratio in population, density, and availability of suitable nest-sites. Of 87 pairs in which age of ♂ known, 72 had red ♂ (2 years and older), 14 1-year ♂, one with aberrantly-plumaged ♂, giving ratio of 5 red ♂♂:1 brown ♂. Most 1-year ♂♂ probably breed, but later in season. (Iovchenko 1986.) At Kariniemi, in population of 7-10 pairs over 8 years, most paired ♂♂ similarly over 2 years old, ratio of old to young ♂♂ in pairs 6:1 (Peiponen 1974); in another Tien Shan study 4:1 (Kovshar' 1979). See also Zimin (1981) and, for further details of population composition and dynamics, Bozhko (1980). BREEDING DISPERSION. Territorial. Apparently some tendency to form neighbourhood groups, but Finnish study found this difficult to gauge owing to patchy habitat distribution, though 39 cases out of 332 (12%) where 3 or more pairs bred within circle of *c.* 3 ha. In 20 cases (6%), distance between 2 simultaneously-occupied nests less than 25 m, in 73 cases (22%) less than 50 m; shortest 2-5 m. (Stjernberg 1979.) In St Petersburg, 3 nests (no interference, all successful) separated by 3 and 15 m (Bozhko 1974), and in Talasskiy Alatau (Tien Shan), where nests usually 50-60 m apart, one case of broods fledging same day from nests 15 m apart (Kovshar' 1979). Further reports of nests 35-150 m apart northern Poland and northern Germany (Haas 1939; Kraatz 1979), 400-700 m central European Russia (Ptushenko and Inozemtsev 1968) and Helgoland, Germany (Barth and Moritz 1988); in Ural valley, 8-40 m between pairs (Levin and Gubin 1985). 2 cases (in Sweden) of 2 nests close together on same branch (Risberg 1968, 1970); significance not clear. Map in Delany *et al.* (1982) shows 23 territories in Nepal mostly contiguous or nearly so (see also discussion of densities, below). ♂ arrives first, e.g. at Kristinestad, average 4·7 days (2-8, *n*=8) before ♀, and, having split away from feeding flock, may then move about and sing from various perches (Stjernberg 1979), in area larger than eventual territory (Risberg 1970; see also below). ♂ eventually followed by ♀ from feeding grounds and as he then flies between song-posts presumably demonstrating (potential) territory; see also subsection 5 in Heterosexual Behaviour (below). ♂ also reacts strongly to song outside territory. (Stjernberg 1979.) Study at Rättvik found only song-post used at a given time is defended before pairing and nest-building, so nesting area often unrelated to song-perches of unpaired ♂; *c.* 50% of pairs leave area each year after pairing and breed elsewhere (Björklund 1989*a*, 1990*a*). In Tien Shan study, some local birds are among first to arrive; soon pair up and pair may then feed within very small area over days, 23 out of 24 birds ringed last third of May breeding within 10–50 m of ringing site; behaviour in early period much as later when breeding (i.e. birds apparently territorial from start) (Iovchenko 1986). Passage birds may be temporarily dispersed as if breeding, this (through disputes that arise from intrusions into occupied territories) affecting general dispersion and territory size (see below); later encroachments are by ♂♂ of local population (Risberg 1970; Bozhko 1980). Immature ♂♂ arriving after older birds have established territories may be forced to occupy (apparently) less favourable peripheral sites (Peiponen 1974). However, at Anklam, 3 out of 5 red-plumaged ♂♂ setting up territories late in season were replaced by 1-year-olds (Müller and Wernicke 1990). ♂ more strictly tied to territory only at start of breeding cycle and during nestling phase (Stjernberg 1979). In Rominter Heide (Poland/Russia), ♂ sang mostly within *c.* 10 m of nest (Steinfatt 1937*a*). However, more recent studies have shown nest not necessarily close to regular song-post (Björklund 1990*a*), and ♂ tends to wander far during incubation (e.g. song noted up to *c.* 3 km from nest) and make only short visits to territory (Stjernberg 1979; Björklund 1990*a*). Territory used mainly for courtship, copulation and nesting; also some foraging, but much of this done outside, with excursions (perhaps longer by ♂) 0·5–700 m (*n* = 20) or up to (1–)2 km away (Stjernberg 1979); also Steinfatt (1937*a*), Haas (1939), Bozhko (1980), Göttgens *et al.* (1985), and Levin and Gubin (1985). ♂♂ regularly overfly others' territories to reach feeding grounds (Kraatz 1979). Territory defended by both sexes, though apparently more by ♂, ♀ being more active in defence of nest (Haas 1939); see however, Antagonistic Behaviour, also Parental Anti-predator Strategies (below). Most fights in Finnish study took place near ♂'s song-post in territory (Stjernberg 1979). Defence maintained even when other conspecific birds are attracted into territory during disturbance; territory abandoned gradually after fledging (Haas 1939). In Tien Shan, ♂ tending fledglings seen to chase another ♂ when it sang within *c.* 50-60 m (Kovshar' 1979). At Kristinestad, following nest failure, 13 out of 15 birds re-nested within 500 m; 1 ♀ (probably still with same mate) started, but did not complete, new nest *c.* 2 km away. Territory perhaps larger than average for other Carduelinae (Stjernberg 1979); not clear whether any contraction once breeding under way. One Finnish territory *c.* 1500 m² (Stjernberg 1979), 1600 m² Ural valley (Levin and Gubin 1985), average 3000 m² (*n* = 7) in northern Poland (Haas 1939), 1600–3600 m² (*n* = 10) on Öland, Sweden (Rodebrand 1975); at Hofors (Sweden), 11050 m² (*n* = 11) (Risberg 1970). Contiguous territories *c.* 200-250 m across (*c.* 3·1–4·9 ha) reported from Yamal peninsula (northern Russia) (Danilov *et al.* 1984) and north-east Russia (Kishchinski 1980). Not clear, however, how much defended and larger figures in particular likely to include feeding area (Bozhko 1980). High densities recorded in taiga in Asiatic part of range, in some montane populations, and locally forest and other habitats further west in range where expansion still in progress (see Distribution) (Bozhko 1980, which see for summary and further references).

In fire-affected pine *Pinus* forest with some birch *Betula*, central Siberia, 200 pairs per km² (Reymers 1966); see also (e.g.) Rogacheva (1988). In Kirgizskiy Alatau, up to 270 pairs per km² (Iovchenko 1986). 23 territories in *c.* 25 ha (92 per km²) recorded in Ladakh (Delany *et al.* 1982). Increase in density with altitude in Caucasus: 0·25–2 pairs and higher up locally 57 birds, per km² (Zhordania and Gogilashvili 1976), though at *c.* 3800 m in Pamirs only 6-7 singing ♂♂ along 3-km stretch (Potapov 1966). Water meadows in Estonia held 300 pairs per km² (Sits 1937). Other data as follows (pairs per km² unless stated otherwise). On north-east coast of Germany, locally 20 ♂♂ (16 paired) along 5 km of hedges (Klafs and Stübs 1987). In northern Poland, up to 9 ♂♂ along 200 m of elder *Sambucus* scrub; in 7·8 ha, 19 pairs per km² or (only suitable nesting habitat considered) 77 (Haas 1939); inland in Poland, 10 in ash *Fraxinus*-alder *Alnus* forest, Białowieża (Tomiałojć *et al.* 1984). On Öland (Sweden), in different years, 13 territories or up to 50 pairs in 17·8 ha of deciduous forest (Fritz 1989, 1991). At Kristinestad (Finland), birds used *c.* 35% of 7 km² (town and surrounding farmland); average over 8 years (based on nests found) 3·4 in total area (8·8 in favoured habitat of coastal woods), 9·8 (21·4) in area used (Stjernberg 1979). Other data from Finland include 120 in Kumo river delta (Haukioja 1968), 28-40 over 8 years in natural park at Kariniemi (Peiponen 1974), and at Siikalahti lake, 17 in undrained area, 12–13 where drained, *C. erythrinus* eventually disappearing (Koskimies 1985). See also Järvinen and Väisänen (1976). In north-west Russia, up to 32·5 in pine plantations, Kareliya (Bozhko 1980) and, in Pskov region, 12·8 in semi-mature forest, 4 in secondary growth on rides, etc., 3·6 in village (Uryadova 1986). 4-5 in Moscow region, 8-9 along Oka river, Ryazan' (Ptushenko and Inozemtsev 1968). Average over 8 years in northern Urals (Russia) 9·9 ± 1·2 (Shutov 1990). Site-fidelity marked, especially in birds that have bred once (Stjernberg 1985). Study at Kristinestad found brown ♂♂ show strong attachment to site where summered or bred in year after hatching. Of older birds, *c.* 80% of returning ♂♂, *c.* 60% of ♀♀ bred less than 500 m from their previous year's nest. Average distance between nests of same individuals in successive years 320 m for ♂♂ (*n* = 23, longest 1800 m), 570 m (significantly longer) for ♀♀ (*n* = 37, longest 3800 m). One ♂ bred within circle of radius 85 m in 6 successive years. (Stjernberg 1979.) At Rättvik, 12 out of 26 ♂♂ ringed in 1985 returned in 1986, 12 out of 30 in 1987, 14 of 31 in 1988; ♂♂ with failed nests returned as often as successful ones (Björklund 1989*b*). Of 36 birds ringed as nestlings and returning to natal area, 34 did so within 5 km of ringing site (average 2 km, *n* = 13), but up to 100 km recorded (Stjernberg 1979). Of 30 ringed as nestlings in Tien Shan, only 1 returned (Kovshar' 1979). For further details, see Peiponen (1974) and Levin and Gubin (1985). ROOSTING. Solitary or communal; nocturnal. In India, in winter, communal (5–10 or up to several tens); sometimes in sugar cane *Saccharum* and mixed with many other species (Gadgil and Ali 1975; Dhindsa and Toor 1981). In Rominter Heide, ♀ roosted on nest from 3rd egg, ♂ accompanying her back, then roosting in nearby tree (Steinfatt 1937*a*). ♀ in another study apparently roosted on nest-rim when young *c.* 12 days old; young perhaps return to nest to roost after fledging (Kraatz 1979). At Markakol' lake (Kazakhstan), large numbers came to eat salt June–July, mainly 05.00–10.00 hrs, fewer around midday and evening (Gavrilov 1968).

2. Generally confiding when breeding (e.g. Steinfatt 1937*a*, Lassey and Wallace 1992), but little information for other times, though migrant flock in Switzerland rather shy and wary (Maurizio 1987). Birds quiet and secretive in post-breeding period, in marked contrast to behaviour (♂♂ especially) before breeding (Thienemann 1903). Indications of excitement (at least in ♂♂)

A

are adoption of sleeked upright-posture, ruffling of crown (Fig A), and spreading or flicking of tail, sometimes while giving alarm-call (see 2 in Voice) (Thienemann 1903; Haas 1939; Göttgens *et al.* 1985). In Sweden, ♂'s only reaction to nearby Carrion Crow *Corvus corone* was to stop singing (Risberg 1970). FLOCK BEHAVIOUR. Winter flocks in India quiet and unobtrusive, feeding on ground, but flying into trees when disturbed (Whistler 1941). ♂♂ feeding together before breeding intermittently fly to shrubs and trees where sing, preen, and eat buds (Stjernberg 1979). SONG-DISPLAY. Whistling-song (see 1a in Voice) given by ♂ mainly for mate attraction according to Björklund (1989*c*), though territorial function suggested by Diesselhorst (1968) and Stjernberg (1979); see Antagonistic Behaviour (below). ♂ also sings vigorously outside territory, and intensely before copulation (Haas 1939; Stjernberg 1979). Study in eastern Germany found little song where density low (pairs isolated) (Dittberner *et al.* 1979). Song-posts used, sometimes over several days, hidden or exposed (Thienemann 1903; Sits 1937); tree-top, bush or overhead wire, at less than 1 m up to *c.* 10 m (Steinfatt 1937*a*; Czikeli and Busch 1974; Kraatz 1979; Björklund 1990*a*). Singing bird (Fig B) has head pointed up at angle, flanks, crown and

B

throat ruffled (see also Antagonistic Behaviour, below), body upright, and bill wide open. After singing a little, usually droops wings to expose (ruffled) rump. Excited bird may flap or flick wings, also move tail up and down and side to side. Quieter twittering (Subsong or similar: see 1b in Voice) given without raising head and with bill closed. (Haas 1939; Reinsch 1965; Czikeli and Busch 1974; Bergmann and Helb 1982; Giebing 1992.) Known to sing in flight, but apparently rare. Reported by Newton (1972) to flutter from one bush-top to another while singing or to fly erratically through bushes. On Helgoland, several Song-flights observed were all just before start of nest-

building; ♂ had head laid back and flew with bat-like wing-beats up to *c.* 30–40 m only between song-posts, usually giving loud, continuous Whistling-song, though 1-year-old ♂ silent in one flight and gave only quiet song in another (H Göttgens). In Poland, during nest-building, ♂ performed Song-flight similar to Greenfinch *Carduelis chloris* or Serin *Serinus serinus* (and presumably thus over wide area) giving loud Whistling-song in flight, twittering from perch after landing (Haas 1939); in Kaliningrad (western Russia), singing noted from bird leaving nest, exceptionally while actively migrating (Sick 1938). No Song-flights, even where nesting in open habitat, noted by Mal'chevski (1959); only 2 records in Tien Shan study (Kovshar' 1979). Will sing at any time of day, in bouts of 1–15 min (or longer): e.g. in Rominter Heide, 8 June (during laying), 03.20–17.19 hrs (more in morning) (Steinfatt 1937*a*; Haas 1939); earliest in south and central Finland 1 hr before sunrise 10 July, latest ½ hr after sunset 5 June (Reinikainen 1939); in Tien Shan, during May–June peak, 04.22–20.30 hrs (mainly 05.00–09.00 or 10.00 hrs and evening after 19.00 hrs), with maximum rate 203 songs in 15 min (Kovshar' 1979); in Ural valley, peak 168 in 15 min (Levin and Gubin 1985). Sings in winter quarters before departure for breeding grounds (Witherby *et al.* 1938), on spring migration (e.g. Sits 1937, Sick 1938, Moritz 1982) and, with varying intensity, more or less from arrival and throughout breeding season, waning in July, but noted up to mid-August, or September (Sick 1938; Reinikainen 1939; Mal'chevski 1959; Bozhko 1974, 1980; Stjernberg 1979; Levin and Gubin 1985; Giebing 1992). Study by Björklund (1990*a*) found singing rates greatly reduced once paired, but all ♂♂ (including bigamist) started to sing again during laying. Peak period in northern Poland coincided with incubation (Haas 1939); rapid decline during nestling period (Steinfatt 1937*a*). In Tien Shan, over 5 years, song-period 83–96 days (Kovshar' 1979). ANTAGONISTIC BEHAVIOUR. (1) General. In Ural valley study, disputes between ♂♂ regular during territory occupation and pair-formation, rare later during nest-building (Levin and Gubin 1985). In defence of territory, uses song (Song-duels quite common early in season), threat-postures, chases, supplanting attacks and aerial fights (Stjernberg 1979). At Rättvik, disputes between 2 unpaired ♂♂ usually short displacements, smaller ♂ always withdrawing; 56% of contests between 2 paired ♂♂ escalated into fight, either bird yielding (whether bigger or smaller than opponent). Most fights between paired and unpaired ♂♂ escalated (60% where unpaired ♂ smaller, 75% where larger); unpaired ♂ withdrew in 78·3% of 23 such contests, suggesting pairing status more important than size in settling contest. (Björklund 1989*a*.) (2) Threat and fighting. Territory-owner may adopt sleeked upright-posture on sighting strange ♂ (Haas 1939). In ♂–♂ confrontations, one or both birds adopt Threat-posture (throat, breast and belly ruffled, in territory-owner often more so) in which breast often lowered against perch and wings drooped and vibrated. Intruder may be subjected to series of supplanting attacks, fleeing (typically with body and crown sleeked) to higher perch, etc., often leads to chase (noted on feeding grounds and throughout breeding season) and fight. (Stjernberg 1979.) During threat-display, one ♂ may sing usually with head ruffled, wings slightly drooped and tail fanned (Haas 1939); call 4 (see Voice) also given while flapping wings (Reinsch 1965). Song-duels occur, birds typically giving longer songs than usual (Stjernberg 1979); chase may ensue (Scheer 1953). In northern Poland, intruders usually chased to edge of territory, territory-owner singing during pursuit; ♀ also seen to chase off strange ♂ attracted by her calls during disturbance (Haas 1939). In Finnish study, chases sometimes over many hundred metres, and participation of ♀ recorded in 56% of 68 such actions to evict intruder (Stjernberg 1979). In aerial fights, 2 ♂♂ (no participation by ♀) may flutter up *c.* 30 m beating wings violently and apparently pecking and clawing at each other; may become locked together, fluttering to ground and continuing fight there (Stjernberg 1979; Björklund 1989*a*). 2 ♀♀ seen collecting nest material in same shrub without aggression (Levin and Gubin 1985). In Sweden, 2 pairs similarly recorded foraging amicably in same large tree, but 2 ♂♂ rarely sang simultaneously in same tree; unpaired ♂♂ commonly attack one another, regularly flying several hundred metres to do so, also react strongly to playback (Björklund 1989*a*, 1990*a*). Other species attacked include White-throat *Sylvia communis* (Risberg 1970), Bluethroats *Luscinia svecica* coming close to nest under construction (Levin and Gubin 1985) and, following playback of territory-owner's song, once each Willow Warbler *Phylloscopus trochilus*, House Sparrow *Passer domesticus*, and Ortolan Bunting *Emberiza hortulana* (Stjernberg 1979). HETEROSEXUAL BEHAVIOUR. (1) General. At Rättvik, ♂♂ arrive 3-4 days before ♀♀ and pairing apparently rapid (within 1-2 days) (Björklund 1990*a*). Most ♀♀ arrive in Kareliya 3-4 weeks after first ♂♂, thus just before onset of breeding (Bozhko 1980). Pair-formation in Talasskiy Alatau mid-May to early June (Kovshar' 1979); in Kirgizskiy Alatau, *c.* 1 month between arrival and start of breeding (Iovchenko 1986). Exact process of pair-formation not known; courtship probably mainly associated with copulation (Stjernberg 1979). (2) Pair-bonding behaviour. Pair-formation presumed to take place over several days. In early stages, each may attack the other suddenly, though this stops within a few days; later in season (when ♀ dominant), ♀ may make mock-attacks on mate. (Stjernberg 1979.) 2-3 ♂♂ recorded attempting to court same ♀ (Kovshar' 1979). Courtship observed on ground, rock, fallen tree or branch, close to nest just completed, around start of laying. Birds usually close together (often within a few cm). ♂ (Fig C, right) assumes

*et al.* 1985.) ♂ seen to make deep bowing movements, then throw head back and give rapid series of song-phrases (Göttgens *et al.* 1985). Standing in front of ♀, ♂ may also perform Pivoting-display: slowly swings body side to side, while vibrating drooped wings (Stjernberg 1979: Fig E); in another description, ♂ moved

E

only neck, also gave Subsong and, at peak excitement, quiet Whistling-song (Mal'chevski 1959; see also Levin and Gubin 1985). 2 records (one of captive bird displaying to other species) of ♂ picking up nest material during courtship (Reinsch 1965; Göttgens *et al.* 1985). Billing noted (Thienemann 1903). ♂ also recorded flying to ♀ and pecking rapidly at her bill and head, then moving in Moth-flight (see subsection 4, below) into vegetation; perhaps initial stages of pre-copulatory behaviour, though seen mainly when pair disturbed during nest-building (Stjernberg 1979). Once paired, ♂ guards mate while she is feeding, etc.; sings, including if loses sight of ♀ (Iovchenko 1986; Björklund 1990*a*). For further descriptions, see Bozhko (1980). (3) Courtship-feeding. ♂ feeds ♀ on or close to nest (Figs F-G).

C

Head-up posture: head (ruffled, especially crown) and tail raised, wings drooped (at times brushing ground) and vibrated; patters on spot or slowly circles (sometimes apparently indifferent) ♀ (Fig C, left); at times, birds circle simultaneously, ♂ in centre, and both may adopt extreme variant of Head-up posture (Fig D). (Mal'chevski 1959; Risberg 1970; Stjernberg 1979; Göttgens

D

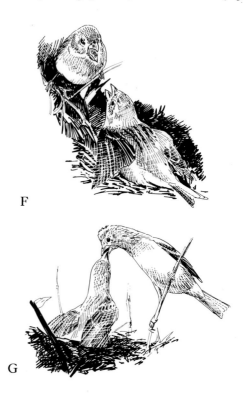

F

G

Not recorded during early pairing stages, but occurs at low rate (e.g. once per hr, 3 times during 4 hrs of morning) during incubation, also after hatching (Steinfatt 1937a; Björklund 1990a; Müller and Wernicke 1990). ♂ may sing while approaching nest, causing ♀ to start wing-shivering (Kraatz 1979). Will also try to lure ♂ to nest with begging-calls (see 6 in Voice). ♀ leaving nest to meet ♂ gave contact-call (see 2 in Voice) and pair flew off together after ♂ had fed her; after hatching, ♀ pattered about on branch and gave begging-calls in posture shown in Fig D, but ♂ finally fed her (and young) at nest (Steinfatt 1937a). (4) Mating. Recorded in morning, on ground, in tree or nest-shrub (Göttgens *et al.* 1985). Of 17 cases in Finnish study, 82% were within 1–5 days of laying (May–June); once 7th day of incubation (presumed polygynous ♂). Initiated by either sex, mostly by ♂ without elaborate ceremony, though seen to adopt Head-up posture and wing-shiver, apparently demonstrating breast to ♀; both sexes once faced each other thus. Soliciting-posture of ♀ much as Head-up, though head raised only slightly; ♀ sometimes calls quietly (presumably begging-calls or similar). ♂ seen to perform Butterfly-flight (head and tail raised, slow, deep wing-beats), sometimes descending like Tree Pipit *Anthus trivialis*; copulation ensued if ♀ followed. Also, from Pivoting-display, ♂ ascended briefly with rapid, shallow wing-beats (Moth-flight), then landed and copulated. For report of both sexes apparently in Pivoting-display, see Thienemann (1908). After copulation, ♀ may preen quietly, or continues to solicit, and ♂, sometimes also wing-shivering close by (Thienemann 1903), makes further mountings or leaves in Butterfly-flight. Recorded copulating (successfully) when attracted to nest by ♀'s alarm-calls during disturbance, other ♂♂ similarly attracted also attempting to mount her. (Stjernberg 1979.) (5) Nest-site selection. Final choice made by ♀ after ♂ has demonstrated several potential sites (Ptushenko and Inozemtsev 1968; Risberg 1970; Stjernberg 1979). In early stages (perhaps still courtship and demonstration of territory), ♂ flies about and sings, ♀ following and sometimes carrying straw (♂ once a leaf), but apparently not inspecting sites carefully or starting to build (Stjernberg 1979). ♀ seen to disappear into shrubs briefly or for up to ½ hr, calling (see 8a in Voice) frequently (Göttgens *et al.* 1985). In Ural valley study, ♀ crouched in potential sites and attempted to deposit material, but ♂ led her into shrub in which nest eventually built (Levin and Gubin 1985). Only ♀ builds, ♂ singing quietly near site, and sometimes accompanying her on collecting trips (up to *c.* 50–100 m away), then also singing and displaying (Korelov *et al.* 1974; Kovshar' 1979; Stjernberg 1979; Levin and Gubin 1985). (6) Behaviour at nest. Laying starts within 1–5 days of nest-completion (Ptushenko and Inozemtsev 1968; Kovshar' 1979; Levin and Gubin 1985). May add material after start of laying (Kovshar' 1979; Stjernberg 1979). ♀ typically furtive in approach and departure from nest; sometimes leaves on hearing song of ♂ who may then escort her off nest and back to it; for further details (e.g. incubation stints), see Steinfatt (1937a) and Breeding. RELATIONS WITHIN FAMILY GROUP. Hatching takes place over 2–3 days (Steinfatt 1937a; Haas 1939). Eggshells eaten or carried away and discarded (Giebing 1992). Young brooded by ♀ (once by ♂ at 1–2 days: Haas 1939) more or less continuously for first 2–3 days (Steinfatt 1937a; Levin and Gubin 1985); though recorded at *c.* 7 (still continuous) or *c.* 10–12 days (close to fledging), older young being otherwise only sheltered from rain or sun (Haas 1939; Kraatz 1979). For further details of brooding (stints, etc.), see Steinfatt (1937a), Ptushenko and Inozemtsev (1968), and Kovshar' (1979). Eyes of chicks open to slit by day 2, fully by day 6 (see Steinfatt 1937a, Stjernberg 1979, and Bozhko 1980, including for physical development). Most food brought by ♂ in early stages. Young start begging on

hearing rustle of vegetation, etc. made by arriving parent. Whole brood normally fed each visit and by regurgitation. ♂ recorded feeding young direct (♀ meanwhile begging from nest-rim), transferring food via ♀, sometimes even receiving food back from ♀ and then feeding nestlings. ♀ seen to mock-feed young when ♂ about to fly down, but usually anyway passes some food received to young immediately or during brooding stint (seeds perhaps further digested) following ♂'s visit. (Steinfatt 1937a; Bates 1938; Kraatz 1979; Levin and Gubin 1985.) Nest kept clean by both sexes (more by ♀), nestling faeces being swallowed more or less up to fledging (Steinfatt 1937a; Kraatz 1979), or left to accumulate on nest-rim during last days (Scheer 1953; Bozhko 1980). For fledging period, see Breeding. Often barely able to flutter about on leaving nest, young tend to remain nearby on hidden perch for first few days, calling persistently (Thienemann 1903; Haas 1939; Kovshar' 1979); leave breeding area within 2–3 weeks (Stjernberg 1979). ANTI-PREDATOR RESPONSES OF YOUNG. Young 12 days old crouched when nest (with ♀ perched on rim) was examined in evening (Kraatz 1979). In one study, not shy of man at 8–9 days, tending rather to beg (Steinfatt 1937a), but will vacate nest prematurely at 9–10 days, creeping into cover and freezing, such behaviour being also recorded at 34 days old (Stjernberg 1979). PARENTAL ANTI-PREDATOR STRATEGIES. (1) Passive measures. Some ♀♀ leave nest when man several metres away, making much commotion and frequently deserting; others typically slip away furtively, but most, especially near hatching, sit tightly, leaving only when nest touched or permitting handling (Nyberg 1932; Sits 1937; Bozhko 1974, 1980; Stjernberg 1979). (2) Active measures: general. Reaction tends to be stronger after hatching. Some ♀♀ silent when disturbed, others give persistent and vigorous alarm-calls (see 1 in Voice). (Bozhko 1974; Stjernberg 1979.) (3) Active measures: against birds. Alarm-calls given for Kestrel *Falco tinnunculus*, also (from cover) for Red-backed Shrike *Lanius collurio* (Risberg 1970; Peiponen 1974; Stjernberg 1979); see Breeding. (4) Active measures: against man. Alarm-calls noted in Tien Shan study even before birds had started breeding (Iovchenko 1986). Both sexes may approach closely and give alarm-calls when man at nest (e.g. Björklund 1990b, Lassey and Wallace 1992), but ♀ in particular often bold and aggressive in defence of eggs or young. Recorded pecking when handled, 'falling' from nest in short, fluttering flight (distraction-lure display of disablement-type), bird returning to repeat performance when not successful, flicking wings and tail or hopping about excitedly, ruffling plumage, assuming threat-posture, giving alarm-calls, sometimes also hissing (see 5 in Voice), and launching repeated attacks, concentrating on head or face, and even following short distance when intruder left. (Sits 1937; Steinfatt 1937a; Haas 1939; Mal'chevski 1959; Bozhko 1974, 1980; Kraatz 1979.) (5) Active measures: against other animals. In Altai (Russia), when snake (Viperidae) at nest, ♀ gave alarm-calls within 1–1·5 m (Irisov 1972).

(Figs by D Nurney: A and F from photographs in Müller and Wernicke 1990; B from photograph by R Siebrasse in *Vogelwelt* 101, 1980; C–E from drawings in Stjernberg 1979; G from photograph in Lind 1952.)  MGW

**Voice.** Distinctive loud song of ♂ well known; repertoire of calls apparently small, but not investigated in detail. Contact- and alarm-calls (see 2, below) given frequently after breeding (Kovshar' 1979), but birds generally quiet in winter, though start to sing before departure for breeding grounds (Witherby *et al.* 1938). Major study of song of nominate *erythrinus* in central Sweden found slight microgeographical variation ('song neighbourhoods') rather than large-scale regional dialects (Björklund 1989*c*, which see for sonagrams), though variation between regions not studied (Bergmann 1983*b*). Other races little known, but probably no fundamental differences from nominate *erythrinus* (as confirmed for *roseatus* by Osmaston 1931 and Diesselhorst 1968). For useful early descriptions allied with musical notation, see Stadler (1926*a*) and Nyhlén (1950); for additional sonagrams, see also Stjernberg (1979), Winkler (1979*a*), Bergmann and Helb (1982), and Bergmann (1983*b*).

CALLS OF ADULTS. (1) Song. (1a) Whistling-song of ♂. Far-carrying, sprightly whistling or piping phrase, characteristically clear, short, melodious, and attractive, comprising 3–5(–7) steeply ascending or descending notes given in stereotyped fashion. Presumed species-specific song (noted in recordings from Denmark, Sweden, Finland, and Germany; see also Bergmann 1983*b*, Björklund 1989*c*) usually of 5 portamento-type notes in 0·7–1·6 s, with regular pattern of 1st and 3rd note ascending, other 3 descending in pitch, and the last (typically descending) note often having a short ascending note preceding and attached to it. This basic pattern shown in Figs I–IV (song in Fig II probably given by 1-year-old ♂). In 20 out of the 22 song-phrases illustrated by Björklund (1989*c*) (see below), 1st note also ascends, last descends, and pattern of ascending and descending notes is maintained, usually in alternation, throughout the phrases. Song sometimes likened to song of Golden Oriole *Oriolus oriolus*, but lacks mellow, fluting quality of that species, being much thinner and clearer. Rendered 'WEEje-wü

WEEja', 'tsitsewitsa' (or 'checheveetsa', like Russian name), 'ste-weedye-vyu', etc. (Bozhko 1980; Bergmann and Helb 1982; Bruun *et al.* 1986; Jonsson 1992; J Hall-Craggs.) For further descriptions, see (e.g.) Sick (1931) and Steinfatt (1937*a*). Figs V–IX show how phrase may be modified within skeletal form and ascending-descending pattern; versatility of individual ♂ illustrated in Figs V–VII, bird even taking over from singing Willow Warbler *Phylloscopus trochilus* and matching its structure and temporal pattern. Phrases suggest 'hui tsoooeeee tseeoo hui-teeooo' (Fig VIII) and 'te tseeooo tserrr teeooo' (Fig IX). Song-phrase may have initial ascending note missing or very weak, final descent maintained, but regular alternation may be lost: see (e.g.) Figs IX–XI, phrase in Fig XI suggesting 'te-tsi d' yoo te tsi d' yoo'. In another recording (Fig XII), 1st note a firm descent balanced by (uniquely in recordings analysed here) long ascending final note. (J Hall-Craggs.) Study of songs of 60 ♂♂ (some over 2–3-year period) in Sweden by Björklund (1989*c*) found each ♂ generally to give only one song-type (very little variation) all season, though some (4 in 3 years) occasionally gave another type (compare song-phrases given by Finnish bird in Figs V–VII). Songs comprised average 5·3 ± SD0·6 units and lasted 0·75–1·29 s, depending on type; frequency range 1·90–6·83 kHz (mean 3·81–4·45 kHz). In 1986, 12 song-types were distributed among 31 ♂♂; in 1987, 10 among 28 ♂♂; in 1988, 15 among 28 ♂♂. Similarity between ♂♂ sharing given song-type very close (some differences of detail). Spatial distribution of song-types fairly constant between years. Among 18 new ♂♂ in 1987 (not present in 1986), only 3 sang new,

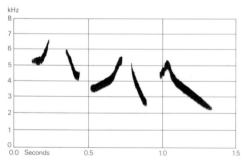

I  P S Hansen  Denmark  June 1971

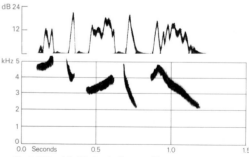

II  J-C Roché  Finland  June 1968

III  P J Sellar  Sweden  May 1971

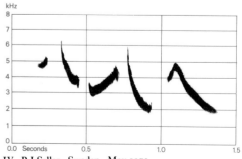

IV  P J Sellar  Sweden  May 1971

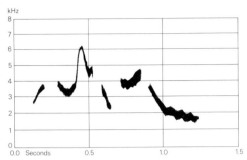

V  P A D Hollom  Finland  June 1988

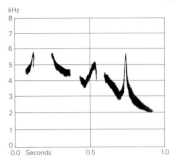

VI  P A D Hollom  Finland  June 1988

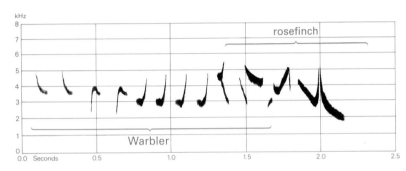

VII  P A D Hollom  Finland  June 1988

VIII  F Jüssi  Estonia  June 1977

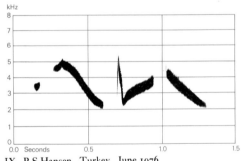

IX  P S Hansen  Turkey  June 1976

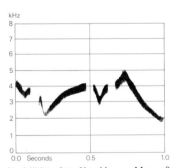

X  Mild (1987)  Kazakhstan  May 1987

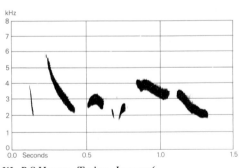

XI  P S Hansen  Turkey  June 1976

independent songs; others matched ♂ already present or gave song given in same area year before (despite absence of 'model'). No obvious differences in song given by 1-year and older ♂♂. Of 10 1-year-old ♂♂, one switched from a foreign to a local song-type, 4 gave variants of other local types, 4 gave perfect copies of dominant song-type where they settled, and one gave versions of song 1b (see below).

(Björklund 1989c.) Similar study in Caucasus based on 363 songs of 27 birds in 5 districts also found each ♂ to give only one song-type; no evidence for dialects (Sultanov and Gumbatova 1986); see also Sultanov (1987a, b). In Swedish study, one record of loud Whistling-song given by ♀ (when disturbed at nest; see also 1b, below), but unlike ♂ in changing song-type between phrases, and none

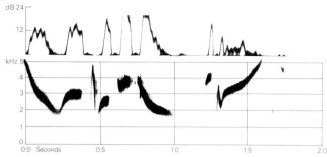

XII P S Hansen Turkey June 1976

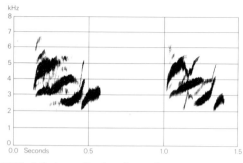

XIII S Carlsson Sweden June 1984

of ♀'s song-types was noted from any ♂ over 3 years (Björklund 1989c); for another report of ♀ giving Whistling-song, see Bozhko (1980). In Austria, ♂ perhaps mimicked end of song of Red-breasted Flycatcher *Ficedula parva* (see Winkler 1979a). (1b) Subsong and similar utterances. Well-spaced, drawn-out notes, some diadic or with harmonics, given in quite long phrase-like structure (Bergmann and Helb 1982). In Swedish study, song similar among all ♂♂: initial units like song 1a, but quieter; gradual diminuendo, ending in very quiet chirping (see below), usually given with bill closed; lasts *c.* 2·5 s (*c.* 1·5 s excluding chirping). Given in song-duels, at first approach of ♀, or as response to human interference (Björklund 1989c; also Stjernberg 1979). See also Sick (1931) and, for song apparently of this kind given by captive bird displaying to other species, Reinsch (1965). Clearly related, if not identical (at least in part) is rapid, quiet, at times squeaky twittering or chirping, less clear than song 1a, but sometimes given as coda to or between loud whistling phrases (Mal'chevski 1959; Ptushenko and Inozemtsev 1968; Bergmann and Helb 1982); said by Dittberner *et al.* (1979) to resemble warbling part of song of Lesser Whitethroat *Sylvia curruca*. Twittering (characteristically including 'hoid') given by most adult ♂♂, but not 1-year-olds according to Haas (1939); in study of *ferghanensis* given only rarely, not regular between Whistling-songs and not to be regarded as integral part of normal advertising song (Kovshar' 1979); twittering also noted from ♀ (Bozhko 1980). (2) Contact- and alarm-calls. Some descriptions suggest 2 different calls exist, others that basically the same call is modified in various ways depending on the context (sexual differences perhaps also present). Sprightly, clear-toned 'ueet', with same tonal quality as in song (Bruun *et al.* 1986), or 'huit' like Chiffchaff *P. collybita*, given when perched or in flight (Göttgens *et al.* 1985); such descriptions suggest single ascending tone such as typically occurs in song, and apparently serving mainly for contact. Calls given by both sexes when disturbed (e.g. near nest) strangled, throaty and ascending in pitch, sounding di- or trisyllabic: 'düi', 'düei', or 'chräi' (Bergmann and Helb 1982). 2 calls of complex structure (diadic, if not triadic) are shown in Fig XIII, the first suggesting rough 'kerrrrweeee' (J Hall-Craggs). Further descriptions:

'JAY-ee' or 'djä-iep' (more emphatic in alarm than as contact-call) reminiscent of call of Greenfinch *Carduelis chloris*, but notably harsher (Steinfatt 1937a; Bruun *et al.* 1986; Jonsson 1992), or Brambling *Fringilla montifringilla* (Mauersberger *et al.* 1982); 'täit' of ♂ said by Thienemann (1903) to be more penetrating, lower pitched and louder than 'huit' given by ♀ (see above); see also Field Characters and, for further suggestion of sexual differences in calls, Lassey and Wallace (1992). (3) Quiet, short 'zik', 'zit', or 'zlit' sounds given before take-off and in flight (Sick 1938; Czikeli and Busch 1974; Bergmann and Helb 1982); used in contact according to Giebing (1992). (4) Harsh, rattling 'rä' or 'grääh' sounds given in antagonistic interactions (Scheer 1953; Reinsch 1965). (5) Hissing sound noted from ♀ staying in nest to protect brood (Bozhko 1980); perhaps a threat-call. (6) Begging-calls of ♀. Quiet and delicate series of 'twie' sounds given to lure ♂ to nest and when begging; with waning excitement (after ♂ had fed ♀ and young) ♀ may change to quiet, conversational 'bett' sounds (Steinfatt 1937a). (7) Quiet call (no further details) given when feeding young (Steinfatt 1937a). (8) Other calls. (8a) A 'dije' confusingly similar to Linnet *C. cannabina* (see call 3b of that species) given by ♀ apparently in excitement during nest-site selection, nest-building and start of laying (H Göttgens); rendering suggests perhaps related to 'twie' begging-call of ♀ (see 6, above), whereas relationship with calls described under 2 (above) not clear. (8b) In Iran, 'didididi' noted from flock (apparently exclusively of *C. erythrinus*) when flushed (Schüz 1959).

CALLS OF YOUNG. Small nestlings give very quiet, high-pitched piping 'ziehd' or similar, changing at *c.* 8–9 days to louder 'srieh' sounds (Steinfatt 1937a; Haas 1939). In Austria, 'dschäb' (like *C. chloris*) given by fledglings, especially when flying after parent (Czikeli and Busch 1974); same call from *ferghanensis* young out of nest (Tien Shan mountains) rendered 'tsit' (Kovshar' 1979).  MGW

**Breeding.** SEASON. Late throughout region. Moscow region (central European Russia): eggs laid late May to mid-June; late-June clutches probably replacements (Ptushenko and Inozemtsev 1968). St Petersburg (north-west Russia): eggs laid from end of May to beginning of July

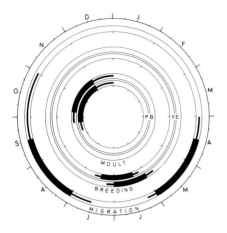

(Bozhko 1974); see also Mal'chevski and Pukinski (1983). Southern Finland: see diagram; peak period early to mid-June (end of May to early July); clutches started after mid-June considered replacements; around 5 days later further north (Reinikainen 1939; Stjernberg 1979, which see for discussion of factors involved). Sweden: peak egg-laying 1st half of June (Risberg 1970; Björklund 1990a). North-east England: eggs laid 1st half of June (Lassey and Wallace 1992). For Scotland, see Mullins (1984b); for France, Pochelon (1992); for Helgoland (north-west Germany), see Barth and Moritz (1988). Northern Turkey: May–July (Hüe and Etchécopar 1970). North-west Kazakhstan: end of May to beginning of July (Levin and Gubin 1985). For review, see Bozhko (1980). SITE. Low in dense bush or young tree, generally well hidden close to trunk; sometimes in tangle of scrub, herbs (etc.), but only very rarely on ground. Of 1162 nests in Finland, 42% in juniper *Juniperus*, 25% spruce *Picea*, 8% currant *Ribes*; conifers preferred, and configuration of branches and twigs as important as cover since nests loosely attached; 82% less than 1·25 m above ground (0–10 m, *n* = 875) (Stjernberg 1979, which see for comparison with other studies); other Finnish studies found 53% in young spruce, *n* = 139 (Reinikainen 1939), and 48% in elder *Sambucus*, *n* = 81 (Peiponen 1974). In central Sweden, of 65 nests, 25% in *Ribes*, 23% willow *Salix*, 17% cherry *Prunus*; 74% less than 1·5 m above ground (Björklund 1990b); see also Risberg (1970). For northern Germany, see Müller (1973); for parks in St Petersburg, see Bozhko (1974). In northern Turkey and Iran, mostly in juniper and hazel *Corylus* (Hüe and Etchécopar 1970). For review, see Bozhko (1980). Nest: untidy, tangled foundation of twigs, stems, and grass, often including dried flowers and Umbelliferae stalks, with inner layer of finer grasses lined with rootlets, plant down, sometimes moss and lichen, and, where available, often large amounts of horsehair (Steinfatt 1937a; Haas 1939; Bozhko 1974; Stjernberg 1979). Range of dimensions in review were outer diameter 8·5–18·0 cm, inner diameter 4·0–7·0, overall height 4·7–12·5, and depth of cup 2·8–5·0 cm (Bozhko 1980); 2 nests

had dry weights 7·5 and 8·0 g (Steinfatt 1937a; Haas 1939). Building: by ♀ only, sometimes accompanied by ♂, usually in morning; takes 3–4 days (2–6); material generally gathered from within 50 m of nest, though can be up to 200 m (Haas 1939; Risberg 1970; Bozhko 1974; Levin and Gubin 1985). In southern Finland, took average 3·9 days (3–7, *n* = 82) (Stjernberg 1979, which see for details of technique). EGGS. See Plate 61. Sub-elliptical, smooth and glossy; light bluish-green with purplish or blackish-brown spots and hairstreaks at broad end, remainder of surface hardly marked; sometimes dark violet-grey undermarkings (Harrison 1975; Makatsch 1976; Bozhko 1980). Nominate *erythrinus*: 20·4 × 14·5 mm (18·0–22·7 × 13·3–15·9), *n* = 125; calculated weight 2·25 g (Schönwetter 1984). *C. e. kubanensis*: 19·9–23·0 × 14·0–16·0 mm (Dementiev and Gladkov 1954; Hüe and Etchécopar 1970). Clutch: 4–6 (3–7). In southern Finland, of 236 full clutches: 4 eggs, 9%; 5, 69%; 6, 21%; 7, 0·4%; average 5·13; early June 5·26 (*n* = 158), mid-June 4·86 (*n* = 66), late June and early July 4·75 (*n* = 12); 5 replacements all smaller than 1st clutches of same ♀♀ (Stjernberg 1979, which see for influence of habitat, density, age of ♀, etc.); see also Peiponen (1974). In St Petersburg, of 26 clutches: 3 eggs, 4%; 4, 35%; 5, 50%; 6, 12%; average 4·7 (Bozhko 1974). Average in Sweden 5·3, *n* = 12 (Risberg 1970), and in Czechoslovakia 4·4, *n* = 9 (Hudec 1983). Eggs laid daily in early morning 0–2(–5) days after nest completion (Steinfatt 1937a; Haas 1939; Stjernberg 1979). One brood (Bozhko 1974, 1980; Stjernberg 1979; Levin and Gubin 1985); 2 recorded in captivity (Giebing 1992). In one case, replacement clutch started 4–6 days after loss of 8-day-old young (Stjernberg 1979). INCUBATION. 11–12(–14) days, by ♀ only; starts with 3rd or 4th egg (Steinfatt 1937a; Risberg 1970; Bozhko 1980). In southern Finland, took average 12·1 days (11–14, *n* = 77) from laying of last egg to hatching of last young; on 9th day, ♀ spent 80% of 15 hrs on eggs in stints of *c*. 90–180 min (Stjernberg 1979). At one nest, ♀ on full clutch sat for average 48·5 min (1–152) and took breaks of 9·2 min (2–19) (Steinfatt 1937a). See also Levin and Gubin (1985) for routine. YOUNG. Fed and cared for by both parents (Haas 1939; Stjernberg 1979, which see for feeding; Bozhko 1980); brooded by ♀; for stints, see Steinfatt (1937a) and Stjernberg (1979). See also Lind (1952) and Björklund (1990a). FLEDGING TO MATURITY. Fledging period 10–13 days; usually leaves nest before able to fly (Steinfatt 1937a; Haas 1939; Risberg 1970). In southern Finland, average 11·7 days (9–15, *n* = 27); young fed by parents, especially ♀, for *c*. 2 weeks after fledging; age of first breeding for ♂♂ 1 year (Stjernberg 1979, which see for development of young). See also Bozhko (1980). BREEDING SUCCESS. In southern Finland, of 620 eggs, 62% hatched and 54% produced fledged young; 24% of eggs lost to predators, 9% to desertion, 5% failed to hatch; 64% of total losses due to predation, most probably by weasel, etc. (Mustelidae) and crows (Corvidae); ♀♀ may desert easily. 273 nests produced 3·0 fledged young per nest, 4·6 per successful nest, $n=175$. (Stjernberg 1979, which see for many details, including comparison between habitats and review of other studies.) Presence of Red-backed Shrike *Lanius collurio* in nesting area in Finland reduces success mainly by causing desertion of eggs and broods (Peiponen 1974). In central Sweden over 4 years, of 246 eggs in 49 clutches, 67·5% hatched and 60·6% produced 6-day-old young (3·0 per nest overall); 37% (31–43) of 65 nests failed completely; losses most likely caused by Mustelidae and domestic cats, which probably take sitting ♀♀ first; nests well hidden from aerial predators and those nearest woodland edge bordering on fields most liable to loss (Björklund 1990a, b). In St Petersburg area, 73% of 214 eggs hatched and 70% produced flying young (3·4 per nest overall); main predators Corvidae, *L. collurio*, and Mustelidae; 44% of 34 nests in city parks failed completely, principally because of human interference causing ♀ to desert eggs (Bozhko 1974, 1980, which see also for Kareliya, north-west Russia). In north-west Kazakhstan, only 23% of 602 eggs resulted in fledged young; nest-failure rate of 60–70% much higher than in western Europe; due mainly to Corvidae, diurnal raptors, martens *Martes*, and stoat *Mustela erminea* (Levin and Gubin 1985). Nest prone to falling as often only loosely anchored (Haas 1939). In Mecklenburg (north-east Germany), 27 young flew from 12 of 18 nests, giving 2·3 young per successful nest, 1·5 per nest overall (Müller 1973); for Helgoland, see Barth and Moritz (1988).                                              BH

**Plumages.** ADULT MALE. In fresh plumage (winter), forehead and crown dull dark rosy-red, tips of feathers narrowly edged pale brown-grey. Supercilium slightly paler rosy-red, extending from rear of eye backwards, merging into mixed grey-brown and dark rosy-red upper side of neck and hindneck. Lore, narrow ring round eye, and broad but rather indistinct stripe over upper ear-coverts dark grey-brown with faint rosy-red speckling; some fine off-white speckling below eye. Upper mantle and lower side of neck dark grey-brown; lower mantle, scapulars, and back dark olive-brown, each feather of lower mantle and scapulars suffused dull rosy-red subterminally (partly concealed) and suffused pinkish-brown on tip; shafts slightly darker brown, especially on mantle. Rump and shorter upper tail-coverts dull rosy-red, feather-tips suffused grey-brown or pink-brown; longer upper tail-coverts dark olive-brown with paler brown or pink-brown fringes. Lower cheek dull rosy-red, partly concealed by brown-grey feather-tips. Chin, throat, and chest pink-red, forming broad bib; feathers narrowly fringed off-white, bib appearing scalloped, widest on chest (shape of bib and amount of fringing as in fresh-plumaged ♂ House Sparrow *Passer domesticus*, but pink-red instead of black); centres of pink-red feathers often deeper rosy-red, especially on chest. Upper belly (at border of chest), side of breast, flank, and thigh pale buff-brown or light grey-brown, merging into white on mid-belly and vent and into cream-white or pale pink-cream on under tail-coverts; belly, side of breast, and upper flank washed pink-buff or rosy-pink to varying extent, pink much less intense than pink-red of chest; some faint dark brown shaft-streaks on upper belly, side of breast, and flank; under tail-coverts with pale brown-grey central marks, partly shining through. Tail dark sepia-brown, outer

webs of central 5 pairs (t1–t5) narrowly fringed rufous-pink or buff-pink, inner webs narrowly and faintly edged white. Flight-feathers, greater upper primary coverts, and bastard wing brown-black, outer webs narrowly but sharply fringed pink-buff or dull pink (faintly pale grey on outer web of p9, on emarginated parts of p6–p8, and on longest feather of bastard wing). Inner webs of tertials brown-black, outer webs grey shading to dark brown-grey on base, partly washed pink-buff or dull pink. Greater and median upper wing-coverts brown-black, greater with broad dull pink outer fringe, median with dull pink tip 2–3 mm wide; lesser coverts dark grey-brown with dull pink-red fringes. Under wing-coverts and axillaries grey-white with varying amount of pink suffusion; shorter coverts along leading edge of wing mottled brown-black and white. *In worn plumage* (late spring and summer), brown-grey feather-fringes of upperparts and whitish fringes of chin to chest worn off, pink and red more exposed; due to abrasion, red brighter, more scarlet or (when heavily worn) orange-scarlet, more glossy. Cap scarlet-red, some dark grey-brown of feather-bases visible; supercilium (backwards from eye) and lower cheek bright pink-red, separated by distinct dull black eye-stripe; hindneck, side of neck, mantle, scapulars, and back olive-brown (when rather fresh) to dull brown-grey (when heavily abraded), partly dotted rosy-red (mainly on lower hindneck and mantle); rump bright rosy-red to ruby-red, more contrasting than in fresh plumage. Chin to chest bright rosy-red or carmine, merging into pink on upper belly and upper flank; belly more extensively white; fringes along outer webs of tail- and flight-feathers paler, light pink-buff or pink-yellow; outer webs of greater coverts and tips of median paler and more contrastingly pink, forming more distinct double wing-bar. ADULT FEMALE. In fresh plumage, entire upperparts olive-brown, slightly browner on cap, slightly more greenish-olive backwards from mantle; feathers of forehead and forecrown with dull black marks on centres, appearing spotted, dark marks more grey and more diffuse on hindcrown, mantle, and scapulars, appearing obscurely streaked; back to upper tail-coverts unstreaked. Side of head and neck greenish olive-brown, mottled pale buff or grey-white at base of both mandibles and on lore, narrowly streaked grey or pale buff behind eye, on cheeks, and on ear-coverts. Chin and upper throat cream-white, partly marked with small triangular brown specks, bordered at side by mottled black-brown malar stripe. Lower throat, chest, side of breast, flank, and thigh light buff-brown or brown-grey, merging into buffish-cream on upper and side of belly and on under tail-coverts, this in turn to white on mid-belly and vent; lower throat and chest with fairly broad and distinct dark olive-grey streaks, narrower and more diffuse light brown-grey streaks on side of breast, flank, and upper belly; centres of tail-coverts light brown-grey, mainly concealed. Tail, flight-feathers, greater upper primary coverts, and bastard wing as adult ♂, but outer fringes pale olive-brown or olive-green, no pink or red. Tertials and greater upper wing-coverts dark brown-grey, fringed light olive-brown or olive-green along outer webs, shading to paler olive-grey towards tips, but latter soon bleaching to off-white; median coverts dark brown-grey with contrasting cream-buff tips 1–2 mm wide, latter soon bleaching to white; lesser coverts dark brown-grey with olive-brown fringes. Under wing-coverts and axillaries light ash-grey, feather-tips washed cream-buff to off-white, coverts along leading edge of wing spotted brown-grey and cream-white. In worn plumage, upperparts duller, greyer, more streaky, brighter and more uniform green-olive on rump only; side of head and neck brown-grey, palest on lore and round eye, faintly streaked grey-white on cheeks and ear-coverts; chin and throat off-white with some dusky specks, sharply bordered by dark brown-grey malar stripe at each side and by gorget of rather sharp dark brown-grey and grey-white streaks across lower throat and upper chest; streaks of gorget gradually more diffuse on lower chest and side of breast, upper belly and flank light brown-grey with ill-defined darker grey streaks; belly extensively off-white; pale fringes of tail- and flight-feathers and tertials paler green-grey or grey-white; terminal part of fringes of greater coverts and tips of median coverts bleached to off-white, sometimes largely worn off when heavily abraded. NESTLING. Down dark grey; dense and *c.* 6-10 mm long on upperparts, scanty and much shorter on side of neck, flank, thigh, and belly (Bozhko 1980, which see for photographs of birds of various ages). JUVENILE. Cap dull buff-brown with narrow dusky olive-brown shaft-streaks. Hindneck and upper mantle buff, feathering rather loose, rather narrowly but sharply streaked dark grey-brown; ground-colour paler than in adult ♀, streaks darker and more sharply defined. Lower mantle and scapulars olive-brown, streaked sooty-grey; borders of streaks rather diffuse, but streaks more distinct than in adult ♀. Back to upper tail-coverts buff- or olive-brown, upper tail-coverts with narrow sooty streaks, less greenish than adult ♀. Side of head and neck and side of breast tawny-buff, densely mottled dark olive-brown; eye-ring uniform buff, conspicuous. Chin and throat off-white, closely marked with contrasting dusky olive-brown spots (less uniform than in adult ♀); chest, belly, and flank buff or isabelline with sharply contrasting dusky olive-brown streaks, each widening at tip of feather (ground-colour paler than in adult ♀, streaks darker, reaching feather-tip). Vent, thigh, and under tail-coverts pale buff or isabelline-white. Tail, flight-feathers, and greater upper primary coverts as in adult ♀ but fringes slightly more buffish, less greenish; greater upper wing-coverts and tertials as adult ♀ but terminal part of outer fringe contrastingly cream or pink-buff (2-4 mm long), remainder of outer web more narrowly fringed olive-green; median coverts with similar cream or pink-buff tip; lesser coverts brownish-olive. Plumage fresh at same time as adult ♀ is heavily worn. FIRST ADULT MALE. Like adult ♀, but juvenile flight-feathers (except sometimes outer primaries), greater upper primary coverts, and sometimes a few tail-feathers or outer greater upper wing-coverts retained. If outer primaries new, these blacker and less frayed on tips than browner neighbouring juvenile ones (though difference sometimes hard to detect); if some outer greater coverts still juvenile, these with distinctly more worn, narrower, and more sharply-defined white tip than those of neighbouring new ones. Retained primary coverts browner and more frayed on tips than those of adult ♀ at same time of year. A few birds obtain some rosy-red or orange-red feathering during post-juvenile moult, e.g. on crown, upper mantle, rump, throat, chest, flank, belly, or wing-coverts, but these uncommon (e.g., 1 bird out of 301 of all ages and sexes examined by Vaurie 1949*b*, only 2 examined by Paludan 1959; but 21% of 33 examined by Ström 1991), and head and body of most birds indistinguishable from adult ♀; differences in colour and intensity of streaking, as cited by, e.g., Bozhko (1980) and Müller and Wernicke (1990), mainly due to difference in wear, olive of upperparts of 1st adult (both ♂ and ♀) fading more rapidly to grey during winter and spring than in adult ♀, underparts and wing-bars more rapidly becoming whiter, dark streaking on head and body becoming more pronounced; once breeding season started, adult ♀♀ fade rapidly also, sometimes becoming more worn and thus greyer and more streaky than some 1st adult ♂♂. In breeding season, ♀♀ of all ages have brood-patch (apparently never so in ♂); retained until later stages of complete post-breeding moult in late autumn or early winter (Noskov 1982). FIRST ADULT FEMALE. Like adult ♀, but part of juvenile feathering retained, as in 1st adult ♂. Often hard to distinguish from adult ♀ when no juvenile tail-feathers or outer greater

coverts retained or when no new outer primaries obtained. Difference in gradual abrasion of head, body, and wing from adult ♀ as in 1st adult ♂; both adult and 1st adult ♀ generally obtain brood-patch, and this only way to separate 1st adult ♀ from 1st adult ♂, apart from song of ♂. Never any red in plumage, unlike some 1st adult ♂♂. SECOND ADULT. Like adult, but some ♂♂ retain some old brown feathers amidst scarlet of head and body, or some old greater or median coverts (white-tipped rather than pink) (Svensson 1992); mantle mostly brown, less tinged red; belly, fringes of flight-feathers, and tips of coverts less tinged pink (Witherby *et al.* 1938).

**Bare parts.** ADULT, FIRST ADULT. Iris hazel-brown to dark brown. Bill steel-grey or bluish leaden-grey, cutting edges and base of lower mandible paler silvery-grey or horn with yellow, pink, mauve, or purplish tinge; in 1st adult in autumn, more horn-brown or dark horn-grey with black-brown culmen, light flesh-brown or grey-brown lower mandible, and paler yellow-horn tip and cutting edges. Mouth rose-red. Leg and foot dull flesh or dull horn-grey to dusky brown, in latter often with flesh tinge; rear of tarsus and soles sometimes dirty white or pink-flesh; leg in adult ♀ occasionally entirely purplish-flesh, livid-flesh, or flesh-yellow. (Sharpe 1888; Williamson 1962; Ali and Ripley 1974; C M Liebregts-Haaker; ZMA.) NESTLING. At hatching, bare skin of body pink; bill white, a dark triangular spot on culmen from day 4; mouth bright carmine (Bozhko 1980). JUVENILE. At fledging, bill dark horn-grey with distinct green tinge at side of lower mandible; cutting edges and traces of gape-flanges yellow, mouth bright red (Heinroth and Heinroth 1924-6). At 1-3 months, bill pale leaden-grey with darker grey culmen ridge, cutting edges and lower mandible with pink or yellowish tinge, gape pink-flesh (ZMA).

**Moults.** ADULT POST-BREEDING. Complete; primaries descendent. Starts immediately on arrival in winter quarters in Indian subcontinent, September-October (Ali and Ripley 1974). In September, moult not started in heavily worn adults and 1-year-olds from Iran, Afghanistan, Baltistan, Ladakh, Kashmir, and Nepal (Vaurie 1949*b*; Diesselhorst 1968; Niethammer 1973; ZMA). A few birds in last stages of moult in early November in northern India; outer 2-4 primaries and 1-2 tail-feathers still growing (primary moult score 40-46), body and wing-coverts mainly new but feathers of head and neck still partly growing (BMNH, ZMA). Moult completed in single bird from Shantung (China) on 8 November and in several from India and Yunnan in December (Vaurie 1949*b*; Svensson 1992). In captivity, moult October-November (Dementiev and Gladkov 1954). Moult of captive birds rapid, completed in 55±8 days; starts with p1 in late October; moult of tail centrifugal, starting with t1 when (p5-)p6 shed, completed with regrowth of t6 when p8(-p10) completed; 1st secondary (s1) shed at same time as p6, last one (s6) completed after regrowth of p9; tertials gradually replaced between loss of p5 and regrowth of p6 (sequence: s8-s9-s7); body starts at same time as shedding of (p2-)p3-p4, head with p5, all completed with regrowth of p9-p10 about December (Noskov 1978, 1982, which see for effect on moult of artificially altering day length). See also Carlotto (1991). POST-JUVENILE. Partial: head, body, lesser, median, and many to (usually) all greater upper wing-coverts, many or all tail-feathers, 1-3 tertials, and frequently (1-)4-5 outer primaries (in 5 of 10 birds examined in detail). Mainly in winter quarters, timing of moult as adult post-breeding or slightly later (start November: Noskov 1978), but (unlike adult) some birds start (with scattered feathers of body) when still in or near breeding area. 10 birds from Ladakh in body moult September (Vaurie 1949*b*; ZMA), and vagrant in

Netherlands had moult completed (including outer primaries) mid-November (ZMA). Some birds still in moult December (Noskov 1982); moult completed in birds examined from January and later (RMNH, ZMA).

**Measurements.** ADULT, FIRST ADULT. Nominate *erythrinus*. Northern and central Europe east to central Siberia, mainly May–July; skins (RMNH, ZMA). Bill (S) to skull, bill (N) to distal corner of nostril; exposed culmen on average 3·8 shorter than bill (S).

| | | | | | |
|---|---|---|---|---|---|
| WING | ♂ | 84·8 (1·65; 32) 82–88 | ♀ | 82·7 (1·44; 13) 80–85 |
| TAIL | | 56·3 (1·95; 33) 53–60 | | 55·4 (2·20; 13) 52–60 |
| BILL (S) | | 14·2 (0·54; 31) 13·3–15·1 | | 14·3 (0·68; 13) 13·1–15·3 |
| BILL (N) | | 7·9 (0·31; 33) 7·4–8·6 | | 7·9 (0·38; 13) 7·3–8·6 |
| TARSUS | | 19·0 (0·60; 29) 18·1–20·0 | | 19·1 (0·59; 13) 18·3–20·0 |

Sex differences significant for wing.

Ages combined in sample above, as wing and tail of brown juvenile, brown 1st adult (with juvenile or partly new flight-feathers and new tail), and red older ♂ hardly differ; thus: wing, red ♂ 84·6 (1·44; 21) 82–87, brown juvenile or 1st adult ♂ 85·3 (1·99; 12) 82–88; tail, red ♂ 56·2 (1·99; 22) 53–60, brown ♂ 56·4 (1·95; 11) 54–59.

Red ♂ of (1) *kubanensis*, northern Caucasus, May–June, and (2) *ferghanensis*, western Kun Lun and Tien Shan mountains (central Asia), May–August (RMNH, ZMA).

| | | | |
|---|---|---|---|
| WING (1) | 86·4 (1·42; 15) 84–89 | (2) | 86·7 (1·38; 21) 84–90 |
| TAIL | 56·2 (1·60; 15) 53–59 | | 58·2 (2·01; 21) 55–62 |
| BILL (S) | 14·5 (0·50; 15) 13·8–15·3 | | 14·7 (0·47; 21) 13·8–15·6 |
| BILL (N) | 8·1 (0·35; 15) 7·5–8·6 | | 8·3 (0·44; 21) 7·4–9·1 |
| TARSUS | 18·9 (0·52; 13) 18·2–19·8 | | 19·2 (0·60; 20) 18·2–20·0 |

Wing of single ♀ *kubanensis* 83·5, bill (S) 15·2, tarsus 19·8; wing of ♀ *ferghanensis* 82·2 (1·72; 5) 80–85, bill (S) 14·4 (0·60; 5) 13·8–15·4, tarsus 19·3 (0·72; 5) 18·4–20·2. Small sample of *roseatus* from northern India and Nepal: wing, ♂ 85·8 (3·07; 4) 83–90, ♀ 81·0 (1·17; 5) 79–83; bill (S), ♂ 14·4 (0·46; 4) 13·8–14·9, ♀ 14·5 (0·69; 5) 13·9–15·3; tarsus, ♂ 18·8 (0·30; 4) 18·5–19·2, ♀ 18·5 (0·55; 5) 17·9–19·0 (RMNH, ZMA).

Wing. Nominate *erythrinus*: (1) Netherlands, live birds, June–August (C M Liebregts-Haaker; ZMA); (2) Yamal peninsula (Danilov *et al.* 1984). *C. e. kubanensis*: (3) Turkey, Caucasus, and northern Iran, April–July (Hesse 1915; Stresemann 1928b; Paludan 1940; Schüz 1959; Rokitansky and Schifter 1971). *C. e. ferghanensis*: (4) Afghanistan, June–July (Paludan 1959). *C. e. roseatus*: (5) Kansu (China) (Stresemann *et al.* 1937). Intermediates between nominate *erythrinus*, *ferghanensis*, and/or *grebnitskii*: (6) Mongolia, May–August (Piechocki and Bolod 1972; Piechocki *et al.* 1982). Adults and 1st adults combined.

| | | | | |
|---|---|---|---|---|
| (1) | ♂ | 84·2 (2·57; 10) 80–87 | ♀ | 82·1 (1·62; 9) 78–83 |
| (2) | | 84·4 ( — ; 7) 81–89 | | — ( — ; 5) 80–85 |
| (3) | | 85·7 (1·61; 25) 82–89 | | 84·6 (2·09; 6) 81–88 |
| (4) | | 84·0 ( — ; 14) 82–88 | | 82·2 (1·48; 5) 80–84 |
| (5) | | 84·7 ( — ; 15) 82–90 | | 81·5 ( — ; 2) 81–82 |
| (6) | | 85·2 (1·72; 8) 82–87 | | 83·5 (1·60; 5) 80–85 |

Nominate *erythrinus*, live birds, Lake Chany (south-west Siberia): juvenile 80·1 (2·84; 11) 76–86 (Havlín and Jurlov 1977). Netherlands, live juveniles, July–September: 80·1 (2·27; 9) 76–83·5 (C M Liebregts-Haaker; ZMA). Live migrants Kashmir, race unknown, August–September: adult ♂ 86·6 (1·82; 101) 82–91, 1st adult ♂ 83·8 (2·18; 61) 78–91, adult and 1st adult ♂ 82·9 (2·20; 141) 77–89, juvenile 83·0 (2·27; 86) 78–88 (P R Holmes and Oxford University Expedition Kashmir 1983). See also Stegmann (1931a) and Vaurie (1949b). For growth of wing of nestling, see Stjernberg (1979).

**Weights.** ADULT, FIRST ADULT. Nominate *erythrinus*: (1) Finland, May–July (Stjernberg 1979); (2) Netherlands, June to early August (C M Liebregts-Haaker; ZMA); (3) Yamal peninsula (Russia) (Danilov *et al.* 1984). *C. e. kubanensis*: (4) Turkey and northern Iran, May–July (Paludan 1940; Schüz 1959; Kumerloeve 1967; Rokitansky and Schifter 1971; ZFMK). *C. e. ferghanensis*: (5) Afghanistan, June–September (Paludan 1959; Niethammer 1973). Nominate *erythrinus* and/or *ferghanensis*: (6) Kazakhstan, May–August (Korelov *et al.* 1974). Intermediates between nominate *erythrinus*, *ferghanensis*, and/or *grebnitskii*: (7) Mongolia, May–August (Piechocki and Bolod 1972; Piechocki *et al.* 1982). *C. e. roseatus*: (8) Nepal, June–September (Diesselhorst 1968). Race unknown: (9) migrants Kashmir, August–September (P R Holmes and Oxford University Expedition Kashmir 1983); (10) India and Nepal, September–April (Diesselhorst 1968; Ali and Ripley 1974).

| | | | | |
|---|---|---|---|---|
| (1) | ♂ | 23·1 (1·3 ; 38) — | ♀ | 23·7 (1·8 ; 87) — |
| (2) | | 22·0 (1·21; 14) 20·0–25·0 | | 23·9 (2·37; 11) 19·5–27·0 |
| (3) | | 23·0 ( — ; 7) 20·2–25·6 | | 21·6 ( — ; 5) 18·4–23·2 |
| (4) | | 22·7 (1·96; 7) 19·1–25 | | — ( — ; — ) |
| (5) | | 21·3 ( — ; 15) 20–24 | | 23·8 (2·14; 6) 21–26 |
| (6) | | 22·2 ( — ; 66) 17·0–28·5 | | 22·1 ( — ; 35) 18·5–29·7 |
| (7) | | 23·0 (2·10; 6) 20–25 | | 23·2 (1·91; 8) 20–25 |
| (8) | | 23·4 (2·20; 8) 21·0–28·0 | | 24·6 ( — ; 2) 23·9–25·3 |
| (9) | | 22·4 (1·55; 203) 19·4–26·5 | | 22·4 (1·63; 140) 18·0–27·0 |
| (10) | | 23·8 ( — ; 10) 19·7–30 | | 23·6 ( — ; 9) 21–26 |

See also Kumar and Tewari (1983). For *grebnitskii*, see Weigold (1926), Dementiev and Gladkov (1954), and Piechocki (1959). For influence of moult on weight, see Noskov (1982).

JUVENILE. Nominate *erythrinus*, Netherlands, late July to October: 22·5 (1·13; 14) 20·5–24 (C M Liebregts-Haaker, ZMA). Belgium, August: 25·0 (De Ruwe *et al.* 1990). Lake Chany (south-west Siberia), late summer: 22·0 (1·49; 11) 20·2–25·3 (Havlín and Jurlov 1977). For growth of nestling, see Stjernberg (1979).

Race unknown. Migrants Kashmir, August–September: 22·0 (1·71; 188) 18·1–28·0 (P R Holmes and Oxford University Expedition Kashmir 1983).

**Structure.** Wing rather long, broad at base, tip bluntly pointed. 10 primaries: p8 longest, p9 0·5–3 shorter, p7 0–1, p6 2–5, p5 8–13, p4 12–17, p3 15–20, p2 17–23, p1 20–26. P10 reduced, a tiny pin concealed below outermost greater upper primary covert, 53–61 shorter than p8, 8–11 shorter than longest upper primary covert. Outer web of p6–p8 emarginated, inner web of p7–p9 with faint notch. Tip of longest tertial reaches to tip of p2–p3. Tail of average length, tip forked; 12 feathers, t1 5·5 (20) 3–8 shorter than t6. Bill short, thick, markedly swollen at base; depth at base 8·4 (15; 7·8) 7·8–9·3, width at base 8·4 (18) 7·7–9·0 (all races); culmen and cutting edges decurved, gonys straight or slightly convex; tip rather sharp, not as blunt as in (e.g.) Bullfinch *Pyrrhula pyrrhula*. Nostrils rather small, rounded, protected by small tufts of bristly feathers projecting from forehead. A few short bristles at base of upper mandible above gape. Tarsus and toes rather short and slender. Middle toe with claw 18·8 (15) 18·0–20·5; outer toe with claw *c.* 71% of middle with claw, inner *c.* 68%, hind *c.* 77%.

**Geographical variation.** Very slight in size, more marked in general body colour. Nominate *erythrinus*, *grebnitskii*, and *roseatus* slightly smaller, *kubanensis* and *ferghanensis* slightly larger, mainly in wing and bill, but difference slight and individual variation large. 2 groups separable on colour of ♂: (1) scarlet-red birds with limited red on mantle, scapulars, and belly in north, from northern and central Europe through Siberia to Kamchatka, and (2) darker rosy-red birds with more extensive red on body in mountain areas from Turkey and Caucasus area to central Asia. In both groups, birds in west paler red in ♂ and with paler olive-brown and less streaked ground-colour in both sexes, birds in east darker and with darker and browner ground-colour (Johansen 1944). In north, where distribution continuous, paler western and central race nominate *erythrinus* runs clinally into darker *grebnitskii* of eastern Asia, boundary between them hard to define, and difference too slight to warrant recognition of intermediate race *diamesa* Stantschinsky, 1929; in south, difference between western *kubanensis* (Turkey to Iran and western Turkmeniya) and eastern *roseatus* (central and eastern Himalayas to south-west and central China) more marked, with intermediate race *ferghanensis* in mountains of west-central Asia separable. *C. e. kubanensis* rather pale; adult ♂ similar to nominate *erythrinus*, but upper belly and flanks pink-red, scarcely contrasting with red of chest, not as pink-white as in ♂ nominate *erythrinus*; mantle and scapulars with slightly more distinct red feather-tips; ♀ slightly paler and greyer than ♀ nominate *erythrinus*, less olive-brown; in *ferghanensis*, red of head and upperparts

of ♂ and brown ground-colour of these in both sexes darker than in *kubanensis*, belly and flank of ♂ slightly deeper and more extensively rosy-red than in ♂ *kubanensis*; ♂ *roseatus* has head and upperparts dark brown with extensive purplish-carmine tinge, mantle and scapulars streaked black-brown, rump and underparts deep carmine, ♀ dark brown above, heavily and broadly streaked black-brown on cap, mantle, scapulars, and underparts. Depth of brown and red of *grebnitskii* and amount of streaking of ♀ roughly intermediate between *roseatus* and *ferghanensis*; red of ♂ carmine, less purplish- or rosy-red than *roseatus*, less bright scarlet than *ferghanensis*; extent of red above and below as in *kubanensis*, but ground-colour of upperparts darker grey-brown. Only a few ♂♂ examined from Altai, Sayan mountains, and south-central Siberia, where nominate *erythrinus*, *ferghanensis*, and *grebnitskii* are in contact; these nearest to *ferghanensis*, but ground-colour of head and underparts of some birds paler (tending towards nominate *erythrinus*), in some darker (tending towards *grebnitskii*). See also Stantschinsky (1929), Stegmann (1931a), Hartert and Steinbacher (1932–8), Kozlova (1939), and Vaurie (1949b, 1959).

Forms superspecies with Purple Finch *C. purpureus* of North America (Mayr and Short 1970). CSR

## *Carpodacus synoicus*  Sinai Rosefinch

Du. Sinaï-roodmus   Fr. Roselin du Sinaï   Ge. Sinaigimpel
Ru. Бледная чечевица   Sp. Camachuelo del Sinaí   Sw. Sinairosenfink

*Pyrrhula synoica* Temminck, 1825

Polytypic. Nominate *synoicus* (Temminck, 1825), Sinai, Israel, Jordan, and extreme north-west Saudi Arabia. Extralimital: *salimalii* (Meinertzhagen, 1938), Afghanistan in northern Hindu Kush between 65 and 69°E; *stolickzae* (Hume, 1874), foothills of eastern Pamir and northern Kun Lun in south-west Tarim basin (China), between Kashgar and Kotan; *beicki* (Stresemann, 1930), Tsaidam basin, north-east Tsinghai, and neighbouring part of Kansu (China).

**Field characters.** 14·5 cm; wing-span 25–27·5 cm. Slightly larger than Scarlet Rosefinch *C. erythrinus* and Pale Rock Sparrow *Carpospiza brachydactyla*; bill proportionately a little finer than in *C. erythrinus*. Medium-sized, unobtrusive but nervous, ground- and cliff-haunting rosefinch, restricted to sand desert and hills. Adult ♂ almost wholly drenched carmine-red and pink; ♀ and juvenile pale fawn, showing only soft streaks. No other striking features except for pale legs; beware confusion of ♀ and juvenile with *C. brachydactyla*. Sexes dissimilar; no seasonal variation. Juvenile separable.

ADULT MALE. Moults: July–September (complete). Surround to bill carmine-red; rest of head, underbody, and rump rose-carmine, with silvery-white streaks and spots most obvious on crown. Rear crown, nape, and mantle grey-brown, washed mauve-red and indistinctly streaked. Wings and tail grey-brown, with indistinctly paler margins and some red wash; wings least-patterned of west Palearctic *Carpodacus*. Bill less bulbous than *C. erythrinus*, with finer tip; grey-horn. Legs pink-brown.

ADULT FEMALE. Palest and least-marked of west Palearctic *Carpodacus*, looking faded and recalling *Carpospiza brachydactyla*. At distance can appear wholly dun-coloured but at close range shows gingery face, isabelline-brown to drab-grey upperparts with indistinct browner streaks on crown and mantle, slightly paler uniform rump, dusky-brown wings with paler-margined coverts and tertials, and pale ochre underparts with obsolete, soft brown streaks on chest and flanks. Slight rose wash to plumage rarely shows in the field. JUVENILE. Resembles ♀ but general colour buff-brown, with distinctly paler, creamier rump and even more indistinct streaks.

In west Palearctic, occurs only in sandstone deserts of southern Levant and north-west Arabia where no possibility of confusion with other *Carpodacus* except rare vagrant *C. erythrinus*, but needs to be separated from similarly sized *Carpospiza brachydactyla* and smaller Trumpeter Finch *Bucanetes githagineus* which share its habitat and can show similar featureless plumage in brief view (see those species). Flight light and bounding, with

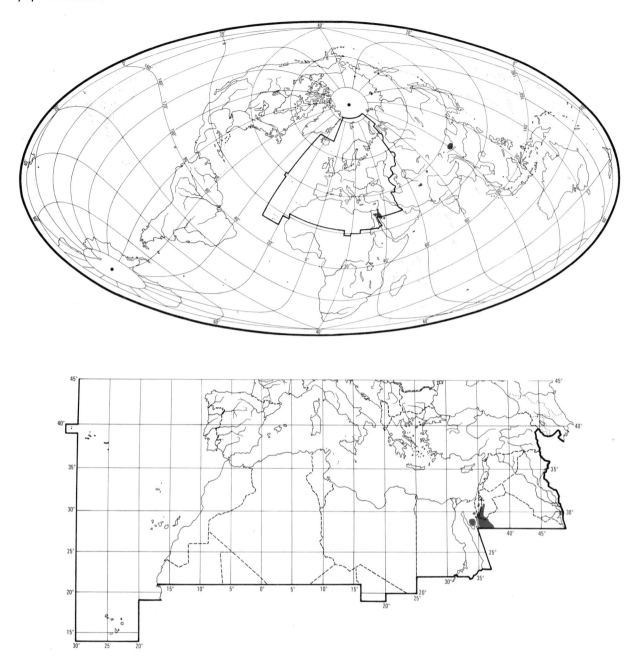

fast ascent. Gait a hop, varied by shuffling. Feeds on ground. Sociable, forming small parties after breeding. Normally nervous, escaping up cliff faces, but can become tame.

Song varied and melodious, ♂ in display also giving buzzing note (beware confusion with *B. githaginea*). Contact- and alarm-call a quiet, high-pitched 'tweet', 'tsweet', or 'ts-tsweet'. Flight-call a fairly rich 'trizp', suggesting Tree Pipit *Anthus trivialis*. Other calls include a frequent 'chig' and 'tieu', latter suggesting Yellowhammer *Emberiza citrinella*.

**Habitat.** In west Palaearctic, narrowly limited to highly arid, warm temperate regions with steep broken gorges, gullies, or crags, minimally vegetated. Avoids open areas (even of desert), grassland, or scrub, especially wooded steppe or vicinity of forests or cultivation. Extensive lowlands are rarely suitable; e.g. preferred sites in Jordan include ruins of Petra and precipices of Wadi Rum (Mountfort 1965). Occurs in south-west Asia up to *c.* 2000 m; in Afghanistan, Paludan (1959) found it at 2600–3050 m in narrow side valleys, especially where they narrowed to form canyons. A ground feeder, never perching on branches (Etchécopar and Hüe 1967). Among west Palearctic *Carpodacus*, ranks as most extreme in habitat choice, and most distinct from typical Fringillidae in its lifestyle.

**Distribution and population.** No changes reported.

JORDAN. 41–80 pairs breeding at Petra, 1983 (Wittenberg 1987).

**Movements.** Some populations sedentary, others dispersive, making chiefly altitudinal movements.

In Israel, more widely dispersed winter than summer, as rains cause extension of suitable habitat. Some populations strictly sedentary; in other populations, some birds make short-distance movements (primarily altitudinal) up to a few tens of km, to lower-lying and more sheltered areas, e.g. some birds breeding in Elat mountains descend to wadis that flow into Arava. Numbers in Elat mountains and eastern Negev markedly higher winter than summer, and apparently include immigrants, e.g. from Sinai, central Negev, and Petra area (Jordan). Movement chiefly October–November in autumn and February–March in spring. (Shirihai in press.)

In Sinai, where breeds at 1000–2000 m in southern and central areas, many birds remain at 1000–1500 m in mild winters; in cold winters most birds move down to wadis and other low-lying districts, including Gulf of Suez area and northern Sinai. With recent increase in human habitations at high altitude, more birds have remained in high areas in winter. (Inbar 1979; Shirihai in press.)

Assumed resident in rest of range (Afghanistan, China) (Schauensee 1984; S C Madge), with no movements reported.     DFV

**Food.** Mainly seeds; also leaves, buds, and fruit. Picks seeds from ground among rocks and plants; feeds also in fruit trees and wormwood *Artemisia* shrubs in flocks of up to 200. (Darom 1979; Inbar 1979; Paz 1987.) Often tamely forages amongst refuse left by tourists, even inside wastepaper-baskets, and accustomed to artificial feeding (e.g.) at St Katherine Monastery in Sinai (eastern Egypt) (Inbar 1979; Paz 1987; P D Housley). Feeds in early morning and late afternoon (except where food provided by man), and very frequently recorded drinking at water sources, up to 30 times per hr (Darom 1979; Inbar 1979; Paz 1987; Wittenberg 1987; Shirihai in press).

In Sinai, seems to feed on what is most abundant over year; seeds of yellow bugloss *Alkanna orientalis* in March, spurge *Euphorbia* in April, mulberry *Morus* in July, fig *Ficus* in August (Darom 1979). In southern Israel, feeds on seeds of *Hammada scoparia* (Chenopodiaceae) and *Ochradenus baccatus* (Resedaceae) (Inbar 1979). In extralimital Afghanistan, fed with Rock Sparrow *Petronia petronia* and sparrows *Passer* on newly-sown cereals (Meinertzhagen 1938).

Young are fed on seeds (Inbar 1979).     BH

**Social pattern and behaviour.** Little known. Most information from Israel and Sinai (Egypt): see Inbar (1979) and Shirihai (1989 and in press).

1. Gregarious all year, though more so outside breeding season. Forms flocks of varying size for feeding, drinking, and roosting. In post-breeding period (June or July to August), flocks up to *c.* 75 recorded (largest typically at water), these comprising adults and juveniles which remain on breeding grounds for a few more months after fledging; during local dispersal, up to 50, while concentrations of several hundred occur in autumn and winter (Roi 1923; Darom 1979; Inbar 1979; Shirihai in press). In Jordan in February, 50–70 birds were mostly dispersed in flocks of up to 8 (Wittenberg 1987). In late winter and spring, water source becomes centre of flock's home-range; some flocks break up for breeding in March–April after arrival on breeding grounds in February–March (Inbar 1979; Shirihai 1989). On most days of year, flocks predominantly ♀♀ and young ♂♂. Non-breeders (see also Bonds, below) remain in pairs or flocks. (Shirihai in press.) BONDS. In Sinai, in April (start of and during breeding season), pairs each often accompanied closely by 1(–2) young ♂♂, perhaps offspring of previous year (Zedlitz 1912); not clear whether this pattern persists during actual breeding attempt. Duration of pair-bond not known. Pair-formation takes place on breeding grounds. In Kansu (China), bond perhaps maintained in winter (Stresemann *et al.* 1937). Both sexes have brood-patch (Zedlitz 1912; Darom 1979), and both incubate according to Shirihai (in press), though ♀ may do most, being perhaps fed by ♂ (Inbar 1979; Shirihai 1989), but share taken by each in care of nestlings and fledglings requires further investigation. Families each containing up to 7 fledglings are presumed to be both parents accompanying their 2 broods (Shirihai in press); see, however, Breeding. Duration of post-fledging care not known. Age of first breeding 1 year according to Paz (1987); study by Shirihai (in press) recorded copulation between 1-year-old ♂ and ♀ of unknown age; however, majority of 1-year-olds (or perhaps only ♂♂) apparently pair up or remain in flocks but do not breed. BREEDING DISPERSION. Solitary and territorial, sometimes in loose neighbourhood groups with minimum 10 m between nests. Territory small, a few tens of m². (Shirihai in press.) No detailed information on use of territory. Density reported only in southern Jordan where *c.* 6–8 pairs per km² (over 10 km²) and, in Wadi ed-Der, probably 15–20 pairs along 1·2 km (Wittenberg 1987). In Israel, marked fluctuations typical in all areas, birds abandoning breeding site if water source dries up (Shirihai in press). In Sinai, 8 birds ringed in spring (not clear what proportion of total ringed) returned in following spring, 7 out of 8 to same sites (Darom 1979). ROOSTING. In Sinai, outside breeding season, communal and nocturnal, in cracks and crevices on steep slopes (Inbar 1979); roosting habits of breeding birds not known. In Kansu, birds roosted in winter in caves and tunnels in earth cliffs (Stresemann *et al.* 1937). Roosting birds emerge at first light. In winter, typically forage in (early) morning and afternoon, drink also in morning from *c.* 09.00 hrs (in summer, if hot, drink several times per day). Seek shade under shrubs or on rock ledges at hottest time of day. Gather for roosting towards sunset. (Darom 1979; Inbar 1979; Shirihai in press.)

2. In Sinai, tame near human settlements, far shyer further away (Siering 1986); at St Katherine Monastery, where regularly fed, birds have become tame, permitting approach within a few metres (Inbar 1979). Further reports of very confiding birds in winter from Israel (P D Housley) and Jordan (Wittenberg 1987). In contrast, described by Hollom *et al.* (1988) as unobtrusive but flighty, and feeding flocks in Sinai would fly off low if approached (Zedlitz 1912). Information on displays and postures

very scanty, but indications of similarity to other *Carpodacus* (see below). FLOCK BEHAVIOUR. Birds emerging from roost give 'chip' calls (see 2a in Voice) and start to feed (Inbar 1979). Move about quietly when foraging, though can be quite vocal (especially when agitated), giving variety of calls (see 2a–c in Voice) reminiscent of sparrow *Passer* (Zedlitz 1912; Hollom *et al.* 1988; Mild 1990). Rapid 'sip' calls (see 2c in Voice) given when disturbed (Inbar 1979). SONG-DISPLAY. In Sinai, ♂ of one pair sang (see 1 in Voice) from rock, apparently to advertise territory (Inbar 1979). In Israel, song-period mainly 2nd half of March and April (Shirihai in press). ANTAGONISTIC BEHAVIOUR. Only report refers to fights accompanied by rapid and loud 'chik-chik' (see 2a in Voice) (Darom 1979). HETEROSEXUAL BEHAVIOUR. Pair-formation apparently takes place on breeding grounds when flocks break up between late March and mid-April (Inbar 1979; Shirihai in press). In apparent courtship-display (noted, together with copulation, in Jordan late April and early May), ♂ adopted Upright-posture similar to Great Rosefinch *C. rubicilla* (see that species): head raised with bill pointed vertically up and breast conspicuously ruffled; gave buzzing call (see 3 in Voice) (Wallace 1984); in another study, ♀ seen to circle ♂ while calling (not described), and ensuing copulation was accompanied by rapid, loud 'chik' calls (see also Scarlet Rosefinch *C. erythrinus*); Upright-posture of ♂ perhaps (to some extent) aggressive in *C. synoicus* (Darom 1979). In Sinai, early April, in flock comprising 1 adult ♂ and 8 ♀♀ (or young ♂♂), adult ♂ and one of the others were perched close together briefly and both adopted Upright-posture, ♀ (or young ♂) with bill pointed up more steeply (M Ullman; see photograph in *Vår Fågelvärld* 1992, 51 (5), V). RELATIONS WITHIN FAMILY GROUP. No information on care of nestlings. Fledge at 14–16 days, and remain near nest for further *c.* 2 weeks after able to fly (Inbar 1979). ANTI-PREDATOR RESPONSES OF YOUNG, PARENTAL ANTI-PREDATOR STRATEGIES. No information.                                    MGW

Voice. Freely used (Shirihai 1989; Jonsson 1992); large winter flocks sometimes highly vocal (Darom 1979). No detailed description or recording of song, but a variety of calls reported, some reminiscent of sparrow *Passer* (see below).

CALLS OF ADULTS. (1) Song. Reported only for ♂, and apparently serves for territorial advertisement. Of fine quality, melodious and varied, and apparently quite loud. (Arnold 1962; Inbar 1979.) (2) Contact- and alarm-calls. Difficult to know how many different calls exist, and how many separate categories are warranted; following classification necessarily provisional. (2a) Variable short sound, with slightly squeaky quality, sometimes harder sounding, recalling sparrow: 'chig', 'chik', '(t)chip', 'tcheeep', or 'tcheup'; 'pch', 'ptchi', and more distinct 'picht'; also 'zick' or tinny 'dschidd'. Given frequently, on ground or in flight, singly or as series at varying volume and rate of delivery: e.g. as quiet contact-call while feeding or before roosting, or as loud and harsh-sounding staccato burst when flushed and otherwise excited (during fights, copulation). (Darom 1979; Inbar 1979; Wallace 1984; Wittenberg 1987; Hollom *et al.* 1988; Mild 1990; Jonsson 1992; M Ullman, D I M Wallace.) Recording of foraging flocks contains a number of calls, most suggesting House Sparrow *P. domesticus*, and presumably used for contact while also expressing varying levels of excitement. Harsher and shriller sounds shown in Fig I (apparent wing-whirr followed by 2 'twee' calls, wing-beats, and single 'chip' call) are probably linked with dispute over food. Calls in Fig II suggest 'chip chip-chee chir chip' and in Fig III

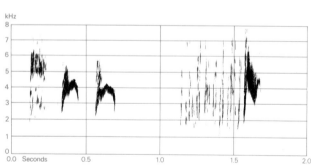

I  Mild (1990)  Israel  December 1988

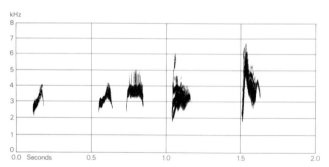

II  Mild (1990)  Israel  December 1988

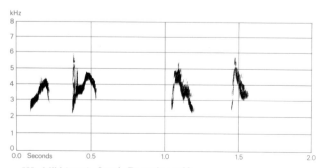

III  Mild (1990)  Israel  December 1988

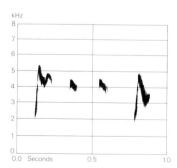

IV  Mild (1990)  Israel  December 1988

'chip chee cheer chip'; delicate 'tsee sip sip tsir' shown in Fig IV ('sip' perhaps related to call 2c or 2e; see below). (J Hall-Craggs, M G Wilson.) Presumed equivalent call of *beicki* (China), given in flight, short 'pit pit' or 'wit wit' (Stresemann *et al.* 1937). Calls 2b–e provisionally kept separate here, though probably some overlap (including with call 2a). (2b) A 'touit', signalling alarm (Hüe and Etchécopar 1970); in Jordan, noted less frequently than call 2a (Wittenberg 1987). Presumably this or related call also rendered high-pitched 'tweet' (Meinertzhagen 1954); single loud and strident sound serving as warning-call (Darom 1979). (2c) High-pitched and metallic 'tswet' or 'tzewt tweet' (quieter version given by ♀) (Shirihai 1989), also rendered quiet 'tsweet' or 'ts-tsweet', given frequently while feeding (Hollom *et al.* 1988). Quiet, at times slightly shriller, sparrow-like piping mentioned by Zedlitz (1912) is presumably also in this category, as are drawn-out 'pseeeh' (Jonsson 1992), and rapid series of 'sip' sounds in alarm (Inbar 1979). (2d) Quite rich 'trizp', resembling call of Tree Pipit *Anthus trivialis*, is reported to be usual flight-call (Hollom *et al.* 1988). (2e) A 'tieu' like Yellowhammer *Emberiza citrinella* (Hollom *et al.* 1988); quiet 'pleu' or 'pee' (Mild 1990) probably the same. (3) Buzzing sound given by ♂ in display (Wallace 1984).

CALLS OF YOUNG. No information.                     MGW

**Breeding.** SEASON. Southern Israel: eggs laid 1st half of April to May (late March to July) (Inbar 1979; Paz 1987; Shirihai in press). Sinai (eastern Egypt): at 1500–2000 m, laying starts 2nd half of April; elsewhere, fledged young recorded 20 April (Siering 1986; Shirihai in press). Jordan: eggs recorded from mid-April, and family parties at end of month (Hollom 1959; Hüe and Etchécopar 1970); see also Wittenberg (1987). SITE. In crevice *c.* 50 cm deep in rock-face, at least 5 m above ground and often much higher in inaccessible cliffs (Hüe and Etchécopar 1970; Inbar 1979; Paz 1987). Also recorded on ground (Shirihai 1989). Nest: basket-shaped structure of delicate to rough stalks and stems, twigs, and long leaves lined with plant fibres, hair, and fur (Inbar 1979; Shirihai 1989). Nests found built on top of each other, over several seasons, perhaps belonged to this species (Darom 1979). Building: no information. EGGS. See Plate 61. Sub-elliptical, smooth and glossy; light blue or blue-green with brown-black speckles at broad end, though these sometimes lacking (Harrison 1975; Inbar 1979). Nominate *synoicus*: 18·0–20·0 × 14·0–14·5 mm, sample size *c.* 6; calculated weight 2·11 g (Hüe and Etchécopar 1970; Inbar 1979; Schönwetter 1984). Clutch: 4–5 (Inbar 1979; Paz 1987; Shirihai in press); possibly 2 broods in Israel as families with 6–7 young seen July–August (Darom 1979; Shirihai in press), but according to Hüe and Etchécopar (1970) clutch can contain 7 eggs. INCUBATION. 13–14 days (Paz 1987; Shirihai in press); *c.* 3 weeks according to Inbar (1979). Possibly by both sexes as ♂ has brood-patch (Darom 1979; Paz 1987). YOUNG. No information. FLEDGING TO MATURITY.

Fledging period 14–16 days (Inbar 1979; Paz 1987; Shirihai in press). Age of first breeding said to be 1 year (Paz 1987). BREEDING SUCCESS. No information.

**Plumages.** (nominate *synoicus*). ADULT MALE. Forehead, chin, and upper throat glossy crimson-red, merging into pale rosy-red or ruby-red on side of head. Crown silvery-pink, feathers somewhat stiffened and glossy, pink strongly contrasting with red on forehead and red above eye, more gradually merging into brown-grey feathers with pink fringes on rear of crown. Hindneck, side of neck, mantle, scapulars, back, and upper tail-coverts light drab-grey, centres of feathers slightly darker brown-grey, tips paler grey, sides suffused pink; rump rather contrastingly uniform rosy-red or rosy-pink. Red of upper throat gradually shades into rosy-pink of chest, belly, vent, and under tail-coverts; some grey-white of feather-bases shining through, especially on vent, pink appearing paler; side of breast and flank greyish-white or cream-white, unstreaked. Tail dark drab-grey to greyish-black, paler drab-grey on central (t1) and outermost pair (t6); outer webs of all feathers with rather narrow and ill-defined pale cream-grey or isabelline edges. Flight-feathers, greater upper primary coverts, and bastard wing dark drab-grey, slightly darker on tips of inner webs of primaries; outer webs indistinctly fringed paler drab-grey, sometimes with faint pink tinge, fringes narrower but paler and more contrasting on basal and middle portions of primaries; tips of flight-feathers faintly fringed pale grey. Tertials and greater upper wing-coverts dark drab-grey, paler drab-grey or isabelline-grey on borders of outer web. Lesser and median upper wing-coverts drab-grey with slightly paler grey fringes, latter sometimes slightly tinged pink. Under wing-coverts and axillaries cream-white, sometimes with slight pink suffusion. In fresh plumage (autumn), red of face slightly duller and pink of head, upperparts, and chest partly concealed below narrow pale grey feather-tips. *In worn plumage* (about late May to July), red of face more glossy carmine; pink of crown more silvery-white, more restricted; remainder of upperparts duller grey-brown, less grey and pink, more closely marked with ill-defined darker streaks; more grey-white of feather-bases of underparts visible between pink; outer fringes of tertials, tail, and primaries bleached to off-white. ADULT FEMALE. Upperparts and side of head and neck drab-grey, feathers of cap, mantle, scapulars, upper tail-coverts, and side of head with dusky brown shaft-streaks and slightly pinkish-brown borders; rump slightly paler and more uniform drab-grey. Ground-colour of underparts pale cream-grey, faintly tinged pink when plumage quite fresh; vent, thigh, and under tail-coverts paler, grey-white; chin, throat, chest, side of breast, and flank marked with narrow dark grey-brown shaft-streaks. Wing and tail as in adult ♂, without any hint of pink; under wing-coverts and axillaries cream-white, grey of feather-bases partly visible. *In worn plumage* (June to early August), upperparts (except rump) duller grey-brown, more distinctly streaked on cap and mantle; underparts greyish-buff with very faint streaking. NESTLING. No information. JUVENILE. Upperparts and side of head and neck light buff-brown, centres of feathers of cap, lower mantle, scapulars, and longer upper tail-coverts slightly darker grey-brown; rump and shorter upper tail-coverts paler, almost uniform cream-buff. Underparts pale buff-brown, some cream-white of feather-bases partly visible, thigh and under tail-coverts off-white; entirely unstreaked, or with a few very faint dusky marks on chest and flank only. Tail, flight-feathers, tertials, greater upper primary coverts, and bastard wing dark drab-grey; outer webs and tips of tail- and flight-feathers with ill-defined pale cream-grey fringe *c.* 1 mm wide, tertials with similar but broader and even more poorly defined outer border, primary coverts with narrow drab-grey edges. Lesser upper wing-coverts buff-brown; median and greater coverts dark drab-grey with rather ill-defined pale cream-brown fringe along tip. Entire plumage rather like adult ♀ but tinged buff or slightly tawny, less drab-grey, dark marks along shafts of feathers very faint or absent; feathering of (especially) rump and vent shorter, more fluffy. *In worn plumage*, upperparts more diluted sandy-brown with faintly darker brown feather-centres, underparts dirty cream-white with faint buff-brown streaks on chin, throat, chest, and side of breast. FIRST ADULT. Like adult ♀, but juvenile tail, flight-feathers, greater upper primary coverts, and bastard wing retained; in autumn, fringes of these slightly more creamy, less pale grey than in adult ♀, tips of feathers sometimes slightly narrower and more pointed, less broad and truncate, especially bastard wing; in spring, feathers browner than in adult, more sepia, less drab-grey, tips frayed (in adult, still smoothly bordered until March–May). Sexes similar, but 1st adult ♂ sometimes with slight pink shade on head or body, like some adult ♀♀.

**Bare parts.** ADULT, FIRST ADULT. Iris brown or dark hazel. Bill in breeding ♂ grey-horn to steel-grey, slightly darker horn-grey or blackish on culmen and tip, in winter and in ♀ paler grey-horn with darker horn culmen, sometimes with pink or yellow tinge on base. Leg grey-horn with pink tinge, purple-pink with grey tinge, or pink-brown, upper surface of toes greyish-purple, dull grey-horn, or dark brown. (BMNH, ZMA, ZFMK.) NESTLING, JUVENILE. No information.

**Moults.** ADULT POST-BREEDING. 27 nominate *synoicus* in moult examined, but all from rather restricted period, 21 July to 26 August (ZFMK). Of these, primary moult score 6·3 (3) 3–10 on 21–31 July, 8·2 (4) 5–13 on 3–10 August, 24·1 (3) 17–28 on 11–15 August, 29·3 (10) 25–42 on 16–19 August, 34·1 (7) 27–42 on 21–26 August; moult thus probably starts mid-July to early August (with peak in last few days of July), completed late August to mid-September (with peak in early September). Tail moult centrifugal, with many feathers growing simultaneously; starts with primary moult score 5–20, completed at primary score 33–41. Body feathering mainly old at primary moult score 1–15, in full moult (body mainly new, head mainly old, many feathers growing) at score 20–35, mainly new (feathering, especially of head and neck, still growing) at score 40–42. Moult completed in birds from 25 September. (C S Roselaar.) In *salimalii*, Afghanistan, part of feathering of body still growing 13 September to 12 October (Paludan 1959); 2 others just completed 26–30 September (Vaurie 1949b). POST-JUVENILE. Partial: head, body, and lesser, median, and greater upper wing-coverts. Of 12 *synoicus* examined, Sinai (ZFMK), 4 birds almost fully juvenile 29 July to 18 August, 5 birds in heavy moult 8–24 August, 3 birds had moult virtually completed 18–26 August (C S Roselaar). In *salimalii*, single ♀ halfway through moult 26–30 September, 2 ♂♂ and 2 ♀♀ just completed then (Vaurie 1949b); another in moult between 13 September and 12 October (Paludan 1959). In *beicki*, China, moult starts between early and late August (Stresemann *et al.* 1937).

**Measurements.** Nominate *synoicus*. Sinai, mainly July–September and March; skins (RMNH, ZFMK). Data of ♂ refer to red adults only, those of ♀ include all ages. Bill (S) to skull, bill (N) to distal corner of nostril; exposed culmen on average 3·6 less than bill (S).

| | | | | | |
|---|---|---|---|---|---|
| WING AD | ♂ | 88·8 (1·58; 31) 86–92 | ♀ | 83·7 (0·79; 10) 82–85 |
| TAIL | | 62·1 (2·43; 13) 58–65 | | 59·2 (1·31; 11) 58–62 |
| BILL (S) | | 13·7 (0·31; 12) 13·3–14·2 | | 13·6 (0·89; 8) 12·7–14·5 |
| BILL (N) | | 7·9 (0·21; 12) 7·5–8·2 | | 8·0 (0·34; 8) 7·5–8·5 |
| TARSUS | | 19·9 (0·49; 12) 19·2–20·4 | | 19·7 (0·56; 11) 18·6–20·3 |

Sex differences significant for wing and tail. Juvenile and 1st adult ♂: wing 86·6 (1·11; 4) 85–88, tail 62·5 (0·41; 4) 62–63 (ZFMK).

Wing. (1) *C. s. salimalii*, northern Afghanistan (Vaurie 1949b; Paludan 1959); (2) *stolickzae*, western Kun Lun mountains (China) (Stresemann 1930; ZMA). (3) *beicki*, north-east Tsinghai and western Kansu (Stresemann *et al.* 1937; see also Vaurie 1972).

| | | | | | |
|---|---|---|---|---|---|
| (1) | ♂ | 97·8 (11) 92–101 | ♀ | 93·0 ( 2) 92–94 |
| (2) | | 93·8 ( 2) 93–94·5 | | 87·8 ( 3) 86–90·5 |
| (3) | | 89·1 (28) 87–91 | | 85·0 (21) 82–90 |

Nominate *synoicus*. Israel: wing 80–81 (Paz 1987). Petra area (Jordan), wing 81–84 (5) (Phillips 1915).

*C. s. salimalii*. Afghanistan, tail: ♂♂ 72, 74 (Vaurie 1949b).

*C. s. stolickzae*. Ken Lun mountains, tail: ♂♂ 67, 71; ♀ 67 (Hartert 1903–10; ZMA).

*C. s. beicki*. North-east Tsinghai and Kansu, wing: 1st adult ♂ 88·4 (13) 86–91 (Stresemann *et al.* 1937).

**Weights.** Nominate *synoicus*. Israel: 17–24 (Paz 1987).

*C. s. salimalii*. Afghanistan, June and September–October: ♂ 20·7 (9) 19–23, ♀♀ 19, 24 (Paludan 1959).

**Structure.** Wing rather long, broad at base, tip bluntly pointed. 10 primaries: p7–p8 longest, p9 1·5–3·5 shorter, p6 0–1·5, p5 2–5, p4 8–11, p3 12–16, p2 15–19, p1 17–21; p10 strongly reduced, 58–65 shorter than p7–p8, 7–12 shorter than longest upper primary covert. Outer web of p5–p8 emarginated, inner web of p6–p9 with notch (sometimes faint). Tip of longest tertial reaches to tip of p1 or to tips of secondaries. Tail fairly long, tip shallowly forked; 12 feathers, t1 4–8 shorter than t6. Bill as in Scarlet Rosefinch *C. erythrinus*, but slightly more slender at base, finer and more sharply pointed on tip; depth at base of nominate *synoicus* c. 7·0–8·0, width c. 6·0–7·0. Tarsus and toes rather short and slender. Middle toe with claw 17·2 (5) 16·5–18; outer toe with claw c. 73% of middle with claw, inner c. 71%, hind c. 81%.

**Geographical variation.** Distinct, involving depth of general colour, extent of pink and red of ♂, and length of wing, tail, and bill. All races well isolated, each occupying markedly restricted range. Nominate *synoicus* from Sinai, north-west Arabia, Israel, and Jordan small; ground-colour of upperparts and side of body rather dark drab-grey, pink and red of adult ♂ extensive, reaching tips of both upper and under tail-coverts. Birds from Petra area (Jordan) perhaps slightly smaller than birds from Sinai, and sometimes separated as *petrae* Phillips, 1915, but difference too slight to warrant recognition. Extent of pink and red of adult ♂♂ of other races more restricted, on upperparts confined to forehead, forecrown, side of crown, and rump, on underparts from chin to upper belly and upper flank; differences between races mainly in saturation of ground-colour, not in extent of red (C S Roselaar). *C. s. beicki* from China about equal in size to nominate *synoicus*, but bill shorter, more conical; ground-colour of ♂ and general colour of ♀ slightly darker grey-brown or earth-brown. *C. s. stolickzae* from south-west border of Tarim basin (China) rather small; ground-colour pale, light sandy drab-brown. *C. s. salimalii* of northern slopes of Hindu Kush in Afghanistan is large, bill fairly slender; ground-colour as *stolickzae*, but much larger in size. CSR

*Carpodacus roseus* **Pallas's Rosefinch**

PLATES 49 and 51 (flight)
[between pages 640 and 641]

DU. Pallas' Roodmus  FR. Roselin rose  GE. Rosengimpel
RU. Сибирская чечевица  SP. Camachuelo de Pallas  Sw. Sibirisk rosenfink

*Fringilla rosea* Pallas, 1776

Monotypic

**Field characters.** 15·5–16 cm; wing-span 25·5–28 cm. Slightly larger than Scarlet Rosefinch *C. erythrinus*, with less bulbous bill, proportionately slightly longer wings, and 10% longer tail. Medium-sized, quite robust but lengthy rosefinch, with rather dark, noticeably streaked mantle. Adult ♂ has rose-pink head, rump, and underbody, with silver feather-tips on crown and throat; brightest double wing-bar and tertial edges of similarly sized west Palearctic *Carpodacus*. ♀ brown, suffused rose-red on face, chest, and rump and well streaked; darkest of similarly sized *Carpodacus*. Sexes dissimilar; some seasonal variation. Juvenile separable.

ADULT MALE. Moults: July–September (complete). Most beautifully plumaged *Carpodacus* occurring in west Palearctic. Head pale rose-red, almost ruby-toned on face and finely streaked or spotted silvery-white on crown, ear-coverts, chin, and throat; has hoary, even sheeny appearance. Underparts ruby- to pink-red, faintly mottled pink-white except on almost white belly and under tail-coverts. Mantle rose-red, strongly streaked blackish-brown, contrasting with long rosy-red, only faintly marked rump. Wings almost black, washed pink on feather-margins and tipped rose-white on median and greater coverts so that bird shows brightest wing-bars of all west Palearctic *Carpodacus*, particularly on median coverts. Tertials black with pink fringes (bleaching to white) creating obvious pale edges. Underwing silvery-white. Tail greyish-black, with rose-red fringes. When worn, silver and ruby tones of plumage increase and pink fringes and tips of feathers become white or disappear. Bill stout but rather straight-edged, thus noticeably more pointed than in *C. erythrinus*; pale grey-horn, with dusky tip in summer. Legs reddish-grey. ADULT FEMALE. Darkest and most streaked of similarly sized *Carpodacus* in west Palearctic. Head and upperparts dusky brown, well streaked on rear crown, hindneck, and mantle, and distinctly suffused rose-red on forehead, face, and rump; rump little streaked. Thin buff eye-ring less striking than in *C. erythrinus*. Underparts suffused rose-red from throat to chest, otherwise brownish-white, finely streaked dark brown from sides of breast to rear flanks. Wings and tail dark brown, with pale ochre-buff margins and tips producing similar but less bright pattern of marks to those of ♂, whole fringes to tertials (not just tips) pale, unlike *C. erythrinus*. When worn, may show hint of silvery marks on forehead, greyer hindneck and mantle, more white-mottled underparts, and paler but narrower wing markings. JUVENILE.

Greyer even than worn ♀, especially below, and more heavily streaked. Red tones restricted to rump. For 1st-year, see Plumages.

Adult ♂ unmistakable with its mantle streaks and bright pale wing-bars lacking in otherwise similar Sinai Rosefinch *C. synoicus* (a most unlikely vagrant in any case). Distinction of ♀ and juvenile from *C. erythrinus*, potentially vagrant or escaped Purple Finch *C. purpureus*, and escaped House Finch *C. mexicanus* not studied in the field, but clear that at least appearance of young, partially red ♂ *C. erythrinus* converges dangerously with ♀ *C. roseus* (see also Plumages of both species). ♀ *C. purpureus* has bold head markings including pale supercilium and strongly and almost completely streaked underparts, but ♀ *C. mexicanus* far less distinctive (reference to skins is recommended for its exclusion). Flight, gait, and behaviour not known to differ from *C. erythrinus*. Forms small flocks in winter when birds spend much time on ground; escapes into bushes or trees.

Call a short, subdued whistle, repeated in song.

**Habitat.** Breeds in east-central Palearctic, in rather dry continental climatic zone, mainly in mountain forests; in Altai in upper part of forest zone, in wide valleys and flat summits of mountain passes, high-mountain shrub thickets, or subalpine meadows. In winter, descends to broad-leaved woods beside mountain rivulets, scrub thickets in small deep dry valleys, and low-growing groves of aspen *Populus* and birch *Betula*. (Dementiev and Gladkov 1954.) In spring stays hidden and in pairs; at other seasons in bushes and trees or in flocks on ground, occupying river floodlands and also gardens and parks (Flint *et al.* 1984).

**Distribution.** Breeds in central and eastern Siberia. Winters in southern parts of breeding range and in Mongolia, north-east China, Korea, and Japan.

Accidental. Denmark: adult ♂, Blåvand, 12 October 1987 (UGS). Czechoslovakia: October 1850. Hungary: December 1850. (Alström and Colston 1991.) Dementiev and Gladkov (1954) listed records in Tatarskaya ASSR (east European Russia) and Ukraine. See Movements.

**Movements.** Partially migratory, also nomadic. Winter range extends far south of Siberian breeding range, reaching east-central China. Data suggest gradual movement southward in varying numbers. Limited information.

In Yakutia, autumn migration recorded from beginning of September to at least 19 October (Vorobiev 1963; Noskov and Gaginskaya 1977); spring passage recorded on Lena river 24 April to 8 May in one year (Reymers 1966). Irregular in winter south of Lake Baykal in Khamar-Daban mountains (where breeds) (Vasil'chenko 1982). Further east, regular on passage on upper Zeya in Amur region, and sometimes winters there (Il'yashenko 1986). In Ussuriland, common on passage, and winters irregularly in small numbers; most observations on edges of woods, roadsides, and in fields. Numbers and timing vary; passage weak in 1959–60, with first arrivals 27 October to 4 November; stronger passage in 1962 peaked last third of October, with few by mid-November; in 1961, numerous only in early December. Spring passage begins end of February; in 1962, highest numbers in late March. (Panov 1973a.) On Sakhalin, present throughout year, with some movement to south of island; birds move down from mountains to river valleys and plains in autumn (Gizenko 1955). In Mongolia (where probably breeds sparsely), recorded in some localities (Hangay, south-west Hentiy) only in winter, October–March (Kozlova 1930, 1933). In China, winters in north-east and east, west to Inner Mongolia and southern Kansu, south to northern Chekiang (Schauensee 1984; Cheng 1987). Common but irregular winter visitor to Korea, also passage migrant; most records January–February (Gore and Won 1971). In Japan, very uncommon and erratic but annual winter visitor to Hokkaido, where occurs widely, and to highlands of northern and central Honshu; accidental further south, and rare migrant on Sea of Japan islands. Numbers vary greatly from year to year. Most records early November to mid-April, with highest numbers January–March. (Brazil 1991.)

Some birds move west and south-west of breeding range, reaching Tomsk (85°E) in most years, and sometimes Barnaul, Semipalatinsk, and Zaysan depression (80–84°E), probably only in years of heaviest snowfall (Johansen 1944; Korelov et al. 1974). Also reported at Tobol'sk (68°12′E), and recorded exceptionally west of Urals (see Distribution). According to Alström and Colston (1991), records furthest west perhaps involve escaped cagebirds; above data show, however, that some westward vagrancy occurs.                                      DFV

**Voice.** See Field Characters.

**Plumages.** ADULT MALE. In fresh plumage (autumn and winter), narrow strip along base of upper mandible, side of head down to lower cheek, hindneck, and side of neck rosy- or ruby-red; crown paler rosy-pink, each feather with pale silvery-grey centre, appearing silvery with pink scalloping. Mantle and scapulars broadly streaked rosy-red and black, feathers tipped buff-white or off-white, partly concealing streaking; outer webs of outer scapulars broadly fringed pale pink or white, in some attitudes showing as pale stripe on each side of upperparts. Back to upper tail-coverts rosy-red, tail-coverts sometimes tinged grey subterminally and with dull black arrow-marks on centres partly visible. Fairly contrasting patch on chin and throat pale silvery-

grey, consisting of somewhat stiffened, glossy, and round-ended feathers, each feather with rosy-pink fringe; remainder of underparts ruby-red, more pink-red on lower flank and under tail-coverts; mid-belly, vent, upper flank, thigh, and bases of tail-coverts pink-white or pure white, faintly streaked dusky on flank, side of lower belly, thigh, and on concealed centres of tail-coverts. Tail greyish-black, outer webs narrowly but sharply fringed rosy-red (except on outermost pair, t6), tips of feathers narrowly fringed white. Flight-feathers, greater upper primary coverts, and bastard wing greyish- or brownish-black, outer webs narrowly fringed dull pink or brown-pink (except on p9, emarginated parts of p6–p8 and longest feather of bastard wing); fringes often paler and whiter near emarginations, slightly browner on secondaries, purer pink-red on shorter feathers of bastard wing. Greater upper wing-coverts and tertials greyish- or brownish-black, fringes along outer webs pale pink or pink-white, widening towards tips, rapidly bleaching to white, especially on tertials; median coverts dull black with broad pale pink or white tips (often forming paler and more conspicuous wing-bar than that formed by tips of greater coverts); lesser upper wing-coverts black-brown with ruby-red fringes. Under wing-coverts and axillaries silvery-white, sometimes partly tinged pale pink; coverts along leading edge of wing blotched black-and-red. *In worn plumage* (about May–July), feathers of crown and throat-patch more contrastingly glistening silvery-white, tinged pale pink, some black bases sometimes partly visible; red of remainder of head, neck, rump and underparts deeper and more glossy ruby-red, less rose-red; pale feather-tips on mantle and scapulars worn off; pale tips of tail- and flight-feathers worn off; outer scapulars, fringes of tertials, and tips of median and greater coverts purer white. Some individual variation: some birds with brown-buff feather-fringes on mantle and scapulars and pink-brown fringes on greater coverts and flight-feathers, other birds have fringes more pink-white; former birds perhaps in 2nd year of life, latter older ones. ADULT FEMALE. Top and side of head pink-red with brown wash; forehead, lore, and front part of upper cheek finely speckled dusky brown; crown, hindneck, ear-coverts, and rear of cheek finely streaked dusky brown; sides of feathers more brown-grey, especially on nape, ear-coverts, and neck, these appearing browner than face. Narrow pale buff-grey eye-ring. Mantle, scapulars, and back broadly streaked brown-black, each feather with limited amount of pink-red at border of black and broad buff-brown or pink-brown fringe; rump and upper tail-coverts contrastingly rosy-red with short brown streaks or arrow-marks, longest upper tail-coverts with large brown centre and dusky shaft-streak. Entire underparts rosy-pink to pink-red, merging into off-white on mid-belly and vent and into tawny or buff-white on side of breast, rear of flank, and thigh; feathers of throat with small slightly glossy silvery-pink centres, under tail-coverts white with pink wash; chest, side of breast, flank, and upper belly marked with rather narrow but distinct dark brown streaks; sometimes a black-brown malar stripe. Tail brown-black, outer webs and tips narrowly fringed buff-white, fringes more pink-brown towards feather-bases. Wing as in adult ♂, but fringes along outer webs of flight-feathers buff-white, those along outer webs and tips of greater coverts and on tips of median coverts cream-buff, narrower than in ♂, without pink, tips of lesser coverts red-brown, not ruby-red, under wing-coverts without pink or red. *In worn plumage* (May–July), forehead sometimes with tiny glossy silvery marks, crown more heavily streaked black, hindneck and mantle greyer with more pronounced streaks, some brown feather-bases sometimes visible on rump, dark streaks on underparts sharper and more distinct, much white of feather-bases often visible among rosy of chest, breast, and flank, pale fringes and tips of tertials and greater and median coverts bleached to white, partly worn off, wing-bars narrow but distinct. JUVENILE. Upperparts dark grey-brown with broad but ill-defined black streaks, rump with faint red tinge. Side of head and underparts light grey with close black streaking. Tail as adult, but tips of inner webs of feathers sloping, less square and broad; outer fringes tawny or pink-brown. Lesser and median upper wing-coverts black-brown, fringes with tawny or pink-brown tips. For remainder of wing, see 1st adult. (Stegmann 1931a.) FIRST ADULT. Like adult ♀, but tail, flight-feathers, tertials, greater upper primary and secondary coverts, and bastard wing still juvenile, tips of tail-feathers and primary coverts often more distinctly pointed than in adult ♀, more worn at same time of year. Head and body of 1st adult ♂ rather like adult ♀, but cap slightly brighter and more extensively rose-red, less tinged brown; dark streaks on mantle and scapulars on average narrower, more extensively bordered pink; rump more uniform rosy-pink, less marked with brown. Often indistinguishable from adult ♀, except by juvenile tail or primary coverts. Head and body of 1st adult ♀ markedly less pink and red; ground-colour of upperparts grey-brown, or underparts buff-white or white, both heavily streaked brown or black; often some traces of red on forehead and side of crown, some traces of pink on rump, throat, chest, upper belly, and upper flank; rump on average more heavily streaked than in adult ♀. In both sexes, cream-buff tips of median coverts narrower than in adult ♀, 1–2 mm wide (in adult ♀, 2–4 mm), those of juvenile greater coverts and tertials *c.* 1 mm wide only, liable to wear off.

**Bare parts.** ADULT, FIRST ADULT. Iris chestnut-brown or dark brown. Bill grey-brown to steel-grey, darker horn-brown to black on culmen and tip; in winter, upper mandible below nostril and base of lower mandible paler and greyer, sometimes tinged yellow at base on 1st autumn. Leg grey-brown, flesh-brown, reddish-grey, dark purplish-grey, or dull flesh-grey; foot darker brown or brown-grey (Sharpe 1888; Hartert 1903–10; BMNH, ZMA.) JUVENILE. Bill dark grey, leg and foot grey, claws black (Nechaev 1977).

**Moults.** ADULT POST-BREEDING. Complete; primaries descendent. Single ♂ from Sakhalin, 29 August, had primary moult score 38 (p1–p7 new, p9–p10 old); t1 new, other tail-feathers growing; some tertials, secondaries, and many feathers of body growing; another ♂, late September, had all flight-feathers new (Browning 1976). In eastern Siberia, moult mid-July to mid-September; individual primary moult scores *c.* 10 on 21 July, *c.* 33 on 16 August, *c.* 17 on 25 August; secondary moult just started in 2 birds from 16 and 26 August, tertials new or growing (Nechaev 1977). Single ♂ from captivity, 23 August, half-way through primary moult (score 24), body in heavy moult (ZFMK). POST-JUVENILE. Partial: head, body, and lesser and median upper wing-coverts. Moult not yet started in birds examined from western Siberia between early July and early September (Johansen 1944), but birds from eastern Siberia in moult from mid-July to late August or early September (Nechaev 1977); starts during last 10 days of August (Dementiev and Gladkov 1954). Moult completed in birds examined from October and later (BMNH, RMNH).

**Measurements.** Central and eastern Siberia, October–May;

skins (BMNH, RMNH, ZFMK, ZMA). Wing and tail of ♂ include those of red adults only; wing and tail of ♀ are of all ages. Bill (S) to skull, bill (N) to distal corner of nostril; exposed culmen on average 4·9 less than bill (S).

| | | | |
|---|---|---|---|
| WING | ♂ 92·2 (1·53; 19) 89–95 | ♀ 88·0 (1·77; 11) 86–91 |
| TAIL | 64·3 (1·82; 20) 61–68 | 60·8 (3·06; 10) 56–64 |
| BILL (S) | 15·2 (0·54; 22) 14·3–16·0 | 14·9 (0·61; 10) 14·0–15·5 |
| BILL (N) | 8·7 (0·36; 21) 7·9–9·2 | 8·7 (0·38; 10) 8·0–9·3 |
| TARSUS | 20·5 (0·64; 21) 19·5–21·8 | 20·6 (0·44; 10) 19·7–21·1 |

Sex differences significant for wing and tail. Wing of 1st adult ♂ 90·6 (2·17; 13) 85–94 (Meise 1934a; BMNH, RMNH, ZMA).

Wing of paler and darker birds (see Geographical Variation) (Browning 1976, which see for other measurements and for 1st adult ♂).

| | | | |
|---|---|---|---|
| PALE | ♂ 89·6 (3·73; 18) 84–100 | ♀ 86·7 (2·63; 14) 83–91 |
| DARK | 89·2 (1·75; 13) 86–92 | 85·1 (1·52; 11) 82–87 |

**Weights.** Eastern Siberia: late May to early July, ♂ 26·6 (10) 25·0–28·5, ♀ (late May) 27·5 (1); moulting, late July and August, ♂ 28·4 (0·76; 4) 27·7–29·3, ♀ 27·4; juvenile on breeding grounds 26·3 (11) 23·6–28·5, during autumn migration 27 (1); 1st-year 26·9 (0·61; 5) 26·0–27·5 (Nechaev 1977). Kazakhstan, winter: ♂ 30·5 (Korelov *et al.* 1974).

**Structure.** Wing rather long, broad at base, tip bluntly pointed. 10 primaries: p8 longest, p9 1–3 shorter, p7 0–1, p6 2–4, p5 9–13, p4 15–19, p3 18–23, p2 20–25, p1 23–27; p10 reduced, a tiny pin; 57–64 shorter than p8, 8–12 shorter than longest upper primary covert. Outer web of p6–p8 emarginated, inner web of (p7-)p8–p9 with notch (sometimes faint). Tip of longest tertial reaches to tip of p1–p2. Tail rather long, tip slightly forked; 12 feathers, t1 4–8 shorter than t6. Bill short, conical; depth at feathering of base of bill *c.* 8·8–9·8, width at base *c.* 8·4–9·4; similar to bill of Scarlet Rosefinch *C. erythrinus* but relatively slightly longer, less bulbous, tip slightly more compressed laterally, tip of upper mandible more sharply pointed. Tarsus and toes short, rather slender. Middle toe with claw 18·0 (5) 17–19; outer toe with claw *c.* 72% of middle with claw, inner *c.* 70%, hind *c.* 78%. Remainder of structure as in *C. erythrinus*.

**Geographical variation.** Slight; involves colour only. Wintering birds are of 2 types: (1) birds with dark crown, rump, and underparts and with white edges to secondaries, wintering in Sakhalin, Japan, and Korea; (2) birds with pale crown, rump, and underparts and with brown-edged secondaries, occurring across entire winter range. Colour differences apparently not related to season or age of specimens, and probably depend on geographic origin, with dark birds likely to originate from east of breeding range, e.g. Sakhalin; pale birds would then be nominate *roseus*, but no name available for dark birds, as type specimen of *sachalinensis* Portenko, 1960, collected late August on eastern Sakhalin, is pale. Further data on breeding range of dark birds is needed, as is confirmation that dark birds are not a colour morph or age-class. (Browning 1976.) Among 40 birds examined (none from further east than Amur Bay and Vladivostok: BMNH, RMNH, ZFMK, ZMA), no paler and darker birds detected; differences in colour found were only those between ages and sexes.

CSR

## *Carpodacus rubicilla* **Great Rosefinch**

PLATES 49 and 51 (flight)
[between pages 640 and 641]

Du. Grote Roodmus    Fr. Roselin tacheté    Ge. Berggimpel
Ru. Большая чечевица    Sp. Camachuelo grande    Sw. Större rosenfink

*Loxia rubicilla* Güldenstädt, 1775

Polytypic. Nominate *rubicilla* (Güldenstädt, 1775), Caucasus area. Extralimital: *severtzovi* Sharpe, 1886, borders of Tarim basin from Hami through Tien Shan, Dzhungarskiy Alatau (probably), eastern Pamirs, and Kun Lun mountains east to Nan Shan and through southern Tibet and Himalayas to Nepal and eastern Tibet; also (perhaps this race) southern Mongolia; *diabolica* (Koelz, 1939), eastern Afghanistan and western Pamirs to Alay mountains; *kobdensis* (Sushkin, 1925), central and south-east Russian Altai east through north-west Mongolia, Tannu Ola, and Sayan mountains to upper Irkut river.

**Field characters.** 20–21 cm; wing-span 34–36·5 cm. Largest finch reaching west Palearctic, 40% larger than Scarlet Rosefinch *C. erythrinus*; approaches small thrush *Turdus* in length and bulk. Very large, long, but plump-bodied rosefinch, with stout bill and strong legs. Plumage rather dark and uniform, particularly at distance; ♂ drenched in crimson, with dull white spots below; ♀'s underbody dark spotted in strongly linear pattern. Flight noticeably slow and undulating. Sexes dissimilar; little seasonal variation. Juvenile separable.

ADULT MALE. Moults: August–September (complete). Head, sides of neck, underparts, and long rump deep rose-red to scarlet, showing at close range silvery-white striations and spots on face and underbody (size of spots increasing with feather size, largest from breast to flanks). When fresh, face may look hoary. Nape and mantle red, shading to dusky brown due to dark grey mottling on scapulars and lower back when fresh. Wings and tail brownish-black, with paler grey centres and dull red margins to feathers not creating striking pattern except for dark centres to tertials and dusky bases to secondaries. Underwing dusky. Bill stout and conical; upper mandible light brown, lower paler, more yellowish. Legs dark brown. ADULT FEMALE. Head and upperparts pale earthy-grey, with blackish foreface (particularly when worn), small dark spots on crown, face, and throat, and blackish streaks on mantle. Underparts pale ashy-buff, palest on belly and under tail-coverts, with dark brown spots on chest becoming short streaks on flanks and belly and all marks forming obvious linear pattern. Wings and tail dark brown, with pale ashy-grey margins strong enough to form indistinct wing-bar on greater coverts and pale fringes to tertials. JUVENILE. Duller, more buff-toned than ♀; best distinguished by much lighter streaks on underparts. Bare parts differ in totally grey-horn bill and paler flesh legs.

Unmistakable. Large size unlikely to be missed, and no other rosefinch occurring in west Palearctic is so uniformly drenched rose-red and pale spotted below in ♂ nor so strongly streaked below in ♀. Flight slow and deeply undulating; flight silhouette long, recalling (with dark plumage) thrush *Turdus*. Gait a hop. A ground bird, escaping into bushes.

Commonest call a soft 'dyyb' or 'dyuit', recalling Bullfinch *Pyrrhula pyrrhula*.

**Habitat.** In west Palearctic, found only in Caucasus, breeding above 2500 m, sometimes at foot of glaciers where birds land on ice and peck snow. Mainly inhabit sunny alpine meadows, hopping about on grass or perching on jutting ledges of rocks or crags. Such habitats, above *Rhododendron* zone, are characterized by piled-up boulders scattered amid stunted vegetation, with small clusters of birch *Betula*, often flanked by large broken rock screes or steep rock faces, or isolated ledges overgrown with creeping *Rhododendron*. Flies freely, usually in small parties, and in snowy winters descends to upper valleys, there occupying thickets of *Viburnum*, etc.

Extralimitally in Asia breeds in mountains up to 5000 m in highest desolate rocky parts of alpine zone, with at best sparse vegetation, feeding mainly on ground and nesting in rock crevices or under stones. Avoids alpine tundra, boggy meadows, and thickets of alpine dwarf birch. Apparently some preference for juniper *Juniperus*. In winter, descends to river valleys only when driven by hard weather, occurring in valley thickets and lower ends of ravines. (Dementiev and Gladkov 1954.)

**Distribution and population.** No changes reported. No data on population.

**Movements.** Altitudinal migrant; extent of movement depends on weather and food availability at high altitude.

In Caucasus (nominate *rubicilla*), where breeds at 3000–3500 m, birds usually remain above 2000 m in winter, feeding in alpine and subalpine zones, especially on steep and windswept slopes with little snow; they descend (predominantly young birds) to valleys at 900–1000 m only after heavy snowfall. Extent of such movements usually no more than 10–15 km, but up to 30–60 km in harsh winters with food shortage, even reaching foothills. Winter records tend to be in south-east rather than north-west.

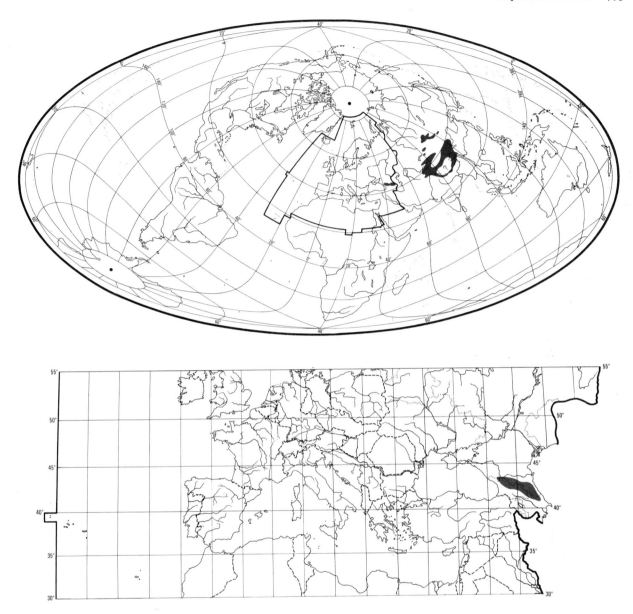

(Loskot 1991.) Record in spring 1865 of 2 ♂♂ 'C. *caucasica*' in Crimea steppe, more than 600 km from breeding range, was attributed to *C. rubicilla* by Kostin (1983), but impossible to verify in absence of skins (Loskot 1991).

Movements of other races have similar altitudinal pattern. In Tuva (south-west of Lake Baykal), Siberian race *kobdensis* regularly encountered below breeding areas in winter, on rocky slopes and in riverine shrubby areas down to 1600 m. In Altai, sometimes descends to 900 m, and recorded down to 500 m in harsh winters. (Loskot 1991.)

In western Tien Shan, *severtzovi* descends to 900–1200 m, occurring fairly regularly in Alma-Ata area; also sometimes descends lower, to foothills near Tashkent, Parkent,

and Yangibazar (Korelov *et al.* 1974; Loskot 1991). In Pamir-Alay, a part of population, predominantly ♂♂, remains above 3600 m, but most birds descend to valleys, usually not below 1100 m; spring return begins at end of March. In lower Zeravshan (western Tadzhikistan), some birds winter on south-facing, snow-free slopes, only returning to breeding grounds at about the end of April. (Abdusalyamov 1977; Loskot 1991.) Widespread in winter in Shakhdara valley (southern Tadzhikistan), occurring chiefly in woods of poplar *Populus* and willow *Salix*, but sometimes even in gardens (Stepanyan 1969*b*). Recorded in winter as high as 4500 m in Tibet, and as low as 1500 m in Himalayas (Gilgit and Sikkim) (Ali and Ripley 1974). In Nepal, usually shows little altitudinal movement; most records at 3660–5000 m, but noted as low as 2650 m in Kali Gandaki valley during particularly cold winter 1982 (Inskipp and Inskipp 1985).                                    DFV

**Food.** Much information extralimital. Diet seeds, flowers, and other plant material; occasionally small insects. Forages on rocky slopes or scree among stunted alpine vegetation, generally at lower altitude than nesting or roosting sites, searching for seeds in cracks and crevices, or on herb-rich meadows with scattered boulders where birds perch to deal with flower- and seed-heads, or among plants on cliff ledges. In spring and summer, feeds almost wholly on ground, very rarely perching in lowest scrub and preferring to pick berries, etc., from ground, but in winter frequently enters bushes (e.g. sea buckthorn *Hippophae*, rose *Rosa*, barberry *Berberis*, juniper *Juniperus*) to take fruits; for most of year forages on, or at edges of, snow. (Sushkin 1938; Dementiev and Gladkov 1954; Neufeldt 1986; Roberts 1992; Loskot 1991.) In Pamirs (Tadzhikistan), in winter, recorded on haystacks and rubbish tips near settlements (Meklenburtsev 1946); in Mongolia, spring, fed around houses and rubbish bins (Piechocki *et al.* 1982), and in Altai (southern Siberia) frequently came to climbers' camps for cereal grains (Kuznetsov 1967). In central Caucasus, birds rapidly descend 1–2 km (0·3–6) in early morning to feeding grounds, returning more slowly in evening. Collects seeds, flowers, and buds from herbs, ground, or snow; small seeds in spring or summer picked from melting snow and swallowed whole; Compositae seeds greatly preferred when unripe and sometimes perches on stem to bend it over to obtain them by pulling away bracts at base of flower-head and extracting them through hole, snipping off pappus before swallowing them; never seen to hold food item under foot; uses hollows and protrusions on stones to hold seed-heads temporarily. Bites ripening Leguminosae seed-pods off at base and picks seeds out of split casing; breaks off stems and unripe fruits of knotgrass *Polygonum* and cinquefoil *Potentilla*, and snips off *Ranunculus* flowers, removing petals in bill and swallowing ovary and stamens. During hot summer days some birds ascend to snowline from alpine belt to feed on grassy slopes until evening. (Loskot 1991, which see for altitudinal change in foraging sites over year.) Pulp of sea buckthorn discarded and only seeds eaten (Stepanyan 1969*b*). In captive study, seeds of berries generally preferred to those of herbs or grasses (Preiser and Massoth 1990, which see for details). In Caucasus, when food abundant will remain for some time in small area, e.g. one ♂ moved 5 m in 2 hrs; in sparser vegetation covers 70–100 m per hr; feeds in bouts of *c.* 1·5–2 hrs before resting; one ♂ made average 23 pecks per min over 4 min. Most seeds eaten are less than 2 mm, though some plant items can be 8 mm in size; takes beetles of mostly 8–10(–12) mm. Unripe seeds of cinquefoil mandibulated for average 16·6 s (8–26, *n* = 23); handling time for seed-head of dandelion *Taraxacum* 48 s (18–72, *n* = 73); usually

extracts *c.* two-thirds of seeds; in optimal conditions deals with 70–75 seed-heads per hr; 180 flowers of milk-vetch *Astragalus* handled in 20 min. (Loskot 1991, which see for many details.)

Diet in west Palearctic includes the following. Invertebrates: flies (Diptera: Trichoceridae), beetles (Coleoptera: Byrrhidae, Curculionidae). Plants: seeds, flowers, buds, etc., of knotweed *Polygonum*, mountain sorrel *Oxyria*, fumitory *Corydalis*, buttercup *Ranunculus*, mouse-ear *Cerastium*, sandwort *Minuartia*, catchfly *Silene*, *Dichodon*, saxifrage *Saxifraga*, Cruciferae, *Sibbaldia*, cinquefoil *Potentilla*, milk-vetch *Astragalus*, vetch *Vicia*, *Oxytropis*, sea buckthorn *Hippophae*, primrose *Primula*, forget-me-not *Myosotis*, speedwell *Veronica*, lousewort *Pedicularis*, bellflower *Campanula*, dandelion *Taraxacum*, goatsbeard *Tragopogon*, grasses (Gramineae), sedge *Carex*. (Boehme 1958; Drozdov and Zlotin 1962; Loskot 1991.)

In central Caucasus, June–July, 16 stomachs contained 1972 items of plant material, of which 69·3% by number fruits (seeds) of *Sibbaldia semiglabra* (Rosaceae), 11·5% fumitory, 8·5% Caryophyllaceae (6% mouse-ear, 2% *Dichodon cerastoides*), 4·2% stamens or ovaries of *Ranunculus crassifolius*, 2·0% unidentified buds, 1·7% seeds of Compositae (almost all dandelion), 1·3% sedge, 1·1% grass; also 23 beetles (22 Curculionidae and 1 Byrrhidae). At end of August, in 10–15 cm of snow, 198 feeding observations made in 2 plots each 50 × 50 m were 49% *Ranunculus caucasicus*, 19% *Taraxacum confusum*, 16% *Polygonum carneum*, 11% *Tragopogon reticulatus*, and 5% *Myosotis alpestris*; *R. caucasicus* clearly preferred, remainder taken according to abundance. In general, preferred families were (in order) Compositae, Rosaceae, Ranunculaceae, Leguminosae, and Caryophyllaceae. (Loskot 1991.) In Kazakhstan, diet consisted of seeds and berries of juniper, sea buckthorn, hawthorn *Crataegus*, avens *Dryas*, liquorice *Glycyrrhiza*, milk-vetch, *Acacia*, and bulbils of bistort *Polygonum viviparum*; main winter food seeds of *Robinia pseudoacacia* (Korelov *et al.* 1974). In Altai, July, ate mainly unripe seeds of juniper (Sushkin 1938); gullets of 2 ♀♀, late July, contained berries of honeysuckle *Lonicera* and unripe Leguminosae seeds (Neufeldt 1986). In Tadzhikistan, preferred seeds in summer were milk-vetch and lyme grass *Elymus* (Abdusalyamov 1977); see also Stepanyan (1969*b*). For 5 winter stomachs from Kirgiziya, see Pek and Fedyanina (1961). In north-west India, very fond of fruit of *Caragana* shrub (Leguminosae), also cultivated peas *Pisum* and cereals (Whistler 1923; Koelz 1937). In central Caucasus, favourite summer food was ovaries (base of future seed-pod) of Leguminosae, especially milk-vetch *Astragalus alpinus* and vetch *Vicia caucasica*; total of 40 species of alpine herbs from 15 families recorded in summer diet, which is characterized by abrupt changes of plants eaten over 1–2 days; insects taken regularly but in small numbers, predominantly large flightless Curculionidae beetles. Other preferred items over summer: at end of June and beginning of July, ovaries of *Primula ruprechtii* and also, though less frequently, buds of *P. algida* and speedwell; towards mid-July, speedwell buds became more important, as well as unripe *Potentilla* fruits, along with dandelion and sandwort; at end of July and throughout August main food was unripe dandelion seeds plus, to lesser extent, flowers of *Polygonum carneum*, and various parts of speedwell, forget-me-not, mouse-ear, saxifrage, and sandwort. (Loskot 1991.) Also in Caucasus, diet in January to early February included sea buckthorn berries and Trichoceridae flies (Drozdov and Zlotin 1962).

In central Caucasus, stomachs contained average 119·6 seeds (37-238, $n = 16$) or 1-3 beetles ($n = 13$). Foraging over 5 m, one bird picked up 7 *Ranunculus* seeds, 4 *Polygonum*, and 2 *Taraxacum*. (Loskot 1991.) Gullet of one bird in Caucasus held 3·5 g of insects (Drozdov and Zlotin 1962).

In central Caucasus, nestlings fed mainly unripe dandelion seeds; adult took *c.* 1 hr to collect 8-10 g, brought to nest in enlarged gullet measuring *c.* 0·7 × 3·0 cm when full (Loskot 1991). In Kazakhstan, grasshoppers, etc. (Orthoptera), and beetles given to young (Korelov *et al.* 1974). Captive brood of 5 received up to 150 beetle larvae per day as well as aphids (Hymenoptera: Aphidoidea) and seeds (Preiser and Massoth 1990, which see for details). BH

**Social pattern and behaviour.** Major study of nominate *rubicilla* in central Caucasus and review by Loskot (1991).

1. Highly sociable, in generally small flocks most of year, rarely 100 or more (Dementiev and Gladkov 1954). Largest flocks (*c.* 150-200) recorded in Caucasus in winter at end of 19th century; 5-20 now generally more usual on feeding grounds, much less frequently *c.* 40-60 (Drozdov and Zlotin 1962; Molamusov 1967; Loskot 1991). 3 local populations studied in Caucasus each comprised 3-5 groups and each of these 5-15 birds (roosting associations: see Roosting, below), increased movement between populations in winter leading to formation of larger flocks. Winter flocks comprise both sexes, but mainly ♀♀ and young ♂♂. (Loskot 1991.) Near Alma-Ata (Kazakhstan) in winter, flocks of 30-50; up to 150 elsewhere (Korelov *et al.* 1974). Single birds and mobile winter flocks (both sexes) up to 10 reported from Tadzhikistan (Stepanyan 1969b; Abdusalyamov 1977), and Mongolia (Piechocki *et al.* 1982); some birds (mainly ♂♂) winter above 3600-3700 m in Pamirs, but most descend to river valleys (Abdusalyamov 1977; Loskot 1991). Flocks of 6-30 also noted in valley bottoms of Pakistan, in winter (Roberts 1992). In Caucasus, remains most of year near glaciers in upper parts of deep river gorges; nesting grounds and roosts separated from feeding grounds by 0·3-0·5(-6) km (altitudinal difference *c.* 100-900 m; see also Breeding Dispersion, below), birds descending to forage there daily. Non-breeders tend to spend virtually all daylight hours on feeding grounds, sometimes (when food abundant) in area of just a few m², but more often unite with neighbouring groups and range over several tens of hectares. From end of May to mid-July in one year, average size of feeding flock 24·3 (3-50, $n = 14$). As slopes became snow-free, focus of activity shifted from *c.* 2600 m at end of May to *c.* 2800-2900 m by mid-July. In late July and August, birds fed practically all day within 0·3-0·5 km of roost. Breeding and some moulting birds in July-August

sometimes solitary, but even these tend to feed for long periods in flocks. Also in breeding season, recorded in pairs and in flocks of varying size: on south slope, July-August, daily average 10·9 birds (3-27, $n = 21$). (Loskot 1991.) Small ♂-only flocks (4-10) occur during incubation and fledging (Dementiev and Gladkov 1954; Korelov *et al.* 1974). In Altai (Mongolia), flocks of 5-6 regularly noted at water 7-10 August (Potapov 1986). One brood in Caucasus joined flock of adults once able to fly well (Loskot 1991). For densities (birds per km²) based on transect counts in central Caucasus, August-September, see Vtorov (1962). In subalpine zone of central Caucasus in winter, 218-293 birds per km², with concentrations in thickets of sea buckthorn *Hippophae* (Drozdov and Zlotin 1962; Voronkova and Ravkin 1974). In Caucasus in summer, birds sometimes associate for feeding with Water Pipit *Anthus spinoletta*, Alpine Accentor *Prunella collaris*, Ring Ouzel *Turdus torquatus*, Red-fronted Serin *Serinus pusillus*, and Scarlet Rosefinch *C. erythrinus* (Loskot 1991). In Tadzhikistan, in September, family parties and single birds recorded in mixed flocks with Snow Finch *Montifringilla nivalis* and Brandt's Rosy Finch *Leucosticte brandti* (Abdusalyamov 1977). In Tibet, ecologically separated from Streaked Great Rosefinch *C. rubicilloides* when breeding, but the 2 species sometimes intermingle in sea buckthorn lower down in winter (Schäfer 1938; Ali and Ripley 1974). BONDS. Mating system apparently essentially monogamous; perhaps some extra-pair copulation, but ringing needed to clarify; other ♂♂ not infrequently tolerated near pair, leading to formation of trios (see, however, Breeding Dispersion, below for defence of ♀) (Loskot 1991). In Pamirs, spring and early summer, trios usually comprised adult ♂, ♀, and immature ♂ (Potapov 1966); similarly in Caucasus study, young ♂ generally tolerated, though not allowed to display very close to ♀. Supernumerary ♂♂ tend to follow pairs especially during ♀'s receptive period (Loskot 1991). Also in Caucasus, mid-July, nest-building ♀ seen to copulate with one of 3 adult ♂♂ displaying (without mutual aggression) nearby; suggested lack of territoriality and possible occurrence of 'polyandry' (Lipkovich 1985, 1986); however, no proof that copulation was not with rightful mate (Loskot 1991). See also below for non-breeders in population. Pair-formation apparently takes place in flocks (see Heterosexual Behaviour, below), and bond close (Schäfer 1938), perhaps extending beyond one breeding season (Loskot 1991). Only ♀ builds nest and incubates; young fed by both sexes (Korelov *et al.* 1974; Loskot 1991), in Tadzhikistan, more by ♀ than ♂ according to Abdusalyamov (1977). Captive nestlings (extra-limital *severtzovi*) fed by both parents, by ♀ only for first 4 days (Preiser and Massoth 1990). Little information on post-fledging care, though in south-east Altai (Russia), 2 broods were each accompanied by ♀ only (Neufeldt 1986). Captive young (fledging at *c.* 16-17 days) started feeding independently at 35 days, but were still fed (by ♂, ♀ having re-laid) for further *c.* 5 days (Preiser and Massoth 1990). Information on age-composition of central Caucasus population as follows. Over 3 years, 35% of 112 birds were adult ♂♂ (probable error *c.* 5%), *c.* 40% 1-year-olds. In winter flock of 60, only 5 adult (red) ♂♂ and, of 12 birds shot, 7 were 1-year ♂♂ (Drozdov and Zlotin 1962). See also Schäfer (1938) for Tibet and Diesselhorst (1968) for Nepal. Despite development of gonads, energetic courtship of ♀♀, and singing, no known case of successful pairing by 1-year-old ♂ (not known whether ♀♀ breed at 1 year); some (sometimes many) adult ♂♂ also remain unpaired and do not breed for unknown reason. (Loskot 1991.) For suggestion that *severtzovi* ♂♂ breed at 1 year, see Schäfer (1938). BREEDING DISPERSION. Solitary, or sometimes apparently a few pairs more clustered (Dementiev and Gladkov 1954; Korelov *et al.* 1974); few nests found and nothing to indicate territoriality, apart, perhaps, from song of ♂. Study in Caucasus found fairly constant (isolated) groups, with little or no movement between them; 3 ♂♂ sang regularly morning and evening 150-200 m apart, and family seen at one of the sites later (Loskot 1991). In typical habitat in Tadzhikistan, pairs usually separated by *c.* 1-2 km, occasionally less (Abdusalyamov 1977). Caucasus study suggested most nests well away from centre of home-range of other birds, so territory defence superfluous; contacts near nest more likely (especially with unpaired ♂♂) where only 0·3-0·5 km between nest and feeding grounds, and ♂ seen to defend radius of 0·5-1 m around ♀. Not known if even nest and its vicinity defended. Only nest located was 70-100 m from roost-sites of another 10 birds and *c.* 1·2 km from main feeding grounds; no other birds seen near this nest over 2 days during laying, and hence no territorial defence, though ♂ sang and preened near nest, including after it was predated, both birds (especially ♂) being seen there for further week. (Loskot 1991.) Copulation observed on feeding grounds, and *en route* to and near nest (Lipkovich 1985, 1986; Loskot 1991). Breeding pairs widely dispersed in Nepal (Diesselhorst 1968), and in Tibet also each pair said to require large area (Schäfer 1938). Little information on density. In alpine zone of Altyn-tu (Altai), only 5 ♂♂ in 10 km² (Sushkin 1938); density also low in summit zone (sparse pine *Pinus* forest and willow *Salix* tundra) of eastern Sayan mountains where 2 birds per km² (Rogacheva 1988), and in Tadzhikistan, highest being 1 pair along *c.* 300 m (Potapov 1966). ROOSTING. Nocturnal and apparently mainly communal. In Caucasus, uses crevices in granite cliffs, roosting throughout year in temperatures near or below 0°C. Summer roost usually located above 2900 m, often on cliffs near large glaciers. Birds seek shelter among rocks from heavy rain or strong wind and, on sunny days, spend hottest period on shady ledges (also reported for Pamirs: Potapov 1966), also preen and sun-bathe; some birds (especially 1-year-old ♂♂) move higher up at this time to forage just below snowline. Composition of roosting flocks changes, especially those of 8-12 birds; some birds evidently use several different roosts, others (both breeders and non-breeders) stick to same one for long period, e.g. in case of one ♂ (recognized by voice) from end of July to mid-August. Feed practically all day in dull, cold conditions. In summer, active *c.* 15½-16 hrs per day; much reduced by mid-winter, but general pattern the same. In July and up to mid-August, leaves feeding grounds when dusk well advanced; in first half of August, returns to roost area *c.* 19.40-20.20 hrs (most 20.10-20.30 hrs). (Loskot 1991.) In Tadzhikistan in summer, birds fly in morning to feeding grounds and to special places to eat salt, shifting thereafter (by midday) to water where they also preen, dust-bathe, etc.; resting period followed by more feeding, then return to rocks for roosting (Abdusalyamov 1977).

2. In Nepal, among shyest of small birds, usually allowing approach only to within *c.* 50 m, and often flying far when disturbed (Diesselhorst 1968). Also found to be very wary in Pamirs (Potapov 1966). Caucasus study found most birds to fly when man *c.* 30-40 m away; sudden sight of man, especially if above birds on slope, elicits Alarm-calls (see 2b in Voice); ♂, late June, permitted gradual approach to *c.* 5-7 m (Loskot 1991), and pair encountered by Stremke (1990) were very confiding, as were those feeding quietly well inside sea buckthorn shrubs in winter, while old ♂♂ in particular much warier on open slope (Drozdov and Zlotin 1962). Further recorded coming into camp to feed in Altai (Kuznetsov 1967), and in Mongolia in winter foraged around houses, while breeding birds (in Altai) still not shy, briefly flying up to high perch, but soon returning to ground to feed (Piechocki *et al.* 1982). In Caucasus, early August, ♀ (perhaps breeding) perching within *c.* 15 m of Kestrel *Falco tinnunculus* gave 'chev-chev' calls (see 2b in Voice) and adopted

horizontal posture with legs and neck extended, head and closed tail raised, crown ruffled, body (apart from belly) sleeked; while calling, made sharp side-to-side movements, rapidly bobbed and flicked wings and tail, also mock-preened; eventually flew off when raptor left (Loskot 1991). FLOCK BEHAVIOUR. Contact-calls (see 2a in Voice) given by birds on ground tend to attract down overflying birds, and much calling (quite often also song from ♂♂) results when these land and join flock; such calls enable birds to locate feeding grounds, even in dense fog. Even during peak courtship period, no sign of aggression among feeding birds: e.g. adult ♂♂ quite often within *c.* 30-50 cm of each other, or several birds on large rock 10-15 cm apart; birds from same group probably know one another. (Loskot 1991.) Similarly, in Nepal, no aggression noted among birds feeding in loose flocks in breeding season (Diesselhorst 1968). Much calling characteristic of birds arriving at roost, others already present responding, sometimes with loud song given by several ♂♂ together, which may lead to conflicts accompanied by threat-calls (presumably 2a-b in Voice) (Loskot 1991). SONG-DISPLAY. ♂ gives loud Advertising-song (see 1a in Voice) from ground or rock; body upright, head raised (even thrown back), crown and throat feathers ruffled (Loskot 1991). Based on study of captive birds, ♂ has head sleeked, wings slightly drooped (and occasionally flapped), tail fanned, neck extended and moved up and down, bird also swaying to left and right and hopping about; loud song given over 10-15 min (Dementiev and Gladkov 1954; Boehme 1958). Subsong (see 1d in Voice) given (with bill closed) occasionally by isolated adult ♂♂ on feeding grounds; usually while hopping about and feeding or while resting on large rocks, and frequently interrupted; bird may also stop and give several phrases of loud Advertising-song (especially as response to song from other ♂♂), then resume quiet Subsong and feeding. In July-August (dawn *c.* 05.00 hrs), song starts *c.* 04.50-05.20 hrs, during 18-21 August 05.50-06.15 hrs. Song given in chorus at roost-site for *c.* 10-30(-50) min after waking, with peak (over *c.* 10-15 min) *c.* 5 min later, most birds then leaving for feeding grounds. Birds give mainly full and loud, often extended, complex phrases (see 1a in Voice), and high level of excitement evident: e.g. 100 songs given in 11 min; singing rate and persistence reduced in poor weather or when only 3-4 ♂♂ in group. On return to roost, song noted when almost dark: e.g. in July, up to 21.05 hrs. (Loskot 1991.) Song-period apparently extends through breeding season, at least to mid-August. Subsong noted from end of June, but mainly July to *c.* 10 August. (Dementiev and Gladkov 1954; Loskot 1991.) ANTAGONISTIC BEHAVIOUR. (1) General. Conflicts between ♂♂ frequent June-July, apparently associated with early and later stages of pair-formation (Loskot 1991); see also Potapov (1966). (2) Threat and fighting. All information from Caucasus study by Loskot (1991). In late June, foraging ♂ gave Subsong and (more than 10 times in 1 hr) adopted Threat-posture (Fig A) and performed Hiss-display: loud Hiss (see 3 in Voice) given while whole body extended and tensed, pivoting from side to side (compare Pivoting-display of, e.g., Goldfinch *Carduelis carduelis*). Hiss-display performed (sometimes *in vacuo*) by 2 other ♂♂, feeding on ground or perched on boulder, July-

A

August; not recorded in grey-plumaged birds; in 4 cases, accompanied actual conflict, once when ♂ approached strange ♀. High level of excitement in ♂ indicated by 3 main variants of Upright-posture all of which indicate aggression, but used in both antagonistic and heterosexual contexts (see Heterosexual Behaviour, below): (a) all plumage sleeked (bird looking tense and unusually slim), neck extended, head and bill pointed up

B

C

(Fig B), often slightly to side (Fig C, left) so that bird being displayed to sees intense red of ear-coverts, and white-spotted red breast and belly; (b) belly ruffled (Fig C, left) with unexpected speed, often followed by preening; (c) head ruffled, contrasting strongly with sleeked body; closed tail often slightly raised and turned to side and wings drooped revealing bright red rump and upper tail-coverts (Fig D, left; see Heterosexual

D

Behaviour, below). ♂ guarding ♀ sometimes has only to move abruptly towards rival (unpaired ♂) and display in Upright-posture, giving threat-call (see 2b in Voice) or loud song in order to discourage closer approach, though once seen to make pecking movements. In more serious conflicts, rival ♂♂ wing up, wing-flick, and Bob up and down, synchronously or alternately; after 12–15 such movements, subordinate ♂ often withdraws a few metres or assumes Submissive-posture (Fig C, right) side-on to dominant bird. For alleged simultaneous bowing by 2 ♂♂ 'bill-to-bill' (doubted by Loskot 1991), see Lipkovich (1986). In 4 cases following Bobbing-display, ♂♂ (once *c.* 1·5 m apart) adopted Threat-posture. 3 confrontations (all on feeding grounds) led to physical contact and fight; occurred, variously, when 2 ♂♂ in dispute over ♀♀, 3rd ♂ joined 2 feeding amicably *c.* 2–3 m apart (leading to Upright-posture, leaps, Bobbing, Bowing, and calls), and between 2 1-year-olds (only conflict of such intensity involving this age-class). All fights aerial, rivals fluttering up *c.* 30–40(–500) cm, one attempting to get above the other, birds pecking and using feet as weapons. Once, when ♀ landed *c.* 2 m from adult ♂ and 2nd ♂ almost simultaneously within *c.* 30–40 cm of 1st, the 2 ♂♂ gave threat-calls in Upright-posture and wing-flicked before flying off in long chase, frequently landing and giving loud song; one eventually sang for *c.* 15 min from cliff, the other returning to ♀. (Loskot 1991.) HETEROSEXUAL BEHAVIOUR. (1) General. In Caucasus study in one year, first signs of behaviour associated with breeding were in last third of June. Process of pair-formation probably differs between first-time breeders and older birds. (Loskot 1991.) (2) Pair-bonding behaviour. In new pair, pair-formation probably takes several days, involving repeated attempts by ♂ to approach ♀ and display before her, until ♀ learns to recognize him and becomes habituated to his close proximity; vocal signals important, including intensive singing by ♂, and pair-members clearly recognize each other's calls. In flocks, 2–3 unpaired ♂♂ sometimes all direct intensive courtship at one ♀. In early stages of pair-formation, ♀ quite often flies off low for *c.* 30 m and lands again when ♂ tries to approach her; usually followed by ♂, so that chase develops, in which ♂ gives rapid song and birds sometimes ascend steeply to *c.* 100 m, flying in zigzags, then descend to ground or rock. Aerial chases recorded mainly end of June and early July (when ♂♂ physiologically more developed than ♀♀). Courtship includes many elements of everyday behaviour, including aggressive. Excited (or alarmed) birds of either sex may change from relaxed posture (body at *c.* 45°, legs bent, belly slightly ruffled, bill horizontal) to Upright-posture with legs extended and crown ruffled. ♂ (in Upright-posture) commonly displays to ♀ by executing 5–7 rapid, pendulum-like leaps (*c.* 30–40 cm) to left and right and 15–20 cm in front of ♀. 1-year-old ♂ had tail slightly raised, performed leaps over *c.* 5 min, and flicked drooped wings abruptly; ♀ was in relaxed posture, but occasionally made pecking movements at ♂; adult *c.* 1·5 m away and assumed Upright-posture, but did not intervene. ♂ will also make Bobbing-Bowing movements, i.e. from Upright- to more bowed posture (which draws attention to brightly coloured rump and upper tail-coverts) and back again; closed tail flicked up almost to vertical and wings flicked outwards (but not above head as reported by Lipkovich 1986). Paired birds quiet when feeding in isolation, not responding even to birds flying over; unpaired birds in contrast highly vocal (including song), and one ♂ seen to perform Soliciting-display (see below) and to run rapidly in circle while calling (see 4 in Voice). (Loskot 1991.) (3) Courtship-feeding. Not recorded. (4) Mating. In Caucasus studies, recorded near nest during building (Lipkovich 1986), and on feeding grounds, in 2 cases 200–400 m from nearest presumed nest-site (Loskot 1991). Noted late

June or mid- to late July, ♀'s receptive period lasting *c.* 2 weeks, at various times 07.30-19.00 hrs. ♂ seen to approach ♀ in Upright-posture (Fig D, left), ♀ then crouching and wing-shivering (Fig D, right), but no reaction from ♂. On another occasion, ♀ approached singing ♂ (after he had returned from chasing another ♂), called (see 4 in Voice) and adopted Soliciting-posture, ♂ then mounting and copulating; both afterwards flew up slope, ♂ singing, also continuing to display and sing, flying from rock to rock while ♀ fed. ♂ (in Upright-posture: Fig D, left) recorded leaping in front of ♀, also wing-flicking, and alternating Bobbing or Bowing with 10-15 mincing steps on extended legs; in one case, ♂ leaped *c.* 1 m after copulating, ♀ then flying off and being pursued by her own and another (1-year-old) ♂. (Loskot 1991.) In study by Lipkovich (1986), 2 other ♂♂ present when apparent pair copulated, and all 3 displayed and chased ♀ afterwards. (5) Nest-site selection and behaviour at nest. No information on selection process, but ♀ likely to play major role as in other Fringillidae (Loskot 1991). When ♀ building, ♂ sang near foot of rock wall (in which nest situated) or accompanied ♀ when she collected material near nest or up to 1 km away (Lipkovich 1985, 1986). Observations of pairs feeding, drinking, and preening together at various times 10.00-17.00 hrs suggest incubating ♀ leaves nest at hottest time of day (Loskot 1991). RELATIONS WITHIN FAMILY GROUP. For nestling phase, no information additional to that given under Bonds (above). In Caucasus, late August, brood of 3 fledglings hid initially in rock crevices. Generally quiet when sated, giving only occasional calls for contact with brood-siblings or in response to other conspecific birds; eventually start to give food-calls 1-2 times per min, increasing when conspecific birds (also *A. spinoletta* and *P. collaris*) fly past, and peaking on hearing contact-calls announcing arrival of parents. (Loskot 1991.) ANTI-PREDATOR RESPONSES OF YOUNG. Hide in crevices after fledging; when able to fly well (with adults), did not permit observer within *c.* 40-50 m (Loskot 1991). PARENTAL ANTI-PREDATOR STRATEGIES. (1) Passive measures. ♂ of pair with eggs flew across to far side of gorge while ♀ stayed near nest and came within 3-4 m of observer. Quiet when feeding young, especially near nest. (Loskot 1991.) (2) Active measures. ♂ sometimes sings when disturbed (e.g. near nest-building ♀). Birds presumed to have nest with young were highly agitated, flying about human intruder and giving Alarm-calls (see 2b in Voice) (Dementiev and Gladkov 1954).

(Figs by D Nurney from drawings and photographs in Loskot 1991.)                                                                        MGW

**Voice.** Only detailed study (including sonagrams) is of nominate *rubicilla* in Caucasus, where vocal signals play important role in communication, dense fog being of frequent occurrence (Loskot 1991). In Nepal, August, *severtzovi* is quiet, not singing (Diesselhorst 1968), but frequent Contact-calls noted in Pakistan in winter (Roberts 1992). Possible geographical variation yet to be investigated. Repertoire overall much as in extralimital Red-breasted Rosefinch *C. puniceus* and songs similar (Gavrilov and Kovshar' 1968; Korelov *et al.* 1974); voice also similar to Streaked Great Rosefinch *C. rubicilloides* (Schäfer 1938; Ali and Ripley 1974).

CALLS OF ADULTS. (1a) Advertising-song of ♂ nominate *rubicilla*. Short flowing phrase comprising 1 or more segments within main frequency range 2-6 kHz somewhat reminiscent of very rapid laugh. Short units like inverted

V or U (i.e. rising and falling in pitch) are frequent (see *severtzovi*, below). Song easily recognizable, but considerable variation (between and within individuals) in number of units, order of delivery, temporal pattern, timbre, and intonation. Commonest (most widespread) song-types comprise 13-17 units and last *c.* 2 s. One such phrase roughly rendered 'vi-fyu-fyu-fyu-fyu-vi-tyuv-yuvyuvyuvyuvyuv-tsi'; stress on first 2 units, middle and especially higher-pitched end-section quieter, sometimes with final flourish. Some song-phrases shorter (*c.* 1·5 s), comprising 7-9 units: e.g. 9 clear-toned fluting whistles ('fyu') delivered at even, relatively slow rate; as reported earlier by Dementiev and Gladkov (1954), first 2-3 units in this phrase especially loud and higher pitched, followed by gradual diminuendo and pitch descent. Another short type is composed of complex units with harmonics delivered at very fast rate, producing trill-like effect, but in several phrase-types accelerando is typical, with shorter pauses between units in final section than between first 1-2 or 5-8. When excited, especially during communal singing near roost or displays near ♀♀ (see Social Pattern and Behaviour), song-phrases of 19 or more units lasting *c.* 3·5-4·0 s, usually 3-4 segments (each of 2-6 loud units) separated by quiet segments of 2-4 merged units. (Loskot 1991.) Song of captive ♂ comprised bell-like 'tschi', louder 'tüt' units, or 'tschüt' followed by 7 'tit' notes (Michaelis 1980). Some songs carry up to 1 km or more (Dementiev and Gladkov 1954). (1b) Song of ♀ nominate *rubicilla*. Short, simple phrase given during nest-building, copulation, and laying (Loskot 1991). (1c) Advertising-song of other races. Brief descriptions suggest no significant differences from nominate *rubicilla* but only 1 recording (of *severtzovi*: see below) available for analysis. In *kobdensis* of Mongolia and Altai (Russia), phrase comprising full whistling notes rising and falling in pitch and similar to calls of Crossbill *Loxia curvirostra* (Sushkin 1938; Mauersberger *et al.* 1982). *C. r. severtzovi*: in Tibet, short, clear, melodious song also likened to *L. curvirostra* (Schäfer 1938); in Nepal, loud, slow and plaintive 'weep', also series of quiet chuckling sounds (Fleming *et al.* 1984); in Pakistan, brief stereotyped song-phrase starting with 2-3 spaced, interrogative whistles, 'twee-twee twe-cho-chush-u' (Roberts 1992). Recording of *severtzovi* contains 5 predominantly tonal song-phrases showing distinct trend to start at high pitch and end at lower pitch (though with some intermediate ascent). Phrase shown in Fig I suggests '(w)hee hoo whee too-too-tu-tu-tu', with some decrease in loudness. In this very small sample, U- or V-shaped units occur as frequently as inverted ones. (J Hall-Craggs.) (1d) Subsong. In Caucasus study, a quiet twittering, comprising predominantly modified units of Advertising-song and main calls interspersed with variety of sounds, including mimicry; individual units and 'trills' (presumably segments) are separated by pauses of up to 0·5 s. Most frequent and accurate imitations are of Ripple-call of Red-fronted Serin *Serinus pusillus*, call of

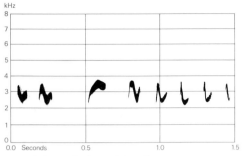

I   T J Roberts Pakistan July 1987

Bullfinch *Pyrrhula pyrrhula*, and 'pink' of Chaffinch *Fringilla coelebs*; less frequently mimics various calls and song-fragments of Alpine Accentor *Prunella collaris*, Ring Ouzel *Turdus torquatus*, Alpine Chough *Pyrrhocorax graculus*, and Goldfinch *Carduelis carduelis*. (Loskot 1991.) (2) Contact- and alarm-calls. (2a) Contact-call. Similar to other *Carpodacus*. Nasal 'kui' or 'kuii' lasting 0·14–0·28 s; sometimes shorter (0·11 s) and more incisive, like modified 'tvi'; also rendered 'vii', 'kvi', 'tvii', 'vit', and 'tvit' (Loskot 1991). Further descriptions: muffled short whistle, suggesting *P. pyrrhula*, but more muffled and coarser (Dementiev and Gladkov 1954); soft 'pyeeu-een pyeeu-een' (Jonsson 1992). Loud 'tooey tooey' noted from *severtzovi* in winter (Roberts 1992) is obviously similar; see also Potapov (1966) and Korelov *et al.* (1974). In Mongolia (*kobdensis*), delicate 'fiu fiu' (Mauersberger *et al.* 1982) or soft 'djü djü' (Piechocki *et al.* 1982); such transcriptions suggest descending calls, but sonagrams in Loskot (1991) also show both ascending and descending calls. In study of nominate *rubicilla*, commonest call, given in variety of contexts: in flight, especially by lone birds, while beating wings (descent phase of undulations silent), at take-off, in conflicts, or by displaying or agitated birds, e.g. ♂ having lost sight of ♀; birds, especially in feeding flocks, always respond to calls, thus usually attracting down conspecifics. (Loskot 1991.) (2b) Alarm-call. Complex call (with harmonics), descending in pitch in final part; frequency range 1·8–8 kHz, duration 0·2 s. Renderings include incisive '(t)chek', 'bzhe(k)', 'chvi', and 'bzhi(i)'; 'chev-chev' given by ♀ near Kestrel *Falco tinnunculus*, and by ♂ in apparent aggressively motivated excitement. (Loskot 1991.) Similar abrupt 'chvi', 'chvik', or 'chik' given by *severtzovi* (Korelov *et al.* 1974). In Mongolia (*kobdensis*), loud 'schütt schütt' noted from ♂ in early June (Mauersberger *et al.* 1982) presumably also belongs here. (2c) Twitter-call. Short twitter noted from flying birds (Dementiev and Gladkov 1954). Apparently same call noted in *severtzovi*, 'twit ping' from birds flying to roost in Tibet (Schäfer 1938); in Pakistan, rapid twittering, lower pitched than in smaller Carduelinae (Roberts 1992). (3) Hissing sound like stream of compressed air given by ♂ in Hiss-display (see Social Pattern and Behaviour) and in conflicts with other ♂♂ (Loskot 1991). (4) Soliciting-call of ♀. Series of 'tiv' sounds, like food-call of young (see below). Once noted from displaying ♂ when flock flying over. (Loskot 1991.)

Calls of Young. No information for nestlings. Brood of 3 recently fledged young (late August) gave loud 'tiu', 'tiv', or more drawn-out 'piiv' sounds; when being fed, calls merged into uninterrupted 'tivtivtivtivtiv' (Loskot 1991).          MGW

**Breeding.** Season. Caucasus: very late due to brief spring–summer period; earliest clutches in central Caucasus not before mid-July, many laid last third of month (Loskot 1991); young found still in nest at end of August (Lipkovich 1986), and well-fledged young recorded mid-August (Dementiev and Gladkov 1954). Kazakhstan: eggs laid from end of June (Potapov 1966; Korelov *et al.* 1974). Tadzhikistan: eggs laid from about mid-July; young still being fed at end of August (Abdusalyamov 1977). Altai: 2–3 weeks earlier than in Caucasus; fledgling seen end of July, and strongly flying young mid-August (Kuznetsov 1967; Neufeldt 1986). Nepal: at 5000 m, newly fledged young recorded late August (Diesselhorst 1968). Site. In cleft or crevice in rock face, or below boulder (Dementiev and Gladkov 1954; Potapov 1966; Lipkovich 1985, 1986; Roberts 1992). In Caucasus, one nest in roomy niche 40 cm deep and 22 × 31 cm wide, protected by overhang, 4·3 m above ground (Loskot 1991); another was 7 m up in crevice (Lipkovich 1985). Accounts of nests in shrubs or low in trees (e.g. Baker 1934, Dementiev and Gladkov 1954, Hüe and Etchécopar 1970) need confirmation. Nest: foundation of thin twigs, stalks, grass, and moss, warmly lined with hair, wool, and in some cases many feathers (Abdusalyamov 1977; Lipkovich 1985; Loskot 1991); 2 nests in Caucasus had outer diameter 12·0 × 16·5 cm and 17·0 cm, inner diameter 7·5 × 9·3 cm and 8·5 cm, overall height 11·3 cm and 14·0 cm, depth of cup 5·8 cm and 3·5 cm; dry weight of one nest 166 g (Lipkovich 1985; Loskot 1991); see Abdusalyamov (1977) for 2 nests in Tadzhikistan. Building: by ♀ only; one nest took *c.* 7 days (Lipkovich 1985); *Saxifraga* stems pulled out by roots, and one nest contained feathers of snowcock *Tetraogallus* collected *c.* 1 km away (Loskot 1991); in captivity took 5 days (Preiser and Massoth 1990). Eggs. See Plate 61. Sub-elliptical to oval, smooth and glossy; intense sky-blue with slight greenish tinge, very sparsely marked at broad end with black specks, blotches, and scrawls (Harrison 1975; Preiser and Massoth 1990; Loskot 1991). *C. r. severtzovi*: 24·0 × 17·0 mm (21·9–28·3 × 16·2–18·7), *n* = 88; calculated weight 3·65 g (Schönwetter 1984); this sample includes 80 eggs described by Baker (1934) which according to Loskot (1991) are not of this species. Clutch: 4–5(–6); in central Asia, 4–6 (Dementiev and Gladkov 1954); in Tadzhikistan, 3 clutches were all of 5 (Abdusalyamov 1977). In central Caucasus, 1 brood and no replacement clutches recorded (Loskot 1991). Captive ♀ started laying 7 days after nest completion, commenced 2nd nest when young 13 days old, and laid 1st egg of 2nd clutch 3 days after 1st brood fledged (Preiser and Massoth 1990). INCUBATION. By ♀ only (Preiser and Massoth 1990; Loskot 1991); in captivity, 16 days starting with 3rd egg (Preiser and Massoth 1990). YOUNG. In Tadzhikistan, ♀ recorded doing more feeding of young than ♂ (Abdusalyamov 1977). In captivity, fed and cared for by both parents, by ♀ only for first 4 days (Preiser and Massoth 1990). FLEDGING TO MATURITY. Fledging period in captivity *c.* 17 days; picked up food 18 days after fledging but still fed by parents for further 5 days at least (Preiser and Massoth 1990). For age of first breeding, see Social Pattern and Behaviour. BREEDING SUCCESS. In Caucasus, nests predated by Alpine Chough *Pyrrhocorax graculus*, and probably also by stoat *Mustela erminea* (Loskot 1991). No further information.

BH

**Plumages.** (nominate *rubicilla*). ADULT MALE. Forehead and crown down to eye glossy carmine-red or dark blood-red, each feather with small white speck subterminally, *c.* 0·5 mm across; feather-base sooty-grey, sometimes partly visible. Hindneck, mantle, scapulars, back, and upper tail-coverts duller carmine-red, sooty-grey of feather-bases shining through or just visible, feathers narrowly fringed dark grey when plumage fresh; rump clearer pure rosy-red or ruby-red. Side of hindneck glossy ruby-red, black of feather-bases partly shining through, partly just visible on lore and just below and behind eye; longer ear-coverts with a silvery-pink sheen. Feathers of entire underparts sooty-grey on bases (darkest on chin and throat), ruby-red or rosy-red on tips; small whitish subterminal specks on chin and throat, gradually larger pale grey ones on chest, side of breast, and flank; dark feather-bases usually exposed on chin and upper throat only (unless plumage heavily worn), pale grey centres and red fringes on chest, side of breast, and flank give underparts scalloped appearance. Rear of flank, vent, and under tail-coverts pink-red with some light grey of feather-bases visible, red less glossy than on remainder of underparts. Tail black with brown or grey tinge; narrow edges along tips and outer webs of feathers light grey, slightly tinged pink on outer web of t1-t5. Flight-feathers, greater upper primary coverts, and bastard wing greyish- or brownish-black, outer webs narrowly and sharply fringed dull pink (except along outer web of p9, on emarginated parts of p5-p8, and on longest feather of bastard wing); tips of flight-feathers narrowly fringed grey-white. Lesser and medium upper wing-coverts sooty-grey with dull carmine red fringe, like scapulars; greater coverts and tertials dark brown-grey to greyish-black with ill-defined rosy-pink fringe along outer web, widening towards feather-tip. Under wing-coverts and axillaries dark grey with rosy-red borders or tips. Occasionally, a few feathers on side of crown, side of breast, flank, or elsewhere brown-grey without red, either newer or older than remainder of plumage; these probably moulted outside moulting season, not indicating immaturity. *In fresh plumage* (autumn), carmine-red of head and body rather dull, partly concealed by grey fringes on mantle and scapulars, silvery tinge on ear-coverts and white specks on cap, chin, and throat inconspicuous; in worn plumage (spring and early summer) red fringes of head and body more glossy scarlet-red, sometimes tinged orange-scarlet; much dark grey of feather-bases visible on hindneck, mantle, scapulars, and from chest downward, much black on cap, face, and throat; grey-white tips of flight-feathers worn off; dull pink fringes along outer webs of flight-feathers, tertials, primary coverts, and

greater coverts bleached to pink-white, but largely worn off when plumage heavily abraded. ADULT FEMALE. Forehead and front part of cheek black with fine pale drab-grey scalloping, gradually merging into less contrastingly light drab-grey and sooty-black streaked crown, ear-coverts, and rear of cheeks; lore almost uniform sooty-grey. Hindneck and side of neck light drab-grey with some faint darker grey streaks; mantle, scapulars, and back light drab-grey with more pronounced though ill-defined dark grey streaks, each streak ending in dull black shaft-streak on tip of feather. Rump pale drab-grey, virtually unstreaked, sometimes with slight sandy-buff tinge; upper tail-coverts darker grey with black shafts and narrow pale drab-grey fringes. Chin and throat black with broad light drab-grey or cream-grey scalloping, merging into cream-grey of remainder of underparts, which are heavily spotted black on chest and side of breast and broadly streaked dull black or sooty-grey on flank, side of belly, and under tail-coverts; lower mid-belly, vent, and thigh uniform greyish-white. Tail as adult ♂, but narrow edges along outer webs and tips of feathers grey without pink tinge, almost white along outer web of t6. Flight-feathers, tertials, greater upper primary coverts, bastard wing, and greater upper wing-coverts as adult ♂, but ground-colour slightly paler sooty-grey and fringes along outer webs without pink; fringes along outer webs and tips of primaries narrow but sharply defined, grey-white, those along primary coverts and secondaries greyer, those along distal half of outer web of greater coverts and tertials broad and pale cream or white, far more contrasting with remainder of wing than pink fringes of adult ♂. Lesser and median coverts like scapulars. Under wing-coverts and axillaries drab-grey with dark shafts; shorter coverts with blackish bases. *In worn plumage* (spring and early summer), pale scalloping of forehead, crown, cheek, chin, and throat largely worn off, face appearing blacker; black streaks on mantle, scapulars, and underparts still heavier, ground-colour of upperparts more dull brown-grey, of underparts dirty white; pale fringes of tail and wing largely or entirely worn off; central pair of tail-feathers and outer web of tertials partly bleached to grey. NESTLING. No information. JUVENILE. Rather like adult ♀, but dark streaks on head and body browner, narrower, less sharply defined; feather-bases on face, side of head and throat much less dark, more sepia, streaking here more similar to that on remainder of body; ground-colour of upperparts slightly more buffy-grey, on underparts more dirty cream to buff-white; breast and belly cream to off-white with faint and narrow dusky streaks, less heavily marked than in adult ♀. Tail as adult ♀ but distal end of inner webs narrower, more sloping to tip, less broad and truncate; wing as in adult ♀ but lesser and median coverts similar to juvenile scapulars; greater coverts with pale grey-buff or cream-grey outer fringe which extends along tip of covert into a slightly darker fringe of equal width on distal part to inner web (in adult ♀, fringe along outer web slightly less well-defined, not extending onto tip of inner web), tip of bastard wing more pointed, less rounded, outer web of longest feather with pale fringe extending to tip (fringe narrower, shorter, or absent in adult ♀). FIRST ADULT. Like adult ♀, but juvenile tail, flight-feathers, tertials, greater primary and secondary coverts, and bastard wing retained; for characters, see juvenile, above. In autumn, shape of tail-feathers and pattern of fringe on greater coverts and bastard wing usually a reliable character for ageing; in spring, when heavily worn, these characters often difficult to detect, but tail and flight-feathers then often browner and more worn than those of adult at same time of year and shape of tip of bastard wing usually reliable. Sexes similar.

**Bare parts.** ADULT, FIRST ADULT. Iris brown or dark brown. Upper mandible light brown or grey-horn, cutting edges and lower mandible dusky yellow, milky-yellow, cream-yellow, light pink-grey, light greenish-horn, light olive-brown or light yellow-horn, tip slightly darker. Leg and foot dark brown, greyish-black, or horn-black, toes deep brown to black. (ZMA.) NESTLING. No information. JUVENILE. Iris brown-black. Bill grey-horn, tinged dark brown on culmen. Leg and foot brownish-flesh, dusky brown when older. (Sharpe 1888.)

**Moults.** ADULT POST-BREEDING. Complete; primaries descendent. Moult August–September (Ali and Ripley 1974); in captivity, late July to September (Dementiev and Gladkov 1954). Moult generally not started by July or early August (Vaurie 1949*b*; Dementiev and Gladkov 1954; ZFMK); moult not started in 3 adults from Ladakh 2–8 July, but single 1-year-old ♀ just started (primary moult score 3) (ZMA). Adult ♂ from Nepal had just started 24 August (Diesselhorst 1968). In Ladakh, 2 ♂♂ near completion of moult 26–27 September (Vaurie 1949*b*); in Caucasus, single ♂ on 19 October (moult score 44) (ZFMK). Mould generally completed October (India, Pakistan, Caucasus, Tien Shan) (Vaurie 1949*b*; RMNH, ZFMK). POST-JUVENILE. Partial: head, body, and lesser and some or all median upper wing-coverts. Probably starts soon after fledging; in July, still in juvenile plumage (Johansen 1944; Dementiev and Gladkov 1954); 3 birds in moult 26–27 September, Nepal (Diesselhorst 1968); moult completed October or later (BMNH, RMNH, ZFMK). See also Loskot (1991).

**Measurements.** Nominate *rubicilla*. Northern Caucasus, October–March; skins (RMNH, ZFMK, ZMA). In this and following samples, data for ♂ include red adults only; ♀ all ages combined, possibly including wrongly sexed 1st adult ♂♂. Bill (S) to skull, bill (N) to distal corner of nostril; exposed culmen on average 6·2 less than bill (S).

| | | | | |
|---|---|---|---|---|
| WING | ♂ 118·9 (1·85; 20) | 116–122 | ♀ 115·1 (2·43; 5) | 112–118 |
| TAIL | 90·3 (2·40; 20) | 86–94 | 86·5 (1·84; 5) | 85–89 |
| BILL (S) | 20·5 (0·97; 20) | 19·0–22·0 | 20·6 (0·64; 5) | 19·6–21·3 |
| BILL (N) | 11·6 (0·66; 20) | 10·9–13·0 | 11·8 (0·17; 5) | 11·5–12·0 |
| TARSUS | 23·9 (0·65; 20) | 22·8–24·8 | 24·8 (0·75; 5) | 24·0–25·5 |

Sex differences significant for wing and tail. Wing: 116·8 (12) 113·5–121·5, ♀ 109·8 (9) 105–112·5 (Dementiev and Gladkov 1954). See also Loskot (1991).

*C. r. severtzovi.* Tien Shan and Kun Lun mountains (central Asia), November–June; skins (RMNH, ZMA).

| | | | | |
|---|---|---|---|---|
| WING | ♂ 115·9 (20·5; 22) | 112–119 | ♀ 111·9 (2·61; 8) | 108–117 |
| TAIL | 85·8 (3·17; 6) | 81–90 | 81·8 (3·28; 8) | 78–87 |
| BILL (S) | 19·4 (0·72; 29) | 18·0–20·4 | 19·2 (0·74; 8) | 18·3–20·2 |
| BILL (N) | 10·9 (0·46; 6) | 10·4–11·6 | 11·3 (0·29; 8) | 10·7–11·6 |
| TARSUS | 23·5 (0·96; 6) | 22·6–24·8 | 23·2 (0·53; 8) | 22·3–24·0 |

Sex differences significant for wing and tail. Wing of 1st adult ♂ 111·0 (2·97; 4) 107–114, adult ♀ 112·4 (3·14; 4) 110–117, 1st adult ♀ 111·5 (2·34; 4) 108–114 (RMNH, ZMA). In Kazakhstan, wing of adult ♂ 110–117 (7), 1st adult ♂ 107–109 (5); tail, adult ♂ 85–91 (7), 1st adult ♂ 81–85 (5) (Korelov *et al.* 1974). See also Loskot (1991).

Wing: (1) Tien Shan, (2) Kun Lun mountains, (3) Ladakh and Kashmir (Vaurie 1949*b*, RMNH, ZMA). (4) Tibet (Vaurie 1972).

| | | | | |
|---|---|---|---|---|
| (1) | ♂ 115·6 (2·10; 30) | 112–119 | ♀ 113·6 (2·56; 4) | 111–117 |
| (2) | 117·2 (0·87; 4) | 116–119 | 110·2 (1·32; 4) | 108–112 |
| (3) | 119·8 (1·42; 12) | 118–122 | 116·7 ( — ; 3) | 114–119 |
| (4) | 12·0 ( — ; 75) | 115–128 | 114·7 ( — ; 43) | 111–121 |

*C. r. kobdensis.* Wing, ♂: north-west Mongolia, 110–119 (21); central Altai, 108–111 (5) (Vaurie 1949*b*) or 115 (RMNH); see also Piechocki *et al.* (1982). See also Loskot (1991).

**Weights.** ADULT, FIRST ADULT. *C. r. severtsovi.* Tien Shan (Kazakhstan): ♂ 45·9 (11) 42·0–48·5, ♀♀ 45, 46·8 (Korelov *et al.* 1974). Nepal, August: adult ♂ 43·2, juveniles 39·5, 40·8, 41·7 (Diesselhorst 1968).

*C. r. kobdensis.* Mongolia, February: ♂ 46·5, ♀ 45·5 (Piechocki *et al.* 1982).

**Structure.** Wing rather long, broad at base, tip bluntly pointed. 10 primaries: p7–p8 longest, p9 and p6 1–4 shorter (on average, p9 2·6 shorter, p6 2·3, *n* = 15), p5 7–11, p4 15–19, p3 20–24, p2 24–29, p1 27–35; p10 minute, a tiny pin concealed below reduced outermost greater upper wing-covert, 74–90 shorter than p7–p8, 10–18 shorter than longest upper primary covert. Outer web of p5–p8 emarginated (sometimes faint on p5), inner web of p6–p9 with notch (often faint on p6–p7). Tip of longest tertial reaches tip of about p1. Tail rather long, tip shallowly forked; 12 feathers, t4–t5 longest, t6 1–3 shorter, t1 3–9 shorter. Bill short, massive, conical; depth at feathering of base 12·5–14·5 in nominate *rubicilla*, width 12·0–12·5, depth 11·5–12·5 in *severtzovi*, width 10·0–12·0; tip of culmen and cutting edges gently decurved. Nostril small, rounded, covered by tuft of short bristly feathers; some longer hair-like feathers at lateral base of upper mandible. Tarsus and toes rather short, but strong. Middle toe with claw, 21·2 (15) 19–23 in nominate *rubicilla*, 19·7 (15) 18·5–21 in *severtzovi*; outer toe with claw *c.* 74% of middle with claw, inner *c.* 69%, hind *c.* 73%.

**Geographical variation.** Pronounced in depth of colour, slight in size. Nominate *rubicilla* from Caucasus area and perhaps eastern Turkey (Kumerloeve 1961) is darkest race: head and underparts of adult ♂ fairly dark blood-red, remainder of upperparts of both sexes and head of ♀ and 1st adult ♂ rather dark greybrown; white spots on head and throat of adult ♂ small, 0·5–1 mm across. *C. r. severtzovi* from mountains surrounding Tarim basin is palest race, adult ♂ rose-red, with spots on head and throat *c.* 2–3 mm across, cap, fore-cheeks, and chin to chest appearing silvery-white with red scalloping; ground-colour of top and side of head and neck and of remainder of upperparts and upperwing much paler drab-grey, suffused pink on mantle, back, upper tail-coverts, and wing, pure rose-pink on rump; underparts paler pink-red; adult ♀ and 1st adult of both sexes much paler light sandy-grey on head, upperparts, and upperwing, narrowly streaked grey-brown on head, mantle, and scapulars; underparts paler cream-white with rather restricted amount of narrow grey-brown streaking; size slightly smaller than in nominate *rubicilla*. *C. r. diabolica* from eastern Afghanistan and western Pamirs north to Alayskiy mountains intermediate in plumage between nominate *rubicilla* and *severtzovi*: head and underparts of adult ♂ dark carmine, but lighter and not as saturated as nominate *rubicilla*, not as rosy as *severtzovi*; white spots on head and throat *c.* 1 mm across; bill on average *c.* 1·5 mm longer than in other races (Vaurie 1949*b*); considered inseparable from *severtzovi* by Loskot (1991). Isolated *kobdensis* from Altai and north-west Mongolia through Sayan mountains to south-west corner of Lake Baykal also intermediate in colour of plumage between nominate *rubicilla* and *severtzovi*, adult ♂ rather dark carmine, near ♂ of nominate *rubicilla*, but white spots on head and throat large, near those of *severtzovi*; ♀ about as dark as ♀ nominate *rubicilla* on upperparts, but slightly browner and with more yellowish rump; rather pale below, near *severtzovi*. Situation in southern part of species' range in central Asia not fully elucidated: birds in southern Himalayas from Chitral (Pakistan) east to Nepal and Lhasa (Tibet) larger than those from north of Karakoram range in eastern Pamirs and Kun Lun mountains (which are inseparable from *severtzovi* of Tien Shan), red of nape of adult ♂ slightly darker and more extensive, ♀ slightly darker on underparts; perhaps separable as *eblis* Koelz, 1939, with wing 118 and over in adult ♂ and 116 and over in ♀ (except for 1 of 15 birds), in Kun Lun and Tien Shan mainly below these values (except for 5 of 42 birds) (Vaurie 1949*b*; RMNH, ZMA). However, overlap apparently larger when more birds examined (Vaurie 1956*b*, 1959), and *eblis* not recognized here, pending further research on extent of variation in *diabolica* and in populations occurring from Himachal Pradesh (northern India) to Nepal.

**Recognition.** ♂ of darker races (nominate *rubicilla* and *kobdensis*) close to Pine Grosbeak *Pinicola enucleator*, but wing of latter has contrasting white markings and upper mandible distinctly heavier and more strongly decurved; ♀ *P. enucleator* quite different in colour, not heavily streaked. In south and east of central Asian range, rather similar to Streaked Great Rosefinch *C. rubicilloides*, and in Tien Shan to pale race *kilianensis* of Rose-breasted Rosefinch *C. puniceus*. In *C. rubicilloides*, adult ♂ distinctly streaked dusky on nape, mantle, and scapulars, ♀ and 1st adult ♂ distinctly streaked dark brown on entire body, unlike any *C. rubicilla*. In *C. puniceus kilianensis*, rosy tinge of adult ♂ more contrasting on rump but less extensive on cap and belly than in *C. rubicilla*; in ♀ and 1st adult ♂, upperparts and chin to breast spotted dark brown, rump contrastingly yellow or orange-yellow (RMNH).

CSR

---

## *Pinicola enucleator* Pine Grosbeak

PLATES 51 (flight) and 52
[between pages 640 and 641]

Du. Haakbek  Fr. Durbec des pins  Ge. Hakengimpel
Ru. Щур  Sp. Camachuelo picogrueso  Sw. Tallbit

*Loxia Enucleator* Linnaeus, 1758

Polytypic. Nominate *enucleator* (Linnaeus, 1758), northern Europe east to valley of lower and middle Yenisey and lowlands north of Altai. Extralimital: *pacata* Bangs, 1913, eastern Siberia east of Yenisey valley and from Altai through Sayan mountains and northern Mongolia to Stanovoy mountains, to western shore of Sea of Okhotsk and Kolyma basin; *kamtschatkensis* Dybowski, 1883, Anadyrland, Kamchatka, and northern shore of Sea of Okhotsk; *sakhalinensis* Buturlin, 1915, Sakhalin, Kuril Islands from Shumshu southward, and Hokkaido; *eschatosa* Oberholser, 1914, north-east USA north to Newfoundland and southern Quebec; *leucura* (Statius Müller, 1776), northern Canada from central Mackenzie east to Labrador; 4–5 races in western North America.

**Field characters.** 18·5 cm; wing-span 30·5–35 cm. Longer than any other finch in west Palearctic except Great Rosefinch *Carpodacus rubicilla*; structure recalls Bullfinch *Pyrrhula pyrrhula* but 15% larger than even northern race of that species. Very large, plump but attenuated finch, with stubby bill shaped like *P. pyrrhula* and secretive behaviour. In all plumages shows somewhat mottled, patchy colours with black wings and striking double white wing-bar, white tertial-fringes, and black tail; wing pattern recalls Two-barred Crossbill *Loxia leucoptera*. Adult ♂ drenched in rose-red, ♀ and juvenile in orange, yellow, and grey. Sexes dissimilar; no seasonal variation. Juvenile separable at close range.

ADULT MALE. Moults: July–September (complete). Head, upperparts to upper tail-coverts, and underparts to lower flanks and belly pale crimson-red, looking uniform at distance. At close range, head shows black surround to bill, black lore, pale grey crescent below dark eye, dusky rear ear-coverts, and dusky nape; upperparts show dusky brown mottling and spotting on mantle, and greyer scapulars and smoky-grey shoulders and upper flanks; grey lower belly and greyish-white vent and under tail-coverts. Wings dusky-black, marked by pink fringes to leading coverts, white tips to median coverts and bolder white tips to greater coverts forming 2 wing-bars, white margins to tertials, and (when fresh) reddish outer fringes to secondaries. Underwing grey. Upper tail-coverts black, fringed dusky-pink; tail black. With wear, crimson plumage brightens to raspberry- or orange-red and white wing marks become narrower. ADULT FEMALE. Plumage pattern as ♂ but crimson replaced by reddish-gold to golden-yellow on head, centre of mantle and chest, and more grey on scapulars, back, rump, and flanks. Spotting of upperparts more isolated, often fainter than ♂. JUVENILE. Resembles neither adult ♂ nor ♀, being rather dark sepia on head and upperparts and ashy-brown on underparts, with only faint yellowish or buff wash on throat and breast. Wings and tail dark brown, with indistinct, narrow double wing-bar formed by yellowish-buff tips to median coverts and whitish-buff tips to greater coverts, dull whitish fringes to tertials, and dull whitish-buff margins to tail-feathers. FIRST-YEAR MALE. Resembles adult ♀ but with duller wings. Most show beginnings of pink plumage on head, mantle, and fore-underparts. FIRST-YEAR FEMALE. Closely resembles adult ♀ but yellow tones on head and forebody paler. Bill shape much as *P. pyrrhula* but with tip of upper mandible appearing as sharp decurved point beyond lower mandible; covered by dark bristles at base; dark grey to horn-brown. Legs dark brown.

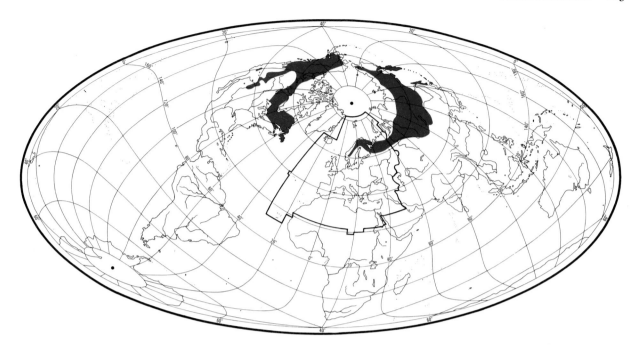

Adult unmistakable, with diagnostic combination of large size, white wing markings, and grey belly. Juvenile puzzling, however, with dull plumage quite unlike adult's. Flight essentially finch-like, but large size, long silhouette, and strength of wing-beats also recall long-tailed thrush, e.g. Fieldfare *Turdus pilaris*; over short distance, progress can appear slow and floating, but in full flight undulations become deep and powerful. Gait a hop, varied by clamber in foliage and some walking steps on ground. Movements deliberate. Flicks up wings and tail when uneasy. Sociable and tame.

Song a fast, crystal-clear series of notes, recalling song of Wood Sandpiper *Tringa glareola*; also described (for Nearctic birds) as a pleasant but short warble, recalling rosefinch *Carpodacus* but less energetically delivered and interspersed with twanging notes. Commonest call a whistled trisyllable, recalling yelp of Greenshank *T. neb-ularia* but less strident, 'tee-tee-tew' or 'tew-tew-tew'; also described (for Fenno-Scandian birds) as a clear, strong, fluting 'peeleeJEEH peeleeJÜ'. Alarm-call of ♀ a musical disyllable, 'chee-vli'. Contact-call a low 'büt büt'.

**Habitat.** In west Palearctic, breeds and largely winters in boreal forests north to treeline, within July isotherms 10–17°C. Uses all species of trees and shrubs, both coniferous and deciduous such as birch *Betula*, alder *Alnus*, willow *Salix*, poplar *Populus*, and juniper *Juniperus*. Predominantly arboreal, descending in summer to take berries from shrubby plants, and sometimes to gather seeds on ground. In Lapland, resorts mainly to drier, pine-clad fellsides and larger wooded knolls on marshes, but locally numerous in thick birchwoods near heads of fjords, show-ing marked preference for slopes and gullies where junipers predominate among scrub. Also inhabits wet thickets and sometimes open partly deforested country. Naturally tame but does not favour human settlements (Bannerman 1953a).

Extralimitally, nests in mountain forests, in Kamchatka to 1240 m (Dementiev and Gladkov 1954). Prefers rather light open forests, pine-clad hillsides and subarctic birch-woods (Newton 1972), but records suggesting thinner distribution in denser forests may be partly influenced by the species' unobtrusive habits.

In winter, migrants often attracted into towns where berried trees or shrubs present. Both geographically and in terms of acceptable habitat is among the most inflexible of west Palearctic birds.

**Distribution.** Has contracted range to north in Finland, but recently spread in southern Norway.

NORWAY. Recent spread in south (VR). FINLAND. Considerable contraction of southern limit of breeding range to north in 20th century (OH). CZECHOSLOVAKIA. Possibly bred in southern Bohemia, 1964 and 1966 (KH).

Accidental. Britain, France, Netherlands, Germany, Denmark, Poland, Austria, Hungary, Switzerland, Italy.

**Movements.** Some populations migratory, most resident and eruptive. Limited information; no long-distance ringing recoveries. Irrupting birds often enter towns to feed (Newton 1972). Invasions are regularly followed by return movements (Markgren and Lundberg 1959).

In northern Europe, some birds remain in breeding areas all year, but most move short distance south or

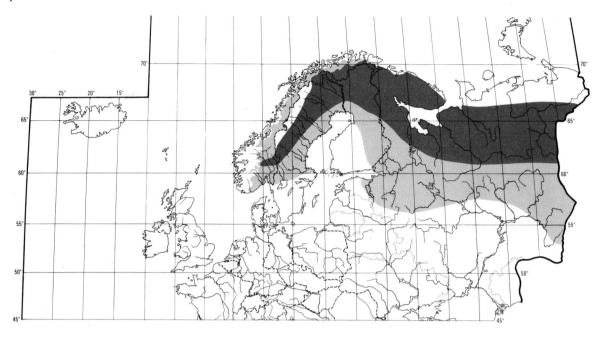

south-west (Newton 1972); data from northern Fenno-Scandia and adjoining Kola peninsula (Russia) show that areas north of arctic circle are vacated, though birds may overwinter only *c.* 100 km further south if sufficient food (often rowan *Sorbus*) available; depart south August–October, returning February–April (Pulliainen 1979*b*; Semenov-Tyan-Shanski and Gilyazov 1991). At Kemi (65°44′N, Finland), numbers wintering depend strongly on *Sorbus* crop, e.g. none wintered in 1943–4 when crop failed, but many when crop good in 1939–40 and 1942–3; birds left city when supply of berries exhausted (Grenquist 1947). Reaches St Petersburg region (western Russia) every year, in greatly fluctuating numbers; usually arrives mid- or late October, later if *Sorbus* crop good further north; present all winter if many berries available, rare in other years; departs end of February or beginning of March (Noskov *et al.* 1981; Mal'chevski and Pukinski 1983).

Irruptions into Europe apparently involve chiefly birds from Russia; numbers in central and southern Sweden (where irruptions most marked) too large to be accounted for by northern Scandinavian populations alone, and distribution of records suggests arrival from east (Markgren 1955; Markgren and Lundberg 1959; Källander *et al.* 1978). Recorded less frequently immediately east in Åland islands (Finland) (Linkola 1960), but presumably unobserved as birds fly at great height (Hildén 1977; Douhan 1978). Occurred in southern Sweden and (chiefly eastern) Denmark in larger numbers in 19th century (Jespersen 1945; Malmberg 1949; Blume 1963). Birds reach central and western Europe only in exceptional years. Since 1900, 11 records in Mecklenburg, eastern Germany (Klafs and

Stübs 1987), 6 in Thüringen, eastern Germany (Knorre *et al.* 1986), 2 in Switzerland (Winkler 1984), and recorded in 14 winters in Poland (Tomiałojć 1990). Rarely recorded in western Germany (Peitzmeier 1955; Bundesdeutscher Seltenheitenausschuss 1989). In Denmark, 9 records 1955–90, in 1975, 1981, 1989, and 1990 (Olsen 1987*a*, 1991*b*; Frich and Nordbjærg 1992). 10 British records up to 1992, all 30 October to 15 May, in eastern Britain from Kent north to Shetland (Scotland) (Dymond *et al.* 1989; Harrop 1992).

Irruption in 1954 reached central and southern Sweden (including Skåne in extreme south) mostly in small flocks, but occasionally over 50 together; movement from 2nd half of October, peaking end of November and beginning of December, some birds remaining until March (Markgren 1955). Exceptional numbers also reached southern Norway (*Br. Birds* 1955, **48**, 134); in Denmark, 23 birds reported, all in east (Blume 1963). One record on Isle of May (Scotland) (Flower *et al.* 1955). In 1956, major influx (involving more than 10 000 birds) restricted almost entirely to northern and central Sweden; very few reached southern Sweden or Norway. Recorded in small numbers from early October, birds remaining mostly in east at first; arrival peaked end of January and beginning of February, when many passed north of Gulf of Bothnia, others crossed central Sweden; they stopped on reaching mountain range, and gradually returned north and east (Markgren and Lundberg 1959). In 1976, numbers unprecedented in Sweden; most records between 59°N and 61°N, but north to *c.* 64°N and south to *c.* 56°50′N; at Torö island, 60 km south of Stockholm, 16 400 flew west between dawn and early afternoon on 13 November, and a further 1000 the following day (Allsopp 1976; Källander *et al.* 1978). In Norway also, 1976 invasion probably largest ever (*Br. Birds* 1977, **70**, 219), and heaviest since 1958 in Finland (Hildén 1977); not reported further south-west. Recorded in various parts of Latvia end of November to January (Vīksne 1983), also in Estonia and Poland (Rute and Baumanis 1986; Tomiałojć 1990). In 1989, influx into central and southern Sweden was largest since 1976; began October, with highest numbers November–December (*Br. Birds* 1990, **83**, 229–30).

In western Russia, marked invasion in 1978; probably came from Urals, as unusually high numbers recorded in Kirov region, west of Urals, late October to early December; in Moscow area, 289 birds recorded 9 December to 25 January; in Kaluga, *c.* 150 km south-west of Moscow, 1st records for 10 years; small groups present in different parts of town from late December to February, with stragglers remaining until end of March (Baranov and Margolin 1983; Shurupov 1985). Winter reports 1978–9 also from Latvia (Rute and Baumanis 1986), and one record in western Germany (*Br. Birds* 1980, **73**, 578). In north-central Russia, birds disperse southward, reaching Tomsk in western Siberia in some years; in 1975–6, large numbers wintered there in parks and gardens (Gyngazov and Milovidov 1977). In Kazakhstan, frequently winters in small numbers in Orenburg in north-west, and occasionally reported from north-east; exceptional elsewhere (Korelov *et al.* 1974). In southern Baykal region further east, local birds sedentary; arrivals from other areas swell winter numbers (Vasil'chenko 1982). Mostly resident in extreme eastern Russia; winter distribution in Kamchatka depends on richness of seed-crop of pine *Pinus pumila*, which varies from place to place; when seed-crop poor, regularly encountered near human habitations (Lobkov 1986). Some birds breeding in Koryak highlands (north of Kamchatka) move down to riverine woods in late September, remaining there mostly until end of March; others remain in subalpine areas (Kishchinski 1980). In Ussuriland, passage-migrant in small numbers and irregular winter visitor (Panov 1973*a*). Present all year throughout Sakhalin; birds tend to move towards south of island in mid-winter, remaining there until beginning of April (Gizenko 1955). Some Japanese birds move to lower levels in autumn and winter, exceptionally moving south from Hokkaido to Honshu; migrants from northern and central Kuril islands winter in southern Kuril islands and occasionally reach Hokkaido; winter records in western Honshu thought to involve birds crossing Sea of Japan (Brazil 1991).

In North America (*leucura*), movements irregular and erratic, sometimes insignificant in numbers, at other times having character of invasions; birds either remain in or near breeding grounds all year, or move southward when food supply becomes scarce; those breeding at high altitude tend to move down to lower levels (Bent 1968). Accidental in Bermuda and Greenland (American Ornithologists' Union 1983). Californian race *californica* rarely descends below 200 m (Gaines 1988).

DFV

**Food.** Buds, shoots, and seeds, especially of spruce *Picea* and other conifers, rowan *Sorbus*, and berry-bearing shrubs; invertebrates in breeding season. Winter numbers, and to some extent movements, probably correlated with berry crop, particularly rowan and, where common, juniper *Juniperus*. Diet rather like that of Bullfinch *Pyrrhula pyrrhula*, but items larger on average; moves to herbs when preferred foods scarce. (Carpelan 1929*b*; Grenquist 1947; Newton 1972; Pulliainen 1979*b*; Mal'chevski and Pukinski 1983; Shurupov 1985.) Commonly enters towns in winter flocks if berries plentiful (Grenquist 1947, which see for effect of temperature on foraging behaviour; Mal'chevski and Pukinski 1983). In spring and summer, feeds more often on or near ground taking fruits of understorey shrubs; in winter, when snow on ground, higher up in trees; winter flocks will remain in same tree for some time if undisturbed, methodically removing all fruits or buds, but (e.g.) in towns may move restlessly from tree to tree (Grenquist 1947; Bent 1968; Newton 1972; Semenov-Tyan-Shanski and Gilyazov 1991). For a bulky bird, a fairly agile and skilful forager in trees, moving slowly and deliberately along thin twigs to reach buds and shoots, sometimes approaching head-down or stretching out from neighbouring perch; uses hooked upper mandible in climbing like crossbill *Loxia* and to grasp buds by tilting head 90° then straightening up to pull bud off; only eats soft kernel, discarding sticky outer scales, and peels soft green spruce cones to get at core; eats pulp and seeds of many berries but rejects skin; often sits quietly for long periods in same position between feeding bouts (Bent 1968; Newton 1972; Massoth 1981; Semenov-Tyan-Shanski and Gilyazov 1991). On herbs, not seen to use feet to hold down seed-head, etc.; Compositae heads bitten open and seeds flung out by shaking; eats flowers, seeds, parts of leaves, and stems if milky. Large insects bitten all over, not beaten; generally discards skins of larvae, etc., and eats snails up to *c.* 1·5 cm whole with shell. (Newton 1972; Massoth 1981.) Frequently catches flying insects, particularly in breeding season, jumping up from branch or making brief pursuit flights (French 1954; Massoth 1981; Blomgren 1983). In North America, frequently comes to feeders in gardens in winter, where particularly fond of sunflower *Helianthus* seeds (Dunn 1990; Walker 1990). In Alaska and in Murmansk region (north-west Russia), recorded feeding on oats *Avena* in or near settlements in snow, in flocks of up to 100 (Cade 1952; Semenov-Tyan-Shanski and Gilyazov 1991). Seen feeding on horse-droppings in Norway, and not infrequently takes spores of moss, notably *Polytrichum* (Blair 1936; Semenov-Tyan-Shanski and Gilyazov 1991). Vagrant in Shetland fed on nucleus of Sitka spruce *P. sitchensis* buds (Harrop 1992). Food stored for transportation in special gular pouches opening on each side of tongue which are only developed during breeding season (French

1954, which see for many details). For diet in captivity, see Bernhoft-Osa (1960), Adkisson (1977a), and, especially, Massoth (1981).

Diet in west Palearctic includes the following. Invertebrates: grasshoppers, etc. (Orthoptera), bugs (Hemiptera: Aphidoidea), larval Lepidoptera (Geometridae), flies (Diptera: Tipulidae, Culicidae), adult and larval Hymenoptera (sawflies Symphyta), adult and larval beetles (Coleoptera: Curculionidae), spiders (Araneae), mites (Acari), snails (Pulmonata). Plants: buds, shoots, seeds, fruits, etc. of juniper *Juniperus*, larch *Larix*, spruce *Picea*, pine *Pinus*, aspen *Populus*, willow *Salix*, birch *Betula*, alder *Alnus*, elm *Ulmus*, maple *Acer*, ash *Fraxinus*, dogwood *Cornus*, rose *Rosa*, rowan *Sorbus*, apple *Malus*, cherry, etc. *Prunus*, raspberry, etc. *Rubus*, *Cotoneaster*, crowberry *Empetrum*, bearberry *Arctostaphylos*, bilberry, etc. *Vaccinium*, cow-wheat *Melampyrum*, guelder rose, etc. *Viburnum*, mugwort *Artemisia*, grasses (Gramineae), sedges (Cyperaceae), rushes (Juncaceae), moss (Musci). (Schuster 1930; Blair 1936; Novikov 1952; Mal'chevski 1959; Turček 1961; Schmidt 1966; Haftorn 1971; Newton 1972; Pulliainen 1974; Popov 1978; Baranov and Margolin 1983; Sabel 1983; Ojanen and Lähdesmäki 1984; Semenov-Tyan-Shanski and Gilyazov 1991.)

In Murmansk region, spring–summer, 7 stomachs contained Tipulidae, beetles (including Curculionidae), sawfly larvae, mites, and seeds of dogwood, bilberry, and crowberry (Novikov 1952). In Moscow area, winter, of 289 feeding observations, 91% on small fruits and apples, 8% rowan berries, 1% larch buds (Shurupov 1985); 3 winter stomachs from south of Moscow contained only apple seeds (Baranov and Margolin 1983). In Finnish Lapland, February–April, 28 of 30 birds fed on spruce buds, 2 on willow buds; spruce buds richer in protein than seeds (Pulliainen 1974, which see for details). In northern Kamchatka peninsula (eastern Russia), 13 gullets and stomachs in June held 55% by volume seeds of dwarf pine *Pinus pumila*, 12% crowberry, 11% alder, 9% *Vaccinium*, 8% birch leaf-buds, 4% *Vaccinium* leaf-buds, 1% invertebrates; 5 samples, August–September, were 98% *P. pumila* seeds; 12, January–March, were 88% *P. pumila*, 11% buds and shoots of willow *Chosenia*, 1% vole (Kishchinski 1980). In Anadyr' region (eastern Russia), contents of 6 stomachs, February–May, included 137 insects, of which 96% by number beetle larvae (Portenko 1939). For other east Russian studies, see Nechaev (1969) and Zonov (1978). In North America, 365 winter stomachs contained 99% plant material, including 24% conifer buds, 17% seeds of snowberry *Symphoricarpos*, 14% *Rubus*, 14% other fruits, 8% herbs, and 6% beechmast *Fagus* and acorns *Quercus*; 16% of contents of 29 summer stomachs invertebrates, mostly Orthoptera, caterpillars, ants (Formicidae), and spiders (Bent 1968). In Russia, main food in early spring is buds and growing tips of trees especially conifers; in summer, berries of *Empetrum*, *Vaccinium*, *Rubus*, etc., herb seeds, and insects; in autumn

into winter, rowan and juniper berries, seeds of trees such as birch, maple, ash, and especially apple, as well as conifers (Popov 1978; Mal'chevski and Pukinski 1983; Shurupov 1985; Semenov-Tyan-Shanski and Gilyazov 1991). One bird, mid-March, Murmansk region, contained 163 whole nuclei of spruce buds (Semenov-Tyan-Shanski and Gilyazov 1991).

Since young are fed regurgitated pulp from gular pouches of adults, often difficult to determine composition, but insects apparently always present. In south European Russia, 16 collar-samples contained plant material and 79 invertebrates, of which 34% by number aphids, 19% beetles (18% larvae), 19% adult Diptera, 17% larval Lepidoptera, 10% spiders, and 1% snails (Mal'chevski 1959). In Murmansk region, 8 collar-samples contained Diptera (Tipulidae and Culicidae), Geometridae caterpillars, sawfly larvae, aphids, spiders, snails, seeds of cowberry *Vaccinium*, and parts of spruce (Semenov-Tyan-Shanski and Gilyazov 1991). See also Pulliainen and Hakanen (1972) and Blomgren (1983). Parents in captive studies fed young wide variety of invertebrates as well as seeds (Bernhoft-Osa 1960; Adkisson 1977; Massoth 1981).　BH

**Social pattern and behaviour.** No comprehensive studies. Account includes data on both Palearctic and Nearctic populations.

1. Gregarious outside breeding season. Migrates south in large numbers in some years (see Movements): flocks during such irruptions small, up to *c.* 50, sometimes several hundred or even several thousand (most data for Sweden: see Faxén 1945, Grenquist 1947, Markgren 1955, Markgren and Lundberg 1959, Douhan 1978, Källander *et al.* 1978). Winter flocks of varying size reported also from Russia: e.g. Kamchatka (Bergman 1935), up to 100 in Khamar-Daban mountains (south of Lake Baykal) in winter 1973–4 (Vasil'chenko 1982). In typically small winter flocks in Nova Scotia (Canada), adult ♂♂ outnumbered by ♀♀ and juveniles (Tufts 1961). Flocks mainly of 5–30 occur at other times of year, including during period after arrival on breeding grounds, and when families unite in flocks for feeding after breeding (e.g. Dementiev and Gladkov 1954 and Semenov-Tyan-Shanski and Gilyazov 1991 for Russia; Carpelan 1929b for Finland). In northern Russia, pairs (as well as small parties and single birds) noted western Taymyr from early April in year with good larch *Larix* and spruce *Picea* cone crop (Krechmar 1966), and Yamal peninsula from mid-May (Danilov *et al.* 1984). Sometimes associates for feeding with crossbills *Loxia* (Bannerman 1953a; Markgren and Lundberg 1959), Bullfinch *Pyrrhula pyrrhula* (Carpelan 1929b; Shurupov 1985; see also Voice), and in Anadyr' (north-east Russia), late February, once with mixed flock of tits *Parus*, Redpolls *Carduelis flammea*, and Nuthatch *Sitta europaea* (Portenko 1939). In Nova Scotia, apparently loath to intermingle with Evening Grosbeak *Hesperiphona vespertina*, with which perhaps some competition (Tufts 1961); see also Antagonistic Behaviour (below). BONDS. No evidence of other than monogamous mating system. In North America, where song noted from February or March, pair-formation perhaps in winter quarters (Adkisson 1981). In Russia, birds initially in flocks after arrival on breeding grounds; form pairs and take up territories from early May (Dementiev and Gladkov 1954). No information on duration of pair-bond. ♀ builds nest and incubates alone (Bernhoft-Osa 1960; Adkisson 1977a; Pulliainen 1979b), though

occasionally some incubation by ♂ according to Løvenskiold (1947). Young fed by both sexes (Davidson 1951), about equally (e.g. Pulliainen and Hakanen 1972, Blomgren 1983). Captive young independent *c.* 3 weeks after fledging at *c.* 15–18 days (Adkisson 1977*a*), or more or less independent (still fed occasionally) at *c.* 40 days old (Massoth 1981). Age of first breeding not known, but old ♂♂ usually predominate among early arrivals on Kola peninsula (north-west Russia) (Semenov-Tyan-Shanski and Gilyazov 1991), and of 40–50 paired ♂♂ seen east Finnmark (Norway) majority were in red plumage, only 2 greenish-yellow (Blair 1936). BREEDING DISPERSION. Territorial, and apparently solitary: no definite reports of neighbourhood groups such as occur in some other Carduelinae. In hills above Kolyma valley (north-east Russia), ♂♂ recorded every *c.* 400–500 m along 5 km, mid-June (Krechmar *et al.* 1978). Little information on size or function of territory. In north-east Utah (USA), one territory *c.* 400 m across; not clear how much of this defended, but strangers tolerated within territory late in season (Bent 1968). In Kareliya (north-west Russia), each pair's territory reported to be 'fairly extensive' (Bannerman 1953*a*). From one nest in Norrbotten (Sweden), ♂ apparently made longer foraging excursions than ♀ who sometimes stayed near nest and foraged there; best foraging site *c.* 200 m from nest (Blomgren 1983); no indication of any territory defence. Captive ♂ chased other birds away from nest (St Quintin 1907). Study in northern Finland showed *P. enucleator* (and other northern taiga species) negatively affected by fragmentation of forest: density 0·4 pairs per km² in virgin pine *Pinus* (absent from managed forest), 1·3 and 2·9 pairs per km² in uniform forest (absent from fragmented forest) (Virkkala 1987). Along Chuna river (Kola peninsula), 0·6–2·1 pairs per km² in June (Semenov-Tyan-Shanski and Gilyazov 1991). In extralimital eastern Russia, forest of pine *Pinus pumila* favoured, e.g. 6·5–8·6 birds per km² in Khamar-Daban mountains (Vasil'chenko 1982; Durnev *et al.* 1984), 11·1 on Vilyuy river (Shmelev and Brunov 1986), 15 on lower Amur river (Babenko 1984*a*); up to 23·5 pairs per km² in Kamchatka, where 7 pairs per km² in forest predominantly of birch *Betula ermani*, and 6·4 pairs per km² in forest dominated by alder *Alnus* (Lobkov 1986). For further densities in Siberia and Altai, see Ravkin (1973, 1984) and Rogacheva (1988). ROOSTING. Nocturnal and, outside breeding season at least, apparently communal. At Kemi (Finland), winter flocks apparently roosted in birch trees by farms and in forest, coming into town from *c.* 08.30 hrs to feed in morning twilight; quiet at dawn, and very inactive around midday; renewed activity from 14.30 hrs and departed for roost again by 15.30–16.00 hrs (Grenquist 1947). Similar pattern at Juneau (Alaska, USA) in winter: feeding in town, roosting (perhaps together with waxwings *Bombycilla* and American Robin *Turdus migratorius*) in surrounding forests (Bent 1968). In Anadyr', large flock seen gathering for roosting 22 May (Portenko 1938). At nest in Finnish Lapland (67°45′N), during period of continuous daylight, ♂ stopped feeding young *c.* 1 hr earlier than ♀, began *c.* 1 hr earlier in morning; break from feeding around midnight, apparently determined by light; average length of feeding pause 6·2 hrs for both sexes (Pulliainen and Hakanen 1972). In Norrbotten, early July, birds feeding young were active one day from 01.55 hrs, ceased another day 20.35 hrs (Blomgren 1983).

2. Resembles *P. pyrrhula* in being generally quiet and unobtrusive, calls not very far-carrying. Well known also as exceptionally confiding, including at nest. (Carpelan 1929*b*; Blair 1935; Stegmann 1936; Davidson 1941; Bruun *et al.* 1986.) When uneasy, flicks wings and tail; breast feathers ruffled in display (Witherby *et al.* 1938; Newton 1972). On hearing alarm-calls (see 2c in Voice), captive birds crouched and froze or adopted

upright-posture, looking about (Massoth 1981). For buzzards *Buteo* and owl *Bubo*, birds seen to freeze in upright-posture and to give alarm-calls. Call 2a ('location call': see Voice) sometimes also expresses alarm: e.g. abbreviated version given by captive birds mobbing cat. (Adkisson 1981.) When handled, may give Distress-call (see 2d in Voice), other conspecific birds then approaching and calling, plumage sleeked, crown ruffled, flicking wings excitedly; very restless, awaiting developments (Massoth 1981), or flutter about observer and give calls 2a and 2c (Adkisson 1981). Lone wild birds often give call 2a spontaneously, approaching captive bird or playback (always more responsive to playback than pairs or flocks); may change from call 2a to contact-call (see 2b in Voice) during approach (Adkisson 1981). FLOCK BEHAVIOUR. At Kemi (Finland), birds typically restless on arrival from and prior to departure for roost; much more so, moving about often without feeding properly, during marked drop in temperature and increasing barometric pressure (Grenquist 1947). Prior to take-off, birds may gather briefly in prominent tree-top; as in (e.g.) *Loxia*, contact-calls are given with increasing volume and rate of delivery, peaking just before typically sudden departure. Contact-calls also given frequently when moving about (at times apparently in synchrony with wing-beats); flying birds, especially if scattered, give frequent location-calls. (Adkisson 1981.) SONG-DISPLAY. Both sexes sing (Newton 1972), though loud advertising-song perhaps given only by ♂, while ♀'s song may be of subsong type (see 1a–b in Voice). Song given from exposed perch (e.g. Stegmann 1936); in Anadyr', early August, ♀♀ seen to sing from top of shrub when no young visible (Portenko 1939). Singing in flight also noted (Bent 1968), sometimes while gliding between trees (Bannerman 1953*a*); no reports of a more elaborate, ritualized song-flight. Frequency and persistence of loud song increase with start of breeding season (Shurupov 1985). In Sweden, singing noted throughout light summer night (Bernhoft-Osa 1960). In northern Fenno-Scandia, song-period from late March to mid-June (Bannerman 1953*a*); similar on Kola peninsula, where song also given (including by young birds) mid-August to mid-October (Semenov-Tyan-Shanski and Gilyazov 1991). See also Massoth (1981) for captive bird. In Koryak mountains (north of Kamchatka), a few ♂♂ singing 26 July to 18 August, but song regular (even at temperatures down to −30°C) mid-September to mid-November and mid-February to end of March, from adults and 1st-years (Kishchinski 1980). For further reports of singing at various times in eastern Russia, see Stegmann (1936), Portenko (1939), Gizenko (1955), Vorobiev (1963), and Danilov *et al.* (1984). ANTAGONISTIC BEHAVIOUR. Captive ♂ dominated 3 ♀♀ at food, threatening them by gaping, Bill-snapping, and hissing (see 3 in Voice) (Massoth 1981); captive ♂♂ will fight persistently if put near each other in breeding season (Pulliainen 1979*b*). In Ontario (Canada), during major invasion 1985–6, small flocks recorded at garden feeders where some interspecific fighting over seeds with *H. vespertina* (Pittaway 1989). HETEROSEXUAL BEHAVIOUR. (1) General. In Russia, pair-formation from early May; inconspicuous, though much song from ♂♂ at this time (Dementiev and Gladkov 1954). (2) In Norrbotten, when pair feeding close together away from nest, ♂ suddenly made off in fast low flight, moving up and down and side to side over *c.* 15 m (Blomgren 1983). In Utah (USA), ♂ and ♀ mostly came to nest together to feed young; except when ♀ brooded, also departed together, either sex initiating departure by diving at mate, then chasing it across meadow (Bent 1968). Courtship in captive birds proceeded as follows: ♂, plumage sleeked, head drawn in, wings slightly drooped, adopted almost horizontal posture; occasionally flicked wings and made jerky movements of slightly raised tail, then also swayed body from side to side. In 2nd, evidently higher-intensity stage, ♂ carried nest-material and assumed upright posture, breast ruffled, head held back, tail moderately fanned and raised; circled ♀ thus, wings held out and vigorously quivering. (Massoth 1981.) See also subsection 4 (below). (3) Courtship-feeding. ♂ feeds ♀ regularly from early in courtship at least up to hatching, also during 1st week of brooding (though some food then presumably destined for young)—on nest or nearby branch, sometimes apparently further away. ♀, sometimes called off nest by ♂, crouches with bill open, wing-shivers, and raises tail. (Davidson 1951; Bent 1968; Adkisson 1981; Blomgren 1983.) According to Davidson (1951), frequent contact-calls are given by both birds; only Begging-calls from ♀ (see 4 in Voice) reported by Adkisson (1981). ♀ recorded continuing to beg for several minutes after ♂'s departure (Adkisson 1981), or following mate into forest after being fed (Blomgren 1983). After fledging, Allofeeding by pair observed in nest-tree (Pulliainen and Hakanen 1972). (4) Mating. Captive pair copulated frequently during 3–6 days of nest-building (Adkisson 1977a). ♀ of another captive pair adopted posture on ground much as described for Courtship-feeding (see above); after copulating, ♂ would fall off ♀, and lie trembling on ground for several seconds, tail spread, wings fluttering, head bent back (Bernhoft-Osa 1956). ♂ of 3rd captive pair first circled mate on ground (see subsection 2, above), and mounted ♀ as soon as she crouched; shortly before and during copulation, ♂ called, ♀ also once ready to copulate (see 5 in Voice); after copulation, ♂ remained wing-shivering for a few seconds by ♀ in posture similar to her soliciting-posture, before moving away (Massoth 1981). (5) Nest-site selection and behaviour at nest. No information additional to that given above. RELATIONS WITHIN FAMILY GROUP. ♂ recorded carrying away eggshells (Bernhoft-Osa 1960; Blomgren 1983). All brooding by ♀ (Adkisson 1977a). In Finnish Lapland, ♀ continued brooding at night up to c. 5 days before fledging, also shaded young from sun by day (Pulliainen and Hakanen 1972). Young fed by both parents by regurgitation (Bernhoft-Osa 1960; Pulliainen and Hakanen 1972). At one Swedish nest, ♂ brought food for first few days while ♀ brooded, ♀ also leaving nest a few times briefly to feed herself, and starting to help ♂ collect food from c. 1 week after hatching (Blomgren 1983). Captive ♂ fed young direct from day 6, previously only via ♀ (Massoth 1981). In Finnish Lapland, if ♀ brooding young when ♂ arrived, ♀ climbed onto rim of nest to allow ♂ to feed young; both often came to nest together and left together (Pulliainen and Hakanen 1972). ♀ of Swedish pair seen to take food from nestlings after ♂ had fed them and swallow it herself; continued begging from ♂ even after going off to feed herself (Blomgren 1983). Nest also kept clean by both sexes, nestling faeces being swallowed (by captive ♀ up to 6 days at least) or carried away, up to fledging in case of captive pair (Pulliainen and Hakanen 1972; Massoth 1981). Young fledge at 13–18 days (includes data for captive birds: see Breeding). ANTI-PREDATOR RESPONSES OF YOUNG. No information. PARENTAL ANTI-PREDATOR STRATEGIES. (1) Passive measures. Neither sex shows any special alarm for man at nest. ♀ typically tame and silent; sits very tightly, allowing close approach; some can even be handled, making no attempt to peck, and returning to nest immediately after release. ♂ will also make close approach. (Davidson 1951; Bannerman 1953a; Vorobiev 1963; Pulliainen 1979b; Blomgren 1983.) On Kola peninsula, adults reported to be temporarily slightly warier after hatching (Semenov-Tyan-Shanski and Gilyazov 1991), and birds tending fledged young may come close and express agitation by wing- and tail-flicking (Witherby et al. 1938). (2) Active measures. In Utah, both adults united to drive off Gray Jays Perisoreus canadensis, ♂ also drove squirrel (Sciuridae) away from nest (Bent 1968).          MGW

**Voice.** Contact- and alarm-calls (see 2a–c, below) given by both sexes all year; song (or Subsong) also noted in winter, but loud song apparently given (by ♂) mainly close to and during breeding season (Naumann 1900; Portenko 1939; Kishchinski 1980; Adkisson 1981). Some reports of singing also by ♀ (Portenko 1939; Bernhoft-Osa 1956 for captive bird). Generally quiet and thus easily overlooked (Blair 1936; Witherby et al. 1938). Major study in North America found relatively small repertoire of loud 'primary song' and 'whisper song' (Subsong), and 4 distinctly different calls (some additional calls for Palearctic birds described below); marked geographical variation, especially in 'location call' (2a), probably also to some extent in song and other adult and juvenile calls, though further investigation needed (Adkisson 1981, which see for details and sonagrams). Extent of possible geographical variation within Palearctic, also between Palearctic and Nearctic populations, not studied. For additional sonagrams, see also Bergmann and Helb (1982). Bill-snapping noted in antagonistic context among captive birds (Massoth 1981).

CALLS OF ADULTS. (1) Song. (1a) Loud advertising-song presumed to be given mainly or exclusively by ♂. Short yodelling phrases (c. 2 s) comprising series of clear, rich and melodious whistling or fluting motifs, including some resembling calls; song-phrases typically quieter at start and at end than in middle (Carpelan 1929b; Bergmann and Helb 1982; Bruun et al. 1986; Jonsson 1982). Loud song variously likened to that of Wood Sandpiper Tringa glareola (Bruun et al. 1986), mellifluous components in song of Redwing Turdus iliacus, or Woodlark Lullula arborea (Portenko 1939). In North America, reported to resemble Purple Finch Carpodacus purpureus (compare sonagrams of both in Robbins et al. 1983), phrase containing 8–19 complex units or motifs delivered at rate of c. 6–7 per s; each unit of given phrase usually differs from all others (see, however, discussion of recordings, below); sequence of units also varies, and songs of given pattern vary in length; not much variation between birds (Borror 1961; Bergmann and Helb 1982). Recording from Canada contains short warbling phrases with warm, mellow timbre, so unvaried as to sound monotonous. Constant variation in pitch proceeds by short ascending and descending glissandi within each unit, but contained overall within a range of one octave (c. 2–4 kHz). No clearly defined temporal patterning: pauses between units mostly so short as to escape notice (c. 40–60 ms), occasionally longer (c. 100–200 ms). Initial crescendo and final diminuendo show well in amplitude traces above Figs I–II. In longer phrase shown in Fig III, pause of c. 200 ms at 2 s is followed by complex unit comprising much shorter sub-units than evident elsewhere in song-phrases, then 2 relatively low-pitched, short, simple tones of type also in Fig I. Recording by S Palmér of Swedish bird temporarily held captive contains shriller, less mellow song-phrases than those of Canadian bird (Figs I–III); differs further in almost always repeating units within phrase (e.g. 3–4 'tui' and 3

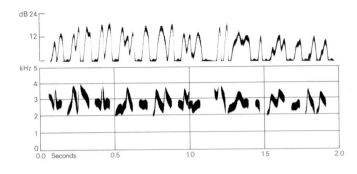

I  W W H Gunn/BBC  Canada  June 1957

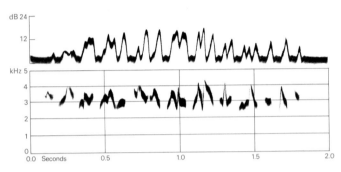

II  W W H Gunn/BBC  Canada  June 1957

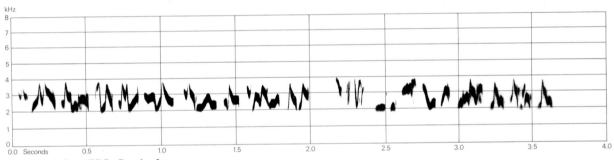

III  W W H Gunn/BBC  Canada  June 1957

'fiu' notes), having generally slower delivery rate, and in lacking persistent ascending and descending glissandi. (J Hall-Craggs.) According to Blair (1936) and Bannerman (1953a), fluting motifs are followed by harsh 'twang' resembling Greenfinch *Carduelis chloris* (reference is perhaps to wheeze in song of *C. chloris*), but no similar description in other sources and no sound of this type in recordings of song analysed. (1b) Subsong. Quiet and continuous chattering, given also in winter when may contain numerous mimicked sounds of other bird species (Bergmann and Helb 1982). In Moscow area (Russia), melodious fluting motifs mixed with fragments of song of other finches (Brambling *Fringilla montifringilla* and various Carduelinae), Meadow Pipit *Anthus pratensis*, Arctic Warbler *Phylloscopus borealis*, and whistles of tits *Parus* (Shurupov 1985). In study of captive pair, song ('twittering phrases') of both sexes essentially similar, but occasional lower-pitched fluty notes (somewhat like Song Thrush *T. philomelos*) interspersed in ♂'s song (Bernhoft-

Osa 1956, 1960); provisionally included in this category, but ♀ perhaps gives loud song (1a) as well. (2) Contact- and alarm-calls. Not always easy to distinguish between contact-calls 2a and 2b in some descriptions, and perhaps some genuine overlap; both often likened to piping sounds of Bullfinch *Pyrrhula pyrrhula*. (2a) Short series (usually 2-3) of attractive piping or fluting notes often descending in pitch overall and with tonal quality of *P. pyrrhula*; subdued, or at times quite loud, perhaps carrying several hundred metres. Renderings include 'chuleewü' and simpler 'puee' (though this could belong in 2c) (Jonsson 1992), 'tuithu' (Bergmann and Helb 1982), 'tee tee', or 'ti ti tew' (Portenko 1939; Newton 1972; Harrop 1992), and clear 'peeleeJEEH peeleeJÜ' (Bruun *et al.* 1986). Recording (Fig IV) contains single descending 'peeooo', similar to some 'location-calls' of simpler structure illustrated by Adkisson (1981). More-complex calls in this category, again resembling some shown in Adkisson (1981), include wholly tonal, ethereal, silken-soft, and quiet 'chliudweee'

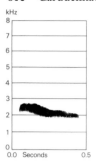

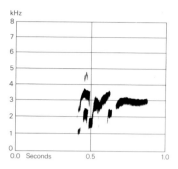

IV  S Carlsson  Sweden  November 1981

V  S Wahlström  Sweden

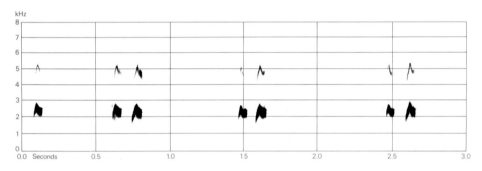

VI  S Wahlström  Sweden

VII  S Carlsson  Sweden  November 1981

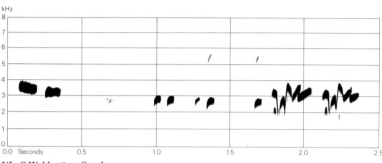

VIII  S Wahlström  Sweden

of remarkable beauty (Fig V), and last 2 complex units in Fig VI which are apparently shorter, more rudimentary forms, suggesting 'chliudwe chliudwe' (for discussion of other units in Fig VI, see call 2b, below). (J Hall-Craggs, M G Wilson.) In North American study, 'location call' was most conspicuous call, given all year and by both sexes, frequently by members of flying flock, less so when feeding; also serves as alarm-call (given when mobbing predator). Length of call (number of units) varies (2-3 types) and considerable discontinuous geographical variation exists, with 2 main categories, birds of taiga and coastal Alaska giving series of ascending and descending whistles (some obviously much as described for Palearctic birds), while western montane and island populations have on average lower-pitched, more complex calls with marked frequency modulation (varying greatly between localities); birds with whistling calls tend not to respond to frequency-modulated calls and vice versa, this apparently promoting assortative flocking in winter when different populations may come into contact. (Adkisson 1981.) (2b)

Low-pitched, quiet, short piping sounds like *P. pyrrhula* (Blair 1936; Dementiev and Gladkov 1954), virtually indistinguishable (even at close range) when given in mixed flocks with *P. pyrrhula* (Shurupov 1985): 'büt büt' (Bruun *et al.* 1986), 'düüb düüb düüb' (Massoth 1981). Presumably this is call (constant soft piping audible only up to *c.* 20 m) given by both sexes during courtship-feeding (Davidson 1951), and 'tjuuu' noted from ♀ being fed (Bernhoft-Osa 1960); for clearly different Begging-call given by ♀ in North America, see call 4 (below). Recording (Fig VII) contains series of extremely quiet, but attractive 'tew' notes, ascending slightly in pitch overall. Contact-calls in another recording similarly suggest 'tew' or 'tiu': 'tiu tiu-tiu tiu-tiu tiu-tiu' (Fig VIII); in Fig VI, after strongly tonal 'see-too' (of unknown significance) contact-calls (up to 1·5 s) suggest piping notes of Oystercatcher *Haematopus ostralegus*. (J Hall-Craggs, M G Wilson.) In North America, contact-call given when perched (including prior to take-off) and in flight obviously similar to that described for Palearctic birds: series of short, clear, fairly low-pitched and highly ventriloquial tones, with differently structured calls in California (Adkisson 1981). (2c) Alarm-calls. Loud excited 'tui-tui-tui' (Massoth 1981); melodious piping 'cheevli cheevli' noted from ♀ tending young, also (from both sexes) subdued chirping sounds audible only at close range (Witherby *et al.* 1938). Recording (Fig IX) contains 5 'chewee'

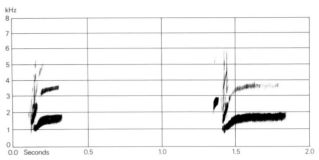

XI S Wahlström Sweden

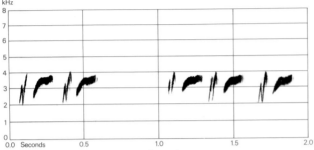

IX S Carlsson Sweden November 1981

X S Wahlström Sweden

calls; presumably at least expressing excitement. Fig X shows apparently related call in which transient-like 'che' of 'chewee' is replaced by loud tone: 'CHEE-wee'. Another call given when disturbed is rather harsh, loud, explosive squeak or 'sneeze-whistle': rendered 'füd tschri-hüid' in which 'tschri' very loud (Bergmann and Helb 1982, which see for sonagram). (J Hall-Craggs, M G Wilson.) North American study found call (apparently elicited mainly by aerial predators) to resemble whistling aerial-predator alarm-call of other woodland passerines, being similarly difficult to locate, but much lower pitched at *c.* 2 kHz; some geographical variation, with pitch tending to descend in California, otherwise steady or with only slight descent (Adkisson 1981). (2d) Distress-call. Screaming or screeching sound given when handled (Adkisson 1981); presumably much as juvenile distress-call (see below). (3) Cat-like hissing noted from captive ♂ threatening conspecific birds in dispute (Massoth 1981). (4) Begging-call of ♀. In North America, loud ringing sound unlike any other call (Adkisson 1981). (5) Calls associated with copulation. High-pitched whispering piping sounds given by both sexes, very high pitched from ♂ shortly before and during copulation (Massoth 1981). (6)

Other calls. Recording of various sounds given by birds in migrant flock includes short and long sighing mews: 'plee' and 'pleeee', tonal and quiet, with pleading quality (Fig XI). Function and possible relationship with calls described above not clear. (J Hall-Craggs, M G Wilson.)

CALLS OF YOUNG. Drawn-out whistling reported as food-call of nestlings, but none noted from captive birds until after fledging (Massoth 1981). In North American study, high-pitched (*c.* 8 kHz) 'seee' given when fed from *c.* 5 days, calls dropping in pitch (to *c.* 2·5 kHz) at fledging and given until independence when adult calls 2a–c appear, these then indistinguishable from adult calls after *c.* 15 weeks, when young give also Subsong (Adkisson 1981, which see for development of adult call 2a); in another study, captive young sang from *c.* 39 days old (Massoth 1981). Renderings of fledgling calls include clear, bell-like 'tee-klee', remarkably ventriloquial and hard to locate (Witherby *et al.* 1938) and, in Japan, persistent loud 'widü widü' (Jahn 1942). Loud, hoarse, squealing or screeching 'kreeeee' given when handled (Witherby *et al.* 1938).                    MGW

Breeding. SEASON. Finland: eggs laid late May to mid-July, mainly 1st half of June with repeat clutches to end of June (Montell 1917; Haartman 1969; Newton 1972; Pulliainen 1979*b*); of 119 clutches, 4% found 21–31 May, 28% 1–10 June, 49% 11–20 June, 12% 21–30 June, 3% 1–10 July, 4% 11–20 July, 1% 21–30 July (Haartman 1969). Kola peninsula (north-west Russia): eggs laid late May (Semenov-Tyan-Shanski and Gilyazov 1991). SITE. Usually against, or close to trunk of pine *Pinus* or spruce *Picea*, and in some areas juniper *Juniperus* or birch *Betula* (Jourdain 1906; Carpelan 1929*b*; Witherby *et al.* 1938; Dementiev and Gladkov 1954). Of 37 nests in north-east Finland, 89% were in spruce, 8% in pine, and 3% in juniper. 82% of 28 nests were against main trunk, 7% 5 cm from trunk, 4% 30 cm, 4% 50 cm, 4% 120 cm (Pulliainen 1979*b*). Nest usually rather low; in conifers 2–4 m (1·2–7) above ground, but nests in birch and shrubs seldom more than 1·75 m high (Bannerman 1953*a*; Newton 1972; Semenov-Tyan-Shanski and Gilyazov 1991). Heights of 37 Finnish nests: 3% less than 1 m, 17% 1–1·9 m, 33% 2–2·9 m, 11% 3–3·9 m, 25% 4–4·9 m, 8% 5–5·9 m, and

3% more than 6 m. Most nests on south side of tree: of 26 nests, 39% south-facing, 15% south-west, 12% south-east, 23% west, and 4% north (Pulliainen 1979b). Nest: like that of Bullfinch *Pyrrhula pyrrhula*, but larger and deeper (Carpelan 1929b; Witherby *et al.* 1938; Davidson 1951; Bannerman 1953a; Newton 1972). Rather untidy, loosely-built foundation of interwoven fine twigs of mainly spruce but also birch, juniper, or pine sometimes of considerable length (Jourdain 1906; Carpelan 1929b; Witherby *et al.* 1938; Davidson 1951; Newton 1972; Blomgren 1983); stiff dead twigs project from all sides of nest (Newton 1972; Pulliainen 1979b; Blomgren 1983); in Finland, sometimes stems of bilberry *Vaccinium* and twinflower *Linnaea*, occasionally roots, shoots, and needles of pine and shoots of crowberry *Empetrum* (Pulliainen 1979b, which see for many details). Rootlets and grass blades, delicate shoots and soft fragments of lichen filaments sometimes woven into outer structure. (Dementiev and Gladkov 1954; Semenov-Tyan-Shanski and Gilyazov 1991.) Cup mainly dry grass, with moss, roots of bilberry, dry spruce twigs, and juniper shoots, lined with very thin roots and *Usnea* lichen, thin grass, moss, and sometimes animal hair (Jourdain 1906; Montell 1917; Pulliainen 1979b; Blomgren 1983; Semenov-Tyan-Shanski and Gilyazov 1991). Average inner diameter of 5 nests in Finland, 7·8 × 7·4 cm (7·0–9·0 × 6·5–7·5); average depth of cup 3·3 cm (2·8–3·8) (Pulliainen 1979b). On Kola peninsula, outer diameter 15·0–22·0 × 16·5–35·0 cm, inner diameter 6·8–7·3 cm, overall height 7·2–11·0 cm, depth of cup 3·0–4·3 cm (Semenov-Tyan-Shanski and Gilyazov 1991). Building: by ♀ only (Witherby *et al.* 1938; Haftorn 1971; Makatsch 1976). EGGS. See Plate 61. Sub-elliptical, often long sub-elliptical; smooth, little or no gloss; variably green-blue, with blotches and spots of black or dark purple-brown and underlying pale violet-grey markings; sometimes freckled all over with small spots, others with a few very large blotches; markings often tend to be concentrated in zone around broad end (Jourdain 1906; Dementiev and Gladkov 1954; Newton 1972; Makatsch 1976). Nominate *enucleator*: 26·0 × 17·7 mm (23·0–29·6 × 16·1–19·1), *n* = 308; calculated weight 4·30 g (Schönwetter 1984). Clutch: 3–4 (2–5), mainly 4 (Montell 1917; Løvenskiold 1947; Haartman 1969; Semenov-Tyan-Shanski and Gilyazov 1991). Of 116 clutches, Finland: 1 egg, 1·7%; 2, 0·9%; 3, 23·3%; 4, 66·4%; 5, 7·8%; average 3·78 (Haartman 1969); average of 23 complete clutches 3·8 (Pulliainen 1979b). Probably 1 brood. Eggs laid daily, occasionally with interval of 2 days between eggs. Apparently no seasonal variation in clutch size, and repeat clutches the same size as 1st clutch (Newton 1972; Pulliainen 1979b). INCUBATION. 13–14 days; by ♀ only (Bernhoft-Osa 1960; Makatsch 1976; Adkisson 1977a; Blomgren 1983), but see Løvenskiold (1947) and Bannerman (1953a) for reports of ♂ occasionally taking over. Probably starts with last egg (St Quintin 1907; Makatsch 1976) although according to Danilov *et al.* (1984) apparently from 3rd egg. Eggs hatch over 1(–2) days (Pulliainen 1979b). YOUNG. Fed and cared for by both parents (Witherby *et al.* 1938; Bernhoft-Osa 1960; Makatsch 1976). ♀ broods nestlings for *c.* 7–10 days after hatching (Pulliainen and Hakanen 1972, which see for details; Blomgren 1983). FLEDGING TO MATURITY. Fledging period 14 (13–18) days (Adkisson 1977a; Pulliainen 1979b; Blomgren 1983). Young do not feed themselves until 3 weeks out of nest (Adkisson 1977a). For development of young, see Pulliainen and Hakanen (1972). Age of first breeding not known. BREEDING SUCCESS. In Finland, of 55 eggs, 89% hatched; 39% of 23 clutches produced fledged young, 30% suffered predation at egg stage, 17% at nestling stage, 4% abandoned during egg stage, and 4% during nestling stage. Main predators apparently crows (Corvidae): in Finland, Siberian Jay *Perisoreus infaustus* (Pulliainen 1979b); on Yamal peninsula (northern Russia), Carrion Crow *Corvus corone* (Danilov *et al.* 1984). MK

**Plumages.** (nominate *enucleator*). ADULT MALE. Nasal bristles, feathering along lateral base of bill, lore, narrow stripe over eye, and small patch behind eye dark grey or black, darkest from nostril to just above eye. Forehead, crown, and hindneck bright rosy-red, darkest at side of crown, each feather with grey centre, light grey shaft-streak, and black-brown subterminal bar, which are mostly concealed, except for traces on mid-crown and hindneck. Remainder of upperparts with red feather-tips, similar to crown and hindneck, but grey of feather-centres more fully exposed and black-brown subterminal bar on each feather more distinct (except on rump), giving scalloped appearance to mantle; red of feather-tips more rosy-pink, especially on rump; rosy-pink tips of scapulars relatively narrow, much grey (on bases) and brown (subterminally) exposed, grey and brown predominating over red (on remainder of upperparts, red predominating over grey, except when plumage heavily worn). Indistinct stripe from gape to below eye pale grey with darker grey and rosy-red mottling, merging into rosy-pink of cheeks, ear-coverts, and side of neck, latter with traces of grey and brown bars. Upper chin dirty grey-white; side of breast, upper flank, rear of flank, lower belly, thigh, vent, and under tail-coverts medium ash-grey, tips of feathers often paler grey-white or cream-grey on rear of flank, vent, and under tail-coverts; remainder of underparts rosy-pink, some grey of feather-centres sometimes shining through or just visible (especially if plumage worn), some brown suffusion subterminally on feathers of chest and at border of grey of side of breast and flank. Tail dark greyish- or fuscous-brown, outer webs and tips of feathers faintly edged off-white, shading to faint pale brown or pink-brown on bases. Flight-feathers dark fuscous-brown, outer webs narrowly and rather indistinctly edged dull pink or pink-brown, off-white along terminal part and tip of primaries, more buff or off-white along outer webs of inner secondaries. Greater upper primary coverts and bastard wing greyish-black, outer webs and tips narrowly edged off-white, except for longest feather of bastard wing. Greater upper wing-coverts and tertials black-brown, terminal halves of outer webs contrastingly bordered by white fringes 1–2 mm wide, fringe *c.* 1 mm wide on tip; white slightly or clearly tinged pale pink, especially on greater coverts and when plumage fresh. Lesser upper wing-coverts dark brown with narrow rufous or red-brown tips, longest narrowly tipped pale pink; median coverts black-brown with contrasting pale pink or pink-white fringe *c.* 2–4 mm wide along tip. Pink or white on tips of median and greater coverts forms distinct double wing-bar. Under wing-coverts and axillaries medium grey, tips of longer feathers tinged pale pink, those of marginal coverts narrowly rosy-red. *In worn plumage* (May–July), red of head and body more ruby-red, crimson, or (when heavily abraded) orange-red; more brown and grey of feather-bases exposed, plumage less uniform than in fresh plumage; grey of flank, lower belly, vent, thigh, and under tail-coverts darker; less washed cream or white; tips and fringes of tertials and median and greater coverts bleached to white, partly worn off but generally prominent, those of tail- and flight-feathers paler also, but sometimes fully worn off. Occasionally, tips of some feathers yellow instead of red; probably due to deficiency in diet during growth (see Adkisson 1977*a*). ADULT FEMALE. Mainly like adult ♂, but red of body replaced by golden-yellow, and no dark brown subterminal bars on mantle, scapulars, back, upper tail-coverts, side of neck, and chest, more extensive dark ash-grey of feather-centres more gradually merging into golden-yellow of feather-tip. Small dark grey or dull black mask, as in adult ♂; crown to upper mantle bright golden- or buff-yellow, somewhat more grey of feather-bases exposed than in adult ♂, remainder of upperparts mainly ash-grey with slightly darker grey feather-centres, only faintly washed golden-yellow to yellow-green on mantle, back, rump, and upper tail-coverts only. Extent of golden-yellow or buff-yellow on side of head and neck and on underparts similar to that of rosy-red of adult ♂, but some more grey of feather-centres usually visible on side of neck, throat, and flank. Wing and tail as adult ♂, but fringes along basal parts of tail-feathers, primaries, and outer secondaries pale grey-green, not pink-brown, tips of lesser coverts and of marginal under wing-coverts yellow-green, not rufous or red, tips and fringes of tertials and of median and greater upper wing-coverts pure white, without pink tinge, underwing without pink tinge. *In worn plumage*, yellow of plumage slightly more glossy, tinged golden-orange, yellow on feather-tips on upperparts backwards from lower mantle virtually worn off, more dark grey of feather-bases visible on head, neck, and chin to belly; grey from belly downwards darker and more uniform; white on tertials and median and greater upper wing-coverts sometimes largely worn off. NESTLING. Down greyish-black (on 1st day) to light brown-grey (at 1 week), dense, *c.* 10–13 mm long; on head, back, rump, upper wing, and patch above thigh (Bernhoft-Osa 1960; RMNH). JUVENILE. Upperparts and side of head and neck dark drab-grey, feathers tinged rusty-olive on forehead and above eye (forming paler supercilium), feather-tips purer and almost uniform ochre to yellow-olive on hindcrown, hindneck, rump, upper tail-coverts, and ear-coverts, faintly tinged olive on feather-tips on remainder. Narrow sooty black line on lore, stripe below lore to below eye buff-white. Chin isabelline, merging to pale olive-buff on lower throat. Chest, side of breast, flank, and thigh drab-grey with buff wash, merging into light grey-buff of mid-belly, vent, and under tail-coverts. Lesser upper wing-coverts grey with slight olive tinge on tips; median coverts dull black, tip with narrow pale buff fringe 1–2 mm wide; greater coverts and tertials dull black with pale yellow-buff or yellow-white outer fringe and tip, 1–2 mm wide. Under wing-coverts and axillaries isabelline-buff. Tail and flight-feathers as adult, but slightly browner, less blackish, especially when worn, outer webs of primaries and tail-feathers narrowly fringed pale buff, pale greenish-yellow, or yellow-white, fringes along terminal part of secondaries broader and whiter. FIRST ADULT MALE. Like adult ♀, but juvenile tail, flight-feathers, tertials, greater primary and secondary coverts, and apparently often median upper wing-coverts retained; white tips and fringes of tertials and of median and greater upper wing-coverts slightly tinged grey, pale buff, or pale yellow in autumn,

less pure white than in adult ♀, less sharply defined but difference negligible once plumage somewhat abraded; tips of inner webs of tail-feathers more sloping, less broad and truncate (see Svensson 1992 for shape), but some birds intermediate in shape. Colour of head and body as in adult ♀, but golden-yellow often partly replaced by golden-orange, orange-rufous, or (occasionally) dull orange-red, especially on top and side of head, rump, and chest; part of feathering sometimes with rosy-pink tips, especially on side of head. Chin and short stripe below lore on average paler than in adult ♀, brown-grey or dirty white. FIRST ADULT FEMALE. Like adult ♀, but part of juvenile feathers retained, extent and character as in 1st adult ♂. Head and body as adult ♀, but golden-green of mantle and belly on average less extensive, grey predominating. Inseparable in plumage from those 1st adult ♂♂ which show no orange or rosy-pink tinges.

**Bare parts.** ADULT, JUVENILE, FIRST ADULT. Iris cinnamon to dark brown. Bill plumbeous-grey, blackish on tip, basal half of lower mandible and cutting edges paler greyish-horn, in juvenile and in 1st autumn with yellow or flesh tinge and sometimes with traces of orange-yellow gape-flanges. Leg and foot dark grey-horn with slight flesh tinge (in juvenile and 1st autumn) to black-brown or slate-black. (RMNH, ZMA.) NESTLING. Mouth deep plum-red (Blair 1936) or bright red (Bernhoft-Osa 1960); gape-flanges yellow (RMNH).

**Moults.** ADULT POST-BREEDING. Complete; primaries descendent. In breeding area, July–September (Johansen 1944; Pyle *et al.* 1987). Starts when feeding young; moult in captive non-breeders in late August and September (Adkisson 1977*a*). In USSR, single ♂ on 22 July had moulted feathers of throat but not yet primaries; another in heavy moult on 31 August (Dementiev and Gladkov 1954). In northern Sweden, single ♂ not started moulting mid-July, but ♀ from same date had shed p1–p2 (primary moult score 3); ♂ from 10 September had primary moult score 29 (p1–p4 new, p8–p10 old), tail in full moult (t1 new, t2–t6 simultaneously growing), body and tertials mainly new, secondaries, and part of nape in moult (RMNH). None of 28 adult birds examined in RMNH and ZMA had arrested secondary moult, unlike Parrot Crossbill *Loxia pytyopsittacus*, which is of similar size. POST-JUVENILE. Partial: head, body, and lesser and variable number of median upper wing-coverts. In breeding area, starting shortly after fledging; in moult July–October (Pyle *et al.* 1987) or August (Witherby *et al.* 1938). Moult completed in bird from 10 September, Finland, and in birds from various localities in October (RMNH, ZFMK, ZMA).

**Measurements.** ADULT, FIRST ADULT. Nominate *enucleator*. Fenno-Scandia and Latvia, whole year; skins (RMNH, ZMA). Bill (S) to skull, bill (N) to distal corner of nostril; exposed culmen on average 5·8 shorter than bill (S).

| | ♂ | ♀ |
|---|---|---|
| WING | 109·4 (1·78; 22) 106–113 | 109·2 (1·15; 11) 107–111 |
| TAIL | 84·8 (2·58; 21) 80–89 | 84·9 (2·73; 11) 80–89 |
| BILL (S) | 21·0 (0·62; 21) 20·1–22·3 | 21·0 (0·66; 11) 20·0–22·0 |
| BILL (N) | 11·9 (0·50; 21) 11·2–13·1 | 11·8 (0·52; 11) 11·0–12·7 |
| TARSUS | 22·2 (0·78; 20) 21·0–23·8 | 22·5 (0·55; 11) 21·9–23·4 |

Sex differences not significant. Ages combined, though wing and tail of 1st adult (with retained juvenile flight-feathers and tail) on average slightly shorter than in older birds; thus, sexes combined: wing, adult 109·9 (1·35; 18) 107–113, 1st adult 108·7 (1·61; 15) 106–113; tail, adult 85·0 (2·92; 17) 80–89, 1st adult 84·6 (2·24; 15) 80–89.

Wing. North-west Russia: ♂ 111·0 (21) 105–116, ♀ 110·0

(19) 102–117 (Danilov *et al.* 1984; Semenov-Tyan-Shanski and Gilyazov 1991). Norway: ♂ 106–113 (11), ♀ 101–111(-115) (9) (Haftorn 1971).

**Weights.** Nominate *enucleator*. Norway, October–December: ♂♂ 56, 63; ♀♀ 53·9 (3) 51·3–56·8; both sexes 55·0 (20) 49–64 (Haftorn 1971; ZMA). Russian Lapland: ♂ 53·3 (16) 47–60, ♀ 53·0 (16) 47–60 (Semenov-Tyan-Shanski and Gilyazov 1991). Northern Russia: ♂ 53·0 (11) 42·5–60, ♀ 54·6 (6) 41·3–63·8 (Dementiev and Gladkov 1954; Danilov *et al.* 1984).

P. e. *pacata*. Single ♀ 47 (Dementiev and Gladkov 1954).

P. e. *kamtschatkensis*. Kuril islands: 46·3 (3) 45·4–47·3 (Nechaev 1969).

North American races. See Van Tyne (1934), Amadon (1943), and Johnston (1963).

**Structure.** Wing rather long, broad at base, tip bluntly pointed. 10 primaries: p8 longest, p9 2–5 shorter, p7 0–1, p6 0·5–4, p5 7–12, p4 15–21, p3 20–26, p2 23–31, p1 26–35; p10 strongly reduced, a tiny pin hidden below reduced outermost greater upper primary covert, 72–80 shorter than p8, 10–16 shorter than longest upper primary covert. Outer web of (p5-)p6-p8 emarginated, inner web of p7-p9 with slight notch. Tip of longest tertial reaches to tip of p1-p2. Tail rather long, tip slightly forked; 12 feathers, t4-t5 longest, t1 4–8 shorter. Bill short, heavy, slightly swollen at base; depth at base 11·3 (30) 10·8–12·0 mm in nominate *enucleator*, width 9·9 (30) 9·3–10·6 mm; culmen strongly decurved, upper mandible ending in small hook, projecting over tip of lower mandible. Cutting edge of upper mandible with shallow tooth-like bulge in middle. Nostril rather small, rounded, covered by broad tuft of feathers projecting over basal side of upper mandible. A few short bristles at gape. Tarsus and toes short, but strong. Middle toe with claw 21·9 (10) 21–23 mm; outer and hind toe with claw both *c.* 75% of middle with claw, inner *c.* 68%.

**Geographical variation.** Slight in Eurasia, rather more pronounced in North America. Involves depth of general colour, size (wing and tail length, weight), and relatively length, depth, and width of bill. Within northern Europe and western Siberia, birds gradually paler towards east, mainly in depth of grey of vent and under tail-coverts; birds of western Siberia sometimes separated as *stschur* Portenko, 1931, but even in western Siberia only half of specimens are paler than typical nominate *enucleator* from Scandinavia, and race therefore not recognized (Johansen 1944); size and bill of birds examined from Tomsk similar to those from Fenno-Scandia. *P. e. pacata* from Altai, Sayan mountains, and Yenisey east to Kolyma basin and Stanovoy mountains and south to northern Mongolia closely similar to nominate *enucleator* in colour and size, but bill on average shorter, to skull 18·5–20·5, to nostril 10–12 (in nominate *enucleator*, 20–22 and 11–13 respectively), appearing more swollen, though actual depth and width of bill at base close to nominate *enucleator*; in some birds, bill rather compressed laterally at tip (Vaurie 1956*b*), but similar birds occur occasionally elsewhere in range (Hartert 1921–2). *P. e. kamtschatkensis* from Anadyrland and shores of Shelekhova bay to Kamchatka has bill still shorter than *pacata*, to nostril 9·8–11·0(-11·5) (Stegmann 1931*a*; Démentieff 1935; Vaurie 1956*b*); bill deeper at base than in *pacata* and nominate *enucleator*, 11·5–13. Further south, Kuril Islands, Sakhalin, and Hokkaido inhabited by *sakhalinensis*, which has width and depth of bill similar to *kamtschatkensis* (depth 11·7–13·7, width 9·5–10·3), but bill length rather similar to nominate *enucleator* (to nostril 11·0–13·5, to skull 20–23) (Hartert 1921–2; Démentieff 1935; BMNH, RMNH). Most North American races distinctly larger than nominate *enucleator*, bill relatively shorter, broader, and deeper, wing and tail virtually black, less brown. In smaller and darker *eschatosa*, ranging from extreme north-east USA to Newfoundland and southern Quebec, wing of 10 birds 112·9 (108–118), tail 85·8 (81–89), bill to skull 19·6 (19·0–20·7), to nostril 10·8 (10·2–11·6), average depth of bill 11·1, average width 9·9. In larger and paler *leucura* from eastern half of northern Canada, wing of 8 birds 117·8 (114–123), tail 93·5 (87–98), bill to skull 20·5 (19·5–21·5), to nostril 11·4 (11·1–11·8), average depth of bill 11·4, average width 10·3 (RMNH, ZMA). For these and other North American races, see Griscom (1934), Adkisson (1977*b*), and Godfrey (1986).    CSR

*Uragus sibiricus* Pallas, 1773   **Long-tailed Rosefinch**

Fr. Roselin à longue queue   Ge. Meisengimpel

PLATE 54
[between pages 640 and 641]

Breeds in valleys and woodland of southern and eastern Siberia, northern Mongolia, northern and central China, south-east Tibet, and Japan. Winters in breeding range, and in west south to Russian and Chinese Turkestan. Recorded as presumed escape in several European countries, e.g. Britain, Sweden, Germany, European Russia. Record from Finland, 25–27 April 1989, formerly accepted as a wild bird (Jännes *et al.* 1990) but now presumed to have been an escape (Jännes 1992).

# *Pyrrhula pyrrhula* **Bullfinch**

Du. Goudvink      Fr. Bouvreuil pivoine      Ge. Gimpel
Ru. Обыкновенный снегирь      Sp. Camachuelo común      Sw. Domherre

*Loxia Pyrrhula* Linnaeus, 1758

Polytypic. Nominate *pyrrhula* (Linnaeus, 1758), northern Eurasia from Scandinavia east to Verkhoyansk mountains, Olekma basin, and Yablonovyi mountains, grading into *cassinii* in eastern Siberia north of Sea of Okhotsk; in south-west to Skåne (Sweden), Bornholm, north-east Poland, Belorussiya, eastern and southern Carpathians, and north-east Yuogoslavia; intermediates with *europoea* which are nearer nominate *pyrrhula* ('*germanica*' C L Brehm, 1831) occur from Sjælland (Denmark) south through eastern half and south of former East Germany (west to Thüringen), Poland (except north-east), Czechoslovakia, Hungary, north-west Rumania, and south-west Germany (west to Bamberg, Nürnberg, Augsburg, and Oberbayern) to Alps, Yugoslavia (except coastal districts, and north-east), Bulgaria, and northern Greece; intermediates which are nearer *europoea* ('*coccinea*' Gmelin, 1789) occur in Jutland, Fyn (Denmark), western half of former East Germany south to Harz, in western Germany west to Kassel, Frankfurt, and Pfalz, in Vosges and Jura (eastern France), in Switzerland north of Alps and Italy south of Alps, and in coastal Yugoslavia; *europoea* Vieillot, 1816, from northern and north-west Germany (south to Rheinland) and Netherlands south to France, except east and Pyrénées; *pileata* MacGillivray, 1837 (synonym: *nesa* Mathews and Iredale, 1917), Britain and Ireland; *iberiae* Voous, 1951, Pyrénées and northern Portugal and Spain; *rossikowi* Derjugin and Bianchi, 1900, northern Turkey and Caucasus south to Azarbaijan (north-west Iran); *murina* Godman, 1866, São Miguel (Azores). Extralimital: *caspica* Witherby, 1908, northern Iran, perhaps grading into *rossikowi* in Lenkoran' area (eastern Transcaucasia); *cassinii* Baird, 1869, Kamchatka and Paramushir (northern Kuril Islands); *rosacea* Seebohm, 1882, western shore of Sea of Okhotsk, Amurland, and Manchuria east to Sakhalin, probably grading into nominate *pyrrhula* further west; *griseiventris* Lafresnaye, 1841, Japan and southern Kuril Islands.

**Field characters.** 14·5–16·5 cm; wing-span 22–29 cm. Size varies, being noticeably large in races or populations occupying highest latitudes and altitudes; typical bird of northern race nominate *pyrrhula* over 15% larger than birds of European races *pileata*, *europoea*, and *iberiae*. Medium-sized to rather large, bull-headed, quite long-winged, and seemingly long-tailed finch, with short bulbous bill and distinctive plumage. ♂ and ♀ share black face and cap, black wings with white wing-bar, white rump, and black tail; ear-coverts and underbody pink in ♂, pale pink-brown in ♀. Juvenile like ♀ but lacks black cap. Call diagnostic. Sexes dissimilar (except in Azores race, *murina*); no seasonal variation. Juvenile separable. 6 races in west Palearctic; of 5 main races, 3 smaller ones in western Europe separable from 2 larger ones to north and east.

(1) North Eurasian race, nominate *pyrrhula*. ADULT MALE. Moults: July–November (complete). Chin, lores, and long crown glossy black, combining to form heavy-looking cap. Nape, mantle, scapulars, and back blue-grey, partially or wholly suffused pink in some (see Plumages). Rump square and pure white, contrasting with back, inner wing-feathers, and tail. Wings predominantly glossy black but blue-grey lesser and median coverts show at shoulder, while broad grey to almost white tips to greater coverts form striking wing-bar; innermost tertial shows grey and pink outer fringe. Upper tail-coverts and tail pure black. Underparts bright pink from ear-coverts down to breast and along flanks; white from lower belly to under tail-coverts. Underwing mainly white. With wear, back

becomes purer grey and pink underparts have initially orange wash replaced by redder tone. Bill massive, forming bulb in front of whole face; black. Eye brown, lost in head cap. Legs dark brown. ADULT FEMALE. Plumage pattern as ♂ but upper- and underparts duller and black areas less glossy. Nape strikingly pale grey, contrasting with buff-grey mantle and back. Bar across greater coverts buffier, less grey and white. Ear-coverts, fore-underparts, and flanks pale pinkish-grey-brown. With wear, back becomes greyer and breast and flanks cleaner and more fawn. JUVENILE. Resembles ♀ but lacks cap, having head concolorous with warm grey-brown back. Rump buff-white. Whole underparts yellowish-brown to buff. Bar across greater coverts fully buff. Can look almost orange on fore-face. Dark brown eye prominent in amorphous face. (2) North-west European race, *europoea*. Averages over 10% smaller than nominate *pyrrhula*. ♂ slightly darker, more red-toned below; ♀ slightly browner above and below. (3) British and Irish race, *pileata*. Slightly smaller than *europoea*. Typical ♂ from England even duller below than *europoea*; ♀ more olive-brown above, with dirty nape, and much browner below. Beware local variation (see Geographical Variation). (4) Iberian race, *iberiae*. Only marginally smaller than *europoea* with distinctive plumage tones. ♂ very rich red below; ♀ much paler and greyer above and below.

Unmistakable. Plumage pattern and colours as unique as voice (see below). Separation of south-migrating bird of northern nominate race from smaller more sedentary birds of western and south-west Europe not difficult; lar-

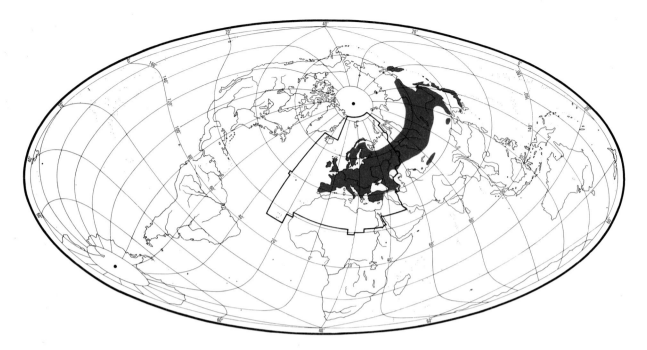

ger size readily evident, particularly in flight when body bulk, wing-span, and tail length are eye-catching. Important to recognize that colour tones change noticeably according to light intensity, and racial identification should not be made on plumage colours alone. Flight at times rather weak for finch, flitting within cover or along leading line—with white rump easy to follow; strong over distance, however, with action developing typical finch undulations and achieving considerable height between woods or on migration. Escape-flight usually very short, to low dense patch of cover (rarely to treetop). Gait a clumsy hop, making bird look uncomfortable and untidy on ground. Feeds more slowly than other finches, visibly manipulating buds and seeds in bill. Has habit of flicking and twisting tail. Less sociable than other finches, with few parties of over 10 birds. Secretive, staying in cover and feeding unobtrusively; parties fond of glades within tall trees.

No loud song but both sexes may pipe a squeaky, wheezy warble. Commonest call from perched or flying bird a rather low-pitched, sad, piped or fluted 'phew', occasionally repeated; hollow in timbre, but carries far.

**Habitat.** Breeds across most of temperate west Palearctic in both oceanic and continental climates, and also summers over much of boreal zone, within July isotherms *c.* 12–21°C (Voous 1960*b*). Arboreal and mainly lowland, but in mountains in south of range ascends to breed up to treeline—in Caucasus up to 2500–2700 m. In USSR, largely in coniferous taiga forest, but also in mixed and broadleaved woods; in Caucasus, in pine forest and tall beechwoods. Elsewhere, prefers mature mixed or coniferous

forest with dense undergrowth. In winter often appears in orchards and towns, preferring gardens, parks, and plantings of sunflower and hemp (Dementiev and Gladkov 1954; Flint *et al.* 1984). In Switzerland, only exceptionally in pure broad-leaved woodland, preferring spruce *Picea* and usually keeping close to dense stands, but some nest in gardens and parks with thick undergrowth, even in large cities (Glutz von Blotzheim 1962). In Britain, breeds mainly in broad-leaved woods, but commonly also in fringing or detached groups of trees, thickets, tall and dense hedgerows, large gardens, yews *Taxus* in churchyards, and even gorse *Ulex* on heaths; where conifers used, some preference for younger plantations or ornamental groups; orchards often visited in spring to pick buds but rarely favoured for nesting (Bannerman 1953*a*; Yapp 1962; Sharrock 1976).

Due to its shyness is more readily observed outside woodland or along its edge, and this may lead to underestimation of use of interior of forests, especially where there are glades, clearings, or regenerating patches. Apparently, in parts especially of western Europe, habitats outside woodland have been increasingly colonized in recent decades, but questionable whether such habitats account for significant fraction of total population. Tall mature trees seem to be less favoured than younger or smaller trees and above all tall and dense shrubs, even in moist fen or riverside situations. Presence of water however is not an evident attraction, and open areas and windy, wet, or exposed situations are avoided. Successful within its limits but shows little tendency to adapt lifestyle or change traditional habitats, beyond quite minor degrees. Even within human settlements normally avoids

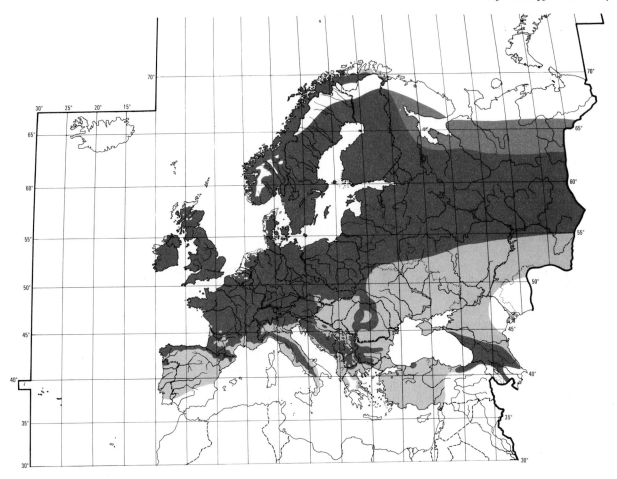

contact with people, and does not usually show itself on ground, or even flying in the open. Only very locally and recently have a few adopted seasonal habit of visiting bird-tables. (Newton 1972.) For *murina*, see Bibby *et al.* (1992).

**Distribution.** Some expansion in north and west of range, mostly not recent. Decrease reported only in endemic Azores race *murina*.

BRITAIN, IRELAND. Probably slow spread to north and west in last 200 years (Sharrock 1976; Hutchinson 1989). NORWAY. Some recent spread in north and west (VR). POLAND. In early 19th century known to breed only in mountains and in north-east; other areas colonized in 2nd half of 19th century or 20th century, expansion proceeding from Silesian mountains (AD, LT). AZORES. São Miguel only. Formerly more widespread; now confined to 650 ha in an eastern valley. Reduction of range due to destruction of native vegetation, to which it is confined. (Brien *et al.* 1982; GLG.)

Accidental. Iceland, Faeroes, Malta, Morocco, Algeria, Tunisia.

**Population.** Increases reported from some northern parts of range, decline (recent, perhaps not long-term) only in Britain. Isolated Azores population has declined, but now less in danger of extinction with recent conservation measures (Brien *et al.* 1982; Bibby *et al.* 1992).

BRITAIN, IRELAND. Estimated *c.* 600 000 pairs, 1968–72 (Sharrock 1976). Britain: decline since mid-1970s; 300 000–350 000 pairs (Marchant *et al.* 1990). Ireland: increasing throughout this century (Hutchinson 1989). FRANCE. 100 000 to 1 million pairs (Yeatman 1976). BELGIUM. Estimated 16 000 pairs, 1973–7; probably increasing (Devillers *et al.* 1988). NETHERLANDS. Estimated 15 000–20 000 pairs (SOVON 1987). GERMANY. Estimated 438 000 pairs (Rheinwald 1992). East Germany: 50 000 ± 25 000 pairs (Nicolai 1993). SWEDEN. Estimated 400 000 pairs (Ulfstrand and Högstedt 1976). FINLAND. Perhaps 200 000–500 000 pairs (Koskimies 1989). Some increase in recent years due to intensified winter feeding and increase of young spruce *Picea* forests favoured by modern forestry (OH). POLAND. Major increase with expansion of range (see Distribution), and probably still increasing (AD, LT). AZORES. Estimated 108 or 133 pairs, 1991 (J A Ramos). Previous estimates 30–40 pairs (Le

Grand 1982), 100 pairs (Bibby and Charlton 1991; Bibby *et al.* 1992).

Survival. Britain: average annual mortality of adult ♂♂ 52 ± SE3%, adult ♀♀ 59 ± SE3% (Dobson 1987). Finland: average annual mortality 47 ± SE6·4% (Haukioja 1969). Oldest ringed bird 17 years 6 months (Rydzewski 1978).

**Movements.** Sedentary to migratory; probably most populations partially migratory. Winters chiefly within breeding range. Most migrants move short or medium distances, but some (apparently chiefly from Russia) move longer distances; in northern and central Europe, no evidence that northern populations move further than southern ones. North European birds move within wider compass than central European birds. Also eruptive migrant; numbers migrating show marked annual fluctuations (e.g. Rendahl 1964, Haftorn 1971, Winkler 1984); no link with particular food source established.

Chiefly sedentary in Britain (*pileata*); winter distribution very similar to summer (Lack 1986). Of 185 ringing recoveries, 1910–60, 93% under 5 km, 6·5% 5–25 km, 0·5% over 25 km; records 1961–74 show significant increase in longer movements: of 1367 recoveries, 77·8% under 5 km, 17% 5–25 km, 5·2% over 25 km. Some evidence to suggest south-east movement in autumn and corresponding return north-west in spring; mean direction was south-east for 19 adults recovered in winter more than 25 km from summer ringing site, and NNW for 9 birds recovered in summer more than 25 km from winter ringing-site. Occurs only irregularly on passage at coastal sites, and absent in many years; in occasional years, small proportion shows eruptive tendency; in 1961–2, marked movement from October included exceptional numbers at Dungeness (Kent) and southward coasting passage at Minsmere (Suffolk); higher than usual numbers also in 1967–8. (Summers 1979.) Of 4 recoveries between southern England and continental Europe (not all assigned to race), 2 in 1961: Huntingdon (April) to Belgium (October, 362 km south-east), and Kent (November) to Versailles, France (November, 298 km south-east); also bird ringed Dorset, April 1977, recovered SSE in Maine-et-Loire (France), January 1978, and bird ringed Netherlands, November 1954, found Kent, February 1960 (Spencer 1961, 1962; Spencer and Hudson 1979). Radio-tracking shows local movements in midwinter; individuals may remain for weeks near good food source, then suddenly move several kms to new site (Lack 1986). No evidence of nominate *pyrrhula* migrating west to southern Britain, though recorded very occasionally (Newton 1972; Summers 1979), but scattered reports in Orkney and Shetland (Scotland), where no breeding occurs, are probably that race (Lack 1986). In Iceland, 37 records (presumably nominate *pyrrhula*) up to 1989, of which 16 since 1979 (Pétursson *et al.* 1992).

In Spain (*iberiae*), birds breeding in valleys or at low

levels are sedentary, those breeding at high levels tend to make altitudinal movements (Noval 1971). No movements reported in Azores (*murina*) (Bannerman and Bannerman 1966; G Le Grand).

Eurasian race nominate *pyrrhula* and west European race *europoea* partially migratory. ♂♂ are more resident than ♀♀, and adults more than juveniles (Møller 1978b; Saurola 1979). In Sweden, nominate *pyrrhula* winters north to 68°21′N; some birds winter in breeding areas, some move south within Sweden, others move further afield (Rendahl 1964; *Rep. Swedish Bird-Ringing* 1987). Many remain to winter in Finland (Koskimies 1989); most of those wintering in southern Finland have moved only short distance: 17 of 20 recoveries in breeding season were within 100 km of wintering site (Saurola 1979). Local movements occur mid-winter, both inland and on coast (notably Helsinki district, December–January) (Haila *et al.* 1986). On Kola peninsula (north-west Russia), most birds depart, but in recent years some have wintered in towns (Semenov-Tyan-Shanski and Gilyazov 1991). In St Petersburg region, many birds leave woods and occur in more open areas; local movements in midwinter are typical; in some years, sharp increase in numbers gives evidence of immigration (Mal'chevski and Pukinski 1983). Widespread ringing shows that a large proportion (perhaps 75%) of French birds (*europoea*) are sedentary, others move up to 500 km; most long-distance recoveries involve birds from north-west moving chiefly south, and birds from eastern France moving chiefly south-west. Individuals from both directions reach Spain. 2 birds ringed in Asturias (northern Spain), winter, and recovered in north-west France, April and July, had crossed or skirted Bay of Biscay, also bird ringed eastern Spain, October, recovered in eastern France, July. (Aubry 1970; Noval 1986.) Winter population in France includes (mostly in east and south) passage and breeding birds from Germany, passage (chiefly) and breeding birds from Switzerland, also birds from Belgium and Luxembourg, and (in northeast) from Poland and Lithuania (Aubry 1970; Yeatman-Berthelot 1991). Southern German populations also partially migratory; 21 recoveries in subsequent winters of juveniles and adults ringed in breeding season, of which 11 long-distance, 230–780 km (Bairlein 1979). In Rheinland (western Germany), disperses locally, occurring in non-breeding areas autumn and winter; some birds make altitudinal movements, though higher levels not entirely vacated (Mildenberger 1984). Local birds in Thüringen (eastern Germany) recovered up to 100 km from ringing-site (Knorre *et al.* 1986). Of 4 recoveries of birds ringed in Belgium in breeding season, 2 at site of ringing and 2 in France, up to 250 km (Lippens and Wille 1972). In Switzerland, birds usually remain in breeding areas, but move down to plains in times of seed scarcity in forests (Schifferli *et al.* 1982). In northern Italy, makes limited altitudinal movements, occurring outside breeding areas only in severest weather (Mingozzi *et al.* 1988). In Greece, where few breed, uncommon but much more widespread in winter (G I Handrinos). Few recorded south of range in Mediterranean region. Birds of unknown origin occur exceptionally on both sides of Strait of Gibraltar; southernmost Moroccan records at Settat (33°04′N), 1976, and Meknès (33°53′N), 1987 (Pineau and Giraud-Audine 1979; Finlayson 1992; *Br. Birds* 1988, 81, 338). No recent records elsewhere in North Africa. In Portugal, rare south of Tajo river (Cary 1973). Only one record since 1909 (7 November 1972) in Malta (Sultana and Gauci 1982), 4 records 1872-1976 in Sicily (Iapichino and Massa 1989), and none south of Turkey in Middle East (Moore and Boswell 1957; Hovel 1987; Flint and Stewart 1992; F E Warr). In Turkey (breeds in north), occasionally recorded more widely across western two-thirds in winter (Beaman 1986). In Russia, vacates extreme north of range (Dementiev and Gladkov 1954), but present all year in western Siberia (Gyngazov and Milovidov 1977). In Volga-Kama area, more common winter than summer, with immigrants from further north joining local populations (Popov 1978). In Kazakhstan (breeds only in extreme north-west), encountered very widely on passage and in winter; common in north, with numbers diminishing southward; rare at Alma-Ata (Korelov *et al.* 1974).

Swedish ringing data show that in autumn birds from Norway and northern Sweden initially migrate south-east and south, and some Finnish birds south-west, through central and southern Sweden; from southern Sweden, most birds head south-west or SSW, but some south or south-east. 13 birds ringed north of 59°N, and recovered over 200 km from ringing site, had moved 596 km on average (231-904 km); 10 birds ringed south of 59°N, and recovered over 200 km from ringing site, had moved 589 km on average (230-1334 km). Foreign recoveries of birds ringed all seasons are west to *c.* 5°E (Belgium), south to *c.* 47°N (Switzerland, Rumania), and east to *c.* 35°E (Russia). Spring recoveries show movement through Sweden both north or north-west and also north-east; of spring birds over 200 km from ringing site, 14 recovered north of 62°N had moved 605 km on average (239-899 km), and 7 south of 62°N had moved 492 km on average (203-1005 km). (*Rep. Swedish Bird-Ringing* 1987.) Birds ringed Finland on autumn passage have dispersed before following March in all southerly directions between west and east (Saurola 1979); also 2 birds ringed Finland in breeding season were recovered 300 and 860 km south-east in winter. Birds from northern and central European Russia, also from Altai area further east, migrate chiefly south-west; longest distance 2980 km: bird ringed Belgium, November 1932, recovered Arkhangel'sk region, May 1933. (Sapetina 1962.) Recoveries involving movement to or from eastern Germany extend further west and south than from Scandinavia, reaching France and Italy; those furthest east are from Austria and Czechoslovakia (Knorre *et al.* 1986; Klafs and Stübs 1987). Of birds ringed October-March in southern Germany, 50 recoveries in breeding season of

which 13 over 100 km, mostly within 400 km, but up to 1600 km; average direction of both short and long-distance recoveries NNE (Bairlein 1979).

Data, especially from Finland and Russia, show that (in accordance with eruptive tendency) individuals winter at greatly differing distances from breeding areas in different years. Of 17 long-distance recoveries of birds ringed Finland October to early April, 1956-66, and found in later winters November-March, 9 were at 550-1000 km, and 8 over 1000 km; 15 were between SSW (Hungary and Poland) and east (Russia) of ringing site (including 2 at 1920 and 2350 km east in Siberia), one was NNE and one NNW (*Finnish Ringing Rep.*). Also 3 birds ringed Sweden, winter, were recovered at 475-1491 km, between NNE and south-east, in later winters (Sapetina 1962; *Rep. Swedish Bird-Ringing*). Reports of winter site-fidelity vary; in Finland, only 8 of 31 recoveries in subsequent winters were within 20 km of winter ringing site (Saurola 1979); in Latvia, 25 of 26 birds ringed in winter and recovered in later winters were in or near ringing-site (Sapetina 1962). Many birds remain at or near same locality throughout winter; individuals recorded up to 104 days in Russia (Sapetina 1962), and up to 109 days in Fenno-Scandia (Rendahl 1964).

Autumn migration begins late, and is fairly brief, mostly October-November (Sapetina 1962; Bairlein 1979; *Rep. Swedish Bird-Ringing* 1987). At Ottenby (south-east Sweden), passage almost entirely October-November, peaking mid- to late October (Enquist and Pettersson 1986). Arrivals in northern Denmark also show mid- to late October peak (Møller 1978a); proportion of ♂♂ slightly lower than of ♀♀ on passage, but slightly higher in winter; proportion of ♂♂ increases in cold years, and proportion of ♀♀ increases in irruptive years (Møller 1978b). Main arrival in Belgium November (Lippens and Wille 1972). On Kola peninsula (north-west Russia), departures very irregular, with most birds leaving October in some years, but many remaining to mid-December in other years (Semenov-Tyan-Shanski and Gilyazov 1991). Study in Schwäbische Alb (south-west Germany), 1969-74, showed that passage begins between mid-September and mid-October and ends mid-December or later, with main movement mid-October to end of November; ♂♂ average 4 days later than ♀♀. In normal years, proportion of ♂♂ in entire movement 31-36%, but 41% in unusually cold autumn 1974 (perhaps temperature initiating migration is lower in ♂♂ than in ♀♀). Among wintering birds, however, ♂♂ predominate 7: 3 over ♀♀. (Gatter 1976.)

Spring migration February-April. Winter visitors leave Switzerland mid-February to March (Winkler 1984) and Germany February-April (Mildenberger 1984; Rutschke 1983; Klafs and Stübs 1987). In northern Denmark, movement peaks early to mid-April (Møller 1978a), and at Ottenby passage late March to April, with stragglers into May; median 10 April (Enquist and Pettersson 1986). Birds return to Kola peninsula usually March or 1st half of April, occasionally from February (Semenov-Tyan-Shanski and Gilyazov 1991). Passage in Moscow region mostly 12 March to 24 April (Ptushenko and Inozemtsev 1968).

In Iran, *caspica* recorded rarely in winter in forests of western Caspian, south of breeding area in northern Azerbaijan (D A Scott). In north-west Caucasus, *rossikowi* makes vertical movements and perhaps disperses more widely (Polivanov and Polivanova 1986).

East Asian races also apparently partially migratory. In Kamchatka peninsula (*cassinii*), southward movement reported September–October and northward movement 2nd half of April; numbers wintering vary; accidental in Japan (Lobkov 1986; Brazil 1991). In Japan, *griseiventris* breeds mainly at high altitude and winters at lower elevations; some short-distance movement southward within Japan (Brazil 1991); winter visitor to Ussuriland in varying numbers, October–April (Panov 1973*a*), and irregular winter visitor to north-east China and Korea (Vaurie 1959; Schauensee 1984).                                        DFV

**Food.** Seeds of fleshy fruits and herbs, buds, and shoots; invertebrates important in diet of young. Forages in woodland, thickets, hedgerows, etc., some moving to open country, large gardens, or orchards in late autumn to early spring; rarely feeds on ground and hardly ever more than *c*. 10 m from cover. In southern England, took widest variety of native fleshy fruits among Fringillidae, virtually ignoring exotic species. In continental Europe, apparently feeds more often on ground and at garden bird-tables than in Britain. (Klehm 1967; Newton 1967*a*, *b*; Greig-Smith and Wilson 1984; Snow and Snow 1988.) In captivity, 70% of feeding postures were clinging to sloping perch, 22% normal perching, 3% leaning forward, 2% clinging to vertical perch, 2% standing on ground, 1% in air; in the wild, 98% of time spent feeding in vegetation; never seen to use feet in feeding or to hang upside-down (Kear 1962; Newton 1967*a*). Generally extracts seeds from fruit on plant, since only large seeds can be picked up from ground because of bill shape; small seed-heads and sometimes small fruits may be bitten off and carried to ground for removal of seeds. In tree or bush, either extracts seeds from fruit *in situ*, leaving skin and pulp hanging, or mandibulates fruit by turning it in bill with tongue, removing pulp against lower mandible and swallowing seeds, sometimes without de-husking; bites into fruits and soft seed-heads much more than other Fringillidae of region and bill well-adapted to this. On ground or herbs, either pulls down seed-head or stands on stem to bend it over, pulling seeds out in bundles and de-husking them or removing pappi by mandibulation; restricted to small, soft seed-heads such as groundsel *Senecio* or sow-thistle *Sonchus* because rounded bill unable to extract seeds one by one and jaw too weak to tackle hard seeds; squeezes Cruciferae seeds out of pods. Also hovers fairly frequently at seed-heads or fruits on outer twigs, or snatches exposed seeds

while flying up from ground. Makes sallies after flying insects. Invertebrates simply crushed in bill and swallowed, small snails sometimes turned in bill like berries between crushing and swallowing. (Doerbeck 1963; Klehm 1967; Newton 1967*a*, *b*; Snow and Snow 1988; M G Wilson.) Of 549 feeding observations in trees and bushes, 7% hovering to pluck off fruits; 46% of 52 observations at honeysuckle *Lonicera* involved hovering (Snow and Snow 1988). For comparison of handling times and selection of ash *Fraxinus* and sunflower *Helianthus* seeds in experiments, see Greig-Smith and Crocker (1986) and Greig-Smith (1987). Feeds in selected trees, at times taking all fruits or buds from some and ignoring others, seeming to prefer those with high fat and low phenol content, extending even to differentiation between varieties of cultivated fruit (Newton 1964*c*; Summers 1982; Greig-Smith and Wilson 1985). Major pest in fruit-growing areas, feeding on flower buds in winter and spring, preferring them to buds of its natural food species at this time of year, thus preventing development of fruit; plum *Prunus* and pear *Pyrus* worst affected, followed by gooseberry and currants *Ribes*, then cherry *Prunus* and apple *Malus*; for many details, see (e.g.) Newton (1964*c*), Summers (1979), and Greig-Smith and Wilson (1984). However, most damage in orchards in southern England perhaps occurs in years of low tree seed crop, especially ash (Newton 1964*c*; Summers 1979, which see for movements related to food supply). Can consume buds at rate of 30 or more per min, systematically moving from outer twigs to trunk, removing all on branch then flying out to start on next one; moves towards centre of orchard from surrounding trees and bushes (Newton 1964*c*, 1972). In England, size of seeds eaten ranged from less than 0·5 mm (willowherb *Epilobium*) to 14 mm (ash) (Newton 1967*b*). Average weight of 9 most commonly taken fleshy fruit seeds was 1·9 mg, about half weight of those taken by Greenfinch *Carduelis chloris* (Snow and Snow 1988). Handling time for ash seeds 20–50 s, *c*. 10 s for de-husking (Greig-Smith and Wilson 1985). Pair fed intermittently over *c*. 2 weeks in very cold weather on meaty bones suspended on verandah of house, pulling off meat 'with evident relish' (Payn 1982). Gut relatively long, associated with proportion of green bud material in diet (Eber 1956, which see for bill morphology, comparison with other seed-eaters, etc.); see also Ziswiler (1965) and Newton (1967*a*).

Diet in west Palearctic includes the following. Invertebrates: aphids (Hemiptera: Aphidoidea), adult and larval Lepidoptera (Tortricidae), flies (Diptera), larval sawflies (Hymenoptera: Symphyta), beetles (Coleoptera: Curculionidae), spiders (Araneae), snails (Pulmonata). Plants: seeds, buds, shoots, etc., of yew *Taxus*, japanese cedar *Cryptomeria*, juniper *Juniperus*, *Thuja*, larch *Larix*, spruce *Picea*, pine *Pinus*, fir *Abies*, willow *Salix*, poplar *Populus*, birch *Betula*, alder *Alnus*, hornbeam *Carpinus*, beech *Fagus*, oak *Quercus*, elm *Ulmus*, mulberry *Morus*, maple etc. *Acer*, ash *Fraxinus*, privet *Ligustrum*, laurel *Laurus*, lilac *Syringa*, holly *Ilex*, lily-of-the-valley tree *Clethra*, mistletoe *Viscum*, hemp *Cannabis*, hop *Humulus*, nettle *Urtica*, knotgrass *Polygonum*, dock *Rumex*, amaranth *Amaranthus*, fat hen *Chenopodium*, orache *Atriplex*, chickweed *Stellaria*, mouse-ear *Cerastium*, corn spurrey *Spergula*, Ranunculaceae (especially buttercup *Ranunculus*), poppy *Papaver*, Cruciferae (especially shepherd's purse *Capsella*, garlic mustard *Alliaria*, charlock *Sinapis*), mignonette *Reseda*, Rosaceae (especially rose *Rosa*, hawthorn *Crataegus*, rowan *Sorbus*, bramble *Rubus*, meadowsweet *Filipendula*, cultivated fruits), Leguminosae, *Oxalis*, flax *Linum*, spurge *Euphorbia*, cranesbill *Geranium*, spindle *Euonymus*, sea buckthorn *Hippophae*, dog's mercury *Mercurialis*, bryony *Bryonia*, willowherb *Epilobium*, Umbelliferae, violet, etc. *Viola*, balsam *Impatiens*, St John's-wort *Hypericum*, dogwood *Cornus*, buckthorn *Rhamnus*, alder buckthorn *Frangula*, Ericaceae, forget-me-not *Myosotis*, Labiatae, nightshade *Solanum*, figwort *Scrophularia*, yellow rattle *Rhinanthus*, plantain *Plantago*, Caprifoliaceae (especially honeysuckle *Lonicera*, guelder rose *Viburnum*), scabious *Scabiosa*, *Knautia*, Compositae (especially dandelion *Taraxacum*, sow-thistle *Sonchus*, groundsel *Senecio*), bluebell *Endymion*, grasses (Gramineae), rushes (Juncaceae), ferns (Filicopsida). (Collinge 1924–7; Schuster 1930; Creutz 1953; Löhrl 1957; Doerbeck 1963; Klehm 1967; Ptushenko and Inozemtsev 1968; Sabel 1983; Guitián Rivera 1985; Snow and Snow 1988; Bibby and Charlton 1991; G Le Grand.) For further details, see (e.g.) Turček (1961), Béress and Molnár (1964), and Newton (1967*a*, *b*).

In open cultivated country near Oxford (southern England), October–April, over 2 years, of 1439 feeding observations, 26% by number seeds of dock, 22% hawthorn buds, 13% nettle seeds, 9% buds of cultivated fruit, 7% fat hen seeds, 6% bramble seeds, 6% ash seeds (only in year of heavy ash crop); in year of poor ash seed crop, winter diet 54% seeds, 46% buds, when crop heavy, 80% and 20%; of 700 observations, May–September in 1 year, 23% seeds of sow-thistle, 12% dandelion, 8% fat hen, 6% dock, 6% chickweed, 5% charlock, 4% invertebrates. In woodland, October–April, over 2 years, of 1779 observations, 28% bramble seeds, 18% hawthorn buds, 14% dock seeds, 12% ash (only in good crop year), 10% nettle, 7% birch (only in good crop year); of 902 observations, May–September in one year, 21% birch seeds, 17% dog's mercury, 6% dock, 6% bramble, 5% meadowsweet, 5% grasses, 5% invertebrates. Fairly selective in choice of food plants over year; in cultivated area, taking account of abundance, in spring preferred seeds of dandelion and chickweed; in summer, charlock, sow-thistle, and buttercup; into autumn, fat hen and dock, then later also nettle and bramble, and towards early spring hedgerow ash seeds and buds of both hawthorn and cultivated fruits; grasses and cereals common in study area but hardly eaten. In woodland, where *c.* 75% of 15 favoured plants scarce or restricted and many common species ignored as seeds

probably too hard or difficult to deal with quickly, preferred items in spring and summer were seeds of dog's mercury, almost ripe seed-heads of dandelion, elm seeds, leaves of blackthorn *Prunus*, oak and willow flowers, and invertebrates; in autumn, seeds of birch, meadowsweet, dock, bramble, privet, and honeysuckle, then into winter also ash. In years when seeds scarce, starts feeding on tree flower-buds around January (80% of diet in February in one year); many moved away from woods to feed in gardens on (e.g.) lilac, *Buddleia*, or *Forsythia*; when seed crop good, only takes buds from around March. Ash seeds apparently rather disliked and only taken in absence of preferred species; unable to survive on buds alone in cold, short days of mid-winter; eats flesh of hawthorn berries because seeds too hard to crack. (Newton 1964*c*, 1967*b*, which see for many details and for diet over year when feeding in orchards.) Also in southern England, when feeding on berries apparently much prefers those of various Caprifoliaceae and *Sorbus* species (Snow and Snow 1988). See also Collinge (1924–7). Bud damage in pear orchard, south-east England, peaked late November and early December and again in February (Greig-Smith and Wilson 1984, which see for many details). In Germany, diet over year very similar to studies above; locally important components include seeds of spruce, maple, *Viola*, nettle, and *Sorbus* (Doerbeck 1963; Klehm 1967). In Moscow region (Russia), of 460 items from 8 stomachs in breeding season, 90% by number seeds (of *Rumex*, lady's mantle *Alchemilla*, *Rubus*, *Viola*), 7% beetles (5% Curculionidae), 2% sawfly larvae, 2% other insects (including caterpillars and Diptera) and spiders. Leaves forest in winter to feed in gardens or fields near forest edge, taking seeds such as *Rumex*, orache, amaranth, and bur-marigold *Bidens*. (Ptushenko and Inozemtsev 1968.) For winter diet in northern Rumania, see Béress and Molnár (1964); for north-west Spain, see Guitián Rivera (1985). On Azores, feeds on seeds, buds, etc., of Japanese cedar, laurel, holly, lily-of-the-valley tree, *Prunus*, knotgrass *Polygonum*, tormentil *Potentilla*, self-heal *Prunella*, figwort, sow-thistle, and young fronds and sporangia of ferns (Bibby and Charlton 1991; Bibby *et al.* 1992; G Le Grand, J A Ramos). In captive trials, *pileata* preferred seeds of hemp, sunflower, and canary grass *Phalaris* (Kear 1962; Ziswiler 1965).

In captivity, average daily intake of pear bud material 1·4–7·0 g dry weight, with energy content 15·5–126·3 kJ, which was insufficient to lay down enough fat to survive overnight during November–January (Summers 1962); average daily consumption of seed kernel 4·8–5·0 g (Kear 1962). See Guitián Rivera (1985) for change over year in daily energy intake per 10 ha in woodland, north-west Spain.

Food for young brought by adults in buccal pouches opening on each side of tongue which develop in breeding season. Adults feed young more invertebrates than they eat themselves, but same kinds of seeds, and large ones may be broken up for very young nestlings. In Oxford area, *c.* 350 collar-samples from 2 broods aged 6-11 days in woodland contained 55% by volume seeds of dog's mercury, 17% elm, 10% spiders, 7% caterpillars, 4% Diptera; 24% of diet invertebrates and diet of both broods identical. Same number of samples from open cultivated area showed marked differences between 2 broods; diet of one was 36% by volume seeds of groundsel, 31% shepherd's purse, 13% chickweed, 7% Diptera, 6% garlic mustard, 2% caterpillars, 2% spiders (13% of diet invertebrates); diet of other was 30% dandelion, 25% caterpillars, 21% elm, 5% dog's mercury, 5% chickweed, 4% shepherd's purse (29% of diet invertebrates). 119 collar-samples from woodland nests showed decline in invertebrates from 38% by volume on days 1-5 to 10% on day 12; 202 samples from cultivated habitat showed decline from 20% to 1%. In woodland, June, caterpillars were 35% of invertebrates in diet, spiders 32%, and in July snails were 35%; in open country, June, caterpillars were 46% of invertebrates, and in July spiders were 41%. Differences between samples probably due to local habitat variations, since adults generally forage near nest, though those in cultivated area sometimes flew up to 1·5 km to woodland to collect preferred caterpillars and dog's mercury seeds. (Newton 1967*b*.) Also in England, 34 nestling stomachs contained mostly small adult and larval Lepidoptera, plus some beetles and Diptera; plant material present in only 7 stomachs (Collinge 1924-7). In Moscow region, of 938 items from 18 collar-samples, 25% by number spiders, 22% *Oxalis* seeds, 12% lady's mantle seeds; among other seeds, spruce and pine most important, while insects and snails accounted for only 2% of diet (Ptushenko and Inozemtsev 1968). In south-west Germany, 24 collar-samples and visual observations at 2 nests yielded 78 spiders, 37-42 caterpillars (Tortricidae), 3 snails, 1 beetle, 1 Lepidoptera pupa, and many seeds, particularly bilberry *Vaccinium*; common seeds in vicinity of nests apparently ignored (Löhrl 1957). 2 other German studies showed only seeds and aphids in diet of young, but invertebrates can be difficult to detect in regurgitated pulp; main components were seeds of dandelion, bilberry, *Viola*, maple, garlic mustard, sow-thistle, and, in suburban gardens, hemp and sunflower (Doerbeck 1963; Klehm 1967).                                        IN, BH

**Social pattern and behaviour.** Well studied in general, including birds in captivity, notably by Hinde (1955-6), Nicolai (1956), and Newton (1972), whose studies form basis of following account; but no thorough study of free-living population of colour-ringed birds, so little can be said about some important aspects.

1. Occurs singly, in pairs, or in small groups (maximum *c.* 20: Goodwin 1985) outside breeding season; groups form mostly at sites where food abundant. Migrates singly or in small scattered flocks. In study in southern England, single birds commonest group category October-February. In March-April, modal group size 2; even-numbered groups with equal sex-ratio commoner than expected by chance, and odd-numbered groups and

those with unequal sex-ratio rarer. Aggressive encounters occurred at high rates in larger groups, but higher than expected in groups of 3 and groups with supernumerary ♂♂. ♂♂ initiated more aggressive encounters than ♀♀, and these directed almost exclusively at other ♂♂. Much of aggression between individuals probably related to mate defence; ♂♂ almost exclusively aggressive to ♂♂, and ♀♀ attacked other ♀♀ more frequently than ♂♂. (Wilkinson 1982.) These interactions indicate pair-formation occurring in groups before start of breeding, and consistent with supposition that established pairs are permanent, persisting through winter (Nicolai 1956). Radio-tracking study, southern England, showed that individuals remain largely within small areas of 1-2 ha for periods of a few weeks, occasionally making brief excursions of a few hundred metres; over longer periods, occasional shifts of home-range may occur (Greig-Smith 1985). BONDS. Monogamous mating system. Pair-bond lasts at least for duration of breeding season; pairs also seen together in winter, apparently faithful for successive breeding seasons, but not confirmed by ringing. ♀ builds nest (accompanied by ♂), incubates, and broods young; both sexes feed young. Age of first breeding 1 year (I Newton). BREEDING DISPERSION. Pairs breed solitarily, but not obviously territorial. Throughout breeding cycle, ♂ may ignore other ♂♂ near nest (Newton 1972). Nests may be close together, in extreme case 2 nests in same bush (Snow 1953), but not known if 2 ♂♂ involved; in another case, 2 ♀♀ incubating in same nest, attended by single ♂ (Dale 1980). Most food collected in vicinity of nest, but some much further away, up to more than 1 km (I Newton). Breeding densities very variable; difficult to assess because of inconspicuous behaviour when breeding, and often thickly vegetated breeding habitat. Following is sample of available data. New Forest, southern England (woodland, heath, and bog mosaic): 1-2 pairs per km² (Glue 1973). Mixed farmland, southern England: average 4·9 pairs per km² (Benson and Williamson 1972). Young conifer plantation, Wales: up to 21 pairs per km² (Insley and Wood 1973). Semi-natural oak *Quercus* forest, Bourgogne (France): 1-4 pairs per km² in coppice-with-standards, 2-7 pairs in shelterwood (Ferry and Frochot 1970). Mature spruce *Picea* forest, Harz mountains (Germany): 3 pairs per km² (Oelke 1992). Primeval forest, Poland: average 2 pairs per km² in both mixed coniferous-deciduous forest, and in pine *Pinus* with bilberry *Vaccinium*, less than 1 in other forest types (Tomiałojć *et al.* 1984). Finland: 1-4 pairs per km² in different forest types, most abundant in mixed forest (Palmgren 1930). Pskov region (north-west European Russia): average 7·1 pairs per km² in semi-mature forest over several years, 3·3 pairs per km² in village (gardens and cemetery) (Uryadova 1986). Bulgaria: up to 68 pairs per km² in different types of pine forest, 23-36 in spruce forest (Simeonov 1975); 10 in oak forest, 13 in beech *Fagus* forest (Petrov 1988). Density of *murina*, Azores, in restricted area of occurrence, estimated 58 birds per km² (J A Ramos). ROOSTING. Solitarily, in pairs, or (outside breeding season) up to 10(-20) birds together; usually in thick foliage of evergreen trees and bushes, also tall thorn trees (Newton 1972; I Newton). In Finland and Russia in winter, regularly in burrows or cavities under snow (Sulkava 1969; Novikov 1972).

2. Differs markedly in many aspects of social behaviour from most other Fringillidae, showing some resemblance to Hawfinch *Coccothraustes coccothraustes*. Most characteristic features include: persistence of sexual behaviour throughout year, with ♀ generally dominant over ♂; unusually intimate relationship between pair-members, with apparently persistent pair-bond (see Bonds, above); lack of display-flight; soft, quiet song, highly variable between individuals (see Voice); lack of flocking and other social behaviour in breeding season (Newton 1972, which see for comparison with other Fringillidae). Generally rather

timid, but not wary or suspicious. Thus easy to stalk, often failing to notice watcher if still, but, when it does notice, flees sooner than would Chaffinch *Fringilla coelebs* or Greenfinch *Carduelis chloris*, especially if on or near ground; always flies into thick cover, unlike other Fringillidae, which, when disturbed by man, fly up and away, or up into tree. (Goodwin 1985.) Especially inconspicuous when breeding (Newton 1972). FLOCK BEHAVIOUR. Little developed. At favoured feeding sites, birds continually arrive and depart throughout day; maintain contact with soft piping call (see 2 in Voice) (I Newton). See also introduction to part 1 (above). SONG-DISPLAY. Given by both sexes, but most commonly by ♂ (Wilkinson 1990). Inconspicuous; song not loud nor given from conspicuous vantage point, has no territorial function, but plays minor part in pair-formation and in ♂'s display to ♀, mainly in early stages of courtship (Nicolai 1956; Newton 1972). In breeding season, ♂ sings especially when alone; thus, when ♀ incubating, ♂'s usual call gradually changes to song. Outside breeding season, song occasionally given by birds in groups, especially around midday. (Nicolai 1956.) 'Directed song', given in presence of ♀, is very low-volume, sometimes evident only from throat movements; 'undirected song', usually given at distance from other individuals, is louder, audible up to *c.* 20 m (Wilkinson 1990); see Voice. In winter quarters in Italy, occasionally sings in early spring (Alexander 1917). ANTAGONISTIC BEHAVIOUR. Most aggression over food, but in spring, while still in groups, ♂ may chase other ♂♂ from vicinity of ♀ (see introduction to part 1, above). In Threat-display, commonly seen at feeding sites, bird crouches, with tail twisted to one side, head feathers sleeked, and bill open and directed towards opponent; utters hoarse, drawn-out braying call (see 7 in Voice). Supplanting attacks also common, where one bird flies at and displaces another from feeding site, then perches in its place; occasionally leads to short chase, or even fight in which combatants flutter in air, bills touching, until one gives way. (I Newton.) If 2 lone ♂♂ or ♀♀ meet by chance, typically fight fiercely for a moment and then separate (Newton 1972). HETEROSEXUAL BEHAVIOUR. (1) General. Sexual behaviour occurs all year. Relationship between members of pair unusually intimate; bond, once established, apparently permanent. Most displays mutual, with both sexes playing active role and adopting similar postures. (Newton 1972, which see for comparison with other finches.) (2) Pair-bonding behaviour. Pairing takes place in flocks (usually small groups), from which pairs then separate. When 2 unacquainted birds of opposite sex meet, Introduction-ceremony takes place. ♀ behaves aggressively, lowering head, ruffling belly feathers, and giving hoarse mono-syllabic call (see 4e in Voice) with bill wide open. ♂ gives way at first, then, unless quite uninterested, flies from perch to perch in tense attitude with wings slightly drooped (raised just out of supporting feathers and lowered by rotation at shoulder), exposing white rump, body feathers ruffled, and tail swivelled towards ♀; lateral body feathers usually especially ruffled so that body appears dorso-ventrally flattened (Fig A). This display often

accompanied by low warble (see 'directed song' in Voice) and by frequent displacement bill-wiping. If ritual attack by ♀ subsides, ♂ approaches in this posture, and behaviour may pass into next stage of courtship, Bill-caressing: with nibbling bill movements, ♂ touches ♀'s bill for a moment with his, then turns away and hops a little to one side (Fig A); repeats procedure, and ♀ may respond with similar behaviour. At full development, display becomes mutual: ♂ and ♀, with ruffled belly feathers and tail twisted round, hop towards each other, touch bills, and simultaneously turn away; repeat display again and again. (Bill-caressing display silent, in contrast to related Fringillidae which give soft calls during similar display.) When pair-bond established, ♀'s threat becomes ritualized into greeting ceremony, with posture less pronounced and single brief call, to which ♂ responds with same behaviour, even less pronounced. (Hinde 1955-6; Nicolai 1956; Newton 1972.) (3) Courtship-feeding. Begins when pair firmly established. Accompanying behaviour very variable, but typically ♂ approaches ♀ in upright posture, feathers usually a little ruffled and wings sometimes a little drooped; opens and closes bill rapidly as he approaches. ♀ adopts begging posture similar to that of dependent young, with body held low and wings fluttered; pivots from side to side between accepting food from ♂, who feeds ♀ from crop. (Hinde 1955-6; Nicolai 1956.) ♀ does not call (Nicolai 1956); sometimes gives Begging-calls (see 9 in Voice) (Hinde 1955-6). Twig-display (Fig B). Begins when time for nest-building approaches. ♂ picks up and offers nest-material to ♀, adopting same posture as in Courtship-feeding (Fig B). ♀ often responds by picking twig;

B

may fly towards ♂, or both may fly off, one behind the other, with nest-material in bills. Display has strong sexual content, and often followed by mating. (Nicolai 1956; Newton 1972.) (4) Mating. Usually takes place in early morning, at roost-site. ♀ regularly searches for nest-material, and then solicits mating when she has found some. ♂ also often picks nest-material and holds it during mating. Soliciting ♀ gives repeated subdued call (see 8 in Voice); crouches with side-to-side body movements and vibrating wings. ♂ approaches in upright posture, with wings a little drooped (rarely, slightly shivered), tail a little raised, belly feathers maximally ruffled, and bill raised; may make up-and-down bobbing movements of whole body, then mounts. Details very variable: often no bobbing, and sometimes all preliminaries omitted. (Hinde 1955-6; Nicolai 1956.) (5) Nest-site selection and behaviour at nest. ♂ takes initiative in choosing site; leads ♀ from site to site, crouching in suitable places and giving special call (see 7 in Voice) to draw ♀'s attention. Sometimes gives same call when standing in half-completed nest. (Newton 1972.) ♀ builds nest, accompanied by ♂ (I Newton). During incubation, ♀ called off nest and fed by ♂; not fed on nest (Bechtold 1967). In exceptional case of 2 ♀♀ incubating combined clutch in same nest, ♂ called one ♀ at a time off nest; on one occasion, ♀ returning to nest fed the other (Dale 1980).

A

RELATIONS WITHIN FAMILY GROUP. ♀ broods nestlings for first few days, fed on nest by ♂; regurgitates for nestlings food which ♂ has brought. As nestlings grow, both parents forage and visit nest together, both regurgitating food directly to young. For first few days, brooding ♀ eats nestlings' faecal sacs; later, faecal sacs carried away by both parents. Nestlings leave nest at short intervals, over 1–2 days; perch near by at first, later move further away into dense cover. When young can fly well, c. 7 days after leaving nest, follow parents to feeding sites. Usually both parents tend young for c. 14 days after leaving nest; then ♀, if continuing to breed, leaves family and builds new nest. When begging, young pivot from side to side, giving hoarse calls (see Voice). ANTI-PREDATOR RESPONSES OF YOUNG. When disturbed by human intruder, nestlings lie down in nest: from day 11, may leave nest prematurely and scatter if disturbed. PARENTAL ANTI-PREDATOR STRATEGIES. Quiet and inconspicuous at nest; otherwise none recorded. (I Newton.)

(Figs by D Nurney from drawings in Newton 1972.)   DWS

Voice. Well studied, in both wild and captive birds, especially by Wilkinson and Howse (1975) and Wilkinson (1990), which see for further sonagrams and for discussion of temporal characteristics, individual variation in calls, and individual recognition based on vocal differences. Most vocalizations used by both sexes, in contrast to most passerines, which show pronounced sexual dimorphism in vocal repertoire, ♀ having smaller repertoire; difference may be related to mutual courtship-displays, strong pair-bond, and lack of territoriality. Large vocal repertoire (c. 17 distinct calls recognized) may be related to need for effective communication in dense vegetation in which vocal signals are usually given. (Wilkinson 1990.) For ability to learn and imitate human music in captivity, see

Thorpe (1955). For further data on vocalizations of caged birds, see Wilkinson and Howse (1975) and Nicolai (1956). Following account based on Wilkinson (1990), with references to other authors where appropriate.

CALLS OF ADULTS. (1) Song (Figs I–II). Distinctive timbre, but neither loud nor very different in quality from other calls, so that existence of full-song (as distinct from sub-song) not always recognized (e.g. Witherby *et al.* 1938). Highly variable between individuals (Newton 1972), but this aspect of variability apparently not studied by sonagraphic analysis. Consists of varied units, many in pairs, often continuing (without marked pauses) for several minutes. Units include call 7 (see below) and those used in 'sequence calls' (see 11, below), also 'trills' and rattles. Analysis of songs of 4 birds gave range of 15–33 distinct units. Given either as 'directed song' (can be considered type of sub-song) as part of courtship-display (see Social pattern and Behaviour), or as 'undirected song' usually at a distance from other individuals. Both categories of song given by both ♂ and ♀, most commonly by ♂. 'Directed song' audible only within a few metres, sometimes only evident from throat movements; 'undirected song' usually louder, audible up to c. 20 m. (Wilkinson 1990.) Geographical differences reported, but not studied by sonagraphic analysis. Song in central Europe ('*germanica*', '*coccinea*') more continuous and distinctly higher pitched than that of nominate *pyrrhula* of northern Europe; differences potentially useful as field character (Nicolai 1956). For study of song using musical notation, see Hoffmann (1926). (2) Contact- and related calls. Most familiar call a piping 'phew', of 2·5–3·0 kHz, pitch nor-

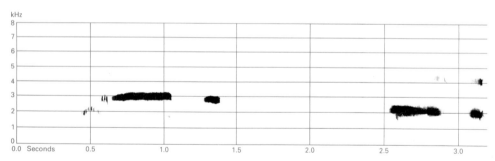

I   A G Field   England   June 1963

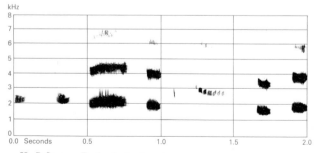

II   R Savage   England   April 1976

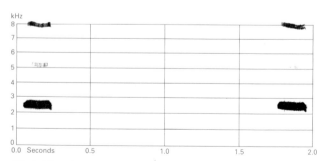

III   J-C Roché   France   May 1978

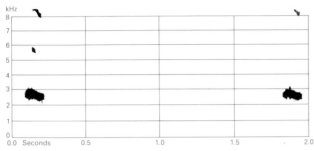

IV  V C Lewis  England  May 1961

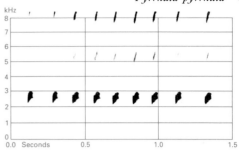

VII  P A D Hollom  England  June 1989

mally descending with time (e.g. Figs III–IV); mean duration 199 ms (105–272) ($n = 20$). Almost certainly individually distinctive; calling bird's mate will respond to it from considerable distance, ignoring calls of other individuals (Nicolai 1956). Given by both sexes throughout year. Birds separated from mates give repeated 'phew' calls which function as long-distance contact-calls. 'Phew' calls also given when startled, and during mobbing of owl; very abrupt 'phew', duration 95 ms, recorded from alarmed ♀ (e.g. Fig V). Following variants recognized: 'phee', sus-

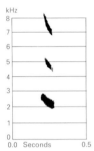

V  J-C Roché  France  June 1984

tained whistle of similar pitch to 'phew' but more protracted (*c.* 700 ms), rarely given and function not known; 'phy', similar in pitch to 'phew' but of shorter duration (*c.* 100 ms), sounding more urgent; often given by startled birds (e.g. on sight of stuffed owl), and may be combined with 'phew' as 'phy-phew'; apparently similar call also recorded from ♀ at nest with week-old young (Fig VI, at 0·7 s). (3) Short calls. Very short (*c.* 10 ms) 'bit' calls, most

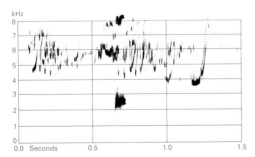

VI  P A D Hollom  England  August 1976

frequently uttered as repeated series (Fig VII); often associated with movement, commonly preceding flight. Occasionally quiet 'bit' calls given by birds feeding in parties, probably as short-distance contact-call. Rate of delivery and loudness variable, apparently reflecting degree of arousal; may be so rapid as to become a 'trill' (bird alerted by hunting Sparrowhawk *Accipiter nisus*; others responded by immediately seeking cover). (4) Calls associated with aggression and other interactions. (4a) Phwhy-call. Intermediate in structure between 'phew' and 'phy' (see above), but differs in more complex structure; sudden drop in pitch, then rises in pitch and terminates with 1–2 cycles of rapid frequency modulation. Given by both sexes, and most commonly heard during bouts of chasing in early spring; may also function as alarm-call. (4b) Whee-call. Intermediate in structure between 'phwhy' and 'chewit' (4c, below); more frequently heard than either of them. Given by both sexes but most often by ♂, during bouts of aggressive chasing. (4c) Chewit-call. Perhaps variant of 'whee' call, differing in being longer, louder, and having abrupt inflection of pitch midway through call. Usually indicative of high state of arousal, during chasing; also occasionally given when alarmed. (4d) Yew-call. Deep, hoarse call, differing from all previous calls in lower pitch and harmonic structure; typically in form of inverted U, with energy concentrated in descending limb. Often given before supplanting-flights and -chases, appearing to function as threat-call. (4e) Open-bill threat-call. Hoarse, monosyllabic 'hhwhore', or 'hhweh' (Nicolai 1956), also 'phee-yore' (hiss preceded by whistled note similar to 'phew'), given with bill wide open, in head-forward threat posture; almost exclusively by ♀ when threatening ♂. (5) Warning-call for flying predators. Drawn-out, penetrating, somewhat plaintive 'dü-dü' (Nicolai 1956, who pointed out that there is no warning-call for ground predators in this pronouncedly arboreal species). (6) Distress-call. 2 types, given by bird when grasped in the hand (and presumably when seized by predator): whistled note beginning at *c.* 1 kHz and rapidly rising, to be maintained at *c.* 2 kHz; also (distress-scream) a drawn-out, noisy call with complex harmonic structure. (7) Braying-call. Specific to ♂♂ in breeding season; usually begins at frequency of *c.* 2·5 kHz, gradually falling to *c.* 2·0 kHz and then rapidly modulated over terminal section. Given by ♂ when indicating nest-sites to mate; also heard from ♂♂ in groups distant from breeding areas. According to Hinde (1955), also given at nest-site before and during nest-building, and a component of song. (8) Soliciting-call of ♀. When soliciting mating, gives subdued, repeated note, becoming more rapid with increasing arousal, 'die—die—die-die-die-diediediediedidi' (Nicolai 1956). (9) Begging-call. Loud, repeated 'hhwhy-hhwhy. . .', accompanied by fluttering of wings; whistled call with rapidly modulated terminal portion, in this resembling Braying-call, but differing in being shorter and having initial rise in pitch. Only heard during incubation from ♀♀ after being called off nest by mate, and so characteristic of this that useful as indicator of presence of nest with eggs in vicinity of caller. Call given by ♀ during courtship-feeding described as thin 'zi zi zi' (Hinde 1955). (10) Nest-approach call and 'ugh'. Series of short calls of low to moderate volume, given by both sexes when approaching nest. Differ from otherwise similar 'bit' and 'phy' calls in being rapidly and roughly modulated. Often followed by lower-pitched 'ugh' (falling in pitch) coincident with landing on nest, which elicits gaping by young. (Nicolai 1956 mentioned low-pitched, 'upwardly modulated' call as call given by ♀ inviting nestlings to beg.) (11) Sequence-calls and Long-calls. Sequence-calls differ from all preceding calls in comprising several different notes given in set pattern; differ from full song in not consisting of more than 6 notes. Notes usually paired, e.g. AB CD, these paired units following each other in invariable sequence. Long-call a simple whistle of similar pitch to 'phew' but intermediate in length between 'phew' and 'phee' (see 2, above); considered together with Sequence-call because it often precedes or is interspersed with bouts of Sequence-calls. Sequence-calls and Long-calls given by solitary birds; not specific to particular context; often follow bout of 'phew' calls, and perhaps function as contact-calls supplementary to 'phew'. Also given in company with conspecifics and occasionally replace song in courtship-displays; heard in all months but most often in early spring, from both sexes but most commonly ♂. (Wilkinson 1990.)

CALLS OF YOUNG. From day of hatching, nestlings give short quiet 'peep' calls, spectrographically of inverted U form with peak at 3·5-4 kHz. Main energy shifts in course of days 1-3 from descending to ascending limb of U. By day 2-3, call shows 2 fundamentals (sound sources), lower one louder and sometimes modulated; call also becomes louder and of longer duration, and peak frequency increases to up to c. 16 kHz. New call appears from day 14: 'tchewy', lower pitched and louder than preceding calls, developed from them by decrease in pitch and concentration of energy in initial portion of lower fundamental. This develops into whistled 'tchew' of older nestlings after further decrease in pitch and loss of terminal modulated portion. 'Satisfied' call, after being fed, a soft, 'trilling' 'trr-trr-trr' (Nicolai 1956). Commonest call of recently fledged young similar to 'tchew' of older

nestlings: a whistled note, sometimes terminated with rapid frequency modulations, functioning as contact-call enabling parents to find hidden young. This call changes gradually into adult 'phew' at c. 6 weeks old. (Wilkinson 1990.) Begging call of fledged young a harsher version of 'tchew' (Wilkinson 1990, which see for further details), also higher-pitched call recorded from fledglings 3 days, 4 days, and 10 days out of nest. For evidence that young birds learn contact-call of ♂ parent, see Nicolai (1959).    DWS

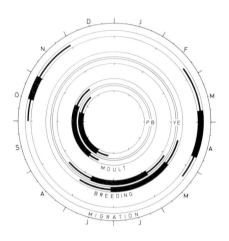

**Breeding.** SEASON. Britain: laying starts late April but mainly early May (Witherby *et al.* 1938; Newton 1964*b*); exceptionally 1st half of April; repeat, 2nd, and possibly 3rd clutches into late August (Greig-Smith and Wilson 1984) and even mid- to late September (Witherby *et al.* 1938; Newton 1972). Fenno-Scandia: May to late July (Witherby *et al.* 1938); in southern Finland (see diagram), eggs laid from end of April, later in north (Haartman 1969). USSR: nest-building starts end of April in south, much later further north (Dementiev and Gladkov 1954). Netherlands: laying starts April and continues to end of August; 86% of 339 clutches started May–July with 2 laying peaks, one in May and another in June–July, corresponding to 1st and 2nd broods; start of laying in April related to temperature (Bijlsma 1982). Germany: eggs laid mainly end of April to end of August (Mildenberger 1984; Klafs and Stübs 1987); in mild years, building may start March (Doerbeck 1963; Mildenberger 1984) with eggs laid early April (Doerbeck 1963; Klehm 1967). Azores: eggs laid June (G Le Grand). SITE. Thick bushes and trees of many kinds, often evergreen, and in particular in dense branches of spruce *Picea* (Géroudet 1964*b*; Klehm 1967; Bijlsma 1982; Preuss 1991*b*). Of 138 nests in Germany, 40% in conifers (spruce 28%, yew *Taxus* 7%), 25% in broad-leaved trees (hornbeam *Carpinus* 7%, beech *Fagus* 6%), 15% in thorny bushes (hawthorn *Crataegus* 9%, bramble *Rubus* 6%), and 11% in climbers (honeysuckle *Lonicera* 5%) (Mildenberger 1984). In Slovakia, where usually nests in conifers, apparently a recent tendency in some areas to nest in earth walls of forest tracks, well hidden by roots (Kaňuščák 1988); also in Switzerland, where one nest sited like that of Wren *Troglodytes troglodytes* in exposed roots on steep embankment (Glutz von Blotzheim 1962). On flat branch or flat tangle of twigs, not in fork, usually well hidden (Klehm 1967; Newton 1972; Preuss 1991*b*). In young conifers, usually 0-10 cm from trunk (Glutz von Blotzheim 1962; Preuss 1991*b*). In taller trees, usually in outer twigs, as growth around trunk too dense (Preuss 1991*b*). Of 134 nests in Germany, 62·7% 0·4-2 m above ground, 21·6% 2-3 m, 10·4% 3-4 m, 3·7% 4-5 m; 1·5% more than 5 m (Mildenberger 1984); average height of 61 nests in eastern Germany 1·1 m (0·7-1·9) (Klehm 1967). of 46 nests in Finland, average height 1·8 m (0·75-5) (Haartman 1969), and in Czechoslovakia 2·5 m (0-20), n = 42 (Hudec 1983). For review of nests on buildings, see Baege (1968). Nest: 2-layered structure similar to that of Hawfinch *Coccothraustes coccothraustes* and Pine Grosbeak *Pinicola enucleator*; loose base of dry twigs, which are often bitten from trees, with some moss and lichen; twigs may be up to 30 cm long so that they project from base (Nicolai 1956; Klehm 1967; Haartman 1969; Newton 1972; Bochenski and Oles 1981; Preuss 1991*b*). Twigs often spruce, birch *Betula*, or honeysuckle, also thick grass stems or roots and occasionally stalks of herbs (Doerbeck 1963; Klehm 1967; Bocheński and Oleś 1981; Preuss 1991*b*). Nest-cup of dry grass and fine rootlets interwoven into foundation and lined with small quantity of fine rootlets and grasses, more rarely lichen, moss and leaves (Dementiev and Gladkov 1954; Nicolai 1956; Doerbeck 1963; Newton 1972; Bocheński and Oleś 1981; Preuss 1991*b*). Various nests had outer diameter 10-19·8 cm, inner diameter 5-8·2 m, overall height 5-7 cm, depth of cup 4-5 cm (Dementiev and Gladkov 1954; Ringleben 1960; Campbell and Ferguson-Lees 1972). Building: by ♀ only, accompanied by ♂ (Nicolai 1956; Klehm 1967; Bocheński and Oleś 1981; Preuss 1991*b*); rarely, ♂ brings some lining (Doerbeck 1963); building occurs in bouts of 10-40 min, which may be followed by hours or even days when nothing is added; usually takes 5-7 days (Schuster 1944; Klehm 1967) but can be built in 2 days (Ringleben 1960; Preuss 1991*a*); first nests of season take longer (Newton 1972). EGGS. See Plate 61. Sub-elliptical, though may be rather pyriform, smooth and slightly glossy. Ground-colour variable, but usually pale blue or clear greenish-blue with dark purple-brown spots, speckles and streaks which tend to be concentrated around broad end; sometimes also paler purple-violet blotches; varieties with reddish markings on white ground-colour also occur. (Witherby *et al.* 1938; Dementiev and Gladkov 1954; Campbell and Ferguson-Lees 1972; Harrison 1975; Makatsch 1976.) Nominate *pyrrhula*: 20·8 × 14·9 mm (17·8-23·2 × 14·0-16·4), n = 238; calculated weight 2·41 g. *P. p. pileata*: 19·5 × 14·5 mm (17·2-22·1 × 13·0-15·6), n = 214; calculated weight 2·14 g. *P. p. europoea*: 19·3 × 14·4 mm

(17·5-21·4 × 12·6-15·7), n = 161; calculated weight 2·10 g. *P. p. murina*: 18·7 × 15·0 mm (18·0-19·5 × 14·4-15·4), n = 3; calculated weight 2·21 g. (Schönwetter 1984.) Clutch: 4-5 (3-7). Southern England: average clutch size declines through season; May 4·8 (n = 209), June 4·6 (n = 64), July 4·3 (n = 59), August 3·8 (n = 21) (Newton 1964*b*). In Netherlands, of 421 full clutches: 3 eggs, 3·8%; 4, 34·0%; 5, 54·6%; 6, 7·1%; 7, 0·5%; average 4·67; clear seasonal variation, with maximum in mid-May declining thereafter, average of 1st clutches 5·2 (n = 12), 2nd 4·8 (n = 12), replacements 4·2 (n = 13) (Bijlsma 1982). Similar pattern in Germany: average 4·85 in May (n = 39), declining to 4·14 in July-August (n = 14) (Mildenberger 1984). Clutch size increases with latitude (Newton 1972); average in Finland 5·7 (n = 24) (Haartman 1969). 2 broods in Britain (Campbell and Ferguson-Lees 1972), and 3 possible in a season where no losses suffered (Witherby *et al.* 1938; Newton 1967*b*). In Germany, 2 broods usual (Nicolai 1956; Mildenberger 1984) but 3 possible under favourable conditions, and including replacements many ♀♀ lay 3-5 clutches per season (Nicolai 1956). In Netherlands, at least 90% of 20 pairs double-brooded, and average time between layings 20-21 days (n = 12) (Bijlsma 1982). 1st eggs in successive clutches of one pair with 3 broods were 33 and 35 days apart (Newton 1972). Apparently 2 and possibly 3 broods in Finland (Haartman 1969). Reportedly only 1 brood on Azores (G Le Grand). One case of replacement nest with 5 eggs 11 days after loss of 8-day-old nestlings (Ringleben 1960). Eggs laid daily in early morning (Doerbeck 1963; Klehm 1967; Preuss 1991*a*). INCUBATION. 12-14(-15) days; by ♀ only (Nicolai 1956; Klehm 1967; Newton 1972; Bijlsma 1982; Preuss 1991*a*). ♀ leaves nest every 50 min for periods of *c.* 10 min to be fed by ♂ (Witherby *et al.* 1938; Nicolai 1956; Doerbeck 1963); may also leave for 30-45 min every 1-2 hrs, more often if food scarce, to feed herself (Schuster 1944; Doerbeck 1963; Bijlsma 1982). Incubation proper begins after 4th or penultimate egg, though sometimes not until last egg (Schuster 1944; Ringleben 1960; Doerbeck 1963; Newton 1972). Hatching asynchronous over 2·5-4 days (Schuster 1944; Steinfatt 1944) and last-hatched chick often dies (Preuss 1991*a*); according to Klehm (1967), hatching synchronous and runts are rare. YOUNG. Fed and cared for by both parents by regurgitation (Doerbeck 1963; Klehm 1967; Newton 1967; Campbell and Ferguson-Lees 1972). ♀ broods until day 6 when young partially feathered, and until then only ♂ brings food which is fed to young mostly by ♀. After day 6, both parents collect food and feed young directly (Pitt 1918; Witherby *et al.* 1938; Doerbeck 1963; Newton 1967*b*; Harrison 1975). For development of young, see Bechtold (1967). FLEDGING TO MATURITY. Fledging period 14-16 days (12-18) (Nicolai 1956; Doerbeck 1963; Klehm 1967; Newton 1967*b*; Bijlsma 1982; Preuss 1991*a*). Fully independent at (10-)15-20 days after fledging (Nicolai 1956; Klehm 1967; Newton 1972; Bijlsma 1982). Fed mostly by ♂ at this time as ♀ often leaves to build new nest (Doerbeck 1963; Newton 1967*b*; Bijlsma 1982; Preuss 1991*a*). First breeding at 1 year old (I Newton). Breeding Success. In Netherlands, of 1321 eggs, 62·5% hatched and 50·0% produced fledged young. Of 279 clutches, 40·8% lost completely, mainly during egg stage. Losses mainly due to predation and desertion; desertion often associated with prolonged bad weather, with predation accounting for 59–75(–90)% of complete failures in this study. Clutches started in April least successful with success increasing throughout season, e.g. breeding success in one year 38·8% in April, 45·6% in May, 52·1% June, 60·4% July, 66·7% August (Bijlsma 1982). In Germany, 73% of 64 eggs hatched and 44% produced fledged young, giving 2·2 per pair overall (*n* = 13 pairs), 4·0 per successful pair (*n* = 7) (Doerbeck 1963). Losses apparently greatest during egg stage (Newton 1972; Bijlsma 1982), although in study by Nicolai (1956) nestling mortality was greater. Main predators Jay *Garrulus glandarius*, weasels, etc. *Mustela*, small rodents, and Carrion Crow *Corvus corone* (Nicolai 1956; Doerbeck 1963; Newton 1964*b*, 1972; Klehm 1967). In woodland, southern England, 15% of clutches started April–May produced fledged young, 50% in June, and 70% July–August; equivalent figures for nearby farmland 44%, 67%, and 84%; predation main cause of failure (Newton 1972). In northern and western Europe, no difference between breeding success of different clutch sizes (4–6), but absolute number of fledged young highest in clutches of 6 at 2·9 (*n* = 16); success highest in years of good conifer seed crop (Bijlsma 1982).                IN, MK, BH

**Plumages.** (nominate *pyrrhula*). Adult Male. Cap down to lower edge of eye, nasal bristles, feathering at base of both mandibles, and chin deep black, slightly glossed oil-blue on cap, sharply contrasting with remainder of body. Nape, upper side of neck, mantle, scapulars, and back uniform medium blue-grey; occasionally, some rosy-pink suffusion to grey of upperparts, varying from small pink patches on a few feathers to dense pink suffusion over entire mantle and scapulars; in Sweden and Finland, 17·6% of 125 ♂♂ show such a variable amount of pink (Voipio 1961). Rump white, contrasting sharply with blue-grey of back and with violet-glossed deep black upper tail-coverts. Side of head down from below eye and from ear-coverts, lower side of neck, and underparts from throat to lower belly and centre of flank uniform rosy-pink; rear of flank, vent, and under tail-coverts contrastingly white. Tail black with slight violet lustre; inner web of t6 occasionally with white spot or stripe subterminally along shaft (in Sweden and Finland, present in 18·4% of 125 ♂♂: Voipio 1961). Flight-feathers, greater upper primary coverts, and bastard wing sooty-grey, edged deep black along outer webs of primaries (narrowly white at emarginations of outer primaries), more broadly fringed black along outer webs of secondaries, black on latter with violet lustre; shorter feathers of bastard wing partly black on base and with grey-white fringe along tip of outer web. Tertials black with violet lustre, inner-most with rosy-pink outer web (rarely, outer web grey). Greater upper wing-coverts black with violet lustre, outermost with white tips 2–4 mm long; pale tips gradually longer towards inner coverts, black of bases virtually concealed on innermost, and white of tips gradually shades to light grey inwards; thus, inner-

most coverts (tertial coverts) with light grey tips of 9–12 mm long, partly fringed white. Median and lesser upper wing-coverts medium blue-grey, like mantle and scapulars. Under wing-coverts and axillaries light grey, almost white on tips of longer secondary coverts and axillaries; basal and middle portion of inner webs of flight-feathers with ill-defined grey-white border. Influence of bleaching and wear limited; when plumage quite fresh, rosy-pink of underparts and lower side of neck tinged orangey, vent and rear of flank with slight pale cream-buff suffusion; when worn (May–July), grey of upperparts slightly duller, less bluish; some dull black of feather-bases sometimes partly exposed on rump; pink of side of head and underparts slightly duller, sometimes contaminated by dirt, some pale grey of feather-centres showing through or partly exposed. Adult Female. Black of face and cap as in adult ♂. Nape medium grey with faint drab suffusion, merging into medium drab-grey or fawn-coloured mantle, scapulars, and back; rump white, sometimes with slight cream suffusion; upper tail-coverts black with violet lustre. Side of head and entire underparts (except for black chin and white under tail-coverts) medium drab-grey with slight vinous-pink cast; drab-grey slightly paler at border of cap below and behind eye (sometimes forming narrow and short white line), gradually merging into grey of nape on side of neck and into white of under tail-coverts on rear of flank and vent. Tail and wing as adult ♂; 26·3% of 72 ♀♀ from Sweden and Finland show some white on inner web of t6 (Voipio 1961); black of tip of longest tertial and of centres of greater coverts sometimes less deep, less glossy, more greyish; pale tips of greater coverts on average perhaps slightly narrower and greyer; outer web of innermost tertial ash-grey with variable amount of rosy-pink suffusion (usually a dot on centre, but sometimes entire web pink or grey). Influence of bleaching and abrasion more marked than in adult ♂, drab or fawn of upperparts gradually worn off in spring, hindneck to back becoming more uniform grey with slight brown cast on lower mantle and scapulars (grey less bluish than in ♂), side of head and underparts more fawn or brown-grey, less drab-grey. Nestling. Tufts of down fairly long and plentiful on upperparts, upperwing, and lower flank; dark grey to blackish-grey (Heinroth and Heinroth 1924–6; Witherby *et al.* 1938). Juvenile. Cap down to lore and middle of eye dark olive-brown or fuscous-brown, sometimes more buff-brown on forehead and lore. Nape to back dark olive-brown or umber-brown, darker and browner than upperparts of adult ♀, less uniform grey on hindneck, but some grey of feather-bases often visible. Rump white, washed buff at border of back and tail-coverts. Upper tail-coverts black, some washed brown on tip. Side of head and neck and chin to side of breast and upper belly umber-brown or greyish buff-brown, remainder of underparts brighter buff-brown or ochre, more cream-buff on vent and under tail-coverts; some bare skin visible on chin and throat, some grey of feather-bases showing through on remainder. Feathering of head and body markedly shorter and looser than in adult, especially on rump and underparts. Tail, flight-feathers, greater upper primary coverts, and tertials as adult, primary coverts sometimes with narrow buff edge when fresh and innermost tertial with broad buff-brown outer fringe; bastard wing and shorter upper primary coverts as adult, but feathers with ill-defined buff-brown or pale grey-buff fringe along outer web and tip (except on longest feather of bastard wing). Lesser and median upper wing-coverts dull grey with buff-brown tips; greater dull grey with broad but ill-defined black border along outer web and buff-brown tip; latter rather pale and *c.* 2 mm wide on outer web of outer coverts, warmer buff and *c.* 4 mm wide on inner greater coverts (in adult, much greyer and wider on innermost, whitish on outermost). Under wing-coverts and axillaries short, fluffy, grey-white, those along leading edges of wing mottled pale grey and ochre. Sexes similar (distinguishable as soon as 1st adult feathers on underparts start to grow, rosy-pink in ♂, vinous-drab-grey in ♀). FIRST ADULT. Like adult, but juvenile tail, flight-feathers, greater upper primary coverts, and usually bastard wing and shorter upper primary coverts still juvenile; most of these similar to adult, but shorter feathers of bastard wing less extensively black than in adult, fringes along outer webs and tips of these and of juvenile shorter primary coverts (if any) brown or buff (in adult, sharply-defined pale grey or whitish). Some outer greater upper wing-coverts often still juvenile, contrasting markedly in colour with new neighbouring feathers: pale tips narrow, pale grey-buff or pink-buff, mainly confined to outer web (in adult, much broader, white or pale grey, more sharply contrasting, and covering entire tip), outer webs duller and less extensively black. In Netherlands, 3·2 (20) 0-5 outer greater coverts retained; in Britain, 2·0 (10) 0-5; in Scandinavia, 3·6 (10) 1-6; in Spain and Italy 1·5 (10) 0-3 (BMNH, RMNH, ZMA); in Oxford area (England), ♂ 1·3 (1·63; 70) 0-4, ♀ 2·8 (1·79; 52) 0-5 (Newton 1966b).

**Bare parts.** ADULT, FIRST ADULT. Iris brown or dark brown. Bill black or greyish-black, in 1st autumn sometimes with faint flesh tinge at base. Leg and foot dark horn-brown with flesh tinge to black-brown, in 1st autumn sometimes horn-brown or light greyish-horn (BMNH, ZMA.) NESTLING. At hatching, skin flesh-pink with yellow tinge, bill pink, leg and foot flesh-colour; mouth pink to bright red with purple-grey or bluish marks at sides, gape-flanges pale or bright yellow. JUVENILE. Iris dark brown. Bill yellow-horn, green-grey, or horn-grey, paler on lower mandible. Leg and foot greyish- or brownish-flesh. (Hartert 1903-10; Heinroth and Heinroth 1924-6; Witherby et al. 1938; RMNH, ZMA.)

**Moults.** ADULT POST-BREEDING. Complete; primaries descendent. In Britain, starts with p1 between mid-July and early September, rarely to late September (on average, 8-31 August in various years); primary moult duration 69±9 days (c. 70-75 days in birds starting early, c. 60 days in late ones); moult of ♂ in known pairs slightly ahead of that of ♀. Secondaries start with s1 at primary moult score 19-25, completed at same time as outer primaries (primary score 45) or up to 1 week later; tail moult centrifugal, t1 shed at primary score 9-19, t6 full-grown at primary score 35-43. (Newton 1966b, 1972; Ginn and Melville 1983). For duration of growth of each flight-feather, see Newton (1967d, 1972); for number of feathers growing at same time, see Newton (1966b). All moult completed early October to mid-November, occasionally early December (Newton 1966b). Moult on average 9 August to 2 November, duration c. 85 days (Newton 1968). In captivity, duration of primary moult 86·2 (4) 84-89 days, secondaries 65·2 (4) 63-66 days (completed up to 9 days later than primaries), total duration of moult in individuals 94·0 (4) 93-95 days (Newton 1967d, which see for sequence of moult of various feather-tracts). In north-west Russia, starts early July to late August, peaking late July; completed early October to early November, mainly 15-25 October; many of 110 birds examined 20 July to 20 August in moult, all 22 from 20 August to 20 September, c. 60% of 137 birds still in moult mid-October (Rymkevich 1990). In Kursk area (south-west European Russia), moult on average c. 10 July to c. 5 October (Dol'nik and Gavrilov 1974). Body moult rapid, parts of bare skin sometimes exposed due to simultaneous loss of feathers (Dementiev and Gladkov 1954). POST-JUVENILE. Partial: head, body, lesser and median upper wing-coverts, variable number of greater upper wing-coverts (see First Adult in Plumages), usually tertials (RMNH,

ZMA), exceptionally shorter upper primary coverts (Flegg and Matthews 1980). In birds hatched May-June, moult starts at age of c. 10 weeks, duration c. 9 weeks; in birds hatched August, starts at c. 4 weeks (sometimes before juvenile tail- and flight-feathers full-grown), duration c. 7 weeks (Newton 1966b, 1972). Moult in Britain and Netherlands starts mid-July to late September (sometimes October), completed mid-September to November(-December) (Newton 1996b; Ginn and Melville 1983; RMNH, ZMA). In central Europe, birds in moult recorded 6 September to 25 October (once 10 November), others not yet started June to 10 September (once 23 September) (Stresemann 1919). In north-west Russia, moult starts mid-July to late August (mainly early August), completed late September to early November (mainly mid-October); of 100 birds examined 20 July to 31 August, 70-85% had started, all 38 birds from September in moult, c. 40% of 158 in 1st third of October, c. 30% of 1105 in 2nd third, c. 20% of 246 in last third, c. 10% of 28 in 1st third of November (Rymkevich 1990). In *murina*, some juvenile feathers present in cap of ♀ from March (no traces of juvenile feathers on head or body in 45 other birds examined March-May) (BMNH, ZFMK).

**Measurements.** ADULT, FIRST ADULT. Nominate *pyrrhula*. Norway and Sweden, all year; skins (BMNH, RMNH, ZMA: A J van Loon, C S Roselaar). Bill (S) to skull, bill (N) to distal corner of nostril; exposed culmen on average 5·3 less than bill (S).

| | ♂ | | ♀ | |
|---|---|---|---|---|
| WING | 93·8 (1·77; 21) | 90-97 | 91·8 (2·27; 13) | 87-95 |
| TAIL | 70·5 (1·95; 20) | 67-75 | 69·6 (1·95; 13) | 66-72 |
| BILL (S) | 15·5 (0·43; 12) | 14·9-16·2 | 15·2 (0·55; 14) | 14·5-16·1 |
| BILL (N) | 8·6 (0·38; 20) | 8·0-9·4 | 8·4 (0·33; 13) | 8·1-9·2 |
| TARSUS | 18·0 (0·50; 19) | 17·3-19·0 | 17·7 (0·62; 12) | 16·9-18·5 |

Sex differences significant for wing.

*P. p. europoea*. Netherlands, all year; skins (RMNH, ZMA: A J van Loon).

| | ♂ | | ♀ | |
|---|---|---|---|---|
| WING | 82·8 (2·10; 50) | 79-87 | 81·4 (1·81; 44) | 78-85 |
| TAIL | 63·7 (1·73; 45) | 60-66 | 62·7 (2·02; 42) | 59-67 |
| BILL (S) | 13·8 (0·47; 24) | 13·1-14·5 | 13·7 (0·58; 19) | 12·9-14·7 |
| BILL (N) | 7·6 (0·33; 42) | 7·0-8·2 | 7·6 (0·34; 40) | 6·9-8·2 |
| TARSUS | 16·6 (0·66; 38) | 15·8-18·1 | 16·7 (0·55; 39) | 15·5-17·7 |

Sex differences significant for wing. In this and other samples, ages combined, though retained juvenile wing and tail of 1st adult on average often shorter than in older birds; thus, in Netherlands: wing, adult ♂ 83·9 (1·93; 18) 81-87, 1st adult ♂ 81·9 (1·60; 18) 79-86, adult ♀ 82·0 (2·67; 8) 77-85, 1st adult ♀ 81·4 (1·35; 16) 79-84; tail, adult ♂ 64·2 (1·48; 46) 61-66, 1st adult ♂ 63·1 (1·78; 17) 60-66, adult ♀ 62·9 (2·55; 8) 59-67, 1st adult ♀ 62·9 (1·43; 14) 60-66 (A J van Loon). For age differences, see also Álbu (1983), and Eck (1985).

*P. p. pileata*. Wing and bill to skull from (1) England, (2) Scotland; other measurements combined, all year; skins (BMNH, RMNH, ZMA: A J van Loon, C S Roselaar).

| | ♂ | | ♀ | |
|---|---|---|---|---|
| WING (1) | 82·3 (2·13; 18) | 79-87 | 80·6 (1·37; 15) | 78-83 |
| (2) | 82·0 (2·02; 14) | 78-86 | 80·0 (1·37; 5) | 78-82 |
| TAIL | 64·0 (1·90; 13) | 60-67 | 62·7 (1·80; 12) | 59-66 |
| BILL (1) | 13·8 (0·62; 14) | 12·9-14·6 | 13·8 (0·25; 10) | 13·4-14·3 |
| (2) | 13·0 (0·47; 12) | 12·2-13·6 | 13·2 (0·39; 6) | 12·7-13·7 |
| BILL (N) | 7·5 (0·31; 12) | 7·0-7·9 | 7·6 (0·26; 11) | 7·2-8·1 |
| TARSUS | 16·7 (0·33; 11) | 16·0-17·1 | 16·8 (0·42; 11) | 16·0-17·4 |

Sex differences not significant. See also Voous (1949), Vaurie (1956c), and Newton (1966a).

*P. p. iberiae*. North-west Spain and Portugal, October-May; skins (BMNH, RMNH, ZMA: A J van Loon).

| | ♂ | | ♀ | |
|---|---|---|---|---|
| WING | 81·2 (1·21; 15) | 78-83 | 79·2 (1·22; 9) | 77-81 |

| | | |
|---|---|---|
| TAIL | 62·0 (1·77; 15)   57–65 | 60·9 (1·56; 9)   59–64 |
| BILL (S) | 13·5 (0·36; 12) 12·9–14·2 | 13·6 (0·47; 8)   12·9–14·3 |
| BILL (N) | 7·6 (0·30; 14)   7·2–8·3 | 7·5 (0·37; 9)   7·0–8·0 |
| TARSUS | 16·5 (0·60; 14) 15·2–17·8 | 16·4 (0·47; 9)   15·7–17·0 |

Sex differences significant for wing. See also Voous (1949, 1951b) and Vaurie (1956c).

*P. p. murina*. Azores, October and March–May; skins (BMNH, RMNH, ZFMK, ZMA: A J van Loon, C S Roselaar).

| | | |
|---|---|---|
| WING | ♂  89·3 (1·46; 29)   87–93 | ♀  87·6 (0·91; 19)   86–90 |
| TAIL | 70·5 (1·58; 10)   68–73 | 70·1 (1·24; 5)   69–72 |
| BILL (S) | 17·3 (0·53; 9) 16·4–18·0 | 16·9 (0·42; 5) 16·4–17·4 |
| BILL (N) | 9·5 (0·33; 10)   8·7–9·8 | 9·4 (0·30; 5)   8·9–9·7 |
| TARSUS | 21·1 (0·61; 10) 19·8–21·7 | 20·4 (1·16; 5) 19·4–22·2 |

Sex differences significant for wing. Exposed culmen: ♂ 12·3 (0·62; 25) 10·5–14·0, ♀ 12·2 (0·40; 15) 11–12·9 (G Le Grand). Azores, live birds (J A Ramos).

| | | |
|---|---|---|
| WING | ♂  90·8 (1·58; 12)   87–93 | ♀  88·8 (1·60; 23)   85–92 |
| TAIL | 74·0 (1·18; 12)   72–75 | 71·1 (1·73; 23)   67–74 |
| CULMEN | 12·6 (0·34; 12) 12·1–13·0 | 12·6 (0·28; 23) 12·0–13·1 |

Wing. Samples mainly from Stresemann (1919, 1928), Cerny (1938), Mayaud (1933c, 1939), Vaurie (1949b, 1956c), Voous (1949), Eck (1985), and birds examined (RMNH, ZMA); some samples are partly from Witherby (1908), Bacmeister and Kleinschmidt (1920), Løppenthin (1935, 1943), Jordans and Steinbacher (1948), Jordans (1950), Niethammer (1950), Pielowski (1963), Kumerloeve (1967a), Scherner (1968), Wolters (1968), Bocheński (1970), Rokitansky and Schifter (1971), Handtke (1975), Bocheński and Oleś (1977), Noval (1981), and BMNH (A J van Loon). Breeding birds, but data of Stresemann (1919) include winter birds of presumed local origin. Nominate *pyrrhula*: (1) Orenburg (south-west Urals), (2) Baltic countries and Pskov (western Russia), (3) Fenno-Scandia, (4) Bornholm island (southern Baltic Sea), (5) north-east Poland (former East Prussia) and Belorussiya, (6) eastern and southern Carpathians (easternmost Slovenia and Rumania). *P. p. 'germanica'* (intermediate, nearer nominate *pyrrhula*): (7) Warsaw (Poland), (8) Krakow area, Beskidy and Krkonoše mountains (southern Poland), and Tatra mountains (Slovenia), (9) Serrahn (central-north of former East Germany), (10) south-east of former East Germany (Erzgebirge and Altenburg to Lausitz), (11) south-west of former East Germany (Thüringer Wald to Erfurt and Jena), (12) south-west Czechia, (13) south-east of former West Germany and neighbouring part of Austria (Böhmer Wald to Oberbayern, Salzburg area, and Gmunden), (14) Alps of southern Bayern, Austria, Switzerland, and northern Italy, (15) Bulgaria. *P. p. 'coccinea'* (intermediate, nearer *europoea*): (16) Jutland and Fyn (Denmark), (17) along northern border between former West and East Germany (Göttingen, Harz, Wolfsburg, and Mecklenburg), (18) south-west Germany (Schwarzwald, Karlsruhe, and Lohr), (19) Pfalz (western Germany) and neighbouring north-east France to Vosges, (20) Jura and northern Switzerland to foot of Alps, (21) central Italy, (22) west-central Germany (Taunus, Westerwald, Marburg). *P. p. europoea*: (23) western France. *P. p. iberiae*: (24) Spain. *P. p. pileata*: (25) Ireland. *P. p. rossikowi*: (26) north-west Turkey, (27) eastern Turkey, Caucasus, Transcaucasia, and Iranian Azarbaijan. *P. p. caspica*: (28) northern Iran.

| | | |
|---|---|---|
| (1) | ♂  93·7 ( – ; 20)   91–98 | ♀  91·5 ( – ; 29)   89–95 |
| (2) | 93·6 (1·64; 24)   90–96 | 92·1 ( – ; 17)   89–96 |
| (3) | 92·5 (1·93; 36)   90–97 | 90·5 (2·35; 14)   86–95 |
| (4) | 91·6 (2·64; 7)   89–96 | 90·5 ( – ; 2)   89–92 |
| (5) | 93·2 (1·81; 15)   90–95 | 91·0 (1·36; 12)   89–93 |
| (6) | 91·8 (2·91; 12)   86–96 | 86·8 ( – ; 3)   85–89 |
| (7) | 91·1 (1·93; 30)   87–96 | 88·9 (2·93; 17)   82–93 |
| (8) | 90·6 (2·31; 30)   86–96 | 90·2 (3·40; 4)   87–95 |

| | | |
|---|---|---|
| (9) | 87·8 ( – ; 17)   85–95 | 86·3 ( – ;   4)   84–87 |
| (10) | 89·1 (1·24; 11)   87–92 | 87·2 (1·68; 7)   85–90 |
| (11) | 89·1 (2·36;   5)   85–92 | 86·2 ( – ;   3)   84–90 |
| (12) | 89·4 (1·50; 10)   87–92 | 87·3 (1·86; 6)   85–90 |
| (13) | 89·1 (2·39; 92)   84–93 | 87·8 (1·84; 49)   84–91 |
| (14) | 89·8 (2·30; 27)   86–93 | 87·0 (2·41; 13)   83–92 |
| (15) | 89·7 (1·70; 13)   87–92 | 86·5 (0·58;   4)   86–87 |
| (16) | 84·6 (1·14;   5)   83–86 | 83·5 ( – ;   2)   83–84 |
| (17) | 84·7 ( – ; 31)   81–89 | 83·3 ( – ; 15)   79–87 |
| (18) | 85·2 (2·05; 20)   82–89 | 85·3 (3·59; 6)   82–92 |
| (19) | 85·6 (1·95; 15)   82–89 | 84·7 (2·86; 8)   81–88 |
| (20) | 85·7 (1·73;   9)   82–88 | 83·5 ( – ;   2)   83–84 |
| (21) | 85·5 ( – ; 14)   82–88 | 83·3 ( – ;   6)   82–88 |
| (22) | 84·5 (1·76; 65)   80–89 | 82·8 (1·81; 83)   78–88 |
| (23) | 80·6 ( – ; 52)   76–85 | 79·3 ( – ; 15)   78–83 |
| (24) | 80·4 ( – ; 39)   79–84 | 78·5 ( – ; 30)   76–81 |
| (25) | 82·0 (1·99; 15)   79–85 | 80·8 (0·84;   5)   80–82 |
| (26) | 85·6 (3·36;   5)   82–91 | 85·0 ( – ;   1)   – |
| (27) | 90·2 ( – ; 20)   87–93 | 88·5 (1·69; 12)   85–91 |
| (28) | 86·8 (1·89;   4)   84–88 | 90·0 ( – ;   1)   – |

Exposed culmen: nominate *pyrrhula*, Orenburg and Pskov, ♂ 10·1 (31) 9·3–10·7, ♀ 9·9 (42) 9·0–10·5; *rossikowi*, Tbilisi and Iranian Azarbaijan, ♂ 11·5 (9) 10·5–12·0, ♀ 11·2 (0·55; 6) 10·5–12·0 (Vaurie 1949b).

**Weights.** ADULT, FIRST ADULT. Nominate *pyrrhula*. (1) Kazakhstan (Korelov *et al.* 1974). Norway: (2) November–March, (3) April–May (Haftorn 1971). (4) Poland, winter (Pielowski 1963). (5) Eastern Germany, October–March (Eck 1985). (6) Czechoslovakia, November–April (Havlín and Havlínova 1974). Intermediates between nominate *pyrrhula* and *europoea*: 'germanica', (7) Warsaw area (Poland), summer (Pielowski 1963). (8) Czechoslovakia, whole year (Havlín and Havlínová 1974, which see for monthly averages); 'coccinea', (9) north-east France, January–March (Bacmeister and Kleinschmidt 1920). *P. p. europoea*. (10) Netherlands, all year (ZMA); (11) Western France, all year (Mountfort 1935; Mayaud 1939). *P. p. pileata*. Oxford area (England): (12) at peak, January–February; (13) at low, July; (14) starting moult or about to do so, August (Newton 1966a, b, 1968b, 1972, which see for influence of moult on weight and many other data).

| | | |
|---|---|---|
| (1) | ♂  31·4 ( – ; 13) 27·5–35·0 | ♀  29·9 ( – ;   6) 27·6–32·7 |
| (2) | 33·1 ( – ; 61) 27·5–38·6 | 33·2 ( – ; 38) 26·3–39·5 |
| (3) | 30·9 ( – ; 24) 27·0–36·2 | 31·2 ( – ; 28) 26·1–35·6 |
| (4) | 29·8 (2·42; 63)   25–35 | 28·9 (1·86; 74)   25–33 |
| (5) | 32·1 (2·26; 23) 29·0–37·2 | 30·5 (2·79; 21) 26·5–35·8 |
| (6) | 34·3 (2·33; 14)   29–38 | 32·7 (3·10; 29) 28·5–40 |
| (7) | 27·1 (1·81; 30)   24–31 | 27·4 (1·58; 17)   24–30 |
| (8) | 30·5 (2·18; 26) 26·2–35·0 | 30·8 (2·88; 24) 26·1–36·0 |
| (9) | 21·3 (0·51;   4) 20·7–21·7 | 20·5 ( – ;   1)   – |
| (10) | 22·5 (4·30; 17) 16·6–26·0 | 22·5 (2·99; 12) 16·0–26·5 |
| (11) | 22·9 ( – ; 26)   18–26 | 21·9 (1·55;   9) 20·0–24·4 |
| (12) | 25·7 (1·8 ; 76)   – | 25·8 (1·5 ; 58)   – |
| (13) | 21·9 (1·2 ; 26)   – | 24·3 (1·7 ; 9)   – |
| (14) | 21·9 (1·4 ; 55)   – | 23·2 (1·5 ; 51)   – |

*P. p. 'germanica'*. Germany: 22–29 (8) (Niethammer 1937). Swiss Alps, March: ♂ 32 (Eck 1985). Southern Yugoslavia, October: ♀ 26 (Makatsch 1950). Bulgaria, July: ♂ 23 (Niethammer 1950).

*P. p. pileata*. For winter weights and fat contents, see Newton and Evans (1966); for fattening, see Newton (1969b).

*P. p. rossikowi*. North-west Asia Minor, July: ♂ 26·6, ♀ 25·2 (Rokitansky and Schifter 1971). North-east Turkey, October–November: ♂♂ 26, 29 (Kumerloeve 1967a).

*P. p. murina.* Average *c.* 30 (Bibby *et al.* 1992).

NESTLING, JUVENILE. For weight increase during growth, see Heinroth and Heinroth (1924-6). At fledging, *pileata* 16-18; in 2nd half of July 20·8 (0·86; 12) (Newton 1966*a, b,* 1972, which see for further data).

**Structure.** Wing rather long, broad at base, tip bluntly pointed. 10 primaries: p7 longest; in nominate *pyrrhula* (*n* = 10), p8 0-4 shorter, p9 2-5, p6 0-1·5, p5 4-7, p4 12-15, p3 15-18, p2 17-20, p1 19-22; in *europoea* and *pileata* (*n* = 10 in each), p8 0-2 shorter, p9 2-6, p6 0-1, p5 3-6, p4 10-14, p3 12-17, p2 15-19, p1 16-21, in *iberiae* (*n* = 9), p1 14-19 shorter; in *murina* (*n* = 3), p8 0-2 shorter, p9 3-6, p6 0-1, p5 2-5, p4 7-12, p3 11-15, p2 13-18, p1 15-20. P10 reduced, minute, pointed, concealed below reduced outermost greater upper primary covert, 54-66 shorter than p7, 7-12 shorter than longest upper primary covert. Outer web of p5-p8 emarginated, inner of p6-p9 with notch (sometimes faint). Tip of longest tertial reaches to about tip of p1. Tail rather long, tip square or very slightly forked; 12 feathers, t1 and t6 0-3 shorter than t4. Tail-coverts long, reaching just beyond middle of tail. Bill markedly short, rounded, and swollen at base; culmen and cutting edges decurved; tip of upper mandible forming small point, extending over tip of lower mandible; gonys slightly convex. Depth of bill at base 9·6 (25) 9·0-10·2 in nominate *pyrrhula*, width at base 10·3 (26) 9·7-11·0; in '*germanica*', depth 9·5 (8) 9·0-9·9, width 9·9 (9) 9·7-10·2; in '*coccinea*', depth 8·7 (7) 8·2-9·4, width 9·6 (7) 9·2-10·0; in *europoea*, depth 8·9 (18) 8·4-9·5, width 9·3 (20) 8·6-9·9; in *pileata* from England, depth 8·8 (21) 8·3-9·4, width 9·2 (21) 8·3-9·9; in *pileata* from Scotland, depth 8·3 (8) 8·0-8·6, width 8·6 (11) 8·4-8·9; in *iberiae*, depth 8·6 (9) 8·3-9·2, width 8·9 (9) 8·5-9·3; in *murina* (*n* = 2), depth 10·7-10·8, width 10·4-10·5; thus, width and depth of bill of *iberiae*, *pileata*, *europoea*, and '*coccinea*' rather small (as is length to nostril: see Measurements), '*germanica*' and nominate *pyrrhula* larger, *murina* distinctly heavier. For internal structure of bill, see Newton (1972). Nostril small, rounded, covered by short tuft of feathers projecting from forehead. Some short bristles at base of upper mandible, projecting obliquely downward over gape. Small gular pouch during breeding (Newton 1972). For weight of various tracts of feathering in several stages of wear, see Newton (1968*b*). Leg and foot short and slender. Middle toe with claw 17·3 (10) 16·5-18·5 in nominate *pyrrhula*, 16·6 (20) 15·5-17·5 in *europoea*, *pileata*, and *iberiae*, 19·2 (2) 18·5-20·0 in *murina*; outer and inner toe with claw both *c.* 71% of middle with claw, hind *c.* 77%.

**Geographical variation.** Marked in size, less so in colour, except in far east. Races *pileata*, *europoea*, and *iberiae* from western Europe small, average wing of ♂ *c.* 81·5 (averages of various populations 80-83, measurements of individuals mainly 78-85); other measurements and bill size of these races also closely similar (see Measurements and Structure). Northern and eastern Europe inhabited by nominate *pyrrhula*, with average wing length of ♂ *c.* 92·5 (averages of various populations 91·5-94·5, measurements of individuals mainly 89-98). For each sex, virtually no overlap in measurements between nominate *pyrrhula* and smaller races, except slightly for bill length, bill depth, and tarsus length. Large and small populations grade clinally into each other in wide zone through central Europe, and opinions differ on whether 1, 2, or no intermediate races should be recognized in transition zone. In west of this zone, '*coccinea*' sometimes recognized, which has slightly longer wing than *europoea*, but which is otherwise closely similar to it in size; typical '*coccinea*' has average wing of ♂ *c.* 85·2, but populations with average wing of ♂ of *c.* 84-87 included in '*coccinea*' here, with range in

individuals mainly 81-89. In east of transition zone, populations have been named '*germanica*': measurements close to nominate *pyrrhula*, but average wing of ♂ in various populations *c.* 88-91 (in typical '*germanica*', 89·1), with range in individuals mainly 85-93. Transition between *europoea* and '*coccinea*' as well as that between '*germanica*' and nominate *pyrrhula* is gradual, but apparently less so between '*coccinea*' and '*germanica*': smaller '*coccinea*' in many localities rather abruptly replaced by larger '*germanica*', especially along foot of Alps ('*coccinea*' in lowlands, '*germanica*' in mountains) and apparently in Germany. Thus, *europoea* and '*coccinea*' should probably be included in single race (named *coccinea*), and '*germanica*' should be merged with nominate *pyrrhula*, but multivariate character analysis may prove that all 4 taxa can be separated and it may be practical to do so, notwithstanding clinal variation.

Differences between small western populations based on colour. *P. p. europoea* on continent from northern Germany and Netherlands to foothills of Pyrénées, Alps, and Vosges in France slightly darker grey on upperparts of ♂ than in nominate *pyrrhula*, medium to dark neutral grey rather than light to medium neutral grey, underparts slightly darker rosy-red, less rosy-pink; tips of outer greater coverts light grey, less whitish; ♀ distinctly darker and browner on upperparts and more vinous (less grey) on underparts. ♂ *pileata* from Britain and Ireland about as dark above as ♂ *europoea*, but more diluted dull pink-red below; ♀ slightly darker and browner above, distinctly so below, especially on flank. ♂ *iberiae* slightly paler grey on upperparts than ♂ *europoea* (but less so than nominate *pyrrhula*), underparts more fiery rosy-red, ♀ intermediate in colour of upperparts and underparts between nominate *pyrrhula* and *europoea* (BMNH, RMNH, ZMA; see also Voous 1951). Within *pileata*, not all populations uniform: birds from Ireland on average paler and greyer than birds from southern England, near *europoea* from western France; ♂♂ from western and central Scotland on average slightly darker grey and ♀♀ paler and more greyish-brown than typical *pileata* from Hertfordshire (Clancey 1947, 1948*b*), and bill smaller (Clancey 1947, 1948*b, e*; Harrison 1958; see also Measurements and Structure), and Scottish birds therefore sometimes separated as '*wardlawi*' Clancey, 1947, but colour differences negligible in birds examined, and difference in bill size rather too small to support recognition of this race. For individual variation in *pileata*, see Harrison (1958).

Birds from Caucasus area separated as *rossikowi*: rather like nominate *pyrrhula* in colour and size, but upperparts of ♂ paler grey, underparts brighter red, less pinkish, ♀ darker and duller brown; bill broader, more swollen on basal half, but more attenuated at tip; depth and width of bill at nostril on average 0·7-0·9 more than in nominate *pyrrhula* (Buturlin 1906; Vaurie 1949*b*). Position of birds breeding north-west Turkey uncertain: smaller in size than *rossikowi*, near '*coccinea*'; colour like birds from central Europe (Jordans and Steinbacher 1948) or deeper red (Rokitansky and Schifter 1971); bill longer and more swollen at base than in '*coccinea*', similar to *rossikowi*. *P. p. caspica* from northern Iran apparently smaller than *rossikowi*, near '*coccinea*', but bill larger; ♂ richer red below than *rossikowi*, ♀ paler and greyer, both sexes purer blue-grey on upperparts.

*P. p. murina* from São Miguel (Azores) markedly different from all other races. Larger in size, bill longer and deeper, and wing relatively shorter and more rounded at tip. Sexes virtually identical in colour, ♂ without red, both sexes without white rump: upperparts grey-brown with marked buff tinge, paler grey with purer buff tinge on rump (♀ slightly greyer than ♂, rump, mantle, and scapulars less intensely buff); sides of head and neck and underparts buffish drab-grey, often with more rufous-buff tinge on ear-coverts, belly, flank, and vent (especially in ♂); tips of greater coverts light grey; outer web of shortest tertial usually orange-pink. Extensive buff tinge on body (including rump) rather similar to that of juvenile of other races, but cap black and bases of rump feathers light grey. Aberrant features probably due to long isolation from other races and *murina* perhaps best considered a separate species.

In eastern Siberia, *cassinii* of Kamchatka similar to nominate *pyrrhula* but underparts of ♀ and upperparts of both sexes paler and purer grey, tips of greater upper wing-coverts broadly whitish; often a white streak on outermost tail-feather, and ♂ frequently with pink-washed upperparts. Further south, situation not fully elucidated (compare, e.g. Stegmann 1931*a*, Hartert and Steinbacher 1932-8, Johansen 1944, Vaurie 1956*c*, 1959, and Stepanyan 1978, 1990). Area from Ob' valley and northern Mongolia east to Sea of Okhotsk and Sea of Japan inhabited by Grey Bullfinch *P. cineracea*, which is rather similar to *P. pyrrhula* but in which pink-red of side of head, tertials, and underparts of ♂

replaced by light grey; sometimes considered a race of *P. pyrrhula* due to occasional hybridization, especially at edges of range, but both occur side-by-side in many places without interbreeding, behaving as good species. Races *rosacea* from far-eastern Russia (including Sakhalin) and Manchuria, and *griseiventris* from central and southern Kuril Islands and Japan, characterized by pink-red of side of head and throat of ♂, contrasting with paler pink (*rosacea*) or grey (*griseiventris*) remainder of underparts; amounts of grey and pink below rather variable in both races. Both usually included in *P. pyrrhula*, but considered a separate species, Japanese Bullfinch *P. griseiventris*, by Stepanyan (1978, 1983, 1990), forming superspecies with *P. pyrrhula*, though *griseiventris* (and *cineracea*) are no more different from *P. pyrrhula* than is *murina*. Bill of *cineracea* and *griseiventris* smaller than in nominate *pyrrhula*.

For distributional history of *P. pyrrhula*, see Stresemann (1919) and Voous (1949).                                              CSR

## *Eophona migratoria* Hartert, 1903  Yellow-billed Grosbeak

Fr. Gros-bec migrateur          Ge. Schwarzschwanz-kernbeisser

An east Palearctic species, breeding in eastern Siberia, Manchuria, and Korea, and wintering mainly in eastern China; also an isolated population resident in southern China. Individuals recorded in Europe—e.g. Britain (P R Colston), Sweden (LR), Germany (Weber 1991)—are thought to have been escapes.

## *Eophona personata* (Temminck and Schlegel, 1848)  Japanese Grosbeak

Fr. Gros-bec masqué          Ge. Maskenkernbeisser

An east Palearctic species, breeding in eastern Siberia, Manchuria, northern China, and Japan, and wintering within and to south and west of breeding range. 2 records, from Sweden (LR) and from Vestfold (Norway), April 1990 (VR), are probably of escapes.

## *Coccothraustes coccothraustes*  Hawfinch

PLATES 53 (flight) and 56
[between pages 640 and 641]

Du. Appelvink          Fr. Gros-bec cassenoyaux          Ge. Kernbeisser
Ru. Дубонос          Sp. Picogordo          Sw. Stenknäck

*Loxia Coccothraustes* Linnaeus, 1758

Polytypic. Nominate *coccothraustes* (Linnaeus, 1758), Eurasia east to Argun valley in western Inner Mongolia (China), south to Iberia, Sardinia, central Italy, northern Yugoslavia and Bulgaria, plains north of Caucasus, and *c.* 40°N in Siberia, grading into *nigricans* in southern Yugoslavia, southern Bulgaria, Greece, and north-west Asia Minor; *buvryi* Cabanis, 1862, north-west Africa; *nigricans* Buturlin, 1908, Crimea and Caucasus mountains, south through Transcaucasia to northern Iran and perhaps Kopet-Dag (south-west Turkmeniya) and in eastern Turkey west through Taurus mountains. Extralimital: *schulpini* Johansen, 1944, Amurland (eastern Siberia) and Manchuria (north-east China) east to Sea of Japan; *japonicus* Temminck and Schlegel, 1848, Sakhalin, southern Kuril Islands, central and southern Kamchatka, and Japan south to Hondo; *humii* Sharpe, 1886, isolated in central Asia from eastern Afghanistan through Pamir-Alay ranges to Tien Shan and Dzhungarskiy Alatau mountains.

**Field characters.** 18 cm; wing-span 29–33 cm. 20% larger than Greenfinch *Carduelis chloris*, with distinctive top-heavy, parrot-like proportions due to massive conical bill on large head, bull-neck, and short, square tail. Very large, huge-billed, big-headed, short-tailed, short-legged finch, bigger than all other common finches of temperate woodlands. Adult plumage warm buff, with bill blue-grey in breeding season, yellow in winter, emphasized by black lore and bib, grey nape, brown back, black flight-feathers boldly panelled white on larger coverts and across primaries, and white-tipped tail. Sexes dissimilar at close range; little seasonal variation. Juvenile separable. 3 races in west Palearctic, with north-west African form distinguishable (see Geographical Variation).

ADULT MALE. Moults: July–October (complete). Border to mandibles, lore, and bib black; fore-crown and deep ear-coverts yellowish-brown; rear crown and nape more chestnut; nape extending below rear face to shoulder pinkish-grey. Mantle and scapulars rich umber-brown; rump yellowish-brown, contrasting with back. Wings strongly patterned: lesser coverts and bases of median almost black; tips of median coverts and outer webs of greater coverts essentially white, forming obvious blaze across wing but with inner feathers washed rufous-brown; flight-feathers when folded black, glossed green and purple, but when extended strikingly marked by white panels on inner webs of primaries which create flashing patch on outer wing, noticeable as blaze on coverts. Tail-feathers black-based, ending in grey- to rufous-brown and then bright white tips (last particularly noticeable from below). Body pinkish-brown, more rufous on flanks and becoming white from belly to under tail-coverts, creating pale patch (obvious from below). With wear, forehead becomes pale buff. Massive bill noticeable even in flight, with almost triangular outline recalling parrot (Psittacidae); almost wholly blue-grey in breeding season, becoming yellowish-horn with dusky tip after autumn moult. Eye amber to reddish-brown, contrasting with black lore and giving intense expression. Legs pink-flesh when breeding, fading to brownish-flesh or light horn in winter. ADULT FEMALE. Indistinguishable from ♂ except at close range when less deeply coloured plumage can be seen. Black on face duller and less extensive, crown and rump greyer, underparts paler and less pink. Pale panel formed by grey fringes to secondaries distinctive, reducing contrast of dull white covert-blaze. JUVENILE. Easily separated from adult by yellow-horn bill, paler, more orange head, with only dusky lore and bib, dull mottled upperparts, and paler more yellowish-buff underparts, strikingly barred brown-black. Differences in ♂ and ♀ wing markings already present.

Unmistakable. Flight rapid with bursts of strong wing-beats propelling bird forward in shooting, then bounding, and soon deeply undulating progress; typically at considerable height (over 50 m) even when only moving between woods. Bird looks like whirring, front-heavy and short-tailed projectile. Shy and secretive, foraging quietly in treetops, but betrays presence by explosive call and remarkable territorial flight, with ♂ following 'roller-coaster' track over wood ending in dramatic plummet. Escape-flight direct to dense canopy foliage. Gait on ground a heavy hop, with erect carriage; along branches and twigs, a twisting, parrot-like waddle, with more level stance. Sociable, forming roving parties and even large flocks outside breeding season. Solitary vagrants in open habitat look remarkably incongruous.

Song poor, a low halting 'deek-waree ree ree' or 'tchee-tchee-turr-wee-wee' with strained quality until more liquid and musical end to phrase; whole phrase somewhat recalls bunting *Emberiza* or Bullfinch *Pyrrhula pyrrhula*, with ending suggesting Goldfinch *Carduelis carduelis*. Commonest call a clipped metallic 'tick', 'pix', or 'tzik', sometimes quickly repeated, used freely in flight (often only clue to bird's sudden passage overhead) and somewhat recalling Robin *Erithacus rubecula* but much more powerful and explosive. In breeding season, also utters harsh, whistling, prolonged 'tzeep' or 'tsip', 'srree' recalling Blackbird *Turdus merula*, and harsh 'chi' recalling Spotted Flycatcher *Muscicapa striata*.

**Habitat.** Breeds in west Palaearctic in lowland and hilly temperate zone, and parts of boreal, Mediterranean, and steppe zones, continental and to lesser extent oceanic, between July isotherms of *c.* 17–25°C. Most characteristically a specialist bird of natural open mixed oak *Quercus* and hornbeam *Carpinus* forest, but extends freely to most other tall deciduous trees which carry large fruits within handling capacity of massive bill, especially beech *Fagus*, ash *Fraxinus*, wych elm *Ulmus*, and sycamore or maple *Acer*. Accordingly mainly found in crowns and forest canopy, liking to perch on topmost twigs. Significant secondary habitats are ribbons of mature trees along rivers or streams or fronting lakes and pools, and similar trees planted in avenues, parks, cemeteries, and large gardens. Occupies mixed broad-leaved–conifer woodlands and forests where broad-leaved predominate, but rarely breeds in conifers in west of range. Ascends freely in mountains to limits of deciduous forest. (Voous 1960b; Newton 1972.)

In Britain, favours moderately hilly country interspersed with patches of mixed woodlands, coppices, thickets, orchards, and hedges, preferring upper canopy for all activities. In winter, feeds more often on ground beneath woodland and orchard trees, and along hedgerows. Breeds up to and above 300 m. (Mountfort 1957.) Special features recorded in Britain are sensitivity to habitat change, often leading to desertion of established haunts, and marked aversion to western regions more subject to oceanic climates, including whole of Ireland (Bannerman 1953a; Mountfort 1957; Sharrock 1972; Parslow 1973).

Berlin apparently marks western fringe of occupancy of mixed pine and oak woodland (Bruch *et al.* 1978). In Switzerland breeds sporadically up to limit of broad-

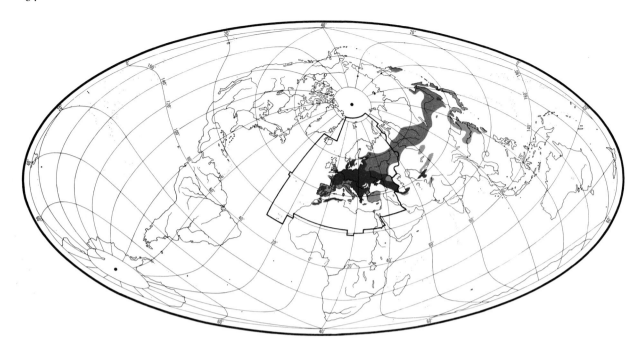

leaved trees (*c.* 1300 m), with preference for oak–hornbeam woods and relative aversion to those with mixture of conifers (Glutz von Blotzheim 1962).

In Portugal, influxes have been noted linked with ripening of pea and bean crops, and Iberian pinewoods are frequented by migrants, though breeders in North Africa prefer cork oaks *Q. suber* and alders *Alnus* growing among pines (Bannerman 1953*a*).

In European USSR, avoids coniferous woods in west, but further east nests in pine forests as well as broad-leaved stands; in Caucasus, will extend into conifers but tends to stay below 300 m which is below conifer zone. In Transcaspia in non-breeding season, found in Atrek desert and on seashore. Elsewhere in Asia occurs at higher altitudes (2200 m in Afghanistan: Paludan 1959) and shows some different habitat preferences, e.g. thickets in gorges and by riversides. (Dementiev and Gladkov 1954.)

Firmly arboreal but also a remarkably aerial species, flying freely above tree height, even up to 300 m. Avoids open ground, and even beneath trees and bushes is clumsy and nervous. Seems to show less interest in water than many Fringillidae, and is little seen on artefacts of any kind, from buildings to overhead wires, masts, or walls. Often appears one of the wariest and most easily disturbed of birds but has in certain urban areas learnt to live at close quarters with people. (Newton 1972.) Rarely perches on low non-woody vegetation, or even on isolated trees. Seems strongly influenced by seasonal local abundance of food plants, especially trees whose fruits are too hard to attract competition from other species. Among terrestrial birds based in temperate lowlands of west Palearctic, exceptional in having maintained apparently almost unchanged a peculiar lifestyle and habitat blend stemming from unique evolutionary development of capabilities.

**Distribution.** No recent changes reported, except for spread in parts of Scandinavia and adjacent Russia.

BRITAIN. Major expansion during 19th and early 20th centuries, followed by erratic fluctuations (Sharrock 1976). IRELAND. Not known to have bred. Regular winter visitor in 19th century; then irregular, few records since 1910, and now apparently more or less accidental (Hutchinson 1989). NORWAY. Spreading. Very few breeding records before 1960; in 1960s to 1980s bred in several new localities in south and south-east. (VR.) SWEDEN. Apparently spreading north; isolated breeding sites (e.g. Sundsvall, *c.* 62°30′N; Umeå, *c.* 64°N) recorded in 1980s (LR). RUSSIA. First recorded breeding Karel'skaya area in 1970s; breeding subsequently recorded up to 62°30′N (Khokhlova *et al.* 1983).

Accidental. Iceland, Faeroes, Madeira, Libya, Jordan.

**Population.** Has recently increased in Netherlands, Scandinavia, and perhaps Finland; no consistent changes reported elsewhere.

BRITAIN. Population has fluctuated this century, apparently erratically; estimated 5000–10 000 pairs, 1968–72 (Sharrock 1976). FRANCE. 1000–10 000 pairs; no evidence of change (Yeatman 1976). BELGIUM. Estimated 7000 pairs; locally some slight increases but no evidence of general change (Devillers *et al.* 1988). NETHERLANDS. Recent increase; estimated 3000–4500 pairs 1979–80, 9000–12 000 pairs 1985–6 (SOVON 1987). GERMANY. Estimated 329 000 pairs (Rheinwald 1992). East Germany:

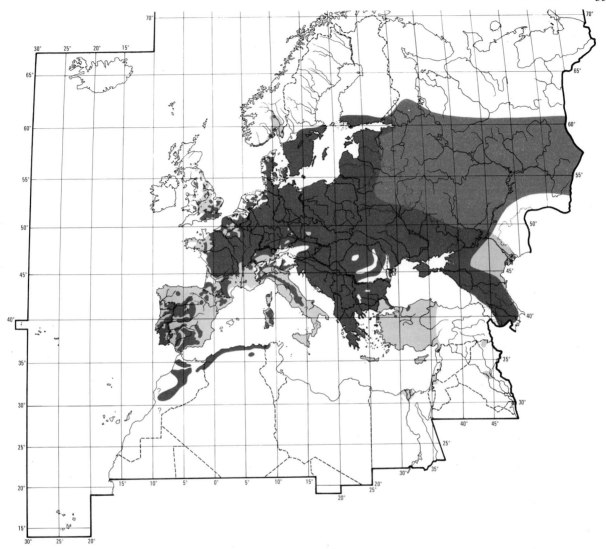

85 000 ± 45 000 pairs (Nicolai 1993). NORWAY. Increasing; estimated 70-100 pairs (VR). SWEDEN. Rough estimate 5000 pairs (Ulfstrand and Högstedt 1976); increasing in north of range and perhaps also in south (LR). FINLAND. Very rare breeder; some dozens of pairs at most (OH). Winter observations suggest increase since 1970s, probably due to mild winters; estimated *c.* 50 pairs (Koskimies 1989).

Survival. Oldest ringed bird 11 years 8 months (Krüger 1979); in captivity *c.* 18 years (Dathe 1971).

**Movements.** Sedentary to migratory; northern populations migrate more than southern ones. Juveniles migrate more than adults, and ♀♀ more than ♂♂. Migration mostly diurnal, but nocturnal also reported. European migrants (nominate *coccothraustes*) head between west and south, wintering chiefly within breeding range; numbers fluctuate markedly from year to year. Makes local feeding movements in small flocks in wide variety of directions; longer movements are probably also associated primarily with food availability. No detailed studies.

British birds mostly resident as far as is known; some migration long suspected. Winter distribution suggests some withdrawal from eastern coastal areas, and from southern Scotland (where few breed regularly). Few ringing recoveries; longest distance recorded 293 km: bird ringed Somerset (south-west England), October 1977, recovered Nottingham (central England), March 1980; all other recoveries within 76 km. (Lack 1986; Mead and Clark 1988.) Marked movement 1988-9, origin unknown; flocks larger than previously recorded in Devon (up to 44 birds), and in Ireland, where rare: at Limerick, flock of 35 from 30 October, increasing to *c.* 95 in February, with last record 6 March; small numbers elsewhere in south (Adams 1989; O'Sullivan and Smiddy 1989). Continental birds probably reach Britain occasionally; a few old records of flocks arriving on east coast, and sporadic records on lightships and at coastal sites, e.g. Isles of Scilly (south-west England), Cape Clear (Ireland), and Lincolnshire (eastern England) in autumn 1969, and Shetland (Scotland) in May 1970 (Bonham and Sharrock 1969; Bonham 1970*b*; Lack 1986). Perhaps birds making east-west coastal movements along Baltic occasionally overshoot to Britain (Mountfort 1957). Bird ringed eastern Germany, March 1962, recovered Shetland, May 1967 (Hudson 1969).

Partially migratory elsewhere in northern and central Europe. Chiefly resident in France, with dispersive feeding movements; local individuals recorded moving from northern to southern France, from north-east France to Portugal, and from eastern France to Italy and Spain (Yeatman-Berthelot 1991). In Belgium, recoveries of local birds are chiefly within 35 km, though some reach Spain (Herroelen 1962; Lippens and Wille 1972). In Rheinland (western Germany), absent from higher levels in winter, but many local birds winter at lower levels (Mildenberger 1984); in Thüringen (eastern Germany), more numerous and widespread winter than summer (Knorre *et al.* 1986). Some birds also remain in southern Sweden (SOF 1990) and Poland (Tomiałojć 1990). Mostly sedentary in Iberia (Bernis 1954; Asensio and Antón 1990).

Many birds move south-west across Europe, and winter chiefly in northern Italy and southern France, also in north-east Spain, central Italy, and Balkans. Others cross Mediterranean, and some winter in Balearic Islands, Sardinia, and Corsica. (Mountfort 1957; Asensio and Antón 1990.) Regular in varying numbers on passage in Malta, and rare winter visitor there; exceptional influx of at least 250, October 1974 (Sultana and Gauci 1982). In North Africa (where local race *buvryi* probably makes only local movements), reports of immigrants, irregular and in small numbers, are widespread along coast; scarce winter visitor to north-west Libya, up to *c.* 100 km from coast (Bundy 1976), also to Tunisia and Algeria (Heim de Balsac and Mayaud 1962). Passage regular at Gibraltar, but presumably very limited, as none detected south of Strait (Pineau and Giraud-Audine 1979; Finlayson 1992); perhaps such birds are of Iberian origin, as central European birds occur mostly in north-east Spain, and above data suggest broad-front movement across western Mediterranean. Further east (including *nigricans*), in central Europe, winters mostly south of 52°N, in southern Ukraine, Crimea, and Caucasus (Dementiev and Gladkov 1954); in Carpathian mountains (western Ukraine) and Crimea, winters especially in years of good beechmast *Fagus* crop (Strautman 1954; Kostin 1983). Occurs more widely and in larger numbers in winter than summer in Turkey (Martins 1989), and scarce (occasionally fairly common) winter visitor to Cyprus (Flint and Stewart 1992). Occasionally recorded in Egypt, south to Cairo; 4 birds seen in autumn over several years in migration study at Bahig on coast (Goodman and Meininger 1989). Irreg-

ular winter visitor to northern and central Israel, with very little evidence of passage further south; most are nominate *coccothraustes*, some are intermediates with *nigricans* (Shirihai in press). Occasional winter visitor to northern Iraq (Moore and Boswell 1957); in Iran, resident breeder in north; some birds move to lower levels in winter (D A Scott).

Ringing data illustrate importance of southern France and northern Italy for wintering, and also show wide range of movement. Longest distance 1950 km, from Prague (Czechoslovakia) to Malaga (southern Spain) (Mountfort 1957). 3 Polish birds ringed in breeding season recovered France, Italy, and Hungary (Krüger 1979). In western Germany, up to 1968, 11 of 16 recoveries of birds ringed as nestlings were at more than 50 km, in Italy (8), eastern France (2), and Poland (1); of 91 other foreign recoveries, 38 in southern France, 33 in northern (mostly) and central Italy, 4 in eastern Spain, 2 in northern France, Switzerland, Poland, and Yugoslavia, and 1 in Shetland, Belgium, Corsica, Balearic Islands, Portugal, Czechoslovakia, Austria, and Rumania (Zink 1969). 139 long-distance recoveries of birds ringed eastern Germany, 1964–76, were in northern (mostly) and central Italy (58), southern France (49), western Germany (15), Czechoslovakia (5), northern France (4), Sweden (2), Switzerland (2), and 1 each from Denmark, Belgium, southern Italy, and Portugal (Krüger 1979). Swedish recoveries 1972–88 were from Denmark (6), France (3), western Germany (3), Italy (2), Lithuania (2), and 1 each from Estonia, Hungary, Czechoslovakia, and eastern Germany (*Rep. Swedish Bird-Ringing*); all 3 recoveries in Lithuania and Estonia, May, were of birds ringed in Sweden January–March, showing that some birds from Baltic states (or further east) winter in Sweden. Birds ringed on Kaliningrad coast (western Russia), April–June, recovered northern Italy (2) and Rheinland (1), October–February (Paevski 1971; Mildenberger 1984). Hungarian birds mostly move less far; of 63 long-distance recoveries, 54 from northern and central Italy, others from Yugoslavia, Austria, Poland, and France (Krüger 1979). Of winter visitors to Spain (including Balearics), 12 were ringed in western Germany (of which 5 in breeding season), 5 in Belgium (2), 5 in Switzerland (1), 4 in Czechoslovakia (1), 3 in France, and 1 in Netherlands and eastern Germany (Asensio and Antón 1990). 2 recoveries on Algerian coast are of birds ringed Switzerland and Germany (Heim de Balsac and Mayaud 1962). Data, e.g. from Yugoslavia and Netherlands (Krüger 1979), show that individuals winter in different areas in different years; 11 birds ringed January–March, 1968–73, at Blekinge (south-east Sweden), recovered October–February in later years in Germany (6), Denmark (2), Italy (2), and France (1) (Strömberg 1975). Apparently no records of winter site-fidelity in successive years; high proportion of recoveries are of birds shot or trapped in Mediterranean area.

Annual numbers on passage and wintering fluctuate

markedly, and influxes occur in some years (Géroudet 1957). At Col de Bretolet (western Switzerland), passage very strong in some years, almost absent in others; wintering numbers vary greatly (Winkler 1984). In Rheinland, passage usually sparse; large flocks occasionally occur, in good years of hornbeam *Carpinus* seed crop (Mildenberger 1984). Of 4 years in Schwäbische Alb (south-west Germany), 1978 (1228 birds) and 1979 (1458) regarded as normal years, 1977 (3664) and 1980 (8042) influx years (Schmid and Gatter 1986). In eastern Belgium numbers also vary, e.g. common in 1964–5, contrastingly rare in 1965–6 (Fouarge and Rappe 1966). In Spain, numbers wintering fluctuate markedly, but have increased overall since 1978 (Asensio and Antón 1990). In Israel, varies from extremely rare to quite common (Shirihai in press). In 1969–70, stronger movement than normal reported from Britain (see above), Netherlands, and northern and southern Germany (Tekke 1971). Varying scale of movement probably linked to fluctuation in food availability (Géroudet 1957), but no direct evidence. In southern Sweden, seed crop of *Carpinus* varies greatly between years and perhaps influences numbers wintering; crop very rich in 1976, but failed in 1977 (Axelsson *et al.* 1977); numbers were unusually high in some other areas in 1977–8: at least 10 reports of wintering in Finland, markedly more than usual, and exceptionally numerous in Switzerland, e.g. flocks of up to 100 at Geneva and 600 at Basle, January–February (*Br. Birds* 1978, **71**, 257).

Birds ringed as nestlings show far greater percentage of long-distance recoveries than those ringed as adults (Mountfort 1957). Also more ♀♀ move long distances than ♂♂; in eastern Germany, of 182 recoveries of ♀♀, 42·3% over 50 km; of 381 recoveries of ♂♂, 25·7% over 50 km (Krüger 1979). Most winter visitors to Spain are ♀♀ (Asensio and Antón 1990).

Inconspicuous on passage, and fewer data than for most other Fringillidae. Swedish data (Strömberg 1975) suggest initial movement west to Denmark by shortest sea crossing, though records from Mediterranean show that longer sea crossings not infrequent. Passage includes high mountains; seen up to 2400 m in Alps (Géroudet 1957; Mingozzi *et al.* 1988). Movement apparently protracted; reported from late August to early December, mostly mid- or late September to early November. In central Netherlands, local dispersal August–November (Bijlsma 1979). In Belgium, regular in east, sometimes in fairly large numbers; departure and passage September to mid-November (Lippens and Wille 1972; Desmet 1981). In St Petersburg region, passage mid-September to mid-November (Noskov *et al.* 1981). At Col de Bretolet, passage begins late August or early September in years of high numbers; main passage from last third of September, peaking 3–17 October and ending early November (Winkler 1984). In northern Italy, passage September to mid-November, chiefly October (Bassini 1970), and mostly mid-October to mid-November in Malta (Sultana and Gauci 1982). Arrives Israel from

end of October, mostly from 2nd week of November (Shirihai in press). Recoveries of immigrants into Spain are 13 October to 15 April (Asensio and Antón 1990), and recorded October to March in Libya and Egypt (Bundy 1976; Goodman and Meininger 1989). Same-season fidelity recorded January–April (2 birds) and September–March (1 bird) in Brandenburg (eastern Germany) (Rutschke 1983), and December to early March in Belgium (Desmet 1981).

Spring movement chiefly February–April. Wintering birds remain in Belgium until end February or beginning March (Herroelen 1962), and inconspicuous passage at Col de Bretolet mid-February to April (Winkler 1984). In Lausitz (eastern Germany), numerous in late winter 1965 (chiefly 5–17 February); retraps were up to 22 days, with few retraps of birds ringed after mid-February; earliest birds were mostly ♂♂, and 1st-years were later than older birds (Creutz 1967). In Denmark, peak passage early to mid-April (Møller 1978a). Wintering birds leave western Ukraine end February to March (Strautman 1963). Arrives in St Petersburg region mostly mid-April to mid-May (Mal'chevski and Pukinski 1983). In Kareliya (north-west Russia, recently colonised), arrives from 1st week of April (Khokhlova *et al.* 1983).

In western Siberia, some birds winter as far north as Tomsk and Krasnoyarsk (*c.* 56°30′N) in years of good rowan *Sorbus* crop; tends also to occur near human settlements (Johansen 1944; Gyngazov and Milovidov 1977). Remarkable summer recovery in western Siberia, of bird ringed *c.* 3800 km south-east in Korea in winter, has established long-distance east–west route (McClure 1974). Other birds move south. In Kazakhstan, winters irregularly in north (regular at Zaysan depression), more frequently, but uncommon, in south, e.g. at Tashkent and Alma-Ata; widespread on passage. Autumn movement late August to October; spring movement gradual, March to May. (Korelov *et al.* 1974.) In lower Ural valley, in one autumn, juveniles arrived 4 days earlier than adults, and predominated August–September, whereas adults predominated in October; in both age groups, ♀♀ began migration before ♂♂ and proportion of ♀♀ slightly higher than ♂♂ (Gavrilov *et al.* 1984). In Mongolia, passage recorded September–October and late April to May (Piechocki *et al.* 1982). In Ussuriland (south-east Russia), some local birds remain for winter; autumn passage begins very early, end of July or early August, with birds flying southward 'at great height'; more marked mid-September to end of October; spring passage chiefly in May, when local birds have already started breeding (Panov 1973a). Locally common but irregular winter visitor to Korea (Austin 1948). In China (breeds in extreme north-east), winters in east as far south as Fukien (*c.* 26°N) (Schauensee 1984). In eastern Hopeh (north-east China), usually present November to end of April; ♂♂ predominate in winter, ♀♀ arriving from south in spring (Hemmingsen 1951; Hemmingsen and Guildal 1968).

*C. c. humei* (breeding south-central Asia) makes altitudinal movements when snowfall heavy, usually end November to December; present at lower levels until April or May (Abdusalyamov 1977). Some birds move south; uncommon and local winter visitor in irregular numbers in northern and western Pakistan, south to northern Baluchistan; arrives October–December and leaves in April (Ali and Ripley 1974). In Japan (*japonicus*), birds breeding in Hokkaido mostly depart, to winter in Honshu and further south; those breeding in northern Honshu make altitudinal movements; migrates mid-October to early December in autumn, and chiefly April in spring, when large flocks occur on islands of Sea of Japan (Brazil 1991).                                        DFV

**Food.** Large hard seeds, buds, and shoots of trees and shrubs; invertebrates, especially caterpillars, in breeding season. In spring and summer, forages mainly in woodland trees; in autumn and winter, in hedges and on ground. (Mountfort 1957; Newton 1967a; Krüger 1979; Desmet 1981.) In captivity, 51% of feeding done while perched normally, 38% on ground, 9% clinging to bent twigs or stems, and 2% stretching forward; in the wild, spent equal time on ground and in vegetation over year; recorded hanging upside-down while feeding on garden peas *Pisum* (Mountfort 1957; Newton 1967a). On ground, searches amongst leaf-litter like Blackbird *Turdus merula* for fallen fruits and seeds and occasionally insect larvae and earthworms (Lumbricidae); small worms eaten whole, large ones cut into 3–4 pieces; in some places, often takes cockchafer beetles *Melolontha* on ground as they emerge as adults. In trees, pulls or bites seed or fruit from stalk and eats it on the spot, or if several fruits collected in small bunch moves to branch or fork to pick seeds out one by one; groups can return to same tree, if undisturbed, over several days until stripped of seeds or buds; in late spring and summer frequently in tree-tops, often in groups, feeding on caterpillars of defoliating moths, principally oak-roller *Tortrix viridana* and winter moth *Operophtera brumata*, which are carefully extracted from rolled-up leaves; birds, especially ♂♂ when young in nest, fly 2–3 km to reach infested trees. Seeds extracted from fleshy fruits by turning in bill, peeling off pulp (which is sometimes eaten) against lower mandible, then seeds cracked in bill; never seen to use feet when dealing with fruits. (Mountfort 1957; Kear 1962; Newton 1967a; Krüger 1979.) Massive bill adapted for splitting large hard seeds to get at kernel; 2 striated knobs at base of each mandible, which develop only in maturity, grip seed tightly so considerable pressure can be brought to bear. Suture of seed placed vertically lengthways along bill by combined movements of head and tongue, so 2 halves fall away to side when de-husked and kernel swallowed whole. Laboratory measurements have shown that force in excess of 50 kg can be employed, large muscles encasing skull providing power. (Sims 1955, which see for details; Ziswiler 1965; Newton 1967a.) Occasionally makes sallies after flying insects, notably *Melolontha*, and sometimes hovers rather awkwardly to pluck (e.g.) beechnuts *Fagus* from trees or at garden feeders of compacted peanuts (Mountfort 1957; Krüger 1979). In Ukraine, small flocks attack walnuts *Juglans* before they harden (Orlov 1955). In general, unable to extract conifer seeds from closed cones, but in Switzerland and Austria small flocks recorded feeding in winter and early spring on wet, soft cones of Arolla pine *Pinus cembra*, often after Nutcracker *Nucifraga caryocatactes* had already tackled them, and in competition with Crossbill *Loxia curvirostra*: perched on twig and pulled seeds from open cones or picked them from ground (Bürkli 1972; Löhrl 1987b). In Germany, flock of over 100 fed in pine *Pinus sylvestris*, ripping off cone scales then landing on ground after 5–10 min to eat fallen seeds (Batt 1993). Picked small snails (Pulmonata) from vegetation and swallowed them after crushing shell in bill (Köhler 1990). Will dig up planted seeds with bill, notably beans *Phaseolus*, and tear open pods to get at peas, often sliding down support wires; on ground, hacks open apples *Malus* and pears *Pyrus* while balanced on top of them, and pulls down grass and herb seed-heads. Readily visits winter bird-tables, especially for seeds of sunflower, and often seen with flocks of feeding thrushes (Turdidae) taking seeds in their pellets and faeces; very frequently reported drinking at water (Mountfort 1957). 68% of seeds eaten heavier then 100 mg, 30% 10–100 mg, 1% 1–10 mg, by far largest seeds in diet of Fringillidae of the region (Newton 1967a). In the wild, 3–6 hornbeam *Carpinus* nuts or 1 beechnut de-husked per min; young birds need more time as bill still soft (Krüger 1979). Captive birds needed c. 3–6 s to open various seeds; took longest to de-husk flax *Linum* and cherry *Prunus*; sunflower preferred and de-husked relatively quickly; took 2–3 times longer than other Fringillidae to handle smaller seeds, optimum size c. 4–5 mm (Kear 1962; Ziswiler 1965, which see for details). Feeding on cherries *P. serotina* on ground, took average c. 1 min to consume kernel (Desmet 1981).

Diet in west Palearctic includes the following. Invertebrates: locusts (Orthoptera: Acrididae), bugs (Hemiptera: Pentatomidae, Aphididae), adult and larval Lepidoptera (Nymphalidae, Pieridae, Tortricidae, Noctuidae, Lymantriidae, Saturniidae, Geometridae), Hymenoptera (Tenthredinidae), beetles (Coleoptera: Silphidae, Lucanidae, Scarabaeidae, Curculionidae), spiders (Araneae), earthworms (Lumbricidae), snails (Pulmonata: Helicidae). Plants: seeds, buds, shoots, etc., of yew *Taxus*, false cypress *Chamaecyparis*, juniper *Juniperus*, *Thuja*, larch *Larix*, spruce *Picea*, pine *Pinus*, fir *Abies*, willow *Salix*, aspen, etc. *Populus*, wingnut *Pterocarya*, walnut *Juglans*, birch *Betula*, alder *Alnus*, hazel *Corylus*, hornbeam *Carpinus*, hop-hornbeam *Ostrya*, beech *Fagus*, oak *Quercus*, elm *Ulmus*, nettle-tree *Celtis*, mulberry *Morus*, laurel *Laurus*, *Robinia*, maple, etc. *Acer*, lime *Tilia*, ash *Fraxinus*, olive *Olea*, privet *Ligustrum*, lilac *Syringa*, mistletoe *Viscum*, hemp *Cannabis*, dock *Rumex*, chickweed

*Stellaria*, traveller's-joy *Clematis*, barberry *Berberis*, radish *Raphanus*, Rosaceae (including apple *Malus*, pear *Pyrus*, cherry, etc. *Prunus*, hawthorn *Crataegus*, rowan *Sorbus*, raspberry, etc. *Rubus*), currant *Ribes*, bean *Phaseolus*, pea *Pisum*, broom *Cytisus*, *Laburnum*, holly *Ilex*, spindle *Euonymus*, dogwood *Cornus*, buckthorn *Rhamnus*, alder buckthorn *Frangula*, grape *Vitis*, ivy *Hedera*, spurge laurel *Daphne*, nightshade *Atropa*, plantain *Plantago*, elder *Sambucus*, honeysuckle *Lonicera*, guelder rose *Viburnum*, Compositae (including dandelion *Taraxacum*, burdock *Arctium*, sunflower *Helianthus*), grasses (Gramineae). (Schuster 1930; Mountfort 1957; Newton 1972; Melde and Melde 1976; Khokhlova *et al.* 1983; Sabel 1983; Köhler 1990; Maestri and Voltolini 1990). For further details, see Turček (1961) and Krüger (1979).

Apparently no quantitative study of adult diet in west Palearctic. Prefers some foods throughout region: seeds of hornbeam, beech, elm, maple, and Rosaceae, especially cherry and other *Prunus*; in Mediterranean area, seeds of olive, nettle-tree, and acorns *Quercus* important, while in central and eastern Asia acorns can form considerable part of diet, availability influencing autumn and winter movements; seeds of sunflower, maize *Zea*, and other crops also taken in large quantities (Mountfort 1957; Krüger 1979). See Turček (1961) for preferences in captivity using seeds of 23 plant species. In Europe, winter to early spring, takes seeds and buds of (e.g.) hornbeam, beech, oak, Rosaceae, and parts of fresh yew tips; in spring and summer, buds and shoots of trees plus catkin stigmas can be main food at times, but otherwise invertebrates, principally caterpillars *Tortrix* and *Operophtera*; in late summer and autumn, cherries, rose hips, hawthorn berries, seeds of hornbeam, beech, and maple, and some larvae taken; in winter, various seeds eaten on ground, then later buds, especially of hornbeam, oak, and *Prunus* (Mountfort 1957; Newton 1967*a*; Melde and Melde 1976; Bijlsma 1979; Krüger 1979). In southern Sweden, apparently very dependent on hornbeam nuts for winter food, size of crop possibly affecting winter movements (Axelsson *et al.* 1977, which see for details); in Switzerland and Austria, Arolla pine seeds can be most important part of winter diet at times (Bürkli 1972; Löhrl 1987*b*), and in Bruges (Belgium) considerable numbers overwintered feeding solely on seeds of *P. serotina* which was common there (Desmet 1981). In Kareliya (north-west European Russia), some birds overwinter if crop of cherry *P. avium* good; this seed most important component in diet all year; in autumn-winter was 79% of 185 feeding observations, 14% spruce seeds, 5% birch, 3% raspberry; still most important into April when buds also major element, as are previous year's cherry seeds taken on ground as snow melts. In some areas in winter, apple and *Cotoneaster* form large part of diet with cherry, and in spring, alongside cherry buds and seeds of alder and aspen; by end of May, none of previous year's cherries remaining. In spring and early summer, feeds also on seeds of elm, lilac, and larch on ground,

moving into elm, maple, and ash trees as new seeds start to ripen. (Khokhlova *et al.* 1983.)

Captive birds ate average 258 hornbeam nuts per day, fresh weight 3·4 g and energy content 23·4 kJ, 495 cherry *P. padus* kernels, 5·5 g dry weight, or 500 mixed seeds (Turček 1961, which see for details). In the wild in southern Sweden, minimum daily requirement in winter estimated 75 hornbeam nuts (Axelsson *et al.* 1977); in Bruges, consumed average 400 *P. serotina* kernels per day (Desmet 1981).

Young fed on regurgitated seed pulp from gullet, and adult and larval invertebrates brought in bill (Mountfort 1957; Newton 1972). In Voronezh region (south European Russia), collar-samples contained 168 items, of which 83% (by number) caterpillars (38% Geometridae, 32% Tortricidae, 4% Noctuidae), 10% sawflies (Symphyta), 5% larval and pupal beetles, remainder Diptera and spiders; several nestlings fed per visit and invertebrates always crushed in bill before being given. Some plant material brought to 1-day-old young. (Mal'chevski 1959.) In eastern England, 70% of items brought to one nest were oak-roller moth caterpillars, sometimes 15 at a time; most of remainder of diet seed pulp (Robertson 1954); in England as a whole, caterpillars of winter moth also major component, along with some adult moths, dismembered *Melolontha* beetles, and spiders (Mountfort 1957). These 2 caterpillar species important in Germany and Belgium, and flesh of cherries also given (Melde and Melde 1976; Bijlsma 1979). Nestlings in Kareliya also fed cherry pulp (Khokhlova *et al.* 1983.) BH

**Social pattern and behaviour.** Fairly well known in spite of very shy habits and inconspicuous behaviour, due mainly to major studies by Mountfort (1956, 1957) in France and England, and Krüger (1979) in Germany.

1. In flocks or smaller groups outside breeding season. Aggregations begin to form at end of June (Germany) and become prominent in July, by coalescing of family parties at feeding sites; but no bonds between them, each family group keeping to itself. Birds become very inconspicuous in August, during moult, keeping to woodland; then from early September small groups begin to form again and flocks grow rapidly, persisting until end of March; of 10-300(-1200) birds. Smaller groups (10-62) of unpaired birds persist into April. In northern Europe, where largely migratory, large flocks hardly ever occur in midwinter, but flocks of 50-150 wintering birds regular in southern Europe. (Krüger 1979.) BONDS. Monogamous mating system. Little information on mate fidelity. Close associations between ♂ and ♀ in winter flocks near Vienna (Austria) suggest maintenance of pair-bond in winter (Bauer 1953). Pairs that have not bred successfully said to remain together until following breeding season if both survive (Krüger 1979), so this probably also true of pairs that have bred successfully. Nestlings fed and nest sanitation performed by both sexes. Young out of nest tended by both parents. Independence of young a slow process, young accompanying parents for some time after they are able to feed themselves (see Relations within Family Group, below). Brood-division usual but does not always occur (see Relations within Family Group). (Krüger 1979.) Age of first breeding 1 year (see Breeding). BREEDING DISPERSION. Solitary, or (especially in

favourable habitats) in small groups. Solitary pairs defend definite boundaries of small territory (*c*. 2000 m²) round nest. In breeding colonies, only small area immediately round nest defended, allowing pairs to nest in close proximity; e.g. 5 nests in woodland within radius of 27 m, nearest 2 only 3·7 m apart; 3 nests in orchard 3·7 m and 7·6 m apart; 7 nests in 163 m of hawthorn *Crataegus* hedge (Mountfort 1957); nests in poplar *Populus* plantation 22–120 m apart (Krüger 1979). Hence breeding densities very variable, depending on size of area surveyed and degree of clumping of nests. Selected breeding densities (pairs per km²) Germany: broad-leaved woodland, 5–33; damp oak *Quercus* and hornbeam *Carpinus* woodland, 20–128 (high values due to special protective measures); beech *Fagus* woodland, 0·5–25; mixed woodland, cemeteries, copses, 3–14; small gardens, town parks with trees, 3–10; poplar plantations with undergrowth, 42–81 (Krüger 1979). Primeval forest, Poland: mixed ash *Fraxinus* and alder *Alnus*, 19–23; oak and hornbeam, 35–54; mixed coniferous and deciduous, 2; pine *Pinus*, 2 (Tomiałojć 1990). Atlas mountains (Morocco): broad-leaved forest, 19·6; scrubby evergreen oak woodland, 1·5 (Thévenot 1982). Defended area around nest prevents interference with mating; only occasionally suffices for foraging, most food being obtained well away from nest (Mountfort 1956; Krüger 1979). Colonial breeding more successful than solitary breeding, apparently because communal defence of combined breeding area is effective against predators (Mountfort 1956, 1957). ROOSTING. Nocturnal, in thick cover. May roost singly, but more usually communally, groups of up to 50 birds being reported. Usually in thick cover, such as yew *Taxus*, holly *Ilex*, or *Rhododendron*, birds if necessary flying long distances to suitable sites. (Mountfort 1957.) ♀ usually roosts on nest at night after laying 3rd egg (Krüger 1979).

2. Extremely wary compared with other Fringillidae, especially so when on ground, taking flight at slightest alarm and 'rocketing' silently up into tree-tops (see Flock Behaviour, below). Hence difficult to observe, and tends to be under-recorded. Avoids open areas, and when crossing open ground flies high and fast, plunging down into shelter of first trees encountered. Heavy head, with associated muscles, has led to increase in size of flight muscles and rapid, rather unmanoeuvrable flight, with steering done mainly by wings rather than by short tail; hence especially liable to accidents from collisions with wires, windows, etc. (Mountfort 1957.) In high-speed flight, whistling sound may be made; not certain whether made by primaries or by modified secondaries, and whether or not it has social significance (Pounds 1972). FLOCK BEHAVIOUR. Winter flocks more closely integrated than in other Fringillidae. Reactions very rapid; first to give alarm in mixed-species flocks. Slight crouch of alarmed bird, with feathers sleeked, is instantly copied by conspecifics, and all spring up almost vertically and disappear into tops of trees. Also the last species to rejoin feeding flocks after an alarm. In feeding flocks, maintains greater individual distance between birds than other Fringillidae. (Mountfort 1957.) Pair-formation, involving displays and flight chases, takes place in early spring, while still in flocks (see Heterosexual Behaviour, below). SONG-DISPLAY. Song of ♂ (see 1 in Voice) never frequent, even where bird is abundant; not loud, not used for defining or defence of territory (see, however, below), and not delivered from special posts. Main function apparently related to pair-formation and courtship, to warn off other ♂♂ and to stimulate mate. During period of pair-formation in winter flocks, ♂♂ may withdraw from flock and sing in nearby trees. ♂♂ also sing from trees near nest, but do not defend them against other ♂♂, who may sing in same tree. Singing bird droops wings and jerks wing-tips and tail in unison with each unit of song. (Mountfort 1956, 1957; Krüger 1979.)

Song-period: in Germany, short period in autumn, after moult, then again from January, but mainly from February (Krüger 1979); in England, from February until 1st or 2nd week of June (early record 16 January; late records 20 July, 16 September) (Mountfort 1957). Most protracted song from dawn until 08.00 hrs; often a brief spell in latter half of afternoon (Mountfort 1957). Wintering birds in Italy sing regularly, from end of January until departure about mid-March (Alexander 1917). ♀♀ sing occasionally, but rather weakly (Mountfort 1957). ANTAGONISTIC BEHAVIOUR. Occurs mainly in winter flocks as reaction by foraging bird to another approaching too closely (see Flock Behaviour, above), or as supplanting attack, initiated by either sex, either over food or simply to assert dominance. Actual fighting rare. Dominance hierarchies probably develop in winter flocks, but not studied. At end of winter, when flocks break up (January–March), persistent chasing of other birds by both ♂♂ and ♀♀ frequent, probably associated with pair-formation (see below). (Mountfort 1957; Krüger 1979.) Threat-display involves following elements: plumage sleeked; head lowered (Fig A); at

A

high intensity, body crouched and horizontally elongated, bill slightly uptilted to display black throat-patch; at highest intensity, head or whole head and body swayed from side to side, bill opened wide (showing pink gape), culminating in lunge forward with wing-tips momentarily twisted upwards over tail and bill snapped shut with loud report (Fig B). During winter conflicts,

B

wings may be held out horizontally, partly spread. Juveniles in aggressive encounters may flick one wing forward horizontally towards opponent (apparently does not occur in adults). Submissive birds raise head, with feathers slightly ruffled (Fig C).

C

Conflict between threat and fleeing may be indicated by slight compression of crown feathers and lowering of head, or head may be lowered with crown sleeked and tail suddenly spread so as to produce white flash. (Mountfort 1957.) HETEROSEXUAL BEHAVIOUR. (1) Pair-bonding behaviour. Pair-formation lasts 2–3 months, during which ♀'s initial aggressive reactions are gradually subdued. Begins in flocks in late winter, during fine weather. ♂♂ begin to trespass repeatedly on individual space of other birds in flock, apparently learning which are ♀♀ from their aggressive reaction. Persistent chasing of other individuals in flight, from January onwards, by both ♂♂ and ♀♀, is characteristic of early stages. Courtship-displays begin later, February–March. Pairs often formed by time breeding territories occupied, in which case pair alternates between visiting territory and rejoining flock; or unpaired ♂♂ alternate between seeking out and visiting prospective territory (where they perch silently, sometimes for long periods, but do not sing) and courting ♀♀ in flock. (Mountfort 1956, 1957; Krüger 1979.) Unpaired ♂♂ may break off twigs and take them to forks in branches; significance not clear (Krüger 1979). ♂'s courtship-display includes following postures and actions. (a) Body held upright, head withdrawn, with feathers of neck ruffled (Fig D). (b) Penguin-walk: ♂ in upright posture, with body almost vertical and wings drooped loosely forwards, walks in front of ♀ with waddling gait (Fig E);

D                    E

body may be swivelled from side to side, and white shoulder-patches displayed. (c) ♂ makes deep bow in front of ♀, bill lowered to almost between legs, grey hind-collar ruffled and prominently displayed (Fig F); (d) ♂ makes frontal approach to

F

♀ with wings half extended and wing-tips dragging along ground; swings round in semicircle, displaying white spots on black primaries. This display evidently variable. Above refers to England (Mountfort 1957). In France, ♂♂ recorded making semicircular turn with only one wing dragged in arc in front of ♀ (Mountfort 1957); in Germany, ♀ (on perch) may slightly open and droop both wings, fan tail, and swing round towards ♀ through arc of *c.* 30° (Krüger 1979). During early stages of pair-formation, ♀ may repulse ♂ with forward lunge and bill-snapping. When pair-formation advanced, birds perform bill-touching display: lean forward with necks outstretched and briefly touch closed bill-tips. (Mountfort 1957.) Courtship-flight occurs, apparently rarely: ♂ flies slowly down towards ♀ in 'moth' flight, with

very rapid, shallow wing-beats (Mountfort 1957), or makes short flight with rapidly beating wings to nearby tree (Krüger 1979). (2) Courtship-feeding. Develops during latter part of pair-formation, as ♀ becomes increasingly used to close proximity of ♂. In preliminary early stages, birds face one another, close together, and rapidly bob heads up and down alternately, with closed bills held horizontally, no food being passed. Later, ♀ begs in crouching attitude, with feathers ruffled, head withdrawn, tail slightly raised, and wings quivering; body rocked from side to side or pivoted, and cheeping call given (see 6 in Voice). ♂ feeds ♀ by regurgitation, both birds twisting heads in opposite directions so that food can be passed (otherwise not possible, owing to massive bills). (Mountfort 1957.) In Germany, rather different aspects emphasized by Krüger (1979). Courtship-feeding occurs from early March. Before breeding begins, ♂ may approach ♀ with sleeked plumage and head stretched forward, and pass food to ♀ who may show no special reaction. When breeding begins, ♂ approaches ♀ without sleeking plumage, and ♀ crouches in begging posture with feathers ruffled and wings vibrating. (Krüger 1979.) ♀ in captivity regularly performed striking 'spread-wing dance', first to ♂ and later to human, with body almost upright and pivoted from side to side, plumage ruffled, tail flattened and spread, and wings held out from body to varying extent, from slightly out to fully spread with leading edge perpendicular to body, at same time making braying call. Not clear whether normal display, but not recorded in wild birds. (Kear 1960.) (3) Mating. Flight-chases described here, apparently an essential element in bringing ♀ into receptive condition for mating, though not immediately followed by mating. ♂ makes supplanting attack on ♀, who retreats in flight. Flight-chase follows, both birds with very fast, twisting flight, often hitting foliage. (Mountfort 1957.) Mating takes place only in nesting territory, usually close to nest (Krüger 1979). ♂ makes many attempts to mate before ♀ receptive; repulsed with bill-snapping. Soliciting ♀ adopts posture similar to courtship-feeding, but fluttering wings more stiffly extended, head and tail more elevated, and tail more widely spread; under tail-coverts noticeably erected, exposing cloaca. (Mountfort 1957.) ♂ approaches in display posture; wings slightly drooped, tail may be raised and fanned (posture apparently variable), before flying onto ♀ (Krüger 1979; see also Wallin 1966). (4) Nest-site selection and behaviour at nest. Nest-site chosen by both members of pair, ♂ usually the more active. Both sexes build; ♂ may pass nest-material to ♀, or build it in himself. (Krüger 1979.) ♀ alone incubates; fed during incubation by ♂, either on nest or near it. ♂ on arriving near nest gives soft call (see 6 in Voice); ♀ begs, in first days of incubation period, with loud calls (see 6 in Voice) and vibrating wings. RELATIONS WITHIN FAMILY GROUP. ♀ alone broods young, for periods during day (up to fledging), and at night. Both parents feed young; ♂ does most, especially in first few days, at first passing food to ♀, who feeds them, later feeding them directly. Faecal sacs removed by both parents, either swallowed or carried 30–50 m away and dropped. (Mountfort 1957; Krüger 1979.) Young cared for by both parents after fledging; families either split, each parent caring for part of brood, or cared for jointly. For first few days, until young fully capable of flight, remain within *c.* 100 m of nest; for first 2 days stay in thick cover, then move up into crowns of trees. If food scarce in immediate vicinity of nest, parents then leave nest area with young; if food locally plentiful and cover good, family may remain in nest area for *c.* 30 days. Independence a slow process; young continue to accompany parents after they are able to feed themselves, and families may still be together in October–November. (Krüger 1979.) ANTI-PREDATOR RESPONSES OF YOUNG. No details. Young sometimes very noisy during last few days in nest

(Mountfort 1957). PARENTAL ANTI-PREDATOR STRATEGIES. In breeding colonies, adults join in concerted defence of nesting area, vigorously mobbing raptors, crows (Corvidae), squirrels *Sciurus*, and other potential predators (Mountfort 1957). At nest, if danger threatens, parent calls to young (see 7 in Voice), causing them to crouch motionless in nest (Melde and Melde 1976). Parent disturbed at nest may drop down close to ground and fly along low before rising to higher perch nearby, probably as form of distraction-display (Krüger 1979).

(Figs by D Nurney from drawings in Mountfort 1957.)

DWS

**Voice.** Song and other calls often neither loud nor of striking quality. Calls other than song are mostly mono-syllabic and of 2 main types: sharp 'tzic', 'zick', 'tvit', 'tvut', etc., and more drawn-out 'seee', 'zee', 'zree', 'zieht', etc. Birds perched together in tree-tops may give varied mixture of both types of call (Figs I-II); significance

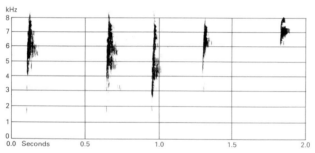

I  P A D Hollom  England  March 1983

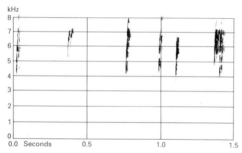

II  P A D Hollom  England  March 1983

uncertain. Distinctions between these calls and their vari-ants need further study, and scheme used below is pro-visional, based mainly on contexts with which calls have been associated. Most characteristic vocalization is Flight-call (see 2, below), by which observer most often made aware of bird's presence. For evidence that voice is lower pitched in Morocco than in northern France, see Chappuis (1969). Aggressive bill-snapping occurs.

CALLS OF ADULTS. (1) Song of ♂. Very variable, both individually and between localities, but no detailed study. Irregular, usually continually altering series of 'tk', 'zick', 'ziet', 'zi', or very high-pitched shrill 'jiis' units, typically ending with more musical 'zie-öh'; often with long pauses between them, thus difficult to recognize as song (Krüger

1979; Bergmann and Helb 1982, which see for sonagram of part of song). Typically in 2 parts: introduction of 2-3 evenly spaced units, followed by one or more longer and slightly more liquid units, with quiet, grating and clicking sounds irregularly interjected; e.g. 'tchee... tchee... tur-wee-wee', 'tsee... tsee-it... tsee-ooo', 'tititit... tsee... tchur-wee' (Mountfort 1957). Based on birds in aviaries, song described as in 3 parts: (a) series of loud, rather high-pitched units, ending in drawn-out 'zieh'; (b) much lower-pitched units with ventriloquial quality; (c) high-pitched 'ziie', as in 1st part (Stadler 1952, which see for musical notation). Subsong, occasionally given by birds when completely relaxed, a series of soft tittering and nasal, squeaky units, 'quee ti-oo eep', etc. (Mountfort 1957); consists of very varied units, more rapidly repeated than full song (Bergmann and Helb 1982, which see for sonagram). ♀♀ sing occasionally but rather weakly (Mountfort 1957). (2) Flight-call. Single, explosive 'tzic' (Mountfort 1957) or 'zieck' (Krüger 1979), repeated at intervals of 2-4 s; in leisurely flight, a 'tzic' accompanying each upward bound (Mountfort 1957). Pairs in flight maintain contact with low-volume 'zieht' (Krüger 1979), this or similar call being used to maintain contact between members of pair when moving from perch to perch (Mountfort 1957). (3) Contact-call between birds in groups similar to Flight-call but much quieter; may be confused with 'tchack' of Blackbird *Turdus merula*, 'tick-ing' of Robin *Erithacus rubecula*, or even 'chicking' of Starling *Sturnus vulgaris*. (4) Alarm-call. Single 'tzic' delivered with great force, bill wide open; given from high vantage point, in favourable conditions audible at *c.* 500 m. A similar call also a sign of aggression. (Mountfort 1957.) Adults accompanying fledged young gave more muffled 'tvut' or double 'tvutut' calls (Fig III), apparently

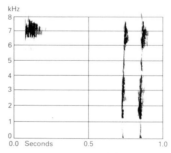

III  J Jackson  England  June 1977

as warning or alarm (J Jackson). (5) Excitement- or Warning-calls. A single 'zick', 'zicke', or 'zick-zick'; or 'zick' repeated in series, rapidly and harshly (Krüger 1979); this last presumably same as rattling 'strreet' given when approaching stuffed Sparrowhawk *Accipiter nisus*, changing to loud, harsh 'zee' when flying away from it (Mountfort 1957). (6) Calls given between members of pair in particular contexts include continuous sharp cheeping 'zit-zit-zit' given by ♀ during courtship-feeding (Hayman

1949); quiet 'tzic-it' used by ♂ when calling ♀ off nest, ♀ sometimes giving soft, rapid titter as food passed to her; wheezy 'zee... zee', given by ♀ as invitation to copulation (Mountfort 1957); soft 'zieht' used as greeting-call by pair-members, stopping when partner arrives (Krüger 1979); very quiet 'quilp' (only just audible at 3 m), given by ♂ to warn ♀ of his approach to nest, also by both sexes as greeting, and to rouse nestlings when they fail to gape. (Mountfort 1957.) Call given by parent announcing arrival at nest also described as short 'xi', which partner on nest answers with disyllabic, harsh 'xickick' (Melde and Melde 1976). (7) Other calls. Parent warning young of danger gives soft 'wuäd', audible for only 3-4 m (Melde and Melde 1976). Distinct vibrant, rather harsh call (Fig IV)

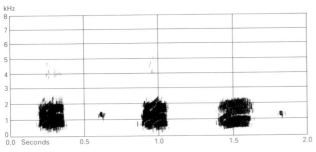

IV  J Jackson  England  June 1977

given by parents immediately before feeding fledged young (J Jackson), and also recorded from birds in flock outside breeding season (P A D Hollom). Caged birds, in both migration periods, give drawn-out 'zieht' calls at night, very like calls of migrating thrushes *Turdus* (Krüger 1979).

CALLS OF YOUNG. Nestlings give whispered, mouse-like cheeping by day 6, changing to clearer 'quist' as they grow (Mountfort 1957); described as very variable 'zrie', 'zirk', 'zieht', 'zierck', or 'ziek', also (presumably the later-developing 'quist') a clear 'zitt zitt' (Krüger 1979). Fledged young give loud 'tchic', 'tchup', and 'tseep' loc-

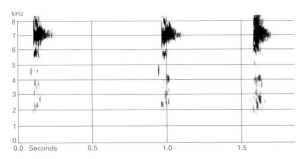

V  J Jackson  England  June 1977

ation calls (Mountfort 1957); rendered 'tziip' (Bergmann and Helb 1982, which see for sonagram); also, when accompanying parents, very high-pitched 'tvit' or 'zit' (Fig V).                                                                    DWS

**Breeding.** SEASON. Britain: eggs laid 1st half of April to end of July, mainly late April to late June (Campbell and Ferguson-Lees 1972; Newton 1972; Harrison 1975); 80% of clutches completed between 1st week of May and 1st week of June ($n=113$) (Mountfort 1957). Central Sweden: 2 weeks later than Britain (Mountfort 1957). Netherlands: beginning of April to mid-August; 71% of clutches in 2nd half of April and in May ($n=62$); in colonies, breeding synchronized so that nearly all clutches completed in May and most eggs hatch May-June (Bijlsma 1979). European USSR: laying begins last days of April or early May (Dementiev and Gladkov 1954); range of 4 weeks in breeding season between north and south USSR (Mountfort 1957). Greece, Italy, and southern Spain: *c.* 3 weeks earlier than Britain, but late clutches at end of May in southern Spain (Mountfort 1957). North Africa (Algeria and Tunisia): eggs laid end of March to end of May (Heim de Balsac and Mayaud 1962). SITE. Prefers old, shrubby trees, especially oak *Quercus* and fruit trees (Campbell 1953; Dementiev and Gladkov 1954; Mountfort 1957; Newton 1972; Bijlsma 1979; Glück 1983; Roberts and Lewis 1988). Of 285 nest-trees in England, 20% apple *Malus*, 16% sycamore *Acer*, 10% pear *Pyrus*, 9% oak *Quercus*, 8% hawthorn *Crataegus*, 6% blackthorn *Prunus*, and 4% birch *Betula* (Mountfort 1957). In Oberlausitz (eastern Germany), 90% of 104 nests in deciduous trees and bushes, 58% in poplar *Populus* and birch, only 7% in oak (Krüger 1979, which see for comparison with other studies). Position of nest varies, though generally well-lit and easy of access; clear preference for horizontal branches, especially in fruit trees; often attached to trunk or on supporting branch close by (Dementiev and Gladkov 1954; Bijlsma 1979; Krüger 1979; Glück 1983), but also in small forks (Mountfort 1957; Campbell and Ferguson-Lees 1972; Bijlsma 1979; Roberts and Lewis 1988) and on leader of small broad-leaved trees (Mountfort 1957; Roberts and Lewis 1988). In south-west Germany, average distance from trunk in fruit trees 1·8 m, and diameter of nest-carrying branch 5·4 cm, $n=28$ (Glück 1983). In orchards, mean height 3-4·5 m, in woodland 6-12 m, $n=218$ (Mountfort 1957); but in USSR, normally 2-2·5 m and as low as 0·6 m (Dementiev and Gladkov 1954). Average height of 93 nests in Netherlands 14 m (7-19) (Bijlsma 1979); in 2 German studies, 4·2 m ($n=28$) in fruit trees (Glück 1983), and 65% of 191 nests less than 5 m above ground (Krüger 1979). Often in cover of ivy *Hedera* (Roberts and Lewis 1988) or honeysuckle *Lonicera* (Verheyen 1957*a*). Nest: bulky foundation of dry twigs, distinct from 2nd layer of thin twiglets and blades of grass in which cup with soft plant matter is shaped (Niethammer 1937; Newton 1972). In England, birch twigs most often used for main structure (Mountfort 1957; Campbell and Ferguson-Lees 1972), strengthened with stouter twigs like oak, interwoven with bark or climbing stems like *Convolvulus*, *Clematis*, or honeysuckle (Mountfort 1957; Campbell and Ferguson-Lees 1972); main fabric of nest twigs is collected from treetops, though exceptionally taken from previous years' nests (Mountfort 1957). Cup shallow, composed of roots and strong grass, twigs, dry moss, and lichen (Verheyen 1946; Mountfort 1957; Campbell and Ferguson-Lees 1972; Krüger 1979). Inner lining of rootlets and grass, rarely hair, and apparently never feathers (Mountfort 1957; Krüger 1979). Average outer diameter 20-22 cm, 9-11 cm for cup itself; inner diameter 7-8 cm; depth of cup 4-5 cm (Campbell 1953; Dementiev and Gladkov 1954; Campbell and Ferguson-Lees 1972; Krüger 1979). Building: ♂ starts nest, ♀ joins ♂ after a few days and completes nest (Wallin 1966; Bijlsma 1979); overall, ♀ has c. 65% of activity, ♂ c. 35%; twigs snipped from trees, other material gathered on ground 4-5(-60) m from nest (Krüger 1979); according to Mountfort (1957), ♂ only accompanies ♀, sometimes carrying material, but never building. Building occurs in bouts of 10-40 min (Newton 1972); takes 3-12 days, replacement nests 2-3 days (Haartman 1969; Krüger 1979). EGGS. See Plate 61. Sub-elliptical, but varying in shape from short to long sub-elliptical, occasionally pyriform, smooth and slightly glossy or non-glossy; light blue or greyish-green, rarely pale buff or grey, sparsely, but usually fairly evenly marked with bold spots and scrawls and paler scribbling of blackish-brown, sometimes concentrated towards broad end. Usually faint underlying spots or streaks of pale ash-grey; considerable variation in colour includes unmarked slate-grey, white or greenish-blue; creamy-white with bold chocolate brown marks, and pure white with a few dark spots at broad end (Witherby et al. 1938; Dementiev and Gladkov 1954; Mountfort 1957; Harrison 1975; Makatsch 1976). Size also varies, and often there is one large, heavily marked egg. Nominate coccothraustes: 24·1 × 17·5 mm (19·8-27·6 × 14·0-19·5), n=723; calculated weight 3·89 g. C. c. nigricans: 23·8 × 17·2 mm (22·0-25·0 × 14·0-18·0), n=18; calculated weight 3·69 g. C. c. buvryi: 23·8 × 17·4 mm (20·1-24·5 × 15·4-17·7), n=20; calculated weight 3·77 g. (Schönwetter 1984.) Clutch: Britain: 4-5 (2-7) (Newton 1972; Harrison 1975). Netherlands: average 4·6; 4·3 (n=15) in solitary pairs, 4·7 (n=26) where population denser; largest April-May (Bijlsma 1979). In Belgium, of 129 completed clutches: 2 eggs, 6%; 3, 19%; 4, 33%; 5, 36%; 6, 5%; average 4·16 (Herroelen 1962). In western Germany, average of 67 completed clutches 4·67 (Mildenberger 1984); see Krüger (1979) for review. Clutch size may increase with latitude: in North Africa, 61% of 13 clutches were of 3 eggs; in England, 76·5% of 149 were of 4-5; in Sweden, 58% of 24 clutches were of 5 (Mountfort 1957). Clutches larger in years of high abundance of defoliating larvae, e.g. Tortrix and Operophtera (Mountfort 1956). Eggs laid daily, in early hours of morning, immediately after nest completion (Krüger 1979). Probably 1 brood; see Mountfort (1957), Krüger (1979), and Mildenberger (1984) for reports of possible 2nd brood; in captivity can produce 3 broods in a year and has attempted 4 (Mountfort 1957). See

Mildenberger (1984) for replacement clutch laid after loss of young. INCUBATION. 11-13 days (9-14); by ♀ only (Witherby et al. 1938; Mountfort 1957; Newton 1972; Harrison 1975). In Oberlausitz, 11·5-13·5 days, n=16 (Krüger 1979). Almost always starts with 3rd egg, young hatching over 2-3 days. Average stint on eggs c. 20 min (4-35) with breaks of 1-10(-77) min (Krüger 1979); started with penultimate egg at all nests watched by Mountfort (1957). YOUNG. Fed and cared for by both parents (Dementiev and Gladkov 1954; Harrison 1975); considerable variation in share of roles; at some nests ♂ feeds more, at others ♀; ♂ can feed young directly immediately after hatching, but usually via ♀ at first (Mountfort 1957; Krüger 1979). ♀ broods right up to fledging (Robertson 1954; Krüger 1979). FLEDGING TO MATURITY. Average fledging period 12·5 days at 19 nests (Mountfort 1957); 12-13 days (Krüger 1979); young independent at c. 30 days old (Bijlsma 1979; Krüger 1979); see also Social Pattern and Behaviour. First breeding at 1 year (Niethammer 1937; Dementiev and Gladkov 1954; C S Roselaar). BREEDING SUCCESS. In Netherlands, 74·5% of 286 eggs produced fledged young; average 4·4 fledged per successful nest (n=48), 3·4 overall (n=62); 77% of 62 layings produced fledged young. Breeding success of solitary pairs c. 66% that of groups (Bijlsma 1979, which see for discussion of varying success), perhaps due to increased susceptibility to predation, especially from Sparrowhawk Accipiter nisus, Jay Garrulus glandarius, Magpie Pica pica, Carrion Crow Corvus corone, and squirrel Sciurus which can be mobbed more effectively by colonies. A. nisus takes both sitting ♀♀ and nestlings which are conspicuous in tall saplings (Mountfort 1956), and newly-fledged birds suffer high predation from A. nisus (Mountfort 1957). In Oberlausitz, 72 breeding attempts produced only 38 fledged young (1·9 per successful attempt, 0·5 overall); of 66 attempts, 11% abandoned during laying, 48% lost eggs, 11% lost young, and 30% produced fledged young; as hatching asynchronous, youngest nestling often dies, and early in season exposed nest easily predated by Red-backed Shrike Lanius collurio, G. glandarius, squirrels, and martens Martes (Krüger 1979). Success highest late in season when full leaf cover gives more protection to nests (Roberts and Lewis 1988).

MK

Plumages. (nominate coccothraustes). ADULT MALE. Short tuft of feathers above nostril and narrow line along lateral base of mandible black, connected with broadly black lore and latter in turn with narrow black eye-ring. Forehead buffish-cinnamon, gradually darkening to rich umber-brown on slightly elongated feathers of side and rear of crown. Side of head from ear-coverts down to lower cheek tawny-cinnamon, darkest below eye. Hindneck, side of neck, and upper mantle light grey, contrasting sharply with hindcrown, slightly less so with side of head. Lower mantle and scapulars burnt umber or dark rufous sepia-brown, merging into cinnamon-brown on back and this in turn to tawny- or rufous-cinnamon on rump and upper tail-coverts. Chin and upper throat black, forming large and contrasting rounded patch, often narrowly bordered white at sides and sometimes along lower border. Lower throat and side of breast backwards to belly and flank light drab-grey with distinct vinous tinge, merging into vinous-white on mid-belly and to pure white on vent and under tail-coverts. Central pair of tail-feathers (t1) cinnamon-brown with ill-defined white tip, some concealed grey and black on base; inner webs of t2–t6 deep black with strongly contrasting white tips, white c. 10–15 mm long on t2, 22·1 (28) 20–26 mm on t6 (ZMA), 20–27 (n = 24) (Herroelen 1962), or 17–28 (Svensson 1992; see also Drost 1940b); outer web of t2–t5 black, merging into rufous-grey on tip, which latter grades into white of tip of inner web; outer web of t6 entirely contrastingly deep black, except sometimes for sharply-defined tiny white triangle on extreme tip. Flight-feathers, greater upper primary coverts, and bastard wing deep black, tips of p7–p8(–p9), fringes along outer webs of primary coverts and bastard wing, and tips of inner webs of secondaries and inner primaries strongly glossed metallic (green-)blue, outer webs of modified secondaries and inner primaries strongly glossed purple or violet (for shape of modified feathers, see Structure, below); middle portion of inner web of outer 4 primaries with strongly contrasting white squarish patch, middle and basal portion of inner 5 primaries largely white (in flight, visible as broad bar on outer wing, both from above and below); basal portions of inner webs of secondaries with broad white border (showing as white bar on inner wing, visible only from below). Tertials cinnamon-brown, merging into burnt umber of scapulars; innermost greater coverts (tertial coverts, gc7–gc9) cinnamon-brown or tawny-buff, partly with dark grey bases; middle 3 greater coverts (gc4–gc6) light buff-grey to silvery-grey with contrasting black inner web; outer 3 greater coverts (gc1–gc3) black with (green-)blue gloss along outer web and tip. Lesser and innermost median upper wing-coverts dull black with narrow grey or cinnamon fringes, central and outer median coverts contrastingly buff-white or off-white, forming crescent over spread wing. Under wing-coverts and axillaries white; longer under primary coverts dark grey, shorter under primary coverts dark grey or dull black with white tips. Influence of bleaching and wear rather limited; in worn plumage (May–July), forehead and mid-crown paler, buff-yellow, less warm cinnamon; lower back and rump sometimes slightly tinged grey; underparts darker vinous-drab, less pale pinkish and grey; tips of outer webs of t1–t5 bleached to off-white (that of t6 still contrastingly black); outer webs of modified secondaries and inner primaries somewhat frayed, gloss more bronzy-purple. ADULT FEMALE. Like adult ♂, but black of line along base of bill, of lore, and of throat-patch slightly duller, less sharply defined, lore more greyish-black; throat-patch marked with narrow pale buff feather-fringes, less broadly rounded at lower end. Forehead and crown olive-brown or buff-brown, less bright tawny-buff on front and less rich umber at side and at rear; lower mantle and scapulars not as dark sepia or burnt umber as adult ♂, more dark cinnamon-brown, rump and upper tail-coverts more olive-yellow, less rufous. Light grey band across hindneck and side of neck slightly tinged buff, continued as ill-defined pale buff band across upper chest (below black of throat); underparts backwards from chest distinctly less vinous than adult ♂, mainly drab-buff with restricted tinge of vinous on flank, mid-belly more extensively pale grey-buff or off-white. Tail as adult ♂, but white on tips of inner webs more restricted, length of patch on t6 17·5 (12) 13–20 mm (ZMA), 12–18 (n = 5) (Herroelen 1962), or 9–21 (Svensson 1992); outer web of t6 fully black or with small grey tip, mainly visible from below. Wing as adult ♂, but outer webs of secondaries broadly fringed light ash-grey (least so on bases of innermost), not glossy black, tips hardly modified; distal half

of outer web of p5–p8 and of feathers of bastard wing and broad fringe along outer web of outer 3 greater upper wing-coverts light ash-grey; some traces of grey on middle portions of inner primaries; white bar on inner webs of flight-feathers and modified inner primaries as in adult ♂, but bar on average slightly narrower. Outer web of longest tertial fringed magenta. In worn plumage (May–July), forehead and rump tinged grey, underparts dull grey-buff, and mid-belly extensively dirty white (but less so than in 1st adult ♀). NESTLING. Down (bluish-)white, long and copious, covering upperparts and upperwing; shorter white down on belly, thigh, and lower flank. Sexing possible once primaries half-grown: outer web of ♂ black, of ♀ grey (Mohr 1974). For development, see Krüger (1979). JUVENILE MALE. Narrow line along lateral base of bill and on upper chin dull black, lore dull grey; eye-ring buff, mixed grey. Remainder of top and side of head yellow-buff, each feather with ill-defined darker buff-brown tip and paler cream base, appearing mottled on cap and slightly streaked on side of head; some yellow mixed in on lower cheek. Hindneck, side of neck and upper mantle mottled buff-brown and pale ash-grey. Lower mantle, scapulars, and back buff-brown, feathers with ill-defined dark grey tips and paler buff bases, appearing mottled or barred; feathering of rump and upper tail-coverts yellow-buff, sometimes with traces of dark grey tips, loose and woolly. Chin and throat sulphur-yellow; chest tawny-buff merging into pale buff or cream-buff on side of breast, flank, upper belly, and side of belly, all feathers with contrasting dark grey bar on tip and off-white base, underparts appearing spotted or barred; mid-belly and vent white with pale buff tinge; under tail-coverts uniform pale tawny-buff. Tail as adult ♂, but outer web of t6 less deep black, fading to grey on tip (especially as seen from below), merging into buff-white on extreme tip (in adult, sharp contrast between black and white on tip); width of t6 (at 5 mm from tip) 7–9 mm (in adult, 8·5–11 mm: Herremans 1990b). Wing as adult ♂ (thus, without grey on outer webs of flight-feathers, bastard wing, and outer greater coverts, or with a trace of grey on greater coverts only), but tip of p8–p9 without gloss (in adult, p9 only); modified flight-feathers on average less broadened on tips, but much variation; white bar on inner webs of primaries on average narrower; upper wing-coverts as in adult ♂, but lesser brown with buff-brown tips, median and middle greater ones tipped pale tawny-buff, inner greater extensively black on bases, outer greater washed ash-grey along outer borders or tips. JUVENILE FEMALE. Like juvenile ♂, but no black along base of bill; lore pale brown-grey; ear-coverts, upper cheek, and (sometimes) chin light drab-grey with rather contrasting yellow stripe over lower cheek, bordered by mottled black malar stripe below (in ♂, warmer yellow-buff of side of head gradually turns into yellow of chin). Tail and upper wing-coverts as in juvenile ♂ (thus, t5 narrower and with outer web less contrastingly black than in adult ♀), flight-feathers as in adult ♀, partly fringed grey (grey more extensive than in adult ♀), but 2–3 outer primaries without gloss on tips (in adult ♀, usually one), secondaries not modified, and inner primaries less so than in adult ♀. FIRST ADULT MALE. Like adult ♂, but juvenile flight-feathers, greater upper primary coverts, and tail retained; gloss and modified tips of flight-feathers as in juvenile ♂; outer web of t6 of juvenile tail distinctly less uniform and less contrasting black on tip, especially as seen from below. Head and body as in adult ♂, but lower back and rump on average more grey- or olive-brown, less buff-brown. Tips of tail-feathers and inner primaries somewhat frayed from about February onwards (in adult, from about June). FIRST ADULT FEMALE. Like adult ♀, but flight-feathers and tail still juvenile, tips of primaries less glossy than in adult, secondaries not modified, and inner primaries less so than in adult ♀; outer web of t6 as in juvenile

♀, less uniform deep black than adult ♂ or ♀. Head and body as in adult ♀, but cap often extensively tinged ash-grey, sometimes scarcely darker grey-brown than side of head and hindneck, especially when worn; back light brown-grey, rump and shorter upper tail-coverts light olive-brown or buff-grey, less tawny than in adult ♀; underparts with no trace of vinous, light brown-grey or buff-grey, extensively dirty white on belly. Ageing characters as given by Herremans (1990*b*) supported by *c.* 140 specimens examined in BMNH, RMNH, and ZMA (C S Roselaar), contra Svensson (1992); pattern on t6 seems best character, as none of birds examined appeared to have replaced outer tail-feathers. Characters given by Drost (1940*b*) and Mayaud (1941) not fully reliable.

**Bare parts.** ADULT, FIRST ADULT. Iris pink-amber, light cinnamon, light red-brown, or deep reddish-chocolate, depending on degree of excitement; grey-white after death. Bill slate-blue in breeding season, paler and bluer at base, more slaty at tip, sides (except at base) and extreme tip steel-black, underside of lower mandible pink or flesh-yellow, ♂ on average brighter and more extensively blue than ♀; in winter, dusky yellowish-horn, greyish-horn, or dull whitish-horn, slightly paler and tinged yellowish or flesh at base, darker horn-brown or brown-grey on extreme tip. Bill turns to bluish and black from late February to early April (sometimes December); base gradually paler and greyer but tip darker in June; basal half light brown, distal half dark grey, and tip black by early August; largely or fully light brown by late August; sandy horn-yellow from September–December. Mouth pale pink with pinkish-blue palate and jet-black throat-patch; tongue bluish-pink. Leg and foot flesh-pink with slight lilac or lavender tinge in breeding season, brightest in ♂, slightly duller in ♀; brown-flesh to light yellowish-horn in winter. NESTLING. At hatching, bare skin reddish-yellow, shading to purplish-pink on belly; bill, leg, and foot yellowish-cream; mouth and tongue deep red, surrounded by livid purple-blue; centre of tongue pink with whitish spurs; gape-flanges and inside tip of lower mandible bright yellow. JUVENILE. Up to age of *c.* 3 weeks, iris neutral violet-grey, thereafter green-grey, grey-brown, hazel, or light red-brown. Bill, leg, and foot as in adult winter, bill tinged pink or yellow-green; mouth pale lilac-pink, tongue red and pink. (Hartert 1903–10, Stresemann 1920; Heinroth and Heinroth 1924–6; Niethammer 1937; Witherby *et al.* 1938; Dementiev and Gladkov 1954; Mountfort 1957; Krüger 1979; RMNH, ZMA.)

**Moults.** ADULT POST-BREEDING. Complete; primaries descendent. In Britain, sample of only 8 in moult examined indicates start with shedding of p1 about early July to mid-August, completing with regrowth of p9–p10 about late September to mid-October; secondaries start 2–3 weeks after p1 (Mountfort 1957; Ginn and Melville 1983). In October–June birds examined, no moult except for ♂ with primary moult score of 38 on 27 October, Spain, and ♂ with some moult on hindneck December, Netherlands; primary moult score of ♂ 5 on 22 July and 42 on 24 September (Netherlands), 48 and 50 on 29 September (Italy) (ZMA). In Germany, moult mainly late July to September (Niethammer 1937); ♀♀ and 1-year-olds in particular sometimes from June (Krüger 1979). In Yugoslavia, 2 had just started 24–30 July, another had not (Stresemann 1920); in Morocco, 4 birds from 16–18 October in moult of head, wing, and tail (Meinertzhagen 1940); in Iran, 2 birds not yet moulting 10–11 July, another just started 25 July (Vaurie 1949*b*). In USSR, nominate *coccothraustes* starts primary moult at end of July, moult advanced mid-August, almost completed late August or early September; *nigricans* moults late July to late September or mid-October

(Dementiev and Gladkov 1954). No pre-breeding moult. POST-JUVENILE. Partial: head, body, lesser, median, and greater upper wing-coverts, usually tertials and rarely t1 or some other tail-feathers. Timing highly variable, depending on fledging date; starts at age of 6–7 weeks (Heinroth and Heinroth 1924–6) or at 10(–13) weeks (Krüger 1979), between late June and early September, completed late August to mid-November (Dementiev and Gladkov 1954; Krüger 1979; RMNH, ZMA). Back replaced first, followed by breast, belly, neck, and head (Mountfort 1957).

**Measurements.** ADULT, FIRST ADULT. Nominate *coccothraustes*. Britain, Netherlands, France, Germany, Italy, and Hungary, all year; skins (RMNH, ZMA: A J van Loon). Bill (S) to skull, bill (N) to distal corner of nostril; exposed culmen on average *c.* 5·1 less than bill (S).

| | | | | | |
|---|---|---|---|---|---|
| WING | ♂ | 103·6 (1·99; 54) 99–110 | ♀ | 101·4 (2·37; 29) 97–105 |
| TAIL | | 57·2 (1·93; 47) 54–61 | | 55·5 (2·42; 25) 51–60 |
| BILL (S) | | 25·3 (0·81; 48) 24·1–27·0 | | 24·4 (0·65; 23) 23·0–25·3 |
| BILL (N) | | 16·5 (0·77; 48) 15·0–17·8 | | 15·7 (0·56; 23) 14·9–16·7 |
| TARSUS | | 21·4 (0·71; 45) 19·6–22·6 | | 20·9 (0·73; 20) 19·6–21·9 |

Sex differences significant, except tarsus. Retained juvenile wing of 1st adult slightly shorter than in older bird; thus, in ♂, adult 104·7 (2·36; 15) 102–110, 1st adult 102·3 (1·77; 16) 100–106 (ZMA).

*C. c. buvryi.* Morocco, Algeria, and Tunisia, January–July; skins (BMNH, RMNH, ZMA: A J van Loon).

| | | | | | |
|---|---|---|---|---|---|
| WING | ♂ | 97·8 (3·06; 16) 94–102 | ♀ | 96·2 (2·26; 14) 92–100 |
| TAIL | | 56·5 (3·10; 15) 54–61 | | 54·4 (1·52; 13) 52–57 |
| BILL (S) | | 24·0 (0·64; 16) 22·7–25·2 | | 22·8 (0·41; 14) 21·9–23·8 |
| BILL (N) | | 15·6 (0·67; 16) 14·6–16·6 | | 15·0 (0·43; 14) 14·3–15·7 |
| TARSUS | | 20·7 (0·76; 15) 19·6–21·9 | | 20·5 (0·41; 13) 19·8–21·0 |

Sex differences significant for tail and bill.

Wing. Nominate *coccothraustes*. (1) Frankfurt am Main (Germany) (Eck 1985). Eastern Germany: all year, (2) adult, (3) juvenile and 1st adult (Creutz 1967); (4) breeding (Eck 1985; see also Bährmann 1976). (5) Corsica and Sardinia (Hartert 1921–2; Bezzel 1957; RMNH, ZMA). (6) Central and southern Iberia (BMNH, RMNH, ZMA). (7) Mongolia (Piechocki and Bolod 1972; Piechocki *et al.* 1982). *C. c. nigricans.* (8) Partly intermediate with nominate *coccothraustes*, southern Yugoslavia and Rumania (Stresemann 1920; ZMA). (9) Caucasus (Dementiev and Gladkov 1954). (10) Northern Iran (Stresemann 1928; Vaurie 1949*b*; Diesselhorst 1962). *C. c. humii.* (11) Central Asia (Dementiev and Gladkov 1954; Paludan 1959).

| | | | | |
|---|---|---|---|---|
| (1) | ♂ | 102·8 (3·07; 30) | ♀ 101·1 (2·84; 35) | — |
| (2) | | 104·8 ( — ; 145) 99–111 | 102·9 ( — ; 81) | 97–108 |
| (3) | | 103·7 ( — ; 37) 97–109 | 102·3 ( — ; 36) | 96–106 |
| (4) | | 104·4 (2·58; 25) — | 102·6 (2·58; 11) | — |
| (5) | | 101·5 (1·41; 6) 99–103 | 98·2 (2·33; 4) | 96–101 |
| (6) | | 101·6 (2·69; 13) 96–105 | 99·3 (2·09; 8) | 96–103 |
| (7) | | 106·0 (1·79; 6) 104–109 | 104·6 (0·98; 6) | 103–106 |
| (8) | | 102·6 (1·99; 7) 100–105 | 101·0 ( — ; 1) | — |
| (9) | | 100·6 ( — ; 37) 96–106 | 99·8 ( — ; 17) | 96–104 |
| (10) | | 103·7 ( — ; 18) 95–110 | 101·7 ( — ; 20) | 97–106 |
| (11) | | 101·1 ( — ; 11) 98–105 | 98·0 ( — ; 4) | 95–102 |

For additional data on various populations, see also Witherby (1928), Johansen (1944), Vaurie (1949*b*, 1959), Jordans (1950), and Mountfort (1957).

**Weights.** ADULT, FIRST ADULT. Nominate *coccothraustes*. Netherlands, Germany, and northern Russia, combined: (1) June–September, (2) November–April (Krohn 1915; Weigold 1926; Danilov *et al.* 1984; Eck 1985; RMNH, ZMA). (3) Belgium (Herroelen 1962). (4) Frankfurt am Main (Germany) (Mountfort

1957). (5) Eastern Germany, mainly February (Creutz 1967, which see also for monthly averages and variation within day). (6) Czechoslovakia, whole year, ages combined (Havlín and Havlínová 1974, which see for monthly variations). (7) Mongolia, June–October (Piechocki and Bolod 1972; Piechocki *et al.* 1982).

| | | | | | | | |
|---|---|---|---|---|---|---|---|
| (1) | ♂ | 56·5 (4·88; 7) | 51·0–64·4 | ♀ | 52·9 ( — ; 3) | 48·5–60·1 |
| (2) | | 57·2 (5·80; 13) | 51·0–69·2 | | 51·9 (7·72; 5) | 46·0–64·9 |
| (3) | | 52·0 ( — ; 8) | 40–66 | | 56·0 ( — ; 3) | 50–61 |
| (4) | | 56·3 ( — ; 30) | 46·3–67·6 | | 54·5 ( — ; 35) | 46·4–63·7 |
| (5) | | 60·2 ( — ; 74) | 48·9–70·8 | | 57·4 ( — ; 59) | 48·8–69·2 |
| (6) | | 58·8 (5·37; 69) | 46·3–72·0 | | 57·0 (4·15; 18) | 50·3–64·9 |
| (7) | | 54·8 (1·83; 6) | 53–57 | | 48·0 (6·30; 7) | 35–54 |

Britain: 54·9 (7) 48·7–59·9 (Jourdain 1935). Exhausted birds, Netherlands: ♂♂ 44·1, 44·5; ♀♀ 36·2, 37·5 (ZMA).

*C. c. nigricans.* Caucasus: ♂ 50·4 (3) 50–51; ♀♀, May, 65, 70 (Dementiev and Gladkov 1954).

*C. c. japonicus.* 50·4 (4·22; 6) 44·5–54·8 (Weigold 1926; Nechaev 1969).

*C. c. humii.* Western Afghanistan, July: ♂ 49 (Paludan 1959). Kazakhstan: ♂ 54·0 (6) 47–61·5, once 83, ♀ 52·5 (5) 50–60 (Korelov *et al.* 1974).

NESTLING, JUVENILE. For growth, see Krüger (1979).

Structure. Wing rather short, broad at base, tip bluntly pointed. 10 primaries: p8 longest, p9 0–2 shorter, p7 0–1, p6 2·5–5, p5 13–18, p4 18–24, p3 23–28, p2 25–31, p1 28–36; p10 reduced, a tiny narrow pin, equal in length or slightly longer than strongly reduced outermost greater upper primary covert under which it is concealed; 63–75 shorter than p8, 11–15 shorter than longest upper primary covert. Outer web of p6–p8 emarginated, inner web of p8 (p7–p9) with notch, often faint. Tips of secondaries and p2–p5 of both sexes modified, shaped like woodman's cleaver: in adult, tips of secondaries square; tip of both webs of secondaries of ♂ and of outer web of p2–p5 of both sexes ending laterally in sharp point; tips of both webs of secondaries of ♀ ending in square hook; tip of inner web of p2–p5 with deeply indented curve; in 1st adult, point on tip of outer web often less long and sharp, especially in ♀, tip of inner web of p2–p5 almost straight or only slightly concave. Tip of longest tertial reaches to tip of p2. Tail short, tip forked; 12 feathers, t1 3–8 shorter than t6. Bill enormous, thick and conical, depth and width at base about equal to length from nostril to tip in population from Britain, central Europe, and northern Italy: depth at base 16·5 (20) 15·8–17·3 in ♂, 15·3 (10) 14·5–15·9 in ♀, width at base 15·7 (20) 15·0–16·5 in ♂, 14·8 (10) 14·2–15·4 in ♀. Culmen and cutting edges slightly decurved, gonys virtually straight; tip of bill blunt, tip of upper mandible often ending in small hook. Inside of bill modified for cracking hard seeds: back of palate (upper mandible) has oblong horny pad and middle of inside of lower mandible

has 2 striated horny bosses; horny sheath covering bill greatly thickened inside mouth (Witherby *et al.* 1938; Mountfort 1957). Nostrils small, rounded; at fledging, obvious cowls or bosses above. Some fine bristles project downward from side of upper mandible at base. Tarsus and toes short, strong. Middle toe with claw 22·3 (10) 21–24; outer toe with claw *c.* 72% of middle with claw, inner *c.* 69%, hind *c.* 74%.

Geographical variation. Slight, largely clinal. Only *buvryi* from North Africa and *humii* from central Asia clearly separable on plumage: *buvryi* mainly because of slightly paler and more diluted colours of head and body and less white on inner web of t6, *humii* because of yellow-ochre mantle and scapulars (less dark fuscous or burnt umber than in other races) and tawny tinge on underparts (less grey or vinous). Many other races described on minor differences in colour, but none separable from nominate *coccothraustes* in fresh plumage; in March, southern populations are already in worn plumage (yellowish crown in ♂ or grey crown in ♀, more rusty-brown mantle and scapulars, greyer back, buffish-olive rump, and browner, less vinous, underparts) when northern birds still quite fresh, this difference in wear explaining much of variation described between various races. Variation in size more pronounced, mainly involving wing length and depth and width of bill at base. Size fairly large in nominate *coccothraustes* from northern and central Europe south to Pyrénées, northern Italy, northern Yugoslavia, northern Rumania, and northern Ukraine; populations become gradually larger and bill even more massive through western Siberia ('*verticalis*' Tugarinov and Buturlin, 1911), to end in *schulpini* from Manchuria and far-eastern Russia; replaced by smaller *japonicus* in Japan and off-lying islands, which has bill on average 1 mm less wide and deep than nominate *coccothraustes* from Europe (see Stejneger 1892, Johansen 1944, and Vaurie 1956c, 1959). Further south in Europe, birds slightly smaller in Iberia, Corsica, Sardinia, southern Italy, southern Balkan countries, Greece, Turkey, Crimea, and Caucasus area to northern Iran; depth of bill at base on average *c.* 15·5 mm in ♂, 14·7 in ♀, width at base *c.* 15·0 in ♂, *c.* 14·5 in ♀. Western birds tend somewhat towards *buvryi*, but nearer nominate *coccothraustes*, and here included in nominate, birds from Balkans eastward named *nigricans*, following traditional view (Johansen 1944; Vaurie 1949b, 1959; Matvejev and Vasić 1973; Stepanyan 1990), though specimens examined suggest split perhaps not warranted, nor is it according to Hartert (1921–2), Stresemann (1928), Mountfort (1957), and Diesselhorst (1962). Isolated *buvryi* and *humii* both still smaller, bill depth at base on average *c.* 14·5 in both sexes of both races, width *c.* 14·0.

For relationships, see Evening Grosbeak *Hesperiphona vespertina*.    CSR

---

## *Hesperiphona vespertina* Evening Grosbeak

PLATES 53 (flight) and 56
[between pages 640 and 641]

DU. Avonddikbek    FR. Gros-bec errant    GE. Abendkernbeisser
RU. Вечерний американский дубонос    SP. Picogordo vespertino    SW. Aftonstenknäck

*Fringilla vespertina* Cooper, 1825. Synonym: *Coccothraustes vespertinus.*

Polytypic. Nominate *vespertina* (Cooper, 1825), north-east USA and southern Canada, west to central Alberta and north-east British Columbia. Extralimital: *brooksi* Grinnell, 1917, western Canada and western USA from north-central British Columbia and south-west Alberta south to about California; *montana* Ridgway, 1873, eastern and central Rockies south to central Arizona and south-central New Mexico; *mexicana* Chapman, 1897, south-east Arizona (USA) south through Sierra Madre Occidental to Oaxaca (Mexico).

**Field characters.** 17·5 cm; wing-span 32·5–34·5 cm. Close in size to Hawfinch *Coccothraustes coccothraustes* but with proportionately rather smaller bill and head. Large, heavy, strong-billed, and short-tailed finch, with form somewhat recalling *C. coccothraustes* but in west Palearctic unique colours of pale bill, dusky and pale yellow head and body, and white-patched black wings and tail. ♂ has distinctive yellow forecrown, with white on flight-feathers restricted to bold patch on inner greater coverts, inner secondaries and tertials; ♀ lacks yellow on head and has greenish body with additional white panel over bases of primaries and bold white tips on central tail-feathers. Call distinctive. Sexes dissimilar; little seasonal variation. Juvenile separable.

ADULT MALE. Moults: July–October (complete), March–May (mainly forehead). Forehead and stripe over eye yellow; crown black; nape, centre of mantle, and fore-underparts brownish-olive, appearing dusky at any distance. Scapulars, rear mantle, back, rump, flanks, and under tail-coverts dull yellow, merging with brownish-olive foreparts. Wings black but inner greater coverts and tertials white, forming bold patch on inner wing. Upper tail-coverts and slightly cleft tail black. Depth of both dusky and yellow tones on body varies, yellow more golden when fresh. Bill cone-shaped, with noticeably broad culmen; pale yellowish-green when breeding, white to pale horn in winter. Legs dull, brown to flesh. ADULT FEMALE. Head and upperparts silver-washed greenish-grey, darkest on crown and palest on rump, where buff or off-white tone creates pale patch. Behind ear-coverts and around shoulder, yellowish areas may show as collar, but buff-grey to pale greenish-grey underbody usually marked only by buff-white chin and upper throat. Wings black, with inner greater coverts increasingly white, outer webs of tertials greyish-white and all but 3 outer primaries narrowly white at bases forming short bar on outer wing. Longest upper tail-coverts black-centred and white-fringed; tail black but with inner webs of feathers broadly tipped white. Bill and legs as ♂. JUVENILE. Resembles ♀ but duller and browner, with olive tinge on head. ♂ and ♀ already show adult wing patterns but white panels less bright. FIRST-WINTER. Resembles adult but distinguished by retained juvenile wing- and tail-feathers; young ♂ also has blackish inner margins to tertials.

Unmistakable. In west Palearctic, only *C. coccothraustes* bears passing resemblance in shape, but differs distinctly in brown-buff body plumage and wholly dark tertials, with white panel on wing brightest towards carpal joint. White-winged Grosbeak *Mycerobas carnipes* of north-east Iran unlikely to stray but is similarly coloured, though plumage pattern very different (see Hollom *et al.* 1988). Flight rather heavy, but swift and strong, with bursts of wing-beats producing bounding undulations; flies high above trees when moving between feeding areas. Gait a hop. Looks compact when perched, recalling *C. coccothraustes*; will feed high in trees. Quiet and undemonstrative.

Song a rather short, uneven, jerky warble. Commonest call a loud, ringing 'cleer', 'cleep', 'clee-ip', or 'chreep', recalling House Sparrow *Passer domesticus* but louder and more strident.

**Habitat.** Breeds in boreal Nearctic coniferous forest, usually tall and mature, but also in mixed forest, including second growth, and at times in willow *Salix* growing beside rivers, or in town gardens and shade trees. In winter, strongly attracted to seeding box elder *Acer negundo*, and to bird-tables with sunflower seeds. Accessible salt is also an attraction. Recent changes have involved major eastward extension of range and more frequent wintering in and around human settlements, including parks and gardens. (Pough 1949; Forbush and May 1955; Godfrey 1979; Johnsgard 1979).

**Distribution.** Breeds in North America, from south-west and north-central British Columbia east to Nova Scotia, south through Rocky Mountains to southern Mexico, and east of the mountains south to Central Minnesota, southern Ontario, northern New York, and Massachusetts. Winters throughout breeding range, sporadically south to lowland areas of southern USA.

Accidental. Britain: ♂, St Kilda (Outer Hebrides), 26 March 1969; ♀, Nethybridge (Highland), 10–25 March 1980 (Dymond *et al.* 1989). Norway: ♂, Østfold, 2–9 May 1973; ♂, Sør-Trøndelag, 17–26 May 1975 (Ree 1976).

**Movements.** Erratic, nomadic, or inconsistently migratory. May arrive and depart regularly for several years, then not appear at all for a year or more. During range expansion to eastern North America, 1920–50, distribution and movements correlated well with widespread planting of box elder *Acer negundo*, on whose fruits it feeds in winter. Since 1950, summer occurrence correlates well with epidemics of spruce budworm *Choristoneura fumiferana*, on whose larvae it feeds. (Bent 1968; A J Erskine.)

Birds start to disperse soon after breeding, appearing in many years in late August or early September from Maine to Minnesota (Bailey 1955; Bull 1974; Janssen 1987), and in southern British Columbia and Washington state (Jewett *et al.* 1953; Campbell *et al.* 1974; Cannings *et al.* 1987). In years with southward irruptions, arrivals extend further as autumn progresses, sometimes reaching mid-latitude states in September, but more usually not before late October (Stewart and Robbins 1958; Sutton 1967; Bohlen 1989). Occasionally flights even reach Mexican Gulf coast states (Lowery 1974; Oberholser 1974), but large numbers seldom go as far as Virginia, and are not regular south of 40°N (Prescott 1991). Often they do not appear in numbers in an area until December or January (e.g. Palmer 1949, Hall 1983). Movements are as likely to be east–west as north–south, as shown by ringing recoveries (see below). In west, short-distance and altitudinal movements are regular (e.g. Grinnell and Miller 1944, Bent 1968, Alcorn 1988).

Most ringing is in winter at sites where artificial food available in cities and towns, and shows rapid turnover of birds. No comprehensive summary of movements shown by ringing available. Examples of spring movements include 80 km east overnight, Ontario to New York (Quilliam 1973), and 1300 km east in 68 and 77 days, Michigan to Maine (Palmer 1949). Recoveries in subsequent years include birds that moved between Saskatchewan and Wisconsin or Michigan (Houston and Street 1959), between Maryland and New York, Massachusetts, or Michigan (Stewart and Robbins 1958), and between New Brunswick and Quebec, Ontario, Massachusetts, Pennsylvania, or Virginia (Christie 1963). Ringing recoveries of birds that had irrupted to Alabama and Arkansas showed movements only to or from north-east, except for one from Minnesota (Imhof 1976; James and Neal 1986). Birds retreat northward chiefly during April, reaching breeding grounds by early or mid-May (Sadler and Myres 1976; Sprague and Weir 1984; Janssen 1987).

Given their nomadic habits and difficulty of confirming nesting, birds probably breed further north than stated in most handbooks (see Erskine and Davidson 1976; Godfrey 1986; Cadman *et al.* 1987). Reports north of mapped or inferred breeding range include birds found in south-east Alaska in winter (Kessel and Gibson 1978), southern Mackenzie District in spring (Scotter *et al.* 1985), and Newfoundland and nearby St-Pierre-et-Miquelon in winter and spring (Etcheberry 1982).    AJE

**Voice.** See Field Characters.

**Plumages.** (nominate *vespertina*). ADULT MALE. Narrow line along base of bill as well as lore and narrow line above and below eye black. Forehead and short supercilium bright yellow or green-yellow, contrasting sharply with velvety-black line along bill-base and with glossy black cap, latter extending over centre and rear of crown. Hindneck and side of head dark fuscous-olive, almost black near corner of mouth, gradually paler dull green-olive on upper mantle, side of neck, chin, throat, and chest, this in turn merging into bright yellow of outer and lower scapulars, lower back, and rump, and through dark yellowish-olive of side of breast and belly into bright yellow or green-yellow of lower flank, vent, and under tail-coverts. Thighs black with yellow mottling. Upper tail-coverts glossy black; tips of under tail-coverts sometimes pale yellow or yellow-white, occasionally with a few black subterminal marks. Tail uniform black. Tertials, inner 3 secondaries (except extreme base), and inner greater upper wing-coverts white, fringed pale yellow when plumage

fresh; remainder of upper wing, including all primaries, primary coverts, bastard wing, and outer secondaries contrastingly deep black, secondaries at border of white ones white on inner web and partly suffused grey or off-white on tips. Under wing-coverts and axillaries bright yellow, shorter coverts with black bases, small coverts along leading edge of wing and primary coverts mainly black. *In fresh plumage* (autumn), yellow of forehead, scapulars, rump, and from belly and flank downwards slightly more golden; if worn (late spring and summer) yellow tinged olive-green, hindneck, mantle and chin to belly duller green-olive or brown-olive with some dull yellow of feather-bases visible, hindneck and side of head duller fuscous-brown, white of inner wing sometimes contaminated by dirt. ADULT FEMALE. Unlike ♀♀ of most other finches, rather boldly patterned, and as this pattern strongly different from that of adult ♂, easily mistaken for different species. Narrow line along base of bill dark grey to dull black, less contrasting than in adult ♂, not extending across chin, but running into bold black malar stripe. Lore and narrow short lines just above and below eye dark grey or dull black, remaining top and side of head medium or dark ash-grey, sometimes with some darker grey feather-centres on forehead and crown. Lower cheek pale ash-grey to off-white, merging into darker grey of side of head but contrasting sharply with black malar stripe. A collar round hindneck and down side of neck bright yellowish-green. Mantle, scapulars, and back medium ash-grey, slightly suffused yellow-green; rump ash-grey with extensive sandy-buff suffusion; upper tail-coverts black, shorter central ones tipped buff, outer and longer ones with bold black spot. Chin and throat white, faintly tinged cream; chest and side of breast down to belly and flank pale cinnamon-buff, slightly tinged grey on chest, slightly cream-yellow elsewhere; mid-belly cream-white, vent and under tail-coverts white with slight cream suffusion. Tail black, tips with large white blotches, terminal 25–30% of inner web of outer 5 pairs all-white. Flight-feathers black, base of inner primaries (p1–p6) white (partly tinged pale yellow when fresh), white extending 8–15 mm beyond longest upper primary coverts, forming contrasting white speculum (thus, unlike adult ♂, both tail and wing with contrasting pattern, not uniform black); tips of outer webs of variable number of flight-feathers with white border; secondaries with broader white border or white drop on tip of outer web and broad white border along inner web and tip, inner secondaries sometimes largely white except for irregular black streak along part of shaft. Inner webs and bases of tertials dull black, outer webs light brown-grey, grading into off-white fringe on tip, centre of longest tertials marbled white to varying extent. Greater upper primary coverts, bastard wing, outer greater and all median and lesser upper wing-coverts black; inner greater coverts and tertial coverts with black base and inner web, largely white or yellow outer web, and varying amount of light brown-grey subterminally (especially on tertial coverts) and white along shaft. Under wing-coverts and axillaries bright lemon-yellow, shorter coverts with black bases (yellow less bright than in adult ♂, black along leading edge of wing less extensive). *In fresh plumage* (autumn), slight yellow tinge to mantle, scapulars, belly, and flanks, and to white of wing; rump tinged warm buff; lower cheek, chin, throat, and under tail-coverts tinged cream; in worn plumage (spring and summer), grey of head and neck darker and duller; rump, belly, and flanks greyer, less buffish, without yellow tinge, underparts sometimes contaminated by soil; when heavily worn, white tips of tail-feathers, primaries, and tertials partly worn off. JUVENILE. Similar to adult ♀, but colours duller, browner, and less saturated; dark malar stripe less distinct, sometimes obsolete; top and side of head tinged olive with dusky line above ear-coverts. In ♂, general body colour yellow-tan; malar stripe distinct; tail black, outer feathers sometimes with indistinct grey-white spot on tip of inner web; upper tail-coverts grey-black with off-white tip or fringe; flight-feathers black, tertials, inner secondaries, and tertial-coverts yellow-white with buff fringe; in ♀, general body colour grey-tan; malar stripe less distinct; inner 6 primaries with yellow-white patch at base and with white fringe along tip of outer web; tertials and inner secondaries mixed black-brown and white or yellow-white; tertial coverts yellow-white, tipped black, lesser and median upper wing-coverts fringed olive; tail with much white on feather-tips, upper tail-coverts grey with white tips. (Dwight 1900; Shaub and Shaub 1953; Bent 1968.) FIRST ADULT MALE. Like adult, but tail, flight-feathers, usually tertials and bastard wing, and greater upper primary coverts still juvenile, primaries and secondaries slightly browner and more worn than those of adult at same time of year; tips of tail-feathers slightly pointed, less truncate than adult; outer tail-feathers and secondaries sometimes with trace of grey-white spot on tip of inner web; primary coverts and primaries slightly paler than neighbouring median and outer greater coverts; *c*. 6 outer primaries and (faintly) primary coverts with narrow yellow or white fringes (in adult, uniform black); inner secondaries dirty white, outer web suffused grey, tip sometimes with black spot; tertials light grey-brown with broad black inner border. FIRST ADULT FEMALE. Like adult ♀, but part of juvenile feathering retained, as in 1st adult ♂; these feathers slightly greyer and more worn than in adult ♀ at same time of year, tips of juvenile tail-feathers more pointed; tertials light brown-grey on outer web, brown-black on inner web; inner secondaries white with much black on inner web and base. See also Yunick (1977a).

**Bare parts.** ADULT, FIRST ADULT. Iris dark brown. Bill pale lime-green, greenish-yellow, olive-yellow, or straw-yellow, tinged greenish-grey or brown at base and tip of upper mandible, at nostril, and at tip of lower mandible; yellow brighter towards cutting edges of both mandibles, but edges proper dull olive or green-grey. In winter, bill horn-colour or bone-colour with light olive or pale yellow at gape, changing to light green or pea-green in late spring and into light bluish-green in June (Bent 1968, which see for details). Leg and foot pink-flesh or dark flesh, more or less strongly tinged grey or brown on tarsus. (RMNH, ZFMK, ZMA.) JUVENILE. In 1-month-old ♀, upper mandible green-grey with greenish tip, lower mandible pinkish, dark grey at gape; in 2-month-old ♂, upper mandible drab or dusky olive, extreme base pale green, foot brown (Bent 1968).

**Moults.** ADULT POST-BREEDING. Complete; primaries descendent. Primarily on breeding grounds, starting late June to August, completed October or (occasionally) up to early November. ADULT PRE-BREEDING. Restricted to limited amount of feathering of head and body (mainly face and throat), March–May, or no moult at all. POST-JUVENILE. Partial: head, body, lesser, median, and greater upper wing-coverts, occasionally tertials. Mainly in breeding area, August–November, staring shortly after fledging when *c*. 1 month old, completed at 3 months; 2nd half of September and October in state of New York (USA). FIRST PRE-BREEDING. Like adult pre-breeding. (Magee 1928, 1930; Grinnell *et al.* 1930; Shaub and Shaub 1953; Bent 1968; Pyle *et al.* 1987; RMNH, ZMA.)

**Measurements.** ADULT, FIRST ADULT. Nominate *vespertina*. Eastern Canada and eastern USA, whole year; skins (BMNH, RMNH, ZFMK, ZMA). Bill (S) to skull, bill (N) to distal corner of nostril; exposed culmen on average 3·3 less than bill (S).

| | | | | |
|---|---|---|---|---|
| WING AD | ♂ 115·0 (2·11; 11) | 111–118 | ♀ 112·2 (0·95; 7) | 110–114 |
| TAIL AD | 64·4 (2·11; 11) | 61–68 | 62·3 (2·36; 7) | 58–67 |
| BILL (S) | 23·2 (1·29; 11) 21·6–24·7 | | 23·2 (1·66; 7) | 21·0–23·2 |
| BILL (N) | 14·7 (1·05; 11) 12·9–16·2 | | 15·2 (0·55; 7) | 14·5–15·9 |
| TARSUS | 21·2 (0·95; 11) 20·3–22·6 | | 22·2 (0·58; 7) | 21·4–22·9 |

Sex differences significant for wing.

**Weights.** Nominate *vespertina*. Pennsylvania (USA): November–December, (1) adult, (2) 1st adult; ages combined, (3) January–March, (4) April, (5) May (Clench and Leberman 1978).

| | | | |
|---|---|---|---|
| (1) | ♂ 62·0 (3·94; 78) 54·8–73·4 | ♀ 59·2 (3·32; 34) 51·5–65·6 | |
| (2) | 59·4 (4·32; 13) 51·6–65·6 | 60·6 (4·27; 25) 54·7–70·6 | |
| (3) | 60·4 (4·14; 445) 46·5–72·1 | 59·2 (4·77; 642) 46·5–72·5 | |
| (4) | 58·6 (4·57; 295) 38·7–86·1 | 57·4 (4·42; 380) 43·2–73·5 | |
| (5) | 65·7 (5·95; 29) 54·9–77·7 | 62·9 (5·49; 32) 52·1–73·2 | |

New Mexico (USA), December: ♂ 53·6, ♀ 52·7 (ZFMK). See also Shaub and Shaub (1953) and (for western race *brooksi*) Grinnell *et al.* (1930).

**Structure.** Wing rather short, broad at base, tip bluntly pointed. 10 primaries: p7–p8 longest or either one 0–1·5 shorter than other; p9 1–3·5 shorter than longest, p6 2–7 shorter, p5 14–19, p4 21–26, p3 25–31, p2 28–35, p1 33–40; p10 strongly reduced; narrow and pointed, 74–80 shorter than longest, 10–17 shorter than longest upper primary covert, 0–3 shorter than reduced outermost upper primary covert. Outer web of p6–p8 emarginated; inner web of p7–p9 with faint notch. Tip of longest tertial reaches tip of p1–p2. Tail rather short, tip slightly forked; (t4–)t5 longest, t1 3–6 shorter, t6 0–3 shorter. Bill short, conical, very heavy; depth at base *c*. 14–17 mm, width at base *c*. 14–15; very similar to bill of Hawfinch *Coccothraustes coccothraustes* (but slightly less deep and wide at base), quite different from bills of Rose-breasted Grosbeak *Pheucticus ludovicianus* and Blue Grosbeak *Guiraca caerulea* (which belong to different family), differing in (e.g.) heavier bill-base, more broadly flattened base of culmen, virtually straight cutting edges, deeper base of upper mandible, and shallower base of lower. Middle toe with claw 22·6 (5) 21–25 mm; outer toe with claw *c*. 70% of middle with claw, inner *c*. 64%, hind *c*. 68%. Remainder of structure as in *C. coccothraustes*, but secondaries normal, without broad truncate tip.

**Geographical variation.** Rather slight, mainly involving size of bill. Bill of nominate *vespertina* from central and eastern Canada and north-east USA rather short and stubby; body of ♀ slightly greyer than in other races. *H. v. montana* from southern Rockies in western USA has longer but deep-based bill, and body of ♀ on average browner; *brooksi* from north-west USA and western Canada perhaps slightly darker than typical *montana*, ♀ less washed yellow, but difference probably due mainly to greater wear and discoloration by soot and grease of skins and better not separated (Phillips *et al.* 1964). *H. v. mexicana* from south-east Arizona and Mexico similar to *montana*, but bill, though long, is less wide and deep at base, and yellow band on forehead of ♂ averages narrower.

Often included in *Mycerobas* (a genus containing 4 species of black-and-yellow grosbeaks from central Asia), or combined with *Mycerobas* to form a subgenus of an enlarged *Coccothraustes* (Paynter 1968), or placed in *Coccothraustes* proper (excluding 4 species of *Mycerobas*, but including *C. coccothraustes* and 2 species of masked grosbeaks *Eophona* of eastern Asia) (Mayr and Short 1970). ♀ in particular shares a number of characters with *C. coccothraustes* and with Yellow-billed Grosbeak *E. migratoria*, but ♂ superficially more like *Mycerobas*. In view of uncertain position, here retained in *Hesperiphona*, together with Hooded Grosbeak *H. abeillei* from Mexico.

CSR

# REFERENCES

ABBOTT, W M (1931) *Irish Nat. J.* 3, 191-2. ABDULALI, H (1947) *J. Bombay nat. Hist. Soc.* 46, 704-8. ABDUSALYAMOV, I A (1973) *Fauna Tadzhikskoy SSR* 19 Ptitsy 2; (1977) 3. Dushanbe. ÅBRO, A (1964) *Sterna* 6, 81-5. ABS, M (1961) *Falke* 8, 370-1; (1964) *Vogelwarte* 22, 173-6; (1966) *Ostrich* suppl. 6, 41-9. ABSHAGEN, K (1963) *Beitr. Vogelkde.* 8, 325-38. ACHARYA, H G (1953) *J. Bombay nat. Hist. Soc.* 50 (1951), 169-70. ACKERMANN, A (1967) *J. Orn.* 108, 430-73. ADAMS, A L (1864) *Ibis* (1) 6, 1-36. ADAMS, R (1989) *Devon Birds* 42, 40-43. ADAMS, R G (1948) *Br. Birds* 41, 210-11. ADAMYAN, M S (1965) *Zool. Zh.* 44, 569-77. ADKISSON, C S (1977a) *Avic. Mag.* 83, 195-8; (1977b) *Wilson Bull.* 89, 380-95; (1981) *Condor* 83, 277-88. ADOLPH, P A (1943) *Br. Birds* 37, 134. ADRET-HAUSBERGER, M (1983) *Z. Tierpsychol.* 62, 55-71; (1984) *Ibis* 126, 372-8; (1989) *Bioacoustics* 2, 137-62. ADRET-HAUSBERGER, M, and GÜTTINGER, H R (1984) *Z. Tierpsychol.* 66, 309-27. ADRET-HAUSBERGER, M, and JENKINS, P F (1988) *Behaviour* 107, 138-56. AERTS, M A P A and SPAANS, A L (1987) *Limosa* 60, 169-74. AHARONI, I (1931) *Beitr. Fortpfl. Vögel* 7, 161-6, 222-6; (1942) *Bull. zool. Soc. Egypt* 4, 13-19. AICHHORN, A (1966) *J. Orn.* 107, 398-9; (1969) *Verh. dt. zool. Ges.* 32, 690-706; (1970) *Ber. Nat.-Med. Ver. Innsbruch* 58, 347-52; (1989) *Egretta* 32, 58-71. AKHMEDOV, K P (1957) *Uchen'iye Zap. Stalinabad Zh. pedagog. Inst.* 1, 101-13. ALATALO, R (1975) *Lintumies* 10, 1-7. ALBERNY, J-C, TANGUY LE GAC, J, and VENANT, H (1965) *Oiseaux de France* 44, 18-25. ÅLBU, T (1983) *Fauna norv. (C) Cinclus* 6, 53-6. ALCORN, J R (1988) *The birds of Nevada.* Fallon. AL-DABBAGH, K Y and JIAD, J H (1988) *Int. Stud. Sparrows* 15, 22-43. ALDRICH, J W (1940) *Ohio J. Sci.* 40, 1-8; (1984) *Orn. Monogr. AOU* 35. ALEKSEEVA, N S (1986) *Ornitologiya* 21, 145. ALERSTAM, T (1988) *Anser* 27, 181-218; (1990) *Bird migration.* Cambridge. ALERSTAM, T and ULFSTRAND, S (1972) *Ornis scand.* 3, 99-139. ALEX, U (1985) *Zool. Abh. Staatl. Mus. Tierkde. Dresden* 41, 200. ALEXANDER, B (1898a) *Ibis* (7) 4, 74-118; (1898b) *Ibis* (7) 4, 277-85. ALEXANDER, C J (1917) *Br. Birds* 11, 98-102. ALEXANDER, W B (1933) *J. Anim. Ecol.* 2, 24-35. ALEXANDER, W B and FITTER, R S R (1955) *Br. Birds* 48, 1-14. ALI, S (1963) *J. Bombay nat. Hist. Soc.* 60, 318-21; (1968) *The book of Indian birds.* Bombay. ALI, M H, RAO, B H K, RAO, M A, and RAO, P S (1982) *J. Bombay nat. Hist. Soc.* 79, 201-4. ALI, S and RIPLEY, S D (1972) *Handbook of the birds of India and Pakistan* 5; (1973a) 8; (1973b) 9; (1974) 10. Bombay. ÅLIND, P (1991) *Calidris* 20, 94-8. AL-JOBORAE, F F (1979) D Phil Thesis. Oxford. ALLARD, H A (1939) *Science* 90, 370-1. ALLAVENA, S (1970) *Riv. ital. Orn.* 40, 460-1. ALLIN, E K (1968) *Br. Birds* 61, 541-5. ALLISON, G W (1975) *Glos. nat. Soc. J.* 26 (7), 84. ALLOUSE, B E (1953) *Iraq nat. Hist. Mus. Publ.* 3. ALLSOPP, K (1976) *Br. Birds* 69, 532-4. ALLSOPP, K and NIGHTINGALE, B (1991) *Br. Birds* 84, 137-45. ALMOND, W E (1946) *Br. Birds* 39, 315. ALONSO, J A, MUÑOZ-PULIDO, R, BAUTISTA, L M, and ALONSO, J C (1991) *Bird Study* 38, 45-51. ALONSO, J C (1984a) *J. Orn.* 125, 209-23; (1984b) *J. Orn.* 125, 339-40; (1984c) *Ardeola* 30, 3-21; (1985a) *Ardeola* 32, 31-8; (1985b) *Ardeola* 32, 405-8; (1985c) *J. Orn.* 126, 195-205; (1986a) *Intl. studies on sparrows* 13, 35-43; (1986b) *Ekol. Polska* 34, 63-73. ALSOP, F J (1973) *Wilson Bull.* 85, 484-5. ALSTRÖM, P and COLSTON, P (1991) *A field guide to the rare birds of Britain and Europe.* London. ALTEVOGT, R, and DAVIS, T A (1980) *J. Bombay nat. Hist. Soc.* 76, 283-90. ALTNER, H (1957) *Orn. Mitt.* 9, 115. ALTNER, H and REGER, K (1959) *Anz. orn. Ges. Bayern* 5, 224-34. ALTRICHTER, K (1974) *Anz. orn. Ges. Bayern* 13, 231-9. ALVAREZ, F (1975) *Doñana Acta Vert.* 1 (2), 67-75. ALVAREZ, F and AGUILERA, E (1988) *Ardeola* 35, 269-75. ALVAREZ, F and ARIAS DE REYNA, L (1975) *Doñana Acta Vert.* 1 (2), 77-95. AMADON, D (1943) *Wilson Bull.* 55, 164-77; (1944) *Auk* 61, 136-7; (1967) *Linnaean News-Letter* 20, 2-3. AMANOVA, M A (1977) *Ekologiya* 1, 99-101. AMAT, J A and OBESO, J R (1989) *Ardeola* 36, 219-24. AMBEDKAR, V C (1972) *J. Bombay nat. Hist. Soc.* 69, 268-82. AMERICAN ORNITHOLOGISTS' UNION (1957) *Checklist of North American birds,* 5th ed. Baltimore; (1983) *Checklist of North American birds,* 6th edn. Lawrence. AMLANER, C J and BALL, N J (1983) *Behaviour* 87, 85-119. AMMERSBACH, R (1960) *Gef. Welt* 84, 81-85. ANANIN, A A and ANANINA, T L (1983) In Kuchin, A P (ed.) *Ptitsy Sibiri,* 163-4. Gorno-Altaysk. ANANIN, A A and FEDOROV, A V (1988) In Korneeva, T M (ed.) *Fauna Barguzinskogo Zapovednika,* 8-33. Moscow. ANDELL, P, EBENMAN, B, EKBERG, B, LARSSON, P-G, NILSSON, L, PERSSON, O, and ÖHRSTRÖM, P (1983) *Anser,* Suppl. 15. ANDERSEN, H H (1989) *Br. Birds* 82, 380-1. ANDERSEN-HARILD, P, BLUME, C A, KRAMSHJ, E, and SCHELDE, O (1966) *Dansk orn. Foren. Tidsskr.* 60, 1-13. ANDERSON, A (1961) *Scott. Nat.* 70, 60-74. ANDERSON, B W (1971) *Condor* 73, 342-7. ANDERSON, B W and DAUGHERTY, R J (1974) *Wilson Bull.* 86, 1-11. ANDERSON, T R (1977) *Condor* 79, 205-8; (1978) *Occ. Pap. Mus. nat. Hist. Kansas* (70), 1-58; (1980) *Proc. int. orn. Congr.* 17, 1162-70; (1984) *Ekol. Polska* 32, 693-707; (1991) In Pinowski, J and Summers-Smith, J D (eds) *Granivorous birds in the agricultural landscape,* 87-93. Warsaw. ANDREEV, A V (1977) *Zool. Zh.* 56, 1578-81; (1982a) *Ornitologiya* 17, 72-82; (1982b) In Gavrilov, V M and Potapov, R L (eds) *Ornithological studies in the USSR* 2, 364-76. Moscow. ANDRÉN, H (1985) *Vår Fågelvärld* 44, 261-8. ANDREW, D G (1969) *Br. Birds* 62, 334-6. ANDREW, R J (1956a) *Behaviour* 10, 179-204; (1956b) *Br. J. anim. Behav.* 4, 125-32; (1956c) *Ibis* 98, 502-5; (1956d) *Br. Birds* 49, 107-11; (1957a) *Behaviour* 10, 255-308; (1957b) *Ibis* 99, 27-42; (1961) *Ibis* 103a, 315-48. ANDRIESCU, C, and ANDRIESCU, I (1972) *Muzeul de Ştiinţele Naturii Dorohoi, Botoşani, Studii şi comunicări 1972,* 205-10. ANDRIESCU, C and CORDUNEANU, V (1972) *Muzeul de Ştiinţele Naturii Dorohoi, Botoşani, Studii şi comunicări 1972,* 199-204. ANGELL-JACOBSEN, B (1980) *Ornis scand.* 11, 146-54. ANGWIN, E (1977) *Bull. E. Afr. nat. Hist. Soc.* Nov-Dec, 131. ANIKIN, V I (1963) *Ornitologiya* 6, 463. ANON (1976) *BTO News* 79, 3; (1985) *Reader's Digest complete book of New Zealand birds.* Sydney; (1986) *Israel Land Nat.* 12, 37; (1987) *Strait of Gibraltar Bird Observatory Spec. Rept.* 1; (1990) *Ardeola* 37 (2), 325-52. ANTAL, L, FERNBACH, J, MIKUSKA, J, PELLE, I, and SZLIVKA, L (1971) *Larus* 23, 73-127. ANTIKAINEN, E (1978) *Savonia* 2, 1-45; (1981) *Ornis fenn.* 58, 72-7. ANTIKAINEN, E, SKARÉN, U, TOIVANEN, J, and UKKONEN, M (1980) *Ornis fenn.* 57, 124-31. APLIN, O V (1911) *Zoologist* (4) 15, 112-13. ARAUJO, J (1975)

*Ardeola* 21 (Espec.), 469–85. ARCHER, G and GODMAN, F M (1961) *Birds of British Somaliland and the Gulf of Aden* 4. Edinburgh. ARDAMATSKAYA, T B (1968) In Ivanov, A I (ed.) *Migratsii Zhivotnykh* 5, 146–52. St Petersburg. ARENDT, E and SCHWEIGER, H (1982) *Publ. Wiss. Film. Sekt. Biol.* 15/34, 3–10. ARFF, E (1962) *Falke* 9, 390. ARHEIMER, O and ENEMAR, A (1974) *Fauna och Flora* 69, 153–64. ARIAS DE REYNA, L M, RECUERDA, P, CORVILLO, M, and CRUZ, A (1984) *Doñana Acta Vert.* 11 (1), 79–92. ARMANI, G C (1985) *Guide des passereaux granivores: Embérizinés.* Paris. ARMITAGE, J (1927) *Br. Birds* 21, 117–19; (1932) *Br. Birds* 26, 206–7; (1933) *Br. Birds* 27, 153–7; (1937) *Br. Birds* 31, 98–100. ARMSTRONG, E A (1954) *Ibis* 96, 1–30. ARNHEM, R and VAN LOMMEL, J (1964) *Gerfaut* 54, 458–65. ARNOLD, E L and ELLIS, J C S (1957) *Br. Birds* 50, 347. ARNOLD, M A (1955) *Br. Birds* 48, 91. ARNOLD, P (1962) *Birds of Israel.* Haifa. ARRIGONI DEGLI ODDI, E (1929) *Ornitologia italiana.* Milan; (1931) *Riv. ital. Orn.* (2) 1, 100–4. ARROYO, B and TELLERÍA, J L (1984) *Ardeola* 30, 23–31. ARTHUR, R W (1963) *Br. Birds* 56, 49–51. ARVIDSSON, B L, BOSTRÖM, U, DAHLÉN, B, DE JONG, A, KOLMODIN, U, and NILSSON, S G (1992) *Ornis svecica* 2, 67–76. ASBIRK, S and FRANZMANN, N-E (1978) In Green, G H and Greenwood, J J D (Eds) *Joint biological expedition to north east Greenland 1974*, 132–42. Dundee; (1979) *Dansk orn. Foren. Tidsskr.* 73, 95–102. ASCHOFF, J and HOLST, D VON (1960) *Proc. int. orn. Congr.* 12, 55–70. ASENSIO, B (1984) *Ardeola* 31, 128–34; (1985a) *Ardeola* 32, 173–8; (1985b) *Ardeola* 32, 179–86; (1986a) *Ardeola* 33, 176–83; (1986b) *Doñana Acta Vert.* 13, 103–10; (1986c) *Suppl. Ric. Biol. Selvaggina* 10, 375–6. ASENSIO, B and ANTÓN, C (1990) *Ardeola* 37, 29–35. ASENSIO, B and CARRASCAL, L M, (1990) *Folia Zool.* 39, 125–30. ASH, J (1949) *Br. Birds* 42, 289. ASH, J S (1964) *Br. Birds* 57, 221–41; (1969) *Ibis* 111, 1–10; (1980) *Proc. Pan-Afr. orn. Congr.* 4, 199–208; (1988) *Sandgrouse* 10, 85–90. ASH, J S and NIKOLAUS, G (1991) *Bull. Br. Orn. Club* 111, 237–9. ASHTON-JOHNSON, J F R (1961) *Ool. Rec.* 35, 49–55. ÅSTRÖM, G (1976) *Zoon Suppl.* 2. ATKINSON, C T, and RALPH, C J (1980) *Auk* 97, 245–52. ATTLEE, H G (1949) *Br. Birds* 42, 85. AUBIN, T (1987) *Behaviour* 100, 123–33. AUBRY, J (1970) *Ann. Zool.-Écol. animale* 2, 509–22. AUEZOVA, O N (1982) In Gvozdev, E V (ed.) *Zhivotnyi mir Kazakhstana i problemy ego okhrany*, 9–11. Alma-Ata. AUSTIN, O L (1948) *Bull. Mus. comp. Zool.* 101 (1). AVERY, G R (1991) *Glos. Nat. Soc. J.* 42, 2–5. AXELL, H (1989) *Acrocephalus* 10, 36–7. AXELSSON, P, KÄLLANDER, H, and NILSON, S (1977) *Anser* 16, 241–6.

BABENKO, V G (1984a) *Ornitologiya* 19, 171–2; (1984b) *Ornitologiya* 19, 172 BACCETTI, N, FRUGIS, S, MONGINI, E, and SPINA, F (1981) *Riv. ital. Orn.* 51, 191–240. BACCHUS, J G (1941) *Br. Birds* 35, 17; (1943) *Br. Birds* 37, 38. BACHKIROFF, Y (1953) *Le moineau steppique au Maroc.* Rabat. BACMEISTER, W and KLEINSCHMIDT, O (1920) *J. Orn.* 68, 1–32. BAEGE, L (1968) *Beitr. Vogelkde.* 14, 81–3. BAER, W (1910) *Orn. Monatsschr.* 35, 401–8. BAEYENS, G (1979) *Ardea* 67, 28–41; (1981a) *Ardea* 69, 69–82; (1981b) *Ardea* 69, 125–39; (1981c) *Ardea* 69, 145–66. BAGG, A M (1943) *Auk* 60, 445. BÄHRMANN, U (1937) *Mitt. Verein: sächs. Orn.* 5, 115–18; (1942) *Beitr. Fortpfl. Vögel* 18, 203–4; (1950) *Syllegomena biologica 1950*, 41–9; (1958) *Vogelwelt* 79, 129–35; (1960a) *Abh. Ber. Mus. Tierkde. Dresden* 25, 71–9; (1960b) *Anz. orn. Ges. Bayern* 5, 510–13; (1960c) *Anz. orn. Ges. Bayern* 5, 573–7; (1963) *Zool.*

*Abh. Staatl. Mus. Tierkde. Dresden* 26, 187–218; (1964) *Zool. Abh. Staatl. Mus. Tierkde. Dresden* 27, 1–9; (1966) *Zool. Abh. Staatl. Mus. Tierkde. Dresden* 28, 221–34; (1967) *Beitr. Vogelkde.* 12, 363–6; (1968a) *Die Elster.* Wittenberg Lutherstadt; (1968b) *Zool. Abh. Staatl. Mus. Tierkde. Dresden* 29, 177–90; (1968c) *Beitr. Vogelkde.* 14, 8–28; (1970a) *Beitr. Vogelkde.* 15, 434–6; (1970b) *Beitr. Vogelkde.* 15, 454–5; (1971a) *Beitr. Vogelkde.* 17, 180–1; (1971b) *Beitr. Vogelkde.* 17, 413–4; (1972) *Beitr. Vogelkde.* 18, 89–122; (1973) *Beitr. Vogelkde.* 19, 153–69; (1976) *Zool. Abh. Staatl. Mus. Tierkde. Dresden* 34, 1–37; (1978a) *Zool. Abh. Staatl. Mus. Tierkde. Dresden* 34, 199–228; (1978b) *Zool. Abh. Staatl. Mus. Tierkde. Dresden* 35, 223–52. BÄHRMANN, U and ECK, S (1975) *Zool. Abh. Staatl. Mus. Tierkde. Dresden* 33, 237–43. BAILEY, A M and NIEDRACH, R J (1967) *Pictorial checklist of Colorado birds.* Denver. BAILEY, A M, NIEDRACH, R J, and BAILY, A L (1953) *Publ. Denver Mus. Nat. Hist.* 9. BAILEY, W (1955) *Birds in Massachusetts: when and where to find them.* South Lancaster. BAIRD, J C (1958) *Bird-Banding* 29, 224–8; (1967) *Bird-Banding* 38, 236–7. BAIRLEIN, F (1979) *Vogelwarte* 30, 1–6; (1988) *Vogelwarte* 34, 237–48. BAKER, A J (1980) *Evolution* 34, 638–53. BAKER, A J, DENNISON, M D, LYNCH, A, and LE GRAND, G (1990b) *Evolution* 44, 981–99. BAKER, A J, PECK, M K, and GOLDSMITH, M A (1990a) *Condor* 92, 76–88. BAKER, C E (1927) *Br. Birds* 20, 200. BAKER, E C S (1926) *The fauna of British India: Birds* 3. London; (1932) *The nidification of birds of the Indian Empire.* 1. London; (1934) *The nidification of birds of the Indian Empire.* 3. London. BAKER, H R and INGLIS, C M (1930) *The birds of southern India.* Madras. BAKER, M C (1975) *Evolution* 29, 226–41. BALANÇA, G (1984a) *Gibier Faune sauvage* 2, 45–78; (1984b) *Gibier Faune sauvage* 3, 37–61; (1984c) *Gibier Faune sauvage* 4, 5–27. BALÁT, F (1963) *Larus* 15, 141–4; (1971) *Zool. Listy* 20, 265–80; (1974) *Zool. Listy* 23, 123–35; (1976) *Zool. Listy* 25, 39–49. BALDA, R P (1980) *Z. Tierpsychol.* 52, 331–46. BALDWIN, P J and MEADOWS, B S (1988) *Birds of Madinat Yanbu Al-Sinaiyah and its hinterlands.* Riyadh. BALDWIN, S P and KENDEIGH, S C (1938) *Auk* 55, 416–67. BALDWIN, S P, OBERHOLSER, H C, and WORLEY, L G (1931) *Sci. Publ. Cleveland Mus. Nat. Hist.* 2, 1–165. BALFOUR, D (1976) *Sterna* 15, 169–73. BALFOUR, E (1968) *Scott. Birds* 5, 89–104. BALPH, M H (1975) *Bird-Banding* 46, 126–30. BALTVILKS, J (1970) *Mat. 7. Pribalt. orn. Konf.* 2, 23–8. BANGJORD, G (1986) *Vår Fuglefauna* 9, 251. BANKIER, A M (1984) *Br. Birds* 77, 121. BANKS, K W, CLARK, H, MACKAY, I R K, MACKAY, S G, and SELLERS, R M (1989) *Ring. Migr.* 10, 141–57; (1991a) *Bird Study* 38, 10–19; (1991b) *Scott. Birds* 16, 57–65. BANKS, R C (1959) *Condor* 61, 96–109; (1964) *Univ. Calif. Publ. Zööl.* 70, 1–123. BANNERMAN, D A (1911) *Ibis* (9) 5, 401–2; (1912) *Ibis* (9) 6, 557–627; (1919) *Ibis* (11) 1, 84–131; (1936) *The birds of tropical West Africa* 4; (1948) *The birds of tropical West Africa* 6; (1949) *The birds of tropical West Africa* 7. London; (1953a) *The birds of the British Isles* 1; Edinburgh; (1953b) *The birds of west and equatorial Africa* 2. Edinburgh; (1956) *The birds of the British Isles* 5. Edinburgh; (1963) *Birds of the Atlantic islands* 1. Edinburgh. BANNERMAN, D A and BANNERMAN, W M (1958) *Birds of Cyprus.* London; (1965) *Birds of the Atlantic islands* 2; (1966) 3; (1968) 4. Edinburgh; (1971) *Handbook of the birds of Cyprus and migrants of the Middle East.* Edinburgh; (1983) *The birds of the Balearics.* London. BANNERMAN, D and PRIESTLEY, J (1952) *Ibis* 94, 654–82.

BANZHAF, E (1937a) *Vogelzug*, 8, *114–8*. BANZHAF, W (1937b) *Ver. orn. Ges. Bayern* 21, 123–36. BAPTISTA, L F (1977) *Condor* 79, 356–70; (1990) *Vogelwarte* 35, 249–56. BARANCHEEV, L M (1963) *Ornitologiya* 6, 173–6. BARANOV, L S and MARGOLIN, E A (1983) *Ornitologiya* 18, 186. BARBA, E and LÓPEZ, J A (1990) *Mediterránea Ser. Biol.* 12, 79–88. BÁRDHARSON, H R (1986) *Birds of Iceland*. Reykjavík. BARDIN, A V (1990) *Trudy Zool. Inst. Akad. Nauk SSSR* 210, 18–34. BARIŞ, S, AKÇAKAYA, R, and BILGIN, C (1984) *Birds of Turkey* 3. BARLOW, J C (1973) *Ornith. Monogr. AOU* 14, 10–23; (1980) *Proc. int. orn. Congr.* 17, 1143–9. BARLOW, J C and POWER, D M (1970) *Canad. J. Zool.* 48, 673–80. BARNARD, C (1979) *New Scientist* 83, 818–20. BARNARD, C J (1980a) *Anim. Behav.* 28, 295–309; (1980b) *Anim. Behav.* 28, 503–11; (1980c) *Behaviour* 74, 114–27. BARNARD, C J and SIBLY, R M (1981) *Anim. Behav.* 29, 543–50. BARNES, J A G (1941) *Br. Birds* 35, 17. BARRAUD, E M (1956) *Br. Birds* 49, 289–97. BARREAU, D, BERGIER, P, and LESNE, L (1987) *Oiseau* 57, 307–67. BARRETT, J H (1947) *Ibis* 89, 439–50. BARRIETY, L (1965) *Bull. Centr. Étud. Sci. Biarritz* 5, 267–71. BARROWCLOUGH, G F (1980) *Auk* 97, 655–68. BARROWS, W B (1889) *US Dept. Agric. Div. Econ. Orn. and Mamm. Bull.* 1. BARTA, Z (1977) *Aquila* 83, 308. BARTELS, M (1931) *Beitr. Fortpfl. Vögel* 7, 129–30. BARTELS, M and BARTELS, H (1929) *J. Orn.* 77, 489–501. BARTH, R and MORITZ, D (1988) *Beitr. Naturkde. Niedersachs.* 41, 118–29. BARTHEL, P H, HANOLDT, W, HUBATSCH, K, KOCH, H-M, KONRAD, V, and LANNERT, R (1992) *Limicola* 6, 265–86. BARTLETT, E (1976) *Br. Birds* 69, 312. BÄSECKE, K (1950) *Vogelwelt* 71, 53; (1955) *Vogelwelt* 76, 187–8; (1956) *Vogelwelt* 77, 190. BASSINI, E (1970) *Ric. Zool. appl. Caccia* 47. BATES, D J (1979) *Scott. Birds* 10, 276–7. BATES, G G and BATES, I M (1989) *Scott. Birds* 15, 132. BATES, G G and WHITAKER, D S (1980) *Scott. Birds* 11, 85–7. BATES, G L (1930) *Handbook of the birds of West Africa*. London; (1934) *Ibis* (13) 4, 685–717; (1936) *Ibis* (13) 6, 531–56; (1937) *Ibis* (14) 1, 786–830. BATES, R S P (1938) *J. Bombay nat. Hist. Soc.* 40, 183–90. BATES, R S P and LOWTHER, E H N (1952) *Breeding birds of Kashmir*. Oxford. BATT, L (1993) *Br. Birds* 86, 133. BATTLORI, X and NOS, R (1985) *Misc. Zool.* 9, 407–11. BAU, A (1902–3) *Z. Ool.* 12, 81–6. BAUER, C-A (1976) *Anser* 15, 221. BAUER, K (1953) *Orn. Mitt.* 5, 224–5. BAUER, K, DVORAK, M, KOHLER, B, KRAUS, E, and SPITZENBERGER, F (1988) *Artenschutz in Österreich*. Vienna. BAUER, K and ROKITANSKY, G (1951) *Arbeiten aus der Biologischen Station Neusiedler See* 4 (1). BAUER, M (1961) *Vogelwelt* 82, 118–19. BAUER, M, HELVERSEN, O VON, HODGE, M, and MARTENS, J (1969). *Catalogus Faunae Graeciae* 2. Thessaloniki. BAUGNEE, J-Y (1988) *Aves* 25, 59–61. BAUMGART, W (1967) *J. Orn.* 108, 341–5; (1978) *Falke* 25, 372–85; (1980) *Falke* 27, 78–85; (1984) *Beitr. Vogelkde.* 30, 217–42. BAUMGART, W and KASPAREK, M (1992) *Zool. Middle East* 6, 13–19. BAUMGART, W and STEPHAN, B (1974) *Zool. Abh. Staatl. Mus. Tierkde. Dresden* 33, 103–38; (1987) *Mitt. zool. Mus. Berlin* 63, Suppl. *Ann. Orn.* 11, 57–95. BÄUMER-MÄRZ, C and SCHUSTER, A (1991) In Glandt, D (ed.) *Der Kolkrabe (Corvus corax) in Mitteleuropa. Metelener Schriftenreihe Naturschutz* 2, 69–81. BAUR, P (1981) *Vögel Heimat* 51, 209. BAUWENS, P, BAUWENS, L, and RAHDER, J (1976) *Limosa* 49, 201–3. BAWTREE, R F (1950) *Br. Birds* 43, 16. BAXTER, E V and RINTOUL, L J (1953) *The birds of Scotland*. BAZELY, D R (1987) *Condor* 89, 190–2. BAZIEV, D K (1976) *Int. Stud. Sparrows* 9, 30–4. BEAMAN, M (ed.) (1978) *Orn. Soc. Tur-key Bird Rep. 1974–5*. Sandy; (1986) *Sandgrouse* 8, 1–41. BEAMAN, M, PORTER, R F, and VITTERY, A (eds) (1975) *Orn. Soc. Turkey Bird Rep. 1970–3*. Sandy. BEAR, A (1991) *Torgos* 9 (2), 41–2, 73. BEAUD, M (1991) *Nos Oiseaux* 41, 249–50. BEAUD, M and SAVARY, L (1987) *Nos Oiseaux* 39, 40–1. BEAUD, P and MANUEL, F (1983) *Nos Oiseaux* 37, 39–41. BEAUDOIN, J-C (1976) *Alauda* 44, 77–90. BEAUFORT, L F DE (1947) *Ardea* 35, 226–30. BECHER, O M (1949) *Country Life* 106, 120. BECHSTEIN, J M (1853) *Cage and chamber-birds*. London. BECHTOLD, I (1967) *Aquila* 73–4, 161–70. BECKER, G B and STACK, J W (1944) *Bird-Banding* 15, 45–68. BEECHER, W J (1978) *Bull. Chigago Acad. Sci.* 11, 269–98. BEHLE, W H (1950) *Condor* 52, 193–219. BEHLE, W H and ALDRICH, J W (1947) *Proc. Biol. Soc. Washington* 60, 69–72. BEKLOVÁ, M (1972) *Zool. Listy* 21, 337–46. BELCHER, C F (1930) *The birds of Nyasaland*. London. BELIK, V P (1981) *Ornitologiya* 16, 151–2. BELL, B D (1967) *Br. Birds* 60, 139; (1968) *Br. Birds* 61, 529–30; (1969) *Br. Birds* 62, 209–18; (1970) *Bird Study* 17, 2169–81. BELL, B D and HORNBY, R J (1969) *Ibis* 111, 402–5. BELL, D G (ed.) (1978) *County Cleveland Bird Rep. 1977*, 36. BELL, T H (1962) *The birds of Cheshire*. Altrincham. BELOPOL'SKI, L (1962) *Loodus. Seltsi Aastaraamat* 55, 227–39. BELOUSOV, Y A (1986) *Tez. Dokl. 1. S'ezda Vsesoyuz. orn. Obshch. 9. Vsesoyuz. orn. Konf.* 1, 75–6. BEL'SKAYA, G S (1963) *Ornitologiya* 6, 464–5; (1974) In Rustamov, A K (ed.) *Fauna i ekologiya ptits Turkmenii* 1, 34–54. Ashkhabad; (1987) *Izv. Akad. Nauk Turkmen. SSR Ser. biol. Nauk* (5), 41–9; (1989) *Izv. Akad. Nauk Turkmen. SSR Ser. biol. Nauk* (3), 53–9. BENEDEN, A VAN (1946) *Alauda* 14, 70–86. BENDER, R O (1949) *Bird-Banding* 20, 180–2. BENDINI, L and SPINA, F (eds) (1990) *Boll. dell'Attività Inanellamento* 3. BENECKE, W (1970) *Falke* 17, 268–9. BENKMAN, C W (1987a) *Ecol. Monogr.* 57, 251–67; (1987b) *Wilson Bull.* 99, 351–68; (1988a) *Ibis* 130, 288–93; (1988b) *Auk* 105, 370–1; (1988c) *Auk* 105, 578–9; (1988d) *Auk* 105, 715–19; (1988e) *Behav. Ecol. Sociobiol.* 23, 167–75; (1989a) *Ornis scand.* 20, 65–8; (1989b) *Auk* 106, 483–5; (1990) *Auk* 107, 376–86; (in press) White-winged Crossbill. In Poole, A, Stettenheim, P, and Gill, F (eds) *The birds of North America*. Philadelphia. BENNETT, C J L (1976) *Cyprus Orn. Soc. (1957) Rep.* 20, 1–63. BENSON, C W (1946) *Ibis* 88, 444–61. BENSON, G B G and WILLIAMSON, K (1972) *Bird Study* 19, 34–50. BENSON, G B G (1967) *Br. Birds* 60, 343. BENSON, H (1890) *Zoologist* (3) 14, 17–18. BENT, A C (1946) *Bull. US natn. Mus.* 191; (1950) *Bull. US natn. Mus.* 197; (1953) *Bull. US natn. Mus.* 203; (1958) *Bull. US Nat. Mus.* 211; (1968) *Bull. US natn. Mus.* 237. BENTZ, P-G (1987) *Vår Fuglefauna* 10, 91–5; (1988) *Vår Fuglefauna* 11, 87–93; (1989) *Vår Fuglefauna* 12, 101–10; (1990) *Fauna och Flora*, 1–9. BENTZ, P-G and GÉNSBØL, B (1988) *Norsk fuglehåndbok*. Tånder. BERCK, K-H (1961–2) *Vogelwelt* 82, 129–73; 83, 8–26. BERCK, K-H and BERCK, U (1976) *Anz. orn. Ges. Bayern* 15, 95–6. BÉRESS, J and MOLNÁR, P (1964) *Aquila* 69–70, 57–70. BERETZK, P and KEVE, A (1971) *Lounais-Hämeen Luonto* 42. BERETZK, P, KEVE, A, and MARIÁN, M (1962) *Acta zool. Acad. Sci. Hung.* 8, 251–71. BERETZK, P, KEVE, A, and MARIÁN, M (1969) *Bonner zool. Beitr.* 20, 50–9. BEREZOVIKOV, N N (1983) *Ornitologiya* 18, 187. BEREZOVIKOV, N N and KOVSHAR', A F (1992) *Russ. Orn. Zh.* 1, 221–6. BERG, A B VAN DEN (1982a) *Dutch Birding* 4, 60–2; (1982b) *Dutch Birding* 4, 136–9; (1988) *Dutch Birding* 10, 92–3; (1989) *List of Dutch bird species 1989*; (1990) *Vogels nieuw in Nederland*

*1990*. BERG, A B VAN DEN, BY, R A DE, and CDNA (1991) *Dutch Birding* 13, 41-57; (1992) *Dutch Birding* 14, 73-90. BERG, A B VAN DEN, and BLANKERT, J J (1980) *Dutch Birding* 2, 33-5. BERG, A B VAN DEN and COTTAAR, F (1986) *Dutch Birding* 8, 57-9. BERG, A B VAN DEN and ROEVER, J W DE (1984) *Dutch Birding* 6, 139-40. BERGER, S B (1968) *Sterna* 8, 157, 159. BERGH, L M J VAN DEN, JAARSVELD, B VAN, and DRIEL, F VAN (1989) *Limosa* 62, 91-2. BERGIER, P (1989) *Newsl. Bahrain nat. Hist. Soc.* 1. BERGMAN, G (1952) *Ornis fenn.* 29, 105-7; (1953) *Acta Soc. Faun. Flor. Fenn.* 69(4), 1-15; (1956) *Ornis fenn.* 33, 61-71. BERGMAN, S (1935) *Zur Kenntnis nordostasiatischer Vögel.* Stockholm. BERGMANN, H-H (1983a) *Cyprus orn. Soc.* (*1969*), *Rep.* 8, 41-54; (1983b) *Vogelkdl. Ber. Niedersachs.* 15, 1-4; (1991) *Gef. Welt* 115, 426-8. BERGMANN, H-H, FABREWITZ, S, GRAUPNER, B, HINRICHS, K, and ZUCCHI, H (1982) *Math.-naturwiss. Unterricht* 35, 172-81. BERGMANN, H-H and HELB, H-W (1982) *Stimmen der Vögel Europas.* Munich. BERGMANN, H-H, ROY, A, and SCHRÖDER, H (1988) *Gef. Welt* 112, 280-4. BERGMANN, H-H, ZIETLOW, S, and HELB, H-W (1984) *J. Orn.* 125, 59-67. BERG-SCHLOSSER, G (1978) *J. Orn.* 119, 111-13. BERNASEK, O (1985) *Gef. Welt* 109, 124-5. BERNDT, R (1960) *Orn. Mitt.* 12, 181. BERNDT, R and DANCKER, P (1960) *Proc. int. orn. Congr.* 12, 97- 109. Helsinki. BERNDT, R and FRANTZEN, M (1987) *Vogelkdl. Ber. Niedersachs.* 19, 93. BERNDT, R and FRIELING, F (1939) *J. Orn.* 87, 593-638. BERNDT, R and FRIELING, H (1944) *Beitr. Fortpfl. Vögel* 22, 68. BERNDT, R and MOELLER, J (1960) *Braunschweig. Heimat* 46, 119-24. BERNDT, R and WINKEL, W (1987) *Vogelwelt* 108, 98-105. BERNHOFT-OSA, A (1956) *Vår Fågelvärld* 15, 245-7; (1959) *Stavanger Mus. Årb. 1959*, 139-42; (1960) *Vår Fågelvärld* 19, 220-3; (1965) *Stavanger Mus. Årb.*, 109-18; (1978) *Vår Fuglefauna* 1, 93-5. BERNIS, F (1933) *Bol. Soc. Esp. Hist. Nat.* 33, 377-84; (1945) *Bol. Soc. Esp. Hist. Nat.* 43, 93-145; (1954) *Ardeola* 1, 11-85; (1989a) *Commun. INIA 53 Ser. Recursos naturales*; (1989b) *Commun. INIA 54 Ser. Recursos naturales.* BERROW, S D, KELLY, T C, and MYERS, A A (1991) *Irish Birds* 4, 393-412. BERRUTI, A and NICHOLS, G (1991) *Birding in Southern Africa* 43, 52-7. BERTHOLD, P (1964) *Vogelwarte* 22, 236-75; (1971) *Vogelwelt* 92, 141-7. BERTHOLD, P and BERTHOLD, H (1987) *Bocagiana* 110, 1-8. BERTHOLD, P and GWINNER, E (1972) *Vogelwarte* 26, 356-7; (1978) *J. Orn.* 119, 338-9. BERTHOLD, P and SCHLENKER, R (1982) *Dutch Birding* 4, 100-2. BESSON, J (1968) *Alauda* 36, 292-3; (1982) *Nos Oiseaux* 36, 289-90. BÉTHUNE, G DE (1961) *Gerfaut* 51, 387-98; (1986) *Oriolus* 52, 38. BETTMANN, H (1969) *Orn. Mitt.* 21, 26-7. BEVEN, G (1946) *Br. Birds* 39, 23; (1947) *Br. Birds* 40, 308-10; (1964) *Lond. Nat.* 43, 86-109. BEZEMER, K W L (1979) *Vogeljaar* 27, 128-30. BEZZEL, E (1957) *Anz. orn. Ges. Bayern* 4, 589-707; (1972) *Vogelwarte* 26, 346-52; (1985) *J. Orn.* 126, 434-9; (1987) *Vogelwelt* 108, 71-2; (1988) *J. Orn.* 129, 71-81. BEZZEL, E and BRANDL, R (1988) *Anz. orn. Ges. Bayern* 27, 45-65. BEZZEL, E and LECHNER, F (1978) *Die Vögel des Werdenfelser Landes.* Greven. BEZZEL, E, LECHNER, F, and RANFTL, H (1980) *Arbeitsatlas der Brutvögel Bayerns.* Greven. BHARUCHA, E K (1989) *J. Bombay nat. Hist. Soc.* 86, 450. BIBBY, C J (1977) *Ring. Migr.* 1, 148-57. BIBBY, C J and CHARLTON, T D (1991) *Açoreana* 7, 297-304. BIBBY, C J, CHARLTON, T D, and RAMOS, J (1992) *Br. Birds* 85, 677-80. BIBBY, C J and LUNN, J (1982) *Biol. Conserv.* 23, 167-186. BIBER, J-P and LINK, R (1974) *Nos Oiseaux* 32, 273-4. BIBER, O (1984) *Orn. Beob.* 81, 1-28. BIDDULPH, C H (1954) *J. Bombay*

*nat. Hist. Soc.* 52, 208-9. BIEDERMANN-IMHOOF, R (1913) *Orn. Monatsber.* 21, 4-6. BIELFELD, H (1980) *Kanarien.* Stuttgart; (1981) *Zeisige, Kardinäle und andere Finkenvögel.* Stuttgart. BIERI, W (1945) *Orn. Beob.* 42, 140-2. BIGGER, W K (1931) *Ibis* (13) 1, 584-5. BIGNAL, E and CURTIS, D J (eds) (1989) *Choughs and land-use in Europe.* Scottish Chough Study Group. BIGNAL, E, MONAGHAN, P, BENN, S, BIGNAL, S, STILL, E, and THOMPSON, P M (1987) *Bird Study* 34, 39-42. BIGOT, L (1966) *Terre Vie* 113, 295-315. BIJLSMA, R G (1979) *Limosa* 52, 53-71; (1982) *Ardea* 70, 25-30. BIJLSMA, R G and MEININGER, P L (1984) *Gerfaut* 74, 3-13. BIJLSMA, R G, RODER, F E DE, and BEUSEKOM, R VAN (1988) *Limosa* 61, 1-6. BILLE, R-P (1978) *Nos Oiseaux* 34, 261; *Nos Oiseaux* 35, 227-31. BILLETT, A E (1989) *Br. Birds* 82, 81. BINFORD, L C (1971) *Californian Birds* 2, 1-10; (1989) *Orn. Monogr. AOU* 43. BIRCH, A (1990) *Birding World* 3, 308-9. BIRD, E G (1935) *Ibis* (13) 5, 438-41. BIRKHEAD, T R (1972) *Br. Birds* 65, 356-7; (1974a) *Br. Birds* 67, 221-9; (1974b) *Ornis scand.* 5, 71-81; (1979) *Anim. Behav.* 27, 866-74; (1982) *Anim. Behav.* 30, 277-83; (1989) *Br. Birds* 82, 583-600; (1991) *The magpies.* London. BIRKHEAD, T R, ATKIN, L, and MØLLER, A P (1987) *Behaviour* 101, 101-38. BIRKHEAD, T R and CLARKSON, K (1985) *Behaviour* 94, 324-32. BIRKHEAD, T R, EDEN, S F, CLARKSON, K, GOODBURN, S F, and PELLATT, J (1986) *Ardea* 74, 59-68. BIRKHEAD, T R and GOODBURN, S F (1989) In Newton, I (ed.) *Lifetime reproduction in birds*, 173-82. London. BIRKHEAD, T R and MØLLER, A P (1992) *Sperm competition in birds.* London. BISWAS, B (1950) *J. zool. Soc. India* 2 (1). BJORDAL, H (1983a) *Fauna norv.* (C) *Cinclus* 6, 105-8; (1983b) *Vår Fuglefauna* 6, 34-6; (1984) *Fauna norv.* (C) *Cinclus* 7, 21-3. BJÖRKLUND, M (1989a) *Behav. Ecol. Sociobiol.* 25, 137-40; (1989b) *Anim. Behav.* 38, 1081-83; (1989c) *Ornis scand.* 20, 255-64; (1990a) *Auk* 107, 35-44; (1990b) *Ibis* 132, 613-17. BJØRNSEN, B (1988) Cand. Sci. Anim. Ecol. Thesis. Bergen Univ. BLAIR, C M G (1961) *Ibis* 103a, 499-502. BLAIR, H M S (1936) *Ibis* (13) 6, 280-308. BLAIR, R H and TUCKER, B W (1941) *Br. Birds* 34, 206-15. BLAKE, C H (1956) *Bird-Banding* 27, 16-22; (1962) *Bird-Banding* 33, 97-9; (1964) *Bird-Banding* 35, 125-7; (1967) *Bird-Banding* 38, , 234; (1969) *Bird-Banding* 40, 133-9. BLANA, E (1970) *Charadrius* 6, 23-5. BLANCHARD, B D (1941) *Univ. Calif. Publ. Zöol.* 46 (1). BLANCHARD, B D and ERICKSON, M M (1949) *Univ. Calif. Publ. Zöol.* 47, 255-318. BLANCO, G, CUEVAS, J A, and FARGALLO, J A (1991) *Ardeola* 38, 91-9. BLANCOU, L (1939) *Oiseau* 9, 410-85. BLANFORD, W T (1876) *Eastern Persia. An account of the journeys of the Persian Boundary Commission, 1870-2* 2: *Zoology and Geology.* London. BLASER, P (1970) *Orn. Beob.* 67, 297-9; (1973) *Orn. Beob.* 70, 186-7; (1974) *Orn. Beob.* 71, 322-3. BLASIUS, R (1886) *Ornis* 2, 437-550. BLATHWAYT, F L (1903) *Zoologist* (4) 7, 26-7. BLEDSOE, A H (1988) *Wilson Bull.* 100, 1-8. BLEM, C R (1975) *Wilson Bull.* 87, 543-9; (1981) *Condor* 83, 370-6. BLOCH, D and SØRENSEN, S (1984) *Yvirlit yvir Føroya fuglar.* Tórshavn. BLOK, A A and SPAANS, A L (1962) *Limosa* 35, 4-16. BLOMGREN, A (1964) *Lavskrika.* Stockholm; (1971) *Br. Birds* 64, 25-8; (1983) *Vår Fågelvärld* 42, 343-6. BLONDEL, J (1963) *Alauda* 31, 22-6; (1969) *Synécologie des passereaux résidents et migrateurs dans le Midi Méditerranéen français.* Marseille; (1979) *Biogéographie et écologie.* Paris. BLONDEL, J and ISENMANN, P (1981) *Guide des oiseaux de Camargue.* Neuchâtel; 750e 764e 768e 687e. BLUME, C A (1963) *Dansk orn. Foren. Tidsskr.* 57, 19-21. BLUME, D (1967) *Aus-*

*drucksformen unserer Vögel*. Wittenberg Lutherstadt. BLU-MEL, H (1976) *Der Grünling*. 2nd edition. Wittenber Lutherstadt; (1982) *Die Rohrammer*. Wittenberg Lutherstadt; (1983a) *Der Grünling*. Wittenberg Lutherstadt; (1983b) *Abh. Ber. Naturkundemus. Görlitz* **56** (4); (1986) *Abh. Ber. Naturkdemus. Görlitz* **59** (3). BLYTH, E (1867) *Ibis* (2) **3**, 1-48. BOAG, P T and RATCLIFFE, L M (1979) *Condor* **81**, 218-9. BOBRETSOV, A V and NEUFELD, N D (1986) *Tez. Dokl. 1. S'ezda Vsesoyuz. orn. Obshch. 9. Vsesoyuz. orn. Konf.* **1**, 85-6. BOCCA, M and MAFFEI, G (1984) *Gli uccelli della Valle d'Aosta*. Aosta. BOCK, W J and MARONY, J J (1978) *Bonn. zool. Beitr.* **29**, 122-47. BOCHEŃSKI, Z (1970) *Acta Zool. Cracov.* **15**, 1-59. BOCHE-ŃSKI, Z and OLEŚ, T (1977) *Acta Zool. Cracov.* **22**, 319-71; (1981) *Acta Zool. Cracov.* **25**, 3-12. BOCK, C E and LEPTHIEN, L W (1976) *Amer. Nat.* **110**, 559-71. BOCK, W J and MORONY, J J (1978) *Bonn. zool. Beitr.* **29**, 122-47. BODDY, M (1979) *Ringers' Bull* **5**, 87; (1981) *Ring. Migr.* **3**, 193-202; (1983) *Ornis scand.* **14**, 299-308; (1984) *Ring. Migr.* **5**, 91-100. BODDY, M and BLACKBURN, A C (1978) *Ring. Migr.* **2**, 27-33. BODDY, M and SELLERS, R M (1983) *Ring. Migr.* **4**, 129-38. BODENSTEIN, G (1953) *Orn. Mitt.* **5**, 72. BOECKER, M (1970) *Bonn. zool. Beitr.* **21**, 183-236. BOEHME, R L (1958) *Uchen. zap. Severo-Osetinsk. gos. ped. Inst.* **23** (1), 111-83. BOERTMANN, D, OLSEN, K M, and PEDERSEN, B B (1986) *Dansk orn. Foren. Tidsskr.* **80**, 35-57. BOERTMANN, D, SØRENSEN, S, and PIHL, S (1986) *Dansk. orn. Foren. Tidsskr.* **80**, 121-30. BOESMAN, P (1992) *Dutch Birding* **14**, 161-9. BOURNE, W R P (1967) *Ibis* **109**, 141-67. BOGLIANI, G (1985) *Riv. ital. Orn.* **55**, 140-50. BOGLIANI, G and BRANGI, A (1990) *Bird Study* **37**, 195-8. BOGORODSKI, Y V (1981) *Ornitologiya* **16**, 153. BOHAC, D (1967) *Angew. Orn.* **2**, 151-2. BOHLEN, H D (1989) *The birds of Illinois*. Bloomington. BÖHMER, A (1973) *Orn. Beob.* **70**, 103-12; (1974) *Orn. Beob.* **71**, 279-82; (1976a) *Orn. Beob.* **73**, 109-36; (1976b) *Orn. Beob.* **73**, 136-40. BÖHNER, J, CHAIKEN, M L, BALL, G F, and MARLER, P (1990) *Hormon. Behav.* **24**, 582-94. BÖHR, H-J (1962) *Bonn. zool. Beitr.* **13**, 50-114. BOIE, F (1866) *J. Orn.* **14**, 1-4. BÖKER, H (1922) *Biol. Centralbl.* **42**, 87-93. BOLAM, G (1912) *Birds of Northumberland and the eastern borders*. Alnwick; (1913) *Wildlife in Wales*. London. BOLLE, C (1856) *J. Orn.* **4**, 17-31; (1857) *J. Orn.* **5**, 305-51; (1858a) *J. Orn.* **6**, 125-51; (1858b) *Naumannia* **8**, 369-93. BOLTON, M (1986) *Distribution of breeding birds in the Algarve in relation to habitat*. In Pullan, R (ed.) *A Rocha Bird Report 1986*, 5-8. BOND, J (1938) *Can. Field-Nat* **52**, 3-5; (1985) *Birds of the West Indies*. London. BOND, L M G (1946) *Bird Notes News* **22**, 32. BONHAM, P F (1970a) *Br. Birds* **63**, 28-32; (1970b) *Br. Birds* **63**, 262-4. BONHAM, P F and SHARROCK, J T R (1969) *Br. Birds* **62**, 550-52. BOOTH, C J (1979) *Scott. Birds* **10**, 261-7; (1986) *Scott. Birds* **14**, 51. BOOTH, C, CUTHBERT, M, and REYNOLDS, P (1984) *The birds of Orkney*. Stromness; (1986) *Orkney Bird Rep. 1985*, 3-57. BOOTH, C J and REYNOLDS, P (1992) *Br. Birds* **85**, 245-6. BORGVALL, T (1952) *Vår Fågelvärld* **11**, 11-15. BOROS, I and HORVÁTH, L (1955) *Acta Zool. Acad. Sci. Hung.* **1**, 43-51. BORRAS, A and SENAR, J C (1986) *Misc. Zool.* **10**, 403-6; (1991) *J. Orn.* **132**, 285-9. BORROR, D J (1961) *Ohio J. Sci.* **61**, 161-74. BORTOLI, L (1973) *Productivity, population dynamics and systematics of granivorous birds*, 249-52. Warsaw. BORTOLI, L and BRUGGERS, R (1976) PNUD/Recherche pour la lutte contre les oiseaux granivores 'Quelea quelea'. SR 258. Dakar. BOS, G, SLIJPER, H J, and TAAPKEN, J (1945) *Limosa* **18**, 56-68. BÖSENBERG, K

(1958) *Falke* **5**, 58-61. BOSSEMA, I (1967) *Levende Nat.* **70**, 86-92; (1979) *Behaviour* **70**, 1-117; (1980) *Corvid Newsl.* **1** (2). BOSSEMA, I and BENUS, R F (1985) *Behav. Ecol. Sociobiol.* **16**, 99-104. BOSSEMA, I and POT, W (1974) *Levende Nat.* **77**, 265-79. BOSSEMA, I, RÖELL, A, BAEYENS, G, ZEEVALKING, H, and LEEVER, H (1976) *Levende Nat.* **79**, 149-66. BOSWALL, J (1970) *Br. Birds*, **63**, 256-7; (1985a) *BBC Wildl.* **3**, 174-9; (1985b) In Campbell and Lack (1985), 599. BOSWELL, C and NAYLOR, P (1957) *Iraq nat. Hist. Mus. Publ.* **13**, 16. BOUARD, R (1983) *Bull. UCAGO* 3-4, 6-15. BOURDELLE, E and GIBAN, J (1950-51) *Bull. stations françaises de Baguage* 7. Paris. BOURNE, W R P (1953) *Br. Birds* **46**, 381-2; (1955) *Ibis* **97**, 508-56; (1957) *Ibis* **99**, 182-90; (1966) *Ibis* **108**, 425-9; (1967) *Ibis* **109**, 141-67; (1986) *Bull. Br. Orn. Club* **106**, 163-70. BOURNE, W R P (1992) *Dutch Birding* **7**, 21-2. BOURNE, W R P and NELDER, J A (1951) *Br. Birds* **44**, 386-7. BOURRILLON, P (1961) *Oiseau* **31**, 247. BOUTET, J-Y and PETIT, P (1987) *Atlas des oiseaux nicheurs d'Aquitaine 1974-1984*. Bordeaux. BOXBERGER, L VON (1930) *Beitr. Fortpfl. Vögel* **6**, 211. BOWDEN, C G R (1987) *Sandgrouse* **9**, 94-7. BOWDEN, C G R and BROOKS, D J (1987) *Sandgrouse* **9**, 111-14. BOYARCHUK, V P (1990) *Vestnik Zool.* 1990 (2), 52-7. BOYD, A W (1932) *Br. Birds* **25**, 278-85; (1933) *Br. Birds* **26**, 273-4; (1934) *Br. Birds* **27**, 259-60; (1935) *Br. Birds* **28**, 347-9; (1949) *Br. Birds* **42**, 213-4; (1951) *A country parish*. London. BOYLE, G (1966) *Br. Birds* **59**, 342. BOZHKO, S I (1974) *Acta Orn.* **14**, 39-57; (1980) *Der Karmingimpel*. Wittenberg Lutherstadt. BOZSKO, S (1977) *Aquila* **83**, 289-90, 305. BRAAE, L (1975) *Dansk orn. Foren. Tidsskr.* **69**, 41-53. BRACK, H (1977) *Bird-Banding* **48**, 370. BRAČKO, F (1992) *Acrocephalus* **13**, 57-8. BRADER, M (1989) *Egretta* **32**, 18-20. BRADSHAW, C (1991) *Birding World* **4**, 354-5; (1991) *Br. Birds* **84**, 310-11; (1992) *Br. Birds* **85**, 653-65. BRANDL, R and BEZZEL, E (1988) *Zool. Anz.* **221**, 411-17; (1989) *Orn. Beob.* **86**, 137-43. BRANDNER, J (1991) *Egretta* **34**, 73-85. BRANDT, H (1960) *Anz. orn. Ges. Bayern* **5**, 597-8; (1962) *Vogelwelt* **83**, 81-2. BRAY, O E, DE GRAZIO, J W, GUARINO, J L, and STREETER, R G (1974) *Inland Bird Banding News* **46**, 204-9. BRAY, O E, ROYALL, W C, GUARINO, J L, and DE GRAZIO, J W (1973) *Bird-Banding* **44**, 1-12. BRAZIER, H, DOWDALL, J F, FITZHARRIS, J E, and GRACE, K (1986) *Irish Birds* **3**, 287-336. BRAZIL, M A (1991) *The birds of Japan*. London. BRAUN, P (1989) *Falke* **36**, 154-5. BRAUNBERGER, C W (1990) *Orn. Mitt.* **42**, 15-16. BREIFE, B, HIRSCHFELD, E, KJELLÉN, N, and ULLMAN, M (1990) *Vår Fågelvärld Suppl.* **13**. BRÉMOND, J-C (1962) *Angew. Orn.* **1**, 49-63. BRENCHLEY, A (1986) *J. Zool. Lond.* **210**, 261-78. BRENNECKE, H-E (1953) *Orn. Mitt.* **5**, 149. BREWER, D (1990) *Br. Birds* **83**, 289-90. BRIAN, M V and BRIAN, A D (1948) *Trans. Herts. nat. Hist. Soc.* **23**, 30-6. BRICHETTI, P (1973) *Riv. ital. Orn.* **43**, 519-649; (1976) *Atlante ornitologico italiano* **2**. Brescia; (1982) *Riv. ital. Orn.* **52**, 3-50; (1983) *Riv. ital. Orn.* **53**, 101-44; (1986) *Riv. ital. Orn.* **56**, 3-39. BRICHETTI, P and COVA, C (1976) *Gli Ucc. Ital.* **1**, 28-31. BRICHETTI, P, CAFFI, M, and GANDINI, S (1992) *Nat. Bresciana* 28. BRICHETTI, P and MASSA, B (1984) *Riv. ital. Orn.***54**, 3-37; (1991) *Riv. ital. Orn.* **61**, 3-9. BRICHOVSKY, A (1968) *Falke* **15**, 170-3. BRIDGMAN, C J (1962) *Br. Birds* **55**, 461-70. BRIEN, Y, BESSEC, A, and LESOUEF, J-Y (1982) *Oiseau* **52**, 87-9. BRIGGS, F S and OSMASTON, B B (1928) *J. Bombay Nat. Hist. Soc.* **32**, 744-61. BRIGHOUSE, U W (1954) *Devon Birds* **7**, 38-41. BRINGELAND, R (1964) *Sterna* **6**, 4-6. BRITISH ORNITHOLOGISTS' UNION (1992) *Ibis* **134**, 211-14. BRIT-

TON, P L (1970) *Ostrich* **41**, 145-90; (ed.) (1980) *Birds of East Africa*. Nairobi. BRITTON, P L, and DOWSETT, R J (1969) *Ostrich* **40**, 55-60. BROAD, R A (1974) *Br. Birds* **67**, 297-301; (1978) *Scott. Birds* **10**, 58-9; (1981) *Br. Birds* **74**, 90-4. BROAD, R A and ODDIE, W E (1980) *Br. Birds* **73**, 402-8; (1982) In Sharrock, J T R and Grant, P J (eds) *Birds new to Britain and Ireland*, 212-16. Calton. BROCKMANN, H J (1980) *Wilson Bull.* **92**, 394-98. BROD, G (1988) *Beih. Veröff. Nat. Land. Bad.-Württ.* **53**, 83-90. BRODIE, E (1985) *Br. Birds* **78**, 244. BROKHOVICH, S A (1990) *Ornitologiya* **24**, 168-9. BROMLEY, F C (1947) *Br. Birds* **40**, 114. BROOD, K and SÖDERQUIST, T (1967) *Vår Fågelvärld* **26**, 266-8. BROOKE, R K (1962) *Ostrich* **33** (1), 23-5; (1973) *Auk* **90**, 206; (1976) *Bull. Br. Orn. Club* **96**, 8-13. BROOKS, D J (1987) *Sandgrouse* **9**, 115-20. BROOKS, D J, EVANS, M I, MARTINS, R P, and PORTER, R F (1987) *Sandgrouse* **9**, 4-66. BROOKS, W S (1968) *Wilson Bull.* **80**, 253-80; (1973) *Bird-Banding* **44**, 13-21. BROSSET, A (1956) *Alauda* **24**, 266-71; (1957a) *Alauda* **25**, 43-50; (1957b) *Alauda* **25**, 224-6; (1961) *Trav. Inst. sci. chérifien Ser. Zool.* **22**, 7-155; (1984) *Alauda* **52**, 81-101. BROSSET, A and ERARD, C (1986) *Les oiseaux des régions forestières du nord-est du Gabon* 1. Paris. BROUN, M (1971) *Auk* **88**, 924-5. BROWN, A W, LEVEN, M R, and PRATO, S R D DA (1984) *Scott. Birds* **13**, 107-111. BROWN, B J (1985) *Br. Birds* **78**, 244. BROWN, J L (1987) *Helping and communal breeding in birds*. Princetown. BROWN, L H (1967) *Ibis* **109**, 275. BROWN, R G B (1959) *Br. Birds* **52**, 98; (1963) *Ibis* **105**, 63-75. BROWN, R H (1924) *Br. Birds* **18**, 122-8; (1942) *Naturalist*, 39-45. BROWN, R N (1974) M Sc Thesis. Alaska Univ. BROWNE, K and BROWNE, E (1956) *Br. Birds* **49**, 241-57. BROWNE, P W P (1960) *Br. Birds* **53**, 575-7; (1981) *Bull. Br. Orn. Club* **101**, 306-10. BROWNING, M R (1976) *Bull. Br. Orn. Club* **96**, 44-7. BROYD, S J (1985) *Br. Birds* **78**, 647-56. BRUCH, A, ELVERS, H, POHL, C, WESTPHAL, D, and WITT, K (1978) *Orn. Ber. Berlin (West)* 3 suppl. Berlin. BRUCH, A and LÖSCHAU, M (1960) *Orn. Mitt.* **12**, 31. BRUCKER, J W (ed.) (1985) *Oxford Orn. Soc. Rep. 1984*. BRUCKER, J W, GOSLER, A G, and HERYET, A R (eds) (1992) *Birds of Oxfordshire*. Newbury. BRUDENELL-BRUCE, P G C (1975) *The birds of New Providence and the Bahama Islands*. London. BRUGGERS, R L and BORTOLI, L (1976) *Terre Vie* **30**, 521-7. BRUSTER, K-H (1988) *Hamburger avifaun. Beitr.* **21**, 189. BRUUN, B (1984) *Courser* 1, 44-6. BRUUN, B, DELIN, H, and SVENSSON, L (1986) *Birds of Britain and Europe*. Twickenham. BRYSON, D K (1947) *Br. Birds* **40**, 209. BUB, H (1953) *Orn. Mitt.* **5**, 6; (1955) *Ibis* **97**, 25-37; (1962) *Falke* **9**, 164-71; (1969) *Vogelwarte* **25**, 134-41; (1976a) *Orn. Mitt.* **28**, 6-12; (1976b) *J. Orn.* **117**, 461; (1977) *Orn. Mitt.* **29**, 55-60; (1978) *Sterna* **17**, 21-3; (1985) *Beitr. Vogelkde.* **31**, 189-213; (1986) *Beitr. Vogelkde.* **32**, 249-65; (1987) *Beitr. Vogelkde.* **33**, 313-25; (1989) *Falke* **36**, 41. BUB, H, HEFT, H, and WEBER, H (1959) *Falke* **6**, 3-9, 48-54. BUB, H and HINSCHE, A (1982) *Hercynia (NF)* **19**, 322-62. BUB, H and KUMERLOEVE, H (1954) *Orn. Mitt.* **6**, 205-12, 225-31. BUB, H and PANNACH, G (1988) *Verh. orn. Ges. Bayern* **24**, 411-65; (1991) *Beitr. Naturk. Niedersachs.* **44**, 272-90; (1992) *Beitr. Naturk. Niedersachs.* **45**, 192-215 BUB, H and PRÄKELT, A (1952) *Beitr. Naturkde. Niedersachs.* 1, 10-2 BUB, H and VRIES, R de (1973) *Das Planberingungs-Programm am Berghänfling (Carduelis f. flavirostris) 1952-70*. Wilhelmshaven. BUBNOV, M A (1956) *Zool. Zh.* **35**, 316-18. BÜCHEL, H P (1974) *Mitt. nat.-forsch. Ges. Luzern* **24**, 73-94; (1983) *Orn. Beob.* **80**, 1-28. BUCKLAND, S T, BELL, M

V, and PICOZZI, N (1990) *The birds of north-east Scotland*. Aberdeen. BUCKLAND, S T and KNOX, A G (1980) *Br. Birds* **73**, 360-1. BUCKNALL, R H (1983) *Scott. Birds* **12**, 191-3. BÜHLER, A (1968) *Orn. Beob.* **65**, 26-8. BUITRON, D (1983a) *Anim. Behav.* **31**, 211-20; (1983b) *Behaviour* **87**, 209-36; (1988) *Condor* **90**, 29-39. BULL, J (1974) *Birds of New York State*. New York. BULLOCK, and DEL-NEVO, A (1983) *Peregrine* **5**, 226-9. BULLOCK, I (1985) *Br. Birds* **78**, 247-8. BULLOCK, I D, DREWETT, D R, and MICKLEBURGH, S P (1983a) *Br. Birds* **76**, 377-401; (1983b) *Irish Birds* **2**, 257-71; (1983c) *Peregrine* **5**, 229-37; (1986) *Nature in Wales* (N.S.) **4** (1985), 46-57. BULLOUGH, W S (1942a) *Ibis* (14) **6**, 225-39; (1942b) *Phil. Trans. Roy. Soc.* (B) **231**, 165-241. BULLOUGH, W S and CARRICK, R (1940) *Nature* **145**, 629. BÜLOW, B von (1990) *Charadrius* **26**, 151-89. BUNDESDEUTSCHER SELTENHEITENAUSSCHUSS (1989) *Limicola* **3**, 157-96; (1991) *Limicola* **5**, 186-220; (1992) *Limicola* **6**, 153-177. BUNDY, G (1976) *The birds of Libya*. London. BUNDY, G, CONNOR, R J, and HARRISON, C J O (1989) *Birds of the Eastern Province of Saudi Arabia*. London. BUNDY, G and MORGAN, J H (1969) *Bull. Br. Orn. Club* **89**, 139-44, 151-9. BUNDY, G and WARR, E (1980) *Sandgrouse* **1**, 4-49. BUNYARD, P F (1932) *Bull. Br. Orn. Club* **52**, 83-4. BURG, G DE (1911) *Les oiseaux de la Suisse* **13**. Berne. BURKE, T and BRUFORD, M W (1987) *Nature* **327**, 149-52. BURKITT, J P (1935) *Br. Birds* **28**, 322-6; (1936) *Br. Birds* **29**, 334-8. BÜRKLI, W (1972) *Orn. Beob.* **69**, 183-4. BURLEIGH, T D (1958) *Georgia birds*. Norman; (1960) *Auk* **77**, 210-5; (1972) *Birds of Idaho*. Caldwell. BURLEIGH, T D and PETERS, H (1948) *Proc. Biol. Soc. Washington* **61**, 111-26. BURNETT, B W (1965) *Countryman* **65**, 130. BURNIER, E (1977) *Alauda* **45**, 238-9. BURNS, D W (1993) *Br. Birds* **86**, 115-20. BURNS, P S (1957) *Bird Study* **4**, 62-71. BURT, E H and HAILMAN, J P (1978) *Ibis* **120**, 153-70. BURTON, J A (1974) *The naturalist in London*. Newton Abbot. BURTON, M (1976) *Daily Telegraph* London 30 Oct 1976; (1979) *Wildlife* **21** (10), 19. BURTON, M and BURTON, R (1977) *Inside the animal world*. London. BUSCHE, G (1970) *Corax* **3**, 51-70; (1980) *Vogelbestände des Wattenmeeres von Schleswig-Holstein*. Heide; (1989) *Vogelwarte* **35**, 11-20; (1991) *Vogelwelt* **112**, 162-76. BUSCHE, G, BOHNSACK, P, and BERNDT, R K (1975) *Corax* **5**, 114-26. BUSCHING, W-D (1988) *Falke* **35**, 42-7. BUSSE, P (1962) *Acta Orn.* **6**, 209-30; (1963) *Acta Orn.* **7**, 189-220; (1965) *Ekol. Polska* (A) **13**, 491-514; (1968) *Notatki Orn.* **9** (1-2), 24-6; (1969) *Acta orn.* **11**, 263-328; (1970) *Notatki Orn.* **11**, 1-15; (1976) *Acta Zool. Cracov.* **21**, 121-261. BUSSE, P and HALASTRA, G (1981) *Acta Orn.* **18** (3), 1-122. BUTLER, A G (1899) *Foreign species in captivity*; (1905) *Ibis* (8) **5**, 301-401. BUTLER, A S (1989) *Br. Birds* **82**, 474-5. BUTLER, R W and CAMPBELL, R W (1987) *Can. Wildl. Serv. Occas. Pap.* **65**. BUTLIN, S M (1959) *Br. Birds* **52**, 387-8. BUTTERFIELD, R (1906) *Zoologist* (4) **10**, 31-3. BUTURLIN, L (1929) *Sistematicheskie zametki o ptitsakh Severnogo Kavkaza*. Makhachkala. BUTURLIN, S A (1906) *Ibis* (8) **6**, 407-27; (1907) *Orn. Monatsber.* **15**, 8-9. BUXTON, E J M (1960) *Ibis* **102**, 127-9. BUZZARD, G G (1989) *Br. Birds* **82**, 620-1. BYARS, T and GALBRAITH, H (1980) *Br. Birds* **73**, 2-5. BYKOVA, L P (1990) In Kurochkin, E N (ed.) *Sovremennaya ornitologiya*, 98-116. Moscow. BYRD, G V (1979) *Elepaio* **39**, 69-70.

CABOT, D (1965) *Irish Nat. J.* **15**, 95-100. CADE, T J (1952) *Condor*, **54**, 363; (1953) *Condor* **55**, 43-4. CADMAN, M D, EAGLES, P F J, and HELLEINER, F M (1987) *Atlas of the* *breeding birds of Ontario.* Waterloo. CADMAN, W A (1947) *Br. Birds* **40**, 209-10. CAIRNS, J (1952) *Malayan Nat. J.* **7**, 106-7. CALDWELL, J A (1949) *Br. Birds* **42**, 288. CALDWELL, L D, ODUM, E P, and MARSHALL, S G (1963) *Wilson Bull.* **75**, 428-34. CALHOUN, J B (1947*a*) *Auk* **64**, 305-6; (1947*b*) *Amer. Nat.* **81**, 203-28. CALVERT, M (1988) *Br. Birds* **81**, 531-2. CAMBI, D and MICHELI, A (1986) *Nat. Bresc. Ann. Mus. Civ. Sc. Nat.* **22**, 103-78. CAMPBELL, B (1953) *Finding nests.* London; (1968) *Countryman* **70** (2), 306-9; (1973) *Forest Comm. Forest Rec.* 86. CAMPBELL, B and FERGUSON-LEES, J (1965) *A field guide to birds' nests.* London. CAMPBELL, B and LACK, E (eds) (1985) *A dictionary of birds.* Calton. CAMPBELL, P O (1972) M Sc Thesis. Massey Univ. CAMPBELL, J W (1936) *Br. Birds* **29**, 306-9; (1936) *Br. Birds* **30**, 209-18. CAMPBELL, R W, SHEPARD, M G, and MACDONALD, B A (1974) *Vancouver birds in 1972.* Vancouver. CAMPINHO, F, LOURENÇO, J, and RODRIGUES, P (1991) *Airo* **2**, 21. CANNELL, P F, CHERRY, J D, and PARKES, K C (1983) *Wilson Bull.* **95**, 621-7. CANNINGS, R A, CANNINGS, R J, and CANNINGS, S G (1987) *Birds of the Okanagan valley, British Columbia.* Victoria. CANO, A and KÖNIG, C (1971) *J. Orn.* **112**, 461-2. CARACO, T and BAYHAM, M C (1982) *Anim. Behav.* **30**, 990-6. CARLO, E A DI (1991) *Sitta* **5**, 35-47. CARLOTTO, L (1991) *Riv. ital. Orn.* **61**, 48-9. CARLSON, T (1946) *Vår Fågelvärld* **5**, 37-38. CARPELAN, J (1929*a*) *Beitr. Fortpfl. Vögel* **5**, 60-3; (1929*b*) *Beitr. Fortpfl. Vögel* **5**, 198-201; (1932) *Beitr. Fortpfl. Vögel* **8**, 56-8. CARR, D (1969) *Br. Birds* **62**, 238. CARRASCAL, L M (1987) *Ardeola* **34**, 193-224; (1988) *Doñana Acta Vert.* **15**, 111-31. CARROLL, C J (1916) *Br. Bird* **9**, 293-4. CARRUTHERS, D (1910) *Ibis* (9) **4**, 436-75. CARY, P (1973) *A guide to birds of southern Portugal.* Lisbon. CASÉN, R and HILDÉN, O (1965) *Ornis fenn.* **42**, 33-5. CASTELLI, M (1988) *Cahiers Éthol. appl.* **8**, 501-82. CASTLE, M E (1977) *Scott. Birds* **9**, 327-34. CASTO, S D (1974) *Wilson Bull.* **86**, 176-7. CATLEY, G P and HURSTHOUSE, D (1985) *Br. Birds* **78**, 482-505. CATZEFLIS, F (1975) *Nos Oiseaux* **33**, 64-5. CAVE, F O and MACDONALD, J D (1955) *Birds of the Sudan.* Edinburgh. CAWKELL, E M (1947) *Br. Birds* **40**, 212; (1949) *Br. Birds* **42**, 85; (1951) *Br. Birds* **44**, 36. CEDERHOLM, G, FLODIN, L-Å, FREDRIKSSON, S, GUSTAFSSON, L, JACOBSSON, S, and PATERSSON, L (1974) *Fauna och Flora* **69**, 134-45. CELLIER, M (1992) D. Phil. Thesis. Oxford Univ. CERNY, W (1938) *Alauda* **10**, 76-90. ČERNÝ, W (1946) *Sylvia* **8**, 13-18. CHAIKEN, M (1986) Ph D Thesis. Rutgers University; (1990) *Develop. Psychobiol.* **23**, 233-46; (1992) *Behaviour* **120**, 139-50. CHAKIR, N (1986) *Ecologie du Bruant striolé, contribution à la biologie et à la dynamique de population à Marrakech.* C.E.A. de biologie générale, Faculté des Sciences de Marrakech.. CHALMERS, M L (1986) *Annotated checklist of the birds of Hong Kong.* Hong Kong. CHALMERS, M L, TURNBULL, M, and CAREY, G J (1991) *Hong Kong Bird Rep. 1990*, 4-63. CHAMBERS, W T H (1867) *Ibis* (2) **3**, 97-104. CHANDLER, C R and MULVIHILL, R S (1988) *Ornis scand.* **19**, 212-16; (1990) *Condor* **92**, 54-61; (1992) *Auk* **109**, 235-41. CHANG, JAMES WAN-FU (1980) *A field guide to the birds of Taiwan.* Tai-pei. CHAPIN, J P (1954) *Bull. Amer. Mus. nat. Hist.* **75B** (4). CHAPMAN, E A and McGEOCH, J A (1956) *Ibis* **98**, 577-94. CHAPMAN, F M (1935) *Auk* **52**, 21-9; (1939) *Auk* **56**, 364-5; (1940) *Auk* **57**, 225-233. CHAPPATTE, B (1980) *Nos Oiseaux* **35**, 345. CHAPPELL, B M A (1946) *Br. Birds* **39**, 352; (1949) *Br. Birds* **42**, 84. CHAPPELLIER, A (1932) *Oiseau* (NS) **2**, 535-42 CHAPPUIS, C (1969) *Alauda* **37**, 59-71; (1976) *Alauda* **44**, 475-95. CHAPPUIS, C, HEIM DE

BALSAC, H, and VIELLIARD, J (1973) *Bonn. zool. Beitr.* **24**, 302-16. CHARLES, J K (1972) Ph D Thesis. Aberdeen Univ. CHARMAN, K (1965) Ph D Thesis. Durham Univ. CHARVOZ, P (1953) *Nos Oiseaux* **22**, 137. CHAVIGNY, J DE and MAYAUD, N (1932) *Alauda* **4**, 304-48, 416-41. CHEESMAN, R E (1919) *Bull. Br. Orn. Club* **40**, 59. CHEKE, A S (1966) *Ibis* **108**, 630-1; (1967) *Ringers Bull.* **3** (2), 7-8; (1973) In Kendeigh, S C and Pinowski, J (eds) *Productivity, population dynamics and systematics of granivorous birds*, 211-12. Warsaw. CHENG, T-H (1964) *China's economic fauna: birds.* Washington; (1987) *A synopsis of the avifauna of China.* Peking. CHERNOV, Y I and KHLEBOSOLOV, E I (1989) In Chernov, Y T (ed.) *Ptitsy v soobshchestvakh tundrovoy zony*, 39-51. Moscow. CHESNEY, M C (1986) *Bird Study* **33**, 196-200; (1987) *Tay Ringing Group Rep. 1984-6*, 26-32. CHETTLEBURGH, M R (1952) *Br. Birds* **45**, 359-64. CHEYLAN, G (1973) *Alauda* **41**, 213-26. CHIA HSIANG-KANG and LI SHI-CHUN (1973) *Acta zool. Sinica* **19**, 190-7. CHILGREN, J D (1977) *Auk* **94**, 677-88; (1978) *Condor* **80**, 222-9. CHIPLEY, R M (1980) In Keast, A and Morton, E S (eds) *Migrant birds in the Neotropics: ecology, behavior, distribution, and conservation*, 309-17 Washington, DC. CHITTENDEN, D E (1973) *Br. Birds* **66**, 121. CHONG, L T (1938) *Contr. biol. Lab. Sci. Soc. China* **12** (Zool. Ser. 9), 183-373. CHRISTEN, W (1984) *Orn. Beob.* **81**, 227-31. CHRISTENSEN, H Ø (1957) *Dansk orn. Foren. Tidsskr.* **51**, 168-75. CHRISTENSEN, N H and ROSENBERG, N T (1964) *Dansk orn. Foren. Tidsskr.* **58**, 13-35. CHRISTIANSEN, A (1935) *Dansk orn. Foren. Tidsskr.* **29**, 22-9. CHRISTIE, D A (1983) *Br. Birds* **76**, 462-3. CHRISTIE, D S (1963) *Saint John Nat. Club Bull.* **4**, 1. CHRISTIE, G H (1927) *Scott. Nat.* **167**, 158-9. CHRISTISON, A F P (1940) *Birds of Northern Baluchistan.* Quetta; (1941) *Ibis* (14) **5**, 531-56. CHRISTOPHERS, S M (1984) *Birds in Cornwall 1983*, 11-96. CHURCHER, P B and LAWTON, J H (1987) *J. Zool. Lond.* **212**, 439-55. CICHOKI, W (1988) *Notatki Orn.* **28**, 97-8. CINK, C L (1976) *Condor* **78**, 103-4. CLANCEY, P A (1938) *Ibis* (14) **2**, 746-54; (1940) *Ibis* (14) **4**, 91-9; (1943) *Bull. Br. Orn. Club* **64**, 27-31; (1945) *Bull. Br. Orn. Club* **66**, 20-1; (1946) *Ibis* **88**, 518-19; (1947) *Bull. Br. Orn. Club* **67**, 76-7; (1948*a*) *Br. Birds* **41**, 115-16; (1948*b*) *Bull. Br. Orn. Club* **68**, 92-4; (1948*c*) *Bull. Br. Orn. Club* **68**, 132-7; (1948*d*) *Bull. Br. Orn. Club* **68**, 137-41; (1948*e*) *Ibis* **90**, 132-4; (1953) *Bull. Br. Orn. Club* **73**, 72; (1954) *Ibis* **96**, 317-18; (1989) *Cimbebasia* **11**, 111-33; (1964) *Bull. Br. Orn. Club* **84**, 110. CLARK, C C and CLARK L (1990) *Wilson Bull.* **102**, 167-9. CLARK, J H (1903) *Auk* **20**, 306-7. CLARK, L and MASON, J R (1988) *Oecologia* **77**, 174-80. CLARKE, G C W (1949) *Country Life* **105**, 1131. CLARKSON, K (1984) Ph D Thesis. Sheffield Univ. CLARKSON, K and BIRKHEAD, T R (1987) *BTO News* **151**, 8-9. CLARKSON, K, EDEN, S F, SUTHERLAND, W J, and HOUSTON, A I (1986) *J. Anim. Ecol.* **55**, 111-21. CLEGG, T M (1962) *Br. Birds* **55**, 88-9. CLEMENTS, F A (1990) *Sandgrouse* **12**, 55-6. CLENCH, M H (1973) *Orn. Monogr. AOU* **14**, 32-33. CLENCH, M H and LEBERMAN, R C (1978) *Bull. Carnegie Mus. nat. Hist.* **5**. CLERGEAU, P (1989) *Oiseau* **59**, 101-15; (1990) *J. Orn.* **131**, 458-60. CLESSE, B, DEWITTE, T, and FOUARGE, J-P (1991) *Aves* **28**, 57-74. CLUNIE, F (1976) *Notornis* **23**, 77. COCHET, P and FAURE, R (1987) *Bièvre* **9**, 83-5. CODD, R B (1947) *Br. Birds* **40**, 210. CODOUREY, J (1966) *Nos Oiseaux* **28**, 177; (1968) *Nos Oiseaux* **29**, 338-41. COE, M and COLLINS, N M (eds) (1986) *Kora: an ecological inventory of the Kora National Reserve, Kenya.* London. COHEN, E (1963) *Birds of Hampshire and the Isle* *of Wight*. Edinburgh. COLE, A (1990) *Devon Birds* **43**, 63-71. COLE, L R and FLINT, P R (1970) *Cyprus Orn. Soc. (1957) Rep.* **17**, 9-94. COLEIRO, C (1989) *Il-Merill* **26**, 1-26. COLEMAN, J D (1972) *Notornis* **19**, 118-39; (1973) *Notornis* **20**, 324-9; (1977) *Proc. New Zealand Ecol. Soc.* **24**, 94-109. COLEMAN, J D and ROBSON, A B (1975) *Proc. New Zealand Ecol. Soc.* **22**, 7-13. COLLAR, N J and STUART, S N (1985) *Threatened birds of Africa and related islands*. Cambridge. COLLETTE, P (1972) *Aves* **9**, 226-40. COLLETTE, P and FOUARGE, J (1978) *Aves* **15**, 19-29. COLLIAS, N E and COLLIAS, E C (1964) *Univ. Calif. Publ. Zool.* **73**, 1-239. COLLINGE, W E (1924-7) *The food of some British wild birds*. York; (1930) *J. Min. Agric.* (May), 151-8. COMTE, A (1926) *Bull. Soc. Zool. Genève* **3**, 34-8. CONDER, P J (1947) *Br. Birds* **40**, 212-3; (1948) *Ibis* **90**, 493-525. CONGREVE, W M (1936) *Ool. Rec.* **16**, 73-8. CONNELL, C E, ODUM, E P, and KALE, H (1960) *Auk* **77**, 1-9. CONNER, R N (1985) *Condor* **87**, 379-88. CONNOR, R J (1965) *Ool. Rec.* **39** (4), 4-9. CONRAD, E (1979) *Regulus* **13**, 83. CONRADS, K (1968) *Vogelwelt Suppl.* **2**, 7-21; (1969) *J. Orn.* **110**, 379-420; (1971) *Vogelwarte* **26**, 169-75; (1976) *J. Orn.* **117**, 438-50; (1977) *Vogelwelt* **98**, 81-105; (1984) *J. Orn.* **125**, 241-4. CONRADS, K and CONRADS, W (1971) *Vogelwelt* **92**, 81-100. CONRADS, K and QUELLE, M (1986) *Limosa* **59**, 67-74. CONSUL, C and ALVAREZ, F (1978) *Doñana Acta Vert.* **5**, 73-88. COOK, A (1975) *Bird Study* **22**, 165-8. COOKE, C H (1947) *Br. Birds* **40**, 308. COOKE, F, ROSS, R K, SCHMIDT, R K, and PAKULAK, A J (1975) *Can. Fld.-Nat.* **89**, 413-22. COOKE, H (1960) *Br. Birds* **53**, 229. COOMBS, C J F (1945) *Br. Birds* **38**, 154; (1946) *Cornwall Bird Watching Pres. Soc. Ann. Rep.* **16**, 49-50; (1960) *Ibis* **102**, 394-419; (1961) *Bird Study* **8**, 32-7, 55-70; (1978) *The crows*. London. COOPER, J and UNDERHILL, L G (1991) *Ostrich* **62**, 1-7. COOPER, J E S (1985) *Ring. Migr.* **6**, 61-5; (1987) *Sussex Bird Rep.* **39**, 88-91. COOPER, J E S and BURTON, P J K (1988) *Ring. Migr.* **9**, 93-4. COOPER, W W (1847) *Zoologist* **5**, 1775. COPETE, J L (1990) *Butll. GCA* **7**, 19-20. CORBIN, K W and SIBLEY, C G (1977) *Condor* **79**, 335-42. CORDERO, P J (1990) *Butll. GCA* **7**, 3-6; (1991) In Pinowski, J, Kavanagh, B P, and Górski, W (eds) *Proc. Int. Symp. Working Group Granivorous Birds, INTECOL*, 111-20. Warsaw. CORDERO, P J and RODRIGUEZ-TEIJEIRO, J D (1990) *Ekol. Polska* **38**, 443-52. CORDERO, P J and SALAET, M A (1987) *Publ. Dept. Zool. Barcelona* **13**, 111-16. CORDERO, P J and SUMMERS-SMITH, J D (1993) *J. Orn.* **134**, 69-77. CORKHILL, P (1973) *Bird Study* **20**, 207-20. CORNISH, A V (1947) *Br. Birds* **40**, 115. CORNWALLIS, L and PORTER, R F (1982) *Sandgrouse* **4**, 1-36. CORNWALLIS, R K and SMITH, A E (1964) *The bird in the hand*. Oxford (BTO guide 6). CORTÉS, J E (1982) *Alectoris* **4**, 26-9. CORTÉS, J E, FINLAYSON, J C, MOSQUERA, M A J, and GARCIA, E F J (1980) *The birds of Gibraltar*. Gibraltar. CORTI, U A (1935) *Bergvögel*. Bern; (1939) *Orn. Beob.* **36**, 121-40; (1949) *Einführung in die Vogelwelt des Kantons Wallis*. Chur; (1952) *Die Vogelwelt der schweizerischen Nordalpenzone*. Chur; (1959) *Die Brutvögel der deutschen und österreichischen Alpenzone*. Chur; (1961) *Die Brutvögel der französischen und italienischen Alpenzone*. Chur. CORTOPASSI, A J and MEWALDT, L R (1965) *Bird-Banding* **36**, 141-69. COULSON, J C (1960) *J. Anim. Ecol.* **29**, 251-71. COUNSILMAN, J J (1971) MA Thesis. Auckland Univ; (1974a) *Notornis* **21**, 318-33; (1974b) *Emu* **74**, 135-48; (1977) *Babbler* **1**, 1-13. COURSE, H A (1941) *Br. Birds* **35**, 154-5. COURTEILLE, C and THÉVENOT, M (1988) *Oiseau* **58**, 320-49. COVERLEY, H W (1933) *Ibis* (13) **3**, 782-5. COWDY,

S (1962) *Br. Birds* **55**, 229-33; (1973) *Bird Study* **20**, 117-20; (1976) *Birds* **6** (4), 30-32. COWIE, R J and HINSLEY, S A (1988) *Bird Study* **35**, 163-8. COX, J (1981) *Wielewaal* **47**, 322-5. COX, S (1984) *A new guide to the birds of Essex*. Ipswich. CRAGGS, J D (1967) *Bird Study* **14**, 53-60; (1976) *Bird Study* **23**, 281-4. CRAIG, A J F K (1983) *Ibis* **125**, 346-52; (1988) *Bonn. zool. Beitr.* **39**, 347-60. CRAMB, A P D (1972) *Br. Birds* **65**, 167. CRAMP, S (1969) In Gooders, J (ed.) *Birds of the world*, 2715-16; (1971) *Ibis* **113**, 244-5; (1985) *The birds of the western Palearctic* **4**. Oxford. CRAMP, S, PARRINDER, E R, and RICHARDS, B A (1957) In London Natural History Society *The birds of the London area since 1900*, 106-17. CRAMP, S and TEAGLE, W G (1952) *Br. Birds* **45**, 433-56. CRASE, F T, DE HAVEN, R W, and WORONECKI, P P (1972) *Bird-Banding* **43**, 197-204. CRAWFORD, R D and HOHMAN, W L (1978) *Bird-Banding* **49**, 201-7. CRAWHALL, E W (1952) *Country Life* **111**, 506. CREUTZ, G (1949) *Zool. Jahrb.* **78**, 133-72; (1953) *Beitr. Vogelkde.* **3**, 91-103; (1961) *Vår Fågelvärld* **20**, 302-18; (1962) *Orn. Mitt.* **14**, 64-6; (1967) *Falke* **14**, 93-6; (1970) *Falke* **17**, 426; (1981) *Der Graureiher*. Wittenberg Lutherstadt; (1988) *Beitr. Vogelkde.* **34**, 61. CREUTZ, G and FLÖSSNER, D (1958) *Beitr. Vogelkde.* **6**, 234-51. CROCQ, C (1974) *Alauda* **42**, 39-50; (1990) *Le casse-noix moucheté*. Paris. CROMBRUGGHE, S DE (1980) *Aves* **17**, 48. CROOK, J H (1963a) *J. Bombay nat. Hist. Soc.* **60**, 1-48; (1963b) *Ibis* **105**, 238-62; (1964a) *Proc. zool. Soc. Lond.* **142**, 217-55; (1964b) *Behaviour Suppl.* **10**, 1-178; (1969) In Hinde, R A (ed.) *Bird vocalizations*, 265-89. CROOK, J H and ALLEN, P M (1960) *J. E. Afr. nat. Hist. Soc.* **23**, 246. CROOK, S (1921) *Br. Birds* **15**, 10-15. CROSBY, M J (1988) *Br. Birds* **81**, 449-52. CROSSNER, K A (1977) *Ecology* **58**, 885-92. CROUCH, D J (1948) *Br. Birds* **41**, 149. CROUSAZ, G DE and LEBRETON, P (1963) *Nos Oiseaux* **27**, 46-61. CROWE, T M, BROOKE, R K, and SIEGFRIED, W R (1980) Abstr. 5th Pan-Afr. Orn. Congr. Malawi. CROZE, H (1970) *Z. Tierpsychol. Suppl.* **5**, 1-85. CRUMB, D W (1985) *Kingbird* **35**, 238-40. CRUON, R, ÉRARD, C, LEBRETON, J-D, and NICOLAU-GUILLAUMET, P (1992) *Alauda* **60**, 57-63. CRUON, R and NICHOLAU-GUILLAUMET, P (1985) *Alauda* **53**, 34-63. CRUZ, C DE LA, LOPE, F DE, and SILVA, E DA (1990) *Ardeola* **37**, 179-95. CRUZ SOLIS, C DE LA (1989) Thesis. Badajoz Univ. CRUZ SOLIS, C DE LA, LOPE REBOLLO, F DE, and SANCHEZ, J M (1992) *Ring. Migr.* **13**, 27-35. CRUZ SOLIS, C DE LA, LOPE REBOLLO, F DE, and SILVA RUBIO, E DA (1991a) *Ardeola* **38**, 101-15. CRUZ SOLIS, C DE LA, LOPE REBOLLO, F DE, and SILVA RUBIO, E DA (1991b) *Ring. Migr.* **12**, 86-90. CSIKI, E (1913) *Aquila* **20**, 375-96; (1914) *Aquila* **21**, 210-29; (1919) *Aquila* **26**, 76-104. CTYROKÝ, P (1987) *Beitr. Vogelkde.* **33**, 141-204; (1989) *Zpravy MOS* **47**, 107-24. CUCCO, M and FERRO, M (1988) *Sitta* **2**, 99-103. CUDWORTH, J (1979) *Br. Birds* **72**, 291-3. CUGNASSE, J-M (1975) *Alauda* **43**, 478-9. CUGNASSE, J-M and RIOLS, C (1987) *Nos Oiseaux* **39**, 57-65. CULLEN, J P (1989) In Bignal, E and Curtis, D J (eds) *Choughs and land-use in Europe*, 19-22. Scottish Chough Study Group. CULLEN, J M, GUITON, P E, HORRIDGE, G A, and PEIRSON, J (1952) *Ibis* **94**, 68-84. CULLEN, P (1978) *Peregrine* **4**, 264-73. CUMMING, I G (1979) *Br. Birds* **72**, 53-9. CURRY-LINDAHL, K (1962) *Vår Fågelvärld* **21**, 161-73. CURTIS, S (1969) *Passenger Pigeon* **31**, 151-9. CUSTER, T W, OSBORN, R G, PITELKA, F A, and GESSAMAN, J A (1986) *Arctic Alpine Res.* **18**, 415-27. CUSTER, T W and PITELKA, F A (1975) *Condor* **77**, 210-12; (1977) *Auk* **94**, 505-25; (1978) *Condor* **80**, 295-301. CUTHILL, I and HINDMARSH, A M (1985)

*Anim. Behav.* **33**, 326-8. CVITANIĆ, A (1980) *Larus* **31-2**, 385-419; (1986) *Larus* **36-7**, 249-52. CVITANIĆ, A and TOLIĆ, R (1988) *Larus* **40**, 129-36.

DAAN, S (1972) *Flora och Fauna* **67**, 211-14. DAANJE, A (1941) *Ardea* **30**, 1-42. DACHSEL, M (1975) *Gef. Welt* **99**, 168. DAHLÉN, B (1988) *Vår Fågelvärld* **47**, 150. DALE, E M S (1924) *Can. Fld.-Nat.* **38**, 119-20. DALE, I H (1980) *Br. Birds* **73**, 480. DALMON, J (1932) *Oiseau* **2** (NS), 339-72. DALZIEL, L, FITCHETT, A, and WYNDE, R M (1986) *Shetland Bird Rep. 1985*, 3-35. DAMSTÉ, P H (1947) *J. exp. Biol.* **24**, 20-35. DANCKER, P (1956) *J. Orn.* **97**, 430-7. DANDL, J (1959) *Aquila* **65**, 175-88. DANFORD, C G (1877-8) *Ibis* (4) **1**, 261-74, (4) **2**, 1-35. DANIELS, D and EASTON, A (1985) *Devon Birds* **38**, 3-8. DANILOV, N N, NEKRASOV, E S, DOBRINSKI, L N, and KOPEIN, K I (1969a) *Int. Stud. Sparrows* **3**, 21-7. (1969b) *Ekol. Polska A* **17**, 489-501. DANILOV, N N, RYZHANOVSKI, V N, and RYAB-ITSEV, V K (1984) *Ptitsy Yamala*. Moscow. DARE, P J (1986) *Bird Study* **33**, 179-89. DAROM, L (1979) *Israel Land Nat.* **5**, 23. DARWIN, C (1872) *The expressions of the emotions in man and animals*. London. DATHE, H (1962) *Orn. Mitt.* **14**, 56. (1971) *Beitr. Vogelkde.* **17**, 83-4. (1983) *Falke* **30**, 204. (ed.) (1986) *Handbuch des Vogelliebhabers* **2**. Berlin. DAVIDSON, A (1951) *Br. Birds* **44**, 346-8. DAVIDSON, A (1954) *A bird watcher in Scandinavia*. London. DAVIDSON, C (1985) *Trans. Norfolk Norwich Nat. Soc.* **27** (2), 98-102 705ghi. DAVIES, M (1988) In Cadbury, C J and Everett, M (eds) *RSPB Cons. Rev.* **2**, 91-4. DAVIS, D E (1959) *Ecology* **40**, 136-9. (1960) *Bird-Banding* **31**, 216-9. DAVIS, G (1928) *Avic. Mag.* **6**, 241-7. (1930) *Avic. Mag.* **8**, 289-94. DAVIS, J (1953) *Condor* **55**, 117-20. (1954) *Condor* **56**, 142-9. (1957) *Condor* **59**, 195-202. DAVIS, M (1951) *Auk* **68**, 529-30. DAVIS, P (1954) *Br. Birds* **47**, 21-3. (1963) *Bird Migration* **2**, 260-4. (1964) *Br. Birds* **57**, 477-501. DAVIS, P and DENNIS, R H (1959) *Br. Birds* **52**, 419-21. DAVIS, P E and DAVIS, J E (1986) *Nat. Wales* **3** (NS) (1984), 44-54. DAVIS, P G (1976) *Ring. Migr.* **1**, 115-6. (1977) *Bird Study* **24**, 127-9. DAVYDOV, A F (1976) *Dokl. Uchast. 2 Vsesoyuz. Konf. Poved. Zhiv.* 87-9. Moscow. DAVYDOVICH, L I and GORBAN', I M (1990) *Ornitologiya* **24**, 147 DAWSON, A (1991) *Ibis* **133**, 312-6. DAWSON, R (1975) *Br. Birds* **68**, 159-60. DAY, D H (1975) *Ostrich* **46**, 192-4. DEAN, F (1947) *Br. Birds* **40**, 191. DEBRU, H (1958) *Oiseau* **28**, 112-22. (1961) *Oiseau* **31**, 100-10. DE BRUN, N (1988) *Ornis Flandriae* **7**, 10. DECKERT, G (1962) *J. Orn.* **103**, 428-86. (1968a) *Beitr. Vogelkde.* **14**, 97-102. (1968b) *Der Feldsperling*. Wittenberg Lutherstadt. (1969) *Beitr. Vogelkde.* **15**, 1-84. (1980) *Beitr. Vogelkde.* **26**, 305-34. DEELDER, C L (1949) *Ardea* **37**, 1-88. (1952) *Ardea* **40**, 63-6. DEHN, W (1990) *Gef. Welt* **114**, 356. DEIGNAN, H G (1945) *US natn. Mus. Bull.* **186**. DEJONGHE, J-F (1984) *Les oiseaux de montagne*. Maisons-Alfort. DEJONGHE, J F and CZAJKOWSKI, M A (1983) *Alauda* **51**, 27-47. DEKEYSER, P L and VILLIERS, A (1950) *Bull. IFAN* **12** (3), 660-99. DELACOUR, J (1929) *Ibis* (12) **5**, 403-29. (1935) *Oiseau* **5** (NS), 377-88. (1943) *Zoologica* **28**, 69-86. (1947) *Birds of Malaysia*. New York. DELAMAIN, J (1912) *Rev. fr. Orn.* **2**, 298-302, 322-5. (1929) *Alauda* **1**, 59-63. DELANY, S, CHADWELL, C, and NORTON, J (1982) In *Univ. Southampton Ladakh Exped. Rep.* (1980), 5-153. Southampton Univ. DELAVELEYE, R (1964) *Gerfaut* **54**, 12-15. DELBOVE, P and FOUILLET, F (1986) *Oiseau* **56**, 77. DELESTRADE, A (1989) In Bignal, E, and Curtis, D J (eds) *Choughs and land-use in Europe*, 70-71.

Scottish Chough Study Group. DELESTRADE, A (1991) In Curtis, D J, Bignal, E M, and Curtis, M A (eds) *Birds and pastoral agriculture in Europe*, 72-5. Scottish Chough Study Group. DELIN, H and SVENSSON, L (1988) *Photographic guide to the birds of Britain and Europe*. London. DELMOTTE, C (1981) *Bull. Rech. Agron. Gembloux* **16**, 99-110. DELMOTTE, C, and DELVAUX, J (1981) *Aves* **18**, 108-18. DELVAUX, J (1983) *Aves* **20**, 174-5. DELVINGT, W (1961) *Gerfaut* **51**, 53-63. (1962) *J. Orn.* **103**, 260-5. DEMARTIS, A M (1987) *Gli uccelli Ital.* **12**, 15-26. DÉMENTIEFF, G (1934) *Oiseau* **4**, 525-9. DÉMENTIEFF, G P (1935) *Alauda* **7**, 153-69. DEMENTIEV, G P and GLADKOV, N A (1954) *Ptitsy Sovetskogo Soyuza* **5**. Moscow. DEMENTIEV, G P, KARTASHEV, N N, and SOLDATOVA, A N (1953) *Zool. Zh.* **32**, 361-75. DEMENTIEV, G P, KARTASHEV, N N, and TASHLIEV, A O (1956) *Trudy Inst. Biol. Akad. Nauk Turkmen. SSR* **4**, 77-119. DEMENTIEV, G P and PTUSHENKO, E (1939) *Ibis* (14) **3**, 507-12. DEMENTIEW, G (1937) *Orn. Monatsber.* **45**, 86-7. DENDALETCHE, C (1991) In Curtis, D J, Bignal, E M and Curtis, M A (eds) *Birds and pastoral agriculture in Europe*, 68-9. Scottish Chough Study Group. DENDALETCHE, C and SAINT-LÈBE, N (1988) *Acta biol. mont.* **8** (no. spéc.), 147-70. (1991) *Acta biol. mont.* **10**, 45-50. DENIS, J V (1981) *N. Amer. Bird Band.* **6**, 88-96. DENISOVA, M N and GOMOLITSKAYA, R D (1967) *Ornitologiya* **8**, 344-5. DENNEMAN, W D (1981) *Vogeljaar* **29**, 194-203. DENNIS, R H (1969) *Br. Birds* **62**, 144-8. DENNY, J (1950) *Br. Birds* **43**, 333. DENSLEY, M (1990) *Br. Birds* **83**, 195-201. DE RUWE, F, DE PUTTER, G, and VANPRAET, J (1990) *Mergus* **4**, 166-70. DESBROSSE, A and ETCHEBERRY, R (1986) *Oiseau* **56**, 291-4. DESFAYES, M (1951) *Nos Oiseaux* **21**, 132. (1969) *Oiseau* **39**, 21-7. DESFAYES, M and PRAZ, J C (1978) *Bonn. zool. Beitr.* **29**, 18-37. DESMET, J (1981) *Gerfaut* **71**, 627-57. DESTRE, R (1984) Doctoral Thesis. Languedoc Univ. DEUNERT, J (1981) *Beitr. Vogelkde.* **27**, 125-6. (1989) *Abh. Ber. Naturkundemus. Görlitz* **63** (2). DEVILLERS, P, ROGGEMAN, W, TRICOT, J, MARMOL, P DEL, KERWIJN, C, JACOB, J-P, and ANSELIN, A (1988) *Atlas des oiseaux nicheurs de Belgique*. Brussels. DEWOLFE, B B (1967) *Condor* **69**, 110-32. DEWOLFE, B B, WEST, G C, and PEYTON, L J (1973) *Condor* **75**, 43-59. DHARMAKUMARSINHJI, R S (1955) *Birds of Saurashtra, India*. Bhavnagar. DHINDSA, M S (1983) *Ibis* **125**, 243-5. (1986) *Indian Rev. Life Sci.* **6**, 101-40. DHINDSA, M S and BOAG, D A (1989a) *Ibis* **132**, 595-602. (1989b) *Ornis scand.* **20**, 76-9. DHINDSA, M S, KOMERS, P E, and BOAG, D A (1989) *Can. J. Zool.* **67**, 228-32. DHINDSA, M S and TOOR, H S (1981) *Indian J. Ecol.* **8**, 156-62. (1990) In Pinowski, J and Summers-Smith, J D (eds) *Granivorous birds in the agricultural landscape*, 217-36. Warsaw. DHONDT, A A, and SMITH, J N M (1980) *Can. J. Zool.* **58**, 513-20. DIAMOND, A W, LACK, P, and SMITH, R W (1977) *Wilson Bull.* **89**, 456-66. DIAMOND, A W and SMITH, R W (1973) *Bird-Banding* **44**, 221-4. DÍAZ, J A and ASENSIO, B (1991) *Bird Study* **38**, 38-41. DI CARLO, E A (1950) *Riv. ital. Orn.* **26**, 55-61. DI CARLO, E A and LAURENTI, S (1991) *Gli uccelli Ital.* **16**, 81-96. DICE, L R (1918) *Condor* **20**, 129-31. DICK, W (1973) *Beitr. Vogelkde.* **19**, 397-405. DICKINSON, E C, KENNEDY, R S, and PARKES, K C (1991) *The birds of the Philippines*. Tring. DICKINSON, B H B and DOBINSON, H M (1969) *Bird Study* **16**, 135-46. DICKINSON, J C (1952) *Bull. Mus. comp. Zool.* **107**, 271-352. DICKSON, R C (1969) *Br. Birds* **62**, 497. (1972) *Br. Birds* **65**, 221-2. DIEN, J (1965) *Hamburger avifaun. Beitr.* **2**, 120-94. DIERSCHKE, J (1989) *Limicola* **3**, 246—51. DIESING, P (1984) *Beitr.*

*Naturk. Niedersachs.* **37**, 196-7. (1987) *Beitr. Naturk. Niedersachs.* **40**, 302. DIESSELHORST, G (1949) *Orn. Ber.* **2**, 1-31. (1950) *Orn. Ber.* **3**, 69-112. (1956) *Vogelwelt* **77**, 190. (1962) *Stuttgarter Beitr. Nturkde* **86**. (1968) *Khumbu Himal* **2**. Innsbrück. (1971*a*) *Anz. orn. Ges. Bayern* **10**, 38-42. (1971*b*) *Vogelwelt* **92**, 201-26. (1986) *Mitt. Zool. Mus. Berlin* **62**, Suppl. *Ann. Orn.* **10**, 3-23. DIESSELHORST, G and POPP, K (1963) *Vogelwelt* **84**, 184-90. DIJKSEN, A J (1976) *Limosa* **49**, 204-6. DIJKSEN, L J (1989) *Limosa* **62**, 48. DIJKSEN, L J and KLEMANN, M (1992) *OSME Bull.* **28**, 21. DILGER, W C (1960) *Wilson Bull.* **72**, 115-32. DIMITROPOULOS, A (1987) *Herptile* **12**, 72-81. DIMOVSKI, A, and MATVEJEV, S (1955) *Arhiv Biol. Nauka Srpsko Biol. Društvo* **7**, 121-138. DINETTI, M and ASCANI, P (1990) *Atlante degli uccelli nidificanti nel comune di Firenze.* Florence. DIRECTIE NMF (1989) *Limosa* **62**, 11-14. DITTBERNER, H and DITTBERNER, W (1971) *Falke* **18**, 418-23. (1992) *Orn. Mitt.* **44**, 123-7. DITTBERNER, H, DITTBERNER, W, and LENZ, M (1969) *Vogelwelt* **90**, 225-33. DITTBERNER, H, DITTBERNER, W, and SADLIK, J (1979) *Falke* **26**, 296-8. DITTBERNER, H and KAGE, J (1991) *Beitr. Vogelkde.* **37**, 239-49. DITTRICH, W (1981) *J. Orn.* **122**, 181-5. DOBBEN, W H VAN (1949) *Ardea* **37**, 89-97. DOBBRICK, L (1931) *Abh. Westfäl. Mus. Naturk.* **2**, 27-33. DOBBRICK, W (1933) *Orn. Monatsber.* **41**, 55-6. DOBRYNINA, I N (1981) *Tez. Doklad. 10 Pribalt. orn. Konf.* **1**, 107-10. (1982) *Ornitologiya* **17**, 181. DOBRYNINA, I N (1986) In Sokolov, V E and Dobrynina, I N (eds) *Kol'tsevanie i mechenie ptits v SSSR, 1979-82, gody*, 35-167. Moscow. DOBSON, A P (1987) *Ornis Scand.* **18**, 122-8. DODSWORTH, P T L (1911) *J. Bombay nat. Hist. Soc.* **21**, 248-9. DOERBECK, F (1963) *Vogelwelt* **84**, 97-114. DOHERTY, P (1992) *Br. Birds* **85**, 595-600. DOHRN, H (1871) *J. Orn.* **19**, 1-10. DOÏCHEV, R L (1973) *Ann. Univ. Sofia Fac. Biol.* **65**, 1-10. DOLGUSHIN, I A (1968) *Trudy Inst. Zool. Akad. Nauk Kaz. SSR* **29**, 15-18. DOLGUSHIN, I A, KORELOV, M N, KUZ'MINA, M A, GAVRILOV, E I, GAVRIN, V F, KOVSHAR', A F, BORODIKHIN, I F, and RODIONOV, E F (1970) *Ptitsy Kazakhstana* **3**. Alma-Ata. DOL'NIK, V R (1975) *Zool. Zh.* **54**, 1048-56. DOL'NIK, V R (1980) *Zool. Zh.* **59**, 91-9. DOL'NIK, V R (1982) *Populyatsionnaya ekologiya zyablika.* Leningrad. DOL'NIK, V R and BLYUMENTHAL, T I (1967) *Condor* **69**, 435-68. DOL'NIK, V R and GAVRILOV, V M (1974) *Ornitologiya* **11**, 110-25. (1975) *Ekol. Polska* **23**, 211-26. (1979) *Auk* **96**, 253-64. (1980) *Auk* **97**, 50-62. DONALD, P F, WILSON, J D, and SHEPHERD, M (in press) *Br. Birds.* DONČEV, S (1958) *Bull. Inst. zool. Acad. Sci. Bulgarie* **7**, 269-313. (1981) *Aquila* **87**, 27-9. DONNER, E (1908) *Orn. Monatsschr.* **33**, 30-8. DONOVAN, J W (1978) *Nat. Wales* **16**, 142-3. (1984) *Br. Birds* **77**, 491. DONTSCHEV, S (1986) *Suppl. alle Ric. di Biologia della Selvaggina* **10**, 117-21. DORKA, U (1986) *Orn. Jahresh. Bad.-Württ.* **2**, 57-71. DORKA, V (1966) *Orn. Beob.* **63**, 165-223. (1973) *Anz. orn. Ges. Bayern* **12**, 114-21. DORKA, V, PFAU, K, and SPAETER, C (1970) *J. Orn.* **111**, 495-6. DORN, J L (1972) M Sc Thesis. Wyoming. DORNBERGER, W (1977) *Auspicium* **6**, 163-74. (1978) *Anz. orn. Ges. Bayern* **17**, 335-7. (1979) *Vogelwarte* **30**, 28-32. (1982) *Voliere* **5**, 59-61. (1983) *Verh. orn. Ges. Bayern* **23**, 501-9. DORNBUSCH, M (1972) *Apus* **2**, 286. (1973) *Falke* **20**, 193-5. (1981) *Beitr. Vogelkde.* **27**, 73-99. DOROGOY, I V (1982) *Ornitologiya* **17**, 119-24. DORSCH, H (1970) *Beitr. Vogelkde.* **15**, 437-51. DOS SANTOS, J R, DOS SANTOS, J N, and PEREIRA, A DE J (1985) *Cyanopica* **3**, 269-308. DOST, H (1957) *Falke* **4**, 101-3. DOUHAN, B (1978) *Vår Fågelvärld* **37**, 33-6. Dow,

D D (1966) *Ontario Bird-Banding* **2**, 1-14. DRESSER, H E (1871-81) *A history of the birds of Europe.* London. DRIVER, R (1957) *Br. Birds* **50**, 397-8. DROST, R (1930) *Vogelzug* **1**, 69-72. (1940*a*) *Orn. Monatsber.* **48**, 61-2. (1940*b*) *Vogelzug* **11**, 65-70. DROZDOV, N N (1965) *Ornitologiya* **7**, 166-99. (1968) *Ornitologiya* **9**, 345-7. DROZDOV, N N and ZLOTIN, R I (1962) *Ornitologiya* **5**, 193-207. DRURY, W H (1961) *Bird-Banding* **32**, 1-46. DRURY, W H and KEITH, J A (1962) *Ibis* **104**, 449-89. DUBALE, M S and PATEL, G (1975) *Pavo* **10**, 8-20. DUBERY, P (1983) *Birds* **9** (6), 71. DUBININ, N P (1953) *Trudy Inst. Lesa Akad. Nauk SSSR* **18**. DUBOIS, P J and COMITÉ D'HOMOLOGATION NATIONAL (1989) *Alauda* **57**, 263-94; (1990) *Alauda* **58**, 245-66. DUBOIS, P and DUHAUTOIS, L (1977) *Alauda* **45**, 285-91. DUBOIS, P J and YÉSOU, P (1986) *Inventaire des espèces d'oiseaux occasionnelles en France.* Paris. (1992) *Les oiseaux rares en France.* Bayonne. DUCHROW, H (1959) *Orn. Mitt.* **11**, 12. DUFF, A G (1979) *Alauda* **47**, 216-17. DUCKWORTH, E (1983) *Bird Life* (May-June), 43. DUIN, G VAN (1992) *Dutch Birding* **14**, 173-6. DUNAJEWSKI, A (1938) *Acta Orn. Mus. Zool. Polonici* **2**, 145-56. DUNLOP, E B (1917) *Br. Birds* **10**, 278-9. DUNN, E H (1990) *Ontario Birds* **7**, 87-91. DUNN, P J (1985) *Br. Birds* **78**, 151-2. (1990) *Br. Birds* **83**, 123-4. DUNN, P O and HANNON, S J (1989) *Auk* **106**, 635-44. DUNNET, G M (1955) *Ibis* **97**, 619-62. (1956) *Ibis* **98**, 220-30. DUNNET, G M, FORDHAM, R A, and PATTERSON, I J (1969) *J. appl. Ecol.* **6**, 459-73. DUPOND, C (1939) *Gerfaut* **29**, 185-203. DUPONT, P-L (1944) *Gerfaut* **34**, 23-8. DUPUY, A (1966) *Oiseau* **36**, 256-68. (1969) *Oiseau* **39**, 225-41. DUPUY, A and JOHNSON, E D H (1967) *Oiseau* **37**, 143. DUQUET, M (1984) *Nos Oiseaux* **37**, 331-40. (1986) *Nos Oiseaux* **38**, 263-8. DUQUET, M and PÉPIN, D (1987) *Nos Oiseaux* **39**, 170-1. DURAND, A L (1963) *Br. Birds* **56**, 157-64. DURANGO, S (1948) *Alauda* **16**, 1-20. DUREL, J (1927) *Oiseau* **8**, 235-43. DURHAM, M E and SELLERS, R M (1984) *Gloucs. Bird Rep.* **22**, 57-60. DURNEV, Y A, SONIN, V D, and SIROKHIN, I N (1984) *Ornitologiya* **19**, 177-8. DURNEV, Y A, SONIN, V D, LIPIN, S I, and SIROKHIN, I N (1991) In: Tsyrenov, V Z (ed.) *Ekologiya i fauna ptits Vostochnoy Sibiri*, 45-54. Ulan-Ude. DWENGER, R (1989) *Die Dohle.* Wittenberg Lutherstadt. DWIGHT, J (1897) *Auk* **14**, 259-72. (1900) *Ann. New York Acad. Sci.* **13**, 73-360. DWIGHT, J and GRISCOM, L (1927) *Amer. Mus. Novit.* **257**. DYBBRO, T (1976) *De danske ynglefugles udbredelse.* Copenhagen. DYER, M I, PINOWSKI, J, and PINOWSKA, B (1977) In Pinowski, J and Kendeigh, S C (eds) *Granivorous birds in ecosystems*, 53-105. London. DYMOND, J N (1991) *The birds of Fair Isle.* Edinburgh. DYMOND, J N, FRASER, P A, and GANTLETT, S J M (1989) *Rare birds in Britain and Ireland.* Calton. DYRCZ, A (1966) *Acta Orn.* **9**, 227-40.

EADES, R A (1984) *Br. Birds* **77**, 616-17. EAGLE CLARKE, W (1904) *Ibis* (8) **4**, 112-42. EARLÉ, R and GROBLER, N J (1987) *First atlas of bird distribution in the Orange Free State.* Bloemfontein. EAST, M (1988) *Ibis* **130**, 294-9. EASTWOOD, E, ISTED, G A, and RIDER, G C (1962) *Proc. Roy. Soc. Lond.* (B) **156**, 242-67. EATON, S W (1957*a*) *Auk* **74**, 229-39; (1957*b*) *Sci. Studies St. Bonaventura Univ.* **19**, 7-36. EBELS, E B (1991) *Dutch Birding* **13**, 86-9. EBER, G (1956) *Biol. Abh.* **13-14**, 1-60. ECK, S (1975) *Zool. Abh. Staatl. Mus. Tierkde. Dresden* **33**, 277-302; (1977) *Falke* **24**, 114-7; (1981) *Zool. Abh. Staatl. Mus. Tierkde Dresden* **37**, 183-207; (1984) *Zool. Abh. Staatl. Mus. Tierkde Dresden* **40**, 1-32; (1985*a*) *Zool. Abh. Staatl. Mus. Tierkde*

Dresden **40**, 79-108; (1985*b*) *Zool. Abh. Staatl. Mus. Tierkde Dresden* **41**, 1-32; (1990) *Zool. Abh. Staatl. Mus. Tierkde Dresden* **46**, 1-55. ECK, S and PIECHOCKI, R (1988) *Zool. Abh. Staatl. Mus. Tierkde Dresden* **43**, 135-41. EDDINGER, C R (1967) *Elepaio* **28**, 1-5, 11-18. EDELSTAM, C (1972) *Vår Fågelvärld*, supp. 7. EDEN, S F (1985) *J. Zool. Lond.* (A) **205**, 325-34; (1987*a*) *Ibis* **129**, 477-90; (1987*b*) *Anim. Behav.* **35**, 608-10; (1987*c*) *Anim. Behav.* **35**, 764-72; (1989) *Ibis* **131**, 141-53. EDGAR, A T (1972) *Notornis* **19** suppl. 89; (1974) *Notornis* **21**, 349-78. EDHOLM, M (1979) *Vår Fågelvärld* **38**, 106-7; (1980) *Vår Fågelvärld* **39**, 102. EDQVIST, T (1945) *Fauna och Flora* **40**, 92-3. EDWARDS, K D and OSBORNE, K C (1972) *Br. Birds* **65**, 203-5. EELLS, M M (1980) *Murrelet* **61**, 36-7. EENS, M (1992) Ph D Thesis. Antwerp Univ. EENS, M and PINXTEN, R (1990) *Ibis* **132**, 618-19. EENS, M, PINXTEN, R, and VERHEYEN, R F (1989) *Ardea* **77**, 75-86. EENS, M, PINXTEN, R, and VERHEYEN, R F (1990) *Bird Study* **37**, 48-52. EENS, M, PINXTEN, R, and VERHEYEN, R F (1991*a*) *Behaviour* **116**, 210-38. EENS, M, PINXTEN, R, and VERHEYEN, R F (1991*b*) *Belg. J. Zool.* **121**, 257-78. EENS, M, PINXTEN, R, and VERHEYEN, R F (1992*a*) *Ibis* **134**, 72-6. EENS, M, PINXTEN, R, and VERHEYEN, R F (1992*b*) *Anim. Behav.* EGGELING, W J (1960) *The Isle of May*. Edinburgh. EGGERMONT, D (1956) *Gerfaut* **46**, 17-21. EHRENROTH, B and JANSSON, B (1966) *Vår Fågelvärld* **25**, 97-105. EHRLICH, P R, DOBKIN, D S, and WHEYE, D (1986) *Auk* **103**, 835. EIFLER, G (1990) *Abh. Ber. Naturkundemus. Görlitz* **64** (2). EIFLER, G and BLÜMEL, H (1983) *Abh. Ber. Natkundemus. Görlitz* **57** (2). EISENHUT, E and LUTZ, W (1936) *Mitt. Vogelwelt* **35**, 1-14. EISENMANN, E (1969) *Bird-Banding* **40**, 144-5. EISFELD, D, STRÖDE, P, and OPHOVEN, E (1991) In Glandt, D (ed.) *Der Kolkrabe (Corvus corax) in Mitteleuropa. Metelener Schriftenreihe Naturschutz* **2**, 41-3. Metelen. EKELÖF, O and KUSCHERT, H (1979) *Corax* **7**, 37-9. EKSTRÖM, S (1952) *Vår Fågelvärld* **11**, 36. ELCAVAGE, P and CARACO, T (1983) *Anim. Behav.* **31**, 303-4. ELEY, C (1991) D Phil Thesis. Sussex Univ. ELGAR, M A and CATTERALL, C P (1982) *Emu* **82**, 109-11. ELGOOD, J H (1982) *The birds of Nigeria*. London. ELGOOD, J H, FRY, C H, and DOWSETT, R J (1973) *Ibis* **115**, 1-45, 375-411. ELISEEV, D O (1985) In Prokofieva, I V (ed.) *Ekologiya ptits i reproduktivnyi period*, 3-10. St Petersburg. ELISEEVA, V I (1961) *Zool. Zh.* **40**, 583-91. ELKINS, N (1985) *Br. Birds* **78**, 51-2 ELLIOT, R D (1985) *Anim. Behav.* **33**, 308-14. ELLIOTT, C C H and JARVIS, M J F (1970) *Ostrich* **41**, 1-117; (1973) *Ostrich* **44**, 34-78. ELLIOTT, S (1991) *Wildlife Sound* **6** (6), 28-31. ELLIS, C R (1966) *Wilson Bull.* **78**, 208-24. ELLISON, A (1910) *Br. Birds* **3**, 300-2. ELMBERG, J (1991*a*) *Br. Birds* **84**, 344-5; (1991*b*) *Vår Fågelvärld* **50** (2), 38-9; (1992) *Birding World* **5**, 193. ELSACKER, L VAN, PINXTEN, R, and VERHEYEN, R F (1988) *Behaviour* **107**, 122-30. ELVERS, H, PFEIFFER, K, and WESTPHAL, D (1974) *Orn. Mitt.* **26**, 83-6. ELY, C A, LATAS, P J, and LOHOEFENER, R R (1977) *Bird-Banding* **48**, 275-6. ELZEN, R VAN DEN (1983) *Girlitze: Biologie, Haltung and Pflege*. Baden-Baden. ELZEN, R VAN DEN, and KÖNIG, C (1983) *Bonn. zool. Beitr.* **34**, 149-96. ELZEN, R VAN DEN, KÖNIG, C, and WOLTERS, H E (1978) *Bonn. zool. Beitr.* **29**, 323-59. EMLEN, S T (1971) *Anim. Behav.* **19**, 407-8. EMMERSON, K, MARTÍN, A, and BACALLADO, J J (1982) *Doñana Acta Vert.* **9**, 408-9. ENA, V (1984*a*) *Alytes* **2**, 144-59; (1984*b*) *Ibis* **126**, 240-9. ENA ALVAREZ, V (1979) Doctoral Thesis. Oviedo Univ. ENCKE, F-W (1965) *Beitr. Vogelkde.* **11**, 153-84. ENEMAR, A (1963) *Acta Univ. Lund* **2** (58), 1-21;

(1964) *Flora och Fauna* **59**, 1-23; (1969) *Vår Fågelvärld* **28**, 230-5. ENEMAR, A, HANSON, S Å, and SJÖSTRAND, B (1965) *Acta Univ. Lund* **2** (5), 1-11. ENEMAR, A and NYSTRÖM, B (1981) *Vår Fågelvärld* **40**, 409-26. ENEMAR, A and SJÖSTRAND, B (1970) *Bull. ecol. Res. Comm. Lund* **9**, 33-7. ENGGIST-DÜBLIN, P VON (1988) *Beih. Veröff. Natursch. Landschaftsplege Bad.-Württ.* **53**, 175-82 ENGLAND, M D (1945*a*) *Br. Birds* **38**, 274; (1945*b*) *Br. Birds* **38**, 315; (1951) *Br. Birds* **44**, 386; (1970) *Br. Birds* **63**, 385-7; (1974) *Br. Birds* **67**, 218. ENNION, E A R and ENNION, D (1962) *Ibis* **104**, 158-68. ENQUIST, M and PETTERSSON, J (1986) *Spec. Rep. Ottenby Bird Obs.* **8**. EPPRECHT, W (1965) *Orn. Beob.* **62**, 118. ERARD, C (1964) *Alauda* **32**, 105-28; (1966) *Alauda* **34**, 102-19; (1970) *Alauda* **38**, 1-26. ERARD, C (1968) *Bulletin du Centre de Recherches sur les Migrations des Mammifères et des Oiseaux* **19**, 3-62. ERARD, C and ETCHÉCOPAR, R-D (1970) *Mém. Mus. natn. Hist. nat.* (A) **66**. ERARD, C, and LARIGAUDERIE, F (1972) *Oiseau* **42**, 81-169. ERDMANN, E (1972) *Vogelk. Ber. Niedersachs.* **4**, 13-14. ERDMANN, G (1985) *Falke* **32**, 84-7. ERIKSEN, J (1990) *Oman Bird News* **9**, 10-13. ERIKSSON, K (1970*a*) *Ann. zool. Fenn.* **7**, 273-82; (1970*b*) *Ornis fenn.* **47**, 52-68; (1970*c*) *Sterna* **9**, 77-90. ERIKSSON, M and HANSSON, J-Å (1973) *Vår Fågelvärld* **32**, 11-22. ERIKSTAD, K E, BLOM, R, and MYRBERGET, S (1982) *J. Wildl. Mgmt* **46**, 109-14. ERLANGER, C F VON (1899) *J. Orn.* **47**, 449-532. ERNST, S (1983*a*) *Falke* **30**, 150-6; (1983*b*) *Naturschutzarb. Sachsen* **25**, 22-6; (1986) *Falke* **33**, 28-9; (1988) *Mitt. Zool. Mus. Berlin* **64**, *Suppl. Ann. Orn.* **12**, 3-50; (1990) *Beitr. Vogelkde.* **36**, 65-108. ERNST, S and THOSS, M (1977) *Falke* **24**, 48-53. ERPINO, M J (1968*a*) *Condor* **70**, 91-2; (1968*b*) *Condor* **70**, 154-65. ERSKINE, A J (1964) *Murrelet*, **45**, 15-22; (1971) *Wilson Bull.* **83**, 352-370; (1977) *Can. Wildl. Serv. Rep. Ser.* **41**; (1985) *Can. Field-Nat.* **99**, 188-95; (1992) *Atlas of breeding birds of the Maritime Provinces*. Halifax. ERSKINE, A J and DAVIDSON, G S (1976) *Syesis* **9**, 1-11. ERZ, W (1968) *Anthus*, **5**, 4-8. ESPMARK, Y (1972) *Fauna och Flora* **67**, 250-3. ESTARRIOL JIMÉNEZ, M (1974) *Ardeola* **20**, 330-1. ESTEN, S R (1931) *Auk* **48**, 572-4. ETCHEBERRY, R (1982) *Les oiseaux de St. Pierre et Miquelon*. Saint-Pierre. ETCHÉCOPAR, R-D (1961) *Oiseau* **31**, 158-9. ETCHÉCOPAR, R-D and HÜE, F (1967) *The birds of North Africa*. Edinburgh; (1983) *Les oiseaux de Chine, de Mongolie et de Corée: passereaux*. Paris. EVANS, A D (1992) *Bird Study* **39**, 17-22. EVANS, G H (1965) *Br. Birds* **58**, 457-61. EVANS, P G H (1980) D Phil Thesis. Oxford; (1986) *Ibis* **128**, 558-61; (1988) *Anim. Behav.* **36**, 1282-94. EVANS, P R (1966) *Ibis* **108**, 183-216; (1969*a*) *Condor* **71**, 316-30; (1969*b*) *J. Anim. Ecol.* **38**, 415-23; (1971) *Ornis fenn.* **48**, 131-2. EVANS, P R, ELTON, R A, and SINCLAIR, G R (1967) *Ornis fenn.* **44**, 33-41. EVANS, S M (1970) *Anim. Behav.* **18**, 762-7. EVERETT, M (1988) *Orn. Soc. Middle East Bull.* **20**, 3-5. EWERT, D N, and LANYON, W E (1970) *Auk* **87**, 362-3. EWIN, J P (1976) *Ibis* **118**, 468-9; (1978) Ph D Thesis, Univ. London. EWINS, P J (1979) *Ornithological observations in Morocco*. Unpubl; (1986) *Scott. Bird News* **2**, 2; (1989) *Br. Birds* **82**, 331. EWINS, P J and DYMOND, J N (1984) *Shetland Bird Rep. 1983*, 48-57. EWINS, P J, DYMOND, J N, and MARQUISS, M (1986) *Bird Study* **33**, 110-6. EXCELL, J, KORKOLAINEN, V, and LINKOLA, P (1974) *Lintumies* **9**, 40-4. EYBERT, M-C (1980) *Oiseau* **50**, 295-7; (1985) Thèse Univ. Rennes I. EYCKERMAN, R, LOUETTE, M, and BECUWE, M (1992) *Zool. Middle East* **6**, 29-37. EYGELIS, Y K (1958) *Vestnik Leningrad. Univ.* **3**, *Biol.* **1**, 108-15; (1961) *Zool. Zh.* **40**, 888-

99; (1964) *Zool. Zh.* **43**, 1517-29; (1965) *Zool. Zh.* **44**, 95-100; (1970) *Zool. Zh.* **49**, 892-7.

FAABORG, J and ARENDT, W J (1984) *J. Fld. Orn.* **55**, 376-8. FAIRHURST, A R (1970) *Br. Birds* **63**, 387; (1974) *Br. Birds* **67**, 215. FAIRON, J (1971) *Gerfaut* **61**, 146-61; (1975) *Gerfaut* **65**, 107-34; (1972) *Gerfaut* **62**, 325-30. FALCONER, D S (1941) *Br. Birds* **35**, 98-104. FALLET, M (1958a) *Schrift. naturwiss. Ver. Schleswig-Holstein* **29**, 39-46; (1958b) *Zool. Anz.* **161**, 178-87. FARINHA, J C (1989) In Bignal, E and Curtis, D J (eds) *Choughs and land-use in Europe*, 89-93. Scottish Chough Study Group. FARNER, D S, DONHAM, R S, MOORE, M C, and LEWIS, R A (1980) *Auk* **97**, 63-75. FARRAND, J (ed.) (1983) *The Audubon Society Master Guide to Birding* **3**. New York. FARRAR, R B (1966) *Auk* **83**, 616-22. FASOLA, M and BRICHETTI, P (1983) *Avocetta* **7**, 67-84. FASOLA, M, PALLOTTI, E, CHIOZZI, G, and BALESTRAZZI, E (1986) *Riv. ital. Orn.* **56**, 172-80. FATIO, V and STUDER, T (1889) *Catalogue des Oiseaux de la Suisse.* Genève. FAWKS, E and PETERSEN, P, JR (1961) *A field list of birds of the tri-city region.* Davenport. FAXÉN, L (1945) *Vår Fågelvärld* **4**, 18-26. FEARE, C J (1974) *J. appl. Ecol.* **11**, 897-914; (1975) *Bull. Br. Orn. Club* **95**, 48-50; (1976) *J. Bombay nat. Hist. Soc.* **73**, 525-7; (1978) *Ann. appl. Biol.* **88**, 329-34; (1980) *Proc. int. orn. Congr.* **17** (2), 1331-6; (1981) In Thresh, J M (ed.) *Pests, pathogens and vegetation*, 393-400. London; (1984) *The Starling*, Oxford; (1986) *Gerfaut* **76**, 3-11; (1991) *Ibis* **133**, 75-9. FEARE, C J (1993) *Wilson Bull.* **105**. FEARE, C J and BURHAM, S E (1978) *Bird Study* **25**, 189-91. FEARE, C J and CONSTANTINE, D A T (1980) *Bird Study* **27**, 119-20. FEARE, C J, FRANSSU, P D DE, and PERIS, S J (1992) *Proc. Vert. Pest Contr. Conf.* **15**, 83-8. FEARE, C J, GILL, E L, MCKAY, H V, and BISHOP, J D (in press) *Ibis*. FEARE, C J and INGLIS, I R (1979) *Ornis scand.* **10**, 42-7. FEARE, C J and McGINNITY, N (1986) *Bird Study* **33**, 164-7. FEARE, C J and MUNGROO, Y (1989) *Bull. Br. Orn. Club* **109**, 199-201. FEARE, C J, DUNNET, G M, and PATTERSON, I J (1974) *J. appl. Ecol.* **11**, 867-96. FEDERSCHMIDT, A (1988) *Ökol. Vögel* **10**, 151-64. FEDIUSCHIN, A V (1927) *J. Orn.* **75**, 490-5. FEDYUSHIN, A V and DOLBIK, M S (1967) *Ptitsy Belorussii.* Minsk. FEENEY, P P, ARNOLD, R W, and BAILEY, R S (1968) *Ibis* **110**, 35-86. FEIJEN, H R (1976) *Limosa* **49**, 28-67. FELLENBERG, W (1986) *Charadrius* **22**, 199-215; (1988a) *Charadrius* **24**, 85-7; (1988b) *Charadrius* **24**, 92-5. FELLENBERG, W and PFENNIG, H G (1986) *Charadrius* **22**, 216-20. FELTON, C (1969a) *Br. Birds* **62**, 80; (1969b) *Br. Birds* **62**, 445-6. FENK, R (1911a) *Orn. Monatsschr.* **36**, 233-44; (1911b) *Orn. Monatsschr.* **36**, 429-38; (1914) *Orn. Monatsber.* **22**, 85-90. FENNELL, J F M, SAGAR, P M, and FENNELL, J S (1985) *Notornis* **32**, 245-53. FENNELL, J F M and STONE, D A (1976) *Ring. Migr.* **1**, 108-14. FERDINAND, L (1991) *Bird voices in the North Atlantic.* Torshavn. FERGMAN, U (1988) *Torgos* **7** (2), 74-9, 101-2. FERGUSON-LEES, I J (1956) *Br. Birds* **49**, 398-400; (1957) *Br. Birds* **50**, 200; (1958) *Br. Birds* **51**, 99-103; (1959) *Br. Birds* **52**, 161-3; (1967) *Br. Birds* **60**, 344-7; (1971) In Gooders, J (ed.) *Birds of the world*, 2753-5. London. FERNANDEZ-CRUZ, M, ARAUJO, J, TEIXEIRA, A M, MAYOL, J, MUNTANER, J, EMMERSON, K W, MARTIN, A, and LE GRAND, G (1985) *Situacion de la avifauna de la Península Ibérica, Baleares y Macaronesia.* Madrid. FERRER, X (1987) *Ardeola* **34**, 110-3. FERRER, X, MOTIS, A, and PERIS, S J (1991) *J. Biogeography* **18**, 631-6. FERRY, C and FROCHOT, B (1970) *Terre Vie* **24**, 153-250. FEY, A

(1982) *Gef. Welt* **106**, 343-4. FFRENCH, R P (1967) *Living bird* **6**, 123-40; (1976) *A guide to the birds of Trinidad and Tobago.* Valley Forge, Pa; (1991) *Glos. nat. Soc. J.* **42**, 16. FICKEN, M S (1965) *Wilson Bull.* **77**, 71-5. FICKEN, M S and FICKEN, R W (1965) *Wilson Bull.* **77**, 363-75. FICKEN, R W, FICKEN, M S, and HAILMAN, J P (1978) *Z. Tierpsychol.* **46**, 43-57. FIEBIG, J (1983) *Mitt. zool. Mus. Berlin* **59**, Suppl. Ann. Orn. **7**, 163-87. FIEBIG, J and JANDER, G (1987) *Mitt. zool. Mus. Berlin* **63** Suppl. Ann. Orn. **11**, 123-35. FINLAYSON, C (1992) *Birds of the Strait of Gibraltar.* London. FINLAYSON, J C and CORTÉS, J E (1987) *Alectoris* **6**. FIRTH, F and FIRTH, F M (1945) *Br. Birds* **38**, 235-6. FISCHER, S, MAUERSBERGER, G, SCHIELZETH, H, and WITT, K (1992) *J. Orn.* **133**, 197-202. FISHER, J (1948) *Agriculture* **55**, 20-3. FISHER, J and HINDE, R A (1949) *Br. Birds* **42**, 347-57. FISHER, R H (1969) *Animals* **12**, 362. FISK, E J (1979) *Bird-Banding* **50**, 224-43, 297-303. FITTER, R S R (1948) *Br. Birds* **41**, 343; (1949) *London's Birds.* London. FITTER, R S R and LOUSLEY, J E (1953) *The natural history of the City.* London. FITTER, R S R and RICHARDSON, R A (1951) *Br. Birds* **44**, 16. FITZHARRIS, J and GRACE, K (1986) *IWC News* **47**, 13. FITZPATRICK, J (1978) *Br. Birds* **71**, 134. FITZWATER, W D (1967) *J. Bombay nat. Hist. Soc.* **64**, 111. FIUCZYNSKI, D (1961) *J. Orn.* **102**, 96-8. FJELD, P E and SONERUD, G A (1988) *Ornis scand.* **19**, 268-74. FJELDSÅ, J (1972) *Norwegian J. Zool.* **20**, 147-55; (1976) *Sterna* **15**, 133-5; (1981) *Dansk orn. Foren. Tidsskr.* **75**, 31-9. FLAXMAN, E W (1983) *Br. Birds* **76**, 352. FLEGG, J J M (1974) *Br. Birds* **67**, 517. FLEGG, J M and MATTHEWS, N J (1980) *Ringers' Bull.* **5**, 95. FLEISCHER, R C and JOHNSTON, R F (1984) *Can. J. Zool.* **62**, 405-10. FLEISCHER, R C, LOWTHER, P E, and JOHNSTON, R F (1984) *J. Fld. Orn.* **55**, 444-56. FLEISCHER, R C and ROTHSTEIN, S I (1988) *Evolution* **42**, 1146-58. FLEISCHER, R C, ROTHSTEIN, S I, and MILLER, L S (1991) *Condor* **93**, 185-9. FLEMING, R L, SR, FLEMING, R L, JR, and BANGDEL, L S (1976) *Birds of Nepal.* Kathmandu. FLIEGE, G (1984) *J. Orn.* **125**, 393-446. FLINT, P R and STEWART, P F (1992) *The birds of Cyprus.* London. FLINT, V E, BOEHME, R L, KOSTIN, Y V, and KUZNETSOV, A A (1984) *A field guide to birds of the USSR.* Princeton. FLOHART, G (1985) *Héron* **3**, 51. FLOWER, W U, WEIR, T, and SCOTT, D (1955) *Br. Birds* **48**, 133-4. FLUX, J E C (1978) *Notornis* **25**, 350-2. FOCKE, E (1966) *Veröff. Übersee Mus. Bremen* A, **3**, 259-64. FOELIK, R F (1970) *Zool. Jhrb. Anat.* **87**, 523-87. FOG, M (1963) *Danish Rev. Game Biol.* **4**, 63-110. FOKKEMA, J, BAKKER, A G, HOLLENGA, D, JUKEMA, J, and RIJPMA, U (1987) *Vanellus* **31**, 130-5. FOLK, Č (1966) *Zool. Listy* **15**, 273-83; (1967a) *Zool. Listy* **16**, 61-72; (1967b) *Zool. Listy* **16**, 379-93; (1968) *Zool. Listy* **17**, 221-36. FOLK, Č and BEKLOVÁ, M (1971) *Zool. Listy* **20**, 357-63. FOLK, Č, HAVLÍN, J, and HUDEC, K (1965) *Zool. Listy* **14**, 143-50. FOLK, Č and NOVOTNÝ, I (1970) *Zool. Listy* **19**, 333-42. FOLK, Č and TOUŠKOVÁ, I (1966) *Zool. Listy* **15**, 23-32. FOLKESTAD, A O (1967) *Sterna* **7**, 343-4; (1978) *Cinclus* **1**, 8-11. FOLLOWS, G (1969) *Ringers' Bull.* **3** (5), 11-12. FONSTAD, T (1981) *Fauna norv.* (C) *Cinclus* **4**, 89-96; (1984) *Oikos* **42**, 314-22. FORBUSH, E H and MAY, J B (1955) *A natural history of American birds.* New York. FORMOSOF, A N (1933) *J. Anim. Ecol.* **2**, 70-81. FORMOZOV, A N, OSMOLOVSKAYA, V I, and BLAGOSKLONOV, K N (1950) *Ptitsy i vrediteli lesa.* Moscow. FORNAIRON, F (1977) *Alauda* **45**, 341-2. FOUARGE, J (1980) *Nos Oiseaux* **35**, 373-5. FOUARGE, J and RAPPE, A (1966) *Aves* **3**, 52-9. FOX, A D, FRANCIS, I S, MADSEN, J, and STROUD, J M (1987) *Ibis* **129**, 541-52.

FOX, A D, FRANCIS, I S, McCARTHY, J P, and McKAY, C R (1992) *Dansk orn. Foren. Tidsskr.* 86, 155–62. FRANCIS, C M and COOKE, F (1990) *J. Fld. Orn.* 61, 404–12. FRANCIS, I S, FOX, A D, McCARTHY, J P, and McKAY, C R (1991) *Ring. Migr.* 12, 28–37. FRANCIS, C M and WOOD, D S (1989) *J. Fld. Orn.* 60, 495–503. FRANDSEN, J (1982) *Birds of the south western Cape.* Cape Town. FRANK, F (1943) *Orn. Monatsber.* 51, 138–9; (1951) *J. Orn.* 93, 61. FRANZ, D, HAND, R, and KAMRAD-SCHMIDT, M (1987) *Anz. Orn. Ges. Bayern* 26, 237–50. FRANZ, J (1949) *Z. Tierpsychol.* 6, 309–29. FRATICELLI, F (1984) *Riv. ital. Orn.* 54, 98. FRATICELLI, F and GUSTIN, M (1987) *Avocetta* 11, 161. FRAZIER, J G, SALAS, S S, and SALEH, M A (1984) *Courser* 1, 17–27. FREDRIKSSON, S, JACOBSSON, S, and SILVERIN, B (1973) *Vår Fågelvärld* 32, 245–51. FREEDMAN, B and SVOBODA, J (1982) *Can. Fld.-Nat.* 96, 56–60. FREITAG, F (1978) *Orn. Mitt.* 30, 255. FRENCH, N R (1954) *Condor* 56, 83–5. FRENDIN, H (1943) *Fauna och Flora* 38, 116–22. FRETWELL, S (1980) In Keast, A and Morton, E S (eds) *Migrant birds in the Neotropics: ecology, behavior, distribution, and conservation,* 517–27. Washington, D C; (1986) In Johnston, R F (ed.) *Current ornithology* 4, 211–42. FREUCHEN, P and SALOMONSEN, F (1959) *The Arctic year.* London. FREY, M (1989a) *Orn. Beob.* 86, 265–89; (1989b) *Orn. Beob.* 86, 291–305. FRICH, A S and NORDBJÆRG, L (1992) *Dansk orn. Foren. Tidsskr.* 86, 107–22. FRIEDMANN, H (1929) *The cowbirds.* Springfield; (1950) *The breeding habits of the weaverbirds: a study in the biology of behaviour patterns.* Washington. FRIEDRICH, W (1974) *Vogelwarte* 27, 223–4. FRINGS, H, FRINGS, M, JUMBER, J, BUSNEL, R-G, GIBAN, J, and GRAMET, P (1958) *Écology* 39, 126–31. FRITH, J H (1957) *Emu* 57, 287–8. FRITZ, Ö (1989) *Calidris* 18, 143–64; (1991) *Calidris* 20, 48–57. FROCHOT, B and PETITOT, F (1964) *Jean le Blanc* 3, 32–40. FRÖDING, L (1987) BSC Thesis. Lund Univ. FROST, M P (1985) *Br. Birds* 78, 50. FROST, R A (1979) *Br. Birds* 72, 595–6; (1986) *Br. Birds* 79, 508–9. FROST LARSEN, K and AAGAARD ANDERSEN, P (1965) *Feltornithologen* 7, 24–5. FRUGIS, S, PARMIGIANI, S, PARMIGIANI, E, and PELLONI, C (1983) *Avocetta* 7, 13–24. FRY, C H (1970) *Ostrich Suppl.* 8, 239–63. FUCHS, E (1964) *Orn. Beob.* 61, 132–7. FUCHS, W (1984) *Vögel Heimat* 55, 42–3. FUGLE, G N, ROTHSTEIN, S I, OSENBERG, C W, and McGINLEY, M A (1984) *Anim. Behav.* 32, 86–93. FUGLE, G N and ROTHSTEIN, S I (1985) *J. Fld. Orn.* 56, 356–68. FUJIMAKI, Y and TAKAMI, M (1986) *Jap. J. Orn.* 35, 67–73. FULLER, R J (1982) *Bird habitats in Britain.* Calton. FULTON, H T (1906) *J. Bombay nat. Hist. Soc.* 16, 44–64. FUYE, M DE LA (1911) *Rev. fr. Orn.* 3, 147–51.

GABRIELSON, I N and LINCOLN, F C (1959) *The birds of Alaska.* Harrisburg. GADGIL, M and ALI, S (1975) *J. Bombay nat. Hist. Soc.* 72, 716–27. GAGINSKAYA, E R (1969) *Vopr. ekol. biotsenol* 9, 37–48. GAILLY, P (1982) *Aves* 19, 13–21, 99–102; (1988) *Alauda* 56, 404. GAINES, D (1988) *Birds of Yosemite and the East Slope.* Lee Vining. GAIT, R P (1947) *Br. Birds* 40, 341–2. GALEA, R (1987) *Il-Merill* 24, 16. GALLACHER, H (1978) *De Spreeuw.* Utrecht. GALLAGHER, M (1989a) *Oman Bird News* 5, 7–8; (1989b) *Oman Bird News* 7, 10–11. GALLAGHER, M and WOODCOCK, M W (1980) *The birds of Oman.* London. GALLAGHER, M D (1977) *J. Oman. Stud. spec. Rep.* 1, 27–58. GALLAGHER, M D and ROGERS, T D (1978) *Bonn. zool. Beitr.* 29, 5–17; (1980) *J. Oman Stud. spec. Rep.* 2, 347–85. GALLARDO, M (1986) *Faune de Provence* 7, 18–29. GALLEGO, S and

BALCELLS, E (1960) *Ardeola* 6, 337–9. GALLOWAY, D (1972) *Br. Birds* 65, 522–6. GAMBLE, R and HAYCOCK, R J (1989) In Bignal, E and Curtis, D J (eds) *Choughs and land-use in Europe,* 39–41. Scottish Chough Study Group. GANGULI, U (1975) *A guide to the birds of the Delhi area.* New Delhi. GANSO, M (1960) *Egretta* 3, 26–31. GANZHORN, J U (1986) *Ökol. Vögel* 8, 49–56. GARCÍA DORY, M A (1983) *Alytes* 1, 411–47. GARLING, M (1941) *Beitr. Fortpfl. Vögel* 17, 51–8; (1943) *Beitr. Fortpfl. Vögel* 19, 165; (1949) *Vogelwelt* 70, 101–4. GARRETT, K, and DUNN, J (1981) *Birds of southern California.* Los Angeles. GARSTANG, W (1923) *Songs of the birds.* London. GARZÓN, J (1969) *Ardeola* 14, 97–130. GASSMANN, H (1989–90) *Voliere* 12, 324–8, 13, 17–20. GASTON, A J (1968) *Ibis* 110, 17–26; (1970) *Bull. Br. Orn. Club* 90, 53–60, 61–6; (1978) *J. Bombay nat. Hist. Soc.* 75, 115–28. GATEHOUSE, A G and MORGAN, M J (1973) *Nat. Wales* 13, 267. GÄTKE, H (1891) *Die Vogelwarte Helgoland.* Braunschweig. GATTER, W (1974) *Vogelwarte* 27, 278–89; (1976) *Vogelwarte* 28, 165–70; (1977) *Verh. orn. Ges. Bayern* 23, 61–9. GATTER, W, KLUMP, G, and SCHÜTT, R (1979) *Vogelwarte* 30, 101–7. GAUCI, C and SULTANA, J (1983) *Il-Merill* 22, 12–16. GAUHL, F (1984) *Vogelwelt* 105, 176–87. GAVRILO, M V (1986) *1. S'ezda Vsesoyuz. orn. Obshch. 9. Vsesoyuz. orn. konf.* 1, 140–1. St Petersburg. GAVRILOV, E I (1962a) *Ibis* 104, 416–17; (1962b) *Trudy nauch.-issled. Inst. Zashch. Rast.* 7, 459–528; (1963) *J. Bombay nat. Hist. Soc.* 60, 301–17; (1965) *Bull. Br. Orn. Club* 85, 112–14; (1968) *Ornitologiya* 9, 343–4; (1972) *Int. Stud. Sparrows* 6, 11–23. GAVRILOV, E I and GISTSOV, A P (1985) *Sezonnye perelety ptits v predgor'yakh zapadnogo Tyan'-Shanya.* Alma-Ata. GAVRILOV, E I, GUBIN, B M, and LEVIN, A S (1984) In Shukurov, E D (ed.) *Migratsii ptits v Azii* 7, 74–96. Frunze. GAVRILOV, E I and KORELOV, M N (1968) *Byull. Mosk. Obshch. Ispyt. Prir. Otd. Biol.* 73 (4), 115–22. GAVRILOV, E I and KOVSHAR', A F (1968) *Trudy Inst. Zool. Akad. Nauk Kazakh. SSR* 29, 41–9. GAVRILOV, E I, NAGLOV, V A, FEDOSENKO, A K, SHEVCHENKO, V L, and TATARINOVA, O M (1968) *Trudy Inst. Zool. Akad. Nauk. Kazakh. SSR* 29, 153–207. GAVRILOV, V M (1979) *Ornitologiya* 14, 158–63. GEBHARDT, E (1944) *Beitr. Fortpfl. Vögel* 20, 98; (1955) *Vogelwelt* 76, 188. GEE, J P (1984) *Malimbus* 6, 31–66. GEE, N G and MOFFETT, L I (1917) *A key to the birds of the lower Yangtse valley.* Shanghai. GEILER, H (1959) *Beitr. Vogelkde.* 6, 359–66. GEISTER, I (1974) *Biol. Vestn. Ljubljana* 22, 71–3. GÉNARD, M and LESCOURRET, F (1986) *Vie Milieu* 36, 27–36; (1987) *Bird Study* 34, 52–63. GEORG, P V (1969) *Bull. Iraq. nat. Hist. Mus.* 4, 21; (1971) *Bull. Iraq nat. Hist. Mus.* 5, 45. GEORG, P V and VIELLIARD, J (1970) *Bull. Iraq nat. Hist. Mus.* 4, 61–85. GEORGE, J C (1976) *J. Bombay nat. Hist. Soc.* 71, 394–404. GEORGE, W G (1968) *Wilson Bull.* 80, 496–7. GÉNSBØL, B (1964) *Feltornithologen* 6, 59–63. GENT, C J (1949) *Br. Birds* 42, 242. GENTZ, K (1970) *Beitr. Vogelkde.* 16, 109–18; (1971) *Falke* 18, 112–18. GERBER, R (1955) *Beitr. Vogelkde.* 5, 36–45; (1956) *Die Saatkrähe.* Wittenberg Lutherstadt; (1963) *Beitr. Vogelkde.* 8, 341–8. GERMOGENOV, N I (1982) In Labutin, Y V (ed.) *Migratsii i ekologiya ptits Sibiri,* 74–87. Novosibirsk. GÉROUDET, P (1951a) *Les passereaux* 1. Neuchâtel; (1951b) *Nos Oiseaux* 21, 1–6, 23–31; (1952) *Nos Oiseaux* 21, 160–8; (1954) *Nos Oiseaux* 22, 145–56; (1955) *Nos Oiseaux* 23, 89–95; (1957) *Les passereaux* 3. Neuchâtel; (1961) *Les passereaux* 1 2nd edn. Neuchâtel; (1964a) *Nos Oiseaux* 27, 251; (1964b) *Nos Oiseaux* 27, 299–303; (1991) *Nos Oiseaux* 41, 119–36. GEUENS, A (1968)

*Wielewaal* **34**, 357-8. GEYER, C (1985) *Aves* **22**, 53-4. GEYR VON SCHWEPPENBURG, H (1918) *J. Orn.* **66**, 121-76; (1920) *Falco* **16**, 17-26; (1930) *Orn. Monatsber.* **38**, 118-21; (1939) *Beitr. Fortpfl. Vögel* **15**, 198-9; (1942a) *Beitr. Fortpfl. Vögel* **18**, 27; (1942b) *Beitr. Fortpfl. Vögel* **18**, 1-5. GHABBOUR, S I (1976) *Int. Stud. Sparrows* **9**, 17-29. GHIGI, A (1932) *Ann. Mus. Civ. Stor. nat. Genova* **55**, 268-92. GHIOT, C (1976) *Gerfaut* **66**, 267-305. GIBAN, J (1947) *Ann. des Épiphyt.* **13** (NS), 19-41. GIBB, J (1951) *Ibis* **93**, 109-27. GIBBONS, D W (1987) *J. Anim. Ecol.* **56**, 403-14. GIBSON-HILL, C A (1950) *Malayan Nat. J.* **5**, 58-75. GIEBING, M (1992) *Voliere* **15**, 209-13. GIEROW, P and GIEROW, M (1991) *Orn. Svec.* **1**, 103-11. GIL, A (1927) *Bol. real Soc. española Hist. nat.* **27**, 81-96. GIL-DELGADO, J A (1981) *Mediterránea* **5**, 97-114. GIL-DELGADO, J A and CATALÁ, M C (1989) *Mediterránea Ser. Biol.* **11**, 121-32. GIL-DELGADO, J A and GÓMEZ, J A (1988) *Doñana Acta Vert.* **15**, 201-14. GILL, E H N (1923) *J. Bombay nat. Hist. Soc.* **29**, 757-68. GILL, E L (1919) *Br. Birds* **13**, 23-5. GILLESPIE, M and GILLESPIE, J A (1932) *Auk* **49**, 96. GILROY, N (1922) *Ool. Rec.* **2**, 76-80. GINN, H B and MELVILLE, D S (1983) *Moult in birds.* Tring. GINN, P J, McILLERON, W G, and MILSTEIN, P LE S (1989) *The complete book of southern African birds.* Cape Town. GIRAUD-AUDINE, M and PINEAU, J (1973) *Alauda* **41**, 317. GIRAUDOUX, P, DEGAUQUIER, R, JONES, P J, WEIGEL, J, and ISENMANN, P (1988) *Malimbus* **10**, 1-140. GISTSOV, A P and GAVRILOV, E I (1984) *Int. Stud. Sparrows* **11**, 22-33. GIZENKO, A I (1955) *Ptitsy Sakhalinskoy oblasti.* Moscow. GLADKOV, N A and ZALETAEV, V S (1962) *Ornitologiya* **5**, 31-4. GLADWIN, T W (1985) *Br. Birds* **78**, 109-10. GLANDT, D (ed.) (1991) *Der Kolkrabe (Corvus corax) in Mitteleuropa. Meteler Schriftenreihe Naturschutz* **2**. Metelen. GLANDT, D and JANSEN, M (1991) In Glandt (1991), 113-16. GLAS, P (1960) *Arch. Néerl. Zool.* **13**, 466-72. GLATTHAAR, R and ZISWILER, V (1971) *Rev. suisse Zool.* **78**, 1222-30. GLAUBRECHT, M (1989) *J. Orn.* **130**, 277-92. GLAUSE, J (1969) *Vogelwelt* **90**, 66. GLAYRE, D (1970) *Nos Oiseaux* **30**, 230-4; (1979) *Nos Oiseaux* **35**, 71-4. GLAYRE, D and MAGNENAT, D (1984) *Oiseaux nicheurs de la haute vallée de l'Orbe.* Geneva. GLIEMANN, L (1973) *Die Grauammer.* Wittenberg Lutherstadt. GLOE, P (1982) *Ökol. Vögel* **4**, 209-11. GLÜCK, E (1978) *J. Orn.* **119**, 336-8; (1980) *Ökol. Vögel* **2**, 43-91; (1982) *Vogelwarte* **31**, 395-422; (1983) *J. Orn.* **124**, 369-92; (1984) *Voliere* **7**, 7-12; (1985) *Ibis* **127**, 421-9; (1986) *Oecologia* **71**, 149-55; (1987) *Oecologia* **71**, 268-72; (1988) *Experientia* **44**, 537-9. GLÜCK, E and MASSOTH, K (1985) *Voliere* **8**, 193-226. GLUE, D E (1973) *Br. Birds* **66**, 461-72. GLUSHCHENKO, F P (1986) *Ornitologiya* **21**, 158-9. GLUTZ VON BLOTZHEIM, U N (1956) *Orn. Beob.* **53**, 36-40; (1962) *Die Brutvögel der Schweiz.* Aarau; (1987) *Orn. Beob.* **84**, 249-74. GNIELKA, R (1978) *Orn. Mitt.* **30**, 81-90; (1986) *Beitr. Vogelkde.* **32**, 235-44. GODFREY, W E (1965) *Auk* **82**, 510-11; (1979) *The birds of Canada.* Ottawa; (1986) *The birds of Canada* rev. ed. Ottawa. GODIN, J, DEGAUQUIER, R, and TONNEL, R (1977) *Héron* **4** (4), 1-26. GODMAN, F DU C (1870) *Natural history of the Azores, or Western Islands.* London; (1872) *Ibis* (3) **2**, 209-24; (1907-10) *A monograph of the Petrels.* London. GOOCH, S, BAILLIE, S R, and BIRKHEAD, T R (1991) *J. appl. Ecol.* **28**, 1068-86. GOODACRE, M J (1959) *Bird Study* **6**, 180-92. GOODBODY, I M (1955) *Scott. Nat.* **67**, 90-7. GOODFELLOW, D J and SLATER, P J B (1990) *Bioacoustics* **2**, 249-51. GOODFELLOW, P (1977) *Birds as builders.* Newton Abbot. GOODMAN, S M (1984) *Bonn.*

*zool. Beitr.* **35**, 39-56. GOODMAN, S M and AMES, P L (1983) *Sandgrouse* **5**, 82-96. GOODMAN, S M and ABDEL MOWLA ATTA, G (1987) *Gerfaut* **77**, 3-41. GOODMAN, S M and MEININGER, P L (eds) (1989) *The birds of Egypt.* Oxford. GOODMAN, S M, MEININGER, P L, and MULLIÉ, W C (1986) *Misc. Publ. Mus. Zool. Univ. Michigan* **172**. GOODMAN, S M and STORER, R W (1987) *Gerfaut* **77**, 109-45. GOODMAN, S M and WATSON, G E (1983) *Bull. Br. Orn. Club* **103**, 101-6. GOODPASTURE, K A (1963) *Bird-Banding* **34**, 191-9; (1972) *Bird-Banding* **43**, 136. GOODWIN, D (1949) *Br. Birds* **42**, 278-87; (1951) *Ibis* **93**, 414-42, 602-25; (1952a) *Behaviour* **4**, 293-316; (1952b) *Br. Birds* **45**, 113-22; (1952c) *Br. Birds* **45**, 364; (1953a) *Br. Birds* **46**, 113; (1953b) *Ibis* **95**, 147-9; (1955a) *Br. Birds* **48**, 181-3; (1955b) *Bull. Br. Orn. Club* **75**, 97-8; (1956) *Ibis* **98**, 186-219; (1960) *Avic. Mag.* **66**, 174-99; (1962) *Ibis* **104**, 564-6; (1964) *Br. Birds* **57**, 82-3; (1965a) *Avic. Mag.* **71**, 76-80; (1965b) *Domestic birds.* London; (1971) *Avic. Mag.* **77**, 88-93; (1975) *Br. Birds* **68**, 484-8; (1982) *Estrildid finches of the world.* London; (1985) *Avic. Mag.* **91**, 143-56; (1986) *Crows of the world.* London; (1987) *Avic. Mag.* **93**, 38-50. GORDON, P, MORLAN, J, and ROBERSON, D (1989) *Western Birds* **20**, 81-7. GORDON, S (1949) *Country Life* **106**, 1239. GORE, M E J (1990) *Birds of The Gambia.* 2nd rev. edn. Tring. GORE, M E J and WON, P-O (1971) *The birds of Korea.* Seoul. GÖRNER, M (1971) *Beitr. Vogelkde.* **17**, 173-4. GÖRNITZ, K (1921) *Falco* **17**, 3; (1922) *Verh. orn. Ges. Bayern* **15**, 134-46; (1927) *J. Orn.* **75**, 58-60. GÓRSKI, W (1976) *Acta Orn.* **16**, 79-116; (1982) *Notatki Orn.* **23**, 3-13; (1988) *Acta Orn.* **24**, 29-62. GOSNELL, H T (1932) *Br. Birds* **26**, 196-7; (1947) *The science of birdnesting.* Wirral. GOTHE, H (1954) *Vogelwelt* **75**, 204-5. GÖTHEL, H (1969) *Falke* **16**, 410-16. GÖTMARK, F (1981) *Vår Fågelvärld* **40**, 47-56; (1982) *Vår Fågelvärld* **41**, 315-22. GÖTMARK, F and ÅHLUND, M (1984) *J. Wildl. Management* **48**, 381-7. GÖTMARK, F, WALLIN, K, JACOBSSON, S, and ALSTRÖM, P (1979) *Vår Fågelvärld* **38**, 201-20. GÖTTGENS, F, GÖTTGENS, H, and KOLLIBAY, F-J (1985) *Beitr. Naturkde. Niedersachs.* **38**, 233-8. GÖTTGENS, H (1989) *Beitr. Naturkde. Niedersachs.* **42**, 148-57. GOTZMAN, J and WISIŃSKI, P (1965) *Przeglad Zool.* **9**, 280-3. GÖWERT, R (1978) *Gef. Welt* **102**, 84-5, 224-5. GRABER, R R and GRABER, J W (1962) *Wilson Bull.* **74**, 74-88. GRABOVSKI, V I (1983) *Zool. Zh.* **62**, 389-98. GRACZYK, R (1961) *Przeglad Zool.* **5**, 241-5. GRACZYK, R and MICHOCKI, M (1975) *Roczniki Akad. Roln. Poznan.* **87**, 79-87. GRAHN, M (1990) *Ornis scand.* **21**, 195-201. GRAMET, P (1956) *Bull. Soc. zool. Fr.* **81**, 207-17; (1973) In Kendeigh, S C and Pinowski, J (eds) *Productivity, population dynamics and systematics of granivorous birds*, 181-94. Warsaw. GRAMET, P and DUBAILLE, E (1983) *Académie d'agriculture de France, Séance du 13 Avril*, 455-64. GRAMPIAN AND TAY RINGING GROUP (1981) *Rep. Grampian Ring. Group* **3**, 61-6. GRANDE, J L G (1986) *Ronda*, March 1986, 37-44. GRANT, C H B and MACKWORTH-PRAED, C W (1944) *Bull. Br. Orn. Club* **64**, 35-40. GRANT, G S (1982) *Elepaio* **42**, 97-8. GRANT, G S and QUAY, T L (1970) *Bird-Banding* **41**, 274-8. GRANT, P J (1968) *Br. Birds* **61**, 176-80; (1970) *Br. Birds* **63**, 153-5. GRANT, P R (1972) *System. Zool.* **21**, 23-30; (1976) *Proc. int. orn. Congr.* **16**, 603-15; (1979) *Biol. J. Linn. Soc.* **11**, 301-32; (1980) *Bonn. zool. Beitr.* **31**, 311-17. GRANVIK, H (1916) *J. Orn.* **64**, 371-8; (1923) *J. Orn. Suppl.*, 1-280. GRÄTZ, D and GRÄTZ, H-P (1985) *Falke* **32**, 301-2. GRAY, A P (1958) *Bird hybrids.* Farnham Royal. GREEN, G H and SUMMERS, R W (1975) *Bird Study*

22, 9-17. GREEN, P T (1980) *Br. Birds* **73**, 358-60; (1981) *Ring. Migr.* **3**, 203-12; (1982a) *Ibis* **124**, 193-6; (1982b) *Ibis* **124**, 320-4. GREENHALGH, M (1965) *Br. Birds* **58**, 511. GREENING, M (1992) *J. Gloucs. Nat. Soc.* **43**, 34. GREENSLADE, D W (1979) *Br. Birds* **72**, 553. GREIG-SMITH, P W (1977) *Bull. Niger. orn. Soc.* **13**, 3-14; (1982) *Ibis* **124**, 529-34; (1985) In Sibly, R M and Smith, R H *Behavioural ecology*, 387-92. Oxford; (1987) *Behaviour* **103**, 203-16. GREIG-SMITH, P W and CROCKER, D R (1986) *Anim. Behav.* **34**, 843-59. GREIG-SMITH, P W and DAVIDSON, N C (1977) *Bull. Br. Orn. Club* **97**, 96-9. GREIG-SMITH, P W and WILSON, G M (1984) *J. appl. Ecol.* **21**, 401-22. GREIG-SMITH, P W and WILSON, M F (1985) *Oikos* **44**, 47-54. GRENQUIST, P (1947) *Ornis fenn.* **24**, 1-10. GREVE, K (1983) *Vogelkde. Ber. Niedersachs.* **15**, 5-10; (1990) *Beitr. Naturkde. Niedersachs.* **43**, 28-37. GREVE, K, and DORNIEDEN-GREVE, R (1982) *Beitr. Naturkde. Niedersachs.* **35**, 127-8. GRIEVE, A (1987) *Br. Birds* **80**, 466-73. GRIMES, L G (1987) *The birds of Ghana.* London. GRIMM, H (1954) *J. Orn.* **95**, 306-18; (1989) *Acta ornithoecol.* **2**, 100-2. GRIMSBY, A and RØER, J E (1992a) *Fauna norv. (C) Cinclus* **15**, 17-24; (1992b) *Vår Fuglefauna* **15**, 90-1. GRINNELL, J (1908) *Univ. Calif. Publ. Zool.* **5**, 1-170; (1928) *Condor* **30**, 185-9. GRINNELL, J, DIXON, J, and LINSDALE, J M (1930) *Univ. Calif. Publ. Zool.* **35**. GRINNELL, J and MILLER, A H (1944) *Pacific Coast Avifauna* **27**. GRINNELL, L I (1943) *Wilson Bull.* **55**, 155-63; (1944) *Auk* **61**, 554-60. GRISCOM, L (1937) *Proc. Boston Soc. nat. Hist.* **41**, 77-209. GRITTNER, I (1941) *Vogelzug* **12**, 56-73. GROBE, D W (1983) *Orn. Mitt.* **35**, 159. GRODZIŃSKI, Z (1971) *Acta Zool. Cracov.* **16**, 735-72; (1976) *Acta Zool. Cracov.* **21**, 465-500; (1980) *Acta Zool. Cracov.* **24**, 375-410. GROEBBELS, F (1960) *Vogelwelt* **81**, 94. GROH, G (1975) *Mitt. Pollichia* **63**, 72-139; (1982) *Mitt. Pollichia* **70**, 217-34; (1988) *Mitt. Pollichia* **75**, 261-87. GROMADZKA, J (1980) *Acta Orn.* **17**, 227-55. GROMADZKA, J and GROMADZKI, M (1978) *Acta Orn.* **16**, 335-64. GROMADZKA, J and LUNIAK, M (1978) *Acta Orn.* **16**, 275-85. GROMADZKI, M (1969) *Ekol. Polska* **17**, 287-311; (1980) *Acta Orn.* **17**, 195-224. GROMADZKI, M and KANIA, W (1976) *Acta Orn.* **15**, 279-321. GROŠELJ, P (1983) *Acrocephalus* **4**, 56-8. GROSS, A O (1921) *Auk* **38**, 1-26, 163-84. GROTE, H (1934a) *Beitr. Fortpfl. Vögel* **10**, 20-6; (1934b) *Orn. Monatsber.* **42**, 17-21; (1943a) *Beitr. Fortpfl. Vögel* **19**, 98-104; (1943b) *J. Orn.* **91**, 136-43; (1937) *Beitr. Fortpfl. Vögel* **13**, 150-1; (1940) *Vogelzug* **11**, 127-9; (1947) *Orn. Beob.* **44**, 84-90. GROTH, J G (1988) *Condor* **90**, 745-60; (1992a) *Auk* **109**, 383-5; (1992b) *Western Birds* **23**, 35-7. GRÜLL, A (1981) *Egretta* **24** Suppl., 39-63; (1988) *Beih. Veröff. Nat. Land. Bad.-Württ.* **53**, 65. GRÜN, G (1975) *Int. Stud. Sparrows* **8**, 24-103. GRÜNKORN, T (1991) In Glandt (1991), 9-15. GRUYS-CASIMIR, E M (1965) *Arch. Néerl. Zool.* **16**, 175-279. GUBIN, B M (1979) *Ornitologiya* **14**, 211-13; (1980) *Ornitologiya* **15**, 111-16. GUBLER, W (1978) *Orn. Beob.* **75**, 279-80. GUÉNIAT, E (1948) *Orn. Beob.* **45**, 81-98. GUERMEUR, Y, HAYS, C, L'HER, M, and MONNAT, J-Y (1973) *Ar Vran* **6**, 199-260. GUERRERO, J, LOPE, F DE, and CRUZ, C DE LA (1989) *Alauda* **57**, 234. GUEX, M-L (1986) *Nos Oiseaux* **38**, 343. GUICHARD, K M (1955) *Ibis* **97**, 393-424. GUILLORY, H D and DESHOTELS, J H (1981) *Wilson Bull.* **93**, 554. GUILLOU, J J (1964) *Alauda* **32**, 196-225; (1981) *Oiseau* **51**, 177-88. GUITIÁN, J (1987) *Ardeola* **34**, 25-35; (1989) *Ardeola* **36**, 73-82. GUITIÁN RIVERA, J (1985) *Ardeola* **32**, 155-72. GULAY, V I (1989) In Konstantinov, V M and Klimov, S M (eds) *Vranovye ptitsy v estestvennykh i antropogennykh landshaftakh* **1**, 53-5. Lipetsk. GURIEV, V N and BASHKOVA, E N (1986) *Tez. Dokl. S'ezda 1. Vsesoyuz. orn. Kongr. 9. Vsesoyuz. orn. Konf.* **1**, 182-3. St Petersburg. GUSH, G H (1975) *Br. Birds* **68**, 342; (1978) *Br. Birds* **71**, 40-1; (1980) *Devon Birds* **33**, 75-80. GÜTTINGER, H R (1970) *Z. Tierpsychol.* **27**, 1011-75; (1974) *J. Orn.* **115**, 321-37; (1976) *Bonn. zool. Beitr.* **27**, 218-44; (1977) *Behaviour* **60**, 304-18; (1978) *J. Orn.* **119**, 172-90; (1981) *Gef. Welt* **105**, 210-22; (1985) *Behaviour* **94**, 254-78. GÜTTINGER, H R and CLAUSS, G (1982) *J. Orn.* **123**, 269-86. GÜTTINGER, H R, WOLFFGRAMM, J, and THIMM, F (1978) *Behaviour* **65**, 241-62. GUYOT, A, LIGOR, J-L, and ROSE, R (1991) *Nos Oiseaux* **41**, 195. GWINNER, E (1964) *Z. Tierpsychol.* **21**, 657-748; (1965a) *J. Orn.* **106**, 145-78; (1965b) *Vogelwarte* **23**, 1-4; (1966) *Vogelwelt* **87**, 129-33. GWINNER, E and KNEUTGEN, J (1962) *Z. Tierpsychol.* **19**, 692-6. GYLLIN, R (1965a) *Fauna och Flora* **60**, 225-74; (1965b) *Vår Fågelvärld* **24**, 420-1; (1967) *Vår Fågelvärld* **26**, 19-29. GYLLIN, R and KÄLLANDER, H (1976) *Ornis scand.* **7**, 113-25; (1977) *Fauna Flora* **72**, 18-24. GYLLIN, R, KÄLLANDER, H and SYLVÉN, M (1977) *Ibis* **119**, 358-1. GYNGAZOV, A M and MILOVIDOV, S P (1977) *Ornitofauna Zapadno-Sibirskoy ravniny.* Tomsk. GYÖRGYPÁL, Z (1981) *Aquila* **87**, 71-8.

HAACK, W and SCHMIDT, G A J (1989) *Vogelkundl. Tageb. Schleswig-Holstein* **16**, 509-29. HAAPANEN, A (1965) *Ann. zool. Fenn.* **2**, 153-96; (1966) *Ann. zool. fenn.* **3**, 176-200. HAAR, H (1975) *Mitt. Abt. Zool. Landesmus. Joanneum* **4**, 105-14. HAARHAUS, D (1968) *Oecologia* **1**, 176-218. HAARTMAN, L VON (1952) *Ornis fenn.* **29**, 73-6; (1969) *Comm. Biol. Soc. Sci. Fenn.* **32**; (1972) *Ornis fenn.* **49**, 15; (1973) *Ornis fenn.* **50**, 49. HAARTMAN, L VON and NUMERS, M VON (1992) *Ornis fenn.* **69**, 65-71. HAAS, G (1939) *Beitr. Fortpfl. Vögel* **15**, 52-62; (1943) *Beitr. Fortpfl. Vögel* **19**, 43-6. HAASE, B J M (1988) *Ardea* **76**, 210. HAASE, E (1975) *Gen. comp. Endocrin.* **26**, 248-52. HACHFELD, B (1979) *Voliere* **2**, 7-11, 79-84; (1983) *Voliere* **6**, 146-9; (1984) *Voliere* **7**, 67. HAENSEL, J (1967) *Beitr. Vogelkde.* **13**, 1-28; (1970) *Beitr. Vogelkde.* **16**, 169-91. HAFFER, J (1977) *Bonn. zool. Monogr.* **10**; (1989) *J. Orn.* **130**, 475-512. HAFSTEINSSON, H T and BJÖRNSSON, H (1989) *Bliki* **8**, 47-9. HAFTORN, S (1952) *Fauna* **5**, 105-141. Oslo; (1971) *Norges Fugler.* Oslo. HAGEN, Y (1942) *Archiv Naturgesch. NF* **11**, 1-132; (1956) *Nytt Magasin for Zoologi* **4**, 107-8. HAGGER, C H E (1961) *Br. Birds* **54**, 291. HAGMANN, J and DAGAN, D (1992) *Orn. Beob.* **89**, 72. HAILA, Y, JÄRVINEN, O, and VÄISÄNEN, R A (1979) *Ornis scand.* **10**, 48-55. HAILA, Y, TIAINEN, J, and VEPSÄLÄINEN, K (1986) *Ornis fenn.* **63**, 1-9. HAILS, C J (1985) *Studies of problem bird species in Singapore: I Sturnidae.* Rep. Parks and Recreation Dept., Ministry of National Development, Singapore. HÁJEK, V (1969) *Vertebr. Zprávy* **1969** (2), 73-6; (1974) *Sylvia* **19**, 145-56. HÁJEK, V and BAŠOVÁ, D (1963) *Zool. Listy* **12**, 115-20. HAKALA, A V K and NYHOLM, E S (1973) *Ornis fenn.* **50**, 46-7. HAKE, M and EKMAN, J (1988) *Ornis scand.* **19**, 275-9. HÅLAND, A (1980) *Viltrapport* **10**, 104-6. HALD-MORTENSEN, P (1970) *Ibis* **112**, 265-6. HALL, B P (1953) *Bull. Br. Orn. Club* **73**, 2-8; (1957) *Bull. Br. Orn. Club* **77**, 44-6. HALL, B P and MOREAU, R E (1970) *An atlas of speciation in African passerine birds.* London. HALL, G A (1981) *J. Fld. Orn.* **52**, 43-9; (1983) *West Virginia birds.* Pittsburgh; (1985) *American birds* **39**, 911-4. HAM, I and ŠOTI, J P (1986) *Larus* **36-37**, 297-303. HAMILTON, S and JOHNSTON, R F (1978) *Auk* **95**, 313-23. HAMILTON,

T H (1958) *Wilson Bull.* **70**, 307-46. HAMILTON-HUNTER, R (1909) *Br. Birds* **3**, 188-9. HAMMER, H (1958) *Falke* **5**, 141. HAMMER, M (1948) *Danish Rev. Game Biol.* **1** (2), 1-59. HANCOX, M (1985) *Scott. Nat.* 1985, 37-40. HANDTKE, K (1975) *Naturk. Jber. Mus. Heineanum* **10**, 33-41. HANDTKE, K and WITSACK, W (1972) *Naturk. Jber. Mus. Heineanum* **7**, 21-41. HANF, B (1887) *Ornis* **3**, 267. HANFORD, D M (1969) *Br. Birds* **62**, 158. HANMER, D B (1989) *Safring News* **18**, 19-30. HANN, H W (1937) *Wilson Bull.* **49**, 145-237. HANSCH, A (1938) *Vogelzug* **9**, 110. HANSEN, L (1950) *Dansk orn. Foren. Tidsskr.* **44**, 150-61. HANSEN, O and BIRKHOLM-CLAUSEN, F (1968) *Dansk orn. Foren. Tidsskr.* **62**, 97. HANSEN, P (1975) *Biophon* **3** (3), 2-5; (1981a) *Nat. Jutlandica* **19**, 107-20; (1981b) *Nat. Jutlandica* **19**, 121-38; (1984) *Ornis scand.* **15**, 240-7; (1985) *Natura Jutlandica* **21**, 209-19. HANSEN, W and OELKE, H (1976) *Beitr. Naturkde. Niedersachs.* **29**, 85-158. HANSKI, I K and HAILA, Y (1988) *Ornis fenn.* **65**, 97-103. HANTZSCH, B (1905) *Beitrag zur Kenntnis der Vogelwelt Islands.* Berlin. HARASZTHY, L (1988) *Magyarország madárvendégei.* Budapest. HARBARD, C (1989) *Songbirds.* London. HARBER, D D (1945a) *Br. Birds* **38**, 211; (1945b) *Br. Birds* **38**, 296. HARDENBERG, J D F (1965) In Busnel, R-G and Giban, J (eds) *Colloque. Le problème des oiseaux sur les aérodromes*, 121-6. Paris. HÁRDI, M (1989) *Orn. Beob.* **86**, 209-17. HARDING, K C (1931) *Auk* **48**, 512-22. HARDY, E (1932) *Br. Birds* **25**, 301; (1946) *A handlist of the birds of Palestine.* Unpubl. MS, Edward Grey Inst., Oxford Univ; (1971) *Br. Birds* **64**, 77-8. HARLOW, R C (1922) *Auk* **39**, 399-410. HARMS, W (1975) *Hamburger avifaun. Beitr.* **13**, 133-44. HARPER, D G C (1984) *Ring. Migr.* **5**, 101-4. HARPUM, J R (1985) *Glos. nat. Soc. J.* **36**, 115-7. HARRIS, A, TUCKER, L, and VINICOMBE, K (1989) *The Macmillan field guide to bird identification.* London. HARRIS, G J (1957) *Br. Birds* **50**, 206-8. HARRIS, G J, PARSLOW, J L F, and SCOTT, R E (1960) *Br. Birds* **53**, 513-8. HARRIS, M P (1962) *Br. Birds* **55**, 97-103; (1974) *A field guide to the birds of Galapagos.* New York. HARRIS, M P, NORMAN, F I, and MCCOLL, R H S (1965) *Br. Birds* **58**, 288-94. HARRISON, C J O (1956) *Avic. Mag.* **62**, 128-41; (1962a) *Proc. zool. Soc. Lond.* **139**, 261-82; (1962b) *Bull. Br. Orn. Club* **82**, 126-32; (1962c) *J. Orn.* **103**, 369-79; (1964) *Ibis* **106**, 462-8; (1965a) *Ardea* **53**, 57-72; (1965b) *Bull. Br. Orn. Club* **85**, 26-30; (1965c) *Behaviour* **24**, 161-209; (1967) *Wilson Bull.* **79**, 22-7; (1975) *A field guide to the nests, eggs and nestlings of European birds.* London; (1976) *Bird Study* **23**, 59; (1978a) *A field guide to the nests, eggs and nestlings of North American birds.* Glasgow; (1978b) *Avic. Mag.* **84**, 80-7; (1982) *An atlas of the birds of the western Palearctic.* London; (1983) *Avic. Mag.* **89**, 163-9. HARRISON, G R, DEAN, A R, RICHARDS, A J, and SMALLSHIRE, D (1982) *The birds of the West Midlands.* Studley. HARRISON, H H (1984) *Wood Warblers' World.* New York. HARRISON, J M (1928) *Br. Birds* **22**, 36-7; (1934) *Ibis* (13) **4**, 396-8; (1937) *Bull. Br. Orn. Club* **57**, 64-5; (1938) *Vogelzug* **9**, 36; (1947a) *Ibis* **89**, 411-8; (1947b) *Ibis* **89**, 664; (1954) *Bull. Br. Orn. Club* **74**, 105-12; (1955) *Bull. Br. Orn. Club* **75**, 6-12, 17-21; (1958) *Bull. Br. Orn. Club* **78**, 9-14, 23-8; (1961) *Bull. Br. Orn. Club* **81**, 96-103, 119-124. HARRISON, J M and PATEFF, P (1933) *Ibis* (13) **3**, 494-521. HARRISON, R (1970) *Br. Birds* **63**, 302-3. HARROP, A (1988) *Bull. Oriental Bird Club* **8**, 31. HARROP, H (1992) *Birding World* **5**, 133-7. HARROP, J M (1970) *Nature in Wales* **12** (2), 65-9. HARS, D (1991) *Héron* **24**, 289-92. HARTBY, E (1968) *Dansk orn. Foren.*

*Tidsskr.* **62**, 205-30. HARTERT, E (1903-10) *Die Vögel der paläarktischen Fauna* **1.** Berlin; (1913) *Novit. Zool.* **20**, 37-76; (1915) *Novit. Zool.* **22**, 61-79; (1918a) *Bull. Br. Orn. Club* **38**, 58-60; (1918b) *Novit. Zool.* **25**, 327-337; (1918c) *Novit. Zool.* **25**, 361; (1921-2) *Die Vögel der paläarktischen Fauna* **3.** Berlin; (1928) *Novit. Zool.* **34**, 197. HARTERT, E and OGILVIE-GRANT, W R (1905) *Novit. Zool.* **12**, 80-128. HARTERT, E and STEINBACHER, F (1932-8) *Die Vögel der paläarktischen Fauna, Ergänzungsband.* Berlin. HARTLEY, I R (1991) Ph D Thesis. Leicester Univ. HARTLEY, I R, SHEPHERD, M, ROBSON, T, and BURKE, T (1993) *Behav. Ecol.* **4** in press. HARTOG, J C DEN (1987) *Zool. Meded. Leiden* **61** (28), 405-19; (1990) *Cour. Forsch.-Inst. Senckenberg* **129**, 159-90. HARTWIG, W (1886) *J. Orn.* **34**, 452-86. HARVEY, P (1990) *Birding World* **3**, 266-7; (1991) *Fair Isle Bird Observatory Rep.* **43**, 15-57. HARWIN, R M (1959) *Ostrich* **30**, 97-104. HARWOOD, N (1959) *Br. Birds* **52**, 166. HASSE, H (1961) *Regulus* **41**, 115-9; (1962) *Vogelwelt* **83**, 173-7; (1963) *Die Goldammer.* Wittenberg Lutherstadt; (1965) *Beitr. Vogelkde.* **10**, 406-7. HAUKIOJA, E (1968) *Ornis fenn.* **45**, 105-13; (1969) *Ornis fenn.* **46**, 171-8; (1970) *Ornis fenn.* **47**, 101-35; (1971a) *Ornis fenn.* **48**, 25-32; (1971b) *Ornis fenn.* **48**, 45-67; (1971c) *Rep. Kevo Subarctic Res. Stat.* **7**, 60-9. HAUKIOJA, E, and REPONEN, J (1969) *Orn. Soc. Pori Ann. Rep.* **2**, 49-51. HAURI, R (1956) *Orn. Beob.* **53**, 28-35; (1957) *Orn. Beob.* **54**, 41-4; (1966) *Orn. Beob.* **63**, 77-85; (1988a) *Orn. Beob.* **85**, 1-79; (1988b) *Orn. Beob.* **85**, 305-7. HAUSBERGER, M and GUYOMARCH, J C (1981) *Biol. Behav.* **6**, 79-98. HAUSBERGER, M and BLACK, J M (1991) *Ethol. Ecol. Evol.* **3**, 337-44. HAUSER, D C (1957) *Wilson Bull.* **69**, 78-90; (1973) *Chat* **37**, 91-5. (1974) *Auk* **91**, 537-63. HAVERSCHMIDT, F (1934) *Beitr. Fortpfl. Vögel* **10**, 73; (1937) *Beitr. Fortpfl. Vögelkde.* **13**, 228-30. HAVLÍN, J (1957) *Zool. Listy* **6**, 247-56; (1976) *Zool. Listy* **25**, 51-63; (1988) *Fol. zool.* **37**, 59-66. HAVLÍN, J and FOLK, C (1965) *Zool. Listy* **14**, 193-208. HAVLÍN, J and HAVLÍNOVÁ, S (1974) *Sylvia* **19**, 89-116. HAVLÍN, J and JURLOV, K T (1977) *Acta Sci. Nat. Brno* **11** (2), 1-50. HAYCOCK, R J and BULLOCK, I D (1982) *Br. Birds* **75**, 91-2. HAYHOW, S J (1984) *Magpie* **3**, 22-9. HAYMAN, R W (1949) *Br. Birds* **42**, 84-5; (1953) *Br. Birds* **46**, 378; (1958) *Br. Birds* **51**, 275. HAZELWOOD, A and GORTON, E (1953) *Bull. Br. Orn. Club* **73**, 1-2; (1955) *Bull. Br. Orn. Club* **75**, 98. HAZEVOET, C J and HAAFKENS, L B (1989) *Nature reserve development and ornithological research in the Republica de Cabo Verde.* ICBP Rep. Amsterdam. HEATHERLEY, F (1910) *Br. Birds* **3**, 234-42. HECKE, P VAN and VERSTUYFT, S (1972) *Gerfaut* **62**, 245-72. HEEB, P A (1991) D Phil Thesis. Oxford Univ. HEGELBACH, J (1980) *Orn. Beob.* **77**, 60; (1984) Ph D Thesis. Zurich Univ. HEGELBACH, J and ZISWILER, V (1979) *Orn. Beob.* **76**, 119-132. HEGNER, R E and WINGFIELD, J C (1986) *Horm. Behav.* **20**, 294-312; (1987) *Auk* **104**, 470-80. HEIDEMANN, J and SCHÜZ, E (1936) *Mitt. Vogelwelt* **35**, 37-44. HEIJ, C J (1986) *Int. Stud. Sparrows* **13**, 28-34. HEIJ, C J and MOELIKER C W (1991) In Pinowski, J and Summers-Smith, J D (eds) *Granivorous birds in the agricultural landscape*, 59-85. Warsaw. HEIM DE BALSAC, H (1929) *Alauda* **1**, 68-77; (1948) *Alauda* **16**, 75-96. HEIM DE BALSAC, H and HEIM DE BALSAC, T (1949-50) *Alauda* **17-18**, 206-21. HEIM DE BALSAC, H and MAYAUD, N (1962) *Les oiseaux du nord-ouest de l'Afrique.* Paris. HEIMERDINGER, M A and PARKES, K C (1966) *Br. Birds* **59**, 315-6. HEINIGER, P H (1991) *Orn. Beob.* **88**, 193-207. HEINRICH, B (1988a) *Condor* **90**, 950-2; (1988b) *Behav. Ecol. Sociobiol.* **23**, 141-56; (1988c) In

Slobodchikoff, C N (ed.) *The ecology of social behaviour*, 285-311. London; (1990) *Ravens in winter*. London. HEINRICH, B and MARZLUFF, J (1992) *Condor* **94**, 549-50. HEINROTH, O and HEINROTH, M (1924-6) *Die Vögel Mitteleuropas* 1. Berlin-Lichterfelde; (1931) *Die Vögel Mitteleuropas*. Ergänzungsband. HEINZE, J and KROTT, N (1979) *Vogelwelt* **100**, 225-7. HEINZEL, H, FITTER, R, and PARSLOW, J (1972) *The birds of Britain and Europe with North Africa and the Middle East*. London. HEINZEL, H and WOLTERS, H E (1970) *J. Orn.* **111**, 497-8. HEISE, G (1970) *Orn. Mitt.* **22**, 144-5. HELB, H-W (1974) *Verhandl. Ges. Ökol.* (1974), 55-8; (1981) *J. Orn.* **122**, 325; (1985) *Behaviour* **94**, 279-323; (1986) In Wüst, W (ed.) *Avifauna Bavariæ* 2. Munich. HELL, P and SOVIŠ, B (1958) *Zool. Listy* 7, 38-56; (1959) *Aquila* **65**, 145-60. HELLMAYR, C E (1929) *Field Mus. nat. Hist. Publ.* **263**, *Zool. Ser.* 17, 27-144; (1935) *Catalogue of birds of the Americas and the adjacent islands* 8; (1938) 11. Chicago. HELLMICH, J (1985) *Orn. Mitt.* **37**, 178-81. HELM, F (1894) *Orn. Monatsschr.* **19**, 239. HELMS, C W (1959) *Wilson Bull.* **71**, 244-53. HELMS, C W, AUSSIKER, W H, BOWER, E B, and FRETWELL, S D (1967) *Condor* **69**, 560-78. HELMS, C W and DRURY, W H (1960) *Bird-Banding* **31**, 1-40. HÉMERY, G and PASCAUD, P-N (1981) *Oiseau* **51**, 1-16. HEMMINGSEN, A M (1951) *Spolia zool. Mus. Haun.* 11; (1958) *Vidensk. Medd. dansk nat. Foren.* **120**, 189-206. HEMMINGSEN, A M and GUILDAL, J A (1968) *Spolia zool. Mus. haun.* 28. HENDERSON, I G (1990) *Proc. int. orn. Congr.* 20 suppl., 377-8; (1991) *Ring. Migr.* **12**, 23-7. HENDY, E W (1939) *Br. Birds* **33**, 162; (1943) *Br. Birds* **35**, 37; (1943) *Somerset birds*. London. HENRICI, P (1927) *Beitr. Fortpfl. Vögel* 3, 7-13. HENRIKSEN, K (1989) *Dansk orn. Foren. Tidsskr.* **83**, 55-9. HENRY, G M (1971) *A guide to the birds of Ceylon*. London. HENS, P A (1931) *Proc. int. orn. Congr.* 7, 439-64. HENS, P A and MARLE, J G van (1933) *Org. Club. Nederl. Vogelk.* 6, 49-58. HENTY, C J (1975) *Br. Birds* **68**, 463-6; (1979) *Bird Study* **26**, 192-4. HENZE, O (1975) *Gef. Welt* **99**, 31-2; (1979) *Falke* **26**, 13-20. HEPWORTH, N M (1946) *Br. Birds* **39**, 84-5. HERMANN, H (1983) *Verh. orn. Ges. Bayern* **23**, 459-77. HERREMANS, L (1973) *Wielewaal* **39**, 185-7. HERREMANS, M (1977) *Wielewaal* **43**, 133-41; (1982) *Gerfaut* **72**, 243-54; (1987) *Oriolus* **53**, 149-53; (1988) *Gerfaut* **78**, 243-60; (1989) *Dutch Birding* **11**, 9-15; (1990a) *Ardea* **78**, 441-58; (1990b) *Ring. Migr.* **11**, 86-9. HERRERA, C M (1980) *Ardeola* **25**, 143-80. HERRLINGER, E (1966) *Egretta* **9**, 55-60. HERRMANN, J (1977) *Orn. Ber. Berlin (West)* **2** (2), 121-38. HERROELEN, P (1962) *Gerfaut* **52**, 173-205; (1967) *Giervalk* **57**, 81-3; (1974) *Wielewaal* **40**, 69-74; (1980) *Ornis Brabant* **85**, 13-15; (1983) *Ornis Flandriae* 1983, 92-3; (1987) *Gerfaut* **77**, 99-104. HESELER, U (1966) *Luscinia* **39**, 69-71. HESS, R (1975) *Orn. Beob.* **72**, 120. HESSE, E (1915) *Orn. Monatsber.* **23**, 112-8. HEUER, J (1986) *Gef. Welt* **110**, 59. HEUGLIN, M T von (1869-74) *Ornitologie Nordost-Afrika's* 1. Kassel. HEWSON, R (1957) *Br. Birds* **50**, 432-4; (1981) *Br. Birds* **74**, 509-12; (1984) *J. appl. Ecol.* **21**, 843-68. HEWSON, R and LEITCH, A F (1982) *Bird Study* **29**, 235-8. HICKLING, R (1983) *Enjoying Ornithology*. Calton. HICKS, L E (1934) *Bird-Banding* 5, 103-18. HIETT, J C and CATCHPOLE, C K (1982) *Anim. Behav.* **30**, 568-74. HILDÉN, O (1969a) *Ornis fenn.* **46**, 93-112; (1969b) *Ornis fenn.* **46**, 179-87; (1972) *Ornis fenn.* **49**, 14-15; (1974a) *Lintumies* 9, 45-51; (1974b) *Ornis fenn.* **51**, 10-35; (1977) *Ornis fenn.* **54**, 170-9; (1988) *Sitta* 2, 21-57. HILDÉN, O and NIKANDER, P J (1986) *Lintumies* **21**, 88-93; (1988) *Lintumies* **23**, 80-5; (1991) *Lintumies* **26**,

104-117. HILL, A (1986) *Orn. Mitt.* **38**, 72-84. HILL, D, TAYLOR, S, THAXTON, R, AMPHLET, A, and HORN, W (1990) *Bird Study* **37**, 133-41. HILL, R A (1976) *Bird-Banding* **47**, 112-4. HILPRECHT, A (1964) *Beitr. Vogelkde.* 10, 177-83; (1965) *Auspicium* 2, 91-118. HILTY, S L and BROWN, W L (1986) *A guide to the birds of Colombia*. Princeton. HIMMER, K H (1967) *Gef. Welt* **91**, 188-92. HINDE, R A (1947) *Br. Birds* **40**, 246-7; (1952) *Behaviour Supp.* 2; (1953) *Behaviour* 5, 1-31; (1954) *Behaviour* 7, 207-32; (1954) *Proc. Roy. Soc. B* **142**, 306-31, 331-58; (1955-6) *Ibis* **97**, 706-45; **98**, 1-23; (1958a) *Anim. Behav.* 6, 211-18; (1958b) *Proc. zool. Soc. Lond.* **131**, 1-48; (1959) *Bird Study* 6, 15-19; (1960) *Proc. Roy. Soc. B* **153**, 398-420. HINDE, R A and STEEL, E (1972) *Anim. Behav.* **20**, 514-25. HINDMARSH, A M (1984) *Behaviour* **90**, 302-24; (1986) *Behaviour* **99**, 87-100. HIRALDO, F and HERRERA, C M (1974) *Doñana, Acta Vertebr.* 1 (2), 149-70. HIRSCHFELD, E (1988) *Birding World* 1, 380. HIRSCHFELD, E, HOLST, O, KJELLEN, N, PERSSON, O, and ULLMAN, M (1983) *Anser Suppl.* 14. HIRSCHFELD, E and SYMENS, P (1992) *Sandgrouse* **14**, 48-51. HIRSCHI, W (1986) *Orn. Beob.* **83**, 145-6. HISS, J-P (1979) *Ciconia* 3, 184-5. HJORT, C and LINDHOLM, C-G (1978) *Oikos* **30**, 387-92. HOBBS, J N (1955) *Emu* **55**, 202. HOEHL, O (1939) *Jber. vogelkd. Beob. Stat. 'Untermain', Frankfurt*, 24-6. HOFER, H (1935) *Verh. zool. bot. Ges. Wien* **85**, 60-87. HOFFMAN, E C (1930) *Bird-Banding* 1, 80-1. HOFFMANN, B (1925) *Mitt. Ver. sächs. Orn.* 1 suppl., 37-43; (1928) *Ver. orn. Ges. Bayern* **18**, 75-107. HOFFMANN, L (ed.) (1955) *Premier compte rendu 1950-54* Station Biol. de la Tour du Valat; (1956) *Deuxième compte rendu 1955* Station Biol. de la Tour du Valat. HOFSHI, H (1985) *M Sc Thesis*. Ben Gurion Univ. HOFSHI, H, GERSANI, M, and KATZIR, G (1987a) *Ostrich* **58**, 156-9; (1987b) *Ibis* **129**, 389-90. HÖGLUND, J (1985) *Ornis fenn.* **62**, 19-22. HOGSTAD, O (1967) *Sterna* 7, 255-60; (1969) *Nytt Mag. Zool.* **17**, 81-91; (1975) *Norw. J. Zool.* **23**, 223-34; (1977) *Sterna* **16**, 19-27; (1982) *Fauna norv. (C) Cinclus* 5, 59-64; (1985) *Ornis fenn.* **62**, 13-18; (1988) *Fauna norv. (C) Cinclus* 11, 27-39. HOGSTAD, O and RØSKAFT, E (1986) *Fauna norv. (C) Cinclus* 10, 7-10. HÖGSTEDT, G (1980a) *Ornis scand.* 11, 110-15; (1980b) *Nature* **283**, 64-6; (1981) *J. Anim. Ecol.* **50**, 219-29. HOHLT, H (1956) *Vogelwelt* **77**, 194. HOLGERSEN, H (1982) *Sterna* **17**, 85-123. HOLIAN, J J and FORTEY, J E (1992) *Br. Birds* **85**, 370-6. HÖLLER, C and TEIXEIRA, A M (1983) *Vogelwarte* **32**, 81-2. HOLLIDAY, S T (1990) *Alectoris* 7, 49-57. HOLLOM, P A D (1940) *Br. Birds* **34**, 86-7; (1955) *Ibis* **97**, 1-17; (1959) *Ibis* **101**, 183-200; (1962) *Br. Birds* **55**, 158-64; (1971) *The popular handbook of British birds*. London. HOLLOM, P A D, PORTER, R F, CHRISTENSEN, S, and WILLIS, I (1988) *Birds of the Middle East and North Africa*. Calton. HOLLOWAY, J (1984) *Fair Isle's garden birds*. Lerwick. HOLLYER, J N (1970) *Br. Birds* **63**, 353-73; (1971) *Br. Birds* **64**, 196-7. HOLMAN, D (1990) *Br. Birds* **83**, 430-2. HOLMAN, D and KEMP, J (1991) *Birding World* 4, 353-4. HOLMAN, D J (1981) *Br. Birds* **74**, 203-4. HOLMAN, D J and MADGE, S C (1982) *Br. Birds* **75**, 547-53. HOLMES, D A and WRIGHT, J O (1969) *J. Bombay nat. Hist. Soc.* **66**, 8-30. HOLMES, P (1982) *Birds of Oxfordshire 1981*, 36-8. HOLMES, P R, HOLMES, H J, and PARR, A J (eds) (1985) *Rep. Oxford Univ. Exped. Kashmir 1983*. Oxford. HOLMSTRÖM, C T (ed.) (1959-62) *Våra Fåglar i Norden*. Stockholm. HOLSTEIN, V (1934) *Dansk orn. Foren. Tidsskr.* **28**, 116-18. HOLTMEIER, F-K (1966) *J. Orn.* **107**, 337-45. HOLYOAK, D (1967a) *Br. Birds* **60**, 52; (1967b) *Bird Study*

14, 61–2; (1967c) *Bird Study* 14, 153–68; (1968) *Bird Study* 15, 147–53; (1970a) *Ibis* 112, 397–400; (1970b) *Bull. Br. Orn. Club.* 90, 40–2; (1971) *Bird Study* 18, 97–106; (1972a) *Bird Study* 19, 59–68; (1972b) *Bird Study* 19, 215–27; (1974a) *Bird Study* 21, 15–20; (1974b) *Bird Study* 21, 117–28. HOLYOAK, D and RATCLIFFE, D A (1968) *Bird Study* 15, 191–7. HOLYOAK, D T and SEDDON, M B (1991) *Alauda* 59, 55–7, 116–20. HOLTMEIER, F K (1966) *J. Orn.* 107, 337–45. HÖLZINGER, J (1987) *Die Vögel Baden-Württembergs* 1. Karlsruhe; (1992a) *Kart. Med. Brutvögel* 7, 3–8; (1992b) *Kart. Med. Brutvögel* 7, 15; (1992c) *Kart. Med. Brutvögel* 7, 17–25. HOMANN, J (1959) *Beitr. Naturkde. Niedersachs.* 12, 58–62. HONGELL, H (1977) *Ornis fenn.* 54, 138. HOOGENDOORN, W B (1991) *Dutch Birding* 13, 104–6. HOPE JONES, P (1980) *Br. Birds* 73, 561–8. HOPKINS, J R (1985) *Br. Birds* 78, 597. HOPPE, R (1976) *Falke* 23, 29–33. HOPSON, A J (1964) *Bull. Nigerian orn. Soc.* 1 (4), 7–15. HORDOWSKI, J (1989) *Notatki orn.* 30 (1–2), 21–36. HORNBUCKLE, J (1984) *Derbyshire Bird Rep.* 1983, 62–3. HORNBY, R J (1971) Ph D Thesis. Nottingham Univ. HORNER, K O (1977) Ph D Thesis. Virginia Polytechnic Inst. State Univ. HORTLING, I (1928) *Ornis fenn.* 5, Supp; (1929) *Ornitologisk handbok.* Helsingfors; (1938) *Mitt. Ver. sächs. Orn.* 5, 219–27. HORTLING, I and BAKER, E C S (1932) *Ibis* (13) 2, 100–27. HORVÁTH, L (1975) *Aquila* 80–1, 310–1; (1976) *Aquila* 82, 37–47; (1977) *Aquila* 83, 91–5. HORVÁTH, L and HÜTTLER, B (1963) *Acta Zool. Sci. Hung.* 9, 271–6. HORVÁTH, L and KEVE, A (1956) *Bull. Br. Orn. Club* 76, 92–5. HOSONO, T (1966a) *Misc. Rep. Yamashina Inst. Orn.* 4, 327–47; (1966b) *Misc. Rep. Yamashina Inst. Orn.* 4, 481–7; (1967a) *Misc. Rep. Yamashina Inst. Orn.* 5, 34–47; (1967b) *Misc. Rep. Yamashina Inst. Orn.* 5, 177–93; (1967c) *Misc. Rep. Yamashina Inst. Orn.* 5, 278–86; (1969) *Misc. Rep. Yamashina Inst. Orn.* 5, 659–75; (1971) *Misc. Rep. Yamashina Inst. Orn.* 6, 231–49; (1973) *Misc. Rep. Yamashina Inst. Orn.* 7, 56–72; (1975) *Misc. Rep. Yamashina Inst. Orn.* 7, 533–49; (1983) *Misc. Rep. Yamashina Inst. Orn.* 15, 63–71; (1989) *Jap. J. Orn.* 37, 103–27. HOUSTON, C S (1963) *Bird-Banding* 34, 94–5. HOUSTON, C S and STREET, M G (1959) *Sask. Nat. Hist. Soc. Spec. Publ.* 2. HOUSTON, D (1974) Unpubl. Rep., Dept. Forestry nat. Res., Edinburgh Univ; (1977a) *J. appl. Ecol.* 14, 1–15; (1977b) *J. appl. Ecol.* 14, 17–29; (1978) *Ann. appl. Biol.* 88, 339–41. HOVEL, H (1960) *Bull. Br. Orn. Club* 80, 75–6; (1987) *Check-list of the birds of Israel with Sinai.* Tel-Aviv. HOWARD, H E (1929) *An introduction to the study of bird behaviour.* HOWARD, D V (1968) *Bird-Banding* 39, 132. HOWARD, D V and DICKINSON HENRY, D (1966) *Bird-Banding* 37, 123. HOWARD, H E (1929) *An introduction to the study of bird behaviour.* Cambridge; (1930) *Territory in bird life.* London. HOWARD, R and MOORE, A (1980) *A complete checklist of the birds of the world.* Oxford. HOWELL, T R, PAYNTER, R A, and RAND, A L (1968) Carduelinae, in *Check-list of birds of the world* 14. Cambridge (Mass.). HOWELLS, V (1956) *A naturalist in Palestine.* London. HUBÁLEK, Z (1976) *Vertebr. Zprávy* 1975–6 (1), 82–6; (1978a) *Vest. Česk. Spol. Zool.* 42, 1–14; (1978b) *Vest. Česk. Spol. Zool.* 42, 15–22; (1980) *Acta orn.* 16, 535–53; (1983) *Acta Sc. nat. Brno (NS)* 17 (1), 1–52. HUBÁLEK, Z and HORÁKOVÁ, M (1988) *Acta Sci. Nat. Brno* 22 (5), 1–44. HUBBARD, J P (1969) *Auk* 86, 393–432; (1970) *Wilson Bull.* 82, 355–69; (1980) *Nemouria* 25, 1–9. HUBER, B (1991) In Glandt (1991), 45–59. HUCKRIEDE, B (1969) *Vogelwarte* 25, 23–5. HUDEC, K (1983) *Ptáci ČSSR* 3 (1). Prague. HUDEC, K and FOLK, C (1961) *Zool. Listy*

10, 305–30. HUDSON, R (1967) *Br. Birds* 60, 423–6; (1969) *Br. Birds* 62, 13–22. HUDSON, R and MARCHANT, J H (1984) *BTO Res. Rep.* 13. HUDSON, W H (1915) *Birds and man.* London. HÜE, F and ETCHÉCOPAR, R-D (1958) *Terre Vie* 105, 186–219; (1970) *Les oiseaux du proche et du moyen orient.* Paris. HUGHES, J (1976) *Br. Birds* 69, 273. HUGHES, S W M (1972) *Br. Birds* 65, 445; (1986) *Br. Birds* 79, 342. HUI, P A and HUI, M (1974) *Vögel Heimat* 45, 122–4. HULTEN, M (1967) *Regulus* 47, 27–31; (1972) *Regulus* 10, 463–5. HUME, A O (1874a) *Stray Feathers* 2, 29–324; (1874b) *Stray Feathers* 2, 467–84. HUME, R A (1975) *Br. Birds* 68, 515–16; (1980) *Br. Birds* 73, 478–9; (1983) *Br. Birds* 76, 90. HUMPHREY, D (1967) *Ool. Rec.* 41, 69–71. HUMPHREYS, G R (1928) *Bull. Br. Ool. Assoc.* 2, 51–7. HUND, K and PRINZINGER, R (1981) *Ökol. Vögel* 3, 261–5. HUNTINGTON, C E (1952) *Syst. Zool.* 1, 149–70. HURRELL, A G (1951) *Br. Birds* 44, 88–9. HURRELL, H G (1956) *Br. Birds* 49, 28–31. HUSAIN, K Z (1964) *Bull. Br. Orn. Club* 84, 9–11. HUSBY, M (1986) *J. Anim. Ecol.* 55, 75–83. HUSSELL, D J T (1972) *Ecol. Monogr.* 42, 317–64; (1985) *Ornis scand.* 16, 205–12. HUSSELL, D J T and HOLROYD, G L (1974) *Can. Fld.-Nat.* 88, 197–212. HUSTINGS, F, POST, F, and SCHEPERS, F (1990) *Limosa,* 63, 103–11. HUT, R M G VAN DER (1985) *Graspieper* 5, 43–64, 103–16. HUT, R M G VAN DER (1986) *Ardea* 74, 159–76. HUTCHINSON, C D (1989) *Birds in Ireland.* Calton. HUTSON, H P W (1945) *Ibis* 87, 456–9; (1954) *The birds about Delhi.* Kirkee. HYTÖNEN, O (1972) *Ardeola* 16, 277.

IANKOV, P N (1983) Ph D Thesis. Inst. Zool. Acad. Sci. Belorussiya Minsk. IAPICHINO, C and MASSA, B (1989) *The birds of Sicily.* Tring. IDZELIS, R F (1986) *Tez. Dokl. 1. S'ezda Vsesoyuz. orn. Obshch. 9. Vsesoyuz. orn. Konf.* 1, 261–2. IJZENDOORN, A L J VAN (1950) *The breeding-birds of the Netherlands.* Leiden. ILANI, G and SHALMON, B (1983) *Israel Land Nat.* 9, 39. IL'YASHENKO, V Y (1986) *Trudy Zool. Inst. Akad. Nauk SSSR* 150, 77–81. IL'YASHENKO, V Y, KALYAKIN, M V, SOKOLOV, E P, and SOKOLOV, A M (1988) *Trudy Zool. Inst. Akad. Nauk SSSR* 182, 70–88. ILYINA, T A (1990) In Kurochkin, E N (ed.) *Sovremennaya ornitologiya 1990,* 48–54. Moscow. IMHOF, T A (1976) *Alabama birds.* Alabama Univ. IMMELMANN, K (1966) *Ostrich Suppl.* 6, 371–9; (1973) *Der Zebrafink.* Wittenberg Lutherstadt. IMMELMANN, K, STEINBACHER, J, and WOLTERS, H E (1965) *Vögel in Käfig und Voliere: Prachtfinken* 1. Aachen. INBAR, R (1971) *Birds of Israel.* Tel Aviv; (1979) *Israel Land Nat.* 5, 20–2. INDYKIEWICZ, P (1988) *Przeglad Zool.* 32, 281; (1990) In Pinowski, J and Summers-Smith, J D (eds) *Granivorous birds in the agricultural landscape,* 95–121. Warsaw. INGLIS, I R, FLETCHER, M R, FEARE, C J, GREIG-SMITH, P W, and LAND, S (1982) *Ibis* 124, 351–5. INGRAM, C (1965) *Bull. Br. Orn. Club* 85, 20. INOZEMTSEV, A A (1962) *Ornitologiya* 5, 101–4. INSKIPP, C and INSKIPP, T (1985) *A guide to the birds of Nepal.* London. INSLEY, H and WOOD, J B (1973) *Nat. Wales* 13, 165–73. ION, I (1971) *Muz. Stiint. Nat. Bacau. Stud. Communic.* 263–76. ION, I and SARACU, S (1971) *Studii şi Comunicaŕri Ştiinţele Naturii, Muz. Judeţ. Suceava,* 2, 271–8. IOVCHENKO, N P (1986) *Trudy Zool. Inst. Akad. Nauk SSSR* 147, 7–24. IRISOV, E A (1972) *Ornitologiya* 10, 334. IRVING, L (1960) *Bull. US natn. Mus.* 217. IRWIN, M P S (1981) *The birds of Zimbabwe.* Salisbury. ISAKOV, Y A and VOROBIEV, K A (1940) *Trudy Vsesoyuz. orn. zapoved. Gassan-Kuli* 1, 5–159. ISENMANN, P (1990) *Nos Oiseaux* 40, 308; (1992) *Alauda* 60, 109–11.

ISHIMOTO, A (1992) *J. Yamashina Inst. Orn.* **24**, 1–12. IVANAUSKAS, T L (1961) *Ekol. Migr. Ptits Pribaltiki*, 329–31. IVANCHEV, V P (1988) *Ornitologiya* **23**, 209–10. IVANITSKI, V V (1984) *Zool. Zh.* **63**, 1374–87; (1985a) *Nauch. Dokl. vyssh. Shk. Biol. Nauki* (9), 50–5; (1985b) *Zool. Zh.* **64**, 1213–23; (1986) *Zool. Zh.* **65**, 387–98; (1991) *Zool. Zh.* **70**, 104–17. IVANOV, A I (1969) *Ptitsy Pamiro-Alaya*. Leningrad. IVANOV, B E (1987) *Ekol. Polska* **35**, 699–721. IVANTER, E V (1962) *Ornitologiya* **5**, 68–85. IVASHCHENKO, A A and KOVSHAR', A F (1972) *Ornitologiya* **10**, 333–4 IVOR, H R (1944) *Wilson Bull.* **56**, 91–104. IZMAYLOV, I V and BOROVITSKAYA, G K (1967) *Ornitologiya* **8**, 192–7.

JACK, J (1974) *Br. Birds* **67**, 356. JÄCKEL, A J (1891) *Systematische Übersicht der Vögel Bayerns*. Munich. JACKSON, F J (1938) *Birds of Kenya colony and the Uganda protectorate* **2**. London. JACKSON, H D (1989) *Bull. Br. Orn. Club* **109**, 100–6. JACOB, J-P (1982) *Aves* **19**, 37–45; (1984) *Aves* **21**, 261. JACOBSEN, J R (1963) *Dansk orn. Foren. Tidsskr.* **57**, 181–220. JACOBY, H, KNÖTZSCH, G, and SCHUSTER, S (1970) *Orn. Beob.* **67** suppl. JAHN, H (1942) *J. Orn.* **90**, 7–302. JAHNKE, W (1955) *Orn. Mitt.* **7**, 51. JAKOBS, B (1959) *Orn. Mitt.* **11**, 121–5. JAKOBSEN, O (1986) *Zool. Middle East* **1**, 32–3. JAKUBIEC, Z (1972a) *Ekol. Polska* **20**, 609–35; (1972b) *Ochrona Przyrody* **17**, 135–52. JALIL, A K (1985) BSc Hons Thesis. National Univ. Singapore. JAMES, D A and NEAL, J C (1986) *Arkansas birds*. Fayetteville. JAMES, P (1986) *Br. Birds* **79**, 299–300. JAMES, P C, BARRY, T W, SMITH, A R, and BARRY, S J (1987) *Ornis scand.* **18**, 310–12. JANES, S W (1976) *Condor* **78**, 409. JÄNNES, H (1992) *Lintumies* **27**, 240–7. JÄNNES, H, NUMMINEN, T, NIKANDER, P J, and PALMGREN, J (1990) *Lintumies* **25**, 254–71. JANSSEN, R B (1983) *Loon* **55**, 64–5; (1987) *Birds in Minnesota*. Minneapolis; (1990) *Loon* **62**, 69–71. JARDINE, D (1991) *Birds in Northumbria 1990*, 103–6. JARDINE, D C (1992a) *Br. Birds* **85**, 619; (1992b) *Scottish Bird Rep. 1990*, 65–9. JARRY, C (1976) *Passer* **12**, 64–75. JARRY, C and LARIGAUDERIE, F (1974) *Oiseau* **44**, 62–71. JÄRVI, E and MARJAKANGAS, A (1991) *Ornis fenn.* **62**, 171. JÄRVINEN, A and PIETIÄINEN, H (1982) *Mem. Soc. Fauna Flora Fenn.* **58**, 21–6. JÄRVINEN, O and VÄISÄNEN, R A (1976) *Ornis fenn.* **53**, 115–18; (1978) *J. Orn.* **119**, 441–9; (1979) *Oikos* **33**, 261–71. JEDRASZKO-DABROWSKA, D, and SZEPIETOWSKA, S B (1987) *Falke* **34**, 337–8. JENNER, H E (1947) *Br. Birds* **40**, 176. JENNI, L (1980) *Orn. Beob.* **77**, 62; (1982) *Orn. Beob.* **79**, 265–72; (1983) *Orn. Beob.* **80**, 136–7; (1985) *Vogelwarte* **33**, 53–63; (1986) *Orn. Beob.* **83**, 267–8; (1987) *Ornis scand.* **18**, 84–94; (1991) *Ornis scand.* **22**, 327–34. JENNI, L and JENNI-EIERMANN, S (1987) *Ardea* **75**, 271–84. JENNI, L and NEUSCHULZ, F (1985) *Orn. Beob.* **82**, 85–106. JENNI, L and SCHAFFNER, U (1984) *Orn. Beob.* **81**, 61–7. JENNING, W (1959) *Vogelwarte* **20**, 35–6. JENNINGS, M C (1980) *Sandgrouse* **1**, 71–81; (1981a) *The birds of Saudi Arabia: a check-list*. Whittlesford; (1981b) *Birds of the Arabian Gulf*. London; (1981c) *J. Saudi Arab. nat. Hist. Soc.* **2** (1), 8–14; (1986a) *Phoenix* **3**, 1–2; (1986b) *Phoenix* **3**, 8–9; (1987) *Phoenix* **4**, 7–8; (1988a) *Phoenix* **5**, 3–4; (1988b) *Fauna of Saudi Arabia* **9**, 457–67; (1992) *Sandgrouse* **14**, 27–33. JENNINGS, M C and AL SALAMA, M I (1989) *Nat. Commiss. Wildl. Cons. Dev. Tech. Rep.* **14**. JENNINGS, M C, AL SALAMA, M I, and FELEMBAN, H M (1988) *Nat. Commiss. Wildl. Cons. Dev. Tech. Rep.* **4**. JENNINGS, M C, AL SHODOUKHI, S A, AL ABBASI, T M, and COLLENETTE, S (1990) *Nat. Commiss. Wildl. Cons. Dev. Tech. Rep.* **19**. JENNINGS, M C, ABDULLA, I A, and MOHAM-

MED, N K (1991) *Nat. Commiss. Wildl. Cons. Dev. Tech. Rep.* **25**. JENNINGS, P (1981d) *Peregrine* **5** (3), 114–19; (1984) *Peregrine* **5** (6), 282. JENTZSCH, M (1988) *Acta ornithoecol.* **1**, 415. JEPSON, P R (1987) *Br. Birds* **80**, 19. JERDON, T C (1877) *The birds of India* **2** (1). Calcutta. JERZAK, L (1988) *Notatki Orn.* **29**, 27–41. JERZAK, L and KAVANAGH, B (1991) *Br. Birds* **84**, 441–3. JESPERSEN, P (1945) *Dansk orn. Foren. Tidsskr.* **39**, 92–8. JESSE, W (1902) *Ibis* (8) **2**, 531–66. JEWETT, S G, TAYLOR, W P, SHAW, W T, and ALDRICH, J W (1953) *Birds of Washington state*. Seattle. JIAD, J H and BUNNI, M K (1965) *J. biol. Sci. Res. Baghdad* **16**, 5–15; (1988) *Bull. Iraq nat. Hist. Mus.* **8**, 145–8. JIRSÍIK, J (1951) *Sylvia* **13**, 125–32. JOENSEN, A H and PREUSS, N O (1972) *Medd. Grøn.* **191** (5). JOHANNSEN, O F (1974) *Vögel Heimat* **45**, 85. JOHANSEN, H (1907) *Orn. Jahrb.* **18**, 198–203; (1944) *J. Orn.* **92**, 1–105. JOHANSSON, L and LUNDBERG, A (1977) *Vår Fågelvärld* **36**, 229–37. JOHN, A W G and ROSKELL, J (1985) *Br. Birds* **78**, 611–37. JOHNSGARD, P A (1979) *Birds of the Great Plains*. Lincoln. JOHNSON, L R (1958) *Iraq nat. Hist. Mus. Publ.* **16**, 1–32. JOHNSON, N K and ZINK, R M (1985) *Wilson Bull.* **97**, 421–35. JOHNSON, N K, ZINK, R M, and MARTEN, J A (1988) *Condor* **90**, 428–45. JOHNSTON, D W (1962) *Auk* **79**, 387–98; (1963) *Wilson Bull.* **75**, 435–46; (1967a) *Bird-Banding* **38**, 211–14. JOHNSTON, D W and DOWNER, A C (1968) *Bird-Banding* **39**, 277–93. JOHNSTON, D W, and HAINES, T P (1957) *Auk* **74**, 447–58. JOHNSTON, R F (1967b) *Int. Stud. Sparrows* **1**, 34–40; (1967c) *Auk* **84**, 275–7; (1969a) *Condor* **71**, 129–39; (1969b) *Auk* **86**, 558–9; (1969c) *Syst. Zool* **18**, 206–31; (1972) *Boll. Zool.* **39**, 351–62; (1973a) *Syst. Zool.* **22**, 219–26; (1973b) *Orn. Monogr. AOU*, **14**, 24–31; (1976) *Occ. Pap. Mus. nat. Hist. Univ. Kansas* **56**, 1–8; (1981) *J. Fld. Orn.* **52**, 127–33. JOHNSTON, R F and FLEISCHER, R C (1981) *Auk* **98**, 503–11. JOHNSTON, R F and KLITZ, W J (1977) In Pinowski, J and Kendeigh, S C (eds) *Granivorous birds in ecosystems*, 15–51. Cambridge. JOHNSTON, R F and SELANDER, R K (1964) *Science* **144**, 548–50; (1966) *System. Zool.* **15**, 357–8; (1971) *Evolution* **25**, 1–28; (1973a) In Kendeigh, S C and Pinowski, J (eds) *Productivity, population dynamics, and systematics of granivorous birds*, 301–26. Warsaw; (1973b) *Amer. Nat.* **107**, 373–90. JOLLET, A (1984) *Oiseau* **54**, 109–30; (1985) *Alauda* **53**, 263–86. JOLLIE, M (1985) *J. Orn.* **126**, 303–5. JONASSON, H (1960) *Ornis fenn.* **37**, 46–51. JONES, A E (1947–8) *J. Bombay nat. Hist. Soc.* **47**, 117–25, 219–49, 409–32. JONES, H (1955) *Br. Birds* **48**, 91. JONES, M B (1950) *Field* **195**, 761. JONES, P J and WARD, P (1977) *Ibis* **119**, 200–3. JONES, R, DAVIS, P G, and MEAD, C J (1975) *Ringers' Bull.* **4**, 99–100. JONG, C DE and SCHILTHUIZEN, M (1987) *Vogeljaar* **35**, 344–5. JONG, F DE (1981) *Vogeljaar* **29**, 261. JONIN, M and LE DEMEZET, M (1972) *Penn ar bed* **8** (NS), 214–18. JONKERS, D A (1983) *Vogeljaar* **31**, 37. JONSSON, L (1978a) *Birds of lake, river, marsh and field*. Harmondsworth; (1978b) *Birds of sea and coast*. Harmondsworth; (1979) *Birds of mountain regions*. Harmondsworth; (1982) *Birds of the Mediterranean and Alps*. London; (1992) *Birds of Europe with North Africa and the Middle East*. London. JÖNSSON, P E (1982) *Anser* **21**, 213–22; (1989) *Anser* **28**, 17–24; (1992) *Anser* **31**, 101–8. JONSSON, P N (1949) *Fauna och Flora* **44**, 76–84. JORDANS, A VON (1923) *Arch. Naturgesch.* **89A** (3), 1–147; (1924) *J. Orn.* **72**, 381–410; (1935) *Mitt. Vogelwelt* **34**, 81–5; (1950) *Syllegomena biol.*, 65–181. JORDANS, A VON and STEINBACHER, J (1948) *Senckenbergiana* **28**, 159–86. JOST, K (1951) *Orn. Mitt.* **3**, 140. JOUARD, H (1930) *Orn. Mon-*

atsber. **38**, 137-9; (1934) *Alauda* **6**, 396-9. JOURDAIN, F C R (1906) *The eggs of European birds*. London; (1915) *Ibis* (10) **3**, 133-69; (1930) *Orn. Monatsber.* **38**, 155; (1935) *Br. Birds* **29**, 148; (1936) *Ibis* (13) **6**, 725-63. JOURDAIN, F C R and LYNES, H (1936) *Ibis* (13) **6**, 39-47. JÓZEFIK, M (1960) *Acta Orn.* **5**, 307-24; (1976) *Acta Orn.* **15**, 339-482. JUANA, A E DE and COMITÉ IBÉRICO DE RAREZAS DE AL SEO (1989) *Ardeola* **36**, 111-23; (1991) *Ardeola* **38**, 149-66. JUKEMA, J (1992a) *Dutch Birding* **14**, 12-14. JUKEMA, J (1992b) *Limosa* **65**, 30-1. JUKEMA, J and FOKKEMA, J (1992) *Limosa* **65**, 67-72. JUNG, E (1966) *Falke* **13**, 408-11; (1968) *Falke* **15**, 238-9. JUNG, K (1955) *Beitr. Naturkde. Niedersachs.* **8**, 44-6. JUNG, N (1975) *Falke* **22**, 194; (1983) *Beitr. Vogelkde.* **29**, 249-73. JUNGE, G C A (1942) *Ardea* **31**, 19-22.

KAATZ, C (1986) *Falke* **33**, 328-31. KAATZ, C and OLBERG, S (1975) *Int. Stud. Sparrows* **8**, 107-16. KADHIM, A-H H, AL-DABBAGH, K Y, MAYSOON, M A-N, and WAHEED, I N (1987) *J. Biol. Sci. Res. Baghdad* **18**, 1-9. KADOCHNIKOV, N P and EYGELIS, Y K (1954) *Zool. Zh.* **33**, 1349-57. KAINADY, P V G (1976) *Bull. Basrah nat. Hist. Mus.* **3**, 107-9. KAISER, W (1983) *Falke* **30**, 17-23. KALCHREUTER, H (1969a) *Auspicium* **3**, 437-57; (1969b) *Anz. orn. Ges. Bayern* **8**, 578-92; (1970) *Vogelwarte* **25**, 245-55; (1971) *Jh. Ges. Naturkde. Württ.* **126**, 284-338. KALDEN, G (1983) *Vogelkdl. Hefte Edertal* **9**, 91-2. KALE, H W, II and MAEHR, D S (1990) *Florida's birds*. Sarasota. KÄLIN, H (1983) *Orn. Beob.* **80**, 296-7. KALINOSKI, R (1975) *Condor* **77**, 375-84. KALITSCH, L VON (1943) *Beitr. Fortpfl. Vögel* **19**, 116-17. KÄLLANDER, H (1982) *Vår Fågelvärld* **41**, 268; (1988) *Ökol. Vögel* **10**, 113-14. KÄLLANDER, H, NILSSON, S G, and SVENSSON, S (1978) *Vår Fågelvärld* **37**, 37-46. KALMBACH, E R (1940) *US Dept. Agric. Tech. Bull.* **711**. KALOTÁS, Z (1986a) *Aquila* **92**, 162-70; (1986b) *Aquila* **92**, 175-239; (1988) *Beih. Veröff. Nat. Land. Bad.-Württ.* **53**, 67-74. KAMIŃSKI, P (1983) *Notatki orn.* **24**, 167-75; (1986) *J. Orn.* **127**, 315-29. KAMIŃSKI, P and KONARZEWSKI, M (1991) *Ekol. Pol.* **32**, 125-39. KANE, C P (1960) *Br. Birds* **53**, 223. KANEKO, Y (1976) *Misc. Rep. Yamashina Inst. Orn.* **8**, 206-12. KANG, N (1989) Ph D Thesis. National Univ. Singapore; (1991) *Malaysiana* **16**, 98-103; (1992) In Priede, I G and Swift, S M (eds) *Wildlife telemetry, remote monitoring and tracking of animals*, 633-41. Chichester. KANIA, W (1981) *Acta Orn.* **18**, 375-418. KANISS, M (1970) *Anz. orn. Ges. Bayern* **9**, 173-4. KAŇUŠČÁK, P (1979) *Zprávy MOS* **37**, 69-97; (1988) *Orn. Mitt.* **40**, 227-9. KAŇUŠČÁK, P and SNAJDAR, M (1972) *Ochrana Fauny* **6**, 128-32. KAPITONOV, V I and CHERNYAVSKI, F B (1960) *Ornitologiya* **3**, 80-97. KAREILA, R (1958) *Ornis fenn.* **35**, 140-50. KARLSSON, J (1983) Dissertation. Lund Univ. KARLSSON, L (1969) *Vår Fågelvärld* **28**, 252. KARPLUS, M (1952) *Ecology* **33**, 129-34. KARR, J R (1976) *Bull. Br. Orn. Club* **96**, 92-6. KARTTUNEN, L, LAAKSONEN, A, and LAPPI, E (1970) *Ornis fenn.* **47**, 30-4. KASPAREK, M (1979) *Ring. Migr.* **2**, 158-9; (1980) *Ökol. Vögel* **2**, 1-36; (1981) *Die Mauser der Singvögel Europas: ein Feldführer;* (1992) *Die Vögel der Türkei: eine Übersicht.* Heidelberg. KATZIR, G (1981) *Ardea* **69**, 209-10; (1983) *Ibis* **125**, 516-23. KAUFMAN, K (1989) *Amer. Birds* **43**, 385-8. KAVANAGH, B P (1987a) *Irish Birds* **3**, 387-94; (1987b) *Br. Birds* **80**, 383; (1988) *Ring. Migr.* **9**, 83-90. KAZAKOV, B A (1976) *Ornitologiya* **12**, 61-7. KAZAKOV, B A and LOMADZE, N K (1984) *Ornitologiya* **19**, 179-80. KAZANTSEV, A N (1967) *Ornitologiya* **8**, 356-7. KEAR, J (1960) *Ibis* **102**,

614-16; (1962) *Proc. zool. Soc. Lond.* **138**, 163-204. KEAST, A and MORTON, E S (eds) (1980) *Migrant birds in the Neotropics: ecology, behavior, distribution, and conservation.* Washington DC. KECK, W N (1934) *J. Exp. Zoöl.* **67**, 315-47. KEIL, W (1973) In Kendeigh, S C and Pinowski, J (eds) *Productivity, population dynamics and systematics of granivorous birds*, 253-62. Warsaw. KEKILOVA, A F (1978) *Tez. Soobshch. 2. Vsesoyuz. Konf. Migr. Ptits.* **1**, 29-30. Alma-Ata. KELLER, M (1979) *Notatki orn.* **20**, 1-16. KELM, H (1936) *Vogelzug* **7**, 67-8. KELM, H and ECK, S (1985) *Zool. Abh. Staatl. Mus. Tierkde. Dresden* **42**, 1-40. KEMPER, H (1964) *Z. angew. Zool.* **51**, 31-47. KEMPER, T (1959) *Auk* **76**, 181-9. KEMPPAINEN, J and KEMPPAINEN, O (1991) *Lintumies* **26**, 20-9. KENDRA, P E, ROTH, R R, and TALLAMY, D W (1988) *Wilson Bull.* **100**, 80-90. KENNEDY, R J (1969) *Br. Birds* **62**, 249-58. KENNEDY, R S (1984) *Bull. Br. Orn. Club* **104**, 149-50. KENNERLEY, P R (1987) *Hong Kong Bird Rep. 1984-5*, 97-111. KEPLER, A K, KEPLER, C B, and DOD, A (1975) *Condor* **77**, 220-1. KERÄNEN, S and SOIKKELI, M (1985) *Ornis fenn.* **62**, 23-4. KESSEL, B (1951) *Bird-Banding* **22**, 16-23; (1957) *Amer. Midl. Nat.* **58**, 257-331. KESSEL, B and GIBSON, D D (1978) *Stud. avian Biol.* **1**. KESSEL, B and SPRINGER, H K (1966) *Condor* **68**, 185-95. KETTERSON, E D, and NOLAN, V (1978) *Auk* **95**, 755-8; (1983) *Wilson Bull.* **95**, 628-35. KEULEMANS, J G (1866) *Nederland. Tijds. Dierk.* **3**, 363-74. KEVE, A (1943) *Anz. Akad. Wiss. Wien, math.-naturwiss.* **80**, 16-20; (1958a) *Bull. Br. Orn. Club* **78**, 88-90; (1958b) *Bull. Br. Orn. Club* **78**, 155-7; (1960) *Proc. int. Orn. Congr.* **12**, 376-95; (1966a) *Lounais-Hämeen Luonto* **23**, 49-52; (1966b) *Riv. ital. Orn.* **36**, 315-23; (1966c) *Aquila* **71-2**, 39-65; (1967a) *Bull. Br. Orn. Club* **87**, 39-40; (1967b) *Ibis* **109**, 120-2; (1969) *Der Eichelhäher.* Wittenberg Lutherstadt; (1970) *Riv. ital. Orn.* **40**, 37-42; (1973) *Zool. Abh. Staatl. Mus. Tierkde. Dresden* **32**, 175-98; (1976a) *Emu* **76**, 152-3; (1976b) *Il-Merill* **17**, 25-6; (1978) *Zool. Abh. Staatl. Mus. Tierkde. Dresden* **34**, 245-73; (1985) *Der Eichelhäher.* Wittenberg Lutherstadt. KEVE, A and DONČEV, S (1967) *Zool. Abh. Staatl. Mus. Tierkde. Dresden* **29**, 1-16. KEVE, A and KOHL, Š (1978) *Nymphaea* **6**, 583-606. KEVE, A and ROKITANSKY, G (1966) *Ann. naturhist. Mus. Wien* **69**, 225-83. KEVE, A and STERBETZ, I (1968) *Falke* **15**, 184-7, 230-3. KEVE-KLEINER, A (1942) *Aquila* **46-9**, 146-224; (1943) *Aquila* **50**, 369-70. KEYMER, I F (1975) *Br. Birds* **68**, 49. KHAIRALLAH, N H (1986) *Bull. Orn. Soc. Middle East* **16**, 16-17. KHAKHLOV, V A (1991) *Ornitologiya* **25**, 214-15. KHAN, R (1986) *Devon Birds* **39**, 63-4; (1987) *Devon Birds* **40**, 41. KHOKHLOV, A N and KONSTANTINOV, V M (1983) In Kuchin, A P (ed.) *Ptitsy Sibiri*, 224-5. Gorno-Altaysk. KHOKHLOVA, N A (1960) *Ornitologiya* **3**, 259-60. KHOKHLOVA, T Y, SAZONOV, S V, and SUKHOV, A V (1983) In Ivanter, E V (ed.) *Fauna i ekologiya ptits i mlekopitayushchikh severo-zapada SSSR*, 41-52. Petrozavodsk. KIDONO, H (1977) *Misc. Rep. Yamashina Inst. Orn.* **9**, 271-9. KIIS, A (1985) *Dansk. orn. Foren. Tidsskr.* **79**, 107-12; (1986) *Ornis scand.* **17**, 80-3. KIIS, A and MØLLER, A P (1986) *Anim. Behav.* **34**, 1251-5. KILHAM, L (1989) *The American Crow and the Common Raven.* Texas. KILIÇ, A and KASPAREK, M (1989) *Birds of Turkey* **8**. KILIN, S V (1983) In Kuchin, A P (ed.) *Ptitsy Sibiri*, 140-1. KILLPACK, M L (1986) *Utah Birds* **2**, 23-4. KILZER, R and BLUM, V (1991) *Atlas der Brutvögel Vorarlbergs.* Wolfurt. KING, B (1969) *Br. Birds* **62**, 201; (1971) *Br. Birds* **64**, 423; (1976a) *Bristol Orn.* **9**, 159; (1976b) *Br. Birds* **69**, 507; (1985a) *Br. Birds* **78**, 401;

(1985*b*) *Br. Birds* 78, 512-13. KING, B (1978) *J. Saudi Arab. nat. Hist. Soc.* 1 (21), 3-24. KING, B and KING, M (1968) *Br. Birds* 61, 316. KING, B and ROLLS, J C (1968) *Br. Birds* 61, 417-18. KING, B, WOODCOCK, M, and DICKINSON, E C (1975) *A field guide to the birds of south-east Asia.* London. KING, J R, BARKER, S, and FARNER, D S (1963) *Ecology* 44, 513-21. KING, J R and FARNER, D S (1959) *Condor* 61, 315-24; (1961) *Condor* 63, 128-42; (1966) *Amer. Nat.* 100, 403-18. KING, J R, FARNER, D S, and MORTON, M L (1965) *Auk* 82, 236-52. KING, J R and MEWALDT, L R (1987) *Condor* 89, 549-65. KING, R (1976*c*) *N. Am. Bird-Bander* 1, 172-3. KING, W B and KEPLER, C B (1970) *Auk* 87, 376-8. KINGTON, B L (1973) *Br. Birds* 66, 231. KINSEY, R N (1972) IIIB Project. Auckland Univ. KINZELBACH, R (1962) *Vogelwelt* 83, 187; (1969) *Bonn. zool. Beitr.* 20, 175-81. KINZELBACH, R and MARTENS, J (1965) *Bonn. zool. Beitr.* 16, 50-91. KIPP, F A (1978) *Vogelwelt* 99, 185-9. KIRNER, O (1964) *Gef. Welt* 88, 29-31, 50-3. KIRSCH, K-W (1992) *Beitr. Naturkde. Niedersachs.* 45, 89-122. KISHCHINSKI, A A (1980) *Ptitsy Koryakskogo nagor'ya.* Moscow. KISHCHINSKI, A A, TOMKOVICH, P S, and FLINT, V E (1983) In Flint, V E and Tomkovich, P S (eds) *Rasprostranenie i sistematika ptits*, 3-76. Moscow. KISLENKO, G S (1974) *Ornitologiya* 11, 381-2. KISS, J B and RÉKÁSI, J (1983) *An. Baustului Stüntele nat.* Timişoara 15, 133-40; (1986) *A Magyar Madártani Egyesület 2 Tudományos Ülése*, 75-82. KISS, J B, RÉKÁSI, J, and STERBETZ, I (1978) *Avocetta* 1 (2), 3-18. KISTYAKOVSKI, O B (1950) *Trudy Inst. Zool. Akad. Nauk USSR* 4, 3-77. KITSON, A R (1979) *Br. Birds* 72, 94-100; (1982) *Br. Birds* 75, 40. KITSON, A R and ROBERTSON, I S (1983) *Br. Birds* 76, 217-25. KIUCHI, K (1988) *Strix* 7, 296. KIZIROĞLU, I, ŞIŞLI, M N, and ALP, Ü (1987) *Vogelwelt* 108, 169-75. KLAAS, C (1967) *Natur Mus. Frankfurt* 97, 29-32. KLAFS, G and STÜBS, J (1977) *Die Vogelwelt Mecklenburgs.* Jena; (1987) *Die Vogelwelt Mecklenburgs* 3rd edn. Jena. KLEHM, K (1967) *Falke* 14, 328-33. KLEIN, R (1988) Diss. Saarbrücken Univ; (1989) *J. Orn.* 130, 361-5. KLEINER, A (1938) *Oiseau* 8, 149-50; (1939*a*) *Aquila* 42-5, 79-140; (1939*b*) *Aquila* 42-5, 141-226; (1939*c*) *Aquila* 42-5, 542-9; (1939*d*) *Bull. Br. Orn. Club* 59, 70-1; (1939*e*) *Orn. Beob.* 36, 117-18; (1939*f*) *Bull. Br. Orn. Club* 60, 11-14. KLEINSCHMIDT, A (1938) *Falco* 34, 49-52. KLEINSCHMIDT, O (1893) *Orn. Jahrb.* 4, 167-219; (1906) *J. Orn.* 54, 78-99; (1909-11) Berajah, Zoographia infinita. Corvus Nucifraga. Halle; (1919) *Falco* 14, 15-17; (1922) *Abh. Ber. Zool. Mus. Dresden* 15 (3), 1; (1940) *Falco* 36, 22-25. KLICKA, J and WINKER, K (1991) *Condor* 93, 755-7. KLIJN, H B (1975) *Bijdr. Dierkde.* 45, 39-49. KLIMANIS, A (1987) *Gef. Welt* 111, 184-6. KLITZ, W J (1973) *Orn. Monogr. AOU* 14, 34-8. KLUIJVER, H N (1933) *Versl. Meded. Plantenziektekd. Dienst* 69; (1935) *Versl. Meded. Plantenziektekd. Dienst* 81; (1945) *Limosa* 18, 1-11. KNAPTON, R W (1979) *Sask. Nat. Hist. Soc. Spec. Publ.* 10. KNAPTON, R W and FALLS, J B (1982) *Can. J. Zool.* 60, 452-9. KNECHT, S (1960) *Anz. orn. Ges. Bayern* 5, 525-56. KNECHT, S and SCHEER, U (1968) *Z. Tierpsychol.* 25, 155-69. KNEIS, P (1977) *Falke* 24, 132-3; (1987) *Beitr. Vogelkde.* 33, 56-8. KNEUTGEN, J (1969) *J. Orn.* 110, 158-60. KNIEF, W (1988) *Beih. Veröff. Nat. Land. Bad.-Württ.* 53, 31-54. KNIGHTS, J C and WALKER, D (1989) *Adjutant* 19, 8-12. KNIJFF, P DE (1977) *Vogeljaar* 25, 268; (1991) *Birding World* 4, 384-91. KNOLLE, F (1990) *Vogelkdl. Ber. Niedersachs.* 22, 70-2; (1991) *Vogelkdl. Ber. Niedersachs.* 23, 48-53. KNORRE, D VON, GRÜN, G, GÜNTHER, R, and

SCHMIDT, K (1986) *Die Vogelwelt Thüringens.* Jena. KNOWLES, R K (1972) *Bird-Banding* 73, 114-17. KNOX, A G (1976) *Bull. Br. Orn. Club* 96, 15-19; (1987*a*) *Br. Birds* 80, 482-7; (1987*b*) In Cameron, E (ed.) *Glen Tanar: its human and natural history*, 64-74; (1988*a*) *Br. Birds* 81, 206-11; (1988*b*) *Ardea* 76, 1-26; (1990*a*) *Br. Birds* 83, 89-94; (1990*b*) *Ibis* 132, 454-66; (1990*c*) *Scott. Birds* 16, 11-18; (1992) *Biol. J. Linn. Soc.* 47, 325-35. KNYSTAUTAS, A (1987) *The natural history of the USSR.* London. KOBUS, D (1967) *Orn. Mitt.* 19, 259. KOCH, W (1914) *Orn. Monatsschr.* 39, 241-57, 273-88. KOELZ, W (1937) *Ibis* (14) 1, 86-104; (1948) *Auk* 65, 444-5; (1949) *Auk* 66, 208-9. KOENIG, A (1888) *J. Orn.* 36, 121-298; (1890) *J. Orn.* 38, 257-488; (1895) *J. Orn.* 43, 113-238, 257-321, 361-457; (1896) *J. Orn.* 44, 101-216; (1905) *J. Orn.* 53, 259-60; (1920) *J. Orn.* 68 suppl., 83-148; (1924) *J. Orn.* 72 suppl; (1926) *J. Orn.* 74 suppl., 152. KOES, R F (1989) *Blue Jay* 47, 104-6. KÖHLER, F (1943) *Fauna och Flora* 38, 3-7. KÖHLER, K-H (1990) *Orn. Mitt.* 42, 129. KOIVUNEN, P, NYHOLM, E S, and SULKAVA, S (1975) *Ornis fenn.* 52, 85-96. KOKHANOV, V D (1982) In Zabrodin, V A (ed.) *Ekologiya i morfologiya ptits na kraynem severo-zapade SSSR*, 124-37. Moscow. KOKHANOV, V D and GAEV, Y G (1970) *Trudy Kandalaksh. gos. Zapoved.* 8, 236-74. KOLBE, H and KOLBE, E (1981) *Falke* 28, 228-31. KOLBE, U (1982) *Falke* 29, 197-201, 209. KOLLIBAY, P (1913) *J. Orn.* 61, 612-17. KOLLMANSPERGER, F (1959) *Bonn. zool. Beitr.* 10, 21-67. KOLTHOFF, K (1932) *Medd. Göteborgs Mus. Zool. Avd.* 59. KOMEDA, S, YAMAGISHI, S, and FUJIOKA, M (1987) *Condor* 89, 835-41. KOMERS, P E (1989) *Anim. Behav.* 37, 256-65. KOMERS, P E and BOAG, D A (1988) *Can. J. Zool.* 66, 1679-84. KOMERS, P E and DHINDSA, M S (1989) *Anim. Behav.* 37, 645-55. KÖNIG, D (1966) *Corax* 1 (17), 203-9. KÖNIGSTEDT, D and MÜLLER, H E J (1988) *Zool. Abh. Staatl. Mus. Tierkde. Dresden* 43, 143-8. KÖNIGSTEDT, D and ROBEL, D (1977) *Zool. Abh. Staatl. Mus. Tierkde. Dresden* 34, 301-18; (1983) *Mitt zool. Mus. Berlin* 59, Suppl. *Ann. Orn.* 7, 127-49. KÖNIGSTEDT, D, ROBEL, D, and GOTTSCHALK, W (1977) *Beitr. Vogelkde.* 23, 347-50. KONRADT, H-U (1968) *Falke* 15, 278-9. KONSTANTINOV, V M, BABENKO, V G, and BARYSHEVA, I K (1982) *Zool. Zh.* 61, 1837-45. KONSTANTINOV, V M, MARGOLIN, V A, and BARANOV, L S (1986) *Tez. Dokl. I. S'ezda Vesesoyuz. orn. Obshch. 9. Vsesoyuz. orn. Konf.* 1, 312-13. KONTOGIANNIS, J E (1967) *Auk* 84, 390-5. KOOIKER, G (1991) *Vogelwelt* 112, 225-36. KORBUT, V V (1981) *Zool. Zh.* 60, 115-25; (1982) *Nauch. dokl. vyssh. Shk. Biol. nauki* (4) 29-33; (1985) *Ornitologiya* 20, 186-9; (1989*a*) *Zool. Zh.* 68 (11), 125-34; (1989*b*) *Zool. Zh.* 68 (12), 88-95. KORELOV, M N, KUZ'MINA, M A, GAVRILOV, E I, KOVSHAR', A F, GAVRIN, V F, and BORODIKHIN, I F (1974) *Ptitsy Kazakhstana* 5. Alma-Ata. KORENBERG, E I, RUDENSKAYA, L V, and CHERNOV, Y I (1972) *Ornitologiya* 10, 151-60. KORHONEN, K (1981) *Ann. zool. fenn.* 18, 165-7. KORPIMÄKI, E (1978) *Ornis fenn.* 55, 93-104. KORODI GÁL, I (1958) *Orn. Mitt.* 10, 66-9; (1965) *Zool. Abh. Staatl. Mus. Tierkde. Dresden* 28, 113-25; (1968) *Falke* 15, 296-301; (1969) *Revista Musedor* 3 (6), 251-5; (1972) *Trav. Mus. Hist. nat. 'Grigore Antipa'* 12, 355-83 KORTE, J de (1972) *Beaufortia* 20, 23-58. KORZYUKOV, A I (1979) *Ornitologiya* 14, 216. KOSHKINA, T V and KISHCHINSKI, A A (1958) *Trudy Kandalaksh. gos. Zapoved.* 1, 79-88. KOSKIMIES, P (1985) *Lintumies* 20, 302-6; (1989) *Distribution and numbers of Finnish breeding birds.* Helsinki. KOSTIN, Y V (1983) *Ptitsy Kryma.* Moscow. KOUKI, J and

Häyrinen, U (1991) *Ornis fenn.* 68, 170-7. Kovačević, J and Danon, M (1952) *Larus* 4-5, 185-217; (1959) *Larus* 11, 111-30. Kovács, G (1981) *Aquila* 87, 49-70. Kováts, L (1973) *Nymphaea* 1, 71-85. Kovshar', A F (1966) *Ptitsy Talasskogo Alatau.* Alma-Ata; (1979) *Pevchie ptitsy v subvysokogor'e Tyan'-Shanya.* Alma-Ata; (1981) *Osobennosti razmnozheniya ptits v subvysokogor'e.* Alma-Ata; (ed.) (1985) *Ptitsy Kurgal'dzhinskogo Zapovednika.* Alma-Ata. Kovshar', A F, Ivashchenko, A A, and Kovshar', V A (1986) *Vestnik Zool.* (5), 36-40. (1987) *Vestnik Zool.* (1), 59-64. Kowschar, A F (1966) *Falke* 13, 48-53. Kozlova, E V (1930) *Ptitsy yugo-zapadnogo Zabaykal'ya severnoy Mongolii i tsentral'noy Gobi.* St Petersburg; (1932) *Trudy Mongol. Komiss. Akad. Nauk SSSR* 3; (1933) *Ibis* (13) 3, 59-87; (1939) *Byull. Mosk. Obshch. Ispyt. prir. Otd. Biol.* 48, 63-70; (1975) *Ptitsy zonal'nykh stepey i pustyn' Tsentral'noy Azii.* St Petersburg. Kraatz, S (1979) *Falke* 26, 299-306. Krabbe, N (1980) Checklist of the birds of Elat. Unpubl. Krägenow, P (1986) *Der Buchfink.* Wittenberg Lutherstadt. Kramer, G (1941) *J. Orn.* 89 suppl., 105-31. Krampitz, H-E (1950) *Vogelwelt* 71, 7-9. Krätzig, H (1936) *Vogelzug* 7, 1-16. Kraus, M and Gauckler, A (1970) *Vogelwelt* 91, 18-23. Krechmar, A V (1966) *Trudy Zool. Inst. Akad. Nauk SSSR* 39, 185-312. Krechmar, A V, Andreev, A V, and Kondratiev, A Y (1978) *Ekologiya i rasprostranenie ptits na severo-vostoke SSSR.* Moscow. Krechmar, A V, Andreev, A V, and Kondratiev, A Y (1991) *Ptitsy severnykh ravnin.* St Petersburg. Kreibig, K (1957) *Falke* 4, 63. Krementz, D G, Nichols, J D, and Hines, J E (1989) *Ecology* 70, 646-55. Kreutzer, M (1979) *Behaviour* 71, 291-321; (1983) *C. R. Acad. Sci. Paris* (3) 297, 71-4; (1985) Thèse Doc. Etat. Univ. Pierre et Marie Curie, Paris; (1987) *J. comp. Psychol.* 101, 382-6; (1990) *Terre Vie* 45, 147-64. Kreutzer, M and Güttinger, H R (1991) *J. Orn.* 132, 165-77. Kricher, J C and Davis, W E (1986) *J. Fld. Orn.* 57, 48-52. Krištín, A (1988) *Folia Zool.* 37, 343-56. Krohn, H (1915) *Orn. Monatsber.* 23, 147-51. Kroodsma, D E (1974a) *Auk* 91, 54-64; (1974b) *Wilson Bull.* 86, 230-6; (1975) *Auk* 92, 66-80; (1981) *Auk* 98, 743-51. Kroymann, B (1965) *Orn. Mitt.* 17, 231-2; (1967) *Vogelwelt* 88, 170-3. Kroymann, B and Girod, R (1980) *BUND Information* 9, 37-40. Kroymann, B and Kroymann, H (1992) *Kart. Med. Brutvögel* 7, 47-8. Krüger, C (1944) *Dansk. orn. Foren. Tidsskr.* 38, 105-14. Krüger, S (1976) *Falke* 23, 283-4; (1979) *Der Kernbeisser.* Wittenberg Lutherstadt. Krull, F, Demmelmeyer, H, and Remmert, H (1985) *Naturwiss.* 72, 197-203. Krüper, T (1875) *J. Orn.* 23, 258-85. Kruseman, G (1942) *Ardea* 31, 302. Kübel, M and Ullrich, B (1975) *J. Orn.* 116, 323-4. Kuenzel, W J and Helms, C W (1974) *Auk* 91, 44-53. Kuenzi, A J, Moore, F R, and Simons, T R (1991) *Condor* 93, 869-83. Kuhk, R (1931 *J. Orn.* 79, 269-78. Kulczycki, A (1973) *Acta Zool. Cracov.* 18, 583-666. Kulczycki, A and Mazur-Gierasińska, M (1968) *Acta Zool. Cracov.* 13, 231-50. Kumar, V and Tewary, P D (1983) *Indian J. Zool.* 13, 25-31. Kumari, E (1958) *J. Orn.* 99, 32-4; (1972) *Soobshch. Pribalt. Kom. Izuch. Migr. Ptits* 7, 58-83. Kumari, E V (1960) *Trudy prolemn. tematich. Soveshch. Zool. Inst. Akad. Nauk SSSR* 9, 119-28. Kumerloeve, H (1957) *Orn. Mitt.* 9, 133; (1961) *Bonn. zool. Beitr.* 12 spec. vol; (1962a) *Bonn. zool. Beitr.* 13, 327-32; (1962b) *Iraq nat. Hist. Mus. Publ.* 20, 1-36; (1963) *Alauda* 31, 110-36, 161-211; (1964a) *Istanbul Üniv. Fakült. Mecmuasi* (B) 27, 165-228; (1964b) *J. Orn.* 105, 307-25; (1965a) *J. Orn.* 106, 112; (1965b) *Alauda* 33, 257-64; (1967a) *Istanbul Üniv.*

*Fakült. Mecmuasi* (B) 32, 79-213; (1967b) *Alauda* 35, 1-19; (1969a) *Ibis* 111, 617-18; (1969b) *Istanbul Üniv. Fen. Fakült. Mecmuasi* (B) 34, 245-312; (1969c) *Orn. Mitt.* 21, 84-5; (1969d) *Alauda* 37, 43-58, 114-34, 188-205; (1970a) *Istanbul Üniv. Fen. Fakült. Mecmuasi* (B) 35, 85-160; (1970b) *Beitr. Vogelkde.* 16, 239-49; (1974) *Orn. Mitt.* 26, 235. Kummer, J (1983) *Beitr. Vogelkde.* 29, 244-5. Kunkel, P (1959) *Z. Tierpsychol.* 16, 302-50; (1961) *Z. Tierpsychol.* 18, 471-89; (1967a) *Behaviour* 29, 237-61; (1967b) *Bonn. zool. Beitr.* 18, 139-68; (1969) *Z. Tierpsychol.* 26, 277-83. Kunz, H (1950) *Orn. Beob.* 47, 1-4. Kunz-Plüss, H (1969) *Orn. Beob.* 66, 22-3. Kuusisto, A P (1927) *Luonnon Ystävä* 31, 106-7. Kuznetsov, A A (1962) *Ornitologiya* 5, 215-42; (1967) *Ornitologiya* 8, 262-6. Kuzyakin, A P and Vtorov, P P (1963) *Ornitologiya* 6, 184-94. Kydyraliev, A (1972) *Ornitologiya* 10, 352-6.

Labitte, A (1937) *Oiseau* 7, 85-104; (1953) *Oiseau* 23, 247-60; (1954) *Oiseau* 24, 197-210; (1955a) *Alauda* 23, 212-16; (1955b) *Oiseau* 25, 57-8. Lack, D (1942-3) *Ibis* (14) 6, 461-84, 85, 1-27. Lack, D (1940) *Condor* 42, 239-41; (1946) *Br. Birds* 39, 258-64; (1948) *Evolution* 2, 95-110; (1954a) *The natural regulation of animal numbers.* Oxford; (1954b) *Br. Birds* 47, 1-15; (1955) *Proc. int. orn. Congr.* 11, 176-8; (1957) *Br. Birds* 50, 10-19. Lack, D and Southern, H N (1949) *Ibis* 91, 607-26. Lack, P (ed.) (1986) *The atlas of wintering birds in Britain and Ireland.* Calton. Lack, P C (1988) *Sitta* 2, 3-20. Łacki, A (1959) *Biul. Inst. Ochr. Ros.* 5, 239-49; (1962) *Acta Orn.* 6, 195-207. Læssøe, and Nørregaard, K (1968) *Dansk orn. Foren. Tidsskr.* 62, 95-6. Laferrère, M (1968) *Alauda* 36, 260-73. Lakhanov, D L (1977) *Trudy Samarkand gos. univ.* NS 324, 33-49. Lakhanov, Z L (1967) *Ornitologiya* 8, 364-66. Lam, C Y and Costin, R (1991) *Hong Kong Bird Rep.* 1990, 123-4. Lamarche, B (1981) *Malimbus* 3, 73-102; (1988) *Études sahariennes et ouest-africaines* 1 (4). Lamba, B S (1963a) *J. Bombay nat. Hist. Soc.* 60, 121-33; (1963b) *Res. Bull.* (NS) *Panjab Univ.* 14, 11-20; (1969) *J. Bombay nat. Hist. Soc.* 65, 777-8. Lambert, K (1965) *Falke* 12, 318. Lambertini, M (1981) *Avocetta* 5, 65-86. Lancum, F H (1928) *Br. Birds* 21, 264. Land, H C (1970) *Birds of Guatemala.* Wynnewood. Land, R and Lewis, P (1986) *Trans. Norfolk Norwich Nat. Soc.* 27, 256-9. Lane, C (1957) *Ibis* 99, 116. Lane, M (1984) *Notornis* 31, 283-4. Lang, E M (1939) *Orn. Beob.* 36, 141-5; (1946a) *Orn. Beob.* 43, 33-43; (1946b) *Orn. Beob.* 43, 117-18; (1948) *Orn. Beob.* 45, 197-205. Lange, G (1960) *J. Orn.* 101, 360. Lang, M, Bandorf, H, Dornberger, W, Klein, H, and Mattern, U (1990) *Ökol. Vögel* 12, 97-126. Langrand, O (1990) *Guide to the birds of Madagascar.* New Haven. Lanner, R M and Nikkanen, T (1990) *Ornis fenn.* 67, 24-7. Lansdown, P and Charlton, T D (1990) *Br. Birds* 83, 240-2. Lansdown, P, Riddiford, N, and Knox, A (1991) *Br. Birds* 84, 41-56. Lanz-Wälchli, H (1953) *Orn. Beob.* 50, 12-20. Lapous, E (1988) *Alauda* 56, 437-8. Larionov, G P, Degtyarev, V G, and Larionov, A G (1991) *Ptitsy Leno-Amginskogo mezhdurech'ya.* Novosibirsk. Larionov, P D (1959) *Zool. Zh.* 38, 253-60. Larionov, V F (1927) *Trudy Lab. exper. Biol. Mosk. Zoo.* 3, 119-137. Larrison, E J, and Sonnenberg, K G (1968) *Washington birds.* Seattle. Larsen, T and Tombre, I (1988) *Vår Fuglefauna* 11, 68-70; (1989) *Fauna norv.* (C) *Cinclus* 12, 3-10. Laskey, A R (1940) *Bird Lore* 42, 25-30. Lassey, P A and Wallace, D I M (1992) *Bird Watching* 79, 84-5. La Touche, J D D (1912) *Bull. Br. Orn. Club* 29, 124-

60; (1920) *Ibis* (11) **2**, 629-71; (1923) *Ibis* (11) **5**, 300-332; (1925-30) *A handbook of the birds of eastern China* **1**. London. LATSCHA, H (1979) *Orn. Mitt.* **31**, 225. LATZEL, G (1968) *Vogelwelt* **89**, 231-2. LATZEL, G and WISNIEWSKI, H-J (1971) *Vogelkde. Ber. Niedersachs.* **3**, 79-81. LAUB-MANN, A (1912a) *Orn. Jahrb.* **23**, 81-8; (1912b) *Verh. Orn. Ges. Bayern* **11**, 164-5. LAUDAGE, C and SCHROETER, W (1982) *Orn. Mitt.* **34**, 168. LAURENT, G and MOUILLARD, B (1939) *Alauda* **11**, 104-74. LAURENT, J-L (1986) *Oiseau* **56**, 263-86. LAVAUDEN, L (1930) *Alauda* **2**, 133-5. LAVIN CASTANEDO, J (1978) *Ardeola* **24**, 245-8. LAWTON, J (1959) *Br. Birds* **52**, 433-4. LAY, H G (1970) *Br. Birds* **63**, 38-9. LEA, D and BOURNE, W R P (1975) *Br. Birds* **68**, 261-83. LEADER, P J (1992) *Hong Kong Bird Rep. 1991*, 127-30. LEBEDEV, V G (1986) *Tez. Dokl. 1. s'ezda Vsesoyuz. orn. Obshch. 9. Vsesoyuz. orn. Konf.* **2**, 16-17. LEBERMAN, R C (1984) *J. Fld. Orn.* **55**, 486-7. LEBEURIER, E (1955) *Oiseau* **25**, 102-43. LEBEURIER, E and RAPINE, J (1937) *Oiseau* **7**, 583-93; (1939) *Oiseau* **9**, 219-32. LEBRETON, J D (1975) *Oiseau* **45**, 65-71; (1976) *Trav. scient. Parc. nat. Vanoise* **7**, 157-61. LEBRETON, P (ed.) (1977) *Atlas ornithologique Rhône-Alpes.* Lyon. LEBRETON, P, TOURNIER, H, and LEBRETON, J D (1976) *Trav. scient. Parc. nat. Vanoise* **7**, 163-243. LECK, C F (1973) *Auk* **90**, 888; (1975) *Bird-Banding* **46**, 201-3. LE DU, R (1935) *Alauda* **7**, 198-209. LEES-SMITH, D T and MADGE, S C (1982) *Bull. Orn. Soc. Middle East* **9**, 5-6. LEEVER, J (1982) *Roek in landbouw.* Zeist. LEFEVER, H and HUBLÉ, J (1955) *Gerfaut* **45**, 299. LEFRANC, N and PFEFFER, J-J (1975) *Alauda* **43**, 103-10. LE GRAND, G (1982) *Acoreana* **6**, 195-211; (1983) *Arquipélago* **4**, 85-116; (1986) *Dutch Birding* **8**, 55-7. LEHIKOINEN, E (1979) *Ornis fenn.* **56**, 24-9. LEHIKOINEN, E, and LAAKSONEN, M (1977) *Ornis fenn.* **54**, 133-4. LEHIKOINEN, E and NIEMELÄ, P (1977) *Lintumies* **12**, 33-44. LEHMANN, E VON (1952) *Vögel Heimat* **22**, 95-6; (1962) *Orn. Mitt.* **14**, 70-1. LEHMANN, H and MERTENS, R (1969) *Ool. Rec.* **43**, 2-16. LEIBL, F and MELCHIOR, F (1985) *Anz. orn. Ges. Bayern* **24**, 125-33. LEICESTER, M (1959) *Emu* **59**, 295-6. LEIN, M R (1978) *Can. J. Zool.* **56**, 1266-83. LEINONEN, A (1978) *Ornis fenn.* **55**, 182-3. LEKAGUL, B and ROUND, P D (1991) *A guide to the birds of Thailand.* Bangkok. LEKAGUL, B, ROUND, P D, and KOMOLPHALIN, K (1985) *Br. Birds* **78**, 2-39. LENNERSTEDT, I (1964) *Fauna och Flora* **59**, 94-123. LENSINK, R, BIJTEL, H J V VAN DEN, and SCHOLS, R M (1989) *Limosa* **62**, 1-10. LENSINK, R and HUSTINGS, F (1991) *Limosa* **64**, 29-30. LEONOVICH, V V (1976) *Ornitologiya* **12**, 87-94; (1983) *Ornitologiya* **18**, 23-32. LEONTIEV, A N (1965) *Ornitologiya* **7**, 478-9. LEONTIEV, A N and PAVLOV, E I (1963) *Ornitologiya* **6**, 165-72. LEVER, C (1987) *Naturalized birds of the world.* Harlow. LEVIN, A S and GUBIN, B M (1985) *Biologiya ptits intrazonaľnogo lesa.* Alma-Ata. LEWIS, A and POMEROY, D (1989) *A bird atlas of Kenya.* Rotterdam. LEWIS, A D (1989) *Scopus* **13**, 129-31. LEWIS, L R (1985) *Newbury Distr. orn. Club ann. Rep. 1984*, 34-9. LEWIS, S (1920) *Br. Birds* **14**, 26-33. LIEDER, K (1987) *Beitr. Vogelkde.* **33**, 46-8. LIEDEKERKE, R DE (1970) *Aves* **7**, 122; (1979) *Nos Oiseaux* **35**, 288. LIEFF, M R and JORDAN, N P (1950) *Br. Birds* **43**, 56. LIEN, L, ÖSTBYE, E, HAGEN, A, KLEMETSEN, A, and SKAR, H J (1970) *Nytt Mag. Zool.* **18**, 245-51. LIGON, J S (1961) *New Mexico birds and where to find them.* Albuquerque. LILFORD, LORD (1866) *Ibis* (2) **2**, 377-92. LILJA, C (1982) *Growth* **46**, 367-87. LIMBRUNNER, A (1987) *Verh. Orn. Ges. Bayern* **24**, 541-2. LINCOLN, G A, RACEY, R A, SHARP, P J, and KLANDORF, H (1980) *J. Zool. Lond.* **190**, 137-53. LIND, E

(1952) *Vår Fågelvärld* **11**, 145-53. LINDGREN, F (1975) *Fauna och Flora* **70** (5), 198-210. LINDNER, C (1906) *Orn. Montsschr.* **31**, 46-65, 105-21; (1907) *Orn. Monatsschr.* **32**, 398-410; (1917) *J. Orn.* **65** (2), 161-5. LINDNER, F (1911) *Orn. Monatsschr.* **36**, 62-72. LINDSTRÖM, Å (1987) *Ornis fenn.* **64**, 50-6; (1989) *Auk* **106**, 225-32; (1990) D Phil Thesis. Lund Univ. LINDSTRÖM, Å and ALERSTAM, T (1986) *Behav. Ecol. Sociobiol.* **19**, 417-24. LINDSTRÖM, Å and NILSSON, J-Å (1988) *Ornis scand.* **19**, 165-6. LINDSTRÖM, Å, OTTOSSON, U, and PETTERSSON, J (1984) *Vår Fågelvärld* **43**, 525-30. LINK, R and RITTER, M (1973a) *Orn. Beob.* **70**, 185-6; (1973b) *Orn. Beob.* **70**, 267-72. LINKE, H (1975) *Orn. Mitt.* **27**, 170-1. LINKOLA, P (1960) *Vår Fågelvärld* **19**, 66-7. LINN, H (1984) *Br. Birds* **77**, 489-90. LINSDALE, J M (1928) *Univ. Calif. Publ. Zool.* **30**, 251-392. LINSDALE, J M and SUMNER, E L (1934) *Condor* **36**, 107-12; (1937) *Condor* **39**, 162-63. LINT, A (1964) *Loodus. Selts. aast.* **56**, 167-88; (1971) *Orn. Kogumik* **5**, 132-63. LIPKOVICH, A D (1985) In Sokolov, V E and Sablina, T B (eds) *Izuchenie i okhrana redkikh i ischezayushchikh vidov zhivotnykh fauny SSSR*, 102-5. Moscow; (1986) In Amirkhanov, A M (ed.) *Ekosistemy ekstremaľnykh usloviy sredy v zapovednikakh RSFSR*, 128-34. Moscow. LIPPENS, L (1968) *Gerfaut* **58**, 3-23. LIPPENS, L and WILLE, H (1972) *Atlas des oiseaux de Belgique et d'Europe occidentale.* Tielt; (1976) *Les oiseaux du Zaïre.* Tielt. LI SHI-CHUN, LIU XI-YUO, TAN YAO-KUANG, and SIEN YAO-HUA (1975) *Acta zool. Sin.* **21**, 71-7. LISITSYNA, T Y and NIKOL'SKI, I D (1979) *Ornitologiya* **14**, 216-19. LITTLE, R (1990) *Br. Birds* **83**, 504-6. LITTLEJOHN, A C (1952) *Fair Isle Bird Obs. Bull.* **1** (8), 21-3, 34. LITUN, V I (1986) *Ornitologiya* **21**, 165. LITUN, V I and PLESSKI, P V (1983) *Ornitologiya* **18**, 64-9. LLOYD-EVANS, L (1948) *Br. Birds* **41**, 213. LLOYD-EVANS, L and NAU, B S (1965) *Rye Meads ann. Rep.* **3**, 23-39. LOBB, M G (1981) *J. RAF orn. Soc.* **12**, 25-7. LOBKOV, E G (1986) *Gnezdyashchiesya ptitsy Kamchatki.* Vladivostok. LOCK, J M (1971) In Gooders, J (ed.) *Birds of the world*, 2676-8. London. LOCKIE, J D (1955) *Ibis* **97**, 341-69; (1956a) *Bird Study* **3**, 180-90; (1956b) *J. Anim. Ecol.* **25**, 421-8; (1959) *Br. Birds* **52**, 332-4. LOCKLEY, A K (1992) *J. Orn.* **133**, 77-82. LOCKLEY, R M (1953) *Br. Birds* **46**, 347-8. LØFALDLI, L (1983) *Vår Fuglefauna* **6**, 183-9. LOFTIN, H (1977) *Bird-Banding* **48**, 253-58. LOFTS, B and MARSHALL, A J (1960) *Ibis* **102**, 209-14. LOFTS, B, MURTON, R K, WESTWOOD, N J, and THEARLE, R J P (1973) *Gen. comp. Endocrin.* **21**, 202-9. LOGMINAS, V (1991) *Lietuvos fauna: Paukščiai* **2**. Vilnius. LÖHRL, H (1950) *Z. Tierpsychol.* **7**, 130-3; (1957) *J. Orn.* **98**, 122-3; (1963) *J. Orn.* **104**, 62-8; (1964) *J. Orn.* **105**, 153-81; (1967) *Vogelwelt* **88**, 148-52; (1970) *Anz. orn. Ges. Bayern* **9**, 185-96; (1978a) *Gef. Welt* **102**, 162-5; (1978b) *Vogelwelt* **99**, 121-31; (1982) *Ökol. Vögel* **4**, 81; (1980) *J. Orn.* **121**, 408; (1987a) *Vogelwelt* **108**, 151-2; (1987b) *Vogelwelt* **108**, 189-90. LÖHRL, B and BÖHRINGER, R (1957) *J. Orn.* **98**, 229-40. LOISON, M (1984) *Aves* **21**, 109. LOMAN, J (1975) *Ornis scand.* **6**, 169-78; (1977) *Oikos* **29**, 294-301; (1980a) *Ekol. pol.* **28**, 95-109; (1980b) *Holarctic Ecol.* **3**, 26-35; (1980c) *Ibis* **122**, 494-500; (1980d) Doctoral Thesis. Lund Univ; (1984) *Ornis scand.* **15**, 183-7; (1985) *Ardea* **73**, 61-75. LOMAN, J and TAMM, S (1980) *Amer. Nat.* **115**, 285-9. LOMBARDO, M P, POWER, H W, STOUFFER, P C, ROMAGNANO, L C, and HOFFENBERG, A S (1989) *Behav. Ecol. Sociobiol.* **24**, 217-23. LONDEI, T and GNISCI, R (1988) *Riv. ital. Orn.* **58**, 59-73. LONDEI, T and MAFFIOLI, B (1989) *Riv. ital. Orn.* **59**, 241-58. LONDON

Natural History Society (1957) *The birds of the London area since 1900*. London. Loneux, M (1988) *Cahiers Ethol. appl.* 8, 337-406. Long, J L (1981) *Introduced birds of the world*. New York. 668efgi. Longstaff, T G (1932) *J. Anim. Ecol.* 1, 119-42. Lönnberg, E (1905) *Arkiv Zool.* 2 (9), 1-23; (1909) *Arkiv Zool.* 5 (9), 1-42; (1918) *Fauna och Flora* 13, 87-8. Looft, V (1965) *Corax* 1 (17), 1-9; (1967) *Corax* 2 (18), 27-31; (1971a) *Corax* 3, 188-96; (1971b) *Corax* 3, 196-9. Lope, F de, Guerrero, J, and Cruz, C de la (1984) *Alauda* 52, 312. Lope, F de, Guerrero, J, Cruz, C de la and Silva, E da (1985) *Alauda* 53, 167-80. Lope, F de, Guerrero, J, García, M E, Cruz, C de la Carretero, J J, Navarro, J A, Silva, E da, and Otano, J (1983) *Alytes* 1, 393-9. Løppenthin, B (1935) *Dansk orn. Foren. Tidsskr.* 29, 15-22; (1943) *Dansk orn. Foren. Tidsskr.* 37, 193-214. Lorenz, K S (1931) *J. Orn.* 79, 67-127; (1932) *J. Orn.* 80, 50-98; (1940) *Z. Tierpsychol.* 3, 278-92; (1952) *King Solomon's ring*. London; (1955) *Ik sprak met viervoeters, vogels en vissen*. Amsterdam; (1963) *On aggression*. London. Lorenz, T (1890) *J. Orn.* 38, 98-100. Loskot, V M (1986a) *Trudy Zool. Inst. Akad. Nauk SSSR* 150, 44-56; (1986b) *Trudy Zool. Inst. Akad. Nauk SSSR* 150, 147-70; (1991) *Trudy Zool. Inst. Akad. Nauk SSSR* 231, 43-116. Lo Valvo, F and Lo Verde, G (1987) *Riv. ital. Orn.* 57, 97-110. Lovari, S (1976a) In Pedrotti, F (ed.) *SOS Fauna*, 189-214. Camerino; (1976b) *Gerfaut* 66, 207-19; (1978) *Gerfaut* 68, 163-76; (1979) *Biol. Behav.* 4, 311-26. Lovaty, F (1985) *Oiseau* 55, 351-7; (1991) *Nos Oiseaux* 41, 99-106. Love, J A and Summers, R W (1973) *Scott. Birds* 7, 399-403. Løvenskiold, H L (1947) *Håndbok over Norges fugler*. Oslo; (1963) *Avifauna svalbardensis*. Oslo. Lowery, G H (1974) *Louisiana birds*. Baton Rouge. Lowery, G H and Monroe, B L (1968) In *Peters' check-list of birds of the world* 14. Lowther, J K (1961) *Can. J. Zool.* 39, 281-92. Lowther, P E (1979a) *Bird-Banding* 50, 160-2; (1979b) *Inland Bird Banding* 51, 23-9; (1988) *J. Fld. Orn.* 59, 51-4. Loxton, R G (1968) *Nat. Wales* 11, 126-30. Lucas, J V and Took, J M E (1974) *Cyprus Orn. Soc. (1957) Rep.* 19, 1-72. Ludewig, G (1989) *Gef. Welt* 113, 361-2. Ludwig, H (1984) *Gef. Welt* 108, 226-8. Ludlow, F (1928) *Ibis* (12) 4, 51-73. Ludlow, F and Kinnear, N B (1933) *Ibis* (13) 3, 658-94. Lukač, G (1988) *Orn. Mitt.* 40, 287-91. Lulav, S (1967) *IUCN Bull.* NS 2, 11. Lundberg, A and Edholm, M (1982) *Br. Birds* 75, 583-5. Lundberg, A, Mattsson, R, Nilsson, B, and Widén, P (1980) *Vår Fågelvärld* 39, 225-30. Lundberg, P (1981) *J. Orn.* 122, 65-72. Lundberg, P and Eriksson, L-O (1984) *Ornis scand.* 15, 105-9. Lundevall, C-F (1950) *Dansk orn. Foren. Tidsskr.* 44, 30-40; (1952) *Kung. Svenska Vetensk. Akad., Avh. Naturskydd.* 7, 1-73. Lundin, A (1962) *Vår Fågelvärld* 21, 81-95. Luniak, M (1977a) *Acta Orn.* 16, 213-40; (1977b) *Acta Orn.* 16, 241-74. Lüps, P, Hauri, R, Herren, H, Märki, H, and Ryser, R (1978) *Orn. Beob.* 75 suppl. Lütgens, H (1955) *Orn. Mitt.* 7, 113. Lüttschwager, H (1926) *Orn. Monatsber.* 34, 41-3. Lyaister, A F and Sosnin, G V (1942) *Materialy po ornitofaune Armyanskoy SSR (Ornis Armeniaca)*. Yerevan. Lye, R J (1948) *Br. Birds* 41, 211. Lykhvar', V P (1983) In Kuchin, A P (ed.) *Ptitsy Sibiri*, 147-8. Gorno-Altaysk. Lynch, J F, Morton, E S, and Van der Voort, M E (1985) *Auk* 102, 714-21. Lynes, H (1924) *Ibis* (11) 6, 648-719; (1926) *Ibis* (12) 2, 346-405; (1930) *Diary of Algerian expedition*. Unpubl. MS, Edward Grey Inst., Oxford Univ. Lyon, B E (1984) M Sc Thesis. Queen's Univ. Kingston. Lyon, B E and Montgomerie, R D (1985) *Behav. Ecol. Sociobiol.* 17, 279-84; (1987) *Ecology* 68, 713-22. Lyon, B E, Montgomerie, R D, and Hamilton, L D (1987) *Behav. Ecol. Sociobiol.* 20, 377-82.

McAtee, W L (1950) *Auk* 67, 247. MacBean, A F (1949) *Scott. Nat.* 61, 176-7. McCabe, T T and McCabe, E B (1933) *Condor* 35, 136-47. McCabe, T T and Miller, A H (1933) *Condor* 35, 192-7. McClure, H E (1974) *Migration and survival of the birds of Asia*. Bangkok. McCracken, D I (1989) In Bignal, E and Curtis, D J (eds) *Choughs and land-use in Europe*, 52-6. Scottish Chough Study Group. MacDonald, D (1965) *Scott. Birds* 3, 235-46; (1968) *Scott. Birds* 5, 177-8. Macdonald, J D (1973) *Birds of Australia*. Sydney. MacDonald, J W (1962) *Bird Study* 9, 147-67; (1963) *Bird Study* 10, 91-101. MacDonald, M (1960) *Birds in my Indian garden*. London. Macdonald, R A and Whelan, J (1986) *Ibis* 128, 540-57. McGillivray, W B (1980) *J. Fld. Orn.* 51, 371-2; (1984) *Can. J. Zool.* 62, 381-5. McGillivray, W B and Johnston, R F (1987) *Auk* 104, 681-7. McGregor, P K (1980) *Z. Tierpsychol.* 54, 285-97; (1981) D Phil Thesis. Oxford Univ; (1983) *Z. Tierpsychol.* 62, 256-60; (1986) *J. Orn.* 127, 37-42. McGregor, P K and Thompson, D B A (1988) *Ornis scand.* 19, 153-9. McGregor, P K, Walford, V R, and Harper, D G C (1988) *Bioacoustics* 1, 107-29. McIntyre, N (1953) *Br. Birds* 46, 377-8. Macke, T (1965) *J. Orn.* 106, 461-2; (1980) *Charadrius* 16, 5-13. McKee, J (1985) *Br. Birds* 78, 150-1. McKendry, W G (1973) *Br. Birds* 66, 400. McKilligan, N G (1980) *Bird Study* 27, 93-100. MacKinnon, J (1988) *Field guide to the birds of Java and Bali*. Yogyakarta. Mackintosh, D R (1941) *Bull. zool. Soc. Egypt* 3, 7-29. Mackowicz, R, Pinowski, J, and Wieloch, M (1970) *Ekol. Polska* 18, 465-501. Mackworth-Praed, C W and Grant, C H B (1960) *Birds of eastern and north eastern Africa* 2. London; (1963) *Birds of the southern third of Africa* 2. London; (1973) *Birds of west central and western Africa* 2. London. McLachlan, G R (1963) *Ostrich* 34, 102-9. McLaren, I A (1981) *Auk* 98, 243-57. McLaughlin, R L and Montgomerie, R D (1985) *Auk* 102, 687-95; (1989a) *Auk* 106, 738-41; (1989b) *Behav. Ecol. Sociobiol.* 25, 207-15. Maclean, G L (1985) *Roberts' birds of Southern Africa*. Cape Town. Macleay, K N G (1960) *Univ. Khartoum nat. Hist. Mus. Bull.* 1. McLennan, J A and MacMillan, B W H (1983) *New Zealand J. agric. Res.* 26, 139-45. MacLeod, I C (1987) *Scott. Bird News* 7, 11. McNair, D B (1988) *Migrant* 59, 45-8. McNeil, R (1982) *J. Fld. Orn.* 53, 125-32. McVean, A and Haddlesey, P (1980) *Ibis* 122, 533-6. Madge, S C, Hearl, G C, Hutchings, S C, and Williams, L P (1990) *Br. Birds* 83, 187-95. Madon, P (1928a) *Les corvidés d'Europe*. Paris; (1928b) *Mém. Soc. Orn. Mamm. France*. 1. Madsen, J (1981) In Fox, A D and Stroud, D A (eds) *Rep. Greenland White-fronted Goose Study 1979 Exped. Eqalungmiut Nunât, west Greenland*, 183-7; (1982) *Dansk orn. Foren. Tidsskr.* 76, 137-45. Madsen, J J (1990) *Dutch Birding* 12, 77. Maes, P (1989) *Oriolus* 55, 66-72. Maestri, F and Voltolini, L (1990) *Riv. ital. Orn.* 60, 99-100. Maestri, F, Voltolini, L, and Lo Valvo, F (1989) *Riv. ital. Orn.* 59, 159-71. Magee, M J (1928) *Bull. NE Bird-Banding Assoc.* 4, 149-52; (1930) *Bird-Banding* 1, 43-5. Magnenat, D (1969) *Nos Oiseaux* 30, 69-70. Magnússon, K G (1986) *Bliki* 5, 1-2. Mahabal, A and Vaidya, V G (1989) *Proc. Indian Acad. Sci. Anim. Sci.* 98, 199-209. Mahé, E (1985) Thèse. Univ. Sci. Tech. Languedoc. Mahéo, R (1969) *Ar Vran* 2, 176-87. Maher, W J (1964)

*Ecology* **45**, 520-8. MAKATSCH, W (1950) *Die Vogelwelt Macedoniens.* Leipzig; (1955) *Aquila* **59-62**, 347-50; (1963) *Zool. Abh. Staatl. Mus. Tierkde. Dresden* **26**, 135-86; (1971) *Zool. Abh. Staatl. Mus. Tierkde. Dresden* **32**, 17-41; (1976) *Die Eier der Vögel Europas* **2**. Radebeul; (1979) *Vögel Heimat* **49**, 163-6; (1981) *Verzeichnis der Vögel der Deutschen Demokratischen Republik.* Leipzig. MALBRANT, R (1952) *Faune du Centre Africain français.* Paris. MAL'CHEVSKI, A S (1959) *Gnezdovaya zhizn' pevchikh ptits.* St Petersburg. MAL'CHEVSKI, A S and PUKINSKI, Y B (1983) *Ptitsy Leningradskoy oblasti i sopredel'nykh territoriy* **2**. St Petersburg. MALCOLM-COE, Y (1981) *Bull. E. Afr. nat. Hist. Soc.* Jan-Feb, 8-11. MALLALIEU, M (1988) *Birds in Islamabad, Pakistan, 1985-7.* MALLOCH, J R (1922) *Auk* **39**, 569-70. MALMBERG, T (1949) *Vår Fågelvärld* **8**, 121-31; (1971) *Ornis scand.* **2**, 89-117. MAMBETZHUMAEV, A M and ABDREIMOV, T (1972) In Reymov, R (ed) *Ekologiya vazhneyshikh mlekopitayushchikh i ptits Karakalpakii,* 200-12. Tashkent. MANN, P, HERLYN, H, and UNTHEIM, H (1990) *Vogelwelt* **111**, 142-55. MANN, S and HOCHBERG, O (1982) *Israel Land Nat.* **8**, 10-14. MANNICHE, A L V (1910) *Medd. Grønland* **45**. MANSFELD, K (1950) *Nachricht. dtsch. Pflanzenschutzdienst* (NF) **4**, 131-6, 147-54, 164-75. MANSON-BAHR, P (1953) *Br. Birds* **46**, 414-15; (1954) *Br. Birds* **47**, 313. MANWELL, C and BAKER, C M A (1975) *Austral. J. Biol. Sci.* **28**, 545-57. MARCHANT, J (1984) *BTO News* **134**, 7-10. MARCHANT, J and WHITTINGTON, P (1988) *BTO News* **157**, 7-10. MARCHANT, J H, HUDSON, R, CARTER, S P, and WHITTINGTON, P (1990) *Population trends in British breeding birds.* Tring. MARCHANT, S (1962) *Bull. Iraq nat. Hist. Mus.* **2** (1), 1-40; (1963a) *Ibis* **105**, 369-98; (1963b) *Ibis* **105**, 516-57. MARCHANT, S and MACNAB, J W (1962) *Bull. Iraq nat. Hist. Inst.* **2** (3), 1-48. MARDER, J (1973) *Comp. Biochem. Physiol.* **45A**, 421-30. MARÉCHAL, P (1986) *Vogeljaar* **34**, 73-81; (1988) *Vogeljaar* **36**, 88. MARFURT, B (1971) *Orn. Beob.* **68**, 245-9. MARION, L and MARION, P (1975) *Bull. Soc. Sci. nat. Ouest France Suppl.* MARISOVA, I V, GORBAN', I M, and DAVIDOVICH, L I (1990) *Mat. dopov. 5. naradi orn. amat. orn. rukhu Zakhid. Ukraïni,* 29-32. MARJAKANGAS, A (1981) *Ornis. fenn.* **58**, 90-1; (1983) *Ornis fenn.* **60**, 89. MARKGREN, G (1955) *Vår Fågelvärld* **14**, 168-77. MARKGREN, G and LUNDBERG, S (1959) *Vår Fågelvärld* **18**, 185-205. MÄRKI, H (1976) *Orn. Beob.* **73**, 67-88. MÄRKI, H and BIBER, O (1975) *Jahrb. Naturhist. Mus. Bern* **5**, 153-64. MARKKOLA, J and VIERIKKO, E (1983) *Aureola* **8**, 147-50. MARLAN, J (1991) *Birding* **23**, 220-3. MARLE, J G VAN (1949) *Bull. Br. Orn. Club* **69**, 118-19. MARLE, J G VAN and HENS, P A (1938) *Limosa* **11**, 86-92. MARLE, J G VAN and VOOUS, K H (1988) *The birds of Sumatra.* Tring. MARLER, P (1956a) *Behav. Suppl.* **5**; (1956b) *Ibis* **98**, 496-501; (1957) *Behaviour* **11**, 13-39. MARLER, P and BOATMAN, D J (1951) *Ibis* **93**, 90-9. MARLER, P and HAMILTON, W J (1966) *Mechanisms of animal behaviour.* New York. MARLER, P and MUNDINGER, P C (1975) *Ibis* **117**, 1-17. MARPLES, B J and GURR, L (1943) *Emu* **43**, 67-71. MARQUARDT, K (1975) *Vögel Heimat* **46**, 56-9. MARQUISS, M (1980) *Ring. Migr.* **3**, 35-6. MARQUISS, M and BOOTH, C J (1986) *Bird Study* **33**, 190-5. MARQUISS, M, NEWTON, I, and RATCLIFFE, D A (1978) *J. appl. Ecol.* **15**, 129-44. MARR, V and KNIGHT, R L (1982) *Murrelet* **63**, 25. MARSHALL, A J and COOMBS, C J F (1957) *Proc. zool. Soc. Lond.* **128**, 545-88. MARSHALL, D and RAE, R (1981) *Rep. Grampian Ring. Group* **3**, 23-35. MARTEN, J A and JOHNSTON, N K (1986) *Condor* **88**,

409-20. MARTENS, J (1972) *Bonn. zool. Beitr.* **23**, 95-121; (1979) *Nat. Mus.* **109**, 337-43. MARTI, C D (1974) *Condor* **76**, 229. MARTÍN, A (1985) *Alauda* **53**, 309; (1987) *Atlas de las aves nidificantes en la isla de Tenerife.* Inst. Estud. Canarios Monogr. **32**. MARTÍN, A, BACALLADO, J J, EMMERSON, K W, and BAEZ, M (1984) *Il Reunión Iberoamer. Cons. Zool. Vert.*, 130-9. MARTIN, A J (1990) *Stour Ring. Group Rep. 1990*, 51-64. MARTIN, C E (1939) *Br. Birds* **33**, 108. MARTIN, G R (1986) *J. comp. Physiol.* **159**, 545-57. MARTIN, J-L (1980) *Problèmes de biogéographie insulaire: les cas des oiseaux nicheurs de Corse.* Thèse doct. Univ. Montpellier. MARTINS, R P (1989) *Sandgrouse* **11**, 1-41. MARZOCCHI, J F (1990) *Contribution à l'étude de l'avifaune du Cap Corse.* Bastia. MASATOMI, H and KOBAYASHI, S (1982) *J. Yamashina Inst. Orn.* **14**, 306-24. MASCHER, J (1952) *Vår Fågelvärld* **11**, 34-6, 37. MASON, C F (1989) *Bird Study* **36**, 145-6. MASON, C W and MAXWELL-LEFROY, H (1912) *Mem. Dept. Agric. India Ent. Ser.* **3**. MASSA, B (1976) *Ric. Biol. Selvag.* **7** suppl., 427-74; (1984) *Riv. ital. Orn.* **54**, 102-3; (1987) *Bull. Br. Orn. Club* **107**, 118-29; (1989) *Bull. Br. Orn. Club* **109**, 196-8. MASSI, A, MESCHINI, E, and ROSELLI, A (1991) *Riv. ital. Orn.* **61**, 63-5. MASSOTH, K (1981) *Voliere* **4**, 115-21; (1989) *Voliere* **12**, 135-9. MAŠTROVIĆ, A (1942) *Die Vögel des Küstenlandes Kroatiens* **1**. Zagreb. MAT, H A and DAVISON, G W H (1984) *Malaysiana* **13**, 231-8. MATHER, J (1986) *The birds of Yorkshire.* London. MATHEW, K L and NAIK, R M (1986) *Ibis* **128**, 260-5. MATHEWS, G F (1864) *Naturalist* **1**, 49-51, 69-71, 88-90. MATHIASSON, S (1972) *Bull. Br. Orn. Club* **92**, 103-6. MATOUŠEK, B (1968) *Act. Rer. Natur. Mus. Nat. Slov. Bratislava* **14** (2), 101-18; (1969) *Act. Rer. Natur. Mus. Nat. Slov. Bratislava* **15** (1), 59-76; (1971) *Act. Rer. Natur. Mus. Nat. Slov. Bratislava* **17** (1), 155-66. MATOUŠEK, B and JABLONSKI, B (1969) *Annot. zool. bot. Bratislava* **63**, 1-9. MATT, D (1983) *Orn. Mitt.* **35**, 216. MATTES, H (1976) *Orn. Beob.* **73**, 247-8; (1978) *Münster. Geogr. Arb.* **2**. MATTES, H and BÜRKLI, W (1979) *Orn. Beob.* **76**, 317-20. MATTES, H and JENNI, L (1984) *Orn. Beob.* **81**, 303-15. MATTHEWS, M, RECKITT, R, MUŽINIĆ, J, and MIKUSKA, J (1988) *Upoznajmo Ptice.* ICBP Migratory Birds Programme, Zagreb. MATVEJEV, S D (1948) *Godišnjaka Biol. Inst. Sarajevu* **1948**, 75-8; (1955) *Acta Mus. Maced. Sci. Nat.* **4**, 1-22; (1976) *Conspectus Avifaunae Balcanicae* **1**. Belgrade. MATVEJEV, S D and VASIĆ, V F (1973) *Catalogus Faunae Jugoslaviae* **4** (3). Aves. Ljubljana. MATZ, W (1967) *Falke* **14**, 130-3. MAU, K-G (1980) *Gef. Welt* **104**, 171-5, 187-9, 213-15, 234-8; (1982) *Gef. Welt* **106**, 215-19, 246-9. MAUERSBERGER, G (1960) In Stresemann, E and Portenko, L A (eds) *Atlas der Verbreitung Palaearktischer Vögel* **1**. Berlin; (1971) *J. Orn.* **112**, 232-33; (1972) *J. Orn.* **113**, 53-9; (1973) *Beitr. Vogelkde.* **19**, 76-7; (1976) *Falke* **23**, 51-5; (1980) *Mitt. zool. Mus. Berlin* **56**, Suppl. Ann. Orn. **4**, 77-164; (1982) *Mitt. zool. Mus. Berlin* **58**, Suppl. Ann. Orn. **6**, 101-13; (1983) *Mitt. zool. Mus. Berlin* **59**, Suppl. Ann. Orn. **7**, 47-83. MAUERSBERGER, G and MÖCKEL, R (1987) *Mitt. zool. Mus. Berlin* **63**, Suppl. Ann. Orn. **11**, 97-111. MAUERSBERGER, G and PORTENKO, L A (1971) In Stresemann, E, Portenko, L A, and Mauersberger, G (eds) *Atlas der Verbreitung Palaearktischer Vögel* **3**. Berlin. MAUERSBERGER, G, WAGNER, S, WALLSCHLÄGER, D, and WARTHOLD, R (1982) *Mitt. zool. Mus. Berlin* **58**, 11-74. MAUNDER, J E and MONTEVECCHI, W A (1982) *A field checklist of the birds of insular Newfoundland.* St John's. MAURIZIO, R (1978) *Vögel Heimat* **48**, 74-9, 104-7; (1987) *Orn. Beob.* **84**, 133-

4. Mautsch, H and Rank, H (1973) *Falke* **20**, 268-72. May, R C (1951) *Br. Birds* **44**, 17. Mayaud, N (1933a) *Alauda* **5**, 192-4; (1933b) *Alauda* **5**, 195-220, 345-82; (1933c) *Alauda* **5**, 453-99; (1936) *Inventaire des oiseaux de France*. Paris; (1939) *Oiseau* **9**, 486-506; (1941) *Arch. suisses Orn.* (1) **12**, 539-43; (1948) *Alauda* **16**, 168-79; (1960) *Alauda* **28**, 287-302 Mayes, W E (1926) *Br. Birds* **20**, 273-4. Mayhoff, H (1911) *Orn. Monatsschr.* **36**, 72-86; (1915) *Verh. orn. Ges. Bayern* **12**, 109-18. Mayo, A L W (1950) *Br. Birds* **43**, 368. Mayr, E (1926) *J. Orn.* **74**, 571-671; (1927) *J. Orn.* **75**, 596-619; (1949) *Ibis* **91**, 304-6; (1963) *Animal species and evolution*. Cambridge, Mass. Mayr, E and Short, L L (1970) *Publ. Nuttall Orn. Club* **9**. Mazzucco, K (1974) *Egretta* **17**, 53-9. Mead, C (1983) *BTO News* **129**, 1; (1986) *BTO News* **145**, 1. Mead, C J (1975) *Ardeola* **21**, 699-732. Mead, C J and Clark, J A (1987) *Ring. Migr.* **8**, 135-200; (1988) *Ring. Migr.* **9**, 169-204; (1989) *Ring. Migr.* **10**, 159-96; (1990) *Ring. Migr.* **11**, 137-76; (1991) *Ring. Migr.* **12**, 139-76; (1993) *Ring. Migr.* in press. Mead, C J and Hudson, R (1983) *Ring. Migr.* **4**, 281-319; (1984) *Ring. Migr.* **5**, 153-92; (1985) *Ring. Migr.* **6**, 125-72. Meade-Waldo, E G (1889a) *Ibis* (6) **1**, 1-13; (1889b) *Ibis* (6) **1**, 503-20; (1893) *Ibis* (6) **5**, 185-207. Mearns, R and Mearns, B (1989) *Scott. Birds* **15**, 179. Medway, Lord and Wells, D R (1976) *The birds of the Malay peninsula* **5**. London. Meek, E R (1984) *Br. Birds* **77**, 160-4. Meese, R J and Fuller, M R (1989) *Ibis* **131**, 27-32. Mehlum, F (1978) *Vår Fuglefauna* **1**, 87-9. Meidell, O (1943) *Nytt Mag. Naturv.* **84**, 1-91. Meiklejohn, M and Meiklejohn, R F (1938) *Br. Birds* **32**, 194. Meiklejohn, M F M (1948) *Ibis* **90**, 76-86; (1950) *Br. Birds* **43**, 264. Meiklejohn, R F (1930) *Ibis* (12) **6**, 560-4; (1936) *Ibis* (13) **6**, 377-8. Meineke, T (1979) *Beitr. Naturkde. Niedersachs.* **32**, 86-93. Meinertzhagen, R (1920) *Ibis* (11) **2**, 132-95; (1921) *Ibis* (11) **3**, 621-71; (1924) *Ibis* (11) **6**, 601-25; (1926) *Novit. Zool.* **33**, 57-121; (1927) *Ibis* (12) **3**, 363-422; (1930) *Nicoll's birds of Egypt*. London; (1935) *Ibis* (13) **5**, 110-51; (1938) *Ibis* (14) **2**, 480-520, 671-717; (1939) *Bull. Br. Orn. Club* **59**, 67-8; (1940) *Ibis* (14) **4**, 106-36, 187-234; (1947) *Bull. Br. Orn. Club* **67**, 90-8; (1949) *Ibis* **91**, 465-82; (1953) *Bull. Br. Orn. Club* **73**, 41-4; (1954) *Birds of Arabia*. Edinburgh; (1959) *Pirates and predators*. Edinburgh. Meininger, P L, Duiven, P, Marteijn, E C L, and Spanje, T M van (1990) *Malimbus* **12**, 19-24. Meininger, P L, Mullié, W C, and Bruun, B (1980) *Gerfaut* **70**, 245-50. Meininger, P L and Sørensen, U G (1984) *Bull. Br. Orn. Club* **104**, 54-7. Meise, K E (1958) *Falke* **5**, 141. Meise, W (1928) *J. Orn.* **76**, 1-203; (1934a) *Zool. Abh. Staatl. Mus. Tierkde. Dresden* **18** (2), 1-86; (1934b) *Orn. Monatsber.* **42**, 9-15; (1936) *J. Orn.* **84**, 631-72; (1938) *Compt. Rendu Congr. Orn. Int.* **9**, 233-48. Meklenburtsev, R N (1946) *Byull. Mosk. Obshch. Ispyt. Prir. Otd. Biol.* **5** (1), 87-110. Melcher, R (1951) *Orn. Beob.* **48**, 122-35. Melchior, E (1975) *Regulus* **11**, 351-63. Melchior, E, Mentgen, E, Peltzer, R, Schmitt, R, and Weiss, J. (1987) *Atlas der Brutvögel Luxemburgs*. Luxembourg. Melde, F and Melde, M (1976) *Falke* **23**, 88-92. Melde, M (1984) *Raben- und Nebelkrähe*. Wittenberg Lutherstadt. Meltofte, H and Fjeldså, J (1989) *Fuglene i Danmark* **2**. Copenhagen. Melville, D S and Round, P D (1984) *Bull. Br. Orn. Club* **104**, 127-38. Mendelssohn, H (1955) *Sal'it* **2**, 27-36. Mengel, R M (1963) *Wilson Bull.* **75**, 201-3; (1964) *Living Bird* **3**, 9-43; (1965) *Orn. Monogr. AOU* **3**. Mentgen, E (1988) *Regulus* **9** suppl., 62-8. Menzbier, M and Suschkin,

P (1913) *Orn. Monatsber.* **21**, 192-3. Menzel, H (1983) *Beitr. Vogelkde.* **29**, 310. Méric, J-D (1973) *Alauda* **41**, 161-3. Merikallio, E (1951) *Proc. int. orn. Congr.* **10**, 484-93. Merkel, F W (1978) *Luscinia* **43**, 163-81; (1980) *Luscinia* **44**, 133-58. Messineo, D J (1985) *Kingbird* **35**, 233-7. Mester, H (1971) *Bonn. zool. Beitr.* **22**, 28-89. Mester, H and Prünte, W (1982) *J. Orn.* **123**, 381-99. Metz, E (1981) *Orn. Mitt.* **33**, 272. Metzmacher, M (1984) *Cah. Ethol. appl.* **3**, 191-214; (1986a) *Gerfaut* **76**, 131-8; (1986b) *Gerfaut* **76**, 317-34; (1986c) *Gerfaut* **76**, 335-42; (1986d) *Oiseau* **56**, 229-62; (1990) In Pinowski, J and Summers-Smith, J D (eds) *Granivorous birds in the agricultural landscape*, 151-68. Mewaldt, L R (1977) *N. Am. Bird Bander* **2** (4), suppl. Mewaldt, L R and King, J R (1977) *Condor* **79**, 445-55; (1978) *Auk* **95**, 168-74; (1986) *J. Fld. Orn.* **57**, 155-67. Mewaldt, L R, Kibby, S S, and Morton, M L (1968) *Condor* **70**, 14-30. Mewes, W and Homeyer, E F von (1886) *Ornis* **2**, 181-288. Mey, E (1988) *Mitt. zool. Mus. Berlin* **64**, Suppl. Ann. Orn. **12**, 79-128. Meyer, D and Schloss, W (1968) *Auspicium* **3**, 33-68. Meyer, R M (1990) *Bird Study* **37**, 199-209. Meyer-Deepen, H (1954) *Gef. Welt* **78**, 154-5. Mezhennyi, A A (1964) *Zool. Zh.* **43**, 1679-87; (1979) In Krechmar, A V and Chernyavski, F B (eds) *Ptitsy severo-vostoka Azii*, 64-7. Vladivostok. Mezzavilla, F and Battistella, U (1987) *Riv. ital. Orn.* **57**, 33-40. Michaelis, H J (1980) *Falke* **27**, 286. Michaelsen, J (1985) *Vår Fuglefauna* **8**, 49-52. Michel, C (1986) *Birds of Mauritius*. Mauritius. Michels, H (1973) *Orn. Mitt.* **25**, 273. Michener, H and Michener, J R (1943) *Condor* **45**, 113-16. Middleton, A L A (1969) *Emu* **69**, 145-54; (1970) *Emu* **70**, 12-16. Mikhaylov, K E (1984) *Ornitologiya* **19**, 205-6. Mikhaylov, K E and Fil'chagov, A V (1984) *Ornitologiya* **19**, 22-9. Mikheev, A V (1939) *Zool. Zh.* **18**, 924-38. Mikkola, K (1979) *Lintumies* **14**, 125-36. Mikkola, K and Koivunen, P (1966) *Orn. fenn.* **43**, 1-12. Mikkonen, A V (1981) *Ornis scand.* **12**, 194-206; (1983) *Ornis scand.* **14**, 36-47; (1984) *Ornis fenn.* **61**, 33-53; (1985a) *Ann. zool. fenn.* **22**, 137-56; (1985b) *Acta Univ. Ouluensis* (A) *Sci. Rer. Nat.* **172** Biol. **24**. Milchev, B (1990) Ph D Thesis. Sofia Univ. Mild, K (1987) *Soviet bird songs* (2 cassettes and booklet). Stockholm; (1990) *Bird songs of Israel and the Middle East* (2 cassettes and booklet). Stockholm. Mildenberger, H (1940) *Beitr. Fortpfl. Vögel* **16**, 77-9; (1968) *Bonn. zool. Beitr.* **19**, 322-8; (1984) *Die Vögel des Rheinlandes* **2**. Düsseldorf. Mildenberger, H and Schulze-Hagen, K (1973) *Charadrius* **9**, 52-7. Miles, P (1968) *Opera Corcontica* **5**, 201-11. Miller, A H (1941) *Univ. Calif. Publ. Zool.* **44**, 173-434; (1942) *Condor* **44**, 185-6. Miller, W de W (1906) *Bull. Am. Mus. nat. Hist.* **22**, 161-83. Millington, R and Harrap, S (1991) *Birding World* **4**, 52-4. Mills, D G H (1982) *Br. Birds* **75**, 290-1. Mills, E D and Rogers, D T (1990) *Wilson Bull.* **102**, 146-50. Milsom, T P and Watson, A (1984) *Scott. Birds* **13**, 19-23. Mingozzi, T (1982) *Riv. ital. Orn.* **52**, 43-5. Mingozzi, T, Boano, G, and Pulcher, C (1988) *Atlante degli uccelli nidificanti in Piemonte e Val d'Aosta 1980-84*. Turin. Mirza, Z B (1973) In Kendeigh, S C and Pinowski, J (eds) *Productivity, population dynamics and systematics of granivorous birds*, 141-50. Warsaw; (1974) *Int. Stud. Sparrows* **7**, 76-87. Mirza, Z B, Kora, A, Sadik, L S, and Dahnous, K (1975) *Int. Stud. Sparrows* **8**, 117-23. Misra, R K and Short, L L (1974) *Condor* **76**, 137-46. Mlíkovský, J (1982) *Vogelwarte* **31**, 442-5. Moberg, A (1949) *Vår Fågelvärld* **8**, 187. Mocci Demartis, A (1973) *Alauda* **41**, 35-62. Model,

N and Otremba, W (1986) *Anz. orn. Ges. Bayern* **24**, 177–9. Mödlinger, P (1977) *Aquila* **83**, 79–89. Moeed, A (1976) *Notornis* **23**, 246–9. Moffat, C B (1943) *Irish Nat. J.* **8**, 54–5. Mohr, R (1967) *J. Orn.* **108**, 484–90; (1974) *J. Orn.* **115**, 106–7. Mojsisovics, A (1886) *Mitt. orn. Ver. Wien Schwalbe* **10**, 113. Moksnes, A (1973) *Norw. J. Zool.* **21**, 113–38. Moksnes, A and Røskaft, E (1987) *Ornis scand.* **18**, 168–72. Molamusov, K T (1967) *Ptitsy tsentral'noy chasti Severnogo Kavkaza*. Nal'chik. Molau, U (1985) *Vår Fågelvärld* **44**, 5–20. Moll, W (1986) *Vogelkdl. Ber. Niedersachs.* **18**, 11–14. Møller, A P (1978a) *Nordjyllands fugle*. Klampenborg; (1978b) *Dansk orn. Foren. Tidsskr.* **72**, 61–3; (1978c) *Dansk orn. Foren. Tidsskr.* **72**, 197–215; (1979) *Dansk orn. Foren. Tidsskr.* **73**, 305–9; (1981a) *Dansk orn. Foren. Tidsskr.* **75**, 69–78; (1981b) *Dutch Birding* **3**, 148–50; (1982a) *Ornis scand.* **13**, 94–100; (1982b) *Ornis scand.* **13**, 239–46; (1983a) *J. Orn.* **124**, 147–61; (1983b) *Ornis fenn.* **60**, 105–11; (1983c) *Ornis scand.* **14**, 81–9; (1985) *J. Orn.* **126**, 405–19; (1987) *Anim. Behav.* **35**, 203–10; (1988) *Ethology* **78**, 321–31. Moltoni, E (1950) *Riv. ital. Orn.* **20**, 75–8; (1951) *Riv. ital. Orn.* **21**, 45–51; (1969) *Riv. ital. Orn.* **39**, 128–57; (1973) *Riv. ital. Orn.* **43** suppl. Moltoni, E and Brichetti, P (1978) *Riv. ital. Orn.* **48**, 65–142. Monaghan, P (1989) In Bignal and Curtis (1989) 4–8, 63–4. Monaghan, P and Bignal, E (1985) *Bull. Br. ecol. Soc.* **16**, 208–10. Monaghan, P, Bignal, E, Bignal, S, Easterbee, N, and McKay, C R (1989) *Scott. Birds* **15**, 114–18. Monaghan, P and Thompson, P M (1984) *Bull. Br. ecol. Soc.* **15**, 145–6. Monk, J F (1954) *Bird Study* **1**, 2–14. Mönke, R (1975) *Beitr. Vogelkde.* **21**, 370–1. Monroe, B L, Jr (1968) *Orn. Monogr. AOU* **7**. Montell, J (1917) *Acta Soc. Fauna Flora fennica* **44** (7). Montgomerie, R D, Cartar, R V, McLaughlin, R L, and Lyon, B (1983) *Arctic* **36**, 65–75. Moody, C (1954) *Br. Birds* **47**, 406. Moore, F R, Kerlinger, P, and Simons, T R (1990) *Wilson Bull.* **102**, 487–500. Moore, A S (1991) *Peregrine* **7**, 47–53. Moore, H J and Boswell, C (1956) *Iraq nat. Hist. Mus. Publ.* **10**; (1957) *Iraq nat. Hist. Mus. Publ.* **12**. Moore, N C (1962) *J. Northants. nat. Hist. Soc.* **34**, 130–2. Morales, J A G (1969) *Ardeola* **13**, 265. Moreau, R E (1931) *Ibis* (13) **1**, 204–8; (1960) *Ibis* **102**, 298–321, 443–71; (1966) *The bird faunas of Africa and its islands*. London; (1967) *Ibis* **109**, 445; (1972) *The Palaearctic–African bird migration systems*. London. Morel, G and Roux, F (1966) *Terre Vie* **113**, 19–72, 143–76; (1973) *Terre Vie* **27**, 523–50. Morel, G J (1980) *Liste commentée des oiseaux du Sénégal et de la Gambie. Suppl.* **1**. Dakar. Morel, G J and Morel, M-Y (1976) *Terre Vie* **30**, 493–520; (1978) *Cah. ORSTOM Sér. Biol.* **13**, 347–58; (1980) *Proc. int. orn. Congr.* **17**, 1150–4; (1990) *Les oiseaux de Sénégambie*. Paris. Morel, M-Y (1964) *Terre Vie* **111**, 436–451; (1966) *Ostrich Suppl.* **6**, 435–42; (1967) *Terre Vie* **114**, 77–82; (1969) *Doc. Sci. Nat. Thesis*. Rennes Univ; (1973) *Mém. Mus. natn. Hist. nat. Paris* (A) *Zool.* **78**, 1–156. Morel, M-Y and Morel, G J (1973a) *Oiseau* **43**, 97–118; (1973b) *Oiseau* **43**, 314–29. Moreno, J M (1988) *Guía de las aves de las Islas Canarias*. Santa Cruz de Tenerife. Morgan, A (1971) *Br. Birds* **64**, 422–3. Morgan, P A and Howse, P E (1973) *Anim. Behav.* **21**, 481–91; (1974) *Anim. Behav.* **22**, 688–94. Morgan, J and Shirihai, H (1992) *Birding World* **5**, 344–7. Moritz, D (1982) *Vogelwelt* **103**, 16–18. Morley, A (1943) *Ibis* **85**, 132–58. Morozov, V V (1984) *Ornitologiya* **19**, 30–40; (1986) *Tez. Dokl. 1. S'ezda Vsesoyuz. orn. Obshch. 9. Vsesoyuz. orn. Konf.* **2**, 83–5; (1987) *Ornitologiya* **22**, 134–47. Morphy,

M J (1965) *Ibis* **107**, 97–100. Morrison, A (1948) *Ibis* **90**, 381–7. Morrison, C M (1977) *Scott. Birds* **9**, 302. Mortensen, P H and Birkholm-Clausen, F (1963) *Dansk orn. Foren. Tidsskr.* **57**, 22–4. Morton, E S (1989) *Wilson Bull.* **101**, 460–2. Morton, G A and Morton, M L (1990) *Condor* **92**, 813–28. Morton, M L and Welton, D E (1973) *Condor* **75**, 184–9. Morton, M L, Horstmann, J L, and Carey, C (1973) *Auk* **90**, 83–93. Morton, M L, King, J R, and Farner, D S (1969) *Condor* **71**, 376–85. Morton, M L and Morton, G A (1987) *Condor* **89**, 197–200. Morton, M L, Orejuela, J E, and Budd, S M (1972) *Condor* **74**, 423–30. Morton, M L and Pereyra, M E (1987) *J. Fld. Orn.* **58**, 6–21. Mosimann, P (1988) *Orn. Beob.* **85**, 179–81. Moss, D (1979) *Wicken Fen Group Rep.* **10**, 12–14. Motis, A (1985) Master's Thesis. Barcelona Univ; (1986) *Historia Natural dels Paisos Catalans.* **12**: *Ocells. Fundació Enciclopedia Catalana,* 336–415. Barcelona; (1987) *Misc. Zool.* **11**, 339–46. Motis, A, Martinez, A, Peris, S, and Ferrer, X (1987) *9th Jorn. Orn. Españolas, Madrid,* poster paper. Motis, A, Mestre, P, and Martínez, A (1983) *Misc. Zool.* **7**, 131–7. Mould, J E M (1974) *Bird Study* **21**, 157–8. Mountfort, G and Ferguson-Lees, I J (1961a) *Ibis* **103a**, 86–109; (1961b) *Ibis* **103a**, 443–71. Mountfort, G R (1935) *Br. Birds* **29**, 145–8; (1956) *Ibis* **98**, 490–5; (1957) *The Hawfinch.* London; (1958) *Portrait of a wilderness.* London; (1962) *Portrait of a river.* London; (1965) *Portrait of a desert.* London. Mountjoy, D J and Lemon, R E (1991) *Behav. Ecol. Sociobiol.* **28**, 97–100. Moyer, J W (1930) *Auk* **47**, 567–8. Moyson, I Y (1973) *Gerfaut* **63**, 257–78. Mühlethaler, F (1952) *Orn. Beob.* **49**, 173–82. Muir, R C (1959) *Br. Birds* **52**, 434–5. Müller, A K (1953) *Anz. Orn. Ges. Bayern* **4**, 76. Müller, H E J (1983) *Falke* **30**, 24–31. Müller, H E J and Wernicke, P (1990) *Falke* **37**, 83–6. Müller, L (1982) *Falke* **29**, 421. Müller, S (1970) *Falke* **17**, 199–203; (1973) *Corax* **4**, 112–30; (1977) *Falke* **24**, 268. Muller, Y (1987) *Acta Oecol.* **8**, 185–9. Mulligan, J A and Olsen, K C (1969) In Hinde, R A (ed) *Bird vocalizations,* 165–84. Cambridge. Mullins, J R (1984a) *Br. Birds* **77**, 26–7; (1984b) *Br. Birds* **77**, 133–5; (1984c) *Br. Birds* **77**, 426. Mulvihill, R S and Chandler, C R (1990) *Auk* **107**, 490–9; (1991) *Condor* **93**, 172–5. Münch, H (1957) *Orn. Beob.* **54**, 194–5. Mundinger, P C (1970) *Science NY* **168**, 480–2; (1979) *Syst. Zool.* **28**, 270–83. Mundy, P J and Cook, A W (1977) *Ostrich* **48**, 72–84. Munkejord, A (1987) *Fauna norv.* (C) *Cinclus* **10**, 73–80. Munkejord, A, Hauge, F, Folkedal, S, and Kvinnesland, A (1985) *Fauna norv.* (C) *Cinclus* **8**, 1–8. Munn, P W (1931) *Novit. Zool.* **37**, 53–132. Muñoz-Pulido, R, Bautista, L M, Alonso, J C, and Alonso, J A (1990) *Bird Study* **37**, 111–14. Munro, I C (1977) *Scott. Birds* **9**, 382. Munro, J H B (1975) *Scott. Birds* **8**, 309–14. Munro, J A and Cowan, I M (1947) *British Columbia Prov. Mus. spec. Publ.* **2**. Muntaner, J, Ferrer, X, and Martínez-Vilalta, A (1983) *Atlas. dels ocells nidificants de Catalunya i Andorra*. Barcelona. Munteanu, D (1966) *Bull. Br. Orn. Club* **86**, 98–100; (1967) *Larus* **19**, 179–203. Murphy, M E and King, J R (1984) *Auk* **101**, 164–7. Murphy, M E, King, J R, and Lu, J (1988) *Can. J. Zool.* **66**, 1403–13. Murphy, R C (1924) *Bull. Amer. Mus. nat. Hist.* **50**, 211–78. Murr, F (1957) *Anz. orn. Ges. Bayern* **4**, 556–8. Murray, B G (1965) *Wilson Bull.* **77**, 122–33; (1989) *Auk* **106**, 8–17. Murray, B G and Jehl, J R (1964) *Bird-Banding* **35**, 253–63. Murray, R D (1978) *Br. Birds* **71**, 318–9. Murton, R K (1971) *Man and birds*. London. Murton, R K and

WESTWOOD, N J (1974) *Ibis* **116**, 298-313. MYCOCK, J (1987) *Orn. Soc. Middle East Bull.* **18**, 1-3. MYLNE, C K (1960) *Br. Birds* **53**, 86-8; (1961) *Br. Birds* **54**, 206-7; (1957) *Br. Birds* **50**, 171-2. MYRCHA, A and PINOWSKI, J (1970) *Condor* **72**, 175-81.

NAGY, E (1943) *Beitr. Fortpfl. Vögel* **19**, 9-13. NAIK, N L and NAIK, R M (1969) *Pavo* **7**, 57-73. NAKAMURA, T, YAMAGUCHI, S, IIJIMA, K, and KAGAWA, T (1968) *Misc. Rep. Yamashina Inst. Orn.* **5**, 313-36. NANKINOV, D N (1974) In Boehme, R L and Flint, V E (eds) *Mat. 6 Vsesoyuz. orn. Konf.* **2**, 92-3. Moscow; (1978) *Acta Orn.* **16**, 285-94; (1984) *Intl. Studies on Sparrows* **11**, 47-70. NARANG, M L and LAMBA, B S (1984) *Rec. zool. Survey India misc. Publ. occ. Pap.* **44**, 1-76. NARDIN, C and NARDIN, G (1985) *Nos Oiseaux* **38**, 113-20. NATORP, O (1931) *J. Orn.* **79**, 338-46; (1940) *Orn. Monatsber.* **48**, 46-8. NAU, B S (1960) *Bird Study* **7**, 185-8. NAUMANN, J A (1900) (ed. Hennicke, C R) *Naturgeschichte der Vögel Mitteleuropas* **3**; (1901) (ed. Hennicke, C R) **4**. NAUROIS, R DE (1969) *Bull. de l'Inst. fond. d'Afrique noire* **31**A, 143-218; (1981) *Bull. de l'Inst. fond. d'Afrique noire* **43**, 202-18; (1986) *Cyanopica* **3**, 533-8; (1988) *Bol. Mus. Mun. Funchal* **40**, 253-73. NAUROIS, R DE and BONNAFFOUX, D (1969) *Alauda* **37**, 93-113. NEALE, J J (1899-1900) *Trans. Cardiff Nat. Soc.* **32**, 1-5; (1901) *Rep. Trans. Cardiff Nat. Soc.* **32**, 49-53. NECHAEV, V A (1969) *Ptitsy yuzhnykh Kuril'skikh ostrovov.* Leningrad; (1974) *Trudy Biol.-pochvenn. Inst. Akad. Nauk SSSR* **17** (120), 120-35; (1975) *Trudy Biol.-pochvenn. Inst. Akad. Nauk SSSR* **29** (132), 114-60; (1977) *Byull. Mosk. Ispyt. Prir. Otd. Biol.* **82** (3), 31-9. NEIN, R (1982) *Beitr. Naturkde. Wetterau* **2**, 154. NELSON, B (1973) *Azraq: desert oasis.* London. NELSON, J B (1950) *Br. Birds* **43**, 293. NELSON, T H (1907) *The birds of Yorkshire* **1**. London. NERO, R W (1951) *Wilson Bull.* **63**, 84-8; (1963) *Sask. nat. Hist. Soc. spec. Publ.* **5**; (1967) *Sask. nat. Hist. Soc. spec. Publ.* **6**. NERO, R W and LEIN, M R (1971) *Sask. Nat. Hist. Soc. Spec. Publ.* **7**. NERUCHEV, V V and MAKAROV, V I (1982) *Ornitologiya* **17**, 125-9. NETHERSOLE-THOMPSON, D (1966) *The Snow Bunting.* Edinburgh; (1975) *Pine Crossbills.* Berkhamsted; (1976) *Scott. Birds* **9**, 147-62. NETHERSOLE-THOMPSON, C and NETHERSOLE-THOMPSON, D (1940) *Br. Birds* **34**, 135; (1943a) *Br. Birds* **37**, 70-4; (1943b) *Br. Birds* **37**, 88-94. NETHERSOLE-THOMPSON, D and WATSON, A (1974) *The Cairngorms.* London. NETHERSOLE-THOMPSON, D and WHITAKER, D (1984) *Scott. Birds* **13**, 87. NEUB, M (1973) *Anz. orn. Ges. Bayern* **12**, 248-55. NEUFELDT, I A (1961) *Zool. Zh.* **40**, 416-26; (1970) *Trudy Zool. Inst. Akad. Nauk SSSR* **47**, 111-81; (1986) *Trudy Zool. Inst. Akad. Nauk SSR* **150**, 7-43. NEUFELDT, I A and LUKINA, E W (1966) *Falke* **13**, 121-5. NEUMANN, O (1907) *Orn. Monatsber.* **15**, 144-6. NEUSCHWANDER, J (1973) *Alauda* **41**, 163-5. NEVES, F I M (1984) *Cyanopica* **3**, 183-96. NEWSTEAD, R (1908) *J. Board Agric. Suppl.* **15** (9). NEWTON, I (1964a) *Bird Study* **11**, 47-68; (1964b) D Phil Thesis. Oxford Univ; (1964c) *J. appl. Ecol.* **1**, 265-79; (1966a) *Br. Birds* **59**, 89-100; (1966b) *Ibis* **108**, 41-67; (1967a) *Ibis* **109**, 33-98; (1967b) *J. Anim. Ecol.* **36**, 721-44; (1967c) *Ibis* **109**, 440-1; (1967d) *Bird Study* **14**, 10-24; (1968a) *Bird Study* **15**, 84-92; (1968b) *Condor* **70**, 323-32; (1969a) *J. Orn.* **110**, 53-61; (1969b) *Physiol. Zoöl.* **42**, 96-107; (1970) In Watson, A (ed.) *Animal populations in relation to their food resources*, 337-57; (1972) *The finches.* London. NEWTON, I, DAVIS, P E, and DAVIS, J E (1982) *J.*

*appl. Ecol.* **19**, 681-706. NEWTON, I and EVANS, P R (1966) *Bird Study* **13**, 96-8. NICE, M M (1937) *Trans. Linn. Soc. New York* **4**; (1938) *Bird-Banding* **9**, 1-11; (1943) *Trans. Linn. Soc. New York* **6**; (1946) *Condor* **48**, 41-2. NICHOLS, J T (1935) *Bird-Banding* **6**, 11-15; (1945) *Bird-banding* **16**, 29-32; (1953) *Bird-Banding* **24**, 16-7. NICHOLSON, E M (1930) *Ibis* (12) **6**, 280-313, 395-428; (1951) *Birds and men.* London. NICHOLSON, E M and NICHOLSON, B D (1930) *J. Ecol.* **18**, 51-66. NICHT, M (1961) *Zool. Abh. Staatl. Mus. Tierkde. Dresden* **26**, 79-99. NICKELL, W P (1968) *Bird-banding* **39**, 107-16. NICOLAI, B (ed.) (1993) *Atlas der Brutvögel Ostdeutschlands.* Jena. NICOLAI, J (1956) *Z. Tierpsychol.* **13**, 93-132; (1957) *J. Orn.* **98**, 363-71; (1959) *J. Orn.* **100**, 39-46; (1960) *Zool. Jb.* **87**, 317-62; (1964) *Z. Tierpsychol.* **21**, 129-204. NICOLL, M J (1909a) *Bull. Br. Orn. Club* **23**, 99-100; (1909b) *Ibis* (9) **3**, 471-84; (1912) *Ibis* (9) **6**, 405-53; (1922) *Ibis* (11) **4**, 688-701. NIEDRACH, R J and ROCKWELL, R B (1939) *The birds of Denver and mountain parks.* Denver. NIELSEN, O K (1979) *Nátúrufr.* **49**, 204-20; 654ce; (1986) Ph D Thesis. Cornell Univ. NIETHAMMER, G (1936) *Beitr. Fortpfl. Vögel* **12**, 161-2; (1937) *Handbuch der deutschen Vogelkunde* **1**. Leipzig; 647i 658b 665b 651j 653j 658gh 674j 678bcfi 679gh; (1942) *Handbuch der deutschen Vogelkunde* **3**. Leipzig; (1943) *J. Orn.* **91**, 167-238; (1950) *Syllegomena biologica*, 267-86. Leipzig; (1953a) *Bonn. zool. Beitr.* **4**, 73-8; (1953b) *J. Orn.* **94**, 282-9; (1955) *Bonn. zool. Beitr.* **6**, 29-80; (1957a) *Bonn. zool. Beitr.* **8**, 230-47; (1957b) *Bonn. zool. Beitr.* **8**, 275-84; (1957c) *J. Orn.* **98**, 363-71; (1958) *J. Orn.* **99**, 431-7; (1961) *Falke* **8**, 367-70; (1962) *Bonn. zool. Beitr.* **13**, 209-15; (1963) *Bonn. zool. Beitr.* **14**, 129-50; (1967) *Ibis* **109**, 117-8; (1969) *J. Orn.* **110**, 205-8; (1971) *J. Orn.* **112**, 202-26; (1973) *Bonn. zool. Beitr.* **24**, 270-84. NIETHAMMER, G and BAUER, K (1960) *Orn. Beob.* **57**, 241-2. NIETHAMMER, VON G, KRAMER, H, and WOLTERS, H E (1964) *Die Vögel Deutschlands.* Frankfurt. NIETHAMMER, G and THIEDE, W (1962) *J. Orn.* **103**, 289-93. NIETHAMMER, G and WOLTERS, H E (1966) *Bonn. zool. Beitr.* **17**, 157-85; (1969) *Bonn. zool. Beitr.* **20**, 351-4. NIGG, F (1974) *Vögel Heimat* **45**, 38. NIGHTINGALE, B and ALLSOPP, K (1990) *Br. Birds* **83**, 541-8; (1991) *Br. Birds* **84**, 316-28; (1992) *Br. Birds* **85**, 636-47. NIKANDER, P J and JÄNNES, H (1992) *Lintumies* **27**, 275-83. NIKOLAUS, G (1987) *Bonn. zool. Monogr.* **25**. NIKOLAUS, G and PEARSON, D (1991) *Ring. Migr.* **12**, 46-7. NILSSON, I (1957) *Vår Fågelvärld* **16**, 50. NILSSON, R (1971) *Vår Fågelvärld* **30**, 124-5. NILSSON, S G (1979) *Ibis* **121**, 177-85. NISBET, I C T (1970) *Bird-Banding* **41**, 207-40. NISBET, I C T, DRURY, W H, and BAIRD, J (1963) *Bird-Banding* **34**, 107-59. NISBET, I C T and SMOUT, T C (1957) *Br. Birds* **50**, 201-4. NITSCHE, G (1980) *Orn. Mitt.* **32**, 274-5. NOË, A (1983) *Anz. orn. Ges. Bayern* **22**, 110-11. NOESKE, A and HOFFMANN, M (1988) *Limicola* **2**, 22-7. NOGALES, M (1990) Doct. Thesis. La Laguna Univ; (1992) *Ecología* **6**, 215-23. NOGGE, F (1973) *Bonn. zool. Beitr.* **24**, 254-69. NOLAN, V and KETTERSON, E D (1983) *Wilson Bull.* **95**, 603-20. NOLAN, V and KETTERSON, E D (1990) *Wilson Bull.* **102**, 469-79. NOLAN, V, KETTERSON, E D, ZIEGENFUS, C, CULLEN, D P, and CHANDLER, C R (1992) *Condor* **94**, 364-70. NOLL, H (1956) *Ber. St. Gall. naturw. Ges.* **75**, 48-52; 651g. NOLTE, W (1930) *J. Orn.* **78**, 1-19. NOORDEN, B VAN (1991) *Limosa* **64**, 69-71. NORDMEYER, A, OELKE, H, and PLAGEMANN, E (1970) *Int. Stud. Sparrows* **4**, 50-4. NORMAN, D, CROSS, D, and COCKBAIN, R (1981) *BTO News* **114**, 9. NORMAN, E J (1966) *Birds Illustr.* **12**, 74-5. NØHR, H (1984) *Vår*

*Fågelvärld* **43**, 241-3. Nørrevang, A and Hartog, J C den (1984) *Cour. Forsch.-Inst. Senckenberg* **68**, 107-34. Norris, R A and Hight, G L (1957) *Condor* **59**, 40-52. Norris, R A and Johnston, D W (1958) *Wilson Bull.* **70**, 114-29. North, C A (1968) Diss. Oklahoma State Univ; (1973) In Hardy, J W and Morton, M L (eds) *Orn. Monogr. AOU* **14**, 79-91. Norton, W J E (1958) *Ibis* **100**, 179-89. Noskov, G A (1975) *Zool. Zh.* **54**, 413-24; (1977) *Zool. Zh.* **56**, 1676-86; (1978) *Ekologiya* **1**, 61-9; (ed.) (1981) *Polevoy vorobey.* Leningrad; (1982) *Orn. Stud. USSR* **2**, 348-63. Noskov, G A and Gaginskaya, A R (1977) *Ornitologiya* **13**, 190-1. Noskov, G A, Rymkevich, T A, Shibdov, A A, and Nankinov, D N (1975) *Vestnik Leningradskogo Univ.* **3** (1), 11-16. Noskov, G A and Shamov, S V (1983) *Soobshch. Pribalt. Kom. Izuch. Migr. Ptits* **14**, 125-9. Noskov, G A and Smirnov, E N (1979) *Nauch. Dokl. vyssh. Shk. Biol. Nauki* 1970 (3), 38-45. Noskov, G A, Zimin, V B, Rezvyi, S P, Rymkevich, T A, Lapshin, N V, and Golovan', V I (1981) In Noskov, G A (ed.) *Ekologiya ptits Priladozh'ya*, 3-86. Leningrad. Nothdurft, W (1972) *Anz. orn. Ges. Bayern* **11**, 185-9. Nothdurft, W, Knolle, F, and Zang, H (1988) *Vogelkdl. Ber. Niedersachs.* **20**, 33-85. Noval, A (1971) *Ardeola* spec. vol., 491-507; (1986) *Guía de las aves de Asturias.* Gijon. Novikov, B G (1938) *Trudy Nauch.-issled. Inst. eksper. Morfogen. Mosk. Univ.* **6**, 485-93. Novikov, G A (1949) *Zool. Zh.* **28**, 461-70; (1952) *Trudy Zool. Inst. Akad. Nauk SSSR* **9**, 1133-54; (1972) *Aquilo (Ser. Zool.)* **13**, 95-7. Novotný, I (1970) *Acta Sci. Nat. Brno* **4**, 1-57. Numerov, A D (1978) *Trudy Okskogo gos. Zapoved.* **14**, 356-7. Nürnberger, F, Siebold, D, and Bergmann, H-H (1989) *Bioacoustics* **1**, 273-86. W Soo06 TNuttall, J (1972) *Naturalist* **923**, 140-1. Nyberg, M (1932) *Ornis fenn.* **9**, 83-5. Nyhlén, G (1950) *Vår Fågelvärld* **9**, 49-63. Nyholm, N E I and Myhrberg, H E (1977) *Oikos* **29**, 336-41. Nyström, B and Nyström, H (1987) *Vår Fågelvärld* **46**, 119-28; (1991) *Ornis Svecica* **1**, 65-8. Nyström, E W (1925) *Ornis fenn.* **2**, 8-13.

Oakes, C (1953) *The birds of Lancashire.* Edinburgh. Oates, E W (1883) *A handbook to the birds of British Burmah* **1**; (ed.) (1890) *Hume's nests and eggs of Indian birds.* London. Oberholser, H C (1919) *Auk* **36**, 549-55; (1974) *The bird life of Texas* **2**. Austin. O'Connor, R J (1973) *Rep. Rye Meads Ringing Group* **6**, 16-27. O'Connor, R J and Shrubb, M (1986) *Farming and birds.* Cambridge. Oehler, J (1977) *Wissenschaftl. Z. Humboldt-Univ. Berlin Math.-Nat. R.* **26**, 425-9; (1978) *Biol. Zbl.* **97**, 279-87. Oelke, H (1992) *Beitr. Naturk. Niedersachs.* **45**, 1-17. Oeser, R (1975) *Beitr. Vogelkde.* **21**, 475-6. Oeser, R E (1984) *Beitr. Vogelkde.* **30**, 162-8. Oddie, W E (1981) *Br. Birds* **74**, 532-3. Odinzowa, N P (1967) *Falke* **14**, 414-5. Odum, E P (1931) *Wilson Bull.* **43**, 316-7; (1949) *Wilson Bull.* **61**, 3-14; (1958) *Bird-Banding* **29**, 105-8. Odum, E P, Connell, C E, and Stoddard, H L (1961) *Auk* **78**, 515-27. Odum, E P and Perkinson, J D (1951) *Physiol. Zool.* **24**, 216-30. Oggier, P A (1986) *Orn. Beob.* **83**, 295-9. Ogilvie, C M (1947) *Br. Birds* **40**, 135-9; (1949) *Br. Birds* **42**, 65-8; (1951) *Br. Birds* **44**, 1-5. Ogilvie, M A and Ogilvie, C C (1984) *Br. Birds* **77**, 368. Ogilvie-Grant, W R (1901) *Ibis* (8) **1**, 518-21. Ogorodnikova, L I and Mironova, V E (1989) In Konstantinov, V M and Klimov, S M (eds.) *Vranovye ptitsy v estestvennykh i antropogennykh landshaftakh* (Materialy II Vsesoyuznogo soveshchaniya) **4**, 12-13. Lipetsk. Ojanen, M and Kyl-

mänen, R (1984) *Ornis fenn.* **61**, 123. Ojanen, M, Orell, M, and Hirvelä, J (1979) *Holarctic Ecol.* **2**, 81-7. Ojanen, M and Lähdesmäki, P (1984) *Aureola* **9**, 14-21. Ólafsson, E (in press) *Náttúrufr.* Okulewicz, J (1989) *Ptaki Śl SKA* **7**, 1-39; (1991) *Ptaki Śląska* **8**, 1-17. Olier, A (1958) *Alauda* **26**, 65-66; (1959) *Alauda* **27**, 205-10. Olioso, G (1972) *Alauda* **40**, 171-4; (1973) *Alauda* **41**, 227-32; (1974) *Alauda* **42**, 502; (1987) *Bièvre* **9**, 1-8. Olivier, G (1949) *Oiseau* **19**, 102-4. Olsen, K (1984) *Vår Fuglefauna* **7**, 233-4. Olsen, K M (1987a) *Dansk orn. Foren. Tidsskr.* **81**, 109-20; (1987b) *Fugle* **7** (2), 20-1; (1989) *Dansk orn. Foren. Tidsskr.* **83**, 97-101; (1991a) *Anser* **30**, 29-40; (1991b) *Dansk orn. Foren. Tidsskr.* **85**, 20-34. Olson, S L (1985) *Bull. Br. Orn. Club* **105**, 29-30. Olsson, C (1988) *Vår Fågelvärld* **47**, 197-9. Olsson, M (1954) *Vår Fågelvärld* **13**, 113, 120. Olsson, V (1960) *Vår Fågelvärld* **19**, 1-19; (1964) *Br. Birds* **57**, 118-23; (1969) *Vogelwarte* **25**, 147-56; (1971) *Br. Birds* **64**, 213-23; (1985) *Vår Fågelvärld* **44**, 269-83. Olsthoorn, H (1987) *Br. Birds* **80**, 117-18. O'Mahony, J (1977) *Br. Birds* **70**, 340. O'Malley, S L C (1993) D Phil Thesis. Leicester Univ. Oordt, G J van, and Damsté, P H (1939) *Acta Brevia Neerlandica* **9**, 140-3. Oren, D C and Smith, N J H (1981) *Wilson Bull.* **93**, 281-282. Orford, N (1973) *Bird Study* **20**, 50-62, 121-6. Orford, N W (1959) *Nature* **184**, 650. Orlov, P P (1955) *Zool. Zh.* **34**, 950-2. Ornithological Society of Japan (1974) Check-list of Japanese birds. Tokyo. Ortlieb, R (1971) *Anz. Orn. Ges. Bayern* **10**, 186-7. Oscar, P (1986) *Larus* **36-7**, 155-65. Osieck, E R (1973) *Vogeljaar* **21**, 274-7. Osmaston, B B (1925) *Ibis* (12) **1**, 663-719; (1927) *J. Bombay Nat. Hist. Soc.* **31**, 975-99. Ostapenko, M M, Kashkarov, D Y, Shernazarov, E, Lanovenko, E N, and Filatov, A K (1980) In Abdusalyamov, I A (ed.) *Migratsii ptits v Azii*, 75-97. Dushanbe. Ostapenko, V A (1981) *Ornitologiya* **16**, 179-80. Ostashchenko, A N, Popov, E A, and Yablonkevich, M L (1985) *Trudy Zool. Inst. Akad. Nauk SSSR* **137**, 119-28. Osthaus, H and Schloss, W (1975) *Auspicium* **6**, 45-89. O'Sullivan, O and Smiddy, P (1988) *Irish Birds* **3**, 609-48; (1989) *Irish Birds* **4**, 79-114; (1991) *Irish Birds* **4**, 423-62. Ottosson, U and Haas, F (1991) *Ornis Svecica* **1**, 113-8. Ottow, B (1912) *Orn. Monatsber.* **20**, 11. Oubron, G (1967) *Organisation commune de lutte anti-acridienne et de lutte antiaviare* (Dakar), 707/403.33.21/LAV. Ouweneel, G L (1970) *Limosa* **43**, 156-8. Owen, D A L (1985) D Phil Thesis. Univ. Oxford; (1989) In Bignal and Curtis (eds), 57-62. Owen, D F (1953) *Br. Birds* **46**, 353-64; (1956) *Bird Study* **3**, 257-65; (1959) *Ibis* **101**, 235-9. Owen, J H (1950) *Br. Birds* **43**, 16.

Paatela, J E (1938) *Ornis fenn.* **15**, 65-9; (1948) *Ornis fenn.* **25**, 21-8. Paccaud, O (1954) *Nos Oiseaux* **22**, 214. Pacheco, F, Alba, F J, García Díaz, E, and Pérez Mellado, V (1977) *Ardeola* **22**, 55-73. Packard, G C (1967a) *Syst. Zool.* **16**, 73-89; (1967b) *Wilson Bull.* **79**, 345-6. Paevski, V A (1968) *Migr. Zhivot.* **5**, 153-60; (1970a) *Mat. 7. Pribalt. orn. Konf.* **1**, 69-75; (1970b) *Zool. Zh.* **49**, 798-9; (1971) *Trudy Zool. Inst. Akad. Nauk SSSR* **50**, 3-110; (1981) *Zool. Zh.* **60**, 109-14; (1985) *Demografiya ptits.* Leningrad. Paevski, V A, and Vinogradova, N V (1974) *Trudy Zool. Inst. Akad. Nauk SSSR* **55**, 186-206. Paevski, V A, Vinogradova, N V, Shapoval, A P, Shumakov, M E, and Yablonkevich, M L (1990) *Trudy Zool. Inst. Akad. Nauk SSSR* **210**, 63-72. Page, J (1988)

*Bat News* **15**, 3. PAIGE, J P (1960) *Ibis* **102**, 520–5. PAIL-LERETS, DE B DE (1934) *Alauda* **6**, 395–6. PAINTIN, A (1991) *Ann. Bull. Soc. Jersiaise* **25**, 422–4. PALMER, R S (1949) *Bull. Mus. comp. Zool.* **102**. PALMÉR, S and BOSWALL, J (1981) *A field guide to the bird songs of Britain and Europe.* Stockholm. PALMER, W (1894) *Auk* **11**, 282–91. PALMGREN, P (1930) *Acta zool. fenn.* **7**,; (1935) *Ornis fenn.* **12**, 107–21; (1936) *Ornis fenn.* **13**, 153–9. PALUDAN, K (1936) *Vidensk. Medd. dansk nat. Foren.* **100**, 247–346; (1938) *J. Orn.* **86**, 562–638; (1940) *Danish Sci. Invest. Iran* **2**, 11–54; (1959) *Vidensk. Medd. dansk nat. Foren.* **122**. PANDOLFI, M (1987) *Riv. ital. Orn.* **57**, 115–6. PANNACH, D (1983) *Beitr. Vogelkde.* **29**, 317–18; (1984) *Beitr. Vogelkde.* **30**, 211–12; (1990) *Abh. Ber. Naturkundemus. Görlitz* **63** (3). PANOV, E N (1973*a*) *Ptitsy yuzhnogo Primor'ya.* Novosibirsk; (1973*b*) In Vorontsov, N N (ed.) *Problemy Evolyutsii* **3**, 261–94; (1989) *Gibridizatsiya i etologicheskaya izolyatsiya u ptits.* Moscow. PANOV, E N and BULATOVA, N S (1972) *Byull. Mosk. Obshch. Ispyt. Prior. Otd. Biol.* **77** (4), 86–94. PANOV, E N and RADZHABLI, S I (1972) In Vorontsov, N N (ed.) *Problemi Evolyutsii* **2**, 263–75. Novosibirsk. PAQUET, A (1979) *Aves* **16**, 162–3. PARKES, K C (1951) *Wilson Bull.* **63**, 5–15; (1952) *Wilson Bull.* **64**, 161–2; (1961) *Wilson Bull.* **73**, 374–9; (1967) *Wilson Bull.* **79**, 456–8; (1978) *Auk* **95**, 682–90; (1988) *J. Fld. Orn.* **59**, 60–2; (1988*b*) *Bird Obs. (Mass.)* **16**, 324–5; (1990) *Wilson Bull.* **102**, 733–4. PARKHURST, R and LACK, D (1946) *Br. Birds* **39**, 358–64. PARKIN, D T (1988) *Proc. Int. Orn. Congr.* **19**, 1652–7, 1669–73. PARKS, G H (1962) *Bird-Banding* **33**, 148–51. PARMELEE, D F and MACDONALD, S D (1960) *Bull. nat. Mus. Canada* **169**. PARMELEE, D F and PARMELEE, J M (1988) *Condor* **90**, 952. PAROVSHCHIKOV, V Y (1962) *Ornitologiya* **4**, 7–10. PARRACK, J D (1973) *The naturalist in Majorca.* Newton Abbot. PARROTT, J, PHILLIPS, J, AND WOOD, V (1987) *Dutch Birding* **9**, 17–19. PARROT, C (1907) *Zool. Jahrb.* **25**, 1–78. PARRY, P E (1948) *Br. Birds* **41**, 344–5. PARSLOW, J L F (1973) *Breeding birds of Britain and Ireland.* Berkhamsted. PARSLOW, J L F and CARTER, M J (1965) *Br. Birds* **58**, 208–14. PARSONS, J and BAPTISTA, L F (1980) *Auk* **97**, 807–15. PARTRIDGE, L and GREEN, P (1987) *Anim. Behav.* **35**, 982–90. PASCUAL, J A (1992) Ph D Thesis, Salamanca Univ. PASCUAL, J A, CALVO, J M, LEHMANN, S, and PERIS, S J (1992) *Congr. nac. Iberoamer. Etología* **4**, 381–7. Cáceres. PASHLEY, D N and MARTIN, R P (1988) *Amer. Birds* **42**, 1164–76. PASPALEVA, M (1965) *Bull. Inst. Zool. Acad. Sci. Bulg.* **19**, 33–7. PASSBURG, R E (1959) *Ibis* **101**, 153–69. PASTEUR, G (1956) *Bull. Soc. Sci. nat. Phys. Maroc* **36**, 165–84. PATEFF, P (1947) *Ibis* **89**, 494–507. PATTERSON, I J (1970) *Br. ecol. Soc. Symp.* **10**, 249–52; (1975) In Baerends, G, Beer, C, and Manning, A (ed.) *Function and evolution in behaviour*, 169–83. Oxford; (1980*a*) *Ardea* **68**, 53–62; (1980*b*) *Br. Birds* **73**, 359–60. PATTERSON, I J, CAVALLINI, P, and ROLANDO, A (1991) *Ornis scand.* **22**, 79–87. PATTERSON, I J, DUNNET, G M, and FORDHAM, R A (1971) *J. appl. Ecol.* **8**, 815–33. PATTERSON, I J and GRACE, E S (1984) *J. Anim. Ecol.* **53**, 559–72. PÄTZOLD, R (1975) *Falke* **22**, 300–4. PAUDTKE, B (1988) *Limicola* **2**, 74. PAYN, W A (1926) *Ool. Rec.* **6**, 70; (1947) *Bull. Br. Orn. Club* **67**, 41–2. PAYN, W H (1982) *Br. Birds* **75**, 134. PAYNE, R B (1972) *Condor* **74**, 485–6; (1973) *Orn. Monogr. AOU* **11**; (1976) *Auk* **93**, 25–38; (1980) *Ibis* **122**, 43–56; (1983) *Anim. Behav.* **31**, 788–805; (1987) *Occas. Papers Mus. Zool. Univ. Michigan* **714**; (1991) In Poole, A, Stettenheim, P and Gill, F B (eds) *Birds of North America* **4**. Philadelphia. PAYNE, R B, PAYNE, L L, and

DOEHLERT, S M (1988) *Ecology* **69**, 104–17. PAYNE, R B, THOMPSON, W L, FIALA, K L, and SWEANY, L L (1981) *Behaviour* **77**, 199–221. PAYNTER, R A (1964) *Condor* **66**, 277–81; (1968) In Peters, J L *Check-list of birds of the world* **14**; (1970) **13**. Cambridge, Mass. PAZ, U (1987) *The birds of Israel.* Tel-Aviv. PAZZUCONI, A (1970) *Riv. ital. Orn.* **40**, 458–9. PEABODY, P B (1907) *Auk* **24**, 271–8. PEACH, W J, GIBSON, T S H, and FOWLER, J A (1987) *Bird Study* **34**, 37–8. PEAKALL, D B (1960) *Bird Study* **7**, 94–102. PEAKE, E (1929) *Br. Birds* **22**, 174–5. PEARCE, A (1978) *Nat. Wales* **16**, 142. PEARSE, T (1940) *Condor* **42**, 124–5. PEARSON, D L (1980) In Keast, A and Morton, E S (eds) *Migrant birds in the Neotropics: ecology, behavior, distribution and conservation*, 273–83. Washington DC. PECK, K M (1989) *Biol. Cons.* **48**, 41–57. PEDERSEN, B B (1980) *Dansk orn. Foren. Tidsskr.* **74**, 127–40. PEDROCCHI, C (1979) *Estudio integrado y multidisciplinario de la dehesa salmantina. 1. Estudio fisiográfico descriptivo.* **3**, 221–49. PEDROLI, J-C (1967) *Nos Oiseaux* **29**, 164. PEIPONEN, V (1957) *Ornis fenn.* **34**, 41–64. PEIPONEN, V A (1962) *Ornis fenn.* **39**, 37–60; (1967) *Ann. zool. fenn.* **4**, 547–59; (1970) *Proc. Helsinki Sympl. UNESCO*, 281–7; (1974) *Ann. zool. fenn.* **11**, 155–65. PEITZMEIER, J (1955) *J. Orn.* **96**, 347–8. PEK, L V and FEDYANINA, T F (1961) In Yanushevich, A I (ed.) *Ptitsy Kirgizii* **3**, 59–118. Frunze. PELLERIN, M (1981) *Compt. Rend. Acad. Sci. Paris* **293**, 713–15; (1982) *Behaviour* **81**, 287–95; (1983) *Behav. Proc.* **8**, 157–63. PEPPER, S R and KENNEDY, H J (1970) *Ornis fenn.* **47**, 35–6. PEPPERBERG, I M (1985) *Auk* **102**, 854–64. PERDECK, A C (1967) *Bird Study* **14**, 129–52; (1970) *Ardea* **58**, 142–70. PÉREZ PADRÓN, F (1981) *Vida Silvestre* **40**, 258–63. PERIS, S J (1979) Doct. Thesis. Univ. Complutense de Madrid; (1980*a*) *Ardeola* **25**, 207–40; (1980*b*) *Doñana Acta Vert.* **7**, 249–60; (1981) *Stud. Oec.* **2**, 155–69; (1983) *J. Orn.* **124**, 78–81; (1984*a*) *Bol. Real Soc. Española Hist. Nat. (Biol.)* **80**, 37–46; (1984*b*) *Ardeola* **31**, 3–16; (1984*c*) *Reunión Iberoamer. Cons. Zool. Vert. Cáceres* **2**, 140–52; (1984*d*) *Salamanca, Revista de Estudios* **11–12**, 175–234; (1988) *Gerfaut* **78**, 101–112; (1989) *Misc. Zool.* **13**, 217–20; (1991) *Ring. Migr.* **12**, 124–5. PERIS, S J, MOTIS, A, and MARTINEZ-VILALTA, A (1987) *Acta VIII BRSE Hist. Nat., Pamplona*, 151–6. PERIS, S, MOTIS, A, MARTINEZ-VILALTA, A, and FERRER, X (1991) *J. Orn.* **132**, 445–9. PERSSON, C (1942) *Vår Fågelvärld* **1**, 27–8. PESCH, A and GÜTTINGER, H-R (1985) *J. Orn.* **126**, 108–10. PESENTI, P G (1945) *Riv. ital. Orn.* **15**, 89–97. PESKOV, V N (1990) *Vestnik Zool.* (6), 62–7. PETER, H (1968) *Egretta* **11**, 58. PETER, H-U and STEIDEL, G (1990) *Acta XX Congr. Int. Orn. Supp. Abstr.* **703**, 384. PETERS, H S and BURLEIGH, T D (1951) *The birds of Newfoundland.* St John's. PETERS, J L (1931) *Auk* **48**, 575–87. PETERS, J L and GRISCOM, L (1938) *Bull. Mus. Comp. Zool.* **80**, 445–78. PETERS, R A (ed.) (1968) *Check-list of birds of the world* **14**. Cambridge. PETERSEN, Æ (1985) *Bliki* **4**, 57–67; (1989) *Bliki* **8**, 56–61. PETERSON, R, MOUNTFORT, G, and HOLLOM, P A D (1983) *A field guide to the birds of Britain and Europe.* London. PETERSON, R T and CHALIF, E L (1973) *A field guide to Mexican birds.* Boston. PETRETTI, F (1979) *Br. Birds* **72**, 290. PETRIDES, G A (1943) *Wilson Bull.* **55**, 193–4. PETROV, T (1982) *Isv. Mus. Yuzh. Bŭlg.* **8**, 21–41; (1988) *Isv. Mus. Yuzh. Bŭlg.* **14**, 25–45. PETTERSSON, Å (1977) *Vår Fågelvärld* **36**, 161–73. PETTERSSON, M (1959) *Nature* **184**, 649–50. PETTITT, R G and BUTT, D V (1949) *Br. Birds* **42**, 327. PÉTURSSON, G and ÓLAFSSON, E (1985) *Bliki* **4**, 13–39; (1986) *Bliki* **5**, 19–46; (1988) *Bliki* **6**, 33–68. PÉT-

URSSON, G, THRÁINSSON, G, and ÓLAFSSON, E (1991) *Bliki* **10**, 15–54; (1992) *Bliki* **11**, 31–63. PEUS, F (1954) *Bonn. zool. Beitr.* **5** suppl. **1**, 1–50. PFEIFFER, S (1928) *Anz. orn. Ges. Bayern* (**12**), 142–3. PFEIFER, S (1953) *Vogelwelt* **74**, 216–17; (1954) *J. Orn.* **95**, 185–6; (1956) *Orn. Mitt.* **8**, 148–9; (1966) *Orn. Mitt.* **18**, 28–30; (1974) *Orn. Mitt.* **26**, 159–60. PFEIFER, S and KEIL, W (1956) *Nachrichtenblatt des Deutschen Pflanzenschutzdienstes* **8** (9), 129–31. PFENNIG, H G (1986) *Charadrius* **22**, 221–6; (1988) *Charadrius* **24**, 88–91. PFIRTER, A (1975) *Vögel Heimat* **45**, 128. PFLUMM, W (1978) *J. Orn.* **119**, 308–24; (1984) *J. Orn.* **125**, 481–2. PFORR, M and LIMBRUNNER, A (1982) *The breeding birds of Europe* **2**. London. PFÜTZNER, W (1988) *Abh. Ber. Naturkundemus: Görlitz* **62** (2). PHILIPSON, W R (1933) *Br. Birds* **27**, 66–71; (1939) *Br. Birds* **32**, 272. PHILLIPS, A R (1966) *Bull. Br. Orn. Club* **86**, 148–59; (1974) *Abstr. Int. Orn. Congr.* **16**, 68–9; (1977) *Bird-Banding* **48**, 110–7; (1991) *The known birds of North and Middle America*. Part II. Denver. PHILLIPS, A R, MARSHALL, J, and MONSON, G (1964) *The birds of Arizona*. Tucson. PHILLIPS, J C (1915) *Auk* **32**, 273–89. PHILLIPS, N J and ROUND, P D (eds) (1975) *Crete Ringing Group Rep. 1973–5*. Aberdeen Univ. PHILLIPS, N R (1982) *Sandgrouse* **4**, 37–59. PHILLIPS, W W A (1953) *Ibis* **95**, 548–9; (1978) *Annotated checklist of the birds of Ceylon (Sri Lanka)*. Colombo. PICOZZI, N (1975a) *Br. Birds* **68**, 409–19; (1975b) *J. Wildl. Mgmt.* **39**, 151–5; (1976) *Ibis* **118**, 254–7; (1982) *Scott. Birds* **12**, 23–4. PIECHOCKI, R (1954) *J. Orn.* **95**, 297–305; (1956) *Falke* **3**, 10–17; (1959) *Zool. Abh. Staatl. Mus. Tierkde. Dresden* **24**, 105–203; (1971a) *Falke* **18**, 4–26, 40–57; (1971b) *Falke* **18**, 94–100. PIECHOCKI, R and BOLOD, A (1972) *Mitt. zool. Mus. Berlin* **48**, 41–175. PIECHOCKI, R, STUBBE, M, UHLENHAUT, K, and SUMJAA, D (1982) *Mitt. zool. Mus. Berlin* **58**, Suppl. Ann. Orn. **6**, 3–53. PIELOWSKI, Z (1963) *Przegląd Zool.* **7**, 71–8. PIELOWSKI, Z and PINOWSKI, J (1962) *Bird Study* **9**, 116–22. PIERSMA, T and BLOKSMA, N (1987) *Bird Study* **34**, 127–8. PIESKER, O (1972) *Beitr. Vogelkde.* **18**, 452. PIKULA, J (1973) *Zool. Listy* **22**, 155–64; (1989) *Folia Zool.* **38**, 167–82. PIKULA, J and FOLK, C (1970) *Zool. Listy* **19**, 261–73. PILCHER, C W T (1986) *Sandgrouse* **8**, 102–6. PINDER, J M (1991) *Br. Birds* **84**, 198. PINEAU, J and GIRAUD-AUDINE, M (1975) *Alauda* **43**, 135–41; (1979) *Trav. Inst. Sci. Rabat Sér. Zool.* **38**. PINOWSKA, B and PINOWSKI, J (1977) *Int. Stud. Sparrows* **10**, 26–41. PINOWSKI, J (1959) *Ekol. Polska A* **7**, 435–82; (1965a) *Bird Study* **12**, 27–33; (1965b) *Bull. Acad. pol. Sci.* **13**, 509–14; (1966) *Ekol. Polska A* **14**, 145–72; (1967a) *Ardea* **55**, 241–8; (1967b) *Ekol. Polska A* **15**, 1–30; (1968) *Ekol. Polska A* **16**, 1–58. PINOWSKI, J and KENDEIGH, S C (eds) (1977) *Granivorous birds in ecosystems*. Cambridge. PINOWSKI, J and PINOWSKA, B (1985) *Ring* **124–5**, 51–6. PINOWSKI, J, TOMEK, T, and TOMEK, W (1973) In Kendeigh, S C and Pinowski, J (eds) *Productivity, population dynamics and systematics of granivorous birds*, 263–73, Warsaw. PINOWSKI, J and WOJCIK, Z (1969) *Falke* **16**, 256–61. PINXTEN, R, EENS, M, VAN ELSACKER, L, and VERHEYEN, R F (1989b) *Bird Study* **36**, 45–8. PINXTEN, R, EENS, M, and VERHEYEN, R F (1989a) *Behaviour* **111**, 234–56; (1990) *J. Orn.* **131**, 141–50; (1991) *Ardea* **79**, 15–30. PINXTEN, R, VAN ELSACKER, L, and VERHEYEN, R F (1987) *Ardea* **75**, 263–9. PIOTROWSKI, S H (ed.) (1991) *Suffolk Birds* **40**, 45–136. PIPER, W H and WILEY, R H (1989) *J. Fld. Orn.* **60**, 73–83; (1991) *J. Fld. Orn.* **62**, 40–5. PITMAN, C R S (1921) *Ool. Rec.* **1**, 15–38, 73–91; (1922) *Ool. Rec.* **2**, 49–57; (1961) *Bull. Br. Orn. Club* **81**, 148–9. PITT, F (1918) *Br. Birds* **12**, 122–31. PITTAWAY, R (1989) *Ontario Birds* **7**, 65–7. PIVAR, G (1965) *Larus* **16–18** (1962–64), 159–280. PLATH, L (1983) *Beitr. Vogelkde.* **29**, 53–4; (1984) *Beitr. Vogelkde.* **30**, 215; (1987) *Orn. Mitt.* **39**, 16; (1988a) *Falke* **35**, 27–8; (1988b) *Falke* **35**, 358–61; (1989) *Falke* **36**, 143–7. PLAYFORD, P F J (1985) *Gwent Bird Rep.* **20**, 26. PLESKE, T (1928) *Birds of the Eurasian trundra*. Boston. PLUCINSKI, A (1970) *Orn. Mitt.* **22**, 3–4. POCHELON, G (1992) *Alauda* **60**, 148. PODARUEVA, V I (1979) In Labutin, Y V (ed.) *Migratsii i ekologiya ptits Sibiri*, 170–2. Yakutsk. PODMORE, R E (1948) *Br. Birds* **41**, 272. PODOL'SKI, A L and SADYKOV, O F (1983) *Ornitologiya* **18**, 176–8. POHL, H (1971a) *Ibis* **113**, 185–93; (1971b) *J. Orn.* **112**, 266–78; (1972) *Naturwissenschaften* **59**, 518; (1974) *Naturwissenschaften* **61**, 406; (1980) *Physiol. Zool.* **53**, 186–209; (1989) In Mercer, J B (ed.) *Thermal Physiology*, 713–18. New York. POKORNY, F and PIKULA, J (1986–7) *J. World Pheasant Assoc.* **12**, 75–80. POKROVSKAYA, I V (1956) *Zool. Zh.* **35**, 96–110; (1963) *Uch. Zap. Leningr. gos. ped. Inst.* **230**, 93–102. POLATZEK, J (1909) *Orn. Jahrb.* **20**, 1–24. POLIVANOV, V M and POLIVANOVA, N N (1986) *Trudy Teberdinsk. gos. zapoved.* **10**, 11–164. POLIVANOVA, N N and POLIVANOV, V M (1977) *Ornitologiya* **13**, 82–90. POLOZOV, S A (1990) *Ornitologiya* **24**, 132–3. PONOMAREVA, T S (1974) *Ornitologiya* **11**, 404–7; (1981) *Ornitologiya* **16**, 16–21; (1983) *Ornitologiya* **18**, 57–63. POPOV, A V (1959) *Ptitsy Gissaro-Karategina*. Dushanbe. POPOV, V A (1978) *Ptitsy Volzhsko-Kamskogo kraya*. Moscow. POPPE, D (1976) *Wielewaal* **42**, 345. PORTENKO, L A (1929) *Ezhegodnik Zool. Muz. Akad. Nauk SSSR* **29**, 37–81; (1930) *Auk* **47**, 205–7; (1937) *Fauna ptits vnepolyarnoy chasti severnogo Urala*. Moscow; (1939) *Trudy Nauch.-issled. inst. polyarn. zemledeliya zhivotnovod. i promysl. khoz.* **5**, 5–211; (1954) *Ptitsy SSSR* **3**; Moscow; (1973) *Ptitsy Chukotskogo poluostrova i ostrova Vrangelya* **2**. Leningrad. PORTENKO, L A and STÜBS, J (1971) In Stresemann, E, Portenko, L A, and Mauersberger, G (eds) *Atlas der Verbreitung palaearktischer Vögel* **3**. Berlin; (1976) In Dathe, H (ed.) *Atlas der Verbreitung palaearktischer Vögel* **5**. Berlin. PORTENKO, L A and VIETINGHOFF-SCHEEL, E VON (1974) In Stresemann, E, Portenko, L A, and Mauersberger, G (eds) *Atlas der Verbreitung palaearktischer Vögel* **4**. Berlin. PORTER, R F (1983) *Sandgrouse* **5**, 45–74. PORTER, S (1941) *Avic. Mag.* (5) **6**, 3–8. POSLAVSKI, A N (1974) *Ornitologiya* **11**, 238–52. POTAPOV, R L (1966) *Trudy Zool. Inst. Akad. Nauk SSSR* **39**, 3–119; (1986) *Trudy Zool. Inst. Akad. Nauk SSSR* **150**, 57–73. POTTER, E F (1970) *Auk* **87**, 692–713. PARNELL, J F, and TEULINGS, R P (1980) *Birds of the Carolinas*. Chapel Hill. POTTI, J and TELLERÍA, J L (1984) *Stud. Oecol.* **5**, 247–58. POUGH, R H (1949) *Audubon bird guide: small land birds*. New York; (1957) *Audubon western bird guide*. New York. POULSEN, H (1949) *Dansk orn. Foren. Tidsskr.* **43**, 256–7; (1950) *Dansk orn. Foren. Tidsskr.* **44**, 96–9; (1954) *Dansk orn. Foren. Tidsskr.* **48**, 32–7; (1956) *Dansk orn. Foren. Tidsskr.* **50**, 267–98; (1958) *Dansk orn. Foren. Tidsskr.* **52**, 89–105; W Soooo6 T(1959) *Z. Tierpsychol.* **16**, 173–8. POUNDS, H E (1950) *Br. Birds* **43**, 333; (1972) *Br. Birds* **65**, 34. POWER, H W, LITOVICH, E, and LOMBARDO, M P (1981) *Auk* **98**, 386–9. POWER, H W, KENNEDY, E D, ROMAGNANO, L C, LOMBARDO, M P, HOFFENBERG, A S, STOUFFER, P C, and McGUIRE, T C (1989) *Condor* **91**, 753–65. PRATO, E S DA, and PRATO, S R D DA (1978) *Ring. Migr.* **2**, 48–9. PRATT, H D, BRUNER, P L, and BERRETT, D G (1987) *The birds of Hawaii and the* *tropical Pacific*. Princeton. PRAVOSUDOV, V V (1984) *Zool. Zh.* 63, 950-3. PRAZ, J-C (1971) *Nos Oiseaux* 31, 11-13. PRAZ, J-C and OGGIER, P-A (1973) *Nos Oiseaux* 32, 109-12; (1976) *Bull. Murithienne* 93, 29-40. PREBLE, E A (1908) *North American Fauna* 27. PREISER, F (1957) *Diss. Landwirtschaftliche Hochschule Hohenheim*. PREISER, E and MASSOTH, K H (1990) *Voliere* 13, 68-71. PREISS, F (1974) *Gef. Welt* 98, 121-2. PRENDERGAST, E D V (1985) *Adjutant* 15, 17-18. PRENDERGAST, E D V and BOYS, J V (1983) *The birds of Dorset*. Newton Abbot. PRESCOTT, D R C (1991) *Condor* 93, 694-700. PRESCOTT, K W (1978) *Inland Bird-Banding News* 50, 163-83; (1981) *Inland Bird-Banding* 53, 39-48; (1986) *N. American Bird Bander* 11, 46-51. PREUSS, D (1991a) *Gef. Welt* 115, 192-6; (1991b) *Gef. Welt* 115, 340-41. PRICAM, R (1957) *Nos Oiseaux* 24, 160-3. PRICE, T D (1979) *J. Bombay Nat. Hist. Soc.* 76, 379-422. PRIEST, C D (1936) *The birds of southern Rhodesia* 4. London. PRIESTLEY, C F (1947) *Br. Birds* 40, 176. PRIGANN, I (1992) *Kart. Med. Brutvögel* 7, 11-12. PRILL, H (1974) *Orn. Rundbrief Mecklenburgs N.F.* 15, 56-9; (1975) *Falke* 22, 92-4. PRIOLO, A (1988) *Riv. ital. Orn.* 58, 105-24. PRINZINGER, R, and HUND, K (1975) *Anz. Orn. Ges. Bayern* 14, 70-8; (1981) *Ökol. Vögel* 3, 249-59. PROKOFIEV, M A (1962) *Ornitologiya* 4, 333-5. PROKOFIEVA, I V (1963a) *Uch. Zap. Leningr. gos. ped. Inst.* 230, 57-69; (1963b) *Uch. Zap. Leningr. gos. ped. Inst.* 230, 71-86. PROMPTOV, A N (1934) *Zool. Zh.* 13 (3), 523-39. PROZESKY, O P M (1974) *A field guide to the birds of southern Africa*. London. PRŶS-JONES, R P (1976) *Bird Study* 23, 294; (1977) D Phil Thesis. Oxford Univ; (1984) *Gerfaut* 74, 15-37. PRZYGODDA, W (1960) *Orn. Mitt.* 12, 21-5; (1969) *Bonn. zool. Beitr.* 20, 69-74. PTUSHENKO, E S and INOZEMTSEV, A A (1968) *Biologiya i khozyaystvennoe znachenie ptits Moskovskoy oblasti i sopredel'nykh territoriy*. Moscow. PUKINSKI, Y B (1969) *Vopr. ekol. biotsenol.* 9, 62-78. PULLIAINEN, E (1971) *Ann. zool. Fenn.* 8, 326-9; (1972) *Ann. zool. Fenn.* 9, 28-31; (1974) *Ann. zool. Fenn.* 11, 204-6; (1979a) *Aquilo Ser. Zool.* 19, 87-96; (1979b) *Ornis fenn.* 56, 156-62. PULLIAINEN, E and HAKANEN, R (1972) *Ornis fenn.* 49, 86-90. PULLAINEN, E, KALLIO, T, and HALLAKSELA, A-M (1978) *Aquilo Ser. Zool.* 18, 23-7. PULLIAINEN, E and PEIPONEN, V (1981) *Ornis fenn.* 58, 109-16. PULLIAINEN, E and SAARI, L (1976) *Ornis fenn.* 53, 46; (1989) *Ornis fenn.* 66, 161-5. PULLIAINEN, E and TUOMAINEN, J (1978) *Ornis fenn.* 55, 180-2. PULMAN, C B (1978) *Br. Birds* 71, 363. PURCHAS, T P G (1975) *Proc. New Zealand ecol. Soc.* 22, 111-12; (1979) *New Zealand J. Zool.* 6, 321-7; (1980) *New Zealand J. Zool.* 7, 557-78. PURRMANN, P (1973) *Falke* 20, 102. PYLE, P, HOWELL, S N G, YUNICK, R P, and DeSANTE, D F (1987) *Identification guide to North American passerines*. Bolinas.

QUAY, W B (1989) *Condor* 91, 660-70. QUÉPAT, N (1875) *Monographie du Cini*. Paris. QUILLIAM, H R (1973) *History of the birds of Kingston, Ontario*. Kingston.

RABØL, J (1969) *Feltornithologen* 11, 123-31. RABOUD, C (1988) *Orn. Beob.* 85, 385-92. RADEMACHER, B (1951) *Z. Pflanzenkrankheiten u. Pflanzenschutz*. 58, 416-26. RADERMACHER, W (1974) *Orn. Mitt.* 26, 182; (1977) *Charadrius* 13, 75-8; (1983) *Orn. Mitt.* 35, 107; (1984) *Charadrius* 20, 183-5. RADFORD, A P (1966) *Br. Birds* 59, 201; (1970a) *Br. Birds* 63, 138; (1970b) *Br. Birds* 63, 428-9; (1971) *Br. Birds* 64, 233-5; (1974) *Br. Birds* 67, 440; (1983) *Br. Birds* 76, 580; (1985) *Br. Birds* 78, 513-14; (1991) *Br. Birds* 84,

153. RADTKE, G A (1986) *Voliere* 9, 152-4. RADZHABLI, S I and PANOV, E N (1972) In Vorontsov, N N (ed.) *Problemy Evolyutsii* 2, 255-62. Novosibirsk. RADZHABLI, S I, PANOV, E N, and BULATOVA, N S (1970) *Zool. Zh.* 49, 1857-63. RAE, R and MARQUISS, M (1989) *Ring. Migr.* 10, 133-40. RAEVEL, P (1981a) *Héron* (3), 63; (1981b) *Héron* (4), 1-10; (1983) *Héron* (4), 105-7. RAEVEL, P and DEROO, S (1980) *Héron* (4), 118. RAEVEL, P and ROUSSEL, F (1983) *Héron* (4), 104-5. RAFFAELE, H A (1989) *A guide to the birds of Puerto Rico and the Virgin Islands*. Princeton. RAHMAN, M K and HUSAIN, K Z (1988) *Bangladesh J. Zool.* 16, 155-7. RAHMANI, A R and D'SILVA, C (1986) *J. Bombay nat. Hist. Soc.* 82, 657. RAINES, R J (1962) *Ibis* 104, 490-502. RAIT KERR, H (1950) *Field* 196, 302. RALPH, C J and PEARSON, C A (1971) *Condor* 73, 77-80. RAMMNER, C (1977) *Falke* 24, 62-7. RAMOS, M A (1988) *Proc. int. orn. Congr.* 19, 251-93. RAND, A L (1944) *Can. Fld.-Nat.* 58, 111-25; (1946) *Nat. Mus. Canada Bull.* 105; (1961a) *Wilson Bull.* 73, 46-56; (1961b) *Bird-Banding* 32, 71-9. RAND, A L and VAURIE, C (1955) *Bull. Br. Orn. Club* 75, 28. RAPPE, A (1965) *Gerfaut* 55, 4-15. RAPPE, A and HERROELEN, P (1964) *Gerfaut* 54, 3-11. RAPPOLE, J H (1983) *J. Fld. Orn.* 54, 152-9. RAPPOLE, J H, MORTON, E S, LOVEJOY, T E, and RUOS, J L (1983) *Nearctic avian migrants in the Neotropics*. US Dept. Interior, Washington DC. RAPPOLE, J H, RAMOS, M A, OEHLENSCHLAGER, R J, WARNER, D W, and BARKAN, C P (1979) In Drawe, D L (ed.) *Proc. First Welder Wildl. Found. Symposium*, 199-214. Sinton. RAPPOLE, J H and WARNER, D W (1980) In Keast, A and Morton, E S (eds) *Migrant birds in the Neotropics: ecology, behavior, distribution, and conservation*, 353-93. Washington DC. RASHKEVICH, N A (1965) *Zool. Zh.* 44, 1532-7. RASPAIL, X (1901) *Bull. Soc. zool. France* 26, 104-9. RATCLIFFE, D A (1962) *Ibis* 104, 13-39. RAVELING, D G (1965) *Bird-Banding* 36, 89-101. RAVELING, D G and WARNER, D W (1965) *Bird-Banding* 36, 169-79. RAVKIN, Y S (1973) *Ptitsy severo-vostochnogo Altaya*. Novosibirsk; (1984) *Prostranstvennaya organizatsiya naseleniya ptits lesnoy zony*. Novosibirsk. RAW, W (1921) *Ibis* (11) 4, 238-64. RAY, K A (1965) *Br. Birds* 58, 439. RAYMOND, T (1988) *Bird Obs. Mass.* 16, 270. RAYNOR, J H (1948) *Br. Birds* 41, 342. REA, A M (1970) *Condor* 72, 230-3. REBECCA, G W (1985) *Scott. Birds* 13, 188; (1986) *Scott. Bird News* 4, 9. REBER, U (1986) *Gef. Welt* 110, 153; (1988) *Gef. Welt* 112, 324. REBOUSSIN, R (1931) *Oiseau* 1, 283-8. REDMAN, N (1993) *Br. Birds* 86, 131-3. REDONDO, T (1991) *J. Orn.* 132, 145-63. REDONDO, T and ARIAS DE REYNA, L (1988) *Anim. Behav.* 36, 653-61. REDONDO, T, ARIAS DE REYNA, L, GONZALEZ-ARENAS, J, RECUERDA, P, and ZUNIGA, J M (1988) *Misc. Zool.* 10, 287-97. REDONDO, T and CARRANZA, J (1989) *Behav. Ecol. Sociobiol.* 25, 369-78. REDONDO, T and EXPOSITO, F (1990) *Ethology* 84, 307-18. REDONDO, T, HIDALGO DE TRUCIOS, S J, and MEDINA, R (1989) *Etología* 1, 19-31. REE, L van and BERG, A B van den (1987) *Dutch Birding* 9, 108-13. REE, V (1976) *Sterna* 15, 179-197; (1977) *Sterna* 16, 113-202. REEB, F (1977) *Alauda* 45, 293-33. REED, T M (1982) *Anim. Behav.* 30, 171-81. REICHENOW, A (1916) *Orn. Monatsber.* 24, 154-5. REID, J C (1979) *Bonn. zool. Beitr.* 30, 357-66. REINIKAINEN, A (1937) *Ornis fenn.* 14, 55-64; (1939) *Ornis fenn.* 16, 73-95. REINSCH, A (1977) *Orn. Mitt.* 29, 190. REINSCH, H H (1960) *Beitr. Vogelkde.* 7, 153-4; (1965) *Beitr. Vogelkde.* 10, 323-5; (1967) *Beitr. Vogelkde.* 12, 293-4. REISER, O (1905) *Materialien zu einer Ornis Balcanica* 3; (1939) 1. Vienna. REISER, O and FÜHRER, L von (1896) *Materialien*

*zu einer Ornis Balcanica* 4. Vienna. RÉKÁSI, J (1976) *Int. Stud. Sparrows* 9, 72-82. RÉKÁSI, J and STERBETZ, I (1975) *Aquila* 80-1, 215-20. RENDAHL, H (1958) *Vogelwarte* 19, 199-203; (1959) *Ark. Zool.* 12, 303-12; (1964) *Vogelwarte* 22, 229-35; (1968) *Ark. Zool.* 22, 225-78. RENDELL, L (1947) *Br. Birds* 40, 49. RENGLIN, P (1975) *Vår Fågelvärld* 34, 59. RENSSEN, T A (1988) *Limosa* 61, 137-44; (1991a) *Limosa* 64, 30; (1991b) In Glandt (1991), 61-6. RESTALL, R L (1975a) *Avic. Mag.* 81, 107-13; (1975b) *Finches and other seed-eating birds*. London. RETTIG, K (1985) *Beitr. Naturk. Niedersachs.* 38, 222-3. RETZ, M (1966a) Diss. Zool. Forschungsinst., Mus. Alexander Koenig (*not a Ph. D.*); (1966b) *Auspicium* 2, 231-47; (1968) *Auspicium* 2, 412-46. REY, E (1907a) *Orn. Monatsschr.* 32, 205-18; (1907b) *Orn. Monatsschr.* 32, 235-46; (1908) *Orn. Monatsschr.* 33, 221-31; (1910) *Orn. Monatsschr.* 35, 305-13. REYMERS, N F (1954) *Zool. Zh.* 33, 1358-62; (1959) *Zool. Zh.* 38, 907-15; (1966) *Ptitsy i mlekopitayushchie yuzhnoy taygi sredney Sibiri*. Moscow. REYNOLDS, A (1974) *Rye Meads Ring. Gr. Rep.* 7, 44-52. RHEINWALD, G (1973) *Charadrius* 9, 58-64; (1982) *Brutvogelatlas der Bundesrepublik Deutschland*. Bonn; (1992) *Die Vögel von Deutschland: Artenliste*. Dachverband Deutscher Avifaunisten. RIBAUT, J-P (1954) *Nos Oiseaux* 22, 225-8. RICHARD, A (1928) *Nos Oiseaux* 9, 97-104. RICHARDS, C E (1981) *Br. Birds* 74, 187-8. RICHARDS, D B and THOMPSON, N S (1978) *Behaviour* 64, 184-203. RICHARDS, M L (1973) *Br. Birds* 66, 365-6. RICHARDS, P R (1974) *Br. Birds* 67, 215; (1976a) *Bird Study* 23, 207-11; (1976b) *Bird Study* 23, 212. RICHARDSON, C (1989) *Oman Bird News* 5, 6-7; (1990) *The birds of the United Arab Emirates*. Warrington. RICHARDSON, S C, PATTERSON, I J, and DUNNET, G M (1979) *J. Anim. Ecol.* 48, 103-10. RICHFORD, A S (1978) D. Phil. Thesis. Oxford Univ. RICHMOND, C W (1895a) *Proc. US. natn. Mus.* 18, 451-503; (1895b) *Proc. US natn. Mus.* 18, 569-91. RICHMOND, W K (1963) *Country Life*, 3 Jan., 14-15. RICHNER, H (1989a) *J. Anim. Ecol.* 58, 427-40; (1989b) *Anim. Behav.* 38, 606-12; (1990) *Ibis* 132, 105-8. RICHNER, H and MARCLAY, C (1991) *Anim. Behav.* 41, 433-38. RICHNER, H, SCHNEITER, P, and STIRNIMANN, H (1989) *Funct. Ecol.* 3, 617-24. RICHTER, H (1958) *Abh. Ber. Staatl. Mus. Tierkde. Dresden* 23, 219-40. RICHTERS, W (1952) *Orn. Mitt.* 4, 193-9. RICKLEFS, R E (1979) *Auk* 96, 10-30; (1984) *Auk* 101, 319-33. RICKLEFS, R E and PETERS, S (1979) *Bird-Banding* 50, 338-48. RICKLEFS, R E and HUSSELL, D J T (1984) *Ornis scand.* 15, 155-61. RIDDIFORD, N and BROOME, T (1983) *Br. Birds* 76, 174-82. RIDDIFORD, N, HARVEY, P V, and SHEPHERD, K B (1989) *Br. Birds* 82, 603-12. RIDDOCH, J (1986) *Scott. Bird News* 3, 2. RIDGELY, R S (1981) *A guide to the birds of Panama*. Princeton. RIDGELY, R S and TUDOR, G (1989) *The birds of South America* 1. Oxford. RIDGWAY, R (1869) *Proc. Acad. Nat. Sci. Philadelphia* 2, 129-33; (1901-11) *Bull. US natn. Mus.* 50 (1-5). RIGGENBACH, H E (1970) *Orn. Beob.* 67, 255-69; (1979) *Orn. Beob.* 76, 153-68. RIMMER, C C (1986) *J. Fld. Orn.* 57, 114-25; (1988) *Condor* 90, 141-56. RINGLEBEN, H (1960) *Vogelwelt* 81, 146-51; (1981) *Vogelk. Ber. Niedersachs.* 13, 73-8. RINNE, U and BAUCH, J (1970) *Luscinia* 41, 16-20. RINNHOFER, G (1965) *Beitr. Vogelkde.* 11, 118-19; (1968) *Falke* 15, 312-14; (1969) *Beitr. Vogelkde.* 14, 324-9; (1972) *Beitr. Vogelkde.* 18, 401-16; (1976) *Falke* 23, 20-1. RION, P (1990) *Aves* 27, 119-28. RIS, H (1957) *Orn. Beob.* 54, 195-7. RISBERG, E L (1968) *Vår Fågelvärld* 27, 173-4; (1970) *Vår Fågelvärld* 29, 77-89. RISBERG, L and RISBERG, B (1975) *Vår Fågelvärld* 34,

139-51. RISING, J D (1969) *Comp. Biochem. Physiol.* 31, 915-25; (1970a) *System. Zool.* 19, 315-51; (1973) In Kendeigh, S C and Pinowski, J (eds) *Productivity, population dynamics, and systematics of granivorous birds*, 327-35. Warsaw; (1973) *Can. J. Zool.* 51, 1267-73; (1983) *Auk* 100, 885-97; (1987) *Evolution* 41, 514-24; (1988) *Wilson Bull.* 100, 183-203. RISING, J D and SHIELDS, G F (1980) *Evolution* 34, 654-62. RISTOW, D, WINK, C, and WINK, M (1986) *Ric. Biol. Selvaggina Suppl.* 10, 285-95. RITTINGHAUS, H (1957) *Vogelwarte* 19, 90-7. ROBBINS, C A (1932) *Auk* 49, 159-65. ROBBINS, C S, BRUUN, B, and ZIM, H S (1983) *Birds of North America*. New York. ROBEL, D (1983) *Beitr. Vogelkde.* 29, 326-7. ROBERT, G (1975) *Gerfaut* 65, 168-9; (1977) *Gerfaut* 67, 101-31. ROBERT, J-C (1979) *Héron* (4), 75-7. ROBERTS, E L (1955) *Br. Birds* 48, 91. ROBERTS, J L (1979) *Country Life* 4 Jan., 32-3. ROBERTS, P (1982) *Bird Study* 29, 155-61; (1983) *Bird Study* 30, 67-72; (1985) *Br. Birds* 78, 217-32. ROBERTS, P J (1989) In Bignal and Curtis (eds), 9-11; (1982) *Br. Birds* 75, 38-40. ROBERTS, S and LEWIS, J (1988) *Gwent Bird Rep.* 23 (1987), 7-10. ROBERTS, T J (1992) *The birds of Pakistan* 2. Karachi. ROBERTSON, A W P (1953) *Br. Birds* 46, 380-1; (1954) *Bird pageant*. London. ROBERTSON, C J R (ed.) (1985) *Reader's Digest complete book of New Zealand Birds*. Sydney. ROBERTSON, H A, WHITAKER, A H, and FITZGERALD, B M (1983) *New Zeal. J. Zool.* 10, 87-97. ROBERTSON, I S (1975a) *Br. Birds* 68, 115; (1975b) *Br. Birds* 68, 453-5. ROER, F (1979) *Prachtfinken*. Berlin. ROBINSON, J C (1990) *Passenger Pigeon* 52, 113-18. ROCHE, J (1958) *Inst. Rech. Sahar. Univ. Alger* 1958, 151-65. RODE, M and LUTZ, K (1991) *Corax* 14, 95-109. RODEBRAND, S (1975) *Calidris* 4, 3-12. RÖDER, J (1991) *Voliere* 14, 204-6. RODRÍGUEZ, L G (1972) *Ardeola* 16, 215-22. RODRÍGUEZ-TEIJEIRO, J D and CORDERO-TAPIA, P J (1983) *Doñana Acta Vert.* 10, 77-90. RÖELL, A (1978) *Behaviour* 64, 1-124; (1979) *Ardea* 67, 123-9. RÖELL, A and BOSSEMA, I (1982) *Behav. Ecol. Sciobiol.* 11, 1-6. ROFSTAD, G (1986) *J. Zool. Lond.* (A) 208, 299-323; (1988) *Ornis scand.* 19, 27-30. ROFSTAD, G and SANDVIK, J (1985) *Ornis scand.* 16, 38-44. ROGACHEVA, E V (1988) *Ptitsy Sredney Sibiri*. Moscow. ROGERS, D T (1965) *Bird-Banding* 36, 115-6. ROGERS, D T, GARCIA, B J, and RÓGEL, B.A (1986a) *Wilson Bull.* 98, 163-7. ROGERS, D T, HICKS, D L, WISCHUSEN, E W, and PARRISH, J R (1982a) *J. Fld. Orn.* 53, 133-8. ROGERS, D T, JR and ODUM, E P (1964) *Auk* 81, 505-13; (1966) *Wilson Bull.* 78, 415-33. ROGERS, M J (1982) *Br. Birds* 75, 387; (1984a) *Br. Birds* 77, 120; (1984b) *Isles of Scilly Bird Rep.* 1983, 11-62. ROGERS, M J and RARITIES COMMITTEE (1979) *Br. Birds* 72, 503-49; (1982b) *Br. Birds* 75, 482-533; (1985) *Br. Birds* 78, 529-89; (1986b) *Br. Birds* 79, 526-88; (1988) *Br. Birds* 81, 535-96; (1989) *Br. Birds* 82, 505-63; (1990) *Br. Birds* 83, 439-96; (1991) *Br. Birds* 84, 449-505; (1992) *Br. Birds* 85, 507-54. ROGERS, T D (1988) *A new list of the birds of Masirah island, Sultanate of Oman*. Muscat. ROGGEMAN, W (1983) *Gerfaut* 73, 451-82. ROHNER, C (1980) *Orn. Beob.* 77, 103-10; (1981) *Orn. Beob.* 78, 1-11. ROHWER, S (1986) *Auk* 103, 281-92. ROHWER, S, KLEIN, W P, and HEARD, S (1983) *Wilson Bull.* 95, 199-208. ROHWER, S and MANNING, J (1990) *Condor* 92, 125-40. ROHWER, S, THOMPSON, C W, and YOUNG, B E (1992) *Condor* 94, 297-300. ROI, O LE (1923) *J. Orn.* 71, 196-252. ROKITANSKY, G (1934) *Falco* 30, 6-8; (1969) *Gef. Welt* 93, 68-9. ROKITANSKY, G and SCHIFTER, H (1971) *Ann. naturhist. Mus. Wien* 75, 495-538. ROLAND, J (1988) *Beih. Veröff. Nat. Land. Bad.-*

*Württ.* **53**, 93- 108. ROLANDO, A and ZUNINO, M (1992) *Ornis scand.* **23**, 201-2. ROLFE, R L (1965) *Br. Birds* **58**, 150-1. ROLFE, R (1966) *Bird Study* **13**, 221-36. ROLLIN, N (1958) *Br. Birds* **51**, 290-303. ROLLS, J C (1973) *Br. Birds* **66**, 169. ROMER, M L R (1947) *Br. Birds* **40**, 176-7. ROOKE, K B (1950) *Br. Birds* **43**, 114-15. ROOS, G (1975) *Anser* **14**, 237-46; (1978) *Anser* **17**, 69-89; (1984) *Anser Suppl.* **13**; (1985) *Anser* **24**, 1-28; (1991) *Anser* **30**, 229-58. ROOT, T (1988) *Atlas of wintering North American birds.* Chicago. ROOTSMÄE, L (1990) *Soobshch. Pribalt. Kom. Izuch. Migr. Ptits* **23**, 123-8. ROOTSMÄE, L and VEROMAN, H (1967) *Orn. Kogumik* **4**, 177-99. RÖRIG, G (1900) *Arb. biol. Abt. Land.-Forstwirtsch. kaiserl. Gesundheitsamt Berlin* **1**, 285-400. ROSELAAR, C S (1976) *Bull. zool. Mus. Univ. Amsterdam* **5**, 13-18; (1991) Lijst van alle vogelsoorten van de wereld. In Perrins, C M (ed.) *Geïllustreerde encyclopedie van de vogels.* Weert. ROSENBERG, E (1953) *Fåglar i Sverige.* Stockholm. ROSENIUS, P (1929) *Svenska fåglar och fågelbon.* Lund. ROSEVEARE, W L (1951) *Br. Birds* **44**, 16-17. RØSKAFT, E (1978) *Vår Fuglefauna* **1**, 207; (1980a) *Fauna norv.* (C) *Cinclus* **3**, 9-15; (1980b) *Fauna norv.* (C) *Cinclus* **3**, 56-9; (1980c) *Vår Fuglefauna* **3**, 5-10; (1981a) *Fauna norv.* (C) *Cinclus* **5**, 5-9; (1981b) *Fauna norv.* (C) *Cinclus* **4**, 76-81; (1983a) *Ornis scand.* **14**, 175-9; (1983b) *Ornis scand.* **14**, 180-7; (1983c) *Fauna norv.* (C) *Cinclus* **6**, 78-80; (1985) *J. Anim. Ecol.* **54**, 255-60; (1987) *Ornis scand.* **18**, 70-1. RØSKAFT, E and ESPMARK, Y (1982) *Ornis scand.* **13**, 38-46; (1984) *Behav. Proc.* **9**, 223-30. RØSKAFT, E, ESPMARK, Y, and JÄRVI, T (1983) *Ornis scand.* **14**, 169-74. ROSS, H A (1980) *Auk* **97**, 721-32. ROSS, W M (1948) *Scott. Nat.* **60**, 147-56. ROST, F (1992) *Anz. Ver. Thür. Orn.* **1**, 41-2. RÖTHING, H (1969) *Vogelwelt* **90**, 146-7. ROTHSCHILD, M (1955) *Proc. int. orn. Congr.* **11** (1954), 611-617; (1957) *Nos Oiseaux* **24**, 1-6; (1960) *Entomologist,* July 1960, 139-40. ROTHSCHILD, W and HARTERT, E (1911) *Novit. Zool.* **18**, 456-550. ROTHSTEIN, S I (1978) *Auk* **95**, 152-60. ROTHSTEIN, S I, VERNER, J, and STEVENS, E (1980) *Auk* **97**, 253-67. RÖTTLER, G (1985a) *Orn. Mitt.* **37**, 35-6; (1985b) *Orn. Mitt.* **37**, 243-4. ROUX, P (1990) *Oiseau* **60**, 16-38. ROUX, P, CHAKIR, N, and LESNE, L (1990) *Bièvre* **11**, 13-20. RØV, N (1975) *Ornis scand.* **6**, 1-14. ROWELL, C H F (1957) *Bird Study* **4**, 33-50. ROWORTH, P C (1983) *Br. Birds* **76**, 351. ROYALL, W C, JR, GUARINO, J L, DE GRAZIO, J W, and GAMMELL, A (1971) *Condor* **73**, 100-6. RUCNER, D (1973a) *Larus* **24**, 168-70. RUCNER, R (1973b) *Larus* **25**, 27-45. RUDAT, V (1976) *Falke* **23**, 316-17; (1984) *Orn. Jber. Mus. Hein.* **8-9**, 77-85. RUDAT, V and RUDAT, W (1971) *Falke* **18**, 387-9; (1978) *Zool. Jb. Syst.* **105**, 386-98. RUDEBECK, G (1950) *Vår Fågelvärld. Suppl.* **1**. RUELLE, M (1987) *Ornithologue* **55**, 46-57. RUELLE, P J and SEMAILLE, R (1982) *Malimbus* **4**, 27-32. RUF, M (1977) *Vögel Heimat* **47**, 146. RUFINO, R (1989) *Atlas das aves que nidificam em Portugal continental.* Lisbon. RUGE, K (1988) *Beih. Veröff. Nat. Land. Bad.-Württ.* **53**, 21-30. RUNDE, O J (1984) *Sterna* **17**, 129-55. RÜPPELL, G (1970) *Publ. Wiss. Sekt. Biol. E 1508/1969.* Göttingen. RÜPPELL, W (1944) *J. Orn.* **92**, 106-32. RUPPRECHT, A L (1990) *Zool. Abh. Staatl. Mus. Tierkde. Dresden* **45**, 151-4. RUSSELL, D N (1971) *Condor* **73**, 369-72. RUSSELL, S M (1964) *Orn. Monogr. AOU* **1**. RUSTAMOV, A K (1954) *Ptitsy pustyny Kara-Kum.* Ashkhabad; (1958) *Ptitsy Turkmenistana* **2**. Ashkhabad; (1977) In Yurlov, K T (ed.) *Migratsii ptits v Azii,* 202-4. Novosibirsk; (1984) In Konstantinov, V M (ed.) *Ekologiya, biotsenoticheskoe i khozyaystvennoe znachenie vrano-*

*vykh ptits,* 114-15. Moscow. RUSTAMOW, A K and SOPYEW, O (1990) *Falke* **37**, 12-15. RUTE, J (1984) *Loodusevaatlusi 1981* (1), 89-94. RUTE, J J and BAUMANIS, J A (1986) In Sokolov, V E and Dobrynina, I N (eds) *Koľtsevanie i mechenie ptits v SSSR, 1979-82,* 23-9. Moscow. RUTHENBERG, H (1968) *Falke* **15**, 406-13. RUTHKE, P (1939a) *Beitr. Fortpfl. Vögel* **15**, 41-50; (1939b) *Orn. Monatsber.* **47**, 181-2; (1971) *Vogelwelt* **92**, 191. RUTSCHKE, E (ed.) (1983) *Die Vogelwelt Brandenburgs.* Jena. RUTTLEDGE, R F (1954) *Br. Birds* **47**, 447; (1966) *Ireland's birds.* London; (1975) *A list of the birds of Ireland.* Dublin. RUTTLEDGE, W (1965) *Br. Birds* **58**, 442-3. RYALL, C (1986) *New Scientist* **1528**, 48-9; (1990) *Scopus* **14**, 14-16. RYALL, C and REID, C (1987) *Swara* **10** (1), 9-11. RYDZEWSKI, W (1960) *Proc. int. orn. Congr.* **12** (2), 641-4; (1978) *Ring* **96-7**, 218-62. RYMKEVICH, T A (1976) *Zool. Zh.* **55**, 1695-1703; (1977) *Ornitologiya* **13**, 67-73; (1979) *Vestnik Leningrad gos. Univ. 3 Biol.* **1**, 37-47; (ed.) (1990) *Lin'ka vorob'inykh ptits Severo-zapada SSSR.* Leningrad. RYVES, B H (1948) *Bird life in Cornwall.* London. RYVES, B H and RYVES, I N M (1934) *Br. Birds* **28**, 2-26. RYZHANOVSKI, V N (1986) *Zool. Zh.* **65**, 1041-50.

SAARELA, S, KLAPPER, B, and HELDMAIER, G (1988) *Abstr. 10th int. Congr. Photobiol.* **40**. Jerusalem. SABEL, K (1963) *Gef. Welt* **87**, 1-4, 41-5; (1965) *Gef. Welt* **89**, 32-4, 49-51; (1983) *Naturgemässe Finkenzucht.* Bassum. SACARRÃO, G F (1968) *Cyanopica* **1**, 37-46. SACARRÃO, G F and SOARES, A A (1975) *Est. Fauna Portug.* **8**, 1-14; (1976) *Arquivos Mus. Bocage* (2) **6**, 1-13. SACHTLEBEN, H (1918) *Arch. Naturgeschichte* 84 (A) **6**, 88-153. SADLER, T S and MYRES, M T (1976) *Prov. Mus. Alberta Nat. Hist. Sect. Occ. Pap.* **1**. SADLIK, J and HAFERLAND, H-J (1981) *Orn. Jber. Mus. Hein.* **5-6**, 77-80. SÆTHER, B-E (1982) *Ornis scand.* **13**, 149-63. SÆTHER, B-E and FONSTAD, T (1981) *Anim. Behav.* **29**, 637-9. SAGE, B L (1957) *Br. Birds* **50**, 353; (1960) *Ardea* **48**, 160-78. SAGE, B and WHITTINGTON, P A (1985) *Bird Study* **32**, 77-81. SAGE, B L and NAU, B S (1963) *Trans. Herts. nat. Hist. Soc.* **25**, 226-44. SAGE, B L and VERNON, J D R (1978) *Bird Study* **25**, 64-86. SAGITOV, A K (1962) *Ornitologiya* **4**, 354-66. SAGITOV, A K and BAKAEV, S (1980) *Ornitologiya* **15**, 142-5. SAGITOV, A K, BELYALOVA, L E, and FUNDUKCHIEV, S E (1990) In Kurochkin, E N (ed.) *Sovremennaya ornitologiya,* 86-97. Moscow. SAINI, H K, DHINDSA, M S, and TOOR, H S (1989) *Gerfaut* **79**, 69-79. SAINO, N and MERIGGI, A (1990) *Ethol. Ecol. Evol.* **2**, 205-14. ST LOUIS, V L and BARLOW, J C (1987) *Wilson Bull.* **99**, 628-41; (1988) *Evolution* **42**, 266-76; (1991) *Wilson Bull.* **103**, 1-12. ST QUINTIN, W H (1907) *Avic. Mag.* **5**, 55-6. SAITO, S-I (1983) *Tori* **32**, 13-20. SALATHÉ, T (1979) *Orn. Beob.* **76**, 247-56; (1987) *Ardea* **75**, 221-9. SALATHÉ, T and RAZUMOVSKY, K (1986) *Terre Vie* **41**, 343-53. SALFELD, D (1963) *Br. Birds* **56**, 221; (1969) *Br. Birds* **62**, 238. SALIKHBAEV, K S and BOGDANOV, A N (1967) *Fauna Uzbekskoy SSR* **2**, *Ptitsy* **4**. Tashkent. SALMANOVA, L M (1986) *Tez. Dokl. 1 S'ezda Vsesoyuz. orn. Obshch. 9 Vsesoyuz. orn. Konf.* **2**, 221-2. SALMEN, H (1982) *Die Ornis Siebenbürgens* **2**. Köln. SALMON, L (1948) *Br. Birds* **41**, 84. SALOMONSEN, F (1928) *Vidensk. medd. Dansk Naturhist. Foren.* **86**, 123-202; (1930a) *Dansk orn. Foren. Tidsskr.* **24**, 9-101; (1930b) *Dansk orn. Foren. Tidsskr.* **24**, 101-4; (1931a) *Ibis* (13) **1**, 57-70; (1931b) *Orn. Monatsber.* **39**, 112-13; (1935) *Zoology of the Faroes* **64** *Aves.* Copenhagen; (1947a) *Dansk orn. Foren. Tidsskr.* **41**, 136-40; (1947b) *Dansk orn. Foren. Tidsskr.* **41**, 216-21;

(1948) *Dansk orn. Foren. Tidsskr.* **42**, 27-8; (1949) *Dansk orn. Foren. Tidsskr.* **43**, 1-45; (1950-1) *Grønlands fugle.* Copenhagen; (1959) *Dansk orn. Foren. Tidsskr.* **53**, 31-9; (1972) *Proc. int. orn. Congr.* **15**, 25-77. SALT, G W (1963) *Proc. int. orn. Congr.* **13**, 905-17. SALT, J R (1984) *Alberta Nat.* **14**, 104. SALT, W R and SALT, J R (1976) *The birds of Alberta.* Edmonton. SALVAN, J (1963) *Oiseau* **33**, 161-2; (1967-9) *Oiseau* **37**, 255-84, **38**, 53-85, 127-50, 249-73, **39**, 38-69. SALZMANN, E (1909) *Orn. Monatsschr.* **34**, 357-67, 400-14; (1911) *Orn. Monatsschr.* **36**, 425-9. SAMCHUK, N D (1971) *Vestnik Zool.* (1), 69-73. SÁNCHEZ-AGUADO, F J (1984) *Ardeola* **31**, 33-45; (1985) *Doñana Acta Vert.* **12**, 197-209; (1986) *Ardeola* **33**, 17-33. SANDBERG, P and WALLENGAARD, R (1987) *Vår Fågelvärld* **46**, 130. SANDEN, W VON (1956) *Orn. Mitt.* **8**, 16. SANTOS JÚNIOR, J R DOS (1968) *Cyanopica* **1**, 9-36. SANTOS, T, SUAREZ, F, and TELLÉRIA, J L (1981) *Proc. 7th int. Conf. Bird Census, 5th Meeting EOAC,* 79-88. León. SAPETINA, I M (1962) *Trudy Okskogo gos. Zapoved.* **4**, 327-36. SAPOZHENKOV, Y F (1962) *Ornitologiya* **5**, 177-82. SAPPINGTON, J N (1977) *Wilson Bull.* **89**, 300-9. SAPSFORD, A (1991) *Calf of Man Bird Obs. Rep. 1990,* 42-61. SARUDNY, N (1904) *Orn. Jahrb.* **15**, 108; (1906) *Orn. Monatsber.* **14**, 47-8; (1907) *Orn. Monatsber.* **15**, 61-3. SARUDNY, N and HÄRMS, M (1912) *J. Orn.* **60**, 592-619; (1914) *Orn. Monatsber.* **22**, 53-4. SAUNDERS, A A (1951) *A guide to bird songs.* Garden City. SAUNDERS, D and SAUNDERS, S (1992) *Br. Birds* **85**, 337-43. SAUNIER, A (1971) *Nos Oiseaux* **31**, 66-7. SAUROLA, P (1977) *Lintumies* **12**, 118-23; (1979) *Lintumies* **14**, 161-7. SAVAGE, E U (1928) *Br. Birds* **22**, 57. SAVIGNI, G and MASSA, R (1983) *Riv. ital. Orn.* **53**, 3-14. SAWLE, C (1988) *Devon Birds* **41**, 76-7. SAXBY, H L (1874) *The birds of Shetland.* Edinburgh. SCHABER, S (1983) *Nos Oiseaux* **37**, 41-2. SCHAETZEN, R de and JACOB, J P (1985) *Aves* **22**, 127-9. SCHÄFER, E (1938) *J. Orn.* **86** suppl; (1991) *Gef. Welt* **115**, 55. SCHÄFER, K J (1968) *Natur Heimat* **28**, 67-72. SCHÄPPER, R (1986) *Orn. Beob.* **83**, 142-5. SCHARLAU, W (1989) *Kart. med. Brutvögel* **3**, 3-23. SCHAUENSEE, R M DE (1966) *The species of birds of South America and their distribution.* Philadelphia; (1984) *The birds of China.* Oxford. SCHAUENSEE, R M DE and PHELPS, W H (1978) *A guide to the birds of Venezuela.* Princeton. SCHEER, G (1952) *Ornis fenn.* **29**, 77-82; (1953) *Anz. orn. Ges. Bayern* **4**, 70-2. SCHEIFLER, H (1968) *Gef. Welt* **92**, 189-90. SCHEKKERMAN, H (1986a) *Dutch Birding* **8**, 89-97; (1986b) *Graspieper* **6**, 110-15; (1989) *Limosa* **62**, 29-34; (1992) *Dutch Birding* **14**, 7-10. SCHELDE, O (1970) *Feltornithologen* **12**, 188. SCHENK, J (1907) *Aquila* **14**, 252-75; (1919) *Aquila* **26**, 129-31; (1929) *Verh. VI int. Orn.-Kongr.,* 250-64; (1934) *Aquila* **38-41**, 121-53. SCHERNER, E R (1968) *Beitr. Naturkde. Niedersachs.* **20**, 120-1; (1969) *Vogelwelt* **90**, 64-5; (1972a) *Vogelwelt* **93**, 41-68; (1972b) *Angew. Orn.* **4**, 35-42; (1972c) *Orn. Mitt.* **24**, 221; (1974) *Vogelwelt* **95**, 41-60; (1979) *Faun. Mitt. Süd. Niedersachsen* **2**, 11-17. SCHIFFERLI, A (1932) *Orn. Beob.* **29**, 66-84; (1953) *Orn. Beob.* **50**, 65-89; (1963) *Proc. int. orn. Congr.* **13**, 468-74; (1992) *Orn. Beob.* **89**, 48-9. SCHIFFERLI, A, GÉROUDET, P, and WINKLER, R (1982) *Verbreitungsatlas der Brutvögel der Schweiz.* Sempach. SCHIFFERLI, A and LANG, E M (1940a) *J. Orn.* **88**, 550-75; (1940b) *Rev. suisse Zool.* **47**, 217-23; (1946) *Orn. Beob.* **43**, 114-17. SCHIFFERLI, L (1977) *Orn. Beob.* **74**, 71-4; (1978) *Orn. Beob.* **75**, 44-7; (1980) *Avocetta* **4**, 49-62; (1981) *Orn. Beob* **78**, 113-15. SCHIFFERLI, L and FUCHS, E (1981) *Orn. Beob.* **78**, 233-43. SCHIFFERLI, L and SCHIFFERLI, A (1980) *Orn. Beob.* **77**, 21-6. SCHIØLER, E L (1922) *Dansk orn.*

*Foren. Tidsskr.* **16**, 1-55. SCHLEGEL, R (1920) *Z. Ool. Orn.* **25**, 29-35. SCHLENKER, R (1976) *Vogelwarte* **28**, 313-14. SCHLEUSSNER, G (1990) *J. Orn.* **131**, 151-5. SCHLEUSSNER, G, DITTAMI, J P, and GWINNER, E (1985) *Physiol. Zool.* **58**, 597-604. SCHLINGER, B A and ADLER, G H (1990) *Wilson Bull.* **102**, 545-50. SCHLÖGEL, N (1987) *Beitr. Vogelkde.* **33**, 65-71. SCHLOSS, W (1984) *Auspicium* **7**, 257-75. SCHMID, H (1991) *Orn. Beob.* **88**, 101-9. SCHMID, T (1974) *Nos Oiseaux* **32**, 274. SCHMID, U (1979) *Vogelkdl. Ber. Niedersachs.* **11**, 45-6. SCHMID, U and GATTER, W (1986) *Vogelwarte* **33**, 335-8. SCHMIDT, E (1960a) *Vogelwarte* **20**, 199-205. SCHMIDT, G (1960b) *Orn. Mitt.* **12**, 3-8; (1962) *Orn. Mitt.* **14**, 33; (1964) *Heimat* **71**, 394-6; (1966) *Vogelwelt* **87**, 154-6. SCHMIDT, G A J (1957) *Orn. Mitt.* **9**, 121-6. SCHMIDT, K (1988) *Beih. Veröff. Nat. Land. Bad.-Württ.* **53**, 191-210. SCHMIDT, O (1991) *Orn. Mitt.* **43**, 3-5. SCHMITT, C and STADLER, H (1914) *Orn. Monatsschr.* **39**, 300-1. SCHMITZ, L (1987) *Aves* **24**, 1-18; (1989) *Aves* **26**, 73-87. SCHNEIDER, W (1957) *Beitr. Vogelkde.* **6**, 43-74; (1964) *Beitr. Vogelkde.* **9**, 455; (1972a) *Der Star.* Wittenberg Lutherstadt; (1972b) *Beitr. Vogelkde.* **18**, 310-46; (1982) *Beitr. Vogelkde.* **28**, 207-21; (1984) *Falke* **31**, 42-3. SCHNELL, F H (1950) *Vogelwelt* **71**, 168. SCHNURRE, O (1959) *Bonn. zool. Beitr.* **10**, 343-50. SCHOENENBERGER, A (1972) *Alauda* **40**, 23-36. SCHÖLL, R W (1959) *J. Orn.* **100**, 439-40; (1960) *Anz. orn. Ges. Bayern* **5**, 591-6. SCHOLS, R (1987) *Limosa* **60**, 119-22. SCHÖLZEL, H (1981) *Orn. Mitt.* **33**, 327-8. SCHÖNBECK, H (1956) *Mitt. Landesmus. Joanneum* **556D**, 68-82. SCHÖNFELD, M and BRAUER, P (1972) *Hercynia* (NF) **1**, 40-68. SCHÖNWETTER, M (1984) *Handbuch der Oologie* **3**. SCHOOF, E (1988) *Vogelkdl. Hefte Edertal* **14**, 41-2. SCHOPPE, R (1986) *Beitr. Naturkde. Niedersachs.* **39**, 44-52. SCHOT, W E M VAN DER (1989) *Dutch Birding* **11**, 170-2. SCHRANTZ, F G (1943) *Auk* **60**, 367-87. SCHREIBER, M (1987) *J. Orn.* **128**, 388. SCHREIBER, R W and SCHREIBER, E A (1984) *Bull. Br. Orn. Club* **104**, 62-8. SCHROETER, W (1982a) *Orn. Mitt.* **34**, 127-8; (1982b) *Orn. Mitt.* **34**, 166-8; (1982c) *Orn. Mitt.* **34**, 171. SCHUBERT, G and SCHUBERT, M (1982) *Falke* **29**, 366-72. SCHUBERT, M (1982) *Stimmen der Vögel Zentralasiens.* (2 LP discs). Eterna. SCHUBERT, P (1988) *Beitr. Vogelkde.* **34**, 69-84. SCHUBERT, W (1977) *Anz. orn. Ges. Bayern* **16**, 45-57. SCHUMANN, H (1956) *Beitr. Naturkde. Niedersachs.* **9**, 94. SCHUPHAN, I and HESELER, U (1965) *Vogelwarte* **23**, 77-9. SCHUSTER, L (1921-3) *J. Orn.* **69**, 153-200, 535-70; **71**, 287-361; (1926) *Beitr. Fortpfl. Vögel* **2**, 55-8; (1930) *J. Orn.* **78**, 273-301; (1944) *Beitr. Fortpfl. Vögel* **20**, 132-33; (1950) *Vogelwelt* **71**, 9-17. SCHÜZ, E (1932) *Orn. Monatsber.* **40**, 123; (1941) *Vogelzug* **12**, 152-63; (1959) *Die Vogelwelt des südkaspischen Tieflandes.* Stuttgart. SCHWAB, A (1969) *Orn. Beob.* **66**, 230-1. SCHWAB, R G and MARSH, R E (1967) *Bird-Banding* **38**, 143-7. SCHWABL, H and FARNER, D S (1989) *Condor* **91**, 108-12. SCHWABL, H, SCHWABL-BENZINGER, I, GOLDSMITH, A R, and FARNER, D S (1988) *Gen. comp. Endocrinol.* **71**, 398-405. SCHWARTZ, P (1964) *Living Bird* **3**, 169-84. SCHWEIGER, H (1959) *J. Orn.* **100**, 350-1. SCLATER, W L and MOREAU, R E (1933) *Ibis* (13) **3**, 399-440. SCOTT, D A, HAMADANI, H M, and MIRHOSSEYNI, A A (1975) *Birds of Iran.* Tehran. SCOTT, R E (1959) *Br. Birds* **52**, 388; (1961) *Bird Study* **8**, 152-4; (1965a) *Bull. Br. Orn. Club* **85**, 66-7; (1965b) *Vår Fågelvärld* **24**, 156-71. SCOTTER, G W, CARBYN, L N, NEILY, W P, and HENRY, J D (1985) *Sask. nat. Hist. Soc. spec. Publ.* **15**. SEAGO, M J (ed.) (1986) *Trans. Norfolk Norwich Nat. Soc.* **27** (4), 274-303; (1987) *Trans. Norfolk Norwich Nat.*

*Soc.* **27** (6), 430-62; (1988) *Trans. Norfolk Norwich Nat. Soc.* **28** (2), 118-51; (1991) *Trans. Norfolk Norwich Nat. Soc.* **29** (2), 123-54. SEALY, S G (1971) *Blue Jay* **29**, 12-16; (1979) *Can. J. Zool.* **57**, 1473-8; (1985) *N. Amer. Bird Bander* **10**, 12-17. SEALY, S G, SEXTON, D A, and COLLINS, K M (1980) *Wilson Bull.* **92**, 114-16. SEARCY, W A (1979) *Condor* **81**, 304-5. SEASTEDT, T R (1980) *Condor* **82**, 232-3. SEASTEDT, T R and MACLEAN, S F (1979) *Auk* **96**, 131-42. SEEL, D C (1960) *Br. Birds* **53**, 303-10; (1964) *Bird Study* **11**, 265-71; (1966) *Bird Study* **13**, 207-9; (1968a) *Ibis* **110**, 129-44; (1968b) *Ibis* **110**, 270-82; (1969) *Ibis* **111**, 36-47; (1970) *Ibis* **112**, 1-14; (1976) *Ibis* **118**, 491-536; (1983) *Bangor Occ. Pap.* **15**. Inst. terr. Ecol., Bangor. SEIFERT, S and SCHÖNFUSS, G (1959) *Beitr. Vogelkde.* **6**, 387-95. SEITZ, E (1964) *Orn. Mitt.* **16**, 212. SELANDER, R K (1967) *Syst. Zool.* **16**, 286-7. SELANDER, R K and GILLER, D R (1960) *Condor* **62**, 202-14. SELANDER, R K and JOHNSTON, R F (1967) *Condor* **69**, 217-58. SELL, M (1984) *Charadrius* **20**, 73-7. SELLERS, R M (1986) *Ring. Migr.* **7**, 99-111. SELLEY, E (1976) *Aquila* **82**, 250. SELLIN, D (1987) *Vogelwelt* **108**, 13-27; (1988) *Beitr. Vogelkde.* **34**, 157-76; (1991) In Glandt (1991), 21-6. SELOUS, F C (1907) *Br. Birds* **1**, 48-51. SEMA, A (1978) *Trudy Inst. Zool. Akad. Nauk Kazakh. SSR Ser. Zool.* **38**, 42-57. SEMENOV-TYAN-SHANSKI, O I and GILYAZOV, A S (1991) *Ptitsy Laplandii.* Moscow. SEMPLE, K R (1971) *Avic. Mag.* **77**, 166-7. SENAR, J C (1983) *Anthropos* **26-7**, 84; (1983) *Misc. Zool.* **7**, 224-6; (1984a) In De Haro, A and Espadaler, X (eds) *Processus d'acquisition précoce. Les communications*, 351-5. Barcelona; (1984b) *Condor* **86**, 213-14; (1985) *Misc. Zool.* **9**, 347-60; (1988) *Ring. Migr.* **9**, 91-2; (1989) *Gerfaut* **79**, 185-7. SENAR, J C and BORRAS, A (1985) *BTO News* **141**, 6. SENAR, J C, BURTON, P J K, and METCALFE, N B (1992) *Ornis scand.* **23**, 63-72. SENAR, J C, CAMERINO, M, and METCALFE, N B (1989) *Behav. Ecol. Sociobiol.* **25**, 141-5. SENAR, J C and COPETE, J L (1990a) *Butll. Gr. Cat. Anell.* **7**, 11-12; (1990b) *Bird Study* **37**, 40-3. SENAR, J C, COPETE, J L, and METCALFE, N B (1990) *Ornis scand.* **21**, 129-32. SENAR, J C, MASÓ, G, and VALLE, M J DEL (1986) *Res. Congr. Nac. Etol.* **8**. Córdoba. SENAR, J C and METCALFE, N B (1988) *Anim. Behav.* **36**, 1549-50. SENGUPTA, S N (1968) *Proc. zool. Soc. Calcutta* **21**, 1-27; (1969) *Auk* **86**, 556; (1973) *J. Bombay nat. Hist. Soc.* **70**, 204-6; (1976) *Proc. Indian natn. Sci. Acad.* **42**, 338-45. SEO, LAND BIRD CENSUS COMMITTEE (1985) *Proc. 9th Int. Conf. Bird Census Atlas Work*, 117-22. Tring. ŠERE, D (1986) *Acrocephalus* **7**, 33-4. SEREBRENNIKOV, M K (1931) *J. Orn.* **79**, 29-56. SERLE, W and MOREL, G J (1977) *A field guide to the birds of West Africa.* London. SERMET, E (1967) *Nos Oiseaux* **29**, 17-20; (1973) *Nos Oiseaux* **32**, 113-15. SEUTIN, G, BOAG, P T, WHITE, B N, and RATCLIFFE, L M (1991) *Auk* **108**, 166-70. SEVESI, A (1939) *Riv. ital. Orn.* (2) **9**, 112-13. SHAPOVAL, A P (1989) In Konstantinov, V M and Klimov, S M (eds) *Vranovye ptitsy v estestvennykh i antropogennykh landshaftakh* (1), 76-8. Lipetsk. SHARPE, R B (1885) *Catalogue of the birds in the British Museum* **10**; (1888) **12**; (1890) **13**. London. SHARROCK, J T R (1963) *Br. Birds* **56**, 221; (1974) *Br. Birds* **67**, 356; (1976) *The atlas of breeding birds in Britain and Ireland.* Tring; (1984) *Br. Birds* **77**, 489. SHAUB, B M (1950) *Bird-Banding* **21**, 105-11. SHAUB, B M and SHAUB, M S (1953) *Bird-Banding* **24**, 135-41. SHAW, G (1990) *Bird Study* **37**, 30-5. SHAW, TSEN-HWANG (1935) *Bull. Fan Mem. Inst. Biol. Zool.* **6**, 65-70. SHEPHERD, M (1992) Ph D Thesis. Nottingham Univ. SHIOVITZ, K A and THOMPSON, W L (1970) *Anim. Behav.*

**18**, 151-8. SHIRIHAI, H (1987) *Dutch Birding* **9**, 152-7; (1989) *Br. Birds* **82**, 52-5; (1992) *Br. Birds*, 289; (in press) *Birds of Israel.* SHIRIHAI, H and ALSTRÖM, P (1990) *Br. Birds* **83**, 262-72. SHIRIHAI, H, JONSSON, A, and SEBBA, N (1987) *Dutch Birding* **9**, 120-2. SHKATULOVA, A P (1962) *Ornitologiya* **4**, 176-81; (1979) *Ornitologiya* **14**, 97-107. SHMELEV, A A and BRUNOV, V V (1986) *Tez. Dokl. 1. S'ezda Vsesoyuz. orn. Obshch. 9 Vesesoyuz. orn. Konf.* **2**, 339-41. SHNITNIKOV, V N (1949) *Ptitsy Semirech'ya.* Moscow. SHOEMAKER, H H (1939) *Auk* **56**, 381-406. SHORT, L L (1969) *Auk* **86**, 84-105. SHORT, L L and HORNE, J F M (1981) *Sandgrouse* **3**, 43-61. SHORT, L L and ROBBINS, C S (1967) *Auk* **84**, 534-43. SHORT, L L and SIMON, S W (1965) *Condor* **67**, 438-42. SHRUBB, M (1979) *The birds of Sussex.* Chichester. SHUFORD, W D (1981) *Amer. Birds* **35**, 264-6. SHUKLA, K K (1981) BSc Hons Thesis. Malaya Univ. SHUKUROV, E D (1986) *Ptitsy elovykh lesov Tyan'-Shanya.* Frunze. SHURUPOV, I I (1985) *Ornitologiya* **20**, 201; (1986) *Tez. Dokl. 1. S'ezda Vsesoyuz. orn. Obshch. 9. Vsesoyuz. orn. Konf.* **2**, 348-9. St Petersburg. SHUTOV, S V (1990) *Zool. Zh.* **69** (5), 93-9. SIBLET, J-P (1988) *Les oiseaux du massif de Fontainbleau et des environs.* Paris. SIBLEY, C G (1950) *Univ. Calif. Publ. Zoöl.* **50**, 109-94; (1954) *Evolution* **8**, 252-90; (1970) *Bull. Peabody Mus. nat. Hist.* **32**. SIBLEY, C G and AHLQUIST, J E (1984) *Auk* **101**, 230-43; (1990) *Phylogeny and classification of birds.* New Haven. SIBLEY, C G, CORBIN, K W, AHLQUIST, J E, and FERGUSON, A (1974) In Wright, C A (ed.) *Biochemical and immunological taxonomy of animals*, 89-176. New York. SIBLEY, C G and SHORT, L L (1959) *Auk* **76**, 443-63; (1964) *Condor* **66**, 130-50. SIBLEY, C G and WEST, D A (1958) *Condor* **60**, 85-104; (1959) *Auk* **76**, 326-38. SICK, H (1931) *Mitt. Ver. sächs. Orn.* **3**, 150-4; (1938) *Beitr. Fortpfl. Vögel* **14**, 176-81; (1939) *Orn. Monatsber.* **47**, 65-71; (1957) *Vogelwelt* **78**, 1-18. SIEGFRIED, W R (1968) *Ostrich* **39**, 105-29. SIERING, M (1986) *Verh. orn. Ges. Bayern* **24**, 319-32. SIIVONEN, L (1963) *Sitzber. Finn. Akad. Wiss.* **1962**, 111-25. SILVOLA, T (1966) *Ornis fenn.* **43**, 60-70. SIMEONOV, S D (1964) *Ann. Univ. Sofia* **56**, 239-75; (1970) *Vogelwelt* **91**, 59-67; (1971) Dissertation. Sofia; (1975) *Ekologiya (Sofia)* **1**, 55-63. SIMEONOV, S and DOICHEV, R (1973) *Ann. Univ. Sofia Fac. Biol.* **65**, 163-71. SIMEONOV, S and PETROV, T (1977) *Ann. Univ. Sofia Fac. Biol.* **71**, 39-47. SIMMONS, K E L (1952) *Br. Birds* **45**, 323-5; (1954) *Ibis* **96**, 478-81; (1957a) *Br. Birds* **50**, 401-24; (1957b) *Ibis* **99**, 178-81; (1960) *Br. Birds* **53**, 11-15; (1961a) *Ibis* **103a**, 37-49; (1961b) *Avic. Mag.* **67**, 124-32; (1963) *Avic. Mag.* **69**, 148; (1966) *J. Zool. Lond.* **149**, 145-62; (1968) *Br. Birds* **61**, 228-9; (1970) *Br. Birds* **63**, 175-7; (1974) *Br. Birds* **67**, 243; (1984) *Br. Birds* **77**, 121; (1985) In Campbell, B and Lack, E (eds) *A dictionary of birds.* 101-5. Calton; (1986a) *The sunning behaviour of birds.* Bristol; (1986b) *Br. Birds* **79**, 595-6. SIMMS, E (1948) *Br. Birds* **41**, 344; (1962) *Br. Birds* **55**, 1-36; (1971) *Woodland birds.* London; (1975) *Birds of town and suburb.* London. SIMON, P, DELMÉE, E, and DACHY, P (1975) *Gerfaut* **65**, 153-64; (1983) *Gerfaut* **73**, 207-11. SIMON, T (1921) *Orn. Beob.* **19**, 4-6. SIMPSON, T (1970a) *Br. Birds* **63**, 177; (1970b) *Br. Birds* **63**, 254-5. SIMROTH, H (1908) *Orn. Monatsschr.* **33**, 61-71. SIMS, R W (1955) *Bull. Br. Mus. (nat. Hist.)* **2**, 369-93. SIMSON, E C L (1958) *Bull. Jourdain Soc.* **4**, 35-40. SINGER, R and YOM-TOV, Y (1988) *Ornis scand.* **19**, 139-44. SITASUWAN, N and THALER, E (1984) *Gef. Welt* **108**, 12-15, 45-7; (1985) *J. Orn.* **126**, 181-93. SITS, E (1937) *Beitr. Fortpfl.*

*Vögel* **13**, 140–3. SITTERS, H P (1982) *Br. Birds* **75**, 105–8; (1985) *Bird Study* **32**, 1–10; (1991) M Sc Thesis. Aberdeen Univ. SKEAD, C J (1967) *Ostrich Suppl.* **7**. SKEAD, D M (1974) *Ostrich* **45**, 189–92; (1977) *Ostrich Suppl.* **12**, 117–31. SKEAD, D M and DEAN, W R J (1977) *Ostrich Suppl.* **12**, 3–42. SKARPHÉTHINSSON, K H, NIELSEN, Ó K, THÓRISSON, S, and PETERSEN, I K (1992) *Bliki* **11**, 1–26. SLAGSVOLD, T (1979a) *Fauna norv.* (C) *Cinclus* **2**, 1–6; (1979b) *Fauna norv.* (C) *Cinclus* **2**, 60–4; (1979c) *Fauna norv.* (C) *Cinclus* **2**, 65–9; (1980) *Fauna norv.* (C) *Cinclus* **3**, 16–35; (1981) *Fauna norv.* (C) *Cinclus* **4**, 47–8; (1984) *Fauna norv.* (C) *Cinclus* **7**, 127–31; (1982a) *Ornis scand.* **13**, 141–4; (1982b) *Ornis scand.* **13**, 165–75; (1985) *Fauna norv.* (C) *Cinclus* **8**, 9–17. SLATER, P J B (1981) *Z. Tierpsychol.* **56**, 1–24; (1983) *Anim. Behav.* **31**, 272–81. SLATER, P J B, CLEMENTS, F A, and GOODFELLOW, D J (1984) *Behaviour* **88**, 76–97. SLATER, P J B and INCE, S A (1979) *Behaviour* **71**, 146–66; (1982) *Ibis* **124**, 21–6. SLATER, P J B and SELLAR, P J (1986) *Behaviour* **99**, 46–64. SLESSERS, M (1970) *Auk* **87**, 91–9. SLOAN-CHESSER, S (1937) *Ool. Rec.* **17**, 83–8. SMART, J H (1978) *Br. Birds* **71**, 86. SMETANA, N M (1980) In Kovshar', A F (ed.) *Biologiya ptits Naurzumskogo gosudarstvennogo zapovednika,* 75–104. Alma-Ata. SMETS, F and DRAULANS, D (1982) *Wielewaal* **48**, 55–7. SMIDDY, P (1986) *Br. Birds* **79**, 251–2. SMITH, A E (ed.) (1951) *Gibraltar Point Bird Rep. 1950;* (1953) *Gibraltar Point Bird Rep. 1952;* (1954) *Gibraltar Point Bird Rep. 1953*. SMITH, A E and CORNWALLIS, R K (1953a) In Smith (1953), 8–29; (1953b) *Br. Birds* **46**, 428–30. SMITH, E C (1974) *Ardea* **62**, 226–35. SMITH, F and BORG, S (1976) *Il-Merill* **17**, 25–6. SMITH, F R (1959) *Br. Birds* **52**, 1–9. SMITH, F R and RARITIES COMMITTEE (1967) *Br. Birds* **60**, 309–38; (1973) *Br. Birds* **66**, 331–60. SMITH, J N M and ZACH, R (1979) *Evolution* **33**, 460–7. SMITH, K, WALDON, J, and WILLIAMS, G (1992) *RSPB Conserv. Rev.* **6**, 40–4. SMITH, K D (1955a) *Ibis* **97**, 65–80; (1955b) *Ibis* **97**, 480–507; (1957) *Ibis* **99**, 1–26, 307–37; (1960) *Ibis* **102**, 536–44; (1962–4) Unpubl. diaries of Moroccan expeditions. Edward Grey Inst., Oxford Univ; (1965) *Ibis* **107**, 493–526; (1968) *Ibis* **110**, 452–92. SMITH, K G (1978) *Western Birds* **9**, 79–81. SMITH, C E (1966a) *Bird-Banding* **37**, 49–51. SMITH, P W (1985) *American Birds* **39**, 255–8. SMITH, R (1991) In Stroud, D A and Glue, D (eds) *Britain's birds in 1989/90: the conservation and monitoring review,* 112–13. BTO/NCC; (1992) *Ring. Migr.* **13**, 43–51. SMITH, V W (1966b) *Ibis* **108**, 492–512; (1967) *Bull. Niger. Orn. Soc.* **4** (13–14), 40. SMYTHIES, B E (1960) *The birds of Borneo.* Edinburgh; (1981) *The birds of Borneo.* 3rd edn. Kuala Lumpur; (1986) *The birds of Burma.* Liss. SNIGIREWSKI, S I (1928) *J. Orn.* **76**, 587–607. SNOW, B and SNOW, D (1988) *Birds and berries.* Calton. SNOW, D W (1952) *Ibis* **94**, 473–98; (1953) *Br. Birds* **46**, 379–80. SNOW, D W and MAYER-GROSS, H (1967) *Bird Study* **14**, 43–52. SNOW, D W, OWEN D F, and MOREAU, R E (1955) *Ibis* **97**, 557–71. SNYDER, L L (1957) *Arctic birds of Canada.* Toronto; (1937) *Can. Fld.-Nat.* **51**, 37–9. SOBANSKI, G G (1979) In Labutin, Y V (ed.) *Migratsii i ekologiya ptits Sibiri,* 46. Yakutsk. SOF (SVERIGES ORNITOLOGISKA FÖRENING) (1990) *Vår Fågelvärld Suppl.* **14**. SOKOLOV, E P (1986a) *Trudy Zool. Inst. Akad. Nauk SSSR* **147**, 71–81. SOKOLOV, E P and LOBKOV, E G (1985) *Ornitologiya* **20**, 33–41. SOKOLOV, L V (1986b) *Zool. Zh.* **65**, 1544–51. SOKOŁOWSKI, J (1962) *Acta Orn.* **7**, 33–67. SOLER, J J and SOLER, M (1991) *Ardeola* **38**, 69–89 SOLER, M (1987) *Doñana Acta Vert.* **14**, 67–81; (1988) *Bird Study* **35**, 69–76; (1989a) *Bird Study* **36**, 73–6; (1989b) *Ardeola* **36**, 3–24; (1989c) In Bignal, E and Curtis, D J (eds) *Choughs and land-use in Europe,* 29–33. Scottish Chough Study Group; (1990) In Pinowski, J and Summers-Smith, J D (eds) *Granivorous birds in the agricultural landscape,* 253–61. Warsaw. SOLER, M, ALCALÁ, N, and SOLER, J J (1990) *Doñana Acta Vert.* **17**, 17–48. SOLER, M and SOLER, J J (1987) *Ardeola* **34**, 3–14; (1990) *Ardeola* **37**, 37–52. SOLLENBERG, P (1959) *Vår Fågelvärld* **18**, 128–31. SOLONEN, T (1985) *Suomen linnusto.* Helsinki. SOMEREN, V D VAN (1958) *A bird watcher in Kenya.* Edinburgh. SOMEREN, V G L VAN (1922) *Novit. Zool.* **29**, 1–246; (1956) *Fieldiana Zool.* **38**. SOMERKOSKI, M (1984) *Lounais-Hameen Luonto* **70**, 47–9. SÖMMER, P (1991) In Glandt (1991), 17–20. SONDBØ, S D (1993) *Birding World* **5**, 458–60. SONDELL, J (1977) *Vår Fågelvärld* **36**, 174–84. SONERUD, G A and BEKKEN, J (1979) *Vår Fuglefauna* **2**, 78–85. SONERUD, G A and FJELD, P E (1987) *Ornis scand.* **18**, 323–5. SOPER, E A (1969) *Br. Birds* **62**, 200–1. SOPER, T (1986) *The bird table book.* Newton Abbot. SOPYEV, O (1965) *Ornitologiya* **7**, 134–41; (1967) *Ornitologiya* **8**, 221–35; (1979) *Izv. Akad. Nauk Turkmen. SSR Ser. Biol. Nauk* (4), 53–7. SORBI, S, ROBBRECHT, G, STEEMAN, C, and WILLE, E (1990) *Aves* **27**, 39–47. SORCI, G, MASSA, B, and CANGIALOSI, G (1971) *Riv. ital. Orn.* **41**, 1–10. SOROKIN, A G (1977) *Ornitologiya* **13**, 210–11. SOUTHERN, H N (1945) *Ibis* **87**, 287. SOUZA, J A de (1991) *Ardeola* **38**, 179–98. SOVON (SAMENWERKENDE ORGANISATIES VOGELONDERZOEK NEDERLAND) (1987) *Atlas van de Nederlandse vogels.* Arnhem. SPAANS, A L (1977) *Ardea* **65**, 83–7. SPAANS, A L, RODENBURG, S, and WOLF, J DE (1982) *Vogeljaar* **30**, 31–5. SPAEPEN, J (1952) *Gerfaut* **42**, 164–214. SPANGENBERG, E P (1965) *Sbor. Trud. zool. Mus. MGU* **9**, 98–202. SPANÒ, S and TRUFFI, G (1986) *Riv. ital. Orn.* **56**, 231–9. SPARKS, J H (1963a) *Ibis* **105**, 558–61; (1963b) *Nature* **200**, 281; (1964) *Anim. Behav.* **12**, 125–36; (1965) *Proc. zool. Soc. Lond.* **145**, 387–403. SPEEK, B J and SPEEK, G (1984) *Thieme's vogeltrekatlas.* Zutphen. SPEICHER, K (1989) *Unser Kanarienvogel.* Stuttgart. SPEIRS, J M (1985) *Birds of Ontario* **2**. Toronto. SPELLMAN, C B, LEMON, R E, and MORRIS, M M J (1987) *Wilson Bull.* **99**, 257–61. SPENCE, B R and CUDWORTH, J (1966) *Br. Birds* **59**, 198–201. SPENCER, K G (1966a) *Naturalist* **91**, 73–80. SPENCER, R (1961) *Br. Birds* **54**, 449–95; (1962) *Br. Birds* **55**, 493–556; (1963) *Br. Birds* **56**, 477–524; (1964) *Br. Birds* **57**, 525–82; (1966b) *Br. Birds* **59**, 441–91; (1967) *Br. Birds* **60**, 429–75; (1969) *Br. Birds* **62**, 393–442; (1972) *Bird Study* **19** *Suppl.* SPENCER, R and GUSH, G H (1973) *Br. Birds* **66**, 91–9. SPENCER, R and HUDSON, R (1978a) *Ring. Migr.* **1**, 189–252; (1978b) *Ring. Migr.* **2**, 57–104; (1979) *Ring. Migr.* **2**, 161–208; (1980) *Ring. Migr.* **3**, 65–108; (1981) *Ring. Migr.* **3**, 213–56; (1982) *Ring. Migr.* **4**, 65–128. SPENCER, R and RARE BREEDING BIRDS PANEL (1988a) *Br. Birds* **81**, 99–125; (1988b) *Br. Birds* **81**, 417–44; (1991) *Br. Birds* **84**, 349–70, 379–92; (1993) *Br. Birds* **86**, 62–90. SPENNEMANN, A (1926) *Trop. Natuur* **15**, 86–9; (1928) *Beitr. Fortpfl. Vögel* **4**, 112–13; (1937) *Beitr. Fortpfl. Vögel* **13**, 120. SPERL, J (1992) *Falke* **39**, 244–5. SPILLNER, W (1973) *Falke* **20**, 166–9; (1975) *Falke* **22**, 276–9. SPITTLE, R J (1950) *Bull. Raffles Mus.* **21**, 184–204. SPJØTVOLL, Ø (1972) *Sterna* **11**, 201–13. SPRAGUE, R T and WEIR, R D (1984) *The birds of Prince Edward County.* Kingston. SPRAY, C J (1978) Ph D Thesis. Aberdeen Univ. SQUIRES, W A (1976) *New Brunswick Museum Monogr. Ser.* **7**. St John. SSOKOLOW, J J (1932) *Orn. Beob.* **30**, 20–4. STAAV, R (1976) *Fauna och Flora* **71**, 202–7; (1983)

*Fauna och Flora* **78**, 265-76. STAAV, R and FRANSSON, T (1987) *Nordens fåglar*. Stockholm. STACHANOW, W S (1931) *J. Orn.* **79**, 315-17. STADIE, C (1983) *Voliere* **6**, 35. STADLER, H (1926a) *Ber. Ver. schles. Orn.* **12**, 22-38; (1926b) *Ber. Ver. schles. Orn.* **12**, 82-94; (1927) *Ber. Ver. schles. Orn.* **13**, 40-9, 117-25; (1931) *Verh. orn. Ges. Bayern* **19**, 331-59; (1956) *Nachr. naturw. Mus. Aschaffenburg* **51**, 1-39. STAGG, A (1974) *J. RAF orn. Soc.* **9**, 17-38; (1985) *The birds of south-west Saudi Arabia*. Riyadh; (1987) *Birds of the Riyadh region*. Riyadh. STAGG, A and WALKER, F (1982) *A checklist of the birds of Tabuk Kingdom of Saudi Arabia*. STAHLBAUM, G (1957) *Vogelwelt* **78**, 127; (1967) *Beitr. Vogelkde.* **13**, 224. STAINTON, J M (1982) *Br. Birds* **75**, 65-86; (1991) *Br. Birds* **84**, 66-7. STANFORD, J K (1954) *Ibis* **96**, 449-73, 606-24. STANTSCHINSKY, W W (1929) *J. Orn.* **77**, 309-15. STAŠAITIS, J N (1982) *Ornitologiya* **17**, 173. ŠŤASTNÝ, K and BEJČEK, V (1991) *Panurus* **3**, 27-36. STEGEMAN, L C (1954) *Auk* **71**, 179-85. STEGEMANN, K-D (1975) *Beitr. Vogelkde.* **21**, 383-4. STEGMANN, B (1928) *Ezhegodnik Zool. Muz. Akad. Nauk SSSR* **28**, 366-90; (1931a) *J. Orn.* **79**, 137-236; (1931b) *Orn. Monatsber.* **39**, 183-4; (1932) *J. Orn.* **80**, 99-114; (1935) *Orn. Monatsber.* **43**, 29-30; (1936) *J. Orn.* **84**, 58-139; (1948) *Zool. Zh.* **27**, 241-4; (1956) *J. Orn.* **97**, 236. STEIN, G (1929) *Orn. Monatsber.* **37**, 7-12. STEINBACHER, F (1930a) *J. Orn.* **78**, 471-87; (1930b) *Aquila* **36-7**, 88-91. STEINBACHER, J (1952) *Bonn. zool. Beitr.* **3**, 23-30; (1954) *Senckenbergiana* **34**, 307-10; (1956) *Senck. biol.* **37**, 213-19. STEINBACHER, J and WOLTERS, H E (1963-5) *Vögel in Käfig and Voliere: Prachtfinken*. Aachen. STEINER, H (1960) *J. Orn.* **101**, 92-112; (1955) *Proc. int. Orn. Congr.* **11**, 350-5. STEINER, H M (1969) *Bonn. zool. Beitr.* **20**, 75-84. STEINFATT, O (1937a) *Beitr. Fortpfl. Vögel* **13**, 210-23; (1937b) *Verh. orn. Ges. Bayern* **21**, 139-54; (1940) *Ber. Ver. schles. Orn.* **25**, 11-22; (1942) *Beitr. Fortpfl. Vögel* **18**, 21-6; (1943) *Beitr. Fortpfl. Vögel* **19**, 68-71; (1944) *Orn. Monatsber.* **52**, 8-16; (1954) *J. Orn.* **95**, 245-62. STEJNEGER, L (1892) *Proc. US natn. Mus.* **15**, 289-359. STENHOUSE, D (1962a) *Ibis* **104**, 250-2; (1962b) *Notornis* **10**, 61-7. STEPANITSKAYA, E V (1987) In Sokolov, V E and Dobrynina, I N (eds) *Kol'tsevanie i mechenie zhivotnykh, 1983-4 gody*, 80-155. Moscow. STEPANOV, E A (1960) *Ornitologiya* **3**, 292-7; (1987) *Ornitologiya* **22**, 118-23. STEPANYAN, L S (1969a) *Nauch. Dokl. vyssh. Shk. biol. Nauki* (2), 22-6; (1969b) *Uchen. zap. Mosk. gos. ped. Inst.* **362**, 176-302; (1970) *Uchen. zap. Mosk. gos. ped. Inst.* **394**, 102-50; (1978) *Sostav i raspredelenie ptits fauny SSSR, Passeriformes*. Moscow; (1983) *Nadvidy i vidy-dvoyniki v avifaune SSSR*. Moscow; (1990) *Konspekt ornitologicheskoy fauny SSSR*. Moscow. STEPANYAN, L S and GALUSHIN, V M (1962) *Ornitologiya* **4**, 200-7. STEPHAN, B (1974) *Falke* **21**, 31; (1982) *Mitt. zool. Mus. Berlin* **58**, Suppl. *Ann. Orn.* **6**, 91-100; (1984) *Mitt. zool. Mus. Berlin* **60**, Suppl. *Ann. Orn.* **8**, 89-96; (1986) *Mitt. zool. Mus. Berlin* **62**, Suppl. *Ann. Orn.* **10**, 25-68. STEPHAN, B and GAVRILOV, E I (1980) *Mitt. zool. Mus. Berlin* **56**, Suppl. *Ann. Orn.* **4**, 25-8. STERBETZ, I (1964) *Angew. Orn.* **2**, 30-6; (1967) *Aquila* **73-4**, 203; (1968) *Aquila* **75**, 151-7; (1971) *Állattani közl.* **58**, 171-2. STEVENS, C J (1945) *Br. Birds* **38**, 295-6. STEVENSON, H (1866) *The birds of Norfolk* **1**. London. STEVENSON, H M (1957) *Wilson Bull.* **69**, 39-77. STEVENSON, J (1950) *Countryman* **41**, 334-5. STEWART, P A (1937) *Auk* **54**, 324-32. STEWART, R E (1952) *Auk* **69**, 50-9. STEWART, R E and ROBBINS, C S (1958) *North American Fauna* **62**. STEWART, W (1927) *Scott. Nat.* **163**, 104-

7. STEWART-HESS, C (1992) *Devon Birds* **45**, 25-6. STIEFEL, A (1976) *Ruhe und Schlaf bei Vögeln*. Wittenberg Lutherstadt. STIEHL, R B (1978) Ph D Thesis. Portland State Univ. STIEHL, R B and TRAUTWEIN, S N (1991) *Wilson Bull.* **103**, 83-92. STILES, F G and CAMPOS, R G (1983) *Condor* **85**, 254-5. STILES, F G and SKUTCH, A F (1989) *A guide to the birds of Costa Rica*. London. STILL, E, MONAGHAN, P, and BIGNAL, E (1987) *Ibis* **129**, 398-403. STINGELIN, A (1935) *Arch. suisses Orn.* **1**, 251-6. STIRRUP, S A and EVERSHAM, B (1984) *Br. Birds* **77**, 491. STISHOV, M S, PRIDATKO, V I, and BARANYUK, V V (1991) *Ptitsy ostrova Vrangelya*. Novosibirsk. STJERNBERG, T (1979) *Acta zool. Fenn.* **157**; (1985) *Proc. int. orn. Congr.* **18**, 743-53. STOBO, W T and MCLAREN, I A (1975) *Proc. Nova Scotia Inst. Sci.* **27**, suppl. 2. STODDARD, H L (1978) *Tall Timbers Res. Stn. Bull.* **21**. STOKOE, R (1949) *Br. Birds* **42**, 359-60. STOLT, B-O (1977) *Zoon* **5**, 57-61; (1987) *Vår Fågelvärld* **46**, 48-53; (1993) *J. Orn.* **134**, 59-68. STOLT, B-O and ÅSTRÖM, G (1975) *Fauna och Flora* **70**, 145-54. STOLT, B O and MASCHER, J W (1971) *Vår Fågelvärld* **30**, 84-90. STONE, W (1937) *Bird Studies at old Cape May* **2**. Philadelphia. STONER, D (1923) *Auk* **40**, 328-30. STORER, R W (1969) *Living Bird* **8**, 127-36. STORER, R W, and ZIMMERMAN, D A (1959) *Occ. Pap. Mus. Zool. Univ. Michigan* **609**, 1-13. STORK, H-J, JÄNICKE, B, and WENDENBURG, U (1976) *Orn. Ber. Berlin* **1**, 295-316. STRACHE, R-R and MADAS, K (1988) *Beitr. Vogelkde.* **34**, 201-2. STRAHM, J (1958) *Nos Oiseaux* **24**, 177-84; (1960) *Nos Oiseaux* **25**, 265-71; (1962) *Nos Oiseaux* **26**, 179-85, 297-303. STRAKA, U (1991) *Egretta* **34**, 34-41. STRAUSS, E (1938a) *Z. Tierpsychol.* **2**, 145-72; (1938b) *Z. Tierpsychol.* **2**, 172-97. STRAUTMAN, F I (1963) *Ptitsy zapadnykh oblastey USSR* **2**. L'vov; (1954) *Ptitsy Sovetskikh Karpat*. Kiev. STRAVINSKI, C and SHCHEPSKI, Y (1972) *Soobshch. Pribalt. Kom. Izuch. Migr. Ptits* **7**, 44-57. STREBEL, S (1991) *Orn. Beob.* **88**, 217-42. STREMKE, D (1990) *Beitr. Vogelkde.* **36**, 10-16. STRESEMANN, E (1910) *Orn. Monatsber.* **18**, 33-9; (1919) *Beitr. Zoogeogr. paläarktischen Region* **1**, 25-56; (1920) *Avifauna Macedonica*. Munich; (1924) *Orn. Monatsber.* **32**, 42-3; (1928a) *Orn. Monatsber.* **36**, 41-2; (1928b) *J. Orn.* **76**, 313-411; (1930) *Orn. Monatsber.* **38**, 17-18; (1935) *Orn. Monatsber.* **43**, 30-1; (1940) *Orn. Monatsber.* **48**, 102-4; (1943a) *Orn. Monatsber.* **51**, 166-8; (1943b) *J. Orn.* **91**, 305-24; (1943c) *J. Orn.* **91**, 448-514; (1956) *J. Orn.* **97**, 44-72. STRESEMANN, E, MEISE, W, and SCHÖNWETTER, M (1937) *J. Orn.* **85**, 375-576. STRESEMANN, E and SCHIEBEL, G (1925) *J. Orn.* **73**, 658-9. STRESEMANN, E and STRESEMANN, V (1966) *J. Orn.* **107** suppl; (1969a) *J. Orn.* **110**, 291-313; (1969b) *J. Orn.* **110**, 475-81; (1970) *Beitr. Vogelkde.* **16**, 386-92. STRIEGLER, R and STRIEGLER, U, and JOST, K-D (1982) *Falke* **29**, 164-70. STROKOV, V V (1962) *Ornitologiya* **5**, 290-9. STRÖM, K (1991) *Ornis Svecica* **1**, 119-20. STRÖMBERG, G (1975) *Fåglar Blekinge* **11**, 232-7. STÜBS, J (1958) *Beitr. Vogelkde.* **5**, 312-14. STUDD, M, MONTGOMERIE, R D, and ROBERTSON, R J (1983) *Can. J. Zool.* **61**, 226-31. STUDER-THIERSCH, A (1969) *Orn. Beob.* **66**, 105-44; (1984) *Orn. Beob.* **81**, 29-44. STYAN, F W (1891) *Ibis* (6) **3**, 316-59. SUDHAUS, W (1969a) *Orn. Mitt.* **21**, 18; (1969b) *Vogelwelt* **90**, 53-9; (1969c) *Vogelwelt* **90**, 234-5. SUDILOVSKAYA, A M (1957) *Byull. Mosk. Obshch. Ispyt. Prir. Otd. Biol.* **57** (3), 19-23. SUEUR, F (1981) *Alauda* **49**, 300-4; (1988) *Oiseau* **58**, 156-8; (1990a) *Oiseau* **60**, 60-2; (1990b) *Oiseau* **60**, 63-5. SUHONEN, J and JOKIMÄKI, J (1988) *Ornis fenn.* **65**, 76-83. SUKHININ, A N (1959) *Trudy Inst. Zool. Parasitol. Akad. Nauk Turkmen. SSR*

4, 69-124. SULKAVA, S (1969) *Aquilo Ser. Zool.* 7, 33-7. SULLIVAN, G A (1976) *Anim. Behav.* 24, 880-8. SULTANA, J and GAUCI, C (1982) *A new guide to the birds of Malta.* Valletta; (1985a) *Il-Merill* 23, 11; (1985b) *Il-Merill* 23, 32-40; (1988) *Il-Merill* 25, 41-52. SULTANOV, E G (1987a) *Izv. Akad. Nauk Azer. SSR Ser. Biol. Nauk* (1), 43-8; (1987b) *Dokl. Akad. Nauk Azer. SSR* 43 (9), 72-5. SULTANOV, E G and GUMBATOVA, S E (1986) *Tez. Dokl. I S'ezda Vsesoyuz. orn. Kongr. 9 Vsesoyuz. orn. Konf.* 2, 265-6. St Petersburg. SUMMERS, D D B (1979) *Br. Birds* 72, 249-63; (1982) *J. appl. Ecol.* 19, 813-19. SUMMERS, R W (1989) *Scott. Birds* 15, 181. SUMMERS, R W and CROSS, S C (1987) *Ring. Migr.* 8, 11-18. SUMMERS, R W, WESTLAKE, G E, and FEARE, C J (1987) *Ibis* 129, 96-102. SUMMERS-SMITH, D (1954a) *Ibis* 96, 116-28; (1954b) *Br. Birds* 47, 249-65; (1955) *Ibis* 97, 296-305; (1956) *Br. Birds* 49, 465-88; (1958) *Ibis* 100, 190-203; (1959) *Br. Birds* 52, 164-5; (1979) *Il-Merill* 20, 18-19; (1980) *Il-Merill* 21, 17-18; (1983) *Br. Birds* 76, 411; (1984) *Br. Birds* 77, 25-6. SUMMERS-SMITH, D and LEWIS, L R (1952) *Bird Notes* 25, 44-8. SUMMERS-SMITH, D and SUMMERS-SMITH, M (1952) *Br. Birds* 45, 75. SUMMERS-SMITH, D and VERNON, J D R (1972) *Ibis* 114, 259-62. SUMMERS-SMITH, J D (1963) *The House Sparrow.* London; (1984a) *Ostrich* 55, 141-6; (1984b) *Bull. Br. Orn. Club* 104, 138-42; (1985) In Campbell, B and Lack, E (eds) *A dictionary of birds,* 555-6; (1988) *The sparrows.* Calton; (1989) *Bird Study* 36, 23-31; (1990) *Phoenix* 7, 17-19; (1990) In Pinowski, J and Summers-Smith, J D (eds) *Granivorous birds in the agricultural landscape,* 11-29. Warsaw; (1992) *In search of sparrows.* London. SUMMERS-SMITH, M (1951) *Br. Birds* 44, 16. SUNDERLIN, M A (1978) *Kingbird* 28, 94. SUNDIN, B (1988) *Vår Fågelvärld* 47, 15. SUORMALA, K (1938) *Ornis fenn.* 15, 16-20. SUSCHKIN, P and STEGMANN, B (1929) *J. Orn.* 77, 386-406. SUSHKIN, P (1913) *Bull. Soc. Imp. Nat. Moskva* 26, 198-400; (1933) *Ibis* (13) 3, 55-8. SUSHKIN, P P (1925) *Proc. Boston Soc. nat. Hist.* 38, 1-55; (1938) *Ptitsy Sovetskogo Altaya* 2. Moscow. SUTTER, E (1946) *Orn. Beob.* 43, 81-5; (1948) *Orn. Beob.* 45, 98-106; (1985) *Proc. int. Orn. Congr.* 18, 1055. SUTTER, E and AMANN, F (1953) *Orn. Beob.* 50, 89-90. SUTTON, G M (1932) *Mem. Carnegie Mus.* Pittsburgh 12 (part 2, sect. 2); (1935) *Cranbrook Inst. Sci. Bull.* 3, 1-36; (1938) *Auk* 55, 1-6; (1967) *Oklahoma birds.* Norman; (1968) *Oklahoma orn. Soc.* 1, 1-7. SUTTON, G M and PARMELEE, D F (1954) *Wilson Bull.* 66, 159-79. SUTTON, R W W and GRAY, J R (1972) In Vittery, A and Squire, J E (eds) *Orn. Soc. Turkey Bird Rep. 1968-9,* 186-205. Sandy. SVÄRDSON, G (1957) *Br. Birds* 50, 314-43. SVENSSON, B W (1978) *Ornis scand.* 9, 66-83. SVENSSON, L (1975) *Vår Fågelvärld* 34, 311-18; (1984a) *Soviet birds.* Stockholm (cassette and booklet); (1984b) *Identificaion guide to European passerines* 3rd edn. Stockholm; (1991) *Birding World* 4, 349-52; (1992) *Identification guide to European passerines* 4th edn. Stockholm. SVENSSON, L O (1973) *Vår Fågelvärld* 32, 46-7. SVENSSON, S (1964) *Vår Fågelvärld* 23, 43-56; (1988) *Vår Fågelvärld* 47, 119-20; (1990) *Baltic Birds* 5 (2), 180-91. SVENSSON, S, CARLSSON, U T, and LILJEDAHL, G (1984) *Ann. zool. fenn.* 21, 339-50. SWANBERG, P O (1951a) *Fauna och Flora* 46, 11-29, 111-36; (1951b) *Proc. int. orn. Congr.* 10, 545-54; (1952) *Br. Birds* 45, 60-1; (1954) *Vår Fågelvärld* 13, 213-40; (1956a) *Ibis* 98, 412-19; (1956b) In Wingstrand, K G (ed.) *Bertil Hanström: zoological papers in honour of his sixty-fifth birthday,* 278-97. Lund; (1969) *Br. Birds* 62, 239-40; (1981) *Vår Fågelvärld* 40, 399-408. SWANN, R L

(1954) *Sterna* 14, 111-12; (1988) *Ring. Migr.* 9, 1-4. SWARTH, H S (1920) *Univ. Calif. Publ. Zool.* 21, 75-224. SWENK, M H (1930) *Wilson Bull.* 42, 81-95. SWINGLAND, I R (1976) *Anim. Behav.* 24, 154-8; (1977) *J. Zool. Lond.* 182, 509-28. SWINHOE, R (1861) *Ibis* (1) 3, 323-45. SWYNNERTON, C F M (1916) *Ibis* (10) 4, 264-94. SYKES, T K (1986) *Br. Birds* 79, 594-5. SYLVESTER, G (1968) *Vogelwelt* 89, 232. SYMENS, P (1990) *Sandgrouse* 12, 3-7. SYMENS, D (1991) *Oriolus* 57, 26-32. SZABÓ, L V (1962) *Aquila* 67-8, 260-1; (1976) *Aquila* 82, 145-54. SZABÓ, L V and GYÖRY, J (1962) *Aquila* 67-8, 141-9. SZCZEPSKI, J P (1970) *Acta Orn.* 12, 103-75; (1976) *Acta Orn.* 15, 145-276. SZEMERE, L (1957) *Aquila* 63-4, 349. SZIJJ, J (1957) *Aquila* 63-4, 71-101. SZIVKA, L (1983) *Larus* 33-5, 141-59. SZYMCZAK, J T (1987a) *J. interdiscip. Cycle Res.* 18, 49-57; (1987b) *J. comp. Physiol. (A) Sens. Neural Behav. Physiol.* 161, 321-7. SZYMCZAK, J T, NAREBSKI, J, and KADZIELA, W (1989) *J. interdiscip. Cycle Res.* 20, 281-8.

TAAPKEN, J (1976) *Vogeljaar* 24, 39. TAAPKEN, J, BLOEM, F, and BLOEM, T (1955a) *Ardea* 43, 145-74; (1955b) *Ardea* 43, 286-9; (1957) *Vår Fågelvärld* 16, 105-12. TACZANOWSKI, L (1873) *J. Orn.* 21, 81-119. TAHON, J, TORREKENS, C, and GIGOT, J (1978) *EPPO Publ.* (B) 84, 83-153. TAIT, W C (1924) *The birds of Portugal.* London. TAITT, M J (1973) *Bird Study* 20, 226-36. TAKEISHI, M (1985) *Orn. Far East Newsl.* 6, 1-2. TALLMAN, D A and ZUSI, R L (1984) *Auk* 101, 155-8. TÁLPEANU, M and PASPALEVA, M (1973) *Oiseaux du delta du Danube.* Bucharest; (1979) *Trav. Mus. Hist. nat. Grigore Antipa* 20, 441-9. TAMM, S (1977) *Behav. Proc.* 2, 293-9. TARANENKO, L I (1979) *Ornitologiya* 14, 198-9. TARASHCHUK, V I (1953) *Ptitsy polezashchitnykh nasazhdeniy.* Kiev. TAST, J (1968) *Ann. zool. fenn.* 5, 159-78; (1970) *Ornis fenn.* 47, 74-82. TAST, J and RASSI, P (1973) *Ornis fenn.* 50, 29-45. TATNER, P (1982a) *J. Zool. Lond.* 197, 559-81; (1982b) *Bird Study* 29, 227-34; (1982c) *Naturalist* 107, 47-58; (1983) *Ibis* 125, 90-107; (1986) *Ring. Migr.* 7, 112-18. TATSCHL, J L (1968) *Auk* 85, 514. TAUCHNITZ, H (1972) *Apus* 2, 245-54. TAVERNER, J H (1958) *Br. Birds* 51, 126. TAVERNER, P A and SUTTON, G M (1934) *Ann. Carnegie Mus.* 23. TAYLOR, D W (1980) *Br. Birds* 73, 39. TAYLOR, D W, DAVENPORT, D L, and FLEGG, J J M (1981) *The birds of Kent.* Meopham. TAYLOR, F (1922) *Br. Birds* 16, 103-4; (1935) *Br. Birds* 29, 102-4. TAYLOR, K (1985) *BTO News* 140, 1; (1986) *BTO News* 142, 9. TAYLOR, M (1987) *The birds of Sheringham.* North Walsham. TAYLOR, W K (1972) *Bird-Banding* 43, 15-19; (1974) *Auk* 91, 485-7; (1976) *Bird-Banding* 47, 72-3. TEBBUTT, C F (1949) *Br. Birds* 42, 242. TEIXEIRA, R M (ed) (1979) *Atlas van de Nederlandse broedvogels.* Deventer. TEKKE, M J (1971) *Limosa* 44, 19-22. TELLERÍA, J L (1981) *La migración de las aves en el Estrecho de Gibraltar* 2. Madrid. TEMMINCK, C (1835) *Manuel d'Ornithologie ou tableau systématique des oiseaux qui se trouvent en Europe.* Paris. TEMPLE LANG, J and DEVILLERS, P (1975) *Gerfaut* 65, 137-52. TENOVUO, R (1963) *Ann. zool. Soc. Vanamo* 25 (5). TENOVUO, R and LEMMETYINEN, R (1970) *Ornis fenn.* 47, 159-66. TERENTIEV, P V (1966) *Trudy Inst. Biol. Sverdlovsk* 51, 35-55; (1970) *Byull. Mosk. Obshch. Ispyt. Prir. Otd. Biol.* 75 (6), 129-134. TERNE, T (1978) *Vår Fågelvärld* 37, 255-6. TERNOVSKI, D V (1954) *Byull. Mosk. Obshch. Ispyt. Prir. Otd. Biol.* 59 (1), 37-40. TERRES, J K (1980) *Encyclopedia of North American birds.* New York. TERRY, J H (1986) *Trans. Herts. nat. Hist. Soc.* 29, 303-12. THALER, E (1977)

*Zool. Garten* (NF) **47**, 241-60. THANNER, R VON (1903) *Orn. Jahrb.* **14**, 211-17; (1910) *Orn. Jahrb.* **21**, 81-101. THESING, G (1987) *Orn. Mitt.* **39**, 320. THÉVENOT, M (1982) *Oiseau* **52**, 21-86, 97-152. THÉVENOT, M, BEAUBRUN, P, BAOUAB, R E, and BERGIER, P (1982) *Docum. Inst. Sci. Rabat* **7**. THÉVENOT, M, BERGIER, P, and BEAUBRUN, P (1981) *Docum. Inst. Sci. Rabat* **6**. THIBAULT, J-C (1983) *Les oiseaux de la Corse.* Ajaccio. THIEDE, W (1982) *Ornis fenn.* **59**, 37-8; (1987) *Orn. Mitt.* **39**, 269-75; (1989) *Orn. Mitt.* **41**, 6-11. THIEDE, W and THIEDE, U (1974) *Vogelwelt* **95**, 88-95. THIELCKE, G (1969) In Hinde, R A (ed.) *Bird vocalizations*, 311-39. Cambridge. THIENEMANN, J (1903) *J. Orn.* **51**, 212-23; (1908) *J. Orn.* **56**, 393-470; (1910) *Orn. Monatsber.* **18**, 66. THIENEMANN, J and SCHÜZ, E (1931) *Vogelzug* **2**, 103-10. THIES, H (1990) *Corax* **13**, 281-308. THIOLLAY, J-M (1985) *Malimbus* **7**, 1-59. THOM, V M (1986) *Birds in Scotland.* Calton. THOMAS, A (1989) In Bignal and Curtis (1989), 23-4. THOMAS, M (1982) *Br. Birds* **75**, 36-7. THOMPSON, C F and FLUX, J E C (1988) *Ornis scand.* **19**, 1-6. THOMPSON, D, EVANS, A, and GALBRAITH, C (1992) *BTO News* **178**, 8-9. THOMPSON, D B A and GRIBBIN, S (1986) *Bull. Br. ecol. Soc.* **17**, 69-75. THOMPSON, D B A and NETHERSOLE-THOMPSON, D (1984) *Br. Birds* **77**, 368. THOMPSON, N S (1969a) *Comm. Behav. Biol.* part A, **3**, 1-5; (1969b) *Comm. Behav. Biol.* Part A, **4**, 269-71; (1982) *Behaviour* **80**, 106-17. THOMPSON, W L (1970) *Auk* **87**, 58-71. THOMPSON, W L and COUTLEE, E L (1963) *Wilson Bull.* **75**, 358-72. THOMSEN, P and JACOBSEN, P (1979) *The birds of Tunisia.* Copenhagen. THOMSON, A L (1949) *Bird migration.* London. THOMSON, A L and SPENCER, R (1954) *Br. Birds* **47**, 361-92. THÖNEN, W (1965) *Orn. Beob.* **62**, 196-7. THÓRISSON, S (1981) *Náttúrufr.* **51**, 145-63. THORNEYCROFT, H B (1966) *Science* **154**, 1571-2; (1975) *Evolution* **29**, 611-21. THORPE, W H (1954) *Nature* **173**, 465-9; (1955) *Ibis* **97**, 247-51; (1956) *Br. Birds* **49**, 389-95; (1958) *Ibis* **100**, 535-70. THOUY, P (1976) *Alauda* **44**, 135-51. TIAINEN, J (1979) In Hildén, O, Tiainen, J, and Valjakka, R (eds) *Muuttolinnut*, 264-71. Helsinki. TIAINEN, J and YLIMAUNU, J (1984) *Lintumies* **19**, 26-9. TICEHURST, C B (1910a) *Br. Birds* **3**, 261-2; (1910b) *Br. Birds* **4**, 70-2; (1915) *Ibis* (10) **3**, 662-9; (1922) *Ibis* (11) **4**, 526-72, 605-62; (1926) *J. Bombay nat. Hist. Soc.* **31**, 368-78; (1926-7) *J. Bombay nat. Hist. Soc.* **31**, 687-711, 862-81, **32**, 64-97; (1932) *A history of the birds of Suffolk.* London; (1940) *Ibis* (14) **4**, 523-5. TICEHURST, C B, BUXTON, P A, and CHEESMAN, R E (1921-2) *J. Bombay nat. Hist. Soc.* **28**, 210-50, 381-427, 650-74, 937-56. TICEHURST, C B and CHEESMAN, R E (1925) *Ibis* (12) **1**, 1-31. TICEHURST, C B, COX, P, and CHEESMAN, R E (1926) *J. Bombay nat. Hist. Soc.* **31**, 91-119. TICEHURST, C B and WHISTLER, H (1933) *Ibis* (13) **3**, 97-112; (1938) *Ibis* (14) **2**, 717-46. TICEHURST, N F (1909) *A history of the birds of Kent.* London. TICHON, M (1989) *Aves* **26**, 57. TIETZE, F (1971) *Falke* **18**, 89-93. TIMMERMANN, G (1938) *Beitr. Fortpfl. Vögel* **14**, 201-6; (1938-49) *Die Vögel Islands.* Reykjavik. TIMMIS, W H (1972) *Avic. Mag.* **78**, 9-11; (1973) *Avic. Mag.* **79**, 3-7. TIMOFÉEFF-RESSOVSKY, N W (1940) *J. Orn.* **88**, 334-40. TINBERGEN, J M (1981) *Ardea* **69**, 1-67. TINBERGEN, L (1934) *Ardea* **23**, 99-100; (1946) *Ardea* **34**, 1-213; (1953) *Br. Birds* **46**, 377. TINBERGEN, N (1939) *Trans. Linn. Soc. New York* **5**. TINNING, P C and TINNING, P A (1970) *Br. Birds* **63**, 83. TINTORI, G (1964) *Nos Oiseaux* **27**, 250-1. TIPPER, R P (1987) *Hong Kong Bird Rep. 1986*, 81-21. TISCHLER, F (1931) *Orn. Monatsber.* **39**, 113-15. TODD, W E C (1963) *Birds of the Labrador peninsula and adjacent areas.* Toronto. TODHUNTER, J F (1987) *Suffolk orn. Group Bull.* **76**, 31-2. TOHMÉ, G and NEUSCHWANDER, J (1978) *Oiseau* **48**, 319-27. TOLSTOY, V A (1986) *Tez. Dokl. 1. S'ezda Vsesoyuz. orn. Obshch. 9. Vsesoyuz. orn. Konf.* **2**, 284-5. TOMBRE-STEEN, I (1991a) *Ornis scand.* **22**, 383-6; (1991b) *Vår Fuglefauna* **14**, 222-3. TOMEK, T and WALIGÓRA, E (1976) *Acta Zool. Cracov.* **21**, 13-30. TOMIAŁOJĆ, L (1967) *Acta Orn.* **10**, 109-56; (1974) *Acta Orn.* **14**, 59-97; (1976a) *Birds of Poland.* Warsaw; (1976b) *Przeglad. Zool.* **20**, 361-4; (1988) *Ring* **12** (134-5), 31; (1990) *Ptaki Polski* 2nd edn. Warsaw. TOMIAŁOJĆ, L and PROFUS, P (1977) *Acta Orn.* **16**, 117-77. TOMIAŁOJĆ, L, WESOŁOWSKI, T, and WALANKIEWICZ, W (1984) *Acta Orn.* **20**, 241-310. TOMKOVICH, P S and MOROZOV, V V (1982) *Ornitologiya* **17**, 173-5. TOMKOVICH, P S and SOROKIN, A G (1983) *Sbor. Trud. Zool. Mus.* **21**, 77-159. TOMLINSON, A G (1917) *J. Bombay nat. Hist. Soc.* **24**, 825-9. TOMPA, F S (1975) *Orn. Beob.* **72**, 181-98; (1976) *Orn. Beob.* **73**, 195-208. TOOK, J M E (1971) *Cyprus orn. Soc. (1957) Rep.* **18**, 40-9; (1973) *Common birds of Cyprus.* Nicosia. TORDOFF, H B (1950) *Wilson Bull.* **62**, 3-4; (1952) *Condor* **54**, 200-3; (1954) *Condor* **56**, 346-58. TORDOFF, H B and DAWSON, W R (1965) *Condor* **67**, 416-22. TORDOFF, H B and MENGEL, R M (1956) *Univ. Kansas Publ. Mus. nat. Hist.* **10**, 1-44. TÖRÖK, J (1990) In Pinowski, J and Summers-Smith, J D (eds) *Granivorous birds in the agricultural landscape*, 199-210. Warsaw. TORRES, J A and LEON, A (1985) *Serv. Publ. Univ. Córdoba España.* TOUPS, J A and JACKSON, J A (1987) *Birds and birding on the Mississippi coast.* Jackson. TOWNSEND, C W (1906) *Auk* **23**, 172-9; (1909) *Auk* **26**, 13-19. TRACY, N (1927) *Br. Birds* **21**, 155. TRANSEHE, N VON (1965) *Die Vogelwelt Lettlands.* Hannover. TRAUTMAN, M B (1940) *Univ. Michigan Mus. Zool. Misc. Publ.* **44**. TRAYLOR, M A (1960) *Nat. Hist. Misc. Chicago Acad. Sci.* **175**, 1-2. TRETTAU, W (1964) *J. Orn.* **105**, 475-82. TRETTAU, W and WOLTERS, H E (1967) *Bonn. zool. Beitr.* **18**, 308-20. TRET'YAKOV, G P (1978) In Kashkarov, D Y (ed.) *Migratsii ptits v Azii*, 126-30. Tashkent. TRICOT, J (1968) *Aves* **5**, 146-56. TRISTRAM, H B (1859) *Ibis* (1) **1**, 22-41; (1864) *Proc. zool. Soc. Lond.*, 444; (1868) *Ibis* (2) **4**, 204-15; (1884) *The survey of western Palestine.* London. TROMMER, G (1971) *Orn. Mitt.* **23**, 170-1. TROTMAN, N (1974) *J. Gloucs. Nat. Soc.* **25**, 358-9. TROTTER, W D C (1970) *Oiseau* **40**, 160-70. TROY, D M (1983) *J. Fld. Orn.* **54**, 146-51; (1984) *Can. J. Zool.* **62**, 2302-6; (1985) *Auk* **102**, 82-96. TROY, D M and BRUSH, A H (1983) *Condor* **85**, 443-6. TROY, D M and SHIELDS, G F (1979) *Condor* **81**, 96-7. TRUSCOTT, B (1944) *Br. Birds* **38**, 74. TRYON, P R and MACLEAN, S F (1980) *Auk* **97**, 509-20. TSCHUSI ZU SCHMIDHOFFEN, V VON (1890) *Orn. Jahrb.* **1**, 65-81. TSUNEKI, K (1960) *Jap. J. Ecol.* **10**, 177-89; (1962) *Mem. Fac. Lib. Arts Fukui Univ. Ser. II Nat. Sci.* **12**, 117-78. TSVELYKH, A N (1993) *Russ. orn. Zh.* **2** (1), 94-6. TUAJEW, D G and WASSILJEW, W I (1974) *Falke* **21**, 18-19. TUCKER, B W (ed.) (1950) *Br. Birds* **43**, 114. TUCKER, M and ROWCLIFFE, J P G (1950) *Br. Birds* **43**, 370. TUCKER, N and TUCKER, L A (1978) *Br. Birds* **71**, 363-4. TUCKER, V R (1980a) *Br. Birds* **73**, 538; (1980b) *Devon Birds* **33**, 55-9. TUFTS, H F (1906) *Auk* **23**, 339-40. TUFTS, R W (1961) *Birds of Nova Scotia.* Halifax; (1986) *Birds of Nova Scotia* 3rd edn. Halifax. TURČEK, F J (1948) *Auk* **65**, 297; (1961) *Ökologische Beziehungen der Vögel und Gehölze.* Bratislava. TURČEK, F J and KELSO, L (1968) *Comm. Behav. Biol.* (A) **1**, 277-97. TURNER, B C (1959a) *Br. Birds* **52**, 129-31; (1959b) *Br. Birds* **52**, 388-

90. Turner, D (1983) *Sunday Express* 21 Aug 1983. Tutman, I (1950) *Larus* 3, 353-60; (1969) *Vogelwelt* 90, 1-8. Tutt, H R (1952) *Ibis* 94, 162-3. Tweedie, M W F (1960) *Common Malayan birds*. London. Tyler, S J (1971) *Br. Birds* 64, 230-1; (1980) *Scopus* 4, 44-5. Tyrberg, T (1987) *Vår Fågelvärld* 46, 375-417; (1988) *Vår Fågelvärld* 47, 378-418; (1990) *Vår Fågelvärld* 49, 389-428; (1991a) *Ornis Svecica* 1, 3-10; (1991b) *Vår Fågelvärld* 50 (6-7), 27-61.

Udvardy, M D F (1956) *Ark. Zool.* (2) 9, 499-505. Uhlig, R (1984) *Beitr. Vogelkde.* 30, 75-6. Ulbricht, H (1975) *Beitr. Vogelkde.* 21, 452-70. Ulfstrand, S (1959) *Vår Fågelvärld* 18, 131-62. Ulfstrand, S and Högstedt, G (1976) *Anser* 15, 1-32. Ulfstrand, S, Roos, G, Alerstam, T, and Österdahl, L (1974) *Vår Fågelvärld* suppl. 8. Ullrich, B (1986) *Orn. Jahresh. Bad.-Württ.* 2, 79-80. Uloth, W (1977) *Falke* 24, 98-9. Umrikhina, G S (1969) *Tez. dokl. V Vsesoyuz. orn. Konf.* 2, 652-5; (1970) *Ptitsy Chuyskoy doliny*. Frunze. Upton, R (1962) *Br. Birds* 55, 592. Urban, E K and Brown, L H (1971) *A checklist of the birds of Ethiopia*. Addis Ababa. Urbánek, B (1959) *Sylvia* 16, 253-61. Uryadova, L P (1986) *Tez. Dokl. 1. S'ezda Vsesoyuz. orn. Obshch 9. Vsesoyuz. Orn. Konf.* 2, 288-90. St Petersburg. Uspenski, S M (1959) *Ornitologiya* 2, 7-15. Ussher, R J (1889) *Zoologist* (3) 13, 180-1.

Vakarenko, V I and Mikhalevich, O A (1986) *Ornitologiya* 21, 150-2. Valeur, P (1946) *Naturen* 9, 270-9. Valverde, J-A (1953) *Nos Oiseaux* 22, 78-82; (1957) *Aves del Sahara Español*. Madrid; (1967) *Monogr. Estac. biol. Doñana* 1, 1-219. Van der Elst, D (1990) *Aves* 27, 73-82. Van der Mueren, E (1980) *Gerfaut* 70, 455-70. Van der Plas, L H W and Wattel, J (1986) *Abstr. Symp. XIX int. orn. Congr. Ottawa*, 516. Van Oss, R M (1950) *Br. Birds* 43, 292-3. Van Tyne, J (1934) *Auk* 51, 529-30. Van Tyne, J and Drury, W H (1959) *Occ. Pap. Mus. Zool. Univ. Michigan* 615. Van Winkel, J (1968) *Wielewaal* 34, 359-60. Vardy, L E (1971) *Condor* 73, 401-14. Varshavski, S N (1977) *Byull. Mosk. Obshch. Ispyt. Prir. Otd. Biol.* 82 (5), 51-7. Varshavski, S N and Shilov, M N (1958) *Zool. Zh.* 37, 1521-30. Vartapetov, L G (1984) *Ptitsy taezhnykh mezhdurechiy Zapadnoy Sibiri*. Novosibirsk. Vásárhelyi, I (1967) *Aquila* 73-4, 196. Vasil'chenko, A A (1982) *Ornitologiya* 17, 130-4. Vaughan, J H (1930) *Ibis* (12) 6, 1-48. Vaughan, R (1953) *Riv. ital. Orn.* 23, 137-42; (1979) *Arctic summer*. Shrewsbury; (1992) *In search of arctic birds*. London. Vauk, G (1961) *Vogelwelt* 82, 179-182; (1964) *Vogelwelt* 85, 113-20; (1968) *Vogelwelt* 89, 142-5; (1970) *Vogelwelt* 91, 11-15; (1972) *Die Vögel Helgolands*. Hamburg; (1973) *Beitr. Vogelkde.* 19, 225-60; (1980) *Z. Jagdwiss.* 26, 93-5. Vaurie, C (1949a) *Amer. Mus. Novit.* 1406; (1949b) *Amer. Mus. Novit.* 1424; (1954a) *Amer. Mus. Novit.* 1658; (1954b) *Amer. Mus. Novit.* 1668; (1954c) *Amer. Mus. Novit.* 1694; (1955) *Amer. Mus. Novit.* 1753; (1956a) *Amer. Mus. Novit.* 1775; (1956b) *Amer. Mus. Novit.* 1786; (1956c) *Amer. Mus. Novit.* 1788; (1956d) *Amer. Mus. Novit.* 1795; (1956e) *Amer. Mus. Novit.* 1805; (1956f) *Amer. Mus. Novit.* 1814; (1957) *Dansk orn. Foren. Tidsskr.* 51, 9-11; (1958a) *Amer. Mus. Novit.* 1898; (1958b) *Ibis* 100, 275-6; (1959) *Birds of the Palearctic fauna: passeriformes*. London; (1972) *Tibet and its birds*. London. Vedum, T V and Tøråsen, A (1988) *Vår Fuglefauna* 11, 83-6. Veger, Z (1968) *Avic. Mag.* 74, 157-9. Veh, M

(1988) *Beih. Veröff. Nat. Land. Bad.-Württ.* 53, 75-82. Veiga, J P (1990) *Anim. Behav.* 39, 496-502. Venables, L S V (1940) *Br. Birds* 33, 334-5; (1949) *Br. Birds* 42, 182. Veprintsev, B N and Zablotskaya, M M (1982) *Akusticheskaya kommunikatsiya tundryanoy chechetki Acanthis hornemanni (Holboell)*. Pushchino. Verbeek, N A M (1972) *J. Orn.* 113, 297-314. Vercauteren, P (1984) *Gerfaut* 74, 327-60. Vere Benson, S (1970) *Birds of Lebanon and the Jordan area*. London. Verheyen, R (1953) *Explor. Parc natn. Upemba Miss. G F de Witte* 19; (1954) *Gerfaut* 44, 324-42; (1955a) *Gerfaut* 45, 5-25; (1955b) *Gerfaut* 45, 173-84; (1956) *Gerfaut* 46, 1-15; (1957a) *Les passereaux de Belgique* 1. Brussels; (1957b) *Gerfaut* 47, 161-70; (1960) *Gerfaut* 50, 101-53. Verheyen, R F (1968) *Gerfaut* 58, 369-93; (1969a) *Gerfaut* 59, 239-59; (1969b) *Gerfaut* 59, 378-84; (1980) In Wright, E N, Inglis, I R, and Feare, C J (eds) *Bird problems in agriculture*, 69-82. Croydon. Veroman, H (1978) *Orn. Kogumik* 8, 253-4. Vertse, A (1943) *Aquila* 50, 142-248. Vertzhutski, B N, Ravkin, Y S, Seryshev, A A, and Verzhutskaya, N V (1979) In Labutin, Y V (ed.) *Migratsii i ekologiya ptits Sibiri*, 127-8. Yakutsk. Vickholm, M, Virolainen, E, and Zetterberg, P (1981) *Ornis fenn.* 58, 133-4. Vidal, A (1991) *Orn. Anz.* 30, 173-5. Vielliard, J (1962) *Oiseau* 32, 74-9. Vierhaus, H and Bruch, A (1963) *J. Orn.* 104, 250. Vieweg, J (1981) *Falke* 28, 205. Vīksne, J (1983) *Ptitsy Latvii*. Riga; (1989) *Latvian breeding bird atlas 1980-84*. Riga. Vincent, A W (1949) *Ibis* 91, 660-88. Vines, G (1981) *Ibis* 123, 190-202. Vinicombe, K E (1988) *Br. Birds* 81, 240-1. Vinogradova, N V, Lyuleeva, D S, Paevski, V A, Popov, E A, and Shumakov, M E (1985) *Trudy Zool. Inst. Akad. Nauk SSSR* 137, 138-54. Vinter, S V and Sokolov, E P (1983) *Trudy Zool. Inst. Akad. Nauk SSSR* 116, 61-71. Virkkala, R (1987) *Ann. zool. fenn.* 24, 281-94; (1988) *Ornis fenn.* 65, 104-13; (1989) *Ann. zool. fenn.* 26, 277-85; (1991) *Ornis fenn.* 68, 193-203. Vleugel, D A (1941) *Ardea* 30, 89-106; (1974) *Alauda* 42, 429-35. Vogt, W (1974) *Orn. Beob.* 71, 320. Voigt, A (1961) *Exkursionsbuch zum Studium der Vogelstimmen* 12th edn. Heidelberg. Voipio, P (1961) *Ornis fenn.* 38, 81-92; (1968) *Ornis fenn.* 45, 10-16; (1969) *Ardea* 57, 48-63. Voisin, R (1963) *Nos Oiseaux* 27, 164-71; (1965) *Nos Oiseaux* 28, 28; (1966) *Bull. Murithienne* 83, 107-12; (1968) *Nos Oiseaux* 29, 286-92. Völker, O (1957) *J. Orn.* 98, 210-14. Volsøe, H (1949) *Dansk orn. Foren. Tidsskr.* 43, 237-42; (1951) *Vidensk. Medd. dansk nat. Foren.* 113. Vondráček, J (1988) *Beih. Veröff. Nat. Land. Bad.-Württ.* 53, 66. Vonk, H and IJzendoorn, E J van (1988) *Dutch Birding* 10, 127-30. Voous, K H (1944) *Ardea* 33, 42-50; (1945) *Limosa* 18, 11-22; (1946) *Gerfaut* 36, 199-202; (1947) *Alauda* 15, 172-6; (1949) *Condor* 51, 52-81; (1950) *Limosa* 23, 281-92; (1951a) *Limosa* 24, 81-91; (1951b) *Limosa* 24, 131-3; (1953) *Beaufortia* 2 (30), 1-41; (1960a) *Limosa* 33, 128-34; (1960b) *Atlas of European birds*. London; (1977) *List of recent Holarctic bird species*. London; (1978) *Br. Birds* 71, 3-10. Vorobiev, G P and Likhatski, Y P (1987) *Ornitologiya* 22, 176-7. Vorobiev, K A (1954) *Ptitsy Ussuriyskogo kraya*. Moscow; (1963) *Ptitsy Yakutii*. Moscow; (1980) *Ornitologiya* 15, 194-6. Vorobiev, V N and Kaganova, O Z (1980) *Ornitologiya* 15, 133-7. Voronkova, K A and Ravkin, E S (1974) *Ornitologiya* 11, 364-6. Vowles, G A and Vowles, R S (1987) *Ring. Migr.* 8, 119-20. Vries, R de (1982) *Seevögel* 3 suppl., 27-33. Vronski, I A (1986) *Tez. Dokl. 1 S'ezda Vsesoyuz. orn. Obshch. 9 Vesesoyuz. orn. Konf.* 1, 136.

VTOROV, P P (1962) *Ornitologiya* 4, 218-33; (1967) *Ornitologiya* 8, 254-61; (1972) *Ornitologiya* 10, 242-7. VTOROV, P P and DROZDOV, N N (1960) *Ornitologiya* 3, 131-8. VUILLEUMIER, F (1977) *Terre Vie* 31, 459-88.

WADE, V E and RYLANDER, M K (1982) *Bird Study* 29, 166. WADEWITZ, O (1976) *Falke* 23, 160-4. WAGNER, U (1981) *Vogelwelt* 102, 32. WAHLMINO, H and PETERSSON, B (1956) *Vår Fågelvärld* 15, 61. WAITE, H W (1948) *J. Bombay nat. Hist. Soc.* 48, 93-117. WAITE, R K (1978) M A Thesis. Keele Univ; (1981) *Z. Tierpsychol.* 57, 15-36; (1984a) *Ornis scand.* 15, 55-62; (1984b) *Behav. Ecol. Sociobiol.* 15, 55-9; (1985a) *Bird Study* 32, 45-9; (1985b) *Proc. int. orn. Congr.* 18, 1189. Moscow; (1986) *Br. Birds* 79, 659-60. WALBRIDGE, G (1978) *Br. Birds* 71, 314-15. WALCHER, A (1918) *Orn. Jahrb.* 29, 51-5. WALFORD, N T (1930) *Br. Birds* 24, 51. WALICZKY, Z, MAGYAR, G, and HRASKÓ, G (1983) *Aquila* 90, 73-9. WALKER, B (1990) *Ontario Birds* 7, 86-7. WALKER, F J (1981a) *Sandgrouse* 2, 33-55; (1981b) *Sandgrouse* 2, 56-85. WALKINSHAW, L H (1948) *Condor* 50, 64-70. WALLACE, D I M (1957) *Br. Birds* 50, 208-9; (1976a) *Br. Birds* 69, 27-33; (1976b) *Br. Birds* 69, 465-73; (1982a) *Br. Birds* 75, 291; (1982b) *Sandgrouse* 4, 77-99; (1983) *Sandgrouse* 5, 1-18; (1984) *Sandgrouse* 6, 24-47. WALLACE, D I M, COBB, F K, and TUBBS, C R (1977) *Br. Birds* 70, 45-9. WALLER, C S (1970) *Br. Birds* 63, 147-9. WALLGREN, H (1954) *Acta zool. fenn.* 84; (1956) *Acta Soc. Fauna Flora fenn.* 71 (4). WALLIN, L (1966) *Vår Fågelvärld* 25, 327-45. WALLIS, H M (1887) *Ibis* (5) 5, 454-5; (1912) *Bull. Br. Orn. Club* 29, 83; (1932) *Bull. Br. Orn. Club* 52, 38-9. WALLSCHLÄGER, D (1983) *Mitt. zool. Mus. Berlin* 59, Suppl. Ann. Orn. 7, 85-116. WALPOLE-BOND, J (1905) *Countryside* 12 Aug; (1932) *Br. Birds* 25, 292-300; (1938) *A history of Sussex birds.* London. WALSH, T A (1976) *Br. Birds* 69, 222. WALTER, H (1965) *J. Orn.* 106, 81-105. WALTER, H and DEMARTIS, A M (1972) *J. Orn.* 113, 391-406. WALTERS, J (1988) *Limosa* 61, 33-40. WALTERS, P M and LAMM, D W (1980) *N. Am. Bird Bander* 5, 15. WARD, N (1987) *Br. Birds* 80, 500-2. WARD, P (1977) *Ann. Rep. Inst. terr. Ecol.* Cambridge, 54-6. WARD, P and POH, G E (1968) *Ibis* 110, 359-63. WARD, P and ZAHAVI, A (1973) *Ibis* 115, 517-34. WARMBIER, N (1973) *Falke* 20, 67. WARNCKE, K (1960) *Vogelwelt* 81, 178-84; (1968) *J. Orn.* 109, 300-2. WARNES, J M (1983) *Scott. Birds* 12, 238-46. WARNES, J M and STROUD, D A (1988) In Bignal, E and Curtis, D J (eds) *Choughs and land-use in Europe,* 46-51. Scottish Chough Study Group. WARREN, D R (1974) *Br. Birds* 67, 440. WARRILOW, G J, FOWLER, J A, and FLEGG, J J M (1978) *Ring. Migr.* 2, 34-7. WASHINGTON, D (1974) *Br. Birds* 67, 213-14. WASSENICH, V (1969) *Regulus* 9, 362-70; (1973) *Regulus* 11, 55. WASSMANN, R (1990) *Vogelkdl. Ber. Niedersachs.* 22, 48. WASYLIK, A and PINOWSKI, J (1970) *Bull. Acad. Pol. Sci.* 18, 29-32. WATERHOUSE, M J (1949) *Ibis* 91, 1-16. WATSON, A (1957a) *Can. Fld.-Nat.* 71, 87-109; (1957b) *Sterna* 2, 65-99; (1963) *Arctic* 16, 101-8; (1989) *Scott. Birds* 15, 178-9; (1992a) *Scott. Birds* 16, 273-5; (1992b) *Scott. Birds* 16, 287. WATSON, A and O'HARE, P J (1980) *Irish Birds* 1, 487-91. WATSON, A and SMITH, R (1991) *Scott. Birds* 16, 53-6. WATSON, D (1972) *Birds of moor and mountain.* Edinburgh. WATSON, G E (1960) *Postilla* 52, 1-15; (1964) Ph D Thesis. Yale Univ. WATT, D J (1986) *J. Fld. Orn.* 57, 105-13. WATT, D J, RALPH, C J, and ATKINSON, C T (1984) *Auk* 101, 110-20. WATTEL, J (1971) *Ostrich* 42, 229. WEAVER, R L (1942) *Wilson Bull.*

54, 183-91; (1943) *Auk* 60, 62-74. WEBER, C (1990a) *Limicola* 4, 222-8; (1990b) *Limicola* 4, 276-84; (1991) *Limicola* 5, 92. WEBER, H (1954) *Orn. Mitt.* 6, 168-70; (1959) *Beitr. Vogelkde.* 6, 351-6; (1971-2) *Falke* 18, 306-14, 19, 16-27. WECHSLER, B (1988a) *Behaviour* 106, 252-64; (1988b) *Behaviour* 107, 267-77; (1989) *Ethology* 80, 307-17. WEHRLE, C M (1989) *Orn. Beob.* 86, 53-68. WEIGOLD, H (1926) *Wiss. Meeresunters.* (NF) 15 (3) article 17. WEINZIERL, H (1961) *Orn. Mitt.* 13, 153. WEISE, C M (1962) *Auk* 79, 161-72. WEISE, R (1992) *Vogelwelt* 113, 47-51. WEISS, I and WIEHE, H (1984) *Orn. Mitt.* 36, 162. WENDLAND, V (1958) *J. Orn.* 99, 203-8. WERNICKE, P (1990) *Beitr. Vogelkde.* 36, 1-9. WERNLI, W (1970) *Vögel der Heimat* 40, 93-109. WEST, D A (1962) *Auk* 79, 399-424. WEST, G C, PEYTON, L J, and SAVAGE, S (1968) *Bird-Banding* 39, 51-5. WEST, M J and KING, A P (1990) *Amer. Sci.* 78, 107-14. WEST, M J, STROUD, A N, and KING, A P (1983) *Wilson Bull.* 95, 635-40. WESTERFRÖLKE, P (1958) *Vogelwelt* 79, 117. WESTERNHAGEN, W VON (1956) *Orn. Mitt.* 8, 169. WESTERSKOV, K (1953) *Notornis* 5, 189-91. WESTERTERP, K (1973) *Ardea* 61, 137-58. WESTERTERP, K, GORTMAKER, W, and WIJNGAARDEN, H (1982) *Ardea* 70, 153-62. WESTIN, L (1973) *Vår Fågelvärld* 32, 44. WESTPHAL, D (1976) *J. Orn.* 117, 70-4; (1981) *Vogelwarte* 31, 94-101. WETHERBEE, O P (1937) *Bird-Banding* 8, 1-10. WETMORE, A (1936) *Smithson. Misc. Coll.* 95 (17); (1949) *J. Washington Acad. Sci* 39, 137-9. WETTON, J H, CARTER, R E, PARKIN, D T, and WALTERS, D T (1987) *Nature* 327, 147-9. WETTON, J H and PARKIN, D T (1991a) *Proc. Roy. Soc. Lond. Biol. Sci.* 245, 227-33; (1991b) *Proc. int. orn. Congr.* 20, 2435-41. WETTSTEIN, O (1959) *J. Orn.* 100, 103-4. WHISTLER, H (1922a) *Ibis* (11) 4, 259-309; (1922b) *J. Bombay nat. Hist. Soc.* 28, 990-1006; (1923) *Ibis* (11) 5, 611-29; (1924) *J. Bombay nat. Hist. Soc.* 30, 177-88; (1941) *Popular handbook of Indian birds.* London; (1945) *J. Bombay nat. Hist. Soc.* 45, 106-22. WHISTLER, H and HARRISON, J M (1930) *Ibis* (12) 6, 453-70. WHITAKER, B (1955) *Br. Birds* 48, 145-6. WHITAKER, J I S (1894) *Ibis* (6) 6, 78-100; (1898) *Ibis* (7) 4, 592-610. WHITAKER, L M (1957) *Wilson Bull.* 69, 195-262. WHITE, C M and WEST, G C (1977) *Oecologia* 27, 227-38. WHITE, C M N (1960) *Ibis* 102, 138-9; (1967) *Bull. Br. Orn. Club* 87, 62-3. WHITE, S J and HINDE, R A (1968) *J. Zool. Lond.* 155, 145-55. WHITEHEAD, C H T (1909) *Ibis* (9) 3, 214-84; (1911) *J. Bombay nat. Hist. Soc.* 20, 776-99. WHITELEY, J D, PRITCHARD, J S, and SLATER, P J B (1990) *Bird Study* 37, 12-17. WHITNEY, B (1983) *Birding* 15, 219-22. WHITTAKER, A (1990) *Br. Birds* 83, 73. WHITTLE, C L (1938) *Bird-Banding* 9, 196-7. WICKLER, W (1961) *Z. Tierpsychol.* 18, 320-42; (1982) *Auk* 99, 590-1. WIECZOREK, P (1975) *Falke* 22, 282. WIEDENFELD, D A (1991) *Condor* 93, 712-23. WIEHE, H (1988) *Orn. Mitt.* 40, 252; (1990) *Orn. Mitt.* 42, 294-6. WIELOCH, M (1975) *Pol. ecol. Stud.* 1, 227-42. WIENS, J A and DYER, M I (1977) In Pinowski, J and Kendeigh, S C (eds) *Granivorous birds in ecosystems,* 205-66. Cambridge. WILDASH, P (1968) *Birds of South Vietnam.* Tokyo. WILD BIRD SOCIETY OF JAPAN (1982) *A field guide to the birds of Japan.* Tokyo. WILDE, J (1962) *Br. Birds* 55, 560-2. WILDER, G D and HUBBARD, H W (1924) *J. N. China Branch Roy. Asiatic Soc.* 55, 156-239; (1938) *Birds of north-eastern China.* Peking. WILKINSON, D M (1988) *Br. Birds* 81, 657-8. WILKINSON, R (1975) Ph D Thesis. Southampton Univ; (1980) *Z. Tierpsychol.* 54, 436-56; (1982) *Ornis scand.* 13, 117-22; (1990) *Bioacoustics* 2, 179-97. WILKINSON, R and HOWSE, P E (1975) *Z. Tierpsychol.*

38, 200-11. WILLCOX, D R C and WILLCOX, B (1978) *Ibis* 120, 329-33. WILLE, H-G (1983) *Orn. Mitt.* 35, 269-73. WILLIAMS, C H and WILLIAMS, C E (1929) *J. Bombay nat. Hist. Soc.* 33, 598-613. WILLIAMS, J G (1941) *Ibis* (14) 5, 245-64; (1963) *A field guide to the birds of East and Central Africa*. London. WILLIAMS, J G and ARLOTT, N (1980) *A field guide to the birds of East Africa*. London. WILLIAMS, L P (1986) *Br. Birds* 79, 423-6. WILLIAMS, M D (1986) (ed.) *Rep. Cambridge orn. Exped. China 1985*. WILLIAMS, M D, CAREY, G J, DUFF, D G, and WEISHU, X (1992) *Forktail* 7, 3-55. WILLIAMS, T D, REED, T M, and WEBB, A (1986) *Scott. Birds* 14, 57-60. WILLIAMS, T S (1946) *Br. Birds* 39, 149-50. WILLIAMS, W M (1989) *Devon Birds* 42, 49-50. WILLIAMSON, F S L and RAUSCH, R (1956) *Condor* 58, 165. WILLIAMSON, K (1939) *Br. Birds* 33, 78; (1953) *Scott. Nat.* 65, 65-94; (1955) *Fair Isle Bird Obs. Bull.* 2, 327-8; (1956) *Dansk orn. Foren. Tidsskr.* 50, 125-33; (1957) *Scott. Nat.* 69, 190-2; (1959) *Peregrine* 3, 8-14; (1961a) *Bird Migr.* 1, 235-40; (1961b) *Bird Migr.* 2, 43-5; (1961c) *Br. Birds* 54, 238-41; (1962) *Br. Birds* 55, 130-1; (1963a) *Bird Migr.* 2, 207-23; (1963b) *Bird Migr.* 2, 252-60, 329-40; (1965) *Fair Isle and its birds*. Edinburgh; (undated) *Observations of the Chough*. Mona's Herald Ltd; (1968) *Q. J. Forestry* 62, 118-31. WILLIAMSON, K and DAVIS, P (1956) *Br. Birds* 49, 6-25. WILLIAMSON, K and SPENCER, R (1960) *Bird Migr.* 1, 176-81. WILLIAMSON, P and GRAY, L (1975) *Condor* 77, 84-9. WILLOUGHBY, E J (1992) *Condor* 94, 295-7. WILMORE, S B (1977) *Crows, jays, ravens and their relatives*. Newton Abbot. WILSON, C W (1883) *Ibis* (5) 1, 575-7. WILSON, P R (1965) B Sc Hons Thesis. Victoria Univ; (1973) Ph D Thesis. Victoria Univ. WILSON, R T (1981) *African J. Ecol.* 19, 285-94. WIMAN, C (1943) *Vår Fågelvärld* 2, 94-5. WINDSOR, R E (1935) *Br. Birds* 29, 126. WINGATE, D B (1958) *Auk* 75, 359-60; (1973) *A checklist and guide to the birds of Bermuda*. Bermuda. WINKELMAN, J E (1981) *Limosa* 54, 81-8. WINKLER, H (1979a) *Egretta* 22, 29-30. WINKLER, K (1979b) *Gef. Welt* 103, 201-3; (1992) *Gef. Welt* 116, 156-7. WINKLER, R (1975) *Bull. Murithienne (Sion)* 92, 48; (1984) *Orn. Beob. Suppl.* 5. WINKLER, R, DAUNICHT, W D, and UNDERHILL, L G (1988) *Orn. Beob.* 85, 245-59. WINKLER, R and JENNI, L (1987) *J. Orn.* 128, 243-6. WINKLER, R and WINKLER, A (1985) *Orn. Beob.* 82, 55-66; (1986) *Orn. Beob.* 83, 76. WINTERBOTTOM, J M (1962) *Ostrich* 33 (2), 43-50; (1975) *Ostrich* 46, 236-50. WIPRÄCHTIGER, P (1987) *Vögel Heimat* 57, 187-8. WIRDHEIM, A and CARLÉN, T (1986) *Fågelstråk*. Halmstad. WITCHELL, C A (1896) *The evolution of bird-song*. London. WITHERBY, H F (1901) *Ibis* (8) 1, 237-78; (1903) *Ibis* (8) 3, 501-71; (1908) *Bull. Br. Orn. Club* 23, 48; (1910) *Ibis* (9) 4, 491-517; (1913) *Br. Birds* 7, 126-39; (1915) *Bull. Br. Orn. Club* 36, 3-4; (1928) *Ibis* (12) 4, 385-436. WITHERBY, H F, JOURDAIN, F C R, TICEHURST, N F, and TUCKER, B W (1938) *The handbook of British birds* 1. London. WITSCHI, E (1936) *Proc. Soc. exp. Biol. Med.* 33, 484-6. WITSCHI, E and MILLER, R A (1938) *J. exp. Zool.* 79, 475-87. WITSCHI, E and WOODS, R P (1936) *J. exp. Zool.* 73, 445-59. WITT, K (1988) *Orn. Ber. Berlin (West)* 13, 119-55; (1989) *Vogelwelt* 110, 142-50. WITTENBERG, G (1970) *Orn. Mitt.* 22, 129-31. WITTENBERG, J (1968) *Zool. Jb. Syst.* 95, 16-146; (1976) *Vogelwarte* 28, 230-2; (1987) *Verh. naturwiss. Ver. Hamburg (NF)* 29, 5-49; (1988) *Beih. Veröff. Nat. Land. Bad.-Württ.* 53, 109-18. WOHL, E (1980) *Mitt. Abt. Zool. Landesmus. Joanneum Graz* 9, 137-40; (1981) *Mitt. Abt. Zool. Landesmus. Joanneum Graz* 10, 81-4; (1985) *Mitt.*

*Abt. Zool. Landesmus. Joanneum Graz* 34, 65-8. WOLFSON, A (1942) *Condor* 44, 237-63; (1945) *Condor* 47, 95-127; (1954a) *Auk* 71, 413-34; (1954b) *J. exp. Zool.* 125, 353-76; (1954c) *Wilson Bull.* 66, 112-18. WOLTERS, H E (1952) *Bonn. zool. Beitr.* 3, 231-88; (1957) *Bonn. zool. Beitr.* 8, 90-129; (1958) *Bonn. zool. Beitr.* 9, 200-7; (1962) *Bonn. zool. Beitr.* 13, 324-6; (1968) *Bonn. zool. Beitr.* 19, 157-64. WON, P-O (1961) *Avi-mammalian Fauna of Korea* 12, 31-139. WON, P-O, WOO, H-C, CHUN, M-Z, and HAM, K-W (1966) *Misc. Rep. Yamashina Inst. Orn.* 4, 445-68. WONG, M (1983) *Wilson Bull.* 95, 287-94. WONTNER-SMITH, C (1939) *Br. Birds* 33, 194. WOOD, D L and WOOD, D S (1972) *Bird-Banding* 43, 182-90. WOOD, H B (1945) *Auk* 62, 455-6. WOODCOCK, M (1980) *Collins handguide to the birds of the Indian sub-continent*. London. WOODFORD, J and LOVESY, F T (1958) *Bird-Banding* 29, 109-10. WOODS, H E (1950) *Br. Birds* 43, 82-3; (1975) *Bird-Banding* 46, 344-6. WOODWARD, I D (1960a) *Nat. Wales* 6, 26-7; (1960b) *Devon Birds* 13, 23-5. WORKMAN, W B (1961) *Br. Birds* 54, 250-1; (1963) *Br. Birds* 56, 52-3. WORTELAERS, F (1950) *Gerfaut* 40, 207-12. WOTKYNS, D B (1962) *Audubon Mag.* 64, 235. WRIGHT, M (1972) *Br. Birds* 65, 260-1. WUNSCH, H (1976) *Gef. Welt* 100, 42-4. WÜST, W (1961) *Anz. orn. Ges. Bayern* 6, 91-2; (1986) *Avifauna Bavariae* 2. Munich. WYDOSKI, R S (1964) *Auk* 81, 542-50. WYNNE-EDWARDS, V C (1927) *Br. Birds* 21, 229-30; (1952) *Auk* 69, 353-91; (1962) *Animal dispersion in relation to social behaviour*. Edinburgh.

XIMENIS, J A (1977) *Ardeola* 22, 111.

YABLONKEVICH, M L, BARDIN, A V, BOL'SHAKOV, K V, POPOV, E A, and SHAPOVAL, A P (1985a) *Trudy Zool. Inst. Akad. Nauk SSSR* 137, 69-97. YABLONKEVICH, M L, BOL'SHAKOV, K V, BULYUK, V N, ELISEEV, D O, EFREMOV, V D, and SHAMURADOV, A K (1985b) *Trudy Zool. Inst. Akad. Nauk SSSR* 137, 11-59. YAKOBI, V E (1979) *Zool. Zh.* 58, 136-7. YAMASHINA, Y (1982) *Birds in Japan*. Tokyo. YANG, S H and SELANDER, R K (1968) *Syst. Zool.* 17, 107-43. YANUSHEVICH, A I, TYURIN, P S, YAKOVLEVA, I D, KYDYRALIEV, A, and SEMENOVA, N I (1960) *Ptitsy Kirgizii* 2. Frunze. YAPP, W B (1951) *J. Anim. Ecol.* 20, 169-72; (1962) *Birds and woods*. London; (1975) *Br. Birds* 68, 342. YARBROUGH, C G, and JOHNSTON, D W (1965) *Wilson Bull.* 77, 175-91. YEATES, G K (1932) *Br. Birds* 26, 30-3; (1934) *The life of the Rook*. London; (1951) *The land of the loon*. London. YEATMAN, L J (1971) *Histoire des oiseaux d'Europe*. Paris; (1976) *Atlas des oiseaux nicheurs de France*. Paris. YEATMAN-BERTHELOT, D (1991) *Atlas des oiseaux de France en hiver*. Paris. YEO, P F (1947) *Br. Birds* 40, 211-12. YEO, V Y Y (1990) B Sc Hons Thesis. Singapore Univ. YÉSOU, P (1983) *Alauda* 51, 161-78. YLIMAUNU, J, YLIMAUNU, O, and YLIPEKKALA, J (1986) *Lintumies* 21, 98-9. YOM-TOV, Y (1974) *J. Anim. Ecol.* 43, 479-98; (1975a) *Bird Study* 22, 47-51; (1975b) *Auk* 92, 778-85; (1976) *Behaviour* 59, 247-51; (1980a) *Ibis* 122, 234-7; (1980b) *Teva Va'aretz* 22, 98-101; (1992) *Bird Study* 39, 111-14. YOM-TOV, Y and AR, A (1980) *Israel J. Zool.* 29, 171-87. YOM-TOV, Y, AR, A, and MENDELSSOHN, H (1978) *Condor* 80, 340-3. YOM-TOV, Y, DUNNET, G M, and ANDERSON, A (1974) *Ibis* 116, 87-90. YOSEF, R (1991) *Sandgrouse* 13, 73-9. YOSEF, R and YOSEF, D (1991) *Wilson Bull.* 103, 518-20. YOUNG, B E (1991) *Condor* 93, 236-50. YOUNG, J G (1984) *Scott. Birds* 13, 88. YOUNG, S (1990) *Br. Birds* 83, 508-9. YTREBERG, N-J (1972) *Norw. J. Zool.* 20, 61-

89. YUNICK, R P (1972) *Bird-Banding* **43**, 38-46; (1976) *Bird-Banding* **47**, 276-7; (1977*a*) *N. Amer. Bird Bander* **2**, 12-13; (1977*b*) *N. Amer. Bird Bander* **2**, 155-6; (1981) *N. Amer. Bird Bander* **6**, 97; (1984) *N. Amer. Bird Bander* **9**, 2-4, 6. YURLOV, K T, CHERNYSHOV, V M, KOSHELEV, A I, SAGITOV, R A, TOTUNOV, V M, KHODKOV, G I, and YURLOV, A K (1977) In Yurlov, K T (ed.) *Migratsii ptits v Azii*, 205-9. Novosibirsk.

ZABLOTSKAYA, M M (1975) *Byull. Mosk. Obshch. Ispyt. Prir. Otd. Biol.* **53** (3), 22-38; (1976*a*) *Dokl. Uchast. 2. Vsesoyuz. Konf. Poved. Zhiv.*, 125-7; (1976*b*) *Dokl. Uchast. 2. Vsesoyuz. Konf. Poved. Zhiv.*, 127-9; (1976*c*) *Dokl. Uchast. 2. Vsesoyuz. Konf. Poved. Zhiv.*, 130-1; (1978*a*) *Zool. Zh.* **57**, 105-13; (1978*b*) *Byull. Mosk. Obshch. Ispyt. Prir. Otd. Biol.* **83** (4), 36-54; (1981) *Akusticheskaya kommunikatsiya obyknovennoy chechetki Acanthis flammea flammea* (*L*). Pushchino; (1982) *Akusticheskaya kommunikatsiya konoplyanki Acanthis cannabina* (*L*). Pushchino. ZAHARONI, M (1991) *Israel Bird Ring. Cent. Ringer's Newsl.* **3**, 21. ZARUDNYI. N A (1911) *Orn. Vestnik.* **3-4**, 298-306 (1916) *Nasha Okhota* **20**, 37-8. ZEDLITZ, O VON (1911) *J. Orn.* **59**, 1-92; (1912) *J. Orn.* **60**, 325-64, 529-69; (1921) *Flora och Fauna* **16**, 275-80; (1925) *Flora och Fauna* **20**, 145-73; (1926) *J. Orn.* **74**, 296-308. ZEIDLER, K (1966) *J. Orn.* **107**, 113-53. ZHORDANIA, R G and GOGILASHVILI, G S (1976) *Acta Orn.* **15**, 323-38. ZHURAVLEV, M N and AFONIN, P V (1982) *Ornitologiya* **17**, 182. ZIEGER, R (1967) *Beitr. Vogelkde.* **13**, 117-24. ZIMIN, V B (1981) In Ivanter, E V (ed.) *Ekologiya nazemnykh pozvonochnykh*

*Severo-zapada SSSR*, 13-31. Petrozavodsk. ZIMKA, J (1968) *Acta Orn.* **11**, 87-102. ZIMMERLI, E (1986) *Vogel Heimat* **56**, 211-12. ZIMMERMAN, J L (1965) *Wilson Bull.* **77**, 55-70. ZIMMERMANN, D (1951) *Orn. Beob.* **48**, 73-111; (1987) *Orn. Beob.* **84**, 66. ZIMMERMANN, R (1907) *Z. Orn. prakt. Geflügelz.* **31**, 2-4, 17-18; (1913) *Orn. Monatsber.* **21**, 112-14; (1931) *Orn. Monatsber.* **39**, 99-102. ZINK, G (1969) *Auspicium* **3**, 195-291; (1981) *Der Zug europäischer Singvögel* **1**; (1985) **4**. Möggingen. ZINK, R M (1982) *Auk* **99**, 632-49. ZINK, R M, DITTMANN, D L, and ROOTES, W L (1991*a*) *Auk* **108**, 578-84. ZINK, R M, DITTMANN, D L, CARDIFF, S W, and RISING, J D (1991*b*) *Condor* **93**, 1016-19. ZINK R M, and KLICKA, J T (1990) *Wilson Bull.* **102**, 514-20. ZINK, R M, ROOTES, W L, and DITTMANN, D L (1991*c*) *Condor* **93**, 318-29. ZINNENLAUF, B (1967) *Orn. Beob.* **64**, 113-18. ZINO, P A (1969) *Bocagiana* **21**, 1-7. ZISWILER, V (1959) *Vjschr. Naturf. Ges. Zürich* **104**, 222-6; (1965) *J. Orn.* **106**, 1-48; (1967*a*) *Orn. Beob.* **64**, 105-10; (1967*b*) *Rev. suisse Zool.* **74** 620-8; (1967*c*) *Zool. Jb. Syst.* **94**, 427-520; (1979) *Rev. suisse Zool.* **86**, 823-31. ZLOTIN, R I (1968) *Ornitologiya* **9**, 158-63. ZONFRILLO, B (1988) *Bull. Br. Orn. Club* **108**, 71-5. ZONOV, G B (1978) In Tolchina, S N (ed.) *Rol' ptits v biotsenozakh Vostochnoy Sibiri*, 168-82. ZUCCHI, H (1975) *Orn. Mitt.* **27**, 171-2. ZUMSTEIN, F (1921) *Verh. orn. Ges. Bayern* **15**, 68-73; (1927) *Beitr. Fortpfl. Vögel* **3**, 181-4. ZUÑIGA, J M (1989) In Bignal, E and Curtis, D J (eds) *Choughs and land-use in Europe*, 65-9. Scottish Chough Study Group. ZYKOVA, L Y and PANOV, E N (1982) *Zool. Zh.* **61**, 1113-16.

# CORRECTIONS

## CORRECTIONS TO VOLUME I

Page 494. *Anas formosa* Baikal Teal. **Distribution.** Column 2, line 1. Delete 'Finland,'.

## CORRECTIONS TO VOLUME IV

Page 129. *Sterna albifrons* Little Tern. **Breeding.** Column 1, line 16. Amend to read '. . . mean 2·0 (N W Orr, E J Wiseman). . .'

## CORRECTIONS TO VOLUME V

Page 236. *Riparia riparia* Sand Martin. Map. Winter range in South America has been omitted.

## CORRECTIONS TO VOLUME VI

Page 168. *Acrocephalus brevipennis* Cape Verde Cane Warbler. German name should be 'Dornrohrsänger'.

Page 168. *Acrocephalus brevipennis* Cape Verde Cane Warbler. **Habitat.** Column 1, last 2 lines. Amend to read '. . . drier areas, but occurs up to *c.* 600 m above sea-level (C J Hazevoet). Like. . .'

Page 338. *Sylvia conspicillata* Spectacled Warbler. West Palearctic map. In Cape Verde Islands, Boavista (large eastern island) should be red and grey.

Page 491. *Sylvia borin* Garden Warbler. **Voice.** Column 1, line 15. Amend to read '. . . Hazevoet 1984. . .'

Page 497. *Sylvia atricapilla* Blackcap. **Distribution.** Line 12. Amend to read '. . . except Sal, Boavista, and Maio; has almost. . .'

Page 509. *Sylvia atricapilla* Blackcap. **Voice.** Column 2, last line. Amend to read '. . . one from France (Fig I). . .'

Page 667. *Regulus calendula* Ruby-crowned Kinglet. **Habitat.** Line 2. Amend to read 'and Rockwell 1939). . .'

Page 687. *Regulus ignicapillus* Firecrest. Map. Madeira should be red and grey.

Page 716. References. Last line. Amend to read '. . . 1-153. Voous, K H (1955). . .'

Page 723. Corrections. Column 1, last line. To read '. . . Myres 1964; insert on p. 1003, column 1, line 11).

Plate 9. Caption. Last line to read '. . . 244): 7 juv/1st ad. . .'

## CORRECTIONS TO VOLUME VII

Page 12. *Muscicapa striata* Spotted Flycatcher. **Distribution.** Accidental: add Cape Verde Islands.

Page 251. *Parus cyanus* Azure Tit. **Social pattern and behaviour.** Line 24. Amend to read '. . . Frank and Voous 1969;. . .'

Page 257. *Parus major* Great Tit. World map. Range shown incorporates Turkestan Tit *P. bokharensis*, which should be excluded (see Geographical Variation, p. 280).

Page 288. *Sitta whiteheadi* Corsican Nuthatch. Map. Breeding range should be shown, coincident with winter range.

Page 292. *Sitta ledanti* Algerian Nuthatch. German name should be 'Kabylenkleiber'.

Page 301. *Sitta europea* Nuthatch. **Movements.** Paragraph 2. Delete last sentence ('Complete vacation. . .'); results from typographical error in original source.

Page 333. *Tichodroma muraria* Wallcreeper. **Social pattern and behaviour.** Column 1, lines 18-19 from foot of page. Amend to read 'of 17 deliveries of food seen during 1 hour's observations at nest . . .'

Page 348. *Certhia familiaris* Treecreeper. World map. Range in North and Central America has been omitted.

Page 428. *Oriolus oriolus* Golden Oriole. **Voice.** Column 2, line 3: amend to read '. . . *glandarius* (Fig VII).'

Page 444. *Lanius cristatus* Brown Shrike. **Distribution.** Paragraph 2, line 2. Amend to read 'October 1985 (Rogers. . .'

Page 502. *Lanius excubitor* Great Grey Shrike. Map. Range in North America has been omitted.

Page 518. *Lanius excubitor* Great Grey Shrike. **Breeding.** Last line. Amend to read 'declined through season (Yosef 1989).'

References. Add:
SELL, M (1974) *Alcedo* 1, 1-15.
SLUYS, R (1982) *J. Orn.* **123**, 175-82.

Plate 9. Caption. Amend to read '. . . Long-tailed Tit (p. 133). . .'

Plate 19. Caption. Line 2: amend to read 'Penduline Tit (p. 377). . .'

# INDEXES

Figures in **bold type** refer to plates

## SCIENTIFIC NAMES

*Acridotheres tristis*, 280, **17**, **18**; eggs, **59**
*albus* (*Corvus*), 195, **14**, **15**
*amandava* (*Amandava*), 427, **31**; eggs, **60**
*Amandava amandava*, 427, **31**; eggs, **60**
*astrild* (*Estrilda*), 420, **31**; eggs, **60**

*bengalus* (*Uraeginthus*), 420
*brachydactyla* (*Carpospiza*), 357, **24**, **27**; eggs, **59**
*Bucanetes githagineus*, 754, **47**, **50**; eggs, **61**
　*mongolicus*, 748, **47**, **50**

*canaria* (*Serinus*), 528, **34**, **39**; eggs, **60**
*cannabina* (*Carduelis*), 604, **40**, **44**; eggs, **60**
*cantans* (*Euodice*), 437, **31**
*carduelis* (*Carduelis*), 568, **36**, **39**; eggs, **60**
*Carduelis cannabina*, 604, **40**, **44**; eggs, **60**
　*carduelis*, 568, **36**, **39**; eggs, **60**
　*chloris*, 548, **35**, **39**; eggs, **60**
　*flammea*, 639, **42**, **44**; eggs, **60**
　*flavirostris*, 625, **41**, **44**; eggs, **60**
　*hornemanni*, 661, **42**, **45**; eggs, **60**
　*pinus*, 604
　*spinus*, 587, **37**, **39**; eggs, **60**
*Carpodacus erythrinus*, 764, **48**, **50**; eggs, **61**
　*roseus*, 789, **49**, **51**
　*rubicilla*, 792, **49**, **51**; eggs, **61**
　*synoicus*, 783, **48**, **51**; eggs, **61**
*Carpospiza brachydactyla*, 357, **24**, **27**; eggs, **59**
*caryocatactes* (Nucifraga), 76, **4**, **5**; eggs, **57**
*chloris* (*Carduelis*), 548, **35**, **39**; eggs, **60**
*citrinella* (*Serinus*), 536, **34**, **39**; eggs, **60**
*coccothraustes* (*Coccothraustes*), 832, **53**, **56**; eggs, **61**
*Coccothraustes coccothraustes*, 832, **53**, **56**; eggs, **61**
*coelebs* (*Fringilla*), 448, **29**, **38**; eggs, **60**
*corax* (*Corvus*), 206, **11**, **15**; eggs, **58**
*corone* (*Corvus*), 172, **14**, **15**; eggs, **58**
Corvidae, 5
*Corvus albus*, 195, **14**, **15**
　*corax*, 206, **11**, **15**; eggs, **58**
　*corone*, 172, **14**, **15**; eggs, **58**

*dauuricus*, 140, **8**, **9**
*frugilegus*, 151, **14**, **15**; eggs, **58**
*monedula*, 120, **7**, **9**; eggs, **57**
*rhipidurus*, 223, **10**, **15**; eggs, **58**
*ruficollis*, 197, **10**, **15**; eggs, **58**
*splendens*, 143, **8**, **15**; eggs, **57**
*cucullatus* (*Ploceus*), 401
*curvirostra* (*Loxia*), 686, **43**, **45**; eggs, **61**
*cyana* (*Cyanopica*), 42, **2**, **5**; eggs, **57**
*Cyanopica cyana*, 42, **2**, **5**; eggs, **57**

*dauuricus* (*Corvus*), 140, **8**, **9**
*domesticus* (*Passer*), 289, **20**, **25**, **26**; eggs, **59**

*enucleator* (*Pinicola*), 802, **51**, **52**; eggs, **61**
*Eophona migratoria*, 832
　*personata*, 832
*erythrinus* (*Carpodacus*), 764, **48**, **50**; eggs, **61**
*Estrilda astrild*, 420, **31**; eggs, **60**
Estrildidae, 409
*Euodice cantans*, 437, **31**
　*malabarica*, 437, **31**

*flammea* (*Carduelis*), 639, **42**, **44**; eggs, **60**
*flavifrons* (*Vireo*), 440, **32**
*flavirostris* (*Carduelis*), 625, **41**, **44**; eggs, **60**
*Fringilla coelebs*, 448, **29**, **38**; eggs, **60**
　*montifringilla*, 479, **29**, **38**; eggs, **60**
　*teydea*, 474, **29**, **38**; eggs, **60**
Fringillidae, 447
*frugilegus* (*Corvus*), 151, **14**, **15**; eggs, **58**

*Garrulus glandarius*, 7, **1**, **5**; eggs, **57**
*githagineus* (*Bucanetes*), 754, **47**, **50**; eggs, **61**
*glandarius* (*Garrulus*), 7, **1**, **5**; eggs, **57**
*graculus* (*Pyrrhocorax*), 95, **6**, **9**; eggs, **57**

*Hesperiphona vespertina*, 847, **53**, **56**
*hispaniolensis* (*Passer*), 308, **21**, **26**; eggs, **59**
*hornemanni* (*Carduelis*), 661, **42**, **45**; eggs, **60**

*iagoensis* (*Passer*), 327, **22**, **26**; eggs, **59**
*infaustus* (*Perisoreus*), 31, **2**, **5**; eggs, **57**

*Lagonosticta senegala*, 411, **30**; eggs, **60**
*leucoptera* (*Loxia*), 672, **43**, **45**; eggs, **61**
*Loxia curvirostra*, 686, **43**, **45**; eggs, **61**

*leucoptera*, 672, **43**, **45**; eggs, **61**
*pytyopsittacus*, 717, **43**, **45**; eggs, **61**
*scotica*, 707, **43**; eggs, **61**
*luteus* (*Passer*), 351, **23**, **27**; eggs, **59**

*malabarica* (*Euodice*), 437, **31**
*manyar* (*Ploceus*), 401, **30**; eggs, **59**
*migratoria* (*Eophona*), 832
*moabiticus* (*Passer*), 320, **22**, **26**; eggs, **59**
*monedula* (*Corvus*), 120, **7**, **9**; eggs, **57**
*mongolicus* (*Bucanetes*), 748, **47**, **50**
*montanus* (*Passer*), 336, , **21**, **27**; eggs, **59**
*montifringilla* (*Fringilla*), 479, **29**, **38**; eggs, **60**
*Montifringilla nivalis*, 386, **28**; eggs, **59**

*nivalis* (*Montifringilla*), 386, **28**; eggs, **59**
*Nucifraga caryocatactes*, 76, **4**, **5**; eggs, **57**

*obsoleta* (*Rhodopechys*), 739, **46**, **50**; eggs, **61**
*olivaceus* (*Vireo*), 444, **32**
*Onychognathus tristramii*, 229, **12**, **18**

*Passer domesticus*, 289, **20**, **25**, **26**; eggs, **59**
　*hispaniolensis*, 308, **21**, **26**; eggs, **59**
　*iagoensis*, 327, **22**, **26**; eggs, **59**
　*luteus*, 351, **23**, **27**; eggs, **59**
　*moabiticus*, 320, **22**, **26**; eggs, **59**
　*montanus*, 336, , **21**, **27**; eggs, **59**
　*simplex*, 331, **23**, **26**; eggs, **59**
Passeridae, 288
*Perisoreus infaustus*, 31, **2**, **5**; eggs, **57**
*personata* (*Eophona*), 832
*petronia* (*Petronia*), 371, **25**, **27**; eggs, **59**
*Petronia petronia*, 371, **25**, **27**; eggs, **59**
　*xanthocollis*, 365, **24**, **27**; eggs, **59**
*philadelphicus* (*Vireo*), 442, **32**
*pica* (*Pica*), 54, **3**, **5**; eggs, **57**
*Pica pica*, 54, **3**, **5**; eggs, **57**
*Pinicola enucleator*, 802, **51**, **52**; eggs, **61**
*pinus* (*Carduelis*), 604
Ploceidae, 400
*Ploceus cucullatus*, 401
　*manyar*, 401, **30**; eggs, **59**
*pusillus* (*Serinus*), 499, **33**, **38**; eggs, **60**
*pyrrhocorax* (*Pyrrhocorax*), 105, **6**, **9**; eggs, **57**
*Pyrrhocorax graculus*, 95, **6**, **9**; eggs, **57**
　*pyrrhocorax*, 105, **6**, **9**; eggs, **57**
*Pyrrhula pyrrula*, 815, **53**, **54**, **55**; eggs, **61**

*pyrrula* (*Pyrrhula*), 815, **53, 54, 55**; eggs, **61**
*pytyopsittacus* (*Loxia*), 717, **43, 45**; eggs, **61**

*quelea* (*Quelea*), 409
*Quelea quelea*, 409

*rhipidurus* (*Corvus*), 223, **10, 15**; eggs, **58**
*Rhodopechys obsoleta*, 739, **46, 50**; eggs, **61**
  *sanguinea*, 729, **46, 50**
*roseus* (*Carpodacus*), 789, **49, 51**
*roseus* (*Sturnus*), 269, **13, 18**; eggs, **59**
*rubicilla* (*Carpodacus*), 792, **49, 51**; eggs, **61**
*ruficollis* (*Corvus*), 197, **10, 15**; eggs, **58**

*sanguinea* (*Rhodopechys*), 729, **46, 50**
*scotica* (*Loxia*), 707, **43**; eggs, **61**
*senegala* (*Lagonosticta*), 411, **30**; eggs, **60**
*serinus* (*Serinus*), 508, **33, 38**; eggs, **60**
*Serinus canaria*, 528, **34, 39**; eggs, **60**
  *citrinella*, 536, **34, 39**; eggs, **60**
  *pusillus*, 499, **33, 38**; eggs, **60**
  *serinus*, 508, **33, 38**; eggs, **60**
  *syriacus*, 521, **33, 38**; eggs, **60**
*sibiricus* (*Uragus*), 814, **54**
*simplex* (*Passer*), 331, **23, 26**; eggs, **59**
*sinensis* (*Sturnus*), 237
*spinus* (*Carduelis*), 587, **37, 39**; eggs, **60**
*splendens* (*Corvus*), 143, **8, 15**; eggs, **57**
Sturnidae, 228
*sturninus* (*Sturnus*), 234, **12, 19**
*Sturnus roseus*, 269, **13, 18**; eggs, **59**
  *sinensis*, 237
  *sturninus*, 234, **12, 19**
  *unicolor*, 260, **17, 19**; eggs, **59**
  *vulgaris*, 238, **16, 19**; eggs, **59**
*synoicus* (*Carpodacus*), 783, **48, 51**; eggs, **61**
*syriacus* (*Serinus*), 521, **33, 38**; eggs, **60**

*teydea* (*Fringilla*), 474, **29, 38**; eggs, **60**
*tristis* (*Acridotheres*), 280, **17, 18**; eggs, **59**
*tristramii* (*Onychognathus*), 229, **12, 18**

*unicolor* (*Sturnus*), 260, **17, 19**; eggs, **59**

*Uraeginthus bengalus*, 420
*Uragus sibiricus*, 814, **54**

*vespertina* (*Hesperiphona*), 847, **53, 56**
*Vireo flavifrons*, 440, **32**
  *olivaceus*, 444, **32**
  *philadelphicus*, 442, **32**
Vireonidae, 439
*vulgaris* (*Sturnus*), 238, **16, 19**; eggs, **59**

*xanthocollis* (*Petronia*), 365, **24, 27**; eggs, **59**

ENGLISH NAMES

Avadavat, Red, 427, **31**; eggs, **60**

Brambling, 479, **29, 38**; eggs, **60**
Bullfinch, 815, **53, 54, 55**; eggs, **61**

Canary, 528, **34, 39**; eggs, **60**
Chaffinch, 448, **29, 38**; eggs, **60**
  Blue, 474, **29, 38**; eggs, **60**
  Canary Islands, 474, **29, 38**; eggs, **60**
Chough, 105, **6, 9**; eggs, **57**
  Alpine, 95, **6, 9**; eggs, **57**
Cordon-bleu, Red-cheeked, 420
Crossbill, 686, **43, 45**; eggs, **61**
  Parrot, 717, **43, 45**; eggs, **61**
  Scottish, 707, **43**; eggs, **61**
  Two-barred, 672, **43, 45**; eggs, **61**
Crow, Carrion, 172, **14, 15**; eggs, **58**
  House, 143, **8, 15**; eggs, **57**
  Pied, 195, **14, 15**

Finch, Citril, 536, **34, 39**; eggs, **60**
  Crimson-winged, 729, **46, 50**
  Desert, 739, **46, 50**; eggs, **61**
  Trumpeter, 754, **47, 50**; eggs, **61**
  Trumpeter, Mongolian, 748, **47, 50**
Firefinch, Red-billed, 411, **30**; eggs, **60**

Goldfinch, 568, **36, 39**; eggs, **60**
Grackle, Tristram's, 229, **12, 18**
Greenfinch, 548, **35, 39**; eggs, **60**
Grosbeak, Evening, 847, **53, 56**
  Japanese, 832
  Pine, 802, **51, 52**; eggs, **61**
Grosbeak, Yellow-billed, 832

Hawfinch, 832, **53, 56**; eggs, **61**

Jackdaw, 120, **7, 9**; eggs, **57**
  Daurian, 140, **8, 9**
Jay, 7, **1, 5**; eggs, **57**
  Siberian, 31, **2, 5**; eggs, **57**

Linnet, 604, **40, 44**; eggs, **60**

Magpie, 54, **3, 5**; eggs, **57**
  Azure-winged, 42, **2, 5**; eggs, **57**
Mynah, Common, 280, **17, 18**; eggs, **59**

Nutcracker, 76, **4, 5**; eggs, **57**

Quelea, Red-billed, 409

Raven, 206, **11, 15**; eggs, **58**

Brown-necked, 197, **10, 15**; eggs, **58**
Fan-tailed, 223, **10, 15**; eggs, **58**
Redpoll, 639, **42, 44**; eggs, **60**
  Arctic, 661, **42, 45**; eggs, **60**
Rook, 151, **14, 15**; eggs, **58**
Rosefinch, Great, 792, **49, 51**; eggs, **61**
  Long-tailed, 814, **54**
  Pallas's, 789, **49, 51**
  Scarlet, 764, **48, 50**; eggs, **61**
  Sinai, 783, **48, 51**; eggs, **61**

Serin, 508, **33, 38**; eggs, **60**
  Red-fronted, 499, **33, 38**; eggs, **60**
  Syrian, 521, **33, 38**; eggs, **60**
Silverbill, African, 437, **31**
  Indian, 437, **31**
Siskin, 587, **37, 39**; eggs, **60**
  Pine, 604
Snow Finch, 386, **28**; eggs, **59**
Sparrow, Dead Sea, 320, **22, 26**; eggs, **59**
  Desert, 331, **23, 26**; eggs, **59**
  House, 289, **20, 25, 26**; eggs, **59**
  Iago, 327, **22, 26**; eggs, **59**
  Pale Rock, 357, **24, 27**; eggs, **59**
  Rock, 371, **25, 27**; eggs, **59**
  Spanish, 308, **21, 26**; eggs, **59**
  Sudan Golden, 351, **23, 27**; eggs, **59**
  Tree, 336, **21, 27**; eggs, **59**
  Yellow-throated, 365, **24, 27**; eggs, **59**
Starling, 238, **16, 19**; eggs, **59**
  Daurian, 234, **12, 19**
  Grey-backed, 237
  Rose-coloured, 269, **13, 18**; eggs, **59**
  Spotless, 260, **17, 19**; eggs, **59**
  Tristram's, 229, **12, 18**

Twite, 625, **41, 44**; eggs, **60**

Vireo, Philadelphia, 442, **32**
  Red-eyed, 444, **32**
  Yellow-throated, 440, **32**

Waxbill, Common, 420, **31**; eggs, **60**
Weaver, Streaked, 401, **30**; eggs, **59**
  Village, 401

NOMS FRANÇAIS

Amarante du Sénégal, 411, **30**; oeufs, **60**
Astrild ondulé, 420, **31**; oeufs, **60**

Bec-croisé bifascié, 672, **43, 45**; oeufs, **61**
  d'Ecosse, 707, **43**; oeufs, **61**
  perroquet, 717, **43, 45**; oeufs, **61**
  des sapins, 686, **43, 45**; oeufs, **61**
Bengali rouge, 427, **31**; oeufs, **60**
Bouvreuil pivoine, 815, **53, 54, 55**; oeufs, **61**

Capucin  bec d'argent, 437, **31**
  bec de plomb, 437, **31**
Cassenoix moucheté, 76, **4**, **5**; oeufs,
  **57**
Chardonneret élégant, 568, **36**, **39**;
  oeufs, **60**
Chocard des Alpes, 95, 6, 9; oeufs, **57**
Choucas de Daourie, 140, **8**, **9**
  des tours, 120, 7, 9; oeufs, **57**
Corbeau brun, 197, **10**, **15**; oeufs, **58**
  familier, 143, 8, **15**; oeufs, **57**
  freux, 151, **14**, **15**; oeufs, **58**
  à queue courte, 223, **10**, **15**; oeufs,
  **58**
Corbeau-pie, 195, **14**, **15**
Cordonbleu à joues rouges, 420
Corneille noir, 172, **14**, **15**; oeufs, **58**
Crave à bec rouge, 105, 6, 9; oeufs, **57**

Durbec des pins, 802, **51**, **52**; oeufs,
  **61**

Etourneau de Daourie, 234, **12**, **19**
  mandarin, 237, **16**, **19**; oeufs, **59**
  roselin, 269, **13**, **18**; oeufs, **59**
  sansonnet, 238
  unicolore, 260, **17**, **19**; oeufs, **59**

Geai des chênes, 7, **1**, **5**; oeufs, **57**
Grand Corbeau, 206, **11**, **15**; oeufs, **58**
Gros-bec cassenoyaux, 832, **53**, **56**;
  oeufs, **61**
  errant, 847, **53**, **56**
  masqué, 832
  migrateur, 832

Linotte à bec jaune, 625, **41**, **44**; oeufs,
  **60**
  mélodieuse, 604, **40**, **44**; oeufs, **60**

Martin triste, 280, **17**, **18**; oeufs, **59**
Mésangeai imitateur, 31, **2**, **5**; oeufs,
  **57**
Moineau blanc, 331, **23**, **26**; oeufs, **59**
  du Cap-Vert, 327, **22**, **26**; oeufs, **59**
  domestique, 289, **20**, **25**, **26**; **oeufs**,
  **59**
  doré, 351, **23**, **27**; oeufs, **59**
  espagnol, 308, **21**, **26**; oeufs, **59**
  friquet, 336, **21**, **27**; oeufs, **59**
  à gorge jaune, 365, **24**, **27**; oeufs, **59**
  de la Mer Morte, 320, **22**, **26**; oeufs,
  **59**
  pâle, 357, **24**, **27**; oeufs, **59**
  soulcie, 371, **25**, **27**; oeufs, **59**

Niverolle alpine, 386, **28**; oeufs, **59**

Pie bavarde, 54, **3**, **5**; oeufs, **57**
  bleue , 42, **2**, **5**; oeufs, **57**
Pinson des arbres, 448 , **29**, **38**; oeufs,
  **60**
  bleu, 474, **29**, **38**; oeufs, **60**
  du Nord, 479, **29**, **38**; oeufs, **60**

Roselin à ailes roses, 729, **46**, **50**
  cramoisi, 764, **48**, **50**; oeufs, **61**
  githagine, 754, **47**, **50**; oeufs, **61**
  de Lichtenstein, 739, **46**, **50**; oeufs,
  **61**
  à longue queue, 814, **54**
  de Mongolie, 748, **47**, **50**
  rose, 789, **49**, **51**
  du Sinai, 783, **48**, **51**; oeufs, **61**
  tacheté, 792, **49**, **51**; oeufs, **61**
Rufipenne de Tristram, 229, **12**, **18**

Serin des Canaries, 528, **34**, **39**; oeufs,
  **60**
  cini, 508, **33**, **38**; oeufs, **60**
  à front d'or, 499, **33**, **38**; oeufs, **60**
  syriaque, 521, **33**, **38**; oeufs, **60**
Sizerin blanchâtre, 661, **42**, **45**; oeufs,
  **60**
Sizerin flammé, 639, **42**, **44**; oeufs, **60**

Tarin des aulnes, 587, **37**, **39**; oeufs,
  **60**
  des pins, 604
Tisserin gendarme, 401
  manyar, 401, **30**; oeufs, **59**
Travailleur à bec rouge, 409

Venturon montagnard, 536, **34**, **39**;
  oeufs, **60**
Verdier d'Europe, 548, **35**, **39**; oeufs,
  **60**
Viréo à gorge jaune, 440, **32**
  à oeil rouge, 444, **32**
  de Philadelphie, 442, **32**

## DEUTSCHE NAMEN

Aaskrähe, 172,, **14**, **15**; eier, **58**
Abendkenbeisser, 847, **53**, **56**
Alpendohle, 95, 6, 9; eier, **57**
Alpenkrähe, 105, 6, 9; eier, **57**

Bergfink, 479, **29**, **38**; eier, **60**
Berggimpel, 792, **49**, **51**; eier, **61**
Berghänfling, 625, **41**, **44**; eier, **60**
Bindenkreutzschnabel, 672, **43**, **45**;
  eier, **61**
Birkenzeisig, 639, **42**, **44**; eier, **60**
Blauelster, 42, **2**, **5**; eier, **57**
Bluthänfling, 604, **40**, **44**; eier, **60**
Blutschnabelweber, 409
Borstenrabe, 223, **10**, **15**; eier, **58**
Buchfink, 448 , **29**, **38**; eier, **60**

Daurischer Star, 234, **12**, **19**
Dohle, 120, 7, 9; eier, **57**

Eichelhäher, 7, **1**, **5**; eier, **57**
Einfarbstar, 260, **17**, **19**; eier, **59**
Elster, 54, 3, **5**; eier, **57**
Elsterdohle, 140, **8**, **9**

Erlenzeisig, 587, **37**, **39**; eier, **60**

Fahlsperling, 357, **24**, **27**; eier, **59**
Feldsperling, 336, **21**, **27**; eier, **59**
Fichtenkreutzschnabel, 686, **43**, **45**;
  eier, **61**
Fichtenzeisig, 604

Gelbkehlsperling, 365, **24**, **27**; eier, **59**
Gelbkehlvireo, 440, **32**
Gimpel, 815, **53**, **54**, **55**; eier, **61**
Girlitz, 508, **33**, **38**; eier, **60**
Glanzkrähe, 143, **8**, **15**; eier, **57**
Grünling, 548, **35**, **39**; eier, **60**

Hakengimpel, 802, **51**, **52**; eier, **61**
Haussperling, 289, **20**, **25**, **26**; **eier**, **59**
Hirtenmaina, 280, **17**, **18**; eier, **59**

Kanariengirlitz, 528, **34**, **39**; eier, **60**
Karmingimpel, 764, **48**, **50**; eier, **61**
Kernbeisser, 832, **53**, **56**; eier, **61**
Kiefernenkreutzschnabel, 717, **43**,
  **45**; eier, **61**
Kolkrabe, 206, **11**, **15**; eier, **58**

Malabarfasänchen, 437, **31**
Mandarinenstar, 237
Manyarweber, 401, **30**; eier, **59**
Maskenkernbeisser, 832
Meisengimpel, 814, **54**
Moabsperling, 320, **22**, **26**; eier, **59**
Mongolengimpel, 748, **47**, **50**

Philadelphiavireo, 442, **32**
Polarbirkenzeisig, 661, **42**, **45**; eier, **60**

Rosengimpel, 789, **49**, **51**
Rosenstar, 269, **13**, **18**; eier, **59**
Rostsperling, 327, **22**, **26**; eier, **59**
Rotaugenvireo, 444, **32**
Rotflügelgimpel, 729, **46**, **50**
Rotstirngirlitz, 499, **33**, **38**; eier, **60**

Saatkrähe, 151, **14**, **15**; eier, **58**
Schildrabe, 195, **14**, **15**
Schmetterlingsfink, 420
Schneefink, 386, **28**; eier, **59**
Schottischer Kreutzschnabel, 707, **43**;
  eier, **61**
Schwarzschwanz-Kernbeisser, 832
Senegalamarant, 411, **30**; eier, **60**
Silberschnäbelchen, 437, **31**
Sinaigimpel, 783, **48**, **51**; eier, **61**
Star, 238, **16**, **19**; eier, **59**
Steinsperling, 371, **25**, **27**; eier, **59**
Stieglitz, 568, **36**, **39**; eier, **60**
Sudangoldsperling, 351, **23**, **27**; eier,
  **59**

Tannenhäher, 76, **4**, **5**; eier, **57**
Textor, 401
Teydefink, 474, **29**, **38**; eier, **60**
Tigerfink, 427, **31**; eier, **60**

Tristramstar, 229, **12**, **18**

Unglückshäher, 31, **2**, **5**; eier, **57**

Weidensperling, 308, **21**, **26**; eier, **59**

Weissflügelgimpel, 739, **46**, **50**; eier, **61**

Wellenastrild, 420, **31**; eier, **60**

Wüstengimpel, 754, **47**, **50**; eier, **61**

Wüstenrabe, 197, **10**, **15**; eier, **58**

Wüstensperling, 331, **23**, **26**; eier, **59**

Zederngirlitz, 521, **33**, **38**; eier, **60**

Zitronengirlitz, 536, **34**, **39**; eier, **60**

# EGG PLATES

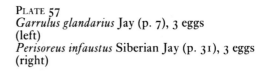

PLATE 57
*Garrulus glandarius* Jay (p. 7), 3 eggs
(left)
*Perisoreus infaustus* Siberian Jay (p. 31), 3 eggs
(right)

*Cyanopica cyanus* Azure-winged Magpie (p. 42), 4 eggs

*Pica pica* Magpie (p. 54), 6 eggs

*Nucifraga caryocatactes* Nutcracker (p. 76), 3 eggs
(left)
*Corvus monedula* Jackdaw (p. 120), 3 eggs
(right)

*Pyrrhocorax graculus* Alpine Chough (p. 95), 3 eggs
(left)
*Pyrrhocorax pyrrhocorax* Chough (p. 105), 3 eggs
(right)

*Corvus splendens* House Crow (p. 143), 5 eggs

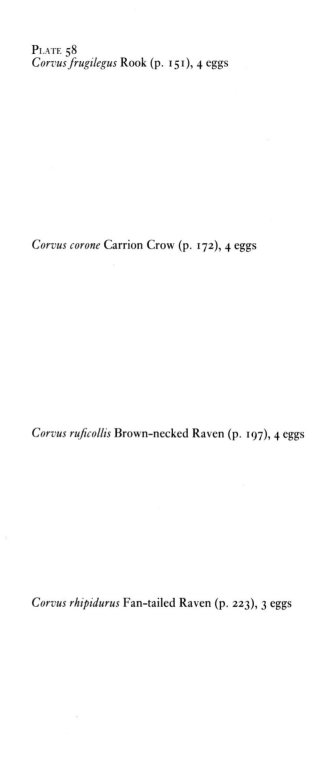

PLATE 58
*Corvus frugilegus* Rook (p. 151), 4 eggs

*Corvus corone* Carrion Crow (p. 172), 4 eggs

*Corvus ruficollis* Brown-necked Raven (p. 197), 4 eggs

*Corvus rhipidurus* Fan-tailed Raven (p. 223), 3 eggs

*Corvus corax* Raven (p. 206), 4 eggs

PLATE 59

*Sturnus vulgaris* Starling (p. 238), 2 eggs
(left)
*Sturnus unicolor* Spotless Starling (p. 260), 2 eggs
(centre)
*Sturnus roseus* Rose-coloured Starling (p. 269), 2 eggs
(right)

*Acridotheres tristis* Common Myna (p. 280), 2 eggs

*Passer domesticus* House Sparrow (p. 289), 4 eggs
(left)
*Passer hispaniolensis* Spanish Sparrow (p. 308), 4 eggs
(right)

*Passer moabiticus* Dead Sea Sparrow (p. 320), 4 eggs
(left)
*Passer iagoensis* Iago Sparrow (p. 327), 4 eggs
(right)

*Passer simplex* Desert Sparrow (p. 331), 4 eggs
(left)
*Passer montanus* Tree Sparrow (p. 336), 4 eggs
(right)

*Carpospiza brachydactyla* Pale Rock Sparrow (p. 357), 2
eggs
(left)
*Petronia xanthocollis* Yellow-throated Sparrow (p. 365), 4
eggs
(right)

*Petronia petronia* Rock Sparrow (p. 371), 4 eggs
(left)
*Ploceus manyar* Streaked Weaver (p. 401), 4 eggs
(right)

*Montifringilla nivalis* Snow Finch (p. 386), 4 eggs

PLATE 60

*Lagonosticta senegala* Red-billed Firefinch (p. 411), 3 eggs
(left)
*Estrilda astrild* Common Waxbill (p. 420), 3 eggs
(centre)
*Amandava amandava* Red Avadavat (p. 427), 3 eggs
(right)

*Fringilla coelebs* Chaffinch (p. 448), 7 eggs

*Fringilla teydea* Blue Chaffinch (p. 474), 4 eggs
(left)
*Fringilla montifringilla* Brambling (p. 479), 4 eggs
(right)

*Serinus pusillus* Red-fronted Serin (p. 499), 4 eggs
(left)
*Serinus serinus* Serin (p. 508), 4 eggs
(right)

*Serinus syriacus* Syrian Serin (p. 521), 4 eggs
(left)
*Serinus canaria* Canary (p. 528), 4 eggs
(right)

*Serinus citrinella* Citril Finch (p. 536), 4 eggs
(left)
*Carduelis carduelis* Goldfinch (p. 568), 4 eggs
(right)

*Carduelis chloris* Greenfinch (p. 548), 4 eggs
(left)
*Carduelis spinus* Siskin (p. 587), 4 eggs
(right)

*Carduelis cannabina* Linnet (p. 604), 4 eggs
(left)
*Carduelis flavirostris* Twite (p. 625), 4 eggs
(right)

*Carduelis flammea* Redpoll (p. 639), 4 eggs
(left)
*Carduelis hornemanni* Arctic Redpoll (p. 661), 4 eggs
(right)

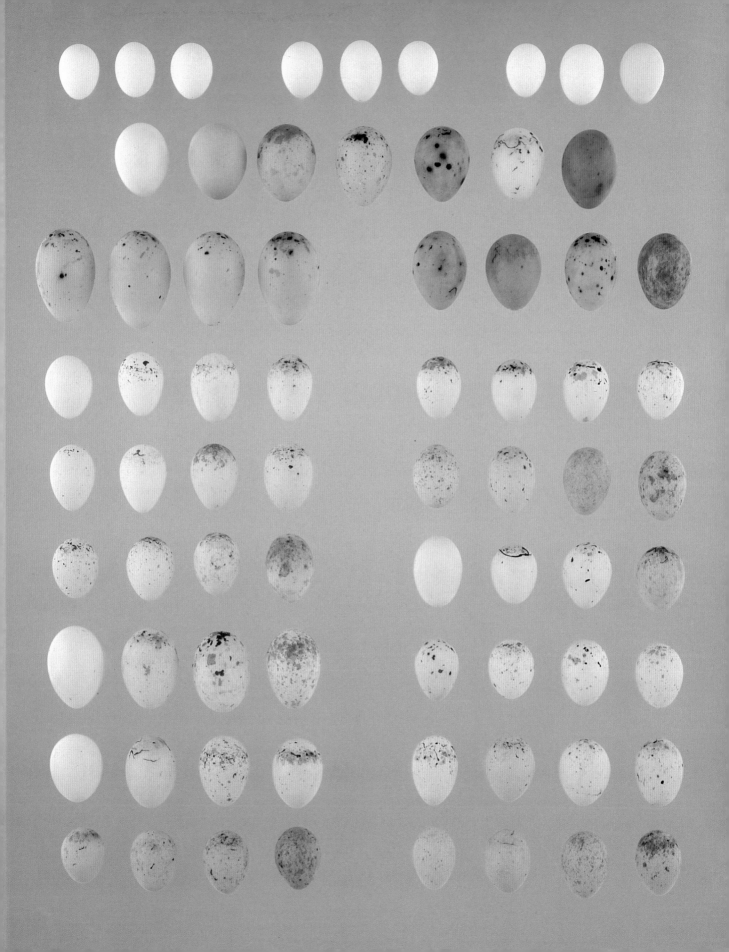

PLATE 61

*Loxia curvirostra* Crossbill (p. 686), 4 eggs
(left)
*Loxia scotica* Scottish Crossbill (p. 707), 4 eggs
(right)

*Loxia pytyopsittacus* Parrot Crossbill (p. 717), 4 eggs
(left)
*Loxia leucoptera* Two-barred Crossbill (p. 672), 4 eggs
(right)

*Rhodospiza obsoleta* Desert Finch (p. 739), 4 eggs
(left)
*Bucanetes githagineus* Trumpeter Finch (p. 754), 4 eggs
(right)

*Carpodacus erythrinus* Scarlet Rosefinch (p. 764), 4 eggs
(left)
*Carpodacus synoicus* Sinai Rosefinch (p. 783), 1 egg
(right)

*Carpodacus rubicilla* Great Rosefinch (p. 792), 4 eggs
(left)
*Pinicola enucleator* Pine Grosbeak (p. 802), 3 eggs
(right)

*Pyrrhula pyrrhula* Bullfinch (p. 815), 5 eggs

*Coccothraustes coccothraustes* Hawfinch (p. 832), 7 eggs

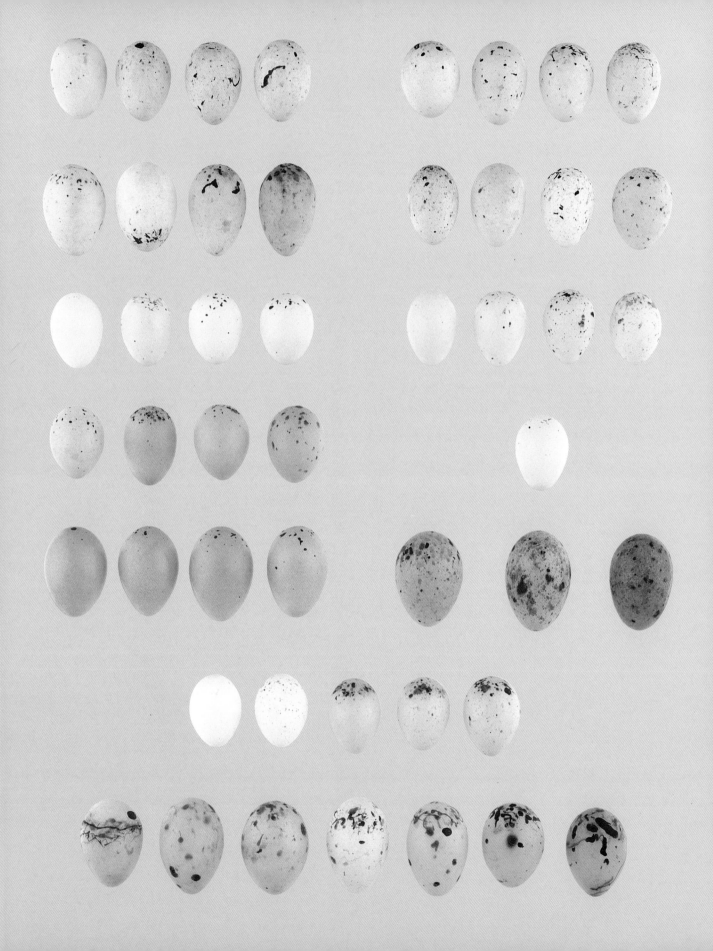

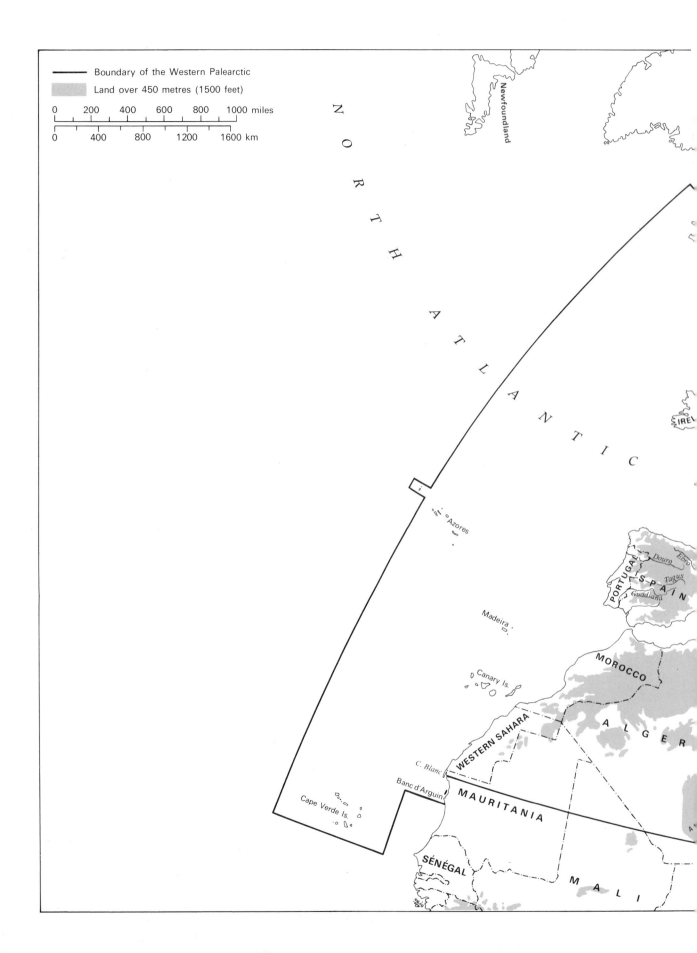

Boundary of the Western Palearctic

Land over 450 metres (1500 feet)

| 0 | 200 | 400 | 600 | 800 | 1000 miles |

| 0 | 400 | 800 | 1200 | 1600 km |

NORTH ATLANTIC

Newfoundland

IREL

Azores

Madeira

Canary Is.

PORTUGAL

SPAIN

Douro

Ebro

Tagus

Guadiana

MOROCCO

ALGER

WESTERN SAHARA

C. Blanc

Banc d'Arguin

MAURITANIA

Cape Verde Is.

SÉNÉGAL

MALI